Ulrich Tietze • Christoph Schenk • Eberhard Gamm
Electronic Circuits

U. Tietze · Ch. Schenk · E. Gamm

Electronic Circuits

Handbook for Design and Application

2nd edition

with 1771 Figures and CD-ROM

 Springer

Dr.-Ing. Ulrich Tietze
Lecturer at the Technical Electronics Institute
University of Erlangen-Nuernberg, Germany

Dr.-Ing. Christoph Schenk
General Manager of Dr. Schenk GmbH
Munic, Germany

Dr.-Ing. Eberhard Gamm
Communications Consultant
Erlangen, Germany

www.tietze-schenk.com
mail@tietze-schenk.com

Translation of
Tietze, U.; Schenk, Ch.: Halbleiter-Schaltungstechnik. 12. edition, 2002

Existing translations:
Polish: Naukowo-Techniczne, Warsaw 1976, 1987, 1996
Hungarian: Müszaki, Budapest 1974, 1981, 1990
Russian: Mir, Moscow 1982, Dodeca Publishin, Moscow 2007
Spanish: Marcombo, Barcelona 1983
Chinese: (bootleg) 1985
English: Springer, Heidelberg 1978, 1991

Springer is a part of Springer Science+Business Media
springer.com
© Springer-Verlag Berlin Heidelberg 2008

ISBN 978-3-540-00429-5

Library of Congress Control Number: 2007936735

Printed on acid-free paper

9 8 7 6 5 4 3 2 1

Preface

The purpose of this book is to help the reader to understand off-the-shelf circuits and to enable him to design his or her own circuitry. The book is written for students, practicing engineers and scientists. It covers all major aspects of analog and digital circuit design. The book is a translation of the current 12th edition of the German bestseller *Halbleiter-Schaltungstechnik*.

Part I describes semiconductor devices and their behavior with respect to the models used in circuit simulation. This part introduces all major aspects of transistor level design (IC-design). Basic circuits are analyzed in five steps: large-signal transfer characteristic, small-signal response, frequency response and bandwidth, noise and distortion. Digital circuits are covered starting with the internal circuitry of gates and flip-flops up to the construction of combinatorial and sequential logic systems with PLDs and FPGAs. Design examples and a short form guide for the digital synthesis tool *ispLever* are included on the CD enclosed.

Part II is dedicated to board level design. The main chapters of this part describe the use of operational amplifiers for signal conditioning including signal amplification, filtering and AD-conversion. Further chapters cover power amplifiers, power supplies and other important functional blocks of analog systems. The chapters are self-contained with a minimum of cross-reference. This allows the advanced reader to familiarize himself quickly with the various areas of applications. Each chapter offers a detailed overview of various solutions to a given requirement. In order to enable the reader to proceed quickly from an idea to a working circuit, we discuss only those solutions we have tested thoroughly by simulation. Many of these simulation examples are included on the CD enclosed.

Part III describes circuits for analog and digital communication over wireless channels. The first chapter is dedicated to transmission channels, scattering parameters and analog and digital modulations. Further chapters treat the architecture of transmitters and receivers, the high frequency behavior of components, circuits for impedance matching, high frequency amplifiers and mixers for frequency conversion.

To support analog circuit design, design examples and a short-form guide for the well known circuit simulator *PSpice* are included on the CD. This package contains libraries with examples of scalable transistors for IC-like design. The library also supports S-parameter and loop-gain simulations. An HTML-based index allows comfortable navigation throughout the simulations.

Our homepage *www.tietze-schenk.com* offers updates, supplements and design examples. We encourage you to use our email-address *mail@tietze-schenk.com* for feedback and comments.

We would like to thank Dr. Merkle at Springer Heidelberg for the administration, Gerhard Büsching for the translation and Danny Lewis at PTP-Berlin for the assembly of this book. In particular we like to thank Dr. Eberhard Gamm for the contribution of the first four chapters of circuit design fundamentals in part I and the chapters of communications in part III. We have added him as a young innovative author.

Erlangen, July 2007 U. Tietze, Ch. Schenk, E. Gamm

Overview

Contents

Part II. General Applications   723

11. Operational Amplifier Applications   725

Part I

Device Models and Basic Circuits

Chapter 1:
Diode

The diode is a semiconductor component with two connections, which are called the *anode* (*A*) and the *cathode* (*K*). Distinction has to be made between discrete diodes, which are intended for installation on printed circuit boards and are contained in an individual case, and integrated diodes, which are produced together with other semiconductor components on a common semiconductor carrier (*substrate*). Integrated diodes have a third connection resulting from the common carrier. It is called the *substrate* (*S*); it is of minor importance for electrical functions.

Construction: Diodes consist of a pn or a metal-n junction and are called pn or Schottky diodes, respectively. Figure 1.1 shows the graphic symbol and the construction of a diode. In pn diodes the p and the n regions usually consist of silicon. Some discrete diode types still use germanium and thus have a lower forward voltage, but they are considered obsolete. In Schottky diodes the p region is replaced by a metal region. This type also has a low forward voltage and is therefore used to replace germanium pn diodes.

In practice the term *diode* is used for the silicon pn diode; all other types are identified by supplements. Since the same graphic symbol is used for all types of diodes with the exception of some special diodes the various types of discrete diodes can be distinguished only by means of the type number printed on the component or the specifications in the data sheet.

Operating modes: A diode can be operated in the *forward, reverse* or *breakthrough mode*. In the following Section these operating regions are described in more detail.

Diodes that are used predominantly for the purpose of rectifying alternating voltages are called *rectifier diodes*; they operate alternately in the forward and reverse region. Diodes designed for the operation in the breakthrough region are called *Zener diodes* and are used for voltage stabilization. The *variable capacitance diodes* are another important type. They are operated in the reverse region and, due to the particularly strong response of the junction capacitance to voltage variations, are used for tuning the frequency in resonant circuits. In addition, there is a multitude of special diodes which are not covered here in detail.

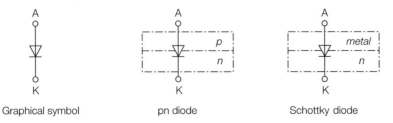

Fig. 1.1. Graphical symbol and diode construction

1.1
Performance of the Diode

The performance of a diode is described most clearly by its characteristic curve. This shows the relation between current and voltage where all parameters are *static* which means that they do not change over time or only very slowly. In addition, formulas that describe the diode performance sufficiently accurately are required for mathematical calculations. In most cases simple equations can be used. In addition, there is a model that correctly reflects the *dynamic performance* when the diode is driven with sinusoidal or pulse-shaped signals. This model is described in Sect. 1.3 and knowledge of it is not essential to understand the fundamentals. The following Sections focus primarily on the performance of silicon pn diodes.

1.1.1
Characteristic Curve

Connecting a silicon pn diode to a voltage $V_D = V_{AK}$ and measuring the current I_D in a positive sense from A to K results in the characteristic curve shown in Fig. 1.2. It should be noted that the positive voltage range has been enhanced considerably for reasons of clarity. For $V_D > 0\,V$ the diode operates in the forward mode, i.e. in the *conducting state*. In this region the current rises exponentially with an increasing voltage. When $V_D > 0.4\,V$, a considerable current flows. If $-V_{BR} < V_D < 0\,V$ the diode is in the reverse-biased state and only a negligible current flows. This region is called the *reverse region*. The *breakthrough voltage* V_{BR} depends on the diode and for rectifier diodes amounts to $V_{BR} = 50\ldots1000\,V$. If $V_D < -V_{BR}$, the diode breaks through and a current flows again. Only Zener diodes are operated permanently in this *breakthrough region*; with all other diodes current flow with negative voltages is not desirable. With germanium and Schottky diodes a considerable current flows in the forward region even for $V_D > 0.2\,V$, and the breakthrough voltage V_{BR} is $10\ldots200\,V$.

In the forward region the voltage for typical currents remains almost constant due to the pronounced rise of the characteristic curve. This voltage is called the *forward voltage V_F*

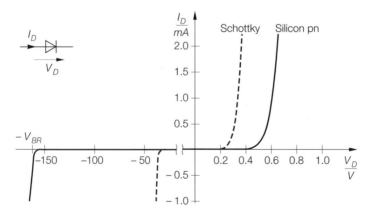

Fig. 1.2. Current-voltage characteristic of a small-signal diode

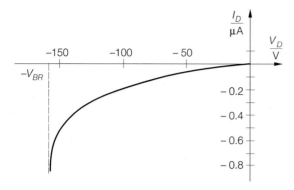

Fig. 1.3. Characteristic curve of a small-signal diode in the reverse region

and for both germanium and Schottky diodes lies at $V_{F,Ge} \approx V_{F,Schottky} \approx 0.3\ldots0.4\,V$ and for silicon pn diodes at $V_{F,Si} \approx 0.6\ldots0.7\,V$. With currents in the ampere range as used in power diodes the voltage may be significantly higher since in addition to the *internal* forward voltage a considerable voltage drop occurs across the spreading and connection resistances of the diode: $V_F = V_{F,i} + I_D R_B$. In the borderline case of $I_D \to \infty$ the diode acts like a very low resistance with $R_B \approx 0.01\ldots10\,\Omega$.

Figure 1.3 shows the enlarged reverse region. The *reverse current* $I_R = -I_D$ is very small with a low reverse voltage $V_R = -V_D$ and increases slowly when the voltage approaches the breakthrough voltage while it shoots up suddenly at the onset of the breakthrough.

1.1.2
Description by Equations

Plotting the characteristic curve for the region $V_D > 0$ in a semilogarithmic form results approximately in a straight line (see Fig. 1.4); this means that there is an exponential relation between I_D and V_D due to $\ln I_D \sim V_D$. The calculation on the basis of semiconductor physics leads to [1.1]:

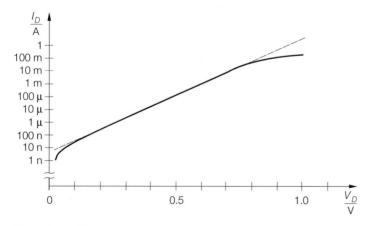

Fig. 1.4. Semilogarithmic representation of the characteristic curve for $V_D > 0$

$$I_D(V_D) = I_S \left(e^{\frac{V_D}{V_T}} - 1 \right) \qquad \text{for } V_D \geq 0$$

For the correct description of a real diode a correction factor is required which enables the slope of the straight line in the semilogarithmic representation to be adapted [1.1]:

$$I_D = I_S \left(e^{\frac{V_D}{nV_T}} - 1 \right) \tag{1.1}$$

Here, $I_S \approx 10^{-12} \ldots 10^{-6}$ A is the *reverse saturation current, $n \approx 1 \ldots 2$* is the *emission coefficient* and $V_T = kT/q \approx 26$ mV is the *temperature voltage* at room temperature.

Even though (1.1) actually applies only to $V_D \geq 0$ it is sometimes used for $V_D < 0$. For $V_D \ll -nV_T$ this results in a constant current $I_D = -I_S$ which is generally much smaller than the current that is actually flowing. Therefore, only the qualitative statement that a small negative current flows in the reverse region is correct. The shape of the current curve as shown in Fig. 1.3 can only be described with the help of additional equations (see Sect. 1.3).

$V_D \gg nV_T \approx 26 \ldots 52$ mV applies to the forward region and the approximation

$$I_D = I_S\, e^{\frac{V_D}{nV_T}} \tag{1.2}$$

can be used. Then the voltage is:

$$V_D = nV_T \ln \frac{I_D}{I_S} = nV_T \ln 10 \cdot \log \frac{I_D}{I_S} \approx 60 \ldots 120\,\text{mV} \cdot \log \frac{I_D}{I_S}$$

This means that the voltage increases by $60 \ldots 120$ mV when the current rises by a factor of 10. With high currents the voltage drop $I_D R_B$ at the spreading resistance R_B must be taken into account, which occurs in addition to the voltage at the pn junction:

$$V_D = nV_T \ln \frac{I_D}{I_S} + I_D R_B$$

In this case it cannot be described in the form $I_D = I_D(V_D)$.

For simple calculations the diode can be regarded as a switch that is opened in the reverse region and is closed in the forward region. Given the assumption that the voltage is approximately constant in the forward region and that no current flows in the reverse region, the diode can be replaced by an ideal voltage-controlled switch and a voltage source with the forward voltage V_F (see Fig. 1.5a). Figure 1.5b shows the characteristic curve of this equivalent circuit which consists of two straight lines:

$$
\begin{aligned}
I_D &= 0 & \text{for } V_D < V_F &\quad \rightarrow \text{switch open (a)} \\
V_D &= V_F & \text{for } I_D > 0 &\quad \rightarrow \text{switch closed (b)}
\end{aligned}
$$

When the additional spreading resistance R_B is taken into consideration, we have:

$$
I_D = \begin{cases}
0 & \text{for } V_D < V_F \quad \rightarrow \text{switch open (a)} \\[2mm]
\dfrac{V_D - V_F}{R_B} & \text{for } V_D \geq V_F \quad \rightarrow \text{switch closed (b)}
\end{cases}
$$

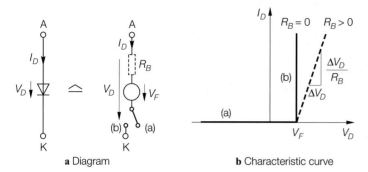

a Diagram   **b** Characteristic curve

Fig. 1.5. Simple equivalent circuit diagram for a diode without (-) and with (- -) spreading resistance

The voltage V_F is $V_F \approx 0.6\,\text{V}$ for silicon pn diodes and $V_F \approx 0.3\,\text{V}$ for Schottky diodes. The corresponding circuit diagram and characteristic curve are shown in Fig. 1.5 as dashed lines. Different cases must be distinguished for both variations, that is, it is necessary to calculate with the switch open *and* closed and to determine the situation in which there is no contradiction. The advantage is that either case leads to linear equations which are easy to solve. In contrast, when using the e function according to (1.1), it is necessary to cope with an implicit nonlinear equation that can only be solved numerically.

Example: Figure 1.6 shows a diode in a bridge circuit. To calculate the voltages V_1 and V_2 and the diode voltage $V_D = V_1 - V_2$ it is assumed that the diode is in the reverse state, that is, $V_D < V_F = 0.6\,\text{V}$ and the switch in the equivalent circuit is open. In this case, V_1 and V_2 can be determined by the voltage divider formula $V_1 = V_b R_2/(R_1 + R_2) = 3.75\,\text{V}$ and $V_2 = V_b R_4/(R_3 + R_4) = 2.5\,\text{V}$. This results in $V_D = 1.25\,\text{V}$, which does not comply with the assumption. Consequently the diode is conductive and the switch in the equivalent circuit is closed; this leads to $V_D = V_F = 0.6\,\text{V}$ and $I_D > 0$. From the nodal equations

$$\frac{V_1}{R_2} + I_D = \frac{V_b - V_1}{R_1} \quad , \quad \frac{V_2}{R_4} = I_D + \frac{V_b - V_2}{R_3}$$

it is possible to eliminate the unknown elements I_D and V_1 by adding the equations and inserting $V_1 = V_2 + V_F$; this leads to:

$$V_2 \left(\frac{1}{R_1} + \frac{1}{R_2} + \frac{1}{R_3} + \frac{1}{R_4} \right) = V_b \left(\frac{1}{R_1} + \frac{1}{R_3} \right) - V_F \left(\frac{1}{R_1} + \frac{1}{R_2} \right)$$

This results in $V_2 = 2.76\,\text{V}$, $V_1 = V_2 + V_F = 3.36\,\text{V}$ and in $I_D = 0.52\,\text{mA}$ by substitution in one of the nodal equations. The initial condition $I_D > 0$ has been fulfilled, that is, there is no contradiction and the solution has been found.

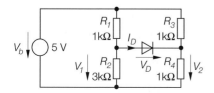

Fig. 1.6. Example for the demonstration of the use of the equivalent circuit of Fig. 1.5

1.1.3
Switching Performance

In many applications the diodes operate alternately in the forward mode and in the reverse mode, for example when rectifying alternating currents. The transition does not follow the static characteristic curve as the parasitic capacitance of the diode stores a charge that builds up in the forward state and is discharged in the reverse state. Figure 1.7 shows a circuit for determining the *switching performance* with an ohmic load ($L = 0$) or an ohmic-inductive load ($L > 0$). Applying a square wave produces the transitions shown in Fig. 1.8.

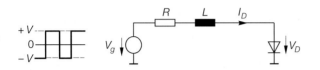

Fig. 1.7. Circuit for determining the switching performance

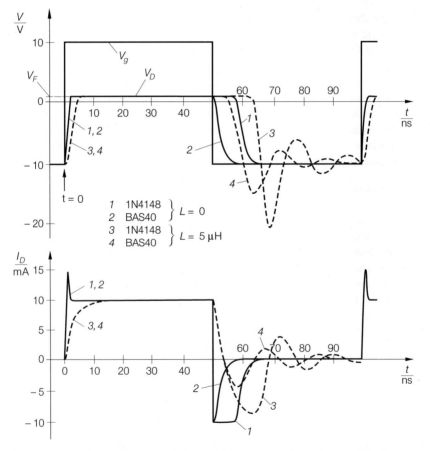

Fig. 1.8. Switching performance of the silicon diode 1N4148 and the Schottky diode BAS40 in the measuring circuit of Fig. 1.7 with $V = 10\,V$, $f = 10\,MHz$, $R = 1\,k\Omega$ and $L = 0$ or $L = 5\,\mu H$

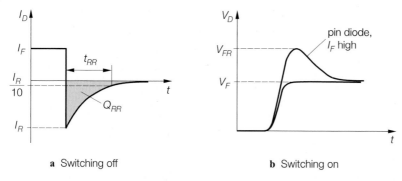

a Switching off   **b** Switching on

Fig. 1.9. Illustration of switching performance

Switching performance with ohmic load: With an ohmic load ($L = 0$) a current peak caused by the charge built up in the capacitance of the diode occurs when the circuit is activated. The voltage rises during this current peak from the previously existing reverse voltage to the forward voltage V_F which terminates the switch-on process. In pin diodes[1] higher currents may cause a voltage overshoot (see Fig. 1.9b) as these diodes initially have a higher spreading resistance R_B at the switch-on point. Subsequently the voltage declines to the static value in accordance with the decrease of R_B. When switching off there is a current in the opposite direction until the capacitance is discharged; then the current returns to zero and the voltage drops to the reverse voltage. Since the capacitance of Schottky diodes is much lower than that of silicon diodes of the same size, their turn-off time is significantly shorter (see Fig. 1.8). Therefore, Schottky diodes are preferred for rectifier diodes in switched power supplies with high cycle rates ($f > 20\,\text{kHz}$), while the lower priced silicon diodes are used in rectifiers for the mains voltage ($f = 50\,\text{Hz}$). When the frequency becomes so high that the capacitance discharge process is not completed before the next conducting state starts, the rectification no longer takes place.

Switching performance with ohmic-inductive load: With an ohmic-inductive load ($L > 0$) the transition to the conductive state takes longer since the increase in current is limited by the inductivity; no current peaks occur. While the voltage rises relatively fast to the forward voltage, the current increases with the time constant $T = L/R$ of the load. During switch-off the current first decreases with the time constant of the load until the diode cuts off. Then, the load and the capacitance of the diode form a series resonant circuit, and the current and the voltage perform damped oscillations. As shown in Fig. 1.8 high reverse voltages may arise which are much higher than the static reverse voltage and consequently require a high diode breakthrough voltage.

Figure 1.9 shows the typical data for *reverse recovery (RR)* and *forward recovery (FR)*. The *reverse recovery time* t_{RR} is the period measured from the moment at which the current passes through zero until the moment at which the reverse current drops to 10 %[2] of its maximum value I_R. Typical values range from $t_{RR} < 100\,\text{ps}$ for fast Schottky diodes to $t_{RR} = 1\ldots 20\,\text{ns}$ for small-signal silicon diodes or $t_{RR} > 1\,\mu\text{s}$ for rectifier diodes. The *reverse recovery charge* Q_{RR} transported during the capacitance discharge corresponds to

---

[1] pin diodes have a nondoped (*intrinsic*) or slightly doped layer between the p and n layers in order to achieve a higher breakthrough voltage.

[2] With rectifier diodes the measurement is sometimes taken at 25 %.

the area below the x axis (see Fig. 1.9a). Both parameters depend on the previously flowing forward current I_F and the cutoff speed; therefore the data sheets show either information on the measuring conditions or the measuring circuit. An approximation is $Q_{RR} \sim I_F$ and $Q_{RR} \sim |I_R| t_{RR}$ [1.2]; this means that in a first approximation the reverse recovery time is proportional to the ratio of the forward and reverse current: $t_{RR} \sim I_F/|I_R|$. However, this approximation only applies to $|I_R| < 3 \ldots 5 \cdot I_F$, in other words, t_{RR} can not be reduced endlessly. In pin diodes featuring a high breakdown voltage, the high cutoff speed may even cause the breakdown to occur far below the static breakdown voltage V_{BR} if the reverse voltage at the diode increases sharply before the weakly doped i-layer is free of charge carriers. With the transition to the forward state the *forward recovery voltage* V_{FR} occurs, which also depends on the actual switching conditions [1.3]; data sheets quote a maximum value for V_{FR}, typically $V_{FR} = 1 \ldots 2.5 \, \text{V}$.

1.1.4
Small-Signal Response

The performance of the diode when controlled by *small* signals around an operating point characterized by $V_{D,A}$ and $I_{D,A}$ is called the *small-signal response*. In this case, the nonlinear characteristic given in (1.1) can be replaced by a tangent to the operating point; with the small-signal parameters

$$i_D = I_D - I_{D,A} \quad , \quad v_D = V_D - V_{D,A}$$

one arrives at:

$$i_D = \left. \frac{d I_D}{d V_D} \right|_A v_D = \frac{1}{r_D} v_D$$

From this the *differential resistance* r_D of the diode is derived:

$$r_D = \left. \frac{d V_D}{d I_D} \right|_A = \frac{n V_T}{I_{D,A} + I_S} \overset{I_{D,A} \gg I_S}{\approx} \frac{n V_T}{I_{D,A}} \tag{1.3}$$

Thus, the equivalent small-signal circuit for the diode consists of a resistance with the value r_D; with large currents r_D becomes very small and an additional spreading resistance R_B must be introduced (see Fig. 1.10).

The equivalent circuit shown in Fig. 1.10 is only suitable for calculating the small-signal response at low frequencies $(0 \ldots 10 \, \text{kHz})$; therefore, it is called the *DC small-signal equivalent circuit*. For higher frequencies it is necessary to use the AC small-signal equivalent circuit given in Sect. 1.3.3.

Fig. 1.10. Small-signal equivalent circuit of a diode

1.1.5
Limit Values and Reverse Currents

The data sheet for a diode shows limit values that must not be exceeded. These are the limit voltages, limit currents and maximum power dissipation. In order to deal with positive values for the limit data the reference arrows for the current and the voltage are reversed in their direction for reverse-biased operation and the relevant values are given with the index R (*reverse*); the index F (*forward*) is used for forward-biased operation.

Limit Voltages

Reaching the *breakthrough voltage* $V_{(BR)}$ or V_{BR} causes the diode to break through in the reverse mode and the reverse current rises sharply. Since the current already increases markedly when approaching the breakthrough voltage, as shown in Fig. 1.3, a *maximum reverse voltage* $V_{R,max}$ is specified up to which the reverse current remains below a limit value in the µA range. Higher reverse voltages are permissible when driving the diode with a pulse chain or a single pulse; they are called the *repetitive peak reverse voltage* V_{RRM} and the *peak surge reverse voltage* V_{RSM}, respectively, and they are chosen so that the diode remains undamaged. The pulse frequency is considered to be $f = 50\,\text{Hz}$ since it is assumed that it will be used as a mains rectifier. Due to the reversed direction of the reference arrow all voltages are positive and are related in the following way:

$$V_{R,max} < V_{RRM} < V_{RSM} < V_{(BR)}$$

Limit Currents

For forward-biased operation a *maximum steady-state forward current* $I_{F,max}$ is specified. It applies to situations in which the diode case is kept at a temperature of $T = 25\,°\text{C}$; at higher temperatures the permissible steady-state current is lower. Higher forward currents are permissible when driving the diode with several pulses or a single pulse; they are called the *repetitive peak forward current* I_{FRM} and the *peak surge forward current* I_{FSM}, respectively, and they depend on the duty cycle or the pulse duration. The currents are related:

$$I_{F,max} < I_{FRM} < I_{FSM}$$

With very short single pulses $I_{FSM} \approx 4 \ldots 20 \cdot I_{F,max}$. The current I_{FRM} is of particular importance for rectifier diodes because of their pulsating periodic current (see Sect. 16.2); in this case the maximum value is much higher than the mean value.

For the breakthrough region a *maximum current-time area* I^2t is quoted which may occur at the breakthrough caused by a pulse:

$$I^2t = \int I_R^2 dt$$

Despite its unit A^2s it is often referred to as the *maximum pulse energy*.

Reverse Current

The *reverse current* I_R is measured at a reverse voltage below the breakthrough voltage and depends largely on the reverse voltage and the temperature of the diode. At room temperature $I_R = 0.01 \ldots 1\,\text{µA}$ for a small-signal silicon diode, $I_R = 1 \ldots 10\,\text{µA}$ for a small-signal Schottky diode and a silicon rectifier diode in the Ampere range and $I_R >$ 10 μA for a Schottky rectifier diode; at a temperature of $T = 150\,°C$ these values are increased by a factor of $20 \ldots 200$.

Maximum Power Dissipation

The power dissipation of the diode is the power converted to heat:

$$P_V = V_D I_D$$

This occurs at the junction or, with large currents, at the spreading resistance R_B. The temperature of the diode increases up to a value at which, due to the temperature gradients, the heat can be dissipated from the junction through the case to the environment. Section 2.1.6 describes this in more detail for bipolar transistors; the same results apply to the diode when P_V is replaced by the power dissipation of the diode. Data sheets specify the *maximum power dissipation* P_{tot} for the situation in which the diode case is kept at a temperature of $T = 25\,°C$; P_{tot} is lower at higher temperatures.

1.1.6
Thermal Performance

The thermal performance of components is described in Sect. 2.1.6 for bipolar transistors; the parameters and conditions described there also apply to the diode when P_V is replaced by the power dissipation of the diode.

1.1.7
Temperature Sensitivity of Diode Parameters

The characteristic curve of a diode is heavily dependent on the temperature; an explicit statement of the temperature sensitivity means for the silicon pn diode [1.1]

$$I_D(V_D, T) = I_S(T) \left(e^{\frac{V_D}{n V_T(T)}} - 1 \right)$$

with:

$$V_T(T) = \frac{kT}{q} = 86.142 \frac{\mu V}{K} T \overset{T=300\,\mathrm{K}}{\approx} 26\,\mathrm{mV}$$

$$I_S(T) = I_S(T_0)\, e^{\left(\frac{T}{T_0}-1\right)\frac{V_G(T)}{n V_T(T)}} \left(\frac{T}{T_0}\right)^{\frac{x_{T,I}}{n}} \quad \text{with } x_{T,I} \approx 3 \qquad (1.4)$$

Here, $k = 1.38 \cdot 10^{-23}$ VAs/K is *Boltzmann's constant*, $q = 1.602 \cdot 10^{-19}$ As is the *elementary charge* and $V_G = 1.12\,V$ is the *gap voltage* of silicon; the low temperature sensitivity of V_G may be ignored. The temperature T_0 with the respective current $I_S(T_0)$ serves as a reference point; usually $T_0 = 300\,K$ is used.

In reverse mode the reverse current $I_R = -I_D \approx I_s$ flows; with $x_{T,I} = 3$ this yields the temperature coefficient of the reverse current:

$$\frac{1}{I_R}\frac{dI_R}{dT} \approx \frac{1}{I_S}\frac{dI_S}{dT} = \frac{1}{nT}\left(3 + \frac{V_G}{V_T}\right)$$

In this region $n \approx 2$ applies to most diodes, resulting in:

$$\frac{1}{I_R}\frac{dI_R}{dT} \approx \frac{1}{2T}\left(3 + \frac{V_G}{V_T}\right) \overset{T=300\,\text{K}}{\approx} 0.08\,\text{K}^{-1}$$

This means that the reverse current doubles with a temperature increase of 9 K and rises by a factor of 10 with a temperature increase of 30 K. In practice there are often lower temperature coefficients; this is caused by surface and leakage currents which are often higher than the reverse current of the pn junction and have a different temperature response.

The temperature coefficient of the current at constant voltage in forward-bias operation is calculated by differentiation of $I_D(V_D, T)$:

$$\frac{1}{I_D}\frac{dI_D}{dT}\bigg|_{V_D=\text{const.}} = \frac{1}{nT}\left(3 + \frac{V_G - V_D}{V_T}\right) \overset{T=300\,\text{K}}{\approx} 0.04\ldots0.08\,\text{K}^{-1}$$

By means of the total differential

$$dI_D = \frac{\partial I_D}{\partial V_D}\,dV_D + \frac{\partial I_D}{\partial T}\,dT = 0$$

the temperature-induced change of V_D at constant current can be determined:

$$\frac{dV_D}{dT}\bigg|_{I_D=\text{const.}} = \frac{V_D - V_G - 3V_T}{T} \overset{\substack{T=300\,\text{K}\\V_D=0.7\,\text{V}}}{\approx} -1.7\,\frac{\text{mV}}{\text{K}} \tag{1.5}$$

This means that the forward voltage decreases when the temperature rises; a temperature increase of 60 K causes a drop in V_D of approximately 100 mV. This effect is used in integrated circuits for measuring the temperature.

These results also apply to Schottky diodes when setting $x_{T,I} \approx 2$ and replacing the gap voltage V_G by the voltage that describes the energy difference between the n and metal regions: $V_{Mn} = (W_{Metal} - W_{n\text{-}Si})/q$; thus $V_{Mn} \approx 0.7\ldots0.8\,\text{V}$ [1.1].

1.2
Construction of a Diode

Diodes are manufactured in a multi-step process on a semiconductor *wafer* that is then cut into small *dies*. On one chip there is either a discrete diode or an *integrated circuit (IC)*, comprising several components.

1.2.1
Discrete Diode

Internal design: Discrete diodes are mostly produced using epitaxial-planar technology. Figure 1.11 illustrates the construction of a pn and a Schottky diode where the active areas are particularly emphasized. Doping is heavy in the n^+ layer, medium in the p layer and low in the n^- layer. The special arrangement of differently doped layers helps to minimize the spreading resistance and to increase the breakthrough voltage. Almost all pn diodes are designed as *pin diodes*, in other words, they feature a middle layer with little or no doping

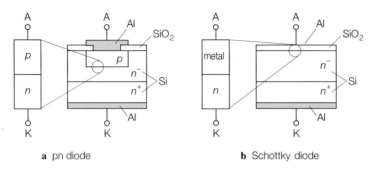

a pn diode   **b** Schottky diode

Fig. 1.11. Construction of a semiconductor chip with one diode

and with a thickness that is roughly proportional to the breakthrough voltage; in Fig. 1.11a this is the n^- layer. For practical purposes diodes are referred to as *pin diodes* only if the lifetime of the charge carriers in the middle layer is very high, thus producing a particular characteristic; this will be described in more detail in Sect. 1.4.2. In Schottky diodes the weakly doped n^- layer is required for the Schottky contact (see Fig. 1.11b); in contrast a junction between metal and a layer of medium or heavy doping produces an inferior diode effect or no effect at all, in which case it behaves rather like a resistor (*ohmic contact*).

Case: To mount a diode in a case the bottom side is soldered to the cathode terminal or connected to a metal part of the case. The anode side is connected to the anode terminal via a fine gold or aluminum *bond wire*. Finally the diode is sealed in a plastic compound or mounted in a metal case with screw connector.

For the various diode sizes and applications there is a multitude of case designs that differ in the maximum heat dissipation capacity or are adapted to special geometrical requirements. Figure 1.12 shows a selection of common models. Power diodes are provided

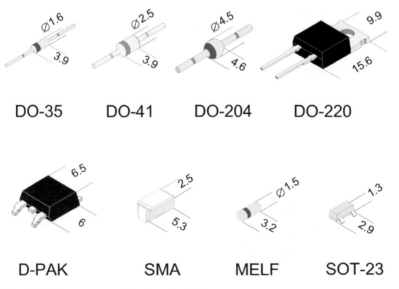

Fig. 1.12. Common cases for discrete diodes

with a heat sink for their installation; the larger the contact surface, the better the heat dissipation. Rectifier diodes are often designed as *bridge rectifiers* consisting of four diodes to serve as full-wave rectifiers in power supply units (see Sect. 1.4.4); the *mixer* described in Sect. 1.4.5 is also made of four diodes. High-frequency diodes require special cases because in the GHz frequency range their electrical performance depends on the case geometry. Often, the case is omitted altogether and the diode chip is soldered or bonded directly to the circuit.

1.2.2
Integrated Diode

Integrated diodes are also produced using epitaxial-planar technology. Here, all connections are located at the top of the chip and the diode is electrically isolated from other components by a reverse-biased pn junction. The active region is located in a very thin layer at the surface. The depth of the chip is called the *substrate* (*S*) and forms a common connection for all components of the integrated circuit.

Internal construction: Figure 1.13 illustrates the design of an integrated pn diode. The current flows from the *p* layer through the pn junction to the n^- layer and from there via the n^+ layer to the cathode; a low spreading resistance is achieved by means of the heavily doped n^+ layer.

Substrate diode: The equivalent circuit diagram in Fig. 1.13 shows an additional substrate diode located between the cathode and the substrate. The substrate is connected to the negative supply voltage so that this diode is always in the reverse mode to act as isolation relative to other components and the substrate.

Differences between integrated pn and Schottky diodes: In principle an integrated Schottky diode can be built like an integrated pn diode by simply omitting the *p* junction at the anode connection. However, for practical applications this is not so easy as different metals must be used for the Schottky diodes and for the component wiring, and in most manufacturing processes for integrated circuits the necessary steps are not intended.

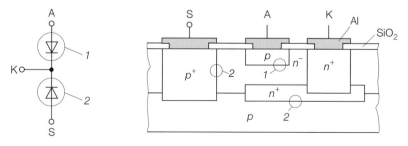

Fig. 1.13. Equivalent circuit and construction of an integrated pn diode with useful diode (1) and parasitic substrate diode (2)

1.3
Model of a Diode

Section 1.1.2 describes the static performance of the diode using an exponential function; but this neglects the breakthrough and the second-order effects in the forward operation. For computer-aided circuit design a model is required that considers all of these effects and, in addition, correctly reflects the *dynamic* performance. The *dynamic small-signal model* is derived from this *large-signal model* by linearization.

1.3.1
Static Performance

The description is based on the ideal diode equation given in (1.1) and also takes other effects into account. A standardized diode model like the the Gummel-Poon model for bipolar transistors does not exist; some of the CAD programs therefore have to use several diode models to describe a real diode with all of its current components. The diode model is almost unnecessary for the design of integrated circuits since here the base-emitter diode of a bipolar transistor is usually used as a diode.

Range of Medium Forward Currents

In pn diodes the *diffusion current* I_{DD} dominates in the range of medium forward currents; this follows from the ideal diode theory and can be described according to (1.1):

$$I_{DD} = I_S \left(e^{\frac{V_D}{n V_T}} - 1 \right) \tag{1.6}$$

The model parameters are the *saturation reverse current* I_S and the *emission coefficient* n. For the ideal diode $n = 1$; for real diodes $n \approx 1 \ldots 2$. This range is called the *diffusion range*.

   In Schottky diodes the emission current takes the place of the diffusion current. But since both current conducting mechanisms lead to the same characteristic curve (1.6) can also be used for Schottky diodes [1.1, 1.3].

Other Effects

With very small and very high forward currents as well as in reverse operation there are deviations from the *ideal* performance according to (1.6):

- High forward currents produce the *high-current effect,* which is caused by a sharp rise in the charge carrier concentration at the edge of the depletion layer [1.1]; this is also referred to as a *strong injection*. This also affects the diffusion current and is described by an extension to (1.6).
- Because of the recombination of charge carriers in the depletion layer a *leakage* or *recombination current* I_{DR} occurs in addition to the diffusion current which is described by a separate equation [1.1].
- The application of high reverse voltages causes the diode to break through. The *breakthrough current* I_{DBR} is also described in an additional equation.

The current I_D thus comprises three partial currents:

$$I_D = I_{DD} + I_{DR} + I_{DBR} \tag{1.7}$$

High-current effect: The high-current effect causes the emission coefficient to rise from n in the medium current range to $2n$ for $I_D \to \infty$; it can be described by an extension to (1.6) [1.4]:

$$I_{DD} = \frac{I_S\left(e^{\frac{V_D}{nV_T}} - 1\right)}{\sqrt{1 + \frac{I_S}{I_K}\left(e^{\frac{V_D}{nV_T}} - 1\right)}} \approx \begin{cases} I_S\, e^{\frac{V_D}{nV_T}} & \text{for } I_S\, e^{\frac{V_D}{nV_T}} < I_K \\ \sqrt{I_S I_K}\; e^{\frac{V_D}{2nV_T}} & \text{for } I_S\, e^{\frac{V_D}{nV_T}} > I_K \end{cases} \tag{1.8}$$

An additional parameter is the *knee-point current* I_K, which marks the beginning of the *high-current region*.

Leakage current: Based on the ideal diode theory the following is applicable to the leakage current [1.1]:

$$I_{DR} = I_{S,R}\left(e^{\frac{V_D}{n_R V_T}} - 1\right)$$

This equation only describes the recombination current accurately enough for forward operation. Setting $V_D \to -\infty$ yields a constant current $I_{DR} = -I_{S,R}$ in the reverse region, while in a real diode the recombination current rises with an increasing reverse voltage. A more accurate description is achieved by taking into account the voltage sensitivity of the width of the depletion layer [1.4]:

$$I_{DR} = I_{S,R}\left(e^{\frac{V_D}{n_R V_T}} - 1\right)\left(\left(1 - \frac{V_D}{V_{Diff}}\right)^2 + 0.005\right)^{\frac{m_J}{2}} \tag{1.9}$$

Additional parameters are the *leakage saturation reverse current* $I_{S,R}$, the *emission coefficient* $n_R \geq 2$, the *diffusion voltage* $V_{Diff} \approx 0.5 \ldots 1\,\text{V}$ and the *capacitance coefficient* $m_J \approx 1/3 \ldots 1/2$.[3] From (1.9) it follows that:

$$I_{DR} \approx -I_{S,R}\left(\frac{|V_D|}{V_{Diff}}\right)^{m_J} \qquad \text{for } V_D < -V_{Diff}$$

The magnitude of the current rises as the reverse voltage increases; its actual curve depends on the capacitance coefficient m_J. In the forward mode the additional factor given in (1.9) has almost no effect since in this case the exponential dependence of V_D is dominant.

Since $I_{S,R} \gg I_S$, the recombination current is larger than the diffusion current at low positive voltages; this region is called the *recombination region*. For

$$V_{D,RD} = V_T\, \frac{n n_R}{n_R - n}\, \ln \frac{I_{S,R}}{I_S}$$

both currents have the same value. With larger voltages the diffusion current becomes dominant and the diode operates in the diffusion region.

---

[3] V_{Diff} and m_J are primarily used to describe the depletion layer capacitance of the diode (see Sect. 1.3.2).

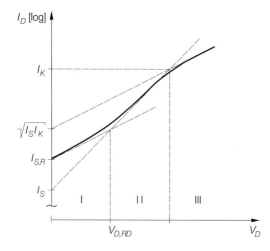

Fig. 1.14. Semi-logarithmic diagram of I_D in forward mode: (I) recombination, (II) diffusion, (III) high-current regions

Figure 1.14 is the semilogarithmic presentation of I_D in the forward region and shows the importance of parameters I_S, $I_{S,R}$ and I_K. In some diodes the emission coefficients n and n_R are almost identical. In such cases the semilogarithmic characteristic curve has the same slope in the recombination and diffusion regions and can be described for both regions using *one* exponential function.[4]

Breakthrough: For $V_D < -V_{BR}$ the diode breaks through; the flowing current can be approximated by an exponential function [1.5]:

$$I_{DBR} = -I_{BR}\, e^{-\frac{V_D+V_{BR}}{n_{BR}V_T}} \tag{1.10}$$

For this, the *breakthrough voltage* $V_{BR} \approx 50\ldots1000\,\mathrm{V}$, the *breakthrough knee-point current* I_{BR} and the *breakthrough emission coefficient* $n_{BR} \approx 1$ are required. For $n_{BR} = 1$ and $V_T \approx 26\,\mathrm{mV}$ the current is:[5]

$$I_D \approx I_{DBR} = \begin{cases} -I_{BR} & \text{for } V_D = -V_{BR} \\ -10^{10} I_{BR} & \text{for } V_D = -V_{BR} - 0.6\,\mathrm{V} \end{cases}$$

Quoting I_{BR} and V_{BR} is not a clear definition since the same curve can be described with different value sets (V_{BR}, I_{BR}); therefore, the model for a certain diode may have different parameters.

Spreading Resistance

The spreading resistance R_B is necessary for the full description of the static performance; according to Fig. 1.15 it is comprized of the resistances of the various layers and it is represented in the model by a series resistor. A distinction has to be made between the *internal diode voltage* V_D' and the *external diode voltage*

$$V_D = V_D' + I_D R_B . \tag{1.11}$$

In the equations for I_{DD}, I_{DR} and I_{DBR} voltage V_D must be replaced by V_D'. The spreading resistance is between $0.01\,\Omega$ for power diodes and $10\,\Omega$ for small-signal diodes.

---

[4] Figure 1.4 shows the characteristic curve of such a diode.
[5] Based on $10V_T \ln 10 = 0.6\,\mathrm{V}$.

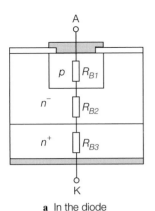

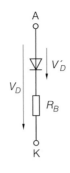

a In the diode   **b** In the model

Fig. 1.15. Spreading resistance of a diode

1.3.2
Dynamic Performance

The response to pulsating or sinusoidal signals is called the *dynamic performance*, and it cannot be derived from the characteristic curves. The reasons for this are the nonlinear *junction capacitance* of the pn or metal-semiconductor junction and the *diffusion charge* that is stored in the pn junction and determined by the *diffusion capacitance*, which is also nonlinear.

Junction Capacitance

A pn or metal–semiconductor junction has a voltage-dependent *junction capacitance* C_J that is influenced by the doping of the adjacent layers, the doping profile, the area of the junction and the applied voltage V'_D. The junction can be visualized as a plate capacitor with the capacitance $C = \epsilon A / d$; where A represents the junction area and d the junction width. A simplified view of the pn junction gives $d(V) \sim (1 - V/V_{Diff})^{m_J}$ [1.1] and thus:

$$C_J(V'_D) = \frac{C_{J0}}{\left(1 - \dfrac{V'_D}{V_{Diff}}\right)^{m_J}} \qquad \text{for } V'_D < V_{Diff} \tag{1.12}$$

The parameters are the *zero capacitance* $C_{J0} = C_J(V'_D = 0)$, the *diffusion voltage* $V_{Diff} \approx 0.5 \ldots 1\,\text{V}$ and the *capacitance coefficient* $m_J \approx 1/3 \ldots 1/2$ [1.2].

For $V'_D \to V_{Diff}$ the assumptions leading to (1.12) are no longer met. Therefore, the curve for $V'_D > f_C V_{Diff}$ is replaced by a straight line [1.5]:

$$C_J(V'_D) = C_{J0} \begin{cases} \dfrac{1}{\left(1 - \dfrac{V'_D}{V_{Diff}}\right)^{m_J}} & \text{for } V'_D \le f_C V_{Diff} \\[4ex] \dfrac{1 - f_C(1 + m_J) + \dfrac{m_J V'_D}{V_{Diff}}}{(1 - f_C)^{(1+m_J)}} & \text{for } V'_D > f_C V_{Diff} \end{cases} \tag{1.13}$$

where $f_C \approx 0.4 \ldots 0.7$. Figure 2.32 on page 70 shows the curve of C_J for $m_J = 1/2$ and $m_J = 1/3$.

Diffusion Capacitance

In forward operation the pn junction contains a stored diffusion charge Q_D that is proportional to the diffusion current flowing through the pn junction [1.2]:

$$Q_D = \tau_T I_{DD}$$

The parameter τ_T is the *transit time*. Differentiation of (1.8) produces the *diffusion capacitance*:

$$C_{D,D}(V_D') = \frac{dQ_D}{dV_D'} = \frac{\tau_T I_{DD}}{n V_T} \frac{1 + \dfrac{I_S}{2 I_K} e^{\frac{V_D'}{n V_T}}}{1 + \dfrac{I_S}{I_K} e^{\frac{V_D'}{n V_T}}} \tag{1.14}$$

For the diffusion region $I_{DD} \gg I_{DR}$ and thus $I_D \approx I_{DD}$, meaning that the diffusion capacitance can be approximated by:

$$C_{D,D} \approx \frac{\tau_T I_D}{n V_T} \frac{1 + \dfrac{I_D}{2 I_K}}{1 + \dfrac{I_D}{I_K}} \overset{I_D \ll I_K}{\approx} \frac{\tau_T I_D}{n V_T} \tag{1.15}$$

In silicon pn diodes $\tau_T \approx 1 \ldots 100$ ns; in Schottky diodes the diffusion charge is negligible, since $\tau_T \approx 10 \ldots 100$ ps.

Complete Model of a Diode

Figure 1.16 shows the complete model of a diode; it is used in CAD programs for circuit simulation. The diode symbols in the model represent the diffusion current I_{DD} and the recombination current I_{DR}; the breakthrough current I_{DBR} is shown as a controlled current source. Figure 1.17 contains the variables and equations. The parameters are listed in Fig. 1.18; in addition the parameter designations used in the circuit simulator *PSpice*[6] are shown. Figure 1.19 indicates the parameter values of some selected diodes taken from the component library of *PSpice*. Parameters not specified are treated differently by *PSpice*:

- A standard value is used:
  $I_S = 10^{-14}$ A, $n = 1$, $n_R = 2$, $I_{BR} = 10^{-10}$ A, $n_{BR} = 1$, $x_{T,I} = 3$, $f_C = 0.5$, $V_{Diff} = 1$ V, $m_J = 0.5$
- The parameter is set to zero: $I_{S,R}$, R_B, C_{J0}, τ_T
- The parameter is set to infinity: I_K, V_{BR}

As a consequence of the values zero and infinite the respective effects are removed from the model [1.4].

---

[6] *PSpice* is an *OrCAD* product.

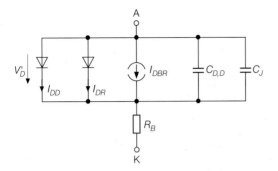

Variable	Designation	Equation
I_{DD}	Diffusion current	(1.8)
I_{DR}	Recombination current	(1.9)
I_{DBR}	Breakthrough current	(1.10)
R_B	Spreading resistance	
C_J	Junction capacitance	(1.13)
$C_{D,D}$	Diffusion capacitance	(1.14)

Fig. 1.16. Full model of a diode

Fig. 1.17. Variables of the diode model

Parameter	PSpice	Designation
Static performance		
I_S	IS	Saturation reverse current
n	N	Emission coefficient
$I_{S,R}$	ISR	Leakage saturation reverse current
n_R	NR	Emission coefficient
I_K	IK	Knee-point current for strong injection
I_{BR}	IBV	Breakthrough knee-point current
n_{BR}	NBV	Emission coefficient
V_{BR}	BV	Breakthrough voltage
R_B	RS	Spreading resistance
Dynamic performance		
C_{J0}	CJO	Zero capacitance of the depletion layer
V_{Diff}	VJ	Diffusion voltage
m_J	M	Capacitance coefficient
f_C	FC	Coefficient for the variation of the capacitance
τ_T	TT	Transit time
Thermal performance		
$x_{T,I}$	XTI	Temperature coefficient of reverse currents account to (1.14)

Fig. 1.18. Parameters in the diode model [1.4]

Parameter	PSpice	1N4148	1N4001	BAS40	Unit
I_S	IS	2.68	14.1	0	nA
n	N	1.84	1.98	1	
$I_{S,R}$	ISR	1.57	0	254	fA
n_R	NR	2	2	2	
I_K	IK	0.041	94.8	0.01	A
I_{BR}	IBV	100	10	10	μA
n_{BR}	NBV	1	1	1	
V_{BR}	BV	100	75	40	V
R_B	RS	0.6	0.034	0.1	Ω
C_{J0}	CJO	4	25.9	4	pF
V_{Diff}	VJ	0.5	0.325	0.5	V
m_J	M	0.333	0.44	0.333	
f_C	FC	0.5	0.5	0.5	
τ_T	TT	11.5	5700	0.025	ns
$x_{T,I}$	XTI	3	3	2	

1N4148 small-signal diode; 1N4001 rectifier diode; BAS40 Schottky diode

Fig. 1.19. Parameters of some diodes

1.3.3
Small-Signal Model

The linear *small-signal model* is derived from the nonlinear model by linearization at an operating point. The *static small-signal model* describes the small-signal response at low frequencies and is therefore called the *DC small-signal equivalent circuit*. The *dynamic small-signal model* also describes the dynamic small-signal response and is required for calculating the frequency response of a circuit; it is called the *AC small-signal equivalent circuit*.

Static Small-Signal Model

Linearization of the static characteristic curve given in (1.11) leads to the small-signal resistance:

$$\left.\frac{dV_D}{dI_D}\right|_A = \left.\frac{dV_D'}{I_D}\right|_A + R_B = r_D + R_B$$

It is made up of the spreading resistance R_B and the *differential resistance* r_D of the inner diode (see Fig. 1.10). Resistance r_D comprises three portions corresponding to the three current components I_{DD}, I_{DR} and I_{DBR}:

$$\frac{1}{r_D} = \left.\frac{dI_D}{dV_D'}\right|_A = \left.\frac{dI_{DD}}{dV_D'}\right|_A + \left.\frac{dI_{DR}}{dV_D'}\right|_A + \left.\frac{dI_{DBR}}{dV_D'}\right|_A$$

The differentiation of (1.6), (1.9) and (1.10) produces complex expressions; for practical purposes the following approximations may be used:

$$\frac{1}{r_{DD}} = \left.\frac{dI_{DD}}{dV_D'}\right|_A \approx \frac{I_{DD,A} + I_S}{nV_T} \frac{1 + \dfrac{I_{DD,A}}{2I_K}}{1 + \dfrac{I_{DD,A}}{I_K}} \overset{I_S \ll I_{DD,A} \ll I_K}{\approx} \frac{I_{DD,A}}{nV_T}$$

$$\frac{1}{r_{DR}} = \left.\frac{dI_{DR}}{dV_D'}\right|_A \approx \begin{cases} \dfrac{I_{DR,A} + I_{S,R}}{n_R V_T} & \text{for } I_{DR,A} > 0 \\[3ex] \dfrac{I_{S,R}}{m_J V_{Diff}^{m_J} |V_{D,A}'|^{1-m_J}} & \text{for } I_{DR,A} < 0 \end{cases}$$

$$\frac{1}{r_{DBR}} = \left.\frac{dI_{DBR}}{dV_D'}\right|_A = -\frac{I_{DBR,A}}{n_{BR}V_T}$$

Thus, the differential resistance r_D is:

$$r_D = r_{DD} \| r_{DR} \| r_{DBR}$$

For operating points that are in the diffusion region and below the high-current region $I_{D,A} \approx I_{DD,A}$ and $I_{D,A} < I_K$;[7] the following approximation can be used:

$$r_D = r_{DD} \approx \frac{nV_T}{I_{D,A}}. \tag{1.16}$$

---

[7] This region is also called the *range of medium forward currents.*

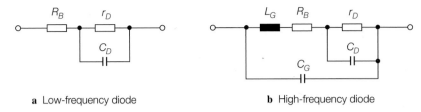

a Low-frequency diode   **b** High-frequency diode

Fig. 1.20. Dynamic small-signal model

This equation corresponds to (1.3) in Sect. 1.1.4. As an approximation it may be used for all operating points in forward mode; in the high-current and recombination regions it provides values that are too low by a factor of $1 \ldots 2$. Setting $n = 1 \ldots 2$ results in:

$$I_{D,A} = 1 \left\{ \begin{array}{c} \mu A \\ mA \\ A \end{array} \right\} \quad \overset{V_T = 26\,\text{mV}}{\Longrightarrow} \quad r_D = 26 \ldots 52 \left\{ \begin{array}{c} k\Omega \\ \Omega \\ m\Omega \end{array} \right\}$$

With small-signal diodes in reverse mode the diffusion resistance is $r_D \approx 10^6 \ldots 10^9\ \Omega$; in the Ampere region of rectifier diodes this value is reduced by a factor of $10 \ldots 100$.

The small-signal resistance in the breakthrough region is required only for Zener diodes since only in Zener diodes an operating point in the breakthrough range is permissible; the resistance is therefore called r_z. For $I_{D,A} \approx I_{DBR,A}$ its value is:

$$r_Z = r_{DBR} = \frac{n_{BR} V_T}{|I_{D,A}|} \tag{1.17}$$

Dynamic Small-Signal Model

Complete model: From the static small-signal model as shown in Fig. 1.10 the dynamic small-signal model according to Fig. 1.20a is derived by adding the junction capacitance and the diffusion capacitance; with reference to Sect. 1.3.2 the following applies:

$$C_D = C_J(V_D') + C_{D,D}(V_D')$$

In high-frequency diodes the additional parasitic influences of the case must be taken into consideration: Figure 1.20b shows the extended model with a case inductivity $L_G \approx 1 \ldots 100\,\text{nH}$ and a case capacitance of $C_G \approx 0.1 \ldots 1$ pF [1.6].

Simplified model: For practical calculations the spreading resistance R_B can be ignored and approximations can be used for r_D and C_D. From (1.15), (1.16) and the estimation $C_J(V_D') \approx 2C_{J0}$ the values for forward operation are:

$$r_D \approx \frac{n V_T}{I_{D,A}} \tag{1.18}$$

$$C_D \approx \frac{\tau_T I_{D,A}}{n V_T} + 2C_{J0} = \frac{\tau_T}{r_D} + 2C_{J0} \tag{1.19}$$

For reverse operation r_D is ignored, that is, $r_D \to \infty$ and $C_D \approx C_{J0}$.

1.4
Special Diodes and Their Application

1.4.1
Zener Diode

A *Zener diode* has a precisely specified breakthrough voltage that is rated for continuous operation in the breakthrough region; it is used for voltage stabilization or limitation. In Zener diodes the breakthrough voltage V_{BR} is called the *Zener voltage* V_Z and amounts to $V_Z \approx 3 \ldots 300\,\text{V}$ in standard Zener diodes. Figure 1.21 shows the graphic symbol and the characteristic for a Zener diode. The current in the breakthrough region is given by (1.10):

$$I_D \approx I_{DBR} = -I_{BR}\, e^{-\frac{V_D+V_Z}{n_{BR} V_T}}$$

The Zener voltage depends on the temperature. The *temperature coefficient*

$$TC = \left. \frac{dV_Z}{dT} \right|_{T=300\,\text{K},\, I_D=\text{const.}}$$

determines the voltage variation at a constant current:

$$V_Z(T) = V_Z(T_0)\,(1 + TC\,(T - T_0)) \quad \text{with } T_0 = 300\,\text{K}$$

When the Zener voltage is below 5 V, the Zener effect dominates with a negative temperature coefficient, while higher voltages produce the avalanche effect with a positive temperature coefficient; typical values are $TC \approx -6 \cdot 10^{-4}\,\text{K}^{-1}$ for $V_Z = 3.3\,\text{V}$, $TC \approx 0$ for $V_Z = 5.1\,\text{V}$ and $TC \approx 10^{-3}\,K^{-1}$ for $V_Z = 47\,\text{V}$.

The differential resistance in the breakthrough region is denoted by r_Z and corresponds to the reciprocal of the slope of the characteristic; from (1.17) it follows that:

$$r_Z = \frac{dV_D}{dI_D} = \frac{n_{BR} V_T}{|I_D|} = -\frac{n_{BR} V_T}{I_D} \approx \frac{\Delta V_D}{\Delta I_D}$$

The differential resistance depends largely on the emission coefficient n_{BR} that reaches a minimum of $n_{BR} \approx 1 \ldots 2$ with $V_Z \approx 8\,\text{V}$ and increases with lower or higher Zener voltages; typical values are $n_{BR} \approx 10 \ldots 20$ for $V_Z = 3.3\,\text{V}$ and $n_{BR} \approx 4 \ldots 8$ for

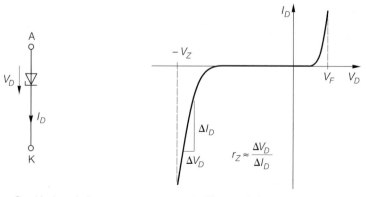

a Graphical symbol   **b** Characteristic

Fig. 1.21. Zener diode

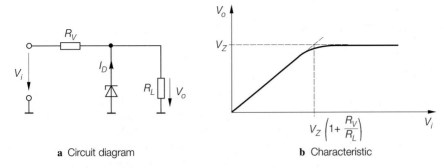

a Circuit diagram   **b** Characteristic

Fig. 1.22. Voltage stabilization with Zener diode

$V_Z = 47\,\text{V}$. The voltage-stabilizing effect of the Zener diode is based on the fact that the characteristic is very steep in the breakthrough region so that the differential resistance is very low; Zener diodes are best suited with $V_Z \approx 8\,\text{V}$ since here the characteristic shows the steepest slope due to the minimum value of n_{BR}. For $|I_D| = 5\,\text{mA}$ the resistance is $r_Z \approx 5\ldots10\,\Omega$ for $V_Z = 8.2\,\text{V}$ and $r_Z \approx 50\ldots100\,\Omega$ for $V_Z = 3.3\,\text{V}$.

Figure 1.22a displays a typical circuit for voltage stabilization. For $0 \le V_o < V_Z$ the Zener diode is reverse-biased and the output voltage is generated by voltage division with resistors R_V and R_L:

$$V_o = V_i \frac{R_L}{R_V + R_L}$$

$V_o \approx V_Z$ applies to the Zener diode in the conductive state. For the characteristic curve shown in Fig. 1.22b this means that:

$$V_o \approx \begin{cases} V_i \dfrac{R_L}{R_V + R_L} & \text{for } V_i < V_Z\left(1 + \dfrac{R_V}{R_L}\right) \\[2ex] V_Z & \text{for } V_i > V_Z\left(1 + \dfrac{R_V}{R_L}\right) \end{cases}$$

In order to render the stabilization effective, the operating point must be in the region in which the characteristic is almost horizontal. From the nodal equation

$$\frac{V_i - V_o}{R_V} + I_D = \frac{V_o}{R_L}$$

differentiation by V_o generates the *smoothing factor*

$$G = \frac{dV_i}{dV_o} = 1 + \frac{R_V}{r_Z} + \frac{R_V}{R_L} \overset{r_Z \ll R_V, R_L}{\approx} \frac{R_V}{r_Z} \tag{1.20}$$

and the *stabilization factor* [1.7]:

$$S = \frac{\dfrac{dV_i}{V_i}}{\dfrac{dV_o}{V_o}} = \frac{V_o}{V_i}\frac{dV_i}{dV_o} = \frac{V_o}{V_i}G \approx \frac{V_o R_V}{V_i r_Z}$$

Example: In a circuit with a supply voltage $V_b = 12\,\text{V} \pm 1\,\text{V}$ a section A is to be provided with the voltage $V_A = 5.1\,\text{V} \pm 10\,\text{mV}$; it requires a current $I_A = 1\,\text{mA}$. One can regard

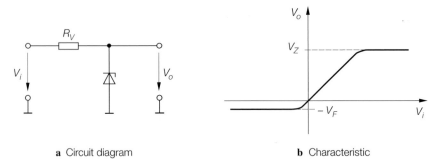

a Circuit diagram   **b** Characteristic

Fig. 1.23. Voltage limitation with Zener diode

this circuit section as a resistor $R_L = V_A/I_A = 5.1\,\mathrm{k\Omega}$ and use the Zener diode circuit in Fig. 1.22 with $V_Z = 5.1\,\mathrm{V}$ if $V_i = V_b$ and $V_o = V_A$. The series resistor R_V must be selected in such a way that $G = dV_i/dV_o > 1\,\mathrm{V}/10\,\mathrm{mV} = 100$; therefore from (1.20) it follows that $R_V \approx Gr_Z \geq 100r_Z$. The nodal equation leads to

$$- I_D = \frac{V_i - V_o}{R_V} - \frac{V_o}{R_L} = \frac{V_b - V_A}{R_V} - I_A$$

and (1.17) leads to $-I_D = n_{BR} V_T/r_Z$; by setting $R_V = Gr_Z$, $G = 100$ and $n_{BR} = 2$ the resistor R_V is:

$$R_V = \frac{V_b - V_A - Gn_{BR}V_T}{I_A} = 1.7\,\mathrm{k\Omega}$$

Then the currents are $I_V = (V_b - V_A)/R_V = 4.06\,\mathrm{mA}$ and $|I_D| = I_V - I_A = 3.06\,\mathrm{mA}$. It can be seen that the Zener diode causes the current to be much higher than the current consumption I_A for the circuit section to be supplied. Therefore, this type of voltage stabilization is suitable only for partial circuits with a low current input. Circuits with a higher current input require a voltage regulator that may be more expensive but, as well as lower power losses, it also offers a better stabilization effect.

The circuit shown in Fig. 1.22a can also be used for voltage limitation. Removing the resistor R_L in Fig. 1.22a leads to the circuit in Fig. 1.23a with the characteristic shown in Fig. 1.23b:

$$V_o \approx \begin{cases} -V_F & \text{for } V_i \leq -V_F \\ V_i & \text{for } -V_F < V_i < V_Z \\ V_Z & \text{for } V_i \geq V_Z \end{cases}$$

In the medium range the diode is reverse-biased, that is, $V_o = V_i$. For $V_i \geq V_Z$ the diode breaks through and limits the output voltage to V_Z. For $V_i \leq -V_F \approx 0.6\,\mathrm{V}$ the diode operates in the forward mode and limits negative voltages to the forward voltage V_F. The circuit in Fig. 1.24a allows a symmetrical limitation with $|V_o| \leq V_Z + V_F$; in the event of limitation one of the diodes is forward-biased and the other breaks through.

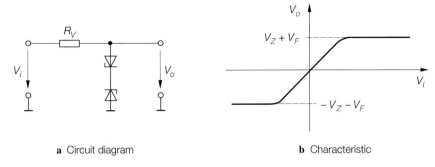

| a Circuit diagram | b Characteristic |

Fig. 1.24. Symmetrical voltage limitation with two Zener diodes

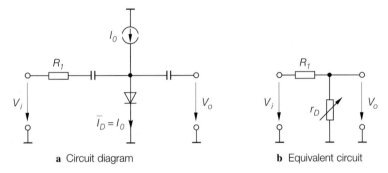

| a Circuit diagram | b Equivalent circuit |

Fig. 1.25. Voltage divider for alternating voltages with pin diode

1.4.2
Pin Diode

In *pin diodes*[8] the life cycle τ of the charge carriers in the nondoped i layer is particularly long. Since a transition from the forward-biased to the reverse-biased mode occurs only after recombination of almost all charge carriers in the i layer, a conductive pin diode remains in the forward mode even with short negative voltage pulses of a pulse duration $t_P \ll \tau$. The diode then acts as an ohmic resistor, with a value that is proportional to the charge in the i layer and thus proportional to the mean current $\overline{I}_{D,pin}$ [1.8]:

$$r_{D,pin} \approx \frac{nV_T}{\overline{I}_{D,pin}} \qquad \text{with } n \approx 1\ldots2$$

On the basis of this property the pin diode may be used with alternating voltages of a frequency $f \gg 1/\tau$ as a *DC-controlled AC resistance*. Figure 1.25 shows the circuit and the small-signal equivalent circuit of a simple variable voltage divider using a pin diode. In high frequency circuits mostly π *attenuators* with three pin diodes are used (see Fig. 1.26); a variable attenuation and a matching of both sides to a certain resistance of usually 50 Ω is then achieved by means of suitable control signals. The capacitances and inductances in Fig. 1.26 result in a separation of the DC and AC circuit paths. Typical pin diodes have $\tau \approx 0.1\ldots5\,\mu s$; this makes the circuit suitable for frequencies $f > 2\ldots100\,\text{MHz} \gg 1/\tau$.

---

[8] Most pn diodes are designed as pin diodes, so that a high reverse voltage is reached across the i layer. The term *pin diode* is used only for diodes with lower impurity concentrations and a correspondingly higher life cycle of the charge carriers in the i layer.

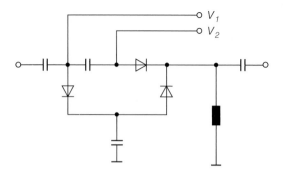

Fig. 1.26. π attenuator with three pin diodes for RF applications

Another important feature of pin diodes is the low junction capacitance due to a relatively thick i layer. This allows pin diodes to be used also for high frequency switches that provide a good off-state attenuation because of the low junction capacitance when the switch is open ($\overline{I}_{D,pin} = 0$). The typical circuit of an RF switch corresponds largely to the attenuator circuit shown in Fig. 1.26 which is designed as a short-series-short-switch with a particularly high off-state attenuation.

1.4.3
Varactor Diodes

Due to the voltage sensitivity of the junction capacitance a diode can be used as a variable capacitor (varactor); in this case the diode is operated in reverse mode and the junction capacitance is controlled by the reverse voltage. Equation (1.12) shows that the region in which the capacitance can be varied depends to a large degree on the capacitance coefficient m_J and increases as m_J increases. A particularly large range of $1 : 3 \ldots 10$ is reached in diodes with *hyperabrupt doping* ($m_J \approx 0.5 \ldots 1$) in which the impurity concentration increases close to the pn border just at the junction to the other region [1.8]. Diodes with this doping profile are called *variable-capacitance diodes* (*tuning diodes, varicap*) and are used predominantly for frequency tuning in LC oscillator circuits. Figure 1.27 shows the graphic symbol of a varactor diode and the curve of the junction capacitance C_J for some typical diodes. Although the curves are similar only diode BB512 shows the particular characteristic of a steeply decreasing junction capacitance. The capacitance coefficient m_J can be derived from the slope in the double logarithmic diagram; therefore Fig. 1.27 also depicts the slopes for $m_J = 0.5$ and $m_J = 1$.

In addition to the curve of the junction capacitance C_J, the quality factor Q is an important measure for the performance of a varactor diode. From the quality definition[9]

$$Q = \frac{|\mathrm{Im}\{Z\}|}{\mathrm{Re}\{Z\}}$$

and the impedance of the diode

$$Z(s) = R_B + \frac{1}{sC_J} \overset{s=j\omega}{=} R_B + \frac{1}{j\omega C_J}$$

---

[9] This quality definition applies to all reactive components.

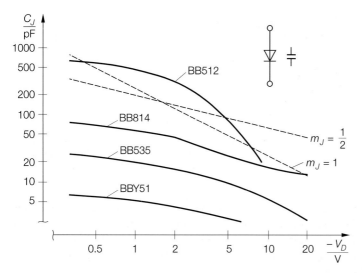

Fig. 1.27. Graphical symbol and capacitance curve of a varactor diode

Q is derived as [1.8]:

$$Q = \frac{1}{\omega C_J R_B}$$

For a given frequency, Q is inversely proportional to the spreading resistance R_B. Therefore, a high performance level is equivalent to a low spreading resistance and corresponds to low losses and a low damping when used in resonant circuits. Typical diodes have a quality factor of $Q \approx 50 \ldots 500$. As it is principally the spreading resistance that is needed for simple calculations and for circuit simulations new data sheets often specify R_B only.

In most cases, the circuits shown in Fig. 1.28 are used for frequency tuning in LC resonant circuits. In the circuit depicted in Fig. 1.28a both the junction capacitance C_J of the diode and the coupling capacitance C_K are connected in series and arranged in parallel with the parallel resonant circuit consisting of L and C. The tuning voltage $V_A > 0$ is provided via the inductivity L_B; with respect to the AC voltage this isolates the resonant circuit from the voltage source V_A and prevents the resonant circuit from being short-circuited by the voltage source. It is essential that $L_B \gg L$ is chosen to ensure that L_B does not affect the resonant frequency. The tuning voltage may also be provided via a resistor which, however, is an additional load to the resonant circuit and thus reduces the

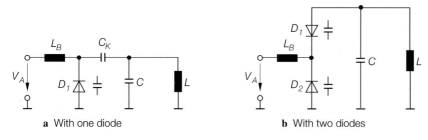

a With one diode   **b** With two diodes

Fig. 1.28. Frequency tuning in LC circuits with varactor diodes

quality of the circuit. The coupling capacitance C_K prevents the voltage source V_A from being short-circuited by the inductance L of the resonant circuit. Provided that $L_B \gg L$, the resonant frequency is:

$$\omega_R = 2\pi f_R = \cfrac{1}{\sqrt{L\left(C + \cfrac{C_J(V_A)\,C_K}{C_J(V_A) + C_K}\right)}} \overset{C_K \gg C_J(V_A)}{\approx} \frac{1}{\sqrt{L\left(C + C_J(V_A)\right)}}$$

The tuning range depends on the characteristic of the junction capacitance and its relation to the resonant circuit capacitance C. The maximum tuning range is achieved with $C = 0$ and $C_K \gg C_J$.

In the circuit depicted in Fig. 1.28b a series connection of two junction capacitances is arranged in parallel to the resonant circuit. Here, too, the inductivity $L_B \gg L$ prevents a high-frequency short-circuit of the resonant circuit by the voltage source V_A. A coupling capacitance is not required since both diodes are in reverse mode so that no DC current can flow into the resonant circuit. In this case, the resonant frequency is:

$$\omega_R = 2\pi f_R = \cfrac{1}{\sqrt{L\left(C + \cfrac{C_J(V_A)}{2}\right)}}$$

Here, again, the tuning range is maximum for $C = 0$; however, only half the junction capacitance is effective so that compared to the circuit shown in Fig. 1.28a either the junction capacitance or the inductance must be twice as high for the same resonant frequency. A material advantage of the symmetrical diode arrangement is the improved linearity with high amplitudes in the resonant circuit; this largely offsets the decrease in the resonant frequency with increasing amplitudes that is caused by the nonlinearity of the junction capacitance [1.3].

1.4.4
Bridge Rectifier

The circuit shown in Fig. 1.29 made up of four diodes is called a *bridge rectifier* and is used for full-way rectification in power supplies and AC voltmeters. Bridge rectifiers for power supplies are divided into high-voltage bridge rectifiers, which are used for direct rectification of the mains voltage and must therefore have a high breakdown voltage ($V_{BR} \geq 350\,\text{V}$), and low-voltage bridge rectifiers, which are used on the secondary side of a line transformer; Sect. 16.5 describes this in more detail. Of the four connections two are marked with $\sim$ and one each with $+$ and $-$.

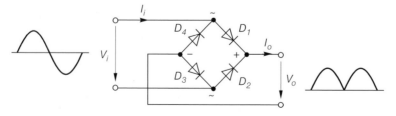

Fig. 1.29. Bridge rectifier

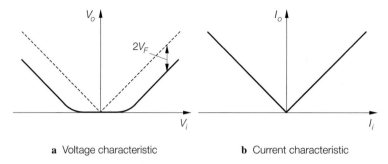

a Voltage characteristic   **b** Current characteristic

Fig. 1.30. Characteristic curves of a bridge rectifier

With a positive input voltage D_1 and D_3 are conductive while D_2 and D_4 are reverse-biased; with a negative input voltage D_2 and D_4 are conductive while D_1 and D_3 are reverse-biased. Since at any given moment the current flows through two conductive diodes, the rectified output voltage is lower (by $2V_F \approx 1.2 \ldots 2\,\text{V}$) than the magnitude of the input voltage:

$$V_o \approx \begin{cases} 0 & \text{for } |V_i| \leq 2V_F \\ |V_i| - 2V_F & \text{for } |V_i| > 2V_F \end{cases}$$

Figure 1.30a shows the voltage characteristic. A peak reverse voltage of $|V_D|_{max} = |V_i|_{max}$, which must be lower than the breakthrough voltage of the diodes, occurs across the diodes in reverse mode.

Unlike the voltages the magnitudes of the currents are in a linear relationship (see Fig. 1.30b):

$$I_o = |I_i|$$

This fact is used in meter rectifiers; the AC voltage to be measured is fed through a voltage-to-current converter and the resulting current is rectified in a bridge rectifier.

1.4.5
Mixer

Mixers are used in communication systems for frequency conversion. There are *passive mixers*, which use diodes or other passive components, and *active mixers*, which use transistors. In the case of passive mixers, the *ring modulator* consisting of four diodes and two transformers with centre tabs is most frequently used. Figure 1.31 shows a ring modulator in *downconverter* configuration with diodes $D_1 \ldots D_4$ and transformers $L_1 - L_2$ and $L_3 - L_4$ [1.9]. The circuit converts the input signal V_{RF} with the frequency f_{RF} by means of the *local oscillator voltage* V_{LO} with a frequency f_{LO} to an *intermediate frequency* $f_{IF} = |f_{RF} - f_{LO}|$. The output voltage V_{IF} is supplied to a resonant circuit in tune with the intermediate frequency in order to strip the signal from additional frequency components generated in the conversion process. The local oscillator provides a sinusoidal or rectangular voltage with an amplitude $\hat{v}_{LO}$; V_{RF} and V_{IF} are sinusoidal voltages of the amplitudes $\hat{v}_{RF}$ and $\hat{v}_{IF}$ respectively. In normal operation $\hat{v}_{LO} \gg \hat{v}_{RF} > \hat{v}_{IF}$ applies; in other words, the voltage of the local oscillator determines which diodes are conductive;

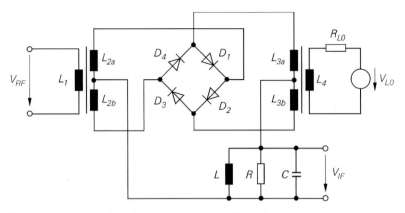

Fig. 1.31. Ring modulator in downconverter configuration

the following applies when using a 1:1 transformer with $L_4 = L_{3a} + L_{3b}$:

$$\left.\begin{array}{r} V_{LO} \geq 2V_F \\ -2V_F < V_{LO} < 2V_F \\ V_{LO} < -2V_F \end{array}\right\} \Rightarrow \left\{\begin{array}{l} D_1 \text{ and } D_2 \text{ are conductive} \\ \text{No diode is conductive} \\ D_3 \text{ and } D_4 \text{ are conductive} \end{array}\right.$$

V_F is the forward voltage of the diodes. Due to their better switching performance Schottky diodes with $V_F \approx 0.3\,\text{V}$ are used exclusively; the current through the diodes is limited by the internal resistance R_{LO} of the local oscillator.

When D_1 and D_2 are conductive a current caused by V_{RF} flows through L_{2a} and $D_1 - L_{3a}$ or $D_2 - L_{3b}$ in the IF resonant circuit; when D_3 and D_4 are conductive, the current flows through L_{2b} and $D_3 - L_{3b}$ or $D_4 - L_{3a}$. The polarity of V_{IF} is different from that of V_{RF} so that the local oscillator and the diodes cause a polarity change at the frequency f_{LO} (see Fig. 1.32). If V_{LO} is a square wave signal with $\hat{v}_{LO} > 2V_F$ the change in polarity occurs suddenly; that is, the ring modulator multiplies the input signal with the square wave signal. The IF filter extracts the desired components with $m = 1, n = -1$ or $m = -1$ and $n = 1$ from the generated frequency components in the form $|mf_{LO} + nf_{RF}|$ with any integer value for m and $n = \pm 1$.

The ring modulator is available as a component with six connections, two each at the RF, LO, and IF sides [1.9]. Furthermore, there are integrated circuits containing only the diodes, and therefore have only four connections. In this context it must be noted that, despite their similarity in form, the mixer and the bridge rectifier differ from one another in terms of the arrangement of the diodes, as shown by a comparison of Figs. 1.31 and 1.29.

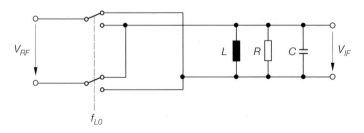

Fig. 1.32. Functioning of a ring modulator

Chapter 2:
Bipolar Transistor

The bipolar transistor is a semiconductor component with three terminals that are known as the *base* (*B*), *emitter* (*E*) and *collector* (*C*). There are discrete transistors that are used for mounting to printed circuit boards and are contained in their own individual case and integrated transistors that are produced together with other semiconductor elements on a common substrate. Integrated transistors feature a fourth connection called the *substrate* (*S*), which represents the common carrier; it is of secondary importance for the transistor's electrical function.

Equivalent circuits with diodes: Bipolar transistors consist of two anti-serially connected pn diodes that have a common p or n region. Fig. 2.1 shows the graphic symbol and the *equivalent diode circuits* of an npn transistor with a common p region and a pnp transistor with a common n region. The equivalent diode circuit diagrams, however, do not correctly reflect the bipolar transistor function but allow an overview of operating modes and illustrate how the type (npn or pnp) and the base terminal of an unknown transistor can be determined with the help of a continuity tester; due to the symmetrical design it is not easy to distinguish the collector from the emitter.

Operating modes: The bipolar transistor is used to amplify and to switch signals and is usually operated in *normal mode (forward region)*, meaning that the emitter diode junction (BE diode) is biased in the forward direction and the collector diode junction (BC diode) is reverse-biased. In some applications the BC diode may also be operated temporarily in the forward region, which is called the *saturation region*. Interchanging the emitter and collector makes the transistor operate in the *reverse region*; this operating mode is advantageous only in exceptional cases. In the *cutoff region* both diodes are nonconductive. Figure 2.2 shows the polarity of voltages and currents for pnp and npn transistors in normal operating mode.

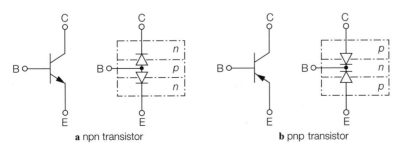

a npn transistor **b** pnp transistor

Fig. 2.1. Graphical symbols and equivalent diode circuits

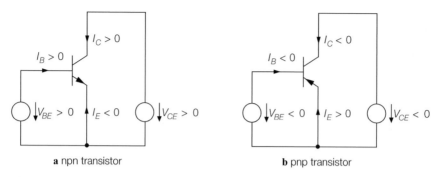

a npn transistor   **b** pnp transistor

Fig. 2.2. Voltages and currents in normal operation

2.1
Performance of a Bipolar Transistor

The easiest way to demonstrate the behavior of a bipolar transistor is to look at its characteristics, which describe the relationship between the currents and voltages in the transistor, provided that all parameters are *static*, that is, not or only slowly variable over time. To reflect its behavior with sufficient accuracy, the calculatory description of the bipolar transistor requires equations. Considering solely its normal operation, which is of particular importance in practical applications, and neglecting any secondary effects makes the equations very simple. But when the proper functionality of a circuit is being evaluated by means of computer simulation it is important to consider the influence of secondary effects as well. For this purpose there are sophisticated models that are capable of correctly demonstrating the *dynamic performance* when applying sinusoidal or pulsed signals. These models are described in Sect. 2.3 and are not required to understand the basics. The following section describes the behavior of an npn transistor; for pnp transistors all voltages and currents have the opposite polarity.

2.1.1
Characteristics

Family of output characteristics: The application of various base–emitter voltages (V_{BE}) to the circuit shown in Fig. 2.2a and measurement of the collector current I_C as a function of the collector–emitter voltage V_{CE} produces the family of output characteristics illustrated in Fig. 2.3. With the exception of a small region close to the I_C-axis the characteristic curves depend only slightly on V_{CE} and the transistor is in normal mode; that is, the BE diode is forward-biased and the BC diode is reverse-biased. Close to the I_C-axis voltage V_{CE} is so small that the BC diode is in the forward region and the transistor enters the saturation region. At this border, which is characterized by the saturation voltage $V_{CE,sat}$, there is a sharp bend in the curves and they run approximately through the origin of the characteristics.

Family of transfer characteristics: In normal mode the collector current I_C essentially depends on V_{BE} alone. Plotting I_C for various V_{CE} values that are typical of normal operation as a function of V_{BE} results in the family of transfer characteristics shown in Fig. 2.4a. The characteristic curves are very close together because of the small degree of dependence on V_{CE}.

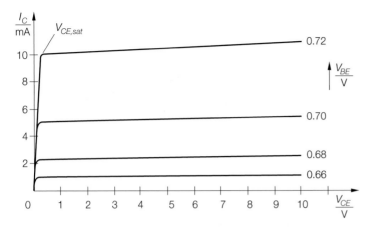

Fig. 2.3. Output characteristics of an npn transistor

Family of input characteristics: A comprehensive description of the behavior of the transistor also requires the family of input characteristics as shown in Fig. 2.4b for which the base current I_B for various values of V_{CE} typical of normal operation are plotted as a function of V_{BE}. Here, too, the dependence on V_{CE} is insignificant.

Current gain: A comparison of the transfer characteristics in Fig. 2.4a and the input characteristics in Fig. 2.4b shows a striking similarity in shape. This means that in normal operating mode the collector current I_C is approximately proportional to the base current I_B. The proportionality constant B is called the *current gain*:

$$B = \frac{I_C}{I_B} \qquad (2.1)$$

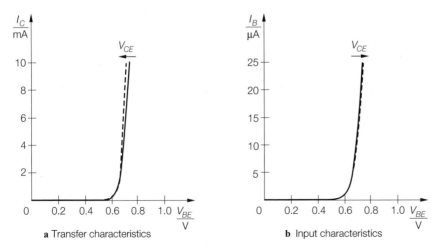

Fig. 2.4. Families of characteristics in normal operation

2.1.2
Description by Way of Equations

The equations required for a mathematical description are based on the fact that a transistor's behavior can be explained for the main part by the characteristic function of the BE diode. The exponential relationship between current and voltage that is typical of a diode is reflected in the transfer and input characteristics of the transistor by the exponential dependence of the currents I_B and I_C on the voltage V_{BE}. Taking the general formulas $I_C = I_C(V_{BE}, V_{CE})$ and $I_B = I_B(V_{BE}, V_{CE})$ as a basis gives us the equations for normal operation [2.1]:

$$I_C = I_S\, e^{\frac{V_{BE}}{V_T}} \left(1 + \frac{V_{CE}}{V_A}\right) \tag{2.2}$$

$$I_B = \frac{I_C}{B} \quad \text{with} \quad B = B(V_{BE}, V_{CE}) \tag{2.3}$$

In this case $I_S \approx 10^{-16} \ldots 10^{-12}\,\text{A}$ is the *saturation reverse current* of the transistor and V_T is the temperature voltage (*the temperature equivalent* of thermal energy); therefore, at room temperature $V_T \approx 26\,\text{mV}$.

Early effect: The dependence on V_{CE} is caused by the *Early effect* and is empirically described by the terms on the right-hand side of (2.2). The basis for this description is the observation that the extrapolated characteristic curves of the output characteristics intersect approximately at one point [2.2]; Fig. 2.5 illustrates this relationship. The constant V_A is called *Early voltage* and is in the range of $V_{A,npn} \approx 30 \ldots 150\,\text{V}$ for npn transistors and $V_{A,pnp} \approx 30 \ldots 75\,\text{V}$ for pnp transistors. Section 2.3.1 explains the Early effect in more detail; the empirical description is sufficient for the normal operation considered here.

Base current and current gain: The base current I_B is related to I_C; this results in the current gain B being the constant of proportionality. This method is chosen because the dependence of the current gain on V_{BE} and V_{CE} can be neglected in many simple calculations; B then represents an independent constant. However, in most cases the dependence on V_{CE} is taken into consideration since it is also caused by the Early effect [2.2]. Therefore:

$$B(V_{BE}, V_{CE}) = B_0(V_{BE}) \left(1 + \frac{V_{CE}}{V_A}\right) \tag{2.4}$$

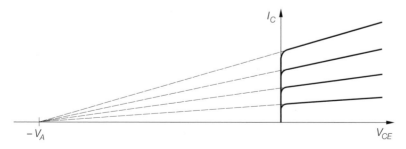

Fig. 2.5. Early effect and Early voltage V_A in the output characteristics

$B_0(V_{BE})$ is the extrapolated current gain for $V_{CE} = 0\,\text{V}$. The extrapolation is necessary because for $V_{CE} = 0\,\text{V}$ the transistor no longer operates in normal mode.

Large-signal equations: Inserting (2.4) into (2.3) leads to the following *large-signal equations* for the bipolar transistor:

$$I_C = I_S\, e^{\frac{V_{BE}}{V_T}} \left(1 + \frac{V_{CE}}{V_A}\right) \tag{2.5}$$

$$I_B = \frac{I_S}{B_0}\, e^{\frac{V_{BE}}{V_T}} \tag{2.6}$$

2.1.3
Characteristic of the Current Gain

The Gummel-plot: The current gain $B(V_{BE}, V_{CE})$ will be examined more closely in the section below. Because of the exponential dependence of the currents I_B and I_C on V_{BE}, the semilogarithmic plot over V_{BE} with V_{CE} as a parameter seems to be the obvious choice. This diagram, which is illustrated in Fig. 2.6, is called the *Gummel plot* and shows that the exponential curves in (2.5) and (2.6) become straight lines if B_0 is assumed to be constant:

$$\ln\left(\frac{I_C}{I_S}\right) = \frac{V_{BE}}{V_T} + \ln\left(1 + \frac{V_{CE}}{V_A}\right)$$

$$\ln\left(\frac{I_B}{I_S}\right) = \frac{V_{BE}}{V_T} - \ln(B_0)$$

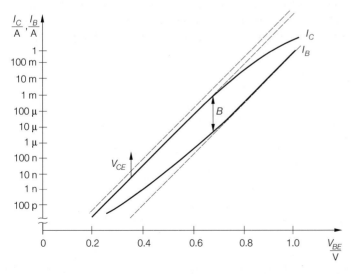

Fig. 2.6. Semilogarithmic plot of the currents I_B and I_C in normal mode (a Gummel plot)

In Fig. 2.6 the straight lines for two values of V_{CE} are shown as broken lines. The current gain B is seen as a shift in the y direction:

$$\ln(B) = \ln\left(\frac{I_C}{I_B}\right) = \ln(B_0) + \ln\left(1 + \frac{V_{CE}}{V_A}\right)$$

The real curves are also plotted in Fig. 2.6. In one large region they correspond to the straight lines, which means that B_0 can be assumed to have a constant value. However, in two regions there are deviations [2.2]:

- In the event of very low collector currents the base current is *higher* than the value for a constant B_0 given in (2.6). This deviation is caused by additional portions in the base current and results in a decrease of B or B_0. The large-signal (2.5) and (2.6) are valid in this region as well.
- With very high collector currents the collector current is *lower* than the value given by (2.5). This deviation is caused by the *high-current effect* and results in a decrease in B or B_0. In this region the large-signal (2.5) and (2.6) are no longer valid since, according to these equations, the decrease in B_0 leads to an increase in I_B and not, as required, to a decrease in I_C. This region is used in power transistors only.

Curve description: For practical purposes the current gain B is expressed as a function of I_C and V_{CE}; that is, $B(V_{BE}, V_{CE})$ is replaced by $B(I_C, V_{CE})$ by utilizing the relationship between I_C and V_{BE} for a constant V_{CE} in order to change the variables. Similarly $B_0(V_{BE})$ is replaced by $B_0(I_C)$. This change in the expression facilitates the dimensioning of circuit components because when setting the operating point, first I_C and V_{CE} are chosen and then the relevant base current is determined by means of $B(I_C, V_{CE})$; this is the procedure for setting the operating point of the basic circuits shown in Sect. 2.4.

Figure 2.7 shows the curves of the current gain B and the differential current gain β versus I_C for two different values of V_{CE}.

$$\beta = \frac{dI_C}{dI_B}\bigg|_{V_{CE}=\text{const.}} \tag{2.7}$$

B is called the *large-signal current gain* and β is the *small-signal current gain*.

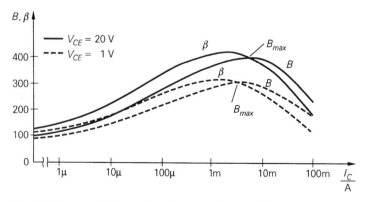

Fig. 2.7. Curves of the large-signal current gain B and the small-signal current gain β in normal operation

These curves are typical of low-power transistors in which the maximum of the current gain is achieved for $I_C \approx 1\ldots10\,\text{mA}$. For power transistors this maximum moves into the ampere range. In practice the transistor is operated in the region of its maximum or to the left of it; that is, with lower collector currents. The region to the right of the maximum is avoided, where possible, since the high-current effect not only reduces B but also lowers the transistor's response time and cutoff frequencies; Sects. 2.3.2 and 2.3.3 explain this in more detail.

The small-signal current gain β is required to describe the small-signal response in the next section. On the basis of (2.7) the following equation

$$\frac{1}{\beta} = \frac{dI_B}{dI_C}\bigg|_{V_{CE}=\text{const.}} = \frac{\partial\left(\dfrac{I_C}{B(I_C, V_{CE})}\right)}{\partial I_C}$$

is derived from the relationship between β and B [2.3]:

$$\beta = \frac{B}{1 - \dfrac{I_C}{B}\dfrac{\partial B}{\partial I_C}}$$

In the region to the left of the maximum B the derivative $(\partial B/\partial I_C)$ is positive, so that $\beta > B$. At the maximum the derivative $(\partial B/\partial I_C) = 0$, so that $\beta = B$. To the right of the maximum B the derivative $(\partial B/\partial I_C)$ is negative and thus $\beta < B$.

Determining the values: When the transistor is operated with a collector current in the region of the maximum of the current gain B, the following approximation can be applied:

$$\boxed{\beta(I_C, V_{CE}) \approx B(I_C, V_{CE}) \approx B_{max}(V_{CE})} \qquad (2.8)$$

Here, $B_{max}(V_{CE})$ is the maximum of B depending on V_{CE}, as shown in Fig. 2.7.

If a transistor data sheet shows the curve for B in the form of a diagram as in Fig. 2.7, then $B(I_C, V_{CE})$ can be taken from the diagram and the approximation given in (2.8) can be used if the β curves are missing. If only one value for B is stated on the data sheet, it can be used as a replacement value for B and β. Typical values are $B \approx 100\ldots500$ for low-power transistors and $B \approx 10\ldots100$ for power transistors. Darlington transistors consist of two transistor elements connected internally so that $B \approx 500\ldots10.000$ is reached, depending on the power rating. Section 2.4.4 describes the Darlington circuit in more detail.

2.1.4
Operating Point and Small-Signal Response

One area of application for bipolar transistors is the linear amplification of signals in *small-signal operation*. The transistor is operated at an operating point A and driven by *small signals* around the operating point. In this case the nonlinear characteristics can be replaced by a line that is tangential to the operating point, and that describes an approximately linear response.

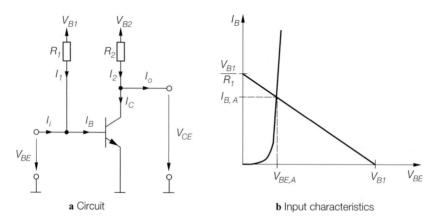

a Circuit   **b** Input characteristics

Fig. 2.8. An example for determining the operating point

Determining the Operating Point

The operating point A is characterized by the voltages $V_{CE,A}$ and $V_{BE,A}$ and the currents $I_{C,A}$ and $I_{B,A}$, and is determined by the external circuitry of the transistor. This determination is called the *operating point adjustment* or *biasing*. As an example the operating point of the simple amplifier circuit shown in Fig. 2.8a is determined. It is adjusted using resistors R_1 and R_2, which are of known value.

Numeric solution: The following system of equation is derived for $I_i = I_o = 0$ from the transistor's large-signal equations and the nodal equations for base and collector connection. This system consists of four equations and four unknowns:

$$
\left.
\begin{aligned}
I_C &= I_C(V_{BE}, V_{CE}) \\
I_B &= I_B(V_{BE}, V_{CE})
\end{aligned}
\right\}
\text{Characteristics of the transistor}
$$

$$
\left.
\begin{aligned}
I_B &= I_1 = \frac{V_{B1} - V_{BE}}{R_1} \\
I_C &= I_2 = \frac{V_{B2} - V_{CE}}{R_2}
\end{aligned}
\right\}
\text{Straight lines under load}
$$

The operating point values $V_{BE,A}$, $V_{CE,A}$, $I_{B,A}$ and $I_{C,A}$ are calculated by solving the equations.

Graphical solution: Besides the numeric solution, a graphic solution is also possible by drawing the straight lines under load in the diagram of the family of characteristics and determining the intersections. As the family of characteristics comprises essentially only one curve as a result of the very low dependence on V_{CE}, there is only one intersection according to Fig. 2.8b so that $V_{BE,A}$ and $I_{B,A}$ can be determined directly. In the diagram of the output characteristics $V_{CE,A}$ and $I_{C,A}$ can be determined from the point at which the straight line intersects the output characteristic for $V_{BE,A}$ (see Fig. 2.9).

Setting the operating point: Both the numeric and the graphic way of determining the operating point are *analytical* procedures; that is, the operating point can be determined when the external circuit components are known. But for circuit design *synthetic*

2.1 Performance of a Bipolar Transistor   41

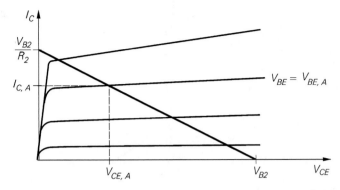

Fig. 2.9. An example for determining the operating point from the output characteristics

procedures are required that are suitable for making decisions on the circuit components necessary to achieve a certain operating point. These procedures are described, together with the basic circuits, in Sect. 2.4.

Small-Signal Equations and Small-Signal Parameters

Small-signal values: When the component is operated close to the operating point the deviations of voltages and currents from the operating point values are called the *small-signal voltages* and *currents*. Per definition:

$$
\begin{aligned}
v_{BE} &= V_{BE} - V_{BE,A} &,\quad i_B &= I_B - I_{B,A} \\
v_{CE} &= V_{CE} - V_{CE,A} &,\quad i_C &= I_C - I_{C,A}
\end{aligned}
$$

Linearization: The characteristics are replaced by their tangents to the operating point; that is, they are *linearized*. For this purpose, a Taylor series expansion is perfomed at the operating point and is interrupted after the linear term:

$$
\begin{aligned}
i_B &= I_B(V_{BE,A} + v_{BE}, V_{CE,A} + v_{CE}) - I_{B,A} \\
&= \left.\frac{\partial I_B}{\partial V_{BE}}\right|_A v_{BE} + \left.\frac{\partial I_B}{\partial V_{CE}}\right|_A v_{CE} + \ldots
\end{aligned}
$$

$$
\begin{aligned}
i_C &= I_C(V_{BE,A} + v_{BE}, V_{CE,A} + v_{CE}) - I_{C,A} \\
&= \left.\frac{\partial I_C}{\partial V_{BE}}\right|_A v_{BE} + \left.\frac{\partial I_C}{\partial V_{CE}}\right|_A v_{CE} + \ldots
\end{aligned}
$$

Figure 2.10 illustrates the linearization using the transfer characteristic as an example; for this purpose the region around the operating point is shown in an enlarged format. The change in current i_C is determined along the characteristic curve from the voltage change v_{BE}; the current change $i_{C,lin}$ is taken from the tangent. $i_C = i_{C,lin}$ can be used with small values of v_{BE}.

Small-signal equations: The partial derivatives at the operating point are called *small-signal parameters*. After introducing specific designations, the *small-signal equations* for the bipolar transistor are:

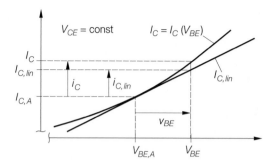

Fig. 2.10. Linearization of the transfer characteristic

$$i_B = \frac{1}{r_{BE}} v_{BE} + g_{m,r} v_{CE} \tag{2.9}$$

$$i_C = g_m v_{BE} + \frac{1}{r_{CE}} v_{CE} \tag{2.10}$$

Small-signal parameters: The *transconductance* g_m characterizes the change in the collector current I_C in response to a change in the base–emitter voltage V_{BE} at the operating point. It can be determined using the diagram of the transfer characteristic according to Fig. 2.4a, from the slope of the tangent at the operating point and thus describes how *steep* the transfer characteristic is at that point. The transconductance can be calculated by differentiating the large-signal (2.5):

$$g_m = \left. \frac{\partial I_C}{\partial V_{BE}} \right|_A = \frac{I_{C,A}}{V_T} \tag{2.11}$$

The *small-signal input resistance* r_{BE} characterizes the change in the base–emitter voltage V_{BE} in response to a change in the base current I_B at the operating point. It can be determined using the input characteristics according to Fig. 2.4b from the inverse value of the slope of the tangent. Use of the following relationship renders the differentiation of the large-signal (2.6) unnecessary:

$$r_{BE} = \left. \frac{\partial V_{BE}}{\partial I_B} \right|_A = \left. \frac{\partial V_{BE}}{\partial I_C} \right|_A \left. \frac{\partial I_C}{\partial I_B} \right|_A$$

This makes it possible to calculate r_{BE} from the transconductance g_m given in (2.11) and the small-signal current gain β according to (2.7):

$$r_{BE} = \left. \frac{\partial V_{BE}}{\partial I_B} \right|_A = \frac{\beta}{g_m} \tag{2.12}$$

The *small-signal output resistance* r_{CE} characterizes the change in the collector–emitter voltage V_{CE} in response to a change in the collector current I_C at the operating point. It can be determined using the output characteristics shown in Fig. 2.3 from

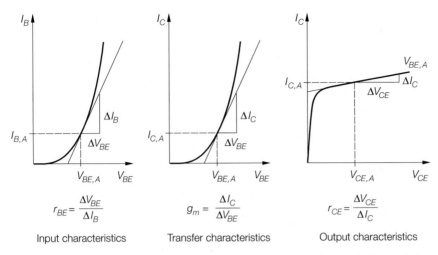

$$r_{BE} = \frac{\Delta V_{BE}}{\Delta I_B} \qquad\qquad g_m = \frac{\Delta I_C}{\Delta V_{BE}} \qquad\qquad r_{CE} = \frac{\Delta V_{CE}}{\Delta I_C}$$

Input characteristics   Transfer characteristics   Output characteristics

Fig. 2.11. Determination of the small-signal parameters from the family of characteristics

the inverse value of the slope of the tangent. It can be calculated by differentiating the large-signal (2.5):

$$r_{CE} = \left.\frac{\partial V_{CE}}{\partial I_C}\right|_A = \frac{V_A + V_{CE,A}}{I_{C,A}} \overset{V_{CE,A} \ll V_A}{\approx} \frac{V_A}{I_{C,A}} \qquad\qquad (2.13)$$

In practice, the approximation given in (2.13) is used.

The *reverse transconductance* $g_{m,r}$ characterizes the change in the base current I_B in response to a change in the collector–emitter voltage V_{CE} at the operating point. It is so small that it can be disregarded. This dependence is already disregarded in the large-signal (2.6); that is, I_B does not depend on V_{CE}:

$$g_{m,r} = \left.\frac{\partial I_B}{\partial V_{CE}}\right|_A \approx 0 \qquad\qquad (2.14)$$

The small-signal parameters may also be derived from the characteristic curves; drawing the tangent to the operating point makes it possible to determine the transconductance values (see Fig. 2.11). In practice, however, this procedure is rarely used due to its low accuracy; furthermore, the characteristics are not usually shown on a transistor's data sheets.

Small-Signal Equivalent Circuit

Assuming $g_{m,r} = 0$, the small-signal (2.9) and (2.10) lead to the *small-signal equivalent circuit* of the bipolar transistor, as shown in Fig. 2.12. When the values $I_{C,A}$, $V_{CE,A}$ and β at the operating point are known, the transistor parameters can be calculated using (2.11), (2.12) and (2.13).

The equivalent circuit diagram is suitable for calculating the small-signal behavior of transistor circuits with low frequencies (0...10 kHz); for this reason it is also called the *DC small-signal equivalent circuit*. Information about the behavior at higher frequencies, the

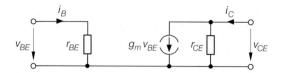

Fig. 2.12. Small-signal equivalent circuit of a bipolar transistor

frequency response and the cutoff frequency of a transistor circuit can only be obtained by using the AC small-signal equivalent circuit, which will be described in Sect. 2.3.3.

Network Matrices

The small-signal equations can also be expressed in matrix notation as:

$$
\begin{bmatrix} i_B \\ i_C \end{bmatrix} =
\begin{bmatrix} \dfrac{1}{r_{BE}} & g_{m,r} \\ g_m & \dfrac{1}{r_{CE}} \end{bmatrix}
\begin{bmatrix} v_{BE} \\ v_{CE} \end{bmatrix}
$$

This corresponds to the conductivity presentation of a network and thus relates to network theory. The conductivity presentation describes a network by the *Y matrix* $\mathbf{Y}_e$:

$$
\begin{bmatrix} i_B \\ i_C \end{bmatrix} = \mathbf{Y}_e
\begin{bmatrix} v_{BE} \\ v_{CE} \end{bmatrix} =
\begin{bmatrix} y_{11,e} & y_{12,e} \\ y_{21,e} & y_{22,e} \end{bmatrix}
\begin{bmatrix} v_{BE} \\ v_{CE} \end{bmatrix}
$$

The subscript e indicates that the transistor is operated in the common-emitter mode which means that the emitter connection is jointly used for the input *and* output ports according to the through connection in the small-signal equivalent circuit shown in Fig. 2.12. Section 2.4 describes the common-emitter operating mode in more detail.

The hybrid presentation based on the *H matrix* $\mathbf{H}_e$ is also commonly used:

$$
\begin{bmatrix} v_{BE} \\ i_C \end{bmatrix} = \mathbf{H}_e
\begin{bmatrix} i_B \\ v_{CE} \end{bmatrix} =
\begin{bmatrix} h_{11,e} & h_{12,e} \\ h_{21,e} & h_{22,e} \end{bmatrix}
\begin{bmatrix} i_B \\ v_{CE} \end{bmatrix}
$$

A comparison reveals the following relationships:

$$
r_{BE} = h_{11,e} = \frac{1}{y_{11,e}} \quad , \quad \beta = h_{21,e} = \frac{y_{21,e}}{y_{11,e}}
$$

$$
g_m = \frac{h_{21,e}}{h_{11,e}} = y_{21,e} \quad , \quad g_{m,r} = -\frac{h_{12,e}}{h_{11,e}} = y_{12,e}
$$

$$
r_{CE} = \frac{h_{11,e}}{h_{11,e}h_{22,e} - h_{12,e}h_{21,e}} = \frac{1}{y_{22,e}}
$$

Range of Validity for Small-Signal Applications

In connection with the small-signal equivalent circuit the question is often raised as to how large the maximum control values around the operating point may become before leaving the region of the small-signal mode. There is no general answer to this question. From a mathematical point of view the equivalent circuit is valid only for an *infinitesimal* – that is, very small – control range. For practical considerations, the nonlinear distortions generated by finite control, which should not exceed an application-specific limit value, are the decisive factor. This limit value is often given as the maximum permissible harmonic *distortion factor*.

Section 4.2.3 describes this in more detail. The small-signal equivalent circuit diagram is derived from the linear term of the interrupted Taylor series expansion. If subsequent terms of the Taylor expansion are taken into account, for constant V_{CE} the small-signal collector current becomes [2.1]:

$$i_C = \left.\frac{\partial I_C}{\partial V_{BE}}\right|_A v_{BE} + \left.\frac{1}{2}\frac{\partial^2 I_C}{\partial V_{BE}^2}\right|_A v_{BE}^2 + \left.\frac{1}{6}\frac{\partial^3 I_C}{\partial V_{BE}^3}\right|_A v_{BE}^3 + \dots$$

$$= \frac{I_{C,A}}{V_T} v_{BE} + \frac{I_{C,A}}{2V_T^2} v_{BE}^2 + \frac{I_{C,A}}{6V_T^3} v_{BE}^3 + \dots$$

For a harmonic input signal with $v_{BE} = \hat{v}_{BE} \cos \omega t$, this leads to:

$$\frac{i_C}{I_{C,A}} = \left[\frac{1}{4}\left(\frac{\hat{v}_{BE}}{V_T}\right)^2 + \dots\right] + \left[\frac{\hat{v}_{BE}}{V_T} + \frac{1}{8}\left(\frac{\hat{v}_{BE}}{V_T}\right)^3 + \dots\right] \cos \omega t$$

$$+ \left[\frac{1}{4}\left(\frac{\hat{v}_{BE}}{V_T}\right)^2 + \dots\right] \cos 2\omega t + \left[\frac{1}{24}\left(\frac{\hat{v}_{BE}}{V_T}\right)^3 + \dots\right] \cos 3\omega t$$

$$+ \dots$$

Polynomes with even or uneven exponents are contained in the brackets. With little input signals – that is, not considering higher exponents – the ratio of the first harmonic with $2\omega t$ to the fundamental with ωt leads to an approximation of the *distortion factor k* [2.1]:

$$k \approx \frac{i_{C,2\omega t}}{i_{C,\omega t}} \approx \frac{\hat{v}_{BE}}{4V_T} \tag{2.15}$$

If k is to be kept below 1 % , then $\hat{v}_{BE} < 0.04V_T \approx 1$ mV. This means that in this case only very small input signals are permissible.

2.1.5
Limit Data and Reverse Currents

For a transistor, various ratings are specified which must not be exceeded. They are classified in terms of limit voltages, limit currents and maximum power dissipation. The following description again applies to npn transistors; for pnp transistors the polarity of all voltages and currents has to be reversed.

Breakdown Voltages

BE diode: At the *emitter–base breakdown voltage* $V_{(BR)EBO}$, the emitter diode breaks through in the cutoff mode. The *addition* "(BR)" means *breakdown*; the subscript *O* indicates that the third terminal, in this case the collector, is *open*. $V_{(BR)EBO} \approx 5\dots7$ V applies to almost all transistors; thus, $V_{(BR)EBO}$ is the smallest of the limit voltages. It is of minor importance, since transistors are rarely operated with negative base–emitter voltages.

BC diode: At the *collector–base breakdown voltage* $V_{(BR)CBO}$ the collector diode breaks through in the cutoff mode. Since the collector diode is reverse-biased in normal mode $V_{(BR)CBO}$ is an important upper limit of the collector–base voltage in practice. In low

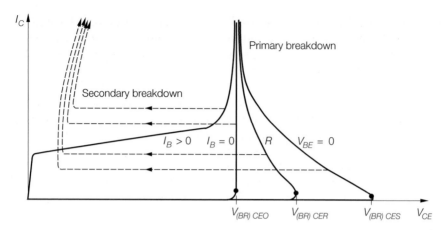

Fig. 2.13. Output characteristics, with breakdown characteristics, of an npn transistor

voltage transistors $V_{(BR)CBO} \approx 20...80\,\text{V}$, while in high voltage transistors $V_{(BR)CBO}$ reaches values of up to $1.300\,\text{V}$. $V_{(BR)CBO}$ is the highest of the transistor limit voltages.

Collector–emitter section: The maximum permissible collector–emitter voltage V_{CE} is of particular importance for practical applications. An overview is provided by the family of output characteristics shown in Fig. 2.13 in which the region around V_{CE} is expanded compared to the output characteristics shown in Fig. 2.3. At a certain collector–emitter voltage a breakdown occurs resulting in a marked increase in the collector current and, in most cases, in the destruction of the transistor. The *breakdown characteristics* shown in Fig. 2.13 are measured with external base circuitry. When measuring the characteristic "$I_B > 0$", a positive base current is produced using a current source. In the region of the *collector–emitter breakdown voltage* $V_{(BR)CEO}$, the current rises significantly and the curve approaches an approximately vertical pattern. $V_{(BR)CEO}$ is the collector–emitter voltage at which the collector current exceeds a certain value due to the breakdown, despite the fact that the base is open; that is, $I_B = 0$. In order to determine $V_{(BR)CEO}$ the characteristic $I_B = 0$ is used; this approaches an approximately vertical pattern at $V_{(BR)CEO}$. To plot the characteristic "R" a resistor is placed between the base and the emitter which increases the breakdown voltage to $V_{(BR)CER}$. In this case, the increase in current that occurs at breakdown results in a drop in the collector–emitter voltage from $V_{(BR)CER}$ to approximately $V_{(BR)CEO}$, causing a branch of the characteristic curve to have a negative slope. The base current I_B is negative. The characteristic "$V_{BE} = 0$", which is measured with a short circuit connection between the base and the emitter, has the same shape. The resulting breakdown voltage $V_{(BR)CES}$ is the highest of the collector–emitter breakdown voltages quoted. The subscript S indicates that the base is *short-circuited*. In general, the voltage levels are:

$$V_{(BR)CEO} < V_{(BR)CER} < V_{(BR)CES} < V_{(BR)CBO}$$

Secondary Breakdown

In addition to the *normal* or *primary breakdown* described so far, there is a *second* or *secondary breakdown*; this is characterized by localized overheating due to a nonhomogenous current distribution (*contraction*) which results in localized melting and thus the destruction of the transistor. The characteristics of secondary breakdown are plotted as broken lines in Fig. 2.13. First, there is the normal breakdown, in the course of which the contraction takes place. Secondary breakdown is characterized by a breakdown of the collector–emitter voltage followed by a marked increase in current. It occurs in power and high-voltage transistors, at high collector–emitter voltages. It is rare in low-power transistors for low voltage applications; here, it is usually primary breakdown that takes place, and – with a suitable current limitation – does not cause the destruction of the transistor.

The characteristics of secondary breakdown cannot be measured under static conditions, since it is an irreversible dynamic process. In contrast, the characteristics of normal breakdown can be measured statically – in other words, with a characteristic curve recorder – provided that the currents are limited and the measuring time is kept short, to prevent overheating and to avoid the region of secondary breakdown.

Limit Currents

A distinction is made between maximum *continuous currents* and maximum *peak currents*. For maximum continuous currents, the data sheets contain no specific identifiers; they are called $I_{C,max}$, $I_{B,max}$ and $I_{E,max}$. The maximum peak currents are called I_{CM}, I_{BM} and I_{EM} in data sheets, and refer to pulsed operation with a predetermined pulse duration and repetition rate; compared to continuous currents they are higher by a factor of 1.2...2.

Cut-Off Currents

For emitter and collector diodes data sheets specify not only the breakdown voltages $V_{(BR)EBO}$ and $V_{(BR)CBO}$ but also the *cutoff currents* I_{EBO} and I_{CBO}, which are measured at a voltage below their relevant breakdown voltage. Similarly, for the collector–emitter portion, the cutoff currents I_{CO} and I_{CES} are specified; these are measured with the base open or short-circuited at a voltage below $V_{(BR)CEO}$ and $V_{(BR)CES}$, respectively. In general, the following applies:

$$I_{CES} < I_{CEO}$$

Maximum Power Dissipation

The *maximum power dissipation* is a particularly important limit value. The dissipated power is the power that the transistor converts to heat:

$$P_V = V_{CE}I_C + V_{BE}I_B \approx V_{CE}I_C$$

This is produced predominantly in the junction of the collector diode. The junction temperature rises to a level at which, due to the temperature gradient, the heat can be dissipated from the junction to the environment via the case; Sect. 2.1.6 describes this in more detail.

The junction temperature must not exceed a material-specific limit, which is 175 °C (345 °F) for silicon; for safety reasons a limit temperature of 150 °C (300 °F) for silicon is used in calculations. For this limit value the maximum power dissipation depends on the construction and installation of the transistor; in the data sheet it is called P_{tot} and is given for two situations:

– Operation in an upright position on a printed circuit board without any additional means of cooling at a free-air temperature of $T_A = 25$ °C (77 °F); the subscript A means *ambient* air.

– Operation at a *case temperature* of $T_C = 25\,°C$ (77 °F); there are no stipulations as to the means of cooling to achieve this case temperature.

The two maximum values are called $P_{V,25(A)}$ and $P_{V,25(C)}$. With small-power transistors designed for upright installation without heat sink only $P_{tot} = P_{V,25(A)}$ is quoted; often the steady state case temperature T_C is stated as well. With power transistors that are only be used with heat sinks, $T_{tot} = P_{V,25(C)}$ is quoted. $T_A = 25\,°C$ (77 °F) or $T_C = 25\,°C$ (77 °F) usually cannot be guaranteed in practice. As P_{tot} decreases with an increase in temperature, the data sheet often provides a *power derating curve* that plots P_{tot} versus T_A or T_C (see Fig. 2.15a). Section 2.1.6 describes the thermal performance in more detail.

Safe Operating Area

The *safe operating area (SOA)* is derived from the limit data in the output characteristics; it is limited by the maximum collector current $I_{C,max}$, the collector–emitter breakdown voltage $V_{(BR)CEO}$, the maximum power dissipation P_{tot} and the boundary of the region of secondary breakdown. Figure 2.14 shows the SOA in a linear plot and a diagram with logarithmic scales on both axes. The linear plot shows hyperbolas for both maximum power dissipation and secondary breakdown [2.2]:

$$\text{Power dissipation:} \quad I_{C,max} = \frac{P_{tot}}{V_{CE}}$$

$$\text{Secondary breakdown:} \quad I_{C,max} \approx \frac{\text{const.}}{V_{CE}^2}$$

In the plot with logarithmic scales on both axes, the hyperbolas become straight lines with a slope of -1 or -2.

For low-power transistors the curve for secondary breakdown remains above the maximum power dissipation even at high voltages; thus, it does not represent the SOA limit. For power transistors, there are additional limit curves for pulsed operation with various

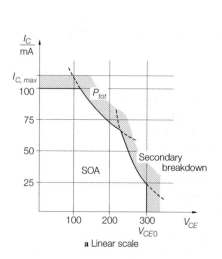

a Linear scale

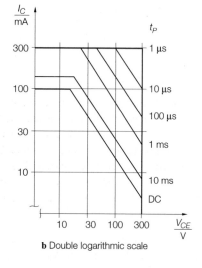

b Double logarithmic scale

Fig. 2.14. Safe operating area (SOA)

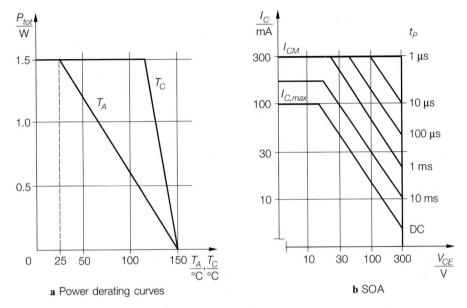

a Power derating curves   **b** SOA

Fig. 2.15. Limiting curves of a high-voltage switching transistor

pulse durations. If the pulse duration is very short and the duty cycle is low, the transistor can be operated at the maximum voltage $V_{(BR)CEO}$ *and* the maximum collector current I_{CM} *at the same time*; in this case, the SOA is a square. For this reason, the transistor can switch loads the power of which is high compared to the maximum power dissipation; Sect. 2.1.6 describes this in more detail.

Figure 2.15b shows the SOA of a high-voltage switching transistor that comes in three different versions with $V_{(BR)CEO} = 160\,\mathrm{V}$, $250\,\mathrm{V}$ or $300\,\mathrm{V}$. The maximum continuous current is $I_{C,max} = 100\,\mathrm{mA}$ and the maximum permissible peak current for one pulse of $1\,\mathrm{ms}$ duration is $I_{CM} = 300\,\mathrm{mA}$. If the pulse duration is shorter than $1\,\mu\mathrm{s}$, the SOA is a square. It allows the switching of loads with a power dissipation of up to $P = V_{(BR)CEO} I_{C,max} = 90\,\mathrm{W} \gg P_{tot} = 1.5\,\mathrm{W}$.

2.1.6
Thermal Performance

The arrangement shown in Fig. 2.16 is used to explain the thermal performance. The elements shown are provided with insulation on the outside and have the temperatures T_1, T_2 and T_3; $C_{th,2}$ is the *thermal capacity* (*thermal storage capacity*) of the centre element. The differences in temperature cause the heat flows P_{12} and P_{23},[1] which can be calculated using the *thermal resistances* $R_{th,12}$ and $R_{th,23}$ of the junctions:

$$P_{12} = \frac{T_1 - T_2}{R_{th,12}} \quad ; \quad P_{23} = \frac{T_2 - T_3}{R_{th,23}}$$

---

[1] In thermodynamics, Θ is the symbol used to detote heat flux. Here, we use P, since in electrical components heat flows are caused by the power dissipation P_V.

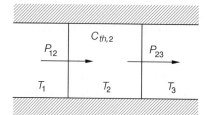

Fig. 2.16. An illustration of the thermal performance

Balancing the heat flows makes it possible to determine the *thermal quantity* $Q_{th,2}$ stored in the centre element and the temperature T_2:

$$Q_{th,2} = C_{th,2}T_2$$

$$\frac{dQ_{th,2}}{dt} = P_{12} - P_{23} \quad \Rightarrow \quad \frac{dT_2}{dt} = \frac{P_{12} - P_{23}}{C_{th,2}}$$

If the temperatures T_1 and T_3 are constant, temperature T_2 changes until $P_{12} = P_{23}$; in this state the heat flows into and out of the centre element balance and T_2 remains constant. If the supplied heat flow P_{12} is constant and the right-hand element represents the *ambient air*, with an ambient temperature of $T_3 = T_A$, the centre element warms up to the temperature $T_2 = T_3 + R_{th,23}P_{23}$; here, too, the steady state is $P_{12} = P_{23}$.

Thermal equivalent circuit: For the thermal performance an electrical equivalent circuit diagram can be used. The parameters *heat flow, thermal resistance, thermal capacity* and *temperature* correspond to the electrical parameters *current, resistance, capacity* and *voltage*. For a transistor the elements *junction* (J), *case* (C), *ambient air* (A) and, if used, *heat sink* (H) are important. The heat dissipation P_V is assigned to the junction as s heat flow; the ambient temperature T_A is constant. The resulting *thermal equivalent circuit* shown in Fig. 2.17 allows the calculation of the time characteristics of the temperatures T_J, T_C and T_H using the known time characteristic of P_V.

Operation without a heat sink: Where no heat sink is used, $R_{th,CH}$, $R_{th,HA}$ and $C_{th,H}$ are replaced by the heat resistance $R_{th,CA}$ between the case and the ambient air. On the data sheet for a transistor, the thermal resistance $R_{th,JA}$ between the junction and ambient air is often quoted with the transistor installed upright on a printed circuit board and operated without a heat sink:

$$R_{th,JA} = R_{th,JC} + R_{th,CA}$$

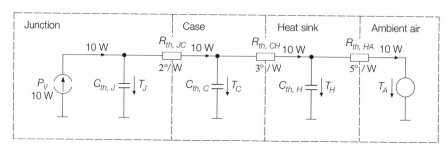

Fig. 2.17. Thermal equivalent circuit of a transistor with a heat sink

Operation with a heat sink: The thermal resistance $R_{th,HA}$ of the heat sink is stated on the data sheet for the heat sink; it depends on the size, design and installation position. The thermal resistance $R_{th,CH}$ depends on the transistor's installation on the heat sink; in order to prevent any loss in the efficiency of the heat sink it must be kept low through a specific heat transfer compound (thermolube). The use of insulating washers for electrical isolation between a transistor and a heat sink may cause $R_{th,CH}$ to become so high that the efficiency of a large heat sink with a low $R_{th,HA}$ value is drastically reduced; the relation $R_{th,CH} < R_{th,HA}$ should always be maintained. Therefore:

$$R_{th,JA} = R_{th,JC} + R_{th,CH} + R_{th,HA}$$

If several transistors are mounted on a common heat sink, the equivalent circuit contains several junctions and cases that are connected to the heat sink *node*.

SMD transistors: Transistors that employ SMD technology dissipate the heat to the circuit board via the connecting wires. The thermal resistance between the junction and the soldering point is called $R_{th,JS}$ on the data sheet; the subscript S indicates the *soldering point*. This results in:

$$R_{th,JA} = R_{th,JS} + R_{th,SA}$$

Thermal Performance in Static Operation

In static operation the power dissipation P_V is constant and depends only on the operating point; this also applies to small-signal operation:

$$\boxed{P_V = V_{CE,A} I_{C,A}} \tag{2.16}$$

The temperature of the junction is:

$$T_J = T_A + P_V R_{th,JA} \tag{2.17}$$

Thus, the maximum permissible *static* power dissipation is:

$$\boxed{P_{V,max(stat)} = \frac{T_{J,limit} - T_{A,max}}{R_{th,JA}}} \tag{2.18}$$

$T_{J,limit} = 150\,°C\ (300\,°F)$ is used for calculations with silicon transistors. $T_{A,max}$ must be given for the specific application and determines the maximum ambient temperature allowed for circuit operation.

The data sheet of a transistor quotes $P_{V,max(stat)}$ as a function of T_A and/or T_C; Fig. 2.15a shows the *power derating curves*. Their down-sloping portion is described by (2.18) when the respective values for T and R_{th} are inserted:

$$P_{V,max(stat)}(T_A) = \frac{T_{J,limit} - T_A}{R_{th,JA}}$$

$$P_{V,max(stat)}(T_C) = \frac{T_{J,limit} - T_C}{R_{th,JC}}$$

Therefore, the thermal resistances $R_{th,JA}$ and $R_{th,JC}$ can also be determined from the negative slope of these curves.

Thermal Performance in Pulsed Operation

In pulsed operation the maximum power dissipation $P_{V,max(puls)}$ may exceed the maximum static power dissipation $P_{V,max(stat)}$ according to (2.18). Using the *pulse duration* t_p, the *repetition rate* $f_W = 1/T_W$ and the *duty factor* $D = t_p f_W$, one can determine the mean power dissipation $\overline{P}_V = D P_{V(puls)}$ from the power dissipation $P_{V(puls)}$; the power dissipation in the off-state is negligible. In the on-state T_J increases, while in the cutoff state it decreases. This results in a sawtooth-shaped curve for T_J. The mean value of $\overline{T}_J$ can be calculated from $\overline{P}_V$ using (2.17). The more important maximum value $T_{J,max}$ depends on the ratio between the pulse parameters t_p and D and the thermal time constant; the latter is determined from the thermal capacities and the thermal resistances. The condition $T_{J,max} < T_{J,limit}$ gives us the maximum power dissipation $P_{V,max(puls)}$.

Determination of the maximum power dissipation in pulsed operation: In practice there are two ways to determine $P_{V,max(puls)}$:

– First, the maximum static power dissipation $P_{V,max(stat)}$ is calculated using (2.18) and from this the value of $P_{V,max(puls)}$; for this purpose the data sheet shows the plot of the ratio $P_{V,max(puls)}/P_{V,max(stat)}$ for several values of D versus t_p (see Fig. 2.18a). With a declining pulse duration t_p the amplitude of the sawtooth-shaped portion of the T_J curve decreases more and more; $t_p \to 0$ results in $\overline{T}_J = T_{J,max}$ and thus:

$$\lim_{t_P \to 0} \frac{P_{V,max(puls)}}{P_{V,max(stat)}} = \frac{1}{D}$$

These limit values can be taken from the vertical axis of the graph in Fig. 2.18a: for $D = 0.5$, and with a very short pulse duration the ratio is $P_{V,max(puls)} = 2P_{V,max(stat)}$, and so on.

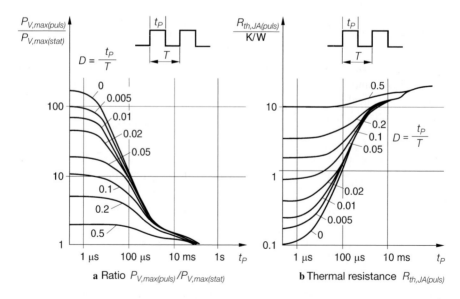

Fig. 2.18. Determination of the maximum power dissipation $P_{V,max(puls)}$

- The data sheet contains values for the thermal resistance in pulsed operation from which $P_{V,max(puls)}$ can be calculated directly:

$$P_{V,max(puls)}(t_P, D) = \frac{T_{J,limit} - T_{A,max}}{R_{th,JA(puls)}(t_P, D)} \tag{2.19}$$

The data sheet shows a plot of $R_{th,JA(puls)}$ for different values of D versus t_P (see Fig. 2.18b).

The two procedures are equivalent. Apart from a constant factor, the ratio $P_{V,max(puls)}/P_{V,max(stat)}$ is basically the reciprocal value of $R_{th,JA(puls)}$:

$$\frac{P_{V,max(puls)}}{P_{V,max(stat)}} = \frac{T_{J,limit} - T_{A,max}}{R_{th,JA(puls)}} \frac{1}{P_{V,max(stat)}} \sim \frac{1}{R_{th,JA(puls)}}$$

2.1.7
Temperature Sensitivity of Transistor Parameters

The characteristics of a bipolar transistor are highly sensitive to temperature. Of particular importance is the temperature-dependent relationship between I_C and V_{BE}. Where the dependence of V_{BE} and of temperature T are explicitly quoted, the current is:

$$I_C(V_{BE}, T) = I_S(T) e^{\frac{V_{BE}}{V_T(T)}} \left(1 + \frac{V_{CE}}{V_A}\right)$$

The reason for the temperature sensitivity of I_C is the temperature sensitivity of the reverse current I_S and the temperature voltage V_T [2.2], [2.4]:

$$V_T(T) = \frac{kT}{q} = 86.142 \frac{\mu V}{K} T$$

$$I_S(T) = I_S(T_0) e^{\left(\frac{T}{T_0} - 1\right)\frac{V_G(T)}{V_T(T)}} \left(\frac{T}{T_0}\right)^{x_{T,I}} \qquad \text{with } x_{T,I} \approx 3 \tag{2.20}$$

Here, $k = 1.38 \cdot 10^{-23}$ VAs/K is *Boltzman's constant*, $q = 1.602 \cdot 10^{-19}$ As is the *elementary charge* and $V_G = 1.12$ V is the *gap voltage* of silicon; the low-temperature sensitivity of V_G is negligible.

The relative change in I_S is derived by differentiation of $I_S(T)$:

$$\frac{1}{I_S}\frac{dI_S}{dT} = \frac{1}{T}\left(3 + \frac{V_G}{V_T}\right) \overset{T=300\,K}{\approx} 0.15\,K^{-1}$$

If the temperature increases by 1 K then I_S rises by 15 %. The change in I_C can be calculated accordingly:

$$\frac{1}{I_C}\frac{dI_C}{dT}\bigg|_{V_{BE}=\text{const.}} = \frac{1}{T}\left(3 + \frac{V_G - V_{BE}}{V_T}\right) \overset{\substack{T=300\,K \\ V_{BE}=0.7\,V}}{\approx} 0.065\,K^{-1}$$

With an increase in temperature of 11 K the collector current I_C doubles its value. Therefore it is not possible to select an operating point A that is not affected by the temperature for small-signal operation by a predetermined value of $V_{BE,A}$; instead, $I_{C,A}$ must be approximately constant with temperature since the small-signal parameters depend on $I_{C,A}$ and not on $V_{BE,A}$ (see Sect. 2.1.4). In cases in which $I_{C,A}$ is almost unresponsive to temperature changes,

$$dI_C = \frac{\partial I_C}{\partial T} dT + \frac{\partial I_C}{\partial V_{BE}} dV_{BE} \equiv 0$$

can be used to calculate the temperature sensitivity of V_{BE}:

$$\left.\frac{dV_{BE}}{dT}\right|_{I_C=\text{const.}} = \frac{V_{BE} - V_G - 3V_T}{T} \overset{\substack{T=300\,\text{K}\\ V_{BE}=0.7\,\text{V}}}{\approx} -1.7\,\frac{\text{mV}}{\text{K}} \tag{2.21}$$

The current gain B is also temperature-responsive [2.2]:

$$B(T) = B(T_0)\, e^{\left(\frac{T}{T_0}-1\right)\frac{\Delta V_{dot}}{V_T(T)}}$$

The voltage ΔV_{dot} is a matter constant and amounts to approx 44 mV for npn transistors made of silicon. Differentiation leads to:

$$\frac{1}{B}\frac{dB}{dT} = \frac{\Delta V_{dot}}{V_T T} \overset{T=300\,\text{K}}{\approx} 5.6 \cdot 10^{-3}\,\text{K}^{-1}$$

In practice a simplified relationship is often used [2.4]:

$$B(T) = B(T_0)\left(\frac{T}{T_0}\right)^{x_{T,B}} \qquad \text{with } x_{T,B} \approx 1.5 \tag{2.22}$$

For the region used in practice the same temperature sensitivity is obtained:

$$\frac{1}{B}\frac{dB}{dT} = \frac{x_{T,B}}{T} \overset{T=300\,\text{K}}{\approx} 5 \cdot 10^{-3}\,\text{K}^{-1} \tag{2.23}$$

With a temperature increase of 1 K, the current gain increases by about 0.5 %. This, however, is of secondary practical importance, since the deviations in current gain resulting from the manufacturing process are much higher. It is meaningful only in differential considerations – such as, for example, in the calculation of the temperature coefficient of a circuit.

2.2
Design of a Bipolar Transistor

In general, the bipolar transistor has an asymmetric design. This clearly enables a distinction to be made between the collector and the emitter and causes the differing behavior in the normal and inverse modes as described below. Discrete and integrated transistors are comprised of more than three regions, with the collector region consisting of at least two sub-regions. Therefore, the type designations "npn" and "pnp" only describe the sequence of the active inner regions. Transistors are produced in a multi-stage process on a *wafer* that is subsequently cut by sawing into little *chips*. Each chip carries either a single transistor or an array of several integrated transistors and other components; that is, an *integrated circuit (IC)*.

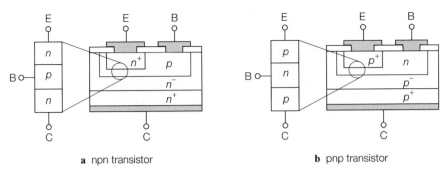

a npn transistor   **b** pnp transistor

Fig. 2.19. Design of a chip with a single epitaxial planar transistor

2.2.1
Discrete Transistors

Internal design: Discrete transistors are manufactured predominantly by using the epi-taxial planar technique. Fig. 2.19 shows the construction principles of npn and pnp transistors with emphasis on the active region. Doping is heavy in the n^+ and p^+ layers, medium in the n and p layers and weak in the n^- and p^- layers. This particular layering of differently doped regions improves the electrical properties of the transistor. The bottom layer of the chip forms the collector while the base and emitter are at the top.

Case: The transistor is mounted in a case by soldering the underside to the collector wire or a metallic part of the case. The two other connections are bonded with fine gold or aluminum wires to the relevant connecting wires. Figure 2.20 shows a low-power and a power transistor after soldering and bonding. Finally the low-power transistor is cast in a plastic material; the case of the power transistor is closed off with a lid.

For the various designs and applications there is a multitude of cases that differ in their maximum heat dissipation capacity and that are adapted to specific geometric requirements. Figure 2.21 shows a selection of the most common models. Power transistors have cases

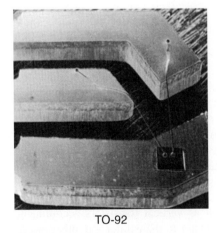

TO-92   TO-3

Fig. 2.20. Connections inside the case

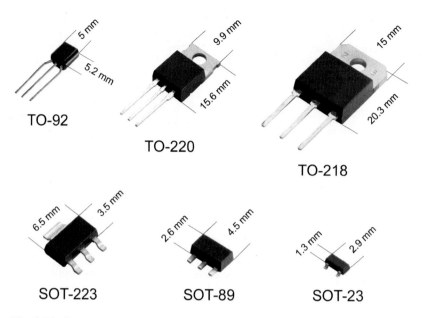

Fig. 2.21. Common case models for discrete transistors

intended for installation in heat sinks; they have a large contact surface to enhance the heat flow. SMD transistors for high power ratings are provided with two collector wires for improved heat conductance to the circuit board. Special case models are used for high-frequency transistors, as the electrical behavior at frequencies in the GHz range is heavily dependent on the geometry; some cases have two emitter connections in order to increase the contact to earth.

Complementary transistors: Due to the separately optimized manufacturing processes for npn and pnp transistors it is not too difficult to produce *complementary* transistors. An npn and a pnp transistor are complementary if their electric characteristics are identical with the exception of the current and voltage signs.

2.2.2
Integrated Transistors

Integrated transistors are also produced using the epitaxial planar technique. Here, too, the collector terminal is placed on the upper side of the chip. The individual transistors are electrically isolated from each other by reverse-biased pn junctions. Only a very thin layer close to the surface forms the active region of the transistors. The depth of the chip is called the *substrate* (S) and represents a fourth connection common to all transistors; it is also arranged on the top. As npn and pnp transistors have to be produced in the same process, the two types deviate considerably in construction and electrical data.

Internal design: Since npn transistors are built as vertical transistors according to Fig. 2.22, the current flows vertically from the collector to the emitter – in other words, its flow is perpendicular to the plane of the chip. In contrast, pnp transistors are usually

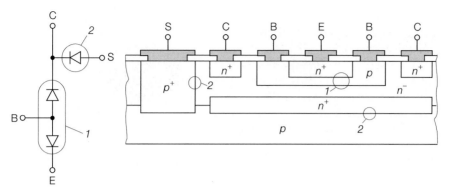

Fig. 2.22. Equivalent diode circuit and design of an integrated vertical npn transistor

constructed as lateral transistors according to Fig. 2.23; the current flows laterally – in other words, parallel to the surface of the chip.

Substrate diodes: The equivalent diode circuits illustrated in Figs. 2.22 and 2.23 show an additional substrate diode arranged between the collector and substrate of the vertical npn transistor or between the base and substrate of the lateral pnp transistor. The substrate is connected to the negative supply voltage so that the diodes are always reverse-biased and act as isolators between the transistors themselves and between the transistors and the substrate.

Differences between vertical and lateral transistors: As the thickness of the base region can be kept thinner in a vertical transistor, the current gain exceeds that of a lateral transistor by a factor of 3...10; the response time and the cutoff frequencies are also significantly higher in the vertical transistor. Therefore, more and more vertical pnp transistors are now being produced. Their design corresponds to that of a vertical npn transistor if the n and p doping are interchanged in every region. Isolation from the substrate is achieved by embedding the transistor in a tub of n-doped material that is connected to the positive supply voltage. In this case, the npn and pnp transistors are called *complementary* – even if their electrical data do not conform as well as is the case in complementary discrete transistors.

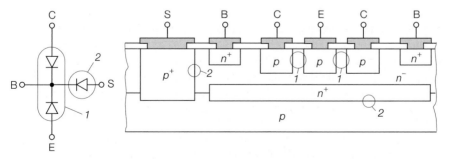

Fig. 2.23. Equivalent diode circuit and design of an integrated lateral pnp transistor

2.3
Models of Bipolar Transistors

Section 2.1.2 described the *static performance* of a bipolar transistor in normal mode by means of large-signal (2.5) and (2.6); secondary effects were not taken into account or were only considered qualitatively as in the description of the current gain in Sect. 2.1.3. Computer-aided circuit design using CAD programs requires a model that allows for all effects, applies to all operating modes and, in addition, correctly reflects the *dynamic performance*. Linearization at the operating point of this *large-signal model* leads to the *dynamic small-signal model* that is necessary for calculating the frequency response of a circuit.

2.3.1
Static Performance

The static performance is demonstrated for an npn transistor; for the pnp transistor all currents and voltages have the opposite polarity. The simplest model of a bipolar transistor is the *Ebers–Moll model*, based on the equivalent diode circuit. This model has only three parameters and describes all primary effects. More accurate modelling necessitates a conversion that first leads to the *transport model* and, after adding other parameters to describe the secondary effects, to the *Gummel–Poon model*; the latter is capable of a very accurate description of the static performance and is used in CAD programs.

Ebers–Moll Model

An npn transistor consists of two pn diodes in anti-serial connection with a common p region. The diodes are called emitter or BE diode and collector or BC diode. The functionality of a bipolar transistor depends on the fact that large portions of the diode currents can be drained through the third connection because of the very thin common-base region. Therefore, the *Ebers–Moll model* in Fig. 2.24 comprises the two diodes of the equivalent diode circuit and two current-controlled current sources that describe the current flow through the base. The controlling factors of the controlled sources are called A_N for normal operation and A_I for inverse operation; typical values are $A_N \approx 0.98...0.998$ and $A_I \approx 0.5...0.9$. The different values for A_N and A_I are a consequence of the asymmetric construction explained in Sect. 2.2.

Basic equations: From the emitter and collector diode currents

$$I_{D,N} = I_{S,N} \left(e^{\frac{V_{BE}}{V_T}} - 1 \right)$$

$$I_{D,I} = I_{S,I} \left(e^{\frac{V_{BC}}{V_T}} - 1 \right)$$

the currents through the terminals can be calculated according to Fig. 2.24 [2.5]:

$$I_C = A_N I_{S,N} \left(e^{\frac{V_{BE}}{V_T}} - 1 \right) - I_{S,I} \left(e^{\frac{V_{BC}}{V_T}} - 1 \right)$$

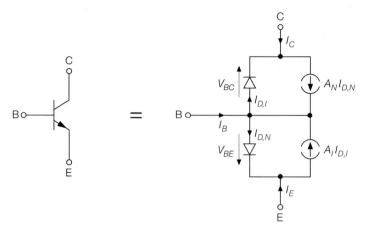

Fig. 2.24. Ebers–Moll model for an npn transistor

$$I_E = -I_{S,N}\left(e^{\frac{V_{BE}}{V_T}} - 1\right) + A_I I_{S,I}\left(e^{\frac{V_{BC}}{V_T}} - 1\right)$$

$$I_B = (1 - A_N)I_{S,N}\left(e^{\frac{V_{BE}}{V_T}} - 1\right) + (1 - A_I)I_{S,I}\left(e^{\frac{V_{BC}}{V_T}} - 1\right)$$

The following equation can be deduced from the theorem on reciprocal networks:

$$A_N I_{S,N} = A_I I_{S,I} = I_S$$

Therefore, the model is completely parameterized by A_N, A_I and I_S.

Normal operation: In normal mode the BC diode is reverse-biased since $V_{BC} < 0$; together with the related controlled source this can be neglected since $I_{D,I} \approx -I_{S,I} \approx 0$. In addition, for $V_{BE} \gg V_T$ the term -1 can be neglected compared with the exponential function, which leads to:

$$I_C = I_S\, e^{\frac{V_{BE}}{V_T}}$$

$$I_E = -\frac{1}{A_N} I_S\, e^{\frac{V_{BE}}{V_T}}$$

$$I_B = \frac{1 - A_N}{A_N} I_S\, e^{\frac{V_{BE}}{V_T}} = \frac{1}{B_N} I_S\, e^{\frac{V_{BE}}{V_T}}$$

Figure 2.25a shows the reduced model with the most significant relationships; in this case, A_N is the *current gain in the common-base circuit* and B_N is the *current gain in the*

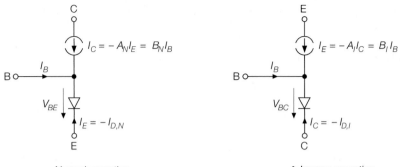

a Normal operation   **b** Inverse operation

Fig. 2.25. Reduced Ebers–Moll model for an npn transistor

common-emitter circuit:[2]

$$A_N = -\frac{I_C}{I_E}$$

$$B_N = \frac{A_N}{1 - A_N} = \frac{I_C}{I_B}$$

Typical values are $A_N \approx 0.98...0.998$ and $B_N \approx 50...500$.

Inverse operation: The reduced model shown in Fig. 2.25b for inverse operation is arrived at in a similar way; the current gains are:

$$A_I = -\frac{I_E}{I_C}$$

$$B_I = \frac{A_I}{1 - A_I} = \frac{I_E}{I_B}$$

Typical values are $A_I \approx 0.5...0.9$ and $B_I \approx 1...10$.

Saturation voltage: The transistor used as a switch is driven from normal operation into saturation; what is interesting here is the achievable minimum collector–emitter voltage $V_{CE,sat}(I_B, I_C)$. This voltage is:

$$V_{CE,sat} = V_T \ln \frac{B_N (1 + B_I) (B_I I_B + I_C)}{B_I^2 (B_N I_B - I_C)}$$

For $0 < I_C < B_N I_B$ the voltage is $V_{CE,sat} \approx 20...200\,\text{mV}$.
   The minimum of $V_{CE,sat}$ is reached when $I_C = 0$:

$$V_{CE,sat}(I_C = 0) = V_T \ln \left(1 + \frac{1}{B_I}\right) = -V_T \ln A_I$$

---

[2] For the current gains a distinction must be made between model parameters and measurable external current gains. In the Ebers–Moll model the model parameters A_N and B_N of normal operation and A_I and B_I of inverse operation are identical to the external current gains; therefore, they may be defined as external currents.

After exchanging emitter and collector, the voltage for $I_E = 0$ when switching from inverse operation into saturation is:

$$V_{EC,sat}(I_E = 0) = V_T \ln\left(1 + \frac{1}{B_N}\right) = -V_T \ln A_N$$

Due to $A_I < A_N < 1$ the following applies: $V_{EC,sat}(I_E = 0) < V_{CE,sat}(I_C = 0)$. Typical values are $V_{CE,sat}(I_C = 0) \approx 2...20\,\mathrm{mV}$ and $V_{EC,sat}(I_E = 0) \approx 0.05...0.5\,\mathrm{mV}$.

Transport Model

An equivalence conversion transforms the Ebers–Moll model into the *transport model* [2.5] shown in Fig. 2.26; it has only one controlled source and it forms the basis for modelling other effects, as described in the next section.

General equations: Using the currents

$$I_{B,N} = \frac{I_S}{B_N}\left(e^{\frac{V_{BE}}{V_T}} - 1\right) \tag{2.24}$$

$$I_{B,I} = \frac{I_S}{B_I}\left(e^{\frac{V_{BC}}{V_T}} - 1\right) \tag{2.25}$$

$$I_T = B_N I_{B,N} - B_I I_{B,I} = I_S\left(e^{\frac{V_{BE}}{V_T}} - e^{\frac{V_{BC}}{V_T}}\right) \tag{2.26}$$

the following currents are obtained from Fig. 2.26:

$$I_B = \frac{I_S}{B_N}\left(e^{\frac{V_{BE}}{V_T}} - 1\right) + \frac{I_S}{B_I}\left(e^{\frac{V_{BC}}{V_T}} - 1\right)$$

$$I_C = I_S\left(e^{\frac{V_{BE}}{V_T}} - \left(1 + \frac{1}{B_I}\right)e^{\frac{V_{BC}}{V_T}} + \frac{1}{B_I}\right)$$

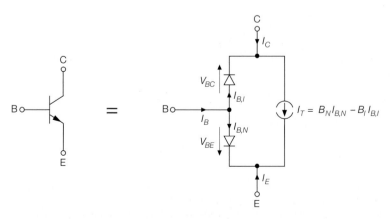

Fig. 2.26. Transport model for a npn transistor

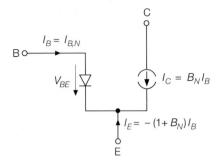

Fig. 2.27. Reduced transport model for normal operation

$$I_E = I_S \left(-\left(1 + \frac{1}{B_N}\right) e^{\frac{V_{BE}}{V_T}} + e^{\frac{V_{BC}}{V_T}} + \frac{1}{B_N} \right)$$

Normal operation: In normal operation, if the reverse currents are neglected, the following equations are obtained:

$$I_B = \frac{I_S}{B_N} e^{\frac{V_{BE}}{V_T}}$$

$$I_C = I_S\, e^{\frac{V_{BE}}{V_T}}$$

Taking the relationship between A_N and B_N into consideration these equations are identical to those of the Ebers-Moll model. Figure 2.27 shows the reduced transport model for normal operation.

Properties: The transport model describes the primary DC current response of the bipolar transistor under the assumption that the emitter and collector diodes are ideal. An important property of the model is that the *transport current* I_T flowing through the base region occurs *separately*; this is not the case in the Ebers–Moll model. As in the Ebers–Moll model three parameters are necessary for the description: I_S, B_N and B_I [2.5].

Other Effects

The transport model can be expanded for a more accurate description of the static performance. The resulting modeled effects have already been described qualitatively in Sects. 2.1.2 and 2.1.3:

- The recombination of the charge carriers in the pn junctions generates additional *leakage currents* in the emitter and collector diodes; these currents add to the base current and have no influence on the transport current I_T.
- With large currents the transport current I_T is smaller than the value resulting from (2.26). This *high current effect* is caused by the significantly increased charge carrier concentration in the base region; this is also called *strong injection*.
- The voltages V_{BE} and V_{BC} influence the effective thickness of the base region and thus also influence the transport current I_T; this is called the *Early effect*.

Leakage currents: In order to take the leakage currents into account, the transport model is expanded by two more diodes with the currents [2.5]:

$$I_{B,E} = I_{S,E} \left(e^{\frac{V_{BE}}{n_E V_T}} - 1 \right) \tag{2.27}$$

$$I_{B,C} = I_{S,C} \left(e^{\frac{V_{BC}}{n_C V_T}} - 1 \right) \tag{2.28}$$

This requires four additional model parameters: The *leakage saturation reverse currents* $I_{S,E}$ and $I_{S,C}$ and the *emission coefficients* $n_E \approx 1.5$ and $n_C \approx 2$.

High-current effect and the Early effect: The influence of the high-current and Early effects on the transport current I_T is reflected by the nondimensional quantity q_B [2.5]:

$$I_T = \frac{B_N I_{B,N} - B_I I_{B,I}}{q_B} = \frac{I_S}{q_B} \left(e^{\frac{V_{BE}}{V_T}} - e^{\frac{V_{BC}}{V_T}} \right) \tag{2.29}$$

General equations: The currents $I_{B,N}$ and $I_{B,I}$ are available from (2.24) and (2.25). Figure 2.28 shows the expanded model. The currents are:

$$I_B = I_{B,N} + I_{B,I} + I_{B,E} + I_{B,C}$$

$$I_C = \frac{B_N}{q_B} I_{B,N} - \left(\frac{B_I}{q_B} + 1 \right) I_{B,I} - I_{B,C}$$

$$I_E = - \left(\frac{B_N}{q_B} + 1 \right) I_{B,N} + \frac{B_I}{q_B} I_{B,I} - I_{B,E}$$

Definition of q_B: The quantity q_B is a measure for the *relative majority carrier charge* in the base and is made up of the quantities q_1, which describes the Early effect, and q_2,

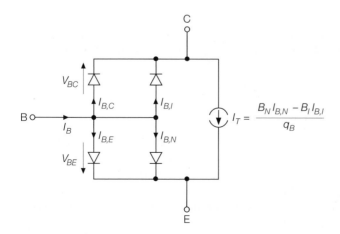

Fig. 2.28. Expanded transport model for a npn transistor

which describes the high-current effect.[3]

$$q_B = \frac{q_1}{2}\left(1 + \sqrt{1 + 4q_2}\right) \tag{2.30}$$

$$q_1 = \frac{1}{1 - \dfrac{V_{BE}}{V_{A,I}} - \dfrac{V_{BC}}{V_{A,N}}}$$

$$q_2 = \frac{I_S}{I_{K,N}}\left(e^{\frac{V_{BE}}{V_T}} - 1\right) + \frac{I_S}{I_{K,I}}\left(e^{\frac{V_{BC}}{V_T}} - 1\right)$$

The *Early voltages* $V_{A,N}$ and $V_{A,I}$ and the *knee-point currents for strong injection* $I_{K,N}$ and $I_{K,I}$ are required as further model parameters. The Early voltages range between 30 V and 150 V, but lower values are possible with integrated and high-frequency transistors. The knee-point currents depend on the size of the transistor; they are in the milliampere range for low power transistors and in the ampere range for power transistors.

Influence of q_B in normal operation: The influence of q_B becomes obvious when the collector current in normal operation is observed. When the reverse currents are ignored, we obtain:

$$I_C = \frac{B_N}{q_B}I_{B,N} = \frac{I_S}{q_B}e^{\frac{V_{BE}}{V_T}} \tag{2.31}$$

– $q_2 \ll 1$ for small and medium currents and thus $q_B \approx q_1$. Due to $V_{BE} \approx 0.6\dots0.8$ V we have $V_{BE} \ll V_{A,I}$ and $V_{BC} = V_{BE} - V_{CE} \approx -V_{CE}$; this results in an approximation for q_1:

$$q_1 \approx \frac{1}{1 + \dfrac{V_{CE}}{V_{A,N}}}$$

From insertion into (2.31) it follows that:

$$I_C \approx I_S\, e^{\frac{V_{BE}}{V_T}}\left(1 + \frac{V_{CE}}{V_{A,N}}\right) \qquad \text{for } I_C < I_{K,N}$$

When considering $V_A = V_{A,N}$ this equation corresponds to the large-signal (2.5) of Sect. 2.1.2.[4]

– $q_2 \gg 1$ for high currents and thus $q_B \approx q_1\sqrt{q_2}$; when using the approximation for q_1 mentioned above this results in:

$$I_C \approx \sqrt{I_S I_{K,N}}\, e^{\frac{V_{BE}}{2V_T}}\left(1 + \frac{V_{CE}}{V_{A,N}}\right) \qquad \text{for } I_C \to \infty$$

---

[3] In the specialist literature (for example [2.5]) another expression for q_B is often found; the expression quoted here is used by *Spice* [2.4, 2.6].

[4] The large-signal equations in Sect. 2.1.2 only apply to normal operation; additional identification by a subscript N is therefore not necessary.

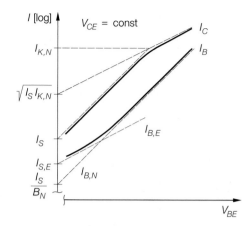

Fig. 2.29. A semilogarithmic plot of currents I_B and I_C in normal operation (a Gummel plot)

Figure 2.29 shows the curves of I_C and I_B in the semilogarithmic plot and illustrate the importance of parameters $I_{K,N}$ and $I_{S,E}$. When the reverse currents are not taken into consideration, I_B is:

$$I_B = \frac{I_S}{B_N} e^{\frac{V_{BE}}{V_T}} + I_{S,E} e^{\frac{V_{BE}}{n_E V_T}} \tag{2.32}$$

Comparing the curves in Fig. 2.29 with the measured values plotted in Fig. 2.6 on page 37 shows that parameters $I_{K,N}$, $I_{S,E}$ and n_E offer a very good description of the true behavior in normal operation; the same is true of the parameters $I_{K,I}$, $I_{S,C}$ and n_C in inverse operation.

Current Gain in Normal Operation

The current gain has been qualitatively described in Sect. 2.1.3 and shown as a diagram on page 38. A full description of B is possible using (2.31) for I_C and (2.32) for I_B:

$$B = \frac{I_C}{I_B} = \frac{B_N}{q_B + B_N \left(\dfrac{q_B}{I_S}\right)^{\frac{1}{n_E}} I_{S,E} I_C^{\left(\frac{1}{n_E} - 1\right)}}$$

$B = B(V_{BE}, V_{CE})$ since I_C and q_B depend on V_{BE} and V_{CE}; the qualitative relationship given in Sect. 2.1.2 is thus made into a quantitative one.

Course of the current gain: The formula $B = B(I_C, V_{CE})$ is better suited in practice but does not allow a full description of B. Instead, three portions can be distinguished:

– For small collector currents the leakage current $I_{B,E}$ is the dominant component of the base current so that $I_B \approx I_{B,E}$; for $q_B \approx q_1$ this leads to

$$B \approx \frac{I_C^{\left(1 - \frac{1}{n_E}\right)}}{I_{S,E} \left(\dfrac{q_1}{I_S}\right)^{\frac{1}{n_E}}} \sim I_C^{\left(1 - \frac{1}{n_E}\right)} \left(1 + \frac{V_{CE}}{V_{A,N}}\right)^{\frac{1}{n_E}}$$

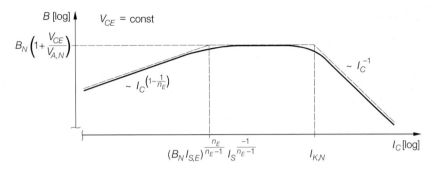

Fig. 2.30. Dependence of the large-signal current gain B on the collector current

Using $n_E \approx 1.5$ results in $B \sim I_C^{1/3}$. In this portion, B is smaller than for medium collector currents and increases with a rising collector current. This portion is called the *leakage current range*.

– $I_B \approx I_{B,N}$ for medium collector currents and thus:

$$B \approx B_N \left(1 + \frac{V_{CE}}{V_{A,N}}\right) \tag{2.33}$$

In this portion, B reaches a maximum and depends only to a small extent on I_C. This portion is called the *normal range*.

– For large collector currents the high-current effect sets in; for $I_B \approx I_{B,N}$ we obtain:

$$B \approx \frac{B_N}{q_B} \approx B_N \frac{I_{K,N}}{I_C} \left(1 + \frac{V_{CE}}{V_{A,N}}\right)^2$$

In this portion B is proportional to the reciprocal value of I_C and thus decreases rapidly with an increasing collector current. This portion is called the *high-current range*.

Figure 2.30 shows a plot of B with both axes plotted using a logarithmic scale; the approximations for the three portions change to straight lines with slopes of $1/3$, 0 and -1. The boundaries of the portions are also shown:

Normal range $\leftrightarrow$ leakage current region : $I_C = \left(B_N I_{S,E}\right)^{\frac{n_E}{n_E-1}} I_S^{\frac{-1}{n_E-1}}$

Normal range $\leftrightarrow$ high current region   : $I_C = I_{K,N}$

Current gain maximum: The maximum value of B at a constant voltage V_{CE} is called $B_{max}(V_{CE})$ (see Fig. 2.7 on page 38 and (2.8)). For transistors with a low leakage current $I_{S,E}$ and a high knee-point current $I_{K,N}$, the normal range is so wide that B is almost tangential to the horizontal approximation given in (2.33). In this case, $B_{max}(V_{CE})$ is determined by (2.33), and the maximum value $B_{0,max}$ extrapolated for $V_{CE} = 0$ is determined by B_N. For transistors with a high leakage current and a low knee-point current the normal range may be very narrow or even nonexistent. In this case B is below the straight line of (2.33) and thus does not come up to the value indicated by this line; therefore $B_{0,max} < B_N$.

Substrate Diodes

Integrated transistors have a substrate diode that lies between the substrate and the collector in the vertical npn transistor and between the substrate and the base in the lateral pnp transistor (see Figs. 2.22 and 2.23). The current flowing through these diodes is expressed using the simple diode equation – for example, for vertical npn transistors:

$$I_{D,S} = I_{S,S} \left(e^{\frac{V_{SC}}{V_T}} - 1 \right) \tag{2.34}$$

A further parameter is the *substrate saturation reverse current* $I_{S,S}$. As these diodes are normally nonconductive, more accurate modelling is not required; the only thing that is important is the fact that a current can flow if a certain – that is, wrong – circuitry is connected to the substrate or the surrounding tub. For lateral pnp transistors it is necessary to replace V_{SC} by V_{SB}.

Spreading Resistances

To describe the static performance fully the spreading resistances must also be taken into account. In Fig. 2.31a a discrete transistor is used to illustrate these resistances:

- The *emitter spreading resistance* R_E is of a low value because of the high doping (n^+) and the small longitudinal/cross section ratio of the emitter region; typical values are $R_E \approx 0.1 \ldots 1\,\Omega$ for low-power transistors and $R_E \approx 0.01 \ldots 0.1\,\Omega$ for power transistors.
- The *collector spreading resistance* R_C is caused predominantly by the low doping (n^-) of a section in the collector region; typical values are $R_C \approx 1 \ldots 10\,\Omega$ for low-power transistors and $R_C \approx 0.1 \ldots 1\,\Omega$ for power transistors.
- The *base spreading resistance* R_B is formed by the *external base spreading resistance* R_{Be} between the base connection and the active base region plus the *internal base spreading resistance* R_{Bi} throughout the active base region. With high currents R_{Bi} has only a small effect as the current flows predominantly in a region close to the base connection due to the *current displacement* (*emitter edge displacement*). The Early

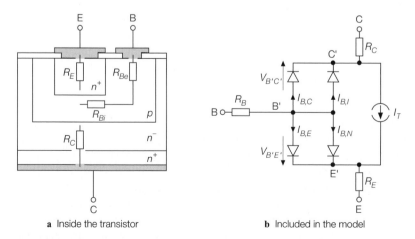

a Inside the transistor   b Included in the model

Fig. 2.31. Spreading resistances in a discrete transistor

effect on the thickness of the base region has an additional influence. These effects can be described by the quantity q_B according to (2.30)[5]:

$$R_B = R_{Be} + \frac{R_{Bi}}{q_B} \qquad (2.35)$$

Consequently, for normal operation the following applies:

$$R_B = \begin{cases} R_{Be} + R_{Bi}\left(1 + \dfrac{V_{CE}}{V_{A,N}}\right) & \text{for } I_C < I_{K,N} \\ R_{Be} & \text{for } I_C \to \infty \end{cases}$$

Typical values are $R_{Be} \approx 10 \ldots 100\,\Omega$ for low-power transistors and $R_{Be} \approx 1 \ldots 10\,\Omega$ for power transistors; the values of R_{Bi} are higher by a factor of $3 \ldots 10$.

Figure 2.31b shows the relevant expanded model. A distinction must now be made between the *external* connections B, C and E and the *internal* connections B', C' and E'; in other words, all diode currents and the transport current I_T no longer depend on V_{BE}, V_{BC} and V_{SC}, but on $V_{B'E'}$, $V_{B'C'}$ and $V_{SC'}$.

Effects of the spreading resistances: The voltages across the spreading resistances are very low in low-power transistors; therefore, the emitter and collector spreading resistances are often neglected. The base spreading resistance cannot be neglected because, even if it is very low, it influences the speed of operation and the cutoff frequencies. The voltage drop at R_B is only $1\,\text{mV}$ for $R_B = 100\,\Omega$ and $I_B = 10\,\mu\text{A}$ which are typical values for low-power transistors; the cutoff frequencies, however, are clearly reduced in most circuits. Therefore, the dependence of R_B on the operating point according to (2.35) must only be taken into account to reflect the dynamic performance exactly.

With power transistors working with high currents all spreading resistances must be taken into account; for $I_B = I_C/B$ and $I_E \approx -I_C$, the voltages are:

$$V_{BE} \approx V_{B'E'} + I_C \left(\frac{R_B}{B} + R_E\right)$$

$$V_{CE} \approx V_{C'E'} + I_C \left(R_C + R_E\right)$$

Here, the external voltages V_{BE} and V_{CE} may deviate sharply from the internal voltages $V_{B'E'}$ and $V_{C'E'}$. If a power transistor is used as a switch in the saturation region with $I_C = 5\,\text{A}$ and $B = 10$, the external voltages are $V_{BE} = 1.5\,\text{V}$ and $V_{CE,sat} = 1.85\,\text{V}$ with $V_{B'E'} = 0.75\,\text{V}$, $V_{C'E',sat} = 0.1\,\text{V}$, $R_B = 1\,\Omega$, $R_E = 0.05\,\Omega$ and $R_C = 0.3\,\Omega$. Due to the spreading resistances the values for V_{BE} and $V_{CE,sat}$ can be relatively high.

2.3.2
Dynamic Performance

The response of a transistor to pulsed or sinusoidal signals is called the *dynamic performance* and cannot be determined from the characteristic curves. This is because of the nonlinear *junction capacitances* of the emitter, the collector and, with integrated transistors, the substrate diode as well as the *diffusion charge*, stored in the base region, which is also described by the nonlinear *diffusion capacitances*.

---

[5] This equation is used by *PSpice* as a standard [2.6]; there is, however, an alternative expression for R_B [2.4, 2.6], which is not described here.

Junction Capacitances

A pn junction has a *junction capacitance* C_J that depends on the doping rate of the adjacent regions, the doping profile, the junction surface and the applied voltage V; a simplified expression is provided by [2.2]:

$$C_J(V) = \frac{C_{J0}}{\left(1 - \dfrac{V}{V_{Diff}}\right)^{m_J}} \qquad \text{for } V < V_{Diff} \tag{2.36}$$

The *zero capacitance* $C_{J0} = C_J(V = 0\,\text{V})$ is proportional to the junction surface and rises with an increase in doping. The *diffusion voltage* V_{Diff} also depends on the doping and increases with it; the range is $V_{Diff} \approx 0.5\ldots 1\,\text{V}$. With the *capacitance coefficient* m_J, the doping profile of the junction is taken into account; $m_J \approx 1/2$ for *abrupt* junctions with a doping jump, while $m_J \approx 1/3$ for *linear* junctions.

The simplifying assumptions leading to (2.36) are no longer fulfilled for $V \to V_{Diff}$. A more accurate calculation shows that (2.36) can be used up to about $0.5\,V_{Diff}$ only; with higher values of V the zero capacitance C_J rises only slightly compared to (2.36). A sufficiently accurate description is reached when the curve of C_J for $V > f_C V_{Diff}$ is replaced by a tangent to point $f_C V_{Diff}$:

$$C_J(V > f_C V_{Diff}) = C_J(f_C V_{Diff}) + \left.\frac{dC_J}{dV}\right|_{V = f_C V_{Diff}} (V - f_C V_{Diff})$$

Insertion leads to [2.4]:

$$C_J(V) = C_{J0} \begin{cases} \dfrac{1}{\left(1 - \dfrac{V}{V_{Diff}}\right)^{m_J}} & \text{for } V \le f_C V_{Diff} \\[3em] \dfrac{1 - f_C(1 + m_J) + \dfrac{m_J V}{V_{Diff}}}{(1 - f_C)^{(1+m_J)}} & \text{for } V > f_C V_{Diff} \end{cases} \tag{2.37}$$

Here, $f_C \approx 0.4\ldots 0.7$. Figure 2.32 shows the curve of C_J for $m_J = 1/2$ and $m_J = 1/3$; the curve resulting from (2.36) is also plotted.

Junction capacitance of a bipolar transistor: In accordance with the pn junctions, discrete transistors feature two junction capacitances and integrated transistors have three:

- The junction capacitance $C_{J,E}(V_{B'E'})$ of the emitter diode with parameters $C_{J0,E}, m_{J,E}$ and $V_{Diff,E}$.
- The junction capacitance $C_{J,C}$ of the collector diode with parameters $C_{J0,C}, m_{J,C}$ and $V_{Diff,C}$. It consists of the *internal* junction capacitance $C_{J,Ci}$ of the active region and the *external* junction capacitance $C_{J,Ce}$ of the regions close to the connections. $C_{J,Ci}$ acts on the internal base B' and $C_{J,Ce}$ acts on the external base B. Parameter x_{CJC} determines the portion of $C_{J,C}$ that acts internally:

$$C_{J,Ci}(V_{B'C'}) = x_{CJC}\, C_{J,C}(V_{B'C'}) \tag{2.38}$$
$$C_{J,Ce}(V_{BC'}) = (1 - x_{CJC})\, C_{J,C}(V_{BC'}) \tag{2.39}$$

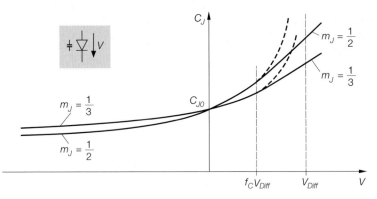

Fig. 2.32. Curves of the junction capacitance C_J for $m_J = 1/2$ and $m_J = 1/3$ according to (2.36) (*broken line*) and (2.37)

In discrete transistors $C_{J,Ce}$ is mostly lower than $C_{J,Ci}$; that is, $x_{CJC} \approx 0.5\ldots1$. For integrated transistors $x_{CJC} < 0.5$.

- In integrated transistors there is an additional junction capacitance $C_{J,S}$ of the substrate diode with parameters $C_{J,S}$, $m_{J0,S}$ and $V_{Diff,S}$. In vertical npn transistors the junction capacitance acts on the internal collector C' – that is, $C_{J,S} = C_{J,S}(V_{SC'})$ – and in lateral pnp transistors it acts on the internal base B' – that is, $C_{J,S} = C_{J,S}(V_{SB'})$.

Extended model: Figure 2.33 shows the static model of an npn transistor extended by the junction capacitances $C_{J,E}$, $C_{J,Ci}$, $C_{J,Ce}$ and $C_{J,S}$; also included are the diffusion capacitances $C_{D,N}$ and $C_{D,I}$, which are described in the next section.

Diffusion Capacitances

A pn junction contains a *diffusion capacitance* Q_D, which is in the first approximation, proportional to the ideal current through the pn junction. In a transistor $Q_{D,N}$ is the diffusion charge of the emitter diode and $Q_{D,I}$ is the diffusion charge of the collector

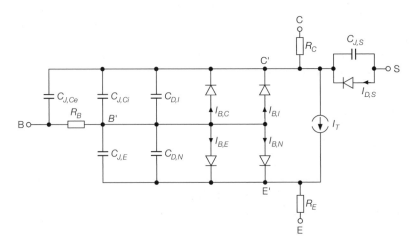

Fig. 2.33. Complete *Gummel–Poon model* of an npn transistor

diode; both are related to the respective portions of the ideal transport current I_T according to (2.26) – that is, to $B_N I_{B,N}$ or $B_I I_{B,I}$ [2.5]:

$$Q_{D,N} = \tau_N B_N I_{B,N} = \tau_N I_S \left(e^{\frac{V_{B'E'}}{V_T}} - 1 \right)$$

$$Q_{D,I} = \tau_I B_I I_{B,I} = \tau_I I_S \left(e^{\frac{V_{B'C'}}{V_T}} - 1 \right)$$

Parameters τ_N and τ_I are called the *transit times*. Differentiation leads to the *diffusion capacitances* $C_{D,N}$ and $C_{D,I}$ [2.5]:

$$C_{D,N}(V_{B'E'}) = \frac{dQ_{D,N}}{dV_{B'E'}} = \frac{\tau_N I_S}{V_T} e^{\frac{V_{B'E'}}{V_T}} \tag{2.40}$$

$$C_{D,I}(V_{B'C'}) = \frac{dQ_{D,I}}{dV_{B'C'}} = \frac{\tau_I I_S}{V_T} e^{\frac{V_{B'C'}}{V_T}} \tag{2.41}$$

Figure 2.33 presents the model with the two capacitances $C_{D,N}$ and $C_{D,I}$.

Normal operation: The diffusion capacitances $C_{D,N}$ and $C_{D,I}$ are parallel to the junction capacitances $C_{J,E}$ and $C_{J,Ci}$ (see Fig. 2.33). In normal operation the collector diffusion capacitance $C_{D,I}$ is very small due to $V_{B'C'} < 0$ and may be ignored with respect to the collector junction capacitance $C_{J,Ci}$ arranged in parallel; therefore, $C_{D,I}$ can be described by a constant transit time $\tau_I = \tau_{0,I}$. For small currents the emitter diffusion capacitance $C_{D,N}$ is smaller than the emitter junction capacitance $C_{J,E}$, while for higher currents it is larger. A more accurate description of the dynamic performance with high currents requires a more detailed model for τ_N.

Current dependence of the transit time: With high currents the diffusion charge increases disproportionately due to the high-current effect. In this region the transit time τ_N is no longer constant but increases with rising currents. There is also an influence from the Early effect, which changes the effective thickness of the base region and thus the stored charge. However, even with the parameters already introduced, $I_{K,N}$ for the high current effect and $V_{A,N}$ for the Early effect, it is not possible to give a satisfactory description; an empirical equation is therefore used [2.6]:

$$\tau_N = \tau_{0,N} \left(1 + x_{\tau,N} \left(3x^2 - 2x^3 \right) 2^{\frac{V_{B'C'}}{V_{\tau,N}}} \right)$$

with

$$x = \frac{B_N I_{B,N}}{B_N I_{B,N} + I_{\tau,N}} = \frac{I_S \left(e^{\frac{V_{B'E'}}{V_T}} - 1 \right)}{I_S \left(e^{\frac{V_{B'E'}}{V_T}} - 1 \right) + I_{\tau,N}} \tag{2.42}$$

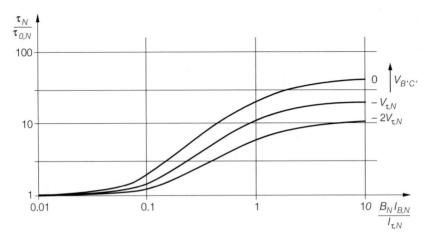

Fig. 2.34. Curves of $\tau_N/\tau_{0,N}$ for $x_{\tau,N} = 40$ and $V_{\tau,N} = 10\,\text{V}$

The new model parameters are the *ideal transit time* $\tau_{0,N}$, the *coefficient for the transit time* $x_{\tau,N}$, the *transit time knee-point current* $I_{\tau,N}$ and the *transit time voltage* $V_{\tau,N}$. Coefficient $x_{\tau,N}$ determines the maximum increase in τ_N for $V_{B'C'} = 0$:

$$\lim_{I_{B,N} \to \infty} \tau_N \bigg|_{V_{B'C'}=0} = \tau_{0,N} \left(1 + x_{\tau,N}\right)$$

For $B_N I_{B,N} = I_{\tau,N}$ half of the maximum increase is reached:

$$\tau_N \big|_{B_N I_{B,N} = I_{\tau,N}, V_{B'C'}=0} = \tau_{0,N} \left(1 + \frac{x_{\tau,N}}{2}\right)$$

When $V_{B'C'}$ decreases by the voltage $V_{\tau,N}$, the increase is only half as much; for $V_{B'C'} = -n V_{\tau,N}$ it is lowered by a factor of 2^n. To make this easier to understand Fig. 2.34 shows the curve for $\tau_N/\tau_{0,N}$ for $x_{\tau,N} = 40$ and $V_{\tau,N} = 10\,\text{V}$.

The increase in τ_N with high currents results in a decrease in the cutoff frequencies and the operating speed of a transistor; these consequences are described in Sect. 2.3.3.

Gummel–Poon Model

Figure 2.33 shows the complete model of an npn transistor; it is called the *Gummel–Poon model* and is used for circuit simulations in CAD programs. Figure 2.35 lists the variables and equations of the model. Figure 2.36 lists the parameters together with the names of the parameters in the *PSpice* circuit simulator[6].

Figure 2.37 lists the parameters of some selected transistors taken from the *PSpice* component library; it only contains the parameters for normal operation in the forward region. Parameters not specified are treated by *PSpice* as follows:

– A standard value is used:
$I_S = 10^{-16}\,\text{A}$, $B_N = 100$, $B_I = 1$, $n_E = 1.5$, $n_C = 2$, $x_{T,I} = 3$, $f_C = 0.5$
$V_{Diff,E} = V_{Diff,C} = V_{Diff,S} = 0.75\,\text{V}$, $m_{J,E} = m_{J,C} = 0.333$, $x_{CJC} = 1$

---

[6] *PSpice* is a *OrCAD* product.

Variable	Name	Equation
$I_{B,N}$	ideal base current of emitter diode	(2.24)
$I_{B,I}$	ideal base current of collector diode	(2.25)
$I_{B,E}$	Base leakage current of emitter diode	(2.27)
$I_{B,C}$	Base leakage current of collector diode	(2.28)
I_T	Collector–emitter transport current	(2.29),(2.30)
$I_{D,S}$	Current of substrate diode	(2.34)
R_B	Base spreading resistance	(2.35)
R_C	Collector spreading resistance	
R_E	Emitter spreading resistance	
$C_{J,E}$	Junction capacitance of emitter diode	(2.37)
$C_{J,Ci}$	Internal junction capacitance of collector diode	(2.37),(2.38)
$C_{J,Ce}$	External junction capacitance of collector diode	(2.37),(2.39)
$C_{J,S}$	Junction capacitance of substrate diode	(2.37)
$C_{D,N}$	Diffusion capacitance of emitter diode	(2.40),(2.42)
$C_{D,I}$	Diffusion capacitance of collector diode	(2.41)

Fig. 2.35. Variables of the Gummel–Poon model

- The parameter is set to zero:
  $I_{S,S}, I_{S,E}, I_{S,C}, R_B, R_C, R_E, C_{J0,E}, C_{J0,C}, C_{J0,S}, m_{J,S}, \tau_{0,N}, x_{\tau,N}$
  $I_{\tau,N}, \tau_{0,I}, x_{T,B}$
- The parameter is set to infinity:
  $I_{K,N}, I_{K,I}, V_{A,N}, V_{A,I}, V_{\tau,N}$

The values zero and infinity have the effect that the respective effects are not modeled [2.6].

PSpice uses an extended version of the Gummel–Poon model that allows the modelling of additional effects (see [2.6]); these effects and the additional parameters are not described here.

2.3.3
Small-Signal Model

By linearization at an operating point the nonlinear Gummel–Poon model becomes a linear *small-signal model*. For practical purposes the operating point is selected so that the transistor operates in forward mode; therefore, the small-signal models described here only apply to this operating mode. Similarly, one can set up small-signal models for other operating modes, although these are of secondary importance.

The *static small-signal model* describes the small-signal characteristics at low frequencies and for that reason is called the *DC small-signal equivalent circuit*. The *dynamic small-signal model* also describes the dynamic small-signal characteristics and is required for calculating the frequency response of circuits; it is called the *AC small-signal equivalent circuit*.

Static Small-Signal Model

Linearization and the small-signal parameters of the Gummel–Poon model: An accurate small-signal model is achieved by linearizing the Gummel–Poon model. Removing the capacitances from Fig. 2.33 and neglecting the reverse currents

Parameter	PSpice	Name
Static performance		
I_S	IS	Saturation reverse current
$I_{S,S}$	ISS	Saturation reverse current of substrate diode
B_N	BF	Ideal current gain in forward mode
B_I	BR	Ideal current gain in reverse mode
$I_{S,E}$	ISE	Leakage saturation reverse current of emitter diode
n_E	NE	Emission coefficient of emitter diode
$I_{S,C}$	ISC	Leakage saturation reverse current of collector diode
n_C	NC	Emission coefficient of collector diode
$I_{K,N}$	IKF	Knee-point current for strong injection in forward mode
$I_{K,I}$	IKR	Knee-point current for strong injection in reverse mode
$V_{A,N}$	VAF	Early voltage in forward mode
$V_{A,I}$	VAR	Early voltage in reverse mode
R_{Be}	RBM	External base spreading resistance
R_{Bi}	—	Internal base spreading resistance ($R_{Bi} = $ RB − RBM)
—	RB	Base spreading resistance (RB $= R_{Be} + R_{Bi}$)
R_C	RC	Collector spreading resistance
R_E	RE	Emitter spreading resistance
Dynamic performance		
$C_{J0,E}$	CJE	Zero capacitance of emitter diode
$V_{Diff,E}$	VJE	Diffusion voltage of emitter diode
$m_{J,E}$	MJE	Capacitance coefficient of emitter diode
$C_{J0,C}$	CJC	Zero capacitance of collector diode
$V_{Diff,C}$	VJC	Diffusion voltage of collector diode
$m_{J,C}$	MJC	Capacitance coefficient of collector diode
x_{CJC}	XCJC	Capacitance distribution in collector diode
$C_{J0,S}$	CJS	Zero capacitance of substrate diode
$V_{Diff,S}$	VJS	Diffusion voltage of substrate diode
$m_{J,S}$	MJS	Capacitance coefficient of substrate diode
f_C	FC	Coefficient for variation of capacitances
$\tau_{0,N}$	TF	Ideal transit time in forward mode
$x_{\tau,N}$	XTF	Coefficient for transit time in forward mode
$V_{\tau,N}$	VTF	Transit time voltage in forward mode
$I_{\tau,N}$	ITF	Transit time current in forward mode
$\tau_{0,I}$	TR	Transit time in reverse mode
Thermal performance		
$x_{T,I}$	XTI	Temperature coefficient of reverse currents (2.20)
$x_{T,B}$	XTB	Temperature coefficient of current gains (2.22)

Fig. 2.36. Parameters of the Gummel–Poon model

($I_{B,I} = I_{B,C} = I_{D,S} = 0$) results in the *static* Gummel–Poon model for forward operation as shown in Fig. 2.38a. The nonlinear variables $I_B = I_{B,N}(V_{B'E'}) + I_{B,E}(V_{B'E'})$ and $I_C = I_T(V_{B'E'}, V_{C'E'})$ are linearized at the operating point A:

$$g_m = \left.\frac{\partial I_C}{\partial V_{B'E'}}\right|_A = \frac{I_{C,A}}{V_T}\left(1 - \frac{V_T}{q_B}\left.\frac{\partial q_B}{\partial V_{B'E'}}\right|_A\right)$$

$$\frac{1}{r_{BE}} = \left.\frac{\partial I_B}{\partial V_{B'E'}}\right|_A = \frac{I_S}{B_N V_T} e^{\frac{V_{B'E',A}}{V_T}} + \frac{I_{S,E}}{n_E V_T} e^{\frac{V_{B'E',A}}{n_E V_T}}$$

Parameter	PSpice	BC547B	BC557B	BUV47	BFR92P	Unit
I_S	IS	7	1	974	0.12	fA
B_N	BF	375	307	95	95	
B_I	BR	1	6.5	20.9	10.7	
$I_{S,E}$	ISE	68	10.7	2570	130	fA
n_E	NE	1.58	1.76	1.2	1.9	
$I_{K,N}$	IKF	0.082	0.092	15.7	0.46	A
$V_{A,N}$	VAF	63	52	100	30	V
R_{Be} [a]	RBM	10	10	0.1	6.2	Ω
R_{Bi} [a]	—	0	0	0	7.8	Ω
[a]	RB	10	10	0.1	15	Ω
R_C	RC	1	1.1	0.035	0.14	Ω
$C_{J0,E}$	CJE	11.5	30	1093	0.01	pF
$V_{Diff,E}$	VJE	0.5	0.5	0.5	0.71	V
$m_{J,E}$	MJE	0.672	0.333	0.333	0.347	
$C_{J0,C}$	CJC	5.25	9.8	364	0.946	pF
$V_{Diff,C}$	VJC	0.57	0.49	0.5	0.85	V
$m_{J,C}$	MJC	0.315	0.332	0.333	0.401	
x_{CJC}	XCJC	1	1	1	0.13	
f_C	FC	0.5	0.5	0.5	0.5	
$\tau_{0,N}$	TF	0.41	0.612	21.5	0.027	ns
$x_{\tau,N}$	XTF	40	26	205	0.38	
$V_{\tau,N}$	VTF	10	10	10	0.33	V
$I_{\tau,N}$	ITF	1.49	1.37	100	0.004	A
$\tau_{0,I}$	TR	10	10	988	1.27	ns
$x_{T,I}$	XTI	3	3	3	3	
$x_{T,B}$	XTB	1.5	1.5	1.5	1.5	

BC547B: npn low-power transistor, BC557B: pnp low-power transistor,
BUV47: npn power transistor, BFR92P: npn high-frequency transistor
[a] With the exception of BFR92P, the base spreading resistances are quoted as flat values while the current-dependent internal portion is not specified. Therefore, there are inaccuracies with high frequencies. More accurate values can be derived from the noise values (see Sect. 2.3.4).

Fig. 2.37. Parameters of some discrete transistors

$$\frac{1}{r_{CE}} = \frac{\partial I_C}{\partial V_{C'E'}}\bigg|_A = \frac{I_{C,A}}{V_{A,N} + V_{C'E',A} - V_{B'E',A}\left(1 + \dfrac{V_{A,N}}{V_{A,I}}\right)}$$

Approximation for small-signal parameters: Small-signal parameters g_m, r_{BE} and r_{CE} are determined with the above equations in CAD programs only. Approximations or other relations are used in practice:

$$g_m = \frac{\partial I_C}{\partial V_{B'E'}}\bigg|_A \approx \frac{I_{C,A}}{V_T}\frac{I_{K,N}+I_{C,A}}{I_{K,N}+2I_{C,A}} \overset{I_{C,A}\ll I_{K,N}}{\approx} \frac{I_{C,A}}{V_T}$$

$$r_{BE} = \frac{\partial V_{B'E'}}{\partial I_B}\bigg|_A = \frac{\partial V_{B'E'}}{\partial I_C}\bigg|_A \frac{\partial I_C}{\partial I_B}\bigg|_A = \frac{\beta}{g_m}$$

$$r_{CE} = \frac{\partial V_{C'E'}}{\partial I_C}\bigg|_A \approx \frac{V_{A,N}+V_{C'E',A}}{I_{C,A}} \overset{V_{C'E',A}\ll V_{A,N}}{\approx} \frac{V_{A,N}}{I_{C,A}}$$

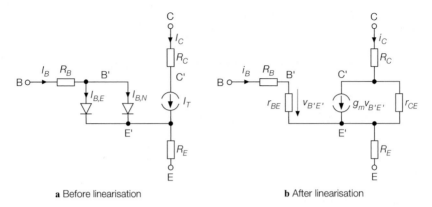

a Before linearisation   **b** After linearisation

Fig. 2.38. Static small-signal model as derived from the static Gummel–Poon model by linearization

The approximations for r_{BE} and r_{CE} are in accordance with (2.12) and (2.13) given in Sect. 2.1.4. In order to determine r_{BE} it is necessary to know the small-signal current gain β or to assume a reasonable value.

The equation for transconductance g_m is derived by the approximate evaluation of the entire expression; (2.11) is thus extended by one term to describe the high-current effect. The high-current effect causes a relative reduction of g_m with high collector currents; that is, to two-thirds of $I_{C,A}/V_T$ for $I_{C,A} = I_{K,N}$ and to half of $I_{C,A}/V_T$ for $I_{C,A} \to \infty$. If the reduction is to remain below 10 %, then $I_{C,A} < I_{K,N}/8$ must be chosen.

DC small-signal equivalent circuit: Figure 2.38b shows the resulting *static small-signal model*. For almost all practical calculations the spreading resistances R_B, R_C and R_E are neglected; this leads to the small-signal equivalent circuit described in Sect. 2.1.4 and shown again in Fig. 2.39a.

If the Early effect is also disregarded ($r_{CE} \to \infty$), the alternative version shown in Fig. 2.39b can also be used in addition to the reduced equivalent circuit of Fig. 2.39a. The following thus applies:

$$r_E = \frac{1}{g_m + \dfrac{1}{r_{BE}}} \approx \frac{1}{g_m} \quad ; \quad \alpha = \frac{\beta}{1+\beta} = g_m\, r_E$$

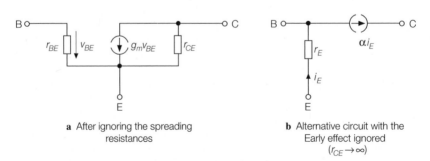

a After ignoring the spreading   **b** Alternative circuit with the
resistances   Early effect ignored
                                                  ($r_{CE} \to \infty$)

Fig. 2.39. Simplified static small-signal models

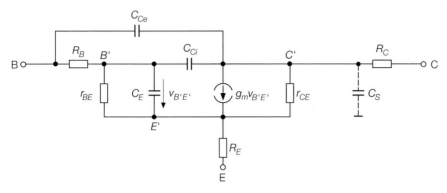

Fig. 2.40. Dynamic small-signal model

This alternative version is obtained by linearization of the reduced Ebers–Moll model according to Fig. 2.25a. It is mentioned here for the sake of completeness because it can be used to advantage only in a few special cases; in many other cases the results may be insufficient if the Early effect[7] is not taken into account.

Dynamic Small-Signal Model

Complete model: Adding the junction and diffusion capacitances to the static small-signal model according to Fig. 2.38b leads to the dynamic small-signal model shown in Fig. 2.40. With regard to Sect. 2.3.2 the following applies:

$$C_E = C_{J,E}(V_{B'E',A}) + C_{D,N}(V_{B'E',A})$$
$$C_{Ci} = C_{J,Ci}(V_{B'C',A}) + C_{D,I}(V_{B'C',A}) \approx C_{J,Ci}(V_{B'C',A})$$
$$C_{Ce} = C_{J,Ce}(V_{BC',A})$$
$$C_S = C_{J,S}(V_{SC',A})$$

The *emitter capacitance* C_E is made up of the emitter junction capacitance $C_{J,E}$ and the diffusion capacitance $C_{D,N}$ for forward operation. The *internal collector capacitance* C_{Ci} reflects the internal collector junction capacitance; the diffusion capacitance $C_{D,I}$ connected in parallel is of a negligible size due to $V_{BC} < 0$. The *external collector capacitance* C_{Ce} and the *substrate capacitance* C_S reflect the respective junction capacitances; the latter exists only in integrated transistors.

Simplified model: In practical calculations the spreading resistances R_E and R_C are ignored; due to its influence on the dynamic performance the base spreading resistance R_B can only be ignored in certain exception cases. In addition, the internal and external collector capacitances are combined into an internal *collector capacitance* C_C; only in integrated transistors with a dominating external portion is it connected externally. The result is the simplified dynamic small-signal model shown in Fig. 2.41 which is used for

---

[7] Some literature shows a variation with an additional resistor r_C between the base and the collector. This results from linearization of the collector–base diode of the Ebers–Moll model, which may not be ignored in this case, and is therefore not used for modelling the Early effect as often assumed. For this reason, this variation is not an equivalent to the simplified model shown in Fig. 2.39a.

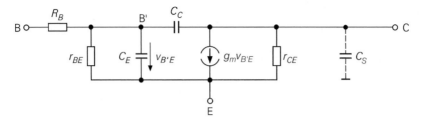

Fig. 2.41. Simplified dynamic small-signal model

the calculations in the following Sections. The *practical* determination of the capacitances C_E and C_C is described in more detail in the next Section.

Cut-Off Frequencies in Small-Signal Operation

With the small-signal model shown in Fig. 2.41 the frequency responses for small-signal current gains α and β and the transadmittance $y_{21,e}$ can be calculated. The respective cutoff frequencies f_α, f_β and f_{Y21e} and the *transit frequency* f_T provide a measure of the bandwidth and the operating speed of the transistor.

Frequency response of the small-signal current gain β: In a common-emitter circuit operating in forward mode and with constant $V_{CE} = V_{CE,A}$ the ratio of the small-signal currents i_C and i_B after Laplace transformation is called the *transfer function of the small-signal current gain* β, with the mathematical symbol $\beta(s)$:

$$\underline{\beta}(s) = \frac{\underline{i}_C}{\underline{i}_B} = \frac{\mathcal{L}\{i_C\}}{\mathcal{L}\{i_B\}}$$

By inserting $s = j\omega$, $\underline{\beta}(s)$ becomes the frequency response $\underline{\beta}(j\omega)$ and, by taking the absolute value, the magnitude frequency response $|\underline{\beta}(j\omega)|$.

In order to determine $\underline{\beta}(s)$, a small-signal current source with current i_B is connected to the base and i_C is then determined. Figure 2.42 shows the corresponding small-signal equivalent circuit diagram; due to $v_{CE} = V_{CE} - V_{CE,A} = 0$ the collector is connected to ground. From the nodal equations

$$\underline{i}_B = \left(\frac{1}{r_{BE}} + s\,(C_E + C_C) \right) \underline{v}_{B'E}$$

$$\underline{i}_C = (g_m - sC_C)\,\underline{v}_{B'E}$$

and $\beta_0 = g_m r_{BE}$:[8]

$$\underline{\beta}(s) = \frac{r_{BE}\,(g_m - sC_C)}{1 + s\,r_{BE}\,(C_E + C_C)} \approx \frac{\beta_0}{1 + s\,r_{BE}\,(C_E + C_C)}$$

The transfer function contains a pole and a zero; the zero can be disregarded due to the very short time constant $C_C g_m^{-1}$. Figure 2.43 illustrates the magnitude fre-

---

[8] The *static* small-signal current gain in a common-emitter circuit which has previously been called β is now named β_0 to distinguish it from the inversely Laplace-transformed $\beta = \mathcal{L}^{-1}\{\underline{\beta}(s)\}$; the subscript zero means that the frequency is zero with the consequence $\beta_0 = |\underline{\beta}(j0)|$.

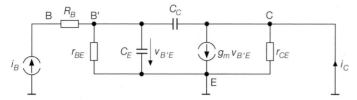

Fig. 2.42. Small-signal equivalent circuit to calculate $\underline{\beta}(s)$

quency response $|\underline{\beta}(j\omega)|$ for $\beta_0 = 100$ with the zero considered; at the β *cutoff frequency*

$$\omega_\beta = 2\pi f_\beta \approx \frac{1}{r_{BE}(C_E + C_C)} \tag{2.43}$$

it is reduced by 3 dB in relation to β_0 [2.7].

Transit frequency: The frequency at which $|\underline{\beta}(j\omega)|$ decreases to a value of one is called the *transit frequency* f_T; therefore [2.7]:

$$\omega_T = 2\pi f_T = \beta_0 \omega_\beta \approx \frac{g_m}{C_E + C_C} \tag{2.44}$$

Due to the approximations in the small-signal model and in calculating $\underline{\beta}(s)$ the transit frequency according to (2.44) is not identical to the real transit frequency of the transistor; therefore it is also called the *extrapolated transit frequency* because it can be derived by extrapolating the declining portion of $|\underline{\beta}(j\omega)|$ similar to a lowpass filter of first degree. Transistor data sheets always state the extrapolated transit frequency.

The transit frequency depends on the operating point; outside the high-current region:

$$g_m = \frac{I_{C,A}}{V_T} \quad , \quad C_E = \frac{\tau_N I_{C,A}}{V_T} + C_{J,E} \quad , \quad C_C = C_{J,C}$$

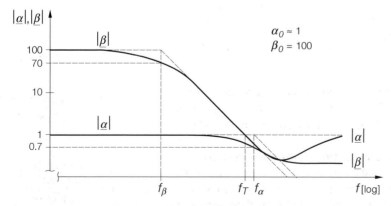

Fig. 2.43. Absolute-value frequency responses $|\underline{\alpha}(j\omega)|$ and $|\underline{\beta}(j\omega)|$

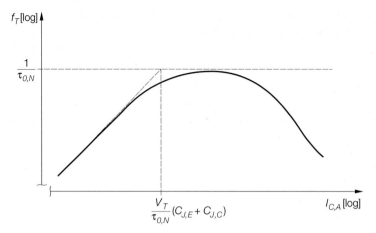

Fig. 2.44. Dependence of the transit frequency on the collector current $I_{C,A}$

This results in [2.7]:

$$\omega_T \approx \frac{1}{\tau_N + \dfrac{I_{C,A}}{V_T}\left(C_{J,E} + C_{J,C}\right)}$$

Figure 2.44 shows the dependence of the transit frequency on the collector current $I_{C,A}$. Three regions can be distinguished:

— With small collector current:

$$\omega_T \approx \frac{I_{C,A}}{V_T\left(C_{J,E} + C_{J,C}\right)} \sim I_{C,A} \qquad \text{for } I_{C,A} < \frac{V_T}{\tau_{0,N}}\left(C_{J,E} + C_{J,C}\right)$$

In this region f_T is approximately proportional to $I_{C,A}$.
— With medium collector currents below the high-current region:

$$\omega_T \approx \frac{1}{\tau_N} \approx \frac{1}{\tau_{0,N}} \qquad \text{for } \frac{V_T}{\tau_{0,N}}\left(C_{J,E} + C_{J,C}\right) < I_{C,A} \ll I_{\tau,N}$$

Here, f_T reaches a maximum and is dependent on $I_{C,A}$ only to a small degree.
— In the high-current region the relation $\omega_T \approx 1/\tau_N$ also applies, but τ_N increases according to (2.42) so that f_T decreases with a rising $I_{C,A}$.

Frequency response of the small-signal current gain α: In a common-base circuit operating in forward mode and with constant $V_{BC} = V_{BC,A}$ the ratio of the small-signal currents i_C and i_E after the Laplace transformation is gcalled the *transfer function of the small-signal current gain α*, with the mathematical symbol $\underline{\alpha}(s)$. In order to determine $\underline{\alpha}(s)$ a small-signal current source with current i_E is connected to the emitter and i_C is determined; the base and the collector are connected to earth, the latter due to $v_{BC} = V_{BC} - V_{BC,A} = 0$. $r_{CE} \to \infty$ and $\alpha_0 = g_m r_E$[9] result in:

$$\underline{\alpha}(s) = -\frac{\underline{i}_C}{\underline{i}_E} = \alpha_0 \frac{1 + s\dfrac{R_B C_C}{\alpha_0} + s^2 \dfrac{r_E C_E R_B C_C}{\alpha_0}}{(1 + s\, r_E C_E)(1 + s R_B C_C)}$$

The transfer function comprises two poles and two zeros; the magnitude frequency response $|\underline{\alpha}(j\omega)|$ is shown in Fig. 2.43 [2.8]. In general, $R_B C_C \ll r_E C_E$, so that the following approximation can be used:

$$\underline{\alpha}(s) \approx \frac{\alpha_0}{1 + s\, r_E C_E}$$

This leads to the α *cutoff frequency*:

$$\boxed{\omega_\alpha = 2\pi f_\alpha \approx \frac{1}{r_E C_E}} \qquad (2.45)$$

Frequency response of transadmittance $y_{21,e}$:   The replacement of the small-signal current source with current i_B in Fig. 2.42 by a small-signal voltage source that provides the voltage v_{BE} and determination of the ratio of i_C and v_{BE} after Laplace transformation lead to the *transfer function of the transadmittance* $y_{21,e}$

$$\underline{y}_{21,e}(s) = \frac{\underline{i}_C}{\underline{v}_{BE}} = \frac{g_m - sC_C}{1 + \dfrac{R_B}{r_{BE}} + sR_B(C_E + C_C)} \approx \frac{g_m}{1 + sR_B(C_E + C_C)}$$

with the *transconductance cutoff frequency*:

$$\boxed{\omega_{Y21e} = 2\pi f_{Y21e} \approx \frac{1}{R_B(C_E + C_C)}} \qquad (2.46)$$

The transconductance cutoff frequency depends on the operating point; however, its dependence on $I_{C,A}$ is not easy to describe since R_B depends on the operating point in a highly nonlinear manner. The general tendency is that the transconductance frequency decreases with an increase in the collector current $I_{C,A}$.

Relationship and meaning of cutoff frequencies: A comparison leads to the following relation:

$$f_\beta < f_{Y21e} < f_T \lesssim f_\alpha$$

The driving of a transistor in a common-emitter circuit from a current source or a from source with an internal resistance $R_i \gg r_{BE}$, is called *current control*; in this case, the cutoff frequency of the circuit is limited to a maximum by the β *cutoff frequency* f_β. The driving of a transistor from a voltage source or from a source with an internal resistance $R_i \ll r_{BE}$ is called *voltage control*; in this case the cutoff frequency of the circuit has a maximum limit because of the *transconductance cutoff frequency* f_{Y21e}. Thus, voltage

---

[9] The *static* small-signal current gain in a common-base circuit which has previously been called α is now named α_0 to distinguish it from the inversely Laplace-transformed $\alpha = \mathcal{L}^{-1}\{\alpha(s)\}$; the subscript zero means that the frequency is zero with the consequence $\alpha_0 = |\underline{\alpha}(j0)|$.

control generally allows a higher bandwidth (see Sect. 2.4.1); this applies in the same way to the common-collector circuit (see Sect. 2.4.2).

The highest bandwidth is reached with the common-base circuit; with the general condition $R_i > r_E$ the transistor is current-controlled and the bandwidth of the circuit is limited to a maximum by α *cutoff frequency* f_α (see Sect. 2.4.3).

Selection of the operating point: Among other things, the bandwidth of a circuit depends on the operating point of the transistor. The common-emitter circuit with current control and the common-base circuit yield the maximum bandwidth when the collector current $I_{C,A}$ is chosen in such a way that the transit frequency f_T is at a maximum. In the common-emitter circuit with voltage control the situation is more complicated; the transconductance cutoff frequency f_{Y21e} decreases with an increasing $I_{C,A}$, but with the same gain the resistance of the circuitry at the collector node is lower so that the bandwidth on the output side increases (see Sect. 2.4.1).

Determination of the small-signal capacitances: The data sheet for a transistor contains the transit frequency f_T and the output capacitance C_{obo} in a common-base circuit (*output, grounded base, open emitter*); C_{obo} corresponds to the collector–base capacitance. Using (2.44) the following can be calculated:

$$C_C \approx C_{obo}$$
$$C_E \approx \frac{g_m}{\omega_T} - C_{obo}$$

Summary of the Small-Signal Parameters

The parameters of the small-signal model shown in Fig. 2.41 can be determined according to Fig. 2.45 using the collector current $I_{C,A}$ in the operating point and the data sheet specifications.

2.3.4
Noise

In resistors and pn junctions there are noise voltages or noise currents the generation of which is attributed to thermal agitation of the charge carriers in the resistors and to discontinuous current flow due to the cross-over of individual charge carriers in the pn junctions.

Noise Densities

Since noise is a stochastic occurrence, it is not possible to calculate in the normal way using voltages and currents. A noise voltage v_r is described by means of the *noise voltage density* $|\underline{v}_r(f)|^2$ and a noise current i_r by means of the *noise current density* $|\underline{i}_r(f)|^2$; the densities reflect the spectral distribution of the effective values v_{reff} or i_{reff}[10]

---

[10] In this case the *unilateral* frequency f with $0 < f < \infty$ is used instead of the *bilateral* angular frequency ω or $j\omega$ with $-\infty < \omega < \infty$ as the frequency variable. $|\underline{v}_r(f)|^2 = 4\pi|\underline{v}_r(j\omega)|^2$ is applicable; here, factor 4π is made up of the factor 2π according to $\omega = 2\pi f$ and the factor 2 for the transfer to the unilateral frequency variable.

Param.	Name	Method of determination
g_m	Transconductance	$g_m = \dfrac{I_{C,A}}{V_T}$ for $V_T \approx 26\,\text{mV}$ at $T = 300\,\text{K}$
(β)	Small-signal current gain	directly from the data sheet *or* indirectly from the data sheet using $\beta \approx B$ *or* a reasonable assumption ($\beta \approx 50\ldots500$)
r_{BE}	Small-signal input resistance	$r_{BE} = \dfrac{\beta}{g_m}$
R_B	Base spreading resistance	Reasonable assumption ($R_B \approx 10\ldots1000\,\Omega$) *or* from the optimum noise figure according to (2.58)
(V_A)	Early voltage	From the slope of the curves in the group of output characteristics (Fig. 2.5) *or* a reasonable assumption ($V_A \approx 30\ldots150\,\text{V}$)
r_{CE}	Small-signal output resistance	$r_{CE} = \dfrac{V_A}{I_{C,A}}$
(f_T)	Transit frequency	From data sheet
C_C	Collector capacitance	From data sheet (for example, C_{obo})
C_E	Emitter capacitance	$C_E = \dfrac{g_m}{2\pi f_T} - C_C$

Fig. 2.45. Small-signal parameters (auxiliary values in parentheses)

$$|\underline{v}_r(f)|^2 = \frac{d(v_{reff}^2)}{df}$$

$$|\underline{i}_r(f)|^2 = \frac{d(i_{reff}^2)}{df}$$

The effective values can be determined from the noise densities by integration [2.9]:

$$v_{reff} = \sqrt{\int_0^\infty |\underline{v}_r(f)|^2 df}$$

$$i_{reff} = \sqrt{\int_0^\infty |\underline{i}_r(f)|^2 df}$$

If the noise densities are constant the noise signal is called *white noise*. A noise signal can be white only in a certain region; for $f \to \infty$ in particular, the noise density must approach zero so that the integrals remain finite.

Transfer of noise densities in circuits: If a noise voltage $v_{r,e}$ with a noise voltage density $|\underline{v}_r(f)|^2$ exists at a point e, it is possible to calculate the resulting noise voltage $v_{r,a}$ with noise voltage density $|\underline{v}_{r,a}(f)|^2$ at a given point a by means of the transfer function $\underline{H}(s) = \underline{v}_{r,a}(s)/\underline{v}_{r,e}(s)$ [2.9]:

$$|\underline{v}_{r,a}(f)|^2 = |\underline{H}(j2\pi f)|^2 \, |\underline{v}_{r,e}(f)|^2$$

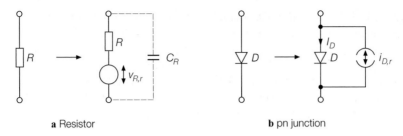

a Resistor   **b** pn junction

Fig. 2.46. Modelling of the noise signal by introducing noise sources

With several noise sources the noise densities at any point can be added if the noise sources are noncorrelated; that is, independent of each other – this is generally the case. For instance, if a noise voltage source has the density $|\underline{v}_r(f)|^2$ and a noise current source has the density $|\underline{i}_r(f)|^2$, the situation at point a can be calculated with the use of $\underline{H}_a(s) = \underline{v}_{r,a}(s)/\underline{v}_r(s)$ and $\underline{Z}_a(s) = \underline{v}_{r,a}(s)/\underline{i}_r(s)$:

$$|\underline{v}_{r,a}(f)|^2 = |\underline{H}_a(j2\pi f)|^2\, |\underline{v}_r(f)|^2 + |\underline{Z}_a(j2\pi f)|^2\, |\underline{i}_r(f)|^2$$

Noise of a resistor: A resistor R generates a noise voltage $v_{R,r}$, with the noise voltage density [2.9]:

$$|\underline{v}_{R,r}(f)|^2 = 4kTR$$

Here, $k = 1.38 \cdot 10^{-23}$ VAs/K is *Boltzmann's constant* and T is the resistor temperature in Kelvin. This noise is called *thermal noise,* since it is caused by the thermal agitation of the charge carriers; the noise voltage density is thus proportional to the temperature. For $R = 1\,\Omega$ and $T = 300$ K, $|\underline{v}_{R,r}(f)|^2 \approx 1.66 \cdot 10^{-20}$ V^2/Hz or $|\underline{v}_{R,r}(f)| \approx 0.13$ nV/$\sqrt{\text{Hz}}$.

Figure 2.46a shows a noise voltage source for noise modelling; the arrow pointing in both directions marks the source as a noise source. It is white noise because of the constant noise voltage density; therefore, the result of the calculated effective value is ∞. However, this result is not correct since for $f \to \infty$ the parasitic capacitance C_R of the resistor must be taken into account; this is shown in Fig. 2.46a. Using the equation

$$\underline{v}'_{R,r}(s) = \frac{\underline{v}_{R,r}(s)}{1 + sRC_R}$$

for the noise voltage $v'_{R,r}$ across the resistor leads to the expression:

$$|\underline{v}'_{R,r}(f)|^2 = \frac{|\underline{v}_{R,r}(f)|^2}{1 + (2\pi f RC_R)^2}$$

Its integration results in a finite effective value [2.10]:

$$v'_{R,reff} = \sqrt{\frac{kT}{C_R}}$$

Noise of a pn junction: A pn junction – that is, an ideal diode – generates a noise current $i_{D,r}$ with a noise current density [2.9]:

$$|\underline{i}_{D,r}(f)|^2 = 2q\,I_D$$

Here, $q = 1.602 \cdot 10^{-19}$ As is the *elementary charge*. The noise current density is proportional to the current I_D flowing through the pn junction. This noise is called *shot noise*. $I_D = 1$ mA results in $|\underline{i}_{D,r}(f)|^2 \approx 3.2 \cdot 10^{-22}$ A^2/Hz or $|\underline{i}_{D,r}(f)| \approx 18$ pA/$\sqrt{\text{Hz}}$.

Figure 2.46b shows a noise current source for noise modelling; here, too, an arrow pointing in both directions characterizes the source as a noise source. It is white noise as in the case of the resistor; the same considerations apply to the effective value – that is, for $f \to \infty$ the capacitance of the pn junction must be taken into consideration.

1/f noise: Resistors and pn junctions produce an additional $1/f$ *noise* with a noise density that is inversely proportional to the frequency. For resistors this portion is usually negligible; for a pn junctions it is

$$|\underline{i}_{D,r(1/f)}(f)|^2 = \frac{k_{(1/f)} I_D^{\gamma_{(1/f)}}}{f}$$

with the experimental constants $k_{(1/f)}$ and $\gamma_{(1/f)} \approx 1...2$ [2.10].

When calculating the effective value the result is ∞ if $f = 0$ is used as the lower limit for the integration. But since in practice this process can be observed for a finite time only it is the reciprocal value of the observation time that is used for the lower limit. With measuring instruments the frequency portions below the reciprocal value of the measuring time are called *drift* instead of noise.

Noise Sources of a Bipolar Transistor

Three noise sources exist in bipolar transistors at an operating point defined by $I_{B,A}$ and $I_{C,A}$ [2.10]:

– Thermal noise of the base spreading resistance with:

$$|\underline{v}_{RB,r}(f)|^2 = 4kT R_B$$

Generally, the thermal noise of other spreading resistances may be ignored.
– Shot noise of the base current with:

$$|\underline{i}_{B,r}(f)|^2 = 2q I_{B,A} + \frac{k_{(1/f)} I_{B,A}^{\gamma_{(1/f)}}}{f}$$

– Shot noise of the collector current with:

$$|\underline{i}_{C,r}(f)|^2 = 2q I_{C,A} + \frac{k_{(1/f)} I_{C,A}^{\gamma_{(1/f)}}}{f}$$

The upper part of Fig. 2.47 shows the small-signal model with the noise voltage source $v_{RB,r}$ and the noise current sources $i_{B,r}$ and $i_{C,r}$.

In shot noise the $1/f$ portion dominates at low frequencies, while at medium and high frequencies it is the white portion that dominates. The frequency for which both portions are of equal size is called the $1/f$ *cutoff frequency* $f_g(1/f)$:

$$f_{g(1/f)} = \frac{k_{(1/f)} I_{C,A}^{(\gamma_{(1/f)}-1)}}{2q} \overset{\gamma_{(1/f)}=1}{=} \frac{k_{(1/f)}}{2q}$$

For $\gamma_{(1/f)} = 1$ the $1/f$ cutoff frequency does not depend on the operating point. Low-noise transistors feature values of $\gamma_{(1/f)} \approx 1.2$ and $f_{g(1/f)}$ increases with higher operating point currents. Typical values are in the range of $f_{g(1/f)} \approx 10$ Hz...10 kHz.

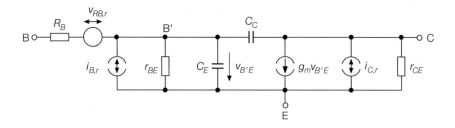

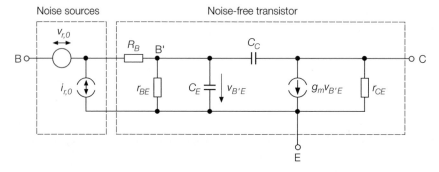

Fig. 2.47. Small-signal model of a bipolar transistor with the original (*top*) and the equivalent noise sources (*bottom*)

Equivalent Noise Sources

To facilitate the calculation of the noise of a circuit the noise sources are converted to the base–emitter path. This leads to the small-signal model shown in the lower part of Fig. 2.47, in which the original noise sources are represented by an *equivalent noise voltage source* $v_{r,0}$ and an *equivalent noise current source* $i_{r,0}$; then the actual transistor is free of noise.

$$|\underline{v}_{r,0}(f)|^2 = |\underline{v}_{RB,r}(f)|^2 + R_B^2\, |\underline{i}_{B,r}(f)|^2 + \frac{|\underline{i}_{C,r}(f)|^2}{|\underline{y}_{21,e}(j2\pi f)|^2}$$

$$|\underline{i}_{r,0}(f)|^2 = |\underline{i}_{B,r}(f)|^2 + \frac{|\underline{i}_{C,r}(f)|^2}{|\beta(j2\pi f)|^2}$$

Using $\beta/g_m = r_{BE} > R_B$, $B \approx \beta \gg 1$ and $\gamma_{(1/f)} = 1$ leads to [2.10]:

$$|\underline{v}_{r,0}(f)|^2 = 2q\,I_{C,A}\left(\left(\frac{1}{g_m^2} + \frac{R_B^2}{\beta}\right)\left(1 + \frac{f_{g(1/f)}}{f}\right) + R_B^2\left(\frac{f}{f_T}\right)^2\right) + 4kT\,R_B$$

$$\tag{2.47}$$

$$|\underline{i}_{r,0}(f)|^2 = 2q\,I_{C,A}\left(\frac{1}{\beta}\left(1 + \frac{f_{g(1/f)}}{f}\right) + \left(\frac{f}{f_T}\right)^2\right) \tag{2.48}$$

In the frequency range $f_{g(1/f)} < f < f_T/\sqrt{\beta}$ the equivalent noise densities are constant; that is, the noise is white noise. For $g_m = I_{C,A}/V_T$:

$$|\underline{v}_{r,0}(f)|^2 = \frac{2kT\,V_T}{I_{C,A}} + 4kT\,R_B + \frac{2q\,R_B^2\,I_{C,A}}{\beta} \tag{2.49}$$

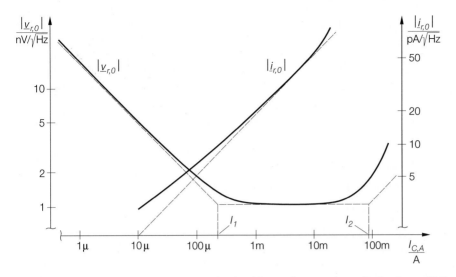

Fig. 2.48. Depencence of the equivalent noise densities on the operating point for $R_B = 60\,\Omega$: an asymptotic curve for $\beta = 100$ (*broken line*) and the actual curve with a β which depends on the operating point and with $\beta_{max} = 100$

$$|\underline{i}_{r,0}(f)|^2 = \frac{2q\,I_{C,A}}{\beta} \tag{2.50}$$

For $f < f_{g(1/f)}$ and $f > f_T/\sqrt{\beta}$ the noise densities increase. For low-noise low-power transistors we have $f_{g(1/f)} \approx 100\,\text{Hz}$ and $f_T/\sqrt{\beta} \approx 10\,\text{MHz}$.

Operating point dependence: Figure 2.48 shows the dependence of the equivalent noise densities on the operating point current $I_{C,A}$ for the frequency range $f_{g(1/f)} < f < f_T/\sqrt{\beta}$. The noise current density $|\underline{i}_{r,0}(f)|^2$ for $\beta = \text{const.}$ is proportional to $I_{C,A}$; this relation is plotted in Fig. 2.48 as an asymptotic curve (broken line). With low and high collector currents the real curve runs above the asymptote due to the decrease in β. A distinction has to be made between three regions for the noise voltage density $|\underline{v}_{r,0}(f)|^2$:

$$|\underline{v}_{r,0}(f)|^2 \approx \begin{cases} \dfrac{2kT\,V_T}{I_{C,A}} & \text{for } I_{C,A} < \dfrac{V_T}{2R_B} = I_1 \\[2ex] 4kT\,R_B & \text{for } \dfrac{V_T}{2R_B} < I_{C,A} < \dfrac{2\beta\,V_T}{R_B} \\[2ex] \dfrac{2q\,R_B^2\,I_{C,A}}{\beta} & \text{for } I_{C,A} > \dfrac{2\beta\,V_T}{R_B} = I_2 \end{cases}$$

The three portions of the curve are shown in Fig. 2.48 as a asymptotic curve (broken line) for $\beta = \text{constant}$. With high collector currents the real curve runs above the asymptote due to the decrease in β.

Equivalent Noise Source and the Noise Figure

The driving of the transistor from a signal generator can be illustrated by the small-signal equivalent circuit diagram with a schematic presentation of the transistor as shown in Fig. 2.49a. The signal generator provides the signal voltage v_g and the noise voltage $v_{r,g}$.

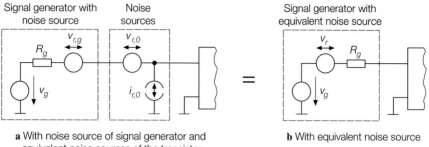

a With noise source of signal generator and   **b** With equivalent noise source
equivalent noise sources of the transistor

Fig. 2.49. Transistor operation with a signal generator

The noise source of the signal generator can be combined with the equivalent noise sources of the transistor into an *equivalent noise source* v_r, as shown in Fig. 2.49b. The noise voltage density is:

$$|\underline{v}_r(f)|^2 = |\underline{v}_{r,g}(f)|^2 + |\underline{v}_{r,0}(f)|^2 + R_g^2|\underline{i}_{r,0}(f)|^2 \tag{2.51}$$

One assumes that the noise of the transistor is caused by the signal generator, and the ratio of the noise density due to the equivalent noise source to the noise density of the signal generator is called the *spectral noise figure* [2.10]:

$$F(f) = \frac{|\underline{v}_r(f)|^2}{|\underline{v}_{r,g}(f)|^2} = 1 + \frac{|\underline{v}_{r,0}(f)|^2 + R_g^2|\underline{i}_{r,0}(f)|^2}{|\underline{v}_{r,g}(f)|^2} \tag{2.52}$$

The *mean noise figure F* indicates the reduction in the *signal-to-noise-ratio (SNR)* caused by the transistor in a frequency interval $f_a < f < f_b$; the signal-to-noise-ratio is the ratio of the powers of the useful signal and the noise. Since the power of a signal is proportional to the square of the effective value, the signal-to-noise-ratio of the signal generator is:

$$SNR_g = \frac{v_{g\text{eff}}^2}{v_{r,g\text{eff}}^2} = \frac{v_{g\text{eff}}^2}{\int_{f_a}^{f_b} |\underline{v}_{r,g}(f)|^2 df}$$

The transistor intensifies the noise density by the spectral noise figure $F(f)$; consequently the signal-to-noise-ratio decreases to:

$$SNR = \frac{v_{g\text{eff}}^2}{\int_{f_a}^{f_b} |\underline{v}_r(f)|^2 df} = \frac{v_{g\text{eff}}^2}{\int_{f_a}^{f_b} F(f)|\underline{v}_{r,g}(f)|^2 df}$$

Thus the mean noise figure is [2.9]:

$$F = \frac{SNR_g}{SNR} = \frac{\int_{f_a}^{f_b} F(f)|\underline{v}_{r,g}(f)|^2 df}{\int_{f_a}^{f_b} |\underline{v}_{r,g}(f)|^2 df}$$

If one assumes that the noise of the signal generator is caused by the thermal noise of the internal resistance R_g so that $|\underline{v}_{r,g}(f)|^2 = 4kTR_g$, this expression can be placed in front of the integral, which leads to:

$$F = \frac{1}{f_b - f_a} \int_{f_a}^{f_b} F(f)df$$

In this case the mean noise figure F is achieved by averaging over the spectral noise figure $F(f)$. Often $F(f)$ is constant for the frequency interval observed; then $F = F(f)$, which is simply called the *noise figure F*.

Noise Figure of a Bipolar Transistor

The spectral noise figure $F(f)$ of a bipolar transistor can be determined by inserting the equivalent noise densities $|\underline{v}_{r,0}(f)|^2$ from (2.47) and $|\underline{i}_{r,0}(f)|^2$ from (2.48) into (2.52). Figure 2.50 shows the plot of $F(f)$ using a numeric example. When $f < f_1 < f_{g(1/f)}$ the 1/f noise dominates and $F(f)$ is inversely proportional to the frequency; for $f > f_2 > f_T/\sqrt{\beta}$ the noise figure $F(f)$ is proportional to f^2.

Insertion from (2.49) and (2.50) into (2.52) yields the noise figure F for $f_{g(1/f)} < f < f_T/\sqrt{\beta}$; in this frequency range all noise densities remain constant – that is, F is not dependent on the frequency:

$$F = F(f) = 1 + \frac{1}{R_g}\left(R_B + \frac{V_T}{2I_{C,A}} + \frac{R_B^2 I_{C,A}}{2\beta V_T}\right) + \frac{I_{C,A}R_g}{2\beta V_T} \tag{2.53}$$

The noise figure is usually given in decibels: $F_{dB} = 10\log F$. Figure 2.51 shows the noise figure of a low-power transistor as a function of the operating point current $I_{C,A}$ for various internal resistances R_g of the signal generator. Figure 2.51a contains the plots for a frequency above the 1/f cutoff frequency $f_{g(1/f)}$; here, (2.53) applies, which means that the noise figure is independent of the frequency. Figure 2.51b contains the plots for

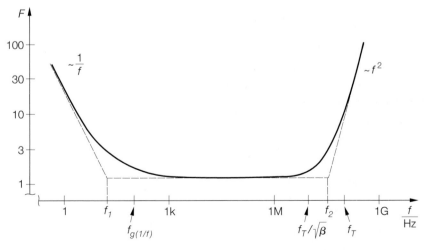

Fig. 2.50. Plots of the spectral noise figure $F(f)$ of a bipolar transistor with $I_{C,A} = 1\,\text{mA}$, $\beta = 100$, $R_B = 60\,\Omega$, $R_g = 1\,\text{k}\Omega$, $f_{g(1/f)} = 100\,\text{Hz}$ and $f_T = 100\,\text{MHz}$

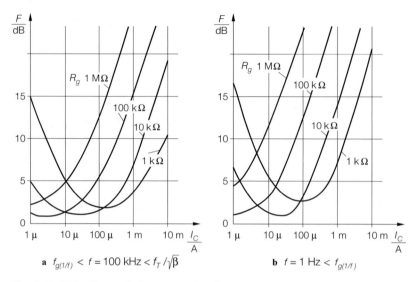

a $f_{g(1/f)} < f = 100\,\text{kHz} < f_T/\sqrt{\beta}$ b $f = 1\,\text{Hz} < f_{g(1/f)}$

Fig. 2.51. Noise figure of a low-power transistor

a frequency below $f_{g(1/f)}$; here, the noise figure depends on the frequency, which means that the plots are true only for the frequencies indicated.

Minimizing the noise figure: We can see in Fig. 2.51a that the noise figure reaches a minimum under certain conditions; the optimum operating point current $I_{C,Aopt}$ can be taken directly from the respective R_g plot. This is presented more clearly in Fig. 2.52, which shows the contours of the noise figure in a log–log diagram of $I_{C,A}$ and R_g. Using

$$\frac{\partial F}{\partial I_{C,A}} = 0$$

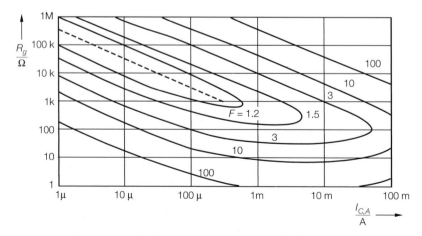

Fig. 2.52. Contours of the noise figure in the $I_{C,A} - R_g$ plane for $R_B = 60\,\Omega$ and $\beta = 100$

the optimum operating point current $I_{C,Aopt}$ can be calculated with (2.53) if R_g is given:

$$I_{C,A\,opt} = \frac{V_T\sqrt{\beta}}{\sqrt{R_g^2 + R_B^2}} \approx \begin{cases} \dfrac{V_T\sqrt{\beta}}{R_B} & \text{for } R_g < R_B \\[3mm] \dfrac{V_T\sqrt{\beta}}{R_g} & \text{for } R_g > R_B \end{cases} \tag{2.54}$$

For low-resistance signal generators with $R_g < R_B$ the optimum operating point current $I_{C,Aopt}$ is determined by R_B, β and V_T; that is, it does not depend on R_g. The current $I_{C,Aopt} \approx 1...50\,\text{mA}$ if $R_B \approx 10...300\,\Omega$ and $\beta \approx 100...400$. However, in practice this case seldom occurs. For signal generators with $R_g > R_B$, the current $I_{C,A\,opt}$ is inversely proportional to R_g; for low-power transistors the following estimation can be used:

$$I_{C,A\,opt} \approx \frac{0.3\,\text{V}}{R_g} \qquad \text{for } R_g \geq 1\,\text{k}\Omega \tag{2.55}$$

This is shown as a broken line in Fig. 2.52.

Similarly, the optimum source resistance R_{gopt} can be determined for a given value of $I_{C,A}$:

$$R_{gopt} = \sqrt{R_B^2 + \frac{\beta\,V_T}{I_{C,A}}\left(\frac{V_T}{I_{C,A}} + 2R_B\right)} \tag{2.56}$$

Three regions can be distinguished:

$$R_{gopt} \approx \begin{cases} \dfrac{V_T\sqrt{\beta}}{I_{C,A}} & \text{for } I_{C,A} < \dfrac{V_T}{2R_B} = I_1 \\[3mm] \sqrt{\dfrac{2\beta\,V_T\,R_B}{I_{C,A}}} & \text{for } \dfrac{V_T}{2R_B} < I_{C,A} < \dfrac{2\beta\,V_T}{R_B} \\[3mm] R_B & \text{for } I_{C,A} > \dfrac{2\beta\,V_T}{R_B} = I_2 \end{cases}$$

The plot of $|\underline{v}_{r,0}(f)|^2$ shown in Fig. 2.48 illustrates these regions, which clearly have identical boundaries. The relation between $I_{C,A}$ and R_{gopt} is illustrated in Fig. 2.53; the optimum internal resistance is $R_{gopt} \sim 1/I_{C,A}$ for low currents at the operating point and $R_{gopt} \sim 1/\sqrt{I_{C,A}}$ for medium currents.

Inserting the optimum operating point current $I_{C,Aopt}$ from (2.54) into (2.53) leads to the following expression for the optimum noise figure:

$$F_{opt} = 1 + \frac{R_B}{R_g} + \frac{1}{\sqrt{\beta}}\sqrt{1 + \left(\frac{R_B}{R_g}\right)^2} \overset{R_g > R_B}{\approx} 1 + \frac{R_B}{R_g} + \frac{1}{\sqrt{\beta}} \tag{2.57}$$

It can be seen that the optimum noise figure of a transistor is determined by the base spreading resistance R_B and the small-signal current gain β. Transistors with a low base spreading resistance and a high small-signal current gain must be used in low-noise circuits; for a high internal resistance R_g it is important to have a high small-signal current gain β, while a low R_g requires a low base spreading resistance R_B. Since β depends on the operating point the absolute minimum of F_{opt} is not achieved with $R_g \to \infty$ as suggested by (2.57) but with a finite value of $R_g \approx 100\,\text{k}\Omega...1\,\text{M}\Omega$ with $I_{C,Aopt} \approx 1\,\mu\text{A}$.

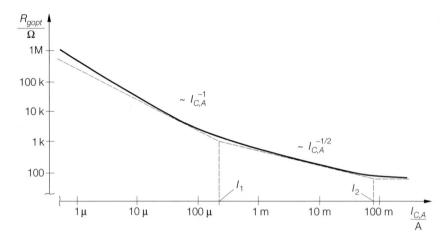

Fig. 2.53. Interdependence of the operating point and the optimum internal resistance R_{gopt} for $R_B = 60\,\Omega$: an asymptotic curve for $\beta = 100$ (*broken line*) and the real curve shape with β, depending on the operating point and $\beta_{max} = 100$

Noise figure in the range of 1/f noise: If $f < f_{g(1/f)}$, substituting (2.47) and (2.48) into (2.52) leads to:

$$F(f) = 1 + \frac{1}{R_g}\left(R_B + \frac{f_{g(1/f)}}{2f}\left(\frac{V_T}{I_{C,A}} + \frac{R_B^2 I_{C,A}}{\beta\, V_T}\right)\right) + \frac{I_{C,A} R_g f_{g(1/f)}}{2\beta\, V_T f}$$

The noise figure increases for $f \to 0$. Likewise in the range of 1/f noise the optimum operating point current $I_{C,Aopt}$ is determined by (2.54); that is, it is not dependent on the frequency. This means that with a given internal resistance R_g the optimum noise figure is achieved with $I_{C,Aopt}$ according to (2.54) at any frequency $f < f_T/\sqrt{\beta}$. In contrast, with a given $I_{C,A}$ the optimum internal resistance R_{gopt}, $(1/f)$ depends on the frequency:

$$R_{gopt,(1/f)} = \sqrt{R_B^2 + \frac{\beta\, V_T}{I_{C,A}}\left(\frac{V_T}{I_{C,A}} + \frac{2R_B f}{f_{g(1/f)}}\right)}$$

In practice R_{gopt}, $(1/f)$ is of minor importance since no broad-band matching is possible due to the dependence on the frequency.

Consequently, the optimum noise figure is:

$$F_{opt,(1/f)} = 1 + \frac{R_B}{R_g} + \frac{f_{g(1/f)}}{\sqrt{\beta}\,f}\sqrt{1 + \left(\frac{R_B}{R_g}\right)^2} \overset{R_g > R_B}{\approx} 1 + \frac{R_B}{R_g} + \frac{f_{g(1/f)}}{\sqrt{\beta}\,f}$$

where $F_{opt,(1/f)}$ increases for $f \to 0$; of particular importance in this case is a high small-signal current gain β.

Noise figure at high frequencies: Taking into account the increase in the equivalent noise densities for $f > f_T/\sqrt{\beta}$ leads to the following equations if $f_{g(1/f)} < f < f_T$:

$$R_{gopt,RF} \approx \sqrt{R_B^2 + \frac{\dfrac{\beta V_T}{I_{C,A}}\left(\dfrac{V_T}{I_{C,A}} + 2R_B\right)}{1 + \beta\left(\dfrac{f}{f_T}\right)^2}}$$

$$F_{opt,RF} \approx 1 + \sqrt{\left(\frac{1}{\beta} + \frac{2R_B I_{C,A}}{\beta V_T} + \left(\frac{R_B I_{C,A}}{\beta V_T}\right)^2\right)\left(1 + \beta\left(\frac{f}{f_T}\right)^2\right)}$$

For $f > f_T/\sqrt{\beta}$, the optimum source resistance $R_{gopt,RF}$ decreases with increasing frequency. Despite the frequency dependence it makes sense to specify $R_{gopt,RF}$ since most radio frequency circuits have a narrow-band characteristic. In such circuits the current $I_{C,A}$ at the operating point needs to be optimized with respect to the gain; that is, it is not available as a free parameter for minimizing the noise figure. Therefore, $F_{opt,RF}$ is given as a function of $I_{C,A}$.

At very high frequencies the noise sources in a transistor are no longer independent. For the equivalent noise densities this induces terms which cause the optimum internal resistance of the signal generator to be no longer real; in this range, the equations shown here provide only approximate values for $R_{gopt,RF}$ and $F_{opt,RF}$.

Notes about minimizing the noise figure: Various aspects should be taken into consideration in order to minimize the noise figure:

– Minimizing the noise figure does not cause an absolute minimum of the noise signal; rather, as can be seen from the definition of the noise figure it is the decrease in the signal noise ratio SNR that is minimized. The minimum absolute noise – that is, the minimum noise density $|\underline{v}_r(f)|^2$ of the equivalent noise source – is achieved for $R_g = 0$ according to (2.51). Which of the parameters needs to be minimized depends on the application involved: in a circuit for signal transmission it is necessary to minimize the noise figure in order to achieve an optimum SNR at the output; on the other hand, in a circuit not intended for signal transfer – for example, a current source to set the operating point – it is necessary to minimize the absolute noise at the output. Therefore, the noise figure is relevant only for signal transmission systems.
– The absolute minimum of the noise figure is achieved with a high internal resistance R_g and a small operating point current $I_{C,A}$. However, the result is true only for $f < f_T/\sqrt{\beta}$. For $I_{C,A} \approx 1\,\mu A$ a typical low-power transistor with a maximum transit frequency of 300 MHz and a maximum small-signal current gain of 400 achieves only $f_T \approx 200\,kHz$ and $\beta \approx 100$; therefore, this consideration only applies if $f < 20\,kHz$. Thus it is not possible to decrease $I_{C,A}$ to an arbitrary low value; its lower limit is determined by the required bandwidth of the circuit.
– In most cases the internal resistance R_g is given and $I_{C,Aopt}$ can be calculated using (2.54) or estimated using (2.55). If, in the event of very high quality requirements, the value determined by these methods should prove unsatisfactory, a transformer as shown in Fig. 2.54 can be used to transform the internal resistance. This method is employed with very low internal resistances since the optimum noise figure according to (2.57)

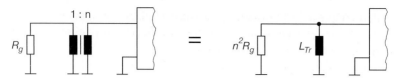

Fig. 2.54. Transforming the internal resistance of a signal generator by use of a transformer

is relatively high in such cases. The transformer converts the internal resistance to a higher value $n^2 R_g$ for which a lower optimum noise figure can be achieved. Due to the inductance L_{Tr} of the transformer the resulting highpass filter has a cutoff frequency $f_{Tr} = n^2 R_g/(2\pi L_{Tr})$; f_{Tr} must be lower than the minimum signal frequency that is of interest.

Example: According to (2.54) $I_{C,Aopt} = 3.3\,\text{mA}$ in a transistor with $\beta = 100$ and $R_B = 60\,\Omega$, if the internal resistance is $R_g = 50\,\Omega$, and $F_{opt} = 2.36 = 3.7\,\text{dB}$ according to (2.57). If we assume that only a minimum operating point current $I_{C,A} = 1\,\text{mA}$ is required due to the necessary bandwidth, then $R_{gopt} = 620\,\Omega$ according to (2.56). By using a transformer with $n = 4$ the internal resistance can be transformed to $n^2 R_g = 800\,\Omega$ and approximately matched to R_{gopt}. Since the optimum is not achieved with an integral value of n, the noise figure has to be determined using (2.53): $F = 1.18 = 0.7\,\text{dB}$. Thus, the use of a transformer in this example yields an increase in the SNR of 3 dB.

– Optimization of the noise figure by matching R_g to R_{gopt} cannot be achieved by additional resistances since these resistances introduce additional noise sources that are not taken into account in the definition of the noise figure according to (2.52); therefore, the formulas for F_{opt}, $I_{C,Aopt}$ and R_{gopt} are not applicable. In any case, with additional resistances the noise figure becomes worse. The matching must take place without additional noise sources being generated. This requirement is fulfilled when a transformer is introduced to transform the internal resistance, as long as the intrinsic noise of the transformer is negligible; in narrow-band applications of radio frequency circuits the matching can be achieved by using LC circuits or strip lines.

Example: In the previous example an attempt is made to match $R_g = 50\,\Omega$ to $R_{gopt} = 620\,\Omega$ using a series resistor $R = 570\,\Omega$. After extending (2.51), the equivalent noise source has a noise density of:

$$|\underline{v}_r(f)|^2 = |\underline{v}_{r,g}(f)|^2 + |\underline{v}_{R,r}(f)|^2 + |\underline{v}_{r,0}(f)|^2 + R_{gopt}^2|\underline{i}_{r,0}(f)|^2$$

and the noise figure can be determined with $|\underline{v}_{r,g}(f)|^2 = 8.28 \cdot 10^{-19}\,\text{V}^2/\text{Hz}$, $|\underline{v}_{R,r}(f)|^2 = 9.44 \cdot 10^{-18}\,\text{V}^2/\text{Hz}$ and $|\underline{v}_{r,0}(f)|^2 = 1.22 \cdot 10^{-18}\,\text{V}^2/\text{Hz}$ from (2.49), and with $|\underline{i}_{r,0}(f)|^2 = 3.2 \cdot 10^{-24}\,\text{A}^2/\text{Hz}$ from (2.50):

$$F(f) = \frac{|\underline{v}_r(f)|^2}{|\underline{v}_{r,g}(f)|^2} = 15.36 = 11.9\,\text{dB}$$

The series resistance increases the noise figure by 8.2 dB compared to the circuit without a transformer and by 11.2 dB compared to the circuit with a transformer.

– To optimize the noise figure it was assumed that the noise of the signal generator is to be attributed to the thermal noise of the internal resistance; that is, $|\underline{v}_{r,g}(f)|^2 = 4kTR_g$.

In general, this is not the case. Optimization of the noise figure by partial differentiation of (2.52) is, however, independent of $|\underline{v}_{r,g}(f)|^2$, as the constant 1 is eliminated by differentiation and the remaining expression is only scaled by $|\underline{v}_{r,g}(f)|^2$. This causes F_{opt} to change, but the corresponding values of R_{gopt} and $I_{C,Aopt}$ remain constant.

Determination of the Base Spreading Resistance

It is possible to determine the base spreading resistance R_B from the optimum noise figure F_{opt} by assuming $f < f_T/\sqrt{\beta}$; insertion into the equation for $F_{opt,RF}$ leads to:

$$R_B \approx \frac{\beta\,V_T}{I_{C,A}}\left(\sqrt{1 - \frac{1}{\beta} + \left(F_{opt} - 1\right)^2} - 1\right) \tag{2.58}$$

This is often done in practice as measuring R_B directly is very complex. For the high-frequency transistor BRF92P, for example, the base spreading resistor evaluates to $R_B \approx 40\,\Omega$ when the values $F_{opt} = 1.41 = 1.5\,\text{dB}$ for $f = 10\,\text{MHz} < f_T/\sqrt{\beta} = 300\,\text{MHz}$, $\beta \approx 100$ and $I_{C,A} = 5\,\text{mA}$ are assumed.

2.4
Basic Circuits

Basic circuits using bipolar transistors: There are three basic circuits in which bipolar transistors can be used: the *common-emitter circuit*, the *common-collector circuit* and the *common-base circuit*. The name reflects the node of the transistor that is connected as a common reference point for the input *and* output of the circuit; this is illustrated in Fig. 2.55.

In many circuits this condition is not met strictly, so a less stringent criterion must be used:

> *The designation reflects the node of the transistor that does not serve either as the input or the output of the circuit.*

Example: Figure 2.56 shows a three-stage amplifier with feedback. The first stage consists of the npn transistor T_1. The base connection serves as the input to the stage with the input voltage V_i across R_1 and the feedback output voltage V_o across R_2, while the collector serves as the output; thus, T_1 is operated in a common-emitter circuit. A difference compared to the stringent criterion is that despite the designation as a common-emitter circuit,

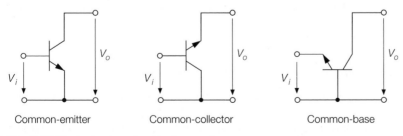

| Common-emitter | Common-collector | Common-base |

Fig. 2.55. Basic circuits using bipolar transistors

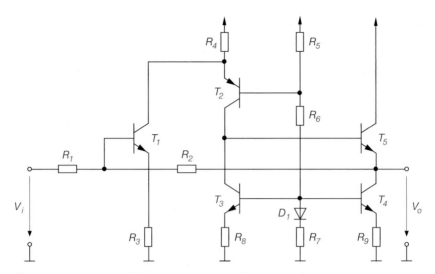

Fig. 2.56. An example of bipolar transistors used in basic configurations

it is not the emitter but the ground that is used as the common reference for the input and output of the stage. The output of the first stage is connected to the input of the second stage, which consists of a pnp transistor T_2. Here, the emitter is the input and the collector the output; therefore, T_2 is operated in a common-base circuit. Again, the base is not used as the reference point. The third stage is made up of the npn transistor T_5. The base is the input while the emitter forms the output of that stage and, at the same time, the output of the entire circuit; consequently T_5 is operated in a common-collector circuit. Transistors T_3 and T_4 serve as current sources and provide the bias currents for T_2 and T_5.

Basic circuits with several transistors: There are several configurations with two or more transistors. They are used so often that they, too, must be regarded as basic circuits which serve, for example, as differential amplifiers or current mirrors; such circuits are described in Sect. 4.1. A special configuration is the *Darlington circuit,* which uses two transistors in such a way that they can be treated as *one* transistor (see Sect. 2.4.4).

Polarity: Since their electrical characteristics are better suited, npn transistors are used predominantly in all these circuits; this applies especially to integrated circuits. In principle, npn and pnp transistors can be interchanged in all configurations by reversing the polarity of the supply voltages, the electrolytic capacitors and the diodes.

2.4.1
Common-Emitter Circuit

Figure 2.57a shows a common-emitter circuit consisting of the transistor, the collector resistance R_C, the supply voltage source V_b and the signal voltage source V_g with the internal resistance R_g. In what follows, we assume that $V_b = 5\,\text{V}$ and $R_C = R_g = 1\,\text{k}\Omega$ so that typical numeric results can be quoted in addition to the formulas.

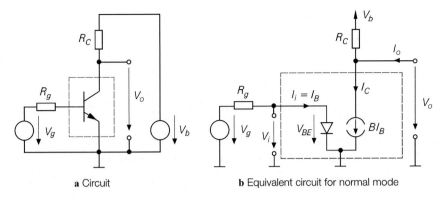

a Circuit   **b** Equivalent circuit for normal mode

Fig. 2.57. Common-emitter circuit

Transfer Characteristic of the Common-Emitter Circuit

The transfer characteristic shown in Fig. 2.58 is obtained by measuring the output voltage V_o as a function of the signal voltage V_g. For $V_g < 0.5\,\text{V}$ the collector current is negligible and $V_o = V_b = 5\,\text{V}$. If $0.5\,\text{V} \leq V_g \leq 0.72\,\text{V}$, a collector current I_C flows that increases with V_g, and the output voltage declines according to $V_o = V_b - I_C R_C$. Up to this point the transistor operates in the normal mode. For $V_g > 0.72\,\text{V}$ the transistor enters the saturation region, which results in $V_o = V_{CE,sat}$.

Normal mode: Figure 2.57b shows the equivalent circuit for normal mode, for which the simplified transport model according to Fig. 2.27 replaces the transistor; thus:

$$I_C = B I_B = I_S\, e^{\frac{V_{BE}}{V_T}}$$

This equation is derived from the basic (2.5) and (2.6) by disregarding the Early effect and assuming a constant large-signal current gain which leads to $B = B_0 = \beta$.

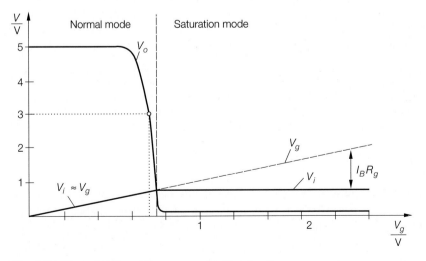

Fig. 2.58. Characteristics of the common-emitter circuit

The formulas for the voltages are:

$$V_o = V_{CE} = V_b + (I_o - I_C) R_C \overset{I_o=0}{=} V_b - I_C R_C \tag{2.59}$$

$$V_i = V_{BE} = V_g - I_B R_g = V_g - \frac{I_C R_g}{B} \approx V_g \tag{2.60}$$

Equation (2.60) is based on the assumption that the voltage drop across R_g can be disregarded if B is large enough and R_g is small enough.

A point located in the centre of the declining segment of the transfer characteristic is selected for the operating point; this allows maximum output signal. When selecting $B = \beta = 400$ and $I_S = 7\,\text{fA}$[11] the output voltage at the operating point, which is indicated in Fig. 2.58, will be for $V_b = 5\,\text{V}$ and $R_C = R_g = 1\,\text{k}\Omega$:

$$V_o = 3\,\text{V} \Rightarrow I_C = \frac{V_b - V_o}{R_C} = 2\,\text{mA} \Rightarrow I_B = \frac{I_C}{B} = 5\,\mu\text{A}$$

$$\Rightarrow V_i = V_{BE} = V_T \ln \frac{I_C}{I_S} = 685\,\text{mV} \Rightarrow V_g = V_i + I_B R_g = 690\,\text{mV}$$

In this case, the voltage drop across R_g is only 5 mV and can be disregarded; therefore, $V_i \approx V_g$ in normal mode according to Fig. 2.58.

In the calculation of the variables a *backwards* approach was selected; in other words, $V_g = V_g(V_o)$ was determined. This method allows the calculation of all values successively without approximation. In contrast, $V_o = V_o(V_g)$ cannot be calculated directly since, due to $I_B = I_B(V_{BE})$, (2.60) only represents an implicit equation for V_{BE} which cannot be solved with respect to V_{BE}; further successive calculations are only possible on the basis of the approximation $V_{BE} \approx V_g$.

Saturation mode: The transistor enters the saturation mode when V_{CE} reaches the saturation voltage $V_{CE,sat}$; then with $V_{CE,sat} \approx 0.1\,\text{V}$:

$$I_C = \frac{V_b - V_{CE,sat}}{R_C} = 4.9\,\text{mA} \Rightarrow I_B = \frac{I_C}{B} = 12.25\,\mu\text{A}$$

$$\Rightarrow V_i = V_{BE} = V_T \ln \frac{I_C}{I_S} = 709\,\text{mV} \Rightarrow V_g = V_i + I_B R_g = 721\,\text{mV}$$

For $V_g > 0.72\,\text{V}$ the transistor operates in the saturation region which means that the collector diode is forward-biased. In this region all variables with the exception of the base current are approximately constant:

$$I_C \approx 4.9\,\text{mA} \quad , \quad V_i = V_{BE} \approx 0.72\,\text{V} \quad , \quad V_o = V_{CE,sat} \approx 0.1\,\text{V}$$

The base current is

$$I_B = \frac{V_g - V_{BE}}{R_g} \approx \frac{V_g - 0.72\,\text{V}}{R_g}$$

and is the sum of the currents through the emitter and the collector diode. In this case the internal resistance R_g has to limit the base current to acceptable values. In Fig. 2.58 it was assumed that $V_{g,max} = 2\,\text{V}$; for $R_g = 1\,\text{k}\Omega$ it follows that $I_{B,max} \approx 1.28\,\text{mA}$ which is a permissible value for low-power transistors.

---

[11] Typical values for the npn low-power transistor BC547B

Small-Signal Response of Common-Emitter Circuits

The response to signals around an operating point A is called the *small-signal response*. The operating point is determined by the operating point values $V_{i,A} = V_{BE,A}$, $V_{o,A} = V_{CE,A}$, $I_{i,A} = I_{B,A}$ and $I_{C,A}$; the operating point as determined above is used as an example: $V_{BE,A} = 685\,\text{mV}$, $V_{CE,A} = 3\,\text{V}$, $I_{B,A} = 5\,\mu\text{A}$ and $I_{C,A} = 2\,\text{mA}$.

For a deeper understanding of the relation between the nonlinear characteristics and the small-signal equivalent circuit we first calculate the small-signal response from the characteristics and then from the small-signal equivalent circuit.

Calculation from the characteristics: The *small-signal voltage gain* corresponds to the slope of the transfer characteristic (see Fig. 2.59); differentiation of (2.59) leads to:

$$A = \left.\frac{\partial V_o}{\partial V_i}\right|_A = -\left.\frac{\partial I_C}{\partial V_{BE}}\right|_A R_C = -\frac{I_{C,A} R_C}{V_T} = -g_m R_C$$

When $g_m = I_{C,A}/V_T = 77\,\text{mS}$ and $R_C = 1\,\text{k}\Omega$, then $A = -77$. This gain is also called the *no-load gain* because it describes operation without any load ($I_o = 0$). Furthermore, it is obvious that the small-signal voltage gain is proportional to the voltage drop $I_{C,A} R_C$ across the collector resistance R_C. Due to $I_{C,A} R_C < V_b$, the possible maximum gain when using an ohmic collector resistance R_C is proportional to the supply voltage V_b.

The *small-signal input resistance* results from the input characteristic:

$$r_i = \left.\frac{\partial V_i}{\partial I_i}\right|_A = \left.\frac{\partial V_{BE}}{\partial I_B}\right|_A = r_{BE}$$

If $r_{BE} = \beta/g_m$ and $\beta = 400$, then $r_i = 5.2\,\text{k}\Omega$.

The *small-signal output resistance* can be determined from (2.59):

$$r_o = \left.\frac{\partial V_o}{\partial I_o}\right|_A = R_C$$

In this case $r_o = 1\,\text{k}\Omega$.

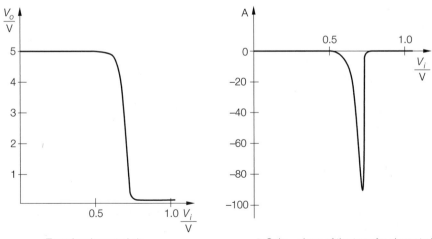

a Transfer characteristic

b Gain = slope of the transfer characteristic

Fig. 2.59. Gain of the common-emitter circuit

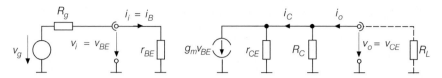

Fig. 2.60. Small-signal equivalent circuit of the common-emitter configuration

The calculation based on the characteristics leads to the small-signal parameters g_m and r_{BE} of the transistor (see Sect. 2.1.4).[12] Therefore, it is common practice to calculate directly from the small-signal equivalent circuit of the transistor without using the characteristics.

Calculation from the small-signal equivalent circuit: Figure 2.60 shows the small-signal equivalent circuit of the common-emitter configuration which is gained by inserting the small-signal equivalent circuit according to Fig. 2.12 or Fig. 2.39a, short-circuiting the DC voltage sources, omitting the DC current sources and changing over to the small-signal values:[13]

$$v_i = V_i - V_{i,A} \quad , \quad i_i = I_i - I_{i,A}$$
$$v_o = V_o - V_{o,A} \quad , \quad i_o = I_o - I_{o,A}$$
$$v_g = V_g - V_{g,A} \quad , \quad i_C = I_C - I_{C,A}$$

Omitting the load R_L leads to the *common-emitter circuit* as shown in Fig. 2.60:

Common-emitter circuit

$$A = \frac{v_o}{v_i}\bigg|_{i_o=0} = -g_m\,(R_C\|r_{CE}) \overset{r_{CE}\gg R_C}{\approx} -g_m R_C \tag{2.61}$$

$$r_i = \frac{v_i}{i_i} = r_{BE} \tag{2.62}$$

$$r_o = \frac{v_o}{i_o} = R_C\|r_{CE} \overset{r_{CE}\gg R_C}{\approx} R_C \tag{2.63}$$

If we take into consideratin that the Early effect has been disregarded, which means that $r_{CE} \to \infty$ was assumed, the results are the same as those determined from the characteristics. If $r_{CE} = V_A/I_{C,A}$ and $V_A \approx 100\,\text{V}$, then $A = -75$, $r_i = 5.2\,\text{k}\Omega$ and $r_o = 980\,\Omega$.

A, r_i and r_o fully describe the common-emitter circuit; Fig. 2.61 presents the corresponding equivalent circuit. The load resistance R_L may be an ohmic resistance or an equivalent element for the input resistance of a circuit connected to the output. In this situation it is important that the operating point is not shifted by R_L; in other words, no or only a minute direct current is allowed to flow through R_L. This will be outlined in more detail together with the method for setting the operating point.

---

[12] The output resistance r_{CE} of the transistor is of no significance here since when the characteristics curves were established the Early effect was omitted; that is, by assuming $r_{CE} \to \infty$.

[13] Changing over to the small-signal values by subtracting the operating point values corresponds to short-circuiting the DC voltage sources or omitting the DC current sources since the operating point values are DC voltages and DC currents.

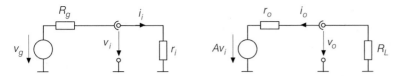

Fig. 2.61. Equivalent circuit with the equivalent parameters A, r_i and r_o

The *small-signal operating gain* is calculated from Fig. 2.61:

$$A_B = \frac{v_o}{v_g} = \frac{r_i}{r_i + R_g} A \frac{R_L}{R_L + r_o} \tag{2.64}$$

It comprises the gain A of the circuit and the voltage divider factors at the input and output. From the assumption that a common-emitter circuit of the same values is connected to the output as a load of $R_L = r_i = 5.2\,\text{k}\Omega$ it follows that $A_B \approx 0.7 \cdot A = -53$.

Maximum gain μ and $\beta - V_A$ product: The gain of the common-emitter circuit reaches its maximum with $R_C \to \infty$; the *maximum gain* is derived from (2.61):

$$\mu = \lim_{R_C \to \infty} |A| = g_m r_{CE} = \frac{I_{C,A}}{V_T} \frac{V_A}{I_{C,A}} = \frac{V_A}{V_T}$$

It is difficult to reach this borderline case with an ohmic collector resistance R_C, since $R_C \to \infty$ causes $R_C \gg r_{CE}$ so that the voltage drop across R_C needs to be much larger than the Early voltage $V_A \approx 100\,\text{V}$ because $I_{C,A} R_C \gg I_{C,A} r_{CE} = V_A$. However, this borderline situation can be achieved by replacing the collector resistance by a constant current source with the current $I_0 = I_{C,A}$; this results in very high small-signal resistances even with low voltages.

In practice μ is very seldom stated because it is only a substitute for the Early voltage V_A. One can summarize by saying that the possible maximum gain of a bipolar transistor is proportional to V_A. With npn transistors $V_A \approx 30...150\,\text{V}$ and thus $\mu \approx 1000...6000$ while in pnp transistors the voltage $V_A \approx 30...75\,\text{V}$ results in $\mu \approx 1000...3000$.

The maximum gain μ is reached only in the no-load condition. In many circuits – especially in integrated circuits – the input resistance of the subsequent stage acts as a load which, in common-emitter and common-collector circuits, is proportional to the current gain β. Thus, the gain actually achieved depends on both V_A *and* β; therefore it is often the $\beta - V_A$ *product* that is used as a quality criterion for bipolar transistors. Typical values range between $1000...60000$.

Non-linearity: Section 2.1.4 describes a relationship between the amplitude of a sinusoidal input signal $\hat{v}_i = \hat{v}_{BE}$ and the *distortion factor* k of the collector current which, in the common-emitter circuit, is identical to the distortion factor of the output voltage v_o (see (2.15) on page 45). Thus $\hat{v}_i < k \cdot 0.1\,\text{V}$; that is, for $k < 1\,\%$ the signal must be $\hat{v}_i < 1\,\text{mV}$. Due to $\hat{v}_o = |A| \hat{v}_i$, the corresponding output amplitude is dependent on the gain A; if $A = -75$ is the numeric example, then $\hat{v}_o < k \cdot 7.5\,\text{V}$.

Temperature sensitivity: The temperature dependence can be seen from (2.21); this shows that with a constant collector current I_C, the base–emitter voltage V_{BE} declines at a rate of $1.7\,\text{mV/K}$. This means that the input voltage must be reduced by $1.7\,\text{mV/K}$ in order to keep the operating point of the circuit $I_C = I_{C,A}$ constant. If, on the other hand, the input voltage is kept at a constant level, then a temperature increase acts like an increase in the input voltage of $dV_i/dT = 1.7\,\text{mV/K}$; the *temperature drift* of the output voltage can thus be calculated using the gain:

$$\left.\frac{dV_o}{dT}\right|_A = \left.\frac{\partial V_o}{\partial V_i}\right|_A \frac{dV_i}{dT} \approx A \cdot 1.7\,\text{mV/K} \tag{2.65}$$

In the numeric example this leads to $(dV_o/dT)|_A \approx -127\,\text{mV/K}$.

We can see that a temperature change of only a few Kelvin results in a marked shift of the operating point; this is accompanied by a change in A, r_i and r_o due to the altered operating point and a change in A and an additional change in A and r_i due to the temperature sensitivity of g_m and/or V_T and β. As, in practice, temperatures may change by 50 K and above, it is essential that the operating point be stabilized; this can be achieved, for example, by *feedback*.

Common-Emitter Circuit with Current Feedback

The nonlinearity and temperature sensitivity of the common-emitter circuit may be reduced by a *current feedback*; for this an *emitter resistor* R_E is introduced (see Fig. 2.62a). Figure 2.63 shows the transfer characteristic $V_o(V_g)$ and the characteristic for V_i and V_E for $R_C = R_g = 1\,\text{k}\Omega$ and $R_E = 500\,\Omega$. If $V_g < 0.5\,\text{V}$, then the collector current is negligible which leads to $V_o = V_b = 5\,\text{V}$. For $0.5\,\text{V} \le V_g \le 2.3\,\text{V}$ there is a collector current I_C that increases with V_g, and the output voltage declines according to $V_o = V_b - I_C R_C$; due to the feedback the characteristic is almost linear in this region. For $V_g \le 2.3\,\text{V}$, the transistor operates in normal forward mode. If $V_g > 2.3\,\text{V}$ the transistor enters the saturation region.

Normal mode: Figure 2.62b shows the equivalent circuit for normal operation. The voltages are:

$$V_o = V_b + (I_o - I_C)\,R_C \overset{I_o=0}{=} V_b - I_C R_C \tag{2.66}$$

$$V_i = V_{BE} + V_E = V_{BE} + (I_C + I_B)\,R_E \approx V_{BE} + I_C R_E \tag{2.67}$$

$$V_i = V_g - I_B R_g \approx V_g \tag{2.68}$$

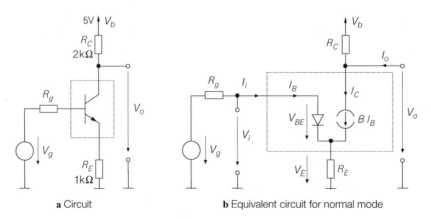

a Circuit   **b** Equivalent circuit for normal mode

Fig. 2.62. Common-emitter circuit with current feedback

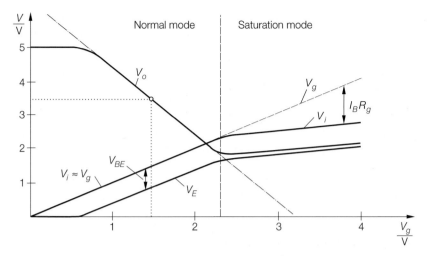

Fig. 2.63. Characteristics of the common-emitter circuit with current feedback

In (2.67) the base current I_B is very small compared to the collector current I_C because $B \gg 1$. It is assumed in (2.68) that the voltage drop across R_g can be disregarded. In (2.67) the current feedback is demonstrated by the fact that the voltage V_{BE} is reduced to $V_{BE} \approx V_i - I_C R_E$ by the collector current I_C, unlike in the common-emitter circuit without negative feedback where $V_{BE} = V_i$ (see (2.60)).

If $0.8\,\text{V} < V_g < 2.2\,\text{V}$, then $V_{BE} \approx 0.7\,\text{V}$; thus, from (2.67) and (2.68) it follows that

$$I_C \approx \frac{V_g - 0.7\,\text{V}}{R_E}$$

which can be substituted in (2.66):

$$V_o \approx V_b - \frac{R_C}{R_E}\left(V_g - 0.7\,\text{V}\right) \tag{2.69}$$

This linear relationship is shown as a broken line in Fig. 2.63 and corresponds closely to the transfer characteristic for $0.8\,\text{V} < V_g < 2.2\,\text{V}$; the difference compared to the transfer characteristic in this range is now only dependent on R_C and R_E. The feedback therefore has the effect that the behavior of the circuit in its first approximation no longer depends on the nonlinear properties of the transistor, but solely on the linear resistances; in addition, inter-component deviations of transistor parameters have virtually no influence.

A point located in the centre of the down-sloping segment of the transfer characteristic is selected for the operating point; this allows maximum output signal. With the sample operating point marked in Fig. 2.63 for $V_b = 5\,\text{V}$, $I_S = 7\,\text{fA}$, $B = \beta = 400$, $R_C = R_g = 1\,\text{k}\Omega$ and $R_E = 500\,\Omega$ we achieve:

$$V_o = 3.5\,\text{V} \;\Rightarrow\; I_C = \frac{V_b - V_o}{R_C} = 1.5\,\text{mA} \;\Rightarrow\; I_B = \frac{I_C}{B} = 3.75\,\mu\text{A}$$

$$\Rightarrow\; V_E = (I_C + I_B)\,R_E = 752\,\text{mV}$$

$$\Rightarrow\; V_i = V_{BE} + V_E = V_T \ln\frac{I_C}{I_S} + V_E = 1430\,\text{mV}$$

$$\Rightarrow V_g = V_i + I_B R_g = 1434\,\text{mV}$$

For $V_o = 3.5\,\text{V}$, (2.69) yields the approximation $V_g \approx 1.45\,\text{V}$.

Saturation mode: The transistor enters the saturation region when V_{CE} reaches the saturation voltage $V_{CE,sat}$; if $V_E \approx V_g - 0.7\,\text{V}$, (2.69) leads to:

$$V_{CE} \approx V_o - V_E = V_b - \left(1 + \frac{R_C}{R_E}\right)(V_g - 0.7\,\text{V})$$

Inserting $V_{CE} = V_{CE,sat} \approx 0.1\,\text{V}$ and solving the equation with respect to V_g leads to $V_g \approx 2.3\,\text{V}$. For $V_g > 2.3\,\text{V}$ the collector diode is forward-biased and a base current flows through the emitter diode and the collector diode; this current increases with V_g and is limited by R_g (see Fig. 2.63). Since the base current flows through R_E the voltages V_i, V_o and V_E cannot be considered to be constant values as in the common-emitter circuit without feedback, but they increase with V_g.

Small-signal response: The *voltage gain A* corresponds to the slope of the transfer characteristic (see Fig. 2.64); it is almost constant in the region of linear approximation according to (2.69). Calculation of A is carried out by using the small-signal equivalent circuit shown in Fig. 2.65. From the nodal equations

$$\frac{v_i - v_E}{r_{BE}} + g_m v_{BE} + \frac{v_o - v_E}{r_{CE}} = \frac{v_E}{R_E}$$

$$g_m v_{BE} + \frac{v_o - v_E}{r_{CE}} + \frac{v_o}{R_C} = i_o$$

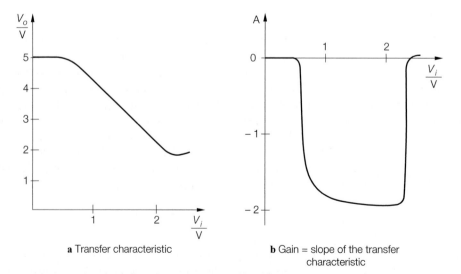

a Transfer characteristic

b Gain = slope of the transfer characteristic

Fig. 2.64. Gain of the common-emitter circuit with negative current feedback

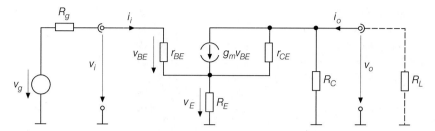

Fig. 2.65. Small-signal equivalent circuit for common-emitter configuration with negative current feedback

for $v_{BE} = v_i - v_E$, it follows that:

$$A = \left.\frac{v_o}{v_i}\right|_{i_o=0} = -\frac{g_m R_C \left(1 - \dfrac{R_E}{\beta\, r_{CE}}\right)}{1 + R_E \left(g_m \left(1 + \dfrac{1}{\beta} + \dfrac{R_C}{\beta\, r_{CE}}\right) + \dfrac{1}{r_{CE}}\right) + \dfrac{R_C}{r_{CE}}}$$

$$\overset{\substack{r_{CE} \gg R_C, R_E \\ \beta \gg 1}}{\approx} \quad -\frac{g_m R_C}{1 + g_m R_E} \overset{g_m R_E \gg 1}{\approx} -\frac{R_C}{R_E}$$

If $g_m R_E \gg 1$, the gain only depends on R_C and R_E. When a load resistance R_L is used the relevant operating gain A_B can be calculated by replacing R_C by the parallel resistors R_C and R_L (see Fig. 2.65). For the sample operating point selected we obtain an *exact* value of $A = -1.927$ with $g_m = 57.7\,\text{mS}$, $r_{BE} = 6.9\,\text{k}\Omega$, $R_C = R_g = 1\,\text{k}\Omega$ and $R_i = 500\,\Omega$; the first approximation yields $A = -1.933$ and the second approximation $A = -2$.

The *input resistance* is:

$$r_i = \left.\frac{v_i}{i_i}\right|_{i_o=0} = r_{BE} + \frac{(1+\beta)\,r_{CE} + R_C}{r_{CE} + R_E + R_C}\,R_E$$

$$\overset{\substack{r_{CE} \gg R_C, R_E \\ \beta \gg 1}}{\approx} \quad r_{BE} + \beta R_E$$

It depends on the load resistance for which in this case the *open-circuit input resistance* is given due to $i_o = 0$ ($R_L \to \infty$). The input resistance for other values of R_L can be calculated by replacing R_C with the two resistors R_C and R_L connected in parallel; by setting $R_L = R_C = 0$ we obtain the *short-circuit input resistance*. However, the dependence on R_L is so slight that it is eliminated by the approximation. For the operating point used as an example $r_{i,o} = 202.1\,\text{k}\Omega$ is the *exact* open-circuit input resistance and $r_{i,s} = 205\,\text{k}\Omega$ the *exact* short-circuit input resistance; the approximation gives us $r_i = 206.9\,\text{k}\Omega$.

The *output resistance* depends on the internal resistance R_g; here we will look at the borderline cases only. The *short-circuit output resistance* applies to a short-circuited input with $v_i = 0$ and $R_g = 0$:

$$r_{o,s} = \left. \frac{v_o}{i_o} \right|_{v_i=0} = R_C \,\|\, r_{CE} \left(1 + \frac{\beta + \dfrac{r_{BE}}{r_{CE}}}{1 + \dfrac{r_{BE}}{R_E}} \right)$$

$$\overset{\substack{r_{CE} \gg r_{BE} \\ \beta \gg 1}}{\approx} R_C \,\|\, r_{CE} \, \frac{\beta R_E + r_{BE}}{R_E + r_{BE}} \overset{r_{CE} \gg R_C}{\approx} R_C$$

If $i_i = 0$ or $R_g \to \infty$, the *open-circuit output resistance* is:

$$r_{o,o} = \left. \frac{v_o}{i_o} \right|_{i_i=0} = R_C \,\|\, (R_E + r_{CE}) \overset{r_{CE} \gg R_C}{\approx} R_C$$

Here, too, the dependence on R_g is so insignificant that it is negligible in practical applications. In our example we have $r_o = R_C = 1\,\mathrm{k\Omega}$.

For $r_{CE} \gg R_C, R_E, \beta \gg 1$ and no load resistance R_L, the following are obtained for the *common-emitter circuit with current feedback*:

Common-emitter circuit with current feedback

$$A = \left. \frac{v_o}{v_i} \right|_{i_o=0} \approx -\frac{g_m R_C}{1 + g_m R_E} \overset{g_m R_E \gg 1}{\approx} -\frac{R_C}{R_E} \tag{2.70}$$

$$r_i = \frac{v_i}{i_i} \approx r_{BE} + \beta R_E = r_{BE}(1 + g_m R_E) \tag{2.71}$$

$$r_o = \frac{v_o}{i_o} \approx R_C \tag{2.72}$$

Comparison with the common-emitter circuit without feedback: A comparison of (2.70) with (2.61) shows that the current feedback reduces the gain approximately by the *feedback factor* $(1 + g_m R_E)$; the input resistance is increased simultaneously by the same factor, as is revealed by a comparison of (2.71) and (2.62).

It is particularly easy to describe the effect of the current feedback by means of the *reduced transconductance*

$$g_{m,red} = \frac{g_m}{1 + g_m R_E} \tag{2.73}$$

The emitter resistance R_E reduces the effective transconductance of the transistor to the value $g_{m,red}$: $A \approx -g_m R_C$ and $r_i = r_{BE} = \beta/g_m$ apply to the common-emitter circuit without negative feedback, while $A \approx -g_{m,red} R_C$ and $r_i \approx \beta/g_{m,red}$ apply to the common-emitter circuit with negative feedback.

Non-linearity: The nonlinearity of the transfer characteristic is reduced markedly by the current feedback. The distortion factor of the circuit can be determined approximately by a series expansion of the characteristic at the operating point. From (2.67):

$$V_i = I_C R_E + V_T \ln \frac{I_C}{I_S}$$

Insertion of the operating point, changing to small-signal values and serial expansion results in

$$v_i = i_C R_E + V_T \ln\left(1 + \frac{i_C}{I_{C,A}}\right)$$

$$= i_C R_E + V_T \frac{i_C}{I_{C,A}} - \frac{V_T}{2}\left(\frac{i_C}{I_{C,A}}\right)^2 + \frac{V_T}{3}\left(\frac{i_C}{I_{C,A}}\right)^3 - \cdots$$

and yields after, inverting the series,

$$\frac{i_C}{I_{C,A}} = \frac{1}{1 + g_m R_E}\left[\frac{v_i}{V_T} + \frac{1}{2\left(1 + g_m R_E\right)^2}\left(\frac{v_i}{V_T}\right)^2 + \cdots\right]$$

For input signal $v_i = \hat{v}_i \cos \omega t$ the ratio of the first harmonic with $2\omega t$ to the fundamental with ωt represents the approximate *distortion factor* k when using low amplitudes; that is, when neglecting higher exponential powers:

$$k \approx \frac{v_{o,2\omega t}}{v_{o,\omega t}} \approx \frac{i_{C,2\omega t}}{i_{C,\omega t}} \approx \frac{\hat{v}_i}{4 V_T \left(1 + g_m R_E\right)^2} \tag{2.74}$$

If a maximum is specified for k, then $\hat{v}_i < 4k V_T(1 + g_m R_E)^2$ must be applied. With $\hat{v}_o = |A|\hat{v}_i$ this leads to the maximum output amplitude. For the numeric example we thus have $\hat{v}_i < k \cdot 93\,\text{V}$ and, if $A \approx -1.93$, $\hat{v}_o < k \cdot 179\,\text{V}$.

A comparison with (2.15) shows that due to the negative feedback the permissible input amplitude $\hat{v}_i$ is increased by the square of the feedback factor $(1 + g_m R_e)$. As the gain is reduced by the feedback factor at the same time, the permissible output amplitude is increased by the feedback factor as long as the distortion factor remains the same and the transistor is not overloaded or driven into saturation; that is, as long as the range of validity of the series expansion is not exceeded. With the same output amplitude the distortion factor is reduced by the feedback factor.

Temperature sensitivity: Since according to (2.21) the base–emitter voltage declines at a rate of $1.7\,\text{mV/K}$, a temperature increase at a constant input voltage has the same effect as an increase in the input voltage of $1.7\,\text{mV/K}$ at a constant temperature. Therefore, the *temperature drift* of the output voltage can be calculated using (2.65). For the numeric example used earlier, we obtain $(dV_o/dT)|_A \approx -3.3\,\text{mV/K}$. For most applications this value is sufficiently low that other measures to stabilize the operating point are not required.

Common-Emitter Circuit with Voltage Feedback

Another type of feedback is voltage feedback; here a portion of the output voltage is returned to the base of the transistor via resistances R_1 and R_2 (see Fig. 2.66a). For a circuit powered by a voltage source V_i[14] the characteristics are as shown in Fig. 2.67 for $R_C = R_1 = 1\,\text{k}\Omega$ and $R_2 = 2\,\text{k}\Omega$. For $V_i < -0.8\,\text{V}$ the collector current is negligible and V_o is determined by the voltage divider formed by R_1 and R_2. If $-0.8\,\text{V} \le V_i \le 1\,\text{V}$, the

---

[14] The common-emitter circuit without feedback according to Fig. 2.57a requires the internal resistance R_g of the signal voltage source in order to limit the base current in the saturation mode; in the present case the base current is limited by R_1, i.e. we may set $R_g = 0$ and use the voltage source $V_i = V_g$ to drive the circuit. This approach is chosen in order to prevent the characteristics from depending on R_g in normal mode.

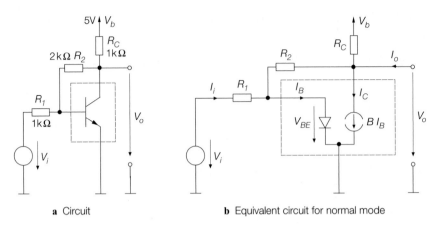

a Circuit   **b** Equivalent circuit for normal mode

Fig. 2.66. Common-emitter circuit with voltage feedback

collector current increases together with V_i and the output voltage decreases accordingly; this segment of the characteristic is almost linear because of the feedback. Up to here the transistor operates in normal mode. It enters the saturation region if $V_i > 1$ V and we have $V_o = V_{CE,sat}$.

Normal mode: Figure 2.66b presents the equivalent circuit for the normal mode. From the nodal equations

$$\frac{V_i - V_{BE}}{R_1} + \frac{V_o - V_{BE}}{R_2} = I_B = \frac{I_C}{B}$$

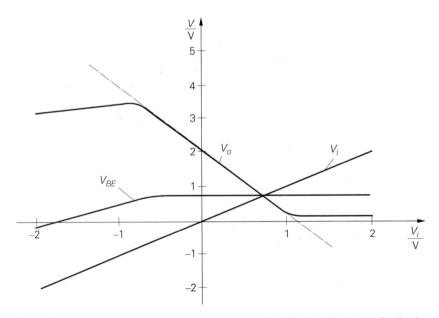

Fig. 2.67. Characteristics of the common-emitter circuit with negative current feedback

$$\frac{V_b - V_o}{R_C} + I_o = \frac{V_o - V_{BE}}{R_2} + I_C$$

For operation without any load – that is, $I_o = 0$ – it follows that:

$$V_o = \frac{V_b R_2 - I_C R_C R_2 + V_{BE} R_C}{R_2 + R_C} \tag{2.75}$$

$$V_i = \frac{I_C R_1}{B} + V_{BE} \left(1 + \frac{R_1}{R_2}\right) - V_o \frac{R_1}{R_2} \tag{2.76}$$

Solving (2.75) with respect to I_C and substituting into (2.76) leads to the following equation if $B \gg 1$ and $B R_C \gg R_2$:

$$V_o \approx \frac{V_b R_2}{B R_C} + \left(1 + \frac{R_2}{R_1}\right) V_{BE} - \frac{R_2}{R_1} V_i \tag{2.77}$$

If $-0.6\,\text{V} \le V_i \le 0.9\,\text{V}$, then $V_{BE} \approx 0.7\,\text{V}$; thus (2.77) describes a linear relation between V_o and V_i, as indicated as a broken line in Fig. 2.67, which corresponds very well to the transfer characteristic. Thus, voltage feedback causes the transfer line in this region to depend only on R_1 and R_2 as a first approximation.

$V_{i,A} = 0\,\text{V}$ is selected for the operating point; this point is located approximately in the middle of the linear segment. In this case it is not possible to successively calculate the operating point values because only implicit equations can be derived from (2.75) and (2.76). However, the operating point can still be determined very accurately with the help of approximations and an iterative approach; this is based on estimated values that become more and more accurate in the course of the calculation. For $R_1 = 1\,\text{k}\Omega$, $R_2 = 2\,\text{k}\Omega$, $B = \beta = 400$, $V_i = 0$ and the estimated value $V_{BE} \approx 0.7$ we obtain from (2.76)

$$V_o = 3\,V_{BE} + I_C \cdot 5\,\Omega \approx 3\,V_{BE} \approx 2.1\,\text{V}$$

From the nodal equation at the output for $V_b = 5\,\text{V}$ and $R_C = 1\,\text{k}\Omega$ it follows that:

$$I_C = \frac{V_b - V_o}{R_C} - \frac{V_o - V_{BE}}{R_2} \approx 2.2\,\text{mA}$$

The estimated value for I_C and $I_S = 7\,\text{fA}$ make it possible to determine V_{BE} more accurately:

$$V_{BE} = V_T \ln \frac{I_C}{I_S} \approx 688\,\text{mV}$$

Repeating the calculation with this new value leads to:

$$V_{BE} \approx 688\,\text{mV} \;\Rightarrow\; V_o \approx 2.07\,\text{V} \;\Rightarrow\; I_C \approx 2.24\,\text{mA}$$

$$\Rightarrow\; I_B = \frac{I_C}{B} \approx 5.6\,\mu\text{A} \;\Rightarrow\; V_i \overset{(2.76)}{\approx} 2.6\,\text{mV} \approx 0$$

These values represent a very accurate solution of (2.75) and (2.76) for $V_i = 0$.

Saturation mode: The transistor enters the saturation region when V_o reaches the saturation voltage $V_{CE,sat}$; inserting $V_o = V_{CE,sat} \approx 0.1\,\text{V}$ and $V_{BE} \approx 0.7\,\text{V}$ into (2.77) leads to $V_i \approx 1\,\text{V}$. For $V_i > 1\,\text{V}$ the collector diode is operated in the forward region.

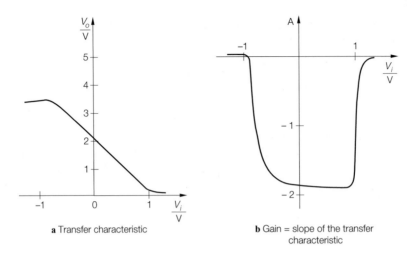

a Transfer characteristic

b Gain = slope of the transfer characteristic

Fig. 2.68. Gain of the common-emitter circuit with negative current feedback

Small-signal response: The *voltage gain A* corresponds to the slope of the transfer characteristic (see Fig. 2.68); it is approximately constant in the region for which the linear approximation according to (2.77) applies. Calculation of *A* is carried out with the aid of the small-signal equivalent circuit shown in Fig. 2.69. From the nodal equations

$$\frac{v_i - v_{BE}}{R_1} + \frac{v_o - v_{BE}}{R_2} = \frac{v_{BE}}{r_{BE}}$$

$$g_m v_{BE} + \frac{v_o - v_{BE}}{R_2} + \frac{v_o}{r_{CE}} + \frac{v_o}{R_C} = i_o$$

and $R_C' = R_C \| r_{CE}$ it follows that:

$$A = \frac{v_o}{v_i}\bigg|_{i_o=0} = \frac{-g_m R_2 + 1}{1 + R_1\left(g_m\left(1 + \frac{1}{\beta}\right) + \frac{1}{R_C'}\right) + \frac{R_2}{R_C'}\left(1 + \frac{R_1}{r_{BE}}\right)}$$

$$\overset{\substack{r_{CE} \gg R_C \\ \beta \gg 1}}{\approx} \frac{-g_m R_2 + 1}{1 + g_m R_1 + \frac{R_1}{R_C} + \frac{R_2}{R_C}\left(1 + \frac{R_1}{r_{BE}}\right)}$$

$$\overset{\substack{r_{BE} \gg R_1 \\ R_1, R_2 \gg 1/g_m}}{\approx} -\frac{R_2}{R_1 + \frac{R_1 + R_2}{g_m R_C}} \overset{g_m R_C \gg 1 + R_2/R_1}{\approx} -\frac{R_2}{R_1}$$

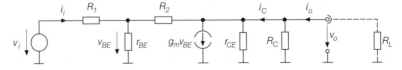

Fig. 2.69. Small-signal equivalent circuit for the common-emitter configuration with negative current feedback

If all conditions are met then A depends only on R_1 and R_2; the last condition means that the gain without feedback – that is, $-g_m R_C$ – must be much higher than the *ideal* gain with feedback – that is, $-R_2/R_1$. If the circuit is operated with a load resistance R_L the operating gain A_B can be calculated by replacing R_C by the parallel resistors R_C and R_L (see Fig. 2.69). For the operating point selected as an example the *exact* value is $A = -1.885$ for $g_m = 86.2\,\mathrm{mS}$, $r_{BE} = 4.6\,\mathrm{k\Omega}$, $r_{CE} = 45\,\mathrm{k\Omega}$, $R_C = R_1 = 1\,\mathrm{k\Omega}$ and $R_2 = 2\,\mathrm{k\Omega}$; while the first approximation is $A = -1.912$, the second approximation is $A = -1.933$ and the third approximation is $A = -2$.

For the *open-circuit input resistance* it follows with $R'_C = R_C \,||\, r_{CE}$ that:

$$r_{i,o} = \left.\frac{v_i}{i_i}\right|_{i_o=0} = R_1 + \frac{r_{BE}\left(R'_C + R_2\right)}{r_{BE} + (1 + \beta)\,R'_C + R_2}$$

$$\overset{\substack{r_{CE} \gg R_C \\ \beta \gg 1}}{\approx} R_1 + \frac{r_{BE}\,(R_C + R_2)}{r_{BE} + \beta R_C + R_2}$$

$$\overset{\beta R_C \gg r_{BE},R_2}{\approx} R_1 + \frac{1}{g_m}\left(1 + \frac{R_2}{R_C}\right)$$

$$\overset{g_m R_C \gg R_2/R_1}{\approx} R_1 + \frac{1}{g_m} \overset{g_m R_1 \gg 1}{\approx} R_1$$

Here $i_o = 0$; that is, $R_L \to \infty$. For other values of R_L the input resistance is calculated by replacing R_C by the parallel arrangement of R_C and R_L. Inserting $R_L = R_C = 0$ leads to the *short-circuit input resistance*:

$$r_{i,s} = \left.\frac{v_i}{i_i}\right|_{v_o=0} = R_1 + r_{BE} \,||\, R_2$$

With the sample operating point the open-circuit input resistance is *exactly* $r_{i,o} = 1034\,\Omega$; the first approximation also provides $r_{i,o} = 1034\,\Omega$, the second approximation $r_{i,o} = 1035\,\Omega$, the third approximation $r_{i,o} = 1012\,\Omega$ and the fourth approximation $r_{i,o} = 1\,\mathrm{k\Omega}$. The short-circuit input resistance is $r_{i,s} = 2.4\,\mathrm{k\Omega}$.

With $R'_C = R_C || r_{CE}$ the *short-circuit output resistance* is:

$$r_{o,s} = \left.\frac{v_o}{i_o}\right|_{v_i=0} = R'_C \,||\, \frac{r_{BE}\,(R_1 + R_2) + R_1 R_2}{r_{BE} + R_1\,(1 + \beta)}$$

$$\overset{\substack{r_{CE} \gg R_C \\ \beta \gg 1}}{\approx} R_C \,||\, \frac{r_{BE}\,(R_1 + R_2) + R_1 R_2}{r_{BE} + \beta R_1}$$

$$\overset{\beta R_1 \gg r_{BE}}{\approx} R_C \,||\, \left(\frac{1}{g_m}\left(1 + \frac{R_2}{R_1}\right) + \frac{R_2}{\beta}\right)$$

For $R_1 \to \infty$ the *open-circuit output resistance* is:

$$r_{o,o} = \left.\frac{v_o}{i_o}\right|_{i_i=0} = R'_C \,||\, \frac{r_{BE} + R_2}{1 + \beta} \overset{\substack{r_{CE} \gg R_C \\ \beta \gg 1}}{\approx} R_C \,||\, \left(\frac{1}{g_m} + \frac{R_2}{\beta}\right)$$

For the sample operating point the short-circuit output resistance is *exactly* $r_{o,s} = 37.5\,\Omega$; the first approximation is also $r_{o,s} = 37.5\,\Omega$, and the second approximation $r_{o,s} = 38.3\,\Omega$.

The open-circuit output resistance is *exactly* $r_{o,o} = 16.2\,\Omega$; the approximation is $r_{o,o} = 16.3\,\Omega$.

In the first approximation the following equations apply to the

Common-emitter circuit with voltage feedback

$$A = \frac{v_o}{v_i}\bigg|_{i_o=0} \approx -\frac{R_2}{R_1 + \dfrac{R_1 + R_2}{g_m R_C}} \overset{g_m R_C \gg 1 + R_2/R_1}{\approx} -\frac{R_2}{R_1} \tag{2.78}$$

$$r_i = \frac{v_i}{i_i} \approx R_1 \tag{2.79}$$

$$r_o = \frac{v_o}{i_o} \approx R_C \,\|\, \left(\frac{1}{g_m}\left(1 + \frac{R_2}{R_1}\right) + \frac{R_2}{\beta}\right) \tag{2.80}$$

Non-linearity: The voltage feedback significantly reduces the nonlinearity of the transfer characteristic. The distortion factor of the circuit can be determined approximately by a series expansion of the characteristic at the operating point. Insertion of the operating point into (2.75) and (2.76) leads to:

$$v_o = \frac{R_C}{R_2 + R_C}\left(-R_2 i_C + V_T \ln\left(1 + \frac{i_C}{I_{C,A}}\right)\right)$$

$$v_i = \frac{R_1}{\beta} i_C + \left(1 + \frac{R_1}{R_2}\right) V_T \ln\left(1 + \frac{i_C}{I_{C,A}}\right) - \frac{R_1}{R_2} v_o$$

From the series expansion and by eliminating i_C it follows for $\beta \gg 1$ and $g_m R_2 \gg 1$ that:

$$v_o \approx -\frac{R_2}{R_1}\left(v_i + \left(\frac{1}{R_2} + \frac{1}{R_C}\right)^2 \left(1 + \frac{R_2}{R_1}\right)\frac{V_T R_2}{2 I_{C,A}^2 R_1} v_i^2 + \cdots\right)$$

For signals $v_i = \hat{v}_i \cos \omega t$ an approximation of the *distortion factor k* can be obtained from the ratio of the first harmonic with $2\omega t$ to the fundamental with ωt when using low amplitudes; in other words, when disregarding higher order terms:

$$k \approx \frac{v_{o,2\omega t}}{v_{o,\omega t}} \approx \frac{\hat{v}_i}{4 V_T} \frac{\dfrac{R_2}{R_1}\left(1 + \dfrac{R_2}{R_1}\right)}{g_m^2 (R_2 \,\|\, R_C)^2}$$

If a maximum value is specified for k then

$$\hat{v}_i < 4k V_T \frac{g_m^2 (R_2 \,\|\, R_C)^2}{\dfrac{R_2}{R_1}\left(1 + \dfrac{R_2}{R_1}\right)}$$

The maximum output amplitude is obtained using $\hat{v}_o = |A|\hat{v}_i$. In the numeric example above the values are as follows: $\hat{v}_i < k \cdot 57\,\text{V}$ and for $A \approx -1.89$ we have $\hat{v}_o < k \cdot 108\,\text{V}$.

Temperature sensitivity: The base–emitter voltage V_{BE} decreases at a rate of 1.7 mV/K according to (2.21). The resulting *temperature drift* of the output voltage can be calculated using the small-signal response by adding a voltage source v_{TD} with $dv_{TD}/dT = -1.7\,\text{mV/K}$ in series with r_{BE} (see Fig. 2.70) and calculating their effect on the output

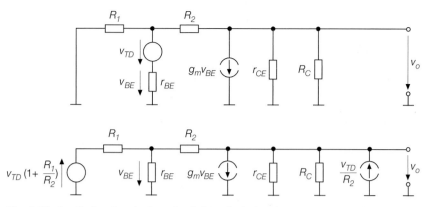

Fig. 2.70. Small-signal equivalent circuit for calculating the temperature drift of the common-emitter circuit with voltage feedback: with voltage source v_{TD} (above) and after shifting the source (below)

voltage. The calculation can be considerably simplified by shifting the voltage source in a suitable manner: replacing the voltage source by two sources in series with R_1 and R_2, converting the latter into two current sources v_{TD}/R_2 at the base and at the collector nodes and then reconverting the source at the base node into a voltage source $v_{TD}R_1/R_2$ leads to the equivalent small-signal circuit diagram shown in Fig. 2.70. Use of the values already defined for A and $r_{o,s}$ results in:

$$\left.\frac{dV_o}{dT}\right|_A = \left(-\left(1 + \frac{R_1}{R_2}\right)A + \frac{r_{o,s}}{R_2}\right)\frac{dv_{TD}}{dT} \approx \left(1 + \frac{R_1}{R_2}\right)A \cdot 1.7\,\frac{\text{mV}}{\text{K}}$$

For the sample operating point we arrive at a temperature drift $(dV_o/dT)|_A \approx -4.8\,\text{mV/K}$ for $A = 1.885$ and $r_o = r_{o,s} = 37.5\,\Omega$.

Operation as a current-to-voltage converter: Short-circuiting resistance R_1 in the common-emitter circuit with voltage feedback and driving the circuit with a current source I_i provides the circuit shown in Fig. 2.71a which operates as a *current-to-voltage converter*; it is also known as a *transimpedance amplifier*.[15] Figure 2.71b shows the characteristics $V_o(I_i)$ and $V_i(I_i)$ for $V_b = 5\,\text{V}$, $R_C = 1\,\text{k}\Omega$ and $R_2 = 2\,\text{k}\Omega$.

From the nodal equations for input and output for the normal mode – that is, $-1.3\,\text{mA} < I_i < 0.2\,\text{mA}$ – it follows that:

$$V_o = \frac{V_b R_2 - I_i B R_2 R_C + V_i (1 + B) R_C}{R_2 + (1 + B) R_C}$$

$$\overset{\underset{\beta R_C \gg R_2}{B \gg 1}}{\approx} \frac{R_2}{B R_C} V_b - R_2 I_i + V_i$$

For $V_i = V_{BE} \approx 0.7\,\text{V}$ the approximation is $V_o \approx 0.72\,\text{V} - 2\,\text{k}\Omega \cdot I_i$.

The small-signal response of the current–voltage converter can be determined from two equations for the common-emitter circuit with voltage feedback. The *transfer resistance*

---

[15] The term *transimpendence amplifier* is also used for operational amplifiers with current feedback and voltage output (CV-OPA).

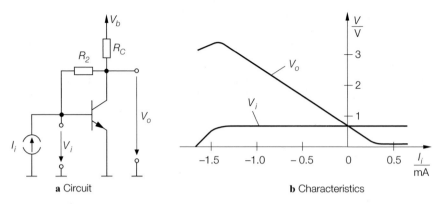

a Circuit   **b** Characteristics

Fig. 2.71. Current-to-voltage converter

(*transimpedance*) takes the place of the gain; from (2.78) it follows that:

$$R_T = \left.\frac{v_o}{i_i}\right|_{i_o=0} = \lim_{R_1 \to \infty} R_1 \left.\frac{v_o}{v_i}\right|_{i_o=0} = \lim_{R_1 \to \infty} R_1 A$$

$$= \frac{-g_m R_2 + 1}{g_m\left(1 + \dfrac{1}{\beta}\right) + \dfrac{1}{R'_C}\left(1 + \dfrac{R_2}{r_{BE}}\right)}$$

$$\underset{\substack{r_{CE} \gg R_C \\ \beta \gg 1}}{\approx} \frac{-g_m R_2 + 1}{g_m + \dfrac{1}{R_C}\left(1 + \dfrac{R_2}{r_{BE}}\right)} \underset{\substack{\beta R_C \gg R_2 \\ g_m R_2 \gg 1}}{\approx} - R_2$$

The *input resistance* can be calculated from the equations for the common-emitter circuit with voltage feedback by setting $R_1 = 0$. The *output resistance* corresponds to the open-circuit output resistance of the common-emitter circuit with voltage feedback. To summarize, the equations for the *current–voltage converter in common-emitter configuration* are:

> *Current-to-voltage converter*
>
> $$R_T = \left.\frac{v_o}{i_i}\right|_{i_o=0} \approx - R_2 \tag{2.81}$$
>
> $$r_i = \frac{v_i}{i_i} \approx \frac{1}{g_m}\left(1 + \frac{R_2}{R_C}\right) \tag{2.82}$$
>
> $$r_o = \frac{v_o}{i_o} \approx R_C \| \left(\frac{1}{g_m} + \frac{R_2}{\beta}\right) \tag{2.83}$$

Setting the Operating Point

Use as a small-signal amplifier requires a stable setting for the transistor's operating point. The operating point should depend as little as possible on the parameters of the transistor since these are temperature-sensitive and subject to variations induced by production processes; important in this respect are the current gain B and the saturation reverse current I_S:

	B	I_S
Temperature coefficient	$+0.5\,\%/\text{K}$	$+15\,\%/\text{K}$
Variance	$-30/+50\,\%$	$-70/+200\,\%$

There are two fundamentally different methods for setting the operating point: *AC coupling* and *DC coupling*.

Setting the Operating Point with AC Coupling: In AC coupling the amplifier or the amplifier stage is connected to the signal source and the load via coupling capacitors (see Fig. 2.72). This makes it possible to set the operating point independent of the DC voltage of the signal source and independent of the load; the charge of the coupling capacitors reflects the voltage difference. Since no DC current can flow through the coupling capacitors one can connect any signal source or load without risking a shift in the operating point. With multi-stage amplifiers the operating point can be set for every stage separately.

Each coupling capacitor forms a highpass filter together with the input or output resistance of the coupled stage and with the signal source or the load. Figure 2.73 shows a portion of the small-signal equivalent circuit of a multi-stage amplifier; the small-signal equivalent circuit according to Fig. 2.61 with parameters A, r_i and r_o is used for every stage.

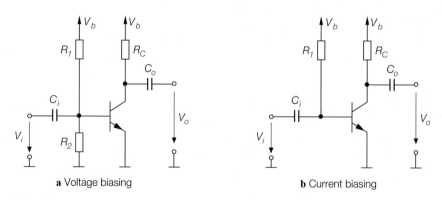

a Voltage biasing   **b** Current biasing

Fig. 2.72. Setting the operating point in the case of AC coupling

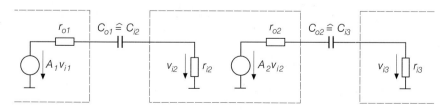

Fig. 2.73. Small-signal equivalent circuit of a multi-stage amplifier to calculate the highpass filters for AC coupling

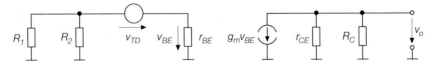

Fig. 2.74. Calculation of the temperature drift for voltage biasing

From the small-signal equivalent circuits it is possible to calculate the cutoff frequencies of the highpass filters. The dimensions of the coupling capacitors must be selected so that the lowest signal frequency of interest is still transmitted in full. DC voltages cannot be transmitted.

In a common-emitter circuit the operating point may be set by voltage or current biasing; $V_{BE,A}$ or $I_{B,A}$ is specified to produce the desired collector current $I_{C,A}$ and thus the desired output voltage $V_{o,A}$. Due to

$$V_{BE,A}(T, C) \; = \; V_T(T) \ln \frac{I_{C,A}}{I_S(T, C)} \quad , \quad I_{B,A}(T, C) \; = \; \frac{I_{C,A}}{B(T, C)}$$

$V_{BE,A}$ and $I_{B,A}$ depend on the temperature T and on the individual component C.

Voltage biasing: For voltage biasing according to Fig. 2.72a the voltage $V_{BE,A}$ is adjusted using the resistances R_1 and R_2. When a current through the resistances is much higher than $I_{B,A}$, the operating point is no longer influenced by a change in $I_{B,A}$. The dependency on the individual component can be eliminated by replacing R_2 by a potentiometer, in order to adjust the operating point. In order to calculate the temperature drift of the output voltage caused by V_{BE} a voltage source v_{TD} with $dv_{TD}/dT = -1.7$ mV/K is introduced in the small-signal equivalent circuit (see Fig. 2.74). It acts like a signal voltage source $v_g = -v_{TD}$ with an internal resistance $R_g = R_1 \| R_2$ as shown by a comparison with Fig. 2.60. Thus:

$$\left. \frac{dV_o}{dT} \right|_A \; = \; - \frac{r_i}{r_i + R_g} A \frac{dv_{TD}}{dT} \; = \; \frac{r_{BE}}{r_{BE} + (R_1 \| R_2)} A \cdot 1.7 \frac{\text{mV}}{\text{K}} \tag{2.84}$$

Example: If $A = -75$ and $R_1 \| R_2 = r_{BE}$, then $(dV_o/dT)|_A \approx -64$ mV/K.

Due to its high temperature drift this method of setting the operating point is not used in practice.

Current biasing: For current biasing according to Fig. 2.72b the base current $I_{B,A}$ is set by resistance R_1:

$$R_1 \; = \; \frac{V_b - V_{BE,A}}{I_{B,A}} \; \approx \; \frac{V_b - 0.7\,\text{V}}{I_{B,A}}$$

If $V_b \gg V_{BE,A}$, then a change of $V_{BE,A}$ has practically no influence on $I_{B,A}$; on the basis of $V_o = V_b - I_C R_C$, it follows that:

$$\left. \frac{dV_o}{dT} \right|_A \approx - R_C \left. \frac{dI_C}{dT} \right|_{I_B=\text{const.}} = - I_B R_C \frac{dB}{dT} = - \frac{I_{C,A} R_C}{V_T} \frac{V_T}{B} \frac{dB}{dT}$$

$$\approx A \frac{V_T}{B} \frac{dB}{dT} \overset{(2.23)}{\approx} A \cdot 0.13 \frac{\text{mV}}{\text{K}} \tag{2.85}$$

Example: If $A = -75$, then $(dV_o/dT)|_A \approx -9.8$ mV/K.

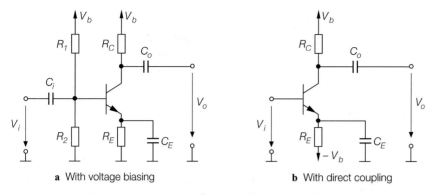

a With voltage biasing   **b** With direct coupling

Fig. 2.75. Setting the operating point by means of DC current feedback

Even though the temperature drift is lower than with voltage biasing, it is still too high for practical applications. Due to the large variance of β it is necessary to replace R_1 by a potentiometer in order to adjust the operating point. For this reason this method of setting the operating point is not used in practice.

Setting the operating point by DC current feedback: The temperature drift is proportional to the gain (see (2.84) and (2.85)); therefore, the stability of the operating point can be improved by reducing the gain. Since the temperature drift is a slow process it is sufficient to reduce the *DC voltage gain* A_{DC}; the *AC voltage gain* A_{AC} may remain unaltered. This can be achieved by a frequency-dependent feedback which is effective only for zero-frequency and frequencies below the lowest signal frequency of interest, while it is totally or partially ineffective for higher frequencies. This is the principle on which the setting of the operating point by means of *DC current feedback* according to Fig. 2.75a is based; here, voltage biasing is combined with a current feedback through resistance R_E. With increasing frequencies the capacitance C_E causes R_E to be short-circuited and thus renders the feedback ineffective for higher frequencies.

The voltage

$$V_{B,A} = \left(I_{C,A} + I_{B,A} \right) R_E + V_{BE,A} \approx I_{C,A} R_E + 0.7 \,\text{V}$$

that is required at the base of the transistor for operation at the operating point is adjusted using R_1 and R_2; the cross-current through the resistances is thus selected to be much higher than $I_{B,A}$, so that the operating point no longer depends on $I_{B,A}$. With the signal source providing a suitable DC voltage and the required base current $I_{B,A}$, the resistances and the coupling capacitor C_i may be omitted and a direct coupling is possible; then $V_{B,A}$ can be tuned by adjusting R_E to the existing DC input voltage. R_E must not be made too small since then the feedback would not be effective and the stability of the operating point would be reduced. For low positive and negative DC input voltages it is possible to achieve direct coupling by an additional negative supply voltage (see Fig. 2.75b).

The temperature drift of the output voltage follows from (2.84) by inserting the values of the common-emitter circuit with current feedback according to (2.70) and (2.71) instead of A and r_i; then $A = A_{DC}$. $r_i \gg R_1 \| R_2$ results in the worst-case scenario:

$$\left. \frac{dV_o}{dT} \right|_A \approx A_{DC} \cdot 1.7 \,\frac{\text{mV}}{\text{K}} \stackrel{g_m R_E \gg 1}{\approx} -\frac{R_C}{R_E} \cdot 1.7 \,\frac{\text{mV}}{\text{K}}$$

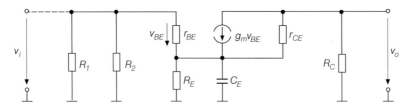

Fig. 2.76. Small-signal equivalent circuit for Fig. 2.75a

In addition, R_E should be as high as possible to achieve a low DC voltage gain A_{DC} and thus a low temperature drift. In practice values in the range of $R_C/R_E \approx 1\ldots 10$ are used.

The frequency response of the gain can be determined by means of the small-signal equivalent circuit shown in Fig. 2.76 or by means of (2.70) through substitution of $R_E\|(1/sC_E)$ for R_E:

$$\underline{A}(s) \approx -\frac{g_m R_C (1+sC_E R_E)}{1+g_m R_E + sC_E R_E} \stackrel{g_m R_E \gg 1}{\approx} -\frac{R_C}{R_E}\frac{1+sC_E R_E}{1+s\dfrac{C_E}{g_m}}$$

Figure 2.77 shows the absolute value of the frequency response $A = |\underline{A}(j2\pi f)|$ and the corner frequencies f_1 and f_2; in this case:

$$\omega_1 = 2\pi f_1 = \frac{1}{C_E R_E} \quad , \quad \omega_2 = 2\pi f_2 \approx \frac{g_m}{C_E}$$

For $f < f_1$ the feedback is fully effective; we have $A \approx A_{DC} \approx -R_C/R_E$. For $f > f_2$ the feedback is ineffective and we have $A \approx A_{AC} \approx -g_m R_C$. The cross-over region lies between these two values. The capacitance C_E must be dimensioned so that f_2 is lower than the lowest signal frequency of interest.

The small-signal equivalent circuit according to Fig. 2.76 further shows that R_1 and R_2 are connected in parallel at the input and must be taken into consideration when calculating the input resistance r_i; for $f > f_2$ the input resistance is:

$$r_i = r_{BE} \| R_1 \| R_2$$

R_1 and R_2 must not be too small since otherwise the input resistance will drop sharply.

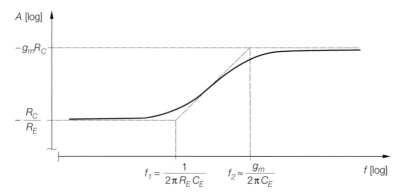

Fig. 2.77. Absolute value of frequency response $A = |\underline{A}(j2\pi f)|$

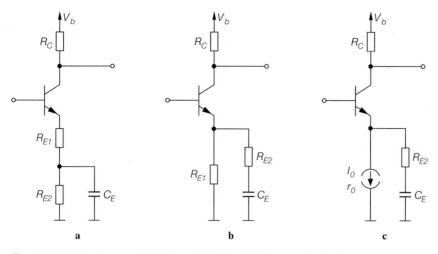

Fig. 2.78. Setting the operating point with DC and AC current feedback

If, for example, in order to reduce the nonlinear distortions, a current feedback is desired for AC voltages, which means for $f > f_2$, and if the AC voltage gain is to be higher than the DC voltage gain, the circuits shown in Fig. 2.78 can be used. Figure 2.79 lists the characteristic parameters.

In the circuit shown in Fig. 2.78c a constant current source with current I_0 and the internal resistance r_0 is used to set the operating point; thus $I_{C,A} \approx I_0$. Due to $r_0 \gg R_C$, the DC voltage gain A_{DC}, and hence the temperature drift caused by the transistor, is very small; in this case the temperature drift of the circuit depends on the temperature drift of the constant current source:

$$\left. \frac{dV_o}{dT} \right|_A \approx -\frac{R_C}{r_0} \cdot 1.7 \, \frac{mV}{K} - R_C \frac{dI_0}{dT} \stackrel{r_0 \gg R_C}{\approx} -R_C \frac{dI_0}{dT}$$

	Fig. 2.75	Fig. 2.78a	Fig. 2.78b and Fig. 2.78c ($R_{E1} = r_0$)
A_{AC}	$-g_m R_C$	$-\dfrac{g_m R_C}{1 + g_m R_{E1}}$	$-\dfrac{g_m R_C}{1 + g_m (R_{E1} \| R_{E2})}$
A_{DC}	$-\dfrac{R_C}{R_E}$	$-\dfrac{R_C}{R_{E1} + R_{E2}}$	$-\dfrac{R_C}{R_{E1}}$
ω_1	$\dfrac{1}{C_E R_E}$	$\dfrac{1}{C_E R_{E2}}$	$\dfrac{1}{C_E (R_{E1} + R_{E2})}$
ω_2	$\dfrac{g_m}{C_E}$	$\dfrac{1}{C_E ((1/g_m + R_{E1}) \| R_{E2})}$	$\dfrac{g_m}{C_E (1 + g_m R_{E2})}$
Assumption	$g_m R_E \gg 1$	$g_m (R_{E1} + R_{E2}) \gg 1$	$g_m R_{E1} \gg 1$

Fig. 2.79. Characteristic parameters of the common-emitter circuit with DC current feedback

Example: A signal with an amplitude $\hat{v}_g = 10\,\text{mV}$ provided by a source with an internal resistance of $R_g = 10\,\text{k}\Omega$ is to be amplified to $\hat{v}_o = 200\,\text{mV}$ and to be supplied to a load $R_L = 10\,\text{k}\Omega$. A lower cutoff frequency $f_L = 20\,\text{Hz}$ and a distortion factor $k < 1\,\%$ are required. The supply voltage is $V_b = 12\,\text{V}$. From (2.74) it follows that with $\hat{v}_i \approx \hat{v}_g = 10\,\text{mV}$ and $k < 0.01$ a current feedback with $g_m R_E > 2.2$ is required; therefore a common-emitter circuit with AC current feedback must be used. The operating gain A_B is derived from (2.64) by inserting for A and r_o the values of the common-emitter circuit with current feedback according to (2.70) and (2.72):

$$A_B = \frac{r_i}{r_i + R_g} A \frac{R_L}{R_L + r_o} \approx -\frac{r_i}{r_i + R_g} \frac{g_m (R_C \| R_L)}{1 + g_m R_E}$$

$A_B = \hat{v}_o/\hat{v}_g = 20$ is required. The attenuation caused by the input resistance r_i cannot be taken into consideration at this point since r_i is not yet known; it is assumed that $r_i \to \infty$. In order to keep the attenuation caused by the output resistance $r_o \approx R_C$ small we select $R_C = 5\,\text{k}\Omega < R_L$. For $g_m R_E > 2.2$ we obtain $R_E = 115\,\Omega \to 120\,\Omega$,[16] $g_m = 21.3\,\text{mS}$ and $I_{C,A} = g_m V_T \approx 0.55\,\text{mA}$. Assuming $B \approx \beta \approx 400$ and $I_S \approx 7\,\text{fA}$ for the transistor, we have $V_{BE,A} \approx 0.65\,\text{V}$, $I_{B,A} \approx 1.4\,\mu\text{A}$ and $r_{BE} \approx 19\,\text{k}\Omega$. To achieve a stable operating point an additional DC current feedback is used according to Fig. 2.78a with $R_{E1} = R_E$ and $R_{E2} = 4.7\,\text{k}\Omega \approx R_C$ (see Fig. 2.80); this results in a DC gain of one and the temperature drift is correspondingly low. The voltage at the base is $V_{B,A} \approx I_{C,A}(R_{E1}+R_{E2})+V_{BE,A} \approx 3.3\,\text{V}$. The voltage divider at the base is intended to cause the current $I_Q = 10 I_{B,A}$; this results in $R_2 = V_{B,A}/I_Q \approx 240\,\text{k}\Omega$ and $R_1 = (V_b - V_{B,A})/(I_Q + I_{B,A}) \approx 560\,\text{k}\Omega$. It is now possible to determine the input resistance: $r_i = R_1 \| R_2 \| (r_{BE} + \beta R_{E1}) \approx 48\,\text{k}\Omega$. With $R_g = 10\,\text{k}\Omega$ a gain reduction by a factor of $1 + R_g/r_i \approx 1.2$ is achieved through r_i. This reduction can be compensated for by changing R_C retrospectively in order to increase the value of $(R_C \| R_L)$ by this factor; the result is $R_C = 6.8\,\text{k}\Omega$. Now all the resistances have been dimensioned (see Fig. 2.80). In a final step the highpass filters caused by the capacitors C_i, C_o and C_E must be rated so that $f_L = 20\,\text{Hz}$; each individual highpass filter cutoff frequency is to be $f_L' = f_L/\sqrt{3} \approx 11\,\text{Hz}$:

$$C_i = \frac{1}{2\pi f_L' (R_g + r_i)} = 250\,\text{nF} \to 270\,\text{nF}$$

$$C_o = \frac{1}{2\pi f_L' (R_C + R_L)} = 860\,\text{nF} \to 1\,\mu\text{F}$$

$$C_E = \frac{1}{2\pi f_L' ((1/g_m + R_{E1}) \| R_{E2})} = 90\,\mu\text{F} \to 100\,\mu\text{F}$$

Use of AC coupling: AC coupling can be used only where no DC voltages are to be transferred; that is, if the amplifier may have the characteristic of a highpass filter. AC amplifiers with very low cutoff frequencies are not used in practice because coupling capacitors with very high values would be required; in such cases, direct coupling is very often required, despite the fact that no DC voltages have to be amplified.

The essential advantage of AC coupling is its independence on the DC voltage at the signal source and the load. The consequence of the highpass characteristic is that the

---

[16] Rounded to standard values.

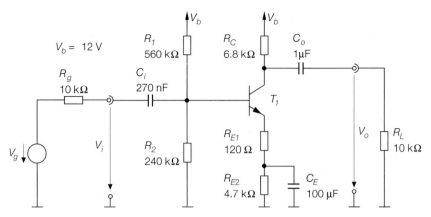

Fig. 2.80. Example of dimensioned components in a common-emitter circuit with DC and AC feedback

temperature drift is noticed only as a shift in the operating point in the respective stage and is not transmitted to the next stage as is the case with direct coupling.

Despite the advantages that AC coupling offers for straight AC amplifiers it is avoided in practice if possible, because of the need for additional capacitors and resistors. This is particularly the case with low-frequency amplifiers since their high capacitances make it necessary to use electrolytic capacitors which are large and expensive and have a high failure rate. AC coupling is widely found in high-frequency amplifiers; here ceramic capacitors in the picofarad range can be used – these are small and comparatively inexpensive. In integrated circuits AC coupling is only found in exceptional cases, because capacitors are very difficult to integrate. Where capacitors are still required, they are often connected externally.

Setting the operating point with DC coupling: In DC coupling, which is also known as *direct* or *galvanic* coupling, the amplifier or the amplifier stage is directly connected to the signal source and the load. This requires adaptation of the DC voltages at the input and the output at the operating point – that is, $V_{i,A}$ and $V_{o,A}$ – to the DC voltage of the signal source and the load. With multi-stage amplifiers the operating point can no longer be set separately for each stage.

In multi-stage amplifiers DC coupling is almost always used in combination with a feedback across all stages; this means that the individual stages are coupled directly and the operating point is set by the feedback. $V_{i,A} = V_{o,A}$ is a frequent requirement which means that the amplifier is not meant to alter the DC voltage component of the signal.

Example: Figure 2.81 shows an amplifier with DC voltage coupling that is made up of two stages in common-emitter configuration and features feedback across both stages. The first stage consists of npn transistor T_1 and resistor R_1, the second stage of pnp transistor T_2 and resistor R_2; the resistors R_3, R_4 and R_5 make up the feedback for setting the operating point and the gain. The amplifier is rated for $V_{i,A} = V_{o,A} = 2.5\,\mathrm{V}$ and $A = 10$. For a common-emitter circuit with an npn transistor working at the operating point, the output voltage is higher than the input voltage while it is lower in a common-emitter circuit with a pnp transistor. Therefore, it is useful to use a pnp transistor in the second stage in order

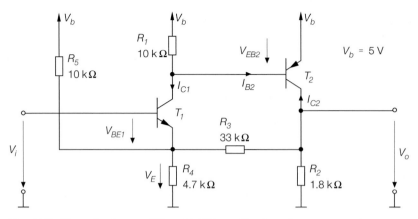

Fig. 2.81. Example of an amplifier with DC coupling featuring two stages in common-emitter configuration and feedback

to meet the requirement $V_{i,A} = V_{o,A}$. The dimensioning of the resistors is not described here.

Calculation of the operating point is based on $V_{o,A} = 2.5\,\text{V}$. Disregarding the current through R_3 leads to $I_{C2,A} \approx -V_{o,A}/R_2 \approx -1.4\,\text{mA}$. If $I_{S2} = 1\,\text{fA}$ and $\beta_2 = 300$,[17] then $V_{EB2,A} = V_T \ln(-I_{C2,A}/I_{S2}) \approx 0.73\,\text{V}$ and $I_{B2,A} \approx -4.7\,\mu\text{A}$. From this it follows that $I_{C1,A} = V_{EB2,A}/R_1 - I_{B2,A} \approx 78\,\mu\text{A}$. From the nodal equation

$$\frac{V_{E,A}}{R_4} = \frac{V_{o,A} - V_{E,A}}{R_3} + \frac{V_b - V_{E,A}}{R_5} + I_{C1,A}$$

at the emitter connection of T_1 we have $V_{E,A} = 1.9\,\text{V}$. If $I_{S1} = 7\,\text{fA}$, then $V_{BE1,A} = V_T \ln(I_{C1,A}/I_{S2}) \approx 0.6\,\text{V}$ which leads to $V_{i,A} = V_{BE1,A} + V_{E,A} \approx 2.5\,\text{V}$. Finally a check must be made as to whether it is permissible to disregard the current through R_3 in the calculation of $I_{C2,A}$: $I_{R3} = (V_{o,A} - V_{E,A})/R_3 \approx 18\,\mu\text{A} \ll |I_{C2,A}|$. This calculation again illustrates the procedure for calculating operating points.

The use of DC coupling: DC coupling is unavoidable in cases in which DC voltages are to be amplified.[18] If possible, the individual stages are also coupled directly in multi-stage AC amplifiers in order to eliminate the coupling capacitors and the additional resistors.

A dizadvantage of DC coupling is that a shift in the operating point caused by a temperature drift in any of the amplifier stages is transferred to the load; if other stages follow then the drift will be amplified further. Therefore, DC coupling requires special measures for drift suppression or circuit configurations that are less prone to drift, such as, for example, differential amplifiers.

Frequency Response and Upper Cut-Off Frequency

The small-signal gain A and operating gain A_B as calculated so far apply only to low signal frequencies; with higher frequencies both these parameters decline as a result of

---

[17] Typical values for a pnp low-power transistor BC557B.

[18] Exceptions are special circuit configurations such as *chopper amplifiers* or amplifiers with switched capacitances in which the DC component of the signal is transferred via a separate path.

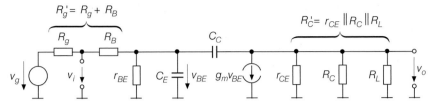

Fig. 2.82. Dynamic small-signal equivalent circuit for the common-emitter circuit without feedback

the capacitances of the transistor. To obtain information on the frequency response and the upper cutoff frequency it is essential to use the dynamic small-signal model of the transistor according to Fig. 2.41 on page 78 as a basis for calculation; this model takes into account the base spreading resistance R_B as well as the emitter capacitance C_E and the collector capacitance C_C.

Common-emitter circuit without feedback: Figure 2.82 shows the dynamic small-signal equivalent circuit of the common-emitter circuit without feedback. The *operating gain* $\underline{A}_B(s) = \underline{v}_o(s)/\underline{v}_g(s)$ with $R'_g = R_g + R_B$ and $R'_C = R_L||R_C||r_{CE}$ is:

$$\underline{A}_B(s) = -\frac{(g_m - sC_C)\,R'_C}{1 + \dfrac{R'_g}{r_{BE}} + s\left(C_E R'_g + C_C\left(R'_g + R'_C + g_m R'_C R'_g\right)\right) + s^2 C_E C_C R'_g R'_C}$$

$$(2.86)$$

Figure 2.83 shows the magnitude of the frequency response with the corner frequencies f_{P1} and f_{P2} of the two poles and the corner frequency f_N of the zero. The zero can be disregarded on the basis of its short time constant $C_C g_m^{-1} = (2\pi f_N)^{-1}$. The two poles are real and far apart. Therefore, the frequency response can be described approximately by

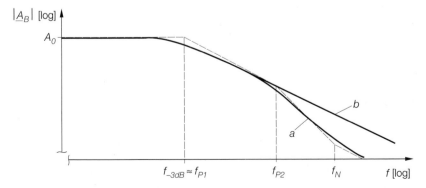

Fig. 2.83. Magnitude frequency response $|\underline{A}_B|$ of the common-emitter circuit: (**a**) calculation with (2.86) and (**b**) approximation (2.88)

a lowpass filter of first order, after the s^2 term in the denominator has been removed.[19] Using the low-frequency gain

$$A_0 = \underline{A}_B(0) = -\frac{r_{BE}}{r_{BE} + R'_g} g_m R'_C \tag{2.87}$$

results in:

$$\underline{A}_B(s) \approx \frac{A_0}{1 + s\left(C_E + C_C\left(1 + g_m R'_C + \frac{R'_C}{R'_g}\right)\right)\left(r_{BE} \| R'_g\right)} \tag{2.88}$$

Figure 2.83 shows the magnitude frequency responses of the approximation given in (2.88) and of the full expression (2.86).

From (2.88) we obtain the approximation for the $-3dB$ *cutoff frequency* f_{-3dB} at which the magnitude of the gain declines by 3 dB:

$$\omega_{-3dB} = 2\pi f_{-3dB} \approx \frac{1}{\left(C_E + C_C\left(1 + g_m R'_C + \frac{R'_C}{R'_g}\right)\right)\left(r_{BE} \| R'_g\right)} \tag{2.89}$$

In most cases we have R'_C, $R'_g \gg 1/g_m$:

$$\boxed{\omega_{-3dB} = 2\pi f_{-3dB} \approx \frac{1}{\left(C_E + C_C g_m R'_C\right)\left(r_{BE} \| R'_g\right)}} \tag{2.90}$$

The upper cutoff frequency depends on the low-frequency gain A_0. Under the assumption that a change in A_0 is induced by a change of R'_C and that all other parameters remain constant, solving (2.87) for R'_C and substituting an expression with two time constants that ar independent of A_0 into (2.89) leads to:

$$\omega_{-3dB}(A_0) \approx \frac{1}{T_1 + T_2|A_0|} \tag{2.91}$$

$$T_1 = (C_E + C_C)\left(r_{BE} \| R'_g\right) \tag{2.92}$$

$$T_2 = C_C\left(R'_g + \frac{1}{g_m}\right) \tag{2.93}$$

Two regions can be distinguished:

– If $|A_0| \ll T_1/T_2$, then $\omega_{-3dB} \approx T_1^{-1}$; that is, the upper cutoff frequency is independent of the gain. For the borderline case of $A_0 \to 0$ and $R_g = 0$ we obtain the maximum upper cutoff frequency:

$$\omega_{-3dB,max} \approx \frac{1}{(C_E + C_C)(r_{BE} \| R_B)} \overset{r_{BE} \gg R_B}{\approx} \frac{1}{(C_E + C_C) R_B}$$

---

[19] This approach corresponds to a well-known procedure from control engineering in which several poles are considered to be one pole with the sum of the time constants: $(1 + sT_1)(1 + sT_2)\dots(1 + sT_n) \approx 1 + s(T_1 + T_2 + \dots + T_n)$. The coefficient of s is the sum of the time constants. The combination is thus achieved by omitting the higher powers of s.

This corresponds to the *transconductance cutoff frequency* ω_{Y21e} (see (2.46)).

– If $|A_0| \gg T_1/T_2$, then $\omega_{\text{-3dB}} \approx (T_2|A_0|)^{-1}$; that is, the upper cutoff frequency is proportional to the reciprocal gain, which leads to a *constant* **gain–bandwidth product** GBW:

$$GBW = f_{\text{-3dB}} |A_0| \approx \frac{1}{2\pi T_2} \tag{2.94}$$

The gain–bandwidth product *GBW* is an important parameter since it represents an absolute upper limit for the product of the gain at low frequencies and of the upper cutoff frequency so that $GBW \geq f_{-3dB}|A_0|$ is true for all values of $|A_0|$.

For $1/g_m \ll R'_g \ll r_{BE}$, (2.89) can be expressed approximately as:

$$\omega_{\text{-3dB}} \approx \frac{1}{R'_g (C_E + C_C (1 + |A_0|))}$$

This shows that, contrary to C_E, the capacitance C_C enters the cutoff frequency enhanced by a factor of $(1 + |A_0|)$. This is known as the *Miller effect* and is caused by the fact that with low frequencies a higher voltage

$$v_{BE} - v_o \approx v_g - v_o = v_g (1 - A_0) = v_g (1 + |A_0|)$$

occurs across C_C, while across C_E only the voltage $v_{BE} \approx v_g$ is present; the approximation $v_g \approx v_{BE}$ results from the condition $r_{BE} \gg R'_g$. The capacitance C_C is also known as the *Miller capacitor* C_M.

In addition to the ohmic resistance the load often has a capacitive component which means that a load capacitance C_L exists in parallel to the load resistance R_L. The influence of C_L can be determined by replacing resistance $R'_C = r_{CE}||R_C||R_L$ by an impedance

$$\underline{Z}_C(s) = R'_C \, || \, \frac{1}{sC_L} = \frac{R'_C}{1 + sC_L R'_C} \tag{2.95}$$

as shown in Fig. 2.84. When inserting $\underline{Z}_C(s)$ into (2.86), neglecting according to (2.88) and determining the time constants T_1 and T_2, it becomes obvious that T_1 does not change, while T_2 can be expressed as:

$$T_2 = \left(C_C + \frac{C_L}{\beta} \right) R'_g + \frac{C_C + C_L}{g_m} \tag{2.96}$$

Due to the load capacitance C_L the gain bandwidth product GBW is reduced in accordance with the increase of T_2 (see (2.94)).

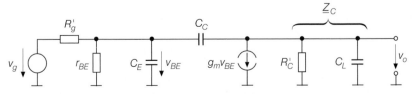

Fig. 2.84. Small-signal equivalent circuit of common-emitter configuration with capacitive load C_L

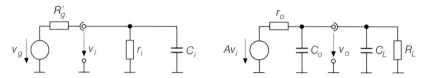

Fig. 2.85. Equivalent circuit with the substitutes A, r_i, r_o, C_i and C_o

Equivalent circuit diagram: The common-emitter circuit can be described approximately by the equivalent circuit shown in Fig. 2.85. It is developed from Fig. 2.61 by adding the *input capacitance* C_i and the *output capacitance* C_o and is suitable for approximately calculating the gain $\underline{A}_B(s)$ and the upper cutoff frequency f_{-3dB}. The formulas for C_i and C_o are based on the premise that the calculation of $\underline{A}_B(s)$ after elimination of the s^2 term in the denominator must lead to (2.88):

$$C_i \approx C_E + C_C\left(1 + |A_0|\right) \tag{2.97}$$

$$C_o \approx C_C \frac{r_{BE}}{r_{BE} + R'_g} \tag{2.98}$$

Both depend on the circuitry at the input and output as A_0 and R'_g depend on R_g and R_L; therefore, they can only be determined once R_g and R_L are known. A, r_i and r_o are given by (2.61)–(2.63) and are independent of the circuitry. The base spreading resistance R_B is regarded as a component of the signal generator's internal resistance: $R'_g = R_g + R_B$.

Where another amplifier stage follows, R_L and C_L are determined by the r_i and C_i of that stage. The equivalent circuit shown in Fig. 2.85 can easily be cascaded by identifying R'_g with r_o, r_i with R_L and C_i with $C_L + C_o$; the base spreading resistance R_B of the subsequent stage which in Fig. 2.85 would be located *between* C_o and C_L is moved to the left side of C_o and combined with r_o without a noticeable error.

Example: $I_{C,A} = 2\,\text{mA}$ was selected for the numeric example used in the common-emitter circuit without feedback according to Fig. 2.57a. For $\beta = 400$, $V_A = 100\,\text{V}$, $C_{obo} = 3.5\,\text{pF}$ and $f_T = 160\,\text{MHz}$ we obtain from Fig. 2.45 on page 83 the small-signal parameters $g_m = 77\,\text{mS}, r_{BE} = 5.2\,\text{k}\Omega, r_{CE} = 50\,\text{k}\Omega, C_C = 3.5\,\text{pF}$ and $C_E = 73\,\text{pF}$. For $R_g = R_C = 1\,\text{k}\Omega$, $R_L \to \infty$ and $R'_g \approx R_g$ it follows from (2.87) that $A_0 \approx -63$, from (2.89) that $f_{-3dB} \approx 543\,\text{kHz}$ and from (2.90) that $f_{-3dB} \approx 554\,\text{kHz}$. It follows from (2.92) that $T_1 \approx 64\,\text{ns}$, from (2.93) that $T_2 \approx 3.55\,\text{ns}$ and from (2.94) that $GBW \approx 45\,\text{MHz}$. With a load capacitance $C_L = 1\,\text{nF}$ we obtain from (2.96) $T_2 \approx 19\,\text{ns}$, from (2.91) $f_{-3dB} \approx 126\,\text{kHz}$ and from (2.94) $GBW \approx 8.4\,\text{MHz}$.

Common-emitter circuit with current feedback: The frequency response and the upper cutoff frequency of the common-emitter circuit with current feedback according to Fig. 2.62a can be derived from the corresponding parameters of the common-emitter circuit without feedback. Figure 2.86a shows a portion of the small-signal equivalent circuit in Fig. 2.82 with the additional resistance R_E of the current feedback; here resistance r_{CE} is disregarded. This portion may be converted to the circuit shown in Fig.2.86b,[20] which

---

[20] This is not an equivalent conversion since it is based on the neglection of one pole in the Y matrix. For any value of R_E, however, the cutoff frequency of this pole is above the transistor transit frequency f_T and thus in a range that cannot be applied to the small-signal model of the transistor anyway; therefore, the conversion is *virtually* equivalent [2.11].

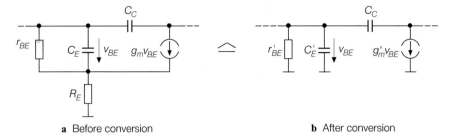

a Before conversion   **b** After conversion

Fig. 2.86. Conversion of the small-signal equivalent circuit of the common-emitter circuit with current feedback

leads back to the original small-signal equivalent circuit of Fig. 2.82; consequently:

$$r'_{BE} = r_{BE} (1 + g_m R_E) \tag{2.99}$$

$$g'_m = \frac{g_m}{1 + g_m R_E} \tag{2.100}$$

$$C'_E = \frac{C_E}{1 + g_m R_E} \tag{2.101}$$

It is thus possible to transform a transistor with resistance R_E of the current feedback into an equivalent transistor without current feedback by replacing r_{BE}, g_m and C_E by r'_{BE}, g'_m and C'_E respectively; here, g'_m corresponds to the *reduced transconductance* $g_{m,red}$ already introduced in (2.73).

The equivalent values can now be inserted into (2.91)–(2.94) for the common-emitter circuit without feedback. We see that T_2 and the gain bandwidth product GBW do not change with the high internal resistances of the signal source – that is, $R'_g \gg 1/g'_m$ – since in this case they are only dependent on R'_g and C_C. Consequently, in the region of $|A_0| > T_1/T_2$ with constant GBW the upper cutoff frequency increases at exactly the same rate as the gain decreases due to the current feedback. Therefore, the upper cutoff frequency can be increased by means of a current feedback at the cost of the gain, while the product of both does not increase.

The influence of load capacitance C_L can be determined from (2.96) by inserting the equivalent value g'_m to replace g_m. With high current feedback small values of C_L already have a comparably high bearing, since T_2 increases rapidly due to $g'_m \ll g_m$; the gain–bandwidth product GBW decreases accordingly.

The common-emitter circuit with current feedback can be described approximately by the equivalent circuit shown in Fig. 2.85. The input capacitance C_i and the output capacitance C_o are determined from (2.97) and (2.98) by inserting the equivalent values r'_{BE} and C'_E for r_{BE} and C_E; A, r_i and r_o are given by (2.70)–(2.72).

Example: $I_{C,A} = 1.5\,\text{mA}$ is selected for the numeric example of the common-emitter circuit with current feedback shown in Fig. 2.62a. For $\beta = 400$, $C_{obo} = 3.5\,\text{pF}$ and $f_T = 150\,\text{MHz}$ the small-signal parameters $g_m = 58\,\text{mS}$, $r_{BE} = 6.9\,\text{k}\Omega$, $C_C = 3.5\,\text{pF}$ and $C_E = 58\,\text{pF}$ are determined from Fig. 2.45 on page 83; r_{CE} is neglected. Conversion to (2.99)–(2.101) provides for $R_E = 500\,\Omega$ the equivalent values $r'_{BE} = 207\,\text{k}\Omega$, $g'_m = 1.93\,\text{mS}$ and $C'_E = 1.93\,\text{pF}$. For $R_g = R_C = 1\,\text{k}\Omega$, $R_L \to \infty$ and $R'_g \approx R_g$ it results in $A_0 \approx -1.93$ from (2.87), $T_1 \approx 5.4\,\text{ns}$ from (2.92), $T_2 \approx 5.3\,\text{ns}$ from (2.93), $f_{-3dB} \approx 10\,\text{MHz}$ from (2.93) and $GBW \approx 30\,\text{MHz}$ from (2.94). With a load capacitance $C_L = 1\,\text{nF}$

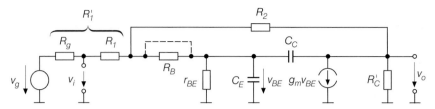

Fig. 2.87. Dynamic small-signal equivalent circuit of the common-emitter circuit with voltage feedback

it follows from (2.96) that $T_2 \approx 526\,\text{ns}$, from (2.91) that $f_{-3dB} \approx 156\,\text{kHz}$ and from (2.94) that $GBW \approx 303\,\text{kHz}$.

A comparison with the example of the common-emitter circuit without feedback on page 126 shows that the gain bandwidth product GBW without load capacitance is identical; therefore, the upper cutoff frequency is reduced by a factor of 30 due to the gain which is 30 times higher. For $C_L = 1\,\text{nF}$ the upper cutoff frequency is about the same despite the difference in gain; in this case the influence of T_2 is predominant and the result for both circuits is $(\omega_{-3dB})^{-1} \approx T_2|A_0| \approx C_L R'_C \approx 1\,\mu\text{s}$.

Common-emitter circuit with voltage feedback: Figure 2.87 shows the small-signal equivalent circuit for the common-emitter configuration with voltage feedback; here $R'_C = r_{CE}||R_C||R_L$ still applies. The calculation of $\underline{A}_B(s)$ is complicated. However, it is possible to use the results from the common-emitter circuit by neglecting – that is, short-circuiting – the base spreading resistance R_B as shown in Fig. 2.87 and substituting into (2.86) the parallel arrangement of C_C and R_2 for C_C and resistance $R'_1 = R_1 + R_g$ for R_g. For R'_1, R_2, $R'_0 \gg 1/g_m$ and $r_{BE} \gg R'_1$, the resulting approximation is sufficiently accurate for practical purposes:

$$A_0 \approx -\frac{R_2}{R'_1 + \dfrac{R_2}{g_m R'_C}} \overset{g_m R'_C R'_1 \gg R_2}{\approx} -\frac{R_2}{R'_1} \tag{2.102}$$

$$\underline{A}_B(s) \approx \frac{A_0}{1 + s\left(\dfrac{C_E}{g_m}\left(1 + \dfrac{R_2}{R'_C}\right) + C_C R_2\right) + s^2 \dfrac{C_E C_C R_2}{g_m}} \tag{2.103}$$

Even though the two poles are not as far apart as in the common-emitter circuit without feedback and the common-emitter circuit with current feedback, the upper cutoff frequency can be estimated with sufficient accuracy by neglecting the s^2 term in the denominator of $\underline{A}_B(s)$:

$$\omega_{-3dB} = 2\pi f_{-3dB} \approx \frac{1}{\dfrac{C_E}{g_m}\left(1 + \dfrac{R_2}{R'_C}\right) + C_C R_2} \tag{2.104}$$

This value depends on A_0. Considering $A_0 \approx -R_2/R_1$, and assuming a change in A_0 induced by a change in R_2 and a constant R'_1, leads to a simple explicit expression with two time constants that are independent of A_0:

$$\omega_{-3dB}(A_0) \approx \frac{1}{T_1 + T_2|A_0|} \tag{2.105}$$

$$T_1 = \frac{C_E}{g_m} \tag{2.106}$$

$$T_2 = \left(\frac{C_E}{g_m R'_C} + C_C\right) R'_1 \tag{2.107}$$

The influence of a load capacitance can be determined by the change in $R'_C \rightarrow \underline{Z}_C(s)$ according to (2.95) similar to the method used in the common-emitter circuit without feedback. Consequently:

$$T_1 = \frac{C_E + C_L}{g_m} \tag{2.108}$$

$$T_2 = \left(\frac{C_E}{g_m R'_C} + C_C\right) R'_1 + \frac{C_L}{g_m} \tag{2.109}$$

Where a strong voltage feedback exists, the two poles of $\underline{A}_B(s)$ may also be complex conjugate; in this case the upper cutoff frequency can be estimated only very roughly using (2.105)–(2.109).

The common-emitter circuit with voltage feedback can also be described approximately by the equivalent circuit shown in Fig. 2.85. The capacitances C_i and C_o result from the requirement that the calculation of $\underline{A}_B(s)$ must lead to (2.103) if the s^2 term in the denominator is removed:

$$C_i = 0$$

$$C_o \approx \left(C_E \left(\frac{1}{R_2} + \frac{1}{R'_C}\right) + C_C g_m\right) (R'_1 \parallel R_2 \parallel r_{BE})$$

Thus, the input impedance is purely ohmic.[21] A, r_i and r_o are given by (2.78)–(2.80).

Example: $I_{C,A} = 2.24$ mA is chosen for the numeric example of the common-emitter circuit with voltage feedback shown in Fig. 2.66a. For $\beta = 400$, $C_{obo} = 3.5$ pF and $f_T = 160$ MHz the small-signal parameters $g_m = 86$ mS, $r_{BE} = 4.6$ kΩ, $C_C = 3.5$ pF and $C_E = 82$ pF are determined from Fig. 2.45 on page 83; r_{CE} is disregarded. For $R_C = R_1 = 1$ kΩ, $R_2 = 2$ kΩ, $R_L \rightarrow \infty$ and $R_g = 0$ from (2.102) we have $A_0 \approx -1.96$, from (2.106) $T_1 \approx 0.95$ ns, from (2.107) $T_2 \approx 4.45$ ns, from (2.105) $f_{-3dB} \approx 16$ MHz and from (2.94) $GBW \approx 36$ MHz. With a load capacitance $C_L = 1$ nF, from (2.108) we obtain $T_1 \approx 12.6$ ns, from (2.109) $T_2 \approx 16.1$ ns, from (2.105) $f_{-3dB} \approx 3.6$ MHz and from (2.94) $GBW \approx 9.9$ MHz.

A comparison with the example of the common-emitter circuit with current feedback on page 127 shows that in both circuits approximately the same upper cutoff frequency is achieved without load capacitance. With a load capacitance of $C_L = 1$ nF the common-emitter circuit with voltage feedback produces an upper cutoff frequency that is 20 times higher; this is caused by the significantly lower output resistance r_o. For this reason voltage feedback should be preferred to the current feedback where large load capacitances are involved.

---

[21] In practical circuit applications there is a parasitic stray capacitance of some pF caused by the physical circuit arrangement.

Summary

The common-emitter circuit can be operated without feedback, with current feedback or with voltage feedback. Figure 2.88 shows the three variations; Fig. 2.89 lists the most important characteristic values.

The gain of the common-emitter circuit without feedback is heavily dependent on the operating point; therefore an accurate and temperature-stable setting of the operating point

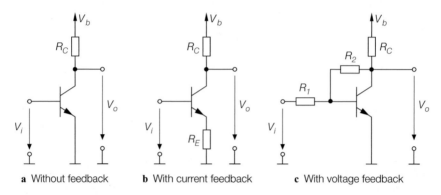

a Without feedback   **b** With current feedback   **c** With voltage feedback

Fig. 2.88. Variations of common-emitter circuits

	Without feedback Fig. 2.88a	With current feedback Fig. 2.88b	With voltage feedback Fig. 2.88c
A	$-g_m R_C$	$-\dfrac{R_C}{R_E}$	$-\dfrac{R_2}{R_1}$
r_i	r_{BE}	$r_{BE} + \beta R_E$	R_1
r_o	R_C	R_C	$g_m\left(1+\dfrac{R_2}{R_1}\right)+\dfrac{R_2}{\beta}$
k	$\dfrac{\hat{v}_i}{4V_T}$	$\dfrac{\hat{v}_i}{4V_T\,(1+g_m R_E)^2}$	$\dfrac{\hat{v}_i\,R_2\,(R_1+R_2)}{4V_T\,(g_m R_1\,(R_2||R_C))^2}$
GBW	$\dfrac{1}{2\pi C_C\left(R_g' + \dfrac{1}{g_m}\right)}$	$\dfrac{1}{2\pi C_C\left(R_g' + \dfrac{1}{g_m'}\right)}$	$\dfrac{1}{2\pi\left(\dfrac{C_E}{g_m R_C'}+C_C\right)R_1'}$
	for $R_g' = R_g + R_B$	for $R_g' = R_g + R_B$ and g_m' according to (2.100)	for $R_1' = R_1 + R_g$ and $R_C' = R_C||R_L$

A Small-signal voltage gain for open circuit,
r_i Small-signal input resistance,
r_o Small-signal output resistance,
k Distortion factor at low amplitudes,
GBW Gain bandwidth product without load capacitance

Fig. 2.89. Characteristic parameters of the common-emitter circuit

is of particular importance. Furthermore, the high degree of dependence on the operating point causes major nonlinear distortions since gain changes with the output voltage. In designs with feedback the first approximation of the gain is determined by two resistances and is thus practically independent of the operating point of the transistor; setting the operating point is less difficult and with the same output signal there are fewer distortions. However, the use of an effective feedback only allows a much lower gain.

With the same collector current the common-emitter circuit with current feedback has the highest input resistance; that is, it provides the lowest load for the signal source. Next in the ranking are the common-emitter circuit without feedback and the common-emitter circuit with voltage feedback. The output resistance of the common-emitter circuit with voltage feedback is significantly lower than that of the other circuit designs; this is advantageous with low-ohmic and capacitive loads.

The gain–bandwidth product is about the same in all configurations when assuming that $R'_g \gg 1/g_m$, $C_E \ll g_m R'_C C_C$ and $R'_g \approx R'_1$. Due to the Miller effect, it depends largely on the collector capacitance C_C.

2.4.2
Common-Collector Circuit

Figure 2.90a shows the common-collector configuration consisting of the transistor, the emitter resistance R_E, the supply voltage source V_b and the signal voltage source V_g with the internal resistance R_g. The following explanations are based on the assumption that $V_b = 5\,\text{V}$ and $R_E = R_g = 1\,\text{k}\Omega$.

Transfer Characteristic of the Common-Collector Circuit

Measurement of the output voltage V_o as a function of the signal voltage V_g results in the transfer characteristic shown in Fig. 2.91. For $V_g < 0.5\,\text{V}$ the collector current is negligible and $V_o = 0\,\text{V}$. For $V_g \geq 0.5\,\text{V}$, there is a collector current I_C that increases with V_g, and the output voltage *follows* the input voltage with a *difference* of V_{BE}; therefore, the common-collector circuit is also known as an *emitter follower*. The transistor operates continuously in normal mode.

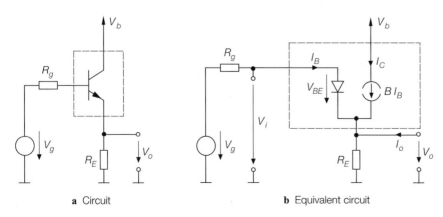

a Circuit   **b** Equivalent circuit

Fig. 2.90. Common-collector circuit

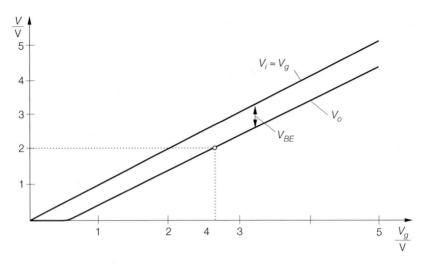

Fig. 2.91. Characteristics of the common-collector circuit

Figure 2.90b shows the equivalent circuit of the common-collector configuration in which the transistor is replaced by the simplified transport model according to Fig. 2.27, with:

$$I_C = B I_B = I_S \, e^{\frac{V_{BE}}{V_T}}$$

From Fig. 2.90b it follows that

$$V_o = (I_C + I_B + I_o) R_E \approx (I_C + I_o) R_E \overset{I_o=0}{=} I_C R_E \qquad (2.110)$$

$$V_i = V_o + V_{BE} \qquad (2.111)$$

$$V_i = V_g - I_B R_g = V_g - \frac{I_C R_g}{B} \approx V_g \qquad (2.112)$$

It is assumed in (2.112) that the voltage drop across R_g can be neglected if B is sufficiently large and R_g is sufficiently small; in (2.110) the base current I_B is neglected.

For $V_i > 1\,\text{V}$ we obtain from (2.111) with $V_{BE} \approx 0.7\,\text{V}$, the approximation:

$$V_o \approx V_i - 0.7\,\text{V} \qquad (2.113)$$

Due to the virtually linear characteristic, the operating point can be chosen from within a large range. The sample operating point shown in Fig. 2.91 is achieved under the assumption that $B = \beta = 400$ and $I_S = 7\,\text{fA}$[22] with $V_b = 5\,\text{V}$, $R_E = R_g = 1\,\text{k}\Omega$ and $I_o = 0$:

$$V_o = 2\,\text{V} \;\Rightarrow\; I_C \approx \frac{V_o}{R_E} = 2\,\text{mA} \;\Rightarrow\; I_B = \frac{I_C}{B} = 5\,\mu\text{A}$$

$$\Rightarrow\; V_i = V_o + V_{BE} = V_o + V_T \ln \frac{I_C}{I_S} = 2.685\,\text{V}$$

$$\Rightarrow\; V_g = V_i + I_B R_g = 2.69\,\text{V}$$

---

[22] Typical values for an npn low-power transistor BC547B.

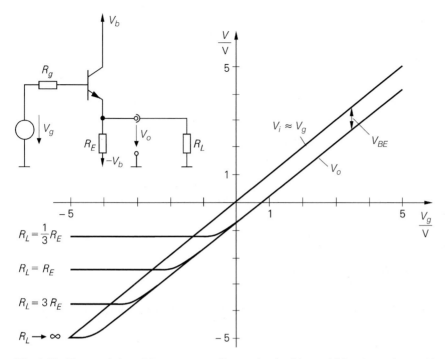

Fig. 2.92. Characteristics of the common-collector circuit with an additional negative supply voltage and the load R_L

Here, the voltage drop across R_g is only 5 mV and can be neglected; therefore in Fig. 2.91 we have $V_i \approx V_g$.

Feeding the common-collector circuit with an additional negative supply voltage $-V_b$ and placing a load R_L between output and earth as shown in Fig. 2.92 enables the generation of negative output voltages. The transfer characteristic then depends on the ratio between resistances R_E and R_L as the minimum output voltage $V_{o,min}$ is determined by the voltage divider formed by R_L and R_E:

$$V_{o,min} = - \frac{V_b R_L}{R_E + R_L}$$

This means that a large signal range is only achieved if $|V_{o,min}|$ is large; this requires $R_L > R_E$. For $V_g < V_{o,min}$ the transistor is in reverse mode because $V_{BE} < 0$ and therefore $V_o = V_{o,min}$. For $V_g \geq V_{o,min}$ the transistor is in forward mode and the characteristic is as shown in Fig. 2.91. In Fig. 2.92 the supply voltages are *symmetrical*; in other words, the positive and negative supply voltages have the same absolute value. This is typical in practice, but, as a general rule, the negative supply voltage can be chosen independent of the positive supply voltage.

Small-Signal Response of Common-Collector Circuits

The performance of the circuit modulated around the operating point A is known as the *small-signal response*. The operating point is determined by the operating point parameters

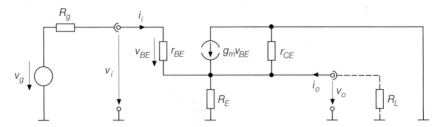

Fig. 2.93. Small-signal equivalent circuit of the common-collector circuit

$V_{i,A}$, $V_{o,A}$, $I_{i,A} = I_{B,A}$ and $I_{C,A}$; as an example we use the operating point as determined above with $V_{i,A} = 2.69\,\text{V}$, $V_{o,A} = 2\,\text{V}$, $I_{B,A} = 5\,\mu\text{A}$ and $I_{C,A} = 2\,\text{mA}$.

The *small-signal voltage gain* corresponds to the slope of the transfer characteristic. As the output voltage follows the input voltage the differentiation of (2.113) leads as expected to the approximation:

$$A = \left.\frac{\partial V_o}{\partial V_i}\right|_A \approx 1$$

A more accurate calculation of A is possible using the small-signal equivalent circuit shown in Fig. 2.93. The nodal equation

$$\frac{v_i - v_o}{r_{BE}} + g_m v_{BE} = \left(\frac{1}{R_E} + \frac{1}{r_{CE}}\right) v_o$$

for $v_{BE} = v_i - v_o$ and $R'_E = R_E \| r_{CE}$ leads to:

$$A = \left.\frac{v_o}{v_i}\right|_{i_o=0} = \frac{\left(1 + \dfrac{1}{\beta}\right) g_m R'_E}{\left(1 + \dfrac{1}{\beta}\right) g_m R'_E + 1}$$

$$\overset{\substack{r_{CE} \gg R_E \\ \beta \gg 1}}{\approx} \frac{g_m R_E}{g_m R_E + 1} \overset{g_m R_E \gg 1}{\approx} 1$$

For $g_m = I_{C,A}/V_T = 77\,\text{mS}$, $\beta = 400$, $R_E = 1\,\text{k}\Omega$ and $r_{CE} = V_A/I_{C,A} = 50\,\text{k}\Omega$ we achieve for the selected operating point the *exact* value and the first approximation $A = 0.987$.

The *small-signal input resistance* is:

$$r_i = \left.\frac{v_i}{i_i}\right|_{i_o=0} = r_{BE} + (1 + \beta) R'_E \overset{\substack{r_{CE} \gg R_E \\ \beta \gg 1}}{\approx} r_{BE} + \beta R_E \overset{g_m R_E \gg 1}{\approx} \beta R_E$$

This depends on the load resistance which in our case is the *open-circuit input resistance* since $i_o = 0$ ($R_L \to \infty$). The input resistance for other values of R_L can be calculated by replacing R_E with the parallel arrangement of R_E and R_L (see Fig. 2.93; with $R_L < R_E$, which is often found in practical applications, it thus depends to a high degree on R_L. For $r_{BE} = \beta/g_m$ and $R_L \to \infty$ the input resistance for the chosen operating point is *exactly* $r_i = 398\,\text{k}\Omega$; the first and second approximations are $r_i = 405\,\text{k}\Omega$ and $r_i = 400\,\text{k}\Omega$, respectively.

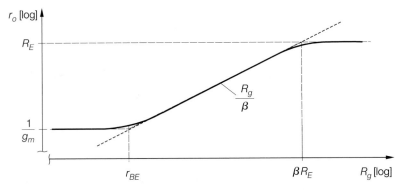

Fig. 2.94. Plot of the small-signal output resistance r_o of the common-collector circuit in its dependence on the internal resistance R_g of the signal generator

The *small-signal output resistance* is:

$$r_o = \frac{v_o}{i_o} = R_E' \parallel \frac{R_g + r_{BE}}{1 + \beta} \overset{\substack{r_{CE} \gg R_E \\ \beta \gg 1}}{\approx} R_E \parallel \left(\frac{R_g}{\beta} + \frac{1}{g_m} \right)$$

It depends on the internal resistance R_g of the signal generator; three ranges can be distinguished:

$$r_o \approx \begin{cases} \dfrac{1}{g_m} & \text{for } R_g < r_{BE} = \dfrac{\beta}{g_m} \\[2mm] \dfrac{R_g}{\beta} & \text{for } r_{BE} < R_g < \beta R_E \\[2mm] R_E & \text{for } R_g > \beta R_E \end{cases}$$

Figure 2.94 shows the plot of r_o in relation to R_g. For $R_g < r_{BE}$ and $R_g > \beta R_E$ the output resistance is constant; that is, independent of R_g. In between there is a range in which the internal resistance R_g is transformed to $r_o \approx R_g/\beta$. Due to this property the common-collector circuit is also known as an *impedance transformer*. A signal source with a subsequent common-collector circuit that operates in the transformation range can be described as an equivalent signal source (see Fig. 2.95); in this case the operating point

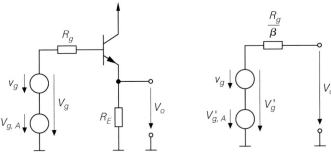

a Circuit with signal source   **b** Equivalent signal source

Fig. 2.95. Common-collector circuit as impedance transformer

voltage of the equivalent signal source is $V'_{g,A} \approx V_{g,A} - 0.7\,\text{V}$ according to (2.113), while the small-signal voltage v_g remains practically unchanged due to $A \approx 1$ and the internal resistance is reduced to R_g/β. For the sample operating point this results *exactly* in $r_o = 15.2\,\Omega$; the approximation is $r_o = 15.3\,\Omega$. From the presentation in ranges for $R_g = 1\,\text{k}\Omega < r_{BE} = 5.2\,\text{k}\Omega$ we obtain an approximation of $r_o \approx 1/g_m = 13\,\Omega$, which means that the circuit is not operating in the transformation range.

For $r_{CE} \gg R_E, \beta \gg 1$ and *no* load resistance R_L we obtain for the common-collector circuit:

Common-collector circuit

$$A = \left.\frac{v_o}{v_i}\right|_{i_o=0} \approx \frac{g_m R_E}{1 + g_m R_E} \overset{g_m R_E \gg 1}{\approx} 1 \tag{2.114}$$

$$r_i = \left.\frac{v_i}{i_i}\right|_{i_o=0} \approx r_{BE} + \beta R_E \overset{g_m R_E \gg 1}{\approx} \beta R_E \tag{2.115}$$

$$r_o = \frac{v_o}{i_o} \approx R_E \,||\, \left(\frac{R_g}{\beta} + \frac{1}{g_m}\right) \tag{2.116}$$

In order to take the load resistance R_L into consideration it is necessary to replace R_E by the parallel arrangement of R_E and R_L in (2.114) and (2.115) (see Fig. 2.93). For $R_g < \beta(R_E\,||\,R_L)$ and $g_m(R_E\,||\,R_L) \gg 1$ this results in:

$$A \approx 1 \quad , \quad r_i \approx \beta(R_E\,||\,R_L) \quad , \quad r_o \approx \frac{R_g}{\beta} + \frac{1}{g_m} \tag{2.117}$$

The relevant equivalent circuit with signal generator and load is shown in Fig. 2.96. One can see that the common-collector circuit features a strong coupling between input and output since in this case, unlike in the common-emitter circuit, the input resistance r_i depends on the load R_L at the output and the output resistance r_o depends on the source resistance R_g of the signal generator at the input.

Figure 2.96 enables the *small-signal operating gain* to be calculated:

$$A_B = \frac{v_o}{v_g} = \frac{r_i}{r_i + R_g} \frac{R_L}{R_L + r_o}$$

In most cases $r_i \gg R_g$ and $R_L \gg r_o$; this results in $A_B \approx 1$.

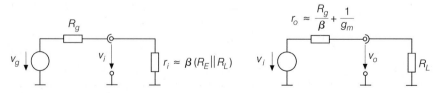

Fig. 2.96. Equivalent circuit with the equivalent parameters r_i and r_o

Non-linearity: The distortion factor of the common-collector circuit can be approximately determined by a series expansion of the characteristic at the operating point. From (2.110) and (2.111) for $I_o = 0$ – that is, $R_L \to \infty$, it follows that:

$$V_i = V_o + V_{BE} = I_C R_E + V_T \ln \frac{I_C}{I_S}$$

For the common-emitter circuit with current feedback we arrive at the same equation; therefore (2.74) also applies to the common-collector circuit. With a load resistance R_L in parallel to R_E from (2.74) it follows that:

$$k \approx \frac{v_{o,2\omega t}}{v_{o,\omega t}} \approx \frac{\hat{v}_i}{4V_T \left(1 + g_m \left(R_E \| R_L\right)\right)^2} \tag{2.118}$$

If a maximum value for k is given, the following must be the case: $\hat{v}_i < 4kU_T(1 + g_m(R_E\|R_L))^2$. In most practical cases $1/g_m \ll R_L \ll R_E$; it is possible to use the approximation:

$$k \approx \frac{\hat{v}_i}{4V_T g_m^2 R_L^2} \tag{2.119}$$

The distortion factor is then inversely proportional to the square of the load resistance which means that it increases significantly when R_L declines. It can be reduced only by an increased transconductance g_m; this means that the operating point current $I_{C,A} = g_m V_T$ must be increased accordingly. $R_L \to \infty$ results in $\hat{v}_i < k \cdot 631$ V for the numeric example. However, assuming $R_L = 100\,\Omega$ results in a much more stringent demand, $\hat{v}_i < k \cdot 6.7$ V. It thus follows from (2.119) that $\hat{v}_i < k \cdot 6.2$ V.

Temperature sensitivity: According to (2.21) the base–emitter voltage V_{BE} decreases at a rate of 1.7 mV/K when the collector current I_C is constant. Since in the common-collector circuit the difference between input and output voltage is just V_{BE} according to (2.111), the *temperature drift* of the output voltage for a constant input voltage is:

$$\frac{dV_o}{dT} = -\frac{dV_{BE}}{dT} \approx 1.7\,\mathrm{mV/K}$$

If we take into consideration that $A \approx 1$ in the common-collector circuit, the same result is achieved with (2.65), which was originally applied to the common-emitter circuit.

Setting the Operating Point

In the common-collector circuit it is much easier to establish a stable operating point for small-signal operation than in the common-emitter circuit, because the characteristic is linear over a much larger range and small deviations from the desired operating point have practically no influence on the small-signal response.[23] The temperature sensitivity and the production-induced variations in the current gain B and the saturation reverse current I_S of the transistor [24] have only a minor effect because with a given collector current $I_{C,A}$ at the operating point the base current $I_{B,A}$, which depends on B, is usually negligibly small and there is only a logarithmic dependence of the base-emitter voltage $V_{BE,A}$ on I_S.

---

[23] Compare Fig. 2.91 on page 132 with Fig. 2.58 on page 97.
[24] Values for temperature sensitivity and variations are given on page 115.

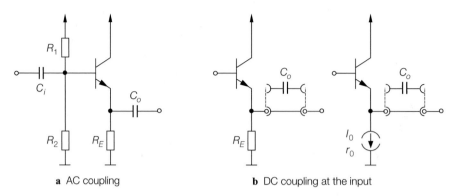

a AC coupling   **b** DC coupling at the input

Fig. 2.97. Setting the operating point

When setting the operating point a distinction is made between *AC coupling* and *DC coupling*. In addition to the *purely* AC or DC coupling in many common-collector circuits a DC coupling at the input is combined with an AC coupling at the output.

Setting the operating point in the case of AC coupling: Figure 2.97a shows AC coupling. The signal source and the load are connected via coupling capacitors and the operating point voltages can be selected independent of the DC voltages of the signal source and of the load; other properties are described on page 115. The voltage

$$V_{B,A} = (I_{C,A} + I_{B,A}) R_E + V_{BE,A} \approx I_{C,A} R_E + 0.7\,\text{V}$$

required at the operating point at the base of the transistor is determined by R_1 and R_2; the current through the resistors is selected to be much higher than the base current $I_{B,A}$ so that the operating point is not influenced by $I_{B,A}$.

In practice *pure* AC coupling is rarely used because in most cases DC coupling is possible at the input at least; this means that resistors R_1 and R_2 and coupling capacitor C_i can be omitted.

Setting the operating point by DC coupling at the input: Figure 2.97b shows the common-collector circuit with DC coupling at the input and DC or AC coupling at the output. The input voltage $V_{i,A}$ at the base of the transistor is given by the output voltage of the signal source when it is assumed that the voltage drop $I_{B,A} R_g$ caused by the base current $I_{B,A}$ through the internal resistance of the signal source can be neglected. With AC coupling at the output the collector current at the operating point can be set by resistance R_E, according to

$$I_{C,A} \approx \frac{V_{i,A} - V_{BE,A}}{R_E} \approx \frac{V_{i,A} - 0.7\,\text{V}}{R_E} \tag{2.120}$$

or by a current source; Fig. 2.97b shows both possibilities. If a current source is used, then $I_{C,A} \approx I_0$; furthermore, in the small-signal calculation the internal resistance r_0 of the current source must be used instead of resistance R_E. In addition, for DC coupling at the output the output current $I_{o,A}$ through the load must be taken into account.

Example: In the example on page 120 a common-emitter circuit is dimensioned for a load $R_L = 10\,\text{k}\Omega$ (see Fig. 2.80 on page 121). Now the circuit should be operated with a load

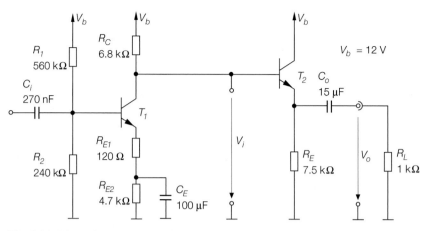

Fig. 2.98. Dimensioned example of a common-collector circuit (T_2) used as an impedance transformer for the common-emitter circuit (T_1)

$R_L = 1\,\text{k}\Omega$. As the output resistance $r_o \approx R_C = 6.8\,\text{k}\Omega$ is larger than R_L, the connection of R_L directly to the output of the common-emitter circuit results in a significant reduction in the operating gain A_B. Therefore, a common-collector circuit functioning as an impedance converter must be added to the output in order to reduce the output resistance and thus preserve the operating gain A_B (see Fig. 2.98). The amplitude of the input voltage to the common-collector circuit is $\hat{v}_i = 200\,\text{mV}$ in accordance with the amplitude at the output of the common-emitter circuit. The common-emitter circuit is dimensioned for a distortion factor $k < 1\,\%$. To make sure that the distortion factor does not rise considerably due to the additional common-collector circuit, the latter is determined to be $k < 0.2\,\%$ for the common-collector circuit. From (2.119) it follows that $g_m > 31\,\text{mS}$ or $I_{C,A} > 0.81\,\text{mA}$; $I_{C,A} = 1\,\text{mA}$ is selected. If $B \approx \beta \approx 400$ and $I_S \approx 7\,\text{fA}$ are assumed values for transistor T_2, then $V_{BE,A} \approx 0.67\,\text{V}$, $I_{B,A} = 2.5\,\mu\text{A}$, $g_m \approx 38.5\,\text{mS}$ and $r_{BE} \approx 10.4\,\text{k}\Omega$. The input voltage $V_{i,A}$ can be determined from the voltage at R_C (see Fig. 2.98):

$$V_{i,A} = V_b - \left(I_{C,A(T1)} + I_{B,A}\right)R_C \approx V_b - I_{C,A(T1)}R_C \approx 8.26\,\text{V}$$

From (2.120) it follows that $R_E \approx 7.59\,\text{k}\Omega \rightarrow 7.5\,\text{k}\Omega$.[25] $I_{B,A}$ causes only a negligible voltage drop of $I_{B,A}R_C \approx 17\,\text{mV}$ across the internal resistance $R_g = R_C$ of the signal source. For the components of the equivalent circuit according to Fig. 2.96 with $R_g \approx R_C$, we obtain from (2.117) $r_i \approx 353\,\text{k}\Omega$ and $r_o \approx 43\,\Omega$. Finally, the highpass filter caused by the capacitance C_o at the output should be dimensioned for $f'_L = 11\,\text{Hz}$:

$$C_o = \frac{1}{2\pi f'_L \left(r_o + R_L\right)} = 13.9\,\mu\text{F} \rightarrow 15\,\mu\text{F}$$

DC coupling at the output by short-circuiting C_o causes a DC voltage $V_{o,A} = V_{i,A} - V_{BE,A} \approx 7.5\,\text{V}$ to occur across R_L, so that an output current $I_{o,A} = -V_{o,A}/R_L \approx -7.5\,\text{mA}$ flows; in this case R_E can be omitted. Due to

$$I_{C,A} = \frac{V_{o,A}}{R_E \| R_L} \approx \frac{V_{i,A} - 0.7\,\text{V}}{R_E \| R_L} \geq 7.5\,\text{mA}$$

the range in which the operating point can be selected is very limited.

---

[25] Rounded to standard values.

Using AC and/or DC coupling: The most important considerations for the use of AC and/or DC coupling are described on pages 120 and 122. Usually the use of DC coupling at the output is complicated by the fact that low-resistance loads cause relatively high DC currents to flow even with low DC voltages at the output.

Frequency Response and Upper Cut-Off Frequency

The small-signal gain A and the operating gain A_B decline with higher frequencies because of the capacitances of the transistor. To obtain any information on the frequency response and the upper cutoff frequency it is necessary to use the dynamic small-signal model of the transistor for calculations; Fig. 2.99 shows the resulting dynamic small-signal equivalent circuit of the common-collector configuration. For $R'_g = R_g + R_B$ and $R'_L = R_L || R_E || r_{CE}$ we obtain for the *operating gain* $\underline{A}_B(s) = \underline{v}_o(s)/\underline{v}_g(s)$:

$$\underline{A}_B(s) \; = \; \frac{1 + \beta + sC_E r_{BE}}{1 + \beta + \dfrac{r_{BE} + R'_g}{R'_L} + sc_1 + s^2 C_E C_C R'_g r_{BE}}$$

$$c_1 \; = \; C_E r_{BE} + (C_E + C_C)\,\frac{r_{BE} R'_g}{R'_L} + C_C R'_g\,(1 + \beta)$$

Consequently, for $\beta \gg 1$ the low-frequency gain is

$$A_0 \; = \; \underline{A}_B(0) \; \approx \; \frac{1}{1 + \dfrac{r_{BE} + R'_g}{\beta R'_L}} \tag{2.121}$$

and thus with the further approximations $R'_L \gg 1/g_m$ and $R'_L \gg R'_g/\beta$ the frequency response is:

$$\underline{A}_B(s) \; \approx \; \frac{A_0 \left(1 + s\,\dfrac{C_E}{g_m}\right)}{1 + s\left(\dfrac{C_E}{g_m}\left(1 + \dfrac{R'_g}{R'_L}\right) + C_C R'_g\right) + s^2\,\dfrac{C_E C_C R'_g}{g_m}} \tag{2.122}$$

Both poles are real and due to

$$f_N \; = \; \frac{g_m}{2\pi C_E} \; > \; f_C$$

the corner frequency at the zero is above the transit frequency f_T of the transistor as can be seen by comparison with (2.44). The frequency response can be described in an approxi-

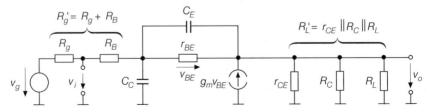

Fig. 2.99. Dynamic small-signal equivalent circuit for the common-collector configuration

mation by a lowpass filter of first order after removing the s^2 term from the denominator and subtracting the linear terms:

$$\underline{A}_B(s) \approx \frac{A_0}{1 + s\left(\dfrac{C_E}{g_m R'_L} + C_C\right) R'_g}$$

This results in an approximation of the upper $-3dB$ *cutoff frequency* f_{-3dB} at which the magnitude of the gain has declined by 3 dB:

$$\omega_{-3dB} = 2\pi f_{-3dB} \approx \frac{1}{\left(\dfrac{C_E}{g_m R'_L} + C_C\right) R'_g} \tag{2.123}$$

Due to $R'_g = R_g + R_B \approx R_g$ this value is proportional to the internal resistance R_g of the signal generator. The maximum upper cutoff frequency is achieved with $R_g \to 0$ and $R'_L \to \infty$

$$\omega_{-3dB,max} \approx \frac{1}{C_C R_B}$$

which is usually higher than the transit frequency f_T of the transistor.

If the load also contains a capacitive portion in addition to the resistive portion – that is, if a load capacitance C_L exists in parallel to the load resistance R_L, we obtain by substituting

$$\underline{Z}_L(s) = R'_L \,||\, \frac{1}{sC_L} = \frac{R'_L}{1 + sC_L R'_L}$$

instead of R'_L:

$$\underline{A}_B(s) \approx \frac{A_0\left(1 + s\dfrac{C_E}{g_m}\right)}{1 + sc_1 + s^2 c_2} \tag{2.124}$$

$$c_1 = \frac{C_E}{g_m}\left(1 + \frac{R'_g}{R'_L}\right) + C_C R'_g + C_L\left(\frac{1}{g_m} + \frac{R'_g}{\beta}\right)$$

$$c_2 = (C_C C_E + C_L (C_C + C_E))\frac{R'_g}{g_m}$$

In this case the poles can be real or complex conjugate. An approximation by a lowpass filter of first order yields a reasonable estimation of the upper cutoff frequency for real poles only:

$$\omega_{-3dB} = 2\pi f_{-3dB} \approx \frac{1}{\left(\dfrac{C_E}{g_m R'_L} + C_C + \dfrac{C_L}{\beta}\right) R'_g + \dfrac{C_L}{g_m}} \tag{2.125}$$

For complex conjugate poles the following estimation must be used:

$$\omega_{-3dB} = 2\pi f_{-3dB} \approx \frac{1}{\sqrt{c_2}} \tag{2.126}$$

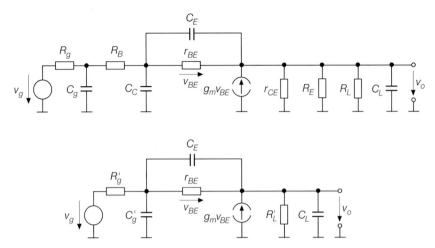

Fig. 2.100. Small-signal equivalent circuit for calculating the range of complex conjugate poles: accurate (top) and simplified (bottom)

From (2.124) it follows that the common-collector circuit is always stable[26]; that is, with complex conjugate poles there is still an oscillation in the step-response, although this does die down again. But in practical applications this circuit may become unstable, which means that a sustained oscillation occurs that, due to overload effects, stabilizes at a certain amplitude and under unfavorable conditions may destroy the transistor. This instability is caused by second order effects that are not taken into consideration by the small-signal equivalent circuit of the transistor used in this case.[27]

Region of complex conjugate poles: For the practical application of a common-collector circuit it is necessary to know for which load capacitances the complex conjugate poles may occur and which circuit design measures can prevent this. For this purpose we look at the small-signal equivalent circuit shown in Fig. 2.100, which is derived from Fig. 2.93 by adding the output capacitance C_g of the signal generator and the load capacitance C_L; due to $R_g \gg R_B$ the elements $R_g - C_g$ and $R_B - C_C$ can be combined into one element consisting of $R'_g = R_g + R_B$ and $C'_g = C_g + C_C$. If the time constants

$$T_g = C'_g R'_g \quad , \quad T_L = C_L R'_L \quad , \quad T_E = \frac{C_E}{g_m} \approx \frac{1}{\omega C} \tag{2.127}$$

and the resistance ratios

$$k_g = \frac{R'_g}{R'_L} \quad , \quad k_S = \frac{1}{g_m R'_L} \tag{2.128}$$

are introduced and C_C is replaced by C'_g, then from (2.124) it follows that:

---

[26] A second order transfer function with positive coefficients in the denominator is stable.

[27] Due to the transit time in the base region of the transistor there is an additional time constant; in the small-signal equivalent circuit of the transistor this effect can be simulated by an inductance in series with the base spreading resistance R_B. This results in a third order transfer function that may become instable with a capacitive load.

$$c_1 = T_E (1 + k_g) + T_g + T_L \left(k_S + \frac{k_g}{\beta} \right)$$
$$c_2 = T_g T_E + T_g T_L k_S + T_L T_E k_g$$

(2.129)

This enables the *quality* to be calculated:

$$Q = \frac{\sqrt{c_2}}{c_1}$$

(2.130)

and under the condition $Q > 0.5$, the range of complex conjugate poles can be determined. In Fig. 2.101 this range is shown for $\beta = 50$ and $\beta = 500$ as a function of the *normalized*

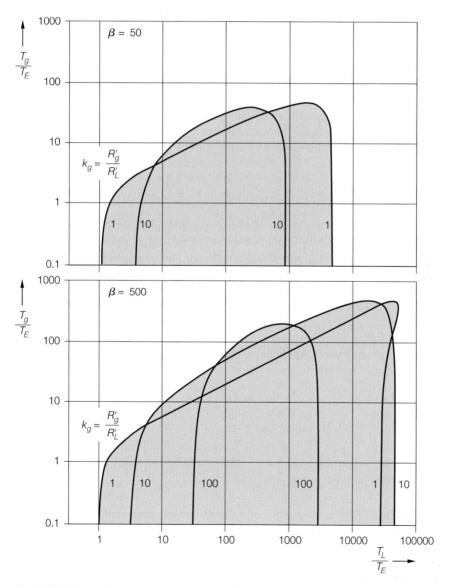

Fig. 2.101. Range of complex conjugate poles for $\beta = 50$ and $\beta = 500$ with gray background

signal source time constant T_g/T_E and the *normalized load time constant* T_L/T_E for various values of k_g; $k_S = 0.01$ is used.

Figure 2.101 shows that no complex conjugate poles occur for very small and very large load capacitances C_L (T_L/T_E small or large) and with a sufficiently large output capacitance C_g of the signal generator (T_g/T_E large). The range of complex conjugate poles is heavily dependent on k_g. The regions for $k_g < 1$ are within the region for $k_g = 1$; no complex conjugate poles occur for $k_g > \beta$. The dependence on k_S has an influence only in the event of a large load capacitance (T_L/T_E large), a high current gain β and small internal resistance R_g of the signal generator; this is responsible for the dent in the right portion of the plot shown in Fig. 2.101 for $\beta = 500$ and $k_g = 1$.

If R_g, C_g, R_L and C_L are given and complex conjugate poles occur, there are four different ways to avoid this range:

1. T_g can be increased in order to leave the range of complex conjugate poles *toward the top*. This requires the insertion of an additional capacitor connected between the input of the common-collector circuit and ground or to a supply voltage; in the small-signal equivalent circuit it is arranged in parallel with C_g and causes an increase in T_g. This method can always be used; for this reason it is very common in practical applications.

2. T_E can be increased in order to leave the range toward the *lower left* if operating close to the left boundary of this range. This requires the use of a *slower* transistor with a higher time constant T_E; that is, a lower transit frequency f_T.

3. T_E can be reduced in order to leave the range toward the *top right* if operating close to the right boundary of this range. This requires the use of a *faster* transistor with a shorter time constant T_E; that is, a higher transit frequency f_T. This method is used, for example, in power supplies because of their high load capacitance as a result of the capacitor at the output which shifts the operating point toward the right boundary; the use of a faster transistor in this case leads to an improved transient response.

4. T_L can be increased in order to leave the range *toward the right* if operating close to the right boundary of this range. This requires an enlargement of the load capacitance C_L by adding an additional capacitor in parallel. This possibility is also used in power supplies; it increases the capacitor at the output accordingly.

The above four methods are indicated in Fig. 2.102. A fifth method is the reduction of T_L, which is seldom used in practical applications since with given values for R_L and C_L it can be achieved only by connecting a resistor in parallel, which means adding an additional load to the output. All of these methods cause a reduction in the upper cutoff frequency. In order to keep this reduction within acceptable limits it is necessary to leave the range of complex conjugate poles *in the shortest way possible*.

Equivalent circuit: The common-collector circuit can be approximately described by the equivalent circuit shown in Fig. 2.103. It is derived from Fig. 2.96 by adding the *input capacitance* C_i, the *output capacitance* C_o and the *output inductance* L_o. C_i, C_o and L_o are obtained under the condition that the calculation of $\underline{A}_B(s)$ leads to (2.124) if both expressions are approximated by a lowpass filter of first order. For the elements of the equivalent circuit this leads to the following equations:

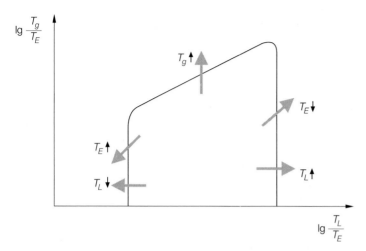

Fig. 2.102. Methods of leaving the range of complex conjugate poles

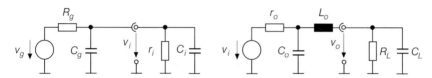

Fig. 2.103. Equivalent circuit with equivalent parameters r_i, r_o, C_i, C_o and L_o

$$r_i = \beta R'_L + r_{BE} \quad , \quad C_i = \frac{C_E r_{BE} + C_L R'_L}{\beta R'_L + r_{BE}}$$

$$r_o = \frac{R'_g}{\beta} + \frac{1}{g_m} \quad , \quad C_o = \frac{\beta C'_g R'_g}{R'_g + r_{BE}}$$

$$L_o = \frac{C_E R'_g}{g_m}$$

This indicates that in addition to the resistances r_i and r_o the capacitances C_i and C_o as well as the inductance L_o are also heavily dependent on the signal source and the load; this means that there is strong coupling between the input and the output.

Example: $I_{C,A} = 2$ mA was selected for the numeric example of Fig. 2.90a. With $\beta = 400$, $V_A = 100$ V, $C_{obo} = 3.5$ pF and $f_T = 160$ MHz we obtain from Fig. 2.45 on page 83 the small-signal parameters $g_m = 77$ mS, $r_{BE} = 5.2$ kΩ, $r_{CE} = 50$ kΩ, $C_C = 3.5$ pF and $C_E = 73$ pF. For $R_g = R_E = 1$ kΩ, $R_L \to \infty$ and $R'_g \approx R_g$ and for $R'_L = R_L \| R_E \| r_{CE} = 980$ Ω it follows from (2.121) that $A_0 = 0.984 \approx 1$ and from (2.123) that $f_{-3dB} \approx 36$ MHz. With a load capacitance $C_L = 1$ nF we achieve using (2.125) $f_{-3dB} \approx 8$ MHz and using (2.126) $f_{-3dB} \approx 5$ MHz. From (2.127) and (2.128) we obtain $T_g = 3.5$ ns, $T_L = 980$ ns, $T_E = 0.95$ ns, $r_g = 0.98$ and $r_S = 0.013$ and thus from (2.129) $c_1 = 20.6$ ns and $c_2 = 979$ (ns)^2. From (2.130) it follows that $Q = 1.52$; that is, there are complex conjugate poles. This result is also achieved by means of Fig. 2.101 since point $T_L/T_E \approx 1000$, $T_g/T_E \approx 4$, $k_g \approx 1$ is within the range of complex conjugate poles; the region for $\beta = 500$ is used because $\beta = 400$. In this case, leaving the range of

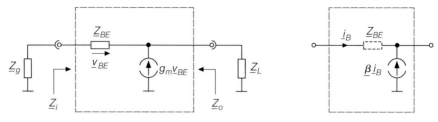

a Simplified small-signal equivalent circuit   **b** Different representation of the transistor

Fig. 2.104. Equivalent circuit for impedance transformation

complex conjugate poles is possible only by increasing T_g to $T_g/T_E \approx 75$; this means that $C'_g \approx 71$ pF is required; that is, a capacitor of $C_g = C'_g - C_C \approx 68$ pF must be connected between the base of the transistor and ground. This causes a reduction in the upper cutoff frequency; from (2.125) we obtain $f_{-3dB} \approx 1.8$ MHz if C_C is replaced by $C'_g = 71$ pF. A smaller C_g could be chosen if a weak complex conjugate pole pair is allowed with resulting overshoots when applying square-wave signals; then the upper cutoff frequency does not drop as much.

Impedance Transformation by the Common-Collector Circuit

The common-collector circuit acts as an impedance transformer. Under static conditions the input resistance r_i depends for the main part on the load and the output resistance r_o depends on the source resistance of the signal generator; for $R_E \gg R_L$ and $R_g \gg r_{BE}$ it follows from (2.117) that $r_i \approx \beta R_L$ and $r_o \approx R_g/\beta$. This property can be generalized. We thus refer to the small-signal equivalent circuit shown in Fig. 2.104a, which is derived from Fig. 2.99 by neglecting R_B, R_E and C_C and by combining r_{BE} and C_E to achieve

$$\underline{Z}_{BE}(s) = r_{BE} \,||\, \frac{1}{sC_E} = \frac{r_{BE}}{1 + sC_E r_{BE}}$$

and by assuming the general generator and load impedances $\underline{Z}_g(s)$ and $\underline{Z}_L(s)$. Instead of the transistor one can also use the circuit shown in Fig. 2.104b with the frequency-dependent small-signal current gain[28]:

$$\underline{\beta}(s) = g_m \underline{Z}_{BE}(s) = \frac{\beta_0}{1 + \dfrac{s}{\omega_\beta}}$$

When calculating the input impedance $\underline{Z}_i(s)$ and the output impedance $\underline{Z}_o(s)$ from Fig. 2.104 we obtain:

$$\underline{Z}_i(s) = \underline{Z}_{BE}(s) + \left(1 + \underline{\beta}(s)\right) \underline{Z}_L(s) \approx \underline{Z}_{BE}(s) + \underline{\beta}(s)\underline{Z}_L(s)$$

$$\underline{Z}_o(s) = \frac{\underline{Z}_{BE}(s) + \underline{Z}_g(s)}{1 + \underline{\beta}(s)} \approx \frac{\underline{Z}_{BE}(s) + \underline{Z}_g(s)}{\underline{\beta}(s)}$$

---

[28] If $C_C = 0$ then $\omega_\beta^{-1} = C_E r_{BE}$, see (2.43); furthermore, $\beta_0 = |\underline{\beta}(j0)| = g_m r_{BE}$.

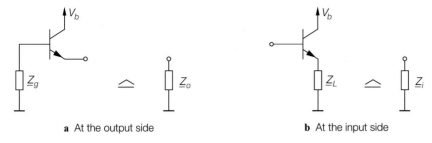

a At the output side   **b** At the input side

Fig. 2.105. Impedance transformation by the common-collector circuit

Figure 2.105 illustrates this relationship. Often it is possible to neglect $\underline{Z}_{BE}(s)$ and to use the simplified transformation formulas:

$$\underline{Z}_i(s) \approx \underline{\beta}(s)\underline{Z}_L(s) \quad , \quad \underline{Z}_o(s) \approx \frac{\underline{Z}_g(s)}{\underline{\beta}(s)}$$

Figure 2.106 shows some selected examples. Particularly noteworthy are the cases of $\underline{Z}_g(s) = sL$ and $\underline{Z}_L(s) = 1/(sC)$ in which the transformation produces a frequency-dependent negative resistance; in this case $\underline{Z}_o(s)$ and $\underline{Z}_i(s)$ are no longer passive and unfavorable circuitry can render the circuit instable. This means for practical purposes that inductances in the base circuit and/or capacitances in the emitter circuit of a transistor may produce unwanted oscillations; an example of this is the common-collector circuit with capacitive load. The RC parallel circuit with the secondary condition $\omega_\beta RC = 1$, as shown in the lower left part of Fig. 2.106 causes a purely ohmic output impedance; here an

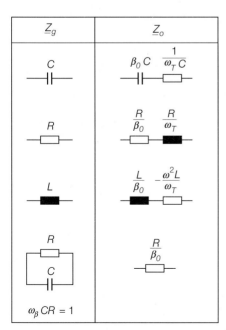

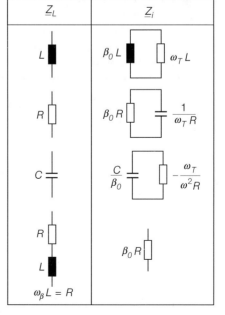

Fig. 2.106. Selected samples of impedance transformation

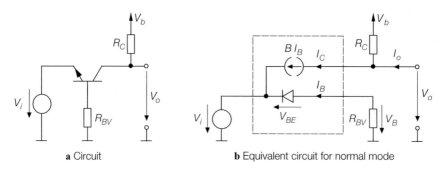

a Circuit   **b** Equivalent circuit for normal mode

Fig. 2.107. Common-base circuit

additional capacitance at the output will not produce complex conjugate poles – in other words, no oscillations will occur.

2.4.3
Common-Base Circuit

Figure 2.107a shows the common-base circuit consisting of the transistor, the collector resistance R_C, the supply voltage source V_b and the signal voltage source V_i.[29] The resistance R_{BV} acts as a limiter for the base current in the event of overload; in normal mode it has no noticeable influence. The following considerations are based on $V_b = 5\,\text{V}$ and $R_C = R_{BV} = 1\,\text{k}\Omega$.

Transfer Characteristic of the Common-Base Circuit

The transfer characteristic shown in Fig. 2.108 is obtained by measuring the output voltage V_o as a function of the signal voltage V_i. For $V_i > -0.5\,\text{V}$ the collector current is negligible and $V_o = V_b = 5\,\text{V}$. The condition $-0.72\,\text{V} \le V_i \le -0.5\,\text{V}$ causes a collector current I_C that increases with a decrease in V_i and the output voltage declines according to $V_o = V_b - I_C R_C$. Up to this point the transistor operates in normal mode. For $V_i < 0.72\,\text{V}$ the transistor enters the saturation region and $V_o = V_i + V_{CE,sat}$.

Normal mode: Figure 2.107b shows the equivalent circuit for the normal mode in which the transistor is replaced by the simplified transport model according to Fig. 2.27 with:

$$I_C = BI_B = I_S e^{\frac{V_{BE}}{V_T}}$$

From Fig. 2.107b it follows that:

$$V_o = V_b + (I_o - I_C)\,R_C \overset{I_o=0}{=} V_b - I_C R_C \tag{2.131}$$

$$V_i = -V_{BE} - I_B R_{BV} = -V_{BE} - \frac{I_C R_{BV}}{B} \approx -V_{BE} \tag{2.132}$$

---

[29] In contrast to the procedure with common-emitter and common-collector circuits, here a voltage source *without* internal resistance is used; from $R_g = 0$ we have $V_i = V_g$ as a comparison with Fig. 2.57b and Fig. 2.90b shows. This approach was selected in order to make the characteristics of normal operation independent of R_g.

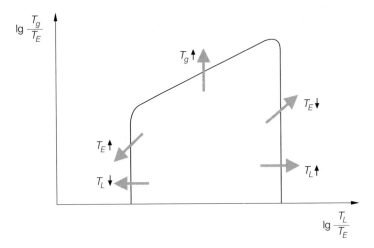

Fig. 2.102. Methods of leaving the range of complex conjugate poles

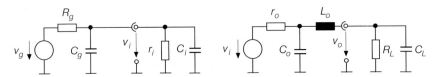

Fig. 2.103. Equivalent circuit with equivalent parameters r_i, r_o, C_i, C_o and L_o

$$r_i = \beta R'_L + r_{BE} \quad , \quad C_i = \frac{C_E r_{BE} + C_L R'_L}{\beta R'_L + r_{BE}}$$

$$r_o = \frac{R'_g}{\beta} + \frac{1}{g_m} \quad , \quad C_o = \frac{\beta C'_g R'_g}{R'_g + r_{BE}}$$

$$L_o = \frac{C_E R'_g}{g_m}$$

This indicates that in addition to the resistances r_i and r_o the capacitances C_i and C_o as well as the inductance L_o are also heavily dependent on the signal source and the load; this means that there is strong coupling between the input and the output.

Example: $I_{C,A} = 2\,\text{mA}$ was selected for the numeric example of Fig. 2.90a. With $\beta = 400$, $V_A = 100\,\text{V}$, $C_{obo} = 3.5\,\text{pF}$ and $f_T = 160\,\text{MHz}$ we obtain from Fig. 2.45 on page 83 the small-signal parameters $g_m = 77\,\text{mS}$, $r_{BE} = 5.2\,\text{k}\Omega$, $r_{CE} = 50\,\text{k}\Omega$, $C_C = 3.5\,\text{pF}$ and $C_E = 73\,\text{pF}$. For $R_g = R_E = 1\,\text{k}\Omega$, $R_L \rightarrow \infty$ and $R'_g \approx R_g$ and for $R'_L = R_L\|R_E\|r_{CE} = 980\,\Omega$ it follows from (2.121) that $A_0 = 0.984 \approx 1$ and from (2.123) that $f_{-3dB} \approx 36\,\text{MHz}$. With a load capacitance $C_L = 1\,\text{nF}$ we achieve using (2.125) $f_{-3dB} \approx 8\,\text{MHz}$ and using (2.126) $f_{-3dB} \approx 5\,\text{MHz}$. From (2.127) and (2.128) we obtain $T_g = 3.5\,\text{ns}$, $T_L = 980\,\text{ns}$, $T_E = 0.95\,\text{ns}$, $r_g = 0.98$ and $r_S = 0.013$ and thus from (2.129) $c_1 = 20.6\,\text{ns}$ and $c_2 = 979\,(\text{ns})^2$. From (2.130) it follows that $Q = 1.52$; that is, there are complex conjugate poles. This result is also achieved by means of Fig. 2.101 since point $T_L/T_E \approx 1000$, $T_g/T_E \approx 4$, $k_g \approx 1$ is within the range of complex conjugate poles; the region for $\beta = 500$ is used because $\beta = 400$. In this case, leaving the range of

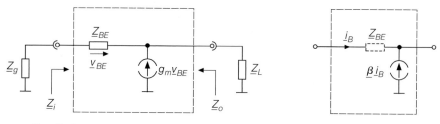

a Simplified small-signal equivalent circuit   **b** Different representation
of the transistor

Fig. 2.104. Equivalent circuit for impedance transformation

complex conjugate poles is possible only by increasing T_g to $T_g/T_E \approx 75$; this means that $C'_g \approx 71\,\text{pF}$ is required; that is, a capacitor of $C_g = C'_g - C_C \approx 68\,\text{pF}$ must be connected between the base of the transistor and ground. This causes a reduction in the upper cutoff frequency; from (2.125) we obtain $f_{-3dB} \approx 1.8\,\text{MHz}$ if C_C is replaced by $C'_g = 71\,\text{pF}$. A smaller C_g could be chosen if a weak complex conjugate pole pair is allowed with resulting overshoots when applying square-wave signals; then the upper cutoff frequency does not drop as much.

Impedance Transformation by the Common-Collector Circuit

The common-collector circuit acts as an impedance transformer. Under static conditions the input resistance r_i depends for the main part on the load and the output resistance r_o depends on the source resistance of the signal generator; for $R_E \gg R_L$ and $R_g \gg r_{BE}$ it follows from (2.117) that $r_i \approx \beta R_L$ and $r_o \approx R_g/\beta$. This property can be generalized. We thus refer to the small-signal equivalent circuit shown in Fig. 2.104a, which is derived from Fig. 2.99 by neglecting R_B, R_E and C_C and by combining r_{BE} and C_E to achieve

$$\underline{Z}_{BE}(s) = r_{BE} \parallel \frac{1}{sC_E} = \frac{r_{BE}}{1 + sC_E r_{BE}}$$

and by assuming the general generator and load impedances $\underline{Z}_g(s)$ and $\underline{Z}_L(s)$. Instead of the transistor one can also use the circuit shown in Fig. 2.104b with the frequency-dependent small-signal current gain[28]:

$$\underline{\beta}(s) = g_m \underline{Z}_{BE}(s) = \frac{\beta_0}{1 + \dfrac{s}{\omega_\beta}}$$

When calculating the input impedance $\underline{Z}_i(s)$ and the output impedance $\underline{Z}_o(s)$ from Fig. 2.104 we obtain:

$$\underline{Z}_i(s) = \underline{Z}_{BE}(s) + \left(1 + \underline{\beta}(s)\right)\underline{Z}_L(s) \approx \underline{Z}_{BE}(s) + \underline{\beta}(s)\underline{Z}_L(s)$$

$$\underline{Z}_o(s) = \frac{\underline{Z}_{BE}(s) + \underline{Z}_g(s)}{1 + \underline{\beta}(s)} \approx \frac{\underline{Z}_{BE}(s) + \underline{Z}_g(s)}{\underline{\beta}(s)}$$

---

[28] If $C_C = 0$ then $\omega_\beta^{-1} = C_E r_{BE}$, see (2.43); furthermore, $\beta_0 = |\underline{\beta}(j0)| = g_m r_{BE}$.

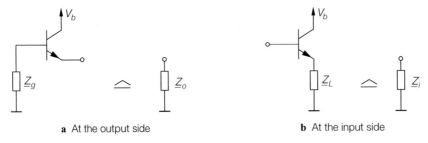

a At the output side   **b** At the input side

Fig. 2.105. Impedance transformation by the common-collector circuit

Figure 2.105 illustrates this relationship. Often it is possible to neglect $\underline{Z}_{BE}(s)$ and to use the simplified transformation formulas:

$$\underline{Z}_i(s) \approx \underline{\beta}(s)\underline{Z}_L(s) \quad , \quad \underline{Z}_o(s) \approx \frac{\underline{Z}_g(s)}{\underline{\beta}(s)}$$

Figure 2.106 shows some selected examples. Particularly noteworthy are the cases of $\underline{Z}_g(s) = sL$ and $\underline{Z}_L(s) = 1/(sC)$ in which the transformation produces a frequency-dependent negative resistance; in this case $\underline{Z}_o(s)$ and $\underline{Z}_i(s)$ are no longer passive and unfavorable circuitry can render the circuit unstable. This means for practical purposes that inductances in the base circuit and/or capacitances in the emitter circuit of a transistor may produce unwanted oscillations; an example of this is the common-collector circuit with capacitive load. The RC parallel circuit with the secondary condition $\omega_\beta RC = 1$, as shown in the lower left part of Fig. 2.106 causes a purely ohmic output impedance; here an

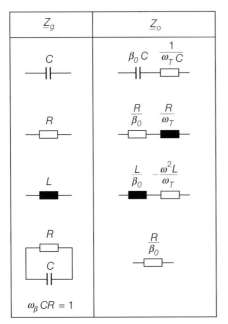

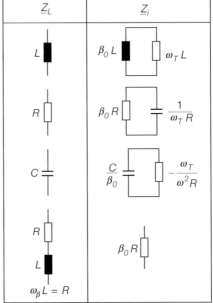

Fig. 2.106. Selected samples of impedance transformation

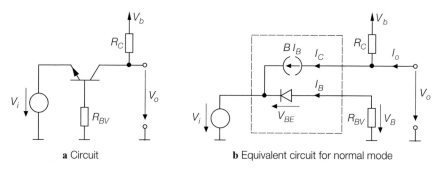

Fig. 2.107. Common-base circuit

additional capacitance at the output will not produce complex conjugate poles – in other words, no oscillations will occur.

2.4.3
Common-Base Circuit

Figure 2.107a shows the common-base circuit consisting of the transistor, the collector resistance R_C, the supply voltage source V_b and the signal voltage source V_i.[29] The resistance R_{BV} acts as a limiter for the base current in the event of overload; in normal mode it has no noticeable influence. The following considerations are based on $V_b = 5\,\text{V}$ and $R_C = R_{BV} = 1\,\text{k}\Omega$.

Transfer Characteristic of the Common-Base Circuit

The transfer characteristic shown in Fig. 2.108 is obtained by measuring the output voltage V_o as a function of the signal voltage V_i. For $V_i > -0.5\,\text{V}$ the collector current is negligible and $V_o = V_b = 5\,\text{V}$. The condition $-0.72\,\text{V} \le V_i \le -0.5\,\text{V}$ causes a collector current I_C that increases with a decrease in V_i and the output voltage declines according to $V_o = V_b - I_C R_C$. Up to this point the transistor operates in normal mode. For $V_i < 0.72\,\text{V}$ the transistor enters the saturation region and $V_o = V_i + V_{CE,sat}$.

Normal mode: Figure 2.107b shows the equivalent circuit for the normal mode in which the transistor is replaced by the simplified transport model according to Fig. 2.27 with:

$$I_C = BI_B = I_S\, e^{\frac{V_{BE}}{V_T}}$$

From Fig. 2.107b it follows that:

$$V_o = V_b + (I_o - I_C)\, R_C \overset{I_o=0}{=} V_b - I_C R_C \tag{2.131}$$

$$V_i = -V_{BE} - I_B R_{BV} = -V_{BE} - \frac{I_C R_{BV}}{B} \approx -V_{BE} \tag{2.132}$$

---

[29] In contrast to the procedure with common-emitter and common-collector circuits, here a voltage source *without* internal resistance is used; from $R_g = 0$ we have $V_i = V_g$ as a comparison with Fig. 2.57b and Fig. 2.90b shows. This approach was selected in order to make the characteristics of normal operation independent of R_g.

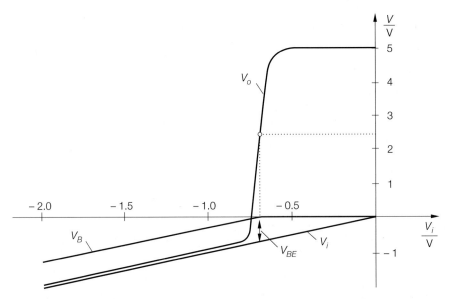

Fig. 2.108. Characteristics of the common-base circuit

It is assumed in (2.132) that the voltage drop across R_{BV} can be neglected if B is sufficiently large and R_{BV} is sufficiently small.

A point in the centre of the declining region of the transfer characteristic is selected for the operating point; this allows maximum output amplitude. When setting $B = \beta = 400$ and $I_S = 7\,\text{fA}$[30] the operating point as shown in Fig. 2.108 under the assumptions $V_b = 5\,\text{V}$ and $R_C = R_{BV} = 1\,\text{k}\Omega$ will be:

$$V_o = 2.5\,\text{V} \;\Rightarrow\; I_C = \frac{V_b - V_o}{R_C} = 2.5\,\text{mA} \;\Rightarrow\; I_B = \frac{I_C}{B} = 6.25\,\mu\text{A}$$

$$\Rightarrow\; V_{BE} = V_T \ln\frac{I_C}{I_S} = 692\,\text{mV} \;\Rightarrow\; V_i = -\,V_{BE} - I_B R_{BV} = -\,698\,\text{mV}$$

In this case the voltage drop across R_{BV} is only 6.25 mV and may be neglected; that is, the voltage at the base of the transistor is $V_B \approx 0$.

Saturation mode: For $V_i < -0.72\,\text{V}$ the transistor enters the saturation region, which means that the collector diode is forward-biased. In this region $V_{CE} = V_{CE,sat}$ and $V_o = V_i + V_{CE,sat}$ and the base current that flows must be limited to acceptable values by resistance R_{BV}:

$$I_B = -\,\frac{V_i + V_{BE}}{R_{BV}} \approx -\,\frac{V_i + 0.72\,\text{V}}{R_{BV}}$$

Transfer characteristic when driven by a current source: It is also possible to drive the circuit by a current source I_i (see Fig. 2.109); for $V_b = 5\,\text{V}$ and $R_C = R_{BV} = 1\,\text{k}\Omega$ the circuit operates as a *current–voltage converter* or a *transimpedance amplifier* for

---

[30] Typical values for an npn low-power transistor BC547B.

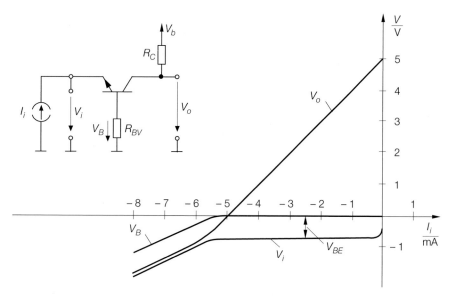

Fig. 2.109. Circuit and characteristics of the common-base circuit when driven by a current source

$-5.5\,\mathrm{mA} \leq I_i \leq 0$:[31]

$$V_o = V_b - I_C R_C = V_b + \frac{B}{1+B} I_E R_C \approx V_b + I_i R_C \tag{2.133}$$

$$V_i = -V_{BE} - I_B R_{BV} \approx -V_{BE} \approx -V_T \ln\left(-\frac{I_i}{I_S}\right) \tag{2.134}$$

Here, $I_i = I_E \approx -I_C$ is used. In this region the transistor operates in normal mode and the transfer characteristic is almost linear. For $I_i > 0$ the transistor is in the reverse region and for $I_i < -5.5\,\mathrm{mA}$ it enters the saturation region.

In most practical applications a common-emitter circuit with open collector or a current mirror is used as input current source; the section on setting the operating point outlines this in more detail.

Small-Signal Response of Common-Base Circuits

The behavior of a circuit near the operating point is called the *small-signal response*. The operating point is determined by the operating point values $V_{i,A}$, $V_{o,A}$, $I_{i,A} = I_{B,A}$ and $I_{C,A}$; the operating point as determined above is used as an example: $V_{i,A} = -0.7\,\mathrm{V}$, $I_{o,A} = 2.5\,\mathrm{V}$, $I_{B,A} = 6.25\,\mu\mathrm{A}$ and $I_{C,A} = 2.5\,\mathrm{mA}$.

The *small-signal voltage gain A* corresponds to the slope of the transfer characteristic. The calculation is based on the small-signal equivalent circuit shown in Fig. 2.110. From the nodal equation

$$\frac{v_o}{R_C} + \frac{v_o - v_i}{r_{CE}} + g_m v_{BE} = 0$$

---

[31] The name *transimpedance amplifier* is also used for operational amplifiers with current feedback and voltage output (CV-OPV).

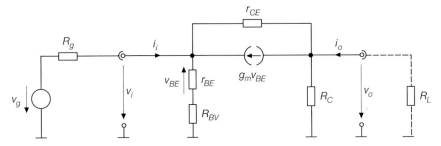

Fig. 2.110. Small-signal equivalent circuit of the common-base circuit

and the voltage division

$$v_{BE} = -\frac{r_{BE}}{r_{BE} + R_{BV}} v_i$$

it follows that:

$$A = \left.\frac{v_o}{v_i}\right|_{i_o=0} = \left(\frac{\beta}{r_{BE} + R_{BV}} + \frac{1}{r_{CE}}\right)(R_C \| r_{CE})$$

$$\overset{\substack{r_{CE} \gg R_C \\ \beta r_{CE} \gg r_{BE}+R_{BV}}}{\approx} \frac{\beta R_C}{r_{BE} + R_{BV}} \overset{r_{BE} \gg R_{BV}}{\approx} g_m R_C$$

The maximum gain is achieved with $R_{BV} = 0$; this requires the base of the transistor to be connected either directly or via a capacitor to earth. The following section on setting the operating point describes this in more detail. When operating with a load resistance R_L, it is possible to calculate the corresponding operating gain A_B by replacing R_C by R_L and R_C connected in parallel (see Fig. 2.110). For $g_m = I_{C,A}/V_T = 96\,\text{mS}$, $\beta = 400$, $r_{BE} = 4160\,\Omega$, $r_{CE} = V_A/I_{C,A} = 40\,\text{k}\Omega$, and $R_{BV} = 1\,\text{k}\Omega$, we obtain the *exact* value and the first approximation $A = 76$; the result $A = 96$ of the second approximation is very inaccurate because the condition $r_{BE} \gg R_{BV}$ is only insufficiently met.

For the *small-signal input resistance* we obtain:

$$r_i = \left.\frac{v_i}{i_i}\right|_{i_o=0} = (r_{BE} + R_{BV}) \| \frac{R_C + r_{CE}}{1 + \dfrac{\beta r_{CE}}{r_{BE} + R_{BV}}}$$

$$\overset{\substack{\beta \gg 1 \\ r_{CE} \gg R_C \\ \beta r_{CE} \gg r_{BE}+R_{BV}}}{\approx} \frac{1}{g_m} + \frac{R_{BV}}{\beta} \overset{r_{BE} \gg R_{BV}}{\approx} \frac{1}{g_m}$$

It depends on the load resistance. Here, r_i is the *open-circuit input resistance* because of $i_o = 0$ ($R_L \to \infty$). The input resistance for other values of R_L is calculated by replacing R_C with R_C and R_L connected in parallel; the *short-circuit input resistance* is obtained by setting $R_L = R_C = 0$. However, the influence of R_L is so insignificant that it is eliminated by the approximation. For the sample operating point we obtain *exactly* $r_i = 13.2\,\Omega$; the approximation is $r_i = 12.9\,\Omega$.

The *small-signal output resistance* is:

$$r_o = \frac{v_o}{i_o} = R_C \| r_{CE} \left(1 + \frac{R_g}{r_{CE}} \frac{\beta r_{CE} + r_{BE} + R_{BV}}{r_{BE} + R_{BV} + R_g}\right)$$

$$\overset{\beta \, r_{CE} \gg r_{BE} + R_{BV}}{\approx} \quad R_C \| r_{CE} \left(1 + \frac{\beta R_g}{r_{BE} + R_{BV} + R_g}\right)$$

$$\overset{r_{CE} \gg R_C}{\approx} \quad R_C$$

This depends on the source resistance R_g of the signal generator. For $R_g = 0$ the *short-circuit output resistance* is

$$r_{o,s} = R_C \| r_{CE}$$

and for $R_g \to \infty$ the *open-circuit input resistance* is:

$$r_{o,o} = R_C \| r_{CE} (1 + \beta) \approx R_C \| \beta \, r_{CE}$$

Since in most practical cases $r_{CE} \gg R_C$, the dependence on R_g can be neglected. In our example we obtain $r_{o,s} = 976\,\Omega$ and $r_{o,o} = 999.94\,\Omega$; the approximation is $r_o = R_C = 1\,\text{k}\Omega$.

For $r_{CE} \gg R_C, \beta r_{CE} \gg r_{BE} + R_{BV}, \beta \gg 1$ and without load resistance R_L we obtain, for the common-base circuit:

Common-base circuit ————

$$A = \left.\frac{v_o}{v_i}\right|_{i_o=0} \approx \frac{\beta R_C}{r_{BE} + R_{BV}} \overset{r_{BE} \gg R_{BV}}{\approx} g_m R_C \qquad (2.135)$$

$$r_i = \frac{v_i}{i_i} \approx \frac{1}{g_m} + \frac{R_{BV}}{\beta} \overset{r_{BE} \gg R_{BV}}{\approx} \frac{1}{g_m} \qquad (2.136)$$

$$r_o = \frac{v_o}{i_o} \approx R_C \qquad (2.137)$$

From a comparison of (2.135)–(2.137) and (2.61)–(2.63) it is clear that the small-signal response of the common-base circuit and the common-emitter circuit without feedback is similar. This similarity is due to the fact that in both circuits the signal generator is connected between the base and the emitter of the transistor and the output signal is taken from the collector. The input circuits are identical when V_g and R_g in Fig. 2.57a on page 97 are replaced by V_i and R_{BV} in Fig. 2.107a and the different polarity of the signal generator is not taken into consideration. The values of the gain are approximately the same but the signs are reversed because of the different polarity of the signal generator. The output resistance is also the same with the exception of a somewhat different influence of r_{CE}. In the common-base circuit the input resistance is lower by approximately the factor β, because instead of the base current i_B it is the emitter current $i_E = -(1+\beta)i_B \approx -\beta i_B$ that serves as the input current. Due to the similarity, the equivalent diagram of the common-emitter circuit shown in Fig. 2.61 on page 101 with the equivalent values A, r_i and r_o can also be used for the common-base circuit.

When a current source is used to drive the circuit, the *transfer resistance* R_T (the *transimpedance*) takes the place of the gain:

$$R_T = \left.\frac{v_o}{i_i}\right|_{i_o=0} = \left.\frac{v_o}{v_i}\right|_{i_o=0} \left.\frac{v_i}{i_i}\right|_{i_o=0}$$

$$= Ar_i = \frac{(\beta \, r_{CE} + r_{BE} + R_{BV}) \, R_C}{(1 + \beta) \, r_{CE} + r_{BE} + R_{BV} + R_C}$$

For $\beta \gg 1$, $r_{CE} \gg R_C$ and $\beta r_{CE} \gg r_{BE} + R_{BV}$, we obtain for the current–voltage converter in the common-base configuration.

> *Current-to-voltage converter in*
> *the common-base configuration*
>
> $$R_T = \left. \frac{v_o}{i_i} \right|_{i_o=0} \approx R_C \tag{2.138}$$

The input and output resistances are according to (2.136) and (2.137).

Non-linearity: If resistance R_{BV} is sufficiently small and a voltage source is used to drive the circuit, then $V_i = -V_{BE}$ (see (2.132)). It follows that $\hat{v}_{BE} \approx \hat{v}_i$ can be used as well as equation (2.15) on page 45 which describes the relation between the amplitude $\hat{v}_{BE}$ of sinusoidal signals and the *distortion factor k* of the collector current, which in the case of a common-base circuit is the same as the distortion factor of the output voltage. Therefore $\hat{v}_i < k \cdot 0.1 \, \text{V}$; that is, if $k < 1\,\%$ then $\hat{v}_i < 1 \, \text{mV}$. Due to $\hat{v}_o = |A| \hat{v}_i$, the corresponding output amplitude depends on the gain A; from the numeric example $A = 76$ it follows that $\hat{v}_o < k \cdot 7.6 \, \text{V}$. When a current source is used to drive the circuit, the distortion factor is very small because of the virtually linear relation between $I_i = I_E$ and I_C.

Temperature sensitivity: According to (2.21) on page 54 the base-emitter voltage V_{BE} decreases at a rate of $1.7 \, \text{mV/K}$ if the collector current I_C is constant. If resistance R_{BV} is sufficiently small and the circuit is driven by a voltage source $V_i \approx -V_{BE}$ (see (2.132)), the input voltage must increase at a rate of $1.7 \, \text{mV/K}$ in order to keep the operating point $I_C = I_{C,A}$ of the circuit constant. If, on the other hand, the input voltage is kept constant, then a temperature increase has the same effect as a decrease in the input voltage by $dV_i/dT = -1.7 \, \text{mV/K}$; the *temperature drift* of the output voltage can thus be calculated by:

$$\left. \frac{dV_o}{dT} \right|_A = \left. \frac{\partial V_o}{\partial V_i} \right|_A \frac{dV_i}{dT} \approx -A \cdot 1.7 \, \text{mV/K}$$

For the numeric example we obtain $(dV_o/dT)|_A \approx -129 \, \text{mV/K}$.

If a current source is used to drive the circuit, then from (2.133) it follows that:

$$\left. \frac{dV_o}{dT} \right|_A = -R_C \left. \frac{dI_C}{dT} \right|_A = -R_C \left(\frac{I_{C,A}}{(1+B) \, B} \frac{dB}{dT} + \frac{B}{1+B} \frac{dI_{i,A}}{dT} \right)$$

In the numeric example, and with an input current independent of temperature, we obtain from (2.23) a temperature drift of $(dV_o/dT)|_A \approx -31 \, \mu\text{V/K}$; in this case only the temperature sensitivity of the current gain B has an effect.

Setting the Operating Point

Operation as a small-signal amplifier requires a stable operating point setting; in this respect, we distinguish between *AC coupling* and *DC coupling*.

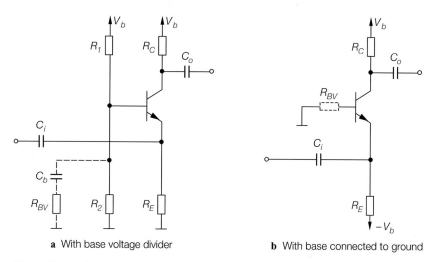

a With base voltage divider   **b** With base connected to ground

Fig. 2.111. Setting the operating point by AC voltage coupling

Setting the operating point by AC coupling: Figure 2.111 shows two versions of AC coupling in which the signal source and the load are connected through coupling capacitors; the other properties are described on page 115. In both versions the operating point is set by the DC current feedback, which is used in the same way in the common-emitter circuit (see Fig. 2.75 on page 117).

In the circuit shown in Fig. 2.111a, the voltage required at the base of the transistor

$$V_{B,A} = \left(I_{C,A} + I_{B,A}\right) R_E + V_{BE,A} \approx I_{C,A} R_E + 0.7\,\text{V}$$

is adjusted with R_1 and R_2; the current through the resistors is chosen to be significantly higher than $I_{B,A}$ to prevent the operating point from being influenced by $I_{B,A}$. The temperature stability of the operating point depends to a high degree on the ratio of the resistances R_C and R_E:

$$\left.\frac{dV_o}{dT}\right|_A \approx -\frac{R_C}{R_E} \cdot 1.7\,\frac{\text{mV}}{\text{K}}$$

R_E must be selected to be as large as possible to minimize the temperature drift; ratios used in practical applications are $R_C/R_E \approx 1\ldots10$. In the small-signal equivalent circuit R_E is connected in parallel to the input resistance r_i, but can be neglected because $R_E \gg r_i = 1/g_m$. The parallel connection of R_1 and R_2 replaces resistance R_{BV} of Fig. 2.107a:[32]

$$R_{BV} = R_1 \parallel R_2$$

The maximum gain is achieved only with a low-resistance base circuit; (2.135) requires the condition $R_{BV} \ll r_{BE}$. In practice, it is seldom possible to make R_1 and R_2 small enough to meet this requirement, since otherwise the current through R_1 and R_2 is too high.

---

[32] In Fig. 2.107a the base connection of the transistor is connected to earth via resistance R_{BV}; R_{BV} may be regarded as the internal resistance of a voltage source with $V = 0$. The equivalent voltage source for the base voltage divider in Fig. 2.111a features the internal resistance $R_1 \parallel R_2$ and the open-circuit voltage $V = V_b R_2/(R_1 + R_2)$.

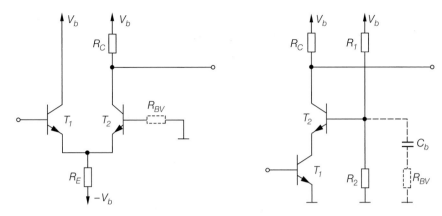

Fig. 2.112. Setting the operating point by DC coupling

Example: If $I_{C,A} = 1\,\text{mA}$ and $\beta = 400$, then $R_{BV} \ll r_{BE} = 10.4\,\text{k}\Omega$; if we select $R_1 = 3\,\text{k}\Omega$ and $R_2 = 1.5\,\text{k}\Omega$ – that is, $R_{BV} = 1\,\text{k}\Omega$ – then for $V_b = 5\,\text{V}$ we obtain a current that is higher than $I_{C,A}$: $I_Q = V_b/(R_1 + R_2) \approx 1.1\,\text{mA}$. However, the requirement that the current must be significantly higher than the base current is already met for $I_Q = 25\,\mu\text{A}$ because $I_{B,A} = I_{C,A}/\beta = 2.5\,\mu\text{A}$.

Therefore, the current must *only* be significantly higher than the base current and the requirement for a low-resistance base circuit is met for AC voltages only by putting a capacitor C_b between the base connection and ground (see Fig. 2.111a);[33] C_b must be selected so that $1/(2\pi f_L C_b) \ll r_{BE}$ is still the case with the smallest signal frequency of interest f_L.

If there is an additional negative supply voltage, the base connection of the transistor can be connected directly to earth (see Fig. 2.111b), and the operating point can be adjusted with R_E:

$$I_{C,A} \approx -I_{E,A} = \frac{V_b - V_{BE,A}}{R_E} \approx \frac{V_b - 0.7\,\text{V}}{R_E}$$

In both versions resistance R_E can be replaced by a current source with current I_0; then $I_{C,A} \approx I_0$. In this case the temperature drift is determined by the temperature drift of the current source.

Setting the operating point by DC coupling: Figure 2.112 shows two versions of DC coupling. In Fig. 2.112a the common-base circuit (T_2) is driven by a common-collector circuit (T_1); this is in fact a voltage control because the common-collector circuit has a low output resistance. The operating point current $I_{C,A}$ is the same in both transistors and, as shown, is adjusted by resistance R_E or a current source. The circuit can be regarded as an asymmetrically operated differential amplifier as can be seen from a comparison with Fig. 4.54c on page 330.

Figure 2.112b shows a *cascode circuit* in which one transistor (T_2) in common-base configuration is driven by a common-emitter circuit (T_1); this is a current control situation. The operating point of the common-base circuit is determined by resistors R_1 and R_2 and

---

[33] Figure 2.111a shows an *additional* resistance R_{BV} to prevent high-frequency oscillations; this will be explained later.

by the operating point of the common-emitter circuit. In Fig. 2.112b the common-emitter circuit is shown symbolically only, since the circuitry required to set the operating point is missing. The cascode circuit is described in more detail in Sect. 4.1.2.

Preventing high-frequency oscillations: Due to the high upper cutoff frequency high-frequency oscillations may occur at the operating point; the circuit then operates as an oscillator. This phenomenon occurs especially in circuits where the base of the transistor is connected to ground either directly or via a capacitor C_b. The reason for this is a parasitic inductance in the base circuit that is caused by transfer time effects in the base region of the transistor and by conductor inductancees. Together with the input capacitance of the transistor and/or the capacitor C_b this parasitic inductance forms a series resonant circuit which, given sufficiently high quality, may run the risk of self-excitation. In order to prevent this the quality of the resonant circuit must be reduced by adding a dumping resistor. This is the purpose of R_{BV}, which is drawn using dotted lines in Figs. 2.111 and 2.112. The resistors used in practical applications are in the range of $10 \ldots 100\,\Omega$ or even higher in exception cases. They should be connected with short leads to ground to keep the inductance low.

Frequency Response and Upper Cut-Off Frequency

Small-signal gain A and operating gain A_B decline with increasing frequency due to the transistor capacitances. Any information obtained on the frequency response and the upper cutoff frequency calculations must be based on the dynamic small-signal model of the transistor.

Voltage source as input signal: Figure 2.113 shows the dynamic small-signal equivalent diagram for the common-base circuit driven by a signal voltage source with source resistance R_g. The accurate calculation of the *operating gain* $\underline{A}_B(s) = \underline{v}_o(s)/\underline{v}_g(s)$ is difficult and leads to complicated expressions. A sufficiently accurate approximation is reached by neglecting resistance r_{CE} and assuming $\beta \gg 1$; for $R'_{BV} = R_{BV} + R_B$, $R'_C = R_C \| R_L$ and the low-frequency gain

$$A_0 = \underline{A}_B(0) \approx \frac{\beta R'_C}{\beta R_g + R'_{BV} + r_{BE}} \tag{2.139}$$

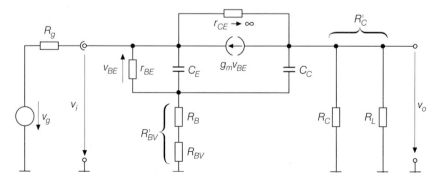

Fig. 2.113. Dynamic small-signal equivalent circuit of the common-base configuration

it follows that:

$$\underline{A}_B(s) \approx A_0 \, \frac{1 + sC_C R'_{BV} + s^2 \dfrac{C_E C_C R'_{BV}}{g_m}}{1 + sc_1 + s^2 c_2}$$

$$c_1 = \frac{C_E r_{BE} \left(R_g + R'_{BV}\right) + C_C \left(R'_{BV} \left(\beta \left(R_g + R'_C\right) + r_{BE}\right) + R'_C \left(\beta R_g + r_{BE}\right)\right)}{\beta R_g + R'_{BV} + r_{BE}}$$

$$c_2 = \frac{C_E C_C \left(R'_{BV} \left(R_g + R'_C\right) + R_g R'_C\right)}{\beta R_g + R'_{BV} + r_{BE}}$$

The transfer function has two real poles and two zeros; the latter are complex conjugate in most cases. The frequency response can be described approximately by a lowpass filter of first order by removing the s^2 terms and subtracting the linear terms:

$$\underline{A}_B(s) \approx \frac{A_0}{1 + s \, \dfrac{C_E r_{BE} \left(R_g + R'_{BV}\right) + C_C R'_C \left(\beta \left(R_g + R'_{BV}\right) + r_{BE}\right)}{\beta R_g + R'_{BV} + r_{BE}}} \tag{2.140}$$

This leads to an approximation for the upper $-3dB$ *cutoff frequency* f_{-3dB} at which the magnitude of the gain has dropped by 3 dB:

$$\omega_{-3dB} \approx \frac{\beta R_g + R'_{BV} + r_{BE}}{C_E r_{BE} \left(R_g + R'_{BV}\right) + C_C R'_C \left(\beta \left(R_g + R'_{BV}\right) + r_{BE}\right)} \tag{2.141}$$

The upper cutoff frequency depends on the low-frequency gain A_0; from (2.139) and (2.141) an expression with two time constants independent of A_0 is derived:

$$\omega_{-3dB}(A_0) \approx \frac{1}{T_1 + T_2 A_0} \tag{2.142}$$

$$T_1 = C_E \, \frac{r_{BE} \left(R_g + R'_{BV}\right)}{\beta R_g + R'_{BV} + r_{BE}} \tag{2.143}$$

$$T_2 = C_C \left(R_g + R'_{BV} + \frac{1}{g_m}\right) \tag{2.144}$$

Here, too, there is a close similarity to the common-emitter circuit as can be seen from a comparison of (2.142)–(2.144) and (2.91)–(2.93). The information regarding the gain bandwidth product *GBW* including (2.94) outlined on page 125 applies here as well.

If the load comprises both a resistive and a capacitive component, which means that there is a load resistance R_L and a load capacitance C_L in parallel, then

$$T_2 = (C_C + C_L) \left(R_g + \frac{1}{g_m}\right) + \left(C_C + \frac{C_L}{\beta}\right) R'_{BV} \tag{2.145}$$

The time constant T_1 does not depend on C_L. The upper cutoff frequency decreases with an increase in T_2.

In an approximation the common-base circuit can be described by the equivalent circuit shown in Fig. 2.85 on page 126. The *input capacitance* C_i and the *output capacitance* C_o

are obtained on the basis of the requirement that a calculation of $\underline{A}_B(s)$ after eliminating the s^2 term must result in (2.140):

$$C_i \approx C_E \frac{r_{BE}\left(R_g + R'_{BV}\right)}{R_g\left(r_{BE} + R'_{BV}\right)} \overset{R'_{BV} \ll R_g,\, r_{BE}}{\approx} C_E$$

$$C_o \approx C_C \frac{\beta\left(R_g + R'_{BV}\right) + r_{BE}}{\beta R_g + R'_{BV} + r_{BE}} \overset{R'_{BV} \ll R_g,\, r_{BE}}{\approx} C_C$$

A, r_i and r_o are determined from (2.135)–(2.137); $R'_{BV} = R_{BV} + R_B$ is used instead of R_{BV}.

Current source as input signal: When a current source is used to drive the circuit it is interesting to know the frequency response of the *transimpedance* $\underline{Z}_T(s)$. On the basis of (2.140) one can describe an approximation by a lowpass filter of first order:

$$\underline{Z}_T(s) = \frac{\underline{v}_o(s)}{\underline{i}_i(s)} = \lim_{R_g \to \infty} R_g \underline{A}_B(s) \approx \frac{R'_C}{1 + s\left(\dfrac{C_E}{g_m} + C_C R'_C\right)} \tag{2.146}$$

The upper cutoff frequency is:

$$\omega_{\text{-3dB}} = 2\pi f_{\text{-3dB}} \approx \frac{1}{\dfrac{C_E}{g_m} + C_C R'_C} \tag{2.147}$$

This result is obtained from (2.141) if $R_g \to \infty$ is assumed. In the event of a capacitive load, C_C must be replaced by $C_L + C_C$.

Comparison with the common-emitter circuit: A comparison of the common-base and the common-emitter circuits is easiest with the equivalent circuits shown in Fig. 2.114; they follow from Fig. 2.85 after inserting the simplified expressions for A_0, r_i, C_i, r_o and C_o. Both circuits are identical on the output side; with the exception of the sign the open-circuit gain is also the same. However, major differences exist in the input circuit. With the common-base circuit both the input resistance and the input capacitance are lower and the latter does not depend on the gain. It follows that the common-base circuit has a much lower input time constant $T_i = C_i r_i$, while the time constant $T_o = C_o r_o = C_C R_C$ on the output side is identical in both circuits. Therefore, the upper cutoff frequency is higher in the common-base circuit, particularly if the time constant at the output side is low and the cutoff frequency depends mainly on the time constant on the input side.

Example: Using the numeric example for the common-base circuit in Fig. 2.107a $I_{C,A} = 2.5\,\text{mA}$ was chosen. For $\beta = 400$, $C_{obo} = 3.5\,\text{pF}$ and $f_T = 160\,\text{MHz}$ we obtain from Fig. 2.45 on page 83 the small-signal parameters $g_m = 96\,\text{mS}$, $r_{BE} = 4160\,\Omega$, $C_C = 3.5\,\text{pF}$ and $C_E = 92\,\text{pF}$. For $R_{BV} = R_C = 1\,\text{k}\Omega$, $R'_{BV} \approx R_{BV}$, $R_L \to \infty$ and $R_g = 0$ it follows from (2.139) that $A_0 \approx 77.5$ and from (2.141) $f_{-3dB} \approx 457\,\text{kHz}$. The comparably low upper cutoff frequency is caused by the resistance R_{BV}. A much higher upper cutoff frequency can be achieved by making R_{BV} smaller or by removing it, as long as this does not cause high-frequency oscillations; this leads to $R'_{BV} \approx R_B$. For $R_B = R_g = 10\,\Omega$, it

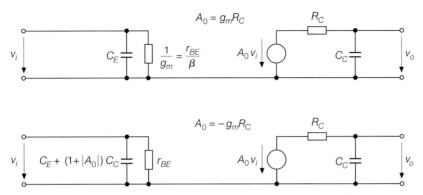

Fig. 2.114. Equivalent circuit of the common-base circuit (top) and the common-emitter circuit (bottom)

follows from (2.139) that $A_0 \approx 49$ and from (2.141) $f_{-3dB} \approx 25.9$ MHz. From (2.143) it follows that $T_1 \approx 0.94$ ns, from (2.144) $T_2 \approx 107$ ps and from (2.94) $GBW \approx 1.5$ GHz. These values are heavily dependent on R_B; if $R_B = 100\,\Omega$ then $A_0 \approx 48$, $f_{-3dB} \approx 6.2$ MHz, $T_1 \approx 5.1$ ns, $T_2 \approx 421$ ps and $GBW = 378$ MHz. For a load capacitance $C_L = 1$ nF and $R_B = 10\,\Omega$ we obtain from (2.145) $T_2 \approx 20.5$ ns, from (2.142) $f_{-3dB} \approx 158$ kHz and from (2.94) $GBW \approx 7.74$ MHz.

If the circuit is driven by a current source and $R_L \rightarrow \infty$, then from (2.146) we obtain $R_T = Z_T(0) \approx R_C = 1$ kΩ and from (2.147) $f_{-3dB} = 35.7$ MHz. In this case, resistance R_{BV} has no effect. For a load capacitance $C_L = 1$ nF we obtain from (2.147) $f_{-3dB} \approx 159$ kHz when C_C is replaced by $C_C + C_L$.

2.4.4
Darlington Circuit

In some applications the current gain of a single transistor is not sufficient; in such cases a Darlington circuit can be used. This consists of two transistors and provides a current gain that is approximately equal to the product of the current gains of the two individual transistors:

$$B \approx B_1 B_2 \tag{2.148}$$

Under the name *Darlington transistor* the Darlington circuit is available as a component in its own case for installation on a circuit board; the connections are called the base, emitter and collector as for a discrete transistor. Of course, the Darlington circuit can be made up of discrete components. Here, the Darlington transistor is an integrated circuit that consists solely of one Darlington circuit.

Figure 2.115 shows the circuit and the graphic symbol for an npn *Darlington transistor* that consists of two npn transistors and a resistor to improve the switching performance. In general, it can be used like an npn transistor. The *pnp Darlington transistor*, which can be used as a pnp transistor, comes in two versions (see Fig. 2.116):

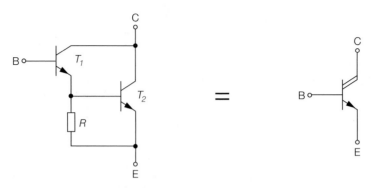

Fig. 2.115. Circuit and graphic symbol of the npn Darlington transistor

– The *normal* pnp Darlington consists of two pnp transistors and is directly complementary to the npn Darlington. It is commonly known as the *pnp Darlington*; that is, without the word *normal*.
– The *complementary* pnp Darlington consists of a pnp and an npn transistor and is indirectly complementary to the npn Darlington as the pnp transistor T_1 determines the polarity; the npn transistor T_2 is only used for additional current amplification.

The current gain of a pnp Darlington is often much lower than that of a comparable npn Darlington because the current gain of the pnp transistor is usually less than that of an npn transistor, which is squared by the Darlington due to the multiplication of the two individual gains. The solution can be the complementary pnp Darlington in which the second pnp transistor is replaced by an npn transistor; thus only one pnp transistor provides the lower current gain.

The following sections describe the npn Darlington, which is used more widely in practice. However, the explanations also apply to the pnp Darlington when the signs of all the currents and voltages are reversed. An exception is the complementary pnp Darlington, which is described separately.

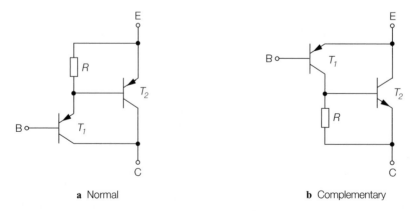

a Normal   **b** Complementary

Fig. 2.116. Circuit of a pnp Darlington transistor

Characteristics of a Darlington Transistor

Figure 2.117 shows the family of output characteristics of an npn Darlington transistor. The diagram looks quite similar to that of an npn transistor, except that the collector–emitter saturation voltage $V_{CE,sat}$ of 0.7…1 V, which is the knee-point of the characteristics, is much higher. For $V_{CE} > V_{CE,sat}$ both T_1 and T_2, and thus the Darlington, operate in normal mode. For $V_{CE} \leq V_{CE,sat}$ transistor T_1 enters the saturation region, while T_2 remains in normal mode; in the Darlington this is also called the saturation mode.

Figure 2.118 shows the region of small collector currents and low collector–emitter voltages. With very small collector currents the voltage across resistance R of the Darlington is so low that T_2 is nonconductive (the bottom characteristic in Fig. 2.118); in this region the current gain corresponds to the current gain of T_1. With a rising collector current T_2 becomes conductive and the current gain increases rapidly; in Fig. 2.118 this is indicated by the fact that a constant increase in I_B produces an ever stronger increase in I_C.

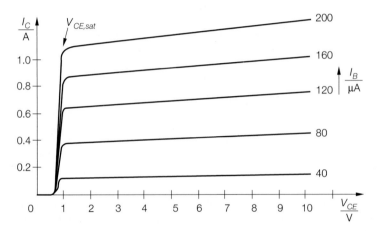

Fig. 2.117. Family of output characteristics of an npn Darlington transistor

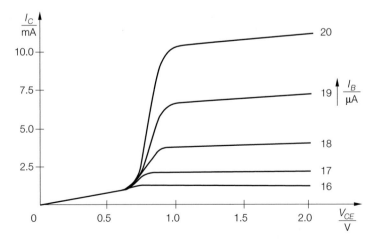

Fig. 2.118. Family of output characteristics for low collector currents

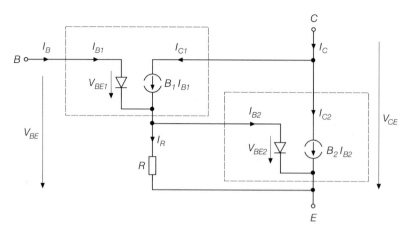

Fig. 2.119. Equivalent circuit of an npn Darlington transistor in normal mode

The family of output characteristics of the pnp Darlington is obtained by inversing the signs. This also applies to the complementary pnp Darlington as there is practically no difference in the family of output characteristics of the two pnp versions. However, there are differences in the input characteristics since the base–emitter junction of the npn and the pnp Darlington consists of two base–emitter transistor junctions, while there is only one in the complementary pnp Darlington; as a result the base-emitter voltage of the complementary pnp Darlington reaches only half the value of that of the normal pnp Darlington at the same current.

Description by Equations

Figure 2.119 shows the equivalent circuit of an npn Darlington transistor in normal mode that combines the equivalent circuits of both transistors with the additional resistance R. The currents are

$$
\begin{aligned}
I_C &= I_{C1} + I_{C2} \\
I_{C1} &= B_1 I_{B1} = B_1 I_B \\
I_{C2} &= B_2 I_{B2} = B_2 (I_{C1} + I_B - I_R)
\end{aligned}
\tag{2.149}
$$

and the base emitter voltage is:

$$
V_{BE} = V_{BE1} + V_{BE2} = V_T \left(\ln \frac{I_{C1}}{I_{S1}} + \ln \frac{I_{C2}}{I_{S2}} \right) = V_T \ln \frac{I_{C1} I_{C2}}{I_{S1} I_{S2}}
$$

Here, I_{S1} and I_{S2} are the saturation reverse currents of T_1 and T_2; in most cases their ratio is $I_{S2} \approx 2...3 I_{S1}$. Medium collector currents produce $V_{BE} \approx 1.2...1.5\,\text{V}$.

Behavior of the Current Gain

Figure 2.120 plots the current gain B versus the collector current I_C. Four regions can be distinguished [2.8]:

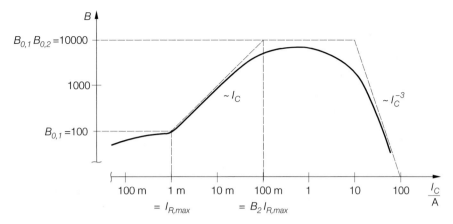

Fig. 2.120. Plot of the current gain of a Darlington transistor

– With small collector currents T_2 is nonconductive and we obtain:[34]

$$B = \frac{I_C}{I_B} = \frac{I_{C1}}{I_{B1}} = B_1 \approx B_{0,1}$$

In this region the current gain of the Darlington corresponds to the current gain of T_1. The borderline for this region can easily be drawn if it is assumed that $V_{BE2} \approx 0.7\,\text{V}$ when T_2 is conductive; the following current flows through resistance R:

$$I_{R,max} \approx \frac{0.7\,\text{V}}{R}$$

Consequently T_2 is nonconductive if $I_C < I_{R,max}$.

– For $I_C > I_{R,max}$ both transistors are conductive; for $I_R = I_{R,max}$ from (2.149) it follows that:

$$I_B = \frac{I_C + B_2 I_{R,max}}{(1 + B_1) B_2 + B_1}$$

Consequently

$$B(I_C) = \frac{I_C}{I_B} = \frac{(1 + B_1) B_2 + B_1}{1 + \dfrac{B_2 I_{R,max}}{I_C}}$$

$$\overset{B_1,B_2 \gg 1}{\approx} \frac{B_1 B_2}{1 + \dfrac{B_2 I_{R,max}}{I_C}} \tag{2.150}$$

This equation describes two regions. If $I_{R,max} < I_C < B_2 I_{R,max}$, then

$$B \approx \frac{B_1 I_C}{I_{R,max}} \approx \frac{B_{0,1} I_C}{I_{R,max}}$$

---

[34] The current gain B_1 and B_2 depend on I_{C1} and I_{C2}, respectively, and thus on I_C; in Fig. 2.120 this dependence is taken into consideration, but is neglected in the calculations due to the assumption that $B_1 \approx B_{0,1}$ and $B_2 \approx B_{0,2}$; that is, B_1 and B_2 are assumed to be constant. This is not the case in the high-current region which is described separately.

In this region the current gain is approximately proportional to the collector current. This property is caused by resistance R as in this region a predominant portion of the collector current I_{C1} flows through resistance R and only a small portion is available as base current for T_2. However, an increase in I_{C1} causes a corresponding increase in I_{B2}, because the current through the resistor R remains approximately constant due to $I_R \approx I_{R,max}$.

- For $I_C > B_2 I_{R,max}$ it follows from (2.150) that:

$$B \approx B_1 B_2 \approx B_{0,1} B_{0,2}$$

This corresponds to (2.148), which was mentioned above. This region is the preferred operating range for the Darlington transistor.

- With a further increase in the collector current, both transistors – first T_2 and then T_1 – enter the high-current region. For

$$B_1 = \frac{B_{0,1}}{1 + \dfrac{I_{C1}}{I_{K,N1}}} \quad , \quad B_2 = \frac{B_{0,2}}{1 + \dfrac{I_{C2}}{I_{K,N2}}}$$

it follows that:

$$B(I_C) = \frac{B_{0,1} B_{0,2}}{1 + \dfrac{I_C}{I_{K,N2}} + \dfrac{I_C}{I_{K,N1} B_{0,2}} \left(1 + \dfrac{I_C}{I_{K,N2}}\right)^2}$$

Here, $I_{K,N1}$ and $I_{K,N2}$ are the knee-point currents for strong injection of T_1 and T_2; in most cases the ratio is $I_{K,N2} \approx 2...3 I_{K,N1}$. In the high-current region the current gain drops rapidly; this becomes obvious when a limit value condition is examined [2.8]:

$$\lim_{I_C \to \infty} B(I_C) = \frac{B_{0,1} I_{K,N1} B_{0,2}^2 I_{K,N2}^2}{I_C^3}$$

In the Darlington the current gain declines at a rate of $1/I_C^3$ for large currents, while for the discrete transistor, it declines at a rate of $1/I_C$.

Small-Signal Response

In order to determine the small-signal response of the Darlington transistor at an operating point A it is also necessary to know the operating point currents $I_{B,A}$ and $I_{C,A}$, and the *internal currents* $I_{C1,A}$ and $I_{C2,A}$, which means that the distribution of the collector current must be known; this gives us the small-signal parameters of the two transistors:

$$g_{m1/2} = \frac{I_{C1/2,A}}{V_T} \quad , \quad r_{BE\,1/2} = \frac{\beta_{1/2}}{g_{m1/2}} \quad , \quad r_{CE1/2} = \frac{V_{A1/2}}{I_{C1/2,A}}$$

The Early voltages are usually about the same so that an Early voltage $V_A \approx V_{A1} \approx V_{A2}$ can be used for calculations. An operating point in the region of large current gain is selected; here $I_{C2,A} \gg I_{C1,A}$, and the approximation $I_{C2,A} \approx I_{C,A}$ can be used, which means that the collector current of the Darlington flows almost completely through T_2.

The upper part of Fig. 2.121 shows the complete small-signal equivalent circuit of a Darlington transistor; it describes the npn and pnp Darlingtons but not the complementary pnp Darlington. However, this comprehensive equivalent diagram is rarely used since, due to its similarity with a discrete transistor, the Darlington can be described with sufficient

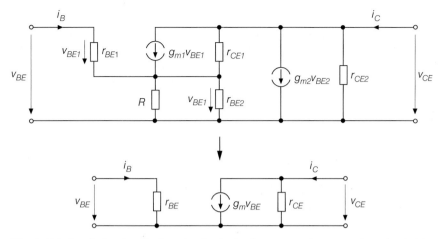

Fig. 2.121. Small-signal equivalent circuit of a Darlington transistor: complete circuit (above) and simplified circuit (below)

accuracy by the equivalent circuit of a discrete transistor (see Fig. 2.121); the parameters g_m, r_{BE} and r_{CE} can be determined either from the characteristics or by conversion using the complete equivalent circuit.[35] For $\beta_1, \beta_2 \gg 1$ the conversion of the parameters provides

$$g_m \approx g_{m1} \frac{1 + g_{m2}(r_{BE2} \| R)}{1 + g_{m1}(r_{BE2} \| R)} \overset{R \gg r_{BE2}}{\approx} \frac{g_{m2}}{2}$$

$$r_{BE} \approx r_{BE1} + \beta_1 (r_{BE2} \| R) \overset{R \gg r_{BE2}}{\approx} 2\, r_{BE1}$$

$$r_{CE} \approx r_{CE2} \| r_{CE1} \frac{1 + g_{m1}(r_{BE2} \| R)}{1 + g_{m2}(r_{BE2} \| R)} \overset{R \gg r_{BE2}}{\approx} \frac{2}{3} r_{CE2}$$

The small-signal current gain is:

$$\beta = g_m r_{BE} \approx \beta_1 \beta_2 \frac{R}{r_{BE2} + R} \overset{R \gg r_{BE2}}{\approx} \beta_1 \beta_2 \qquad (2.151)$$

The requirement $R \gg r_{BE2}$ is fulfilled when the current through resistance R can be neglected because of $I_{B2} \gg I_R$; consequently:

$$I_{C2,A} \approx I_{C,A} \quad , \quad I_{C1,A} \approx \frac{I_{C,A}}{B_2}$$

For this purpose the Darlington must be operated in the region of maximum power gain B; that is, the condition $I_{C,A} \gg B_2 I_{R,max}$ must be met (see Fig. 2.120). The following equations apply to the region of maximum current gain of the *Darlington transistor*:

---

[35] This is not an equivalent transformation, as the conversion introduces an additional resistance between base and collector, although this can be neglected.

Darlington transistor

$$g_m \approx \frac{g_{m2}}{2} \approx \frac{1}{2}\frac{I_{C,A}}{V_T} \tag{2.152}$$

$$r_{BE} = \frac{\beta}{g_m} \approx 2\frac{\beta_1\beta_2 V_T}{I_{C,A}} \tag{2.153}$$

$$r_{CE} \approx \frac{2}{3}r_{CE2} \approx \frac{2}{3}\frac{V_A}{I_{C,A}} \tag{2.154}$$

Similarly, for the complementary pnp Darlington:

$$g_m \approx g_{m1}\left(1 + g_{m2}\left(r_{BE\,2} \,||\, R\right)\right) \quad\overset{R \gg r_{BE\,2}}{\approx}\quad g_{m2}$$

$$r_{BE} = r_{BE\,1}$$

$$r_{CE} = r_{CE2} \,||\, \frac{r_{CE1}}{1 + g_{m2}\left(r_{BE\,2} \,||\, R\right)} \quad\overset{R \gg r_{BE\,2}}{\approx}\quad \frac{1}{2}r_{CE2}$$

Equation (2.151) also applies. For the region of maximum power gain of the complementary Darlington transistor, the following applies:

Complementary Darlington transistor

$$g_m \approx g_{m2} \approx \frac{I_{C,A}}{V_T} \tag{2.155}$$

$$r_{BE} = \frac{\beta}{g_m} \approx \frac{\beta_1\beta_2 V_T}{I_{C,A}} \tag{2.156}$$

$$r_{CE} \approx \frac{1}{2}r_{CE2} \approx \frac{1}{2}\frac{V_A}{I_{C,A}} \tag{2.157}$$

Switching Performance

The Darlington transistor is very often used as a switch; due to the high current gain, it is possible to switch high-load currents with comparably low control currents. Switching off the load is particularly critical: transistor T_1 blocks relatively fast, while transistor T_2 remains conductive until the charge stored in the base is drained through resistance R. Thus a short turn-off time is achieved only with a sufficiently low resistance R (see Fig. 2.122). On the other hand, a small resistance R lowers the current gain. A compromise must therefore be found; Darlingtons used in switching applications use smaller resistances than Darlingtons used for general purposes.

In addition to the two transistors and resistance R, Darlington transistors used in switching applications are provided with three additional diodes; Fig. 2.123 shows the complete circuit diagram of such an npn Darlington. In order to shorten the turn-off time the base current can be inverted; diodes D_1 and D_2 then limit the reverse voltage at the base–emitter junctions. Diode D_3 acts as a free-wheeling diode for inductive loads.

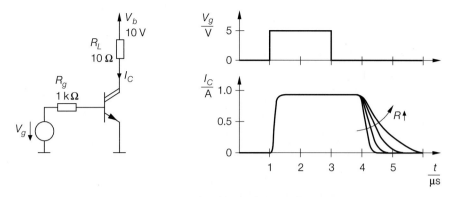

Fig. 2.122. Switching performance of a Darlington transistor

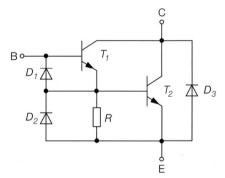

Fig. 2.123. Design of an npn Darlington for switching applications

Chapter 3:
Field Effect Transistor

The field effect transistor (*FET*) is a semiconductor component with three terminals, known as the *gate* (*G*), *source* (*S*) and *drain* (*D*). There are discrete transistors that are used for mounting on printed circuit boards, and are contained in their own housings, and integrated field effect transistors that are produced together with other semiconductor elements on a common *substrate*. Integrated field effect transistors feature a fourth terminal called the substrate or *bulk* (*B*), which results from the common substrate.[1] This terminal also exists internally in discrete transistors, where it is not connected to the outside but to the source terminal.

Mode of operation: In the field effect transistor, a control voltage between the gate and the source is used to control the conductivity of the drain–source junction without the flow of a control current; that is, they are controlled in a Watt-less fashion. Two different effects are utilized:

- In the *MOSFET* (*metal oxide semiconductor FET* or *insulated gate FET, IGFET*) the gate is isolated from the channel by an oxide layer (SiO$_2$) (see Fig. 3.1); this means that the control voltage can have either polarity without a current flowing. The control voltage influences the charge carrier density in the *inversion layer* beneath the gate, which forms a conductive *channel* between the drain and the source, thus enabling current flow. Without the inversion layer, at least one of the pn junctions between the source and the substrate or the drain and the substrate is always reverse-biased and no current can flow. Depending on the doping of the channel, there are "*normally on*" (*depletion*) or "*normally off*" (*enhancement*) MOSFETs; for $V_{GS} = 0$ a drain current flows through depletion MOSFETs but not enhancement MOSFETs. In addition to the gate, the substrate B also has a slight controlling effect; this is described in more detail in Sect. 3.3

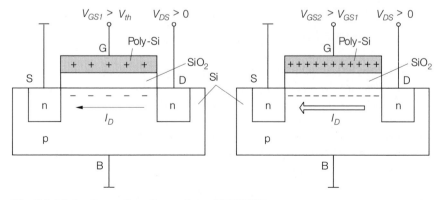

Fig. 3.1. Mode of operation of an n-channel MOSFET

---

[1] In a bipolar transistor, this terminal is called the *substrate* (*S*); since in a field effect transistor *S* stands for source, *bulk* (*B*) is used for the substrate.

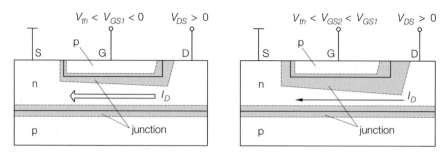

Fig. 3.2. Mode of operation of an n-channel junction FET

– In the *junction FET* (*JFET* or *noninsulated gate FET, NIGFET*) the control voltage influences the junction width of a pn junction operated in reverse mode. This influences the cross-sectional area and thus the conductivity of the channel between the drain and the source (see Fig. 3.2). As the gate is not isolated from the channel, it is possible to use the pn junction in forward mode; but since this mode eliminates the advantage of Watt-less control, it is not used in practice. In the *MESFET* (*metal semiconductor FET*) a metal semiconductor junction (a Schottky junction) is used instead of a pn junction; the functional principle is the same as for normal junction FETs. JFETs and MESFETs are *depletion* FETs; that is, a drain current flows at a control voltage of $V_{GS} = 0$.

It follows from Figs. 3.1 and 3.2 that MOSFETs and junction FETs are generally symmetrical, which means that drain and source may be interchanged. But most discrete FETs are not exactly symmetrical in their design, and in discrete MOSFETs the internal connection between substrate and source makes the terminals noninterchangeable.

Both MOSFETs and junction FETs are available in n-channel and p-channel designs, which means that there are a total of six types of field effect transistors; Fig. 3.3 shows the graphic symbols together with a simplified plot of their characteristics. The polarities stated in Fig. 3.4 are used in normal operation for the voltages V_{GS} and V_{DS}, the drain current I_D and the *threshold voltage* V_{th}.[2]

3.1
Behavior of a Field Effect Transistor

The behavior of a field effect transistor is most easily explained by means of its characteristic curves. They describe the relations between the currents and voltages in the transistor in the event of all values being *static*; in other words, not or only very slowly changing over time. Field effect transistor calculations require simple equations that give a sufficiently accurate description of the behavior of the field effect transistor. In order to check the functionality of a circuit by computer simulation, however, the influence of secondary effects must be taken into consideration. For this purpose, sophisticated models are available that also reflect the *dynamic behavior* when controlled with sinusoidal or pulsed signals. These models are described in Sect. 3.3, but are not needed for a basic understanding. The

---

[2] The threshold voltage V_{th} is usually used in relation to MOSFETs; in the case of junction FETs, the *pinch-off voltage* V_P replaces V_{th}. For the sake of a uniform designation, V_{th} is used for all FETs.

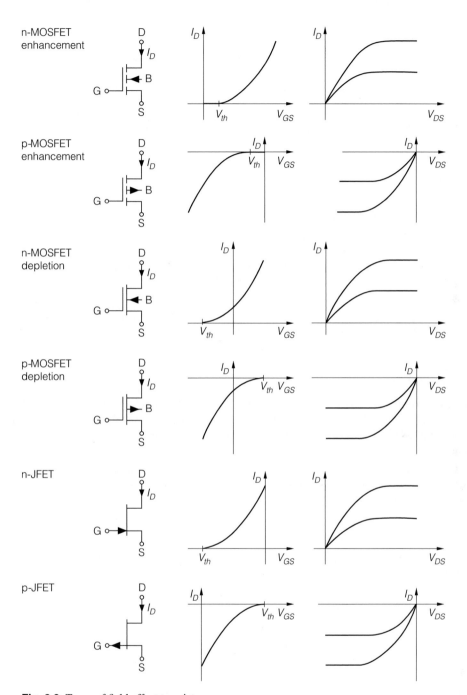

Fig. 3.3. Types of field effect transistors

Type	n-channel	p-channel
MOSFET, enhancement	$V_{th} > 0$ $V_{GS} > V_{th}$ $V_{DS} > 0$ $I_D > 0$	$V_{th} < 0$ $V_{GS} < V_{th}$ $V_{DS} < 0$ $I_D < 0$
MOSFET, depletion	$V_{th} < 0$ $V_{GS} > V_{th}$ $V_{DS} > 0$ $I_D > 0$	$V_{th} > 0$ $V_{GS} < V_{th}$ $V_{DS} < 0$ $I_D < 0$
Junction FET	$V_{th} < 0$ $V_{th} < V_{GS} < 0$ $V_{DS} > 0$ $I_D > 0$	$V_{th} > 0$ $0 < V_{GS} < V_{th}$ $V_{DS} < 0$ $I_D < 0$

Fig. 3.4. Polarity of voltages and currents in normal mode

text below primarily describes the behavior of an n-channel enhancement MOSFET; for p-channel FETs, the polarity of all voltages and currents is reversed.

3.1.1
Characteristic Curves

Family of output characteristics: When one applies different gate–source voltages V_{GS} to an n-channel FET and measures the drain current I_D as a function of the drain–source voltage V_{DS}, one arrives at the family of output characteristics illustrated in Fig. 3.5. These characteristics are basically identical for all n-channel FETs, with the exception of the gate–source voltages V_{GS} for the individual characteristics, which are different for the three n-channel types. A drain current flows only if V_{GS} is higher than the threshold voltage V_{th}. Two regions have to be distinguished:

- When $V_{DS} < V_{DS,po} = V_{GS} - V_{th}$, the FET operates in the *ohmic region* (*triode region*); this designation was chosen because for $V_{DS} = 0$ the characteristics run almost linearly through the origin, so that their behavior is identical to that of an ohmic resistance. When approaching the limiting voltage $V_{DS,po}$, the slope of the characteristics declines until they run almost horizontally when $V_{DS} = V_{DS,po}$.
- If $V_{DS} \geq V_{DS,po}$, then the characteristics are almost horizontal; this region is known as the *pinch-off region* (*saturation region*).[3]

If $V_{GS} < V_{th}$, then no current flows and the FET operates in the *cutoff region*.

Pinch-off region: The pinch-off region of the MOSFET is due to the decline of the charge carrier concentration in the channel, which causes the channel to be *pinched off*; with an increasing voltage V_{DS} this happens first on the drain side, since the voltage between the gate and the channel is the lowest:

---

[3] The term *saturation region* is not a very fortunate choice, since *saturation* has an entirely different meaning in respect to bipolar transistors. The term *pinch-off region* is more neutral and is to be given priority over the term *saturation region*, which is sometimes found in the literature.

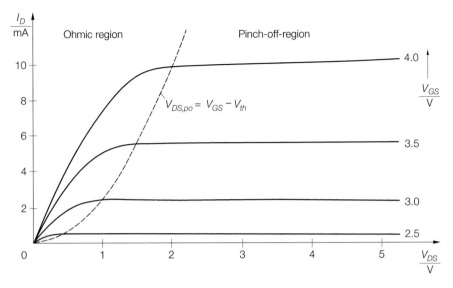

Fig. 3.5. Family of output characteristics of an n-channel field effect transistor

$$V_{GD} = V_{GS} - V_{DS} < V_{GS} \qquad \text{with } V_{DS} > 0$$

The pinch-off effect occurs exactly at the point at which V_{GD} becomes smaller than V_{th}; the borderline between the ohmic region and the pinch-off region can thus be described by:

$$V_{GD} = V_{GS} - V_{DS,po} \equiv V_{th} \quad \Rightarrow \quad V_{DS,po} = V_{GS} - V_{th}$$

A drain current still flows through the channel, as the charge carriers can cross the pinched-off region, but an increase in the voltage V_{DS} only has a slight effect on the nonpinched portion of the channel; thus the drain current remains almost constant. The slight effect of V_{DS} in the pinched-off region is known as *channel-length modulation* and results in a slight increase in the drain current with rising voltage V_{DS}. In the cutoff region the channel is also pinched off at the source side because $V_{GS} < V_{th}$; in this case, the flow of current is no longer possible. Figure 3.6 shows the distribution of the charge carriers in the channel for the three regions.

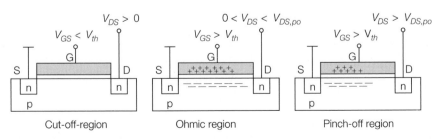

Fig. 3.6. Charge carrier distribution in the channel of the MOSFET

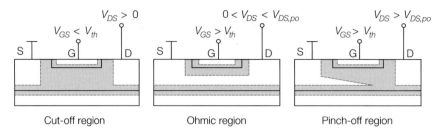

Fig. 3.7. Expansion of the junctions in the junction FET

In the junction FET, the pinch-off effect is due to the fact that the junctions are in contact with one another and close off the channel; with an increasing voltage V_{DS} this happens first on the drain side, because this is where the voltage is highest across the junction. At the border between the ohmic and the pinch-off region we have $V_{DS,po} = V_{GS} - V_{th}$, as with the MOSFET. Here too, the drain current continues, since the charge carriers can cross the pinched-off region. But a further increase of V_{DS} has only a small effect. Figure 3.7 shows the expansion of the junctions in the three regions.

Family of transfer characteristics: In the pinch-off region the drain current I_D essentially depends only on V_{GS}. Plotting I_D for various values of V_{DS} as a function of V_{GS} that belong to the pinch-off region produces the family of transfer characteristics shown in Fig. 3.8. Besides the characteristic of the enhancement MOSFET, the diagram also shows those of the depletion MOSFET and the junction FET; apart from a shift along the V_{GS}-axis they have an identical shape. Due to their small dependence on V_{DS}, the individual characteristics are very close together. No current flows when $V_{GS} < V_{th}$, as in this case the channel is pinched off over its entire length.

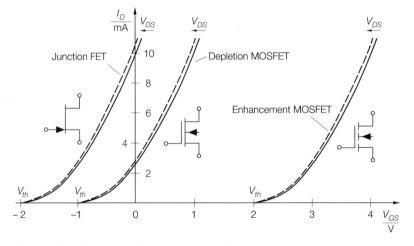

Fig. 3.8. Transfer characteristics of n-channel field effect transistors

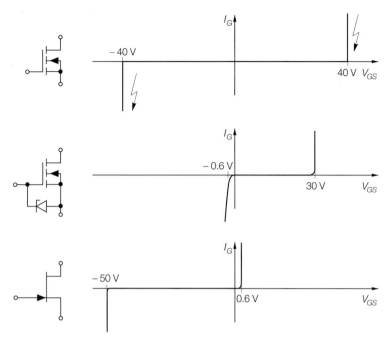

Fig. 3.9. Input characteristics of n-channel field effect transistors

Input characteristics: A complete description also requires the input characteristics as shown in Fig. 3.9, in which the gate current I_G is plotted as a function of V_{GS}. In the normal mode all field effect transistors feature a negligible very small gate current, or none at all. In a MOSFET without over-voltage protection, a gate current only flows when an over-voltage causes an oxide breakthrough; this destroys the MOSFET. Therefore, many MOSFETs protect their gate–source junction against over-voltage using an internal Zener diode, so that the Zener diode characteristic is part of the input characteristic. In the junction FET the pn junction is used in forward mode for $V_{GS} > 0$ and the resulting gate current corresponds to the forward current of a diode; but where $V_{GS} < 0$ the current only starts to flow if the absolute value of the voltage rises to a sufficiently high level to cause a breakthrough of the pn junction.

3.1.2
Description by Equations

On the basis of an ideal charge distribution in the channel, we can calculate the drain current $I_D(V_{GS}, V_{DS})$; the equations for the junction FET and the MOSFET are different, but an approximation by way of a simple equation causes no major errors [3.1]:

$$
I_D = \begin{cases}
0 & \text{for } V_{GS} < V_{th} \\
K V_{DS} \left(V_{GS} - V_{th} - \dfrac{V_{DS}}{2} \right) & \text{for } V_{GS} \geq V_{th},\, 0 \leq V_{DS} < V_{GS} - V_{th} \\
\dfrac{K}{2} (V_{GS} - V_{th})^2 & \text{for } V_{GS} \geq V_{th},\, V_{DS} \geq V_{GS} - V_{th}
\end{cases}
$$

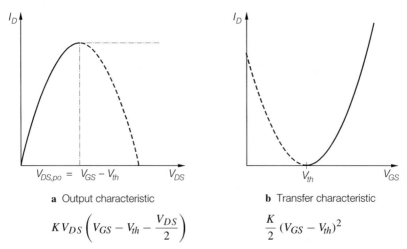

a Output characteristic

$$KV_{DS}\left(V_{GS} - V_{th} - \frac{V_{DS}}{2}\right)$$

b Transfer characteristic

$$\frac{K}{2}(V_{GS} - V_{th})^2$$

Fig. 3.10. Equations for an n-channel FET

The first equation describes the cutoff region, while the second is for the ohmic and the third for the pinch-off regions. The *transconductance coefficient K* is a measure of the slope of the transfer characteristic and is described in more detail below.

Curve description: The equation for the ohmic region is quadratic in V_{DS} and therefore describes a parabola in the family of output characteristics (see Fig. 3.10a). The peak of the parabola is at $V_{DS,po} = V_{GS} - V_{th}$, that is, at the borderline of the pinch-off region, and the validity of this equation ends here, since it only applies for $0 \leq V_{DS} < V_{DS,po}$. For $V_{DS} \geq V_{DS,po}$ we have to use the equation for the pinch-off region, which does not depend on V_{DS} and thus describes parallels to the V_{DS}-axis; Fig. 3.10a shows the related characteristic as a dot–dash line.

The equation for the pinch-off region is quadratic in V_{GS} and therefore describes a parabola in the family of transfer characteristics (see Fig. 3.10b). The peak of the parabola is at $V_{GS} = V_{th}$; the validity of the equation starts here and in n-channel FETs it applies to $V_{GS} > V_{th}$ only.

All equations only apply in the first quadrant of the output characteristics; that is, for $V_{DS} \geq 0$.[4] In a symmetrical FET the characteristics in the third quadrant are symmetric to those in the first quadrant; this is particularly true for integrated FETs. The equations can also be applied to the third quadrant if the drain and source are interchanged; that is, using V_{GD} and V_{SD} instead of V_{GS} and V_{DS}, respectively.[5] Discrete MOSFETs, especially power MOSFETs, are designed asymmetrically and display in the third quadrant a behavior that is different from that in the first quadrant (see Sect. 3.2).

In order to simplify the description, we will use abbreviations for the operating regions of the n-channel FET in the text below:

---

[4] Figure 3.5 depicts this region only.

[5] Because $V_{SD} = -V_{DS}$, one can also use $-V_{DS}$.

$$
\left.
\begin{aligned}
&\text{CR : cutoff region} \\
&\text{OR : ohmic region} \\
&\text{PR : pinch-off region}
\end{aligned}
\right\}
\Rightarrow
\left\{
\begin{aligned}
& V_{GS} < V_{th} \\
& V_{GS} \geq V_{th},\, 0 \leq V_{DS} < V_{GS} - V_{th} \\
& V_{GS} \geq V_{th},\, V_{DS} \geq V_{GS} - V_{th}
\end{aligned}
\right.
\tag{3.1}
$$

Also, taking the influence of channel-length modulation [3.2] into account and complementing the equation for the gate current leads to the *large-signal equations* for a field effect transistor:

$$
I_D =
\begin{cases}
0 & \text{CR} \\[2ex]
K\, V_{DS} \left(V_{GS} - V_{th} - \dfrac{V_{DS}}{2} \right)\left(1 + \dfrac{V_{DS}}{V_A} \right) & \text{OR} \\[2ex]
\dfrac{K}{2} (V_{GS} - V_{th})^2 \left(1 + \dfrac{V_{DS}}{V_A} \right) & \text{PR}
\end{cases}
$$

$$
\tag{3.2}
$$
$$
\tag{3.3}
$$

$$
I_G =
\begin{cases}
0 & \text{MOSFET} \\[2ex]
I_{G,S}\left(e^{\frac{V_{GS}}{V_T}} - 1 \right) & \text{junction FET}
\end{cases}
$$

$$
\tag{3.4}
$$

Transconductance coefficient: The *transconductance coefficient K* is a measure of the slope of the transfer characteristic of a FET. For n-channel MOSFETs, the following applies:

$$
K = K'_n \frac{W}{L} = \mu_n C'_{ox} \frac{W}{L}
\tag{3.5}
$$

Here, $\mu_n \approx 0.05 \ldots 0.07 \, \mathrm{m^2/Vs}$ is the *mobility*[6] of the charge carriers in the channel and C'_{ox} is the *capacitance per unit area of the gate oxide*; W is the width and L is the length of the gate (see Fig. 3.11). Together with the underlying silicon layer, the gate forms a plate capacitor of surface $A = WL$ and a plate distance according to the *oxide thickness* d_{ox}:

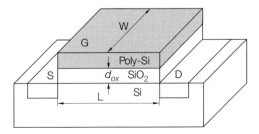

Fig. 3.11. Geometric dimensions of a MOSFET

---

[6] The mobility depends on the doping of the channel and is significantly lower than the mobility in undoped silicon ($\mu_n \approx 0.14 \, \mathrm{m^2/Vs}$).

$$C_{ox} = \epsilon_{ox} \frac{A}{d_{ox}} = \epsilon_0 \epsilon_{r,ox} \frac{W L}{d_{ox}} = C'_{ox} W L$$

With the *dielectric constant* $\epsilon_0 = 8.85 \cdot 10^{-12}$ As/Vm, the *relative dielectric constant* $\epsilon_{r,ox} = 3.9$ for silicon dioxide (SiO$_2$) and $d_{ox} \approx 40 \dots 100$ nm, we arrive at a capacitance per unit area $C'_{ox} \approx 0.35 \dots 0.9 \cdot 10^{-3}$ F/m^2 and the *relative transconductance coefficient*:[7]

$$K'_n = \mu_n C'_{ox} \approx 20 \dots 60 \frac{\mu A}{V^2}$$

The transconductance coefficient K is obtained from (3.5) by multiplying by the factor W/L, which is a measure of the size of the MOSFET. Typical values for discrete transistors are $L \approx 1 \dots 5\,\mu$m and from $W \approx 10$ mm in small-signal MOSFETs up to $W > 1$ m [8] in power MOSFETs; thus K ranges from approximately 40 mA/V^2 up to 50 A/V^2.

In p-channel MOSFETs, the mobility of the charge carriers in the channel is $\mu_p \approx 0.015 \dots 0.03$ m^2/Vs and is thus lower by about a factor of 2–3 than in n-channel MOS-FETs; this leads to $K'_p \approx 6 \dots 20\,\mu$A/V^2.

In junction FETs, K also depends on the geometric dimensions.[9] An accurate illustration is not provided here; refer to reference [3.1]. For small-signal applications with $K \approx 0.5 \dots 10$ mA/V^2, junction FETs are almost always discrete transistors.

Alternative description: For junction FETs, a different description of the characteristics is widely used. One defines

$$I_{D,0} = \frac{K V_{th}^2}{2}$$

If channel-length modulation is neglected, this leads to the following in the pinch-off region:

$$I_D = I_{D,0} \left(1 - \frac{V_{GS}}{V_{th}}\right)^2$$

On the basis of this definition, we have $I_{D,0} = I_D(V_{GS} = 0)$, which means that the transfer characteristic intersects the y-axis at $I_D = I_{D,0}$. In principle, all FETs with $V_{th} \neq 0$ can be described in this fashion; in the case of enhancement FETs, in which the transfer characteristic intersects the y-axis in the cutoff region only, $I_{D,0}$ is taken from $V_{GS} = 2V_{th}$.

Channel-length modulation: The dependence of the drain current on V_{DS} in the pinch-off region is caused by *channel-length modulation* and is empirically described by the term on the right-hand side of (3.3). In order to achieve a continuous changeover from the ohmic to the pinch-off region, this term must also be complemented in (3.2) [3.2]. This description is based on the observation that the extrapolated curves of the family of output characteristics intersect approximately at one point; Fig. 3.12 illustrates this relation. With

---

[7] K'_n is inversely proportional to d_{ox}, so that continuing miniaturization causes ever-increasing values; for example, $K'_n \approx 100 \dots 120\,\mu$A/V^2 in 3.3 V CMOS circuits.

[8] Section 3.2 describes how to arrive at these high values for W.

[9] The transconductance coefficient of a junction FET is usually called β in the literature; here we use K to maintain uniform identification and to avoid confusion with the current gain β of a bipolar transistor.

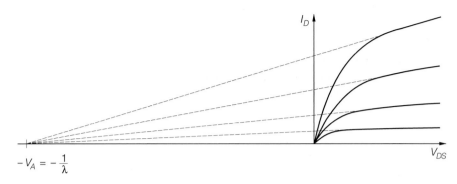

Fig. 3.12. Channel-length modulation and the Early voltage

reference to the bipolar transistor, the constant V_A is called the *Early voltage*: for MOSFETs it lies at $V_A \approx 20 \ldots 100$ V and for junction FETs it is $V_A \approx 30 \ldots 200$ V. The *channel-length modulation parameter* is often used instead of the Early voltage:

$$\lambda = \frac{1}{V_A} \tag{3.6}$$

We have $\lambda \approx 10 \ldots 50 \cdot 10^{-3} \, \text{V}^{-1}$ for MOSFETs and $\lambda \approx 5 \ldots 30 \cdot 10^{-3} \, \text{V}^{-1}$ for junction FETs.

For integrated MOSFETs with small geometric dimensions, this empirical description is very inaccurate. Very complicated equations are required in this case, to describe the *short-channel effect* that occurs. A wide number of models that describe this effect in various ways are used in CAD programs for designing integrated circuits (see Sect. 3.3).

3.1.3
Field Effect Transistor as an Adjustable Resistor

A field effect transistor operated in the ohmic region can be used as an adjustable resistor (see Fig. 3.13a); in this case the control voltage $V_{ctl} = V_{GS}$ changes the resistance of the drain source junction. Differentiation of (3.2) leads to:

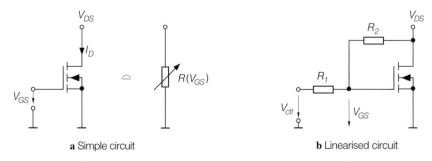

a Simple circuit   **b** Linearised circuit

Fig. 3.13. FET as an adjustable resistor

$$\frac{1}{R(V_{GS})} = \frac{\partial I_D}{\partial V_{DS}}\bigg|_{OR} = K\,(V_{GS} - V_{th} - V_{DS})\left(1 + \frac{2V_{DS}}{V_A}\right) + \frac{K\,V_{DS}^2}{2V_A}$$

However, due to the dependence on V_{DS} the resistance is not linear. Of particular interest is the *on resistance* $R_{DS,on}$ for signals around the point $V_{DS} = 0$:

$$R_{DS,on} = \frac{\partial V_{DS}}{\partial I_D}\bigg|_{V_{DS}=0} = \frac{1}{K\,(V_{GS} - V_{th})} \tag{3.7}$$

As the characteristics around $V_{DS} = 0$ are almost linear, the turn-on resistance $R_{DS,on}$ is independent of V_{DS}, and the FET acts as an adjustable linear resistor when modulated with small amplitudes.

The linearity can be improved by not feeding the control voltage directly to the gate but by adding half of the drain–source voltage to it in advance; this can be achieved by using the circuit shown in Fig. 3.13b, with a voltage divider consisting of two high-resistance components $R_1 = R_2$ in the MΩ range that generates

$$V_{GS} = \frac{V_{DS}R_1 + V_{ctl}R_2}{R_1 + R_2} \overset{R_1=R_2}{=} \frac{V_{DS} + V_{ctl}}{2}$$

Inserting this expression into (3.2) leads to

$$I_D = K\,V_{DS}\left(\frac{V_{ctl}}{2} - V_{th}\right)\left(1 + \frac{V_{DS}}{V_A}\right)$$

Consequently:

$$\frac{1}{R(V_{ctl})} = K\left(\frac{V_{ctl}}{2} - V_{th}\right)\left(1 + \frac{2V_{DS}}{V_A}\right) \overset{V_{DS} \ll V_A}{\approx} K\left(\frac{V_{ctl}}{2} - V_{th}\right)$$

A dependence on V_{DS} remains but is, however, much smaller than that of the simple circuit shown in Fig. 3.13a, a fact that is illustrated by the curves in Fig. 3.14. The nonlinearity can be further reduced by finely tuning the voltage divider. Optimum values

$$\frac{R_1}{R_2} = \frac{V_A - 2V_{ctl} + 2V_{th}}{V_A - 2V_{th}}$$

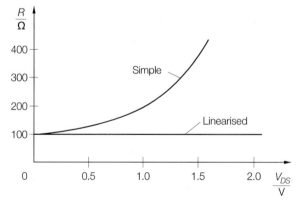

Fig. 3.14. Comparison of the resistance curves for $K = 5\,\text{mA/V}^2$, $V_{th} = 2\,\text{V}$, $V_A = 100\,\text{V}$ and $V_{GS} = 4\,\text{V}$ and $V_{ctl} = 8\,\text{V}$

are calculated by performing the previous calculation without the assumption $R_1 = R_2$; however, they are influenced by the control voltage V_{ctl}, which means that exact linearization is achieved for a certain control voltage only. If $K = 5\,\mathrm{mA/V^2}$, $V_{ctl} = 2\,\mathrm{V}$, $V_A = 100\,\mathrm{V}$ and $V_{ctl} = 8\,\mathrm{V}$, then $R(V_{ctl} = 8\,\mathrm{V}) = 100\,\Omega$ and $R_1/R_2 = 0.917$.

3.1.4
Operating Point and Small-Signal Behavior

One field of field effect transistor application is the linear amplification of signals in *small-signal* circuits. Here, the field effect transistor is modulated with *small* signals around an operating point. In this case, the characteristics can be replaced by their tangents to the operating point.

Operating Point

The operating point A is characterized by the voltages $V_{DS,A}$ and $V_{GS,A}$ and the current $I_{D,A}$, and is determined by external circuitry. For the appropriate use as an amplifier, the operating point must be in the pinch-off region. For the six FET types, Fig. 3.15 shows the

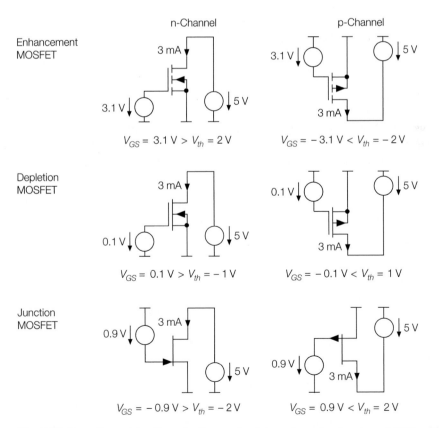

Fig. 3.15. Operating point settings for $I_{D,A} = 3\,\mathrm{mA}$ in n-channel and p-channel FETs with $K = 5\,\mathrm{mA/V^2}$

setting of the operating point and the polarity of voltages and currents; for n-channel FETs, according to the family of characteristics shown in Fig. 3.8 on page 174, a threshold voltage $V_{th} = -2/-1/2$ V and a transconductance coefficient $K = 5$ mA/V^2 are assumed. The chosen example of a current $I_{D,A} = 3$ mA is achieved with $V_{GS,A} = V_{th} + 1.1$ V: [10]

$$I_D \approx \frac{K}{2} (V_{GS} - V_{th})^2 = 2.5 \ \frac{\text{mA}}{\text{V}^2} \cdot 1.1 \ \text{V}^2 \approx 3 \ \text{mA}$$

In p-channel FETs the threshold voltage V_{th} has the opposite sign, producing $I_D = -3$ mA at $V_{GS,A} = V_{th} - 1.1$ V. Procedures for adjusting the operating point are described in Sect. 3.4.

Small-Signal Equations and Small-Signal Parameters

Small-signal values: When the transistor is operated around the operating point, the voltage and current deviations from the values at the operating point are called the *small-signal voltages* and *small-signal currents*. We can define:

$$v_{GS} = V_{GS} - V_{GS,A} \quad , \quad v_{DS} = V_{DS} - V_{DS,A} \quad , \quad i_D = I_D - I_{D,A}$$

Linearization: The characteristics are replaced by the tangents to the operating point; that is, they are *linearized*. For this purpose, a Taylor series expansion is performed at the operating point and is interrupted after the linear term:

$$i_D = I_D(V_{GS,A} + v_{GS}, V_{DS,A} + v_{DS}) - I_{D,A}$$

$$= \left. \frac{\partial I_D}{\partial V_{GS}} \right|_A v_{GS} + \left. \frac{\partial I_D}{\partial V_{DS}} \right|_A v_{DS} + \dots$$

Small-signal equations: The partial derivatives at the operating point are called the *small-signal parameters*. The introduction of specific designations produces *small-signal equations* for the field effect transistor:

$$i_G = 0 \tag{3.8}$$

$$i_D = g_m v_{GS} + \frac{1}{r_{DS}} v_{DS} \tag{3.9}$$

Small-signal parameters in the pinch-off region: The *transconductance* g_m describes how the drain current I_D changes through the gate–source voltage V_{GS} at the operating point. It can be determined from the slope of the tangent to the operating point in the family of transfer characteristics according to Fig. 3.8, and thus it indicates how *steep* the transfer characteristic is at the operating point. Differentiation of the large-signal (3.3) leads to:

$$g_m = \left. \frac{\partial I_D}{\partial V_{GS}} \right|_A = K \left(V_{GS,A} - V_{th} \right) \left(1 + \frac{V_{DS,A}}{V_A} \right) \overset{V_{DS,A} \ll V_A}{\approx} K \left(V_{GS,A} - V_{th} \right) \tag{3.10}$$

By definition, the slope is proportional to the *transconductance coefficient K*. Figure 3.16 presents the curves for n-channel FETs with $K = 5$ mA/V^2; the corresponding transfer

---

[10] The Early effect is neglected.

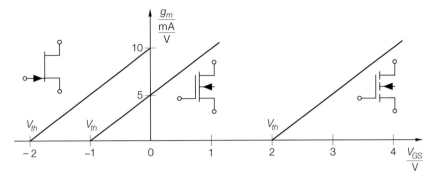

Fig. 3.16. Transconductance of n-channel FETs, with transfer characteristics from Fig. 3.8 ($K = 5\,\text{mA}/\text{V}^2$)

characteristics are shown in Fig. 3.8 on page 174. There are straight lines with the x-axis segment V_{th} and the slope K:

$$K = \frac{\partial g_m}{\partial V_{GS}} = \frac{\partial^2 I_D}{\partial V_{GS}^2}$$

By solving (3.3) for $V_{GS} - V_{th}$ and substituting in (3.10), the transconductance g_m can also be described as a function of the drain current $I_{D,A}$:

$$g_m = \left.\frac{\partial I_D}{\partial V_{GS}}\right|_A = \sqrt{2K I_{D,A}\left(1 + \frac{V_{DS,A}}{V_A}\right)} \overset{V_{DS,A} \ll V_A}{\approx} \sqrt{2K I_{D,A}} \tag{3.11}$$

In contrast to the bipolar transistor, where only the collector current $I_{C,A}$ is required to calculate the transconductance, for the field effect transistor the drain current $I_{D,A}$ and the transconductance coefficient K must be known; the dependence on V_A is low. In practice, the approximation given in (3.11) is used. Data sheets specify the transconductance for a certain drain current instead of K. The value of K can then be determined from the transconductance:

$$K \approx \frac{g_m^2}{2I_{D,A}}$$

The *small-signal output resistance* r_{DS} describes the changes of the drain–source voltage V_{DS} caused by a changing drain current I_D at the operating point. It can be determined from the reciprocal value of the slope of the tangent in the family of output characteristics according to Fig. 3.5. Differentiation of the large-signal (3.3) leads to:

$$r_{DS} = \left.\frac{\partial V_{DS}}{\partial I_D}\right|_A = \frac{V_A + V_{DS,A}}{I_{D,A}} \overset{V_{DS,A} \ll V_A}{\approx} \frac{V_A}{I_{D,A}} \tag{3.12}$$

The approximation given in (3.12) is usually used in practice.

Small-signal parameters in the ohmic region: For the ohmic region, $V_{DS} \ll V_A$; differentiation of (3.2) thus leads to:

$$g_{m,OR} \approx K\, V_{DS,A}$$

$$r_{DS,OR} \approx \frac{1}{K\left(V_{GS,A} - V_{th} - V_{DS,A}\right)}$$

The transconductance and the output resistance in the ohmic region are smaller than in the pinch-off region; clearly, lower levels of amplification can therefore be achieved.

Small-Signal Equivalent Circuit

The *small-signal equivalent circuit* shown in Fig. 3.17 is derived from the small-signal (3.8) and (3.9). Taking the drain current $I_{D,A}$ at the operating point as a basis, the parameters can be determined from (3.11) and (3.12).

This equivalent circuit diagram is suitable for calculating the small-signal response to low frequencies (0 . . . 10 kHz); for this reason, it is called the *DC small-signal equivalent circuit*. The response of the circuit to higher frequencies can only be described by means of the AC small-signal equivalent circuit described in Sect. 3.3.3.

Four-Pole Matrices

Small-signal equations can also be written in matrix form:

$$\begin{bmatrix} i_G \\ i_D \end{bmatrix} = \begin{bmatrix} 0 & 0 \\ g_m & \dfrac{1}{r_{DS}} \end{bmatrix} \begin{bmatrix} v_{GS} \\ v_{DS} \end{bmatrix}$$

This corresponds to the admittance form of characterizing a four-terminal network and thus leads to the four-pole theory. The admittance form describes the four-terminal circuit by the *y matrix* $\mathbf{Y}_s$:

$$\begin{bmatrix} i_G \\ i_D \end{bmatrix} = \mathbf{Y}_s \begin{bmatrix} v_{GS} \\ v_{DS} \end{bmatrix} = \begin{bmatrix} y_{11,s} & y_{12,s} \\ y_{21,s} & y_{22,s} \end{bmatrix} \begin{bmatrix} v_{GS} \\ v_{DS} \end{bmatrix}$$

The subscript S indicates that the FET is operated in common-source configuration; that is, the source terminal is used for both the input *and* output port according to the through connection shown in the small-signal equivalent circuit in Fig. 3.17. Section 3.4.1 describes the common-source circuit in more detail.

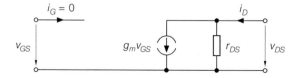

Fig. 3.17. Small-signal equivalent circuit for a field effect transistor

Since $I_G = 0$, the voltage V_{GS} only depends on the circuit connected to the gate and the equation $v_{GS} = v_{GS}(i_G, v_{DS})$ thus does not exist, which means that a hybrid representation with the **H** matrix, as for the bipolar transistor, is not possible for the field effect transistor.

Range of Validity for Small-Signal Applications

It has yet to be clarified how large the maximum signal around the operating point may be to remain within the small-signal operating mode. In practical terms, the linear distortions, which must not exceed an application-specific limit value, are crucial. This limit value is often given in form of a maximum permissible *distortion factor*. Section 4.2.3 describes this in more detail. The small-signal equivalent circuit is derived from the Taylor series expansion that is interrupted after the linear term. If subsequent terms of the Taylor series expansion are taken into account and channel-length modulation ($V_A \to \infty$) is neglected, the small-signal drain current is:

$$
i_D = \left. \frac{\partial I_D}{\partial V_{GS}} \right|_A v_{GS} + \frac{1}{2} \left. \frac{\partial^2 I_D}{\partial V_{GS}^2} \right|_A v_{GS}^2 + \frac{1}{6} \left. \frac{\partial^3 I_D}{\partial V_{GS}^3} \right|_A v_{GS}^3 + \dots
$$

$$
= \sqrt{2K I_{D,A}} \, v_{GS} + \frac{K}{2} v_{GS}^2
$$

Due to the parabolic shape of the characteristic, the series ends after the second term. For harmonic signals with $v_{GS} = \hat{v}_{GS} \cos \omega t$, this leads to:

$$
i_D = \frac{K}{4} \hat{v}_{GS}^2 + \sqrt{2K I_{D,A}} \, \hat{v}_{GS} \cos \omega t + \frac{K}{4} \hat{v}_{GS}^2 \cos 2\omega t
$$

The *distortion factor k* is determined from the ratio of the first harmonic with $2\omega t$ to the fundamental with ωt:

$$
k = \frac{i_{D,2\omega t}}{i_{D,\omega t}} = \frac{\hat{v}_{GS}}{4} \sqrt{\frac{K}{2I_{D,A}}} = \frac{\hat{v}_{GS}}{4 \left(V_{GS,A} - V_{th} \right)} \tag{3.13}
$$

It is inversely proportional to $\sqrt{I_{D,A}}$ or $V_{GS,A} - V_{th}$, so that with the same signal it declines with an increasing drain current. In discrete transistors, $V_{GS,A} - V_{th} \approx 1 \dots 2\,\text{V}$; with $\hat{v}_{GS} < 40 \dots 80\,\text{mV}$, this results in a distortion factor of $k < 1\%$. From a comparison with (2.15) on page 45, it can be noted that with the same distortion factor the possible input signal of the FET is much higher than that of the bipolar transistor, for which $k < 1\%$ can only be achieved with $\hat{v}_{BE} < 1\,\text{mV}$.

3.1.5
Maximum Ratings and Leakage Currents

Various limit data that must not be exceeded are specified for field effect transistors. These are divided up into limit voltages, limit currents and maximum dissipation. Once again, n-channel MOSFETs will be examined here; for p-channel MOSFETs, the voltage and current signs have to be reversed.

Breakthrough Voltages

Gate breakthrough: At the *gate–source breakthrough voltage* $V_{(BR)GS}$, the gate oxide of the MOSFET breaks through on the source side; while at the *drain–gate breakthrough voltage* $V_{(BR)DG}$, the gate oxide breaks through on the drain side. This breakthrough is irreversible and will destroy the MOSFET if it is not protected by Zener diodes. Therefore, discrete MOSFETs without Zener diodes must be protected against static charges, and must not be touched until potential equalization has been carried out.

The gate–source breakthrough is symmetrical, which means that it is independent of the polarity of the gate–source voltage; for this reason, data sheets specify a plus-or-minus value, such as $V_{(BR)GS} = \pm 20\,\text{V}$, or the absolute value of the breakthrough voltage. Typical values are $|V_{(BR)GS}| \approx 10 \ldots 20\,\text{V}$ for MOSFETs in integrated circuits and $|V_{(BR)GS}| \approx 10 \ldots 40\,\text{V}$ for discrete transistors.

The drain region of symmetrically designed MOSFETs is the same as the source region, which means that $|V_{(BR)DG}| = |V_{(BR)GS}|$; this is particularly true of MOSFETs in integrated circuits. In asymmetrically designed MOSFETs, $|V_{(BR)DG}|$ is significantly higher than $|V_{(BR)GS}|$, as a large portion of the voltage drops over a weakly doped layer between the channel and the drain terminal (see Sect. 3.2). On data sheets this voltage is called $V_{(BR)DGR}$ or V_{DGR}, since the measurement is taken with a resistance R between gate and source; the size of the resistance is stated. Since this is the breakthrough of the junction between the substrate and the weakly doped portion of the drain region, a drain–source breakthrough occurs at the same time; therefore, the same value is usually quoted for $V_{(BR)DG}$ as for the drain–source breakthrough voltage $V_{(BR)DSS}$ that is described in the following paragraphs.

In case of the junction FET, $V_{(BR)GSS}$ is the breakthrough voltage of the gate–channel diode; it is measured with the short-circuited drain source path, that is, with $V_{DS} = 0$, and it is negative in n-channel junction FETs and positive in p-channel junction FETs. Typical values of $V_{(BR)GSS}$ for n-channel FETs range from approximately $-50\,\text{V}$ to $-20\,\text{V}$. Also specified are the breakthrough voltages $V_{(BR)GSO}$ and $V_{(BR)GDO}$ on the source and drain sides, respectively; the subscript O indicates that the third terminal is *open*. These voltages are normally the same: $V_{(BR)GSS} = V_{(BR)GSO} = V_{(BR)GDO}$. Since the voltages V_{GS} and V_{DS} are of opposite polarity in the junction FET, $V_{GD} = V_{GS} - V_{DS}$ is the voltage of the highest absolute value, so that $V_{(BR)GDO}$ is of particular importance for practical purposes. Contrary to the MOSFET, the junction FET is not destroyed by the breakthrough as long as the current is limited and overheating is prevented.

Drain–source breakthrough: At the *drain–source breakthrough voltage* $V_{(BR)DSS}$, the junction between the drain region and the substrate of a MOSFET breaks through; this causes a current to flow from the drain region to the substrate and from there through the pn junction between the substrate and the source, which is operated in forward mode, or to the source via a connection between the substrate and the source that exists in discrete transistors. Figure 3.18a shows the characteristics for the breakthrough of a power MOSFET; with high currents in particular, it starts slowly and is reversible as long as the current is limited and overheating does not occur. For the enhancement n-channel MOSFET, $V_{(BR)DSS}$ is measured with short-circuited gate and source; that is, $V_{GS} = 0$, the additional subscript S stands for *short-circuited*. With depletion n-channel MOSFETs, the transistor is reverse-biased by a negative voltage $V_{GS} < V_{th}$. The corresponding drain–source breakthrough voltage is also called $V_{(BR)DSS}$; here, the subscript S indicates the

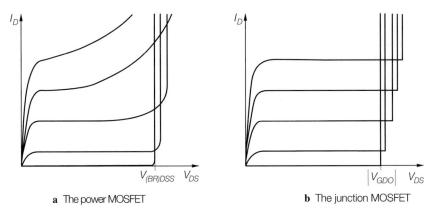

a The power MOSFET   **b** The junction MOSFET

Fig. 3.18. Output characteristics of discrete FETs at breakthrough

small-signal short-circuit, which means biasing the gate with a voltage source that has a negligibly low internal resistance. The values range from $V_{(BR)DSS} \approx 10 \ldots 40\,\mathrm{V}$ in integrated FETs to $V_{(BR)DSS} = 1000\,\mathrm{V}$ in discrete FETs for switching applications.

Junction FETs feature no direct breakthrough between drain and source because of the homogeneity of this region. Here, with the channel pinched off and the drain–source voltage increasing, it is the junction between the drain and the gate that breaks through when the above-mentioned breakthrough voltage $V_{(BR)GDO}$ is reached. Figure 3.18b shows the characteristics for the breakthrough of a small-signal junction FET; it occurs suddenly.

Maximum Currents

Drain current: For the drain current one distinguishes between the maximum *continuous current* and the maximum *peak current*. Data sheets have no specific designation for the maximum continuous current; here, it is referred to as $I_{D,max}$. The maximum peak current is called I_{DM} [11] on data sheets and is quoted for a pulsed current with a given pulse duration and repetition rate; it is higher than the maximum continuous current by a factor of $2 \ldots 5$.

For junction FETs the *drain saturation current* I_{DSS} [12] is specified instead of the maximum continuous current $I_{D,max}$; it is measured with $V_{GS} = 0$ in the pinch-off region and thus represents the maximum possible drain current in normal operation.

Backward diode: Due to the connection between the source and the substrate in discrete MOSFETs, there is a backward diode between the source and the drain (see Sect. 3.2). Two currents are specified for this diode: the maximum continuous current $I_{S,max}$ and the maximum peak current I_{SM}. Due to the component construction, they are of the same magnitude as the corresponding drain currents $I_{D,max}$ and I_{DM}, so that the backward diode can be used without limitations as a freewheeling or commutating diode.

Gate current: For junction FETs the maximum gate current $I_{G,max}$ in forward mode is also specified; typical values are $I_{G,max} \approx 5 \ldots 50\,\mathrm{mA}$. However, this information is of minor importance, since the channel gate diode is usually operated in reverse mode.

---

[11] For MOSFETs in switching applications, $I_{D,puls}$ is often used instead of I_{DM}.

[12] I_{DSS} is also known as $I_{D,S}$ and corresponds to $I_{D,0} = I_D(V_{GS} = 0)$ described for junction FETs in Sect. 3.1.2.

Leakage Currents

Drain current: A small *drain–source leakage current* I_{DSS} flows in enhancement MOS-FETs with short-circuited gate and source; it corresponds to the leakage current of the drain substrate junction and, as such, is very temperature sensitive. Typically, $I_{DSS} < 1\,\mu A$ in integrated MOSFETs and discrete MOSFETs, while $I_{DSS} = 1\ldots 100\,\mu A$ for discrete MOSFETs with currents in the ampere range. In depletion MOSFETs, I_{DSS} is also measured in the cutoff region; this requires a gate–source voltage of $V_{GS} < V_{th}$.

Note that the current I_{DSS} is also specified for junction FETs, but has an entirely different meaning here. In MOSFETs, I_{DSS} is the *minimum* drain current that also flows in the cutoff region and occurs in switching applications as a leakage current across the open switch; in junction FETs, I_{DSS} is the *maximum* drain current in the pinch-off region. Despite the different meanings, it is specified on data sheets using the same designation.

Maximum Power Dissipation

The dissipation of a transistor is the power that is converted into heat:

$$P_V = V_{DS} I_D$$

This mainly takes place in the channel and leads to an increase in the channel temperature. Due to the temperature difference, the heat can be dissipated to the environment via the housing. However, the temperature in the channel must not exceed a material-specific limit value of 175 °C for silicon; for safety reasons, a limit value of 150 °C is used for calculations in practice. The related maximum dissipation of discrete transistors depends on the given construction and installation method; the data sheet specifies the total dissipation P_{tot} for two situations:

– Operation in an upright mounting on the circuit board without further cooling measures, at an ambient temperature (*free-air temperature*) of $T_A = 25\,°C$.
– Operation with a *case temperature* of $T_C = 25\,°C$.

The two maximum values are known as $P_{V,25(A)}$ and $P_{V,25(C)}$. Only $P_{tot} = P_{V,25(A)}$ is specified for small-signal FETs that are used for upright mounting without a heat sink, while for power MOSFETs that are designed exclusively for installation with heat sinks only $P_{tot} = P_{V,25(C)}$ is specified. However, the conditions $T_A = 25\,°C$ or $T_C = 25\,°C$ cannot be guaranteed in real applications. Since P_{tot} decreases as the temperature increases, data sheets often show a *power derating curve* that plots P_{tot} versus T_A or T_C (see Fig. 3.19a). Section 2.1.6 on page 49 describes the temperature response of the bipolar transistor in detail; naturally, the results apply equally to FETs.

Safe Operating Area

The *safe operating area* (*SOA*) can be derived from the family of output characteristics using the limit data; the SOA is determined by the maximum drain current $I_{D,max}$, the drain–source breakthrough voltage $V_{(BR)DSS}$, the maximum dissipation P_{tot} and the $R_{DS,on}$ limit. Figure 3.19b shows the SOA plot as a log–log diagram in which both the hyperbola of the maximum power dissipation, represented by $V_{DS} I_D = P_{tot}$, and the $R_{DS,on}$ limit with $V_{DS} = R_{DS,on} I_D$ appear as straight lines. It follows that the maximum continuous current $I_{D,max}$ can be calculated from P_{tot} and $R_{DS,on}$:

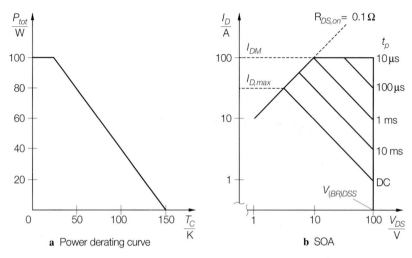

Fig. 3.19. Limiting curves of a MOSFET in switching applications

$$I_{D,max} = \sqrt{\frac{P_{tot}}{R_{DS,on}}}$$

For FETs to be used in switching applications, additional limiting curves for pulsed operation with various pulse durations are shown. If the pulse duration is very short and the duty factor is small, the FET can be operated with a maximum voltage $V_{(BR)DSS}$ *and the maximum drain current* I_{DM} *at the same time*; in this case, the SOA shows a rectangle. A FET enables loads with a power dissipation of up to $P = V_{(BR)DSS}I_{D,max}$ to be switched. This maximum switching capacity is high compared to the maximum dissipation P_{tot}; from Fig. 3.19 it follows that $P = V_{(BR)DSS}I_{D,max} = 100\,\text{V} \cdot 30\,\text{A} = 3\,\text{kW} \gg P_{tot} = 100\,\text{W}$.

3.1.6
Thermal Behavior

The thermal behavior of components is described in Sect. 2.1.6 using the example of the bipolar transistor; the values and relations mentioned there apply likewise to a FET if P_V is replaced by the power dissipation of the FET.

3.1.7
Temperature Sensitivity of FET Parameters

MOSFETs and junction FETs show different temperature responses and must therefore be considered separately in this respect.

MOSFET

In case of the MOSFET, the threshold voltage V_{th} and the transconductance coefficient K are temperature-dependent; the temperature coefficient of the drain current in an n-channel MOSFET in the pinch-off region is derived by differentiation of (3.3):

$$\frac{1}{I_D}\frac{dI_D}{dT} = \frac{1}{K}\frac{dK}{dT} - \frac{2}{V_{GS} - V_{th}}\frac{dV_{th}}{dT} \tag{3.14}$$

From (3.5) and the temperature sensitivity of the mobility related to the reference point T_0 [3.1], it follows that

$$\mu(T) = \mu(T_0)\left(\frac{T_0}{T}\right)^{m_\mu} \qquad \text{with } m_\mu \approx 1.5$$

which means that the transconductance coefficient declines with an increase in temperature:

$$\frac{1}{K}\frac{dK}{dT} = -\frac{m_\mu}{T} \overset{T=300\,\text{K}}{\approx} -5 \cdot 10^{-3}\,\text{K}^{-1}$$

The threshold voltage [3.1] is

$$V_{th} = V_{FB} + V_{inv} + \gamma\sqrt{V_{inv}}$$

where V_{FB} is the *flat-band voltage*, V_{inv} is the *inversion voltage* and γ is the *substrate control factor*. The flat-band voltage depends on the design of the gate and is not important in this context, while the other parameters are described in Sect. 3.3. V_{FB} and γ are not temperature-dependent; therefore

$$\frac{dV_{th}}{dT} = \left(1 + \frac{\gamma}{2\sqrt{V_{inv}}}\right)\frac{dV_{inv}}{dT}$$

Typical values are as follows: $V_{inv} \approx 0.55\ldots0.8\,\text{V}$, dV_{inv}/dT ranges from approximately $-2.3\,\text{mV/K}$ to $-1.7\,\text{mV/K}$ and $\gamma \approx 0.3\ldots0.8\,\sqrt{\text{V}}$. Hence:

$$\frac{dV_{th}}{dT} \approx -3.5\ldots-2\,\frac{\text{mV}}{\text{K}}$$

As the temperature coefficients of K and V_{th} are negative, the temperature coefficient of the drain current may be positive or negative because of the subtraction in (3.14) depending on the operating point. Therefore there is a *temperature compensation point TC* at which the temperature coefficient becomes zero; for n-channel MOSFETs, solving (3.14) leads to:

$$V_{GS,TC} = V_{th} + 2\,\frac{\dfrac{dV_{th}}{dT}}{\dfrac{1}{K}\dfrac{dK}{dT}} \approx V_{th} + 0.8\ldots1.4\,\text{V}$$

$$I_{D,TC} \approx K \cdot 0.3\ldots1\,\text{V}^2$$

Figure 3.20a shows the transfer characteristic of an n-channel MOSFET at the temperature compensation point. For p-channel MOSFETs the values are: $V_{GS,TC} = V_{th} - 0.8\ldots1.4\,\text{V}$ and $I_{D,TC} = -K \cdot 0.3\ldots1\,\text{V}^2$.

These values are for integrated MOSFETs with single diffusion. In contrast, discrete MOSFETs are almost always designed with double diffusion (see Sect. 3.2); in discrete MOSFETs, $dV_{th}/dT \approx -5\,\text{mV/K}$ and thus:

$$V_{GS,TC(DMOS)} \approx V_{th} + 2\,\text{V}$$

$$I_{D,TC(DMOS)} \approx K \cdot 2\,\text{V}^2$$

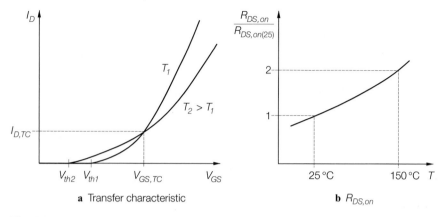

a Transfer characteristic   **b** $R_{DS,on}$

Fig. 3.20. Temperature response of an n-channel MOSFET

In practical applications, most n-channel MOSFETs are used with $V_{GS} > V_{GS,TC}$; the temperature coefficient in this region is negative, that is, the drain current decreases when the temperature increases. This *negative thermal relation* enables thermally stable operation without any circuit modifications. Bipolar transistors, however, require negative electrical feedback in order to prevent positive thermal feedback due to temperature-based current increases that can raise the temperature in the transistor and lead to its destruction.

Of particular importance in the ohmic region is the turn-on resistance $R_{DS,on}$. Differentiation of (3.7) leads to:

$$\frac{1}{R_{DS,on}} \frac{dR_{DS,on}}{dT} = \frac{1}{V_{GS} - V_{th}} \frac{dV_{th}}{dT} - \frac{1}{K} \frac{dK}{dT}$$

$$\overset{V_{GS} \gg V_{th}}{\approx} -\frac{1}{K} \frac{dK}{dT} \approx 5 \cdot 10^{-3} \, \text{K}^{-1}$$

This means that $R_{DS,on}$ almost doubles in value if the temperature increases from 25 °C to 150 °C; Fig. 3.20b shows the resulting curve for $R_{DS,on}$.

Junction FET

Equation (3.14) also applies to n-channel junction FETs. The transconductance coefficient K is proportional to the conductivity σ of the channel; and since $\sigma \sim \mu$, the temperature coefficient is the same as for the MOSFET:

$$\frac{1}{K} \frac{dK}{dT} \approx -5 \cdot 10^{-3} \, \text{K}^{-1}$$

The threshold voltage V_{th} is formed from a temperature-independent portion and the *diffusion voltage* V_{Diff} of the pn junction between the gate and the channel. This leads to:

$$\frac{dV_{th}}{dT} = \frac{dV_{Diff}}{dT} \approx -2.5 \ldots -1.7 \, \text{mV/K}$$

Consequently, the temperature compensation point of an n-channel junction FET is:

$$V_{GS,TC(Jfet)} \approx V_{th} + 0.7 \ldots 1 \, \text{V}$$
$$I_{D,TC(Jfet)} \approx K \cdot 0.25 \ldots 0.5 \, \text{V}^2$$

The transfer characteristic is like that of the MOSFET with the exception of a shift in direction of V_{GS}; the turn-on resistance $R_{DS,on}$ is the same as that of the MOSFET.

3.2
Construction of the Field Effect Transistor

Simple types of MOSFETs and junction FETs are symmetrical. The illustrations in Figs. 3.1 and 3.2 basically reflect this simple construction, which is predominantly used in integrated circuits; for this reason, we will first look at integrated transistors.

3.2.1
Integrated MOSFETs

Construction: Figure 3.21 shows the principle of construction of n-channel and p-channel MOSFETs on a common semiconductor substrate; the drain, gate, source and bulk terminals are marked with respective subscripts. In n-channel MOSFETs the p-doped semiconductor substrate serves as the bulk, with terminal B_n. The p-channel MOSFET requires an n-doped bulk region and must therefore be produced in an n-doped recession; B_p is the corresponding bulk terminal. The drain and source regions have high n-doping in the n-channel MOSFET and high p-doping in the p-channel MOSFET. The gates are made of polysilicon and are insulated against the channel beneath by the thin *gate oxide*. In the outer areas the insulation between the semiconductor regions and the aluminum conductors of the metal-coated layer is achieved by way of the much thicker *field oxide*. Since polysilicon is a relatively good conductor, the leads to the gate can be made of polysilicon; for this reason the metal coating on the gates, as shown in Fig. 3.21, is not essential.

The term *MOS* (*metal-oxide semiconductor*) goes back to the time when the gate was made of a metal (aluminum) instead of polysilicon. The modern MOSFET with a polysilicon gate should be correctly called an *SOS* (*semiconductor-oxide semiconductor*), but the familiar designation has been retained.

CMOS: Circuits designed according to Fig. 3.21 are called *CMOS circuits* (*complementary metal-oxide semiconductor circuits*) because they contain *complementary* MOSFETs. NMOS and PMOS circuits, the obsolete predecessors of CMOS circuits, consisted

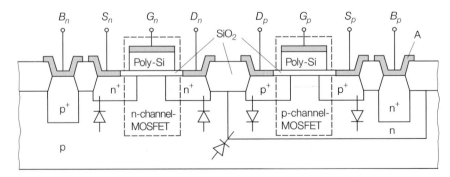

Fig. 3.21. Construction of an n-channel and a p-channel MOSFET in an integrated CMOS circuit

only of n-channel or p-channel MOSFETs, as the name indicates; p-doped or n-doped wafers were used for this purpose and a recession for the other type was not required.

Bulk diodes: In the sequence of layers in a CMOS circuit there are several pn junctions which must be operated in reverse mode; they are shown as diodes in Fig. 3.21. In order for the diodes to block between the drain or source regions and the bulk regions underneath, the respective voltages are $V_{SB} \geq 0$ and $V_{DB} \geq 0$ in the n-channel MOSFET and $V_{SB} \leq 0$ and $V_{DB} \leq 0$ in the p-channel MOSFET: the subscript B denotes the respective bulk region; that is, B_n for the n-channel and B_p for the p-channel MOSFET. Furthermore, $V_{Bn} \leq V_{Bp}$ is necessary in order for the diode between the bulk regions to be nonconductive. Thus it follows that all diodes are reverse-biased if B_n is connected to the negative and B_p to the positive supply voltage of the circuit; all other voltages then lie in between.

Latch-up: In addition to the diodes, the CMOS circuit also has a parasitic thyristor formed by the sequence of layers and the connections $B_n - S_n$ and $B_p - S_p$; Fig. 3.22 shows a simplified equivalent circuit for the thyristor, which is made up of two bipolar transistors and two resistors. The bipolar transistors result from the layer sequence, and R_n and R_p are the spreading resistances of the comparatively high-resistance bulk regions. Normally, these transistors are nonconductive since the bases are connected to the emitters via R_n or R_p, and no current flows through the bulk regions; the thyristor is nonconductive. In the event of under- or over-voltages at one of the inputs of the CMOS circuit, currents will flow to the bulk regions through the protective diodes described in Sect. 7.4.6. This can cause the voltage drop across R_p or R_n to become so high that one of the transistors is forward-biased. The current that then flows causes a voltage drop across the other resistor so that the second transistor also conducts and, in turn, its current keeps the first transistor conductive. Thus we have a positive feedback that short-circuits the supply voltage V_b: the thyristor has fired. This fault situation is called *latch-up* and it almost always destroys the circuit. Modern CMOS circuits feature a high *latch-up* resistance that is achieved by a suitable arrangement of the regions and specific circuitry at the inputs. One specialty is CMOS circuits with *dielectric insulation*, which means that the individual MOSFETs are formed in separate recessions that are insulated by an oxide layer; there is no thyristor formation and the circuit is safe from *latch-up*.

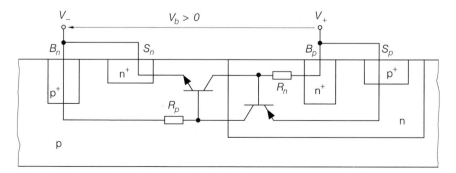

Fig. 3.22. A parasitic thyristor in an integrated CMOS circuit

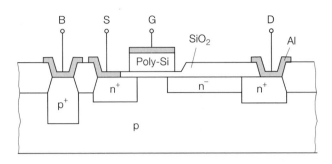

Fig. 3.23. An n-channel MOSFET for high drain–source voltages

MOSFETs for higher voltages: Due to the fact that $K \sim W/L$, the transconductance coefficient of a MOSFET is inversely proportional to the channel length L, which one then tries to keep as short as possible by reducing the distance between the drain and source regions. This, however, causes a reduction in the drain–source breakthrough voltage. If the breakthrough voltage is to be high despite the short channel length, then a weakly doped drift region, across which a major part of the drain–source voltage drops, is introduced between the channel and the drain terminal; this is shown in the example of an n-channel MOSFET in Fig. 3.23. The breakthrough voltage is approximately proportional to the length of the drift region; this is why integrated high-voltage MOSFETs require a larger area on the wafer.

3.2.2
Discrete MOSFETs

Construction: Unlike integrated MOSFETs, discrete MOSFETs are usually built vertically; that is, the drain terminal is on the underside of the substrate. Figure 3.24 shows a three-dimensional section through a *vertical MOSFET*. The weakly doped drift layer with n^- doping does not run laterally on the surface, as in the integrated MOSFET as seen in Fig. 3.23, but vertically; this saves space on the surface while guaranteeing a comparatively high breakthrough voltage that corresponds to the thickness of the n^- region. Again, the channel runs along the surface beneath the gate. The p-doped bulk region is not formed by the substrate, but by diffusion in the n^- substrate and is connected to the source via a p^+ contact region. Since the n^+ source regions are also generated by diffusion, these MOSFETs are also known as *double diffused MOSFETs (DMOS)*.

The *cellular* construction is also visible in Fig. 3.24. A vertical MOSFET consists of a two-dimensional parallel circuitry of small cells, whose source regions are connected to the surface by a metal coating that covers the entire area of the source. The cells are driven by a common polysilicon gate arranged beneath the source metal coating, in the form of a grid that is connected to the outer gate terminal at the edge of the wafer only; the underside serves as the common-drain terminal. This design facilitates a very high channel width W on a small surface and thus a high transconductance coefficient $K \sim W$. Thus it is possible to achieve a channel width of $W = 0.2\ \mu\text{m}$ for a wafer with a surface area of $2 \times 2\ \text{mm}^2$ and a cell size of $20 \times 20\ \mu\text{m}^2$ with $W_{Cell} = 20\ \mu\text{m}$; if $L = 2\ \mu\text{m}$ and $K'_n \approx 25\ \mu\text{A}/\text{V}^2$, then $K = K'_n W/L = 2.5\ \text{A}/\text{V}^2$. Since, with an n-fold reduction of the geometric size of the cells, their number increases by the factor n^2 but the width W per cell is only reduced by a factor of n, any miniaturization results in an increase in the channel width per unit area.

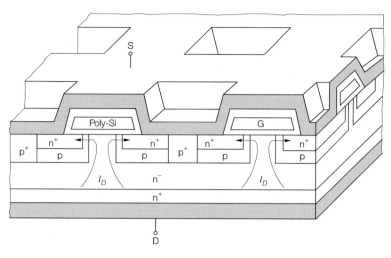

Fig. 3.24. Construction of an n-channel DMOS-FET

Parasitic elements: Due to the specific construction of vertical MOSFETs, there are several parasitic elements, which are shown in Fig. 3.25 together with the resulting equivalent circuit diagram:

– The large overlap of gate and source results in a high *external* gate–source capacitance C_{GS}. This is usually higher than the *internal* gate–source capacitance, which is described in more detail in Sect. 3.3.2.
– The overlap between the gate and the n^- region results in a relatively high *external* gate–drain capacitance C_{GD}, which adds to the *internal* drain–gate capacitance; a detailed description of the latter is found in Sect. 3.3.2.

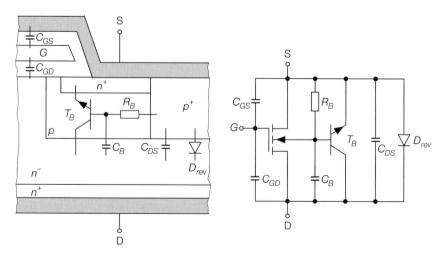

Fig. 3.25. Parasitic elements and an equivalent circuit for an n-channel DMOS-FET

- The drain–source capacitances C_{DS} and C_B lie between the bulk region and the drain region; C_{DS} is located directly between drain and source, while C_B is in series with the spreading resistance R_B of the bulk region.
- Due to the sequence of layers, there is also a bipolar transistor T_B whose base is connected to the emitter via the spreading resistance R_B; T_B is therefore normally reverse-biased. If the drain–source voltage increases very rapidly, then the current $I = C_B\,dV_{DS}/dt$ through C_B increases, causing a high voltage drop across R_B, so that T_B becomes conductive. To prevent this from happening when DMOS power switches are turned off, the slew rate of V_{DS} must be limited by suitable control conditions or by a turn-off relief circuit.
- Between the source and the drain there is a *reverse diode* D_{rev}, which is forward-biased when the drain–source voltage is negative. It can be used as a freewheeling diode when switching inductive loads, but it leads to unwanted currents in push-pull circuits due to their long reverse recovery time t_{RR}, which is determined by the design of the MOSFET.

Characteristics of vertical power MOSFETs: The characteristics of a vertical power MOSFET deviate from the simple large-signal characteristics of (3.2) and (3.3); Fig. 3.26 shows the differences in the transfer and output characteristics:

- With high currents, the influence of parasitic resistances in the source terminal becomes noticeable. In this case the external gate–source voltage V_{GS} across the terminals is comprised of the internal gate–source voltage and the voltage drop across the source resistance R_S; this linearizes the transfer characteristic for high currents (see Fig. 3.26a).
- In vertical MOSFETs the pinch-off voltage $V_{DS,po}$ is higher than $V_{GS} - V_{th}$ due to an additional voltage drop in the drift region. This voltage drop can be described by a nonlinear drain resistance, and it results in a shift in the output characteristics (see Fig. 3.26b).

Equations to describe this effect are presented in Sect. 3.3.1.

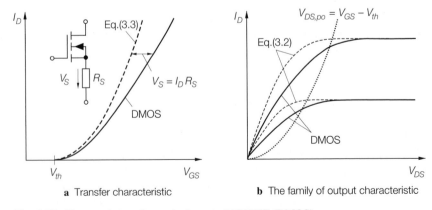

a Transfer characteristic   **b** The family of output characteristic

Fig. 3.26. Characteristics of a vertical power MOSFET (DMOS)

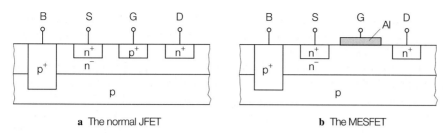

a The normal JFET   **b** The MESFET

Fig. 3.27. Construction of junction FETs

3.2.3
Junction FETs

Figure 3.27 shows the construction of a *normal* n-channel junction FET, with a pn junction between the gate and the channel, and an n-channel MESFET with a metal semiconductor junction (a Schottky junction) between the gate and the channel. In integrated junction FETs, the substrate terminals B are connected to the negative supply voltage to enable the pn junctions between the substrate and the n^- channel regions always to be operated in reverse mode. Furthermore, every FET must be surrounded by a closed p^+ ring, so that the channel regions of the individual FETs are insulated from each other. In discrete junction FETs, the substrate may also be connected to the gate; this provides a controlling effect for the substrate/channel junction in addition to that of the gate/channel junction. The vertical construction used in the MOSFET and the bipolar transistor is not possible in the case of the junction FET.

3.2.4
Cases

Discrete MOSFETs and discrete junction FETs are contained in the same cases as bipolar transistors; Fig. 2.21 on page 56 shows the most common cases. MOSFETs are available in all power ratings and thus in all case sizes. Junction FETs are always small-signal transistors in small cases; an exception is the power MESFET for high-frequency power amplifiers, which comes in a special high-frequency housing for surface mounting. There are also junction FETs with a separate bulk terminal, in cases with four terminals. Dual-gate MOSFETs also require cases with four terminals; these are always high-frequency transistors in special high-frequency cases.

3.3
Models of Field Effect Transistors

Section 3.1.2 describes the *static behavior* of a field effect transistor by way of large-signal (3.2)–(3.4), and with secondary effects neglected. More accurate models that take these effects into consideration and, furthermore, correctly describe the *dynamic response* are required for computer-aided circuit design. The *dynamic small-signal model* that is necessary for calculating the frequency response of circuits is derived from this *large-signal model* by linearization.

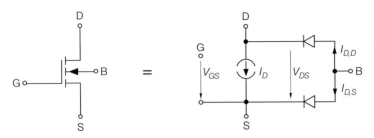

Fig. 3.28. Large-signal equivalent circuit for an n-channel MOSFET

3.3.1
Static Behavior

Contrary to the bipolar transistor, for which the Gummel–Poon model has proven its general applicability, for FETs there are numerous models that have their advantages and disadvantages for the various applications, and can become rather complex. The *level-1 MOSFET* model[13], which is available in almost all CAD programs for circuit simulation, is described below. It is very suitable for discrete transistors with a comparably large channel length and width, but not for integrated MOSFETs, with their small dimensions typical of large-scale integrated circuits. Here, the more elaborate *level-2* and *level-3* models or the *BSIM* models[14] are necessary; they also take into account the *short-channel, narrow-channel* and *sub-threshold effects*. Only a qualitative description of these effects is provided here.

Junctions FETs have their own model, the static response of which corresponds to that of the level-1 MOSFET model, despite the fact that CAD programs often use other parameters or other designations for parameters with the same meaning; this will be discussed in more detail at the end of this section.

Level-1 MOSFET Model

The n-channel MOSFET consists of a p-doped substrate (bulk), the n-doped regions for the drain and the source, an insulated gate and an inversion channel located between the drain and the source. This is reflected in the large-signal equivalent circuit diagram with a controlled current source for the channel and two diodes for the pn junctions between bulk and drain and bulk and source, as shown in Fig. 3.28.

Drain current: The level-1 model uses (3.2) and (3.3) in combination with (3.5); from

$$V_{DS,po} = V_{GS} - V_{th} \tag{3.15}$$

and $K = K'_n W/L$, we obtain:

---

[13] This name is used in circuit simulators of the *Spice* family; for example, *PSpice* from *OrCAD*. In the literature it is often called the *Shichman–Hodges model*, because major portions are taken from a publication by H. Shichman and D.A. Hodges.

[14] The *BSIM* models (*Berkeley short-channel IGFET model*) were developed at the University of California, Berkeley, and are presently considered the most advanced models for short-channel MOSFETs.

$$
I_D = \begin{cases}
0 & \text{for } V_{GS} < V_{th} \\[2mm]
\dfrac{K'_n W}{L} V_{DS} \left(V_{GS} - V_{th} - \dfrac{V_{DS}}{2} \right) \left(1 + \dfrac{V_{DS}}{V_A} \right) & \text{for } V_{GS} \geq V_{th}, \\[1mm]
& 0 \leq V_{DS} < V_{DS,po} \\[2mm]
\dfrac{K'_n W}{2L} (V_{GS} - V_{th})^2 \left(1 + \dfrac{V_{DS}}{V_A} \right) & \text{for } V_{GS} \geq V_{th}, \\[1mm]
& V_{DS} \geq V_{DS,po}
\end{cases}
\tag{3.16}
$$

The parameters are the *relative transconductance coefficient* K'_n, the *channel width* W, the *channel length* L and the *Early voltage* V_A. The *mobility* μ_n and the *oxide thickness* d_{ox} can be given as an alternative to K'_n; then [3.1]:

$$
K'_n = \frac{\mu_n \epsilon_0 \epsilon_{r,ox}}{d_{ox}}
\tag{3.17}
$$

With $\mu_n = 0.05 \ldots 0.07 \, \text{m}^2/\text{Vs}$, $\epsilon_0 = 8.85 \cdot 10^{-12} \, \text{As/Vm}$ und $\epsilon_{r,ox} = 3.9$, it follows that:

$$
K'_n \approx 1700 \ldots 2400 \, \frac{\mu A}{V^2} \cdot \frac{1}{d_{ox}/\text{nm}}
$$

In discrete MOSFETs the oxide layer thickness $d_{ox} \approx 40 \ldots 100 \, \text{nm}$, while it is reduced to 15 nm in large-scale integrated CMOS circuits.

Threshold voltage: The threshold voltage V_{th} is the gate–source voltage above which the inversion channel is generated beneath the gate. As the channel is in the substrate region, the inversion and thus the threshold voltage depend on the gate–substrate voltage V_{GB}. This effect is known as the *substrate effect* and is determined by the doping of the substrate. Since the form $V_{th} = V_{th}(V_{GB})$ is not very illustrating, the source is taken as the reference point, as in V_{GS} and V_{DS}, and replaces $V_{GB} = V_{GS} - V_{BS}$ by the bulk–source voltage V_{BS}. Consequently [3.1]:

$$
V_{th} = V_{th,0} + \gamma \left(\sqrt{V_{inv} - V_{BS}} - \sqrt{V_{inv}} \right)
\tag{3.18}
$$

The parameters are the *zero threshold voltage* $V_{th,0}$, the *substrate control factor* $\gamma \approx 0.3 \ldots 0.8 \, \sqrt{V}$ and the *inversion voltage* $V_{inv} \approx 0.55 \ldots 0.8 \, V$. Figure 3.29 shows the plot of V_{th} versus V_{BS} for $V_{th,0} = 1 \, V$, $\gamma = 0.55 \, \sqrt{V}$ and $V_{inv} = 0.7 \, V$;[15] this requires $V_{BS} \leq 0$ so that the bulk source diode is in reverse mode.

The substrate effect is noted especially in integrated circuits, where all n-channel MOSFETs have a common substrate region and are operated with different bulk–source voltages depending on the operating point; integrated MOSFETs of the same geometric dimensions thus have different characteristics if operated with different bulk–source voltages. In the discrete MOSFET, with its internal connection between source and substrate, this effect does not occur; here, $V_{BS} = 0$ and $V_{th} = V_{th,0}$.

As an alternative to γ and V_{inv}, the *substrate doping density* N_{sub} and the oxide thickness d_{ox} may be given. Consequently [3.1]:

$$
\gamma = \frac{\sqrt{2q\epsilon_0 \epsilon_{r,Si} N_{sub}}}{C'_{ox}} = \sqrt{\frac{2q\epsilon_{r,Si} N_{sub}}{\epsilon_0}} \, \frac{d_{ox}}{\epsilon_{r,ox}}
\tag{3.19}
$$

---

[15] γ and V_{inv} are determined by using (3.19) and (3.20) for $N_{sub} = 10^{16} \, \text{cm}^{-3}$ and $d_{ox} = 32 \, \text{nm}$.

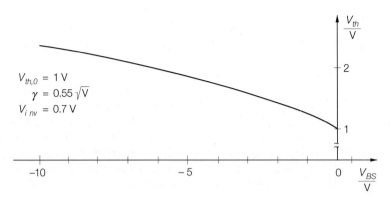

Fig. 3.29. Dependence of the threshold voltage V_{th} on the bulk–source voltage V_{BS} (substrate effect)

$$V_{inv} = 2V_T \ln \frac{N_{sub}}{n_i} \tag{3.20}$$

Inserting the constants $q = 1.602 \cdot 10^{-19}$ As, $\epsilon_0 = 8.85 \cdot 10^{-12}$ As/Vm, $\epsilon_{r,ox} = 3.9$ and $\epsilon_{r,Si} = 11.9$, as well as $V_T = 26\,\text{mV}$ and $n_i = 1.45 \cdot 10^{10}\,\text{cm}^{-3}$ for $T = 300\,\text{K}$, we obtain:

$$\gamma \quad \approx \quad 1.7 \cdot 10^{-10}\,\sqrt{V} \cdot \sqrt{N_{sub}/\text{cm}^{-3}} \cdot d_{ox}/\text{nm}$$

$$V_{inv} \quad \underset{T=300\,\text{K}}{\approx} \quad 52\,\text{mV} \cdot \ln \frac{N_{sub}}{1.45 \cdot 10^{10}\,\text{cm}^{-3}}$$

Typical values are $N_{sub} \approx 1 \ldots 7 \cdot 10^{16}\,\text{cm}^{-3}$ for integrated circuits and $N_{sub} \approx 5 \cdot 10^{14} \ldots 10^{16}\,\text{cm}^{-3}$ for discrete MOSFETs.

Substrate diodes: Due to the construction of a MOSFET, there are *substrate diodes* between bulk and source and bulk and drain; Fig. 3.28 shows the location and polarity of these diodes in the equivalent circuit diagram for an n-channel MOSFET. The currents through these diodes are described by the diode equations

$$I_{D,S} = I_{S,S} \left(e^{\frac{V_{BS}}{nV_T}} - 1 \right) \tag{3.21}$$

$$I_{D,D} = I_{S,D} \left(e^{\frac{V_{BD}}{nV_T}} - 1 \right) \tag{3.22}$$

with *saturation reverse currents* $I_{S,S}$ and $I_{S,D}$ and the *emission factor* $n \approx 1$.

The *reverse current density* J_S and the *edge current density* $J_{S,SW}$ may be given as an alternative to $I_{S,S}$ and $I_{S,D}$; with the areas A_S and A_D and the edge length l_S and l_D for the source and drain regions, it follows that:

$$I_{S,S} = J_S A_S + J_{S,SW} l_S \tag{3.23}$$

$$I_{S,D} = J_S A_D + J_{S,SW} l_D \tag{3.24}$$

In particular, this is used in CAD programs for the design of integrated circuits; in this case, J_S and $J_{S,SW}$ are parameters of the MOS process and are the same for all n-channel

MOSFETs. After the sizes of the individual MOSFETs have been determined, only the areas and edge lengths need to be determined; the CAD program then calculates $I_{S,S}$ and $I_{S,D}$.

In normal operation, the bulk terminal of an n-channel MOSFET has a lower, or at most the same, potential as the drain and source; thus $V_{BS}, V_{BD} \leq 0$ and the diodes are operated in the reverse-biased mode. In discrete MOSFETs, with their internal connection between source and bulk, this condition is automatically met as long as $V_{DS} > 0$. In integrated circuits the common bulk terminal of the n-channel MOSFET is connected to the negative supply voltage, so that the diodes are always reverse-biased. In the case of smaller MOSFETs, the reverse currents $I_{D,S} \approx -I_{S,S}$ and $I_{D,D} \approx -I_{S,D}$ are in the pA region, while in power MOSFETs they are in the μA region; they can usually be neglected.

Other effects: There is a multitude of other effects that are not taken into consideration by the level-1 model. The most important are briefly outlined below [3.2]:

– For short channel lengths L, the region beneath the channel is highly restricted by the depletion layers of the bulk–source and the bulk–drain diodes. Space charges in that area are increasingly compensated by charges in the source and drain regions, resulting in a reduction of the gate charge; this causes the threshold voltage V_{th} to drop. This is known as the *short-channel effect* and it depends on the voltages V_{BS} and V_{BD} or $V_{DS} = V_{BS} - V_{BD}$. With an increasing drain–source voltage, the threshold voltage decreases with a corresponding rise in the drain current; the slope of the output characteristics in the pinch-off region thus depends on V_{DS}. The description of these effects in the level-2/3 and the BSIM models can therefore be regarded as *extended channel-length modulation* which, in this case, is no longer modeled by the Early voltage V_A or the channel-length modulation parameter λ, but by the threshold voltage

$$V_{th} = V_{th,0} + \gamma \left((1 - f(L, V_{DS}, V_{BS})) \sqrt{V_{inv} - V_{BS}} - \sqrt{V_{inv}} \right)$$

The function $f(L, V_{DS}, V_{BS})$ is detailed in reference [3.3]. Figure 3.30a shows the dependence of the threshold voltage on the channel length in an integrated MOSFET.
– With a shrinking channel width W, the charge at the edges of the channel increases compared to the charge within the channel and must be taken into consideration. It is compensated by the charge of the gate, which causes the threshold voltage V_{th} to increase. This is known as the *narrow-channel effect* and it can be described by expanding the threshold voltage equation:

$$V_{th} = V_{th,0} + \gamma \left(\ldots \right) + k \frac{V_{inv} - V_{BS}}{W}$$

The factor k is described in more detail in reference [3.3]. Figure 3.30b shows the dependence of the threshold voltage on the channel width in an integrated MOSFET.
– Free charges exist in the channel area even without the inversion channel; this allows a small drain current to flow even with a voltage below the threshold voltage V_{th}. This effect is called the *sub-threshold effect* and the current is called the *sub-threshold current*. In this *sub-threshold region*, the characteristic is an exponential curve, and in the range of

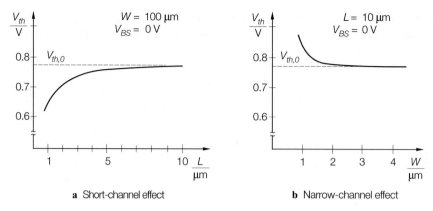

a Short-channel effect   **b** Narrow-channel effect

Fig. 3.30. Influence of the geometric dimensions on the threshold voltage

the threshold voltage it changes to the characteristic of the pinch-off region;

$$
I_D = \begin{cases}
2K \left(\dfrac{n_V V_T}{e} \right)^2 e^{\frac{V_{GS}-V_{th}}{n_V V_T}} \left(1 + \dfrac{V_{DS}}{V_A} \right) & \text{for } V_{GS} < V_{th} + 2n_V V_T \\[3mm]
\dfrac{K}{2} (V_{GS} - V_{th})^2 \left(1 + \dfrac{V_{DS}}{V_A} \right) & \text{for } V_{GS} \geq V_{th} + 2n_V V_T
\end{cases}
\tag{3.25}
$$

Here, $n_V \approx 1.5\dots2.5$ is the *emission factor in the sub-threshold region*. The changeover takes place at $V_{GS} \approx V_{th} + 3\dots5 \cdot V_T \approx V_{th} + 78\dots130\,\text{mV}$. Figure 3.31 shows a plot of the drain current in the range of the threshold region on both linear and logarithmic scales; the latter shows the exponential sub-threshold current as a straight line. In integrated MOS circuits for battery-operated units, the MOSFETs are often operated in this region; this allows the current consumption to be markedly reduced at the expense of circuit bandwidth.

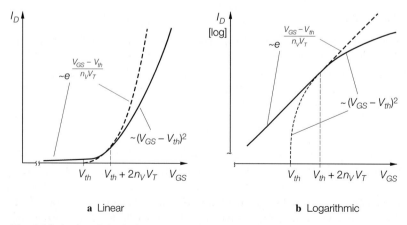

a Linear   **b** Logarithmic

Fig. 3.31. A plot of the drain current in the sub-threshold region

p-channel MOSFET: The characteristics of a p-channel MOSFET are obtained by mirroring the families of output and transfer characteristics of an n-channel MOSFET at their origin. In the equations, these mirrored conditions cause a change in the polarity of all voltages and currents; from

$$V_{DS,po} = V_{GS} - V_{th} < 0$$

it follows that:

$$I_D = \begin{cases} 0 & \text{for } V_{GS} > V_{th} \\[2ex] -\dfrac{K_p' W}{L} V_{DS} \left(V_{GS} - V_{th} - \dfrac{V_{DS}}{2} \right) \left(1 - \dfrac{V_{DS}}{V_A} \right) & \text{for } V_{GS} \leq V_{th}, \\ & V_{DS,po} < V_{DS} \leq 0 \\[2ex] -\dfrac{K_p' W}{2L} (V_{GS} - V_{th})^2 \left(1 - \dfrac{V_{DS}}{V_A} \right) & \text{for } V_{GS} \leq V_{th}, \\ & V_{DS} \leq V_{DS,po} \end{cases}$$

$$V_{th} = V_{th,0} - \gamma \left(\sqrt{V_{inv} + V_{BS}} - \sqrt{V_{inv}} \right)$$

The parameters γ and V_{inv} are also determined for the p-channel MOSFET using (3.19) and (3.20). The Early voltage V_A is positive in both the n channel and the p channel MOSFET; the relative transconductance coefficient is also positive:

$$K_p' = \frac{\mu_p \epsilon_0 \epsilon_{r,ox}}{d_{ox}}$$

Here, $\mu_p = 0.015 \ldots 0.025 \text{ m}^2/\text{Vs}$. The currents in the substrate diodes are:

$$I_{D,S} = -I_{S,S} \left(e^{-\frac{V_{BS}}{n V_T}} - 1 \right)$$

$$I_{D,D} = -I_{S,D} \left(e^{-\frac{V_{BD}}{n V_T}} - 1 \right)$$

Spreading Resistances

For each terminal, there is a spreading resistance that is made up of the resistance of the relevant region and the contact resistance of the metal zone. Figure 3.32a shows the resistances R_G, R_S, R_D and R_B in a sample integrated n-channel MOSFET. In CAD programs for circuit simulations, these resistances can be entered directly or by using the *sheet resistance* R_{sh} and the multipliers n_{RG}, n_{RS}, n_{RD} and n_{RB}:

$$\begin{bmatrix} R_G \\ R_S \\ R_D \\ R_B \end{bmatrix} = R_{sh} \begin{bmatrix} n_{RG} \\ n_{RS} \\ n_{RD} \\ n_{RB} \end{bmatrix} \tag{3.26}$$

In this case, the sheet resistance is a property of the MOS process and is identical for all n-channel MOSFETs in a given integrated circuit. Typical values are $R_{sh} \approx 20 \ldots 50 \, \Omega$ in n-channel MOSFETs and $R_{sh} \approx 50 \ldots 100 \, \Omega$ in p-channel MOSFETs.

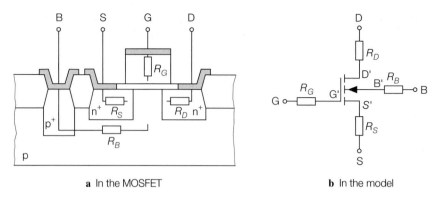

a In the MOSFET   **b** In the model

Fig. 3.32. Spreading resistances in an integrated n-channel MOSFET

Figure 3.32b shows the extended model. We must distinguish between the *external* terminals G, S, D and B and the *internal* terminals G', S', D' and B', which means that the drain current I_D and the diode currents $I_{D,S}$ and $I_{D,D}$ depend on the internal voltages $V_{G'S'}$, $V_{D'S'}$, ...

Vertical Power MOSFETs

The special properties of vertical power MOSFETs (DMOS FETs) have already been described in Sect. 3.2.2; Fig. 3.26 on page 196 presents the relevant characteristics. The displacement of the transfer characteristic in Fig. 3.26a is caused by the source resistance R_S. If $I_G = 0$, it follows from Fig. 3.32b that:

$$V_{GS} = V_{G'S'} + I_D R_S = V_{th} + \sqrt{\frac{2I_D}{K\left(1 + \dfrac{V_{D'S'}}{V_A}\right)}} + I_D R_S$$

$$\overset{V_A \to \infty}{\approx} V_{th} + \sqrt{\frac{2I_D}{K}} + I_D R_S \tag{3.27}$$

This equation is used for parameter extraction; on the basis of at least three pairs of values (V_{GS}, I_D) in the pinch-off region, it is possible to determine the three parameters V_{th}, K and R_S.[16]

The relations in the family of output characteristics shown in Fig. 3.26b are more complicated. It is possible to describe the displacement by a resistance in the drain line but, in contrast to the linear drain resistance R_D in Fig. 3.32b, this resistance is nonlinear. This is caused by the *conductivity modulation* in the drift region, which means that the conductivity in the drift region increases with the current due to an increase in the charge carrier density. The voltage drop V_{Drift} is approximately [3.4]:

$$V_{Drift} = V_0 \left(\sqrt{1 + 2\frac{I_D}{I_0}} - 1\right) \tag{3.28}$$

---

[16] In practice, a large number of pairs of values are used and the parameters are determined by means of an *orthogonal projection*.

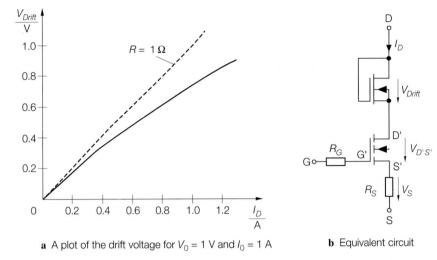

a A plot of the drift voltage for $V_0 = 1$ V and $I_0 = 1$ A

b Equivalent circuit

Fig. 3.33. Drift voltage of a vertical power MOSFET

Here, V_0 and I_0 are the drift region parameters. Figure 3.33a shows the plot of the drift voltage versus I_D for a MOSFET with $V_0 = 1$ V and $I_0 = 1$ A. For small currents the drift region acts like a linear resistor with $R = V_0/I_0$; in Fig. 3.33a, $R = 1\,\Omega$. For higher currents, the conductivity increases and the voltage drop is less than when $R = 1\,\Omega$.

The characteristic given by (3.28) corresponds to that of a depletion MOSFET with the gate connected to the drain; if $V_{GS} = V_{DS}$ and $V_{th} < 0$, then

$$I_D = K\,V_{DS}\left(V_{GS} - V_{th} - \frac{V_{DS}}{2}\right) \overset{\substack{V_{GS}=V_{DS}=V_{Drift} \\ V_{th}<0}}{=} K\,|V_{th}|\,V_{Drift} + \frac{1}{2}\,K\,V_{Drift}^2$$

which can be solved to give:

$$V_{Drift} = |V_{th}|\left(\sqrt{1 + \frac{2I_D}{K\,|V_{th}|^2}} - 1\right)$$

A comparison with (3.28) shows that a depletion MOSFET with $V_{th} = -V_0$ and $K = I_0/V_0^2$ can be used to model the drift region; this leads to the equivalent circuit shown in Fig. 3.33b, which features a depletion MOSFET in place of resistance R_D as compared to Fig. 3.32b.

Junction FETs

The model of a junction FET is derived from the model of a MOSFET by eliminating the isolated gate, renaming the bulk gate and inserting $\beta = K/2$ in the equations; the result is

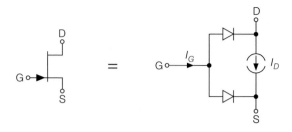

Fig. 3.34. Large-signal equivalent circuit for an n-channel JFET

the equivalent circuit diagram shown in Fig. 3.34, with the equations:

$$
I_D = \begin{cases}
0 & \text{for } V_{GS} < V_{th} \\[2mm]
2\beta V_{DS}\left(V_{GS} - V_{th} - \dfrac{V_{DS}}{2}\right)\left(1 + \dfrac{V_{DS}}{V_A}\right) & \text{for } V_{GS} \geq V_{th}, \\[1mm]
 & 0 \leq V_{DS} < V_{GS} - V_{th} \\[2mm]
\beta\,(V_{GS} - V_{th})^2\left(1 + \dfrac{V_{DS}}{V_A}\right) & \text{for } V_{GS} \geq V_{th}, \\[1mm]
 & V_{DS} \geq V_{GS} - V_{th}
\end{cases}
$$

$$
I_G = I_S\left(e^{\frac{V_{GS}}{nV_T}} + e^{\frac{V_{GD}}{nV_T}} - 2\right) \tag{3.29}
$$

The parameters are the *threshold voltage* V_{th}, the *JFET transconductance coefficient* β, the *Early voltage* V_A, the *saturation leakage current* I_S and the *emission coefficient n*.

As in the MOSFET, additional spreading resistances are allowed in the drain and source lines; the corresponding parameters are R_S and R_D. In the JFET model there is no gate resistance: however, this must be added externally for circuit simulations with CAD programs if an accurate description of the high-frequency response is required.

Unlike the MOSFET model, the JFET model is not scalable; that is, there are no geometric dimensions such as channel length and width. The JFET model is simple but not very accurate.

3.3.2
Dynamic Behavior

The response to pulsed and sinusoidal signals is called the *dynamic behavior* and cannot be determined from the characteristics. This is due to the capacitances between the different regions of the MOSFET, as shown in Fig. 3.35. These may be classified into three groups:

- *Channel capacitances* $C_{GS,ch}$ and $C_{GD,ch}$, which describe the capacitive effect between the gate and the channel. They are active only with an existing channel; that is, when the MOSFET is in the conductive state. Without a channel, there is a capacitance $C_{GB,ch}$ between gate and bulk which is a component part of the *gate–bulk capacitance* C_{GB}. The channel capacitances are linear in the pinch-off region, but nonlinear in the ohmic region.
- The linear *overlap capacitances* $C_{GS,ov}$, $C_{GD,ov}$ and $C_{GB,ov}$ result from the geometric overlap of the gate and the source, drain and bulk regions. $C_{GB,ov}$ results from the overlap of the gate and the bulk at the sides of the channel and is part of C_{GB}.

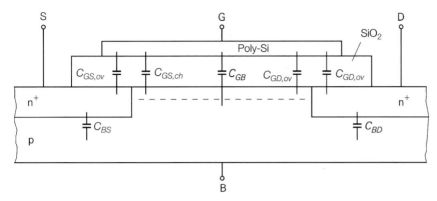

Fig. 3.35. Capacitances of an n-channel MOSFET

– The nonlinear *junction capacitances* C_{BS} and C_{BD} are caused by the pn junctions between bulk and source and between bulk and drain.

In combination, these produce a total of five capacitances:

$$
\begin{aligned}
C_{GS} &= C_{GS,ch} + C_{GS,ov} \\
C_{GD} &= C_{GD,ch} + C_{GD,ov} \\
C_{GB} &= C_{GB,ch} + C_{GB,ov}
\end{aligned}
\tag{3.30}
$$

as well as C_{BS} and C_{BD}.

Channel capacitances: The gate together with the underlying channel form a plate capacitor with an *oxide capacity:*

$$
C_{ox} = \epsilon_{ox} \frac{A}{d_{ox}} = \epsilon_0 \epsilon_{r,ox} \frac{WL}{d_{ox}}
\tag{3.31}
$$

In the cutoff region, that is, without a channel, this capacitance acts between gate and bulk; consequently,

$$
\left.
\begin{aligned}
C_{GS,ch} &= 0 \\
C_{GD,ch} &= 0 \\
C_{GB,ch} &= C_{ox}
\end{aligned}
\right\} \quad \text{for } V_{G'S'} < V_{th}
\tag{3.32}
$$

In the ohmic region the channel extends from the source to the drain region, and the oxide capacitance is distributed along the channel according to the charge distribution. For $V_{D'S'} = 0$, the channel is symmetrical, resulting in $C_{GS,ch} = C_{GD,ch} = C_{ox}/2$. For $V_{D'S'} > 0$, the channel is asymmetrical, resulting in $C_{GS,ch} > C_{GD,ch}$. Thus, the capacitances depend on $V_{D'S'}$ and $V_{G'S'}$, and may be described by the following equations as an approximation [3.3]:

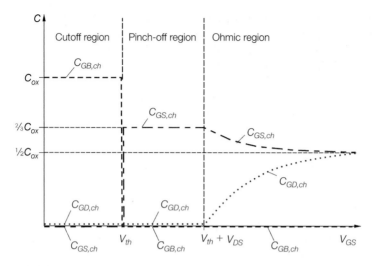

Fig. 3.36. Schematic presentation of the channel capacitances for an n-channel MOSFET. In a real MOSFET, the transitions are smooth

$$
\begin{aligned}
C_{GS,ch} &= \frac{2}{3}\, C_{ox}\left(1 - \left(\frac{V_{G'S'} - V_{th} - V_{D'S'}}{2\,(V_{G'S'} - V_{th}) - V_{D'S'}}\right)^{2}\right) \\[2mm]
C_{GD,ch} &= \frac{2}{3}\, C_{ox}\left(1 - \left(\frac{V_{G'S'} - V_{th}}{2\,(V_{G'S'} - V_{th}) - V_{D'S'}}\right)^{2}\right) \\[2mm]
C_{GB,ch} &= 0
\end{aligned}
\quad
\left.\begin{aligned} \\ \end{aligned}\right\}
\;
\begin{aligned}
&\text{for } V_{G'S'} \geq V_{th}, \\
&V_{D'S'} < V_{G'S'} - V_{th}
\end{aligned}
\quad (3.33)
$$

In the pinch-off region the channel is pinched-off at the drain side so that the connection between the channel and the drain region no longer exists; thus $C_{GD,ch} = 0$. Only $C_{GS,ch}$ is left as a channel capacitance [3.3]:

$$
\left.\begin{aligned}
C_{GS,ch} &= \frac{2}{3}\, C_{ox} \\[1mm]
C_{GD,ch} &= 0 \\[1mm]
C_{GB,ch} &= 0
\end{aligned}\right\}
\quad \text{for } V_{G'S'} \geq V_{th},\ V_{D'S'} \geq V_{G'S'} - V_{th}
\quad (3.34)
$$

Figure 3.36 shows the curves for the three capacitances. Note that the similarity to a plate capacitor only exists where the charge distribution is uniform; only in this case does $C_{GS,ch} + C_{GD,ch} + C_{GB,ch} = C_{ox}$. This is always true for the cutoff region, but only for $V_{D'S'} = 0$ in the ohmic region and never in the pinch-off region.

As Fig. 3.36 shows, the capacitance model provides an abrupt transition from $C_{GB,ch}$ to $C_{GS,ch}$ at the boundary between the cutoff and the pinch-off regions, causing the total capacitance to jump from C_{ox} to $2C_{ox}/3$. This is only a very rough presentation of the true relationship in this region. In a real MOSFET, the transitions are always smooth; the corresponding curves are indicated by broken lines in Fig. 3.36.[17]

---

[17] A relatively simple description of this transition is given in reference [3.3]. An additional problem is charge conservation, consideration of which requires further modification of the equations; a correspondingly expanded model is used in *PSpice* from *OrCAD* [3.5].

Overlap capacitances: As the gate is usually larger than the channel[18], that is, wider than the channel width W and longer than the channel length L, overlaps occur at the edges, producing the *overlap capacitances* $C_{GS,ov}$, $C_{GD,ov}$ and $C_{GB,ov}$. But it is not possible to obtain these capacitances from the related overlap areas by means of the plate capacitive formula, as the field and charge distributions along the margins are not uniform. For this reason, the *capacitances* per unit length $C'_{GS,ov}$, $C'_{GD,ov}$ and $C'_{GB,ov}$ are specified, which are determined by way of measurement or by field simulation; consequently:

$$
\begin{aligned}
C_{GS,ov} &= C'_{GS,ov}\, W \\
C_{GD,ov} &= C'_{GD,ov}\, W \\
C_{GB,ov} &= C'_{GB,ov}\, L
\end{aligned}
\tag{3.35}
$$

$C'_{GB,ov}$ contains portions of both sides and therefore needs to be multiplied by the single channel length only. $C'_{GS,ov} = C'_{GD,ov}$ and $C_{GS,ov} = C_{GD,ov}$ is the case in MOSFETs of symmetrical design; but these values deviate from each other in high-voltage MOSFETs with an additional drift region.

In vertical power MOSFETs the gate source overlap capacitance $C_{GS,ov}$ is particularly large because the whole area of the source metal coating covers the gate grid beneath (see Fig. 3.24 on page 195 or C_{GS} in Fig. 3.25 on page 195). This additional portion of the overlap capacitance depends on W and L; however, in cases in which MOSFETs of different sizes consist of a different number of identical cells, it depends on W only; L is then identical for all MOSFETs.

Junction capacitances: pn junctions between bulk and source and between bulk and drain have a voltage-dependent *junction capacitance*, C_{BS} or C_{BD} respectively, which are determined by the doping, the area of the junction and the applied voltage. The description is like that of a diode; from (1.13) on page 19 follows:

$$
C_{BS}(V_{B'S'}) = \frac{C_{J0,S}}{\left(1 - \dfrac{V_{B'S'}}{V_{Diff}}\right)^{m_J}} \qquad \text{for } V_{B'S'} \le 0
\tag{3.36}
$$

$$
C_{BD}(V_{B'D'}) = \frac{C_{J0,D}}{\left(1 - \dfrac{V_{B'D'}}{V_{Diff}}\right)^{m_J}} \qquad \text{for } V_{B'D'} \le 0
\tag{3.37}
$$

with the *zero capacitances* $C_{J0,S}$ and $C_{J0,D}$, the *diffusion voltage* V_{Diff} and the *capacitance coefficient* $m_J \approx 1/3 \ldots 1/2$.

As alternatives to $C_{J0,S}$ and $C_{J0,D}$, one can use the *junction capacitance per unit area* C'_J, the *sidewall capacitance per unit length* $C'_{J,SW}$, the *sidewall diffusion voltage* $V_{Diff,SW}$ and the *sidewall capacitance coefficient* $m_{J,SW}$; for the areas A_S and A_D and the sidewall length l_S and l_D of the source and drain regions, this results in:

$$
C_{BS} = \frac{C'_J A_S}{\left(1 - \dfrac{V_{B'S'}}{V_{Diff}}\right)^{m_J}} + \frac{C'_{J,SW} l_S}{\left(1 - \dfrac{V_{B'S'}}{V_{Diff,SW}}\right)^{m_{J,SW}}} \qquad \text{for } V_{B'S'} \le 0
\tag{3.38}
$$

---

[18] In order to allow the formation of a continuous channel, the size of the gate must be *at least* as large as the channel region.

$$CBD = \frac{C'_J A_D}{\left(1 - \dfrac{V_{B'D'}}{V_{Diff}}\right)^{m_J}} + \frac{C'_{J,SW} l_D}{\left(1 - \dfrac{V_{B'D'}}{V_{Diff,SW}}\right)^{m_{J,SW}}} \qquad \text{for } V_{B'D'} \le 0 \qquad (3.39)$$

CAD programs make use of this for the design of integrated circuits; in this case, C'_J, $C'_{J,SW}$, V_{Diff}, $V_{Diff,SW}$, m_J and $m_{J,SW}$ are parameters of the MOS process and are the same for all n-channel MOSFETs. If the parameters of the individual MOSFETs are given, then only the areas and the sidewall length have to be determined; from these, the CAD program calculates C_{BS} and C_{BD}.

Here, the range for the use of these equations is limited to $V_{B'S'} \le 0$ and $V_{B'D'} \le 0$. If $V_{B'S'} > 0$ and $V_{B'D'} > 0$, then the pn junctions are forward biased and the diffusion capacitance must be considered in addition to the junction capacitance. In other words, a complete capacitance model, as for a diode, must be used (see Sect. 1.3.2 on page 19); this includes the *transit time* τ_T as an additional parameter that is required for calculating the diffusion capacitance. CAD programs use a complete capacitance model for every pn junction.

Level-1 MOSFET Model

Figure 3.37 shows the complete level-1 model of an n-channel MOSFET; it is used in CAD programs for circuit simulation. Figure 3.38 gives an overview of the model's variables and equations. The parameters used by the model are listed in Fig. 3.38; the figure also contains the parameter designations used in the circuit simulator *PSpice*.[19]

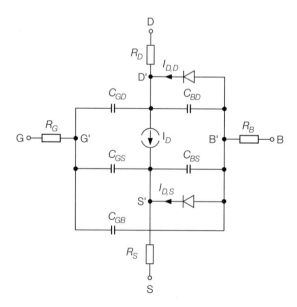

Fig. 3.37. Level-1 MOSFET model for an n-channel MOSFET

---

[19] *PSpice* is an *OrCAD* product.

Variable	Name	Equation(s)
I_D	Ideal drain current	(3.16)
$I_{D,S}$	Current of bulk source diode	(3.21),(3.23)
$I_{D,D}$	Current of bulk drain diode	(3.22),(3.24)
R_G	Gate spreading resistance	
R_S	Source spreading resistance	(3.26)
R_D	Drain spreading resistance	
R_B	Bulk spreading resistance	
C_{GS}	Gate–source capacitance	
C_{GD}	Gate–drain capacitanc	(3.30)–(3.35)
C_{GB}	Gate–bulk capacitance	
C_{BS}	Bulk–source capacitance	(3.36) bzw. (3.38)
C_{BD}	Bulk–drain capacitance	(3.37) bzw. (3.39)

Fig. 3.38. Variables of the level-1 MOSFET model

There are four different parameter types:

– *Process parameters (P)*: these parameters characterize the MOS process and are the same for all n-channel MOSFETs and all p-channel MOSFETs, respectively, contained in one integrated circuit.
– *Scalable process parameters (PS)*: these parameters also characterize the MOS process, but need scaling according to the geometric data of the given MOSFET.
– *Scaling parameters (S)*: these are the geometric data of the MOSFET in question. From these parameters and from the scalable process parameters, the effective parameters of a particular MOSFET are determined; for example $K = K'_n W/L$.
– *Effective parameters (E)*: these parameters apply to a MOSFET of a given size.

Figure 3.40 lists parameter values of typical NMOS and CMOS processes.

Some of the model characteristics can be specified as scalable or effective parameters; this is the case with spreading resistances, for example, which can be specified as the scalable parameters $n_{RG}, \ldots, n_{RB}$ and R_{sh} or as effective parameters $R_G, \ldots, R_B$.

The oxide thickness d_{ox} also influences the dynamic response, because it is required to determine the channel capacitances; it is quoted only once in Fig. 3.39. The parameters K'_n and γ need not be specified, because they can be calculated from d_{ox}, μ_n, V_{inv} and N_{sub}; in turn, V_{inv} can be calculated from N_{sub}. In the event of contradictory information, the direct specification has priority over the calculated value.

Discrete MOSFETs: While the nonscalable Gummel–Poon model applies equally to discrete and integrated bipolar transistors, the scalable level-1 MOSFET model applies, strictly speaking, only to integrated MOSFETs of the most simple design; discrete MOS-FETs designed as vertical DMOS FETs and integrated MOSFETs with a drift junction display different behavior in some respects. However, it has been found that these MOS-FETs can be described approximately by the level-1 model if some of the parameters are used in a different sense; these parameters lose their original meanings and may take on values that seem absurd with respect to semiconductor physics. Figure 3.41 contains the level-1 parameters of some DMOS FETs. The substrate control factor γ can be disregarded since the source and bulk are connected to each other; furthermore, the channel-length modulation is neglected; that is, the parameter λ is omitted.

Parameter	PSpice	Name	Type
Geometric data			
W	W	Channel width	S
L	L	Channel length	S
A_S	AS	Source region area	S
l_S	PS	Source region sidewall length	S
A_D	AD	Drain region area	S
l_D	PD	Drain region sidewall length	S
n_{RG}	NRG	Multiplier for gate spreading resistance	S
n_{RS}	NRS	Multiplier for source spreading resistance	S
n_{RD}	NRD	Multiplier for drain spreading resistance	S
n_{RB}	NRB	Multiplier for bulk spreading resistance	S
Static behavior			
K_n'	KP	Relative transconductance coefficient	PS
$V_{th,0}$	VTO	Zero threshold voltage	P
γ	GAMMA	Substrate control factor	P
λ	LAMBDA	Channel-length modulation parameter	P
V_A	-	Early voltage ($V_A = 1/\lambda$)	P
d_{ox}	TOX	Oxide thickness	P
μ_n	UO	Charge carrier mobility in cm^2/Vs	P
V_{inv}	PHI	Inversion voltage	P
N_{sub}	NSUB	Substrate doping density in cm^{-3}	P
J_S	JS	Reverse current density of bulk diodes	PS
$J_{S,SW}$	JSSW	Sidewall current density of bulk diodes	PS
n	N	Emission coefficient of bulk diodes	P
$I_{S,S}$	IS	Saturation reverse current of bulk source diode	E
$I_{S,D}$	IS	Saturation reverse current of bulk drain diode	E
R_{sh}	RSH	Sheet resistance	PS
R_G	RG	Gate spreading resistance	E
R_S	RS	Source spreading resistance	E
R_D	RD	Drain spreading resistance	E
R_B	RB	Bulk spreading resistance	E
Dynamic behavior			
C_J'	CJ	Junction capacitance per unit length	PS
m_J	MJ	Capacitance coefficient of bulk diodes	P
V_{Diff}	PB	Diffusion voltage of bulk diodes	P
$C_{J,SW}'$	CJSW	Sidewall capacitance per unit length	PS
$m_{J,SW}$	MJSW	Sidewall capacitance coefficient	P
$V_{Diff,SW}$	PBSW	Sidewall diffusion voltage	P
f_C	FC	Coefficient for capacitance curve	P
$C_{J0,S}$	CBS	Zero capacitance of bulk source diode	E
$C_{J0,D}$	CBD	Zero capacitance of bulk drain diode	E
$C_{GS,ov}'$	CGSO	Gate–source overlap capacitance	PS
$C_{GD,ov}'$	CGDO	Gate–drain overlap capacitancet	PS
$C_{GB,ov}'$	CGBO	Gate–bulk overlap capacitance	PS
τ_T	TT	Transit time for substrate diodes	P
Model selection			
– –	LEVEL	LEVEL = 1 selects the level-1 model	– –

Fig. 3.39. Parameters of level-1 MOSFET model

Parameter	PSpice	NMOS		CMOS		Unit
		Enhancement	Depletion	n-channel	p-channel	
K'_n, K'_p	KP	37	33	69	23.5	$\mu A/V^2$
$V_{th,0}$	VTO	1.1	−3.8	0.73	−0.75	V
γ	GAMMA	0.41	0.92	0.73	0.56	$\sqrt{V}$
λ	LAMBDA	0.03	0.01	0.033	0.055	V^{-1}
V_A	-	33	100	30	18	V
d_{ox}	TOX	55	55	25	25	nm
μ_n	UO	590	525	500	170	cm^2/Vs
V_{inv}	PHI	0.62	0.7	0.76	0.73	V
N_{sub}	NSUB	0.2	1	3	1.8	$10^{16}/cm^3$
R_{sh}	RSH	25	25	25	45	Ω
C'_J	CJ	110	110	360	340	$\mu F/m^2$
m_J	MJ	0.5	0.5	0.4	0.5	
V_{Diff}	PB	0.8	0.8	0.9	0.9	V
$C'_{J,SW}$	CJSW	500	500	250	220	pF/m
$m_{J,SW}$	MJSW	0.33	0.33	0.2	0.2	
$V_{Diff,SW}$	PBSW	0.8	0.8	0.9	0.9	V
f_C	FC	0.5	0.5	0.5	0.5	
$C'_{GS,ov}$	CGSO	160	160	300	300	pF/m
$C'_{GD,ov}$	CGDO	160	160	300	300	pF/m
$C'_{GB,ov}$	CGBO	170	170	150	150	pF/m

Fig. 3.40. Parameters of NMOS and CMOS processes

Parameter	PSpice	BSD215	IRF140	IRF9140	Unit
W	W	$540\,\mu$	0.97	1.9	m
L	L	2	2	2	μm
K'_n, K'_p	KP	20.8	20.6	10.2	$\mu A/V^2$
$V_{th,0}$	VTO	0.95	3.2	−3.7	V
d_{ox}	TOX	100	100	100	nm
μ_n	UO	600	600	300	cm^2/Vs
V_{inv}	PHI	0.6	0.6	0.6	V
I_S	IS	125	1.3	10^{-5}	pA
R_G	RG	−	5.6	0.8	Ω
R_S	RS	0.02	0.022	0.07	Ω
R_D	RD	25	0.022	0.06	Ω
R_B	RB	370	−	−	Ω
$C'_{GS,ov}$	CGSO	1.2	1100	880	pF/m
$C'_{GD,ov}$	CGDO	1.2	430	370	pF/m
$C_{J0,D}$	CBD	5.35	2400	2140	pF
m_J	MJ	0.5	0.5	0.5	
V_{Diff}	PB	0.8	0.8	0.8	V
f_C	FC	0.5	0.5	0.5	
τ_T	TT	−	142	140	ns

BSD215, n-channel small-signal FET; IRF140, n-channel power FET;
IRF9140, p-channel power FET.

Fig. 3.41. Parameters of some DMOS FETs

Where accuracy ranks higher, a *macromodel* must be used which, in addition to the actual MOSFET model, includes other components to model specific characteristics. One example is the static equivalent circuit for a DMOS FET, shown in Fig. 3.33b, for which an additional MOSFET is used to model the nonlinear drain resistance. Similar expansions are needed to describe the dynamic behavior of a DMOS FET, but there is no uniform equivalent circuit diagram.

Even though in some cases the level-2 and level-3 models use different equations, their parameters are largely identical. They use the following additional parameters [3.3]:

- *Level-2 Model*: UCRIT, UEXP and VMAX for the voltage dependence of the mobility, and NEFF to describe the channel charge.
- *Level-3 Model*: THETA, ETA and KAPPA for empirical modelling of the static behavior.
- *Both Models*: DELTA for modelling the narrow-channel effect and XQC for the charge distribution in the channel.

Both models describe channel-length modulation with the help of additional parameters; this renders the channel-length modulation parameter λ unnecessary.

Junction FET Model

Figure 3.42 shows the model of an n-channel junction FET. It is derived from the level-1 model of an n-channel MOSFET by eliminating the gate terminal together with its related components and by renaming the bulk gate. The variables and equations are summarized in Fig. 3.43. Figure 3.44 lists the parameters.

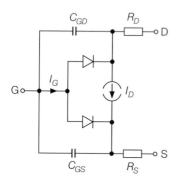

Fig. 3.42. Model of an n-channel junction FET

Variable	Name	Equation
I_D	Ideal drain current	(3.29)
I_G	Gate current	
R_S	Source spreading resistance	
R_D	Drain spreading resistance	
C_{GS}	Gate–source capacitance	(3.36) with $C_{BS} \rightarrow C_{GS}$
C_{GD}	Gate–drain capacitance	(3.37) with $C_{BD} \rightarrow C_{GD}$

Fig. 3.43. Variables of the junction FET model

Parameter	PSpice	Name
Static behavior		
β	BETA	JFET transconductance coefficient
V_{th}	VTO	Threshold voltage
λ	LAMBDA	Channel-length modulation parameter ($\lambda = 1/V_A$)
I_S	IS	Saturation reverse current of diodes
n	N	Emission coefficient of diodes
R_S	RS	Source spreading resistance
R_D	RD	Drain spreading resistance
Dynamic behavior		
$C_{J0,S}$	CGS	Zero capacitance of gate source diode
$C_{J0,D}$	CGD	Zero capacitance of gate drain diode
V_{Diff}	PB	Diffusion voltage of diodes
m_J	M	Capacitance coefficient of diodes
f_C	FC	Coefficient for capacitance variations

Fig. 3.44. Parameters of the junction FET model

3.3.3
Small-Signal Model

A linear *small-signal model* can be derived from the level-1 MOSFET model by linearization at an operation point. In practice, the operating point is selected so that the FET operates in the pinch-off region; the small-signal models described here are thus applicable to this operating mode only.

The *static small-signal model* describes the small-signal response at low frequencies and is therefore also known as the *DC small-signal equivalent circuit*. The *dynamic small-signal model* further describes the dynamic small-signal response and is needed to calculate the frequency response of a circuit; it is also known as the *AC small-signal equivalent circuit*.

Static Small-Signal Model in the Pinch-Off Region

Small-signal parameters of the level-1 MOSFET model: Omitting the capacitances and neglecting the leakage currents ($I_{D,S} = I_{D,D} = 0$) in Fig. 3.37 leads to the *static* level-1 model shown in Fig. 3.45a; the spreading resistances R_G and R_B are disregarded because no current can flow through these paths. Linearizing the large-signal (3.16) und (3.18) at the operating point A leads to:

$$g_m = \left.\frac{\partial I_D}{\partial V_{G'S'}}\right|_A = \frac{K_n' W}{L}\left(V_{G'S',A} - V_{th}\right)\left(1 + \frac{V_{D'S',A}}{V_A}\right)$$

$$g_{m,B} = \left.\frac{\partial I_D}{\partial V_{B'S'}}\right|_A = \left.\frac{\partial I_D}{\partial V_{th}}\right|_A \frac{dV_{th}}{dV_{BS}}$$

$$= \frac{\gamma}{2\sqrt{V_{inv} - V_{B'S',A}}} \frac{K_n' W}{L}\left(V_{G'S',A} - V_{th}\right)\left(1 + \frac{V_{D'S',A}}{V_A}\right)$$

$$\frac{1}{r_{DS}} = \left.\frac{\partial I_D}{\partial V_{D'S'}}\right|_A = \frac{1}{V_A}\frac{K_n' W}{2L}\left(V_{G'S',A} - V_{th}\right)^2$$

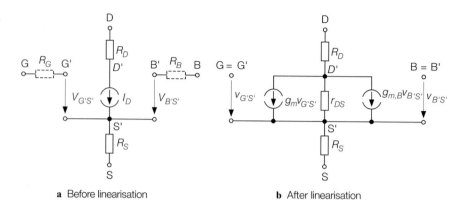

a Before linearisation   **b** After linearisation

Fig. 3.45. Generating the static small-signal model by linearization of the static level-1 MOSFET model

Approximations for small-signal parameters: The small-signal parameters g_m, $g_{m,B}$ and r_{DS} are determined in CAD programs solely by the equations indicated above; practical applications use the following approximations, which are obtained by back-substitution of $I_{D,A}$, reference from $g_{m,B}$ to g_m, the assumption that $V_{D'S',A} \ll V_A$ and by inserting $K = K'_n W/L$:

$$g_m = \left.\frac{\partial I_D}{\partial V_{G'S'}}\right|_A = \sqrt{2K\,I_{D,A}\left(1 + \frac{V_{D'S',A}}{V_A}\right)} \overset{V_{D'S',A} \ll V_A}{\approx} \sqrt{2K\,I_{D,A}} \qquad (3.40)$$

$$g_{m,B} = \left.\frac{\partial I_D}{\partial V_{B'S'}}\right|_A = \frac{\gamma\, g_m}{2\sqrt{V_{inv} - V_{B'S',A}}} \qquad (3.41)$$

$$r_{DS} = \left.\frac{\partial V_{D'S'}}{\partial I_D}\right|_A = \frac{V_A + V_{D'S',A}}{I_{D,A}} \overset{V_{D'S',A} \ll V_A}{\approx} \frac{V_A}{I_{D,A}} \qquad (3.42)$$

The approximations for g_m and r_{DS} correspond with (3.11) and (3.12) in Sect. 3.1.4. An additional small-signal parameter is the *substrate transconductance* $g_{m,B}$, which is only effective with a small-signal voltage $v_{BS} \neq 0$ between source and bulk.

Small-signal parameters in the sub-threshold region: In many integrated CMOS circuits with a particularly low current consumption, the MOSFETs are operated in the sub-threshold region. In this region the drain current I_D shows an exponential dependence on V_{GS} according to (3.25); in this case, the transconductance is:

$$g_m = \frac{I_{D,A}}{n_V V_T} \qquad \text{for } V_{GS} < V_{th} + 2n_V V_T \qquad (3.43)$$

Equations (3.41) and (3.42) for $g_{m,B}$ and r_{DS} also apply to the sub-threshold region. For $n_V \approx 2$, the borderline of the sub-threshold region is at $V_{GS} \approx V_{th} + 4V_T \approx V_{th} + 100\,\text{mV}$ or $I_D \approx 2K\,(n_V V_T)^2 \approx K \cdot 0.005\,\text{V}^2$. The transconductance is continuous; in other words,

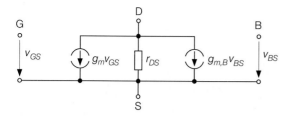

Fig. 3.46. Simplified static small-signal mode

at the transition both (3.40) and (3.43) yield the same result:

$$\sqrt{2K\,I_{D,A}} \quad \overset{I_{D,A}=2K(n_V V_T)^2}{=} \quad \frac{I_{D,A}}{n_V V_T}$$

DC small-signal equivalent circuit: Figure 3.45b shows the resulting static small-signal model. The spreading resistances R_S and R_D are disregarded in almost all practical calculations; this leads to the small-signal equivalent circuit shown in Fig. 3.46, which is derived from the small-signal equivalent circuit described in Sect. 3.1.4 by adding the controlled source with the substrate transconductance $g_{m,B}$.

Small-signal equivalent circuit for junction FETs: Figure 3.46 also applies to junction FETs if the controlled source with the substrate transconductance is removed; the small-signal parameters are obtained using (3.29):

$$g_m \;=\; 2\sqrt{\beta\,I_{D,A}\left(1+\frac{V_{D'S',A}}{V_A}\right)} \;\overset{V_{D'S',A}\ll V_A}{\approx}\; 2\sqrt{\beta\,I_{D,A}} \;=\; \frac{2}{|V_{th}|}\sqrt{I_{D,0}I_{D,A}}$$

$$r_{DS} \;=\; \frac{V_{D'S',A}+V_A}{I_{D,A}} \;\overset{V_{D'S',A}\ll V_A}{\approx}\; \frac{V_A}{I_{D,A}}\,,$$

where $I_{D,0} = I_D(V_{GS}=0) = \beta\,V_{th}^2$. By virtue of the relation $K = 2\beta$, this leads to the same equations as those used for the MOSFET.

Dynamic Small-Signal Model in the Pinch-Off Region

Complete model: Adding the channel, overlap and junction capacitances to the static small-signal model of Fig. 3.45b leads to the dynamic small-signal model for the pinch-off region as shown in Fig. 3.47; with respect to Sect. 3.3.2 it follows that:

$$\begin{aligned}
C_{GS} &= C_{GS,ch}+C_{GS,ov} = \frac{2}{3}C'_{ox}W L + C'_{GS,ov}W \\
C_{GD} &= C_{GD,ov} = C'_{GD,ov}W \\
C_{GB} &= C_{GB,ov} = C'_{GB,ov}L \qquad\qquad (3.44)\\
C_{BS} &= C_{BS}(V_{B'S',A}) \\
C_{BD} &= C_{BD}(V_{B'D',A})
\end{aligned}$$

For this:

$$C'_{ox} \;=\; \frac{\epsilon_0\epsilon_{r,ox}}{d_{ox}} \qquad\qquad (3.45)$$

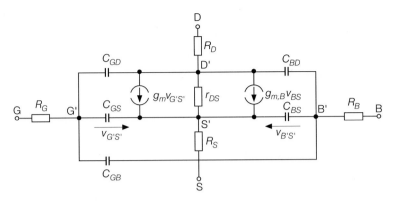

Fig. 3.47. Dynamic small-signal mode

The *gate–source capacitance* C_{GS} comprises the channel capacitance in the pinch-off region and the gate–source overlap capacitance; it is determined by the geometric dimensions alone and is independent of the operating point voltages as long as one remains in the pinch-off region. The *gate–drain capacitance* C_{GD} and the *gate–bulk capacitance* C_{GB} are pure overlap capacitances and therefore are not influenced by the operating point, while the junction capacitances C_{BS} and C_{BD} depend on the operating point voltages $V_{B'S',A}$ and $V_{B'D',A}$.

Simplified model: In practical applications, the spreading resistances R_S, R_D and R_B are neglected; the gate resistance R_G may not be disregarded, since together with C_{GS} it forms a lowpass filter in the gate circuit and must be taken into account when calculating the dynamic behavior of the basic circuit. The gate–bulk capacitance C_{GB} is effective only in MOSFETs with a very narrow channel width W and thus may also be neglected. This leads to the simplified small-signal model shown in Fig. 3.48, which is used for calculating the frequency response of basic circuits.

The source and bulk of discrete MOSFETs are usually connected to each other; this eliminates the source with the substrate transconductance $g_{m,B}$ and the bulk–source capacitance C_{BS}. In this case, the bulk–drain capacitance is arranged between the drain and the source and renamed C_{DS}. The result is the small-signal model shown in Fig. 3.49a, which resembles the small-signal model of a bipolar transistor, as can be seen from a comparison with Fig. 3.49b. On the basis of these similarities, the results of the small-signal

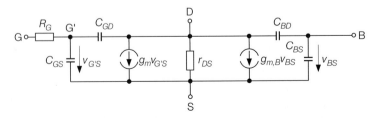

Fig. 3.48. Simplified dynamic small-signal model

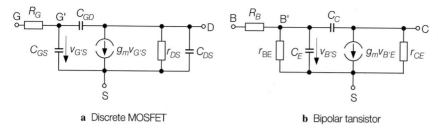

Fig. 3.49. A comparison of the dynamic small-signal models for discrete MOSFET and bipolar transistor

calculations can be used by exchanging the relevant variables, assuming $r_{BE} \to \infty$ and inserting:[20]

$$r_{CE} \doteq \frac{r_{DS}}{1 + s C_{DS} r_{DS}}$$

This model can also apply to integrated MOSFETs if the source and bulk in the small-signal equivalent circuit are combined and connected to the small-signal ground.

Cutoff Frequencies in Small-Signal Operation

The *transconductance cutoff frequency* f_{Y21s} and the *transit frequency* f_T can be calculated by means of the small-signal model. Since both cutoff frequencies are determined for $V_{BS} = 0$ and for $V_{DS} = $ const. (i.e. $v_{DS} = 0$), it is possible to apply the small-signal model of Fig. 3.49a, and in addition to remove r_{DS} and C_{DS}.

Transconductance cutoff frequency: The ratio of the Laplace-transformed small-signal current i_D and the small-signal voltage v_{GS} of the common-source circuit operated in the pinch-off region with constant $V_{DS} = V_{DS,A}$ is called the *transadmittance* $\underline{y}_{21,s}(s)$; from the small-signal equivalent circuit shown in Fig. 3.50a, it follows that:

$$\underline{y}_{21,s}(s) = \frac{i_D}{\underline{v}_{GS}} = \frac{\mathcal{L}\{i_D\}}{\mathcal{L}\{v_{GS}\}} = \frac{g_m - s C_{GD}}{1 + s (C_{GS} + C_{GD}) R_G}$$

with the *transconductance cutoff frequency*

$$\omega_{Y21s} = 2\pi f_{Y21s} \approx \frac{1}{R_G (C_{GS} + C_{GD})} \tag{3.46}$$

The transconductance cutoff frequency does not depend on the operating point as long as one remains in the pinch-off region.

Transit frequency: The *transit frequency* f_T is the frequency at which the absolute value of the small-signal current gain is reduced to 1 when the circuit is operated in the pinch-off

---

[20] In a common-source or common-drain circuit, C_{DS} is located between the output of the circuit and the small-signal ground, and has the effect of a capacitive load (see Sect. 3.4.1 or Sect. 3.4.2); thus it can be alternatively assumed that $r_{CE} \doteq r_{DS}$ and $C_L \doteq C_L + C_{DS}$.

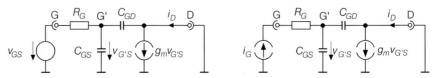

a For calculating the transconductance cut-off frequency   **b** For calculating the transit frequency

Fig. 3.50. Small-signal equivalent circuits for calculating the cutoff frequencies

region and $V_{DS} = V_{DS,A}$ is constant:

$$\left.\frac{|i_D|}{|i_G|}\right|_{s=j\omega_T} \equiv 1$$

From the small-signal equivalent circuit shown in Fig. 3.50b, it follows that

$$\frac{i_D}{i_G} = \frac{g_m - sC_{GD}}{s\,(C_{GS} + C_{GD})}$$

and thus:

$$\omega_T = 2\pi f_T \approx \frac{g_m}{C_{GS} + C_{GD}} \tag{3.47}$$

The transit frequency is proportional to the transconductance g_m and increases with the operating point current because $g_m \sim \sqrt{I_{D,A}}$.

Relationship and meaning of the cutoff frequencies: Comparing the cutoff frequencies leads to the relationship:

$$f_T = f_{Y21s} g_m R_G \overset{g_m R_G < 1}{<} f_{Y21s}$$

Driving a FET in common-source configuration using a voltage source or a source with a small internal resistance is called *voltage control*; in this case, the cutoff frequency of the circuit is limited to a maximum by the transconductance cutoff frequency f_{Y21s}. Driving the FET using a current source or a source with a high internal resistance is called *current control*; in this case, the cutoff frequency of the circuit is limited to a maximum by the *transit frequency* f_T. Thus, voltage control generally provides a higher bandwidth than current control.

Determining the small-signal capacitances from the cutoff frequencies: If the data sheet for a FET specifies the transit frequency f_T, the reverse capacitance C_{rss} (*reverse, grounded source, gate shorted*) and the output capacitance C_{oss} (*output, grounded source, gate shorted*), the capacitances of the equivalent circuit shown in Fig. 3.49a can be determined with the help of (3.47):

$$C_{GS} \approx \frac{g_m}{\omega_T} - C_{rss}$$

$$C_{GD} \approx C_{rss}$$

$$C_{DS} \approx C_{oss} - C_{rss}$$

If the transconductance cutoff frequency f_{Y21s} is also known, then the gate resistance can be calculated:

$$R_G = \frac{f_T}{g_m f_{Y21s}}$$

Summary of the Small-Signal Parameters

For discrete FETs, the parameters according to Fig. 3.51 for the small-signal model shown in Fig. 3.49a can be determined from the drain current $I_{D,A}$ at the operating point and the specifications of the data sheet. Often, the *Y parameters of the common-source circuit* are also specified; if $\omega \ll \omega_{Y21s}$, then R_G can be neglected, which leads to:

$$\mathbf{Y}_s(j\omega) = \begin{bmatrix} y_{11,s}(j\omega) & y_{12,s}(j\omega) \\ y_{21,s}(j\omega) & y_{22,s}(j\omega) \end{bmatrix} = \begin{bmatrix} g_{11} + jb_{11} & g_{12} + jb_{12} \\ g_{21} + jb_{21} & g_{22} + jb_{22} \end{bmatrix}$$

$$\stackrel{R_G \to 0}{\approx} \begin{bmatrix} j\omega(C_{GS} + C_{GD}) & -j\omega C_{GD} \\ g_m - j\omega C_{GD} & 1/r_{DS} + j\omega(C_{DS} + C_{GD}) \end{bmatrix}$$

Consequently:

$$g_m \approx g_{21} \,,\; r_{DS} \approx \frac{1}{g_{22}} \,,\; C_{GD} \approx -\frac{b_{12}}{\omega} \,,\; C_{GS} \approx \frac{b_{11} + b_{12}}{\omega} \,,\; C_{DS} \approx \frac{b_{22} + b_{12}}{\omega}$$

Divided into a real (g_{ij}) and an imaginary (b_{ij}) portion, the Y parameters are usually specified for various frequencies or as curves plotted versus the frequency, and apply to the given operating point. The method described here for determining the small-signal parameters provides a sufficiently high level of accuracy only for relatively low frequencies ($f \leq 10\,\mathrm{MHz}$). With higher frequencies ($f > 100\,\mathrm{MHz}$), the spreading resistance R_G and the conductor inductances of the terminals become effective; it is then no longer possible to determine the small-signal parameters from the Y parameters in a simple manner. Conversion to other operating points is possible as an approximation, by maintaining the values for the capacitances but converting the parameters g_m and r_{DS}:

$$\frac{g_{m1}}{g_{m2}} = \sqrt{\frac{I_{D,A1}}{I_{D,A2}}} \,,\qquad \frac{r_{DS1}}{r_{DS2}} = \frac{I_{D,A2}}{I_{D,A1}}$$

The determination of the small-signal parameters for a discrete FET is more complicated than for a bipolar transistor. In the case of a bipolar transistor, it is possible *without specific data* to determine the transconductance as the most important parameter from the simple relation $g_m = I_{C,A}/V_T$; but for a FET the transconductance coefficient K is required, which is usually not even specified on the data sheet.[21]

The small-signal parameters for integrated MOSFETs can be determined more easily and more accurately, since the process parameters and scaling values are generally

---

[21] This is surprising, since K is *the* specific value of a FET; in contrast, the current gain B or β is always specified as *the* specific value of a bipolar transistor.

Parameters	Name	Method of determination
(K)	Transconductance coefficient	From the transfer characteristic for small currents (R_S is not yet effective): $$K = \frac{2I_D}{(V_{GS} - V_{th})^2} \overset{\text{Jfet: } V_{GS}=0}{=} \frac{2I_{D,0}}{V_{th}^2}$$ or from the transconductance: $$K = \frac{g_m}{V_{GS} - V_{th}}$$ or from a pair of values (I_D, g_m): $$K = \frac{g_m^2}{2I_D}$$
g_m	Transconductance	$g_m = \sqrt{2K\,I_{D,A}} = \frac{2}{|V_{th}|}\sqrt{I_{D,0}I_{D,A}}$
(V_A)	Early voltage	From the slope of the output characteristics (Fig. 3.12) or reasonable assumption ($V_A \approx 20 \ldots 200\,\text{V}$)
r_{DS}	Output resistance	$r_{DS} = \dfrac{V_A}{I_{D,A}}$
(f_T)	Transit frequency	From the data sheet
(f_{Y21s})	Transconductance cutoff frequency	From the data sheet
R_G	Gate spreading resistance	$R_G = \dfrac{f_T}{g_m f_{Y21s}}$ or reasonable assumption ($R_G \approx 1 \ldots 100\,\Omega$)
C_{GD}	Gate–drain capacitance	From the data sheet: $C_{GD} \approx C_{rss}$
C_{GS}	Gate–source capacitance	$C_{GS} \approx \dfrac{g_m}{2\pi f_T} - C_{GD}$
C_{DS}	Drain–source capacitance	From the data sheet: $C_{DS} \approx C_{oss} - C_{rss}$

Fig. 3.51. Method of determining small-signal parameters for a discrete FET (auxiliary parameters in parentheses)

known; it is only necessary to evaluate (3.40)–(3.45). Figure 3.52 outlines the procedure for determining the parameters for the small-signal model shown in Fig. 3.48.

3.3.4
Noise

The basics for describing noise and calculating the noise figure are outlined in Sect. 2.3.4 on page 82, using the example of a bipolar transistor. The same procedure can be applied to the field effect transistor using the relevant noise sources.

Parameters	Name	Method of determination
(d_{ox}, W, L, A_S, A_D)	Geometric dimensions	Oxide thickness, channel width, channel length, area of the source and drain regions
(C_{ox})	Oxide capacitance	$C_{ox} = \dfrac{\epsilon_0 \epsilon_{r,ox} W L}{d_{ox}} \approx 3.45 \cdot 10^{-11} \dfrac{\text{F}}{\text{m}^2} \cdot \dfrac{W L}{d_{ox}}$
(K)	Transconductance coefficient	$K = K'_n \dfrac{W}{L} = \mu_n C'_{ox} \dfrac{W}{L}$; p-Kanal: K'_p, μ_p
g_m	Transconductance	$g_m = \sqrt{2K I_{D,A}}$
$g_{m,B}$	Substrate transconductance	$g_{m,B} = \dfrac{\gamma g_m}{2\sqrt{V_{inv} - V_{BS,A}}}$
r_{DS}	Output resistance	$r_{DS} = \dfrac{V_A}{I_{D,A}} = \dfrac{1}{\lambda I_{D,A}}$
R_G	Gate spreading resistance	From the geometry: $R_G = n_{RG} R_{sh}$ or reasonable assumption ($R_G \approx 1 \ldots 100\,\Omega$)
C_{GS}	Gate–source capacitance	$C_{GS} = \dfrac{2}{3} C_{ox} + C'_{GS,ov} W \approx \dfrac{2}{3} C_{ox}$
C_{GD}	Gate–drain capacitancet	$C_{GD} = C'_{GD,ov} W$
C_{BS}, C_{BD}	Bulk capacitances	$C_{BS} \approx C'_J A_S$ bzw. $C_{BD} \approx C'_J A_D$

Fig. 3.52. Method of determining small-signal parameters for an integrated MOSFET (auxiliary parameters in parentheses)

The noise figure of a field effect transistor is calculated below; first, we look at the noise sources and the correlation between the noise sources. In general, a simplified description is sufficient for practical applications and is given below.

Noise Sources in Field Effect Transistors

In a FET, the following noise sources occur at the operating point in the pinch-off region determined by $I_{D,A}$:

– Thermal noise of the gate spreading resistance:

$$|\underline{v}_{RG,r}(f)|^2 = 4kTR_G$$

The thermal noise of the other spreading resistances can usually be neglected.
– Thermal noise and $1/f$ noise of the channel, with:

$$|\underline{i}_{D,r}(f)|^2 = \frac{8}{3} kT g_m + \frac{k_{(1/f)} I_{D,A}^{\gamma(1/f)}}{f} = \frac{8}{3} kT g_m \left(1 + \frac{f_{g(1/f)}}{f}\right)$$

The thermal portion in $|\underline{i}_{D,r}(f)|^2$ is lower than the thermal noise of an ohmic resistance $R = 1/g_m$ with $|\underline{i}_{R,r}(f)|^2 = 4kT/R$ because in the pinch-off region the channel is

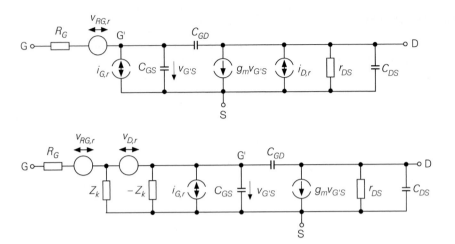

Fig. 3.53. Small-signal model of a FET with the original (*top*) and equivalent (*bottom*) noise sources

neither homogenous nor in thermal equilibrium. Furthermore, there is 1/f noise with the experimental parameters $k_{(1/f)}$ und $\gamma_{(1/f)} \approx 1$. At low frequencies, the 1/f portion is predominant, while at medium and high frequencies the thermal portion dominates. Making the portions equal leads to the *1/f cutoff frequency:*

$$f_{g(1/f)} = \frac{3}{8} \frac{k_{(1/f)} I_{D,A}^{(\gamma_{(1/f)}-1/2)}}{kT\sqrt{K}} \overset{\gamma_{(1/f)}=1}{=} \frac{3}{8} \frac{k_{(1/f)}}{kT} \sqrt{\frac{I_{D,A}}{K}}$$

This increases with a rising operating point current. For the MOSFET it is approximately $k_{(1/f)} \sim 1/L^2$, which means that the 1/f noise decreases as the channel length increases; since MOSFETs in integrated circuits are scaled in correspondence with the current at the operating point ($I_{D,A} \sim K \sim W/L$) a large MOSFET has less $1/f$ noise than a small one with the same current or the same transconductance. Typical values are $f_{g(1/f)} \approx$ 100 kHz ... 10 MHz for MOSFETs and $f_{g(1/f)} \approx$ 10 Hz ... 1 kHz for junction FETs.
 – Induced gate noise, with:

$$|\underline{i}_{G,r}(f)|^2 = \frac{4}{3} kT \, g_m \left(\frac{f}{f_T}\right)^2$$

This noise current is also caused by the thermal noise in the channel, which is transmitted to the gate by the capacitive coupling between the gate and the channel. Therefore, the noise current sources $i_{G,r}$ and $i_{D,r}$ are not independent but *correlated*. This correlation must be taken into account when calculating the noise figure.

The upper portion of Fig. 3.53 shows the small-signal model with noise sources $v_{RG,r}$, $i_{G,r}$ and $i_{D,r}$.

Equivalent Noise Sources

The noise source $i_{D,r}$ is converted to the input to facilitate the calculation of noise. This leads to the small-signal model shown in the lower part of Fig. 3.53, in which the noise current source $i_{D,r}$ is replaced by the *equivalent noise voltage source* $v_{D,r}$, with[22]

$$|\underline{v}_{D,r}(f)|^2 = \frac{|\underline{i}_{D,r}(f)|^2}{|\underline{y}_{21,s}(j2\pi f)|^2} = \frac{8}{3}\frac{kT}{g_m}\left(1 + \frac{f_{g(1/f)}}{f} + \left(\frac{f}{f_{Y21s}}\right)^2\right)$$

and the impedances $\underline{Z}_k$ and $-\underline{Z}_k$ to describe the correlation. The *correlation impedance*[23]

$$\underline{Z}_k \approx -\frac{j\sqrt{2}f_T}{g_m f}$$

only influences the source $v_{D,r}$; for all other sources and signals, the parallel arrangement of $\underline{Z}_k$ und $-\underline{Z}_k$ is ineffective.

Dependence on the operating point: The noise voltage density of the equivalent source $v_{D,r}$ is inversely proportional to the transconductance; in other words, it declines when the transconductance increases, such that $g_m = \sqrt{2K I_{D,A}}$ leads to $|\underline{v}_{D,r}(f)|^2 \sim 1/\sqrt{I_{D,A}}$. For the noise current source $i_{G,r}$, this means that $|\underline{i}_{G,r}(f)|^2 \sim g_m \sim \sqrt{I_{D,A}}$, so that the noise current density increases with the transconductance. The noise of the gate resistance R_G is independent of the operating point.

Example: For a FET with $K = 0.5\,\mathrm{mA/V^2}$, $R_G = 100\,\Omega$ and $f_T = 100\,\mathrm{MHz}$ we obtain, for an operating point current $I_{D,A} = 1\,\mathrm{mA}$, the transconductance $g_m = 1\,\mathrm{mA/V}$ and thus, in the medium frequency range, that is, $f_{g(1/f)} < f < f_{Y21s}$, the frequency-independent noise voltage densities $|\underline{v}_{RG,r}(f)| = 1.3\,\mathrm{nV/\sqrt{Hz}}$ and $|\underline{v}_{D,r}(f)| = 3.3\,\mathrm{nV/\sqrt{Hz}}$ and the noise current density $|\underline{i}_{G,r}(f)| = 2\,\mathrm{pA/\sqrt{Hz}} \cdot f/f_T$, which is proportional to the frequency; for $f = 1\,\mathrm{kHz}$, $|\underline{i}_{G,r}(f)| = 0.02\,\mathrm{fA/\sqrt{Hz}}$. For JFETs, the 1/f cutoff frequency is relatively low, with $f_{g(1/f)} \approx 100\,\mathrm{Hz}$; thus the calculated value for $|\underline{v}_{D,r}(f)|$ applies to a relatively wide frequency range of $f_{g(1/f)} \approx 100\,\mathrm{Hz}$ to $f_{Y21s} = f_T/(g_m R_G) = 1\,\mathrm{GHz}$. In contrast, $f_{g(1/f)} \approx 1\,\mathrm{MHz}$ applies to MOS-FETs; due to the 1/f noise we thus achieve with $f = 1\,\mathrm{kHz}$ a much higher value of $|\underline{v}_{D,r}(f)| = 105\,\mathrm{nV/\sqrt{Hz}}$ compared to the JFET.

Noise Figure of FETs

When using the FET with a signal source with internal resistance R_g and thermal noise voltage density $|\underline{v}_{r,g}(f)|^2 = 4kTR_g$, all noise sources can be combined into one *equivalent noise source* v_r; then [3.6]:

$$|\underline{v}_r(f)|^2 = |\underline{v}_{r,g}(f)|^2 + |\underline{v}_{RG,r}(f)|^2 + \left(1 + \frac{(R_g + R_G)^2}{|\underline{Z}_k|^2}\right)|\underline{v}_{D,r}(f)|^2$$

$$+ (R_g + R_G)^2|\underline{i}_{G,r}(f)|^2$$

---

[22] The cutoff frequency f_{Y21s} is only considered in the conversion of the thermal portion of $i_{D,r}$; the 1/f portion is negligible in this frequency range.

[23] Different data exist for $\underline{Z}_k$; reference [3.6] specifies $\underline{Z}_k \approx -1.39jf_T/(g_m f)$. Instead of 1.39, we use here the factor $\sqrt{2} \approx 1.41$. Here, squaring produces the factor 2 instead of 1.93.

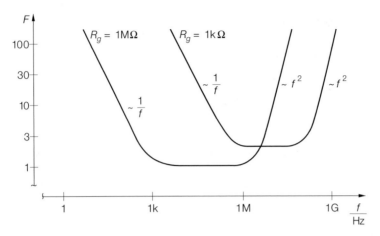

Fig. 3.54. Noise figure curve for a MOSFET with $g_m = 1\,\mathrm{mA/V}$, $R_G = 100\,\Omega$, $f_T = 100\,\mathrm{MHz}$ and $f_{g(1/f)} = 1\,\mathrm{MHz}$ for $R_g = 1\,\mathrm{k}\Omega$ und $R_g = 1\,\mathrm{M}\Omega$

$$\overset{R_g \gg R_G}{\approx}\quad |\underline{v}_{r,g}(f)|^2 + |\underline{v}_{RG,r}(f)|^2 + \left(1 + \frac{R_g^2}{|Z_k|^2}\right)|\underline{v}_{D,r}(f)|^2$$

$$+ R_g^2 |\underline{i}_{G,r}(f)|^2$$

This leads to the *spectral noise figure*:

$$F(f) = \frac{|\underline{v}_r(f)|^2}{|\underline{v}_{r,g}(f)|^2} = 1 + \frac{|\underline{v}_{RG,r}(f)|^2 + \left(1 + \frac{R_g^2}{|Z_k|^2}\right)|\underline{v}_{D,r}(f)|^2 + R_g^2|\underline{i}_{G,r}(f)|^2}{|\underline{v}_{r,g}(f)|^2}$$

Substituting and limiting to $f < f_{Y21s}$ yields:

$$F(f) = 1 + \frac{R_G}{R_g} + \frac{2}{3}\left(\frac{1}{g_m R_g}\left(1 + \frac{f_{g(1/f)}}{f}\right) + g_m R_g \left(\frac{f}{f_T}\right)^2\right) + \frac{1}{3}g_m R_g \frac{f_{g(1/f)}f}{f_T^2}$$

The contribution of the last term is negligibly low, because $f_{g(1/f)} \ll f_T$.

Figure 3.54 shows the noise figure curve for a MOSFET where $R_g = 1\,\mathrm{k}\Omega$ and $R_g = 1\,\mathrm{M}\Omega$; three areas can be distinguished:

– For medium frequencies, the noise figure is approximately constant:

$$\boxed{F \approx 1 + \frac{R_G}{R_g} + \frac{2}{3}\frac{1}{g_m R_g} \overset{R_G \ll 1/g_m}{\approx} 1 + \frac{2}{3}\frac{1}{g_m R_g} \overset{R_g \gg 1/g_m}{\approx} 1} \qquad (3.48)$$

With the gate resistance R_G small and the source resistance R_g high compared to the inverse value of the transconductance, we obtain the optimum noise factor $F = 1$ in this area.

– With low frequencies the 1/f noise dominates; the noise figure in this area is inversely proportional to the frequency:

$$F(f) \approx \frac{2}{3}\frac{1}{g_m R_g}\frac{f_{g(1/f)}}{f}$$

The boundary of the area for medium frequencies is at:

$$f_1 \overset{R_G \ll 1/g_m}{\approx} \frac{f_{g(1/f)}}{1 + \frac{3}{2}g_m R_g} \overset{R_g \gg 1/g_m}{\approx} \frac{2}{3}\frac{f_{g(1/f)}}{g_m R_g}$$

The noise figure and the cutoff frequency f_1 are inversely proportional to the source resistance R_g; therefore the 1/f portion of the noise figure declines as the source resistance increases (see Fig. 3.54).

- With high frequencies, the noise figure increases proportionally with the square of the frequency:

$$F(f) \approx \frac{2}{3}g_m R_g \left(\frac{f}{f_T}\right)^2$$

The boundary of the area for medium frequencies is at:

$$f_2 \approx f_T\sqrt{1 + \frac{3}{2}\frac{1}{g_m R_g}} \overset{R_g \gg 1/g_m}{\approx} \sqrt{\frac{3}{2}}\frac{f_T}{\sqrt{g_m R_g}} \approx \frac{f_T}{\sqrt{g_m R_g}}$$

The noise figure increases with an increasing source resistance R_g, while the cutoff frequency f_2 decreases accordingly (see Fig. 3.54).

For JFETs, the 1/f cutoff frequency, and thus also the 1/f portion of the noise figure, is lower than in MOSFETs by three or four orders of magnitude; therefore, with source resistances in the MΩ range the 1/f portion is virtually unnoticeable, since the cutoff frequency f_1 drops below 1 Hz.

Minimizing the noise figure: Under certain conditions, the noise figure is reduced to a minimum. With a given source resistance R_g, the optimum transconductance and thus the optimum drain current at the operating point can be determined by evaluating:

$$\frac{\partial F(f)}{\partial g_m} = 0$$

It must be noted that according to (3.47) the transit frequency f_T is proportional to the transconductance: $f_T = g_m/(2\pi C)$ with $C = C_{GS} + C_{GD}$; inserting this leads to:

$$F(f) = 1 + \frac{R_G}{R_g} + \frac{2}{3}\frac{1}{g_m}\left(\frac{1}{R_g}\left(1 + \frac{f_{g(1/f)}}{f}\right) + 4\pi^2 C^2 R_g f\left(f + \frac{f_{g(1/f)}}{2}\right)\right)$$

It is obvious that $F(f)$ declines with an increasing transconductance; this means that there is no optimum, so that low-noise FET amplifiers must be operated with the highest possible transconductance or the highest possible drain current.

The *optimum source resistance* R_{gopt} can be calculated by way of

$$\frac{\partial F(f)}{\partial R_g} = 0$$

and by limiting to $f_{g(1/f)} < f < f_{Y21s}$:

$$R_{gopt}(f) \approx \frac{f_T}{g_m f}\sqrt{1 + \frac{3}{2}g_m R_G} \overset{g_m R_G \ll 1}{\approx} \frac{f_T}{g_m f} = \frac{1}{2\pi f (C_{GS} + C_{GD})}$$

Wide-band matching is not possible due to the frequency dependence of R_{gopt}. Inserting R_{gopt} in $F(f)$ leads to the *optimum spectral noise figure* $F_{opt}(f)$. An approximation is [3.6]:

$$F_{opt}(f) \approx 1 + \frac{R_G}{R_g} + \frac{4}{3}\frac{f}{f_T} \overset{R_g \gg R_G}{\approx} 1 + \frac{4}{3}\frac{f}{f_T}$$

Simplified Description

A simplified description is usually sufficient for practical applications. For this purpose, the gate spreading resistance and the correlation between the channel noise and the induced gate noise are neglected, and the frequency range is limited to medium frequencies; that is, $f_{g(1/f)} < f < f_{Y21s}$. An increase in the gate noise current density by a factor of 2 is assumed in order to achieve an approximate compensation of the portion induced by the correlation; for $f_{g(1/f)} < f < f_{Y21s}$ it follows that:

$$|\underline{i}_{D,r}(f)|^2 \approx \frac{8}{3}kT\,g_m$$

$$|\underline{i}_{G,r}(f)|^2 \approx \frac{8}{3}kT\,g_m\left(\frac{f}{f_T}\right)^2$$

With $\underline{y}_{21,s}(s) \approx g_m$, this leads to the equivalent noise sources:

$$|\underline{v}_{r,0}(f)|^2 = \frac{|\underline{i}_{D,r}(f)|^2}{|\underline{y}_{21,s}(j2\pi f)|^2} \approx \frac{8}{3}\frac{kT}{g_m}$$

$$|\underline{i}_{r,0}(f)|^2 = |\underline{i}_{G,r}(f)|^2 \approx \frac{8}{3}kT\,g_m\left(\frac{f}{f_T}\right)^2$$

Figure 3.55 shows the simplified small-signal model with equivalent noise sources. The noise figure is:

$$F(f) = 1 + \frac{|\underline{v}_{r,0}(f)|^2 + R_g^2|\underline{i}_{r,0}(f)|^2}{|\underline{v}_{r,g}(f)|^2} \approx 1 + \frac{2}{3}\left(\frac{1}{g_m R_g} + g_m R_g\left(\frac{f}{f_T}\right)^2\right)$$

With the exception of the missing $1/f$ portion, this yields the same values as in the detailed calculation (see Fig. 3.54). From $(\partial F)/(\partial R_g) = 0$, it follows for the optimum source resistance R_{gopt} and the optimum noise figure F_{opt} that:

$$R_{gopt}(f) \approx \frac{f_T}{g_m f} = \frac{1}{2\pi f\,(C_{GS} + C_{GD})}$$

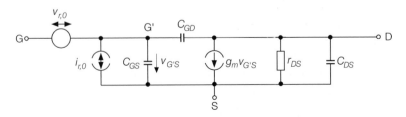

Fig. 3.55. Simplified small-signal model of a FET with equivalent noise sources

$$F_{opt}(f) \approx 1 + \frac{4}{3}\frac{f}{f_T}$$

These values coincide with the approximations of the detailed calculation, because the assumption of a higher gate noise current density exactly replaces the portion caused by the correlation.

Comparison of the Noise Figure for FETs and Bipolar Transistors

A FET achieves the ideal noise figure $F = 1$ with high-resistance sources and medium frequencies. Furthermore, the high $1/f$ noise of a MOSFET becomes comparatively less influential with high-resistance sources, since in this condition the induced gate noise dominates and offsets the 1/f portion of the channel noise; this is illustrated in the example of Fig. 3.54: even though the 1/f cutoff frequency is at 1 MHz, the 1/f region of the noise figure for $R_g = 1\,\mathrm{M}\Omega$ starts below 1 kHz. In JFETs, the 1/f noise is virtually insignificant. Due to these properties, with high-resistance sources the FET is clearly superior to the bipolar transistor. Therefore, amplifiers for high-resistance sources, such as receivers for photodiodes, use a FET in the input stage; JFETS are preferred because of their low 1/f noise.

With low-resistance sources, the noise figure of a FET is higher than that of a bipolar transistor; in addition, the maximum gain is much lower. According to (3.48), a low noise figure requires a high transconductance and thus a correspondingly high quiescent current. Since the transconductance of the FET increases in proportion to the root of the quiescent current, this approach to reducing the noise figure is not very effective. For MOSFETs, the high 1/f noise has its full effect with low-resistance sources and causes the noise figure to increase dramatically at low frequencies (see Fig. 3.54).

3.4
Basic Circuits

Basic circuits using field effect transistors: There are three basic circuits in which FETs can be used: the *common-source circuit*, the *common-drain circuit* and the *common-gate circuit*. The name reflects the terminal of the FET that is connected as a common reference point for the input and output of the circuit; Fig. 3.56 illustrates this relationship using an enhancement n-channel MOSFET.

In many circuits this criterion does not apply in a strict sense, so that a less stringent criterion must be used:

> *The designation reflects the terminal of the FET that is connected to form neither the input nor the output of the circuit.*

The substrate or bulk terminal has no influence on the classification of the basic circuits, but influences their behavior. It is connected to the source terminal in discrete MOSFETs and to ground or to a supply voltage source (= small-signal ground) in integrated circuits; for the common-source circuit, both versions are identical, since in this case the source terminal is connected to (small-signal) ground.

Basic circuits with several FETs: There are several configurations using two or more FETs. They are so common that they too must be regarded as basic circuits, which serve,

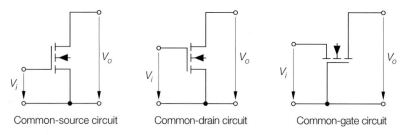

Common-source circuit   Common-drain circuit   Common-gate circuit

Fig. 3.56. Basic circuit configurations for field effect transistors

for example, as differential amplifiers or current mirrors; such circuits are described in Sect. 4.1.

Polarity: n-channel MOSFETs are preferred in all of these configurations, as they feature a higher transconductance coefficient than p-channel MOSFETs with the same channel dimensions due to the higher mobility of charge carriers. Furthermore, enhancement MOS-FETs are more common than depletion MOSFETs; this is particularly true for integrated circuits. With respect to the small-signal response, there is no principal difference between depletion MOSFETs and JFETs on the one hand and enhancement MOSFETs on the other; only their operating point setting is different. All circuits can also be designed using the respective p-channel FETs; but then the polarity of all supply voltages, electrolytic capacitors and diodes must be reversed.

3.4.1
Common-Source Circuit

Figure 3.57a shows the common-source circuit, consisting of the MOSFET, the drain resistance R_D, the supply voltage source V_b and the signal voltage source V_g with the internal resistance R_g. In what follows, we assume $V_b = 5\,\text{V}$, $R_D = 1\,\text{k}\Omega$ and $K = 4\,\text{mA/V}^2$, $V_{th} = 1\,\text{V}$ for the MOSFET.

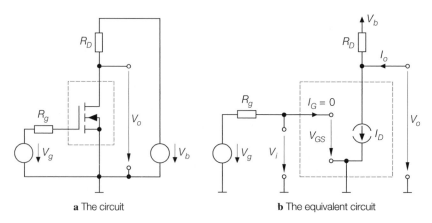

a The circuit   **b** The equivalent circuit

Fig. 3.57. Common-source circuit

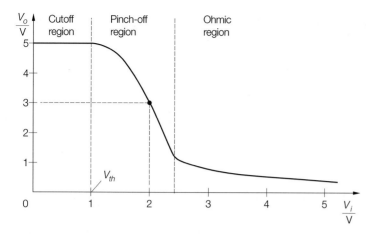

Fig. 3.58. Transfer characteristic in the common-source circuit

Transfer Characteristic of the Common-Source Circuit

The transfer characteristic shown in Fig. 3.58 is obtained by measuring the output voltage V_o as a function of the signal voltage V_g. With $V_g < V_{th} = 1\,\text{V}$, no drain current flows and the output voltage is $V_o = V_b = 5\,\text{V}$. With $V_g \geq 1\,\text{V}$, a drain current I_D flows that increases with V_g, and the output voltage declines accordingly; with $1\,\text{V} \leq V_g \leq 2.4\,\text{V}$, the MOSFET operates in the pinch-off region and with $V_g > 2.4\,\text{V}$ it operates in the ohmic region. The substrate effect that occurs in integrated MOSFETs is not influential in common-source circuits, since the substrate or bulk terminal *and* the source terminal are connected to ground; that is, it always follows that $V_{BS} = 0$.

Operation in the pinch-off region: Figure 3.57b shows the equivalent circuit. Disregarding the Early effect leads to:

$$I_D = \frac{K}{2}(V_{GS} - V_{th})^2$$

With $V_g = V_i = V_{GS}$, the output voltage is:

$$V_o = V_{DS} \overset{I_o = 0}{=} V_b - I_D R_D = V_b - \frac{R_D K}{2}(V_i - V_{th})^2 \tag{3.49}$$

The internal resistance R_g has no effect on the characteristic of a MOSFET due to $I_G = 0$; it only influences the dynamic behavior. But in JFETs there are gate leakage currents in the pA and nA range that cause voltage drops across high internal resistances that can no longer be neglected; MOSFETs are thus preferred for sources with $R_g > 10\,\text{M}\Omega$.

A point at the centre of the declining segment of the transfer characteristic is selected as the operating point; this allows maximum output signals. Setting $V_b = 5\,\text{V}$, $R_D = 1\,\text{k}\Omega$, $K = 4\,\text{mA/V}^2$ and $V_{th} = 1\,\text{V}$ for the operating point shown as an example in Fig. 3.58 results in:

$$V_o = 3\,\text{V} \Rightarrow I_D = \frac{V_b - V_o}{R_D} = 2\,\text{mA} \Rightarrow V_i = V_{GS} = V_{th} + \sqrt{\frac{2I_D}{K}} = 2\,\text{V}$$

Boundary of the ohmic region: With $V_o = V_{o,po} = V_{DS,po}$, the MOSFET reaches the boundary of the ohmic region. For $V_{DS,po} = V_{GS} - V_{th}$ and $V_i = V_{GS}$, the relation $V_o = V_i - V_{th}$ follows; inserting this into (3.49) leads to

$$V_{o,po} = \frac{1}{R_D K}\left(\sqrt{1 + 2V_b R_D K} - 1\right) \overset{2V_b R_D K \gg 1}{\approx} \sqrt{\frac{2V_b}{R_D K}} - \frac{1}{R_D K}$$

and $V_{i,po} = V_{o,po} + V_{th}$. With the numeric example we obtain $V_{o,po} = 1.35\,\mathrm{V}$ and $V_{i,po} = 2.35\,\mathrm{V}$.

If the supply voltage is given, the product $R_D K$ must be increased in order to reduce $V_{o,po}$ and to increase the output voltage range. In practice, the control range is always narrower than in the common-emitter configuration because a bipolar transistor can be controlled down to $V_{CE,sat} \approx 0.1\,\mathrm{V}$ almost independent of the external circuitry.

Small-Signal Response in Common-Source Circuits

The response to signals around an operating point A is called the *small-signal response*. The operating point is determined by the operating point values $V_{i,A} = V_{GS,A}$, $V_{o,A} = V_{DS,A}$ and $I_{D,A}$, and must be located in the pinch-off region in order to reach a useful gain; as an example, we use the operating point established above, with $V_{GS,A} = 2\,\mathrm{V}$, $V_{DS,A} = 3\,\mathrm{V}$ and $I_{D,A} = 2\,\mathrm{mA}$.

Figure 3.59 shows the small-signal equivalent circuit for the common-source configuration that is obtained by inserting the small-signal equivalent circuit for the FET according to Fig. 3.17 or Fig. 3.46 and changing over to small-signal values. The source, with its substrate transconductance $g_{m,B}$ shown in Fig. 3.46 is omitted because $V_{BS} = v_{BS} = 0$.

Without the load resistance R_L it follows from Fig. 3.59 for the *common-source circuit* that:

common source circuit

$$A = \left.\frac{v_o}{v_i}\right|_{i_o=0} = -g_m (R_D \| r_{DS}) \overset{r_{DS} \gg R_D}{\approx} -g_m R_D \tag{3.50}$$

$$r_i = \frac{v_i}{i_i} = \infty \tag{3.51}$$

$$r_o = \frac{v_o}{i_o} = R_D \| r_{DS} \overset{r_{DS} \gg R_D}{\approx} R_D \tag{3.52}$$

If $K = 4\,\mathrm{mA/V^2}$ and $V_A = 50\,\mathrm{V}$, then $g_m = \sqrt{2K\,I_{D,A}} = 4\,\mathrm{mS}$, $r_{DS} = V_A/I_{D,A} = 25\,\mathrm{k\Omega}$, $A = -3.85$ and $r_o = 960\,\Omega$. For comparison, at the same operating point of the common-emitter circuit described in Sect. 2.4.1, with $I_{C,A} = I_{D,A} = 2\,\mathrm{mA}$ and

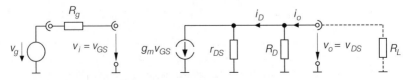

Fig. 3.59. Small-signal equivalent circuit for the common-source circuit

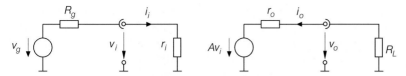

Fig. 3.60. Equivalent circuit with the equivalent values A, r_i and r_o

$R_C = R_D = 1\,\mathrm{k\Omega}$, the gain is $A = -75$. The reason for the low gain of the MOSFET is the low transconductance at the same current: $g_m = 4\,\mathrm{mA/V}$ for the MOSFET and $g_m = 77\,\mathrm{mA/V}$ for the bipolar transistor.

The parameters A, r_i and r_o provide a complete description of the common-source circuit; Fig. 3.60 shows the corresponding equivalent circuit. The load resistance R_L may be an ohmic resistance or an equivalent element for the input resistance of a circuit connected to the output. It is important that the operating point is not shifted by R_L, which means that no or only a negligible DC current may flow through R_L.

Using Fig. 3.60, it is possible to calculate the *small-signal operating gain*:

$$A_B = \frac{v_o}{v_g} = \frac{r_i}{r_i + R_g}\, A\, \frac{R_L}{R_L + r_o} \overset{r_i \to \infty}{=} A\, \frac{R_L}{R_L + r_o} \tag{3.53}$$

It consists of the circuit gain A and the voltage divider factor at the output.

Maximum gain: From (3.50), we obtain the *maximum gain* for $R_D \to \infty$:

$$\mu = \lim_{R_D \to \infty} |A| = g_m\, r_{DS} \approx \sqrt{\frac{2K}{I_{D,A}}}\, V_A = \frac{2V_A}{V_{GS} - V_{th}}$$

This extreme situation is very difficult to obtain with an ohmic drain resistance R_D, because the consequence of $R_D \to \infty$ is $R_D \gg r_{DS}$, so that $I_{D,A} R_D \gg I_{D,A} r_{DS} = V_A$, which means that the voltage drop across R_D must be much higher than the Early voltage $V_A \approx 50\,\mathrm{V}$. The maximum gain is reached when a constant current source with $I_K = I_{D,A}$ is used instead of R_D.

The maximum gain depends on the operating point; it declines with an increasing current or an increasing voltage $V_{GS} - V_{th}$. In order to reach the maximum gain, a MOSFET with the highest possible transconductance coefficient K must be used and operated with the current $I_{D,A}$ as low as possible. The maximum value μ_{max} is achieved in the *sub-threshold region* with $V_{GS} - V_{th} < 100\,\mathrm{mV}$; in this region, the transfer characteristic rises exponentially (see (3.25)) and $\mu_{max} \approx V_A/(2V_T) \approx 400 \ldots 2000$. In practical applications, MOSFETs are often operated close to the *temperature compensation point* $V_{GS,TC} \approx V_{th} + 1\,\mathrm{V}$ (see Sect. 3.1.7); then $\mu \approx 40 \ldots 200$.

Nonlinearity: In Sect. 3.1.4 the *distortion factor* k of the drain current is calculated for sinusoidal signals with $\hat{v}_i = \hat{v}_{GS}$ (see (3.13) on page 185); in the common-source circuit it is identical to the distortion factor of the output voltage v_o. This means that $\hat{v}_i < 4k\,(V_{GS,A} - V_{th})$; that is, for $k < 1\%$ we need $\hat{v}_i < (V_{GS,A} - V_{th})/25$. Using $V_{GS,A} - V_{th} = 1\,\mathrm{V}$ in our numeric example, we obtain $\hat{v}_i < 40\,\mathrm{mV}$. Since $\hat{v}_o = |A|\hat{v}_i$, the corresponding output amplitude depends on the gain A; with $A = -3.85$ in the above example, we obtain $\hat{v}_o < 4k|A|\,(V_{GS,A} - V_{th}) = k \cdot 15.4\,\mathrm{V}$. For comparison, in the common-emitter circuit in Sect. 2.4.1 $\hat{v}_o < k \cdot 7.5\,\mathrm{V}$, which means that the common-source circuit achieves a higher output amplitude for the same distortion factor.

The common-source configuration is particularly suitable for use in amplifiers with bandpass filter properties; for example, transmitters, receivers and IF amplifiers in radio transmission systems. In such amplifiers, the square distortions are not relevant since the generated sum and difference frequencies are outside the passband of the bandpass filters: f_1, f_2 in the passband $\Rightarrow f_1 - f_2$, $f_1 + f_2$ outside the passband. In contrast, the cubic distortions may contain portions with $2f_1 - f_2$ and $2f_2 - f_1$, which may be within the passband. However, in FETs the cubic distortions are very low due to the almost square characteristic. For this reason, high-frequency MOSFETs and GaAs MESFETs in common-source configuration *without negative feedback* are preferred in modern transmitter output stages. In FETs too, the negative feedback does lead to a reduction in the distortion factor, since the comparably high square distortions decrease, but the cubic distortions increase.

Temperature sensitivity: It follows from (3.49) and (3.14) that:

$$\left.\frac{dV_o}{dT}\right|_A = -R_D \left.\frac{dI_D}{dT}\right|_A = -I_{D,A}R_D\left(\frac{1}{K}\frac{dK}{dT} - \frac{2}{V_{GS,A} - V_{th}}\frac{dV_{th}}{dT}\right)$$

$$\approx I_{D,A}R_D \cdot 10^{-3}\,\text{K}^{-1}\left(5 - \frac{4\ldots7\,\text{V}}{V_{GS,A} - V_{th}}\right)$$

The result from our numeric example is that $(dV_o/dT)|_A$ ranges from approximately ≈ -4 to $+2\,\text{mV/K}$. The temperature drift is low, as the MOSFET operates close to the *temperature compensation point* (see Sect. 3.1.7).

A temperature drift comparison between common-source and common-emitter circuits makes sense only in respect to the gain; $(dV_o/dT)|_A$ ranges from approximately -1 to $+0.5\,\text{mV/K} \cdot |A|$ for the common-source circuit and $(dV_o/dT)|_A \approx -1.7\,\text{mV/K} \cdot |A|$ for the common-emitter circuit. This suggests that with the same gain the drift is lower in the common-source circuit, especially if the operating point is located close to the temperature compensation point.

Common-Source Circuit with Current Feedback

Current feedback can reduce the nonlinearity and the temperature sensitivity of a common-source circuit; for this purpose, a *source resistance* R_S is added (see Fig. 3.61a). The transfer characteristic and the small-signal response depend on the circuitry at the bulk terminal. It is connected to the source in discrete MOSFETs and to the most negative supply voltage, to ground in this example, in integrated circuits; Fig. 3.61a therefore features a switch in the bulk terminal.

Figure 3.62 shows the transfer characteristic of a discrete MOSFET ($V_{BS} = 0$) and of an integrated MOSFET ($V_B = 0$) for $R_D = 1\,\text{k}\Omega$ and $R_S = 200\,\Omega$. The borderline drawn between the pinch-off region and the ohmic region applies to the discrete MOSFET.

Operation in the pinch-off region: Figure 3.61b shows the equivalent circuit diagram; for $I_o = 0$ the voltages in the pinch-off region are:

$$V_o = V_b - I_D R_D = V_b - \frac{R_D K}{2}(V_{GS} - V_{th})^2 \tag{3.54}$$

$$V_i = V_{GS} + V_S = V_{GS} + I_D R_S \tag{3.55}$$

For the sample operating point shown in Fig. 3.62 we obtain, with $V_b = 5\,\text{V}$, $K = 4\,\text{mA/V}^2$, $R_D = 1\,\text{k}\Omega$ and $R_S = 200\,\Omega$, for the discrete MOSFET:

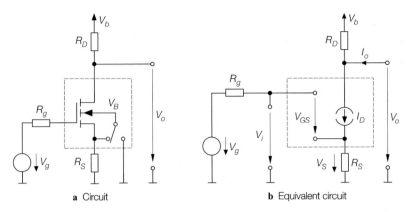

a Circuit   **b** Equivalent circuit

Fig. 3.61. Common-source circuit with current feedback

$$V_o = 3.5\,\text{V} \implies I_D = \frac{V_b - V_o}{R_D} = 1.5\,\text{mA} \implies V_S = I_D R_S = 0.3\,\text{V}$$

$$\implies V_{GS} = V_{th} + \sqrt{\frac{2I_D}{K}} = 1.866\,\text{V} \implies V_i = V_{GS} + V_S = 2.166\,\text{V}$$

For the integrated MOSFET it is necessary to take into consideration the dependence of the threshold voltage on V_{BS}, according to (3.18) on page 199. For the MOSFET we assume $V_{th,0} = 1\,\text{V}$, $\gamma = 0.5\,\sqrt{\text{V}}$ and $V_{inv} = 0.6\,\text{V}$; consequently:

$$V_{BS} = -V_S = -0.3\,\text{V}$$

$$\implies V_{th} = V_{th,0} + \gamma \left(\sqrt{V_{inv} - V_{BS}} - \sqrt{V_{inv}} \right) \approx 1.087\,\text{V}$$

$$\implies V_{GS} = V_{th} + \sqrt{\frac{2I_D}{K}} = 1.953\,\text{V} \implies V_i = V_{GS} + V_S = 2.253\,\text{V}$$

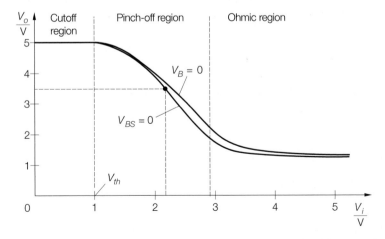

Fig. 3.62. Transfer characteristic of the common-source circuit with current feedback for a discrete MOSFET ($V_{BS} = 0$) and an integrated MOSFET ($V_B = 0$); the borderline between the pinch-off and ohmic regions applies to the discrete MOSFET

Small-signal response: The small-signal equivalent circuit shown in Fig. 3.63 is used for calculations. The nodal equation

$$g_m v_{GS} + g_{m,B} v_{BS} + \frac{v_{DS}}{r_{DS}} + \frac{v_o}{R_D} = 0$$

leads, for $v_{GS} = v_i - v_S$ and $v_{DS} = v_o - v_S$, to the *gain*:

$$A = \left. \frac{v_o}{v_i} \right|_{i_o=0} = -\frac{g_m R_D}{1 + \dfrac{R_D}{r_{DS}} + \left(g_m + g_{m,B} + \dfrac{1}{r_{DS}} \right) R_S}$$

$$\overset{r_{DS} \gg R_D, 1/g_m}{\approx} -\frac{g_m R_D}{1 + \left(g_m + g_{m,B} \right) R_S}$$

$$\overset{v_{BS}=0}{=} -\frac{g_m R_D}{1 + g_m R_S} \overset{g_m R_S \gg 1}{\approx} -\frac{R_D}{R_S}$$

For a discrete MOSFET, that is, without a substrate effect ($v_{BS} = 0$), and a high feedback ($g_m R_S \gg 1$), the gain only depends on R_D and R_S. However, because of the low maximum gain of a MOSFET the feedback cannot be high, as otherwise the gain will become too low; therefore the condition $g_m R_S \gg 1$ is rarely met in practice. When using a load resistance R_L, the operating gain A_B can be calculated by replacing R_D by the parallel arrangement of R_D and R_L (see Fig. 3.63). At the chosen sample operating point, the gain of the discrete MOSFET with $g_m = 3.46$ mS, $r_{DS} = 33$ kΩ, $R_D = 1$ kΩ and $R_S = 200\,\Omega$ is *exactly* $A = -2.002$; the two first approximations yield $A = -2.045$ and the third approximation cannot be used because $g_m R_S < 1$. For the integrated MOSFET, we assume $\gamma = 0.5\,\sqrt{V}$ and $V_{inv} = 0.6$ V; it follows from (3.41) that $g_{m,B} = 0.91$ mS and we have the *exact* value $A = -1.812$; the first approximation yields $A = -1.846$.

The *input resistance* is $r_i = \infty$ and the *output resistance*:

$$r_o = R_D \| r_{DS} \left(1 + \left(g_m + g_{m,B} + \frac{1}{r_{DS}} \right) R_S \right) \overset{r_{DS} \gg R_D}{\approx} R_D$$

For $r_{DS} \gg R_D, 1/g_m$ and with no load resistance R_L, we obtain for the:

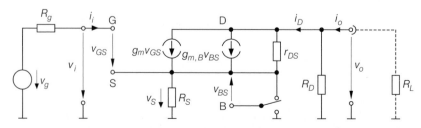

Fig. 3.63. Small-signal equivalent circuit for the common-source circuit with current feedback

common-source circuit with current feedback

$$A = \frac{v_o}{v_i}\bigg|_{i_o=0} \approx -\frac{g_m R_D}{1 + (g_m + g_{m,B}) R_S} \overset{v_{BS}=0}{=} -\frac{g_m R_D}{1 + g_m R_S} \tag{3.56}$$

$$r_i = \infty \tag{3.57}$$

$$r_o = \frac{v_o}{i_o} \approx R_D \tag{3.58}$$

Comparison with the common-source circuit without feedback: A comparison of (3.56) and (3.50) shows that *feedback* reduces the gain approximately by the *factor* $(1 + (g_m + g_{m,B})R_S)$ or $(1 + g_m R_S)$.

This negative feedback effect can easily be described by means of the *reduced transconductance*:

$$g_{m,red} = \frac{g_m}{1 + (g_m + g_{m,B}) R_S} \overset{v_{BS}=0}{=} \frac{g_m}{1 + g_m R_S} \tag{3.59}$$

The effective transconductance is reduced to $g_{m,red}$ by the source resistance R_S: thus $A \approx -g_m R_D$ for the common-source circuit without feedback and $A \approx -g_{m,red} R_D$ for the common-source circuit with current feedback.

Nonlinearity: The current feedback also reduces the nonlinearity of the transfer characteristic. The distortion factor of the circuit can be approximately determined by a series expansion of the characteristic at the operating point. From (3.55) it follows that:

$$V_i = V_{GS} + I_D R_S = V_{th} + \sqrt{\frac{2I_D}{K}} + I_D R_S$$

By inserting the values for the operating point, changing to the small-signal values and carrying out a series expansion, we achieve with (3.18), for $V_{BS} = -V_S = -I_D R_S$

$$v_i = \gamma \sqrt{V_{inv} + I_{D,A} R_S} \left(\sqrt{1 + \frac{R_S i_D}{V_{inv} + I_{D,A} R_S}} - 1 \right)$$

$$+ \sqrt{\frac{2I_{D,A}}{K}} \left(\sqrt{1 + \frac{i_D}{I_{D,A}}} - 1 \right) + R_S i_D$$

$$= \frac{1}{g_m} \left(\left(1 + (g_m + g_{m,B}) R_S\right) i_D + \frac{1}{4} \left(\frac{g_{m,B} R_S^2}{V_{inv} + I_{D,A} R_S} + \frac{1}{I_{D,A}} \right) i_D^2 + \cdots \right)$$

and by inverting the series:

$$i_D = \frac{g_m}{1 + (g_m + g_{m,B}) R_S} \left(v_i + \frac{v_i^2}{4} \frac{\frac{g_m}{I_{D,A}} + \frac{g_m g_{m,B} R_S^2}{V_{inv} + I_{D,A} R_S}}{\left(1 + (g_m + g_{m,B}) R_S\right)^2} + \cdots \right)$$

For input signals with $v_i = \hat{v}_i \cos \omega t$, the *distortion factor k* is calculated approximately from the ratio of the first harmonic with $2\omega t$ to the fundamental with ωt at a low amplitude, that is, by neglecting the higher powers:

$$k \approx \frac{u_{a,2\omega t}}{u_{a,\omega t}} \approx \frac{i_{D,2\omega t}}{i_{D,\omega t}} \approx \frac{\hat{v}_i}{8} \frac{\dfrac{g_m}{I_{D,A}} + \dfrac{g_m g_{m,B} R_S^2}{V_{inv} + I_{D,A} R_S}}{\left(1 + \left(g_m + g_{m,B}\right) R_S\right)^2}$$

$$\overset{v_{BS}=0}{=} \frac{\hat{v}_i}{4\left(V_{GS,A} - V_{th}\right)\left(1 + g_m R_S\right)^2} \tag{3.60}$$

The last expansion uses $g_m/I_{D,A} = 2/(V_{GS,A} - V_{th})$. With the numeric example, we have $\hat{v}_i < k \cdot 11.5$ V and, if $A \approx -2$, $\hat{v}_o < k \cdot 23$ V.

A comparison with (3.13) shows that due to the feedback the permissible input amplitude $\hat{v}_i$ is increased by the square of the feedback factor $(1 + g_m R_S)$. But since the feedback simultaneously reduces the gain by the feedback factor, the permissible output amplitude is increased by the feedback factor if the distortion factor is the same. If the output amplitude is the same, then the distortion factor is reduced by the feedback factor.

A comparison with the common-emitter circuit with current feedback, described in Sect. 2.4.1, shows that with the same gain ($A \approx -2$) and the same operating point current ($I_{D,A} = I_{C,A} = 1.5$ mA) the distortion factor is higher: $k \approx \hat{v}_o/(23$ V) in the common-source circuit and $k \approx \hat{v}_o/(179$ V) in the common-emitter circuit. The reason for this is the low maximum gain of the MOSFET which, for the same circuit gain, results in a lower feedback factor and therefore in a higher distortion factor. For very low operating point currents, the maximum gain of the MOSFET increases and the distortion factor decreases; under these conditions the same values are achieved as with the common-emitter circuit.

A special situation arises for cubic distortions. Due to the almost square characteristic of the MOSFET without feedback, these distortions are very low and increase with the feedback, while the dominating square distortions and thus the distortion factor k decrease when the feedback increases. Figure 3.64 shows the dependence of the distortion factor k

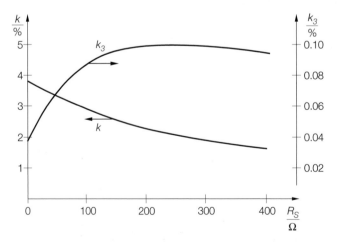

Fig. 3.64. A plot of the distortion factor k and the cubic distortion factor k_3 over the feedback resistance R_S with a constant amplitude at the output for the circuit shown in Fig. 3.61a

and the cubic distortion factor k_3 on the feedback resistance R_S with a constant amplitude at the output. The data for this plot were determined by simulation with *PSpice*.

Temperature sensitivity: Compared to the common-source circuit without feedback, the temperature drift of the output voltage is reduced by the feedback factor:

$$\left.\frac{dV_o}{dT}\right|_A \approx \frac{I_{D,A}R_D}{1+\left(g_m+g_{m,B}\right)R_S} \cdot 10^{-3}\,\text{K}^{-1}\left(5-\frac{4\ldots7\,\text{V}}{V_{GS,A}-V_{th}}\right)$$

For the numeric example, we obtain the result that $(dV_o/dT)|_A$ ranges from approximately $-3\,\text{mV/K}$ to $0.4\,\text{mV/K}$.

Common-Source Circuit with Voltage Feedback

In the common-source circuit with voltage feedback, according to Fig. 3.65a, a portion of the output voltage is fed back to the gate of the FET via the resistors R_1 and R_2; Fig. 3.65b shows the plot of the corresponding characteristic for $V_b = 5\,\text{V}$, $R_D = R_1 = 1\,\text{k}\Omega$, $R_2 = 6.3\,\text{k}\Omega$ and $K = 4\,\text{mA/V}^2$.

Operation in the pinch-off region: From the nodal equations

$$\frac{V_b - V_o}{R_D} + I_o = I_D + \frac{V_o - V_{GS}}{R_2}$$

$$\frac{V_{GS} - V_i}{R_1} = \frac{V_o - V_{GS}}{R_2}$$

for the operation without load, that is, with $I_o = 0$, it follows that:

$$V_o = \frac{V_b R_2 - I_D R_D R_2 + V_{GS} R_D}{R_2 + R_D} \overset{R_2 \gg R_D}{\approx} V_b - I_D R_D \tag{3.61}$$

$$V_i = \frac{V_{GS}(R_1 + R_2) - V_o R_1}{R_2} \tag{3.62}$$

The calculation of the operating point is based on (3.61). Replacing I_D with the equation for the pinch-off region leads to a quadratic equation in V_{GS}, which enables calculation

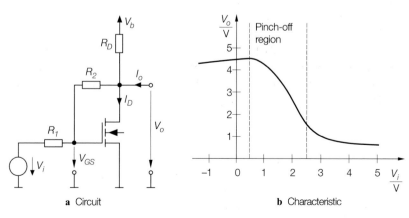

a Circuit   **b** Characteristic

Fig. 3.65. Common-source circuit with voltage feedback

of the operating point voltage $V_{GS,A}$ for a given output voltage $V_{o,A}$. As an alternative, an approximation can be used in which the current through the feedback resistance R_2 is neglected; with $V_{o,A} = 2.5\,\mathrm{V}$ we obtain:

$$V_{o,A} = 2.5\,\mathrm{V} \;\Rightarrow\; I_{D,A} \approx \frac{V_b - V_{o,A}}{R_D} \approx 2.5\,\mathrm{mA}$$

$$\Rightarrow\; V_{GS,A} = V_{th} + \sqrt{\frac{2 I_{D,A}}{K}} \approx 2.12\,\mathrm{V} \;\Rightarrow\; V_{i,A} \approx 2.06\,\mathrm{V} \tag{3.62}$$

Small-signal response: The calculation is carried out by means of the small-signal equivalent circuit shown in Fig. 3.66. From the nodal equations

$$\frac{v_i - v_{GS}}{R_1} + \frac{v_o - v_{GS}}{R_2} = 0$$

$$g_m v_{GS} + \frac{v_o - v_{GS}}{R_2} + \frac{v_o}{r_{DS}} + \frac{v_o}{R_D} = i_o$$

it follows, with $R'_D = R_D \,||\, r_{DS}$, that:

$$A = \left.\frac{v_o}{v_i}\right|_{i_o=0} = \frac{-g_m R_2 + 1}{1 + g_m R_1 + \dfrac{R_1 + R_2}{R'_D}} \;\;\overset{\substack{r_{DS} \gg R_D \\ R_1, R_2 \gg 1/g_m}}{\approx}\;\; -\frac{R_2}{R_1 + \dfrac{R_1 + R_2}{g_m R_D}}$$

If the gain without feedback is much higher than the feedback factor, that is, $g_m R_D \gg 1 + R_2/R_1$, then $A \approx -R_2/R_1$; however, due to the low maximum gain of the FET this condition is rarely met. Operating the circuit with a load resistance R_L allows calculation of the relevant operating gain A_B by replacing R_D with the parallel arrangement of R_D and R_L (see Fig. 3.66). For the chosen sample operating point we obtain, with $g_m = 4.47\,\mathrm{mS}$, $r_{DS} = 20\,\mathrm{k\Omega}$, $R_D = R_1 = 1\,\mathrm{k\Omega}$ and $R_2 = 6.3\,\mathrm{k\Omega}$, an *exact* value of $A = -2.067$; the approximation yields $A = -2.39$.

The *open-circuit input resistance* is with $R'_D = R_D \,||\, r_{DS}$:

$$r_{i,o} = \left.\frac{v_i}{i_i}\right|_{i_o=0} = R_1 + \frac{R_2 + R'_D}{1 + g_m R'_D} \;\;\overset{r_{DS} \gg R_D \gg 1/g_m}{\approx}\;\; R_1 + \frac{1}{g_m}\left(1 + \frac{R_2}{R_D}\right)$$

This applies to $i_o = 0$; that is, $R_L \to \infty$. For other values of R_L the input resistance can be calculated by replacing R_D with the parallel arrangement of R_D and R_L. For the sample operating point selected we obtain an *exact* value of $r_{i,o} = 2.38\,\mathrm{k\Omega}$; the approximation yields $r_{i,o} = 2.63\,\mathrm{k\Omega}$.

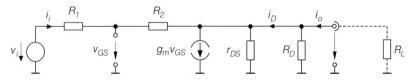

Fig. 3.66. Small-signal equivalent circuit for the common-source circuit with voltage feedback

The *short-circuit output resistance* is with $R'_D = R_D \| r_{DS}$:

$$r_{o,s} = \left.\frac{v_o}{i_o}\right|_{v_i=0} = R'_D \| \frac{R_1 + R_2}{1 + g_m R_1} \overset{\substack{r_{DS} \gg R_D \\ R_1 \gg 1/g_m}}{\approx} R_D \| \frac{1}{g_m}\left(1 + \frac{R_2}{R_1}\right)$$

For $R_1 \to \infty$, this leads to the *open-circuit output resistance*:

$$r_{o,o} = \left.\frac{v_o}{i_o}\right|_{i_i=0} = R'_D \| \frac{1}{g_m} \overset{r_{DS} \gg R_D \gg 1/g_m}{\approx} \frac{1}{g_m}$$

For the sample operating point selected, we obtain *exact* values of $r_{o,s} = 556\,\Omega$ and $r_{o,o} = 181\,\Omega$, while the approximation yields $r_{o,s} = 602\,\Omega$ and $r_{o,o} = 223\,\Omega$.

The summarized equations for the *common-source circuit with voltage feedback* are:

---

common-source circuit with voltage feedback

$$A = \left.\frac{v_o}{v_i}\right|_{i_o=0} \approx -\frac{R_2}{R_1 + \dfrac{R_1 + R_2}{g_m R_D}} \tag{3.63}$$

$$r_i = \left.\frac{v_i}{i_i}\right|_{i_o=0} \approx R_1 + \frac{1}{g_m}\left(1 + \frac{R_2}{R_D}\right) \tag{3.64}$$

$$r_o = \left.\frac{v_o}{i_o}\right|_{v_i=0} \approx R_D \| \frac{1}{g_m}\left(1 + \frac{R_2}{R_1}\right) \tag{3.65}$$

Operation as a current–to–voltage converter: Removing the resistance R_1 and driving the circuit with a current source I_i leads to the circuit shown in Fig. 3.67a, which acts as a *current-to-voltage converter*; it is also known as a *transimpedance amplifier*.[24] Figure 3.67b shows the characteristic for $V_b = 5\,\text{V}$, $R_D = 1\,\text{k}\Omega$ and $R_2 = 6.3\,\text{k}\Omega$.

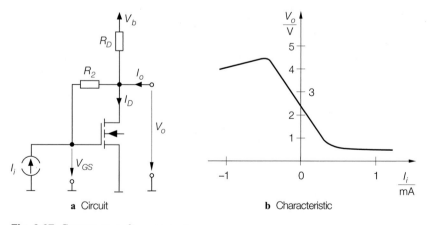

a Circuit   **b** Characteristic

Fig. 3.67. Current–to–voltage converter

---

[24] The term *transimpedance amplifier* is also used for operational amplifiers with current input and voltage output (CV-OPA).

It follows from Fig. 3.67a that:

$$V_o = V_b + (I_i + I_o - I_D) R_D \overset{I_o=0}{=} V_b + I_i R_D - \frac{K R_D}{2} (V_{GS} - V_{th})^2 \tag{3.66}$$

$$I_i = \frac{V_{GS} - V_o}{R_2} \tag{3.67}$$

Inserting the equations into one another leads to a quadratic equation in V_o and I_i, the resolution of which is very extensive. When assuming $|I_i R_D| \ll V_b - V_o$ and defining V_o, V_{GS} can be calculated from (3.66) and I_i from (3.66); for $V_{th} = 1\,\text{V}$, $K = 4\,\text{mA/V}^2$, $R_D = 1\,\text{k}\Omega$ and $R_2 = 6.3\,\text{k}\Omega$, we obtain:

$$V_o = 2.5\,\text{V} \;\Rightarrow\; V_{GS} \approx V_{th} + \sqrt{\frac{2 (V_b + I_i R_D - V_o)}{K R_D}} \overset{|I_i R_D| \ll V_b - V_o}{\approx} 2.12\,\text{V}$$

$$\Rightarrow I_i = \frac{V_{GS} - V_o}{R_2} \approx -60\,\mu\text{A} \quad \text{und} \quad I_D = \frac{K}{2} (V_{GS} - V_{th})^2 \approx 2.509\,\text{mA}$$

With the iterative approach, one inserts the last value of I_i into (3.66) and calculates the new values of V_{GS} and I_i; the next iteration yields $V_{GS} \approx 2.105\,\text{V}$, $I_i \approx -63\,\mu\text{A}$ and $I_D \approx 2.44\,\text{mA}$, which represent almost the exact results.

The small-signal response of the current–to–voltage converter can be derived from the equations for the common-source circuit with voltage feedback. The *transfer resistance* (*transimpedance*) R_T now takes the place of the gain; with $R'_D = R_D \,\|\, r_{DS}$, we obtain:

$$R_T = \left.\frac{v_o}{i_i}\right|_{i_o=0} = \lim_{R_1 \to \infty} R_1 \left.\frac{v_o}{v_i}\right|_{i_o=0} = \lim_{R_1 \to \infty} R_1 A$$

$$= R'_D \frac{1 - g_m R_2}{1 + g_m R'_D} \overset{\substack{g_m R_2 \gg 1 \\ r_{DS} \gg R_D}}{\approx} - R_2 \frac{g_m R_D}{1 + g_m R_D}$$

The *input resistance* can be calculated using the equations for the common-source circuit with voltage feedback by inserting $R_1 = 0$, and the *output resistance* corresponds to the open-circuit output resistance.

The summarized equations for the *current–to–voltage converter in common-source configuration* are:

---

current–to–voltage converter in common-source configuration

$$R_T = \left.\frac{v_o}{i_i}\right|_{i_o=0} \approx - R_2 \frac{g_m R_D}{1 + g_m R_D} \tag{3.68}$$

$$r_i = \left.\frac{v_i}{i_i}\right|_{i_o=0} \approx \frac{1}{g_m}\left(1 + \frac{R_2}{R_D}\right) \tag{3.69}$$

$$r_o = \frac{v_o}{i_o} \approx R_D \,\|\, \frac{1}{g_m} \tag{3.70}$$

At the sample operating point selected, and for $I_{D,A} = 2.44\,\text{mA}$, $K = 4\,\text{mA/V}^2$, $R_D = 1\,\text{k}\Omega$ and $R_2 = 6.3\,\text{k}\Omega$, we obtain the values $R_T \approx -5.14\,\text{k}\Omega$, $r_i \approx 1.65\,\text{k}\Omega$ and $r_o \approx 185\,\Omega$.

The current–to–voltage converter is mostly used in photodiode receivers; the receiving diode is operated in reverse-biased mode and therefore acts like a current source with a very high internal resistance. The current–to–voltage converter transforms the current i_i into the voltage $v_o = R_T i_i$. Due to the high internal resistance of the diode, the noise of the circuit is mainly caused by the input noise current of the FET and the thermal noise of the feedback resistance R_2; the low input noise current of the FET, which is particularly low compared to the bipolar transistor, results in a notably low noise figure.

Setting the Operating Point

Operation as a small-signal amplifier requires a stable setting for the operating point. The operating point should depend as little as possible on the parameters of the FET, since they are temperature-sensitive and subject to variations caused by the production process. The temperature sensitivity of the FET can, of course, be kept low by selecting an operating point close to the temperature compensation point, but the variations in the threshold voltage caused by the production process are considerable, especially in discrete FETs; fluctuations in the range of $\pm 0.5\,\text{V} \ldots \pm 1\,\text{V}$ are common.

Due to their low gain compared to common-emitter circuits, small-signal amplifiers in common-source configuration with discrete FETs are used in exceptional cases only; these include amplifiers for very high-resistance signal sources and the current–to–voltage converter described above.

Setting the operating point for AC voltage coupling: In AC voltage coupling, the amplifier is connected to the signal source and the load via coupling capacitors. For voltage amplifiers the voltage setting with DC feedback is generally used as shown in Fig. 3.68a; this corresponds to the operating point setting in common-emitter circuits shown in Fig. 2.75a. The variations of Figs. 2.75b and 2.78 can also be used with the FET; here, the extremely high input resistance of the FET is only effective with direct coupling at the input, since otherwise the voltage divider at the input determines the input resistance of the circuit.

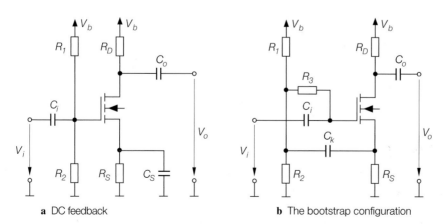

a DC feedback   **b** The bootstrap configuration

Fig. 3.68. Setting the operating point with current feedback

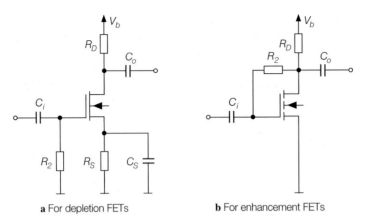

a For depletion FETs   **b** For enhancement FETs

Fig. 3.69. Special circuits for setting the operating point

A special case is current feedback with the *bootstrap* shown in Fig. 3.68b, which reduces the voltage drop across R_3 by feeding the signal back to the voltage divider so that the input resistance increases accordingly: $r_i \approx R_3 (1 + g_m R_S)$. This circuit only works efficiently with high feedback ($g_m R_S \gg 1$) and is therefore predominantly used in drain circuits (see Sect. 3.4.2).

In addition, there are special circuits that are suitable for setting the operating point of depletion FETs only. Since these can be operated with $V_G = 0$, it is possible to remove the resistance R_1 from Fig. 3.68a, leading to the circuit shown in Fig. 3.69a; the same applies to the bootstrap configuration. From the condition $V_{GS} = - I_D R_S$ and from the equation for the pinch-off region, we obtain the calculation:

$$ R_S = \frac{|V_{th}|}{I_{D,A}} \left(1 - \sqrt{\frac{2I_{D,A}}{K V_{th}^2}} \right) = \frac{|V_{th}|}{I_{D,A}} \left(1 - \sqrt{\frac{I_{D,A}}{I_{D,A(max)}}} \right) $$

Here, $I_{D,A(max)} = K V_{th}^2/2$ is the maximum possible operating point current. If the FET is to be operated at the temperature compensation point with $V_{GS,TC} \approx V_{th} + 1\,\text{V}$, then $I_{D,A} \approx K \cdot 0.5\,\text{V}^2$ and, consequently,

$$ R_S = \frac{2|V_{GS,TC}|}{K \left(V_{GS,TC} - V_{th} \right)^2} \approx \frac{|V_{GS,TC}|}{K \cdot 0.5\,\text{V}^2} \qquad \text{for } V_{GS,TC} \leq 0 $$

Enhancement MOSFETs can be operated in the pinch-off region with $V_{GS} = V_{DS}$ (see Fig. 3.69b); since there is no gate current, or only a very small one, resistance R_2 can be so high that the voltage feedback caused by R_2 is negligible; in this case, the input resistance can be determined from (3.69) on page 242.

The properties, advantages and disadvantages of AC voltage coupling are described in detail on page 120 in respect to the common-emitter circuit.

Setting the operating point with DC voltage coupling: In DC voltage coupling, also known as *direct* or *galvanic* coupling, the amplifier is connected directly to the signal source and the load. The DC voltages at the input and output of the amplifier must be adapted to the DC voltages of the signal source and load; therefore it is not possible, for multi-stage amplifiers, to adjust the operating point for each stage separately.

For DC voltage amplifiers, DC voltage coupling is mandatory. The same applies to integrated amplifiers, since it is usually not possible to achieve the values of coupling capacitances that are required in integrated circuits and external coupling capacitances are unwanted. In multi-stage amplifiers, DC voltage coupling is almost always combined with a feedback across all stages, in order to achieve a defined and thermally stable operating point.

Frequency Response and Cutoff Frequency

The small-signal gain A as calculated so far only applies to low signal frequencies; with higher frequencies, the magnitude of the gain decreases due to the FET capacitances. To calculate the frequency response and the cutoff frequency one must use the dynamic small-signal model of the FET according to Fig. 3.48; here, in addition to the capacitances C_{GS}, C_{GD}, C_{BS} and C_{BD}, the gate spreading resistance R_G is also taken into consideration.

For discrete FETs without a bulk terminal it is possible to use the simple small-signal model according to Fig. 3.49a, which corresponds to the small-signal model of the bipolar transistor in most respects. As the cutoff frequency can only be calculated approximately, using the simple small-signal model for integrated FETs also causes no major problems, but allows the results for bipolar transistors to be transferred to FETs if the following replacements are made:

$$R_B \rightarrow R_G , \ r_{BE} \rightarrow \infty , \ r_{CE} \rightarrow r_{DS} , \ C_E \rightarrow C_{GS} , \ C_C \rightarrow C_{GD}$$

Common-source circuit without feedback: Figure 3.70 shows the dynamic small-signal equivalent circuit for the common-source circuit without feedback. With $R'_g = R_g + R_G$ and $R'_D = R_L \| R_D \| r_{DS}$, the operating gain $\underline{A}_B(s) = \underline{v}_o(s)/\underline{v}_g(s)$ can be expressed by:

$$\underline{A}_B(s) = -\frac{(g_m - sC_{GD})R'_D}{1 + sc_1 + s^2c_2} \tag{3.71}$$

$$c_1 = C_{GS}R'_g + C_{GD}\left(R'_g + R'_D + g_m R'_D R'_g\right) + C_{DS}R'_D$$

$$c_2 = (C_{GS}C_{GD} + C_{GS}C_{DS} + C_{GD}C_{DS})R'_g R'_D$$

As with the common-emitter circuit, the frequency response can be described approximately by a first-degree lowpass filter, if the zero is neglected and the s^2 term is removed from the denominator. With the low-frequency gain

$$A_0 = \underline{A}_B(0) = -g_m R'_D \tag{3.72}$$

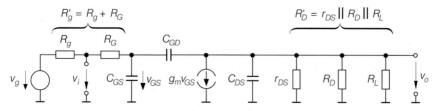

Fig. 3.70. Dynamic small-signal equivalent circuit for the common-source circuit without feedback

it follows that:

$$\underline{A}_B(s) \approx \frac{A_0}{1 + s\left(C_{GS}R'_g + C_{GD}\left(R'_g + R'_D + g_m R'_D R'_g\right) + C_{DS}R'_D\right)} \tag{3.73}$$

This gives an approximation for the $-3\ dB$ *cutoff frequency* f_{-3dB} at which the magnitude of the gain has dropped by 3 dB:

$$\omega_{-3dB} = 2\pi f_{-3dB} \approx \frac{1}{C_{GS}R'_g + C_{GD}\left(R'_g + R'_D + g_m R'_D R'_g\right) + C_{DS}R'_D} \tag{3.74}$$

In most cases $R'_D, R'_g \gg 1/g_m$; this leads to:

$$\omega_{-3dB} = 2\pi f_{-3dB} \approx \frac{1}{C_{GS}R'_g + C_{GD}g_m R'_D R'_g + C_{DS}R'_D} \tag{3.75}$$

As for the common-emitter circuit, it is possible to describe the cutoff frequency by means of the low-frequency gain A_0 and two time constants that are independent of A_0; from (3.74) it follows that:

$$\omega_{-3dB}(A_0) \approx \frac{1}{T_1 + T_2|A_0|} \tag{3.76}$$

$$T_1 = (C_{GS} + C_{GD})\,R'_g \tag{3.77}$$

$$T_2 = C_{GD}R'_g + \frac{C_{GD} + C_{DS}}{g_m} \tag{3.78}$$

Two regions can be distinguished:

- For $|A_0| \ll T_1/T_2$, $\omega_{-3dB} \approx T_1^{-1}$, which means that the cutoff frequency is independent of the gain. The maximum cutoff frequency is achieved for the borderline case $A_0 \to 0$ and $R_g = 0$:

$$\omega_{-3dB,max} = \frac{1}{(C_{GS} + C_{GD})\,R_G}$$

This corresponds to the *transconductance cutoff frequency* ω_{Y21s} (see (3.46)).
- For $|A_0| \gg T_1/T_2$, $\omega_{-3dB} \approx (T_2|A_0|)^{-1}$, which means that the cutoff frequency is proportional to the inverse value of the gain, leading to a constant *gain-bandwidth-product, GBW*:

$$GBW = f_{-3dB}\,|A_0| \approx \frac{1}{2\pi\,T_2} \tag{3.79}$$

The gain–bandwidth product *GBW* is an important parameter, as it represents the absolute upper limit for the product of the low-frequency gain and the cutoff frequency; that is, for all values of $|A_0|$, $GBW \geq f_{-3dB}|A_0|$.

If a load capacitance C_L is connected to the output of the circuit, (3.74)–(3.79) can be used to calculate the related values for f_{-3dB}, T_1, T_2 and *GBW* by replacing C_{DS} by $C_{DS} + C_L$. T_2 is then:

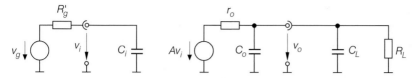

Fig. 3.71. Equivalent circuit with the equivalent parameters A, r_o, C_i and C_o

$$T_2 = C_{GD} R'_g + \frac{C_{GD} + C_{DS} + C_L}{g_m} \qquad (3.80)$$

Equivalent circuit diagram: The common-source circuit can be approximately described by the equivalent circuit shown in Fig. 3.71. It is derived from Fig. 3.60 by adding the *input capacitance* C_i and the *output capacitance* C_o, but only produces an approximate calculation for the gain $\underline{A}_B(s)$ and the cutoff frequency f_{-3dB}. C_i and C_o are determined from the premise that after removing the s^2 term from the denominator, the calculation of $\underline{A}_B(s)$ must lead to (3.73):

$$C_i \approx C_{GS} + C_{GD}(1 + |A_0|) \qquad (3.81)$$

$$C_o \approx C_{GD} + C_{DS} \qquad (3.82)$$

The input capacitance C_i depends on the circuitry at the output, because A_0 depends on R_L. The fact that C_{GD} influences C_i with the factor $(1 + |A_0|)$ is known as the *Miller effect* and C_{GD} is called the *Miller capacitance*. A and r_o are given by (3.50) and (3.52), and are independent of the circuitry. The gate spreading resistance R_G is regarded as a portion of the internal resistance of the signal source: $R'_g = R_g + R_G$.

Example: $I_{D,A} = 2\,\text{mA}$ was selected for the numeric example used in the common-source circuit without feedback, according to Fig. 3.57a. For $K = 4\,\text{mA/V}^2$, $V_A = 50\,\text{V}$, $C_{oss} = 5\,\text{pF}$, $C_{rss} = 2\,\text{pF}$, $f_{Y21s} = 1\,\text{GHz}$ and $f_T = 100\,\text{MHz}$, we obtain from Fig. 3.51 on page 222 the small-signal parameters $g_m = 4\,\text{mS}$, $r_{DS} = 25\,\text{k}\Omega$, $R_G = 25\,\Omega$, $C_{GD} = 2\,\text{pF}$, $C_{GS} = 4.4\,\text{pF}$ and $C_{DS} = 3\,\text{pF}$. For $R_g = R_D = 1\,\text{k}\Omega$, $R_L \to \infty$ and $R'_g \approx R_g$, it follows from (3.72) that $A_0 \approx -3.85$, from (3.74) that $f_{-3dB} \approx 8.43\,\text{MHz}$, and from (3.75) that $f_{-3dB} \approx 10.6\,\text{MHz}$. It follows from (3.77) that $T_1 \approx 6.4\,\text{ns}$, from (3.78) that $T_2 \approx 3.25\,\text{ns}$, and from (3.79) that $GBW \approx 49\,\text{MHz}$. With a load capacitance $C_L = 1\,\text{nF}$ we obtain from (3.80) $T_2 \approx 253\,\text{ns}$, from (3.76) $f_{-3dB} \approx 162\,\text{kHz}$ and from (3.79) $GBW \approx 630\,\text{kHz}$.

A comparison with the values for the common-emitter circuit described on page 126 makes sense only with regard to the gain–bandwidth product, as the low-frequency gains are very different. It can be seen that for the common-source circuit without capacitive load approximately the same GBW is reached as for the common-emitter circuit. However, if a capacitive load is present, the GBW of the common-source circuit is markedly lower; namely, in the borderline case of large load capacitances it is reduced by exactly the ratio of the transconductances, as can be seen by a comparison of (3.80) and (2.96) on page 125. In practice, this means that:

Due to the low transconductance of the FETs, the common-source circuit is poorly suited for driving capacitive loads.

Common-source circuit with current feedback: The frequency response and the cutoff frequency of the common-emitter circuit with current feedback, according to Fig. 3.61a, can be derived from the corresponding parameters of the common-source circuit without feedback. For this purpose, we shall use the conversion of the small-signal equivalent circuit already carried out for the common-emitter circuit with current feedback (see Fig. 2.86 on page 127). Figure 3.72 shows the small-signal equivalent circuit for the common-source configuration with current feedback before and after conversion; here, the small-signal parameters are converted into the equivalent values of a FET without current feedback:

$$
\begin{bmatrix} g'_m \\ C'_{GS} \\ C'_{DS} \\ 1 \\ r'_{DS} \end{bmatrix} = \frac{1}{1 + (g_m + g_{m,B})\, R_S} \cdot \begin{bmatrix} g_m \\ C_{GS} \\ C_{DS} \\ 1 \\ r_{DS} \end{bmatrix} \overset{v_{BS}=0}{=} \frac{1}{1 + g_m R_S} \cdot \begin{bmatrix} g_m \\ C_{GS} \\ C_{DS} \\ 1 \\ r_{DS} \end{bmatrix} \tag{3.83}
$$

The transconductance g'_m corresponds to the *reduced transconductance* $g_{m,red}$ that has already been introduced in (3.59). The gate–drain capacitance C_{GD} remains unchanged.

Now one can insert the equivalent values into (3.72) and (3.76)–(3.78) or (3.80) for the common-source circuit without feedback; with $R'_g = R_g + R_G$ and $R'_D = r'_{DS} \| R_D \| R_L$, it follows:

$$
\omega_{-3dB}(A_0) \approx \frac{1}{T_1 + T_2|A_0|} \tag{3.84}
$$

$$
T_1 = \left(C'_{GS} + C_{GD} \right) R'_g \tag{3.85}
$$

$$
T_2 = C_{GD} R'_g + \frac{C_{GD} + C'_{DS} + C_L}{g'_m} \tag{3.86}
$$

$$
A_0 = -g'_m R'_D \tag{3.87}
$$

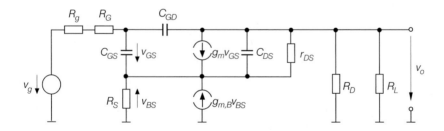

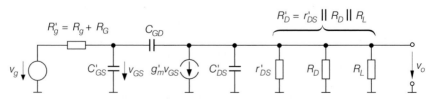

Fig. 3.72. Dynamic small-signal equivalent circuit for the common-source circuit with current feedback prior to (*above*) and after (*below*) conversion

From (3.86) it follows that a strong current feedback has a comparable high effect even for a small capacitance C_L, since T_2 increases relatively strongly due to $g'_m < g_m$; the gain–bandwidth product GBW declines accordingly.

Example: $I_{D,A} = 1.5\,\text{mA}$ is selected for the numeric example of the common-source circuit with current feedback, as shown in Fig. 3.61a. For $K = 4\,\text{mA/V}^2$ and $V_A = 50\,\text{V}$, we obtain from Fig. 3.51 on page 222 the parameter $g_m = 3.46\,\text{mS}$ and $r_{DS} = 33.3\,\text{k}\Omega$. The parameters $R_G = 25\,\Omega$, $C_{GD} = 2\,\text{pF}$, $C_{GS} = 4.4\,\text{pF}$ and $C_{DS} = 3\,\text{pF}$ are taken from the example on page 247,[25] and r_{DS} is disregarded. For $R_S = 200\,\Omega$, the conversion to (3.83) results in the equivalent values $g'_m = 2.04\,\text{mS}$, $C'_{GS} = 2.6\,\text{pF}$, $C'_{DS} = 1.77\,\text{pF}$ and $r'_{DS} = 56.3\,\text{k}\Omega$. If $R_g = R_D = 1\,\text{k}\Omega$ and $R_L \to \infty$, then $R'_D = R_D \,||\, r_{DS} = 983\,\Omega$ and $R'_g = R_g + R_G = 1025\,\Omega$, and it follows from (3.87) that $A_0 \approx -2$, from (3.85) that $T'_1 \approx 4.7\,\text{ns}$, from (3.86) that $T_2 \approx 4.9\,\text{ns}\,(C_L = 0)$, from (3.76) that $f_{-3dB} \approx 11\,\text{MHz}$ and from (3.79) that $GBW \approx 32.5\,\text{MHz}$. With the load capacitance $C_L = 1\,\text{nF}$, we obtain from (3.86) $T_2 \approx 494\,\text{ns}$, from (3.76) $f_{-3dB} \approx 160\,\text{kHz}$ and from (3.79) $GBW \approx 322\,\text{kHz}$.

Common-source circuit with voltage feedback: Figure 3.73 shows the small-signal equivalent circuit in which the gate resistance R_G of the FET is neglected. The results from the common-emitter circuit with voltage feedback can be transferred to the common-source circuit with voltage feedback if one takes into consideration the fact that the capacitance C_{DS} acts like a load capacitance. From (2.102), it follows, with $R'_1 = R_1 + R_g$ and $R'_D = r_{DS} \,||\, R_D \,||\, R_L$, that:

$$A_0 \approx -\frac{R_2}{R'_1 + \dfrac{R'_1 + R_2}{g_m R'_D}} \overset{g_m R'_D \gg 1 + R_2/R'_1}{\approx} -\frac{R_2}{R'_1} \tag{3.88}$$

and, from (2.105)–(2.107):

$$\omega_{-3dB}(A_0) \approx \frac{1}{T_1 + T_2|A_0|} \tag{3.89}$$

$$T_1 = \frac{C_{GS} + C_{DS} + C_L}{g_m} \tag{3.90}$$

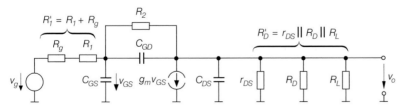

Fig. 3.73. Small-signal equivalent circuit for the common-source configuration with voltage feedback

[25] Strictly speaking, these parameters must be calculated from C_{rss}, f_T and f_{Y21s} with the aid of Fig. 3.51. However, since the dependences of these parameters on the operating point are generally unknown, one makes use of the fact that the capacitances and the gate spreading resistance are essentially geometrically scaled; that is, they depend only on the geometric dimensions of the FET and not on the operating point.

$$T_2 = \left(\frac{C_{GS}}{g_m R'_D} + C_{GD} \right) R'_1 + \frac{C_{DS} + C_L}{g_m} \tag{3.91}$$

Strong voltage feedback can cause complex conjugate poles; in this case, the cutoff frequency can be estimated only very roughly by (3.89)–(3.91).

The common-source circuit with voltage feedback can also be approximately described by the equivalent circuit shown in Fig. 3.71; by analogy to the common-emitter circuit with voltage feedback, and by taking into consideration the additional capacitance C_{DS} that occurs at the output, this leads to:

$$C_i = 0$$

$$C_o \approx \left(C_{GS} \left(\frac{1}{R_2} + \frac{1}{R'_D} \right) + C_{GD} g_m \right) (R'_1 \| R_2) + C_{DS}$$

Therefore, the input impedance is a purely ohmic resistance. A, r_i and r_o are given by (3.63)–(3.65).

Example: $I_{D,A} = 2.5\,\text{mA}$ was selected for the numeric example used in the common-source circuit with voltage feedback, according to Fig. 3.65; for $K = 4\,\text{mA/V}^2$ and $V_A = 50\,\text{V}$, we obtain, from Fig. 3.51 on page 222, $g_m = 4.47\,\text{mS}$ and $r_{DS} = 20\,\text{k}\Omega$. The parameters $R_G = 25\,\Omega$, $C_{GD} = 2\,\text{pF}$, $C_{GS} = 4.4\,\text{pF}$ and $C_{DS} = 3\,\text{pF}$ are adopted from the example on page 247. For $R_D = R_1 = 1\,\text{k}\Omega$, $R_2 = 6.3\,\text{k}\Omega$, $R_L \to \infty$, $r_{DS} \gg R_D$ and $R_g = 0$, we obtain $R'_D \approx R_D = 1\,\text{k}\Omega$ und $R'_1 = R_1 = 1\,\text{k}\Omega$; therefore, it follows from (3.88) that $A_0 \approx -2.6$, from (3.90) that $T_1 \approx 1.66\,\text{ns}$, from (3.91) that $T_2 \approx 3.66\,\text{ns}$, from (3.89) that $f_{\text{-3dB}} \approx 14\,\text{MHz}$ and from (3.79) that $GBW \approx 43\,\text{MHz}$. With a load capacitance $C_L = 1\,\text{nF}$, we obtain from (3.90) $T_1 \approx 225\,\text{ns}$, from (3.91) $T_2 \approx 227\,\text{ns}$, from (3.89) $f_{\text{-3dB}} \approx 195\,\text{kHz}$ and from (3.79) $GBW \approx 700\,\text{kHz}$.

Current-to-voltage converter: Figure 3.74 shows the small-signal equivalent circuit for the current–to–voltage converter of Fig. 3.67a; with $R'_D = R_D \| R_L \| r_{DS}$, and removing the s^2 term from the denominator, it follows that

$$\underline{Z}_T(s) = \frac{v_o(s)}{\underline{i}_i(s)} \approx -\frac{g_m R'_D R_2}{1 + g_m R'_D} \cdot \frac{1}{1 + s \left(\dfrac{C_{GS}(R_2 + R'_D) + C_{DS} R'_D}{1 + g_m R'_D} + C_{GD} R_2 \right)}$$

and thus:

$$\omega_{\text{-3dB}} = 2\pi f_{\text{-3dB}} \approx \frac{1}{\dfrac{C_{GS}(R_2 + R'_D) + C_{DS} R'_D}{1 + g_m R'_D} + C_{GD} R_2}$$

For $r_{DS} \gg R_D \gg 1/g_m$ and $R_L \to \infty$, we have:

$$\omega_{\text{-3dB}} = 2\pi f_{\text{-3dB}} \approx \frac{1}{\dfrac{C_{GS}}{g_m} \left(1 + \dfrac{R_2}{R_D} \right) + \dfrac{C_{DS}}{g_m} + C_{GD} R_2} \tag{3.92}$$

A load capacitance C_L is taken into consideration by replacing C_{DS} by $C_L + C_{DS}$.

Example: $I_{D,A} = 2.44\,\text{mA}$ is selected for the current–to–voltage converter shown in Fig.. 3.67a; for $K = 4\,\text{mA/V}^2$ and $V_A = 50\,\text{V}$, it follows that $g_m = 4.42\,\text{mS}$ and

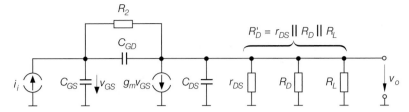

Fig. 3.74. Small-signal equivalent circuit for the current–to–voltage converter

$r_{DS} = 20.5\,\text{k}\Omega$. The parameters $R_G = 25\,\Omega$, $C_{GD} = 2\,\text{pF}$, $C_{GS} = 4.4\,\text{pF}$ and $C_{DS} = 3\,\text{pF}$ are adopted from the example on page 247. With $R_D = 1\,\text{k}\Omega$, $R_2 = 6.3\,\text{k}\Omega$, $R_L \to \infty$ and $r_{DS} \gg R_D$, we obtain from (3.92) $f_{-3dB} \approx 7.75\,\text{MHz}$.

Summary

The common-source circuit can be operated without feedback, with current feedback or with voltage feedback. Figure 3.75 shows the three variations and Fig. 3.76 lists the most important characteristic values. The common-source circuit with voltage feedback is rarely used, since it is not possible to make use of the high input resistance of the FET.

In practice, the common-source circuit without feedback and the common-source circuit with current feedback are used only if a high input resistance or a low noise figure with high-resistance sources is required. In all other cases, the common-emitter circuit is superior due to the larger maximum gain, the much higher transconductance of the bipolar transistor for the same operating current and the lower noise figure for low-resistance sources.

An important role is played by the common-source circuit in integrated CMOS circuits, because here no bipolar transistors are available. This applies in particular to large-scale integrated *mixed analog/digital circuits* (*mixed-mode ICs*) that contain only a few analog components in addition to many digital ones, and it means that they can be produced in the comparably simple and reasonably priced CMOS digital process. However, there is greater tendency toward BICMOS processes, which enable the production of MOSFETs *and* bipolar transistors.

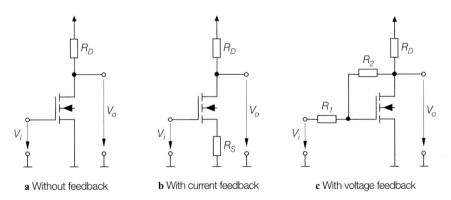

a Without feedback   **b** With current feedback   **c** With voltage feedback

Fig. 3.75. Variations of the common-source circuit

	Without feedback Fig. 3.75a	With current feedback Fig. 3.75b	With voltage feedback Fig. 3.75c
A	$-g_m R_D$	$-\dfrac{g_m R_D}{1 + g_m R_S}$	$-\dfrac{R_2}{R_1 + \dfrac{R_1 + R_2}{g_m R_D}}$
r_i	∞	∞	R_1
r_o	R_D	R_D	$R_D \parallel \dfrac{1}{g_m}\left(1 + \dfrac{R_2}{R_1}\right)$

A: small-signal voltage gain; r_i: small-signal input resistance; r_o: small-signal output resistance

Fig. 3.76. Characteristic parameters of the common-source circuit

3.4.2
Common-Drain Circuit

Figure 3.77a shows the common-drain circuit, which consists of the MOSFET, the source resistance R_S, the supply voltage source V_b and the signal voltage source V_g with the internal resistance R_g. The transfer characteristic and the small-signal response depend on the circuitry at the bulk terminal, which is connected to the source in discrete MOSFETs and to the most negative supply voltage (here, ground) in integrated MOSFETs. Assumptions for the following examination are that $V_b = 5\,\text{V}$ and $R_S = R_g = 1\,\text{k}\Omega$; in addition, $K = 4\,\text{mA/V}^2$ and $U_{th} = 1\,\text{V}$ for the discrete MOSFET and $K = 4\,\text{mA/V}^2$, $V_{th} = 1\,\text{V}$, $\gamma = 0.5\,\sqrt{\text{V}}$ and $V_{inv} = 0.6\,\text{V}$ for the integrated MOSFET.

Transfer Characteristic of the Common-Drain Circuit

The transfer characteristic shown in Fig. 3.78 is obtained by measuring the output voltage V_o as a function of the signal voltage V_g. For $V_g < V_{th} = 1\,\text{V}$, there is no drain current and $V_o = 0$. If $V_g \geq 1\,\text{V}$, a drain current I_D flows which increases with U_g, and the output voltage *follows* the input voltage with a *distance* of V_{GS}; therefore, the common-drain circuit is also known as the *source follower*. The FET always operates in the pinch-off

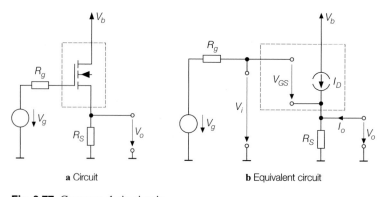

a Circuit   **b** Equivalent circuit

Fig. 3.77. Common-drain circuit

region as long as the signal voltage remains below the supply voltage, or exceeds it by a maximum of V_{th}.

Figure 3.77b shows the equivalent circuit for the common-drain configuration. For $V_g \geq V_{th}$ and $I_o = 0$, the voltages are:

$$V_o = I_D R_S \tag{3.93}$$

$$V_i = V_o + V_{GS} = V_o + \sqrt{\frac{2I_D}{K}} + V_{th} \tag{3.94}$$

Equation (3.94) uses (3.3) solved for V_{GS} for a current in the pinch-off region; the Early effect is neglected. Inserting (3.93) into (3.94) leads to:

$$V_i = V_o + \sqrt{\frac{2V_o}{K R_S}} + V_{th} \tag{3.95}$$

This equation applies to both discrete and integrated MOSFETs, but with integrated MOS-FETs the threshold voltage V_{th} depends on the bulk–source voltage V_{BS} due to the substrate effect; if $V_B = 0$ then $V_{BS} = -V_o$, so that applying (3.18) leads to:

$$V_i = V_o + \sqrt{\frac{2V_o}{K R_S}} + V_{th,0} + \gamma \left(\sqrt{V_{inv} + V_o} - \sqrt{V_{inv}} \right) \tag{3.96}$$

Due to the approximately linear characteristic, the operating point can be selected over a wide range; for the operating point marked on the characteristic curve of the discrete MOSFET in Fig. 3.78, the voltages are:

$$V_o = 2\,\text{V} \;\Rightarrow\; I_D = \frac{V_o}{R_S} = 2\,\text{mA} \;\Rightarrow\; V_{GS} = \sqrt{\frac{2I_D}{K}} + V_{th} = 2\,\text{V}$$

$$\Rightarrow\; V_i = V_o + V_{GS} = 4\,\text{V}$$

With $V_o = 2\,\text{V}$ for the integrated MOSFET, we obtain from (3.96) $V_i = 4.42\,\text{V}$.

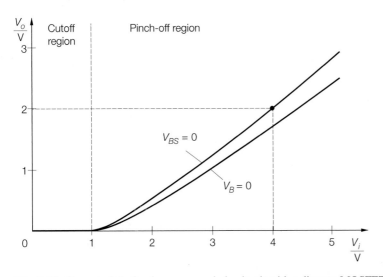

Fig. 3.78. Characteristic for the common-drain circuit with a discrete MOSFET ($V_{BS} = 0$) and an integrated MOSFET ($V_B = 0$)

Small-Signal Response of the Common-Drain Circuit

The response to signals around an operating point A is called the *small-signal response*. The operating point is determined by the operating point values $V_{i,A}$, $V_{o,A}$ and $I_{D,A}$; the operating point as determined above is used as an example: $V_{i,A} = 4\,\text{V}$, $V_{o,A} = 2\,\text{V}$ and $I_{D,A} = 2\,\text{mA}$.

The upper section of Fig. 3.79 shows the small-signal equivalent circuit for the common-drain circuit in its original state. Redrawing this circuit and combining elements in parallel leads to the small-signal equivalent circuit shown in the lower part of Fig. 3.79, with:

$$R'_S = \begin{cases} R_S \,||\, r_{DS} & \text{for the discrete MOSFET}(v_{BS} = 0) \\[2mm] R_S \,||\, r_{DS} \,||\, \dfrac{1}{g_{m,B}} & \text{for the integrated MOSFET}(v_{BS} = -v_o) \end{cases}$$

With integrated MOSFETs, the current source with its substrate transconductance $g_{m,B}$ acts like a resistor, since the control voltage v_{BS} is equal to the voltage at the current source. It is possible to change from the integrated to the discrete MOSFET with the limitation $v_{BS} = 0$; this sets $g_{m,B} = 0$ in the equations.[26]

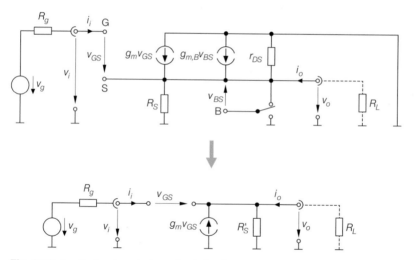

Fig. 3.79. Small-signal equivalent circuit for the common-drain circuit

The nodal equation $g_m\,v_{GS} = v_o/R'_S$ leads, for $v_{GS} = v_i - v_o$, to the *small-signal gain*:

$$A = \left.\frac{v_o}{v_i}\right|_{i_o=0} = \frac{g_m R'_S}{1 + g_m R'_S} \overset{r_{DS} \gg 1/g_m}{\approx} \frac{g_m R_S}{1 + (g_m + g_{m,B}) R_S} \overset{v_{BS}=0}{=} \frac{g_m R_S}{1 + g_m R_S}$$

For $K = 4\,\text{mA/V}^2$, $\gamma = 0.5\,\sqrt{\text{V}}$, $V_{inv} = 0.6\,\text{V}$ and $I_{D,A} = 2\,\text{mA}$, it follows from Fig. 3.51 or Fig. 3.52 that $g_m = 4\,\text{mS}$ and $g_{m,B} = 0.62\,\text{mS}$; for $R_S = 1\,\text{k}\Omega$ we thus obtain $A \approx 0.8$

---

[26] $g_{m,B} = 0$ as a limiting condition would not be correct, as the substrate transconductance of the discrete MOSFET is not zero but has no effect because $v_{BS} = 0$; therefore, $v_{BS} = 0$ is a correct limitation and $g_{m,B} = 0$ is its effect in the equations.

when using a discrete MOSFET and $A \approx 0.71$ when using an integrated MOSFET. Due to the relatively low transconductance, the gain is clearly below 1.

The *small-signal input resistance* is $r_i = \infty$ and the *small-signal output resistance* is:

$$r_o = \frac{v_o}{i_o} = \frac{1}{g_m} \| R_S' \overset{r_{DS} \gg 1/g_m}{\approx} \frac{1}{g_m} \| \frac{1}{g_{m,B}} \| R_S \overset{v_{BS}=0}{=} \frac{1}{g_m} \| R_S$$

For the numeric example, this results to $r_o \approx 200\,\Omega$ when using a discrete MOSFET and $r_o \approx 178\,\Omega$ when using an integrated MOSFET.

With $r_{DS} \gg 1/g_m$ and *without* load resistance R_L, the formulas for the *common-drain configuration* are:

> **common-drain configuration**
>
> $$A = \frac{v_o}{v_i}\bigg|_{i_o=0} \approx \frac{g_m R_S}{1 + (g_m + g_{m,B}) R_S} \overset{v_{BS}=0}{=} \frac{g_m R_S}{1 + g_m R_S} \qquad (3.97)$$
>
> $$r_i = \frac{v_i}{i_i}\bigg|_{i_o=0} = \infty \qquad (3.98)$$
>
> $$r_o = \frac{v_o}{i_o} \approx \frac{1}{g_m} \| \frac{1}{g_{m,B}} \| R_S \overset{v_{BS}=0}{=} \frac{1}{g_m} \| R_S \qquad (3.99)$$

With a load resistance R_L, it is necessary to replace R_S by the parallel arrangement of R_S and R_L in (3.97).

Maximum gain in integrated circuits: The *maximum gain* A_{max} is reached when the source resistance R_S is replaced by an ideal current source. In integrated circuits the maximum gain is:

$$A_{max} = \lim_{R_S \to \infty} A \overset{r_{DS} \gg 1/g_m}{\approx} \frac{g_m}{g_m + g_{m,B}} \overset{\overset{(3.41)}{V_{BS}=-V_o}}{=} \frac{1}{1 + \dfrac{\gamma}{2\sqrt{V_{inv} + V_o}}}$$

For the numeric example with $\gamma = 0.5\,\sqrt{V}$, $V_{inv} = 0.6\,V$ and $V_{o,A} = 2\,V$, we obtain $A_{max} = 0.87$. With a discrete FET, we have $A_{max} = 1$.

Nonlinearity: The distortion factor of the common-drain circuit can be approximately determined by a series expansion of the characteristic in the operating point. As (3.94), which determines the characteristic, also applies to the common-source circuit with current feedback, we can use (3.60):

$$k \approx \frac{\hat{v}_i}{8} \frac{\dfrac{g_m}{I_{D,A}} + \dfrac{g_m g_{m,B} R_S^2}{V_{inv} + I_{D,A} R_S}}{\left(1 + (g_m + g_{m,B}) R_S\right)^2} \overset{v_{BS}=0}{=} \frac{\hat{v}_i}{4 (V_{GS,A} - V_{th}) (1 + g_m R_S)^2} \qquad (3.100)$$

For the numeric example, we obtain $\hat{v}_i < k \cdot 100\,V$ when using a discrete MOSFET and $\hat{v}_i < k \cdot 85.5\,V$ when using an integrated MOSFET.

Temperature sensitivity: The following formula applies:

$$\frac{dV_o}{dT}\bigg|_A = \frac{dV_o}{dV_{GS}}\bigg|_A \frac{dV_{GS}}{dT}\bigg|_A \overset{dV_{GS}=dV_i}{=} A \frac{dV_{GS}}{dT}\bigg|_A \overset{dV_{GS}=dI_D/g_m}{=} \frac{A}{g_m}\frac{dI_D}{dT}\bigg|_A$$

Inserting A according to (3.97) and dI_D/dT according to (3.14) on page 190, and using the typical values, leads to:

$$\frac{dV_o}{dT}\bigg|_A \approx \frac{I_{D,A}R_S}{1 + (g_m + g_{m,B})R_S} \cdot 10^{-3}\,\mathrm{K}^{-1}\left(\frac{4\ldots 7\,\mathrm{V}}{V_{GS,A} - V_{th}} - 5\right)$$

For discrete MOSFETs, we set $g_{m,B} = 0$. Using a discrete MOSFET in the numeric example, we obtain the result that $(dV_o/dT)|_A$ ranges from approximately $-0.4\,\mathrm{mV/K}$ to $+0.8\,\mathrm{mV/K}$; the temperature drift is somewhat lower when using an integrated MOSFET.

Setting the Operating Point

The operating point is set in the same way as in the common-collector circuit; Fig. 2.97 on page 138 shows some examples. While in enhancement n-channel MOSFETs the output voltage $V_{o,A}$ is always below the input voltage $V_{i,A}$ due to $V_{GS,A} > V_{th} > 0$ and $V_{o,A} = V_{i,A} - V_{GS,A}$, the output voltage of depletion n-channel MOSFETs can also be higher. The relation $V_{i,A} \leq V_{o,A}$ always exists in n-channel junction FETs due to $V_{GS,A} \leq 0$.

A special situation is shown in Fig. 3.80 with depletion n-channel MOSFETs and a current source instead of the source resistance R_S; in this case, $V_{i,A} = V_{o,A}$, which is independent of the threshold voltage as long as the transconductance coefficient and the threshold voltage are the same for both MOSFETs. This property can be used in circuits with discrete components when using matched MOSFETs; although the threshold voltages vary within tolerances, they are still almost identical. This principle cannot be applied in integrated circuits, as the threshold voltages depend on the MOSFETs source voltages due to the substrate effect.

The circuit shown in Fig. 3.80a can be used with junction FETs only within certain limits, because at their operating point $V_{GS,A} = 0$, so that the gate–channel diode of the junction FET can become conductive with a sudden increase in the input voltage; it is therefore necessary to use the circuit shown in Fig. 3.80b, in which $V_{GS,A} = -I_{D,A}R$.

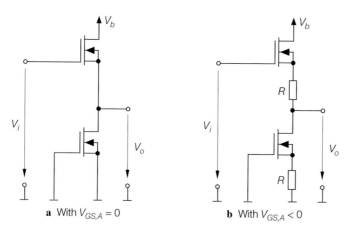

a With $V_{GS,A} = 0$ **b** With $V_{GS,A} < 0$

Fig. 3.80. Setting the operating point with $V_{i,A} = V_{o,A}$

Resistance R results in an increase of the output resistance and for this reason should not be too high.

Frequency Response and Cutoff Frequency

The small-signal gain A and the operating gain A_B of the common-drain circuit decline as the frequency increases, because of the capacitances of the FET. To obtain any information on the frequency response and the cutoff frequency, it is essential to use the dynamic small-signal model of the FET; Fig. 3.81 shows the resulting dynamic small-signal equivalent circuit for the common-drain configuration. With $R'_g = R_g + R_G$ and $R'_L = R_L \| R_S \| r_{DS} \| 1/g_{m,B}$, it follows, for the *operating gain* $\underline{A}_B(s) = \underline{v}_o(s)/\underline{v}_g(s)$, that:

$$\underline{A}_B(s) = \frac{1 + s\,\dfrac{C_{GS}}{g_m}}{1 + \dfrac{1}{g_m R'_L} + sc'_1 + s^2 c'_2}$$

$$c'_1 = \frac{C_{GS} + C_{DS}}{g_m} + (C_{GS} + C_{GD})\,\frac{R'_g}{g_m R'_L} + C_{GD} R'_g$$

$$c'_2 = (C_{GS}C_{GD} + C_{GS}C_{DS} + C_{GD}C_{DS})\,\frac{R'_g}{g_m}$$

The zero can be neglected, because the cutoff frequency

$$f_N = \frac{g_m}{2\pi\,C_{GS}} > f_T$$

is higher than the transit frequency f_T of the FET, as shown by comparison with (3.47). With the low-frequency gain

$$A_0 = \underline{A}_B(0) = \frac{g_m R'_L}{1 + g_m R'_L} \tag{3.101}$$

it follows that:

$$\underline{A}_B(s) \approx \frac{A_0}{1 + sc_1 + s^2 c_2} \tag{3.102}$$

$$c_1 = \frac{(C_{GS} + C_{DS})\,R'_L + C_{GS}R'_g}{1 + g_m R'_L} + C_{GD}R'_g \tag{3.103}$$

$$c_2 = \frac{(C_{GS}C_{GD} + C_{GS}C_{DS} + C_{GD}C_{DS})\,R'_L R'_g}{1 + g_m R'_L} \tag{3.104}$$

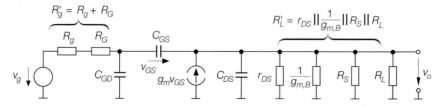

Fig. 3.81. Dynamic small-signal equivalent circuit for the common-drain circuit

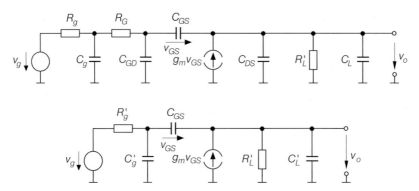

Fig. 3.82. Small-signal equivalent circuit for calculating the region of complex conjugate poles: complete (*above*) and simplified (*below*)

The *quality* of the poles can thus be specified:

$$Q = \frac{\sqrt{c_2}}{c_1} \tag{3.105}$$

With $Q \leq 0.5$, the poles are real; with $Q > 0.5$, they are complex conjugate.

In the case of real poles, the frequency response can be approximately described by way of a first-degree lowpass filter by removing the s^2 term from the denominator:

$$\underline{A}_B(s) \approx \frac{A_0}{1 + sc_1} \overset{g_m R_L' \gg 1}{\approx} \frac{A_0}{1 + s\left(\dfrac{C_{GS} + C_{DS}}{g_m} + \left(\dfrac{C_{GS}}{g_m R_L'} + C_{GD}\right) R_g'\right)}$$

This provides an approximation for the $-3\,dB$ *cutoff frequency* f_{-3dB} at which the absolute gain has dropped by 3 dB:

$$\omega_{-3dB} = 2\pi f_{-3dB} \approx \frac{1}{c_1} \overset{g_m R_L' \gg 1}{\approx} \frac{1}{\dfrac{C_{GS} + C_{DS}}{g_m} + \left(\dfrac{C_{GS}}{g_m R_L'} + C_{GD}\right) R_g'} \tag{3.106}$$

In the case of complex conjugate poles, that is, $Q > 0.5$, the following approximation can be used:

$$\omega_{-3dB} = 2\pi f_{-3dB} \approx \frac{1}{\sqrt{c_2}} \tag{3.107}$$

For $Q = 1/\sqrt{2}$ it produces the exact result, while it is too high for $0.5 < Q < 1/\sqrt{2}$ and too low for $Q > 1/\sqrt{2}$.

If there is a load capacitance C_L, it is in parallel with C_{DS} and can be taken into account by replacing C_{DS} with $C_L + C_{DS}$.

Region of complex conjugate poles: For practical applications of the common-drain circuit, it is important to know for which load capacitances complex conjugate poles occur and which circuit design measures can prevent this. In what follows, we will look at the small-signal equivalent circuit shown in Fig. 3.82, which is derived from Fig. 3.79 by adding the capacitance C_g of the signal generator and the load capacitance C_L. The RC elements R_g-C_g and R_G-C_{GD} can be combined to give one element with $R'_g = R_g + R_G$ and $C'_g = C_g + C_{GD}$, since $R_g \gg R_G$; on the output side $C'_L = C_L + C_{DS}$. After introducing the time constants

$$T_g = C'_g R'_g \quad , \quad T_L = C'_L R'_L \quad , \quad T_{GS} = \frac{C_{GS}}{g_m} \approx \frac{1}{\omega_C} \tag{3.108}$$

and the resistance ratios

$$k_g = \frac{R'_g}{R'_L} \quad , \quad k_S = \frac{1}{g_m R'_L} \tag{3.109}$$

it follows from (3.103) and (3.104) that:

$$
\begin{aligned}
c_1 &= \frac{T_{GS}(1 + k_g) + T_L k_S}{1 + k_S} + T_g \\[2mm]
c_2 &= \frac{T_g T_{GS} + T_g T_L k_S + T_L T_{GS} k_g}{1 + k_S}
\end{aligned}
\tag{3.110}
$$

Using the formula

$$Q = \frac{\sqrt{c_2}}{c_1} > 0.5$$

it is possible to determine the region of the complex conjugate poles. This region is shown in Fig. 3.83 as a function of the *normalized signal source time constant* T_g/T_{GS} and the *normalized load time constant* T_L/T_{GS} for various values of k_g and the typical value $k_S = 0.2$. It is clear that no complex conjugate poles occur with very small and very large load capacitances C_L (T_L/T_{GS} small or large), or with a sufficiently large output capacitance C_g of the signal generator (T_g/T_{GS} large). The region of complex conjugate poles is strongly dependent on k_g.

If the influence of the FET parameters on the time constants T_g and T_L and on the factors k_g and k_S is neglected and the resistances R_S and R_L are combined, the circuit shown in Fig. 3.84 is produced; the time constants and the resistance ratios are:

$$T_g \approx R_g C_g \quad , \quad T_L \approx R_L C_L \quad , \quad T_{GS} \approx \frac{1}{\omega_C}$$

$$k_g \approx \frac{R_g}{R_L} \quad , \quad k_S \approx \frac{1}{g_m R_L}$$

If R_g, C_g, R_L and C_L are given and complex conjugate poles occur, there are four different methods of avoiding this range:

1. One can increase T_g in order to leave the range of complex conjugate poles *toward the top*. This requires the insertion of an additional capacitor from the input of the common-drain circuit to ground or to a supply voltage; in the small-signal equivalent circuit, this is arranged in parallel to C_g and causes an increase in T_g. This method can always be applied; it is therefore often found in practical applications.
2. One can increase T_{GS} in order to leave the range toward the *lower left* when operating close to the left boundary of this range. This requires the use of a *slower* FET with a larger time constant T_{GS}; that is, a lower transit frequency f_T.

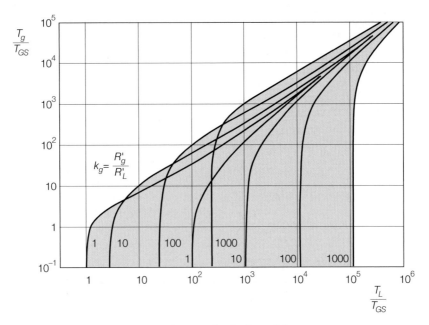

Fig. 3.83. Region of complex conjugate poles for $k_S = 0.2$

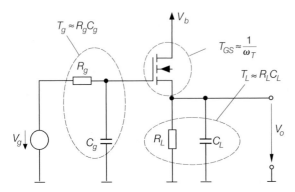

Fig. 3.84. Circuit for the approximate calculation of the time constants

3. One can reduce T_{GS} in order to leave the range toward the *top right* if operating close to the right border of this range. This requires the use of a *faster* FET with a smaller time constant T_{GS}; that is, a higher transit frequency f_T.
4. One can increase T_L in order to leave the range *toward the right* if operating close to the right border of this range. This requires an enlargement of the load capacitance C_L by adding an additional capacitor in parallel.

Figure 3.85 indicates the four methods. The fifth method is the reduction of T_L which is rarely used in practical applications since, with given values for R_L and C_L, it can be achieved only by connecting a resistor in parallel, which then puts an additional load on the output. All of these methods cause a reduction in the lower cutoff frequency. In order to

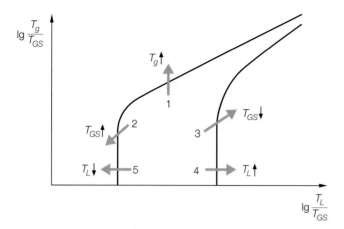

Fig. 3.85. Possibilities for leaving the complex conjugate pole region

keep this reduction within acceptable limits, it is necessary to leave the range of complex conjugate poles *by the shortest route possible*.

Example: $I_{D,A} = 2\,\text{mA}$ is selected for the numeric example, according to Fig. 3.77. For $K = 4\,\text{mA/V}^2$, $V_A = 50\,\text{V}$, $C_{oss} = 5\,\text{pF}$, $C_{rss} = 2\,\text{pF}$, $f_{Y21s} = 1\,\text{GHz}$ and $f_T = 100\,\text{MHz}$, the parameters $g_m = 4\,\text{mS}$, $r_{DS} = 25\,\text{k}\Omega$, $R_G = 25\,\Omega$, $C_{GD} = 2\,\text{pF}$, $C_{GS} = 4.4\,\text{pF}$ and $C_{DS} = 3\,\text{pF}$ are determined from Fig. 3.51 on page 222. For $R_g = R_S = 1\,\text{k}\Omega$ and $R_L \to \infty$, we obtain $R'_g = R_g + R_G = 1025\,\Omega$, $R'_L = R_L||R_S||r_{DS} = 960\,\Omega$ and thus, from (3.101), $A_0 = 0.793$ and, from (3.106), the approximation $f_{-3dB} \approx 31.4\,\text{MHz}$. A more precise calculation with the help of (3.103)–(3.105) results in $c_1 = 4.45\,\text{ns}$, $c_2 = 5.69\,\text{ns}^2$ and $Q \approx 0.54$; this means that complex conjugate poles exist, and (3.107) provides the approximation $f_{-3dB} \approx 67\,\text{MHz}$, which must be considered to be too high due to $0.5 < Q < 1/\sqrt{2}$. For a load capacitance $C_L = 1\,\text{nF}$, we obtain, from (3.108) and (3.109), $T_g = 2.05\,\text{ns}$, $T_L = 960\,\text{ns}$, $T_{GS} = 1.1\,\text{ns}$, $k_g = 1.07$ und $k_S = 0.26$, so that (3.110) provides $c_1 = 202\,\text{ns}$ and $c_2 = 1305\,(\text{ns})^2$; from (3.105) it follows that $Q = 0.179$, that is, the poles are real, and from (3.106) we obtain $f_{-3dB} \approx 788\,\text{kHz}$. The indication of real poles can also be obtained without calculating c_1, c_2 and Q on the basis of Fig. 3.83, since point $T_L/T_{GS} \approx 1000$, $T_g/T_{GS} \approx 2$, $k_g \approx 1$ is not in the region of complex conjugate poles.

3.4.3
Common-Gate Circuit

Figure 3.86 shows a common-gate circuit that consists of the MOSFET, the drain resistance R_D, the supply voltage source V_b, the signal voltage source V_i [27] and the gate series resistance R_{GV}; the latter has no effect on the transfer characteristic but influences the frequency response and the bandwidth. The transfer characteristic and the small-signal response depend on the circuitry at the bulk terminal, which is connected to the source in discrete MOSFETs and to the negative supply voltage in integrated MOSFETs. Due to the fact that the common-gate circuit shown in Fig. 3.86 is operated with a negative input

---

[27] A voltage source *without* an internal resistance R_g is used in this case, so that the characteristics are independent of R_g.

voltage, the bulk terminal of the integrated MOSFET must be connected to an additional negative supply voltage V_B that is lower than the minimum input voltage; this ensures that the bulk–source diode is reverse-biased. In what follows, we assume that $V_b = 5\,\text{V}$, $V_B = -5\,\text{V}$ and $R_D = R_{GV} = 1\,\text{k}\Omega$; in addition, we assume that $K = 4\,\text{mA/V}^2$ and $V_{th} = 1\,\text{V}$ for the discrete MOSFET and $K = 4\,\text{mA/V}^2$, $V_{th,0} = 1\,\text{V}$, $\gamma = 0.5\,\sqrt{\text{V}}$ and $V_{inv} = 0.6\,\text{V}$ for the integrated MOSFET.

Transfer Characteristic of the Common-Gate Circuit

The transfer characteristics of a discrete MOSFET ($V_{BS} = 0$) and of an integrated MOSFET ($V_B = -5\,\text{V}$) are obtained by measuring the output voltage V_o as a function of the signal voltage V_i.

For $-2.7\,\text{V} < V_i < -V_{th} = -1\,\text{V}$, the discrete MOSFET operates in the pinch-off region; using $V_{GS} = -V_i$ and neglecting the Early effect leads to:

$$V_o = V_b - I_D R_D = V_b - \frac{K R_D}{2}(V_{GS} - V_{th})^2 \tag{3.111}$$

$$V_i = -V_{GS} - I_G R_{GV} \overset{I_G=0}{=} -V_{GS} \tag{3.112}$$

By inserting (3.112) into (3.111) for the transfer characteristic, it follows that:

$$V_o = V_b - \frac{K R_D}{2}(-V_i - V_{th})^2 = V_b - \frac{K R_D}{2}(V_i + V_{th})^2 \tag{3.113}$$

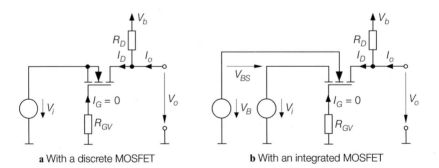

a With a discrete MOSFET   **b** With an integrated MOSFET

Fig. 3.86. Common-gate circuit

At the sample operating point shown in Fig 3.87, we obtain:

$$V_o = 2.5\,\text{V} \;\Rightarrow\; I_D = \frac{V_b - V_o}{R_C} = 2.5\,\text{mA}$$

$$\Rightarrow\; V_{GS} = V_{th} + \sqrt{\frac{2 I_D}{K}} = 2.12\,\text{V} \;\Rightarrow\; V_i = -V_{GS} = -2.12\,\text{V}$$

Transfer characteristic when driven by a current source: This circuit may also be driven by a current source I_i (see Fig. 3.88). With $I_i < 0$, the circuit then operates as a *current–to–voltage converter* or a *transimpendance amplifier*:

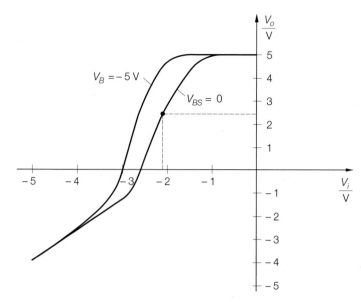

Fig. 3.87. Characteristics of a discrete MOSFET ($V_{BS} = 0$) and of an integrated MOSFET ($V_B = -5\,\text{V}$) in the common-gate circuit

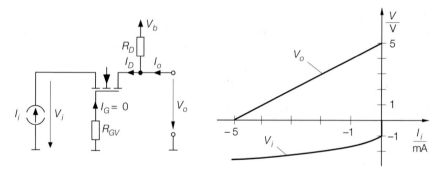

Fig. 3.88. Circuit and characteristic of the common-gate circuit driven by a current source

$$V_o = V_b - I_D R_D \overset{I_D=-I_i}{=} V_b + I_i R_D \tag{3.114}$$

$$V_i = -V_{GS} = -V_{th} - \sqrt{\frac{2I_D}{K}} \overset{I_D=-I_i}{=} -V_{th} - \sqrt{-\frac{2I_i}{K}} \tag{3.115}$$

In practice, a common-source circuit with an open drain or a current mirror is used for driving the current; this will be described in more detail in the section on setting the operating point.

Small-Signal Response of the Common-Gate Circuit

The response to signals around an operating point A is called the *small-signal response*; as an example, we use the operating point determined above with $V_{i,A} = -2.12\,\text{V}$, $V_{o,A} = 2.5\,\text{V}$ and $I_{D,A} = 2.5\,\text{mA}$.

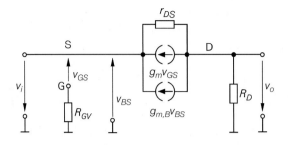

Fig. 3.89. Small-signal equivalent circuit for the common-gate circuit

Figure 3.89 shows the small-signal equivalent circuit for the common-gate circuit. A changeover from the integrated to the discrete MOSFET is based on the limitation of $v_{BS} = 0$; the equations then use $g_{m,B} = 0$.[28] From the nodal equation

$$\frac{v_o}{R_D} + \frac{v_o - v_i}{r_{DS}} + g_m v_{GS} + g_{m,B} v_{BS} = 0$$

it follows, for $v_i = -v_{GS} = -v_{BS}$, that:

$$A = \left.\frac{v_o}{v_i}\right|_{i_o=0} = \left(g_m + g_{m,B} + \frac{1}{r_{DS}}\right)(R_D \| r_{DS})$$

$$\overset{r_{DS} \gg R_D, 1/g_m}{\approx} \left(g_m + g_{m,B}\right) R_D \overset{v_{BS}=0}{=} g_m R_D$$

For $I_{D,A} = 2.5\,\text{mA}$, $K = 4\,\text{mA/V}^2$ and $V_A = 50\,\text{V}$, we obtain from Fig. 3.51 on page 222 the values $g_m = 4.47\,\text{mS}$ and $r_{DS} = 20\,\text{k}\Omega$; for discrete MOSFETs, by inserting $g_{m,B} = 0$ and $R_D = 1\,\text{k}\Omega$, we obtain the *exact* value $A = 4.3$ and in the first approximation $A = 4.47$; while for integrated MOSFETs we obtain a slightly higher gain, because $g_{m,B} > 0$.

For the *small-signal input resistance*, it follows that:

$$r_i = \left.\frac{v_i}{i_i}\right|_{i_o=0} = \frac{R_D + r_{DS}}{1 + \left(g_m + g_{m,B}\right) r_{DS}} \overset{r_{DS} \gg R_D, 1/g_m}{\approx} \frac{1}{g_m + g_{m,B}} \overset{v_{BS}=0}{=} \frac{1}{g_m}$$

This depends on the load resistance which, in this case, is the *open-circuit input resistance* because $i_o = 0$ ($R_L \to \infty$). The input resistance for other values of R_L is calculated by replacing R_D with the parallel arrangement of R_D and R_L; inserting $R_L = R_D = 0$ leads to the *short-circuit input resistance*. However, the dependence on R_L is so low that it is eliminated by the approximation. For the operating point taken as an example, we obtain, for the discrete MOSFET, an *exact* value of $r_i = 232\,\Omega$; the approximation is $r_i = 224\,\Omega$.

For the *small-signal output resistance*, it follows:

$$r_o = \frac{v_o}{i_o} = R_D \| \left[\left(1 + \left(g_m + g_{m,B}\right) R_g\right) r_{DS} + R_g\right] \overset{r_{DS} \gg R_D}{\approx} R_D$$

This depends on the internal resistance R_g of the signal generator; with $R_g = 0$, the *short-circuit output resistance* is

$$r_{o,s} = R_D \| r_{DS}$$

---

[28] $g_{m,B} = 0$ as a limiting condition would not be correct, as the substrate transconductance of the discrete MOSFET is not zero, but has no effect because $v_{BS} = 0$; therefore, $v_{BS} = 0$ is a correct limitation and $g_{m,B} = 0$ is its effect in the equations.

and with $R_g \to \infty$ the *open-circuit output resistance* is:

$$r_{o,o} = R_D$$

In most practical cases $r_{DS} \gg R_D$, and the dependence on R_g can be neglected. In our example, we obtain $r_{o,s} = 952\,\Omega$ and $r_{o,o} = 1\,k\Omega$.

With $r_{DS} \gg R_D$, $1/g_m$ and without the load resistance R_L, it follows for the *common-gate circuit*:

---

common-gate circuit

$$A = \left.\frac{v_o}{v_i}\right|_{i_o=0} \approx \left(g_m + g_{m,B}\right) R_D \overset{v_{BS}=0}{=} g_m R_D \tag{3.116}$$

$$r_i = \left.\frac{v_i}{i_i}\right|_{i_o=0} \approx \frac{1}{g_m + g_{m,B}} \overset{v_{BS}=0}{=} \frac{1}{g_m} \tag{3.117}$$

$$r_o = \frac{v_o}{i_o} \approx R_D \tag{3.118}$$

When driven by a signal source with an internal resistance R_g and a load resistance R_L, the *operating gain* is:

$$A_B = \frac{r_i}{r_i + R_g} A \frac{R_L}{r_o + R_L} \approx \frac{g_m \left(R_D \,||\, R_L\right)}{1 + \left(g_m + g_{m,B}\right) R_g} \overset{v_{BS}=0}{=} \frac{g_m \left(R_D \,||\, R_L\right)}{1 + g_m R_g} \tag{3.119}$$

When the circuit is driven by a current source, the *transfer resistance* R_T (the transimpedance) takes the place of the gain; for the *current–to–voltage converter in common-base configuration* this leads to:

---

current–to–voltage converter in common-base configuration

$$R_T = \left.\frac{v_o}{i_i}\right|_{i_o=0} = \left.\frac{v_o}{v_i}\right|_{i_o=0} \left.\frac{v_i}{i_i}\right|_{i_o=0} = A r_i = R_D \tag{3.120}$$

The input and output resistances are given by (3.117) and (3.118), respectively.

Nonlinearity: If the circuit is driven by a voltage source, then $\hat{v}_{GS} = \hat{v}_i$ and (3.13) on page 185 can be used: this describes the relationship between the amplitude $\hat{v}_{GS}$ of a sinusoidal signal with small amplitude and the *distortion factor* k of the drain current, which in the common-gate circuit is the same as the distortion factor of the output voltage. Consequently, $\hat{v}_i < 4k \left(V_{GS,A} - V_{th}\right)$. When driven by a current source the circuit operates linear, which means that the distortion factor is zero.

Temperature sensitivity: The common-gate circuit has the same temperature drift as the common-source circuit without feedback, because in both configurations there is a constant input voltage between gate and source and the output voltage is given by $V_o = V_b - I_D R_D$. This leads to:

$$\left.\frac{dV_o}{dT}\right|_A = -R_D \left.\frac{dI_D}{dT}\right|_A \approx I_{D,A} R_D \cdot 10^{-3}\,K^{-1} \left(5 - \frac{4\ldots 7\,V}{V_{GS,A} - V_{th}}\right)$$

Setting the Operating Point

The operating point setting is achieved in the same way as in the common-base circuit; Fig. 3.90 shows the voltage-driven and the current-driven versions, which correspond to the circuits shown in Fig. 2.112. In the voltage-driven version as shown in Fig. 3.90a, the common-drain circuit (T_1) is used to drive the common-gate circuit (T_2); this forms a differential amplifier with an asymmetrical input and output. In the current-driven version according to Fig. 3.90b, a common-source circuit (T_1) is used to provide the driving signal; this is also known as a *cascode circuit*. The voltage divider consisting of R_1 and R_2 acts as the gate series resistance, with $R_{GV} = R_1 \| R_2$.

Frequency Response and Cutoff Frequency

The small-signal gain A and the operating gain A_B of the common-gate circuit decline with an increasing frequency due to the capacitances of the FET. To obtain any information on the frequency response and the cutoff frequency, it is essential to use the dynamic small-signal model of the FET for the calculation.

Driving by a voltage source: Obtaining an exact calculation for the *operating gain* $\underline{A}_B(s) = \underline{v}_o(s)/\underline{v}_g(s)$ is a difficult process and leads to complex expressions. A sufficiently accurate approximation is reached by neglecting the resistance r_{DS} and the capacitance C_{DS}; the latter only exists in discrete MOSFETs. Additional parameters that occur in the integrated MOSFETs are the substrate transconductance $g_{m,B}$ and the bulk capacitances C_{BS} and C_{BD}; they are neglected here. This leads to the simplified small-signal equivalent circuit for the discrete *and* the integrated MOSFET as shown in Fig. 3.91, which is to a large extent similar to the small-signal equivalent circuit for the common-base configuration shown in Fig. 2.113. Therefore, the results of the common-base circuit can be transferred to the common-gate circuit by inserting the corresponding small-signal parameters in (2.139) and (2.140) and taking the limit as $\beta \to \infty$; with $R'_{GV} = R_{GV} + R_G$ and $R'_D = R_D \| R_L$ and the low-frequency gain

$$A_0 = \underline{A}_B(0) \approx \frac{g_m R'_D}{1 + g_m R_g} \tag{3.121}$$

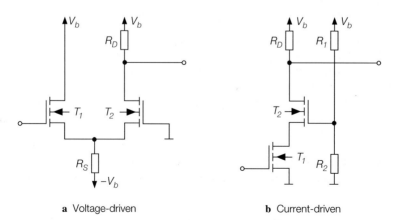

a Voltage-driven   **b** Current-driven

Fig. 3.90. Setting the operating point in a common-gate circuit

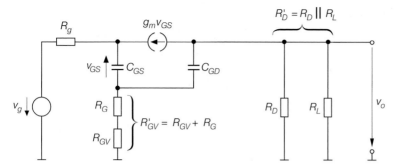

Fig. 3.91. Simplified dynamic small-signal equivalent circuit for the common-gate configuration

an approximation for the frequency response by a first-degree lowpass filter is obtained:

$$\underline{A}_B(s) \approx \frac{A_0}{1 + s \dfrac{C_{GS}\left(R_g + R'_{GV}\right) + C_{GD} R'_D \left(1 + g_m \left(R_g + R'_{GV}\right)\right)}{1 + g_m R_g}} \tag{3.122}$$

This leads to an approximation for the *−3 dB cutoff frequency* f_{-3dB}:

$$\omega_{-3dB} \approx \frac{1 + g_m R_g}{C_{GS}\left(R_g + R'_{GV}\right) + C_{GD} R'_D \left(1 + g_m \left(R_g + R'_{GV}\right)\right)} \tag{3.123}$$

From (3.121) and (3.123), two time constants that are independent of the low-frequency gain A_0 can be calculated:[29]

$$\omega_{-3dB}(A_0) \approx \frac{1}{T_1 + T_2 A_0} \tag{3.124}$$

$$T_1 = C_{GS} \frac{R_g + R'_{GV}}{1 + g_m R_g} \tag{3.125}$$

$$T_2 = C_{GD} \left(R_g + R'_{GV} + \frac{1}{g_m}\right) \tag{3.126}$$

The explanation concerning the gain-bandwidth product *GBW*, including (3.79) given on page 246, also applies to the common-gate circuit.

If there is a load capacitance C_L in parallel with the load resistance R_L, then

$$T_2 = C_{GD} \left(R_g + R'_{GV} + \frac{1}{g_m}\right) + C_L \left(R_g + \frac{1}{g_m}\right) \tag{3.127}$$

C_L has no influence on the time constant T_1.

Driving by a current source: When a current source is used to drive the circuit, the frequency response of the *transimpedance* $\underline{Z}_T(s)$ is interesting; on the basis of (3.122), one can reach an approximation by a first-degree lowpass filter:

---

[29] It is assumed that a change in A_0 is caused by a variation in R'_D; the time constants are thus independent of A_0 if they are independent of R'_D.

$$\underline{Z}_T(s) = \frac{\underline{v}_o(s)}{\underline{i}_i(s)} = \lim_{R_g \to \infty} R_g \underline{A}_B(s) \approx \frac{R'_D}{1 + s \left(\dfrac{C_{GS}}{g_m} + C_{GD} R'_D \right)} \tag{3.128}$$

In this case, the cutoff frequency is:

$$\omega_{\text{-3dB}} = 2\pi f_{\text{-3dB}} \approx \frac{1}{\dfrac{C_{GS}}{g_m} + C_{GD} R'_D} \tag{3.129}$$

In the event of a capacitive load, it is necessary to replace C_{GD} by $C_L + C_{GD}$.

Example: $I_{D,A} = 2.5\,\text{mA}$ is taken as the numeric example of the common-gate circuit shown in Fig. 3.86a. The small-signal parameters of the MOSFET can be taken from the example on page 250: $g_m = 4.47\,\text{mS}$, $R_G = 25\,\Omega$, $C_{GD} = 2\,\text{pF}$ and $C_{GS} = 4.4\,\text{pF}$. For $R_D = 1\,\text{k}\Omega$, $R_L \to \infty$, $r_{DS} \gg R_D$ and $R_g = R_{GV} = 0$, we obtain $R'_D = R_D = 1\,\text{k}\Omega$ and $R'_{GV} = R_G = 25\,\Omega$; from (3.121) it follows that $A_0 \approx 4.47$ and from (3.123) that $f_{\text{-3dB}} \approx 68\,\text{MHz}$. The cutoff frequency strongly depends on R_{GV}; with $R_{GV} = 1\,\text{k}\Omega$ the frequency drops to $f_{\text{-3dB}} \approx 10\,\text{MHz}$.

When the circuit is driven by a current source and $R_L \to \infty$, then it follows from (3.128) that $R_T = \underline{Z}_T(0) \approx R_D = 1\,\text{k}\Omega$ and from (3.129) that $f_{\text{-3dB}} \approx 53\,\text{MHz}$. The resistance R_{GV} is of no influence in this case.

Chapter 4:
Amplifiers

Amplifiers are important elements in analog signal processing. They amplify an input signal of low amplitude to such a degree that it can be used to drive a subsequent unit. For example, a microphone signal has to be amplified in several stages from the microvolt (μV) range to the volt (V) range in order to feed a loudspeaker. Similarly, the signals of thermocouples, photodiodes, magnetic reading heads, receiving antennas and many other signal sources can only be processed after suitable amplification. Since digital circuits such as microprocessors and digital signal processors (DSP) are increasingly being used in the processing and evaluation of complex signals, a typical signal processing chain usually consists of the following elements or stages:

1. A sensor for converting a physical unit such as pressure (microphone), temperature (thermocouple), light (photodiode) or electromagnetc field (antenna) into an electrical signal.
2. One or more amplifiers to amplify and filter the signal.
3. An analog-to-digital (A/D) converter for digitizing the signal.
4. A microprocessor, DSP or other digital circuit for processing the digitized signal.
5. A digital-to-analog (D/A) converter to produce an analog output signal.
6. One or more amplifiers to amplify and filter the signal to such a degree that it can be used to drive an actuator.
7. An actuator to convert the signal into a physical unit such as pressure (loudspeaker), temperature (heating rod), light (incandescent lamp) or electromagnetic field (transmitting antenna).

Figure 4.1 illustrates the seven stages of a signal processing chain; one of the symbols shown in Fig. 4.2 is used for the amplifiers.

The amplifiers of stage 2 process comparatively small signals and are therefore known as *small-signal amplifiers*; in most cases, their output power is below 1 mW. In contrast,

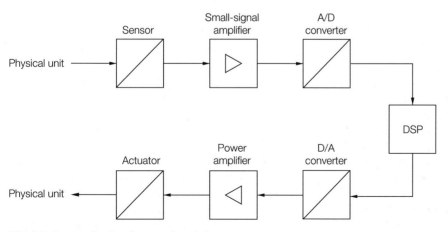

Fig. 4.1. Stages of a signal processing chain

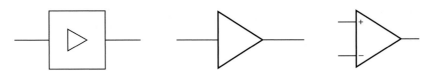

Fig. 4.2. Amplifier symbols

stage 6 requires *power amplifiers* that operate in a power range between a few mW (head-sets, remote controls, etc.) to several kW (large loudspeaker systems, radio stations, etc.). Power amplifiers are described in Chap. 15.

To filter the signals, passive filters are increasingly being replaced by active filters that also contain amplifiers. Therefore, it is not possible to clearly distinguish between the elements *amplifier* and *filter*, as every amplifier also acts as a filter due to its limited bandwidth and each active filter may produce signal amplification. Active filters are covered in Chap. 13.

Another distinctive feature is the frequency range of an amplifier. In the lower cut-off frequency f_L we differentiate between *DC amplifiers* and *AC amplifiers*, and in the upper cutoff frequency f_U between *low-frequency (LF) amplifiers* and *high-frequency (HF) amplifiers*. With regard to the bandwidth $B = f_U - f_L$, we differentiate between *broadband amplifiers* and *narrowband* or *tuned amplifiers*. With respect to the upper limit frequency, another distinction is often made between *audio* or *audio frequency amplifiers (AF amplifiers)*, *video amplifiers*, *intermediate frequency amplifiers (IF amplifiers)* and *radio frequency (RF) amplifiers*. While the division into AC and DC amplifiers results directly from the design of AC or DC voltage coupling, the distinction between LF and HF amplifiers is not defined; often, 1 MHz is used as a limit value. Similarly, there is no definition for broadband and narrowband amplifiers; the latter are usually characterized by the mid-frequency $f_M = (f_U + f_L)/2$ and the bandwidth $B = f_U - f_L$. In narrowband amplifiers, the bandwidth is less than one tenth of the mid-frequency: $B < f_M/10$.

Despite this multitude of amplifier types, the circuit design used is almost identical, because all amplifiers are based on standard transistor circuits that amplify DC voltages. However, a distinction is made in respect to the coupling at the input and the output and between the individual stages of multi-stage amplifiers: DC amplifiers use direct coupling (*DC coupling* or *galvanic coupling*), AC amplifiers use capacitive coupling with coupling capacitances (*AC coupling*) and narrowband amplifiers use selective coupling with LC resonant circuits, ceramic resonators or surface acoustic wave filters. Figure 4.3 shows the type of coupling and the frequency responses of the amplifiers mentioned, together with the parameters f_U, f_L, f_M and B.

Likewise, the distinction between LF and HF amplifiers is not so much the result of the circuit design but, rather, of the transit frequencies of the transistors used. The quiescent currents at the operating point also play a decisive role, since for small currents the transit frequencies are approximately proportional to the quiescent current. Thus, a differential amplifier that achieves a cutoff frequency of 10 MHz at a quiescent current of 1 mA may yield a cutoff frequency of only $100 \ldots 300$ kHz at a quiescent current of $10\,\mu$A.

One speciality is *operational amplifiers*, which are very important as general-purpose DC amplifiers for low frequencies. Operational amplifiers are used almost exclusively in standard applications. Only if particular demands cannot be met by standard operational

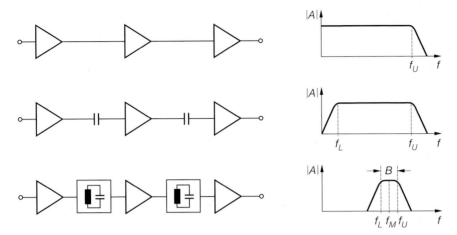

Fig. 4.3. Coupling types and the frequency response of DC amplifiers (*top*), AC amplifiers (*centre*) and narrowband amplifiers (*bottom*)

amplifiers, or those contained in amplifier libraries[1], may the amplifier circuit be made up of discrete transistors, or a special integrated circuit be produced. Operational amplifiers are described in Chap. 5.

4.1
Circuits

Amplifiers encompass one or more amplifier stages, where each stage is made up of one or more coupled basic circuits with bipolar transistors or field effect transistors. Further transistors may additionally be required to set the operating point. Going back to basic circuits in many cases allows the use of the equations determined in Sects. 2.4 and 3.4.

Characteristics of transistors: The circuits below are described using bipolar transistors *and* enhancement MOSFETs as far as this is possible and suitable; depletion MOSFETs and JFETs are used in exception cases only. The basic equations, (2.2) and (2.3), or (3.3) and (3.4), are used to calculate the characteristics and operating points:

$$\text{npn transistor:} \quad I_C = I_S \, e^{\frac{V_{BE}}{V_T}} \left(1 + \frac{V_{CE}}{V_A}\right) \quad , \quad I_B = \frac{I_C}{B}$$

$$\text{n-channel MOSFET:} \quad I_D = \frac{K}{2}\,(V_{GS} - V_{th})^2 \left(1 + \frac{V_{DS}}{V_A}\right) \quad , \quad I_G = 0$$

With MOSFETs the substrate effect must also be taken into account; (3.18) applies to the n-channel MOSFET:

$$V_{th} = V_{th,0} + \gamma \left(\sqrt{V_{inv} - V_{BS}} - \sqrt{V_{inv}}\right)$$

---

[1] Predefined modules available from module libraries are used as far as possible when designing integrated circuits.

Scaling: The presentation is based on integrated circuit design, which makes particular use of the almost unlimited *scaling capabilities* of transistors. In bipolar transistors the reverse saturation current I_S is scaled by varying the emitter surface, while in MOSFETs the transconductance coefficient K is scaled by varying the channel width/length ratio W/L. In MOSFETs it is usually the channel width W that is scaled, while the channel length L remains constant.[2]

Scaling is generally done according to the quiescent currents at the operating point: $I_S \sim I_{C,A}$ or $W \sim K \sim I_{D,A}$ ($L = $ const.); this makes the current density equal in all transistors. The result is that all npn transistors operate with the same base-emitter voltage $V_{BE,A}$ at the operating point, with the exception of a small deviation caused by the Early effect:

$$V_{BE,A} \approx V_T \ln \frac{I_{C,A}}{I_S} \overset{I_{C,A} \sim I_S}{=} \text{const.} \approx 0.7\,\text{V}$$

With MOSFETs the conditions are more complicated because of the substrate effect: two MOSFETs with the same current density – if the Early effect is ignored – only have an identical gate-source voltage $V_{GS,A}$ if the bulk-source voltages are the same:

$$V_{GS,A} \approx V_{th}(V_{BS,A}) + \sqrt{\frac{2I_{D,A}}{K}} \overset{\substack{I_{D,A} \sim K \sim W \\ V_{BS,A} = \text{const.}}}{=} \text{const.}$$

Normalization: The parameters of the various transistors are normalized to the size of a reference transistor; it is given the *relative size* 1. Consequently, a bipolar transistor of size 5 has five times the reverse saturation current I_S and a MOSFET of size 5 has five times the transconductance coefficient K than a transistor of size 1.

Often, the smallest transistor of a certain technology is used as the reference transistor; in this case, there are only relative sizes, which are greater than or equal to one. For bipolar transistors, the reference transistor has the smallest emitter surface and is thus the smallest in terms of *electrical* properties – that is, with regard to I_S – and in geometric terms. For MOSFETs there is an additional degree of freedom, since the channel width W *and* the channel length L may be freely selected. Since the short-channel and narrow-channel effects are both undesirable in analog circuits, W and L must not be below certain technology-dependent values: $W \geq W_{min}$ and $L \geq L_{min}$. Selecting $W = W_{min}$ and $L = L_{min}$ results in the geometrically smallest MOSFET, which represents reference transistors with the relative size 1. Larger MOSFETs are created by increasing W and keeping $L = L_{min}$ constant. However, one may also keep $W = W_{min}$ constant and enlarge L; this creates a MOSFET that is electrically – that is, with regard to $K \sim W/L$ – smaller but geometrically larger than the reference transistor. Therefore, it is necessary to distinguish between the *electrical size* and the *geometric size*. In the following description, the term *size* always refers to the electrical size. Proportionally enlarging W and L results in a MOSFET of the same size; this method is, however, only used in exceptional cases, because of the higher spatial requirements.[3] Figure 4.4 uses bipolar transistors of sizes 1

---

[2] In digital circuits, the channel lengths are mostly $0.2 \ldots 0.5\,\mu\text{m}$, while analog circuits usually have channel lengths of more than $1\,\mu\text{m}$. This is due to the fact that the Early voltage V_A, and thus the maximum gain, rise with an increasing channel length.

[3] MOSFETs that are of the same electrical size but larger in terms of geometry generally feature lower noise and higher Early voltage; the capacitances, on the other hand, increase.

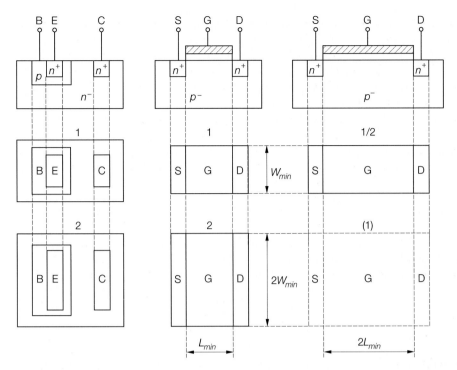

Fig. 4.4. Scaling and normalization of bipolar transistors and MOSFETs

and 2 and n-channel MOSFETs of sizes 1, 2 and 1/2 to illustrate transistor scaling and normalization.

Complementary transistors: Most bipolar technologies only allow lateral pnp transistors, which have significantly poorer electrical properties than those of vertical npn transistors; this is particularly true for the current gain and the transit frequency. These technologies only use npn transistors in the signal path of an amplifier if possible; pnp transistors are only used for current sources or in common-collector or common-base circuits, as in these applications their poorer properties are less noticeable. Vertical pnp transistors with similar properties may be available in special complementary technologies, but here too the npn transistors have somewhat better characteristics. The differences between vertical and lateral bipolar transistors are described in more detail in Sect. 2.2.

MOS technologies are predominantly complementary technologies, i.e. CMOS. Here, n-channel and p-channel MOSFETs with comparable properties are available. However, the relative transconductance coefficient K'_p of p-channel MOSFETs is 2 ... 3 times lower than the relative transconductance coefficient K'_n of n-channel MOSFETs. This means that a p-channel MOSFET with the same channel length L must have a channel width W that is 2 ... 3 times wider than in the n-channel MOSFET in order to achieve the same transconductance coefficient $K = K'_{n/p} W/L$. But all that this means is that just the static characteristics are almost the same. The dynamic characteristics of the p-channel MOSFET are inferior, because the larger dimensions result in higher capacitances. Therefore, n-channel MOSFETs are preferred. If both the static and the dynamic characteristics are to be almost the same, then W and L for the n-channel MOSFET must be enlarged by a factor

Name	Parameter	PSpice	npn	pnp	Unit
Reverse saturation current	I_S	IS	1	0.5	fA
Current gain	B	BF	100	50	
Early voltage	V_A	VAF	100	50	V
Base spreading resistance	R_B	RBM	100	50	Ω
Emitter capacitance	$C_{J0,E}$	CJE	0.1	0.1	pF
Collector capacitance	$C_{J0,C}$	CJC	0.2	0.5	pF
Substrate capacitance	$C_{J0,S}$	CJS	1	2	pF
Transit time	$\tau_{0,N}$	TF	100	150	ps
Maximum transit frequency	f_C		1.3	0.85	GHz
Typical quiescent current	$I_{C,A}$		100	− 100	μA

Fig. 4.5. Parameters of bipolar transistors with (relative) size 1

Name	Parameter(s)	PSpice	n-channel	p-channel	Unit
Threshold voltage	V_{th}	VTO	1	− 1	V
Relative transconductance coefficient	K'_n, K'_p	KP	30	12	μA/V^2
Mobility [4]	μ_n, μ_p	UO	500	200	cm^2/Vs
Oxide thickness	d_{ox}	TOX	57.5	57.5	nm
Gate capacitance per unit length	C'_{ox}		0.6	0.6	fF/μm^2
Bulk capacitance per unit length	C'_J	CJ	0.2	0.2	fF/μm^2
Gate–drain capacitance	$C'_{GD,ov}$	CGDO	0.5	0.5	fF/μm
Early voltage	V_A		50	33	V
Channel length modulation	λ	LAMBDA	0.02	0.033	V^{-1}
Substrate control factor	γ	GAMMA	0.5	0.5	$\sqrt{V}$
Inversion voltage	V_{inv}	PHI	0.6	0.6	V
Channel width	W	W	3	7.5	μm
Channel length	L	L	3	3	μm
Transconductance coefficient	K		30	30	μA/V^2
Typical transit frequency [5]	f_C		1.3	0.5	GHz
Typical quiescent current	$I_{D,A}$		10	− 10	μA

Fig. 4.6. Parameters of MOSFETs with (relative) size 1

of between $\sqrt{2}$ and $\sqrt{3}$, to make the surfaces and thus the capacitances approximately comparable to those of the p-channel MOSFET; this has no influence on the electrical size of the n-channel MOSFET. As this method reduces the transit frequency of the n-channel MOSFET to that of the p-channel MOSFET, it is only used if special symmetric characteristics are required.

---

[4] The mobility is quoted as in *Spice* in cm^2/Vs (UO=500 bzw. UO=200).
[5] The transit frequency is proportional to $V_{GS} - V_{th}$ or $\sqrt{I_{D,A}}$; it is given here for $V_{GS} - V_{th} = 1$ V, which is a typical value in analog circuits.

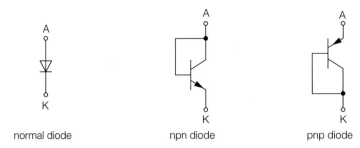

normal diode   npn diode   pnp diode

Fig. 4.7. Bipolar diodes in integrated circuits

The circuits explained below are described on the basis of complementary bipolar and CMOS technologies; the most important transistor parameters are listed in Figs. 4.5 and 4.6.

Effects of manufacturing tolerances: In bipolar technology, npn and pnp transistors are produced in separate steps. For npn transistors, the tolerances of one manufacturing step have approximately the same effect on all npn transistors, and the parameters of all npn transistors are also affected in the same way. This has the particular consequence that a manufacturing tolerance for the reverse saturation currents has no influence on the dimensional proportions caused by scaling: an npn transistor of size 5 always has five times the reverse saturation current than an npn transistor of size 1. The same is true for pnp transistors. However, the dimensional proportions are not constant between npn and pnp transistors. The ratio of reverse saturation currents between an npn and a pnp transistor of size 1 may thus vary significantly. The same considerations also apply to n-channel and p-channel MOSFETs in CMOS technology; in this case, the transconductance coefficients in particular are affected.

Diodes: In integrated circuits, diodes are realized with the help of transistors. An npn or pnp transistor with a short-circuited base-collector junction is used for a bipolar diode (see Fig. 4.7). This particular diode is called *transdiode* and is required in particular for current scaling as described below; a collector or emitter diode is not suitable for this purpose. Furthermore, a distinction must be made between npn and pnp diodes, since they have different parameters. Scaling is done in the same way as in transistors; that is, an npn diode of size 5 corresponds to an npn transistor of size 5 with a short-circuited base-collector junction.

An important application of diodes is current-to-voltage conversion according to Fig. 4.9a, in which the diodes provide an *indication* for the current:

$$I = I_{S,D}\left(e^{\frac{V}{V_T}} - 1\right) \quad \Rightarrow \quad V = V_T \ln\left(\frac{I}{I_{S,D}} + 1\right) \overset{I \gg I_{S,D}}{\approx} V_T \ln\frac{I}{I_{S,D}}$$

Here, $I_{S,D}$ is the reverse saturation current of the diode. Feeding this voltage to the base-emitter junction of a transistor with reverse saturation current $I_{S,D}$ leads to the following equation, provided that the transistor is operating in normal mode and the base current is negligible:

$$I_C \approx I_{S,T}\, e^{\frac{V_{BE}}{V_T}} \overset{V_{BE}=V}{=} I_{S,T}\, e^{\ln\frac{I}{I_{S,D}}} = I\,\frac{I_{S,T}}{I_{S,D}}$$

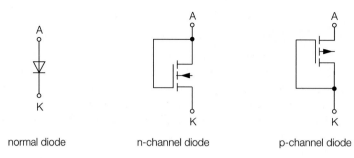

normal diode   n-channel diode   p-channel diode

Fig. 4.8. FET diodes in integrated circuits

This means that the current is scaled in proportion to the reverse saturation currents. Defined scaling is achieved only if an npn diode is combined with an npn transistor or a pnp diode is combined with a pnp transistor; then the ratio of the reverse saturation currents is defined by the dimensional proportions.

In MOS circuits the *FET diodes* shown in Fig. 4.8 may be used. The current-to-voltage conversion according to Fig. 4.9b is:

$$I = \frac{K_D}{2}(V_{GS} - V_{th})^2 \quad \Rightarrow \quad V = V_{th} + \sqrt{\frac{2I}{K_D}}$$

Here, K_D is the transconductance coefficient of the FET diode. Feeding this voltage to the gate-source junction of a MOSFET with the transconductance coefficient K_M leads to the following equation, provided that the MOSFET is operated in the pinch-off region:

$$I_D \approx \frac{K_M}{2}(V_{GS} - V_{th})^2 \overset{V_{GS}=V}{=} I\,\frac{K_M}{K_D}$$

Even in this case, an n-channel FET diode must be combined with an n-channel MOSFET and a p-channel FET diode with a p-channel MOSFET in order to define the current scaling by the dimensional proportions.

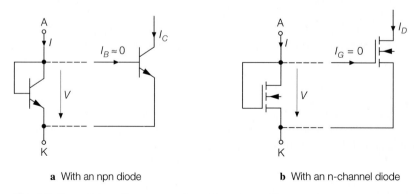

a With an npn diode   **b** With an n-channel diode

Fig. 4.9. Current–to–voltage conversion and current scaling

4.1.1
Current Sources and Current Mirrors

A *current source* supplies a constant output current and is used predominantly for setting the operating point (biasing). A *current mirror* provides an amplified or attenuated copy of the input current at the output; that is, it operates as a current-controlled current source. Any current mirror may also be used as a current source by keeping the input current constant; in this respect, the current source is a special application of the current mirror.

Basics of the current source

The output characteristic of a bipolar transistor and a MOSFET are almost horizontal over a large region (see Fig. 2.3 on page 35 and Fig. 3.5 on page 173); in this region, the collector and drain currents are practically independent of the collector-emitter or the drain-source voltage. Therefore, a single transistor can be used as a current source by feeding a constant input voltage and using the collector or drain connection as the output:

$$
I_o = \begin{cases} I_C(V_{BE}, V_{CE}) & \approx & I_C(V_{BE}) & \overset{V_{BE}=\text{const.}}{=} & \text{const.} \\[2mm] I_D(V_{GS}, V_{DS}) & \approx & I_D(V_{GS}) & \overset{V_{GS}=\text{const.}}{=} & \text{const.} \end{cases}
$$

Stable operation also necessitates negative current feedback in order to keep the output current constant despite variations in the transistor parameters caused by manufacturing tolerances or temperature drift. This results in the circuits shown in Fig. 4.10. A load must be connected to the output of the current source to allow the flow of current I_o; in Fig. 4.10, resistance R_L represents the load.

Output current: The current source with a bipolar transistor according to Fig. 4.10a is described by the following mesh equation:

$$
V_B = V_{BE} + V_R = V_{BE} + (I_C + I_B) R_E \overset{I_C \gg I_B}{\approx} V_{BE} + I_C R_E
$$

With $I_C = I_o$, it follows

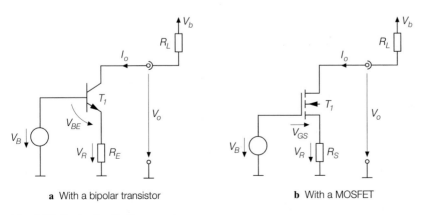

a With a bipolar transistor   **b** With a MOSFET

Fig. 4.10. Basic circuit of a current source

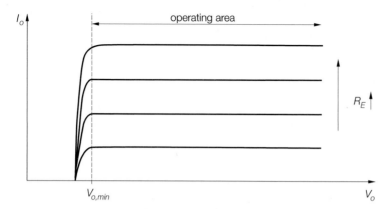

Fig. 4.11. Output characteristics of a current source with a bipolar transistor

$$I_o \approx \frac{V_B - V_{BE}}{R_E} \overset{V_{BE} \approx 0.7\,\text{V}}{\approx} \frac{V_B - 0.7\,\text{V}}{R_E}$$

The influence of V_{BE} can be reduced by making V_B sufficiently high; the borderline case $V_B \gg V_{BE}$ results in $I_o \approx V_B/R_1$. On the other hand, V_B must not be too high, as this reduces at the output voltage range. The current source only functions correctly if transistor T_1 operates in normal mode; this requires $V_{CE} > V_{CE,sat}$ and thus

$$V_o = V_R + V_{CE} > V_R + V_{CE,sat} = V_B - V_{BE} + V_{CE,sat}$$

Output characteristic: Plotting the output current I_o versus V_o for several values of R_E produces the family of characteristics shown in Fig. 4.11, with the minimum output voltage:

$$V_{o,min} = V_B - V_{BE} + V_{CE,sat} \overset{\substack{V_{CE,sat} \approx 0.2\,\text{V} \\ V_{BE} \approx 0.7\,\text{V}}}{\approx} V_B - 0.5\,\text{V}$$

For $V_o > V_{o,min}$ and $V_B = $ const., the circuit functions as a current source. Hereafter, $V_{o,min}$ will be referred to as the useful limit.

Output resistance: Besides the output current I_o and the voltage limit $V_{o,min}$, the output resistance

$$r_o = \left. \frac{\partial V_o}{\partial I_o} \right|_{V_B = \text{const.}}$$

in the operating range is also of particular interest; for an ideal current source this value is $r_o = \infty$, and therefore it should be as high as possible in any real current source. The finite output resistance is caused by the Early effect and can be calculated using the small-signal equivalent circuit. Since the circuit in Fig. 4.10a largely corresponds to the common-emitter circuit with current feedback shown in Fig. 2.62a on page 102, the previous result can be used once the resistances have been changed to $R_g = 0$ and $R_C \to \infty$;[6] this leads to

---

[6] In the common-emitter circuit with current feedback, R_C is assumed to be an inherent part of the circuit and is thus included in the calculation of the output resistance; in the current source, however, it is the output resistance of the collector without other circuit components that is of interest. Resistance R_C is *removed* by setting $R_C \to \infty$.

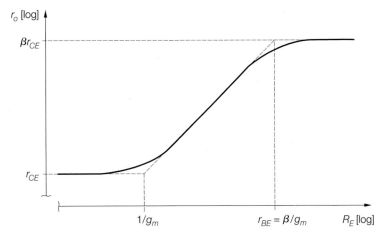

Fig. 4.12. Output resistance of a current source with a bipolar transistor at a constant output current

$$r_o = \frac{v_o}{i_o}\bigg|_{V_B=\text{const.}} \overset{r_{CE}\gg r_{BE}}{\approx} r_{CE}\left(1 + \frac{\beta R_E}{R_E + r_{BE}}\right) \qquad (4.1)$$

The use of $\beta \gg 1$ and $r_{BE} = \beta/g_m$ leads to:

$$r_o \approx \begin{cases} r_{CE}\,(1 + g_m R_E) & \text{for } R_E \ll r_{BE} \\ \beta\, r_{CE} & \text{for } R_E \gg r_{BE} \end{cases}$$

Figure 4.12 shows the curve of r_o versus R_E with a constant output current.

Inserting $r_{CE} = V_A/I_o$, $g_m = I_o/V_T$, $r_{BE} = \beta\, V_T/I_o$ and $V_R \approx I_o R_E$ shows the dependence of the output resistance on the output current:

$$r_o \approx \begin{cases} \dfrac{V_A}{I_o} + \dfrac{V_A}{V_T}\, R_E & \text{for } V_R \ll \beta\, V_T \\[2ex] \dfrac{\beta\, V_A}{I_o} & \text{for } V_R \gg \beta\, V_T \end{cases}$$

The maximum output current is achieved if a voltage drop V_R across the feedback resistance of more than $\beta\, V_T \approx 2.6\,\text{V}$ is selected. This results in a constant $I_o r_o$ product:

$$I_o r_o \approx \beta\, V_A \overset{\substack{V_A \approx 30...200\,\text{V} \\ \beta \approx 50...500}}{\approx} 1.5 \ldots 100\,\text{kV}$$

The product of the Early voltage V_A and the current gain β is thus a crucial parameter for evaluating the use of bipolar transistors in current sources.

Current source with a MOSFET: With $I_o = I_D$, the voltage for the MOSFET current source in Fig. 4.10b is:

$$V_B = V_R + V_{GS} = I_o R_S + V_{GS} = I_o R_S + V_{th} + \sqrt{\frac{2I_o}{K}}$$

Calculating the output current $I_o = I_D$ is rather complex, since MOSFETs do not allow a simple approximation for V_{GS}, such as $V_{BE} \approx 0.7\,\text{V}$ for bipolar transistors. However, for

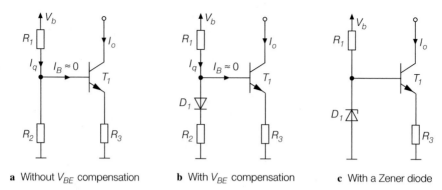

a Without V_{BE} compensation   **b** With V_{BE} compensation   **c** With a Zener diode

Fig. 4.13. Simple current sources for discrete circuits

discrete MOSFETs it is possible to define I_o and V_B in order to calculate R_S:

$$R_S = \frac{V_B - V_{th}}{I_o} - \sqrt{\frac{2}{K I_o}}$$

This cannot be done exactly with integrated MOSFETs, since the threshold voltage does not remain constant due to the substrate effect.

Since the MOSFET must be operated in the pinch-off region (it is only in this region that the output characteristics are almost horizontal), the output voltage limit is $V_{o,min} = V_R + V_{DS,po}$; since $V_{DS,po} > V_{CE,sat}$, the limit is higher than at bipolar transistors. A comparison with the common-source circuit with current feedback defines the output resistance as:

$$r_o = \left. \frac{v_o}{i_o} \right|_{V_B=const.} \overset{r_{DS} \gg 1/g_m}{\approx} r_{DS} \left(1 + \left(g_m + g_{m,B} \right) R_S \right) \overset{g_m \gg g_{m,B}}{\approx} r_{DS} \left(1 + g_m R_S \right) (4.2)$$

It is lower than in bipolar transistors due to the lower Early voltage and the lower transconductance. Therefore, discrete circuits almost exclusively use current sources with bipolar transistors.

Simple current sources for discrete circuits

Figure 4.13 shows the three discrete current sources that are most commonly used. For $I_q \gg I_B \approx 0$, the currents in the circuit of Fig. 4.13a are:

$$\left. \begin{aligned} I_q &\approx \frac{V_b}{R_1 + R_2} \\ I_q R_2 &\approx I_o R_3 + V_{BE} \end{aligned} \right\} \Rightarrow I_o \approx \frac{1}{R_3} \left(\frac{V_b R_2}{R_1 + R_2} - V_{BE} \right) \quad \text{with } V_{BE} \approx 0.7 \, \text{V}$$

The output current depends on the temperature, since V_{BE} is influenced by the temperature:

$$\frac{d I_o}{dT} = -\frac{1}{R_3} \frac{d V_{BE}}{dT} \approx \frac{2 \, \text{mV/K}}{R_3}$$

The temperature sensitivity can be reduced by increasing R_3 to enhance the feedback effect; but this also requires adaptation of R_1 and R_2 in order to keep the output current constant.

In the circuit of Fig. 4.13b the temperature sensitivity is reduced by compensating V_{BE} by the voltage at the diode; with $V_D \approx V_{BE}$ and $I_q \gg I_B \approx 0$, it follows that

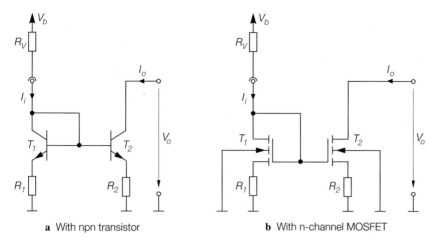

a With npn transistor   **b** With n-channel MOSFET

Fig. 4.14. Simple current mirror

$$\left.\begin{array}{l} I_q \approx \dfrac{V_b - V_D}{R_1 + R_2} \\[2mm] I_q R_2 \approx I_o R_3 \end{array}\right\} \Rightarrow I_o \approx \dfrac{(V_b - V_D)\, R_2}{(R_1 + R_2)\, R_3} \quad \text{with } V_D \approx 0.7\,\text{V}$$

The temperature sensitivity is:

$$\frac{dI_o}{dT} = -\frac{R_2}{(R_1 + R_2)\, R_3}\frac{dV_D}{dT} \approx \frac{2\,\text{mV/K}}{R_3}\frac{R_2}{R_1 + R_2} \approx 2\,\text{mV/K} \cdot \frac{I_o}{V_b - V_D}$$

Compared to the circuit in Fig. 4.13a, this value is lower by the factor $1 + R_1/R_2$ and it becomes zero when R_1 is replaced by a (temperature-independent) current source with current I_q.[7]

The current in the circuit of Fig. 4.13c is:

$$I_o \approx \frac{V_Z - V_{BE}}{R_3} \approx \frac{V_Z - 0.7\,\text{V}}{R_3}$$

Here, V_Z is the breakdown voltage of the Zener diode. The temperature sensitivity depends on the temperature coefficient of the Zener diode. To keep the temperature sensitivity very low, a normal diode can be connected in series as in Fig. 4.13b to compensate V_{BE}; then:

$$I_o \approx \frac{V_Z}{R_3}$$

Only the temperature coefficient of the Zener diode has any influence. The minimum temperature sensitivity is achieved with $V_Z \approx 5 \ldots 6\,\text{V}$.

Simple current mirror

The simplest current mirror consists of two transistors T_1 and T_2 and two optional resistances R_1 and T_2 for current feedback (see Fig. 4.14); as it has no specific name, it will be referred to here as the *simple current mirror*. An additional resistance R_V allows a constant reference current to be set; the current mirror thus becomes a current source.

---

[7] The transition to the current source is achieved by making $R_1 \to \infty$; at the same time this necessitates that $V_b \to \infty$ to keep the output current constant.

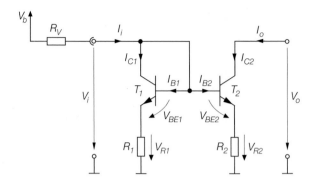

Fig. 4.15. Currents and voltages of the npn current mirror

npn current mirror: Figure 4.15 shows the currents and voltages of the simple current mirror with npn transistors, which is simply referred to as the *npn current mirror*. The mesh equation for the base-emitter junction and the feedback resistances leads to:

$$(I_{C1} + I_{B1}) R_1 + V_{BE\,1} \; = \; (I_{C2} + I_{B2}) R_2 + V_{BE\,2} \tag{4.3}$$

In the normal operating range, both transistors are in normal mode and can be described by the basic (2.2) and (2.3):

$$I_{C1} = I_{S1}\, e^{\frac{V_{BE\,1}}{V_T}} \qquad , \qquad I_{B1} = \frac{I_{C1}}{B}$$

$$I_{C2} = I_{S2}\, e^{\frac{V_{BE\,2}}{V_T}} \left(1 + \frac{V_{CE2}}{V_A}\right) \quad , \quad I_{B2} = \frac{I_{C2}}{B} \tag{4.4}$$

For T_1, the Early effect is ignored, since $V_{CE1} = V_{BE\,1} \ll V_A$. From Fig. 4.15, it thus follows that:

$$I_i \; = \; I_{C1} + I_{B1} + I_{B2} \quad , \quad I_o \; = \; I_{C2} \tag{4.5}$$

npn current mirror without feedback: Under the assumption of $R_1 = R_2 = 0$, (4.3) reads $V_{BE\,1} = V_{BE\,2}$. Inserting (4.4) and (4.5), while taking into account the fact that $V_{CE2} = V_o$, leads to the *current ratio*:

$$k_I \; = \; \frac{I_o}{I_i} \; = \; \frac{1}{\dfrac{I_{S1}}{I_{S2}}\left(1 + \dfrac{1}{B}\right)\dfrac{V_A}{V_A + V_o} + \dfrac{1}{B}} \tag{4.6}$$

For $V_o \ll V_A$, it follows that

$$\boxed{k_I \; = \; \frac{I_o}{I_i} \; \approx \; \frac{1}{\dfrac{I_{S1}}{I_{S2}}\left(1 + \dfrac{1}{B}\right) + \dfrac{1}{B}} \; \overset{B \gg 1, I_{S2}/I_{S1}}{\approx} \; \frac{I_{S2}}{I_{S1}}} \tag{4.7}$$

If the Early voltage V_A and the current gain B are sufficiently high, and the size ratio I_{S2}/I_{S1} of the transistors is significantly lower than the current gain B, then the current ratio k_I corresponds approximately to the ratio of the transistor dimensions. If both transistors are

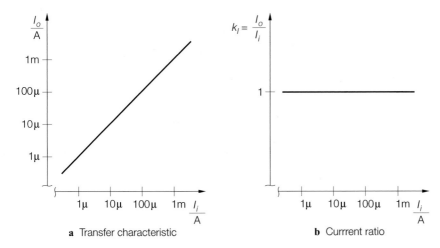

a Transfer characteristic   **b** Currrent ratio

Fig. 4.16. The transfer characteristic of a current mirror with $I_{S1} = I_{S2}$

of the same size, then $I_{S1} = I_{S2}$ and thus:

$$k_I = \cfrac{1}{\left(1 + \cfrac{1}{B}\right) \cfrac{V_A}{V_A + V_o} + \cfrac{1}{B}} \overset{V_o \ll V_A}{\approx} \cfrac{1}{1 + \cfrac{2}{B}} \overset{B \gg 1}{\approx} 1 \qquad (4.8)$$

Figure 4.16 shows the transfer characteristic and the current ratio of a current mirror with $I_{S1} = I_{S2}$; that is, $k_I \approx 1$. One can see that the current mirror features a linear characteristic across several decades. However, with very low and very high currents the current gain drops significantly and the transfer characteristic is no longer linear; this region is not shown in Fig. 4.16.

Output characteristic: In current mirrors not only the current ratio is of interest; indeed, the operating range and the small-signal output resistance in the operating range are also of particular interest. Useful in this respect is the family of output characteristics that show I_o as a function of V_o using I_i as a parameter; usually only one characteristic is plotted for the intended quiescent current $I_i = I_{i,A}$. Figure 4.17 shows the output characteristic of an npn current mirror with $k_I = 1$ for $I_i = 100\,\mu\text{A}$; the characteristic of the n-channel current mirror in Fig. 4.17 will be explained later. The characteristic corresponds to the output characteristic of transistor T_2. For $V_o > V_{CE,sat}$, transistor T_2 operates in normal mode; only in this *operating range* does the current mirror have the calculated current ratio. For $V_o \leq V_{CE,sat}$, transistor T_2 enters the saturation region and the current decreases. The minimum output voltage $V_{o,min}$ is an important parameter; in the npn current mirror it is:[8]

$$V_{o,min} = V_{CE,sat} \approx 0.2\,\text{V}$$

The output resistance corresponds to the reciprocal value of the slope of the output characteristic in the operating range. If only the approximations for the current gain are carried out in (4.6) and the Early voltage is maintained, we obtain the following value within the operating range:

---

[8] A relatively high value of $V_{CE,sat} \approx 0.2\,\text{V}$ is assumed for the collector–emitter saturation voltage, since for this voltage the output characteristic of the transistor ought to be as horizontal as possible.

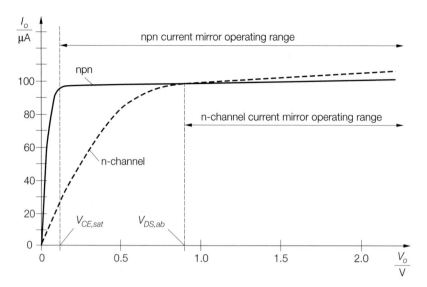

Fig. 4.17. Output characteristics of an npn and an n-channel current mirror for $R_1 = R_2 = 0$

$$k_I = \frac{I_o}{I_i} \approx \frac{I_{S2}}{I_{S1}}\left(1 + \frac{V_o}{V_A}\right)$$

This gives us the following value for the *small-signal output resistance*:

$$r_o = \left.\frac{\partial V_o}{\partial I_o}\right|_{I_i=\text{const.}} = \frac{V_o + V_A}{I_o} \overset{V_o \ll V_A}{\approx} \frac{V_A}{I_o} = \frac{V_A}{I_{C2}} = r_{CE2}$$

Usually, the output resistance is calculated by using the small-signal equivalent circuit; this will be explained in more detail further below.

npn current mirror with feedback: Feedback resistances can be used to stabilise the current ratio and to increase the output resistance. Without feedback resistance the current ratio only depends on the ratio of the transistor dimensions, but with feedback resistances the ratio of the resistances R_2/R_1 also has an influence. Inserting (4.4) into (4.3) and neglecting the Early effect results in:

$$\left(1 + \frac{1}{B}\right)R_1 I_{C1} + V_T \ln\frac{I_{C1}}{I_{S1}} = \left(1 + \frac{1}{B}\right)R_2 I_{C2} + V_T \ln\frac{I_{C2}}{I_{S2}} \tag{4.9}$$

This equation cannot be solved easily, as the collector currents have a linear *and* a logarithmic effect on the result. With sufficiently large resistances the linear terms dominate, so that:

$$R_1 I_{C1} \approx R_2 I_{C2} \tag{4.10}$$

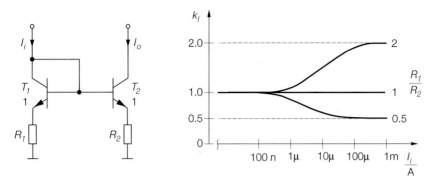

Fig. 4.18. Current dependence of the current ratio for transistors of equal size ($I_{S2}/I_{S1} = 1$) and different R_1/R_2 values

With (4.5), it follows:

$$k_I = \frac{I_o}{I_i} \approx \frac{R_1}{R_2 + \dfrac{R_1 + R_2}{B}} \overset{B \gg 1 + R_1/R_2}{\approx} \frac{R_1}{R_2} \qquad (4.11)$$

In this case, the current ratio depends solely on the ratio of the resistances and no longer on the dimensions of the transistors.

For integrated current mirrors, the ratio of the resistances is usually chosen to correspond to the dimensional ratio of the transistors:

$$\frac{I_{S2}}{I_{S1}} \approx \frac{R_1}{R_2}$$

In this case, the resistances have almost no influence on the current ratio but only cause the output resistance to increase; this will be described in more detail below. For current mirrors driven across a large current range, this condition is in fact mandatory, as the ratio of linear and logarithmic terms in (4.9) depends on the current: for small currents, the current ratio is determined by I_{S2}/I_{S1}, and for large currents by R_1/R_2. Figure 4.18 shows this dependence using a current mirror with transistors of equal size ($I_{S2}/I_{S1} = 1$) and different values for R_1/R_2. A constant current ratio is achieved only in the case of $I_{S2}/I_{S1} = R_1/R_2$.

With current mirrors made of discrete components feedback resistances must always be used, as the tolerances of discrete transistors are so high that the ratio I_{S2}/I_{S1} is practically undefined, even for transistors of the same type; this means that it is essential to adjust the current ratio using the resistances. The required minimum resistance value can be determined by differentiating both sides of (4.9) with respect to the actual current and presuming that the influence of the terms with the resistances dominates:

$$\left(1 + \frac{1}{B}\right) R_1 \gg \frac{V_T}{I_{C1}} \quad , \quad \left(1 + \frac{1}{B}\right) R_2 \gg \frac{V_T}{I_{C2}}$$

It follows:

$$V_{R1} = \left(1 + \frac{1}{B}\right) R_1 I_{C1} \gg V_T \quad , \quad V_{R2} = \left(1 + \frac{1}{B}\right) R_2 I_{C2} \gg V_T$$

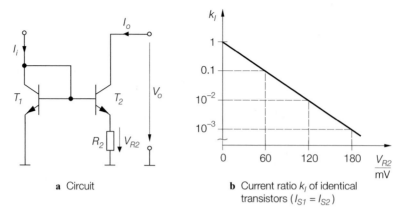

a Circuit

b Current ratio k_I of identical transistors ($I_{S1} = I_{S2}$)

Fig. 4.19. Widlar current mirror

V_{R1} and V_{R2} represent the voltages across resistances R_1 and R_2 (see Fig. 4.15). As both conditions are equivalent due to (4.10), and an approximate factor of 10 is required to meet this condition, one must select

$$V_{R1} \approx V_{R2} \geq 10\,V_T \approx 250\,\text{mV} \tag{4.12}$$

to ensure that the current ratio is only influenced by the resistances. In current mirrors driven across a large current range, the condition given in (4.12) can usually not be met across the entire range; in this case, the current ratio for decreasing currents is increasingly determined by the unknown ratio I_{S2}/I_{S1}.

The feedback reduces the operating range, since the output voltage limit $V_{o,min}$ is increased by a value equal to the voltage across the resistances:

$$V_{o,min} = V_{CE,sat} + V_{R2} \geq 0.2\,\text{V} + 0.25\,\text{V} = 0.45\,\text{V}$$

For this reason, the resistance values cannot be freely chosen.

Operation as a current source: The simple npn current mirror can be operated as a current source by adding resistance R_V as shown in Fig. 4.15; this makes the input current constant. From $V_i = V_{BE1} + V_{R1}$ and $V_b = V_i + I_i R_V$, it follows:

$$V_b = I_i R_V + (I_{C1} + I_{B1})\, R_1 + V_{BE1}$$

If the base currents of the transistors are neglected and we assume that $V_{BE} \approx 0.7\,\text{V}$, then:

$$I_i \approx \frac{V_b - V_{BE1}}{R_V + R_1} \approx \frac{V_b - 0.7\,\text{V}}{R_V + R_1}$$

The output current is $I_o = k_I I_i$.

Widlar current mirror: Where very low current ratios are required, any adjustment by means of the dimensional ratio of the transistors is unfavorable, as T_2 can only be reduced in size down to the basic size, causing T_1 to become very large. In this situation, one can use the *Widlar current mirror* shown in Fig. 4.19a, which only features the feedback resistance R_2; from (4.9), where $R_1 = 0$ and $B \gg 1$, it follows:

$$V_T \ln \frac{I_{C1}}{I_{S1}} = R_2 I_{C2} + V_T \ln \frac{I_{C2}}{I_{S2}}$$

The current ratio with $I_i \approx I_{C1}$ and $I_o \approx I_{C2}$ is:

$$k_I = \frac{I_o}{I_i} \approx \frac{I_{C2}}{I_{C1}} = \frac{I_{S2}}{I_{S1}} e^{-\frac{V_{R2}}{V_T}} \quad \text{with } V_{R2} = R_2 I_{C2} \qquad (4.13)$$

This value depends exponentially on the ratio V_{R2}/V_T and decreases by a factor of 10 if V_{R2} increases by $V_T \ln 10 \approx 60\,\text{mV}$; Fig. 4.19b illustrates this for identical transistors; that is, for $I_{S1} = I_{S2}$. From (4.13) it can also be seen that the Widlar current mirror is only suitable for constant currents, due to the high current dependency of the current ratio.

It would now seem that the same procedure could be used to realize very high current ratios by only using resistance R_1 in Fig. 4.14a. Theoretically, this is correct, but it is not feasible in practice since the higher current at the output naturally requires a larger transistor. This *inverted* Widlar current mirror can only be used if the current ratio is high enough to warrant the use of a Widlar current mirror and the output current is so low that a transistor of size 1 can also be used at the output; this case is, however, very rare.

Example: An output current $I_o = 10\,\mu\text{A}$ is to be derived from an input current $I_i = 1\,\text{mA}$. As in our sample technology a transistor of size 1 according to Fig. 4.5 is rated for a current of $100\,\mu\text{A}$, we select size 10 for T_1 and the minimum size 1 for T_2; thus, $I_{S2}/I_{S1} = 0.1$. For the desired current ratio $k_I = I_o/I_i = 0.01$, this means that the exponential factor in (4.13) must also be 0.1; this leads to $V_{R2} = V_T \ln 10 \approx 60\,\text{mV}$ and $R_2 = V_{R2}/I_o \approx 6\,\text{k}\Omega$.

Three-transistor current mirror: A low current gain of the transistors has a negative effect on the current ratio of the simple current mirror. Especially with large current ratios, the base current of the output transistor can increase so much that the current ratio clearly deviates from the dimensional ratio of the transistors. This means that the current ratio no longer depends solely on the geometric sizes, but also to an increasing degree on the current gain that is affected by imminent tolerances. This can be overcome by the *three-transistor current mirror* shown in Fig. 4.20a, in which the base current for transistors T_1 and T_2 is supplied via an additional transistor T_3. This, in turn, contributes to the input current I_i

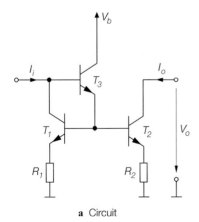

a Circuit

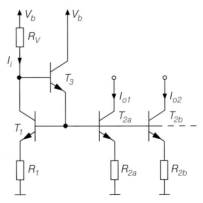

b Use in a current source bank

Fig. 4.20. Three-transistor current mirror

with its very low base current, which significantly reduces the dependence on the current gain.

Without feedback resistances – that is, with $R_1 = R_2 = 0$ – one arrives at $V_{BE\,1} = V_{BE\,2}$, and by ignoring the Early effect at:

$$\frac{I_{C2}}{I_{C1}} = \frac{I_{S2}}{I_{S1}}$$

Inserting the nodal equations

$$I_i = I_{C1} + I_{B3} \quad , \quad I_{B1} + I_{B2} = I_{C3} + I_{B3} \quad , \quad I_o = I_{C2}$$

leads to the current ratio where $I_{B1} = I_{C1}/B$, $I_{B2} = I_{C2}/B$ and $I_{B3} = I_{C3}/B$:

$$k_I = \frac{B^2 + B}{\dfrac{I_{S1}}{I_{S2}}\left(B^2 + B + 1\right) + 1} \overset{B \gg 1}{\approx} \frac{I_{S2}}{I_{S1}} \tag{4.14}$$

For $I_{S1} = I_{S2}$, it follows:

$$k_I = \frac{1}{1 + \dfrac{2}{B^2 + B}} \overset{B \gg 1}{\approx} 1$$

A comparison with (4.8) on page 283 shows that instead of the error term $2/B$, only the error term $2/(B^2 + B) \approx 2/B^2$ occurs. Reducing the error by factor B corresponds exactly to the current gain of T_3. With feedback resistances, one obtains the very same result if the resistances are selected according to the transistor sizes: $I_{S2}/I_{S1} = R_1/R_2$.

Operation as current source: The three-transistor current mirror is predominantly used in *current source banks* according to Fig. 4.20b; in this case, several output transistors are connected to a common reference path. This generates several output currents which can be scaled as required by the dimensional and resistance ratios, and which have a fixed relationship to one another. Since the sum of the base currents of the output transistors may become very high, T_3 must be used to create an additional current gain. From Fig. 4.20b, where $V_{BE} \approx 0.7\,\text{V}$, it follows:

$$I_i \approx \frac{V_b - V_{BE\,3} - V_{BE\,1}}{R_V + R_1} \approx \frac{V_b - 1.4\,\text{V}}{R_V + R_1}$$

Current source banks of this type are chiefly used as quiescent current sources in integrated circuits.

n-channel current mirror: Figure 4.21 shows the currents and voltages in a simple current mirror with n-channel MOSFETs, otherwise known as the *n-channel current mirror*. In the normal operating range, both MOSFETs operate in the pinch-off region and the basic (3.3) applies:

$$I_{D1} = \frac{K_1}{2}(V_{GS1} - V_{th})^2$$

$$I_{D2} = \frac{K_2}{2}(V_{GS2} - V_{th})^2\left(1 + \frac{V_{DS2}}{V_A}\right) \tag{4.15}$$

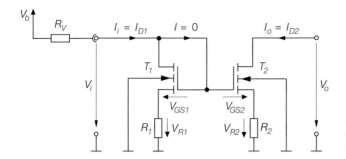

Fig. 4.21. Currents and voltages in the n-channel current mirror

In T_1, the Early effect is neglected due to $V_{DS1} = V_{GS1} \ll V_A$. Since no gate current flows in MOSFETs, the currents at the input and the output correspond to the drain currents:

$$I_i = I_{D1} \quad , \quad I_o = I_{D2} \tag{4.16}$$

The following mesh equation is also obtained from Fig. 4.21:

$$I_{D1} R_1 + V_{GS1} = I_{D2} R_2 + V_{GS2} \tag{4.17}$$

n-channel current mirror without feedback: Where $R_1 = R_2 = 0$ and $V_{DS2} = V_o$, the current ratio can be derived from (4.15)–(4.17):

$$k_I = \frac{I_o}{I_i} = \frac{K_2}{K_1}\left(1 + \frac{V_o}{V_A}\right) \overset{V_o \ll V_A}{\approx} \frac{K_2}{K_1} \tag{4.18}$$

If the Early voltage V_A is sufficiently high, the current ratio is only influenced by the ratio of the MOSFET sizes.

The output characteristic of the n-channel current mirror together with the output characteristic of an npn current mirror of the same design is shown in Fig. 4.17 on page 284. What can be seen here is that the operating range of the n-channel current mirror is narrower due to $V_{o,min} = V_{DS,po} > V_{CE,sat}$. The output voltage limit, however, is not constant but depends on the size of the MOSFETs, since:

$$V_{o,min} = V_{DS,po} = V_{GS} - V_{th} \overset{V_{DS,po} \ll V_A}{\approx} \sqrt{\frac{2 I_D}{K}}$$

It is therefore possible to lower the output voltage limit by increasing the sizes of the MOSFETs. In integrated analog circuits the operating points are usually given by $V_{GS} - V_{th} \approx 1\,\text{V}$; consequently, $V_{o,min} \approx 1\,\text{V}$. In order to achieve an output voltage limit of $V_{o,min} \approx 0.1 \ldots 0.2\,\text{V}$, as is the case in an npn current mirror, the MOSFET dimensions would have to be increased by a factor of $25 \ldots 100$. In practice, this is only possible in exceptional cases, because the gate capacitance increases by the same factor and the transit frequency is reduced accordingly; when used as a current source, the higher output capacitance is problematic.

n-channel current mirror with feedback: Calculating the current ratio is not easily done in this case, as the voltages across resistances R_1 and R_2 not only exert an influence on (4.17), but also cause a shift in the threshold voltages due to the substrate effect; this is due to $V_{BS1} = -V_{R1}$ and $V_{BS2} = -V_{R2}$. If both voltages are identical, the substrate effect has the same influence on both MOSFETs and both threshold voltages increase to the same degree; this requires the resistances to be chosen according to the sizes of the MOSFETs:

$$\frac{K_2}{K_1} = \frac{R_1}{R_2}$$

In this case, one arrives at the same current ratio as with the n-channel current mirror without feedback.

The feedback increases the output resistance of the current mirror; this will be explained in more detail below. In contrast, the output voltage limit is increased by the voltage drop across the resistances:

$$V_{o,min} = V_{DS2,po} + V_{R2} = V_{DS2,po} + I_{D2}R_2 \overset{I_{D2}=I_o}{=} \sqrt{\frac{2I_o}{K_2}} + I_o R_2$$

Operation as a current source: The simple n-channel current mirror can be operated as a current source by adding resistance R_V as shown in Fig. 4.21; this makes for a constant input current. From $V_i = V_{GS1} + V_{R1}$ and $V_b = V_i + I_i R_V$, it follows

$$I_i = \frac{V_b - V_{GS1}}{R_V + R_1}$$

The output current is $I_o = k_I I_i$.

Output resistance: The output current of a current mirror should depend on the input current alone and not on the output voltage; this means that the *small-signal output resistance*

$$r_o = \left.\frac{\partial V_o}{\partial I_o}\right|_{I_i=\text{const.}} = \left.\frac{v_o}{i_o}\right|_{i_i=0}$$

should be as high as possible. This value can be determined from the slope of the output characteristic in the operating range or with the help of the small-signal equivalent circuit. As the definition directly implies, the input is supplied by an ideal current source: $I_i = $ const. or $i_i = 0$. Strictly speaking, this is the *open-circuit output resistance*. In the small-signal equivalent circuit, the open-circuit condition at the input is expressed by the fact that the input is *open*; that is, with no circuitry. In practical applications, however, a true open circuit condition never exists at the input, although the difference between the real output resistance and the open-circuit output resistance is generally negligibly low.

The npn current mirror corresponds to the small-signal equivalent circuit shown in Fig. 4.22; for the transistors, the small-signal equivalent circuit of Fig. 2.12 on page 44 is used. The left section with transistor T_1 and resistor R_1 can be combined to give one resistance R_g:[9]

$$R_g = R_1 + \frac{1}{g_{m1} + \dfrac{1}{r_{BE1}} + \dfrac{1}{r_{CE1}}} \approx R_1 + \frac{1}{g_{m1}}$$

This results in almost the same small-signal equivalent circuit as in the case of the common-emitter circuit with current feedback, as a comparison with Fig. 2.65 on page 105 reveals;

---

[9] The controlled source $g_{m1} v_{BE\,1}$ acts as a resistance $1/g_{m1}$ since the control voltage $v_{BE\,1}$ is equal to the voltage at the source.

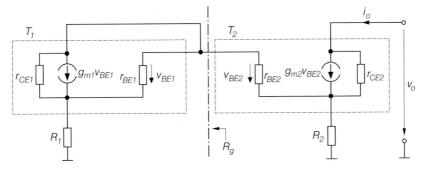

Fig. 4.22. Small-signal equivalent circuit of an npn current mirror

only resistance R_C and source v_g are omitted. Therefore, the output resistance of the current mirror can be derived from the short-circuit output resistance of the common-emitter circuit with current feedback:

$$
r_o = r_{CE2}\left(1 + \frac{\beta + \dfrac{r_{BE2}+R_g}{r_{CE2}}}{1 + \dfrac{r_{BE2}+R_g}{R_2}}\right) \underset{\substack{r_{CE2}>r_{BE2}+R_g \\ \beta \gg 1}}{\approx} r_{CE2}\left(1 + \frac{\beta R_2}{R_2 + r_{BE2}+R_g}\right)
$$

If $r_{BE2} \gg 1/g_{m1}$, replacing R_g leads to:

$$
\boxed{r_o = \left.\frac{v_o}{i_o}\right|_{i_i=0} \approx r_{CE2}\left(1 + \frac{\beta R_2}{R_1 + R_2 + r_{BE2}}\right)} \tag{4.19}
$$

Here, $r_{CE2} = V_A/I_o$ and $r_{BE2} = \beta V_T/I_o$.

Three special cases can be derived:

$$
r_o \approx \begin{cases} r_{CE2} & \text{for } R_2 = 0 & \rightarrow \text{ without feedback} \\ r_{CE2}\,(1 + g_{m2}R_2) & \text{for } R_1, R_2 \ll r_{BE2} & \rightarrow \text{ weak feedback} \\ \beta\, r_{CE2} & \text{for } R_2 \gg R_1, r_{BE2} & \rightarrow \text{ strong feedback} \end{cases}
$$

Weak feedback uses $g_{m2} = \beta/r_{BE2}$ and strong feedback uses $\beta \gg 1$. With strong feedback the output resistance is the highest output resistance that can be achieved with a bipolar transistor with feedback.[10] In practical applications, it is usually achieved by using a current source instead of R_2; an example of this is the *cascode current mirror,* which will be described in more detail below.

The small-signal equivalent circuit shown in Fig. 4.23 is used for calculating the output resistance of an n-channel current mirror; only the output with T_2 and R_2 is shown, because there is no connection to the input side of the current mirror due to the isolated gate. For MOSFETs the small-signal equivalent circuit shown in Fig. 3.17 on page 184 is used. A comparison with Fig. 3.63 on page 236 shows that the small-signal equivalent circuit of the n-channel current mirror corresponds to that of the common-source circuit with current feedback if resistance R_D is removed and the bulk is connected to ground. Thus,

---

[10] The use of amplifiers or positive feedback can achieve even higher output resistances, but only with very accurate matching.

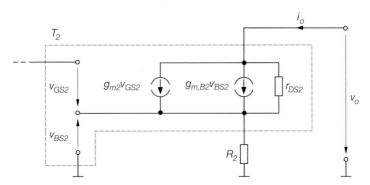

Fig. 4.23. A small-signal equivalent circuit for calculating the output resistance of an n-channel current mirror

this enables us to derive the output resistance; for $g_{m2} \gg 1/r_{DS2}$, we obtain the following:

$$r_o = \frac{v_o}{i_o}\Big|_{i_i=0} \approx r_{DS2}\left(1 + \left(g_{m2} + g_{m,B2}\right)R_2\right) \tag{4.20}$$

In this case, $r_{DS2} = V_A/I_o$.

Two special cases can be derived:

$$r_o \approx \begin{cases} r_{DS2} & \text{for } R_2 = 0 & \rightarrow \text{ without feedback} \\ r_{DS2}g_{m2}R_2 & \text{for } R_2, 1/g_{m,B2} \gg 1/g_{m2} & \rightarrow \text{ strong feedback} \end{cases}$$

In contrast to the npn current mirror, the output resistance of the n-channel current mirror has no upper limit: if $R_2 \rightarrow \infty$, then $r_o \rightarrow \infty$.

Figure 4.24 shows a comparison between the output resistances of an npn current mirror and an n-channel current mirror with $k_I = 1$ at a current of $I_o = 100\,\mu A$. Without feedback, the output resistance of the npn current mirror is usually higher than that of the n-channel current mirror; this is caused by the higher Early voltage of the npn transistors. In the low feedback range, the npn current mirror features $r_o \approx r_{CE2}g_{m2}R_2$ and the n-channel current mirror features $r_o \approx r_{DS2} \ldots r_{DS2}g_{m2}R_2$; the advantage of the npn current mirror is even stronger here, since not only the higher Early voltage but also the significantly higher transconductance of the npn transistors have an influence. With strong feedback, the output resistance of the npn current mirror approaches the maximum value $r_o = \beta\, r_{CE2}$, while in the n-channel current mirror with $r_o \approx r_{DS2}g_{m2}R_2$ it continues to rise. For an output current of $I_o = 100\,\mu A$, an ohmic feedback resistance of up to $R_2 \approx 10\,k\Omega$ can be used; the voltage across the resistances then remains below $V_{R2} \approx I_o R_2 = 100\,\mu A \cdot 10\,k\Omega = 1\,V$. Should one, however, try to realise $R_2 = 10\,M\Omega$ with an ohmic resistance, the voltage across $V_{R2} \approx I_o R_2 = 1000\,V$. For this reason, higher feedback resistances must be realized by using current sources.

Two important facts can be derived from Fig. 4.24:

– In the npn current mirror, the borderline to the region of strong feedback is reached with $R_2 = r_{BE2} = \beta/g_{m2}$; any further increase in R_2 results in no further noticeable improvement. In this case, the voltage drop across R_2 is:

$$V_{R2} = I_o R_2 = I_o\frac{\beta}{g_m} = I_o\frac{\beta\, V_T}{I_o} = \beta\, V_T \stackrel{\beta\approx100}{\approx} 2.6\,V$$

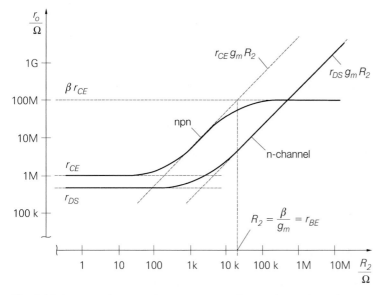

Fig. 4.24. Output resistances of an npn current mirror and an n-channel current mirror with the current ratio $k_I = 1$, $I_i = I_o = 100\,\mu\text{A}$ and $R_1 = R_2$

This suggests that the maximum output resistance can be achieved with an ohmic feedback resistance if an output voltage limit of $V_{o,min} \approx V_{R2} + V_{CE,sat} \approx 2.8\,\text{V}$ is accepted. With lower current gains, the voltage limit is reduced accordingly.

– Due to the significantly lower transconductance of the MOSFETs in n-channel current mirrors, higher feedback resistances must be used in order to achieve output resistances that are as high as those in npn current mirrors; in this case, R_2 must be replaced by a current source, which means that the simple current mirror is converted to a cascode current mirror.

Current mirror with cascode

Where particularly high output resistances are required, either a very high ohmic resistance or a current source must be used for the feedback in simple current mirrors. However, the use of high-value ohmic resistances is generally not feasible due to the rapid increase in the output voltage limit $V_{o,min}$, so that it becomes necessary to use a current source. Since current sources are commonly realized by means of current mirrors, in the simplest case this leads to the *current mirror with cascode* as shown in Fig. 4.25, in which, according to the basic circuit of Fig. 4.10 on page 277, the feedback resistance R_E or R_S is replaced by a simple current mirror that consists of T_1 and T_2. On the output side this results in a series connection of a common-emitter or common-source circuit (T_2) and a common-base or common-gate circuit (T_3), which is called a *cascode circuit* (see Sect. 4.1.2).

It is important in this respect to note the difference between the *current mirror with cascode* described here and the *cascode current mirror* described in the next section. Both use a cascode circuit at the output, but different methods are applied for setting the operating point: in the current mirror with cascode an *external* voltage source V_B is used to set the operating point, while in the cascode current mirror the necessary voltage is generated internally.

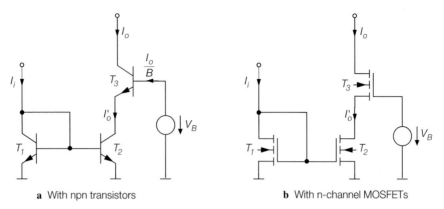

a With npn transistors   **b** With n-channel MOSFETs

Fig. 4.25. Current mirror with a cascode

npn current mirror with cascode: The current ratio k_I of the npn current mirror with cascode shown in Fig. 4.25a can be calculated with the help of the current ratio of the simple current mirror; for the current mirror that consists of T_1 and T_2 we obtain, from (4.6):

$$\frac{I_o'}{I_i} = \frac{1}{\frac{I_{S1}}{I_{S2}}\left(1+\frac{1}{B}\right) + \frac{1}{B}}$$

The Early effect has no influence, since T_2 is driven with the almost constant collector-emitter voltage $V_{CE2} = V_B - V_{BE3} \approx V_B - 0.7\,\text{V}$. Inserting

$$I_o' = I_o + \frac{I_o}{B}$$

leads to:

$$k_I = \frac{I_o}{I_i} = \frac{1}{\frac{I_{S1}}{I_{S2}}\left(1+\frac{1}{B}\right)^2 + \frac{1}{B} + \frac{1}{B^2}} \overset{B\gg1}{\approx} \frac{I_{S2}}{I_{S1}} \tag{4.21}$$

For $I_{S1} = I_{S2}$, it follows that:

$$k_I = \frac{1}{1 + \frac{3}{B} + \frac{2}{B^2}} \overset{B\gg1}{\approx} \frac{1}{1 + \frac{3}{B}} \approx 1$$

The current ratio only depends on the dimensional ratio of transistors T_1 and T_2; T_3 has no influence. As k_I is independent of the output voltage V_o, the first approximation of the output resistance is infinite.

n-channel current mirror with cascode: In the n-channel current mirror with cascode shown in Fig. 4.25b, $I_o = I_o'$; together with (4.18), this leads to

$$k_I = \frac{I_o}{I_i} = \frac{K_2}{K_1} \tag{4.22}$$

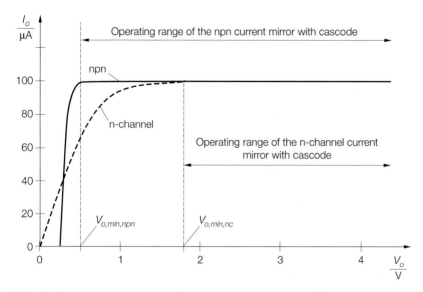

Fig. 4.26. Output characteristics of an npn and an n-channel current mirror with cascode

Here too, the current ratio only depends on the dimensional ratios of MOSFETs T_1 and T_2.

Output characteristics: Figure 4.26 shows the output characteristics of an npn and an n-channel current mirror with cascode. The characteristic of the npn current mirror with cascode is practically horizontal for $V_o > V_{o,min,npn}$, indicating that the output resistance is very high. With $V_{CE,sat} \approx 0.2\,\mathrm{V}$ and $V_{BE} \approx 0.7\,\mathrm{V}$, the output voltage limit is:

$$V_{o,min,npn} = V_B - V_{BE3} + V_{CE3,sat} \approx V_B - 0.5\,\mathrm{V}$$

$V_{CE2} > V_{CE2,sat}$ must be the case to ensure that T_2 operates in normal mode; consequently,

$$V_B = V_{CE2} + V_{BE3} > V_{CE2,sat} + V_{BE3} \approx 0.9\,\mathrm{V}$$

For the borderline case $V_B = 0.9\,\mathrm{V}$, the output voltage limit is $V_{o,min,npn} = 2V_{CE,sat} \approx 0.4\,\mathrm{V}$. The characteristic is inflected below the voltage limit.

For the n-channel cascode current mirror with $V_o > V_{o,min,nc}$, the characteristic is also horizontal; here, the output voltage limit is:

$$V_{o,min,nc} = V_B - V_{GS3} + V_{DS3,po} = V_0 - V_{th3}$$

Here, $V_{DS3,po} = V_{GS3} - V_{th3}$. To ensure that T_2 works in the pinch-off region, $V_{DS2} > V_{DS2,po}$ must be the case; thus:

$$V_B = V_{DS2} + V_{GS3} > V_{DS2,po} + V_{GS3} = V_{GS2} - V_{th2} + V_{GS3}$$

Here, $V_{DS2,po} = V_{GS2} - V_{th2}$. Typical values are $V_{th} \approx 1\,\mathrm{V}$ and $V_{GS} \approx 1.5\ldots2\,\mathrm{V}$; this results in $V_B \approx 2\ldots3\,\mathrm{V}$ and $V_{o,min,nc} \approx 1\ldots2\,\mathrm{V}$. For $I_{D2} = I_{D3} = I_o$ and

$$V_{GS} \approx V_{th} + \sqrt{\frac{2I_D}{K}}$$

the dependence of the output voltage limit on the output current and the sizes of the MOSFETs is:

$$V_{o,min,nc} = V_{GS2} - V_{th2} + V_{GS3} - V_{th3} = \sqrt{2I_o}\left(\frac{1}{\sqrt{K_2}} + \frac{1}{\sqrt{K_3}}\right)$$

This means that the voltage limit can be reduced by increasing the size of the MOSFETs; but the result is only influenced by the square root of the size.

Below the output voltage limit, T_3 enters the ohmic region. The current, however, is determined by T_2 and thus remains approximately constant; the output resistance, on the other hand, is markedly reduced. A further reduction in the output voltage also drives T_2 into the ohmic region and the characteristic becomes identical to the output characteristic of T_2.

Output resistance: The output resistance of the npn current mirror with cascode is calculated by inserting the small-signal parameters of T_3 and r_{CE2} in place of R_E in (4.1):

$$r_o = r_{CE3}\left(1 + \frac{\beta\, r_{CE2}}{r_{CE2} + r_{BE3}}\right)$$

If $r_{CE2} \approx r_{CE3} = V_A/I_o$, $r_{CE2} \gg r_{BE3}$ and $\beta \gg 1$, then:

$$r_o = \left.\frac{v_o}{i_o}\right|_{i_i=0} \approx \beta\, r_{CE3} \tag{4.23}$$

According to (4.2), for the n-channel current mirror with cascode we obtain:

$$r_o = r_{DS3}\left(1 + \left(g_{m3} + g_{m,B3}\right)r_{DS2}\right)$$

If $r_{DS2} = r_{DS3} = V_A/I_o$ and $g_{m3}r_{DS2} \gg 1$, then

$$r_o = \left.\frac{v_o}{i_o}\right|_{i_i=0} \approx \left(g_{m3} + g_{m,B3}\right)r_{DS3}^2 \approx \mu_3\, r_{DS3} \tag{4.24}$$

Cascode current mirror

Another way to increase the output resistance is to connect two simple current mirrors in series, as shown in Fig. 4.27. This is called a *cascode current mirror*, by analogy with the cascode circuit described in Sect. 4.1.2. It is closely related to the current mirror with cascode in Fig. 4.25. However, the cascode current mirror needs no external voltage source and therefore is also called the *self-biased cascode current mirror*. It also differs from the current mirror with cascode in the output voltage limit and output resistance.

npn cascode current mirror: The current ratio of the npn cascode current mirror shown in Fig. 4.27a can be calculated using the current ratio of the simple current mirror; according to (4.6), the ratio for the current mirror consisting of T_1 and T_2 is:

$$\frac{I_o'}{I_i'} = \frac{1}{\dfrac{I_{S1}}{I_{S2}}\left(1 + \dfrac{1}{B}\right) + \dfrac{1}{B}}$$

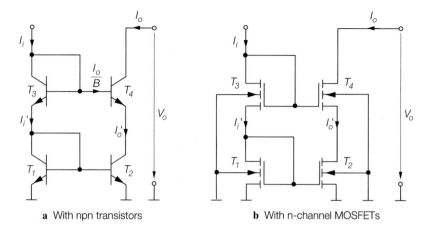

a With npn transistors   **b** With n-channel MOSFETs

Fig. 4.27. Cascode current mirror

The Early effect has no noticeable influence here, since T_2 is driven by the almost constant collector-emitter voltage $V_{CE2} = V_{BE1} + V_{BE3} - V_{BE4} \approx 0.7\,\text{V}$.

$$I_i = I_i' + \frac{I_o}{B} \quad , \quad I_o' = I_o + \frac{I_o}{B}$$

leads to:

$$k_I = \frac{I_o}{I_i} = \frac{1}{\dfrac{I_{S1}}{I_{S2}}\left(1 + \dfrac{1}{B}\right)^2 + \dfrac{2}{B} + \dfrac{1}{B^2}} \overset{B \gg 1}{\approx} \frac{I_{S2}}{I_{S1}} \tag{4.25}$$

For $I_{S1} = I_{S2}$, the current ratio is:

$$k_I = \frac{1}{1 + \dfrac{4}{B} + \dfrac{2}{B^2}} \overset{B \gg 1}{\approx} \frac{1}{1 + \dfrac{4}{B}} \approx 1$$

The current ratio only depends on the ratio of the sizes of transistors T_1 and T_2; T_3 and T_4 have no influence. As k_I is independent of the output voltage V_o, the first approximation of the output resistance is infinite.

n-channel cascode current mirror: In the n-channel cascode current mirror in Fig. 4.27b, $I_i = I_i'$ and $I_o = I_o'$; together with (4.18), this results in:

$$k_I = \frac{I_o}{I_i} = \frac{K_2}{K_1} \tag{4.26}$$

Here too, the current ratio depends solely on the dimensional ratio of MOSFETs T_1 and T_2.

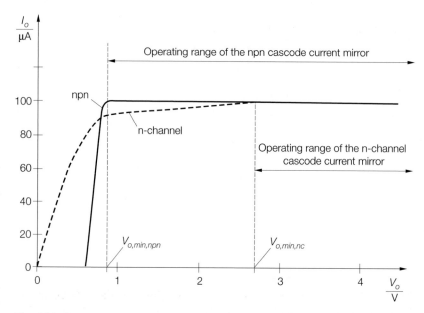

Fig. 4.28. Output characteristics of an npn and an n-channel cascode current mirror

Output characteristics: Figure 4.28 shows the output characteristics of an npn and an n-channel current mirror with cascode. The characteristic of the npn current mirror with cascode is practically horizontal for $V_o > V_{o,min,npn}$, indicating that the output resistance is very high. With $V_{CE,sat} \approx 0.2\,\text{V}$ and $V_{BE} \approx 0.7\,\text{V}$, the output voltage limit is:

$$V_{o,min,npn} = V_{BE1} + V_{BE3} - V_{BE4} + V_{CE4,sat} \approx 0.9\,\text{V}$$

It is higher than that of the current mirror with cascode, which reaches $V_{o,min,npn} \approx 0.4\,\text{V}$ at the minimum voltage V_B.

For $V_o > V_{o,min,nc}$, the characteristic of the n-channel cascode current mirror is also horizontal; the output voltage limit is:

$$V_{o,min,nc} = V_{GS1} + V_{GS3} - V_{GS4} + V_{DS4,po} = V_{GS1} + V_{GS3} - V_{th4}$$

Here, $V_{DS4,po} = V_{GS4} - V_{th4}$. Typical values are $V_{th} \approx 1\,\text{V}$ and $V_{GS} \approx 1.5\ldots 2\,\text{V}$; this results in $V_{o,min,nc} \approx 2\ldots 3\,\text{V}$. If we assume that all MOSFETs have the same threshold voltage V_{th}, that is, if we neglect the substrate effect, then for $I_{D1} = I_{D3} = I_i$ and

$$V_{GS} \approx V_{th} + \sqrt{\frac{2I_D}{K}}$$

the dependence of the output voltage limit on the input current and the sizes of the MOSFETs is:

$$V_{o,min,nc} \approx V_{th} + \sqrt{2I_i}\left(\frac{1}{\sqrt{K_1}} + \frac{1}{\sqrt{K_3}}\right)$$

This suggests that the voltage limit can be reduced by increasing the sizes of the MOSFETs; but only the square root of the size influences the result. The lower limit is given by $V_{o,min,nc} = V_{th}$ and can be approximately reached only with very large MOSFETs. Below

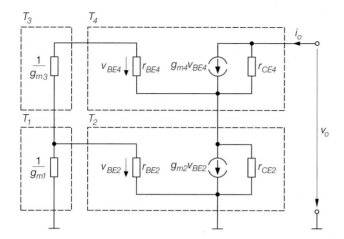

Fig. 4.29. Small-signal equivalent circuit of an npn cascode current mirror

the output voltage limit, transistor T_4 enters the ohmic region. The current, however, is influenced by T_2 and thus remains approximately constant, while the output resistance is markedly reduced. If the output voltage is further increased, T_2 also enters the ohmic region and the characteristic becomes identical to the output characteristic of T_2.

Output resistance: The small-signal equivalent circuit shown in Fig. 4.29 is used for calculating the output resistance of the npn cascode current mirror. The following relationships exist:

$$r_{CE2} \approx r_{CE4} = \frac{V_A}{I_o} \quad , \quad g_{m2} \approx g_{m4} = \frac{I_o}{V_T}$$

$$r_{BE2} \approx r_{BE4} = \frac{\beta V_T}{I_o} \quad , \quad g_{m1} \approx g_{m3} \approx \frac{I_i}{V_T} = \frac{I_o}{k_I V_T}$$

Here, V_A is the Early voltage, V_T is the temperature voltage, β is the small-signal current gain of the transistors and k_I is the current ratio of the current mirror. Calculating the output resistance with $k_I \ll \beta$ leads to:

$$r_o = \left.\frac{v_o}{i_o}\right|_{i_i=0} \approx r_{CE4}\left(1 + \frac{\beta}{1+k_I}\right) \approx \frac{\beta r_{CE4}}{1+k_I} \tag{4.27}$$

The output resistance of the cascode current mirror is higher than that of the simple current mirror by the factor $\beta/(1+k_I)$. The maximum possible output resistance βr_{CE} cannot be reached, since T_4 has an influence on the reference path and the voltage v_{BE2} via the base-emitter junction (see Fig. 4.29); therefore, current $g_{m2}v_{BE2}$ depends on the output voltage and the output resistance of T_2 is smaller than r_{CE2}.

In n-channel cascode current mirrors, there is no such influence on the reference path. Thus it is possible to calculate the output resistance using (4.20), by replacing R_2 by r_{DS2}:

$$r_o = r_{DS4}\left(1 + (g_{m4} + g_{m,B4})r_{DS2}\right)$$

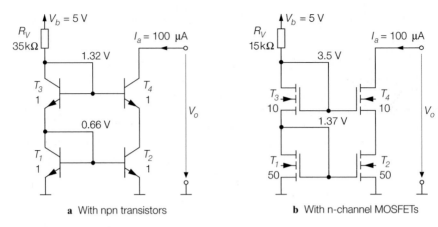

a With npn transistors   **b** With n-channel MOSFETs

Fig. 4.30. An example of a cascode current source

If $r_{DS2} = r_{DS4} = V_A/I_o$ and $g_{m4}r_{DS2} \gg 1$, then:

$$r_o = \left.\frac{v_o}{i_o}\right|_{i_i=0} \approx \left(g_{m4} + g_{m,B4}\right)r_{DS4}^2 \approx \mu_4 \, r_{DS4} \tag{4.28}$$

Example: An npn and an n-channel current source with an output current $I_o = 100\,\mu\text{A}$ is to be dimensioned for the highest possible output resistance and the lowest possible output capacitance. The demand for a high output resistance r_o requires the use of a cascode current mirror, while the demand for a low output capacitance necessitates the use of very small output transistors. With regard to the current ratio, there are contradictory requirements: on the one hand, it should be as large as possible so that only a low input current $I_i = I_o/k_I$ is necessary; on the other hand, it should be as small as possible to make the output resistance of the npn cascode current mirror as high as possible. $k_I \approx 1$ is selected for both current mirrors.

The circuit for the resulting npn cascode current mirror is shown in Fig. 4.30a. Here, transistors of size 1 are used, which are rated for a collector current of $100\,\mu\text{A}$ according to Fig. 4.5; the other parameters are $I_S = 1\,\text{fA}$, $B = \beta = 100$ and $V_A = 100\,\text{V}$. According to (4.25), the current ratio for $I_{S1} = I_{S2} = I_{S3} = I_{S4} = I_S$ is

$$k_I \approx \frac{1}{1 + \dfrac{4}{B}} = \frac{1}{1.04} \approx 0.96$$

and the input current is $I_i = I_o/k_I \approx 104\,\mu\text{A}$. Since the collector currents of the transistors are almost identical, the same base-emitter voltage V_{BE} can be calculated:

$$V_{BE} \approx V_T \ln \frac{I_o}{I_S} = 26\,\text{mV} \cdot \ln \frac{100\,\mu\text{A}}{1\,\text{fA}} \approx 660\,\text{mV}$$

The series resistor R_V is:

$$R_V = \frac{V_b - V_{BE\,1} - V_{BE\,3}}{I_i} \approx \frac{V_b - 2V_{BE}}{I_i} = \frac{3.68\,\text{V}}{104\,\mu\text{A}} \approx 35\,\text{k}\Omega$$

For $r_{CE4} = V_A/I_o = 100\,\text{V}/100\,\mu\text{A} = 1\,\text{M}\Omega$, the output resistance is:

$$r_o \approx \frac{\beta\, r_{CE4}}{1 + k_I} \approx \frac{\beta\, r_{CE4}}{2} \approx 50\,\text{M}\Omega$$

The output voltage limit is $V_{o,min} = V_{BE} + V_{CE,sat} \approx 0.9\,\text{V}$.

The resulting circuit for the n-channel cascode current mirror is shown in Fig. 4.30b. MOSFETs of size 10 according to Fig. 4.6 are used for T_3 and T_4, since size 1 is rated for a drain current of $10\,\mu\text{A}$ and in this case $100\,\mu\text{A}$ is required. MOSFETs of size 10 could also be used for T_1 and T_4; but here MOSFETs of size 50 are used to achieve a reduction of the voltage limit $V_{o,min}$. As the output capacitance is mainly dependent on T_4, the size of T_1 and T_2 has almost no bearing. $K = 30\,\mu\text{A}/\text{V}^2$ for size 1, $\gamma = 0.5\,\sqrt{\text{V}}$, $V_{inv} = 0.6\,\text{V}$ and $V_A = 50\,\text{V}$ can be taken from Fig. 4.6. The current ratio is $k_I = 1$; consequently, $I_i = I_o = 100\,\mu\text{A}$. For the MOSFETs:

$$K_1 = K_2 = 50\,K = 1.5\,\frac{\text{mA}}{\text{V}^2} \quad , \quad K_3 = K_4 = 10\,K = 300\,\frac{\text{uA}}{\text{V}^2}$$

In T_1 and T_2 the substrate effect has no influence, since $V_{BS1} = V_{BS2} = 0$; thus $V_{th1} = V_{th2} = V_{th,0}$ and:

$$V_{GS1} = V_{GS2} = V_{th,0} + \sqrt{\frac{2I_i}{K_1}} = 1\,\text{V} + \sqrt{\frac{200\,\mu\text{A}}{1.5\,\text{mA}/\text{V}^2}} \approx 1.37\,\text{V}$$

In contrast, the voltages in T_3 and T_4 are

$$V_{th3} = V_{th4} = V_{th,0} + \gamma\left(\sqrt{V_{inv} - V_{BS3}} - \sqrt{V_{inv}}\right)$$

$$\overset{V_{BS3}=V_{GS1}}{=} 1\,\text{V} + 0.5\,\sqrt{\text{V}} \cdot \left(\sqrt{1.97\,\text{V}} - \sqrt{0.6\,\text{V}}\right) \approx 1.31\,\text{V}$$

and:

$$V_{GS3} = V_{GS4} = V_{th3} + \sqrt{\frac{2I_i}{K_3}} \approx 1.31\,\text{V} + \sqrt{\frac{200\,\mu\text{A}}{300\,\mu\text{A}/\text{V}^2}} \approx 2.13\,\text{V}$$

Consequently, the series resistor is:

$$R_V = \frac{V_b - V_{GS1} - V_{GS3}}{I_i} \approx \frac{5\,\text{V} - 1.37\,\text{V} - 2.13\,\text{V}}{100\,\mu\text{A}} \approx 15\,\text{k}\Omega$$

For $r_{DS2} = r_{DS4} = V_A/I_o = 500\,\text{k}\Omega$ and

$$g_{m4} = \sqrt{2K_4 I_o} = \sqrt{2 \cdot 300\,\mu\text{A}/\text{V}^2 \cdot 100\,\mu\text{A}} \approx 245\,\frac{\mu\text{A}}{\text{V}}$$

$$g_{m,B4} = \frac{\gamma\, g_{m4}}{2\sqrt{V_{inv} - V_{BS4}}} \overset{V_{BS4}=-V_{GS2}}{=} \frac{0.5\,\sqrt{\text{V}} \cdot g_{m4}}{2\sqrt{1.97\,\text{V}}} \approx 44\,\frac{\mu\text{A}}{\text{V}^2}$$

the output resistance is:

$$r_o \approx \left(g_{m4} + g_{m,B4}\right) r_{DS4}^2 \approx 289\,\frac{\mu\text{A}}{\text{V}} \cdot (500\,\text{k}\Omega)^2 \approx 72\,\text{M}\Omega$$

The output voltage limit is:

$$V_{o,min} = V_{GS1} + V_{GS3} - V_{th4} \approx 1.37\,\text{V} + 2.13\,\text{V} - 1.31\,\text{V} \approx 2.2\,\text{V}$$

This means that with an operating voltage of 5 V, almost half of the operating voltage is lost.

The n-channel cascode current source has a higher output resistance which, however, comes with a disproportionately high output voltage limit, despite the fact that a reduction has already been made by increasing T_1 and T_2. If the same voltage limit as that for the npn cascode current source is required, one can only use a simple n-channel current source, which has a significantly lower output resistance of $r_o = r_{DS2} = 500\,\text{k}\Omega$; the npn cascode current source is thus superior by a factor of 100.

Furthermore, a comparison of the cascode current mirror with the simple current mirror with feedback is very interesting, provided that the output voltage limit is identical. The npn cascode current mirror features a voltage limit of $V_{o,min} = V_{BE} + V_{CE,sat}$, which is $V_{BE} \approx 0.7\,\text{V}$ higher than that of the simple npn current mirror without feedback; therefore, feedback can be complemented with $R_2 = V_{BE}/I_o \approx 7\,\text{k}\Omega$ in order to achieve the same voltage limit. The output resistance of the simple npn current mirror is thus:

$$r_o \approx r_{CE2}\left(1 + g_m R_2\right) = \frac{V_A}{I_o}\left(1 + \frac{I_o}{V_T}\frac{V_{BE}}{I_o}\right) \approx \frac{V_A V_{BE}}{V_T I_o} \approx 27\,\text{M}\Omega < 50\,\text{M}\Omega$$

In this way, the output resistance of the simple npn current mirror is smaller than that of the npn cascode current mirror, but only by a factor of 2; thus output resistances of the same magnitude are achieved in practice with both variants. In simple n-channel current mirrors, the voltage $V_{GS2} \approx 1.37\,\text{V}$ of the n-channel cascode current mirror is available for the feedback resistances if the same output voltage limits are also to be achieved here; consequently, $R_2 \approx 13.7\,\text{k}\Omega$ and:

$$r_o = r_{DS2}\left(1 + \left(g_m + g_{m,B}\right)R_2\right) \approx \left(g_m + g_{m,B}\right)R_2 r_{DS2}$$

$$\approx 289\,\frac{\mu\text{A}}{\text{V}} \cdot 13.7\,\text{k}\Omega \cdot 500\,\text{k}\Omega \approx 2\,\text{M}\Omega \ll 72\,\text{M}\Omega$$

The output resistance of the simple n-channel current mirror with feedback is thus significantly smaller than that of the n-channel cascode current mirror.

Wilson current mirror

In addition to the cascode current mirror, the *Wilson current mirror* shown in Fig. 4.31a, for which only three transistors are required, can also be used where high output resistances are needed. Compared to other current mirrors, the Wilson current mirror has one special characteristic; namely, the very low influence of the current gain on the current ratio when bipolar transistors are used. The Wilson current mirror is therefore a precision current mirror. It can certainly be made up of MOSFETs, but this will not improve the accuracy, due to the lack of a gate current in MOSFETs; its only advantage would be the high output resistance.

npn Wilson current mirror: In our calculations, we take advantage of the fact that the Wilson current mirror contains a simple npn current mirror. With the currents I_i' and I_o', the following applies:

$$\frac{I_o'}{I_i'} = \frac{1}{\dfrac{I_{S2}}{I_{S1}}\left(1 + \dfrac{1}{B}\right) + \dfrac{1}{B}}$$

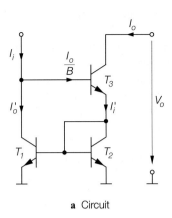

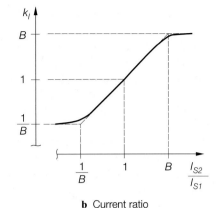

a Circuit   **b** Current ratio

Fig. 4.31. Wilson current mirror with npn transistors

Using

$$I_i = I_o' + \frac{I_o}{B} \quad , \quad I_i' = I_o + \frac{I_o}{B}$$

the current ratio is calculated:

$$k_I = \frac{I_o}{I_i} = \frac{B\left(\dfrac{I_{S2}}{I_{S1}} + \dfrac{1}{B+1}\right)}{\dfrac{I_{S2}}{I_{S1}} + B + \dfrac{1}{B+1}} \stackrel{B \gg 1}{\approx} \frac{B\,\dfrac{I_{S2}}{I_{S1}} + 1}{\dfrac{I_{S2}}{I_{S1}} + B} \qquad (4.29)$$

The size of transistor T_3 has no influence on k_I. Figure 4.31b shows the curve of k_I versus the size ratio I_{S2}/I_{S1}.

For $I_{S1} = I_{S2}$, we obtain:

$$k_I = \frac{1}{1 + \dfrac{2}{B^2 + 2B}} \stackrel{B \gg 1}{\approx} \frac{1}{1 + \dfrac{2}{B^2}}$$

Here, the error amounts only to $2/B^2$, compared to $2/B$ for the simple current mirror or $4/B$ for the cascode current mirror. The error of the three-transistor current mirror is also $2/B^2$, but only with the provision that all three transistors have the same current gain; but since T_3 in Fig. 4.20a carries a much lower current, in practice its current gain is lower than that of the other transistors. This is different in the Wilson current mirror, where $I_{S1} = I_{S2}$; that is, all transistors carry approximately the same current and all transistors have the maximum current gain, provided that they are correctly dimensioned. The fact that the error is smallest in the Wilson current mirror with $I_{S2}/I_{S1} = 1$ can be seen from the symmetry of the curve in Fig. 4.31b.

Output characteristic: The output characteristic of the Wilson current mirror corresponds to that of the cascode current mirror (see Fig. 4.28 on page 298); the output voltage

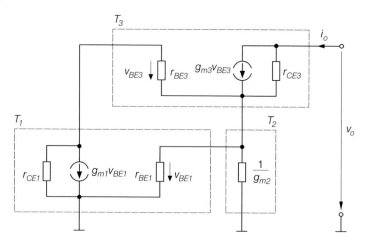

Fig. 4.32. Small-signal equivalent circuit of the Wilson current mirror

limit is also the same:

$$V_{o,min} = V_{BE} + V_{CE,sat} \approx 0.9\,\text{V}$$

Output resistance: The small-signal equivalent circuit shown in Fig. 4.32 is used to calculate the output resistance of the Wilson current mirror. The following equations apply:

$$r_{CE3} = \frac{V_A}{I_o} \quad , \quad r_{CE1} \approx \frac{V_A}{I_i} = \frac{k_I V_A}{I_o} = k_I r_{CE3}$$

$$g_{m2} \approx g_{m3} = \frac{I_o}{V_T} \quad , \quad g_{m1} \approx \frac{I_i}{V_T} = \frac{I_o}{k_I V_T} = \frac{g_{m3}}{k_I}$$

$$r_{BE3} = \frac{\beta V_T}{I_o} = \frac{\beta}{g_{m3}} \quad , \quad r_{BE1} \approx \frac{\beta V_T}{I_i} \approx \frac{k_I \beta V_T}{I_o} = \frac{k_I \beta}{g_{m3}}$$

Here, V_A is the Early voltage, V_T is the temperature voltage, β is the small-signal current gain of the transistors and k_I is the current ratio of the current mirror. For $\beta \gg 1$, the output resistance can be calculated as:

$$r_o = \left. \frac{v_o}{i_o} \right|_{i_i=0} \approx r_{CE3} \left(1 + \frac{\beta}{1 + k_I} \right) \approx \frac{\beta r_{CE3}}{1 + k_I} \overset{k_I=1}{=} \frac{\beta r_{CE3}}{2} \qquad (4.30)$$

A comparison with (4.27) shows that the output resistance of the Wilson current mirror is the same as that of the npn cascode current mirror.

Dynamic behavior

When a current mirror is used for signal transmission, not only the output resistance is of interest. Indeed, the frequency response of the current ratio and the step response at large signals are also interesting. However, a general calculation of the frequency responses is very complex, and the results are difficult to interpret because of the large number of parameters. Therefore, the basic dynamic response of current mirrors is described on the

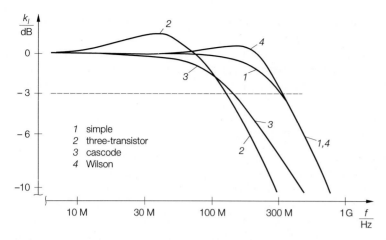

Fig. 4.33. Frequency responses of npn current mirrors with $k_I = 1$, with a small-signal short-circuit at the output

basis of simulation results. Four npn current mirrors are compared: the simple, the three-transistor, the cascode and the Wilson current mirrors, each with $k_I = 1$ and $I_o = 100\,\mu A$. Figure 4.33 shows the frequency responses for a small-signal short circuit at the output ($V_{o,A} = 5\,V$ or $v_o = 0$), while Fig. 4.34 shows the responses to a jump from $I_o = 10\,\mu A$ to $I_o = 100\,\mu A$.

It can be seen that the simple current mirror features the best dynamic characteristics, because it performs like a first-order lowpass filter. The Wilson current mirror has a somewhat higher cutoff frequency due to its complex conjugate poles, but at the cost of a step response, which has an overshoot of about 15%. The cutoff frequency of the cascode current mirror is lower than that of the simple current mirror by a factor of 2.5; consequently, the settling time is prolonged accordingly. The three-transistor current mirror shows the poorest response; it has the lowest cutoff frequency and an overshoot of more than 20%.

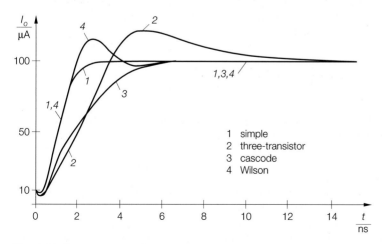

Fig. 4.34. Step responses of npn current mirrors

This is caused by the low quiescent current of transistor T_3 in Fig. 4.20a, which results in a correspondingly low transit frequency.

The numeric values of the cutoff frequency, the settling time and the overshoot naturally depend on the parameters of the transistors used. Other parameters would, of course, produce other values, but the relations among the various current mirrors are almost identical.

Other current mirrors and current sources

After very high output resistances have already been achieved with cascode and Wilson current mirrors, the main aim of other versions is to reduce the output voltage limit $V_{o,min}$. It may be possible to reduce the voltage limits of the cascode and the Wilson current mirrors slightly by increasing the size of the transistors excessively, but this method is ineffective and expensive due to the disproportionately high space requirements in integrated circuits. For this reason, current mirrors with $V_{o,min} \approx 2\,V_{CE,sat}$ or $V_{o,min} \approx 2\,V_{DS,po}$ have been developed.

Cascode current mirror with bias voltage: The *current mirror with bias voltage* shown in Fig. 4.35a is achieved by replacing transistor T_3 in the cascode current mirror of Fig. 4.27a on page 297 by a voltage source with $V_{CE,sat}$. From the equation $V_{CE,sat} + V_{BE1} = V_{CE2,sat} + V_{BE4}$ and $V_{BE1} \approx V_{BE4}$, it follows that $V_{CE2,sat} \approx V_{CE,sat}$ and thus:

$$V_{o,min} = V_{CE2,sat} + V_{CE4,sat} = 2\,V_{CE,sat} \approx 0.4\,V$$

With a constant input current – that is, operating the circuit as a current source – the bias voltage can be generated with a resistor (see Fig. 4.35b); if the base current of T_4 is neglected, then:

$$R_1 \approx \frac{V_{CE2,sat}}{I_i}$$

The current ratio and the output resistance remain almost unaltered (see (4.25) and (4.27)). Since the collector-emitter voltages of T_1 and T_2 are no longer practically identical as in the case of the cascode current mirror, the current ratio shows a slight dependence on the Early voltage of the transistors.

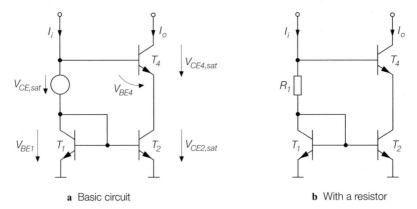

a Basic circuit   **b** With a resistor

Fig. 4.35. Cascode current mirror with bias voltage

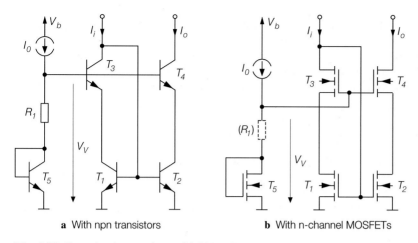

a With npn transistors   **b** With n-channel MOSFETs

Fig. 4.36. Cascode current mirror with bias voltage path

The same procedure can be used in n-channel cascode current mirrors, according to Fig. 4.27b. The following thus apply:

$$V_{o,min} = V_{DS2,po} + V_{DS4,po} = \sqrt{2I_o}\left(\frac{1}{\sqrt{K_2}} + \frac{1}{\sqrt{K_4}}\right)$$

and:

$$R_1 = \frac{V_{DS2,po}}{I_i}$$

The bias voltage may also be generated in a separate *bias voltage path* (see Fig. 4.36); in Fig. 4.36a the following must be the case:

$$V_V \approx V_{BE5} + I_0 R_1 > V_{CE2,sat} + V_{BE4}$$

and in Fig. 4.36b

$$V_V = V_{GS5} + I_0 R_1 > V_{DS2,po} + V_{GS4}$$

As the bias voltage is generated separately, the circuits, unlike that shown in 4.35b, may be operated with variable input currents – that is, as current mirrors – provided that they are dimensioned such that the conditions mentioned above are also met for the maximum current; in other words, with a maximum V_{BE4} or V_{GS4}. The circuits also function without transistor T_3; but then the collector-emitter or the drain-source voltages of T_1 and T_2 are no longer identical, and the current ratio shows a slight dependence on the Early voltage of the transistors. With the use of MOSFETs, R_1 can be omitted if I_0 is selected to be high enough and the size of T_5 small enough to make $V_{GS5} > V_{DS2,po} + V_{GS4}$.

Double cascode current mirror: Fig. 4.37a shows the *npn double cascode current mirror*: compared to the cascode current mirror, the collector of T_4 is connected to the bias voltage V_b and a second cascode with T_5 and T_6 is added. When T_5 and T_6 are operated with $V_{CE} > V_{CE,sat}$, the current ratio is

$$k_I = \frac{I_o}{I_i} \approx \frac{I_{S5}}{I_{S1}}$$

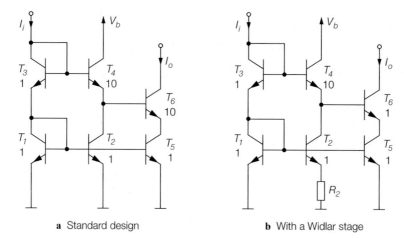

a Standard design   **b** With a Widlar stage

Fig. 4.37. npn double cascode current mirror

and the output resistance is:

$$r_o = \left.\frac{v_o}{i_o}\right|_{i_i=0} \approx \beta \, r_{CE6} = \frac{\beta \, V_A}{I_o}$$

Contrary to the cascode current mirror, there is no factor $(1+k_I)$ in the denominator, since any influence on the reference path by T_6 is prevented by T_4.

The sizes of the transistors can be selected such that T_5 is operated with $V_{CE5} \approx V_{CE,sat}$ and the following output voltage limit is reached:

$$V_{o,min} = V_{CE5,sat} + V_{CE6,sat} = 2\,V_{CE,sat} \approx 0.4\,\text{V}$$

Using the equation

$$V_{BE1} + V_{BE3} = V_{BE4} + V_{CE5} + V_{BE6}$$

under the conditions

$$I_{C1} \approx I_{C3} \approx I_i$$

$$I_{C4} \approx I_{C2} \approx I_i \frac{I_{S2}}{I_{S1}}$$

$$I_{C5} \approx I_{C6} = I_o = k_I I_i$$

and $V_{BE} \approx V_T \ln(I_C/I_S)$, we obtain:

$$V_{CE5} \approx V_T \ln \frac{I_{S4}I_{S6}}{k_I I_{S2} I_{S3}}$$

For the dimensional ratios in Fig. 4.37a, it follows:

$$V_{CE5} \approx V_T \ln \frac{10 \cdot 10}{1 \cdot 1 \cdot 1} = V_T \ln 100 \approx 26\,\text{mV} \cdot 4.6 \approx 120\,\text{mV}$$

Although the voltage is below the saturation voltage $V_{CE,sat} \approx 0.2\,\text{V}$ that has been assumed so far, it is sufficient in most practical applications. This becomes obvious if we consider the output resistance and the current ratio as a function of V_{CE5} (see Fig. 4.38): for $V_{CE} \approx 120\,\text{mV}$ the current ratio is almost one and the output resistance $r_o \approx 30\,\text{M}\Omega$

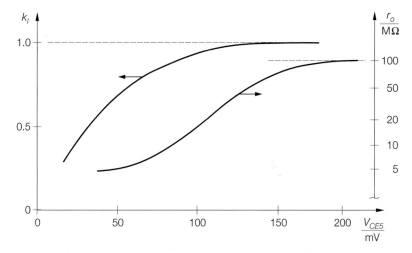

Fig. 4.38. Dependence of the current ratio k_I and the output resistance r_o on V_{CE5} in the npn double cascode current mirror

amounts to one third of the maximum possible value. Making $V_{CE} = 200\,\text{mV}$ will produce better values, but requires size 50 for T_4 and T_6:

$$V_{CE5} \approx V_T \ln \frac{50 \cdot 50}{1 \cdot 1 \cdot 1} = V_T \ln 2500 \approx 200\,\text{mV}$$

On account of their high spatial requirements, transistors of this size are only used in integrated circuits if it is absolutely essential for the function of the circuit. Generally, T_4 and T_5 are both of the same size, as this reduces the spatial requirements for the required value of V_{CE5} to a minimum.

A drawback of the circuit shown in Fig. 4.37a is the high output capacitance caused by the size of T_6. In order to reduce the size of T_6 by a factor of 10 to size 1, it is necessary to either increase the size of T_4 by a factor of 10 to size 100 or to reduce current $I_{C4} \approx I_{C2}$ by a factor of 10. The latter option is achieved by reducing T_2 by a factor of 10 or, if this is not feasible because T_2 is already at the minimum size, by increasing the size of all of the other transistors. If the current mirror is to be used as a current source, current I_{C2} may also be reduced by providing T_2 with a feedback resistance; this leads to the *double cascode current mirror with Widlar stage* shown in Fig. 4.37b.

In Fig. 4.37a, the collector of T_4 can also be used as an additional output; I_{C4} is thus the output current of a cascode current mirror with $k_I \approx I_{S2}/I_{S1}$, and I_{C6} is the output current of the double cascode current mirror with $k_I \approx I_{S5}/I_{S1}$.

Fig. 4.39 shows the *n-channel double cascode current mirror*. When T_5 and T_6 are driven with $V_{DS} > V_{DS,po}$, the current ratio is

$$k_I = \frac{I_o}{I_i} \approx \frac{K_5}{K_1}$$

and the output resistance is:

$$r_o = \left.\frac{v_o}{i_o}\right|_{i_i=0} \approx \left(g_{m6} + g_{m,B6}\right) r_{DS6}^2$$

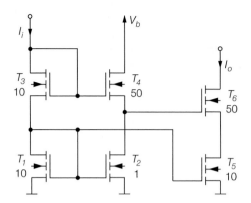

Fig. 4.39. n-channel double cascode current mirror

Neglecting the substrate transconductance $g_{m,B6}$ and using $g_{m6} = \sqrt{2K_6I_o}$ and $r_{DS6} = V_A/I_o$ leads to:

$$r_o \overset{g_{m,B6} \ll g_{m6}}{\approx} V_A^2 \sqrt{\frac{2K_6}{I_o^3}}$$

For the circuit in Fig. 4.39, where $K_6 = 50K = 1.5\,\text{mA/V}^2$, $V_A = 50\,\text{V}$ and $I_o = 100\,\mu\text{A}$, we obtain an output resistance of $r_o \approx 140\,\text{M}\Omega$.

The output voltage limit is minimum when T_5 is driven with $V_{DS5} = V_{DS5,po}$:

$$V_{o,min} = V_{DS5,po} + V_{DS6,po}$$

From the equation

$$V_{GS1} + V_{GS3} = V_{GS4} + V_{DS5} + V_{GS6}$$

under the conditions

$$V_{GS} = V_{th} + \sqrt{2I_D/K}$$

and $I_{D1} = I_{D3} = I_i$, $I_{D2} = I_{D4} = I_i K_2/K_1$ and $I_{D5} = I_{D6} = I_o = I_i K_5/K_1$, we obtain:

$$V_{DS5} = V_{th1} + V_{th3} - V_{th4} - V_{th6}$$
$$+ \sqrt{\frac{2I_o}{K_6}}\left(\sqrt{\frac{K_1 K_6}{K_3 K_5}} + \sqrt{\frac{K_6}{K_5}} - \sqrt{\frac{K_2 K_6}{K_4 K_5}} - 1\right)$$

For the circuit in Fig. 4.39, where $\Delta V_{th} = V_{th1} + V_{th3} - V_{th4} - V_{th6}$, we obtain:

$$V_{DS5} \approx \Delta V_{th} + \sqrt{\frac{2I_o}{K_6}}\left(\sqrt{5} + \sqrt{5} - \sqrt{0.1} - 1\right) \overset{\substack{K_6 = 1.5\,\text{mA/V}^2 \\ I_o = 100\mu\text{A}}}{\approx} \Delta V_{th} + 1.15\,\text{V}$$

Voltage ΔV_{th} combines the differences in the threshold voltages caused by the substrate effect; this voltage is always negative and cannot be calculated directly. Simulation with *PSpice* yields $\Delta V_{th} \approx -0.3\,\text{V}$ and $V_{DS5} = 0.85\,\text{V}$. It thus follows that:

$$V_{DS5} > V_{DS5,po} = \sqrt{\frac{2I_{D5}}{K_5}} = \sqrt{\frac{2I_o}{K_5}} \approx 0.82\,\text{V}$$

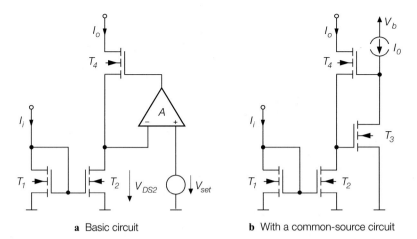

a Basic circuit   **b** With a common-source circuit

Fig. 4.40. Controlled n-channel cascode current mirror

For $V_{DS6,po} = V_{GS6} - V_{th6} = \sqrt{2I_0/K_6} \approx 0.37\,\text{V}$, the output voltage limit is $V_{o,min} = V_{DS5,po} + V_{DS6,po} \approx 1.2\,\text{V}$. A further reduction of $V_{o,min}$ is achieved by increasing the sizes of MOSFETs T_1, T_2 and T_5 proportionally; this decreases $V_{DS5,po}$ in accordance with the increase in K_5.

Controlled cascode current mirror: Removing MOSFET T_3 from the cascode current mirror in Fig. 4.27b and adjusting the gate voltage of T_4 by means of a control amplifier leads to the *controlled cascode current mirror* shown in Fig. 4.40a; with a sufficiently high gain A for the control amplifier, the gate voltage of T_4 is adjusted such that $V_{DS2} \approx V_{set}$. For $V_{set} \approx V_{DS2,po}$, the circuit represents a current mirror with minimum outut voltage limit $V_{o,min}$.

If a simple common-source circuit is used as the control amplifier, we obtain the circuit shown in Fig. 4.40b; voltage V_{set} occurs as the gate-source voltage of T_3 at the operating point:

$$V_{set} = V_{GS3} = V_{th3} + \sqrt{\frac{2I_0}{K_3}}$$

In general, all MOSFETs are operated with $V_{GS} < 2V_{th}$ and $V_{DS,po} = V_{GS} - V_{th} < V_{th}$; in this case, $V_{set} = V_{GS3} > V_{DS2,po}$; that is, T_2 operates in the pinch-off region. If V_{set} is to be kept small in order to achieve the lowest possible output voltage limit, then current I_0 must be small and MOSFET T_3 large; however, this makes the bandwidth of the control amplifier very narrow. In practice, a suitable compromise between voltage range and bandwidth must be found for each application.

The output resistance is calculated according to the small-signal equivalent circuit in Fig. 4.41:

$$r_o = \left.\frac{v_o}{i_o}\right|_{i_i=0} \approx r_{DS4}\left(1 + \left(g_{m4}(1+A) + g_{m,B4}\right)r_{DS2}\right) \overset{\substack{r_{DS2}=r_{DS4} \\ A\gg1}}{\approx} A g_{m4} r_{DS4}^2$$

The output resistance is thus larger than that of the cascode current mirror by a factor equal to the gain A. If a simple common-source circuit according to Fig. 4.40b is used as the

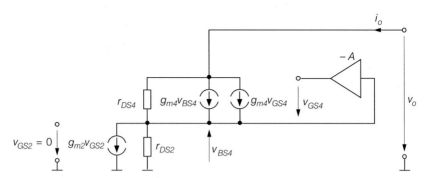

Fig. 4.41. Small-signal equivalent circuit of the controlled n-channel cascode current mirror

control amplifier, then $A = g_{m3}r_{DS3} = \sqrt{2K_3/I_0}\, V_A$; for $I_0 = 10\,\mu A$, $K_3 = 30\,\mu A/V^2$ (T_3 of size 1) and $V_A = 50\,V$, we obtain $A \approx 120$. This leads to output resistances in the GΩ range.

In principle, the controlled cascode current mirror can also be designed with npn transistors, but in this case it is not possible to use a simple common-emitter circuit as the control amplifier. This is due to the fact that correct functioning requires the input resistance $r_{i,amp}$ of the control amplifier to be larger than the output resistance of T_2 (r_{DS2} for MOSFET or r_{CE2} for bipolar transistor). This condition is automatically met in the case of MOSFETs, while bipolar transistors require complicated circuitry to achieve an input resistance $r_{i,amp}$ that is sufficiently high. Similar conditions exist at the output: with MOSFETs, transistor T_4 puts no load on the control amplifier, which can therefore have a high-resistance output, while with bipolar transistors the input resistance of T_4 requires a low-resistance amplifier output. A bipolar control amplifier must therefore consist of several stages. With an ideal amplifier ($r_{i,amp} = \infty$ and $r_{o,amp} = 0$), the same output resistance can be realized as in the controlled n-channel cascode current mirror: $r_o \approx A g_{m4}r_{CE4}^2$.

Current mirror for discrete circuits

In discrete circuits, we cannot rely on the given sizes of the transistors, as even within transistors of the same type the reverse saturation currents and the transconductance co-efficients may widely vary.[11] Therefore, it is essential to use feedback resistances and to adjust the current ratio by means of these resistances. Due to the higher Early voltage and the lower output voltage limit, bipolar transistors are used almost exclusively.

4.1.2
Cascode circuit

When calculating the cutoff frequencies of the common-emitter and common-source circuit according to Sects. 2.4.1 and 3.4.1, the *Miller effect* is a limiting factor. This effect is caused

---

[11] In the computer-aided design of discrete circuits, it is necessary to take into account the fact that in simulation all transistors of the same type have the same data, as the simulation always uses the same model. Therefore, the insensitivity toward parameter variations must be demonstrated by selective parameter variations in *individual* transistors. This can be done using, for example, the *Monte Carlo analysis*, which stochastically varies certain parameters.

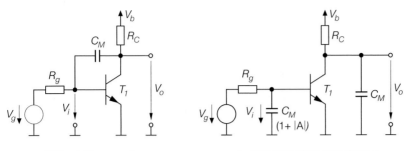

a With a Miller capacitance   **b** With equivalent capacitances

Fig. 4.42. Miller effect in a common-emitter circuit

by the voltage drop across the *Miller capacitance* C_M connected between the base and the collector or the gate and the drain:

$$v_i - v_o = v_i - A v_i \overset{A<0}{=} v_i(1 + |A|) = -v_o\left(1 + \frac{1}{|A|}\right) \overset{|A|\gg1}{\approx} -v_o$$

Here, $A < 0$ is the gain of the common-emitter or common-source circuit. Therefore, the Miller capacitance affects the input side with factor $(1 + |A|)$ and the output side with factor $(1 + 1/|A|) \approx 1$; Fig. 4.42 illustrates this, using a common-emitter circuit as an example.[12] The equivalent input capacitance $C_M(1 + |A|)$ and the internal resistance R_g of the signal source form a lowpass filter with a relatively low cutoff frequency; this significantly reduces the cutoff frequency of the circuit for medium and, in particular, high internal resistances. In bipolar transistors the collector capacitance C_C acts as the Miller capacitance, compared with the gate-drain capacitance C_{GD} in FETs.

This problem is overcome by the *cascode circuit*, in which a common-emitter (CE) and a common-base (CB) or a common-source (CS) and a common-gate (CG) circuit are connected in series; Fig. 4.43 shows the resulting circuits. At the operating point, the same current flows through both transistors if the base current of T_2 is ignored in the npn cascode circuit: $I_{C1,A} \approx I_{C2,A} \approx I_0$ or $I_{D1,A} = I_{D2,A} = I_0$. In the npn cascode circuit with

$$A = \frac{v_o}{v_i} = A_{CE}\frac{r_{i,CB}}{r_{o,CE} + r_{i,CB}}A_{CB}$$

$$= -g_{m1}r_{CE1}\frac{1/g_{m2}}{r_{CE1} + 1/g_{m2}}g_{m2}R_C \overset{r_{CE1}\gg1/g_{m2}}{\approx} -g_{m1}R_C$$

the same gain is obtained as in the simple common-emitter circuit. On the other hand, the operational gain of the common-emitter circuit in the cascode only amounts to:

$$A_{B,CE} \approx -g_{m1}r_{i,CB} = -g_{m1}/g_{m2} \approx -1$$

Therefore, the equivalent input capacitance is $C_M(1 + |A|) \approx 2C_M$; in other words, the Miller effect is avoided. No Miller effect occurs in the common-base circuit of the cascode, because the base of T_2 is at a constant voltage; the collector capacitance of T_2 thus only influences the output. These characteristics apply equally to the n-channel cascode circuit. However, the transconductances g_{m1} and g_{m2} are only identical if the MOSFETs are of the same size: $K_1 = K_2$.

---

[12] Please note that the voltages in Fig. 4.42 are large-signal voltages, but only the small-signal portion influences the calculated result.

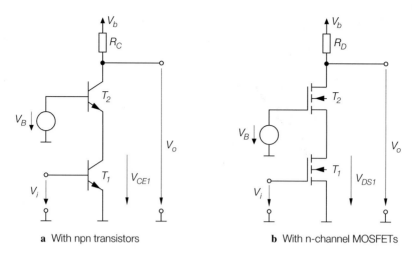

a With npn transistors   **b** With n-channel MOSFETs

Fig. 4.43. Cascode circuit

A voltage source V_B is required to set the operating point (see Fig. 4.43). Voltage V_B must be selected such that:

$$V_{CE1} = V_B - V_{BE2} > V_{CE1,sat} \quad \text{or} \quad V_{DS1} = V_B - V_{GS2} > V_{DS1,po}$$

T_1 thus operates in the normal mode or pinch-off region. Consequently[13]:

$$V_B > \begin{cases} V_{CE1,sat} + V_{BE2} \approx 0.8\ldots 1\,\text{V} \\ V_{DS1,po} + V_{GS2} = V_{GS1} - V_{th1} + V_{GS2} \approx 2\ldots 3\,\text{V} \end{cases}$$

V_B is selected as close to the lower limit as possible, in order to have a maximum output voltage range. In the npn cascode circuit, the voltage drop across two diodes is often used, that is, $V_B \approx 1.4\,\text{V}$ – if the resultant lower voltage range is acceptable.

Small-signal response of the cascode circuit

Cascode circuit with a simple current source: In integrated circuits, current sources are used instead of resistances R_C and R_D; Fig. 4.44 shows the resulting circuits when a simple current source is used. In this case the gain depends on the output resistances r_{oC} and r_{oS} of the cascode and the current source:

$$A = -g_{m1}(r_{oC} \| r_{oS})$$

The output resistance of the cascode corresponds to the output resistance of a current mirror with cascode (see (4.23) and (4.24)):[13]:

$$r_{oC} \approx \begin{cases} \beta_2 r_{CE2} \\ (g_{m2} + g_{m,B2})\, r_{DS2}^2 \quad \overset{g_{m2} \gg g_{m,B2}}{\approx} \quad g_{m2} r_{DS2}^2 \end{cases}$$

---

[13] The values for the npn and the n-channel cascode are given in one equation above one another, following after a brace.

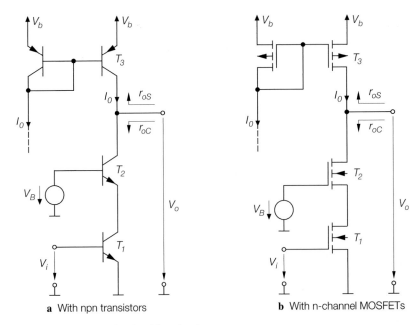

a With npn transistors   **b** With n-channel MOSFETs

Fig. 4.44. Cascode circuit with a simple current source

The output resistances for the simple current source are $r_{oS} = r_{CE3}$ or $r_{oS} = r_{DS3}$. This leads to the following equations for the cascode circuit with a simple current source:

cascode circuit with a simple current source

$$A = \frac{v_o}{v_i}\bigg|_{i_o=0} = -g_{m1}(r_{oC} \| r_{oS}) \overset{r_{oS} \ll r_{oC}}{\approx} \begin{cases} -g_{m1}r_{CE3} \\ -g_{m1}r_{DS3} \end{cases} \tag{4.31}$$

$$r_i = \frac{v_i}{i_i} = \begin{cases} r_{BE1} \\ \infty \end{cases} \tag{4.32}$$

$$r_o = \frac{v_o}{i_o}\bigg|_{v_i=0} = r_{oS} \| r_{oC} \overset{r_{oS} \ll r_{oC}}{\approx} \begin{cases} r_{CE3} \\ r_{DS3} \end{cases} \tag{4.33}$$

For the npn cascode with $g_{m1} \approx I_0/V_T$ and $r_{CE3} \approx V_{A,pnp}/I_0$, it follows:

$$A \approx -\frac{V_{A,pnp}}{V_T} \tag{4.34}$$

Here, $V_{A,pnp}$ is the Early voltage of the pnp transistor T_3 and V_T is the temperature voltage. For the n-channel cascode with $g_{m1} = \sqrt{2K_1 I_0}$ and $r_{DS3} = V_{A,pC}/I_0$, this leads to:

$$A \approx -V_{A,pC}\sqrt{\frac{2K_1}{I_0}} = -\frac{2V_{A,pC}}{V_{GS1} - V_{th,nC}} \tag{4.35}$$

Here, $V_{A,pC}$ is the Early voltage of the p-channel MOSFETs and $V_{th,nC}$ is the threshold voltage of the n-channel MOSFETs. If npn and pnp transistors or n-channel and p-channel

MOSFETs have the same Early voltage, the magnitude of the gain corresponds to the maximum gain μ of the common-emitter or common-source circuit:

$$|A| \approx \mu = \begin{cases} g_m \, r_{CE} = \dfrac{V_A}{V_T} \approx 1000 \ldots 6000 \\[3mm] g_m \, r_{DS} = \dfrac{2V_A}{V_{GS} - V_{th}} \approx 40 \ldots 200 \end{cases}$$

Here again, the low transconductance of the MOSFET compared to the bipolar transistor shows its negative effect.

Cascode circuit with a cascode current source: The gain increases further if the output resistance r_{oS} is enhanced by the use of a current source with cascode to:

$$r_{oS} \approx \begin{cases} \beta_3 r_{CE3} \\ (g_{m3} + g_{m,B3})\, r_{DS3}^2 \end{cases} \overset{g_{m3} \gg g_{m,B3}}{\approx} \quad g_{m3} r_{DS3}^2$$

This leads to the following equations for the *cascode circuit with a cascode current source* shown in Fig. 4.45:

cascode circuit with a cascode current source

$$A = \left. \frac{v_o}{v_i} \right|_{i_o = 0} = -g_{m1} r_o \approx \begin{cases} -g_{m1}\left(\beta_2 r_{CE2} \,\|\, \beta_3 r_{CE3}\right) \\ -g_{m1}\left(g_{m2} r_{DS2}^2 \,\|\, g_{m3} r_{DS3}^2\right) \end{cases} \qquad (4.36)$$

$$r_o = \left. \frac{v_o}{i_o} \right|_{v_i = 0} = r_{oS} \,\|\, r_{oC} \approx \begin{cases} \beta_2 r_{CE2} \,\|\, \beta_3 r_{CE3} \\ g_{m2} r_{DS2}^2 \,\|\, g_{m3} r_{DS3}^2 \end{cases} \qquad (4.37)$$

The input resistance r_i is determined by (4.32).

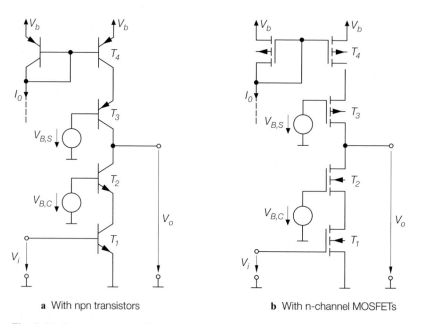

a With npn transistors   b With n-channel MOSFETs

Fig. 4.45. Cascode circuit with a cascode current source

Strictly speaking, the name *cascode circuit with a cascode current source* is not correct, since in Fig. 4.45 a current mirror with a cascode is used as a current source rather than a cascode current source; the correct term *cascode circuit with a current source with a cascode* is, however, too long-winded. If a *true* cascode current mirror is used for the current source, then the gain of the npn cascode is reduced by a factor of 2/3 since, according to (4.27), for a current ratio $k_I = 1$ the cascode current mirror has an output resistance of only $r_{oS} = \beta_3 r_{CE3}/2$ instead of $r_{oS} = \beta_3 r_{CE3}$ in the current mirror with a cascode. For the n-channel cascode, the two versions are equivalent.

In cascode circuits with bipolar transistors, replacing the small-signal parameters leads to

$$A \approx -\frac{1}{V_T \left(\dfrac{1}{\beta_{npn} V_{A,npn}} + \dfrac{1}{\beta_{pnp} V_{A,pnp}} \right)} \tag{4.38}$$

and for the cascode circuit with MOSFETs of the same size ($K_1 = K_2 = K_3 = K$):

$$A \approx -\frac{2K}{I_D \left(\dfrac{1}{V_{A,nC}^2} + \dfrac{1}{V_{A,pC}^2} \right)} = -\frac{4}{(V_{GS} - V_{th})^2 \left(\dfrac{1}{V_{A,nC}^2} + \dfrac{1}{V_{A,pC}^2} \right)} \tag{4.39}$$

If the Early voltages and current gains of the npn and pnp transistors and the Early voltages of the n- and p-channel MOSFETs are the same, then:

$$|A| \approx \begin{cases} \dfrac{\beta\, g_m r_{CE}}{2} = \dfrac{\beta\, V_A}{2V_T} \stackrel{\beta \approx 100}{\approx} 50.000 \ldots 300.000 \\[3mm] \dfrac{g_m^2 r_{DS}^2}{2} = 2\left(\dfrac{V_A}{V_{GS} - V_{th}} \right)^2 \approx 800 \ldots 20.000 \end{cases}$$

It is therefore possible to achieve a gain in the region of $10^5 = 100$ dB with *one* npn cascode circuit; in contrast, the n-channel cascode circuit yields a maximum of about $10^4 = 80$ dB.

Operational gain: The high gain of the cascode circuit is a result of the high output resistance of the cascode and the current source:

$$r_o = r_{oC} \,||\, r_{oS}$$

For $\beta = 100$, $V_A = 100$ V and $I_C = 100\,\mu$A, the output resistance of the npn cascode circuit with a cascode current source is $r_o = \beta\, r_{CE}/2 = 50$ MΩ, and for $K = 300\,\mu$A/V^2, $V_A = 50$ V and $I_D = 100\,\mu$A, the output resistance of the n-channel cascode circuit with a cascode current source is $r_o = g_m\, r_{DS}^2/2 = 31$ MΩ; the same values are assumed for the npn and pnp transistors, as well as for the n-channel and p-channel transistors.

With load R_L, the operational gain

$$A_B = A \frac{R_L}{r_o + R_L} = -g_m (r_o \,||\, R_L)$$

only reaches a value close to A if the value of R_L is comparable to that of r_o. In most cases, an additional amplifier stage with input resistance $r_{i,n}$ is connected to the output of the cascode circuit. If, in a CMOS circuit, the subsequent stage is a common-source or common-drain configuration, then the cascode circuit achieves the maximum operational gain $A_B = A$ without any special measures, since $R_L = r_{i,n} = \infty$. Bipolar circuits, on the other hand, require one or several common-collector circuits for impedance conversion;

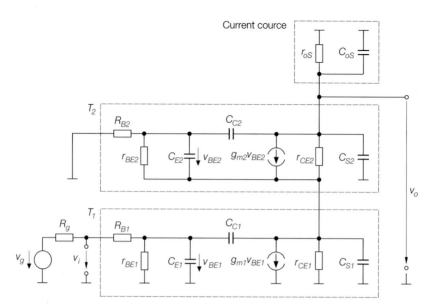

Fig. 4.46. Full small-signal equivalent circuit diagram of an npn cascode circuit

but the output resistance of each common-collector circuit is $r_o \approx R_g/\beta$, which means that each collector circuit reduces the output resistance by the factor β. For $\beta = 100$ and $r_o = 50\,\text{M}\Omega$, the output resistance is reduced to $r_o \approx 500\,\text{k}\Omega$ with one common-collector circuit and $r_o \approx 5\,\text{k}\Omega$ with two common-collector circuits. In many operational amplifiers, a cascode circuit with a cascode current source and three subsequent complementary common-collector circuits is used; this results in $A \approx 2 \cdot 10^5$ and $r_o \approx 50\,\Omega$.

Frequency Response and Cutoff Frequency of the Cascode Circuit

npn cascode circuit: Fig. 4.46 shows the full small-signal equivalent circuit of an npn cascode circuit with transistors T_1 and T_2 and the current source. The small-signal model from Fig. 2.41 on page 78 is used for the transistors, but here the substrate capacitance C_S is also taken into account. The current source is described by the output resistance r_{oS} and the output capacitance C_{oS}. To calculate the frequency response, the small-signal equivalent circuit is simplified as follows:

- The base spreading resistance R_{B2} of transistor T_2 is ignored.
- Resistances r_{CE1}, r_{CE2} and r_{oS} are replaced by the output resistance r_o, which has already been calculated (see (4.33) for use in a simplified current source or (4.37) for use in a current source with cascode).
- Capacitances C_{oS} and C_{S2} are combined to give C'_o.
- Resistances R_g and R_{B1} are combined to give R'_g
- The controlled source $g_{m2}v_{BE2}$ is replaced by two equivalent sources.

This results in the simplified small-signal equivalent circuit shown in the upper part of Fig. 4.47. Modification of the drawing leads to the equivalent circuit shown in the lower

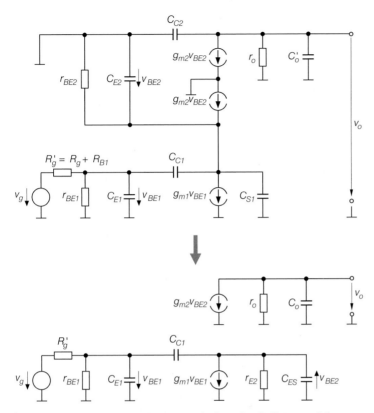

Fig. 4.47. A simplified small-signal equivalent circuit diagram of the npn cascode circuit

part of Fig. 4.47, where:

$$C_o = C_{C2} + C_o' = C_{C2} + C_{S2} + C_{oS} = C_{C2} + C_{S2} + C_{C3} + C_{S3}$$

$$C_{ES} = C_{E2} + C_{S1}$$

$$r_{E2} = 1/g_{m2} \| r_{BE\,2}$$

The simplification is almost equivalent. Only a minor error is caused by neglecting R_{B2}.

Splitting the small-signal equivalent circuit of Fig. 4.47 into input and output portions shows that the cascode circuit has no feedback; this eliminates the Miller effect. The frequency response is the product of the frequency responses $\underline{A}_1(s) = \underline{v}_{BE2}(s)/\underline{v}_g(s)$ and $\underline{A}_2(s) = \underline{v}_o(s)/\underline{v}_{BE2}(s)$:

$$\underline{A}_B(s) = \frac{\underline{v}_o(s)}{\underline{v}_g(s)} = \frac{\underline{v}_o(s)}{\underline{v}_{BE2}(s)} \frac{\underline{v}_{BE2}(s)}{\underline{v}_g(s)} = \underline{A}_2(s)\underline{A}_1(s) \qquad (4.40)$$

Without any load, the frequency response on the output side is:

$$\underline{A}_2(s) = \frac{\underline{v}_o(s)}{\underline{v}_{BE2}(s)} = -\frac{g_{m2}r_o}{1 + sC_o r_o}$$

On the input side, the small-signal equivalent circuit of the cascode circuit corresponds to that of a common-emitter configuration with an ohmic-capacitive load ($R_L = r_{E2}$,

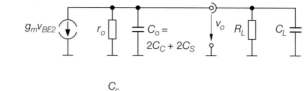

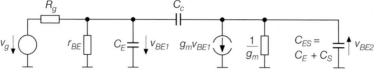

Fig. 4.48. A simplified small-signal equivalent circuit of the npn cascode circuit, with identical small-signal parameters for all transistors and an ohmic-capacitive load

$C_L = C_{ES}$), as a comparison with Fig. 2.82 on page 123 shows. If R'_C is replaced by $r_{E2}/(1 + sC_{ES}r_{E2})$, (2.86) on page 123 results in

$$\underline{A}_1(s) = \frac{g_{m1}r_{E2}}{1 + \dfrac{R'_g}{r_{BE\,1}}} \; \frac{1 - s\,\dfrac{C_{C1}}{g_{m1}}}{1 + sc_1 + s^2c_2}$$

$$c_1 = (C_{E1} + C_{C1}(1 + g_{m1}r_{E2}))\left(R'_g \| r_{BE\,1}\right) + \frac{C_{C1}r_{E2}r_{BE\,1}}{R'_g + r_{BE\,1}} + C_{ES}r_{E2}$$

$$c_2 = (C_{E1}C_{C1} + C_{E1}C_{ES} + C_{C1}C_{ES})\left(R'_g \| r_{BE\,1}\right)r_{E2}$$

where the polarity of $v_{BE\,2}$ is taken into account.

In this case $g_{m1} \approx g_{m2} \approx 1/r_{E2}$, since both transistors are operating with almost identical currents; consequently, $g_{m1}r_{E2} \approx 1$. By neglecting the zero, the s^2 term in the denominator and the middle term in c_1, one arrives at an approximation by a first-order lowpass filter:

$$\underline{A}_1(s) \approx \frac{r_{BE\,1}}{R'_g + r_{BE\,1}} \; \frac{1}{1 + s\left((C_{E1} + 2C_{C1})\left(R'_g \| r_{BE\,1}\right) + \dfrac{C_{ES}}{g_{m1}}\right)}$$

The small-signal equivalent circuit shown in Fig. 4.48 is achieved with $R'_g = R_g + R_{B1} \approx R_g$, an ohmic-capacitive load and on the assumption of identical small-signal parameters for all transistors. By combining $\underline{A}_1(s)$ and $\underline{A}_2(s)$ according to (4.40), again ignoring the s^2 term and replacing r_o by $r_o \| R_L$ and C_o by $C_o + C_L$, an approximation for the frequency response of the cascode circuit is achieved:

$$\underline{A}_B(s) \approx \frac{A_0}{1 + s\left((C_E + 2C_C)R_1 + \dfrac{C_E + C_S}{g_m} + (2C_C + 2C_S + C_L)R_2\right)}$$

$$\approx \frac{A_0}{1 + s((C_E + 2C_C)R_1 + (2C_C + 2C_S + C_L)R_2)} \tag{4.41}$$

$$A_0 = \underline{A}_B(0) = -\frac{\beta R_2}{R_g + r_{BE}} \tag{4.42}$$

$$R_1 = R_g \| r_{BE}$$

$$R_2 = r_o \| R_L$$

In (4.41), the approximation $R_1, R_2 \gg 1/g_m$ is used. The *-3 dB cutoff frequency* is:

$$\omega_{-3dB} = 2\pi f_{-3dB} \approx \frac{1}{(C_E + 2C_C)\left(R_g \| r_{BE}\right) + (2C_C + 2C_S + C_L)\left(r_o \| R_L\right)}$$

(4.43)

The cutoff frequency depends on the low-frequency gain A_0. On the assumption that a change in A_0 is caused by a change in $R_2 = r_o \| R_L$ and that all other values remain constant, a description with two time constants that are independent of A_0 is achieved by solving (4.42) for R_2 and inserting it in (4.43):

$$\omega_{-3dB}(A_0) = \frac{1}{T_1 + T_2|A_0|} \tag{4.44}$$

$$T_1 = (C_E + 2C_C)\left(R_g \| r_{BE}\right) \tag{4.45}$$

$$T_2 = (2C_C + 2C_S + C_L)\left(\frac{R_g}{\beta} + \frac{1}{g_m}\right) \tag{4.46}$$

Due to the high gain, $|A_0| \gg T_1/T_2$ is generally the case; consequently:

$$\omega_{-3dB} \approx \frac{1}{T_2|A_0|}$$

Thus, the cutoff frequency is inversely proportional to the gain, which leads to a constant *gain-bandwidth product* (*GBW*):

$$GBW = f_{-3dB}|A_0| \approx \frac{1}{2\pi\,T_2} \tag{4.47}$$

Two special cases are of particular interest:

– If an ohmic collector resistance R_C is used instead of a current source, the output capacitance $C_{oS} = C_C + C_S$ of the current source is eliminated and thus:

$$T_2 = (C_C + C_S + C_L)\left(\frac{R_g}{\beta} + \frac{1}{g_m}\right)$$

– If the cascode circuit is made up of discrete transistors, the substrate capacitances C_S are eliminated and thus:

$$T_2 = \left(\frac{R_g}{\beta} + \frac{1}{g_m}\right) \cdot \begin{cases} (C_C + C_L) & \text{with collector resistance } R_C \\ (2C_C + C_L) & \text{with current source} \end{cases}$$

Comparison of an npn cascode circuit and a common-emitter circuit: A meaningful comparison of the frequency response of cascode and common-emitter circuits is only possible on the basis of the gain-bandwidth product, since the gain factors with (a) collector resistance R_C, (b) a simple current source and (c) a cascode current source differ and the cutoff frequency is generally lower for higher gains. The gain-bandwidth product *GBW*, on the other hand, is independent of the gain. In what follows, the time constant T_2 will be compared instead of GBW, since this is easier to illustrate (see (4.47)): a smaller time constant T_2 gives a higher GBW and thus a higher cutoff frequency at a given gain.

In discrete circuits with a collector resistance, the time constant T_2 of the common-emitter (CE) circuit is calculated using (2.96) on page 125:[14]

$$T_{2,CE} = \left(C_C + \frac{C_L}{\beta}\right) R_g + \frac{C_C + C_L}{g_m} \overset{C_L=0}{=} C_C \left(R_g + \frac{1}{g_m}\right)$$

In the cascode circuit (CC), T_2 is calculated using (4.46) with $C_S = 0$; that is, eliminating the substrate capacitance, which does not exist in discrete transistors:

$$T_{2,CC} = (C_C + C_L)\left(\frac{R_g}{\beta} + \frac{1}{g_m}\right) \overset{C_L=0}{=} C_C\left(\frac{R_g}{\beta} + \frac{1}{g_m}\right)$$

It can be seen that, especially in the case of a high generator resistance R_g and a low load capacitance C_L, the cascode circuit yields a significantly smaller time constant and thus a higher GBW than the common-emitter circuit. If the generator resistance is very low ($R_g < 1/g_m$) or the load capacitance very high ($C_L > \beta C_C$), the cascode is not advantageous.

In an integrated circuit with current sources, it is necessary to modify the time constant of the common-emitter circuit by taking into consideration the substrate capacitance C_S of the transistor and the capacitance $C_{oS} = C_C + C_S$ of the current source. These act like an additional load capacitance and can therefore be taken into account by replacing C_L with $C_C + 2C_S + C_L$:

$$T_{2,CE} = \left(C_C + \frac{C_C + 2C_S + C_L}{\beta}\right) R_g + \frac{2C_C + 2C_S + C_L}{g_m}$$

Equation (4.46) is used for the cascode circuit:

$$T_{2,CC} = (2C_C + 2C_S + C_L)\left(\frac{R_g}{\beta} + \frac{1}{g_m}\right)$$

For $\beta \gg 1$, this leads to

$$T_{2,CE} \approx T_{2,CC} + C_C R_g \tag{4.48}$$

Here too, the cascode circuit offers a lower time constant and thus a higher GBW. But since $C_S \gg C_C$ almost always applies to integrated circuits, the gain in GBW achieved with the use of a cascode circuit instead of a common-emitter circuit is clearly less than in discrete circuits, even with a high generator resistance R_g and without a load capacitance C_L; a typical factor is between 2 and 3. Therefore, in practice it is very often the higher gain of the cascode circuit – especially in combination with a current source with cascode – and not the higher cutoff frequency that is decisive.

Finally, the circuits shown in Fig. 4.49 will be compared. For very high frequencies the corresponding responses are not shown, since they deviate from the asymptote due to the neglected zero and pole, so that it is not possible to use the GBW to calculate the cutoff frequency. The calculation of the low-frequency gain is based on parameters $\beta = 100$ and $V_A = 100\,\text{V}$ for npn and pnp transistors and on $R_g = 0$ and $R_L \to \infty$. The gain of the cascode circuit with a simple current source is $|A| = V_A/V_T = 4000 = 72\,\text{dB}$ and the gain of the cascode circuit with a cascode current source is $|A| = \beta V_A/(2V_T) = 200000 =$

---

[14] Based on $R'_g = R_g + R_B \approx R_g$.

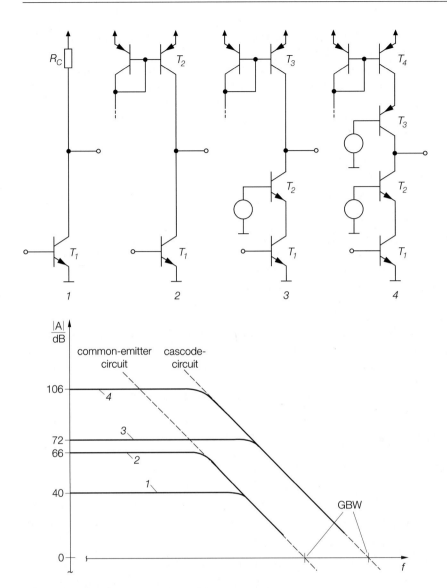

Fig. 4.49. Comparison of circuits and frequency responses

106 dB. By way of comparison, $|A| = V_A/(2V_T) = 2000 = 66$ dB in the common-emitter circuit with a simple current source;[15] $|A| = 100 = 40$ dB is considered to be a typical value for the common-emitter circuit with a collector resistance. A circuit comparison

---

[15] With an *ideal* current source, the common-emitter circuit provides its maximum gain $|A| = \mu = V_A/V_T$. The use of a simple current source with a transistor with the same parameters reduces the output resistance from r_{CE} to $r_{CE} \parallel r_{CE} = r_{CE}/2$; this cuts the gain in half. In a common-emitter circuit with a cascode current source (not shown in Fig.. 4.49), the output resistance of the current source is negligible; therefore, the gain $|A| = V_A/V_T$ has the same value as in the cascode circuit with a simple current source.

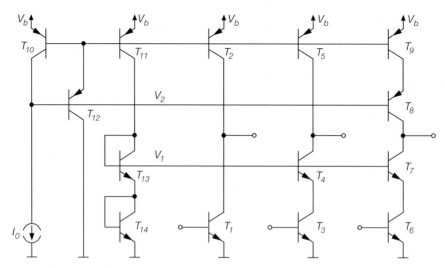

Fig. 4.50. An example of the common-emitter and cascode circuits (all transistors of size 1)

shows that the characteristics that continuously improve from circuit 1 to circuit 4 are achieved by means of additional transistors.

Example: Circuits 2, 3 and 4 in Fig. 4.49 are operated with a quiescent current $I_0 = 100\,\mu\text{A}$ and an operating voltage $V_b = 5\,\text{V}$; Fig. 4.50 shows the circuits with the additions required for setting the operating point:

– The common-emitter circuit with a simple current source (T_1 and T_2).
– The cascode circuit with a simple current source ($T_3 \ldots T_5$).
– The cascode circuit with a cascode current source ($T_6 \ldots T_9$).

The quiescent current is adjusted with a three-transistor current mirror ($T_{10} \ldots T_{12}$) which, together with transistors T_2, T_5 and T_9, forms a current source bank that mirrors the reference current I_0 to four outputs. The current through transistor T_{11} is fed through transistors T_{13} and T_{14}, which operate as diodes, and generates the bias voltage $V_1 = 2V_{BE} \approx 1.4\,\text{V}$ for transistors T_4 and T_7. The bias voltage for transistor T_8 can be taken from the three-transistor current mirror: $V_2 = V_b - 2V_{BE} \approx V_b - 1.4\,\text{V} = 3.6\,\text{V}$. In the most simple case, the current source with reference current I_0 can be realized with a resistance $R = V_2/I_0 \approx 3.6\,\text{V}/100\,\mu\text{A} = 36\,\text{k}\Omega$.

If we neglect the base currents, then $I_{C,A} \approx I_0 = 100\,\mu\text{A}$ for transistors $T_1 \ldots T_9$; thus $g_m = I_{C,A}/V_T \approx 3.85\,\text{mS}$. With the parameters shown in Fig. 4.5 on page 274 for the npn transistors, it follows that $r_{BE,npn} = \beta_{npn}/g_m \approx 26\,\text{k}\Omega$ and $r_{CE,npn} = V_{A,npn}/I_{C,A} \approx 1\,\text{M}\Omega$; for the pnp transistors it follows that $r_{CE,pnp} = V_{A,pnp}/I_{C,A} \approx 500\,\text{k}\Omega$. For junction capacitances, the approximation

$$C_S(U) \approx \begin{cases} C_{S0} & \text{in the reverse region} \\ 2C_{S0} & \text{in the forward region} \end{cases}$$

is used instead of (2.37) on page 69; this means that the voltages at the junction capacitances no longer have to be determined, which is otherwise the case for evaluating (2.37). The collector and substrate diodes operate in the reverse region; therefore:

$$C_C \approx C_{J0,C} \quad , \quad C_S \approx C_{J0,S} \tag{4.49}$$

The parameters from Fig. 4.5 lead to $C_{C,npn} \approx 0.2\,\text{pF}$, $C_{C,pnp} \approx 0.5\,\text{pF}$, $C_{S,npn} \approx 1\,\text{pF}$ and $C_{S,pnp} \approx 2\,\text{pF}$. The emitter capacitance is the combination of the emitter junction capacitance in the forward region and the diffusion capacitance:

$$C_E = C_{J,E} + C_{D,N} \approx 2C_{J0,E} + \frac{\tau_{0,N} I_{C,A}}{V_T} \tag{4.50}$$

For the npn transistors, we obtain $C_E \approx 0.6\,\text{pF}$.

The circuits are to be operated by a signal source with $R_g = 10\,\text{k}\Omega$ and without any load ($R_L \rightarrow \infty$, $C_L = 0$). For the cascode circuit with a cascode current source, this leads to

$$A_0 = -\frac{\beta_{npn} \left(\beta_{npn} r_{CE,npn} \| \beta_{pnp} r_{CE,pnp}\right)}{R_g + r_{BE,npn}} \approx -56.000$$

and for the cascode circuit with a simple current source it leads to:

$$A_0 = -\frac{\beta_{npn} \left(\beta_{npn} r_{CE,npn} \| r_{CE,pnp}\right)}{R_g + r_{BE,npn}} \approx -1400$$

Equation (4.46) applies to both cascode circuits (CC):

$$T_{2,CC} = \left(C_{C,npn} + C_{C,pnp} + C_{S,npn} + C_{S,pnp}\right)\left(\frac{R_g}{\beta_{npn}} + \frac{1}{g_m}\right) \approx 1.3\,\text{ns}$$

From (2.87) and (4.48) for the common-emitter (CE) circuit with a simple current source, it follows:

$$A_0 = -\frac{r_{BE,npn}}{R_g + r_{BE,npn}} g_m \left(r_{CE,npn} \| r_{CE,pnp}\right)$$

$$= -\frac{\beta_{npn} \left(r_{CE,npn} \| r_{CE,pnp}\right)}{R_g + r_{BE,npn}} \approx -900$$

$$T_{2,CE} \approx T_{2,CC} + R_g C_{C,npn} \approx 3.3\,\text{ns}$$

From (4.47), it follows that the gain-bandwidth product for the cascode circuit is $GBW \approx 122\,\text{MHz}$, and for the common-emitter circuit $GBW \approx 48\,\text{MHz}$. With a load capacitance $C_L = 10\,\text{pF}$, the time constants are $T_{2,CC} \approx 4.9\,\text{ns}$ and $T_{2,CE} \approx 6.9\,\text{ns}$; for the cascode circuit this results in $GBW \approx 32\,\text{MHz}$ and for the common-emitter circuit $GBW \approx 23\,\text{MHz}$. This shows that the advantage of the cascode circuit becomes unimportant with an increasing load capacitance and for:

$$C_L \left(\frac{R_g}{\beta} + \frac{1}{g_m}\right) \gg C_C R_g$$

The only remaining advantage is the higher gain.

In discrete circuits, the advantage of the cascode circuit is greater due to the lack of substrate capacitances. For $R_g = 10\,\text{k}\Omega$ and no load ($R_L \rightarrow \infty$, $C_L = 0$), with $C_{S,npn} = C_{S,pnp} = 0$ and with the other parameters identical, we obtain $T_{2,CC} \approx 0.25\,\text{ns}$ and $T_{2,CE} \approx 2.25\,\text{ns}$. Thus the discrete cascode circuit with $GBW \approx 637\,\text{MHz}$ achieves a value in the range of the transit frequency of the transistors, while the discrete common-emitter circuit only achieves $GBW \approx 71\,\text{MHz}$. However, with a load capacity, the advantage of the discrete cascode circuit diminishes rapidly.

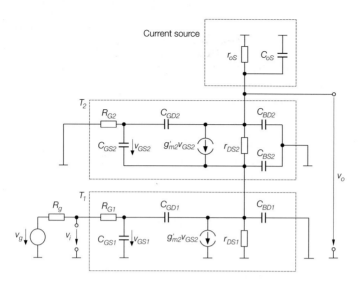

Fig. 4.51. Full small-signal equivalent circuit of an n-channel cascode circuit

n-channel cascode circuit: Figure 4.51 shows the full small-signal equivalent circuit of an n-channel cascode circuit with MOSFETs T_1 and T_2 and a current source. The small-signal model according to Fig. 3.48 on page 218 is used for the MOSFETs; this model does not contain the controlled sources with substrate transconductances $g_{m,B1}$ and $g_{m,B2}$, since:

- For T_1, the source $g_{m,B1} v_{BS1}$ is ineffective because $v_{BS1} = 0$.
- For T_2, the controlled sources $g_{m2} v_{GS2}$ and $g_{m,B2} v_{BS2}$ can be combined to give one source with $g'_{m2} = g_{m2} + g_{m,B2}$.[16]

The current source is described by the output resistance r_{oS} and the output capacitance C_{oS}. A comparison with the small-signal equivalent circuit for the npn cascode circuit in Fig. 4.46 yields, in addition to the usual correlations of parameters ($R_B = R_G, r_{BE} \to \infty$, $C_E = C_{GS}$, etc.), the following correlations:

$$C_{S1} = C_{BD1} + C_{BS2} \quad , \quad C_{S2} = C_{BD2}$$

This enables the results of the npn cascode circuit to be transferred to the n-channel cascode circuit; from (4.43) with $R_g, R_L \gg 1/g_m$, it follows

$$\omega_{\text{-3dB}} = 2\pi f_{\text{-3dB}} \approx \frac{1}{(C_{GS} + 2C_{GD}) R_g + (2C_{GD} + 2C_{BD} + C_L)(r_o \parallel R_L)} \tag{4.51}$$

and, from (4.44)–(4.46)

$$\omega_{\text{-3dB}}(A_0) = \frac{1}{T_1 + T_2 |A_0|} \tag{4.52}$$

$$T_1 = (C_{GS} + 2C_{GD}) R_g \tag{4.53}$$

---

[16] Under static conditions $v_{GS2} = v_{BS2}$, since there is no DC voltage drop across R_{G2}. The same also applies to the dynamic condition, since R_{G2} is ignored in the course of the calculation.

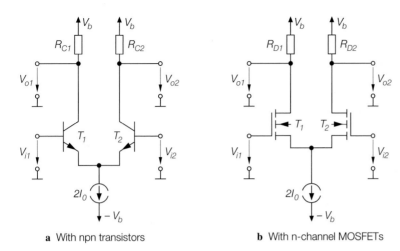

a With npn transistors   **b** With n-channel MOSFETs

Fig. 4.52. Basic circuit of a differential amplifier

$$T_2 = \frac{2C_{GD} + 2C_{BD} + C_L}{g_{m1}} \tag{4.54}$$

with the low-frequency gain:

$$A_0 = \underline{A}_B(0) = -g_{m1}(r_o \| R_L) \tag{4.55}$$

In the n-channel cascode circuit, the low-frequency gain and the time constant T_2 are not influenced by the internal resistance R_g of the signal source due to the infinitely high input resistance ($r_i = \infty$).

4.1.3
Differential Amplifier

The *differential amplifier* is a symmetrical amplifier with two inputs and two outputs. It consists of two common-emitter or two common-source circuits, where the emitter or source connections are connected to a common current source; Fig. 4.52 shows the basic circuit. In general, the differential amplifier is operated with a positive and a negative supply voltage that are often – but not necessarily – symmetrical, as shown in Fig. 4.52. If only one positive or negative supply voltage is available, ground can be used as the second supply voltage; this will be described in more detail below. In integrated differential amplifiers with MOSFETs, the bulk connections of the n-channel MOSFETs are connected to the negative supply voltage and those of the p-channel MOSFETs to the positive supply voltage, while in discrete MOSFETs the bulk connections are connected to the source of the given MOSFET.

Due to the current source, the sum of the currents remains constant:[17]

$$2I_0 = \begin{cases} I_{C1} + I_{B1} + I_{C2} + I_{B2} \approx I_{C1} + I_{C2} & \text{with } B = I_C/I_B \gg 1 \\ I_{D1} + I_{D2} \end{cases}$$

---

[17] The upper and lower lines behind the brace apply to the npn and the n-channel differential amplifiers, respectively.

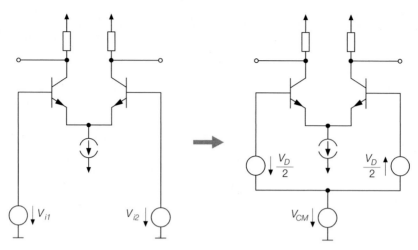

Fig. 4.53. Replacement of the input voltages V_{i1} and V_{i2} by the common-mode voltage V_{CM} and the differential voltage V_D

For the explanations set out below, the following assumption is made: $R_{C1} = R_{C2} = R_C$ and $R_{D1} = R_{D2} = R_D$. Furthermore, the input voltages V_{i1} and V_{i2} are replaced by the symmetric *common-mode voltage* V_{CM} and the skew-symmetric *differential voltage* V_D:

$$V_{CM} = \frac{V_{i1} + V_{i2}}{2} \quad , \quad V_D = V_{i1} - V_{i2} \tag{4.56}$$

Thus:

$$V_{i1} = V_{CM} + \frac{V_D}{2} \quad , \quad V_{i2} = V_{CM} - \frac{V_D}{2} \tag{4.57}$$

Figure 4.53 shows the replacement of V_{i1} and V_{i2} by the symmetric voltage V_{CM} and the skew-symmetric voltage V_D; according to (4.57), the latter leads to two sources with voltage $V_D/2$.

Common-mode and differential gain: For the same input voltages ($V_{i1} = V_{i2} = V_{CM}$, $V_D = 0$), the operation is symmetrical and the current of the current source is split up into equal portions through both transistors:

$$I_{C1} = I_{C2} \overset{B \gg 1}{\approx} I_0 \quad \text{or} \quad I_{D1} = I_{D2} = I_0$$

The output voltages are thus:

$$V_{o1} = V_{o2} \approx V_b - I_0 R_C \quad \text{or} \quad V_{o1} = V_{o2} = V_b - I_0 R_D$$

The variations of the common-mode voltage V_{CM} are called common-mode voltage range and do not change the current distribution as long as the transistors and the current source are not overloaded; this means that in common-mode operation the output voltages remain constant. The *common-mode gain*

$$A_{CM} = \left.\frac{d V_{o1}}{d V_{CM}}\right|_{V_D=0} = \left.\frac{d V_{o2}}{d V_{CM}}\right|_{V_D=0} \tag{4.58}$$

is zero under ideal conditions. In practice, it is slightly negative: A_{CM} ranges between approximately $A_{CM} \approx -10^{-4}...-1$. This is caused by the finite internal resistance of real current sources; the explanation of the small-signal response will go into this in more detail.

For skew-symmetric signals with a differential voltage V_D, the current distribution changes; this also alters the output voltages. This kind of signal is called *differential signal* and the related gain is called the *differential gain*:

$$A_D = \left. \frac{dV_{o1}}{dV_D} \right|_{V_{CM}=\text{const.}} = - \left. \frac{dV_{o2}}{dV_D} \right|_{V_{CM}=\text{const.}} \tag{4.59}$$

This value is negative: it is in the range of $A_D \approx -10...-100$ when ohmic resistances R_C and R_D are used, as in Fig. 4.52, and in the range of $A_D \approx -100 ... -1000$ when current sources are used instead of the resistances.

The ratio of the differential gain and the common-mode gain is called the *common-mode rejection ratio* (*CMRR*):

$$G = \frac{A_D}{A_{CM}} \tag{4.60}$$

Ideally, $A_{CM} \rightarrow -0$ and thus $G \rightarrow \infty$. In reality, differential amplifiers achieve $G \approx 10^3 ... 10^5$ depending on the internal resistance of the current source.[18] The range for G is not as wide as could be assumed on the basis of the extremes of A_{CM} and A_D; this is due to coupling between A_{CM} and A_D, which creates an upper and lower limit for G.

Characteristics of the differential amplifier: On account of its behavior, the central characteristic of the differential amplifier can be defined as follows:

> The differential amplifier amplifies the differential voltage between the two inputs independent of the common-mode voltage, as long as the latter does not exceed a permissible range.

This means that, within a permissible range, the output voltages do not depend on the common-mode voltage V_{CM} but are influenced by the current of the current source. Thus, the operating point for small-signal operation is also largely independent on V_{CM}. A variation of V_{CM} may cause other voltages to vary, but the output voltages and the currents that determine the operating point remain almost constant. This characteristic distinguishes the differential amplifier from all other amplifiers discussed so far, and facilitates both the setting of the operating point and the coupling of multi-stage amplifiers; special circuits for adjusting the DC voltage level or coupling capacitors are not required.

Another advantage of the differential amplifier is the suppression of changes in both branches caused by temperature variations, since these behave similarly to common-mode signals; only the temperature sensitivity of the current source, should this exist, affects the output voltages. In addition, component tolerances are effectively rejected in integrated circuits because of the equidirectional tolerances of the transistors and resistors of the differential amplifier, which are located close to one another.

---

[18] In the differential amplifiers examined here, G is positive, while A_{CM} and A_D are negative. There are, however, cases in which A_{CM} and A_D do not have the same polarity; therefore, sometimes only the magnitude of G is quoted even though G is a signed parameter.

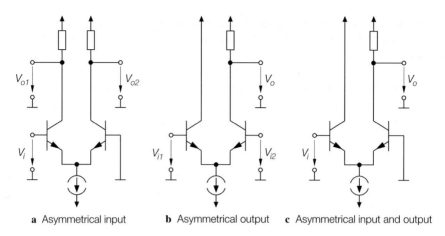

a Asymmetrical input   **b** Asymmetrical output   **c** Asymmetrical input and output

Fig. 4.54. Asymmetrical operation of an npn differential amplifier

Asymmetrical operation: A differential amplifier can be operated asymmetrically by applying a constant voltage to one of the inputs, using one output only or combining both; Fig 4.54 illustrates these three options, using the npn differential amplifier as an example.

In Fig. 4.54a, a constant voltage is applied to input 2 – here ground. In this case, we obtain:

$$A_1 = \frac{dV_{o1}}{dV_{i1}}\bigg|_{V_{i2}=\text{const.}} = \frac{dV_{o1}}{dV_D}\frac{dV_D}{dV_{i1}}\bigg|_{V_{i2}=\text{const.}} + \frac{dV_{o1}}{dV_{CM}}\frac{dV_{CM}}{dV_{i1}}\bigg|_{V_{i2}=\text{const.}}$$

$$= A_D + A_{CM} = A_D\left(1 + \frac{1}{G}\right) \overset{G\gg1}{\approx} A_D$$

$$A_2 = \frac{dV_{o2}}{dV_{i1}}\bigg|_{V_{i2}=\text{const.}} = \frac{dV_{o2}}{dV_D}\frac{dV_D}{dV_{i1}}\bigg|_{V_{i2}=\text{const.}} + \frac{dV_{o2}}{dV_{CM}}\frac{dV_{CM}}{dV_{i1}}\bigg|_{V_{i2}=\text{const.}}$$

$$= -A_D + A_{CM} = -A_D\left(1 - \frac{1}{G}\right) \overset{G\gg1}{\approx} -A_D$$

A sufficiently high common-mode rejection produces output signals of the same amplitude in phase opposition; this circuit is therefore used for converting a single-ended signal into a differential signal.

In Fig. 4.54b, only output 2 is used; output 1 could also be used as an alternative. The common-mode gain and the differential gain can be calculated from (4.58) and (4.59) by setting $V_o = V_{o2}$ or $V_o = V_{o1}$, depending on the output used. Since $A_D < 0$, the version shown in Fig. 4.54b is noninverting with $V_o = V_{o2}$, but inverting with $V_o = V_{o1}$. This circuit is used for converting a differential signal into a single-ended signal.

In Fig. 4.54c, only input 1 and output 2 are used; with respect to the gain A_2, which has already been calculated, this leads to:

$$A = \frac{dV_o}{dV_i} = \frac{dV_{o2}}{dV_{i1}}\bigg|_{V_{i2}=\text{const.}} = A_2 = -A_D + A_{CM} \overset{G\gg1}{\approx} -A_D$$

This circuit can be regarded as a series connection of a common-collector and a common-base circuit. It has a high cutoff frequency, as there is no common-emitter circuit and thus no Miller effect.

Transfer Characteristics of the npn Differential Amplifier

Figure 4.55 shows a circuit with the voltages and currents required to calculate the characteristic curves, where $V_{CM} = 0$. When using transistors of the same size – that is, with the same reverse saturation current I_S – and neglecting the Early effect, we obtain:

$$I_{C1} = I_S e^{\frac{V_{BE1}}{V_T}} \quad , \quad I_{C2} = I_S e^{\frac{V_{BE2}}{V_T}}$$

If the base currents are ignored, the circuit gives us:

$$I_{C1} + I_{C2} = 2I_0 \quad , \quad V_D = V_{BE1} - V_{BE2}$$

The ratio of the collector currents is:

$$\frac{I_{C1}}{I_{C2}} = e^{\frac{V_{BE1}}{V_T}} e^{-\frac{V_{BE2}}{V_T}} = e^{\frac{V_{BE1} - V_{BE2}}{V_T}} = e^{\frac{V_D}{V_T}}$$

Inserting this in $I_{C1} + I_{C2} = 2I_0$ and solving the equation for I_{C1} and I_{C2} leads to

$$I_{C1} = \frac{2I_0}{1 + e^{-\frac{V_D}{V_T}}} \quad , \quad I_{C2} = \frac{2I_0}{1 + e^{\frac{V_D}{V_T}}}$$

With

$$\frac{2}{1 + e^{-x}} = \frac{1 + e^{-x} + 1 - e^{-x}}{1 + e^{-x}} = 1 + \frac{1 - e^{-x}}{1 + e^{-x}} = 1 + \tanh \frac{x}{2}$$

we obtain

$$I_{C1} = I_0 \left(1 + \tanh \frac{V_D}{2V_T} \right) \quad , \quad I_{C2} = I_0 \left(1 - \tanh \frac{V_D}{2V_T} \right) \tag{4.61}$$

and thus for

$$V_{o1} = V_b - I_{C1} R_C \quad , \quad V_{o2} = V_b - I_{C2} R_C$$

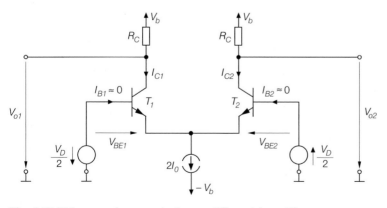

Fig. 4.55. Voltages and currents in the npn differential amplifier

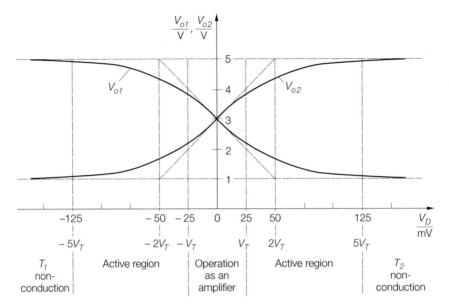

Fig. 4.56. Transfer characteristics of the npn differential amplifier shown in Fig 4.55, where $V_b = 5\,\text{V}$, $R_C = 20\,\text{k}\Omega$ and $I_0 = 100\,\mu\text{A}$

the transfer characteristics of the npn differential amplifier are:

$$
\begin{aligned}
V_{o1} &= V_b - I_0 R_C \left(1 + \tanh\frac{V_D}{2V_T}\right) \\
V_{o2} &= V_b - I_0 R_C \left(1 - \tanh\frac{V_D}{2V_T}\right)
\end{aligned}
\tag{4.62}
$$

Figure 4.56 shows the plot of the characteristics for $V_b = 5\,\text{V}$, $R_C = 20\,\text{k}\Omega$ and $I_0 = 100\,\mu\text{A}$ as a function of the differential voltage V_D, where $V_{CM} = 0$. The slope of the characteristic for $V_D = 0$ is:

$$
\left.\frac{dV_{o1}}{dV_D}\right|_{V_D=0} = -\left.\frac{dV_{o2}}{dV_D}\right|_{V_D=0} = -\frac{I_0 R_C}{2V_T} \approx -\frac{2\,\text{V}}{52\,\text{mV}} \approx -38
$$

This corresponds to the differential gain at the operating point ($V_D = 0$, $V_{CM} = 0$).

The active portion of the characteristic is in the region $|V_D| < 5V_T \approx 125\,\text{mV}$. For $|V_D| > 5V_T$, the differential amplifier is overloaded; in this case, the current of the current source flows almost entirely (over 99%) through one of the two transistors, while the other is in reverse mode. For $V_D < -5V_T$, T_1 is nonconductive and output 1 reaches its maximum output voltage $V_{o,max} = V_b$, while output 2 provides the minimum output voltage $V_{o,min} = V_b - 2I_0 R_C$. For $V_D > 5V_T$, T_2 is nonconductive.

Operating point in small-signal operation: Operation as an amplifier only makes sense in the range $|V_D| < V_T \approx 25\,\text{mV}$: outside this range, the slope of the characteristics decreases; the gain drops and the distortions increase. $V_D = 0$ is chosen for the operating point, and in this case the following applies:

$$V_D = 0 \Rightarrow V_{o1} = V_{o2} = V_b - I_0 R_C \Rightarrow V_{o1} - V_{o2} = 0$$

This shows that in terms of the output differential voltage $V_{o1} - V_{o2}$ the differential amplifier operates as a *true* DC amplifier; that is, without offset. Furthermore, it should be noted that no common-mode voltage V_{CM} is required when selecting the operating point; this voltage can be freely selected within a permissible range.

Common-mode signal range: Using the transistor equations for normal mode in the calculations presented above implied that none of the transistors enters the saturation region. Furthermore, an ideal current source without saturation was assumed. In this case, the characteristics are practically independent of the common-mode voltage V_{CM}; a minor common-mode gain induced by the internal resistance of the current source only causes changes in the mV range. The permissible input voltage range is determined with the help of Fig. 4.57. Two conditions must be met:

- The collector–emitter voltages V_{CE1} and V_{CE2} must be higher than the saturation voltage $V_{CE,sat}$. From Fig. 4.57, it follows

$$V_{CE1} = V_{o1} + V_{BE1} - V_{i1} \quad , \quad V_{CE2} = V_{o2} + V_{BE2} - V_{i2}$$

For $V_{CE} > V_{CE,sat} \approx 0.2$ V, $V_{BE} \approx 0.7$ V and the minimum output voltage $V_{o,min} = V_b - 2I_0 R_C$, it follows:

$$\max\{V_{i1}, V_{i2}\} < V_b - 2I_0 R_C - V_{CE,sat} + V_{BE} \approx V_b - 2I_0 R_C + 0.5 \text{ V}$$

- V_0 must not drop below the lower voltage limit $V_{0,min}$ of the current source; that is, $V_0 > V_{0,min}$ is essential. From Fig. 4.57, it follows

$$V_0 = V_{i1} - V_{BE1} - (-V_b) = V_{i2} - V_{BE2} - (-V_b)$$

Since in normal mode at least one of the transistors is conductive and operated with $V_{BE} \approx 0.7$ V, we obtain:

$$\min\{V_{i1}, V_{i2}\} > V_{0,min} + (-V_b) + V_{BE} \approx V_{0,min} + (-V_b) + 0.7 \text{ V}$$

When using a simple npn current mirror as the current source, $V_{0,min} = V_{CE,sat} \approx 0.2$ V and $\min\{V_{i1}, V_{i2}\} > (-V_b) + 0.9$ V.

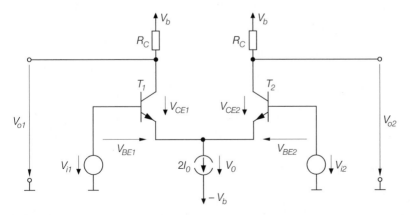

Fig. 4.57. A circuit for calculating the permissible input voltage range of an npn differential amplifier

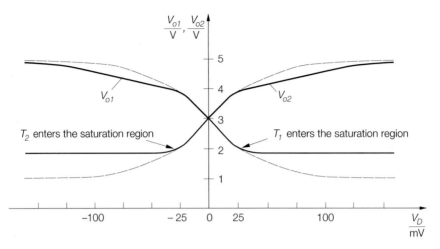

Fig. 4.58. Transfer characteristics of the npn differential amplifier shown in Fig. 4.55, where $V_b = 5\,\text{V}$, $R_C = 20\,\text{k}\Omega$ and $I_0 = 100\,\mu\text{A}$, and where the transistors enter the saturation region ($V_{CM} = 2.5\,\text{V}$)

The permissible input voltage range is usually quoted for pure common-mode operation; that is, $V_{i1} = V_{i2} = V_{CM}$ and $V_D = 0$. In this case, the minimum and maximum operators are eliminated and the *common-mode voltage range* is:[19]

$$V_{0,min} + (-V_b) + V_{BE} \; < \; V_{CM} \; < \; V_b - 2I_0R_C - V_{CE,sat} + V_{BE} \qquad (4.63)$$

For the circuit in Fig. 4.55 with $V_b = 5\,\text{V}$, $(-V_b) = -V_b = -5\,\text{V}$, $R_C = 20\,\text{k}\Omega$, $I_0 = 100\,\mu\text{A}$ and when a simple npn current mirror is used with $V_{0,min} = V_{CE,sat}$, we obtain a common-mode voltage range of $-4.1\,\text{V} < V_{CM} < 1.5\,\text{V}$. If this range is exceeded, the characteristic changes; Fig. 4.58 illustrates this for $V_{CM} = 2.5\,\text{V}$. Since the saturation of one transistor changes the current distribution, it also has an effect on the characteristic of the other branch.

Within the range $|V_D| < 25\,\text{mV}$, the characteristic remains the same; use as an amplifier is still possible even though the common-mode range is exceeded. The reason for this apparent contradiction is that the common-mode range was defined to be the region in which full output signal without saturation is possible. If we restrict ourselves to just a part of the characteristic, the common-mode range is larger. In the borderline case of an infinitesimal small differential voltage, it is sufficient that no saturation occurs with $V_D = 0$. The minimum output voltage is thus $V_{o,min} \approx V_b - I_0R_C$ instead of $V_{o,min} = V_b - 2I_0R_C$. This leads to the *common-mode voltage range in small-signal operation*:

$$V_{0,min} + (-V_b) + V_{BE} \; < \; V_{CM} \; < \; V_b - I_0R_C - V_{CE,sat} + V_{BE} \qquad (4.64)$$

For the values already mentioned, the voltage range of the circuit in Fig. 4.55 is $-4.1\,\text{V} < V_{CM} < 3.5\,\text{V}$. This means that the situation shown in Fig. 4.58, with $V_{CM} = 2.5\,\text{V}$, is still within the small-signal common-mode range.

---

[19] This causes an error, because a differential voltage of at least $5V_T$ is required to reach the minimum output voltage; therefore, one should actually use $\max\{V_{i1}, V_{i2}\} = V_{CM} + V_{D,max}/2$ and $\min\{V_{i1}, V_{i2}\} = V_{CM} - V_{D,max}/2$. As the maximum differential voltage $V_{D,max}$ is application-specific but very low ($V_{D,max} < V_T$) in amplifiers, it is ignored in this case.

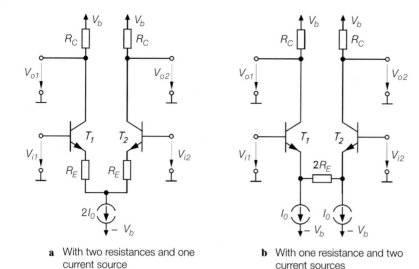

a With two resistances and one current source   **b** With one resistance and two current sources

Fig. 4.59. npn differential amplifier with current feedback

npn differential amplifier with current feedback: The differential amplifier can be provided with current feedback in order to improve its linearity; Fig. 4.59 shows two options that are equivalent in terms of the transfer characteristics. Figure 4.59a uses two resistances R_E and one current source. Without differential signal there is a voltage drop $I_0 R_E$ across both resistors; thus, the lower limit of the common-mode voltage range is increased by this value. In Fig. 4.59b only one resistance is required, which without differential signal carries no current. This means that the common-mode range is not reduced, but two current sources are required.

Figure 4.60 shows the characteristics for $V_b = 5\,\text{V}$, $R_C = 20\,\text{k}\Omega$, $I_0 = 100\,\mu\text{A}$ and several values of R_E; the latter are related to the transconductance of the transistors at the operating point $V_D = 0$:

$$g_m = \frac{I_0}{V_T} \approx \frac{1}{260\,\Omega}\,, \quad g_m R_E = 0\,/\,2\,/\,5 \;\Rightarrow\; R_E = 0\,/\,520\,/\,1300\,\Omega$$

With increasing feedback, the characteristics flatten out and become almost linear over a longer portion. This means that the differential gain is reduced but remains approximately constant over a larger region. The distortions, expressed by the distortion factor, decline as feedback increases.

A direct calculation of the characteristics is not possible. An approximation is possible for high feedback by assuming that the base–emitter voltages remain approximately constant. If the base currents are neglected, the differential voltage for both circuits in Fig. 4.59 is:

$$V_D = V_{i1} - V_{i2} \quad = \quad V_{BE\,1} + I_{C1} R_E - V_{BE\,2} - I_{C2} R_E$$
$$\underset{V_{BE\,1} \approx V_{BE\,2}}{\approx} \quad (I_{C1} - I_{C2})\,R_E$$

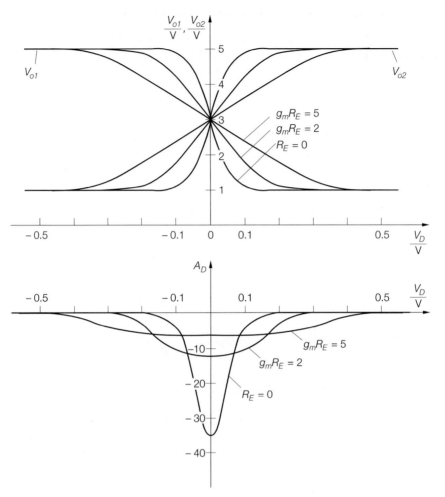

Fig. 4.60. Characteristics and differential gain of an npn differential amplifier with current feedback ($V_b = 5\,\text{V}$, $R_C = 20\,\text{k}\Omega$, $I_0 = 100\,\mu\text{A}$)

Inserting $I_{C1} + I_{C2} = 2I_0$ and solving for I_{C1} and I_{C2} in the range $0 \leq I_{C1}, I_{C2} \leq 2I_0$ leads to

$$I_{C1} \approx I_0 + \frac{V_D}{2R_E} \quad , \quad I_{C2} \approx I_0 - \frac{V_D}{2R_E} \qquad \text{for } |V_D| < 2I_0R_E$$

and consequently to:

$$\left.\begin{aligned} V_{o1} &= V_b - I_{C1}R_C \approx V_b - I_0R_C - \frac{R_C}{2R_E}V_D \\ V_{o2} &= V_b - I_{C2}R_C \approx V_b - I_0R_C + \frac{R_C}{2R_E}V_D \end{aligned}\right\} \quad \text{für } |V_D| < 2I_0R_E \qquad (4.65)$$

Within the active region, the characteristics are almost linear.

Characteristics of the n-Channel Differential Amplifier

Figure 4.61 shows the circuit with the voltages and currents required for the calculation of the characteristics for $V_{CM} = 0$. For MOSFETs of the same size – that is, with the same transconductance coefficient K – and ignoring the Early effect, we obtain the currents:

$$I_{D1} = \frac{K}{2}(V_{GS1} - V_{th})^2 \quad , \quad I_{D2} = \frac{K}{2}(V_{GS2} - V_{th})^2$$

The threshold voltages of the two MOSFETs are identical, since they are both provided with the same bulk–source voltage because their sources are connected to each other. From the circuit, it follows

$$I_{D1} + I_{D2} = 2I_0 \quad , \quad V_D = V_{GS1} - V_{GS2}$$

Further calculation is more complex than for the npn differential amplifier. From the equation

$$V_D = V_{GS1} - V_{GS2} = \sqrt{\frac{2I_{D1}}{K}} - \sqrt{\frac{2I_{D2}}{K}}$$

the term with I_{D2} is isolated on one side. Then both sides are squared, $I_{D2} = 2I_0 - I_{D1}$ is inserted and, after substituting $x = \sqrt{I_{D1}}$, the equation is solved for x by means of the formula for quadratic equations; by squaring, we obtain I_{D1} and $I_{D2} = 2I_0 - I_{D1}$:

$$\left.\begin{aligned} I_{D1} &= I_0 + \frac{V_D}{2}\sqrt{2KI_0 - \left(\frac{K V_D}{2}\right)^2} \\[2mm] I_{D2} &= I_0 - \frac{V_D}{2}\sqrt{2KI_0 - \left(\frac{K V_D}{2}\right)^2} \end{aligned}\right\} \quad \text{für } |V_D| < 2\sqrt{\frac{I_0}{K}} \qquad (4.66)$$

Outside the range of validity of (4.66), the entire current from the current source flows through one of the MOSFETs, while the other MOSFET is nonconductive. With $V_{o1} = V_b - I_{D1}R_D$ and $V_{o2} = V_b - I_{D2}R_D$, the transfer characteristics of the n-channel differential amplifier are as follows:

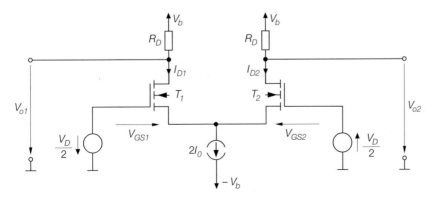

Fig. 4.61. Voltages and currents of the n-channel differential amplifier

$$V_{o1} = V_b - I_0 R_D - \frac{V_D R_D}{2}\sqrt{2K\,I_0 - \left(\frac{K\,V_D}{2}\right)^2}$$

$$V_{o2} = V_b - I_0 R_D + \frac{V_D R_D}{2}\sqrt{2K\,I_0 - \left(\frac{K\,V_D}{2}\right)^2} \qquad\Biggr\}\; \text{für } |V_D| < 2\sqrt{\frac{I_0}{K}} \qquad (4.67)$$

Outside the range of validity of (4.67), one output has the maximum output voltage $V_{o,max} = V_b$ while the other output has the minimum output voltage $V_{o,min} = V_b - 2I_0 R_D$.

A comparison of (4.67) and the corresponding (4.62) for the npn differential amplifier shows that the characteristics of the n-channel differential amplifier *also* depend on the size of the MOSFETs, expressed by the transconductance coefficient K; on the other hand, the size of the bipolar transistors, expressed by the reverse saturation current I_S, has no influence on the characteristic of the npn differential amplifier. Therefore, it is possible to selectively adjust the characteristic of the n-channel differential amplifier by scaling the MOSFETs while keeping the external circuitry the same; in the npn differential amplifier, this is only possible with current feedback. The typical parameter for adjusting the characteristic according to (4.67) is the voltage:

$$V_{DM} = 2\sqrt{\frac{I_0}{K}} \qquad (4.68)$$

This indicates the active segment of the characteristic according to the condition $|V_D| < V_{DM}$. Since $V_D = 0$ at the operating point, the current distribution is $I_{D1} = I_{D2} = I_0$ and at the same time $V_{GS1} = V_{GS2} = V_{GS,A}$, insertion into the characteristic of the MOSFET leads to the alternative expression:

$$V_{DM} = \sqrt{2}\left(V_{GS,A} - V_{th}\right)$$

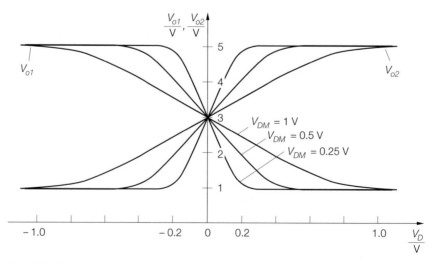

Fig. 4.62. Transfer characteristic of the n-channel differential amplifier shown in Fig. 4.61, where $V_b = 5\,\text{V}$, $R_D = 20\,\text{k}\Omega$ and $I_0 = 100\,\mu\text{A}$.

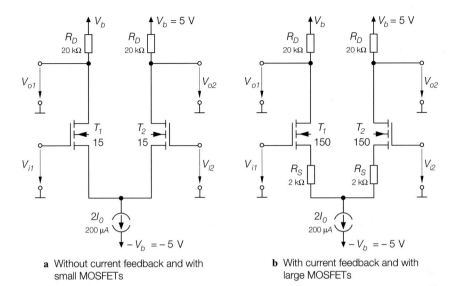

a Without current feedback and with small MOSFETs

b With current feedback and with large MOSFETs

Fig. 4.63. A comparison of n-channel differential amplifiers of the same differential gain with and without current feedback

Figure 4.62 shows the characteristic for $V_b = 5\,\text{V}$, $R_D = 20\,\text{k}\Omega$, and $I_0 = 100\,\mu\text{A}$, and for $K = 0.4\,/\,1.6\,/\,6.4\,\text{mA/V}^2$ or $V_{DM} = 1\,/\,0.5\,/\,0.25\,\text{V}$. A comparison with Fig. 4.60 shows that a variation in the size of the MOSFETs in the n-channel differential amplifier achieves an effect similar to that in the npn differential amplifier with current feedback; the characteristics become less steep if the size of the MOSFETs in the n-channel differential amplifier decreases or if the feedback in the npn differential amplifier increases (R_E is larger). Consequently, the n-channel differential amplifier yields better linearity with smaller MOSFETs, while larger MOSFETs lead to an increase in the differential gain.

Common-mode voltage range: The *common-mode voltage range* can be calculated from (4.63) and (4.64) by replacing V_{BE} with $V_{GS} = V_{th} + \sqrt{2I_D/K}$ and $V_{CE,sat}$ with $V_{DS,po} = V_{GS} - V_{th}$:

$$V_{0,min} + (-V_b) + V_{th} + \sqrt{\frac{4I_0}{K}} < V_{CM} < V_b - 2I_0R_D + V_{th} \qquad (4.69)$$

The *common-mode voltage range in small-signal operation* is:

$$V_{0,min} + (-V_b) + V_{th} + \sqrt{\frac{2I_0}{K}} < V_{CM} < V_b - I_0R_D + V_{th} \qquad (4.70)$$

Here, $V_{0,min}$ is the voltage limit of the current source. It is not possible to determine the limits directly since, due to the substrate effect, the threshold voltage V_{th} depends on the bulk–source voltage V_{BS}, which in turn depends on V_{CM}. For an estimation, one can ignore the substrate effect and insert $V_{th} = V_{th,0}$.

n-channel differential amplifier with current feedback: Current feedback can also be used in the n-channel differential amplifier to improve the linearity. However, the question here is whether this produces a better result with the same gain than does a reduction in the size of the MOSFETs, as discussed in the previous paragraph. For this purpose, the circuits

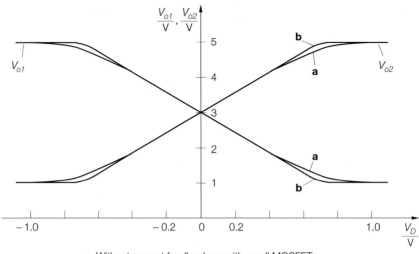

a Without current feedback an with small MOSFETs
b With current feedback and with large MOSFETs

Fig. 4.64. Characteristics of the differential amplifiers shown in Fig. 4.63

shown in Fig. 4.63 will be compared; their characteristics, and thus their differential gains, are identical in the region of the operating point $V_D = 0$; Fig. 4.64 presents the corresponding characteristics. It is clear that the circuit with current feedback and larger MOSFETs offers better linearity; however, the spatial requirements are significantly greater due to the MOSFETs – which are ten times as large – and the necessary feedback resistances, and the bandwidth is significantly narrower than in the circuit without current feedback, due to the larger capacitances of the MOSFETs.

Differential Amplifier with an Active Load

In integrated circuits, current sources are used instead of ohmic collector or drain resistances because with the same, and often even a reduced, spatial requirement they produce a clearly higher differential gain. An npn differential amplifier is used to illustrate the circuits that are commonly used.

Differential amplifier with symmetric output: The circuit in Fig. 4.65a uses two current sources with current I_0 instead of the collector resistances; with regard to (4.61), the output currents are:[20]

$$I_{o1} = I_{C1} - I_0 = I_0 \tanh \frac{V_D}{2V_T} \quad , \quad I_{o2} = I_{C2} - I_0 = -I_0 \tanh \frac{V_D}{2V_T}$$

At the operating point $V_D = 0$, there is no current at either output. The outputs must be connected to a load so that output currents can actually flow without the transistors and current sources entering the saturation region. The output voltages are not defined without external circuitry.

---

[20] As the differential amplifier as a whole represents a current node, the conditions of Kirchhoff's law must be met. In the following equations and in Fig. 4.65, this is only the case if the base currents are ignored.

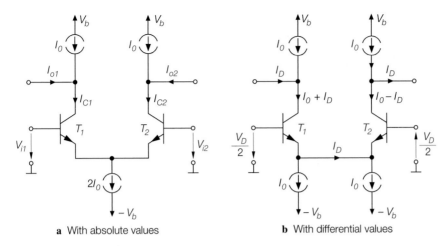

Fig. 4.65. npn differential amplifier with an active load

In order to illustrate the current distribution, the differential current

$$I_D = I_0 \tanh \frac{V_D}{2V_T} \tag{4.71}$$

is shown in the circuit of Fig. 4.65b. The current source $2I_0$ in the emitter branch is split up into two current sources for reasons of symmetry; this causes the current in the cross-connection to be exactly equal to the differential current I_D. The differential current flows from input 1 through T_1, the emitter cross-connection and T_2 to the output 2; it thus flows through the differential amplifier. This means that the current consumption remains constant as long as no transistor and no current source enters the saturation region and the condition $|I_D| < I_0$ is met, or provided that the current *supplied* at one output is *taken up* at the other output.

Differential amplifier with asymmetric output: The circuit of Fig. 4.65a can also be used where an asymmetric output is required, if the output that is not required is connected to the operating voltage V_b and the related current source is removed. A better alternative, which prevails in practical applications, is shown in Fig. 4.66a. In this case, the current sources are replaced by a current mirror, so that the current of the output that has been removed is mirrored to the remaining output:

$$I_o = I_{C2} - I_{C4} \overset{I_{C4} \approx I_{C1}}{\approx} I_{C2} - I_{C1} = -2I_0 \tanh \frac{V_D}{2V_T}$$

At the operating point $V_D = 0$, there is no current at the output. Here again, the output must be connected to a load such that the output current can flow without driving T_2 or T_4 into the saturation region. Figure 4.66b shows the circuit with the differential current I_D. The current of the negative bias voltage source remains constant, while that of the positive source changes by $2I_D$.

Current sources and current mirrors: In essence, all of the circuits described in Sect. 4.1.1 can be used to obtain the current sources in Figs. 4.65 and 4.66; in practice, simple current mirrors or cascode current mirrors are mostly used as current sources. The current mirror of Fig. 4.66 may also be made up differently; since the current ratio

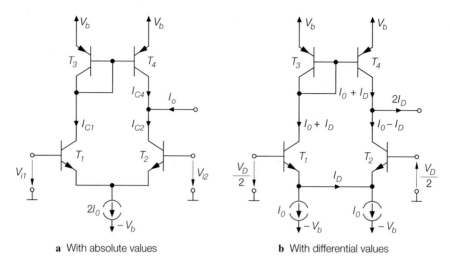

a With absolute values   **b** With differential values

Fig. 4.66. npn differential amplifier with an asymmetric output

should deviate as little as possible from 1, a three-transistor mirror or a Wilson current mirror is often used.

The chosen current source and current mirror have only a negligible influence on the output currents; just the small-signal output resistance changes. This will be explained in more detail in the description of the small-signal behavior.

Offset Voltage of a Differential Amplifier

So far, it has been assumed that the voltages and currents at the operating point $V_D = 0$ are exactly symmetrical. In practice, however, this is not the case, due to the unavoidable tolerances. Furthermore, some circuits are asymmetrical, which means that consideration of just the effects that have been neglected until now leads to a nonsymmetrical current distribution. One example of this is the differential amplifier with asymmetric output in Fig. 4.66, in which the current distribution at $V_D = 0$ is asymmetrical due to the fact that the current ratio of the current mirror deviates from 1 very slightly.

The asymmetry is characterized by the *offset voltage* V_{off}.[21] This specifies the differential voltage that must be applied in order to make the output voltages equal or – in the case of asymmetric outputs – to achieve a certain setpoint value:

$$V_D = V_{off} \Rightarrow V_{o1} = V_{o2} \text{ or } V_o = V_{o,set} \tag{4.72}$$

The relevant current distribution can be symmetrical, although this must not necessarily be the case. In the transfer characteristics, the offset voltage causes a shift in the direction of V_D; Fig. 4.67 illustrates this for $V_{off} > 0$.

As already mentioned, the offset voltage consists of a systematic portion caused by circuit asymmetries and a stochastic portion caused by tolerances. Therefore, in practice a likely range (e.g. 99% probability) is often quoted for the offset voltage.

---

[21] The offset voltage is often called V_O (index O). As this symbol can easily be confused with V_0 (index zero), it is called V_{off} here, for simplicity.

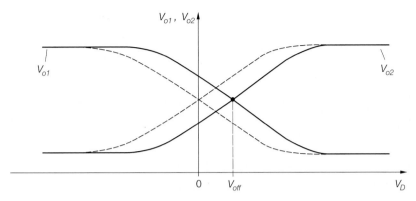

Fig. 4.67. Transfer characteristics where an offset voltage exists

The offset voltage can be calculated if very acurate equations are used for the transistors and upper and lower limits are inserted for all parameters; however, the actual calculation is very complex. It is easier to measure the offset voltage or to determine it by circuit simulation; this is done on the basis of the circuits shown in Fig. 4.68. By feeding the output differential voltage $V_{o1} - V_{o2}$ back to input 1, the output voltages are made approximately equal and the voltage at the input is $V_{i1} \approx V_{off}$. On the one hand, the circuit provides no true differential signal, but, on the other hand, the actual common-mode voltage $V_{CM} \approx V_{off}/2$ has almost no influence on the result, due to the high common-mode rejection.

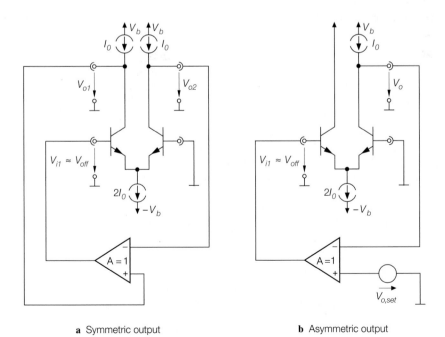

a Symmetric output   **b** Asymmetric output

Fig. 4.68. Circuit for measuring the offset voltage

When measuring the offset voltage, a normal operational amplifier should not be used as a control amplifier, since the differential amplifier causes an additional loop gain, which results in instability in the circuit even with compensated operational amplifiers. The most suitable type is an instrumentation amplifier with a gain of $A = 1$ and a cutoff frequency $f_{g,amp}$ that remains below the cutoff frequency f_g of the differential amplifier by a value at least equal to the differential gain A_D: $f_{g,amp} < f_g/A_D$; this guarantees stable operation. In circuit simulation, a voltage-controlled voltage source with $A = 1$ may be used as a control amplifier; should stability problems occur, A has to be reduced.

Small-Signal Behavior of the Differential Amplifier

The response to input signals around an operating point A is called the *small-signal behavior*. The operating point is characterized by the input voltages $V_{i1,A}$ and $V_{i2,A}$ or $V_{D,A}$ and $V_{CM,A}$, the output voltages $V_{o1,A}$ and $V_{o2,A}$, and the collector and drain currents of the transistors. The following description assumes that the offset voltage is zero; the operating point is thus:

$$V_{D,A} = 0 \quad , \quad V_{o1,A} = V_{o2,A}$$

It is assumed that the common-mode voltage $V_{CM,A}$ remains within the common-mode voltage range and has no influence on the current distribution.

Equivalent circuits for differential and common-mode operation: If the current source in the emitter or source branch of a differential amplifier is split up into two equivalent current sources, the differential amplifier is completely symmetrical; this is shown in the example of an npn differential amplifier in Fig. 4.69. With regards to the changes of the currents and voltages in the plane of symmetry when working at the operating point, the following can be observed:

– With a sufficiently low amplitude, the skew-symmetric differential signal causes a skew-symmetric change of all currents and voltages. Consequently, all voltages in the plane of symmetry remain constant; in Fig. 4.69a this applies to the voltage V_0 at the emitter connections of the transistors. Since a constant voltage can be replaced by a voltage source, the resulting equivalent circuit is as shown in Fig. 4.69a (bottom): the differential amplifier is split up into two common-emitter circuits and the current sources are eliminated. The voltage sources V_0 are ideal and are short-circuited in the conversion to the small-signal equivalent circuit. Therefore, the emitter connections of the transistors are connected to the small-signal ground in the small-signal equivalent circuit.
– The symmetrical common-mode signal produces a symmetrical change in all currents and voltages. This means that all currents flowing through the plane of symmetry are zero; in Fig. 4.69b this applies to current I in the emitter cross-connection. The fact that a dead wire can be removed leads to the equivalent circuit shown in the lower part of Fig. 4.69b: here too, the differential amplifier consists of two common-emitter circuits. The current sources I_0 each represent one *half* of the original current source; Fig. 4.70 explains the conversion from an ideal to a real current source and its splitting into two current sources. In the small-signal equivalent circuit, the current sources are omitted and the negative supply voltage coincides with the small-signal ground.

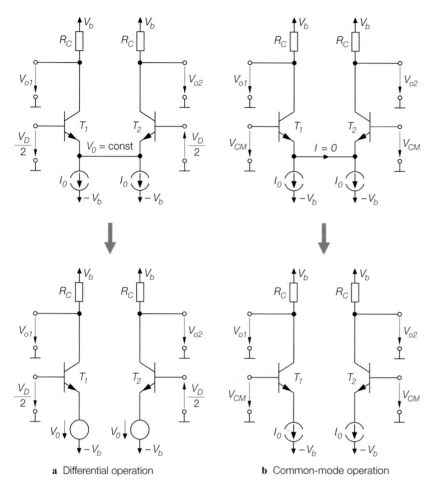

a Differential operation   **b** Common-mode operation

Fig. 4.69. The operation of an npn differential amplifier at the operating point

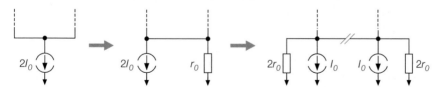

Fig. 4.70. Conversion from an ideal to a real current source and the split into two equivalent current sources

This reduces the npn differential amplifier to a common-emitter circuit, for which the results from Sect. 2.4.1 can be used. The same applies to the n-channel differential amplifier; it is split into equivalent common-source circuits and the results from Sect. 3.4.1 can be used.

The division into separate equivalent circuits for differential and common-mode operation is an application of the *Bartlett symmetry theorem* – which, however, applies to linear circuits only. Strictly speaking, the small-signal equivalent circuit should be used for the differential amplifier in order to allow this theorem to be applied. However, the restriction to linear circuits is only required for differential operation, as here the characteristics of the components are modulated skew-symmetrically, starting from the operating point, which only results in skew-symmetric changes in the case of linear characteristics. In contrast, in common-mode operation the characteristics are modulated symmetrically, which leads to symmetrical changes even in the case of nonlinear characteristics. Therefore, the theorem may also be applied to nonlinear circuits if the differential operation is limited to the range in which the characteristics are almost linear; in the npn differential amplifier, this is the range of $|V_D| < V_T$. This approach was selected because the breakdown of a differential amplifier into two portions is easier to understand from the original circuit than from the small-signal equivalent circuit.

Differential amplifier with resistances: Figure 4.71 shows the circuit of an npn differential amplifier together with the small-signal equivalents for the equivalent common-emitter circuits with differential and common-mode operation which are obtained by linearization of the subcircuits in Fig. 4.69 and by inserting the current source according to Fig. 4.70. For the small-signal values where $V_{D,A} = 0$, it follows:

$$v_{i1} = V_{i1} - V_{i1,A} = V_{i1} - V_{CM,A} \quad , \qquad v_{o1} = V_{o1} - V_{o1,A}$$

$$v_D = V_D - V_{D,A} = V_D \qquad\qquad , \qquad v_{CM} = V_{CM} - V_{CM,A}$$

It is clear that the small-signal equivalent circuit for differential operation corresponds to that of a common-emitter circuit without feedback, and the small-signal equivalent circuit for common-mode operation to that of a common-emitter circuit with current feedback. With common-mode operation the output resistance $2r_0$ of the split current source acts as a feedback resistance. Figure 4.72 shows the corresponding small-signal equivalent circuits of an n-channel differential amplifier.

From the small-signal equivalent circuit for differential operation, the following parameters can be calculated: the *differential gain* A_D, the *differential output resistance* $r_{o,D}$ and the *differential input resistance* $r_{i,D}$:

$$A_D = \left.\frac{v_{o1}}{v_D}\right|_{\substack{i_{o1}=i_{o2}=0 \\ v_{CM}=0}} = \left.\frac{v_o}{2v_i}\right|_{i_o=0} = \frac{1}{2}A_{CE/CS} \tag{4.73}$$

$$r_{o,D} = \left.\frac{v_{o1}}{i_{o1}}\right|_{\substack{v_{o1}=-v_{o2} \\ v_D=v_{CM}=0}} = \left.\frac{v_o}{i_o}\right|_{v_i=0} = r_{o,CE/CS} \tag{4.74}$$

$$r_{i,D} = \left.\frac{v_D}{i_{i1}}\right|_{v_{CM}=0} = \frac{2v_i}{i_i} = 2\,r_{i,CE/CS} \tag{4.75}$$

Here, it is important that the input voltage in the small-signal equivalent circuit for differential operation is $v_D/2$ and not v_D; this is why the differential amplifier has only half the gain but twice the input resistance of the equivalent common-emitter (CE) or common-source (CS) circuit.

The following parameters can be calculated from the small-signal equivalent circuit for common-mode operation: the *common-mode gain* A_{CM}, the *common-mode output resistance* $r_{o,CM}$ and the *common-mode input resistance* $r_{i,CM}$:

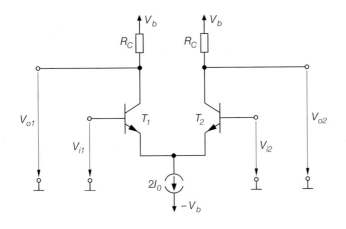

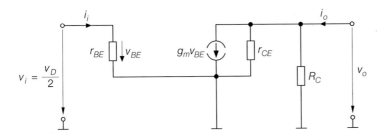

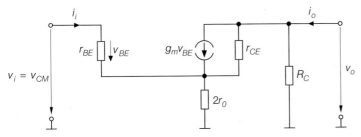

Fig. 4.71. npn differential amplifier with collector resistances: the circuit (*above*) and the small-signal equivalent common-emitter circuits for differential operation (*centre*) and common-mode operation (*below*)

$$A_{CM} = \left.\frac{v_{o1}}{v_{CM}}\right|_{\substack{i_{o1}=i_{o2}=0 \\ v_D=0}} = \left.\frac{v_o}{v_i}\right|_{i_o=0} = A_{CE/CS} \tag{4.76}$$

$$r_{o,CM} = \left.\frac{v_{o1}}{i_{o1}}\right|_{\substack{v_{o1}=v_{o2} \\ v_D=0,v_{CM}=0}} = \left.\frac{v_o}{i_o}\right|_{v_i=0} = r_{o,CE/CS} \tag{4.77}$$

$$r_{i,CM} = \frac{v_{CM}}{i_{i1}} = \frac{v_i}{i_i} = r_{i,CE/CS} \tag{4.78}$$

Here, the differential amplifier provides the same values as the equivalent common-emitter or common-source circuits. It should be noted that the small-signal parameters

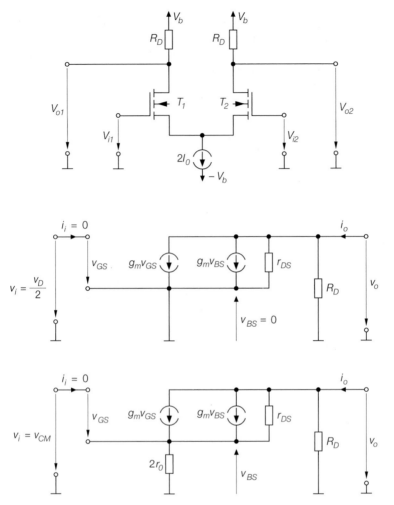

Fig. 4.72. n-channel differential amplifier with drain resistances: the circuit (*above*) and the small-signal equivalent common-source circuits for differential operation (*centre*) and common-mode operation (*below*)

in (4.76)–(4.78) are from a different small-signal equivalent circuit than those in (4.73)–(4.75); for example, $A_D = A_{CM}/2$ is *not* a consequence of (4.73) and (4.76).

Pure differential or common-mode operation must exist when measuring or simulating these parameters. This not only applies to the input, where this is expressed by parameters v_D and v_{CM}, but also to the output. As no specific differential and common-mode parameters are defined for the output, the secondary conditions $v_{o1} = -v_{o2}$ and $v_{o1} = v_{o2}$ must be used to characterize the differential and common-mode operation. The consequence is that the definitions of the differential and the common-mode output resistances are only distinguished by the secondary conditions and not by the small-signal parameters. v_{o1}/i_{o1} is formed for both output resistances; the difference is produced by the different operation of the second output.

In the npn differential amplifier, the output resistances depend on the internal resistance R_g of the signal source, as is the case in the common-emitter circuit. As the latter is generally smaller than the input resistances, calculations can be restricted to the short-circuit output resistances without causing major errors; therefore, $r_{o,D}$ and $r_{o,CM}$ are given for the secondary condition $v_D = v_{CM} = 0$. This dependence does not exist in the n-channel differential amplifier due to the isolated gate connections of the MOSFETs; here, R_g is not *visible* at the output.

The results for the common-emitter circuit in Sect. 2.4.1 and for the common-source circuit in Sect. 3.4.1 lead to the following equations for the *differential amplifier with resistances*:[22]

differential amplifier with resistance

$$A_D = \left. \frac{v_{o1}}{v_D} \right|_{i_{o1}=i_{o2}=0} = \begin{cases} -\dfrac{g_m}{2}\left(R_C \parallel r_{CE}\right) & \overset{r_{CE} \gg R_C}{\approx} & -\dfrac{1}{2}\, g_m R_C \\[2ex] -\dfrac{g_m}{2}\left(R_D \parallel r_{DS}\right) & \overset{r_{DS} \gg R_D}{\approx} & -\dfrac{1}{2}\, g_m R_D \end{cases} \tag{4.79}$$

$$r_{o,D} = \left. \frac{v_{o1}}{i_{o1}} \right|_{v_{o1}=-v_{o2}} = \begin{cases} R_C \parallel r_{CE} & \overset{r_{CE} \gg R_C}{\approx} & R_C \\[2ex] R_D \parallel r_{DS} & \overset{r_{DS} \gg R_D}{\approx} & R_D \end{cases} \tag{4.80}$$

$$r_{i,D} = \frac{v_D}{i_{i1}} = \begin{cases} 2r_{BE} \\[1ex] \infty \end{cases} \tag{4.81}$$

$$A_{CM} = \left. \frac{v_{o1}}{v_{CM}} \right|_{i_o=0} \approx \begin{cases} -\dfrac{R_C}{2r_0} \\[2ex] -\dfrac{g_m R_D}{2\left(g_m + g_{m,B}\right)r_0} & \overset{g_m \gg g_{m,B}}{\approx} & -\dfrac{R_D}{2r_0} \end{cases} \tag{4.82}$$

$$r_{o,CM} = \left. \frac{v_{o1}}{i_{o1}} \right|_{v_{o1}=v_{o2}} = \begin{cases} R_C \parallel \beta r_{CE} \approx R_C \\ R_D \parallel 2g_m\, r_{DS}r_0 \approx R_D \end{cases} \tag{4.83}$$

$$r_{i,CM} = \frac{v_{CM}}{i_{i1}} = \begin{cases} 2\beta r_0 + r_{BE} \approx 2\beta r_0 \\ \infty \end{cases} \tag{4.84}$$

$$G = \frac{A_D}{A_{CM}} \approx \begin{cases} g_m r_0 \\ \left(g_m + g_{m,B}\right)r_0 & \overset{g_m \gg g_{m,B}}{\approx} & g_m r_0 \end{cases} \tag{4.85}$$

Here, the following equations were used: (2.61)–(2.63) on page 100, (2.70)–(2.72) on page 106, (3.50)–(3.52) on page 232 and (3.56)–(3.58) on page 237; $R_E = 2r_0$ is inserted into (2.70) and $R_S = 2r_0$ and $2g_m\, r_0 \gg 1$ are inserted into (3.56).

The common-mode gain of the n-channel differential amplifier with integrated MOS-FETs depends on the common-mode voltage $V_{CM,A}$ at the operating point, because the bulk–source voltage V_{BS} and the substrate transconductance $g_{m,B}$ depend on $V_{CM,A}$. But since $V_{BS} > 0$ according to Fig. 4.73, the substrate transconductance is lower in the n-

---

[22] The results for the npn and the n-channel differential amplifiers are presented following braces. The upper values apply to the npn differential amplifier and the lower values to the n-channel differential amplifier.

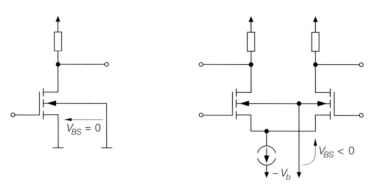

Fig. 4.73. Bulk–source voltage V_{BS} in the common-source circuit and in the n-channel differential amplifier

channel differential amplifier than in the common-source circuit and may therefore be ignored in most practical applications. In the differential amplifier with discrete MOS-FETs, however, $V_{BS} = 0$, so that in the relevant equations the substrate transconductance can be set to $g_{m,B} = 0$. In the following considerations, the substrate transconductance is generally ignored.

Basically, any of the circuits described in Sect. 4.1.1 can be used to realize the current source; the output resistance r_0 has a significant influence on the common-mode gain and the common-mode rejection ratio. In practice, a simple current mirror is usually used.

Basic equations of a symmetric differential amplifier: The differential gain can be described by means of the differential output resistance and the common-mode gain by means of the common-mode output resistance; the *basic equations of a symmetric differential amplifier*, derived from (4.79)–(4.85), are:

$$A_D = \left. \frac{v_{o1}}{v_D} \right|_{i_{o1}=i_{o2}=0} = -\frac{1}{2} g_m\, r_{o,D} \tag{4.86}$$

$$A_{CM} = \left. \frac{v_{o1}}{v_{CM}} \right|_{i_{o1}=i_{o2}=0} \approx -\frac{r_{o,CM}}{2r_0} \tag{4.87}$$

$$G = \frac{A_D}{A_{CM}} \approx g_m\, r_0 \frac{r_{o,D}}{r_{o,CM}} \overset{r_{o,D} \approx r_{o,CM}}{\approx} g_m\, r_0 \tag{4.88}$$

If the output resistances $r_{o,D}$ and $r_{o,CM}$ are almost identical, as is the case in the differential amplifier with resistances, then the common-mode rejection only depends on the transconductance of the transistors and the output resistance r_0 of the current source.

Current feedback, as shown in Fig. 4.59 on page 335 or in Fig. 4.63b on page 339, can be taken into account simply by replacing the transconductance g_m with the *reduced transconductance*:

$$
g_{m,red} = g'_m = \begin{cases} \dfrac{g_m}{1 + g_m R_E} \\[4mm] \dfrac{g_m}{1 + (g_m + g_{m,B}) R_S} \end{cases} \overset{g_m \gg g_{m,B}}{\approx} \dfrac{g_m}{1 + g_m R_S} \tag{4.89}
$$

The differential gain thus drops accordingly. The common-mode gain remains the same because the feedback resistance in the small-signal equivalent circuit for common-mode operation is connected in series to the output resistance r_0 of the current source, and can be ignored since $r_0 \gg R_E, R_S$. Thus the common-mode rejection $G = A_D/A_{CM}$ is lower with current feedback.

Differential amplifiers with simple current sources: Figure 4.74 shows an npn and an n-channel differential amplifier, both with simple current sources instead of resistances. In the small-signal equivalent circuit and in the equations, the resistances are replaced by the output resistance of the simple current source: $R_C \rightarrow r_{CE3}$ for the npn differential amplifier and $R_D \rightarrow r_{DS3}$ for the n-channel differential amplifier. For *differential amplifiers with simple current sources*, this leads to:

differential amplifier with simple current source

$$
A_D = \left.\dfrac{v_{o1}}{v_D}\right|_{i_{o1}=i_{o2}=0} = -\dfrac{1}{2} g_{m1} r_{o,D}
$$

$$
r_{o,D} = \left.\dfrac{v_{o1}}{i_{o1}}\right|_{v_{o1}=-v_{o2}} \approx \begin{cases} r_{CE1} \| r_{CE3} & \overset{r_{CE1}\approx r_{CE3}}{\approx} \dfrac{r_{CE3}}{2} \\[3mm] r_{DS1} \| r_{DS3} & \overset{r_{DS1}\approx r_{DS3}}{\approx} \dfrac{r_{DS3}}{2} \end{cases} \tag{4.90}
$$

$$
A_{CM} = \left.\dfrac{v_{o1}}{v_{CM}}\right|_{i_{o1}=i_{o2}=0} \approx -\dfrac{r_{o,CM}}{2r_0}
$$

$$
r_{o,CM} = \left.\dfrac{v_{o1}}{i_{o1}}\right|_{v_{o1}=v_{o2}} \approx \begin{cases} \beta_1 r_{CE1} \| r_{CE3} \approx r_{CE3} \\[2mm] 2g_{m1} r_{DS1} r_0 \| r_{DS3} \approx r_{DS3} \end{cases} \tag{4.91}
$$

$$
G = \dfrac{A_D}{A_{CM}} \approx g_{m1} r_0 \dfrac{r_{o,D}}{r_{o,CM}} \overset{\substack{r_{CE1}\approx r_{CE3} \\ r_{DS1}\approx r_{DS3}}}{\approx} \dfrac{g_m r_0}{2} \tag{4.92}
$$

The input resistances $r_{i,D}$ and $r_{i,CM}$ remain unaltered; in other words, (4.81) and (4.84) also apply to the differential amplifier with current sources.

After inserting $g_{m1} = I_0/V_T$, $r_{CE1} = V_{A,npn}/I_0$ and $r_{CE3} = V_{A,pnp}/I_0$, the differential gain for npn differential amplifiers with simple current sources is:

$$
A_D = -\dfrac{1}{2V_T \left(\dfrac{1}{V_{A,npn}} + \dfrac{1}{V_{A,pnp}} \right)} \tag{4.93}
$$

Here, $V_{A,npn}$ and $V_{A,pnp}$ are the Early voltages of the transistors; the temperature voltage at $T = 300\,\text{K}$ is $V_T \approx 26\,\text{mV}$. The transistor parameters and the quiescent current I_0

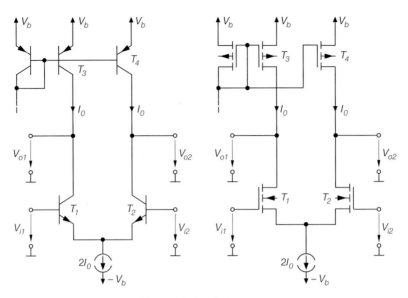

Fig. 4.74. Differential amplifiers with simple current source

have no influence on the differential gain. From Fig. 4.5, the values for the transistor are $V_{A,npn} = 100\,\mathrm{V}$ and $V_{A,pnp} = 50\,\mathrm{V}$; consequently, $A_D = -640$.

For n-channel differential amplifiers with simple current sources, where $g_{m1} = \sqrt{2K_1 I_0}$, $r_{DS1} = V_{A,nC}/I_0$ and $r_{DS3} = V_{A,pC}/I_0$, the differential gain is as follows:

$$A_D = -\sqrt{\frac{K_1}{2I_0}}\,\frac{1}{\dfrac{1}{V_{A,nC}} + \dfrac{1}{V_{A,pC}}} = -\frac{1}{(V_{GS1} - V_{th1})\left(\dfrac{1}{V_{A,nC}} + \dfrac{1}{V_{A,pC}}\right)} \tag{4.94}$$

Here, $V_{A,nC}$ and $V_{A,pC}$ are the Early voltages of the MOSFETs. In this case, the differential gain also depends on the sizes of the MOSFETs T_1 and T_2, expressed by the transconductance coefficient K_1; this gain increases when the size of the MOSFETs is increased. From Fig. 4.6, the values for the MOSFETs are $V_{A,nC} = 50\,\mathrm{V}$ and $V_{A,pC} = 33\,\mathrm{V}$; for the typical value $V_{GS1} - V_{th1} = 1\,\mathrm{V}$, it follows that $A_D = -20$.

Differential amplifiers with cascode current sources: The differential gain can be increased by using current sources with cascode or cascode current sources[23] instead of the simple current sources; Fig. 4.75 shows the resulting circuits when using current sources with cascode. Strictly speaking, the name *differential amplifier with cascode current sources* is not quite correct, but it is preferred to the longer designation *differential amplifier with current sources with cascode*.

When using the current sources with a cascode, the output resistance of the current source increases from r_{CE3} or r_{DS3} to:

$$r_{oS} \approx \begin{cases} \beta_3 r_{CE3} \\ (g_{m3} + g_{m,B3})\,r_{DS3}^2 & \overset{g_{m3} \gg g_{m,B3}}{\approx} \quad g_{m3} r_{DS3}^2 \end{cases}$$

---

[23] The difference is shown in Fig. 4.25 on page 294 and Fig. 4.27 on page 297.

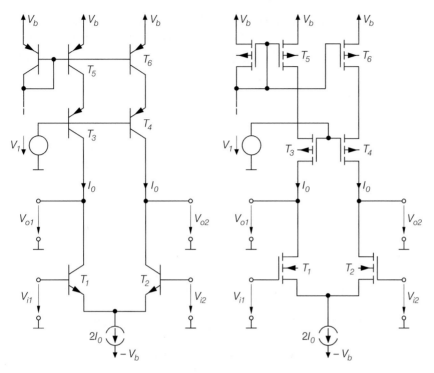

Fig. 4.75. Differential amplifiers with cascode current sources

This leads to the following equations for the *differential amplifier with cascode current sources:*

---

differential amplifier with cascode current sources

$$A_D = \left.\frac{v_{o1}}{v_D}\right|_{i_{o1}=i_{o2}=0} = -\frac{1}{2}g_{m1}r_{o,D}$$

$$r_{o,D} = \left.\frac{v_{o1}}{i_{o1}}\right|_{v_{o1}=-v_{o2}} \approx \begin{cases} r_{CE1} \| \beta_3 r_{CE3} \approx r_{CE1} \\ r_{DS1} \| g_{m3}r_{DS3}^2 \approx r_{DS1} \end{cases} \tag{4.95}$$

$$A_{CM} = \left.\frac{v_{o1}}{v_{CM}}\right|_{i_{o1}=i_{o2}=0} \approx -\frac{r_{o,CM}}{2r_0}$$

$$r_{o,CM} = \left.\frac{v_{o1}}{i_{o1}}\right|_{v_{o1}=v_{o2}} \approx \begin{cases} \beta_1 r_{CE1} \| \beta_3 r_{CE3} \\ 2g_{m1}r_{DS1}r_0 \| g_{m3}r_{DS3}^2 \end{cases} \tag{4.96}$$

$$G = \frac{A_D}{A_{CM}} \approx g_{m1}r_0 \frac{r_{o,D}}{r_{o,CM}} \tag{4.97}$$

Here, the common-mode output resistance $r_{o,CM}$ is typically larger than the differential output resistance $r_{o,D}$ by a factor of $20\ldots200$; as compared to the differential amplifier with resistances, this reduces the common-mode rejection accordingly:

$$G \approx \frac{g_{m1}r_0}{20\ldots200}$$

For npn differential amplifiers with cascode current sources, the differential gain is obtained by inserting $g_{m1} = I_0/V_T$ and $r_{CE1} = V_{A,npn}/I_0$:

$$A_D = -\frac{V_{A,npn}}{2V_T} = -\frac{\mu}{2} \tag{4.98}$$

Here, $\mu = V_A/V_T$ represents the maximum gain of a bipolar transistor introduced in connection with the common-emitter circuit. $V_{A,npn} = 100\,\text{V}$ leads to a differential gain of $A_D = -1920$, as compared to $A_D = -640$ in the npn differential amplifier with simple current sources.

For the n-channel differential amplifier with cascode current sources, where $g_{m1} = \sqrt{2K_1I_0}$ and $r_{DS1} = V_{A,nC}/I_0$, it follows that:

$$A_D = -\sqrt{\frac{K_1}{2I_0}}\,V_{A,nC} = -\frac{V_{A,nC}}{V_{GS1} - V_{th1}} = -\frac{\mu}{2} \tag{4.99}$$

Here, μ represents the maximum gain of the MOSFETs introduced in connection with the common-source circuit. For $V_{A,nC} = 50\,\text{V}$ and $V_{GS1} - V_{th1} = 1\,\text{V}$, the differential gain is $A_D = -50$, as compared to $A_D = -20$ in the n-channel differential amplifier with simple current sources.

The differential amplifier with cascode current sources is used whenever the pnp or p-channel transistors have a clearly lower Early voltage than the npn or n-channel transistors. In this case, the simple current sources only yield an insufficient gain.

Cascode differential amplifier: An additional increase in the differential gain with a simultaneous rise in the gain–bandwidth product is achieved by converting the differential amplifier into a cascode differential amplifier. This requires a symmetric extension of the cascode circuits shown in Fig. 4.45 on page 316; Fig. 4.76 shows the resulting circuits. The advantages of the cascode circuit are described in Sect. 4.1.2, and also apply to the cascode differential amplifier.

In Fig. 4.76, current sources with a cascode are used to achieve the highest possible differential gain. However, if the sole goal is an increase in the gain–bandwidth product, simple current sources may also be used; transistors T_5 and T_6 can then be omitted. However, the higher differential gain is generally of more importance than an increase in the gain–bandwidth product. This particularly applies to the n-channel differential amplifier which reaches only a comparatively low differential gain without the cascode stages in both the differential amplifier *and* the current sources.

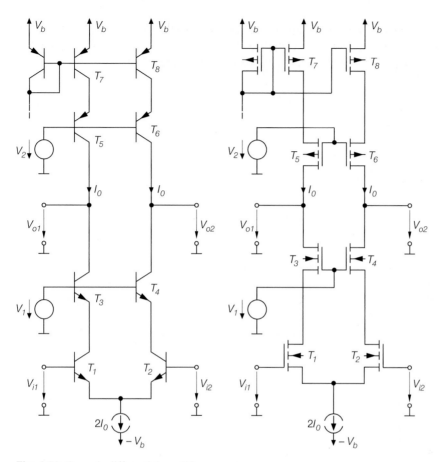

Fig. 4.76. Cascode differential amplifier

For the *cascode differential amplifier*, it follows from (4.36) and (4.37) that:

Cascode differential amplifier

$$A_D = \left.\frac{v_{o1}}{v_D}\right|_{i_{o1}=i_{o2}=0} = -\frac{1}{2}g_{m1}r_{o,D}$$

$$r_{o,D} = \left.\frac{v_{o1}}{i_{o1}}\right|_{v_{o1}=-v_{o2}=0} \approx \begin{cases} \beta_3 r_{CE3} \parallel \beta_5 r_{CE5} \\ g_{m3}r_{DS3}^2 \parallel g_{m5}r_{DS5}^2 \end{cases} \qquad (4.100)$$

$$A_{CM} = \left.\frac{v_{o1}}{v_{CM}}\right|_{i_{o1}=i_{o2}=0} \approx -\frac{r_{o,CM}}{2r_0}$$

$$r_{o,CM} = \left.\frac{v_{o1}}{i_{o1}}\right|_{v_{o1}=v_{o2}=0} \approx \begin{cases} \beta_3 r_{CE3} \parallel \beta_5 r_{CE5} \\ g_{m5}r_{DS5}^2 \end{cases} \qquad (4.101)$$

$$G = \frac{A_D}{A_{CM}} \approx g_{m1}r_0 \frac{r_{o,D}}{r_{o,CM}}$$

With common-mode operation of the n-channel cascode differential amplifier, the output resistance at the drain connection of T_3 increases to $2g_{m1}g_{m3}r_{DS3}^2r_0$ and can be ignored. In the npn cascode differential amplifier, the maximum output resistance $\beta_3 r_{CE3}$ at the collector of T_3 is already achieved with differential operation; it is not possible to increase it any further.

For the npn cascode differential amplifier, the insertion of the small-signal parameters leads to

$$A_D \approx -\frac{1}{2V_T \left(\dfrac{1}{\beta_{npn} V_{A,npn}} + \dfrac{1}{\beta_{pnp} V_{A,pnp}} \right)} \tag{4.102}$$

and for the n-channel cascode differential amplifier with MOSFETs of the same size – that is, the same transconductance coefficients K – we obtain:

$$A_D \approx -\frac{K}{I_D \left(\dfrac{1}{V_{A,nC}^2} + \dfrac{1}{V_{A,pC}^2} \right)} = -\frac{2}{(V_{GS} - V_{th})^2 \left(\dfrac{1}{V_{A,nC}^2} + \dfrac{1}{V_{A,pC}^2} \right)} \tag{4.103}$$

With the bipolar transistors from Fig. 4.5, the differential gain is $A_D \approx -38500$ and with the MOSFETs from Fig. 4.6 it is $A_D \approx -1500$.

If the Early voltage and the current gain of the npn and pnp transistors and the Early voltages of the n-channel and p-channel MOSFETs have the same magnitude, then:

$$|A_D| \approx \begin{cases} \dfrac{\beta\, g_m r_{CE}}{4} = \dfrac{\beta\, V_A}{4V_T} \overset{\beta \approx 100}{\approx} 25.000 \ldots 150.000 \\[4mm] \dfrac{g_m^2 r_{DS}^2}{4} = \left(\dfrac{V_A}{V_{GS} - V_{th}} \right)^2 \approx 400 \ldots 10.000 \end{cases}$$

Therefore, it is possible to achieve a differential gain in the region of $10^5 = 100\,\text{dB}$ with *one* npn cascode differential amplifier; in contrast, the maximum differential gain with an n-channel cascode differential amplifier is approximately $10^4 = 80\,\text{dB}$.

Differential amplifier with a current mirror: The use of a current mirror produces a differential amplifier with an asymmetric output; Fig. 4.77a shows a simple circuit design that has already been presented in Fig. 4.66 on page 342 and studied with regard to the large-signal performance. The use of the cascode current mirror in the cascode differential amplifier results in the circuit shown in Fig. 4.77b. The current ratio of the current mirror must be $k_I = 1$ (in practice, $k_I \approx 1$).

The small-signal parameters can be derived easily if the following properties are considered:

- The current mirror doubles the output current in the case of differential operation (see Fig. 4.66); this increases the differential gain by a factor of 2.
- In common-mode operation the currents change proportionally and are subtracted at the output by the current mirror. With ideal subtraction by an ideal current mirror, the output voltage remains constant; consequently, $A_{CM} = 0$. Real current mirrors generate a small common-mode gain.
- The output resistance r_o corresponds to the differential output resistance $r_{o,D}$ of the corresponding symmetric circuit.

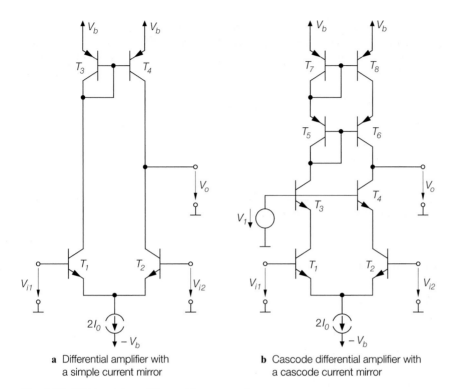

a Differential amplifier with
a simple current mirror

b Cascode differential amplifier with
a cascode current mirror

Fig. 4.77. Differential amplifiers with current mirrors

This leads to the *basic equations for an asymmetric differential amplifier with a current mirror*:

$$A_D = \left.\frac{v_{o1}}{v_D}\right|_{i_o=0} = -g_m r_o \tag{4.104}$$

$$A_{CM} = \left.\frac{v_{o1}}{v_{CM}}\right|_{i_o=0} \approx 0 \tag{4.105}$$

$$G = \frac{A_D}{A_{CM}} \to \infty \tag{4.106}$$

The output resistance of the *differential amplifier with a simple current mirror* is

$$r_o = \left.\frac{v_{o1}}{i_{o1}}\right|_{v_D=0} \approx \begin{cases} r_{CE2} \,\|\, r_{CE4} \\ r_{DS2} \,\|\, r_{DS4} \end{cases} \tag{4.107}$$

and for the *cascode differential amplifier with a cascode current mirror*:

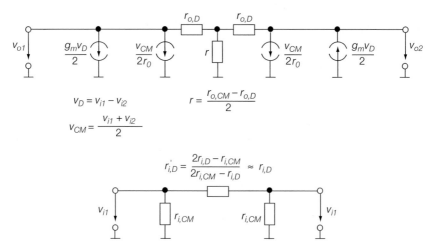

Fig. 4.78. Equivalent circuit of a differential amplifier

$$r_o = \left. \frac{v_{o1}}{i_{o1}} \right|_{v_D=0} = \begin{cases} \beta_4 r_{CE4} \, || \, \dfrac{\beta_6 r_{CE6}}{2} \\[2ex] g_{m4} r_{DS4}^2 \, || \, g_{m6} r_{DS6}^2 \end{cases} \tag{4.108}$$

In the npn cascode differential amplifier with a cascode current mirror, it should be noted that the output resistance of the cascode current mirror with $k_I = 1$ is only half of the output resistance of a current source with a cascode (see (4.23) and (4.27)).

Equivalent circuit: With the help of the small-signal parameters of a differential amplifier, we obtain the equivalent circuit shown in Fig. 4.78. On the input side, it comprises a π network with three resistances, to simulate the input resistances $r_{i,D}$ and $r_{i,CM}$ of the npn differential amplifier; the resistances are omitted in the n-channel differential amplifier. Since the resistances $r_{i,CM}$ also affect the differential operation, the cross-resistance must have the value

$$r'_{i,D} = \frac{2r_{i,D} r_{i,CM}}{2r_{i,CM} - r_{i,D}}$$

This ensures that the effective differential input resistance is $r_{i,D}$. In practice, $r_{i,CM} \gg r_{i,D}$ and thus $r'_{i,D} \approx r_{i,D}$. On the output side, a T network consisting of three resistances simulates the actual output resistances. The T network has the advantage that the differential output resistance has a direct influence, which is important in practical applications, and that for $r_{o,D} = r_{o,CM}$ the resistance r enters the short-circuited state. Two current sources controlled by the differential voltage v_D and the common-mode voltage v_{CM} are connected to each output; the corresponding transconductance values are $g_m/2$ for differential operation and $1/(2r_0)$ for common-mode operation.

Nonlinearity: The series expansion of the characteristics allows us to determine the approximate value of the distortion factor of a differential amplifier. For the npn differential amplifier, by transition to the small-signal values it follows from (4.62) that:

$$v_{o1} = -I_0 R_C \tanh \frac{v_D}{2V_T} = -I_0 R_C \left[\frac{v_D}{2V_T} - \frac{1}{3} \left(\frac{v_D}{2V_T} \right)^3 + \cdots \right]$$

Inserting $v_D = \hat{v}_D \cos \omega t$ leads to:

$$v_{o1} = -I_0 R_C \left[\left(\frac{v_D}{2V_T} - \frac{v_D^3}{32V_T^3} + \cdots \right) \cos \omega t - \left(\frac{v_D^3}{96V_T^3} - \cdots \right) \cos 3\omega t + \cdots \right]$$

For small amplitudes ($v_D < 2V_T$), an approximation of the *distortion factor of the npn differential amplifier without current feedback* results from the ratio of the amplitudes at $3\omega t$ and ωt:

$$k \approx \frac{1}{48} \left(\frac{\hat{v}_D}{V_T} \right)^2 \tag{4.109}$$

For $V_T = 26\,\mathrm{mV}$ and where the maximum distortion factor is given, we obtain:

$$\hat{v}_D < V_T \sqrt{48k} = 180\,\mathrm{mV} \cdot \sqrt{k}$$

The following must be the case: $\hat{v}_D < 18\,\mathrm{mV}$ for $k < 1\%$. This makes the npn differential amplifier considerably more linear than the common-emitter circuit, for which an amplitude of only $\hat{v}_i < 1\,\mathrm{mV}$ is permissible for $k < 1\%$. Furthermore, in order to reduce the distortion factor, the amplitude must be reduced proportionally to the root of the distortion factor, and not linearly as is the case in the common-emitter circuit.

The calculation is only valid as long as no overload occurs at the output; this was implied by the assumption of an ideal tanh characteristic. However, the gain of most differential amplifiers with current sources is so high that a differential operation of as little as a few millivolts causes the output to be overloaded; this is particularly the case in cascode differential amplifiers. In this case, the differential amplifier works virtually linearly up to the point of overload of the output and the distortion factor is correspondingly low. As soon as overload at the output sets in, however, the distortion factor increases rapidly.

The differential voltage of the npn differential amplifier with current feedback is:

$$V_D = V_{BE1} + I_{C1}R_E - V_{BE2} - I_{C2}R_E = V_{BE1} - V_{BE2} + (I_{C1} - I_{C2}) R_E$$

If V_D is replaced by $V_D' = V_{BE1} - V_{BE2}$ it follows from (4.61) that

$$I_{C1} - I_{C2} = 2I_0 \tanh \frac{V_D'}{2V_T}$$

Insertion and transition to the small-signal values leads to:

$$v_D = v_D' + 2I_0 R_E \tanh \frac{v_D'}{2V_T}$$

It follows from (4.62) that:

$$v_{o1} = -I_0 R_C \tanh \frac{v_D'}{2V_T}$$

Series expansion and the elimination of v_D' results in

$$v_{o1} = -\frac{I_0 R_C}{I_0 R_E + V_T} \left(v_D - \frac{V_T v_D^3}{12 (I_0 R_E + V_T)^3} + \cdots \right)$$

which allows us to calculate the *distortion factor of an npn differential amplifier with current feedback*:

$$k \approx \frac{V_T v_D^2}{48 \left(I_0 R_E + V_T\right)^3} \overset{g_m = I_0/V_T}{=} \frac{1}{48 \left(1 + g_m R_E\right)^3} \left(\frac{\hat{v}_D}{V_T}\right)^2 \tag{4.110}$$

As the feedback factor $(1 + g_m R_E)$ has a cubic influence on the distortion factor but only a linear influence on the differential gain, the distortions drop quadratically with the feedback factor when the output amplitude is kept constant. Thus, the linearizing effect of the current feedback is much stronger in the differential amplifier than in the common-emitter circuit, in which the distortions at the output only drop linearly with the feedback factor when the output amplitude is constant.

With the same approach in the n-channel differential amplifier, we obtain the *distortion factor of an n-channel differential amplifier*:

$$k \approx \frac{K \hat{v}_D^2}{64 I_0 \left(1 + \sqrt{2 K I_0} R_S\right)^3} \overset{g_m = \sqrt{2 K I_0}}{=} \frac{K \hat{v}_D^2}{64 I_0 \left(1 + g_m R_S\right)^3} \overset{R_S = 0}{=} \frac{K \hat{v}_D^2}{64 I_0} \tag{4.111}$$

Here too, the feedback factor $(1 + g_m R_S)$ has a cubic effect. In contrast to the npn differential amplifier, the size of the MOSFETs influences the results in the form of the transconductance coefficient K. Without feedback ($R_S = 0$) the distortion factor increases linearly with the size of the MOSFETs ($k \sim K$), but it reduces when the negative feedback is high ($k \sim 1/\sqrt{K}$ for $g_m R_S \gg 1$). The equations are only valid under the condition that the output is not overloaded.

Relating the distortion factor to the amplitude $\hat{v}_o$ at the output of a differential amplifier with resistances and demanding a certain differential gain leads to an expression that is helpful for dimensioning the circuit. We shall therefore look at the differential amplifiers with current feedback shown in Fig. 4.79, which enter the state of a corresponding differential amplifier without current feedback when $R_E = 0$ or $R_S = 0$. For the npn differential amplifier, we obtain:

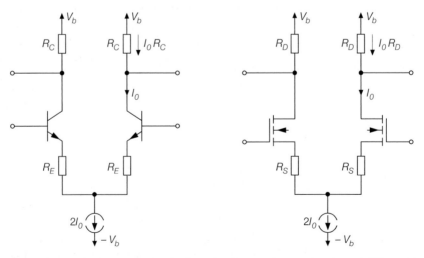

Fig. 4.79. Circuits for comparing the distortion factors of npn and n-channel differential amplifiers

$$
\left. \begin{aligned}
k_{npn} &\approx \frac{1}{48\,(1 + g_m R_E)^3} \left(\frac{\hat{v}_D}{V_T} \right)^2 \\[2mm]
|A_D| &\approx \frac{\hat{v}_o}{\hat{v}_D} = \frac{g_m R_C}{1 + g_m R_E}
\end{aligned} \right\} \quad \Rightarrow \quad k_{npn} \approx \frac{|A_D|\,V_T\,\hat{v}_o^2}{6\,(I_0 R_C)^3}
$$

Here, $I_0 R_C$ is the voltage drop across the collector resistance (see Fig. 4.79a). For the n-channel differential amplifier:

$$
\left. \begin{aligned}
k_{nC} &\approx \frac{K\,v_D^2}{64\,I_0\,(1 + g_m R_S)^3} \\[2mm]
|A_D| &\approx \frac{\hat{v}_o}{\hat{v}_D} = \frac{g_m R_D}{1 + g_m R_S}
\end{aligned} \right\} \quad \Rightarrow \quad k_{nC} \approx \frac{|A_D|\,(V_{GS} - V_{th})\,\hat{v}_o^2}{32\,(I_0 R_D)^3}
$$

Here, $I_0 R_D$ is the voltage drop across the drain resistance (see Fig. 4.79b). In both differential amplifiers, the distortion factor is proportional to the third power of the voltage drop across resistances R_C and R_D. As this voltage drop must be selected as a function of the supply voltage V_b, the distortion factor increases almost cubically when V_b decreases: halving the supply voltage thus results in an eight-fold distortion factor. The feedback resistances R_E and R_S are not shown explicitly, as their values are in a fixed relation to R_C and R_D, respectively, due to the presupposed constant differential gain. From the ratio

$$
\frac{k_{nC}}{k_{npn}} \approx \frac{3}{16} \frac{V_{GS} - V_{th}}{V_T} \overset{V_T = 26\,\mathrm{mV}}{=} \frac{V_{GS} - V_{th}}{140\,\mathrm{mV}}
$$

it follows that the distortion factor of an npn differential amplifier is usually lower than that of an n-channel differential amplifier with the same differential gain.

Example: In the description of the n-channel differential amplifier with current feedback, the characteristics of the circuits shown in Fig. 4.63 on page 339 were compared (see Fig. 4.64). The outcome was that the characteristics of the differential amplifier without current feedback are less linear than that of the differential amplifier with current feedback. This result can be verified by means of the distortion factor approximations. Both circuits have the same quiescent current and the same differential gain; that is, the same output amplitude and the same input amplitude $\hat{v}_D$. For the differential amplifier without feedback, where $I_0 = 100\,\mu\mathrm{A}$, $K = 15 \cdot 30\,\mu\mathrm{A/V}^2 = 0.45\,\mathrm{mA/V}^2$ (size 15) and $\hat{v}_D = 0.5\,\mathrm{V}$, the distortion factor is $k \approx 1.76\%$; for the differential amplifier with feedback, where $K = 150 \cdot 30\,\mu\mathrm{A/V}^2 = 4.5\,\mathrm{mA/V}^2$ (size 150) and $R_S = 2\,\mathrm{k}\Omega$, and with the same values for all other parameters, the distortion factor is $k \approx 0.72\%$. This confirms the previous result.

Setting the Operating Point

The operating point of the differential amplifier is essentially set by the current source $2I_0$. This determines the quiescent currents of the transistors and thus the small-signal response; only in the differential amplifier with resistances can the resistances be treated as additional freely selectable parameters. The operating point voltages play a minor role in the differential amplifier as long as all bipolar transistors are in normal mode at the operating point and all MOSFETs operate in the pinch-off region. This requirement is met in full when the common-mode voltage V_{CM} remains within the *common-mode input voltage range*; this has already been explained in connection with the characteristics. The

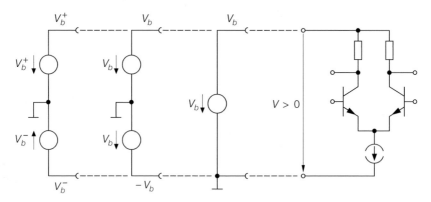

Fig. 4.80. Supply voltages for differential amplifiers: general, symmetrical and unipolar

common-mode voltage range depends on the construction of the differential amplifier, as well as on the supply voltages and the required output amplitude.

Supply voltages: In general, a differential amplifier is provided with the two supply voltages V_b^+ and V_b^-; where $V_b^+ > V_b^-$. The voltage difference $V_b^+ - V_b^-$ must be at least large enough to keep all transistors in the normal or pinch-off region, and small enough to ensure that the maximum permissible voltages are not exceeded in any transistor. Theoretically, all combinations that meet these conditions are possible, but in practice two cases are particularly common:

– A symmetrical voltage supply, with $V_b^+ > 0$ and $V_b^- = -V_b^+$. In this case, the supply voltage terminals are usually called V_b and $-V_b$. Examples: ± 12 V, ± 5 V.
– A unipolar voltage supply, with $V_b^+ > 0$ and $V_b^- = 0$. Here, the terminal V_b^- is connected to ground. The terminal V_b^+ is usually called V_b. Examples: 12 V; 5 V; 3.3 V.

Figure 4.80 shows a comparison of the general case and the two practical solutions. For unipolar voltage supply only one supply voltage source is required.

Common-mode voltage range: The common-mode voltage range of a differential amplifier with a unipolar voltage supply is entirely within the range of positive voltages; that is, at the operating point $V_{CM} > 0$ must be the case. For a symmetrical voltage supply with a sufficiently high voltage V_b, both $V_{CM} = 0$ and $V_{CM} < 0$ are possible, because the common-mode voltage extends across positive and negative voltages. Therefore, the inputs of a differential amplifier with symmetrical voltage supply may be connected directly to a signal source without a DC voltage; a particular feature is that one input may be connected to ground, as was done in the differential amplifiers with asymmetric input shown in Fig. 4.54 on page 330.

Differential amplifier with resistances: Fig. 4.81a shows the common method of setting the operating point in a differential amplifier with resistances, taking an npn differential amplifier as an example. Current $2I_0$ is derived from the reference current I_1 by an npn current mirror, the current ratio being $k_I = 2I_0/I_1$. Current I_1 can be most easily adjusted with a resistance R_1. Voltage V_0 at the output of the current mirror must not drop below the lower limit $V_{0,min}$ – in the simple current mirror $V_{CE,sat}$ or $V_{DS,po}$ – this represents a lower limit of the common-mode voltage range.

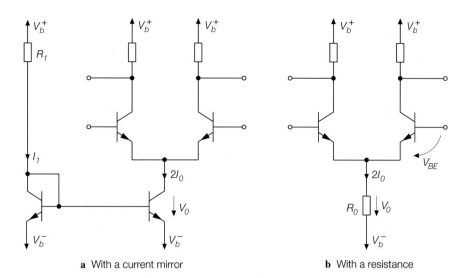

a With a current mirror   **b** With a resistance

Fig. 4.81. Common methods for setting the operating point in npn differential amplifiers with resistances

For only slight variations in the common-mode voltage, the current source can be replaced by a resistance (see Fig. 4.81b):

$$R_0 = \frac{V_0}{2I_0} = \frac{V_{CM} - V_{BE} - V_b^-}{2I_0}$$

In this case, the common-mode rejection is comparatively low because resistance R_0 is usually significantly lower than the output resistance r_0 of a real current source.

Differential amplifier with current sources: Fig. 4.82 uses an npn differential amplifier to show the common method of setting the operating point in differential amplifiers with simple or cascode current mirrors. As in the differential amplifier with resistances, the current mirror $2I_0$ is realized by an npn current mirror with the current ratio $k_I = 2I_0/I_1$. A pnp current mirror with two outputs is used for the current sources at the output; here, the same reference current I_1 is used, which leads to a current ratio of $k_I = I_0/I_1$. Here again, a resistance R_1 is the simplest method of adjusting current I_1. Voltage V_1 for the cascode stage is adjusted to $V_b^+ - 2V_{EB} \approx V_b^+ - 1.4\,\text{V}$ by the two pnp transistor diodes.

Cascode differential amplifier: The cascode differential amplifier with cascode current sources requires two auxiliary voltages; Fig. 4.83 shows an npn cascode differential amplifier as an example of a typical circuit. The current is adjusted in the same way as in the differential amplifier with current sources. Voltage V_2 for the pnp cascode stage is again set to $V_b^+ - 2V_{EB} \approx V_b^+ - 1.4\,\text{V}$ by two pnp transistor diodes. Voltage V_1 for the npn cascode stage is provided via a voltage divider that consists of resistances R_1 and R_2 and a collector circuit for impedance conversion; the current of the collector circuit is set by an additional current source. The value of voltage V_1 affects the voltage range at the input and at the output: a relatively high voltage V_1 provides a larger common-mode voltage range at the input and a lower voltage range at the output; a lower voltage has the opposite effect.

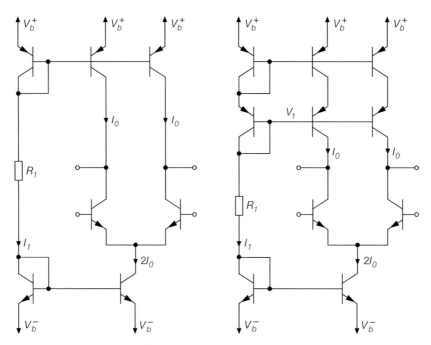

Fig. 4.82. A common method for setting the operating point in npn differential amplifiers with current sources

Differential amplifier with folded cascode: In the ideal case, the input and output volt-age ranges should cover the entire range of the supply voltages. The differential amplifier with folded cascode shown in Fig. 4.84 is very close to this ideal. It is evolved from the normal cascode differential amplifier by folding the cascode stage and the output current sources downward and adding two more current sources. The input and the output can now be modulated almost over the entire range of the supply voltages; a particular conse-quence of this is that the output voltages may be lower than the input voltages. However, the small-signal response remains the same. In practice, an asymmetrical output is usually used, by replacing the output current sources with a cascode current mirror; this results in the circuit shown in Fig. 4.85, which is predominantly used as an input stage in operational amplifiers because of its wide voltage range and its high differential gain and common-mode rejection. Resistance R_1 is often replaced by the reference current source described in Sect. 4.1.5, to make the quiescent currents independent of the supply voltages.

Output voltage control: In all symmetrical differential amplifiers with current sources, the output voltages at the operating point are undefined without external circuits. The reason for this is the small difference between the currents of npn and pnp transistors or n-channel and p-channel transistors, which cause the outputs to reach either the upper or lower voltage limits. With low-resistance loads at the outputs, the operating point is defined by the loads; they absorb the transistor differential currents. If, however, high-

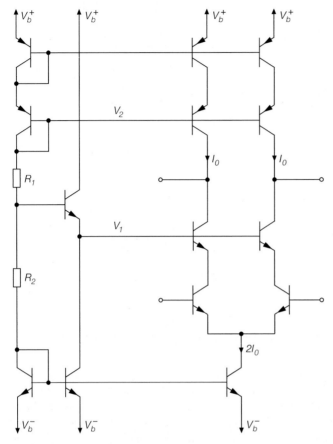

Fig. 4.83. A common method for setting the operating point in an npn cascode differential amplifier with cascode current sources

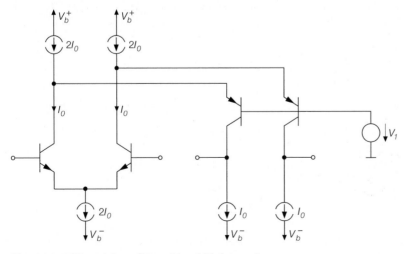

Fig. 4.84. Differential amplifier with a folded cascode

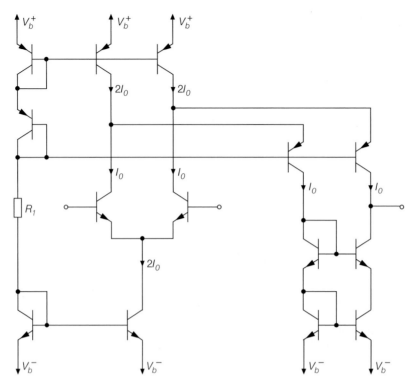

Fig. 4.85. A common configuration of a differential amplifier with a folded cascode and an asymmetrical output

resistive loads are connected, the output voltages must be controlled in order to prevent overload in the circuit; this is done by suitably controlling either current source $2I_0$ or the two output current sources I_0.

If common-collector or common-drain stages are connected to the outputs for impedance conversion, the current source $2I_0$ can be controlled by adjusting the quiescent currents of these circuits via resistances that are connected to the reference path of the current source; Fig. 4.86 illustrates this method using an npn differential amplifier with npn common-collector circuits as an example. For $R_2 = R_3$, the outputs at the operating point are:

$$V_{o,A} = V_b^- + V_{BE\,7} + I_1 R_2 = V_b^- + V_{BE\,7}\left(1 + \frac{R_2}{2R_4}\right) \qquad \text{with } V_{BE\,7} \approx 0.7\,\text{V}$$

This presupposes that the current mirror T_7, T_8 has the current ratio 2, as in the noncontrolled mode. Alternatively, resistance R_4 can be omitted and the operating point can be adjusted with the current ratio k_I of the current mirror T_7, T_8; it thus follows:

$$k_I\,(I_0 + 2I_1) \equiv 2I_0 \;\Rightarrow\; I_1 = I_0\left(\frac{1}{k_I} - \frac{1}{2}\right)$$

The output voltages are related to the supply voltage V_b^-, which is particularly disadvantageous in circuits with variable supply voltages. This is remedied in the version shown in Fig. 4.87, in which the output voltages relate to supply voltage V_b^+ and where the pnp current sources are controlled. Here, the following applies:

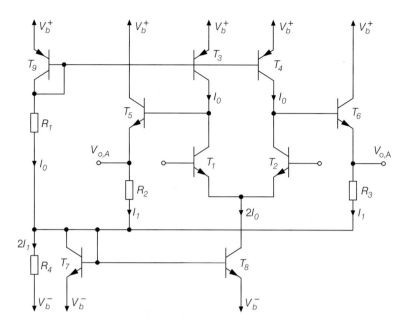

Fig. 4.86. Control of the output voltages in a differential amplifier with common-collector circuits (in relation to the supply voltage V_b^-)

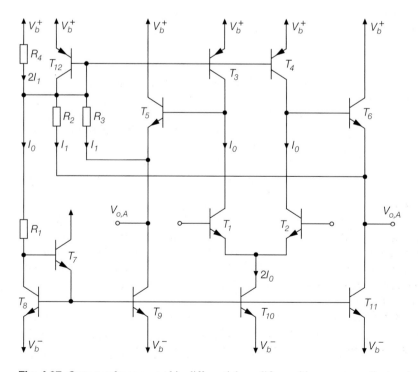

Fig. 4.87. Output voltage control in differential amplifiers with common-collector circuits (in relation to the supply voltage V_b^+

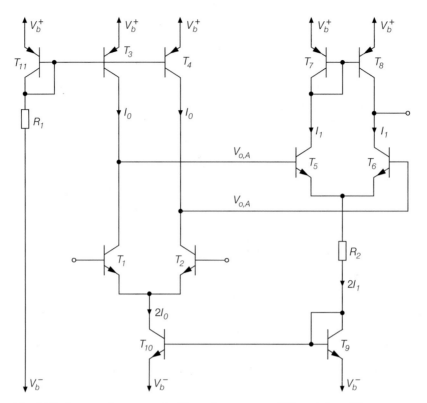

Fig. 4.88. Output voltage control with a subsequent npn differential amplifier

$$V_{o,A} \;=\; V_b^+ - V_{EB12} - I_1 R_2 \;=\; V_b^+ - V_{EB12}\left(1 + \frac{R_2}{2R_4}\right) \qquad \text{with } V_{EB12} \approx 0.7\text{ V}$$

Again, resistance R_4 can be omitted and the operating point can be adjusted using the current ratio k_I of current mirrors T_{12}, T_3 and T_{12}, T_4:

$$I_1 \;=\; \frac{I_0}{2}\left(\frac{1}{k_I} - 1\right)$$

Here, condition $k_I < 1$ must be met; that is, T_{12} must be larger than T_3 and T_4.

In both versions resistances, R_2 and R_3 must not be too small, as they present a load to the output and thus reduce the differential gain. Therefore, in differential amplifiers with very high output resistances it is often necessary to connect two common-collector circuits in series before the resistances can be connected. In the corresponding circuits with MOSFETs, however, a single common-drain circuit prevents any effect on the differential gain as a result of the resistances.

The same method can be used when another npn differential amplifier is connected instead of the common-collector circuits; Fig. 4.88 shows the corresponding circuit. With the current ratio k_I of the current mirror T_9, T_{10}, the parameters are:

$$I_1 \;=\; \frac{I_0}{k_I} \quad , \quad V_{o,A} \;=\; V_b^- + V_{BE9} + 2I_1 R_2 + V_{BE5}$$

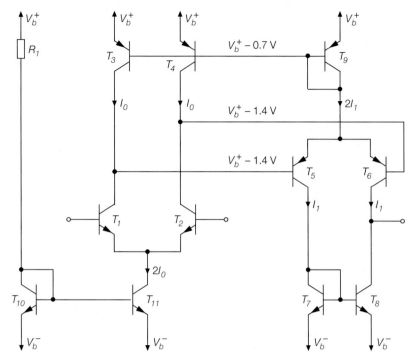

Fig. 4.89. Output voltage control with a subsequent pnp differential amplifier

If a pnp differential amplifier is connected, the circuit shown in Fig. 4.89 can be used, in which the pnp current sources are controlled without additional resistances; here, the voltage is

$$V_{o,A} = V_b^+ - V_{EB9} - V_{EB5} \approx V_b^+ - 1.4\,\text{V}$$

and with the current ratio k_I of the current mirrors T_9, T_3 and T_9, T_4 the current is:

$$I_1 = \frac{I_0}{2k_I}$$

In this version the gain of the control loop is very high and may have to be limited by current feedback resistances in the current mirrors, which means that resistances corresponding to the current ratio have to be inserted on the emitter side of T_3, T_4 and T_9. Such circuits are used especially in precision operational amplifiers

All methods of controlling the output voltages increase the common-mode rejection, because they compensate the unidirectional change of the output voltages caused by the common-mode voltage. Therefore, operational amplifiers provided with the circuit shown in Fig. 4.89 have a particularly high common-mode rejection and, due to the two differential amplifiers, a particularly high differential gain.

Frequency Response and Cutoff Frequency of Differential Amplifiers

The differential and common-mode gains as calculated so far apply to low signal frequencies; with higher frequencies, the capacitances of the transistors must be taken into account and the dynamic small-signal models must be used for calculating the frequency

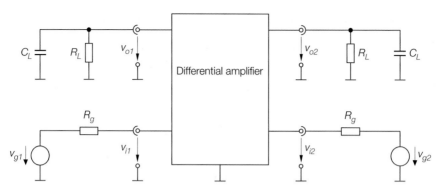

Fig. 4.90. A circuit for determining the frequency response

response. In differential amplifiers we must distinguish between the frequency response of the differential gain and the frequency response of the common-mode gain; the quotient of both is the frequency response of the common-mode rejection.

Since the frequency response depends on the external circuitry, the operational gain has to be considered; in other words, the internal resistances R_g of the signal sources and the load impedances, which consist of the load resistance R_L and the load capacitance C_L, must be taken into account (see Fig. 4.90). The small-signal voltages v_{g1} and v_{g2} of the signal sources are replaced by the *differential voltage* $v_{g,D}$ and the *common-mode voltage* $v_{g,CM}$ in the usual way:

$$v_{g,D} = v_{g1} - v_{g2} \quad , \quad v_{g,CM} = \frac{v_{g1} + v_{g2}}{2} \tag{4.112}$$

This allows us to define the *operational differential gain* $\underline{A}_{B,D}(s)$, the *operational common-mode gain* $\underline{A}_{B,CM}(s)$ and the *operational common-mode rejection* $\underline{G}_B(s)$:

$$\underline{A}_{B,D}(s) = \left. \frac{\underline{v}_{o1}(s)}{\underline{v}_{g,D}(s)} \right|_{v_{g,CM}=0} \tag{4.113}$$

$$\underline{A}_{B,CM}(s) = \left. \frac{\underline{v}_{o1}(s)}{\underline{v}_{g,CM}(s)} \right|_{v_{g,D}=0} \tag{4.114}$$

$$\underline{G}_B(s) = \frac{\underline{A}_{B,D}(s)}{\underline{A}_{B,CM}(s)} \tag{4.115}$$

In what follows, the prefix *operational* is omitted for the sake of simplicity.

The symmetry characteristics are utilized again to calculate the frequency responses. In this way, we can convert the symmetrical differential amplifier back to the corresponding common-emitter, common-source or cascode circuits. This is not possible with asymmetric differential amplifiers with current mirrors because of the asymmetry; in addition, the frequency response of the current mirror has to be taken into account. For the calculation of static parameters, an ideal current mirror was assumed; this meant that the results of the symmetric differential amplifier could simply be transferred to the asymmetric differential amplifier. This procedure may be used here as well, since current mirrors generally have a very high cutoff frequency; an ideal frequency response is thus assumed for the current mirror. The cutoff frequencies of a symmetric and an asymmetric differential amplifier of the same construction are thus identical.

Frequency response and the cutoff frequency of the differential gain: The frequency response of the differential gain can be approximately described by a first-order lowpass filter:

$$\underline{A}_{B,D}(s) \approx \frac{A_0}{1 + \dfrac{s}{\omega_g}} \tag{4.116}$$

If the internal resistance R_g of the signal source and the load resistance R_L are taken into account, then the operational gain A_0 at low frequencies is:

$$A_0 = \underline{A}_{B,D}(0) = A_B = \frac{r_{i,D}}{r_{i,D} + 2R_g} A_D \frac{R_L}{r_{o,D} + R_L} \tag{4.117}$$

The *-3 dB cutoff frequency* f_{-3dB}, at which the magnitude of the gain is 3 dB lower, is derived from (4.116): $\omega_{-3dB} \approx \omega_g$. This can be described by means of the low-frequency gain A_0 and two time constants:

$$\omega_{-3dB} = 2\pi f_{-3dB} = \frac{1}{T_1 + T_2|A_0|} \overset{|A_0| \gg T_1/T_2}{\approx} \frac{1}{T_2|A_0|} \tag{4.118}$$

For $|A_0| \gg T_1/T_2$, the cutoff frequency is inversely proportional to the magnitude of the gain A_0 and the result is a constant *gain–bandwidth product* (*GBW*):

$$GBW = f_{-3dB}|A_0| \approx \frac{1}{2\pi T_2} \tag{4.119}$$

The time constants T_1 and T_2 for the different types of differential amplifiers are described in the following sections:

2.4.1   Common-emitter circuit: (2.92), (2.96), (2.99)–(2.101)   pp. 124ff.
3.4.1   Common-source circuit: (3.77), (3.80), (3.83)   pp. 246ff.
4.1.2   Cascode circuit   (4.45), (4.46), (4.53), (4.54)   pp. 321 and 326

Figure 4.91 contains an overview of the case in which the capacitances of the npn and pnp transistors and the n-channel and p-channel MOSFETs are identical. In order to differentiate, for time constant T_2 all capacitances with factor 2 must be replaced by the sum of the corresponding values:

$$2C_C \to C_{C,npn} + C_{C,pnp} \ , \ \ 2C_S \to C_{S,npn} + C_{S,pnp}$$
$$2C_{GD} \to C_{GD,nC} + C_{GD,pC} \ , \ \ 2C_{BD} \to C_{BD,nC} + C_{BD,pC}$$

All other capacitances relate to the npn transistors in the npn differential amplifier or to the n-channel MOSFETs in the n-channel differential amplifier; this also applies to the capacitances shown with factor 2 in the time constant T_1.

Some of the equations in Fig. 4.91 have been modified, as a comparison with the original calculated forms will show:

npn	Time constants
With resistances	$T_1 = (C_E + C_C)(R_g \| r_{BE})$ $T_2 = \left(C_C + \dfrac{C_S + C_L}{\beta}\right) R_g + \dfrac{C_C + C_S + C_L}{g_m}$
With resistances and current feedback	$T_1 = (C'_E + C_C)(R_g \| r'_{BE})$ $T_2 = \left(C_C + \dfrac{C_S + C_L}{\beta}\right) R_g + \dfrac{C_C + C_S + C_L}{g'_m}$ with $g'_m = g_m/(1 + g_m R_E)$, $C'_E = C_E/(1 + g_m R_E)$ and $r'_{BE} = r_{BE}(1 + g_m R_E)$
With current sources	$T_1 = (C_E + C_C)(R_g \| r_{BE})$ $T_2 = \left(C_C + \dfrac{C_C + 2C_S + C_L}{\beta}\right) R_g + \dfrac{2C_C + 2C_S + C_L}{g_m}$
With a cascode	$T_1 = (C_E + 2C_C)(R_g \| r_{BE})$ $T_2 = (2C_C + 2C_S + C_L)\left(\dfrac{R_g}{\beta} + \dfrac{1}{g_m}\right)$

n-channel	Time constants
With resistances	$T_1 = (C_{GS} + C_{GD}) R_g$ $T_2 = C_{GD} R_g + \dfrac{C_{GD} + C_{BD} + C_L}{g_m}$
With resistances and current feedback	$T_1 = (C'_{GS} + C_{GD}) R_g$ $T_2 = C_{GD} R_g + \dfrac{C_{GD} + C_{BD} + C_L}{g'_m}$ with $g'_m \approx g_m/(1 + g_m R_S)$ and $C'_{GS} \approx C_{GS}/(1 + g_m R_S)$
With current sources	$T_1 = (C_{GS} + C_{GD}) R_g$ $T_2 = C_{GD} R_g + \dfrac{2C_{GD} + 2C_{BD} + C_L}{g_m}$
With a cascode	$T_1 = (C_{GS} + 2C_{GD}) R_g$ $T_2 = \dfrac{2C_{GD} + 2C_{BD} + C_L}{g_m}$

Fig. 4.91. Time constants for the cutoff frequency of the differential gain

- The extrinsic base resistance and the gate resistance have been neglected; that is, $R'_g = R_g + R_B$ and $R'_g = R_g + R_G$ have been replaced by R_g.
- For npn differential amplifiers, the basic equations for the common-emitter circuit have been expanded by adding the substrate capacitance C_S; furthermore, C_L has been replaced by $C_L + C_S$, as the substrate capacitance has the effect of a load capacitance.
- In the basic equations for the common-source circuit for n-channel differential amplifiers, the drain–source capacitance C_{DS}, which only occurs in discrete MOSFETs, has been replaced by the bulk–drain capacitance C_{BD}.

Bipolar transistor	MOSFET
$g_m = \dfrac{\beta}{r_{BE}} = \dfrac{I_{C,A}}{V_T}$ (with $\beta \approx B$)	$g_m = \sqrt{2K\,I_{D,A}} = \sqrt{2\mu C'_{ox}\,I_{D,A}\dfrac{W}{L}}$
$C_E \approx g_m\,\tau_{0,N} + 2C_{J0,E}$	$C_{GS} \approx \dfrac{2}{3}C_{ox} = \dfrac{2}{3}C'_{ox}WL$
$C_C \approx C_{J0,C}$	$C_{GD} = C'_{GD,ov}W$
$C_S \approx C_{J0,S}$	$C_{BD} \approx C'_J A_D$ (A_D : Drain area)

Fig. 4.92. Small-signal parameters of integrated bipolar transistors and MOSFETs

With current feedback, some parameters are transformed by the feedback factor; Fig. 4.91 only reflects this for the differential amplifier with resistances, but the same can be done for other types.

The small-signal parameters of integrated bipolar transistors and MOSFETs, which are required to evaluate the time constants, are listed in Fig. 4.92; these are taken from Fig. 2.45 on page 83 (without C_E and C_C), (4.49) and (4.50) on page 325 and Fig. 3.52 on page 223. For the junction capacitances C_C, C_S and C_{BD}, the corresponding zero capacitance $C(U = 0)$ is used without regard for the actual reverse voltage; the true capacitance is lower.

The frequency responses of the differential gain are illustrated in Fig. 4.93. The values of the low-frequency gain apply to npn differential amplifiers; the values for corresponding n-channel differential amplifiers are about ten times lower. Differential amplifiers with simple and cascode current sources achieve a higher differential gain than the differential amplifier with resistances, but feature a lower gain–bandwidth product due to the additional capacitances of the current source transistors. The cascode differential amplifier with cascode current sources offers both the highest differential gain and the highest gain–bandwidth product.

The differential gain and the gain–bandwidth product of the differential amplifier with a simple current mirror are approximately twice as large as those of corresponding symmetric differential amplifiers; therefore, both circuits have the same cutoff frequency. This also applies to the n-channel cascode amplifier with a cascode current mirror. In the npn cascode differential amplifier with a cascode current mirror, the gain–bandwidth product is also about twice as high as that of the npn cascode differential amplifier with a cascode current mirror, but the differential gain is only slightly higher, due to the low output resistance of the cascode current mirror compared to the cascode current source; the cutoff frequency is therefore higher. The frequency responses of differential amplifiers with current mirrors are not shown in Fig. 4.93 for the sake of clarity.

Frequency response of the common-mode gain: The small-signal equivalent circuit of an npn differential amplifier with resistances, as shown in Fig. 4.94, is used for calculating the frequency response of the common-mode gain; it is derived from the static small-signal equivalent circuit for common-mode gain shown in Fig. 4.71 on page 347 by changing from the static to the dynamic small-signal model of the transistor. C_0 is the output capacitance of the current source, of which only half the value has an effect due to the split. The equivalent circuit for common-mode gain differs from the equivalent circuit for differential

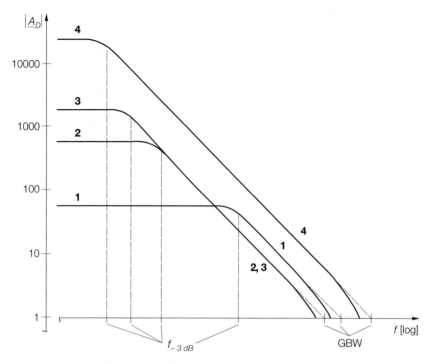

1: Differential amplifier with resistances
2: Differential amplifier with simple current sources
3: Differential amplifier with cascode current sources
4: Cascode differential amplifier with cascode current sources

Fig. 4.93. Frequency responses of the differential gain (the numeric values apply to npn differential amplifiers)

gain only by the impedance of the current source, which causes frequency-dependent current feedback; the approximate frequency response of the common-mode gain can thus be calculated from the frequency response of the differential gain by replacing the transconductance g_m by the reduced transconductance:

$$g_{m,red}(s) \;=\; \frac{g_m}{1 + g_m \left(2\,r_0 \,\|\, \dfrac{2}{s\,C_0}\right)} \;\overset{g_m r_0 \gg 1}{\approx}\; \frac{1 + sC_0 r_0}{2r_0 \left(1 + s\,\dfrac{C_0}{2g_m}\right)}$$

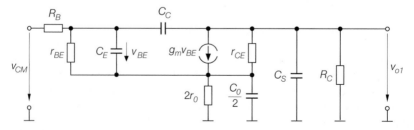

Fig. 4.94. Dynamic small-signal equivalent circuit of an npn differential amplifier with resistances, with common-mode input signal

As the common-mode signal provides the full common-mode voltage at each input, the result must be multiplied by two. Taking the output resistances into account, it follows from (4.116):

$$\underline{A}_{B,CM}(s) \approx 2\underline{A}_{B,D}(s) \frac{g_{m,red}(s)r_{o,CM}}{g_m r_{o,D}} \approx \frac{A_0 r_{o,CM}}{g_m r_0 r_{o,D}} \frac{1 + sC_0 r_0}{\left(1 + s\dfrac{C_0}{2g_m}\right)\left(1 + \dfrac{s}{\omega_g}\right)}$$

If one inserts the common-mode rejection

$$G = \frac{g_m r_0 r_{o,D}}{r_{o,CM}}$$

and replaces the time constant $C_0 r_0$ by the *cutoff frequency of the common-mode rejection*

$$\omega_{g,G} = 2\pi f_{g,G} = \frac{1}{C_0 r_0} \tag{4.120}$$

then:

$$\underline{A}_{B,CM}(s) \approx \frac{A_0}{G} \frac{1 + \dfrac{s}{\omega_{g,G}}}{\left(1 + \dfrac{s}{2G\omega_{g,G}}\right)\left(1 + \dfrac{s}{\omega_g}\right)} \tag{4.121}$$

$$\underline{G}_B(s) \approx G \frac{1 + \dfrac{s}{2G\omega_{g,G}}}{1 + \dfrac{s}{\omega_{g,G}}} \tag{4.122}$$

Figure 4.95 shows the frequency responses $|\underline{A}_{B,D}|$, $|\underline{A}_{B,CM}|$ and $|\underline{G}_B|$ for the cases $f_{g,G} < f_g$ and $f_{g,G} > f_g$.

The case $f_{g,G} < f_g$ is typical for differential amplifiers with resistances or with simple current sources. The magnitude of the common-mode gain increases in the region between the common-mode cutoff frequency $f_{g,G}$ and the cutoff frequency f_g, remains constant above f_g and at high frequencies is twice as high as the magnitude of the differential gain. The common-mode rejection decreases from the common-mode cutoff frequency $f_{g,G}$ to higher frequencies at the rate of 20 dB per decade, and approaches $1/2$ at high frequencies.

The situation $f_{g,G} > f_g$ mostly occurs in cascode differential amplifiers that have a relatively low cutoff frequency f_g because of their very high low-frequency gain; f_g is still very low even if the gain–bandwidth product is high. The common-mode gain decreases between the cutoff frequency f_g and the common-mode cutoff frequency $f_{g,G}$, is constant above $f_{g,G}$ and at high frequencies is twice as high as the magnitude of the differential gain. The common-mode rejection is the same as for $f_{g,G} < f_g$.

The simplified derivation of the frequency response of the common-mode gain makes understanding easier, but leads to inaccuracies:

- Due to the frequency-dependent feedback in common-mode operation, the cutoff frequency f_g is different than in differential operation. In most circuits this effect is insignificant, but in some it is quite considerable; it then creates an additional pole and an additional zero in the common-mode rejection. In the differential amplifier with resistances, this creates a region in which the common-mode rejection decreases at a rate of 40 dB per decade, and in the differential amplifier with cascode current mirrors it

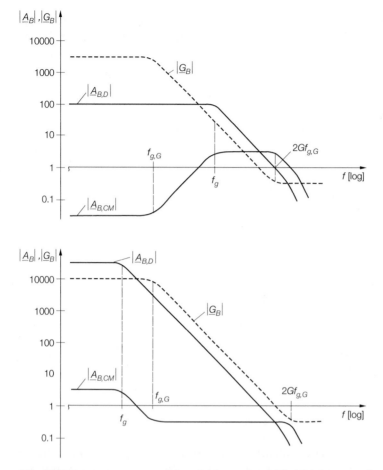

Fig. 4.95. Frequency responses $|\underline{A}_{B,D}|$, $|\underline{A}_{B,CM}|$ and $|\underline{G}_B|$ for the cases $f_{g,G} < f_g$ (*above*) and $f_{g,G} > f_g$ (*below*)

creates a region in which the common-mode rejection increases; Fig. 4.96 illustrates these special cases.

– In the npn differential amplifier, the differential and the common-mode portions of the input signal are attenuated differently because of the different input resistances in differential and in common-mode operation. Thus, especially in the case of high-resistance sources, the low-frequency value of the operational common-mode rejection $\underline{G}_B(s)$ does not correspond to the common-mode rejection G, but is reduced by a value equal to the ratio of the voltage divider factors:

$$\frac{\dfrac{r_{i,CM}}{r_{i,CM} + 2R_g}}{\dfrac{r_{i,D}}{r_{i,D} + 2R_g}} \overset{R_g \ll r_{i,CM}}{\approx} 1 + \frac{2R_g}{r_{i,D}}$$

This effect is not seen with low-resistance sources where $R_g \ll r_{i,D}$.

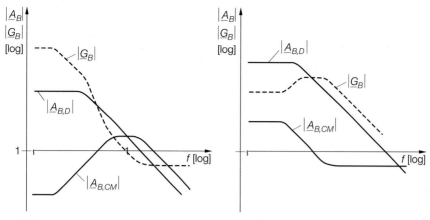

a A differential amplifier with resistances

b A differential amplifier with cascode current sources

Fig. 4.96. Frequency responses $|\underline{A}_{B,D}|$, $|\underline{A}_{B,CM}|$ and $|\underline{G}_B|$

Example: The various npn and n-channel differential amplifiers are compared below. All circuits are designed for a unipolar supply voltage $V_b = 5\,\text{V}$ and an output voltage $V_{o,A} = 2.5\,\text{V}$. The parameters for bipolar transistors are taken from Fig. 4.5 on page 274 and those for MOSFETs from Fig. 4.6 on page 274. The quiescent current is $I_0 = 100\,\mu\text{A}$ in npn differential amplifiers and $I_0 = 10\,\mu\text{A}$ in n-channel differential amplifiers. Bipolar transistors of size 1 are generally used per $100\,\mu\text{A}$ quiescent current; this corresponds to the typical value quoted in Fig. 4.5. According to Fig. 4.6, MOSFETs of size 1 would also be sufficient, but the resulting gate–source voltage of $|V_{GS}| \approx 1.8\dots2\,\text{V}$ ($|V_{BS}| = 0\dots1\,\text{V}$) is too high for the given supply voltage of 5 V; therefore n-channel MOSFETs of size 5 ($V_{GS} \approx 1.4\dots1.6\,\text{V}$) and p-channel MOSFETs of size 2 ($V_{GS} \approx -1.6\dots-1.8\,\text{V}$) are used per $10\,\mu\text{A}$ quiescent current. Since the dimensional ratio of n-channel and p-channel MOSFETs of size 1 is exactly 2/5, all MOSFETs, with the exception of the MOSFET in the current source, have the same geometric size:

$$W = 15\,\mu\text{m} \quad , \quad L = 3\,\mu\text{m}$$

The common-mode voltage at the input is $V_{CM,A} = 1\,\text{V}$ in npn differential amplifiers and $V_{CM,A} = 2\,\text{V}$ in n-channel differential amplifiers; thus the current sources in the common-emitter or common-source branch are operated just above their signal limits.

Figure 4.97 shows the differential amplifiers with resistances; the collector and drain resistances are chosen such that the desired output voltage $V_{o,A} = 2.5\,\text{V}$ is reached:

$$\left.\begin{array}{c} R_C \\ R_D \end{array}\right\} = \frac{V_b - V_{o,A}}{I_0} = \left\{\begin{array}{l} 25\,\text{k}\Omega \\ 250\,\text{k}\Omega \end{array}\right.$$

In contrast, the operating point is not automatically adjusted in the differential amplifiers with simple current sources and simple current mirrors shown in Fig. 4.98. As, in general, the collector and drain currents of transistors T_1 and T_3 or T_2 and T_4 are not exactly the same at the desired operating point, the transistor with the larger current enters the saturation or the pinch-off region; this situation overloads the outputs. In an integrated circuit the actual operating point depends on the external components at the outputs and an operating point control circuit, should this exist; the latter is described in more detail

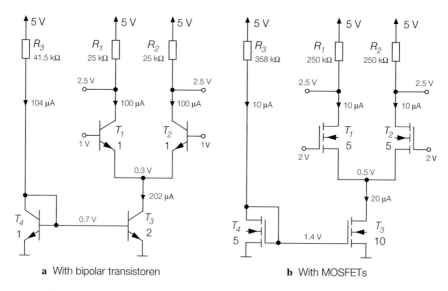

a With bipolar transistoren   **b** With MOSFETs

Fig. 4.97. Example: differential amplifiers with resistances

in the section on setting the operating point in differential amplifiers. During the circuit simulation the desired operating point can be set by, for example, connecting the outputs to a voltage source with $V_{o,A}$ via very high inductances (e.g. $L = 10^9$ H); this keeps the DC output potential at $V_{o,A}$, while for AC voltage we have an open circuit due to the inductances that present a very high impedance even at low frequencies. This method must be used for all differential amplifiers with current sources and current mirrors. For the differential amplifiers in this example, an operating point of $V_{o,A} = 2.5$ V is assumed, but the necessary external circuitry or operating point control circuit is not illustrated.

Auxiliary voltages are required to set the operating point of the cascode transistors in the differential amplifiers with cascode current sources in Fig. 4.99 and the cascode differential amplifiers with cascode current sources and cascode current mirrors as shown in Figs. 4.100 and 4.101, respectively; the generation of these voltages is detailed in Sect. 4.1.5.

The small-signal parameters of the transistors can be determined with the help of the information in Fig. 4.92 on page 373 and the parameters in Figs. 4.5 on page 274 and 4.6 on page 274, and on the basis of the quiescent currents and the transistor sizes. By means of the following equations, it is possible to calculate the gain, the output and input resistances of differential amplifiers for differential and common-mode operation:

With resistances:	(4.79)–(4.85)
With simple current sources:	(4.90)–(4.92)
With a simple current mirror:	(4.90), (4.104)–(4.106)
With a cascode current mirror:	(4.95)–(4.97)
Cascode with current sources:	(4.100), (4.101)
Cascode with current mirror:	(4.100), (4.104)–(4.106)

The operational differential gain A_0 is determined from (4.117), the time constants T_1 and T_2 from Fig. 4.91, the gain–bandwidth product (GBW) from (4.119), the -3 dB cutoff

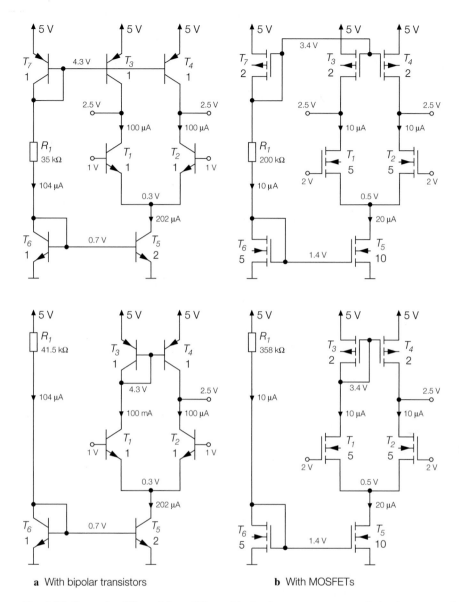

a With bipolar transistors   **b** With MOSFETs

Fig. 4.98. Example: differential amplifiers with simple current sources and simple current mirrors

frequency f_{-3dB} from (4.118) and the cutoff frequency $f_{g,G}$ of the common-mode rejection from (4.120).

When calculating the small-signal parameters of the npn transistors, the slight differences in the quiescent currents of the individual transistors are neglected; that is, $|I_{C,A}| \approx I_0 \approx 100\,\mu\text{A}$ is used. Consequently,

npn: $g_m = 3.85\,\text{mS}$, $\beta = 100$, $r_{BE} = 26\,\text{k}\Omega$, $r_{CE} = 1\,\text{M}\Omega$,
$\quad\quad C_E = 0.6\,\text{pF}$, $C_C = 0.2\,\text{pF}$, $C_S = 1\,\text{pF}$

pnp: $\beta = 50$, $r_{CE} = 500\,\text{k}\Omega$, $C_C = 0.5\,\text{pF}$, $C_S = 2\,\text{pF}$

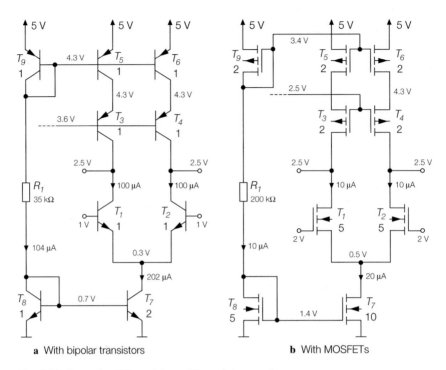

a With bipolar transistors   **b** With MOSFETs

Fig. 4.99. Example: differential amplifiers with cascode current sources

For the current source, $r_0 = V_{A,npn}/(2I_0) = 500\,\text{k}\Omega$. The output capacitance C_0 of the current source is the sum of the substrate capacitance and the collector capacitance of the current source transistors. Since the transistors are of size 2, both partial capacitances are twice as high as in the other npn transistors; therefore $C_0 = 2(C_S + C_C) = 2.4\,\text{pF}$. From (4.120), the cutoff frequency of the common-mode rejection is thus $f_{g,G} = 133\,\text{kHz}$. The resulting values for the npn differential amplifiers are listed in Fig. 4.102. In differential amplifiers with a current mirror, the values for common-mode operation are determined with the help of circuit simulation; they are quoted in parentheses.

With $I_0 = 10\,\mu\text{A}$, the values for MOSFETs are:

$$\text{n-channel:}\ K = 150\,\mu\text{A/V}^2\,,\ g_m = 54.8\,\mu\text{S}\,,\ r_{DS} = 5\,\text{M}\Omega\,,$$
$$C_{GS} = 18\,\text{fF}\,,\ C_{GD} = 7.5\,\text{fF}\,,\ C_{BD} = 17\,\text{fF}$$
$$\text{p-channel:}\ K = 60\,\mu\text{A/V}^2\,,\ g_m = 34.6\,\mu\text{S}\,,\ r_{DS} = 3.3\,\text{M}\Omega\,,$$
$$C_{GD} = 7.5\,\text{fF}\,,\ C_{BD} = 17\,\text{fF}$$

This assumes that the drain areas are 5 μm long and 2 μm wider than the channel width W. Consequently,

$$A_D = (15 + 2) \cdot 5\,\mu\text{m}^2 = 85\,\mu\text{m}^2 \Rightarrow C_{BD} = C_J' A_D = (0.2 \cdot 85)\,\text{fF} = 17\,\text{fF}$$

For the current source, $r_0 = V_{A,nC}/(2I_0) = 2.5\,\text{M}\Omega$. The output capacitance C_0 of the current source consists of the bulk–drain and the gate–drain capacitances of the current source MOSFET and the bulk–source capacitances of MOSFETs T_1 and T_2; the latter have the same size as the bulk–drain capacitances because of the symmetrical configuration.

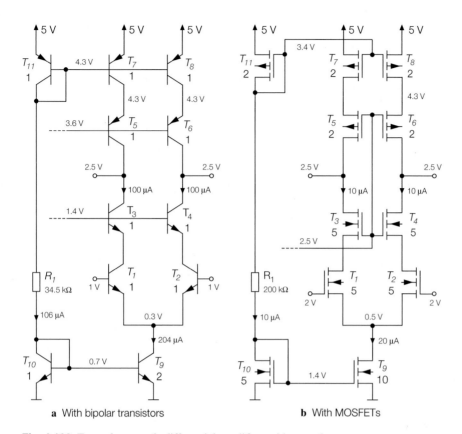

a With bipolar transistors   **b** With MOSFETs

Fig. 4.100. Example: cascode differential amplifiers with cascode current sources

With the drain area of the current source MOSFET $A_D = (32 \cdot 5)\,\mu\text{m}^2 = 160\,\mu\text{m}^2$, the output capacitance is:

$$C_O = C'_J A_D + 2C_{GD} + 2C_{BD} = (0.2 \cdot 160 + 2 \cdot 7.5 + 2 \cdot 17)\,\text{fF} = 83\,\text{fF}$$

The cutoff frequency of the common-mode rejection is thus $f_{g,G} = 767\,\text{kHz}$. The resulting values for the n-channel differential amplifiers are listed in Fig. 4.103. Again, the values for common-mode operation in differential amplifiers with current mirror are determined by circuit simulation.

A comparison of the values for npn and n-channel differential amplifiers shows that the differential gain in npn differential amplifiers is approximately ten times higher than that of the corresponding n-channel differential amplifiers; only in cascode differential amplifiers is the difference less. It must be taken into consideration that n-channel MOSFETs were chosen that are five times as large as the size that the quiescent current calls for; this increases the differential gain by a factor of $\sqrt{5}$. The reason for the lower differential gain in the n-channel differential amplifiers is the lower maximum gain of the MOSFETs. The situation improves with the MOSFETs of the cascode differential amplifiers, as their output resistance rises without any limit with increasing current feedback, while in bipolar transistors it is limited to $\beta\, r_{CE}$. This means that with additional cascode stages the differ-

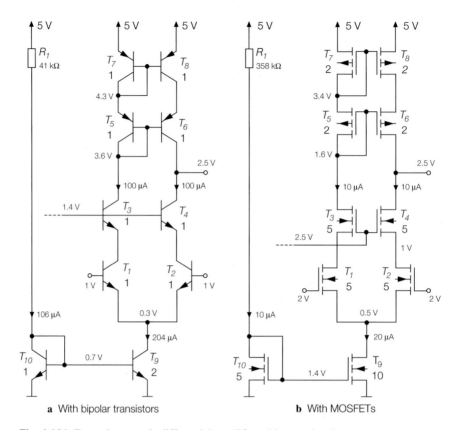

Fig. 4.101. Example: cascode differential amplifier with cascode mirror

ential gain of an n-channel cascode differential amplifier may be increased almost without any limitation.

Usually, additional amplifier stages are connected to the outputs of a differential amplifier. To ensure that the differential gain is fully maintained, the input resistances of these stages must be higher than the output resistances of the differential amplifier. In CMOS circuits this condition is met automatically, because of the isolated gate connections of the MOSFETs, which means that the maximum operational gain $A_{B,D} = A_D$ is achieved without additional measures. In bipolar circuits, on the other hand, each output must be provided with an impedance converter with one or more common-collector circuits, to reduce the output resistances to a value below the input resistance of the next stage. Impedance converters are explained in more detail in Sect. 4.1.4.

Due to the significant variations in gain, a meaningful comparison of the cutoff frequencies of the differential amplifiers described here is only possible on the basis of the gain–bandwidth product. Due to the very low capacitances of the integrated MOSFETs, and despite their low quiescent currents, n-channel differential amplifiers achieve higher values than npn-differential amplifiers. Since the input capacitances of the subsequent amplifier stages are also very low, this advantage of integrated circuits is maintained to its full extent. However, if higher load capacitances are connected or exist outside an integrated circuit, npn differential amplifiers reach a higher gain–bandwidth product, due

npn	RES	SCS	SCM	CCS	CWCS	CWCM	Unit
Gain, output and input resistances							
A_D	-47	-641	-1282	-1851	$-38{,}500$	$-42{,}800$	–
$A_{D,dB}$	33	56	62	65	92	93	dB
A_{CM}	-0.025	-0.5	(-0.008)	-20	-20	(-0.8)	–
$A_{CM,dB}$	-32	-6	(-42)	26	26	(-2)	dB
G	1880	1282	$(160{,}000)$	93	1925	$(54{,}000)$	–
G_{dB}	65	62	(104)	39	66	(95)	dB
$r_{o,D}$	24.4	333	333	962	20{,}000	11{,}100	kΩ
$r_{o,CM}$	25	498	–	20{,}000	20{,}000	–	kΩ
$r_{i,D}$			26				kΩ
$r_{i,CM}$			100				MΩ
Frequency response and cutoff frequency with $R_g = 10\,\text{k}\Omega$, $R_L = \infty$, $C_L = 0$							
A_0	-34	-463	-926	-1337	$-27{,}800$	$-30{,}900$	–
$A_{0,dB}$	31	53	59	63	89	90	dB
T_1	5.67	5.67	2.84	5.67	7.10	3.55	ns
T_2	2.41	3.31	1.66	3.31	1.33	0.67	ns
GBW	66	48	96	48	120	240	MHz
f_{-3dB}	1800	103	103	36	4.3	7.7	kHz
$f_{g,G}$			133				kHz

RES: with resistances (Fig. 4.97a)
SCS: with simple current sources (Fig. 4.98a)
SCM: with a simple current mirror (Fig. 4.98a)
CCS: with cascode current sources (Fig. 4.99a)
CWCS: cascode with current sources (Fig. 4.100a)
CWCM: cascode with current mirror (Fig. 4.101a)

Fig. 4.102. Small-signal parameters of npn differential amplifiers (simulated values in parentheses)

to the higher transconductance of the bipolar transistors. This can be seen from the time constant T_2 in Fig. 4.91, for the borderline case of high load capacitances C_L:

$$\lim_{C_L \to \infty} T_2 = \begin{cases} C_L \left(\dfrac{R_g}{\beta} + \dfrac{1}{g_m} \right) & \text{npn differential amplifier} \\[2ex] \dfrac{C_L}{g_m} & \text{n-channel differential amplifier} \end{cases}$$

If we assume load capacitances of $C_L = 100\,\text{pF}$ for the npn differential amplifiers and $C_L = 10\,\text{pF}$ for the n-channel differential amplifiers, which makes the ratio of the quiescent current to the load capacitance equal in both cases, the gain–bandwidth product is $GBW \approx 4.4\,\text{MHz}$ for the npn differential amplifier and $GBW \approx 870\,\text{kHz}$ for the n-channel differential amplifier. Here too, it must be taken into account that the chosen n-channel MOSFETs are already five times larger than would have been necessary due to the quiescent current; this increases the transconductance and thus the gain–bandwidth product by a factor of $\sqrt{5}$ for a capacitive load.

n-Kanal	RES	SCS	SCM	CCS	CWCS	CWCM	Unit
Gain, output and input resistances							
A_D	−6.5	−55	−110	−135	−8110	−16,220	−
$A_{D,dB}$	16	35	41	42	78	84	dB
A_{CM}	−0.05	−0.67	(−0.005)	−59	−75	(−0.035)	−
$A_{CM,dB}$	−26	−3	(−46)	35	38	(−29)	dB
G	130	82	(22,000)	2.3	108	(460,000)	−
G_{dB}	42	38	(87)	7	40	(113)	dB
$r_{o,D}$	0.238	2	2	4.93	296	296	MΩ
$r_{o,CM}$	0.25	3.3	−	296	376	−	MΩ
$r_{i,D}$				∞			Ω
$r_{i,CM}$				∞			Ω
Frequency response and cutoff frequency with $R_g = 100\,\text{k}\Omega$, $R_L = \infty$, $C_L = 0$							
A_0	−6.5	−55	−110	−135	−8110	−16,220	−
$A_{0,dB}$	16	35	41	42	78	84	dB
T_1	2.55	2.55	1.28	2.55	3.30	1.65	ns
T_2	1.20	1.64	0.82	1.64	0.58	0.29	ns
GBW	133	97	194	97	275	550	MHz
f_{-3dB}	15,000	1700	1700	700	34	34	kHz
$f_{g,G}$				767			kHz

RES: with resistances (Fig. 4.97b)
SCS: with simple current sources (Fig. 4.98b)
SCM: with a simple current mirror (Fig. 4.98b)
CCS: with cascode current sources (Fig. 4.99b)
CWCS: cascode with current sources (Fig. 4.100b)
CWCM: cascode with current mirror (Fig. 4.101b)

Fig. 4.103. Small-signal parameters of n-channel differential amplifiers (simulated values in parentheses)

Summary

Due to its particular characteristics, the differential amplifier is one of the most important circuits in integrated circuit engineering. It is found not only in amplifiers but also in comparators, ECL logic circuits, voltage regulators, active mixers and numerous other circuit designs. It has a special ranking among amplifier circuits, primarily on account of the almost unlimited choice of common-mode voltage levels at the input, which means that any signal source with a DC voltage within the common-mode signal range can be connected; voltage dividers for setting the operating point or coupling capacitances are not required. The differential amplifier is therefore a true DC voltage amplifier. As it virtually amplifies the differential signal alone, it is *the* choice for a feedback control circuit, since by subtraction it determines the deviation and subsequently amplifies it; in other words, it combines the operational blocks *subtractor* and *control amplifier* of an automatic control circuit. It thus forms the basis of the operational amplifier. In this respect, the differential amplifier is the *simplest* operational amplifier, and the operational amplifier is the *better* differential amplifier.

4.1.4
Impedance Converters

The output resistance of an amplifier stage with a high voltage gain is generally very high and must be reduced by means of an impedance converter before other amplifier stages or load resistances can be connected without loss of gain. Single-stage or multi-stage common-collector and common-drain circuits are used as impedance converters.

Single-Stage Impedance Converter

Figure 4.104 shows the simplest type of a single-stage impedance converter, with a common-collector or common-drain circuit (T_1) and a current mirror (T_2,T_3) for adjusting the operating point; resistance R_g represents the output resistance of the previous stage. The output resistance can be calculated from (2.116) and (3.99):

$$
r_o = \begin{cases}
\dfrac{R_g}{\beta} + \dfrac{1}{g_m} \overset{g_m R_g \gg \beta}{\approx} \dfrac{R_g}{\beta} & \text{Common-collector circuit} \\[3mm]
\dfrac{1}{g_m + g_{m,B}} \overset{g_m \gg g_{m,B}}{\approx} \dfrac{1}{g_m} & \text{Common-drain circuit}
\end{cases}
\tag{4.123}
$$

Common-collector circuit: With a high-resistance signal source, the output resistance of a common-collector circuit depends solely on the internal resistance R_g and the current gain β; the quiescent current I_0 has no effect as long as $g_m R_g \gg \beta$. A guidance value for selecting the quiescent current can be derived from this, where $g_m = I_0/V_T$ and $g_m R_g \approx 10\beta$:

$$
I_0 \approx \frac{10\beta\, V_T}{R_g} \overset{\beta \approx 100}{\approx} \frac{26\,\text{V}}{R_g}
\tag{4.124}
$$

A higher quiescent current is usually required with high-resistance signal sources, as otherwise the bandwidth of the circuit becomes too narrow; the reason for this is the decreasing transit frequency of a transistor at small currents. If impedance conversion by a factor of β is insufficient, a multi-stage impedance converter is required. In low-resistance signal

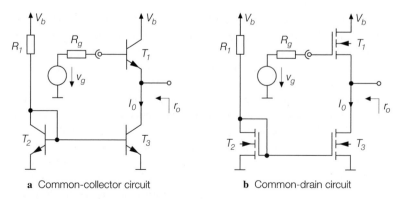

a Common-collector circuit   **b** Common-drain circuit

Fig. 4.104. Single-stage impedance converter

sources with $g_m R_g \ll \beta$, the transconductance of the transistor determines the output resistance:

$$r_o \approx \frac{1}{g_m} = \frac{V_T}{I_0} \approx \frac{26\,\mathrm{mV}}{I_0}$$

Common-drain circuit: The behavior of the common-drain circuit with high-resistance signal sources is completely different. Here, the output resistance only depends on the transconductance:

$$r_o \approx \frac{1}{g_m} = \frac{1}{\sqrt{2K\,I_0}} = \frac{V_{GS} - V_{th}}{2I_0} \tag{4.125}$$

For the MOSFETs in Fig. 4.6 on page 274, a typical quiescent current of $10\,\mu\mathrm{A}$ times the size of the MOSFET produces the values $V_{GS} - V_{th} \approx 0.8\,\mathrm{V}$ and $r_o \approx 0.4\,\mathrm{V}/I_0$. For small output resistances, large MOSFETs with a correspondingly high input capacitance are required; with high-resistance signal sources, this reduces the bandwidth significantly. If the bandwidth is insufficient, a multi-stage impedance converter is required.

Output voltage: At the operating point of both circuits, the output voltage is lower than the input voltage by a value equal to a base–emitter or gate–source voltage. A pnp common-collector circuit or a p-channel common-drain circuit could be used as an alternative; in this case, the output voltage at the operating point is higher than the input voltage. However, pnp transistors generally have a lower current gain than npn transistors, and p-channel MOSFETs are larger in size than n-channel MOSFETs with the same transconductance coefficient; that is, they have higher capacitances.

Multi-stage Impedance Converters

Multi-stage impedance converters are required if:

- The impedance conversion of a common-collector circuit is insufficient.
- The capacitances of a common-drain circuit with the required output resistance become so high that the bandwidth is insufficient.

Figure 4.105 shows a two-stage impedance converter with the current mirror required to set the operating point. The optimum rating for a multi-stage impedance converter requires the quiescent currents and transistor sizes to be selected using optimum values.

Multi-stage common-collector circuit: In multi-stage common-collector circuits, the quiescent current of each stage can be selected by means of (4.124). According to this calculation, the quiescent current should increase by the current gain β from one stage to the next, as the effective internal resistance of the signal source decreases at each stage by a factor of β; this would give an optimum impedance conversion with high-resistance signal sources. But since each stage has to provide the base current for the next stage and the base current should be significantly lower than the quiescent current of the previous stage, a quiescent current ratio of approximately $B/10 \approx \beta/10$ is used in practice; this means that the quiescent current of each stage is ten times higher than the base current of the next stage. As very high-resistance signal sources often require the quiescent current of the first stage to be higher than necessary according to (4.124), a quiescent current ratio of $B/10$ in two-stage common-collector circuits is also advantageous in this respect. Therefore, for two-stage common-collector circuits, (4.124) is used to select the quiescent

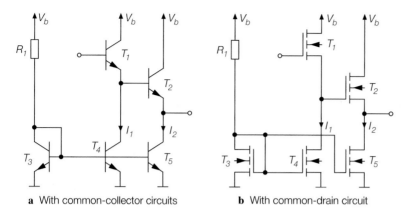

a With common-collector circuits   **b** With common-drain circuit

Fig. 4.105. Two-stage impedance converters

current I_2 of the second stage; the effective source resistance at this point is R_g/β. The quiescent currents of both stages are thus:

$$I_2 \approx \frac{10\beta^2 V_T}{R_g} \overset{\beta\approx100}{\approx} \frac{2600\,\text{V}}{R_g} \quad , \quad I_1 \approx \frac{10I_2}{B} \overset{B\approx\beta\approx100}{\approx} \frac{260\,\text{V}}{R_g} \tag{4.126}$$

In a third stage, the quiescent current would be $I_3 = I_2 B/10$.

Example: A signal source with $R_g = 2.6\,\text{M}\Omega$ is to be connected to a low-resistance load via a two-stage common-collector circuit with the quiescent current ratio $B/10$; here, $B \approx \beta \approx 100$. From (4.126), we obtain $I_2 = 1\,\text{mA}$ and $I_1 = 100\,\mu\text{A}$. At the output of the second stage, the effective internal resistance of the signal source has decreased to $R_g/\beta^2 \approx 260\,\Omega$. For a third stage with quiescent current $I_3 = 10\,\text{mA}$, it follows that $g_m R_g = I_3 R_g/V_T = 100$; in other words, the condition $g_m R_g \gg \beta$ is no longer met. Therefore, the output resistance must be calculated without the approximation in (4.123): $r_o = R_g/\beta + 1/g_m = (2.6 + 2.6)\,\Omega = 5.2\,\Omega$.

Darlington circuit: A Darlington transistor may also be used to obtain a two-stage common-collector circuit; the only requirements here are to combine transistors T_1 and T_2 in Fig. 4.105a into one Darlington transistor and to remove transistor T_4. In this case, the quiescent current of T_1 corresponds with the base current of T_2. In practice, however, the bandwidth achieved is usually insufficient, since the transit frequency of T_1 is very low due to the low quiescent current.

Multi-stage common-drain circuit: According to (4.125), the output resistance of the common-drain circuit depends solely on the quiescent current; therefore, the output resistance of a multi-stage common-drain circuit depends solely on the quiescent current of the last stage. However, the quiescent currents of the other stages influence the bandwidth, as each stage is loaded with the input capacitance of the subsequent stage. The optimum choice for quiescent currents is explained for a two-stage common-drain circuit; Fig. 4.106 shows the circuit and the corresponding small-signal equivalent circuit. The output resis-

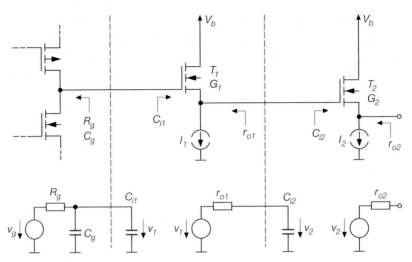

Fig. 4.106. Two-stage common-drain circuit: circuit (*above*) and the small-signal equivalent circuit (*below*)

tances and input capacitances depend on the sizes[24] G_1 and G_2 of the MOSFETs T_1 and T_2:

$$r_{o1} = \frac{r_o'}{G_1} \quad , \quad r_{o2} = \frac{r_o'}{G_2} \quad , \quad C_{i1} = C_i' G_1 \quad , \quad C_{i2} = C_i' G_2$$

Here, r_o' and C_i' apply to a MOSFET of size 1. The small-signal equivalent circuit in Fig. 4.106 allows us to determine the time constants

$$T_1 = R_g \left(C_g + C_{i1} \right) = R_g \left(C_g + C_i' G_1 \right) \quad , \quad T_2 = r_{o1} C_{i2} = \frac{r_o' C_i' G_2}{G_1}$$

and the -3 dB cutoff frequency:

$$\omega_{-3dB} = 2\pi f_{-3dB} \approx \frac{1}{T_1 + T_2} = \frac{1}{R_g C_g + R_g C_i' G_1 + \dfrac{r_o' C_i' G_2}{G_1}} \tag{4.127}$$

The cutoff frequency decreases as the size G_2 increases. An optimum value for size G_1 is reached under the condition $\partial(T_1 + T_2)/\partial G_1 = 0$:

$$G_{1,opt} = \sqrt{\frac{r_o' G_2}{R_g}} = G_2 \sqrt{\frac{r_{o2}}{R_g}} \tag{4.128}$$

It can be seen that the optimum size ratio G_1/G_2 depends on the current ratio R_g/r_{o2}. The square root shows that the transformation takes place to the same extent in both stages. The same approach is used when dealing with three-stage or multi-stage common-drain circuits. The optimum sizes for a circuit made up of n stages are:

$$G_{i,opt} = G_n \left(\frac{r_{o,n}}{R_g} \right)^{\frac{n-i}{n}} \quad \text{for } i = 1 \dots n - 1 \tag{4.129}$$

---

[24] In this context, *size* means the electrical and not the geometric size; that is, $G \sim K$.

Example: A load resistance $R_L = 1\,\text{k}\Omega$ is to be connected to a signal source with $R_g = 2\,\text{M}\Omega$ and $C_g = 20\,\text{fF}$ via an impedance converter. To keep the signal attenuation low at the output, an output resistance $r_o = 100\,\Omega$ is selected. For the typical value $V_{GS} - V_{th} \approx 0.8\,\text{V}$ for the MOSFETs in Fig. 4.6, the necessary quiescent current can be determined using (4.125) on page 386:

$$I_0 = \frac{V_{GS} - V_{th}}{2r_o} = \frac{0.4\,\text{V}}{100\,\Omega} = 4\,\text{mA}$$

The required MOSFET size is $G = 4\,\text{mA}/10\,\mu\text{A} = 400$. The input capacitance of a common-drain circuit can be calculated using (3.106) on page 258, by taking into account the capacitance related to R'_g and inserting $R'_L = 1/g_{m,B}$:

$$C_i = C_{GS} \frac{g_{m,B}}{g_m} + C_{GD} \overset{g_{m,B}/g_m \approx 0.2}{\approx} 0.2 \cdot C_{GS} + C_{GD}$$

With the parameters in Fig. 4.6 on page 274, for an n-channel MOSFET of size 1 with $W = L = 3\,\mu\text{m}$ and a quiescent current of $10\,\mu\text{A}$, we obtain:

$$r'_o \approx \frac{1}{g_m} = \frac{1}{\sqrt{2K\,I_0}} = \frac{1}{\sqrt{2 \cdot 30\,\mu\text{A}/\text{V}^2 \cdot 10\,\mu\text{A}}} \approx 40\,\text{k}\Omega$$

$$C'_i \approx 0.2 \cdot \frac{2C'_{ox}\,W L}{3} + C'_{GD,ov}\,W = 0.72\,\text{fF} + 1.5\,\text{fF} \approx 2.2\,\text{fF}$$

A MOSFET of size 400 thus has an input capacitance of $C_i = 400 \cdot 2.2\,\text{fF} = 880\,\text{fF}$. With this MOSFET connected directly to the signal source, the time constant is $T = R_g(C_g + C_i) = 1.8\,\mu\text{s}$ and the cutoff frequency is $f_{-3dB} = 1/(2\pi\,T) \approx 88\,\text{kHz}$. For a two-stage common-drain circuit, the optimum size for the MOSFET of the first stage is determined from (4.128):

$$G_{1,opt} = G_2 \sqrt{\frac{r_o}{R_g}} = 400 \cdot \sqrt{\frac{100\,\Omega}{2\,\text{M}\Omega}} = 2\sqrt{2} \approx 3$$

From (4.127), we obtain a cutoff frequency of $f_{-3dB} \approx 2.5\,\text{MHz}$. Therefore, the use of a two-stage common-drain circuit features a bandwidth that is 28 times higher than that of a single-stage.

Output voltage: In a two-stage npn common-collector circuit, the output voltage at the operating point is lower than the input voltage by a value equal to $2\,V_{BE} \approx 1.4\,\text{V}$. In a two-stage common-drain circuit, the voltage offset of $2\,V_{GS} \approx 3 \dots 4\,\text{V}$ is so high that an input voltage of at least $4 \dots 5\,\text{V}$ is required when utilizing the operation limit of the current source of about $1\,\text{V}$. With impedance converters that consist of more than two stages, the voltage offset increases further. As an alternative, one may design one or more stages as pnp common-collector or p-channel common-drain circuits; this compensates the base–emitter or gate–source voltage fully or partially. Figure 4.107 shows an example of a two-stage impedance converter with $V_{i,A} \approx V_{o,A}$.

Complementary Impedance Converters

With low-resistance or high capacitive loads, the use of complementary impedance converters is preferred. The essential configuration of these converters is described below, followed by an explanation of their advantages.

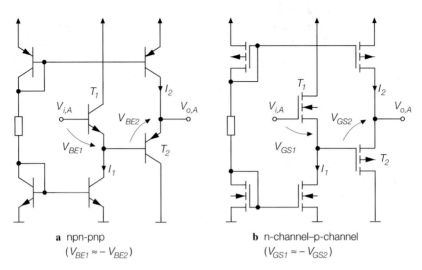

a npn-pnp

$(V_{BE1} \approx - V_{BE2})$

b n-channel–p-channel

$(V_{GS1} \approx - V_{GS2})$

Fig. 4.107. Two-stage impedance converters with $V_{i,A} \approx V_{o,A}$

Figure 4.108 shows the basic circuit of a single-stage complementary impedance converter with bipolar transistors and with MOSFETs. The quiescent currents must be adjusted using bias voltage sources, the practical aspects of which will be described later. At the operating point, the input and output voltages are identical; that is, there is no offset voltage. For reasons of symmetry, the circuits are shown with a symmetrical voltage supply; however, a unipolar voltage supply is also possible.

Complementary impedance converters offer the advantage of providing a high output current in both directions; Fig. 4.109 illustrates this with a sudden change of the input voltage by comparing a complementary and a simple common-collector circuit. In the complementary common-collector circuit, the output current is supplied via an active common-collector circuit in both directions and can therefore become very high; in this case, the other common-collector circuit is nonconductive. In the case of a sudden step-

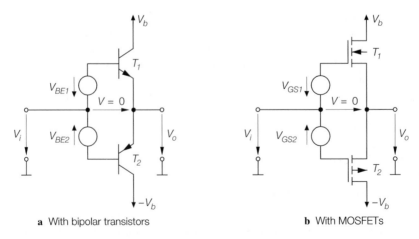

a With bipolar transistors

b With MOSFETs

Fig. 4.108. Basic circuit of a single-stage complementary impedance converter

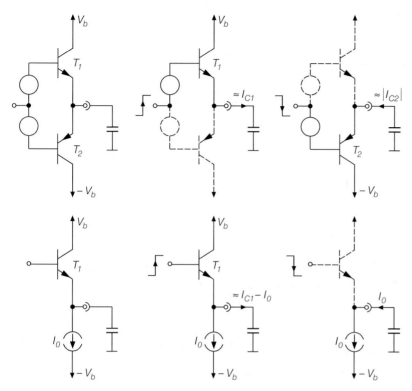

Fig. 4.109. Comparison of a complementary and a simple common-collector circuit with regards to their response to a sudden change in the input voltage

down in the input voltage in the simple common-collector circuit, the output current is supplied by the current source; in other words, it is limited to the quiescent current. For this reason, complementary impedance converters are used whenever simple impedance converters would require a disproportionately high quiescent current.

Single-stage complementary impedance converters: The use of transistor and MOS-FET diodes to realize the bias voltage sources in Fig. 4.108 leads to the circuits shown in Fig. 4.110. At the operating point, the input and output voltages are identical if the size ratios T_1/T_3 and T_2/T_4 are the same; in terms of quiescent currents, T_3,T_1 and T_4,T_2 act as current mirrors with the current ratio:

$$k_I \approx \frac{I_{S1}}{I_{S3}} = \frac{I_{S2}}{I_{S4}} \quad \text{or} \quad k_I = \frac{K_1}{K_3} = \frac{K_2}{K_4}$$

Here $I_{S1} \dots I_{S4}$ are the reverse saturation currents of the bipolar transistors and $K_1 \dots K_4$ are the transconductance coefficients of the MOSFETs. The relation of the quiescent currents is:

$$I_1 = k_I I_0$$

The circuit shown in Fig. 4.110a may be regarded as one npn and one pnp common-collector circuit connected in parallel; the output resistance is thus:

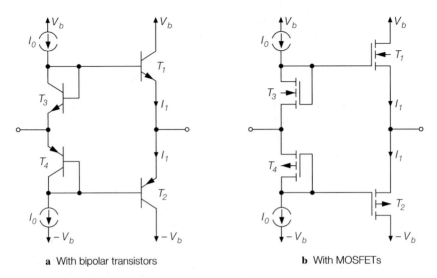

a With bipolar transistors   **b** With MOSFETs

Fig. 4.110. Single-stage complementary impedance converters

$$r_o \approx \frac{1}{2}\left(\frac{R_g}{\beta_1}+\frac{R_g}{\beta_2}+\frac{1}{g_m}\right) \overset{g_m R_g \gg \beta_1,\beta_2}{\approx} \frac{R_g}{2}\left(\frac{1}{\beta_1}+\frac{1}{\beta_2}\right) \overset{\beta_1=\beta_2=\beta}{=} \frac{R_g}{\beta} \qquad (4.130)$$

The differential resistance of transistor diodes T_3 and T_4 is ignored, as it is much lower than R_g. The transconductances of transistors T_1 and T_2 are the same: $g_m = I_1/V_T$. Similarly, the circuit in Fig. 4.110b can be regarded as one n-channel and one p-channel common-drain circuit connected in parallel, so that:

$$r_o = \frac{1}{g_{m1}} \,\|\, \frac{1}{g_{m2}} = \frac{1}{g_{m1}+g_{m2}} \overset{g_{m1}=g_{m2}=g_m}{=} \frac{1}{2g_m} = \frac{1}{2\sqrt{2K\,I_1}} \qquad (4.131)$$

Two-stage complementary common-collector circuit: Replacing transistor diodes T_3 and T_4 in Fig. 4.110a by common-collector circuits which cause an impedance conversion in addition to generating the bias voltages, without any additional measures, leads to the two-stage complementary common-collector circuit shown in Fig. 4.111. It should be noted that the npn transistor diode T_3 is replaced by a pnp common-collector circuit and the pnp transistor diode T_4 is replaced by an npn common-collector circuit. The current mirrors T_5,T_7 and T_6,T_8 are already used for the current sources. The circuit can be regarded as a parallel circuit consisting of a pnp-npn (T_3,T_1) and an npn-pnp common-collector circuit (T_4,T_2).

The easiest way to adjust the quiescent currents and select the transistor sizes is to follow the same approach as used in the simple complementary common-collector circuit; the current ratio for the same size ratio of the npn and pnp transistors is:

$$k_I \approx \frac{I_{S1}}{I_{S4}} = \frac{I_{S2}}{I_{S3}}$$

It follows that $I_2 = k_I I_1$. The general calculation is described in the next section. However, the input and output voltages at the operating point are not equal, since the transistor diodes have been replaced with common-collector circuits of opposite polarities, and the base–

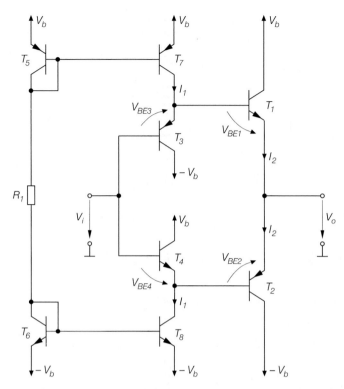

Fig. 4.111. Two-stage complementary common-collector circuit

emitter voltages of npn and pnp transistors of the same size are different for the same current. The offset voltage can be minimized by scaling the transistors accordingly.

The general calculation of the quiescent currents and the offset voltage is based on the following equation:

$$V_{BE\,3} + V_{BE\,1} - V_{BE\,2} - V_{BE\,4} = 0$$

The base–emitter voltages are:

$$V_{BE} = \begin{cases} V_T \ln \dfrac{I_C}{I_S} & \text{npn transistor} \\[3mm] -V_T \ln \dfrac{-I_C}{I_S} & \text{pnp transistor} \end{cases}$$

Inserting and dividing by V_T leads to:

$$-\ln \frac{-I_{C3}}{I_{S3}} + \ln \frac{I_{C1}}{I_{S1}} + \ln \frac{-I_{C2}}{I_{S2}} - \ln \frac{I_{C4}}{I_{S4}} = 0$$

Neglecting the base currents allows us to insert $-I_{C2} = I_{C1} \approx I_2$ and $-I_{C3} = I_{C4} \approx I_1$; it thus follows that

$$\ln \frac{I_{S3} I_{S4} I_2^2}{I_{S1} I_{S2} I_1^2} \approx 0$$

and, consequently:

$$k_I = \frac{I_2}{I_1} \approx \sqrt{\frac{I_{S1}I_{S2}}{I_{S3}I_{S4}}} = \sqrt{g_{npn}g_{pnp}} \quad \text{with } g_{npn} = \frac{I_{S1}}{I_{S4}}, \ g_{pnp} = \frac{I_{S2}}{I_{S3}} \quad (4.132)$$

Here, g_{npn} is the size ratio of the npn transistors T_1 and T_4; g_{pnp} is the size ratio of the pnp transistors T_2 and T_3.

In general, the same size ratios and the same sizes for T_1 and T_2 are chosen – for example, size 10 for T_1 and T_2 and size 1 for T_3 and T_4 – then $k_I \approx g_{npn} = g_{pnp} = 10$ and $I_2 \approx 10\,I_1$. Factor 10 is typical for practical applications because here, as is the case in simple multi-stage common-collector circuits, a quiescent current ratio of about $B/10$ is used and $B \approx \beta \approx 100$ is a typical value for integrated transistors.

The *offset voltage* is $V_{off} = V_{i,A} - V_{o,A}$; from Fig. 4.111, it follows:

$$V_{off} = V_{BE\,1} + V_{BE\,3} \approx V_T \ln \frac{I_2}{I_{S1}} - V_T \ln \frac{I_1}{I_{S3}} = V_T \ln \frac{I_{S3}I_2}{I_{S1}I_1}$$

When equal size ratios and the same sizes for T_1 and T_2 are selected, then $k_I = I_2/I_1 \approx g_{npn} = g_{pnp}$; consequently,

$$V_{off} \approx V_T \ln \frac{I_{S2}}{I_{S1}} = V_T \ln \frac{I_{S3}}{I_{S4}} = V_T \ln \frac{I_{S,pnp}}{I_{S,npn}}$$

Here, $I_{S,npn}$ and $I_{S,pnp}$ are the reverse saturation currents of npn and pnp transistors of the same size; for example, size 1. The value for the transistors in Fig. 4.5 on page 274 is $I_{S,npn} = 2\,I_{S,pnp}$; consequently, $V_{off} = V_T \cdot \ln 0.5 \approx -18\,\text{mV}$.

The offset voltage becomes zero when the reverse saturation currents of transistors T_1 and T_2 reach the same level. For the transistors in Fig. 4.5, this means that T_2 must be twice as large as T_1 and – in order to maintain the same size ratios – must be twice the size of T_4. In practice, this causes a significant reduction in the offset voltage; typical values are in the range of a few millivolts. The reason for the remaining offset voltage is the asymmetrical current distribution caused by the differing current gains of npn and pnp transistors. This could also be eliminated by:

– Slightly adapting the size of T_1 or T_2.
– Slightly increasing T_8 until the absolute values of the collector currents in T_3 and T_4 are the same; then the relatively high base current of T_2, which is due to the lower current gain of the pnp transistors, is also provided by the lower current mirror T_6, T_8.

Despite these measures, the offset voltage of the circuit cannot be reduced as much as in the circuit shown in Fig. 4.110a, because here the offset voltage depends on the ratio of the reverse saturation currents of the pnp and npn transistors, which varies in practice due to production tolerances.

The two-stage complementary common-collector circuit can be regarded as a series connection of two single-stage complementary common-collector circuits, which means that the output resistance can be calculated by applying (4.130) on page 392 twice.

Two-stage complementary common-drain circuit: The two-stage complementary impedance converter in Fig. 4.111 can also be made up of MOSFETs (see Fig. 4.112). In this case, the size ratios must be determined by means of circuit simulation, because the MOSFETs $T_1 \ldots T_4$ operate with different, initially unknown, bulk–source voltages, and therefore have different threshold voltages due to the substrate effect. To a first approximation, the substrate effect may be neglected and the size ratios can be chosen according

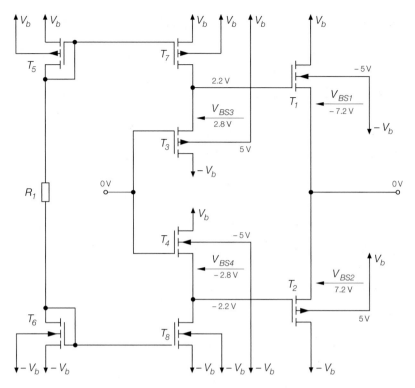

Fig. 4.112. Two-stage complementary common-drain circuit

to the optimum for the two-stage common-drain circuit (see (4.128) on page 388). The quiescent current and the size of the MOSFETs in the second stage can be calculated using (4.131) on page 392, by giving the value desired for the output resistance.

Since the bulk–source voltages change during operation, the quiescent current of the second stage also varies. Here too, it is necessary to ensure, by circuit simulation, that the circuit meets the given requirements in the desired operation range. The quiescent current is usually largest when the input voltage is approximately in the middle of the supply voltage range, and it decreases when it approaches one of the supply voltages. In contrast, the quiescent currents of the first stage remain constant, as they are determined by the current mirror.

4.1.5
Circuits for Setting the Operating Point

In integrated circuits, the operating point is set in most cases by injecting the quiescent currents by means of current sources or current mirrors. Therefore, the setting of a stable operating point requires, first and foremost, reference current sources that are insensitive to temperature variations and independent of the supply voltage. On the other hand, reference voltage sources are seldom required; for example, the auxiliary voltages necessary to adjust the operating point of cascode stages can be generated without particularly complex circuit designs and without stringent demands on stability. The most important reference current sources are described below, followed by an examination of the circuits used for current distribution.

V_{BE} Reference Current Source

In this reference current source, the approximately constant base–emitter voltage V_{BE} of a bipolar transistor is used as the reference value; Fig. 4.113 shows the basic circuit. Transistor T_1 receives its base current I_{B1} via resistance R_2. The collector current $I_{C1} = BI_{B1}$ increases until the voltage across feedback resistance R_1 is high enough to make T_2 conductive, thus preventing a further increase in I_{B1} and I_{C1}. Neglecting the base current and assuming an approximately constant base–emitter voltage of $V_{BE2} \approx 0.7\,\text{V}$ results in a reference current:

$$I_{ref} = I_{C1} \approx \frac{V_{BE2}}{R_1} \approx \frac{0.7\,\text{V}}{R_1}$$

In the first approximation, this value does not depend on current I_2 and therefore does not depend on the supply voltage V_b.

Characteristics: Figure 4.114 shows the characteristic of the V_{BE} reference current source with $R_1 = 6.6\,\text{k}\Omega$ and $R_2 = 36\,\text{k}\Omega$. For $V_b > 1.4\,\text{V}$, the current is approximately constant; only in this region does the circuit act as a current source.

When calculating the characteristic, the dependence of the base–emitter voltage V_{BE2} on the current $I_{C2} \approx I_2$ must be taken into account:

$$I_2 \approx I_{C2} = I_{S2}\left(e^{\frac{V_{BE2}}{V_T}} - 1 \right) \quad \Rightarrow \quad V_{BE2} \approx V_T \ln\left(\frac{I_2}{I_{S2}} + 1 \right)$$

Here, I_{S2} is the reverse saturation current of T_2 and V_T is the temperature voltage; at room temperature, $V_T \approx 26\,\text{mV}$. The reference current is:

$$\boxed{I_{ref} \approx \frac{V_T}{R_1} \ln\left(\frac{I_2}{I_{S2}} + 1 \right) \overset{I_2 \gg I_{S2}}{\approx} \frac{V_T}{R_1} \ln \frac{I_2}{I_{S2}}} \tag{4.133}$$

For

$$I_2 = \frac{V_b - V_{BE1} - V_{BE2}}{R_2} \approx \frac{V_b - 1.4\,\text{V}}{R_2}$$

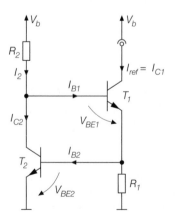

Fig. 4.113. Basic circuit of a V_{BE} reference current source

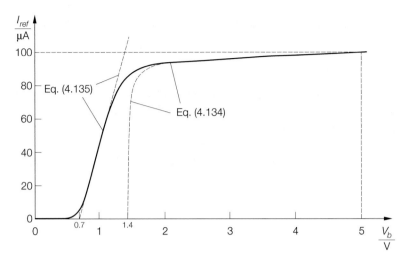

Fig. 4.114. Characteristic of a V_{BE} reference current source with $R_1 = 6.6\,\mathrm{k}\Omega$ and $R_2 = 36\,\mathrm{k}\Omega$

the reference current is:

$$I_{ref} \approx \frac{V_T}{R_1} \ln \frac{V_b - 1.4\,\mathrm{V}}{I_{S2}R_2} \qquad \text{for } V_b > 1.4\,\mathrm{V} \tag{4.134}$$

For $V_b < 1.4\,\mathrm{V}$, T_2 is in reverse mode; it thus follows from $V_b = (I_{C1} + I_{B1})R_1 + V_{BE1} + I_{B1}R_2$ that:

$$I_{ref} = I_{C1} \approx \frac{V_b - 0.7\,\mathrm{V}}{R_1 + \dfrac{R_1 + R_2}{B}} \qquad \text{for } V_b < 1.4\,\mathrm{V} \tag{4.135}$$

The approximations given in (4.134) and (4.135) are plotted in Fig. 4.114.

V_{BE} reference current source with a current mirror: A significant improvement of the circuit behavior is achieved when current feedback is introduced by means of a current mirror; Fig. 4.115 shows the circuit with a simple current mirror. Current I_2 is no longer adjusted by a resistor, but is derived from the reference current. In standard circuits, all transistors are of the same size; in this case, the current mirror has a current ratio $k_I \approx 1$, so that $I_2 \approx I_{ref}$. Inserting this into (4.133) leads to the transcendental equation:

$$I_{ref} \approx \frac{V_T}{R_1} \ln \left(\frac{I_{ref}}{I_{S2}} + 1 \right)$$

The solution to this equation depends solely on V_T, R_1 and I_{S2}, and no longer on the supply voltage V_b. In practice, a very low dependence remains due to the Early voltage of the transistors, which is ignored in this context.[25] As the current I_2 is now also stabilized, the base–emitter voltage V_{BE2} can be regarded as constant and the following approximation

---

[25] When the Early effect is taken into account, the calculation shows that the Early factor $1 + V/V_A$ only enters the argument of the logarithm and its effect is thus attenuated by a factor of $20\ldots 30$; this leads to an output resistance which is the same as that reached in a cascode circuit.

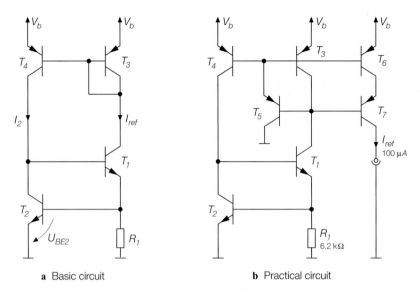

a Basic circuit   **b** Practical circuit

Fig. 4.115. V_{BE} reference current source with a current mirror

can be used:

$$\boxed{I_{ref} \approx \frac{V_{BE\,2}}{R_1}} \tag{4.136}$$

The practical configuration of the V_{BE} reference current source with a current mirror is shown in Fig. 4.115b. Transistor T_5 expands the current mirror T_3, T_4 to form a three-transistor current mirror, in which T_6 is an additional output for the reference current. The additional output must be provided with a cascode stage T_7 to ensure that the independence of the supply voltage is not impaired by the Early effect of T_6. To make certain that the desired reference current is obtained at the output, R_1 must be somewhat smaller than calculated in (4.136), to compensate the current losses caused by the various base currents. Figure 4.116 shows the resulting characteristics for $R_1 = 6.2\,\mathrm{k\Omega}$ at room temperature ($T = 27\,°C$) and at the limits of the temperature range for general applications ($T = 0 \ldots 70\,°C$).

Temperature sensitivity: One disadvantage of the V_{BE} reference current source is the relatively high temperature sensitivity, caused by the temperature dependence of the base–emitter voltage. From (2.21) on page 54, it follows that $dV_{BE}/dT \approx -1.7\,\mathrm{mV/K}$; this causes a current change of

$$\frac{dI_{ref}}{dT} = \frac{1}{R_1}\frac{dV_{BE\,2}}{dT} \approx -\frac{1.7\,\mathrm{mV/K}}{R_1} \tag{4.137}$$

and a *temperature coefficient* of:

$$\frac{1}{I_{ref}}\frac{dI_{ref}}{dT} = \frac{1}{V_{BE\,2}}\frac{dV_{BE\,2}}{dT} \overset{V_{BE\,2}\approx0.7\,\mathrm{V}}{\approx} -2.5\cdot10^{-3}\,\mathrm{K}^{-1}$$

This means that a temperature increase of 4 K causes a reference current reduction of 1%.

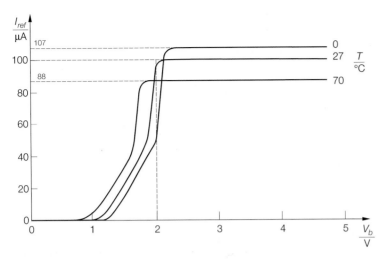

Fig. 4.116. Characteristics of the V_{BE} reference current source in Fig. 4.115

Starting circuit: In addition to the desired operating point, the V_{BE} reference current source has another operating point at which all transistors are at zero current. Whether this second operating point is stable or instable depends on the leakage currents of the transistors; these depend on the production process used and are disregarded in most simulation models. If lateral pnp transistors are used to make up current mirror $T_3 \ldots T_5$, the relatively high leakage current of T_4, which is caused by its large surface, is high enough to provide a sufficiently high starting current for T_1. In this case, there is no stable operating point with zero current. Otherwise, a separate starting circuit is required, to provide the starting current that is then cut off when the desired operating point has almost been reached.

Figure 4.117 shows a simple and commonly used starting circuit [4.1, 4.2]. It consists of diodes $D_1 \ldots D_4$ designed as transistor diodes and the resistances R_3. Diodes $D_1 \ldots D_3$

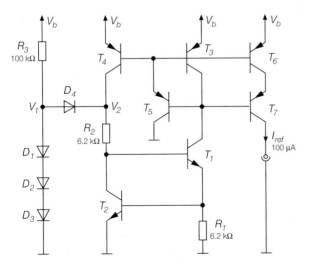

Fig. 4.117. V_{BE} reference current source with a starting circuit

and resistance R_3 form a simple reference voltage source of $V_1 = 3\,V_{BE} \approx 2.1$ V, which provides T_1 with a starting current by means of D_4 and resistance R_2. Resistance R_2 is rated such that the starting collector current of T_4 causes voltage V_2 to increase until D_4 enters the blocking state at the desired operating point. When selecting

$$R_2 \approx \frac{V_{BE}}{I_{ref}} \approx R_1$$

the voltage at the desired operating point is $V_1 = V_2$; D_4 is thus in the blocking state. Resistance R_3 must be selected small enough to make the starting current sufficiently high even with the supply voltage at a minimum; on the other hand, it must not be too small to prevent the cross-current through diodes $D_1 \ldots D_3$ from becoming too high at a maximum supply voltage.

Example: The V_{BE} reference current source in Fig. 4.117 is to be rated for a reference current of $I_{ref} = 100\,\mu$A. For the npn transistors in Fig. 4.5 on page 274, $V_{BE} \approx V_T \ln I_{ref}/I_S \approx 0.66$ V; (4.136) thus leads to $R_1 \approx 6.6$ kΩ. Circuit simulation is used for fine-tuning to $R_1 = 6.2$ kΩ. For the starting circuit, this leads to $R_2 = R_1 = 6.2$ kΩ. The value for resistance R_3 can be in a wide range; here, it is selected such that at a maximum supply voltage of $V_b = 12$ V the current in the starting circuit is lower than the reference current: $R_3 \approx (V_b - 3\,V_{BE})/I_{ref} \approx 100$ kΩ.

PTAT Reference Current Source

Replacing the V_{BE} reference current source T_1,T_2 in Fig. 4.115a with a Widlar current mirror leads to the *PTAT reference current source* shown in Fig. 4.118. *PTAT* stands for *proportional to absolute temperature* and indicates that the current is proportional to the absolute temperature in Kelvin. This means that, unlike the V_{BE} reference current source, the PTAT reference current source has a positive temperature coefficient.

The following equation is derived from Fig. 4.118:

$$V_{BE\,2} = V_{BE\,1} + (I_{C1} + I_{B1})\,R_1 \overset{I_{ref} = I_{C1} \gg I_{B1}}{\approx} V_{BE\,1} + I_{ref}\,R_1$$

For $V_{BE} = V_T \ln I_C/I_S$, $I_{C1} = I_{ref}$ and $I_{C2} \approx I_2$, this leads to:

$$V_T \ln \frac{I_2}{I_{S2}} \approx V_T \ln \frac{I_{ref}}{I_{S1}} + I_{ref}\,R_1$$

The current mirror T_3,T_4 normally has a current ratio of $k_I \approx 1$; thus, $I_2 \approx I_{ref}$. By inserting this into the previous equation and solving for I_{ref}, we obtain:

$$\boxed{I_{ref} \approx \frac{V_T}{R_1} \ln \frac{I_{S1}}{I_{S2}} \quad \text{for } I_{S1} > I_{S2} \text{ und } k_I \approx 1} \tag{4.138}$$

Since I_{ref} must be positive, the limitation $I_{S1} > I_{S2}$ is necessary in (4.138); this means that T_1 must be larger than T_2. Usually, I_{ref} and $I_{S1}/I_{S2} \approx 4 \ldots 10$ are specified to calculate R_1.

The PTAT reference current source also has a second operating point at zero current, which has to be eliminated by a starting circuit. Figure 4.118b shows one possible circuit, which has already been used and described in connection with the V_{BE} reference current source. However, here, resistance R_2 must be higher than in the V_{BE} reference current

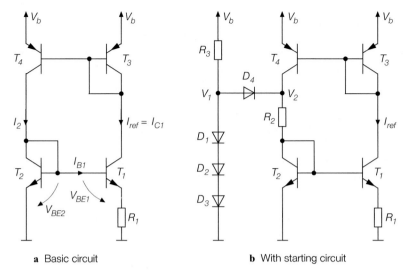

a Basic circuit   **b** With starting circuit

Fig. 4.118. PTAT reference current source

source, in order to render voltage V_2 sufficiently high at the desired operating point; the guidance value here is $I_{ref} R_2 \approx 2 V_{BE} \approx 1.4\,\text{V}$.

To ensure that the PTAT reference current source provides a current independent of the supply voltage, it is necessary to add cascode stages to eliminate the Early effect of transistors T_1 and T_4 and an output. Figure 4.119 shows a practical solution, which has the following additional features compared to Fig. 4.118b:

- T_5 is added to current mirror T_3,T_4 to form a three-transistor current mirror and the cascode stage T_6 is added at the output.

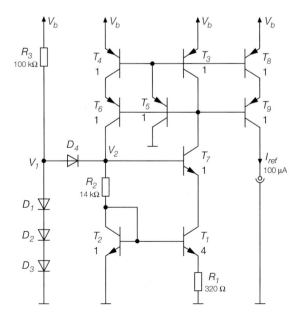

Fig. 4.119. Practical configuration of a PTAT reference current source

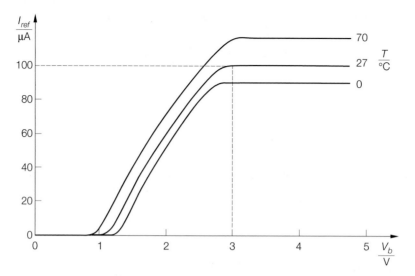

Fig. 4.120. Characteristics of the PTAT reference current source in Fig. 4.119

- Transistor T_1 is provided with cascode stage T_7, which uses voltage V_2 of the starting circuit as the base bias voltage.
- Transistor T_8 and the corresponding cascode stage T_9 are used to provide the reference current output node.

Figure 4.120 shows the resulting characteristics for various temperatures.

Controlled PTAT reference current source: Figure 4.121 shows the principle of a controlled PTAT reference current source; here, the PTAT current according to (4.138) is adjusted not with a current mirror, but with two control amplifiers A1 and A2.
    If both control amplifiers have high-resistance inputs and a gain A, then:

$$V_1 = A(V_0 - V_4)$$
$$V_2 = V_3 - I_{C2}R_2$$
$$V_3 = A(V_0 - V_2)$$
$$V_4 = V_3 - I_{C1}R_2$$

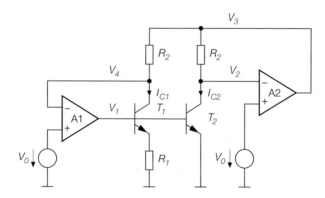

Fig. 4.121. Controlled PTAT reference current source

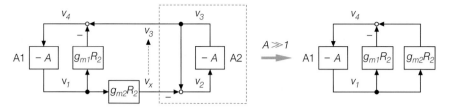

Fig. 4.122. Equivalent control circuit of the controlled PTAT reference current source

If the gain A is sufficiently high and the circuit is stable, then the parameters at the operating point are $V_2 = V_4 = V_0$ and $I_{C1} = I_{C2} = I_{ref}$; the latter only applies to the PTAT current according to (4.138), because of the common-base voltage V_1. The stability is checked using the small-signal equivalent circuit; this leads to

$$v_1 = -Av_4$$

$$v_2 = v_3 - i_{C2}R_2 = v_3 - g_{m2}R_2v_1$$

$$v_3 = -Av_2$$

$$v_4 = v_3 - i_{C1}R_2 = v_3 - g_{m1}R_2v_1$$

with the transconductances:

$$g_{m1} = \frac{I_{ref}}{V_T + I_{ref}R_1} \quad , \quad g_{m2} = \frac{I_{ref}}{V_T} > g_{m1}$$

Figure 4.122 shows the equivalent control circuit for the static situation; due to

$$\frac{v_3}{v_x} = \frac{A}{1+A} \overset{A\gg1}{\approx} 1$$

the circuit with control amplifier A2 can be replaced by a direct connection. This means that transistor T_1 with $g_{m1}R_2$ provides positive feedback and transistor T_2 with $g_{m2}R_2$ negative feedback for control amplifier A1; since $g_{m2} > g_{m1}$, the circuit is *statically stable*. Dynamic stability must be ensured by way of frequency response compensation for the two control amplifiers; this will be outlined in more detail below.

Figure 4.123 shows a practical configuration for the controlled PTAT reference current source. One common-emitter circuit (T_3,T_5) with a subsequent common-collector circuit (T_4,T_6) is used for each of the control amplifiers; an additional nonlinear level converter (R_6,T_7), which linearizes the circuit with regard to large signals, is contained in control amplifier A1. The voltages V_0 correspond to the base–emitter voltages of transistors T_3 and T_5 at the operating point: $V_2 \approx V_4 \approx V_0 \approx 0.7\,\text{V}$. This means that transistors T_1 and T_2 are operated with a constant collector voltage; the Early effect thus has no influence on the reference current. Decoupling is done by connecting additional transistors to voltage V_1; this is indicated on the far left of Fig. 4.123. Here too, the decoupling transistors may need a cascode stage to eliminate their Early effect. Figure 4.124 shows the characteristics for various temperatures.

Frequency response compensation is required for both control amplifiers to ensure that the circuit is dynamically stable; this is done by capacitances C_1 and C_2. The component dimensions are determined by circuit simulation. For this purpose, a transient analysis is performed in which a current source injects a short current pulse into node V_1; this

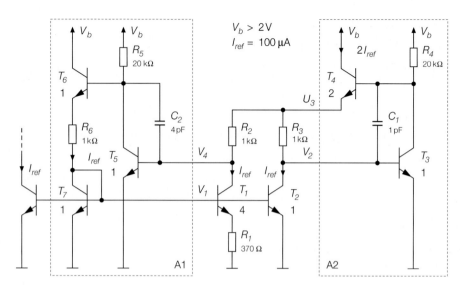

Fig. 4.123. Practical configuration of a controlled PTAT reference current source

allows us to evaluate the pulse response at the different nodes and select the capacitances accordingly. Without compensation the circuit is usually unstable; in such cases, the circuit simulation produces undamped oscillation.

Temperature sensitivity: Since the current of the PTAT reference current source is proportional to the temperature voltage V_T, the influence of the temperature sensitivity is:

$$V_T = \frac{kT}{q} \quad \Rightarrow \quad \frac{dV_T}{dT} = \frac{k}{q} \approx 86 \, \mu V/K$$

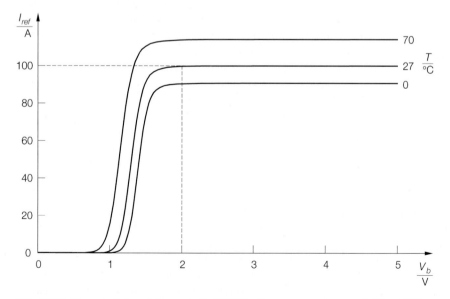

Fig. 4.124. Characteristics of the controlled PTAT reference current source in Fig. 4.123

This causes a current change of

$$\frac{d I_{ref}}{dT} = \frac{1}{R_1} \ln \frac{I_{S1}}{I_{S2}} \frac{dV_T}{dT} \approx \frac{86 \,\mu V/K}{R_1} \ln \frac{I_{S1}}{I_{S2}} \qquad (4.139)$$

and a *temperature coefficient* of:

$$\frac{1}{I_{ref}} \frac{d I_{ref}}{dT} = \frac{1}{V_T} \frac{dV_T}{dT} = \frac{1}{T} \stackrel{T=300\,K}{=} 3.3 \cdot 10^{-3} \, K^{-1}$$

With a temperature increase of 3 K, the reference current rises by 1%. This makes the temperature sensitivity of the PTAT reference current source even higher than that of the V_{BE} reference current source; its polarity, however, is reversed.

Use in bipolar amplifiers: Despite its high temperature sensitivity, the PTAT reference current source is used as the reference source for quiescent currents in bipolar amplifiers. In such cases the temperature sensitivity may even be an advantage, as the gain of bipolar amplifier stages without current feedback is proportional to the transconductance $g_m = I_{C,A}/V_T$ of the transistors; with $I_{C,A} \sim I_{ref} \sim V_T$ the transconductance, and thus the gain, remains constant.

Temperature-Independent Reference Current Source

Adding the currents of a V_{BE} and a PTAT reference current source and selecting the values such that

$$\frac{d I_{ref}}{dT}\bigg|_{V_{BE}-Ref.} + \frac{d I_{ref}}{dT}\bigg|_{PTAT-Ref.} = 0$$

leads to the temperature-independent reference current source shown in Fig. 4.125. The left portion of the circuit corresponds to the PTAT reference current source in Fig. 4.119.

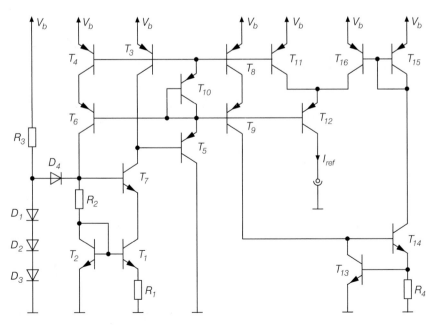

Fig. 4.125. Temperature-independent reference current source

The transistor diode T_{10} has been added to allow the bases of the pnp cascode transistors to be connected to the emitter of T_5; this reduces the errors caused by the base currents. The V_{BE} reference current source T_{13}, T_{14} is connected to the original output T_8, T_9; in this case, it is already supplied with a stabilized current and feedback via a current mirror is not required. With T_{11}, T_{12}, the PTAT reference current source has an additional output at which the current of the V_{BE} reference current source is added via the current mirror T_{15}, T_{16}. Equations (4.136)–(4.139) determine the ratio of the currents

$$\frac{I_{ref,VBE}}{V_{BE}}\frac{dV_{BE}}{dT} + \frac{I_{ref,PTAT}}{V_T}\frac{dV_T}{dT} = 0 \quad \Rightarrow \quad \frac{I_{ref,VBE}}{I_{ref,PTAT}} = -\frac{V_{BE}}{V_T}\frac{\dfrac{dV_T}{dT}}{\dfrac{dV_{BE}}{dT}} \approx 1.3$$

and the reference current:

$$I_{ref} = I_{ref,VBE} + I_{ref,PTAT} \approx 2.3 \cdot I_{ref,PTAT} \approx 1.77 \cdot I_{ref,VBE}$$

A reference current of $I_{ref} = 100\,\mu A$ results in $I_{ref,PTAT} \approx I_{ref}/2.3 \approx 43\,\mu A$ and $I_{ref,VBE} \approx I_{ref}/1.77 \approx 57\,\mu A$.

Reference Current Sources in MOS Circuits

The V_{BE} reference current source in Fig. 4.113 may also be realized with MOSFETs; it is then called the V_{GS} reference current source [4.2]. When operated in the square region of the characteristic, the stabilization of the current is comparatively poor. A far better response is achieved by making the MOSFETs so large that they operate in the sub-threshold region; then the characteristic follows an exponential curve and the MOSFETs respond approximately like bipolar transistors. According to (3.25) on page 202, for operation in the sub-threshold region the following must be the case:

$$|V_{GS} - V_{th}| < 2n_V V_T \overset{n_V \approx 1.5...2.5}{\approx} 3...5 \cdot V_T$$

This makes the MOSFETs very large, even with small currents. One disadvantage of this is the dependence on the threshold voltage V_{th}, which varies due to production tolerances.

The PTAT reference current source may also be realized with MOSFETs in the sub-threshold region; here, when calculating the current voltage V_T is replaced by $n_V V_T$, since for MOSFETs in the sub-threshold region the following applies:

$$I_D \sim e^{\frac{V_{GS}-V_{th}}{n_V V_T}} \quad \text{with } n_V \approx 1.5...2.5$$

On the other hand, the shift by the threshold voltage V_{th} has no influence on the current, but merely causes a shift in the operating point voltages.

In general, reference current sources with MOSFETs have far poorer properties than bipolar reference current sources. This is the reason why integrated circuits that have to satisfy high demands in terms of accuracy and temperature response are mostly produced using bipolar technology.

Pinch-Off Current Sources

Where accuracy and temperature sensitivity are not the main concern, the pinch-off current sources shown in Fig. 4.126 can be employed, all of which use the constant drain current

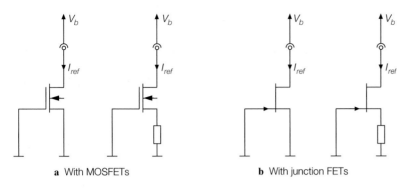

a With MOSFETs   **b** With junction FETs

Fig. 4.126. Pinch-off current sources

of a depletion FET in the pinch-off region as the reference current; We can thus operate with $V_{GS} = 0$, or in the case of current feedback with a resistance with $V_{GS} < 0$.

The pinch-off current sources with junction FETs in Fig. 4.126b are realized in integrated circuits by means of a *pinch resistor*. This is an integrated high-resistance resistor, which is pinched off as the voltage increases. Since its basic construction corresponds to that of a junction FET, its characteristic behavior is practically identical. The disadvantages are the high production tolerances, which are typically in the range of $\pm 30\%$ [4.1].

Setting the Operating Point in Integrated Amplifier Circuits

In integrated circuits, the operating point is mainly set by means of current sources for adjusting the quiescent current and auxiliary voltages for cascode stages; the current sources are usually set up as current source banks with a common reference.

Bipolar circuits: Figure 4.127 shows a typical circuit for setting the operating point in bipolar amplifier circuits. It comprises of a PTAT reference current source ($T_1 \dots T_8$) with a starting circuit ($D_1 \dots D_5$) as well as an npn (T_9) and a pnp collector circuit (T_{11}) with the relevant current sources (T_{10}, T_{12}) to provide the auxiliary voltages V_1 and V_2 for the cascode stages; the transistor diode D_6 represents a simple method of producing additional auxiliary voltages. As the PTAT reference current source offers not only decoupling at the current mirror $T_3 \dots T_6$ but also decoupling at the Widlar current mirror T_1, T_2, the Widlar current mirror is, in contrast to Fig. 4.119, expanded by T_8 to form a three-transistor current mirror to reduce the error caused by the base currents; this, however, makes an additional transistor diode necessary in the starting circuit, to raise the starting voltage accordingly. Resistance R_3 represents a p-channel pinch resistance. This is nothing unusual, as in most cases resistances rated around $100\,k\Omega$ can only be produced in this form. The fact that with large voltages pinch resistances act as constant current sources (see the section on pinch-off current sources) is of advantage here, since this limits the current in the starting circuit; likewise, the production tolerances have no negative effect, since the current in the starting circuit may vary by almost one order of magnitude without restricting the function.

Simple current sources or current sources with a cascode of any current ratio can be connected to the decoupled signals and the auxiliary voltages; Fig. 4.127 exemplifies this with each current source with cascode. Other auxiliary voltages, such as voltage V_3, can

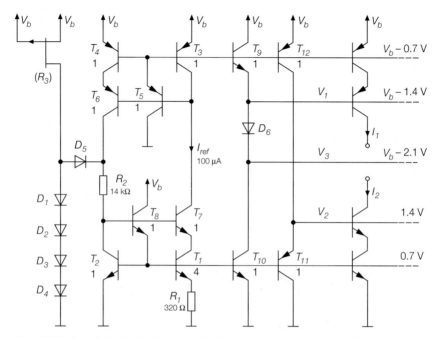

Fig. 4.127. A typical circuit with a PTAT reference current source to set the operating point in bipolar amplifier circuits (numeric example for $I_{ref} = 100\,\mu A$ at $V_b > 3.5\,V$, using the bipolar transistors in Fig. 4.5 on page 274)

be easily generated with transistor diodes; common-collector circuits, as with V_1 and V_2, must be used where higher currents are needed.

MOS circuits: Figure 4.128 shows a typical circuit for setting the operating point in MOS amplifier circuits. It comprises a V_{GS} reference current source (T_1, T_2) with current mirror (T_3, T_4) and starting circuit (T_5, T_6), as well as a decoupling with an auxiliary voltage generator ($T_8 \ldots T_{12}$). The starting circuits supplies the starting current via T_5, which is turned off by T_6 after the reference current becomes active. The depletion MOSFET T_7 serves as the quiescent current source (the pinch-off current source) for T_6; its current must be smaller than the reference current to be able to turn the starting circuit off via T_6. The size of T_7 depends on the threshold voltage of the depletion MOSFET, which is determined by the given manufacturing process.

A circuit with this configuration only makes sense if the production tolerances of resistance R_1 and the threshold voltage of T_2 are lower than the tolerance of the threshold voltage of T_7; otherwise, it is better to use the current of the pinch-off current source T_7 as the reference current.

4.2
Properties and Parameters

The properties of an amplifier are described by the characteristic parameters. These are based on the characteristic curves of the amplifier. The small-signal parameters (e.g. the

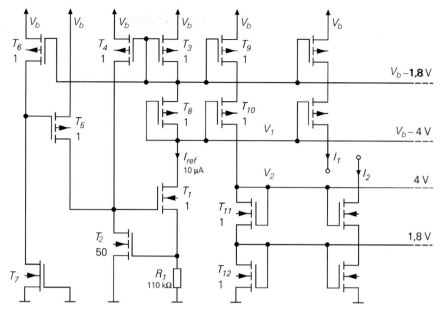

Fig. 4.128. A typical circuit using a V_{GS} reference current source to set the operating point in MOS amplifier circuits (numeric example for $I_{ref} = 10\,\mu A$ at $V_b > 7\,V$, using the MOSFETs in Fig. 4.6 on page 274)

gain) are determined by linearization at the operating point, and the nonlinear parameters (e.g. the distortion factor) are determined by series expansion. As direct calculation of the characteristics is often not possible, it may be necessary to rely on measurements or circuit simulations.

4.2.1
Characteristics

An amplifier with one input and one output is usually described by two families of characteristics; using the parameters of Fig. 4.129, the currents are:

$$I_i = f_I(V_i, V_o)$$
$$I_o = f_O(V_i, V_o)$$

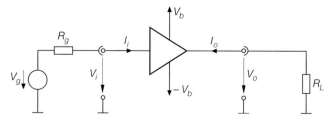

Fig. 4.129. Voltages and currents in an amplifier with one input and one output

In most amplifiers, the effect that the output has on the input is negligibly small in the interesting region; that is, the input characteristic is almost independent of the output voltage. This leads to:

$$I_i = f_I(V_i) \tag{4.140}$$
$$I_o = f_O(V_i, V_o) \tag{4.141}$$

With an open output, this leads to the *open-circuit transfer characteristic*:

$$I_o = f_O(V_i, V_o) = 0 \quad \Rightarrow \quad V_o = f_{Tr}(V_i) \tag{4.142}$$

This is often simply referred to as the *transfer characteristic*.

If the amplifier is operated with a signal source with an internal resistance R_g and a load R_L, then according to Fig. 4.129:

$$I_i = \frac{V_g - V_i}{R_g} \quad , \quad I_o = -\frac{V_o}{R_L} \tag{4.143}$$

The straight lines described by these equations are known as the *source* and *load characteristics*. Insertion into (4.140) and (4.141) leads to the nonlinear set of equations

$$\begin{aligned} V_g &= V_i + R_g f_I(V_i) \\ 0 &= V_o + R_L f_O(V_i, V_o) \end{aligned} \tag{4.144}$$

and thus the *operational transfer characteristic*:

$$V_o = f_{Tr,B}(V_g) \tag{4.145}$$

It is only possible in exceptional cases to solve (4.142) and the set given in (4.144), as well as to determine the operational transfer characteristic. In practice, circuit simulation programs are used, which solve the equations in the course of a *DC analysis* for several points and plot the characteristics graphically. If the characteristics of an amplifier are available in a graphic plot, the set given in (4.144) can also be solved graphically by drawing the straight lines according to (4.143) into the family of input or output characteristics and determining the intersections.

Example: Applying the transport model in Fig. 2.26 on page 61 to the common-emitter circuit shown in Fig. 4.130 leads to:

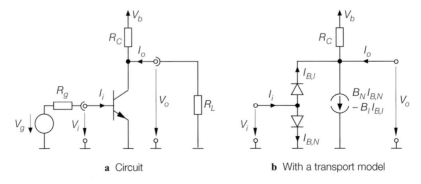

a Circuit   **b** With a transport model

Fig. 4.130. Example: the common-emitter circuit

$$I_i = f_I(V_i, V_o) = I_{B,N} + I_{B,I}$$

$$= \frac{I_S}{B_N}\left(e^{\frac{V_i}{V_T}} - 1\right) + \frac{I_S}{B_I}\left(e^{\frac{V_i - V_o}{V_T}} - 1\right)$$

$$I_o = f_O(V_i, V_o) = \frac{V_o - V_b}{R_C} + B_N I_{B,N} - (1 + B_I) I_{B,I}$$

$$= \frac{V_o - V_b}{R_C} + I_S\left(e^{\frac{V_i}{V_T}} - 1\right) - \frac{1 + B_I}{B_I} I_S\left(e^{\frac{V_i - V_o}{V_T}} - 1\right)$$

For practical applications, only the region of normal transistor operation, $V_o > V_{CE,sat} \approx$ 0.2 V, is of interest; in this region, the output voltage has no effect on the input characteristic. If the reverse currents are neglected, it follows that:

$$I_i = f_I(V_i) = \frac{I_S}{B_N} e^{\frac{V_i}{V_T}}$$

$$I_o = f_O(V_i, V_o) = \frac{V_o - V_b}{R_C} + I_S e^{\frac{V_i}{V_T}}$$

The characteristics are shown in Fig. 4.131. Here, the open-circuit transfer characteristic can be calculated straight forward:

$$f_O(V_i, V_o) = 0 \quad \Rightarrow \quad V_o = f_{Tr}(V_i) = V_b - I_S R_C e^{\frac{V_i}{V_T}}$$

The straight-line source characteristic in Fig. 4.131a and the straight-line load characteristic in Fig. 4.131b are determined for $V_g = 1\,\mathrm{V}$, $R_g = 100\,\mathrm{k\Omega}$ and $R_L = 10\,\mathrm{k\Omega}$. The intersections provide $V_i(V_g = 1\,\mathrm{V}) \approx 0.69\,\mathrm{V}$ and $V_o(V_i = 0.69\,\mathrm{V}) \approx 1\,\mathrm{V}$. Thus, one

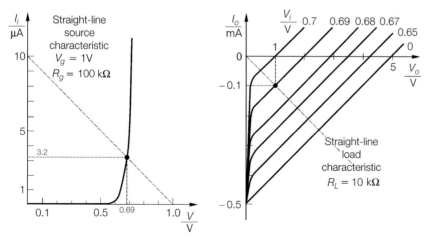

a Input characteristic   b Family of output characteristics

Fig. 4.131. Characteristics of the common-emitter circuit in Fig. 4.130, with $V_b = 5\,\mathrm{V}$ and $R_C = 10\,\mathrm{k\Omega}$

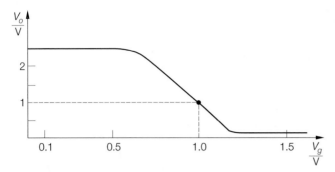

Fig. 4.132. Operational transfer characteristic of the common-emitter circuit in Fig. 4.130, with $V_b = 5\,\text{V}$, $R_C = 10\,\text{k}\Omega$, $R_g = 100\,\text{k}\Omega$ and $R_L = 10\,\text{k}\Omega$

point of the operational transfer characteristic is known: $V_o(V_g = 1\,\text{V}) \approx 1\,\text{V}$. For other values of V_g, the characteristic can be determined point for point. A circuit simulation program uses the same basic method of solving the set given in (4.144) numerically for the V_g values given by the user; the result is shown in Fig. 4.132.

4.2.2
Small-Signal Characteristics

The small-signal characteristics describe the quasi-linear behavior of an amplifier when driven with low amplitudes around one operating point; this mode of operation is called *small-signal operation.*

Operating Point

The operating point A is characterized by the voltages $V_{i,A}$ and $V_{o,A}$ and by the currents $I_{i,A}$ and $I_{o,A}$:

$$I_{i,A} = f_I(V_{i,A}) \quad , \quad I_{o,A} = f_O(V_{i,A}, V_{o,A})$$

In general, the operating point depends on the signal source and the load. One exception are amplifiers with AC coupling via coupling capacitances or transformers, which allow the operating point to be set independent of the signal source and the load. However, for calculating the small-signal parameters, it is of no importance how the operating point is adjusted.

Small-Signal Values

In the analysis of the small-signal characteristics, only the deviations from the operating point are considered. These are described by the small-signal values:

$$v_i = V_i - V_{i,A} \quad , \quad i_i = I_i - I_{i,A}$$
$$v_o = V_o - V_{o,A} \quad , \quad i_o = I_o - I_{o,A}$$

As the operating point values $V_{i,A}$, $I_{i,A}$, $V_{o,A}$ and $I_{o,A}$ normally correspond to the DC components of V_i, I_i, V_o and I_o, the small-signal values have no DC component; that is, they have zero mean value.

Example:

$$V_i = V_0 + v_1 \cos \omega_1 t + v_2 \cos \omega_2 t \quad \Rightarrow \quad \begin{cases} V_{i,A} = V_0 \\ v_i = v_1 \cos \omega_1 t + v_2 \cos \omega_2 t \end{cases}$$

Linearization

Insertion of the small-signal values into the characteristics given in (4.140) and (4.141) and series expansion at the operating point lead to[26]:

$$I_i = I_{i,A} + i_i = f_I(V_{i,A} + v_i)$$

$$= f_I(V_{i,A}) + \left.\frac{\partial f_I}{\partial V_i}\right|_A v_i + \frac{1}{2}\left.\frac{\partial^2 f_I}{\partial V_i^2}\right|_A v_i^2 + \frac{1}{6}\left.\frac{\partial^3 f_I}{\partial V_i^3}\right|_A v_i^3 + \cdots$$

$$I_o = I_{o,A} + i_o = f_O(V_{i,A} + v_i, V_{o,A} + v_o)$$

$$= f_O(V_{i,A}, V_{o,A}) + \left.\frac{\partial f_O}{\partial V_i}\right|_A v_i + \left.\frac{\partial f_O}{\partial V_o}\right|_A v_o$$

$$+ \frac{1}{2}\left.\frac{\partial^2 f_O}{\partial V_i^2}\right|_A v_i^2 + \frac{1}{2}\left.\frac{\partial^2 f_O}{\partial V_i \partial V_o}\right|_A v_i v_o + \frac{1}{2}\left.\frac{\partial^2 f_O}{\partial V_o^2}\right|_A v_o^2 + \cdots$$

With sufficiently small input signals, the series expansion can be aborted after the linear term; this results in linear relations between the small-signal values:

$$i_i = \left.\frac{\partial f_I}{\partial V_i}\right|_A v_i$$

$$i_o = \left.\frac{\partial f_O}{\partial V_i}\right|_A v_i + \left.\frac{\partial f_O}{\partial V_o}\right|_A v_o$$

The transition to these linear equations is called *linearization at the operating point*.

Small-Signal Parameters

The partial derivations resulting from the linearization, which are evaluated at operating point A, are called the *small-signal parameters*.

– The *small-signal input resistance* r_i:

$$r_i = \frac{v_i}{i_i} = \left(\left.\frac{\partial f_I}{\partial V_i}\right|_A\right)^{-1} \tag{4.146}$$

– The *small-signal output resistance* r_o:

$$r_o = \left.\frac{v_o}{i_o}\right|_{v_i=0} = \left(\left.\frac{\partial f_O}{\partial V_o}\right|_A\right)^{-1} \tag{4.147}$$

This is also called the *short-circuit output resistance*, since the input is short-circuited with regard to small signals ($v_i = 0$). In practice, this means that a voltage source with sufficiently low internal resistance, which keeps the input voltage constant at $V_{i,A}$, is connected to the input.

---

[26] A partial differentiation of the input characteristic f_I is also used in what follows; this indicates that f_I is generally dependent on a second variable (V_o).

- The *small-signal gain A*:

$$A = \frac{v_o}{v_i}\bigg|_{i_o=0} = -\frac{\partial f_O}{\partial V_i}\bigg|_A \left(\frac{\partial f_O}{\partial V_o}\bigg|_A\right)^{-1} \tag{4.148}$$

This is also called the *open-circuit gain*, as there is no load connected to the output; that is, it is open with regard to small signals ($i_o = 0$). The gain can also be calculated from the open-circuit transfer characteristic given in (4.142):

$$A = \frac{df_{Tr}}{dV_i}\bigg|_A$$

- The *transconductance g_m*:

$$g_m = \frac{i_o}{v_i}\bigg|_{v_o=0} = \frac{\partial f_O}{\partial V_i}\bigg|_A \tag{4.149}$$

This is of minor importance in amplifiers that have a low-resistance output (small r_o) and thus provide an output voltage, but it plays an important role in transistors and amplifiers with high-resistance outputs (high r_o). From a comparison of (4.147) and (4.148), it follows:

$$g_m = -\frac{A}{r_o} \quad \text{of} \quad A = -g_m \, r_o \tag{4.150}$$

This means that one of the parameters A, r_o and g_m is redundant.

Small-Signal Equivalent Circuit for an Amplifier

With the help of the small-signal parameters, the *small-signal equivalent circuits* shown in Fig. 4.133 are derived using the following equations:

$$i_i = \frac{v_i}{r_i} \tag{4.151}$$

$$v_o = A\,v_i + i_o r_o \quad \text{or} \quad i_o = g_m\,v_i + \frac{v_o}{r_o} \tag{4.152}$$

When operating the amplifier with a signal source with an internal resistance R_g and a load R_L, the *small-signal operational gain* can be obtained from the small-signal equivalent circuit in Fig. 4.134:

$$A_B = \frac{v_o}{v_g} = \frac{r_i}{R_g + r_i}\,A\,\frac{R_L}{r_o + R_L} \overset{A=-g_m r_o}{=} -\frac{r_i}{R_g + r_i}\,g_m\,\frac{r_o R_L}{r_o + R_L} \tag{4.153}$$

Here, $v_g = V_g - V_{g,A}$ is the small-signal voltage of the signal source. The small-signal operational gain comprises the open-circuit gain A and the voltage divider factors at the input and the output; when using the transconductance g_m to describe the circuit, the factor at the output side changes over to the parallel connection of r_o and R_L. The small-signal operational gain may also be calculated from the operational transfer characteristic given in (4.145):

$$A_B = \frac{df_{Tr,B}}{dV_g}\bigg|_A$$

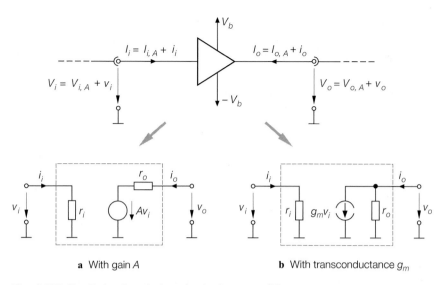

a With gain A **b** With transconductance g_m

Fig. 4.133. Small-signal equivalent circuits for an amplifier

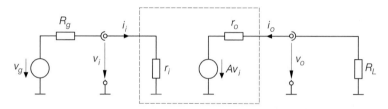

Fig. 4.134. Small-signal equivalent circuit for an amplifier with a signal source and a load

Example: For the common-emitter circuit in Fig. 4.130a on page 410, the following characteristics were determined:

$$I_i = f_I(V_i) = \frac{I_S}{B_N} e^{\frac{V_i}{V_T}} \quad , \quad I_o = f_O(V_i, V_o) = \frac{V_o - V_b}{R_C} + I_S e^{\frac{V_i}{V_T}}$$

For $V_g = 1\,\text{V}$, $R_g = 100\,\text{k}\Omega$ and $R_L = R_C = 10\,\text{k}\Omega$, we obtained $V_i \approx 0.69\,\text{V}$ and $V_o \approx 1\,\text{V}$. This point is now used as the operating point; for $I_S = 1\,\text{fA}$, $B_N = 100$ and $V_T = 26\,\text{mV}$, it follows:

$$V_{i,A} \approx 0.69\,\text{V} \ , \ I_{i,A} = f_I(V_{i,A}) \approx 3\,\mu\text{A}$$

$$V_{o,A} \approx 1\,\text{V} \ , \ I_{o,A} = -\frac{V_{o,A}}{R_L} \approx -100\,\mu\text{A}$$

The upper portion of Fig. 4.135 shows the circuit with the operating point values.
    From (4.146) for

$$\left.\frac{\partial f_I}{\partial V_i}\right|_A = \left.\frac{I_S}{V_T B_N} e^{\frac{V_i}{V_T}}\right|_A = \left.\frac{I_i}{V_T}\right|_A = \frac{I_{i,A}}{V_T} \approx \frac{3\,\mu\text{A}}{26\,\text{mV}} \approx 0.115\,\text{mS}$$

we obtain an input resistance of $r_i \approx 8.7\,\text{k}\Omega$; similarly, for

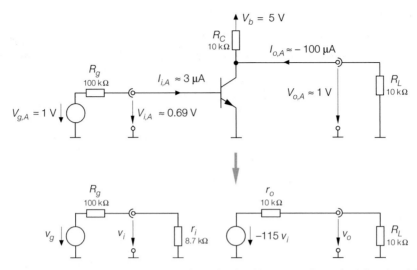

Fig. 4.135. Example: the common-emitter circuit with an operating point (*above*) and the resulting small-signal equivalent circuit (*below*)

$$\left.\frac{\partial fo}{\partial V_o}\right|_A = \frac{1}{R_C} = 0.1\,\text{mS}$$

it follows from (4.147) that the output resistance is $r_o = R_C = 10\,\text{k}\Omega$, and from (4.149) for

$$\left.\frac{\partial fo}{\partial V_i}\right|_A = \left.\frac{I_S}{V_T}\,e^{\frac{V_i}{V_T}}\right|_A \approx \frac{300\,\mu\text{A}}{26\,\text{mV}} \approx 11.5\,\text{mS}$$

we obtain the transconductance of $g_m \approx 11.5\,\text{mS}$. The gain A can be determined from g_m and r_o using (4.150): $A = -g_m r_o \approx -115$. The lower portion of Fig. 4.135 shows the resulting small-signal equivalent circuit. From (4.153), the operational gain is thus $A_B \approx -4.6$; this corresponds to the slope of the operational transfer characteristic in Fig. 4.132 on page 412 at the operating point shown.

Amplifier with Reverse Gain

In some amplifiers, the effect of the output on the input is not negligible.[27] In this case, the input current is also influenced by the output voltage:

$$I_i = f_I(V_i, V_o) \tag{4.154}$$

Two further small-signal parameters can be determined by linearization:

– The *reverse gain* A_r:

$$A_r = \left.\frac{v_i}{v_o}\right|_{i_i=0} = -\left.\frac{\partial f_I}{\partial V_o}\right|_A \left(\left.\frac{\partial f_I}{\partial V_i}\right|_A\right)^{-1} \tag{4.155}$$

---

[27] Here, only the *static reactive effect* will be looked at; but due to parasitic capacitances many amplifiers have *dynamic* reactive effects that become evident at higher frequencies.

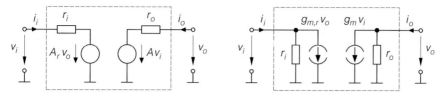

a With gains A and A_r **b** With transconductances g_m and $g_{m,r}$

Fig. 4.136. Small-signal equivalent circuits for an amplifier with reverse gain

– The *reverse transconductance* $g_{m,r}$:

$$g_{m,r} = \left. \frac{i_i}{v_o} \right|_{v_i=0} = \left. \frac{\partial f_I}{\partial V_o} \right|_A \tag{4.156}$$

From a comparison of (4.146) and (4.155), it follows:

$$g_{m,r} = -\frac{A_r}{r_i} \qquad \text{or} \qquad A_r = -g_{m,r}r_i \tag{4.157}$$

This means that one of the parameters A_r, r_i and $g_{m,r}$ is redundant.

Figure 4.136 shows the small-signal equivalent circuits for amplifiers with reverse gain; the parameters are:

$$v_i = A_r v_o + i_i r_i \qquad \text{or} \qquad i_i = g_{m,r} v_o + \frac{v_i}{r_i} \tag{4.158}$$

$$v_o = A v_i + i_o r_o \qquad \text{or} \qquad i_o = g_m v_i + \frac{v_o}{r_o} \tag{4.159}$$

In this case, input resistance r_i is also known as the *short-circuit input resistance*, as it is determined with the output short-circuited for small signals ($v_o = 0$). Furthermore, gain A is also called the *forward gain* and transconductance S the *forward transconductance* if the difference with regard to the related reverse parameter is to be emphasized.

In addition to the two small-signal equivalent circuits shown in Fig. 4.136, there are a further two circuits, since the input or output can be shown in terms of the given gain or the given transconductance. The two mixed forms are, however, very seldom used. These four possible presentations must not be confused with the four four-pole representations using the Y, Z, H or P matrices, since here the controlled sources are always voltage-controlled; therefore, the four small-signal equivalent circuits are variations of the Y representation with:

$$y_{11} = \frac{1}{r_i} \quad , \quad y_{12} = g_{m,r} \quad , \quad y_{21} = g_m \quad , \quad y_{22} = \frac{1}{r_o}$$

The small-signal equivalent circuit in Fig. 4.136b corresponds to the common Y representation. The three other small-signal equivalent circuits are obtained by converting the current source into an equivalent voltage source in either the input circuit or the output circuit, or both; this causes the transconductances g_m and $g_{m,r}$ to change into the gains A and A_r, respectively.

For operation with a signal source with internal resistance R_g and a load R_L, the operational gain A_B can be calculated directly with the help of the small-signal equivalent circuit in Fig. 4.137; the result is a complex expression that allows no insight into the relationships. The procedure is therefore broken down into three steps:

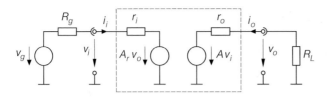

Fig. 4.137. Small-signal equivalent circuit for calculating the operational gain of an amplifier with reverse gain

- First, the operational gain is calculated for operation with an ideal signal voltage source; that is, $R_g = 0$:

$$A_{B,0} = \left.\frac{v_o}{v_g}\right|_{R_g=0} = \frac{v_o}{v_i} = A\,\frac{R_L}{r_o + R_L} \tag{4.160}$$

This comprises the open-circuit gain A and the voltage divider factor at the output and is independent of the reverse gain A_r. Subscript 0 of $A_{B,0}$ indicates that $R_g = 0$.

- Then, the *operational input resistance* $r_{i,B}$ is calculated:

$$r_{i,B} = \frac{v_i}{i_i} = \frac{r_i}{1 - A_r A\,\dfrac{R_L}{r_o + R_L}} = \frac{r_i}{1 - A_r A_{B,0}} \tag{4.161}$$

In amplifiers with reverse gain ($A_r \neq 0$), this value depends on the load R_L; in amplifiers without reverse gain ($A_r = 0$), we obtain $r_{i,B} = r_i$.

- With the help of the operational input resistance, the voltage divider factor at the input, and thus the operational gain, can be calculated:

$$A_B = \frac{r_{i,B}}{R_g + r_{i,B}}\,A_{B,0} = \frac{r_{i,B}}{R_g + r_{i,B}}\,A\,\frac{R_L}{r_o + R_L} \tag{4.162}$$

Inserting $r_{i,B} = r_i$ leads to the operational gain for amplifiers without reverse gain according to (4.153).

Therefore, it is possible to treat an amplifier with reverse gain like an amplifier without reverse gain provided that the operational input resistance $r_{i,B}$ is used instead of the input resistance r_i. This is the reason why Sects. 2.4 and 3.4, on the calculation of basic transistor circuits, show how the input resistance can be calculated for a given load R_L provided that such a dependence – that is, reverse gain – exists; this replaces the calculation of the reverse gain A_r or the reverse transconductance $g_{m,r}$. The operational gain of basic transistor circuits can thus be calculated using (4.160)–(4.162) by replacing r_o with the short-circuit output resistance $r_{o,s}$ and $r_{i,B}$ with the input resistance r_i for operation with a load R_L:

$$r_o = r_{o,s} \quad , \quad r_{i,B} = r_i(R_L)$$

While the gain A presents no interpretation problems, specifications for the input and output resistances generally require the operating conditions to be taken into consideration; these relationships are listed in Fig. 4.138.

Calculating the Small-Signal Parameters with the Help of the Small-Signal Equivalent Circuit

For complex circuits, the characteristics f_I and f_O cannot be quoted as a whole; thus, it is no longer possible to calculate the small-signal parameters by differentiating the characteristic

Operating conditions	Input resistance	Output resistance
General operation	$r_{i,B} = \dfrac{r_i}{1 - A_r A \dfrac{R_L}{r_o + R_L}}$	$r_{o,B} = \dfrac{r_o}{1 - A_r A \dfrac{R_g}{r_i + R_g}}$
Short circuit	$r_{i,s} = r_{i,B}\big|_{R_L=0} = r_i$	$r_{o,s} = r_{o,B}\big|_{R_g=0} = r_o$
Open circuit	$r_{i,o} = r_{i,B}\big|_{R_L=\infty}$	$r_{o,o} = r_{o,B}\big|_{R_g=\infty}$
	$= \dfrac{r_i}{1 - A_r A}$	$= \dfrac{r_o}{1 - A_r A}$

Fig. 4.138. Input and output resistances for the amplifier in Fig. 4.137, for different operating conditions. Note that r_i and r_o are short-circuit resistances *by definition*

curves according to (4.146)–(4.149). However, if the operating point as expressed by all voltages and currents is known or can be determined approximately, the components can be linearized individually and the small-signal parameters can be calculated from the resultant small-signal equivalent circuit; for each component, the relevant small-signal circuit is used. Figure 4.139 shows this method in comparison with the procedure that uses the characteristics. Specifications from the circuit are required to calculate the operating point, to select the relevant small-signal equivalent circuits and develop the small-signal equivalent circuit of the actual circuit.

In practice, the method employing the small-signal equivalent circuit of the actual circuit is used exclusively. Likewise, circuit simulation programs can only use this method, since they can only perform numeric calculations; these programs are not capable of drawing up, reformulating and differentiating equations in a continuous fashion. However, some programs (e.g. *PSpice*) can calculate individual points of the characteristic of a circuit and illustrate them in a numerically differentiated form. These representations are useful when looking at the small-signal characteristics at the operating point. The numeric differentiation, however, may cause significant errors in characteristic curve sections with very low or very high slopes.

Example: Fig. 4.140 again shows the common-emitter circuit from Fig. 4.130a; the only nonlinear component is the transistor. The small-signal equivalent circuit of the circuit is derived by inserting the small-signal equivalent circuit of the transistor. The transistor parameters β and V_A and the collector current $I_{C,A}$ at the operating point are required to calculate the parameters g_m, r_{BE} and r_{CE}; for $\beta = 100$, $V_A = 100\,\text{V}$ and $I_{C,A} = 300\,\mu\text{A}$, we obtain:

$$g_m = \frac{I_{C,A}}{V_T} = \frac{300\,\mu\text{A}}{26\,\text{mV}} \approx 11.5\,\text{mS} \quad , \quad r_{BE} = \frac{\beta}{g_m} = \frac{100}{11.5\,\text{mS}} \approx 8.7\,\text{k}\Omega$$

$$r_{CE} = \frac{V_A}{I_{C,A}} = \frac{100\,\text{V}}{300\,\mu\text{A}} \approx 333\,\text{k}\Omega$$

From a comparison with Fig. 4.133b on page 415, we obtain $r_i = r_{BE} \approx 8.7\,\text{k}\Omega$, $r_o = r_{CE} \| R_C \approx 9.7\,\text{k}\Omega$, $g_m \approx 11.5\,\text{mS}$ (here the transconductance of the amplifier corresponds with the transconductance of the transistor) and $A = -g_m r_o \approx -112$.

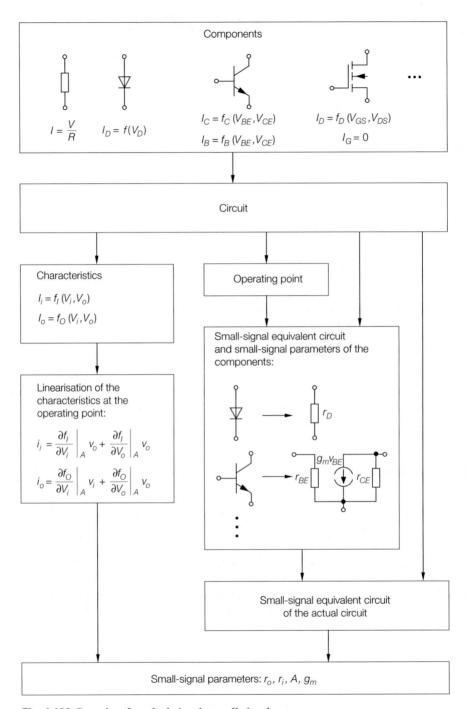

Fig. 4.139. Procedure for calculating the small-signal parameters

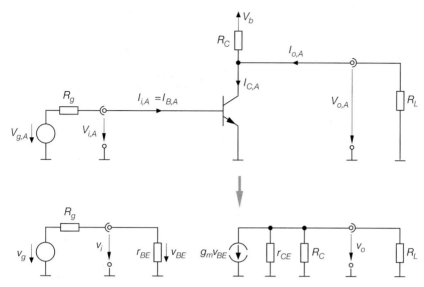

Fig. 4.140. Example: common-emitter circuit with an operating point (*top*) and the resulting small-signal equivalent circuit when using the small-signal equivalent circuit of the transistor (*bottom*)

The values for A and r_o deviate slightly from the values in Fig. 4.135, as the small-signal equivalent circuit of the transistor also takes into account the Early effect, which is represented by resistance r_{CE} and which was ignored when calculating the values from the characteristics.

Series Connection of Amplifiers

Several amplifiers can be combined to make one amplifier by connecting them in series. As an amplifier generally consists of several basic transistor circuits connected in series, this method is also used for calculating the characteristic values of individual amplifiers, by treating the individual basic circuits as partial amplifiers.

Combining basic transistor circuits is generally a complex task, as some basic circuits have a reverse gain that cannot be neglected; however, this rarely applies to amplifiers that consist of several basic circuits, since the series connection of basic circuits has no reverse gain as long as it contains one basic circuit without reverse gain.

Series connection of amplifiers without reverse gain: Several amplifiers without reverse gain connected in series can be combined into one single amplifier. The following applies to n amplifiers:

- The input resistance corresponds to the input resistance of the first amplifier: $r_i = r_{i1}$.
- The output resistance corresponds to the output resistance of the last amplifier: $r_o = r_{o(n)}$.
- The gain corresponds to the product of the individual gains and the voltage divider factors between each pair of subsequent amplifiers:

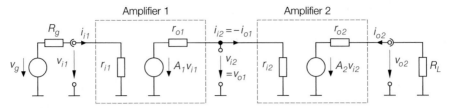

Fig. 4.141. Example: series connection of two amplifiers without reverse gain

$$A = \prod_{i=1}^{n} A_{(i)} \cdot \prod_{i=1}^{n-1} \frac{r_{i(i+1)}}{r_{o(i)} + r_{i(i+1)}} \tag{4.163}$$

– The operating gain is calculated using (4.153):

$$A_B = \frac{r_i}{R_g + r_i} \; A \; \frac{R_L}{r_o + R_L} \tag{4.164}$$

$$= \prod_{i=1}^{n} A_{(i)} \cdot \prod_{i=0}^{n} \frac{r_{i(i+1)}}{r_{o(i)} + r_{i(i+1)}} \qquad \text{with } r_{o,(0)} = R_g, \, r_{i(n+1)} = R_L$$

Here, the voltage divider factors at the input ($i = 0$) and output ($i = n$) must be added.

Example: The small-signal parameters of the two amplifiers without reverse gain connected in series in Fig. 4.141 are

$$r_i = r_{i1} \quad , \quad r_o = r_{o2} \quad , \quad A = A_1 \frac{r_{i2}}{r_{o1} + r_{i2}} A_2$$

and the operating gain is:

$$A_B = \frac{r_i}{R_g + r_i} \; A \; \frac{R_L}{r_o + R_L} = \frac{r_{i1}}{R_g + r_{i1}} \; A_1 \; \frac{r_{i2}}{r_{o1} + r_{i2}} \; A_2 \; \frac{R_L}{r_{o2} + R_L}$$

Series connection of amplifiers with reverse gain: It is a very complex task to determine the small-signal parameters of amplifiers with reverse gain connected in series. In contrast, the calculation of the operating gain A_B is simple: if the operating input resistances $r_{i,B(i)}$ are used instead of the input resistances $r_{i(i)}$, it is possible to proceed as with amplifiers without reverse gain connected in series and use (4.164). The operating input resistances can be determined *backwards*: The operating input resistance of the last amplifier depends on the load R_L; this in turn forms the load of the second-last amplifier and so on. For n amplifiers, the following applies:

$$R_L \; \to \; r_{i,B(n)}(R_L) \; \to \; r_{i,(n-1)}(r_{i,B(n)}) \; \to \; \cdots \; \to \; r_{i,B1}(r_{i,B2})$$

In general, this backwards calculation can only be performed with numeric values, as inserting several formulas into one another soon leads to extremely comprehensive expressions. In relation to this, it should be noted that the dependence on R_L means that the gain A cannot be calculated using (4.163); the operating input resistances are not defined in this case.

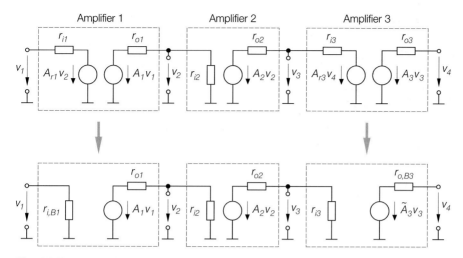

Fig. 4.142. Conversion of amplifiers with reverse gain connected in series into one amplifier without reverse gain

Series connection of at least one amplifier without reverse gain: As mentioned above, a chain of amplifiers in series has no reverse gain if *at least* one amplifier in the series has no reverse gain. In such cases, it is possible to determine the small-signal parameters A, r_i and r_o by successively converting the amplifiers with reverse gain into amplifiers without reverse gain; Fig. 4.142 shows an example of this. The procedure is based on the fact that an amplifier with reverse gain arranged in front of or behind an amplifier without reverse gain can be converted into an amplifier without reverse gain; the successive application of this rule effectively removes the reverse gain from all amplifiers.

First, we will look at *amplifier 1* in Fig. 4.142. It is arranged in front of the *amplifier 2* without reverse gain and is therefore operated with a defined load; that is, r_{i2}. This allows the operating input resistance $r_{i,B1} = r_{i,B1}(r_{i2})$ to be calculated and the conversion to be performed.

Amplifier 3 in Fig. 4.142 is arranged behind the *amplifier 2* without reverse gain and is therefore operated with a defined internal resistance of the signal source; that is, r_{o2}. This allows the operating output resistance $r_{o,B3} = r_{o,B3}(r_{o2})$ to be calculated. In addition, it is necessary to change the gain of the voltage-controlled voltage source from A_3 to:[28]

$$\tilde{A}_3 = A_3 \, \frac{r_{o,B3}}{r_{o3}}$$

Example: Figure 4.143 shows a three-stage amplifier with one common-emitter circuit with voltage feedback at the input and one at the output (T_1 and T_3) and a common-emitter circuit with current feedback in between (T_2). The common-emitter circuits with voltage feedback have a reverse gain that cannot be neglected and which is mainly caused by the resistances R_{21} and R_{23}; the reverse gain of common-emitter circuit with current feedback,

---

[28] This formula results when converting the values, but can also be derived by observing the short-circuit current at the output: with a short at the output ($v_4 = 0$) the current $A_3 v_3 / r_{o3}$ flows in the original amplifier and current $\tilde{A}_3 v_3 / r_{o,B3}$ flows in the converted amplifier; since in both cases v_3 is the same due to $A_{r3} v_4 = 0$, the equality of the currents leads to the formula for the gain factors.

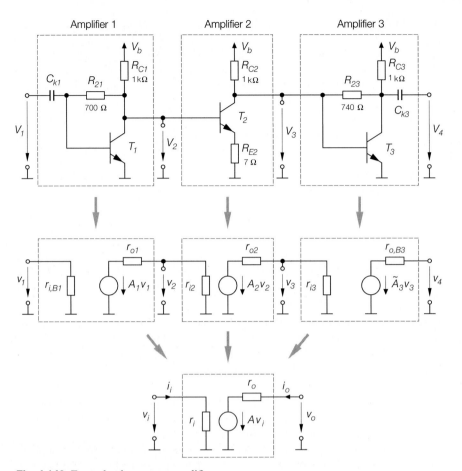

Fig. 4.143. Example: three-stage amplifier

on the other hand, is practically zero. The small-signal values A, r_i and r_o are determined below.

The supply voltage is $V_b = 1.7\,\text{V}$; this causes a quiescent collector current of $1\,\text{mA}$ in all three transistors. If $\beta = 100$ and $V_A = 100\,\text{V}$, then $g_m = I_C/V_T = 38\,\text{mS}$ and $r_{BE} = \beta V_T/I_C = 2.6\,\text{k}\Omega$; compared to the resistances in the circuit, the collector-emitter resistance $r_{CE} = V_A/I_C = 100\,\text{k}\Omega$ can be neglected.

First, the parameters of the common-emitter circuit with current feedback are determined:

– From (2.70), we obtain the gain:

$$A_2 = -\frac{g_m R_{C2}}{1 + g_m R_{E2}} = -30$$

– From (2.71), we obtain the input resistance:

$$r_{i2} = r_{BE} + \beta R_{E2} = 3.3\,\text{k}\Omega$$

– From (2.72), we obtain the output resistance:

$$r_{o2} = R_{C2} = 1\,\text{k}\Omega$$

The common-emitter circuits with voltage feedback lack the resistance R_1 that exists in the basic circuit of Fig. 2.66; therefore, it is necessary to provide the formulas for this situation first:

– From the deduction for gain A, the formulas can be used with the conditions that $r_{CE} \gg R_C$, $\beta \gg 1$ and $R_1 = 0$:

$$A = \frac{-g_m R_2 + 1}{1 + \dfrac{R_2}{R_C}}$$

– The short-circuit input resistance for $R_1 = 0$ is taken from the deduction for the input resistance:

$$r_i = r_{i,s} = r_{BE} \parallel R_2$$

– The operating input resistance $r_{i,B}$ corresponds to the open-circuit input resistance in the deduction if R_C is replaced by R_C and R_L connected in parallel (see Fig. 2.69); if $r_{CE} \gg R_C$, $\beta \gg 1$, $\beta R_C \gg r_{BE}$, $\beta R_C \gg R_2$, and $R_1 = 0$, then:

$$r_{i,B} = \frac{1}{g_m} \left(1 + \frac{R_2}{R_C \parallel R_L} \right)$$

– From the deduction for the short-circuit output resistance, the formula can be used with the conditions that $r_{CE} \gg R_C$, $\beta \gg 1$ and $R_1 = 0$:

$$r_o = R_C \parallel R_2$$

– The same formula is used with $R_1 = R_g$ to calculate the operating output resistance, since in this case the internal resistance R_g replaces the missing resistance R_1:

$$r_{o,B} = R_C \parallel \frac{r_{BE}\left(R_g + R_2\right) + R_g R_2}{r_{BE} + \beta R_g}$$

These formulas yield the following values for the first common-emitter circuit with voltage feedback with $R_2 = R_{21} = 700\,\Omega$ and $R_C = R_{C1} = 1\,k\Omega$:

$$A_1 = -15 \quad , \quad r_{i,B1}(R_L = r_{i2}) = 50\,\Omega \quad , \quad r_{o1} = 412\,\Omega$$

For the second common-emitter circuit with voltage feedback with $R_2 = R_{23} = 740\,\Omega$ and $R_C = R_{C3} = 1\,k\Omega$, the values are:

$$A_3 = -15.6 \quad , \quad r_{i3} = 576\,\Omega \quad , \quad r_{o3} = 425\,\Omega$$

$$r_{o,B3}(R_g = r_{o2}) = 49\,\Omega \quad , \quad \tilde{A}_3 = -1.8$$

All elements of the small-signal equivalent circuit shown in the centre of Fig. 4.143 are now determined and the series connection can be combined:

$$A = A_1 \frac{r_{i2}}{r_{o1} + r_{i2}} A_2 \frac{r_{i3}}{r_{o2} + r_{i3}} \tilde{A}_3 = -263$$

$$r_i = r_{i,B1} = 50\,\Omega$$

$$r_o = r_{o,B3} = 49\,\Omega$$

This represents an amplifier that is matched to $50\,\Omega$ on both sides. If we use a $50\,\Omega$ signal source and a $50\,\Omega$ load, then the voltage dividers at the input and output have the factor $1/2$;

consequently, the operating gain is $A_B = A/4 = -66$. Circuit simulation with *PSpice* yields $r_i = r_o = 50\,\Omega$ and $A_B = -61$.

It should be noted that the gain is achieved by the first two stages; while the third stage, together with the voltage divider factor between the second and the third stage, provides attenuation. The third stage only serves as an impedance converter from $r_{o2} = 1\,\mathrm{k}\Omega$ to $r_{o,B3} = 50\,\Omega$; it is necessary to use a common-emitter circuit with voltage feedback, as a directly coupled npn common-collector circuit cannot be used due to the low DC output voltage of the second stage ($V_{3,A} \approx 0.7\,\mathrm{V}$). Also, the circuit is to be produced in HF semiconductor technology, in which no pnp transistors are available that are fast enough.

This detailed example shows that a multi-stage amplifier can be calculated exactly using the method described here.[29] However, it also shows that the parameters in the small-signal equivalent circuit must be calculated very carefully, and that it may be necessary to revert to the full equations of the basic transistor circuits.

4.2.3
Nonlinear Parameters

In relation to the small-signal parameters, the question arises as to how high the maximum signals around the operating point may become before leaving the small-signal mode. From a mathematical point of view, the small-signal equivalent circuit is only valid for *infinitesimal* – that is, arbitrarily small – signals. In practice, the nonlinear distortions that increase disproportionately with an increasing amplitude are crucial and should not exceed an application-specific limiting value.

The nonlinear response of an amplifier is described by the *distortion factor*, the *compression point* and the *intercept points*. These can be calculated from the coefficients of the series expansion for the transfer characteristic. If it is not possible to calculate a closed representation of the transfer characteristic, the transfer characteristic must be measured or determined by means of circuit simulation.

Series Expansion at the Operating Point

Figure 4.144 shows a nonlinear amplifier with its operating transfer characteristic $V_o = f_{Tr,B}(V_g)$. The corresponding series expansion (*Taylor series*) at the operating point is [4.3]:

$$V_o = V_{o,A} + v_o = f_{Tr,B}(V_g) = f_{Tr,B}(V_{g,A} + v_g)$$

$$= f_{Tr,B}(V_{g,A}) + \left.\frac{df_{Tr,B}}{dV_g}\right|_A v_g + \frac{1}{2}\left.\frac{d^2 f_{Tr,B}}{dV_g^2}\right|_A v_g^2$$

$$+ \frac{1}{6}\left.\frac{d^3 f_{Tr,B}}{dV_g^3}\right|_A v_g^3 + \frac{1}{24}\left.\frac{d^4 f_{Tr,B}}{dV_g^4}\right|_A v_g^4 + \cdots$$

This leads to the small-signal parameters:

---

[29] Differences to circuit simulation are the result of the approximations $\beta \gg 1$ and $r_{CE} \gg R_C$; calculation without these approximations yields exactly the same values as the simulation.

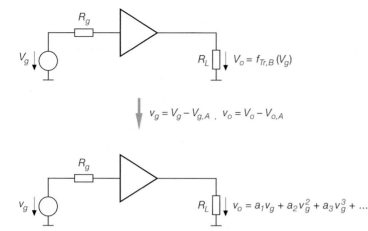

$$V_o = f_{Tr,B}(V_g)$$

$$v_g = V_g - V_{g,A} \ , \ v_o = V_o - V_{o,A}$$

$$v_o = a_1 v_g + a_2 v_g^2 + a_3 v_g^3 + \cdots$$

Fig. 4.144. Nonlinear amplifier (*top*) and series expansion at the operating point (*bottom*)

$$v_o = \left.\frac{df_{Tr,B}}{dV_g}\right|_A v_g + \frac{1}{2}\left.\frac{d^2 f_{Tr,B}}{dV_g^2}\right|_A v_g^2 + \frac{1}{6}\left.\frac{d^3 f_{Tr,B}}{dV_g^3}\right|_A v_g^3 + \frac{1}{24}\left.\frac{d^4 f_{Tr,B}}{dV_g^4}\right|_A v_g^4 + \cdots$$

$$= \sum_{n=1\ldots\infty} a_n v_g^n \quad \text{with} \quad a_n = \frac{1}{n!}\left.\frac{d^n f_{Tr,B}}{dV_g^n}\right|_A \tag{4.165}$$

The coefficients $a_1, a_2, \ldots$ are known as the *coefficients of the Taylor series*. Coefficient a_1 corresponds to the small-signal operating gain A_B and is nondimensional; all of the other coefficients are dimensional parameters:

$$[a_n] = \frac{1}{V^{n-1}} \quad \text{for } n = 2\ldots\infty$$

Example: For the common-emitter circuit in Fig. 4.135 on page 416, it is comparatively simple to calculate the series expansion of the operating transfer characteristic; we start with the series expansion of the input equation

$$V_g = I_i R_g + V_i = I_B R_g + V_{BE} = \frac{I_C R_g}{B} + V_T \ln\frac{I_C}{I_S}$$

at the operating point:

$$v_g = \frac{i_C R_g}{B} + V_T \ln\left(1 + \frac{i_C}{I_{C,A}}\right)$$

$$= \left(\frac{I_{C,A} R_g}{B} + V_T\right)\frac{i_C}{I_{C,A}} - \frac{V_T}{2}\left(\frac{i_C}{I_{C,A}}\right)^2 + \frac{V_T}{3}\left(\frac{i_C}{I_{C,A}}\right)^3 - \cdots$$

The use of

$$i_C = -\frac{v_o}{R_C \| R_L}$$

and $V_k = I_{C,A}(R_C \| R_L)$ leads to

$$v_g = -\left(\frac{I_{C,A} R_g}{B} + V_T\right)\frac{v_o}{V_k} - \frac{V_T}{2}\left(\frac{v_o}{V_k}\right)^2 - \frac{V_T}{3}\left(\frac{v_o}{V_k}\right)^3 - \cdots$$

If we use the values $R_C = R_L = 10\,\text{k}\Omega$, $R_g = 100\,\text{k}\Omega$, $I_{C,A} = 300\,\mu\text{A}$, $B = 100$ and $V_T = 26\,\text{mV}$, then

$$v_g = -0.2173\,v_o - \frac{5.78\,v_o^2}{10^3\,\text{V}} - \frac{2.57\,v_o^3}{10^3\,\text{V}^2} - \frac{1.28\,v_o^4}{10^3\,\text{V}^3} - \frac{0.685\,v_o^5}{10^3\,\text{V}^4}$$

and, after inversion:

$$v_o = -4.6\,v_g - \frac{0.563\,v_g^2}{\text{V}} + \frac{v_g^3}{\text{V}^2} - \frac{2\,v_g^4}{\text{V}^3} + \frac{4\,v_g^5}{\text{V}^4} - \cdots$$

Consequently:

$$a_1 = -4.6 \quad , \quad a_2 = -\frac{0.563}{\text{V}} \quad , \quad a_3 = \frac{1}{\text{V}^2} \quad , \quad a_4 = -\frac{2}{\text{V}^3} \quad , \quad a_5 = \frac{4}{\text{V}^4}$$

Output Signal with a Sinusoidal Input

With a signal

$$v_g = \hat{v}_g \cos \omega t$$

the terms v_g^n in (4.165) provide the desired output signal (*useful signal*)

$$v_{o,us} = \hat{v}_o \cos \omega t = a_1 \hat{v}_g \cos \omega t$$

as well as portions of the multiple of ω:

$$v_o = \sum_{n=1\ldots\infty} a_n v_g^n = \sum_{n=1\ldots\infty} a_n \hat{v}_g^n \cos^n \omega t$$

$$= \left(\frac{a_2 \hat{v}_g^2}{2} + \frac{3a_4 \hat{v}_g^4}{8} + \frac{5a_6 \hat{v}_g^6}{16} + \cdots \right) \qquad \text{DC component}$$

$$+ \left(a_1 + \frac{3a_3 \hat{v}_g^2}{4} + \frac{5a_5 \hat{v}_g^4}{8} + \frac{35a_7 \hat{v}_g^6}{64} + \cdots \right) \hat{v}_g \cos \omega t \qquad \text{fundamental wave}$$

$$+ \left(\frac{a_2}{2} + \frac{a_4 \hat{v}_g^2}{2} + \frac{15a_6 \hat{v}_g^4}{32} + \cdots \right) \hat{v}_g^2 \cos 2\omega t \qquad \text{first harmonic wave}$$

$$+ \left(\frac{a_3}{4} + \frac{5a_5 \hat{v}_g^2}{16} + \frac{21a_7 \hat{v}_g^4}{64} + \cdots \right) \hat{v}_g^3 \cos 3\omega t \qquad \text{second harmonic wave}$$

$$+ \left(\frac{a_4}{8} + \frac{3a_6 \hat{v}_g^2}{16} + \cdots \right) \hat{v}_g^4 \cos 4\omega t \qquad \text{third harmonic wave}$$

$$+ \left(\frac{a_5}{16} + \frac{7a_7 \hat{v}_g^2}{64} + \cdots \right) \hat{v}_g^5 \cos 5\omega t \qquad \text{fourth harmonic wave}$$

$$+ \cdots$$

$$= \sum_{n=0\ldots\infty} b_n \hat{v}_g^n \cos n\omega t \qquad \text{with } b_n = (\cdots)_n \qquad (4.166)$$

The b_n coefficients are obtained by converting the terms $\cos^n \omega t$ into the form $\cos n\omega t$ and by sorting them according to frequencies. We can see that the *even* coefficients $a_2, a_4, \ldots$ generate a DC component b_0 – that is, a shift in the operating point; with the amplitudes used in practical applications the shift is minor and can be neglected. Furthermore, at even-numbered multiples of the frequency ω the even coefficients produce additional portions. Similarly, at *odd-numbered* multiples of frequency ω the odd coefficients $a_3, a_5, \ldots$ produce additional portions. The odd coefficients have an effect on the amplitude of the useful signal; therefore, with higher amplitudes the operating gain is no longer constant.

The portion at the frequency ω is called the *fundamental wave*. The other portions are known as *harmonic waves* and are numbered according to their order: the first harmonic wave at 2ω, the second harmonic wave at 3ω and so on. Alternatively, the portions can also be called *harmonics*: the first harmonic at ω, the second harmonic at 2ω and so on.

In practice, amplitudes are used at which the harmonic waves are significantly smaller than the fundamental. In this case, only the first term of the expressions in parentheses in (4.166) must be taken into account; that is, the coefficients b_n are approximately constant and no longer depend on the input amplitude $\hat{v}_g$, but only on the coefficient a_n of the characteristic:

$$b_n \approx \frac{a_n}{2^{n-1}} \qquad \text{for } n = 1 \ldots \infty \tag{4.167}$$

For the amplitudes of the fundamental and the harmonic waves, it follows:

$$\hat{v}_{o(FW)} = |b_1|\hat{v}_g \approx |a_1|\hat{v}_g$$

$$\hat{v}_{o(1.HW)} = |b_2|\hat{v}_g^2 \approx \left|\frac{a_2}{2}\right| \hat{v}_g^2 \tag{4.168}$$

$$\hat{v}_{o(2.HW)} = |b_3|\hat{v}_g^3 \approx \left|\frac{a_3}{4}\right| \hat{v}_g^3$$

$$\vdots$$

Thus the amplitude of the fundamental shows a linear increase with the input amplitude, while the amplitudes of the harmonic waves increase disproportionately. A requirement for the approximation is the condition:

$$\hat{v}_{o(FW)} \gg \hat{v}_{o(1.HW)}, \hat{v}_{o(2.HW)}, \ldots$$

Insertion of the coefficients leads to

$$|b_1|\hat{v}_g \gg |b_2|\hat{v}_g^2, |b_3|\hat{v}_g^3, |b_4|\hat{v}_g^4, |b_5|\hat{v}_g^5, \ldots$$

and, solving this for $\hat{v}_g$,

$$\hat{v}_g \ll \left|\frac{b_1}{b_2}\right|, \sqrt{\left|\frac{b_1}{b_3}\right|}, \sqrt[3]{\left|\frac{b_1}{b_4}\right|}, \sqrt[4]{\left|\frac{b_1}{b_5}\right|}, \ldots$$

$$\hat{v}_g \ll \min_n \sqrt[n-1]{\left|\frac{b_1}{b_n}\right|} \stackrel{(4.167)}{=} 2 \min_n \sqrt[n-1]{\left|\frac{a_1}{a_n}\right|} \tag{4.169}$$

Example: With (4.167) and the coefficients $a_1, \ldots, a_5$ on page 428, we obtain for the common-emitter circuit in Fig. 4.135:

$$b_1 \approx a_1 = -4.6 \quad , \quad b_2 \approx \frac{a_2}{2} = -\frac{0.282}{V} \quad , \quad b_3 \approx \frac{a_3}{4} = \frac{0.25}{V^2}$$

$$b_4 \approx \frac{a_4}{8} = -\frac{0.25}{V^3} \quad , \quad b_5 \approx \frac{a_5}{16} = \frac{0.25}{V^4}$$

All other coefficients also have the magnitude 0.25. From (4.169) we obtain, for the amplitude:

$$\hat{v}_g \ll min\,(16.3\,V\,;\ 4.3\,V\,;\ 2.6\,V\,;\ 2\,V\,;\ \ldots) = 1\,V$$

The minimum is reached for $n \to \infty$. For $\hat{v}_g = 100\,mV$, (4.168) yields the fundamental of $\hat{v}_{o(FW)} \approx 460\,mV$, the first harmonic wave of $\hat{v}_{o(1.HW)} \approx 2.82\,mV$ and the second harmonic wave of $\hat{v}_{o(2.HW)} \approx 0.25\,mV$.

Range of Validity for the Series Expansion

The operating transfer characteristic has been described by the polynomial in (4.165) for a limited range only. This range depends on the number of terms taken into account, but ends at the latest when the saturation limits are reached, since from here on the characteristic is approximately horizontal and can no longer be described by a polynomial. In most cases, the active range cannot be described either when the saturation limits have almost been reached, so that (4.165) only covers a more or less large range around the operating point. Figure 4.145 shows this range of the operating transfer characteristic, using a common-emitter circuit as an example.

Distortion Factor

For sinusoidal signals, the *distortion factor k* is used as a measure of the nonlinear distortions:

> *The distortion factor k describes the ratio of the effective value of all harmonic waves in one signal to the effective value of the total signal.*

For a sinusoidal signal without harmonic waves, $k = 0$.

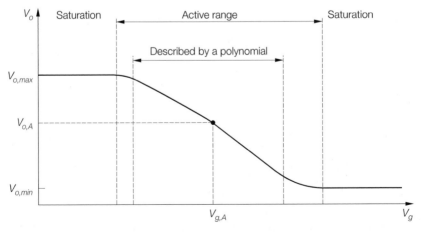

Fig. 4.145. Validity range of the series expansion for the operating transfer characteristic

If the relation between the amplitude and the effective value ($v_{\text{eff}}^2 = \hat{v}^2/2$) is taken into account, then (4.166) leads to:

$$k = \sqrt{\frac{\displaystyle\sum_{n=2\ldots\infty} \frac{1}{2}\left(b_n\hat{v}_g^n\right)^2}{\displaystyle\sum_{n=1\ldots\infty} \frac{1}{2}\left(b_n\hat{v}_g^n\right)^2}} = \sqrt{\frac{\displaystyle\sum_{n=2\ldots\infty} b_n^2\hat{v}_g^{2n}}{\displaystyle\sum_{n=1\ldots\infty} b_n^2\hat{v}_g^{2n}}} \tag{4.170}$$

The DC component b_0 is neglected. For small signals with a low distortion factor, the harmonic waves can be neglected when calculating the effective value of the total signal. Consequently:

$$k \approx \frac{\sqrt{\displaystyle\sum_{n=2\ldots\infty} b_n^2\hat{v}_g^{2n}}}{b_1\hat{v}_g}$$

Often, not all of the harmonic waves are transferred in systems with filters; *partial distortion factors*

$$k_n = \left|\frac{b_n\hat{v}_g^n}{b_1\hat{v}_g}\right| = \left|\frac{b_n}{b_1}\right|\hat{v}_g^{n-1} \qquad \text{for } n = 2\ldots\infty$$

that describe the ratio of the effective values of individual harmonic waves to the fundamental wave are thus quoted. This allows the distortion factor k to be calculated from the partial distortion factors:

$$k = \sqrt{\frac{\displaystyle\sum_{n=2\ldots\infty} k_n^2}{1 + \displaystyle\sum_{n=2\ldots\infty} k_n^2}} \stackrel{k_n \ll 1}{\approx} \sqrt{\sum_{n=2\ldots\infty} k_n^2} \tag{4.171}$$

From (4.166), we obtain:

$$k_2 = \left|\frac{b_2}{b_1}\right|\hat{v}_g = \left|\frac{\dfrac{a_2}{2} + \dfrac{a_4\hat{v}_g^2}{2} + \dfrac{15a_6\hat{v}_g^4}{32} + \cdots}{a_1 + \dfrac{3a_3\hat{v}_g^2}{4} + \dfrac{5a_6\hat{v}_g^4}{8} + \cdots}\right|\hat{v}_g \approx \left|\frac{a_2}{2a_1}\right|\hat{v}_g$$

$$k_3 = \left|\frac{b_3}{b_1}\right|\hat{v}_g^2 = \left|\frac{\dfrac{a_3}{4} + \dfrac{5a_5\hat{v}_g^2}{16} + \dfrac{21a_7\hat{v}_g^7}{64} + \cdots}{a_1 + \dfrac{3a_3\hat{v}_g^2}{4} + \dfrac{5a_6\hat{v}_g^4}{8} + \cdots}\right|\hat{v}_g^2 \approx \left|\frac{a_3}{4a_1}\right|\hat{v}_g^2$$

$$k_4 = \left|\frac{b_4}{b_1}\right|\hat{v}_g^3 \approx \left|\frac{a_4}{8a_1}\right|\hat{v}_g^3$$

$$\vdots$$

$$k_n = \left|\frac{b_n}{b_1}\right|\hat{v}_g^{n-1} \approx \left|\frac{a_n}{2^{n-1}a_1}\right|\hat{v}_g^{n-1} \qquad \text{for } n = 2\ldots\infty \tag{4.172}$$

It can be seen that for small amplitudes the nth partial distortion factor only depends on the coefficients a_1 and a_n, and increases with the power $(n-1)$ of the input amplitude. At medium amplitudes other components become influential and cause a deviating response. At very high amplitudes, the amplifier becomes fully saturated; in this case the output provides a square-wave signal with:

$$
k_n = \begin{cases} 0 & \text{for } n = 2, 4, 6, \ldots \\[2mm] \dfrac{1}{n} & \text{for } n = 3, 5, 7, \ldots \end{cases}
$$

The distortion factor is $k \approx 0.48$. In practice, saturation is usually not exactly symmetrical, so that the even-numbered partial distortion factors do not approach zero.

Example: With the coefficients a_n from page 428, we obtain the following partial distortion factors for the common-emitter circuit in Fig. 4.135:

$$
k_2 \approx \frac{0.061\,\hat{v}_g}{V} \quad , \quad k_3 \approx \frac{0.054\,\hat{v}_g^2}{V^2} \quad , \quad k_4 \approx \frac{0.054\,\hat{v}_g^3}{V^3} \quad , \quad k_5 \approx \frac{0.054\,\hat{v}_g^4}{V^4}
$$

Figure 4.146 shows the curves of $k_2 \ldots k_5$. In the quasi-linear region (I), the partial distortion factors are in accordance with (4.172); in the graph with two logarithmic scales, the powers of $\hat{v}_g$ become straight lines with the respective slopes shown. In the region of weak saturation (II), the partial distortion factors increase markedly. With a further increase in saturation, the output signal shows segments where some partial distortion factors approach zero; in Fig. 4.146 this is the case for $\hat{v}_g \approx 0.2$ V and $\hat{v}_g \approx 0.5$ V. In the region of strong saturation (III), the output signal is almost rectangular; the distortion factors are $k_3 \approx 1/3$, $k_5 \approx 1/5$ and $k_2, k_4 \to 0$.

It follows from Fig. 4.146 that the distortion factor k in the quasi-linear region roughly corresponds to the partial distortion factor k_2:

$$
k \approx k_2 \approx \left| \frac{a_2}{2a_1} \right| \hat{v}_g
$$

All other partial distortion factors are clearly lower. In circuits with a symmetric characteristic ($a_2 = 0$) $k_2 = 0$; here, in the quasi-linear region the following applies:

$$
k \approx k_3 \approx \left| \frac{a_3}{4a_1} \right| \hat{v}_g^2
$$

An example of this is the differential amplifier.

Compression Point

The odd coefficients of the series expansion also affect the amplitude of the fundamental wave (see (4.166)); this makes the effective operating gain of the circuit dependent on the amplitude:

$$
A'_B(\hat{v}_g) = b_1 = a_1 + \frac{3a_3}{4}\hat{v}_g^2 + \frac{5a_5}{8}\hat{v}_g^4 + \frac{35a_7}{64}\hat{v}_g^6 + \cdots
$$

With an increasing amplitude, the absolute value of the operating gain may initially increase ($a_3/a_1 > 0$, *gain expansion*) or decrease ($a_3/a_1 < 0$, *gain compression*), starting from $|A_B| = |a_1|$. When saturation sets in, the value always declines and approaches zero as saturation increases. This region is not covered by the series expansion.

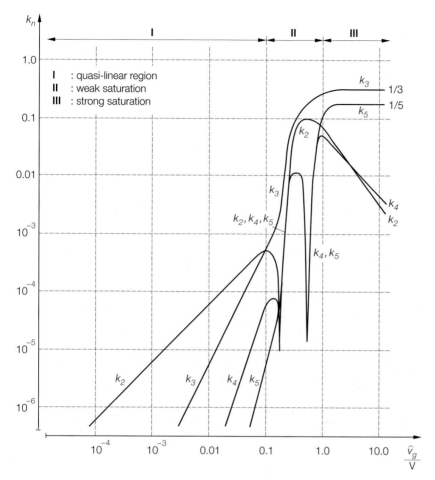

Fig. 4.146. A plot of the partial distortion factors $k_2 \dots k_5$ for the common-emitter circuit in Fig. 4.135

For amplifiers, the *1 dB compression point* is quoted as a characteristic figure for the onset of saturation:

> *The 1 dB compression point describes the amplitude at which the operating gain drops to* 1 dB *below the small-signal operating gain due to the onset of saturation.*

We can distinguish between the *input compression point* $\hat{v}_{g,comp}$ with

$$\left| A'_B(\hat{v}_{g,comp}) \right| = 10^{-1/20} \cdot |A_B| \approx 0.89 \cdot |A_B| \tag{4.173}$$

and the *output compression point*:

$$\hat{v}_{o,comp} = 10^{-1/20} \cdot |A_B| \hat{v}_{g,comp} \approx 0.89 \cdot |A_B| \hat{v}_{g,comp} \tag{4.174}$$

In practice, both are determined by measurement or by circuit simulation. Figure 4.147 shows a diagram of the magnitude of the operating gain for an amplifier with and without *gain expansion*.

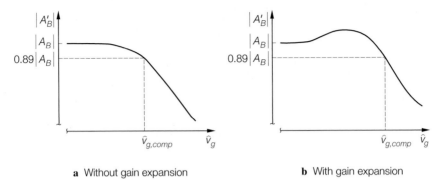

a Without gain expansion   **b** With gain expansion

Fig. 4.147. Magnitude of the operating gain, including the 1 dB compression point

Example: Circuit simulation for the common-emitter circuit in Fig. 4.135 yields $\hat{v}_{g,comp} \approx 0.3$ V and $\hat{v}_{o,comp} \approx 1.2$ V.

Intermodulation and Intercept Points

In systems with bandpass filters, the harmonic distortions described by the distortion factors are usually of no importance, as they are usually outside the passband of the filters; consequently, there are no distortions in the passband if the system is driven with one sinusoidal signal (*single-tone operation*). However, if two or more sinusoidal signals that are within the passband are fed, then some of the distortion products are within the passband. These components are known as *intermodulation distortions*, and result from the fact that the operation of a nonlinear characteristic of order N with a multi-tone signal of frequencies $f_1, f_2, \ldots, f_m$ generates not only the harmonics $nf_1, nf_2, \ldots, nf_m$ ($n = 1 \ldots N$), but also composite products of the frequencies

$$\pm n_1 f_1 \pm n_2 f_2 \pm \cdots \pm n_m f_m \qquad \text{with } n_1 + n_2 + \ldots + n_m \leq N$$

some of which are within the passband [4.4, 4.5].

In practice, a two-tone signal is used with frequencies f_1 and f_2, which are close to one another in the centre of the passband, and with equal amplitudes; with $f_1 < f_2$, the following portions occur due to the powers $n = 1 \ldots 5$:

$$
\begin{aligned}
n = 1 &\Rightarrow f_1, \ f_2 \\
n = 2 &\Rightarrow 2f_1, \ 2f_2, \ f_2 - f_1 \\
n = 3 &\Rightarrow 3f_1, \ 3f_2, \ 2f_1 + f_2, \ 2f_1 - f_2, \ 2f_2 + f_1, \ 2f_2 - f_1 \\
n = 4 &\Rightarrow 4f_1, \ 4f_2, \ 3f_1 + f_2, \ 3f_1 - f_2, \ \ldots \\
n = 5 &\Rightarrow 5f_1, \ 5f_2, \ \ldots, \ 3f_1 - 2f_2, \ 3f_2 - 2f_1, \ \ldots
\end{aligned}
$$

Figure 4.148 shows the portions of two-tone operation compared to one-tone operation. One can see that the portions caused by the odd powers for

$$2f_1 - f_2, \ 2f_2 - f_1, \ 3f_1 - 2f_2, \ 3f_2 - 2f_1$$

are within the passband. Inserting

$$v_g = \hat{v}_g \left(\cos \omega_1 t + \cos \omega_2 t \right)$$

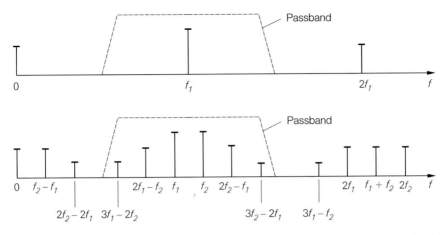

Fig. 4.148. Portions resulting from feeding a single-tone signal (*top*) and a two-tone signal (*bottom*) to a system with a characteristic of fifth order

into the series given by (4.165) on page 427 leads to:

$$
v_o = \left(a_1 + \frac{9a_3\hat{v}_g^2}{4} + \frac{25a_5\hat{v}_g^4}{4} + \frac{1225a_7\hat{v}_g^6}{64} + \cdots \right) \hat{v}_g \cos \omega_1 t \qquad f_1
$$

$$
+ \left(a_1 + \frac{9a_3\hat{v}_g^2}{4} + \frac{25a_5\hat{v}_g^4}{4} + \frac{1225a_7\hat{v}_g^6}{64} + \cdots \right) \hat{v}_g \cos \omega_2 t \qquad f_2
$$

$$
+ \left(\frac{3a_3}{4} + \frac{25a_5\hat{v}_g^2}{8} + \frac{735a_7\hat{v}_g^4}{64} + \cdots \right) \hat{v}_g^3 \cos(2\omega_1 - \omega_2)t \qquad 2f_1 - f_2
$$

$$
+ \left(\frac{3a_3}{4} + \frac{25a_5\hat{v}_g^2}{8} + \frac{735a_7\hat{v}_g^4}{64} + \cdots \right) \hat{v}_g^3 \cos(2\omega_2 - \omega_1)t \qquad 2f_2 - f_1
$$

$$
+ \left(\frac{5a_5}{8} + \frac{245a_7\hat{v}_g^2}{64} + \cdots \right) \hat{v}_g^5 \cos(3\omega_1 - 2\omega_2)t \qquad 3f_1 - 2f_2
$$

$$
+ \left(\frac{5a_5}{8} + \frac{245a_7\hat{v}_g^2}{64} + \cdots \right) \hat{v}_g^5 \cos(3\omega_2 - 2\omega_1)t \qquad 3f_2 - 2f_1
$$

$$
+ \cdots
$$

$$
= \sum_{n=0\ldots\infty} c_{2n+1}\hat{v}_g^{2n+1} \cos\left[(n+1)\,\omega_1 - n\omega_2 \right] t
$$

$$
+ \sum_{n=0\ldots\infty} c_{2n+1}\hat{v}_g^{2n+1} \cos\left[(n+1)\,\omega_2 - n\omega_1 \right] t \qquad (4.175)
$$

$$
+ \cdots
$$

$$
\text{with}\ \ c_{2n+1} = (\cdots)_{2n+1}
$$

The sum is of practical relevance only in so far as the portions are still within the passband. At low amplitudes, the coefficients c_n are approximately constant:

$$c_1 \approx a_1 \quad , \quad c_3 \approx \frac{3a_3}{4} \quad , \quad c_5 \approx \frac{5a_5}{8} \quad , \quad \ldots$$

It follows:

$$c_{2n+1} \approx \frac{2n+1}{2^{n+1}} a_{2n+1} \qquad \text{for } n = 1, \ldots, \infty \tag{4.176}$$

Intermodulation: Distortions within the passband are called *intermodulation products*:

> *In multi-tone operation, those distortions in the passband whose frequencies consist of at least two signal frequencies are known as intermodulation or intermodulation products.*

The portions at $2f_1 - f_2$ and $2f_2 - f_1$ are called *intermodulation of third order* (IM3) and those at $3f_1 - 2f_2$ and $3f_2 - 2f_1$ are called *intermodulation of fifth order* (IM5). In general, the following applies:

> *The distortions at frequencies $(n+1)f_1 - nf_2$ and $(n+1)f_2 - nf_1$ are called intermodulation of the order $2n+1$.*

In practice, only the dominant portions IM3 and IM5 are of interest, since according to their order, the amplitudes of intermodulation products depend on the input amplitude.

For the amplitudes of the useful signal and the intermodulations, we obtain:

$$\hat{v}_{o,us} = |c_1|\hat{v}_g \approx |a_1|\hat{v}_g$$

$$\hat{v}_{o,IM3} = |c_3|\hat{v}_g^3 \approx \left|\frac{3a_3}{4}\right|\hat{v}_g^3 \tag{4.177}$$

$$\hat{v}_{o,IM5} = |c_5|\hat{v}_g^5 \approx \left|\frac{5a_5}{8}\right|\hat{v}_g^5$$

$$\vdots$$

Intermodulation ratio: The abbreviations $IM3$ and $IM5$ are also used to refer to the intermodulation ratio:

> *The ratio of the useful signal amplitude to the amplitude of a certain intermodulation product is known as the intermodulation ratio.*

Use of the amplitudes from (4.177) leads to:

$$IM3 = \frac{\hat{v}_{o,us}}{\hat{v}_{o,IM3}} = \left|\frac{c_1}{c_3}\right|\frac{1}{\hat{v}_g^2} \approx \left|\frac{4a_1}{3a_3}\right|\frac{1}{\hat{v}_g^2} \tag{4.178}$$

$$IM5 = \frac{\hat{v}_{o,us}}{\hat{v}_{o,IM5}} = \left|\frac{c_1}{c_5}\right|\frac{1}{\hat{v}_g^4} \approx \left|\frac{8a_1}{5a_5}\right|\frac{1}{\hat{v}_g^4} \tag{4.179}$$

In practice, the intermodulation ratios are usually quoted in decibels (dB):

$$IM3_{dB} = 20\,\text{dB} \cdot \log IM3 \quad , \quad IM5_{dB} = 20\,\text{dB} \cdot \log IM5$$

In terms of meaning, the intermodulation ratios correspond to the partial distortion factors of one-tone operation, because the intermodulation ratios describe the relationship between the useful signal and the distortion product; while the partial distortion factors describe the relationship between the distortion product and useful signal. Thus the inverse values of the intermodulation ratios can be conceived as the multi-tone partial distortion factors.

Intercept points: In order to characterize the intermodulation products by a value that is independent of amplitude $\hat{v}_g$, amplitudes are obtained that have intermodulation ratios with the *theoretical* value one; for this purpose, the approximations for small amplitudes according to (4.178) and (4.179) are extrapolated beyond their calculatory range of validity. The resulting amplitudes are called *intercept points (IP)*:

Intercept points have an input and output amplitude at which the extrapolated amplitude of a certain intermodulation product has the same value as the extrapolated amplitude of the useful signal.

One distinguishes between *input intercept points (input IP; IIP)*

$$\hat{v}_{g,IP3} = \hat{v}_g\bigg|_{IM3=1} \overset{(4.178)}{=} \sqrt{\left|\frac{4a_1}{3a_3}\right|} \tag{4.180}$$

$$\hat{v}_{g,IP5} = \hat{v}_g\bigg|_{IM5=1} \overset{(4.179)}{=} \sqrt[4]{\left|\frac{8a_1}{5a_5}\right|} \tag{4.181}$$

and *output intercept points (output IP, OIP)*:

$$\hat{v}_{o,IP3} = |a_1|\hat{v}_{g,IP3} \quad , \quad \hat{v}_{o,IP5} = |a_1|\hat{v}_{g,IP5} \tag{4.182}$$

The latter exceed the input intercept points by a value equal to the small-signal operating gain ($|a_1| = |A_B|$) and are often simply called *intercept points IP3 and IP5*, without explicit reference to the output.

Figure 4.149 shows the amplitude curve of the useful signal $\hat{v}_{o,us} = |c_1|\hat{v}_g$ and the intermodulation products $\hat{v}_{o,IM3} = |c_3|\hat{v}_g^3$ and $\hat{v}_{o,IM5} = |c_5|\hat{v}_g^5$ as a function of the input amplitude $\hat{v}_g$, in the form of a diagram with two logarithmic axes. With small amplitudes, this results in straight lines with slopes equal to 1 for $\hat{v}_{o,us}$, 3 for $\hat{v}_{o,IM3}$ and 5 for $\hat{v}_{o,IM5}$. The intercept points *IP3* and *IP5* are determined by extrapolation, as the points at which the straight lines are intercepted. Furthermore, examples of the intermodulation ratios *IM3* and *IM5* and the compression point are indicated.[30]

With the help of the intercept points, we can calculate the amplitudes of the intermodulation products and the intermodulation ratios for any given input and output amplitudes in the quasi-linear region. According to the approximations in (4.177), the amplitudes of the intermodulation products, when referenced to the input intercept points, are:

$$\hat{v}_{o,IM3} \approx \left|\frac{3a_3}{4}\right|\hat{v}_g^3 = \left|\frac{3a_3}{4a_1}\right||a_1|\hat{v}_g^3 \overset{(4.180)}{=} \frac{|a_1|\hat{v}_g^3}{\hat{v}_{g,IP3}^2}$$

---

[30] Due to $b_1 \neq c_1$, two-tone operation of the amplifier results in a different compression point than one-tone operation [see (4.166) and (4.175)]; $c_1 \approx b_1 \approx a_1$ is true for small amplitudes only. Therefore, diagrams such as that in Fig. 4.149 usually present the curve of the intermodulation products in two-tone operation and the curve of the useful portion in one-tone operation. This has no bearing on the intercept points, as they are determined by means of extrapolated values.

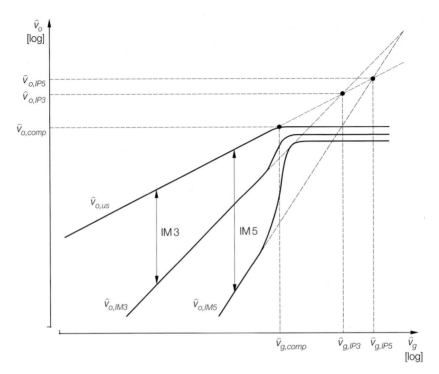

Fig. 4.149. Intercept points at the input ($\hat{v}_{g,IP3}, \hat{v}_{g,IP5}$) and at the output ($\hat{v}_{o,IP3}, \hat{v}_{o,IP5}$) and the intermodulation ratios $IM3$ and $IM5$

$$\hat{v}_{o,IM5} \approx \left| \frac{5a_5}{8} \right| \hat{v}_g^5 = \left| \frac{5a_5}{8a_1} \right| |a_1| \hat{v}_g^5 \overset{(4.181)}{=} \frac{|a_1| \hat{v}_g^5}{\hat{v}_{g,IP5}^4}$$

With reference to the output intercept points, and taking $\hat{v}_{o,us} = |a_1| \hat{v}_g$ and (4.182) into consideration, we obtain:

$$\hat{v}_{o,IM3} \approx \frac{|a_1| \hat{v}_g^3}{\hat{v}_{g,IP3}^2} = \frac{\left(|a_1| \hat{v}_g \right)^3}{\left(|a_1| \hat{v}_{g,IP3} \right)^2} = \frac{\hat{v}_{o,us}^3}{\hat{v}_{o,IP3}^2}$$

$$\hat{v}_{o,IM5} \approx \frac{|a_1| \hat{v}_g^5}{\hat{v}_{g,IP5}^4} = \frac{\left(|a_1| \hat{v}_g \right)^5}{\left(|a_1| \hat{v}_{g,IP5} \right)^4} = \frac{\hat{v}_{o,us}^5}{\hat{v}_{o,IP5}^4}$$

In general, the following applies:

$$\hat{v}_{o,IMn} \approx \frac{|a_1| \hat{v}_g^n}{\hat{v}_{g,IPn}^{n-1}} = \frac{\hat{v}_{o,us}^n}{\hat{v}_{o,IPn}^{n-1}} \tag{4.183}$$

From the approximations in (4.178) and (4.179), and taking $\hat{v}_{o,us} = |a_1| \hat{v}_g$ and (4.182) into consideration, we obtain the intermodulation ratios:

$$IM3 \approx \left| \frac{4a_1}{3a_3} \right| \frac{1}{\hat{v}_g^2} \overset{(4.180)}{=} \frac{\hat{v}_{g,IP3}^2}{\hat{v}_g^2} = \frac{\left(|a_1| \hat{v}_{g,IP3} \right)^2}{\left(|a_1| \hat{v}_g \right)^2} = \frac{\hat{v}_{o,IP3}^2}{\hat{v}_{o,us}^2}$$

$$IM5 \approx \left| \frac{8a_1}{5a_5} \right| \frac{1}{\hat{v}_g^4} \overset{(4.181)}{=} \frac{\hat{v}_{g,IP5}^4}{\hat{v}_g^4} = \frac{\left(|a_1| \hat{v}_{g,IP5} \right)^4}{\left(|a_1| \hat{v}_g \right)^4} = \frac{\hat{v}_{o,IP5}^4}{\hat{v}_{o,us}^4}$$

In general, the following applies:

$$IMn \approx \left(\frac{\hat{v}_{g,IPn}}{\hat{v}_g} \right)^{n-1} = \left(\frac{\hat{v}_{o,IPn}}{\hat{v}_{o,us}} \right)^{n-1} \tag{4.184}$$

Example: Using (4.180)–(4.182) and the coefficients a_n on page 428, the intercept points for the common-emitter circuit in Fig. 4.135 are:

$$\hat{v}_{g,IP3} = 2.5\,\text{V} \Rightarrow \hat{v}_{o,IP3} = 11.4\,\text{V} \quad , \quad \hat{v}_{g,IP5} = 1.2\,\text{V} \Rightarrow \hat{v}_{o,IP5} = 5.4\,\text{V}$$

These are always clearly higher than the actual amplitudes. For a two-tone signal with $\hat{v}_g = 100\,\text{mV}$, (4.177) yields $\hat{v}_{o,us} = 460\,\text{mV}$, $\hat{v}_{o,IM3} \approx 0.7\,\text{mV}$ and $\hat{v}_{o,IM5} \approx 0.024\,\text{mV}$; with (4.178), $IM3 \approx 610$; and with (4.179), $IM5 \approx 19000$.

Series Connection of Amplifiers

If two amplifiers are connected in series as in Fig. 4.150, their characteristics

$$v_{o1} = a_{1,1}v_{g1} + a_{2,1}v_{g1}^2 + a_{3,1}v_{g1}^3 + \cdots$$
$$v_{o2} = a_{1,2}v_{g2} + a_{2,2}v_{g2}^2 + a_{3,2}v_{g2}^3 + \cdots$$

can be used to derive the characteristic for series connection by insertion:

$$\begin{aligned} v_o &= a_1 v_g + a_2 v_g^2 + a_3 v_g^3 + \cdots \\ &= a_{1,1}a_{1,2}v_g + \left(a_{1,2}a_{2,1} + a_{1,1}^2 a_{2,2} \right) v_g^2 \\ &\quad + \left(a_{1,2}a_{3,1} + 2a_{1,1}a_{2,1}a_{2,2} + a_{1,1}^3 a_{3,2} \right) v_g^3 + \cdots \end{aligned} \tag{4.185}$$

Note that in all values $x_{n,m}$, subscript n signifies the power within the series and subscript m the number of the amplifier.

The characteristic of amplifier 1 corresponds to the operating transfer characteristic for operation with a signal source with internal resistance R_g and a load corresponding to the input resistance r_{i2} of amplifier 2; with regard to the small-signal parameters, this leads to:

$$a_{1,1} = A_{B1} = \frac{r_{i1}}{R_g + r_{i1}} A_1 \frac{r_{i2}}{r_{o1} + r_{i2}}$$

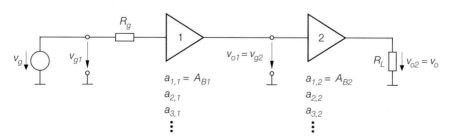

Fig. 4.150. Two amplifiers connected in series

By way of contrast, amplifier 2 is operated with an ideal signal voltage source, since the voltage v_{g2} is fed directly to the amplifier input (see Fig. 4.150); therefore, $R_g = 0$ and:

$$a_{1,2} = A_{B2} = A_2 \frac{R_L}{r_{o2} + R_L}$$

Distortion factor in series connections: For series connections, the partial distortion factors are obtained using (4.172):

$$k_2 \approx \left| \frac{a_2}{2a_1} \right| \hat{v}_g \quad , \quad k_3 \approx \left| \frac{a_3}{4a_1} \right| \hat{v}_g^2 \quad , \quad \ldots$$

If we assume that all harmonic distortions add up – in other words, that all terms in parentheses in (4.185) have the same sign – and we take into account the fact that $\hat{v}_{g2} \approx |a_{1,1}| \hat{v}_{g1}$, then the partial distortion factors of the series connection can be expressed by the partial distortion factors of amplifier 1

$$k_{2,1} \approx \left| \frac{a_{2,1}}{2a_{1,1}} \right| \hat{v}_{g1} \quad , \quad k_{3,1} \approx \left| \frac{a_{3,1}}{4a_{1,1}} \right| \hat{v}_{g1}^2 \quad , \quad \ldots$$

and by the partial distortion factors of amplifier 2:

$$k_{2,2} \approx \left| \frac{a_{2,2}}{2a_{1,2}} \right| \hat{v}_{g2} \approx \left| \frac{a_{1,1} a_{2,2}}{2a_{1,2}} \right| \hat{v}_{g1}$$

$$k_{3,2} \approx \left| \frac{a_{3,2}}{4a_{1,2}} \right| \hat{v}_{g2}^2 \approx \left| \frac{a_{1,1}^2 a_{3,2}}{4a_{1,2}} \right| \hat{v}_{g1}^2 \quad , \quad \ldots$$

$$k_2 \approx k_{2,1} + k_{2,2}$$
$$k_3 \approx k_{3,1} + k_{3,2} + 2k_{2,1}k_{2,2}$$
$$k_4 \approx k_{4,1} + k_{4,2} + 2k_{3,1}k_{2,2} + 3k_{2,1}k_{3,2} + k_{2,1}^2 k_{2,2}$$

$$\vdots$$

If all of the partial distortion factors are much smaller than one, then the products of the partial distortion factors are negligible:

$$k_2 \approx k_{2,1} + k_{2,2} \quad , \quad k_3 \approx k_{3,1} + k_{3,2} \quad , \quad k_4 \approx k_{4,1} + k_{4,2} \quad , \quad \ldots$$

Thus, the partial distortion factors in series connections can be obtained from the sum of the partial distortion factors of both amplifiers. This result can be extended to cover the series connection of any number of amplifiers:

The partial distortion factors of a series connection of several amplifiers correspond approximately to the sum of the corresponding partial distortion factors of the individual amplifiers.

The equation for the series connection of M amplifiers is:

$$k_n \approx \sum_{m=1\ldots M} k_{n,m} \qquad (4.186)$$

If a compensation of harmonic waves occurs in a series connection, then the partial distortion factors of the series are smaller than the sum; therefore, the sum can be considered to be an estimate of the upper limit (the *worst case*).

For the general case, it is not possible to describe a simple relationship between the total distortion factor k of a series connection as calculated using (4.171) on page 431 from the partial distortion factors and the distortion factors of the individual amplifiers. However, in practice, one partial distortion factor is usually dominant, so that $k \approx k_2$ or – in the case of symmetrical characteristics – $k \approx k_3$; in this case, (4.186) can be used and the distortion factor of the series connection can be evaluated by adding up the distortion factors of the individual amplifiers.

Intercept points of series connections: For the input intercept point *IIP3* of the series connection, it follows from (4.180) and (4.185) that:

$$\frac{1}{\hat{v}^2_{g,IP3}} = \left| \frac{3a_3}{4a_1} \right| = \left| \frac{3a_{3,1}}{4a_{1,1}} + \frac{3a^2_{1,1}a_{3,2}}{4a_{1,2}} + \frac{3a_{2,1}a_{2,2}}{2a_{1,2}} \right|$$

If we assume that the first two terms have the same sign and the third term is negligible, because the denominator is the product of two comparably small values $a_{2,1}$ and $a_{2,2}$, this expression can be presented with the help of the intercept point of amplifier 1

$$\hat{v}_{g1,IP3} = \sqrt{\left| \frac{4a_{1,1}}{3a_{3,1}} \right|} = \sqrt{\left| \frac{4A_{B1}}{3a_{3,1}} \right|}$$

and that of amplifier 2

$$\hat{v}_{g2,IP3} = \sqrt{\left| \frac{4a_{1,2}}{3a_{3,2}} \right|} = \sqrt{\left| \frac{4A_{B2}}{3a_{3,2}} \right|}$$

as:

$$IIP3: \qquad \frac{1}{\hat{v}^2_{g,IP3}} \approx \frac{1}{\hat{v}^2_{g1,IP3}} + \frac{|A_{B1}|^2}{\hat{v}^2_{g2,IP3}}$$

With

$$\hat{v}_{o1,IP3} = |A_{B1}| \hat{v}_{g1,IP3} \quad , \quad \hat{v}_{o2,IP3} = |A_{B2}| \hat{v}_{g2,IP3}$$

this leads to the output intercept point *OIP3*:

$$OIP3: \qquad \frac{1}{\hat{v}^2_{o,IP3}} \approx \frac{1}{|A_{B2}|^2 \hat{v}^2_{o1,IP3}} + \frac{1}{\hat{v}^2_{o2,IP3}}$$

The intercept point IP5 is obtained in the same way:

$$IIP5: \qquad \frac{1}{\hat{v}^4_{g,IP5}} \approx \frac{1}{\hat{v}^4_{g1,IP5}} + \frac{|A_{B1}|^4}{\hat{v}^4_{g2,IP5}}$$

$$OIP5: \qquad \frac{1}{\hat{v}^4_{o,IP5}} \approx \frac{1}{|A_{B2}|^4 \hat{v}^4_{o1,IP5}} + \frac{1}{\hat{v}^4_{o2,IP5}}$$

Using the equation for parallel circuits,

$$\frac{1}{c} = \frac{1}{a} + \frac{1}{b} \quad \Rightarrow \quad c = a \,\|\, b$$

we obtain:

$$IIP3 : \qquad \hat{v}_{g,IP3}^2 \approx \hat{v}_{g1,IP3}^2 \parallel \left(\frac{\hat{v}_{g2,IP3}}{|A_{B1}|} \right)^2$$

$$OIP3 : \qquad \hat{v}_{o,IP3}^2 \approx \left(|A_{B2}| \hat{v}_{o1,IP3} \right)^2 \parallel \hat{v}_{o2,IP3}^2$$

$$IIP5 : \qquad \hat{v}_{g,IP5}^4 \approx \hat{v}_{g1,IP5}^4 \parallel \left(\frac{\hat{v}_{g2,IP5}}{|A_{B1}|} \right)^4$$

$$OIP5 : \qquad \hat{v}_{o,IP5}^4 \approx \left(|A_{B2}| \hat{v}_{o1,IP5} \right)^4 \parallel \hat{v}_{o2,IP5}^4$$

It can be seen that with the help of the operating gains A_{B1} and A_{B2} the intercept points of the amplifiers can be converted to the input or output of the series connection and can be *connected in parallel* to the second and fourth power.

This result can be extended to apply to the series connections of any number of amplifiers:

> *The input intercept point IIPn of amplifiers in series connection can be determined by converting the intercept points of the individual amplifiers to the input by means of the operating gains and by connecting them in parallel with the power $(n - 1)$. In the same way, the output intercept point OIPn can be determined by conversion to the output.*

Operating Situation When Determining the Nonlinear Parameters

Here, nonlinear parameters are determined on the basis of the operating transfer characteristic; that is, for amplifier operation with a signal source with an internal resistance R_g and a load R_L. These parameters are thus related to a certain operating situation and are therefore not properties of the amplifier alone. This procedure is in accordance with practical methods, as parameters such as the distortion factor and intercept points are always established for a certain circuitry. This external circuitry is stated on the amplifier's data sheet. There are two typical operating situations that often occur:

- In low-frequency amplifiers, the input impedance is often much higher than the internal resistance of typical signal sources ($r_i \gg R_g$) and the output resistance is much smaller than the load ($r_o \ll R_L$). In this case, the voltage division at the input and the output is negligible, and the operating gain A_B corresponds to the open-circuit gain A. Since $v_g \approx v_i$, it is of no importance whether the nonlinear parameters refer to v_g or v_i.
- High-frequency amplifiers are operated in a matched mode; that is, $R_g = r_i = r_o = R_L = Z_W$, where Z_W is the characteristic impedance of the lines used. Typical values are $Z_W = 50\,\Omega$ and $Z_W = 75\,\Omega$ for coaxial cables and $Z_W = 110\,\Omega$ for *twisted-pair* lines. With an individual amplifier, this divides the signal amplitude in half by voltage division at the input and output. Consequently:

$$A_B = \frac{A}{4} \quad , \quad v_i = \frac{v_g}{2}$$

If the nonlinear parameters are not to be related to v_g but to v_i, it is necessary to use the corresponding equations $(2v_i)^n$ instead of v_g^n. This applies accordingly to the first amplifier of a series connection; but for each subsequent amplifier the voltage division at the output must also be taken into account:

$$A_{B(i)} = \frac{A_{(i)}}{2} \quad , \quad v_{i(i)} = v_{g(i)} = v_{o(i-1)} \qquad \text{for } i \geq 2$$

4.2.4
Noise

The basics for discussing the noise and calculating the noise figure are described in Sect. 2.3.4, using a bipolar transistor as an example. Here, the results are converted to amplifiers in general. For an explanation of the term *noise density*, we recommend that you read Sect. 2.3.4 on page 82.

Noise Sources and Noise Densities of an Amplifier

Semiconductor amplifiers are made up of transistors and resistances that each contain one or more noise sources. It is possible to convert all noise sources to the amplifier input and to combine them into one noise voltage source $v_{r,0}$ and one noise current source $i_{r,0}$ (see Fig. 4.151); this makes the actual amplifier noiseless. The noise sources $v_{r,0}$ and $i_{r,0}$ are also known as *equivalent noise sources*, since they *equivalently* describe the noise performance of the amplifier. Calculation of the related noise densities $|\underline{v}_{r,0}(f)|^2$ and $|\underline{i}_{r,0}(f)|^2$ is usually rather elaborate; in practice, they are determined by measurement or by circuit simulation. Often, the squares of the absolute values with the units

$$\left[|\underline{v}_{r,0}(f)|^2 \right] = \frac{V^2}{Hz} \quad , \quad \left[|\underline{i}_{r,0}(f)|^2 \right] = \frac{A^2}{Hz}$$

are replaced by the absolute values $|\underline{v}_{r,0}(f)|$ and $|\underline{i}_{r,0}(f)|$, with the units:

$$\left[|\underline{v}_{r,0}(f)| \right] = \frac{V}{\sqrt{Hz}} \quad , \quad \left[|\underline{i}_{r,0}(f)| \right] = \frac{A}{\sqrt{Hz}}$$

In the medium-frequency range, the noise densities are approximately constant; in other words, independent of the frequency. This is known as *white noise* and the corresponding frequency range is called the *white noise region*. The noise densities increase due to the $1/f$ *noise* at low frequencies and the decreasing gain at high frequencies. An exception to this is the noise current density of a FET, which increases in proportion to the frequency across the entire frequency range. Figure 4.152 shows the typical noise density curves of an amplifier with bipolar transistors; in this diagram, the white noise region with $|\underline{v}_{r,0}(f)| = 1.1\,\text{nV}/\sqrt{Hz}$ and $|\underline{i}_{r,0}(f)| = 1.8\,\text{pA}/\sqrt{Hz}$ extends from 5 kHz to 50 MHz.

Equivalent Noise Source and the Noise Figure

Operation with a signal generator results in the equivalent circuit shown in Fig. 4.153a; here, v_g is the signal voltage and $v_{r,g}$ is the noise voltage of the signal generator. The noise

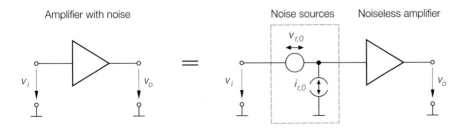

Fig. 4.151. Noise sources of an amplifier

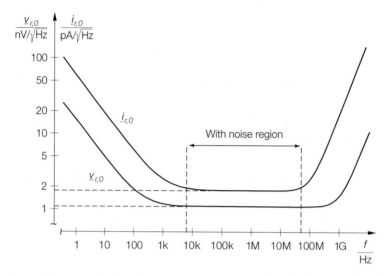

Fig. 4.152. Typical noise density curves of an amplifier with bipolar transistors

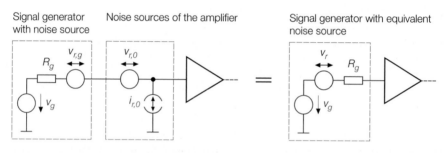

Signal generator with noise source

Noise sources of the amplifier

Signal generator with equivalent noise source

a With the noise source of the signal generator and the equivalent noise sources of the amplifier

b With an equivalent noise source

Fig. 4.153. Operation with a signal source

source of the signal generator and the noise sources of the amplifier can be combined into one *equivalent noise source* (see Fig. 4.153b). Consequently:

$$|\underline{v}_r(f)|^2 = |\underline{v}_{r,g}(f)|^2 + |\underline{v}_{r,0}(f)|^2 + R_g^2 |\underline{i}_{r,0}(f)|^2 \tag{4.187}$$

It is assumed that the amplifier noise originates from the signal generator, and the ratio of the noise density of the equivalent noise source to the noise density of the signal generator is known as the *spectral noise figure* [4.6]:

$$F(f) = \frac{|\underline{v}_r(f)|^2}{|\underline{v}_{r,g}(f)|^2} = 1 + \frac{|\underline{v}_{r,0}(f)|^2 + R_g^2 |\underline{i}_{r,0}(f)|^2}{|\underline{v}_{r,g}(f)|^2} \tag{4.188}$$

In other words:

The noise density of the equivalent noise source representing the noise from the signal generator and from the amplifier is higher than the noise density of the signal generator by a value equal to the spectral noise figure $F(f)$. Thus the *spectral noise figure describes the factor by which the noise already existing in the signal generator is increased as a result of the amplifier noise. This means that the noise density at the amplifer output is higher than the noise density at the output of a noiseless amplifier with the same gain by a value equal to the spectral noise figure. Therefore, a noiseless amplifier has a spectral noise figure of one.*

In order to prevent the noise figure from depending on the properties of one particular signal generator, an *ideal signal generator* with a noise density that corresponds to the thermal noise density of the internal resistance R_g is used for calculations [4.7]:

$$|\underline{v}_{r,g}| = \sqrt{4kTR_g} = 0.13 \, \frac{\text{nV}}{\sqrt{\text{Hz}}} \cdot \sqrt{\frac{R_g}{\Omega}} \tag{4.189}$$

The noise density in real signal generators is usually much higher.

When quoting the *noise figure F* in practice, we are referring to the spectral noise figure for the frequency range that is of interest for the intended application. If the noise figure in this region is not constant, then, strictly speaking, the mean noise figure must be calculated using the integral given in (4.233). In this book, we follow the general convention and use the term $F(f)$ only in cases in which the frequency dependence of the noise figure is expressly being referred to; noise densities are treated accordingly.

The relationship between the noise densities, the internal resistance R_g of the signal generator and the noise figure is illustrated in Fig. 4.154; the graph shows separate curves for the portions $|\underline{v}_{r,g}|$, $|\underline{v}_{r,0}|$ and $R_g|\underline{i}_{r,0}|$ that originate in the white noise region of the equivalent noise source for the amplifier shown in Fig. 4.152. On the two logarithmic axes, these portions have the slopes 0, 1/2 and 1, respectively:

$$|\underline{v}_{r,0}| = \text{const.} \sim R_g^0 \quad , \quad |\underline{v}_{r,g}| \sim R_g^{1/2} \quad , \quad R_g|\underline{i}_{r,0}| \sim R_g$$

In this plot, the noise figure F corresponds to the ratio of the noise density $|\underline{v}_r|$ of the equivalent noise source and the noise density $|\underline{v}_{r,g}|$ of the signal generator; this is shown

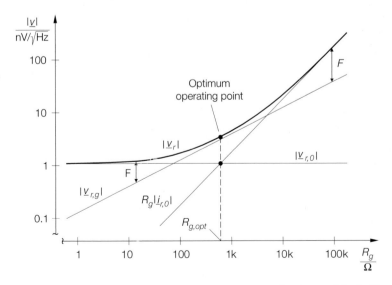

Fig. 4.154. Noise density of the equivalent noise source for the amplifier in Fig. 4.152 within the white noise region

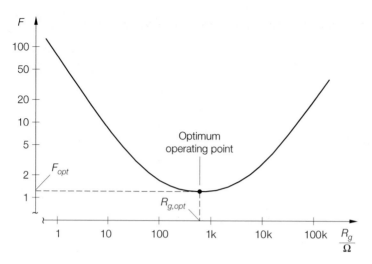

Fig. 4.155. Noise figure of the amplifier in Fig. 4.152 in the white noise region

on a separate diagram in Fig. 4.155. Due to the different slopes, there is always one point at which the noise figure reaches a minimum; this point is indicated on the curves of Figs. 4.154 and 4.155 as the *optimum operating point*. The corresponding internal resistance is known as the *optimum source resistance* R_{gopt}.

A fundamental property can be derived from the curves in Fig. 4.154:

> *If operated with an internal resistance that is clearly below the optimum source resistance, the noise density of the equivalent noise source depends primarily on the noise voltage density of the amplifier; similarly, if operated with an internal resistance that is clearly above the optimum source resistance, then the noise density depends primarily on the noise current density of the amplifier. This is also true for the noise figure.*

Therefore:

$$R_g \ll R_{gopt} \Rightarrow |\underline{v}_r| \approx |\underline{v}_{r,0}| \quad , \quad R_g \gg R_{gopt} \Rightarrow |\underline{v}_r| \approx R_g |\underline{i}_{r,0}|$$

No general statement can be made for operation with an internal resistance in the region of R_{gopt}, since in this case the ratio of the noise densities of the amplifier to the noise density of the signal generator is decisive.

An amplifier has low noise if there is a region in which the portions $|\underline{v}_{r,0}|$ and $R_g |\underline{i}_{r,0}|$ caused by the amplifier are clearly below the noise density $|\underline{v}_{r,g}|$ of the signal generator. The borderline case is where the noise density of the signal generator is equal to the sum of the noise densities of the amplifier:

$$|\underline{v}_{r,g}|^2 \overset{!}{=} |\underline{v}_{r,0}|^2 + R_g^2 |\underline{i}_{r,0}|^2$$

The resulting noise figure is $F = 2$; therefore $F < 2$ is often defined as a condition for a low-noise amplifier.

Optimum Noise Figure and the Optimum Source Resistance

The optimum operating point is characterized by the *optimum noise figure* F_{opt} and the *optimum source resistance* R_{gopt} (see Fig. 4.155). The values are calculated by inserting the noise density of the ideal signal generator into (4.188). With the condition

$$\frac{\partial F}{\partial R_g} = 0$$

the minimum value for the noise figure can be determined; at first, this leads to the optimum source resistance

$$R_{gopt}(f) = \frac{|\underline{v}_{r,0}(f)|}{|\underline{i}_{r,0}(f)|} \tag{4.190}$$

from which the optimum noise figure can be determined by inserting this into (4.188):

$$F_{opt}(f) = 1 + \frac{|\underline{v}_{r,0}(f)|\,|\underline{i}_{r,0}(f)|}{2kT} \tag{4.191}$$

Both values are dependent on the frequency; therefore optimum wide-band operation is not usually possible.[31]

The white noise region: In the white noise region, the frequency dependence can be neglected; it thus follows:

$$R_{gopt} = \frac{|\underline{v}_{r,0}|}{|\underline{i}_{r,0}|} \tag{4.192}$$

$$F_{opt} = 1 + \frac{|\underline{v}_{r,0}|\,|\underline{i}_{r,0}|}{2kT} \tag{4.193}$$

Here, $|\underline{v}_{r,0}|$ and $|\underline{i}_{r,0}|$ are the noise densities in the white noise region. The quantity equation for F_{opt} is:

$$F_{opt} \overset{T=300\,K}{=} 1 + 0.12 \cdot \frac{|\underline{v}_{r,0}|}{nV/\sqrt{Hz}} \cdot \frac{|\underline{i}_{r,0}|}{pA/\sqrt{Hz}}$$

For the amplifier in Fig. 4.152, the values are $|\underline{v}_{r,0}| = 1.1\,nV/\sqrt{Hz}$ and $|\underline{i}_{r,0}| = 1.8\,pA/\sqrt{Hz}$; it follows that $R_{gopt} = 610\,\Omega$ and $F_{opt} = 1.24$.

At the optimum operating point, the absolute noise source quantities of the amplifier are of the same value: $|\underline{v}_{r,0}| = R_{gopt}|\underline{i}_{r,0}|$. This relationship is shown in Fig. 4.154: the corresponding straight lines intersect at the optimum operating point.

Operation using a nonoptimum source resistance: In practice, the amplifier cannot normally be operated using the optimum source resistance, since the signal source is given and $R_g \neq R_{gopt}$; in this case, the noise figure can be calculated using (4.188), provided

---

[31] In a wide-band amplifier that is to be used in the region of the $1/f$ and the high-frequency noise, the *mean noise figure* must be optimized and not the *spectral noise figure*; to do so, the minimum value of the integral (4.233) must be determined. This may then produce *optimum* operation of the amplifier, but this is not the case *at any frequency*.

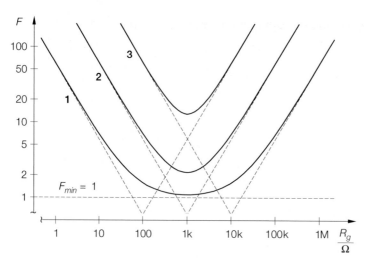

Fig. 4.156. A plot of the noise figures of three amplifiers with different optimum noise figure

that $|\underline{v}_{r,0}|$ and $|\underline{i}_{r,0}|$ are known. It is also possible to start with F_{opt} and R_{gopt}; from (4.192) and (4.193), it follows:

$$|\underline{v}_{r,0}|^2 \;=\; 2kT R_{gopt}\left(F_{opt}-1\right) \quad , \quad |\underline{i}_{r,0}|^2 \;=\; \frac{2kT}{R_{gopt}}\left(F_{opt}-1\right)$$

and by inserting this into (4.188), we obtain:

$$F \;=\; 1 + \frac{1}{2}\left(F_{opt}-1\right)\left(\frac{R_g}{R_{gopt}} + \frac{R_{gopt}}{R_g}\right) \tag{4.194}$$

$R_g = R_{gopt}$ leads by definition to $F = F_{opt}$; if $R_g \neq R_{gopt}$, then $F > F_{opt}$.

It should be noted that the increase in the noise figure is not only due to the ratio of the resistances, but also to the optimum noise figure:

> *An amplifier with lower noise not only has a lower optimum noise figure but also a wider minimum. In the borderline case of a noiseless amplifier, the minimum becomes infinitely wide; that is, $F = 1$ for all values of R_g.*

Figure 4.156 illustrates this for three amplifiers with different optimum noise figures: amplifier 1 with $F_{opt} = 1.12$, amplifier 2 with $F_{opt} = 2.2$ and amplifier 3 with $F_{opt} = 13$.[32]

For $R_g \ll R_{gopt}$, the noise figure is almost only dependent on $|\underline{v}_{r,0}|$, while for $R_g \gg R_{gopt}$ it is almost only dependent on $|\underline{i}_{r,0}|$:

$$F \approx \begin{cases} \dfrac{|\underline{v}_{r,0}|^2}{4kT R_g} \;=\; \dfrac{1}{2}\left(F_{opt}-1\right)\dfrac{R_{gopt}}{R_g} & \text{für } R_g \ll R_{gopt} \\[3mm] \dfrac{R_g\,|\underline{i}_{r,0}|^2}{4kT} \;=\; \dfrac{1}{2}\left(F_{opt}-1\right)\dfrac{R_g}{R_{gopt}} & \text{für } R_g \gg R_{gopt} \end{cases}$$

This relationship has already been pointed out in one of the emphasized notes. The corresponding asymptotes are shown in Fig. 4.156.

---

[32] For amplifier 1 it was assumed that $|\underline{v}_{r,0}| = 1\,\text{nV}/\sqrt{\text{Hz}}$ and $|\underline{i}_{r,0}| = 1\,\text{pA}/\sqrt{\text{Hz}}$. The values for amplifiers 2 and 3 are higher by the factors $\sqrt{10}$ and 10, respectively.

A note on the selection or dimensioning of amplifiers: The parameters F_{opt} and R_{gopt} play an important role in the optimum operation of a given amplifier. On the other hand, making a choice between several standard amplifiers must always be done on the basis of the noise figure F for the given source resistance R_g: the optimum amplifier is the one that has the lowest noise figure when operated with the given source resistance. The *operating noise figure* is calculated using (4.188) and (4.189):

$$F = 1 + \frac{|\underline{v}_{r,0}(f)|^2 + R_g^2 |\underline{i}_{r,0}(f)|^2}{4kTR_g} \overset{T=300\,\mathrm{K}}{=} 1 + \frac{|\underline{v}_{r,0}(f)|^2 + R_g^2 |\underline{i}_{r,0}(f)|^2}{1.656 \cdot 10^{-20}\,\mathrm{V^2/\Omega} \cdot R_g}$$

Here, the parameters F_{opt} and R_{gopt} are only relevant because, according to (4.194), they can be used to calculate the operating noise figure. Therefore, the best choice is *not* the amplifier with the lowest noise figure F_{opt} or the amplifier whose optimum source resistance R_{gopt} best corresponds to the given source resistance R_g; usually, neither produces an optimum result. Accordingly, when dimensioning an integrated amplifier it is the operating noise figure with the given source resistance that should be used as a criterion for optimization; the optimizations F_{opt} or R_{gopt} as individual parameters are useless. There is only one exception: if an impedance transformation of R_g to R_{gopt} is made by means of a transformer, then $F = F_{opt}$ can be achieved; this allows the amplifier with the lowest noise figure F_{opt} to be selected provided that the required transformation ratio R_g/R_{gopt} can be realized.

Noise Figure of Amplifiers Connected in Series

The noise figure of a series of amplifiers may be calculated from the noise figures of the individual amplifiers. For this purpose, we first calculate the noise figure for two amplifiers connected in series, as shown in Fig. 4.157, and then generalize. In Fig. 4.157, the amplifiers are presented in the form of the equivalent circuit of an amplifier without reverse gain with the equivalent noise sources $v_{r,0}$ and $i_{r,0}$; this equivalent circuit also applies to amplifiers with reverse gain if the operating input resistances are used instead of the input resistances (see Sect. 4.2.2).

Calculating the noise figure: As the noise densities of the equivalent noise sources are related to the source voltages and not to the input voltages of the amplifiers, it is not possible to use the operating gain values $A_{B1} = v_2/v_g$ and $A_{B2} = v_o/v_2$ in this case. Instead, we must use the *noise operating gains*

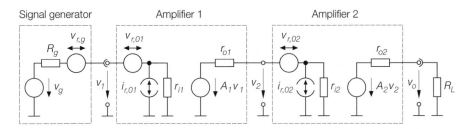

Fig. 4.157. Equivalent circuit for calculating the noise figure of two amplifiers connected in series

$$A_{B,r1} = \frac{A_1 v_1}{v_g} = \frac{r_{i1}}{R_g + r_{i1}} A_1$$

$$A_{B,r2} = \frac{A_2 v_2}{A_1 v_1} = \frac{r_{i2}}{r_{o1} + r_{i2}} A_2$$

and the *load factor:*

$$k_L = \frac{R_L}{r_{o2} + R_L}$$

It thus follows:

$$A_B = \frac{v_o}{v_g} = A_{B,r1} A_{B,r2} k_L$$

The noise operating gains are made up of the voltage divider factor at the input and the open-circuit gain, and indicate the gain from one source voltage to the next.

At first, all noise sources are converted to the output of the series connection by means of the relevant gain; without a signal voltage ($v_g = 0$) it follows:

$$v_o = \left(v_{r,g} + v_{r,01} + R_g i_{r,01} \right) A_{B,r1} A_{B,r2} k_L + \left(v_{r,02} + r_{o1} i_{r,02} \right) A_{B,r2} k_L$$

The voltage of the equivalent noise source for the series connection is obtained by conversion to the signal generator:

$$v_r = \frac{v_o}{A_B} = v_{r,g} + v_{r,01} + R_g i_{r,01} + \frac{v_{r,02} + r_{o1} i_{r,02}}{A_{B,r1}} \qquad (4.195)$$

Since all noise sources are independent, the noise density of the equivalent noise source is:

$$|\underline{v}_r|^2 = |\underline{v}_{r,g}|^2 + |\underline{v}_{r,01}|^2 + R_g^2 |\underline{i}_{r,01}|^2 + \frac{|\underline{v}_{r,02}|^2 + r_{o1}^2 |\underline{i}_{r,02}|^2}{A_{B,r1}^2}$$

For the noise figure of the series connection, it follows:

$$F = \frac{|\underline{v}_r|^2}{|\underline{v}_{r,g}|^2} = 1 + \frac{|\underline{v}_{r,01}|^2 + R_g^2 |\underline{i}_{r,01}|^2}{|\underline{v}_{r,g}|^2} + \frac{|\underline{v}_{r,02}|^2 + r_{o1}^2 |\underline{i}_{r,02}|^2}{A_{B,r1}^2 |\underline{v}_{r,g}|^2} \qquad (4.196)$$

The noise figure of the first amplifier is:

$$F_1 = 1 + \frac{|\underline{v}_{r,01}|^2 + R_g^2 |\underline{i}_{r,01}|^2}{|\underline{v}_{r,g}|^2} = 1 + \frac{|\underline{v}_{r,01}|^2 + R_g^2 |\underline{i}_{r,01}|^2}{4kT R_g}$$

For the second amplifier, R_g is replaced by r_{o1}. To calculate the noise figure, we assume an ideal signal generator with thermal noise from the internal resistance. The thermal noise

$$|\underline{v}_{r,o1}|^2 = 4kT r_{o1}$$

must therefore be used as a reference value. This does *not* mean that resistance r_{o1} in Fig. 4.157 produces thermal noise, but only that the noise figure of the second amplifier is determined for an internal signal generator resistance of the value r_{o1}. Therefore:

$$F_2 = 1 + \frac{|\underline{v}_{r,02}|^2 + r_{o1}^2 |\underline{i}_{r,02}|^2}{|\underline{v}_{r,o1}|^2} = 1 + \frac{|\underline{v}_{r,02}|^2 + r_{o1}^2 |\underline{i}_{r,02}|^2}{4kT r_{o1}}$$

By inserting the noise figures F_1 and F_2 into (4.196), we obtain:

$$F = F_1 + \frac{F_2 - 1}{A_{B,r1}^2} \frac{r_{o1}}{R_g}$$

which by generalization leads to the *noise figure of a series connection of n amplifiers*:

noise figure of a series connection of n amplifiers

$$F = F_1 + \frac{F_2 - 1}{A_{B,r1}^2} \frac{r_{o1}}{R_g} + \frac{F_3 - 1}{A_{B,r1}^2 A_{B,r2}^2} \frac{r_{o2}}{R_g} + \cdots$$

$$= F_1 + \sum_{i=2}^{n} \left(\frac{F_{(i)} - 1}{\displaystyle\prod_{k=1}^{i-1} A_{B,r(k)}^2} \frac{r_{o(i-1)}}{R_g} \right) \tag{4.197}$$

$$A_{B,r(k)} = \frac{r_{i(k)}}{r_{o(k-1)} + r_{i(k)}} A_{(k)} \qquad \text{with } r_{o0} = R_g \tag{4.198}$$

The noise figure of the first amplifier directly affects the noise figure of the series connection; for all subsequent amplifiers, the *additional noise figure*

$$F_Z = F - 1 \tag{4.199}$$

which is rated according to the inverse square of the *previous* operating noise gain and the ratio of the source resistances, enters into the result. A minimum noise figure is achieved mainly by optimizing the first amplifier:

- Minimizing the noise figure F_1.
- Maximizing the noise operating gain $A_{B,r1}$ by maximizing A_1 and r_{i1}.
- Minimizing the output resistance r_{o1}.

The last item is of particular importance for small internal resistances R_g, since the factor $1/A_{B,r1}^2$ can be overcompensated by the factor r_{o1}/R_g; the noise figure F_2 dominates in this case. If the second amplifier also features a high-noise operating gain, then the contributions of the subsequent amplifiers are negligible.

Expression by means of the available power gain: In radio-frequency engineering, the *available power gain* G_A is quoted instead of the gain; this indicates the ratio of the *available power* at the amplifier output to the available power of the signal generator. The available power of the signal generator is:[33]

$$P_{A,g} = \frac{v_g^2}{4R_g}$$

This is composed of the source voltage and the internal resistance. Correspondingly, the available power at the output of an amplifier with gain A is:

$$P_{A,amp} = \frac{(Av_i)^2}{4r_o} = \left(\frac{r_i}{R_g + r_i} \right)^2 A^2 \frac{v_g^2}{4r_o}$$

---

[33] Here, we use r.m.s. *values*; that is, $P = u^2/R$; a distinction between DC and AC voltages is therefore not required.

Consequently:

$$GA = \frac{P_{A,amp}}{P_{A,g}} = \left(\frac{r_i}{R_g + r_i}\right)^2 A^2 \frac{R_g}{r_o} = A_{B,r}^2 \frac{R_g}{r_o} \tag{4.200}$$

Insertion into (4.197) leads to:

$$F = F_1 + \frac{F_2 - 1}{G_{A1}} + \frac{F_3 - 1}{G_{A1}G_{A2}} + \cdots = F_1 + \sum_{i=2}^{n} \left(\frac{F_{(i)} - 1}{\prod\limits_{k=1}^{i-1} G_{A(k)}} \right) \tag{4.201}$$

The following relationship should be noted:

$$G_{A1}G_{A2}\cdots G_{A(i-1)} = A_{B,r1}^2 \frac{R_g}{r_{o1}} A_{B,r2}^2 \frac{r_{o1}}{r_{o2}} \cdots A_{B,r(i-1)}^2 \frac{r_{o(i-2)}}{r_{o(i-1)}}$$

$$= A_{B,r1}^2 A_{B,r2}^2 \cdots A_{B,r(i-1)}^2 \frac{R_g}{r_{o(i-1)}}$$

Equation (4.201) is often quoted in the form

$$F = F_1 + \frac{F_2 - 1}{G_1} + \frac{F_3 - 1}{G_1 G_2} + \cdots$$

with the power gains $G_{(i)}$ not specified in detail. Since a number of different power gains are used in radio-frequency engineering, we expressly emphasize that, in general, the available power gain G_A must be used; only in cases where all amplifiers are matched are all power gains equal and the general term "power gain" can be used without further specification.

Minimum noise figure for a series connection of amplifiers: In a series connection of matched amplifiers of the *same characteristic impedance*, the order of the amplifiers may be changed without affecting the gain or the noise figures of the individual amplifiers. In order to find a criterion for obtaining the minimum noise figure, two amplifiers with the available power gains G_{A1} and G_{A2} and noise figures F_1 and F_2 will be reviewed: the noise figures of the two possible series connections are:

$$F_{12} = F_1 + \frac{F_2 - 1}{G_{A1}} \quad , \quad F_{21} = F_2 + \frac{F_1 - 1}{G_{A2}}$$

From the condition $F_{12} < F_{21}$, by separating the variables it follows:

$$\frac{F_1 - 1}{1 - \dfrac{1}{G_{A1}}} < \frac{F_2 - 1}{1 - \dfrac{1}{G_{A2}}}$$

The factor

$$M = \frac{F - 1}{1 - \dfrac{1}{G_A}} \tag{4.202}$$

is known as the *noise measure* [4.8]. Consequently, the amplifiers must be arranged according to the noise measures: the amplifier with the lowest noise measure must be placed at the front and that with the highest noise measure at the end.

Equivalent noise sources: The equivalent noise sources of the series connection in Fig. 4.157 can be determined by means of the noise voltage of the equivalent noise source; inserting $A_{B,r1}$ into (4.195) and arranging the terms into groups with and without R_g leads to:

$$v_r = \underbrace{v_{r,g} + v_{r,01} + \frac{v_{r,02} + r_{o1}i_{r,02}}{A_1}}_{v_{r,0}} + R_g \underbrace{\left(i_{r,01} + \frac{v_{r,02} + r_{o1}i_{r,02}}{A_1 r_{i1}} \right)}_{i_{r,0}}$$

The equivalent noise sources are interdependent, as the noise sources of the second amplifier enter into the noise voltage source $v_{r,0}$ *and* the noise current source $i_{r,0}$. Since calculations with interdependent noise sources are rather complex, we will refrain from using this model, but we nevertheless wish to emphasize an important relationship: the equivalent noise sources of a multi-stage amplifier, which are regarded as basic transistor circuits connected in series, are only approximately independent if the contribution of the second and any subsequent stage can be neglected in at least one of the two equivalent noise sources.

Optimization of the Noise Figure

The following description is limited to the optimization of the noise figure in the white noise region; in this case, the noise densities are independent of the frequency. An optimization outside this range is much more complex, as this requires the minimization of integral equations; in practice, this is solely done numerically. However, optimization in the white noise region generally results in an improvement in the region of $1/f$ or high-frequency region as well; the results thus correspond in terms of *tendency*.

The task of optimization depends on the surrounding conditions. The user of integrated circuits usually faces the task of amplifying the signal of a given source with as little noise as possible. He must select an amplifier that obtains the lowest possible noise figure for the given internal resistance R_g; he can use either (4.188) or (4.194), depending on whether the data sheet specifies $|\underline{v}_{r,0}|$ and $|\underline{i}_{r,0}|$ or F_{opt} and R_{gopt}. In the low-frequency range VV operational amplifiers are usually used; their noise properties are described in Sect. 5.2.8. For video and high-frequency applications, the task is easier in so far as matching – that is, fixed sources and load resistances – is used: $R_g = 75\,\Omega$ or $R_g = 50\,\Omega$. Data sheets usually specify the noise figures for this operating mode so that the user can make a direct comparison. Specific amplifiers are available for special applications such as photodiode receivers.

Designers of integrated circuits approach the optimization task from a different angle. They must choose a suitable type of technology, a suitable circuit and a suitable operating point for the circuit in order to achieve an optimum result. In what follows, we will describe the basic conditions that influence this three-step selection process.

Noise sources in integrated circuits: The noise of integrated circuits is generated by the transistors and the ohmic resistances; capacitances and inductances are noise-free. Figure 4.158 shows the noise sources of a bipolar transistor, a MOSFET and an ohmic resistance.

Ohmic resistance: For ohmic resistances, we prefer to use the model of a noise current source with the noise density:

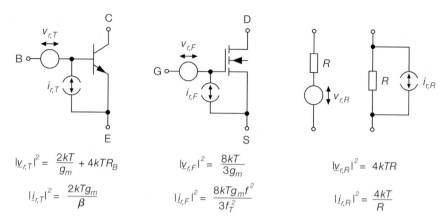

Fig. 4.158. Noise sources of a bipolar transistor, a MOSFET and an ohmic resistance

$$|\underline{i}_{r,R}|^2 = \frac{4kT}{R}$$

However, this method only determines the thermal noise of an ideal resistance; real resistances in integrated circuits may have a clearly higher noise density, depending on their construction.

Bipolar transistors: According to (2.49) and (2.50), the following applies to a bipolar transistor in the white noise region; that is, for $f_{g(1/f)} < f < f_C/\sqrt{\beta} \approx f_C/10$:

$$|\underline{v}_{r,T}|^2 = \frac{2kT V_T}{I_{C,A}} + 4kT R_B = \frac{2kT}{g_m} + 4kT R_B$$

$$|\underline{i}_{r,T}|^2 = \frac{2q I_{C,A}}{\beta} = \frac{2kT g_m}{\beta}$$

Here, $g_m = I_{C,A}/V_T$ and $V_T = kT/q$. We shall limit our analysis to small and medium currents; this eliminates the third term of (2.49). It is obvious that a low-noise bipolar transistor must have a low base spreading resistance R_B and a high current gain β. While the current gain is determined by the technology, the base spreading resistance can be influenced by scaling: in general, R_B is inversely proportional to the size of the transistor. It is therefore possible to reduce the noise voltage density in the medium current range by enlarging the transistor. But this is at the cost of the bandwidth, since the capacitances of the transistor increase while the transconductance remains unchanged.

Figure 4.159 shows the noise figure of a bipolar transistor with $\beta = 100$ and $R_B = 10\,\Omega$ in the $I_{C,A}$–R_g plane. Inserting the noise densities into (4.192) and (4.193) leads to:[34]

$$R_{gopt,T} = \frac{\sqrt{\beta}}{g_m}\sqrt{1 + 2g_m R_B} \overset{R_B \to 0}{\approx} \frac{\sqrt{\beta}}{g_m} \overset{\beta \approx 100}{\approx} \frac{10}{g_m} = \frac{0.26\,\text{V}}{I_{C,A}}$$

$$F_{opt,T} = 1 + \frac{1}{\sqrt{\beta}}\sqrt{1 + 2g_m R_B} \overset{R_B \to 0}{\approx} 1 + \frac{1}{\sqrt{\beta}} \overset{\beta \approx 100}{\approx} 1.1$$

---

[34] The relationships $|\underline{v}_{r,T}|^2 = |\underline{v}_{r,0}(f)|^2$ and $|\underline{i}_{r,T}|^2 = |\underline{i}_{r,0}(f)|^2$ apply since $v_{r,T}$ and $i_{r,T}$ are the equivalent noise sources of the bipolar transistor.

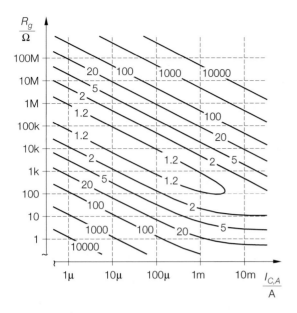

Fig. 4.159. Noise figure of a bipolar transistor with $\beta = 100$ and $R_B = 10\,\Omega$

However, this optimum only applies if g_m and/or $I_{C,A}$ are regarded as given variables while R_g varies. The relationship is different if R_g is given and the optimum quiescent current $I_{C,A}$ is to be determined; in this case:

$$I_{C,A\,opt}(R_g) = \frac{V_T\sqrt{\beta}}{\sqrt{R_g^2 + R_B^2}}$$

$$F_{opt,T}(R_g) = 1 + \frac{R_B}{R_g} + \frac{1}{\sqrt{\beta}}\sqrt{1 + \left(\frac{R_B}{R_g}\right)^2}$$

This optimum has already been determined in Sect. 2.3.4 [see (2.54) and (2.57)]. In practice, it is more significant, because R_g is usually given and can only be matched by means of a transformer or a resonance transformer. By contrast, it is easy to alter $I_{C,A}$.

The difference between $F_{opt,T}$ and $F_{opt,T}(R_g)$ is caused by the base spreading resistance; for $R_B = 0$, the values are equal. To illustrate this, Fig. 4.160 shows the noise figure of a bipolar transistor with $\beta = 100$ and $R_B = 100\,\Omega$, together with the curves for $F_{opt,T}$ and $F_{opt,T}(R_g)$. It is clear that $F_{opt,T}$ represents the optimum in the R_g direction and that $F_{opt,T}(R_g)$ is the optimum in the $I_{C,A}$ direction. If $R_g < R_B$, then $F_{opt,T}(R_g)$ increases rapidly; for $R_g = 1\,\Omega$, the noise figure is 100. Figure 4.160 also shows that there is an upper limit for the quiescent current:

$$I_{C,A\,opt}(R_g \to 0) = \frac{V_T\sqrt{\beta}}{R_B} \stackrel{\beta \approx 100}{\approx} \frac{0.26\,\mathrm{V}}{R_B}$$

A higher quiescent current is pointless.

For currents in the range of $I_{C,A} = 10\,\mu\mathrm{A}\ldots 1\,\mathrm{mA}$, $R_{gopt,T} \approx 26\,\mathrm{k}\Omega\ldots 260\,\Omega$. For higher quiescent currents, the base spreading resistance has to be taken into account. The value $\sqrt{\beta}R_B \approx 10R_B$ can be regarded as the lower limit for $F_{opt,T}(R_g)$; the portions in $F_{opt,T}(R_g)$ that are caused by β and R_g are then of approximately the same size and

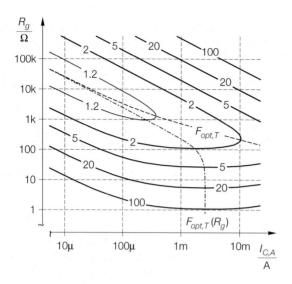

Fig. 4.160. Optimum noise figures $F_{opt,T}$ and $F_{opt,T}(R_g)$ of a bipolar transistor with $\beta = 100$ and $R_B = 100\,\Omega$

$F_{opt,T}(R_g) \approx F_{opt,T}$ still applies. The upper limit for $R_{gopt,T}$ depends on the required bandwidth; $I_{C,A}$ cannot be reduced as one wishes, since in the range of very small currents the transit frequency f_T decreases proportionally to $I_{C,A}$ (see Fig. 2.44 on page 80). The limiting current, below which the transit frequency declines, is:

$$I_{C,A} = \frac{V_T}{\tau_{0,N}}\left(C_{J,E} + C_{J,C}\right) \approx \frac{V_T}{\tau_{0,N}}\left(2C_{J0,E} + C_{J0,C}\right)$$

For an npn transistor of size 1 with the parameters given in Fig. 4.5, the limit current is $100\,\mu A$; consequently, $R_{gopt} = 5.7\,k\Omega$ and $F_{opt} = 1.22$. In pnp transistors the base spreading resistance is usually lower than in npn transistors; with low-resistance sources in particular, this produces a lower noise figure. However, even in complementary technology, the transit frequency of a pnp transistor is lower than that of an npn transistor. In technologies where only lateral pnp transistors are available, the transit frequency of the pnp transistors is lower than that of npn transistors by up to three orders of magnitude.

Example: A wide-band transistor BFR93 with $\beta = 95$ and $R_B = 15\,\Omega$ is to be operated by a signal source with $R_g = 50\,\Omega$. This results in $I_{C,A\,opt}(R_g) = 5\,mA$ and $F_{opt,T}(R_g) = 1.3$. The parameters at this operating point are $g_m = 192\,mS$, $R_{gopt,T} = 132\,\Omega$ and $F_{opt,T} = 1.27$; in other words, the optimum for this *operating point* is not quite achieved. If, by mistake, we evaluate the equation for $R_{gopt,T}$ with the condition $R_{gopt,T} = 50\,\Omega$, then we obtain $g_m = 1.17\,S$ and $I_{C,A} = 30.5\,mA$; this results in $F_{opt,T} = 1.62$. This example shows that for low-resistance signal sources it is necessary to optimize in the $I_{C,A}$ direction; optimizing in the R_g direction leads to a higher noise figure.

MOSFETs: For MOSFETs, we use the simplified description in Sect. 3.3.4:

$$|\underline{v}_{r,F}|^2 = \frac{8kT}{3g_m} = \frac{8kT}{3\sqrt{2K\,I_{D,A}}}$$

$$|\underline{i}_{r,F}(f)|^2 = \frac{8kT\,g_m}{3}\left(\frac{f}{f_T}\right)^2 = \frac{8kT\sqrt{2K\,I_{D,A}}}{3}\left(\frac{f}{f_T}\right)^2$$

Here, $g_m = \sqrt{2K I_{D,A}}$. This description applies up to the transconductance cutoff frequency f_{Y21s}, which is usually higher than the transit frequency f_T. The noise current density depends on the frequency; in other words, there is no region of white current noise. But when we examine the basic circuits, we will note that additional noise sources that cover the current noise of the MOSFET in a more or less wide range are caused by the external circuitry; this makes the equivalent noise current density of the circuit independent of the frequency in this range.

Inserting the noise densities into (4.192) and (4.193) leads to:[35]

$$R_{gopt,F}(f) = \frac{f_T}{g_m f}$$

$$F_{opt,F}(f) = 1 + \frac{4}{3}\frac{f}{f_T}$$

For $f \ll f_T$ we have $F_{opt,F} \to 1$, which means that at the optimum operating point a MOSFET is practically free of noise. In narrow-band applications, the frequency dependency can be neglected and the centre frequency can be used for f. In contrast, for wide-band applications the mean noise current density

$$|\underline{i}_{r,F}[f_L, f_U]|^2 = \frac{1}{f_U - f_L} \int_{f_L}^{f_U} |\underline{i}_{r,F}(f)|^2 df$$

$$= \frac{8kT g_m}{f_T^2} \frac{f_U^3 - f_L^3}{f_U - f_L} \stackrel{f_U \gg f_L}{\approx} 8kT g_m \left(\frac{f_U}{f_T}\right)^2$$

must be used in the range between the lower cutoff frequency f_L and the upper cutoff frequency f_U; for $f_U \gg f_L$, this leads to:

$$R_{gopt,F}[f_L, f_U] \approx \frac{f_T}{\sqrt{3} g_m f_U} = R_{gopt,F}(f)\big|_{f=\sqrt{3} f_U}$$

$$F_{opt,F}[f_L, f_U] \approx 1 + \frac{4}{\sqrt{3}}\frac{f_U}{f_T} = F_{opt,F}(f)\big|_{f=\sqrt{3} f_U}$$

Thus, we only have to insert $f = \sqrt{3} f_U$ in order to achieve optimum values for a wide-band application. Here too, $F_{opt,F} \to 1$, since in many applications the upper cutoff frequency f_U is at least 100 times lower than the transit frequency. From the equations for $R_{gopt,F}$, one must not conclude that the matching to a source can be optimized by the proper choice of the transit frequency; the noise figure has a minimum value for $f_T \to \infty$ not only in optimum operation but, indeed, for *any* given operation. Figure 4.161 shows the noise figure of a MOSFET for

$$V_{GS} - V_{th} = 1\,\text{V} \Rightarrow g_m = \frac{2I_{D,A}}{V_{GS} - V_{th}} = \frac{2I_{D,A}}{1\,\text{V}}$$

and $f_U/f_T = 100$ in the $I_{D,A}$–R_g plane. Due to the diagonal plots across the entire plane, it is not necessary to distinguish between optimization in the R_g and $I_{D,A}$ directions, which means that the equation for $R_{gopt,F}$ can be solved for g_m and can be used to calculate the optimum transconductance.

---

[35] The relationships $|\underline{v}_{r,F}|^2 = |\underline{v}_{r,0}(f)|^2$ and $|\underline{i}_{r,F}(f)|^2 = |\underline{i}_{r,0}(f)|^2$ apply since $v_{r,F}$ and $i_{r,F}$ are the equivalent noise sources of the MOSFET.

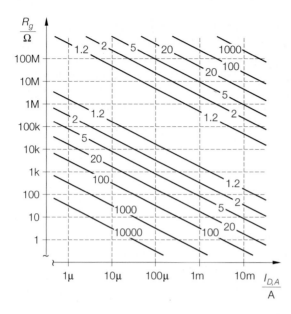

Fig. 4.161. Noise figure of a MOSFET for $V_{GS} - V_{th} = 1$ V and $f_U/f_T = 100$

Unlike the bipolar transistor, the MOSFET offers an additional degree of freedom, since the transconductance not only depends on the current $I_{D,A}$, but also on the size of the MOSFET, which is expressed by the transconductance coefficient K: $g_m = \sqrt{2K I_{D,A}}$. It is only the transconductance that enters the noise density; in other words, the choice of $I_{D,A}$ and K does not affect $R_{gopt,F}$ and $F_{opt,F}$ in the *white noise region*. However, the *width* of this range depends on this choice, since it influences the 1/f cutoff frequency. With the given area $A = WL$, we prefer to increase L at the expense of W, since:

$$f_{g(1/f)} \sim k_{(1/f)} \sqrt{\frac{I_{D,A}}{K}} \sim \frac{1}{L^2} \sqrt{\frac{I_{D,A} L}{W}} = I_{D,A}^{1/2} W^{-1/2} L^{-3/2}$$

This causes K to decline. The minimum 1/f noise is thus achieved with MOSFETs that are geometrically large but electrically small, and that are operated using high currents. Here, as in the case for the bipolar transistor, there is a conflict with the bandwidth:

$$f_T \sim \frac{g_m}{C} \sim \frac{\sqrt{2K I_{D,A}}}{A} \sim \frac{\sqrt{\frac{W}{L}} \sqrt{I_{D,A}}}{WL} = I_{D,A}^{1/2} W^{-1/2} L^{-3/2}$$

It follows that $f_{g(1/f)} \sim f_T$, which means that a reduction in the 1/f cutoff frequency will also reduce the transit frequency.

Comparison of bipolar transistor and MOSFET: In the white noise region, if we assume that the transconductances are the same, the noise voltage densities of a bipolar transistor and a MOSFET are almost equal:

$$\frac{|\underline{v}_{r,T}|^2}{|\underline{v}_{r,F}|^2} = \frac{3}{4} \frac{g_{m,F}}{g_{m,T}} \overset{g_{m,T}=g_{m,F}}{=} \frac{3}{4}$$

If, on the other hand, we assume that the currents are the same, then the noise voltage density of the bipolar transistor is clearly lower, due to the higher transconductance:

$$\frac{|\underline{v}_{r,T}|^2}{|\underline{v}_{r,F}|^2} \overset{I_{C,A}=I_{D,A}}{=} \frac{3}{2} \frac{V_T}{V_{GS,A} - V_{th}} \overset{V_{GS,A}-V_{th}\approx 1\,\mathrm{V}}{\approx} \frac{1}{25}$$

Equal bandwidth, which is of particular interest in practice, occurs between these two borderline cases. Since the capacitances of a MOSFET are usually lower than those of a bipolar transistor, the transconductance required for the given gain–bandwidth product is also lower; therefore, we usually have $g_{m,F} < g_{m,T}$ and $I_{D,A} > I_{C,A}$. Under comparable conditions, the noise voltage density of a MOSFET is therefore somewhat higher than that of a bipolar transistor. An entirely different picture is found in the region of the 1/f noise; here, the noise voltage density of a MOSFET is significantly higher due to the 1/f cutoff frequency, which can be higher as much as four orders of magnitude.

Contrary to the noise voltage density, the noise current density of a MOSFET in the region of lower and medium frequencies is clearly lower than that of a bipolar transistor:

$$\frac{|\underline{i}_{r,T}|^2}{|\underline{i}_{r,F}(f)|^2} \overset{f<f_{T,T}/\sqrt{\beta}}{=} \frac{3}{4\beta} \frac{g_{m,T}}{g_{m,F}} \left(\frac{f_{T,F}}{f}\right)^2 \overset{g_{m,T}=g_{m,F}}{=} \frac{3}{4\beta} \left(\frac{f_{T,F}}{f}\right)^2 \overset{f\to 0}{\longrightarrow} \infty$$

The condition $f < f_{T,T}/\sqrt{\beta} \approx f_{T,T}/10$ must be met, since we have restricted the discussion to the white noise region. For $f > f_{T,T}/\sqrt{\beta}$, the noise current density of a bipolar transistor increases in proportion to $(f/f_{T,T})^2$ (see (2.48)); therefore, the noise current densities in this region are almost the same if the transit frequencies are the same.

A basic finding for the white noise region can be derived from the relations of the noise densities:

> *Under comparable conditions, the noise voltage density of a MOSFET is somewhat higher, but the noise current density is considerable lower, than the corresponding noise densities of a bipolar transistor. It follows that the optimum source resistance of a MOSFET is clearly higher than that of a bipolar transistor. Therefore, a lower noise figure is achieved with bipolar transistors for low-resistance sources and with MOSFETs for high-resistance sources.*

This leads to the question of the limit – that is, the source resistance – for which a bipolar transistor and a MOSFET achieve the same noise figure. From the condition

$$|\underline{v}_{r,T}|^2 + R_g^2 |\underline{i}_{r,T}|^2 = |\underline{v}_{r,F}|^2 + R_g^2 |\underline{i}_{r,F}(f)|^2$$

it follows:

$$R_{g,T\leftrightarrow F} = \sqrt{\frac{|\underline{v}_{r,F}|^2 - |\underline{v}_{r,T}|^2}{|\underline{i}_{r,T}|^2 - |\underline{i}_{r,F}(f)|^2}} \overset{|\underline{i}_{r,T}|^2\gg|\underline{i}_{r,F}(f)|^2}{\approx} \sqrt{\frac{|\underline{v}_{r,F}|^2 - |\underline{v}_{r,T}|^2}{|\underline{i}_{r,T}|^2}}$$

The limit is independent of the frequency as long as the noise current density of the MOSFET can be neglected. The following must therefore apply:

$$\left(\frac{f_U}{f_{T,F}}\right)^2 \ll \frac{1}{\beta} \approx \frac{1}{100}$$

In other words, the upper cutoff frequency f_U must be at least 30 times lower than the transit frequency of the MOSFET. By inserting the noise densities, we obtain:

$$R_{g,T\leftrightarrow F} \overset{|\underline{i}_{r,T}|^2\gg|\underline{i}_{r,F}(f)|^2}{\approx} \sqrt{\frac{\beta}{g_{m,T}}\left(\frac{4}{3g_{m,F}} - \frac{1}{g_{m,T}}\right)} \qquad (4.203)$$

It follows:

$$g_{m,T} > \frac{3}{4} g_{m,F}$$

Otherwise, the MOSFET is generally superior in the white noise region.

In practice, the limit $R_{g,T \leftrightarrow F}$ is seldom of interest, since general limitations are caused by the technology employed, which cannot be chosen with regard to the noise figure of *one* amplifier alone; it is of greater interest to determine the region in which the *required* noise figure is achieved. Solving (4.194) for R_g results in a quadratic equation with the solution:

$$R_{g,l/u} = R_{gopt} \left(\frac{F-1}{F_{opt}-1} \pm \sqrt{\left(\frac{F-1}{F_{opt}-1}\right)^2 - 1} \right)$$

For $F > F_{opt}$, the lower limit is achieved with $R_{g,l} < R_{gopt}$ and the upper limit with $R_{g,u} > R_{gopt}$; for $F = F_{opt}$, we have $R_{g,l} = R_{g,u} = R_{gopt}$ and no solution exists for $F < F_{opt}$. Furthermore, $R_{g,l} R_{g,u} = R_{gopt}^2$. For

$$F - 1 > 2\left(F_{opt} - 1\right)$$

it is possible to perform a series expansion of the square root and interrupt the series after the linear term;[36]

$$R_{g,l} \approx \frac{R_{gopt}}{2} \frac{F_{opt}-1}{F-1} \quad , \quad R_{g,u} \approx 2 R_{gopt} \frac{F-1}{F_{opt}-1}$$

The results for a bipolar transistor are:

$$R_{g,lT} \approx \left(\frac{1}{2g_m} + R_B\right) \frac{1}{F-1} \quad , \quad R_{g,uT} \approx \frac{2\beta}{g_m}(F-1)$$

The base spreading resistance R_B only influences the lower limit. The values for a MOSFET in wide-band applications are:

$$R_{g,lF} \approx \frac{2}{3g_m(F-1)} \quad , \quad R_{g,uF} \approx \frac{F-1}{2g_m}\left(\frac{f_T}{f_U}\right)^2$$

Only the upper solution $R_{g,uF}$ depends on the frequency.

Optimum operating point: For both the bipolar transistor and the MOSFET, the optimum source resistance in the white noise region depends primarily on the transconductance; in principle, this is also the case for the 1/f and the high-frequency noise, as is shown in the relevant equations in Sects. 2.3.4 and 3.3.4. The operating point plays a predominant role in the optimization of the noise figure. Bipolar transistors allow no margin due to the relationship $g_m = I_{C,A}/V_T$; that is, an optimum collector current $I_{C,A\,opt}(R_g)$ exists for every source resistance. MOSFETs, on the other hand, allow variations in the ratio of the transconductance coefficient K to the drain current $I_{D,A}$, since $g_m = \sqrt{2K I_{D,A}}$.

In practice, it is not usually possible to select the collector or drain current at the operating point solely on the basis of noise considerations, since conflicting demands exist with regard to the bandwidth, the impedance level and – due to the increasing miniaturization and portability of modern systems – the power consumption. The tendency is favorable: with increasing frequencies, the resistances must be made lower due to the unavoidable capacitances, which leads to a reduction of the source resistances in the circuits; furthermore,

---

[36] For $a > 2$, the following applies: $\sqrt{a^2 - 1} \approx a - 1/(2a)$.

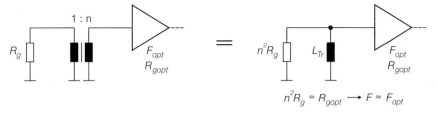

Fig. 4.162. Noise matching with a transformer

the transconductance of the transistors must be increased, which leads to a reduction in their optimum source resistances, so that they generally follow the tendency of the source resistances in the circuit.

In what follows, we will discuss two methods of noise matching by impedance transformation that are used in highly demanding fields such as radio receiving systems.

Noise matching with a transformer: If no DC voltage gain is needed and the demands with respect to noise are particularly high, then a transformer can be used for noise matching; it will transform the value of the internal resistance R_g to a value in the region of the optimum source resistance R_{gopt}. This method is used with very small internal resistances ($R_g < 50\,\Omega$) in particular, since amplifiers with a correspondingly low optimum source resistance are not available. Figure 4.162 shows the transformation of R_g to $n^2 R_g$ by means of a 1:n transformer; a numeric example is presented at the end of Sect. 2.3.4.

The lower cutoff frequency is determined by the inductance of the transformer:

$$f_L \;=\; \frac{n^2 R_g}{2\pi L_{Tr}}$$

Therefore, in LF applications it is necessary to use transformers with a high inductance and correspondingly large dimensions, which is impractical in most applications; however, transformers for the frequency range from 1 MHz to 1 GHz are available as SMD components, with a volume of $0.1 \ldots 0.5\,\mathrm{cm}^3$.

Noise matching with a resonance transformer: For high frequency and a low bandwidth, a special resonance transformer can be used instead of the standard transformer. Very often, a π element with two capacitances and one inductance, known as a *Collins filter* or a *Collins transformer*, is used. An RF amplifier usually contains two resonance transformers; one at the output for power matching and one at the input for power *or* noise matching. Figure 4.163 shows one version with discrete components and one with strip lines. Dimensioning is discussed in more detail in Sect. 27.2.7.

Equivalent Noise Sources of Basic Circuits

So far, we have only taken the equivalent noise sources of the transistor into account when calculating the noise figure. This is correct for the ideal situation in which the noise sources of the ohmic resistances and current sources belonging to the circuit can be neglected. Besides, we have only looked at one transistor. The following calculations apply to the equivalent noise sources of basic circuits, the cascode configuration and the differential amplifier, with consideration of the required resistances and current sources.

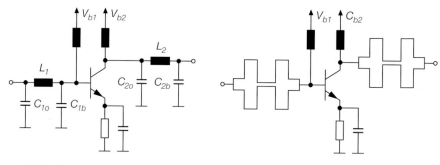

a With inductances and capacitances   **b** With strip lines

Fig. 4.163. Noise matching at the input and power matching at the output using Collins filters

Procedure for calculating the equivalent noise sources: Every noise source of an amplifier can be converted into an equivalent noise voltage source and an equivalent noise current source at the amplifier input. This is done in four steps:

– Calculation of gain $A = v_o/v_i$ using a driving voltage from an ideal voltage source ($v_i = v_g$), and calculation of the transimpedance $R_T = v_o/i_i$ using a drive signal from an ideal current source ($i_i = i_g$). Since $v_i = i_i r_i$, we have $R_T = A r_i$; this is of importance since in the basic circuits we have only calculated A and r_i, and not R_T.
– Calculation of the short-circuit output voltage

$$v_{o,S} = A_{S,x} v_{r,x} \quad \text{or} \quad v_{o,S} = R_{S,x} i_{r,x} \qquad \text{for } v_i = 0$$

and the open-circuit output voltage

$$v_{o,O} = A_{O,x} v_{r,x} \quad \text{or} \quad v_{o,O} = R_{O,x} i_{r,x} \qquad \text{for } i_i = 0$$

for each noise source $v_{r,x}$ and/or $i_{r,x}$.
– Calculation of the equivalent noise voltage

$$v_{r,0x} = \frac{A_{S,x} v_{r,x}}{A} \quad \text{or} \quad v_{r,0x} = \frac{R_{S,x} i_{r,x}}{A}$$

and the equivalent noise current

$$i_{r,0x} = \frac{A_{O,x} v_{r,x}}{R_T} \quad \text{or} \quad i_{r,0x} = \frac{R_{O,x} i_{r,x}}{R_T}$$

for each noise source $v_{r,x}$ and/or $i_{r,x}$.
– Calculation of the noise densities of the equivalent noise sources:

$$v_{r,0} = \sum_x v_{r,0x} \Rightarrow |\underline{v}_{r,0}|^2 = \sum_x |\underline{v}_{r,0x}|^2$$

$$i_{r,0} = \sum_x i_{r,0x} \Rightarrow |\underline{i}_{r,0}|^2 = \sum_x |\underline{i}_{r,0x}|^2$$

This assumes that the noise sources $v_{r,x}$ and/or $i_{r,x}$ are independent; thus the equivalent noise sources $v_{r,0x}$ and/or $i_{r,0x}$ are also independent and the noise densities can be added.

Figure 4.164 shows the first three steps of this procedure, using the example of a noise current source $i_{r,x}$.

1: Calculation of gain A

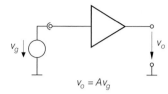

$$v_o = A v_g$$

1: Calculation of transimpedance R_T

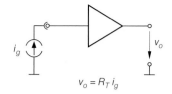

$$v_o = R_T i_g$$

2: Calculation of the short-circuit
    output voltage for noise source $i_{r,x}$

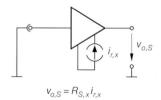

$$v_{o,S} = R_{S,x} i_{r,x}$$

2: Calculation of the open-circuit
    output voltage for noise source $i_{r,x}$

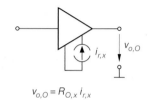

$$v_{o,O} = R_{O,x} i_{r,x}$$

3: Calculation of the equivalent
    noise voltage $v_{r,0x}$

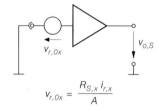

$$v_{r,0x} = \frac{R_{S,x} i_{r,x}}{A}$$

3: Calculation of the equivalent
    noise current $i_{r,0x}$

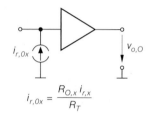

$$i_{r,0x} = \frac{R_{O,x} i_{r,x}}{R_T}$$

a Equivalent noise voltage source   **b** Equivalent noise current source

Fig. 4.164. Method for calculating the equivalent noise sources for a noise current source $i_{r,x}$

In general, every noise source contributes to the equivalent noise voltage source and the equivalent noise current source; therefore, strictly speaking, the equivalent noise sources are always dependent. However, the ratio of the values is mostly such that each noise source contributes significantly to only one equivalent noise source, while its contribution to the other equivalent noise source is negligibly low; thus the equivalent noise sources are practically independent.

Common-emitter circuit with current feedback: Figure 4.165a shows a common-emitter circuit with current feedback and resistances for setting the operating point. For $R_E = 0$, we obtain a common-emitter circuit without feedback; in other words, this situation is also included. To calculate the equivalent noise sources we use the small-signal equivalent circuit in Fig. 4.165b, which contains the noise sources of the transistors and the resistances. The collector–emitter resistance r_{CE} of the transistor can be disregarded and is therefore not shown. The base spreading resistance R_B of the transistor is also omitted in the small-signal equivalent circuit. However, it is contained in the noise voltage density $|v_{r,T}|^2$ and is thus included in the following calculations when the noise densities are

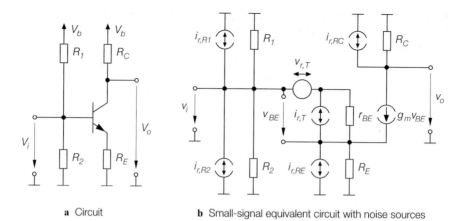

a Circuit   **b** Small-signal equivalent circuit with noise sources

Fig. 4.165. Common-emitter circuit with current feedback

inserted. Resistances R_1 and R_2 are combined to obtain $R_b = R_1 \parallel R_2$; this means that the noise currents $i_{r,R1}$ and $i_{r,R2}$ are also combined into one noise current $i_{r,Rb}$.

The gain and the transimpedance can be taken from (2.70) and (2.71); if we take the influence of R_b on the input resistance r_i into account, then:

$$A = -\frac{g_m R_C}{1 + g_m R_E}$$

$$R_T = A r_i \overset{r_i = R_b \parallel (r_{BE} + \beta R_E)}{=} -\frac{\beta R_C R_b}{R_b + r_{BE} + \beta R_E}$$

It is rather complex to calculate the short-circuit and the open-circuit voltages for the noise sources; here, we shall only give the result of the conversion to the input:

$$v_{r,0} = v_{r,T} + \left(i_{r,T} + i_{r,RE}\right) R_E + i_{r,RC}\left(R_E + \frac{1}{g_m}\right)$$

$$i_{r,0} = \frac{v_{r,T}}{R_b} + i_{r,T}\left(1 + \frac{R_E}{R_b}\right) + i_{r,Rb} + \frac{i_{r,RE} R_E}{R_b} + i_{r,RC}\left(\frac{1}{\beta} + \frac{R_E + 1/g_m}{R_b}\right)$$

The typical size ratios ($g_m R_C \gg 2$, $g_m R_b \gg 1/2$, $R_b \gg R_E$, $g_m R_E \ll 2\beta$) result in:

$$v_{r,0} \approx v_{r,T} + i_{r,RE} R_E$$

$$i_{r,0} \approx i_{r,T} + i_{r,Rb}$$

In this case the equivalent noise sources are independent, as none of the noise sources enters both equivalent noise sources. The noise source of the collector resistance R_C has no influence; only with very small values for R_C does it become noticeable. For the equivalent noise densities of the common-emitter circuit with current feedback, this leads to:

$$|\underline{v}_{r,0}|^2 \approx |\underline{v}_{r,T}|^2 + 4kT R_E = \frac{2kT}{g_m} + 4kT\left(R_B + R_E\right) \tag{4.204}$$

$$|\underline{i}_{r,0}|^2 \approx |\underline{i}_{r,T}|^2 + \frac{4kT}{R_b} \overset{R_b \gg 2r_{BE}}{\approx} |\underline{i}_{r,T}|^2 = \frac{2kT g_m}{\beta} \tag{4.205}$$

The current feedback primarily affects the equivalent noise voltage density. By way of contrast, the internal resistance R_b of the base voltage divider only influences the noise current density; this influence is negligible for $R_b \gg 2r_{BE}$. It should be noted that (4.204) uses the sum of the feedback resistance R_E and the base spreading resistance R_B; therefore, all noise equations for bipolar transistors can be used if we replace R_B with $R_B + R_E$. However, R_E is not usually an independent parameter, but is correlated to the transconductance via the loop gain $k_E = g_m R_E$; in this case, the calculation of the optimum quiescent current leads to:

$$
I_{C,A\,opt}(R_g) \;=\; \frac{V_T\sqrt{\beta}}{\sqrt{R_g^2 + R_B^2}}\sqrt{1+2k_E} \;\overset{\substack{\beta\approx100\\R_g>R_B}}{\approx}\; \frac{0.26\,\mathrm{V}}{R_g}\sqrt{1+2k_E}
$$

Due to the current feedback, the optimum noise figure increases to:

$$
F_{opt,T}(R_g) \;=\; 1+\frac{R_B}{R_g}+\frac{1}{\sqrt{\beta}}\sqrt{1+\left(\frac{R_B}{R_g}\right)^2}\sqrt{1+2k_E}
$$

Therefore, no current feedback is used if optimization is based on noise alone. In practice, it is often the amplitude limit, given by an permissible distortion factor or intermodulation ratio, that must be optimized. The potential gain in amplitude due to the linearizing effect of the current feedback may be higher than the loss caused by the increased noise. An example of this kind of application is the receiving amplifier of a mobile telephone, which may receive extremely varying input signals depending on the distance to the transmitter; here, it is necessary to *sacrifice* sensitivity in order to enable high input signals to be processed with low intermodulation.

Common-source circuit with current feedback: The equivalent noise densities of the common-source circuit with current feedback in Fig. 4.166 correspond to those of the common-emitter circuit with current feedback; but here the equivalent noise current density is determined by the gate voltage divider, since the noise current density of the MOSFET is so low that it can be ignored. If $R_b = R_1 \| R_2$ and the substrate transconductance is disregarded, then:

$$
|\underline{v}_{r,0}|^2 \;\approx\; |\underline{v}_{r,F}|^2 + 4kT\,R_S \;\overset{k_S=g_m R_S}{=}\; \frac{8kT}{3g_m}\left(1+\frac{3}{2}k_S\right) \tag{4.206}
$$

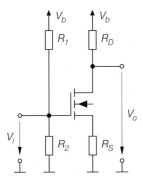

Fig. 4.166. Common-source circuit with current feedback

$$|\underline{i}_{r,0}|^2 \approx |\underline{i}_{r,F}(f)|^2 + \frac{4kT}{R_b} \approx \frac{4kT}{R_b} \tag{4.207}$$

Here, $k_S = g_m R_S$ is the loop gain. It is necessary to select R_b as high as possible to prevent the noise figure from declining considerably in the region of high source resistances.

In the common-source circuit direct optimization is not possible, as there are no opposed values; here, it is necessary to make g_m and R_b as high as possible and the loop gain k_S as low as possible. In LF applications the minimum for the noise figure is not very pronounced. If k_S is given by the permissible distortion factor, then the transconductance should be increased until the noise figure drops below the desired limit value. This may be achieved by using the lower limit $R_{g,lF}$ for the source resistance and taking into consideration the additional factor

$$1 + \frac{3}{2}k_S$$

from (4.206); if $R_g = R_{g,lF}$, then:

$$g_m = \frac{2}{3R_g\,(F-1)}\left(1 + \frac{3}{2}k_S\right)$$

With increasing frequencies the minimum becomes more pronounced; in this case, we use the equation for the optimum source resistance in wide-band applications – that is, $R_{gopt,F}[f_L,\,f_U]$ – and, likewise, take into consideration the additional factor:

$$g_m = \frac{f_T}{\sqrt{3}R_g\,f_U}\left(1 + \frac{3}{2}k_S\right)$$

Common-emitter circuit with voltage feedback: Figure 4.167 shows the common-emitter circuit with voltage feedback and the corresponding small-signal equivalent circuit with all noise sources. In this circuit the resistance R_1 shown in Fig. 2.66 on page 108 is omitted, since the source resistance R_g takes over the function of R_1; however, R_g belongs to the signal source and therefore does not contribute to the noise of the circuit. An *additional* resistance R_1 is undesirable, as it reduces the gain and increases the noise figure.

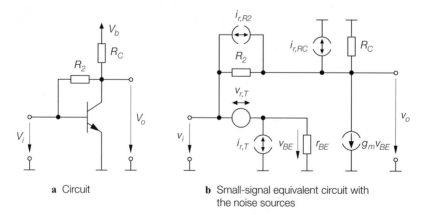

a Circuit

b Small-signal equivalent circuit with the noise sources

Fig. 4.167. Common-emitter circuit with voltage feedback

Calculation of the equivalent noise sources yields:

$$v_{r,0} = \frac{g_m R_2 v_{r,T} + R_2 \left(i_{r,R2} + i_{r,RC}\right)}{g_m R_2 - 1}$$

$$i_{r,0} = i_{r,T} + \frac{g_m v_{r,T} + g_m R_2 i_{r,R2} + \left(1 + \dfrac{R_2}{r_{BE}}\right) i_{r,RC}}{g_m R_2 - 1}$$

For typical size ratios ($g_m R_C \gg 2$ and $g_m R_2 \gg 2$), we obtain:

$$v_{r,0} \approx v_{r,T}$$

$$i_{r,0} \approx i_{r,T} + i_{r,R2}$$

In this case, the equivalent noise sources are independent. The noise source of the collector resistance R_C has no influence; only with very small values of R_C does it become noticeable. Thus the equivalent noise densities of the common-emitter circuit with voltage feedback are:

$$|\underline{v}_{r,0}|^2 \approx |\underline{v}_{r,T}|^2 = \frac{2kT}{g_m} + 4kT R_B \tag{4.208}$$

$$|\underline{i}_{r,0}|^2 \approx |\underline{i}_{r,T}|^2 + \frac{4kT}{R_2} = \frac{2kT g_m}{\beta} + \frac{4kT}{R_2} \stackrel{g_m R_2 \gg 2\beta}{\approx} \frac{2kT g_m}{\beta} \tag{4.209}$$

If $g_m R_2 \gg 2\beta$ or $R_2 \gg 2r_{BE}$, the noise current source of the feedback resistance R_2 is negligible; the noise densities of the circuit thus correspond to those of the transistor and the noise equations for the bipolar transistor can be used for optimization.

As in the case of the common-emitter circuit with current feedback, the criterion for selecting the feedback resistance R_2 is the permissible distortion factor. The voltage feedback, however, offers the advantage that the distortion factor can *also* be influenced by the collector resistance R_C without affecting the noise figure. From the optimum collector current

$$I_{C,A\,opt} \stackrel{R_g > R_B}{\approx} \frac{V_T \sqrt{\beta}}{R_g}$$

the relationship $R_g = \sqrt{\beta}/g_m$ is obtained; by inserting it into the equation for the distortion factor of the common-emitter circuit with voltage feedback, and taking into consideration the fact that $R_2 \gg R_1$, $R_1 = R_g$ and $\hat{v}_i = \hat{v}_g$, we obtain:[37]

$$k \approx \frac{\hat{v}_g}{4\beta V_T} \left(1 + \frac{R_2}{R_C}\right)^2$$

Therefore, it is possible to meet the demand for a minimum noise figure – that is, $R_2 \gg 2r_{BE}$ – and at the same time achieve a low distortion factor if the chosen collector resistance R_C is sufficiently high; the use of a low-noise current source instead of R_C is therefore optimum. As a load connected to the output is parallel to R_C for small signals, the circuit must be followed by an impedance converter; that is, a common-collector circuit (see Fig. 4.168a). Here, the feedback signal is picked up at the output of the common-collector circuit.

---

[37] Here, the parameters v_g and R_g correspond to v_i and R_1 of the common-emitter circuit with voltage feedback in Sect. 2.4.1.

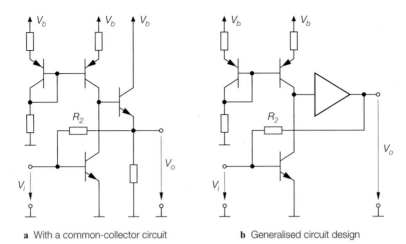

a With a common-collector circuit   **b** Generalised circuit design

Fig. 4.168. A practical design for a low-noise common-emitter circuit with voltage feedback

Figure 4.168b shows the generalized circuit design of a low-noise common-emitter circuit with voltage feedback, in which the common-collector circuit is replaced by an amplifier with a high input and a low output resistance. This circuit is of great importance in practice, since it features minimum noise and low distortions, and allows easy and stable adjustment of the operating point. It is ideal for low and medium source resistances, and is only outmatched by the corresponding circuit with MOSFETs, where high source resistances are involved. An important field of use is optical detectors for fibre-optic transmission systems. Even though photodetector diodes are high-resistance elements, the operating frequencies are so high that the noise current densities of bipolar transistors and MOSFETs are approximately the same; bipolar transistors are preferred because of the higher transconductance. A comparative overview of this and other optical broadband receiver circuits is contained in reference [4.9].

The common-source circuit with voltage feedback: The equivalent noise densities of the common-source circuit with voltage feedback correspond to those of the common-emitter circuit with voltage feedback; but here the equivalent noise current density is determined by the feedback resistance, since the noise current density of the MOSFET is negligibly low:

$$|\underline{v}_{r,0}|^2 \approx |\underline{v}_{r,F}|^2 = \frac{8kT}{3g_m} \tag{4.210}$$

$$|\underline{i}_{r,0}|^2 \approx |\underline{i}_{r,F}(f)|^2 + \frac{4kT}{R_2} \approx \frac{4kT}{R_2} \tag{4.211}$$

The practical design is shown in Fig. 4.168; here, however, it is not essential that an impedance converter in the form of a common-drain circuit follows, as a common-source circuit also features a high input resistance. This circuit is ideal for high-resistance sources provided that the upper cutoff frequency remains far below the transit frequency. This circuit is preferred in optical receivers for frequencies around 10 MHz; Fig. 4.169 shows the corresponding circuit. The feedback resistance R_2, together with the capacitance C_D of the photodiode and the gate–source capacitance C_{GS}, form a lowpass filter that limits the bandwidth; therefore, in practical applications R_2 must be chosen according to the

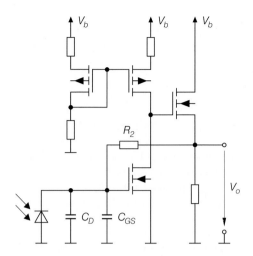

Fig. 4.169. A practical design for an optical detector circuit with a photodiode

required bandwidth. Since photodiodes are high-resistance components, only the noise current density has an influence; by means of the sensitivity of the diode, the noise current density is converted to a corresponding illuminance, known as the *noise equivalent power* (NEP) [4.7].

Common-collector and common-drain circuits: Figure 4.170 shows the basic design of both a common-collector and a common-drain circuit used as impedance converters. The noise sources of the common-collector circuit are:

$$v_{r,0} = \frac{g_m R_E v_{r,T} + R_E \left(i_{r,T} + i_{r,RE} \right)}{g_m R_E + 1} \overset{g_m R_E \gg 1}{\approx} v_{r,T} + \frac{i_{r,T} + i_{r,RE}}{g_m}$$

$$i_{r,0} = i_{r,T} + \frac{i_{r,RE}}{\beta}$$

If $g_m R_E \gg 2$ and $\beta \gg 1$, then the equivalent noise densities are:

$$|\underline{v}_{r,0}|^2 \approx |\underline{v}_{r,T}|^2 = \frac{2kT}{g_m} + 4kT R_B \tag{4.212}$$

$$|\underline{i}_{r,0}|^2 \approx |\underline{i}_{r,T}|^2 = \frac{2kT g_m}{\beta} \tag{4.213}$$

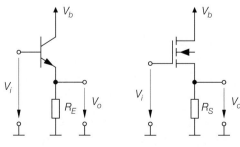

a Common-collector circuit

b Common-drain circuit

Fig. 4.170. Impedance converters

Accordingly, for the common-drain circuit we obtain:

$$|\underline{v}_{r,0}|^2 \approx |\underline{v}_{r,F}|^2 = \frac{8kT}{3g_m} \tag{4.214}$$

$$|\underline{i}_{r,0}|^2 \approx |\underline{i}_{r,F}[f_L, f_U]|^2 = 8kT g_m \left(\frac{f_U}{f_T}\right)^2 \tag{4.215}$$

The equivalent noise densities in both circuits are approximately equal to those of the respective transistor.

Although the external circuitry of a bipolar transistor or MOSFET used as an impedance converter has almost no influence on the equivalent noise densities, both of these circuits are used as input stages in low-noise amplifiers in exceptional cases only, because they have no gain and the noise densities of the subsequent stages thus fully affect the input as well.

Common-base and common-gate circuits: Figure 4.171 shows a common-base and a common-gate circuit with AC coupling, which is typical for high-frequency applications. For AC signals a short-circuited base or gate voltage divider – or, as shown in Fig. 4.171, a negative supply voltage – is used to adjust the quiescent current. The noise sources of the common-base circuit are:

$$v_{r,0} = v_{r,T} + \frac{i_{r,RC}}{g_m}$$

$$i_{r,0} = \frac{v_{r,T}}{R_E} + i_{r,T} + i_{r,RE} + \frac{g_m R_E + 1}{g_m R_E} i_{r,RC}$$

If $g_m R_C \gg 2$ and $g_m R_E \gg 1$, then:

$$|\underline{v}_{r,0}|^2 \approx |\underline{v}_{r,T}|^2 = \frac{2kT}{g_m} + 4kT R_B \tag{4.216}$$

$$|\underline{i}_{r,0}|^2 \approx |\underline{i}_{r,T}|^2 + \frac{4kT}{R_E} + \frac{4kT}{R_C} = 4kT \left(\frac{1}{2r_{BE}} + \frac{1}{R_E} + \frac{1}{R_C}\right) \tag{4.217}$$

Unlike in the common-emitter circuit, here the collector resistance R_C also makes a contribution, as no attenuation by the factor β occurs in the conversion from $i_{r,RC}$ to the emitter. When using a voltage divider at the base, its internal resistance R_b acts like an

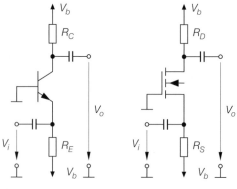

a Common-base circuit   **b** Common-gate circuit

Fig. 4.171. Common-base and common-gate circuits

additional base spreading resistance, so that the equivalent noise voltage density increases accordingly; the same applies to the resistance R_{BV} in Sect. 2.4.3.

The corresponding values for the common-gate circuit are:

$$|\underline{v}_{r,0}|^2 \approx |\underline{v}_{r,F}|^2 = \frac{8kT}{3g_m} \tag{4.218}$$

$$|\underline{i}_{r,0}|^2 \approx |\underline{i}_{r,F}(f)|^2 + \frac{4kT}{R_S} + \frac{4kT}{R_D} \approx 4kT\left(\frac{1}{R_S} + \frac{1}{R_D}\right) \tag{4.219}$$

The noise current density of the MOSFET is negligible.

Current source: For a current source, the noise current density at the output is of interest; it should be so low that the noise figure of the circuit to which the current source is introduced shows no or very little increase. Usually, a current source is used instead of a high-value resistor; for example, a collector or drain resistor. In the small-signal equivalent circuit of a circuit, this is shown by the output resistance r_o and a noise current source; the noise current density $|\underline{i}_{o,r}|^2$ is significantly greater than that of a corresponding ohmic resistor:

$$|\underline{i}_{o,r}|^2 \gg \frac{4kT}{r_o}$$

Therefore, a current source may drastically increase the noise figure of a circuit, even if an ohmic resistance at the same location had no notable influence.

Figure 4.172 shows the circuit and the small-signal equivalent circuit of a current source based on a simple current mirror with current feedback. A strict analysis yields the result

$$i_o = \frac{\beta}{r_{BE\,2} + \beta R_2 + r_i}\left[i_{r,RV}r_i + g_{m1}r_i\,\frac{v_{r,T1} + i_{r,R1}R_1}{1 + g_{m1}R_1} + v_{r,T2}\right.$$

$$\left. + i_{r,T2}\,(r_i + R_2) + i_{r,R2}R_2\right]$$

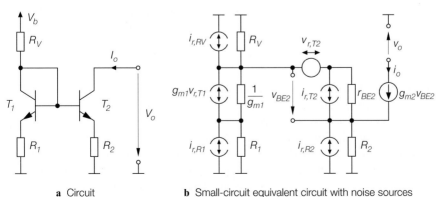

a Circuit   **b** Small-circuit equivalent circuit with noise sources

Fig. 4.172. A simple current mirror used as a current source

where

$$r_i = \frac{R_V (1 + g_{m1} R_1)}{1 + g_{m1} (R_V + R_1)}$$

is the internal resistance of the left branch. Here, our observations are restricted to the case of cross-symmetrical transistors and resistors with the current ratio:

$$k_I = \frac{g_{m2}}{g_{m1}} = \frac{R_1}{R_2}$$

Furthermore, we assume that $R_V \gg 1/g_{m1} + R_1$; thus $r_i \approx 1/g_{m1} + R_1$ and:

$$i_o \approx \frac{\beta}{\beta + k_I} \frac{g_{m2}}{1 + g_{m2} R_2} \left[v_{r,T1} + v_{r,T2} + \left(k_I i_{r,R1} + i_{r,R2} + (1 + k_I) i_{r,T2} \right) R_2 \right]$$

In general, the current ratio is much smaller than the current gain: $k_I \ll \beta$. Consequently:

$$|\underline{i}_{o,r}|^2 \approx \left(\frac{g_{m2}}{1 + g_{m2} R_2} \right)^2 \left[(1 + k_I) \left(|\underline{v}_{r,T2}|^2 + 4kT R_2 \right) + (1 + k_I)^2 |\underline{i}_{r,T2}|^2 R_2^2 \right]$$

An approximation for this expression can be reached by examining the extreme situations without feedback ($R_2 = 0$) and with high feedback ($g_{m2} R_2 \gg 1$); this provides accurate results for the extreme situations and only deviates very slightly from the exact value in the region $g_{m2} R_2 \approx 1$:

$$|\underline{i}_{o,r}|^2 \approx \frac{1 + k_I}{\dfrac{1}{g_{m2}^2 |\underline{v}_{r,T2}|^2} + \dfrac{1}{|\underline{i}_{r,R2}|^2}} + (1 + k_I)^2 |\underline{i}_{r,T2}|^2$$

$$\approx \frac{4kT (1 + k_I)}{\dfrac{2}{g_{m2} (1 + 2g_{m2} R_{B2})} + R_2} + (1 + k_I)^2 \frac{2kT g_{m2}}{\beta} \qquad (4.220)$$

The noise current density declines with increasing feedback, since the denominator of the first term increases with R_2; the lower limit is determined by the second term. For practical applications, a current ratio of $k_I = 1$ is selected in order to prevent an unnecessary increase in the noise current density. A ratio of $k_I < 1$ may be selected for highly demanding applications.

For $k_I \leq 1$, both terms in (4.220) have approximately the same size for $g_{m2} R_2 \approx \beta$, which means that a further increase in R_2 has very little effect; the voltage drop across R_2 is:

$$I_{C2} R_2 = g_{m2} R_2 V_T \overset{g_{m2} R_2 \approx \beta}{\approx} \beta V_T \overset{\beta \approx 100}{\approx} 2.6 \,\text{V}$$

In practice, it is seldom possible to select the voltage drop across R_2 to be so high that the lower limit is reached; therefore, especially in circuits with very low supply voltages, no low-noise current sources can be realized.

Without feedback or with low feedback, the noise voltage densities of the transistors are of dominating influence. In this case, large transistors can be used to keep the base spreading resistance, and thus the noise voltage densities, low; however, this causes an increase in the output capacitance. With medium and high feedback, the influence of the noise voltage densities of the transistors is low; the transistor size can then be chosen according to the normal scaling.

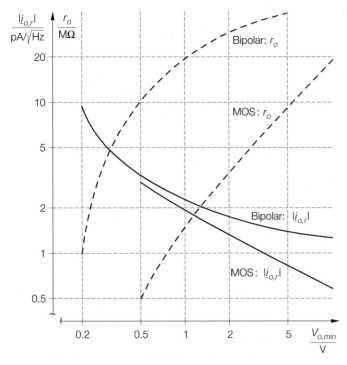

Fig. 4.173. Noise current densities and output resistances of a bipolar and a MOS current source with $I_o = 100\,\mu\text{A}$ as a function of the output voltage limit $V_{o,min}$ ($V_{CE,sat} = 0.2\,\text{V}$, $V_{DS,po} = 0.5\,\text{V}$)

For a current source with MOSFETs, we obtain:

$$|\underline{i}_{o,r}|^2 \approx (1+k_I)\,\frac{4kT}{\dfrac{3}{2g_{m2}}+R_2} + (1+k_I)^2\,8kT\,g_{m2}\left(\frac{f_U}{f_T}\right)^2$$

$$\overset{f_U \ll f_T}{\approx} (1+k_I)\,\frac{4kT}{\dfrac{3}{2g_{m2}}+R_2} \tag{4.221}$$

With $f_U \ll f_T$ there is practically no lower limit since, on account of the low transconductance of the MOSFET, the feedback resistance R_2 must be extremely high so that the terms reach the same value; the voltage drop across R_2 is then unacceptably high.

A comparison of the noise current densities of a bipolar and a MOS current source only makes sense on the basis of identical output voltage limits $V_{o,min}$. Figure 4.173 shows a comparison of the noise current densities and the output resistances of a bipolar and a MOS current source with $I_o = 100\,\mu\text{A}$ for the transistors in Figs. 4.5 and 4.6; it is assumed that $V_{CE,sat} = 0.2\,\text{V}$ and $V_{DS,po} = V_{GS} - V_{th} = 0.5\,\text{V}$. The output voltage limits of the MOS current source can be further reduced by increasing the MOSFET size, although to halve $V_{DS,po}$ the MOSFET has to be quadrupled in size. Thus, in the range of $V_{o,min} \approx 0.5\ldots2\,\text{V}$, which is of interest for practical applications, the noise current density of the MOS current source is only slightly lower than that of the bipolar current

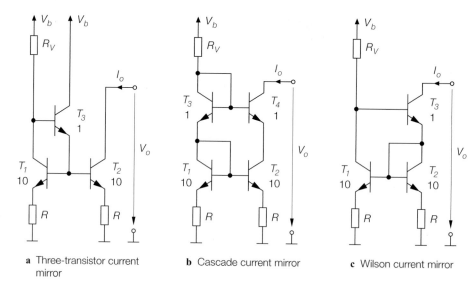

a Three-transistor current mirror

b Cascade current mirror

c Wilson current mirror

Fig. 4.174. Other current sources

source; this is the case even after the MOSFETs have been changed in size, which means that in terms of noise the MOS current source is always slightly better. The bipolar current source, however, has a significantly higher output resistance; for this reason it is always to be preferred in situations that require a high output resistance.

Figure 4.174 shows other current sources with bipolar transistors, on the basis of three-transistor, cascode and Wilson current mirrors. The noise current densities differ only in terms of their lower limit at high feedback: for the three-transistor current mirror this limit is lower than in the current source with a simple current mirror, while in the other two current mirrors the limit is higher than in the current source with a simple current mirror. The differences are small and of no significance in practice, as such high feedbacks can only seldom be used in practice. Without feedback or with low feedback, the noise current density again depends on the size of transistors T_1 and T_2, while the sizes of transistors T_3 and T_4 have no noticeable influence. For this reason, one can optimize the current sources with cascode or Wilson current mirrors without feedback by selecting large transistors for T_1 and T_2 to minimize the noise current density and keeping the other transistors small to minimize the output capacitance (see Fig. 4.174b,c). With the same output voltage limit, the output resistance of the current source with cascode or Wilson current mirrors is higher than for the current source with a simple current mirror; however, the noise current density is also higher. Therefore, one must check which parameter is more important for the given application.

Common-emitter and common-source circuits with a current source: Figure 4.175 shows the circuit design and small-signal equivalent circuit of a common-emitter circuit with current source; $i_{o,r}$ is the noise current source of the current source and r_o is the output resistance of the circuit. One can see that the source $i_{o,r}$ affects the output in the same way as the controlled source $g_{m3}v_{BE3}$ of transistor T_3; therefore, its current can be converted to an equivalent input voltage by means of transconductance g_{m3} and to an equivalent input current by means of current gain β_3. For the equivalent noise densities of the circuit, it

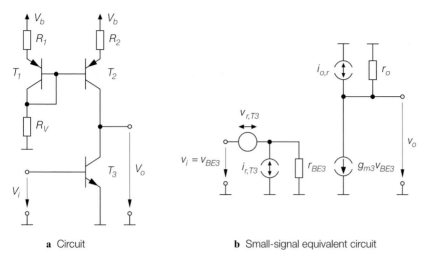

a Circuit   **b** Small-signal equivalent circuit

Fig. 4.175. Common-emitter circuit with a current source

follows:

$$|\underline{v}_{r,0}|^2 = |\underline{v}_{r,T3}|^2 + \frac{|\underline{i}_{o,r}|^2}{g_{m3}^2} \stackrel{g_{m3}R_2 \gg 2+2k_I}{\approx} |\underline{v}_{r,T3}|^2 \tag{4.222}$$

$$|\underline{i}_{r,0}|^2 = |\underline{i}_{r,T3}|^2 + \frac{|\underline{i}_{o,r}|^2}{\beta_3^2} \approx |\underline{i}_{r,T3}|^2 \tag{4.223}$$

The equivalent noise current density is hardly increased by the current source; this is also the case without current feedback ($R_1 = R_2 = 0$). In contrast, a current feedback with $g_{m3}R_2 \gg 2 + 2k_I$ is required in order to prevent the equivalent noise voltage density from increasing significantly; without current feedback ($R_1 = R_2 = 0$) and under the assumption that the base spreading resistances R_{B2} and R_{B3} are equal, it follows:

$$|\underline{v}_{r,0}|^2 \approx (2 + k_I)\,|\underline{v}_{r,T3}|^2 \stackrel{k_I=1}{=} 3\,|\underline{v}_{r,T3}|^2$$

This also applies to the common-source circuit.

If the common-emitter or common-source circuit already includes current feedback through resistance R_E or R_S, then the reduced transconductance

$$g_{m,red} = \frac{g_m}{1 + g_m R_E} \quad \text{or} \quad g_{m,red} = \frac{g_m}{1 + g_m R_S}$$

must be used to convert the noise current density of the current source; this increases its influence. However, the noise voltage density is also increased by resistances R_E and R_S; here, the noise of the current source for $R_2 \gg (1 + k_I)R_E$ or $R_2 \gg (1 + k_I)R_S$ is negligible.

Common-collector and common-drain circuits with a current source: Here, the same conditions apply as for the common-emitter and common-source circuits; that is, the equivalent noise current density is hardly increased. However, current feedback of the current source is required in order to maintain the equivalent noise voltage density. In practice, however, current sources without current feedback are usually used, since common-collector

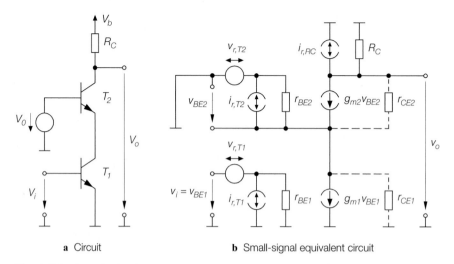

Fig. 4.176. Cascode circuit

and common-drain circuits are generally used as impedance converters if high source resistances are involved; in this case, the noise figure depends primarily on the equivalent noise current density, so that an increase in the equivalent noise voltage density by a factor of three has almost no effect on the noise figure.

Cascode circuit: Figure 4.176 shows the circuit and the small-signal equivalent circuit of a cascode circuit with bipolar transistors. Since the small-signal characteristics of cascode and common-emitter circuits are essentially the same, one can use the equations for the common-emitter circuit; in particular, the noise of the collector resistance R_C can be disregarded. The noise sources of transistor T_2 are also negligible:

- The noise voltage source $v_{r,T2}$ has virtually no effect, as the current is impressed by T_1 and there is very little change in the current distribution due to the collector–emitter resistances r_{CE1} and r_{CE2}.
- The noise current source $i_{r,T2}$ affects the output in the same way as source $g_{m1}v_{BE\,1}$ and is therefore converted to the input by means of current gain β; this can be neglected compared to $i_{r,T1}$.

For the cascode circuit, it follows:

$$|\underline{v}_{r,0}|^2 \approx |\underline{v}_{r,T1}|^2 \tag{4.224}$$

$$|\underline{i}_{r,0}|^2 \approx |\underline{i}_{r,T1}|^2 \tag{4.225}$$

For versions with current feedback or a base voltage divider, the equations of the corresponding common-emitter circuit can be used. Similarly, the equations for the common-source circuit apply to the cascode circuit with MOSFETs.

Differential amplifier: Figure 4.177 shows the circuit design and the small-signal equivalent circuit of a differential amplifier with bipolar transistors and a resistor for setting the quiescent current. For the sake of clarity, the small-signal equivalent circuit of the transistors is shown in the form of a block diagram.

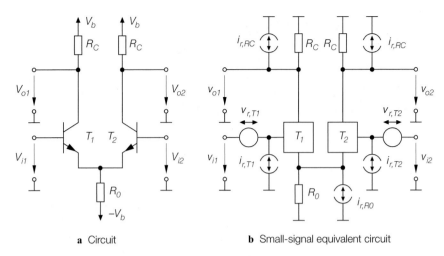

a Circuit   **b** Small-signal equivalent circuit

Fig. 4.177. Differential amplifier

The differential amplifier can be reduced to a common-emitter circuit; this method has already been utilized in Sect. 4.1.3. It follows that the noise of the collector resistances can also be neglected in the case of the differential amplifier. The two noise voltage sources $v_{r,T1}$ and $v_{r,T2}$ are combined to form one equivalent noise voltage source

$$|\underline{v}_{r,0}|^2 = |\underline{v}_{r,T1}|^2 + |\underline{v}_{r,T2}|^2 \tag{4.226}$$

which is arranged in front of one of the two inputs. This is possible regardless of the external circuitry, as the noise voltage sources enter directly into the equivalent noise source. By way of contrast, the contribution of the noise current sources depends on the source resistances of the two inputs; this is the reason why – provided that the transistors are equal – an equivalent noise current source with

$$|\underline{i}_{r,01}|^2 = |\underline{i}_{r,02}|^2 = |\underline{i}_{r,T1}|^2 = |\underline{i}_{r,T2}|^2 \tag{4.227}$$

is connected to *both* inputs. The influence of the noise current source $i_{r,R0}$ depends on the circuitry at the output and is discussed separately; in most cases it can be neglected.

The noise equivalent circuit shown in Fig. 4.178 is obtained with two signal sources; for the equivalent noise source, it thus follows:

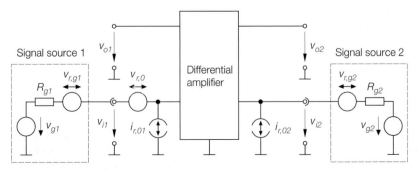

Fig. 4.178. Noise equivalent circuit of a differential amplifier

$$v_r = v_{r,g1} + v_{r,g2} + v_{r,0} + i_{r,01}R_{g1} + i_{r,02}R_{g2}$$

and for the noise figure:

$$F = \frac{|\underline{v}_r|^2}{|\underline{v}_{r,g1}|^2 + |\underline{v}_{r,g2}|^2} = 1 + \frac{|\underline{v}_{r,0}|^2 + R_{g1}^2\,|\underline{i}_{r,01}|^2 + R_{g2}^2\,|\underline{i}_{r,02}|^2}{4kT\left(R_{g1} + R_{g2}\right)} \tag{4.228}$$

Two modes of operation are very often found in practical applications:

- Symmetrical operation with $R_{g1} = R_{g2} = R_g$:

$$F = 1 + \frac{|\underline{v}_{r,0}|^2 + 2R_g^2\,|\underline{i}_{r,01}|^2}{8kT\,R_g} = 1 + \frac{|\underline{v}_{r,T1}|^2 + R_g^2\,|\underline{i}_{r,T1}|^2}{4kT\,R_g}$$

  This mode corresponds to the common-emitter circuit. The noise densities of the differential amplifier may be higher by a factor of 2 due the use of two transistors, however, this is compensated by the fact that the noise density of both signal sources is also increased by a factor of 2.

- Asymmetrical operation with $R_{g1} = R_g$ and $R_{g2} = 0$:

$$F = 1 + \frac{|\underline{v}_{r,0}|^2 + R_g^2\,|\underline{i}_{r,01}|^2}{4kT\,R_g} = 1 + \frac{2\,|\underline{v}_{r,T1}|^2 + R_g^2\,|\underline{i}_{r,T1}|^2}{4kT\,R_g}$$

  In this mode the noise voltage density is higher than in the common-emitter circuit by a factor of 2; the optimum source resistance and the noise figure thus increase accordingly.

Any direct comparison of the differential amplifier and the common-emitter circuit must be based on *one* single signal source with source resistance R_g which is used to drive the common-emitter circuit in the same way as the differential amplifier in symmetrical and in asymmetrical operation. In the case of asymmetrical operation with double the noise voltage density, the comparison can be made directly. However, in the case of symmetrical operation, the comparison must be made with $R_{g1} = R_{g2} = R_g/2$; here, both noise densities are increased by a factor of 2. As a consequence, the common-emitter circuit is most favorable, followed by the differential amplifier in asymmetric mode, while the differential amplifier with symmetric operation is the least favorable solution.

In asymmetrical mode, the input that is not used is not normally connected directly to the small-signal ground, but via a resistance equivalent to the source resistance R_g; this compensates the voltage drop caused by the base current of the transistors. In low-noise circuits, this resistance must be short-circuited for small signals by a capacitance connected in parallel, so that the noise current source at this input has no influence. In practice, this may sometimes cause undesirable oscillations; this can be prevented by connecting a small resistance of $10 \ldots 100\,\Omega$ in series, which is not short-circuited by the capacitance.

The noise current source $i_{r,R0}$ or the noise current source of a current source used instead of R_0 has the same effect on both outputs of the differential amplifier, as its current is divided equally to the transistors according to the quiescent current; thus this has no influence if the output signal is processed differentially. When using only one output, only half the noise current, and thus a quarter of the noise current density, is effective; but in an asymmetrical output with a current mirror according to Fig. 4.66 the full noise current is effective. In practice, the influence of $i_{r,R0}$ can be disregarded, as R_0 is usually much higher than the transconductance resistance of the transistors: $g_m R_0 \gg 1$. The same applies to the use of a current source with feedback if the feedback resistance meets the same condition.

All of the above considerations apply equally well to a differential amplifier with MOSFETs.

Mean Noise Figure and the Signal-to-Noise Ratio

The use of the *spectral noise figure* $F(f)$ allows us to calculate the *mean noise figure* F. This indicates the loss in signal-to-noise ratio (SNR) caused by the amplifier in an application-specific frequency interval $f_L < f < f_U$. Sometimes $\overline{F}$ is used as a symbol for the mean noise figure to prevent any confusion with the spectral noise figure. We use $F(f)$ for the spectral noise figure and F for the mean noise figure.

Signal-to-noise ratio: The SNR is the ratio of the useful signal power to the noise power:

$$SNR = \frac{P_{us}}{P_n} \tag{4.229}$$

P_n is the noise power in the frequency interval $f_L < f < f_U$. Since the power of a signal is proportional to the square of its effective value (the root-mean-square value), the signal-to-noise ratio of the signal generator is:

$$SNR_g = \frac{v_{geff}^2}{v_{r,geff}^2} = \frac{v_{geff}^2}{\int_{f_L}^{f_U} |\underline{v}_{r,g}(f)|^2 df} \tag{4.230}$$

The amplifier increases the noise density by the spectral noise figure $F(f)$; the signal-to-noise ratio at the amplifier input is thus

$$SNR_i = \frac{v_{geff}^2}{v_{reff}^2} = \frac{v_{geff}^2}{\int_{f_L}^{f_U} |\underline{v}_r(f)|^2 df} = \frac{v_{geff}^2}{\int_{f_L}^{f_U} F(f) |\underline{v}_{r,g}(f)|^2 df} \tag{4.231}$$

which is lower than the signal-to-noise ratio of the signal generator.

Mean noise figure: The mean noise figure corresponds to the relation of the signal-to-noise ratios [2.9]:

$$F = \frac{SNR_g}{SNR_i} = \frac{\int_{f_L}^{f_U} F(f) |\underline{v}_{r,g}(f)|^2 df}{\int_{f_L}^{f_U} |\underline{v}_{r,g}(f)|^2 df} \tag{4.232}$$

Usually, it is quoted in decibels (dB):

$$F_{dB} = 10 \log F$$

If $SNR_{i,dB} = 10 \log SNR_i$ and $SNR_{g,dB} = 10 \log SNR_g$, then:

$$F_{dB} = SNR_{g,dB} - SNR_{i,dB}$$

When operated with an ideal signal generator with the frequency-independent noise density $|\underline{v}_{r,g}(f)|^2 = 4kTR_g$, this expression can be moved in front of the integral, so that:

$$F = \frac{1}{f_U - f_L} \int_{f_L}^{f_U} F(f) df \tag{4.233}$$

In this case the mean noise figure F is obtained by averaging across the spectral noise figure $F(f)$. Often, $F(f)$ is constant in the given frequency interval; thus $F = F(f)$, which is generally referred to as *the* noise figure F.

Use of weighting filters: In some applications a *weighting filter* with the transfer function $\underline{H}_B(s)$ is used for determining the noise power; the signal-to-noise ratios are then calculated with the *weighted noise density:*

$$|\underline{v}_{r(B),g}(f)|^2 = |\underline{H}_B(j2\pi f)|^2 |\underline{v}_{r,g}(f)|^2$$

Thus:

$$SNR_{B,g} = \frac{v_{geff}^2}{\displaystyle\int_{f_L}^{f_U} |\underline{H}_B(j2\pi f)|^2 |\underline{v}_{r,g}(f)|^2 df}$$

$$SNR_{B,i} = \frac{v_{geff}^2}{\displaystyle\int_{f_L}^{f_U} F(f) |\underline{H}_B(j2\pi f)|^2 |\underline{v}_{r,g}(f)|^2 df}$$

This is used if the noise causes stronger interference in certain ranges of the given frequency interval than in other ranges. The weighting filter, whose absolute frequency response is proportional to the disturbing effect of the noise, provides more meaningful values for the SNRs. A *weighted noise figure* can also be introduced, but this is not common.

A typical use for a weighting filter is to determine the signal-to-noise ratio of an audio amplifier. Such amplifiers usually operate in the frequency range $20\,Hz < f < 20\,kHz$. Since the human ear is particularly sensitive to noise in the range of $1\,kHz < f < 4\,kHz$, a weighting filter is used that enhances this range and suppresses other frequency ranges. This filter is known as an *A filter*; the SNR is thus quoted in *decibel* A or dBA.

Bandwidth of the amplifier: The bandwidth of an amplifier must at least cover the given frequency interval of the useful signal $f_L < f < f_U$ to provide equal amplification; this means that the noise is also equally amplified in this range. In practice, the bandwidth is usually wider than required; that is, the amplifier also amplifies the ranges $f < f_L$ and $f > f_U$ with its operating gain $\underline{A}_B(s)$. These ranges do not contain any useful signal, but only the noise of the signal generator and the amplifier. This means that *without any limitation of the frequency range* the amplifier provides the noise power at its output:

$$P_{n,o} = \int_0^\infty |\underline{A}_B(j2\pi f)|^2 |\underline{v}_r(f)|^2 df$$

$$= \int_0^\infty |\underline{A}_B(j2\pi f)|^2 F(f) |\underline{v}_{r,g}(f)|^2 df \tag{4.234}$$

As the useful signal is amplified with the useful gain $\underline{A}_{B,us}$, which is assumed to be constant in the range $f_L < f < f_U$, the signal-to-noise ratio at the amplifer output without limitation of the frequency range is:

$$SNR_o = \frac{|\underline{A}_{B,us}|^2 v_{geff}^2}{P_{n,o}} = \frac{|\underline{A}_{B,us}|^2 v_{geff}^2}{\displaystyle\int_0^\infty |\underline{A}_B(j2\pi f)|^2 F(f) |\underline{v}_{r,g}(f)|^2 df} \tag{4.235}$$

This is lower than the SNR obtained using (4.231), since the total noise is taken into account, rather than just the portion in the range $f_L < f < f_U$.

In practice, the noise power $P_{n,o}$ plays a major role, as it may be considerably higher than the useful power if the bandwidth of the amplifier is correspondingly large; as a

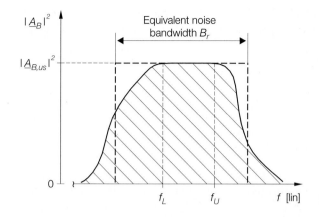

Fig. 4.179. Equivalent noise bandwidth of an amplifier

consequence, the following components of the signal processing chain are primarily driven, and in some cases even overdriven, by the amplified noise.

The signal-to-noise ratio SNR_o is only of importance if the noise outside the range $f_L < f < f_U$ is transmitted to the output of the signal processing chain and actually causes interference here. In this event, the weighting filter comes into play: the signal-to-noise ratio SNR_i is obtained using (4.231) by introducing an additional ideal bandpass filter with the lower cutoff frequency f_L and the upper cutoff frequency f_U into (4.235).

Equivalent noise bandwidth: If the noise densities of the signal generator and the amplifier are almost constant within the transfer bandwidth, which means that the noise figures are also almost constant, then the noise power at the amplifier output is:

$$P_{n,o} \approx F \, |\underline{v}_{r,g}|^2 \int_0^\infty |\underline{A}_B(j2\pi f)|^2 \, df \; = \; F \, |\underline{v}_{r,g}|^2 \, |\underline{A}_{B,us}|^2 \, B_r$$

The bandwidth

$$B_r \;=\; \frac{\displaystyle\int_0^\infty |\underline{A}_B(j2\pi f)|^2 \, df}{|\underline{A}_{B,us}|^2} \tag{4.236}$$

is known as the *equivalent noise bandwidth*. It indicates the bandwidth of an ideal filter with a gain of $\underline{A}_{B,us}$ which has the same noise power $P_{n,o}$ as the amplifier. This means that the area under the plot of the squared magnitude $|\underline{A}_B(j2\pi f)|^2$ is replaced by a rectangle of the same area with height $|\underline{A}_{B,us}|^2$ and width B_r (see Fig. 4.179).

In the following example, we will calculate the equivalent noise bandwidth of an amplifier with a transfer function that corresponds to a lowpass filter of first order:

$$\underline{A}_B(s) \;=\; \frac{A_0}{1 + \dfrac{s}{\omega_g}} \qquad \Rightarrow \qquad |\underline{A}_B(j2\pi f)|^2 \;=\; \frac{A_0^2}{1 + \left(\dfrac{f}{f_g}\right)^2}$$

For $|\underline{A}_{B,us}|^2 = A_0^2$, it follows:

$$B_r \;=\; \int_0^\infty \frac{1}{1 + \left(\dfrac{f}{f_g}\right)^2} \, df \;=\; \left[f_g \arctan \frac{f}{f_g} \right]_0^\infty \;=\; \frac{\pi}{2} \, f_g \approx 1.57 \cdot f_g$$

Order	B_r/f_{-3dB}	
	Multiple pole	Butterworth
1	$\pi/2 = 1.57$	$\pi/2 = 1.57$
2	1.22	1.11
3	1.15	1.05
4	1.13	1.03
5	1.11	1.02

Fig. 4.180. Equivalent noise bandwidth B_r for lowpass filters

The equivalent noise bandwidth of a first-order lowpass filter is higher than the cutoff frequency f_g by a factor of 1.57, which in this case corresponds to the 3 dB cutoff frequency f_{-3dB}. Figure 4.180 includes the factors for lowpass filters of higher orders for cases with multiple poles and for a Butterworth characteristic with a maximum flat magnitude frequency response. In this and most other cases found in practice, the equivalent noise bandwidth is larger than the 3 dB bandwidth; however, it may also be smaller.

The equivalent noise bandwidth is not often used in practical calculations; instead, the 3 dB bandwidth is used and the slight error that occurs especially with higher orders is accepted. For measuring noise densities, however, the equivalent noise bandwidth is of high importance; the frequency range of interest *is scanned* with a bandpass filter of narrow bandwidth and, with the help of the equivalent noise bandwidth, the noise power at the output is converted into the noise density to be determined.

Chapter 5:
Operational Amplifiers

An operational amplifier is a multi-stage DC-coupled amplifier realized as integrated circuit. It is available as an individual component or as a library component for the design of larger integrated circuits. There is essentially no difference between a normal amplifier and an operational amplifier. Both are used to provide voltage or power gain. However, whereas the characteristics of a normal amplifier are governed by its internal design, an operational amplifier is constructed such that its mode of operation can be determined primarily by external feedback circuitry. In order to make this possible, operational amplifiers are designed as direct voltage-coupled amplifiers with a high gain. To render any additional measures for setting the operating point unnecessary, the input and output potential must be 0 V. As a rule, this requires two operating voltage sources: one positive and one negative. Amplifiers of this kind were at one time used exclusively in analog computers and for performing mathematical operations such as addition and integration – hence the term "operational amplifier."

5.1
General

Operational amplifiers are available in a wide variety of monolithic integrated circuits and differ little from discrete transistors in terms of size and price. As their characteristics are ideal in many respects, they are much easier to use than discrete transistors. The advantage of the classical operational amplifier is its high accuracy at low frequencies. But for many applications it is too slow. For this reason, versions with a modified architecture were developed that feature desirable high-frequency characteristics. Today, there are virtually no fields of application in which discrete transistors offer advantages. The detailed description of the internal design of operational amplifiers provided in this chapter is intended only to explain certain properties and characteristics of integrated circuits. Nowadays, circuit design on a transistor level is only needed for the development of specific integrated circuits.

Today, a vast selection of operational amplifiers is available; they differ not only in their data but also in their basic construction. Four families can be distinguished, whose internal design and effect on the parameters will be described in the sections below. Section 5.6 compares the four different versions in order to establish their common features and their differences.

Circuit calculations are performed on models that are explained in respect to the internal design. It is, of course, not possible to explain every single transistor, since this would render the circuit analysis too complicated. Instead, macromodels that allow an easy interpretation of the entire circuit are used. Depending on the effect to be analyzed, only the relevant section of the circuit is modeled in more detail. In many cases, the calculation of operational amplifier circuits is so simple that the quickest way to do it is by hand. With the aid of macromodels, the behavior of a circuit can be studied in more detail using simulation programs such as *PSpice*. This approach provides information on the capability of the

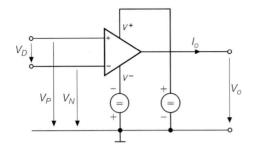

Fig. 5.1. Terminals of an operational amplifier

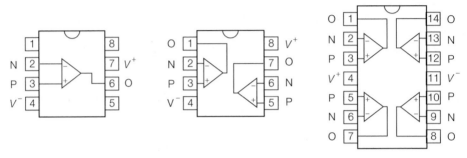

Fig. 5.2. Typical terminal configuration of operational amplifiers in dual inline cases, seen from top

circuit as early as in the design phase. The hardware construction is not carried out until the simulation results are satisfactory.

Figure 5.1 shows the graphic symbol for the operational amplifier. It has two inputs – one inverting and the other noninverting – and one output.

The ideal operational amplifier only amplifies the differential voltage $V_D = V_P - V_N$ between the input terminals. The noninverting input is called the P-input and is identified by the plus (+) sign in the graphic symbol; the inverting input is called the N-input and is identified by the minus (−) sign. For the power supply of the operational amplifier, there are two supply voltage terminals which provide operating voltages that are positive and negative in respect to ground in order to allow quiescent input and output potentials of 0 V. The operational amplifier has no ground terminal, even though the input and output voltages are related to ground. Typical supply voltages are ±15 V for general applications; but today's tendency is to use voltages of ±5 V and there is a trend to reduce the voltages further. Typical pin-outs for operational amplifiers are illustrated in Fig. 5.2. As many circuits require more than one operational amplifier, there are dual and quadruple operational amplifiers on the market to save space and money.

5.1.1
Types of Operational Amplifier

As shown in Fig. 5.3, there are four different types of operational amplifier. The differences are the high-resistive or low-resistance inputs and outputs. All four types have high-resistive noninverting inputs.

In the *standard operational amplifier* (the voltage feedback operational amplifier) the inverting input also has high resistance, which means that it is voltage controlled. Its output

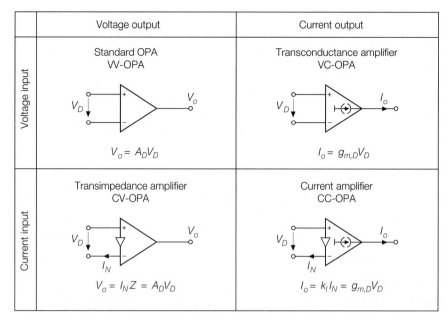

Fig. 5.3. Graphic symbols and transfer equations for the four types of operational amplifier

acts like a voltage source with a low internal resistance; that is, a low-resistance output. For this reason, standard operational amplifiers are also known as VV operational amplifiers. The first V (voltage) indicates the voltage control at the (inverting) input, while the second V indicates the voltage source at the output. At one time, only this type existed; even today, it is still the most important type and it holds the highest market share. The output voltage

$$V_o = A_D V_D = A_D(V_P - N_N) \tag{5.1}$$

is identical to the amplified input voltage difference; A_D is the differential gain. Values of $A_D = 10^4 ... 10^6$ are necessary to create circuits with high negative feedback. The transfer characteristics of an *ideal* VV operational amplifier are shown in Fig. 5.4a. The differential gain

$$A_D = \left. \frac{dV_o}{dV_D} \right|_b \tag{5.2}$$

corresponds to the slope of the curve in the operating or bias point. It is obvious that fractions of 1 mV are sufficient to obtain the maximum output swing. The linear operating range $V_{o,min} < V_o < V_{o,max}$ is called the output voltage swing. After this limit is reached, an increase in V_D causes no further increase in V_o; that is, the amplifier is overdriven. The literature often specifies an ideal operational amplifier that features a differential gain of $A_D = \infty$; here, we do not wish to share this point of view, since it makes a thorough understanding more difficult.

Unlike the normal operational amplifier, however, the operational *transconductance amplifier* also has high-resistance output. This acts like a current source, whereby the current is controlled by the input voltage difference V_D. Therefore, its graphic symbol shows a current source symbol at the output (see Fig. 5.3). This is an operational amplifier

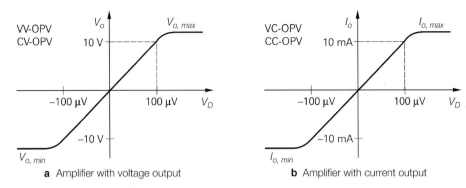

a Amplifier with voltage output   **b** Amplifier with current output

Fig. 5.4. Transfer characteristics of operational amplifiers

with a voltage-controlled inverting input and an output that acts like a current source. The transconductance amplifier is thus also known as a VC-OPA. The output current

$$I_o = g_{m,D}V_D = g_{m,D}(V_P - V_N) \tag{5.3}$$

is proportional to the input voltage difference. The differential transconductance

$$g_{m,D} = \left.\frac{dI_o}{dV_D}\right|_b \tag{5.4}$$

indicates the degree of output current increase with a rising input voltage. The differential transconductance generally corresponds to the transconductance of a transistor, and here too a transistor is used to determine the transconductance. The term "transconductance amplifier" is based on the fact that the transconductance $g_{m,D}$ governs the behavior of the amplifier. Figure 5.4b shows the typical transfer characteristic of a VC operational amplifier. It can be seen that small voltage differences are sufficient to achieve the maximum output current swing.

The two operational amplifiers with current input in Fig. 5.3 have a low-resistance inverting input; that is, they are current controlled. At first glance this seems to be a drawback, but for high frequencies it provides major advantages, as we will see below, since:

– The internal signal path is shortened and the oscillation tendency reduced.
– The gain of the OPA can be adapted to the specific demand.

The *transimpedance amplifier* (current feedback amplifier) shown in Fig. 5.3 has a current-controlled inverting input and a voltage source at the output; it is thus a CV-OPA. The output voltage

$$V_o = A_D V_D = I_N Z \tag{5.5}$$

can be calculated either from the differential gain – as with the standard OPA – or from the input current I_N and an internal impedance Z in the megaohm range. Due to this characteristic impedance Z, the CV-OPA is also known as the transimpedance amplifier.

The *current amplifier* (diamond transistor, drive-R amplifier) has a current-controlled input like the CV-OPA and a current-controlled output like the VC-OPA. Therefore, it is known as a CC-OPA. The transfer response

$$I_o = g_{m,D}V_D = k_I I_N \tag{5.6}$$

is determined by the transconductance. However, it is usually easier to use the current transfer factor

$$k_I = \left. \frac{dI_o}{dI_N} \right|_b \tag{5.7}$$

for calculations, because it is in the range $k_I = 1 \ldots 10$ depending on the type. The current amplifier is also known as a diamond transistor (with the trade name of Burr Brown) as in many respects it behaves like an ideal transistor (as will be shown in Sect. 5.5).

5.1.2
Principle of Negative Feedback

The VV operational amplifier is used as an example to explain the principle of negative feedback. An operational amplifier with negative feedback can be regarded as a closed-control circuit for which the rules of automatic control engineering apply. Figure 5.5 shows the design of a general closed-control circuit. The target value is derived from the reference variable using the reference variable shaper, given here as multiplication by k_F. The actual value is derived from the output variable, using the controller given here, as multiplication by k_R. In the controlled system, the difference between the target and the actual value is multiplied by A_D. The equation for the system deviation,

$$V_D = k_F V_e - k_R V_i$$

leads to the definitions

$$k_F = \left. \frac{V_D}{V_i} \right|_{V_o=0} \quad \text{and} \quad k_R = -\left. \frac{V_D}{V_a} \right|_{V_i=0} \tag{5.8}$$

The gain of the control loop in Fig. 5.5 can be calculated from $V_o = A_D V_D$ and (5.1.2):

$$A = \frac{V_o}{V_i} = \frac{k_F A_D}{1 + k_R A_D} \overset{k_R A_D \gg 1}{\approx} \frac{k_F}{k_R} \tag{5.9}$$

In an operational amplifier circuit, the operational amplifier represents the controlled system. The reference variable shaper and the controller are formed by the operational amplifier's external circuitry. The subtraction is achieved either by the differential input of the differential amplifier or by the external circuitry.

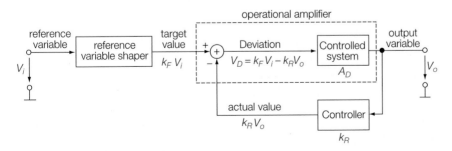

Fig. 5.5. General control loop

Noninverting Amplifier

In the general control loop shown in Fig. 5.5, if the setpoint value is identical to the reference variable and the controller is a voltage divider, then the circuit is a noninverting amplifier, as shown in Fig. 5.6. For a qualitative investigation of the transient response, the input voltage is changed from zero to a positive value V_i. In the first instant the output voltage, and hence the feedback voltage, is still zero. Therefore, the voltage at the amplifier input is $V_D = V_i$. Since this voltage is amplified with the high differential gain A_D, V_o and thus the feedback voltage $k_R V_o$ rapidly assume positive values; this leads to a reduction in V_D. The fact that a change in the output voltage counteracts a change in the input voltage is typical of the negative feedback principle. This suggests that a stable condition will ultimately be reached.

In order to calculate the steady state values, it is assumed that the output voltage rises until it reaches the same level as the amplified input voltage difference:

$$V_o = A_D V_D = A_D (V_P - k_R V_o)$$

This leads to the *voltage gain*

$$A = \frac{V_o}{V_i} = \frac{A_D}{1 + k_R A_D} = \begin{cases} \dfrac{1}{k_R} & \text{for } k_R A_D \gg 1 \\ A_D & \text{for } k_R A_D \ll 1 \end{cases} \tag{5.10}$$

where

$$\boxed{g = k_R A_D} \tag{5.11}$$

represents the *loop gain*. If the loop gain $g \gg 1$, then the 1 in the denominator of (5.10) can be disregarded, thus giving the gain of the feedback circuit:

$$A = \frac{V_o}{V_i} = \frac{1}{k_R} = 1 + \frac{R_N}{R_1} \tag{5.12}$$

In this case it is determined by the external circuitry only and not by the amplifier. This approximation can be derived directly from the circuit because a large loop gain results in $V_D = 0$; that is, $V_N = V_i$. For the negative feedback voltage divider, it follows:

$$V_i = \frac{R_1}{R_1 + R_N} V_o \quad \Rightarrow \quad A = \frac{V_o}{V_i} = 1 + \frac{R_N}{R_1}$$

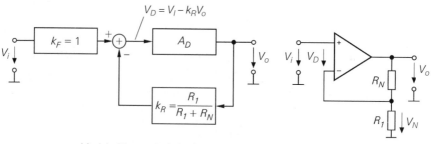

$$V_D = V_I - k_R V_o$$

a Model of the control circuit   b Noninverting amplifier

Fig. 5.6. Control engineering model of the noninverting amplifier, using the VV-OPA as an example

This leads to the most important rule for calculating operational amplifier circuits:

> *The output voltage of an operational amplifier takes on a value at which the input voltage difference is zero.*

A condition for this is that the loop gain is high and that true negative feedback takes place, with no positive feedback; otherwise a Schmitt trigger occurs, as described in Sect. 6.5.2 on page 601. If the loop gain $g \ll 1$, then $A = A_D$ as seen in (5.10); in this case, the gain is not affected by the negative feedback.

A useful method for calculating the loop gain results from (5.11) and (5.12) if $g \gg 1$:

$$g = k_R A_D = \frac{A_D}{A} \tag{5.13}$$

In order to prevent the error caused by the approximation in (5.12) from exceeding $1^0\!/_{00}$, a loop gain of $g = 1000$ is required. If the gain of the negative feedback circuit is to be $A = 100$, the required differential gain can be calculated from (5.13): $A_D = gA = 1000 \cdot 100 = 10^5$. This explains why the differential gain of the operational amplifier must be as high as possible. Four different gains must be distinguished:

A_D the open-loop gain of the amplifier
A the closed-loop gain of the amplifier
g the loop gain, $g = A_D/A$
k_R the feedback factor β

The literature sometimes specifies an additional gain – noise gain, which is the inverse value of the feedback factor; this represents the gain determined by the external circuitry. The loop gain may also be exemplified: we make $V_i = 0$ and interrupt the loop at the input of the external circuitry as shown in Fig. 5.7a; then we introduce a test signal V_S at this point and measure the signal at the other side of the interruption; in other words, at the amplifier output. As can be seen in Fig. 5.6,

$$V_o = -k_R A_D V_S = -g V_S \tag{5.14}$$

The test signal is amplified with gain $g = k_R A_D$ while passing through the open loop. The loop can also be interrupted at the inverting input and a test signal introduced there (see Fig. 5.7b). In this case, the gain is initially A_D, followed by k_R; the loop gain is once again $g = k_R A_D$.

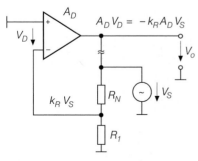

a Interruption at the output   **b** Interruption at the input

Fig. 5.7. Illustration of loop gain

The loop gain can also be measured in a closed loop. To this end, a voltage V_i is fed into the input of the circuit and measured as V_N and V_D according to Fig. 5.6b. The ratio of these two voltages represents the loop gain:

$$\frac{V_N}{V_D} = \frac{k_R V_o}{V_D} = \frac{k_R V_o}{V_o/A_D} = k_R A_D = g \tag{5.15}$$

Inverting Amplifier

In addition to the circuitry shown in Fig. 5.6, there is a second fundamental method of making an operational amplifier to an amplifier with negative feedback. Here, the feedback must, of course, come from the output toward the *inverting* input too, in order to achieve negative and no positive feedback. However, it is also possible to supply the input voltage to the base point of the feedback voltage divider instead of the noninverting input. This leads to the circuit shown in Fig. 5.8. Inserting k_F and k_R into (5.9) results in

$$A = \frac{V_o}{V_i} = \frac{-k_F A_D}{1 + k_R A_D} = \frac{-\dfrac{R_N}{R_1 + R_N} A_D}{1 + \dfrac{R_1}{R_1 + R_N} A_D} \overset{k_R A_D \gg 1}{\approx} -\frac{R_N}{R_1} \tag{5.16}$$

This means that this is an inverting amplifier. This can also be seen in the circuit itself when a positive input voltage is applied. Since this voltage reaches the inverting input via R_1, the output voltage becomes negative. In the ideal operational amplifier with $A_D = \infty$, the output voltage becomes negative until $V_D = 0$; thus, this is also called a *virtual ground*. In order to obtain the output voltage, the nodal equation is applied to the inverting input, leading to

$$\sum_{I=0} = \frac{V_i}{R_1} + \frac{V_o}{R_N} = 0$$

This equation can directly be solved for V_o:

$$V_o = -\frac{R_N}{R_1} V_i \quad \Rightarrow \quad A = \frac{V_o}{V_i} = -\frac{R_N}{R_1}$$

Compared to the noninverting amplifier in Fig. 5.5, the voltage gain here is negative and its value is reduced by 1. Of course, it is also possible to calculate the gain of the circuit shown

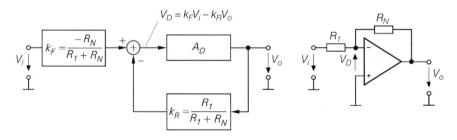

$$V_D = k_F V_i - k_R V_o$$

a Model of the control circuit

b Inverting amplifier

Fig. 5.8. Circuitry of an operational amplifier as an inverting amplifier, using the VV-OPA as an example. The values shown for k_F and k_R are in accordance with the definitions in (5.8)

in Fig. 5.8, for a finite differential gain A_D. In this case, one must take into consideration the fact that $V_D \neq 0$.

From

$$\frac{V_E + V_D}{R_1} + \frac{V_o + V_D}{R_N} = 0$$

and $V_o = A_D V_D$, it follows:

$$A = \frac{V_o}{V_i} = -\frac{R_N A_D}{R_1 A_D + R_N + R_1} = k_F \frac{A_D}{1 + k_R A_D} \tag{5.17}$$

The loop gain determines the deviation from the ideal performance. For $g = k_R A_D \gg 1$ it follows:

$$A = \frac{k_F A_D}{k_R A_D} = \frac{k_F}{k_R} = \frac{-R_N/(R_1 + R_N)}{R_1/(R_1 + R_N)} = -\frac{R_N}{R_1} \tag{5.18}$$

In the simplest configuration, the external circuitry consists of a voltage divider only, as shown in Figs. 5.6 and 5.8. The use of an RC network produces an integrator, a differentiator, or an active filter. Nonlinear components such as diodes can also be used in the external circuitry to form exponential functions or logarithms. These applications are described in Sect. 11.7 on page 739. The explanations here are limited to the simplest ohmic feedback.

5.2
Normal Operational Amplifier (VV-OPA)

This section explains the various ways to design an operational amplifier so that the user understands the practical consequences that result from the internal design. The purpose is not to recommend building operational amplifiers from discrete transistors and resistors. This would not only result in greater costs, but also clearly poorer data.

An amplifier that is to be used as an operational amplifier must meet several requirements, which determine its internal design. Nowadays, as shown in Fig. 5.3, there are four different types of operational amplifier on the market, each of which comes in several variations depending on the given field of application. All types comply with the following general requirements:

– Direct voltage coupling.
– Differential input.
– Zero input and output voltage operating point

Operational amplifiers can be built with bipolar or field effect transistors, or with a combination of both. The following description is generally based on bipolar transistors. Differential amplifiers are mostly used for the input stage, because this compensates for both the base–emitter voltages and the temperature sensitivity.

The use of npn transistors for the amplification renders the output potential of the amplifier stage positive to the input potential. To reduce the output operating point to zero, it is essential to shift the potential downwards in a stage of the amplifier. This can be achieved by two fundamentally different methods, which are shown in Fig. 5.9.

– Zener diodes, as shown in Fig. 5.9a, cause virtually no attenuation of the desired signal, due to their low dynamic internal resistance. However, sufficient current must flow

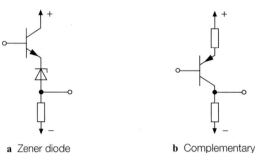

a Zener diode   **b** Complementary
                        transistors   **Fig. 5.9.** Methods for potential shift

through the Zener diode to insure that its noise does not reach too high a level. Consequently, these devices should only be used after emitter followers. Another disadvantage
is their fixed potential shift, which does not adapt itself to the operating voltage. However, it is advantageous that only npn transistors are required, which makes this method
particularly suitable for high-frequency amplifiers.

– Complementary transistors, as shown in Fig. 5.9b, represent the easiest and most elegant method of compensating the potential shift of the amplifier stage with that of the
subsequent stage. In most cases pnp transistors are used in the current mirror configuration, as shown in Fig. 4.66 on page 342. A disadvantage here, however, is the fact that
the transit frequencies of pnp transistors in integrated circuits are often clearly poorer.
Only the more modern, more complex production processes enable the manufacture of
equivalent pnp transistors.

5.2.1
Principle

VV operational amplifiers have a voltage-controlled (in other words, high-resistance) input
and a low-resistance output. Therefore, it is convenient to use a differential amplifier at
the input and an emitter follower at the output. This creates the simplest VV operational
amplifier, as shown in Fig. 5.10. The circuit was only extended by an additional Zener diode
at the output to bring the quiescent output potential down to 0 V. Operational amplifiers
should meet the following three requirements:

– Common-mode voltage: almost up to the operating voltages.
– Output voltage swing: almost up to the operating voltages.
– Differential gain: as large as possible: $A_D = 10^4 \ldots 10^6$.

In Fig. 5.10, the positive limit of the common-mode voltage (see Sect. 4.1.3 on
page 333) is reached at $V_N = V_P = V_{Gl} = 7.5\,\mathrm{V}$, because otherwise the collector–
base diode of T_2 becomes conductive. The negative limit is given by the current source I_0.
If one assumes a minimum voltage drop of 1 V, the emitter potential of the differential amplifier may drop to $-14\,\mathrm{V}$. This results in a minimum common-mode voltage of $-13.4\,\mathrm{V}$.
Consequently, the limits of the common-mode voltage can be expressed in the inequality:
$-13.4\,\mathrm{V} < V_{Gl} < +7.5\,\mathrm{V}$.

The positive limit of the output voltage is reached when the transistor T_2 becomes
nonconductive; the base potential of the emitter follower then rises to 15 V and the output

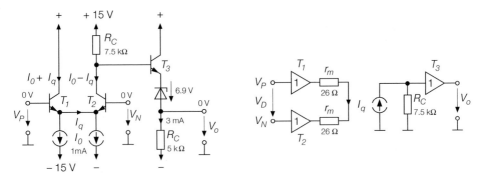

Fig. 5.10. Simple operational amplifier with dimensioning examples, and the resulting quiescent voltages and quiescent currents. The emitter current is divided into two parts to allow the current difference I_q to be indicated. In this and the following circuits, it is just as good to use a common emitter current of $I_k = 2I_0$

voltage to $+7.5$ V. The lower control limit is determined by T_2, because its collector potential cannot fall below 0 V, since otherwise the collector–base diode becomes conductive. At this point the related output voltage is -7.5 V. For the output swing, this leads to the inequality -7.5 V $< V_o < +7.5$ V. This range becomes even more unfavorable when a positive common-mode voltage is applied. For $V_{GI} = 5$ V, the negative output swing is even limited to -2.5 V.

The model for the differential gain of the operational amplifier also shown in Fig. 5.10 is equal to that of the differential amplifier if one takes into consideration the fact that the emitter follower features a voltage gain of almost 1 and the Zener diode causes no attenuation. Therefore, we can use the model shown in Fig. 5.10 for calculating the voltage gain. If a differential input voltage V_D is applied, this voltage also lies between the transconductance resistances of the input transistors, producing a current

$$I_q = \frac{V_D}{2\,r_m} = \frac{1}{2}\frac{I_C}{V_T}\,V_D = \frac{1}{2}\frac{1\,\text{mA}}{26\,\text{mV}}\,V_D = 19\frac{\text{mA}}{\text{V}}\,V_D$$

At the collector resistance R_C this current causes a voltage change that is identical to the output voltage, due to the almost constant potential shift between collector of T_2 and output:

$$V_o = I_q R_C = \frac{1}{2}\frac{I_C R_C}{V_T}\,V_D = \frac{V_{RC}}{2\,V_T}\,V_D = \frac{7.5\,\text{V}}{2\cdot 26\,\text{mV}}\,V_D = 144\cdot V_D$$

One can see that the common-mode and the output voltage range, as well as the differential gain of the circuit, are not even close to reaching the desired values; the operational amplifier shown in Fig. 5.10 therefore needs to be improved in all respects. A marked improvement is achieved by replacing the Zener diode used for potential shift with a current mirror made of pnp transistors. Figure 5.11 illustrates this variation. Here, the common-mode control is clearly improved, as the collector potential of T_2 is close to the positive operating voltage: -13.4 V $< V_{GI} < +14.4$ V. At the same time the output swing is improved, since the emitter follower at the output allows a swing close to the positive and negative operating voltages. If a minimum voltage drop of 1 V is required for the current sources, then we achieve -14 V $< V_o < +13.8$ V.

The differential gain of the circuit shown in Fig. 5.11 can easily be calculated if one takes into account the fact that the current mirror T_2, T_3 only reverses the direction of the

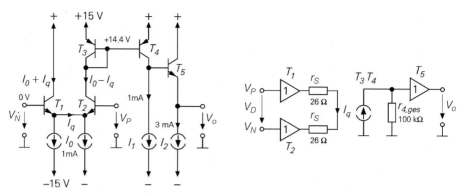

Fig. 5.11. Operational amplifier with a current mirror for potential shift

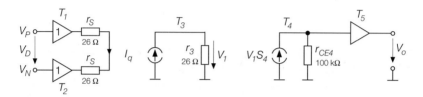

Fig. 5.12. Model of the amplifier shown in Fig. 5.11, with two amplifier stages

collector current of T_2. The necessary load resistance consists of the parallel arrangement of all resistances $r_{4,ges}$ connected to the collector of T_4. In the event of ideal current sources only r_{CE4} needs to be considered, because the input resistance of the emitter follower is infinitely high in the no-load condition:

$$A_D = \frac{r_{CE4}}{2\, r_S} = \frac{1}{2} \frac{I_{C2}}{V_T} \frac{V_A}{I_{C2}} = \frac{1}{2} \mu = \frac{1}{2} \frac{100\,\text{V}}{26\,\text{mV}} = 1923 \qquad (5.19)$$

The differential gain can, however, also be calculated on the basis of two amplifier stages, as illustrated in Fig. 5.12. The first stage is formed by the differential amplifier, with the transdiode T_3 as collector resistance. It has the gain

$$A_2 = \frac{r_3}{2\, r_S} = \frac{1}{2} \frac{I_{C2}}{V_T} \frac{V_T}{I_{C3}} = \frac{1}{2} \qquad (5.20)$$

since both collector currents are equal. The second amplifier stage is formed by transistor T_4, operated in common-emitter configuration with the voltage gain

$$A_4 = S_4 r_{CE4} = \frac{I_{C4}}{V_T} \frac{V_A}{I_{C4}} = \mu = \frac{100\,\text{V}}{26\,\text{mV}} = 3846 \qquad (5.21)$$

assuming that the internal resistances of the current sources are infinite. Thus, in accordance with (5.19), the differential gain of the entire operational amplifier is $A_D = 1923$.

5.2.2
Multipurpose Amplifiers

The operational amplifier shown in Fig. 5.11 is far from being an ideal multi-purpose operational amplifier because of its gain of only $A_D \approx 2000$. There are two methods for substantially increasing the voltage gain:

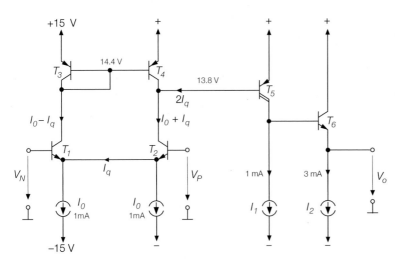

Fig. 5.13. Operational amplifier with two stages for voltage amplification. The values shown are examples of quiescent potentials and quiescent currents

- By increasing the internal resistance at the collector of T_4. This node is the high-impedance point of the circuit, which means that it is the point of the highest ohmic resistance in the signal path. Its impedance determines the voltage gain and the cutoff frequency of the circuit. The internal resistance of this high-impedance point can be increased by using cascode circuits; this method is used in the wide-band amplifiers described in Sect. 5.2.6.
- By amplifying the voltage in two stages. This method is used in the multipurpose amplifiers that are outlined below.

As we have seen, the circuit in Fig. 5.11 can be regarded as an amplifier with two amplifier stages. Due to its low collector resistance, the differential amplifier with the transdiode load T_3 only provides a gain of $1/2$. This means one must increase the collector resistance of T_2 in order to enhance the voltage gain. In Fig. 5.13, the current source T_4 has been introduced for this purpose. It is very useful to complement this current source with T_3 to make it a current mirror; in this way the current changes in T_1 are used to double the current changes at the output of the differential amplifier and thus the differential gain. An even more important advantage of the current mirror is the fact that the quiescent current of T_4 always has the appropriate value, regardless of the quiescent current I_0. The tolerances of I_0 thus have no influence on the offset voltage of the differential amplifier. This makes it easier to integrate the circuit. To prevent the second amplifier stage from compromising the high internal resistance at the collector of T_2, a Darlington configuration with its high input resistance must be used for T_5, as described in Sect. 2.4.4 on page 159.

Most integrated multipurpose amplifiers are based on the principle shown in Fig. 5.13. However, the input differential amplifier is realized by means of several npn and pnp transistors, which act together like a pnp differential amplifier. In this case, an npn transistor must be used in the second stage for potential shift. In Fig. 5.14, one can see that this leads to a circuit that is exactly complementary to that in Fig. 5.13. In integrated circuits much smaller quiescent currents are used. The collector currents of the differential amplifier are only $10\,\mu A$. In integrated operational amplifiers the final stage is always designed as a

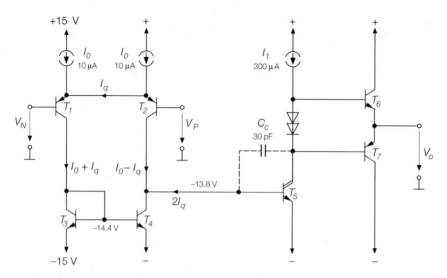

Fig. 5.14. Operational amplifier of the 741 class. This circuit illustrates the principle only; due to technological restrictions, the differential amplifier is made up of an array of transistors. Capacitance C_c serves to correct the frequency response; its effect is described in Sect. 5.2.7. Current $2I_q$ is not the base current of T_5, but the signal current that determines the voltage gain at this point

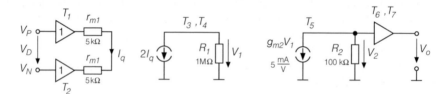

Fig. 5.15. Model of an operational amplifier of the 741 class

complementary emitter follower, with the aim of maintaining positive and negative output currents that are high compared to the quiescent current.

The model shown in Fig. 5.15 can be used to calculate the differential gain of the operational amplifier. The transistors T_1 and T_2 of the input differential amplifier are represented by voltage followers. The emitters are connected via the transconductance resistances $r_{m1} = 1/g_{m1}$. The current I_q indicates how much the current through one transistor is increased and the current through the other transistor is decreased when an input signal is applied: $I_q = V_D/2r_{m1}$. This current reaches the output of the differential amplifier through the current mirror and produces a voltage across the existing internal resistance. This voltage can be determined:

$$V_1 = -2I_q R_1 = -2R_1 \frac{V_D}{2r_{m1}} = -\frac{1\,\text{M}\Omega}{5\,\text{k}\Omega} V_D = -200 \cdot V_D$$

For the parameters shown in the model, the differential amplifier has a voltage gain of $A_1 = V_1/V_D = -200$ and the transconductance

$$g_{m1} = \frac{2I_o}{V_D} = \frac{1}{r_{m1}} = 0.2\frac{\text{mA}}{\text{V}} .$$

The Darlington circuit T_5 amplifies the voltage V_1 and provides the output current $g_{m2}V_1$ that produces a voltage drop across the internal resistance R_2. This voltage can be determined as:

$$V_2 = -g_{m2}V_1 R_2 = -5\frac{\text{mA}}{\text{V}} \cdot 100k\,\Omega \cdot V_1 = -500 \cdot V_1$$

For the parameters shown in the model, the second amplifier stage has a gain of $A_5 = -500$. Under the assumption that the emitter follower has a voltage gain of 1 at its output, we obtain an overall gain for the model of

$$A_D = A_1 A_2 = (-200) \cdot (-500) = 10^5$$

5.2.3
Operating Voltages

So far, we have used a *symmetrical* operating voltage of ±15 V. Normal operational amplifiers such as those described here then have a common-mode control range and an output swing of approximately ±13 V. This is shown in Fig. 5.16a. The limitation results from the internal construction with a minimum voltage drop of 2 V. Of course, 15 V can be added to both operating voltages without the operational amplifier noticing this, since there is no ground connection. This case is illustrated in Fig. 5.16b. Here, the operational amplifier can be operated with *one single* supply voltage. This, however, causes a 15 V shift in positive direction in the common-mode control range and the output swing, so that the quiescent input and output potentials of 0 V can no longer be reached; this leads to $2\,V < V_{CM}, V_o < 28$ V. This causes the operational amplifier to lose an important property that renders its application so simple: zero quiescent input and output potential. This can be remedied by generating an additional positive auxiliary potential of $+15$ V as a reference for all other voltages;[1] but this still makes a second voltage source necessary, resulting in a number of disadvantages. If, however, it is known from the start that no negative common mode and output voltages will occur, one can drive the operational amplifier with an *asymmetrical* operating voltage in order to increase the positive control range. For the example shown in Fig. 5.16c, this results in a control voltage of $-1\,V < V_{CM}$, $V_o < +25$ V.

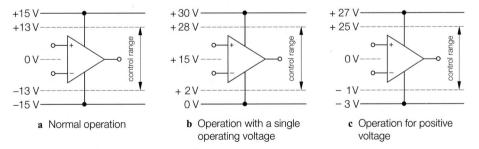

a Normal operation   **b** Operation with a single operating voltage   **c** Operation for positive voltage

Fig. 5.16. Influence of the operating voltages on the common-mode control range and the output swing.

---

[1] A circuit designed for this purpose is the rail-splitter TLE2426, from Texas Instruments.

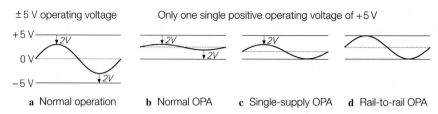

±5 V operating voltage   Only one single positive operating voltage of +5 V

a Normal operation   **b** Normal OPA   **c** Single-supply OPA   **d** Rail-to-rail OPA

Fig. 5.17. Control range of operational amplifiers when driven with low operating voltages

Operational amplifiers intended for a nominal operating voltage of ±15 V can also usually be driven with ±5 V, but then the control range is reduced to ±3 V – as shown in Fig. 5.17a – if one again assumes a minimum voltage drop of 2 V. There is a growing desire to supply operational amplifiers with a single operating voltage of not more than +5 V, or even +3.3 V, since in most cases this voltage is already available for the supply of digital circuits. Most multipurpose amplifiers are not specified to operate at such low voltages. Even if they were operated at +5 V, they would be of little use because the control range would be reduced to $2\,\mathrm{V} < V_{CM}, V_o < 3\,\mathrm{V}$, as shown in Fig. 5.17b. For this reason, *single-supply amplifiers* have been developed, whose common-mode control range and output swing include the negative operating voltage, as shown in Fig. 5.17c. Even with a negative operating voltage of 0 V, they allow quiescent input and output potentials as low as 0 V. Indeed, there are operational amplifiers whose common-mode control range and output swing reach to both the negative and the positive operating voltage. Such amplifiers are known as *rail-to-rail* amplifiers; their control range is shown in Fig. 5.17d.

5.2.4
Single-Supply Amplifiers

The classic single-supply amplifier is model LM324, the basic design of which is shown in Fig. 5.18. This circuit is related to the multipurpose amplifier shown in Fig. 5.14, but it features some modifications to allow a control range down to negative operating voltages:

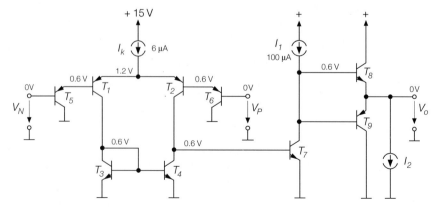

Fig. 5.18. Basic circuit of the single-supply amplifier LM324. The potentials shown apply to a control voltage equal to the negative operating voltage; in other words, zero potential in this case

- The emitter followers T_5 and T_6 have been added to shift the emitter potential of the differential amplifier upward by 0.6 V. This keeps the collector–emitter voltage of the differential amplifier at 0.6 V even in the most critical case of 0 V input voltage, as shown here.
- The second amplifier stage T_7 is shown here as a simple common-emitter configuration, in order to achieve a quiescent base potential of 0.6 V. The Darlington circuit shown in Fig. 5.14 would result in a quiescent potential of 1.2 V, so that with an input voltage of 0 V, transistor T_2 would enter the saturation region.
- The current source I_2 is added to allow an output swing down to almost 0 V. Of course, for output voltages under 0.6 V transistor T_9 is nonconductive, so that in this region the output can only accept currents lower than I_2.

Phase Reversal

If a single-supply amplifier such as that shown in Fig. 5.18 is driven in a range below the negative operating voltage, the collector–emitter voltage of the transistors in the differential amplifier T_1, T_2 is still at a level of 0.6 V. This value is clearly above the saturation voltage of $V_{CE,sat} = 0.2$ V. For this reason, the common-mode voltage may even be 0.4 V below the negative operating voltage. For even more negative common-mode voltages, transistor T_2 enters the saturation region and its base–collector diode becomes forward biased. Then the emitter of T_6 is connected to the base of T_7 and the inverting gain of T_2 changes to a noninverting signal transfer. With a further voltage drop at the P-input, transistor T_7 becomes nonconductive and the output voltage increases until it reaches the positive limit of the control range. This effect is known as *phase reversal*. The consequence of this effect in practical applications is shown in Fig. 5.19 for a noninverting amplifier that amplifies a sinusoidal alternating voltage. Due to the limitation caused by the final stage, the output voltage cannot become negative. However, it is not clipped to 0 V as one might expect, but due to the phase reversal jumps to the positive control limit if the input voltage drops below the reversal voltage $V_r \approx -1$ V. Phase reversal can have a very negative effect in single-supply amplifiers if sufficiently negative input voltages are applied. This may be prevented; for example, by using a Schottky diode at the input, which becomes conductive at -0.5 V. The better solution, however, would be to use operational amplifiers, which, due to their circuit design, do not cause phase reversal.

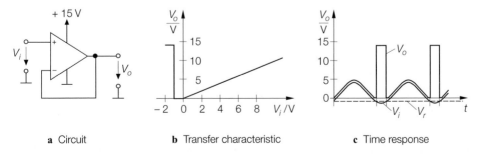

a Circuit   **b** Transfer characteristic   **c** Time response

Fig. 5.19. Effect of phase reversal

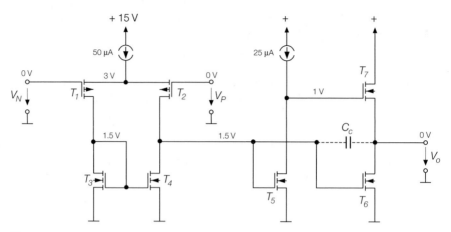

Fig. 5.20. Single-supply CMOS operational amplifier of the TLC series. The substrates of the n-channel FETs are connected to zero potential, and the substrates of the p-channel FETs to the positive operating voltage

CMOS operational amplifiers do not suffer from phase reversal, since the isolated gate electrodes do not allow any current flow. An effect similar to the conductive mode of the base–collector diode in a bipolar transistor does not exist for the MOSFET. A common circuit is illustrated in Fig. 5.20. A comparison with Fig. 5.18 shows that the design is very similar. The p-channel FETs T_1 and T_2 form the differential amplifier. Both output signals are combined by the current mirrors T_3 and T_4 and transferred to the second amplifier stage T_5. The source follower T_7 serves as an impedance converter. The only difference is the operating mode of T_6. It does not act as a complementary source follower, but in the common-source configuration it amplifies the signal, as does T_5. This enables this transistor to reduce the output voltage to 0 V when T_7 is nonconductive, as shown in the example of Fig. 5.20. Therefore, the current source required at the output of LM324 is not needed in this case. Here, the threshold voltage of all MOSFETs is $|V_{th}| = 1$ V. For the p-channel MOSFETS this value is raised to 2.5 V due to the substrate effect because all bulk-electrodes of these MOSFETs are connected to the positive supply voltage. Therefore the bulk-source voltage of the input transistors amounts $V_{BS} = 12$ V. This effect is quite useful, because sufficient drain–source voltage remains, even with a common-mode voltage, down to the negative operating voltage as shown in the example. As this effect does not exist in bipolar transistors, the LM324 shown in Fig. 5.18 requires the additional transistors T_5 and T_6 for level shift.

5.2.5
Rail-to-Rail Amplifiers

Rail-to-rail amplifiers are special operational amplifiers that allow the common-mode control range to extend not only to the negative operating voltage, as is the case with single-supply amplifiers, but also to the positive operating voltage. In Fig. 5.20 the enhancement MOSFETs provide a potential shift that is sufficient to allow common-mode control down to the negative operating voltage. However, common-mode control up to the positive operating voltage is not possible, since this would require the source potential of the differential amplifier to increase above the positive operating voltage. The rail-to-rail

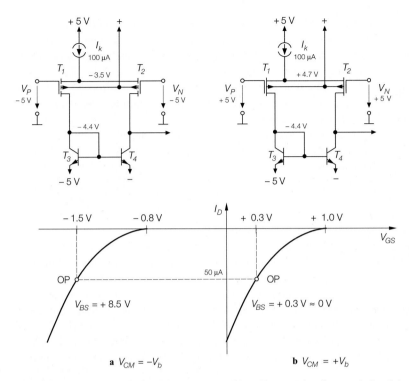

Fig. 5.21. Rail-to-rail CMOS differential amplifier. The transfer characteristic of the MOSFETs is shown for both extreme situations. OP marks the operating point. This principle is used, for example, in the LMC6484

amplifier with this configuration becomes possible due to the use of MOSFETs which – as previously seen – are in enhancement mode at the negative control limit, but which change to the depletion mode at the positive control limit. The – usually rather disturbing – substrate effect is used to shift the threshold voltage. This method is shown in Fig. 5.21.

In Fig. 5.21b, the maximum positive common-mode control voltage is $V_{BS} \approx 0$ and the transistors are in depletion mode; this leaves a voltage drop of 0.3 V for the current source. With a maximum negative common-mode voltage, the bulk–source voltage of $V_{BS} = 8.5$ V causes a shift in the threshold voltage, as explained in Sect.3.3.1:[2]

$$V_{th} = V_{th,0} - \gamma(\sqrt{V_{inv} + V_{BS}} - \sqrt{V_{inv}})$$

$$= 1\,\text{V} - 0.8\sqrt{\text{V}}(\sqrt{0.6\,\text{V} + 8.5\,\text{V}} - \sqrt{0.6\,\text{V}})$$

$$= 1\,\text{V} - 1.8\,\text{V} = -0.8\,\text{V}$$

At the operating point, this leads to a voltage of $V_{GS} = -1.5$ V, as shown in Fig. 5.21a. For the differential amplifier, this leaves a voltage of $V_{DS} = -0.9$ V, which is sufficient to remain above the pinch-off voltage:

$$V_{DS} < V_{DS,\,po} = V_{GS} - V_{th} = -1.5\,\text{V} + 0.8\,\text{V} = -0.7\,\text{V}$$

---

[2] In practice, special MOSFETs with a higher substrate effect are used. The substrate control factor is therefore assumed to be $\gamma = 0.8\sqrt{V}$, which is higher than the usual value of $\gamma = 0.6\sqrt{V}$.

This method is not practicable for bipolar transistors because they offer a fixed base-emitter voltage of 0.6 V. They require the use of two complementary single-supply differential amplifiers, one of which can be driven up to the positive and the other down to the negative operating voltage, and the combination of their output signals. This method is shown in Fig. 5.22. The differential amplifier consisting of pnp transistors T_1, T_2 allows input signals down to the negative operating voltage, while becoming nonconductive at common-mode voltages close to the positive operating voltage; in this region, the npn differential amplifier T_3, T_4 connected in parallel takes over. The subsequent transistors $T_5 - T_8$ combine the output signals of the differential amplifiers, so that all current changes produced enhance the gain: the output current at the collectors of T_6 and T_8 is $4I_q$. If one of the two differential amplifiers fails because the common-mode voltages are close to the operating voltages, the transconductance of the rail-to-rail input stage, and thus the voltage gain, halve in value. However, this effect is not noticeable in a circuit with negative feedback, in which the gain is determined by the external circuitry.

In principle, it is possible to use the voltage V_2 as the output voltage of the operational amplifier. Transistors $T_9 - T_{12}$ then form a conventional complementary emitter follower. But this would not provide a rail-to-rail output; the limits for the output voltages would be approximately 1 V below the operating voltages. A rail-to-rail output stage can only be realized with complementary transistors whose emitters are connected to the operating voltages. In the common-emitter configuration, the resultant output voltage swing almost reaches the operating voltages. In this case, the minimum voltage drop reaches a collector–emitter saturation voltage $V_{CE,sat}$ of T_{15} or T_{16}, which for low currents amounts to only a few millivolts.

The two output transistors are more difficult to drive in this configuration for the following reasons:

– Throughout the entire control range, a constant quiescent current must be achieved via the output-stage transistors.
– When controlling or loading the output, the current must be increased through one of the transistors and decreased through the other.
– Output currents that are high compared to the quiescent current must be possible, so that the output-stage transistors are operated in class-AB mode.

There are several different circuit designs for driving the output transistors. The circuit shown in Fig. 5.22 is very reliable and requires no specific technological tricks; this makes it easier to understand. In order to accurately control the output transistors T_{15} and T_{16}, they are expanded to current mirrors by adding transistors T_{13} and T_{14}.

In the analysis of the output stage, we start with the assumption $R_3 = \infty$. Driving the voltage followers $T_9 - T_{12}$ with a positive signal causes a current to flow through R_2 that increases the collector current of T_{11} and decreases that of T_{12}. The current difference is identical to the current flowing through R_2. This is even the case if the current through R_2 is so high that one of the transistors becomes nonconductive. For this reason, the output current difference of T_{11}, T_{12} is proportional to the current through R_2. To allow high output currents, the areas of the output transistors have been multiplied in size by a factor of ten ($A = 10$).

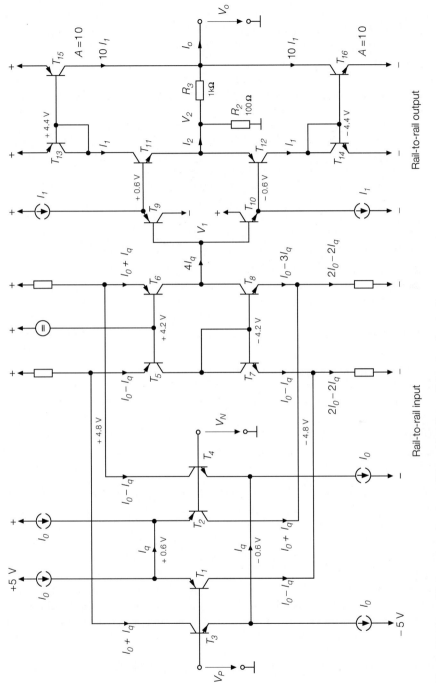

Fig. 5.22. Example of an operational amplifier with rail-to-rail input and output. Quiescent current: for example, $I_0 = 10\ \mu A$ and $I_1 = 100\ \mu A$

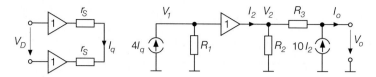

Fig. 5.23. Model of a rail-to-rail operational amplifier for analyzing the output stage

Due to its common-emitter configuration, the rail-to-rail output stage features a high output resistance and thus is a current output. In order to achieve a VV operational amplifier with a low-resistance output, internal negative voltage feedback was realized with resistors R_2 and R_3 in the output stage. The rail-to-rail output stage is a CC operational amplifier: this circuit design is described in more detail in Sect. 5.5.

The behavior of the rail-to-rail output stage is explained most easily using the model in Fig. 5.23. The situation may seem somewhat confusing, as the output stage represents a current-controlled current source with negative voltage feedback. To analyze the circuit, one can apply the nodal equations to both nodes of the output stage:

$$I_2 - \frac{V_2}{R_2} + \frac{V_o - V_2}{R_3} = 0 \tag{5.22}$$

$$\frac{V_2 - V_o}{R_3} + 10\, I_2 - I_o = 0 \tag{5.23}$$

For $I_o = 0$, this leads to the open-circuit gain of the output stage:

$$V_o = \frac{11\, R_2 + 10\, R_3}{11\, R_2}\, V_2 = \frac{111}{11}\, V_2 \approx 10\, V_2 \qquad \text{for } R_3 = 10\, R_2$$

From (5.22) and (5.23), where $V_2 = 0$, this leads to an output resistance:

$$r_o = -\frac{V_o}{I_o} = \frac{1}{11}\, R_3 = \frac{1}{11}\, 1\,\text{k}\Omega \approx 100\,\Omega$$

This means that the resistance is low compared to the output resistance of the output transistors that amounts to $r_{CE} = V_A/(10 I_1) = 100\,\text{V}/(10 \cdot 100\mu\text{A}) = 100\,\text{k}\Omega$.

5.2.6
Wide-Band Operational Amplifiers

In wide-band amplifiers, a single amplifier stage is preferred to bring the full voltage gain, since this usually eliminates the need for a frequency response correction that would in turn affect the bandwidth. However, the maximum gain achievable with one bipolar transistor is limited:

$$A_D = \mu = g_m r_{CE} = \frac{I_C}{V_T} \cdot \frac{V_A}{I_C} = \frac{V_A}{V_T} = \frac{100\,\text{V}}{26\,\text{mV}} \approx 4000$$

This is insufficient even for a universal wide-band operational amplifier. A higher voltage gain may be achieved by increasing the internal resistance above r_{CE}. This is possible with the cascode circuit shown in Fig. 5.24a. Unlike Fig. 5.11, here the additional transistor T_6 together with T_4 forms the cascode circuit. Its output resistance has already been determined in Sect. 4.1.1 on page 294; according to (4.23) its value is $r_o = \beta_{CE}$. This means that in

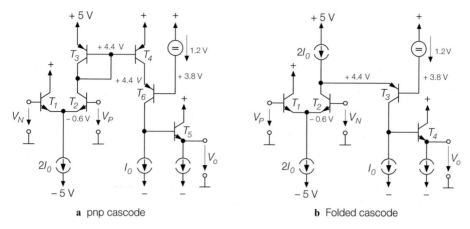

Fig. 5.24. Cascode circuit for increasing the differential gain A_D

this case the output resistance is higher by the current gain β than that of a single transistor. We thus obtain an open-circuit gain in the cascode circuit for $\beta = 100$:

$$A_D = \beta g_m r_{CE} = \beta \mu = \beta \frac{V_A}{V_T} = \beta \frac{100\,\text{V}}{26\,\text{mV}} \approx 400.000$$

This method allows us to build a high-gain operational amplifier with a single amplifier stage. The circuit shown in Fig. 5.24a may be simplified by omitting the current mirror. In Fig. 5.24b, this eliminates the phase shift caused by the current mirror without degrading the voltage gain. Only the polarity of the gain changes, but this can be compensated by interchanging the inputs.

The realization of this principle in practice is shown in Fig. 5.25. In order to make the circuit symmetrical, the two output signals of the differential amplifier are used and combined by the current mirror T_7, T_8. To prevent a reduction of the high internal resistance at the collector of T_4, a cascode current mirror according to Fig. 4.27 is used. Its output resistance is $\beta r_{CE}/2$.

As the high-impedance node (the collector of T_4) has a very high internal resistance due to the cascode circuit, a simple emitter follower is not sufficient to serve as an impedance converter. The emitter followers T_9 and T_{10} not only cause additional impedance transformation, but also generate the bias voltage necessary for the complementary emitter followers T_{11} and T_{12}.

The model shown in Fig. 5.26 can be used to calculate the voltage gain. The maximum voltage gain is achieved for $R_E = 0$. This causes a current in the input differential amplifier of

$$I_q = \frac{V_D}{2\,r_m} = \frac{I_0}{2\,V_T}\,V_D$$

At the high-impedance node of the circuit, which is represented here by $R_2 = \beta r_{CE} \parallel \beta r_{CE}/2 = \beta r_{CE}/3$, this current causes a voltage drop:

$$V_o = 2\,I_q R_2 = \frac{I_0}{V_T}\frac{\beta}{3}\frac{V_A}{I_0}\,V_D = \frac{\beta}{3}\frac{V_A}{V_T}\,V_D \overset{\substack{\beta=100\\V_A=100\,\text{V}}}{=} 1.3 \cdot 10^5\,V_D$$

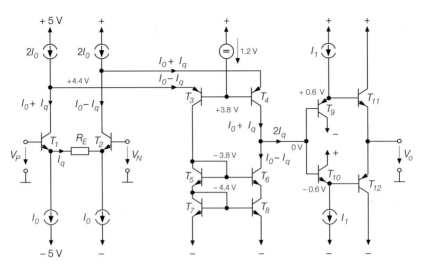

Fig. 5.25. Operational amplifier with a complementary cascode differential amplifier (folded cascode). This principle is used, for example, in the AD797 from Analog Devices, the OP640 from Burr Brown, and the LT1363 from Linear Technology

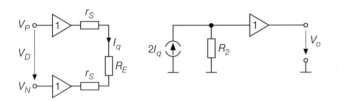

Fig. 5.26. Model of the operational amplifier shown in Fig. 5.25 for calculating the differential gain

A differential gain of $A_D = 1.3 \cdot 10^5$ follows. The transconductance of the differential amplifier, and thus the voltage gain, can be reduced by means of the emitter resistance R_E. This has the advantage that the bandwidth of the input stage is increased and the tendency toward oscillations in the circuit, due to feedback, is reduced.

A precondition for satisfactory high-frequency amplifiers is the production of pnp transistors with good high-frequency properties that correspond to npn transistors. Therefore, the circuits shown here use vertical pnp transistors in p wells, which are more expensive to produce.

To insure equally steep rising and falling edges of a signal, one can use the push–pull principle, which is based on inversely controlled transistors so that both the positive and the negative edge cause a current increase in one half of the circuit. In order to apply this principle to wide-band amplifiers, one can symmetrically complement the circuit shown in Fig. 5.25, which results in the push–pull operational amplifier shown in Fig. 5.27.

At the quiescent point, the same current I_0 flows through transistors T_5 and T_6. As soon as a positive differential voltage V_D is supplied, the collector current of T_5 increases by I_q, while the collector current of T_6 decreases by the same amount. Therefore, current $I_q = V_D/R_E$ has the same value as in the previous circuit, shown in Fig. 5.25.

One problem that is common to all circuits discussed so far is the constant emitter current of the input differential amplifier, because it is the maximum current $I_{q,max} = I_0$ with which the circuit capacitances at the high-impedance node can be charged and discharged

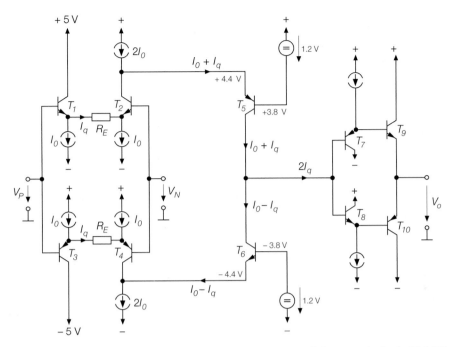

Fig. 5.27. Push–pull operational amplifier. This principle is used, for example, in the EL2038 from Elantec and the HFA0001 from Harris

resulting in a limited slew rate. One can naturally make the quiescent currents sufficiently high, but this also increases the power dissipation of the operational amplifier. However, a criterion for the quality of a given circuit design is whether a high bandwidth can be achieved despite low quiescent currents. A satisfactory circuit is one that can provide high charge currents for the parasitic capacitances even at low quiescent currents. In complementary emitter followers, this principle is common in class AB operation. A remarkable development in the area of wide-band amplifiers is that of a differential amplifier suitable for class AB operation.

An operational amplifier with a differential amplifier in AB mode at the input is shown in Fig. 5.28. Transistors $T_1 - T_8$ form two voltage followers that are connected via the resistance R_E; they form the differential amplifier. Here, the emitter current $I_q = V_D/R_E$ is not limited; it rises continuously with the voltage difference at the input. The output signals are decoupled by the current mirror T_9, T_{11} or T_{10}, T_{12}. With $I_q > I_0$, the entire current flows through the upper signal path (see Fig. 5.28). For this reason, the charge of capacitance C_c at the high-impedance node can be reversed as quickly as desired if the voltage difference at the input is high. Here, the slew-rate limitation that occurs in all of the operational amplifiers discussed so far (see Sect. 5.2.7) is practically nonexistent.

The model shown in Fig. 5.29 will be used to explain the functional principle of the amplifier. The voltage followers at the inputs are coupled via the resistance R_E and the two output resistances $r_m = 1/g_m$ are in series. This results in

$$I_q = \frac{V_D}{R_E + 2r_m}$$

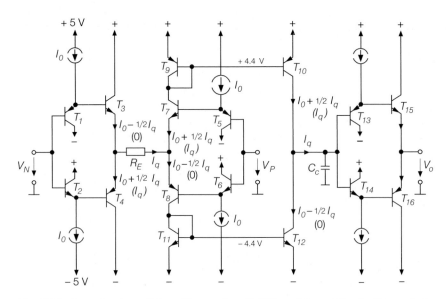

Fig. 5.28. Operational amplifier with a push–pull differential amplifier in AB mode (*current on demand*). The values quoted in parentheses reflect the conditions with a high current $I_q > I_0$. This principle is used, for example, in the OP 467 models from Analog Devices, the LT1819 from Linear Technology, and the LM7171 from National

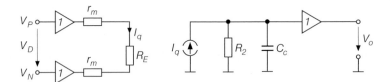

Fig. 5.29. Model to explain the mode of operation of the operational amplifier shown in Fig. 5.28

This current is mirrored to the output and produces a voltage drop V_o at the resistance R_2, which represents the parallel arrangement of all impedances occurring at the output of the current mirror. This voltage drop can be determined as:

$$V_o = I_q R_2 = \frac{V_D R_2}{R_E + 2r_m}$$

This voltage drop is transferred to the output via a voltage follower. The differential gain of the circuit is thus

$$A_D = \frac{V_o}{V_D} = \frac{R_2}{R_E + 2r_m} = \frac{R_2}{R_{E,\,ges}}$$

The highest voltage gain is reached when $R_E = 0$. With $R_2 = r_{CE}/2$ we obtain:

$$A_D = \frac{1}{2}\frac{R_2}{r_m} = \frac{1}{2}\frac{I_C}{V_T}\frac{1}{2}\frac{V_A}{I_C} = \frac{1}{4}\frac{V_A}{V_T} = \frac{1}{4}\mu \approx 1000$$

With a simple common-emitter current mirrors (T_{10}, T_{12}), no higher voltage gain can be expected. But here too, it can be increased significantly by using a cascode current mirror as seen in Fig. 5.25, since this causes the internal resistance at the high-impedance node to increase by the current gain β.

It may be somewhat unusual to build a differential amplifier from two voltage followers, and to transfer the output signal by means of two complementary current mirrors from the supply terminals, but this arrangement can also be beneficial in other situations – for example, in rail-to-rail output stages, as shown in Fig. 5.22.

There is also another way of interpreting the functional principle of the voltage followers shown in Fig. 5.28: transistors T_3 and T_7 form an npn differential amplifier and transistors T_4 and T_8 form a pnp differential amplifier, whose inputs are connected in parallel and whose outputs are combined. This illustrates the relationship with the push–pull operational amplifier in shown Fig. 5.27. Both circuits use complementary differential amplifiers to generate output signals in phase opposition, which are then amplified in the subsequent stage. The circuit shown in Fig. 5.27 limits the output current of the differential amplifier to $2I_0$. The current of the operational amplifier in Fig. 5.28 is, however, not limited; for this reason, it is essential to use the current mirrors here; otherwise, the advantage of the unlimited output currents of the differential amplifier would be lost.

5.2.7
Frequency Compensation

Basic Principles

When using an operational amplifier as an voltage amplifier, the feedback, as shown in Fig. 5.30, must always take place between the output and the *inverting* input, so that negative feedback is achieved. Direct (positive) feedback is undesirable, because it may produce a latching circuit with bistable behavior.

Operational amplifiers such as model 741, shown in Fig. 5.14, are multi-stage amplifiers, where every stage has lowpass characteristic. The model in Fig. 5.31 shows the most important lowpass filters of the amplifiers. The differential amplifier has the lowest cutoff frequency of $f_{c1} = 10\,\text{kHz}$, as it is operated with very low currents and the effective resistance at the collector is very high. The cutoff frequency of the second amplifier stage is clearly higher due to the higher currents, and amounts to $f_{c2} = 100\,\text{kHz}$. In low-cost tech-

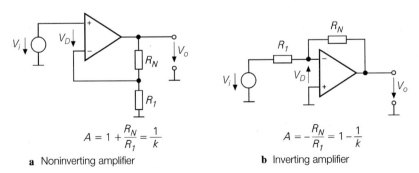

$$A = 1 + \frac{R_N}{R_1} = \frac{1}{k}$$

a Noninverting amplifier

$$A = -\frac{R_N}{R_1} = 1 - \frac{1}{k}$$

b Inverting amplifier

Fig. 5.30. Comparison of noninverting and inverting amplifiers. For $V_i = 0$, both circuits are identical. Then $V_D = \frac{R_1}{R_1 + R_N} V_o = k V_o$ (k is used here instead of k_R)

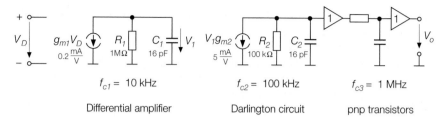

$f_{c1} = 10\,\text{kHz}$ $f_{c2} = 100\,\text{kHz}$ $f_{c3} = 1\,\text{MHz}$

Differential amplifier   Darlington circuit   pnp transistors

Fig. 5.31. The three most important cutoff frequencies in operational amplifiers of the 741class

nologies, the quality of pnp transistors is greatly inferior to that of the npn types; therefore, pnp transistors cause a third lowpass filter with a cutoff frequency of $f_{c3} = 1\,\text{MHz}$.

Each lowpass filter gives rise to a reduction in the gain by 20 *dB per decade* above the cutoff frequency, and causes an additional phase lag which, at the cutoff frequency, amounts to 45° and thereafter increases up to 90°, as described in Sect. 29.3.1 on page 1488. Figure 5.32 shows the resulting Bode diagram. The gain reduction by 20 *dB per decade* starts at f_{c1}, with a phase shift of 45°. From f_{c2} onward, this value falls by 40 dB per decade and the phase shift is as much as 135°; this value is a combination of 90° from the first lowpass filter and 45° from the second. Due to the third lowpass filter, above f_{c3} the gain drops at a rate of -60 *dB per decade* and the phase shift rises asymptotically to $-270°$. At a frequency of (here) $f_{180} = 300\,\text{kHz}$, the phase lag passes $-180°$. At this point the functionality of the inputs interchanges and the negative feedback turns into positive feedback.

Whether or not this circuit oscillates at this frequency depends on whether or not the oscillating condition is met:

$$\underline{g} = \underline{k}\,\underline{A_D} \equiv 1 \;\Rightarrow\; \begin{cases} |\underline{g}| = |\underline{k}||\underline{A_D}| \equiv 1 & \text{Amplitudecondition} \\ \phi(\underline{k}\underline{A_D}) \equiv 0°, 360°, \dots & \text{Phasecondition} \end{cases} \tag{5.24}$$

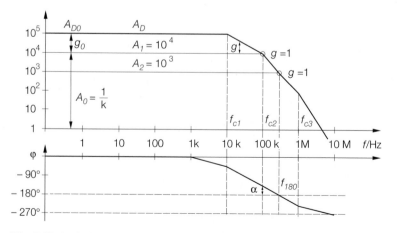

Fig. 5.32. Bode diagram of a noncompensated operational amplifier of the 741 class

This condition comprises two parts: the amplitude condition and the phase condition. An oscillation with constant amplitude arises only if both conditions are met. This is the case in Fig. 5.32 with a feedback, to achieve a gain of $A_2 = 1,000$. At the frequency f_{180}, the loop gain is then $|\underline{k}|A_D| = 1$. The loop gain can be taken directly from the Bode diagram. Since $g = kA_D = A_D/A$ becomes $lg\,g = lg\,A_D - lg\,A$ in the logarithmic diagram, the (logarithmic) loop gain is identical to the distance between the differential gain and the negative feedback gain.[3] From Fig. 5.32 one can see that this distance becomes smaller as the frequency increases and reaches zero at the intersection with the set gain; at these points, $g = 1$.

If $kA_D > 1$ and the phase condition is met, an oscillation with an increasing amplitude occurs. In this case, the oscillation amplitude increases until the amplifier is overdriven. If $kA_D < 1$, then there is a damped oscillation. For an amplifier, this is the only case of interest. In our example, it occurs when the gain set by the negative feedback exceeds 1,000; for example, $A_1 = 10,000$. For the frequency f_{180} it follows that $g = kA_D = 1/10$; the loop gain is thus below the oscillating point by a factor of ten. This is also called a *gain margin* of ten; this means that the loop gain can be increased by a factor of ten before an undamped oscillation occurs.

It is more common to specify the phase shift distance to $-180°$, when the amplitude condition $g = kA_D = 1$ is met. The value

$$\alpha = 180° - \varphi(f_k) \tag{5.25}$$

is called *phase margin*. It states the angle by which the phase shift may increase before an undamped oscillation occurs at the *critical frequency* f_k at which the amplitude condition is met. In the Bode diagram of Fig. 5.32, it is marked by circles.

The phase margin is a particularly useful parameter for judging the damping effect and the tendency toward oscillation of a given system. Figure 5.33 shows the transient responses for various phase margins and the corresponding frequency responses. It is clear that a declining phase margin causes move overshoot in the step response and an increasing peaking of the frequency response. The aperiodic limit is reached at a 90° phase margin: there is no overshoot, but there is clearly a longer rise time and a drastically reduced

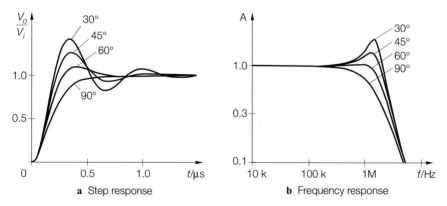

Fig. 5.33. Step response and frequency response for several phase margins α. Overshoot in the time interval corresponds to peaking in the frequency range

---

[3] In this abbreviated form, $g = |\underline{g}|$ and $A = |\underline{A}|$.

bandwidth. A particularly favorable response in both time and frequency is reached at a phase margin of $\alpha = 60°$.

The explanation of frequency compensation will, however, not be based on $\alpha = 60°$ but on the more simple case of $\alpha = 45°$. Here, the critical frequency f_k, at which the amplitude condition $|g| = 1$ is met, is the same as the second cutoff frequency at which the phase margin is 45°. As the gain of the operational amplifier in the frequency range between f_{c1} and f_{c2} is inversely proportional to the frequency, it follows;

$$f_{c1} = \frac{f_{c2}}{g_0} \tag{5.26}$$

This relationship is also evident in Fig. 5.32, and leads to the rule of frequency compensation:

> *The first cutoff frequency must be below the second cutoff frequency by a value equal to the loop gain g_0.*

In order to achieve a phase margin of $\alpha = 60°$ the first cutoff frequency has to be halved additionally.

In the Bode diagram of the noncompensated gain, as shown in Fig. 5.32, the phase margin of 45° is achieved at a gain of $A_1 = 10,000$. When the negative feedback increases, the phase margin decreases. Reducing the gain to $A_2 = 1,000$ causes the amplifier to start oscillating, since the phase margin is zero. An amplifier without frequency compensation allows only poor negative feedback, since it otherwise oscillates.

General Frequency Compensation

For universal frequency compensation, the frequency response is modified so that the amplifier remains stable even in the event of full negative feedback, $A_3 = 1$ in Fig. 5.34. In order to achieve a phase margin of 45°, the gain at f_{c2} must have dropped to $A_D = 1$, so that in this case the second cutoff frequency is equal to the transit frequency, which is defined by $A_D = 1$. From (5.26) it follows:

$$f_{c1} = \frac{f_{c2}}{g_0} = \frac{f_T}{A_{D0}} = \frac{100\,\text{kHz}}{10^5} = 1\,\text{Hz}$$

This situation is depicted in Fig. 5.34. In order to lower the cutoff frequency f_{c1} from 10 kHz to 1 Hz, it is necessary to increase capacitance C_1 in Fig. 5.31, which determines the frequency, from 16 pF to 160 nF. A capacitance of this size cannot be realized in integrated circuits; it must either be connected externally or its value must be reduced by circuit design tricks to enable integration. Generally, the condition

$$f_{c1} = \frac{1}{2\pi R_1 C_1} = \frac{f_T}{A_{D0}} \quad \text{and} \quad A_{D0} = g_{m1} R_1 g_{m2} R_2 \tag{5.27}$$

can be used to calculate the compensation capacity:

$$C_1 = \frac{A_{D0}}{2\pi R_1 f_T} = \frac{g_{m1} g_{m2} R_2}{2\pi f_T} = \frac{0.2\frac{\text{mA}}{\text{V}} \cdot 5\frac{\text{mA}}{\text{V}} \cdot 100\,\text{k}\Omega}{2\pi \cdot 100\,\text{kHz}} = 160\,\text{nF}$$

As shown in Fig. 5.34, the frequency compensation reduces the gain, not the phase shift. This moves the critical frequency f_k into a range where phase margin is $\alpha \geq 45°$.

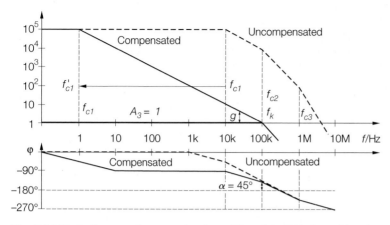

Fig. 5.34. Bode diagram of an operational amplifier of the 741 class with universal frequency compensation

Pole Splitting

Pole splitting utilizes the Miller effect in order to reduce the compensation capacity down to a value that can be integrated. In the common-emitter circuit, the collector–base capacitance represents the Miller capacity. It results in a significantly narrowed bandwidth, since it has the effect of an input capacitance that is increased by the voltage gain (refer to Sect. 2.4.1 on page 125). This otherwise disadvantageous effect can be of benefit here, as it enables the use of a lower compensation capacitance. The Miller capacitor C_c is shown in the model shown in Fig. 5.35. For a voltage gain of 500, a capacitance of only $160\,\text{nF}/500 = 320\,\text{pF}$ is sufficient.

Utilization of the Miller effect has a second advantage: since the Miller capacitor induces negative voltage feedback, the output resistance of the amplifier stage is also reduced. In our example, this results in an increase of the cutoff frequency f_{c2} from $100\,\text{kHz}$ to $10\,\text{MHz}$; it thus exceeds the third cutoff frequency of $1\,\text{MHz}$. As the cutoff frequency

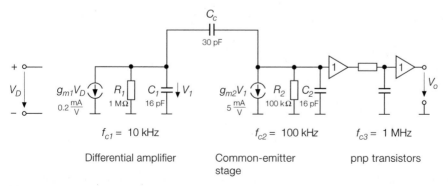

Fig. 5.35. Frequency compensation using a Miller capacitor

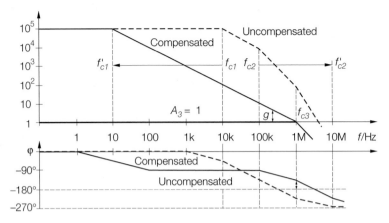

Fig. 5.36. Frequency compensation by means of pole splitting

is increased, this is a particularly desirable case. As a consequence, the cutoff frequency

$$f_{c1} = \frac{f_{c3}}{g_0} = \frac{f_{c3}}{A_{D0}} = \frac{1\,\text{MHz}}{10^5} = 10\,\text{Hz}$$

can be increased by a factor of ten. This is also reflected in the Bode diagram shown in Fig. 5.36. The necessary compensation capacitance is reduced by an additional factor of ten and amounts to only $C_c \approx 30\,\text{pF}$. A capacitance of this size can easily be realized in integrated circuits; it has already been included in Fig. 5.14.

This method of reducing the cutoff frequency (f_{c1}) and at the same time increasing another frequency (f_{c2}), which is used here to shift the two limit frequencies further apart, is known as *pole splitting*.

In general, the compensation capacitance can be calculated by taking into account the fact that a capacitance $A_2 C_c$ is arranged in parallel to C_1 (Fig. 5.35):

$$f_{c1} = \frac{1}{2\pi R_1 (C_1 + g_{m2} R_2 C_c)} \approx \frac{1}{2\pi R_1 g_{m2} R_2 C_c} = \frac{f_T}{A_{D0}}$$

Consequently,

$$C_c = \frac{A_{D0}}{2\pi R_1 g_{m2} R_2 f_T} = \frac{g_{m1} R_1 g_{m2} R_2}{2\pi R_1 g_{m2} R_2 f_T} = \frac{g_{m1}}{2\pi f_T} \tag{5.28}$$

The transconductance g_{m1} can be determined from the constant current of the differential amplifier, if one takes into account the fact that the transconductance of a differential amplifier is half that of the transistors:

$$C_c = \frac{I_0/2}{2\pi V_T f_T} = \frac{I_0}{4\pi V_T f_T} = \frac{10\,\mu\text{A}}{4\pi \cdot 26\,\text{mV} \cdot 1\,\text{MHz}} = 30\,\text{pF}$$

Adapted Frequency Compensation

It is convenient to work with amplifiers that have general frequency compensation, because this eliminates the need to give frequency compensation any further thought. Such

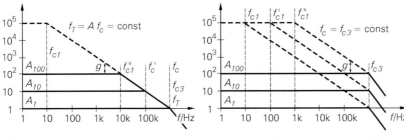

a General frequency compensation.
Constant gain bandwith product $GBW = f_T$

b Adapted frequency compensation.
Constant cutoff frequency

Fig. 5.37. Comparison of general and adapted frequency compensation for the gains of $A_{min} = 1$, 10, and 100

amplifiers provide a good transient response for any external circuitry and remain stable as long as the negative feedback does not produce an additional phase lag. However, using them to produce a gain $A > 1$ gives away bandwidth. This disadvantage can be avoided by adapted frequency compensation. For a gain of $A_0 = 10$ we obtain, from (5.26):

$$f_{c1} = \frac{f_{c3}}{g_0} = \frac{f_{c3}}{kA_{D0}} = \frac{f_{c3}}{A_{D0}}A_0 = \frac{1\,\text{MHz}}{10^5}\,10 = 100\,\text{Hz}$$

This means the first cutoff frequency can be increased by the gain A_0, compared to the case of general frequency compensation. This makes the cutoff frequency of the amplifier with negative feedback *constantly equal to the second cutoff frequency*, as can be seen from Fig. 5.37. The phase margin is always 45°. In contrast, general frequency compensation provides a *constant gain–bandwidth product*; that is, the bandwidth decreases by the same factor as the gain increases.

With operational amplifiers of the 741 class, the adapted frequency compensation can only be used up to a cutoff frequency of $f_{g1} = 100\,\text{Hz}$; this results in a compensation capacitance of $C_c = 3\,\text{pF}$. A higher decompensation renders the pole splitting ineffective, and the second cutoff frequency is again reduced to $100\,\text{kHz}$.

The terminals of the compensation capacitor are critical. Therefore, in more recent operational amplifiers they are no longer accessible to the user. Instead, for some types of operational amplifier fully compensated and partially compensated terminals are available (for example, for a minimum gain of $A_{min} = 2, 5$, or 10).

Slew Rate

Besides reducing the bandwidth and the loop gain, frequency compensation has yet another disadvantage: the maximum speed the output voltage can change – the *slew rate* – is limited to a relatively low value. The reason for this can easily be seen in the equivalent circuit shown in Fig. 5.38.

If, in the event of overdrive, only T_2 is conductive, then $I_1 = 2I_0$. If only T_1 is conductive, then the entire current flows through the current mirror, resulting in $I_1 = -2I_0$. The charge current of C_c is limited to the maximum output current of the differential

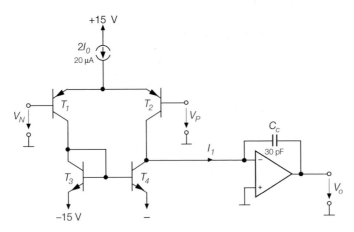

Fig. 5.38. Model to explain the slew rate, using a 741 class amplifier as an example. The second amplifier stage with the Miller capacitor is shown here as integrator

amplifier: $I_{1max} = \pm 2I_0 = \pm 20\ \mu A.$[4] As the compensation capacity carries the full output voltage, $I = C\dot{V}$ leads to the slew rate SR

$$SR = \left.\frac{dV_o}{dt}\right|_{max} = \frac{I_{1max}}{C_c} = \frac{2I_0}{C_c} = \frac{20\ \mu A}{30\ pF} = 0.6\ \frac{V}{\mu s} \tag{5.29}$$

This means that the maximum change in the output voltage within $1\ \mu s$ is $0.6\ V$. A rectangular-wave signal with an output amplitude of $\pm 20\ V$ has a rise- or falltime of

$$\Delta t = \frac{\Delta V_o}{SR} = \frac{20V}{0.6\ V/\mu s} = 33\ \mu s$$

Likewise, a sinusoidal input voltage swing can produce no faster change at any point than that allowed by the slew rate. If we take the output voltage to be $V_o = \hat{V}_o \sin \omega t$, then the maximum rate of voltage change occurs at the point of zero crossing resulting in

$$SR = \frac{dV_o}{dt} = \hat{V}_o \omega = 2\pi f \hat{V}_o \tag{5.30}$$

This allows us to calculate the frequency up to which an undistorted sinusoidal signal with full output amplitude is possible:

$$f_p = \frac{SR}{2\pi \hat{V}_o} = \frac{0.6\ V/\mu s}{2\pi \cdot 10\ V} = 10\ kHz \tag{5.31}$$

This value is known as the *power bandwidth*, because it represents the frequency up to which the full output power is available. In amplifiers of the 741 class it is only $f_p = 10\ kHz$, as can be seen, even though the small-signal bandwidth is $f_T = 1\ MHz$. According to (5.30), the maximum output voltage swing decreases when the frequency f is increased:

$$\hat{V}_o = \frac{SR}{2\pi f} \tag{5.32}$$

As shown in Fig. 5.39, an amplifier of the 741 class achieves the full output swing up to $10\ kHz$. At $100\ kHz$ this drops to $1\ V$, and at $1\ MHz$ as low as $0.1\ V$.

---

[4] The maximum current of the second amplifier stage, shown here as an integrator, is also limited, but at $300\ \mu A$ it is significantly higher and therefore has no limiting effect.

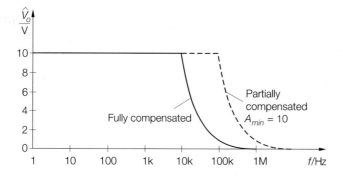

Fig. 5.39. Dependence of the output swing on the frequency in an amplifier of the 741 class

When the output signal exceeds the slew rate limitation, the curve segments can be replaced by straight lines that represent the slope of the slew rate. This is shown in Fig. 5.40. When the slew rate is considerably exceeded the output signal assumes a triangular shape and, with the exception of the frequency, has little in common with the undistorted signal.

In order to improve the slew rate, one could assume on the basis of (5.29) that it increases as the current I_0 increases. However, in order to examine this we also have to take into account the current dependence of C_c, as shown in (5.28):

$$SR = \frac{2I_0}{C_c} = 2\pi f_T \frac{2I_0}{g_{m1}} \tag{5.33}$$

For the given transit frequency, the slew rate increases as the current I_0 increases at a given transconductance. For bipolar transistors, however, the ratio I_0/S_1 is constant, since the transconductance is proportional to I_0:

$$\frac{2I_0}{S_1} = \frac{2I_0}{2I_0/4V_T} = 4V_T \approx 100\,\mathrm{mV}$$

For the slew rate, it follows:

$$SR = \frac{4 \cdot 2I_0 V_T}{2I_0} 2\pi f_T = 8\pi V_T f_T = 8\pi \cdot 26\,\mathrm{mV} \cdot 1\,\mathrm{MHz} = 0.6\,\frac{\mathrm{V}}{\mu\mathrm{s}} \tag{5.34}$$

Therefore the slew rate does not depend on the current I_0, since the required compensation capacity increases at the same rate as I_0. However, current feedback of the input differential amplifier allows the current I_0 to be increased with a constant transconductance; this is often utilized in wide-band operational amplifiers. Favorable values are also achieved with operational amplifiers by using field effect transistors at the input, as they have a much lower transconductance as bipolar transistors, at the same current I_0.

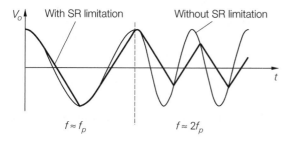

Fig. 5.40. Effect of the slew rate on a sinusoidal output signal. *Left*, slightly exceeding the power bandwidth; *right*, a signal with double the frequency

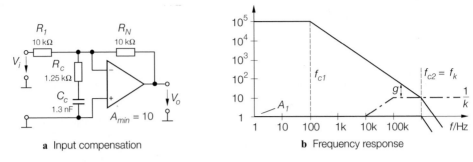

a Input compensation   **b** Frequency response

Fig. 5.41. Use of a partially compensated amplifier with $A_{min} = 10$ at a gain of $A_{min} = 1$

Since partially compensated operational amplifiers have a clearly lower compensation capacity, their slew rate is correspondingly higher. For this reason, it may be desirable to operate them below the minimum gain A_{min}. This requires additional frequency compensation to maintain an acceptable transient response. However, this cannot be realized by increasing the internal compensation, because it is not externally accessible and provides no advantage compared to a fully compensated amplifier. However, in any case, the loop gain can be reduced by reducing the negative feedback at the input. The additional RC network shown in Fig. 5.41 serves this purpose. If an amplifier that is compensated internally for $A_{min} = 10$ is to be used for a gain of $A = 1$, the feedback signal must be reduced to 1/10:

$$V_N = \frac{R_1 || R_k}{R_N + R_1 || R_c} \, V_o \equiv \frac{1}{10} \, V_o$$

R_c must then have the value

$$R_c = \frac{R_1 R_N}{9 R_1 - R_N} = 1.25 \, \text{k}\Omega$$

The capacitor connected in series renders the attenuation ineffective at low frequencies, in order to maintain the full loop gain in this frequency range. A sufficiently high capacity must be selected to insure that no significant phase lag occurs at the critical frequency f_k. This leads to the condition:

$$C_c = \frac{10}{2\pi \cdot R_c f_k} = \frac{10}{2\pi \cdot 1.25 \, \text{k}\Omega \cdot 1 \, \text{MHz}} = 1.3 \, \text{nF}$$

The way in which the input compensation functions can be seen from the Bode diagram in Fig. 5.41. At high frequencies, only $k = 1/10$ of the output signal is fed back, so that k has the same value as is usually the case for a gain of $A = 1/k = 10$. Naturally, the input noise of the amplifier increases with this gain; for this reason, $1/k$ is also known as the *noise gain*. The input compensation reduces the loop gain without increasing the gain of the circuit.

Frequency compensation at the input can support interior compensation, but can never replace it; it is therefore used in partially compensated amplifiers. Input compensation does not work well with noninverting amplifiers, as it depends on the source resistance which is in series with R_c.

Capacitive Load

If a capacitive load C_L is connected to the output of an operational amplifier, it forms, together with the output resistance r_o, an additional lowpass with a cutoff frequency f_{cC}, as shown in Fig. 5.42. Operational amplifiers with a simple emitter follower at the output have output resistances (of the open amplifier) in the range of $r_o \approx 1\ \mathrm{k\Omega}$, while in Darlington circuits and in RF operational amplifiers this value is mostly below 100Ω. If the load capacitance is low ($C_L < 100\ \mathrm{pF}$), the additional cutoff frequency f_{cC} is higher than the second cutoff frequency of the amplifier; the phase margin then decreases only slightly. For higher load capacitances, the additional cutoff frequency drops below the second cutoff frequency; this case is shown in Fig. 5.43. As one can see, the phase shift above f_{cC} becomes so large that the circuit would oscillate at lower closed loop gains. In order to achieve stable operating conditions in spite, additional frequency compensation is required.

As standard operational amplifiers usually have internal compensation, it is not possible to subsequently reduce the minimum cutoff frequency f_{c1}. But with the aid of input compensation it is possible to carry out additional compensation by external circuitry. This method has already been demonstrated in Fig. 5.41. In the example shown in Fig. 5.43, the gain was still 1,000 for a second cutoff frequency $f_{cC} = 1\ \mathrm{kHz}$, making it necessary to reduce the loop gain by this factor. Such a great attenuation is not practicable with input compensation.

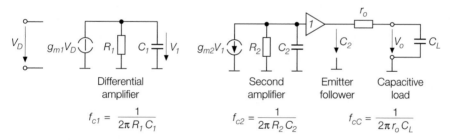

$$f_{c1} = \frac{1}{2\pi R_1 C_1} \qquad f_{c2} = \frac{1}{2\pi R_2 C_2} \qquad f_{cC} = \frac{1}{2\pi r_o C_L}$$

Fig. 5.42. Operational amplifier with capacitive load. The cutoff frequencies of a fully compensated amplifier of the 741 class are $f_{c1} = 10\ \mathrm{Hz}$ and $f_{c2} = 1\ \mathrm{MHz}$

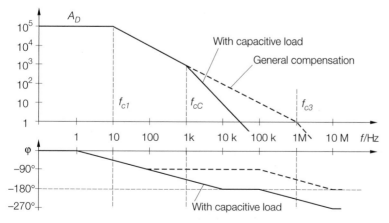

Fig. 5.43. Effect of a capacitive load on a fully compensated operational amplifier

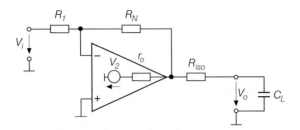

Fig. 5.44. Isolation resistance for phase compensation with a capacitive load

It is better to connect an isolation resistance in series with the capacitive load, as shown in Fig. 5.44. For high frequencies at which the load capacitor acts as a short-circuit, the output of the amplifier is connected solely to a voltage divider formed by r_o and R_{iso}, which produces no phase lag. From the Bode diagram in Fig. 5.45 it can be seen that, compared to Fig. 5.43, the phase does not change up to 1 kHz, but above this frequency it approaches the nonloaded situation. At the critical frequency $f_k = f_{c2} = 100$ kHz, the phase margin is $90°$; this determines the transient response of the circuit, whereby it is of no importance that the phase margin is smaller at lower frequencies. Here we have the specific case of the phase margin decreasing with less feedback: for a gain of $A = 10$, the critical frequency is 10 kHz; the phase margin at this point is only $45°$.

An example will serve to explain the dimensioning of the components. An amplifier with an open-circuit output resistance of $r_o = 1$ kΩ is to be loaded at the output with a capacity of $C_L = 160$ nF. This results in a cutoff frequency of

$$f_{cC} = \frac{1}{2\pi r_o C_L} = \frac{1}{2\pi\ 1\,\text{k}\Omega\ 160\,\text{nF}} = 1\,\text{kHz} \tag{5.35}$$

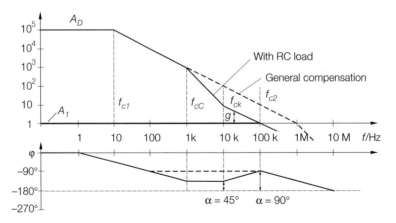

Fig. 5.45. Reversal of the phase shift above f_{ck} by way of isolation resistance

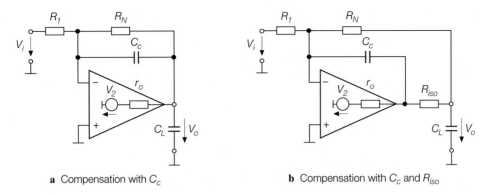

a Compensation with C_c **b** Compensation with C_c and R_{iso}

Fig. 5.46. Phase reversal with C_c at capacitive loads

In order to reduce the phase shift caused by the load, when the critical frequency $f_{c2} = 100\,\text{kHz}$ is reached, we take $f_{ck} = 10\,\text{kHz}$ according to Fig. 5.45. It then follows from (5.35)

$$R_{iso} = \frac{1}{2\pi f_{ck} C_L} = \frac{f_{cC}}{f_{ck}} r_o = \frac{1\,\text{kHz}}{10\,\text{kHz}}\, 1\,\text{k}\Omega = 100\,\Omega \tag{5.36}$$

To achieve as much bandwidth as possible, one can reduce R_{iso} slightly. Due to the reduced phase margin this leads to a gain peaking as shown in Fig. 5.33, which can compensate the gain reduction caused by the lowpass filter $R_{iso} C_L$ within a certain frequency range.

According to Fig. 5.44, the use of an isolation resistance is disadvantageous in many applications, as the load is not operated from a low source resistance. The conventional circuitry can then be expanded by the capacitance C_c shown in Fig. 5.46a. This can compensate the phase lag caused by the load. Its size should be continually increased until the desired transient response or frequency response is achieved.

In very difficult cases, one can insert an additional isolation resistance as shown in Fig. 5.46b. In order to realize the proper output voltage V_o at the load capacitor the amplifier produces a leading voltage V_1. Feeding this voltage back via the compensation capacitance C_c enhances the stabilizing effect.

Internal Load Compensation

In order to make operational amplifiers as user-friendly as possible, manufacturers attempt to provide general internal compensation for capacitive loads. The philosophy behind this is to enhance the existing compensation for capacitive loads. The $R_c C_{c2}$ network, which bypasses the emitter follower at the output in Fig. 5.47, is used for this purpose. With a low load, there is practically no voltage drop across r_o; the RC network thus has no effect. With a high load, the additional compensation capacitance C_{c2} is almost in parallel to C_{c1}. This approach helps to prevent the operational amplifier from oscillating when used with capacitive loads; however, in most cases the transient response is not satisfying and additional measures are required.

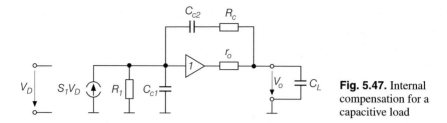

Fig. 5.47. Internal compensation for a capacitive load

Two-Pole Frequency Compensation

For frequency compensation with capacitive loads, it was shown in Fig. 5.45 that in a certain frequency range the phase margin may be small. This is acceptable, as only the phase margin at the critical frequency f_k determines the transient response. It is therefore possible to reduce the gain of an operational amplifier simultaneously with *two* lowpass filters for phase compensation and accept the fact that the phase margin will drop to almost zero in the relevant frequency range. The advantage of 2-pole frequency compensation can be seen in Fig. 5.48. Here, we can see that due to the sharp rolloff the loop gain is clearly higher than it is with 1-pole compensation.

2-pole compensation reduces the first and second cutoff frequencies of the amplifier to an *identical* value, which in this example is $f'_{c1} = 1\,\text{kHz}$. This effect is achieved by the two capacitances C_{c1} and C_{c2} in Fig. 5.49. Consequently, the phase shift at this frequency

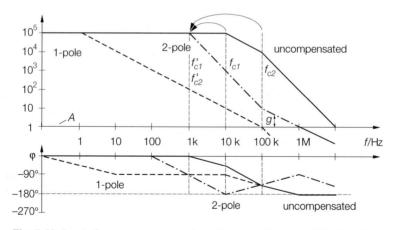

Fig. 5.48. 2-pole frequency compensation of an operational amplifier. 1-pole correction without pole splitting is also shown for comparison

$$f'_{c1} = \frac{1}{2\pi R_1 (C_1 + C_{c1})} = 1\,\text{kHz} \qquad f'_{c2} = \frac{1}{2\pi R_2 (C_2 + C_{c2})} = 1\,\text{kHz}$$

Fig. 5.49. Model for 2-pole frequency compensation

is already $-90°$ and increases to $-180°$. In order to maintain a sufficient phase margin around the critical frequency, a zero point is introduced by R_{c2}, which shifts back the phase of *one* lowpass filter. This reduces the phase shift; the phase margin then reaches 45° and more in the range from 100 kHz to 1 MHz. The amplifier is thus correctly compensated for gains in the range of $A = 1 - 10$. For less negative feedback with a gain of $A = 1,000$, the critical frequency is 10 kHz; but at this point the phase margin is almost zero and the transient response is correspondingly poor.

In Fig. 5.48, one can see that the 2-pole compensation creates a higher loop gain without degrading the frequency response of the feedback circuit. The nonlinear distortions are also reduced to the same extent as the loop gain increases. This technology is therefore particularly interesting in electro-acoustics to insure low nonlinear distortions. However, bipolar compensation impairs the transient response: there is an overshooting and undershooting tendency in the range of some percent, which decays very slowly. And since bipolar frequency compensation must be tailored to each individual application, it is rarely used. It is more convenient to use a faster operational amplifier with 1-pole standard compensation. This technology is applied only in cases in which it is essential to keep the nonlinear distortions as low as possible even up to high frequencies.

5.2.8
Parameters of Operational Amplifiers

The most important parameters of operational amplifiers are listed in Fig. 5.50. These parameters, together with their influence on the noninverting and inverting amplifiers, are described below.

The **μA741** and **TLC272** amplifiers are older standard types that feature no particularly high-quality data compared to newly developed types. The reason for their widespread use in large numbers even today is that they are very low-priced. The TLC272 is built exclusively from n-channel and p-channel enhancement MOSFETs. Its maximum operating voltage is therefore limited to 16 V. As its common-mode control range and output swing reach the negative operating voltage, it is known as a single supply amplifier. Due to the MOS input its input currents are extremely low and the input resistances are correspondingly high. These values are usually not determined by the chip but by the case and the PCB.

The **OP177** is an operational amplifier that is suitable for achieving a particularly high precision. On the one hand, its offset voltage is very low and can be totally neglected in most applications and the user must insure that the thermal voltages at the soldering point cause no major errors. On the other hand, this type of amplifier features extremely high values for differential gain and common-mode rejection, which can be assumed to be infinite.

The **AD797** is a particularly low-noise amplifier for audio applications. Its noise voltage density of $1\,\text{nV}/\sqrt{\text{Hz}}$ is at the limit of technical feasibility. Its noise current, however, is not lower than in normal operational amplifiers. For this reason, the AD797 is particularly advantageous with low-resistance sources (see page 537). The gain–bandwidth product of 110 MHz seems to be unnecessarily high for audio applications. However, this is the precondition for a high loop gain and thus low distortions. Up to 20 kHz, the distortions remain 120 dB below the useful signal.

Parameter	Symbol	Standard amplifiers		Special amplifiers		
		μA 741 (bipolar)	TLC 272 (MOS)	OP 177 (precision)	AD 797 (low–noise)	LM 7171 (fast)
Differential gain	A_D	10^5	$4 \cdot 10^4$	$\mathbf{10^7}$	$2 \cdot 10^7$	$2 \cdot 10^4$
Common mode rejection	CMR	$3 \cdot 10^4$	$2 \cdot 10^4$	$\mathbf{10^7}$	10^7	$2 \cdot 10^5$
Offset voltage	V_O	1 mV	1 mV	$\mathbf{10\,\mu V}$	25 μV	1 mV
Offset voltage drift	$\Delta V_O / \Delta \vartheta$	6 μV/K	2 μV/K	$\mathbf{0.1\,\mu V/K}$	0.2 μV/K	35 μV/K
Input quiescent current	I_B	80 nA	$\mathbf{1\,pA}$	1 nA	250 nA	3 μA
Offset current	I_O	20 nA	0.5 pA	0.3 nA	100 nA	0.1 μA
Offset current drift	$\Delta I_O / \Delta \vartheta$	0.5 nA/K		3 nA/K	1 nA/K	1 μA/K
Differential input resistance	r_D	1 MΩ	1 TΩ	50 MΩ	7.5 kΩ	3 MΩ
Common mode resistance	r_{CM}	1 GΩ	1 TΩ	200 GΩ	100 MΩ	40 MΩ
Common mode control range	$V_{CM\,max}$	±13 V	0...14 V	±13 V	±12 V	±13 V
Input noise voltage density	$V_{nd} \cdot \sqrt{Hz}$	13 nV	25 nV	10 nV	$\mathbf{1\,nV}$	14 nV
Input noise current density	$I_{nd} \cdot \sqrt{Hz}$	2 pA	$\mathbf{1\,fA}$	0.3 pA	2 pA	2 pA
Maximum output current	$I_{o\,max}$	±20 mA	±20 mA	±20 mA	±20 mA	$\mathbf{\pm 100\,mA}$
Output swing	$V_{o\,max}$	±13 V	0...13 V	±14 V	±13 V	±13 V
Output resistance	r_o	1 kΩ	200 Ω	60 Ω	300 Ω	15 Ω
3 dB bandwidth	f_{gA}	10 Hz	50 Hz	0.06 Hz	5 Hz	$\mathbf{10\,kHz}$
Gain bandwidth product	f_T	1 MHz	2 MHz	0.6 MHz	110 MHz	$\mathbf{200\,MHz}$
Slew rate	dV_o/dt	0.6 V/μs	5 V/μs	0.3 V/μs	20 V/μs	$\mathbf{3000\,V/\mu s}$
Power bandwidth	f_p	10 kHz	100 kHz	5 kHz	300 kHz	$\mathbf{50\,MHz}$
Operating voltage	V_b	±15 V	0/+15 V	±15 V	±15 V	±15 V
Operating current	I_b	1.7 mA	1.4 mA	1.6 mA	8 mA	7 mA
Circuit in Fig.		5.14	5.20		5.25	5.28

Fig. 5.50. Parameters of operational amplifiers

The **LM7171** is a particularly fast operational amplifier that can be used up to 200 MHz. This can be seen from the high bandwidth and slew rate. But this is at the cost of poor DC data; the offset voltage drift and the input quiescent currents are high, while the differential gain is low.

For the calculation and design of operational amplifier circuits, it is generally possible to carry out a precise analysis of the circuit with all of its error sources. But it is easier to initially assume that the operational amplifier to be ideal and then calculate the deviations caused by the individual parameters of the real operational amplifier.

Differential and Common-Mode Gain

The output voltage of an operational amplifier is a function of the differential and common-mode gain: $V_o = f(V_D, V_{Gl})$. This leads to the total differential:

$$dV_o = \frac{\partial V_o}{\partial V_D} dV_D + \frac{\partial V_o}{\partial V_{CM}} dV_{CM} \tag{5.37}$$

The differential quotients are:

Differential gain $\qquad A_D = \dfrac{\partial V_o}{\partial V_D}$ $\qquad\qquad$ (5.38)

Common-mode gain $\qquad A_{CM} = \dfrac{\partial V_o}{\partial V_{CM}}$ $\qquad\qquad$ (5.39)

The voltage difference V_D applied to the inputs of an operational amplifier is amplified with the differential gain and transferred to the output. The slope of the transfer characteristic in Fig. 5.51a represents the differential gain. Due to the high differential gain, differential voltages below 1 mV are sufficient to exceed the possible output swing.

Applying the voltage V_{CM} to both inputs produces a common-mode signal, which would require a zero output voltage in an ideal operational amplifier. For a real operational amplifier there is a common-mode gain that is usually around 1, which is several orders of magnitude lower than the differential gain.

With the definitions of (5.38) and (5.39), it follows from (5.37)

$$dV_o = A_D \, dV_D + A_{CM} \, dV_{CM} \tag{5.40}$$

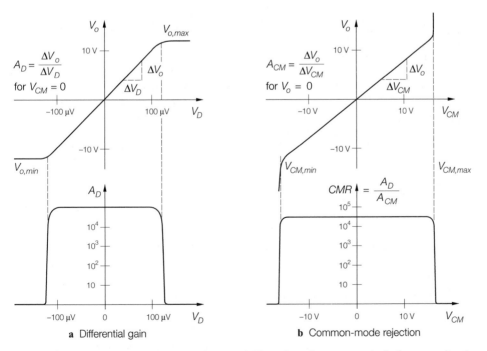

Fig. 5.51. Differential and common-mode control. The values shown are typical of an operational amplifier of the 741 class.

Since the transfer characteristics are approximately linear within the permissible control limits, (5.40) also applies in large-signal situations:

$$V_o = A_D V_D + A_{CM} V_{CM}$$

This equation can be solved for V_D; at the same time, the common-mode gain can be replaced by the more conventional common-mode rejection $CMR = A_D/A_{CM}$:

$$V_D = \frac{V_o}{A_D} - \frac{V_{CM}}{CMR} = \begin{cases} V_o/A_D & \text{for } V_{CM} = 0 \\ -V_{CM}/CMR & \text{for } V_o = 0 \end{cases} \qquad (5.41)$$

This represents the common definition of the differential gain

$$A_D = \left.\frac{\partial V_o}{\partial V_D}\right|_{dV_{CM}=0} = \left.\frac{V_o}{V_D}\right|_{V_{CM}=0} \qquad (5.42)$$

and an additional definition for the common-mode rejection:[5]

$$CMR = \frac{A_D}{A_{CM}} = \frac{\partial V_o}{\partial V_D} \cdot \frac{\partial V_{CM}}{\partial V_o} = \left.\frac{\partial V_{CM}}{\partial V_D}\right|_{dV_o=0} = \left.\frac{V_{CM}}{V_D}\right|_{V_o=0} \qquad (5.43)$$

The relationships between the common-mode and the differential voltages are obtained by applying a differential voltage at a certain common-mode voltage that is so high that it causes a zero output voltage. This is the voltage required to compensate the effect of the common-mode control. The slope of this function is the common-mode rejection, which is shown in Fig. 5.51b. From the abrupt drop in the common-mode rejection, one can clearly see the limits of the common-mode control capability. The limit defined by the circuitry in Fig. 5.14 is due to the fact that one of the transistors in the differential amplifier or the corresponding current source enters the saturation region. A comparison of Figs. 5.51a and 5.51b shows that the differential gain and the common-mode rejection look very similar.

In (5.41) one can see that the input voltage V_D consists of two parts: the first is determined by the output swing and the second is added during the common-mode control. As A_D and CMR are generally very high, within the linear operating region voltage V_D is usually very low, with values in the millivolt range. In order to take the effects of the finite differential gain and the common-mode rejection into account, it is easiest to use the models shown in Fig. 5.52. From the condition of an ideal operational amplifier that the output voltage assumes a value at which the input voltage difference becomes zero, we can calculate the output voltage:

$$V_i - \frac{V_o}{A_D} + \frac{V_{CM}}{CMR} = kV_o$$

Since, in a noninverting amplifier, the common-mode voltage

$$V_{CM} = (V_P + V_N)/2 \approx V_P = V_i \qquad (5.44)$$

is practically equal to V_i, the gain is

$$A = \frac{V_o}{V_i} = \frac{A_D}{1 + kA_D}\left(1 + \frac{1}{CMR}\right) \approx \frac{A_D}{1 + kA_D} \approx \frac{1}{k} = 1 + \frac{R_N}{R_1} \qquad (5.45)$$

---

[5] For the common-mode gain and common-mode rejection, only the absolute value is specified; therefore the sign is of no importance. To avoid the impression that other effects could be compensated by common-mode rejection, it is always good practice to use the less favorable sign.

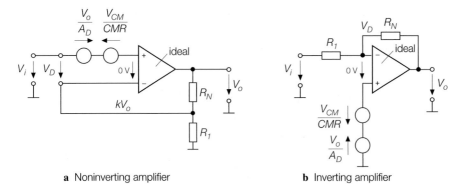

a Noninverting amplifier   **b** Inverting amplifier

Fig. 5.52. Effect of the finite differential gain and the common-mode rejection on the gain

The deviation from the ideal behavior caused by the finite differential gain is

$$\frac{\Delta A}{A} = \frac{A_{id} - A}{A_{id}} = \frac{\dfrac{1}{k} - \dfrac{A_D}{1 + kA_D}}{1/k} = \frac{1}{1 + kA_D} \approx \frac{1}{g} \qquad (5.46)$$

Thus, the relative deviation from the ideal behavior is the same as the reciprocal value of the loop gain; therefore, it is generally very low. Production tolerances and the changes in differential gain due to temperature variations are also reduced by the same factor.

The inverting amplifier in Fig. 5.52b has a common-mode voltage $V_{CM} = V_{D/2} \ll V_i$. This means that the finite common-mode rejection CMR has no influence on the voltage gain; the gain is thus as defined in (5.17).

Offset Voltage

The transfer characteristic of a real operational amplifier does not pass through zero, but is shifted by the *input offset voltage*; this is illustrated in Fig. 5.53. The offset voltage is usually in the millivolt range; high-quality operational amplifiers even feature offset voltages in the microvolt range, as shown in Fig. 5.50. But even a low offset voltage can cause the amplifier to be overdriven if both inputs are connected to ground (that is, $V_D = 0$); this can also be seen in Fig. 5.53. The reason for this is the high differential gain that amplifies even small offset voltages to such a degree that the output swing is exceeded.

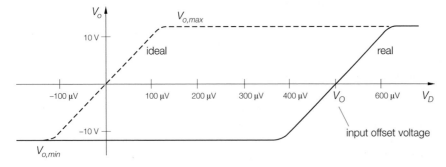

Fig. 5.53. Effect of the offset voltage on the transfer characteristic of an operational amplifier

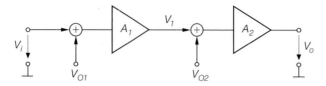

Fig. 5.54. Model of the influence of offset voltages in multi-stage amplifiers

However, operational amplifiers are not usually operated in open-circuit mode but with negative feedback; then, the error caused by the offset voltage is only amplified with the closed loop gain as the input signal. The effect is similar to that achieved with the signal voltage source connected in series. If this small error still causes disturbance, the offset voltage can be brought down to zero. Some operational amplifiers have special terminals for connecting a trimmpoti to adjust the offset voltage. However, it is often better to use a type with so small an offset voltage that the problem does not arise. Type OP177 in Fig. 5.50 illustrates how low the offset voltage may be. It is usually much cheaper for the operational amplifier to be balanced by the manufacturer rather than by the user, since this would require not only a trimmer but also a measuring station, a technician, and adjustment instructions.

The offset voltage is caused by various factors. In addition to the tolerances when matching the input transistors, the input amplifier and the subsequent circuitry also have asymmetries and tolerances. The greatest influence, however, is the input stage. This is exemplified by the model of a two-stage amplifier shown in Fig. 5.54, where the offset voltage is supplied to the input of each stage. This leads to the output voltage:

$$V_o = (V_1 + V_{O2}) A_2 = [(V_i + V_{O1}) A_1 + V_{O2}]A_2$$
$$= A_1 A_2 V_i + A_1 A_2 V_{O1} + A_2 V_{O2}$$

In order to determine the offset voltage of the entire circuit in relation to the input, one sets $V_o = 0$ and calculates the input voltage for this situation:

$$V_i (V_o = 0) = V_O = -V_{O1} - \frac{1}{A_1} V_{O2} \tag{5.47}$$

The input offset voltage of the first stage affects to full amount – whereas the offset of the second stage is reduced by the factor $1/A_1$. The goal is therefore to make the gain of the first stage as high as possible.

With the offset voltage reduced to zero, only its dependence on temperature, time, and the operating voltage can be noticed:

$$dV_O(\vartheta, t, V_b) = \frac{\partial V_O}{\partial \vartheta} d\vartheta + \frac{\partial V_O}{\partial t} dt + \frac{\partial V_O}{\partial V_b} dV_b \tag{5.48}$$

Here, $\partial V_O / d\vartheta$ is the temperature drift, with typical values of $3 - 10 \,\mu\text{V/K}$. The long-term drift $\partial V_O / \partial t$ is several microvolts per month. This can be regarded as a low-frequency portion of the noise. The supply voltage rejection ratio $\partial V_O / \partial V_b$ characterizes the influence that the supply voltage variations have on the offset voltage. It has a value of $10 - 100 \,\mu\text{V/V}$. To keep this contribution to the offset voltage low, the supply voltage should be regulated.

According to Fig. 5.53, the transfer characteristic of an operational amplifier with an offset voltage within the linear control range is:

$$V_o = A_D(V_D - V_O) \tag{5.49}$$

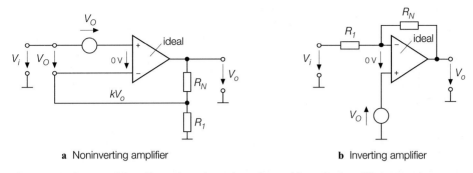

a Noninverting amplifier **b** Inverting amplifier

Fig. 5.55. Influence of the offset voltage in noninverting and inverting amplifiers

In order to reduce the output quiescent potential to zero, it is necessary to either zero the offset voltage or apply a voltage $V_D = V_O$ to the input. This results in the rule:

The offset voltage is the voltage required at the input to achieve a zero output voltage.

To examine the effect of the offset voltage in circuits with negative feedback, use of the equivalent circuits in Fig. 5.55 is recommended. If $V_i = 0$, then both circuits are identical. This leads to an offset voltage at the output:

$$V_o(V_i = 0) = -\left(1 + \frac{R_N}{R_1}\right) V_O \qquad (5.50)$$

This complies with the voltage gain of the noninverting amplifier. Thus, the offset voltage of a noninverting amplifier is amplified like the input voltage, while in an inverting amplifier this is only an approximation.

Input Currents

The input quiescent current of an operational amplifier corresponds to the base or gate current of the input transistors. Its value depends on the current used to drive the input transistors. In multipurpose amplifiers with bipolar transistors at the input, operated with collector currents of $10\,\mu A$, the input quiescent currents may be $0.1\,\mu A$. In wide-band amplifiers with collector currents up to $1\,mA$, the input currents may be several microamperes. With Darlington circuits at the input, the input quiescent current is in the nanoampere range. The lowest input quiescent currents are found in operational amplifiers with field effect transistors at the input. Here, they often amount to a few picoamperes only.

Since the input transistors are operated with constant collector currents, the base currents are constant as well at common mode input signals; therefore, the inputs represent constant current sources. In reality, the input currents are similar, but not identical. This is the reason why data sheets specify the mean *input quiescent current*

$$I_B = \frac{1}{2}(I_P + I_N) \qquad (5.51)$$

and the *input offset current*

$$I_O = |I_P - I_N| \qquad (5.52)$$

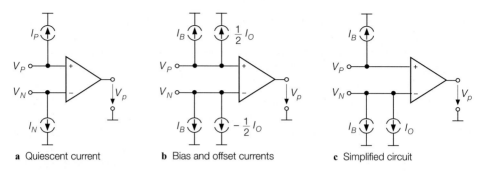

a Quiescent current   **b** Bias and offset currents   **c** Simplified circuit

Fig. 5.56. Converting the input currents to bias and offset currents

The input currents can be calculated from these definitions:

$$I_N = I_B \pm I_O/2 \quad or \quad I_P = I_B \mp I_O/2 \tag{5.53}$$

This relationship is illustrated in Fig. 5.56. To simplify matters, the offset current may be added to one of the input currents; the resultant error is usually small, since $|I_O| \ll |I_B|$.

The influence that input currents have on the amplifier circuits is calculated with the help of Fig. 5.57 and the simplified model in Fig. 5.56c. According to Fig. 5.57a, the output voltage is

$$V_o = \left(1 + \frac{R_N}{R_1}\right) V_i + I_B \left(R_N - \frac{R_g (R_1 + R_N)}{R_1}\right) + \frac{I_O}{2} \left(R_N + \frac{R_g (R_1 + R_N)}{R_1}\right) \tag{5.54}$$

If the input resistances are adjusted according to the relationship

$$R_g = \frac{R_N R_1}{R_N + R_1} = R_N \| R_1 \tag{5.55}$$

then the effect of I_B is eliminated and (5.54) is simplified:

$$V_o = \left(1 + \frac{R_N}{R_1}\right) V_i + I_O R_N$$

Only the error caused by the offset current remains; this is usually small compared to the input quiescent current, as seen in Fig. 5.50. The amount of the offset current varies from

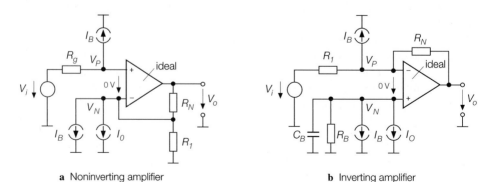

a Noninverting amplifier   **b** Inverting amplifier

Fig. 5.57. Effect of the input currents

amplifier to amplifier, as does its polarity. In principle one could adjust the offset current in the same way as the offset voltage, but it is better to dimension the circuitry so low-ohmic that the resulting error remains negligible. Furthermore, as for the offset voltage, the offset current is also temperature sensitive; the offset current drift indicates the rate at which it changes with temperature.

For the inverting amplifier shown in Fig. 5.57b, the noninverting input usually carries ground potential. The input current thus causes an offset of $I_N R_N$ at the output. Here too, this error can be compensated by connecting the noninverting input indirectly to ground via resistance $R_B = R_N \| R_1$, so that the total resistances of both inputs are equal. The error $I_O R_N$ caused by the offset current is the only remaining error. To prevent resistance R_B from generating additional noise, the alternating voltages are shorted by capacitance C_B.

We have shown that the error caused by the input currents rises proportionally to the dimensions of the resistors connected. Therefore, these resistors should be dimensioned as low as possible so that this error remains small. If you must use high feedback resistors, an operational amplifier with sufficiently low input currents should be selected. There are very significant differences, as can be seen in Fig. 5.50.

Input Resistances

As in the case of the differential amplifier, the operational amplifier has two input resistances, the differential resistance and the much larger common-mode resistance. The influence that the common-mode resistance may have on the noninverting amplifier can be taken from the equivalent circuit in Fig. 5.58a. The resistors are connected between input and ground; that is, they are in parallel to the inputs and therefore they are not affected by the negative feedback. The common-mode resistance at the noninverting input causes an attenuation whereas the resistance at the inverting input results in an increase in gain. After matching the internal resistances of both inputs to $R_g = R_N \| R_1$, the effect is fully compensated. Due to their very high resistance common mode resistance has little influence.

In order to examine the effect of the differential input resistances, it is necessary to take a real operational amplifier with finite gain and common-mode rejection. We thus look at Fig. 5.58b and calculate the current flowing through the differential input resistance. According to (5.41), the current is:

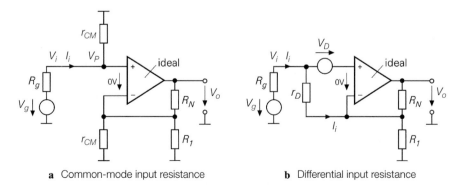

a Common-mode input resistance   **b** Differential input resistance

Fig. 5.58. Effect of the differential and common-mode input resistances in noninverting amplifiers

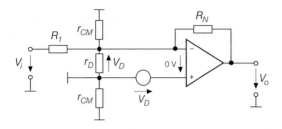

Fig. 5.59. Input resistance in an inverting amplifier

$$I_i = \frac{V_D}{r_D} = \left(\frac{V_o}{A_D} + \frac{V_{CM}}{CMR}\right)\frac{1}{r_D}$$

For $V_o = V_i/k$, $V_{CM} = V_i$, and $g = kA_D$, we have a contribution to the input resistance due to r_D:

$$r_D' = \frac{V_i}{I_i} = r_D\,\frac{g\,CMR}{g + CMR} = \begin{cases} g\,r_D & \text{for} \quad CMR \gg g \\ CMR\,r_D & \text{for} \quad g \gg CMR \end{cases} \tag{5.56}$$

Therefore, the differential input resistance is significantly increased by the negative feedback, as the differential voltage V_D, which is only a fraction of the input voltage V_i. The resulting input resistance of the noninverting amplifier is thus $r_i = r_{CM}||r_D'$; since both portions are very high, we obtain input resistances in the gigaohm range, even with operational amplifiers with bipolar transistors.

A differential voltage can, of course, also be caused by the offset voltage, so that $V_D = V_O$. In this case, a constant current $I_i = V_O/r_d$ flows through r_D, causing a constant offset at the output. The output voltage caused by I_i can be calculated according to Fig. 5.58b:

$$\Delta V_o = -\left(R_N + \frac{R_1 + R_N}{R_1}R_g\right)\frac{V_O}{r_D}$$

Furthermore, even if the resistances are balanced, the offset is not compensated, because both terms have the same sign. In the balanced situation of $R_g = R_N||R_1$, the equation is simplified:

$$\Delta V_o = -\frac{2\,R_N}{r_D}\,V_O \tag{5.57}$$

The situation is much easier for the inverting amplifier shown in Fig. 5.59. Since the differential voltage V_D is in the millivolt range, the inverting input represents a virtual ground here. Therefore, resistance R_1 acts as if connected to a real ground. The input resistance of the circuit is thus identical to R_1. It is not influenced measurably by the differential and the common-mode input resistances of the amplifier. However, R_1 is usually in the range of $1 - 100\,\text{k}\Omega$, and is thus several orders of magnitude lower than the input resistance of a noninverting amplifier.

Output Resistance

As Fig. 5.50 indicates, in terms of output resistance, real operational amplifiers are far from ideal. However, the output resistance is reduced by negative feedback: the output voltage reduction induced by the load is fed back to the inverting input via the voltage divider

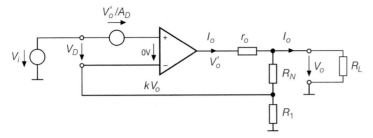

Fig. 5.60. Model for calculating the output resistance

R_N, R_1 in Fig. 5.60. This increases V_D and counteracts the original decrease of the output voltage.

The model shown in Fig. 5.60 is used for quantitative analysis. Neglecting the current through the negative feedback voltage divider, the output voltage is calculated from

$$V_i - \frac{V_o'}{A_D} = kV_o \quad \text{and} \quad V_o' = V_o + I_o r_o$$

which leads to the output voltage:

$$V_o = \frac{A_D V_i - I_o r_o}{1 + kA_D} \approx \frac{V_i}{k} - \frac{I_o r_o}{g}$$

In addition to the normal output voltage, there is also a reduction caused by the current, but this is reduced by the loop gain. Consequently, the output resistance is:

$$r_o' = -\frac{dV_o}{dI_o} = \frac{r_o}{g} \tag{5.58}$$

This means that the negative feedback reduces the output voltage by the loop gain.

An Example: Static Errors

The size of the various static errors can be demonstrated using a numeric example. For this purpose, we use the noninverting amplifier in Fig. 5.61, whose gain has been adjusted to 10 by resistors R_N and R_1. In order to make the various errors clearly visible, an operational amplifier with relatively poor data is used; to this end we have chosen the data of μA741 from Fig. 5.50. The input voltage source provides a voltage of 1 V. With a gain of $A = 10$ an ideal operational amplifier produces an output voltage of $V_o = 10$ V. The deviations caused by the various nonideal characteristics will be calculated.

If a differential gain of $A_D = 10^5$ is taken into account, the voltage error calculated for the input, according to Fig. 5.52, will be

$$\frac{V_o}{A_D} = \frac{10\,\text{V}}{10^5} = 100\,\mu\text{V}$$

At a gain of 10, this results in an output voltage error of 1 mV. The error induced by the common-mode control is:

$$\frac{V_{CM}}{CMR} = \frac{1\,\text{V}}{3 \cdot 10^4} = 33\,\mu\text{V}$$

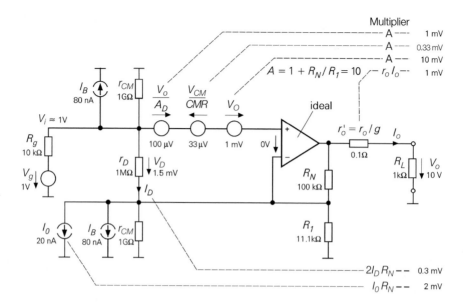

Fig. 5.61. Static error of a noninverting amplifier with a closed loop gain $A = 10$ for an operational amplifier of the 741 class

This error also undergoes a 10-fold gain and at the output amounts to 0.33 mV. According to Fig. 5.55, the effect of the offset voltage can be regarded similarly: a voltage of 1 mV at the input produces an error of 10 mV at the output.

With matched input resistances, the output quiescent current I_B has no effect. The offset current causes an error of:

$$\Delta V_o = I_O R_N = 20\,\text{nA} \cdot 100\,\text{k}\Omega = 2\,\text{mV}$$

If the source had no internal resistance, it would be possible to insert a 10 kΩ resistance with a capacitor in parallel to suppress unwanted noise. If the input resistances are not matched, the input quiescent current must be taken into account, which would then produce a voltage error of $I_B R_N = 8\,\text{mV}$.

The common-mode input resistances are obviously very high compared to all other resistances, so that they seldom produce an error. Here, their effect is cancelled out because the input resistances are matched. But the differential input resistance produces an error that is caused by the current:

$$I_D = \frac{V_D}{r_D} = \frac{1.5\,\text{mV}}{1\,\text{M}\Omega} = 1.5\,\text{nA} \tag{5.59}$$

According to (5.57), this current induces an output voltage error of

$$\Delta V_o = 2 \cdot R_N \frac{V_D}{r_D} = 2 \cdot R_N I_D = 2 \cdot 1.5\,\text{nA} \cdot 100\,\text{k}\Omega = 0.3\,\text{mV}$$

The error caused by the output resistance must also be evaluated. Assuming that the output is loaded with the resistance $R_L = 1\,\text{k}\Omega$, the output current is $I_o = 10\,\text{V}/1\,\text{k}\Omega = 10\,\text{mA}$.[6] Across the output resistance that is transformed according to (5.58), there is a voltage drop of

$$\Delta V_o = \frac{r_o}{kA_D} I_o = \frac{1\,\text{k}\Omega}{10^4} 10\,\text{mA} = 1\,\text{mV}$$

When calculating the error, we have not taken the sign into consideration. Due to errors caused by the differential gain and the output resistance, the output voltage becomes smaller. But the signs of the offset voltage, the offset current, and the common-mode rejection are not defined; it is therefore impossible to say with which sign they influence the output voltage. The magnitude of the individual errors is more important: in this example, no single error exceeds 1% of the output voltage. Most disturbing is the $10\,\text{mV}$ error caused by the offset voltage, as it is independent of the output voltage value. With an output voltage of $100\,\text{mV}$, this amounts to 10%. For this reason, the offset voltage is an important parameter when choosing an operational amplifier.

Bandwidth

Operational amplifier as a lowpass filter: Now that we have seen that a frequency-corrected operational amplifier behaves approximately like a first-order lowpass filter, its frequency response can be simply described as:

$$\underline{A}_D = \frac{A_{D0}}{1 + j\dfrac{f}{f_g}} \tag{5.60}$$

The differential gain of the open amplifier is usually very high; it often reaches values of $A_{D0} = 10^5$ and above, as shown in Fig. 5.62. On the other hand, the cutoff frequency of the open amplifier is usually very low and often amounts to only $f_g = 10\,\text{Hz}$.

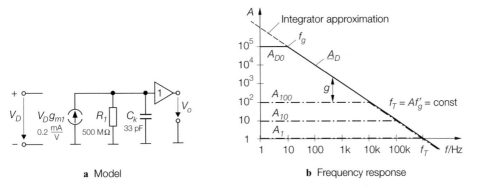

a Model   **b** Frequency response

Fig. 5.62. Frequency-corrected operational amplifier as a first-order lowpass filter for calculating the frequency response of a circuit with negative feedback. An amplifier of the 741 class is taken as an example. f_g' is the cutoff frequency of the amplifier with feedback.

---

[6] For a standard operational amplifier, this is a relatively high current, which is not far from the maximum output current of 20 mA. Such high currents should only be allowed if they are unavoidable, since the operational amplifier becomes heated up by the power dissipation. The offset voltage drift and offset current drift then cause additional errors.

According to (5.10), the frequency response of the circuit with negative feedback is:

$$\underline{A} = \frac{\underline{A}_D}{1 + k\underline{A}_D} = \frac{1/k}{1 + \dfrac{1}{k\underline{A}_D}} \tag{5.61}$$

Inserting (5.60) leads to:

$$\underline{A} = \frac{A_{D0}}{1 + kA_{D0}} \frac{1}{1 + j\dfrac{f}{f_g\,(1 + kA_{D0})}} \overset{kA_{D0} \gg 1}{\approx} \frac{1/k}{1 + j\dfrac{f}{kf_T}} \tag{5.62}$$

A comparison of the right-hand sides of (5.61) and (5.62) shows that one can use a simplified frequency response of an open amplifier:

$$\underline{A}_D = -j\frac{f_T}{f} \tag{5.63}$$

This represents the frequency response of an integrator; for this reason it is also known as the *integrator approximation* of the operational amplifier. As shown in Fig. 5.62b, only low frequencies lead to a deviation from the exact frequency response of the open amplifier: here, the gain approaches infinity while the actual gain approaches A_{D0}. Insertion of the integrator approximation (5.63) into (5.61) results in:

$$\underline{A} = \frac{\underline{A}_D}{1 + k\underline{A}_D} = \frac{A_0}{1 + j\dfrac{A_0}{f_T}f} = \begin{cases} A_0 = 1/k & \text{für } f \ll f'_g \\ \underline{A}_D = -jf_T/f & \text{für } f \gg f'_g \end{cases} \tag{5.64}$$

where $A_0 = 1/k$ is the gain defined by the negative feedback. In this way, we obtain the result of (5.62) with less calculation and without further approximation. Thus, up to the cutoff frequency $f'_g = f_T/A_0 = f_T k$, the gain of the circuit with negative feedback has a value determined by the feedback; above the cutoff frequency the gain is the same as in the open amplifier. This is illustrated in Fig. 5.62b, which includes the frequency responses for different closed loop gains. It follows that the product of closed loop gain and bandwidth is constant as can be seen in the figure:

$$\boxed{f_T = A_0 \cdot f'_g = \text{GBW} = \text{Gain BandWidth product}}$$

Limitation by external circuitry: An unexpected limitation of the bandwidth may be caused by parasitic capacitances of the negative feedback resistances. This effect is shown in Fig. 5.63. Each resistor has a parasitic capacitance that depends only on the mechanical construction and not on the resistance value. Thus, for high frequencies, the gain of the circuit

$$A_{HF} = 1 + \frac{C_1}{C_N} = 2$$

is independent of the resistors. For $R_N = R_1$, a frequency-corrected voltage divider exists; the gain is thus always $A = 2$ for all frequencies. To achieve a frequency-corrected voltage divider in other cases the time constants must be identical:

$$R_1 C_1 = R_N C_N \quad \Rightarrow \quad C_1 = \frac{R_N}{R_1} C_N = \frac{100\,\text{k}\Omega}{11.1\,\text{k}\Omega}\,1\,\text{pF} = 9\,\text{pF} \tag{5.65}$$

In this case, it is necessary to connect another capacitor of 8 pF in parallel to C_1.

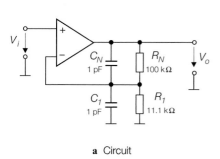

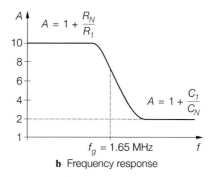

a Circuit   **b** Frequency response

Fig. 5.63. Cutoff frequency caused by parasitic capacitances of the feedback resistances with an ideal amplifier

Noise

The noise of an operational amplifier can be described by the noise voltage and noise current densities related to the input, as is the case for discrete transistors. Typical values are listed in Fig. 5.50 on page 524. These have to be multiplied by the root of the bandwidth in order to calculate the noise voltage and the noise current:

$$V_n = V_{nd}\sqrt{B} \quad \text{or} \quad I_n = I_{nd}\sqrt{B} \tag{5.66}$$

Resistances also produce noise; their noise power

$$P_n = 4\,kTB \tag{5.67}$$

is independent of the value of the resistance. The value k represents Boltzmann's constant and T is the absolute temperature; at room temperature we have $4kT = 1.6 \times 10^{-20}$ W s. This can be used to calculate the noise voltage:

$$V_n = \sqrt{PR} = \sqrt{4kTBR} = 0.13\,\text{nV}\sqrt{\frac{B}{\text{Hz}}}\sqrt{\frac{R}{\Omega}} \tag{5.68}$$

A $10\,\text{k}\Omega$ resistance has a noise voltage density of $V_{nd} = 13\,\text{nV}/\sqrt{\text{Hz}}$. Figure 5.64 shows all the noise voltage sources of an operational amplifier configured as a noninverting amplifier. It can be seen that every resistance is a noise voltage source, and that the noise voltage of the operational amplifier acts like the offset voltage and the noise current like the input quiescent current. In the internal resistance R_g of the signal source, the input noise current of the amplifier generates a noise voltage $I_n R_g$ which, together with the noise of R_g and the voltage noise of the amplifier, is amplified in the same way as the useful signal. The noise of resistance R_1 is amplified with the gain of 9 of the inverting amplifier. The noise current at the inverting input produces a voltage drop across R_N. In order to calculate the resulting noise voltage at the amplifier's output, the individual noise components cannot be simply added. Since these are uncorrelated noise sources, the squares of the noise components must be added:

$$V_{n,\,tot} = \sqrt{\sum V_n^2} \tag{5.69}$$

This has the consequence that smaller contributions have almost no influence on the result. Thus, in this example, the resulting noise voltage density $V_{no,tot} = 353\,\text{nV}/\sqrt{\text{Hz}}$. To

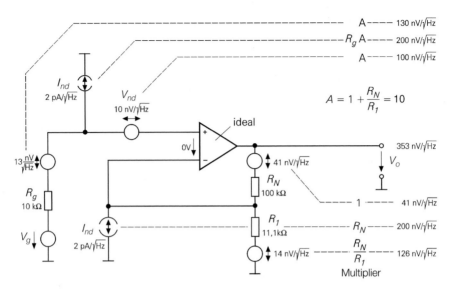

Fig. 5.64. Noise sources of a noninverting amplifier, taking the 741 class as an example. The closed loop gain of the circuit is $A = 10$

calculate the noise voltage, the bandwidth must be taken into consideration. This requires multiplication by $\sqrt{B}$ and a correction factor of $\pi/2 = 1.57$, which takes into account the fact that the noise above the cutoff frequency will not suddenly become zero, but will decline in the same way as a first-order lowpass filter (see Fig. 4.180 on page 482). For a bandwidth of $B = 100\,\text{kHz}$, the example in Fig. 5.64 gives an output noise voltage

$$V_{no,\,tot} = \frac{\pi}{2}\,\frac{353\,\text{nV}}{\sqrt{\text{Hz}}}\,\sqrt{100\,\text{kHz}} = 175\,\mu\text{V} \tag{5.70}$$

In order to reduce the noise, it is necessary to configure a circuit with lower resistance values and to use an operational amplifier with less voltage and current noise. Reducing the resistances in Fig. 5.64 by a factor of 100 causes noise voltage reduction by a factor of ten. Using the model AD797 amplifier, which has a noise voltage density of only $1\,\text{nV}/\sqrt{\text{Hz}}$, we obtain a noise voltage of only $V_{no,tot} = 11\,\mu\text{V}$ at the output for the same bandwidth.

Since it already exists at the amplifier input, the internal resistance of the input voltage source R_g represents a lower noise limit. This value can be calculated using (5.68). For the purpose of comparison, one can convert the noise voltage at the output of the amplifier to the value at the input by dividing it by the gain. The noise of the amplifier, including its circuitry, is obtained by taking all resistors including the generator resistor and the amplifier in Fig. 5.64 into account. This allows a comparison to be made in Fig. 5.65a as to how much the amplifier contributes to the total noise. The voltage noise of the amplifier is dominant in the case of low source resistances, while at high source resistances it is the current noise that produces a noise voltage across R_g, which dominates. Since this voltage is proportional to R_g, it rises in the logarithmic graph with double the slope of the resistance noise R_g which rises with the square root of R_g according to (5.68). In order to reduce the noise at low generator resistances, an amplifier with low voltage noise must be used. For the purpose of comparison, we chose the AD797 amplifier, whose value $1\,\text{nV}/\sqrt{\text{Hz}}$ is only $1/10$ of the voltage noise. It is obvious that source resistances in the

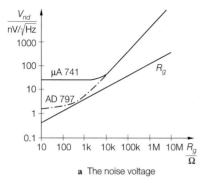

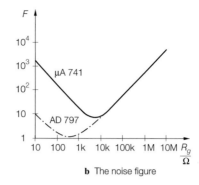

a The noise voltage   **b** The noise figure

Fig. 5.65. Dependence of the noise voltage and the noise figure on the source resistance, using models μA741 and AD797 as examples

range of 500 Ω bring us close to the theoretical limits. For high generator resistances, the low noise voltage offers no advantage. In this case, an amplifier with low current noise is more favorable.

It is obvious from Fig. 5.65a that it is not the absolute noise value that determines the quality of an amplifier, but the factor by which the circuit with amplifier produces more noise than the generator resistance itself.

> *The noise figure indicates by which factor the noise power with a real amplifier is higher than that of an ideal – that is, noise-free – amplifier at constant source resistance noise.*

It is calculated most easily from the relation

$$F = \left(\frac{\text{total noise voltage}}{\text{noise voltage of the source resistance}} \right)^2 \tag{5.71}$$

Due to the ratio it makes no matter whether the quotient is taken at the input or the output of the amplifier. You only have to translate all noise voltages with the gain to the output or the input. For the example in Fig. 5.64 we obtain a noise figure

$$F = \left(\frac{(353/10)\,\text{nV}/\sqrt{\text{Hz}}}{13\,\text{nV}/\sqrt{\text{Hz}}} \right)^2 = \left(\frac{353\,\text{nV}/\sqrt{\text{Hz}}}{(13 \cdot 10)\,\text{nV}/\sqrt{\text{Hz}}} \right)^2 = 7.4$$

The interdependence of the noise figure and source resistance is shown in Fig. 5.65b. As can be seen, there is a distinct minimum; the optimum noise figure is achieved with an optimum generator resistance:

$$R_{g,\,opt} = \frac{V_{nd}}{I_{nd}} = \begin{cases} 10\,\text{nV}/2\,\text{pA} = 5 \ \ \text{k}\Omega \quad \text{for} \quad \mu\text{A741} \\ 1\,\text{nV}/2\,\text{pA} = 0.5\,\text{k}\Omega \quad \text{for} \quad \text{AD797} \end{cases} \tag{5.72}$$

There are systematic differences in the noise characteristic between the various technologies used in the construction of an input differential amplifier. A comparison is shown in Fig. 5.66. Operational amplifiers with bipolar transistors at the input have the lowest noise voltage, which in good models amounts to only $1\,\text{nV}/\sqrt{\text{Hz}}$. Junction FETs at the input result in clearly higher noise voltages, even in high-quality models. CMOS operational amplifiers have the highest noise voltage but the lowest current noise at high frequencies. Junction FETs are superior at low frequencies.

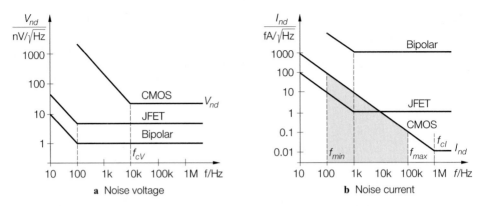

Fig. 5.66. Voltage and current noise of a low-noise operational amplifier with bipolar transistors, with junction FETs and MOSFETs at the input

Below a certain frequency there is an increase in both the voltage and current noise, as shown in Fig. 5.66. As the noise density is inversely proportional to the frequency, this noise is called $1/f$ noise. The frequency at which it enters the white noise is clearly higher in CMOS operational amplifiers than in types with bipolar transistors or junction FETs at the input. Data sheets usually specify the noise density in the region of white noise; this is the region in which the noise density is independent of the frequency. In order to evaluate the contribution of the noise voltage in the $1/f$ region, we require integration over the noise density; this leads to:

$$V_n = \sqrt{\int_{f_{min}}^{f_{max}} V_{nd}^2 \left(\frac{f_{cV}}{f} + 1\right) df} = V_{nd} \sqrt{f_{cV} \ln \frac{f_{max}}{f_{min}} + (f_{max} - f_{min})} \qquad (5.73)$$

$$I_n = \sqrt{\int_{f_{min}}^{f_{max}} I_{nd}^2 \left(\frac{f_{cI}}{f} + 1\right) df} = I_{nd} \sqrt{f_{cI} \ln \frac{f_{max}}{f_{min}} + (f_{max} - f_{min})} \qquad (5.74)$$

Here, f_{max} and f_{min} are the cutoff frequencies of the region of interest and f_{cV} and f_{cI} are the cutoff frequencies of the $1/f$ noise. They are shown in Fig. 5.66 as examples of the current noise of a CMOS operational amplifier. For the frequency range from 100 Hz to 100 kHz, the noise current is:

$$I_n = 0.01 \frac{fA}{\sqrt{Hz}} \sqrt{1\,MHz \ln \frac{100\,kHz}{100\,Hz} + (100\,kHz - 100\,Hz)} = 26fA$$

5.3
Transconductance Amplifier (VC-OPA)

The operational transconductance amplifier (OTA) differs from a conventional amplifier due to its high-resistive output; its output acts like a current source, as illustrated in the overview in Fig. 5.63. Any VV operational amplifier can be converted into a VC operational amplifier by omitting the emitter follower at the output.

5.3.1
Internal Construction

The simplest circuit for a VC operational amplifier is achieved by taking the operational amplifier shown in Fig. 5.11 and removing the emitter follower. The result is the circuit shown in Fig. 5.67. Here, the characteristic value is the transconductance, which can be calculated from the model:

$$g_D = \frac{I_q}{V_D} = \frac{I_{os}}{V_D} = \frac{1}{2\,r_m} = \frac{1}{2}g_m = \frac{1}{2}\frac{I_0}{V_T} \tag{5.75}$$

With no load at the output, the current I_q flows through the output resistance r_{CE4} and generates the open-circuit voltage gain

$$\frac{V_o}{V_D} = \frac{I_q}{V_D}r_{CE4} = g_D\,r_{CE4} = \frac{1}{2}\frac{I_0}{V_T}\frac{V_A}{I_0} = \frac{1}{2}\frac{100\,\text{V}}{26\,\text{mV}} = 1923$$

which is identical to that of the VV operational amplifier in Fig. 5.11. The amount of the voltage gain with a connected load depends very much on the load resistance, since here the output resistance of $100\,\text{k}\Omega$ is much higher than in a VV operational amplifier, where it is $1\,\text{k}\Omega$ or less. For high-resistive loads that permit a high gain, the VC operational amplifier behaves almost identically to the VV operational amplifier.

In a practical realization as in Fig. 5.68, both output currents of the input differential amplifier are used to control the output. This not only leads to doubled output currents but also to markedly improved zero stability, as the quiescent currents I_0 cancel each other out at the output.

A special feature is the fact that the user can control the current I_0. The easiest way to do this is to use a resistor R_{cont} with a voltage drop that is 0.6 V lower than the negative operating voltage. A nominal current $I_{cont} = 0.5\,\text{mA}$ therefore requires a resistance of

$$R_{cont} = \frac{14.4\,\text{V}}{0.5\,\text{mA}} = 28.8\,\text{k}\Omega$$

The maximum output current is thus $I_{o,max} = I_{cont} = 2I_0 = 0.5\,\text{mA}$. The transconductance of the circuit according to (5.75), and thus the voltage gain, can be controlled by adjusting the current I_{cont}. If a load resistance R_L that is small compared to the output resistance of the circuit, the output voltage is

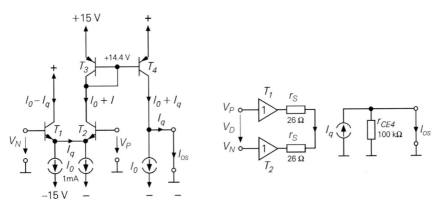

Fig. 5.67. Simple VC operational amplifier. The values shown are for $I_0 = 1\,\text{mA}$

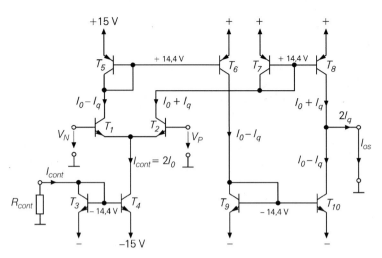

Fig. 5.68. Schematic circuit for the CA3080 from Harris. It is also known as an *operational transconductance amplifier (OTA)* or g_m-cell

$$V_o = g_D R_L V_D = \frac{R_L}{2\,V_T} I_{cont} V_D$$

This characteristic can be utilized to multiply two voltages by making current I_{cont} proportional to a second input voltage. A more comprehensive description of such circuits is contained in Sect. 11.8.2 on page 754.

Nowadays, model CA3080 is of no practical significance because of its outdated technology and the small output currents. But the MAX436 from Maxim or the OPA615 from Burr Brown are modern successors that operate in push–pull AB mode and thus provide high output currents. The circuit in Fig. 5.69 is obtained by taking the VV operational amplifier shown in Fig. 5.28 and removing the impedance converter at the output stage. This circuit has the particular advantage of working even with currents $I_q > 2\,I_0$ when the upper or lower half of the circuit is without current.

The newer types enable the user to reduce the transconductance by means of an external emitter resistance R_E. Figure 5.70 shows how the resistance is connected in series with the transconductance resistances of the input transistors. If we then take into consideration the fact that the current mirrors of the MAX436 have a transformation ratio of $k_I = 8$, the transconductance of the circuit becomes

$$g_D = \frac{I_{os}}{V_D} = \frac{k_I I_q}{V_D} = \frac{k_I}{2\,r_m + R_E} \approx \frac{k_I}{R_E}$$

Calculation of the voltage gain of the circuit takes the load resistance into account:

$$A = g_D\,(R_L||r_o) = \frac{k_I}{R_E}\,(R_L||r_o) \overset{r_o \gg R_L}{\approx} k_I\,\frac{R_L}{R_E}$$

As the gain can be adjusted to any value due to the current feedback via R_E, no additional negative voltage feedback is required.

One can see that all of the VC operational amplifiers shown here use transistors in common-emitter configuration at their output in order to achieve a high output resistance. For this reason, no special circuitry is required to realize a rail-to-rail output, as is the case

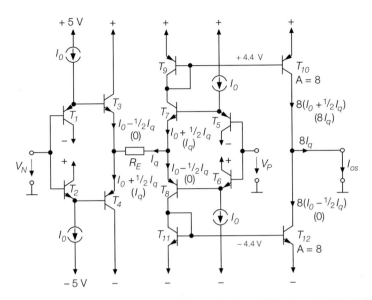

Fig. 5.69. Example of a modern VC operational amplifier (e.g., the MAX436). The circuit is also known as a wide-band transconductance amplifier (WTA). The values in parentheses represent the case $I_q > I_0$

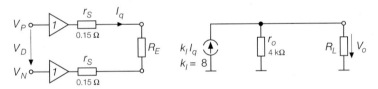

Fig. 5.70. Model of the MAX436 amplifier

for the VV operational amplifiers described in Sect. 5.2.5. However, some of the standard VC operational amplifiers use Wilson current mirrors that require a minimum voltage drop of 0.8 V, as described in Sect. 4.1.1 on page 303.

5.3.2
Typical Applications

VC operational amplifiers are particularly suitable for driving coaxial cables. It is assumed that their output resistance is high compared to the characteristic impedance of the cable. This allows the cable to be terminated at either end by the characteristic resistance R_t connected in parallel. The tranconductance of the amplifier is controlled by current feedback through R_E. The output voltage is:

$$V_o = \frac{1}{2} I_o R_t = \frac{k_I}{2} \frac{R_t}{R_E} V_i$$

In order to make $V_o = V_i$, it is necessary to use a negative current feedback resistance of $R_E = k_I R_t/2$. The advantage of the parallel termination is the fact that the voltage at the coaxial cable is the same as the output voltage of the amplifier. With low operating voltages

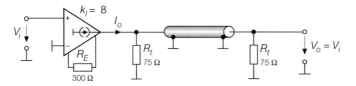

Fig. 5.71. Use of a VC operational amplifier to drive coaxial cables

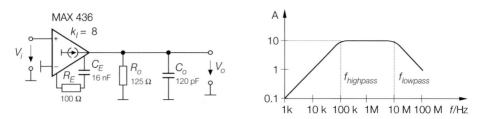

Fig. 5.72. Passive bandpass filter with decoupled cutoff frequencies

this is particularly advantageous compared to series terminals with low-resistance outputs, since the amplifier must then produce double the voltage. The fact that the amplifier must provide twice the amount of current is usually no problem at low operating voltages.

Another typical application is the bandpass filter shown in Fig. 5.72. Here too, the VC operational amplifier uses an emitter resistance with a defined transconductance. In contrast to the circuits described so far, this circuit features a complex emitter resistance to form a highpass filter. The RC network at the output acts as a lowpass filter. The two cutoff frequencies are decoupled by the amplifier:

$$f_{highpass} = \frac{1}{2\pi R_E C_E} \qquad\qquad f_{lowpass} = \frac{1}{2\pi R_o C_o}$$

At medium frequencies, the gain is $A = k_I R_o / R_E$. The capacitive load at the output is not critical, as the circuit has no voltage feedback. But even with voltage feedback, VC operational amplifiers are not affected by capacitive loads, because the high impedance point with the lowest cutoff frequency is at the output. The load capacitance lowers the cutoff frequency and thus improves the circuit stability.

5.4
Transimpedance Amplifier (CV-OPA)

A transimpedance amplifier differs from a conventional operational amplifier in that its inverting input has low resistance; this input is thus current-controlled, as shown in the overview in Fig. 5.3. For this reason, transimpedance amplifiers are also called CV operational amplifiers.

5.4.1
Internal Design

The most simple type of CV amplifier is shown in Fig. 5.73b. For comparison, a normal VV amplifier is illustrated in Fig. 5.73a. Transistor T_1 of the differential amplifier in the VV operational amplifier can be regarded as an impedance converter for the inverting input. This is omitted in the CV operational amplifier, resulting in a low-resistance inverting input. However, the emitter–base voltage of T_2 must then be compensated at some other point. This is done by the pnp transistor T_1 at the noninverting input in Fig. 5.73b. In the CV operational amplifier, the npn emitter follower at the emitter of T_2 is replaced by a pnp emitter follower at the base of T_2. Transistors T_1 and T_2 form a voltage follower that leads from the noninverting input to the inverting input. Its output resistance is $r_S = 1/g_m$. The signal at the noninverting input is transferred to the inverting input by two-step impedance conversion. In this way, the voltage difference between the inputs becomes zero due to the circuit configuration and not only as a result of external feedback, as is the case in VV operational amplifies. Therefore the additional small amplifier symbol points from the noninverting to the inverting input of the graphic symbol in Fig. 5.74.

The current control of the inverted input is somewhat unusual. If current I_q flows in Fig. 5.73b, then the current through T_2 increases by I_q. This increase is transmitted by the current mirror and current I_q remains after removing current I_0. This current is not the base current of T_5, which may be neglected in this case, but the current generating the voltage gain at the internal circuit resistance r_{CE}. The functional principle is easy to understand using the model shown in Fig. 5.74. The resulting output voltage is:

$$V_o = I_q Z = \frac{V_D}{r_m} Z$$

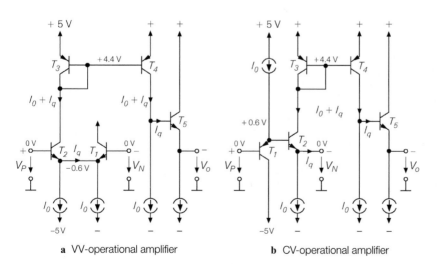

a VV-operational amplifier   **b** CV-operational amplifier

Fig. 5.73. Comparison of a VV operational amplifier and a CV operational amplifier. Current I_q is not the base current of T_5, but the signal current that controls the voltage gain at this point. The emitter current of the differential amplifier is divided into two halves in order to depict the current difference I_q.

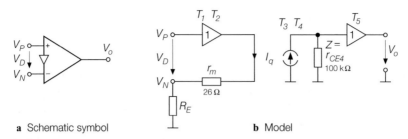

a Schematic symbol   **b** Model

Fig. 5.74. Schematic symbol and model of a CV operational amplifier. The values shown apply to $I_0 = 1\,\text{mA}$

For an unloaded output, this results in a voltage gain:

$$A_D = \frac{V_o}{V_D} = \frac{Z}{r_m} = \frac{r_{CE4}}{r_m} = \frac{V_A}{V_T} = \frac{100\,\text{V}}{26\,\text{mV}} = 3846$$

Here, Z is the transimpedance after which the amplifier is named. The higher Z is, the higher is the differential gain. In terms of circuit arrangement, we have the internal resistance at the node with the highest impedance, the collector of T_4 in this case. For the negative feedback, the CV operational amplifier feeds a portion of the output voltage back to the inverting input via a voltage divider, as is the case with VV operational amplifiers. But here the internal resistance shown as r_m in the model in Fig. 5.74 reduces the open loop

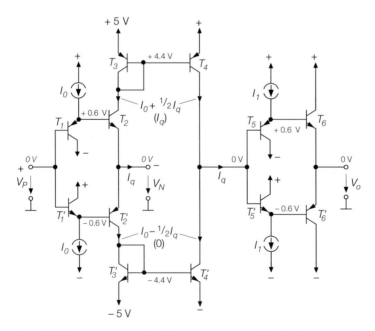

Fig. 5.75. Practical design of a CV operational amplifier in AB push–pull mode.
The values in parentheses show the situation for large currents $I_q > I_0$. Operational amplifiers with complementary current mirrors are called Nelson amplifiers

voltage gain of the operational amplifier:

$$A_{operating} = \frac{V_o}{V_P} = \frac{Z}{R_E + r_m} \tag{5.76}$$

From Fig. 5.73, one can see that current I_q can reach high positive values that are transferred to the output by the current mirror. The negative currents, however, must not be higher than I_0, as otherwise transistor T_1, and consequently the current mirror, becomes nonconductive. In order to enable the circuit to process large signal currents of either polarity with low quiescent currents, practicable amplifiers are symmetrically complemented as shown in Fig. 5.75 and operated in AB push–pull mode. The circuit then corresponds to the VV operational amplifier in AB mode as shown in Fig. 5.28; only the impedance converter at the inverting input has been omitted. CV operational amplifiers are always arranged for the AB push–pull mode (*current on demand*). The principle of the CV operational amplifier was first used in hybrid circuits by Comlinear. Of course, these amplifiers were expensive and therefore were only used in special applications. This type of amplifier only found widespread use after the monolithic types became available at a price that did not exceed that of normal operational amplifiers; model EL2030 from Elantec then became the industrial standard. This, however, presumes a technology that also allows the production of pnp transistors with equivalent good high-frequency characteristics as the npn transistors.

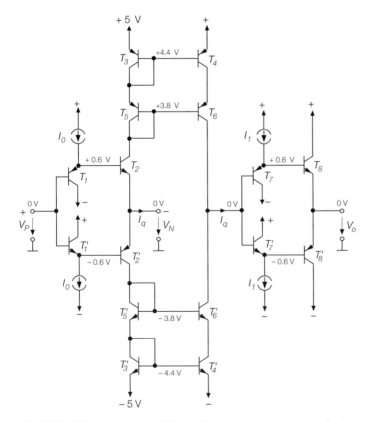

Fig. 5.76. CV operational amplifier with cascode current mirrors for increased voltage gain

VV operational amplifier

Without negative feedback

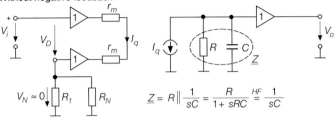

$$\underline{Z} = R \parallel \frac{1}{sC} = \frac{R}{1+sRC} \overset{HF}{=} \frac{1}{sC}$$

Output voltage $\qquad \underline{V}_o = I_q \underline{Z} = \dfrac{V_i}{2r_m} \underline{Z} = A_D V_i$

Gain $\qquad \underline{A}_D = \dfrac{\underline{V}_o}{\underline{V}_i} = \dfrac{\underline{Z}}{2r_m} = \dfrac{R/2r_m}{1+sRC}$

Case distinction $\quad \underline{A}_D = \begin{cases} A_{D0} = \dfrac{R}{2r_m} & \text{for } f \ll f_c \\[2mm] \dfrac{1}{2s\,r_m C} & \text{for } f \gg f_c \end{cases}$

Cutoff frequency $\quad f_c = \dfrac{1}{2\pi RC}$

Transit frequency $\quad f_T = \dfrac{1}{4\pi r_m C}$

With negative feedback

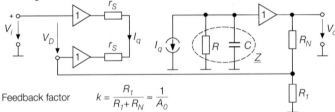

Feedback factor $\qquad k = \dfrac{R_1}{R_1 + R_N} \approx \dfrac{1}{A_0}$

Gain $\qquad \underline{A} = \dfrac{\underline{V}_o}{\underline{V}_i} = \dfrac{\underline{A}_D}{1 + k\,\underline{A}_D} = \dfrac{1/k}{1 + 2r_m/(k\cdot\underline{Z})} \approx \dfrac{1/k}{1 + 2s\,r_m C/k}$

Loop gain

Case distinction $\quad \underline{A} = \begin{cases} A_0 \approx \dfrac{1}{k} = 1 + \dfrac{R_N}{R_1} & \text{for } f \ll f_c \\[2mm] \dfrac{1}{2s\,r_m C} = -j\dfrac{f_T}{f} & \text{for } f \gg f_c \end{cases}$

$g_0 = \dfrac{A_{D0}}{A_0} \sim \dfrac{1}{A_0}$

Cutoff frequency

Transit frequency $\qquad f_T = \dfrac{1}{4\pi r_m C} = \text{const}$

$f_c = \dfrac{k}{4\pi r_m C} \approx \dfrac{f_T}{A_0}$

Frequency response

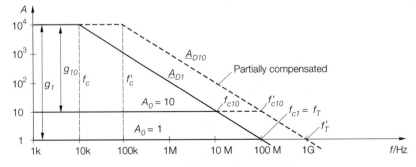

Fig. 5.77. Comparison of the VV and CV operational amplifiers

CV operational amplifier

Without negative feedback

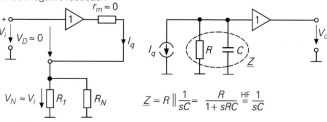

$$\underline{Z} = R \parallel \frac{1}{sC} = \frac{R}{1+sRC} \overset{HF}{=} \frac{1}{sC}$$

Output voltage $\quad \underline{V}_o = I_q \underline{Z} = \dfrac{V_i}{R_1 \parallel R_N} \underline{Z}$

Gain $\quad \underline{A}_D = \dfrac{V_o}{V_i} = \dfrac{\underline{Z}}{R_1 \parallel R_N} = \dfrac{R/(R_1 \parallel R_N)}{1+sRC}$

Case distinction $\quad \underline{A}_D = \begin{cases} A_{DO} = \dfrac{R}{R_1 \parallel R_N} & \text{for } f \ll f_c \\[3mm] \dfrac{1}{s\,(R_1 \parallel R_N)C} & \text{for } f \gg f_c \end{cases}$

Cutoff frequency $\quad f_c = \dfrac{1}{2\pi\,RC}$

Transit frequency $\quad f_T = \dfrac{1}{2\pi\,(R_1 \parallel R_N)C}$

With negative feedback

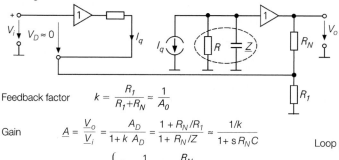

Feedback factor $\quad k = \dfrac{R_1}{R_1 + R_N} \approx \dfrac{1}{A_0}$

Gain $\quad \underline{A} = \dfrac{V_o}{V_i} = \dfrac{\underline{A}_D}{1+k\,\underline{A}_D} = \dfrac{1+R_N/R_1}{1+R_N/\underline{Z}} \approx \dfrac{1/k}{1+s\,R_N C}$

Loop gain

Case distinction $\quad \underline{A} = \begin{cases} A_0 \approx \dfrac{1}{k} = 1 + \dfrac{R_N}{R_1} & \text{for } f \ll f_c \\[3mm] \dfrac{1}{s\,k\,R_N C} = \dfrac{1}{s\,(R_1 \parallel R_N)C} & \text{for } f \gg f_c \end{cases}$

$g_0 = k\,A_{DO} = \dfrac{R}{R_1} = \text{const}$

Cutoff frequency

Transit frequency $\quad f_T = \dfrac{1}{2\pi\,(R_1 \parallel R_N)C} = \dfrac{A_0}{2\pi\,R_N C} \sim A_0$

$f_c = \dfrac{k}{2\pi\,R_n C} = \text{const}$

Frequency response

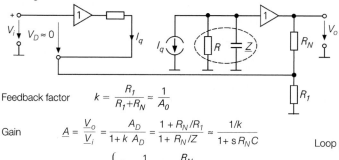

Fig. 5.77. Comparison of the VV and CV operational amplifiers

The voltage gain resulting from (5.76) is usually insufficient, as it is further reduced by the external resistance R_E formed by the feedback voltage divider. In order to increase the voltage gain, it is common practice to increase the internal resistance at the high-impedance point. This is equivalent to increasing the transimpedance Z. As in the VV operational amplifier shown in Fig. 5.25, cascode current mirrors can be used as shown in Fig. 5.76. According to (4.27), this increases the internal resistance at the high-impedance point by the current gain β of the transistors. The differential gain also rises by this factor: (5.76):

$$A_{operating} = \frac{Z}{R_E + r_m} = \frac{1}{2}\frac{\beta\, r_{CE}}{R_E + r_m} \qquad (5.77)$$

The factor of 1/2 accounts for the fact that two current sources are switched in parallel at the high-impedance point. A drawback of the cascode current sources is the fact that the common-mode and output control range is reduced by 0.6 V. At an operating voltage of ± 5V, this gives a value of only ± 3.6 V.

5.4.2
Frequency Response

Transimpedance amplifiers are only used where high bandwidth or short rise times are of importance. Lately, wide-band VV operational amplifiers have also become available; they are produced with the same technology and operate in AB mode as shown in Fig. 5.28. The two amplifier types have been placed side by side in Fig. 5.77, to illustrate the differences. The major distinction becomes visible in the model: in CV operational amplifiers there is no impedance converter at the inverting input. Therefore, the transconductance of the input stage is determined by the resistance at the inverting input:

$$g_m = \frac{I_q}{V_i} = \frac{1}{r_m + R_E} = \frac{1}{r_m + R_1 || R_N}$$

For this reason, the resistors R_1 and R_N must be taken into account when analyzing the open circuit. In practice $r_m \ll R_1 || R_N$, which means that the effect of r_m may be neglected. This means that voltage $V_D \approx 0$ and voltage V_i is generated across R_1 and R_N. Therefore, the gain of a CV operational amplifier is clearly lower than that of a comparable VV operational amplifier with the same resistance R at the high-impedance point. As R is usually of the order of magnitude of 1 MΩ, it only influences the gain at low frequencies. The parasitic capacitances C, of a few picofarad, already cause a drop in gain at medium frequencies. This can be accounted for in the calculation by using $\underline{Z}$, the parallel arrangement of R and C. With high frequencies it is the capacitance that determines the characteristics, so that the resistance need not to be taken into account; this facilitates the calculation. One can see that the cutoff frequencies of both amplifiers are the same. The transit frequencies, however, are different: while that of the VV operational amplifier is given by the inner circuit design, that of the CV operational amplifier depends on the external circuitry.

When analyzing the CV operational amplifier with negative feedback, it is essential to take the current at the inverting input into consideration and not to assume the feedback voltage divider to be without load, as it is the case with VV operational amplifiers. Kirchhoff's first law is applied to the inverting input to calculate the voltage gain:

$$\frac{V_o - V_i}{R_N} - \frac{V_i}{R_1} + \frac{V_o}{Z} = 0$$

For low frequencies, this yields the same result as for the VV operational amplifier, as shown by the comparison in Fig. 5.77. It seems strange that the current at the inverting input does not alter the result. The reason for this is that the current I_q is small, since for a resistance of $R = 1\,M\Omega$ even an output voltage of 5 V only requires a current of $I_q = 5\,\mu A$.

The cutoff frequency of the VV operational amplifier with negative feedback is inversely proportional to the set gain; the gain–bandwidth product is constant equal to the transit frequency as shown in Fig. 5.77. In CV operational amplifiers with negative feedback, a cutoff frequency that is independent of the adjusted gain can be achieved by keeping R_N constant and setting the gain by means of R_1. For in this case the loop gain remains constant: decreasing R_1 in order to heighten the gain increases the open-circuit gain by lowering the current feedback to the same degree. For this reason, manufacturers usually quote the optimum value for R_N at which the loop gain just reaches the level at which a favorable transient response is achieved. This value of R_N is already incorporated into some types.

If resistance R_N of a CV operational amplifier is kept constant and the gain is set by means of R_1, the following *deviations* from the VV operational amplifier can be observed, as expressed in the formulas of Fig. 5.77:

- The bandwidth of the feedback circuit is independent of the gain selected.
- The loop gain of the feedback circuit is independent of the gain selected.
- The transit frequency of the feedback circuit is proportional to the gain selected.

A matched frequency response correction is also possible in VV operational amplifiers (see Fig. 5.37); however, here it would be necessary to change the correction capacitance together with the gain, but this is only possible by exchanging the OPAmp. Under this condition the cutoff frequency remains here also constant, as shown in Fig. 5.77.

Two properties in which both circuits are *equal* are demonstrated by the formulas:

- The bandwidth of the feedback circuit is increased by the loop gain, compared to that of the open amplifier, by a value equal to the loop gain.
- The transit frequency of the circuit is not altered by the feedback.

These conditions are illustrated by the frequency–response plots in Fig. 5.77.

5.4.3
Typical Applications

In order to allow the feedback of the CV operational amplifier to control the gain through negative current feedback, it must consist of ohmic resistances. The circuit becomes unstable if R_N or R_1 is replaced by a capacitor. This is the reason why an integrator or differentiator cannot be realized by means of a CV operational amplifier. Therefore, they are mainly used as amplifiers with high bandwidth; for example, as video amplifiers. They can be used as inverting or noninverting amplifiers as shown in Fig. 5.78. Resistance R_N determines the loop gain and is thus given by the amplifier to a large degree. Resistance R_1 determines the voltage gain and is of rather low impedance for higher gains. As in all inverting amplifier, resistance R_1 represents the input resistance; the noninverting mode is therefore usually preferred.

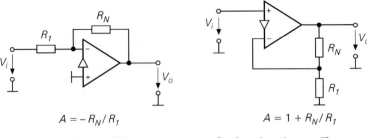

a An inverting amplifier

b A noninverting amplifier

Fig. 5.78. Using CV operational amplifiers as amplifiers. **a** An inverting amplifier

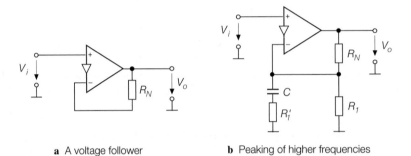

a A voltage follower

b Peaking of higher frequencies

Fig. 5.79. CV operational amplifiers as noninverting amplifiers

It is important that resistance R_N holds its optimum value even for gain $A = 1$ and that it must not be reduced to zero (see Fig. 5.79a). Only resistance R_1 may be omitted here. To counteract the reduction in gain that occurs around the cutoff frequency, the gain can be increased by the additional $R_1' C$ element shown in Fig. 5.79b. However, one should check whether it is possible to achieve the same effect with a higher loop gain that results from low resistances in the feedback voltage divider.

5.5
The Current Amplifier (CC-OPA)

The CC operational amplifier differs from the CV operational amplifier in the same way as the VC operational amplifier differs from the VV operational amplifier; that is, in that the impedance converter is omitted at the output.

5.5.1
The Internal Design

The basic circuit is shown in Fig. 5.80a. From a comparison with the CV amplifier in Fig. 5.73, one can see that the only difference is the lack of the emitter follower at the output. Assuming that the inverting input is connected to ground, the circuit can be divided into two parts:

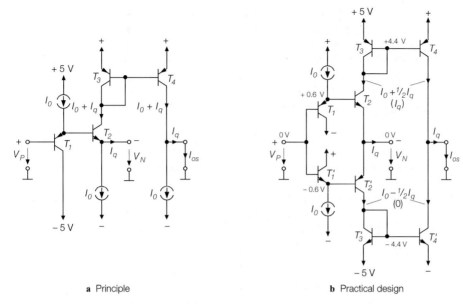

a Principle   **b** Practical design

Fig. 5.80. Internal design of CC operational amplifiers

– Transistor T_2, in common-emitter connection, with base-emitter voltage compensation by T_1.
– The current mirror formed by transistors T_3 and T_4.

Since the current flowing at the inverted input (the emitter of T_2), is transferred to the output, the CC operational amplifier is also called a *current amplifier*. The output current of the circuit shown in Fig. 5.80 is equal to the input current; this means that the current amplification factor is one. Higher amplification factors may be achieved by providing the current mirror with a gain; values of up to $k_I = 8$ are available.

Thus, the entire operational amplifier is no more than an expanded transistor. This is the reason why two graphic symbols are used for CC operational amplifiers, as shown in Figs. 5.81a,b. When they are used in circuits that are also common for VV operational amplifiers, the graphic symbol for the operational amplifier is preferred. However, the CC operational amplifier can also be used like a transistor, in which case the symbol for the

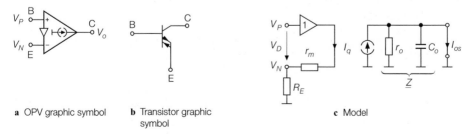

a OPV graphic symbol   **b** Transistor graphic symbol   **c** Model

Fig. 5.81. Graphic symbols and a model for CC operational amplifiers

transistor is more familiar. The CC operational amplifier – the current amplifier – and the simple transistor have a lot in common:

– The collector current has the same (absolute) value as the emitter current.
– The input resistance is high at the base and low at the emitter.
– The output resistance at the collector is high.

But there are also differences, which make it easier to use a CC operational amplifier than a transistor:

– The collector current has the opposite polarity because of the current mirror.
– The base–emitter voltage is zero: $V_{BE} = 0$ due to compensation by T_1.
– The emitter and collector currents can have both polarities.
– The operating point is set internally.

For these reasons, the CC operational amplifier performs like an ideal transistor. Therefore the manufacturer Burr Brown calls it a *diamond transistor*.

Due to its short internal signal path, the CC operational amplifier has particular advantages for high frequencies. For this reason it is constructed in AB push–pull mode (*current on demand*), to allow high output currents even with small quiescent currents. A circuit used in practice is shown in Fig. 5.80b. A comparison with the basic circuit shows that the current sources are replaced by a complementary circuit.

The model given in Fig. 5.81c shows the high-resistance noninverting and low-resistance inverting inputs. The output has a high resistance. The dominating lowpass filter is at the output. Its cutoff frequency depends on the load connected. One can see that the short-circuit current at the output is equal to the current at the inverting input. The short-circuit transconductance of the circuit is equal to the transconductance of transistor T_2:

$$g_m = \frac{I_{os}}{V_D} = \frac{1}{r_m} \tag{5.78}$$

In practical applications, there is usually a resistance R_E at the inverting input, which reduces the transconductance. The operational transconductance of the circuit is then:

$$g_{m,op} = \frac{I_{os}}{V_P} = \frac{1}{r_m + R_E} \tag{5.79}$$

This allows the open-circuit voltage gain to be calculated:

$$A_{op} = \frac{V_o}{V_P} = g_{m,op} \, R_{o,tot} = \frac{R_{o,tot}}{r_m + R_E} = \frac{R_{o,tot}}{R_{E,tot}} \tag{5.80}$$

5.5.2
Typical Applications

In most applications the behavior of the CC operational amplifier is determined by the current feedback at the inverting input; voltage feedback is used in special cases only.

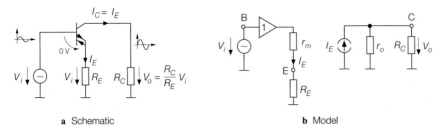

a Schematic   **b** Model

Fig. 5.82. Common-emitter circuit of a CC operational amplifier

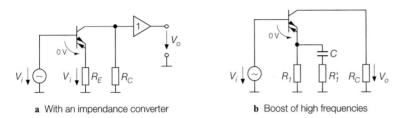

a With an impendance converter   **b** Boost of high frequencies

Fig. 5.83. Expansion of a CC operational amplifier in common-emitter configuration

Applications with Current Feedback

Common-emitter circuit: Since a CC operational amplifier behaves largely like a transistor, it makes sense to use it in the three basic circuits. The common-emitter circuit is shown in Fig. 5.82. If the transconductance resistance r_m is neglected, then $V_{BE} = 0$. The emitter current is then $I_E = V_i/R_E$. Since the collector current has the same value, it follows that the output voltage is $V_o = R_C V_i = V_i R_C/R_E$.

The voltage gain can be calculated exactly using the model of the CC operational amplifier in Fig. 5.81 with the load resistance R_C. The resulting emitter current in Fig. 5.82 is

$$I_E = \frac{V_i}{r_m + R_E}$$

This current flows in the output circuit, where it causes the voltage drop

$$V_o = I_E (r_o \| R_C) = \frac{r_o \| R_C}{r_m + R_E} V_i \approx \frac{R_C}{R_E} V_i$$

If a load resistance is connected to the output, this must be taken into account when calculating the voltage gain. It is easiest to combine it with a collector resistance and to calculate with $R_{C,res}$. In order to prevent the load resistance from affecting the voltage gain, it is possible to insert a voltage follower (a CC operational amplifier in common-collector configuration), as shown in Fig. 5.83a. This possibility should be remembered in all applications for CC operational amplifiers. Since the operating point is set internally – as with any operational amplifier – the collector resistance is connected to ground and not to the operating voltage. This is the reason why circuits that for a normal transistor only represent the small-signal equivalent work with a diamond transistor.

To counteract a drop in gain at high frequencies, the effective emitter resistance can be decreased in this frequency range by adding an additional RC element in parallel, as shown in Fig. 5.83b.

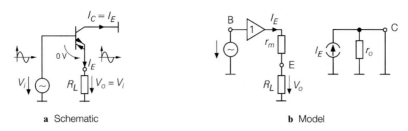

a Schematic   **b** Model

Fig. 5.84. Common-collector circuit of a CC operational amplifier

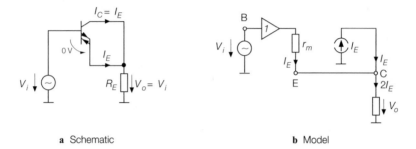

a Schematic   **b** Model

Fig. 5.85. Use of the collector current in the CC operational amplifier as an emitter follower

Common-collector circuit: In the common-collector circuit of Fig. 5.84, the output signal is available at the emitter. The collector carries constant potential. As the operating point is set internally, the collector – which is not required – is set to zero potential. If we assume the approximation $V_{BE} = 0$, then it is obvious that the voltage gain is $A = 1$. Here, the emitter resistance is not required functionally; it can therefore be regarded as a load resistance.

The model in Fig. 5.84 is most suitable for a more accurate calculation of the voltage gain. It shows that a voltage divider with the transconductance resistance exists, whereby the voltage gain is

$$A = \frac{V_o}{V_i} = \frac{R_L}{r_m + R_L} \approx 1$$

The model also shows that the collector current flows to ground and is unused. This allows us to remove the current mirrors in Fig. 5.80 when using the CC operational amplifier in common-collector configuration. What remains is the complementary Darlington circuit in AB mode.

However, the collector current can also be utilized in practice by connecting the collector to the emitter, as shown in Fig. 5.85. This doubles the output current, as the collector current of the CC operational amplifier has the same polarity as the emitter current. To calculate the voltage gain we again use the model and apply Kirchhoff's current law to the collector:

$$2\frac{V_i - V_o}{r_m} - \frac{V_o}{R_E} = 0 \quad \Rightarrow \quad V_o = \frac{R_E}{R_E + r_m/2} V_i$$

As can be seen, this method splits the output resistance in half.

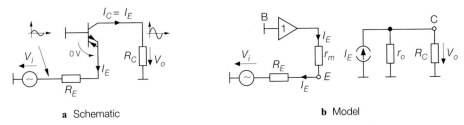

a Schematic   **b** Model

Fig. 5.86. Common-base circuit of a CC operational amplifier

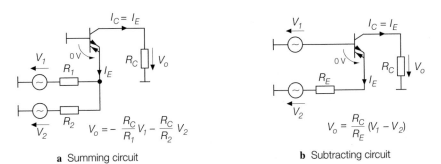

a Summing circuit   **b** Subtracting circuit

Fig. 5.87. Common base CC operational amplifier as a summing amplifier

Common-base circuit: In the common-base circuit, the input signal is fed to the emitter via a resistor, and the collector current generates the amplified output signal across the collector resistance, as shown in Fig. 5.86. If we assume an ideal CC operational amplifier with $V_{BE} = 0$, the emitter current is $I_E = V_i/R_i$; across the collector resistance, this current generates the voltage

$$V_o = I_C R_C = -\frac{R_C}{R_e} V_i$$

The model in Fig. 5.86 is best suited for an exact analysis and yields:

$$V_o = -I_E (r_o\|R_C) = -\frac{r_o\|R_C}{r_m + R_E} V_i \approx -\frac{R_C}{R_E} V_i$$

This is the same result as in the common-emitter circuit but with a negative sign. Compared to the simple transistor, the common-emitter and the common-base configurations of the CC operational amplifier produce the opposite polarity for the voltage amplification.

Since in the common-base circuit the emitter carries zero potential via a low resistance r_m, currents at this point can be added interaction-free, as at the summing point of a VV operational amplifier. This possibility is shown in Fig. 5.87a. It is also possible to combine the common-emitter and common-base circuits to form the subtractor shown in Fig. 5.87b.

Differential amplifier: A differential amplifier can be built from two CC operational amplifiers, as shown in Fig. 5.88. It is very similar to the conventional differential amplifier with current feedback presented in Fig. 4.59 on page 335. However, as the operating point is set internally, it does not require an emitter current source, and the collector resistances are connected to ground. If we assume an ideal CC operational amplifier with $V_{BE} = 0$, then the cross-current can be calculated directly:

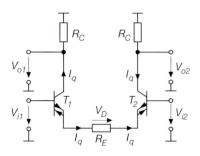

Fig. 5.88. Differential amplifier made up of two CC operational amplifiers

$$I_q = \frac{V_{i1} - V_{i2}}{R_E} = \frac{V_D}{R_E}$$

Since the collector current has the same value, the output voltages are:

$$V_{o1} = I_q R_C = \frac{R_C}{R_E} V_D \quad \text{and} \quad V_{o2} = -I_q R_C = -\frac{R_C}{R_E} V_D$$

The model shown in Fig. 5.89 can be used for an accurate calculation of the voltage gain. It allows the transconductance resistances r_m to be taken into account when calculating the cross-current:

$$I_q = \frac{V_D}{R_E + 2r_m}$$

The output resistances are in parallel to the collector resistances, so that the output voltages are:

$$V_{o1} = \frac{R_C || r_o}{R_E + 2r_m} V_D \quad \text{and} \quad V_{o2} = -\frac{R_C || r_o}{R_E + 2r_m} V_D$$

The differential amplifier has already been used in applications for the VC operational amplifier in Figs. 5.71 and 5.72. The external emitter resistance R_E determines the transconductance of the circuit. The short-circuit transconductance is:

$$g_m = \frac{I_{a1}}{V_D} = \frac{1}{R_E + 2r_m}$$

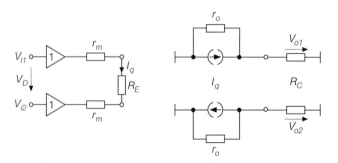

Fig. 5.89. Model of a differential amplifier made up of two CC operational amplifiers

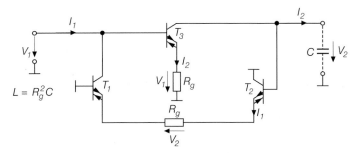

Fig. 5.90. Gyrator realized with CC operational amplifiers as voltage-controlled current sources

Gyrator: Voltage-controlled current sources are particularly suitable for realizing gyrators, as they enable the direct realization of the required transfer equations:

$$I_1 = \frac{1}{R_g} V_2 \qquad I_2 = \frac{1}{R_g} V_1 \qquad\qquad (5.81)$$

These formulas can be derived directly from Fig. 5.90 if we assume that $V_{BE} = 0$ and
$I_B = 0$. To achieve the correct sign for the current, it suffices to place a simple CC
operational amplifier in the signal path from left to right; while in the opposite direction a
differential amplifier is necessary, according to Fig. 5.88. In order to achieve high-quality
gyrators, the current sources must have high output resistances. Model OPA615 (Burr
Brown) is particularly well suited for this purpose, as it contains a cascode current mirror
at the output with an output resistance in the megaohm region.
    If one connects a capacitor C to one side of the gyrator, the inductance $L = R_G^2 C$ is
produced at the other side according to (12.21) on page 782. Connecting a capacitor to
both sides produces an oscillating circuit that can be damped with a parallel resistor. The
resulting circuit is identical to the filter in Fig. 5.93 if an input signal is fed to the base of
T_1 and the output signal is moved from the emitter of T_2 in Fig. 5.90. It is astonishing that
both filters, which are totally different in approach, produce the same circuit.

Integrator: Driving a capacitor with a voltage-controlled current source results in an
integrator. This is the operating principle of the integrator shown in Fig. 5.91. The CC
operational amplifier forms the voltage-controlled current source; it supplies current $I_C =
V_i/R$ if the transconductance resistor is neglected. The voltage across the capacitor is thus

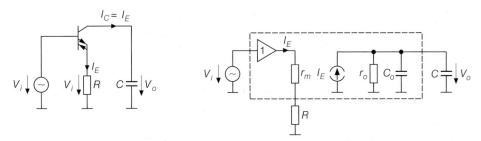

Fig. 5.91. CC operational amplifier used as an integrator

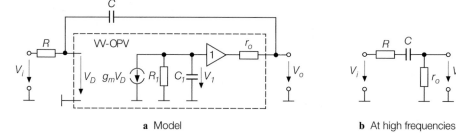

a Model   **b** At high frequencies

Fig. 5.92. VV operational amplifier in integrator configuration

$$V_o = \frac{1}{C} \int I_C \, dt = \frac{1}{RC} \int V_i \, dt$$

Of course, it is also possible to calculate the transfer function if the complex resistance of the capacitor is considered

$$\underline{V}_o = \frac{\underline{I}_C}{sC} = \frac{\underline{V}_i}{sRC}$$

In order not to affect the operation of the circuit, it is necessary to tap the voltage at the capacitor without a load; this normally requires an additional impedance converter.

The model shown in Fig. 5.91 can be used to study the effects of the real properties of a CC operational amplifier used as integrator. Here too, the transconductance resistance r_m is in series with the external emitter resistance R and can be taken into account if the external resistance is reduced accordingly.

The dominating lowpass filter $r_o C_o$ limits the lower cutoff frequency of the integrator to $f_{low} = 1/2\pi r_o(C + C_o)$. This limitation is common to all integrators, since with low frequencies in ideal integrators the gain would reach infinity. The parasitic capacitance C_o presents no limitation as it is parallel to the integration capacitance C. To take this into account, one can make the external integration capacitance C accordingly smaller. Thus, on the basis of the model, the bandwidth is not limited toward high frequencies. Of course, due to secondary effects, the CC integrator also has an upper cutoff frequency; however, this value is in the very high frequency range.

By way of comparison, the features of the VV operational amplifier are much more disadvantageous. Figure 5.92a shows the model of a VV operational amplifier that is configured as an integrator. Above the transit frequency of the opamp $f_T = g_m/(2\pi C)$ the voltage becomes $V_1 = 0$; the output resistor r_o is thus at zero potential. In this case, the model can be simplified as shown in Fig. 5.92b. The integration capacitance now acts as a coupling capacitance and transfers the input signal to the output instead of short-circuiting it. In this frequency range, the circuit operates solely as a voltage divider according to $V_o = V_i r_o/(r_o + R)$.

Filter: Since CC operational amplifiers can be used to achieve integrators with very good properties for high frequencies, they are particularly suitable for active high-frequency filters based on integrators (see Sect. 13.11 on page 831). An example of a combined bandpass/lowpass filter of second order is shown in Fig. 5.93. It consists of two CC integrators and one voltage follower. In contrast to the integrator filters with VV operational

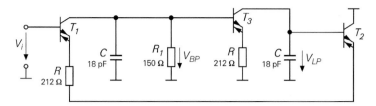

Fig. 5.93. Active second-order high-frequency filter, with bandpass or lowpass output. The dimensions shown result in a resonance or cutoff frequency of 30 MHz with Q-factor of $1/\sqrt{2}$ (Butterworth). When dimensioning the circuit, the transconductance resistances of $r_m = 10\,\Omega$ and the circuit capacitances of 6 pF in parallel to the integration capacitors were taken into account. The circuit is very accurate up to more than 300 MHz

amplifiers, the CC integrators are not inverting. Therefore, no inverter is required in the filter loop. The transfer function can be derived from the circuit:

$$\frac{\underline{V}_{LP}}{\underline{V}_i} = \frac{1}{1 + sCR^2/R_1 + s^2C^2R^2} \tag{5.82}$$

$$\frac{\underline{V}_{BP}}{\underline{V}_i} = \frac{sRC}{1 + sCR^2/R_1 + s^2C^2R^2} \tag{5.83}$$

$$f_r = \frac{1}{2\pi RC} \qquad Q = \frac{R_1}{R}$$

We can see from the formulas that the resonance frequency and the Q-factor can be set independently of one another.

Applications with Voltage Feedback

Voltage feedback in CC operational amplifiers is possible by feeding a portion of the output voltage back to the inverting input via a voltage divider, as shown in Fig. 5.94a. The resulting circuit is typical for noninverting amplifiers in VV operational amplifiers. However, the difference is that the feedback voltage divider is loaded with the input current. If we draw the circuit diagram using the transistor symbol, as in Fig. 5.94b, then we can see that current feedback exists at the same time. The feedback from the collector to the emitter is a negative feedback since, unlike in a simple transistor, here the collector current

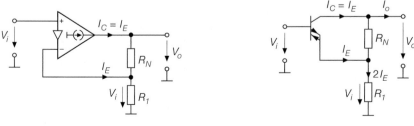

a With the OPA symbol   b With the transistor symbol

Fig. 5.94. CC operational amplifier with additional voltage feedback. This configuration is called direct feedback

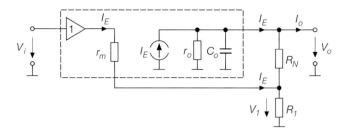

Fig. 5.95. Model of a CC operational amplifier for calculating the voltage gain and the bandwidth with voltage feedback

is inverted. As the feedback loop follows the shortest path possible – that is, there is no impedance converter at the input or at the output – this is also known as *direct feedback*.

To calculate the voltage gain, we assume an ideal CC operational amplifier with $V_D = V_{BE} = 0$. With no load at the output, the current $2I_E$ flows through resistor R_1. The emitter current is $I_E = 2V_i/2R_1$. The output voltage can now be calculated:

$$V_o = V_i + I_E R_N = V_i + \frac{R_N}{2R_1} V_i = \left(1 + \frac{R_N}{2R_1}\right) V_i$$

The formula is very similar to that for the VV operational amplifier, apart from the factor of two in the denominator. The formula is based on the assumption that for an unloaded output, $I_o = 0$. A load causes a reduction in the gain. Should this be a problem, a voltage follower can be connected at the output.

In order to calculate for the influence of the transconductance resistance and the output resistance in the CC operational amplifier, the model shown in Fig. 5.95 is used. If Kirchhoff's first law is applied to the emitter and the collector, then:

$$\frac{V_i - V_1}{r_m} + \frac{V_o - V_1}{R_N} - \frac{V_1}{R_1} = 0$$

$$\frac{V_i - V_1}{r_m} + \frac{V_o - V_1}{R_N} - \frac{V_o}{r_o} = 0$$

Consequently, the accurate value of the open-circuit voltage gain is

$$A = \frac{V_o}{V_i} = \frac{1 + \dfrac{R_N}{2\,R_1}}{1 + \dfrac{1}{2r_o}\left(R_N + r_m + \dfrac{r_m R_N}{R_1}\right) + \dfrac{r_m}{2R_1}} \overset{\substack{r_o \to \infty \\ r_m = 0}}{=} 1 + \frac{R_N}{2R_1}$$

The bandwidth of the circuit shown in Fig. 5.94 is best calculated on the basis of the model shown in Fig. 5.95. For the sake of simplicity, we presume an ideal CC operational amplifier – with the exception of capacitance C_o – which means that $r_S = 0$ and $r_o = \infty$. Again, for circuit analysis Kirchhoff's first law can be applied to the emitter and the collector:

$$I_E + \frac{V_o - V_i}{R_N} - \frac{V_i}{R_1} = 0$$

$$I_E - \frac{V_o - V_i}{R_N} - V_o s C_o = 0$$

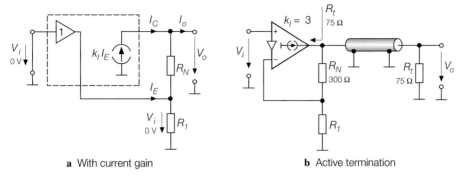

a With current gain   **b** Active termination

Fig. 5.96. CC operational amplifier for active termination

This leads to the voltage gain:

$$\underline{A} = \frac{V_o}{V_i} = \frac{1 + \dfrac{R_N}{2R_1}}{1 + sR_NC_o/2}$$

This expression includes the known low-frequency gain. The cutoff frequency results from the condition that the imaginary part in the denominator must be 1; it follows:

$$\omega_g = 2\pi f_g = \frac{2}{R_NC_o} = \frac{1}{r_{out}C_o}$$

Since C_o acts as a load capacitance which, together with the output resistance of the circuit, forms a lowpass filter with the output resistance becomes $r_{out} = R_N/2$.

The output resistance of the circuit is neither as high as that of the amplifier itself, due to the feedback voltage divider at the output, nor as low as in VV operational amplifiers with negative feedback, since here the loop gain is lower. To calculate the output resistance, we again start with the ideal CC operational amplifier in Fig. 5.96a and determine the relation between the output current and the output voltage for the short-circuited input $V_i = 0$. We also want to look at CC operational amplifiers with a current gain of $k_I = I_C/I_E$, which is higher than 1. The output current is:

$$I_o = I_C - \frac{V_o}{R_N} = k_I I_E - \frac{V_o}{R_N} = -(k_I + 1)\frac{V_o}{R_N}$$

This leads to an output resistance of

$$r_{out} = -\frac{V_o}{I_o} = \frac{R_N}{k_I + 1} \tag{5.84}$$

Resistance R_N is thus actively reduced by the amplifier. Using resistance R_N, the output resistance of the circuit can be adjusted to any given value; R_1 can then be used to select the voltage gain independently. For this reason, CC operational amplifiers are also called *drive-R-amplifiers*.

To drive a transmission line with a characteristic impedance of $R_t = 75\,\Omega$, a resistance

$$R_N = R_t(k + 1) = 4R_t = 4 \cdot 75\,\Omega = 300\,\Omega$$

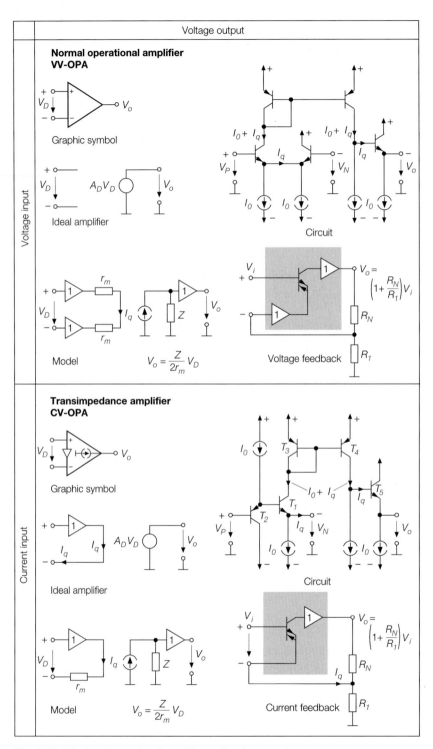

Fig. 5.97. Matrix of operational amplifiers: Circuit comparison

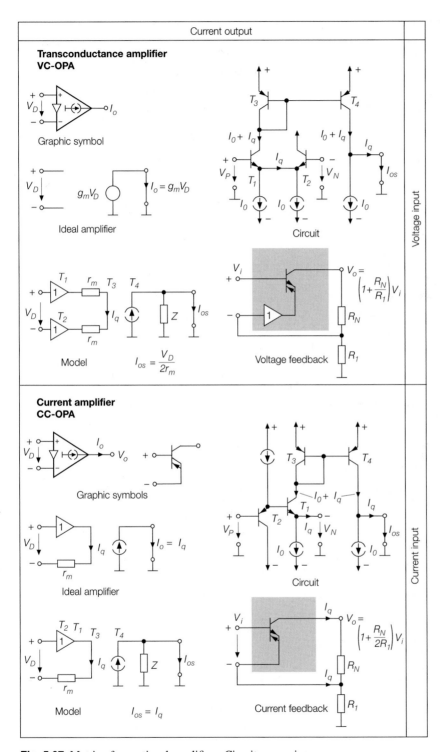

Fig. 5.97. Matrix of operational amplifiers: Circuit comparison

Voltage output	
Common name	**Normal operational amplifier**
Systematic name	VV operational amplifier
Function as controlled source	Voltage-controlled voltage source VCVS
Feedback/output description	Voltage feedback, voltage output VFVO
Type of feedback	Voltage feedback
Applications	Amplifier for low frequencies
Advantages	Low offset voltage Low drift High precision at low frequencies
Disadvantages	Unsuitable for high frequencies Stability problems with capacitive and inductive loads
Typical example	OP177 (Analog Devices)
Offset voltage	10 µV ☺
Offset voltage drift	0.1 µV/K ☺
Input current	1 nA ☺
Large-signal bandwidth	5 kHz ☹
Slew rate	0.3 V/µs ☹

(Voltage input)

Common name	**Transimpendance amplifier**
Systematic name	CV operational amplifier
Function as controlled source	Current-controlled voltage source, CCVS
Feedback/output description	Current feedback, voltage output, CFVO
Type of feedback	Current feedback
Applications	Line drivers
Advantages	High bandwidth High slew rate
Disadvantages	Stability problems with capacitive and inductive loads
Typical example	AD8024
Offset voltage	2 mV ☹
Offset voltage drift	1.5 µV/K ☺
Input current	1 µA ☺
Large-signal bandwidth	350 MHz ☺
Slew rate	2400 V/µs ☺

(Current input)

Fig. 5.98. Matrix of operational amplifiers: comparison of characteristics

Current output		
Common name	**Transconductance amplifier**	
Systematic name	VC operational amplifier	
Function as controlled source	Voltage-controlled current source, VCCS	
Feedback/output description	Voltage feedback, current output, VFCO	
Type of feedback	Voltage feedback	
Applications	Drivers for capacitive loads	
Advantages	Low offset voltage Low drift Good transient response with capacitive loads	Voltage input
Disadvantages	Load must be known for calculation	
Typical example	MAX436 (Maxim)	
Offset voltage	0.3 mV ☺	
Offset voltage drift	4 µV/K ☹	
Input current	1 µA ☹	
Large-signal bandwidth	200 MHz ☺	
Slew rate	850 V/µs ☺	

Common name	**Current amplifier**	
Systematic name	CC operational amplifier	
Function as controlled source	Current-controlled current source, CCCS	
Feedback/output description	Current feedback, current output, CFCO	
Type of feedback	Current feedback	
Applications	Active filter for hight frequencies, current drivers for magnetic heads, laser diodes, line drivers	Current input
Advantages	High bandwidth High slew rate	
Disadvantages	Load must be known for calculation	
Typical example	OPA615 (Burr Brown)	
Offset voltage	8 mV ☹	
Offset voltage drift	40 µV/K ☺	
Input current	0.3 µA ☺	
Large-signal bandwidth	500 MHz ☺	
Slew rate	3000 V/µs ☺	

Fig. 5.98. Matrix of operational amplifiers: comparison of characteristics

is required according to (5.84) when using a CC-OPAmp with a current gain of $k_I = 3$. This example is illustrated in Fig. 5.96b. For the MAX436 with $k_i = 8$, a resistance of $R_N = 675\,\Omega$ would be necessary. We can see that the power loss in the termination resistance is clearly lower in active termination than in passive termination by means of parallel or series resistances. While in passive termination the power loss is equal to the power output, here only $1/(1 + k_I)$ of the power fed into the line is lost in the resistance.

The method of reducing the output resistance of a high-resistive amplifier to a defined value using *direct feedback* has already been used in the rail-to-rail amplifier. The conformity of the rail-to-rail output stage in Fig. 5.22 with the CC operational amplifier in Fig. 5.80 is apparent.

5.6
Comparison

An overview is given of the common features and differences between the four different operational amplifiers. The important properties are listed in Figs. 5.97 and 5.98. In the graphic symbols, the current source symbol of types with current output is an indication of a high-resistance output with an impressed output current. The types with current input have an amplifier symbol between the inputs, to indicate a high-resistance noninverting input and a low-resistance inverting input.

Every operational amplifier may be regarded as a controlled source that describes the ideal amplifier, whereby amplifiers with low-resistance outputs are voltage sources and those with high-resistance outputs are current sources. A high-resistance (inverting) input forms a voltage-controlled input and a low-resistance input a current-controlled input. The descriptions of their function as controlled sources in Fig. 5.98 provide the two-letter designations used so far for any of the four types of operational amplifiers. From the systematic presentation it is obvious that no other types can exist; each and every circuit fits into the matrix of the four operational amplifiers.

The models shown in Fig. 5.97 describe the most important real characteristics of operational amplifiers. If the element Z is realized as a parallel connection of a resistor and a capacitor the frequency response can also be modeled. This has been used to calculate the cutoff frequencies of the various types.

The circuit diagrams show simple realizations of the examples already discussed. Operational amplifiers with a voltage input feature a differential amplifier at the input, while those with a current input have a voltage follower with compensated base–emitter voltage. Those with voltage outputs feature an emitter follower at the output that does not exist in types with current outputs.

A particularly instructive comparison is achieved by presenting the CC operational amplifier, the most simple of the four amplifier types, as a (diamond) transistor and realizing the remaining three types by adding impedance converters. This makes it clear that the CV operational amplifier requires a voltage follower at its output, the VC operational amplifier a voltage follower at its inverting input, and the VV operational amplifier both simultaneously. For this reason, it is possible to realize all four types of operational amplifiers with CC operational amplifiers, as shown by a comparison of Figs. 5.97 and 5.99.

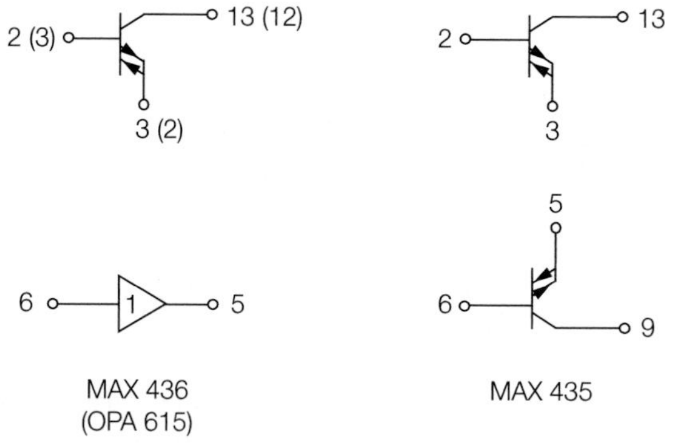

2 (3) o———|◄   o 13 (12)

3 (2)

2 o———|◄   o 13

3

5

6 o———|1|▷———o 5

MAX 436
(OPA 615)

6 o———|◄   o 9

MAX 435

Fig. 5.99. Comparison of standard amplifiers with current output: illustrations of the pin numbers of the dual inline case

For the purpose of comparison, the four operational amplifiers are presented as non-inverting amplifiers in Fig. 5.97. As the feedback always leads to the inverting input it determines the kind of feedback: If the inverting input is a voltage input (high impedance) a voltage feedback amplifier results; if it is a current input (low impedance) a current feedback amplifier results even though voltage feedback exists at the same time. Pure current feedback is achieved by simply connecting the inverting input to ground via a resistor (see Fig. 5.82–5.93). The voltage gain for voltage feedback shown in Fig 5.97 is the same in all cases, except for the CC operational amplifier, which has a factor of two in the denominator. The feedback output description in Fig. 5.98 also leads to the typical systematic short names for the operational amplifiers.

The feedback loops in Fig. 5.97 show that the longest path exists in the VV operational amplifier and the shortest path in the CC operational amplifier. This is the reason why, in CC operational amplifiers, high frequencies cause least phase lag and thus the fewest stability problems. For this reason, the CC operational amplifier is particularly suitable for high frequencies. This difference is shown in Fig. 5.100. Even though both circuits require the same amplifier and have a low output resistance, the CC operational amplifier features a shorter feedback loop than the transimpedance amplifier where the impedance converter is outside the feedback loop.

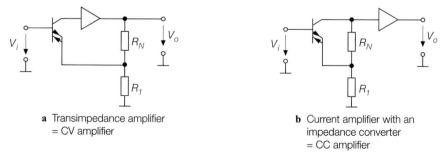

a Transimpedance amplifier = CV amplifier

b Current amplifier with an impedance converter = CC amplifier

Fig. 5.100. Comparison of a CV operational amplifier and a CC operational amplifier with an impedance converter

5.6.1
Practical Implementation

Many parasitic effects cannot be detected by circuit simulation. In particular, these include the inductances caused by wiring, as they depend on the printed circuit board layout. Only few simulation programs are capable of extracting these parameters from the layout and automatically taking them into account during simulation (post-layout simulation). This is not required for low-frequency circuits, but it becomes more and more important with increasing frequencies above 1 MHz. Above 30 MHz, even the inductance of the cases of integrated circuits play an important role. For this reason, SMD components are especially advantageous for high frequencies, as their parasitic inductances are significantly lower on account of their small dimensions. The most important issues to be observed in the practical use of operational amplifiers are described below.

Blocking operating voltages: Operating voltages must be well blocked. Naturally, the operating voltage leads have an inductance that increases with their length. As shown in Fig. 5.101a, these inductances are short-circuited by capacitors in order to prevent any voltage drop. A precondition is that the ground lead inductance of the capacitor is small in respect to the supply inductance. One way to achieve this is with a close ground network, or – even better – with a ground plane that has gaps for pins only. Capacitors also widely vary in their high-frequency response. An electrolytic capacitor has a low resistance low frequencies on account of its high capacitance. However, its resistance increases with higher frequencies due to its parasitic inductance. To achieve low resistances even with these frequencies, ceramic capacitors, with low inductance, are connected in parallel.

Tendency to oscillate: The circuit may oscillate, especially with capacitive loads or when driving an amplifier below A_{min}. The cause of this, however, may also turn out to be an unfavorable layout or insufficient blocking of the operating voltages. Often, the amplitude is low and the frequency high, so that the oscillation does not become directly apparent. An indication of oscillation is often the inaccurate operation of the circuit with DC voltages. Therefore it is always recommended to use an oscilloscope in order to make sure that the circuit is working properly. However, what must be borne in mind is the fact that the input of an oscilloscope represents a capacitive load that contributes to the tendency of the operational amplifier to oscillate. For this reason, the oscilloscope should never be connected via a coaxial cable or a 1:1 probe, but only via a 1:10 probe whose

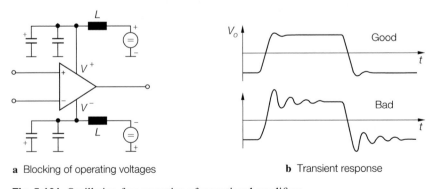

a Blocking of operating voltages   **b** Transient response

Fig. 5.101. Oscillation-free operation of operational amplifiers

capacitance is usually only a few picofarad. Grounding should be carried out via a short lead connected to a location close to the test point.

Damping: Having established that no amplifier of a given board oscillates, the next step is to insure that the amplifier is operated far from any oscillating condition. On the one hand, oscillations may occur if temperature or load changes; on the other hand, there is a general desire for a well-damped transient response. It is therefore beneficial to apply a square-wave signal and observe the output signals on an oscilloscope. This gives a good impression of the circuit damping. An example of a good and a bad square-wave response are shown in Fig. 5.101b.

Feedback resistances: With VV operational amplifiers, there is a considerable degree of freedom as regards dimensioning the feedback resistances. On the one hand, the resistances should be selected to be so low that no significant error occurs due to the input currents of the operational amplifier and the resistance noise. On the other hand, the resistances should be selected to be so high that their contribution to the current consumption and the heating of the operational amplifier remains low. The parasitic capacitances of the resistances must also be taken into account. The input of operational amplifiers also contains capacitances, which may cause undesired lowpass filter effects in the feedback loop. The resistances should therefore be made as high as the dynamic performance allows. If high resistances are required, then low capacitances should connected in parallel to maintain the desired gain with higher frequencies (see Fig. 5.63). In the CV operational amplifiers in Fig. 5.77, the feedback resistance R_N determines the loop gain and thus the transient response; therefore, its size is usually given by the manufacturer. The gain is determined by the series resistance R_1; its size must selected according to the given application.

Dissipation: The operating voltage should be selected to be as low as possible, to keep the dissipation of the circuit low. One must consider whether an output control range of ± 10 V, which was common in the past, is really required, since this means that operating voltages of between ± 12 V and ± 15 V are required. Operating voltages of ± 5 V, or even a single operating voltage of 3.3 V, are often sufficient when rail-to-rail amplifiers are used. Other significant differences exist with regard to the current consumption of operational amplifiers: it ranges from a few microamperes to several milliamperes. Amplifiers with a higher current consumption usually have a higher bandwidth. Therefore, one should not use faster operational amplifiers than are needed for the given task.

Cooling: Higher output currents require additional cooling of the operational amplifier. As long as the dissipation is around 1 W, the PC board can be used as heat sink via a few square centimeters of metal-plated.

Overdriving: Overdriving an amplifier usually means that internal transistors are in saturation mode and the capacitor for frequency response correction charges. After an amplifier has been overdriven, it often takes some time until it returns to normal operation. Overdriving should therefore be avoided if possible. Where this is not possible, *clamping amplifiers* should be used, since these need almost no recovery time due to their internal construction.

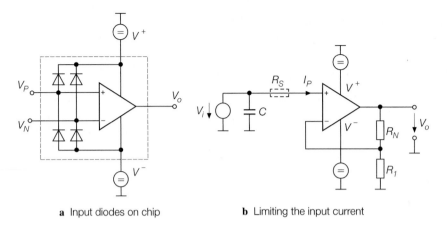

a Input diodes on chip   **b** Limiting the input current

Fig. 5.102. Overcurrents at the inputs

Input protection: The input voltages of an integrated circuit must not exceed the operating voltages, since otherwise the parasitic diodes shown in Fig. 5.102a become conductive. Often, the maximum permissible currents are around as little as 10 mA. The moment after turn-off, when the operating voltages drop to zero, is particularly critical, because at this instant the maximum input voltage is only ±0.6 V. If a charged capacitor is connected to the input, dangerous high discharge currents may flow through the diodes. The same situation occurs if a correspondingly high input signal remains. In both cases, the protective resistance R_S shown in Fig. 5.102b is very useful for limiting the current.

5.6.2
Types

There are many types of operational amplifier, that are optimized for a variety of different applications. To provide the reader with an overview on which data can be expected, some of the typical devices for the various applications were set out in Fig. 5.2. The list of manufacturers shows which companies are particularly active in this field. Of course, each of them offers many more types of operational amplifier, whose data sheets are available via the Internet. For details of the website addresses, see Sect. 29.8 on page 1518.

The offset voltage and the quiescent input current have been given as criteria for comparing the accuracy of DC voltages. This is followed by the gain–bandwidth product and the slew rate, which characterize their high frequency properties. The gain–bandwidth product has a clear meaning: it shows the bandwidth of an amplifier for the gain $A = 1$. The slew rate allows the power bandwidth to be calculated; this is the frequency up to which the full output amplitude is available. According to (5.31), this frequency is:

$$f_p = \frac{SR}{2\pi \widehat{V}_o}$$

The values for the minimum or maximum operating voltage indicate which types are suitable for low operating voltages and which types can provide high output amplitudes. The figures stated here are the minimum and maximum voltages between the positive and a negative operating voltage terminal for which the chip is specified. Since operational amplifiers have no ground connection, it is up to the user to decide how the voltage is distributed between positive and negative operating voltages (see Sect. 5.2.5). Symmetrical operating voltages of $\pm 5\,V$ or $\pm 15\,V$ are normally used. At rail-to-rail output (RRO) amplifiers the output voltage can nearly reach the supply voltages. Thus, appreciable output amplitudes can be obtained even at low supply voltages.

The maximum common-mode voltage and output voltage of normal operational amplifiers is approximately 2 V within the operating voltages. At rail-to-rail input-output (RRIO) amplifiers the common-mode voltage may also reach the supply voltages and even exceed it by 200–300 mV. They are especially adapted for single supply operation.

Often, two or four amplifiers, rather than just one single operational amplifier, are contained in one case. Figure 5.103 shows how many amplifiers are contained in one case. We have preferred to list two-fold or four-fold operational amplifiers. Often, single operational amplifiers are available under a similar model designation. Where the current consumption is quoted, this always refers to one amplifier.

Universal types: These have no specific electrical properties; but the old standard types, such as the 741 and 324 are particularly low-priced. From the value of the input current it is possible to determine the technology used at the input of the differential amplifier: for bipolar transistors this value is in the nanoampere range, while for field effect transistors it is in the picoampere range.

Precision types: The most important preconditions for high accuracy with DC voltages and low frequencies are a low offset voltage and a high differential gain. A low quiescent input current is, of course, also desirable; the error that it causes can be reduced by low-resistance dimensioning of the feedback resistances (see Fig. 5.61 on page 534). Due to the low offset voltage, one must insure that the unavoidable thermal voltages do not affect the circuit. It is important that the corresponding points of the circuit are at the same temperature level, so that the thermal voltages compensate each other as much as possible. In critical cases one may use special solder or make wire connections by thermal compression.

Low-noise types: The types listed here are the operational amplifiers on the market that have the lowest noise. While amplifiers with bipolar transistors reach noise voltage densities of $1\,nV/\sqrt{Hz}$, the best operational amplifiers with FETs at the input have values five times as high. Nevertheless, they are advantageous for high-resistance sources due to their noise current density, which is lower by three orders of magnitude. In any case, feedback resistances should be set as low as possible in order to keep the noise voltages caused by the noise current of the amplifier and the inherent noise of the resistors as low as possible (see Fig. 5.64 on page 538).

Rail-to-rail output amplifiers (RRO):   At these amplifiers the output voltage can nearly reach the supply voltages. So appreciable output amplitudes can be obtained even at low supply voltages. Often the common-mode range of these amplifiers includes the negative supply voltage. Therefore these Amplifiers can be used as single supply amplifiers (see Sect. 5.2.4 on page 498).

Type	Manufacturer	Offset-voltage	Bias current	Gain-bandwidth product	Slew rate	Operating voltage min/max	OPs per case	Special feature
VV-OPAmps: Low cost universal types								
...741	Many	1 mV	80 nA	1.5 MHz	0.6 V/µs	6/36 V	1	high voltage
...324	Many	2 mV	45 nA	1 MHz	0.6 V/µs	3/32 V	4	single supply
AD8034	Analog D.	1 mV	1 pA	80 MHz	80 V/µs	5/24 V	2	fast
AD8058	Analog D.	1 mV	0.5 µA	325 MHz	1000 V/µs	3/12 V	2	damned fast
AD8604	Analog D.	1 mV	0.2 pA	8 MHz	5 V/µs	3/ 6 V	4	RRIO
AD8619	Analog D.	0.5 mV	0.2 pA	0.4 MHz	0.1 V/µs	2/ 5 V	4	RRIO
AD8674	Analog D.	20 µV	5 nA	10 MHz	4 V/µs	8/36 V	4	$V_{nd} = 3\,\mathrm{nV}/\sqrt{\mathrm{Hz}}$
ADA4851	Analog D.	0.6 mV	2 µA	130 MHz	375 V/µs	3/10 V	4	RRO
OP177	Analog D.	20 µV	1 nA	0.6 MHz	0.3 V/µs	6/36 V	1	Low offset
LTC6242	Lin. Tech.	50 µV	1 pA	18 MHz	10 V/µs	3/ 6 V	4	$V_{nd} = 7\,\mathrm{nV}/\sqrt{\mathrm{Hz}}$
MAX4094	Maxim	30 µV	20 nA	0.5 MHz	0.2 V/µs	3/ 6 V	4	RRIO
MAX4351	Maxim	1 mV	8 µA	200 MHz	490 V/µs	9/11 V	2	RRO, fast
MAX4495	Maxim	0.3 mV	0.2 µA	5 MHz	3 V/µs	5/12 V	4	RRO
MAX9916	Maxim	0.2 mV	1 pA	1 MHz	0.5 V/µs	2/ 6 V	2	RRIO
LF356	National	1 mV	30 pA	5 MHz	12 V/µs	8/36 V	1	robust
LMC6034	National	1 mV	40 fA	1.4 MHz	1 V/µs	5/15 V	4	low bias
LMC6484	National	0.1 mV	20 fA	1.5 MHz	1 V/µs	3/15 V	4	RRIO
LMH6646	National	1 mV	0.4 µA	55 MHz	22 V/µs	3/12 V	2	RRIO
OPA2244	Texas I.	1 mV	10 nA	0.3 MHz	0.1 V/µs	2/36 V	2	$I_6 = 40\,\mu\mathrm{A}$
OPA4134	Texas I.	0.5 mV	5 pA	8 MHz	20 V/µs	5/36 V	4	low distortion
OPA4343	Texas I.	2 mV	0.2 pA	5 MHz	6 V/µs	3/ 6 V	4	RRIO
TLC084	Texas I.	0.4 mV	3 pA	10 MHz	16 V/µs	5/16 V	4	single supply
TLC274	Texas I.	1 mV	0.1 pA	2 MHz	3 V/µs	3/16 V	4	single supply

Fig. 5.103. Examples for OPAmps, typical data

Type	Manufacturer	Offset-voltage	Bias current	Gain-bandwidth product	Slew rate	Operating voltage min/max	OPs per case	Special feature
VV-OPAmps: Precision types								
AD8574	Analog D.	1 μV	10 pA	1.5 MHz	0.4 V/μs	3/ 5 V	4	RRIO
AD8675	Analog D.	10 μV	0.5 nA	10 MHz	2.5 V/μs	6/32 V	1	RRO
OP177	Analog D.	20 μV	1 nA	0.6 MHz	0.3 V/μs	6/36 V	1	low cost
LT1028	Lin. Tech.	10 μV	25 nA	75 MHz	15 V/μs	8/32 V	1	$V_{nd} = 1\,nV/\sqrt{Hz}$
LT1218	Lin. Tech.	25 μV	30 nA	0.3 MHz	0.1 V/μs	3/30 V	1	RRIO
LT1469	Lin. Tech.	50 μV	10 nA	90 MHz	22 V/μs	4/32 V	2	fast
LT1882	Lin. Tech.	30 μV	0.2 nA	1 MHz	0.3 V/μs	3/36 V	4	RRO
LT2079	Lin. Tech.	30 μV	6 nA	0.2 MHz	0.1 V/μs	5/40 V	4	single supply
LTC2052	Lin. Tech.	0.5 μV	8 pA	3 MHz	2 V/μs	3/11 V	4	autozero
LMC2001	National	0.5 μV	6 pA	6 MHz	5 V/μs	3/ 6 V	4	autozero
LMP7701	National	40 μV	0.2 pA	2.5 MHz	0.9 V/μs	3/12 V	4	RRIO
OPA4227	Texas I.	10 μV	3 nA	8 MHz	2 V/μs	5/36 V	4	$V_{nd} = 3\,nV/\sqrt{Hz}$
OPA4277	Texas I.	20 μV	3 nA	1 MHz	0.8 V/μs	4/36 V	4	high gain
OPA2334	Texas I.	1 μV	70 pA	2 MHz	1.6 V/μs	3/ 6 V	2	autozero
OPA4727	Texas I.	15 μV	10 pA	20 MHz	30 V/μs	4/13 V	4	$V_{nd} = 6\,nV/\sqrt{Hz}$

Type	Manufacturer	Offset-voltage	Bias current	Gain-bandwidth product	Slew rate	Operating voltage min/max	OPs per case	Special feature V_{nd}/I_{nd} per $\sqrt{Hz}$
VV-OPAmps: Low noise types								
AD743	Analog D.	250 μV	150 pA	5 MHz	3 V/μs	9/36 V	1	3 nV/7 fA
AD797	Analog D.	25 μV	300 nA	8 MHz	20 V/μs	10/36 V	1	1 nV/2 pA
AD8066	Analog D.	400 μV	2 pA	145 MHz	180 V/μs	5/24 V	2	7 nV/0.6 fA
AD8099	Analog D.	100 μV	60 nA	500 MHz	1300 V/μs	10/12 V	1	1 nV/3 pA
AD8513	Analog D.	80 μV	25 pA	8 MHz	20 V/μs	9/36 V	4	8 nV/2 fA
AD8652	Analog D.	100 μV	1 pA	50 MHz	40 V/μs	3/ 6 V	2	5 nV/4 fA
AD8674	Analog D.	20 μV	3 nA	10 MHz	4 V/μs	8/36 V	4	3 nV/0.3 pA
EL2125	Intersil	200 μV	22 μA	220 MHz	230 V/μs	5/30 V	1	0.8 nV/2.4 pA

Fig. 5.103. Examples for OPAmps, typical data

VV-OPAmps: Low noise types (cont.)

Type	Manufacturer	Offset-voltage	Bias current	Gain-bandwidth product	Slew rate	Operating voltage min/max	OPs per case	Special feature
LT1113	Lin. Tech.	500 μV	300 pA	6 MHz	4V/μs	8/36 V	2	5 nV/10 fA
LT1115	Lin. Tech.	75 μV	40 nA	70 MHz	15V/μs	8/36 V	1	1 nV/1 pA
LTC6242	Lin. Tech.	50 μV	1 pA	18 MHz	10V/μs	3/ 6 V	4	7 nV/0.6 fA
LMC6001	National	200 μV	25 fA	1 MHz	1V/μs	5/14 V	1	22 nV/0.1 fA
LMH6626	National	700 μV	50 nA	90 MHz	300V/μs	4/6 V	2	1 nV/2 pA
LMP7712	National	20 μV	50 fA	14 MHz	8V/μs	4/11 V	2	6 nV/10 fA
OPA4227	Texas I.	10 μV	3 nA	8 MHz	2V/μs	5/36 V	4	3 nV/0.4 pA
OPA627	Texas I.	40 μV	1 pA	16 MHz	55V/μs	9/36 V	1	5 nV/2 fA
OPA656	Texas I.	250 μV	2 pA	500 MHz	290V/μs	9/12 V	1	7 nV/1 fA
OPA686	Texas I.	400 μV	10 μA	300 MHz	600V/μs	8/12 V	1	1 nV/2 pA

VV-OPAmps: Rail-to-Rail Output (RRO)

Type	Manufacturer	Offset-voltage	Bias current	Gain-bandwidth product	Slew rate	Operating voltage min/max	OPs per case	Special feature
AD8054	Analog D.	2 mV	2 μA	150 MHz	150V/μs	3/10 V	4	low cost
AD8625	Analog D.	50 μV	0.25 pA	5 MHz	5V/μs	5/26 V	4	precision
OP481	Analog D.	500 μV	3 nA	100 kHz	0.25V/μs	3/12 V	4	$I_b = 3\,\mu A$
LT1635	Lin. Tech.	300 μV	2 nA	175 kHz	45V/ms	1.2/14 V	1	$V_{ref} = 0.2$V
LT6012	Lin. Tech.	20 μV	20 pA	0.3 MHz	90V/ms	3/36 V	4	precision
LT6232	Lin. Tech.	100 μV	5 μA	200 MHz	60V/μs	3/12 V	4	low noise
MAX4220	Maxim	4 mV	5 μA	200 MHz	600V/μs	3/11 V	4	fast
MAX4254	Maxim	100 μV	1 pA	3 MHz	0.3V/μs	3/ 6 V	4	low noise
MAX4472	Maxim	500 μV	200 pA	9 kHz	2V/ms	2/ 6 V	4	$I_b = 0.8\,\mu A$
MAX4478	Maxim	70 μV	1 pA	10 MHz	3V/μs	3/ 6 V	4	low noise
LMC6442	National	1 mV	5 fA	10 kHz	4V/ms	2/11 V	2	$I_b = 1\,\mu A$
LMH6643	National	1 mV	2 μA	130 MHz	130V/μs	3/12 V	2	$I_o = 75$ mA
LMV651	National	100 μV	80 nA	12 MHz	3V/μs	3/ 6 V	1	$I_b = 110\,\mu A$
OPA4336	Texas I.	100 μV	10 pA	100 kHz	30V/ms	2/ 6 V	4	$I_b = 20\,\mu A$
OPA3355	Texas I.	2 mV	3 pA	450 MHz	360V/μs	3/ 6 V	3	fast

Fig. 5.103. Examples for OPAmps, typical data

VV-OPAmps: Rail-to-rail Input and Output (RRIO)

Type	Manufacturer	Offset-voltage	Bias current	Gain-bandwidth product	Slew rate	Operating voltage min/max	OPs per case	Special feature
AD8040	Analog D.	1.6 mV	0.7 µA	125 MHz	60 V/µs	3/12 V	4	fast
AD8062	Analog D.	1 mV	4 µA	300 MHz	600 V/µs	3/ 8 V	2	dammed fast
AD8527	Analog D.	1 mV	0.2 µA	7 MHz	8 V/µs	2/ 6 V	2	low voltage
AD8618	Analog D.	25 µV	0.2 pA	24 MHz	12 V/µs	3/ 6 V	4	low offset
OP484	Analog D.	65 µV	60 nA	3 MHz	2 V/µs	3/36 V	4	high voltage
OP496	Analog D.	300 µV	10 nA	0.3 MHz	0.3 V/µs	3/12 V	4	$I_b = 60$ µA
LT1496	Lin. Tech.	200 µV	0.3 nA	3 kHz	1 V/ms	2/36 V	2	$I_b = 1.2$ µA
LT1636	Lin. Tech.	100 µV	5 nA	0.2 MHz	70 V/ms	3/44 V	2	$I_b = 55$ µA
LT1639	Lin. Tech.	200 µV	15 nA	1.2 MHz	0.5 V/µs	3/44 V	4	$V_{CM} > V_S$
LT1679	Lin. Tech.	35 µV	2 nA	20 MHz	6 V/µs	3/36 V	4	low noise
LT6204	Lin. Tech.	100 µV	1 µA	90 MHz	24 V/µs	3/12 V	4	low noise
LTC1152	Lin. Tech.	1 µV	10 pA	0.7 MHz	0.5 V/µs	3/14 V	1	autozero
MAX4196	Maxim	250 µV	50 nA	5 MHz	2 V/µs	3/ 6 V	4	high I_o
MAX9916	Maxim	200 µV	1 pA	1 MHz	0.5 V/µs	2/ 6 V	2	$I_b = 20$ µA
LM8261	National	700 µV	1 µA	21 MHz	12 V/µs	3/30 V	1	high voltage
LMC6484	National	100 µV	20 fA	1.5 MHz	1 V/µs	3/15 V	4	low bias
LMH6646	National	1 mV	400 nA	55 MHz	22 V/µs	3/12 V	2	fast
LMV982	National	1 mV	15 nA	1.4 MHz	0.4 V/µs	2/ 5 V	2	low voltage
LPV511	National	200 µV	300 pA	27 kHz	8 V/ms	3/12 V	1	$I_b = 0.9$ µA
OPA2349	Texas I.	2 mV	1 pA	70 kHz	20 V/ms	2/ 6 V	2	$I_b = 1$ µA
OPA4340	Texas I.	500 µV	10 pA	6 MHz	6 V/µs	3/ 5 V	4	low distortion
OPA4350	Texas I.	150 µV	0.5 pA	35 MHz	22 V/µs	3/ 6 V	4	low noise
TLV2764	Texas I.	500 µV	3 pA	500 kHz	0.2 V/µs	2/ 4 V	4	$I_b = 20$ µA

Fig. 5.103. Examples for OPAmps, typical data

Type	Manufacturer	Offset-voltage	Bias current	Gain-bandwidth product	Slew rate	Operating voltage min/max	OPs per case	Special feature
VV-OPAmps: High bandwidth								
AD829	Analog D.	0.2 mV	3 μA	120 MHz	230 V/μs	10/36 V	1	high voltage
AD8021	Analog D.	0.4 mV	7 μA	490 MHz	120 V/μs	5/24 V	1	$I_{nd} = 2nV\sqrt{Hz}$
AD8036	Analog D.	2 mV	4 μA	240 MHz	1500 V/μs	6/12 V	1	clamping
AD8039	Analog D.	0.5 mV	0.4 μA	350 MHz	425 V/μs	3/12 V	2	$I_b = 1mA$
AD8045	Analog D.	0.2 mV	0.2 μA	1000 MHz	1300 V/μs	3/12 V	1	low distortion
ADA4851	Analog D.	0.6 mV	2 μA	130 MHz	370 V/μs	3/12 V	4	RRO
ISL55004	Intersil	1 mV	0.6 μA	220 MHz	300 V/μs	5/30 V	4	
LT1224	Lin. Tech.	0.5 mV	4 μA	45 MHz	400 V/μs	5/30 V	1	C Load stable
LT1365	Lin. Tech.	0.5 mV	1 μA	70 MHz	1000 V/μs	5/30 V	4	
LT1812	Lin. Tech.	0.4 mV	1 μA	100 MHz	750 V/μs	5/12 V	1	
LT1817	Lin. Tech.	0.2 mV	2 μA	220 MHz	1500 V/μs	5/12 V	4	
LT1819	Lin. Tech.	0.2 mV	2 μA	400 MHz	2500 V/μs	5/12 V	2	low distortion
MAX4413	Maxim	0.4 mV	1.6 μA	500 MHz	220 V/μs	3/ 6 V	2	
MAX4418	Maxim	0.5 mV	1.3 μA	400 MHz	200 V/μs	3/ 5 V	4	RRO
MAX4454	Maxim	0.4 mV	0.8 μA	200 MHz	240 V/μs	3/ 5 V	4	
LM6172	National	0.4 mV	1.2 μA	100 MHz	3000 V/μs	10/32 V	2	$I_b = 0.6 mA$
LM7171	National	0.2 mV	3 μA	200 MHz	4100 V/μs	10/32 V	1	$I_b = 2.3 mA$
LMH6609	National	0.8 mV	2 μA	900 MHz	1400 V/μs	6/12 V	1	
LMH6655	National	1 mV	5 μA	250 MHz	200 V/μs	5/12 V	2	
LMH6658	National	1 mV	5 μA	270 MHz	700 V/μs	5/12 V	2	
OPA643	Texas I.	2 mV	20 μA	800 MHz	1000 V/μs	9/11 V	1	$V_{nd} = 2nV\sqrt{Hz}$
OPA657	Texas I.	0.1 mV	1 pA	1600 MHz	700 V/μs	9/12 V	1	FET input
OPA698	Texas I.	2 mV	3 μA	450 MHz	1100 V/μs	9/12 V	1	clamping
OPA842	Texas I.	0.3 mV	20 μA	400 MHz	400 V/μs	9/12 V	1	
THS4271	Texas I.	5 mV	6 μA	1400 MHz	1000 V/μs	5/16 V	1	

Fig. 5.103. Examples for OPAmps, typical data

Type	Manufacturer	Offset-voltage	Bias current	Gain-bandwidth product	Slew rate	Operating voltage min/max	OPs per case	Special feature
VV-OPAmps: Differential output								
AD8137	Analog D.	5 mV	0.3 µA	110 MHz	450 V/µs	3/12 V	1	$I_b = 2.6\,\text{mA}$
AD8138	Analog D.	1 mV	4 µA	320 MHz	1200 V/µs	3/11 V	1	low distortion
AD8139	Analog D.	0.2 mV	2 µA	410 MHz	800 V/µs	9/12 V	1	$V_{nd} = 2\,\text{nV}/\sqrt{\text{Hz}}$
AD8390	Analog D.	1 mV	4 µA	60 MHz	300 V/µs	10/24 V	1	$I_o = 400\,\text{mA}$
LT1994	Lin. Tech.	1 mV	18 µA	70 MHz	65 V/µs	3/12 V	1	
LTC1992	Lin. Tech.	0.3 mV	2 pA	4 MHz	2 V/µs	3/11 V	1	
THS4120	Texas I.	3 mV	1 pA	100 MHz	50 V/µs	3.3 V	1	RRO
THS4130	Texas I.	0.2 mV	2 µA	150 MHz	50 V/µs	5/30 V	1	$V_{nd} = 1.3\,\text{nV}/\sqrt{\text{Hz}}$
THS4140	Texas I.	1 mV	5 µA	160 MHz	450 V/µs	5/30 V	1	
THS4150	Texas I.	1 mV	7 µA	150 MHz	650 V/µs	5/30 V	1	
THS4502	Texas I.	1 mV	4 µA	370 MHz	2800 V/µs	5/15 V	1	
THS4509	Texas I.	0.5 mV	8 µA	1900 MHz	6600 V/µs	3/ 5 V	1	damned fast

Fig. 5.103. Examples for OPAmps, typical data

Type	Manufacturer	Offset-voltage	Bias current	Gain-bandwidth product	Slew rate	Operating voltage min/max	OPs per case	Special feature
VV-OPAmps: High output voltage								
PA78*	Apex	8 mV	9 pA	1 MHz	350 V/μs	10/ 350 V	1	$I_0 = 150$ mA
PA85*	Apex	0.5 mV	5 pA	100 MHz	1000 V/μs	30/ 450 V	1	$I_0 = 200$ mA
PA89*	Apex	0.5 mV	5 pA	10 MHz	16 V/μs	150/1200 V	1	$I_0 = 75$ mA
PA90*	Apex	0.5 mV	200 pA	100 MHz	300 V/μs	80/ 400 V	1	$I_0 = 200$ mA
PA93*	Apex	2 mV	200 pA	12 MHz	50 V/μs	80/ 400 V	1	$I_0 = 8$ A
PA240	Apex	25 mV	50 pA	3 MHz	30 V/μs	100/ 350 V	1	$I_0 = 60$ mA
OPA452	Texas I.	1 mV	7 pA	1.8 MHz	8 V/μs	20/ 80 V	1	$I_0 = 50$ mA
VV-OPAmps: High output current								
PA17*	Apex	5 mV	10 pA	2 MHz	50 V/μs	30/ 200 V	1	$I_0 = 50$ A
PA19*	Apex	0.5 mV	10 pA	100 MHz	900 V/μs	30/ 80 V	1	$I_0 = 3$ A
PA35	Apex	1.5 mV	1 μA	0.6 MHz	1 V/μs	5/ 40 V	2	$I_0 = 1.7$ A
PA45	Apex	5 mV	20 pA	4.5 MHz	27 V/μs	30/ 150 V	1	$I_0 = 5$ A
PA52*	Apex	5 mV	10 pA	3 MHz	50 V/μs	50/ 200 V	1	$I_0 = 40$ A
PA93*	Apex	2 mV	200 pA	12 MHz	50 V/μs	80/ 400 V	1	$I_0 = 8$ A
LM12	National	2 mV	150 nA	0.7 MHz	9 V/μs	20/ 60 V	1	$I_0 = 10$ A
LM675	National	1 mV	200 nA	5.5 MHz	8 V/μs	16/ 60 V	1	$I_0 = 3$ A
OPA2544	Texas I.	1 mV	15 pA	1.4 MHz	8 V/μs	20/ 70 V	2	$I_0 = 2$ A
OPA548	Texas I.	3 mV	500 nA	1 MHz	6 V/μs	8/ 60 V	1	$I_0 = 3$ A
OPA549	Texas I.	1 mV	100 nA	1 MHz	10 V/μs	8/ 60 V	1	$I_0 = 8$ A
OPA561	Texas I.	1 mV	10 pA	17 MHz	50 V/μs	7/ 15 V	1	$I_0 = 1.2$ A
OPA569	Texas I.	0.5 mV	1 pA	1.2 MHz	1 V/μs	3/ 6 V	1	$I_0 = 2$ A

*Hybrid circuit (probably expensive)

Fig. 5.103. Examples for OPAmps, typical data

Type	Manufacturer	Offset-voltage	Bias current	Gain-bandwidth product	Slew rate	Operating voltage min/max	OPs per case	Special feature
CV-OPAmps: Transimpedance amplifiers								
AD815	Analog D.	5 mV	2 µA	120 MHz	900 V/µs	8/34 V	2	$I_o = 500\,mA$
AD8005	Analog D.	5 mV	0.5 µA	270 MHz	1500 V/µs	5/12 V	1	$I_b = 0.4\,mA$
AD8009	Analog D.	2 mV	50 µA	1000 MHz	5500 V/µs	8/12 V	1	fast
AD8010	Analog D.	5 mV	6 µA	230 MHz	800 V/µs	9/12 V	1	$I_o = 200\,mA$
AD8012	Analog D.	2 mV	3 µA	350 MHz	2200 V/µs	3/12 V	2	$I_b = 1\,mA$
AD8024	Analog D.	2 mV	1 µA	350 MHz	2400 V/µs	5/24 V	4	
ADA4861	Analog D.	0.1 mV	1 µA	730 MHz	650 V/µs	5/12 V	3	
EL5367	Intersil	0.5 mV	0.7 µA	1000 MHz	6000 V/µs	5/12 V	3	fast
EL5462	Intersil	1.5 mV	0.5 µA	500 MHz	4000 V/µs	5/12 V	4	$I_b = 1.5\,mA$
LT1207	Lin. Tech.	3 mV	2 µA	60 MHz	900 V/µs	10/30 V	2	$I_o = 250\,mA$
LT1210	Lin. Tech.	3 mV	2 µA	35 MHz	900 V/µs	10/30 V	1	$I_o = 1100\,mA$
LT1399	Lin. Tech.	1.5 mV	10 µA	300 MHz	800 V/µs	4/12 V	3	
MAX4119	Maxim	1 mV	3.5 µA	270 MHz	1200 V/µs	6/11 V	4	
MAX4187	Maxim	1.5 mV	1 µA	270 MHz	450 V/µs	6/11 V	4	$I_b = 1\,mA$
MAX4223	Maxim	0.5 mV	2 µA	1000 MHz	1000 V/µs	6/11 V	1	
MAX4226	Maxim	0.5 mV	2 µA	250 MHz	1100 V/µs	6/11 V	2	
CLC502	National	0.5 mV	10 µA	150 MHz	800 V/µs	6/12 V	1	clamping
LMH6702	National	1 mV	6 µA	1700 MHz	3100 V/µs	5/12 V	1	low distortion
LMH6715	National	2 mV	5 µA	480 MHz	1300 V/µs	9/12 V	2	
LMH6725	National	1 mV	2 µA	370 MHz	600 V/µs	4/12 V	4	
OPA2691	Texas I.	1.3 mV	30 µA	280 MHz	2100 V/µs	5/11 V	2	$I_b = 1\,mA$
OPA2694	Texas I.	3 mV	4 µA	800 MHz	17000 V/µs	9/11 V	2	$I_o = 150\,mA$
OPA684	Texas I.	1.5 mV	5 µA	120 MHz	650 V/µs	5/12 V	1	
THS3001	Texas I.	1 mV	1 µA	420 MHz	6500 V/µs	9/32 V	2	$I_b = 1.7\,mA$
THS3061	Texas I.	0.7 mV	2 µA	300 MHz	7000 V/µs	10/32 V	2	$I_o = 140\,mA$
THS3202	Texas I.	0.7 mV	13 µA	2000 MHz	9000 V/µs	6/15 V	1	damned fast
THS6012	Texas I.	2 mV	3 µA	315 MHz	1300 V/µs	9/32 V	2	$I_o = 500\,mA$

Fig. 5.103. Examples for OPAmps, typical data

Type	Manufacturer	Offset-voltage	Bias current	Gain-bandwidth product	Slew rate	Operating voltage min/max	OPs per case	Output current
VC-OPAmps: Transconductance amplifiers								
CA3080	Intersil	1 mV	2 μA	110 kHz	8 V/μs	4/30 V	1	0.5 mA
LT1228	Lin. Tech.	0.5 mV	0.4 μA	80 MHz	600 V/μs	4/30 V	1	1 mA
MAX436	Maxim	0.3 mV	1 μA	200 MHz	800 V/μs	9/11 V	1	20 mA
OPA660	Texas I.	7 mV	2 μA	700 MHz	3000 V/μs	9/11 V	1	15 mA

Type	Manufacturer	Offset-voltage	Bias current	Gain-bandwidth product	Slew rate	Current transfer ratio	OPs per case	Special feature
CC-OPAmps: Current amplifiers; $V_s = 9/11$ V								
MAX435	Maxim	0.3 mV	1 μA	275 MHz	850 V/μs	4	2	10 mA
MAX436	Maxim	0.3 mV	1 μA	275 MHz	850 V/μs	8	1	20 mA
OPA615	Texas I.	8 mV	0.3 μA	500 MHz	3000 V/μs	1	1	20 mA
OPA660	Texas I.	7 mV	2 μA	700 MHz	3000 V/μs	1	1	15 mA

Fig. 5.103. Examples for OPAmps, typical data

Some models feature a current consumption of only a few microamperes. They are particularly useful for battery operation; quite often, an on–off switch is not needed. However, the bandwidth and the slew rate also decline with the current consumption, as the overview shows.

Rail-to-rail input-output amplifiers (RRIO):   The operational amplifiers listed here can be used up to the operating voltages at the input and output. While the output voltage only reaches the proximity of the operating voltage (especially under load), common-mode voltages that exeed the supply voltages by several hundreds of millivolts, are usually permissible. Rail-to-rail amplifiers are particularly useful with low operating voltages, as they offer the maximum possible voltage sewing. The minimum operating voltage given indicates how low an operating voltage is allowed to be. It should be noted that most rail-to-rail amplifiers cannot be operated with ± 15 V, and some not even with ± 5 V.

High bandwidth: We can see that there is a multitude of VV operational amplifiers that are faster than amplifiers of the 741 class by up to three orders of magnitude. This, however, usually results in unfavorable DC data: a high offset voltage, a high input current, a low differential gain, and high current consumption. Most wide-band operational amplifiers are manufactured in a low voltage process for operating voltages of ± 5 V, since it is easier to make high-frequency transistors. However, the comparison shows that there are other high-frequency amplifiers that are suitable for operation with ± 15 V.

Most operational amplifiers require a relatively long recovery time to return to normal operation after overdrive. Models with a clamping output are preferred, where overdriving cannot be prevented. In this case, internal additions prevent transistors from entering the saturation region; this reduces the recovery time to a few nanoseconds. Furthermore a positive and negative output voltages limit may be set by special voltage limiting pins. This can also protect subsequent circuits, such as A/D converters, against overdrive.

Differential output amplifiers: Newly there is a family of OPAmps that offer complementary outputs. They are especially suitable for driving symmetrical loads, for instance AD-converters. Here the common-mode output voltage can be controlled by an additional input so that both output voltages are always positive. Furthermore these amplifiers are the precondition for fully symmetrical circuits.

High output voltage: There are relatively few operational amplifiers that can provide high output voltages and allow correspondingly high operating voltages, since the standard manufacturing processes do not support these features. For operating voltages exceeding 100 V, the relatively expensive hybrid circuits are usually used. An exception here is model PA240.

High output currents: High output currents cause high power dissipation in operational amplifiers. Therefore, operating voltages should not be unnecessarily high and amplifiers should be effectively cooled. The overview shows that, as a rule, power amplifiers are slow and have low slew rates; but there are exceptions to this rule. Here too, high-quality products can only be produced in hybrid technology.

CV operational amplifiers   are very similar to conventional wide-band amplifiers and perform accordingly. But despite the same technology and current consumption, they have a higher slew rate and larger bandwidth than equivalent VV operational amplifiers. The main difference for the user is that here only ohmic feedback is possible. Correspondingly high operating currents are required to achieve large bandwidths. At amplifiers with specially low current consumption the supply current is given under remarks. In order to keep power dissipation within limits, operating voltages of $\pm 5\,$V are normally used; models for higher operating voltages are an exception here. Since low-resistance loads normally have to be driven at high frequencies, the maximum output currents are generally higher than 20 mA. This value is specified for models that allow particularly high output currents.

VC operational amplifiers   have been available for some time. The first-generation models, like the CA3080, provide data that no longer satisfy modern demands: they are too slow and their output currents are too low. The OPA615 is CC operational amplifiers that can be expanded to form VC operational amplifiers by means of a voltage follower (see Fig. 5.97). In that case and in the MAX436, the transconductance of the operational amplifier can be reduced to any given value using an external resistance. This is the prerequisite for operation with pure current feedback, as seen in Figs. 5.82–5.91. Furthermore, all models allow the current consumption, and thus the maximum output current, to be set by means of an external resistance.

CC operational amplifiers   are the most versatile operational amplifiers for high frequencies. The fact that they are not widely used is mainly due to the habit of thinking in terms of voltage rather than current. The advantages of CC operational amplifiers have been explained in the example of the integrator filter; see Fig. 5.93. Figure 5.103 also specifies the current gain, which is determined by the transfer ratio of the current mirror at the output (see Fig. 5.80 on page 553). Here, it is also possible to set the quiescent current using an external resistance.

Classification

The technology determines the input data and thus the accuracy of the DC voltage. This becomes obvious when the quiescent input currents and offset voltages of operational amplifiers are plotted, as in Fig. 5.104. Here we can see that the operational amplifiers with FET differential amplifiers at the input have the lowest input currents but high offset voltages. Amplifiers with an automatic zero offset have especially low offset voltages.

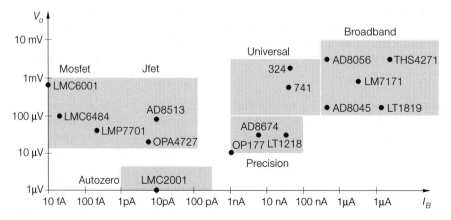

Fig. 5.104. Offset voltage and input bias current of various operational amplifier technolgies

Some of the precision amplifiers with bipolar transistors at the input have also very low offset voltages but significantly higher quiescent input currents. Most unfavorable are the DC data of wide-band operational amplifiers: they feature high offset voltages as well as high input currents.

When comparing the operational amplifiers in terms of their noise levels, two clearly distinguishable groups can be seen, as illustrated in Fig. 5.105: operational amplifiers with field effect transistors at the input show clearly less current noise due to their low input current than opamps with bipolar transistor input. This is the reason why fet input opamps are advantageous at high-resistance sources. On the other hand, one can see that operational amplifiers with bipolar transistors have lower voltage noise than those with FETs. For this reason, these are more advantageous for low-resistance sources (see Fig. 5.64).

The bandwidth–current diagram in Fig. 5.106 can be used to compare the dynamic responses of operational amplifiers. In order to enlarge the bandwidth of an operational amplifier, the transistors must be driven with higher currents; the bandwidth should therefore be proportional to the current. However, technology with small parasitic capacitances can help you achieve large bandwidths even with medium currents. Correspondingly, circuits operating in current on demand AB mode (see Fig. 5.28) offer a larger bandwidth than circuits in conventional A mode at the same quiescent current. The circuits near the top line in Fig. 5.106 have a more favorable technology and/or circuit design with regard to the frequency response than the models at the bottom right. From Fig. 5.106 one can see that significant differences exist with respect to both the current at constant bandwidth and the bandwidth at constant current. If, on the other hand, a line is drawn along which the bandwidth is proportional to the current, then all operational amplifiers of the same technology or circuit design will be located on a line. This shows that high-quality operational amplifiers have a bandwidth: current ratio above 100 MHz/mA, while older models do not even reach a level of 1 MHz/mA.

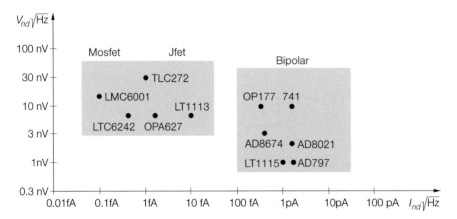

Fig. 5.105. Comparison of noise voltages and noise currents

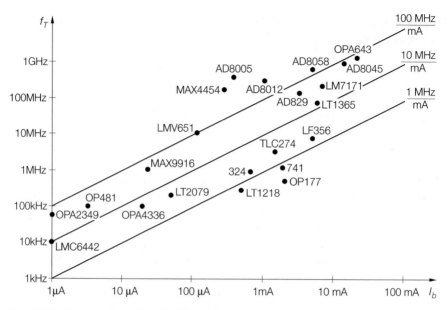

Fig. 5.106. A comparison of bandwidth and supply current

Chapter 6:
Latching Circuits

6.1
Transistor as Switch

In the case of linear circuits, we set the collector quiescent potential between V^+ and $V_{CE\,sat}$, thus enabling them to be driven about this operating point. The characteristic feature of linear circuits is that the swing is kept so small that the output voltage is a linear function of the input voltage. Consequently, the output voltage must not attain the positive or negative limits of the swing, as this would result in distortion. With *digital* circuits, on the other hand, only two operating states are employed. We are only interested in whether a voltage is greater than a specified value V_H or less than a specified value $V_L < V_H$. If the voltage exceeds V_H, it is referred to as being in the H (high) state, and if it is below V_L, it is said to be in the L (low) state.

The absolute values of the levels V_H and V_L depend entirely on the circuit design. For an unambiguous interpretation, steady-state levels between V_H and V_L must not occur. The circuit design implications of this will now be discussed with reference to the level inverter in Fig. 6.1. The circuit must exhibit the following characteristics:

$$\text{For } V_i \leq V_L \;\rightarrow\; V_o \geq V_H$$

and

$$\text{for } V_i \geq V_H \;\rightarrow\; V_o \leq V_L.$$

This relationship must hold good, even under worst-case conditions; that is, for $V_i = V_L$, V_o must not be lower than V_H, and for $V_i = V_H$, V_o must not be higher than V_L. This condition can only be satisfied by selecting suitable values for V_H, V_L, R_C, and R_B. The following worked example should serve to indicate a possible approach.

When the transistor in Fig. 6.1 is turned off, the output voltage under no-load conditions is equal to V^+. Let us assume that the lowest output load resistance is $R_V = R_C$; in this case, V_o is consequently equal to $V^+/2$. This is therefore the lowest output voltage in the H state. To be on the safe side, we specify $V_H < V^+/2$ for a supply voltage of V^+5 V; for example, $V_H = 1.5$ V. In accordance with the above-mentioned requirement, the input voltage must be in state L for $V_0 \geq V_H$. V_L is therefore defined as the highest input voltage for which the transistor is just certain to remain in the blocking state. For a silicon transistor, we can therefore take a value of 0.4 V if the device is at ambient temperature. Consequently, we select $V_L \leq 0.4$ V. Having determined the two levels V_H and V_L in this way, we must now select component values for the circuit in such a way that an output voltage $V_o \leq V_L$ is obtained for $V_i = V_H$. Even under worst-case conditions, we require a certain safety margin, namely that the output voltage shall remain below $V_L = 0.4$ V for $V_i = V_H = 1.5$ V. The collector resistance R_C is chosen low enough to insure that the switching times are sufficiently short, but without making the current drain unnecessarily high. Typically, we select $R_C = 5\,\text{k}\Omega$. We must now select a value for R_B that insures that the output voltage falls below the value $V_L = 0.4$ V for an input voltage of $V_i = 1.5$ V. For this to occur, a collector current of $I_C \approx V^+/R_C = 1$ mA has to flow. Transistors for this kind of application normally have a current gain of $B = 100$. The base current required

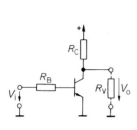

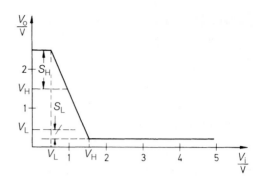

Fig. 6.1. Transistor as an inverter

Fig. 6.2. Transfer characteristic for $R_V = R_C$
S_L: L noise margin, S_H: H noise margin

is therefore $I_{B\,min} = I_C/B = 10\mu A$. In order to insure that the transistor is driven into saturation, we select $I_B = 100\ \mu A$; that is, it is ten times overdriven. We thus obtain

$$R_B = \frac{1.5\,V - 0.6\,V}{100\,\mu A} = 9\,k\Omega$$

Figure 6.1 shows the transfer characteristic for these parameters.

For $V_i = V_L = 0.4$ V, the output voltage $V_o = 2.5$ V at full load ($R_V = R_C$) and is therefore 1 V above the minimum value $V_H = 1.5$ V required. We now specify an *H noise margin* $S_H = V_o - V_H$ for $V_i = V_L$. In our example, it is 1 V. Similarly, we can define an *L noise margin* $S_L = V_L - V_o$ for $V_i = V_H$. In Fig. 6.1 it is identical to the voltage difference between V_L and the collector–emitter saturation voltage $V_{CE\,sat} \approx 0.2$ V, and has a value of $S_L = 0.4\,V - 0.2\,V = 0.2V$. The noise margins are a measure of the reliability of the circuit performance. Their general definition is:

$$\left.\begin{array}{l} S_H = V_o - V_H \\ S_L = V_L - V_o \end{array}\right\} \text{ for a worst-case condition at the input}$$

If we wish to improve the L noise margin, it is necessary to increase V_L, as voltage $V_o(V_i = V_H) \approx V_{CE\,sat}$ cannot be reduced much further. For this purpose, we can insert one or more diodes in front of the base, as in Fig. 6.3a. Resistor R_2 serves as a drain for the collector–base reverse current, thereby insuring that the transistor turns off reliably. Another possibility is to simply insert a preceding voltage divider, as shown in Fig. 6.3b or Fig. 6.3c.

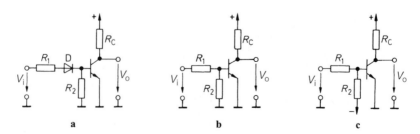

Fig. 6.3. Methods of improving the L noise margin

The output loading capability (fan-out) of the inverter in Fig. 6.1 is low. No more than two identical inputs can be connected to one output; otherwise, the output voltage would fall below 2.5 V in the H state.

Dynamic Characteristics

When using a transistor as a switch, the main parameter of interest is the switching time. Within the square-wave response we differentiate between various periods, as shown in Fig. 6.4.

We can see that the storage time t_S is considerably greater than the other switching times. It is incurred when a previously saturated transistor ($V_{CE} = V_{CE\,sat}$) is turned off. If, for the conducting transistor, V_{CE} is greater than $V_{CE\,sat}$, the storage time is considerably reduced. We exploit this fact in high-speed switches and prevent $V_{CE\,sat}$ from being reached. Digital circuits operating on this principle are known as *unsaturated logic circuits*. Their design will be discussed for the relevant circuits in Sect. 7.4.5.

The time behavior of digital circuits is generally characterized by the propagation delay time t_{pd}:

$$t_{pd} = \frac{1}{2}(t_{pd\,L} + t_{pd\,H})$$

where $t_{pd\,L}$ is the time difference between the 50% value of the input edge and the 50% value of the falling output edge. $t_{pd\,H}$ is the corresponding time difference for the rising output edge. This is illustrated in Fig. 6.5.

In the circuit shown in Fig. 6.1, we saw that the H level was far below the supply voltage and was a function of the load. In order to avoid this, an emitter follower can be connected, as shown in Fig. 6.6.

If T_1 is off, the output current flows via the emitter follower T_2, thus keeping the current in collector resistor R_C low. If T_1 becomes conducting, its collector potential falls to low levels. With a resistive output load, the output voltage likewise falls. With capacitive loading, the circuit must pick up the capacitor discharge current. As transistor T_2 is blocking in this case, diode D is inserted to allow the discharge current to flow via conducting transistor T_1 However, this increases the output voltage to about 0.8 V in the L state.

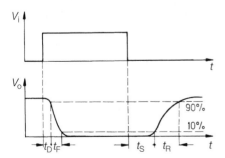

Fig. 6.4. Square-wave response of an inverter
t_S: Storage time
t_R: Rise time
t_D: Delay time
t_F: Fall time

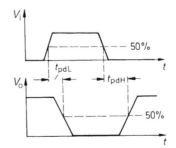

Fig. 6.5. Defining the propagation delay time
t_{pd}: propagation delay time

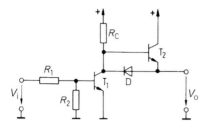

Fig. 6.6. Push–pull output stage for digital circuits

6.2
Latching Circuits Using Saturated Transistors

Latching circuits are positive-feedback digital circuits. They differ from positive-feedback linear circuits (oscillators) in that their output voltage does not vary continuously, but only jumps back and forth between two fixed values. This *switching process* can be initiated in various ways:

A *bistable* circuit changes the output state only changes when switching is initiated by an input signal. With a *flip-flop* (latch), a short pulse suffices, whereas a *Schmitt trigger* requires a sustained input signal.

A *monostable* circuit has a *single* stable state. Its other state is only stable for a fixed length of time, which is determined by the circuit component values. When this time has elapsed, the circuit automatically returns to the stable state. It is therefore also known as a time interval switch, monostable multivibrator, or one-shot.

An *astable* circuit has no stable state, but constantly oscillates between states without external triggering. It is therefore also known as a multivibrator.

The three types of circuit can be implemented using the basic configuration shown in Fig. 6.7. The only difference is in the construction of the two coupling devices CD1 and CD2, as listed in Fig. 6.8.

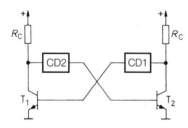

Fig. 6.7. Basic latching configuration, using saturated transistors

Type	Name	CD1	CD2
Bistable	Flip-flop, Schmitt trigger	R	R
Monostable	One-shot	R	C
Astable	Multivibrator	C	C

Fig. 6.8. Coupling networks required for the various types of latching circuit

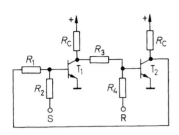

Fig. 6.9. Positive-feedback circuit comprising two inverters

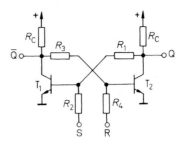

Fig. 6.10. *RS* flip-flop

6.2.1
Bistable Circuits

Flip-Flops

As shown in Fig. 6.9, a bistable circuit can be implemented by connecting two inverters in series with the coupling divices R_1, R_2 and R_3, R_4 providing positive feedback. We can see that the two inverters have equal status, and so the symmetrical circuit diagram given in Fig. 6.10 is generally preferred.

The mode of operation is as follows: a positive voltage at the set input S renders T_1 conducting. This causes its collector potential to fall, thereby reducing the base current of T_2, whose collector potential increases. This increase causes the base current of T_1 to rise via resistor R_1. The steady-state condition is therefore reached when the collector potential of T_1 has fallen to the saturation voltage. T_2 then turns off and T_1 is maintained ON via resistor R_1, Consequently, the voltage at the S-input can be made zero again at the end of the switching process without inducing any further changes. The flip-flop can be returned to its original state by applying a positive voltage pulse to the reset input R. When the two input voltages are zero, the flip-flop retains the last state assumed. This characteristic allows it to be used as a data *memory*.

If the two input voltages are simultaneously changed to the H state, both transistors become conducting during this time. However, in this case the base currents are supplied only by the control voltage sources and not by the adjacent transistor, as the two collector potentials are low. Consequently, this state is not stable. If the two control voltages are again made zero, this causes the two collector potentials to rise simultaneously. However, as there is never complete symmetry, one collector potential rises somewhat faster than the other. Due to the positive feedback, this difference is amplified, so that a stable state is eventually reached in which one transistor is OFF and the other is ON. However, as it is impossible to predict definitively which of the two stable states the flip-flop will assume,

R	S	Q	$\overline{Q}$
H	H	(L)	(L)
H	L	L	H
L	H	H	L
L	L	Q_{-1}	$\overline{Q}_{-1}$

Fig. 6.11. Truth table for the *RS* flip-flop Q_{-1} and $\overline{Q}_{-1}$ mean that the flip-flop remains in the last state

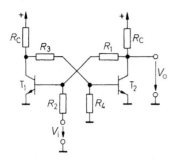

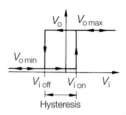

Fig. 6.12. Schmitt trigger

Fig. 6.13. Transfer characteristic of the Schmitt trigger

the input state $R = S = H$ is logically impermissible. If it is avoided, the output states are always complementary. This process is summarized by the truth table in Fig. 6.11.

Schmitt Trigger

The *RS* flip-flop described above is caused to change state by applying a positive voltage pulse to the base of the currently nonconducting transistor in order to turn it on. Another possibility consists of only using one input and switching it off with a *negative* input voltage. A flip-flop that operates in this way is known as a *Schmitt trigger*. Its simplest circuit is shown in Fig. 6.12.

When the input voltage exceeds the upper trigger threshold $V_{i\,ON}$, the output voltage jumps to the positive saturation limit $V_{o\,max}$. It only returns to zero when the input voltage falls below the lower trigger threshold $V_{i\,OFF}$. This characteristic allows the Schmitt trigger to be used as a square-wave converter. By way of example, Fig. 6.14 shows how a sine wave is converted into a square wave. Due to positive feedback, the change from one state to the other is instantaneous, even though the input voltage changes only slowly.

The transfer characteristic is illustrated in Fig. 6.13. The voltage difference between the turn-on and turn-off levels is termed the *hysteresis*. This is smaller the more the difference between $V_{o\,max}$ and $V_{o\,min}$ is reduced, or the greater the attenuation in the voltage divider R_1, R_2 is. Any attempt to reduce the hysteresis will adversely affect the positive feedback in the Schmitt trigger and could result in it ceasing to be bistable. For $R_1 \to \infty$, the circuit becomes a conventional two-stage amplifier.

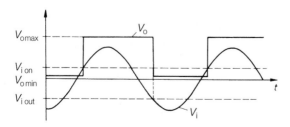

Fig. 6.14. Schmitt trigger as a square-wave converter

6.2.2
Monostable Circuits

A one-shot is basically an *RS* flip-flop in which one of the two feedback resistors is replaced by a capacitor, as shown in Fig. 6.15. As the capacitor blocks DC, transistor T_2 is ON and transistor T_1 is OFF under steady-state conditions.

A positive input pulse turns on transistor T_1, causing its collector potential to jump from its steady-state value V^+ to zero. This jump is coupled by the capacitor RC to the base of T_2, causing its base potential to go from $0.6\,V$ to $-V^+ + 0.6\,V \approx -V^+$, and T_2 turns off. T_1 is maintained ON via feedback resistor R_1, even if the input voltage has already returned to zero as shown in Fig. 6.16.

Capacitor C is charged via resistor R connected to V^+. As described in Chap. 29.3.2, the base potential of T_2 increases in accordance with the relation

$$V_{B2}(t) \approx V^+(1 - 2e^{-t/RC}) \tag{6.1}$$

Transistor T_2 remains in the blocking state until V_{B2} has risen to approximately $+0.6\,V$. We can obtain the time t_{ON} required for this sequence by substituting $V_{B2} \approx 0$ in (6.1), giving

$$t_{ON} \approx RC \ln 2 \approx 0.7\,RC \tag{6.2}$$

When this time has elapsed, transistor T_2 begins to conduct again; that is, the circuit flips back to its stable state. Figure 6.16 shows the relevant voltage waveforms.

The output returns to its initial state within the defined turn-on time even if the input pulse is longer than the turn-on time. In this case, transistor T_1 remains ON until the input pulse disappears, and the positive feedback has no effect. T_2 does not therefore begin to conduct instantaneously, but only in accordance with the rate of rise of V_{B2}.

When the switching cycle is complete, capacitor C must be charged via R_C. If the capacitor is not fully charged until the next turn-on pulse occurs, the subsequent turn-on time is reduced. In order to reduce the on-time by less than 1%, T_1 must remain OFF for a recovery time of at least $5\,R_C \cdot C$.

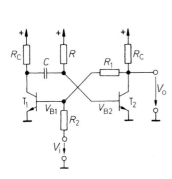

Fig. 6.15. One-shot *Turn-on time:*
$t_{ON} = RC \ln 2$

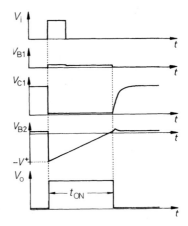

Fig. 6.16. Voltage waveforms

The supply voltage of the circuit should not be choosen above 5 V, as this could cause the emitter–base breakdown voltage of T_2 to be exceeded when T_1 becomes conducting. This effect would reduce the turn-on time as a function of the supply voltage.

6.2.3
Astable Circuits (Multivibrators)

If the second feedback resistor of a one-shot is also replaced by a capacitor, as in Fig. 6.17, the two states are each stable only for a limited period of time. The circuit therefore continuously oscillates between the two states therefore the name: multivibrator. From (6.2), the switching times are given by

$$t_1 = R_1 C_1 \ln 2$$

and

$$t_2 = R_2 C_2 \ln 2$$

The voltage waveforms are shown in Fig. 6.18. As we can see, t_1 is the OFF-time of T_1 and t_2 is the OFF-time of T_2. The circuit therefore always changes state when the opposite nonconducting transistor is turned on.

The circuit designer has little margin in selecting the values of resistors R_1 and R_2. On the one hand, they must be small compared to βR_C so that sufficient current flows through them to drive the conducting transistor into saturation. On the other hand, they must be large compared to R_C so that the capacitors can be charged up to the supply voltage before the next cycle. This condition can be expressed as

$$R_C \ll R_1, \ R_2 \ll \beta R_C$$

As with the one-shot in Fig. 6.15, the supply voltage must not be selected larger than 5 V in order not to exceed the emitter–base breakdown voltage.

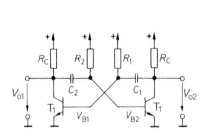

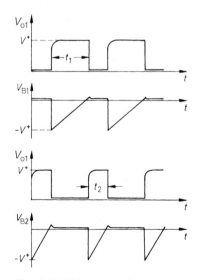

Fig. 6.17. Multivibrator
Switching times: $t_1 = R_1 C_1 \ln 2$
$t_2 = R_2 C_2 \ln 2$

Fig. 6.18. Voltage waveforms

It is possible that the multivibrator in Fig. 6.17 will not start to oscillate. For example, if one output is short-circuited, both transistors go into saturation, and this condition continues to obtain even when the short circuit is removed.

At frequencies below 100 Hz, the capacitors are unwieldy and large; and at frequencies above 10 kHz, the storage times of the transistors become a noticeable problem. Consequently, the circuit in Fig. 6.17 is of little practical use. The precision circuits with comparators (Sect. 6.5.3) are preferable for low-frequency applications, whereas the emitter-coupled multivibrators in Sect. 6.3.2 tend to be used for high-frequency applications.

6.3
Latching Circuits with Emitter-Coupled Transistors

6.3.1
Emitter-Coupled Schmitt Trigger

A noninverting amplifier can also be implemented by using a differential amplifier. By applying positive feedback through a resistive voltage divider, we obtain the emitter-coupled Schmitt trigger shown in Fig. 6.19. Both of its trigger thresholds are positive.

By selecting suitable component values for the circuit, we can cause current I_k to switch from one transistor to the other when the circuit changes state, without the transistors becoming saturated. This eliminates the storage time t_S during switchover, and high switching frequencies can be achieved. This principle is known as "unsaturated logic."

6.3.2
Emitter-Coupled Multivibrator

Due to the elimination of the storage times, considerably higher frequencies can be achieved with emitter-coupled multivibrators than using saturated transistors. A suitable circuit is shown in Fig. 6.21.

To explain how the circuit operates, let us assume that the amplitude of the AC voltages present is small at all points in the circuit; say, $V_{pp} \approx 0.5\,\text{V}$. When T_1 is OFF, its collector potential is equal to the supply voltage, resulting in an emitter potential of $V^+ - 1.2\,\text{V}$ at T_2. Its emitter current is $I_1 + I_2$. In order to produce the required amplitude of oscillation at R_1, the value $R_1 = 0.5\,\text{V}/(I_1 + I_2)$ must therefore be selected. This gives us an emitter

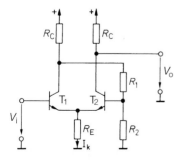

Fig. 6.19. Emitter-coupled Schmitt trigger

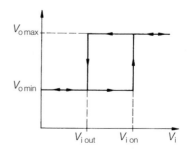

Fig. 6.20. Transfer characteristic

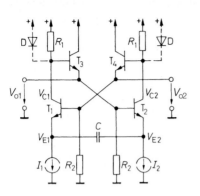

Fig. 6.21. Emitter-coupled multivibrator

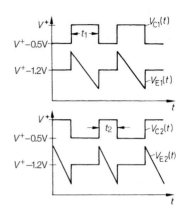

Fig. 6.22. Voltage waveforms

potential of $V^+ - 1.1$ V at T_4 under this operating condition. As long as T_1 is OFF, the current from the left current source flows via capacitor C, causing the emitter potential of T_1 to fall at a rate of

$$\frac{\Delta V_{E1}}{\Delta t} = -\frac{I_1}{C}$$

T_1 begins to conduct when its emitter potential has fallen to $V^+ - 1.7$ V. The base potential of T_2 then falls by 0.5 V, turning the device off, and its collector potential rises to V^+. The base potential of T_1 rises with it via emitter follower T_4, causing the emitter potential of T_1 to jump to $V^+ - 1.2$ V. This step is transferred via capacitor C to the emitter of T_2, producing a potential increase from $V^+ - 1.2$ V to $V^+ - 0.7$ V across this device.

During the OFF-time of T_2, current 2 flows via capacitor C and causes the emitter potential of T_2 to fall at a rate of

$$\frac{\Delta V_{E2}}{\Delta t} = -\frac{I_2}{C}$$

Transistor T_2 remains OFF until its emitter potential has fallen from $V^+ - 0.7$ V to $V^+ - 1.7$ V, giving a switching time of

$$t_2 = \frac{1\,\text{V}\cdot C}{I_2} \quad \text{or, more generally,} \quad t_2 = 2\left(1 + \frac{I_1}{I_2}\right)R_1 C \tag{6.3}$$

Similarly, we obtain

$$t_1 = \frac{1\,\text{V}\cdot C}{I_1} \quad \text{or, more generally,} \quad t_1 = 2\left(1 + \frac{I_2}{I_1}\right)R_1 C \tag{6.4}$$

The voltage waveforms for the circuit are shown in Fig. 6.22. We can see that by selecting $V_{pp} = 0.5$ V, none of the transistors is driven into saturation. This circuit allows frequencies of more than 100 MHz to be achieved with no great cost or complexity.

The circuit is particularly suitable to extend to a voltage controlled oscillator VCO. For this purpose, we select the currents $I_1 = I_2 = I$ and control them with a common control voltage. In order to insure that the oscillation amplitude at R_1 remains constant, a diode can be connected in parallel as shown by the dashed lines in Fig. 6.21. The oscillation frequency then becomes

$$f = \frac{1}{t_1 + t_2} = \frac{I}{4U_D C}$$

where U_D is the forward voltage of the diodes.

Emitter-coupled multivibrators are available as monolithic integrated circuits.

IC types:

TTL	SN 74 LS 624...629	$f_{max} =$	20 MHz (Texas Instruments)
CMOS	LTC 1799	$f_{max} =$	33 MHz (Linear Technology)
CMOS	LTC 6905	$f_{max} =$	170 MHz (Linear Technology)

6.4
Latching Circuits Using Gates

Latching circuits can be implemented not only using transistors, but also using integrated logic devices (gates), as described in Chap. 7.4. Readers who are unfamiliar with basic logic functions should therefore read that chapter first.

6.4.1
Flip-Flops

Let us return to the flip-flop shown in Fig. 6.10. Transistor T_1 is ON when a positive voltage is dropped across resistor R_1 OR resistor R_2. If we also take into account the inverter function produced by the transistor, we can see that components R_1, R_2, T_1, and R_C form a NOR gate. The same applies to the other half of the circuit. Inserting the appropriate circuit symbols, we obtain the circuit diagram shown in Fig. 6.23, with the associated truth table (Fig. 6.24).

6.4.2
One-Shot

The circuit shown in Fig. 6.25 provides a simple means of generating short pulses with a duration of just a few gate propagation delay times. As long as the input variable $x = 0$, a logic 0 will be produced at the output of the AND gate. If $x = 1$, the AND element produces a logic 1 until the signal has passed through the inverter chain. When the input signal returns to zero, the AND condition is not satisfied.

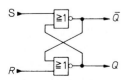

Fig. 6.23. Flip-flop with NOR gates

R	S	Q	$\overline{Q}$
0	0	Q_{-1}	$\overline{Q}_{-1}$
0	1	1	0
1	0	0	1
1	1	(0)	(0)

Fig. 6.24. Truth table

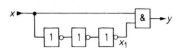

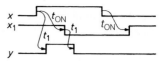

Fig. 6.25. One-shot for short switching times
Turn-on time: t_{ON} = sum of inverter propagation delay times

Fig. 6.26. Waveform diagram.
t_1 = propagation delay time of the AND gate

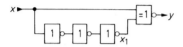

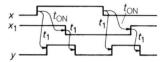

Fig. 6.27. Two-edge-triggered one-shot
Turn-on time: $t_{ON} = 3t_{pd}$

Fig. 6.28. Waveform diagram
t_1 = propagation delay time of exclusive NOR gate

The timing diagram is shown in Fig. 6.26. The duration of the output pulse is equal to the delay in the inverter chain. This can be specified by an appropriate number of gates, taking care to insure that there is an odd number of gates. As we can see from Fig. 6.26, for this one-shot the trigger signal must be present at least for the duration of the output pulse.

To obtain longer switching times, the delay chain becomes unmanageably long. In this case, it is preferable to employ integrated one-shots (monostable multivibrators), whose switching times are determined by an external *RC* network.

IC types:
TTL   74 LS 121…123, 422, 423   (Texas Instruments)

If the AND gate in Fig. 6.25 is replaced by an exclusive-NOR gate, we obtain a one-shot that produces an output pulse on each edge of the input signal. Figure 6.27 shows the relevant circuit, and Fig. 6.28 the associated waveform pattern. Under steady-state conditions, the inputs of the exclusive-NOR gate are complementary and the output signal is zero. If the input variable x changes its state, temporarily identical input signals appear at the exclusive-NOR gate due to the delay through the inverters. During this time, the output signal y becomes logic 1.

6.4.3
Multivibrator

A simple multivibrator comprising two inverters is shown in Fig. 6.29. In order to explain how it works, let us assume that signal x is in the H state and that y is therefore in the L state. This causes capacitor C to charge up via resistor R until potential V exceeds the switching level V_S of gate G_1. x then changes to the L state and y to the H state, causing potential V to go positive by the amplitude of the output signal. The capacitor then discharges via resistor R until the voltage falls below the switching level.

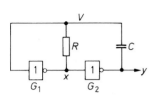

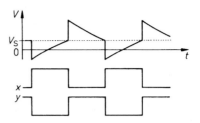

Fig. 6.29. A multivibrator comprising two inverters *Cycle time:* $T = 2 \ldots 3RC$

Fig. 6.30. Waveform diagram *Switching level:* V_S

The voltage waveform is shown in Fig. 6.30. Assuming that the switching level lies halfway between the output levels, the cycle time is given by

$$T = 2RC \ln 3 \approx 2.2 RC$$

In general, this assumption is only approximately satisfied in practical circuits. Additional deviations are caused by the fact that the input of gate G_1 loads the RC network. In the case of low-power Schottky TTL circuits, there is very little choice for resistor R: $R = 1.0 \, k\Omega \ldots 3.9 \, k\Omega$.

When using CMOS gates, a high value can be selected for resistor R, thereby achieving relatively large cycle times. However, in this case a series resistor is required at the input of gate G_1 in order to minimize the load on the RC network. This load would be caused by the protection circuit at the input of G_1 becoming conducting as long as V exceeds the supply voltage or falls below ground potential.

A circuit in which this problem does not arise is shown in Fig. 6.31. In this circuit, capacitor C is charged via resistor R to the switch-off level of the Schmitt trigger and then discharged to the switch-on level. We can see from Fig. 6.32 that the voltage across the capacitor oscillates between the two trigger levels. When using low-power Schottky TTL circuits, R must be selected sufficiently low to insure that it can pull the input below the switch-on level for the input current that is flowing. Recommended values are between 220 Ω and 680 Ω. These limitations do not apply to CMOS Schmitt triggers.

Particularly high frequencies above 50 MHz can be achieved by employing ECL gates. If positive feedback is applied to a line receiver (e.g., the MC 10116), we obtain a Schmitt trigger that can be configured as a multivibrator, similar to Fig. 6.31. The external circuitry and internal design are shown in Figs. 6.33 and 6.34 respectively.

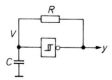

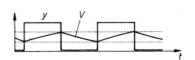

Fig. 6.31. Multivibrator using a Schmitt trigger
Cycle time: (TTL) $T = 1.4 \ldots 1.8RC$
           (5 V-CMOS) $T = 0.5 \ldots 1.0RC$

Fig. 6.32. Waveform diagram

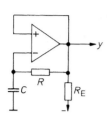

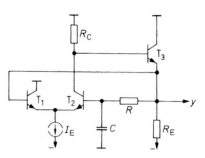

Fig. 6.33. Multivibrator with an ECL line receiver

Cycle time: $T \approx 3RC$

Fig. 6.34. Internal design of the line-receiver multivibrator

6.5
Latching Circuits Using Comparators

6.5.1
Comparators

An operational amplifier without feedback, as shown in Fig. 6.35, represents the basic circuit of a comparator. Its output voltage is given by

$$V_o = \begin{cases} V_{o\,max} & \text{for } V_1 > V_2 \\ V_{o\,min} & \text{for } V_1 < V_2 \end{cases}$$

The corresponding transfer characteristic is shown in Fig. 6.36. Due to the high gain, the circuit responds to very small voltage differences $V_1 - V_2$. It is thus suitable for the comparison of two voltages, and operates with high accuracy.

At the zero crossing of the input voltage difference, the output voltage does not immediately reach the saturation level because the transition is limited by the slew rate. For frequency-compensated standard operational amplifiers, it can be as low as $1\,\text{V/\textmu s}$. A voltage rise from $-12\,\text{V}$ to $+12\,\text{V}$ therefore takes $24\,\text{\textmu s}$. An additional delay is incurred due to the recovery time needed after the amplifier has been saturated.

As the amplifier possesses no negative feedback, it also does not require any frequency compensation. Its omission can improve the slew rate and recovery time by a factor of about 20.

Considerably shorter response times can be attained when using special comparator amplifiers, which are designed for use without feedback and have especially small recovery times. However, the gain and hence the accuracy of the threshold are somewhat lower than those of operational amplifiers. Usually, the amplifier output is directly connected to a level-

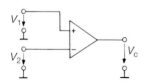

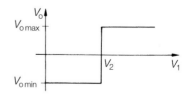

Fig. 6.35. Operational amplifier as a comparator

Fig. 6.36. Transfer characteristic

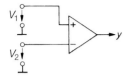 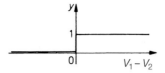

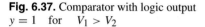

Fig. 6.37. Comparator with logic output $y = 1$ for $V_1 > V_2$

Fig. 6.38. Transfer characteristic

Type	Manufacturer	Comparators per package	Output	Consumption per unit	Switching time
ADCMP 371	Analog Dev.	1	CMOS	12 µW	4 µs
ADCMP 573	Analog Dev.	1	PECL	140 mW	0,15 ns
LT 1443	Lin. Tech	4	CMOS	6 µW	12 µs
LT 1720	Lin. Tech	2	TTL	12 mW	4 ns
MAX 919	Maxim	1	CMOS	2 µW	25 µs
MAX 978	Maxim	4	CMOS	3 mW	20 ns
MAX 9012	Maxim	2	TTL	5 mW	5 ns
MAX 9031	Maxim	4	CMOS	180 µW	230 ns
MAX 9144	Maxim	4	TTL	500 µW	40 ns
MAX 9602	Maxim	4	PECL	85 mW	0,5 ns
LM 339	National	4	TTL	8 mW	600 ns
LMC 6717	National	2	CMOS	2 µW	12 µs
LMV 761	National	1	CMOS	1 mW	120 ns
TLV 3704	Texas I.	4	CMOS	5 µW	50 µs

Fig. 6.39. Examples of comparators

shift circuit, which permits compatible operation with logic circuits. Its application and characteristics are shown in Figs. 6.37 and 6.38. A number of commonly used comparators are listed in Fig. 6.39.

Window Comparator

A window comparator can determine whether or not the value of the input voltage lies between two reference voltage levels. Figure 6.40 shows that two comparators are used to decide whether the input voltage is above a lower *and* below an upper reference voltage. When this condition arises, both comparator amplifiers produce a logical 1 and the output becomes $y = 1$. The characteristics in Fig. 6.40 illustrate the operation.

6.5.2
Schmitt Trigger

The Schmitt trigger is a comparator, for which the positive and negative transitions of the output occur at different levels of the input voltage. Their difference is characterized by the hysteresis ΔV_i. As described in earlier sections, Schmitt triggers can be realized by transistors, but in the following section some designs involving comparators are discussed.

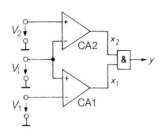

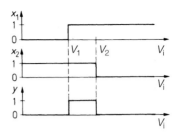

Fig. 6.40. Window comparator $y = 1$ for $V_1 < V_i < V_2$

Fig. 6.41. Variables as a function of the input voltage

Inverting Schmitt Trigger

In the Schmitt trigger of Fig. 6.42, the hysteresis is produced by a positive feedback of the comparator, via the voltage divider R_1, R_2. If a large negative voltage V_i is applied, $V_o = V_{o\,max}$. At the P-input, the potential is then given as

$$V_{P\,max} = \frac{R_1}{R_1 + R_2} V_{o\,max}$$

If the input voltage is changed toward positive values, V_o does not change at first; only when V_i reaches the value $V_{P\,max}$ does the output voltage reduce, and therefore also V_P. The difference $V_D = V_P - V_N$ becomes negative. Due to the positive feedback, V_o falls very quickly to the value $V_{o\,min}$. The potential V_P assumes the value

$$V_{P\,min} = \frac{R_1}{R_1 + R_2} V_{o\,min}$$

V_D is negative and large, this resulting in a stable state. The output voltage changes to $V_{o\,max}$ only when the input voltage has reached $V_{P\,min}$. The corresponding transfer characteristic is shown in Fig. 6.43.

The circuit has two stable states only if the loop gain is $g = \frac{A_D R_1}{R_1 + R_2} > 1$.

Figure 6.44 shows the output of the Schmitt trigger for a sinusoidal input voltage.

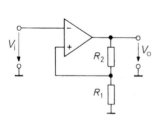

Fig. 6.42. Inverting Schmitt trigger

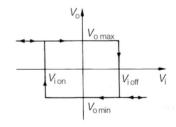

Fig. 6.43. Transfer characteristic

$$\text{Switch} - \text{on level:} \qquad V_{i\,ON} = \frac{R_1}{R_1 + R_2} V_{o\,min}$$

$$\text{Switch} - \text{off level:} \qquad V_{i\,OFF} = \frac{R_1}{R_1 + R_2} V_{o\,max}$$

$$\text{Hysteresis:} \qquad \Delta V_i = \frac{R_1}{R_1 + R_2} (V_{o\,max} - V_{o\,min})$$

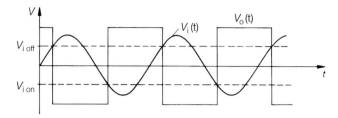

Fig. 6.44. Voltage waveform of an inverting Schmitt trigger for a sinusoidal input voltage

Noninverting Schmitt Trigger

The input signal for the Schmitt trigger in Fig. 6.42 can also be applied to the low end of the feedback voltage divider when, at the same time, the inverting input is grounded. The new configuration is the noninverting Schmitt trigger shown in Fig. 6.45.

If a large positive input voltage V_i is applied, $V_o = V_{o\,max}$. When reducing V_i, V_o does not change until V_p crosses zero. This is the case for the input voltage

$$V_{i\,\text{OFF}} = -\frac{R_1}{R_2}V_{o\,max}$$

The output voltage jumps to $V_{o\,min}$ as soon as V_i reaches or falls below this value. The transition is initiated by V_i but is then determined only by the positive feedback via R_2. The new state is stable until V_i returns to the level

$$V_{i\,\text{ON}} = -\frac{R_1}{R_2}V_{o\,min}$$

Figure 6.47 depicts the time function of the output voltage for a sinusoidal input. Since, at the instant of transition, $V_P = 0$, the formulas for the trigger level have the same form as those for the inverting amplifier.

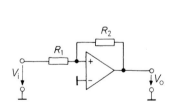

Fig. 6.45. Noninverting Schmitt trigger

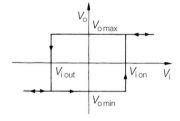

Fig. 6.46. Transfer characteristic

Switch-on level: $V_{i\,\text{ON}} = -\dfrac{R_1}{R_2}V_{o\,min}$

Switch-off level: $V_{i\,\text{OFF}} = -\dfrac{R_1}{R_2}V_{o\,max}$

Hysteresis: $\Delta V_i = \dfrac{R_1}{R_2}(V_{o\,max} - V_{o\,min})$

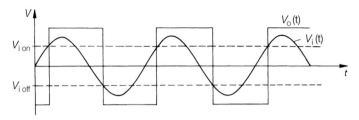

Fig. 6.47. Voltage waveform of a noninverting Schmitt trigger for a sinusoidal input voltage

Precision Schmitt Trigger

In the Schmitt triggers described, the switching levels are not as precise as one would expect from operational amplifier circuits. This is because the trigger levels are influenced by the imprecisely defined output voltages $V_{o\,max}$ or $V_{o\,min}$. This drawback can be overcome, as shown in Fig. 6.48, by using two comparators that compare the input signal with the required switching levels. They then set an *RS* flip-flop if the upper trigger level is exceeded and reset it if the lower trigger level is undershot. The mode of operation is illustrated in Fig. 6.49.

The precision Schmitt trigger shown in Fig. 6.48 can be implemented particularly simply using the dual comparator NE 521, as the latter already incorporates the two NAND gates required. For low frequencies there is yet another single-chip solution using an NE 555 timer, which will be described in greater detail in the next section.

6.5.3
Multivibrators

If an inverting Schmitt trigger is configured such that the output signal is fed to the input with a delay, we obtain a multivibrator of the type shown in Fig. 6.50.

When the potential at the N-input exceeds the trigger level, the circuit changes state, and the output voltage jumps to the opposite output limit. The potential at the N-input attempts to follow the transition in the output voltage, but the capacitor prevents it from changing rapidly. The potential therefore rises or falls gradually until the other trigger level is reached. The circuit then flips back to its initial state. The voltage wave-shapes are

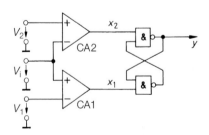

Fig. 6.48. Precision Schmitt trigger

Switch-on level: $\quad V_{i\,ON} = V_2$
Switch-off level: $\quad V_{i\,OFF} = V_1$ $\Big\}$ for $V_2 > V_1$

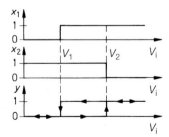

Fig. 6.49. Variables as a function of the input voltage

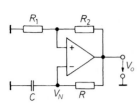

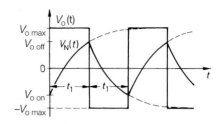

Fig. 6.50. Multivibrator using an operational amplifier
Cycle time: $T = 2RC \ln(1 + 2R_1/R_2)$

Fig. 6.51. Waveform of the multivibrator

shown in Fig. 6.51. From Fig. 6.42, the trigger levels for $V_{o\,max} = -V_{o\,min} = V_o$ are given by

$$V_{i\,ON} = -\alpha V_o$$

and

$$V_{i\,OFF} = \alpha V_o$$

where $\alpha = R_1/(R_1 + R_2)$.

The differential equation for V_N can be taken directly from the circuit diagram, as

$$\frac{dV_N}{dt} = \frac{\pm V_o - V_N}{RC}$$

With the initial condition $V_N(t = 0) = V_{i\,ON} = -\alpha V_o$, we obtain the solution

$$V_N(t) = V_o\left[1 - (1 + \alpha)e^{-\frac{t}{RC}}\right]$$

The switching level $V_{i\,OFF} = \alpha V_{max}$ is reached after a time

$$t_1 = RC \ln\frac{1 + \alpha}{1 - \alpha} = RC \ln\left(1 + \frac{2R_1}{R_2}\right).$$

The period is therefore

$$T = 2t_1 = 2RC \ln\left(1 + \frac{2R_1}{R_2}\right). \tag{6.5}$$

For $R_1 = R_2$

$$T = 2RC \ln 3 \approx 2.2RC.$$

Multivibrator Using a Precision Schmitt Trigger

The frequency stability of the multivibrator in Fig. 6.50 can be improved by incorporating the precision Schmitt trigger of Fig. 6.48. The resultant circuit is shown in Fig. 6.52. The enclosed section represents integrated timer NE 555, which offers the simplest solution for low-frequency applications. Depending on the external circuitry, it can be operated as an astable circuit (Fig. 6.52), a one-shot (Fig. 6.54), or a precision Schmitt trigger (Fig. 6.48).

The internal voltage divider R defines the trigger levels as $\frac{1}{3}V^+$ and $\frac{2}{3}V^+$, but these can be adjusted within certain limits using terminal 5. When the capacitor potential exceeds the upper threshold, $\overline{R}$ goes low. The output of the flip-flop assumes the H-state and transistor

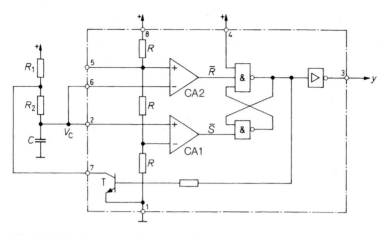

Fig. 6.52. Multivibrator using a timer circuit

$Period:$ $T = (R_1 + 2R_2)C \ln 2 \approx 0.7(R_1 + 2R_2)C$

T is turned on. Capacitor C is then discharged by the resistor R_2 until the lower threshold $\frac{1}{3}V^+$ is reached, this process requiring the time

$$t_2 = R_2 C \ln 2 \approx 0.693 R_2 C$$

On reaching the lower threshold, $\overline{S}$ goes low and the *flip-flop* resumes its former state. Its output goes low and transistor T is turned off. Charging of the capacitor takes place via the series connection of R_1 and R_2. The time interval needed to reach the upper trigger level is

$$t_1 = (R_1 + R_2)C \ln 2 \approx 0.693(R_1 + R_2)C$$

Hence the frequency is:

$$f = \frac{1}{t_1 + t_2} = \frac{1}{(R_1 + 2R_2)C \ln 2} \approx \frac{1.44}{(R_1 + 2R_2)C}$$

The wave-shapes of the signals y and V_C are shown in Fig. 6.53. The reset input 4 allows interruption of the oscillation.

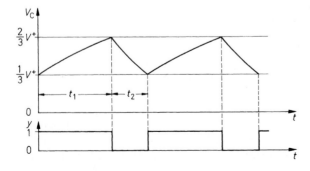

Fig. 6.53. Voltage waveforms for a timer as an astable circuit

When supplying a voltage to pin 5, the trigger thresholds can be shifted. In this way, the charging time t_1 and therefore the frequency of the multivibrator can be varied. A change in the potential $V_5 = \frac{2}{3}V^+$ by ΔV_5 results in a relative frequency shift of

$$\frac{\Delta f}{f} \approx -3.3 \cdot \frac{R_1 + R_2}{R_1 + 2R_2} \cdot \frac{\Delta V_5}{V^+}$$

As long as the voltage deviation is not too large, the frequency modulation is reasonably linear.

6.5.4
One-Shots

The timer 555 is also useful for generating single pulses (one-shot), with pulse times between few microseconds and several minutes. The required external connections are shown in Fig. 6.54.

When the capacitor potential exceeds the upper trigger threshold, the flip-flop is re-set; that is, the output voltage resumes the L-state. Transistor T becomes conducting and discharges the capacitor. As the lower comparator is no longer connected to the capacitor, this state remains unchanged until the flip-flop is set by an L-pulse at trigger input 2. The ON-time t_1 is equal to the time required by the capacitor potential to rise from zero to the upper threshold $\frac{2}{3}V^+$. It is given by

$$t_1 = R_1 C \ln 3 \approx 1.1 R_1 C.$$

If a new trigger pulse occurs during this time interval, the flip-flop remains set and the pulse is ignored. Figure 6.55 shows the signals involved.

Discharging of capacitor C at the end of t_1 is not as fast as could be wished, as the collector current of the transistor is limited. The discharge time is known as the *recovery time*. If a trigger pulse occurs during this interval, the ON-time is reduced and is therefore no longer precisely defined.

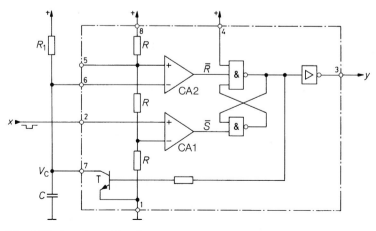

Fig. 6.54. One-shot with a timer
ON-*time*: $t_1 = R_1 C \ln 3 \approx 1.1 R_1 C$

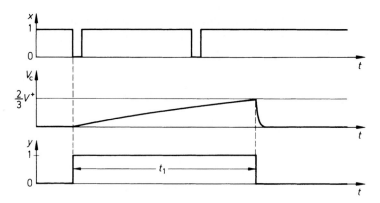

Fig. 6.55. Signals for the one-shot

Retriggerable Timer

There are cases in which the ON-time is not to be counted from the first pulse of a pulse train, as in the previous circuit, but from the last pulse of the train. Circuits that have this characteristic are termed "retriggerable timers." The appropriate connection of the timer 555 is shown in Fig. 6.56, where it is used only as a precision Schmitt trigger.

When the capacitor potential exceeds the upper trigger threshold, the flip-flop is reset the output y assumes the L-state. The capacitor will not discharge, as transistor T is not connected. The capacitor potential therefore rises to V^+, this being the stable state. The capacitor must be discharged by a sufficiently long positive trigger pulse applied to the base of the external transistor T'. The flip-flop is set by the lower comparator and the output y assumes the H-state. If a new trigger pulse occurs before the end of the ON-time, the capacitor is discharged again and the output remains in the H-state. It flips back only if no new trigger pulse occurs for a time interval of at least

$$t_1 = R_1 C \ln 3 \approx 1.1 R_1 C$$

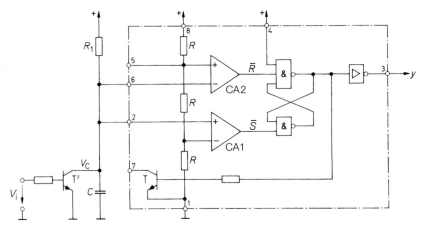

Fig. 6.56. Retriggerable timer
ON-*time:* $t_1 = R_1 C \ln 3 \approx 1.1 R_1 C$

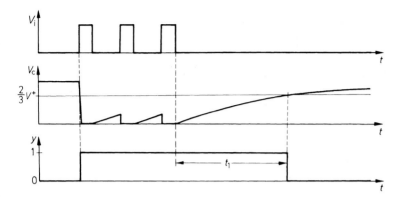

Fig. 6.57. Signals in the retriggerable timer

The circuit is therefore also called the "missing pulse detector." The signals within the circuit are shown in Fig. 6.57 for several consecutive trigger pulses.

Chapter 7:
Logic Families

Although, at first sight, digital equipment appears to be relatively complicated, its design is based on the simple concept of the repeated use of a small number of basic logic circuits. We can work out how these basic logic elements have to be linked by applying purely formal methods to the problem. This approach is based on Boolean algebra which, when applied specifically to digital circuit design, is known as computational algebra. In the next few subsections, we shall summarize the basics of computational algebra.

7.1
Basic Logic Functions

Unlike a variable in conventional algebra, a logic variable can only assume two discrete values (binary variable), generally referred to as logic or Boolean one and logic or Boolean zero, for which the symbols 0 and 1 are used. There is no risk of confusion with the numbers 0 and 1, as it is always clear from the context whether a number or a logic value is meant.

There are three basic relations that link logic variables: conjunction, disjunction, and negation. Using the mathematical signs of numeric algebra,

Conjunction: $\quad y = x_1 \wedge x_2 = x_1 \cdot x_2 = x_1 x_2$
Disjunction: $\quad y = x_1 \vee x_2 = x_1 + x_2$
Negation: $\quad y = \bar{x}$

A number of theorems relating these operations are listed below:

Commutative law:

$$x_1 x_2 = x_2 x_1 \tag{7.1a}$$

$$x_1 + x_2 = x_2 + x_1 \tag{7.1b}$$

Associative law:

$$x_1 (x_2 x_3) = (x_1 x_2) x_3 \tag{7.2a}$$

$$\begin{aligned} x_1 + (x_2 + x_3) \\ = (x_1 + x_2) + x_3 \end{aligned} \tag{7.2b}$$

Distributive law:

$$x_1 (x_2 + x_3) = x_1 x_2 + x_1 x_3 \tag{7.3a}$$

$$\begin{aligned} x_1 + x_2 x_3 \\ = (x_1 + x_2)(x_1 + x_3) \end{aligned} \tag{7.3b}$$

Absorption law:

$$x_1 (x_1 + x_2) = x_1 \tag{7.4a}$$

$$x_1 + x_1 x_2 = x_1 \tag{7.4b}$$

Tautology:

$$xx = x \tag{7.5a}$$

$$x + x = x \tag{7.5b}$$

Law of negation

$$x\bar{x} = 0 \tag{7.6a}$$

$$x + \bar{x} = 1 \tag{7.6b}$$

Double negation:
$$\overline{(\overline{x})} = x \qquad (7.7)$$

De Morgans law:

$$\overline{x_1 x_2} = \overline{x}_1 + \overline{x}_2 \qquad (7.8a)$$

$$\overline{x_1 + x_2} = \overline{x}_1 \overline{x}_2 \qquad (7.8b)$$

Operations with 0 and 1:

$$x \cdot 1 = x \qquad (7.9a)$$

$$x + 0 = x \qquad (7.9b)$$

$$x \cdot 0 = 0 \qquad (7.10a)$$

$$x + 1 = 1 \qquad (7.10b)$$

$$\overline{0} = 1 \qquad (7.11a)$$

$$\overline{1} = 0 \qquad (7.11b)$$

Many of these formulas are already familiar from ordinary algebra. However, (7.3b), (7.4a,b), (7.5a,b), and (7.10b) do not apply to algebraic numbers, and the term "negation" is not used at all in connection with numbers. Expressions such as $2x$ and x^2 do not occur in switching algebra, due to tautology.

If one compares the equations on the left with those on the right, the important principle of duality becomes apparent: if conjunction and disjunction and 0 and 1 are interchanged in any identity, an identity on the other side is obtained.

Using (7.9)–(7.11), it is possible to work out the conjunction and disjunction for all the possible values of variables x_1 and x_2. Figure 7.1 shows the function or truth table for conjunction and Fig. 7.2 the truth table for disjunction.

We can see from Fig. 7.1 that y is only 1 if x_1 *and* x_2 are 1. Consequently, conjunction is also known as the AND operation. In the case of disjunction, y is always 1 if x_1 *or* x_2 is 1. This operation is therefore also known as the OR operation. Both of these logic operations can be extended to apply to any number of variables.

The question now is how to implement these logic operations using electrical circuits. As logic variables can only assume two discrete values, the only candidates are circuits that possess two clearly distinguishable operating states. The simplest means of representing a logic variable is a switch, as shown in Fig. 7.3. An open switch can now be defined as representing a logic "zero" and a closed switch a logic "one." Switch S therefore represents variable x if it is closed for $x = 1$. It represents variable $\overline{x}$ if it is open for $x = 1$.

Let us first determine which logic function is obtained when two switches, x_1 and x_2, are connected in series, as in Fig. 7.4. The value of dependent variable y is characterized by whether the resulting switch arrangement between the terminals is open or closed. As we can see, current can only flow if x_1 and x_2 are closed; that is, both 1. The series connection therefore represents an AND operation. Similarly, an OR operation is obtained by connecting switches in parallel.

x_1	x_2	y
0	0	0
0	1	0
1	0	0
1	1	1

Fig. 7.1. Truth table for conjunction (AND) $y = x_1 x_2$

x_1	x_2	y
0	0	0
0	1	1
1	0	1
1	1	1

Fig. 7.2. Truth table for disjunction (OR) $y = x_1 + x_2$

Fig. 7.3. Representation of a logic variable by a switch

Fig. 7.4. An AND circuit

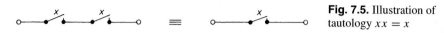

Fig. 7.5. Illustration of tautology $xx = x$

Using this switching logic, we can now verify the theorems stated above. We shall illustrate this by taking tautology as an example. In Fig. 7.5, both sides of (7.5a) have been realized using switch arrangements. We can see that the given identity is satisfied, because two switches connected in series and simultaneously opened and closed have the same effect as a single switch.

As we saw in Sect. 6.1, another way of representing logic variables is by electrical voltages. To the two distinct levels H = high and L = low we can now assign the logic states 1 and 0. By making H = 1 and L = 0, we obtain what is known as "positive logic." Conversely, we can also make H = 0 and L = 1, this being termed "negative logic."

The basic logic functions can be implemented by appropriate electronic circuits with one or more inputs and a single output. These are generally known as "gates." The voltage levels at the inputs and the types of logic operation determine the output level. As a logic function can be implemented electronically in a variety of ways, to simplify matters circuit symbols have been introduced which only show the logic function and give no indication of the internal design. These symbols are given in Figs. 7.6–7.8. The complete standard can, for instance, be found in IEC 60617-12 and IEEE Std 91-1984. To facilitate the understanding of old-style circuit diagrams, the symbols formerly used are shown in Figs. 7.9–7.11.

In digital circuits, we are not concerned with the voltage as a physical quantity, but merely with the logic state that it represents. Instead of denoting the input and output signals by V_1, V_2, and so on, they are therefore expressed directly using the logic variables shown.

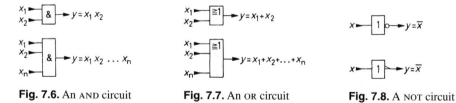

Fig. 7.6. An AND circuit   **Fig. 7.7.** An OR circuit   **Fig. 7.8.** A NOT circuit

Figs. 7.6–7.8. Circuit symbols to IEC 60617-12 and IEEE Std 91-1984

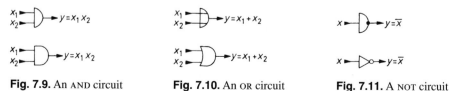

Fig. 7.9. An AND circuit   **Fig. 7.10.** An OR circuit   **Fig. 7.11.** A NOT circuit

Figs. 7.9–7.11. Old-style circuit symbols

7.2
Construction of Logic Functions

In digital circuit design, the task is generally presented in the form of a function table, also known as a truth table. The objective is then to find a logic function that satisfies this truth table. The next step is to reduce this function to its simplest form, so that it can be implemented by a suitable combination of basic logic circuits. The logic function is generally given in the *disjunctive normal* form (sum-of-products, standard product terms, canonical products). The procedure is done in 3 steps:

1) We find all the rows in the truth table in which the output variable y is 1.
2) From each of these rows, we form the logical product (conjunction) of all the input variables, substituting x_i if the relevant variable is 1 and using $\overline{x}_i$ if the variable is 0. We thus obtain as many product terms as there are rows with $y = 1$.
3) We obtain the required function by forming the logical sum (disjunction) of all the product terms found.

By way of an example, let us consider the truth table in Fig. 7.12. In rows 3, 5, and 7, we have $y = 1$. We must therefore form the logical products of these rows.

$$\text{Row 3:} \quad K_3 = \overline{x}_1 x_2 \overline{x}_3,$$
$$\text{Row 5:} \quad K_5 = x_1 \overline{x}_2 \overline{x}_3,$$
$$\text{Row 7:} \quad K_7 = x_1 x_2 \overline{x}_3$$

The required function is now obtained as the sum of the products (disjunction of the conjunctions):

$$y = K_3 + K_5 + K_7,$$
$$y = \overline{x}_1 x_2 \overline{x}_3 + x_1 \overline{x}_2 \overline{x}_3 + x_1 x_2 \overline{x}_3$$

This is the disjunctive normal form of the required logic function. To simplify, we now apply (7.3a) and obtain

$$y = [\overline{x}_1 x_2 + x_1 (\overline{x}_2 + x_2)]\overline{x}_3$$

Equations 7.6b) and (7.9a) yield a further simplification:

$$y = (\overline{x}_1 x_2 + x_1)\overline{x}_3$$

From (7.3b),

$$y = (x_1 + x_2)(x_1 + \overline{x}_1)\overline{x}_3$$

Row	x_1	x_2	x_3	y
1	0	0	0	0
2	0	0	1	0
3	0	1	0	1
4	0	1	1	0
5	1	0	0	1
6	1	0	1	0
7	1	1	0	1
8	1	1	1	0

Fig. 7.12. Example of a truth table

By again applying (7.6b) and (7.9a), we finally obtain the simple result

$$y = (x_1 + x_2)\bar{x}_3$$

If the output variable y has more "ones" than "zeros," a large number of product terms is obtained. Simplification can be performed in this case by considering the complemented (negated, barred) output variable $\bar{y}$ instead of y. There are sure to be fewer "ones" than "zeros" for this negated variable. Consequently, by expressing the logic function for the complemented variable $\bar{y}$, we obtain from the start a smaller number of product terms; that is, a simpler function. It only needs to be negated at the end to obtain the required function for y. To do this, we merely interchange the operations $(+)$ and $(\cdot)$, and complement all the variables and constants individually.

7.2.1
Karnaugh Map

The Karnaugh map provides an important means of obtaining a logic function in its simplest form. It is basically another version of the truth table. However, the values of the input variables are not simply written one below the other, but arranged at the horizontal and vertical sides of an area that is divided into squares like a chessboard. If the number of input variables is even, half are written horizontally and half vertically. If the number of input variables is odd, one side will have one more variable than the other.

The various combinations of input function values must be arranged in such a way that only *one* variable changes from one square to the next. In the squares themselves are entered the values of the output variables y associated with the input variable values written at the sides. Figure 7.13 shows once again the truth table of the AND function for two input variables, and Fig. 7.14 shows the corresponding Karnaugh map.

As the Karnaugh map is merely a simplified version of the truth table, it can be used for obtaining the disjunctive normal form of the associated logic function in the manner described above. Its advantage is that it makes it easier to spot possible simplifications. We shall illustrate this by means of the example in Fig. 7.15.

To write the disjunctive normal form, we must first, as described above, form the logical product of all the input variables for each cell containing a 1. For the cell in the top left-hand corner, we obtain

$$K_1 = \bar{x}_1\bar{x}_2\bar{x}_3\bar{x}_4$$

For the cell immediately to the right,

$$K_2 = \bar{x}_1 x_2 \bar{x}_3 \bar{x}_4$$

x_1	x_2	y
0	0	0
0	1	0
1	0	0
1	1	1

Fig. 7.13. Truth table for the AND function

x_2 \ x_1	0	1
0	0	0
1	0	1

Fig. 7.14. Karnaugh map for the AND function

x_1	x_2	x_3	x_4	y
0	0	0	0	1
0	0	0	1	1
0	0	1	0	1
0	0	1	1	1
0	1	0	0	1
0	1	0	1	0
0	1	1	0	0
0	1	1	1	0
1	0	0	0	1
1	0	0	1	0
1	0	1	0	1
1	0	1	1	1
1	1	0	0	0
1	1	0	1	0
1	1	1	0	1
1	1	1	1	1

$x_3 x_4$ \\ $x_1 x_2$	00	01	11	10
00	1 (B)	1	0	1 (A)
01	1	0	0	0
11	1 (D)	0	1	1 (C)
10	1	0	1	1 (A)

Fig. 7.15. Truth table with a corresponding Karnaugh map

If we finally form the sum of all the products, one possible expression is

$$K_1 + K_2 = \overline{x}_1\overline{x}_2\overline{x}_3\overline{x}_4 + \overline{x}_1 x_2 \overline{x}_3 \overline{x}_4$$

This can be simplified to

$$K_1 + K_2 = \overline{x}_1\overline{x}_3\overline{x}_4(\overline{x}_2 + x_2) = \overline{x}_1\overline{x}_3\overline{x}_4$$

This illustrates the general simplification rule for the Karnaugh map.

If a rectangle or square with 2, 4, 8, 16, ... cells contains all "ones," the logical product of the entire block can be obtained directly *by only taking into account the input variables that possess a constant value in all the cells of the block.*

Therefore, in our example we obtain for pair B the logical product

$$K_B = \overline{x}_1\overline{x}_3\overline{x}_4$$

which corresponds to the function given above. It is also possible to block together those cells located at the left and right extremities of a row, or at the top and bottom of a column. For the block of four D in Fig. 7.15, we obtain

$$K_D = \overline{x}_1\overline{x}_2$$

Similarly, for the block of four C, the logical product is given by

$$K_C = x_1 x_3$$

This still leaves the 1 in the top right-hand corner. As shown, it can be combined with the 1 at the bottom of that column to form a pair K_A. Another possibility would be to combine it with the 1 at the other end of the same row. However, the simplest solution is to note that there is a 1 at each corner of the Karnaugh map. These can be combined to form a block of four, and we obtain

$$K'_A = \overline{x}_2\overline{x}_4$$

For the disjunctive normal form, we now obtain the already considerably simplified result:

$$y = K'_A + K_B + K_C + K_D,$$
$$y = \overline{x}_2\overline{x}_4 + \overline{x}_1\overline{x}_3\overline{x}_4 + x_1x_3 + \overline{x}_1\overline{x}_2$$

7.3
Extended Functions

In the preceding discussion, we have shown that every logic function can be represented by a suitable combination of the basic functions OR, AND, and NOT. We shall now consider a number of derived functions that occur so frequently in circuit design that they have been given names of their own. Their truth tables and circuit diagrams are shown in Fig. 7.16.

The NOR and NAND functions are the complements of the OR and AND functions, respectively: NOR = not or; NAND = not and. Thus:

$$x_1 \text{ NOR } x_2 = \overline{x_1 + x_2} = \overline{x}_1\overline{x}_2, \tag{7.12}$$
$$x_1 \text{ NAND } x_2 = \overline{x_1x_2} = \overline{x}_1 + \overline{x}_2 \tag{7.13}$$

In the case of the *equivalence function*, $y = 1$ if the two input variables are the same. By writing the disjunctive normal form, we obtain from the truth table

$$y = x_1 \text{ EQUIV } x_2 = \overline{x}_1\overline{x}_2 + x_1x_2$$

The *antivalence (nonequivalence) function* is a complemented equivalence function for which y is 1 if the input variables are different. Written in the disjunctive normal form:

$$y = x_1 \text{ ANTIV } x_2 = \overline{x}_1x_2 + x_1\overline{x}_2$$

The truth table reveals another meaning of the antivalence function: it coincides with the OR function in all values except in the case in which all the input variables are 1. It is therefore also known as the exclusive-or (EXOR) function. Similarly, the equivalence function may also be termed the exclusive-NOR function (EXNOR).

When using integrated circuits, it is sometimes preferable to implement functions using only NAND or NOR gates because they are easy to be implemented. For this purpose, the functions are modified such that only the desired logic operations occur. A simple method

Input variables x_1 x_2	$y = x_1 + x_2$ $= x_1 \text{ OR } x_2$	$y = x_1 \cdot x_2$ $= x_1 \text{ AND } x_2$	$y = \overline{x_1 + x_2}$ $= x_1 \text{ NOR } x_2$	$y = \overline{x_1 \cdot x_2}$ $= x_1 \text{ NAND } x_2$	$y = x_1 \oplus x_2$ $= x_1 \text{ EXOR } x_2$ $= x_1 \text{ ANTIV } x_2$	$y = \overline{x_1 \oplus x_2}$ $= x_1 \text{ EXNOR } x_2$ $= x_1 \text{ ÄQUIV } x_2$
0 0	0	0	1	1	0	1
0 1	1	0	0	1	1	0
1 0	1	0	0	1	1	0
1 1	1 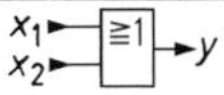	1 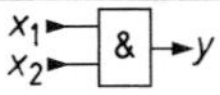	0 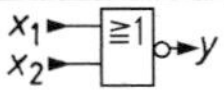	0 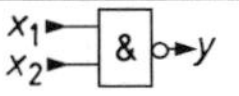	0 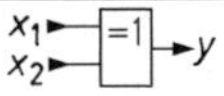	1 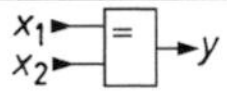

Fig. 7.16. Extended functions derived from the AND or OR functions

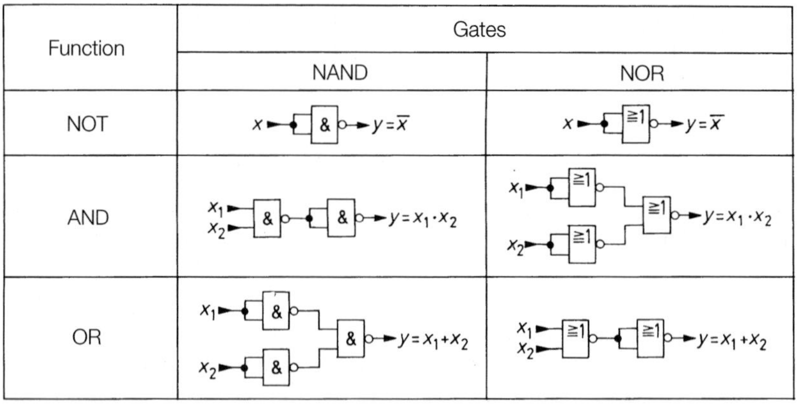

Function	Gates	
	NAND	**NOR**
NOT	$x \longrightarrow \boxed{\&} \!\!\circ \longrightarrow y = \overline{x}$	$x \longrightarrow \boxed{\geq 1} \!\!\circ \longrightarrow y = \overline{x}$
AND	$\begin{array}{c} x_1 \\ x_2 \end{array} \longrightarrow \boxed{\&} \!\!\circ \longrightarrow \boxed{\&} \!\!\circ \longrightarrow y = x_1 \cdot x_2$	$x_1 \longrightarrow \boxed{\geq 1}\!\!\circ \longrightarrow \boxed{\geq 1}\!\!\circ \longrightarrow y = x_1 \cdot x_2 \quad x_2 \longrightarrow \boxed{\geq 1}\!\!\circ$
OR	$x_1 \longrightarrow \boxed{\&}\!\!\circ \quad x_2 \longrightarrow \boxed{\&}\!\!\circ \longrightarrow \boxed{\&}\!\!\circ \longrightarrow y = x_1 + x_2$	$\begin{array}{c} x_1 \\ x_2 \end{array} \longrightarrow \boxed{\geq 1}\!\!\circ \longrightarrow \boxed{\geq 1}\!\!\circ \longrightarrow y = x_1 + x_2$

Fig. 7.17. Implementation of basic functions using NOR and NAND gates

of doing this is to replace the basic functions by NAND and NOR operations. For the AND function,

$$x_1 x_2 = \overline{\overline{x_1 x_2}} = \overline{x_1 \text{ NAND } x_2},$$
$$x_1 x_2 = \overline{\overline{\overline{x_1}} \, \overline{\overline{x_2}}} = \overline{\overline{x_1} + \overline{x_2}} = \overline{x_1} \text{ NOR } \overline{x_2}$$

This yields the possible implementations shown in Fig. 7.17 .

$$x_1 + x_2 = \overline{\overline{\overline{x_1}} + \overline{\overline{x_2}}} = \overline{\overline{x_1}\,\overline{x_2}} = \overline{x_1} \text{ NAND } \overline{x_2},$$
$$x_1 + x_2 = \overline{\overline{x_1 + x_2}} = \overline{x_1 \text{ NOR } x_2}$$

7.4
Circuit Implementation of the Basic Functions

In the preceding subsections, we have manipulated logic circuits without concerning ourselves with their internal design. This approach is justified because, nowadays, digital circuitry is based virtually exclusively on integrated circuits which, apart from their power supply connections, possess only the inputs and outputs mentioned.

For implementation of the individual basic operations, a wide variety of circuit technologies exists, with differing characteristics in terms of power consumption, supply voltage, H and L level, gate propagation delay, and fanout. In order to make an appropriate choice, it is necessary to have at least a basic understanding of the internal design of these circuits. We have therefore summarized the principal families of circuits below.

When ICs are interconnected, a large number of gate inputs are often connected to one output. The number of inputs of the same family of circuits that can be connected without falling below the guaranteed noise margin is characterized by *the fanout*. A fanout of ten therefore means that ten gate inputs can be connected. If the fanout is inadequate, a *buffer* is used instead of a standard gate.

It is a characteristic of a gate that a specific output state is associated with each input state. As described in Chap. 6, these states can be designated by H and L, depending on whether the voltage is greater than V_H or less than V_L. The operation of a gate can be described by a level table, as shown in Fig. 7.18. However, this does not determine which

V_1	V_2	V_o
L	L	H
L	H	H
H	L	H
H	H	L

Fig. 7.18. Example of a level table (function table)

x_1	x_2	y
0	0	1
0	1	1
1	0	1
1	1	0

Fig. 7.19. Truth table for positive logic: the NAND function

x_1	x_2	y
1	1	0
1	0	0
0	1	0
0	0	1

Fig. 7.20. Truth table for negative logic: the NOR function

logic function the gate will implement, as we have not yet defined the assignment of levels to logic states. Although this assignment is arbitrary, it is advisable to adopt a uniform system within the equipment. The level/state assignment

$$H \cong 1, \qquad L \cong 0$$

is termed "positive logic" and produces, in our example, the truth table shown in Fig. 7.19, which may be identified as that of the NAND operation. The assignment

$$H \cong 0, \qquad L \cong 1$$

is known as "negative logic." In our example, it results in the truth table shown in Fig. 7.20; that is, the NOR operation.

One and the same circuit can therefore represent either a NOR or a NAND circuit, depending on the type of logic selected. The logic functions of digital circuits are normally described in positive logic. In negative logic, the operations are reversed:

NOR ⇔ NAND,

OR ⇔ AND,

NOT ⇔ NOT

7.4.1
Resistor-Transistor Logic (RTL)

RTL circuits are the IC implementation of inverter circuits with saturated transistors, as shown in Fig. 6.10, for example. If one input voltage of the RTL gate in Fig. 7.21 is in the H state, the relevant transistor is turned on and the output goes low (L). We therefore obtain a NOR operation in positive logic. The relatively low-valued base resistors insure that the transistors are turned fully on even for low current gains. However, the fanout

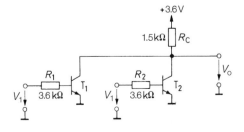

Fig. 7.21. RTL NOR gate, type MC717

Power dissipation :	P_V	$= 5\,\text{mW}$
Gate propagation delay :	t_{pd}	$= 25\,\text{ns}$

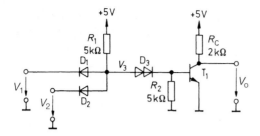

Fig. 7.22. DTL NAND gate, type MC849
Power dissipation: 15 mW   *Gate propagation delay:* $t_{pd} = 25$ ns

is consequently low. The following circuits are considerably better in this respect and, nowadays, RTL circuits are no longer used.

7.4.2
Diode-Transistor Logic (DTL)

In the DTL circuit in Fig. 7.22, the base current for the output transistor is injected via resistor R_1 if input diodes D_1 and D_2 are blocking; in other words, when all of the input voltages are in the H state. In this case, transistor T_1 is turned on and the output voltage goes low. We thus obtain a NAND operation in positive logic. If the same NAND gates are reconnected to the output, the output voltage in the H state is not loaded by the inputs. It therefore assumes, in the H state, the value V^+. Due to the large gate propagation delay caused by saturation of the transistors, DTL circuits are no longer used.

7.4.3
High-Level Logic (HLL)

Modified DTL circuits are available for use in equipment in which high noise levels cannot be avoided. In these circuits, the double diode D_3 is replaced by a Zener diode, as shown in Fig. 7.23. This increases the switching level at the input to approximately 6 V, and a noise margin of 5 V is obtained for a 12 V supply voltage. In order to increase the fanout,

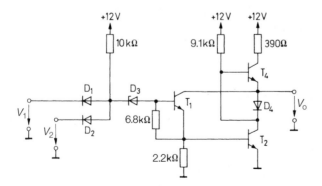

Fig. 7.23. HLL NAND gate, type FZH 101 A
Power dissipation: 180 mW;   *Gate propagation delay:* $t_{pd} = 175$ ns

HLL circuits have a push–pull stage as shown in Fig. 6.6. The switching time is artificially increased by using low-speed transistors, and it can be increased still further by means of an external capacitor. As a result, short spikes have no effect even if their amplitude is greater than the noise margin. HLL circuits are also known as low-speed logic circuits (LSL).

7.4.4
Transistor-Transistor Logic (TTL)

Basically, TTL gates operate in exactly the same way as DTL gates. The only difference is in the design of the diode gate and amplifier. With the standard TTL gate shown in Fig. 7.24, the diode gate is replaced by transistor T_1, which incorporates several emitters. If all of the input levels are in the H state, the current from R_1 flows via the forward-biased base–collector diode of the input transistor to the base of T_2, turning it on. If one input is at low potential, the relevant base–emitter diode becomes conducting and takes over the base current of T_2. This turns T_2 off and the output potential goes high.

In TTL circuits, the amplifier consists of drive transistor T_2 and a push–pull output stage (a totem-pole circuit). When T_2 is conducting, T_3 is also ON and T_4 is OFF. The output is at L and transistor T_3 can accept high currents originating, for example, from the connected gate inputs. (In the L state, a current flows from the inputs!)

When T_2 is OFF, T_3 is also OFF. In this case, T_4 is turned on and delivers an H signal to the output. The transistor operated as an emitter follower can then supply high output currents, and thus rapidly charge up load capacitances. Standard TTL circuits, as shown in Fig. 7.24, are no longer used due to the gate propagation delay caused by the saturation of the transistors.

One method to avoid saturation consists of connecting a Schottky diode in parallel with the collector-based (Fig. 7.25). When the transistor is conducting, it provides voltage feedback to prevent the collector–emitter voltage from falling below about 0.3 V. A TTL gate employing "Schottky transistors" of this type is shown in Fig. 7.26; this is actually a simplified representation of a low-power Schottky TTL gate. A comparison with the standard TTL gate in Fig. 7.24 shows that the values of the circuit resistors are higher by a factor of five. The power consumption is therefore lower by a factor of five, being only 2 mW. Nevertheless, the gate propagation delay is no greater, being only 10 ns. The input

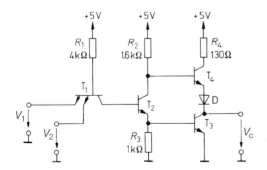

Fig. 7.24. Standard TTL NAND gate, type 7400
Power dissipation: 10 mW;   *Gate propagation delay:* $t_{pd} = 10$ ns

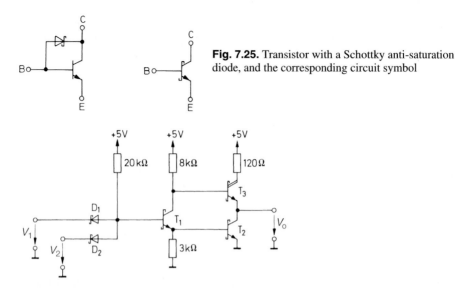

Fig. 7.25. Transistor with a Schottky anti-saturation diode, and the corresponding circuit symbol

Fig. 7.26. Low-power Schottky TTL gate, type 74LS00
Power dissipation: 2 mW   *Gate propagation delay:* $t_{pd} = 10$ ns

diode gate, as in DTL circuits, consists of separate diodes. The diode D required in the output stage for level shifting (Fig. 7.24) is replaced here by the Darlington transistor T_3.

The transfer characteristic of the low-power Schottky TTL inverter (NOT operation) is shown in Fig. 7.27. We can see that the switching level is around 1.1 V at the input. The specified tolerance limits are well exceeded: at the maximum permissible L level at the input of 0.8 V, an H level of at least 2.4 V must be present at the output. For the minimum H level at the input of 2.0 V, the L level at the output must be no more than 0.4 V.

Open-Collector Outputs

The problem sometimes arises that a large number of gate outputs must be logically linked. For 20 outputs, for instance, a gate with 20 inputs would be required, with 20 individual leads leading to them. This complexity can be avoided by using gates with an *open-collector output*. As shown in Fig. 7.28, their output stage consists merely of an npn transistor whose

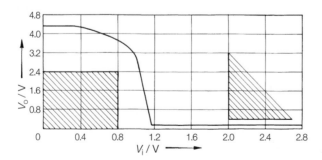

Fig. 7.27. Transfer characteristic of a low-power Schottky TTL inverter.
Hatched areas: tolerance limits

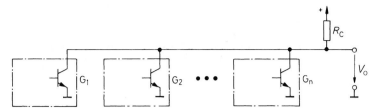

Fig. 7.28. Logical linking of open-collector gate outputs

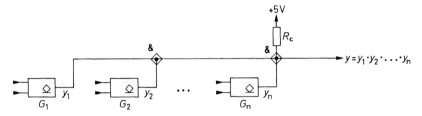

Fig. 7.29. Representation of a wired-AND operation with logic symbols. The ◇-symbol in the gates denotes an open-collector output that is active low

emitter is connected to ground. Outputs of this kind can simply be paralleled – unlike the push–pull output stages otherwise used – and provided with a common-collector resistor as shown in Fig. 7.28.

The output potential therefore only goes high if all the outputs are high. Consequently, in positive logic an AND operation is produced. On the other hand, we can see that the output voltage then goes low if one or more of the outputs assumes the L state. We therefore have an OR operation in negative logic. As the operation is realized by the external wiring, it is referred to as a wired-AND or wired-OR circuit. As the gates are at low impedance in the L state only, they are also known as active-low outputs. The wired-AND operation is represented using logic symbols as shown in Fig. 7.29.

An OR operation can also be implemented using open-collector outputs by applying the wired-AND operation to the complemented variables. De Morgan's law states that:

$$y_1 + y_2 + \ldots + y_n = \overline{\overline{y}_1 \cdot \overline{y}_2 \cdot \ldots \cdot \overline{y}_n}$$

The corresponding circuit is shown in Fig. 7.30.

A disadvantage of using open-collector outputs is that the output voltage rises more slowly than with push–pull outputs, because the circuit capacitances can only charge up via resistor R_C. In this respect, open-collector TTL gates have the same disadvantages as

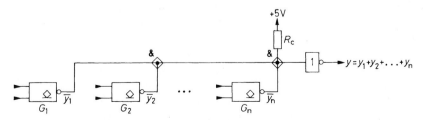

Fig. 7.30. Wired-OR circuit with open-collector outputs

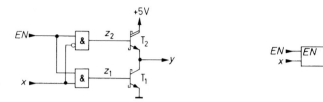

Fig. 7.31. Inverter with a tristate output

Fig. 7.32. Circuit symbol for an inverter with a tristate output

the RTL circuits in Fig. 7.21. There, the logical linking can likewise be interpreted as a wired-AND operation.

Tristate Outputs

There is another important application in which circuit simplification can be achieved by paralleling gate outputs; namely, when any one of several gates connected to a signaling line is to determine the logic state. This is then referred to as a *bus system*.

This end can also be attained using open-collector gates, as shown in Fig. 7.29, by placing all of the outputs, apart from one, in the high-impedance H state. However, the main disadvantage of the low rate of rise can be avoided in this particular application by using gates with a *tristate* output instead of gates with an open-collector output. The tristate output is a genuine push–pull output with the additional property that it can be placed in a high-impedance state using a special control signal. This state is known is the Z state.

The basic circuit implementation is shown in Fig. 7.31. When the *enable* signal $EN = 1$, the circuit operates as a normal inverter: for $x = 0$, $z_1 = 0$ and $z_2 = 1$; that is, T_1 is OFF and T_2 is ON. For $x = 1$, T_1 is turned on and T_2 is turned off. However, if the control variable $EN = 0$, the signals $z_1 = z_2 = 0$, and both output transistors are OFF. This is the high-impedance Z state.

There are many realizations for TTL circuits that differ in speed, supply voltage and propagation delay. The various Schottky TTL families are listed in Fig. 7.46.

7.4.5
Emitter-Coupled Logic (ECL)

We saw in Fig. 4.56 that, in a differential amplifier with an input voltage difference of about ± 100 mV, current I_k can he completely switched from one transistor to the other. The amplifier therefore possesses two defined switching states, namely $I_C = I_k$ or $I_C = 0$. It is therefore also known as a current switch. If, for this switch mode, suitably low-resistance components are selected to insure that the change in voltage across the collector resistors remains sufficiently low, the conducting transistor can be prevented from being driven into saturation.

Figure 7.33 shows a typical ECL gate. Transistors T_2 and T_3 form a differential amplifier. A constant potential $V_{ref} = -1.3$ V is applied to the base of T_3 via the voltage divider. If all of the input voltages are in the L state, transistors T_1 and T_2 are turned off. In this case, the emitter current flows via transistor T_3, producing a voltage drop across R_2. Output voltage V_{o1} is therefore in the L state, and V_{o2} in the H state. When at least one

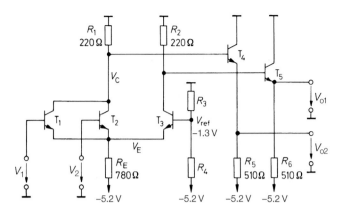

Fig. 7.33. ECL NOR-OR gate, type MC10102. The emitter resistors R_5 and R_6 are not incorporated in the IC and must be connected externally if required

Power dissipation per gate:	$P_{dg} = 25\,\text{mW}$
Power dissipation R_5, R_6, each:	$P_{dR} = 30\,\text{mW}$
Gate propagation delay:	$t_{pd} = 2\,\text{ns}$

input level goes high, the output states are reversed. Positive logic gives an OR operation for V_{o1} and a NOR operation for V_{o2}.

We shall now examine the potentials within the circuit. We will assume that the base–emitter voltage of a conducting transistor will be 0.8 V. When transistor T_3 is OFF, there is hardly no voltage drop across R_2. Consequently, in this case the emitter potential of T_5 is -0.8 V. This is the output H level. If this level is applied, for example, to the base of T_2, the emitter potential is

$$V_E = -0.8\,\text{V} - 0.8\,\text{V} = -1.6\,\text{V}$$

If T_1 or T_2 is conducting the collector current of 4.6 mA will produce a voltage drop of 1 V across R_2. This results in an output voltage of $V_L = -1.8\,V$.

V_{ref} must now be selected such that the input transistors are sure to be ON at an input voltage of $V_H = -0.8\text{V}$ and OFF at an input voltage of $V_L = -1.8\text{V}$. This condition can best be satisfied by setting V_{ref} halfway between V_H and V_L; that is, at about $-1.3\,\text{V}$. The complete transfer characteristic is shown in Fig. 7.34. We can see that the switching level is $-1.3\,\text{V}$. At the maximum permissible input L level of $-1.5\,\text{V}$, an H level of at least $-1.0\,\text{V}$ must be produced at the NOR output. At the lowest input H level of $-1.1\,\text{V}$, the L level at the output must not exceed $-1.65\,\text{V}$.

In contrast to other logic circuit families, the input voltage of ECL in the H state is tightly constrained at the upper limit. If it exceeds $-0.8\,\text{V}$, the relevant input transistor will be driven into saturation. This can be seen from the bend in the transfer characteristic for the NOR output, at an input voltage of $-0.4\,\text{V}$. As the voltage increases further, the collector potential V_C increases with the emitter potential due to the saturation of transistor T_2, and therefore output voltage V_{o2} also increases.

We can see from Fig. 7.34 that the logic levels are much closer to zero potential than to the negative supply voltage ($-5.2\,\text{V}$). Moreover, the magnitude of the supply voltage does not affect the H level, as this is determined only by the base–emitter voltage of the emitter followers. Therefore it was usual in the beginning of ECL-logic to define the positive supply voltage as ground and use a negative emitter supply. However one can shift all

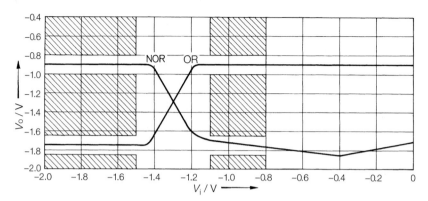

Fig. 7.34. Transfer characteristic of an ECL gate of the MC10000 series
Hatched areas: tolerance limits

voltages by 5.2 V in positive direction. If ECL-circuits are operated this way one speaks of Positive-ECL shortly PECL-circuits. Then the positive supply voltage is usually reduced to 5.0 V or even 3.3 V in order to use the supply voltages used in the remaining circuits.

Of all the logic families, ECL circuits have the smallest gate propagation delays. Indeed, they are even faster than Schottky TTL circuits, which are also operated unsaturated. The difference is that the collector–emitter voltage across the conducting transistors is higher – never less than 0.6 V. Not only does this provide a greater margin to the saturation voltage, but it also results in a lower collector–base junction capacitance.

Another reason for the high speed of ECL circuits lies in the small signal amplitudes of only 1 V involved in switching. The unavoidable switching capacitances can therefore be charged rapidly. The low output resistance of the emitter followers also promotes fast switching times. From (2.117), this is given by

$$r_o \approx 1/g_m \; = \; V_T/I_C \; = \; 26\,\text{mV}/7.7\,\text{mA} = 3.4\,\Omega$$

The high speed of ECL circuits is obtained at the expense of high power dissipation. For a gate in the MC10.000 series, this can be as much as 25 mW. To this must be added the power dissipation in the emitter resistors. For an average output voltage of $-1.3\,\text{V}$, there is a power dissipation of 30 mW in a 510 Ω emitter resistor; that is, more than in the entire gate. For this reason, emitter resistors will only be connected to the outputs used. This dissipation in the emitter resistors can be reduced to 10 mW if instead of connecting 510 Ω resistors to the $-5.2\,\text{V}$ supply, 50 Ω resistors are used on an additional supply voltage of $V_{TT} = -2\,\text{V}$. However, the associated cost and complexity is only justifiable for extensive ECL circuitry. Additional care must be taken to insure that the $-2\,\text{V}$ supply voltage is generated with high efficiency in the power supply. Otherwise, the problem of power dissipation is merely shifted from the circuit to the power supply. For this reason, it is impractical to produce the $-2\,\text{V}$ from the $-5.2\,\text{V}$ using a series regulator.

Wired-OR Operation

By connecting ECL outputs in parallel, it is possible – as with open-collector outputs – to implement a logic operation. This possibility is illustrated in Fig. 7.35. As the H level is predominant when the emitter followers are connected in parallel (active high), we obtain an OR operation in positive logic. The advantage of a wired- OR operation using

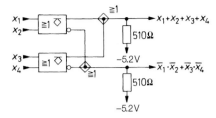

$x_1 + x_2 + x_3 + x_4$

$510\,\Omega$

$-5.2\,V$

$\overline{x_1} \cdot \overline{x_2} + \overline{x_3} \cdot \overline{x_4}$

$510\,\Omega$

$-5.2\,V$

Fig. 7.35. Wired-OR operation for ECL circuits. The ⌀ symbol in the gates denotes an open-emitter output which is active high

ECL circuits is that the speed is not reduced. We not only save one gate, but also one propagation delay time.

To summarize, let us enumerate once more the main considerations for using ECL gates in high-speed logic circuits:

1) They exhibit the shortest gate propagation delays.
2) Their power consumption is independent of the switching state. No current or voltage spikes occur during switching. Consequently, high-frequency noise injected into the printed circuit board remains low.
3) The balanced outputs allow noise-immune differential signal transmission over comparatively large distances (see Sect. 7.5).

A list of the various ECL families is given in Fig. 7.46 below.

7.4.6
Complementary MOS Logic (CMOS)

CMOS logic circuits constitute a family characterized by extremely low power consumption. Figure 7.36 shows an inverter circuit. It is noticeable that the circuit consists exclusively of enhancement-type MOSFETS. The source electrode of the n-channel FET is connected to ground and that of the p-channel FET to the supply voltage V_{DD}. The two FETs therefore operate in source connection and amplify the input voltage on an inverting basis, with one transistor constituting the pull-up resistance for the other.

The pinch-off voltage of the two MOSFETs is about 1.5 V. For a supply voltage of 5 V, at least one of the two MOSFETs is therefore ON. If we make $V_i = 0$, the p-channel FET T_2 is ON and the n-channel FET T_1 is OFF. The output voltage is equal to V_{DD}. For $V_i = V_{DD}$, T_2 is OFF and T_1 is ON. The output voltage becomes zero. We can see that under steady-state conditions no current flows through the circuit. It is only during switching that a small crossover current flows, as long as the input voltage is in the range $|V_p| < V_i < V_{DD} - |V_p|$. The crossover current I_{DD} and the transfer characteristic are plotted in Fig. 7.37.

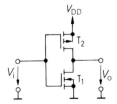

Fig. 7.36. CMOS inverter

$(V_{DD} = 5\,V)$	Standard	High speed
Type	74C04	74HC04
Power dissipation	0.3 μW/kHz	0.5 μW/kHz
Gate propagation delay	90 ns	10 ns

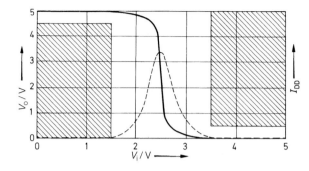

Fig. 7.37. Transfer characteristic and crossover current of a CMOS gate for a 5 V supply voltage
Hatched areas: tolerance limits
Broken line: current drain I_{DD}

The logic levels depend on the supply voltage selected. The permissible supply voltage range for CMOS circuits is very large. In the case of silicon gate circuits it is between 3 V and 6 V, while for metal gate circuits it is as much as 3–15 V. For reasons of symmetry, the switching level is always at half the supply voltage. Consequently, for a supply voltage of 5 V the H level must be above 3.5 V, as we can see from Fig. 7.37. In order to drive a CMOS gate input with a TTL gate output, an additional pull-up resistor is therefore required. On the other hand, HCT (high-speed CMOS TTL) circuits are fully TTL-compatible, because they have input transistors with an adapted pinch-off voltage.

The current consumption of the CMOS gate comprises three parts:

1. A low reverse current of only a few microamperes that flows if the input voltage is at zero level or equal to V_{DD}.
2. A cross current that flows temporarily through both transistors whenever the state of the input signal changes.
3. The principal contribution occurs during charging and decharging the transistor capacitances C_T.

During the charging process the energy $\frac{1}{2}C_T V_{DD}^2$ is stored in the transistor capacitances C_T; simultaneously the same amount of energy is converted to heat in the charging FET. During discharging the energy stored in the capacitance is is converted to heat in the discharging FET. The energy $W = C_T V_{DD}^2$ is therefore converted to heat during a L-H-L cycle. This results in a dissipation power of

$$P_V = W/t = W \cdot f = C_T \cdot V_{DD}^2 \cdot f$$

Since the losses caused by the cross current are also proportional to the frequency they can also be taken into account by defining a power dissipation capacity C_{PD} as described by the formula

$$C_{PD} = \frac{P_{V\,\text{ges}}}{V_{DD}^2 \cdot f}$$

This is slightly higher than the pure transistor capacitances C_T.

The potential of open CMOS inputs is *undefined*. The open inputs must therefore be connected to ground or V_{DD}. This is necessary even for unused gates, because otherwise an input potential can occur at which an undefined but high crossover current flows through the two transistors, resulting in unexpectedly high power dissipation.

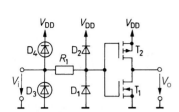

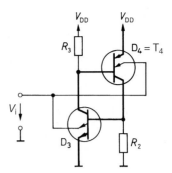

Fig. 7.38. Input protection circuit for CMOS gates

Fig. 7.39. Parasitic thyristor caused by the junction insulation of the MOSFET

Precautions When Employing CMOS Circuits

The gate electrodes of MOSFETs are highly sensitive to static charges. The inputs of MOS integrated circuits are therefore protected by diodes, as shown in Fig. 7.38. Nevertheless, careful handling is required.

These protective diodes, however, introduce a further constraint that affects the use of CMOS circuits. Due to the junction insulation of the two MOSFETs T_1 and T_2, a parasitic thyristor is produced between the supply voltage terminals by the protection diodes, as shown in Fig. 7.39. This thyristor normally has no effect, as transistors T_3 and T_4 are in the OFF state. Their reverse currents are drained off via resistors R_2 and R_3. However, if one of the protective diodes acting as an additional emitter, thyristor T_3, is forward biased, T_4 may fire. This will cause the two transistors to turn on, short-circuiting the supply voltage, and the high currents produced will destroy the chip. In order to prevent this "latch-up" effect, the input voltage must not fall below the ground potential or exceed the supply voltage. If this cannot be avoided, the current flowing via the protective diodes should at least be limited to values of 1 to 100 mA, depending on the technology. A simple series resistor is generally adequate for this purpose. The parasitic thyristor will also be fired if a voltage that exceeds the supply voltage range is applied to the output V_o.

CMOS Gates

Figure 7.40 shows a CMOS NOR gate which operates on the same principle as the inverter described above. In order to insure that the pull-up resistance is high when one of the input voltages assumes the H state, a suitable number of p-channel FETs must be connected in series. By replacing the parallel circuit with a series arrangement, the NOR gate becomes the NAND gate shown in Fig. 7.41.

Transmission Gates

In Sect. 7.1 we saw that logic operations can also be implemented using switches. This possibility is also utilized in MOS technology, as it frequently results in circuit simplification. The resultant component is known as a transmission gate and is used in addition to the conventional gates. Its circuit symbol and equivalent circuit are shown in Fig. 7.42. The way it operates is that input and output are either connected via a low ON resistance

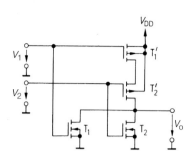

Fig. 7.40. CMOS NOR gate

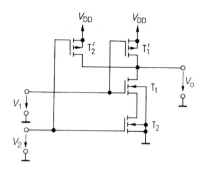

Fig. 7.41. CMOS AND gate

of a conducting MOSFET or isolated. As, in effect, the two terminals are interchangeable, the signal can be transmitted in both directions with minimal delay.

Whereas the logic level in conventional gates is always regenerated, no level regeneration occurs here. The noise margin therefore deteriorates as the number of interconnected transmission gates is increased. Consequently, they are only used in conjunction with conventional gates.

Circuit implementation in CMOS technology is shown in Fig. 7.43. The actual switch is formed by the two complementary MOSFETs, T_1 and T_2. The drive arrangement consists of the inverter producing complementary gate potentials. When $V_{contr.} = 0$, $V_{GN} = 0$ and $V_{GP} = V_{DD}$. This causes the two MOSFETs to be turned off, provided that the signal voltages V_1 and V_2 lie within the range zero to V_{DD}. If, on the other hand, we make $V_{contr.} = V_{DD}$, then $V_{GN} = V_{DD}$ and $V_{GP} = 0$. In this case there is always at least one MOFSFET conducting, as long as V_1 and V_2 are within the permissible signal voltage range.

As we shall see in Chap. 17.2.1, this configuration can also be used as an analog switch. It differs from the transmission gate in that the gate electrodes of T_1 and T_2 are not controlled with logically complementary signals, but driven with signals of opposite polarity. This makes it possible to switch positive and negative signal voltages.

Due to their low power requirement and wide supply voltage range, CMOS circuits are particularly suitable for battery-operated equipment. The various CMOS families are listed in Fig. 7.46.

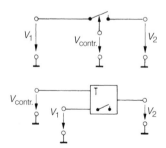

Fig. 7.42. Circuit symbol and operation of a transmission gate

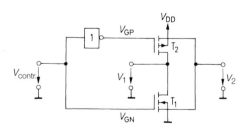

Fig. 7.43. Internal design of a transmission gate

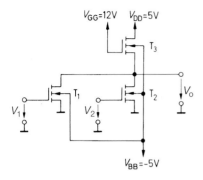

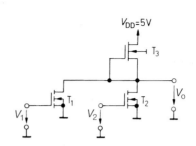

Fig. 7.44. Standard circuit for the NMOS NOR gate

Fig. 7.45. NMOS NOR gate with a depletion load

7.4.7
NMOS Logic

The feature of NMOS ICs is that they consist exclusively of n-channel MOFSFETS. They are therefore particularly easy to manufacture and, for this reason, they are mainly used in large-scale integrated (LSI) circuits.

The NMOS NOR gate shown in Fig. 7.44 is a close relative of the RTL NOR gate shown in Fig. 7.21. For technological reasons, a MOSFET is used instead of the ohmic pull-up resistor. As with the input FETs an enhancement type is employed. To make it conduct, a high gate potential V_{GG} must be applied. If the output voltage in the H state is to rise to the drain potential V_{DD}, the auxiliary potential V_{GG} must be selected to be higher than V_{DD} by at least the pinch-off voltage. In addition, a negative substrate bias V_{BB} is often required, in order to turn off the input FETs reliably and reduce the junction capacitances.

As we can see from Fig. 7.44, T_1, operates as a source follower for V_{GG}. The internal resistance r_o therefore has the value $1/g_m$. In order to achieve the high resistance values required, it is given a considerably lower transconductance than the input FETs.

The positive auxiliary voltage V_{GG} can be dispensed with by using a depletion-type MOSFET for T_3. This possibility is illustrated in Fig. 7.45, in which T_3 is operated as a constant current source as in Fig. 4.126. However, the input FETs must always be of the enhancement type, as the control voltage would otherwise have to be negative, whereas the output voltage is always positive. Direct coupling of such gates would therefore be impossible.

Using ion implantation, it is possible to integrate depletion and enhancement type MOSFETs on the same chip. The negative auxiliary voltage can be eliminated by selecting suitable pinch-off voltages, or can be generated from the positive supply voltage using a voltage converter incorporated in the device.

NMOS logic has no more importance because all MOS chips are made in CMOS technology today.

7.4.8
Summary

Figure 7.46 provides a list of the most commonly used logic families. In each case, the data refer to a single logic gate. We can see that each technology is available in various

Family	Nominal voltage	Propagation delay	Power for one gate	5 V TTL compatibility	Technology
7400	5 V	10 ns	10 mW	✓	Bipolar
74LS00	5 V	10 ns	2 mW	✓	Bipolar
74ALS00	5 V	8 ns	1 mW	✓	Bipolar
74ABT00	5 V	3 ns	2 mW + 0.5mW/MHz	✓	BiCMOS
74HC00	5 V	10 ns	0.5 mW/MHz	–	CMOS
74HCT00	5 V	10 ns	0.5 mW/MHz	✓	CMOS
74AC00	5 V	6 ns	0.8 mW/MHz	–	CMOS
74ACT00	5 V	6 ns	0.8 mW/MHz	✓	CMOS
74LVT00	3.3 V	3 ns	1 mW+0.2 mW/MHz	✓	BiCMOS
74LVC00	3.3 V	4 ns	0.1 mW/MHz	✓	CMOS
74AVC00	2.5 V	2 ns	0.2 mW/MHz	–	CMOS
74AUC00	1.8 V	2 ns	0.1 mW/MHz	–	CMOS

Fig. 7.46a. Some families of the 7400-series. At CMOS circuits the power is calculated from the power dissipation capacitance C_{PD} according $P = f \cdot C_{PD} \cdot V_{dd}^2$

Family	Nominal voltage	Propagation delay	Power for one gate	Technology
MC 10,000	5.2 V	2000 ps	35 mW	Bipolar
MC 100,000	4.5 V	750 ps	50 mW	Bipolar
MC 10E000 MC 100E000	5 V	350 ps	50 mW	Bipolar
MC 10EP000 MC 100EP000	3.3/5 V	200 ps	60/85 mW	Bipolar

Fig. 7.46b. Some ECL-families. Power includes one output resistor of 50 Ω connected to $V_{TT} = V_{CC} - 2V$ resulting in 10 mW average power dissipation

versions, which differ in terms of power dissipation and propagation delay. Clearly, the newer families, possess a noticeably higher speed at lower power. The reason for this is that they are dielectrically insulated and therefore have smaller switching capacitances than the older junction-insulated types.

The power consumptions of the logic circuit families vary greatly. We can see from Fig. 7.47 that the CMOS circuits perform well at low frequencies. However, above 1 MHz there is little difference in power dissipation between low-power Schottky and CMOS circuits. It is noticeable that in this frequency range the power consumption of TTL circuits also rises. The reason for this is that a crossover current flows through the totem-pole output stage at each switching cycle, which significantly increases the power consumption at high frequencies. ECL circuits do not have this drawback. Consequently, apart from being more expensive, ECL circuits offer advantages at frequencies above 300 MHz.

Digital ICs will only function properly with a well-designed power supply arrangement. All of the logic circuit families generate high-frequency current pulses on the supply lines during switching. As all of the signals are referred to ground potential, low-resistance and low-inductance grounding of all ICs is required. This requirement is best satisfied on a printed circuit board, by means of a ground layer. At frequencies above 50 MHz, it is advisable to metallize one side of the board completely as a ground layer and only cut out

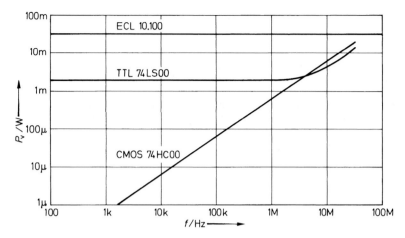

Fig. 7.47. Power dissipation versus frequency

the terminals (see the next section). In order to prevent the current pulses from contaminating the supply voltage during switching, the latter must be fed to the ICs with very low resistance and inductance. If the ground connection is made well, interference can be prevented by smoothing the supply voltage with capacitors. For this purpose, 10 . . . 100 nF ceramic capacitors are used. Electrolytic capacitors are unsuitable due to their poor high-frequency performance. Depending on the requirements, one capacitor is assigned to 2 . . . 5 chips.

7.5
Connecting Lines

So far, we have assumed that the digital signals are transmitted undistorted from one to another. However, at signals with low rise- and fall-times the effect of the connecting lines is not negligible. As a rule of thumb, a simple connecting wire is no longer adequate if the delay on that wire attains the order of magnitude of the rise time of the signal. Consequently, the maximum length for such connections is approximately 10 cm per nanosecond rise time.

If this length is exceeded, severe pulse deformations, reflections, and more or less damped oscillations occur. This problem can be overcome by using lines of defined characteristic impedance (coaxial cable, microstrip lines), which are terminated in their characteristic (surge) impedance. This is generally between 50 and 300 Ω.

Microstrip lines can be produced by fabricating all the connecting tracks on the underside of a circuit board and fully metallizing the component side, while providing small clearances for the insulation of the component terminals. In this way, all of the connecting tracks on the underside become microstrip lines. If the circuit board used has a relative permittivity $\varepsilon_r = 5$ and a thickness $d = 1.2$ mm, we obtain a surge impedance of 75 Ω for a conductor track width of $w = 1$ mm; see Chap. 24.2.1.

For connections from one board to another, coaxial cables can be used. However, they have the major disadvantage that they are difficult to run via multi-pin connectors. It is much

Fig. 7.48. Data transmission via an unsymmetrically driven twisted-pair line

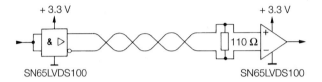

Fig. 7.49. Data transmission via a symmetrically driven twisted-pair line

Fig. 7.50. Data transmission via a symmetrically driven twisted-pair line with LVDS-chips (Low Voltage Differential Signaling)

simpler to run the signal via a simple insulated twisted-wire pair, which can be connected to two adjacent pins of a normal multi-pin connector. If these twisted-pair lines have approximately 100 turns per meter, a characteristic impedance of about $110\,\Omega$ is obtained.

The simplest method of transmitting signals on a twisted-pair line is shown in Fig. 7.48. Due to the low-impedance termination required, the sending-end gate must be able to deliver a correspondingly high output current. Such gates are available in IC form as "line drivers" (buffers). It is advisable to use a Schmitt trigger gate as a receiver to reshape the signal edges.

Signal transmission that is unsymmetrical about ground, as illustrated in Fig. 7.48, is relatively sensitive to external disturbances, such as voltage spikes on the ground wire. In larger systems, *symmetrical* signal transmission as shown in Fig. 7.49 is therefore preferable. Here, complementary signals are transmitted on the two wires of the twisted-pair line and a comparator is used as a receiver. In this mode of operation, the information is contained in the polarity of the difference voltage, rather than in its absolute value. A noise voltage therefore only causes a common-mode input, which remains ineffective due to the common mode rejection of the comparator.

When forming the complementary signals, it must be insured that there is no time delay between the two signals. Consequently, a special circuit with complementary outputs must be used with TTL circuits instead of a simple inverter.

Complementary outputs of this type are inherent in ECL gates. They are therefore particularly well suited for symmetrical data transmission. In order to exploit their high speed capability, a simple differential amplifier with an ECL-compatible output is used as a comparator. It is known as a *line receiver.* The relevant circuit arrangement is shown in Fig. 7.50.

Chapter 8:
Combinatorial Circuits

A combinatorial circuit describes an arrangement of digital circuitry without variable memory. The output variables y_i are clearly defined by the input variables x_i according to the block diagram in Fig. 8.1, whereas in *sequential logic circuits* the output variables are also dependent not only on the current state of the system but also on its history.

The description of a combinatorial circuit, i.e. the assignment of output variables to input variables uses truth tables or Boolean equations. For the realization of combinatorial circuits one thinks primarily in terms of using gates. However, as shown in Fig. 8.2, this is not the only and usually not even the best method. If the zeros and ones were statistically distributed in the truth tables as e.g. in a program code, then the logical function would be very complex. In this case it is advantageous to store the truth tables in the form of a table on a ROM (see Chap. 10).

If there are only a few ones in the truth table, then the number of product terms in the logical functions will be correspondingly low. But they may be simple even with many ones if the regularity of the pattern they follow is sufficiently as e.g. in the function $y_j = x_i$. For this reason it is always worth testing whether the logical functions can be simplified. As a manual process this is cumbersome with both the Boolean algebra and the Karnaugh map. In the era of computer-aided circuit design, one therefore employs a simplifier for this task. Only if this results in a few very simple functions is its realization using individual gates such as e.g. from the 7400 family practical.

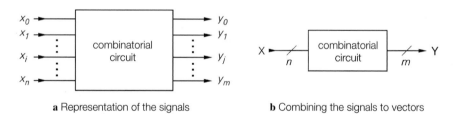

a Representation of the signals   **b** Combining the signals to vectors

Fig. 8.1. Block diagram of a combinational circuit

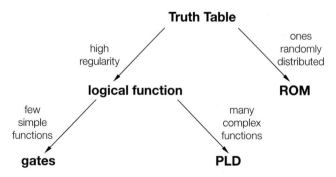

Fig. 8.2. Implementing combinational circuits

When many complicated functions have to be realized one is soon confronted with the notorious TTL grave. In such a case the use of programmable logic devices (PLD) is of great advantage, since they allow the realization of even the most complicated functions by means of a single chip in view of the fact that components containing more than 100,000 gates are available. In principle, with PLDs the logical operations are realized in the same way as with the use of discrete gates. The only difference is that all gates needed may be contained on a single chip and that the required connections of the gates can be configured by programming the PLD (see Sect. 10.4 on page 711).

8.1
Number Representation

As digital circuits can only process binary – that is, two-valued – variables, the number representation must be converted from the conventional decimal system to a binary system. The various methods of doing this are summarized in the following sections.

8.1.1
Positive Integers in Straight Binary Code

The simplest form of representation is the straight binary code. The digit positions are arranged in ascending powers of 2. For the straight binary representation of an N-digit number, we therefore have:

$$Z_N = z_{N-1} \cdot 2^{N-1} + z_{N-2} \cdot 2^{N-2} + \ldots + z_1 \cdot 2^1 + z_0 2^0 = \sum_{i=0}^{N-1} z_i 2^i$$

As in the decimal system, we simply write the digit sequence $\{z_{N-1} \ldots z_0\}$ and think in terms of multiplying the digits by the relevant power of 2 and adding.

Example:
$$15253_{dec} = \underline{1\ 1\ 1\ 0\ 1\ 1\ 1\ 0\ 0\ 1\ 0\ 1\ 0\ 1} \quad \text{Straight binary}$$
$$2^{13} \qquad\qquad\qquad\qquad 2^0 \quad \text{Weight}$$

Octal Code

As can be seen, the straight binary representation is difficult to read. One therefore uses an abbreviated notation by condensing binary numbers into groups of three and writing them as decimal digits. As the resulting digits are arranged in powers of $2^3 = 8$, this is known as octal code.

Example:	3	5	6	2	5	Octal
$15253_{dec} =$	0 1 1	1 0 1	1 1 0	0 1 0	1 0 1	Dual
	2^{12}	2^9	2^6	2^3	2^0	
	8^4	8^3	8^2	8^1	8^0	Weight

Hexadecimal Code

Another commonly used abbreviated notation is obtained by combining binary digits into groups of four. As the resulting digits are arranged in powers of $2^4 = 16$, this is known as hexadecimal, or simply hex, code. Each digit can assume values between 0 and 15. Since we only have ten decimal digits, the numbers "ten" to "fifteen" are represented by the letters A to F.

Example:	3	B	9	5	Hex
$15253_{dec} =$	0 0 1 1	1 0 1 1	1 0 0 1	0 1 0 1	Dual
	2^{12} 16^3	2^8 16^2	2^4 16^1	2^0 16^0	Weight

8.1.2
Positive Integers in BCD Code

Straight binary numbers are unsuitable for numeric input and output, as we are accustomed to calculating in the decimal system. Binary-coded decimal (BCD) notation has therefore been introduced, in which each individual decimal digit is represented by a binary number; in other words, by the corresponding straight binary number. In this case we have, for example:

	1	5	2	5	3	Dec
$15253_{dec} =$	0 0 0 1	0 1 0 1	0 0 1 0	0 1 0 1	0 0 1 1	BCD
	10^4	10^3	10^2	10^1	10^0	Weight

A decimal number encoded in this way could more precisely be termed a BCD number in 8421 code, or a natural BCD number. The individual decimal digits can also be represented by other binary combinations of four or more digits. As the 8421 BCD code is the most commonly used, it is often known simply as BCD code. We shall adopt this convention and draw the reader's attention to any deviations from natural BCD code.

Numbers between 0 and 15_{dec} can be represented using a four-digit straight binary number (a tetrad). As only ten combinations are used in BCD code, this form of representation requires more bits than straight binary code.

8.1.3
Binary Integers of Either Sign

Signed-Magnitude Representation

A negative number can be characterized quite simply by placing a sign bit s in front of the highest-order digit. Zero means "positive" and one means "negative." An unambiguous interpretation is only possible if a fixed word length has been agreed upon.

Example for an 8-bit word length:

$$+118_{dec} = \boxed{0} \quad 1 \quad 1 \quad 1 \quad 0 \quad 1 \quad 1 \quad 0_2$$

$$-118_{dec} = \boxed{1} \quad 1 \quad 1 \quad 1 \quad 0 \quad 1 \quad 1 \quad 0_2$$

$$(-1)^s \quad 2^6 \quad 2^5 \quad 2^4 \quad 2^3 \quad 2^2 \quad 2^1 \quad 2^0$$

Two's-Complement Representation

The disadvantage of signed-magnitude representation is that positive and negative numbers cannot be added simply. An adder must be switched over to subtraction mode when a minus sign occurs. With two's-complement representation, this is unnecessary.

With 2's-complement representation, the most significant bit is given a negative weight. The rest of the number is represented in normal binary form. Once again, a fixed word length must be agreed upon, so that the most significant bit is unambiguously defined. For a positive number, the most significant bit is 0. For a negative number, the most significant bit must be 1, because only this position has a negative weight.
Example for an 8-bit word length:

$$+118_{dec} = \boxed{0} \quad \underbrace{1 \quad 1 \quad 1 \quad 0 \quad 1 \quad 1 \quad 0}_{B_N}$$

$$-118_{dec} = \boxed{1} \quad \underbrace{0 \quad 0 \quad 0 \quad 1 \quad 0 \quad 1 \quad 0}_{X} = -128 + 10$$

$$-2^7 \quad 2^6 \quad 2^5 \quad 2^4 \quad 2^3 \quad 2^2 \quad 2^1 \quad 2^0 \qquad \text{Weight}$$

The transition from a positive to a negative number of equal magnitude is of course somewhat more difficult than with signed-magnitude representation. Let us assume that the binary number B_N has word length N without the sign bit. The sign digit position therefore has the weight -2^N. The number $-B_N$ is therefore obtained in the form

$$-B_N = -2^N + X$$

The positive remainder X is therefore

$$X = 2^N - B_N$$

This expression is known as the *two's-complement* B_N to $B_N^{(2)}$. It can easily be calculated from B_N. For this purpose, we consider the largest number that can he represented in binary form using N digit positions. It has the value

$$1111\ldots \cong 2^N - 1$$

If we subtract any binary number, B_N, from this number, we obviously obtain a binary number that is produced by negation of all the digits. This number is known as the one's-complement $B_N^{(1)}$ to B_N. We obtain

$$B_N^{(1)} = 2^N - 1 - B_N = \underbrace{2^N - B_N}_{B_N^{(2)}} - 1 = B_N^{(2)} - 1$$

and

$$B_N^{(2)} = B_N^{(1)} + 1 \tag{8.1}$$

The 2's-complement of a binary number is therefore the result of negation of all the digits and the addition of 1.

It can easily be demonstrated that it is not necessary to deal with the sign digit separately, but that, to change the sign, it is possible merely to form the 2's-complement of the entire number including the sign digit. For binary numbers in 2's-complement representation, the following relationship therefore holds:

$$-B_N = B_N^{(2)} \tag{8.2}$$

This relation applies to the case in which we likewise consider only N digits in the result and disregard the overflow digit.

Example of an eight-digit binary number in 2's-complement representation:

$118_{dec}=$ 0 1 1 1 0 1 1 0

1's-complement: 1 0 0 0 1 0 0 1

           +   1

2's-complement: 1 0 0 0 1 0 1 0$= -118_{dec}$

Reconversion:

1's-complement: 0 1 1 1 0 1 0 1

         +   1

           0 1 1 1 0 1 1 0$= +118_{dec}$

Sign Extension

If we wish to expand a positive number to a larger word length, we simply add to the leading zeros. In 2's-complement, a different rule applies: we have to extend the sign bit.

Example: 8 bit   16 bit

118_{dec} $= 0 1 1 1 0 1 1 0 = 0 0 0 0 0 0 0 0 0 1 1 1 0 1 1 0$

-118_{dec} $= 1 0 0 0 1 0 1 0 = \underbrace{1 1 1 1 1 1 1 1}_{\text{sign extension}} 1 0 0 0 1 0 1 0$

The proof is simple. For an N-digit negative number, the sign bit has the value -2^{N-1}. If we extend the word length by one bit, we have to insert an additional leading "one". The added sign digit has the value -2^N. The old sign digit changes its value from -2^{N-1} to $+2^{N-1}$. The two together therefore have the value:

$$-2^N + 2^{N-1} = -2 \cdot 2^{N-1} + 2^{N-1} = -2^{N-1}$$

Therefore, the value remains unchanged.

Decimal	2's-complement								Offset binary							
	b_7	b_6	b_5	b_4	b_3	b_2	b_1	b_0	b_7	b_6	b_5	b_4	b_3	b_2	b_1	b_0
127	0	1	1	1	1	1	1	1	1	1	1	1	1	1	1	1
1	0	0	0	0	0	0	0	1	1	0	0	0	0	0	0	1
0	0	0	0	0	0	0	0	0	1	0	0	0	0	0	0	0
−1	1	1	1	1	1	1	1	1	0	1	1	1	1	1	1	1
−127	1	0	0	0	0	0	0	1	0	0	0	0	0	0	0	1
−128	1	0	0	0	0	0	0	0	0	0	0	0	0	0	0	0

Fig. 8.3. Relationship between 2's-complement and offset binary representation

Offset Binary

Some circuits can only process positive numbers. Therefore, they always interpret the most significant digit as being positive. In such cases, we define the midpoint of the number range as being represented as zero (offset binary representation).

Using an eight-digit positive binary number, the range 0–255_{dec} can be represented; while using an eight-digit 2's-complement number, we have the range from -128_{dec} to $+127_{dec}$. To change to offset binary representation, we shift the number range by adding 128 to $0 - 255$. Numbers above 128 are therefore to be treated as positive, and numbers under 128 as negative. In this case, the range midpoint "128" means zero. The addition of 128 can simply be performed by negating the sign bit in the 2's-complement notation. The number range is shown in Fig. 8.3.

8.1.4
Fixed-Point Binary Numbers

Like a decimal fraction, a binary fraction is defined such that the weights to the right of the point are interpreted as negative powers of 2.

Example:

$225.8125_{dec} =$	1	1	1	0	0	0	0	1	, 1	1	0	1
	2^7	2^6	2^5	2^4	2^3	2^2	2^1	2^0	2^{-1}	2^{-2}	2^{-3}	2^{-4}

In general, a fixed number of digits after the point is stipulated; hence the term "fixed-point binary digit." Negative fixed-point numbers are given in signed-magnitude form.

When specifying a defined number of digits, it is possible, by multiplying by the reciprocal of the lowest power of 2, to produce integers that can be processed in the notations described. For the numeric output, the multiplication is again reversed.

8.1.5
Floating-Point Binary Numbers

By analogy with floating-point decimal numbers,

$$Z_{10} = M \cdot 10^E$$

a floating-point binary number is defined as

IEEE format	Word length	Sign S	Exponent		Mantisse	
			length E	range	length M	precision
Single	32 bit	1 bit	8 bit	$2^{\pm127} \approx 10^{\pm38}$	23 bit $\widehat{=}$ 7 dec. places	
Double	64 bit	1 bit	11 bit	$2^{\pm1023} \approx 10^{\pm308}$	52 bit $\widehat{=}$ 16 dec. places	
Internal	80 bit	1 bit	15 bit	$2^{\pm16383} \approx 10^{\pm4932}$	64 bit $\widehat{=}$ 19 dec. places	

Fig. 8.4. Specifications of the IEEE-floating-point formats

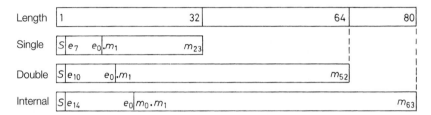

Fig. 8.5. Comparison of floating-point formats

$$Z_2 = M \cdot 2^E$$

where M is the mantissa and E is the exponent.

Example:

225.8125	decimal, fixed point
$= 2.258125 \ \mathrm{E} \ 2$	decimal, floating point
$= 11100001.1101$	straight binary, fixed point
$= 1.11000011101 \ \mathrm{E} \ 0111$	straight binary, floating point

For computation with floating-point numbers, the notation specified in *IEEE Standard P 754* is universally employed nowadays. This notation is used not only in mainframe computers, but also in PCs and even in some signal processors, and is in many cases supported by the corresponding arithmetic hardware. The user can choose between two precision formats: 32-bit single precision and double precision with 64 bits. Internally, computation is performed with 80-bit precision. These three formats are shown in Figs. 8.4 and 8.5. There are three distinct parts in each format: the sign bit S, the exponent E, and the mantissa M. The word lengths of the exponent and mantissa are functions of the precision selected.

In the IEEE Standard, the mantissa M is specified by the digits $m_0, m_1, m_2, \ldots$ Usually, the mantissa is normalized to $m_0 = 1$:

$$M = 1 + m_1 \cdot 2^{-1} + m_2 \cdot 2^{-2} + \ldots = 1 + \sum_{i=1}^{k} m_i 2^{-i},$$

Its absolute value is therefore $1 \leq M < 2$. The digit $m_0 = 1$ is only specified for the internal notation; otherwise, it is hidden and must be restored for the calculation.

The exponent E is specified in IEEE format as an offset binary number, so that positive and negative values can be defined. For the calculation, an offset amounting to half the range must be subtracted, namely:

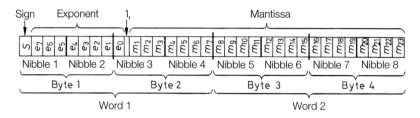

Fig. 8.6. Segmentation of a 32-bit floating-point number

$$2^7 - 1 = \quad 127 \quad \text{offset for single precision,}$$
$$2^{10} - 1 = \quad 1\,023 \quad \text{offset for double precision,}$$
$$2^{14} - 1 = 16\,383 \quad \text{offset for internal precision.}$$

The sign of the number is determined by the sign bit S, thus producing a signed-magnitude representation. The value of an IEEE number can therefore be calculated in the following manner:

$$Z = (-1)^S \cdot M \cdot 2^{E - \text{offset}}$$

Taking the example of IEEE 32-bit single precision, we shall now examine this in somewhat greater detail. The segments of a word are shown in Fig. 8.6. The most significant bit is the sign bit S, followed by 8 bits for the exponent and 23 bits for the mantissa. The MSB (most significant bit) of the mantissa $m_0 = 1$ is hidden; the point is to the left of m_1. The weight of m_1 is therefore $1/2$.

A 32 bit floating-point number can be split up into two words of 16 bits, 4 bytes, or 8 nibbles each. It can therefore be expressed by eight hex characters. A number of examples are given in Fig. 8.7. The normalized number NOR_1 has an exponent of 127; after subtracting the offset from 127, we obtain a multiplier of $2^0 = 1$. The noted value of the mantissa is 0.75. This, together with the hidden 1, produces the specified value $+1.75$. In the second example, NOR_2, a negative number has been selected; in this case, $S = 1$. The number 10 in the third example is represented in normalized form as $10 = 2^3 \cdot 1.25$. We arrive at the given hex representation in the usual way, by organizing the bit string into groups of four and using the associated hex symbols. Unfortunately, the hex representation of IEEE numbers is very involved, because the first symbol contains the sign and part of the exponent, and the third symbol a mixture of exponent and mantissa.

A couple of special cases are also listed in Fig. 8.7. The largest number that can be represented in 32-bit IEEE format is

$$\begin{aligned} \text{NOR}_{max} &= 2^{254-127}(1 + 1 - 2^{-23}) \\ &= 2^{127}(2 - 2^{-23}) \approx 2^{128} \approx 3.4 \cdot 10^{38} \end{aligned}$$

The exponents 0 and 255 are reserved for exceptions. The exponent 255 is interpreted in conjunction with the mantissa $M = 0$ as $\pm\infty$, depending on the sign. If the exponent and the mantissa are both 0, the number is defined as $Z = 0$. In this case, the sign is irrelevant.

NOR_1	$= 3\,\text{F}\,\text{E}\,0\,0\,0\,0\,0_{\text{Hex}} =$	0	$\underbrace{0\,1\,1\,1\,1\,1\,1\,1}_{127}$,	$\underbrace{1\,1\,0\,0\ldots 0}_{0.75}$	$= +1.75$

$$\text{NOR}_1 = 3\,\text{F}\,\text{E}\,0\,0\,0\,0\,0_{\text{Hex}} = 0 \quad \underbrace{0\,1\,1\,1\,1\,1\,1\,1}_{127} \;,\; \underbrace{1\,1\,0\,0\ldots 0}_{0.75} \quad = +1.75$$

$$\text{NOR}_2 = \text{B}\,\text{F}\,\text{B}\,0\,0\,0\,0\,0_{\text{Hex}} = 1 \quad \underbrace{0\,1\,1\,1\,1\,1\,1\,1}_{127} \;,\; \underbrace{0\,1\,1\,0\ldots 0}_{0.375} \quad = -1.375$$

$$\text{NOR}_3 = 4\,1\,2\,0\,0\,0\,0\,0_{\text{Hex}} = 0 \quad \underbrace{1\,0\,0\,0\,0\,0\,1\,0}_{130} \;,\; \underbrace{0\,1\,0\,0\ldots 0}_{0.25} \quad = +10$$

$$\text{NOR}_{max} = 7\,\text{F}\,7\,\text{F}\,\text{F}\,\text{F}\,\text{F}_{\text{Hex}} = 0 \quad \underbrace{1\,1\,1\,1\,1\,1\,1\,0}_{254} \;,\; \underbrace{1\,1\,1\,1\ldots 1}_{1 - 2^{-23}} \quad = +2^{127}(2 - 2^{-23})$$

$$\text{INF} = 7\,\text{F}\,8\,0\,0\,0\,0\,0_{\text{Hex}} = 0 \quad \underbrace{1\,1\,1\,1\,1\,1\,1\,1}_{255} \;,\; \underbrace{0\,0\,0\,0\ldots 0}_{0} \quad = +\infty$$

$$\text{ZERO} = 0\,0\,0\,0\,0\,0\,0\,0_{\text{Hex}} = \times \quad \underbrace{0\,0\,0\,0\,0\,0\,0\,0}_{0} \;,\; \underbrace{0\,0\,0\,0\ldots 0}_{0} \quad = 0$$

Fig. 8.7. Examples of normalized numbers and exceptions in 32-bit floating-point format

8.2
Multiplexer – Demultiplexer

Multiplexer are circuits that connect one of several data sources to a single output. Which of the sources is selected must be determined by an address. The inverse circuit operation which distributes data to several outputs according to their address is known as demultiplexer. In both circuits, addressing the selected input or output is performed by a 1-out-of-n decoder, described below.

8.2.1
1-of-n Decoder

A 1-of-n decoder is a circuit with n outputs and ld n address inputs. The outputs y_i are numbered from 0 to $(n - 1)$. An output therefore goes to "one" precisely when the input binary number A is identical to the number i of the relevant output. Figure 8.9 shows the

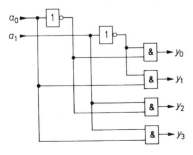

Fig. 8.8. Circuit for a 1-of-4 decoder

A	a_1	a_0	y_3	y_2	y_1	y_0
0	0	0	0	0	0	1
1	0	1	0	0	1	0
2	1	0	0	1	0	0
3	1	1	1	0	0	0

Fig. 8.9. Truth table for a 1-of-4 decoder
$y_0 = \bar{a}_0\bar{a}_1$, $y_1 = a_0\bar{a}_1$, $y_2 = \bar{a}_0 a_1$, $y_3 = a_0 a_1$

truth table for a 1-of-4 decoder. The variables a_0 and a_1 represent the binary code of the number A. The sum of the products (disjunctive normal form) of the recoding functions can be taken directly from the truth table. Figure 8.8 shows the corresponding implementation.

When using monolithic integrated circuits, NAND functions are often chosen rather than AND functions, so that the output variables are complemented. For further IC types, see the following section on demultiplexers.

Type of IC:	TTL	CMOS
10 outputs	74 LS 42	4028

8.2.2
Demultiplexer

A demultiplexer can be used to distribute input d to various outputs. It represents an extension of the 1-of-n decoder. The addressed output does not go to "one," but assumes the value of input data d. Figure 8.10 illustrates the principle by means of switches, while Fig. 8.11 shows its implementation using gates. If we make $d = 1$, the multiplexer operates as a 1-of-n decoder. Commonly used demultiplexers are listed in Fig. 8.12.

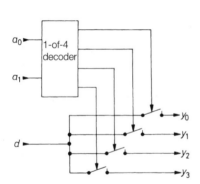

Fig. 8.10. Principle operation of a demultiplexer

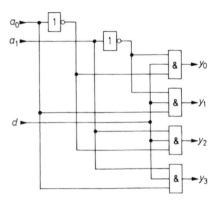

Fig. 8.11. Demultiplexer circuit
$y_0 = \bar{a}_0\bar{a}_1 d$, $y_1 = a_0\bar{a}_1 d$, $y_2 = \bar{a}_0 a_1 d$, $y_3 = a_0 a_1 d$

outputs	TTL	ECL	CMOS
16	74 LS 154		4514
8	74 LS 138	10162	74 HC 138
8	74 ALS 538[1]		40 H 138
2 × 4	74 LS 139	10172	74 HC 139
2 × 4	74 ALS 539[1]		4555

[1] output polarity selectable

Fig. 8.12. Integrated demultiplexers

8.2.3
Multiplexer

The opposite of a demultiplexer is a multiplexer. Starting from the circuit in Fig. 8.10, it can be implemented by swapping the outputs and input to give the basic circuit shown in Fig. 8.13. This provides a particularly simple illustration of the mode of operation: a 1-of-n decoder selects from n inputs the one whose number coincides with the address and switches it to the output. The corresponding gate implementation is shown in Fig. 8.14.

In CMOS technology, a multiplexer can be implemented using both gates and analog switches (transmission gates). When analog switches are employed, signal transmission is bidirectional. In this case, therefore, the multiplexer is identical to the demultiplexer, as comparison of Figs. 8.10 and 8.13 will show. The circuit is then known as an analog multiplexer/demultiplexer.

The OR operation required in multiplexers can also be implemented using a wired- OR connection. This possibility is shown for open-collector outputs in Fig. 8.15. In positive logic, this connection results in an AND operation, therefore it is necessary to use the complemented signals – as in Fig. 7.30.

In order to overcome the disadvantage associated with open-collector outputs, namely the higher switching time, tristate outputs can be connected in parallel, with only one being activated at a time. This alternative is shown in Fig. 8.16.

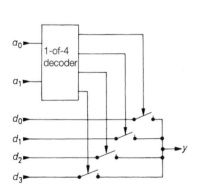

Fig. 8.13. Basic multiplexer operation

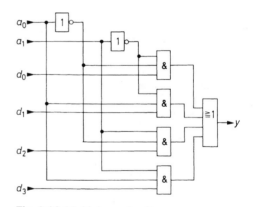

Fig. 8.14. Multiplexer circuit
$$y = \bar{a}_0\bar{a}_1 d_0 + a_0\bar{a}_1 d_1 + \bar{a}_0 a_1 d_2 + a_0 a_1 d_3$$

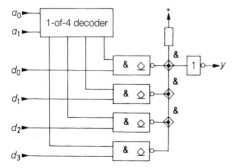

Fig. 8.15. Multiplexer with open-collector gates

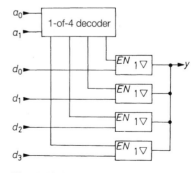

Fig. 8.16. Multiplexer with tristate gates

Inputs	TTL	ECL	CMOS digital	CMOS analog
16	74 LS 150		4515	4067
2 × 8				4097
8	74 LS 151	10164	4512	4051
2 × 4	74 LS 153	10174	4539	4052
8 × 2	74 LS 604			
4 × 2	74 LS 157	10159	4519	4066

Fig. 8.17. Integrated multiplexers: "CMOS analog" means a multiplexer/demultiplexer with transmission gates

Although the possible implementations of the OR operation shown in Figs. 8.15 and 8.16 are not employed in integrated multiplexers, they are useful if the signal sources of the multiplexer are spatially distributed. Arrangements of this kind are found in bus systems of computers. Some commonly used multiplexers are listed in Fig. 8.17.

8.3
Priority Decoder

The 1-of-n code can be converted to binary code by using *a priority decoder*. At its outputs a binary number appears that corresponds to the highest input number, which is logic 1. The value of the lower-index input variables is irrelevant; hence the name *priority decoder*. This property enables the circuit to convert not only the 1-of-n code but also a sum code in which not just one variable is 1, but also the less significant bits as it is the matter at parallel AD-converters in Chap. 18.10.1. The truth table for the priority decoder is shown in Fig. 8.18.

IC types:
1-of-10 code: SN 74147 (TTL)
1-of-8 code, extendable: SN 74148 (TTL); MC 10165 (ECL);
                            MC 14532 (CMOS)

J	x_9	x_8	x_7	x_6	x_5	x_4	x_3	x_2	x_1	y_3	y_2	y_1	y_0
0	0	0	0	0	0	0	0	0	0	0	0	0	0
1	0	0	0	0	0	0	0	0	1	0	0	0	1
2	0	0	0	0	0	0	0	1	×	0	0	1	0
3	0	0	0	0	0	0	1	×	×	0	0	1	1
4	0	0	0	0	0	1	×	×	×	0	1	0	0
5	0	0	0	0	1	×	×	×	×	0	1	0	1
6	0	0	0	1	×	×	×	×	×	0	1	1	0
7	0	0	1	×	×	×	×	×	×	0	1	1	1
8	0	1	×	×	×	×	×	×	×	1	0	0	0
9	1	×	×	×	×	×	×	×	×	1	0	0	1

Fig. 8.18. Truth table for a priority decoder: × $\hat{=}$ don't care

8.4
Combinatorial Shift Register (Barrel Shifter)

For many arithmetic operations, a bit pattern must be shifted by one or more binary digits. This operation is usually carried out by a shift register, as described in Sect. 9.5. A single clock pulse results in a shift by one bit. There is a disadvantage, however, in that a sequential controller is necessary to organize the loading of the bit pattern into the shift register and the subsequent shifting by a given number of binary digits.

The same operation may be carried out without a clocked by employing a combinatorial network involving multiplexers, as illustrated in Fig. 8.19. For this reason, the unclocked shift registers involved are termed combinatorial or asynchronous shift registers or barrel shifter that are mainly used in signal processors. If, in Fig. 8.19, the address $A = 0$ is applied, then $y_3 = x_3$, $y_2 = x_2$, and so on, but if $A = 1$, then $y_3 = x_2$, $y_2 = x_1$, $y_1 = x_0$, and $y_0 = x_1$, due to the wiring arrangement of the multiplexers. The bit pattern X therefore appears at the output left-shifted by one digit. As with a normal shift register, the MSB is lost. If multiplexers with N inputs are used, a shift of 0, 1, 2, ..., $(N - 1)$ bits can be executed. For the example shown in Fig. 8.19, $N = 4$; the corresponding function table is shown in Fig. 8.20.

If the loss of MSBs is to be avoided, the shift register may be extended by adding identical elements, as illustrated in Fig. 8.21. For the chosen example, where $N = 4$, a 5-bit number X can be shifted in this way by a maximum of 3 bits without loss of information. The shifted number then appears at outputs $y_3 - y_7$.

The circuit shown in Fig. 8.19 can also be operated as a ring shifter if the extension inputs $x_{-1} \ldots x_{-3}$ are connected to the inputs $x_1 \ldots x_3$, as shown in Fig. 8.22.

IC Types:

16-bit (TTL):   SN 74 AS 897   from Texas Instruments

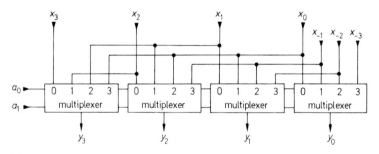

Fig. 8.19. Barrel shifter realized with multiplexers

a_1	a_0	y_3	y_2	y_1	y_0
0	0	x_3	x_2	x_1	x_0
0	1	x_2	x_1	x_0	x_{-1}
1	0	x_1	x_0	x_{-1}	x_{-2}
1	1	x_0	x_{-1}	x_{-2}	x_{-3}

Fig. 8.20. Function table for the barrel shifter

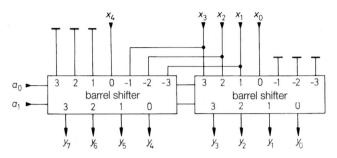

Fig. 8.21. Extended barrel shifter

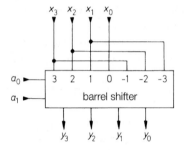

Fig. 8.22. Ring shifter

8.5
Digital Comparators

Comparators check two numbers, A and B, against one another, the relations of interest being $A = B$, $A > B$, and $A < B$. We shall first consider comparators that determine whether two binary numbers are equal (identity comparators). The criterion for this is that all corresponding bits of the two numbers are identical. The comparator will produce a logic 1 at its output if the numbers are equal, and otherwise a logic 0. In the simplest case, the two numbers consist of only one bit each; to compare them, the EQUIV operation (the exclusive NOR gate) may be used. Two N-bit numbers are compared bit by bit using an EQUIV circuit for each binary digit, and the outputs are combined by an AND gate, as shown in Fig. 8.23.

IC Types:

2×8 inputs: SN 74 LS 688 (TTL) from Texas Instruments

Comparators have a wide range of application if, in addition to indicating equality, they can also determine which of two numbers is the larger. Such circuits are known as magnitude comparators. To enable a comparison of the magnitudes of two numbers, their

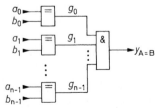

Fig. 8.23. Identity comparator for two N-bit numbers

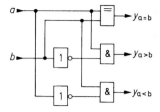

Fig. 8.24. 1-bit magnitude comparator

a	b	$y_{a>b}$	$y_{a=b}$	$y_{a<b}$
0	0	0	1	0
0	1	0	0	1
1	0	1	0	0
1	1	0	1	0

Fig. 8.25. Truth table for a 1-bit magnitude comparator

codes must be known, for the following, we assume that both numbers are straight binary coded, in other words that

$$A = a_N \cdot 2^N + a_{N-1} \cdot 2^{N-2} + \ldots + a_1 \cdot 2^1 + a_0 \cdot 2^0 .$$

The simplest case is again that of comparing two single-bit numbers. The formulation of the logic functions is based on the truth table of Fig. 8.25. From these, we can directly obtain the circuit shown in Fig. 8.24.

The following algorithm is used for comparing numbers that consist of more than one bit: To begin with, the most significant bit (MSB) of A is compared with the MSB of B. If they are different, these bits are sufficient to determine the result. If they are equal, the next lower significant bit must compared, and so on. If the identity variable of digit i is denoted by g_i, as in Fig. 8.23, the magnitude comparison of an N-digit number is given by the general relation

$$y_{A>B} = a_{N-1} \cdot \overline{b}_{N-1} + g_{N-1} \cdot a_{N-2} \cdot \overline{b}_{N-2} + \ldots$$
$$+ g_{N-1} \cdot g_{N-2} \cdot \ldots \cdot g_1 \cdot a_0 \cdot \overline{b}_0$$

IC Types:
for 5-digit comparison: MC10166 (ECL)
for 8-digit comparison: SN 74 LS 682–689 (TTL).

The circuits can be cascaded serially or in parallel; the serial method is shown in Fig. 8.26. When the three most significant bits are the same, the outputs of comparator 1 determine the result, as they are connected to the LSB inputs of comparator 2 (LSB = least significant bit).

When comparing many-bit numbers it is better to employ parallel cascading, as shown in Fig. 8.27, since the propagation delay time is shorter.

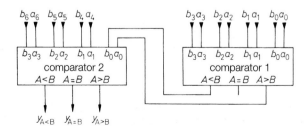

Fig. 8.26. Serial magnitude comparator

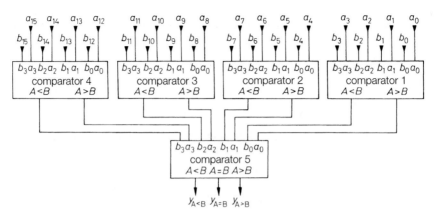

Fig. 8.27. Parallel magnitude comparator

8.6
Adders

8.6.1
Half-Adder

Adders are circuits that give the sum of two binary numbers. We shall first describe adders for straight binary numbers. The simplest case is the addition of two single-bit numbers. To devise the logic circuit, all possible cases must first be investigated so that a logic function table can be compiled. If two single-bit numbers A and B are to be added, the following cases can occur:

$$0 + 0 = 0,$$
$$0 + 1 = 1,$$
$$1 + 0 = 1,$$
$$1 + 1 = 10$$

If both A and B are 1, a carry to the next higher bit is obtained. The adder must therefore have two outputs, one for the sum and one for the carry to the next higher bit. The truth table shown in Fig. 8.29 can be deduced by expressing the numbers A and B by the logic variables a_0 and b_0. The carry is represented by the variable c_1, and the sum by the variable s_0.

By setting up the canonical products, the Boolean functions

$$c_1 = a_0 b_0 \quad \text{and} \quad s_0 = \bar{a}_0 b_0 + a_0 \bar{b}_0 = a_0 \oplus b_0$$

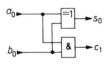

a_0	b_0	s_0	c_1
0	0	0	0
0	1	1	0
1	0	1	0
1	1	0	1

Fig. 8.28. Circuit for the half-adder

Fig. 8.29. Truth table for the half-adder

are obtained. The carry thus represents an AND operation, and the sum an exclusive OR operation. A circuit that implements both operations is known as a half-adder and is shown in Fig. 8.28.

8.6.2
Full-Adder

If two straight binary numbers of more than one digit are to be added, the half-adder can only be used for the LSB. For all other binary digits, not two but three bits must be added, as the carry from the next lower binary digit must be included. In general, each bit requires a logic circuit with three inputs, a_i, b_i, and c_i, and two outputs, s_i and c_{i+1}. Such circuits are called full-adders and can be implemented as shown in Fig. 8.30, using two half-adders. Their truth table is given in Fig. 8.31.

For each bit, a full-adder is required, but for the LSB a half-adder is sufficient. Figure 8.32 shows a circuit suitable for adding two 4-bit numbers, A and B. Such circuits are

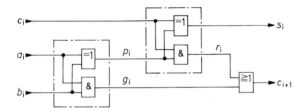

Fig. 8.30. Full-adder.
$$s_i = a_i \oplus b_i \oplus c_i$$
$$c_{i+1} = a_i b_i + a_i c_i + b_i c_i$$

Input			Internal			Output		Decimal
a_i	b_i	c_i	p_i	g_i	r_i	s_i	c_{i+1}	Σ
0	0	0	0	0	0	0	0	0
0	1	0	1	0	0	1	0	1
1	0	0	1	0	0	1	0	1
1	1	0	0	1	0	0	1	2
0	0	1	0	0	0	1	0	1
0	1	1	1	0	1	0	1	2
1	0	1	1	0	1	0	1	2
1	1	1	0	1	0	1	1	3

Fig. 8.31. Truth table for the full-adder

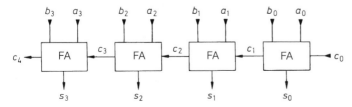

Fig. 8.32. 4-bit adder with ripple carry

available as ICs, but usually a full-adder is also used for the LSB, in order to enable the circuit to be extended as required (SN 74 LS 83).

8.6.3
Look-Ahead Carry Logic

The computing time of the adder in Fig. 8.32 is considerably longer than that of the individual stages, because the carry c_4 can assume its correct value only after c_3 has been determined. The same applies to all the previous carries (ripple carry). To shorten the computing time for the addition of many-bit straight binary numbers, a look-ahead carry generator (parallel or simultaneous carry logic) can be used. In this method, all carries are determined directly from the input variables. From the truth table in Fig. 8.31, the general relation for the carry of stage i can be deduced:

$$c_{i+1} = \underbrace{a_i b_i}_{g_i} + \underbrace{(a_i \oplus b_i)}_{p_i} c_i \tag{8.3}$$

The quantities g_i and p_i introduced for brevity appear as intermediate variables in the full-adder of Fig. 8.30. Their calculation therefore requires no additional complexity. These variables can be interpreted as follows: the quantity g_i indicates whether or not the input combination a_i, b_i results in a carry in stage i, and is therefore called the generate variable. The quantity p_i indicates whether the input combination causes a carry from the next lower-order stage to be absorbed or passed on. It is therefore called the propagate variable. From (8.3), we obtain successively the individual carries

$$
\begin{aligned}
c_1 &= g_0 + p_0 c_0, \\
c_2 &= g_1 + p_1 c_1 = g_1 + p_1 g_0 + p_1 p_0 c_0, \\
c_3 &= g_2 + p_2 c_2 = g_2 + p_2 g_1 + p_2 p_1 g_0 + p_2 p_1 p_0 c_0, \\
c_4 &= g_3 + p_3 c_3 = g_3 + p_3 g_2 + p_3 p_2 g_1 + p_3 p_2 p_1 g_0 + p_3 p_2 p_1 p_0 c_0
\end{aligned}
\tag{8.4}
$$

$$\vdots \qquad \vdots$$

It can be seen that the expressions become ever more complicated, but that they can be computed from the auxiliary variables within 2 propagation delay times: one for the products and one for the sum.

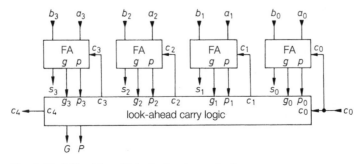

Fig. 8.33. 4-bit adder with look-ahead carry logic

Figure 8.33 shows the block diagram of a 4-bit adder with look-ahead carry logic. The equations given in (8.4) are implemented in the carry generator. The complete circuit is available on a single chip.

IC Types:

TTL: SN 74 LS 181; SN 74 S 281; SN 74 LS 381; SN 74 LS 382; SN 74 LS 681

Adder networks for more than 4 bits can be realized by cascading several 4-bit blocks. The carry c_4 would then be applied as c_0 to the next block up. However, this method is somewhat inconsistent, because the carry is parallel-processed within the blocks but serially processed between the blocks.

To obtain short computation times, the carries from block to block must therefore also be parallel-processed. The relationship for c, in (8.4) is thus reconsidered:

$$c_4 = \underbrace{g_3 + p_3 g_2 + p_3 p_2 g_1 + p_3 p_2 p_1 g_0}_{G} + \underbrace{p_3 p_2 p_1 p_0}_{P} c_0 \tag{8.5}$$

To abbreviate this, the block-generate variable G and the block-propagate variable P are introduced, and

$$c_4 = G + P c_0$$

is obtained. The form of this equation is the same as (8.3). Within the individual 4-bit adder blocks, only the additional auxiliary variables G and P need be computed; when these are known, the algorithm given in (8.4), and used for the bit-to-bit carries, can also be used for the carries from block to block. The result is the block diagram given in Fig. 8.34 for a 16-bit adder with look-ahead carry logic. The carry logic is identical to that of the 4-bit adder in Fig. 8.33. It can be obtained as a separate IC. When performing a 16-bit addition with TTL circuits, the computation time is 36 ns; for Schottky-TTL circuits, it is reduced to 19 ns.

IC carry blocks:
For four digits: SN 74182 (TTL), MC 10179 (ECL), MC 14582 (CMOS)
For eight digits: SN 74LS 882 (TTL)

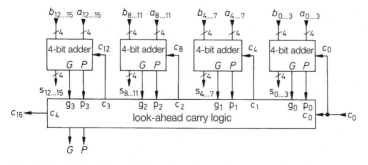

Fig. 8.34. 16-bit adder with look-ahead carry logic on two levels

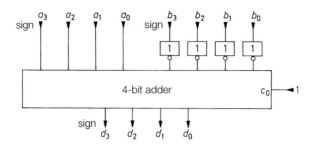

Fig. 8.35. Subtraction of two's complement numbers $D = A - B$

8.6.4
Subtraction

The subtraction of two numbers can be reduced to an addition, since

$$D = A - B = A + (-B) \tag{8.6}$$

If the numbers are represented in 2's-complement, for a specified word length N, we can derive, from (8.2), the simple relation

$$-B_N = B_N^{(2)} = B_N^{(1)} + 1$$

The difference is therefore

$$D_N = A_N + B_N^{(2)}$$

To calculate the difference, we therefore have to form the 2's-complement of B_N and add it to A_N. For this purpose, we must negate all the digits of B_N (1's-complement) and add 1, as stated in (8.1). The addition of A_N and 1 can be performed by one and the same adder, by utilizing the carry input. This results in the 4-bit circuit shown in Fig. 8.35.

In order to insure that the difference D_N appears in the correct 2's-complement notation, A_N and B_N must likewise be entered in this format; that is, for positive numbers the highest bits a_3 and b_3 must be 0.

The 181-series integrated adders described in Sect. 8.6.3 have control inputs that enable the input numbers to be complemented. They are therefore also suitable as subtractors. As further control inputs can be used to select logic operations for the input variables, these devices are generally known as arithmetic logic units (ALUs).

8.6.5
Two's-Complement Overflow

When two positive N-bit binary numbers are added, the result can be an $(N+1)$-bit number. This overflow is recognizable from the fact that, from the most significant bit, a carry is produced.

In 2's-complement notation, the leftmost digit position is reserved for the sign. When two negative numbers are added, a carry into the overflow position will systematically occur, as the sign bit of the two numbers is 1. When processing 2's-complement numbers of either sign, the occurrence of a carry into the overflow position does not therefore necessarily mean that an overflow has taken place.

An overflow can be detected as follows: when two positive numbers are added, the result must also be positive. If the sum is out of range, a carry into the overflow position

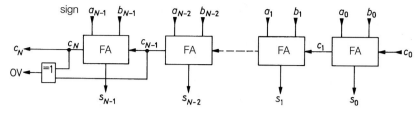

Fig. 8.36. Calculating the 2's-complement overflow OV

occurs; that is, the result becomes "negative". This indicates a positive overflow. Similarly, a negative overflow is present if a "positive" result is obtained when two negative numbers are added. When a positive and a negative number are added, no overflow can occur, as the magnitude of the difference is then smaller than the numbers entered.

The occurrence of a 2's-complement overflow can be easily detected by comparing the carry c_{N-1} with the carry c_N (Fig. 8.36). An overflow has taken place precisely when these two carries are different. This case is decoded by the exclusive-OR gate. This output is available on the 4-bit arithmetic unit SN 74 LS 382.

8.6.6
Addition and Subtraction of Floating-Point Numbers

When processing floating-point numbers, the mantissa and exponent must be handled separately. For addition, it is first necessary to adjust the exponents so that they are the same. To do this, we take the difference of the exponents and shift the mantissa associated with the smaller exponent by the corresponding number of bits to the right (see Sect. 8.4). Both numbers then have the same – namely, the larger – exponent. It is passed on to the output via the multiplexer shown in Fig. 8.37. The two mantissas can now he added or subtracted, generally producing a nonnormalized result; that is, the leading 1 in the mantissa is not

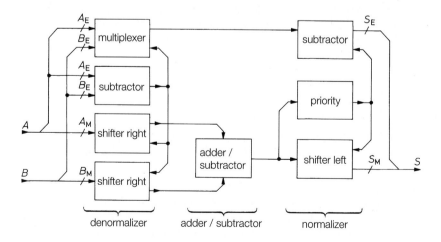

Fig. 8.37. Arrangement for adding/subtracting floating-point numbers A and B (indices: E = exponent, M = mantissa)

in the correct position. To normalize the result, the leftmost 1 in the mantissa is located using a priority decoder (see Sect. 8.3). The mantissa is then shifted to the left by the corresponding number of bits and the exponent is reduced accordingly.

8.7
Multipliers

8.7.1
Multiplication of Fixed-Point Numbers

Multiplication of two straight binary numbers is best illustrated by an example. The product $13 \cdot 11 = 143$ is to be calculated:

$$
\begin{array}{r}
1\,1\,0\,1 \quad \cdot \quad 1\,0\,1\,1 \\
\hline
1\,1\,0\,1 \\
+ \qquad 1\,1\,0\,1 \\
+ \qquad 0\,0\,0\,0 \\
+ \quad \;1\,1\,0\,1 \\
\hline
1\,0\,0\,0\,1\,1\,1\,1
\end{array}
$$

This calculation is particularly easy, because only multiplications by either 1 or 0 occur. The product is obtained by consecutive shifting of the multiplicand to the left by 1 bit at a time, and by adding or not adding, depending on whether the corresponding multiplier bit is 1 or 0. The individual bits are processed consecutively, and this method is therefore known as serial multiplication.

Multiplication can be implemented by combining a shift register and an adder, although such a circuit would require a sequential controller. If the adders are suitably staggered and interconnected, the shifting process can be carried out by wiring. This method requires a large number of adders, but the shift register and the sequential controller are no longer needed. The main advantage, however, is the considerably reduced computation time since, instead of the time-consuming clock control, only gate propagation delays are incurred.

Figure 8.38 shows a suitable circuit for a combinatorial 4×4 bit multiplier. For the additions, we can usefully employ the SN 74 LS 381 chip, whereby addition can be activated and deactivated by the mode control input, giving

$$
S = \begin{cases} A + 0 & \text{für } m = 0 \\ A + B & \text{für } m = 1 \end{cases}
$$

The multiplier is applied bit by bit to the control inputs m. The multiplicand is fed in parallel to the four addition inputs b_0 to b_3.

To begin with, we assume that the number $K = 0$. We then obtain the expression

$$
S_0 = X \cdot y_0
$$

at the output of the first element, corresponding to the first partial product in the above multiplication algorithm. The LSB of S_0 represents the LSB of the product P; it is transferred directly to the output. The more significant bits of S_0 are added in the second element to the expression $X \cdot y_1$. The resulting sum is the subtotal of the partial products in the first and second line of the multiplication algorithm. The LSB of this sum gives the second lowest bit of P and is therefore transferred to the output p_1. The subsequent subtotals are

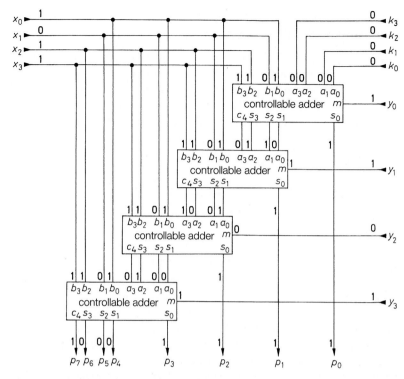

Fig. 8.38. Multiplier for two 4-bit numbers. The values entered refer to the example $13 \cdot 11 = 143$.
Result: $P = X \cdot Y + K$

treated accordingly. We have entered the numbers of the above example in Fig. 8.38 to demonstrate this process.

The additional inputs $k_0 - k_3$ can be used to add a 4-bit number K to the total product P, so that the multiplier function becomes

$$P = X \cdot Y + K .$$

The method of expansion for larger numbers can now be understood. For each additional bit of the multiplier Y, a further arithmetic unit is added at the bottom left of the circuit. If the multiplicand X is to be increased, the word length is enlarged by cascading an appropriate number of arithmetic units at each stage.

In the multiplication method described, the new partial product is always added to the previous subtotal. This technique requires the fewest elements and results in straightforward and easily extensible circuitry. The computing time can be shortened if as many summations as possible are carried out simultaneously, and if the individual subtotals are added afterwards by a fast adder circuit. Several procedures are available, which differ only in the adding sequence (the Wallace Tree). Another way of reducing the computing time is to use the Booth algorithm. The multiplier bits are combined into pairs, thereby halving the number of adders required, and the computing time is reduced accordingly.

8.7.2
Multiplication of Floating-Point Numbers

To multiply floating-point numbers, the mantissas of both numbers have to be multiplied and their exponents added, as shown in Fig. 8.39. During this process, an overflow may occur in the mantissa. The result can be renormalized by shifting the mantissa one place to the right and increasing the exponent by 1. In this case, denormalization as used with the floating-point adder in Fig. 8.37 is not required; the complexity lies in the multiplier.

Formerly many hardware multiplier chips have been available. Now the multipliers shown are part of CPUs and signal processors like those in Fig. 19.59.

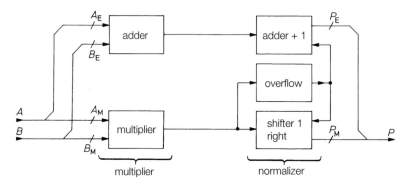

Fig. 8.39. Multiplication of floating-point numbers (indices: E exponent, M = mantissa)

Chapter 9:
Sequential Logic Systems

A sequential logic system is an arrangement of digital circuits that can carry out logic operations and, in addition, store the states of individual variables. It differs from a combinatorial logic system in that the output variables y_j are not only dependent on the input variable x_i, but also on the previous history, which is represented by the state of flip-flops. In what follows, we shall first discuss the design and operation of integrated flip-flops.

9.1
Integrated Flip-Flops

We described simple transistor flip-flops in Sect. 6.2.1. We shall now demonstrate the operation of flip-flops with reference to gates. This approach defines their basic operation irrespective of the particular technology employed.

9.1.1
Transparent Flip-Flops

By connecting two NOR gates in a feedback arrangement, as shown in Fig. 9.1, we obtain a flip-flop that has complementary outputs Q and $\overline{Q}$ and two inputs S (Set) and R (Reset).

If the complementary input state $S = 1$ and $R = 0$ is applied, we have

$$\overline{Q} = \overline{S + Q} = \overline{1 + Q} = 0$$

and

$$Q = \overline{R + \overline{Q}} = \overline{0 + 0} = 1$$

The two outputs therefore assume complementary states. Similarly, for $R = 1$ and $S = 0$, the opposite output state is obtained. If we make $R = S = 0$, the old output state is retained. This explains why RS flip-flops are used as memories. When $R = S = 1$, the two outputs simultaneously become zero; however, the output state is no longer defined when R and S simultaneously become zero. Consequently, the input state $R = S = 1$ is generally avoided. The switching states are summarized in the truth table shown in Fig. 9.2, with which we are already familiar from the transistor circuit in Fig. 6.10 on page 591.

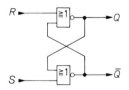

S	R	Q	$\overline{Q}$
0	0	Q_{-1}	$\overline{Q}_{-1}$
0	1	0	1
1	0	1	0
1	1	(0)	(0)

Fig. 9.1. *RS* flip-flop consisting of NOR gates

Fig. 9.2. Truth table for an *RS* flip-flop

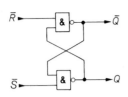

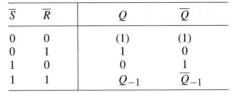

$\overline{S}$	$\overline{R}$	Q	$\overline{Q}$
0	0	(1)	(1)
0	1	1	0
1	0	0	1
1	1	Q_{-1}	$\overline{Q}_{-1}$

Fig. 9.3. *RS* flip-flop consisting of NAND gates

Fig. 9.4. Truth table for an *RS* flip-flop comprising NAND gates

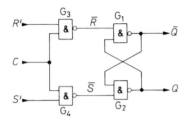

Fig. 9.5. Statically clocked *RS* flip-flop

In Sect. 7.1 on page 611 we showed that a logic equation does not change if all the variables are negated and the arithmetic operations (+) and (·) are interchanged. Applying this rule here, we arrive at the *RS* flip-flop comprising NAND gates shown in Fig. 9.3, which has the same truth table as that shown in Fig. 9.2. However, note that the input variables are now $\overline{R}$ and $\overline{S}$. As we shall be using the *RS* flip-flop comprising NAND gates frequently, we have given its truth table for input variables $\overline{R}$ and $\overline{S}$ in Fig. 9.4.

Clocked RS Flip-Flops

We frequently require an *RS* flip-flop that only reacts to the input state at a specific moment in time, which is determined by an additional clock variable C. Figure 9.5 shows a statically clocked *RS* flip-flop of this kind. If $C = 0$, then $\overline{R} = \overline{S} = 1$. In this case, the flip-flop stores the old state. For $C = 1$, we obtain

$$R = R' \quad \text{and} \quad S = S'$$

The flip-flop then behaves like a normal *RS* flip-flop.

Clocked D Flip-Flops

We shall now examine how the value of a logic variable D can be stored using the flip-flop shown in Fig. 9.5. We have seen that $Q = S$ if complementary input states are applied and $C = 1$. In order to store the value of a variable D, we therefore need only make $S = D$ and $R = \overline{D}$. The inverter G_5 in Fig. 9.6 is used for this purpose. In the resulting data latch, $Q = D$ as long as clock $C = 1$. This may also be seen from the truth table in Fig. 9.7. Due to this property, the clocked data latch is also known as a transparent D flip-flop. If we make $C = 0$, the existing output state is stored.

We can see that NAND gate G_4 in Fig. 9.6 acts as an inverter for D when $C = 1$. Inverter G_5 can therefore be omitted, producing the practical implementation of a D latch shown in Fig. 9.8. The circuit symbol is given in Fig. 9.9.

IC types:
74 LS 75 (TTL);   10133 (ECL);   4042 (CMOS)

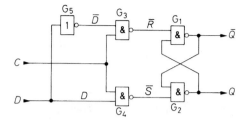

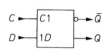

C	D	Q
0	0	Q_{-1}
0	1	Q_{-1}
1	0	0
1	1	1

Fig. 9.6. Transparent D flip-flop (D latch)

Fig. 9.7. Truth table for the transparent D-flip-flop

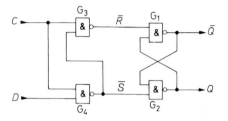

Fig. 9.8. Practical implementation of a transparent D flip-flop

Fig. 9.9. Circuit symbol for a transparent D flip-flop

9.1.2
Flip-Flops with Intermediate Storage

For many applications, such as counters and shift registers, transparent flip-flops are unsuitable. In these cases, flip-flops are required that temporarily store the input state and only transfer it to the output when the inputs are inhibited. There is no clock state where the flip-flops are transparent: this means where data are forewarded from the input to the output. They therefore comprise two flip-flops: the master flip-flop at the input and the slave flip-flop at the output.

Two-Edge-Triggered Flip-Flops

Figure 9.10 shows a master–slave flip-flop of this kind. It consists of two statically clocked RS flip-flops of the type shown in Fig. 9.5. The two flip-flops are mutually inhibited by complementary clock signals. Gate G_{15} is used for clock inversion. As long as clock $C = 1$,

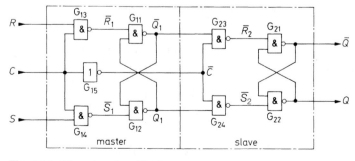

Fig. 9.10. RS master–slave flip-flop

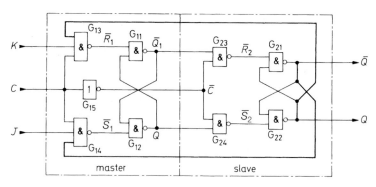

Fig. 9.11. *JK* master–slave flip-flop

the input information is read into the master. The output state remains unchanged, because the slave is disabled.

When the clock goes to zero, the master is disabled, thereby freezing the state that was present immediately prior to the negative-going edge of the clock signal. The slave is simultaneously enabled, thus transferring the state of the master to the output. Data transmission therefore occurs on the negative-going edge; however, there is no clock state in which the input data have a direct effect on the output, as is the case with transparent flip-flops.

The input combination $R = S = 1$ necessarily results in undefined behavior, because inputs $\overline{S}_1$, $\overline{R}_1$ in the master simultaneously go from 00 to 11 when clock C goes to zero. In order to make use of this input combination, the complementary output data are additionally applied to the input gates. The feedback circuit shown in heavy type in Fig. 9.11 is used for this purpose. The external inputs are then designated J and K respectively. For $J = K = 1$ at each clock cycle complementary data are applied to the master flip-flop. So this input combination is not longer forbidden but results in toggling the flip-flop. This can also be seen in the truth table in Fig. 9.13. This is the same as dividing the frequency by two, as shown in Fig. 9.12. Consequently, JK master–slave flip-flops provide a particularly simple means of constructing counters.

However, because of the feedback, operation of the JK flip-flop is subject to an important *limitation*: the truth table in Fig. 9.13 only applies if the state at the JK inputs remains unchanged as long as clock $C = 1$. This is because, unlike the RS master–slave flip-flop in Fig. 9.10, the master–slave flip-flop here can only change state once but cannot change back, as one of the two input NAND gates is always disabled by the feedback. Failure to observe this limitation is a frequent source of errors in digital circuits.

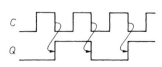

J	K	Q
0	0	Q_{-1} (unchanged)
0	1	0
1	0	1 $\quad (Q = J)$
1	1	$\overline{Q}_{-1}$ (inverted)

Fig. 9.12. *JK* master–slave flip-flop as a frequency divider ($J = K = 1$)

Fig. 9.13. Output state of a *JK* master–slave flip-flop after a (010) clock cycle

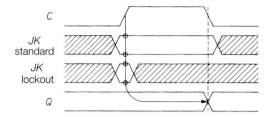

Fig. 9.14. Timing diagram of the input and output signals of JK master–slave flip flops

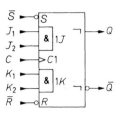

Fig. 9.15. Circuit symbol of a JK master–slave flip-flop

Special types of JK master–slave flip-flops are available which are not subject to this limitation. They are provided with data lockout: the input state is read only during the positive-going edge. Immediately after this edge, the two input gates are disabled and no longer react to changes in the input states [10.1]. This is shown in Fig. 9.14. Whereas with normal JK flip-flops the J and K inputs must not change as long as clock $C = 1$, with a data lockout JK flip-flop they must remain constant only during the positive-going edge of the clock signal. The common feature of both flip-flops is that the information read in on the positive-going edge of the clock signal does not appear at the output until the negative-going edge. Due to this delay, the circuit symbol in Fig. 9.15 additionally has a delay sign at the outputs.

JK flip-flops frequently have several J and K inputs leading to an internal AND gate. The internal J and K variables are then only 1 when all of the respective J and K inputs are 1.

In addition to the JK inputs, the JK flip-flops additionally possess Set and Reset inputs, which operate independently of the clock—that is, asynchronously. This enables master slave flip-flops to be set or cleared. The RS inputs have priority over the JK inputs. In order to allow clock-controlled operation, there must be $R = S = 0$ or $\overline{R} = \overline{S} = 1$.

Typical IC types:

	TTL	ECL	CMOS
Standard	7476	10135	4027
Data lockout	74 LS 111		

Single-Edge-Triggered Flip-Flops

Flip-flops with intermediate storage can also be implemented by connecting two transparent D flip-flops (Fig. 9.8) in series and clocking them with complementary signals. This produces the circuit shown in Fig. 9.16. As long as clock $C = 0$, the master follows the input signal and we have $Q_1 = D$. Meanwhile, the slave stores the old state. When the clock goes to 1, the data D present at that instant is frozen in the master and transferred to the slave, and thus to the Q output. The information present at the D input on the positive-going edge of the clock signal is therefore instantaneously transmitted to the Q output. The state of the D input has no effect for the rest of the time. This can also be seen from Fig. 9.17. Instead of waiting for the negative-going edge, as in the JK flip-flop with data lockout, the input value appears at the output immediately. For this reason, the circuit symbol in Fig. 9.18 also has no delay symbols. This constitutes a significant advantage, in that the entire clock cycle is now available for forming the new D signal. If JK flip-flops

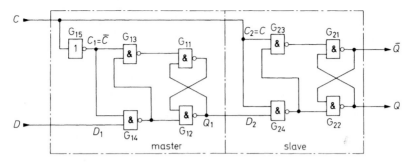

Fig. 9.16. Single-edge-triggered D flip-flop

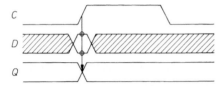

Fig. 9.17. Timing diagram for the input and output signals in the single-edge-triggered D flip-flop

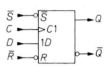

Fig. 9.18. Circuit symbol for the single-edge-triggered D flip-flop

are used, this process must take place while the clock is zero; that is, with symmetrical clock pulses in half the time.

Examples of IC types:
74 LS 74 (TTL);   10131 (ECL);   4013 (CMOS)

Single-edge-triggered D flip-flops can also be operated as toggle flip-flops. For this purpose, we make $D = \overline{Q}$, as shown in Fig. 9.19. The output state therefore inverts at each positive-going edge of the clock signal. This is illustrated in Fig. 9.20. If transparent D flip-flops were used, an oscillation would be obtained while clock $C = 1$, instead of a frequency division. This is caused by the transparent propagation of the signal through the circuit, resulting in a signal inversion after every propagation delay time.

It is also possible to make the inversion dependent on a control variable by providing feedback from either $\overline{Q}$ or Q to the D input via a multiplexer. The latter is controlled by the toggle input T shown in Fig. 9.21. The same mode of operation is possible using the JK flip-flop shown in Fig. 9.22, with interconnected JK inputs.

Multipurpose flip-flops can be obtained by additionally providing synchronous data input. The multiplexer can then be given another input preceding the D input. This additional input is selected via Load input L as shown in Fig. 9.23. If $L = 1$, then $y = D$ and therefore, after the next clock signal, $Q = D$. When $L = 0$, the circuit operates in exactly

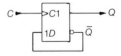

Fig. 9.19. Single-edge-triggered D flip-flop as a frequency divider

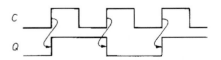

Fig. 9.20. Timing pattern in the frequency divider

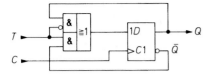

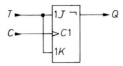

Fig. 9.21. With D flip-flop   **Fig. 9.22.** With JK flip-flop

Figs. 9.21 and **9.22.** Controllable toggle flip-flops

$$Q = \begin{cases} Q_{-1} \\ \overline{Q}_{-1} \end{cases} \quad \text{for} \quad \begin{aligned} T &= 0 \\ T &= 1 \end{aligned}$$

the same way as that in Fig. 9.21. The mode of operation of this multifunction flip-flop is summarized in Fig. 9.25.

The same behavior can also be obtained using a JK flip-flop, as shown in Fig. 9.24, when $L = 1$, $J = D$, or $K = \overline{D}$. Therefore, after the next clock signal, $Q = D$. When $L = 0$, we have $J = K = T$; the circuit then operates like in Fig. 9.22 as toggle flip-flop. In the case of JK flip-flops, it must be remembered that the data have to be present before the positive-going edge of the clock signal, but only appear after the output on the negative-going edge. With normal JK flip-flops (as in Fig. 9.11), it must also be insured that the J and K inputs do not change as long as $C = 1$. During this time, the L, T, and D inputs must therefore also remain unchanged.

Due to their versatility, the multifunction flip-flops in Figs. 9.23 and 9.24 constitute the basic building blocks of counters.

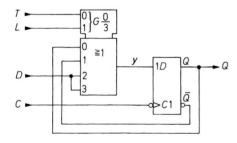

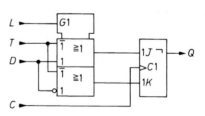

Fig. 9.23. With D flip-flop   **Fig. 9.24.** With JK flip-flop

Figs. 9.23 and **9.24.** Multifunction flip-flops
$T = \text{Toggle}, \quad L = \text{Load}, \quad D = \text{Data}, \quad C = \text{Clock}$

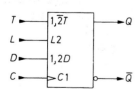

Fig. 9.25. Circuit symbol for a
multifunction flip-flop

L	T	Q
0	0	Q_{-1}
0	1	Q_{-1}
1	0	D
1	1	D

Fig. 9.26. Function table for a multifunction
flip-flop

9.2
Straight Binary Counters

Counters are an important group of sequential logic systems. A counter may be any circuit which, within certain limits, has a defined relationship between the number of input pulses and the state of the output variables. As each output variable can have only two values, for n outputs, there are 2^n possible output combinations, although often only some of these are used. It is not important which number is assigned to which combination, but it is useful to choose a representation that can subsequently be easily processed. The simplest circuits are obtained for straight binary notation.

Figure 9.28 shows the relationship between the number, Z, of input pulses and the values of output variables z_i, for a four-bit straight binary counter. If this table is read from top to bottom, two change conditions can be recognized:

① an output variable z_i always changes state when the next lower value z_{i-1} changes from 1 to 0;

② an output variable z_i always changes state when all lower variables $z_{i-1} \ldots z_0$ have the value 1 and a new pulse arrives.

These conditions can also be seen in the timing diagram in Fig. 9.27. Pattern ① is the basis of an asynchronous counter (ripple counter), whereas pattern ② yields the synchronous counter.

Occasionally, counters are required whose output state is reduced by 1 for each count pulse. The operational principle of such a *down-counter* can also be inferred from the table in Fig. 9.28 by reading it from the bottom up. It follows that:

①a an output variable z_i of a down-counter changes state whenever the next lower variable z_{i-1} changes from 0 to 1;

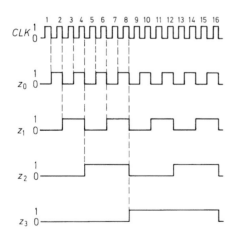

| Z | z_3 | z_2 | z_1 | z_0 |
---	2^3	2^2	2^1	2^0
0	0	0	0	0
1	0	0	0	1
2	0	0	1	0
3	0	0	1	1
4	0	1	0	0
5	0	1	0	1
6	0	1	1	0
7	0	1	1	1
8	1	0	0	0
9	1	0	0	1
10	1	0	1	0
11	1	0	1	1
12	1	1	0	0
13	1	1	0	1
14	1	1	1	0
15	1	1	1	1
16	0	0	0	0

Fig. 9.27. Output states of a straight binary up-counter, as a function of time

Fig. 9.28. State table of a straight binary counter

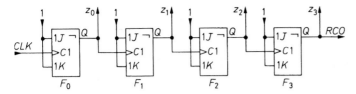

Fig. 9.29. Asynchronous straight binary up-counter
CLK = Clock RCO = Ripple Carry Output

2a) an output variable z_i of a down-counter always changes state when all lower variables $z_{i-1} \ldots z_0$ have the value 0 and a new clock pulse arrives.

9.2.1
Asynchronous Straight Binary Counters

A straight binary *asynchronous* (ripple) counter can be implemented by arranging toggle flip-flops in a chain, as shown in Fig. 9.29, and by connecting each clock input C to the output Q of the previous flip-flops. If the circuit is to be an up-counter, the flip-flops must change their output states when their clock inputs C change from 1 to 0. Edge-triggered flip-flops are therefore required; for example, JK master–slave flip-flops where $J = K = 1$. The counter may be extended to any size. Using this principle, one can count up to 1,023 with only ten flip-flops.

Flip-flops triggered by the positive-going edge of the clock pulse can also be employed; for example, single-edge-triggered D flip-flops. If they are connected in the same way as in Fig. 9.29, down-counter operation is obtained. For up-counter operation, their clock pulse must be inverted. This is achieved by connecting each clock input to the $\overline{Q}$-output of the previous flip-flop.

Every counter is also a frequency divider. The frequency at the output of flip-flop F_0 is half the counter frequency. A quarter of the input frequency appears at the output of F_1, an eighth at the output of F_2, and so on. This property of frequency division can be seen clearly in Fig. 9.27.

IC-types:

Length	TTL	ECL	CMOS
4 bit	74 LS 93	10178	
7 bit			4024
8 bit	74 LS 393		
24 bit			4521
30 bit	74 LS 292		

9.2.2
Synchronous Straight Binary Counters

It is characteristic of an *asynchronous* counter that the clock pulse is applied only to the input of the first flip-flop, while the remaining flip-flops are connected to the previous outputs. This means that the input signal of the last flip-flop does not arrive until all of the preceding stages have changed state. Each change of the output states z_0–z_n is therefore

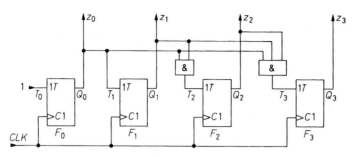

Fig. 9.30. Synchronous straight binary counter

delayed by the set-up time of the previous flip-flops. For long chains and high counter frequencies, this may result in z_n changing with a delay of one or more clock cycles. After the last clock pulse, it is therefore necessary to wait for the delay time of the entire counter chain before the result can be evaluated. If evaluation of the counter state is required during counting, the period of the clock pulse must not be smaller than the delay time of the counter chain.

Synchronous counters do not have these drawbacks, as the clock pulses are applied *simultaneously* to all clock inputs C. In order that the flip-flops do not all change state at every clock pulse, controlled toggle flip-flops – as shown in Figs. 9.21 and 9.22 – are used, which only change state when the control variable $T = 1$. In accordance with Fig. 9.28, a flip-flop of a straight binary counter may only change state when all the lower-order flip-flops are 1. To bring this about, we make $T_0 = 1$, $T_1 = z_0$, $T_2 = z_0 \cdot z_1$ and $T_3 = z_0 \cdot z_1 \cdot z_2$. The AND gates required for this purpose are shown in Fig. 9.30.

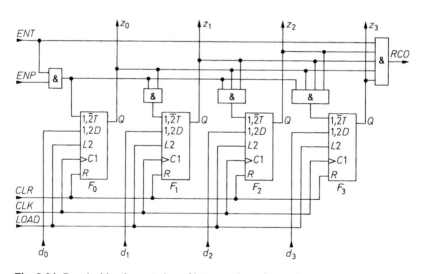

Fig. 9.31. Practical implementation of integrated synchronous counters.

$$ENT = \text{Enable } T \qquad\qquad ENP = \text{Enable } P$$
$$CLR = \text{Clear} \qquad\qquad\qquad CLK = \text{Clock}$$
$$RCO = \text{Ripple Carry Output}$$

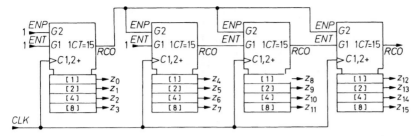

Fig. 9.32. Cascading of synchronous counter stages.
CT = Content

Integrated synchronous counters have yet more inputs and outputs, whose function and application will be described in further detail with reference to Fig. 9.31. The entire counter can be initialized using the Clear input CLR ($Z = 0$). It can be set to any number $Z = D$ via the Load input. Whereas the Clear input mostly operates asynchronously like any Reset input, both synchronous and asynchronous types are available for the load process.

Large (multiple-bit) counters can be implemented by cascading several four-bit counter stages. The stages are connected via the ripple carry output RCO and the enable input ENT, which can be used to inhibit the entire counter stage and the carry output. The latter must therefore go to 1 when a count of 1111 is reached and all the lower-order stages likewise produce a carry. For this to occur, the logic operation

$$RCO = ENT \cdot z_0 \cdot z_1 \cdot z_2 \cdot z_3$$

must be performed in each counter stage. The corresponding output gate is shown in Fig. 9.31.

To cascade the counter stages, it is merely necessary to connect the ENT input of a stage to the RCO output of the next lower-order stage. However, as the delays are cumulative due to the cascaded AND operations, multiple-bit counters are subject to a reduction in the maximum possible counting frequency. In this case, it is preferable to perform the required AND operations in parallel in each counter stage. To do this, the lowest-order stage is omitted from the serial RCO–ENT operation, and the enabling of the higher-order stages is controlled in parallel via the ENP inputs. In this way, the parallel AND operation can be implemented without external gates, as shown in Fig. 9.32.

Typical IC types:

Length	Reset	TTL	ECL	CMOS
4 bit	asynchronous	74 LS 161 A		4161
4 bit	synchronous	74 LS 163 A	10136	4163
8 bit	synchronous	74 LS 590		

9.2.3
Up–Down Counters

A distinction is drawn between up–down counters with one clock input and a second input which determines the mode of counting, and those with two clock inputs, one for incrementing the count, the other for decrementing it.

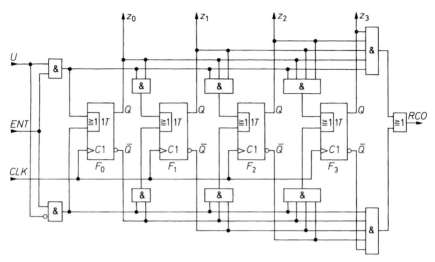

Fig. 9.33. Binary counter with up–down control $U = \begin{cases} 1 & \text{up} \\ 0 & \text{down} \end{cases}$

Counters with Up–Down Control

As may be seen from Fig. 9.28, the switching condition for a down counter is that a flip-flop must change state if all the lower-order bits are zero. In order to decode this, the up-counter logic used in Fig. 9.31 can be connected to the $\overline{Q}$ outputs. In the case of the counter with up–down control shown in Fig. 9.33, either the top part of the up-counter logic or the bottom part of the down-counter logic is enabled via up–down control input U.

A carry into the next higher counter stage can occur in two cases, namely when the count is 1111 during "up" operation ($U = 1$), or when the count is 0000 during "down" operation. The carry variable is therefore given by

$$RCO = [z_0 z_1 z_2 z_3 U + \overline{z}_0 \overline{z}_1 \overline{z}_2 \overline{z}_3 \overline{U}] ENT$$

This variable is applied to the enable input of the next counter stage, as shown in Fig. 9.32. The carry is always interpreted with correct sign if the counting direction is changed over simultaneously for all the counters.

Typical IC types:

Length	TTL	ECL	CMOS
4 bit	74 LS 191	10136	4516
8 bit	74 AS 867		

Counters with Separate Up and Down Inputs

Figure 9.34 shows a counter with two clock inputs for counting up and down respectively. In the previous circuits, the clock signal was fed to all of the flip-flops. Those flip-flops not intended to change state were disabled via control input T. In the case of the counter shown in Fig. 9.34, the clock pulses are prevented from reaching particular flip-flops. An "up" clock signal CUP is only applied to the clock inputs of those flip-flops whose predecessors are at 1. Similarly, a "down" clock signal CDN is only fed to those flip-flops for which all preceding outputs are at 0.

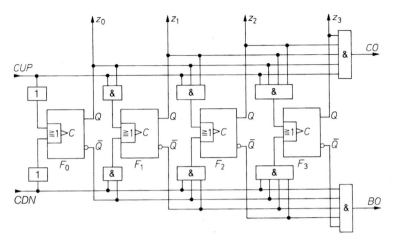

Fig. 9.34. Straight binary counter with clock-up and clock-down inputs. $F_0 \dots F_3$ are toggle flip-flops

CUP = Clock Up	CDN = Clock Down
CO = Carry Output	BO = Borrow Output

As the flip-flops that are to change state receive their clock pulses at virtually the same time, the flip-flops for the more significant bits change state simultaneously with those for the less significant ones. The circuit therefore operates as a synchronous counter. The AND gates at the output determine the carry for up-counter operation and that for down-counter operation. It is possible to connect another identical counter that is in itself synchronous but delayed with respect to the first; that is, operating asynchronously. This mode of operation is termed "semisynchronous."

IC type:
four-bit: 74 LS 193 (TTL)

Coincidence Cancellation

The interval between two count pulses and their duration must not be smaller than the set-up time t_{su} of the counter, or the second pulse would be incorrectly processed. Counters with only one clock input can therefore count at a maximum possible frequency off $f_{max} = 1/2t_{su}$. For the counter shown in Fig. 9.34, the situation is more complicated. Even if the counter frequencies at the up-clock and at the down-clock input are considerably lower than f_{max}, the interval between an up- and a down-clock pulse may, in asynchronous systems, be smaller than t_{su}. Such close or even coinciding pulses result in a spurious counter state. This can be avoided only by preventing these pulses from reaching the counter inputs. The state of the counter then remains unchanged, as would also be the case after one up- and one down-clock pulse.

Such a coincidence cancellation circuit can, for example, be designed as in Fig. 9.35, where one-shots are used [9.2]. The one-shots (monostable multivibrators) M_1 and M_2 convert the counter pulses CUP and CDN into the signals x_{UP} and x_{DN}, each having a defined length t_1. Their trailing edges are used to trigger the two one-shots M_4 and M_5, which in turn generate the output pulses. Gate G_1 decides whether the normalized input pulses x_{UP} and x_{DN} overlap. If this is the case, a positive-going edge appears at its output,

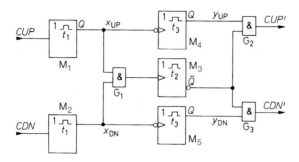

Fig. 9.35. Coincidence cancellation circuit

which triggers one-shot M_3. Both output gates G_2 and G_3 are then disabled for a time t_2, and no pulses can appear at the output. In order that pulses are safely suppressed, the relationship must hold:

$$t_2 > t_1 + t_3$$

Time t_3 defines the duration of the output pulses. The interval between them is shortest just before coincidence is detected; that is, $\Delta t = t_1 - t_3$. For correct operation of the counter, the additional conditions

$$t_3 > t_{su} \quad \text{und} \quad t_1 - t_3 > t_{su}$$

must therefore be fulfilled. The shortest permissible ON-times of the one-shots are thus $t_3 = t_{su}$, $t_1 = 2t_{su}$ and $t_2 = 3t_{su}$. The maximum counter frequency at the two inputs of the coincidence detector is then

$$f_{max} = \frac{1}{t_2} = \frac{1}{3t_{su}}$$

The coincidence cancellation circuit therefore reduces f_{max} by a factor of 1.5. The "anti-race clock generator" in the 40110 counter (CMOS) operates on this principle.

Subtraction Method

A considerably more elegant method consists of counting the up and down pulses in separate counters and then subtracting the results, as shown in Fig. 9.36. Counter pulses that coincide then produce no unwanted effects. A further advantage is that the simpler logic circuitry of an up-counter inherently permits a higher clock frequency.

The carry bit of the subtractor cannot be used to indicate the mathematical sign, as a positive difference would be misinterpreted as being negative if one of the two counters has

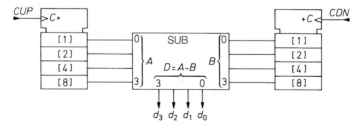

Fig. 9.36. Straight binary up–down counter that is insensitive to coincident clock pulses

an overflow and starts at zero. However, the result is obtained with the correct sign if the difference is interpreted as a four-digit 2's-complement number. Bit d_3 therefore exhibits the correct sign, provided that the difference does not exceed the permissible range of -8 to $+7$.

9.3
BCD Counters

9.3.1
Asynchronous BCD Counters

The table in Fig. 9.28 shows that a three-bit counter can count up to seven, and a four-bit counter up to 15. In a counter for straight BCD numbers, a four-bit straight binary counter used as a decade counter is required for each decimal digit. This decade counter differs from the normal straight binary counter in that it is reset to zero after every tenth count pulse, and it produces a carry. This carry bit controls the decade counter for the next higher decimal digit.

With BCD counters, a decimal display of the count is achieved much more easily than for the straight binary counter, as each decade can be separately decoded and displayed as a decimal digit.

In straight BCD code, each decimal digit is represented by a four-bit straight binary number, the bit weightings of which are $2^3, 2^2, 2^1$ and 2^0. It is therefore also known as the 8421 code. The state table of a decade counter employing the 8421 code is shown in Fig. 9.38. By definition, it must be identical with that in Fig. 9.28 up to the number 9, but the number $10 = 10_{dec.}$ is again represented by 0000. The associated timing diagram for the output variables is shown in Fig. 9.37.

Obviously, additional logic circuitry is required to reset the counter at every tenth input pulse. However, gates may be saved by using JK flip-flops with several J and K inputs, as shown in Fig. 9.39. In contrast to the normal straight binary counter shown in Fig. 9.29, the circuit operates as follows: flip-flop F_1 may not change state at the tenth counting pulse,

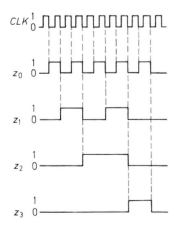

Fig. 9.37. Timing diagram of the output states of an 8421-code counter

Z	z_3	z_2	z_1	z_0
	2^3	2^2	2^1	2^0
0	0	0	0	0
1	0	0	0	1
2	0	0	1	0
3	0	0	1	1
4	0	1	0	0
5	0	1	0	1
6	0	1	1	0
7	0	1	1	1
8	1	0	0	0
9	1	0	0	1
10	0	0	0	0

Fig. 9.38. State table for 8421 code

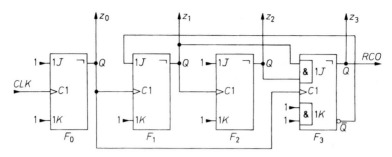

Fig. 9.39. Asynchronous BCD counter

even though z_0 changes from 1 to 0. From Fig. 9.29, we deduce a simple criterion for this case: z_1 must be kept at 0 if z_3 is 1 prior to the clock signal. To achieve this, the J input of F_1 is connected to $\overline{z}_3$. The condition that z_2 must remain 0 at the tenth pulse is therefore automatically satisfied.

The second difference with respect to a straight binary counter is that z_3 changes from 1 to 0 at the tenth pulse. However, if the clock input of F_3 were connected to z_2 as in a normal straight binary counter, z_3 would be unable to change after the eighth counting pulse, since flip-flop F_1 is disabled by the feedback signal. The clock input of F_1, must therefore be connected to the output of the flip-flop that is not disabled by the feedback signal, in this case z_0.

On the other hand, the J inputs must be controlled so that they prevent flip-flop F_3 from changing state with every clock of z_0. Figure 9.38 indicates that z_3 must not go to 1 unless both z_1 and z_2 are 1 prior to the clock signal. This may be achieved by connecting the two J inputs of F_3 to z_1 and z_2 respectively. Then, at the eighth counting pulse, $z_3 = 1$. Since z_1 and z_2 become zero simultaneously, z_3 resumes the state $z_3 = 0$ as soon as possible; that is, at the tenth counting pulse, when z_0 has its next transition from 1 to 0. Figure 9.38 indicates that this is precisely the right instant.

IC types:
4 bit   74 LS 90 (TTL)   10138 (ECL)
2 × 4 bit   74 LS 390 (TTL)

9.3.2
Synchronous BCD Counters

The synchronous decade counter in Fig. 9.40 has largely similar circuitry to the synchronous straight binary counter in Fig. 9.31. As with the asynchronous decade counter, two additional features are again required to insure that, at the transition from $9_{dec} = 1001_2$ to $0_{dec} = 0000_2$, flip-flop F_3 changes state and not flip-flop F_1. The disabling of F_1 is achieved in Fig. 9.40 via the feedback path of $\overline{Q}_3$ and the change of state of F_3 by additionally decoding the 9 at the toggle control input.

Examples of synchronous BCD counters: 74 LS 160 (TTL);   4160 (CMOS);
with up–down control: 74 LS 190 (TTL);   10137 (ECL);   4510 (CMOS);
with an up–down clock input: 74 LS 192 (TTL)

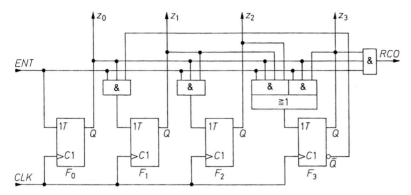

Fig. 9.40. Synchronous BCD counter

9.4
Presettable Counters

Presettable counters are circuits that produce an output signal when the number of input pulses equals a predetermined number M. The output signal can be used to trigger any desired process and is employed to stop the counter or reset it to its initial state. If the counter is allowed to continue counting after reset, it operates as a modulo-M counter, the counting cycle of which is determined by the preselected number M.

The most obvious method of implementing a presettable counter consists of comparing the count Z with the preselected number M, as in Fig. 9.41. For this purpose, we can use an identity comparator, as described in Sect. 8.5 on page 648. If $Z = M$ after M clock pulses, y becomes 1 and the counter is cleared ($Z = 0$). The equality signal y is present for the duration of the clearing process. With an asynchronous CLR input, this time only amounts to a few gate propagation delays. For this reason a synchronous clear input is preferable; then the equality signal is present for precisely one clock period. The counter in Fig. 9.41 therefore returns to zero after $M + 1$ clock pulses. It thus represents a modulo-$(M + 1)$ counter.

The comparator in Fig. 9.41 can be dispensed with by using the LOAD inputs generally provided in synchronous counters (Fig. 9.31). The circuits in Figs. 9.42 and 9.43 make use of this possibility. The counter in Fig. 9.42 is loaded with the number $P = Z_{max} - M$. After M clock pulses, the maximum count Z_{max} is therefore reached, which is internally decoded and produces a carry $RCO = 1$. If this output is connected to the LOAD input as shown in Fig. 9.42, the preset number P is reloaded with clock pulse $M + 1$. Once

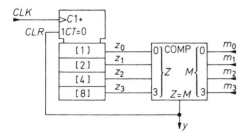

Fig. 9.41. Modulo-$(M + 1)$ counter with comparator

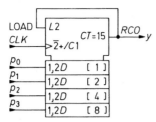

Fig. 9.42. Modulo-($M + 1$) counter with a parallel input of $P = Z_{max} - M$ for $Z = 15$

Fig. 9.43. Modulo-($M + 1$)-counter with a parallel input of M for $Z = 0$, using a down-counter

again, we have a modulo-($M + 1$) counter. For straight binary counters, the number P is particularly easy to determine: it is equal to the 1's-complement of M (see Sect. 8.1.3).

The counter in Fig. 9.43 is loaded with the preset number M itself. It then counts down to zero. At zero, a carry RCO is generated (see Fig. 9.33), which can be used to reload the counter.

9.5
Shift Registers

Shift registers are chains of flip-flops that allow data applied to the input to be advanced by one flip-flop with each clock pulse. After passing through the chain, the data are available at the output with a delay, but are otherwise unchanged.

9.5.1
Basic Circuit

The shift-register principle is illustrated in Fig. 9.44. On the first clock pulse, the information, D, present at the input is read into flip-flop F_1. On the second clock pulse, it is passed on to flip-flop F_2; simultaneously, new information is read into flip-flop F_1. As an example, Fig. 9.45 illustrates the mode of operation for a four-bit shift register. We can see that the shift register is filled serially with input data after four clock pulses. These are then available in parallel at the four flip-flop outputs $Q_1 \dots Q_4$, or they can be extracted serially once more at output Q_4 on subsequent clock pulses. All flip-flops with intermediate storage can be used. Transparent flip-flops are unsuitable, because the information applied to the input would immediately pass through all flip-flops to the last flip-flop when the clock goes to 1 for the first time.

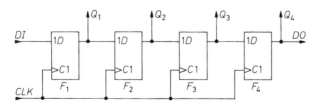

Fig. 9.44. Simplest version of a four-bit shift register.
DI = Data Input
DO = Data Output
CLK = Clock

CLK	Q_1	Q_2	Q_3	Q_4
1	D_1	—	—	—
2	D_2	D_1	—	—
3	D_3	D_2	D_1	—
4	$\mathbf{D_4}$	$\mathbf{D_3}$	$\mathbf{D_2}$	$\mathbf{D_1}$
5	D_5	D_4	D_3	$\mathbf{D_2}$
6	D_6	D_5	D_4	$\mathbf{D_3}$
7	D_7	D_6	D_5	$\mathbf{D_4}$

Fig. 9.45. Function table for a four-bit shift register

9.5.2
Shift Registers with Parallel Inputs

If, as in Fig. 9.46, a multiplexer is connected in front of each D input, it is possible to switch over to parallel data input via the Load input. On the next clock pulse, data d_1, to d_4, are loaded in parallel and appear at $Q_1 \ldots Q_4$. This allows not only *serial-to-parallel conversion but also parallel-to-serial conversion.*

A shift register with parallel load inputs can also be operated as a bidirectional shift register. For this purpose, the parallel load inputs are connected to the output of the next flip-flop on the right. If $LOAD = 1$, the data are shifted from right to left.

Typical IC types:

Length	TTL	ECL	CMOS
4 bit	74 LS 194 A	10141	40194
8 bit	74 LS 164, 299		4014
16 bit	74 LS 673		4006

9.6
Processing of Asynchronous Signals

Sequential logic circuits can be realized either in asynchronous or synchronous – that is, clocked – mode. Asynchronous operation normally requires simpler circuits but creates a number of problems: it must be insured that the spurious transitions (hazards) which may

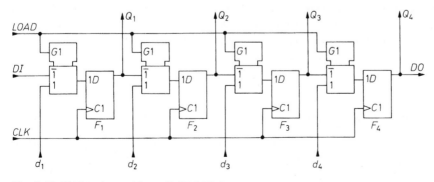

Fig. 9.46. Shift register with parallel LOAD inputs

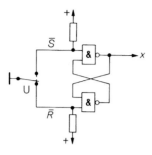

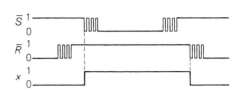

Fig. 9.47. Switch debouncing   **Fig. 9.48.** Timing diagram

appear temporarily because of the difference in propagation delay times are not decoded as valid states. In synchronous system, the conditions are far more simple. Any transition within the system can only take place at the triggering edge of a clock pulse. The clock pulse therefore indicates when the system is in the steady-state condition. It is advisable to construct the system so that all changes consistently occur on one edge of the clock pulse. If, for instance, all the circuits are triggered by the rising edge, the system is certain to be in the steady-state condition at the next rising edge of the clock if the clock frequency is not too high.

As a rule, external data fed to the system are not synchronized with its clock. In order that they may be processed synchronously, they must be synchronized by special circuits, some examples of which are described below.

9.6.1
Debouncing of Mechanical Contacts

If a mechanical switch is opened or closed, vibrations usually generate a pulse train. A counter then detects an undefined number of pulses instead of the single pulse intended. One way of avoiding this is to use mercury-wetted contacts, although this is rather expensive. A simple method of electronic debouncing by means of an RS flip-flop is illustrated in Fig. 9.47. When the switch U is in its lower position (break contact) $\overline{R} = 0$ and $\overline{S} = 1$; that is, $x = 0$. When the switch is operated, a pulse train initially occurs at the $\overline{S}$ input, because the break contact is opened. Since $\overline{R} = \overline{S} = 1$, this being the storing condition, the output remains unchanged. After the complete opening of the break contact, a pulse train is generated by the opposite make contact. With the very first pulse, $\overline{R} = 1$ and $\overline{S} = 0$, the flip-flop is set to $x = 1$. This state is stored during the bouncing that follows. The flip-flop changes to its off state only when the lower break contact is touched again. The timing diagram in Fig. 9.48 illustrates this behavior.

9.6.2
Edge-Triggered RS Flip-Flops

A flip-flop with RS inputs is set as long as $S = 1$ and reset as long as $R = 1$. Both inputs must not be 1 simultaneously. To achieve this, we can generate short R or S pulses. A simpler possibility is shown in Fig. 9.49. Here, the input signals are fed to the inputs of positive-edge-triggered D flip-flops. This insures that only the instant of the positive-going

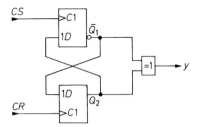

Fig. 9.49. Edge-triggered RS flip-flop
$CS =$ Clock Set $CR =$ Clock Reset

edge is important, and the rest of the clock pulse is immaterial. When a positive-going Set edge occurs, $Q_1 = Q_2$. This results in the exclusive-OR operation

$$y = \overline{Q_1 \oplus Q_2} = 1$$

When a positive-going Reset edge occurs, $Q_2 = \overline{Q_1}$. In this case, $y = 0$. Output y therefore behaves like the Q output of an RS flip-flop.

However, the time characteristics of the input signals are once again subject to a limitation: the positive-going input edges must not occur simultaneously. They must be separated in time by at least the propagation delay time plus the data setup time. In TTL circuits of the 74 LS series, this amounts to approximately 20 ns. If the input pulse edges occur simultaneously, the output signal is inverted.

9.6.3
Pulse Synchronization

The simplest method of synchronizing pulses employs D-type flip-flops. As shown in Fig.9.50, the external unsynchronized signal x is applied to the D-input, and the system clock Φ to the C-input. In this manner, the state of the input variable x is monitored and transferred to the output on the positive-going edge of the clock pulse. As the input signal can also change during the positive-going edge of the clock pulse, metastable states may occur in flip-flop F_1. Additional flip-flop F_2 has therefore been provided to prevent errors occurring in output signal y. Flip-flops that operate according this principle are called "metastable resistant" as the type 74 ACT 11478.

Figure 9.51 shows a typical timing diagram. Any pulse too short to be registered by the leading edge of a clock pulse is ignored. This case is also shown in Fig. 9.51. If such short pulses are not to be lost, they must be read into an intermediate store before being transferred to the D flip-flop. The D flip-flop F_1 in Fig. 9.52 serves this purpose. It is set asynchronously via the $\overline{S}$-input when x becomes 1. With the next positive-going clock edge, $y = 1$. If, at this moment, x has already returned to zero, flip-flop F_1 is reset by the

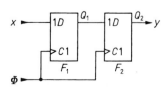

Fig. 9.50. Synchronization circuit

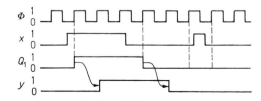

Fig. 9.51. Timing diagram

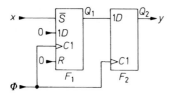

Fig. 9.52. Detection of short pulses

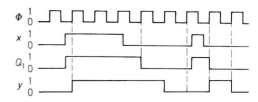

Fig. 9.53. Timing diagram

same edge. A short pulse x is thus prolonged until the next clock edge occurs and cannot therefore be lost. This property may also be seen in the example in Fig. 9.53.

9.6.4
Synchronous One-Shot

It is possible, using the circuit in Fig. 9.54, to generate a pulse that is synchronized with the clock. The pulse length equals one clock period and is independent of the length of the trigger signal x.

If x changes from 0 to 1, $Q_1 = 1$ at the positive-going edge of the next clock pulse; that is, $y = 1$. On the subsequent leading edge, $\overline{Q}_2$ becomes 0 and y becomes 0 again. This state remains unchanged until x has been zero for at least one clock period. Short trigger pulses that are not registered by the leading edge of a clock pulse are lost, as with the synchronizing circuit in Fig. 9.50. If they too are to be considered, an additional flip-flop, as shown in Fig. 9.52, must store the pulses until they are transferred to the main flip-flop. The timing diagram in Fig. 9.55 shows an example of operation.

A synchronous one-shot for ON-times longer than one clock period can be realized quite simply by using a synchronous counter, as is shown in Fig. 9.56. If the trigger variable x is at 1, the counter is loaded in the parallel-in mode on the next clock pulse. The following clock pulses are used to count to the maximum output state Z_{max}. At this number, the carry output $RCO = 1$. The counter is then inhibited via count-enable input ENP; the output variable y is 0. The ordinary enable input ENT cannot be employed for this purpose, as it not only affects the flip-flops but also the RCO directly, and this would result in an unwanted oscillation.

A new cycle is started by parallel read-in. Immediately after loading, RCO becomes zero and y is then 1. The feedback from RCO to the AND gate at the x-input prevents a new loading process unless the counter has reached the state Z_{max}. By this time, x should have

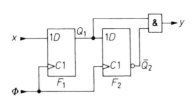

Fig. 9.54. Generation of a single, but clock-synchronous, pulse

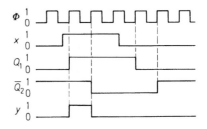

Fig. 9.55. Timing diagram

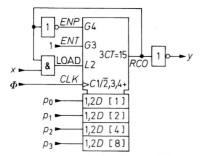

Fig. 9.56. Synchronous one-shot

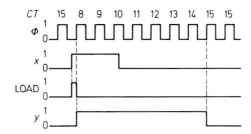

Fig. 9.57. Timing diagram. CT = Content

returned to 0; if not, the counter is loaded again – that is, it is operating as a modulo-$(M+1)$ counter, as shown in Fig. 9.42.

The timing diagram is shown in Fig. 9.57 for an ON-time of seven clock pulses. If a four-bit straight binary counter is employed, it must, for this particular ON-time, be loaded with $P = 8$. The first clock pulse is needed for the loading process and the remaining six pulses for counting up to 15.

9.6.5
Synchronous Edge Detector

A synchronous edge detector gives an output signal that is synchronized with the clock pulse whenever the input variable x has changed. For the implementation of such an arrangement, we consider the one-shot circuit shown in Fig. 9.54. It produces an output pulse whenever x changes from 0 to 1. In order that a pulse is also obtained at the transition from 1 to 0, the AND gate must be replaced by an exclusive-OR gate, producing the circuit shown in Fig. 9.58. Its characteristics are illustrated by the timing diagram in Fig. 9.59.

9.6.6
Synchronous Clock Switch

The problem often arises of how to switch the clock on and off without interrupting the clock pulse generator. In principle, an AND gate could be used for this purpose, but this would result in the first and the last pulse being of undefined length if the switching signal

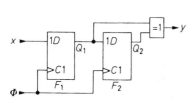

Fig. 9.58. Synchronous edge detector

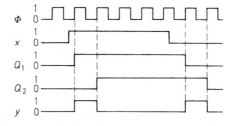

Fig. 9.59. Timing diagram

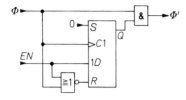

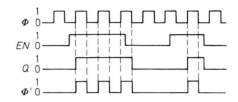

Fig. 9.60. Synchronous clock switch   **Fig. 9.61.** Timing diagram

is not clock-synchronized. This effect can be avoided by employing a D-type flip-flop for the synchronization, as is shown in Fig. 9.60. If $EN = 1$, at the next leading pulse edge $Q = 1$ and therefore $\Phi' = 1$. The first pulse of the switched clock Φ' always has the full length because of the edge-triggering property.

The leading pulse edge cannot be used to switch the control flip-flop off since Q would go to zero a moment after the transition, which would result in a very short output pulse. The flip-flop is therefore cleared asynchronously via the reset input when EN and Φ are 0, this being achieved by the NOR gate at the R-input. As is obvious from Fig. 9.61, only full-length clock pulses can reach the output of the AND gate.

9.7
Systematic Design of Sequential Circuits

9.7.1
State Diagram

To enable the systematic design of sequential circuits, it is necessary to obtain a clear description of the problem. The starting point is the block diagram in Fig. 9.62.

In contrast to a combinatorial logic system, the output variables y_i depend not only on the input variables x_i but also on the previous history of the system. All of the system logic variables affecting the transition to the next state, apart from the input variables, are called state variables z_n. To insure that they can become effective on the arrival of the next clock signal, they are stored in the state variable memory for one clock pulse.

The number of input variables x_i is called the input vector:

$$X = \{x_1, x_2 \ldots x_l\}$$

The number of output variables y_j is called the output vector:

$$Y = \{y_1, y_2 \ldots y_m\}$$

The number of state variables z_n is called the state vector:

$$Z = \{z_1, z_2 \ldots z_n\}$$

We shall denote the various states through which the sequential logic system passes by S_z. To simplify the notation, the state vector is preferably read as a straight binary number and the corresponding decimal number is simply written as a subscript.

The new state $S(t_{k+1})$ is determined both by the old state $S(t_k)$ and by the input variables (qualifiers) x_i. The sequence in which the states occur can therefore be influenced using the qualifiers X. The appropriate assignment is made by a combinatorial logic system: if the old

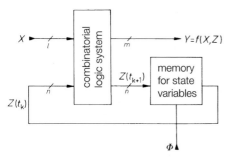

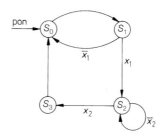

Fig. 9.62. Basic arrangement of a sequential logic system called Mealy machine. If Y depends of Z only its a Moore machine.

Input vector: X
Output vector: Y
State vector: Z
Clock: Φ

Fig. 9.63. Example of a state diagram
State 0: Initial state
State 1: Branching state
State 2: Wait state
State 3: Temporary state

state vector $Z(t_k)$ is applied to its inputs, the new state vector $Z(t_{k+1})$ appears at its output. The corresponding system state must obtain until the next clock pulse. Consequently, the state vector $Z(t_{k+1})$ must not be transmitted to the outputs of the flip-flops until the next clock pulse. For this reason, edge-triggered flip-flops must be used.

There are a few important special types of sequential logic circuit. For example, a special case arises when the state variables can be used directly as outputs. A second simplification occurs when the sequence of states is always the same, in which case no input variables are required. We have made use of these simplifications for the counters.

A general description of the state sequence is provided by a state diagram, as shown in Fig. 9.63. Each state S_Z of the system is illustrated by a circle. The transition from one state to another is shown by an arrow. The symbol on the arrow indicates under which condition a transition is to occur. For the example in Fig. 9.63, state $S(t_k) = S_1$, is followed by state $S(t_{k+1}) = S_2$ if $x_1 = 1$. For $x_1 = 0$, however, $S(t_{k+1}) = S_0$. An unmarked arrow denotes an unconditional transition.

For a *synchronous* sequential circuit, there is an additional condition that a transition will only occur at the next clock pulse edge, and not at the precise moment at which the transition condition is fulfilled. As this restriction applies to all transitions in the system, it is usually not entered in the state diagram, but indicated in the description. We deal below only with synchronous sequential circuits, as their design is less problematic.

If the system is in state S_Z and no transition condition is fulfilled that might lead out of this state, the system remains in state S_Z. This obvious fact can sometimes be emphasized by entering an arrow that starts and ends at S_Z (the wait state). Such a case is illustrated at state S_2 in Fig. 9.63.

When the power supply is switched on, a sequential circuit must be set to a defined initial state. This is the "pon" condition ("power on"). Its signal is produced by a special logic circuit; it is 1 for a short time after switch-on of the supply and is otherwise 0. This signal is generally used to clear the state variable memory by applying it to the RESET inputs of the flip-flops.

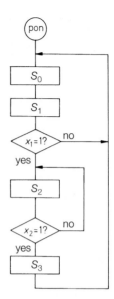

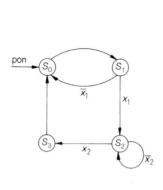

Fig. 9.64. Conversion of a state diagram to a flow diagram

The operation of a sequential circuit can also be represented by a flowchart, as shown for the same example in Fig. 9.64. This representation suggests the implementation of a sequential circuit by a microcomputer program.

9.7.2
Example for a Programmable Counter

We shall demonstrate the design process for a counter, the counting cycle of which is either 0, 1, 2, 3 or 0, 1, 2, depending on whether the control variable x is 1 or 0. The appropriate state diagram is given in Fig. 9.65. As the system can assume four stable states, we require

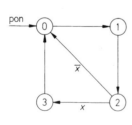

	$Z(t_k)$		$Z(t_{k+1})$		
x	z_1	z_0	z_1'	z_0'	y
0	0	0	0	1	0
0	0	1	1	0	0
0	1	0	0	0	1
0	1	1	0	0	0
1	0	0	0	1	0
1	0	1	1	0	0
1	1	0	1	1	0
1	1	1	0	0	1

ROM address   ROM contents

Fig. 9.65. State diagram of a counter with a programmable counting cycle

$$Counting\ cycle = \begin{cases} 3 & \text{for } x = 0 \\ 4 & \text{for } x = 1 \end{cases}$$

Fig. 9.66. Truth table for the state diagram shown in 9.65

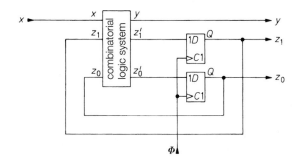

Fig. 9.67. Sequential circuit for implementing the programmable counter

two flip-flops for storage of the state vector Z, which consists of two variables, z_0 and z_1. Since these variables immediately indicate the state of the counter, they are simultaneously used as output variables. In addition, a carry y should be produced when the counter state is 3 for $x = 1$, or 2 for $x = 0$.

We thus obtain the circuit shown in Fig. 9.67, with the truth table given in Fig. 9.66. The left-hand side of the table shows all the possible bit combinations of the input and state variables. The state diagram in Fig. 9.65 shows which is the next system state for each combination. This is shown on the right-hand side of the table. The respective values of the carry bit y are also added.

If a ROM (read-only memory) is used to realize the combinatorial system, the truth table of Fig. 9.66 can be directly employed to program the memory, the state and input variables being used as address variables. The new value Z' of the state vector Z and the output variable y are stored at the appropriate addresses. Hence, to implement our example, we require a ROM for eight words of three bits.

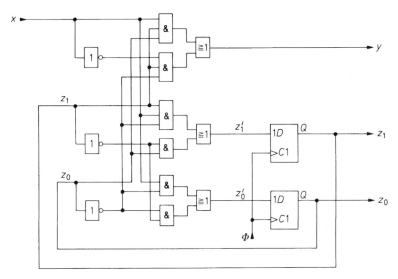

Fig. 9.68. Programmable counter using a combinatorial system consisting of gates. This circuit would also result as a configuration of a PLD

The truth table shown in Fig. 9.66 supplies the Boolean functions:

$$z_1' = z_0\bar{z}_1 + x\bar{z}_0 z_1,$$
$$z_0' = \bar{z}_0\bar{z}_1 + x\bar{z}_0,$$
$$y = \overline{x}\overline{z}_0 z_1 + x z_0 z_1$$

Figure 9.68 shows the realization of this combinatorial system by means of gates. However one should not use discrete gate because of the multitude of chips required. A PLD (programmable logic device) would be preferable because with it a single chip realization is possible that contains the registers for the state variables also. The application of a ROM or PLD also has another decisive advantage, namely flexibility: the ROM or PLD need only be reprogrammed to provide a circuit with different properties, without further changes.

The use of gates for a sequential circuit is thus recommended only in certain simple cases; for instance, in the standard counters described in the previous sections.

9.8
Dependency notation

The new standard for digital circuit symbols (IEC 60617-12 and IEEE Std 91-1984) does not merely involve replacing the earlier round symbols by square ones. It also constitutes a significant advance by introducing the so-called dependency notation which allows complex circuits to be represented in a readily comprehensible manner.

The underlying concept is the use of precisely defined labeling rules extending beyond the gate symbol itself to indicate how specific variables affect other variables. Controlling terminals are differentiated from controlled terminals. A controlled terminal can in turn also act as a controlling terminal for other terminals.

Various types of dependency have been standardized. These are denoted by specific letters as shown in Fig. 9.69. The relevant letter is written inside the circuit symbol next to the controlling terminal. The letter is followed by an identification number which is also entered at all the terminals affected by the relevant operation.

Symbol	Meaning
G	AND
V	OR
N	Exclusive-OR (controllable negation)
Z	Unaltered transmission
C	Clock, Time
S	Set
R	Reset
EN	Enable
M	Mode
L	Load
T	Toggle
A	Addres
CT	Content (e.g. of a counter)

Fig. 9.69. Dependency notation symbols

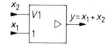

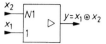

Fig. 9.70. AND-operation

Fig. 9.71. OR-operation

Fig. 9.72. Exclusive-OR-operation

Abb 9.70–9.72. Dependency notation demonstrated by the example of a driver

By way of example, Fig. 9.70 shows the extension of a driver gate to form and AND gate using the dependency notation. Similarly, Figs. 9.71 and 9.72 show the extension to OR or EXOR gates.

A terminal can be controlled simultaneously by several other terminals. In this case the various identification numbers are separated by commas, as in Fig. 9.73. The relevant operations must be carried out successively from left to right.

As an example, Fig. 9.74 shows how a control terminal acts on several other terminals. A bar over the identification number indicates that the variable in question must be linked with the negated control variable.

As shown in Fig. 9.75, several terminals can be combined to form one control variable. In this case the identification number is a straight binary number resulting from the weighting written inside the brace. The number range in question is entered after the function symbol. The notation $\frac{0}{3}$ means 0 to 3. In the example, input x_0 is only effective if control inputs a_0 and a_1 represent the straight binary number 0.

From the examples given so far, it is clear that controlled inputs are designated only by identification numbers. However, there are cases in which a mnemonic designation of a terminal is desirable for other reasons, e.g. D for data. In such cases the identification numbers are placed before the designation letter.

Figure 9.76 gives an example of the use of various modes (M) and the effect and control action of a content (CT). The example shown is of an up-down counter with parallel loading inputs. Depending on mode, the clock CLK has various effects.

The notation 2,4+ at the clock input means that the count is incremented (+) when mode 2 is present $(LOAD = 0, UP = 1)$ and $ENABLE = 1$. Similarly, in mode 0 the count is decremented. The condition for this is 0,4-. The various modes of a terminal are simply written alongside one another, separated by obliques.

In the third mode, the clock initiates parallel data transfer at the D inputs (loading). The notation 1,5 D means that the parallel loading process in mode 1 is taking place in

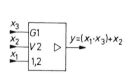

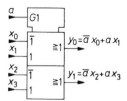

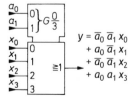

Fig. 9.73. Multiple control of an input

Fig. 9.74. Control of several inputs, using two 2-to-1 multiplexers as an example

Fig. 9.75. Control block with several control variables, using a 4-to-1 multiplexer as an example

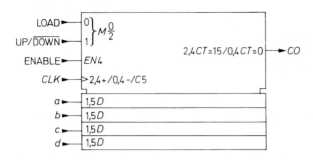

Fig. 9.76. Description of several operating modes using the example of an up-down counter with synchronous parallel loading inputs

synchrony with the clock. Similarly the notation $1D$ would mean a clock-independent (i.e. asynchronous) data transfer.

The carry output CO is controlled by the counter content. It is "1" if the content is 15 when counting up ($2,4\,CT = 15$) or if the content "0" when counting down ($0,4\,CT = 0$).

Chapter 10:
Semiconductor Memories

Semiconductor memories fall into two main categories, as shown in Fig. 10.1: *table memories* and *function memories*. With table memories, an address A is defined in the range

$$0 \leq A \leq n = 2^N - 1$$

The word width of the address is between $N = 5$ and $N = 22$, depending on the size of memory. Data can be stored at each of the 2^N addresses. The data word width is $m = 1 - 16$ bits. Figure 10.2 shows an example for $N = 3$ address bits and $m = 2$ data bits.

The memory capacity $K = m \cdot 2^N$ is specified in bits, and also in bytes (K/8) for data word widths of 8 or 16 bits. When using several memory chips, both the address space and the word width can be increased by any amount. This will allow any tables such as truth tables, computer programs, or results of measurements (numbers) to be stored.

Function memories store logic functions instead of tables. Each variable of a truth table can be expressed as a logic function. Written in standard product terms, the logic function of variable d_0 in Fig. 10.2 becomes

$$d_0 = \bar{a}_2\bar{a}_1\bar{a}_0 d_{00} + \bar{a}_2\bar{a}_1 a_0 d_{10} + \cdots + a_2 a_1 a_0 d_{70}$$

If d_0 contains no regularity, and the zeros and ones are therefore statistically distributed, we obtain $n/2$ – in this case, four – nonvanishing product terms. This situation occurs, for instance, when programs are being stored. In this case, the implementation of the logic function is more complex than its storage in a table.

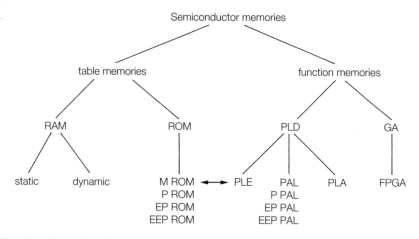

Fig. 10.1. Categories of commonly used semiconductor memories

RAM = Random Access Memory	PLD = Programmable Logic Device
ROM = Read Only Memory	PLA = Programmable Logic Array
M = Mask programmed	PAL = Programmable Array Logic
P = Programmable	GA = Gate Array
EP = Erasable and Programmable	FPGA = Field Programmable Gate Array
EEP = Electrically Erasable and Programmable	

	A			D		
	a_2	a_1	a_0	d_1	d_0	
0	0	0	0	d_{01}	d_{00}	D_0
1	0	0	1	d_{11}	d_{10}	D_1
2	0	1	0	d_{21}	d_{20}	D_2
3	0	1	1	d_{31}	d_{30}	D_3
4	1	0	0	d_{41}	d_{40}	D_4
5	1	0	1	d_{51}	d_{50}	D_5
6	1	1	0	d_{61}	d_{60}	D_6
7	1	1	1	d_{71}	d_{70}	D_7

Fig. 10.2. Table of a memory
Address word width $N = 3\,\text{bit}$
Data word width $m = 2\,\text{bit}$

However, if a truth table is used as the starting point, extensive simplification is often possible for the logic function due to the underlying regularity. One such case is when there are only very few "ones." For example, if in the function d_0 only $d_{70} = 1$, we only require a single conjunction $d_0 = a_2 a_1 a_0$. Another case is when the logic functions can be simplified using Boolean algebra. For example, if $d_0 = a_1$ in Fig. 10.2, an extremely simple function is obtained, even though it contains four "ones." In such cases, the use of function memories generally produces much better solutions than storage in a table.

Table memories are subdivided into two distinct categories, namely RAMs and ROMS. RAM is a general designation for read–write memories. The contents of the memory can be both read and written during normal operation. The abbreviation actually stands for Random Access Memory. "Random access" means that any data word within the memory can be accessed at any time, in contrast to shift register memories, in which data can only be read from the memory in the same sequence in which they were written into it. As shift register memories are no longer very important, the term RAM has become a generic term for memories with read–write capabilities. This is somewhat misleading, in that ROMs also allow random access to any data word.

ROM is an abbreviation for Read-Only Memory. This designation identifies memory ICs that retain their contents when there is no supply voltage, even without backup batteries. In normal operation, data are only read from such memories but not written into them. Special equipment is normally required for writing the data, the storage procedure being referred to as programming. The sub-categories listed in Fig. 10.1 differ in the type of programming employed, which is described in greater detail below.

10.1
Random Access Memories (RAMs)

10.1.1
Static RAMs

A RAM is a memory device in which data can be stored at a specified address and subsequently read out from that address (random access). For technological reasons, the individual memory cells are not arranged linearly but in a square matrix. To select a particular memory cell, address A is decoded, as shown in Fig. 10.3, by a row and column decoder.

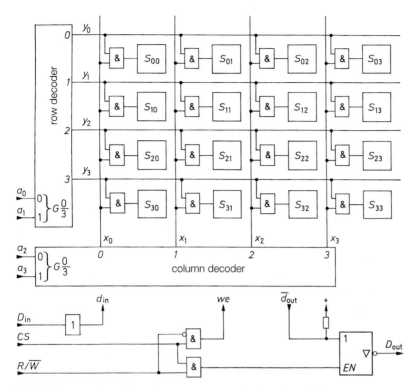

Fig. 10.3. Internal structure of a RAM: an example showing 16-bit memory capacity

D_{in}	=	Data input	
CS	=	Chip Select	
we	=	write enable	
D_{out}	=	Data output	
$R/\overline{W}$	=	Read/Write	

In addition to its address inputs, a RAM has an extra data input D_{in}, a data output D_{out}, a *read/write* pin $R/\overline{W}$, and a chip select (CS) or chip enable (CE) pin. The latter is used for multiplexing more than one memory operated via a common data line (a bus system). When $CS = 0$, the data output D_{out} assumes high impedance and thus has no effect on the data line. To allow this, the data output is always implemented as a tristate gate.

During the write process $(R/\overline{W} = 0)$, the output gate is likewise switched to high impedance by an additional logic operation. This allows D_{in} to be connected to D_{out}, enabling data to be transmitted in both directions via the same line (a bidirectional bus system).

Another logic gate prevents a switchover to the write state $(we = 1)$, if the chip is not selected $CS = 0$. This prevents data from being written accidentally until the relevant memory has been selected.

Figure 10.3 shows the logic operations mentioned above. Lines d_{in}, d_{out}, and we (write enable) are connected to each memory cell internally, as illustrated schematically in Fig. 10.4. Data should only be written into the memory cell when address condition $x_i = y_i = 1$ is satisfied and also $we = 1$. This logic operation is performed by gate G_1. The contents of the memory cell must only reach the output if the address condition is satisfied. This operation is performed by gate G_2, which has an open-collector output. When the cell is not addressed, the output transistor is off. The outputs of all the cells are

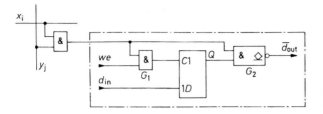

Fig. 10.4. Equivalent logic circuit for the structure of a memory cell

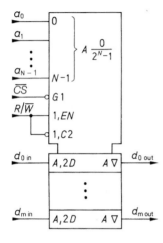

Fig. 10.5. RAM circuit symbol

internally wire-ANDed together and connected to the memory output D_{out} via the tristate gate shown in Fig. 10.3.

Unless the supply voltage is switched off, the memory contents are retained until they are modified by a write command. Such memories are referred to as "static," to distinguish them from "dynamic" memories, in which the contents have to be refreshed at regular intervals to prevent them from being lost.

The circuit symbol for a RAM is shown in Fig. 10.5. As we can see, there are N address inputs. These are decoded by the address decoder in such a way that the precise memory cell is selected (out of 2^N) that corresponds to the address applied. The read–write signal $R/\overline{W}$ is only activated when chip select $CS = 1$. The tristate output is therefore activated for $R/\overline{W} = 1$ and $CS = 1$; for $R/\overline{W} = 0$ it is high impedance. For this reason, the data input and output can be internally interconnected in the memory IC. This produces a bidirectional data port whose direction of operation is determined by the $R/\overline{W}$ signal.

Frequently, not just a single bit but an m-digit word is stored at an address as shown in Fig. 10.2. The storage of entire words may be seen as an extension of the block diagram in Fig. 10.3 in the third dimension. The additional bits are then stacked in additional memory layers above the base layer shown; their control lines x, y, and we are connected in parallel, and their data lines form the input or output word.

Timing Considerations

For satisfactory operation of the memory, a number of timing conditions must be observed. Figure 10.6 shows the sequence of a write operation. To prevent the data from being written into the wrong cell, the write command must not be applied until a certain time has elapsed

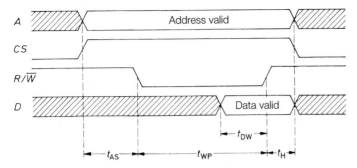

Fig. 10.6. Timing of a write operation
t_{AS} : Address Setup Time
t_{WP} : Write Pulse Width
t_{DW} : Data Valid to End of Write Time
t_H : Hold Time

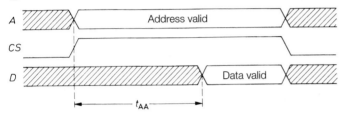

Fig. 10.7. Timing of a read operation
t_{AA} : Address Access Time

after definition of the address. This time is called the address setup time t_{AS}. The duration of the write pulse must not be less than the minimum value t_{WP}, (the write pulse width). The data are read in at the end of the write pulse. They must be valid – that is, stable – for a minimum period of time prior to this. This time is called t_{DW} (Data Valid to End of Write). In a number of memories, the data and addresses must also be present for a further time t_H after the end of the write pulse (the Hold Time). As can be seen from Fig. 10.6, the time required to execute a write operation is expressed as

$$t_W = t_{AS} + t_{WP} + t_H$$

This is referred to as the Write Cycle Time.

The read operation is shown in Fig. 10.7. After the address is applied, it is necessary to wait for time t_{AA} until the data at the output are valid. This time is referred to as the Address Access Time, or simply the Access Time. A list of some static RAMs is given in Fig. 10.8.

10.1.2
Dynamic RAMs

As we wish to maximize the number of cells in a memory, every effort must be made to implement them as simply as possible. They normally consist of just a few transistors; in the case of static CMOS RAMs, a six-transistor cell is normally used. In the simplest case, even the flip-flop is omitted and replaced by a single MOSFET and a capacitor

Capacity	Organization	Type	Manufacturer voltage	Supply Frequency	Acc. time[2]	Case
1 Mbit	128 k × 8	CY7C1018	Cypress	3.3 V	10 ns	SOJ 32
	32 k × 32	CY7C1214	Cypress	3.3 V	100 MHz	TQFP 100
	64 k × 16	CY62126	Cypress	3.3 V	50 ns	TSOP 44
	32 k × 32	IDT71V432	IDT	3.3 V	100 MHz	TQFP 100
	64 k × 16	IDT71V016	IDT	3.3 V	10 ns	SOJ 32
	128 k × 8	M5MV108	Renesas	3.3 V	70 ns	TSOP 32
	128 k × 8	DS1245W[1]	Maxim	3.3 V	100 ns	DIL32
4 Mbit	256 k × 16	CY7C1041	Cypress	3.3 V	10 ns	SOJ 44
	256 k × 16	CY62146	Cypress	3.3 V	55 ns	BGA 48
	256 k × 16	IDT71V416	IDT	3.3 V	10 ns	SOJ 44
	1 M × 4	IDT71V428	IDT	3.3 V	10 ns	SOJ 32
	256 k × 18	IDT71V3578	IDT	3.3 V	150 MHz	TQFP 100
	256 k × 16	R1RW0416	Renesas	3.3 V	12 ns	TSOP 44
	256 k × 16	R1LV0416	Renesas	2.5 V	55 ns	TSOP 44
	256 k × 16	M68AR256	ST	1.8 V	70 ns	BGA 48
	512 k × 8	DS1350W[1]	Maxim	3.3 V	70 ns	DIL34
16 Mbit	1 M × 16	CY7C1061	Cypress	3.3 V	8 ns	SOP 54
	1 M × 16	CY62167	Cypress	1.8 V	55 ns	BGA 48
	512 k × 36	CY7C1316	Cypress	1.8 V	250 MHz	BGA 165
	1 M × 18	IDT71P72804	IDT	1.8 V	250 MHz	BGA 165
	1 M × 18	M5M5V5A36	Renesas	3.3 V	8 ns	QFP 100
	1 M × 16	R1LV1616	Renesas	3.3 V	45 ns	TSOP 48
	2 M × 8	DS1270W[1]	Maxim	3.3 V	100 ns	DIL36
32 Mbit	2 M × 18	HM66AEB18202	Renesas	1.8 V	300 MHz	BGA 165
	2 M × 16	R1WV3216	Renesas	3.3 V	85 ns	TSOP 52

[1] Containing lithium battery; data retention 5–10 years
[2] If access time is specified the RAM is asynchronous; if a frequency is specified the RAM is synchronous.

Fig. 10.8. Examples of SRAMs

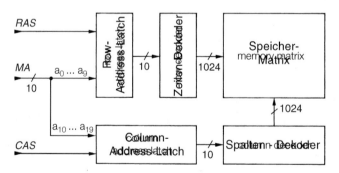

Fig. 10.9. Address decoding in a dynamic 1 Mbit memory.

RAS : Row Address Strobe (simultaneously Chip Enable)
CAS : Column Address Strobe

Capacity	Organization	Type	Manufacturer	Supply voltage	Clock freqency	Interface[1]	Case
512 Mbit	64 M × 8	HBY395512800	Infinion	3.3 V	143 MHz	SDRAM	54TSOP2
	64 M × 8	HBY25D512800	Infinion	2.5 V	166 MHz	DDR	66TSOP2
	64 M × 8	HBY18TD12800	Infinion	1.8 V	266 MHz	DDR2	60TFBGA
	64 M × 8	MT48LC64M8A2P	Micron	3.3 V	133 MHz	SDRAM	54TSOP2
	64 M × 8	MT47H64M8BT	Micron	1.8 V	200 MHz	DDR2	92FBGA
	64 M × 8	K4H510838C	Samsung	2.5 V	200 MHz	DDR	66TSOP2
1 Gbit	128 M × 8	MT47H128M8BT	Micron	1.8 V	200 MHz	DDR2	92FBGA
	128 M × 8	K4S1G0732B	Samsung	3.3 V	133 MHz	SDRAM	54TSOP2
	128 M × 8	K4H1G0838M	Samsung	2.5 V	200 MHz	DDR	66TSOP2
	128 M × 8	K4T1G084QM	Samsung	1.8 V	200 MHz	DDR2	68FBGA

[1] SDRAM = Synchronous Dynamic RAM, DDR = Double Date Rate

Fig. 10.10. Examples for DRAM-chips

Capacity	Organization	Type	Manufacturer	Chips Number	Chips Organization	Interface	Case
256 MB	32 M × 64	HYS64D32301HU	Infineon	4	32 M × 16	SDRAM	184 DIMM
	32 M × 64	MT4VDDT3264AG	Micron	4	32 M × 16	SDRAM	184 DIMM
	32 M × 64	M368L3324BTM	Samsung	4	32 M × 16	DDR	184 DIMM
512 MB	64 M × 64	HYS64D64300HU	Infineon	8	64 M × 8	DDR	184 DIMM
	64 M × 72	MT18LSDF6472	Micron	18[1]	128 M × 4	SDRAM	168 DIMM
	64 M × 64	M368L6423FTN	Samsung	16	32 M × 8	DDR	184 DIMM
1 GB	128 M × 64	HYS64D128021GB	Infineon	16	64 M × 8	DDR	200 SODIMM
	128 M × 64	MT16VDDT12864AG	Micron	16	64 M × 8	SDRAM	184 DIMM
	128 M × 64	M368L2923BTM	Samsung	16	64 M × 8	DDR	184 DIMM
	128 M × 64	M470T2953BSO	Samsung	16	64 M × 8	DDR2	200 SODIMM
	128 M × 72	M378T2953BGO	Samsung	18[1]	64 M × 8	DDR2	240 DIMM
2 GB	256 M × 64	M368L5623MTN	Samsung	16	128 M × 8	DDR	184 DIMM

[1] extra bits for error correction ECC

Fig. 10.11. Examples for DRAM modules

to store 1 bit as a charge. This makes a single-transistor cell possible. However, as the charge is only retained for a short time, the capacitor must be recharged at regular intervals (approximately every 2–70 ms). This operation is known as *refresh*, and the memories are called *dynamic RAMs, DRAMs*.

This disadvantage is offset by several advantages. Dynamic memories can provide about four times more storage capacity on the same printed circuit board area, with the same current drain and at the same cost.

To save on pins, with dynamic memories the address is entered in two stages and buffered in the IC. The block diagram of a 1 Mbit RAM is shown in Fig. 10.9. In the first step, address bits a_0–a_9 are stored in the row-address latch with the *RAS* signal. In the second step, address bits a_{10}–a_{19} are loaded into the column-address latch with the *CAS* signal. This makes it possible to accommodate a 1 Mbit memory in an 18-pin package. Figure 10.10 gives some examples for DRAM chips, in Fig. 10.11 some modules are listed.

Dynamic RAM Controllers

To operate dynamic RAMs, additional circuitry is required. For a normal memory access, the address has to be loaded into the RAM in two consecutive steps. To avoid loss of data, it is necessary to call up all the row addresses at least once within (usually) 64 ms. The column-addresses need not to be selected because a whole row is refreshed with one RAS-cycle. If the contents of memory are not read out cyclically, extra circuitry is required to effect cyclic addressing between normal memory accesses. These circuits are referred to as "dynamic RAM controllers." Figure 10.12 shows the block diagram for a circuit of this type.

For normal memory access, the externally applied address is stored in the row and column address latch when address strobe $AS = 1$, which indicates that the address is valid. Then the DRAM-controller initiates a memory access cycle by forwarding the row address a_0–a_9 via the memory address lines MA and activating the RAS-strobe $RAS = 1$. Subsequently, the column address a_{10}–a_{19} is produced and also read into memory with the column address strobe via the same MA lines. This sequence is illustrated in Fig. 10.13. The next memory access may not occur immediately, but only after a "precharge time," which is of the same order of magnitude as the address access time.

In order to perform the refresh, the 1024 row addresses a_0–a_9 must be applied once every 8 ms. For a refresh cycle time of 100 ns, a total of some 100 μs is required for this purpose. Memory availability is thereby reduced by less than 2%. Three different methods can be used to organize the refresh:

1) *Burst refresh.* In this mode, normal operation is interrupted after 8 ms and a refresh is performed for all memory cells. In many cases, however, it is undesirable for the memory to be inaccessible for 100 μs.

2) *Cycle stealing.* To avoid disabling the memory for a continuous 100 μs period, the refresh process can be subdivided and spread out over 8 ms: if the status of the refresh

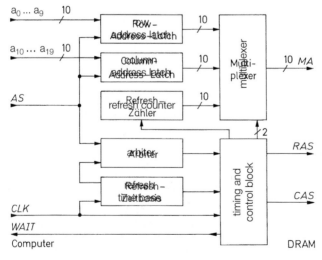

Fig. 10.12. Design of a DRAM controller for 1 Mbit RAMs
AS = Address Strobe RAS = Row Address Strobe
MA = Memory Address CAS = Column Address Strobe

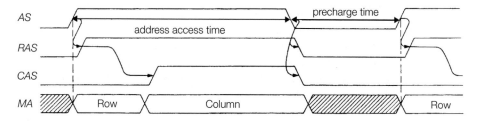

Fig. 10.13. Address input timing sequence for a DRAM

counter is incremented by one every 8 μs, then after $1024 \cdot 8\,\mu s \approx 8\,ms$ all of the row addresses will have been applied once, as required. With *cycle stealing* the processor is stopped every 8 μs for one cycle and a refresh step is performed. For this purpose, the refresh time base shown in the block diagram in Fig. 10.12 divides the frequency of the clock signal *CLK*, so that the timing and control block receives a refresh command every 8 μs.

When a refresh cycle is initiated, the status of the refresh counter is transferred via the multiplexer to the memory and the RAS signal is temporarily set to 1. The counter is then incremented by 1. During the refresh cycle, the memory user is inhibited by a wait signal. This means that the ongoing process is stopped every 8 μs for 100 ns; that is, slowed down likewise, by 2%.

3) *Transparent, or hidden, refresh.* With this method, a refresh step is also performed every 8 μs, but the refresh controller is synchronized to the CPU in such a way that, instead of inhibiting the memory user, the refresh is performed at the precise instant at which the CPU cannot access the memory. This means that no time is lost. If any overlapping of an external access with the refresh cycle cannot be totally eliminated, an additional priority decoder (arbiter) can be employed, as shown in Fig. 10.12. It acknowledges an external request with a wait signal until the current refresh cycle is complete and then executes the request.

10.2
RAM Expansions

10.2.1
Two-Port Memories

Two-port memories are special RAMs that allow two independent processes to access common data. This enables data to be exchanged between the two processes. To be able to do this, the two-port memory must have two separate sets of address, data, and control lines, as shown in Fig. 10.14. Implementation of this principle is subject to limitations, since it is basically impossible to write into the same memory cell simultaneously from both ports.

The "Read-While-Write" memories overcome this problem by only reading from one of the two ports and only writing at the other. Figure 10.15 shows that memories of this type have two separate address decoders, which allow *simultaneous* writing to one address while reading from another.

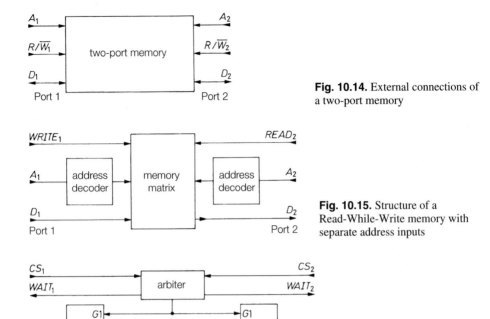

Fig. 10.14. External connections of a two-port memory

Fig. 10.15. Structure of a Read-While-Write memory with separate address inputs

Fig. 10.16. Two-port memory with standard RAMs

If reading and writing are to take place at both ports of a two-port memory, an access conflict can generally only be avoided by preventing simultaneous memory access. To do this, the address, data, and control lines can be made available via multiplexers to the port accessed, as shown in Fig. 10.16. In many cases, the two processes accessing the memory are synchronized to prevent simultaneous access. If this is not possible, a priority decoder (arbiter) can be used which, in the event of access overlap, temporarily stops one of the two processes by a wait signal. Some integrated two-port memories are listed in Fig. 10.17. However, their capacities are limited. In order to implement large two-port memories, it is advisable to use normal RAMs in conjunction with a dual-port RAM controller.

10.2.2
RAMs as Shift Registers

RAMs can be operated as shift registers if the addresses are applied cyclically. The counter shown in Fig. 10.18 is used for this purpose. For each address, the stored data are first read out and new data are then read in. The timing diagram is shown in Fig. 10.19. The positive-going edge of the signal increments the counter. If the CLK signal is simultaneously used as the $R/\overline{W}$ signal, the memory contents are then read out and stored in the output flip-flop on the negative-going edge. While $CLK = 0$, the new data D_{in} are written into the memory cell that has just been read out. In this case, the minimum clock cycle is shorter

Capacity	Organi- zation	Type	Manu- facturer	Supply voltage	Acc. time[1] frequency	Case
64 kbit	4 k × 16	IDT70V24	IDT	3.3 V	15 ns	100TQFP
128 kbit	8 k × 16	IDT70V25	IDT	3.3 V	15 ns	100TQFP
	8 k × 16	CY7C025	Cypress	3.3 V	15 ns	100TQFP
1 Mbit	64 k × 16	IDT70V28	IDT	3.3 V	15 ns	100TQFP
	64 k × 16	IDT70V9289	IDT	3.3 V	83 MHz	108TQFP
	128 k × 8	CY7C009V	Cypress	3.3 V	15 ns	100TQFP
4 Mbit	128 k × 36	CY7C0852V	Cypress	3.3 V	167 MHz	172BGA
	128 k × 36	IDT70V659	IDT	3.3 V	10 ns	256BGA
	128 k × 36	IDT70V3599	IDT	3.3 V	166 MHz	256BGA
18 Mbit	512 k × 36	IDT70T3539	IDT	2.5 V	166 MHz	256BGA
	512 k × 36	IDT70T653	IDT	2.5 V	10 ns	256BGA
36 Mbit	1024 k × 36	IDT70T3509	IDT	3.3 V	133 MHz	256BGA

[1] If access time is specified the RAM is asynchronous, if a frequency is specified the RAM is synchronous.

Fig. 10.17. Examples of two-port memories

than the sum of the read and write cycle times, since the address remains constant. It is equal to what is known as the "Read-Modify-Write Cycle Time."

The difference between this type of shift register and the normal type (see Sect.9.5) is that only the address, which acts as a pointer to the fixed data, is shifted, not the data themselves. The advantage of this method is that normal RAMs can be employed, and these are obtainable with memory capacities far greater than those of normal shift registers. Even dynamic RAMs can be used without a refresh controller if all raw addresses are scanned within the refresh time.

At high shift frequencies, slow low-cost RAMs can be used if several data bits are processed in parallel, and if a serial-to-parallel converter is provided at the input and a parallel-to-serial converter at the output, in order to obtain the required shift frequency.

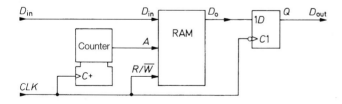

Fig. 10.18. RAM operated as a shift register

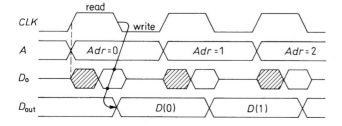

Fig. 10.19. Timing diagram for a RAM shift register

10.2.3
First-In-First-Out Memories (FIFO)

A FIFO is a special type of shift register. The common feature is that the data appear at the output in the same order as they were read in: the first word read in is also the first one read out. With a FIFO, as opposed to a shift register, this process can take place completely asynchronously; that is, the read-out clock is independent of the read-in clock. FIFOs are therefore used for linking asynchronous systems.

Operation is very similar to that of a delay line: the data do not move at a fixed rate from input to output, but only remain in the register long enough for all the previous data to be read out. This is shown schematically in Fig. 10.20. With first-generation FIFOs, the data were actually shifted through a register chain, as illustrated in Fig. 10.20. On entry, the data were passed on to the lowest free memory location and shifted onward from there to the output by the read clock. One disadvantage of this principle was the long fall-through time. This is particularly noticeable when the FIFO is empty, as the input data then have to pass through all the registers before being available at the output. This means that even the smallest FIFOs exhibit fall-through times of several microseconds. Other disadvantages include the complex shift logic and the large number of shift operations, thereby precluding a current-saving implementation in CMOS technology.

To overcome these drawbacks, in the second-generation FIFOs it is no longer the data that are shifted, but merely two pointers that specify the input and output addresses in a RAM. This is illustrated in Fig. 10.21. The input counter points to the first free address A_{in}, and the output counter to the last occupied address A_{out}. Both pointers therefore rotate during ongoing data input and output.

The distance between the two pointers indicates how full the FIFO is. When $A_{in} - A_{out} = A_{max}$, the FIFO is full. No more data must then be entered, as this would mean overwriting data that have not yet been read out. When $A_{in} = A_{out}$, the FIFO is empty. No

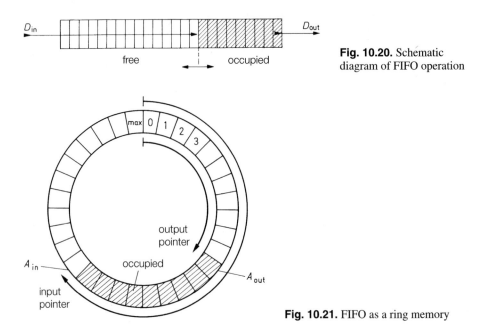

Fig. 10.20. Schematic diagram of FIFO operation

Fig. 10.21. FIFO as a ring memory

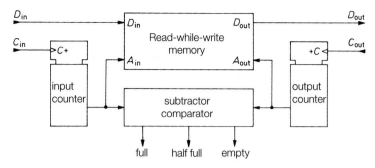

Fig. 10.22. FIFO implementation with read-while-write memory

data must now be read out, as this would mean receiving old data for a second time. An overflow or empty condition can only be avoided if the average data rates for input and output are identical. To achieve this, it is necessary to monitor the occupancy of the FIFO, and to attempt to control the source or sink in such a way that the FIFO is on average half full. The FIFO can then accommodate short-term fluctuations, assuming that it has a sufficient storage capacity.

The design of a FIFO is shown in Fig. 10.22. It is similar to the RAM shift register in Fig. 10.18. Read-while-write memories with separate address inputs (see Fig. 10.15) are particularly suitable here, as reading and writing can occur asynchronously. The more recent FIFOs, examples of which are listed in Fig. 10.23, operate on this principle.

Capacity	Organization	Type	Manufacturer	Supply voltage	Acc. time[1] frequency	Case
128 kbit	16 k × 9	CY7C4261V	Cypress	3.3 V	100 MHz	32PLCC
	16 k × 9	IDT72V06	IDT	3.3 V	15 ns	32PLCC
	8 k × 18	IDT72T1865	IDT	2.5 V	225 MHz	144BGA
	4 k × 36	IDT72T3665	IDT	2.5 V	225 MHz	208BGA
1 Mbit	128 k × 9	CY7C4291V	Cypress	3.3 V	100 MHz	32PLCC
	128 k × 9	IDT72T1895	IDT	2.5 V	225 MHz	144BGA
	64 k × 18	IDT72T1895	IDT	2.5 V	225 MHz	144BGA
	32 k × 36	IDT72T3695	IDT	2.5 V	225 MHz	208BGA
4 Mbit	512 k × 9	IDT72T2113	IDT	3.3 V	166 MHz	100BGA
	256 k × 18	IDT72T18115	IDT	2.5 V	225 MHz	240BGA
	128 k × 36	IDT72V36110	IDT	3.3 V	166 MHz	144BGA
9 Mbit	512 k × 18	IDT72T18125	IDT	2.5 V	255 MHz	240BGA
Bidirectional						
256 kbit	2 × 4 k × 36	CY7C43662AV	Cypress	3.3 V	133 MHz	120TQFP
	2 × 4 k × 36	IDT72V3662	IDT	3.3 V	100 MHz	120TQFP
1 Mbit	2 × 16 k × 36	CY7C43682AV	Cypress	3.3 V	133 MHz	120TQFP
	2 × 16 k × 36	IDT72V3682	IDT	3.3 V	100 MHz	120TQFP
4 Mbit	2 × 64 k × 36	IDT72V36102	IDT	3.3 V	100 MHz	120TQFP

[1] The access time is specified the FIFO is asynchronous, if a frequency is specified the FIFO is synchronous.

Fig. 10.23. Examples of FIFOs

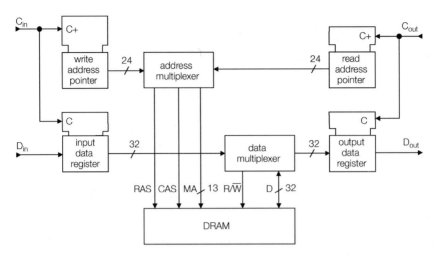

Fig. 10.24. FIFO implementation with standard DRAMs using the sequential flow controller IDT72T6360

FIFO Implementation Using Standard RAMs

For the implementation of large FIFOs, it is advisable to use standard RAMs, thereby providing the maximum memory capacity at lowest cost. This method is shown in Fig. 10.24. For writing data is stored at the address of the write address pointer. It is transferred to the memory chip by the address multiplexer. To read data the address of the read address pointer is applied to the memory. Data is buffered by the input and output data register and transferred to and from the memory by the data multiplexer. The control logic required to operate a RAM as a FIFO can be obtained as an integrated circuit known as FIFO RAM controller or a sequential flow controller.

4 M … 16 M words, 166 MHz, 2.5/3.3 V CMOS: IDT72T6360

10.2.4
Error Detection and Correction

When data are stored in RAMs, two different types of error can occur: permanent errors and transient errors. The permanent errors (hard errors) are caused by faults in the ICs themselves or in the associated controller circuits. The transient errors (soft errors) only occur randomly and are therefore not reproducible. They are mainly caused by α-radiation of the package, which may not only discharge memory capacitors in dynamic RAMs, but also cause flip-flops in static RAMs to change state. Transient errors can also result from noise pulses generated inside or outside the circuit.

The occurrence of memory errors can have far-reaching consequences. Thus a single error in a computer memory might not only produce an incorrect result, but even cause the program to crash completely. Methods have therefore been developed to indicate the occurrence of errors. In order to do this, one or more check bits must be processed in addition to the actual data bits: the more check bits used, the greater the number of errors that can be detected or even corrected.

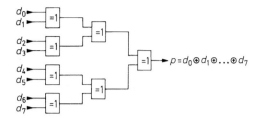

Fig. 10.25. Parity generator for even parity with eight inputs.
IC types: 9 bit: SN74LS658

Parity Bit

The simplest method of error detection consists of using a parity bit p that is added to the data word D. Even or odd parity can be defined. For an even parity check, the parity bit is set to zero if the number of ones in the data word is even. It is set to one if this number is odd. This means that the total number of ones transmitted in a data word including parity bits is always even or, for odd parity, always odd.

The even-parity bit can also be interpreted as the sum (modulo-2) of the data bits. This checksum can be calculated as the exclusive-OR of the data bits.

The implementation of a parity generator is shown in Fig. 10.25. The exclusive-OR gates can be in any sequence. It is chosen such that the sum of the delay times involved remains as small as possible.

For error detection purposes, the parity bit is stored together with the data bits. When the data are read out, the parity can then be regenerated as shown in Fig. 10.26 and compared with the stored parity bit by an exclusive-OR operation. If they differ, an error has occurred and the error output becomes $e = 1$. This allows each single-bit error to be detected. However, no correction is possible, since the bit containing the error cannot be located. If *several* bits contain errors, an odd number of errors can be detected, whereas an even number cannot.

Hamming Code

The principle of the Hamming code consists of using several check bits in order to refine error detection, to the point at which it is possible not just to detect single-bit errors but also to pinpoint their location. Once the error bit in a binary code is located, it can be corrected by complementing it.

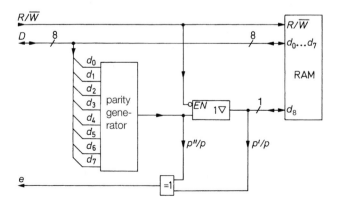

Fig. 10.26. Data memory with parity checking (using eight-bit data words as an example). The parity bit d_8 must be stored as an additional data bit

Number of data bits	m	$1 - 4$	$5 - 11$	$12 - 26$	$27 - 57$	$58 - 120$	$121 - 247$
Number of check bits	k	3	4	5	6	7	8

Fig. 10.27. Minimum number of check bits required to detect and correct a single-bit error

The question of how many check bits are required for this purpose is easily answered: with k check bits, 2^k different bit locations can be identified. With m data bits, the resultant total word length is $m + k$. An additional check bit combination is required to indicate whether the data word received is correct. This yields the condition

$$2^k \geq m + k + 1 .$$

The most important practical solutions are listed in Fig. 10.27. It can be seen that the relative proportion of check bits is smaller the greater the word length.

We shall now examine the procedure for determining the check bits, using a 16-bit word as an example. Figure 10.27 shows that to safeguard 16 bits we need five check bits; that is, a total word length of 21 bits. In accordance with Hamming, the individual check bits are evaluated as parity bits for different parts of the data word. In our example, we therefore require five parity generators. Their inputs are allocated to data bits in such a way that each data bit is connected to at least two of the five generators. If a data bit is now read incorrectly, there is a difference only between those parity bits affected by that particular data bit. Using this method, we therefore obtain a 5-bit error word E, the syndrome word, instead of a parity error bit. This word can assume 32 different values, which allows us to pinpoint the error bit. It can be seen that the identification of a single-bit error is unique only if a different parity bit combination is selected for each data bit location. If a difference in just *one* parity bit is detected, only the parity bit *itself* can be in error, since the parity bit combination chosen means that, for an incorrect data bit, at least two parity bits have to differ. If all of the data bits and parity bits are read without error, the calculated parity bits match those stored and the syndrome word becomes $E = 0$.

An example of the assignment of the five parity bits to the individual data bits is given in Fig. 10.28. This shows that data bit d_0 affects parity bits p_0 and p_1, data bit d_1, affects parity bits p_0 and p_2, and so on. As required, each data bit affects a different combination of parity bits. To simplify the circuitry, the combinations have been distributed in such a way that each parity generator has eight inputs.

During reading ($R/\overline{W} = 1$) the syndrome generator in Fig. 10.29 compares the stored parity word P' with the parity word P'' calculated from data D'. If errors occur, the

Pariy bits	Data bits d_i															
	0	1	2	3	4	5	6	7	8	9	10	11	12	13	14	15
p_0	×	×	×	×							×	×	×		×	
p_1	×				×	×	×				×	×	×	×		
p_2		×			×			×	×		×			×	×	×
p_3			×			×		×		×		×		×	×	×
p_4			×	×	×		×		×	×			×			×

Fig. 10.28. Example of parity bit generation using Hamming code for a 16-bit word

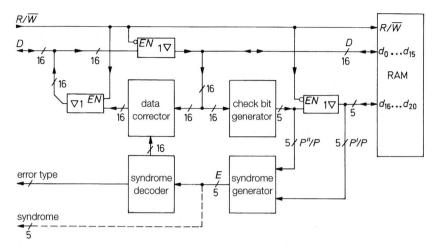

Fig. 10.29. Data memory with error correction (using 16-bit data words as an example)

syndrome word becomes $E = P' \oplus P'' \neq 0$. The syndrome decoder then defines which data bit must be corrected and causes the bit in question to be inverted in the data corrector.

The operation of the syndrome generator will be explained more precisely with reference to Fig. 10.30. Depending on the syndrome word e_0-e_4, three types of error can be identified: data errors d_0-d_{15}, check bit errors p_0-p_4, and multiple errors. The latter type, however, are not completely detected when a Hamming matrix of minimum size is used, and cannot be corrected.

The particular advantage of memories with error correction is that the occurrence of memory errors can be registered, while they remain ineffective as a result of the correction procedure. However, in order to derive the maximum benefit from this, a number of factors have to be taken into consideration: the probability of noncorrectable multiple errors must be minimized. For this reason, a separate memory IC should be used for each data bit d_0-d_{15} and each check bit p_0-p_4. Otherwise, several data bits would be simultaneously falsified in the event of total failure of a memory chip. In addition, it is necessary for each error that is detected to be eliminated as quickly as possible. Consequently, a program should be interrupted when an error is detected in the computer memory and an error service program should be executed. This must first establish whether this error is a transient error that can be rectified by writing the corrected data word back into memory and then reading

Syndrome word	No error	Data error						Check bit error					Multiple error			
		d_0	d_1	d_2	...	d_{14}	d_{15}	p_0	p_1	p_2	p_3	p_4			...	
e_0	0	1	1	1	...	1	0	1	0	0	0	0	0	1	... 0	1
e_1	0	1	0	0		0	0	0	1	0	0	0	1	0	1	1
e_2	0	0	1	0		1	1	0	0	1	0	0	1	0	1	1
e_3	0	0	0	1		1	1	0	0	0	1	0	0	1	1	1
e_4	0	0	0	1		0	1	0	0	0	0	1	0	0	1	1

Fig. 10.30. Table of the syndrome words and their significance

it out again. If the error persists, it is a permanent error. In this case, the syndrome word is read out, as this allows the memory IC involved to be located, and the IC number – together with the frequency of failure – are listed in a table. This table can then be scanned at regular intervals so that defective chips can be replaced. This enables the reliability of a memory with EDC (Error Detection and Correction) to be continually increased.

10.3
Read-Only Memories (ROMs)

The term ROM refers to table memories that are normally only read. They are therefore suitable for storing tables and programs. Their advantage is that the memory content is retained when the supply voltage is disconnected. Their disadvantage is that putting data into the table is more difficult than with RAMs. The categories shown in Fig. 10.1 (MROM, PROM, EPROM, and EEPROM) differ with respect to the input procedure.

10.3.1
Mask-Programmed ROMs (MROMs)

With these devices, the memory content is entered during the final manufacturing stage, using a specific metallization mask. This process is only cost-effective for large production quantities (from approximately 1,000,000 upward) and generally requires several months for implementation. MROMs are for instance used in hand-held calculators.

10.3.2
Programmable ROMs (PROMs)

A PROM is a read-only memory whose content is programmed-in by the user. The programmable components are formerly fuses, which are implemented in the ICs by means of exceptionally thin metallization links. Diodes are also used, which can be shorted by overloading them in the reverse-bias direction. The latest programmable elements for PROMs are special MOSFETs with an additional "floating gate." This is charged up during programming, causing a change of the pinch-off voltage of the MOSFET. As the floating gate has all-round SiO_2 insulation, it can be guaranteed to retain its charge for 10 years.

We shall now describe the internal design of a PROM, using the example of the fuse-type PROM in Fig. 10.31. For technological reasons, the individual memory cells are not arranged linearly, but as a square matrix exactly as with RAMs. A particular memory cell is addressed by applying a logical 1 to the appropriate column and row connection. For this purpose, the address vector $A = (a_0 \ldots a_{n-1})$ must be decoded accordingly. The column and row decoders used operate as 1-of-x decoders.

The memory cell selected is activated by the AND gate at the intersection of the selected row or column line. The ORing of all the memory cell outputs produces output signal D. In order to obviate the need for a gate with 2^n outputs, wired-OR logic is used. In the case of open-collector outputs, this can be implemented by wired-ANDing the negated signals. This method has already been described in Fig. 7.30.

In its basic state, each memory cell addressed generates output signal $D = 1$. To program a zero, the fusible link at the output of the desired cell is blown. For it one selects

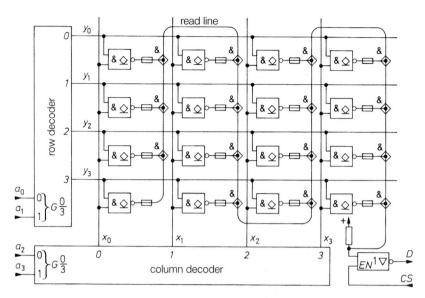

Fig. 10.31. Internal structure of a PROM: an example showing 16-bit memory capacity

the address of the corresponding cell, turning on the output transistor of the NAND gate. A powerful current pulse, which is just large enough to blow the fusible link at the NAND gate output, is then injected into the read line. A timing sequence precisely defined by the manufacturer must be observed. Special programming devices are therefore used, which can be tailored to the particular type of memory.

In the case of PROMs, not just 1 bit but an entire 4-bit or 8-bit word is usually stored at an address. These devices therefore possess a corresponding number of data outputs. Specifying a memory capacity as, say, 1 k × 8 bits means that the memory contains 1,024 8-bit words. The contents are specified in the form of a programming table. By way of an example, Fig. 10.32 shows its organization for a 32 × 8-bit PROM. The circuit symbol for a PROM is like that for the RAM in Fig. 10.5. Here, the $R/\overline{W}$ input becomes the read line as programming input and the data inputs are omitted.

10.3.3
UV-Erasable PROMs (EPROMs)

An EPROM (erasable PROM) is a PROM that can not only once be user-programmed, but that can also be erased using ultraviolet light. A MOSFET that incorporates an additional floating gate is used exclusively as the memory element. This is charged up during programming, thereby changing the pinch-off voltage of the transistor. With EPROMs, however, this charge can be erased again in about 20 minutes by irradiation with UV light. For this purpose, the package is provided with a quartz glass window. Due to the additional complexity involved in producing this package, EPROMs are more expensive than non-window PROMs, even though they are realized using the same technology. Consequently, EPROMs are useful during the developmental stage of new equipment, but the equivalent PROMs are preferred for mass production.

Inputs					Outputs							
x_4	x_3	x_2	x_1	x_0	d_7	d_6	d_5	d_4	d_3	d_2	d_1	d_0
0	0	0	0	0								
0	0	0	0	1								
0	0	0	1	0								
0	0	0	1	1								
$\approx$	$\approx$	$\approx$	$\approx$	$\approx$	$\approx$	$\approx$	$\approx$	$\approx$	$\approx$	$\approx$	$\approx$	$\approx$
1	1	1	1	0								
1	1	1	1	1								

Fig. 10.32. Example of a programming table for a PROM containing 32 8-bit words

EPROMs are programmed on a word-by-word basis; that is, for the usual 8-bit organization one byte at a time. Even in the case of older EPROMs (e.g., the 2716; 2 k × 8 bit), the programming procedure is still simple. One applies a programming voltage of $V_{pp} = 25$ V, together with the required address and the bit pattern to be programmed. Then a programming command that lasts for 50 ms is applied to store the data. Programming can then be terminated or the process can be repeated for another address, using the associated bit pattern. In the case of a 2-kbyte EPROM, the programming of the entire device takes about 2 min. However, it would take almost 2 hours to program a 128 kbyte memory. As this is clearly not tolerable, it was found to be necessary to modify the technology and programming algorithms for larger EPROMs. All fast programming algorithms are based on the fact that most of the cells of an EPROM can be programmed in considerably less than 50 ms. However, as "slower" bytes occur from time to time, it is impossible to generally reduce the programming time. Instead, a variable programming pulse length is employed.

The "fast" or "intelligent" programming algorithm that is generally used nowadays is shown in Fig. 10.33. The programming voltage, $V_{pp} = 12.5$ V, is applied and the supply voltage is raised to $V_{CC} = 6$ V. The higher supply voltage speeds up the programming process, as the transistors assume lower impedance, and this also constitutes the worst-case condition for verification purposes. The address then becomes $A = 0$ and the associated data are applied. Now follows the procedure for programming this byte. For this purpose, an auxiliary counter is set to $n = 0$. A 1 ms programming command is then issued. After the auxiliary counter has been incremented, the memory content is read out to check whether programming has already been successful. If not, up to another 24 programming commands are issued. If the byte has still not been programmed, the chip is deemed to be defective. Normally, only a few programming pulses are required. However, it is still not certain that the floating gate has sufficient charge to last for 10 years. In order to make sure, the charge is tripled. This is done by overprogramming for $3n \cdot 1$ ms.

The first byte is thus programmed and the process can be repeated for the next address, using new data. At the end of programming, we switch back to the read mode and once more check that the entire memory content is in order. The fast programming algorithm reduces the programming time for a 1 Mbit EPROM from around 2 h to less than 10 min. By reducing the programming pulse duration to 100 µs, times of less than 1 min can be achieved with some EPROMs.

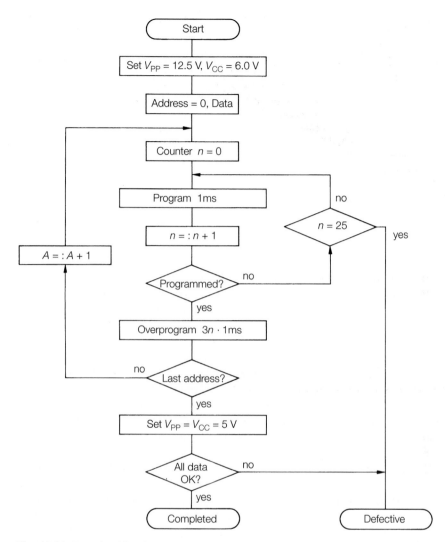

Fig. 10.33. Fast algorithm for programming EPROMS

10.3.4
Electrically Erasable PROMs (EEPROMs)

The term EEPROM (electrically erasable PROM) refers to a PROM which, unlike the EPROM, can also be erased *electrically*. With the more recent types, the voltage converter for generating the programming voltage and the timer for determining the programming pulse duration are incorporated in the memory chip. In order to program a byte, just the address and the data need to be applied. If programming is then initiated by a write command, the EEPROM stores the address and data internally, and immediately releases the address and data lines. The subsequent process takes place autonomously on the chip. The old byte is first erased and the new byte is then programmed. This process is internally monitored to insure that the programmed charge is adequate. The process lasts for 1–

Capacity	Organi- zation	Type	Manu- facturer	Acc. time[2] frequency	Inter-[1] face	Case
256 kbit	32 k × 8	HN58V256AT	Renesas	120 ns	parallel	32 TSOP
	32 k × 8	HN58X25256TIE	Renesas	400 kHz	I²C	8 SOP
	32 k × 8	HN58X25256FPIE	Renesas	5 MHz	SPI	8 SOP
	32 k × 8	S524ADOXF1	Samsung	400 kHz	I²C	8 DIP
	32 k × 8	M24256W	ST	400 kHz	I²C	8 SOP
	32 k × 8	M95256W	ST	5 MHz	SPI	8 SOP
1 Mbit	128 k × 8	HN58V1001	Renesas	250 ns	parallel	32 SOP
	128 k × 8	HN58W241000	Renesas	1 MHz	I²C	8 SOP

[1] Beside the parallel interface serial interfaces by I²C bus and SPI bus are usual.
[2] In the case of serial interfaces the maximum serial clock frequency is specified.

Fig. 10.34. Examples of EEPROMs. Supply voltage is 3.3 V for all types. A single byte can be written or erased

10 ms; that is, it has the same order of magnitude as with EPROMs. Some EEPROMs are capable of storing not just one byte but an entire "page" containing 16–64 bytes in one programming process. To do this, the page is read into an internal RAM before the programming command is issued. This gives effective programming times of 30 µs per byte.

However, despite these simple, fast erase and write procedures, one should not be tempted to use an EEPROM as a RAM. The number of possible write cycles is in fact limited: no byte must be written more than $10^4 \ldots 10^6$ times (depending on the type). With a programming time of 1 ms, the end of the operational life of a byte or a page may be reached in as little as 10 s if continuous programming is performed.

Some examples of EEPROMs are listed in Fig. 10.34. With many EEPROMs, as with most memories, the power dissipation is reduced when they are not selected by $CS = 0$. The smallest power dissipation obviously occurs when the supply voltage is disconnected completely. This does not result in the data being lost – as in the case of all ROMs – but the access time after application of the supply voltage is increased due to the transient response of the read amplifiers. For this reason, it is inadvisable to switch on the supply voltage only when the memory is accessed.

Flash EEPROMs are intermediates between EPROMs and EEPROMs. Like EEP-ROMs, they can be erased electrically, but not byte by byte, as in the case of EPROMS – here, the entire block of 1 . . . 32 kbyte is erased at once. This is the reason for their name. They are erased much more simply than EPROMs: a single erase pulse lasting for a few seconds is required. It is not necessary to take the package out of the circuit and put it into an eraser unit for 20 min. Flash EEPROM technology is simpler than that for standard EEPROMs. Correspondingly, large integration densities can therefore be achieved, consistent with low prices. Some examples are given in Fig. 10.35

A comparison of the write and read performance of the various ROM types with that of RAMs is shown in Fig. 10.36. We can see that the strength of RAMs is their fast write and read processes, which can be repeated any number of times. With all the ROM variants, writing is subject to more or less severe limitations, although all ROMs have the advantage of retaining their contents even in the absence of a supply voltage. This characteristic can be achieved for RAMs by adding a buffer battery. As we can see from Fig. 10.8, the current drain of many CMOS RAMs is generally lower than the self-discharge of a battery.

Capacity	Organi-zation	Type	Manu-facturer	Supply voltage	Access time	Case
64 Mbit	8 M × 8	K9F6408VOC	Samsung	3.3 V	50 ns	44TSOP
	8 M × 8	M8M29PL64LM	Spansion	3.3 V	90 ns	48TSOP
	4 M × 16	MBM29BS64LF	Spansion	1.8 V	70 ns	60BGA
	4 M × 16	M58CR064C	ST	1.8 V	100 ns	56FBGA
256 Mbit	32 M × 8	K9F5608QOC	Samsung	1.8 V	50 ns	63BGA
	32 M × 8	NAND256W3A	ST	1.8 V	50 ns	48TSOP
	32 M × 8	TC58DVM82A1FT	Toshiba	3.3 V	50 ns	48TSOP
	16 M × 16	GE28F256K3C120	Intel	1.8 V	110 ns	64BGA
	16 M × 16	K9F5616QOC	Samsung	1.8 V	50 ns	63BGA
	16 M × 16	S29WS256N	Spansion	1.8 V	70 ns	84BGA
512 Mbit	64 M × 8	S29GL512N	Spansion	3.3 V	90 ns	56TSOP
	64 M × 8	TC58DVM92A1FT	Toshiba	3.3 V	50 ns	48TSOP
	32 M × 16	K9K1216QOC	Samsung	1.8 V	60 ns	63BGA
	32 M × 16	NAND512W3A	ST	1.8 V	50 ns	48TSOP
2048 Mbit	256 M × 8	TC58NVG153BFT	Toshiba	3.3 V	50 ns	48TSOP
	256 M × 8	K9F2G08QOM	Samsung	1.8 V	50 ns	48TSOP
	128 M × 16	K9F2G16QOM	Samsung	1.8 V	50 ns	48TSOP
4096 Mbit	512 M × 8	K9K4G08QOM	Samsung	1.8 V	50 ns	48TSOP

Fig. 10.35. Examples of Flash EEPROMs. Only a whole block can be written or erased. The block size amounts 1 ... 32 kbyte depending on type. Flash EEPROMs are mainly used in memory cards for USB-sticks and memory cards

	RAM	ROM			
		MROM	PROM	EPROM	EEPROM
Write					
No.	any	once	once	... 100 times	$10^4 ... 10^6$ times
Time	10 ... 200 ns	months	minutes	minutes	milliseconds
Read					
No.	any	any	any	any	any
time	10 ... 200 ns	approx. 100 ns	10 ... 300 ns	30 ... 300 ns	30 ... 300 ns

Fig. 10.36. Comparison of RAMs and ROMs in terms of their write and read performance

Hence, data retention of 10 years can be insured also at RAMs by using appropriate backup batteries.

10.4
Programmable Logic Devices (PLDs)

PLDs are used for storing logic functions. The categories in Fig. 10.1 show that there are three variants: PLEs, PALs, and PLAs. The differences between them are in respect of programming flexibility. PALs (programmable array logic) are the easiest to program. They are therefore particularly popular and are available in a wide range of designs. PLAs (programmable logic arrays) are basically more flexible, but it is more complicated to

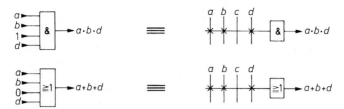

Fig. 10.37. Simplified representation of the AND and OR operations. The crosses indicate which input is connected. An unconnected input has no effect, since it is 1 for the AND operation and 0 for the OR operation

program them. They have therefore ceased to be of major importance. Gate Arrays consist of a see of gates or primitive logical functions the connection of which is configured by programming. GAs are programmed by the manufacturer, therefore they are used with large quantitites only. Very popular are the Field Programmable Gate Arrays FPGAs because they are user programmable like PLDs.

When logic functions are implemented using the standard product terms, it is first necessary to generate all AND (product) terms of the input variables and then form the sum of the products. In order to be able to show these operations clearly, we use the simplified representation in Fig. 10.37. The internal design of PLAs and PALs is shown in Fig. 10.38. The input variables and the intersecting inputs of AND gates form a matrix that enables all the required logical products to be formed. In a corresponding second matrix, the outputs of the AND gates are ORed together in order to form the required logical sums. This requires only one OR gate per output variable. In the case of a PLA both matrices are user-programmable. In the case of a PAL the OR matrix is fixed by the manufacturer, and only the AND matrix can be programmed.

A PROM can also be understood as a function memory if the address decoder, which has a truth table like that shown in Fig. 10.39, is interpreted as an AND matrix. For every address applied, only a single AND operation is 1, namely that corresponding to the address applied. There are therefore $n = 2^N$ product terms, whereas the PLAs and PALs have substantially fewer. The OR-matrix represents the truth-table here.

PROMs designed for implementing logic functions are also known as PLEs (programmable logic elements). The differences become apparent by considering the example in Fig. 10.38. All of the connections that are not required for these functions have been programmed "open". Figure 10.38 shows that all the required logical products are formed in the AND matrices of the PLA and PAL. In the case of the PLA, it is even possible to use a product, which is required several times, twice in the OR matrix. This freedom is not available in the case of the (simple) PAL, as whose OR matrix is not programmable.

With a PROM, in each case the particular product line corresponds to the input combination that is 1. Consequently, in the OR matrix it is necessary to program connections for all combinations that are 1 in the truth table. We can see from Fig. 10.39 that a PROM is the image of the truth table, whereas the PLA and PAL represent the logic functions. A PROM can be used to store any kind of truth table, whereas only a limited number of products and sums are available in a PLA or PAL. For this reason, it is not possible to realize any truth table in a PLA or PAL, but only those that can be converted into simple logic functions. This requires utmost simplification of the functions using Boolean algebra and, if necessary, transformation from AND into OR operations using De Morgan's Law,

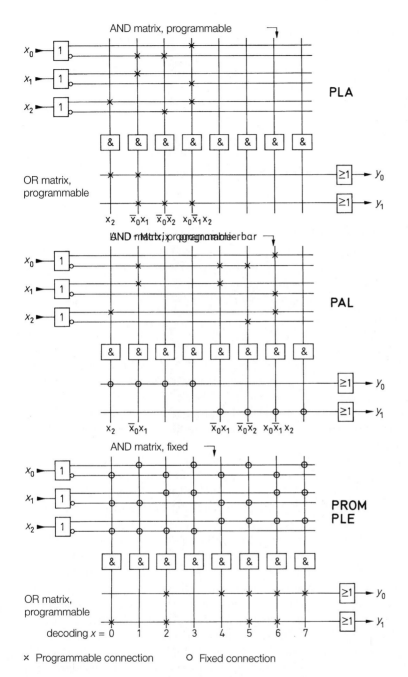

× Programmable connection   ○ Fixed connection

Fig. 10.38. Comparison of the structures of PLA, PAL, and PROM/PLE devices.
An example for $y_0 = x_2 + \bar{x}_0 x_1$ and $y_1 = \bar{x}_0 x_1 + \bar{x}_0 \bar{x}_2 + x_0 \bar{x}_1 x_2$

in order to utilize the PALs as efficiently as possible. Nowadays, this is no longer done
manually, but using computer-aided design programs, which can be run on any PC. Their
application is described in greater detail in Sect. 10.4.2.

Z	x_2	x_1	x_0	y_1	y_0
0	0	0	0	1	0
1	0	0	1	0	0
2	0	1	0	1	1
3	0	1	1	0	0
4	1	0	0	0	1
5	1	0	1	1	1
6	1	1	0	1	1
7	1	1	1	0	1

$$y_0 = x_2 + \overline{x}_0 x_1$$
$$y_1 = \overline{x}_0 x_1 + \overline{x}_0 \overline{x}_2 + x_0 \overline{x}_1 x_2$$

Fig. 10.39. Example of a truth table and its logic functions for the example in Fig. 10.38

10.4.1
Programmable Logic Array (PAL)

PALs are the most important representatives of programmable logic devices (PLDs). They are available in a wide range of variants, all of which are based on the principle shown in Fig. 10.38. The differences are in the implementation of the output circuit. The most commonly used variants are listed in Fig. 10.40. Each different type is designated by the relevant letter shown.

The high (H) output represents the basic type shown in Fig. 10.38. In the case of the low (L) type, the output is negated.

The sharing (S) output has features in common with the PLAs. Here, the OR matrix is also partly programmable: two adjacent OR gates can share the AND operations that are available to them. This makes it possible to use a product term twice in PAL.

With many PALs, an output can also be used as an input or programmed as a bidirectional port (B). This is the purpose of the tristate gate at the output, whose ENABLE is itself a logic function.

An important application of PALs is in sequential logic systems. In order to obviate the need for additional chips, the required registers (R) are incorporated into the PALs. They have a common clock terminal to enable the construction of synchronous systems. In addition, the output signals are generally fed back internally to the AND matrix, thereby eliminating external feedback circuitry (see Fig. 9.62) and saving on pins.

Using the optimum PAL for each application would require a large number of different types – as shown in Fig. 10.40. In order to reduce the variety of types, PALs with a programmable output structure are common today. One such variable "macrocell" (V) is also shown in Fig. 10.40. It is built around a multiplexer that can be used to select any one of four different operating modes. These are defined by programming the function bits f_0 and f_1. The different operating modes are listed in Fig. 10.41. Bit f_0 determines whether or not the output is negated. Bit f_1 switches between combinatorial and registered mode. It also determines, via a second multiplexer, whether feedback is taken from the output or from the register. We can see that most output structures of PALs can be implemented in this way using a single type.

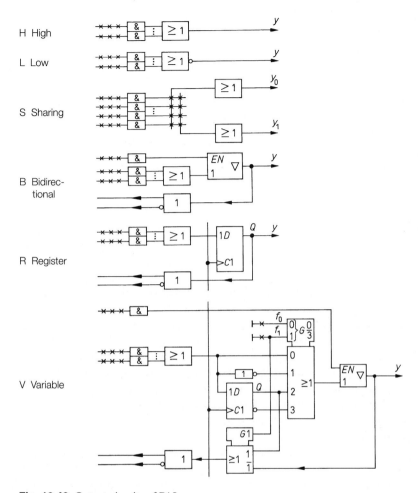

Fig. 10.40. Output circuits of PALs

f_1	f_0	Type	Output	Feedback
0	0	H	Function	Output
0	1	L	Function, negated	Output
1	0	R	Register	Register
1	1	R	Register, negated	Register

Fig. 10.41. Operating modes of the variable macrocell

10.4.2
Computer-Aided PLD Design

In order to "personalize" a PAL, it is first necessary to specify which connections are to be programmed, and then to carry out the programming in a second step. Now that software packages are available which will run on any PC, PAL design is no longer a manual process. The various design phases are shown in Fig. 10.42. There are usually various input formats, of which only the most commonly used will be presented here. The logic function or truth

Input   Design   Verification

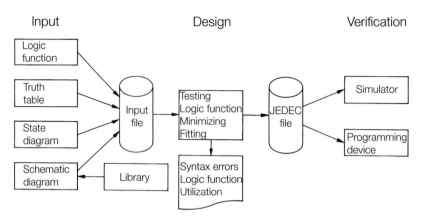

Fig. 10.42. Computer-aided PLD design. You can use ispLEVER with a tutorial in Chap. 29.2 on page 1459

table is entered using a text editor. When designing sequential logic systems, it is also possible to start from a state diagram and specify the transition conditions.

A particularly effective method is to enter the schematic diagram. For this, a library can be used, in which the most common TTL functions are already defined as macros. In addition to gates and flip-flops, the library also provides multiplexers, demultiplexers, adders, comparators, counters and shift registers. This is not only useful for converting an old design that incorporates TTL devices into a PLD design, but it also simplifies the design of new circuits in which the TTL devices are used merely as conceptual models. The input method is supported by a schematic editor.

After input by whatever method, all the data are converted to logic functions and a syntax check is performed. The logic functions are then minimized in accordance with the rules of Boolean algebra. However, this does not yet guarantee that they fit the relevant PLD in their optimum form. For fitting, AND operations, for instance, are converted to OR operations using De Morgan's Law. The programming data (the fuse map) are finally stored in a standard format, the *JEDEC* file.

For the programming process a programmer is connected to the PC and the JEDEC file is downloaded. Newer programmers are controlled entirely by the PC and have no controls of their own. Manually entering the programming data in form of a fuse map is outdated and no longer practical due to the large data quantity. Before choosing a new PLD type it is reasonable to check whether it is supported by the programmer. It is not sufficient to have a universal programmer with software control for the voltages and currents of each pin if the corresponding programming software is not available. These problems do not exist with newer PLDs which are in-circuit or in-system programmable. They are provided with an interface that can be connected directly to the PC. Here, the programming hardware is located on the PLD. This not only renders the programmer or a corresponding socket adapter unnecessary, but also allows to program and update the PLD on the circuit board.

Even though most PLDs can be erased, it is useful to check *prior* to programming whether or not the designed circuit has the desired properties. For this purpose, a *functional simulator* is used to determine whether the output functions respond to the input signals and the clock pulses as intended. The resulting time diagrams are similar to those on the screen of a logic analyser; this, however, does not take the signal delay times into consideration.

Manufacturer	Synthesis	Simulation	Timing
Cadence	Build Gates	NC-VHDL NC-Sim NC-Verilog Verilog-XL	
Expemplar	Leonardo Spectrum		
Mentor Graphics	Precision Leonardo Spectrum	ModelSim	
Model Technology		ModelSim	
Synopsys	Design Compiler FPGA Compiler FPGA Express	Scirocco VCS VSS	Prime Time
Synplicity	Synplify		

Fig. 10.43. Third party support for PLD- and FPGA-design

To test the dynamic response requires a *timing simulator*, which is much more complicated since it must take into account the data and architecture of the particular PLD. At this point you may see whether the fitter needs to cycle twice through the PLD in order to generate a complicated function.

Almost every manufacturer offers a software package for circuit design with their own PLDs. Often, program versions with limited functionality can be obtained free of charge. As a programming language they use HDL or VHDL. For graphic input they offer schematic often in conjunction with a TTL-library.

Programs that are independent of the manufacturer, as listed in Fig. 10.43, offer the great advantage that the manufacturer can be changed without having to re-enter the design. Also beneficial is the fact that they use the same user interface for products of different manufacturers. This, however, requires a device fitter of the respective manufacturer which maps the design in accordance with the architecture of the used module.

10.4.3
Survey of Types Available

Today when designing digital circuits with more than a few 7400 gates it is advantageous to use PLDs for the implementation. The simple PALs listed in Fig. 10.44 already contain 300 to 500 gates. Even if only half of them can be used, a single chip design is already possible with it in many cases. A whole conglomeration of TTLs can be replaced by a single chip. Keeping the number of chips low offers several advantages:

– PCBs can be kept small, saving space and money,
– one PLD is usually less expensive than the total of the components otherwise required
– the reliability increases, since the internal PLD connections are more reliable than those on the PCB
– design changes can often be performed by simply reprogramming the PLD.

All PLDs listed in this survey feature configurable output cells of type V = variable as shown in Fig. 10.40 on page 715. The most important parameter describing the architecture

Type	Manufacturer	Gate equivalent	Architecture			Pins
			Input	Macro cell	Matrix	
16 V 8	La, At, Cy	300	10	8	16×64	20
20 V 8	La, At, Cy	310	12	8	20×64	24
22 V 10	La, At, Cy	400	12	10	22×130	24
26 V 12	La	500	14	12	26×160	28

Manufacturer: AT = Atmel, Cy = Cypress, La = Lattice

Fig. 10.44. Standard PALs in flash EEPROM technology. There are several speed versions: they range between 5 and 25 ns for the delay from input to output. The basic parts do not support in system programming

is the number of macro cells. This is a measure of the number of logical functions that can be formed. Each input of a PLD and each macro cell feed one signal to the AND matrix. This results in the relation:

$$\left.\begin{array}{l} \text{Number of } \textit{inputs} \\ \text{in the AND matrix} \end{array}\right\} = \left\{\begin{array}{l} \text{Number of} \\ \text{inputs + macro cells} \end{array}\right.$$

The number of matrix outputs specifies the possible product terms of the PLD. If these are distributed uniformly to the macro cells the following relation applies:

$$\left.\begin{array}{l} \text{Maximum number of} \\ \text{product terms per function} \end{array}\right\} = \frac{\text{Number of matrix outputs}}{\text{Number of macro cells}}$$

In the 16 V 8 chip with 64 matrix outputs this is $64/8 = 8$. This is sufficient for most functions, but not necessarily for every case. In order to allow implementation of very complicated functions it is sometimes possible to share the product terms of the neighbour (*product term sharing*) or to use *expander product terms* which can be assigned to any function. In this respect they then behave similar to PLAs.

The architecture column in Fig. 10.44 also specifies the number of macro cells in addition to the number of terminals that can be used only as inputs. The output can also be configured as an input, as can be seen for the case of the variable output cell. The architecture of a PAL can be seen from its type designation:

Number of inputs to the matrix   **V**   *Number of macro cells*

22 V 10 is the most commonly used type. There are two options:

– "in system programming" i.e. programming logic integrated in the chip
– "zero power", i.e. the static power consumption is zero.

Large PLDs are realized by a matrix of small PLDs on one chip that can be connected by a programmable interconnect matrix as shown in Fig. 10.45. The product-term allocator of the device fitter determines which product term is realized by what PLD and how the PLDs are interconnected. If a pinout is specified it determines the choice. These composite PLDs are called complex PLDs shortly CPLDs. Some types are collected in Fig. 10.46.

Most PLDs show a surprisingly high power dissipation even in the static state, even though they are CMOS circuits. This is caused by pull-up resistors in wired AND functions. But there are also PLDs with a power consumption that is proportional to the frequency. These consume practically no current in the static state. In Fig. 10.45 these are identified

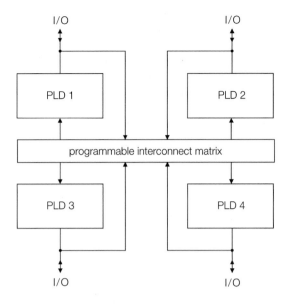

I/O   I/O

PLD 1   PLD 2

programmable interconnect matrix

PLD 3   PLD 4

I/O   I/O

Fig. 10.45. CPLD consisting of 4 PLDs

Manu-facturer	Family	Gate equivalent	Macro cells	Supply voltage	RAM bits	Pins
Altera	MAX3000	600−10 k	32−512	2.5−3.3 V	—	44−256
	MAX7000	600−10 k	32−512	2.5−5 V	—	44−256
Cypress	Delta 39k	16 k−300 k	512−3072	2.5−3.3 V	64 k−384 k	208−676
	Ultra 37k		32−512	3.3−5 V	—	44−400
Lattice	MACH4000Z[1]	1 k−10 k	32−256	1.8 V	—	48−176
	MACH4A	1 k−20 k	32−512	1.8−5 V	—	44−256
	MACH5000	5 k−40 k	128−1024	2.5−3.3 V	—	100−676
	XPLD5000	75 k−300 k	256−1024	1.8−3.3 V	128 k−512 k	208−672
Xilinx	Coolrunner II[1]	750−12 k	32−512	1.8 V	—	44−208
	XPAL3[1]	750−12 k	32−512	3.3 V	—	44−208
	XC9500	800−7 k	36−288	2.5−5 V	—	44−352

[1] Zero power: nearly no static power consumption

Fig. 10.46. Examples for CPLD-familes. All types are in system programmable

as "zero power" types. The current consumption these types is especially low, even during operation.

If a few bits have to be stored in a sequential logic this can be achieved by using the registers of the macro cells. But for larger data quantities this method is not practical; then an external RAM must be connected. With some of the newer CPLDs this is not necessary, since RAMs are available on chip; the storage capacity is shown in Fig. 10.46.

10.4.4
User Programable Gate Arrays

One group of logic circuits that can also be programmed by the user is that if the *Field Programmable Gate Arrays* (FPGAs). They contain many Configurable Logic Blocks, CLBs that can be connected by column- and row-interconnect busses as shown in Fig. 10.47. The

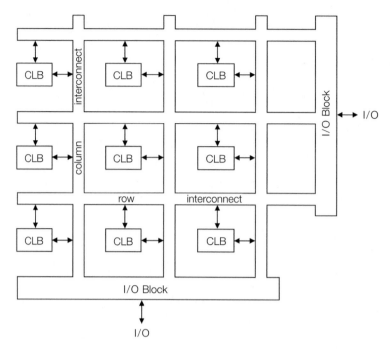

Fig. 10.47. Structure of a FPGA. CLB = Configurable Logic Block

CLBs consist of configurable gates and flip-flops like macrocells in PLDs. Their structure is much simpler than that of macrocells but their number is much greater as can be seen in Fig. 10.48. FPGAs can therefore be used for complex designs. They are particularly advantageous when the architecture of a PLD is not suitable for the desired functionality.

For the circuit designer there is little difference in designing a PLD or FPGA. Some manufacturers offer the same design software for PLDs and FPGAs. But the *device fitter* is much more complex for FPGAs. In a first step it must *place* the functions in the CLBs and in a second step it must *route* the connections. Which connections are time-critical can be specified, since there are usually several data paths of different delay times available and short connections are faster than those forming several corners. Here, a timing simulation which takes account of the wiring (*Post Layout Simulation*) is essential.

In most FPGAs the configuration of the routing is stored in a RAM on the chip. During power-up this is automatically loaded via a serial interface from a separate EPROM. Fundamentally different is the programming of FPGAs in *antifuse* technology. Here, the connections are made by programming. Thus it is the opposite of the *Fusible Link* technology used in PROMs in which the connection is interrupted by the programming process. As with a fusible link, a programmable link – antifuse – cannot be erased.

FPGAs are often used for digital signal processing. For these applications some manufacturers offer pre-programmed multipliers, adders, and RAM blocks on the chips that can be fitted in the design.

Manu-facturer	Family	Gates max	CLBs	RAM bits max	Program-ming	specific features
Actel	ProASIC	75 k−1 M	3 k−56 k	2 k−200 k	Flash	PLL
	Axcelerator	125 k−2 M	672−10 k	18 k−300 k	Anti	PLL
Altera	Cyclone II		5 k−68 k	120 k−1 M	RAM	Mult + PLL
	Startix		10 k−80 k	1 M−7 M	RAM	Add + Mult + PLL
	Apex II	2 M−5 M	16 k−67 k	500 k−1 M	RAM	PLL
Lattice	ORCA4	200 k−900 k	5 k−16 k	74 k−150 k	RAM	
	ECXP	140 k−1 M	2 k−15 k	92 k−414 k	Flash	
	ECP2		6 k−68 k	55 k−1 M	RAM	Add + Mult + PLL
Quicklogic	Eclipse II	47 k−320 k	128−1.5 k	9 k−55 k	Anti	PLL, ZP
	PolarPro	75 k−1 M	512−8 k	37 k−100 k	Anti	PLL, ZP, FiFo
Xilinx	Spartan 3	50 k−5 M	1.7 k−75 k	72 k−1.8 M	RAM	Mult
	Virtex 4		14 k−200 k	800 k−6 M	RAM	Add + Mult

Fig. 10.48. Examples for FPGAs.

Programming: Flash = flash EEPROM on chip with in system programming
                          Anti   = Antifuse = programmable connection, irreversible
                          RAM = configuration RAM on chip. Must be loaded after power-up
Specific features: Mult = wired multiplier blocks on chip
                          Add   = wired Adder on chip
                          PLL = phase locked loop for clock generation on chip
                          ZP   = zero power = nearly no static power consumption

Part II

General Applications

Chapter 11:
Operational Amplifier Applications

Most analog signal processing is done today with circuits using operational amplifiers because opamps are available with good data for little money. Most signals that must be processed in electronic circuits arise in analog form and are needed after processing in analog form also. Therefore analog signal processing is first choice.

The accuracy of analog processing is limited to 0.1 to 1%. If higher accuracy is needed, digital processing is advantageous. The same is true if the required processing is complex or nonlinear or needs memory. To use the advantages of digital signal processing, the signals must be digitized before digital processing and converted back to analog signals afterwards. Therefore, Analog-Digital Converters (ADCs) and Digital-Analog Converters (DACs) as described in Chap. 18 are needed. At the input of ADCs and the output of DACs, the analog signal must be amplified and filtered. Therefore, analog signal processing is necessary also in this case.

In the following sections, the most important families of operational circuits are classified and described. They are circuits for the four fundamental arithmetic operations, for differential and integral operations, and for the synthesis of transcendental or any other chosen functions. In order to illustrate the operating principles of these circuits as clearly as possible, we initially assume ideal characteristics of the operational amplifiers involved. When using real operational amplifiers, restrictions and additional conditions must be observed in the choice of the circuit parameters, and these are treated thoroughly in Chap. 5. We want to discuss in more detail only those effects that play a special role in the performance of the particular circuit.

11.1
Summing Amplifier

When connected as an inverting amplifier, an operational amplifier can be used for the addition of several voltages. As Fig. 11.1 indicates, the input voltages are connected via series resistors to the N-input of the operational amplifier. Since this node represents virtual ground, Kirchhoff's current law (KCL) directly yields the relation for the output voltage:

$$\frac{V_1}{R_1} + \frac{V_2}{R_2} + \cdots + \frac{V_n}{R_n} + \frac{V_o}{R_N} = 0$$

If a DC voltage is added to the signal voltage in the manner described, the inverting summing amplifier can also be used as an amplifier with a wide-range zero adjustment.

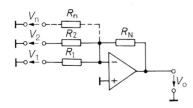

Fig. 11.1. Inverting summing amplifier

Output voltage:

$$-V_o = \frac{R_N}{R_1}V_1 + \frac{R_N}{R_2}V_2 + \cdots + \frac{R_N}{R_n}V_n$$

11.2
Subtracting Circuits

11.2.1
Reduction to an Addition

A subtraction operation can be reduced to the problem of an addition by inverting the signal to be subtracted. This requires the circuit shown in Fig. 11.2. The operational amplifier OA1 inverts the input voltage V_2; the output voltage is then

$$V_o = \alpha_P V_2 - \alpha_N V_1 \tag{11.1}$$

If both resistance ratios are equal

$$\alpha_P = \alpha_N = \alpha$$

then the output voltage is the amplified difference

$$V_o = \alpha(V_2 - V_1) = A_D(V_2 - V_1).$$

If the resistance ratios are not equal because of tolerances, we get an outpout voltage even if both input voltages are equal

$$V_1 = V_2 = V_{CM} \tag{11.2}$$

resulting in

$$V_o = \alpha_P V_{CM} - \alpha_N V_{CM} = V_{CM}(\alpha_P - \alpha_N).$$

With

$$\alpha_P = \alpha + \frac{1}{2}\Delta\alpha \quad \text{and} \quad \alpha_N = \alpha - \frac{1}{2}\Delta\alpha$$

we obtain

$$V_o = V_{CM}\Delta\alpha = V_{CM}A_{CM}.$$

From this we can calculate the common mode rejection

$$G = \frac{A_D}{A_{CM}} = \frac{\alpha}{\Delta\alpha} = \frac{1}{\text{tolerance}}. \tag{11.3}$$

The common-mode rejection ratio thus equals the reciprocal of the relative matching of the individual gains. If resistors with a tolerance of $1\% = 0.01$ are used, a common-mode rejection of $G = 1/0.01 = 100$ can be expected.

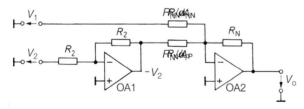

Fig. 11.2. Subtracting circuit using a summing amplifier

Output voltage: $V_o = \alpha_D(V_2 - V_1)$
Condition for coefficients: $\alpha_N = \alpha_P = \alpha$

11.2.2
Subtraction Using a Single Operational Amplifier

To calculate the output voltage of the subtracting amplifier in Fig. 11.3, we may use the principle of superposition. We therefore write

$$V_o = k_1 V_1 + k_2 V_2$$

For $V_2 = 0$, the circuit is an inverting amplifier, where $V_o = -\alpha_N V_1$. It follows $k_1 = -\alpha_N$. For $V_1 = 0$, the circuit represents a noninverting amplifier that has a voltage divider connected at its input. The potential

$$V_P = \frac{R_P}{R_P + R_P/\alpha_P} V_2 = \frac{\alpha_P}{1 + \alpha_P} V_2$$

is thus amplified by the factor $(1 + \alpha_N)$, and this results in the output voltage

$$V_o = \frac{\alpha_P}{1 + \alpha_P} (1 + \alpha_N) V_2$$

If both resistor ratios are the same – that is, if $\alpha_N = \alpha_P = \alpha$ – it follows:

$$V_o = \alpha V_2$$

and that $k_2 = \alpha$. We now obtain the output voltage for the general case, in the form

$$V_o = \alpha (V_2 - V_1)$$

Should the ratios of the resistors at the P- and N-inputs not be precisely equal to α, the circuit will not evaluate the precise difference between the input voltages. In this case,

$$V_o = \frac{1 + \alpha_N}{1 + \alpha_P} \alpha_P V_2 - \alpha_N V_1$$

To calculate the common-mode rejection ratio we again use the formulation of (11.3), and obtain

$$G = \frac{A_D}{A_{CM}} = \frac{1}{2} \cdot \frac{(1 + \alpha_N)\alpha_P + (1 + \alpha_P)\alpha_N}{(1 + \alpha_N)\alpha_P - (1 + \alpha_P)\alpha_N}$$

With $\alpha_N = \alpha - \frac{1}{2}\Delta\alpha$ and $\alpha_P = \alpha + \frac{1}{2}\Delta\alpha$, the expression may be rewritten and expanded into a series. Neglecting higher-order terms, we obtain

$$G \approx (1 + \alpha) \frac{\alpha}{\Delta\alpha} \approx A_D \frac{1}{\text{tolerance}} \tag{11.4}$$

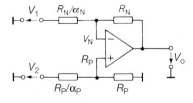

Fig. 11.3. Subtracting circuit using a single amplifier (a subtracting amplifier)

Output voltage: $\qquad$ $V_o = \alpha (V_2 - V_1)$

Condition for coefficients: $\qquad$ $\alpha_N = \alpha_P = \alpha$

For constant α, the common-mode rejection ratio is inversely proportional to the tolerance of the resistor ratios. If the resistor ratios are identical, $G = \infty$, although this applies to ideal operational amplifiers only. In order to obtain a particularly high common-mode rejection ratio under real conditions, R_P may be varied slightly. In this way, $\Delta\alpha$ can be adjusted and the finite common-mode rejection ratio of the operational amplifier can be compensated for.

Equation (11.4) also shows that the common-mode rejection ratio for a given resistor matching tolerance $\Delta\alpha/\alpha$ is approximately proportional to the chosen differential gain $A_D = \alpha$. This is a great improvement over the previous circuit.

This may be best illustrated by an example. Two voltages of about 10 V are to be subtracted one from the other. Their difference is no more than 100 mV. This value is to be amplified and is to appear at the output of the subtraction amplifier as a voltage of 5 V, with an accuracy of 1%. The differential gain must therefore be $A_D = 50$. The absolute error at the output must be smaller than $5\,V \cdot 1\% = 50\,mV$. If we assume the favorable case of the common-mode gain representing the only source of error, we then find it necessary to limit the common-mode gain to

$$A_{CM} \le \frac{50\,mV}{10\,V} = 5 \cdot 10^{-3}$$

that is,

$$G \ge \frac{50}{5 \cdot 10^{-3}} = 10^4 \hateq 80\,dB$$

For the subtracting amplifier in Fig. 11.3, this requirement can be met by a relative resistor matching tolerance of $\Delta\alpha/\alpha = 0{,}5\%$, as follows from (11.4). For the subtraction circuit of Fig. 11.2, however, (11.3) yields a maximum tolerable mismatch of 0.01%.

Figure 11.4 shows an expansion of the subtracting amplifier for any number of additional summing and/or subtracting inputs. The determining factor for the proper functioning of the circuit is that the specified coefficient condition is satisfied.

If this is not achieved with the coefficients α_i and α_i' specified, the voltage 0 can be added or subtracted using the missing coefficient required to satisfy the equation.

In order to deduce the relationships given below the caption to Fig. 11.4, we apply Kirchhoff's current law to the N-input:

$$\sum_{i=1}^{m} \frac{V_i - V_N}{\left(\dfrac{R_N}{\alpha_i}\right)} + \frac{V_o - V_N}{R_N} = 0$$

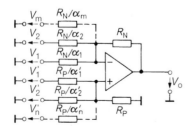

Fig. 11.4. Multiple-input subtracting amplifier

Output voltage: $\qquad V_o = \sum_{i=1}^{n} \alpha_i' V_i' - \sum_{i=1}^{m} \alpha_i V_i$

Condition for coefficients: $\qquad \sum_{i=1}^{n} \alpha_i' = \sum_{i=1}^{m} \alpha_i$

Hence

$$\sum_{i=1}^{m} \alpha_i V_i - V_N \left[\sum_{i=1}^{m} \alpha_i + 1 \right] + V_o = 0$$

Similarly, we obtain for the P-input

$$\sum_{i=1}^{n} \alpha'_i V'_i - V_P \left[\sum_{i=1}^{n} \alpha'_i + 1 \right] = 0$$

With $V_N = V_P$ and the coefficient condition

$$\sum_{i=1}^{m} \alpha_i = \sum_{i=1}^{n} \alpha'_i \qquad (11.5)$$

the subtraction of the two equations results in

$$V_o = \sum_{i=1}^{n} \alpha'_i V'_i - \sum_{i=1}^{m} \alpha_i V_i$$

For $n = m = 1$, the multiple-input subtracting amplifier becomes the basic circuit of Fig. 11.3.

The inputs of computing circuits represent loads to the signal voltage sources. The output resistances of the latter have to be sufficiently low to minimize the computing errors. If the sources themselves are feedback amplifiers, this condition is generally well satisfied. When other signal sources are used, it may become necessary to employ impedance converters connected to the inputs. These converters often take the form of noninverting amplifiers, and the resulting subtracting circuits are then called instrumentation amplifiers. They are commonly used in the field of measurement, and are dealt with extensively in Chap. 20.1.2 on page 1032.

11.3
Bipolar-Coefficient Circuit

The circuit shown in Fig. 11.5 allows the multiplication of the input voltage by a constant factor, the value of which can be set between the limits $+n$ and $-n$ by the potentiometer R_2. If the slider of the potentiometer is positioned as far to the right as possible, then $q = 0$ and the circuit operates as an inverting amplifier with gain $A = -n$. In this case, the resistor $R_1/(n-1)$ is ineffective, since there is no voltage across it.

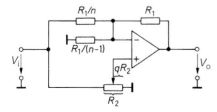

Fig. 11.5. Bipolar-coefficient circuit

Output voltage: $V_o = n(2q - 1)V_i$

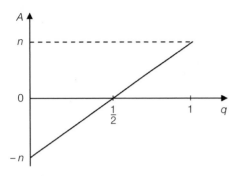

Fig. 11.6. Gain as a function of q. For $q = 1/2$ the gain is $A = 0$

For $q = 1$, the full input voltage V_i is at the P-input. The voltage across R_1/n is therefore zero, and the circuit operates as a noninverting amplifier that has the gain

$$A = 1 + \frac{R_1}{R_1/(n-1)} = +n$$

For intermediate positions, the gain is

$$A = n(2q - 1)$$

It is thus linearly dependent on q and can be easily adjusted – for instance, by means of a calibrated multi-turn potentiometer. The factor n determines the range of the coefficient. The smallest value is $n = 1$; in this case, the resistor $R_1/(n-1)$ may be omitted.

11.4
Integrators

One of the most important applications of the operational amplifier in analog computing circuits is as an integrator. Its output voltage can be expressed in the general form

$$V_o(t) = K \int_0^t V_i(\tilde{t})d\tilde{t} + V_o(t = 0)$$

11.4.1
Inverting Integrator

The inverting integrator shown in Fig. 11.7 differs from the inverting amplifier in that the feedback resistor R_N is replaced by the capacitor C. The output voltage is then expressed by

$$V_o = \frac{Q}{C} = \frac{1}{C}\left[\int_0^t I_C(\tilde{t})d\tilde{t} + Q_0\right]$$

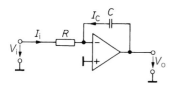

Fig. 11.7. Inverting integrator

Output voltage: $\quad V_o = -\dfrac{1}{RC}\displaystyle\int_0^t V_i(\tilde{t})d\tilde{t} + V_{o0}$

where Q_0 is the charge on the capacitor at the beginning of the integration ($t = 0$). As $I_C = -V_i/R$, it follows:

$$V_o = -\frac{1}{RC}\int_0^t V_i(\tilde{t})d\tilde{t} + V_{o0}$$

The constant V_{o0} represents the initial condition: $V_{o0} = V_o(t = 0) = Q_0/C$. It has to be set to a defined value by the additional measures described in the next section.

Let us now look at two special cases. If the input voltage V_i is constant, the output voltage is

$$V_o = -\frac{V_i}{RC}t + V_{o0}$$

which increases linearly with time. This circuit is thus very well suited to the generation of triangular and sawtooth voltages.

If V_i is a cosinusoidal alternating voltage $v_i = \widehat{V}_i \cos \omega t$, the output voltage becomes

$$V_o(t) = -\frac{1}{RC}\int_0^t \widehat{V}_i \cos \omega \tilde{t} d\tilde{t} + V_{o0} = -\frac{\widehat{V}_i}{\omega RC}\sin \omega t + V_{o0}$$

The amplitude of the alternating output voltage is therefore inversely proportional to the angular frequency ω. When the amplitude–frequency response is plotted using log–log coordinates, the result is a straight line with a slope of -6 dB per octave. This characteristic is a simple criterion for determining whether a circuit behaves like an integrator.

The behavior in the frequency domain can also be determined directly with the help of complex calculus:

$$\underline{A} = \frac{\underline{V_o}}{\underline{V_i}} = -\frac{\underline{Z_C}}{R} = -\frac{1}{s\,RC} \tag{11.6}$$

Hence, it follows for the ratio of the amplitudes:

$$\frac{\widehat{V}_o}{\widehat{V}_i} = |\underline{A}| = \frac{1}{\omega RC}$$

as shown before.

With regard to the frequency compensation, it must be noted that, unlike all of the circuits that have been discussed previously, the feedback network causes a phase shift. This means that the feedback factor becomes complex:

$$\underline{k} = \frac{\underline{V_N}}{\underline{V_o}}\bigg|_{V_i=0} = \frac{s\,RC}{1+s\,RC} \tag{11.7}$$

For high frequencies, $\underline{k}$ approaches $\underline{k} = 1$ and the phase shift becomes zero. Therefore, in this frequency range the same conditions obtain as for a unity-gain inverting amplifier (see

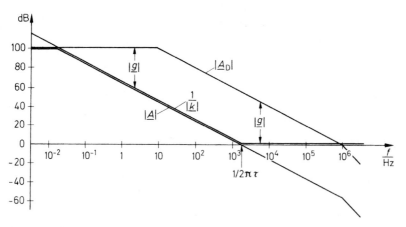

Fig. 11.8. Frequency response of the loop gain g

Chap. 5). The frequency compensation that is necessary in the latter case must therefore also be used for the integrator circuit. Internally compensated amplifiers are normally designed for this application, and are therefore suitable for use as integrators.

The frequency range that is usable for integration can be seen in Fig. 11.8, to give a typical example. The integration time constant chosen is $\tau = RC = 100\,\mu s$. It is apparent that by making this choice, a maximum loop gain of $|g| = |k\,A_D| \approx 600$ is attained, which corresponds to an output accuracy of about $1/|g| \approx 0.2\%$. In contrast to that of the inverting amplifier, the output accuracy falls not only at high, but also at low, frequencies.

For the real operational amplifier, the input bias current I_B and the offset voltage V_O may be very troublesome, as their effects accumulate with time. If the input voltage V_i is reduced to zero, the capacitor carries the error current

$$\frac{V_o}{R} + I_B$$

This results in a change in output voltage

$$\frac{dV_o}{dt} = \frac{1}{C}\left(\frac{V_o}{R} + I_B\right) \tag{11.8}$$

An error current of $I_B = 1\,\mu A$ causes the output voltage to rise at a rate of 1 V/s if $C = 1\,\mu F$. Equation (11.8) indicates that, for a given time constant, the contribution of the input bias current is smaller, the larger is the value of C chosen. The contribution of the offset voltage

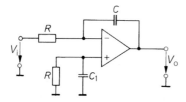

Fig. 11.9. Integrator with input bias current compensation. Capacitor C_1, shorts noise voltages at the P-input

remains constant. Because there is a limit to the size of C, it is important to at least make certain that the effect of I_B does not exceed that of V_0. This is the case if

$$I_B < \frac{V_0}{R} = \frac{V_0 C}{\tau}$$

If a time constant of $\tau = 1\,\text{s}$ has to be achieved with a capacitance of $C = 1\,\mu\text{F}$, an operational amplifier that has an offset voltage of $V_O = 1\,\text{mV}$ should possess an input bias current smaller than

$$I_B = \frac{1\,\mu\text{F} \cdot 1\,\text{mV}}{1\,\text{s}} = 1\,\text{nA}$$

Operational amplifiers with bipolar input transistors rarely have such low input bias currents. Their undesirable effects can be reduced by not connecting the P-input to ground directly, but via a resistor that also has the value R. The voltage $I_B R$ is then dropped across both resistors and the error current through capacitor C is therefore zero. The only remaining error current is now the difference between the input bias currents – that is, the offset current – which, however, is generally small in comparison.

In the case of FET operational amplifiers, the input bias current is usually negligible. They are therefore preferred if the integration time constants are large, even though their offset voltages are often much larger than for operational amplifiers with bipolar input transistors.

Leakage currents through the capacitor may be a further source of error. As electrolytic capacitors have leakage currents that are of the order of microamperes, they cannot be used as integration capacitors. It is therefore necessary to use foil capacitors, which makes capacitances of over $10\,\mu\text{F}$ very bulky.

11.4.2
Initial Condition

An integrator can often be used only if its output voltage $V_o(t = 0)$ can be set independently of the input voltage. Using the circuit shown in Fig. 11.10, it is possible to stop integration and set the initial condition.

If the switch S_1 is closed and S_2 is open, the circuit operates like that in Fig. 11.7; the voltage V_1 is integrated. If switch S_1 is now opened, the charging current becomes zero in the case of an ideal integrator, and the output voltage remains at the value that it had at the time of switching. This may be of use if we want to interrupt computation; for example, in order to read the output voltage at leisure. To set the initial condition, S_1 is left open and S_2 is closed. The integrator becomes an inverting amplifier, with an output voltage of

$$V_o = -\frac{R_N}{R_2} V_2$$

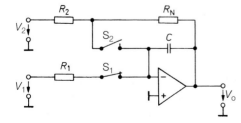

Fig. 11.10. Integrator that has three modes of operation: integrate, hold, and set initial condition

Initial condition : $V_o(t = 0) = -\dfrac{R_N}{R_2} V_2$

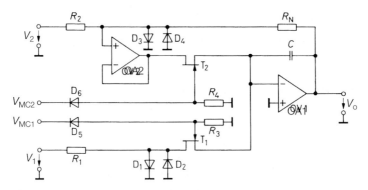

Fig. 11.11. Integrator with electronic mode control

Initial condition : $V_o(t = 0) = -\dfrac{R_N}{R_2} V_2$

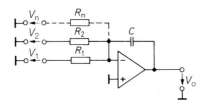

Fig. 11.12. Summing integrator

Output voltage:

$$V_o = -\frac{1}{C} \int_0^t \left(\frac{V_1}{R_1} + \frac{V_2}{R_2} + \cdots + \frac{V_n}{R_n} \right) d\tilde{t} + V_{o\,0}$$

However, the output assumes this voltage only after a certain delay, which is determined by the time constant $R_N C$.

Figure 11.11 shows one possibility for replacing the switches by electronic components. The two FETs T_1 and T_2 replace the switches g_{m1} and g_{m2} shown in Fig. 11.10. They are conducting (ON) if the corresponding mode control signal voltage is greater than zero. For sufficiently negative control voltages they are in the OFF state. The precise operation of the FET switches, and of the diodes $D_1 - D_6$, is described in more detail in Chap. 17.2.1.

The voltage follower OA2 reduces the delay time constant for setting the initial condition, from the value $R_N C$ to the small value of $R_{DS\,on} C$.

11.4.3
Summing Integrator

Just as the inverting amplifier can be extended to become a summing amplifier, so an integrator can be developed into a summing integrator as shown in Fig. 11.12. The relationship given for the output voltage can be derived directly by applying KCL to the summing point.

11.4.4
Noninverting Integrator

For integration without polarity reversal, an inverting amplifier can be added to the integrator. Another solution is shown in Fig. 11.13. The circuit basically consists of a lowpass filter as the integrating element. A NIC, having an internal resistance of $-R$, is connected in

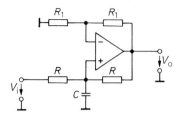

Fig. 11.13. Noninverting integrator

Output voltage:

$$V_o = \frac{2}{RC} \int_0^t V_i(\tilde{t})d\tilde{t} + V_{o0}$$

parallel with the filter and simultaneously acts as an impedance converter (see Chap. 12.5). To calculate the output voltage, we apply KCL to the P-input and obtain

$$\frac{V_o - V_P}{R} + \frac{V_i - V_P}{R} - C\frac{dV_P}{dt} = 0$$

Hence, with $V_P = V_N = \frac{1}{2}V_o$ we arrive at the result

$$V_o = \frac{2}{RC} \int_0^t V_i(\tilde{t})d\tilde{t}$$

It is to be noted that the input voltage source must have a low impedance; otherwise, the stability condition for the NIC is not fulfilled. The operational amplifier evaluates differences between large quantities, and therefore this integrator does not have the same precision as the basic circuit shown in Fig. 11.7.

11.5
Differentiators

11.5.1
Basic Circuit

If the resistor and capacitor of the integrator in Fig. 11.7 are interchanged, we obtain the differentiator shown in Fig. 11.14. The application of KCL to the summing point yields the relationship

$$C\frac{dV_i}{dt} + \frac{V_o}{R} = 0,$$

$$V_o = -RC\frac{dV_i}{dt} \tag{11.9}$$

Thus, for sinusoidal alternating voltages $V_i = \widehat{V}_i \sin \omega t$, we obtain the output voltage

$$V_o = -\omega RC \widehat{V}_i \cos \omega t$$

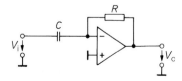

Fig. 11.14. Differentiator

Output voltage: $V_o = -RC\dfrac{dV_i}{dt}$

For the ratio of the amplitudes, it follows:

$$\frac{\widehat{V}_o}{\widehat{V}_i} = |\underline{A}| = \omega\,RC \tag{11.10}$$

When the frequency response of the gain is plotted using log–log coordinates, the result is a straight line with a slope +6 dB per octave. In general, a circuit is said to behave as a differentiator in a particular frequency range if, in that range, the amplitude–frequency response rises at a rate of 6 dB per octave.

The behavior in the frequency domain can also be determined directly with the help of complex calculus:

$$\underline{A} = \frac{V_o}{\underline{V}_i} = -\frac{R}{\underline{Z}_C} = -s\,RC \tag{11.11}$$

Hence

$$|\underline{A}| = \omega\,RC$$

in accordance with (11.10).

11.5.2
Practical Implementation

The practical implementation of the differentiator circuit shown in Fig. 11.14 presents certain problems, since the circuit is prone to oscillations. These are caused by the feedback network, which – at higher frequencies – gives rise to a phase lag of 90°, as the feedback factor is

$$\underline{k} = \frac{V_D}{V_o} = \frac{1}{1 + s\,RC} \tag{11.12}$$

This lag is added to the phase lag of the operational amplifier, which in the most favorable case is already 90°. The remaining phase margin being zero, the circuit is therefore unstable. This instability can be overcome if the phase shift of the feedback network at high frequencies is reduced by connecting a resistor R_1 in series with the differentiating capacitor, as shown in Fig. 11.15. This measure need not necessarily reduce the usable frequency range, since the reduction in loop gain limits the satisfactory operation of the differentiator at higher frequencies.

For the cutoff frequency f_1 of the lowpass element $R_1\,C$, it is advisable to choose the value for which the loop gain becomes unity. To find this value, we consider a fully compensated amplifier, the amplitude–frequency response of which is shown as a dashed

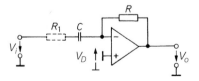

Fig. 11.15. Practical design of a differentiator

Output voltage: $\quad V_o = -RC\,\dfrac{dV_i}{dt}\quad$ for $\quad f \ll \dfrac{1}{2\pi\,R_1 C}$

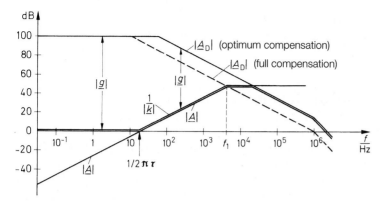

Fig. 11.16. Example for the frequency response of the loop gain $f_1 = \sqrt{f_T/2\pi\tau}$ where $\tau = RC$ and f_T is the unity gain bandwidth

line in the example of Fig. 11.16. The phase margin at frequency f_l is then approximately $45°$. Since, in the vicinity of frequency f_1, the amplifier has a feedback factor of less than unity, an increase in the phase margin can be obtained by reducing the frequency compensation, and hence a transient behavior of near-critical damping can be approached.

To optimize the compensation capacitor C_k, a triangular voltage is applied to the input of the differentiator and C_k is reduced so that the rectangular output voltage is optimally damped.

11.5.3
Differentiator with High Input Impedance

The input impedance of the differentiator described exhibits capacitive behavior, which in some cases can lead to difficulties; for example, an operational amplifier circuit used as an input voltage source can easily become unstable. The differentiator shown in Fig. 11.17 is better in this respect. Its input impedance does not fall below the value of R, even at high frequencies.

The operation of the circuit is best illustrated as follows. Alternating voltages of low frequency are differentiated by the RC network at the input. In this frequency range, the operational amplifier corresponds to a noninverting amplifier, and has a gain of $\underline{A} = 1$.

Alternating voltages of high frequency pass the input RC element unchanged and are differentiated by the feedback amplifier. If both time constants are equal, the effects of differentiation at low frequencies and at high frequencies overlap and produce a smooth changeover.

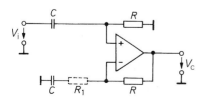

Fig. 11.17. Differentiator with high input impedance

Output voltage: $V_o = RC \dfrac{dV_i}{dt}$

Input impedance: $|\underline{Z}_i| \geq R$

With regard to stabilization against likely oscillations, the same principles apply as for the previous circuit. The damping resistor R_1 is shown by the dashed line in Fig. 11.17.

11.6
Solving Differential Equations

There are many problems that can be described most easily in the form of differential equations. One obtains the solution by using the analog computing circuits described above to model the differential equation, and by measuring the resulting output voltage. In order to avoid stability problems, the differential equation is transformed in such a way that only integrators are required, rather than differentiators.

We will illustrate this method using the example of a linear second-order differential equation:

$$y'' + k_1 y' + k_0 y = f(x) \tag{11.13}$$

In the first step, the independent variable x is replaced by the time variable t:

$$x = \frac{t}{\tau}$$

Since

$$y' = \frac{dy}{dt} \cdot \frac{dt}{dx} = \tau \dot{y} \quad \text{and} \quad y'' = \tau^2 \ddot{y}$$

the differential equation (11.13) becomes

$$\tau^2 \ddot{y} + k_1 \tau \dot{y} + k_0 y = f(t/\tau) \tag{11.14}$$

In the second step, the equation is solved for the undifferentiated quantities:

$$k_0 y - f(t/\tau) = -\tau^2 \ddot{y} - k_1 \tau \dot{y}$$

Thirdly, the equation is multiplied throughout by the factor $-1/\tau$ and integrated:

$$-\frac{1}{\tau} \int [k_0 y - f(t/\tau)] dt = \tau \dot{y} + k_1 y \tag{11.15}$$

In this way, an expression is formed on the left-hand side of (11.15), which can be computed by a simple summing integrator. Its output voltage is termed the state variable, z_n, where n is the order of the differential equation; here, $n = 2$. Therefore,

$$z_2 = -\frac{1}{\tau} \int [k_0 y - f(t/\tau)] dt \tag{11.16}$$

In this equation, the output variable y is initially taken as being known.

By inserting (11.16) in (11.15), we arrive at

$$z_2 = \tau \dot{y} + k_1 y \tag{11.17}$$

This differential equation is now treated in the same way as (11.14), and therefore we obtain

$$z_2 - k_1 y = \tau \dot{y},$$
$$-\frac{1}{\tau} \int [z_2 - k_1 y] dt = -y \tag{11.18}$$

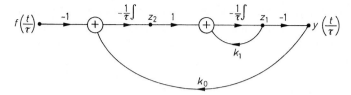

Fig. 11.18. Signal flow graph for solving the differential equation

$$\tau^2 \ddot{y} + k_1 \tau \dot{y} + k_0 y = f\left(\frac{t}{\tau}\right)$$

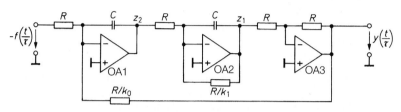

Fig. 11.19. Practical analog computing circuit

The left-hand side represents the state variable z_1:

$$z_1 = -\frac{1}{\tau} \int [z_2 - k_1 y] dt \qquad (11.19)$$

This expression is formed by a second summing integrator. Substitution in (11.18) gives the equation for the output:

$$y = -z_1 \qquad (11.20)$$

Since there are no longer any derivatives, the procedure is terminated. The last equation, (11.20), provides the missing relation for the output variable y, which had initially been taken as known.

The operations that are necessary for solving the differential equation – (11.16), (11.19), and (11.20) – can be represented clearly with the aid of a signal flow graph, as shown in Fig. 11.18. The appropriate analog computing circuit is shown in Fig. 11.19. In order to save an additional inverting amplifier for the computation of the expression $-k_1/y$ in (11.19), we make use of the fact that, from (11.20), $z_1 = -y$.

11.7
Function Networks

The problem often arises that a function, $V_2 = f(V_1)$ is to be calculated, where f is a nonlinear function of V_1, so that for example

$$V_2 = V_A \log \frac{V_1}{V_B}$$

or

$$V_2 = V_A \sin \frac{V_1}{V_B}$$

There are three possibilities for realizing such relationships. One can either make use of a physical effect that complies with the function required, or one can approximate the function by a piecewise linear approximation or by a power series expansion. Below, we give some examples of these methods.

11.7.1
Logarithm

A logarithmic amplifier must produce an output voltage that is proportional to the logarithm of the input voltage. It is therefore possible to make use of the diode characteristic

$$I_A = I_S \left(e^{\frac{V_{AC}}{nV_T}} - 1 \right) \tag{11.21}$$

In this equation, I_S is the saturation leakage current, $V_T = kT/q \approx 26\,\text{mV}$ is the thermal voltage, and n is a correction factor between 1 and 2. For the forward-biased diode, when $I_A \gg I_S$, (11.21) can be approximated with good accuracy by

$$I_A = I_S e^{\frac{V_{AC}}{nV_T}} \tag{11.22}$$

Hence, the forward voltage is

$$V_{AC} = nV_T \ln \frac{I_A}{I_S} \tag{11.23}$$

which is the required logarithmic function. The simplest way of using this relationship for computation of the logarithm is shown in Fig. 11.20, where a diode is incorporated into the feedback loop of an operational amplifier. This amplifier converts the input voltage V_i to a proportional current $I_A = V_i/R_1$. At the same time, the voltage $V_o = -V_{AC}$ appears at its low-impedance output. Therefore

$$V_o = -nV_T \ln \frac{V_i}{I_S R_1} = -nV_T \ln 10 \lg \frac{V_i}{I_S R_1} \tag{11.24}$$

or

$$V_o = -(1\ldots 2) \cdot 60\,\text{mV} \lg \frac{V_i}{I_S R_1} \quad \text{at room temperature.}$$

The usable range is limited by two effects. The diode possesses a parasitic series resistance, across which a considerable voltage is present at high currents, leading to errors in the computation of the logarithm. In addition, the correction factor n is current-dependent. A satisfactory accuracy can therefore only be achieved over an input voltage range of one or two decades.

The unfavorable effect of the varying correction factor n can be eliminated by replacing the diode D by a transistor T, as shown in Fig. 11.21. In accordance with (2.2), if $I_C \gg I_{CS}$, for the collector current we write

$$I_C = I_S e^{V_{BE}/V_T} \tag{11.25}$$

and hence

$$V_{BE} = V_T \ln I_C/I_S \tag{11.26}$$

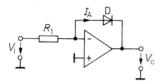

Fig. 11.20. Diode logarithmic amplifier

$$V_o = -nV_T \ln \frac{V_i}{I_S R_1} \quad \text{for} \quad V_i > 0$$

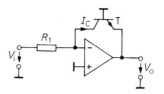

Fig. 11.21. Transistor logarithmic amplifier

$$V_o = -V_T \ln \frac{V_i}{I_{CS} R_1} \quad \text{for} \quad V_i > 0$$

The output voltage of the transistor logarithmic amplifier shown in Fig. 11.21 is therefore:

$$V_o = -V_{BE} = -V_T \ln \frac{V_i}{I_S R_1}$$

As well as eliminating the correction factor n, the circuit shown in Fig. 11.21 has two further advantages. First, no distortion due to the collector–base leakage current occurs, as $V_{CB} = 0$. In addition, the magnitude of the current gain does not affect the result, as the base current flows away to ground. When suitable transistors are employed, the collector current can be varied from the picoampere to the milliampere region; that is, over nine decades. However, operational amplifiers with very low input currents are needed to exploit this range to the full.

Since the transistor T increases the loop gain by its own voltage gain, the circuit is prone to oscillations. The voltage gain of the transistor stage can be reduced quite simply by connecting a resistor R_E between the emitter and the amplifier output, as shown in Fig. 11.22. This limits the voltage gain of the transistor by means of current feedback to the value R_1/R_E. The resistance R_E must not, of course, give rise to voltage saturation of the operational amplifier at the largest possible output current. The capacitor C can further improve stability of the circuit by differentiate action in the feedback. It must be noted, however, that the upper cutoff frequency decreases proportionally to the current because of the nonlinear transistor characteristic.

Enhanced performance is achieved if the transistor is operated from a high-impedance current source. The loop gain is then $g_m \cdot R_1$, where g_m is the transconductance of the drive circuit. As it is independent of the collector current, frequency compensation can be optimized over the entire current range. Operational amplifiers which have a current output – for example, CA 3080 is available as integrated "transconductance amplifier." The disadvantage of these types, however, is that they have a relatively large input bias current.

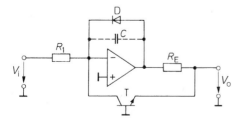

Fig. 11.22. Practical implementation of a logarithmic amplifier

$$V_o = -V_T \ln \frac{V_i}{I_S R_1} \quad \text{for} \quad V_i > 0$$

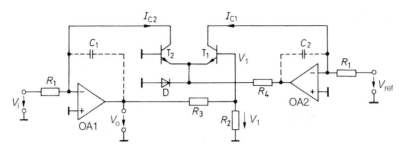

Fig. 11.23. Temperature-compensated logarithmic amplifier

$$V_o = -V_T \cdot \frac{R_3 + R_2}{R_2} \ln \frac{V_i}{V_{\text{ref}}} \quad \text{for } V_i, V_{\text{ref}} > 0$$

Diode D in Fig. 11.22 prevents the operational amplifier from being overdriven in the event of negative input voltages. This insures that transistor T is not damaged by an excessively high emitter–base reverse voltage and it shortens the recovery time.

One disadvantage of the logarithmic amplifier is its strong temperature dependence. The reason for this is that V_T and I_{CS} vary markedly with temperature. For a temperature rise from 20 °C to 50 °C, V_T increases by 10%, while the reverse current I_S multiplies tenfold. The influence of the reverse current can be eliminated by computing the difference between two logarithms. We employ this principle in Fig. 11.23, where the differential amplifier stage T_1, T_2 is used to find the logarithm. In order to examine the operation of the circuit, we determine the current sharing in the differential amplifier stage. From Kirchhoff's voltage law (KVL), it follows:

$$V_1 + V_{BE2} - V_{BE1} = 0$$

The transfer characteristics of the transistors may be written as

$$I_{C1} = I_S e^{\frac{V_{BE1}}{V_T}}$$

$$I_{C2} = I_S e^{\frac{V_{BE2}}{V_T}}$$

and therefore

$$\frac{I_{C1}}{I_{C2}} = e^{\frac{V_1}{V_T}} \tag{11.27}$$

From Fig. 11.23, we get the additional equations

$$I_{C2} = \frac{V_i}{R_1} \qquad I_{C1} = \frac{V_{\text{ref}}}{R_1} \qquad V_1 = \frac{R_2}{R_3 + R_2} V_o,$$

if R_2 is not chosen to be too large. By substitution, we obtain the output voltage

$$V_o = -V_T \frac{R_3 + R_2}{R_2} \ln \frac{V_i}{V_{\text{ref}}} \tag{11.28}$$

The value of R_4 does not appear in the result. Its resistance is chosen so that the voltage across it is smaller than the maximum possible output voltage swing of amplifier OA2.

Logarithmic amplifiers that provide an output voltage of 1 V per decade are frequently required. To determine the sizes of R_2 and R_3 for this special case, we rewrite (11.28) in the form

$$V_o = -V_T \frac{R_3 + R_2}{R_2} \cdot \frac{1}{\lg e} \cdot \lg \frac{V_i}{V_{\text{ref}}} = -1\,\text{V}\,\lg \frac{V_i}{V_{\text{ref}}}$$

With $V_T = 26\,\text{mV}$, the resulting condition is

$$\frac{R_3 + R_2}{R_2} = \frac{1\,\text{V} \cdot \lg e}{V_T} \approx 16.7$$

If we select $R_2 = 1\,k\Omega$, then $R_3 = 15.7\,k\Omega$.

With regard to the frequency compensation of both amplifiers, the same argument holds as for the previous circuit. C_1 and C_2 are the additional compensation capacitors. The temperature effect of V_T can be offset by letting resistor R_2 have a positive, or R_3 a negative, temperature coefficient of $0.3\% K^{-1}$. A realization is found in the MAX 4206 from Maxim and the LOG 112 from Burr-Brown. Another solution is to maintain the differential amplifier at constant temperature using a transistor array with two additional transistors. One of these is then used as a temperature sensor and the other as a heater. A suitable transistor array is, for example, the MAT 04 from Analog Devices.

11.7.2
Exponential Function

Figure 11.24 shows an exponential function amplifier whose design is analogous to that of the logarithmic amplifier in Fig. 11.21. When a negative voltage is applied to the input, the current flowing through the transistor is given by (11.25),

$$I_C = I_S e^{\frac{V_{BE}}{V_T}} = I_S e^{-\frac{V_i}{V_T}},$$

and the output voltage is therefore

$$V_o = I_C R_1 = I_S R_1 e^{-\frac{V_i}{V_T}}.$$

As with the logarithmic amplifier shown in Fig. 11.23, the temperature stability can be improved by using a differential amplifier. The appropriate circuit is represented in Fig. 11.25. Again, from (11.27),

$$\frac{I_{C1}}{I_{C2}} = e^{\frac{V_1}{V_T}}$$

From Fig. 11.25 we deduce the equations:

$$I_{C1} = \frac{V_o}{R_1} \qquad I_{C2} = \frac{V_{\text{ref}}}{R_1} \qquad V_1 = \frac{R_2}{R_3 + R_2} V_i$$

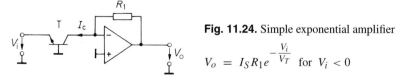

Fig. 11.24. Simple exponential amplifier

$$V_o = I_S R_1 e^{-\frac{V_i}{V_T}} \quad \text{for } V_i < 0$$

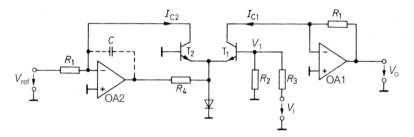

Fig. 11.25. Temperature-compensated exponential amplifier

$$V_o = V_{ref} e^{\frac{R_2}{R_3+R_2} \cdot \frac{V_i}{V_T}} \quad \text{for } V_{ref} > 0$$

By substitution, we obtain the output voltage

$$V_o = V_{ref} e^{\frac{R_2}{R_3+R_2} \cdot \frac{V_i}{V_T}} \tag{11.29}$$

It can be seen that I_S no longer appears in the result if the transistors are well matched. The resistor R_4 limits the current through the transistors T_1 and T_2, and its resistance does not affect the result as long as operational amplifier OA2 is not saturated.

A particularly important design is obtained if the output voltage increases by a factor of 10 for a 1 V increase in the input voltage. The required condition may be derived from (11.29):

$$V_o = V_{ref} \cdot 10^{\frac{R_2}{R_3+R_2} \cdot \frac{V_i}{V_T} \cdot \lg e} = V_{ref} \cdot 10^{\frac{V_i}{1\,V}}$$

Consequently, with $V_T = 26\,\text{mV}$,

$$\frac{R_3 + R_2}{R_2} = \frac{1\,V \cdot \lg e}{V_T} \approx 16{,}7$$

– the same component values as for the logarithmic amplifier shown in Fig. 11.23.

The exponential amplifiers described above enable the computation of expressions of the form

$$y = e^{ax}$$

Since

$$b^{ax} = e^{\ln b^{ax}} = e^{ax \ln b}$$

exponential functions to any base b can be computed according to

$$y = b^{ax}$$

by amplifying the input signal x by the factor $\ln b$, and by applying the result to the input of an exponential function amplifier.

11.7.3
Computation of Power Functions Using Logarithms

The computation of power expressions of the form

$$y = x^a$$

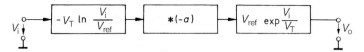

Fig. 11.26. A general power-function network

$$V_o = V_{\text{ref}} \left(\frac{V_i}{V_{\text{ref}}} \right)^a \quad \text{for } V_i > 0$$

can be performed for $x > 0$ by means of logarithmic and exponential function amplifiers because

$$x^a = (e^{\ln x})^a = e^{a \ln x}$$

The basic arrangement for such a circuit is shown in Fig. 11.26. The equations mentioned apply to the logarithmic-function amplifier of Fig. 11.23 and the exponential-function amplifier of Fig. 11.25, where $R_2 = \infty$ and $R_3 = 0$. We therefore obtain the output voltage

$$V_o = V_{\text{ref}} e^{\dfrac{a V_T \ln \dfrac{V_i}{V_{\text{ref}}}}{V_T}} = V_{\text{ref}} \left(\frac{V_i}{V_{\text{ref}}} \right)^a$$

The logarithm and the exponential function can be obtained using a single integrated circuit if so-called multifunction converters are used, such as the AD 538 from Analog Devices.

Involution (raising to the power) by means of logarithms is in principle defined for positive input voltages only. However, from a mathematical point of view, bipolar input signals are also permitted for whole-number exponents a. This case can be realized by using the multipliers described in Sect. 11.8.

11.7.4
Sine and Cosine Functions

The output of a sine-function network should approximate the expression

$$V_o = \widehat{V}_o \sin \left(\frac{\pi}{2} \cdot \frac{V_i}{\widehat{V}_i} \right) \tag{11.30}$$

within the range $-\widehat{V}_i \leq V_i \leq +\widehat{V}_i$. For small input voltages

$$V_o = \widehat{V}_o \cdot \frac{\pi}{2} \cdot \frac{V_i}{\widehat{V}_i}$$

It is advisable to choose a value for $\widehat{V}_o$ so that near the origin, $V_o = V_i$. This is the case for

$$\widehat{V}_o = \frac{2}{\pi} \cdot \widehat{V}_i \tag{11.31}$$

For small input voltages, the sine-function network must accordingly have unity gain, whereas at higher voltages the gain must decrease. Figure 11.27 represents a circuit that fulfills these conditions, based on the principle of *piecewise approximation*.

For small input voltages, all the diodes are reverse biased, and $V_o = V_i$, as required. If V_o rises above V_i, diode D_2 becomes forward biased. V_o then increases more slowly

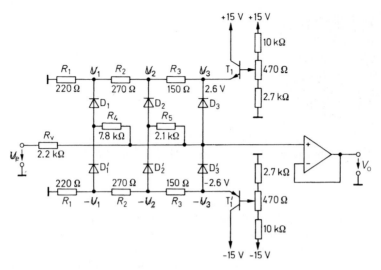

Fig. 11.27. Sine function network with $2n = 6$ breakpoints

$$V_o \approx \frac{2}{\pi} \cdot \widehat{V}_i \sin\left(\frac{\pi}{2}\frac{V_i}{\widehat{V}_i}\right) \quad \text{for } \widehat{V}_i = 5\,\text{V}$$

than V_i, because of the voltage divider formed by R_v and R_4. If V_o becomes larger than V_2, the output of the network is additionally loaded with R_5, so that the rise in voltage is slowed down even more. Diode D_3 finally produces the horizontal tangent at the top of the sine curve. Diodes $D_1' - D_3'$ have the corresponding effects for the negative part of the since function. Considering that diodes do not become conducting suddenly, but have an exponential characteristic, low harmonic distortions of V_o can be obtained with only a small number of diodes.

In order to determine the parameters of the network, we begin by choosing the breakpoints of the approximation curve. It can be shown that the first n odd harmonies disappear if $2n$ breakpoints are assigned to the following values of the input voltage:

$$V_{ik} = \pm \frac{2k}{2n+1}\widehat{V}_i, \qquad 0 < k \leq n \tag{11.32}$$

According to (11.30) and (11.31), the corresponding output voltages are

$$V_{ok} = \pm \frac{2}{\pi}\widehat{V}_i \sin \frac{\pi k}{2n+1}, \qquad 0 < k \leq n \tag{11.33}$$

Therefore, the slope of the line segment above the kth breakpoint is given as

$$m_k = \frac{V_{o(k+1)} - V_{ok}}{V_{i(k+1)} - V_{ik}} = \frac{2n+1}{\pi}\left[\sin \frac{\pi(k+1)}{2n+1} - \sin \frac{\pi k}{2n+1}\right] \tag{11.34}$$

For the highest breakpoint, when $k = n$, the slope becomes zero, as was stipulated earlier in the qualitative description. The slope m_0 must be chosen to be equal to unity.

For reasons of symmetry, no even harmonics appear. With the r.m.s. values of the odd harmonics present in the waveform, we obtain a theoretical distortion factor of 1.8% if $2n = 6$ breakpoints are chosen, this being reduced to 0.8% for $2n = 12$. However, as real diode characteristics do not have sharp breakpoints, the actual distortion is considerably lower. This is illustrated by the following example.

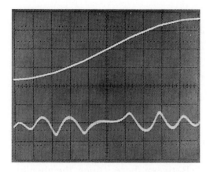

Fig. 11.28. Output voltage and the error voltage (amplified 50 times) as a function of the input voltage Vertical: 2 V per division; Horizontal: 1 V per division

A voltage that has a triangular wave-shape, with a peak value of $\widehat{V}_i = 5\,\text{V}$, is to be converted into a sinusoidal voltage. According to (11.31), the amplitude of the latter must be 3.18 V, so that the line segment around the origin has a unit slope. For the approximation, we want to use $2n = 6$ breakpoints. Following (11.33), these must appear at the output voltages ± 1.4, ± 2.5, and $\pm 3.1\,\text{V}$. For real diodes we assume that a sizeable current only flows for forward voltages of more than 0.5 V. The diode bias voltages must then be reduced by this amount. We thus obtain the voltages $V_1 = 0.9\,\text{V}$, $V_2 = 2.0\,\text{V}$, and $V_3 = 2.6\,\text{V}$, which define the values for the voltage divider chain R_1, R_2, R_3 shown in Fig. 11.27. The emitter-followers T_1 and T'_1 serve as low-impedance sources for V_3 and, simultaneously, as temperature compensation for the forward voltages of the diodes.

From (11.34), we obtain, for the slopes of the three segments, $m_1 = 0.78$, $m_2 = 0.43$, and $m_3 = 0$. We choose $R_v = 2.2\,\text{k}\Omega$. From

$$m_1 = \frac{R_4}{R_v + R_4}$$

and disregarding the internal resistance of the divider chain, we obtain $R_4 = 7.8\,\text{k}\Omega$. The slope of the second segment is

$$m_2 = \frac{(R_5 \| R_4)}{R_v + (R_5 \| R_4)}$$

and thus $R_5 = 2.1\text{k}\Omega$.

For fine adjustment of the network, it is advisable to use a notch filter for the fundamental (see Sect. 13.9) and to display of the remaining error voltage on an oscilloscope screen. The optimum is attained when the peaks of the deviation curve have the same height, as can be seen in the oscillogram shown in Fig. 11.28. The distortion factor measured for this case was 0.42% – clearly below the theoretical value for ideal diodes.

Power Series Expansion

Another method for the approximation of a sine function is to use a power series, since

$$\sin x = x - \frac{x^3}{3!} + \frac{x^5}{5!} - + \cdots$$

To keep the number of components low, the series is truncated after the second term and this results in an error. If the range of values of the argument is now limited to $-\frac{\pi}{2} \leq x \leq \frac{\pi}{2}$, the error can be minimized by changing the coefficients slightly. If we select

$$\sin x \approx y = 0.9825\,x - 0.1402\,x^3 \tag{11.35}$$

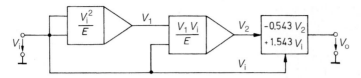

Fig. 11.29. Approximation of the sine function by a power series

$$V_o \approx \widehat{V_i} \sin\left(\frac{\pi}{2} \cdot \frac{V_i}{\widehat{V_i}}\right) \quad \text{for } \widehat{V_i} = E$$

the error becomes zero for $x = 0$, ± 0.96, and $\pm\pi/2$. Between these values, the absolute error is less than 0.57% of the amplitude. The harmonic distortion is 0.6%. It can be reduced to 0,25% by a slight variation of the coefficients, and is therefore somewhat smaller than for the piecewise approximation method using 2×3 breakpoints. The lack of breakpoints is particularly advantageous when the signal is to be differentiated.

For a practical circuit, we define

$$x = \frac{\pi}{2} \cdot \frac{V_i}{\widehat{V_i}} \quad \text{and} \quad y = \frac{V_o}{\widehat{V_o}}$$

Furthermore, we select $\widehat{V_i} = \widehat{V_o}$ and thus obtain, from (11.35),

$$V_o = 1.543\, V_i - 0.543 \frac{V_i^3}{\widehat{V_i}^2} \approx \widehat{V_i} \sin\left(\frac{\pi}{2} \frac{V_i}{\widehat{V_i}}\right)$$

The block diagram for this operation is represented in Fig. 11.29; the input voltage amplitude $\widehat{V_i}$ is equal to the computing unit E for the multipliers. We shall discuss the analog multipliers required in the next section.

Differential Amplifiers

Another method of approximating a sine wave is based on the fact that the function $\tanh x$ has a similar shape for small x. This function can be easily generated using a differential amplifier stage, as shown in Fig. 11.30. It was shown in Sect. 11.7.1 that, for a differential amplifier, using (11.27),

$$\frac{I_{C1}}{I_{C2}} = e^{\frac{V_i}{V_T}} \quad \text{and} \quad I_{C1} + I_{C2} \approx I_k$$

Therefore,

$$I_{C1} - I_{C2} = \frac{e^{\frac{V_i}{V_T}} - 1}{e^{\frac{V_i}{V_T}} + 1} I_k = I_k \tanh \frac{V_i}{2V_T} \tag{11.36}$$

The operational amplifier forms the difference between the two collector currents, such that

$$V_o = R_2(I_{C1} - I_{C2})$$

It follows:

$$V_o = I_k R_2 \tanh \frac{V_i}{2V_T} \tag{11.37}$$

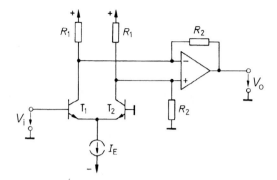

Fig. 11.30. Approximation of the sine function, using a differential amplifier

$$V_o \approx I_k R_2 \cdot \sin\left(\frac{\pi}{2} \frac{V_i}{\widehat{V_i}}\right) \quad \text{for} \ \ \widehat{V_i} = 2.8\,V_T \approx 73\,\text{mV}$$

This function can be interpreted as approximating the sine function

$$V_o \ = \ \widehat{V_o} \sin\left(\frac{\pi}{2} \cdot \frac{V_i}{\widehat{V_i}}\right) \quad \text{for} \quad -\frac{\pi}{2} \leq x \leq \frac{\pi}{2}$$

The quality of the sine approximation is dependent on the peak value $\widehat{V_i}$ chosen. For $\widehat{V_i} = 2.8\,V_T \approx 73\,\text{mV}$, the error becomes minimal and $\widehat{V_o}$ is $0.86\,I_k R_2$. However, the error is still 3%. It can be reduced to 0.02% by providing the differential amplifier with two additional appropriately biased transistors. This is the operating principle of the AD 639 from Analog Devices, which can be used to produce all the other trigonometric functions as well as the sine function.

Cosine Function

The cosine function can be generated for values $0 \leq x \leq \pi$ by means of the sine function networks that have previously been described. The input voltage V_i, which should be between zero and $V_{i\,max}$, is first converted to an auxiliary voltage:

$$V_1 \ = \ V_{i\,max} - 2V_i \tag{11.38}$$

As can be seen in Fig. 11.31, this equation is already a linear approximation of the cosine function. For the necessary rounding-off of the curve near the maximum and the minimum, we apply V_1 to the input of a sine function network. As is obvious from Fig. 11.32, the addition of a summing amplifier is all that is needed to convert a sine function network into a cosine function network.

Simultaneous Generation of the Sine and Cosine Functions

With the networks described so far, sine and cosine functions can be generated over a half-period. In cases in which the range of the argument has to be a full period or more, we initially generate triangular functions as a linear approximation, and we use sine networks to round off the peaks. The shape of the required triangular voltages is represented in Fig. 11.33.

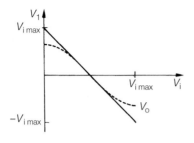

Fig. 11.31. Shape of the auxiliary voltage for generating the cosine function (dashed line)

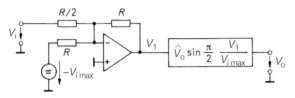

Fig. 11.32. Generation of a cosine function by means of a sine function network

$$V_o = \widehat{V}_o \cos\left(\pi \frac{V_i}{V_{i\,max}}\right) \text{ for }$$

$$0 \leq V_i \leq V_{i\,max}$$

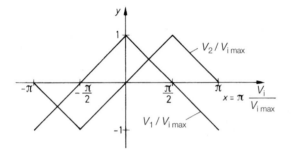

Fig. 11.33. Shape of the auxiliary voltages for generating the sine and cosine functions for

$$-\pi \leq x \leq \pi$$

Voltage V_1 approximates the cosine function. For $V_i > 0$, it is identical to the voltage V_1 in Fig. 11.31. For $V_i < 0$, it is symmetrical about the y-axis. We can therefore use (11.38) by replacing V_i by $|V_i|$, and obtain

$$V_1 \;=\; V_{i\,max} - 2|V_i| \tag{11.39}$$

The relationships for the sine function are somewhat more complicated, since we must differentiate between three cases:

$$V_2 = \begin{cases} -2(V_i + V_{i\,max}) & \text{for} & -\,V_{i\,max} \leq V_i \leq -\tfrac{1}{2}V_{i\,max} & \text{(11.42a)} \\[2mm] 2V_i & \text{for} & -\tfrac{1}{2}V_{i\,max} \leq V_i \leq \tfrac{1}{2}V_{i\,max} & \text{(11.42b)} \\[2mm] -2(V_i - V_{i\,max}) & \text{for} & \tfrac{1}{2}V_{i\,max} \leq V_i \leq V_{i\,max} & \text{(11.42c)} \end{cases}$$

Such functions are best implemented using the general precision function network, which is described below.

11.7.5
Arbitrary Function Networks

In Fig. 11.27, a diode network was used for the piecewise linear approximation of functions. Calculation of the circuit parameters is only possible up to an approximation, because the forward voltage of the diodes and the loading of the voltage divider chains must be taken

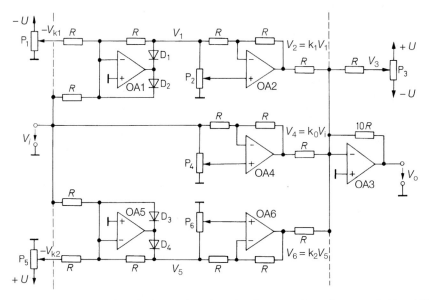

Fig. 11.34. Arbitrary function network. U = unit reference voltage, for example $U = 10\,\text{V}$

into account. Furthermore, the sign of the slope of each linear segment is already defined by the structure of the network. Therefore, such a circuit can only be optimized for one particular function, and its parameters cannot easily be changed.

Figure 11.34, on the other hand, represents a circuit that allows the breakpoint and slope of each individual segment to be set precisely, using a separate potentiometer. The part of the circuit formed by operational amplifiers OA1 and OA2 permits a segment for positive input voltages to be formed, while operational amplifiers OA5 and OA6 are effective for negative input voltages. Amplifier OA4 determines the slope about the origin. The circuit can be extended for any number of segments by adding further sections that are identical to those mentioned.

Amplifiers OA2, OA4, and OA6 are connected as bipolar-coefficient circuits, as in Fig. 11.5 for $n = 1$. Their gain can be adjusted to values between $-1 \le k \le +1$ by the associated potentiometers; and their output voltages are added by the summing amplifier OA3. An additional DC voltage can be added by means of potentiometer P_3.

Near zero input voltage, only amplifier OA4 contributes to the output voltage:

$$V_4 = k_0 V_i$$

Both voltages V_1 and V_5 are zero in this case, because diodes D_1 and D_4 are reverse biased, and amplifiers OA1 and OA5 have a feedback path via the conducting diodes D_2 and D_3.

When the input voltage becomes greater than V_{k1}, diode D_1 is forward biased, and we obtain

$$V_1 = -(V_i - V_{k1}) \quad \text{for} \quad V_i \ge V_{k1} \ge 0$$

Amplifier OA1 therefore operates as a half-wave rectifier, with a positive bias voltage V_{k1}. Operational amplifier OA5 behaves correspondingly for negative input voltages:

$$V_5 = -(V_i - V_{k2}) \quad \text{for} \quad V_i \le V_{k2} \le 0$$

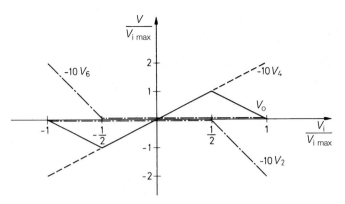

Fig. 11.35. Voltage components for generating voltage V_o in Abb. 11.33

Hence, we obtain the general relationship for the slope of the output voltage V_o as

$$m = \frac{\Delta V_o}{\Delta V_i} = 10 \cdot \begin{cases} -k_0 + k_1 + \cdots + k_m & \text{for} \quad V_i > V_{k\,m} > 0 \\ -k_0 + k_1 & \text{for} \quad V_i > V_{k\,1} > 0 \\ -k_0 & \text{for} \quad V_{k\,2} < V_i < V_{k\,1} \\ -k_0 + k_2 & \text{for} \quad V_i < V_{k\,2} < 0 \\ -k_0 + k_2 + \cdots + k_n & \text{for} \quad V_i < V_{kn} < 0 \end{cases} \qquad (11.41)$$

As an example, we shall demonstrate the implementation of the voltage wave-shape $V_2/V_{i\,max}$ in Fig. 11.33. A positive breakpoint at $V_{k\,1} = \frac{1}{2} V_{i\,max}$ and a negative break-point at $V_{k\,2} = -\frac{1}{2} V_{i\,max}$ are required. According to (11.40b), the slope of the segment through the origin must have the value $m = +2$; therefore, $k_0 = -0.2$. Above the positive breakpoint, the slope must be -2. For this region, we take, from (11.41),

$$m = 10(-k_0 + k_1)$$

and therefore obtain $k_1 = -0.4$, and correspondingly $k_2 = -0.4$. The shapes of the output voltage functions that result from this process are shown in Fig. 11.35.

Even if no calibrated potentiometers are available, the network output can be given the desired shape in a simple way, using the following procedure. Initially, all the breakpoint voltages and slopes are set to their maximum values and the input voltage is made zero. This insures that $|V_i| < |V_{ki}|$. Only the zeroing potentiometer P_3 affects the output; it is used to adjust the output voltage $V_o(V_i = 0)$ to the desired value. In the next step, V_i is made equal to V_{k1} and P_4 is set so that $V_o(V_i = V_{k1})$ assumes the level required. The factor k_0 is now defined. P_1 is then adjusted to a point at which the output voltage just begins to change; this occurs when the setting of P_2 corresponds to V_{k1}. Now, V_i is set to the value of the next higher breakpoint (or to the end of the range, if there is no higher breakpoint), and P_2 is adjusted so that V_o, attains the desired value. In this way, k_1 is defined. The remaining breakpoints and slopes are dealt with in the same manner.

In cases in which no calibrated potentiometers are needed for the adjustment of the segment slopes, the circuit may be simplified. One can replace the bipolar-coefficient circuits by simple potentiometers that are connected to a multiple-input subtracting amplifier, as shown in Fig. 11.36. The subtracting amplifier consists of operational amplifiers OA2 and OA3, and is based on the principle of Fig. 11.2.

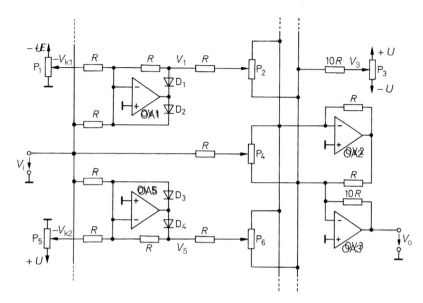

Fig. 11.36. Simplified arbitrary function network

11.8
Analog Multipliers

So far, we have described circuits for addition, subtraction, differentiation, and integration. Multiplication could be carried out only if a constant factor was involved. Below, we deal with the most important principles for the multiplication and division of two variable voltages.

11.8.1
Multipliers with Logarithmic Amplifiers

Multiplication and division can be reduced to an addition and subtraction of logarithms:

$$\frac{xy}{z} = \exp[\ln x + \ln y - \ln z]$$

The function can be implemented by using three logarithmic amplifiers, one exponential function amplifier, and one adder/subtractor circuit. The latter can be eliminated by using the inputs of the differential amplifier for the exponential function amplifier in Fig. 11.25 to perform the subtraction, and by considering the fact that the terminal for the reference voltage can be used as an additional signal input.

The logarithmic amplifiers shown in Fig. 11.37 produce the expressions

$$V_1 = -V_T \ln \frac{V_y}{I_S R_1} \quad \text{and} \quad V_2 = -V_T \ln \frac{V_z}{I_S R_1}$$

The exponential function generator therefore provides an output voltage of the form

$$V_o = V_x e^{\frac{V_2 - V_1}{V_T}} = \frac{V_x V_y}{V_z}$$

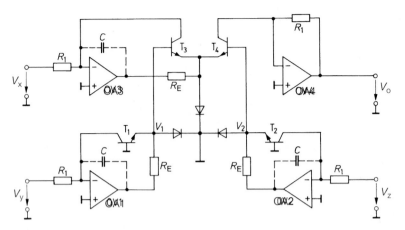

Fig. 11.37. Multiplication by means of logarithms

$$V_o = \frac{V_x V_y}{V_z} \quad \text{for} \quad V_x, V_y, V_z > 0$$

We can see that, in this case, not only the reverse saturation currents I_S but also voltage V_T are eliminated, thereby obviating the need for temperature compensation. However, it is essential that the four transistors have the same characteristics and are at the same temperature. They must therefore be integrated on one chip.

An inherent disadvantage of this method is that all input voltages must be positive, and may not even be zero. Such multipliers are called one-quadrant multipliers.

Multipliers such as those shown in Fig. 11.37 can be implemented using multifunction converters such as the AD 538 (Analog Devices).

11.8.2
Transconductance Multipliers

As shown in Chap. 2.11 on page 42, the transconductance of a transistor is defined as

$$g_m = \frac{dI_C}{dV_{BE}} = \frac{I_C}{V_T}$$

and is therefore proportional to the collector current.

The variation of the collector current is then proportional to the product of the variation in the input voltage and the quiescent collector current. This property is made use of for multiplication in the differential amplifier shown in Fig. 11.38.

The operational amplifier evaluates the difference between the collector currents:

$$V_o = R_z(I_{C2} - I_{C1}) \tag{11.42}$$

Applying a negative voltage V_y and setting V_x to zero, the currents through both transistors are equal, and the output voltage remains zero. If V_x, is made positive, the collector current through T_1 rises and that of T_2 falls; the output voltage is negative. Correspondingly, V_o becomes positive when V_x is negative. The resulting difference in collector currents is greater, the larger the emitter current, i.e. the higher the value $|V_y|$. It can therefore be assumed that V_o is at least approximately proportional to $V_x \cdot V_y$. For a more precise

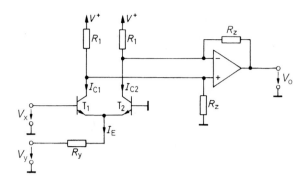

Fig. 11.38. Basic circuit of a transconductance multiplier

$$V_o \approx \frac{R_z}{R_y} \cdot \frac{V_x V_y}{2V_T} \quad \text{for } V_y < 0$$

calculation, we determine the current sharing within the differential amplifier stage. As was shown in Sect. 11.7.4, (11.36) states that

$$I_{C1} - I_{C2} = I_E \tanh \frac{V_x}{2V_T} \tag{11.43}$$

A power series expansion up to the fourth order gives

$$I_{C1} - I_{C2} = I_E \left(\frac{V_x}{2V_T} - \frac{V_x^3}{24V_T^3} \right) \tag{11.44}$$

Hence,

$$I_{C1} - I_{C2} \approx I_E \cdot \frac{V_x}{2V_T} \quad \text{for } |V_x| \ll V_T \tag{11.45}$$

If $|V_y| \gg V_{BE}$, then

$$I_E \approx -\frac{V_y}{R_y}$$

Substitution in (11.45) gives, in conjunction with (11.42), the result

$$V_o \approx \frac{R_z}{R_y} \cdot \frac{V_x V_y}{2V_T} \tag{11.46}$$

If the error in (11.46) is not to exceed 1%, the voltage V_x must be $|V_x| < 0.35\,V_T \approx 9\,\text{mV}$. Because of the small value of V_x, transistors T_1 and T_2 must be closely matched to prevent the offset voltage drift from affecting the result.

For the correct operation of the circuit, it is necessary that U_y is always negative, while the voltage U_x, may have either polarity. Such a multiplier is called a two-quadrant multiplier.

There are several properties of the transconductance multiplier shown in Fig. 11.38 that can be improved. In deducing the output equation (11.46), we had to use the approximation that $|V_y| \gg V_{BE} \approx 0.6\,\text{V}$. This condition can be dropped if resistor R_y is replaced by a controlled current source for which I_E is proportional to V_y.

A further disadvantage of the circuit shown in Fig. 11.38 is that $|V_x|$ must be limited to small values in order to minimize the error. This can be avoided by not applying V_x, directly but, rather, its logarithm.

An expansion to a four-quadrant multiplier – that is, a multiplier for input voltages of either polarity – is possible if a second differential amplifier stage is connected in parallel,

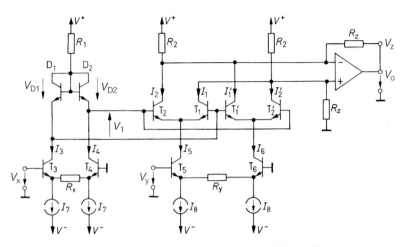

Fig. 11.39. Four-quadrant transconductance multiplier (Gilbert cell)

$$V_o = \frac{2R_z}{R_x R_y} \cdot \frac{V_x V_y}{I_7} \quad \text{for } I_7 > 0$$

the emitter current of which is controlled by V_y, in opposition to that of the first transistor pair.

All these aspects are considered in the four-quadrant transconductance multiplier shown in Fig. 11.39. The differential amplifier stage T_1, T_2 is the same as that of Fig. 11.38. It is supplemented symmetrically by the differential amplifier T_1', T_2'. Transistors T_5, T_6 form a differential amplifier with current feedback. The collectors represent the outputs of two current sources that are controlled by V_y simultaneously but in opposition, as required:

$$I_5 = I_8 + \frac{V_y}{R_y}, \qquad I_6 = I_8 - \frac{V_y}{R_y} \tag{11.47}$$

For the difference between the collector currents of the two differential amplifier stages T_1, T_2 and T_1', T_2' we obtain, by analogy with the previous circuit,

$$I_1 - I_2 = I_5 \tanh \frac{V_1}{2V_T} = \left(I_8 + \frac{V_y}{R_y} \right) \tanh \frac{V_1}{2V_T} \tag{11.48}$$

$$I_1' - I_2' = I_6 \tanh \frac{V_1}{2V_T} = \left(I_8 - \frac{V_y}{R_y} \right) \tanh \frac{V_1}{2V_T} \tag{11.49}$$

As before, the operational amplifier evaluates the difference between the collector currents according to

$$\Delta I = (I_2 + I_1') - (I_2' + I_1) = (I_1' - I_2') - (I_1 - I_2) \tag{11.50}$$

By subtracting (11.48) from (11.49), it follows:

$$\Delta I = -\frac{2V_y}{R_y} \tanh \frac{V_1}{2V_T} \tag{11.51}$$

where V_y may now have either polarity. By expanding this expression into a series, we can see that the same approximation to multiplication is involved as for the previous circuit.

We shall now examine the relationship between V_1 and V_x. Two transistors are connected as diodes (transdiodes), D_1 and D_2, and these are used to form the logarithm of the input signals:

$$V_1 = V_{D2} - V_{D1} = V_T \ln \frac{I_4}{I_{CS}} - V_T \ln \frac{I_3}{I_{CS}}$$

Hence,

$$V_1 = V_T \ln \frac{I_4}{I_3} = V_T \ln \frac{I_7 - \dfrac{V_x}{R_x}}{I_7 + \dfrac{V_x}{R_x}} \qquad (11.52)$$

Substitution in (11.51) gives the current difference:

$$\Delta I = \frac{2 V_x V_y}{R_x R_y I_7} \qquad (11.53)$$

From this, the operational amplifier configured as a current subtractor forms the output voltage

$$V_o = \Delta I R_z = \frac{2 R_z}{R_x R_y I_7} \cdot V_x V_y = \frac{V_x V_y}{U} \qquad (11.54)$$

where $U = R_x R_y I_7 / 2 R_z$ is the computing unit. This is usually chosen to be 10 V. Good temperature compensation is attained, since V_T cancels out. Either (11.53) or (11.54) is obtained without recourse to power expansion, and therefore a considerably larger range of input voltages V_x, is permissible. The limits of the input range are reached when one of the transistors in the controlled current source is turned off. Therefore,

$$|V_x| < R_x I_7 \quad \text{and} \quad |V_y| < R_y I_8$$

If the currents I_7 are controlled by a further input voltage V_7, simultaneous division and multiplication is possible. However, the usable range for I_7 is limited, because I_7 influences all the quiescent potentials within the multiplier and also the permissible range for V_x.

A simpler way of achieving division is to open the connection between V_o and V_z, and to link the voltages V_y and V_o, instead. Because of the resulting feedback, the output voltage assumes a value such that $\Delta I = V_z / R_z$. Therefore, from (11.53),

$$\Delta I = \frac{2 V_x V_y}{R_x R_y I_7} = \frac{V_z}{R_z}$$

Thus the new output voltage is

$$V_o = V_y = \frac{R_x R_y I_7}{2 R_z} \cdot \frac{V_z}{V_x} = U \frac{V_z}{V_x} \qquad (11.55)$$

However, stability is only guaranteed if V_x, is negative; otherwise, the negative feedback becomes positive. The signal V_z, on the other hand, can have either polarity, and therefore the circuit is a two-quadrant divider. The limitation on the sign of the denominator is not peculiar to this arrangement, but is common to all divider circuits.

Transconductance multipliers operating on the principle shown in Fig. 11.40 are available as monolithic integrated circuits. The achievable accuracy is 0.1% referred to computing unit U; that is, for a computing unit of $U = 10$ V. As we shall see in Sect. 11.8.5, the simple types require four trimmers in order to achieve this degree of accuracy. High

IC Type	Manufacturer	Accuracy		Bandwidth	
		without adjustment	with adjustment	1%	3 dB
MPY 100	Burr Brown	0.5 %	0.35%	35 kHz	0.5 MHz
MPY 600	Burr Brown	1 %	0.5 %		60 MHz
AD 534	Analog Dev.	0.25%	0.1 %	70 kHz	1 MHz
AD 633	Analog Dev.	1 %	0.1 %	100 kHz	1 MHz
AD 734	Analog Dev.	0.1 %		1000 kHz	10 MHz
AD 834	Analog Dev.	2 %			500 MHz
AD 835	Analog Dev.		0.1 %	15 MHz	250 MHz
MLT 04*	Analog Dev.	2 %	0.2 %		8 MHz

* 4 multipliers on 1 chip

Fig. 11.40. Transconductance multipliers

accuracy types are already laster trimmed by the manufacturer, so that external adjustment is generally unnecessary.

The 3 dB bandwidth is of the order of 1 MHz and beyond, at which frequency the computing error is already 30%. As a deviation of this magnitude is unacceptable in the majority of applications, a better reference point is the frequency at which the output voltage is reduced by 1%.

Transconductance Dividers with Improved Accuracy

We have mentioned two methods of division, one using the logarithmic multiplier (Fig. 11.37) and the other using the transconductance multiplier described above. For a division, a basic problem arises in the region of zero input, as the output voltage is then chiefly determined by the input offset error. This error is particularly large for the transconductance multiplier, since in the input log-amplifier a positive constant (i.e., I_7 in (11.52)) is added to the input signal to avoid a change of polarity in the argument. The conditions are considerably more favorable if the circuit in Fig. 11.37 is used for the division. However, only one quadrant is here available.

The advantages of the two methods – two-quadrant division and good accuracy near zero input – can be combined. This is achieved not by adding a constant to the argument of the logarithm (to avoid the change in sign) but by adding to the numerator a quantity proportional to the denominator.

The divider output should conform to the expression

$$V_o = U \frac{V_x}{V_z}$$

Assuming that $V_z > 0$ and $|V_x| < V_z$, two auxiliary voltages that are always positive,

$$V_1 = V_z - \frac{1}{2} V_x \qquad V_2 = V_z + \frac{1}{2} V_x \qquad (11.56)$$

can be generated. The logarithms of these two voltages are computed according to the block diagram of Fig. 11.41, each by means of the simple logarithmic amplifier shown in

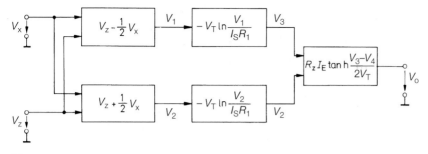

Fig. 11.41. Two-quadrant transconductance divider

$$V_o = \frac{R_z I_E}{2} \cdot \frac{V_x}{V_z} \quad \text{for } V_z > 0 \text{ and } |V_x| \leq V_z$$

Fig. 11.21. Using a differential amplifier stage, as in Fig. 11.38, the hyperbolic tangent of the difference between the output voltages V_3 and V_4 is calculated so that

$$V_o = R_z I_E \tanh \frac{V_T \ln(V_2/V_1)}{2V_T}$$

Therefore, using (11.56),

$$V_o = \frac{R_z I_E}{2} \cdot \frac{V_x}{V_z}$$

With this method, an accuracy of 0.1% of the computing unit U can be obtained over a dynamic range of $1 : 1,000$.

11.8.3
Multipliers Using Electrically Controlled Resistors

A voltage can be multiplied by a constant using a simple voltage divider. Analog multiplication is possible if, by employing closed-loop control, if one insures that this constant is proportional to a second input voltage.

The principle of such a circuit is illustrated in Fig. 11.42. The arrangement contains two identical coefficient elements K_x and K_z, the output voltages of which are proportional to their input voltages. Their constant of proportionality k can be controlled by voltage V_1. Due to the feedback via K_z, the output voltage V_1 of the operational amplifier assumes a level such that $kV_z = V_y$, and this results in $k = V_y/V_z$. If the voltage V_x, is applied to the second coefficient element K_x, its output voltage becomes

$$V_o = kV_x = \frac{V_x V_y}{V_z}$$

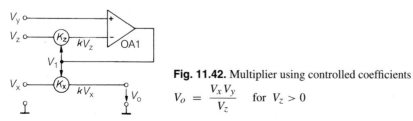

Fig. 11.42. Multiplier using controlled coefficients

$$V_o = \frac{V_x V_y}{V_z} \quad \text{for } V_z > 0$$

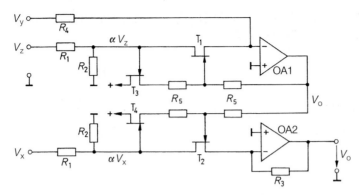

Fig. 11.43. Multiplier with FETs as controlled resistors

$$V_o = \frac{R_3}{R_4} \cdot \frac{V_x V_y}{V_z} \quad \text{for} \quad V_z > 0, \; V_y < 0$$

Voltage V_z, must be larger than zero, so that the negative feedback does not become positive. Whether voltage V_y is allowed to go both positive and negative depends on the design of the coefficient elements. If the latter permits bipolar coefficients, V_y may also be bipolar.

The voltage V_x, can be bipolar in any case. It has the additional advantage that it is not transferred through balancing amplifier OA1. Consequently, very high bandwidths can be achieved for V_x.

FETs can be employed as electrically controlled resistors, as in the circuit shown in Fig. 11.43. Amplifier OA1 operates as a controller for adjusting the coefficients. Its output voltage causes the resistance R_{DS} to vary, so that

$$\frac{\alpha V_z}{R_{DS}} + \frac{V_y}{R_4} = 0$$

Hence,

$$R_{DS} = -\alpha R_4 \frac{V_z}{V_y}$$

The output voltage of operational amplifier OA2 is

$$V_o = -\alpha \frac{R_3}{R_{DS}} V_x = \frac{R_3}{R_4} \cdot \frac{V_x V_y}{V_z}$$

In order that FETs may be operated as resistors, the voltage across them must be kept below approximately 0.5 V. Voltage dividers R_1, R_2 provide the necessary attenuation. The resistors R_5 make the voltage–current characteristic of the FET more linear, as is described in Sect. 3.1.3. In order to prevent any reactive effect of the control voltage V_1 on the input signals V_z and V_x, the circuit incorporates the two additional source followers, T_3 and T_4 respectively. The magnitude of their gate–source voltage is irrelevant, as it is controlled by operational amplifier OA1. It is merely essential for them to be well matched. Double FETs should therefore be used.

To insure negative feedback in the control circuit, V_z must be positive. As only positive coefficients can be realized using the simple coefficient elements shown in Fig. 11.43, V_y must always be negative to allow balancing. However, V_x, can have either polarity.

In order to achieve a high degree of accuracy, FETs T_1 and T_2 should be well matched over a wide range of resistances. This requirement can only be met using monolithic dual FETs.

11.8.4
Adjustment of Multipliers

A multiplier should conform to the expression

$$V_o = \frac{V_x V_y}{U}$$

where U is the computing unit; for example, $-10\,$V. In practice, there is a small offset voltage superposed on any terminal voltage. Therefore, in general,

$$V_o + V_{o0} = \frac{1}{U}(V_x + V_{x0})(V_y + V_{y0})$$

Thus,

$$V_o = \frac{V_x V_y}{U} + \frac{V_y V_{x0} + V_x V_{y0} + V_{x0} V_{y0}}{U} - V_{o0} \tag{11.57}$$

The product V_x, V_y must be zero whenever V_x or V_y is zero. This is only possible if the parameters V_{x0}, V_{y0}, and V_{o0} become zero independently. Therefore, three trimmers are essential for compensating the offset voltages. A suitable trimming procedure is as follows. First, V_x is made zero. Then, according to (11.57),

$$V_o = \frac{V_y V_{x0} + V_{x0} V_{y0}}{U} - V_{o0}$$

When voltage V_y is now varied, the output voltage also changes because of the term $V_y V_{x0}$. The nulling circuit for V_x is adjusted in such a way that, despite variation of V_y, a constant output voltage is obtained; V_{x0} is then zero.

In a second step, V_y is nulled and V_x is varied. In the same way as above, the offset of V_y can now be compensated. Thirdly, V_x, and V_y are made zero and the third trimmer is adjusted such that the output offset V_{o0} becomes zero.

A fourth trim potentiometer may often be necessary for adjusting the constant of proportionality, U, to the desired value.

11.8.5
Expansion to Four-Quadrant Multipliers

There are cases in which one- and two-quadrant multipliers have to be operated with input voltages of a polarity for which they are not designed. The most obvious remedy would then be to invert the polarity of the input and output of the multiplier whenever the impermissible polarity combination occurs. However, this method involves a large number of components and is not particularly fast. It is more convenient to add constant voltages V_{xk} and V_{yk} to the input voltages V_x and V_y, so that the resulting input voltages remain positive. Then, for the output voltage, we get

$$V_3 = \frac{(V_x + V_{xk})(V_y + V_{yk})}{U}$$

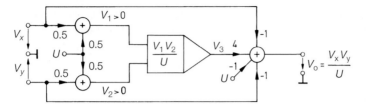

Fig. 11.44. Expansion of a one-quadrant multiplier to a four-quadrant multiplier

Hence,

$$\frac{V_x V_y}{U} = V_3 - \frac{V_{xk}}{U} V_y - \frac{V_{yk}}{U} V_x - \frac{V_{xk} V_{yk}}{U}$$

It follows that a constant voltage – and also two voltages, each of which is proportional to an input voltage – must be subtracted from the output voltage of the multiplier.

The block diagram of the resulting arrangement is shown in Fig. 11.44. The constant voltage and coefficients ($V_{xk} = 0.5\,U$) are selected such that the range of control is fully exploited. If the input voltage V_x is within $-U \le V_x \le +U$, the range for the voltage $V_1 = 0.5\,V_x + 0.5\,U$ is $0 \le V_1 \le U$ and $V_2 = 0.5\,V_y + 0.5\,U$ is $0 \le V_2 \le U$. Therefore, the output voltage obtained is:

$$V_o = 4 \cdot \frac{\frac{1}{2}(V_x + U) \cdot \frac{1}{2}(V_y + U)}{U} - V_x - V_y - U = \frac{V_x V_y}{U}$$

11.8.6
Multiplier as a Divider or Square Rooter

Figure 11.45 illustrates a method by which a multiplier without a division input can be used as a divider. Because of negative feedback, the output voltage of the operational amplifier finds a level such that

$$\frac{V_o V_z}{U} = V_x$$

Thus, the circuit evaluates the quotient $V_o = U V_x / V_z$, but only as long as $V_z > 0$. For negative denominators, the feedback is positive.

A multiplier can be employed as a square-rooter by operating it as a squarer and inserting it in the feedback loop of an operational amplifier, as shown in Fig. 11.46. The output voltage settles at a level such that

$$\frac{V_o^2}{U} = V_i, \quad \text{hence} \quad V_o = \sqrt{U V_i}$$

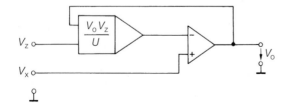

Fig. 11.45. Multiplier used as a divider

$$V_o = U \frac{V_x}{V_z} \quad \text{for } V_z > 0$$

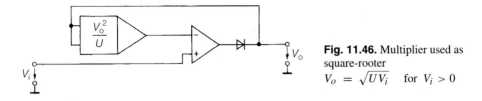

Fig. 11.46. Multiplier used as square-rooter

$$V_o = \sqrt{U V_i} \quad \text{for } V_i > 0$$

Correct operation is insured only for positive input and output voltages. Difficulties may arise if the output becomes momentarily negative; for example, at switch-on. In such a case, the squarer causes a phase inversion in the feedback loop so that positive feedback occurs, and the output voltage becomes more negative until it reaches the negative level of output saturation. The circuit is then said to be in "latch-up" and is inoperable. Therefore, a diode must be used to insure that the output voltage cannot become negative.

11.9
Transformation of Coordinates

Cartesian as well as polar coordinates play an important role in many technical applications. Therefore, in this section we discuss some circuits that allow transformations from one coordinate system to the other.

11.9.1
Transformation from Polar to Cartesian Coordinates

In order to implement the transformation equations

$$\begin{aligned} x &= r \cos \varphi, \\ y &= r \sin \varphi \end{aligned} \tag{11.58}$$

by means of an analog computing circuit, the coordinations must be expressed as voltages. We let

$$\varphi = \pi \frac{V_\varphi}{U} \quad \text{for} \quad -U \le V_\varphi \le +U$$

The range of the argument is therefore defined as being between $\pm\pi$. We define the normalized coordinates:

$$x = \frac{V_x}{U}; \qquad y = \frac{V_y}{U}; \qquad r = \frac{V_r}{U}$$

Thus, (11.58) can be rewritten:

$$V_x = V_r \cos\left(\pi \frac{V_\varphi}{U}\right), \qquad V_y = V_r \sin\left(\pi \frac{V_\varphi}{U}\right) \tag{11.59}$$

To generate the sine and cosine functions for the range of argument $\pm\pi$, we employ the network described in Sect. 12.7.4 and, in addition, two multipliers. The complete circuit is represented in the block diagram of Fig 11.47.

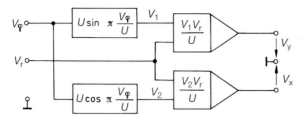

Fig. 11.47. Transformation from polar to Cartesian coordinates

$$V_x = V_r \cos\left(\pi\frac{V_\varphi}{U}\right); \qquad V_y = V_r \sin\left(\pi\frac{V_\varphi}{U}\right)$$

11.9.2
Transformation from Cartesian to Polar Coordinates

Inversion of the transformation equation (11.58) yields

$$r = \sqrt{x^2 + y^2} \quad \text{or} \quad V_r = \sqrt{V_x^2 + V_y^2}, \tag{11.60}$$

and

$$\varphi = \arctan\frac{y}{x} \quad \text{or} \quad V_\varphi = \frac{U}{\pi}\arctan\frac{V_y}{V_x} \tag{11.61}$$

respectively. The magnitude, V_r, of the vector can be computed according to the block diagram shown in Fig. 11.48, using two squarers and one square-rooter. A more simple circuit, which also has a larger input voltage range, can be deduced by applying a few more mathematical operations. From (11.60),

$$V_r^2 - V_y^2 = V_x^2,$$
$$(V_r - V_y)(V_r + V_y) = V_x^2$$

Hence,

$$V_r = \frac{V_x^2}{V_r + V_y} + V_y$$

The implicit equation for V_r can be implemented by means of a multiplier with a division input, as shown in Fig. 11.49. The summing amplifier S_1 evaluates the expression

$$V_1 = V_r + V_y$$

and, therefore,

$$V_2 = \frac{V_x^2}{V_r + V_y}$$

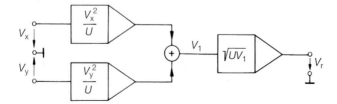

Fig. 11.48. Computation of the vector magnitude

$$V_r = \sqrt{V_x^2 + V_y^2}$$

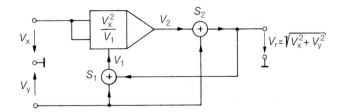

Fig. 11.49. Simplified circuit for computing the vector magnitude

In order to obtain V_r, this voltage V_2 is added to the input voltage V_y using the summing amplifier S_2.

The fact that voltage V_y must always be positive may be easily explained by reference to the special case $V_x = 0$. We thus have $V_2 = 0$ and $V_r = V_y$. This is the correct solution only for positive values of V_y. In addition, as practical dividers cannot handle a sign change in the denominator, it is necessary to form the absolute value for bipolar values of V_y using, for example, the circuit shown in Fig. 20.20. This does not limit the vector calculation, as the intermediate variable V_y^2 is positive in each case.

The simplest implementation of a vector meter is one in which multiplication and division are performed via logarithms, because both operations can be performed using a single circuit, as shown in Fig. 11.37. However, in this case it is also necessary to form the absolute value of V_x.

This is not required when using transconductance multipliers, as these generally allow four-quadrant operation. However, in this case separate circuits are required for multiplication and division. It is advisable, as shown in Fig. 11.50, to perform division before multiplication, as the dynamic range would otherwise be reduced due to the presence of the variable V_x^2.

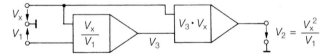

Fig. 11.50. Transconductance multipliers to compute the vector magnitude according to the method in Fig. 11.49

Chapter 12:
Controlled Sources and Impedance Converters

In linear network synthesis, not only passive components are used, but also idealized active elements such as controlled current and voltage sources. In addition, idealized converter circuitry, such as the negative impedance converter (NIC), the gyrator, and the circulator, is often employed. In the following sections, we describe the most common ways of implementing these circuits.

12.1
Voltage-Controlled Voltage Sources

A voltage-controlled voltage source is characterized by having an output voltage V_2 that is proportional to the input voltage V_1. It is therefore nothing more than a voltage amplifier. Ideally, the output voltage should be independent of the output current and the input current should be zero. Hence, the transfer characteristics are:

$$I_1 = 0 \cdot V_1 + 0 \cdot I_2 = 0,$$
$$V_2 = A_v V_1 + 0 \cdot I_2 = A_v V_1$$

In practice, the ideal source can only be approximated. Considering that the reaction of the output on the input is usually negligibly small, the equivalent circuit of a real source is as shown in Fig. 12.1, and its transfer characteristics are:

$$I_1 = \frac{1}{r_i} V_1 + 0 \cdot I_2 \qquad (12.1)$$
$$V_2 = A_v V_1 - r_o I_2$$

The internal voltage source shown is assumed to be ideal. The input resistance is r_i and the output resistance is r_o.

Voltage-controlled voltage sources of low output resistance and of defined, but adjustable, gain have already been described in Chap. 5 in the form of inverting and noninverting amplifiers. They are shown for the sake of completeness in Figs. 12.2 and 12.3. It is easy to obtain output resistances of far less than 1 Ω and therefore to approach the ideal behavior fairly closely. It should be noted, however, that the output impedance is somewhat inductive; in other words, it rises with increasing frequency (see Chap. 5).

The input resistance of an electrometer amplifier is very high. At low frequencies, one easily attains values in the gigaohm range, and hence very nearly ideal conditions. The high (incremental) input resistance must not lead us to overlook the additional errors that may arise due to the constant input bias current I_B, particularly when the output resistance of the signal source is high. In critical cases, an amplifier with FET input should be employed.

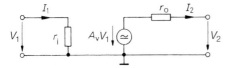

Fig. 12.1. Low-frequency equivalent circuit of a voltage-controlled voltage source

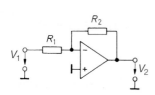

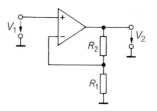

Fig. 12.2. Inverting amplifier as voltage-controlled voltage source

Fig. 12.3. Electrometer amplifier as voltage-controlled voltage source

Ideal
transfer characteristic: $V_2 = -\dfrac{R_2}{R_1} V_1$

Ideal
transfer characteristic: $V_2 = \left(1 + \dfrac{R_2}{R_1}\right) V_1$

Input imdepance: $\underline{Z}_i = R_1$

Input impedance: $\underline{Z}_i = r_{CM} \left\| \dfrac{1}{j\omega C} \right.$

Output impedance: $\underline{Z}_o = \dfrac{r_o}{g}$

Output impedance: $\underline{Z}_o = \dfrac{r_o}{g}$

For low-resistance signal sources, the inverting amplifier circuit in Fig. 12.2 can be used, as its low input resistance R_1 then causes no error. The advantage is that no inaccuracies can arise due to common-mode signals.

12.2
Current-Controlled Voltage Sources

The equivalent circuit of a current-controlled voltage source, shown in Fig. 12.4, is identical to that of the voltage-controlled voltage source in Fig. 12.1. The only difference between the two sources is that the input current is now the controlling signal. It should be influenced by the circuit as little as possible, a condition fulfilled in the ideal case when $r_i = 0$. Disregarding the effect of the output on the input, the transfer characteristics are:

$$
\begin{aligned}
V_1 &= r_i I_1 + 0 \cdot I_2 & \quad V_1 &= 0 \\
V_2 &= R I_1 - r_o I_2 & \quad \Rightarrow \quad V_2 &= R I_1 \\
&\text{(real)} & \text{(ideal, } & r_i = r_o = 0)
\end{aligned}
\tag{12.2}
$$

When implementing this circuit as shown in Fig. 12.5, we use the fact that the summing point of an inverting amplifier represents virtual ground. Because of this, the input resistance is low, as required. The output voltage becomes $V_2 = -R I_1$ if the input bias current of the amplifier is negligible compared with I_1. If very small currents I_1 are to be used as control signals, an amplifier with FET input must be employed. Additional errors may occur due to the offset voltage—and these will increase if the output resistance R_g of the signal source is low, since the offset voltage is amplified by a factor of $(1 + R/R_g)$.

For the output impedance of the circuit, the same conditions hold as for the previous circuit, where the loop gain $\underline{g}$ is dependent on the output resistance R_g of the signal source

$$
\underline{Z}_o = \frac{r_o}{\underline{g}} \quad \text{with } \underline{g} = \underline{k}\,\underline{A}_D = \frac{R_g}{R + R_g}\underline{A}_D
$$

A current-controlled voltage source with a floating input is discussed in Sect. 20.2.1.

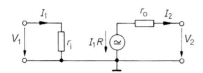

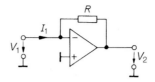

Fig. 12.4. Low-frequency equivalent circuit of a current-controlled voltage source

Fig. 12.5. Current-controlled voltage source

Ideal transfer characteristic: $V_2 = -RI_1$

Input impedance: $\underline{Z}_i = \dfrac{R}{A_D}$

Output impedance: $\underline{Z}_o = \dfrac{r_o}{g}$

12.3
Voltage-Controlled Current Sources

The purpose of voltage-controlled current sources is to impress on a load a current I_2 that is independent of the output voltage V_2 and is determined only by the control voltage V_1. Therefore,

$$I_1 = 0 \cdot V_1 + 0 \cdot V_2$$
$$I_2 = g_m V_1 + 0 \cdot V_2 \tag{12.3}$$

In practice, these conditions can be fulfilled only approximately. Taking into account that the effect of the output on the input is normally very small indeed, the equivalent circuit of a real current source becomes that shown in Fig. 12.6. The transfer characteristics are then:

$$I_1 = \frac{1}{r_i} V_1 + 0 \cdot V_2,$$
$$I_2 = g_m V_1 - \frac{1}{r_o} V_2 \tag{12.4}$$

For $r_i \to \infty$ and $r_o \to \infty$, one obtains the ideal current source. The parameter g_m is the forward transconductance or transfer conductance.

12.3.1
Current Sources for Floating Loads

In inverting and electrometer amplifiers, the current through the feedback resistor is $I_2 = V_1/R_1$ and is therefore independent of the voltage across the feedback resistor. Both circuits can thus be used as current sources if the load R_L is inserted in place of the feedback resistor, as shown in Figs. 12.7 and 12.8.

The same conditions obtain for the input impedance as for the corresponding voltage-controlled voltage sources in Figs. 12.2 and 12.3.

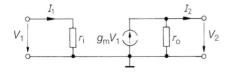

Fig. 12.6. Low-frequency equivalent circuit of a voltage-controlled current source

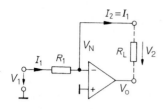

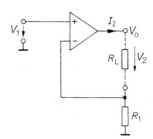

Fig. 12.7. Inverting amplifier as a voltage-controlled current source

Fig. 12.8. Electrometer amplifier as a voltage-controlled current source

Ideal
transfer characteristic: $I_2 = V_1/R_1 = I_1$

Ideal
transfer characteristic: $I_2 = V_1/R_1$

Input impedance: $\underline{Z}_i = R_1$

Input impedance: $\underline{Z}_i = r_{CM} \left\| \dfrac{1}{j\omega C_{CM}} \right.$

Output impedance: $\underline{Z}_o = A_D R_1 \left\| -jR_1 \dfrac{f_T}{f} \right.$

Output impedance: $\underline{Z}_o = A_D R_1 \left\| -jR_1 \dfrac{f_T}{f} \right.$

For a finite open-loop gain A_D of the operational amplifier, the output resistance assumes finite values only, as the potential difference $V_D = V_P - V_N$ does not remain precisely zero. To determine the output resistance, we take the following relationships from Fig. 12.7,

$$I_1 = I_2 = \frac{V_1 - V_N}{R_1} \qquad V_N = -\frac{V_o}{A_D} \qquad V_2 = V_N - V_o$$

and obtain

$$I_2 = \frac{V_1}{R_1} - \frac{V_2}{R_1(1 + A_D)} \approx \frac{V_1}{R_1} - \frac{V_2}{A_D R_1}$$

and therefore, for the output resistance,

$$r_o = -\frac{\partial V_2}{\partial I_2} = A_D R_1 \tag{12.5}$$

which is thus proportional to the differential gain of the operational amplifier.

Since the open-loop gain A_D of a frequency-compensated operational amplifier has a fairly low cutoff frequency (e.g., $f_{cA} \approx 10\,\text{Hz}$ for the 741 type), one must take into account that A_D is complex even at low frequencies. Equation (12.5) must then be rewritten in its complex form:

$$\underline{Z}_o = \underline{A}_D R_1 = \frac{A_D}{1 + j\frac{\omega}{\omega_{cA}}} R_1 \approx -jR_1 \frac{f_T}{f} \tag{12.6}$$

This output impedance may be represented by a parallel connection of a resistor r_o and a capacitor C_o, as the following rearrangement of (12.6) shows:

$$\underline{Z}_o = \frac{1}{\dfrac{1}{A_D R_1} + \dfrac{j\omega}{A_D R_1 \omega_{cA}}} = r_o \left\| \dfrac{1}{j\omega C_o} \right. , \qquad \text{where} \tag{12.7}$$

$$r_o = A_D R_1 \quad \text{and} \quad C_o = \frac{1}{A_D R_1 \omega_{cA}} = \frac{1}{2\pi R_1 f_T}.$$

With an operational amplifier that has $A_D = 10^5$ and $f_T = 1\,\text{MHz}$, one obtains, for $R_1 = 1\,k\Omega$, the result $r_o = 100\,M\Omega$ and $C_o = 159\,\text{pF}$. For a frequency of $10\,\text{kHz}$, the magnitude of the output impedance is reduced to $100\,k\Omega$.

The same considerations apply to the output impedance of the circuit shown in Fig. 12.8. As far as their electrical data are concerned, the two current sources in Figs. 12.7 and 12.8 are well suited for many applications. However, they have a serious technical disadvantage: the load R_L must be floating – that is, it must not be connected to a fixed potential – otherwise the amplifier output or the N-input is short-circuited. The following circuits overcome this restriction.

12.3.2
Current Sources for Grounded Loads

The principle of the current source in Fig. 12.9 is based on the fact that the output current is measured as the voltage drop across R_1. The output voltage of the operational amplifier finds a value such that this voltage is equal to a given input voltage. In order to determine the output current, we apply KCL to the N-input, the P-input, and the output. Thus,

$$\frac{V_o - V_N}{R_2} - \frac{V_N}{R_3} = 0 \qquad\qquad \frac{V_1 - V_P}{R_1 + R_2} + \frac{V_2 - V_P}{R_3} = 0$$

$$\frac{V_o - V_2}{R_1} + \frac{V_P - V_2}{R_3} - I_2 = 0$$

Since $V_N = V_P$, we obtain the output current:

$$I_2 = \frac{V_1}{R_1} + \frac{R_2^2 - R_3^2}{R_1 R_3 (R_2 + R_3)} V_2$$

We can see that the output current for $R_2 = R_3$ is independent of the output voltage. The output resistance then becomes $r_o = \infty$ and the output current is $I_2 = V_1/R_1$. In practice, resistance R_1 is made low so that the voltage across it remains of the order of a volt. Resistance R_2 is selected large in comparison with R_1 so that the operational amplifier and the voltage source V_1 are not unnecessarily loaded. The output resistance of the current source can, at low frequencies, be adjusted to infinity even for a real operational amplifier, by slightly varying R_3. The internal resistance R_g of the controlling voltage source is

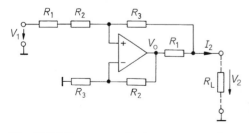

Fig. 12.9. Voltage-controlled current source for grounded loads

Output current: $\qquad I_2 = \dfrac{V_1}{R_1} \qquad$ for $\quad R_2 = R_3$

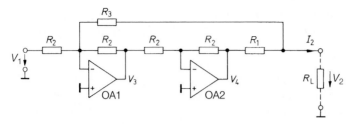

Fig. 12.10. Voltage-controlled current source without a common-mode voltage

Output current: $I_2 = \dfrac{V_1}{R_1}$ for $R_3 = R_2 - R_1$

connected in series with R_1 and R_2. In order to avoid an incorrect result, it must be negligibly small.

The circuit can also be designed as a current source with *negative output resistance*. For this purpose, we select $R_3 < R_2$ and obtain

$$r_o = -\frac{V_2}{I_2} = \frac{R_1 R_3(R_2 + R_3)}{R_3^2 - R_2^2} < 0$$

An alternative for the same application is shown in Fig. 12.10, where both opamps are operated as inverting amplifiers having no common-mode signal. In addition, the loading of the controlling source is independent of U_2 and hence of the load resistance R_L. To determine the output current, we use KVL for the circuit and obtain

$$V_4 = -V_3 = V_1 + \frac{R_2}{R_3} V_2$$

Applying KCL to the output gives

$$\frac{V_4 - V_2}{R_1} - \frac{V_2}{R_3} = -I_2 = 0$$

Eliminating V_4, we obtain

$$I_2 = \frac{V_1}{R_1} + \frac{R_2 - R_3 - R_1}{R_1 R_3} V_2$$

The output current becomes independent of the output voltage when the condition

$$R_3 = R_2 - R_1$$

is fulfilled.

12.3.3
Precision Current Sources Using Transistors

Simple single-ended current sources employ a bipolar transistor or a field effect transistor to supply loads that have one terminal connected to a constant potential. Such circuits have been described in the introductory Chap. 4.1.1 on page 277.

Their disadvantage is that the output current is affected by V_{BE} or V_{GS} and therefore cannot be precisely defined. An operational amplifier can be used to eliminate this effect. Figure 12.11 shows the relevant circuits for a bipolar transistor and for a FET. The output voltage of the operational amplifier finds a value such that the voltage across the resistance

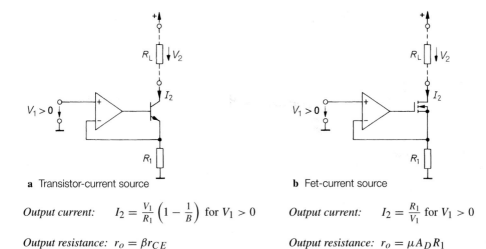

a Transistor-current source   **b** Fet-current source

Output current: $I_2 = \frac{V_1}{R_1}\left(1 - \frac{1}{B}\right)$ for $V_1 > 0$ *Output current:* $I_2 = \frac{R_1}{V_1}$ for $V_1 > 0$

Output resistance: $r_o = \beta r_{CE}$ *Output resistance:* $r_o = \mu A_D R_1$

Fig. 12.11. A current source using a transistor

R_1 equals V_1. Obviously, this holds for positive voltages only, as otherwise the transistors are OFF. Since the current through R_1 is V_1/R_1, the load current is:

for the bipolar transistor: $I_2 = \dfrac{V_1}{R_1}\dfrac{B}{1+B} \approx \dfrac{V_1}{R_1}\left(1 - \dfrac{1}{B}\right)$

and for the FET: $I_2 = \dfrac{V_1}{R_1}$

The difference between these currents is due to the fact that, in the bipolar transistor, part of the emitter current escapes through the base. As the current transfer ratio B is dependent on V_{CE}, the current I_B also changes with the output voltage V_2. As shown in Sect. 4.1.1 on page 277, this effect limits the output resistance to the value βr_{CE}, even if the operational amplifier is assumed to be ideal.

The effect of the finite current transfer ratio can be reduced if the bipolar transistor is replaced by a Darlington circuit. It can be virtually eliminated by using a FET, because the gate current is extremely small. The output resistance of the circuit in Fig. 12.11b is limited only by the finite gain of the operational amplifier. It can be determined by the following relationships, obtained directly from the circuit for $V_1 = $ constant:

$$dV_{DS} \approx -dV_2$$
$$dV_{GS} = dV_G - dV_S = -A_D R_1 dI_2 - R_1 dI_2 \approx -A_D R_1 dI_2$$

Using the basic equation (3.9),

$$dI_2 = g_m dV_{GS} + \frac{1}{r_{DS}} dV_{DS}$$

we obtain for the output resistance

$$r_o = -\frac{dV_2}{dI_2} = r_{DS}(1 + A_D g_m R_1) \approx \mu A_D R_1 \qquad (12.8)$$

It is thus greater by a factor $\mu = g_m \cdot r_{DS} \approx 50$ than that of the corresponding current source in Fig. 12.8, which uses an operational amplifier without a field effect transistor.

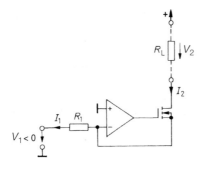

Fig. 12.12. Inverting FET current source

Output current: $I_2 = -\dfrac{V_1}{R_1} = I_1$ for $V_1 < 0$

Output resistance: $r_o = \mu A_D R_1$

Using the same values as in the example given for the circuit in Fig. 12.8, we obtain the very high output resistance of approximately $5\,G\Omega$. Because of the frequency dependence of the open-loop gain A_D, this value holds only for frequencies below the cutoff frequency f_{cA} of the operational amplifier. For higher frequencies, we must take into account the fact that the differential gain is complex and, instead of (12.8), obtain the output impedance

$$\underline{Z}_o = \underline{A}_D \mu R_1 = \frac{A_D}{1 + j\dfrac{\omega}{\omega_{cA}}} \mu R_1 = \frac{1}{\dfrac{1}{\mu A_D R_1} + \dfrac{j\omega}{\mu A_d R_1 \omega_{cA}}} = r_0 \left\| \frac{1}{j\omega C_o} \right. \quad (12.9)$$

A comparison with (12.6) and (12.7) shows that this impedance is equivalent to connecting a resistance $r_o = \mu A_D R_1$ in parallel with a capacitance $C_o = 1/\mu A_D R_1 \omega_{cA}$. For the practical example mentioned, we obtain $C_o = 3\,pF$. The drain-gate capacitance of the FET, which is of the order of a few picofarads, will appear in parallel.

The circuit shown in Fig. 12.11b can be modified by connecting the input voltage directly to R_1 and by grounding the P-input terminal. Figure 12.12 shows the resulting circuit. To insure that the FET is not turned off, V_1 must always be negative. In contrast to the circuit in Fig. 12.11b, the control voltage source is loaded by the current $I_1 = I_2$.

When a current source is required, the output current of which flows in the opposite direction to that in the circuit of Fig. 12.11b, the n-channel FET is simply replaced by a

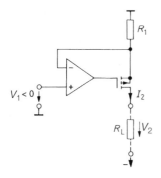

Fig. 12.13. A current source using a p-channel FET

Output current: $I_2 = -\dfrac{V_1}{R_1}$ for $V_1 < 0$

Output resistance: $r_o = \mu A_D R_1$

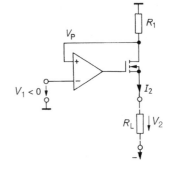

Fig. 12.14. A current source with a quasi-p-channel FET

Output current: $I_2 = -\dfrac{V_1}{R_1}$ for $V_1 < 0$

Output resistance: $r_o = A_D R_1$

p-channel FET, as shown in the circuit given in Fig. 12.13. If no p-channel FET is available, the arrangement in Fig. 12.14 can also be used. In contrast to the previous circuits, the source terminal here serves as the output. However, this does not affect the output current, since it is controlled, as before, by the voltage across R_1. Here the inputs of the opamp must be exchanged because the FET acts as an inverting amplifier. As the source electrode is the current output the amplifier has to transform the low output resistance $r_o = 1/g_m$ of the source follower. Another disadvantage of the arrangement is that the output of the opamp has to follow the load voltage V_2 and therefore it needs a large output voltage sewing. In the previous circuits the opamp must only compensate the voltage drop of the gate-source voltage.

Transistor Current Sources for Bipolar Output Currents

One disadvantage of all the current sources previously described is that they can supply only output currents of one polarity. By combining the circuits in Figs. 12.11b and 12.13, we obtain the current source shown in Fig. 12.15, which can supply currents of either polarity. For zero control voltage, $V_{P1} = \frac{3}{4}V^+$ and $V_{P2} = \frac{3}{4}V^-$. In this case,

$$I_2 = I_{D1} - I_{D2} = \frac{V^+}{4R_1} + \frac{V^-}{4R_1} = 0 \quad \text{for} \quad V^+ = -V^-$$

For positive input voltages V_1, current I_{D2} increases by $V_1/4R_1$, whereas I_{D1} decreases by the same amount. Therefore, one obtains a negative output current

$$I_2 = -\frac{V_1}{2R_1}$$

For negative input voltages, I_{D2} decreases while I_{D1} becomes larger, this resulting in a positive output current.

The circuit has a rather poor zero stability. This is because the output current is itself the difference between the two relatively large currents, I_{D1} and I_{D2}, which are additionally affected by changes in the supply voltages.

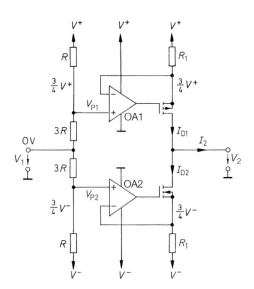

Fig. 12.15. A bipolar FET current source (the potentials entered are quiescent values)

Output current: $I_2 = -\dfrac{V_1}{2R_1}$

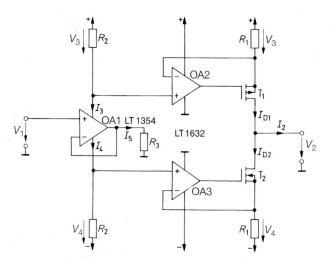

Fig. 12.16. A bipolar FET current source for large output currents

Output current: $\quad I_2 = -\dfrac{R_2}{R_1 R_3} V_1$

The circuit shown in Fig. 12.16 is considerably better in this respect. It differs from the former circuit in that it uses a different kind of control. The two output stages are controlled by the currents I_3 and I_4, which flow in the supply terminals of the amplifier OA1. For the drain currents,

$$I_{D1} = \frac{V_3}{R_1} = \frac{R_2}{R_1} I_3, \qquad\qquad I_{D2} = \frac{V_4}{R_1} = \frac{R_2}{R_1} I_4 \qquad\qquad (12.10)$$

The output stages therefore operate as current mirrors. Hence, the output current is:

$$I_2 = I_{D1} - I_{D2} = \frac{R_2}{R_1}(I_3 - I_4) \qquad\qquad (12.11)$$

Amplifier OA1 operates as a voltage follower. Therefore, the voltage across resistor R_3 is equivalent to the input voltage V_1. Thus, its current is:

$$I_5 = V_1/R_3 \qquad\qquad (12.12)$$

For further processing of this signal, use is made of the fact that the operational amplifier can be regarded as a current node for which, by applying KCL, the sum of the currents must equal zero. As the input currents are negligible and as there is usually no ground connection, the following relationship holds with very good accuracy:

$$I_5 = I_3 - I_4 \qquad\qquad (12.13)$$

Substitution in (12.12) and (12.11) yields the output current:

$$I_2 = \frac{R_2}{R_1 R_3} V_1 = \frac{V_1}{R_1} \quad \text{for} \quad R_2 = R_3$$

For zero input, $I_5 = 0$ and $I_3 = I_4 = I_Q$, where I_Q is the quiescent current flowing in the supply leads of the amplifier OA1. It is small in comparison with the maximum possible output current, I_5, of the amplifier. For a positive input voltage, $I_3 \approx I_5 \gg I_4$.

The output current I_2 is then supplied virtually only from the upper output stage, whereas the lower stage is disabled. The inverse is true for a negative input voltage difference. The circuit is therefore of the class AB push–pull type. Since the quiescent current in the output stage is

$$I_{D1Q} = I_{D2Q} = \frac{R_2}{R_1} I_Q \qquad (12.14)$$

and is small relative to the maximum output current. The output current at zero input signal is now determined by the difference of small quantities. This results in a very good zero-current stability. A further advantage is the high efficiency of the circuit, this being of special interest if the circuit is to be designed for high output currents. For this reason, the device selected for OA1 is an operational amplifier with a low quiescent current drain I_Q.

In the case of the circuit shown in Fig. 12.16, it is particularly advisable to use power MOSFETS. As they are of the enhancement type, their gate potentials are within the supply voltage range, thereby obviating the need for positive and negative auxiliary voltages for operational amplifiers OA2 and OA3 respectively. However, amplifiers must be used whose common-mode input and output voltage range extends to the positive or negative supply voltage. Therefore one should use rail-to-rail amplifiers for OA2 and OA3 as in the example in Fig. 12.16.

If resistor R_3 in Fig. 12.16 is not grounded but is connected to the output of a second voltage follower, the input voltage difference will determine the output current.

12.3.4
Floating Current Sources

In the previous sections, we have discussed two types of current sources. Neither load terminal in the circuits of Figs. 12.7 and 12.8 may be connected to a fixed potential. Such a load is called "off-ground" or "floating," and is illustrated in Fig. 12.17a. For this kind of operation, the load may in practice consist only of passive elements, since for active loads there is normally a connection to ground via the supply. Grounded loads can be supplied by a current source based on the arrangement shown in Fig. 12.17b. Its practical design is shown in Figs. 12.9–12.16.

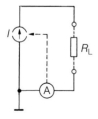
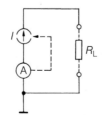
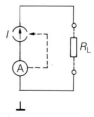

a Current source for floating loads   **b** Current source for grounded loads (single-ended current source)   **c** Floating current source for any load

Fig. 12.17. Different types of loads

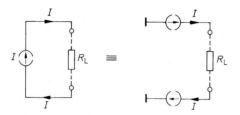

Fig. 12.18. Implementation of a floating current source by two single-ended current sources

If one or other load terminal is to be connected to any desired potential without the current being affected, a floating current source is required. It can be constructed using two grounded current sources which supply equal but opposite currents, as shown in Fig. 12.18. It can be realized especially easy with the MAX435.

12.4
Current-Controlled Current Sources

The equivalent circuit of the current-controlled current source in Fig. 12.19 is identical to that of the voltage-controlled current source shown in Fig. 12.6. The only difference is that the input current is now the controlling signal and should be affected as little as possible by the circuit. This is the case for the ideal condition in which $r_i = 0$. When the effect of the output on the input is neglected, the transfer characteristics are:

$$\begin{aligned} V_1 &= r_i I_1 + 0 \cdot V_2 \quad \Rightarrow \quad V_1 = 0 \\ I_2 &= A_I I_1 - \frac{1}{r_o} \cdot V_2 \quad \Rightarrow \quad I_2 = A_I I_1 \end{aligned} \tag{12.15}$$

In Figs. 12.7 and 12.12, we show two voltage-controlled current sources of finite input resistance. They can be operated as current-controlled current sources that have virtually ideal characteristics if the resistor R_1 is made zero; then, $I_2 = I_1$.

Current-controlled current sources allowing polarity reversal of the output currents are of particular interest. They are called current mirrors (also see Fig. 4.14 on page 281), and one example is shown in Fig. 12.20. It is based on the voltage-controlled current source shown in Fig. 12.11b, and the current-to-voltage conversion is effected by the additional resistor R_1. However, this results in nonideal conditions for the input resistance.

The maximum freedom in specifying the circuit parameters is obtained if a circuit from Sect. 12.2 is used for the current-to-voltage conversion and one of the voltage-controlled current sources described in Sect. 12.3 is connected in series.

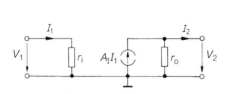

Fig. 12.19. Principle of a current controlled current source. In an ideal circuit $r_i = 0$ and $r_o = \infty$

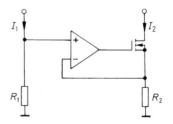

Fig. 12.20. Current mirror

Output current: $I_2 = \dfrac{R_1}{R_2} I_1$

12.5
NIC (Negative Impedance Converter)

There are cases in which negative resistances or voltage sources that have negative internal resistances are required. By definition, the resistance $R = +V/I$, if the arrows for the current and the voltage point in the same direction. If the voltage V across and the current I through a two-terminal network have opposite signs, the quotient V/I then becomes negative. Such a network is said to have a negative resistance. Negative resistances can, in principle, be implemented only by the active circuits known as NICs. There are two types: the UNIC, which reverses the polarity of the voltage without affecting the direction of the current; and the INIC, which reverses the current without changing the polarity of the voltage. The implementation of the INIC is particularly simple. Its ideal transfer characteristics are:

$$V_1 = V_2 + 0 \cdot I_2 \qquad I_1 = 0 \cdot V_2 - I_2 \tag{12.16}$$

These equations can be implemented as shown in Fig. 12.21, by a voltage-controlled voltage source and a current-controlled current source. However, both functions can also be performed by a single operational amplifier, as shown in Fig. 12.22.

For the ideal operational amplifier, $V_P = V_N$, and therefore $V_1 = V_2$, as required. The output potential of the amplifier has the value

$$V_o = V_2 + I_2 R$$

Hence, the current at port 1 is, as required,

$$I_1 = \frac{V_2 - V_o}{R} = -I_2$$

For this deduction, we have tacitly assumed stability of the circuit. However, since it simultaneously employs positive and negative feedback, the validity of this assumption must be examined separately in each case. To do so, we determine what proportion of the output voltage affects the P-input and the N-input respectively. Figure 12.23 shows the INIC in a general application, where R_1 and R_2 are the internal resistances of the circuits connected to it. The feedback of the voltage

$$V_P = V_o \frac{R_1}{R_1 + R}$$

is positive, and that of

$$V_N = V_o \frac{R_2}{R_2 + R}$$

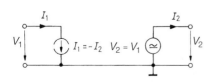

Fig. 12.21. Model for an INIC, using controlled sources

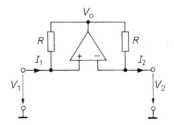

Fig. 12.22. INIC using a single amplifier

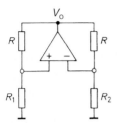

Fig. 12.23. INIC in a general application

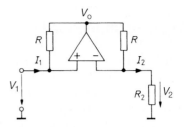

Fig. 12.24. Implementation of negative resistances

Negative resistance: $\dfrac{V_1}{I_1} = -R_2$

is negative. The circuit is stable if the positive-feedback voltage V_P is smaller than V_N; that is, if

$$R_1 < R_2$$

The circuit shown in Fig. 12.24 illustrates the use of the INIC for producing negative resistances. When a positive voltage is applied to port 1, $V_2 = V_1$ becomes positive according to (12.16), and hence I_2 is also positive. From (12.16),

$$I_1 = -I_2 = -\frac{V_1}{R_2}$$

A negative current thus flows into port 1 even though a positive voltage has been applied. Port 1 therefore behaves as if it were a negative resistance with the value

$$\frac{V_1}{I_1} = -R_2 \tag{12.17}$$

This arrangement is stable as long as the internal resistance R_1 of the circuit connected to port 1 is smaller than R_2. For this reason, the arrangement is stable even under short-circuit conditions. It is also possible to have a negative resistance which is stable at open-circuit, by reversing the INIC; that is, by connecting the resistance R_2 to port 1.

Since (12.16) also holds for alternating currents, we can replace the resistance R_2 by a complex impedance $\underline{Z}_2$ and in this way obtain any desired negative impedance.

The INIC can also be operated as a voltage source with negative output resistance. A voltage source that has a no-load voltage V_0 and a source resistance r_o supplies, at load I, an output voltage $V = V_0 - I r_o$. For normal voltage sources, r_o is positive, this resulting in a reduction of V under load. For a voltage source with negative output resistance, however, V increases with load. The circuit shown in Fig. 12.25 exhibits this behavior. It follows from the circuit that

$$V_2 = V_1 = V_0 - I_1 R_1$$

Then, since $I_1 = -I_2$,

$$V_2 = V_0 + I_2 R_1$$

Here, the INIC is connected in such a way that the voltage source is stable under no-load conditions.

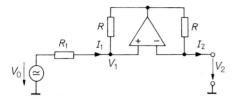

Fig. 12.25. Voltage source with negative output resistance

Output voltage: $V_2 = V_0 + I_2 R_1$

Output resistance: $r_o = -\dfrac{dV_2}{dI_2} = -R_1$

Negative resistances can be connected in series and in parallel just like conventional resistors, and the same laws apply. For example, a voltage source with negative output resistance can be used to compensate for the resistance of a long line so that, at the end of the line, the voltage V_0 is obtained with zero source resistance.

12.6
Gyrator

The gyrator is a converter circuit by which any impedance can be converted into its dual-transformed counterpart; for example, a capacitance can be changed into an inductance. The graphic symbol for a gyrator is shown in Fig. 12.26. The ideal transfer characteristics are:

$$
\begin{aligned}
I_1 &= 0 \cdot V_1 + \frac{1}{R_g} V_2 \\
I_2 &= \frac{1}{R_g} V_1 + 0 \cdot V_2
\end{aligned}
\tag{12.18}
$$

where R_g is the gyration resistance. Hence, the current at one port is proportional to the voltage at the other port. For this reason, the gyrator can be constructed from two voltage-controlled current sources that have high input and output resistances, as shown schematically in Fig. 12.27. A practical realization is shown in Fig. 5.90 on page 559.

Another method of implementing a gyrator is based on the combination of two INICs and is illustrated in Fig. 12.28. To determine the transfer characteristics, we apply KCL to the P- and N-inputs of OA1 and OA2 and obtain:

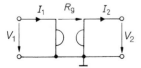

Fig. 12.26. Symbol for the gyrator

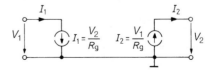

Fig. 12.27. Implementation of a gyrator using two voltage-controlled current sources

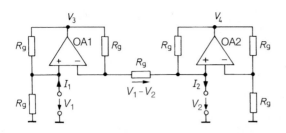

Fig. 12.28. Gyrator using two INICs

For node P_1:
$$\frac{V_3 - V_1}{R_g} - \frac{V_1}{R_g} + I_1 = 0$$

For node N_1:
$$\frac{V_3 - V_1}{R_g} + \frac{V_2 - V_1}{R_g} = 0$$

For node P_2:
$$\frac{V_4 - V_2}{R_g} + \frac{V_1 - V_2}{R_g} - I_2 = 0$$

For node N_2:
$$\frac{V_4 - V_2}{R_g} - \frac{V_2}{R_g} = 0$$

By eliminating V_3 and V_4, the transfer characteristics become

$$I_1 = \frac{V_2}{R_g} \quad \text{and} \quad I_2 = \frac{V_1}{R_g},$$

which are the desired relationships as given in (12.18).

Some applications of the gyrator are described below. In the first example, a resistor R_2 is connected to the right-hand port. Since the arrows for I_2 and V_2 point in the same direction, $I_2 = V_2/R_2$, following Ohm's law. Insertion of this relationship into the transfer characteristics gives

$$V_1 = I_2 R_g = \frac{V_2 R_g}{R_2} \quad \text{and} \quad I_1 = \frac{V_2}{R_g}$$

Port 1 therefore behaves like a resistance that has the value

$$R_1 = \frac{V_1}{I_1} = \frac{R_g^2}{R_2} \tag{12.19}$$

and thus it is proportional to the reciprocal of the load resistance connected to port 2.

The conversion of resistances is also valid for impedances and, according to (12.19), gives

$$\underline{Z}_1 = \frac{R_g^2}{\underline{Z}_2} \tag{12.20}$$

This relationship indicates an interesting application of the gyrator: if a capacitor with value C_2 is connected to one side, the impedance measured on the other side is

$$\underline{Z}_1 = R_g^2 \cdot j\omega C_2 \stackrel{!}{=} j\omega L_1$$

which is the inductance

$$L_1 = R_g^2 \cdot C_2 \tag{12.21}$$

The importance of the gyrator is due to the fact that it can be used to emulate large low-loss inductances. The appropriate circuit is depicted in Fig. 12.29. The two free terminals of

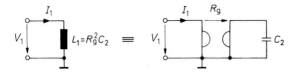

Fig. 12.29. Emulation of an inductance

the gyrator behave, according to (12.21), as if an inductance $L_1 = R_g^2 C_2$ were connected between them. With $C_2 = 1\mu F$ and $R_g = 10\,k\Omega$, we obtain $L_1 = 100\,H$.

If a capacitor C_1 is connected in parallel with the inductance L_1, the result is a parallel resonant circuit that can be used to construct high-Q „L" C-filters.

The Q-factor of the parallel resonant circuit, for $C_1 C_2$, is well suited to describe the deviation of a real gyrator from the ideal behavior, and is also called the Q-factor of the gyrator. The losses of a real gyrator can be represented by two equivalent resistances R_1, connected in parallel to the two ports. For the circuit involving current sources shown in Fig. 12.27, these resistances are given by the parallel connection of the input resistance of one source to the output resistance of the other. For the circuit involving INICs shown in Fig. 12.28, the equivalent resistances are defined by the matching tolerances of the resistors. The equivalent circuit of a parallel-resonant circuit incorporating a real gyrator with losses is shown in Fig. 12.30a. When the conversion equation (12.20) is applied to the right-hand side, the transformed equivalent circuit in Fig. 12.30b is obtained. From this, according to Sect. 29.3.7, the gyrator Q-factor can be determined as $Q = R_1/2R_g$.

This relationship, however, is only valid for low frequencies, as the Q-factor is very sensitive to phase displacements between the current and voltage of the transfer characteristics (12.18). A first order approximation is

$$Q(\varphi) = \frac{1}{\dfrac{1}{Q_0} + \varphi_1 + \varphi_2}$$

where Q_0 is the low-frequency limit value of the Q-factor. The terms φ_1 and φ_2 are the phase displacements between current $\underline{I}_1$ and voltage $\underline{V}_2$, and current $\underline{I}_2$ and voltage $\underline{V}_1$ respectively, at the resonant frequency of the circuit. For lagging phase angles, the Q-factor rises with an increasing resonant frequency. For $|\varphi_1 + \varphi_2| \geq 1/Q_0$, the circuit becomes unstable, and an oscillation occurs at the resonant frequency of the circuit. For leading phase angles, the Q-factor decreases with an increasing resonant frequency.

Not only two-terminal, but also four-terminal networks can be converted using gyrators. For this purpose, the four-terminal network to be converted is connected between two gyrators with identical gyration resistances, as shown in Fig. 12.31. The dual-transformed

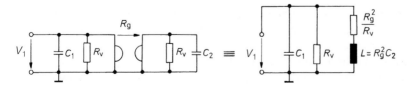

a Emulation of a resonant circuit using a lossy gyrator

b Equivalent circuit of a lossy resonant circuit

Fig. 12.30. Q-factor of a gyrator

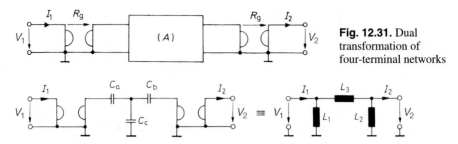

Fig. 12.31. Dual transformation of four-terminal networks

Fig. 12.32. Example of dual transformation

Transformation equations: $L_1 = R_g^2 C_a,\quad L_2 = R_g^2 C_b,\quad L_3 = R_g^2 C_c$

counterpart of the middle two-port then appears between the two outer ports. To deduce the transfer characteristics, the product of the chain matrices is calculated. The four-terminal network to be transformed has the chain matrix:

$$(A) = \begin{pmatrix} A_{11} & A_{12} \\ A_{21} & A_{22} \end{pmatrix}$$

From (12.18), we obtain the following relationship for the gyrator:

$$\begin{pmatrix} V_1 \\ I_1 \end{pmatrix} = \underbrace{\begin{pmatrix} 0 & R_g \\ 1/R_g & 0 \end{pmatrix}}_{(A_g)} \begin{pmatrix} V_2 \\ I_2 \end{pmatrix} \tag{12.22}$$

The chain matrix $(\overline{A})$ of the resulting four-terminal network is then:

$$(\overline{A}) = (A_g)(A)(A_g) = \begin{pmatrix} A_{22} & A_{21} \cdot R_g^2 \\ A_{12}/R_g^2 & A_{11} \end{pmatrix} \tag{12.23}$$

which is the matrix of the dual-transformed middle two-port.

As an example, Fig. 12.32 shows how a circuit of three inductors can be replaced by a dual circuit containing three capacitors.

If a capacitor is externally connected in parallel with each of the inductances L_1 and L_2, we obtain an inductance-coupled bandpass filter consisting exclusively of capacitors. If C_a and C_b are short-circuited, a floating inductance L_3 is obtained.

12.7
Circulator

A circulator is a circuit with three or more ports, the graphic symbol for which is shown in Fig. 12.33. It has the characteristic that a signal applied to one of the terminals is transferred in the direction of the arrow. If a terminal is open, the signal passes unchanged, whereas at a short-circuited terminal the polarity of the signal voltage is inverted. If a resistor $R = R_g$ is connected between one terminal and ground, the signal voltage appears across this resistor and, in this case, will not be passed on to the next terminal.

A circuit that has these properties is shown in Fig. 12.35. It can be seen that it consists of three identical stages, one of which is shown again in Fig. 12.34. Several cases can be distinguished:

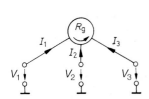

Fig. 12.33. Symbol for a circulator

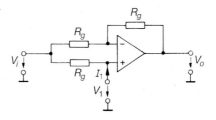

Fig. 12.34. One stage of a circulator

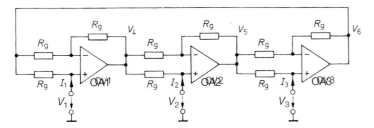

Fig. 12.35. Implementation of a circulator with VV-Opamps

- If terminal 1 is left open, $I_1 = 0$ and $V_P = V_i = V_N$. Hence, no current flows through the feedback resistor, and $V_o = V_i$.
- If terminal 1 is short-circuited, $V_1 = 0$, and the circuit acts as an inverting amplifier with unity gain. In this case, we obtain the output voltage $V_o = -V_i$.
- If a resistor $R_i = R_g$ is connected to terminal 1, the circuit operates as a subtracting amplifier for two equal input voltages V_i, and the voltage V_o is therefore zero.
- If V_1 is made zero and a voltage V_1 is applied to terminal 1, the circuit behaves like a noninverting amplifier that has a gain of 2, and we obtain the output voltage $V_o = 2V_1$.

From these characteristics, the operation of the circuit in Fig. 12.35 can be easily understood. Let us assume that voltage V_1 is applied to port 1, that a resistor R_g is connected at port 2, and that port 3 is left open. We already know that the output voltage of OA2 becomes zero. OA3 has unity gain because of the open terminal 3 and its output voltage is therefore also zero. OA1 consequently operates as a noninverting amplifier with a gain of 2, so that its output voltage is $2V_1$. Half of this voltage (V_1) appears at terminal 2, since this is terminated in R_g. Other special cases can be analyzed in an identical manner.

If a more general case is to be considered, the transfer characteristics of the circulator are used to determine the properties of the circuit. For this purpose, we apply KCL to the P- and N-inputs:

P-inputs

$$\frac{V_6 - V_1}{R_g} + I_1 = 0$$

$$\frac{V_4 - V_2}{R_g} + I_2 = 0$$

$$\frac{V_5 - V_3}{R_g} + I_3 = 0$$

N-inputs

$$\frac{V_6 - V_1}{R_g} + \frac{V_4 - V_1}{R_g} = 0$$

$$\frac{V_4 - V_2}{R_g} + \frac{V_5 - V_2}{R_g} = 0$$

$$\frac{V_5 - V_3}{R_g} + \frac{V_6 - V_3}{R_g} = 0$$

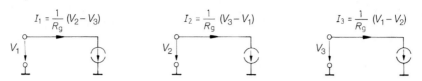

Fig. 12.36. A circulator composed of voltage-controlled current sources

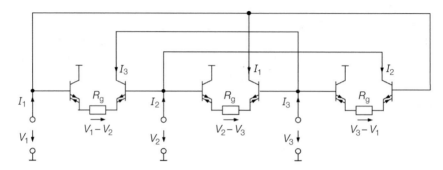

Fig. 12.37. Realization of a circulator with CC-Opamps

Eliminating $V_4 - V_6$, the transfer characteristics become

$$I_1 = \frac{1}{R_g}(V_2 - V_3) \quad I_2 = \frac{1}{R_g}(V_3 - V_1) \quad I_3 = \frac{1}{R_g}(V_1 - V_2) \tag{12.24}$$

It is obvious from (12.24) that a circulator can also be implemented by three voltage-controlled current sources with differential input, as shown in Fig. 12.36. A practical realization with CC-operational amplifiers is shown in Fig. 12.37.

Figure 12.38 shows a circulator employed as an active hybrid set for telephone circuits. It consists of a circulator that has three ports, all of which are terminated in the transfer resistance R_g, The signal from the microphone is relayed to the exchange and does not reach the speaker. The signal from the exchange is transferred to the speaker, but not to the microphone. The cross-talk attenuation is largely determined by the degree to which the terminating resistances are matched.

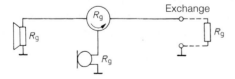

Fig. 12.38. A circulator used as a telephone hybrid set

Chapter 13:
Active Filters

13.1
Basic Theory of Lowpass Filters

Simple lowpass and highpass filters are discussed in Sects. 29.3.1 and 29.3.2, the circuit of the simplest lowpass filter being shown again in Fig. 13.1. The ratio of the output voltage to the input voltage can be expressed using (29.3.1) as

$$\underline{A}(j\omega) = \frac{\underline{V}_o}{\underline{V}_i} = \frac{1}{1 + j\omega RC}$$

and is called the frequency response of the circuit. Replacing $j\omega$ by $j\omega + \sigma = s$ gives the transfer function:

$$A(s) = \frac{L\{V_o(t)\}}{L\{V_i(t)\}} = \frac{1}{1 + s\,RC}$$

This is the ratio of the Laplace-transformed output and input voltages for signals of any time dependence. On the other hand, the transition from the transfer function $A(s)$ to the frequency response $\underline{A}(j\omega)$ for sinusoidal input signals is made by setting σ to zero.

In order to present the problem in a more general form, it is useful to normalize the complex frequency variable s by defining

$$s_n = \frac{s}{\omega_c}$$

Hence, for $\sigma = 0$

$$s_n = \frac{j\omega}{\omega_c} = j\frac{f}{f_c} = j\omega_n$$

The circuit shown in Fig. 13.1 has the cutoff frequency $f_c = 1/2\pi RC$. Therefore, $s_n = s\,RC$ and

$$A(s_n) = \frac{1}{1 + s_n} \tag{13.1}$$

For the absolute value of the transfer function – that is, for the amplitude ratio with sinusoidal input signals – we obtain with $\omega_n = \omega/\omega_c = f/f_c$

$$|\underline{A}(j\omega_n)|^2 = \frac{1}{1 + \omega_n^2}$$

If $\omega_n \gg 1$ – that is, $f \gg f_c$ – then $|\underline{A}| = 1/\omega_n$; this corresponds to a reduction in gain of 20 dB per frequency decade.

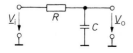

Fig. 13.1. The simplest passive lowpass filter

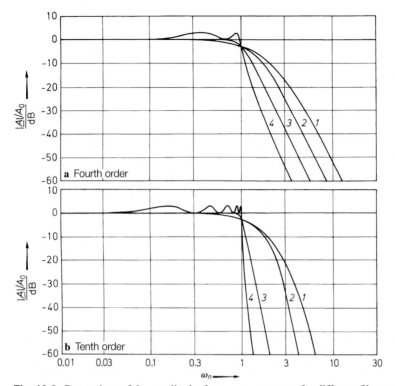

Fig. 13.2. Comparison of the amplitude–frequency responses for different filter types.
Curve 1: Lowpass filter with critical damping. *Curve 2*: Bessel lowpass filter.
Curve 3: Butterworth lowpass filter. *Curve 4*: Chebyshev lowpass filter with 3 dB ripple

If a sharper cutoff is required, N lowpass filters can be connected in series. The transfer function is then expressed in the form:

$$A(s_n) = \frac{1}{(1 + \alpha_1 s_n)(1 + \alpha_2 s_n) \ldots (1 + \alpha_n s_n)} \tag{13.2}$$

where the coefficients $\alpha_1, \alpha_2, \alpha_3, \ldots$ are real and positive. For $\omega_n \gg 1$, $|\underline{A}| \sim 1/\omega_n^N$ is proportional to $1/\omega^N$; the gain therefore falls off at $N \cdot 20$ dB per decade. It can be seen that the transfer function possesses N real negative poles. This is characteristic of Nth-order passive RC lowpass filters. If decoupled lowpass filters of identical cutoff frequencies are cascaded, then

$$\alpha_1 = \alpha_2 = \alpha_3 = \ldots = \alpha = \sqrt{\sqrt[N]{2} - 1}$$

this being the condition for which critical damping occurs. Each individual lowpass filter then has a cutoff frequency that is a factor $1/\alpha$ higher than that of the filter as a whole.

The transfer function of a lowpass filter has the general form:

$$A(s_n) = \frac{A_0}{1 + c_1 s_n + c_2 s_n^2 + \ldots + c_n s_n^N} \tag{13.3}$$

where $c_1, c_2, \ldots, c_N$ are positive and real. The order of the filter is equal to the highest power of s_n. It is advantageous for filter design if the denominator polynomial is written

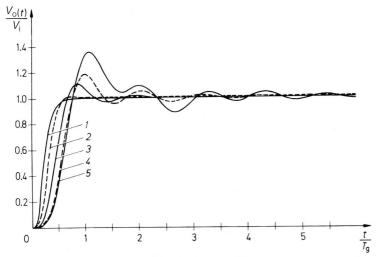

Fig. 13.3. Step response of fourth-order lowpass filters.
Curve 1: Lowpass filter with critical damping. *Curve 2*: Bessel lowpass filter.
Curve 3: Butterworth-Tiefpass. *Curve 4*: Chebyshev lowpass filter with 0.5 dB ripple.
Curve 5: Chebyshev lowpass filter with 3 dB ripple.

in factored form. If complex poles are also permitted, a separation into linear factors as shown in (13.2) is no longer possible, and a product of quadratic expressions is obtained:

$$A(s_n) = \frac{A_0}{(1 + a_1 s_n + b_1 s_n^2)(1 + a_2 s_n + b_2 s_n^2) \dots} \tag{13.4}$$

where a_i and b_i are positive and real. For odd orders N, the coefficient b_1 is zero.

There are several different theoretical aspects with respect to which the frequency response can be optimized. Any such aspect leads to a different set of coefficients a_i and b_i. As will be seen, conjugate complex poles arise. They cannot be realized by passive RC elements, as a comparison with (13.2) shows. One way of implementing conjugate complex poles is to use LRC networks. For high frequencies, the design of the necessary inductances usually presents no difficulties, but in the low-frequency range large inductances are often required. These are unwieldy and have poor electrical properties. However, the use of inductances at low frequencies can be avoided by the addition of active elements (e.g., operational amplifiers) to the RC network. Such circuits are called active filters.

Let us first compare the most important optimized frequency responses, the technical realizations of which are discussed in the following sections.

Butterworth lowpass filters have an amplitude–frequency response that is flat for as long as possible and falls off sharply just before the cutoff frequency. Their step response shows a considerable overshoot, which increases for higher-order filters.

Chebyshev lowpass filters have an even steeper roll-off above the cutoff frequency. In the passband, however, the gain varies with a ripple of constant amplitude. For a given order, the decrease above the cutoff frequency is steeper the larger the permitted ripple. The overshoot in the step response is even greater than for the Butterworth filters.

Bessel lowpass filters have the optimum square-wave response. The prerequisite for this is that the group delay is constant over the largest possible frequency range; in other words, that the phase shift in this frequency range is proportional to the frequency. The amplitude–

	Order				
	2	4	6	8	10
Critical damping					
Normalized rise time t_r/T_c	0.344	0.342	0.341	0.341	0.340
Normalized delay time t_d/T_c	0.172	0.254	0.316	0.367	0.412
Overshoot %	0	0	0	0	0
Bessel					
Normalized rise time t_r/T_c	0.344	0.352	0.350	0.347	0.345
Normalized delay time t_d/T_c	0.195	0.329	0.428	0.505	0.574
Overshoot %	0.43	0.84	0.64	0.34	0.06
Butterworth					
Formalized rise time t_r/T_c	0.342	0.387	0.427	0.460	0.485
Normalized delay time t_d/T_c	0.228	0.449	0.663	0.874	1.084
Overshoot %	4.3	10.8	14.3	16.3	17.8
Chebyshev 0.5 dB *ripple*					
Normalized rise time t_r/T_c	0.338	0.421	0.487	0.540	0.584
Normalized delay time t_d/T_c	0.251	0.556	0.875	1.196	1.518
Overshoot %	10.7	18.1	21.2	22.9	24.1
Chebyshev 1 dB *ripple*					
Normalized rise time t_r/T_c	0.334	0.421	0.486	0.537	0.582
Normalized delay time t_d/T_c	0.260	0.572	0.893	1.215	1.540
Overshoot %	14.6	21.6	24.9	26.6	27.8
Chebyshev 2 dB *ripple*					
Normalized rise time t_r/T_c	0.326	0.414	0.491	0.529	0.570
Normalized delay time t_d/T_c	0.267	0.584	0.912	1.231	1.555
Overshoot %	21.2	28.9	32.0	33.5	34.7
Chebyshev 3 dB *ripple*					
Normalized rise time t_r/T_c	0.318	0.407	0.470	0.519	0.692
Normalized delay time t_d/T_c	0.271	0.590	0.912	1.235	1.557
Overshoot %	27.2	35.7	38.7	40.6	41.6

Fig. 13.4. Comparison of lowpass filters. The rise time and delay time are normalized to the reciprocal cutoff frequency $T_c = 1/f_c$

frequency response of Bessel filters does not fall off as sharply as that of Butterworth or Chebyshev filters.

Figure 13.2 shows the amplitude–frequency responses of the four described filter types for the fourth and tenth orders. It can be seen that the Chebyshev lowpass filter has the most abrupt transition from the passband to the stopband. This is advantageous, but it has the side effect of a ripple in the amplitude–frequency response in the passband. As this ripple is gradually reduced, the behavior of the Chebyshev filter approaches that of a Butterworth filter. Both kinds of filter show a considerable overshoot in the step response, as can be seen in Fig. 13.3. Bessel filters, on the other hand, have only a negligible overshoot. Despite their unfavorable amplitude–frequency response, they will always be used where a good step response is important. A passive RC lowpass filter exhibits no overshoot; however, the relatively small improvement over the Bessel filter involves a considerable deterioration in the amplitude–frequency response. In addition, the corners in the step response are much rounder than for the Bessel filter. The table in Fig. 13.4 compares the rise times, delay times, and overshoots. The rise time is the time in which the output signal rises from 10% to 90% of its final-state value. The delay time is that in which the output signal increases from 0 to 50% of the final-state value.

It can be seen that the rise time does not depend to any great extent on the order; nor does it depend on the type of the filter. Its value is approximately $t_r = 1/3 f_g$, as shown in Sect. 29.3.1. On the other hand, as the order increases, so the delay time and the overshoot increase. The Bessel filters are an exception in that the overshoot decreases for orders higher than four.

It will be seen later that the same circuit allows the implementation of these filter types for a particular order; the values of resistances and capacitances determine the type of filter. In order that the circuit parameters can be defined, the frequency responses of the individual filter types must be known for each order. We shall therefore discuss these in more detail in the following sections.

13.1.1
Butterworth Lowpass Filters

From (13.3), the absolute value of the gain of an nth-order lowpass filter has the general form:

$$|\underline{A}|^2 = \frac{A_0^2}{1 + k_2\omega_n^2 + k_4\omega_n^4 + \cdots + k_{2n}\omega_n^{2N}} \tag{13.5}$$

Odd powers of ω do not occur, since the square of $|\underline{A}|^2$ must be an even function.

Below the cutoff frequency of the Butterworth lowpass filter, the function $|\underline{A}|^2$ must be maximally flat. Since for this range $\omega_n < 1$, this condition is best fulfilled if $|\underline{A}|^2$ is dependent only on the highest power of ω_n. The reason for this is that, for $\omega_n < 1$, the lower powers of ω_n contribute most to the denominator and therefore to the decrease in gain. Hence, for Butterworth-filters

$$|\underline{A}|^2 = \frac{A_0^2}{1 + k_{2n}\omega_n^{2N}}$$

The coefficient k_{2N} is defined by the "normalizing condition"; namely, that the gain at $\omega_n = 1$ is reduced by 3 dB. Thus

$$\frac{A_0^2}{2} = \frac{A_0^2}{1 + k_{2N}}$$
$$k_{2N} = 1$$

Therefore, the square of the gain of Nth-order Butterworth lowpass filters is given by

$$|\underline{A}|^2 = \frac{A_0^2}{1 + \omega_n^{2N}} \tag{13.6}$$

To implement a Butterworth lowpass filter, a circuit must be designed in which the square of the gain has the form given above. However, the circuit analysis initially gives the complex gain $\underline{A}$, rather than the square of the gain, $|\underline{A}|^2$. It is therefore necessary to know the value of the complex gain involved in (13.6). This is found by calculating the absolute

N	
1	$1 + s_n$
2	$1 + \sqrt{2}s_n + s_n^2$
3	$1 + 2s_n + 2s_n^2 + s_n^3 = (1 + s_n)(1 + s_n + s_n^2)$
4	$1 + 2{,}613s_n + 3{,}414s_n^2 + 2{,}613s_n^3 + s_n^4 = (1 + 1{,}848s_n + s_n^2)(1 + 0{,}765s_n + s_n^2)$

Fig. 13.5. Butterworth polynomials

value of (13.3) and by comparing the coefficients with those of (13.6). In this way, the desired coefficients $c_1, \ldots, c_n$ can be defined. The denominators of (13.3) are then the Butterworth polynomials, the first four orders of which are shown in Fig. 13.5.

It is possible to determine the poles of the transfer function analytically. By combining the conjugate complex poles, we immediately obtain the coefficients, a_i and b_i, of the quadratic expressions in (13.4):

even order N:

$$a_i = 2\cos\frac{(2i-1)\pi}{2N} \quad \text{for} \quad i = 1\ldots\frac{N}{2},$$

$$b_i = 1$$

odd order N:

$$a_1 = 1, \qquad a_i = 2\cos\frac{(i-1)\pi}{N} \quad \text{for} \quad i = 2\ldots\frac{N+1}{2},$$

$$b_1 = 0 \qquad b_i = 1$$

The coefficients of the Butterworth polynomials up to tenth order are shown in Fig. 13.14.

It can be seen that the first-order Butterworth lowpass filter is a passive lowpass filter that has the transfer function of (13.1). The higher Butterworth polynomials possess conjugate complex zeros. A comparison with (13.2) shows that such denominator polynomials cannot be implemented by passive RC networks, because in the case of the latter, all of the zeros are real. In such cases, the only choice is to use LRC circuits, with all their disadvantages, or active RC filters. The frequency response of the gain is shown in Fig. 13.6.

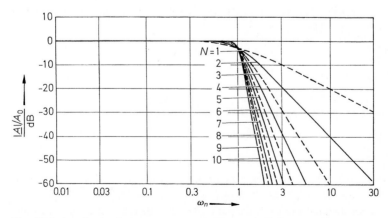

Fig. 13.6. Amplitude–frequency response of Butterworth lowpass filters

13.1.2
Chebyshev Lowpass Filters

At low frequencies, the gain of a Chebyshev lowpass filter has the value A_0, but it varies below the cutoff frequency, having a predetermined ripple. Polynomials that have a constant ripple within a defined range (an equal ripple) are the Chebyshev polynomials,

$$T_N(x) = \begin{cases} \cos(N \arccos x) & \text{for } 0 \le x \le 1 \\ \cosh(N \operatorname{Arcosh} x) & \text{for } x > 1, \end{cases}$$

the first four of which are shown in Fig. 13.7. For $0 \le x \le 1$, $|T(x)|$ oscillates between 0 and 1; for $x > 1$ rises steadily. In order to obtain the equation for a lowpass filter from the Chebyshev polynomials, one defines

$$|\underline{A}|^2 = \frac{kA_0^2}{1 + \varepsilon^2 T_N^2(x)} \tag{13.7}$$

The constant k is chosen such that, for $x = 0$, the square of the gain $|\underline{A}|^2$ becomes A_0^2; that is, $k = 1$ for odd N and $k = 1 + \varepsilon^2$ for even N. The factor a is a measure of the ripple, and is given by

$$\frac{A_{max}}{A_{min}} = \sqrt{1 + \varepsilon^2}$$

and

$$\left.\begin{aligned} A_{max} &= A_0\sqrt{1 + \varepsilon^2} \\ A_{min} &= A_0 \end{aligned}\right\} \quad \text{for even orders}$$

and

$$\left.\begin{aligned} A_{max} &= A_0 \\ A_{min} &= A_0/\sqrt{1 + \varepsilon^2} \end{aligned}\right\} \quad \text{for odd orders}$$

The appropriate values are listed for different ripples in Fig. 13.8. In principle, the complex gain can be calculated from $|\underline{A}|^2$ and hence the coefficients of the factored form can be determined. However, it is possible to derive the poles of the transfer function directly from those of the Butterworth filters. By combining the conjugate complex poles, the coefficients a_i, and b_i in (13.4) are determined:

even order N:

$$\left.\begin{aligned} b_i' &= \frac{1}{\cosh^2 \gamma - \cos^2 \dfrac{(2i-1)\pi}{2N}} \\[2ex] a_i' &= 2b_i' \cdot \sinh \gamma \cdot \cos \dfrac{(2i-1)\pi}{2N} \end{aligned}\right\} \quad \text{for } i = 1 \ldots \frac{N}{2}$$

N	
1	$T_1(x) = x$
2	$T_2(x) = 2x^2 - 1$
3	$T_3(x) = 4x^3 - 3x$
4	$T_4(x) = 8x^4 - 8x^2 + 1$

Fig. 13.7. Chebyshev polynomials

Parameter	Ripple			
	0.5 dB	1 dB	2 dB	3 dB
A_{max}/A_{min}	1.059	1.122	1.259	1.413
k	1.122	1.259	1.585	1.995
ε	0.349	0.509	0.765	0.998

Fig. 13.8. Some Tschebyscheff parameters

odd order N:

$$b'_1 = 0$$
$$a'_1 = 1/\sinh \gamma$$

$$\left. \begin{array}{l} b'_i = \dfrac{1}{\cosh^2 \gamma - \cos^2 \dfrac{(i-1)\pi}{N}} \\[2em] a'_i = 2b'_i \cdot \sinh \gamma \cdot \cos \dfrac{(i-1)\pi}{N} \end{array} \right\} \quad \text{for } i = 2 \dots \dfrac{N+1}{2}$$

where $\gamma = \dfrac{1}{N} \operatorname{Arsinh} \dfrac{1}{\varepsilon}$.

If the coefficients a'_i and b'_i found in this way are used to replace a_i and b_i in (13.4), Chebyshev filters are obtained. However, s_n is then not normalized with respect to the 3 dB cutoff frequency ω_c but, rather, to the frequency ω_x which the gain assumes the value A_{min} for the last time.

For an easy comparison of the different filter types, it is useful to normalize s_n to the 3 dB cutoff frequency. The variable s_n is replaced by αs_n and the normalizing constant α is determined such that the gain, for $s_n = j$, has the value $1/\sqrt{2} \cong -3$ dB. The quadratic expressions in the denominator of the complex gain are then

$$(1 + a'_i \alpha s_n + b'_i \alpha^2 s_n^2)$$

Hence, by comparing the coefficients with those of (13.4),

$$a_i = \alpha a'_i \quad \text{and} \quad b_i = \alpha^2 b'_i$$

The coefficients a_i and b_i are shown in the table of Fig. 13.14 up to the tenth order, and for ripple values of 0.5, 1, 2, and 3 dB. The frequency response of the gain is shown in Fig. 13.9 for ripple values of 0.5 and 3 dB. Figure 13.10 makes possible a direct comparison of fourth-order Chebyshev filters that have different amounts of ripple. It can be seen that the differences in the frequency response in the stopband are very small, and that they become even smaller for higher orders. It is also obvious that even the Chebyshev filter response that has a small ripple of 0.5 dB emerges from the passband much more steeply than that of the Butterworth filter.

The transition from the passband to the stopband can be made even steeper. To accomplish this, zeros are introduced into the amplitude–frequency response above the cutoff frequency. One way of optimizing the design is to give the amplitude–frequency response a constant ripple in the stopband as well. Such filters are called *Cauer and Elliptic* filters.

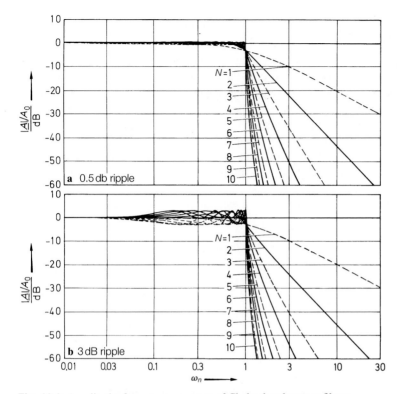

Fig. 13.9. Amplitude–frequency response of Chebyshev lowpass filters

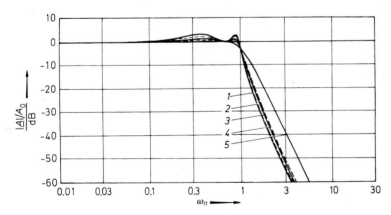

Fig. 13.10. Comparison of fourth-order Chebyshev lowpass filters
Ripple: *Curve 1*: 3 dB. *Curve 2*: 2 dB. *Curve 3*: 1 dB. *Curve 4*: 0.5 dB. *Curve 5*: A fourth-order Butterworth lowpass filter for comparison

Their transfer function differs from the ordinary lowpass filter equation in that the numerator is a polynomial, instead of the constant A_0. For this reason, "steepened" lowpass filters cannot be designed using the simple circuits of Sect. 13.4. However, in Sect. 13.11 we discuss a universal filter with which any numerator polynomial can be implemented.

13.1.3
Bessel Lowpass Filters

As previously shown, Butterworth and Chebyshev lowpass filters have a considerable overshoot in their step response in Figs. 13.3/13.4. An ideal square-wave response is achieved by filters that have a frequency-independent group delay; that is, that have a phase shift that is proportional to the frequency. This behavior is best approximated by Bessel filters, which are sometimes also called Thomson filters. The approximation consists of selecting the coefficients such that the group delay below the cutoff frequency $\omega_n = 1$ is dependent as little as possible on ω_n. This procedure is equivalent to a Butterworth approximation of the group delay; that is, the optimization of a maximally flat group delay.

From (13.4), with $s_n = j\omega_n$, the gain of a second-order lowpass filter is given by

$$\underline{A} = \frac{A_0}{1 + a_1 s_n + b_1 s_n^2} = \frac{A_0}{1 + ja_1\omega_n - b_1\omega_n^2}$$

Therefore the phase shift is

$$\varphi = -\arctan\frac{a_1\omega_n}{1 - b_1\omega_n^2} \tag{13.8}$$

The group delay is defined

$$t_{gr} = -\frac{d\varphi}{d\omega}$$

To simplify further calculations, we introduce the normalized group delay

$$T_{gr} = t_{gr}\,\omega_c = 2\pi\,t_{gr}\cdot f_c = 2\pi\frac{t_{gr}}{T_c} \tag{13.9a}$$

where T_c is the reciprocal of the cutoff frequency. We thus obtain

$$T_{gr} = -\omega_c\cdot\frac{d\varphi}{d\omega} = -\frac{d\varphi}{d\omega_n} \tag{13.9b}$$

and, using (13.8)

$$T_{gr} = \frac{a_1(1 + b_1\omega_n^2)}{1 + (a_1^2 - 2b_1)\omega_n^2 + b_1^2\omega_n^4} \tag{13.9c}$$

In order to find the Butterworth approximation of the group delay, we use the fact that:

$$T_{gr} = a_1\cdot\frac{1 + b_1\omega_n^2}{1 + (a_1^2 - 2b_1)\omega_n^2}\quad\text{for}\quad\omega_n \ll 1$$

This expression becomes independent of ω_n if the coefficients of ω_n^2 in the numerator and denominator are identical. The condition for this is that

$$b_1 = a_1^2 - 2b_1$$

or

$$b_1 = \frac{1}{3}a_1^2 \tag{13.10}$$

N	
1	$1 + s_n$
2	$1 + s_n + \frac{1}{3}s_n^2$
3	$1 + s_n + \frac{2}{5}s_n^2 + \frac{1}{15}s_n^3$
4	$1 + s_n + \frac{3}{7}s_n^2 + \frac{2}{21}s_n^3 + \frac{1}{105}s_n^4$

Fig. 13.11. Bessel polynomials

A second relationship is derived from the normalizing condition, $|\underline{A}|^2 = \frac{1}{2}$ for $\omega_n = 1$:

$$\frac{1}{2} = \frac{1}{(1 - b_1)^2 + a_1^2}$$

Hence, using (13.10),

$$a_1 = 1.3617$$
$$b_1 = 0.6180$$

For higher filter orders, the corresponding calculation becomes difficult, since a system of nonlinear equations arises. Using a different concept it is possible to define the coefficients c_i in (13.3) by a recursion formula:

$$c_1' = 1,$$
$$c_i' = \frac{2(N - i + 1)}{i(2N - i + 1)} c_{i-1}'$$

The denominators of (13.3) obtained in such a way are the Bessel polynomials, and are shown in Fig. 13.11 up to the fourth order. However, it should be noted that, in this representation, the frequency s_n is not normalized with respect to the 3 dB cutoff frequency, but to the reciprocal of the group delay for $\omega_n = 0$. This is of little use for the design of lowpass filters. We have therefore recalculated the coefficients c_i for the 3 dB cutoff frequency, as in the previous section, and in addition we have broken the denominator down into quadratic expressions. The coefficients a_i and b_i of (13.4) thus obtained are listed in Fig. 13.14 for up to the tenth order. The frequency response of the gain is plotted in Fig. 13.12.

In order to demonstrate the amount of phase distortion of other filters in comparison with the Bessel filters, in Fig. 13.13 we have illustrated the frequency response of the phase shift and of the group delay for fourth-order filters. These curves can best be calculated from the factored transfer function in (13.4) by adding together the phase shifts of each individual second-order filter stage and by adding the individual group delays. For a filter of a given order, the following relations are derived from (13.8) and (13.9c):

$$\varphi = -\sum_i \arctan \frac{a_i \omega}{1 - b_i \omega^2}$$

and

$$T_{gr} = \sum_i \frac{a_i (1 + b_i \omega_n^2)}{1 + (a_i^2 - 2b_i)\omega_n^2 + b_i^2 \omega_n^4}$$
$$T_{gr0} = \sum_i a_i$$

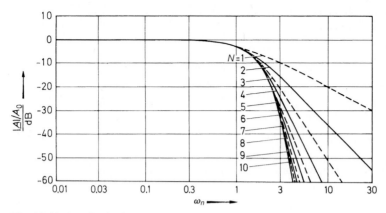

Fig. 13.12. Amplitude–frequency response of Bessel lowpass filters

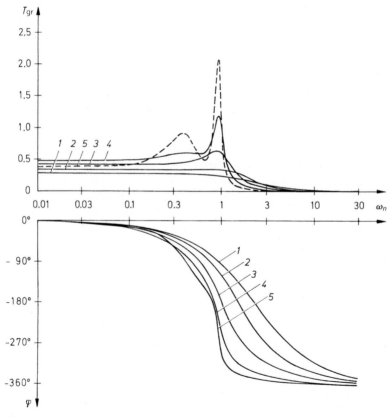

Fig. 13.13. Comparison of the frequency response of the group delay and the phase shift of fourth-order filters
Curve 1: Lowpass filter with critical damping. *Curve 2*: Bessel lowpass filter.
Curve 3: Butterworth lowpass filter. *Curve 4*: Chebyshev lowpass filter with 0.5 dB ripple.
Curve 5: Chebyshev lowpass filter with 3 dB ripple

N	i	a_i	b_i	f_{ci}/f_c	Q_i
Critically damped filters					
1	1	1.0000	0.0000	1.000	–
2	1	1.2872	0.4142	1.000	0.50
3	1	0.5098	0.0000	1.961	–
	2	1.0197	0.2599	1.262	0.50
4	1	0.8700	0.1892	1.480	0.50
	2	0.8700	0.1892	1.480	0.50
5	1	0.3856	0.0000	2.593	–
	2	0.7712	0.1487	1.669	0.50
	3	0.7712	0.1487	1.669	0.50
6	1	0.6999	0.1225	1.839	0.50
	2	0.6999	0.1225	1.839	0.50
	3	0.6999	0.1225	1.839	0.50
7	1	0.3226	0.0000	3.100	–
	2	0.6453	0.1041	1.995	0.50
	3	0.6453	0.1041	1.995	0.50
	4	0.6453	0.1041	1.995	0.50
8	1	0.6017	0.0905	2.139	0.50
	2	0.6017	0.0905	2.139	0.50
	3	0.6017	0.0905	2.139	0.50
	4	0.6017	0.0905	2.139	0.50
9	1	0.2829	0.0000	3.534	–
	2	0.5659	0.0801	2.275	0.50
	3	0.5659	0.0801	2.275	0.50
	4	0.5659	0.0801	2.275	0.50
	5	0.5659	0.0801	2.275	0.50
10	1	0.5358	0.0718	2.402	0.50
	2	0.5358	0.0718	2.402	0.50
	3	0.5358	0.0718	2.402	0.50
	4	0.5358	0.0718	2.402	0.50
	5	0.5358	0.0718	2.402	0.50

Fig. 13.14. Filter coefficients

N	i	a_i	b_i	f_{ci}/f_c	Q_i
Bessel filters					
1	1	1.0000	0.0000	1.000	–
2	1	1.3617	0.6180	1.000	0.58
3	1	0.7560	0.0000	1.323	–
	2	0.9996	0.4772	1.414	0.69
4	1	1.3397	0.4889	0.978	0.52
	2	0.7743	0.3890	1.797	0.81
5	1	0.6656	0.0000	1.502	–
	2	1.1402	0.4128	1.184	0.56
	3	0.6216	0.3245	2.138	0.92
6	1	1.2217	0.3887	1.063	0.51
	2	0.9686	0.3505	1.431	0.61
	3	0.5131	0.2756	2.447	1.02
7	1	0.5937	0.0000	1.684	–
	2	1.0944	0.3395	1.207	0.53
	3	0.8304	0.3011	1.695	0.66
	4	0.4332	0.2381	2.731	1.13
8	1	1.1112	0.3162	1.164	0.51
	2	0.9754	0.2979	1.381	0.56
	3	0.7202	0.2621	1.963	0.71
	4	0.3728	0.2087	2.992	1.23
9	1	0.5386	0.0000	1.857	–
	2	1.0244	0.2834	1.277	0.52
	3	0.8710	0.2636	1.574	0.59
	4	0.6320	0.2311	2.226	0.76
	5	0.3257	0.1854	3.237	1.32
10	1	1.0215	0.2650	1.264	0.50
	2	0.9393	0.2549	1.412	0.54
	3	0.7815	0.2351	1.780	0.62
	4	0.5604	0.2059	2.479	0.81
	5	0.2883	0.1665	3.466	1.42

Fig. 13.14. Filter coefficients (continued)

N	i	a_i	b_i	f_{ci}/f_c	Q_i
Butterworth filters					
1	1	1.0000	0.0000	1.000	–
2	1	1.4142	1.0000	1.000	0.71
3	1	1.0000	0.0000	1.000	–
	2	1.0000	1.0000	1.272	1.00
4	1	1.8478	1.0000	0.719	0.54
	2	0.7654	1.0000	1.390	1.31
5	1	1.0000	0.0000	1.000	–
	2	1.6180	1.0000	0.859	0.62
	3	0.6180	1.0000	1.448	1.62
6	1	1.9319	1.0000	0.676	0.52
	2	1.4142	1.0000	1.000	0.71
	3	0.5176	1.0000	1.479	1.93
7	1	1.0000	0.0000	1.000	–
	2	1.8019	1.0000	0.745	0.55
	3	1.2470	1.0000	1.117	0.80
	4	0.4450	1.0000	1.499	2.25
8	1	1.9616	1.0000	0.661	0.51
	2	1.6629	1.0000	0.829	0.60
	3	1.1111	1.0000	1.206	0.90
	4	0.3902	1.0000	1.512	2.56
9	1	1.0000	0.0000	1.000	–
	2	1.8794	1.0000	0.703	0.53
	3	1.5321	1.0000	0.917	0.65
	4	1.0000	1.0000	1.272	1.00
	5	0.3473	1.0000	1.521	2.88
10	1	1.9754	1.0000	0.655	0.51
	2	1.7820	1.0000	0.756	0.56
	3	1.4142	1.0000	1.000	0.71
	4	0.9080	1.0000	1.322	1.10
	5	0.3129	1.0000	1.527	3.20

Fig. 13.14. Filter coefficients (continued)

N	i	a_i	b_i	f_{ci}/f_c	Q_i
Chebyshev filters, 0.5 dB ripple					
1	1	1.0000	0.0000	1.000	–
2	1	1.3614	1.3827	1.000	0.86
3	1	1.8636	0.0000	0.537	–
	2	0.6402	1.1931	1.335	1.71
4	1	2.6282	3.4341	0.538	0.71
	2	0.3648	1.1509	1.419	2.94
5	1	2.9235	0.0000	0.342	–
	2	1.3025	2.3534	0.881	1.18
	3	0.2290	1.0833	1.480	4.54
6	1	3.8645	6.9797	0.366	0.68
	2	0.7528	1.8573	1.078	1.81
	3	0.1589	1.0711	1.495	6.51
7	1	4.0211	0.0000	0.249	–
	2	1.8729	4.1795	0.645	1.09
	3	0.4861	1.5676	1.208	2.58
	4	0.1156	1.0443	1.517	8.84
8	1	5.1117	11.9607	0.276	0.68
	2	1.0639	2.9365	0.844	1.61
	3	0.3439	1.4206	1.284	3.47
	4	0.0885	1.0407	1.521	11.53
9	1	5.1318	0.0000	0.195	–
	2	2.4283	6.6307	0.506	1.06
	3	0.6839	2.2908	0.989	2.21
	4	0.2559	1.3133	1.344	4.48
	5	0.0695	1.0272	1.532	14.58
10	1	6.3648	18.3695	0.222	0.67
	2	1.3582	4.3453	0.689	1.53
	3	0.4822	1.9440	1.091	2.89
	4	0.1994	1.2520	1.381	5.61
	5	0.0563	1.0263	1.533	17.99

Fig. 13.14. Filter coefficients (continued)

N	i	a_i	b_i	f_{ci}/f_c	Q_i
Chebyshev filters, 1 dB *ripple*					
1	1	1.0000	0.0000	1.000	–
2	1	1.3022	1.5515	1.000	0.96
3	1	2.2156	0.0000	0.451	–
	2	0.5442	1.2057	1.353	2.02
4	1	2.5904	4.1301	0.540	0.78
	2	0.3039	1.1697	1.417	3.56
5	1	3.5711	0.0000	0.280	–
	2	1.1280	2.4896	0.894	1.40
	3	0.1872	1.0814	1.486	5.56
6	1	3.8437	8.5529	0.366	0.76
	2	0.6292	1.9124	1.082	2.20
	3	0.1296	1.0766	1.493	8.00
7	1	4.9520	0.0000	0.202	–
	2	1.6338	4.4899	0.655	1.30
	3	0.3987	1.5834	1.213	3.16
	4	0.0937	1.0423	1.520	10.90
8	1	5.1019	14.7608	0.276	0.75
	2	0.8916	3.0426	0.849	1.96
	3	0.2806	1.4334	1.285	4.27
	4	0.0717	1.0432	1.520	14.24
9	1	6.3415	0.0000	0.158	–
	2	2.1252	7.1711	0.514	1.26
	3	0.5624	2.3278	0.994	2.71
	4	0.2076	1.3166	1.346	5.53
	5	0.0562	1.0258	1.533	18.03
10	1	6.3634	22.7468	0.221	0.75
	2	1.1399	4.5167	0.694	1.86
	3	0.3939	1.9665	1.093	3.56
	4	0.1616	1.2569	1.381	6.94
	5	0.0455	1.0277	1.532	22.26

Fig. 13.14. Filter coefficients (continued)

N	i	a_i	b_i	f_{ci}/f_c	Q_i
Chebyshev filters, 2 dB ripple					
1	1	1.0000	0.0000	1.000	–
2	1	1.1813	1.7775	1.000	1.13
3	1	2.7994	0.0000	0.357	–
	2	0.4300	1.2036	1.378	2.55
4	1	2.4025	4.9862	0.550	0.93
	2	0.2374	1.1896	1.413	4.59
5	1	4.6345	0.0000	0.216	–
	2	0.9090	2.6036	0.908	1.78
	3	0.1434	1.0750	1.493	7.23
6	1	3.5880	10.4648	0.373	0.90
	2	0.4925	1.9622	1.085	2.84
	3	0.0995	1.0826	1.491	10.46
7	1	6.4760	0.0000	0.154	–
	2	1.3258	4.7649	0.665	1.65
	3	0.3067	1.5927	1.218	4.12
	4	0.0714	1.0384	1.523	14.28
8	1	4.7743	18.1510	0.282	0.89
	2	0.6991	3.1353	0.853	2.53
	3	0.2153	1.4449	1.285	5.58
	4	0.0547	1.0461	1.518	18.69
9	1	8.3198	0.0000	0.120	–
	2	1.7299	7.6580	0.522	1.60
	3	0.4337	2.3549	0.998	3.54
	4	0.1583	1.3174	1.349	7.25
	5	0.0427	1.0232	1.536	23.68
10	1	5.9618	28.0376	0.226	0.89
	2	0.8947	4.6644	0.697	2.41
	3	0.3023	1.9858	1.094	4.66
	4	0.1233	1.2614	1.380	9.11
	5	0.0347	1.0294	1.531	27.27

Fig. 13.14. Filter coefficients (continued)

N	i	a_i	b_i	f_{ci}/f_c	Q_i
Chebyshev filters, 3 dB *ripple*					
1	1	1.0000	0.0000	1.000	–
2	1	1.0650	1.9305	1.000	1.30
3	1	3.3496	0.0000	0.299	–
	2	0.3559	1.1923	1.396	3.07
4	1	2.1853	5.5339	0.557	1.08
	2	0.1964	1.2009	1.410	5.58
5	1	5.6334	0.0000	0.178	–
	2	0.7620	2.6530	0.917	2.14
	3	0.1172	1.0686	1.500	8.82
6	1	3.2721	11.6773	0.379	1.04
	2	0.4077	1.9873	1.086	3.46
	3	0.0815	1.0861	1.489	12.78
7	1	7.9064	0.0000	0.126	–
	2	1.1159	4.8963	0.670	1.98
	3	0.2515	1.5944	1.222	5.02
	4	0.0582	1.0348	1.527	17.46
8	1	4.3583	20.2948	0.286	1.03
	2	0.5791	3.1808	0.855	3.08
	3	0.1765	1.4507	1.285	6.83
	4	0.0448	1.0478	1.517	22.87
9	1	10.1759	0.0000	0.098	–
	2	1.4585	7.8971	0.526	1.93
	3	0.3561	2.3651	1.001	4.32
	4	0.1294	1.3165	1.351	8.87
	5	0.0348	1.0210	1.537	29.00
10	1	5.4449	31.3788	0.230	1.03
	2	0.7414	4.7363	0.699	2.94
	3	0.2479	1.9952	1.094	5.70
	4	0.1008	1.2638	1.380	11.15
	5	0.0283	1.0304	1.530	35.85

Fig. 13.14. Filter coefficients (continued)

13.1.4
Summary of the Theory

We have seen that the transfer functions of all lowpass filters have the form

$$A(s_n) = \frac{A_0}{\prod_i (1 + a_i s_n + b_i s_n^2)} \qquad (13.11)$$

The order N of the filter is determined by the highest power of the frequency variable s, respectively, s_n in (13.11) when the denominator is expanded. It defines the slope of the asymptote of the amplitude–frequency response as having the value $-N \cdot 20$ dB per decade. The rest of the amplitude–frequency response curve for a particular order is determined by the type of filter. Of special interest are the Butterworth, Chebyshev, and Bessel filters, all of which have different values for the coefficients a_i and b_i in (13.11). The values of the coefficients are summarized in Fig. 13.14 for up to the tenth order. In addition, the 3 dB cutoff frequency of each individual filter stage in respect to the whole filter is given by the ratio

$$\frac{f_{ci}}{f_c} = \frac{\sqrt{2}}{2b_i}\sqrt{2b_i - a_i^2 + \sqrt{a_i^4 b_i + 8b_i^2}}$$

Although this value is not needed for the design, it is useful for checking the correct operation of the individual filter stages. Also listed are the pole-pair quality factors Q_i of the individual filter stages. By analogy with the Q-factors of the bandpass filters in Sect. 13.6.1, they are defined as $Q_i = \sqrt{b_i}/a_i$. The larger the pole-pair Q-factor, the larger is the transient oscillation of the step response. Filters with real poles have Q-factors of $Q \leq 0.5$.

The frequency response of the gain, phase shift, and group delay can be calculated using the coefficients a_i and b_i of the factored transfer function:

$$|\underline{A}|^2 = \frac{A_0^2}{\prod_i \left[1 + (a_i^2 - 2b_i)\omega_n^2 + b_i^2\omega_n^4\right]} \tag{13.12}$$

$$\varphi = -\sum_i \arctan \frac{a_i\omega_n}{1 - b_i\omega_n^2} \tag{13.13}$$

$$T_{gr} = \sum_i \frac{a_i(1 + b_i\omega_n^2)}{1 + (a_i^2 - 2b_i)\omega_n^2 + b_i^2\omega_n^4} \tag{13.14}$$

13.2
Lowpass/Highpass Transformation

In the logarithmic representation, the amplitude–frequency response of a lowpass filter is transformed into that of the analogous highpass filter response by plotting its mirror image about the cutoff frequency; that is, by replacing ω_n by $1/\omega_n$ and s_n by $1/s_n$. The cutoff frequency remains the same, and A_0 changes to A_∞. Equation (13.11) then becomes

$$A(s_n) = \frac{A_\infty}{\prod_i \left(1 + \dfrac{a_i}{s_n} + \dfrac{b_i}{s_n^2}\right)} = \frac{A_\infty s_n^2}{\prod_i \left(b_i + a_i s_n + s_n^2\right)} \tag{13.15}$$

In the time domain, the performance cannot be transformed, as the step response shows a basically different behavior. This can be seen in Fig. 13.15, where an oscillation about the final-state value occurs even in highpass filters that have critical damping. The analogy with the corresponding lowpass filters still applies inasmuch as the transient oscillation decays more slowly the higher the pole-pair Q-factors are.

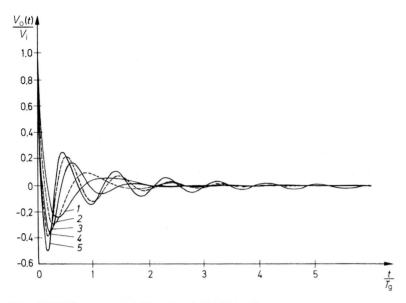

Fig. 13.15. Step response of fourth-order highpass filters
Curve 1: Highpass filter having critical damping. *Curve 2*: Bessel highpass filter.
Curve 3: Butterworth highpass filter. *Curve 4*: Chebyshev highpass filter with 0.5 dB ripple.
Curve 5: Chebyshev highpass filter with 3 dB ripple

13.3
Realization of First-Order Lowpass and Highpass Filters

According to (13.11), the transfer function of a first-order lowpass filter has the general form

$$A(s_n) = \frac{A_0}{1 + a_1 s_n} \tag{13.16}$$

It can be implemented by the simple RC network shown in Fig. 13.1. From Sect. 13.1, it follows for this circuit:

$$A(s_n) = \frac{1}{1 + s\,RC} = \frac{1}{1 + \omega_c RC s_n}$$

The low-frequency gain is defined by the value $A_0 = 1$, but the parameter a_1 can be chosen freely. Its value is found by comparing the coefficients

$$RC = \frac{a_1}{2\pi f_c}$$

As can be seen from the table in Fig. 13.14, all filter types of first order are identical and have the coefficient $a_1 = 1$. When higher-order filters are implemented by cascading filter stages of lower orders, first-order filter stages may be required for which $a_1 \neq 1$. The reason for this is that individual filter stages have, as a rule, a cutoff frequency that is different from that of the filter as a whole, namely $f_{c1} = f_c/a_1$.

The simple RC network shown in Fig. 13.1 has the disadvantage that its properties change when it is loaded. Therefore an impedance converter (buffer) is usually connected

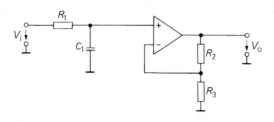

Fig. 13.16. A first-order lowpass filter with an impedance converter

$$A(s_n) = \frac{(R_2 + R_3)/R_3}{1 + \omega_c R_1 C_1 s_n}$$

in series. If it is given voltage gain A_0, the low-frequency gain can then be chosen freely. An appropriate circuit is presented in Fig. 13.16.

In order to arrive at the corresponding highpass filter, the variable s_n in (13.16) must be replaced by $1/s_n$. In the circuit itself, the conversion is achieved by simply exchanging R_1 and C_1.

Somewhat simpler circuits for first-order lowpass and highpass filters are obtained if the filtering network is included in the feedback loop of the operational amplifier. The corresponding lowpass filter is shown in Fig. 13.17. For the actual design, the cutoff frequency, the low-frequency gain A_0, which in this case is negative, and the capacitance C_1 need to be defined. Comparing the coefficients to those of (13.16), it follows:

$$R_2 = \frac{a_1}{2\pi f_c C_1} \quad \text{and} \quad R_1 = -\frac{R_2}{A_0}$$

Figure 13.18 shows the corresponding highpass filter. By comparing coefficients with (13.15), it follows:

$$R_1 = \frac{1}{2\pi f_c a_1 C_1} \quad \text{and} \quad R_2 = -R_1 A_\infty$$

The transfer functions given for the previous circuits apply only to the range of frequency for which the open-loop gain of the operational amplifier is large with respect to the absolute value of $\underline{A}$. This condition is difficult to fulfill for high frequencies, as the magnitude of the open-loop gain falls at a rate of 6 dB per octave because of the necessary frequency compensation. For a standard operational amplifier of the 741-type at 10 kHz, $|\underline{A}_D|$ is only about 100.

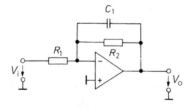

Fig. 13.17. First-order lowpass filter with an inverting amplifier

$$A(s_n) = -\frac{R_2/R_1}{1 + \omega_c R_2 C_1 s_n}$$

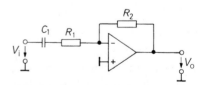

Fig. 13.18. First-order highpass filter with an inverting amplifier

$$A(s_n) = \frac{-s_n R_2/R_1}{\dfrac{1}{R_1 C_1 \omega_c} + s_n}$$

13.4
Realization of Second-Order Lowpass and Highpass Filters

According to (13.11), the transfer function of a second-order lowpass filter has the general form

$$A(s_n) = \frac{A_0}{1 + a_1 s_n + b_1 s_n^2} \qquad (13.17)$$

As can be seen from the table in Fig. 13.14, the optimized transfer functions of the second and higher orders have conjugate complex poles because $Q > 0.5$. Such transfer functions cannot be implemented by passive RC networks, as discussed in Sect. 13.1. One possible way of realizing these circuits is to use inductances, and this is demonstrated in the following example.

13.4.1
LRC Filters

The conventional implementation of second-order filters involves using LRC networks, as shown in Fig. 13.19. Comparison of the coefficients with those of (13.17) yields

$$R = \frac{a_1}{2\pi f_c C} \quad \text{and} \quad L = \frac{b_1}{4\pi^2 f_c^2 C}$$

From Fig. 13.14, the coefficients of a second-order Butterworth lowpass filter are $a_1 = 1.414$ and $b_1 = 1.000$. For a given cutoff frequency of $f_c = 10\,\text{Hz}$ and a capacitance of $C = 10\,\mu\text{F}$, the remaining design parameters are $R = 2.25\,\text{k}\Omega$ and $L = 25.3\,\text{H}$. Obviously, such a filter is extremely difficult to implement because of the size of the inductance. However, this can be avoided by emulating the inductance with an active RC circuit. The gyrator shown in Fig. 12.32 is useful for this purpose, although its implementation involves a considerable number of components.

The desired transfer functions can be put into practice much more simply without inductance emulation by connecting suitable RC networks around operational amplifiers.

13.4.2
Filters with Multiple Negative Feedback

An active second-order RC lowpass filter is shown in Fig. 13.20. By comparing the coefficients with those of (13.17), we obtain the relations

$$A_0 = -R_2/R_1$$
$$a_1 = \omega_c C_1 \left(R_2 + R_3 + \frac{R_2 R_3}{R_1} \right)$$
$$b_1 = \omega_c^2 C_1 C_2 R_2 R_3$$

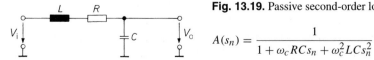

Fig. 13.19. Passive second-order lowpass filter

$$A(s_n) = \frac{1}{1 + \omega_c R C s_n + \omega_c^2 L C s_n^2}$$

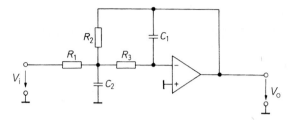

Fig. 13.20. Active second-order lowpass filter with multiple negative feedback

$$A(s_n) = -\frac{R_2/R_1}{1 + \omega_c C_1 \left(R_2 + R_3 + \dfrac{R_2 R_3}{R_1} \right) s_n + \omega_c^2 C_1 C_2 R_2 R_3 s_n^2}$$

For the actual specification, the values of the resistors R_1 and R_3, for example, can be predetermined; the parameters R_2, C_1, and C_2 can then be calculated from the above equations. Such a determination is possible for all positive values of a_1 and b_1, so that any desired type of filter can be realized. The gain at zero frequency, A_0, is negative. At low frequencies, therefore, the filter inverts the signal.

In order to achieve the desired frequency response, the circuit elements must not have too large a tolerance. This requirement is easy to fulfill for resistors, as they can be obtained off the shelf with 1% tolerance, in the E 96 standard series. The situation is different with capacitors, which, as a rule, are only available off the shelf in the E 6 series. It is therefore advantageous in filter design to predetermine the capacitors and calculate the values for the resistors. We therefore solve the design equations for the resistances and obtain:

$$R_2 = \frac{a_1 C_2 - \sqrt{a_1^2 C_2^2 - 4 C_1 C_2 b_1 (1 - A_0)}}{4\pi f_c C_1 C_2}$$

$$R_1 = \frac{R_2}{-A_0}$$

$$R_3 = \frac{b_1}{4\pi^2 f_c^2 C_1 C_2 R_2}$$

In order that the value for R_2 will be real, the condition

$$C_2 \geq \frac{4 b_1 (1 - A_0)}{a_1^2} C_1$$

must be fulfilled. The most favorable design is obtained if the ratio C_2/C_1 is chosen not much larger than is prescribed by this condition. The filter parameters are relatively insensitive to the tolerances of the components, and therefore the circuit is particularly suited to the realization of filters having high Q-factors.

13.4.3
Filter with Single Positive Feedback

Active filters can also be designed using amplifiers with positive feedback. However, the gain must be fixed at a precise value by an internal negative feedback (a "controlled source"). The voltage divider R_3, $(\alpha - 1) R_3$ shown in Fig. 13.21 provides this negative

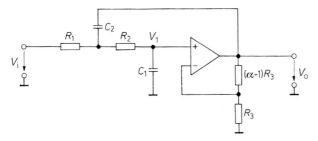

Fig. 13.21. Active second order lowpass filter with single positive feedback

$$A(s_n) = \frac{\alpha}{1 + \omega_c\left[C_1(R_1 + R_2) + (1 - \alpha)R_1C_2\right]s_n + \omega_c^2 R_1 R_2 C_1 C_2 s_n^2}$$

feedback and determines the internal gain as having the value $V_o/V_1 = \alpha$. The positive feedback is provided by capacitor C_2.

Dimensioning can be substantially simplified by certain specializations right from the start. One possible specialization is to select the internal gain $\alpha = 1$. This makes $(\alpha - 1)R_3 = 0$, and both R_3 resistances can be omitted. Such operational amplifiers, with a unit feedback factor, are available as integrated voltage followers. A simple impedance converter or buffer – for instance, an emitter or source follower – is often sufficient. In this way, filters in the MHz range can also be realized. For the special case in which $\alpha = 1$, the transfer function is given by

$$A(s_n) = \frac{1}{1 + \omega_c C_1(R_1 + R_2)s_n + \omega_c^2 R_1 R_2 C_1 C_2 s_n^2}$$

Defining C_1 and C_2 and comparing the coefficients with those of (13.17) gives

$$A_0 = 1$$

$$R_{1/2} = \frac{a_1 C_2 \mp \sqrt{a_1^2 C_2^2 - 4b_1 C_1 C_2}}{4\pi f_c C_1 C_2}$$

In order to arrive at values that are real, the condition

$$C_2/C_1 \geq 4b_1/a_1^2$$

must be fulfilled. As for the filter with multiple negative feedback, the most favorable design is obtained here also if the ratio C_2/C_1 is chosen to be not much larger than is prescribed by this condition. The MAX270 family from Maxim contains two filters of second order with Butterworth characteristics. The cutoff frequency can be externally adjusted by a 7 bit word in 128 steps in the range $1 \ldots 25$ kHZ.

Another interesting specialization is achieved when using identical resistors and identical capacitors, i.e. $R_1 = R_2 = R$ and $C_1 = C_2 = C$. In order to realize the different filter types it is then necessary to vary the internal gain α. The transfer function is then

$$A(s_n) = \frac{\alpha}{1 + \omega_c RC(3 - \alpha)s_n + (\omega_c RC)^2 s_n^2}$$

	Critical	Bessel	Butterworth	3 dB Chebyshev	Undamped
α	1.000	1.268	1.586	2.234	3.000

Fig. 13.22. Internal gain for single positive feedback (second-order)

By comparing the coefficients with those in (13.17), we obtain the relations

$$RC = \frac{\sqrt{b_1}}{2\pi f_c}, \qquad \alpha = A_0 = 3 - \frac{a_1}{\sqrt{b_1}} = 3 - \frac{1}{Q_1}$$

As can be seen, the internal gain α is dependent only on the pole-pair Q-factor and not on the cutoff frequency f_c. The value of α therefore determines the type of filter. Insertion of the coefficients given in Fig. 13.14 for a second-order filter results in the values for α given in Fig. 13.22. For $\alpha = 3$, the circuit produces a self-contained oscillation at the frequency $f = 1/2\pi RC$. It can be seen that the adjustment of the internal gain becomes more difficult as one approaches the value $\alpha = 3$. For the Chebyshev filter in particular, a very precise adjustment is therefore necessary. This is a disadvantage compared with the previous filters. A considerable advantage, however, is the fact that the type of filter is solely determined by α and is not dependent on R and C. Hence, the cutoff frequency of this filter circuit can be changed particularly easily; for example, by using a dual-gang potentiometer for the two identical resistors R_1 and R_2 shown in Fig. 13.21.

If the resistors and capacitors are interchanged, one arrives at the *highpass filter* shown in Fig. 13.23. To simplify the design process, we choose the special case in which $\alpha = 1$ and $C_1 = C_2 = C$. Comparing the coefficients with those of (13.15) gives

$$A_\infty = 1$$

$$R_1 = \frac{1}{\pi f_c C a_1}$$

$$R_2 = \frac{a_1}{4\pi f_c C b_1}$$

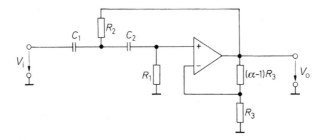

Fig. 13.23. An active second-order highpass filter with single positive feedback

$$A(s_n) = \frac{V_o}{V_i} = \frac{\alpha s_n^2}{\dfrac{1}{R_1 R_2 C_1 C_2 \omega_c^2} + \dfrac{R_2(C_1 + C_2) + R_1 C_2(1 - \alpha)}{R_1 R_2 C_1 C_2 \omega_c} s_n + s_n^2}$$

13.5
Realization of Higher-Order Lowpass and Highpass Filters

In cases in which the filter characteristic is insufficiently steep, filters of higher orders must be employed. For this purpose, first- and second-order filters are cascaded, thereby multiplying the frequency responses of the individual filters. It would, however, be wrong to cascade two second-order Butterworth filters, for example, to obtain a fourth-order Butterworth filter. The resulting filter would have a different cutoff frequency and also a different filter characteristic. The coefficients of the individual filters must therefore be chosen such that the product of the frequency responses gives the desired optimized filter type.

To simplify the design of the individual filters, we have factored the polynomials of the different filter types. The coefficients a_1 and b_1 of the individual filter stages are given in Fig. 13.14. Each filter section can be implemented by one of the second-order filters described previously. It is merely necessary to replace the coefficients a_1 and b_1 by a_i and b_i. To calculate the circuit parameters from the given formulas, the desired cutoff frequency of the *resulting total filter* must be inserted. As a rule, the individual filter stages possess cutoff frequencies that are different from that of the filter as a whole, as can be seen in Fig. 13.14. Odd-order filters contain a factor in which $b_i = 0$. The corresponding filter stage can be implemented by one of the first-order filters described if a_1 is replaced by a_i. In this case also, the cutoff frequency f_c of the total filter must be inserted. Because of the defined value of a_i, this filter stage automatically has the cutoff frequency f_{ci} given in Fig. 13.14.

In principle, the sequence in which the individual filter stages are cascaded is not significant, as the resulting frequency response is the product of the individual stages which remains the same. In practice, however, there are several design considerations governing the best sequence of the filter stages. One such aspect is the permissible voltage swing, for which it is useful to arrange the filter stages according to their cutoff frequencies, with the one having the lowest cutoff frequency at the input; otherwise, it may become saturated while the output of the second stage is still below the maximum permissible voltage swing. The reason for this is that the filter stages that have the higher cutoff frequencies also invariably have the higher pole-pair Q, and therefore show a rise in gain in the vicinity of their cutoff frequency. This can be seen in Fig. 13.24, where the amplitude–frequency responses of a tenth-order 0.5 dB Chebyshev lowpass filter and of its five individual stages are shown. It is obvious that the permissible voltage swing is highest if the filter stages that have low cutoff frequencies are at the input of the filter cascade.

Another aspect that may have to be considered for a suitable arrangement of the filter stages is noise. In this case, just the reverse sequence is the most favorable, as then the filters with the lower cutoff frequencies at the end of the filter chain reduce the noise introduced by the input stages.

The design process is demonstrated for a third-order Bessel lowpass filter. It is to be constructed from the first-order lowpass filter shown in Fig. 13.16 and the second-order lowpass filter shown in Fig. 13.21, for which the special case of $\alpha = 1$ (described in Sect. 13.4.3) is chosen. The low-frequency gain is defined as unity. To achieve this, the impedance converter in the first-order filter stage must have a gain of $\alpha = 1$. The resulting circuit is represented in Fig. 13.25.

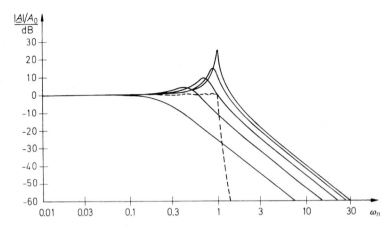

Fig. 13.24. Amplitude–frequency responses of a tenth-order Chebyshev filter with 0.5 dB ripple and of its five individual filter stages

The desired cutoff frequency is $f_c = 100\,\text{Hz}$. For the calculation of the first filter stage, we predetermine $C_1 = 100\,\text{nF}$ and obtain, according to Sect. 13.3, with the coefficients from Fig. 13.14:

$$R_{11} = \frac{a_1}{2\pi f_c C_{11}} = \frac{0.7560}{2\pi \cdot 100\,\text{Hz} \cdot 100\,\text{nF}} = 12.03\,\text{k}\Omega$$

For the second filter stage, we set $C_{22} = 100\,\text{nF}$ and obtain, in accordance with Sect. 13.4.3, the condition for C_{21}:

$$C_{21} \leq C_{22}\frac{a_2^2}{4b_2} = 100\,\text{nF} \cdot \frac{(0.9996)^2}{4 \cdot 0.4772} = 52.3\,\text{nF}$$

We choose the nearest standard value $C_{21} = 47\,\text{nF}$ and arrive at

$$R_{21/22} = \frac{a_2 C_{22} \mp \sqrt{a_2^2 C_{22}^2 - 4b_2 C_{21}C_{22}}}{4\pi f_c C_{21}C_{22}}$$

$$R_{21} = 11{,}51\,\text{k}\Omega, \qquad R_{22} = 22{,}33\,\text{k}\Omega$$

For third-order filters, it is possible to omit the first operational amplifier. The simple lowpass filter of Fig. 13.1 is then connected in front of the second-order filter. Due to the mutual loading of the filters, a different method of calculation must be employed, which is considerably more difficult than in the decoupled case. Figure 13.26 shows such a circuit, which has the same characteristics as that shown in Fig. 13.25.

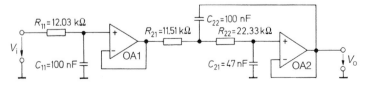

Fig. 13.25. Third-order Bessel lowpass filter with a cutoff frequency of $f_c = 100\,\text{Hz}$

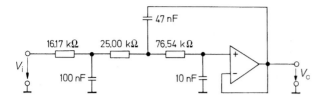

Fig. 13.26. Simplified third-order Bessel lowpass filter with a cutoff frequency of $f_c = 100\,\mathrm{Hz}$

13.6
Lowpass/Bandpass Transformation

In Sect. 13.2 it was shown how, by transforming the frequency variable, a given lowpass frequency response could be converted to the corresponding highpass frequency response. The frequency response of a bandpass filter can be created using a very similar transformation; that is, by replacing the frequency variable s_n in the lowpass transfer function by the expression

$$s_n \rightarrow \frac{1}{\Delta\omega_n}\left(s_n + \frac{1}{s_n}\right) \tag{13.18}$$

By means of this transformation, the amplitude response of the lowpass filter in the range $0 \leq \omega_n \leq 1$ is converted into the pass range of a bandpass filter between the center frequency $\omega_n = 1$ and the upper cutoff frequency $\omega_{n,\,max}$. On a logarithmic frequency scale, it also appears as a mirror image below the center frequency. The lower cutoff frequency is then $\omega_{n,\,min} = 1/\omega_{n,\,max}$. This process is illustrated in Fig. 13.27.

The normalized bandwidth $\Delta\omega_n = \omega_{n,\,max} - \omega_{n,\,min}$ can be chosen freely. The described transformation results in the bandpass filter having the same gain at $\omega_{n,\,min}$ and $\omega_{n,\,max}$ as the corresponding lowpass filter at $\omega_n = 1$. If the lowpass filter, as in the table given in Fig. 13.14, is normalized with respect to the 3 dB cutoff frequency, $\Delta\omega_n$ represents the normalized 3 dB bandwidth of the bandpass filter. Since $\Delta\omega_n = \omega_{n,\,max} - \omega_{n,\,min}$ and $\omega_{n,\,max} \cdot \omega_{n,\,min} = 1$, we obtain, for the normalized 3 dB cutoff frequencies,

$$\omega_{n,\,max/\,min} = \frac{1}{2}\sqrt{(\Delta\omega_n)^2 + 4} \pm \frac{1}{2}\Delta\omega_n \,.$$

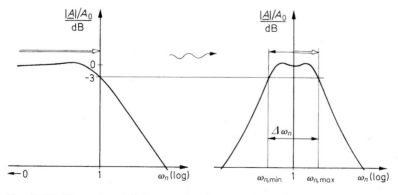

Fig. 13.27. Illustration of the lowpass/bandpass transformation

13.6.1
Second-Order Bandpass Filters

The simplest bandpass filter is obtained by applying the transformation equation (13.18) to a first-order lowpass filter, where

$$A(s_n) = \frac{A_0}{1 + s_n}$$

Therefore, the transfer function of the second-order bandpass filter results

$$A(s_n) = \frac{A_0}{1 + \dfrac{1}{\Delta\omega_n}\left(s_n + \dfrac{1}{s_n}\right)} = \frac{A_0 \Delta\omega_n s_n}{1 + \Delta\omega_n s_n + s_n^2} \tag{13.19}$$

The interesting parameters of bandpass filters are the gain A_r at the resonant frequency f_r, and the quality factor Q. It follows directly from the given transformation characteristic that $A_r = A_0$. This can be easily verified by making $\omega_n = 1$; that is, $s_n = j$ in (13.19). Since A_r is real, the phase shift at resonant frequency is zero.

As for a resonant circuit, the Q-factor is defined as the ratio of the resonant frequency f_r to the bandwidth B. Therefore,

$$Q = \frac{f_r}{B} = \frac{f_r}{f_{max} - f_{min}} = \frac{1}{\omega_{n,max} - \omega_{n,min}} = \frac{1}{\Delta\omega_n} \tag{13.20}$$

Inserting this in (13.19) gives

$$A(s_n) = \frac{(A_r/Q)s_n}{1 + \dfrac{1}{Q}s_n + s_n^2} \tag{13.21}$$

This equation is the transfer function of a second-order bandpass filter; it allows direct identification of all parameters of interest.

With $s_n = j\omega_n$, we obtain from (13.21) the frequency response of the amplitude and the phase:

$$|\underline{A}| = \frac{(A_r/Q)\omega_n}{\sqrt{1 + \omega_n^2\left(\dfrac{1}{Q^2} - 2\right) + \omega_n^4}} \tag{13.22} \qquad \varphi = \arctan\frac{Q(1 - \omega_n^2)}{\omega_n} \tag{13.23}$$

These two functions are shown in Fig. 13.28 for the Q-factors 1 and 10.

13.6.2
Fourth-Order Bandpass Filters

The amplitude–frequency response of second-order bandpass filters becomes more peaked the larger is the Q-factor selected. However, there are many applications where the curve must be as flat as possible in the region of the resonant frequency, but must also have a steep transition to the stopband. This optimization problem may be solved by applying the lowpass/bandpass transformation to higher-order lowpass filters. It is then possible to choose freely not only the bandwidth $\Delta\omega_n$, but also the most suitable type of filter.

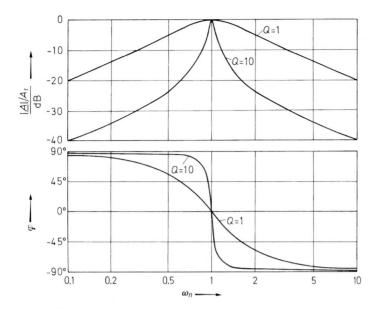

Fig. 13.28. Frequency response of the amplitude and the phase of second-order bandpass filters with $Q = 1$ and $Q = 10$ or $\Delta\omega_n = 1$ and $\Delta\omega_n = 0.1$

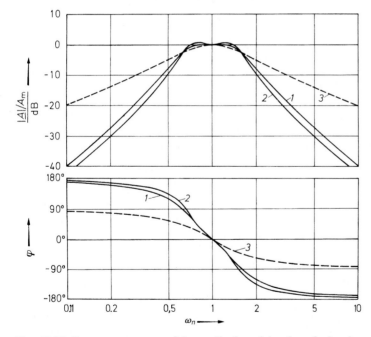

Fig. 13.29. Frequency response of the amplitude and the phase for bandpass filters with a bandwidth of $\Delta\omega_n = 1$
Curve 1: A fourth-order Butterworth bandpass filter. *Curve 2*: A fourth-order Chebyshev bandpass filter with a 0.5 dB ripple. *Curve 3*: A second-order bandpass $Q = 1$ for comparison

The application of the lowpass/bandpass transformation to lowpass filters of second order is particularly important. It results in a mathematical description of fourth-order bandpass filters. Such filters are investigated below. By inserting the transformation equation (13.18) in the second-order lowpass equation (13.17), we obtain the bandpass transfer function:

$$A(s_n) = \frac{s_n^2 A_0 (\Delta \omega_n)^2 / b_1}{1 + \dfrac{a_1}{b_1} \Delta \omega_n s_n + \left[2 + \dfrac{(\Delta \omega_n)^2}{b_1} \right] s_n^2 + \dfrac{a_1}{b_1} \Delta \omega_n s_n^3 + s_n^4} \tag{13.24}$$

It can be seen that the asymptotes of the amplitude–frequency response at low and high frequencies have a slope of ± 40 dB per decade. At the center frequency $\omega_n = 1$, the gain is real and has the value $A_m = A_0$.

In Fig. 13.29 we have plotted the frequency response of the amplitude and the phase of a Butterworth bandpass filter, and of a 0.5 dB Chebyshev bandpass filter, both of which have a normalized bandwidth $\Delta \omega_n = 1$. The frequency response of a second-order bandpass filter with the same bandwidth is shown for comparison.

As in the case of the lowpass filters, we shall simplify the design process by splitting the denominator into quadratic factors for two bandpass filters with symmetrical resonant frequencies:

$$A(s_n) = \frac{s_n^2 A_m (\Delta \omega_n)^2 / b_1}{\left[1 + \dfrac{\alpha s_n}{Q_i} + (\alpha s_n)^2 \right] \left[1 + \dfrac{1}{Q_i} \left(\dfrac{s_n}{\alpha} \right) + \left(\dfrac{s_n}{\alpha} \right)^2 \right]} \tag{13.25}$$

By multiplying out and comparing the coefficients with those in (13.24), we arrive at the equation for α:

$$\alpha^2 + \left[\frac{\alpha \Delta \omega_n a_1}{b_1 (1 + \alpha^2)} \right]^2 + \frac{1}{\alpha^2} - 2 - \frac{(\Delta \omega_n)^2}{b_1} = 0 \tag{13.26}$$

This can easily be solved numerically for any particular application with the aid of a pocket calculator. Having determined α, the pole-pair quality factor Q_i of an individual filter stage is:

$$Q_i = \frac{(1 + \alpha^2) b_1}{\alpha \Delta \omega_n a_1} \tag{13.27}$$

There are two possible ways of implementing the filter, depending on how the numerator is factored. Splitting it up into a constant factor and a factor that contains s_n^2 yields the cascade connection of a highpass and a lowpass filter. This design is useful for realizing large bandwidths $\Delta \omega_n$.

For smaller bandwidths, $\Delta \omega_n \lesssim 1$, it is preferable to use a cascade connection of two second-order bandpass filters, which are not tuned to precisely the same center frequency. This method is called "staggered tuning." For the design of the individual bandpass filter stages, we split the numerator of (13.25) into two factors that contain s_n, and obtain

$$A(s_n) = \frac{(A_r / Q_i)(\alpha s_n)}{1 + \dfrac{\alpha s_n}{Q_i} + (\alpha s_n)^2} \cdot \frac{(A_r / Q_i)(s_n / \alpha)}{1 + \dfrac{1}{Q_i} \left(\dfrac{s_n}{\alpha} \right) + \left(\dfrac{s_n}{\alpha} \right)^2} \tag{13.28}$$

Comparing the coefficients with those of (13.25) and (13.21), we obtain the parameters of the two individual bandpass filters:

$$
\begin{array}{c|ccc}
 & f_r & Q & A_r \\
\hline
\text{1st filter stage} & f_m/\alpha & Q_i & Q_i \Delta\omega_n \sqrt{A_m/b_1} \\
\text{2nd filter stage} & f_m \cdot \alpha & Q_i & Q_i \Delta\omega_n \sqrt{A_m/b_1}
\end{array}
\qquad (13.29)
$$

where f_m is the center frequency of the resulting bandpass filter and A_m is the gain at this frequency. The factors α and Q_i are given by (13.26) and (13.27).

The determination of the parameters of the individual filter stages will be demonstrated by means of an example. A Butterworth bandpass filter is required, with a center frequency of 1 kHz and a bandwidth of 100 Hz. The gain at the center frequency is required to be $A_m = 1$. To begin with, we take the coefficients of a second-order Butterworth lowpass filter from the table in Fig. 13.14: $a_1 = 1.4142$ and $b_i = 1.000$. As $\Delta\omega_n = 0.1$, (13.26) gives $\alpha = 1.0360$. Equation (13.27) yields $Q_i = 14.15$ and, from (13.29), $A_r = 1.415$, $f_{r1} = 965$ Hz, and $f_{r2} = 1.036$ kHz.

13.7
Realization of Second-Order Bandpass Filters

The cascade connection of a highpass and a lowpass filter of first order, as shown in Fig. 13.30, gives a bandpass filter with the transfer function:

$$
A(s) = \frac{1}{1 + \dfrac{s\,RC}{\alpha}} \cdot \frac{1}{1 + \dfrac{1}{\alpha s\,RC}} = \frac{\alpha s\,RC}{1 + \dfrac{1+\alpha^2}{\alpha}s\,RC + (s\,RC)^2}
$$

With the resonant frequency $\omega_r = 1/RC$, we obtain the normalized form. Comparison of the coefficients with those of (13.21) gives the Q-factor:

$$
Q = \frac{\alpha}{1 + \alpha^2}
$$

For $\alpha = 1$, $Q_{max} = \frac{1}{2}$, which is the highest Q-factor that can be achieved by cascading first-order filters. In this case the cutoff frequencies of the lowpass and the highpass are the same. For higher Q-factors, the denominator of (13.21) must have complex zeros, but

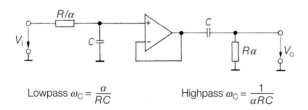

Lowpass $\omega_C = \dfrac{\alpha}{RC}$ Highpass $\omega_C = \dfrac{1}{\alpha RC}$

Fig. 13.30. Bandpass filter consisting of a first-order lowpass filter and a highpass filter

$$
A(s_n) = \frac{\alpha s_n}{1 + \dfrac{1+\alpha^2}{\alpha}s_n + s_n^2}
$$

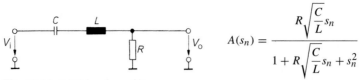

Fig. 13.31. LRC bandpass filter

such a transfer function can only be implemented by LRC circuits or by special active RC circuits, which are discussed below.

13.7.1
LRC Bandpass Filter

A common method of designing selective filters that have high quality factors is the use of resonant circuits. Figure 13.31 shows such a circuit, the transfer function of which is

$$A(s) = \frac{s\,RC}{1 + s\,RC + s^2 LC}$$

With the resonant frequency $\omega_r = 1/\sqrt{LC}$, we obtain the normalized expression as given in Fig. 13.31. Comparison of the coefficients with those of (13.21) gives

$$Q = \frac{1}{R}\sqrt{\frac{L}{C}} \quad \text{and} \quad A_r = 1$$

For high frequencies, the inductances required can be easily implemented and suffer little loss. In the low-frequency range, the inductances become unwieldy and have poor electrical performance. If, for example, a filter having the resonant frequency $f_r = 10\,\text{Hz}$ is to be implemented using the circuit shown in Fig. 13.31, the inductance $L = 25.3\,\text{H}$ is required if a capacitance of $10\,\mu\text{F}$ is selected. As has already been shown in Sect. 13.4.1 for the lowpass and highpass filters, such inductances can be emulated; for example, by using gyrators. In most cases, the desired transfer function of (13.21) can be put into practice much more easily by inserting suitable RC networks into the feedback loop of an operational amplifier.

13.7.2
Bandpass Filter with Multiple Negative Feedback

The principle of multiple negative feedback can also be applied to bandpass filters. The appropriate circuit is shown in Fig. 13.32. As can be seen by comparison with (13.21), the coefficient of s_n^2 must be unity. Therefore, the resonant frequency is given by

$$f_r = \frac{1}{2\pi C}\sqrt{\frac{R_1 + R_3}{R_1 R_2 R_3}} \tag{13.30}$$

Inserting this relation into the transfer function and comparing the remaining coefficients

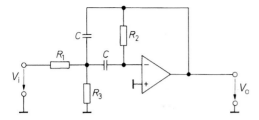

Fig. 13.32. Bandpass filter with multiple negative feedback

$$A(s_n) = \frac{-\dfrac{R_2 R_3}{R_1 + R_3} C \omega_r s_n}{1 + \dfrac{2 R_1 R_3}{R_1 + R_3} C \omega_r s_n + \dfrac{R_1 R_2 R_3}{R_1 + R_3} C^2 \omega_r^2 s_n^2}$$

with those of (13.21) gives the other parameters:

$$-A_r = \frac{R_2}{2 R_1} \tag{13.31}$$

$$Q = \frac{1}{2} \sqrt{\frac{R_2(R_1 + R_3)}{R_1 R_3}} = \pi R_2 C f_r \tag{13.32}$$

It can be seen that the gain, the Q-factor, and the resonant frequency can be freely chosen. From (13.32), we obtain the bandwidth of the filter,

$$B = \frac{f_r}{Q} = \frac{1}{\pi R_2 C}$$

which is independent of R_1 and R_3. On the other hand, it can be seen from (13.31) that A_r is not dependent on R_3. It is therefore possible to vary the resonant frequency by use of R_3 without affecting the bandwidth B and the gain A_r.

If resistor R_3 is omitted, the filter remains functional, but the Q-factor becomes dependent on A_r. This is because, with $R_3 \to \infty$, (13.32) becomes

$$-A_r = 2Q^2$$

In order that the loop gain of the circuit should be very much larger than unity, the open-loop gain of the operational amplifier must be large compared to $2Q^2$. Using resistor R_3, high Q-factors can be attained even for a low gain A_r. As can be seen in Fig. 13.32, the low gain is due only to the fact that the input signal is attenuated by the voltage divider R_1, R_3. Therefore, in this case the open-loop gain of the operational amplifier must also be large compared to $2Q^2$. This requirement is particularly severe: as it must be met at around the resonant frequency, it determines the choice of the operational amplifier, especially for applications at higher frequencies.

The determination of the circuit parameters is shown by the following example. A selective filter is required, having a resonant frequency of $f_r = 10\,\text{Hz}$ and a quality factor of $Q = 100$. Therefore, the cutoff frequencies lie at approximately 9.95 Hz and 10.05 Hz. The gain at the resonant frequency is required to be $A_r = -10$. One of the parameters can be freely chosen – for example, $C = 1\,\mu\text{F}$ – and the remainder must be calculated. To begin with, from (13.32),

$$R_2 = \frac{Q}{\pi f_r C} = 3.18\,\text{M}\Omega$$

Hence, from (13.31),

$$R_1 = \frac{R_2}{-2A_r} = 159\,\text{k}\Omega$$

The resistance R_3 is given by (13.30) as

$$R_3 = \frac{-A_r R_1}{2Q^2 + A_r} = 79.5\,\Omega$$

The open-loop gain of the operational amplifier must, at the resonant frequency, still be large compared to $2Q^2 = 20\,000$.

One advantage of the circuit is that it has no tendency to oscillate at the resonant frequency, even if the circuit elements do not quite match their theoretical values. Obviously, this is true only if the operational amplifier is correctly frequency compensated; otherwise, high-frequency oscillations can occur.

13.7.3
Bandpass Filter with Single Positive Feedback

The application of single positive feedback results in the bandpass circuit shown in Fig. 13.33. The negative feedback via the resistors R_1 and $(k - 1)R_1$ fixes the internal gain at the value k. Comparison of the coefficients with those of (13.21) yields the equations given for determining the circuit parameters.

A disadvantage is that Q and A_r cannot be chosen independently of one another. The advantage, however, is that the Q-factor may be altered by varying k without at the same time changing the resonant frequency.

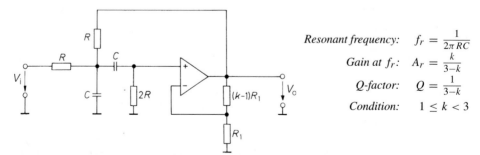

Resonant frequency: $f_r = \dfrac{1}{2\pi RC}$

Gain at f_r: $A_r = \dfrac{k}{3-k}$

Q-factor: $Q = \dfrac{1}{3-k}$

Condition: $1 \le k < 3$

Fig. 13.33. Bandpass filter with single positive feedback

$$A(s_n) = \frac{kRC\omega_r s_n}{1 + RC\omega_r(3 - k)s_n + R^2 C^2 \omega_r^2 s_n^2}$$

For $k = 3$, the gain is infinite, and an undamped oscillation occurs. The adjustment of the internal gain therefore becomes more critical the closer it approaches the value 3.

13.8
Lowpass/Bandstop Filter Transformation

Selective rejection of a particular frequency requires a filter whose gain is zero at the resonant frequency and rises to a constant value at higher and lower frequencies. Such filters are called *rejection filters*, *bandstop filters*, or *notch filters*. To characterize the selectivity, the rejection quality factor is defined as $Q = f_r/B$, where B is the 3 dB bandwidth. The larger the Q-factor of the filter, the more steeply the gain fails off in the vicinity of the resonant frequency f_r.

As in the case of the bandpass filter, the amplitude–frequency response of the band-rejection filter can be derived from the frequency response of a lowpass filter by using a suitable frequency transformation. To accomplish this, the variable s_n is replaced by the expression

$$s_n \rightarrow \frac{\Delta\omega_n}{s_n + \dfrac{1}{s_n}} \tag{13.33}$$

where $\Delta\omega_n = 1/Q$ is the normalized 3 dB bandwidth. By means of this transformation, the amplitude of the lowpass filter in the range $0 \leq \omega_n \leq 1$ is converted to the pass band of the band-rejection filter between $0 \leq \omega_n \leq \omega_{n,c1}$. In addition, it appears as a mirror image above the resonant frequency, when plotted on a logarithmic scale. At the resonant frequency $\omega_n = 1$, the transfer function is zero. As with the bandpass filter, the order of the filter is doubled by the transformation. It is of particular interest to apply the transformation to a first-order lowpass filter. This results in a notch filter of second order, which has the transfer function:

$$A(s_n) = \frac{A_0(1 + s_n^2)}{1 + \Delta\omega_n s_n + s_n^2} = \frac{A_0(1 + s_n^2)}{1 + \dfrac{1}{Q}s_n + s_n^2} \tag{13.34}$$

From this, we obtain the relations for the amplitude–frequency response and the phase–frequency response:

$$|A| = \frac{A_0|(1 - \omega_n^2)|}{\sqrt{1 + \omega_n^2\left(\dfrac{1}{Q^2} - 2\right) + \omega_n^4}}, \qquad \varphi = \arctan\frac{\omega_n}{Q(\omega_n^2 - 1)}$$

The corresponding curves are shown in Fig. 13.34 for the rejection quality factors 1 and 10.

The denominator of (13.34) is identical to that of (13.21) for bandpass filters. It follows from (13.21) that a maximum Q-factor of only $Q = \frac{1}{2}$ can be attained with passive RC circuits; for higher Q-factors, LRC networks or special active RC circuits must be employed.

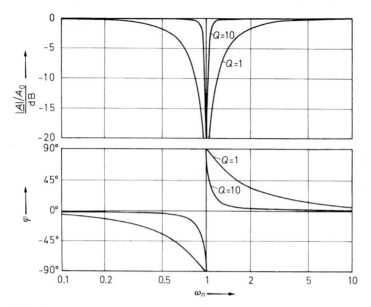

Fig. 13.34. Frequency response of the amplitude and the phase for second-order bandstop filters with $Q = 1$ and $Q = 10$

13.9
Realization of Second-Order Bandstop Filters

13.9.1
LRC Bandstop Filter

A well-known method of implementing rejection filters involves using series resonant circuits, as shown in Fig. 13.35. At the resonant frequency, the arrangement represents a short-circuit and the output voltage is zero. The transfer function of the circuit is

$$A(s) = \frac{1 + s^2 LC}{1 + s\,RC + s^2 LC}$$

Hence, the resonant frequency is $\omega_r = 1/\sqrt{LC}$, and we obtain the normalized form as given in Fig. 13.35. The rejection quality factor is found by comparing the coefficients with those of (13.35):

$$Q = \frac{1}{R}\sqrt{\frac{L}{C}}$$

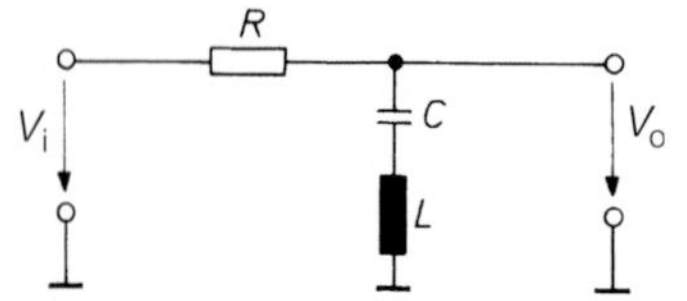

$$A(s_n) = \frac{1 + s_n^2}{1 + R\sqrt{\dfrac{C}{L}}\,s_n + s_n^2}$$

Fig. 13.35. *LRC* bandstop filter

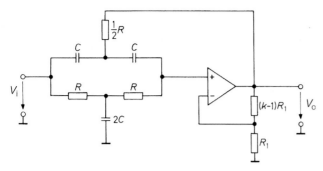

Fig. 13.36. Active parallel-T bandstop filter

$$A(s_n) = \frac{k(1 + s_n^2)}{1 + 2(2 - k)s_n + s_n^2}$$

Resonant frequency: $\quad f_r = \dfrac{1}{2\pi RC}$

Gain: $\quad A_0 = k$

Rejection Q-factor: $\quad Q = \dfrac{1}{2(2 - k)}$

Condition: $\quad 1 \le k < 2$

This holds for lossless inductors only, because only then the output voltage falls to zero. In addition, the same limitations for the use of inductances apply here as for bandpass filters.

13.9.2
Active Parallel-T Bandstop Filter

As shown in Sect. 29.3.6, the parallel-T filter represents a passive RC rejection filter. From (29.3.24), its rejection Q-factor is $Q = 0.25$, which can be increased by incorporating the parallel-T filter in the feedback loop of an amplifier. One possible implementation is shown in Fig. 13.36.

For high and low frequencies, the parallel-T filter transfers the input signal unchanged. The output voltage of the impedance converter is then $k\underline{V}_i$. At the resonant frequency, the output voltage is zero. In this case, the parallel-T filter behaves as if resistor $R/2$ were connected to ground. Therefore, the resonant frequency $f_r = 1/2\pi RC$, remains unchanged.

From the transfer function, the filter data given below Fig. 13.36 can be directly deduced. If the voltage follower has unit gain, $Q = 0.5$. For an increase in gain, Q rises toward infinity as k approaches 2.

A precondition for the correct operation of the circuit is the precise adjustment of the resonant frequency and the gain of the parallel-T filter. This is difficult to achieve for higher Q-factors, since varying one resistance always affects both parameters simultaneously. The active Wien–Robinson bandstop filter is more favorable in this respect.

13.9.3
Active Wien–Robinson Bandstop Filter

As shown in Sect. 29.3.5, the Wien–Robinson bridge also behaves like a notch filter. However, its Q-factor is not much higher than that of the parallel-T filter, but it can also be increased to any desired value by incorporating the filter in the feedback loop of an

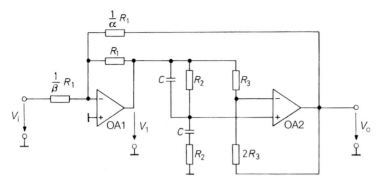

Fig. 13.37. An active Wien–Robinson bandstop filter

$$A(s_n) = -\frac{\dfrac{\beta}{1+\alpha}(1+s_n^2)}{1+\dfrac{3}{1+\alpha}s_n+s_n^2}$$

Resonant frequency: $f_r = \dfrac{1}{2\pi R_2 C}$

Gain: $A_0 = -\dfrac{\beta}{1+\alpha}$

Rejection Q-factor: $Q = \dfrac{1+\alpha}{3}$

operational amplifier. The corresponding circuit is shown in Fig. 13.37, and its transfer function is obtained from the relation for the Wien–Robinson bridge,

$$\underline{V}_o = \frac{1+s_n^2}{1+3s_n+s_n}\underline{V}_1$$

which directly yields the filter data given below Fig. 13.37. For the actual design of the circuit, the parameters f_r, A_0, Q, and C are defined and the remaining parameters are then

$$R_2 = \frac{1}{2\pi f_r C}, \quad \alpha = 3Q-1 \quad \text{and} \quad \beta = -3A_0Q$$

In order to tune the filter to the resonant frequency, the capacitors C are changed in steps and the two resistors R_2 can be varied continuously using potentiometers. If the resonant frequency is not fully suppressed due to a slight mismatch of the bridge components, the final adjustment can be made by slightly varying resistor $2R_3$.

13.10
Allpass Filters

13.10.1
Basic Principles

The filters discussed so far are circuits for which the gain and phase shift are frequency dependent. In this section, we examine circuits for which the gain remains constant and only the phase shift is dependent on the frequency. These are called allpass filters, and they are used for phase correction and signal delay.

Initially, we show how the frequency response of an allpass filter can be derived from the frequency response of a lowpass filter. To do this, the constant factor A_0 in the numerator of (13.11) is replaced by the conjugate complex denominator and, in this way, constant unit gain and phase shift doubling are obtained:

$$A(s_n) = \frac{\prod\limits_i (1 - a_i s_n + b_i s_n^2)}{\prod\limits_i (1 + a_i s_n + b_i s_n^2)} = \frac{\prod\limits_i \sqrt{(1 - b_i \omega_n^2)^2 + a_i^2 \omega_n^2} \, e^{-j\alpha}}{\prod\limits_i \sqrt{(1 - b_i \omega_n^2)^2 + a_i^2 \omega_n^2} \, e^{+j\alpha}} \tag{13.35}$$

$$= 1 \cdot e^{-2j\alpha} = e^{j\varphi}$$

where

$$\varphi = -2\alpha = -2 \sum_i \arctan \frac{a_i \omega_n}{1 - b_i \omega_n^2} \tag{13.36}$$

The use of allpass filters for signal delay is of particular interest. Constant gain, a condition that is always fulfilled by allpass filters, is one prerequisite for undistorted signal transfer. The second prerequisite is that the group delay of the circuit is constant for all frequencies considered. The filters that best fulfill this condition are Bessel lowpass filters, for which the group delay is Butterworth-approximated. Therefore, in order to obtain a "Butterworth allpass filter," the Bessel coefficients must be inserted into (13.35).

It is advisable, however, to renormalize the frequency responses thus obtained, as the 3 dB cutoff frequency of the lowpass filters is meaningless in this case. For this reason, we recalculate the coefficients a_1 and b_1 so that the group delay at $\omega_n = 1$ is reduced to $1/\sqrt{2}$ of its low-frequency value. The coefficients obtained in this way are shown in Fig. 13.38 for filters of up to the tenth order.

The group delay is the time interval by which the signal is delayed in the allpass filter. According to the definition given in (13.9b), it can be determined from (13.36),

$$T_{gr} = t_{gr} \cdot \omega_c = 2\pi \, t_{gr} \cdot f_c = -\frac{d\varphi}{d\omega_n} = 2 \sum_i \frac{a_i(1 + b_i \omega_n^2)}{1 + (a_i^2 - 2b_i)\omega_n^2 + b_i^2 \omega_n^4} \tag{13.37}$$

and therefore at low frequencies it has the value

$$T_{gr\,0} = 2 \sum_i a_i,$$

which is also given for each order in Fig. 13.38. In addition, the pole-pair quality factor $Q_i = \sqrt{b_i}/a_i$ is given. As it is unaffected by the renormalization, it has the same values as for the Bessel filters.

To enable the correct operation of the individual filter stages to be checked, we have also shown the ratio f_i/f_c in Fig. 13.38. Here, f_i is the frequency at which the phase of the particular filter stage approaches the value $-180°$ for a second-order, or $-90°$ for a first-order, filter stage. This frequency is considerably easier to measure than the cutoff frequency of the group delay.

The frequency response of the group delay is shown in Fig. 13.39 for allpass filters of first to tenth order.

The following example shows the steps in the design of an allpass filter. A signal that has a frequency spectrum of 0–1 kHz is to be delayed by $t_{gr\,0} = 2$ ms. In order that the phase distortion is not too large, the cutoff frequency of the allpass filter should be $f_c \geq 1$ kHz. From (13.37), it is therefore necessary that

$$T_{gr\,0} \geq 2 \, \text{ms} \cdot 2\pi \cdot 1 \, \text{kHz} = 12.566.$$

N	i	a_i	b_i	f_i/f_c	Q_i	$T_{gr\,0}$
1	1	0.6436	0.0000	1.554	–	1.2872
2	1	1.6278	0.8832	1.064	0.58	3.2556
3	1	1.1415	0.0000	0.876	–	5.3014
	2	1.5092	1.0877	0.959	0.69	
4	1	2.3370	1.4878	0.820	0.52	7.3752
	2	1.3506	1.1837	0.919	0.81	
5	1	1.2974	0.0000	0.771	–	9.4625
	2	2.2224	1.5685	0.798	0.56	
	3	1.2116	1.2330	0.901	0.92	
6	1	2.6117	1.7763	0.750	0.51	11.5579
	2	2.0706	1.6015	0.790	0.61	
	3	1.0967	1.2596	0.891	1.02	
7	1	1.3735	0.0000	0.728	–	13.6578
	2	2.5320	1.8169	0.742	0.53	
	3	1.9211	1.6116	0.788	0.66	
	4	1.0023	1.2743	0.886	1.13	
8	1	2.7541	1.9420	0.718	0.51	15.7607
	2	2.4174	1.8300	0.739	0.56	
	3	1.7850	1.6101	0.788	0.71	
	4	0.9239	1.2822	0.883	1.23	
9	1	1.4186	0.0000	0.705	–	17.8656
	2	2.6979	1.9659	0.713	0.52	
	3	2.2940	1.8282	0.740	0.59	
	4	1.6644	1.6027	0.790	0.76	
	5	0.8579	1.2862	0.882	1.32	
10	1	2.8406	2.0490	0.699	0.50	19.9717
	2	2.6120	1.9714	0.712	0.54	
	3	2.1733	1.8184	0.742	0.62	
	4	1.5583	1.5923	0.792	0.81	
	5	0.8018	1.2877	0.881	1.42	

Fig. 13.38. Allpass filter coefficients for a maximally flat group delay

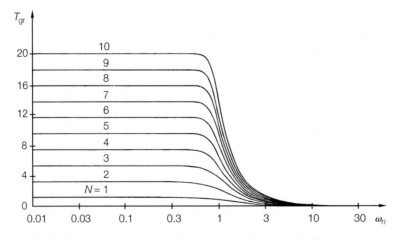

Fig. 13.39. Frequency response of the group delay for orders 1–10

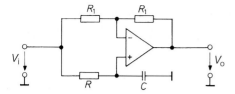

Fig. 13.40. First-order allpass filter

$$A(s_n) = \frac{1 - s\,RC}{1 + s\,RC} = \frac{1 - RC\omega_c s_n}{1 + RC\omega_c s_n}$$

From Fig. 13.38, it can be seen that a filter of at least seventh order is needed, for which $T_{gr0} = 13.6578$. To make the group delay exactly 2 ms, the cutoff frequency must be selected, in accordance with (13.37), as

$$f_c = \frac{T_{gr\,0}}{2\pi \cdot t_{gr\,0}} = \frac{13.6578}{2\pi \cdot 2\,\text{ms}} = 1.087\,\text{kHz}$$

13.10.2
Realization of First-Order Allpass Filters

The circuit in Fig. 13.40 exhibits a gain of +1 at low frequencies and a gain of −1 at high frequencies; in other words, the phase shift changes from 0 to −180°. The circuit is an allpass filter if the magnitude of the gain is also unity for the middle frequency range. To examine this, we consider the transfer function shown in Fig. 13.40. The absolute value of the gain is indeed constant and unity. Comparing the coefficients with those of (13.35) gives

$$RC = \frac{a_1}{2\pi f_c}$$

Using (13.37), the low-frequency value of the group delay is therefore

$$t_{gr\,0} = 2RC$$

The first-order allpass filter shown in Fig. 13.40 is very well suited for use as a phase shifter over a wide range of phase delays. By varying the value of resistor R, the phase delay can be adjusted to values between 0 and −180° without affecting the amplitude. The phase shift is

$$\varphi = -2\arctan(\omega RC)$$

13.10.3
Realization of Second-Order Allpass Filters

The second-order allpass filter transfer function can be implemented, for example, by subtracting the output voltage of a bandpass filter from its input voltage. The transfer function of this circuit is then

$$A(s_n') = 1 - \frac{\dfrac{A_r}{Q}s_n'}{1 + \dfrac{1}{Q}s_n' + s_n'^2} = \frac{1 + \dfrac{1 - A_r}{Q}s_n' + s_n'^2}{1 + \dfrac{1}{Q}s_n' + s_n'^2}$$

It can be seen that, for $A_r = 2$, the transfer function of an allpass filter is obtained. It is normalized not to the cutoff frequency of the allpass, but to the resonant frequency of the bandpass filter. For a correct normalization, we take

$$\omega_c = \beta\omega_r$$

and obtain:

$$s'_n = \frac{s}{\omega_r} = \frac{\beta s}{\omega_c} = \beta s_n$$

Hence, the transfer function is:

$$A(s_n) = \frac{1 - \dfrac{\beta}{Q}s_n + \beta^2 s_n^2}{1 + \dfrac{\beta}{Q}s_n + \beta^2 s_n^2}$$

Comparing the coefficients with those of (13.35) yields

$$a_1 = \frac{\beta}{Q} \quad \text{and} \quad b_1 = \beta^2$$

The data of the required bandpass filter are therefore

$$A_r = 2$$
$$f_r = f_c/\sqrt{b_1}$$
$$Q = \sqrt{b_1}/a_1 = Q_1$$

As an example, let us consider implementation using the bandpass filter shown in Fig. 13.32. As the Q-factors are relatively small, resistor R_3 may be omitted and, instead, the gain adjusted by resistor R/α in Fig. 13.41. The component values are obtained by comparing the coefficients of the transfer function with those of (13.35):

$$R_1 = \frac{a_1}{4\pi f_c C}, \quad R_2 = \frac{b_1}{\pi f_c C a_1} \quad \text{and} \quad \alpha = \frac{a_1^2}{b_1} = \frac{1}{Q_1^2}$$

From the transfer function, a further application of the circuit in Fig. 13.41 can be deduced. If $2R_1 - \alpha R_2 = 0$ a bandstop filter circuit is obtained.

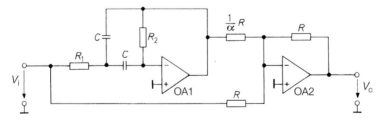

Fig. 13.41. Second-order allpass filter

$$A(s_n) = -\frac{1 + (2R_1 - \alpha R_2)C\omega_c s_n + R_1 R_2 C^2 \omega_c^2 s_n^2}{1 + 2R_1 C\omega_c s_n + R_1 R_2 C^2 \omega_c^2 s_n^2}$$

13.11
Adjustable Universal Filters

As shown previously, the transfer function of a second-order filter element has the general form

$$A(s_n) = \frac{d_0 + d_1 s_n + d_2 s_n^2}{c_0 + c_1 s_n + c_2 s_n^2} \qquad (13.38)$$

The filter families described so far can be deduced from (13.38) by assigning special values to the coefficients of the numerator:

lowpass filter:	$d_1 = d_2 = 0$;	
highpass filter:	$d_0 = d_1 = 0$;	
bandpass filter:	$d_0 = d_2 = 0$;	
bandstop filter:	$d_1 = 0,$	$d_0 = d_2$;
allpass filter:	$d_0 = c_0,$	$d_1 = -c_1, \quad d_2 = c_2$

The numerator coefficients may have either sign, whereas the coefficients of the denominator must always be positive for reasons of stability. The pole-pair Q-factor is defined by the denominator coefficients

$$Q_i = \frac{\sqrt{c_0 c_2}}{c_1} \qquad (13.39)$$

Filter with Arbitrary Coefficients

In the previous sections we have shown specific and, if possible, simple circuits for each type of filter. Sometimes, however, a single circuit is required to realize all of the described kinds of filters, as well as the more general types in (13.38), with any numerator coefficients. This problem can be solved by using the circuit shown in Fig. 13.42. This circuit has the added advantage that the individual coefficients can be set independently of one another, as each coefficient is determined by only one circuit component. In the transfer function (13.40) ω_0 is the normalizing frequency and $\tau = RC$ is the time constant of the two integrators. The coefficients k_i and l_i are resistance ratios and therefore are always positive. If the sign of a numerator coefficient is to be changed, the input voltage of the corresponding resistor must be inverted by an additional amplifier.

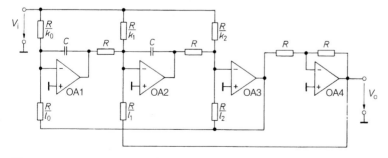

Fig. 13.42. Second-order universal filter with independently adjustable coefficients

$$A(s_n) = \frac{k_0 - k_1 \omega_0 \tau s_n + k_2 \omega_0^2 \tau^2 s_n^2}{l_0 + l_1 \omega_0 \tau s_n + l_2 \omega_0^2 \tau^2 s_n^2} \qquad (13.40)$$

To realize a higher-order filter, the number of integrators can be raised accordingly. However, it is usually easier to split the filter into second-order stages and cascade them.

The design process for the circuit is illustrated by the following example: a second-order allpass filter is required, the group delay curve of which is maximally flat and has, at low frequencies, the value 1 ms. From the table in Fig. 13.38 we take the values $a_1 = 1.6278$, $b_1 = 0.8832$, and $T_{gr\,0} = 3.2556$. Equation (13.9a) gives the cutoff frequency:

$$\omega_c = \frac{T_{gr\,0}}{t_{gr\,0}} = \frac{3.2556}{1\,\text{ms}} = 3.26\,\text{kHz}$$

We select $\tau = 1$ ms and, by comparing the coefficients of (13.40) and (13.35), and using $\omega_0 = \omega_c = 3.26\,\text{kHz}$, we obtain the values

$$l_0 = k_0 = 1 \quad l_1 = k_1 = \frac{a_1}{\omega_0 \tau} = 0.500 \quad l_2 = k_2 = \frac{b_1}{(\omega_0 \tau)^2} = 0.0833$$

Such a low value of l_2 is difficult to obtain in practice. However, it increases more rapidly than the other coefficients when τ is reduced. We therefore choose $\tau = 0.3$ ms and obtain

$$l_0 = k_0 = 1 \qquad l_1 = k_1 = 1.67 \qquad \text{and} \qquad l_2 = k_2 = 0.926$$

Filter with Adjustable Parameters

For some applications of the bandpass filter, it is desirable that the resonant frequency, the Q-factor, and the gain at resonant frequency can be set independently. As a comparison of (13.41) and (13.21) shows, the two coefficients l_1 and k_1 should be simultaneously adjustable to enable the Q-factor to be set without affecting the gain. Figure 13.43 shows a circuit in which any such dependence is eliminated.

The interesting feature of this circuit is that it operates as a selective filter, a blocking filter, a lowpass filter or a highpass filter, depending on the output used. For calculating the filter parameters we obtain the following relations by inserting $\tau = RC$ for the integration time constant:

$$V_{BS} = -V_{BP} - \frac{R_1}{R_2} V_i \qquad\qquad V_{HP} = -\frac{R_3}{R_1} V_{TP} - \frac{R_3}{R_4} V_{BS}$$

$$V_{BP} = -V_{HP}/s\tau \qquad\qquad V_{TP} = -V_{BP}/s\tau$$

The transfer functions are obtained by eliminating any three of the four output voltages. Dimensioning is obtained by the coefficient comparison with (13.11), (13.15), (13.21) and (13.34). This is particularly simple if we set $\tau \cdot \omega_c = 1$; i.e. selecting $RC = 1/2\pi f_c$:

Lowpass filter	Highpass filter	Bandpass, bandstop filter
given: R_1	given: R_1	given: R_1
$R_3 = R_1/b_i$	$R_3 = R_1 b_i$	$R_3 = R_1$
$R_4 = R_1/a_i$	$R_4 = R_1 b_i/a_i$	$R_4 = R_1 Q$
$R_2 = R_1 a_i/A_0$	$R_2 = R_1 a_i/A_\infty$	$R_2 = -R_1/A$

We can see from these equations that R_3 and R_4 determine the filter type for highpass and lowpass filters, and R_2 determines the gain. For any given type of filter, the cutoff frequency and the gain can be adjusted independently.

Even for bandpass and band-rejection filter operation, the resonant frequency, gain, and Q-factor can be varied without one affecting the other. This is because the resonant

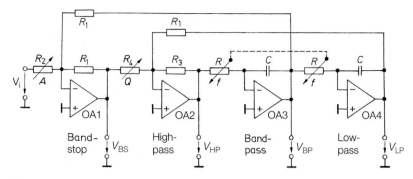

Fig. 13.43. Universal filter of second order with independently adjustable parameters. State Variable Filter, Biquad. Integration time constant $\tau = RC$

$$\frac{V_{LP}}{V_i} = \frac{\dfrac{R_1^2}{R_2 R_4}}{1 + \dfrac{R_1}{R_4}\tau\,\omega_c s_n + \dfrac{R_1}{R_3}\tau^2\omega_c^2 s_n^2}$$

(Lowpass filter)

$$\frac{V_{HP}}{V_i} = \frac{\dfrac{R_1 R_3}{R_2 R_4}s_n^2}{\dfrac{R_3}{R_1\tau^2\omega_c^2} + \dfrac{R_3}{R_4\tau\omega_c}s_n + s_n^2}$$

(Highpass filter)

$$\frac{V_{BP}}{V_i} = \frac{-\dfrac{R_1^2}{R_2 R_4}\tau\,\omega_r s_n}{1 + \dfrac{R_1}{R_4}\tau\,\omega_r s_n + \dfrac{R_1}{R_3}\tau^2\omega_r^2 s_n^2}$$

(Bandpass filter)

$$\frac{V_{BS}}{V_i} = \frac{-\dfrac{R_1}{R_2}\left(1 + \dfrac{R_1}{R_3}\tau^2\omega_r^2 s_n^2\right)}{1 + \dfrac{R_1}{R_4}\tau\,\omega_r s_n + \dfrac{R_1}{R_3}\tau^2\omega_r^2 s_n^2}$$

(Bandstop filter)

frequency is determined solely by the product $\tau = RC$. As these variables do not occur in the equations for A and Q, the frequency can be varied without affecting A and Q. These two parameters can be set independently using resistors R_2 and R_4.

Also when operated as a bandpass or bandstop filter the resonance frequency, the gain and the quality can be varied without mutually affecting each other, since the resonant frequency is determined solely by the product $\tau = RC$. Since these parameters do not appear in the equations for A and Q the frequency can be varied without changing A and/or Q. These two parameters can be adjusted by the resistances R_2 and R_4 independently of each other.

As can be seen in Fig. 5.92 on page 560, integrators with VV operational amplifiers are not very suitable for high frequencies. CC integrators are much better. Figure 5.93 on page 561 shows an example.

Universal filters are available as integrated circuits requiring only a few external resistors to determine the filter type and the cutoff frequency. Some examples are listed in Fig. 13.44. By comparison with the very popular SC filters described in Sect. 13.12, the continuous filters offer the advantage that they need no clock and therefore have no *clock noise*.

Electronic Control of Filter Parameters

As the resistors R have large values for low filter frequencies, it may in this case be advantageous to replace them by fixed resistors connected to voltage dividers. The voltage

Type	Manufac-turer	Filter type	Filter order	Cutoff frequency max.	Dynamic range	Specialty
UAF 42	Burr Br.	Biquad	1×2	100 kHz		1 OA extra
LT 1568	Lin. Tech.	Butterworth	2×2	10 MHz	92 dB	RRIO
LTC 1560-1	Lin. Tech.	Cauer	5	1 MHz	75 dB	8 pin case
LTC 1562	Lin. Tech.	4 Biquads	4×2	150 kHz	118 dB	RRIO
LTC 1563	Lin. Tech.	2 Biquads	2×2	300 kHz	80 dB	
LTC 1563-2	Lin. Tech.	Butterworth	4	300 kHz	80 dB	
LTC 1564	Lin. Tech.	Butterworth	8	150 kHz	120 dB	dig. prog.
LTC 1565-31	Lin. Tech.	Lin. Phase	7	650 kHz	80 dB	8 pin case
MAX 270	Maxim	Butterworth	2×2	25 kHz	96 dB	dig. prog.
MAX 274	Maxim	4 Biquads	4×2	150 kHz	86 dB	
MAX 275	Maxim	2 Biquads	2×2	300 kHz	89 dB	

Fig. 13.44. Integrated active filters

divider can then be realized by low-resistance potentiometers. This method may also be used for the resistors R_1 and R_2.

If a filter parameter is to be voltage controlled, the voltage divider can be replaced by an analog multiplier, where the control voltage is connected to the second input, as shown in Fig. 13.45. The effective resistance is then

$$R_{equiv} = R_0 \cdot \frac{U}{V_{cont}}$$

where V_{cont} is the controlling voltage. If two such circuits are inserted instead of the two frequency-determining resistors R, the resonant frequency of the bandpass filter becomes

$$f_r = \frac{1}{2\pi R_0 C} \cdot \frac{V_{cont}}{U}$$

and is proportional to the control voltage.

For numeric control of filter parameters – for example, via a computer – digital-to-analog converters DACS can be used instead of analog multipliers. These provide an output voltage that is proportional to the product of the applied number and the reference voltage:

$$V_o = V_{ref} \frac{Z}{Z_{max} + 1}$$

The types preferred for use in filters are those whose reference voltage can assume any positive or negative value. Consequently, the multiplying DACs with CMOS switches described in Sect. 18.4 are particularly suitable for this purpose, but as they possess considerable resistor tolerances, they cannot simply be inserted as series resistors in Fig. 13.43. However, the effect of the absolute resistance value can be eliminated by providing a following operational amplifier with a feedback path via a resistor incorporated in the DAC.

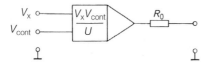

Fig. 13.45. Multiplier as a controlled resistor

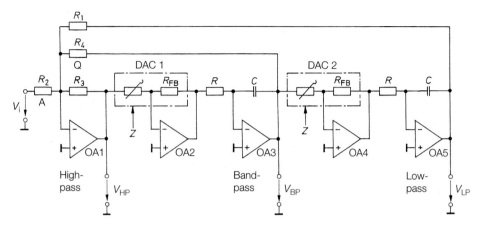

Fig. 13.46. Universal filter with digitally adjustable frequency. The integration time constant $\tau = RC(Z_{max} + 1)/Z$

$$\frac{V_{LP}}{V_i} = \frac{-\dfrac{R_1}{R_2}}{1 + \dfrac{R_1}{R_4}\tau\omega_c s_n + \dfrac{R_1}{R_3}\tau^2\omega_c^2 s_n^2}$$

(Lowpass filter)

$$\frac{V_{HP}}{V_i} = \frac{s_n^2 R_1/R_2}{\dfrac{R_3}{R_1\tau^2\omega_c^2} + \dfrac{R_3}{R_4\tau\omega_c}s_n + s_n^2}$$

(Highpass filter)

$$\frac{V_{BP}}{V_i} = \frac{-\dfrac{R_1}{R_2}\tau\omega_r s_n}{1 + \dfrac{R_1}{R_4}\tau\omega_r s_n + \dfrac{R_1}{R_3}\tau^2\omega_c^2 s_n^2}$$

(Bandpass filter)

The resulting circuit for digital frequency adjustment is shown in Fig. 13.46. The two integrators are preceded by a DAC. The resulting integration time constant is

$$\tau = RC(Z_{max} + 1)/Z \tag{13.41}$$

If the number Z equals the maximum value Z_{max} – that is, all the bits are one – virtually the same resonant frequency is obtained as in the circuit shown in Fig. 13.43.

In comparison with Fig. 13.43, the arrangement of the feedback loops is modified somewhat, because the DACs in conjunction with the associated operational amplifiers and following integrators form a *noninverting integrator.* However, the resulting transfer functions are quite similar. It is particularly simple to select component values if we select $\tau\omega_c = 1$; that is, $f_c = 1/2\pi\tau$:

Lowpass filter: given R_1	Highpass filter: given R_1	Bandpass: given R_1
$R_3 = R_1/b_i$	$R_3 = R_1 b_i$	$R_3 = R_1$
$R_4 = R_1/a_i$	$R_4 = R_3/a_i$	$R_4 = R_1 Q$
$R_2 = -R_1/A_0$	$R_2 = -R_3/A_\infty$	$R_2 = -R_1 Q/A_r$

Substituting the integration time constant in (13.41), we can see that the cutoff and resonant frequencies become proportional to the number Z:

$$f_c = \frac{1}{2\pi\tau} = \frac{1}{2\pi RC} \cdot \frac{Z}{Z_{max} + 1}$$

The outputs of the D/A converters must have a large dynamic range in order to be able to adjust the frequency over a wide band. To prevent any DC errors in the circuit, it is necessary to use operational amplifiers with a low offset voltage. Suitable opamps can be found in Fig. 5.103 on page 574. Suitable DACs include the AD 7528 (8-bit) or the AD 7537 (12-bit) from Analog Devices, as these incorporate two D/A converters with a common computer interface.

A considerably simpler way of implementing a frequency adjustable filter is to use switched capacitor filters of the type described in Sect. 13.12. But they possess some restrictions resulting from switching interference.

13.12
Switched Capacitor Filters

13.12.1
Principle

The active filters described above require an active component in the form of an operational amplifier, as well as passive elements in the form of capacitors and resistors. Normally, filters with a variable cutoff frequency are only realized by varying the capacitors or resistors (see Fig. 13.46). However, it is also possible to simulate a resistor by means of a switched capacitor. The principle involved is shown in Fig. 13.47.

If the switch connects the capacitor to the input voltage, capacitor C receives the charge $Q = C_S \cdot V$. In the other switch position, the capacitor delivers the same charge again. In each switching period it therefore transfers the charge $Q = C_S \cdot V$ from the input to the output of the circuit. This produces an average current flow of $I = C_S \cdot V/T_S = C_S \cdot V \cdot f_S$. Comparing this relation with Ohm's Law, the basic equivalence between the switched capacitor and an ohmic resistor can be expressed in the form

$$I = V/R_{\text{equiv}} = V \cdot C_S \cdot f_S \quad \text{with} \quad R_{\text{equiv}} = 1/(C_S \cdot f_S)$$

Note the proportional relationship between the switching frequency and the equivalent conductance. It is this property that is utilized in switched capacitor (SC) filters.

13.12.2
SC Integrator

The switched capacitor can replace the ohmic resistor in the conventional integrator shown in Fig. 13.48. The result is the SC integrator of Fig. 13.49. In this circuit the integration time constant

$$\tau = C \cdot R_{\text{equiv}} = \frac{C}{C_S \cdot f_S} = \frac{\eta}{2\pi f_S} \tag{13.42}$$

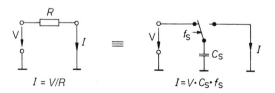

$$I = V/R$$

$$I = V \cdot C_S \cdot f_S$$

Fig. 13.47. Equivalence of the switched capacitor and an ohmic resistor

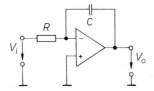

Fig. 13.48. Inverting integrator in RC technology

$$\tau = R \cdot C \qquad \frac{\underline{V}_o}{\underline{V}_i} = -\frac{1}{\tau \cdot s}$$

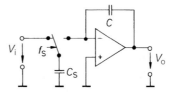

Fig. 13.49. Inverting integrator in SC technology

$$\tau = \frac{1}{f_S} \cdot \frac{C}{C_S} = \frac{1}{f_S} \frac{\eta}{2\pi} \qquad \frac{\underline{V}_o}{\underline{V}_i} = -\frac{1}{\tau \cdot s}$$

can be set by the switching frequency f_S. The capacitance ratio $C/C_S = \eta/2\pi$ is permanently preset by the manufacturer; the parameter η can be obtained from the data sheet, and is generally between 50 and 200.

However, the use of switched capacitors has yet more advantages: the implementation of a noninverting integrator in conventional technology requires an inverting integrator with a preceding or following voltage inverter. With the SC integrator, the polarity of the input voltage can be changed simply by connecting the capacitor, which has been charged up to the input voltage to be sampled, with *reversed* terminals to the operational amplifier input during the following charge transfer phase. Reversal of the terminals can be effected as shown in Fig. 13.50, using an additional changeover switch S_2 that switches simultaneously with S_1.

The charging and discharging of the capacitor C_S does not occur instantaneously, but exponentially due to the unavoidable resistances in the switches. Instantaneous charging and discharging would also be completely undesirable, neither the input voltage source nor the operational amplifier could supply the required currents. On the other hand, these parasitic resistances also determine the maximum switching frequency, as a complete charge–discharge cycle is impossible at too high a switching frequency.

13.12.3
First-Order SC Filter

The two basic SC integrators circuits can be extended by a feedback resistor in parallel to C to produce a first-order lowpass filter similar to that shown in Fig. 13.17. However, a different basic structure is generally selected for the monolithic version. This consists of an

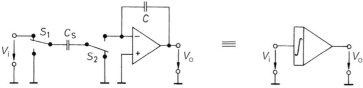

Fig. 13.50. Noninverting integrator in SC technology and its circuit symbol

$$V_o = +f_S \frac{C_S}{C} \int V_i dt = \frac{1}{\tau} \int V_i dt \qquad \frac{\underline{V}_o}{\underline{V}_i} = \frac{f_S}{s} \cdot \frac{C_S}{C} = \frac{1}{\tau \cdot s}$$

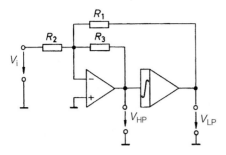

$$\tau = \frac{C}{C_S f_S} = \frac{\eta}{2\pi f_S}$$

$$\frac{V_{LP}}{V_i} = \frac{-R_1/R_2}{1 + \dfrac{\tau \omega_c R_1}{R_3} \cdot s_n}$$

$$\frac{V_{HP}}{V_i} = \frac{-s_n R_3/R_1}{\dfrac{R_3}{\tau \omega_c R_1} + s_n}$$

Fig. 13.51. First-order highpass and lowpass filter

SC-type integrator preceded by an additional summing amplifier, with three supplementary resistors then connected as shown in Fig. 13.51. This arrangement simultaneously provides a highpass and a lowpass filter.

To determine the component values, we simply select $f_S/f_c = \eta$. From this follows $\tau \omega_c = 1$. Coefficient matching yields the component values:

Lowpass filter: given: R_1	Highpass filter: given: R_1
$R_3 = R_1/a_1$	$R_3 = R_1 a_1$
$R_2 = -R_1/A_0$	$R_2 = -R_3/A_\infty$

In the case of first-order filters, for which a, 1 in accordance with Fig. 13.14, we therefore have $R_3 = R_1$. The gains of the lowpass and highpass filters are therefore equal, and thus the circuit represents complementary highpass and lowpass filters.

13.12.4
Second-Order SC Filters

Second-order SC filters are mainly designed as "biquad" structures, as shown in Fig. 13.46. As noninverting integrators are once again employed, we obtain the same structure and identical transfer functions (monolithic IC universal filters always contain this biquad structure). Unlike in the case of the RC filter, the integration time constant τ as formulated in (13.42) is determined by the switching frequency f_S selected.

To determine the transfer function, we derive the following relations from the circuit in Fig. 13.52:

$$V_{HP} = -\frac{R_3}{R_2} V_i - \frac{R_3}{R_4} V_{BP} - \frac{R_3}{R_1} V_{LP}$$

$$V_{BP} = \frac{1}{\tau s} V_{HP} \qquad V_{TP} = \frac{1}{\tau s} V_{BP}$$

From these, we can calculate the specified transfer function for the individual filters. If the switching frequency is again made equal to the ηth multiple of the cutoff frequency (or resonant frequency), we obtain $\tau \omega_c = 1$, and the following design equations:

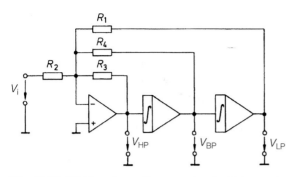

Integration constant

$$\tau = \frac{C}{C_S F_S} = \frac{\eta}{2\pi f_S}$$

Fig. 13.52. SC biquad to realize second-order highpass, lowpass, and bandpass filters

$$\frac{V_{LP}}{V_i} = \frac{-R_1/R_2}{1 + \dfrac{R_1 \tau \omega_c}{R_4} s_n + \dfrac{R_1 \tau^2 \omega_c^2}{R_3} s_n^2}$$

(Lowpass)

$$\frac{V_{HP}}{V_i} = \frac{-s_n^2 R_3/R_2}{\dfrac{R_3}{R_1 \tau^2 \omega_c^2} + \dfrac{R_3}{R_4 \tau \omega_c} s_n + s_n^2}$$

(Highpass)

$$\frac{V_{BP}}{V_i} = \frac{-s_n \tau \omega_r R_1/R_2}{1 + \dfrac{R_1 \tau \omega_r}{R_4} s_n + \dfrac{R_1 \tau^2 \omega_r^2}{R_3} s_n^2}$$

(Bandpass)

Lowpass filter:	Highpass filter:	Bandpass filter:
given: R_1	given: R_1	given: R_1
$R_3 = R_1/b_1$	$R_3 = R_1 b_1$	$R_3 = R_1$
$R_4 = R_1/a_1$	$R_4 = R_3/a_1$	$R_4 = R_1 Q$
$R_2 = -R_1/A_0$	$R_2 = -R_3/A_\infty$	$R_2 = -R_1 Q/A_r$

When the component values of one filter type have been defined, the other two do not, of course, necessarily have the same filter parameters. For the cutoff frequencies (or the resonant frequency), the following relation applies:

$$f_{c\,LP}/\sqrt{b_1} = f_{r\,BP} = f_{c\,HP}\sqrt{b_1}$$

As $b_1 = 1$ for second-order filters, the three frequencies coincide. In this case, the gains are given by

$$A_0 = A_r/Q = A_\infty$$

As a design example, we will calculate the component values of a second-order lowpass filter with a cutoff frequency $f_c = 1\,\text{kHz}$, a gain in the pass band of $A_0 = -1$, and a Butterworth characteristic. From Fig. 13.14, $a_1 = 1.4142$ and $b_1 = 1$. We select $R_2 = 10\,\text{k}\Omega$ and thus obtain $R_1 = R_3 = 10\,\text{k}\Omega$ and $R_4 = 7.15\,\text{k}\Omega$. For $\eta = 100$, the switching frequency must be $f_S = 100\,\text{kHz}$. For these values, we additionally obtain a highpass filter with $A_\infty = -1$ and a Butterworth characteristic, as well as a bandpass filter with $A_r = -0.707$ and $Q = 0.707$.

Higher-order SC filters can be produced by cascading. The coefficients of the filter sections must then be selected from Fig. 13.14.

13.12.5
Implementation of SC Filters with ICs

SC filters are of course implemented not with discrete components but using integrated circuits that contain capacitors and operational amplifiers in addition to the switches. This not only is simpler for the user but also offers significant advantages, as will be shown below.

SC filter ICs employ the two-switch arrangement illustrated in Fig. 13.50, because this arrangement compensates for the effect of stray capacitances. The changeover switches are realized in the form of a transmission gate and are driven by an internal clock generator that provides nonoverlapping timing signals. This insures that no charge is lost during switching.

As we can see, the capacitance ratio C/C_s together with the switching frequency f_s determine the integration time constant. The basic advantage of an IC implementation is that capacitance ratios with 0.1% tolerance can be produced. The use of monolithic SC filters therefore provides well-reproducible accuracies. In addition, the time constant is temperature-invariant, as the two capacitors exhibit identical temperature dependence if both are integrated on the same chip. Reproducible time constants, which are otherwise difficult and costly to achieve in IC technology, can be easily provided using SC devices. To achieve this, the only ratio of the two capacitances must be appropriately selected.

13.12.6
General Considerations for Using SC Filters

Despite the clearly superior characteristics of modern SC circuit design, the use of these components is subject to certain limitations, as these are actually sampling systems. Every time Shannon's sampling theorem is violated, one has to contend with unwanted mixing products in the base frequency band (aliasing). Consequently, the input system must not contain any frequency components above half the switching frequency f_S. In order to insure this, analog filtering is generally required at the input, which must introduce sufficient attenuation (some $70\ldots90\,\text{dB}$) at $\frac{1}{2}f_S$. As the typical sampling frequency of SC filter ICs is approximately $50\ldots100$ times the cutoff frequency, a second-order analog filter that acts as an anti-aliasing filter is normally adequate for this purpose.

The output signal of an SC filter always has a staircase waveform, as the output voltage only changes at the switching instant. It therefore contains spectral components associated with the switching frequency. Consequently, an analog smoothing filter must also be provided at the output at high requirements.

13.12.7
A Survey of Available Types

The SC filters that are available nowadays mainly contain complete functional blocks comprising SC integrators, summing circuits, and the associated (controllable) oscillators for clock generation. Their arrangement on the chip is either permanently fixed by masks (filters with fixed characteristics) or the components can be combined by the user as required (universal filters with variable characteristics). As the universal filters require external circuitry, the IC package involved must have more pins. The number of pins thus limits the complexity of the filter and, therefore, universal filter ICs are for low orders only.

Type	Manufac- turer	Filter type	Filter order	Cutoff frequency max	Dynamic range	Specialty
Universal filter						
LTC 1060	Lin. Tech.	2 Biquads	2 × 2	15 kHz	70 dB	
LTC 1064	Lin. Tech.	4 Biquads	4 × 2	100 kHz	80 dB	
LTC 1067	Lin. Tech.	2 Biquads	2 × 2	17 kHz	80 dB	rail to rail
LTC 1069	Lin. Tech.	4 Biquads	4 × 2	25 kHz	80 dB	
LTC 1264	Lin. Tech.	4 Biquads	4 × 2	200 kHz	80 dB	
MAX 262	Maxim	2 Biquads	2 × 2	75 kHz	80 dB	μC-prog.
MAX 266	Maxim	2 Biquads	2 × 2	100 kHz	80 dB	pin-prog.
MAX 7490	Maxim	2 Biquads	2 × 2	40 kHz	80 dB	
LMF 100	National	2 Biquads	2 × 2	100 kHz	80 dB	
Lowpass filter						
LTC 1069-1	Lin. Tech.	Cauer	8	140 kHz	70 dB	8 pin case
LTC 1069-7	Lin. Tech.	Lin. Phas.	8	140 kHz	70 dB	8 pin case
LTC 1164-5	Lin. Tech.	Butterworth	8	20 kHz	75 dB	Bessel, switch
LTC 1164-6	Lin. Tech.	Lin. Phas.	8	20 kHz	75 dB	Cauer, switch
LTC 1264-7	Lin. Tech.	Lin. Phas.	8	200 kHz	70 dB	
LTC 1569-7	Lin. Tech.	Lin. Phas.	10	300 kHz	80 dB	8 pin case
MAX 291	Maxim	Butterworth	8	25 kHz	70 dB	8 pin case
MAX 292	Maxim	Bessel	8	25 kHz	70 dB	8 pin case
MAX 293	Maxim	Cauer	8	25 kHz	70 dB	8 pin case
MAX 7418	Maxim	Cauer	5	30 kHz	80 dB	8 pin case
MAX 7419	Maxim	Bessel	5	30 kHz	80 dB	8 pin case
MAX 7420	Maxim	Butterworth	5	30 kHz	80 dB	8 pin case
Bandpass filter						
MAX 268	Maxim		2 × 2	75 kHz	80 dB	pin-prog.

Fig. 13.53. Examples of monolithic SC filters

However, dual types in a single package are generally available, which can then be simply cascaded, thus allowing fourth-order filters to be realized on a single chip.

Switching by means of the clock frequency produces background noise in the filters, which limits the signal-to-noise ratio to about 70 . . . 90 dB (Fig. 13.53). This constitutes a disadvantage compared with "continuous" *RC* filters.

Most manufacturers provide filter design programs free of charge which can be downloaded from their homepage. This is the most convenient method for dimensioning freely reconfigurable filters.

Chapter 14:
Signal Generators

In this chapter we shall describe circuits that generate sinusoidal signals. In the case of LC oscillators, the frequency is determined by a tuned circuit, in the case of crystal-controlled oscillators a piezoelectric crystal is used, and with the Wien and differential-equation oscillators, RC networks are the frequency-determining components. The function generators primarily produce a triangular signal, which can be converted into sinusoidal form using a suitable function network.

14.1
LC Oscillators

The simplest method of generating a sine wave is to use an amplifier to eliminate the damping of an LC resonant circuit. In the following section, we deal with some of the basic aspects of this method.

14.1.1
Condition for Oscillation

The principle of an oscillator circuit is shown in Fig. 14.1. The amplifier multiplies the input voltage by the gain $\underline{A}$, and thereby causes a parasitic phase shift a between $\underline{V}_2$ and $\underline{V}_1$. The load resistance R_L and a frequency-dependent feedback network – for example, a resonant circuit – are connected to the amplifier output. The voltage feedback is therefore $\underline{V}_3 = \underline{k}\,\underline{V}_2$ and the phase shift between $\underline{V}_3$ and $\underline{V}_2$ is denoted by β.

To establish whether the circuit can produce oscillations, the feedback loop is opened. An additional resistor R_i is introduced at the output of the feedback network, representing the input resistance of the amplifier. An alternating voltage $\underline{V}_1$ is applied to the amplifier and $\underline{V}_3$ is measured. The circuit is capable of producing oscillations if the output voltage is the same as the input voltage. Hence, the necessary condition for oscillation is:

$$\underline{V}_1 = \underline{V}_3 = \underline{k}\,\underline{A}\,\underline{V}_1$$

The loop gain must therefore be

$$\underline{g} = \underline{k}\,\underline{A} = 1 \tag{14.1}$$

from which two conditions can be deduced; that is,

$$|\underline{g}| = |\underline{k}| \cdot |\underline{A}| = 1 \tag{14.2}$$

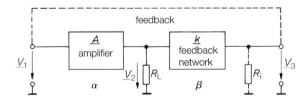

Fig. 14.1. Basic arrangement of an oscillator

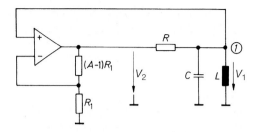

Fig. 14.2. Principle of an LC oscillator

and

$$\alpha + \beta = 0, \ 2\pi, \dots \tag{14.3}$$

Equation (14.2) is the *amplitude condition*, which states that a circuit can oscillate only if the amplifier eliminates the attenuation due to the feedback network. The *phase condition* of (14.3) states that an oscillation can arise only if the output voltage is in phase with the input voltage. Details of the oscillation – for example, the frequency and waveform – can only be obtained if we have additional information on the feedback network. To this end, let us consider the LC oscillator in Fig. 14.2 as an example.

The noninverting amplifier multiplies the voltage $V_1(t)$ by the gain A. As the output resistance of the amplifier is low, the resonant circuit is damped by the parallel resistor R. To calculate the feedback voltage, we apply KCL to node 1 and obtain

$$\frac{V_2 - V_1}{R} - C\dot{V}_1 - \frac{1}{L}\int V_1 dt = 0$$

As $V_2 = AV_1$, it follows:

$$\ddot{V}_1 + \frac{1 - A}{RC}\dot{V}_1 + \frac{1}{LC}V_1 = 0 \tag{14.4}$$

This is the differential equation of a damped oscillation. To abbreviate, we define

$$\gamma = \frac{1 - A}{2RC} \quad \text{and} \quad \omega_0^2 = \frac{1}{LC}$$

and therefore

$$\ddot{V}_1 + 2\gamma \dot{V}_1 + \omega_0^2 V_1 = 0$$

the solution of which is given by

$$V_1(t) = V_0 \cdot e^{-\gamma t} \sin(\sqrt{\omega_0^2 - \gamma^2}\, t) \tag{14.5}$$

One must differentiate between three cases:

1) $\gamma > 0$; that is, $A < 1$.
   The amplitude of the AC output voltage decreases exponentially; the oscillation is damped.
2) $\gamma = 0$; that is, $A = 1$.
   The result is a sinusoidal oscillation with the frequency $\omega_0 = \frac{1}{\sqrt{LC}}$ and with constant amplitude.
3) $\gamma < 0$; that is, $A > 1$.
   The amplitude of the AC output voltage rises exponentially.

With (14.2) we have the necessary condition for an oscillation. This can now be described in more detail: For $A = 1$, a sinusoidal output voltage of constant amplitude and the frequency

$$\omega = \omega_0 = \frac{1}{\sqrt{LC}}$$

is obtained. With reduced feedback the amplitude falls exponentially, while with increased feedback the amplitude rises exponentially. To insure that the oscillation builds up after the supply has been switched on, the gain A must initially be larger than unity. The amplitude then rises exponentially until the amplifier begins to saturate. Because of the saturation, A decreases until $|\underline{g}| = |\underline{k}| \cdot \underline{A}|$ reaches the value 1; the output, however, is then no longer sinusoidal. If a sinusoidal output is required, an additional gain control circuit must insure that $A = 1$ before the amplifier saturates. For the high-frequency range, resonant circuits with high Q-factors are usually simple to implement. The voltage of the resonant circuit is still sinusoidal, even if the amplifier saturates. For this frequency range, additional amplitude control is usually not required, and the voltage across the resonant circuit is then taken as the output voltage.

14.1.2
Meissner Oscillator

The feature of a Meissner circuit is that the feedback is provided by a transformer. A capacitor C, together with the transformer primary winding, forms the frequency-determining resonant circuit. Three Meissner oscillators, each in common-emitter connection, are shown in Figs. 14.3–14.5. At the resonant frequency,

$$\omega_0 = \frac{1}{\sqrt{LC}}$$

the amplified input voltage appears at the collector with maximum amplitude and with a phase shift of $180°$. Part of this alternating voltage is fed back via the secondary winding. To fulfill the phase condition, the transformer must effect a further phase inversion of $180°$. This is achieved by AC-grounding the secondary winding at the end that has the same voltage polarity as the collector end of the primary winding. The dots on the two

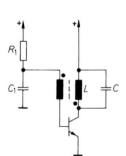

Fig. 14.3. Biasing by a constant base current

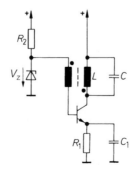

Fig. 14.4. Biasing by current feedback

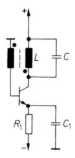

Fig. 14.5. Current feedback for a negative supply voltage

windings indicate which winding ends have the same polarity. The turns ratio is selected such that the magnitude of the loop gain $\underline{k}\,\underline{A}$ at the resonant frequency is always larger than unity. Oscillation then begins when the supply is switched on, its amplitude rising exponentially until the transistor saturates. Saturation reduces the mean value of the gain until $|\underline{k}\,\underline{A}| = 1$ and the amplitude of the oscillation remains constant. Two saturation effects can be distinguished, that at the input and that at the output. The output saturation arises when the collector–base junction is forward biased. This is the case for the circuits shown in Figs. 14.3 and 14.5 when the collector potential goes negative. The maximum amplitude of the oscillation is therefore $\hat{V}_C = V^+$. The maxima of the collector potential are then $\hat{V}_{CE\,max} = 2V^+$. This affects the choice of the transistor. For the circuit in Fig. 14.4, the maximum amplitude is smaller than V^+, the reduction being due to the Zener voltage.

With heavy feedback, input saturation can also occur. Large input amplitudes arise, which are rectified at the emitter–base junction. The capacitor C_1 is therefore charged, and the transistor conducts only during the positive peaks of the AC input voltage.

In the circuit of Fig. 14.3, a few oscillations may be sufficient to charge the capacitor C_1 to such a high negative voltage that the oscillation stops altogether. It restarts only after the base potential has risen to +0.6 V with the relatively large time constant $R_1 C_1$. In this case, a sawtooth voltage appears across C_1. This arrangement has therefore often been used as a sawtooth generator, such a circuit being known as a *blocking oscillator*.

To prevent the circuit from operating in the blocking oscillator mode, the input saturation must be reduced by choosing fewer turns at the base side. In addition, the resistance of the base biasing circuit should be kept as low as possible [14.1]. This is difficult to achieve for the circuit shown in Fig. 14.3, since the base current would then be excessively high. Biasing by series feedback, as shown in Figs. 14.4 and 14.5, is therefore preferable.

14.1.3
Hartley Oscillator

The Hartley oscillator resembles a Meissner oscillator. The only difference is that the transformer is replaced by a tapped winding (an auto-transformer). The inductance of this winding, together with the parallel-connected capacitor, determines the resonant frequency.

A Hartley oscillator in common-emitter connection is shown in Fig. 14.6. An alternating voltage is applied to the base via capacitor C_2; it is in phase opposition to the collector voltage, so that positive feedback occurs. The amplitude of the feedback voltage can be

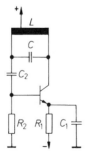

Fig. 14.6. Hartley oscillator in common-emitter connection

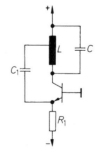

Fig. 14.7. Hartley oscillator in common-base connection

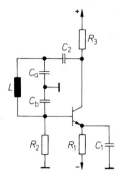

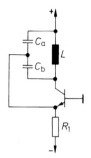

Fig. 14.8. Colpitts oscillator in common-emitter connection

Fig. 14.9. Colpitts oscillator in common-base connection

adjusted to the required value by appropriate positioning of the tap. As with the Meissner oscillator in Fig. 14.5, the collector quiescent current is determined by the series feedback resistor R_1.

For the Hartley oscillator in Fig. 14.7, the transistor is operated in common-base connection. Therefore, a voltage must be taken from the inductor L by means of capacitor C_1, in phase with the collector voltage.

14.1.4
Colpitts Oscillator

A characteristic of the Colpitts circuit is the capacitive voltage divider, which determines the fraction of the output voltage that is fed back. The series connection of the capacitors acts as the oscillator capacitance; that is,

$$C = \frac{C_a C_b}{C_a + C_b}$$

The common-emitter circuit of Fig. 14.8 corresponds to the circuit in Fig. 14.6, but requires an additional collector resistor R_3 for applying the positive supply voltage.

The common-base connection is again much simpler, as can be seen in Fig. 14.9. It corresponds to the Hartley oscillator of Fig. 14.7.

14.1.5
Emitter-Coupled LC Oscillator

A simple way of realizing an oscillator is to use a differential amplifier, as shown in Fig. 14.10. As the base potential of T_1 is in phase with the collector potential of T_2, positive feedback can be attained by directly connecting the two terminals. The loop gain

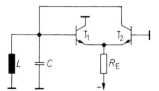

Fig. 14.10. Emitter-coupled oscillator

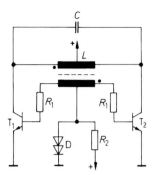

Fig. 14.11. Push–pull oscillator with inductive feedback

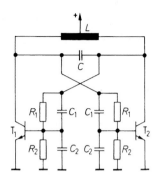

Fig. 14.12. Push–pull oscillator with capacitive feedback

is proportional to the transconductance of the transistors. It can be adjusted over a wide range by varying the emitter current. As the transistors are operated at $V_{CB} = 0$, the amplitude of the output voltage is limited to about 0.5 V.

The amplifier of the emitter-coupled oscillator, together with an output stage and amplitude control, is available as an IC (MC 1648 from On Semiconductor). It is suitable for frequencies of up to about 1 GHz.

14.1.6
Push–Pull Oscillators

Push–pull circuits are used in power amplifiers in order to achieve higher output powers and better efficiency. For the same reason, these circuits can also be employed for the design of oscillators. One such design, consisting basically of two Meissner oscillators in which the transistors T_1 and T_2 are alternately conducting, is shown in Fig. 14.11.

As the base potential of one transistor is in phase with the collector potential of the other, the secondary winding normally required for phase inversion can be omitted. This version is shown in Fig. 14.12. The positive feedback is provided by the capacitive voltage dividers C_1 and C_2. The parallel resistive voltage dividers provide the bias.

In addition to providing greater output power, both circuits also generate fewer harmonics than the single-ended oscillators.

The most suitable amplifier for an oscillator with a parallel resonant circuit is a voltage controlled current source. For this purpose we use a CC-Opamp in Fig. 14.13. It does not load the resonant circuit because it offers both a high input and output resistance. The losses in the series resistance of the inductor R are compensated by an appropriate transconductance of the amplifier with R_1. In order to reduce distortions by hard clipping

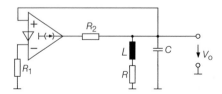

Fig. 14.13. Push–pull oscillator using a CC-operational amplifier

of the amplifier output R_2 is added. The internal structure of a CC-Opamp in Fig. 5.80 shows a push-pull output stage.

14.2
Crystal Oscillators

The frequency of the LC oscillators described is not sufficiently constant for many applications, as it depends on the temperature coefficients of the capacitance and inductance of the resonant circuit. Considerably more stable frequencies can be achieved by using quartz crystals. Such a crystal can be excited by electric fields to vibrate mechanically and, when provided with electrodes, behaves electrically like a resonant circuit that has a high Q-factor. The temperature coefficient of the resonant frequency is very small. The frequency stability that can be attained by a crystal oscillator is of the order of

$$\frac{\Delta f}{f} = 10^{-6} \ldots 10^{-10}$$

14.2.1
Electrical Characteristics of a Quartz Crystal

The electrical behavior of a quartz crystal can be described by the equivalent circuit in Fig. 14.14. The two parameters C and L are well defined by the mechanical properties of the crystal. The resistance R is small and characterizes the damping. C_0 represents the value of the capacitance formed by the electrodes and leads. Typical values for a 4 MHz crystal are

$$L = 100\,\text{mH}, \qquad R = 100\,\Omega$$
$$C = 0.015\,\text{pF}, \qquad C_0 = 5\,\text{pF}$$

giving a Q-factor of

$$Q = \frac{1}{R}\sqrt{\frac{L}{C}} = 26000$$

To calculate the resonant frequency, we initially determine the impedance of the quartz crystal. From Fig. 14.14, and neglecting R,

$$\underline{Z}_q = \frac{1 + s^2 LC}{s(C_0 + C) + s^3 LCC_0} \tag{14.6}$$

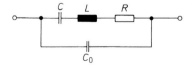

Fig. 14.14. Equivalent circuit of a quartz crystal

Fig. 14.15. Tuning the series resonant frequency

It can be seen that there is a frequency for which $\underline{Z}_q = 0$ and another for which $\underline{Z}_q = \infty$. The quartz crystal therefore has a series and a parallel resonance. To calculate the series resonant frequency f_s, the numerator of (14.6) is set to zero, and thus

$$f_S = \frac{1}{2\pi\sqrt{LC}} \tag{14.7}$$

The parallel resonant frequency is calculated by setting the denominator to zero:

$$f_P = \frac{1}{2\pi\sqrt{LC}}\sqrt{1 + \frac{C}{C_0}} \tag{14.8}$$

As can be seen, the series resonant frequency is dependent only on the well-defined product LC, whereas the parallel resonant frequency is influenced by the electrode capacitance C_0, which is far more susceptible to variations.

The frequency of a quartz oscillator must often be adjustable within a small range. This can be achieved by simply connecting a capacitor C_S in series with the quartz crystal, as shown in Fig. 14.15; C_S must be large compared to C.

To calculate the shift in the resonant frequency, we determine the impedance of the series connection. Using Eq. (14.6), we obtain

$$\underline{Z}'_q = \frac{C + C_0 + C_S + s^2 LC(C_0 + C_S)}{sC_S(C_0 + C) + s^3 LCC_0C_S} \tag{14.9}$$

Setting the numerator to zero gives the new series resonant frequency:

$$f'_S = \frac{1}{2\pi\sqrt{LC}}\sqrt{1 + \frac{C}{C_0 + C_S}} = f_S\sqrt{1 + \frac{C}{C_0 + C_S}} \tag{14.10}$$

By expanding this into a power series, we arrive at the approximation

$$f'_S = f_S\left[1 + \frac{C}{2(C_0 + C_S)}\right]$$

The relative shift in frequency is therefore

$$\frac{\Delta f}{f} = \frac{C}{2(C_0 + C_S)}$$

The parallel resonant frequency is not changed by C_S, as the poles of (14.9) are independent of C_S. A comparison of (14.10) and (14.8) shows that, for $C_S \to \infty$, the series resonant frequency cannot be raised to a value higher than that of the parallel resonant frequency.

14.2.2
Fundamental Frequency Oscillators

In the Pierce oscillator shown in Fig. 14.16, the crystal in conjunction with capacitors C_S and C_1 forms a series resonant circuit with a series capacitance of

$$\frac{1}{C_{S\,ges}} = \frac{1}{C_S} + \frac{1}{C_1}$$

The resonant circuit is excited via the collector. Assuming that the current in the oscillatory circuit is large compared to the excitation current, antiphase signals will be produced at C_1 and C_S, resulting in positive feedback.

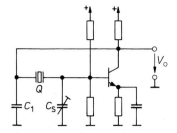

Fig. 14.16. Pierce oscillator with an amplifier in common-emitter connection

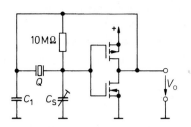

Fig. 14.17. Pierce oscillator with a CMOS inverter as an amplifier

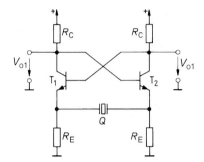

Fig. 14.18. Emitter-coupled crystal oscillator

Nowadays, the amplifiers are usually CMOS inverters. The resulting circuit is shown in Fig. 14.17. Not only does it require fewer components, but it also imposes little damping on the crystal, as its input resistance is high. The resistor fixes the operating point at $V_i = V_o \approx \frac{1}{2} V_b$. Its resistance may be very high, as virtually no input current flows.

The crystal-controlled oscillator shown in Fig. 14.18 operates in the same way as the emitter-coupled multi-vibrator shown in Fig. 6.21. The amount of positive feedback can be adjusted via the transconductance of the transistors, using the emitter resistors. It is selected such that the circuit begins to oscillate reliably, but is not excessively overdriven. The output voltage difference, and hence the current flowing through the crystal, are then virtually sinusoidal. An automatic gain control device of this kind is incorporated, for example, in type MC 12061.

A precision crystal oscillator that allows grounded crystals to be employed is shown in Fig. 14.19. In order not to impair the Q-factor of the crystal, the circuit must be driven at the

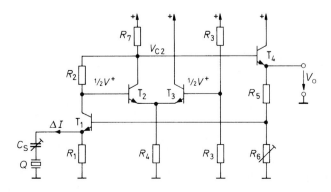

Fig. 14.19. Precision crystal oscillator

Type	Manufacturer	Output	Maximum frequency
74 LS 320	Texas Instruments	TTL	20 MHz
74 LS 624	Texas Instruments	TTL	20 MHz

Fig. 14.20. IC crystal oscillators for external quartz crystals

lowest possible impedance (series resonance). Emitter follower T_1 is used for this purpose. The current ΔI flowing through the crystal is translated into a voltage $\Delta V_{C2} = \Delta I R_2$ in transistor T_2, which is configured as a current–voltage converter. Positive feedback is provided via the emitter follower T_4 and the base of T_1. The reduced transconductance of T_1 and hence the loop gain of the circuit is at its maximum at the series resonant frequency of the crystal. Attenuator R_5, R_6 is adjusted such that the AC voltage across the crystal is only a few tens of millivolts. The power dissipation in the crystal is then so small that the frequency stability is unimpaired. It is preferable to use an electrically controllable attenuator – for example, a transconductance multiplier – which is set to the correct value using an amplitude control circuit. This also insures reliable start-up of the oscillator, and the output voltage has a good sinusoidal waveform. Complete crystal-oscillators are offered in a wide variety in the frequency range from 32 kHz to 50 MHz. Therefore the chips in Fig. 14.20 are seldom used today.

14.2.3
Harmonic Oscillators

Crystals for frequencies above 30 MHz are difficult to manufacture. To obtain high frequencies of this kind with crystal accuracy, one can either stabilize an LC oscillator via a PLL (see Sect. 22.4.5) using a low-frequency crystal, or excite a crystal at a harmonic of its characteristic frequency.

If we examine the frequency response of the crystal impedance shown in Fig. 14.21, we can see that it also possesses resonance points at odd-numbered harmonics. However, the circuits considered so far are unsuitable for operating a crystal at a harmonic. For such a circuit, we require an amplifier that provides the maximum gain close to the desired frequency. This can be achieved by using an additional LC resonant circuit.

If the positive feedback of the Hartley oscillator shown in Fig. 14.7 is obtained via a crystal, we obtain the circuit shown in Fig. 14.22. The LC resonant circuit is tuned to the required harmonic. The gain is then a maximum for this frequency, and the crystal will

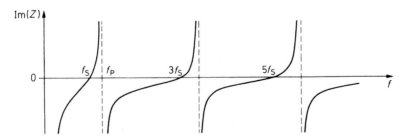

Fig. 14.21. Typical impedance-frequency response of a quartz crystal (imaginary part)

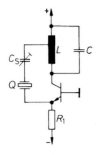

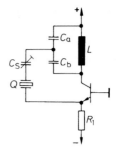

Fig. 14.22. Hartley oscillator with a crystal   **Fig. 14.23.** Colpitts oscillator with a crystal

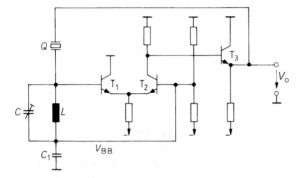

Fig. 14.24. Emitter-coupled crystal-controlled oscillator. Frequencies of up to 100 MHz or more can be achieved using ECL line receiver 10.116

tend to be excited at the corresponding harmonic. The suitably modified Colpitts oscillator of Fig. 14.9 is shown in Fig. 14.23.

A harmonic oscillator can also be implemented using the emitter-coupled oscillator in Fig. 14.10. For this purpose, a crystal is inserted in the positive feedback loop, as shown in Fig. 14.24. This arrangement provides positive feedback with the required crystal harmonic at the resonant frequency of the LC resonant circuit. The simplest way of implementing the high-frequency amplifier required is to use an ECL gate. In this case, a line receiver provides a particularly effective solution, as the reference potential V_{BB} is accessible. If the resonant circuit is connected as shown in Fig. 14.24, the amplifier will be optimally biased. Capacitor C_1 serves merely to short-circuit V_{BB} at high frequencies. The resultant output voltage is very nearly sinusoidal. If a square-wave ECL signal is required, it is merely necessary to connect another line receiver at the output.

14.3
Wien–Robinson Oscillator

In the low-frequency range, LC oscillators are less satisfactory because the inductances and capacitances required become unpleasantly large. Consequently, oscillators are preferred in which RC networks are used to determine the frequency.

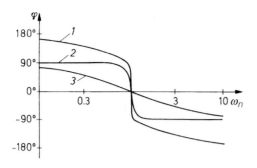

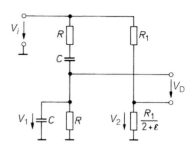

Fig. 14.25. Frequency response of the phase shift
Curve *1*: Wien–Robinson bridge for $\varepsilon = 0.01$
Curve *2*: Resonant circuit for $Q = 10$
Curve *3*: Passive bandpass filter $Q = \frac{1}{3}$

Fig. 14.26. Detuned Wien–Robinson bridge

In principle, an *RC* oscillator can be realized by replacing the resonant circuit in Fig. 14.2 by a passive *RC* bandpass filter. However, the maximum attainable Q-factor would then be limited to 0.5, as shown in Sect. 13.1. Consequently, the frequency stability of the resulting sine wave would be poor, as can be seen from the frequency response of the phase shift shown in Fig. 14.25. The phase shift of a passive lowpass filter with $Q = \frac{1}{3}$ is 27° at half the resonant frequency. In the event of the amplifier causing a phase shift of $-27°$, the circuit would oscillate at half the resonant frequency, as the total phase shift would then be zero. For good frequency stability, a feedback network is thus required for which the gradient of the phase-frequency response at zero phase is as steep as possible. High-Q resonant circuits and the Wien–Robinson bridge have this characteristic. However, the output voltage of the Wien–Robinson bridge is zero at the resonant frequency and the bridge is therefore suitable as a feedback network only under certain conditions. When employed in oscillators, it must be slightly detuned, as shown in Fig. 14.26, where ε is positive and small compared to unity.

The frequency dependence of the phase shift of a detuned Wien–Robinson bridge can be determined qualitatively: At high and low frequencies, $\underline{V}_1 = 0$, and thus $\underline{V}_D \approx -\frac{1}{3}\underline{V}_i$ and the phase shift is $\pm 180°$. At the resonant frequency, $\underline{V}_1 = \frac{1}{3}\underline{V}_i$ and

$$\underline{V}_D = V_1 - V_2 = \left(\frac{1}{3} - \frac{1}{3+\varepsilon}\right)\underline{V}_i \approx \frac{\varepsilon}{9}\underline{V}_i$$

that is, $\underline{V}_D$ is in phase with $\underline{V}_i$. To determine quantitatively the shape of curve 1 in Fig. 14.25, we first calculate the transfer function:

$$\frac{\underline{V}_D}{\underline{V}_i} = -\frac{1}{3+\varepsilon} \cdot \frac{1 - \varepsilon s_n + s_n^2}{1 + 3s_n + s_n^2}$$

and get the frequency response of the phase

$$\varphi = \arctan \frac{(3+\varepsilon)\omega_n(1 - \omega_n^2)}{3\varepsilon\omega_n^2 - (1 - \omega_n^2)^2}$$

which is shown in Fig. 14.25 for $\varepsilon = 0.01$. It can be seen that the phase shift of the detuned Wien–Robinson bridge increases to $\pm 90°$ within a very narrow frequency range. It is narrower the smaller is the value of ε. As far as this aspect is concerned, the Wien–Robinson bridge is comparable to high-quality resonant circuits, and an additional advantage is that

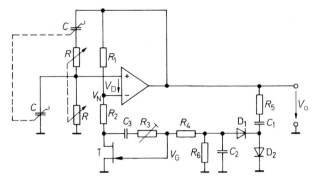

Fig. 14.27. Simple Wien–Robinson oscillator

Resonant frequency: $f_r = 1/2\pi RC$

the phase shift is not limited to $\pm 90°$, but to $\pm 180°$. Harmonics are thereby strongly damped. One disadvantage of the Wien–Robinson bridge is that the attenuation at the resonant frequency becomes higher the smaller is the value of ε. In general, the attenuation at resonant frequency is

$$\frac{\hat{V}_D}{\hat{V}_i} = k \approx \frac{\varepsilon}{9}$$

and, in our example, $k \approx \frac{1}{900}$. To fulfill the amplitude condition of an oscillator, the amplifier must compensate for the attenuation. Such an oscillator circuit is represented in Fig. 14.27.

If the amplifier has open-loop gain A_D, the detuning factor ε must have the value $\varepsilon = 9k = 9/A_D$ to fulfill the amplitude condition $kA_D = 1$. If ε is slightly larger, the oscillation amplitude increases until the amplifier begins to saturate. If ε is smaller, there will be no oscillation. However, it is impossible to adjust the resistors R_1 and $R_1/(2 + \varepsilon)$ with the required precision. Therefore, one of the two resistances must be controlled automatically, depending on the output amplitude. This is the purpose of the field effect transistor T in Fig. 14.27. As shown in Sect. 3.1.3, the channel resistance R_{DS} behaves like a controllable ohmic resistor if V_{DS} remains sufficiently small. To insure that V_{DS} does not become too large, only part of V_N is applied to the FET and the rest appears at R_2. The sum of R_2 and R_{DS} must have the value $R_1/(2 + \varepsilon T)$. The smallest possible value of R_{DS} is $R_{DS\,on}$, and hence R_2 must be smaller than

$$R_2 < \frac{1}{2}R_1 - R_{DS\,on}$$

When the supply voltage is switched on, V_G is initially zero and therefore $R_{DS} = R_{DS\,on}$. If the above design condition is fulfilled, the sum of R_2 and R_{DS} is then smaller than $\frac{1}{2}R_1$. At the resonant frequency of the Wien–Robinson bridge, there is therefore a relatively large signal V_D. As a consequence, oscillation begins and the amplitude increases. The output voltage is rectified by the voltage doubler D_1, D_2. The gate potential thereby becomes negative and R_{DS} increases. The amplitude of the output voltage rises until

$$R_{DS} + R_2 = \frac{R_1}{2 + \varepsilon} = \frac{R_1}{2 + \dfrac{9}{A_D}}$$

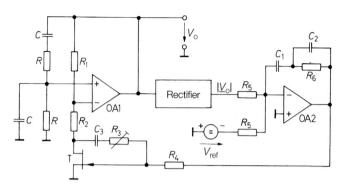

Fig. 14.28. Wien–Robinson oscillator with precise amplitude control

Amplitude: $\quad \hat{V}_o = \dfrac{\pi}{2} V_{\text{ref}}$

The distortion factor of the output voltage is mainly dependent on the linearity of the FET output characteristic, which can be greatly improved if part of the drain–source voltage is added to the gate potential, as shown in Fig. 3.13. This is the purpose of the two resistors R_3 and R_4. The capacitor C_3 insures that no direct current will flow into the N-input of the operational amplifier, as this would cause an output offset. In practice, $R_3 \approx R_4$ but, by fine adjustment of R_3, the distortion factor can be reduced to a minimum and values below 0.1% can be achieved.

If R is made adjustable, the frequency can be continuously controlled. Good matching of the two resistors is important, since the greater their mismatch, the more efficient the amplitude control must be. The maximum value of R should be low enough to insure that there is no noticeable voltage arising from the input bias current of the operational amplifier. On the other hand, R must not be too low, or the output will be overloaded. To adjust the frequency within a range of 1 : 10, fixed resistors with the value $R/10$ are connected in series with the potentiometers R. If, in addition, switches are used to select different values for the capacitances C, a range of output frequencies from 10 Hz to 1 MHz can be covered by this circuit. To insure that the amplitude control does not cause distortion even at the lowest frequency, the charge and discharge time constants, R_5C_1 and R_6C_2 respectively, must be larger than the longest oscillation period by a factor of at least 10.

The data of the field effect transistor T determine the output amplitude. The stability of the output voltage amplitude is not particularly good, since some deviation in amplitude is required to effect a noticeable change in the resistance of the FET. It can be improved by amplifying the gate voltage, and such a circuit is shown in Fig. 14.28.

The rectified AC output voltage is applied to OA2, which is configured as a modified PI-controller (Fig. 22.7). This adjusts the gate potential of the FET in such a way that the mean value of $|\underline{V}_o|$ equals V_{ref}. The controller time constant must be large compared to the oscillation period, or the gain will change even within a single cycle, resulting in considerable distortion. Therefore, a pure PI-controller cannot be used; it is better to have a modified type in which a capacitor is connected in parallel with R_6. The alternating voltage at R_6 is thereby short-circuited even at the lowest oscillator frequency. The proportional controller action comes into effect only below this frequency.

14.4
Differential-Equation Oscillators

Low-frequency oscillations can be generated by programming operational amplifiers to solve the differential equation of a sine wave. From Sect. 14.1.1,

$$\ddot{V}_o + 2\gamma \dot{V}_o + \omega_0^2 V_o = 0 \tag{14.11}$$

the solution of which is

$$V_o(t) = \hat{V}_o e^{-\gamma t} \sin(\sqrt{\omega_0^2 - \gamma^2}t) \tag{14.12}$$

Since operational amplifiers are better suited to integration than differentiation, the differential equation is rearranged by integrating it twice:

$$V_o + 2\gamma \int V_o dt + \omega_0^2 \iint V_o dt^2 = 0$$

This equation can be implemented using two integrators and an inverting amplifier, and several arrangements are available. One of these, which is particularly suitable for an oscillator, is shown in Fig. 14.29. For this circuit, the damping is given by $\gamma = -\alpha/20RC$ and the resonant frequency $f_0 = 1/2\pi RC$. Hence, from (14.12), the output voltage is:

$$V_o(t) = \hat{V}_o e^{\frac{\alpha}{20RC}t} \sin\left(\sqrt{1 - \frac{\alpha^2}{400}}\frac{t}{RC}\right) \tag{14.13}$$

It is evident that the damping of the oscillation is adjusted by α. At the right-hand stop of potentiometer P, $\alpha = 1$. At the left-hand stop, $\alpha = -1$; in the middle, $\alpha = 0$. The damping can thus be varied between positive and negative values. For $\alpha = 1$, the oscillation amplitude is increased, within 20 cycles, by the factor e; for $\alpha = -1$, it is decreased by the factor $1/e$. For $\alpha = 0$, the oscillation is undamped, although this is the case only for ideal conditions. In practice, for $\alpha = 0$, a slightly damped oscillation will occur, and to achieve constant amplitude, α must be given a small positive value. The adjustment is so critical that the amplitude can never be kept constant over a longer period, and one must therefore introduce automatic amplitude control. As for the Wien–Robinson oscillator shown in Fig. 14.28, the amplitude at the output can be measured by a rectifier and α controlled, depending on the difference between this amplitude and a reference voltage. As shown previously, the controller time constant must be large compared to the period of

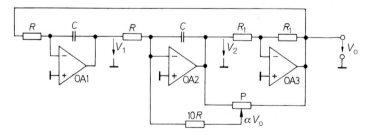

Fig. 14.29. Second-order differential equation for sine-wave generation. The circuit can also be seen as a biquad-filter with $Q = \infty$

Resonant frequency: $f_0 = 1/2\pi RC$

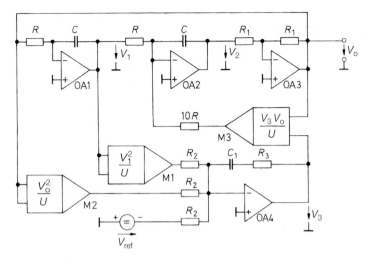

Fig. 14.30. Differential-equation oscillator and supplementary circuit for precise amplitude control. U is the compting unit of the multipliers for instance $U = 10\,\text{V}$

Frequency: $f_0 = 1/2\pi RC,$ *Amplitude:* $\hat{V}_o^2 = U V_{\text{ref}}$

oscillation to insure that amplitude control causes no distortion. This requirement becomes increasingly difficult to fulfill for frequencies below 10 Hz.

The difficulties arise from the fact that a whole cycle of the oscillation must be allowed to occur before the correct amplitude is known. Such problems can be eliminated if the amplitude is measured at every instant of the oscillation period. This is possible for the circuit in Fig. 14.29 where, in the case of an undamped oscillation,

$$V_o = \hat{V}_o \sin \omega_0 t \quad \text{and} \quad V_1 = -\frac{1}{\tau} \int V_o dt = \hat{V}_o \cos \omega_0 t$$

The amplitude can now be determined at any instant by calculation of the vector magnitude

$$V_o^2 + V_1^2 = \hat{V}_o^2 (\sin^2 \omega_0 t + \cos^2 \omega_0 t) = \hat{V}_o^2 \tag{14.14}$$

It is obvious that the expression $V_o^2 + V_1^2$ is dependent only on the amplitude of the output signal and not on its phase. A pure DC voltage is therefore obtained, which requires no filtering and which can be directly compared with the reference voltage.

A differential-equation oscillator whose amplitude is controlled in this way is shown in Fig. 14.30. Analog multipliers M_1 and M_2 form the squares of V_1 and V_o respectively. To these two portions, the reference voltage at the summation point of automatic gain control (AGC) amplifier OA4 is now added, giving an output voltage V_3 such that

$$\frac{V_1^2}{U R_2} + \frac{V_o^2}{U R_2} - \frac{V_{\text{ref}}}{R_2} = 0$$

Applying (14.14), this is the case for an amplitude of $\hat{V}_o^2 = U V_{\text{ref}}$. The time constant of the control amplifier is determined by RC network $R_3 C_1$. The design of the circuit is described in Chap. 22.

Voltage $V_o V_3 / U$ is present at the output of multiplier M_3. This voltage is applied to resistor $10R$ instead of potentiometer P in Fig. 14.29, making $\alpha = V_3 / U$. If the amplitude increases, $\hat{V}_o^2 > UV_{ref}$ and both V_3 and α become negative; that is, the oscillation is damped. If the amplitude decreases, V_3 becomes positive and the oscillation is exited.

Apart from providing a good method of controlling the amplitude, sine-wave generation by solving a second-order differential equation offers a further advantage in that it allows virtually ideal frequency modulation. If this is to be accomplished for LC oscillators, the value of L or C must be varied. However, this will alter the energy of the oscillator and hence the oscillation amplitude, and parametric effects arise. For the differential-equation method, however, the resonant frequency can be changed by varying the two resistors R without affecting the oscillator energy.

As each of the two resistors is connected to virtual ground, multipliers connected in front of them can be used to modulate the frequency. The multipliers then produce the output voltages:

$$V_o' = \frac{V_{St}}{U}V_o \quad \text{and} \quad V_1' = \frac{V_{St}}{U}V_1$$

This is equivalent to an increase in the resistance R by the factor U/V_{St}, so that the resonant frequency is given by

$$f_0 = \frac{1}{2\pi RC} \cdot \frac{V_{St}}{U}$$

that is, it is proportional to the control voltage V_1.

The frequency can also be controlled digitally by connecting D/A converters instead of analog multipliers in front of the integrators. This produces the same arrangement as for the digitally tunable filter in Fig. 13.46. In this way, frequency bands of 1 : 100 can be covered with a high degree of accuracy. In order to keep the damping of the oscillator as constant as possible over such a large frequency range, it is advisable to connect a small capacitance in parallel with input resistor R_1 at OA3. This will compensate for the increased attenuation due to the phase lag of the operational amplifier at higher frequencies.

14.5
Function Generators

As we have seen, the amplitude control involved in the generation of low-frequency sine waves is rather cumbersome. It is much easier to use a Schmitt trigger and an integrator to generate a triangular alternating voltage. A sine wave can then be produced if the sine function network of Sect. 11.7.4 is employed. Since, with this method, a triangular wave, a square wave, and a sinusoidal wave are obtained simultaneously, circuits based on this principle are called function generators. The block diagram of such a circuit is shown in Fig. 14.31.

The principle consists in applying a constant voltage to an integrator. This voltage is either positive or negative depending on the direction in which the integrator output voltage is to be changed. If the integrator output voltage reaches the switch-on or switch-off level of the following Schmitt trigger, the sign at the integrator input is inverted. This produces a triangular voltage at the output, which oscillates between the trigger levels.

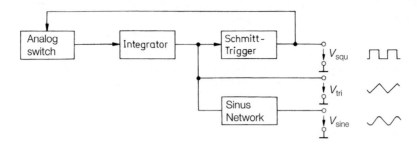

Fig. 14.31. Block diagram of a function generator

14.5.1
Basic Arrangement

Two approaches are possible, which differ in the way in which integration is implemented. In the circuit shown in Fig. 14.32, $+V_i$ or $-V_i$ is applied to an integrator, depending on the position of the analog switch. With the circuit in Fig. 14.33, current $+I_i$ or $-I_i$ is impressed on capacitor C via an analog switch. This also results in a rise or fall in the linear voltage. In order not to falsify the triangular voltage across the capacitor due to loading, an impedance converter is generally required. The advantage of the method in Fig. 14.33, however, is that it is easier to implement the impedance converter and current switch for higher frequencies.

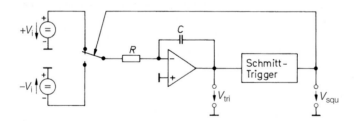

Fig. 14.32. A function generator with an integrator

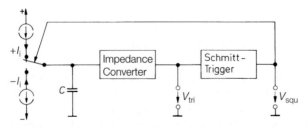

Fig. 14.33. A function generator with constant current sources

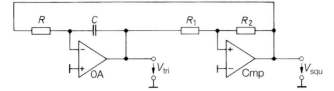

Fig. 14.34. Simple function generator

$$Frequency \ f = \frac{R_2}{4R_1} \cdot \frac{1}{RC}, \qquad Amplitude: \ \hat{V}_{tri} = \frac{R_1}{R_2} V_{squ \, max}$$

14.5.2
Practical Implementation

The simplest method of implementation is to start from the principle illustrated in Fig. 14.32, and to use the output voltage of the Schmitt trigger itself as the input voltage for the integrator. The resulting circuit is shown in Fig. 14.34. The Schmitt trigger supplies a constant output voltage, which is integrated by the integrator. If the integrator output voltage reaches the trigger level of the Schmitt trigger, the voltage V_{squ} to be integrated instantaneously changes sign, causing the integrator output to reverse direction until the other trigger level is reached. In order to insure that the absolute values of the positive and negative slopes are the same, the comparator must have a symmetrical output voltage $\pm V_{squ \, max}$. In accordance with Sect. 6.5.2, the amplitude of the triangular voltage is given by

$$\hat{V}_{tri} = \frac{R_1}{R_2} V_{squ \, max}$$

The oscillation period is equal to four times the time required by the integrator to change from zero to $\hat{V}_{tri}$. It is therefore

$$T = 4\frac{R_1}{R_2} RC$$

An example of a practical implementation of the circuit principle of Fig. 14.33 is shown in Fig. 14.35. The controlled current switch consists of transistors $T_1 - T_3$. As long as control signal $x = L$, the capacitor is discharged via T_1 with current I. If the triangular voltage falls below -1 V, the precision Schmitt trigger from Fig. 6.48 changes state, and $x = H$. This turns T_3 off and current source T_2 is turned on. The latter supplies twice as much current as T_1, namely $2I$. This means that capacitor C is charged with current I and T_1 needs not be turned off.

If the triangular voltage exceeds the upper trigger level of $+1$ V, the Schmitt trigger reverts to the state $x = L$, and capacitor C is discharged again.

The dual comparator NE 521 from On semiconductor is particularly suitable for implementing the precision Schmitt trigger, as it already contains the two gates required. This comparator additionally features particularly short switching times of only some 8 ns, enabling frequencies of up to several MHz to be generated. The impedance converter shown in Fig. 14.33 is only required if the triangular voltage is to be loaded. The subsequent comparators place virtually no load on the triangular voltage.

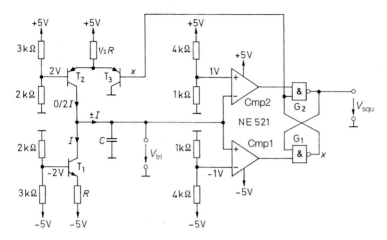

Fig. 14.35. High-speed function generator with a current switch and a precision comparator

$$Frequency:\ f = \frac{I}{4\hat{V}_{tri} \cdot C} = \frac{0.6}{RC}, \qquad Amplitude:\ \hat{V}_{tri} = 1\ V$$

14.5.3
Function Generators with a Controllable Frequency

Using the principle illustrated in Fig. 14.32, the frequency can be controlled quite easily by varying the voltages $+V_i$ and $-V_i$. An example of a function generator of this type is shown in Fig. 14.36. The voltages $+V_i$ or $-V_i$ are available at low impedance at the outputs of OA1 and OA2. These voltages are applied to the integrator input via transistor T_1 or transistor T_2, depending on the switching state of the Schmitt trigger. If the output voltages of the comparator are greater than $\pm V_i$, the two transistors operate as saturated emitter followers and therefore only have a voltage drop of a few millivolts, as described in Sect. 17.2.3.

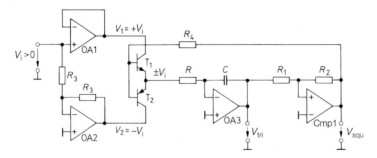

Fig. 14.36. Function generator with a controllable frequency

$$Frequency\ f = \frac{R_2}{4R_1} \cdot \frac{1}{RC} \cdot \frac{V_i}{V_{squ\,max}}, \qquad Amplitude:\ \hat{V}_{tri} = \frac{R_1}{R_2} V_{squ\,max}$$

The Schmitt trigger once again determines the amplitude of the triangular signal, which is given by

$$\hat{V}_{tri} = \frac{R_1}{R_2} V_{squ\,max}$$

The rate of change of the triangular voltage is:

$$\frac{\Delta V_{tri}}{\Delta t} = \pm\frac{V_i}{RC}$$

The period is equal to four times the time required by the integrator to go from zero to $\hat{V}_{tri}$. The frequency is therefore

$$f = \frac{V_i}{4RC\hat{V}_{tri}} = \frac{R_2}{4R_1} \cdot \frac{1}{RC} \cdot \frac{V_i}{V_{squ\,max}}$$

in other words, it is proportional to the input voltage V_i and so the circuit is suitable for use as a voltage–frequency converter. If we select

$$V_i = V_{i0} + \Delta V_i$$

we obtain a linear frequency modulation.

If accuracy and stability of the amplitude and frequency are essential, care must be taken to insure that they are not dependent on $V_{squ\,max}$. This is easily achieved by using a precision Schmitt trigger, as shown in Fig. 14.35. However, we then need an additional amplifier to generate the bipolar signals that are required for driving T_1 and T_2. In this case, it is simpler to replace the bipolar transistors by CMOS analog switches.

Variable Duty Cycle

In order to generate a square-wave voltage with a variable duty cycle, we can use a comparator to compare a symmetrical triangular voltage with a DC voltage. The relationships are somewhat more difficult if not only the square wave but also the triangular voltage is to be asymmetrical in form, as in Fig. 14.37.

One solution is to use the circuit shown in Fig. 14.36, applying two different absolute values for the potentials V_1 and V_2. The rise and fall times of the triangular voltages between $\pm\hat{V}_D$ are therefore

$$t_1 = \frac{2RC\hat{V}_{tri}}{V_1}, \qquad t_2 = \frac{2RC\hat{V}_{tri}}{|V_2|}$$

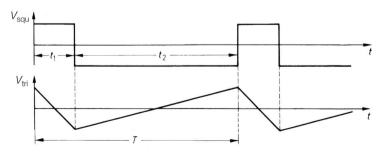

Fig. 14.37. Voltage waveform for a duty factor of $t_1/T = 20\%$

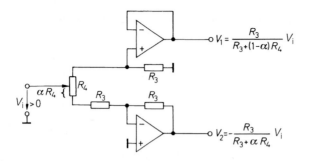

$$V_1 = \frac{R_3}{R_3 + (1-\alpha)R_4} V_i$$

$$V_2 = -\frac{R_3}{R_3 + \alpha R_4} V_i$$

Fig. 14.38. Extension for the variable duty ratio

Type	Manufacturer	Maximum frequency
MAX 038	Maxim	10 MHz
XR-205	Exar	4 MHz
XR-2206	Exar	1 MHz

Fig. 14.39. Integrated function generators with square-, triangle- and sine-output

If we now want to change the symmetry without changing the frequency, we have to increase the absolute value of one potential and reduce that of the other, so that

$$T = t_1 + t_2 = 2RC\hat{V}_{tri}\left(\frac{1}{V_1} + \frac{1}{|V_2|}\right) \tag{14.15}$$

remains constant. This condition can also be easily satisfied by using the drive circuit shown in Fig. 14.38. Its output potentials are given by

$$\frac{1}{V_1} + \frac{1}{|V_2|} = \frac{1}{V_i R_3}[R_3 + (1-\alpha)R_4 + R_3 + \alpha R_4] = \frac{1}{V_i R_3}[2R_3 + R_4]$$

As required, this expression holds irrespective of the duty cycle selected. By substitution in (14.15), we obtain the frequency:

$$f = \frac{R_3}{2RC[2R_3 + R_4]} \cdot \frac{V_i}{\hat{V}_{tri}}$$

The duty cycle t_1/T or t_2/T can be set between

$$\frac{R_3}{2R_3 + R_4} \quad \text{and} \quad \frac{R_3 + R_4}{2R_3 + R_4}$$

using potentiometer R_4. With $R_4 = 3R_3$, we obtain values between 20% and 80%.

Function generators that not only produce triangular and square-wave voltages but also contain a sine-wave function network are available in IC form. Some types are collected in Fig. 14.39. The use of these circuits constitutes the simplest method of implementing function generators. However, the signal quality and the usable frequency range are limited.

14.5.4
Simultaneously Producing Sine and Cosine Signals

The easy amplitude stabilization of function generators can also be utilized for the simultaneous generation of sine and cosine signals. Taking the triangular signal of a function generator, its changes in sign can be determined by means of a comparator. The resulting

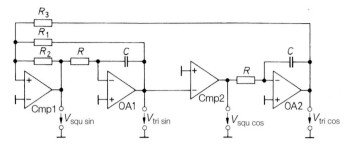

Fig. 14.40. Function generator for producing triangular and square-wave signals that are phase-shifted by 90°

$$Frequency \ \ f = \frac{R_2}{4R_1} \frac{1}{RC}, \qquad Amplitude: \ \hat{V}_{tri} = \frac{R_1}{R_2} V_{squ \, max}$$

square wave is shifted by 90° with respect to the square wave at the input of the integrator. Using a second integrator, the resulting square-wave signal can be converted to a triangular signal, which is also phase-shifted by 90° with respect to the original triangular signal.

A simple version of this principle is illustrated in Fig. 14.40. Operational amplifier OA1 and comparator Cmp1 constitute a function generator of the type shown in Fig. 14.34. Comparator Cmp2 produces the phase-shifted square-wave signal and integrator OA2 the associated triangular signal.

However, the circuit would not operate without a feedback path via R_3: integrator OA2 would run away due to the unavoidable symmetry and offset errors. This can be avoided by inserting an additional resistor R_3. Voltage $V_{tri \, sin}$ can be shifted via this resistor toward positive or negative values, thereby also allowing the duty factor of $V_{squ \, cos}$ to be varied. The feedback via R_3 virtually cancels out the DC voltage superimposed at output $V_{tri \, cos}$.

It is not immediately apparent why the triangular voltage present at output $V_{tri \, cos}$, and fed back via R_3, does not impair the operation of function generator Cmp1, OA1. The reason may be seen in Fig. 14.41, which shows that the triangular voltage $V_{tri \, cos}$ is zero at the peak values of $V_{tri \, sin}$ and therefore does not alter the switching instant of the comparator Cmp1. This can only happen due to a superimposed DC voltage.

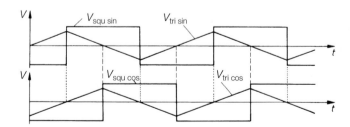

Fig. 14.41. Time diagram of triangular and square-wave voltages shifted by 90°

Chapter 15:
Power Amplifiers

Power amplifiers are designed to provide large output powers, with the voltage gain playing only a minor role. Normally, the voltage gain of a power output stage is near unity and the power gain is thus mainly due to the current gain of the circuit. The output voltage and current must be able to assume positive and negative values. Power amplifiers with unidirectional output current are known as power supplies. They are discussed in Chap. 16.

15.1
Emitter Follower as a Power Amplifier

The operation of the emitter follower has already been described in Sect. 2.4.2. Here, we define some of the parameters that are of particular interest for its application as a power amplifier. First, we calculate the load resistance for which the circuit in Fig. 15.1 delivers maximum power without distortion. If the output is negative, R_L carries some of the current flowing through R_E. The limit for control of the output voltage is reached when the current through the transistor becomes zero. This is the case for the output voltage

$$V_{o\,min} = -\frac{R_L}{R_E + R_L} \cdot V_b$$

If the output voltage is to be controlled sinusoidally around $0\,V$, its amplitude must not exceed the value

$$\hat{V}_{o\,max} = \frac{R_L}{R_E + R_L} \cdot V_b$$

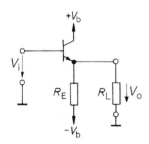

Fig. 15.1. Emitter follower as power amplifier

Voltage gain:	$A \approx 1$
Current gain if load is matched to internal resistance:	$A_i = \frac{1}{2}\beta$
Load resistance if matched:	$R_L = R_E$
Output power for matched load and full sinusoidal output swing	$P_{L\,max} = \frac{V_b^2}{8R_E}$
Maximum efficiency:	$\eta\,max \frac{P_{L\,max}}{P_{tot}} = 6.25\%$
Maximum dissipation of the transistor:	$P_T = \frac{V_b^2}{R_E} = 8P_{L\,max}$

The power delivered to R_L is then

$$P_L = \frac{1}{2} \frac{\hat{V}_{o\,max}^2}{R_L} = \frac{V_b^2 R_L}{2(R_E + R_L)^2}.$$

With $\frac{dP_L}{dR_L} = 0$ it follows that, for $R_L = R_E$, the maximum output power,

$$P_{L\,max} = \frac{V_b^2}{8R_E}$$

is attained. This result is surprising in that one would normally expect maximum power output when the load resistance equals the output resistance r_o of the voltage source. However, this is only the case for a constant open-circuit voltage. Here, the open-circuit voltage is not constant, as it must be reduced when R_L is small.

In the next step, we determine the power consumption within the circuit for any output voltage amplitude and any load resistance. For a sinusoidal voltage, the power

$$P_L = \frac{1}{2} \frac{\hat{V}_o^2}{R_L}$$

is supplied to the load resistance R_L. The power dissipation of the transistor is

$$P_T = \frac{1}{T} \int_0^T (V_b - V_o(t)) \left(\frac{V_o(t)}{R_L} + \frac{V_o(t) + V_b}{R_E} \right) dt.$$

For $V_o(t) = \hat{V}_o \sin \omega t$, it follows:

$$P_T = \frac{V_b^2}{R_E} - \frac{1}{2} \hat{V}_o^2 \left(\frac{1}{R_L} + \frac{1}{R_E} \right).$$

The power dissipation of the transistor is largest for a zero input and output signal. Similarly, the power in R_E is given by

$$P_E = \frac{V_b^2}{R_E} + \frac{1}{2} \frac{\hat{V}_o^2}{R_E}.$$

The circuit therefore draws from the power supplies a total power of

$$P_{tot} = P_L + P_T + P_E = 2 \frac{V_b^2}{R_E}$$

This is a surprising result, since it shows that the total power of the circuit is independent of the drive voltage and of the load, and that it remains constant as long as the circuit is not overdriven. The efficiency η_{max} is defined as the ratio of the maximum obtainable output power to the power consumption at full voltage swing. The results for $P_{L\,max}$ and P_{tot} give $\eta_{max} = \frac{1}{16} = 6.25\%$. The following characteristics are typical for this circuit:

1) The current through the transistor is never zero.
2) The total power consumption is independent of the input drive voltage and the load.

These are the characteristics of class-A operation.

15.2
Complementary Emitter Followers

The output power of the emitter follower shown in Fig. 15.1 is limited in that the resistor R_E restricts the maximum output current. A considerably higher output power and a better efficiency can be attained if R_E is replaced by a second emitter follower, as shown in Fig. 15.2.

15.2.1
Complementary Class-B Emitter Follower

For positive input voltages, T_1 operates as an emitter follower and T_2 is reverse biased; and vice versa for negative drive. The transistors thus carry the current alternately, each for half a period. Such a mode of operation is known as *push–pull class-B operation*. For $V_i = 0$, both transistors are turned off and therefore no quiescent current flows in the circuit. The current taken from the positive or negative power supply is thus the same as the output current. The circuit therefore has a considerably better efficiency than the normal emitter follower. A further advantage is that, at any load, the output can be driven between $\pm V_b$ as the transistors do not limit the output current. The difference between the input and output voltages is determined by the base–emitter voltage of the current-carrying transistor. As it changes only very slightly with the load, $V_o \approx V_i$, irrespective of the load current. The output power is inversely proportional to the resistance R_L and has no extremum, and therefore no matching is required between the load resistance and any internal circuit resistance. The maximum power is determined instead by the permissible peak currents and the maximum power dissipation of the transistors, which, for full sinusoidal drive, is given by

$$P_L = \frac{\hat{V}_o^2}{2R_L}.$$

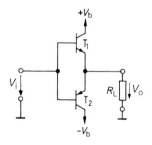

Fig. 15.2. Complementary emitter follower

Voltage gain:	$A \approx 1$
Current gain:	$A_i = \beta$
Output power at full sinusoidal output swing:	$P_L = \dfrac{V_b^2}{2R_L}$
Efficiency at full sinusoidal output swing:	$\eta_{max} = \dfrac{P_L}{P_{tot}} = 78.5\%$
Maximum dissipation of one transistor:	$P_{T1} = P_{T2}\dfrac{V_b^2}{\pi^2 R_L} = 0.2P_L$

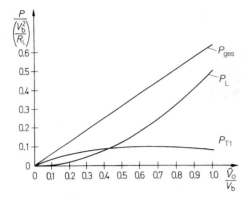

Fig. 15.3. Power and loss curves for the complementary emitter follower as a function of the output amplitude

We now determine the power dissipation P_{T1} of transistor T_1. As the circuit is symmetrical, it is identical to that of T_2:

$$P_{T1} = \frac{1}{T} \int_{0}^{T/2} (V_b - V_o(t)) \frac{V_o(t)}{R_L} dt.$$

For $V_o(t) = \hat{V}_o \sin \omega t$, this is

$$P_{T1} = \frac{1}{R_L} \left(\frac{\hat{V}_o V_b}{\pi} - \frac{\hat{V}_o^2}{4} \right).$$

Hence, the efficiency of the circuit is

$$\eta = \frac{P_L}{P_{tot}} = \frac{P_L}{2P_{T1} + P_L} = \frac{\pi}{4} \cdot \frac{\hat{V}_o}{V_b} \approx 0.785 \frac{\hat{V}_o}{V_b}.$$

It is therefore proportional to the output amplitude and, for a full output swing ($\hat{V}_o = V_b$), it attains a value of $\eta_{max} = 78.5\%$.

The power dissipation of the transistors reaches its maximum not at the full output voltage swing, but at

$$\hat{V}_o = \frac{2}{\pi} V_b \approx 0.64 \, V_b.$$

This follows from the extremum condition:

$$\frac{d P_{T1}}{d \hat{V}_o} = 0.$$

In this case, the power dissipation for each transistor is

$$P_{T \, max} = \frac{1}{\pi^2} \frac{V_b^2}{R_L} \approx 0.1 \frac{V_b^2}{R_L}.$$

The curves for the output power, the power dissipation, and the total power, as functions of the relative output voltage swing $\hat{V}_o / V_b$, are given in Fig. 15.3.

We can see that the power consumption,

$$P_{tot} = 2P_{T1} + P_L = \frac{2V_b}{\pi R_L} \hat{V}_o \approx 0.64 \frac{V_b}{R_L} \hat{V}_o$$

is proportional to the output amplitude. This is characteristic of *class-B operation*.

As described above, only one transistor carries current at any given time. However, this only applies if the period time of the input signal is large with respect to the storage time of the transistors. Some time is needed for the transistors to change from the ON to the OFF state. If the period of the input voltage is smaller than this time, both transistors may conduct simultaneously. Very high crossover currents then flow through both transistors from $+V_b$ to $-V_b$, and will destroy them instantaneously. Oscillations at this critical frequency can occur in amplifiers with feedback or if the emitter followers are loaded by a capacitor. For protection of the transistors, a current limiter should therefore be incorporated.

15.2.2
Complementary Class-AB Emitter Followers

Figure 15.4 represents the transfer characteristic $V_o = V_o(V_i)$ for push–pull class-B operation, as described for the previous circuit. Near zero voltage, the current in the forward-biased transistor becomes very small and the transistor impedance increases. The output voltage at the load therefore remains nearly zero, as indicated by a region in the characteristic near the origin. This gives rise to distortion of the output voltage, known as *crossover distortion*. If a small quiescent current flows through the transistors, their impedance near the origin is reduced and the transfer characteristic shown in Fig. 15.5 is obtained. The transfer characteristic of each individual emitter follower is shown by dashed lines. It can be seen that the crossover distortion is considerably reduced. If the quiescent current is made as large as the maximum output current, the resulting mode of operation would be called push–pull class-A, by analogy with Sect. 17.1. However, the crossover distortion is already greatly reduced if only a fraction of the maximum output current is permitted to flow as a quiescent current. This mode is called push–pull class-AB operation, and its crossover distortion is so small that it can be easily reduced to tolerable values by means of feedback.

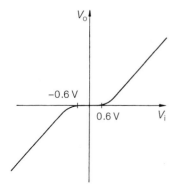

Fig. 15.4. Crossover for push–pull class-B operation

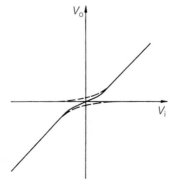

Fig. 15.5. Crossover for push–pull class-AB operation

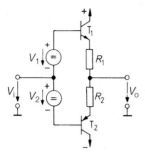

Fig. 15.6. Realization of class-AB operation using two auxiliary voltages

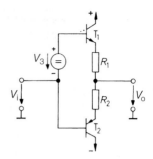

Fig. 15.7. Realization of class-AB operation using a single auxiliary voltage

Additional distortion may arise if positive and negative voltages are amplified at different gains. This is the case if the complementary emitter followers are driven from a high-impedance source and the transistors have different current transfer ratios. If strong feedback is undesirable, the transistors must be selected such that they have identical current transfer ratios.

The basic circuit for the realization of class-AB operation in shown in Fig. 15.6. To obtain a small quiescent current, a DC voltage of about 1.4 V is applied between the base terminals of T_1 and T_2. If the two voltages V_1 and V_2 are the same, the quiescent potential of the output is approximately equal to that of the input. The bias voltage can also be supplied by a single voltage source, $V_3 = V_1 + V_2$, as represented in Fig. 15.7. In this case, the potential difference between the output and the input is about 0.7 V.

The main problem with class-AB operation is keeping the required quiescent current constant over a wide range of temperatures. As the transistors get warmer, the quiescent current increases, which may itself further increase the temperature and finally lead to destruction of the transistors. This effect is known as positive thermal feedback. The increase in the quiescent current can be avoided if the voltages V, and V_2 are each reduced by 2 mV for every degree of temperature rise. For this purpose, diodes or thermistors can be mounted on the heat sinks of the power transistors.

However, temperature compensation is never quite perfect, as the temperature difference between the junction and the case is usually considerable. Therefore, additional stabilization is required in the form of resistors R_1 and R_2, which provide current feedback. This becomes more effective as the resistances chosen increase. As the resistors are connected in series with the load, they reduce the available output power and must therefore be selected to be small compared to the load resistance. This dilemma can be avoided by using Darlington circuits, as will be shown in Sect. 15.3.

15.2.3
Generation of the Bias Voltage

One method of providing a bias voltage is shown in Fig. 15.8. The voltage of $V_1 = V_2 \approx 0.7$ V across each of the diodes D_1 and D_2 just allows a small quiescent current to flow through the transistors T_1 and T_2. To attain a higher input resistance, the diodes can be replaced by emitter followers, this resulting in the circuit shown in Fig. 15.9.

Figure 15.10 represents a driver arrangement that allows adjustment of the bias voltage and its temperature coefficient over a wide range. Feedback is applied to transistor T_3 by

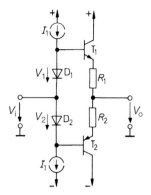

Fig. 15.8. Bias voltage generation by diodes

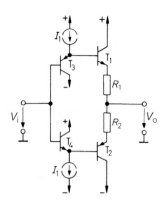

Fig. 15.9. Bias voltage generation by transistors

means of voltage divider R_5, R_6. For a negligible base current, its collector–emitter voltage has the value

$$V_{CE3} = V_{BE3}\left(1 + \frac{R_5}{R_6}\right)$$

To obtain the desired temperature coefficient, R_5 is, in practice, a resistor network that contains an NTC resistor mounted on the heat sink of the output transistors. In this way, the quiescent current can to a large extent be made temperature-independent, even though the temperature of the case is lower than that of the output transistor junctions.

In circuits that use diodes for bias voltage generation, no current can flow from the input into the base of the output transistors. The base current for the output transistors must therefore be supplied from the constant-current sources. The constant current I_1 must be larger than the maximum base current of T_1 and T_2, so that diodes D_1 and D_2 (or transistors T_3 and T_4) will not he turned off before the maximum permissible output voltage swing is reached. For this reason, it would be inadvisable to replace the constant-current sources by resistors, as this would cause the current to decrease with a rising output voltage.

The most favorable driver circuit would be one that supplies a larger base current for an increasing output voltage, and such a circuit is represented in Fig. 15.11. The FETs T_3

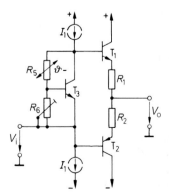

Fig. 15.10. Generation of a bias voltage that has an adjustable temperature coefficient

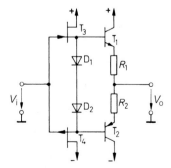

Fig. 15.11. Bias voltage generation by FETs

and T_4 operate as source followers. The difference in their source voltages settles, due to series feedback, at a value of about 1.4 V. FETs that have a large saturation drain current $I_{D\,sat}$ are suitable for this purpose.

15.3
Complementary Darlington Circuits

With the circuits described so far, output currents of up to a few hundred milliamperes can be obtained. For higher output currents, transistors that have higher current gain β must be used. They can be made up of two transistors if they are operated in a Darlington connection. Such circuits and their parameters have already been discussed in Sect. 2.4.4. The basic circuit of a Darlington power amplifier is shown in Fig. 15.12, where the transistor pairs T_1, T_1' and T_2, T_2' are Darlington connected.

For the implementation of push–pull class AB, the adjustment of the quiescent current presents problems, as four temperature-dependent base–emitter voltages must now be compensated. These difficulties can be avoided by allowing the quiescent current to flow only through the driver transistors T_1 and T_2. The output transistors then become conducting only for larger output currents. To achieve this, the bias voltage V_1 is selected such that a voltage of approximately 0.4 V appears across each of resistors R_1 and R_2; thus $V_1 \approx 2(0.4\,\text{V} + 0.7\,\text{V}) = 2.2\,\text{V}$. Then, for zero input, the output transistors carry virtually no current, even at higher junction temperatures.

At higher output currents, the base–emitter voltages of the output transistors rise to about 0.8 V. This limits the current through R_1 and R_2 to double the quiescent value, and therefore most of the emitter current of the driver transistors is available as base current for the output transistors.

Resistors R_1 and R_2 also discharge the base of the output transistors. The lower their resistance, the faster the output transistors can be turned off. This is particularly important when the input voltage changes polarity, as one transistor can become conducting before the other is turned off. In this way, a large crossover current can flow through both output transistors, and the resulting *secondary breakdown* will destroy them immediately. This effect determines the attainable large-signal bandwidth.

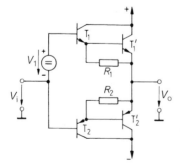

Fig. 15.12. Complementary Darlington pairs

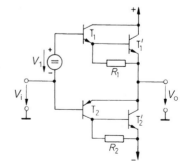

Fig. 15.13. Quasi-complementary Darlington pairs

For integrated circuits it is preferable to use power transistors npn. In such cases, the Darlington circuit T_2, T_2' shown in Fig. 15.12 is replaced by the complementary Darlington connection, as described in Sect. 2.4. The resulting circuit, shown in Fig. 15.13, is known as a *quasi-complementary* power amplifier. To arrive at the same quiescent current conditions as for the previous circuit, a voltage of about 0.4 V is impressed across resistor R_1. Voltage V_1 must then be $0.4\,V + 2 \cdot 0.7V = 1.8\,V$. The quiescent current flows via T_2 and R_2 to the negative supply. If $R_1 = R_2$, a bias voltage of $0.4\,V$ is also obtained for T_2''. The purpose of resistors R_1 and R_2 is the same as for the previous circuit; that is, to discharge the base of the output current transistors. This output structure is found in most integrated power amplifiers.

15.4
Complementary Source Followers

The major advantage of power MOSFETs over bipolar power transistors is that they can be turned on and off much more quickly. Whereas bipolar power transistors have switching times of between 100 ns and 1 μs, the corresponding range for MOSFETs is 10–100 ns. It is therefore preferable to employ power MOSFETs in output stages for frequencies above 100 kHz up to 1 MHz.

As power MOSFETs possess large drain–gate and gate–source capacitances, possibly of a few hundreds of picofarads, it is advantageous to operate them as source followers. This prevents the drain–gate capacitance from being dynamically increased due to the Miller effect, and the gate–source capacitance is even considerably reduced by bootstrapping.

Figure 15.14 shows the basic circuit diagram for a complementary source follower. The two auxiliary voltages V_1 are used, as in the case of the bipolar transistor shown in Fig. 15.6, to set the required quiescent current. For $V_1 = V_{th}$, no quiescent current flows at all: class-B operation is established. However, in order to minimize the transfer distortion, a quiescent current is usually desired to flow by selecting $V_1 > V_{th}$. The magnitude of the quiescent current is stabilized by current feedback via resistors R_1, R_2. The voltage V_1 results from the transfer characteristic of the MOSFETs; that is,

$$V_1 = I_D R_1 + V_{th} + \sqrt{\frac{2I_1}{K_1}}$$

The resulting voltages are markedly higher than for bipolar transistors, as the pinch-off voltage V_{th} of power MOSFETs is between 1 V and 4 V. A simple means of producing the required bias is to replace the emitter followers T_3, T_4 in Fig. 15.9 by source followers. The resultant circuit is shown in Fig. 15.15. Here, T_3 produces a bias voltage of

$$V_1 = V_{th} + \sqrt{\frac{2I_3}{K_3}}$$

If the low-power MOSFETs T_3 and T_4 are fabricated in the same process as the power MOSFETs T_1 T_2, and therefore have the same pinch-off voltages, the maximum quiescent current for $R_1 = R_2 = 0$ is

$$I_1 = \frac{K_1}{K_3} I_3 = \frac{A_1}{A_3} I_3$$

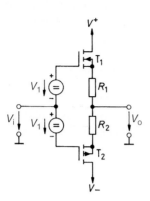

Fig. 15.14. Principle of a complementary source follower

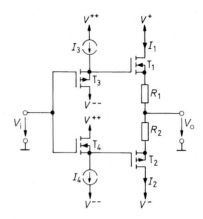

Fig. 15.15. Bias generation for complementary source followers. Examples of transistor types from International Rectifier:

T_3: IRFD 112 T_1: IRF 531
T_4: IRFD 9122 T_2: IRF 9531

Thus it depends only on the ratio of the areas of the transistors on the chip. The current can be reduced using R_1, R_2. The currents I_3, I_4 are selected to be sufficiently large to charge the input capacitor of the source followers T_1, T_2 at the highest frequency that occurs.

Operation of the drive circuit generally requires a supply voltage that is up to 10 V higher than for the output stage. Otherwise, the maximum achievable output voltage may be as much as 10 V below the supply voltage, which results in unacceptably poor efficiency.

15.5
Current Limitation

Due to their low output resistance, power amplifiers can easily be overloaded and therefore easily destroyed. Consequently, it is advisable to limit the output current to a defined maximum value by means of an additional control circuit. The various possibilities are exemplified by the simple complementary emitter follower shown in Fig. 15.8. A particularly simple circuit is shown in Fig. 15.16. Limitation takes effect when multiple diode D_3 or D_4 begins to conduct, because this prevents any further increase in the voltage dropped across R_1 or R_2 The maximum output current is therefore

$$I_{o\,max}^+ = \frac{V_{D3} - V_{BE1}}{R_1} = \frac{0.7\,\text{V}}{R_1}(n_3 - 1),$$

$$I_{o\,max}^- = -\frac{V_{D4} - |V_{BE2}|}{R_2} = -\frac{0.7\,\text{V}}{R_2}(n_4 - 1).$$

where n_3 and n_4 are the number of diodes used for D_3 and D_4 respectively.

Another method of current limitation is shown in Fig. 15.17. If the voltage drop across R_1 or R_2 exceeds a value of about 0.7 V, transistor T_3 or T_4 begins to conduct, thereby preventing any further rise in the base current of T_1 or T_2.

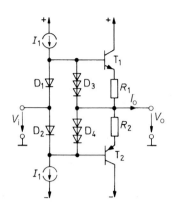

Fig. 15.16. Current limitation with diodes

$I_{o\,max} = \pm 1.4\,\text{V}/R_{1,2}$

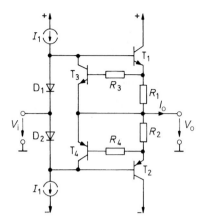

Fig. 15.17. Current limitation with transistors

$I_{o\,max} = \pm 0.7\,\text{V}/R_{1,2}$

This control arrangement limits the output current to a maximum value of

$$I_{o\,max}^{+} \approx \frac{0.7\,\text{V}}{R_1} \quad \text{bzw.} \quad I_{o\,max}^{-} \approx \frac{0.7\,\text{V}}{R_2}$$

The advantage here is that the base-emitter voltages of the current sensing transistors affects the maximum output current and not the varying base-emitter voltages of the power transistors. Resistors R_3 and R_4 are used to protect the current sensing transistors from excessively large base-current peaks.

In the event of a short-circuit, current $I_{o\,max}$ flows through T_1 and T_2 for half a period, while the output voltage is zero. The power dissipation in the power transistors is therefore

$$P_{T1} = P_{T2} \approx \frac{1}{2} V_b I_{o\,max}.$$

As a comparison with Sect. 15.2 shows, this is five times the dissipation during normal operation. Consequently, the ratings of the power transistors and heat sinks must be such that the circuits in Figs. 15.16 and 15.17 are short-circuit proof.

Amplitude-Dependent Current Limitation

The need for additional short-circuit protection at the output stage can be obviated by only allowing resistive loads of a defined value R_L to be connected. It can then be assumed that, for small output voltages, only small output currents will flow. The current limitation does therefore not have to be defined by the maximum current $I_{o\,max} = V_{o\,max}/R_L$, but the output current can be limited to $I_{o\,max} = V_o/R_L$; in other words, it is a function of the output voltage. The maximum current under short-circuit conditions ($V_o = 0$) can therefore be selected to be correspondingly low.

In order to make the current limit dependent on the output voltage, the bias voltage applied to transistors T_3 and T_4 in Fig. 15.18 is increased as the output voltage increases. Resistors R_5 and R_6, which have high values compared to R_3 and R_4, are used for this purpose. For low output voltages, we therefore obtain the same current limit as in Fig. 15.17.

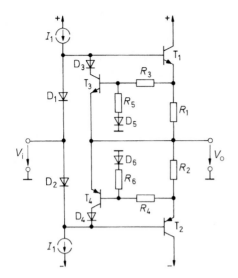

Fig. 15.18. Voltage-dependent current limitation

$$|I_{o\,max}| = \frac{0.7\,\text{V}}{R_{1.2}} + \frac{R_{3.4}}{R_{5.6}} \cdot \frac{V_o}{R_{1.2}}$$

Fig. 15.19. Current limit and output current characteristics for a resistive load

For higher positive output voltages, an additional voltage drop of $V_o R_3 / R_5$ appears across R_3, thereby raising the current limit to

$$I_{o\,max}^+ \approx \frac{0.7\,\text{V}}{R_1} + \frac{R_3}{R_5}\frac{V_o}{R_1}$$

Diode D_5 prevents transistor T_3 from receiving a positive bias when the output voltages are negative, which could result in it being turned on unintentionally. Diode D_3 prevents the collector–base diode of T_3 from conducting when a larger voltage is dropped across R_2 in the event of negative output voltages. Otherwise, the drive circuit would be subject to additional loading. Similar considerations apply to the negative current limitation due to T_4.

These characteristics are graphically illustrated in Fig. 15.19. Using this form of voltage-dependent current limitation, it is therefore possible to utilize fully the safe operating area (SOA) of the power transistors. Consequently, this method is also known as SOA current limitation.

15.6
Four-Quadrant Operation

A power output stage is subject to the most severe operating conditions when a constant current limit $I_{o\,max}^+$ and $I_{o\,max}^-$ is required for any given positive and negative output voltage. Such requirements invariably arise if there is no resistive load present, but a load that can feed energy back to the output stage. Loads of this kind include capacitors, inductors, and electric motors. In this case, it is necessary to use current-limiting arrangements shown in Figs. 15.16 or 15.17. The critical operating condition for the negative output-stage

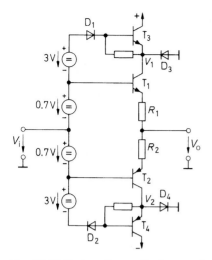

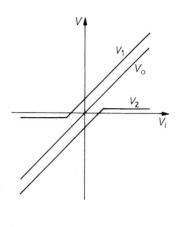

Fig. 15.20. Push–pull stage for four-quadrant operation

Fig. 15.21. Characteristics of the output voltage and auxiliary potentials V_1, and V_2

transistor T_2 occurs when the load feeds the current-limit value $I_{o\,max}^-$ into the circuit while the output voltage is $V_o = V_{o\,max} \approx V^+$. Current $I_{o\,max}^-$ then flows through T_2 at a voltage of $V_{CE2} \approx 2V^+$, resulting in a power dissipation of $P_{T2} = 2V^+ \cdot I_{o\,max}^-$. However, for a voltage of $2\,V^+$, secondary breakdown limits the majority of bipolar transistors such that they can only be loaded to a fraction of their thermal rating. Therefore, it is generally necessary to connect several power transistors in parallel or, preferably, to employ power MOSFETs, which are not subject to secondary breakdown.

One method of halving the voltage across the output-stage transistors is shown in Fig. 15.20. The basic idea is to control the collector potentials of T_1 and T_2 together with the input voltage. For positive input voltages, we obtain

$$V_1 = V_i + 0.7\,\text{V} + 3\,\text{V} - 0.7\,\text{V} - 0.7\,\text{V} = V_i + 2.3\,\text{V}.$$

Transistor T_1 is therefore being operated well outside saturation. For negative input voltages, diode D_3 takes over the current, and we have $V_1 = -0.7\,\text{V}$. If the input voltage falls to $V_i = V_{i\,min} \approx V^-$, a voltage of only $V_{CE1\,max} \approx V^-$ is dropped across T_1. Likewise, the maximum voltage across T_3 is no larger. It occurs when $V_i = 0$ and is given by $V_{CE3\,max} \approx V^+$. The maximum power dissipated in T_1 and T_3 is therefore $P_{max} = V^+ \cdot I_{o\,max}^+$. Consequently, not only is the maximum collector–emitter voltage halved, but also the power dissipation. For the negative side, T_2, T_4 produce corresponding results due to the symmetry of the circuit. To make this clear, the characteristics of V_1 and V_2 are plotted in Fig. 15.21.

15.7
Design of a Power Output Stage

To illustrate the design process for a power output stage in more detail, we use the circuit shown in Fig. 15.22 and determine the rating of its components for an output power of 50 W. The circuit is based on the power amplifier of Fig. 15.12.

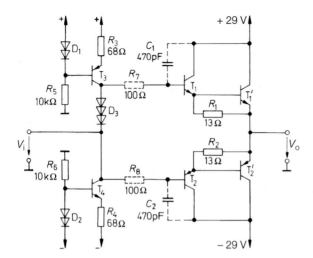

Fig. 15.22. Power output stage for a 50 W sinusoidal output

The amplifier is required to supply a load of $R_L = 5\,\Omega$ with a power of 50 W at sinusoidal output. The amplitude of the output voltage is then $\hat{V}_o = 22.4\,\text{V}$, and that of the current is $\hat{I}_o = 4.48\,\text{A}$. To determine the supply voltage, we calculate the minimum voltage across T_1', T_1, T_3, and R_3. For the base–emitter voltages of T_1 and T_1', at $I_{o\,max}$, we must allow about 2 V in total. A diode forward voltage is dropped across R_3; that is, 0.7 V. The collector–emitter voltage of T_3 should not fall below 0.9 V at full input drive. The output stage is intended for operation from an unregulated supply, the voltage of which may fall by about 3 V at full load. We therefore obtain the no-load supply voltage:

$$V_b = 22.4\,\text{V} + 2\,\text{V} + 0.7\,\text{V} + 0.9\,\text{V} + 3\,\text{V} = 29\,\text{V}.$$

For reasons of symmetry, the negative supply voltage must have the same value. The maximum ratings of transistors T_1', and T_2' can now be determined. The maximum collector current is 4.48 A. To be on the safe side, we choose $I_{C\,max} = 10\,\text{A}$. The maximum collector–emitter voltage occurs at full output swing and is $V_b + \hat{V}_o = 51.4\,\text{V}$. We choose the reverse collector–emitter voltage to be $V_{CER} = 80\,\text{V}$. Using the relation

$$P_T = 0.1\frac{V_b^2}{R_L}$$

from Sect. 15.2.1, we obtain $P_{T\,1'} = P_{T\,2'} = 17\,\text{W}$. The relation between the power dissipation and the thermal resistance is according to (2.18) on page 51

$$P_{D\,max} = \frac{\vartheta_J - \vartheta_A}{R_{th\,JC} + R_{th\,CA}}.$$

The maximum junction temperature ϑ_J is normally 175 °C for silicon transistors. The ambient temperature ϑ_A within the housing of the amplifier should not exceed 55 °C. The thermal resistance between the heat sink and air is assumed to be $R_{th\,CA} = 4\,\text{K/W}$. Hence, the value of the thermal resistance between the junction and the transistor case must be no larger than

$$17\,\text{W} = \frac{175\,°\text{C} - 55\,°\text{C}}{\dfrac{4\,\text{K}}{\text{W}} + R_{th\,JC}},$$

that is,

$$R_{th\,JC} = \frac{3.1\,\text{K}}{\text{W}}.$$

Instead of the thermal resistance $R_{th\,JC}$, the maximum power dissipation P_{D25} at a case temperature of 25 °C is often given in the specifications. It can be calculated from:

$$P_{D25} = \frac{\vartheta_J - 25\,°\text{C}}{R_{th\,JC}} = \frac{150\,\text{K}}{3.1\,\text{K/W}} = 48\,\text{W}.$$

The transistors selected in this way are assumed to have a current transfer ratio of 30 at the maximum output current. We can therefore determine the data of the driver transistors T_1 and T_2. Their maximum collector current is

$$\frac{4.48\,\text{A}}{30} = 149\,\text{mA},$$

although this value applies to low frequencies only. For frequencies above $f_c \approx 20\,\text{kHz}$, the current transfer ratio of audio power transistors falls markedly. When the current rises steeply, the driver transistor must therefore momentarily supply the largest proportion of the output current. To obtain the largest possible bandwidth, we choose $I_{C\,max} = 1\,\text{A}$. Transistors within this range of collector currents, which have gain–bandwidth products in the region of 50 MHz, are still reasonably priced.

We have shown in Sect. 15.3 that it is useful to allow the quiescent current to flow only through the driver transistors, and to have a voltage of about 400 mV across resistors R_1 and R_2. This is the purpose of the three silicon diodes D_3, which have a total forward voltage of about 2.1 V. We select a quiescent current of approximately 30 mA to keep the crossover distortion reasonably small. Thus

$$R_1 = R_2 = \frac{400\,\text{mV}}{30\,\text{mA}} = 13\,\Omega.$$

The power dissipation of the driver transistors is, at zero input voltage, $30\,\text{mA} \cdot 29\,\text{V} \approx 0.9\,\text{W}$, and at maximum input it is still 0.75 W. A small power transistor in a TO-5 case with cooling fins is obviously sufficient. A value of 100 is usual for the current transfer ratio of such a transistor. The maximum base current is then:

$$I_{B\,max} = \frac{1}{100}\left(\frac{4.48\,\text{A}}{30} + \frac{0.8\,\text{V}}{13\,\Omega}\right) \approx 2\,\text{mA}.$$

The current through the constant current sources T_3 and T_4 must be large compared to this value, and we select 10 mA.

Emitter followers are prone to unwanted oscillations in the region of the transit frequency of the output transistors. These oscillations can be damped by additionally loading the output by a series RC element (approximately $1\,\Omega$; $0.22\,\mu\text{F}$). However, this also reduces the efficiency at higher frequencies. Another method of damping that may also be used in addition to that described above is to provide series resistors in the base lead of the driver transistors, in conjunction with an increased collector–base capacitance. If $R_7 = R_8 = 100\,\Omega$, as shown in Fig. 15.22, the voltage across these resistors remains below 0.2 V. The achievable output voltage swing is therefore not significantly reduced.

15.8
Driver Circuits with Voltage Gain

The power amplifiers described above have a certain amount of crossover distortion in the region of zero output voltage, but this can be largely eliminated by feedback. The output stage is connected to a preamplifier stage, and negative feedback is applied across both stages. Figure 15.23 shows a simple possibility. The output stage drive is supplied via current source T_3 which, in conjunction with T_7, forms a current mirror for I_{C6}. The differential amplifier T_5 T_6 provides the required voltage gain. Its effective collector resistance is relatively high, being produced by the paralleled current-source internal resistances T_3, T_4 and the input resistances of the emitter followers T_1, T_2.

The whole arrangement has feedback via resistors R_7 and R_8, and therefore has a voltage gain of $A = 1 + R_8/R_7$. In order to produce sufficient loop gain, A must not be selected to be too large, practicable values being between 5 and 30.

If we only wish to amplify AC voltages, the zero stability of the circuit can be improved by connecting a coupling capacitor in series with R_7, thereby reducing the DC voltage gain to unity. Most power amplifier ICs operate on this principle.

Wideband Drive Circuit

A larger drive circuit bandwidth can be obtained by driving both current sources T_3, T_4 in opposite directions and operating them in common-base mode. This results in the circuit shown in Fig. 15.24, which is similar in some respects to the wideband operational amplifier shown in Fig. 5.27. However, as a power amplifier requires no differential input, half of the push–pull differential stage has been omitted and replaced by the push–pull stage within the operational amplifier. The latter stabilizes the quiescent potential. The overall circuit behaves like an inverting operational amplifier, with a feedback path via R_{15}, and R_{16}. Its gain is therefore $A = -R_{16}/R_{15}$.

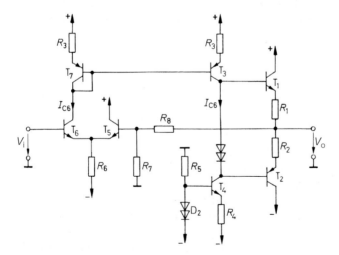

Fig. 15.23. A simple driver circuit with voltage gain

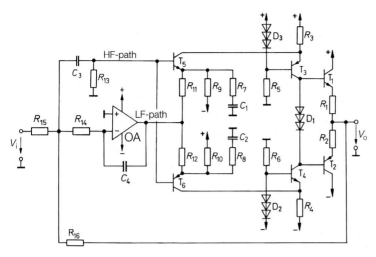

Fig. 15.24. Broadband power amplifier (HF = high frequency, LF = low frequency)

To begin the dimensioning of the circuit, the collector currents of the transistors $T_3 - T_6$ are defined. We choose 10 mA. A current of 20 mA must then flow through resistors R_3 and R_4. A voltage of 1.4 V is dropped across R_3 and R_1. Hence

$$R_3 = R_4 = \frac{1.4\,\text{V}}{20\,\text{mA}} = 70\,\Omega.$$

The output quiescent potential of the operational amplifier is determined by the offset voltage of the power output stage and is close to zero. Hence, with no input drive, the current through resistors R_{11}, and R_{12} is virtually zero. The collector currents of T_5 and T_6 must therefore flow through resistors R_9 and R_{10}. With supply potentials of $\pm 15\,\text{V}$, it follows:

$$R_9 = R_{10} \approx \frac{15\,\text{V}}{10\,\text{mA}} = 1.5\,\text{k}\Omega.$$

To attain the full swing of the current sources T_3 and T_4, the collector currents of T_5 and T_6 must be controlled between zero and 20 mA. These values should be reached for a full output swing of the operational amplifier. Thus, for resistors R_{11} and R_{12},

$$R_{11} = R_{12} \approx \frac{10\,\text{V}}{10\,\text{mA}} = 1\,\text{k}\Omega.$$

The operational amplifier OA is configured as an integrator. Its gain is defined by the external circuitry, and is selected such that it is markedly below the open-loop gain of the operational amplifier. If we select $R_{14} = 10\,\text{k}\Omega$ and $C_4 = 160\,\text{pF}$, for instance, the gain is unity at a frequency of 100 kHz. The lower cutoff frequency of the highpass filter C_3, R_{13} in the high-frequency path must have a lower value; for example, 1 kHz.

The total gain of the circuit can be set by resistors R_{15} and R_{16} to values between 1 and 10. A higher gain is inadvisable, as the loop gain in the high-frequency path then becomes too low. The open-loop gain of the high-frequency path can be varied by means of resistors R_7 and R_8. They are adjusted so as to obtain the desired transient response for the whole circuit. For the operational amplifier, the internal standard frequency compensation is sufficient. To avoid oscillations in the VHF range, it may be necessary to insert resistors in the base leads of the transistors.

15.9
Boosting the Output Current
of Integrated Operational Amplifiers

The output current of integrated operational amplifiers is normally limited, the maximum being about 20 mA. There are many applications for which about ten times this current is required, but where the number of additional components must be kept to a minimum. In such cases, the power output stages described here may be used. For low signal frequencies, the number of components can be reduced by the use of push–pull class-B emitter followers. However, due to the finite slew rate of the operational amplifier, noticeable crossover distortion occurs even with feedback. It can be reduced considerably by inserting a resistance R_1 as in Fig. 15.25, which bypasses the emitter followers in the region of zero voltage. In this case, the slew rate required of the amplifier is reduced from infinity to a value that is $1 + R_1/R_L$ times that of the rate of change of the output voltage.

The arrangement shown in Fig. 15.26 has the same characteristics as the previous circuit. Here, however, the output transistors are controlled by the supply terminals of the operational amplifier. If we make $R_2 = 0$, this, together with the output transistors of the operational amplifier, results in two complementary Darlington connections.

At small output currents, the two output transistors T_1 and T_2 are turned off. The operational amplifier then supplies the whole output current. At larger output currents, transistors T_1 and T_2 alternately become conducting and supply the largest proportion of the output current. The contribution of the operational amplifier remains limited to approximately $0.7\,V/R_1$.

The circuit has an advantage over the previous one in that the quiescent current of the operational amplifier causes biasing of the base–emitter junctions of the power output transistors. The values of the resistors R_1 are such that the bias is about 400 mV. This considerably reduces the range of crossover without the need for a quiescent current in the output transistors, the stabilization of which would require additional measures.

Using the voltage divider R_2, R_3, the output stage can provide an additional voltage gain of $1 + R_2/R_3$. This enables the output voltage swing of the amplifier to be increased nearly to the supply voltage only limited by the saturation voltage of T_1 or T_2. The likelihood of oscillation within the complementary Darlington circuits is also reduced.

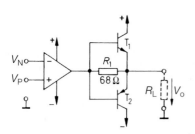

Fig. 15.25. A current booster with complementary emitter followers

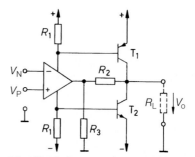

Fig. 15.26. A current booster with complementary common-emitter circuits

Chapter 16:
Power Supplies

Every electronic circuit requires a power supply that provides one or more DC voltages. For larger power requirements, batteries are not economical. The DC voltage is therefore obtained from the AC line supply by transformation and subsequent rectification. The DC voltage thus obtained usually has considerable ripple, and changes in response to variations in the line voltage and the load. Therefore, a voltage regulator is often connected to the rectifier to keep the DC output voltage constant and counteract these variations. The following two sections describe ways of providing the unregulated DC voltage; regulator circuits will be dealt with later.

16.1
Properties of Power Transformers

The internal resistance R_i of the power transformer plays an important part in the design of rectifier circuits. It can be calculated from the rating of the secondary winding $V_{\sec n}$, $I_{\sec n}$, and from the loss factor L, which is defined as the ratio of the no-load voltage $V_{\sec 0}$ to the nominal voltage $V_{\sec n}$:

$$L = \frac{V_{\sec 0}}{V_{\sec n}}; \qquad 1 < L < 1.5 \tag{16.1}$$

Hence, the relation for the internal ohmic resistance is:

$$R_i = \frac{V_{\sec 0} - V_{\sec n}}{I_{\sec n}} = \frac{V_{\sec n}(L-1)}{I_{\sec n}} \tag{16.2}$$

We define a nominal load $R_N = V_{\sec n}/I_{\sec n}$ and obtain, from (16.2),

$$R_i = R_N(l-1) \tag{16.3}$$

The data of the M-core transformers normally used are listed in the table of Fig. 16.1; the corresponding data for toroidal transformers are given in Fig. 16.2.

As toroidal transformers are more difficult to wind, they are significantly more expensive, particularly for low powers. However, this is offset by the advantage of minimal magnetic flux leakage, a higher magnetizing reactance, and therefore a lower magnetizing current and small no-load losses.

The values in Figs. 16.1 and 16.2 are based on a line voltage $V_{p\,\mathrm{rms}} = 230\,\mathrm{V}$ at $f_l = 50\,\mathrm{Hz}$ and a maximum flux density of $\hat{B} = 1.2\,\mathrm{T}$. Should the line voltage deviate slightly from this value, w_1 must be recalculated proportionally to $V_{p\,\mathrm{rms}}$ and d_1 to $1/\sqrt{V_{p\,\mathrm{rms}}}$. If the line frequency is $f_l = 60\,\mathrm{Hz}$, $\hat{B}$ reduces to $1\,\mathrm{T}$, and the parameters w_1 and d_1 in Figs. 16.1 and 16.2 include a safety margin.

Core-type (lateral-length)	Rated power	Loss factor	Number of primary	Primary wire gauge	Normalized number of secondary	Normalized secondary wire
	P_n	L	w_1	d_1	w_2/V_2	$d_2/\sqrt{I_2}$
[mm]	[W]			[mm]	[1/V]	[mm/$\sqrt{A}$]
M 42	4	1.31	4716	0.09	28.00	0.61
M 55	15	1.20	2671	0.18	14.62	0.62
M 65	33	1.14	1677	0.26	8.68	0.64
M 74	55	1.11	1235	0.34	6.24	0.65
M 85a	80	1.09	978	0.42	4.83	0.66
M 85b	105	1.06	655	0.48	3.17	0.67
M 102a	135	1.07	763	0.56	3.72	0.69
M 102b	195	1.05	513	0.69	2.45	0.71

Fig. 16.1. Typical data of M-core supply transformers for a primary voltage of $V_{p\,rms} = 230\,\text{V}$ at $f_l = 50\,\text{Hz}$

Outer diameter approx.	Rated power	Loss factor	Number of primary turns	Primary wire gauge	Normalized number of secondary turns	Normalized secondary wire gauge
D	P_n	L	w_1	d_1	w_2/V_2	$d_2/\sqrt{I_2}$
[mm]	[W]			[mm]	[1/V]	[mm/$\sqrt{A}$]
60	10	1.18	3500	0.15	19.83	0.49
61	20	1.18	2720	0.18	14.83	0.54
70	30	1.16	2300	0.22	12.33	0.55
80	50	1.15	2140	0.30	11.25	0.56
94	75	1.12	1765	0.36	9.08	0.58
95	100	1.11	1410	0.40	7.08	0.60
100	150	1.09	1100	0.56	5.42	0.61
115	200	1.08	820	0.60	4.00	0.62
120	300	1.07	715	0.71	3.42	0.63

Fig. 16.2. Typical data of toroidal core supply transformers for a primary voltage of $V_{p\,rms} = 230\,\text{V}$ at 50 Hz

16.2
Power Rectifiers

16.2.1
Half-Wave Rectifier

The easiest way to rectify an AC voltage is to charge a capacitor via a diode, as shown in Fig. 16.3. If the output is not loaded, the capacitor C is charged during the positive half-cycle to the peak value $V_{o0} = \sqrt{2}\,V_{\sec 0\,\text{rms}} - V_D$, where V_1 is the forward voltage of the diode. The peak reverse voltage of the diode occurs when the transformer voltage is at its negative maximum, and therefore has the value of $\sqrt{2}\,V_{\sec 0\,\text{rms}}$.

When a load resistance R_L is connected to the DC output, it discharges the capacitor C for as long as the diode is reverse biased. Only when the no-load voltage of the transformer exceeds that of the output by the amount V_D is the capacitor recharged. The voltage reached by recharging depends on the internal resistance R_i of the transformer. The shape of the output voltage at steady state is shown in Fig. 16.4. Owing to an unfavorable ratio of the

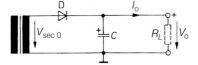

Fig. 16.3. Single-phase half-wave rectifier

No-load output voltage: $\quad\quad\quad\quad\quad\quad V_{o0} = \sqrt{2}V_{\text{sec 0 rms}} - V_D$

On-load output voltage (C infinitely large) : $\quad V_{o\infty} = V_{o0}\left(1 - \sqrt{\dfrac{R_i}{R_L}}\right)$

Peak reverse voltage: $\quad\quad\quad\quad\quad\quad V_{pr} = 2\sqrt{2}V_{\text{sec 0 rms}}$

Mean diode current: $\quad\quad\quad\quad\quad\quad \bar{I}_D = I_o$

Repetitive peak current: $\quad\quad\quad\quad\quad I_{Dp} = \dfrac{V_{oD}}{\sqrt{R_i R_L}}$

Ripple voltage (peak-to-peak): $\quad\quad\quad V_{\text{r pp}} = \dfrac{I_o}{Cf_l}\left(1 - \sqrt[4]{\dfrac{R_i}{R_L}}\right)$

Lowest value of output voltage: $\quad\quad V_{o\,min} \approx V_{o\infty} - \dfrac{2}{3}V_{\text{r pp}}$

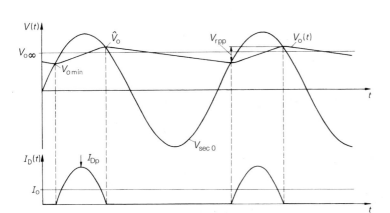

Fig. 16.4. Voltage and current waveforms for a single-phase half-wave rectifier

recharge and discharge times, the output voltage is considerably reduced even for small load currents, and for this reason the circuit is unsuitable for use in power supplies.

16.2.2
Bridge Rectifier

The ratio of the recharge and discharge times can be greatly improved by charging the capacitor C during the positive and negative half-cycles. This is achieved by means of the bridge rectifier shown in Fig. 16.5.

During the recharge period, the diodes connect negative terminal of the transformer to ground, and the positive terminal to the output. The repetitive peak reverse voltage of the diodes is identical to the no-load output voltage:

$$V_{o0} = \sqrt{2}V_{\text{sec 0 rms}} - 2V_D \quad\quad\quad\quad\quad\quad (16.4)$$

and is only half that of the half-wave rectifier.

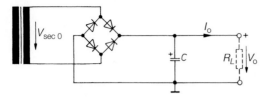

Fig. 16.5. Bridge rectifier

No-load output voltage:	V_{o0}	$= \sqrt{2}V_{\text{sec 0 rms}} - 2V_D$
On-load output voltage:	$V_{o\infty}$	$= V_{o0}\left(1 - \sqrt{\dfrac{R_i}{2R_L}}\right)$
Peak reverse voltage:	V_{pr}	$= 2\sqrt{2}V_{\text{sec 0 rms}}$
Mean diode current:	$\bar{I}_D$	$= \dfrac{1}{2}I_o$
Repetitive peak current:	I_{Dp}	$= \dfrac{V_{o0}}{\sqrt{2R_i R_L}}$
Ripple voltage (peak-to-peak):	$V_{\text{r pp}}$	$= \dfrac{I_o}{2Cf_l}\left(1 - \sqrt[4]{\dfrac{R_i}{2R_L}}\right)$
Lowest value of output voltage:	$V_{o\,min}$	$\approx V_{o\infty} - \dfrac{2}{3}V_{\text{r pp}}$
Rated transformer power:	P_N	$= (1.2\ldots2)V_{o\infty} \cdot I_o$

To calculate the voltage reduction at load, we initially assume an infinitely large storage capacitor C. The output voltage is then a pure DC voltage, which we define as $V_{o\infty}$. The more the output voltage decreases due to the load, the longer is the recharge time. Steady state is reached when the incoming charge of the capacitor equals the outgoing charge; in other words, that supplied to the load. Hence,

$$V_{o\infty} = V_{o0}\left(1 - \sqrt{\frac{R_i}{2R_L}}\right) \tag{16.5}$$

where $R_L = V_{o\infty}/I_o$ is the load resistance. The deduction of this equation is based on calculations involving the approximation of sine waves by parabolas and is omitted here because of its complexity. As comparison with the half-wave circuit in Fig. 16.3 shows, in this bridge rectifier only half the internal resistance of the transformer is responsible for the voltage drop at load.

To dimension the rectifier correctly, the currents must be known. As no DC current flows through the capacitor, the mean forward current of each bridge arm is half of the output current. As the forward voltage is only slightly dependent on the current, the power dissipation of a single diode is given by

$$P_D = \frac{1}{2}V_D I_o$$

During every recharge period a peak current I_{Dp} flows, the value of which may be many times that of the output current:

$$I_{Dp} = \frac{\hat{V}_{\text{sec 0}} - 2V_D - V_{o\infty}}{R_i} = \frac{V_{o0} - V_{o\infty}}{R_i}$$

Using (16.5), it follows:

$$I_{Dp} = \frac{V_{o0}}{\sqrt{2R_i R_L}}$$

It can be seen that the internal resistance R_i of the AC voltage source significantly influences the peak current. If R_i is very small, it may be necessary to series-connect a resistance or inductance so as not to exceed the maximum peak current of the rectifier. This must be taken into account particularly for direct rectification of the line voltage; that is, without a transformer. Full-wave rectification is also better in this respect, as the peak current is reduced by a factor of $\sqrt{2}$.

The rms value of the pulsating charging current is larger than its mean value. Therefore, the DC power must always be kept smaller than the rated power of the transformer for a resistive load, so as not to exceed the thermal transformer rating. The DC power is determined by the power supplied to the load ($I_o V_{o\infty}$) and the losses in the rectifier (approximately $2V_D I_o$). The rated power of the transformer must therefore be selected as

$$P_n = \alpha I_o (V_{o\infty} + 2V_D) \approx \alpha I_o V_{o\infty} \tag{16.6}$$

where α is the form factor allowing for the increased rms value of the current. For full-wave rectification, $\alpha \approx 1.2$. However, it is advisable not to operate at the thermal limit as in (16.6), but to overrate the transformer by using a higher value for α, thereby producing a greater efficiency. The disadvantage of an increased space requirement can be minimized by employing toroidal transformers. Even if they are considerably overrated, the no-load losses remain low.

For a finite capacitance C, a superposed ripple voltage appears at the output. It can be calculated from the charge supplied by the capacitor during the discharge time t_d,

$$V_{rpp} = \frac{I_o t_d}{C}$$

Using (16.5),

$$t_d \approx \frac{1}{2}\left(1 - \sqrt[4]{\frac{R_i}{2R_L}}\right) T_l$$

where $T_L = 1/f_l$ is the reciprocal of the AC supply frequency. Hence

$$V_{rpp} = \frac{I_o}{2Cf_l}\left(1 - \sqrt[4]{\frac{R_i}{2R_L}}\right) \tag{16.7}$$

The lowest instantaneous value of the output voltage is of special interest. It is approximately

$$V_{o\,min} \approx V_{o\infty} - \tfrac{2}{3}V_{rpp} \tag{16.8}$$

The dimensioning of a power rectifier is best illustrated by an example. A DC supply is required, having a minimum output voltage of $V_{o\,min} = 30\,\mathrm{V}$ for an output current of $I_o = 1\,\mathrm{A}$, and a ripple of $V_{rpp} = 3\,\mathrm{V}$.

To begin with, we obtain, from (16.8),

$$V_{o\infty} = V_{o\,min} + \tfrac{2}{3}V_{rpp} = 32\,\mathrm{V}$$

and from (16.6) and $\alpha = 1.5$ the rated power of the transformer:

$$P_n = \alpha I_o(V_{o\infty} + 2V_D) = 1.5\,\mathrm{A}(32\,\mathrm{V} + 2\,\mathrm{V}) = 51\,\mathrm{W}$$

It can be seen from the table in Fig. 16.2 that a toroidal type with $D = 80\,\mathrm{mm}$, having a loss factor of $L = 1.15$, can be used. We now need to know the internal resistance of the transformer; however, it is dependent on the rated voltage, the value of which is not yet known. For its determination, the system of nonlinear equations given in (16.3)–(16.5) must be solved. This is best done by iteration. We set the initial value of $V_{\mathrm{sec\,n\,rms}} \approx V_{o\,min} = 30\,\mathrm{V}$. Using (16.3), it follows:

$$\begin{aligned} R_i &= R_n(L - 1) = \frac{V^2_{\mathrm{sec\,n\,rms}}}{P_n}(L - 1) \\ &= \frac{(30\,\mathrm{V})}{51\,\mathrm{W}} \cdot (1.15 - 1) = 2.65\,\Omega \end{aligned}$$

Hence, using (16.4) and (16.5),

$$\begin{aligned} V_{o\infty} &= (\sqrt{2}V_{\mathrm{sec\,n\,rms}} \cdot L - 2V_D)\left(1 - \sqrt{\frac{R_i}{2R_L}}\right) \\ &= (\sqrt{2} \cdot 30\,\mathrm{V} \cdot 1.15 - 2\,\mathrm{V})\left(1 - \sqrt{\frac{2.65\,\Omega}{2 \cdot 32\,\mathrm{V}/1\,\mathrm{A}}}\right) \approx 37.3\,\mathrm{V} \end{aligned}$$

The voltage is about 5 V higher than that initially required. For the first iteration, we decrease the rated transformer voltage by this amount and, correspondingly, obtain

$$R_i = 1.84\,\Omega \quad \text{and} \quad V_{o\infty} = 32.1\,\mathrm{V}$$

which is already the desired value for the output voltage. The design parameters for the transformer are therefore

$$V_{\mathrm{sec\,n\,rms}} \approx 25\,\mathrm{V}; \qquad I_{\mathrm{sec\,n\,rms}} = \frac{P_n}{V_n} \approx 2\,\mathrm{A}$$

Figure 16.2 gives the winding data for a primary voltage of 220 V/50 Hz:

$$\begin{aligned} w_1 &= 2140 & d_1 &= 0.30\,\mathrm{mm}, \\ w_2 &= 11.25\frac{1}{\mathrm{V}} \cdot 25\,\mathrm{V} = 281, & d_2 &= 0.56\frac{\mathrm{mm}}{\sqrt{\mathrm{A}}}\sqrt{2\,\mathrm{A}} = 0.79\,\mathrm{mm} \end{aligned}$$

The capacitance of the storage capacitor is given by (16.7) as

$$\begin{aligned} C &= \frac{I_o}{2V_{\mathrm{r\,pp}}f_l}\left(1 - \sqrt[4]{\frac{R_i}{2R_L}}\right) \\ &= \frac{1\,\mathrm{A}}{2 \cdot 3\,\mathrm{V} \cdot 50\,\mathrm{Hz}}\left(1 - \sqrt[4]{\frac{1.84\,\Omega}{2 \cdot 32\,\Omega}}\right) \approx 2000\,\mu\mathrm{F} \end{aligned}$$

The no-load output voltage is 39 V. The capacitor must be rated for at least this voltage.

The calculation for transformers that have several secondary windings is the same as that above. For P_n, the rated power of the corresponding secondary winding must be inserted. The total power is the sum of the individual powers of the secondary windings. This determines the choice of the core and therefore the loss factor L.

16.2.3
Center-Tap Rectifier

Full-wave rectification can also be achieved by rectifying two antiphase AC voltages on a half-wave basis. This principle is illustrated by the center-tap circuit shown in Fig. 16.6. As we can see from the data given, the advantages of the bridge circuit are retained.

An additional advantage is that the current need only flow through one diode at a time, rather than through two as in the case of the bridge circuit. As a result, the voltage drop caused by the forward voltages of the diodes is halved. On the other hand, the internal resistance of the transformer is doubled, as each part-winding must be rated for half the output power, thereby further increasing the voltage drop. The ratio of the output voltage to the forward voltage of the diode will dictate which effect predominates. The center-tap circuit is better for low output voltages, and the bridge rectifier circuit for high output voltages.

Double Center-Tap Rectifier

The negative half-cycles, which remain unutilized in the circuit in Fig. 16.6, can be rectified in a second center-tap circuit using diodes of opposite polarity, thereby simultaneously producing a negative DC voltage. This method of generating voltages that are balanced to ground is shown in Fig. 16.7. An IC bridge rectifier can be used to provide the four diodes required. The nominal transformer rating must again be 1.2...2.0 times the DC output power.

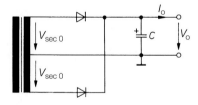

Fig. 16.6. Center-tap circuit

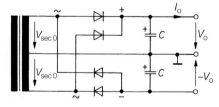

Fig. 16.7. Center-tap circuit for symmetrical output voltages

No-load output voltage:	$V_{o\,0}$	$= \sqrt{2}V_{\text{sec }0} - V_D$
On-load output voltage:	$V_{o\,\infty}$	$= V_{o\,0}\left(1 - \sqrt{\dfrac{R_i}{2R_L}}\right)$
Peak reverse voltage:	V_{pr}	$= 2\sqrt{2}V_{\text{sec }0}$
Mean forward current:	$\bar{I}_D$	$= \dfrac{1}{2}I_o$
Repetitive peak current:	I_{Dp}	$= \dfrac{V_{o\,0}}{\sqrt{2R_i R_L}}$
Ripple voltage:	$V_{\text{r pp}}$	$= \dfrac{I_o}{2Cf_l}\left(1 - \sqrt[4]{\dfrac{R_i}{2R_L}}\right)$
Minimum output voltage:	$V_{o\,\text{min}}$	$\approx V_{o\,\infty} - \dfrac{2}{3}V_{\text{r pp}}$

16.3
Linear Voltage Regulators

Electronic circuits generally require a DC voltage that is accurate to within 5–10% of a specified value. This tolerance must be maintained over the entire range of line voltage, load current, and temperature variations. The ripple voltage must not exceed the millivolt range. For these reasons, the output voltage of the rectifier circuits described is not directly usable as a supply voltage for electronic circuits, but requires stabilization and smoothing by a following voltage regulator.

The principal characteristics of a voltage regulator are:

1) The output voltage and its tolerance.
2) The maximum output current and the short-circuit current.
3) The minimum voltage drop required by the voltage regulator to maintain the output voltage – this is termed "dropout voltage" in data sheets.
4) Suppression of input voltage variations; that is, line regulation.
5) Counteraction of load current variations; that is, load rejection.

16.3.1
Basic Regulator

Output voltage variations due to supply voltage and load current fluctuations can be reduced by inserting a controlled series resistance, this method being known as series regulation.

The simplest series regulator is an emitter follower, with its base connected to a reference voltage source. The reference voltage can, for instance, be obtained from the unstabilized input voltage V_i using a Zener diode, as shown in Fig. 16.8. Other possibilities will be discussed in Sect. 16.4. The output voltage is

$$V_o = V_{\text{ref}} - V_{BE}$$

The extent to which the voltage varies with load is related to the output resistance:

$$r_o = -\frac{\partial V_o}{\partial I_o} = \frac{1}{g_m} = \frac{V_T}{I_o}$$

With $V_T \approx 26\,\text{mV}$, we obtain approximately $0.3\,\Omega$ for $I_o = 100\,\text{mA}$.

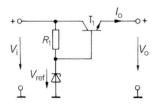

Fig. 16.8. Voltage stabilization using an emitter follower

Output voltage: $V_o = V_{\text{ref}} - V_{BE}$

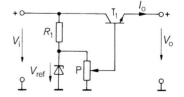

Fig. 16.9. Additional circuitry for output voltage adjustment

$0 \leq V_o \leq V_{\text{ref}} - V_{BE}$

Input voltage variations are compensated by the low differential resistance r_Z of the Zener diode. The output voltage variation is given by

$$\Delta V_o = \Delta V_{\text{ref}} = \frac{r_Z}{R_1 + r_Z}\Delta V_i \approx \frac{r_Z}{R_1}\Delta V_i$$

This represents a regulation of the output voltage between 1% and 10% of the input voltage variation, depending on the component values selected.

If an adjustable output voltage is required, a portion of the reference voltage can be tapped off at a potentiometer, as shown in Fig. 16.9. The potentiometer resistance selected must be small compared to r_{BE}, so that the circuit output resistance is not increased appreciably.

16.3.2
Voltage Regulators with a Fixed Output Voltage

The simple circuits shown in Figs. 16.8 and 16.9 are largely inadequate, or fail to meet the requirements that voltage regulators must satisfy. Consequently, IC voltage regulators contain a voltage control amplifier, a reference voltage source, and several additional modules to protect the power transistor. These are shown in the block diagram of Fig. 16.10.

The current-limiting circuit monitors the voltage drop across current-sensing resistor R. The safe operating area (SOA) of the power transistor is monitored in an additional block. If the voltage drop across the power transistor increases, the current limit is reduced accordingly.

A thermal protection device monitors the chip temperature and reduces the output voltage if hazardous overheating is likely to occur. The diodes realize an analog AND-gate and insure that the output voltage is determined by the lowest of the four correcting variables. The amplifier holds the output voltage at the nominal value only as long as no limit value is exceeded.

The practical implementation of a 7800-series IC voltage regulator is shown in Fig. 16.11. The requirements placed on the amplifier are not particularly stringent, as an emitter follower alone already constitutes an effective voltage regulator. Consequently, it is sufficient to have a simple differential amplifier T_3, T_4 operating in conjunction with Darlington circuit T_1, as a power operational amplifier. It acts as a noninverting amplifier

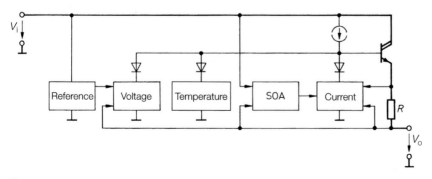

Fig. 16.10. A schematic of an IC voltage regulator

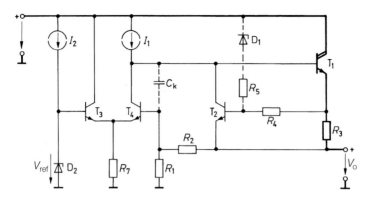

Fig. 16.11. A 7800-series IC voltage regulator

$$V_o = \left(1 + \frac{R_2}{R_1}\right) V_{\text{ref}}; \qquad I_{o\,max} = \frac{0.6\,\text{V}}{R_3}$$

via voltage divider R_1, R_2 in the feedback path and, at the output, produces an amplified reference voltage of

$$V_o = (1 + R_2/R_1)V_{\text{ref}}$$

Transistor T_2 has a current-limiting function. If the voltage drop across R_3 reaches 0.6 V, T_2 is turned on, thereby reducing the output voltage. Due to the feedback path produced, the output voltage is adjusted so that the voltage drop across R_3 is stabilized to the value 0.6 V. This is equivalent to a constant output current of

$$I_{o\,max} = 0.6\,\text{V}/R_3$$

Under these conditions, the output voltage is determined by load resistor R_L, in accordance with $V_o = I_{o\,max} R_L$.

When the maximum current is reached, the power dissipation in output transistor T_1 is given by

$$P = I_{o\,max}(V_i - V_o)$$

In the event of a short-circuit at the output, it is much higher than during normal operation $P = I_{o\,max} V_i$, since the output voltage then falls below the nominal value to zero. In order to prevent this increased power dissipation, the current limit can be reduced as the output voltage decreases. This produces the foldback characteristic shown in Fig. 16.12.

A marked increase in power dissipation may also occur if the input voltage V_1 is increased, as in this case the difference $V_i - V_o$ increases likewise. Consequently, the best

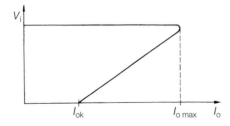

Fig. 16.12. Foldback current limiting

way of protecting output transistor T_1 is to match the current limit $I_{o\,max}$ to the voltage difference $V_i - V_o$. Resistor R_5 and Zener diode D_1, shown by the dashed line in Fig. 16.11, are used for this purpose.

If the potential difference $V_i - V_o$ is less than the Zener voltage V_Z of diode D_1, no current flows through resistor R_5. Consequently, the current limit in this case remains $0.6\,V/R_3$. If the potential difference exceeds the value V_Z, voltage divider R_5, R_4 causes a positive base–emitter bias to be applied to transistor T_2. As a result, the latter will be turned on in response to a correspondingly smaller voltage drop across R_3.

Capacitor C_k provides the frequency compensation required for stability. To provide additional stabilization, it is generally necessary to connect capacitors with approximately $100\,\mu F$ to ground at the input and output.

16.3.3
Voltage Regulators with an Adjustable Output Voltage

In addition to the fixed voltage regulators described above, adjustable types are also available (78 G series). In the latter type, the voltage divider R_1, R_2 is omitted and the amplifier input is brought out as shown in Fig. 16.13. These regulators therefore have four terminals. By connecting voltage divider R_1, R_2 externally, any output voltage between $V_{\mathrm{ref}} \approx 5$ V $\leq V_o < V_i - 3\,V$ can be selected.

Adjustable voltage regulators with only three terminals can be realized by connecting the negative supply voltage of the OA to the regulated output instead of ground. In order to make this difference clear, Fig. 16.13 shows a 78 G series adjustable voltage regulator with four terminals alongside a 317 series adjustable voltage regulator with three terminals (Fig. 16.14). Here, the reference voltage source is not connected to ground, but to the midpoint of the feedback voltage divider. The output voltage therefore increases until voltage V_{ref} is dropped across R_2. The input voltage difference of the operational amplifier is then zero.

The output of the voltage regulator in Fig. 16.14 must not be in open-circuit, as this would prevent the amplifier supply current loop from being closed. It is therefore advisable to select low values for voltage divider $R_1 R_2$, typically $R_2 = 240\,\Omega$; a current of 5 mA therefore flows for a reference voltage of $V_{\mathrm{ref}} = 1.25\,V$. Consequently, the current of

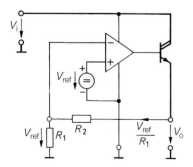

Fig. 16.13. Adjustable voltage regulator with four terminals (78 G series)

$$V_o = \left(1 + \frac{R_2}{R_1}\right) V_{\mathrm{ref}}; \quad V_{\mathrm{ref}} = 5\,V$$

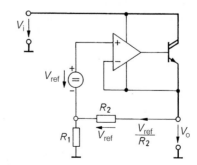

Fig. 16.14. Adjustable voltage regulator with three terminals (317 series)

$$V_o = \left(1 + \frac{R_1}{R_2}\right) V_{\mathrm{ref}}; \quad V_{\mathrm{ref}} = 1.25\,V$$

approximately 100 μA flowing out of the reference voltage source cannot appreciably alter the voltage drop across R_1.

16.3.4
A Voltage Regulator with a Reduced Dropout Voltage

As we can see from Fig. 16.11, the minimum voltage drop between the input and output of the voltage regulator is made up of the voltage drop of 0.6 V across current-sensing resistor R_3, the Darlington circuit's base–emitter voltage of 1.6 V, and the minimum voltage drop of about 0.3 V across current source I_1. The minimum voltage drop (the dropout voltage) is therefore 2.5 V. This is particularly troublesome for regulating low output voltages: in the case of a 5 V regulator, at least 50% of the output power is dissipated. As an additional voltage drop is required to compensate for line and load changes, an even higher power dissipation results, which is generally just as large as the output power.

Removal of the heat generated frequently poses problems. Although IC voltage regulators are provided with thermal protection, this means that the maximum output current is reduced accordingly if cooling is inadequate. It is therefore important to keep the minimum voltage drop as small as possible. This can be achieved in the circuit shown in Fig. 16.11 by operating the current source I_1 from an auxiliary voltage a few volts above the input voltage.

A method without auxiliary voltage is to use a pnp transistor as the power transistor, as shown in Fig. 16.15. Here, the minimum voltage drop across the voltage regulator is equal to the saturation voltage of the power transistor T_1. It can be held below 0.5 V if an appropriately high base current is applied. However, in order to provide the required base currents for T_1, a Darlington pair should not be used, as the minimum voltage drop would be increased by an emitter–base voltage. Transistor T_2 is therefore operated in a common-emitter connection. Current feedback via R_3 limits the maximum output current and simultaneously improves the stability of the regulation circuit. A drawback of this principle is the fact that because of the low current gain of the pnp transistors, the current consumption rises steeply with an increasing output current. This current does not contribute to the output, but is lost since the base current of T_1 flows to ground via T_2. Particularly disturbing is the rise in the current consumption when reaching the minimum voltage drop, since the current flowing through T_2 at this point reaches its maximum. These problems can be bypassed by replacing the pnp power transistor by a p-channel power MOSFET, such as the model Max 1658.

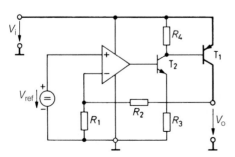

Fig. 16.15. Voltage regulator with a low dropout voltage; $V_o = \left(1 + \frac{R_2}{R_1}\right) V_{\text{ref}}$

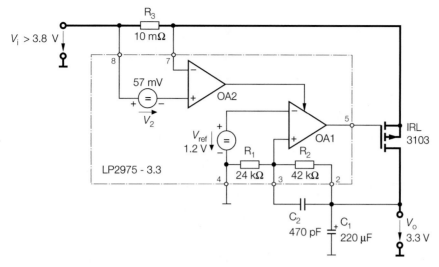

Fig. 16.16. Implementation of a voltage regulator with a low dropout voltage

Output voltage:	V_o	$= 3.3\,\text{V}$
Minimum input voltage:	$V_{i\,min}$	$= 3.8\,\text{V}$
Maximum output current:	$I_{o\,max}$	$= 5\,\text{A}$
Minimum voltage drop:	ΔV_{min}	$= 0.5\,\text{V}$
Short-circuit current:	$I_{o\,short}$	$= 5.7\,\text{A}$

For high currents it is necessary to use discrete power transistors. The control circuitry can be constructed from commercially available operational amplifiers. A very simple solution is achieved when using the voltage regulator LP2975 from National. Its internal construction and external circuitry are shown in Fig. 16.16. The operational amplifier OA1 together with the reference voltage source forms the voltage regulator circuit. Since the external power MOSFET represents an inverting amplifier in common-source connection, the negative feedback signal must be fed to the non-inverted input of the operational amplifier.

Because of the large capacitance C_1 the dominating lowpass filter is located at the output of the voltage regulator circuit. To still obtain a high stability we suggest using capacitance C_2 to turn the phase back in the critical frequency range.

Operational amplifier OA2 is used for current limitation. When the voltage drop across R_3 reaches the built-in current reference of $V_2 = 57\,\text{mV}$, the current regulator becomes active and prevents the output current from rising further.

16.3.5
A Voltage Regulator for Negative Voltages

If a floating input source is available, the voltage regulators described above can also be used for stabilizing negative output potentials. The resulting circuit is shown in Fig. 16.17. We can see that it will operate only if the unstabilized voltage source is floating because the voltage regulator or the output voltage would otherwise be shorted. This problem arises, for instance, if we use the simplified circuit (Fig. 16.7) for simultaneously generating a positive and a negative supply voltage. As the center tap is grounded, the negative supply

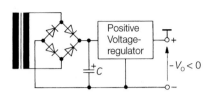

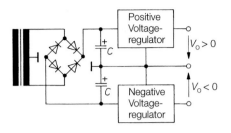

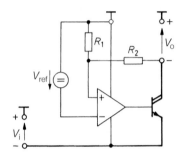

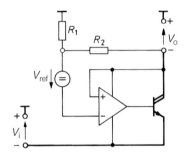

Fig. 16.17. Stabilizing a negative voltage

Fig. 16.18. Stabilizing two voltages balanced to ground

Fig. 16.19. The 7900 family

$$V_o = -\left(1 + \frac{R_2}{R_1}\right) V_{ref}$$

Fig. 16.20. The 337 family

$$V_o = -\left(1 + \frac{R_1}{R_2}\right) V_{ref}$$

potential cannot be stabilized as in Fig. 16.17. In this case voltage regulators for negative output voltages are required, as shown in Fig. 16.18. If IC types complementary to the 7800 or 317 series are used, the power transistor is operated in common-emitter configuration, as this results in an easily fabricated npn transistor. The mode of operation of the circuits shown in Figs. 16.19 and 16.20 thus corresponds to the voltage regulator with reduced dropout voltage in Fig. 16.15. For this reason, the negative voltage regulator ICs have a significantly lower dropout voltage than the corresponding positive voltage regulators.

16.3.6
Symmetrical Division of a Floating Voltage

The problem often arises, especially in battery-operated equipment, of obtaining two regulated balanced-to-ground voltages from a floating unstabilized voltage source. To solve this problem, the sum of the two voltages can be stabilized to the desired value using one of the circuits previously described. A second circuit is then required to insure that the voltage is split in the correct ratio. In principle, we could use a voltage divider with its tap grounded. The division of the voltage is kept constant if the internal resistance of the voltage divider is low, but the loss in the voltage divider is then considerably increased. It is therefore better to use a high resistance voltage divider and two power transistors. Only the transistor connected to the DC bus carrying the smaller load current is turned on at any one time. The relevant circuit is shown in Fig. 16.21.

The voltage divider formed by the two resistors R_1 halves the voltage V_i. It may have a high internal resistance, as its only load is the input bias current of the operational amplifier.

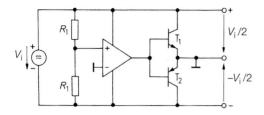

Fig. 16.21. Balancing a floating voltage

If the tap of the voltage divider is at zero potential, the voltage V_i is split into a positive and a negative voltage in the ratio 1:1, as required. The operational amplifier therefore compares the tap potential with the ground potential and adjusts its output voltage so that their difference becomes zero. Negative feedback is provided as follows: if, for instance, the positive output is loaded more than the negative output, the positive output voltage falls, thereby reducing the potential at the P-input of the operational amplifier. Because of the high gain, the amplifier output potential reduces even further, so that T_1 is turned off and T_2 is turned on. This counteracts the assumed voltage dip at the positive output. Under steady-state conditions, the current through T_2 is just large enough to insure that the two output voltages share the load equally. Transistors T_1 and T_2 therefore operate as shunt regulators, only one of which is conducting at any one time.

If the load is only slightly unbalanced, the output stage of the operational amplifier can be used directly, instead of transistors T_1 and T_2. The amplifier output is then simply connected to ground.

16.3.7
Voltage Regulator with Sensor Terminals

The resistance R_w of the connecting wires between the voltage regulator and the load, including possible contact resistances, may cancel out the low output resistance of the regulator. This effect can be eliminated by incorporating the unwanted resistances in the feedback loop; that is, by measuring the output voltage as near to the load as possible. This is the purpose of the sensor terminals S^+ and S^- in Fig. 16.22. The resistors in the sensing leads produce no errors, as only small currents flow in them.

The four-wire regulation method described can also be implemented using IC voltage regulators if the ground or voltage sensor terminal is externally accessible. Suitable types include the 78 G, 79 G, L 200, or LT 1087.

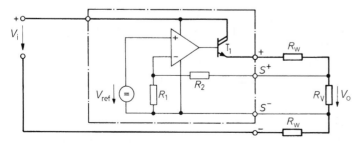

Fig. 16.22. Voltage stabilization at the load

16.3.8
Bench Power Supplies

The output voltage of the voltage regulators described so far can be adjusted only within a certain range $V_o \geq V_{ref}$. The current limit only serves to protect the voltage regulator and is therefore fixed at I_{max}.

A bench power supply must have an output voltage and a current limit, both of which are linearly adjustable between zero and the maximum value. A suitable circuit is shown in Fig. 16.23. Voltage regulation is provided by operational amplifier OA1, which is operated as an inverting amplifier. The output voltage

$$V_o = -\frac{R_2}{R_1} V_{ref\,1}$$

is proportional to the variable resistance R_2. The voltage can also be controlled by varying $V_{ref\,1}$. The output current flows from the floating unregulated power voltage source V_L via Darlington transistor T_1 through the load and via current-sensing resistor R_5 back to the source.

The voltage across R_5 is therefore proportional to the output current I_o. It is compared with a second reference voltage $V_{ref\,2}$ by operational amplifier OA2, which is operated as an inverting amplifier. As long as

$$\frac{I_o R_5}{R_4} < \frac{V_{ref\,2}}{R_3}$$

V_{P2} remains positive. The output voltage of OA2 therefore goes to its positive limit, and diode D_2 is reverse biased. In this operating condition, voltage regulation is therefore unaffected. If the output current reaches the limit

$$I_{o\,max} = \frac{R_4}{R_5 R_3} V_{ref\,2},$$

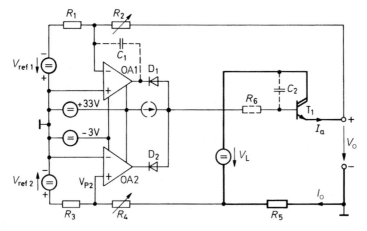

Fig. 16.23. A bench power supply with a fully adjustable output voltage and a current limit
$$V_o = -\frac{R_2}{R_1} V_{ref\,1}; \qquad I_{o\,max} = \frac{R_4}{R_5 R_3} V_{ref\,2}$$

then $V_{P2} = 0$. The output voltage of OA2 falls, and diode D_2 becomes forward biased. This causes the base potential of the Darlington pair to fall; that is, current regulation comes into effect. Amplifier OA1 tries to prevent the fall in output voltage by raising its output potential to the maximum, thereby turning off diode D_1, and current regulation is unimpaired. Therefore the two diodes act as an analog and gate.

In power supplies whose output voltage can be adjusted to zero, exceptionally high power dissipation may occur. In order to be able to achieve $V_{o\,max}$, the unstabilized voltage V_L must be greater than $V_{o\,max}$. Maximum power dissipation in T_1 occurs when the maximum output current $I_{o\,max}$ is allowed to flow at low output voltage levels. It is then approximately $V_{o\,max} \cdot I_{o\,max}$; that is, just as high as the maximum available output power. For this reason, it is preferable, when comparatively high powers are involved, to use switched-mode regulators in the output stage. This is because their power dissipation remains small even if the voltage drop is large.

16.3.9
IC Voltage Regulators

Apart from a small number of voltage regulators for special applications, these devices can be subdivided into two main families: the 7800 or 317 series (see Fig. 16.24). Both categories also include negative voltage regulators. Whereas types with adjustable output voltage are the exception in the 7800 series, all the 317-series types are adjustable and have only three terminals.

We can see that the dropout voltage for all types is 2 V or more. This is particularly troublesome for 5 V regulators that have to handle large currents, as the power dissipation in the voltage regulator is then in excess of 40% of the output power. Consequently, the power supply efficiency is only 25%, which means that three times the power delivered is converted into heat. One solution is to use voltage regulators with a reduced dropout voltage. Nevertheless, by using suitably rated discrete components as – for example, in Fig. 16.16 – even a 3.3 V power supply can be made to operate at over 50% efficiency.

Another way of minimizing losses is to use switching regulators, as we shall describe in Sect. 16.5.

16.4
Reference Voltage Generation

Every voltage regulator requires a reference voltage with which the output voltage is compared. The stability of the output voltage is only as good as that of the reference. In this section, we shall therefore examine various aspects of reference voltage generation in greater detail.

16.4.1
Zener Diode References

The simplest method of generating a reference voltage is to apply the unstabilized input voltage to a Zener diode via a series resistor, as in Fig. 16.25. The quality of the stabilization is characterized by the suppression of input voltage variations (line regulation) $\Delta V_i / \Delta V_{\text{ref}}$,

Type	Manu-facturer	Input voltage min/max	Output voltage min/max	Output current	Drop-out voltage	Special feature
Older universal types						
78xx	various	+7/35 V	+5/24 V	1 A	2 V	fixed output
79xx	various	−7/35 V	−5/24 V	1 A	1.5 V	fixed output
317	various	+7/40 V	+1.2/37 V	1.5 A	2 V	adj. output
337	various	−7/40 V	+1.2/32 V	1.5 A	1.5 V	adj. output
High current LDOs						
LT1084	Lin. Tech	+1.3/30 V	+1.3/25 V	5 A	1 V	
LT1529	Lin. Tech	+5/15 V	+4/14 V	3 A	0.6 V	$I_q = 50\,\mu A$
LT1581	Lin. Tech	+2.5/6 V	+1.3/5 V	10 A	0.4 V	aux. Input
LT1963A	Lin. Tech	+2/20 V	+1.2/20 V	1.5 A	0.4 V	fast response
LM2940	National	+7/26 V	+5/15 V	1 A	0.5 V	fixed output
LM2990	National	−6/26 V	−5/15 V	1 A	0.6 V	fixed output
LP3853	National	+2.5/7 V	+1.8/5 V	3 A	0.4 V	fast response
LP3872	National	+2.5/7 V	+1.8/5 V	1.5 A	0.4 V	fast response
TPS78601	Texas Inst.	+2.8/5.5 V	+1.2/5.5 V	1.5 A	0.4 V	low noise
Low current LDOs						
LT1175	Lin. Tech	−5/20 V	−3.3/15 V	0.5 A	0.5 V	$I_q = 45\,\mu A$
LT3021	Lin. Tech	+1/10 V	+0.2/9 V	0.5 A	0.16 V	$I_q = 120\,\mu A$
MAX1658	Maxim	+3/16 V	+1.2/15 V	0.4 A	0.5 V	$I_q = 30\,\mu A$
MAX5023	Maxim	+6/65 V	+2.5/11 V	0.15 A	1 V	$I_q = 60\,\mu A$
LP2951	National	+2/30 V	1.3/29 V	0.1 A	0.4 V	$I_q = 75\,\mu A$
LP3981	National	+3/6 V	+2.5/3.3 V	0.3 A	0.13 V	$I_q = 100\,\mu A$
Special Types						
LT1185	Lin. Tech	−4/30 V	−2.5/25 V	0–3 A	0.8 V	$I_{o\,max}$ adjustable
L200	ST	+5/40 V	+1.2/36 V	0–1.5 A	2 V	$I_{o\,max}$ adjustable
LT3014	Lin. Tech	+3/80 V	+1.2/60 V	20 mA	0.35 V	$I_q = 7\,\mu A$
LR8	Supertex	+12/450 V	+1.2/440 V	15 mA	5 V	$V_i = \ldots 450\,V$

Fig. 16.24. Examples for linear voltage regulators

which is usually given in decibels. For the circuit in Fig. 16.25,

$$\frac{\Delta V_i}{\Delta V_{\text{ref}}} = 1 + \frac{R}{r_Z} \approx \frac{R}{r_Z} = 10 \ldots 100$$

where r_Z is the dynamic resistance of the Zener diode at the operating point selected. In a first-order approximation, r_Z is inversely proportional to the current flowing in the diode. Increasing the series resistance R for a given input voltage will not therefore produce an improvement in stabilization. An important aspect to consider when defining the diode current is the noise in the Zener voltage, which increases markedly at low currents. The resistance R is selected such that an adequate diode current will still flow at minimum input voltage and maximum output current.

Considerably improved stabilization can be achieved if the series resistor R is replaced by a current source, as shown in Fig. 16.26. The simplest method is to use a FET current source, as this has only two terminals (see Fig. 4.126). Stabilization factors of up to 10,000 can then be achieved.

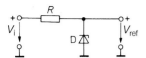

Fig. 16.25. Voltage stabilization using a Zener diode

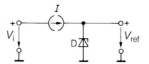

Fig. 16.26. Improving the stabilization using a constant current source

Another way of operating the Zener diode with a constant current is to connect it to the stabilized output voltage, rather than to the unregulated input voltage. As shown in Fig. 16.27, we generate an output voltage

$$V_{\text{ref}} = \left(1 + \frac{R_2}{R_1}\right) V_Z,$$

which is higher than Zener voltage V_Z. Constant current $I_Z = (V_{\text{ref}} - V_Z)/R_3$ then flows through R_3. In this case, line regulation is primarily determined by the supply ripple rejection $D = \Delta V_b/\Delta V_O$ of the operational amplifier, where V_O is the operational amplifier offset voltage. Using the relations

$$\Delta V_O = \Delta V_P - \Delta V_N, \quad \Delta V_P = \frac{r_Z}{r_Z + R_3} \Delta V_{\text{ref}}, \quad \Delta V_N = \frac{R_1}{R_1 + R_2} \Delta V_{\text{ref}}$$

and $\Delta V_b = \Delta V_i$, we obtain

$$\frac{\Delta V_i}{\Delta V_{\text{ref}}} \approx D \left(\frac{r_Z}{r_Z + R_3} - \frac{R_1}{R_1 + R_2}\right) \approx |D| \frac{R_1}{R_1 + R_2} \approx |D|$$

Values of around 10,000 are achieved. Even if the input voltage ripple amounts 10 V, the output voltage will then vary only 1 mV.

Considerably larger variations may occur due to changes in temperature. The temperature coefficient of the Zener voltage is in the order of $\pm 1 \cdot 10^{-3}$/K. For small Zener voltages, it is negative and for larger ones, positive. Its typical characteristic is plotted in Fig. 16.28. We can see that the temperature coefficient is at its smallest for Zener voltages around 6 V. For larger Zener voltages, it can be reduced by connecting forward-biased diodes in series. Although discrete components of this kind are available as *reference diodes,* in most cases IC reference voltage sources containing reference diodes are used, as shown in Fig. 16.27. Temperature coefficients up to 10^{-6}/K $\widehat{=} 1$ ppm are achieved.

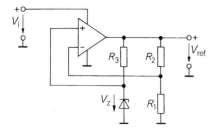

Fig. 16.27. Operating the Zener diode from the regulated voltage

$$V_{\text{ref}} = \left(1 + \frac{R_2}{R_1}\right) V_Z$$

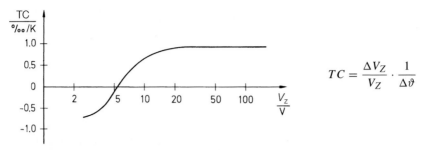

$$TC = \frac{\Delta V_Z}{V_Z} \cdot \frac{1}{\Delta \vartheta}$$

Fig. 16.28. The temperature coefficient as a function of the Zener voltage

16.4.2
Bandgap Reference

In principle, the forward voltage of a diode or the base–emitter voltage of a bipolar transistor can also be used as a voltage reference. However, the temperature coefficient of $-2\,\text{mV/K}$ at $0.6\,\text{V}$ is rather high. It can be compensated for by adding a voltage with a temperature coefficient of $+2\,\text{mV/K}$. The characteristic feature of the circuit in Fig. 16.29 is that this voltage is generated by a second transistor. Transistors T_1 and T_2 are driven by different collector currents; $I_{C2} > I_{C1}$. From the transfer characteristic, we obtain a voltage drop across R_1 of

$$\Delta V_{\text{BE}} = \Delta V_{\text{BE2}} - \Delta V_{\text{BE1}} = V_T \ln \frac{I_{C2}}{I_{C1}}$$

It is therefore proportional to V_T and, because $V_T = kT/e_0$, it is also proportional to the absolute temperature T. A correspondingly larger voltage is dropped across resistor R_2, as not only current $I_{C1} = \Delta V_{BE}/R_1$ but also current I_{C2} flows through this resistor. The operational amplifier causes $V_D = 0$ such that $I_{C2} = nI_{C1}$. Hence

$$V_{\text{Temp}} = R_2(I_{C1} + I_{C2}) = R_2 \frac{\Delta V_{\text{BE}}}{R_1}(1 + n)$$

$$= V_T \frac{R_2}{R_1}(1 + n) \ln n = A V_T$$

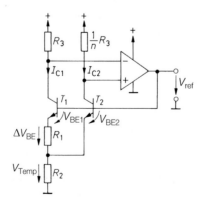

Fig. 16.29. Bandgap reference

$$V_{\text{ref}} = V_{\text{BG}} \approx 1.205\,\text{V}$$

$$V_{\text{Temp}} \approx 2\frac{\text{mV}}{\text{K}}T$$

We now have the possibility of achieving any gain factor $A = (1+n)\ln n \, R_2/R_1$ by selecting suitable values of n and R_2/R_1. Thus, for V_{Temp} we obtain a temperature coefficient of $+2\,\text{mV/K}$ if we select $A \approx 23$, because

$$\frac{\mathrm{d}V_{\text{Temp}}}{\mathrm{d}T} = A \cdot \frac{\mathrm{d}V_T}{\mathrm{d}T} = A \frac{k}{e_0} = A\frac{V_T}{T}$$

$$= 23 \cdot \frac{26\,\text{mV}}{300\,\text{K}} = +2\frac{\text{mV}}{\text{K}}$$

According to (2.21) on page 54 the theoretical value for the temperature coefficient of a bipolar transistor is

$$\frac{\mathrm{d}V_{\text{Temp}}}{\mathrm{d}T} \approx -2\frac{\text{mV}}{\text{K}}$$

where $V_{BG}, = E_g/e_0 = 1.205\,\text{V}$ is the bandgap voltage of silicon, E_g being the bandgap. The temperature coefficient of the output voltage $V_{ref} = V_{\text{temp}} + V_{BE\,2}$ is therefore zero if

$$\boxed{V_{\text{ref}} = AU_T + V_{\text{BE}} = V_{\text{BG}} = 1.2\,\text{V}}$$

This is a more precise and at the same time simpler adjustment criterion than setting the gain A to some calculated value.

For the discrete circuit design shown in Fig. 16.29, we obtain useful component values if $I_{C2} = 10I_{C1}$. In this case, $R_1 = R_2$ is suitable. In order to achieve good matching of T_1 and T_2, a double transistor is necessary, such as an LM 394.

Discrete-component bandgap references are only of interest in special cases, as a wide variety of IC versions is available (see Fig. 16.31). As transistors T_1 and T_2 in Fig. 16.29 are sometimes operated with identical collector currents, different current densities must then be achieved by connecting several transistors in parallel for T_1.

The significant advantage as compared to reference diodes is that bandgap references can be operated at lower voltages, which can be as low as the bandgap voltage $V_{BD} \approx 1.2\,\text{V}$. Reference diodes, on the other hand, require voltages of 6.4 V and above. Moreover, reference voltages of any value can be produced using bandgap references, when only a portion of the operational amplifier output voltage is fed back to the base terminals, as in Fig. 16.30. In this case, the actual reference voltage source T_1, T_2 is operated from the regulated output voltage. This provides significantly improved line regulation, as shown in Fig. 16.27.

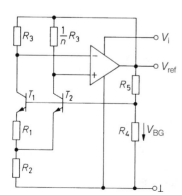

Fig. 16.30. Operating the reference transistors from the regulated voltage

$$V_{\text{ref}} = \left(1 + \frac{R_5}{R_4}\right) V_{\text{BG}}$$

With many IC bandgap references, it is permissible to connect the output to the supply voltage, or the corresponding connection is already established internally. Only two terminals are then brought out of the circuit, and it can be used like a Zener diode.

As voltage V_{Temp} is proportional to the absolute temperature, it can be used for temperature measurement (see Sect. 21.1.5). In many circuits $V_{temp} = 2\,mV/K$ is brought out to the terminals for this purpose.

16.4.3
Types

We have listed a number of commonly used reference voltage sources in Fig. 16.31. Note the tight tolerances and the low temperature coefficients of many types. These are achieved by laser adjustment of the relevant resistors during manufacture. However, the specified values only give a rough indication, as all of the circuits are available in various precision categories.

The two-terminal types behave like Zener diodes. Consequently, their current must never reach zero. Some three-terminal types incorporate a simple emitter follower at the output. This enables them to carry high currents at the output, but accept only low input currents. Other types incorporate complementary emitter followers at the output. Therefore, they can also accept high currents.

All voltage regulators and many AD und DA converters have built-in reference voltage source so that often a separate reference voltage source is not required.

Type	Manu-facturer	Reference voltage	Tolerance	Temperature coefficient	Output current
2 Terminal shunt regulators					
AD1580	Analog Dev.	1.2 V	1%	100 ppm/K	50 µA ... 10 mA
LT1004	Lin. Tech.	1.2 ... 2.5 V	0.3%	20 ppm/K	10 µA ... 10 mA
MAX6006	Maxim	1.2 ... 2 V	0.2%	50 ppm/K	1 µA ... 2 mA
MAX6138	Maxim	1.2 ... 5 V	0.1%	25 ppm/K	65 µA ... 15 mA
LM4050	National	2 ... 10 V	0.1%	50 ppm/K	60 µA ... 15 mA
TL431	Texas Inst.	2.5 V	2%	30 ppm/K	1 mA ... 100 mA
3 Terminal series regulators					
AD1582	Analog Dev.	2.5 ... 5 V	0.1%	50 ppm/K	−5 ... 5 mA
ADR420	Analog Dev.	2 ... 5 V	0.05%	3 ppm/K	0 ... 10 mA
ISL60007	Intersil	1.2 ... 2.5 V	0.05%	10 ppm/K	−7 ... 7 mA
LT1790	Lin. Tech.	1.2 ... 5 V	0.1%	25 ppm/K	−10 ... 10 mA
MAX6018	Maxim	1.2 ... 2 V	0.2%	50 ppm/K	−1 ... 1 mA
MAX6033	Maxim	2.5 ... 5 V	0.04%	7 ppm/K	0 ... 15 mA
MAX6126	Maxim	2 ... 5 V	0.02%	3 ppm/K	−10 ... 10 mA
LM4120	National	1.8 ... 5 V	0.2%	50 ppm/K	−5 ... 5 mA
LM4140	National	1 ... 4 V	0.1%	3 ppm/K	0 ... 7 mA
REF30xx	Texas Inst.	1.2 ... 4 V	0.2%	50 ppm/K	0 ... 25 mA
REF31xx	Texas Inst.	1.2 ... 4 V	0.2%	15 ppm/K	−10 ... 10 mA

Fig. 16.31. Examples for reference voltage sources

16.5
Switched-Mode Power Supplies

The power supplies incorporating linear series regulators described hitherto are subject to three basic loss factors: the line transformer, the rectifier, and the regulating transistor. The efficiency $\eta = P_{output}/P_{input}$ is in most cases only 25 − 50%. The power dissipation,

$$P_{loss} = P_{input} - P_{output} = \left(\frac{1}{\eta} - 1\right) P_{output}$$

can therefore be up to three times as large as the power output. Consequently, there is not only a large power loss but also an associated cooling problem.

The losses in the series regulator can be substantially reduced by replacing the continuously controlled transistor by a switch, as shown in Fig. 16.32. In order to obtain the required DC output voltage, a lowpass filter is additionally required to provide time averaging. The magnitude of the output voltage can be determined in this case by the duty cycle with which the switch is closed. If an LC lowpass filter is used, the regulator no longer possesses an inherently lossy element. As the switching regulator described is connected on the secondary side of the power transformer, these circuits are also known as *secondary switched-mode power supplies*.

The losses in the power transformer are, of course, not reduced by the switching regulator. However, they can be brought down by transforming a high-frequency alternating voltage instead of the line voltage. For this purpose, the line voltage is directly rectified as shown in Fig. 16.33, and an alternating voltage with a frequency in the range 20 kHz...2 MHz is generated using a switching regulator.

As the number of turns required on the power transformer decreases with the switching frequency, the copper losses can be greatly reduced. The secondary voltage is rectified, filtered, and then fed directly to the load. The duty cycle of the switches on the primary side is varied to regulate the DC voltage.

Circuits of this kind are known as *primary switched-mode power supplies*. Their efficiency can be 60–80%. An additional advantage is their compactness and the low weight of the HF transformer.

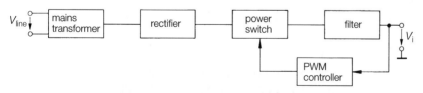

Fig. 16.32. Secondary switching regulator

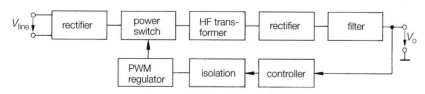

Fig. 16.33. Primary switching regulator (HF: high frequency)

If we compare the two basic circuits shown in Figs. 16.32 and 16.33, we can see that in both cases a switch is used to generate an AC voltage whose duty cycle determines the output voltage. Whereas, with the secondary regulator, line isolation is provided by a normal 50 Hz power transformer, with the primary regulator this function is performed by an HF transformer. Consequently, the switches in the primary regulator are at line potential and must be qualified for line voltages. In this case, the regulator comprises two sections, one that is at the line potential and controls the switches, and another that is at the output potential and measures the output voltage. Both sections must be suitably insulated from each other and fulfill the public regulations.

In spite of these problems and the associated circuit complexity, primary switching power supplies are to be preferred due to their high efficiency. Secondary switching power supplies are mainly used as low-power DC/DC converters.

16.6
Secondary Switching Regulators

The three basic types of DC/DC converters are shown in Figs. 16.34–16.36. Each one consists of three components: a power switch S, a storage choke L, and a smoothing capacitor C. However, each of the three circuits delivers a different output voltage. In the case of the circuit shown in Fig. 16.34, the switch produces an AC voltage whose average value lies between the input voltage and zero, depending on the duty cycle.

With the circuit in Fig. 16.35, $V_o, = V_i$ if the switch remains permanently in the upper position. If the switch is moved to the lower position, energy is stored in the choke and additionally delivered to the output when the switch is returned to the upper position. The output voltage is therefore higher than the input voltage.

In the circuit shown in Fig. 16.36, energy is stored in the choke as long as the switch is in the left-hand position. When it switches to the right, the choke current direction does not change and the capacitor is charged to negative values (if the input voltage is positive).

In the circuit shown in Fig. 16.34, current flows continuously into the reservoir capacitor. The circuit is therefore also known as a *forward converter*. This is not true of Figs. 16.35 and 16.36, because there the capacitor is not recharged as long as the choke is being charged. These circuits are known as *flyback converters*.

16.6.1
Step-Down Converters

The toggle switch can be realized by two simple switches or by using only one switch and a diode for the other path. This results in the step-down converter (the buck regulator)

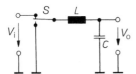
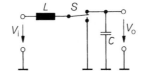
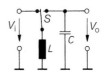

Fig. 16.34. Step-down converter $0 \leq V_o \leq V_i$

Fig. 16.35. Step-up converter $V_o \geq V_i$

Fig. 16.36. Inverting converter $V_o < 0$ for $V_i > 0$

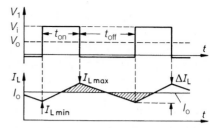

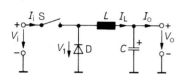

Fig. 16.37. Step-down converter with a simple switch

$$V_o = \frac{t_{on}}{T} V_i \quad \text{for} \quad I_o \geq I_{o\,min}$$

Fig. 16.38. Current and voltage waveforms

$$\bar{I}_i = \frac{t_{on}}{T} I_o \qquad \bar{I}_L = I_o$$

shown in Fig. 16.37. As long as the switch is closed, $V_1 = V_i$. When it opens, the choke current retains its direction and V_1 falls until the diode is turned on; that is, virtually to zero potential. This can be seen from the timing diagram shown in Fig. 16.38.

The time characteristic of the reactor current is deduced from the law of induction:

$$V_L = L \cdot \frac{dI_L}{dt} \tag{16.9}$$

During the ON-time t_{on}, a voltage of $V_L = V_i - V_o$ is applied across the choke; during the OFF-time t_{off}, a voltage of $V_L = -V_o$ is present. Using Eq. (16.9), the rate of change of the current is given by

$$\Delta I_L = \frac{1}{L}(V_i - V_o)t_{on} = \frac{1}{L}V_o t_{off} \tag{16.10}$$

From this, we can calculate the output voltage:

$$V_o = \frac{t_{on}}{t_{on} + t_{off}} V_i = \frac{t_{on}}{T} V_i = p\,V_1 \tag{16.11}$$

where $T = t_{on} + t_{off} = 1/f$ is the period and $p = t_{on}/T$ is the duty cycle[1]. We can see that the output voltage is the arithmetic mean of V_1, as we would expect.

The circuit behaves quite differently if the output current I_o becomes smaller than

$$I_{o\,min} = \frac{1}{2}\Delta I_L V_o = \left(1 - \frac{V_o}{V_i}\right) \tag{16.12}$$

The choke current then reaches zero during the OFF-time of the switch, the diode is turned off, and the voltage across the choke becomes zero, as shown in Fig. 16.39 (discontinuous operation). To calculate the output voltage, we shall assume that the circuit provides lossless operation. The average input power must therefore be equal to the output power:

$$V_i \bar{I}_e = V_o I_o \tag{16.13}$$

The current flowing through the choke rises during t_{on}, from zero to the value $I_L = V_L t_{on}/L$. The arithmetic mean of the input current is therefore

$$\bar{I}_i = \frac{t_{on}}{T} \cdot \frac{1}{2}I_L = \frac{t_{on}^2}{2TL}V_L = \frac{T}{2L}(V_i - V_o)p^2 \tag{16.14}$$

---

[1] Sorry, the commonly used letter for the duty cycle η is spent for the efficiency already.

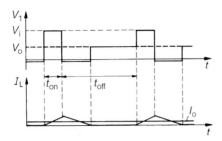

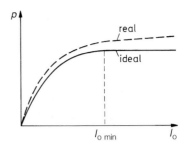

Fig. 16.39. Current and voltage waveforms in the step-down converter for output currents of less than

$$I_{o\,min} = \frac{T}{2L} V_o \left(1 - \frac{V_o}{V_i}\right)$$

Fig. 16.40. Duty cycle $p = t_{on}/T$ as a function of the output current I_o at constant output voltage V_o

Substitution into (16.13) gives the output voltage and duty cycle, respectively:

$$V_o = \frac{V_i^2 p^2 T}{2LI_o + V_i p^2 T} \qquad p = \sqrt{\frac{2L}{T}\frac{V_o}{V_i(V_i - V_o)}}\sqrt{I_o} \qquad (16.15)$$

In order to prevent the output voltage from rising in the event of low currents ($I_o < I_{o\,min}$), p must be reduced accordingly. This is shown schematically in Fig. 16.40. We can see that very short switching times must be achieved in this region. For currents higher than $I_{o\,min}$, the duty cycle remains constant in accordance with Eq. (16.11). However, this only applies to a lossless circuit. Otherwise, p must also be increased – albeit by a much smaller amount – as the output current increases above $I_{o\,min}$, in order to keep the output voltage constant.

Design Considerations

If possible, the inductance of the storage choke is selected to be large enough to prevent the current from falling below $I_{o\,min}$. From (16.12), it therefore follows:

$$L = T\left(1 - \frac{V_o}{V_i}\right)\frac{V_o}{2I_{o\,min}} \qquad (16.16)$$

The maximum current flowing through the storage choke, and consequently through the switch and diode, is therefore $I_{L\,max} = I_o + I_{o\,min}$. This still leaves the parameter with a period of $T = 1/f$. In order to allow a small inductance to be used, the frequency f is selected to be as high as possible. However, the problem then arises that at high frequencies the switching transistor becomes more costly and the drive circuit more complex. In addition, the dynamic switching losses increase proportionally with the frequency. For these reasons, switching frequencies between 20 kHz and 2 MHz are preferred.

The smoothing capacitor C determines the output voltage ripple. The charge current is $I_C = I_L - I_o$. The charge applied and removed during one cycle therefore corresponds to the hatched area in Fig. 16.38. For the ripple, we therefore obtain

$$\Delta V_o = \frac{\Delta Q_C}{C} = \frac{1}{C}\cdot\frac{1}{2}\cdot\left(\frac{t_{on}}{2} + \frac{t_{off}}{2}\right)\cdot\frac{\Delta I_L}{2} = \frac{T}{8C}\Delta I_L$$

Using (16.10) and (16.16), the smoothing capacitance is given by

$$C = \left(1 - \frac{V_o}{V_i}\right)\frac{T^2 V_o}{8L\Delta V_o} = \frac{T I_{o\,min}}{4\Delta V_o} \tag{16.17}$$

When selecting the smoothing capacitor, care must be taken to insure that it has the lowest possible series resistance and series inductance. In order to achieve this, it is usual to connect several ceramic capacitors in parallel.

16.6.2
Generating the Switching Signal

The switching signal is generated using two modules: a pulsewidth modulator and a regulator with voltage reference. The block diagram is shown in Fig. 16.41.

The pulsewidth modulator comprises a sawtooth generator and a comparator. The comparator closes the switch as long as voltage V_R is greater than the triangular voltage. The resultant control voltage, V_{ctl}, is shown in Fig. 16.42 for V_R ranging from the lower limit to the upper limit. The resulting duty cycle,

$$p = \frac{t_{on}}{T} = \frac{V_R}{\hat{V}_{ST}}$$

is therefore proportional to V_R.

The subtractor takes the difference between the reference voltage and the weighted output voltage $V_{ref} - kV_o$. The control amplifier increases V_R until this difference becomes zero. The output then has the value $V_o = V_{ref}/k$.

An example should serve to illustrate the design process for a switching regulator. Let us assume that an output voltage of 5 V at a maximum current of 5 A is required.

The minimum output current is to be 0.3 A, and the input voltage is approximately 15 V. A suitable switching regulator for this application would be the LM 2678-5 from National. The resulting circuit is shown in Fig. 16.43. Apart from the LC output filter only a few external resistors and capacitors are required for the operation of the integrated circuit. The switching regulator is to be operated at a frequency of 250 kHz, resulting in a period of $T = 4\mu s$. From Eq. (16.11), we obtain a switch-on time of

$$t_{on} = T\frac{V_o}{V_i} = 4\,\mu s\,\frac{5\,V}{20\,V} = 1.3\,\mu s$$

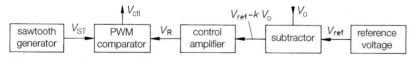

Fig. 16.41. Control unit design of a switching regulator

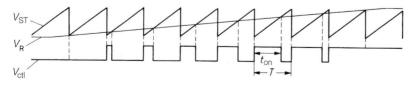

Fig. 16.42. Function of the pulsewidth modulator

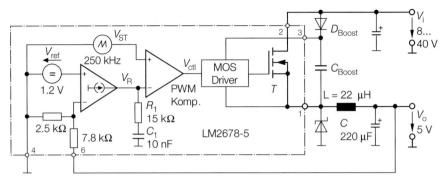

Fig. 16.43. An example of a step-down converter $I_{o\,max} = 5\,\text{A}$

The inductance of the smoothing choke follows from (16.16):

$$L = T\left(1 - \frac{V_o}{V_i}\right)\frac{V_o}{2I_{o\,min}} = 4\,\mu s\left(1 - \frac{5\,\text{V}}{15\,\text{V}}\right)\frac{5\,\text{V}}{2\cdot 0.3\,\text{A}} = 22\,\mu\text{H}$$

If the output ripple is to have a magnitude of 10 mV the value of the smoothing capacitor is determined from (16.17)

$$C = T\frac{I_{o\,min}}{4\Delta V_o} = 4\,\mu s\,\frac{0.3\text{A}}{4\cdot 10\,\text{mV}} = 30\,\mu\text{F}$$

In the calculation of the filter capacity, the equivalent series resistance (ESR) and the equivalent series inductance (ESL) have not been taken into account. To still reach an acceptable level of output ripple requires a distinctly higher capacitance. In the practical realisations several small capacitances are connected in parallel, since this always gives lower values of the series resistance and the series inductance than with one large capacitor.

The output voltage remains constant even if the output current is lower than the value of $I_{o\,min} = 0.3\,\text{A}$. In this case the control amplifier reduces the duty factor through the comparator as in Fig. 16.40. Problems will arise if the required ON-time is shorter than the minimum achievable ON-time of transistor T. In this case the output voltage increases so greatly in response to a single turn-on pulse of T that the transistor is then turned off for several cycles. This results in very unsmooth operation.

The desired control response is achieved by connecting the on-chip network $R_1 C_1$ to the output of the the high-impedance operational amplifier. It must be considered that the voltage regulating circuit of switching regulators is prone to instability. This has two causes: firstly, the switching regulator is a sampling system with an average dead time equal to half the period; secondly, the output filter represents a second-order lowpass filter which produces a phase lag of up to 180°. For these reasons it is advisable to insure that the control amplifier produces no phase lag at high frequencies. Resistor R_1 in Fig. 16.43 is used for this purpose.

The static circuit losses are mainly due to the voltage drops in the power circuit. The smoothing choke can easily be dimensioned so that the resistive losses are low. The losses due to the voltage drop across the power switch formed by transistor T and diode D then remain.

The output current flows through T during t_{on} and through D during t_{off}. For the case of a voltage drop of 0.7 V across the transistor or the diode with an output current of 5 A, a power dissipation of 3.5 W results. The maximum efficiency is therefore

$$\eta = \frac{P_{output}}{P_{input}} = \frac{25\,W}{25\,W + 3.5\,W} = 88\%$$

This does not include the switching losses, which are not insignificant at high switching frequencies. In addition, the efficiency is further diminished by the current consumption of the switching regulator itself. Since this contribution does not depend on the output current, the efficiency is diminished especially at low output currents. To operate the n-channel MOSFET in the resistive region requires a gate potential that is positive compared with the input voltage. For this purpose the MOS gate driver generates an auxiliary voltage that is positive compared with the source potential. The boost capacitor transfers this alternating output voltage to the MOS driver. As an alternative the manufacturer could have used a p-channel MOSFET which has twice as high an on-resistance for the same chip area.

16.6.3
Step-Up Converters

Figure 16.44 shows a practical implementation of the step-up converter (boost regulator) presented in Fig. 16.35, and Fig. 16.45 gives the voltage and current waveforms. Once again, the relations required for the design parameters of the circuit can be derived from the rise or fall in the choke current I_L during the two states of switch S. These relations are given below the diagrams. The smallest output voltage is $V_o = V_i$. For a lossless circuit, this is obtained when switch S is continuously open.

Here too, the specified output voltage is only achieved if the choke current does not become zero. If the output value falls below the minimum $I_{o\,min}$, the ON-time must be reduced, as shown in Fig. 16.40, in order to prevent a rise in the output voltage. This case is shown by the dashed lines in Fig. 16.45. The switching signal is generated in precisely the same way as for the step-down converter.

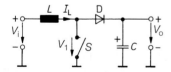

Fig. 16.44. Step-up converter

$$V_o = \frac{T}{t_{off}} V_i \text{ for } I_o > I_{o\,min}$$

$$I_{o\,min} = \frac{V_i^2}{V_o}\left(1 - \frac{V_i}{V_o}\right) \cdot \frac{T}{2L}$$

$$L = \frac{V_i^2}{V_o}\left(1 - \frac{V_i}{V_o}\right) \cdot \frac{T}{2I_{o\,min}}$$

$$C \approx \frac{T I_{o\,max}}{\Delta V_o}$$

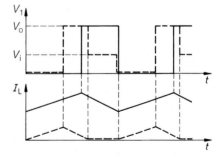

Fig. 16.45. Voltage and current waveforms in the step-up converter Dashed lines: for $I_o < I_{o\,min}$

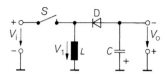

Fig. 16.46. Inverting converter

$$V_o = \frac{t_{on}}{t_{off}} V_i \text{ for } I_o > I_{o\,min}$$

$$I_{o\,min} = \frac{V_i^2 \, V_o}{(V_i + V_o)^2} \cdot \frac{T}{2L}$$

$$L = \frac{V_i \, V_o}{(V_i^2 + V_o)^2} \cdot \frac{T}{2\,L}$$

$$C \approx \frac{T I_{o\,max}}{\Delta V_o}$$

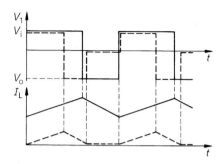

Fig. 16.47. Voltage and current waveforms in the inverting converter. Dashed lines: for $I_o < I_{o\,min}$

16.6.4
Inverting Converter

The inverting converter (inverting regulator) and associated waveforms are shown in Figs. 16.46 and 16.47.

As we can see, the capacitor is charged to a negative voltage via the diode during the OFF-phase. The relations given are again deduced from the fact that the reactor current changes are equal during the ON- and OFF-times.

If the output current falls below the value $I_{o\,min}$, the reactor current occasionally becomes zero. In order to keep the output voltage constant in this case, the ON-time must again be reduced, as shown in Fig. 16.40. This is represented by the dashed lines in Fig. 16.47.

16.6.5
Charge Pump Converter

If the current requirement is low, there is a simple method of inverting a voltage. For this purpose, the input voltage is converted into an AC voltage using switch S_1, as shown in Fig. 16.48. This alternating voltage is made "floating" by capacitor C_1 and then rectified again, this time by switch S_2. In the switch position shown, C_1 is charged up to the input voltage: $V_1 = V_i$. The two switches then change over, causing voltage $-V_1$, to be applied to C_1, and the latter is charged up to the voltage $V_o = -V_1 = -V_i$ after several switching cycles.

Rectification of the output voltage does not necessarily require a controlled switch, but can also be performed by two diodes, as shown in Fig. 16.49. Depending on which potential is applied to the rectifier and on the polarity of the diodes, voltage V_i can be added to or subtracted from that potential. However, the disadvantage of rectification using diodes is that the output voltage is reduced by the two forward voltage drops. Consequently, integrated circuit voltage inverters employ CMOS switches for rectification. Due to the on resistances the voltage drop is proportional to the output current.

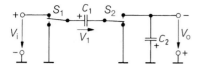

Fig. 16.48. Voltage inverter employing the charge pump principle

$$V_o = -V_i$$

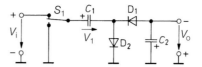

Fig. 16.49. Simplified arrangement for rectifying the output voltage

$$V_o = -(V_i - 2V_D)$$

Additional losses are incurred due charge changes of the capacitors. However, these depend only on the size of the voltage difference produced, which under steady-state conditions can easily be minimized by selecting suitably high-value capacitors.

16.6.6
Integrated Switching Regulators

In Fig. 16.50 charge-pump voltage converters are collected which only need external capacitors. They are especially appropriate for low voltages and low currents. Examples for integrated switching regulators that need a storage choke are listed in Fig. 16.51. Even types with internal power transistors can provide remarkable output power. For high output currents types for external power transistors are available. In order to reduce the power loss in the rectifying diode an internal or external power transistor is used for synchronous rectification.

Type	Manu-facturer	Input voltage min/max	Output voltage min/max	Output current	Quiescent current	Switching frequency
LTC 1751	Lin. Tech.	2.5/5.5 V	5 V	100 mA	20 μA	800 kHz
LTC 3200	Lin. Tech.	2.7/4.5 V	5 V	100 mA	350 μA	2000 kHz
LTC 3215	Lin. Tech.	2.9/4.4 V	...4.4 V	...500 mA	300 μA	900 kHz
LTC 3251	Lin. Tech.	2.7/5.5 V	0.9/1.6 V	500 mA	35 μA	1000 kHz
MAX 682	Maxim	2.7/5.5 V	5 V	250 mA	7000 μA	1000 kHz
MAX 889	Maxim	2.7/5.5 V	$-V_i$	200 mA	5500 μA	1000 kHz
MAX 1680	Maxim	2.0/4.4 V	$-V_i/2V_i$	125 mA	5000 μA	250 kHz
MAX 1759	Maxim	1.6/5.5 V	3.3 V	100 mA	180 μA	1500 kHz
LM 2663	National	1.5./5.5 V	$-V_i/2V_i$	200 mA	1300 μA	150 kHz
LM 2750	National	2.7/5.6 V	5 V	120 mA	5000 μA	1700 kHz
LM 2770	National	2.7/5.5 V	1.2/1.5 V	250 mA	55 μA	700 kHz
LM 2781	National	1.8/5.5 V	$-V_i$	50 mA	700 μA	210 kHz
LM 3354	National	2.5/5.5 V	1.8/5 V	90 mA	375 μA	1000 kHz
TPS 60110	Texas Inst.	2.7/5.4 V	5 V	300 mA	60 μA	300 kHz
TPS 60120	Texas Inst.	1.8/3.6 V	3.3 V	200 mA	60 μA	300 kHz

Fig. 16.50. Examples for charge pump converters without inductors

Type	Manu-facturer	Input voltage min/max	Output voltage min/max	Output current	Switching frequency	Power switch
Step-down converters, Buck converters (Fig. 16.37)						
LT 1374	Lin. Tech.	6/25 V	1.2/24 V	3 A	500 kHz	internal
LT 1616	Lin. Tech.	3.6/25 V	1.3/24 V	0.6 A	1400 kHz	internal
LT 1956	Lin. Tech.	5.5/50 V	1.2/25 V	1.5 A	500 kHz	internal
LTC 3406	Lin. Tech.	2.5/5.5 V	0.6/5 V	0.6 A	1500 kHz	internal
LTC 3411	Lin. Tech.	2.6/5.5 V	0.8/5 V	1.2 A	4000 kHz	internal
LTC 3717	Lin. Tech.	5/28 V	1/14 V		200 kHz	2 N
MAX 1572	Maxim	2.6/5.5 V	0.8/2.5 V	0.8 A	2000 kHz	internal
MAX 1830	Maxim	3/5.5 V	1.1/5 V	3 A	1000 kHz	internal
MAX 1973	Maxim	2.6/5.5 V	1/2.5 V	1 A	1400 kHz	internal
MAX 5035	Maxim	7.5/76 V	1.3/50 V	1 A	125 kHz	internal
LM 2590	National	5/60 V	1.2/57 V	1 A	150 kHz	internal
LM 2678	National	8/40 V	1.2/37 V	5 A	260 kHz	internal
LM 2737	National	2.2/16 V	0.6/12 V		1000 kHz	2 N
LM 5010	National	8/75 V	2.5/70 V	1 A	1000 kHz	internal
TPS 430000	Texas Inst.	1.8/9 V	0.8/5 V		1000 kHz	1 N 1 P
TPS 54350	Texas Inst.	4.5/20 V	0.9/12 V	3 A	500 kHz	internal
TPS 54610	Texas Inst.	3/6 V	0.9/5 V	6 A	550 kHz	internal
TPS 62200	Texas Inst.	2.5/6 V	0.7/5 V	0.3 A	1000 kHz	internal
Step-up converters, Boost converters (Fig. 16.44)						
LT 1615	Lin. Tech.	1/15 V	1.2/34 V	0.1 A	variable	internal
LT 1935	Lin. Tech.	2.3/16 V	1.3/38 V	0.6 A	1200 kHz	internal
LT 3467	Lin. Tech.	2.4/16 V	1.3/40 V	0.4 A	1300 kHz	internal
LTC 3400	Lin. Tech.	0.9/6 V	2.5/5 V	0.1 A	1200 kHz	internal
MAX 1522	Maxim	2.5/5.5 V	1.2/100 V		variable	1 N
MAX 1709	Maxim	0.7/5 V	2.5/5.5 V	4 A	500 kHz	internal
MAX 1896	Maxim	2.6/5.5 V	2.6/13 V	0.3 A	1400 kHz	internal
LM 2623	National	0.9/14 V	1.2/14 V	0.3 A	1000 kHz	internal
LM 2698	National	2.2/12 V	2.2/17 V	0.4 A	1250 kHz	internal
LM 2733	National	2.7/14 V	2.7/20 V	0.3 A	1600 kHz	internal
Inverting converters, flyback converters, Cuk converters (Fig. 16.46)						
LT 1931	Lin. Tech.	2.6/16 V	−1/−12 V	0.2 A	1200 kHz	internal
MAX 1846	Maxim	3/16 V	−2/−100 V		300 kHz	1 P
LM 2611	National	2.7/14 V	−1/−12 V	0.1 A	1400 kHz	internal

Fig. 16.51. Examples for non-isolated (secondary) switching converters using inductors

16.7
Primary Switching Regulators

Primary switching regulators fall into two categories, namely single-ended and push–pull converters. As single-ended types generally require only one power switch, the number of components involved is low. However, their use is limited to low-power applications. For powers in excess of 100 W, push–pull converters are preferred, even though they require two power switches.

16.7.1
Single-Ended Converters

The single-ended converter shown in Fig. 16.52 represents the simplest practical primary switching regulator. It is similar to the flyback converter in Fig. 16.46, except that the storage choke has been replaced by a transformer. As long as power switch S is closed, energy is stored in the transformer. This energy is transferred to smoothing capacitor C when the switch opens. The resulting relation for the output voltage is the same as for the circuit shown in Fig. 16.46. The only difference is that the output voltage is now reduced by the winding ratio W of the transformer, where $W = w_1/w_2$ (see Sect. 18.7.3).

The waveform of the voltage across the switch is plotted in Fig. 16.53. When the switch opens, the voltage rises until diode D is turned on; that is, to $V_{S\,max} = V_i + W V_o$. In order to prevent it from becoming too high, we make the ON-time $t_{on} \leq 0.5T$, which means that $V_{S\,max} \leq 2V_i$. As a DC voltage of

$$V_i = 230\,\text{V} \cdot \sqrt{2} = 325\,\text{V}$$

is produced when rectifying the 230 V AC line, a voltage of $V_{S\,max} = 650\,\text{V}$ is present at the power switch. The voltages actually present are even higher due to the unavoidable leakage inductance.

The current characteristic is also shown in Fig. 16.53. As long as the switch is closed, the rise of the current is $\Delta I = V_i t_{on}/L$. When the switch opens, the diode is turned on and the current falls accordingly $\Delta I = W U_o t_{off}/L$, producing the output voltage indicated. However, the transformer inductance must be large enough to insure that the current does not fall to zero during the OFF-time.

One disadvantage of the circuit is that the transformer has to provide not only AC line isolation and the required stepping-down of the voltage, but must simultaneously act as a storage choke. Due to the DC biasing which occurs, it has to be considerably overrated. A better solution is to keep the transformer free of any DC component and to use a separate storage choke. All of the following circuits operate on this principle.

In the case of the single-ended converter in Fig. 16.54, the primary and secondary windings have the same polarity. Consequently, energy is transferred to the output via diode D_2, as long as the power switch is closed. The circuit is therefore a forward converter. The voltage characteristics are shown in Fig. 16.54. As long as the power switch is closed,

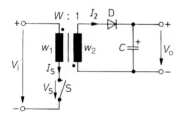

Fig. 16.52. Single-ended flyback converter

$$V_o = \frac{t_{on}}{t_{off}} \cdot \frac{V_i}{W} \quad \text{for} \quad I_o > I_{o\,min}$$

$$V_{S\,max} = V_i \left(1 + \frac{t_{on}}{t_{off}}\right)$$

$$W = w_1/w_2$$

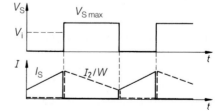

Fig. 16.53. Voltage and current waveforms for $I_o > I_{o\,min}$

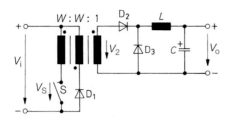

Fig. 16.54. Single-ended forward converter

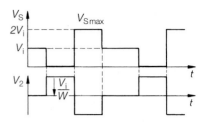

Fig. 16.55. Voltage waveforms

$$V_o = \frac{t_{on}}{T} \cdot \frac{V_i}{W} \text{ for } I_o > I_{o\,min}$$

$$V_{S\,max} = 2V_i$$

the input voltage V_i is present at the primary winding and therefore voltage $V_2 = V_i/W$ is present at the secondary winding. When switch S opens, D_2 is turned off and the current through storage choke L is carried by diode D_3. The conditions on the secondary side are therefore precisely the same as for the forward converter in Fig. 16.37. Consequently, apart from the factor W, we obtain the same relations for the output voltage, and the same considerations apply to the design procedure for the storage chokes and the smoothing capacitor.

At the instant at which the power switch is turned off, diode D_2 also becomes reverse biased. Without further action, the energy stored in the transformer would then generate an extremely high-amplitude voltage spike. In order to prevent this, the transformer is provided with a third winding that has the same number of turns as the primary winding, but a smaller cross-section. For the given polarity, diode D_1 then becomes conducting when the induced voltage equals the input voltage. In this way, the voltage across the power switch is limited to $V_{S\,max} = 2V_i$. In addition, the same energy is fed back to the input voltage source during the OFF-time as was stored in the transformer during the ON-time. In this way, the transformer is operated without DC magnetization.

16.7.2
Push–Pull Converters

With circuits of this type, the DC input voltage is converted to AC form by an inverter that comprises at least two power switches. This AC voltage is stepped down in an HF transformer and subsequently rectified.

In the circuit shown in Fig. 16.56, the period T is subdivided into four time periods. Initially, switch S_1 is closed, which means that diode D_1 is ON and voltage $V_3 = V_i/W$ is present at storage choke L. Switch S_1 then reopens and all of the voltages at the transformer fall to zero. Diodes D_1 and D_2 then each carry half of the choke current.

In the next time period, switch S_1 remains open, whereas switch S_2 closes. This turns on D_2, which likewise transfers voltage $V_3 = V_i/W$. When S_2 reopens, all of the voltages at the transformer become zero once more, as during the second time period. The relevant voltage waveforms are shown in Fig. 16.57.

The secondary side of the circuit therefore operates in basically the same way as the forward converter in Fig. 16.37. However, energy is now transferred to the storage choke

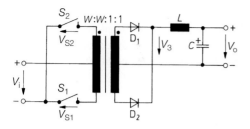

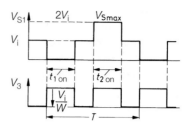

Fig. 16.56. Parallel-fed push–pull converter
Chips: MAX 845, MAX 5069

Fig. 16.57. Voltage waveforms

$$V_o = 2\frac{t_{on}}{T} \cdot \frac{V_i}{W} \quad \text{with} \quad \frac{t_{on}}{T} < 0.5$$

$$V_{S\,max} = 2V_i$$

twice during period T, due to full-wave rectification. Consequently, $\frac{1}{2}T$ instead of T must be substituted in the forward converter equations.

Due to the balanced mode of operation, the transformer operates without direct current. However, this only applies if the ON-times of the power switches are precisely equal; that is, $t_{1\,on} = t_{2\,on} = t_{on}$. This condition must be fulfilled when drive is applied to the switches; otherwise, the transformer will be driven into saturation, the currents will become high, and the switches will be destroyed. For the same reason, it is necessary to prevent one switch from not closing at all during a cycle. However, these conditions are taken into account in the majority of IC drive circuits for push–pull switching regulators. The drive arrangement for the power switches is simple here due to the fact that their two negative terminals are at the same potential.

In the case of the push–pull converter shown in Fig. 16.58, an AC voltage is produced by connecting one end of the primary winding alternately to the positive or negative terminal of the input voltage, while the other is at $\frac{1}{2}V_i$. The power switches are again driven alternately. The voltage waveforms in Fig. 16.59 are the same as for the previous circuit. The only difference is that the amplitude is halved, a feature that is particularly advantageous for switch selection.

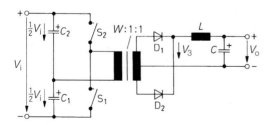

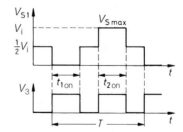

Fig. 16.58. Push–pull converter in a half-bridge
configuration. Chips: MAX 256

Fig. 16.59. Voltage waveforms

$$V_o = \frac{t_{on}}{T} \cdot \frac{V_i}{W} \quad \text{with} \quad \frac{t_{on}}{T} < 0.5$$

$$V_{S\,max} = V_i$$

A further advantage of the circuit is that the transformer is always DC-free due to capacitive coupling, even if the ON-times of the two switches are unequal. In this case, only the DC voltage across capacitors C_1 and C_2 is slightly displaced. However, a disadvantage is that the negative terminals of the power switches are at quite different potentials, which makes the drive arrangement more complex.

16.7.3
High-Frequency Transformers

Storage chokes are commercially available in a wide variety of types. Various manufacturers offer types rated from $1\,\mu\mathrm{H}$ to $10\,\mathrm{mH}$ and from $0.1\,\mathrm{A}$ to $60\,\mathrm{A}$. There is therefore little necessity for the user to wind them him- or herself. However, this is not the case with high-frequency transformers. Here, the user would be lucky to find a ready-made transformer with the appropriate turns ratio. Consequently, if only small quantities are required, the user generally has to calculate the transformer data and also wind the transformers him- or herself.

According to the law of electromagnetic induction, the voltage induced in a transformer is given by

$$V = w\dot{\Phi} = w \cdot A_c \cdot \dot{B},\qquad(16.18)$$

where Φ is the magnetic flux, B is the magnetic induction, and is A_c the cross-sectional area of the core between the two coils. If the number of turns on the primary side is w_1, it follows from Eq. (16.18) that

$$w_1 = \frac{V_1}{A_c \cdot \dot{B}} = \frac{V_1}{A_c} \cdot \frac{\Delta t}{\Delta B}$$

With $\Delta B = \hat{B}$, the minimum number of turns results from the permissible peak value of the magnetic induction, $\hat{B}$ and from the maximum value of

$$\Delta t = t_{on\,max} = p_{max} \cdot T = p_{max}/f = 1/2f$$

Hence

$$w_1 = \frac{V_1}{2A_c\hat{B} \cdot f}\qquad(16.19)$$

We can see that the required number of turns is inversely proportional to the frequency. Consequently, the power that can be transferred for a given core, and thus for a given winding area, is proportional to the frequency.

$$w_2 = w_1\frac{V_2}{V_1} = \frac{w_1}{W}\qquad(16.20)$$

The magnetizing and copper losses can generally be kept negligibly low. The wire gauge depends on the currents to be handled. Current densities of up to $CD = 5\ldots7\,\mathrm{A/mm^2}$ are permissible in terms of thermal requirements. However, if the copper losses are to be minimized, lower values should be adopted. The wire diameter is given by

$$d = 2\sqrt{\frac{I}{\pi \cdot CD}}\qquad(16.21)$$

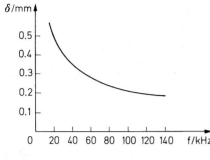

Fig. 16.60. Skin effect: skin depth as a function of frequency

Core type (lateral length) [mm]	Maximum power at 20 kHz [W]	Magnetic cross-section A_c [mm^2]	Inductance factor A_L [µH]
EC 35	50	71	2.1
EC 41	80	106	2.7
EC 521	130	141	3.4
EC 70	350	211	3.9

Fig. 16.61. Ferroxcube cores for high-frequency transformers

Recommended maximum induction: $\hat{B} = 200\,\text{mT} = 2\,\text{kG}$

Inductance: $L = A_L \cdot w^2$

However, due to the *skin effect,* at higher frequencies the current no longer flows uniformly through the entire cross-section, but only at the wire surface. For the skin depth (drop to $1/e$) of the current, one can use the empirical formula

$$\delta = 2.2\,\text{mm}/\sqrt{f/\text{kHz}}\,. \tag{16.22}$$

We can see from Fig. 16.60 how the skin depth reduces with increasing frequency. For this reason, it is inadvisable to select the wire diameter to be greater than twice the skin depth. In order that the required cross-sections can still be achieved, litz wire composed of fine, separately insulated strands can be used. It is also preferable to employ ribbon cable or correspondingly thin copper foils.

The principal characteristics of a number of ferroxcube EC cores are listed in Fig. 16.61. The maximum power is only a rough guideline. If the wire diameter is substantially over-sized in order to minimize the losses, it is possible that the next larger size of core will be required in order to provide sufficient winding space.

16.7.4
Power Switches

The aspects discussed in this section apply to the power switches of all switching regulators. The components that we shall consider here are bipolar transistors and power MOSFETs. The use of IGBTs is only of interest when high powers in the kilowatt range are involved; consequently, these devices will not be discussed here. If we look at the safe operating area (SOA) of power transistors, we can see that there are virtually no power transistors that can handle 100 W at high voltage levels. However, when these devices are used as high-speed switches, there are a number of exceptions, as shown in Fig. 16.62.

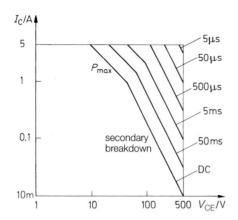

Fig. 16.62. Safe operating area SOA of a bipolar transistor used as a switch

DC power up to	$V_{CE} = 50$ V :	50 W
DC power up to	$V_{CE} = 500$ V :	5 W
Pulse power for 5 μs	$V_{CE} = 500$ V :	2500 W

We can see that the power dissipation and secondary breakdown can be exceeded, albeit briefly, and in extreme cases (for a few microseconds) it is even permissible for $V_{CE\,max}$ and $I_{C\,max}$ to be applied simultaneously. It is therefore possible to use a 50 W transistor to switch several kilowatts, a characteristic that is utilized in switched-mode power supplies.

However, there is a second reason for switching the transistors on and off quickly: a switch only operates in a lossless condition if the transition from the OFF-state to the ON-state and vice versa is instantaneous. Otherwise, so-called switching losses occur. These are greater the longer the switching process lasts. As they occur each time the transistor switches, they are proportional to the switching frequency.

In addition, with most switching regulators it is also preferable to provide short ON-times t_{on}, in order to insure orderly operation even at low load currents $I_o < I_{o\,min}$. For this purpose, it is essential to switch the transistor off rapidly. Consequently, we have to avoid the problem of the storage time of bipolar transistors by preventing them from going into saturation during the conducting phase ($V_{CE} > V_{CE\,sat}$). These two cases are compared in Fig. 16.63. We can see that a slight increase in the voltage drop across the conducting transistor must be tolerated in order to eliminate the storage time.

The basic arrangement for a bipolar transistor operated as a power switch is shown in Fig. 16.64. In order to turn on the transistor, switch S is moved to the upper position, allowing a large base current to flow via resistor R_1. This causes the collector current to rise rapidly, and a short fall time is produced. When the collector potential falls below the base potential, the Schottky diode is turned on, preventing the transistor from being driven into saturation. The major portion of the current through R_1 is now diverted via the diode to the collector, and the remaining base current immediately assumes the value required by the transistor at this operating point. In order to turn the transistor off, it is not sufficient to cut off the base current. Its direction must be reversed in order to remove the charge of the base junction. If this is to be effected rapidly, a large negative base current is required, its magnitude being determined by resistor R_2 in Fig. 16.64.

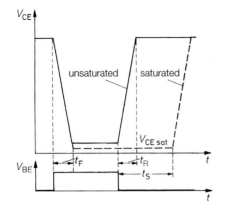

Fig. 16.63. Switching times of a bipolar transistor with and without saturation

t_F : *Fall time*
t_R : *Rise time*
t_S : *Storage time*

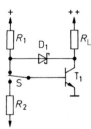

Fig. 16.64. Base drive for short switching times

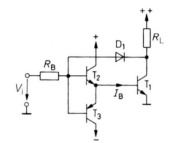

Fig. 16.65. Practical base drive arrangement

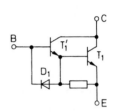

Fig. 16.66. Darlington pair with a speed-up diode

One possible realization is shown in Fig. 16.65. Complementary emitter follower T_2, T_3 provides the required base currents. Resistor R_B limits the base current. Antisaturation diode D_1 insures that the collector potential remains higher than the base potential.

If the power switches have to be used to switch high currents, the required base currents become unmanageably large. In this case, a Darlington circuit can be used, as shown in Fig. 16.66. However, the Darlington configuration has a higher saturation voltage. An important requirement to insure rapid turn-off is that the Darlington circuit contains the speed-up diode D_1 for removal of the base charge, since transistor T_1 cannot otherwise be actively turned off.

Power MOSFETs used as power switches offer considerable advantages: they have no secondary breakdown, no storage time, and they can be turned on and off at least a factor of 10 more quickly than comparable bipolar transistors. Therefore they are to be preferred. However, the use of MOSFETs should not lead to the erroneous assumption that they can be controlled without power. This may best be explained by reference to Fig. 16.67, in which capacitors C_1 and C_2 represent the parasitic capacitances of the power MOSFET. If we now increase the gate voltage from $0\,\mathrm{V}$ to $10\,\mathrm{V}$, the MOSFET is turned on and its drain potential falls from $325\,\mathrm{V}$ to about zero. The associated charge variation in the two capacitors is:

$$\Delta Q = 500\,\mathrm{pF} \cdot 10\,\mathrm{V} + 50\,\mathrm{pF} \cdot 325\,\mathrm{V} = 5\,\mathrm{nC} + 16\,\mathrm{nC} = 21\,\mathrm{nC}$$

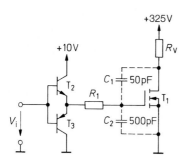
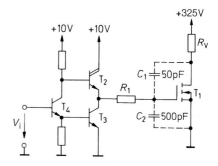

Fig. 16.67. Drive circuit for a power MOSFET with a complementary emitter follower

Fig. 16.68. Drive circuit for a power MOSFET with a totem pole circuit

For the gate potential to rise in 100 ns, a current of $I = 21\,\text{nC}/100\,\text{ns} = 210\,\text{mA}$ is required. The gate current is therefore of the same order of magnitude as the base current of the bipolar transistors. The only difference is that the gate current only flows at the switching instant. In order to switch power MOSFETs on and off rapidly, low-impedance drivers are therefore required. Figure 16.67 shows a complementary emitter follower and Fig. 16.68 a totem pole output stage of the type commonly used in TTL gates. Being easier to implement in monolithic technology, it is therefore preferred in driver ICs. In terms of their drive circuitry, power MOSFETs have the advantage of not requiring negative voltage sources, as is the case with bipolar transistors.

16.7.5
Generating the Switching Signals

The switching signals for single-ended converters can be generated using a pulsewidth modulator, as described earlier in Sect. 16.6.2. However, push–pull converter operation requires two alternately activated pulsewidth modulated outputs. To generate these signals, in the pulsewidth modulator in Fig. 16.41 a toggle flip-flop is added, giving the circuit shown in Fig. 16.69. It changes state on each negative-going edge of the sawtooth signal, thereby enabling one or the other of the AND gates. The waveform diagram is shown in Fig. 16.70. We can see that two signals from the sawtooth generator are required to produce a complete pulse cycle at the output. Its frequency must therefore be twice as high as that at which the

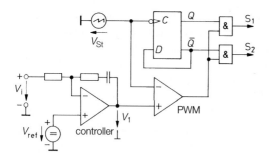
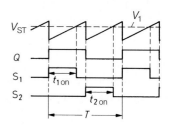

Fig. 16.69. Pulsewidth modulator for push–pull converters

Fig. 16.70. Signal waveforms

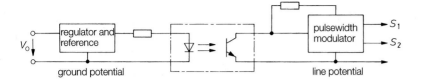

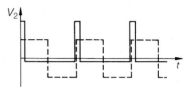

Fig. 16.71. Line isolation using an opto-coupler for the analog regulator signal

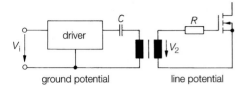

Fig. 16.72. Line isolation using a pulse transformer

Fig. 16.73. Pulse amplitude as a function of the ON-time

HF transformer is to be operated. As the maximum ON-time at its output cannot exceed 50%, this circuit insures that the two power switches can never be simultaneously ON.

An additional problem for controlling primary switching regulators is AC line isolation in the drive circuit, as can be seen in Fig. 16.33. This circuit has the function of monitoring the output voltage of the power supply and providing the switching signals for the power switches, which are at line potential. It therefore requires isolation, which should be provided either in respect of the output signal of the regulator in Fig. 16.69 or in respect of switching signals S1, S2. An opto-coupler can be used to isolate the control voltage, as shown in Fig. 16.71. The regulator then additionally compensates for the nonlinearity of the opto-coupler.

Electrical isolation of the switching signals is additionally complicated when the two power switches are at different potentials – for example, as in Fig. 16.58. As well as opto-couplers, pulse transformers can also be used to provide isolation. Opto-couplers have the disadvantage of being unable to transfer the necessary drive power for the power switches. An auxiliary power supply at the potential of the power switches is therefore required. Pulse transformers, on the other hand, allow the drive power to he transferred directly. The relevant circuit is particularly simple if power MOSFETs are used. The pulse transformer can simply be inserted between driver and MOSFET, as shown in Fig. 16.72. The coupling capacitor keeps the transformer free of any DC component. However, it should be noted that the gate signal amplitude is a function of the ON-time, as the arithmetic mean of V_2 is zero. This effect is illustrated in Fig. 16.73. For this reason, ON-times of over 50% cannot be achieved without further measures; fortunately, they are rarely required.

16.7.6
Loss Analysis

There are three types of losses that determine the efficiency of a switching regulator. The *static losses* are due to the current consumption of the pulsewidth modulator and the drivers, and are compounded by the ON-state power losses of the power switches and the output rectifier. These losses are independent of the switching frequency. The *dynamic losses* occur

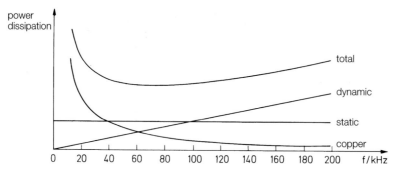

Fig. 16.74. Losses versus frequency in a switching regulator

Static losses:	Current drain of the drive circuit
	on-state power losses of the switches
	on-state power losses of the diodes
Dynamic losses:	Switching losses of the switches
	Magnetizing losses
	Damping of overshoots
Copper losses:	HF transformer
	Choke

as switching losses in the power switches and as magnetic losses in the HF transformer and in the choke. They are approximately proportional to the switching frequency. The *copper losses* in the HF transformer and in the choke result from the voltage drop across the ohmic resistance of the windings. As fewer windings are required at higher frequencies, as stated in (16.19), these losses are inversely proportional to the frequency as long as the skin effect can be neglected. Figure 16.74 shows the three loss sources as a function of frequency. An advisable operating range is between 20 kHz and 200 kHz. Although for high-frequency operation the magnetic components are lighter and smaller, the dynamic losses are so predominant in this range that the overall losses increase.

The overshoots that occur when the power switches are turned off are an additional problem. They are due to the voltages across the leakage inductance of the HF transformer and other circuit inductances. In order to minimize them, all the leads in the power circuit must be kept as short as possible. Nevertheless, during rapid switching, high overshoots may occur even if the leakage inductances are small. To give a numerical example:

$$V = L_{\text{leakage}} \frac{\Delta I}{\Delta t} = 100 \,\text{nH} \frac{1 \,\text{A}}{100 \,\text{ns}} = 100 \,V$$

This voltage is added to the regular voltage of the power switch. In order to prevent damage to the power switches, an additional snubber network is required, although this also causes additional dynamic losses.

16.7.7
IC Drive Circuits

A number of commonly used switching regulators are listed in Fig. 16.75. The control devices for push–pull converters have two outputs that switch alternately. However, these can also be used in single-ended converters if one output is left uncommitted. If limiting the ON-time to 50% causes problems, it is also possible to OR the two outputs together.

Type	Manu-facturer	Switching frequency	Power switch	Special feature
Flyback regulators, off line				
ICE3A2656	Infineon	100 kHz	int. 650 V	family with 32 types
NCP1002	On Semi.	100 kHz	int. 700 V	family with 53 types
TOP204	Power Int.	100 kHz	int. 700 V	family with 44 types
VIPer20	ST	100 kHz	inf. 700 V	family with 41 types
Flyback controller				
LT1737	Lin. Tech	200 kHz	1 N-chan.	primary output sensing
LTC3803	Lin. Tech	200 kHz	1 N-chan.	
MAX5021	Maxim	260 kHz	1 N-chan.	
LM5000	National	600 kHz	int. 80 V	
LM5025	National	300 kHz	1 N-chan.	active clamp
UCC2891	Texas Inst.	500 kHz	1 N-chan.	active clamp
Push-pull controller				
LT3439	Lin. Tech	200 kHz	int. 35 V	slew-rate controlled
LT3723	Lin. Tech	500 kHz	2 N-chan.	rectifier driver
LTC3722	Lin. Tech	250 kHz	4 N-chan.	full bridge, ZVS
LM5030	National	200 kHz	2 N-chan.	

Fig. 16.75. Examples for transformer-based (isolated) switching regulators. *ZVS* = Zero Voltage Switching

The integrated circuits for transformer-based switching regulators in Fig. 16.75 can be divided in 3 groups:

- Flyback controllers with integrated high voltage power MOSFETs. They offer the simplest way to build offline power supplies for low power applications below 50 W. They are especially suitable to build wall power supplies.
- Flyback controllers for external power transistors. Here an arbitrary power transistor can be added that is matched the voltage and current requirements.
- Push-pull controller for two power transistors that are alternately switched on. They are adapted to half- and full-bridge regulators. Normally high voltage floating switch drivers are here needed; some examples are given in Fig. 16.76.

The high power drivers in Fig. 16.76 are suitable for controlling power MOSFETs that need high peak and low continuous gate currents. For half- and full-bridge regulators the upper transistors are floating. They need floating gate drivers which sustain the full intermediate link voltage. For this application drivers with internal high voltage MOSFETs for signal transmission are the cheapest solution. However the floating switch must always be positive with respect to the control input. When full line isolation is needed drivers based on optocouplers or transformers must be applied.

In Fig. 16.76 some drivers for active rectifiers are also added. They are useful if high power is needed at low voltages. Here in the rectifier the usual schottky diodes with on voltages of 0.4–0.7 V can be replaced by power MOSFETs the on voltages of which can be as low as 0.1–0.3 V if low on resistance types are used. In order to generate the appropriate

Type	Manu-facturer	Driver	Output current peak	High/low isolation	Special features
Low voltage drivers					
IR2121	Intern. Rect.	1 low	1.5 A	—	current limiting
EL7202	Intersil	2 low	2 A	—	low supply current
HIP2100	Intersil	1 half	2 A	100 V	matched delays
IXDD430	Ixys	1 low	30 A	—	$V_{DD} = 8.5 - 35\,\text{V}$
LTC1693	Lin. Tech.	2 low	2 A	100 V	2 isolated drivers
MAX5062	Maxim	1 half	2 A	100 V	HIP2100 pinout
LM5100	National	1 half	1.5 A	100 V	HIP2100 pinout
LM5110	National	2 low	3 A	—	split supply
TPS2834	Texas Inst.	1 half	2 A	12 V	deadtime adj.
UCC27223	Texas Inst.	1 half	3 A	12 V	deadtime adj.
UCC37325	Texas Inst.	2 low	4 A	—	low overshoot
Isolated drivers for line voltages					Coupling
HCPL3120	Agilent	1 high	2 A	±1200 V	Optocoupler
ADuM1100	Analog Dev.	1 high	20 mA	±1500 V	Transf. internal
FNA7380	Fairchild	1 half	0.1 A	+600 V	MOSFETs
IR2125	Intern. Rec.	1 high	1 A	+500 V	MOSFETs
IR2184	Intern. Rec.	1 half	1.5 A	+600 V	MOSFETs
HIP2500	Intersil	1 half	2 A	+500 V	MOSFETs
IXBD4410/11	Ixys	1 half	2 A	± any	Transf. external
Active rectifier drivers					
LTC1698	Lin. Tech.	2	1 A	—	primary sync.
LTC3900	Lin. Tech.	2	2 A	—	primary sync.
STSR2P	ST	2	2 A	—	secondary sync.

Fig. 16.76. Examples for MOS-drivers. low = low side driver; high = high side driver; half = half bridge driver = 1 low side + 1 high side driver

gate signal an active rectifier driver is needed that is synchronized to the timing of the controller on the primary side.

Chapter 17:
Analog Switches and Sample-and-Hold Circuits

An analog switch is designed to switch a continuous input signal on and off. When the switch is in the ON-state, the output voltage must be as close to the input voltage as possible; when the switch is OFF, it must be zero. The principal characteristics of an analog switch are defined by the following parameters:

- the forward attenuation (the ON-state resistance),
- the reverse attenuation (the OFF-state current),
- the analog voltage range,
- the switching times.

17.1
Principle

There are several switch arrangements that fulfill the above requirements. They are represented in Fig. 17.1 as mechanical switches.

A single-throw series switch is shown in Fig. 17.1a. As long as its contact is closed, $V_o = V_i$. On opening the switch, the output voltage becomes zero, although this only applies to no-load conditions. For capacitive loads, the output voltage will only fall to zero slowly because of the finite output resistance $r_o = R$.

The single-throw short-circuiting switch shown in Fig. 17.1b overcomes this difficulty. However, in the ON-state – that is, when the contact is open – the circuit possesses a finite output resistance $r_o = R$.

The double-throw series/short-circuiting switch in Fig. 17.1c combines both advantages, and has a low output resistance in both states. The forward attenuation is low and the reverse attenuation is high. However, the fact that the output is short-circuited in the OFF-state may also cause problems – for example, if the output voltage is to be stored in a capacitor, as in the sample-and-hold circuits of Sect. 17.4. In this case, switch S_3 can be added, as shown in Fig. 17.2. When the switch is open, the input signal that is capacitively coupled via S_3 is short-circuited by S_2; however, the output remains at high impedance due to S_3. This arrangement therefore behaves like the series switch in Fig. 17.1a, but has a much better reverse attenuation for high frequencies.

Extending this principle to several inputs, we obtain the arrangement shown in Fig. 17.3. One of the four switches is closed at any one time, which means that the output voltage is

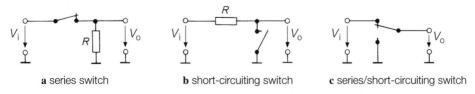

a series switch   **b** short-circuiting switch   **c** series/short-circuiting switch

Fig. 17.1. Switch configurations

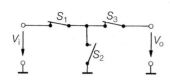

Fig. 17.2. Series switch with improved reverse attenuation

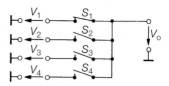

Fig. 17.3. Analog multiplexer–demultiplexer

equal to the particular input voltage selected. This arrangement is therefore also known as an analog *multiplexer*.

By inverting the arrangement, an input voltage can be distributed to several outputs, thus providing an *analog demultiplexer* function. The corresponding circuits for digital signals have already been described in Sect. 8.2 on page 643.

17.2
Electronic Switches

Field effect transistors, diodes, or bipolar transistors are used to implement the switches. They possess quite different characteristics, and specific advantages and disadvantages. They do, however, have the same basic arrangement, which is shown in Fig. 17.4. In most cases, TTL-compatible control signals are required. These are amplified by a power gate followed by a level converter, which generates the voltages that are required for opening or closing the switch.

17.2.1
FET Switch

As we saw in Sect. 3.1.3, a FET behaves like an ohmic resistor whose value can be varied by several orders of magnitude using the gate–source voltage U_{GS}. This behavior makes it extremely useful as a series switch (see Fig. 17.5). The FET is turned off if we apply a control voltage that is negative with respect to the input voltage by at least the threshold $V_{C \, off} \le V_i + V_{th}$.

To make the junction-FET conduct, the voltage V_{GS} must he zero. This condition is not so easy to fulfill, as the source potential is not constant. A solution to this problem is shown in Fig. 17.6, where diode D becomes reverse biased if V_C is made larger than the most positive input voltage, and therefore $V_{GS} = 0$, as required.

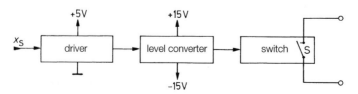

Fig. 17.4. Drive arrangement for a switch

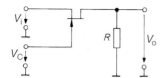

Fig. 17.5. FET series switch

$V_{C\,on} = V_i$

$V_{C\,off} \le V_i + V_{th}$

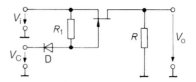

Fig. 17.6. Simplified drive arrangement

$V_{C\,on} = V_{i\,max}$

$V_{C\,off} \le V_i + V_{th}$

For sufficiently negative control voltages, diode D is forward biased and the FET is turned off. In this mode, a current flows from the input voltage source via resistor R_1, into the control circuit. This can usually be tolerated, as the output voltage in this case is zero. However, this effect becomes troublesome if the input voltage is connected to the switch via a coupling capacitor, as the latter becomes charged to a negative voltage during the OFF phase.

These problems do not arise if a MOSFET is used for switching. An n-channel MOS-FET can be made to conduct by applying a control voltage that is larger than the most positive input voltage. No current flows from the gate to the channel, so that diode D and resistor R_1 are no longer necessary. To insure that the input voltage range is as large as possible, it is better to use, instead of a single MOSFET, a CMOS switch consisting of two complementary MOSFETs connected in parallel, as shown in Fig. 17.7.

To turn the switch ON, V^+ is applied to the gate of n-channel MOSFET T_1, and that of p-channel MOSFET T_2 is connected to ground. Around the mid-range of voltage V_i, both MOSFETs are therefore conducting. If the input voltage increases to higher positive values, V_{GS1} is reduced, making T_1 have a higher impedance. However, this has no adverse effect, as the absolute value of $U_{GS\,2}$ increases simultaneously. This makes T_2 go to low impedance, and vice versa for small input voltages. This is illustrated in Fig. 17.8, in which we see that the input voltage can assume any value between 0 and V^+.

With standard CMOS switches, neither the control voltage nor the analog signals must be outside this range, because this could result in the destruction of the switches due to *latch-up*. In this case, the channel substrate diode becomes conducting and floods the substrate with charge carriers that could fire the parasitic thyristor shown in Fig. 7.39 on

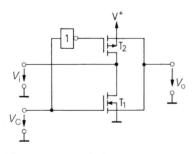

Fig. 17.7. Transmission gate

$V_{C\,on} = V^+$

$V_{C\,off} = 0\,V$

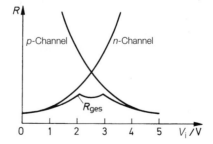

Fig. 17.8. FET resistance versus input voltage for

$V_C = V_{C\,on} = V^+ = 5\,V$

Type	Manu-facturer	Function	On resistance	Voltage range	Power dissipation	Response time
Without level translator, unprotected						
74 HC 4016	Philips	4 × SPST	65 Ω	0...12 V	10 µW	10 ns
74 HC 4053	Philips	3 × SPDT	60 Ω	0...12 V	10 µW	20 ns
74 HC 4066	Philips	4 × SPST	35 Ω	0...12 V	10 µW	20 ns
MAX 4522	Maxim	4 × SPST	60 Ω	0...12 V	10 µW	10 ns
SD 5000	Vishay	4 × SPST	30 Ω	± 10 V	10 µW	1 ns
Fast switching (≤ 100 ns)						
HI 201 HS	Intersil	4 × SPST	30 Ω	± 15 V	120 mW	30 ns
74 HC 4316	Philips	4 × SPST	65 Ω	± 5 V	10 µW	20 ns
DG 611	Vishay	4 × SPST	18 Ω	± 5 V	20 µW	15 ns
Low power (≤ 100 µW) and low on resistance (≤ 100 Ω)						
ADG 511	Analog D.	4 × SPST	30 Ω	± 20 V	20 µW	200 ns
DG 403	Intersil	2 × SPDT	30 Ω	± 15 V	20 µW	150 ns
DG 411	Intersil	4 × SPST	30 Ω	± 15 V	30 µW	150 ns
LT 221	Lin. Tech.	4 × SPST	70 Ω	± 15 V	10 µW	250 ns
MAX 351	Maxim	4 × SPST	22 Ω	± 15 V	35 µW	150 ns
DG 405	Vishay	4 × SPST	20 Ω	± 15 V	10 µW	100 ns
Low on resistance (≤ 100 Ω)						
ADG 211	Analog D.	4 × SPST	60 Ω	± 15 V	10 µW	200 ns
ADG 333	Analog D.	4 × SPDT	20 Ω	± 15 V	1 µW	150 ns
ADG 451	Analog D.	4 × SPST	5 Ω	± 15 V	20 µW	60 ns
MAX 4602	Maxim	4 × SPST	2 Ω	± 15 V	1 µW	180 ns
CDG 271	Vishay	4 × SPST	32 Ω	± 15 V	150 mW	50 ns
High voltage (≥ ± 30 V)						
HV 348	Supertex	2 × SPST	35 Ω	± 50 V	10 mW	500 ns
High off attenuation (≥ 40 dB @ 100 MHz)						
HI 222	Intersil	2 × SPST	35 Ω	± 15 V	75 mW	90 ns
MAX 4545	Maxim	2 × SPST	50 Ω	± 5 V	1 µW	100 ns
DG 540	Vishay	4 × SPST	30 Ω	± 6 V	60 mW	30 ns

Fig. 17.9. Examples for analog switches in CMOS technology. Most other switch configurations are also available with corresponding factory numbers

page 629, short-circuiting the supply voltage. If it cannot be guaranteed that the relevant values will stay within the safe input voltage range, a resistor must be connected at the input to limit the current.

Because of these problems, most integrated CMOS switches are provided with additional protection – that is, current-limiting – structures or are manufactured with *dielectric insulation*. In this case, an oxide layer is used as the insulator to the substrate, instead of a pn junction. Consequently, CMOS components with dielectric insulation are not subject to latch-up effects, but their manufacturing process is considerably more expensive.

A number of commonly used CMOS switches and multiplexers are listed in Fig. 17.9 and 17.10. The 74 HC types are normal, extremely low-cost CMOS gates, but they are prone to latch-up and have only a limited voltage range. The other types are protected against this effect and can therefore be used without any difficulty. The manufacturers listed also offer a wide range of other types, of which just a few examples are given.

Type	Manu-facturer	Function	On resistance	Voltage range	Power dissipation	Response time	Data latch
Fast switching ($\leq$ 100 ns), unprotected							
74 HC 4051	Philips	1×8	$60\,\Omega$	$\pm 5\,$V	$10\,\mu$W	20 ns	no
74 HC 4052	Philips	2×4	$60\,\Omega$	$\pm 5\,$V	$10\,\mu$W	20 ns	no
74 HC 4053	Philips	3×2	$60\,\Omega$	$\pm 5\,$V	$10\,\mu$W	20 ns	no
Low power ($\leq$ 100 μW)							
DG 406	Maxim	1×16	$80\,\Omega$	$\pm 15\,$V	$20\,\mu$W	300 ns	no
DG 408	Maxim	1×8	$80\,\Omega$	$\pm 15\,$V	$20\,\mu$W	200 ns	no
DG 685	Vishay	1×8	$55\,\Omega$	$\pm 15\,$V	$10\,\mu$W	160 ns	yes
High input protection ($\geq \pm 30$ V)							
MAX 378	Maxim	1×8	$2\,$kΩ	$\pm 15\,$V	$2\,$mW	300 ns	no
DG 458	Intersil	1×8	$80\,\Omega$	$\pm 15\,$V	$5\,$mW	200 ns	no
High voltage ($\geq \pm 30$ V)							
HV 22816	Supertex	1×8	$22\,\Omega$	$\pm 80\,$V	$2\,$mW	$4\,\mu$s	yes
High off attenuation ($\geq$ 40 dB @ 100 MHz)							
MAX 310	Maxim	1×8	$150\,\Omega$	$\pm 12\,$V	$1\,$mW	300 ns	no
DG 536	Vishay	1×16	$55\,\Omega$	$0 \ldots 10\,$V	$75\,\mu$W	200 ns	yes
DG 538	Vishay	2×4	$45\,\Omega$	$\pm 6\,$V	$10\,$mW	200 ns	yes
Universal types							
ADG 408	Analog D.	1×8	$80\,\Omega$	$\pm 15\,$V	$2\,$mW	200 ns	no
ADG 526	Analog D.	1×16	$280\,\Omega$	$\pm 15\,$V	$10\,$mW	200 ns	yes
DG 408	Intersil	1×8	$80\,\Omega$	$\pm 15\,$V	$7\,$mW	200 ns	no
MAX 308	Maxim	1×8	$60\,\Omega$	$\pm 15\,$V	$300\,\mu$W	200 ns	no

Fig. 17.10. Examples of analog multiplexers in CMOS technology. Most other switch configurations are also available with corresponding factory numbers

The typical reverse currents of the switches are between 0.1 nA and 1 nA at room temperature. These values double for every 10 degree increase in temperature and can therefore be as much as 100 nA. The switching times are between 100 ns and 300 ns.

17.2.2
Diode Switch

Diodes are also suitable for use as switches because of their low forward resistance and high blocking resistance. If a positive control voltage is applied to the circuit shown in Fig. 17.11, diodes D_5 and D_6 become reverse biased. The impressed current I then flows through branches D_1, D_4 and D_2, D_3, from one current source to the other. The potentials V_1, and V_2 thereby assume the values

$$V_1 = V_i + V_D, \qquad V_2 = V_i - V_D$$

The output voltage is then:

$$V_o = V_1 - V_D = V_2 + V_D = V_i$$

if the forward voltages V_D are the same. Should this not be the case, an offset voltage occurs.

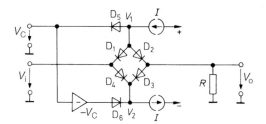

Fig. 17.11. A series switch using diodes

If the control voltage is made negative, the two diodes D_5, D_6 become forward biased, and the diode bridge is turned off. This means that the output is doubly disconnected from the input and the midpoint is at constant potential. The analog switch therefore has a high reverse attenuation as shown in Fig. 17.2.

By employing this principle, switching times of less than 1 ns can be achieved if fast-switching diodes are used. Suitable types include the Schottky diode quartet 5082-2813 from Hewlett-Packard.

Rapid switching naturally requires correspondingly fast drive signals. An example of a suitable drive circuit is shown in Fig. 17.12. This consists of a bridge circuit made up of four constant-current sources, $T_1 \dots T_4$. The upper two are switched on alternately by the drive signal. When T_1 is on, a current of magnitude I flows through the diode bridge, causing it to conduct. When T_2 is on, the diode bridge is reverse biased. In order to insure that, in this case, current sources T_2 and T_3 are not driven into saturation, the reverse voltages are limited by transistors T_5 and T_6. These also insure that the diode bridge driver is at low impedance during reverse-biased operation. Good reverse attenuation is then achieved due to the reduction of capacitive feedthrough.

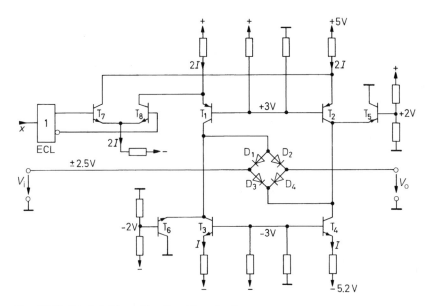

Fig. 17.12. Diode bridge with a fast drive circuit

The amplitude of the analog signal must be smaller than the maximum control voltage across the diode bridge. For the values specified, it must be limited to $\pm 2.7\,\text{V}$.

17.2.3
Bipolar Transistor Switch

To investigate the suitability of a bipolar junction transistor for use as a switch, we examine its output characteristic curves around the origin, an expanded view of which is given in Fig. 17.13 for small positive and negative collector–emitter voltages.

The first quadrant contains the familiar output characteristics shown in Fig. 2.3. If the voltage V_{CE} is made negative without changing the base current, the output characteristics of the third quadrant are obtained. In this reverse-mode polarity, the current gain of the transistor is considerably reduced and is about $\frac{1}{30}\beta$. The maximum permissible collector–emitter voltage in this mode is the breakdown voltage V_{EB0}, since the base–collector junction is forward biased and the base–emitter junction is reverse biased. This type of operation is known as reverse-region operation, and the accompanying current gain is termed the reverse current gain ratio β_r. The collector current is zero for a collector–emitter voltage of about 10...50 mV. If the base current exceeds a few millamperes, this *offset voltage* increases steeply; for small base currents, it remains constant over a large range.

The offset voltage can be reduced considerably by insuring that the transistor is in reverse-region operation when the output current crosses zero. To achieve this, the collector and the emitter must be interchanged. The resulting output characteristics are shown in Fig. 17.14. At larger output currents, virtually the same curves are obtained as for the normal operation in Fig. 17.13, if V_{CE} is still measured at the correct polarity (collector-

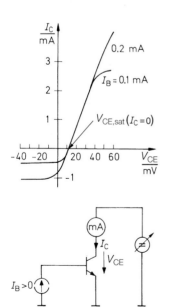

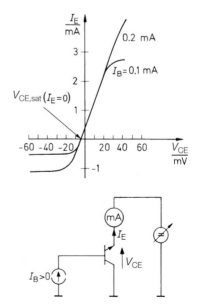

Fig. 17.13. Complete output characteristics of a transistor in common-emitter connection, with the associated test circuit

Fig. 17.14. Complete output characteristics for an interchanged emitter and collector, with the associated test circuit

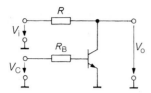

Fig. 17.15. Bipolar transistor as a short-circuiting switch

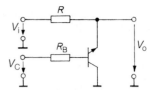

Fig. 17.16. Short-circuiting switch in reverse operation

to-emitter). The reason for this is that the emitter current, which is now the output current, is very nearly the same as the collector current.

Near the origin, however, a major difference arises in that the base current can no longer be neglected with respect to the output current. If, for normal operation, the output current is made zero, the emitter current is identical to the base current – that is, is not zero – and an offset voltage of $10 - 50\,\text{mV}$ appears at the output. If the collector and the emitter are interchanged and the output current is made zero, the collector current becomes the base current. The collector–base junction is then forward biased (reverse operation). The offset voltage for this mode of operation is usually about one-tenth of that in normal operation, but it is also positive since, for the circuit shown in Fig. 17.14, $V_o = -V_{CE}$. Typical values for the offset voltage are between 1 and 5 mV, and it is therefore desirable to operate transistor switches with an interchanged collector and emitter. If the emitter current is kept small, the transistor operates almost exclusively in the reverse mode.

Short-Circuiting Switch

Figures 17.15 and 17.16 show how a transistor can be used as a short-circuiting switch. In the circuit of Fig. 17.15, the transistor is operating in normal mode, whereas in Fig. 17.16 it is in reverse-region mode. To obtain a sufficiently low transistor resistance, the base current must be in the milliampere range. The collector current in Fig. 17.15, and the emitter current in Fig. 17.16, should not be much larger, in order to insure that the offset voltage remains small.

Series Switch

A bipolar transistor used as a series switch is shown in Fig. 17.17. A negative control voltage must be applied to turn off the transistor. It must be more negative than the most negative value of the input voltage, but it also has a limit, since the control voltage may not be more negative than $-V_{EB0} \approx -6\,\text{V}$.

To render the transistor conducting, a positive control voltage is applied which is larger than the input voltage by a value of $\Delta V = I_B R_B$. The collector–base junction is then forward biased and the transistor operates as a switch in reverse-region mode. The disadvantage is that the base current flows into the input voltage source, and unless the internal resistance of the source is kept very small, large errors may occur.

If this condition can be fulfilled, the circuit is particularly suitable for positive input voltages, as the ON-state emitter current is positive. The offset voltage is therefore reduced and even becomes zero for a particular emitter current, as can be seen in Fig. 17.14. In this mode of operation, the circuit is known as a saturated emitter follower, since for control

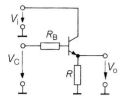

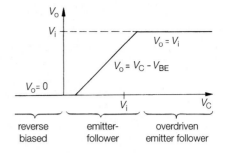

Fig. 17.17. Saturated emitter follower as a series switch

Fig. 17.18. Transfer characteristics for positive input voltages

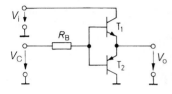

Fig. 17.19. Series/short-circuiting switch

voltages between zero and V_i, it operates as an emitter follower for V_C. This is illustrated in Fig. 17.18, by the transfer characteristics for positive input voltages.

Series/Short-Circuiting Switch

If the saturated emitter follower shown in Fig. 17.17 is combined with the short-circuiting switch shown in Fig. 17.16, a series/short-circuiting switch is obtained, which has a low offset voltage for both modes of operation. It has the disadvantage that complementary control signals are required. The control arrangement is particularly simple if a complementary emitter follower is used, as shown in Fig. 17.19. It is saturated in both the ON and OFF states if $V_{C\,max} > V_i$ and $V_{C\,min} < 0$. Due to the low output resistance, a fast switchover of the output voltage, between zero and V_i, is possible. A practical implementation of this was shown in the case of the function generator in Fig. 14.36 on page 862.

17.2.4
Differential Amplifier Switch

The gain of a differential amplifier is proportional to the transconductance, which is in turn proportional to the collector current. Consequently, the differential gain can be made zero by cutting off the emitter current. Figure 17.20 shows how this principle can be applied to a differential amplifier used as an analog switch.

If the control voltage is made negative, diode D is turned off and the differential amplifier carries the emitter current $I_k = I$. If the output voltage is taken from the collectors, we obtain:

$$V_o = g_m R_c V_i = \frac{I_c}{V_T} R_c V_i = \frac{1}{2V_T} I_k R_c V_i$$

If the control voltage is made positive, the diode takes over the current I and the transistors are turned off; $I_k = 0$. Although this causes the two output potentials to increase to V^+, the output voltage difference V_o becomes zero.

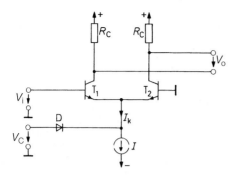

Fig. 17.20. Differential amplifier used as a switch

$$V_o = \begin{cases} 0 & \text{for } V_C = +1\,\text{V} \\ g_m R_C V_i & \text{for } V_C = -1\,\text{V} \end{cases}$$

Figure 17.21 shows how this principle can be employed to design an analog switch for low frequencies. As long as the input voltage $V_D = 0$, the control current I_C is equally divided between the two transistors of the differential amplifier, and current I flows in all the current mirrors. The output current becomes zero. If a positive input voltage is applied, the collector current of T_2 increases by $\Delta I = \frac{1}{2} g_m V_D$ and that of T_1 decreases by the same amount. The output current is therefore:

$$I_o = 2\Delta I = g_m V_D = \frac{I_c}{V_T} V_D = \frac{I_C}{2V_T} V_D$$

If the control current I_C is made zero, all the transistors are turned off and the output current also becomes zero.

Amplifiers that operate on this principle are known as *transconductance amplifiers*. They are available in IC form; for example, the CA 3060 or CA 3280 from Intersil. They can also be used as operational amplifiers if the control current remains constant. If the control current is made proportional to a second input voltage, they can also be used as analog multipliers.

Figure 17.22 shows how the principle illustrated in Fig. 17.20 can be used to design a switch for high frequencies. Here, the two differential amplifiers T_1, T_2 and T_3, T_4 employ common-collector resistors R_1. However, only one pair is in operation at a time: when the

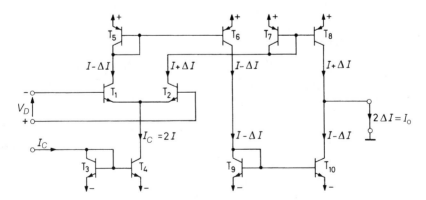

Fig. 17.21. Transconductance amplifier used as a switch

$$I_o = \begin{cases} 0 & \text{for } I_C = 0 \\ I_C V_D / 2V_T & \text{for } I_C > 0 \end{cases}$$

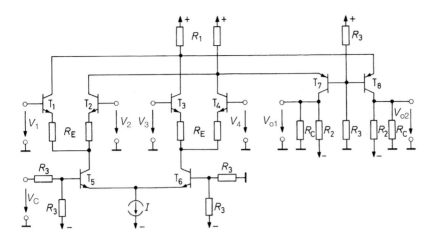

Fig. 17.22. Wideband multiplexer

$$V_{o1} = -V_{o2} = \begin{cases} A(V_1 - V_2) \text{ for } V_C = 1\,\text{V} \\ A(V_3 - V_4) \text{ for } V_C = -1\,\text{V} \end{cases} \qquad A = \tfrac{1}{2} g_{m,\text{red}}(R_C \| R_2)$$

$$g_{m,\text{red}} = g_m/(1 + g_m R_E)$$

control voltage is positive, the differential amplifier T_1/T_2 receives current I; for negative control voltages, current I flows in the pair T_3/T_4. This arrangement has an advantage over that in Fig. 17.20 in that the output potentials remain constant during switching.

We therefore have a device that can be used to switch from one input voltage, $V_{i1} = V_1 - V_2$, to another, $V_{i2} = V_3 - V_4$. If we make $V_3 = V_2$ and $V_4 = V_1$ by connecting the relevant inputs, then $V_{i2} = -V_{i1}$ and we have a polarity changer.

The circuit can be designed as a wideband amplifier, like the complementary cascode differential amplifier shown in Fig. 5.25. The current feedback resistors R_E and cascode circuits T_7, T_8 are used for this purpose. By selecting suitable component values, bandwidths of 100 MHz or more can he achieved. The circuit can therefore be used, for instance, as a modulator, demodulator, or phase detector for telecommunications, and as a channel (beam) chopper in wideband oscilloscopes.

Integrated circuits that employ this principle include the OPA 676 from Burr Brown or the AD 539 from Analog Devices for high-bandwidth applications, or the AD 630 from Analog Devices for high-precision circuits.

17.3
Analog Switches Using Amplifiers

If analog switches are combined with operational amplifiers, a number of special characteristics can be obtained. In the following sections, the switches themselves will only be shown symbolically. The CMOS types listed in Fig. 17.9 and 17.10 are the most suitable for practical implementation.

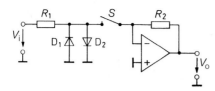

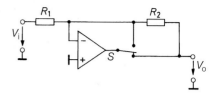

Fig. 17.23. Switching high voltages at low switch voltages

Fig. 17.24. Switching high voltages with high precision

$$V_o = \begin{cases} 0 \\ -V_i R_2/(R_1 + r_{\text{DS on}}) \end{cases}$$

$$V_o = \begin{cases} 0 \\ -V_i R_2/R_1 \end{cases}$$

17.3.1
Analog Switches for High Voltages

In the circuit shown in Fig. 17.23, the operational amplifier operates as an inverting amplifier. When the switch is open, the voltage across it is limited by diodes D_1 and D_2 to $\pm 0.7\,\text{V}$. When the switch is closed, both terminals are at ground potential, as they are connected to the summing point. In this case, the circuit operates as an inverting amplifier. The diodes have no effect, as virtually no voltage is dropped across them. The circuit gain can therefore be selected using R_1 and R_2 such that the operational amplifier is not overdriven even at the highest input voltages.

Another method of switching large voltages is illustrated in Fig. 17.24. In the switch position shown, the operational amplifier again operates as an inverting amplifier. The advantage in this case is that the switch is inserted in the feedback loop and, as a result, its ON-state resistance has no effect on the gain. However, it is necessary to use switches whose analog voltage range is identical to the maximum output voltage swing of the operational amplifier.

When the switch is changed over, the output is connected via R_2 to the summing point; that is, to zero potential.

17.3.2
Amplifier with Switchable Gain

In the circuit shown in Fig. 17.25, the gain of a noninverting amplifier can be switched using an analog multiplexer. Depending on which switch of the multiplexer is closed, any gain factor $A \geq 1$ can be realized by selecting suitable component values for the voltage divider chain. The main advantage of this circuit is that the switches of the analog multiplexer can be operated without current. This means that their ON-state resistance does not affect the output voltage. An IC amplifier that operates on this principle is the AD 526 from Analog Devices. Its gain can be switched between values of 1 and 16.

With the circuit shown in Fig. 17.26, the sign of the gain can be reversed using switch S. When the switch is in the lower position, the circuit operates as an inverting amplifier that provides a gain of $A = -1$.

When the switch is in the upper position, $V_p = V_i$. The output voltage therefore assumes a value such that no voltage is dropped across R_1. This occurs when $V_o = V_i$. The amplifier therefore operates as a noninverting amplifier. The circuit is then very similar to the bipolar coefficient network shown in Fig. 11.5.

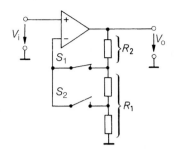

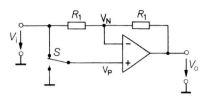

Fig. 17.25. Noninverting amplifier with switchable gain

$$V_o = (1 + R_2/R_1)V_i$$

Fig. 17.26. Inverting/noninverting amplifier

$$V_o = \begin{cases} V_i & \text{for } S = \text{up} \\ -V_i & \text{for } S = \text{down} \end{cases}$$

17.4
Sample-and-Hold Circuits

17.4.1
Basic Principles

The output voltage of a sample-and-hold circuit should follow the input voltage when it is in the ON-state. In this mode, it therefore behaves like an analog switch. In the OFF-state, however, the output voltage must not become zero, but the voltage at the instant of turn-off must be stored. This is why sample-and-holds are also known as track-and-hold circuits.

The basic arrangement for a sample-and-hold circuit is shown in Fig. 17.27. The central component is the capacitor C, which performs the storage function. When switch S is closed, the capacitor is charged up to the input voltage. In order to insure that the input voltage source is not loaded, an impedance converter is employed. This is implemented by voltage follower OA1. It must be capable of delivering high output currents in order to be able to charge and discharge the storage capacitor quickly.

When switch S is open, the voltage across capacitor C must be kept constant for as long as possible. Consequently, a voltage follower is connected after it, to eliminate loading of the capacitor. In addition, the switch must have a high OFF-state resistance and the capacitor a low leakage current.

The main nonideal characteristics of a sample-and-hold circuit are given in Fig. 17.28. When the switch is closed by the sample command, the output voltage does not instantaneously increase to the value of the input voltage, but only at a defined maximum *slew rate*. It is primarily determined by the maximum current of the impedance converter OA1. This is followed by a settling time, whose duration is determined by the damping due to the impedance converter and the ON-state resistance of the switch. The *acquisition time*

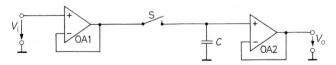

Fig. 17.27. Schematic diagram of a sample-and-hold circuit

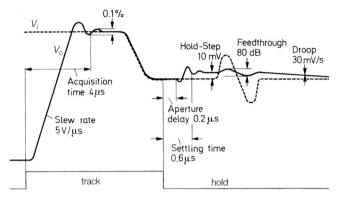

Fig. 17.28. Specifications of a sample-and-hold circuit, showing typical values for an LF 398 with 1 nF capacitor. The duration of the tracking phase must he at least equal to the acquisition time

t_{Ac} is defined as the time that elapses after the start of the track command until the output voltage is equal to the input voltage within the specified tolerance. If the charging of the storage capacitor is determined solely by the ON-state resistance R_S of the switch, the acquisition time can be calculated from the charging function of an RC network and the required accuracy of acquisition. Thus

$$t_{AC} = R_S \cdot C \cdot \begin{cases} 4.6 & \text{for } 1\% \\ 6.9 & \text{for } 0.1\% \end{cases}$$

It is therefore shorter the smaller the value of C selected.

During transition to the hold state, it takes a while for the switch to open. This is known as the *aperture delay* t_{Ap}. It is not usually constant, but tends to vary, often as a function of the particular value of the input voltage. These fluctuations are termed *aperture jitter* Δt_{Ap}.

In general, the output voltage does not now remain at the stored value, but there is a small voltage change ΔV_o (the *hold step*) with subsequent settling. This is due to the fact that, when the circuit is switched off, a small charge is coupled by the drive signal via the switch capacitance C_S into storage capacitor C. The resultant hold step is given by

$$\Delta V_o = \frac{C_s}{C} \Delta V_C$$

where ΔV_C *is* the amplitude of the drive signal. This effect is smaller the larger is the value selected for C.

Another nonideal characteristic is the *feedthrough*. This results from the fact that the input voltage has an effect on the output even though the switch is open. This effect is mainly caused by the capacitive voltage divider formed by the capacitance of the open switch with the storage capacitor.

The most important variable in the store condition is the droop (the hold decay). This is mainly determined by the input current of the impedance converter at the output and the reverse current of the switch. For a discharge current I_L, we have

$$\frac{\Delta V_o}{\Delta t} = \frac{I_L}{C}$$

In order to minimize the discharge current, an FET-input amplifier is used for OA2.

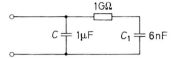

Fig. 17.29. Equivalent circuit of a capacitor. In this example, a 1 µF capacitor with a Mylar dielectric is used

As we can see, all of the characteristics in the hold state are better the larger the value selected for C, whereas during the tracking operation small values of C are desirable. Consequently, a compromise has to be found, depending on the application.

We have hitherto assumed that the hold capacitor possesses ideal characteristics. It is also possible to find capacitors that have virtually no leakage current. Nevertheless, a voltage change can occur in the hold state due to charge storage in the dielectric. This effect may be explained by reference to the equivalent circuit shown in Fig. 17.29. Capacitor C_1 represents the charge stored in the dielectric. It initially remains unchanged in the event of a hold step and only varies slowly. If the sampling time is short, the charge required for this purpose is taken from capacitor C during the hold phase (dielectric absorption). In the case of a hold step of magnitude V, this produces a subsequent voltage change of

$$\Delta V = \frac{C_1}{C} V$$

that is, 0.6% in the example shown in Fig. 17.29. The size of this effect depends on the dielectric used. Teflon, polystyrene, and polypropylene are good in this respect; on the other hand, polycarbonate, Mylar, and most ceramic dielectric materials are poor.

17.4.2
Practical Implementation

The fastest sample-and-hold circuits can be designed on the principle illustrated in Fig. 17.27, if the diode bridge of Fig. 17.12 is used as a switch and the circuits described in Figs. 4.111 and 4.112 are used as voltage followers.

A higher degree of accuracy can be achieved using an overall feedback arrangement, as shown in Fig. 17.30. When the switch is closed, the output potential V_1 of amplifier OA1 assumes a value such that $V_o = V_i$. This eliminates offset errors due to OA2 or the switch. Diodes D_2 and D_3 are nonconducting in this operating condition, as only a small voltage $V_1 - V_o$ is dropped across them, which is precisely equal to the offset voltage.

If the switch is opened, the output voltage remains constant. Resistor R_2 and diodes D_2, D_3 prevent amplifier OA1 from being overdriven in this operating condition. This is an important consideration, because any overdriving is followed by a considerable recovery time, which is added to the settling time.

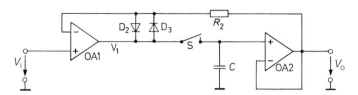

Fig. 17.30. Sample-and-hold circuit with overall feedback

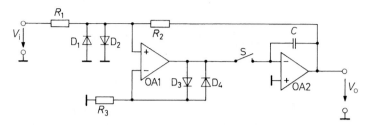

Fig. 17.31. Sample-and-hold circuit with an integrator as a storage device

This is the principle employed by type LF 398 which, being inexpensive, represents the most commonly used sample-and-hold circuit for general applications.

Sample-and-Hold Circuit with Integrator

Instead of a grounded capacitor with voltage follower, an integrator can also be used as an analog storage device. This possibility is illustrated in Fig. 17.31, in which the series switch is connected to the summing point, this resulting in a simple drive arrangement.

When the switch is closed, the output voltage assumes the value $V_o = -V_i R_2/R_1$ due to the negative feedback. As in the previous circuit, amplifier OA1 reduces the settling time and eliminates the offset voltage of FET amplifier OA2.

When the switch is opened, the current flowing through the storage capacitor becomes zero and the output voltage remains constant. In this case, overall feedback is ineffective. Instead, diodes $D_1 - D_4$ come into play and limit the output voltage of OA1 to $\pm 1.2\,V$, thereby preventing it from being overdriven.

With high-speed sample-and-hold circuits, amplifier OA1 is generally omitted. The circuit is then similar to the integrator shown in Fig. 11.10.

A selection of sample-and-hold circuits is listed in Fig. 17.32. With the LF 398, the hold capacitor is connected externally. This makes it possible – within certain limits – to match the properties to the particular application.

Type	Manu-facturer	Hold capacitor	Setting time	Accuracy	Slew rate	Droop	Tech-nology
LF 398	various	10 nF	20 μs	10 bit	0,5 V/μs	3 mV/s	Bifet
LF 398	**various**	**1 nF**	**4 μs**	**10 bit**	**5 V/μs**	**30 mV/s**	**Bifet**
AD 585	Analog Dev.	100 pF*	3 μs	12 bit	10 V/μs	0,1 V/s	Bipolar
SHC 5320	Burr Brown	100 pF*	1.5 μs	12 bit	45 V/μs	0,1 V/s	Bipolar
AD 781	**Analog Dev.**	*	**0.6 μs**	**12 bit**	**60 V/μs**	**10 mV/s**	**Bimos**
AD 682[2]	Analog Dev.	*	0.6 μs	12 bit	60 V/μs	10 mV/s	Bimos
AD 684[4]	Analog Dev.	*	0.6 μs	12 bit	60 V/μs	10 mV/s	Bimos
AD 783	**Analog Dev.**	*	**0.2 μs**	**12 bit**	**50 V/μs**	**20 mV/s**	**Bimos**
LF 6197	**National**	**10 pF***	**0.2 μs**	**12 bit**	**145 V/μs**	**0,6 V/s**	**Bifet**
AD 9100	**Analog Dev.**	**22 pF***	**16 ns**	**12 bit**	**850 V/μs**	**1 kV/s**	**Bipolar**
RTH 050	**Rockwell**	*	**0.2 ns**	**8 bit**	**40 kV/μs**		**GaAs**

* internal hold capacitor   [2] double S&H.   [4] quad S&H.

Fig. 17.32. Examples for integrated sample and hold circuits. The S&H circuits needed for analog-digital converters are today normally included on the ADC chips (Sampling ADC).

Chapter 18:
Digital-Analog and Analog-Digital Converters

To display or process a voltage digitally, the analog signal must be translated into numeric form. This task is performed by an analog-to-digital converter (A/D converter, or ADC). The resultant number Z will generally be proportional to the input voltage V_i:

$$Z = V_i / V_{\text{LSB}}$$

where V_{LSB} is the voltage unit for the least significant bit; that is, the voltage for $Z = 1$.

To convert a number back into a voltage, a digital-to-analog converter (D/A converter, or DAC) is used, whose output voltage is proportional to the numeric input; that is,

$$V_o = V_{\text{LSB}} \cdot Z$$

18.1
Sampling Theorem

A continuous input signal can be converted into a series of discrete values by using a sample-and-hold circuit for sampling the signal at equidistant instants $t_\mu = \mu T_s$, where $f_s = 1/T_s$ is the sampling rate. It is obvious from Fig. 18.1 that a staircase function arises, and that the approximation to the continuous input function is better the higher the sampling rate. However, as circuit complexity increases markedly with higher sampling rates, it is essential to keep the latter as low as possible. The question is now: What is the lowest sampling rate at which the original signal can still be reconstructed *error-free*; that is, without loss of information. This theoretical limit is defined by the sampling theorem (the Nyquist criterion), which we shall now discuss.

In order to obtain a simpler mathematical description, the staircase function shown in Fig. 18.1 is replaced by a series of Dirac impulse functions, as illustrated in Fig. 18.2:

$$\tilde{V}_i(t) = \sum_{\mu=0}^{\infty} V_i(t_\mu) T_s \delta(t - t_\mu) \tag{18.1}$$

Their impulse area $V_i(t_\mu) \cdot T_s$ is represented by an arrow. The arrow must not be mistaken for the height of the impulse, as a Dirac function is, by definition, an impulse with infinite height but zero width, although its area has a finite value. This area is often misleadingly

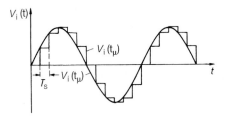

Fig. 18.1. Example of an input signal $V_i(t)$ and sampled values $V_i(t_\mu)$

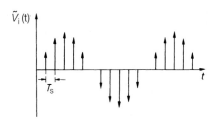

Fig. 18.2. Representation of the input signal by a Dirac impulse sequence

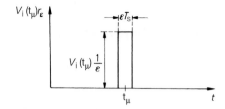

Fig. 18.3. Approximation of a Dirac impulse by a finite voltage pulse

known as the impulse amplitude. The characteristics of the impulse function are shown by Fig. 18.3, where the Dirac impulse function is approximated by a rectangular pulse r_ε; the limit of the approximation is

$$V_i(t_\mu)T_s\delta(t - t_\mu) = \lim_{\varepsilon \to 0} V_i(t_\mu)r_\varepsilon(t - t_\mu) \tag{18.2}$$

To examine the information contained in the impulse function sequence represented by (18.1), we consider its spectrum. By applying the Fourier transformation to (18.1), we obtain

$$\tilde{X}(jf) = T_s \sum_{\mu=0}^{\infty} V_i(\mu T_s)e^{-2\pi j \mu f/f_s} \tag{18.3}$$

It can be seen that this spectrum is a periodic function, the period being identical to the sampling frequency f_s. When this periodic function is Fourier analyzed, it can be shown that the spectrum $|\tilde{X}(jf)|$ is, for $-\frac{1}{2}f_s \leq f \leq \frac{1}{2}f_s$, identical to the spectrum $|\tilde{X}(jf)|$ of the original waveform. Thus it still contains all of the information, although only a few values of the function were sampled.

There is only one restriction, and this is explained with the help of Fig. 18.4. The original spectrum reappears unchanged only if the sampling rate is chosen such that consecutive bands do not overlap. According to Fig. 18.4, this is the case for

$$\boxed{f_s > 2f_{max}} \tag{18.4}$$

this condition being known as the sampling theorem.

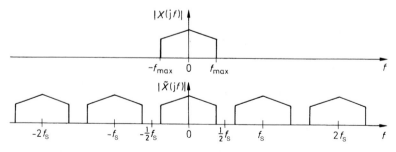

Fig. 18.4. Spectrum of the input voltage before (*upper diagram*) and after (*lower diagram*) sampling

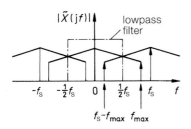

Fig. 18.5. Overlapping of spectra if the sampling frequency is too low

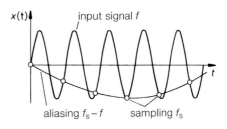

Fig. 18.6. Aliasing due to an excessively low sampling frequency for $f \lesssim f_s$

Recovery of the Analog Signal

From Fig. 18.4, the condition for recovery of the analog signal can be directly deduced. The frequencies above $\frac{1}{2} f_s$ in the spectrum must be cut off by a lowpass filter. The lowpass filter must therefore be designed so as to have zero attenuation at f_{max} and infinite attenuation at $\frac{1}{2} f_s$ and above.

To summarize: the original waveform can be recovered from the sampled values of a continuous band-limited function of time as long as the condition $f_s \geq 2 f_{max}$ is fulfilled. To insure this, the sampled values must be transformed into a sequence of Dirac impulse functions, which is in turn applied to an ideal lowpass filter with $f_c = f_{max}$.

If the sampling rate is lower than the frequency demanded by the sampling theorem, spectral components arise which have the difference frequency $(f_s - f) < f_{max}$. They cannot be suppressed by a lowpass filter and are present as a beat frequency in the output signal (aliasing). These conditions are illustrated in Fig. 18.5. We can see that the spectral components of the input signal above $\frac{1}{2} f_s$ are not simply lost, but are mirrored back into the wanted band. The highest signal frequency f_{max} therefore reappears as the lowest mirror frequency $f_s - f_{max} < \frac{1}{2} f_s$ in the baseband of the output spectrum. Figure 18.6 shows these relationships for an input signal in the time domain whose spectrum contains just one spectral line at $f \lesssim f_s$. We can see from this how a signal with a beat frequency of $f_s - f$ is produced.

18.1.1
Practical Aspects

For a practical realization, the problem arises that a real system is unable to generate Dirac impulse functions. The impulses must thus be approximated, as shown in Fig. 18.3, by a finite amplitude and a finite time interval, thereby abandoning the limit concept of (18.2). By inserting (18.2) into (18.1), we obtain, for finite ε, the approximated impulse sequence:

$$\tilde{V}_i'(t) = \sum_{\mu=0}^{\infty} V_i(t_\mu) r_\varepsilon(t - t_\mu) \tag{18.5}$$

The Fourier transformation yields the spectrum:

$$\tilde{X}'(jf) = \frac{\sin \pi \varepsilon T_s f}{\pi \varepsilon T_s f} \cdot \tilde{X}(jf) \tag{18.6}$$

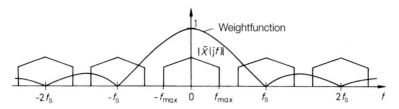

Fig. 18.7. Transition from the spectrum of a Dirac impulse sequence to the spectrum of the staircase function by means of the weighting function $|\sin \pi f/f_s)/(\pi f/f_s)|$

which is the same as for the Dirac impulse sequence, except for a superposed weighting function that causes attenuation of the higher-frequency components. The case of the staircase function is particularly interesting, as the pulse width εT_s is here identical to the sampling interval T_s. The spectrum is then given by:

$$\tilde{X}'(jf) = \frac{\sin(\pi f/f_s)}{\pi f/f_s} \cdot \tilde{X}(jf) \tag{18.7}$$

The magnitude of the weighting function is represented in Fig. 18.7, along with the symbolic spectrum of the Dirac impulse functions. At half the sampling rate, an attenuation of 0.64 is obtained.

The example in Fig. 18.8 should serve to illustrate a possible approach for selecting the sampling rate and the input or output filters. Consider an input spectrum for a music signal in the range $0 \le f \le f_{max} = 16\,\text{kHz}$, which is to be sampled and reconstructed with the utmost fidelity. In this case, it is insignificant whether 16 kHz components actually occur with full amplitude; rather, the linear frequency response should indicate that constant gain is required in this range.

Even if it can be insured that no tones above 16 kHz are present, this does not automatically mean that the spectrum at the sampler input is limited to 16 kHz. A typical source of broadband interference is amplifier noise. For this reason, it is always advisable to provide the input lowpass filter shown in Fig. 18.8. This is designed to limit the input spectrum to half the sampling rate in order to prevent aliasing. Its cutoff frequency must be at least f_{max} in order to maintain the true input signal. On the other hand, it is desirable for it to reject completely a frequency $\frac{1}{2}f_s$ and above that is only slightly higher, in order to allow the lowest possible value to be used for the sampling rate, the point being that the circuit complexity of the A/D or D/A converters and digital filter increases with the sampling frequency. On the other hand, the complexity of the lowpass filter increases with greater filter cutoff sharpness and stop-band attenuation. It is therefore always necessary to find a compromise between the complexity of the lowpass filter on the one hand and that of the converters and digital filter on the other. In the example with $f_{max} = 16\,\text{kHz}$, one could, for example, select $\frac{1}{2}f_s = 22\,\text{kHz}$; in other words, using a sampling rate of $f_s = 44\,\text{kHz}$.

Sampling causes the band-limited input signal to be continued periodically to f_s, as shown in Fig. 18.8. Consequently, the baseband $0 \le f \le \frac{1}{2}f_s$ must be extracted again following D/A conversion. As a staircase function is obtained at the output of the D/A converter, the $(\sin x)/x$ weighting as expressed in (18.7) must also be taken into account.

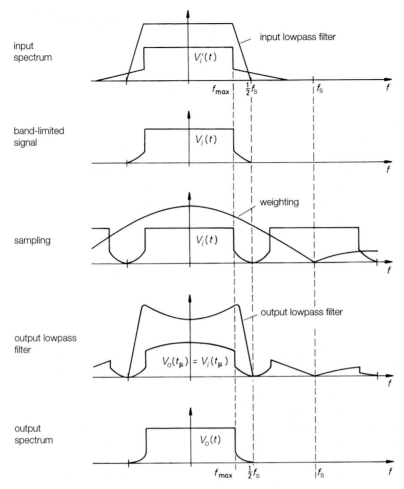

Fig. 18.8. Reconstruction of the input spectrum in a digital system with AD-conversion and following DA-conversion

The equalization required for this purpose can either be provided in the digital domain or performed in the output lowpass filter. The latter possibility is illustrated in Fig. 18.8. However, the main purpose of the output filter is to extract the baseband $0 \leq f \leq \frac{1}{2}f_s$ from the spectrum: at f_{max} it must still exhibit full passband characteristics, whereas the higher frequencies above $\frac{1}{2}f_s$ should attenuate completely. We can see that, in terms of filter steepness, the same problems arise as with the input filter. Consequently, in order to implement the filter, it is again necessary to provide an adequate margin between f_{max} and $\frac{1}{2}f_s$.

The problems of implementing the input or output filter can be mitigated by employing a markedly higher sampling rate – raising it by a factor of two or four, for example. Although this *oversampling* naturally (see page 979) increases the complexity of the A/D and D/A converters, the sampling rate can be reduced again to the value specified by the sampling theorem by inserting a digital lowpass filter after the A/D converter. This avoids high data rates for transmission or storage. Prior to D/A conversion, intermediate values can be recalculated using a digital interpolator, allowing oversampling again and enabling a simple output lowpass filter to be used.

18.2
Resolution

When an analog signal is converted into a digital quantity that has a finite number of bits, a systematic error is incurred due to the limited resolution; this is known as the quantization error. If the number Z is reconverted to a voltage by a D/A converter, the quantization error gives rise to superimposed noise. Figure 18.9 shows the test setup. According to Fig. 18.10, it is $\pm \frac{1}{2} V_{LSB}$; that is, it corresponds to half the input voltage step required to change the least significant bit.

One can show that the noise voltage has the value

$$V_{n\,rms} = \frac{V_{LSB}}{\sqrt{12}} \tag{18.8}$$

For a full sinusoidal swing, the rms output signal voltage for an N-bit converter is determined by

$$V_{s\,rms} = \frac{1}{\sqrt{2}} \cdot \frac{1}{2} \cdot 2^N \cdot V_{LSB}$$

Hence, the signal-to-noise ratio is

$$SNR = 20\,dB\,\lg\frac{V_{s\,rms}}{V_{n\,rms}} = N \cdot 6\,dB + 1.8\,dB \approx N \cdot 6\,dB \tag{18.9}$$

Inversely you can measure the *SNR* and calculate the resolution:

$$N = \frac{SNR}{6\,db} \quad or \quad n = 2^N - 1 = \frac{V_{s,rms}}{V_{n,rms}}$$

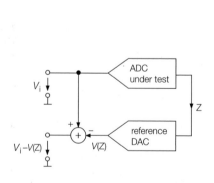
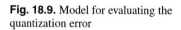

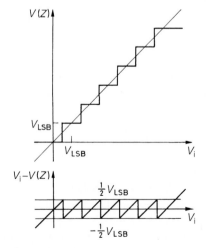

Fig. 18.9. Model for evaluating the quantization error

Fig. 18.10. Quantization noise of an ideal A/D converter

18.3
Principles of D/A Conversion

The purpose of a DAC is to convert a digital number into a proportional voltage. There are basically three methods of conversion:

1) the parallel method,
2) the weighting method,
3) the counting method.

These three methods are shown schematically in Fig. 18.11. With the parallel method (Fig. 18.11a), a voltage divider is used to provide all the possible levels of output voltage. The switch to which the required output voltage level is assigned is then closed by a 1-of-n decoder.

With the weighting method in Fig. 18.11b, a switch is assigned to each bit. The output voltage is then added up via appropriately weighted resistors.

The counting method in Fig. 18.11c requires just a single switch that is opened and closed periodically. Its duty cycle is set using a presettable counter, in such a way that the arithmetic mean of the output voltage assumes the desired value.

A comparison of the three methods shows that the parallel method requires Z_{max} switches, the weighting method ld Z_{max} switches and the counting method a single switch. Due to the large number of switches, the parallel method is rarely used. Likewise, the counting method is seldom used, its main disadvantage being that the output voltage can only change slowly due to the lowpass filter required.

On the other hand, DACs employing the weighting method are widely used, and we shall now describe the various ways in which they can be implemented. Two methods of realizing the switches have become standard: CMOS circuits use the transmission gates shown in Fig. 17.7; while in bipolar circuits, constant currents are generated and switched using diodes or differential amplifiers, as in Fig. 17.20.

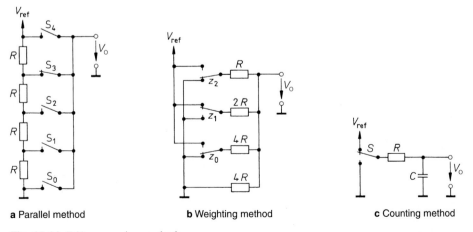

a Parallel method   **b** Weighting method   **c** Counting method

Fig. 18.11. D/A conversion methods

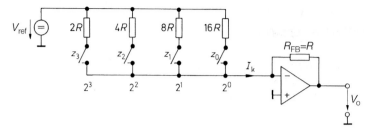

Fig. 18.12. Principle of a D/A converter

$$V_o = -V_{ref}\frac{Z}{16}; \qquad I_k = \frac{V_{ref}}{R}\cdot\frac{Z}{16}$$

18.4
D/A Converters in CMOS Technology

18.4.1
Summation of Weighted Currents

A simple circuit for converting a straight binary number to a voltage that is proportional to it is shown in Fig. 18.12. The resistors are selected such that, when the appropriate switch is closed, a current flows through them, which is equivalent to the relevant binary weight. The switches must therefore always be closed if a logical "1" appears in the relevant bit position. Due to the operational-amplifier feedback via resistor R_{FB}, the summing point remains at zero potential. The current components are therefore added together without affecting one another. If the switch controlled by z_0 is closed, the output voltage is

$$V_o = V_{LSB} = -V_{ref}\frac{R_{FB}}{16\,R} = -\frac{1}{16}V_{ref}$$

In general,

$$V_o = -\frac{1}{2}V_{ref}\,z_3 - \frac{1}{4}V_{ref}\,z_2 - \frac{1}{8}V_{ref}\,z_1 - \frac{1}{16}V_{ref}\,z_0$$

giving

$$V_o = -\frac{1}{16}V_{ref}(8\,z_3 + 4\,z_2 + 2\,z_1 + z_0) = -V_{ref}\frac{Z}{Z_{max}+1} \qquad (18.10)$$

18.4.2
D/A Converters with Double-Throw Switches

A disadvantage of the above D/A converter is that the voltages across the switches depends on their state. As long as the switches are open, they are at V_{ref} potential; when closed, they are at zero potential. As a result, the charges of the stray capacitances of the switch must be reversed every time the switch is operated. This disadvantage can be avoided if double-throw switches are used, as shown in Fig. 18.13, to connect the resistors either to the summing point or to ground. The current through each resistor therefore remains constant. This has a further advantage over the previous circuit in that the load of the reference voltage source is constant and its internal resistance need not be zero. The input

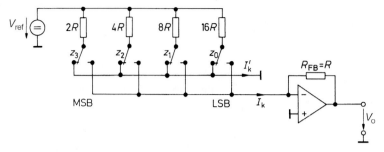

Fig. 18.13. D/A converter with double-throw switches

$$I_k = \frac{V_{ref}}{R}\frac{Z}{Z_{max}+1}; \quad I_k' = \frac{V_{ref}}{R}\cdot\frac{Z_{max}-Z}{Z_{max}+1}; \quad V_o = -V_{ref}\frac{Z}{Z_{max}+1}$$

resistance of the network, and thus the load resistance of the reference voltage source, is given by

$$R_i = 2R \parallel 4R \parallel 8R \parallel 16R = \frac{16}{15}R$$

18.4.3
Ladder Network

When fabricating integrated D/A converters, the implementation of accurate resistances of widely differing values is extremely difficult. The weighting of the bits is therefore often effected by successive voltage division, using a ladder network as shown in Fig. 18.14 consisting of the series resistances R_s and the parallel resistances R_p. The basic element of such a ladder network is the loaded voltage divider in Fig. 18.15, which is required to have the following characteristics: if it is loaded by a resistor R_2, its input resistance R_1

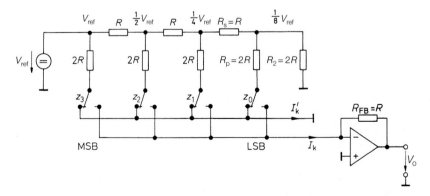

Fig. 18.14. D/A converter with a ladder network. This is the commonly used CMOS circuit

$$V_o = -V_{ref}\frac{Z}{Z_{max}+1}$$

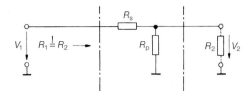

Fig. 18.15. Element of the ladder network

must also assume the value R_2. At this load, the attenuation $\alpha = V_2/V_1$ along the ladder element must have a predetermined value. With these two conditions, we obtain

$$R_s = \frac{(1-\alpha)^2}{\alpha}R_p \quad \text{and} \quad R_2 = \frac{(1-\alpha)}{\alpha}R_p \tag{18.11}$$

In the case of straight binary code, $\alpha = 0.5$. We choose $R_p = 2R$, and obtain

$$R_s = R \quad \text{and} \quad R_2 = 2R \tag{18.12}$$

in accordance with Fig. 18.14.

The reference voltage source in Fig. 18.14 is loaded by the constant resistance

$$R_i = 2R \,||\, 2R = R$$

The output voltage of the summing amplifier is

$$V_o = -R_{\text{FB}}I_k$$
$$= -V_{\text{ref}}\frac{R_{\text{FB}}}{16R}(8\,z_3 + 4\,z_2 + 2\,z_1 + z_0) = -V_{\text{ref}}\frac{Z}{Z_{max}+1} \tag{18.13}$$

The D/A converter in Fig. 18.14 only requires resistors of size R, if the $2R$ resistors are realized by two resistors connected in series. This arrangement is therefore ideally suitable for fabrication in monolithic IC form. Although the required matching tolerances for the resistors can be easily achieved, their absolute values cannot be precisely specified. Consequently, tolerances up to $\pm 50\%$ are common. Of course, the currents I_k or I'_k may also deviate by correspondingly large amounts. In order to obtain tight output voltage tolerances despite this tolerance, feedback resistor R_{FB} is also integrated. This cancels out the absolute value of R from (18.13) for the output voltage. For this reason, the internal feedback resistor should always be used for current-to-voltage conversion, and never an external one.

18.4.4
Inverse Operation of a Ladder Network

Sometimes the ladder network is also operated with exchanged input and output (Fig. 18.16), as no summing amplifier is then required. However, one must then accept the drawbacks, mentioned earlier, of a high voltage swing across the switches and a variable loading of the reference voltage source.

To calculate the output voltage, we need to know the relationship between the applied voltages V_i and the associated node voltages V'_i. For this purpose, we use the superposition principle; in other words, we set all the injected voltages, apart from the voltage V_i in question, equal to zero and add the individual components. If we terminate the network on the right and on the left with resistance $R_L = R_2 = 2R$, we obtain, as required, a load of $R_2 = 2R$ at each node to the right and the left. This gives us the voltage components

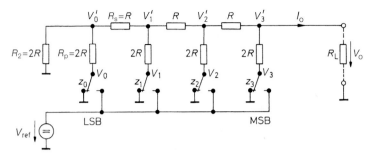

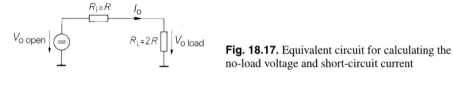

Fig. 18.16. Inversely operated ladder network. This circuit is used in converters with voltage output without operational amplifier

$$V_o = V_{ref} \frac{R_L}{R + R_L} \cdot \frac{Z}{Z_{max} + 1} = V_{ref} \frac{R_L}{R + R_L} \cdot \frac{Z}{16}$$

Fig. 18.17. Equivalent circuit for calculating the no-load voltage and short-circuit current

$\Delta V_i' = \frac{1}{3} \Delta V_i$, and by adding the correspondingly weighted components we obtain the output voltage:

$$V_o = \frac{1}{3} \left(V_3 + \frac{1}{2} V_2 + \frac{1}{4} V_1 + \frac{1}{8} V_0 \right) = \frac{2 V_{ref}}{3} \cdot \frac{Z}{16} \tag{18.14}$$

As the internal resistance of the network, irrespective of the set number Z, has a constant value of

$$R_i = R_2 \parallel R_p = (1 - \alpha) R_p = R \tag{18.15}$$

the weighting is retained even if the load resistance R_L does not possess the initially specified value $R_2 = 2R$. From the equivalent circuit diagram shown in Fig. 18.17, we can calculate the no-load voltage and the short-circuit current directly using (18.14):

$$V_{o\,open} = V_{ref} \frac{Z}{16} = V_{ref} \frac{Z}{Z_{max} + 1}; \quad I_{o\,short} = \frac{V_{ref}}{R} \cdot \frac{Z}{16} = \frac{V_{ref}}{R} \cdot \frac{Z}{Z_{max} + 1} \tag{18.16}$$

18.5
A Ladder Network for Decade Weighting

The ladder network in Fig. 18.14 can be extended to any length if longer straight-binary numbers are to be converted. For the conversion of BCD numbers, the method is modified somewhat, as shown in Fig. 18.18. Each decimal place (decade) is converted by a 4-bit D/A converter, as shown in Fig. 18.13 or 18.14, and the individual converters are connected to a ladder network. This introduces, from stage to stage, an attenuation of $\alpha = \frac{1}{10}$. In (18.11), resistance R_p must then be replaced by the input resistance R_i of the D/A converter stages, so that the coupling resistors are $R_s = 8.1 R_i$ and the terminating resistor is $R_2 = 9 R_i$, as indicated in Fig. 18.18. In this manner, each input voltage for a D/A converter stage

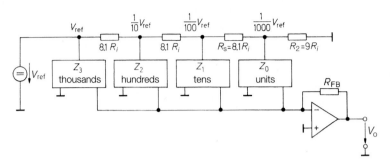

Fig. 18.18. Ladder network for decade weighting

is one-tenth of that for the previous stage. For the example of four decades, the output voltage

$$V_o = -\frac{V_{ref}}{16}\left(Z_3 + \frac{1}{10}Z_2 + \frac{1}{100}Z_1 + \frac{1}{1000}Z_0\right)$$

is obtained if, for each decade, a ladder network as shown in Fig. 18.14 is used.

18.6
D/A Converters in Bipolar Technology

With D/A converters employing bipolar technology, it is easy to implement constant-current sources that individually contribute to the total output current. The principle is illustrated in Fig. 18.19. The currents are weighted according to the significance of the associated bit position. Depending on whether the relevant binary digit is 1 or 0, the associated current flows to the output or is diverted to ground. The busbar for current I_k need not necessarily be at ground potential, as the current of the current sources is not a function of the voltage. However, this only applies within the output voltage range for constant-current operation (compliance voltage range). Consequently, an ohmic load resistance can be used, which need not be connected to ground or to virtual ground.

Simple transistor current sources of the type shown in Fig. 4.18 are used to generate the constant currents. If all of the base potentials are made equal and all of the emitter resistors are connected to V^-, the latter must be inversely proportional to the significance of the associated binary digit. This causes tolerancing problems, even in the bipolar pro-

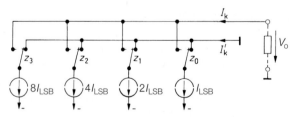

Fig. 18.19. D/A converter with switched current sources

$$V_o = -R_L \cdot I_{LSB} \cdot Z; \quad I_k = I_{LSB} \cdot Z; \quad I'_k = I_{LSB}(Z_{max} - Z)$$

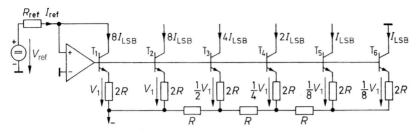

Fig. 18.20. Generation of weighted constant currents. This is the commonly used circuit employing bipolar technology

$$I_{\text{ref}} = \frac{V_{\text{ref}}}{R_{\text{ref}}} = 8I_{\text{LSB}}; \quad V_1 = I_{\text{ref}} \cdot 2R = \frac{2R}{R_{\text{ref}}} V_{\text{ref}}$$

cess. Consequently, a ladder network is again used for current flow division (Fig. 18.20). The current source bank $T_1 - T_6$ is at equal base potential, which is established via the operational amplifier in such a way that current $I_{\text{ref}} = V_{\text{ref}}/R_{\text{ref}}$ flows via reference transistor T_1. This is the case when $V_1 = 2R \cdot I_{\text{ref}}$. If the emitter–base voltages of the other transistors are identical to that of T_1, we obtain the same voltage drops across the emitter resistors and therefore the required weighting of the currents.

However, identical emitter–base voltages do not occur, even if the transistors are completely identical, as the currents are not the same. From the transfer characteristic in Eq. (2.2), we obtain

$$V_{\text{BE}} = V_T \ln \frac{I_C}{I_{C0}}$$

Consequently, the voltage increases by 18 mV if the collector current doubles. To avoid any resultant error, all of the transistors are operated using the same collector current density. For this purpose, a sufficient number of transistors are connected in parallel to insure that only current I_{LSB} flows through each one. In integrated circuits, this is taken into account by using transistors with correspondingly larger areas for the higher currents.

The $2R$ termination of the ladder network shown in Fig. 18.20 must not be connected here to ground, but a point must be selected that is at emitter potential. This is generated by the otherwise unused transistor T_6. For simplicity, its emitter can also be connected in parallel with T_5 and the two emitter resistors combined to form a single resistor of value R.

Another method of D/A conversion using switched current sources is shown in Fig. 18.21. Here, identical currents are generated, which appear at the output after being weighted by a ladder network. The arrangement corresponds to the inversely operated ladder network shown in Fig. 18.16. The resistors $2R$ providing the attenuation within the chain must be connected to ground, as they would have no effect connected in series with the constant current sources. On the other hand, the attenuation in the chain is unchanged by connecting a current source, as – theoretically at least – the latter has an infinitely high internal resistance.

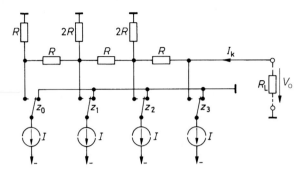

Fig. 18.21. D/A converter with an inversely operated ladder network. This circuit is used in video converters

Short-circuit current ($R_L = 0$): $I_{k\,short} = \dfrac{I\,Z}{8} = 2I\,\dfrac{Z}{Z_{max} + 1}$

Output voltage: $V_o = I_{k\,short}(2R \parallel R_L)$

18.7
D/A Converters for Special Applications

18.7.1
Processing Signed Numbers

When describing D/A converters, we have hitherto assumed that positive numbers are involved which have to be converted into positive or negative voltages, depending on the circuit. We shall now examine ways of producing bipolar output voltages using the D/A converters described. The conventional representation of binary numbers of either sign is in two's-complement notation (see Sect. 8.1.3). In this way, the range from -128 to $+127$ can be represented with 8 bits, as shown again in Fig. 18.22.

To enter data into the D/A converter, the number range is shifted to 0–255 by adding 128. Numbers above 128 are deemed to be positive, and those below negative. In this case, the mid-scale number 128 denotes zero. This characterization of signed numbers by purely positive numbers is known as offset binary representation. The addition of 128 can be performed simply by negation of the sign bit (see Fig. 18.22).

Decimal	Two's complement								Offset binary								Analog	
	v_z	z_6	z_5	z_4	z_3	z_2	z_1	z_0	z_7	z_6	z_5	z_4	z_3	z_2	z_1	z_0	$-V_1/V_{LSB}$	V_o/V_{LSB}
127	0	1	1	1	1	1	1	1	1	1	1	1	1	1	1	1	255	127
126	0	1	1	1	1	1	1	0	1	1	1	1	1	1	1	0	254	126
1	0	0	0	0	0	0	0	1	1	0	0	0	0	0	0	1	129	1
0	0	0	0	0	0	0	0	0	1	0	0	0	0	0	0	0	128	0
-1	1	1	1	1	1	1	1	1	0	1	1	1	1	1	1	1	127	-1
-127	1	0	0	0	0	0	0	1	0	0	0	0	0	0	0	1	1	-127
-128	1	0	0	0	0	0	0	0	0	0	0	0	0	0	0	0	0	-128

Fig. 18.22. Processing negative numbers in D/A converters. $V_{LSB} = V_{ref}/256$

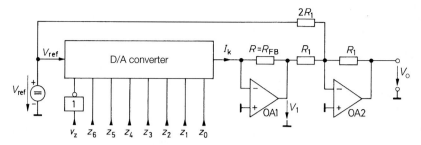

Fig. 18.23. D/A converter with a bipolar output

$$V_o = V_{ref} \frac{Z}{256} \quad \text{for} \quad -128 \le Z \le 127$$

In order to obtain an output voltage of correct sign, the addition of the offset is achieved by subtracting $128 V_{LSB} = \frac{1}{2} V_{ref}$. The summing operational amplifier OA 2 in Fig. 18.23 is used for this purpose, providing the output voltage:

$$V_o = -V_1 - \frac{1}{2} V_{ref} = V_{ref} \frac{Z + 128}{256} - \frac{1}{2} V_{ref} = V_{ref} \frac{Z}{256} \quad (18.17)$$

Its magnitude is listed together with voltage V_1 in Fig. 18.22.

The zero stability of the circuit shown in Fig. 18.23 can be improved by using the complementary output current I'_k, instead of using the fixed reference voltage for subtraction of the offset. In the case of two's-complement number 0, which actually corresponds to 128 in offset binary, we have

$$I_k = 128 I_{LSB} \quad \text{and} \quad I'_k = 127 I_{LSB}$$

Therefore, if we add an I_{LSB} to I'_k and subtract the result from I_k, we obtain the correct zero point. This method is illustrated in Fig. 18.24. As before, operational amplifier OA 1 converts current I_k into the output voltage. To eliminate errors, the amplifier is fed back via DAC internal resistor R_{FB}. Operational amplifier OA 2 inverts the sum of I_{LSB} and I'_k, and adds this current into the summing point of OA 1. The absolute values of the two

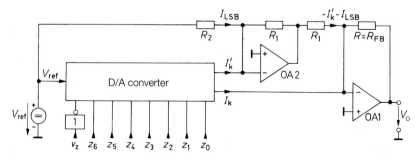

Fig. 18.24. Bipolar D/A converter with improved zero stability. By using both output currents the output voltage is doubled

$$V_o = V_{ref} \frac{Z}{128} \quad \text{for} \quad -128 \le Z \le 127$$

resistors R_1 are irrelevant, they must only have the same value. Current I_{LSB} is added via resistor R_2. If $I_{LSB} = V_{ref}/(256R)$, it follows:

$$R_2 = \frac{V_{ref}}{I_{LSB}} = 256R$$

To calculate the output voltage, we only need to add the currents at the summing point of OA 1 and obtain

$$V_o = R\left[\underbrace{\frac{V_{ref}}{R}\frac{Z+128}{256}}_{I_k} - \underbrace{\frac{V_{ref}}{R}\frac{255-(Z+128)}{256}}_{I'_k} - \underbrace{\frac{V_{ref}}{R}\frac{1}{256}}_{I_{LSB}}\right] = V_{ref}\frac{Z}{128} \qquad (18.18)$$

18.7.2
Multiplying D/A Converters

As we have seen, D/A converters provide an output voltage that is proportional to the input number Z and the reference voltage V_{ref}; in other words, they form the product $Z \cdot V_{ref}$. For this reason, types that allow the reference voltage to be varied are also known as *multiplying* D/A converters.

With circuits realized using bipolar technology, the reference voltage can only assume positive values, as the current sources in Fig. 18.20 would otherwise be turned off. With CMOS types, on the other hand, positive and negative reference voltages are permissible. If circuits such as those shown in Figs. 18.23 and 18.24, which allow positive and negative numbers to be converted with their correct signs, are used, even *four-quadrant multiplication* is possible.

18.7.3
Dividing D/A Converters

A D/A converter can also be operated in such a way that it *divides* by the input number. To achieve this, it is inserted in the feedback loop of an operational amplifier, as shown in Fig. 18.25. This means that the reference voltage V_{ref} is set such that $I_k = -V_i/R_{FB}$. Using the converter equation

$$I_k = \frac{V_{ref}}{R} \cdot \frac{Z}{Z_{max}+1}$$

we obtain the output voltage:

$$V_o = V_{ref} = I_k R\frac{Z_{max}+1}{Z} = -V_i \cdot \frac{R}{R_{FB}} \cdot \frac{Z_{max}+1}{Z} = -V_i \cdot \frac{Z_{max}+1}{Z} \qquad (18.19)$$

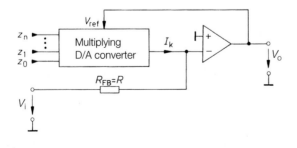

Fig. 18.25. A dividing D/A converter
$$V_o = -V_i \cdot \frac{R}{R_{FB}} \cdot \frac{Z_{max}+1}{Z}$$

This simple method of performing division frequently obviates the need for analog or digital division, with its associated cost and complexity, if some accuracy is required.

18.7.4
D/A Converter as Function Generator

The output voltage V_o of the usual D/A converter is proportional to the applied number Z according to $V_o = aZ$. If, instead of the proportional function, any other relationship $V_o = f(Z)$ is to be realized, the function $X = f(Z)$ must be generated by a digital function network and then applied to a D/A converter.

If no stringent requirements are imposed on the accuracy, there is a much simpler solution: the binary number Z is used to control an analog multiplexer. We apply analog input values, each of which is assigned to the appropriate binary number. For each analog value, a separate switch is needed, and the attainable resolution is therefore limited to about 16 steps.

A possible implementation is shown in Fig. 18.26. In contrast to the usual D/A converters, only one of the switches S_0, to S_7, is closed at any time. The values of the output voltage function are thus given by the expression

$$V_o(Z) = \begin{cases} +V_{\text{ref}} \dfrac{R_N}{R_Z} & \text{for } Z = 0 \ldots 3 \\[3mm] -V_{\text{ref}} \dfrac{R_N}{R_Z} & \text{for } Z = 4 \ldots 7 \end{cases}$$

An important application of this principle is the digital generation of sine waves (e.g., in modems). A simple and widely used method of generating signals of different frequencies, all synchronized to a common time base, is to employ frequency division. However, a serious drawback for use in analog systems is that the signals obtained are square waves. Sine waves can he produced by filtering the fundamental with a lowpass or bandpass filter, but these filters must always be tuned to the appropriate frequency.

The D/A converter described avoids these problems in that it allows the frequency-independent generation of sine waves. According to Fig. 18.27, we require a digital input signal representing a rising and falling sequence of equidistant numbers. This input signal corresponds to the triangular wave-shape for sine-wave generation by an analog function network, as described in Sect. 11.7.4 on page 745 and Sect. 14.5 on page 859.

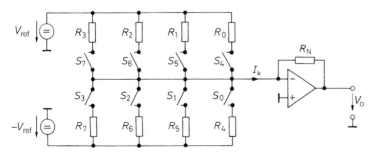

Fig. 18.26. D/A converter for any desired function

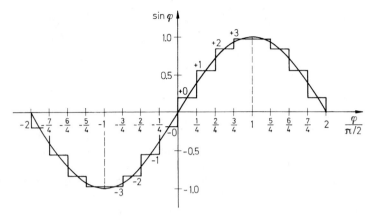

Fig. 18.27. Approximation of a sine wave in 16 steps

If signed-magnitude representation is chosen for the binary numbers, a number se-quence that has the desired properties can be easily generated using a cyclic straight-binary counter. The most significant bit represents the sign. The second significant bit effects a change of counting direction for all lower bits by complementing the corresponding out-puts with the aid of exclusive-OR gates. These bits represent the magnitude. When a 4-bit straight-binary counter is used, the circuit shown in Fig. 18.28 is obtained. The number sequence generated is listed in Fig. 18.29. With a 3-bit number at the input of the analog multiplexer, four positive steps, $+0, 1, 2,$ and $3,$ of the sine function and, correspondingly, four negative steps $- 0, - 1, - 2,$ and $- 3,$ are selected. If the steps are distributed as shown in Fig. 18.27, the function values shown in Fig. 18.29 are obtained, and the appropriate resistors can be determined. As the chosen quantization is rather coarse, it is sufficient to select the nearest standard resistance values.

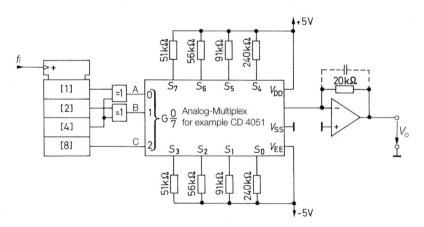

Fig. 18.28. Generating a continuous sine wave

$$V_o = 2\,\text{V} \sin 2\pi \frac{f_i}{16} t$$

Z	Counter-outputs				Multiplexer-inputs			Switch closed	Step number	Output voltage
	z_3	z_2	z_1	z_0	C	B	A			$V_o/\hat{V}_o$
0	0	0	0	0	0	0	0	S_0	+0	0.20
1	0	0	0	1	0	0	1	S_1	+1	0.56
2	0	0	1	0	0	1	0	S_2	+2	0.83
3	0	0	1	1	0	1	1	S_3	+3	0.98
4	0	1	0	0	0	1	1	S_3	+3	0.98
5	0	1	0	1	0	1	0	S_2	+2	0.83
6	0	1	1	0	0	0	1	S_1	+1	0.56
7	0	1	1	1	0	0	0	S_0	+0	0.20
8	1	0	0	0	1	0	0	S_4	−0	−0.20
9	1	0	0	1	1	0	1	S_5	−1	−0.56
10	1	0	1	0	1	1	0	S_6	−2	−0.83
11	1	0	1	1	1	1	1	S_7	−3	−0.98
12	1	1	0	0	1	1	1	S_7	−3	−0.98
13	1	1	0	1	1	1	0	S_6	−2	−0.83
14	1	1	1	0	1	0	1	S_5	−1	−0.56
15	1	1	1	1	1	0	0	S_4	−0	−0.20

Fig. 18.29. Number sequences and the resulting voltages

Since in a complete period each step occurs twice, the sine wave is approximated by a total of 16 steps. Correspondingly, the input frequency f_i of the counter must be 16 times that of the sine wave.

The staircase in the output signal in Fig. 18.27 contains some harmonics. To get a poor sinus a lowpass filter can be added. With variable frequencies a switched capacitor filter is advantageous (Sect. 13.12 on page 836) because its cutoff frequency follows the clock frequency.

18.8
Accuracy of DA Converters

18.8.1
Static Errors

The *zero point error* of a DA converter is determined by the leakage currents flowing through the open switches.

The *full-scale error* is determined by the ON-state resistances of the switches and the accuracy of the feedback resistor R_{FB}. Both errors can be largely eliminated by trimming.

Nonlinearity, on the other hand, cannot be eliminated by adjustment. It is defined as the amount by which a step is larger or smaller than 1 LSB, under worst-case conditions. Figure 18.30 illustrates a nonlinearity of $\pm \frac{1}{2}$ LSB. The critical case occurs at mid-scale: if only the most significant bit is a 1, the current flows via a single switch. If the number is reduced by 1, the total current of all the lower-order switches is reduced by only I_{LSB}.

If the linearity error is greater than 1 LSB, the trend is reversed. The output voltage then falls at the mid-scale point although the number is increased by 1. A serious error of this kind is termed a *monotony error*, an example of which is shown in Fig. 18.31. Most D/A converters are designed such that their nonlinearity does not exceed $\pm \frac{1}{2}$ LSB, as the least significant bit would otherwise be meaningless.

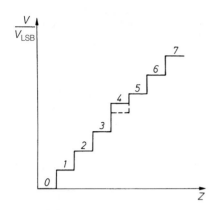

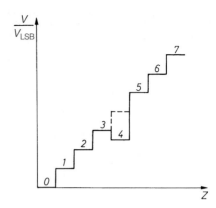

Fig. 18.30. D/A converter with a nonlinearity of $\pm \frac{1}{2}$ LSB

Fig. 18.31. D/A converter with a nonlinearity of $1\frac{1}{2}$ LSB and an associated monotony error

18.8.2
Dynamic Characteristics

The settling time is defined as the time that it takes for the output signal to reach the steady-state value with an accuracy of $\pm \frac{1}{2}$ LSB after the number Z has changed from 0 to Z_{max}. Only then is the analog signal available with the accuracy provided by the resolution of the D/A converter. Defining the settling time with reference to $\pm \frac{1}{2}$ LSB means, of course, that D/A converters with the same time constant but higher resolution settle more slowly than those with lower resolution.

With many D/A converters, a current is initially formed which can be converted into a voltage, as required, in a following operational amplifier. In this case, the settling time of the operational amplifier, which is usually much greater than that of the D/A converter, is also added. In order to achieve short settling times for the voltage it is advantageous to use converters that need no operational amplifier. For ladder networks, the only option is to use the inversely operated ladder network shown in Fig. 18.16. Types with current sources like in Fig. 18.21 can all generate a voltage across an ohmic load resistance. In order to achieve bandwidths in the 100 MHz range, it is advisable to use D/A converters whose output currents are so large that they can produce the required amplitudes across load resistors of 50 Ω or 75 Ω.

Unwanted interference pulses (*glitches*) may also occur at the transition from one input number to the other. In most cases, these are due only to a small extent to capacitive feedthrough of the binary drive signals to the output. Large glitches occur if the switches in

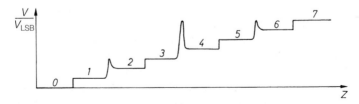

Fig. 18.32. Typical glitches and the large mid-scale glitch

Type	Manu-facturer	Channels	Setting time	Supply voltage min/max		Internal reference	Inter-face	Output
8 bit								
AD5346	Analog Dev.	8	6 µs	2.5/	5.5 V	—	parallel	RRO
AD7304	Analog Dev.	4	2 µs	3/	±5 V	—	SPI	RRO
MAX5515	Maxim	4	8 µs	2.7/	5.25 V	—	I²C	RRO
MAX5820	Maxim	2	4 µs	2.7/	5.5 V	—	I²C	RRO
MAX5594	Maxim	8	2 µs	2.7/	5.25 V	—	SPI	RRO
TLV5624	Texas Inst.	1	1 µs	2.7/	5.5 V	1.24 V	SPI	Spg.
TLV5627	Texas Inst.	4	3 µs	2.7/	5.5 V	—	SPI	RRO
TLV6629	Texas Inst.	8	1 µs	2.7/	5.5 V	—	SPI	Spg.
12 bit								
AD5348	Analog Dev.	8	8 µs	2.5/	5.5 V	—	parallel	RRO
AD5381	Analog Dev.	40	6 µs	4.5/	5.5 V	1.25 V	SPI	RRO
AD5391	Analog Dev.	16	6 µs	4.5/	5.5 V	1.25 V	SPI	RRO
LTC2620	Lin. Tech.	8	7 µs	2.5/	5.5 V	—	SPI	RRO
LTC2621	Lin. Tech.	1	7 µs	2.5/	5.5 V	—	SPI	RRO
MAX5306	Maxim	8	5 µs	2.7/	5.5 V	—	SPI	RRO
MAX5322	Maxim	2	10 µs	5/	±15 V	—	SPI	bipolar
MAX5532	Maxim	2	660 µs	1.8/	5.5 V	—	SPI	RRO
MAX5590	Maxim	8	2 µs	2.7/	5.25 V	—	SPI	RRO
DAC7554	Texas Inst.	4	5 µs	2.7/	5.5 V	—	SPI	RRO
DAC7571	Texas Inst.	1	10 µs	2.7/	5.5 V	—	I²C	RRO
DAC7574	Texas Inst.	4	8 µs	2.7/	5.5 V	—	I²C	RRO
16 bit								
AD5662	Analog Dev.	1	8 µs	2.7/	5.5 V	—	SPI	RRO
AD5764	Analog Dev.	4	8 µs	±12/	±15 V	5 V	SPI	bipolar
LTC2600	Lin. Tech.	8	7 µs	2.5/	5.5 V	—	SPI	RRO
LTC2601	Lin. Tech.	1	7 µs	2.5/	5.5 V	—	SPI	RRO
MAX5616	Maxim	2		±12/	±15 V	—	SPI	bipolar
MAX5732	Maxim	32	20 µs	4.7/	5.5 V	—	SPI	RRO
DAC7642	Texas Inst.	2	10 µs	+5 V/	±5 V	—	parallel	bipolar
DAC7731	Texas Inst.	1	5 µs	±12/	±15 V	10 V	SPI	bipolar
DAC8532	Texas Inst.	2	10 µs	2.7/	5.5 V	—	SPI	RRO
DAC2574	Texas Inst.	4	10 µs	2.7/	5.5 V	—	I²C	RRO

Fig. 18.33. Examples for low-frequency DACs with voltage outputs. Interface: SPI = serial peripheral interface. I²C = I²C-bus interface (serial too). Output: RRO = rail to rail output

the D/A converter do not change state simultaneously. The most critical point again occurs at mid-scale: if the most significant bit (MSB) is a 1, the current flows via one switch only. If the number is reduced by 1, the switch for the MSB opens and all of the others close. If the MSB switch opens before the other switches have closed, the output signal briefly goes to zero. However, if the MSB switch opens slightly late, the output signal momentarily assumes its full-scale value. In this way, unwanted pulses with an amplitude equal to half the range may occur. An example of the switches closing more rapidly than they open is shown in Fig. 18.32.

As the glitches are short pulses, they can be reduced using a lowpass filter at the output. As a result of this, however, they become correspondingly longer, the voltage–time area – in other words, the glitch energy – remaining constant.

Type	Manu-facturer	Sample rate	Supply voltage	Power dissipation	Internal reference	Glitch energy	Inter-face
8 bit							
AD9709*	Analog Dev.	125 MS/s	+5 V	380 mW	1.2 V	5 pVs	CMOS
MAX5186*	Maxim	40 MS/s	+3 V	24 mW	1.2 V	50 pVs	CMOS
12 bit							
AD9735	Analog Dev.	1200 MS/s	+3.3 V	380 mW	1.2 V		LVDS
AD9752	Analog Dev.	125 MS/s	+3.3 V	185 mW	1.2 V	5 pVs	CMOS
LTC 1666	Lin. Tech.	50 MS/s	±5 V	180 mW	2.5 V	5 pVs	CMOS
MAX5886	Maxim	500 MS/s	+3.3 V	230 mW	1.2 V	1 pVs	LVDS
DAC2902*	Texas Inst.	125 MS/s	+3.3 V	310 mW	1.2 V	2 pVs	CMOS
16 bit							
AD9726	Analog Dev.	600 MS/s	+3.3 V	575 mW	1.2 V		LVDS
AD9786	Analog Dev.	500 MS/s	+3.3 V	1250 mW	1.2 V		PECL
LTC1668	Lin. Tech.	80 MS/s	±5 V	180 mW	1.2 V	5 pVs	CMOS
MAX5888	Maxim	500 MS/s	+3.3 V	130 mW	1.2 V	1 pVs	LVDS
DAC5687*	Texas Inst.	500 MS/s	+3.3 V	1250 mW	1.2 V		CMOS

* 2 DACs on the chip

Fig. 18.34. Examples for high speed DACs with current outputs

Glitches can also be eliminated by connecting a sample-and-hold circuit at the output. This can be made to hold during the glitch phase, thereby blanking the glitch. Sample-and-hold circuits specially designed for this purpose are known as *deglitchers*.

A simpler solution, however, is to use low-glitch D/A converters. These generally have an internal edge-triggered data memory for the number Z in order to insure that the control signals are applied to all of the switches simultaneously. The parallel method is also sometimes used for the more significant, critical bits, because this method is intrinsically glitch-free.

Examples for normal DACs with output amplifiers are given in Fig. 18.33. DACs for high conversion rates can be found in Fig. 18.34. Here the output voltage depends on the load resistance that should be matches to the characteristic impedance of the connection line.

18.9
Principles of A/D Conversion

The purpose of an A/D converter (ADC) is to transform an analog input voltage into a proportional digital number. There are three basically different conversion methods that we have distinguished at the D/A converters already:

the parallel method   (a word at a time),
the weighting method   (a digit at a time),
the counting method   (a level at a time).

The parallel (flash) method compares the input voltage with n reference voltages simultaneously and determines between which two reference levels the value of the input voltage lies. The resulting number is thus obtained in a single operation. However, the circuitry involved is very extensive, as a separate comparator is required for each possible number. For a range of measurement from 0 to 255, $n = 255$ comparators are needed.

With the weighting (successive approximation) method, the end result is not obtained in a single operation; instead, one bit of the corresponding straight-binary number is determined at a time. The input voltage is first checked against the most significant bit and the most significant bit is determined. Then the second bit is evaluated: This process is repeated until all bits are processed. For a result with 8 bit accuracy 8 steps are required.

The simplest method is the counting method. It involves counting how often the reference voltage of the least significant bit must be added to arrive at the input voltage. The number of operations is the required result. If the maximum number to be represented is 256, a maximum of 256 operational steps must be performed to obtain the result.

18.10
Design of A/D Converters

18.10.1
Parallel Converter

The construction of a parallel converter for 3 bits is illustrated in Fig. 18.35. With three bits, eight different numbers including zero can be represented, for which seven comparators

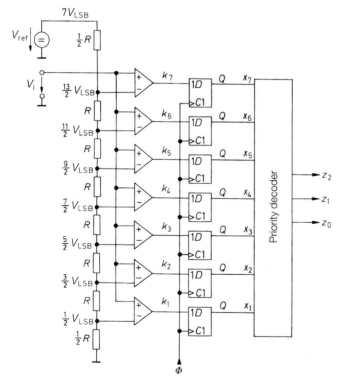

Fig. 18.35. Parallel A/D converter

$$Z = \frac{V_i}{V_{LSB}} = 7\frac{V_i}{V_{ref}} = Z_{max}\frac{V_i}{V_{ref}}$$

Input voltage	Comparator states							Straight-binary number			Decimal equivalent
V_i/V_{LSB}	k_7	k_6	k_5	k_4	k_3	k_2	k_1	z_2	z_1	z_0	Z
0	0	0	0	0	0	0	0	0	0	0	0
1	0	0	0	0	0	0	1	0	0	1	1
2	0	0	0	0	0	1	1	0	1	0	2
3	0	0	0	0	1	1	1	0	1	1	3
4	0	0	0	1	1	1	1	1	0	0	4
5	0	0	1	1	1	1	1	1	0	1	5
6	0	1	1	1	1	1	1	1	1	0	6
7	1	1	1	1	1	1	1	1	1	1	7

Fig. 18.36. States of the variables in the parallel A/D converter as a function of the input voltage

are required. The seven associated equally spaced reference voltages can be generated by means of a voltage divider from a single reference voltage source.

For an input voltage that has a value 3 V_{LSB}, the comparators 1–3 produce ones and the comparators 4–7 zeros. A logic circuit is therefore required to convert these comparator states to the number $Z = 3$. The comparator output states and the corresponding straight-binary numbers are listed in Fig. 18.36. A comparison with Fig. 8.18 on page 646 shows that the required conversion can be carried out by the priority decoder described in Sect. 8.3.

However, the priority decoder must not be connected directly to the outputs of the comparators, since totally erroneous straight-binary numbers may arise if the input voltage is not constant. The example of a change from 3 to 4 – that is, in straight-binary code from 011 to 100 – illustrates this. If the most significant bit changes before the two other bits because of a shorter propagation delay in the priority decoder, the number 111 (i.e., 7_{dec}) occurs temporarily. This is equivalent to an error of half the range. The result of an A/D conversion is usually transferred to a memory, and there is therefore a certain probability that this erroneous number will be stored. The use of a sample-and-hold circuit can prevent this effect, as it holds the input voltage constant during the conversion process. However, high speed sample and holds are expensive. Therefore a *digital sample and hold* is usual in parallel converters.

The D flip-flops at the output of each comparator in Fig. 18.35 are used for this purpose, thus insuring that the priority decoder receives constant input signals for one complete clock

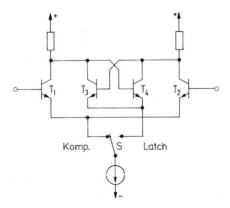

Fig. 18.37. A comparator input with a storage flip-flop

Type	Manu facturer	Sample rate	Supply voltage	Power dissipation	Internal reference	Aperture jitter	Inter- face
8 bit							
MAX104	Maxim	1000 MS/s	±5 V	5250 mW	2.5 V	0.5 ps	PECL
MAX106	Maxim	600 MS/s	±5 V	5250 mW	2.5 V	0.5 ps	PECL
MAX108	Maxim	1500 MS/s	±5 V	5250 mW	2.5 V	0.5 ps	PECL
TLC5540	Texas Inst.	40 MS/s	±5 V	85 mW	—		CMOS

Fig. 18.38. Examples for flash converters

period. Steady-state data is therefore available at the priority decoder output prior to the arrival of the next clock edge.

The sampling instant is essentially determined by the trigger edge of the clock signal, although the actual sampling is performed a little earlier because of the comparator delay. The delay differences is therefore determine as the aperture jitter. In order to achieve the low values the signal delay from the analog input to the storage devices should be as low as possible. On most types, therefore, the storage element is incorporated in the comparator and inserted directly after the analog input. The resulting input circuit of a comparator of this type is shown in Fig. 18.37.

If switch S is set to the left, transistors T_1, T_2 operate as a comparator. When the switch is changed over, comparator T_1, T_2 is deactivated and flip-flop T_3, T_4 is activated instead. The flip-flop then stores the state of the comparator. For this purpose, it is not even necessary for the comparator to have already changed state completely. Since the flip-flop is likewise designed as a differential amplifier, differences of a few millivolts decide whether the flip-flop assumes one state or the other. In this way, the aperture jitter can be reduced to picoseconds.

Figure 18.38 lists a number of A/D converters that employ flash converter. Today mainly 8 bit converters are offered. If higher accuracy is needed additional bits can be generated by averaging. For one additional bit 4 samples must be averaged. This technique is used in high speed oscilloscopes to get up to 11 bits from an 8 bit quantizer if the maximum sample rate is not needed.

The linearity of A/D converters at low signal frequencies is equal to a resolution of $\pm \frac{1}{2}$ LSB, or even $\pm \frac{1}{4}$ LSB in some cases. However, at high signal frequencies, nonlinearity increases, with the result that the least significant, or even the two least significant bits, becomes unusable. The quantizing noise increases correspondingly by 6 or 12 dB, in accordance with (18.9).

18.10.2
Two Step Converters

A disadvantage of flash converters is that the number of comparators required rises exponentially with the word length. For a 10-bit converter, for example, a total of $2^{10} - 1 = 1,023$ comparators are needed. This number can be considerably reduced if at first the five most significant bits are parallel-converted, as can be seen in the block diagram of Fig. 18.39. The result is the coarsely quantized value of the input voltage. A D/A converter is used to produce the appropriate analog voltage, which is then subtracted from the input voltage. The remainder is digitized by a second 5-bit A/D converter.

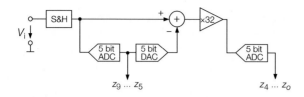

Fig. 18.39. Two step ADC. It requires only $2 \times 32 = 64$ comparators for a 10 bit converter instead of 1023

If the difference between the coarse value and the input voltage is amplified by a factor of 32, two A/D converters with the same input voltage range can be employed. There are, however, different requirements for the accuracy of the two converters: the accuracy of the first 5-bit converter must be as high as that of a 10-bit converter; otherwise, the calculated difference is meaningless.

However, parallel A/D converters that provide such a high degree of linearity are not obtainable; nor can they be implemented for higher signal frequencies. Consequently, the difference signal exceeds the fine range and overdrives the second A/D converter, resulting in serious errors in the output signal (missing codes).

This problem can be overcome by reducing the gain for the difference signal from 32 to 16, as shown in Fig. 18.40. Bit z_5 is then formed both by the coarse and the fine quantizer. If the fine signal now exceeds the range due to errors of the coarse quantizer, the coarse value can be increased or lowered by 1 using z_5'. This allows coarse quantizer linearity errors to be corrected to within $\pm \frac{1}{2}$ LSB. The coarse quantizer linearity need not be better than the resolution, a requirement which is not as stringent as that for the circuit in Fig. 18.40. It is merely necessary for the D/A converter to have full 10-bit accuracy. For successful error correction, the coarse and fine quantizer ranges must overlap by at least one bit. So as not to reduce the resolution of the circuit as a whole, the fine quantizer must therefore have one additional bit.

The coarse and fine values must, of course, be formed by the same input voltage $V_i(t_j)$ in each case. However, due to the transit through the first stage, a time delay is introduced. Consequently, with this method the input voltage must be kept constant using an analog sample-and-hold circuit until the complete number is formed. This constitutes a serious disadvantage as compared with the purely parallel method.

The method can be extended to multi-step converters. In Fig. 18.41 a 3-step converter is presented. Here in each step 5 bit converters are used. In order to get a range overlap for error correction the intermediate amplifiers have a gain of only 16 as in Fig. 18.40.

Here in *each* step a sample and hold circuit is added. The consequence is that the most significant bits of the first input sample V_{i1} will be quantized in step 1 after the first sample. With the second sample the remainder stored in the second S&H for step 2 that forms the intermediate bits. At the same time step 1 quantizes the most significant bits of the next input

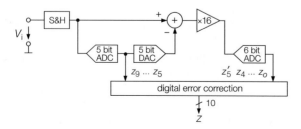

Fig. 18.40. Two step ADC with error correction

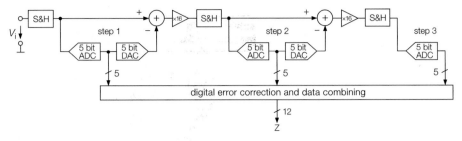

Fig. 18.41. 3 step pipeline ADC with 3 overlap bits

Sample	Step 1	Step 2	Step 3
1	V_{i1}		
2	V_{i2}	V_{i1}	
3	V_{i3}	V_{i2}	V_{i1}

Fig. 18.42. Quantizing 3 successive input voltages V_{i1}, V_{i2}, V_{i3} in a 3 step ADC with pipeline

sample V_{i2}. With the third sample the remainder of V_{i1} stored in the third S&H for step 3 that forms the least bits. At the same time step 1 quantizes the most significant bits of V_{i3} and step 2 quantizes the intermediate bits if V_{i2}. The process is illustrated in Fig. 18.42. The sampling frequency of the converter is only determined by the conversion time of one step. But it needs 3 samples until the first input voltage V_{i1} is quantized and available at the output. Then with each sample a further conversion is available. This latency is typical for pipelined circuits. Without the intermediate S&H circuits the maximum sampling frequency would be only 1/3 because all signals up to the last step must settle before a new sample can be taken.

The two-step converter was introduced to reduce the number of comparators in flash converters. The ultimate reduction of comparators is achieved if only 1 bit is used in one stage because then only one comparator is needed per stage. Then the number of comparators is equal to the number of bits N. The first conversion takes N samples, all following only one. Some examples for commercially available chips are given in Fig. 18.43.

Type	Manu-facturer	Converters on chip	Sample rate	Supply voltage	Power dissipation	Inter-face
8 bit						
AD9283	Analog Dev.	1	50 MS/s	3.3 V	90 mW	CMOS
AD9480	Analog Dev.	1	250 MS/s	3.3 V	590 mW	LVDS
MAX1198	Maxim	2	100 MS/s	3.3 V	264 mW	CMOS
ADC081000	National	1	1000 MS/s	1.9 V	1450 mW	LVDS
ADC08D1500	National	2	1500 MS/s	1.9 V	1900 mW	LVDS
ADS831	Texas Inst.	1	80 MS/s	5 V	275 mW	CMOS

Fig. 18.43. Examples for pipeline AD-converters. All types have an on-chip reference

Type	Manu-facturer	Converters on chip	Sample rate	Supply voltage	Power dissipation	Inter-face
12 bit						
AD9235	Analog Dev.	1	65 MS/s	3.3 V	300 mW	CMOS
AD9430	Analog Dev.	1	210 MS/s	3.3 V	1300 mW	LVDS
LTC2220	Lin. Tech.	1	170 MS/s	3.3 V	890 mW	LVDS
LTC2226	Lin. Tech.	1	25 MS/s	3.3 V	75 mW	CMOS
LTC2294	Lin. Tech.	2	80 MS/s	3.3 V	422 mW	CMOS
MAX1127	Maxim	4	65 MS/s	1.8 V	563 mW	LVDS
MAX1215	Maxim	1	250 MS/s	1.8 V	1000 mW	LVDS
MAX1420	Maxim	1	60 MS/s	3.3 V	220 mW	CMOS
ADC12DL066	National	2	66 MS/s	3.3 V	686 mW	CMOS
ADS5221	Texas Inst.	1	65 MS/s	3.3 V	285 mW	CMOS
ADS5273	Texas Inst.	8	70 MS/s	3.3 V	1100 mW	LVDS
ADS5520	Texas Inst.	1	125 MS/s	3.3 V	578 mW	CMOS
14 bit						
AD9245	Analog Dev.	1	80 MS/s	3.3 V	366 mW	CMOS
LTC2246	Lin. Tech.	1	25 MS/s	3.3 V	75 mW	CMOS
LTC2249	Lin. Tech.	1	80 MS/s	3.3 V	222 mW	CMOS
LTC2255	Lin. Tech.	1	125 MS/s	3.3 V	395 mW	CMOS
LTC2299	Lin. Tech.	2	80 MS/s	3.3 V	444 mW	CMOS
ADS5500	Texas Inst.	1	125 MS/s	3.3 V	578 mW	CMOS
16 bit						
AD9446	Analog Dev.	1	100 MS/s	3.3 V	2300 mW	CMOS
LTC2208	Lin. Tech.	1	130 MS/s	3.3 V	1250 mW	CMOS

Fig. 18.43. (cont.) Examples for pipeline AD-converters. All types have an on-chip reference

18.10.3
Successive Approximation

The basic design of an A/D converter employing successive approximation is shown in Fig. 18.44. The comparator compares the sampled input value with the output voltage of the D/A converter. When measurement starts, the number Z is set to zero. The most significant bit (MSB) is then set to 1 and the comparator checks to ascertain whether the

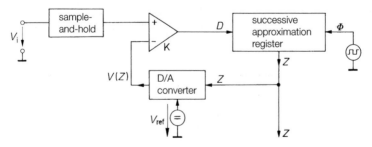

Fig. 18.44. Successive approximation A/D converter

$$Z = (Z_{max} + 1) \frac{V_i}{V_{ref}}$$

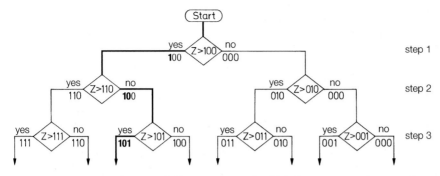

Fig. 18.45. Flowchart for the weighing sequence. The thick line represents the example.

input voltage is greater than $V(Z)$. If so, the bit remains set. If not, it is reset again. Thus the MSB is "weighed." The process is then repeated for each additional bit until, finally, the least significant bit (LSB) is established. In this way, a number is produced in the register. This number is converted in the DAC into a voltage corresponding to V_i within the resolution V_{LSB}, giving

$$V(Z) = V_{ref} \frac{Z}{Z_{max}+1} \overset{!}{=} V_i \quad \text{that is,} \quad Z = (Z_{max}+1)\frac{V_i}{V_{ref}} \tag{18.20}$$

If the input voltage changes during the conversion time, a sample-and-hold circuit is required to buffer the sampled values, so that all the digits are formed from the same input voltage value $V_i(t_j)$. If no sample-and-hold is present, an error may occur that is equal to the input voltage change during the conversion period.

The flowchart for the first three weighing steps is shown in Fig. 18.45. It can be seen that, in each step, a decision is made as to whether the relevant bit is 1 or 0. The previously determined bits remain unchanged.

The timing diagram for the weighing process is shown in Fig. 18.46a for the voltage $V(Z)$, and in Fig. 18.46b for the number Z. Each bit is set on a trial-and-error basis. If, as

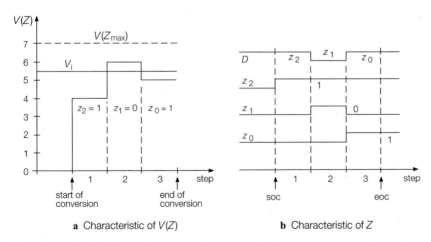

a Characteristic of $V(Z)$ **b** Characteristic of Z

Fig. 18.46. Timing diagram for a successive approximation converter according to the path in Fig. 18.45 for a 3 bit converter

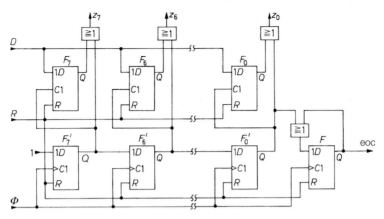

Fig. 18.47. Typical implementation of a successive approximation register (SAR)

a result, the input voltage is exceeded, the bit is reset again. In this example, conversion is therefore complete after 3 weighing steps.

The conversion is controlled by the successive approximation register (SAR). The basic mode of operation will now be discussed, with reference to Fig. 18.47. At the start of the conversion, the reset signal R is used to clear all of the flip-flops. In the shift register F_7' to F_0' a single 1 is shifted one position to the right on each clock pulse, causing bits z_7 to z_0 to be set in turn on a trial-and-error basis. The particular weighing result is stored in latch flip-flops $F_7 - F_0$ by reading the relevant comparator state D. Only the flip-flop whose associated bit is currently being tested is enabled via the C-input at any one time.

When the least significant bit z_0 has also been established, the last flip-flop F of the shift register is set. This indicates end of conversion eoc. Because of the OR gate at the D-input, it retains this state even if further clock pulses are applied. It, together with the result, is not cleared until the next conversion begins.

The truth table for the successive approximation register is shown in Fig. 18.48. As we can see, all of the outputs are cleared with the reset signal. At each step Φ the decision D of the comparator is stored in the relevant position and the next less significant bit is weighed. The truth table illustrates the operation of the shift register. After eight steps, "1" has arrived at the end of conversion output and the conversion is done. The result Z is then

Φ	R	D	z_7	z_6	z_5	z_4	z_3	z_2	z_1	z_0	eoc
0	1	0	0	0	0	0	0	0	0	0	0
1	0	D_7	1	0	0	0	0	0	0	0	0
2	0	D_7	D_7	1	0	0	0	0	0	0	0
3	0	D_6	D_7	D_6	1	0	0	0	0	0	0
4	0	D_5	D_7	D_6	D_5	1	0	0	0	0	0
5	0	D_4	D_7	D_6	D_5	D_4	1	0	0	0	0
6	0	D_3	D_7	D_6	D_5	D_4	D_3	1	0	0	0
7	0	D_2	D_7	D_6	D_5	D_4	D_3	D_2	1	0	0
8	0	D_1	D_7	D_6	D_5	D_4	D_3	D_2	D_1	1	0
9	0	D_0	D_7	D_6	D_5	D_4	D_3	D_2	D_1	D_0	1

Fig. 18.48. Truth table for the successive approximation register

Type	Manu-facturer	Input channels*	Sample rate	supply voltage	Power dissipation	Reference internal	Inter-face
8 bit							
AD7908	Analog Dev.	8 s	1000 kS/s	3.3 V	6 mW	—	SPI
MAX1039	Maxim	12 s	188 kS/s	3.3 V	1.2 mW	2.05 V	I^{2}C
MAX1115	Maxim	1 s	110 kS/s	3.3 V	0.6 mW	2.05 V	SPI
MAX1118	Maxim	2 s	100 kS/s	3.3 V	0.6 mW	—	SPI
ADS7888	Texas Inst.	1 s	1000 kS/s	3.3 V	4 mW	—	SPI
12 bit							
AD7457	Analog Dev.	1 s	100 kS/s	3.3 V	1 mW	—	SPI
AD7490	Analog Dev.	16 s	1000 kS/s	3.3 V	5 mW	—	SPI
AD7928	Analog Dev.	8 s	1000 kS/s	3.3 V	6 mW	—	SPI
LTC1851	Lin. Tech.	8 s	1250 kS/s	5 V	40 mW	2.5 V	12 bit
LTC1860	Lin. Tech.	1 s	250 kS/s	5 V	4.5 mW	—	SPI
LTC1863	Lin. Tech.	8 s	200 kS/s	5 V	6.5 mW	2.5 V	SPI
MAX1239	Maxim	12 s 6 d	100 kS/s	3.3 V	2.2 mW	2.05 V	SPI
MAX1254	Maxim	8 s 4 d	100 kS/s	3.3 V	5 mW	2.5 V	SPI
MAX1287	Maxim	1 s	150 kS/s	3.3 V	0.7 mW	—	SPI
MAX1295	Maxim	6 s 3 d	265 kS/s	3.3 V	6 mW		12 bit
ADS7886	Texas Inst.	1 s	1000 kS/s	3.3 V	11 mW	—	SPI
16 bit							
AD7621	Analog Dev.	1 s	3000 kS/s	3.3 V	65 mW	2.05 V	SPI/16 bit
AD7677	Analog Dev.	1 s	1000 kS/s	5 V	115 mW	—	SPI/16 bit
AD7683	Analog Dev.	1 s	100 kS/s	3.3 V	2 mW	—	SPI
AD7686	Analog Dev.	1 s	500 kS/s	5 V	21 mW	—	SPI
LTC1867	Lin. Tech.	8 s	200 kS/s	5 V	6.5 mW	2.5 V	SPI
MAX1162	Maxim	1 s	200 kS/s	3.3 V	8 mW	—	SPI
MAX1165	Maxim	1 s	165 kS/s	5 V	13 mW	4.1 V	16 bit
MAX1168	Maxim	8 s	200 kS/s	5 V	13 mW	4.1 V	SPI
ADS8323	Texas Inst.	1 s	500 kS/s	5 V	85 mW	2.5 V	16 bit
ADS8325	Texas Inst.	1 s	100 kS/s	3.3 V	4.5 mW	—	SPI
ADS8371	Texas Inst.	1 s	250 kS/s	5 V	130 mW	—	8/16 bit
ADS8412	Texas Inst.	1 s	2000 kS/s	5 V	175 mW	4.1 V	8/16 bit

* s = single ended, d = differential.

Fig. 18.49. Examples for successive aproixamtion (SAR) AD-converters.
Most 3.3 V types can be operated with a 5 V supply at increased power consumption

available in parallel form. However, it can also be obtained in serial form at the comparator output.

Some examples for successive approximation converters are given in Fig. 18.49.

18.10.4
Counting Method

A/D conversion using the counting method requires the least circuit complexity, but the conversion time is considerably longer than with the other methods – generally between 1 ms and 1 s. However, this is adequate for slowly changing signals, such as those involved in temperature measurement, and it is also fast enough for digital voltmeters, as there is a limit to how quickly the result can be read off. The counting method can be implemented in various ways. The principal techniques are discussed below, the most important being the dual-slope and $\Delta\Sigma$ method, as it allows maximum accuracy to be achieved with minimum circuit complexity.

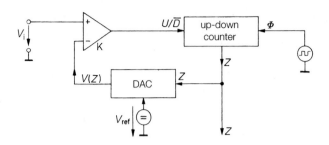

Fig. 18.50. Tracking A/D converter

$Z = (Z_{max} + 1)V_i / V_{ref}$

Compensation Converters

The compensating A/D converter shown in Fig. 18.50 is closely related to the successive approximation (SA) type shown in Fig. 18.44. The basic difference is that an up–down counter is used instead of the SA register.

The comparator compares the input voltage V_i, with the compensating voltage $V(Z)$. If the difference is positive, it causes the counter to count upward, and vice versa. This means that the compensating voltage rises or falls until it has reached the level of the input voltage, and then follows the latter as it changes. For this reason, the circuit is also known as a *tracking A/D converter*.

One drawback with the simple circuit in Fig. 18.50 is that, as the clock is never switched off, the counter never stops, but always oscillates by 1 LSB around the input voltage. If this causes problems, the simple comparator can be expanded to form a window comparator. This allows the clock to be inhibited when the compensating voltage $V(Z)$ is within $\pm \frac{1}{2} V_{LSB}$ of the input voltage V_i.

The significant reduction in control logic as compared to the weighting method is achieved at the expense of considerably longer conversion times, as the compensating voltage only changes in steps of size V_{LSB}. However, if the input voltage only changes slowly, a short settling time can again be achieved, as approximation is performed continuously because of the tracking characteristic, rather than always starting at zero as with the weighting method.

Single-Slope Converter

The single slope A/D converter shown in Fig. 18.51 does not require a DAC. The principle here is that the input voltage is initially converted into a proportional time interval using the sawtooth generator in conjunction with the window comparator K_1, K_2, and G_1.

The sawtooth voltage can be raised from negative to positive values in accordance with the relation

$$V_S = \frac{V_{ref}}{\tau} t - V_0$$

Logic $y = 1$ is present at the EQUIVALENCE gate G_1 only for as long as the sawtooth voltage is between the limits 0 and V_i. The corresponding time interval is $\Delta t = \tau V_i / V_{ref}$. This is measured by counting the oscillations of the crystal-controlled oscillator. If the counter is zeroed at the start of the measurement, the count after the upper comparator threshold has been exceeded is

$$Z = \frac{\Delta t}{T} = \tau f \frac{V_i}{V_{ref}} \tag{18.21}$$

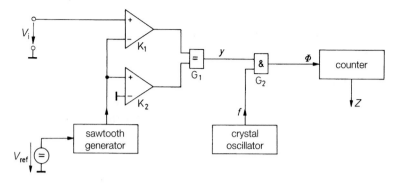

Fig. 18.51. Single-slope A/D converter. $Z = \tau \cdot f \cdot V_i / V_{\text{ref}}$

If a negative measurement voltage is applied, the sawtooth voltage will first cross the measurement voltage and then pass through zero. This sequence therefore enables the sign of the measured voltage to be determined. The measuring time is the same; it is purely a function of the magnitude of the measurement voltage. After each conversion the counter must be reset to zero and the sawtooth voltage adjusted to its negative initial value. In order to retain the result, the old count is normally stored until a new one is available.

As (18.21) indicates, the tolerance of time constant τ directly affects the measuring accuracy. Since it is determined by an RC network, it is subject to the temperature and long-term drift of the capacitor. Consequently, an accuracy of better than 1% is difficult to achieve.

Dual-Slope Converter

With this method, not only the reference voltage but also the input voltage is integrated. In the inactive state, switches S_1 and S_2 in Fig. 18.52 are open, while switch S_3 is closed. As a result, the integrator output voltage is zero.

At the start of the conversion cycle, the counter is cleared, switch S_3 is opened, and S_1 is closed, causing the input voltage V_I to be integrated. If it is positive, the integrator output becomes negative, as shown in Fig. 18.53, and the comparator enables the clock generator. The end of the cycle is reached when the counter overflows after $Z_{max} + 1$ clock pulses and is thus again at zero. Then the reference voltage is integrated by opening S_1 and closing S_2.

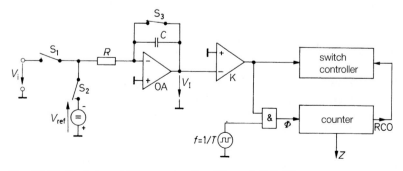

Fig. 18.52. Dual-slope A/D converter. $Z = (Z_{max} + 1)V_i / V_{\text{ref}}$

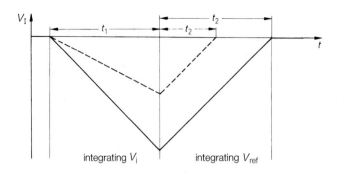

Fig. 18.53. Integrator output voltage for two input voltages in a dual-slope converter

As it is negative the integrator voltage now rises. The second integration interval is finished if V_I has reached zero. Then the comparator goes to zero and the counter is stopped. The counter reading equals the number of clock pulses during the time t_2, therefore the reading is proportional to the input voltage.

The relation between the input voltage V_I and the result Z can be calculated if one keeps in mind that the integration starts at and ends at $V_I = 0$. From

$$V_I = -\frac{1}{RC} \int_0^{t_1} V_i \, dt - \frac{1}{RC} \int_0^{t_2} V_{\text{ref}} \, dt \overset{!}{=} 0 \tag{18.22}$$

follows, if V_i is taken to be constant

$$-\frac{1}{RC} V_i t_1 - \frac{1}{RC} V_{\text{ref}} t_2 = 0$$

With

$$t_1 = (Z_{max} + 1)T \quad \text{and} \quad t_2 = ZT \tag{18.23}$$

results

$$-\frac{1}{RC} V_1 (Z_{max} + 1)T - \frac{1}{RC} V_{\text{ref}} ZT = 0$$

One recognizes that the time constant RC and the clock period T can be reduced from the equation:

$$V_i (Z_{max} + 1) + V_{\text{ref}} Z = 0$$

Solving for Z renders the result:

$$Z = (Z_{max} + 1) \frac{V_i}{V_{\text{ref}}} \tag{18.24}$$

This equation reveals the salient feature of the dual-slope method; namely, that neither the clock frequency $1/T$ nor the integration time constant $\tau = RC$ have any effect on the result. The only requirement is that the clock frequency must be constant during the time $t_1 + t_2$. This short-term stability can be achieved using simple clock generators. For these reasons, accuracies of $0.01\% = 100\,\text{ppm}$ can be achieved by this method with cheap components.

As we have seen, it is not the instantaneous value of the measurement voltage that affects the result, but its average value over measuring time t_1. Consequently, alternating voltages are attenuated more heavily the higher their frequency. Alternating voltages whose frequencies are equal to an integral multiple of $1/t_1$ are rejected completely. It is therefore advisable to adjust the clock generator frequency such that t_1 is equal to the period of the AC supply voltage or a multiple thereof. This will eliminate any hum.

As the dual-slope method allows a high degree of accuracy and noise rejection to be achieved with minimum circuit complexity, it is preferable to use this type of converter in digital voltmeters. In this application, the relatively long conversion times do not present any problem.

The counter in Fig. 18.52 need not be a straight-binary counter. The mode of operation is identical if a BCD counter is employed. This possibility is utilized in digital voltmeters, as no binary-to-decimal conversion of the measured value is then required.

18.10.5
Oversampling

In any AD converter the effective resolution can be raised by using a higher sampling frequency than needed and a following digital lowpass filter that limits the bandwidth to the used range as shown in Fig. 18.54. By oversampling the quantizing noise $V_n^2 = V_{LSB}^2/12$ is distributed over a larger frequency range so that the quantizing noise density

$$V_n'^2 = \frac{V_{LSB}^2}{12 \cdot OSR \cdot (f_s/2)} = \frac{V_{LSB}^2}{6 \cdot f_{OSR}} \tag{18.25}$$

is lowered as shown in Fig. 18.55. If the frequency range is subsequently limited to the required signal bandwidths $f_c = \frac{1}{2} f_S$ the quantizing noise (double hatched area) is reduced. But this method is not very efficient because a fourfold oversampling is needed to halve the quantizing noise V_n' corresponding to additional 1 bit of resolution as can be seen in Fig. 18.57.

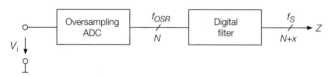

Fig. 18.54. Getting additional bits by oversampling. f_S = sampling frequency; $f_{OSR} \cdot f_S$ = oversampling frequency

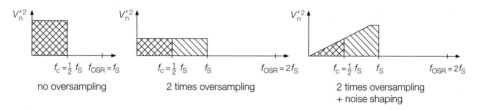

Fig. 18.55. Effect of oversampling on quantizing noise density. Hatched area: quantizing noise that is the same in all three cases. Double hatched area: noise within signal bandwidth. f_c = signal bandwidth, f_S = sampling frequency, f_{OSR} = oversampling frequency

Delta-Sigma Converter

The advantage of a delta-sigma converter is that the $\Delta\Sigma$ modulator in Fig. 18.56 not only performs oversampling but also noise shaping. This is shown in Fig. 18.55 also. Here the quantizing noise is not equally distributed but shifted to higher frequencies where it is effectively removed by the following digital lowpass filter.

The profit of additional bits with first order noise shaping is shown in Fig. 18.57 in comparison to oversampling without noise shaping. At fourfold oversampling only 1 bit is obtained by conventional oversampling whereas 3 bits are obtained by noise shaping oversampling.

The internal construction of a $\Delta\Sigma$ converter is shown in Fig. 18.58. The core is the integrator that ensures that the difference between the input signal and the reconstructed

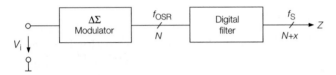

Fig. 18.56. Delta–sigma converter.
f_S = sampling frequency; f_{OSR} = OSR $\cdot$ f_S = oversampling frequency

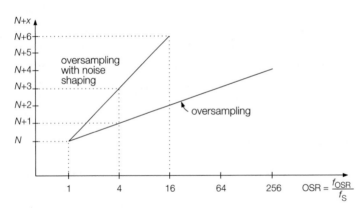

Fig. 18.57. Generation of additional bits by oversampling.
Oversampling ratio = oversampling frequency/conversion frequency. OSR = f_{OSR}/f_S

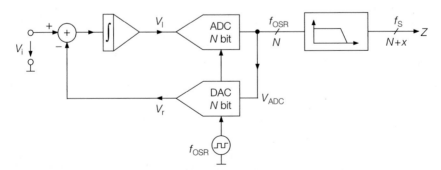

Fig. 18.58. Configuration of a first order $\Delta\Sigma$-converter

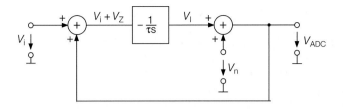

Fig. 18.59. Noise modell for an $\Delta\Sigma$-converter

signal from the DAC will be near zero. The noise transfer function can be calculated from the model in Fig. 18.59. The input signal has lowpass characteristic

$$\frac{V_{ADC}}{V_i} = -\frac{1}{1+\tau s}$$

with a cutoff frequency $f_c = 1/(2\pi\tau)$. The integration time constant τ must be chosen not to limit the signal bandwidth. The noise transfer function has highpass characteristic

$$\frac{V_{ADC}}{V_n} = \frac{\tau s}{1+\tau s} \approx \tau s \quad \text{for} \quad \tau s \ll 1$$

that results in the noise shaping characteristic shown in Fig. 18.55. You can even use a chain of two or three integrators to improve the noise shaping and to get additional bits. But additional feedback paths are needed to stabilize the loop.

The most simple $\Delta\Sigma$ modulator with an $N = 1$ bit ADC and DAC is shown in Fig. 18.60. The 1 bit ADC consists of a comparator and a clocked flip-flop. If the integrator output voltage V_I is negative the flip-flop will be set ($Q = 1$) and the negative refernce voltage will be switched on by the 1 bit DAC causing the integrator to move in positive direction. Like the dual-slope converter it consists of an integrator followed by a comparator. Here the input voltage V_i is always connected to the integrator. In order to measure the input voltage, reference pulses of opposite sign are applied to the integrator so that its output remains near zero. This process is shown in Fig. 18.61. A compensation

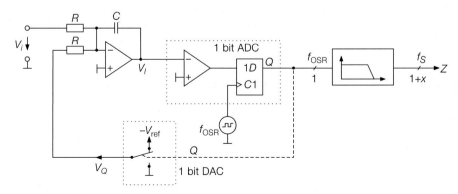

Fig. 18.60. Most simple realization of a $\Delta\Sigma$-converter

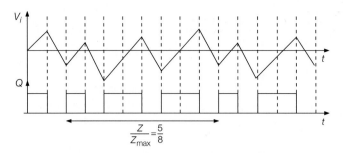

Fig. 18.61. Signals in the simple $\Delta\Sigma$-modulator. Example for $V_i = \frac{5}{8}V_{ref}$ and $N = 3$ bit. The flip-flop is set for one clock if the integrator output voltage is negative at the triggering clock edge

pulse is issued if the integrator output is negative at the triggering clock edge. Its duration is one clock period.

From the number of the compensation pulses required to balance the input charge the reading can be calculated. The current trough of the input resistance must be as high as the average current from the reference source:

$$\frac{V_i}{R} + \frac{V_{ref}}{R}\frac{Z}{Z_{max}} \stackrel{!}{=} 0 \tag{18.26}$$

Hence we get the result

$$Z = \frac{V_i}{V_{ref}}Z_{max} \tag{18.27}$$

The maximum number Z_{max} is the number of clock pulses in which the number of compensation pulses is counted. In the example in Fig. 18.61 we have chosen $Z_{max} = 8$ in order to get a 3 bit conversion. With the supposed input voltage $V_i = \frac{5}{8}V_{ref}$ a number of 5 compensation pulses are required during 8 clock cycles.

The main difference to the dual-slope converter is that the charge is not balanced once at a measurement but with a frequency that is higher by a factor of 2^N. The advantages of the $\Delta\Sigma$ converter are:

- only a simple analog antialiasing filter is required at the input because of the high oversampling frequency
- high increase of resolution by noise shaping oversampling
- high linearity
- good for integration

Some common $\Delta\Sigma$ converters are collected in Fig. 18.62. Some converters with 7-segment interface follow in Fig. 18.63.

Type	Manu-facturer	Resolution	Sample rate	Supply voltage	Power dissipation	Inter-face
AD1870	Analog Dev.	16 bit	44 kS/s	5 V	260 mW	serial
AD1871	Analog Dev.	24 bit	96 kS/s	5 V	260 mW	serial
AD7714	Analog Dev.	24 bit	5–1000 S/s	3.3 V	4 mW	SPI
AD7782	Analog Dev.	24 bit	20 S/s	3.3 V	4 mW	SPI
HI7190	Intersil	24 bit	10–2000 S/s	5 V	325 mW	SPI
LTC2412	Lin. Tech.	24 bit	8 S/s	3.3 V	0.7 mW	SPI
LTC2444	Lin. Tech.	24 bit	13–7000 S/s	5 V	40 mW	SPI
MAX1402	Maxim	18 bit	20–4800 S/s	5 V	1 mW	SPI
MAX1415	Maxim	16 bit	20–200 S/s	3.3 V	1 mW	SPI
MAX1417	Maxim	3.5 digit	5 S/s	3.3 V	2 mW	SPI
MAX1499	Maxim	4.5 digit	5 S/s	3.3 V	2 mW	SPI
ADS1224	Texas Instr.	24 bit	120 S/s	3.3 V	1 mW	serial
ADS1255	Texas Instr.	24 bit	3–30,000 S/s	5 V	35 mW	SPI
ADS1271	Texas Instr.	24 bit	50–105 kS/s	5 V	90 mW	SPI
ADS1605	Texas Instr.	26 bit	5000 kS/s	5 V	560 mW	16 bit

Fig. 18.62. $\Delta\Sigma$-AD-converters with computer interface

Type	Manu-facturer	Tech-nique	Resolution	Sample rate	Supply voltage	Power	Inter-face
TCL7136	Intersil	2 Slope	3.5 digits	3 S/s	9 V	1 mW	LCD
TC7117	Microchip	2 Slope	3.5 digits	3 S/s	5 V	10 mW	LED
TC7126	Microchip	2 Slope	3.5 digits	3 S/s	9 V	4.5 mW	LCD
TC7129	Microchip	2 Slope	4.5 digits	2 S/s	9 V	4.5 mW	LCD
MAX1491	Maxim	$\Delta\Sigma$	3.5 digits	2.5 S/s	3.3 V	3 mW	LCD
MAX1495	Maxim	$\Delta\Sigma$	4.5 digits	2.5 S/s	3.3 V	3 mW	LCD
MAX1497	Maxim	$\Delta\Sigma$	3.5 digits	5 S/s	3.3 V	2 mW	LED
MAX1499	Maxim	$\Delta\Sigma$	4.5 digits	5 S/s	3.3 V	2 mW	LED

Fig. 18.63. AD-converters for seven-segment displays in digital voltmeters

18.11
Errors in AD-Converters

18.11.1
Static Errors

In addition to the systematic quantization noise, errors arise from nonideal circuitry. When the mid-values of the steps are joined, as shown in Fig. 18.64, a straight line is obtained which passes through the origin and has a slope of 1. For a real A/D converter, this line misses the origin (offset error) and its slope is different from 1 (gain error). The gain error gives rise to a *relative* deviation of the output quantity from the desired value, which is constant over the whole range of operation. The offset error gives rise to a constant *absolute* deviation. Both errors can usually be eliminated by adjusting the zero-point and the gain. The remaining deviation is then due to drift and nonlinearity only.

A linearity error that exceeds the quantization error is incurred whenever the steps are of different heights. To determine this linearity error, the offset and gain are adjusted and the maximum deviation of the input voltage from the ideal straight line is measured.

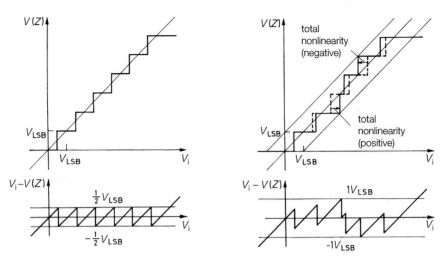

Fig. 18.64. Quantization error of an ideal A/D converter

Fig. 18.65. Transfer characteristic of an A/D converter that has linearity errors

This value, which is reduced by the systematic quantization error of $\frac{1}{2}V_{\text{LSB}}$, is the *total nonlinearity*. It is usually quoted as a percentage of the LSB voltage unit. For the example shown in Fig. 18.65, the total nonlinearity is $\pm \frac{1}{2} V_{\text{LSB}}$.

A further measure of the linearity error is the *differential nonlinearity*, which indicates by how much the widths of the individual steps deviate from the desired value V_{LSB}. If this deviation is larger than V_{LSB}, a number is skipped (missing code). For even larger deviations, the number Z may even decrease for an increasing input voltage (*monotonicity error*).

18.11.2
Dynamic Errors

A/D converter applications fall into two categories: those in digital voltmeters and those in signal-processing circuits. Their use in digital voltmeters is based on the assumption that the input voltage remains constant during the conversion process. For signal-processing applications, however, the input voltage changes continually. For digital processing, samples must be taken from the alternating voltage by means of a sample-and-hold circuit. The samples are then A/D converted. It has been shown in Sect. 18.1 that the resulting number sequence {Z} represents the continuous input signal without loss of information only if the *sampling theorem* (the *Nyquist theorem*) *is* satisfied. The sampling frequency, f_s, must therefore be at least twice the highest signal frequency f_{max}. Consequently, the conversion time of the A/D converter and the settling time of the sample-and-hold circuit must together be less than $1/(2 f_{max})$. In order to meet this requirement without introducing excessive complexity, the signal bandwidth is limited to the lowest value possible. It is therefore necessary to insert a preceding lowpass filter.

In order to judge the attainable accuracy, the properties of the A/D converter and of the sample-and-hold circuit must, in this application, be considered together. For example, it is pointless to operate a 12-bit A/D converter in conjunction with a sample-and-hold circuit that does not settle to $1/4,096 \approx 0.025\%$ of full scale within the allowed time.

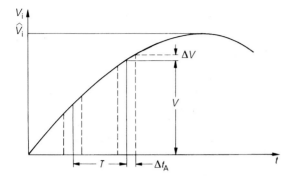

Fig. 18.66. Effect of aperture jitter Δt_A. $T =$ sampling period

Another dynamic error is incurred due to *aperture jitter*. Due to the aperture time t_A of the sample-and-hold circuit, the measured value is with a certain delay. If the aperture time is constant, every measured value will be delayed by the same time, and so equidistant sampling will still be insured. However, if the aperture time varies by the aperture jitter Δt_A, as shown in Fig. 18.66, a measurement error is introduced which is equal to the voltage change ΔV in this time. To calculate the maximum error ΔV, the input signal is taken to be a sine wave with the highest frequency f_{max} for which the system is designed. The maximum slope is at the origin:

$$\frac{dV}{dt}\bigg|_{t=0} = \hat{V}\omega_{max}$$

Hence, the amplitude error is:

$$\Delta V = \hat{V}\omega_{max}\Delta t_A$$

If the error is to be smaller than the smallest conversion voltage V_{LSB} of the A/D converter, the condition for the aperture jitter must be:

$$\Delta t_A < \frac{V_{LSB}}{\hat{V}\omega_{max}} = \frac{V_{LSB}}{\frac{1}{2}V_{max}\omega_{max}} \tag{18.28}$$

It is difficult to fulfill this condition for high signal frequencies, as the following example illustrates. For an 8 bit converter, $V_{LSB}/V_{max} = 1/255$. For a maximum signal frequency of 10 MHz, the aperture jitter must, according to (18.28), be smaller than 125 ps.

18.12
Comparison of AD-Converters

In order to compare the different methods, we have listed their main characteristics in Fig. 18.67. Figure 18.68 shows the resolution and frequency range in which these methods are implemented.

Approach	Number of steps	Number of reference voltages	Characteristics
Parallel method	1	$n = 2^N$	complex, fast
Pipeline method	1	$N = \mathrm{ld}\, n$	fast
Weighting method	$N = \mathrm{ld}\, n$	$N = \mathrm{ld}\, n$	
Counting method	$n = 2^N$	1	simple, slow

Fig. 18.67. Comparison of the different approaches to A/D conversion. With the pipeline method only the first conversion takes N steps all following conversions take only 1 step.

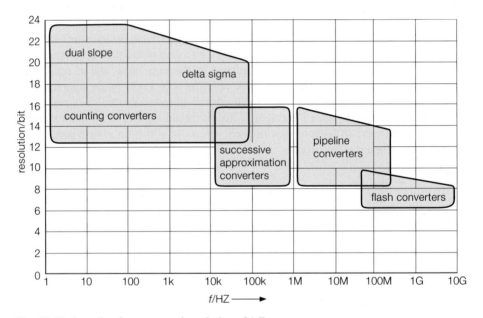

Fig. 18.68. Sampling frequency and resolution of A/D converters

Chapter 19:
Digital Filters

In Chapter 13, several transfer functions are discussed and their realization by active filters is described. The processed signals are voltages which, in turn, are continuous functions of time. The circuits are made up of resistors, capacitors, and amplifiers.

Recently, the trend has been toward signal processing by digital rather than analog circuits. The advantages are high accuracy and consistency in the results, and a lower sensitivity to disturbances. The high number of digital components required is a disadvantage, however, but in view of the increase in integration of digital circuitry, it is becoming less important.

Sequences of discrete numbers are processed instead of continuous signals, and the circuit elements are memories and arithmetic circuits. The transition from an analog to a digital filter raises three questions:

1) How can a sequence of discrete numeric values be derived from the continuous input voltage without loss of information?
2) How must this numeric sequence be processed in order to obtain the required transfer function?
3) How can the output values be converted back into a continuous voltage?

The embedding of a digital filter in an analog environment is shown schematically in Fig. 19.1. At sampling instants t_μ, the sample-and-hold circuit extracts voltages $V_i(t_\mu)$ from the input signal $V_i(t)$ and holds them constant for one sampling interval. In order to prevent any information loss from occurring during sampling, the input signal must be band-limited to half the sampling rate, in accordance with the sampling theorem. Consequently, a lowpass filter is generally required at the input.

The ADC converts the time-discrete voltage sequence $V_i(t_\mu)$ into a numeric sequence $x(t_\mu)$. The x values are usually N bit binary numbers, where N determines the magnitude of the quantizing noise (see (18.9)). Typical sampling frequencies and resolutions are listed in Fig. 19.2.

The digital filter shown in Fig. 19.1 produces the filtered number sequence $y(t_\mu)$. In order to convert it back into a voltage, a DAC is used, which delivers an amplitude- and time-discrete staircase signal at its output. To convert this to a continuous voltage, a following lowpass smoothing filter is required.

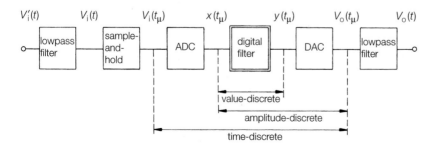

Fig. 19.1. A digital filter in an analog environment

Signal	Sampling frequency	Resolution
Telephone Speech	8 kHz	12 bit
CD music	44.1kHz	16 bit
Digital TV*	10.4MHz	8 bit

* For the luminance-signal at standard resolution

Fig. 19.2. Usual sampling frequencies and word lengths for digital signal processing

19.1
Digital Transfer Function

As we have seen in Chap. 13, analog filters can be implemented using integrators, adders, and coefficient networks. A digital filter is obtained by replacing the integrators by delay elements. The latter can be implemented by, for example, shift registers that are used to shift the input function samples through at the sampling rate f_s. The simplest case is that of delay by one time interval T_s; a delay element of this kind is shown schematically in Fig. 19.3.

19.1.1
Time Domain Analysis

The number sequence $\{x(t_\mu)\} = \{x_\mu\}$ is given and may be used as sampled values with a word length of 8, 16 or 32 bits. These are shifted into a register using a corresponding number of parallel-clocked flip-flops. The output sequence in Fig. 19.3 $\{y(t_\mu)\} = \{y_\mu\}$ represents the input sequence shifted by one clock period T_s. We therefore have

$$y(t_\mu) = x(t_{\mu-1}) \tag{19.1}$$

19.1.2
Frequency Domain Analysis

To examine the frequency response, the sinusoidal sequence $x(t_\mu) = \hat{x} \sin \omega t_\mu$ is applied to the input. If the system is linear, a sinusoidal sequence appears at the output. As for analog filters, the ratio of the amplitudes is equivalent to the magnitude of the transfer function for $s = j\omega$. The linearity of a digital filter is indicated by the linearity of the difference equation. According to (19.1), the filter in Fig. 19.3 is therefore linear.

The transfer function may be inferred from the circuit with the help of complex calculus, as for analog filters. This requires that the frequency response of a delay element be known. With the harmonic input sequence

$$x(t_\mu) = \hat{x} e^{j\omega t_\mu}$$

the harmonic output sequence

$$y(t_\mu) = x(t_{\mu-1})$$
Time Domain

$$Y(z) = z^{-1} X(z) = e^{-j2\pi f/f_a} X(z)$$
Frequency Domain

Fig. 19.3. Representation of a delay element

$$y(t_\mu) = \hat{x}e^{j\omega(t_\mu - T_s)} = \hat{x}e^{j\omega t_\mu} \cdot e^{-j\omega T_s} = x(t_\mu)e^{-j\omega T_s}$$

is obtained, and with $j\omega = s$, the transfer function

$$A(s) = \frac{L\{y(t_\mu)\}}{L\{x(t_\mu)\}} = e^{-j\omega T_s} = e^{-sT_s} \tag{19.2}$$

It is a periodic function, the period being $f = f_s = 1/T_s$, where f_s is the sampling – that is, the clock – frequency. To abbreviate this,

$$z^{-1} = e^{-sT_s} = e^{-j2\pi f/f_s} \tag{19.3}$$

which results, together with (19.2), in the transfer function:

$$\tilde{A}(z) = z^{-1} \tag{19.4}$$

This is the frequency-domain description of the delay element shown in Fig. 19.3.

It was mentioned in Chap. 13 that the transfer function $A(s)$ describes the relationship between the output signal and any desired time-dependent input signal if Laplace transforms according to

$$L\{y(t)\} = A(s) \cdot L\{x(t)\} \tag{19.5}$$

are used. This relationship also holds for a digital system. Using the converted transfer function of (19.4), the relation for number sequences can be simplified, since

$$Z\{y(t_\mu)\} = \tilde{A}(z) \cdot Z\{x(t_\mu)\} \tag{19.6}$$

where

$$Z\{x(t_\mu)\} = X(z) = \sum_{\mu=0}^{\infty} x(t_\mu)z^{-\mu} \tag{19.7}$$

is the Z-transform of the input sequence. The output sequence is obtained by the corresponding reverse transform. Because of this property, $\tilde{A}(z)$ is called the *digital transfer function*.

From this, we can calculate the analog transfer function or the quantities derived from it, such as the magnitude, phase, and group delay. For the delay element, it follows from

$$\tilde{A}(z) = \frac{Y(z)}{X(z)} = z^{-1} \quad \text{with} \quad z^{-1} = e^{-j\omega T_s},$$

that

$$A(j\omega) = z^{-1} = e^{-j\omega T_s} = \cos \omega T_s - j \sin \omega T_s.$$

The magnitude is therefore

$$|\underline{A}(j\omega)| = \sqrt{\cos^2 \omega T_s + \sin^2 \omega T_s} = 1$$

the phase

$$\varphi = \arctan \frac{-\sin \omega T_s}{\cos \omega T_s} = \arctan(-\tan \omega T_s) = -\omega T_s = -2\pi \frac{f}{f_s}$$

and the group delay:

$$t_{gr} = -\frac{d\varphi}{d\omega} = T_s.$$

An Example of a Lowpass Filter

Digital filters may be easily described using the relations for a delay element. The numeric value $x(t_\mu) - \beta_1 y(t_\mu)$ is present at the input of the memory in Fig. 19.4 at instant t_μ. This value appears at the memory output one clock period later. The values of the output sequence are therefore given by the relation

$$y(t_{\mu+1}) = x(t_\mu) - \beta_1 y(t_\mu)$$

This *difference equation* represents the analogon to the differential equation for a continuous system. It can be used as a recursive formula to calculate the output sequence by specifying an initial value $y(t_0)$. As an example, we select $y(t_0) = 0$ and calculate the step response for $\beta_1 = -0.75$. This is plotted in Fig. 19.5. We can see that the circuit exhibits a lowpass characteristic.

The frequency response of the lowpass filter used in this example can be calculated in the same way as for the delay element. From the diagram on the right in Fig. 19.4, we take

$$Y(z) = [X(z) - \beta_1 Y(z)]z^{-1}$$

The digital transfer function is therefore

$$\tilde{A}(z) = \frac{Y(z)}{X(z)} = \frac{z^{-1}}{1 + \beta_1 z^{-1}}$$

To calculate the frequency response, we put

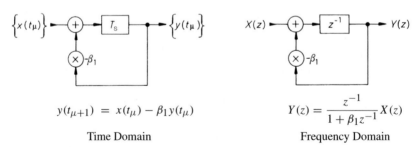

$$y(t_{\mu+1}) = x(t_\mu) - \beta_1 y(t_\mu)$$

Time Domain

$$Y(z) = \frac{z^{-1}}{1 + \beta_1 z^{-1}} X(z)$$

Frequency Domain

Fig. 19.4. Example of a recursive first-order digital filter

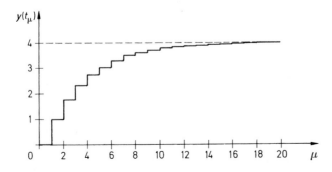

Fig. 19.5. Step response of the digital filter in Fig. 19.4 for $\beta_1 = -0.75$ with an input step from 0 to 1

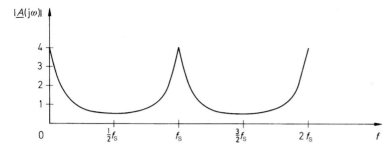

Fig. 19.6. Amplitude–frequency response of the digital filter in Fig. 19.4 for $\beta_1 = -0.75$

$$z^{-1} = e^{-j\omega T_s} = \cos T_s - j \sin \omega T_s$$

and obtain:

$$\underline{A}(j\omega) = \frac{1}{\beta_1 + e^{j\omega T_s}} = \frac{1}{\beta_1 + \cos \omega T_s + j \sin \omega T_s}$$

With $\omega T_s = 2\pi f/f_s$, the resulting magnitude is

$$|\underline{A}(j\omega)| = \frac{1}{\sqrt{(\beta_1 + \cos 2\pi f/f_s)^2 + (\sin 2\pi f/f_s)^2}}$$

We can see from Fig. 19.6 that it is periodic with f_s and symmetrical about $\frac{1}{2}f_s$. This characteristic is shared by all digital filters. The frequency range above $\frac{1}{2}f_s$ however, cannot be used, as this would violate the sampling theorem.

An interesting special case arises for $\beta_1 = -1$. The magnitude of the transfer function can now be simplified in accordance with $\cos^2 x + \sin^2 x = 1$:

$$|\underline{A}(j\omega)| = \frac{1}{\sqrt{2 - 2\cos 2\pi f/f_s}} = \frac{1}{\sqrt{4(\sin \pi f/f_s)^2}} = \frac{1}{2 \sin \pi f/f_s}.$$

Hence, for low frequencies $f \ll f_s$, we obtain, with $\sin x \approx x$,

$$|\underline{A}(j\omega)| = \frac{f_s}{2\pi f} \sim \frac{1}{f};$$

in other words, the frequency response of an integrator. The resulting circuit is the usual arrangement for an adder with memory or accumulator.

19.2
Basic Structures

Lattice filters apart, there are three arrangements for implementing digital filters. These are illustrated in Figs. 19.7–19.9. All three possess the same transfer functions if the filter coefficients α_k and β_k are used at the locations shown.

We can see from Figs. 19.7–19.9 that, in addition to the delay elements, the filters require multipliers that multiply the variables by the constant filter coefficients, and adders that add two or three numbers. The structure shown in Fig. 19.7 is the most commonly used, as in this case each multiplier–accumulator stage (MAC) is separated from the next by a delay element. As a result, a complete clock period is available for these operations.

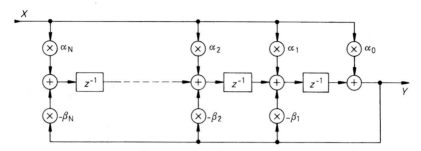

Fig. 19.7. Digital filter with distributed adders

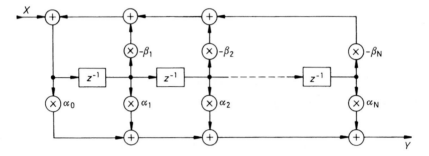

Fig. 19.8. Digital filter with a global adder at the output and input

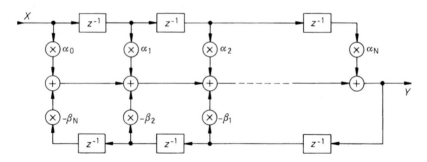

Fig. 19.9. Digital filter with a single global adder at the output

The delay elements here produce a "pipeline" structure. In the other two circuits, many variables have to be added in a single clock period. Although this does not require more adders, it does require more computing time.

In the circuit shown in Fig. 19.8, it can be seen that the input signal for the delay chain is derived from the input signal X and all of the weighted intermediate values. Accordingly, the output signal is the weighted sum of all the intermediate values. The adders can therefore be combined to form two global adders: one at the input and one at the output.

The circuit in Fig. 19.9 has a single global adder at the output. It adds both the delayed and weighted input signal as well as the delayed and weighted output signal. For this purpose, an additional delay chain is required, which is connected to the output. However, the additional complexity involved is minimal.

The number of filter sections determines the order N of the filter. One delay element (two in Fig. 19.9) and two coefficient multipliers are required for each section, and three numbers must be added. Only the first and last sections are somewhat simpler.

We shall analyze the circuits taking Fig. 19.7 as an example. The difference equation is of the form

$$y(t_N) = \sum_{k=0}^{N} \alpha_k x_{N-k} - \sum_{k-1}^{N} \beta_k y_{N-k} \tag{19.8}$$

For the transfer function, the circuit yields the relation

$$Y(z) = X(z) \sum_{k=0}^{N} \alpha_k z^{-k} - Y(z) \sum_{k=1}^{N} \beta_k z^{-k} .$$

Hence, the transfer function is

$$A(z) = \frac{Y(z)}{X(z)} = \frac{\sum\limits_{k=0}^{N} \alpha_k z^{-k}}{1 + \sum\limits_{k=1}^{N} \beta_k z^{-k}} \tag{19.9}$$

$$A(z) = \frac{\alpha_0 + \alpha_1 z^{-1} + \alpha_2 z^{-2} + \ldots + \alpha_{N-1} z^{-(N-1)} + \alpha_N z^{-N}}{1 + \beta_1 z^{-1} + \beta_2 z^{-2} + \ldots + \beta_{N-1} z^{-(N-1)} + \beta_N z^{-N}}$$

To calculate the complex frequency response, we again put

$$z^{-1} = e^{-j\omega T_s} = \cos \omega T_s - j \sin \omega T_s$$

It is also advisable to normalize all of the frequencies to the sampling rate $f_s = 1/T_s$. The normalized frequency variable f_n is therefore

$$f_n = \frac{f}{f_s} \quad \text{or} \quad \omega T_s = 2\pi f_n \tag{19.10}$$

In order not to violate the sampling theorem, the following must apply:

$$0 \le f \le \frac{1}{2} f_s \quad \text{or} \quad 0 \le f_n \le \frac{1}{2}$$

Thus, for the magnitude of the complex frequency response, it follows from (19.9) that

$$|\underline{A}(j\omega)| = \frac{\sqrt{\left[\sum\limits_{k=0}^{N} \alpha_k \cos 2\pi k f_n\right]^2 + \left[\sum\limits_{k=0}^{N} \alpha_k \sin 2\pi k f_n\right]^2}}{\sqrt{\left[\sum\limits_{k=0}^{N} \beta_k \cos 2\pi k f_n\right]^2 + \left[\sum\limits_{k=0}^{N} \beta_k \sin 2\pi k f_n\right]^2}} \tag{19.11}$$

The coefficient β_0 is always 1. The multiplier for β_0 can therefore be omitted in all of the filter structures.

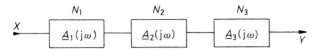

Fig. 19.10. Cascading of filter stages

$$A_{\text{tot}} = A_1 \cdot A_2 \cdot A_3$$
$$|A_{\text{tot}}| = |A_1| \cdot |A_2| \cdot |A_3|$$
$$N_{\text{tot}} = N_1 + N_2 + N_3$$

Equation (19.11) *is* extremely useful, as it allows us to calculate the frequency response of any digital filter if the filter coefficients are known. All of the frequency responses in this section have been calculated in this way.

An extension of the variants shown in Figs. 19.7–19.9 *is* the cascade structure, whereby several filter stages are connected in series (see Fig. 19.10). The resultant frequency response of the filter as a whole is therefore the product of the individual filter stage frequency responses. To design the filter stages, the transfer function to be implemented is factored. Thus, an Nth-order filter (highest power z^{-N}) is broken down into a number of filter stages which, when recombined, possess the order N. The order of the filter stages selected is basically irrelevant, but it must not be lower than $N_i = 2$ in the case of IIR filters, because it will otherwise not be possible to realize the normally occurring conjugate complex poles of the transfer function. We can see that the order of the filter as a whole is retained in the cascade structure. An obvious advantage of this is that the lower-order filter stages are generally easier to design and verify. These properties are utilized in the analog filters of Chap. 13 and in the recursive filters of Sect. 19.5.

19.3
Design Analysis of FIR Filters

The coefficients β_k in the digital filters (see Figs. 19.7–19.9) determine the amount of feedback. If they are all made zero, there is no feedback and the output signal is merely the weighted sum of the input signal and its delays. Filters of this kind are known as *nonrecursive filters, transversal filters*, or *finite impulse response filters*(FIR). The term "FIR" means that the impulse response is of finite length ($N+1$ values). The circuits shown in Figs. 19.7–19.9 then become the simplified circuits shown in Figs. 19.11 and 19.12.

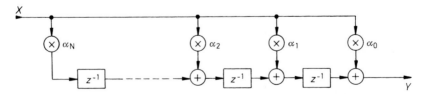

Fig. 19.11. FIR filter with distributed adders

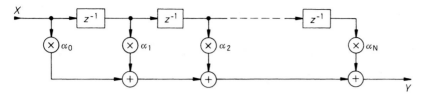

Fig. 19.12. FIR filter with a global adder at the output

19.3.1
Basic Equations

The elimination of the coefficients β_k also simplifies the transfer equations. The difference equation is

$$y_N = \alpha_0 x_N + \alpha_1 x_{N-1} + \ldots + \alpha_{N-1} x_1 + \alpha_N x_0, \qquad y_N = \sum_{k=0}^{N} \alpha_k x_{N-k} \qquad (19.12)$$

For the transfer function, we obtain

$$Y(z) = \left[\alpha_0 + \alpha_1 z^{-1} + \alpha_2 z^{-2} + \ldots + \alpha_{N-1} z^{-(N-1)} + \alpha_N z^{-N} \right] X(z)$$

$$\tilde{A}(z) = \frac{Y(z)}{X(z)} = \sum_{k=0}^{N} \alpha_k z^{-k}. \qquad (19.13)$$

Inserting Euler's relation,

$$z^{-1} = e^{-j2\pi f_n} = \cos 2\pi f_n - j \sin 2\pi f_n, \qquad (19.14)$$

the complex frequency response

$$\underline{A}(j\omega) = \sum_{k=0}^{N} \alpha_k e^{-j2\pi k f_n} \qquad (19.15)$$

is obtained. This relation can be simplified if the coefficients are symmetrical:

$$\alpha_{N-k} = \alpha_k \qquad \text{even symmetry} \qquad (19.16)$$
$$\alpha_{N-k} = -\alpha_k \qquad \text{odd symmetry} \qquad (19.17)$$

Two terms with coefficients that are equal in absolute value can then be combined and a common phase factor can be factored out. Equation (19.15) is then simplified:

for even symmetry,

$$\underline{A}(j\omega) = e^{-j\pi N f_n} \sum_{k=0}^{N} \alpha_k \cos \pi (N - 2k) f_n \qquad (19.18a)$$

for odd symmetry,

$$\underline{A}(j\omega) = e^{-j\pi N f_n} \sum_{k=0}^{N} \alpha_k \sin \pi (N - 2k) f_n. \qquad (19.18b)$$

For odd symmetry, the middle coefficient must disappear in even-order filters; in other words, $\alpha_{\frac{1}{2}N} = 0$. We thus obtain an expression in terms of the magnitude $B(\omega)$ and the phase $e^{j\varphi}$, of the form

$$A(j\omega) = \begin{cases} B(\omega)e^{-j\pi N f_n} & \text{for even symmetry} \\ B(\omega)je^{-j\pi N f_n} & \text{for odd symmetry} \end{cases}$$

In order to calculate the magnitude, we only need to take account of the sum in (19.18). The phase shift follows from the exponential function

$$\varphi = \begin{cases} -\pi N f_n & \text{for even symmetry} \\ -\pi N f_n + \pi/2 & \text{for odd symmetry} \end{cases} \tag{19.19}$$

In both cases, we can see the *linear phase* behavior that is exactly fulfilled for any symmetrical coefficients.

The group delay is obtained from the definition:

$$t_{gr} = -\frac{d\varphi}{d\omega} = -\frac{d\varphi}{df_n} \cdot \frac{df_n}{d\omega} = -\frac{T_s}{2\pi} \cdot \frac{d\varphi}{df_n} \tag{19.20}$$

Hence, by differentiating (19.19),

$$t_{gr} = \frac{1}{2} N T_s \tag{19.21}$$

Consequently, it is frequency-invariant. Delay distortion cannot therefore occur with symmetrical FIR filters. This is one of their chief advantages, and it is the reason why FIR filters are only designed with symmetrical coefficients. The design procedures and examples given in this chapter all produce FIR filters with a constant group delay. Choosing symmetrical coefficients results in another advantage: the calculation of the filter algorithm is simplified because each coefficient multiplier can be used for 2 coefficients.

19.3.2
Simple Examples

In order to become familiar with the behavior and design analysis of FIR filters, it is useful to examine a few simple examples.

The First-Order FIR Filter

The circuit shown in Fig. 19.13 is a first-order FIR filter ($N = 1$). As we can see, it is a lowpass filter. Its DC voltage gain is $|A(f_n = 0)| = 1$. This may also be perceived directly from the circuit: if a unit sequence $x_\mu = 1$ is applied to the input, then $y_\mu = \alpha_0 + \alpha_1 = 0.5 + 0.5 = 1$. This characteristic can be generalized:

> The DC voltage gain of an FIR filter is equal to the sum of all of the filter coefficients.

At the highest signal frequency permitted by the sampling theorem – that is,

$$f = \frac{1}{2} f_s \quad \text{bzw.} \quad f_n = \frac{1}{2}$$

a unit input sequence is produced in which the values $+1$ and -1 occur alternately: $\{X_\mu\} = \{+1, -1, +1, -1, \ldots\}$. Consequently, in Fig. 19.12 the output signal $Y = +0.5 - 0.5 = -0.5 + 0.5 = 0$; that is, it is constantly zero. This is also apparent from the amplitude–frequency response. This characteristic can also be generalized:

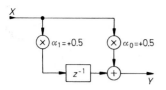
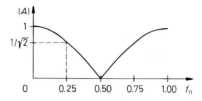

Fig. 19.13. First-order lowpass filter/interpolator

$$\tilde{A}(z) = 0.5(1 + z^{-1})$$
$$\underline{A}(j\omega) = 0.5(1 + \cos 2\pi f_n - j \sin 2\pi f_n)$$
$$|\underline{A}(j\omega)| = |\cos \pi f_n|$$

$$f_{c,n} = 0.25$$
$$\varphi = -\pi f_n$$
$$t_{gr} = 0.5 T_s$$

> The gain of an FIR filter at half the sampling frequency is equal to the sum of the coefficients alternately weighted with $+1$ and -1.

If all of the filter coefficients are multiplied by the same factor, the effect is the same as if the input signal were multiplied by that factor. From this, we can derive the general rule:

> If all of the coefficients of an FIR filter are multiplied by the same factor, only the basic gain of the filter will be modified, but its filter characteristic will remain unchanged.

To calculate the cutoff frequency in Fig. 19.13, we put

$$|\underline{A}(j\omega)| = \cos \pi f_{c,n} = 1/\sqrt{2}$$

and obtain $f_{c,n} = \frac{1}{4} = \frac{1}{4} f_s$.

If an input sequence $x(t_\mu)$, which is equal to 1 only on one clock and is otherwise 0, is applied to an FIR filter, initially the coefficient $y(t_\mu) = \alpha_0$ is obtained at the output, followed by $y(t_{\mu+1}) = \alpha_1$; in other words, the coefficients occur consecutively. To generalize:

> The unit-impulse response of an FIR filter is the sequence of its coefficients. It is $N + 1$ values long.

A first-order highpass filter is shown in Fig. 19.14. We can see from the coefficients $\alpha_0 = +0.5$ and $\alpha_1 = -0.5$ that their sum is zero. This also results in zero DC voltage gain. With an input sequence of $+1, -1, \ldots$ (the highest signal frequency), the same sequence will appear at the output. The gain is therefore unity. The cutoff frequency of the highpass filter, like that of the lowpass filter, is $f_c = \frac{1}{4} f_s$.

The two examples described also reveal the linear phase behavior and the resulting constant group delay. The lowpass filter can also be used for averaging, as may be seen from the coefficients. Similarly, the highpass filter can be used as a differentiator, since for low frequencies,

$$|\underline{A}(j\omega)| = \sin \pi f_n \approx \pi f_n$$

that is, the gain is proportional to the frequency.

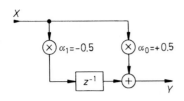

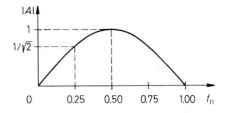

Fig. 19.14. First-order highpass filter/differentiator

$$
\begin{aligned}
\tilde{A}(z) &= 0.5(1 - z^{-1}) \\
\underline{A}(j\omega) &= 0.5(1 - \cos 2\pi f_n + j \sin 2\pi f_n) \\
|\underline{A}(j\omega)| &= |\sin \pi f_n|
\end{aligned}
\qquad
\begin{aligned}
f_{c,n} &= 0.25 \\
\varphi &= -\pi(0.5 - f_n) \\
t_{gr} &= 0.5T_s
\end{aligned}
$$

Second-Order FIR Filter

A second-order lowpass filter, or interpolator, is shown in Fig. 19.15. It may be seen that here the argument of the cosine is twice as large, and consequently, so is the phase shift and the group delay. The corresponding second-order highpass filter (differentiator) is shown in Fig. 19.16. The fact that the sum of the coefficients is zero immediately indicates that we are dealing here with a highpass filter. If we add the coefficients weighted with $+1$ or -1, we can show that the gain at $\frac{1}{2}f_s$ is unity. The Bode-plots for the first- and second-order filters are shown in Figs. 19.17 and 19.18.

For $\alpha_1 = 0$, a band-stop filter (Fig. 19.19) with a center frequency of $f_r = \frac{1}{4}f_s$ is obtained. We can see here that the values at the output cancel each other out when two consecutive input values are $+1$ and the next two are -1. Accordingly, band-stop filters with a low resonant frequency can be realized, likewise, by making (in a comparatively long delay chain) the first and last coefficients $\alpha_0 = \alpha_N = 0.5$ and setting all the others to zero.

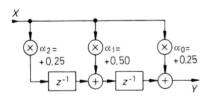

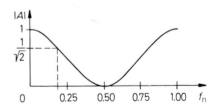

Fig. 19.15. Second-order lowpass filter/interpolator

$$
\begin{aligned}
\tilde{A}(z) &= 0.25 + 0.5z^{-1} + 0.25z^{-2} \\
|\underline{A}(j\omega)| &= 0.5 + 0.5 \cos 2\pi f_n \\
f_{c,n} &= \frac{1}{2\pi} \arccos\left(\sqrt{2} - 1\right) = 0.182
\end{aligned}
\qquad
\begin{aligned}
\varphi &= -2\pi f_n \\
t_{gr} &= T_s
\end{aligned}
$$

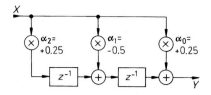

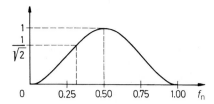

Fig. 19.16. Second-order highpass filter/differentiator

$$\tilde{A}(z) = 0.25 - 0.5z^{-1} + 0.25z^{-2}$$
$$|\underline{A}(j\omega)| = 0.5 - 0.5\cos 2\pi f_n$$
$$f_{c,n} = \frac{1}{2\pi}\arccos\left(1 - \sqrt{2}\right) = 0.318$$

$$\varphi = -2\pi f_n$$
$$t_{gr} = T_s$$

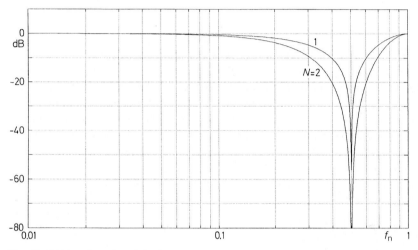

Fig. 19.17. Log–log plot of the frequency responses of the lowpass filters used as examples

$N = 1$ Fig. 19.13 : $\alpha_0 = \alpha_1 = +0.5$
$N = 2$ Fig. 19.15 : $\alpha_0 = \alpha_2 = +0.25$ $\qquad \alpha_1 = +0.5$

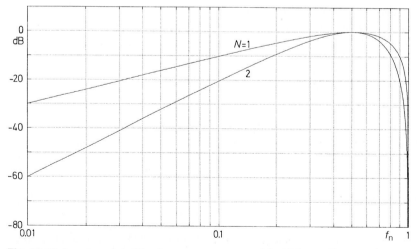

Fig. 19.18. Log–log plot of the frequency responses of the highpass filters used as examples

$N = 1$ Fig. 19.14 : $\alpha_0 = +0.5$ $\qquad\qquad \alpha_1 = -0.5$
$N = 2$ Fig. 19.16 : $\alpha_0 = \alpha_2 = +0.25$ $\qquad \alpha_1 = -0.5$

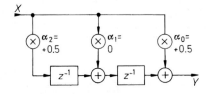

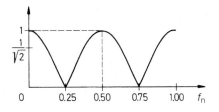

Fig. 19.19. Second-order band-stop filter

$$\tilde{A}(z) = 0.5 + 0.5z^{-2} \qquad\qquad \varphi = -2\pi f_n$$
$$|A(j\omega)| = |\cos 2\pi f_n| \qquad\qquad t_{gr} = T_s$$
$$f_{n,r} = 0.25 \qquad\qquad\qquad Q = f_{n,r}/B = 1$$

19.3.3
Calculating the Filter Coefficients

Two methods are commonly used to calculate the coefficients of FIR filters: the *window method* and the *Remez Exchange Algorithm*. The latter is a numeric method of Chebyshev approximation of a given gain tolerance scheme. It provides a minimal number of coefficients and therefore produces particularly efficient circuits. The advantage of the window method is that it allows a clear understanding of the mode of operation, while at the same time being less computationally intensive. We shall now describe how this method is used to calculate the filter coefficients.

FIR filters possess a particularly distinctive impulse response. If a unit impulse conforming to the sequence

$$\{x(kT_s)\} = \begin{cases} 1 & \text{for } k = 0 \\ 0 & \text{otherwise} \end{cases} \tag{19.22}$$

is applied to the input, we obtain, in accordance with Figs. 19.11 and 19.12 or (19.12), the unit-impulse response

$$\{y(kT_s)\} = \alpha_0, \alpha_1, \alpha_2, \ldots \alpha_N = \{\alpha_k\}; \tag{19.23}$$

that is, the sequence of the filter coefficients.

On the other hand, it can be shown that the impulse response of a system represents the inverse Fourier transform of its frequency response $A_w(j\omega)$, in accordance with

$$y(t) = \int_{-\infty}^{+\infty} A_w(j\omega)e^{j\omega t} d\omega . \tag{19.24}$$

With discrete-time systems, the frequency response is periodic with $f_s = 1/T_s$, and the time can be specified as multiples of the sampling period: $t = kT_s$. This simplifies (19.24) to

$$y(kT_s) = \int_{-1/2f_s}^{+1/2f_s} A_w(jf)e^{j2\pi f kT_s} df . \tag{19.25}$$

The required filter coefficients are obtained by equating (19.23) and (19.25) and specifying the desired frequency response $A_w(j\omega)$.

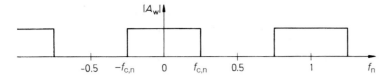

Fig. 19.20. Desired frequency response of an ideal lowpass filter and its periodic continuation

Of particular interest is the case, shown in Fig. 19.20, of the ideal lowpass filter with cutoff frequency $f_{c,n} = f_c/f_s$ and a gain of 1 in the passband and 0 in the stop band. If a constant group delay $t_{gr} = \frac{1}{2}NT_s$ is additionally required, $A_w(jf)$ can also be expressed as a delay function:

$$\underline{A}_w(jf) = \begin{cases} e^{-j\pi f NT_s} & \text{for } -f_c \le f \le f_c \\ 0 & \text{otherwise} \end{cases} \tag{19.26}$$

Inserting this ideal frequency response in (19.25), we obtain

$$\alpha_{kr} = \int_{-f_c}^{f_c} e^{-j\pi f NT_s} e^{j2\pi f k T_s} df = \int_{-f_{c,n}}^{f_{c,n}} e^{j\pi F(2k-N)} df_n$$

$$\boxed{\alpha_{kr} = 2f_{c,n} \frac{\sin(2k-N)\pi f_{c,n}}{(2k-N)\pi f_{c,n}} \quad \text{for} \quad k = 0, 1, 2 \ldots N} \tag{19.27}$$

These are the filter coefficients that are being sought, but we have only their raw values – hence the subscript "r." They must be modified such that the desired cutoff frequency or gain is achieved precisely. For this reason, we shall simplify by omitting factor $2f_{c,n}$, which is common to all coefficients of (19.27). This simplification is allowed since factor $2f_{c,n}$ will eventually be replaced during the necessary normalization of the gain. To the value (sin 0)/0, which occurs for even order – in other words, of odd coefficient numbers – we assign the following limit value:

$$\lim_{x \to 0} \frac{\sin x}{x} = 1.$$

Since, in practice, only finite orders N can be realized, the sequence α_{kr}, must be terminated. This can be interpreted – as in Fig. 19.21 – as multiplication by a square window. This, of course, means that we have an incomplete approximation to the desired frequency response. In Fig. 19.22 we can see a marked deviation from the desired ideal frequency response and a poor stop-band attenuation. This situation can be greatly improved by using, instead of the square window, a window that gradually reduces the coefficients toward the edge. Commonly used window functions are

    the Hamming window,   the Hanning window,
    the Blackman window,   and the Kaiser window.

We shall use the Hamming window, as it provides good results with minimal computing effort. The Hamming function

$$\boxed{W_k = 0.54 - 0.46\cos\frac{2\pi k}{N} \quad \text{for} \quad k = 0, 1, 2 \ldots N} \tag{19.28}$$

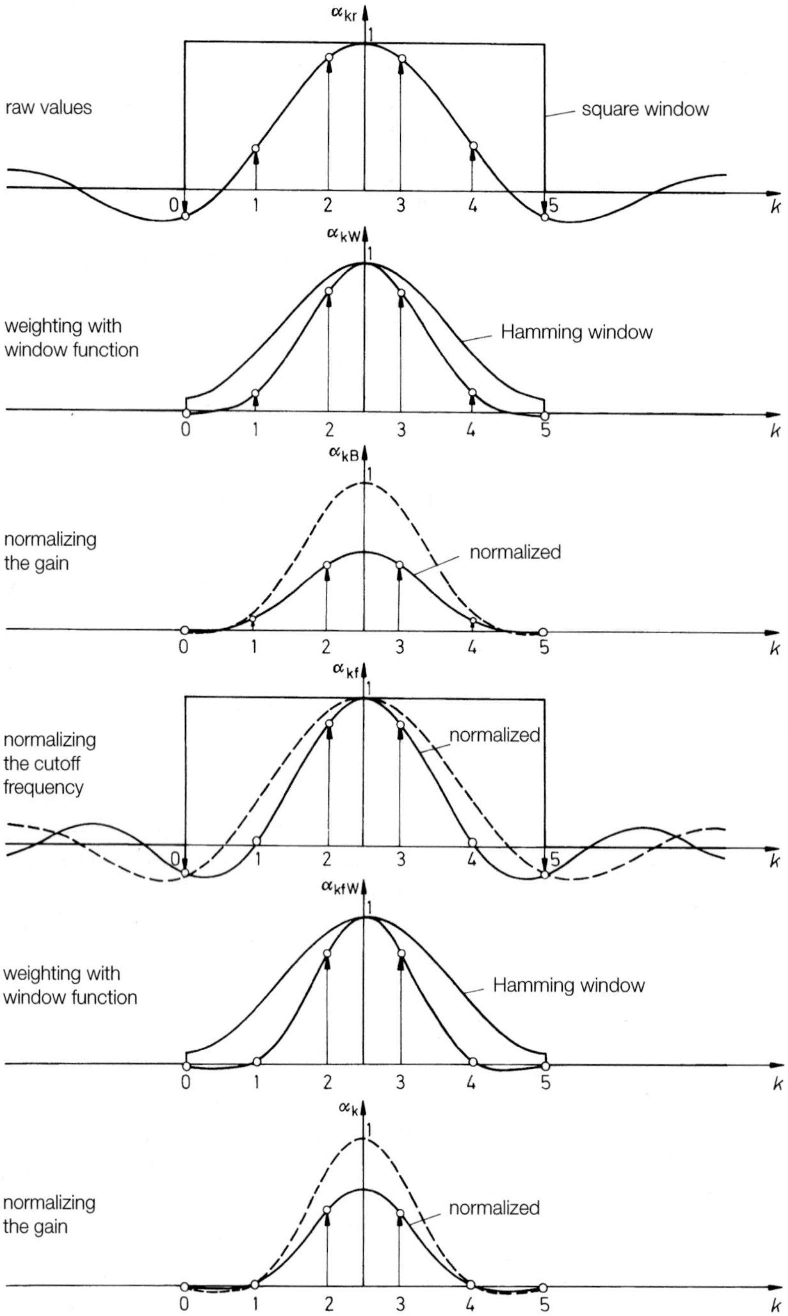

Fig. 19.21. Steps in calculating the filter coefficients based on the example of a fifth-order lowpass filter that has a cutoff frequency $F_{c,n} = 0.25$

Result: $\alpha_0 = \alpha_5 = -0.00979$, $\alpha_1 = \alpha_4 = +0.00979$, $\alpha_2 = \alpha_3 = 0.5000$

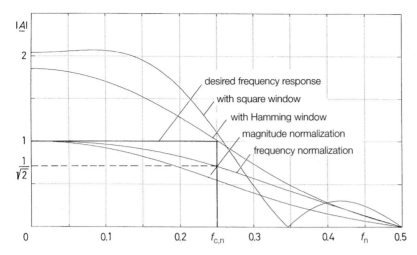

Fig. 19.22. The effect of the individual steps for coefficient calculation on the frequency response of a filter, based on the example of a fifth-order lowpass filter with $f_{c,n} = 0.25$

is plotted in Fig. 19.21. Its boundary values are

$$W(k = 0) \ = \ W(k = N) \ = \ 0.08$$

and in the center it has the value $W(k = \frac{1}{2}) = 1$. The filter coefficients evaluated using this window function are given below Fig. 19.21, and the resulting frequency response is shown in Fig. 19.22. We can see that the unwanted ripple is largely eliminated and the stop-band attenuation is increased.

The low-frequency gain must now be normalized to 1. This is done by dividing each coefficient by the sum of all the coefficients. This step is likewise shown in Figs. 19.21 and 19.22.

The resulting filter does not yet possess the desired cutoff frequency. As the value of $f_{c,n}$ used in (19.22) only yields an approximate solution, its value in this equation must therefore be corrected somewhat in order to achieve a gain of $1/\sqrt{2}$ at the desired cutoff frequency. In this case, it must be increased to $f'_{c,n} = 0.32$. However, to do this it will be necessary to repeat the entire design process using the modified value of F_c. This inevitably entails iteration that must be performed several times, resulting in a lowpass filter that is normalized in terms of gain and cutoff frequency, as illustrated in Figs. 19.21 and 19.22.

The lowpass filters shown in Fig. 19.23 for normalized cutoff frequencies of $f_{c,n} = 0.25, 0.1$, and 0.025 have been designed using this method. The filter coefficients have been tabulated in Fig. 19.24. The lower the cutoff frequency f_c compared to the sampling rate f_s – that is, the smaller $f_{c,n} = f_c/f_s$ – the higher is the lowest order N for which a solution exists. If $f_{c,n} = 0.025$, then $N = 27$ is the lowest order, whereas for $f_{c,n} = 0.01$ it would be $N = 65$.

Here, we have limited our considerations to odd-order lowpass filters. These have two advantages: on the one hand, their frequency response has a zero at $f_n = 0.5$, whereas the responses of the even-order lowpass filters exhibit a (relative) maximum at that point. This results in a considerably enhanced band-stop characteristic, particularly in the case of lower-order lowpass filters with $f_{c,n} = 0.25$. On the other hand, they have an even number of coefficients and therefore integrated FIR filters can generally be better utilized.

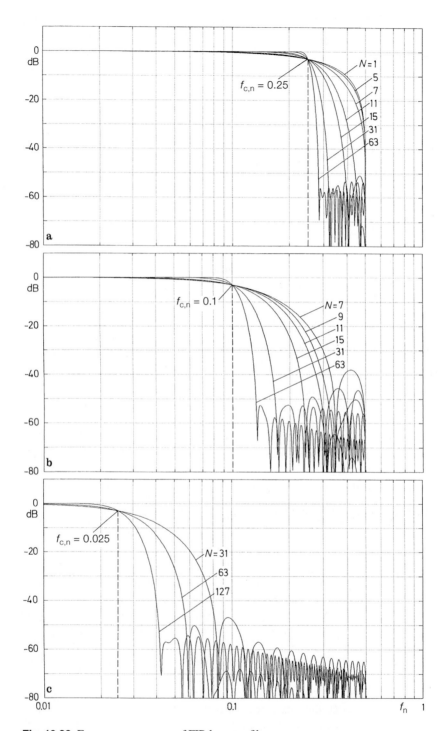

Fig. 19.23. Frequency responses of FIR lowpass filters

Cutoff frequency $f_{c,n} = 0.25$

$N = 1$

$\alpha_0 = \alpha_1 = +0.50000$

$N = 5$

$\alpha_0 = \alpha_5 = -0.00979$	$\alpha_1 = \alpha_4 = +0.00979$	$\alpha_2 = \alpha_3 = +0.50000$

$N = 7$

$\alpha_0 = \alpha_7 = +0.00343$	$\alpha_1 = \alpha_6 = -0.03171$	$\alpha_2 = \alpha_5 = +0.03171$
$\alpha_3 = \alpha_4 = +0.49657$		

$N = 11$

$\alpha_0 = \alpha_{11} = -0.00203$	$\alpha_1 = \alpha_{10} = +0.01056$	$\alpha_2 = \alpha_9 = +0.00010$
$\alpha_3 = \alpha_8 = -0.07531$	$\alpha_4 = \alpha_7 = +0.07734$	$\alpha_5 = \alpha_6 = +0.48934$

$N = 15$

$\alpha_0 = \alpha_{15} = +0.00152$	$\alpha_1 = \alpha_{14} = -0.00561$	$\alpha_2 = \alpha_{13} = -0.00175$
$\alpha_3 = \alpha_{12} = +0.02812$	$\alpha_4 = \alpha_{11} = -0.01076$	$\alpha_5 = \alpha_{10} = -0.09143$
$\alpha_6 = \alpha_9 = +0.09879$	$\alpha_7 = \alpha_8 = +0.48112$	

$N = 31$

$\alpha_0 = \alpha_{31} = +0.00074$	$\alpha_1 = \alpha_{30} = -0.00182$	$\alpha_2 = \alpha_{29} = -0.00083$
$\alpha_3 = \alpha_{28} = +0.00404$	$\alpha_4 = \alpha_{27} = +0.00088$	$\alpha_5 = \alpha_{26} = -0.00898$
$\alpha_6 = \alpha_{25} = +0.00024$	$\alpha_7 = \alpha_{24} = +0.01753$	$\alpha_8 = \alpha_{23} = -0.00429$
$\alpha_9 = \alpha_{22} = -0.03102$	$\alpha_{10} = \alpha_{21} = +0.01441$	$\alpha_{11} = \alpha_{20} = +0.05303$
$\alpha_{12} = \alpha_{19} = -0.03900$	$\alpha_{13} = \alpha_{18} = -0.10015$	$\alpha_{14} = \alpha_{17} = +0.12815$
$\alpha_{15} = \alpha_{16} = +0.46707$		

$N = 63$

$\alpha_0 = \alpha_{63} = +0.00036$	$\alpha_1 = \alpha_{62} = -0.00078$	$\alpha_2 = \alpha_{61} = -0.00036$
$\alpha_3 = \alpha_{60} = +0.00105$	$\alpha_4 = \alpha_{59} = +0.00041$	$\alpha_5 = \alpha_{58} = -0.00158$
$\alpha_6 = \alpha_{57} = -0.00045$	$\alpha_7 = \alpha_{56} = +0.00239$	$\alpha_8 = \alpha_{55} = +0.00044$
$\alpha_9 = \alpha_{54} = -0.00356$	$\alpha_{10} = \alpha_{53} = -0.00030$	$\alpha_{11} = \alpha_{52} = +0.00513$
$\alpha_{12} = \alpha_{51} = -0.00006$	$\alpha_{13} = \alpha_{50} = -0.00714$	$\alpha_{14} = \alpha_{49} = -0.00075$
$\alpha_{15} = \alpha_{48} = +0.00968$	$\alpha_{16} = \alpha_{47} = -0.00190$	$\alpha_{17} = \alpha_{46} = -0.01283$
$\alpha_{18} = \alpha_{45} = +0.00372$	$\alpha_{19} = \alpha_{44} = +0.01677$	$\alpha_{20} = \alpha_{43} = -0.00650$
$\alpha_{21} = \alpha_{42} = -0.02179$	$\alpha_{22} = \alpha_{41} = +0.01074$	$\alpha_{23} = \alpha_{40} = +0.02852$
$\alpha_{24} = \alpha_{39} = -0.01743$	$\alpha_{25} = \alpha_{38} = -0.03839$	$\alpha_{26} = \alpha_{37} = +0.02898$
$\alpha_{27} = \alpha_{36} = +0.05549$	$\alpha_{28} = \alpha_{35} = -0.05326$	$\alpha_{29} = \alpha_{34} = -0.09716$
$\alpha_{30} = \alpha_{33} = +0.14016$	$\alpha_{31} = \alpha_{32} = +0.45891$	

Fig. 19.24a. Coefficients for FIR filters with a cutoff frequency of $f_{c,n} = 0.25$; that is, $f_c = 0.25 f_s$. Order $N = 3$ does not exist here, as the two coefficients α_0 and α_3 vanish

In order to arrive at the simplest solution for realizing lowpass filters, we can ask the question: How must the cutoff frequency $f_{c,n}$ be chosen to insure that, in (19.27), as many filter coefficients as possible become zero? Two special cases of this kind are shown in Fig. 19.25. If we make $f_{c,n} = \frac{1}{2}$, all of the coefficients apart from the middle one with the value $\alpha(\frac{1}{2}N) = 1$ vanish in even-order filters (odd number of coefficients). The resulting filter is an allpass filter, and therefore cannot be used as a lowpass filter.

If the cutoff frequency is halved, *half-band filters* with $f_{c,n} = \frac{1}{4}$ are produced. If this condition is inserted into (19.27), we obtain:

$$\alpha_{kr} = \frac{\sin(2k - N)\pi/4}{(2k - N)\pi/4} \tag{19.29}$$

As we can see from Fig. 19.25, an appreciable simplification is again obtained for even order (an odd number of coefficients), as every second coefficient becomes zero. In order

Cutoff frequency $f_{c,n} = 0.1$

$N = 7$

$\alpha_0 = \alpha_7 = +0.00976$	$\alpha_1 = \alpha_6 = -0.04966$	$\alpha_2 = \alpha_5 = +0.16442$
$\alpha_3 = \alpha_4 = +0.27616$		

$N = 11$

$\alpha_0 = \alpha_{11} = -0.00470$	$\alpha_1 = \alpha_{10} = -0.00605$	$\alpha_2 = \alpha_9 = +0.00818$
$\alpha_3 = \alpha_8 = +0.07006$	$\alpha_4 = \alpha_7 = +0.17404$	$\alpha_5 = \alpha_6 = +0.25848$

$N = 15$

$\alpha_0 = \alpha_{15} = -0.00101$	$\alpha_1 = \alpha_{14} = -0.00521$	$\alpha_2 = \alpha_{13} = -0.01269$
$\alpha_3 = \alpha_{12} = -0.01214$	$\alpha_4 = \alpha_{11} = +0.01830$	$\alpha_5 = \alpha_{10} = +0.08914$
$\alpha_6 = \alpha_9 = +0.17962$	$\alpha_7 = \alpha_8 = +0.24399$	

$N = 31$

$\alpha_0 = \alpha_{31} = -0.00165$	$\alpha_1 = \alpha_{30} = -0.00146$	$\alpha_2 = \alpha_{29} = -0.00037$
$\alpha_3 = \alpha_{28} = +0.00225$	$\alpha_4 = \alpha_{27} = +0.00593$	$\alpha_5 = \alpha_{26} = +0.00823$
$\alpha_6 = \alpha_{25} = +0.00548$	$\alpha_7 = \alpha_{24} = -0.00461$	$\alpha_8 = \alpha_{23} = -0.01979$
$\alpha_9 = \alpha_{22} = -0.03195$	$\alpha_{10} = \alpha_{21} = -0.02944$	$\alpha_{11} = \alpha_{20} = -0.00261$
$\alpha_{12} = \alpha_{19} = +0.04987$	$\alpha_{13} = \alpha_{18} = +0.11780$	$\alpha_{14} = \alpha_{17} = +0.18175$
$\alpha_{15} = \alpha_{16} = +0.22058$		

$N = 63$

$\alpha_0 = \alpha_{63} = +0.00065$	$\alpha_1 = \alpha_{62} = +0.00086$	$\alpha_2 = \alpha_{61} = +0.00073$
$\alpha_3 = \alpha_{60} = +0.00022$	$\alpha_4 = \alpha_{59} = -0.00061$	$\alpha_5 = \alpha_{58} = -0.00148$
$\alpha_6 = \alpha_{57} = -0.00194$	$\alpha_7 = \alpha_{56} = -0.00150$	$\alpha_8 = \alpha_{55} = +0.00001$
$\alpha_9 = \alpha_{54} = +0.00223$	$\alpha_{10} = \alpha_{53} = +0.00418$	$\alpha_{11} = \alpha_{52} = +0.00464$
$\alpha_{12} = \alpha_{51} = +0.00272$	$\alpha_{13} = \alpha_{50} = -0.00144$	$\alpha_{14} = \alpha_{49} = -0.00639$
$\alpha_{15} = \alpha_{48} = -0.00973$	$\alpha_{16} = \alpha_{47} = -0.00909$	$\alpha_{17} = \alpha_{46} = -0.00343$
$\alpha_{18} = \alpha_{45} = +0.00593$	$\alpha_{19} = \alpha_{44} = +0.01532$	$\alpha_{20} = \alpha_{43} = +0.01986$
$\alpha_{21} = \alpha_{42} = +0.01560$	$\alpha_{22} = \alpha_{41} = +0.00177$	$\alpha_{23} = \alpha_{40} = -0.01790$
$\alpha_{24} = \alpha_{39} = -0.03551$	$\alpha_{25} = \alpha_{38} = -0.04135$	$\alpha_{26} = \alpha_{37} = +0.02742$
$\alpha_{27} = \alpha_{36} = +0.00903$	$\alpha_{28} = \alpha_{35} = +0.06348$	$\alpha_{29} = \alpha_{34} = +0.12467$
$\alpha_{30} = \alpha_{33} = +0.17761$	$\alpha_{31} = \alpha_{32} = +0.20829$	

Fig. 19.24b. Coefficients for FIR filters with a cutoff frequency of $f_{c,n} = 0.1$; that is, $f_c = 0.1 f_s$. No solution exists here for order $N < 7$

to arrive at a practicable filter design, the coefficients have still to be evaluated with a window. Using the Hamming window in (19.28), we obtain, with $\alpha_{kw} = \alpha_{kr} \cdot W_k$,

$$\alpha_{kw} = \frac{\sin(2k - N)\pi/4}{(2k - N)\pi/4} \left(0.54 - 0.46 \cos \frac{2\pi k}{N} \right) \quad \text{for } k = 0, 1, 2 \ldots N \qquad (19.30)$$

If filter coefficients for all k have been calculated, it is now merely necessary to normalize them by dividing them by their sum in order to obtain the final coefficients. An iterative process is no longer necessary. Consequently, these filter coefficients may be calculated on a pocket calculator. The resulting cutoff frequencies are of course not precisely $f_{c,n} = \frac{1}{4}$, as they have not been normalized; but a normalization is ruled out, as the advantage of every second coefficient vanishing would be lost. The frequency responses of a number of half-band filters are shown in Fig. 19.26 and a coefficient table is given in Fig. 19.27. It may be seen from Fig. 19.26 that the $-6\,$dB cutoff frequencies approach $f_{c,n} = 0.25$ more accurately the higher the order; that is, they approach the "half band." The $-3\,$dB cutoff frequencies normally specified are therefore lower; their precise values are additionally given in Fig. 19.27. The imprecise values for $f_{c,n}$ nevertheless allow any cutoff frequencies to be realized by selecting the sampling frequency accordingly: $f_s = f_c/f_{c,n}$

Cutoff frequency $f_{c,n} = 0.025$

$N = 31$

$\alpha_0 = \alpha_{31} = +0.00077$	$\alpha_1 = \alpha_{30} = -0.00132$	$\alpha_2 = \alpha_{29} = +0.00236$
$\alpha_3 = \alpha_{28} = +0.00417$	$\alpha_4 = \alpha_{27} = +0.00698$	$\alpha_5 = \alpha_{26} = +0.01095$
$\alpha_6 = \alpha_{25} = +0.01613$	$\alpha_7 = \alpha_{24} = +0.02244$	$\alpha_8 = \alpha_{23} = +0.02968$
$\alpha_9 = \alpha_{22} = +0.03754$	$\alpha_{10} = \alpha_{21} = +0.04559$	$\alpha_{11} = \alpha_{20} = +0.05335$
$\alpha_{12} = \alpha_{19} = +0.06033$	$\alpha_{13} = \alpha_{18} = +0.06606$	$\alpha_{14} = \alpha_{17} = +0.07012$
$\alpha_{15} = \alpha_{16} = +0.07222$		

$N = 63$

$\alpha_0 = \alpha_{63} = -0.00005$	$\alpha_1 = \alpha_{62} = -0.00022$	$\alpha_2 = \alpha_{61} = -0.00042$
$\alpha_3 = \alpha_{60} = -0.00068$	$\alpha_4 = \alpha_{59} = -0.00101$	$\alpha_5 = \alpha_{58} = -0.00141$
$\alpha_6 = \alpha_{57} = -0.00188$	$\alpha_7 = \alpha_{56} = -0.00241$	$\alpha_8 = \alpha_{55} = -0.00295$
$\alpha_9 = \alpha_{54} = -0.00344$	$\alpha_{10} = \alpha_{53} = -0.00383$	$\alpha_{11} = \alpha_{52} = -0.00403$
$\alpha_{12} = \alpha_{51} = -0.00395$	$\alpha_{13} = \alpha_{50} = -0.00350$	$\alpha_{14} = \alpha_{49} = -0.00259$
$\alpha_{15} = \alpha_{48} = -0.00115$	$\alpha_{16} = \alpha_{47} = +0.00089$	$\alpha_{17} = \alpha_{46} = +0.00356$
$\alpha_{18} = \alpha_{45} = +0.00689$	$\alpha_{19} = \alpha_{44} = +0.01084$	$\alpha_{20} = \alpha_{43} = +0.01536$
$\alpha_{21} = \alpha_{42} = +0.02036$	$\alpha_{22} = \alpha_{41} = +0.02573$	$\alpha_{23} = \alpha_{40} = +0.03131$
$\alpha_{24} = \alpha_{39} = +0.03694$	$\alpha_{25} = \alpha_{38} = +0.04243$	$\alpha_{26} = \alpha_{37} = +0.04759$
$\alpha_{27} = \alpha_{36} = +0.05227$	$\alpha_{28} = \alpha_{35} = +0.05618$	$\alpha_{29} = \alpha_{34} = +0.05928$
$\alpha_{30} = \alpha_{33} = +0.06143$	$\alpha_{31} = \alpha_{32} = +0.06252$	

Fig. 19.24c. Coefficients for FIR filters with a cutoff frequency of $f_{c,n} = 0.025$; that is, $f_c = 0.025\,f_s$. No solution exists for orders $N < 27$

Half-band Filters

We can see in Fig. 19.25 that half of the values become zero only for an odd number of coefficients. Consequently, only half-band filters with an even order are used. It is also apparent that the boundary coefficients (α_0 and α_8 in the example) vanish for all orders that are divisible by four. They are therefore particularly useful, as we can obtain two additional filter orders without an additional multiplication. Even the two delay elements

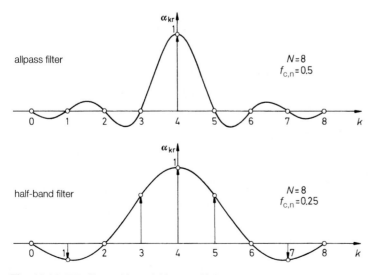

Fig. 19.25. FIR filter with vanishing coefficients

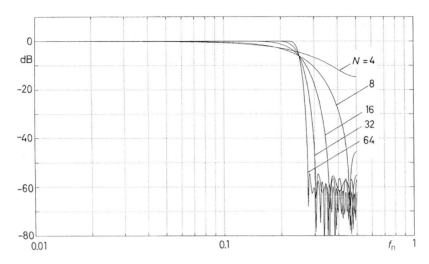

Fig. 19.26. Frequency responses of half-band filters

$N = 4$ 3 coefficients $f_{c,n} = 0.205$
$\alpha_0 = \alpha_4 = 0$ $\alpha_1 = \alpha_3 = +0.20371$ $\alpha_2 = +0.59258$

$N = 8$ 5 coefficients $f_{c,n} = 0.199$
$\alpha_0 = \alpha_8 = 0$ $\alpha_1 = \alpha_7 = -0.02266$ $\alpha_2 = \alpha_6 = 0$ $\alpha_3 = \alpha_5 = +0.27398$
$\alpha_4 = +0.49737$

$N = 16$ 9 coefficients $f_{c,n} = 0.225$
$\alpha_0 = \alpha_{16} = 0$ $\alpha_1 = \alpha_{15} = -0.00524$ $\alpha_2 = \alpha_{14} = 0$ $\alpha_3 = \alpha_{13} = +0.02321$
$\alpha_4 = \alpha_{12} = 0$ $\alpha_5 = \alpha_{11} = -0.07611$ $\alpha_6 = \alpha_{10} = 0$ $\alpha_7 = \alpha_9 = +0.30770$
$\alpha_8 = +0.50087$

$N = 32$ 17 coefficients $f_{c,n} = 0.238$
$\alpha_0 = \alpha_{32} = 0$ $\alpha_1 = \alpha_{31} = -0.00189$ $\alpha_2 = \alpha_{30} = 0$ $\alpha_3 = \alpha_{29} = +0.00386$
$\alpha_4 = \alpha_{28} = 0$ $\alpha_5 = \alpha_{27} = -0.00824$ $\alpha_6 = \alpha_{26} = 0$ $\alpha_7 = \alpha_{25} = +0.01595$
$\alpha_8 = \alpha_{24} = 0$ $\alpha_9 = \alpha_{23} = -0.02868$ $\alpha_{10} = \alpha_{22} = 0$ $\alpha_{11} = \alpha_{21} = +0.05072$
$\alpha_{12} = \alpha_{20} = 0$ $\alpha_{13} = \alpha_{19} = -0.09802$ $\alpha_{14} = \alpha_{18} = 0$ $\alpha_{15} = \alpha_{17} = +0.31594$
$\alpha_{16} = +0.50071$

$N = 64$ 33 coefficients $f_{c,n} = 0.244$
$\alpha_0 = \alpha_{64} = 0$ $\alpha_1 = \alpha_{63} = -0.00084$ $\alpha_2 = \alpha_{62} = 0$ $\alpha_3 = \alpha_{61} = +0.00110$
$\alpha_4 = \alpha_{60} = 0$ $\alpha_5 = \alpha_{59} = -0.00158$ $\alpha_6 = \alpha_{58} = 0$ $\alpha_7 = \alpha_{57} = +0.00235$
$\alpha_8 = \alpha_{56} = 0$ $\alpha_9 = \alpha_{55} = -0.00344$ $\alpha_{10} = \alpha_{54} = 0$ $\alpha_{11} = \alpha_{53} = +0.00490$
$\alpha_{12} = \alpha_{52} = 0$ $\alpha_{13} = \alpha_{51} = -0.00681$ $\alpha_{14} = \alpha_{50} = 0$ $\alpha_{15} = \alpha_{49} = +0.00927$
$\alpha_{16} = \alpha_{48} = 0$ $\alpha_{17} = \alpha_{47} = -0.01243$ $\alpha_{18} = \alpha_{46} = 0$ $\alpha_{19} = \alpha_{45} = +0.01650$
$\alpha_{20} = \alpha_{44} = 0$ $\alpha_{21} = \alpha_{43} = -0.02192$ $\alpha_{22} = \alpha_{42} = 0$ $\alpha_{23} = \alpha_{41} = +0.02944$
$\alpha_{24} = \alpha_{40} = 0$ $\alpha_{25} = \alpha_{39} = -0.04076$ $\alpha_{26} = \alpha_{38} = 0$ $\alpha_{27} = \alpha_{37} = +0.06025$
$\alpha_{28} = \alpha_{36} = 0$ $\alpha_{29} = \alpha_{35} = -0.10408$ $\alpha_{30} = \alpha_{34} = 0$ $\alpha_{31} = \alpha_{33} = +0.31785$
$\alpha_{32} = +0.50039$

Fig. 19.27. Coefficients for the half-band filters

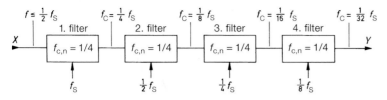

Fig. 19.28. Use of half-band filters in a cascade structure with undersampling.
Number of MAC (multiply and accumulate) operations to calculate on the output value:

$$(N+1) + \frac{1}{2}(N+1) + \frac{1}{4}(N+1) + \ldots = (N+1)\left(1 + \frac{1}{2} + \frac{1}{4} + \ldots\right) = 2(N+1)$$

associated with the two vanishing boundary coefficients $\alpha_0 = \alpha_N = 0$ can be dispensed with.

Half-band filters can be used to advantage in cascade arrangements, as shown in Fig. 19.28. Identical filter blocks are used which already possess a high stop-band attenuation at half the sampling frequency. It is therefore possible, without significantly violating the sampling theorem, to operate the second filter block at half the sampling rate. This reduces the computation by half. In the third and fourth filter blocks, the sampling frequency is again halved in each case. The cutoff frequency of the entire filter is therefore halved with each additional filter block; this is also shown in Fig. 19.28. In this way, cut-off frequencies can be realized which are well below the sampling frequency and whose implementation would otherwise involve considerable complexity.

Highpass Filters

Calculation of the filter coefficients of highpass filters can be related to the design of lowpass filters. To do this, we use the addition theorem of Fourier transformation, which states that an addition in the frequency domain corresponds to an addition in the time domain. Figure 19.29 shows how this statement can be used to design highpass filters. We can see that a highpass filter is produced in the frequency domain by subtracting a lowpass filter from an allpass filter. The associated filter coefficients are therefore obtained by subtracting the coefficients of the lowpass filter from those of the allpass filter, as shown on the right-hand side of the figure. Of course, the coefficients must once again be weighted using a window, and the magnitude of the gain normalized to 1 at $f_n = 0.5$ and to $1\sqrt{2}$ at $f_n = f_{n,c}$.

However, it is apparent that odd-order highpass filters designed using this method have a zero at $f_n = 0.5$, which makes their performance unsatisfactory. Therefore, only even-order filters – that is, those with an odd number of coefficients – have been taken into account in the coefficients tables in Fig. 19.30 and the frequency responses in Fig. 19.31.

Bandpass Filters and Band–Stop Filters

A bandpass filter can be implemented by subtracting the frequency responses of two low-pass filters from one another, as shown in Fig. 19.32. In order to obtain a band-stop filter, the frequency response of a bandpass filter can be subtracted from that of an allpass filter. The coefficients of the required filter can therefore be obtained in each case by subtraction of the relevant sets of coefficients.

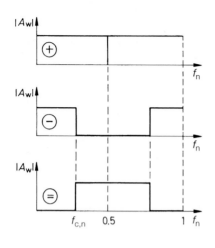

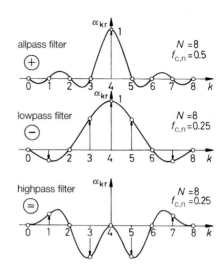

Fig. 19.29. Design of highpass filters

Cutoff frequency $f_{c,n} = 0.25$

$N = 1$
$\alpha_0 = -\alpha_1 = +0.5000$

$N = 6$
$\alpha_0 = \alpha_6 = -0.00009$	$\alpha_1 = \alpha_5 = -0.05091$	$\alpha_2 = \alpha_4 = -0.25163$
$\alpha_3 = +0.60528$		

$N = 10$
$\alpha_0 = \alpha_{10} = -0.00162$	$\alpha_1 = \alpha_9 = +0.01114$	$\alpha_2 = \alpha_8 = +0.03079$
$\alpha_3 = \alpha_7 = -0.05152$	$\alpha_4 = \alpha_6 = -0.27968$	$\alpha_5 = +0.58179$

$N = 14$
$\alpha_0 = \alpha_{14} = +0.00113$	$\alpha_1 = \alpha_{13} = -0.00587$	$\alpha_2 = \alpha_{12} = -0.01005$
$\alpha_3 = \alpha_{11} = +0.02291$	$\alpha_4 = \alpha_{10} = +0.05852$	$\alpha_5 = \alpha_9 = -0.04623$
$\alpha_6 = \alpha_8 = -0.29895$	$\alpha_7 = +0.55709$	

$N = 30$
$\alpha_0 = \alpha_{30} = +0.00053$	$\alpha_1 = \alpha_{29} = -0.00188$	$\alpha_2 = \alpha_{28} = -0.00136$
$\alpha_3 = \alpha_{27} = +0.00375$	$\alpha_4 = \alpha_{26} = +0.00407$	$\alpha_5 = \alpha_{25} = -0.00732$
$\alpha_6 = \alpha_{24} = -0.01026$	$\alpha_7 = \alpha_{23} = +0.01213$	$\alpha_8 = \alpha_{22} = -0.02267$
$\alpha_9 = \alpha_{21} = -0.01739$	$\alpha_{10} = \alpha_{20} = +0.04475$	$\alpha_{11} = \alpha_{19} = +0.02213$
$\alpha_{12} = \alpha_{18} = +0.09366$	$\alpha_{13} = \alpha_{17} = -0.02541$	$\alpha_{14} = \alpha_{16} = -0.31369$
$\alpha_{15} = +0.52709$		

$N = 62$
$\alpha_0 = \alpha_{62} = +0.00025$	$\alpha_1 = \alpha_{61} = -0.00082$	$\alpha_2 = \alpha_{60} = -0.00038$
$\alpha_3 = \alpha_{59} = +0.00104$	$\alpha_4 = \alpha_{58} = +0.00064$	$\alpha_5 = \alpha_{57} = -0.00146$
$\alpha_6 = \alpha_{56} = -0.00110$	$\alpha_7 = \alpha_{55} = +0.00209$	$\alpha_8 = \alpha_{54} = -0.00184$
$\alpha_9 = \alpha_{53} = -0.00291$	$\alpha_{10} = \alpha_{52} = -0.00297$	$\alpha_{11} = \alpha_{51} = +0.00389$
$\alpha_{12} = \alpha_{50} = +0.00457$	$\alpha_{13} = \alpha_{49} = -0.00500$	$\alpha_{14} = \alpha_{48} = -0.00680$
$\alpha_{15} = \alpha_{47} = +0.00620$	$\alpha_{16} = \alpha_{46} = -0.00981$	$\alpha_{17} = \alpha_{45} = -0.00744$
$\alpha_{18} = \alpha_{44} = -0.01387$	$\alpha_{19} = \alpha_{43} = +0.00866$	$\alpha_{20} = \alpha_{42} = +0.01938$
$\alpha_{21} = \alpha_{41} = -0.00982$	$\alpha_{22} = \alpha_{40} = -0.02713$	$\alpha_{23} = \alpha_{39} = +0.01085$
$\alpha_{24} = \alpha_{38} = +0.03879$	$\alpha_{25} = \alpha_{37} = -0.01170$	$\alpha_{26} = \alpha_{36} = -0.05873$
$\alpha_{27} = \alpha_{35} = +0.01235$	$\alpha_{28} = \alpha_{34} = +0.10304$	$\alpha_{29} = \alpha_{33} = -0.01275$
$\alpha_{30} = \alpha_{32} = -0.31713$	$\alpha_{31} = +0.51315$	

Fig. 19.30a. Filter coefficients of FIR highpass filters with a cutoff frequency of $f_{c,n} = 0.25$; that is, $f_c = 0.25 f_s$

Cutoff frequency $f_{c,n} = 0.1$

$N = 12$

$\alpha_0 = \alpha_{12} = -0.01015$	$\alpha_1 = \alpha_{11} = -0.01925$	$\alpha_2 = \alpha_{10} = -0.04453$
$\alpha_3 = \alpha_9 = -0.08090$	$\alpha_4 = \alpha_8 = -0.11882$	$\alpha_5 = \alpha_7 = -0.14737$
$\alpha_6 = +0.84203$		

$N = 30$

$\alpha_0 = \alpha_{30} = -0.00160$	$\alpha_1 = \alpha_{29} = -0.00200$	$\alpha_2 = \alpha_{28} = -0.00212$
$\alpha_3 = \alpha_{27} = -0.00117$	$\alpha_4 = \alpha_{26} = +0.00185$	$\alpha_5 = \alpha_{25} = +0.00723$
$\alpha_6 = \alpha_{24} = +0.01375$	$\alpha_7 = \alpha_{23} = +0.01836$	$\alpha_8 = \alpha_{22} = +0.01674$
$\alpha_9 = \alpha_{21} = +0.00479$	$\alpha_{10} = \alpha_{20} = -0.01960$	$\alpha_{11} = \alpha_{19} = -0.05505$
$\alpha_{12} = \alpha_{18} = -0.09628$	$\alpha_{13} = \alpha_{17} = -0.13521$	$\alpha_{14} = \alpha_{16} = -0.16308$
$\alpha_{15} = +0.82679$		

$N = 62$

$\alpha_0 = \alpha_{62} = +0.00048$	$\alpha_1 = \alpha_{61} = +0.00082$	$\alpha_2 = \alpha_{60} = +0.00096$
$\alpha_3 = \alpha_{59} = +0.00079$	$\alpha_4 = \alpha_{58} = +0.00023$	$\alpha_5 = \alpha_{57} = -0.00070$
$\alpha_6 = \alpha_{56} = -0.00176$	$\alpha_7 = \alpha_{55} = -0.00254$	$\alpha_8 = \alpha_{54} = -0.00252$
$\alpha_9 = \alpha_{53} = -0.00134$	$\alpha_{10} = \alpha_{52} = +0.00099$	$\alpha_{11} = \alpha_{51} = +0.00390$
$\alpha_{12} = \alpha_{50} = +0.00629$	$\alpha_{13} = \alpha_{49} = -0.00689$	$\alpha_{14} = \alpha_{48} = +0.00475$
$\alpha_{15} = \alpha_{47} = -0.00020$	$\alpha_{16} = \alpha_{46} = -0.00683$	$\alpha_{17} = \alpha_{45} = -0.01292$
$\alpha_{18} = \alpha_{44} = -0.01572$	$\alpha_{19} = \alpha_{43} = -0.01296$	$\alpha_{20} = \alpha_{42} = +0.00392$
$\alpha_{21} = \alpha_{41} = +0.00984$	$\alpha_{22} = \alpha_{40} = +0.02439$	$\alpha_{23} = \alpha_{39} = +0.03417$
$\alpha_{24} = \alpha_{38} = +0.03350$	$\alpha_{25} = \alpha_{37} = +0.01835$	$\alpha_{26} = \alpha_{36} = -0.01208$
$\alpha_{27} = \alpha_{35} = -0.05455$	$\alpha_{28} = \alpha_{34} = -0.10217$	$\alpha_{29} = \alpha_{33} = -0.14584$
$\alpha_{30} = \alpha_{32} = -0.17650$	$\alpha_{31} = +0.81246$	

Fig. 19.30b. Filter coefficients of FIR highpass filters with a cutoff frequency of $f_{c,n} = 0.1$; that is, $f_c = 0.1 f_s$

Cutoff frequency $f_{c,n} = 0.025$

$N = 48$

$\alpha_0 = \alpha_{48} = -0.00271$	$\alpha_1 = \alpha_{47} = -0.00288$	$\alpha_2 = \alpha_{46} = -0.00332$
$\alpha_3 = \alpha_{45} = -0.00404$	$\alpha_4 = \alpha_{44} = -0.00503$	$\alpha_5 = \alpha_{43} = -0.00628$
$\alpha_6 = \alpha_{42} = -0.00778$	$\alpha_7 = \alpha_{41} = -0.00951$	$\alpha_8 = \alpha_{40} = -0.01144$
$\alpha_9 = \alpha_{39} = -0.01353$	$\alpha_{10} = \alpha_{38} = -0.01557$	$\alpha_{11} = \alpha_{37} = -0.01811$
$\alpha_{12} = \alpha_{36} = -0.02050$	$\alpha_{13} = \alpha_{35} = -0.02291$	$\alpha_{14} = \alpha_{34} = -0.02530$
$\alpha_{15} = \alpha_{33} = -0.02762$	$\alpha_{16} = \alpha_{32} = -0.02983$	$\alpha_{17} = \alpha_{31} = -0.03189$
$\alpha_{18} = \alpha_{30} = -0.03376$	$\alpha_{19} = \alpha_{29} = -0.03541$	$\alpha_{20} = \alpha_{28} = -0.03680$
$\alpha_{21} = \alpha_{27} = -0.03791$	$\alpha_{22} = \alpha_{26} = -0.03872$	$\alpha_{23} = \alpha_{25} = -0.03921$
$\alpha_{24} = +0.96062$		

Fig. 19.30c. Filter coefficients of FIR highpass filters with a cutoff frequency of $f_{c,n} = 0.025$; that is, $f_c = 0.025 f_s$

Another method of obtaining the filter coefficients of bandpass and band-stop filters consists of multiplying together the transfer functions $\tilde{A}(z)$ of a corresponding highpass and lowpass filter. The implementation can then be performed either from the individual filters in a cascade arrangement or, after multiplying out, in a continuous arrangement.

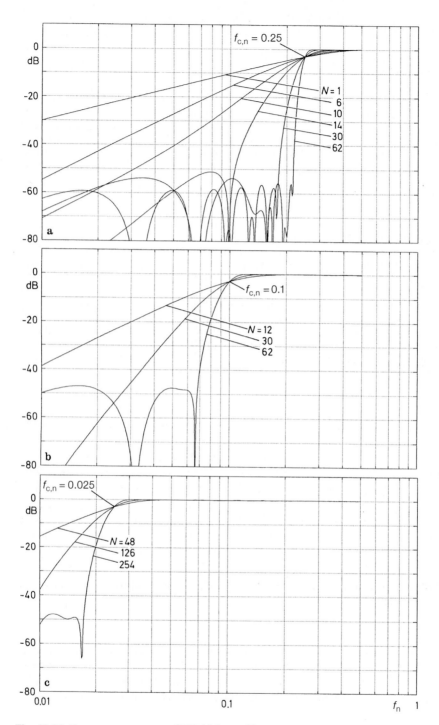

Fig. 19.31. Frequency responses of FIR highpass filters

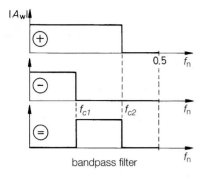

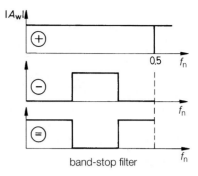

Fig. 19.32. Design of bandpass or band-stop filters

19.4
Realization of FIR Filters

To implement FIR filters, (19.12) on page 995 must be used to calculate the output values,

$$y(t_N) = \sum_{k=0}^{N} \alpha_k x(t_{N-k})$$

as the sum of the N last input values weighted with the coefficients. This operation can be performed either in parallel (i.e., in one step) or serially (i.e., in N steps). In the former case, considerable hardware complexity is involved, and in the latter a considerable amount of time, as shown in Fig. 19.33. If we assume, for example, 10 ns for the basic operation – that is, multiplication and addition (MAC-operation) – sampling frequencies of 100 MHz can be achieved with parallel processing; otherwise, only the Nth part thereof is considered.

To calculate (19.12), all the coefficients and the last N sampling values must of course be available in memory. In both cases, this requires a memory for $2N + 1$ values.

The required word length w of the data x is determined by the signal-to-quantizing-noise ratio, which is in the order of $w \cdot 6\,\text{dB}$. The word length available for the coefficients determines the accuracy to which the calculated coefficients can be realized. They are normally selected to be at least as large as the data word. After multiplication, this produces words of double the word length; that is, $2w$. When calculating the sum, the word length can increase by one bit in each step; that is, to $2w + N$. However, the actual increase is smaller, as the majority of the coefficients $\alpha_k \ll 1$. Nevertheless, a rounding down to smaller word length is generally unavoidable if circuit complexity is to be kept within acceptable limits.

Processing	Multipliers	Adders	Computing time	Memory
Parallel	$N + 1$	N	1 clock period	$2N + 1$
Serial	1	1	$N + 1$ clock periods	$2N + 1$

Fig. 19.33. Estimated requirements for Nth-order FIR filters with parallel or serial processing

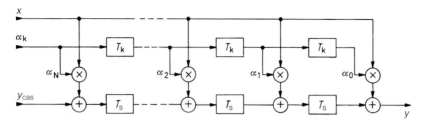

Fig. 19.34. Design of integrated FIR filters employing the parallel method

19.4.1
Realization of FIR Filters Using the Parallel Method

The structure shown in Fig. 19.11 is particularly suitable for realizing FIR filters by the parallel method, as an entire clock period is available for an MAC operation. Multiplication of the input sequence by the filter coefficients can, in principle, be carried out by parallel multipliers. Their multiplicand is defined by connecting bit by bit to 0 or 1, depending on the value of the coefficients. It would also be possible to work out the multiplication table for each coefficient and store it in a look-up table.

However, both methods are now outdated, as FPGAs with signal processing blocks are used. They are highly complex circuits that contain a large number of parallel multipliers, adders, and memories. An additional shift register for storing the coefficients is shown in Fig. 19.34. The coefficients are read into the shift register once the supply voltage has been switched on; the filter is then configured. The coefficients may also be exchanged during operation in order to make the filter characteristic adaptive. This facility is used in echo cancellation, for example. The coefficient input can also be used as a second signal input. In this case, the arrangement calculates the cross-correlation function of the input signals. The additional input y_{cas} allows similar devices to be cascaded so as to raise the filter order.

19.4.2
Realization of FIR Filters Using the Serial Method

The serial realization of FIR filters is derived from the basic structure shown in Fig. 19.12, with a global adder at the output. A shift register is used to store the coefficients, as shown in Fig. 19.35. It is then possible to replace all of the multipliers and adders by a single MAC, as shown in Fig. 19.36. To calculate the output value, the multiplier inputs are shifted

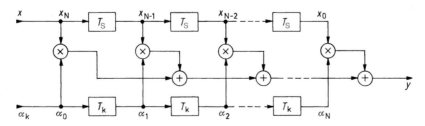

Fig. 19.35. FIR filter with a global adder at the output

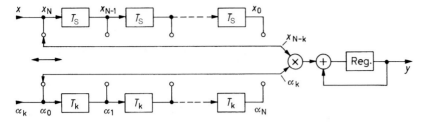

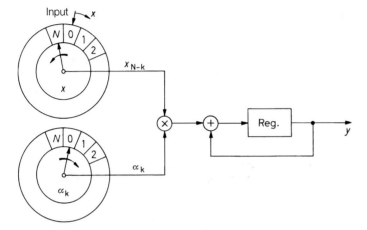

Fig. 19.36. Serial calculation of the sub-products

Fig. 19.37. Serial calculation of y. To calculate the output value, the two output pointers are rotated once and all of the sub-products are added. Then the next value x is read in

once through all the stages and the resultant sub-products are added together. The two shift registers are implemented as FIFOs (see Sect. 10.2.3). It is not necessary to shift the data physically; rather, it is only the relevant input and output pointers that are stepped forward. This is illustrated in Fig.19.37. A number of freely programmable signal processors are listed in Fig. 19.50 on page 1028.

19.5
Design of IIR Filters

Recursive filters are also known as infinite impulse response (IIR) filters, as – at least theoretically – their impulse response possesses an infinite number of nonzero sampling values. Their basic structure and transfer functions, which have already been discussed in Sect. 19.2, apply to digital filters in general.

19.5.1
Calculating the Filter Coefficients

Two methods in particular are commonly used to calculate the filter coefficients, the Yule-walk Algorithm and the bilinear transformation. The Yulewalk Algorithm approximates a given tolerance scheme in the frequency domain by means of a minimum number of filter coefficients. It thus provides coefficients for a minimized IIR filter, and therefore represents the analogon to the Remez Exchange Algorithm for FIR filters. We shall now describe the bilinear transformation in greater detail, because it is less computationally intensive and therefore facilitates understanding of the principles involved.

The bilinear transformation is based on the frequency response of an analog filter and attempts to model it as accurately as possible using an IIR filter. However, this is not directly possible, as the transfer function of a digital filter can only be utilized up to half the sampling frequency $\frac{1}{2} f_s$ and must be periodic beyond that. For this reason, the amplitude–frequency response of the analog filter in the range $0 \leq f \leq \infty$ is mapped into the range $0 \leq f' \leq \frac{1}{2} f_s$ of the digital filter and continued periodically. A transformation that possesses this characteristic is

$$f = \frac{f_s}{\pi} \tan \frac{\pi f'}{f_s} \tag{19.31}$$

For $f \to \infty$, f' tends to $\frac{1}{2} f_s$, as required. For $f' \ll f_s$, we have $f \approx f'$. The compression of the frequency axis is therefore smaller the higher is the clock frequency f_s with respect to the frequency range of interest.

In order to be able to employ normalized frequencies as with the analog filters, we normalize all the frequencies to the sampling frequency:

$$f_n = f/f_s \quad \text{or} \quad f_{c,n} = f_c/f_s \tag{19.32}$$

Equation (19.31) therefore becomes:

$$f_n = \frac{1}{\pi} \tan \pi f_n' . \tag{19.33}$$

To illustrate the transformation of the frequency axis, in Fig. 19.38 we have plotted the amplitude–frequency response of a second-order Chebyshev lowpass filter. We can see that the typical passband characteristic is retained, although the cutoff frequency is shifted. In order to avoid this effect, we introduce a shift-factor l into (19.33) for frequency mapping. We select this factor such that the cutoff frequency is retained in the transformation; that is, $f_{c,n} = f'_{c,n}$:

$$f = f_{c,n} \underbrace{\cot \pi f_{c,n}}_{l} \tan \pi f_n' \tag{19.34}$$

The resultant frequency response curve is shown in Fig. 19.39. We interpret the formally introduced quantity f_n' as a new frequency variable f_n and denote the transformed frequency response by $\underline{A}'(j\omega_n)$. We can see that this provides a good approximation to the analog filter characteristic.

The transformed frequency response $A'(j f_n)$ now possesses a shape which can be implemented using a digital filter. To calculate the digital transfer function $\tilde{A}(z)$, we now require the transformation equation for the complex frequency variable s_n. With $s_n = j\omega_n = j f_n/f_{c,n}$, it follows from (19.34) that

$$s = l \cdot j \tan \pi f_n$$

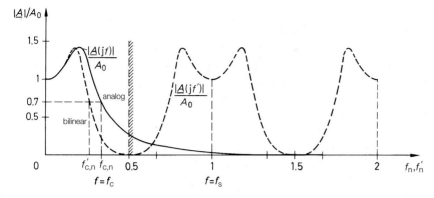

Fig. 19.38. Producing a periodic amplitude frequency response. Second-order Chebyshev characteristic with 3 dB ripple as example. Cutoff frequency: $f_{c,n} = 0.3$, i.e. $f_c = 0.3\,f_s$. Linear plot.

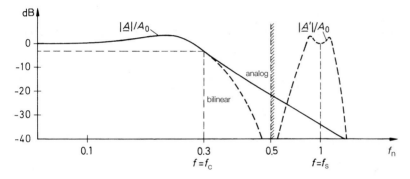

Fig. 19.39. Matching the cutoff frequency. Second-order Chebyshev characteristic with a 3 dB ripple as an example. Cutoff frequency: $f_{c,n} = 0.3$, i.e., $f_c = 0.3\,f_s$. Logarithmic plot

Using the mathematical transformation

$$j \tan x = -\tanh(-jx) = \frac{1 - e^{-2jx}}{1 + e^{-2jx}}$$

and the definition of $z^{-1} = e^{-2\pi j f_n}$, we therefore obtain

$$s = l\,\frac{1 - e^{-2\pi j f_n}}{1 + e^{-2\pi j f_n}} = l\,\frac{1 - z^{-1}}{1 + z^{-1}} \quad \text{with} \quad l = \cot \pi f_{c,n} \tag{19.35}$$

This relation is known as the bilinear transformation.

To summarize, an analog filter function can be transformed into a digital function as follows: in the analog transfer function $A(s_n)$, the normalized frequency variable s_n is replaced by $1(z - 1)/(z + 1)$ and a transfer function $A(z)$ is obtained which can be implemented by a digital filter. The amplitude–frequency response is therefore very similar in shape to that of the analog filter. The characteristic is compressed on the f−axis so that the value $|\underline{A}(j\infty)|$ appears at frequency $\frac{1}{2}\,f_s$. The resultant deviations are smaller the larger is the value of f_s compared to the frequency range $0 < f < f_{max}$ of interest.

However, the phase–frequency response is changed to a far greater extent. Consequently, the statements relating to analog filters cannot be directly applied to digital filters. It would be of no avail, for example, to approximate the amplitude–frequency response of a Bessel filter, because the linearity of the phase delay would be lost. If a filter with linear phase is required, it is preferable to use an FIR filter.

To calculate the filter coefficients of IIR filters, we insert into the frequency response of the linear filter,

$$A(s_n) = \frac{d_0 + d_1 s_n + d_2 s_n^2 + \ldots}{c_0 + c_1 s_n + c_2 s_n^2 + \ldots} = \frac{\sum_{k=0}^{N} d_k s_n^k}{\sum_{k=0}^{N} c_k s_n^k} \tag{19.36}$$

the bilinear transformation (19.35)

$$s_n = l\frac{1 - z^{-1}}{1 + z^{-1}}.$$

A comparison of coefficients with the general frequency response (19.9) of an IIR filter,

$$A(z) = \frac{\alpha_0 + \alpha_1 z^{-1} + \alpha_2 z^{-2} + \ldots}{1 + \beta_1 z^{-1} + \beta_2 z^{-2} + \ldots} = \frac{\sum_{k=0}^{N} \alpha_k z^{-k}}{1 + \sum_{k=1}^{N} \beta_k z^{-k}}$$

then yields the required filter coefficients α_k and β_k.

19.5.2
IIR Filters in a Cascade Structure

As in the case of analog filters, the simplest method of implementing IIR filters is to cascade first- and second-order blocks. In this case, the values tabulated in Fig. 13.14 for analog filters can also be used to calculate the filter coefficients. The recalculation of the filter coefficients is therefore given below in more detail.

First-Order IIR Filters

The structure of a first-order IIR filter shown in Fig. 19.40 is derived from Fig. 19.7 for the case $N = 1$. From the first-order *analog* transfer function

$$A(s_n) = \frac{d_0 + d_1 s_n}{c_0 + c_1 s_n} \tag{19.37}$$

we obtain, by using the bilinear transformation, the digital transfer function

$$\tilde{A}(z) = \frac{\alpha_0 + \alpha_1 z^{-1}}{1 + \beta_1 z^{-1}} \tag{19.38}$$

with the coefficients

$$\alpha_0 = \frac{d_0 + d_1 l}{c_0 + c_1 l}; \quad \alpha_1 = \frac{d_0 - d_1 l}{c_0 + c_1 l}; \quad \beta_1 = \frac{c_0 - c_1 l}{c_0 + c_1 l} \tag{19.39}$$

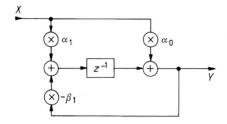

Fig. 19.40. First-order IIR filter

$$\tilde{A}(z) = \frac{Y}{X} = \frac{\alpha_0 + \alpha_1 z^{-1}}{1 + \beta_1 z^{-1}}$$

Applying these general equations to a lowpass filter,

$$A(s_n) = \frac{A_0}{1 + a_1 s_n} \Rightarrow \tilde{A}(z) = \alpha_0 \frac{1 + z^{-1}}{1 + \beta_1 z^{-1}} \tag{19.40}$$

$$\alpha_0 = \alpha_1 = \frac{A_0}{1 + a_1 l}; \quad \beta_1 = \frac{1 - a_1 l}{1 + a_1 l} \tag{19.41}$$

Similarly, for a highpass filter,

$$A(s_n) = \frac{A_\infty}{1 + a_1 \dfrac{1}{s_n}} = \frac{A_\infty s_n}{a_1 + s_n} \Rightarrow \tilde{A}(z) = \alpha_0 \frac{1 - z^{-1}}{1 + \beta_1 z^{-1}}$$

$$\alpha_0 = -\alpha_1 = \frac{A_\infty l}{a_1 + l}; \quad \beta_1 = \frac{a_1 - l}{a_1 + l} \tag{19.42}$$

By way of example, we shall calculate the coefficients for a first-order highpass filter for phone applications. Its cutoff frequency will be $f_c = 100\,\mathrm{Hz}$, and the bandwidth of the input signal will be $3.4\,\mathrm{kHz}$. We select $f_s = 10\,\mathrm{kHz}$ and obtain the normalized cutoff frequency

$$f_{c,n} = f_c/f_s = 100\,\mathrm{Hz}/10\,\mathrm{kHz} = 0.01$$

This yields the normalization factor (19.35)

$$l = \cot \pi f_{c,n} = \cot(0.01 \cdot \pi) = 31.82$$

Hence for $a_1 = 1$ we obtain from (19.42),

$$\alpha_0 = -\alpha_1 = 0.9695 \quad \text{and} \quad \beta_1 = -0.9391$$

The digital transfer function is therefore

$$\tilde{A}(z) = \frac{\alpha_0 + \alpha_1 z^{-1}}{1 + \beta_1 z^{-1}} = \frac{0.9695 - 0.9695 z^{-1}}{1 - 0.9391 z^{-1}}$$

The ratio of the sampling frequency to the cutoff frequency is determined by the parameters selected and has the value 100. The cutoff frequency is therefore proportional to the sampling frequency. It can therefore be simply controlled using the sampling frequency. This characteristic is peculiar to all digital filters. The only other filters exhibiting this property are the switched-capacitor types described in Sect. 13.12.

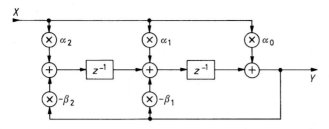

Fig. 19.41. Second-order IIR filter $\tilde{A}(z) = \dfrac{Y}{X} = \dfrac{\alpha_0 + \alpha_1 z^{-1} + \alpha_2 z^{-1}}{1 + \beta_1 z^{-1} + \beta_2 z^{-2}}$

Second-Order IIR

A second-order IIR filter obtained by particularizing Fig. 19.7 on page 992 is shown in Fig. 19.41. Into the linear transfer function,

$$A(s_n) = \frac{d_0 + d_1 s_n + d_2 s_n^2}{c_0 + c_1 s_n + c_2 s_n^2}$$

we insert the bilinear transformation as defined in (19.35) and obtain

$$\tilde{A}(z) = \frac{\alpha_0 + \alpha_1 z^{-1} + \alpha_2 z^{-2}}{1 + \beta_1 z^{-1} + \beta_2 z^{-2}} \tag{19.43}$$

with the coefficients

$$\alpha_0 = \frac{d_0 + d_1 l + d_2 l^2}{c_0 + c_1 l + c_2 l^2}; \qquad \alpha_1 = \frac{2(d_0 - d_2 l^2)}{c_0 + c_1 l + c_2 l^2}; \qquad \alpha_2 = \frac{d_0 - d_1 l + d_2 l^2}{c_0 + c_1 l + c_2 l^2};$$

$$\beta_1 = \frac{2(c_0 - c_2 l^2)}{c_0 + c_1 l + c_2 l^2}; \qquad \beta_2 = \frac{c_0 - c_1 l + c_2 l^2}{c_0 + c_1 l + c_2 l^2}$$

From this, we can calculate the following second-order filters with $l = \cot \pi f_{c,n}$:

Lowpass filter (19.44):

$$A(s_n) = \frac{A_0}{1 + a_1 s_n + b_1 s_n^2} \quad \Rightarrow \quad \tilde{A}(z) = \alpha_0 \frac{1 + 2 z^{-1} + z^{-2}}{1 + \beta_1 z^{-1} + \beta_2 z^{-2}}$$

$$\alpha_0 = \frac{A_0}{1 + a_1 l + b_1 l^2}; \qquad \beta_1 = \frac{2(1 - b_1 l^2)}{1 + a_1 l + b_1 l^2}; \qquad \beta_2 = \frac{1 - a_1 l + b_1 l^2}{1 + a_1 l + b_1 l^2}$$

Highpass filter (19.45):

$$A(s_n) = \frac{A_\infty s_n^2}{b_1 + a_1 s_n + s_n^2} \quad \Rightarrow \quad \tilde{A}(z) = \alpha_0 \frac{1 - 2 z^{-1} + z^{-2}}{1 + \beta_1 z^{-1} + \beta_2 z^{-2}}$$

$$\alpha_0 = \frac{A_\infty l^2}{b_1 + a_1 l + l^2}; \qquad \beta_1 = \frac{2(b_1 - l^2)}{b_1 + a_1 l + l^2}; \qquad \beta_2 = \frac{b_1 - a_1 l + l^2}{b_1 + a_1 l + l^2}$$

Bandpass filter (19.46):

$$A(s_n) = \frac{A_r s_n / Q}{1 + s_n / Q + s_n^2} \quad \Rightarrow \quad \tilde{A}(z) = \alpha_0 \frac{1 - z^{-2}}{1 + \beta_1 z^{-1} + \beta_2 z^{-2}}$$

$$\alpha_0 = \frac{l A_r / Q}{1 + l / Q + l^2}; \qquad \beta_1 = \frac{2(1 - l^2)}{1 + l / Q + l^2}; \qquad \beta_2 = \frac{1 - l / Q + l^2}{1 + l / Q + l^2}$$

Bandstop filter (19.47):

$$A(s_n) = \frac{A_0(1 + s_n^2)}{1 + s_n/Q + s_n^2} \quad \Rightarrow \quad \tilde{A}(z) = \frac{\alpha_0 + A_0\beta_1 - z^{-1} + \alpha_0 z^{-2}}{1 + \beta_1 z^{-1} + \beta_2 z^{-2}}$$

$$\alpha_0 = \frac{A_0(1 + l^2)}{1 + l/Q + l^2}; \quad \beta_1 = \frac{2(1 - l^2)}{1 + l/Q + l^2}; \quad \beta_2 = \frac{1 - l/Q + l^2}{1 + l/Q + l^2}$$

We shall now discuss the design procedure with the aid of an example. We require a second-order Chebyshev lowpass filter with a 0.5 dB ripple and a 3 dB cutoff frequency of $f_c = 100\,\text{Hz}$. The analog signal will have a bandwidth of 3.4 kHz and be sampled at $f_s = 10\,\text{kHz}$. This gives a normalized cutoff frequency of $f_{c,n} = 0.01$ and a normalizing factor of $l = 31.82$. From the table in Fig. 13.14, we can obtain $a_1 = 1.3614$ and $b_1 = 1.3827$. This produces the continuous transfer function

$$A(s_n) = \frac{1}{1 + 1.3614 s_n + 1.3827 s_n^2}$$

Using (19.44), we obtain from this the digital transfer function

$$\tilde{A}(z) = 6.923 \cdot 10^{-4} \frac{1 + 2z^{-1} + z^{-2}}{1 - 1.937 z^{-1} + 0.9400 z^{-2}}$$

As a second example, we will design a bandpass filter. The sampling frequency will be 10 kHz as before. The resonant frequency will be $f_r = 1\,\text{kHz}$. Hence $f_{c,n} = 1\,\text{kHz}/10\,\text{kHz} = 0.1$. For $Q = 10$, the continuous transfer function, in accordance with (13.24), for $A_r = 1$ is of the form

$$A(s_n) = \frac{0.1 s_n}{1 + 0.1 s_n + s_n^2}$$

Using $l = \cot \pi f_{r,n} = 3.078$ and (19.46), we obtain the digital transfer function for $Q = 10$ and $f_{r,n} = 0.1$

$$\tilde{A}(z) = -2.855 \cdot 10^{-2} \frac{1 - z^{-2}}{1 - 1.572 z^{-1} + 0.9429 z^{-2}}$$

Similarly, for $Q = 100$ and $f_{r,n} = 0.1$ we obtain

$$\tilde{A}(z) = -2.930 \cdot 10^{-3} \frac{1 - z^{-2}}{1 - 1.613 z^{-1} + 0.9941 z^{-2}}$$

We will now consider the case $Q = 10$ and $f_{r,n} = 0.01$. This gives

$$\tilde{A}(z) = -3.130 \cdot 10^{-3} \frac{1 - z^{-2}}{1 - 1.990 z^{-1} + 0.9937 z^{-2}}$$

We can see that as Q increases or the resonant frequency $f_{c,n}$ decreases, the coefficient $\alpha_0 \to 0$, whereas $\beta_2 \to 1$ and $\beta_1 \to 2$. The information on the filter characteristic is therefore to be found in the very small deviation with respect to 1 or -2. This means that there is an increasing accuracy requirement on the filter coefficients, which results in a correspondingly large word length in the filter. In order to minimize the circuit complexity, the sampling rate must therefore not be selected to be any higher than necessary.

19.6
Realization of IIR Filters

19.6.1
Construction from Simple Building Blocks

We shall now demonstrate the procedure for arriving at the simplest possible circuit, using the example of the first-order highpass filter from Sect. 19.5.2. There, we have already calculated the digital transfer function for a highpass filter with a cutoff frequency of $f_c = 100\,\text{Hz}$ at a sampling rate of $f_s = 10\,\text{kHz}$; that is, $f_{c,n} = 0.01$:

$$\tilde{A}(z) = \frac{\alpha_0 + \alpha_1 z^{-1}}{1 + \beta_1 z^{-1}} = \frac{0.9695 - 0.9695 z^{-1}}{1 - 0.9391 z^{-1}}$$

The corresponding circuit is shown in Fig. 19.42. We can see that the three coefficients are close to 1. The numerator coefficients α_0 and α_1 can be rounded to 1 without any appreciable error, as they only determine the gain. This does not apply to coefficient β_1, whose deviation from 1 determines the filter cutoff frequency. However, in this case simplification is possible by the transformation

$$-\beta_1 = 1 - \beta_1' = 0.9361 = 1 - 0.0609$$

where $\beta_1' = 1 + \beta_1$ is the deviation from unity. This coefficient possesses substantially fewer significant digits than β_1. The nearest power of two is $2^{-4} = 0.0625$. The binary arithmetic can be greatly reduced by rounding β_1' to this value, as a multiplication by 2^{-4} only represents a shift by four digits, which can be implemented by appropriate wiring. From (19.42), the resulting shift in the cutoff frequency is given by

$$l = \frac{1 - \beta_1}{1 + \beta_1} = \frac{2 - \beta_1'}{\beta_1'} = \frac{2 - 2^{-4}}{2^{-4}} = 31.$$

Thus with (19.35) we get $f_{c,n} = 0.0103$; that is, the cutoff frequency increases to $f_c = 103\,\text{Hz}$.

We simplify further by rounding the numerator coefficients to $\alpha_0 = -\alpha_1 = 1$, and use (19.42) to obtain, for high frequencies ($f \approx \frac{1}{2} f_s$), the gain

$$A_\infty = \alpha_0 \frac{1 + l}{l} = 1 \cdot \frac{1 + 31}{31} = 1.032$$

This small deviation is also acceptable. The resulting simplified arrangement is shown in Fig. 19.43. We can see that in simple filters it is possible to reduce the circuit considerably by slightly modifying the design objective.

The practical implementation is shown in Fig. 19.44 for an input word length of 4 bits. In order to be able to represent positive and negative numbers, we have selected the two's-complement notation introduced in Sect. 8.1.3. The highest-order bit is therefore the sign bit. As we can perform the multiplication by shifting, only adding circuits are required. For this purpose, we use 4-bit arithmetic circuits of type SN 74 LS 382. These can also be operated as subtractors by controlling appropriate inputs. In this way, the computation of the two's-complement of the coefficients $\alpha_1 = -1$ and $-\beta_1 = 1 - 2^{-4}$ can be carried out in the adder.

The two arithmetic circuits IC 8 and IC 9 form the expression

$$r = -\beta_1 y = y - 2^{-4} y$$

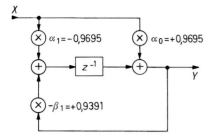

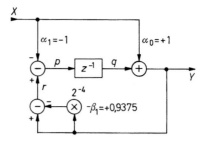

Fig. 19.42. First-order IIR highpass filter

$$\tilde{A}(z) = \frac{0.9695 - 0.9695z^{-1}}{1 - 0.9391z^{-1}}$$

$$f_{c,n} = f_c/f_s = 0.01$$

$$A(f = 0.5f_s) = 1$$

Fig. 19.43. Highpass filter with simplified coefficients

$$\tilde{A}(z) = \frac{1 - z^{-1}}{1 - (1 - 2^{-4})z^{-1}}$$

$$f_{c,n} = f_c/f_s = 0.0103$$

$$A(f = 0.5f_s) = 1.032$$

Multiplication of y by 2^{-4} is achieved by connecting y, displaced by four digits, to the subtractor. This increases the word length from 4 bits to 8 bits.

The sign bit v_y must be connected to *all* of the vacated digits, so that multiplication of y by 2^{-4} can be carried out correctly for both positive and negative values of y (sign extension).

Arithmetic unit IC 2 performs the subtraction $r - x$ at the input of Fig. 19.43, and IC 5 the addition $x + q$ at the output. The delay by one clock period is achieved by ICs 3 and 4, each of which contains four single-edge-triggered D-flip-flops. The flip-flops in IC 1 are used to synchronize the input signal.

The exclusive-OR gates in ICs 6 and 7 provide latch-up protection: as we have already seen in Fig. 8.3, a jump from $+127$ to -128 would occur if the positive number range was exceeded, as the highest-order bit is read as the sign. This unwanted sign change may destabilize the filter in the overdriven state and it may not return to normal operation. This corresponds precisely to the latch-up effect in analog circuitry. One way of preventing it is to set the numbers at the adder output to $+127$ for positive overflow and to -128 for negative overflow. For this purpose, the positive and negative overflows would have to be decoded separately.

However, a distinction between these two cases is unnecessary if the outputs are complemented in the event of an overflow. This produces the characteristic shown in Fig. 19.45. To implement it, exclusive-OR gates are connected to the outputs f_i of those arithmetic units at which an overflow can occur, as shown in Fig. 19.44. This produces a complementation when $OV = 1$. Arithmetic units 74LS382 have an advantage over standard types 74LS181 in that the overflow variable OV is directly available and does not have to be generated externally.

The operation of the digital highpass filter is clearly shown by the step response in Fig. 19.46.

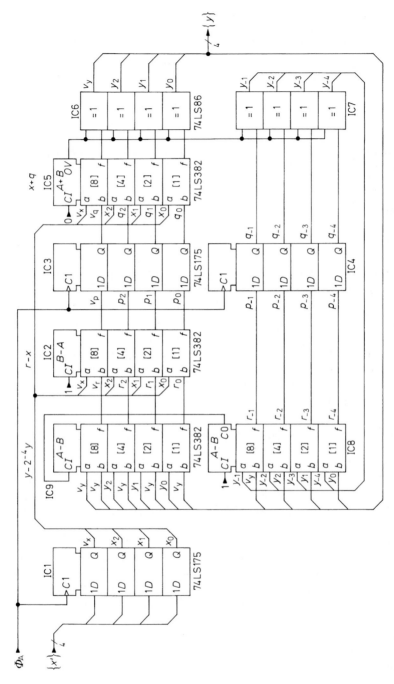

Fig. 19.44. Circuit implementation of a first order digital IIR highpass filter with a word length of 8 bit internal and 4 bit external

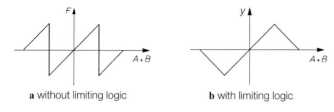

a without limiting logic   **b** with limiting logic

Fig. 19.45. Overflow characteristics of the arithmetic units

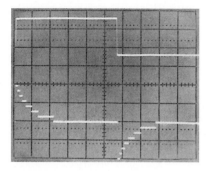

Fig. 19.46. Step response of the digital highpass filter in Fig. 19.44 for a maximum input swing

19.6.2
Design Using LSI Devices

There are three methods of implementing IIR filters using LSI circuitry:

1) using CPLDs or FPGAs programmed as filter,
2) using FIR filters,
3) using programmable signal processors.

Programmable digital hardware circuits like CPLDs and FPGAs can be configured as digital filters. For this task the hardware description language VHDL is advantageous because the adders and multipliers required can be defined with a single command line. FPGAs with predefined DSP-block are especially efficient for high speed processing and little usage of resources. Examples for such devices are given in Fig. 10.48 on page 721.

An IIR filter can be constructed from two FIR filters. It is possible to start with the basic structure shown in Fig. 19.8, with global adders at the input and output, and to double the delay chain. This results in the circuit shown in Fig. 19.47, in which we can identify the two FIR filters. Whether the FIR filters employ a global adder at the output, as shown here, or distributed adders, as shown in Fig. 19.11, is unimportant. The result in both cases is the same if the coefficients are arranged accordingly.

The basic structure shown in Fig. 19.9 can also be broken down into two FIR filters if the global adder at the output is split into two sections. In Fig. 19.48 we can see that this produces two FIR filters whose partial effects (i.e., results) can be combined using an additional adder.

Single-chip signal processors are the most suitable devices for the serial method of implementation of digital filters, as they possess the required data memories in addition to a parallel multiplier with accumulator. The calculation of the filter output sequence can be programmed at machine level in Assembler, as on a microprocessor. Additionally, the programming is now being supported in higher-level languages such as "C." For program-

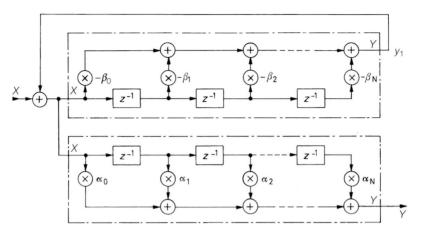

Fig. 19.47. Implementation of an IIR filter with a global adder at the input and the output, from two FIR filters and one additional adder

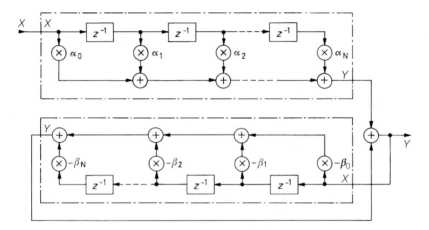

Fig. 19.48. Implementation of an IIR filter using a single global adder at the output, from two FIR filters and one additional adder

ming, it is best to start from the basic structure with a global adder at the output, according to Fig. 19.48. As shown in Fig. 19.49, the new value of the output sequence can then be calculated in accordance with (19.8),

$$Y_N = \sum_{k=0}^{N} \alpha_k x_{N-k} - \sum_{k=1}^{N} \beta_k y_{N-k}$$

by weighting all input and feedback signals with the relevant coefficients and subsequently accumulating them. To do this, the tap is shifted along the delay chain and the particular coefficient α_k or β_k is selected. When the entire chain has been run through once, the new function value y_N is available. Then the contents of the two shift registers can be shifted forward by one clock, in order to calculate a further function value y in the next pass. The data are, of course, not shifted physically: only the pointers addressing the values x_k, y_k, α_k, and β_k rotate.

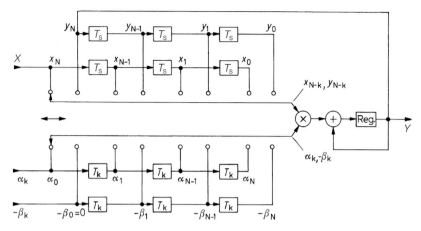

Fig. 19.49. Serial implementation of an IIR filter with a single global adder at the output, preferably using a programmable signal processor. Data X is shifted with the sample clock T_s. Coefficients are shifted in once before start of filtering.

Using a signal processor, it is just as easy to make design calculations for IIR filters in cascade form. For this purpose, the order of the filter in Fig. 19.49 is reduced to $N = 2$ and we calculate

$$y_2 = \alpha_0 x_2 + \alpha_1 x_1 + \alpha_2 x_0 - \beta_1 y_1 - \beta_2 y_0$$

using a small subroutine. The entire filter is therefore obtained by repeatedly calling up the program for a second-order filter and exchanging the relevant data or coefficient sets.

Figure 19.50 provides an overview of a number of more recent signal processors. The preferred number notations are 16-bit fixed-point numbers for universal applications or 32-bit floating-point numbers for high accuracy and dynamic range. The data word length of the accumulator is generally more than twice that size, in order to insure that the rounding errors have no effect on the result. The majority of signal processors have high-speed data and program memories on the chip. These should be used where possible, because each external memory access results in the insertion of wait states even if the access times are short.

The time taken to perform a multiplication and accumulation (MAC) operation determines how rapidly a signal processor can process fitter algorithms, as these consist virtually exclusively of MAC operations. For an Nth-order FIR filter, $N + 1$ MAC operations are required for each sampled value, whereas an IIR filter requires $2N + 1$. In most of the more recent signal processors, a MAC operation is performed in a single machine cycle.

19.7
Comparison of FIR and IIR Filters

If we compare the structure of the IIR filters shown in Figs. 19.7–19.9 with that of the FIR filters shown in Figs. 19.11 and 19.12, we can see that IIR filters of the same order require approximately twice as many MAC operations as FIR filters. However, they possess a higher selectivity than FIR filters with the same number of MAC operations. This is illustrated by the example shown in Fig. 19.51. Generally speaking, the required order for

Type	Manu-facturer	Data word	Internal memory data/progr.	Multiply accumulate MAC	Clock frequency	Power dissipation
ADSP 2188	Analog Dev.	16 fix	112/144 kB	80 M/s	80 MHz	50 mW
ADSP 21369	Analog Dev.	32 fl	256/768 kB	800 M/s	400 MHz	
ADSPTS 203	Analog Dev.	32 fix,fl	512 kB	4000 M/s	500 MHz	2000 mW
ADSPBF 533	Analog Dev.	16 fix	64/32 kB	1200 M/s	600 MHz	
DSP 56311	Freescale	24 fix	384 kB	150 M/s	150 MHz	220 mW
DSP 56858	Freescale	16 fix	48/80 kB	120 M/s	120 MHz	260 mW
MC56F 8014	Freescale	16 fix	4/16 kB	32 M/s	32 MHz	140 mW
MC56F 8367	Freescale	16 fix	32/512 kB	60 M/s	60 MHz	375 mW
TMS320C 2811	Texas Inst.	16 fix	36/256 kB	150 M/s	150 MHz	640 mW
TMS320VC 5416	Texas Inst.	16 fix	256/32 kB	120 M/s	120 MHz	230 mW
TMS320C 6203	Texas Inst.	16 fix	512/384 kB	600 M/s	300 MHz	980 mW
TMS320C 6455	Texas Inst.	16 fix	2048 kB	8000 M/s	1000 MHz	
TMS320C 6727	Texas Inst.	32 fl	256/384 kB	600 M/s	300 MHz	
Pentium 4	Intel	32 fl	1024 kB		3000 MHz	100 W
ARM 1020*	ARM	16 fix	32/32 kB	400 M/s	400 MHz	280 mW
MIPS32-24*	MIPS	32 fix	32/32 kB	600 M/s	600 MHz	360 mW

* Available as Intellectual Property IP.

Fig. 19.50. Examples for Digital Signal Processors DSPs

an FIR filter is, for the same performance, more than twice as high as for an IIR filter. In Sect. 19.5.2 we have shown that a lowpass filter with a low cutoff frequency of $f_{c,n} = 0.01$ can be implemented using a first-order IIR filter. For an FIR filter with this cutoff frequency, an order of at least $N = 65$ would have been required since, to a first approximation, a complete cycle at the cutoff frequency $f_{c,n}$ is to be weighted with the coefficients; that is,

$$N \geq 1/f_{c,n} = f_s/f_c .$$

However, this characteristic must be set against a number of important advantages of FIR filters. We have seen that FIR filters can easily be used to implement linear phase behavior – that is, a constant group delay – with precision. All of the FIR filters mentioned in this section possess this characteristic; that is, they do not produce any phase distortion.

As FIR filters possess no feedback path, they are also stable for any coefficient values. IIR filters, like analog filters, tend to oscillate more the higher is their pole Q-factor or the lower is their cutoff frequency compared to the sampling frequency (see Sect. 19.5.2). In order to insure that no significant deviations from the calculated frequency response are incurred, the coefficients of IIR filters must be realized with considerably more precision than for FIR filters; this requires a larger word length. In addition, with IIR filters the rounding errors due to limited computational accuracy frequently result in limit cycles. These are periodic oscillations in the lowest-order bits, which are particularly disturbing in the case of small input signals. The various advantages and disadvantages are listed in Fig. 19.52.

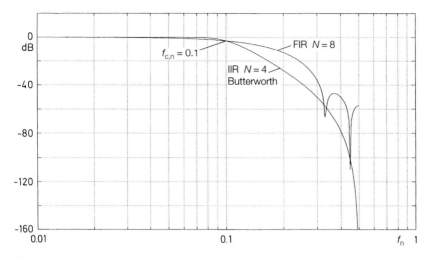

Fig. 19.51. Comparison of an eighth-order FIR lowpass filter with a fourth-order IIR lowpass filter

Feature	FIR filter	IIR filter
selectivity	low	**high**
order required	high	**low**
no. of MAC operations	many	**few**
memory requirement	high	**low**
linear phase	**easily provided**	rarely possible
constant group delay	**easily provided**	rarely possible
stability	**absolute**	limited
required word length	**reasonable**	high
required coefficient accuracy	**reasonable**	high
limit cycles	**none**	present
adaptive filter	**possible**	rarely possible

Fig. 19.52. A comparison of FIR and IIR filters

Chapter 20:
Measurement Circuits

In the previous chapters, a number of methods for processing analog and digital signals have been described. Many applications, however, require that even electrical signals must be conditioned before they can be processed in analog computing circuits or A/D converters. In such cases, measurement circuits are needed which have a low-resistance single-ended output; that is, produce a ground-referenced output voltage.

20.1
Measurement of Voltage

20.1.1
Impedance Converter

If the signal voltage of a high-impedance source is to be measured without affecting the load conditions, the noninverting amplifier (follower-with-gain, electrometer amplifier) shown in Fig. 5.58 can be employed for impedance conversion. It must be noted, however, that the high-impedance input is very sensitive to noise arising from capacitive stray currents in the input lead. Therefore the lead is usually screened, but this results in considerable capacitive loading of the source to ground with $30 \ldots 100\,\mathrm{pF/m}$. For an internal source resistance of $1\,\mathrm{G\Omega}$, for instance, and a capacitance of the coax cable of $100\,\mathrm{pF}$, an upper cutoff frequency of only $1.6\,\mathrm{Hz}$ is obtained.

The capacitance is not constant but can change; for example, when the lead is moved. This gives rise to an additional problem in that this variation produces very large noise voltages. If, for instance, the lead is charged up to $10\,\mathrm{V}$, a change in capacitance of 1% results in a voltage step of $100\,\mathrm{mV}$.

These disadvantages can be avoided if a noninverting amplifier is employed to keep the voltage low between the inner conductor and the shield. The shield is then connected not to ground but to the buffer output, as shown in Fig. 20.1. In this manner, the effect of the lead capacitance is reduced by the open-loop gain of the operational amplifier. Since now only the offset voltage of the amplifier appears across the lead capacitance, the noise also is eliminated to a large extent.

Increasing the Output Voltage Swing

The maximum permissible supply voltage of standard integrated operational amplifiers is usually $\pm 18\,\mathrm{V}$. The attainable output voltage swing is thereby limited to values of about $\pm 15\,\mathrm{V}$. This limit can be raised by allowing the operational amplifier supply potentials to

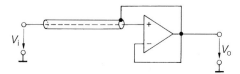

Fig. 20.1. Reduction of shield capacitance and stray current noise by allowing the shield potential to follow the measuring potential: guard drive

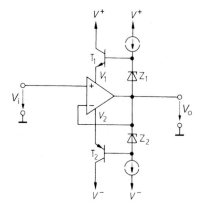

Fig. 20.2. A voltage follower for large input voltages

follow the input voltage. This is achieved by using the bootstrap circuit shown in Fig. 20.2, that has two emitter followers to stabilize the potential differences $V_1 - V_o$ and $V_o - V_2$ to a value $V_Z - 0.7\,\mathrm{V}$. The maximum output voltage swing is thus no longer determined by the operational amplifier, but by the permissible voltage across the emitter followers and the current sources.

20.1.2
Measurement of Potential Difference

The measurement of potential differences involves amplifying the voltage difference

$$V_D = V_2 - V_1$$

while minimizing the effect of the superimposed common-mode voltage

$$V_{CM} = \frac{1}{2}(V_2 + V_1)$$

It is frequently the case that differential voltages in the millivolt range have common-mode voltages of $10\,\mathrm{V}$ or more superimposed on them. The quality of a subtractor is therefore characterized by its common-mode rejection:

$$G = \frac{A_D}{A_{CM}} = \frac{V_o/V_D}{V_o/V_{CM}} = \frac{V_{CM}}{V_D}$$

In the example given, G must be greater than $10\,\mathrm{V}/1\,\mathrm{mV} = 10^4$. Particular problems arise if the superimposed common-mode voltage exhibits high values or high frequencies. There are three different methods of amplifying voltage differences:

- operational amplifiers configured as subtractors;
- differential amplifiers with feedback, and subtraction using switched; capacitors
- subtractor with operational amplifier circuitry.

Subtractor with Operational Amplifier Circuitry

In principle, the subtractor shown in Fig. 11.3 can be used to measure potential differences. However, the potentials to be measured often cannot be loaded with the input resistance of the subtractor because they possess considerable internal resistance. Use of the additional

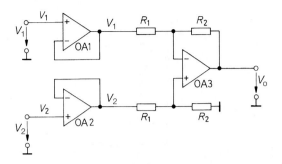

Fig. 20.3. Subtractor with preceding impedance converters

$$V_o = \frac{R_2}{R_1}(V_2 - V_1)$$

voltage followers shown in Fig. 20.3 renders the operation of the subtractor independent of the internal resistances of the potentials to be measured.

A higher common-mode rejection can be achieved by shifting the voltage gain to the impedance converters and giving the subtractor unit gain. This variant is shown in Fig. 20.4. For $R_1 = \infty$, OA1 and OA2 operate as voltage followers; in this case, the circuit is virtually identical to the previous one.

The circuit has the further advantage that the differential gain can be adjusted by varying a single resistor. It can be seen in Fig. 20.4 that the potential difference $V_2 - V_1$ appears across resistor R_1. Hence,

$$V_2' - V_1' = \left(1 + \frac{2R_2}{R_1}\right)(V_2 - V_1)$$

This difference is transferred to the single-ended output by the subtracting amplifier OA3.

For a purely common-mode input drive ($V_1 = V_2 = V_{CM}$), $V_1' = V_2' = V_{CM}$. The common-mode gain of OA1 and OA2 is therefore unity, irrespective of the differential gain chosen. Using Eq. (11.4), we thus obtain the common-mode rejection ratio in the following form:

$$G = \left(1 + \frac{2R_2}{R_1}\right)\frac{2\alpha}{\Delta\alpha}$$

where $\Delta\alpha/\alpha$ is the relative matching tolerance of the resistors R_3.

With the instrumentation amplifier in Fig. 20.4, one operational amplifier can be eliminated at the expense of circuit symmetry. Electrometer amplifier OA2 in Fig. 20.5 has the gain $1 + R_1/R_2$. OA1 amplifies potential V_2 by $1 + R_2/R_1$ and simultaneously adds

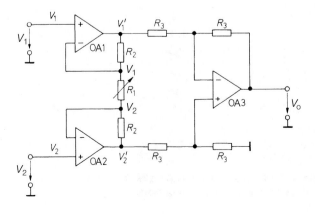

Fig. 20.4. Instrumentation amplifier

$$V_o = \left(1 + \frac{2R_2}{R_1}\right)(V_2 - V_1)$$

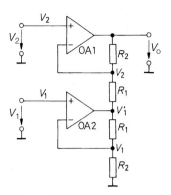

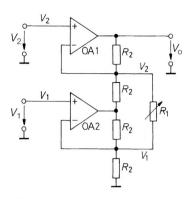

Fig. 20.5. Unbalanced instrumentation amplifier

$$V_o = \left(1 + \frac{R_2}{R_1}\right)(V_2 - V_1)$$

Fig. 20.6. Subtractor with an adjustable gain

$$V_o = 2\left(1 + \frac{R_2}{R_1}\right)(V_2 - V_1)$$

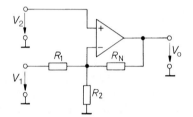

Fig. 20.7. Subtractor with a single high-impedance input

$$V_o = \left(1 + \frac{R_N}{R_1} + \frac{R_N}{R_2}\right)V_2 - \frac{R_N}{R_1}V_1$$

voltage V_1' injected at the base point with weighting $-R_2/R_1$. As a result, the two input potentials are amplified by $1 + R_2/R_1$. If the circuit is modified as shown in Fig. 20.6, the gain can again be set by a single resistor.

For some applications, it is acceptable to use a subtractor with only one high-impedance input. In this case, only a single operational amplifier is required, as shown in Fig. 20.7. However, the transfer equation reveals the limitation that the gain for V_2 is always less than that for V_1 in absolute-value terms (although this is no disadvantage, for example, in the case of the gain and zero offset of sensor signals). An interesting special case arises for $R_N = R_1 = R$ and $R_2 = \infty$; we then obtain the output voltage $V_o = 2V_2 - V_1$.

Subtractors for High Voltages

To subtract high voltages, we can use the circuit shown in Fig. 11.3. The high-voltage operational amplifiers required in this case can often be dispensed with by making $R_1 \gg R_2$; typical component values are shown in Fig. 20.8. The input resistance now becomes so large that the voltage followers can often be omitted. At the same time, the input voltages present at the subtractor are so low that no high-voltage operational amplifier is required. In the example, input voltages of over $200\,\text{V}$ can be applied for a common-mode input voltage range of $\pm10\,\text{V}$.

A disadvantage of selecting these circuit parameters is that they result in subtractors with a gain of $A = R_2/R_1 \ll 1$. Although it is possible to insert a second amplifier to amplify the voltage difference by the required factor, a simpler solution is to use the circuit shown in Fig. 20.9. Here, the attenuation of high input voltages and the gain can

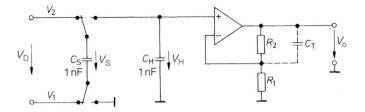

Fig. 20.11. Subtractor in switched-capacitor technology. Example: LTC 2053

$$V_o = \left(1 + \frac{R_2}{R_1}\right)(V_2 - V_1)$$

This voltage can be amplified as required in the following electrometer amplifier, as no further subtraction is required in this case.

The subtraction accuracy is almost entirely determined by the stray capacitances of the switches. In order to minimize their effect, sufficient large capacitors must be used. With the Linear Technology LTC 2053, a common-mode rejection ratio of 120 dB $\cong 10^6$ can be achieved – and this not only for DC voltages, but also at higher frequencies.

The circuit contains three lowpass filters that limit the bandwidth. The first lowpass filter is produced when storage capacitor C_S is charged. The ON-state resistance of the two switches and the internal resistance of the source determine the charge time constant.

A second lowpass filter is produced during the charge transfer to holding capacitor C_H. If the voltage $V_H = 0$, it increases in the first step to $\frac{1}{2}V_D$, in the second to $\frac{3}{4}V_D$, in the third to $\frac{7}{8}V_D$, and so on. The resulting time constant is therefore approximately 2 switch cycles. In order to minimize parasitic charge transfer during switching, low switching frequencies of around 3 kHz $\cong$ 330 µs are selected. Consequently, the circuit can only process low-frequency difference signals, the upper limit being 300 Hz. Superimposed common-mode voltages and ripple voltages of up to 20 kHz have no effect.

A third lowpass filter is provided by capacitor C_T. It is inserted to limit the amplifier bandwidth to the frequency range for which the amplifier is used in order to reduce out of range noise. Examples of IC instrumentation amplifiers are listed in Fig. 20.12.

20.1.3
Isolation Amplifiers

Using the instrumentation amplifiers described above, voltages in the range 10–200 V can be processed, depending on the circuit principle employed. There are many applications, however, in which the voltage to be measured is superimposed on a considerably higher common-mode voltage of perhaps 1000 V. To deal with such high potentials, the measuring circuit is split, as shown in Fig. 20.13, into two electrically isolated units. Electrical isolation may also be required for safety reasons; for example, in most medical applications. The transmitter unit operates at the measuring potential (floating ground), and the receiver unit at system ground. The transmitter must have its own floating supply, which is connected to the floating ground. Although the floating ground is electrically isolated from system ground, it must not be forgotten that there is still some capacitive coupling, which is due mainly to the capacitance C_S of the supply transformer. This is indicated in Fig. 20.13. To keep the coupling low, it is advisable not to use a mains supply transformer but a

Type	Manufacturer	Gain	Offset voltage	Supply voltage	Circuit figure	Pecularity
AD 620	Analog D.	1...1000	50 µV	±2/±18 V	20.10	cheap
AD 623	Analog D.	1...1000	100 µV	±3/± 6 V	20.4	RRO
AD 624	Analog D.	1...1000	25 µV	±6/±18 V	20.5	precision
AD 628	Analog D.	0.01...100	1000 µV	±2/±18 V	20.8	high V_{CM}
AD 8205	Analog D.	50	2000 µV	+ 5 V	20.10	high V_{CM}
AD 8221	Analog D.	1...1000	50 µV	±2/±18 V	20.10	high CMRR at RF
AD 8230	Analog D.	10...1000	5 µV	±4/± 8 V	20.11	autozero
AD 8553	Analog D.	1...10000	25 µV	+2/+ 5 V	20.4	autozero
LT 1101	Lin. Tech.	10, 100	50 µV	±3/±18 V	20.5	low power
LT 1102	Lin. Tech.	10, 100	200 µV	±5/±18 V	20.5	high speed
LT 1167	Lin. Tech.	1...10, 000	20 µV	±2/±18 V	20.4	precision
LT 1190	Lin. Tech.	1...10	800 µV	±2/±18 V	20.9	high V_{CM}
LTC 1100	Lin. Tech.	100	2 µV	±3/ ±9 V	20.5	autozero
LTC 2053	Lin. Tech.	1...1000	10 µV	±2/± 5 V	20.11	autozero
INA 103	Texas I.	1...100	50 µV	±9/±25 V	20.4	low noise
INA 106	Texas I.	10	50 µV	±5/±18 V	20.8	cheap
INA 110	Texas I.	1...500	50 µV	±6/±18 V	20.4	high speed
INA 116	Texas I.	1...1000	2000 µV	±5/±18 V	20.4	low bias current
INA 118	Texas I.	1...10, 000	20 µV	±2/±18 V	20.4	low offset voltage
INA 122	Texas I.	5...10, 000	100 µV	±2/±18 V	20.6	low power, RRO
INA 128	Texas I.	5...10, 000	50 µV	±2/±18 V	20.4	low offset voltage
INA 131	Texas I.	100	25 µV	±2/±18 V	20.4	accurate, cheap
INA 141	Texas I.	10, 100	20 µV	±2/±18 V	20.4	low offset voltage
INA 148	Texas I.	1	1000 µV	±2/±18 V	20.9	high V_{CM}
PGA 204	Texas I.	1...1000	50 µV	±5/±18 V	20.4	digit. gain sel.
PGA 207	Texas I.	1...10	1000 µV	±5/±18 V	20.4	digit. gain sel.

Fig. 20.12. Examples for instrumentation amplifiers. All amplifiers can be operated from a single supply voltage

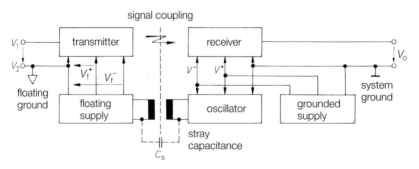

Fig. 20.13. Principle of an isolation amplifier

high-frequency transformer for about 100 kHz, fed from a separate oscillator. In this way, coupling capacitances $C_S < 10$ pF can be achieved.

If both test points have a high internal resistance, even the reduced stray capacitance current may produce a considerable voltage drop in the floating ground loop. In such a case, it is advisable to connect the floating ground terminal to a third point, and to determine the potential difference between the two measuring points using the instrumentation amplifier

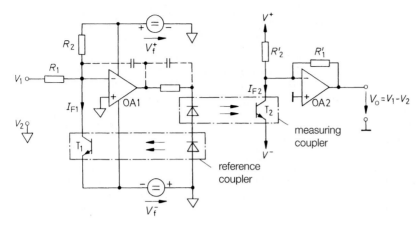

Fig. 20.14. Opto-electronic transmission of an analog signal. A suitable matched dual optocoupler is the HCNR 200 from Agilent.

shown in Fig. 20.4. The measuring signals then carry no current; and the stray current has no influence because it flows through the floating ground connection which is no longer a signal path. The remaining common-mode voltage with respect to floating ground can usually be kept low if the floating ground is connected to a suitable potential within the circuit that is being tested.

The question is how to transmit the analog signal from the transmitter to the receiver with electrical isolation. Three methods exist: transformers, opto-couplers, and capacitors. For transmission by transformers or capacitors, the signal must be modulated by a carrier that offers enough bandwidth (amplitude modulation or pulse-width modulation), whereas opto-couplers enable the direct transmission of DC signals. When high accuracy is required, the analog signal can be digitized at floating-ground potential, and the digital values subsequently transmitted by opto-couplers or capacitors to the receiver unit. Nonlinearities of the transmission path are eleminated this way.

A method of optical analog transmission is presented in Fig. 20.14. In order to compensate for the linearity error of the opto-coupler, the operational amplifier OA1 controls the current though the light-emitting diodes in such a way that the photoelectric current in the reference receiver T_1 equals a desired value. The feedback loop is closed via the reference coupler, so that

$$I_{F1} = \frac{V_f^+}{R_2} + \frac{V_1 - V_2}{R_1}$$

As the photoelectric current cannot change polarity, a constant current V_f^+/R_2 is superimposed on both sides to enable the transmission of bipolar input signals. If the two opto-couplers are well matched, we obtain on the receiver side the current $I_{F2} = I_{F1}$, and thus the output voltage

$$V_o = \frac{R_1'}{R_1}(V_1 - V_2) \quad \text{for} \quad \frac{V_f^+}{R_2} = \frac{V^+}{R_2'}$$

Isolation amplifiers with transformer, opto-, or capacitor coupling are available as ready-made modules. A number of such types are listed in Fig. 20.15. Among the most user-

Type	Manufacturer	Signal transmission	Isolated power	Power bandwidth	Isolation voltage	Pecularity
AD 202	Analog D.	transformer	for input	3 kHz	750 V	cheap
AD 210	Analog D.	transformer	input+output	20 kHz	2500 V	3 port isolation
AD 215	Analog D.	transformer	for input	120 kHz	1500 V	fast
ISO 103	Texas I.	capacitor	for input	10 kHz	1500 V ⎱	complementary
ISO 113	Texas I.	capacitor	for input	10 kHz	1500 V ⎰	power supply
ISO 122	Texas I.	capacitor	external	3 kHz	1500 V	cheap
ISO 124	Texas I.	capacitor	external	32 kHz	1500 V	cheap
HCPL 7510	Agilent	optocoupler	external	15 kHz	1500 V	iso: 15 kV/μs
HCPL 788J	Agilent	digit. opto	external	3 kHz	1500 V	iso: 15 kV/μs

Fig. 20.15. Examples for isolation amplifiers. Examples for isolated power supplies are the DCP 02-series from Texas Instruments or the HPR100-series from Power Convertibles

friendly are the types that already incorporate the required DC converter. An external voltage converter is only worthwhile if it can be used to drive several isolation amplifiers whose floating ground is at the same potential. In the types with a built-in voltage converter, the floating power supply is also available to the user; for example, for driving a preceding instrumentation amplifier or a sensor. An all-purpose device is the Analog Devices AD 210, in which the receiver circuit also operates from a floating supply. Consequently, the receiver signal ground can in this case be isolated from the power supply ground and, as there are therefore three mutually isolated ground terminals, this arrangement is termed "three-port isolation."

20.2
Measurement of Current

20.2.1
Floating Zero-Resistance Ammeter

Section 12.2 describes a current-to-voltage converter that is almost ideally suited for the measurement of currents because of its extremely low input resistance. However, as the input represents virtual ground, only currents to ground can be measured.

Floating ammeters can be realized by the instrumentation amplifier in shown Fig. 20.4 if its two inputs are connected to a measuring shunt. The advantage of the low input resistance is then lost, but if the shunt is incorporated in the feedback loop of the input amplifiers, as illustrated in Fig. 20.16, a floating ammeter is obtained that has virtually zero input resistance.

Since the input voltage differences of both operational amplifiers is zero the potential difference across the inputs 1 and 2 becomes zero, too. If a current I flows into terminal 1, the feedback causes the output potential of OA2 to have the value

$$V_2 = V_i - I R_1 \tag{20.1}$$

With $V_N = V_1$, we obtain

$$V_1 = V_2 + \left(1 + \frac{R_2}{R_2'}\right)(V_i - V_2) = V_i + \frac{R_1 R_2}{R_2'} I \tag{20.2}$$

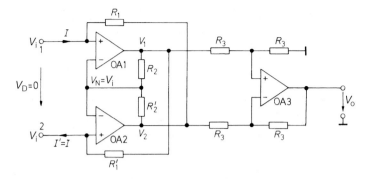

Fig. 20.16. Floating zero-resistance ammeter
$V_o = 2RI$ for $R_1 = R_1' = R_2 = R_2' = R$

Hence, the current leaving terminal 2 is given by

$$I' = \frac{V_1 - V_i}{R_1'} = \frac{R_1 R_2}{R_1' R_2'} I \tag{20.3}$$

If both inputs are to behave like those of a floating circuit, there must be $I' = I$. Hence we have the condition

$$\frac{R_1}{R_1'} = \frac{R_2'}{R_2} \tag{20.4}$$

The subtracting amplifier OA3 computes the difference $V_1 - V_2$. Its output voltage, using (20.1) and (20.2), is therefore

$$V_o = R_1 \left(1 + \frac{R_2}{R_2'}\right) I \tag{20.5}$$

that is, it is proportional to the current I.

20.2.2
Measurement of Current at High Potentials

The permissible common-mode voltage of the previous circuit is limited to values between the supply voltages. For the measurement of currents at higher potentials, the simple circuit of Fig. 12.5 can be employed if it is connected to the floating ground terminal of an isolation amplifier rather than to system ground. Its output voltage is referenced to system ground with the help of the isolation amplifier.

The required circuitry may be reduced quite considerably if a voltage drop of 1...2 V can be tolerated for the measurement of current (e.g., in the anode circuit of high-voltage tubes). In such cases, the current to be measured is made to flow through the light-emitting diode of an opto-coupler, so that a floating supply is no longer needed. For linearization of the transfer characteristic, a reference opto-coupler may be used on the receiver side, as demonstrated in Fig. 20.17. Its input current I_2 is controlled by the operational amplifier in

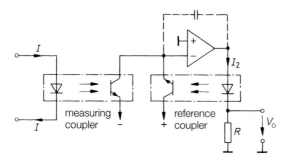

Fig. 20.17. Simple isolation amplifier for current measurement $V_o = RI$

such a manner that the photoelectric currents of the reference and the measuring coupler cancel each other out. If the couplers are well matched,

$$I_2 = I$$

This current can be measured as a voltage across the grounded resistor R.

20.3
AC/DC Converters

Various quantities are used to characterize alternating voltages: the arithmetic mean absolute value, the root-mean-square (rms) value and the positive and negative peak value.

20.3.1
Measurement of the Mean Absolute Value

To obtain the absolute values of an alternating voltage, a circuit is required in which the sign of the gain changes with the polarity of the input voltage; in other words, its transfer characteristic must have the shape represented in Fig. 20.18.

A full-wave rectifier of this kind can be realized by use of a diode bridge. However, the accuracy of such a circuit is limited due to the forward voltages of the diodes. This effect can be avoided if the bridge rectifier is operated from a controlled current source; a simple solution is shown in Fig. 20.19. The operational amplifier is employed as a voltage-controlled current source, in accordance with Fig. 12.8. Hence, independently of the diode forward voltage,

$$I_A = \frac{|V_i|}{R}$$

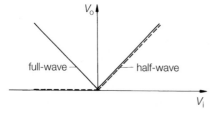

Fig. 20.18. Characteristics of a half-wave and a full-wave rectifier

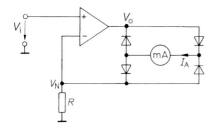

Fig. 20.19. Full-wave rectifier for floating meters
$I_A = |V_i|R$

To display the mean value of this current, a moving-coil ammeter can be used, for instance, and this method is therefore often employed in analog multimeters.

For output potentials in the range $-2V_D < V_o < 2V_D$, the amplifier has no feedback, as none of the diodes conduct. While V_o changes from $2V_D$ to $-2V_D$, V_N remains constant, this causing a delay time within the control loop. Because of this, any phase shift can be incurred in the control loop, depending on the frequencies involved, so that stabilization of the operational amplifier is particularly difficult. To reduce the delay time, amplifiers must be chosen that have a fast slew rate, and the frequency compensation must be stronger than for linear feedback.

Full-Wave Rectifier with a Single-Ended Output

In the previous circuit, the load (i.e., the moving-coil instrument) had to be floating. However, if the signal is to be processed further – for example, digitized – a ground-referenced output voltage is needed; this can be derived from the current I_A by, for example, using a floating current-to-voltage converter. However, Fig. 20.20 shows a simpler method.

Let us first consider the operation of OA1. For positive input voltages, it operates as an inverting amplifier since V_2 is negative; that is, diode D_1 is forward biased and D_2 is reverse biased. Hence, $V_1 = -V_i$. For negative input voltages, V_2 is positive; D_1 is OFF and D_2 is conducting, thereby applying feedback to the amplifier and preventing OA1 from saturating, so that the summing point remains as zero voltage. Since D_1 is reverse biased, V_1 is also zero. Therefore

$$V_1 = \begin{cases} -V_i & \text{for } V_i \geq 0 \\ 0 & \text{for } V_i \leq 0 \end{cases} \tag{20.6}$$

Amplifier OA1 thus operates as an inverting half-wave rectifier.

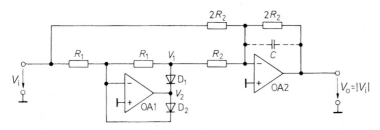

Fig. 20.20. Full-wave rectifier with a single-ended output

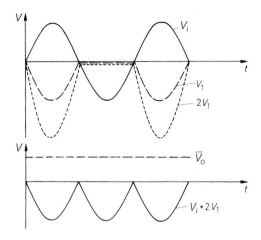

Fig. 20.21. Operation for a sinusoidal input voltage

The extension to a full-wave rectifier is effected by amplifier OA2. It computes the expression

$$V_o = -(V_i + 2V_1) \tag{20.7}$$

which, using Eq. (20.6), becomes

$$V_o = \begin{cases} V_i & \text{for } V_i \geq 0 \\ -V_i & \text{for } V_i \leq 0 \end{cases} \tag{20.8}$$

This is the desired characteristic of a full-wave rectifier. The operation is illustrated in Fig. 20.21.

Amplifier OA2 can be extended by the addition of capacitor C to become a first-order lowpass filter. If the filter cutoff frequency is chosen to be small compared to the lowest signal frequency, a smooth DC voltage is obtained at the output, and has the value

$$V_o = |\overline{V_i}|$$

As in the previous circuit, amplifier OA1 must have a fast slew rate to keep the delay time of the changeover from one diode to the other as short as possible.

Rectification by Sign Reversal

We can see from Eq. (20.8) that with a full-wave rectifier, $A = +1$ for positive voltages and $A = -1$ for negative voltages is realized. This function can also be implemented directly by using an amplifier whose gain can be switched from $+1$ to -1, and by controlling the changeover using the sign of the input voltage. This principle is illustrated in Fig. 20.22. If the input voltages are positive, the noninverting input of the amplifier is used; if they are negative, the comparator changes the switch over to the inverting input.

For amplifier A, it would of course be impossible to use an operational amplifier without external circuitry, as its gain $A_D \gg 1$. However, a suitable solution is provided by the circuit shown in Fig. 17.26 on page 941, where the gain can be switched between $+1$ and -1 using switch S. For higher frequencies, the wideband multiplexer in Fig. 17.22 on page 939 is more suitable, and Fig. 20.23 shows how this circuit can be operated as a rectifier. The input amplifier is connected to the input voltage in such a way that signals

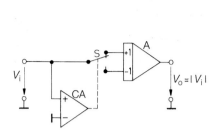

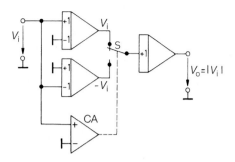

Fig. 20.22. Rectification by sign reversal

Fig. 20.23. Practical implementation of rectification by gain switching

of opposite polarity are produced. Depending on which input amplifier is selected by the comparator, the output voltage $+V_i$ or $-V_i$ is obtained.

This method of rectification is practicable because there are ICs that operate on this principle, such as the Analog Devices AD 630, which also contains the required comparator. However, at high frequencies the delay due to the comparator introduces appreciable errors, as the delayed changeover then becomes a critical factor.

Wideband Full-Wave Rectifier

A differential amplifier provides an inverting and a noninverting output, and can therefore be used as a fast full-wave rectifier. In this application, two parallel-connected emitter followers T_3 and T_4 are used, as shown in Fig. 20.24, to connect the more positive collector potential to the output. The Zener diode compensates for the collector quiescent potential to make the output quiescent potential zero.

The same principle of full-wave rectification can be very well utilized with CC operational amplifiers. The circuit in Fig. 20.25 is based on the differential amplifier of Fig. 5.88 on page 558. From the complementary output currents the respective positive current is routed through D_3 or D_4 to resistance R_2. Diodes D_1 and D_2 conduct negative currents to ground to prevent saturation of the amplifiers. Schottky diodes can be used for the four diodes in bridge-rectifier configuration. A capacitor can be connected in parallel to R_2 to obtain the mean value. Amplifier T_3 serves as an impedance converter.

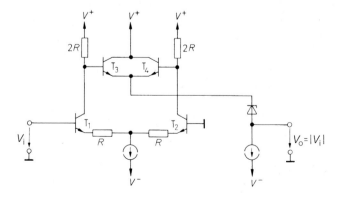

Fig. 20.24. Wideband full-wave rectifier

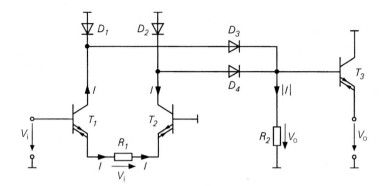

Fig. 20.25. Differential amplifier of CC operational amplifiers for full-wave rectification

$$V_o = |I|R_2 = \frac{R_2}{R_1}|V_i|$$

20.3.2
Measurement of the rms Value

Whereas the arithmetic mean absolute value (the mean modulus) is defined as

$$\overline{|V|} = \frac{1}{T}\int_0^T |V|dt \tag{20.9}$$

the definition of the root-mean-square value (the rms value) is given as

$$V_{\text{rms}} = \sqrt{\overline{(V^2)}} = \sqrt{\frac{1}{T}\int_0^T V^2 dt} \tag{20.10}$$

where T is the measuring interval, which must be large compared to the longest period contained in the signal spectrum. In this manner, a reading is obtained that is independent of T. For strictly periodic functions, averaging over one period is sufficient to yield the correct result.

For sinusoidal voltages,

$$V_{\text{rms}} = \widehat{V}/\sqrt{2}$$

so that the rms value can be determined simply by measuring the peak value. For other wave-shapes, this method would produce errors, particularly for highly peaked voltages; that is, for wave-shapes that have large *crest factors* $\widehat{V}/V_{\text{rms}}$.

The errors become smaller if the measurement of rms values is reduced to that of the mean absolute values. For a sinusoidal voltage,

$$\overline{|V|} = \frac{\widehat{V}}{T}\int_0^T |\sin \omega t|dt = \frac{2}{\pi}\widehat{V} \tag{20.11}$$

and, with $V_{\text{rms}} = \widehat{V}/\sqrt{2}$,

$$V_{\text{rms}} = \frac{\pi}{2\sqrt{2}}\overline{|V|} \approx 1.11 \cdot \overline{|V|} \tag{20.12}$$

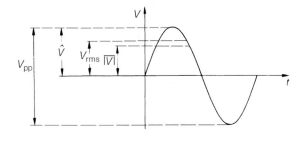

Fig. 20.26. Relative magnitude of peak value, rms value and mean absolute value of sinusoidal signal

Signal	rms value	Reading
Sinusoidal	$V_{rms} = \dfrac{\pi}{2\sqrt{2}}\overline{|V|} = 1.11\,\overline{|V|}$	exact
DC and square wave	$V_{rms} = \overline{|V|}$	11% too large
Triangular wave	$V_{rms} = \dfrac{2}{\sqrt{3}}\overline{|V|} = 1.15\,\overline{|V|}$	4% too small
White Noise	$V_{rms} = \sqrt{\dfrac{\pi}{2}}\overline{|V|} = 1.25\,\overline{|V|}$	11% too small

Fig. 20.27. Errors in the rms value for different waveforms if the mean absolute value is calibrated for sinusoidal signals

The relative magnitude of the individual values is demonstrated in Fig. 20.26. The *form factor* of 1.11 is incorporated in the calibration of most available mean-absolute value meters; for sinusoidal signals, therefore, they show the rms value although they actually measure the mean-absolute value. For other wave-shapes, this modified reading produces varying deviations from the true rms value as shown in Fig. 20.27.

Measurement of True rms Values

For a true, wave-shape-independent measurement of an rms value, either the definition given in (20.10) can be employed, or the power can be measured.

The operation of the circuit shown in Fig. 20.28 is based on (20.10). In order to evaluate the mean value of the squared input voltage, a simple first-order lowpass filter is used, the cutoff frequency of which is low compared to the lowest signal frequency.

One drawback of the circuit is that its minimum input signal must be relatively large; if, for instance, a voltage of 10 mV is applied to the input and the computing unit $U = 10\,\mathrm{V}$ as usual, a voltage of only $10\,\mu\mathrm{V}$ is obtained at the output of the squarer. This, however, will be drowned by the noise of the square-rooter.

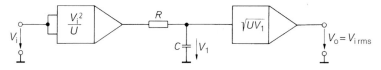

Fig. 20.28. Measurement of true rms values by analog computing circuits

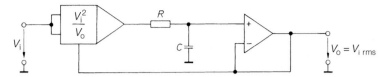

Fig. 20.29. Circuit for true rms measurement, having an improved input voltage range

In this respect, the circuit shown in Fig. 20.29 is preferable, since the square-rooting operation at the output is replaced by division at the input. The voltage at the output of the lowpass filter is thus

$$V_o = \overline{\left(\frac{V_i^2}{V_o}\right)} \tag{20.13}$$

At steady-state, V_o is constant; hence,

$$V_o = \frac{\overline{(V_i^2)}}{V_o} \quad \text{that is} \quad V_o = \sqrt{(\overline{V_i^2})} = V_{\text{rms}}$$

An advantage of this method is that the input voltage V_i is not multiplied by the factor V_i/U which, for low input voltages, is small with respect to unity, but by the factor V_i/V_o, which is close to unity. The available input range is therefore considerably larger. However, the precondition for this is that the division V_i/V_o can be carried out sufficiently accurately even for small signals. Divider circuits based on logarithmic operations are best suited for this purpose, as described in Sect. 11.8.1.

The implicit equation (20.13) can therefore be solved using the principle illustrated in Fig. 20.30. Before taking the logarithm, we must first form the absolute value of the input voltage. Squaring is performed simply by multiplying the logarithm by two. To divide by V_o, the logarithm of the output voltage is subtracted.

The practical implementation of this principle is shown in Fig. 20.31. The full-wave rectified input signal is produced at the summing point of OA2. The latter forms the logarithm of the input voltage. The voltage doubling required for squaring is effected by the two transistors T_1 and T_2 connected in series:

$$V_2 = -2V_T \ln \frac{V_i}{I_{CO}R} = -V_T \ln \left(\frac{V_i}{I_{CO}R}\right)^2$$

OA4 takes the logarithm of the output voltage:

$$V_4 = -V_T \ln \frac{V_o}{I_{CO}R}$$

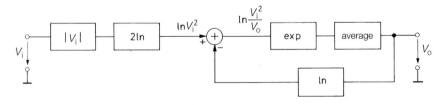

Fig. 20.30. Computation of true rms values via logarithms

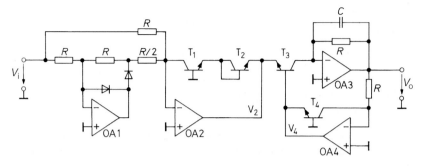

Fig. 20.31. Practical method of true rms calculation.

Output voltage: $V_o = \sqrt{\overline{V_i^2}} = V_{\mathrm{rms}}$

Type	Manufacturer	Technology	Accuracy	Bandwidth
AD 637	Analog Devices	bipolar	0.1%	80 kHz
AD 736	Analog Devices	bipolar	0.3%	30 kHz
AD 8361	Analog Devices	bipolar	0.5 dB	2.5 GHz
LTC 1966	Linear Tech.	CMOS	0.1 %	1 kHz
MX 536	Maxim	bipolar	0.2 %	45 kHz

Fig. 20.32. Chips for true rms conversion

The voltage $V_4 - V_2$ across T_3, implementing the exponential function, produces the output voltage

$$V_o = I_{CS} R \exp \frac{V_4 - V_2}{V_T} = \frac{V_i^2}{V_o} \qquad (20.14)$$

Using capacitor C for averaging, the same output voltage is therefore produced as expressed by (20.13)

Transistors T_1–T_4 must be of monolithic IC form to insure that they possess identical characteristics – as was assumed in the above calculation. It is even possible to integrate the operational amplifiers and resistors on the same chip, as shown in Fig. 20.32.

Thermal Conversion

The rms value of an AC voltage is defined as that DC voltage which will produce the same average power in a resistor; that is,

$$\overline{V_i^2}/R = V_{\mathrm{rms}}^2/R$$

The rms value of an AC voltage V_i can therefore be determined by increasing a DC voltage V_{rms} across a resistor R until the latter is exactly as hot as that heated by V_i. This is the principle on which thermal rms measurement is based. Essentially, any method of temperature measurement can be employed (see Sect. 21.1). A particularly useful one is to use temperature sensors, which can be fabricated together with the heating resistors as an IC. Consequently, diodes are mainly used nowadays as temperature sensors, as illustrated in Fig. 20.33.

Resistor R_1 is heated by the input voltage, and resistor R_2 by the output voltage. The latter increases until the difference between the two diode voltages is zero; in other words,

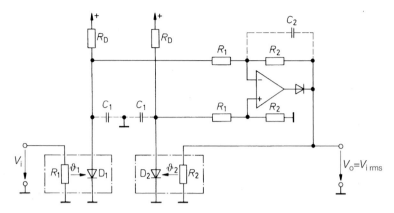

Fig. 20.33. Measurement of rms values by thermal conversion

both temperatures are the same. In this case, the operational amplifier configured as a subtractor with a lowpass filter is used as the control amplifier. The capacitors C_1 keep high-frequency signals away from the operational amplifier.

The diode at the amplifier output prevents resistor R_2 being heated by a negative voltage, as this would result in latch-up of the circuit due to positive thermal feedback.

Since the thermal power is proportional to the square of V_o, the loop gain is also proportional to V_o^2, giving a nonlinear step response: the switch-off time constant is considerably larger than the switch-on time constant. A considerable improvement is achieved with an additional square-function AC feedback circuit.

As resistors R_1 and R_2 are usually low-value components ($50\,\Omega$) in order to achieve a large bandwidth, correspondingly high currents are required to provide the drive. Consequently, an emitter follower is generally inserted at the output of the control amplifier. A preamplifier or impedance converter at the input would have to be of considerably more complex design, since it would not only have to provide high current peaks of several $100\,\text{mA}$, but also be capable of handling the bandwidth of the AC input signal. Therefore the input signal is directly connected to R_1 and preamplifiers are not applied.

In order to obtain accurate measurements, the two measuring couples must have good matching characteristics. An IC that meets these requirements is the LT 1088 from Linear Technology, with which accuracies of 1% can be achieved up to 100 MHz. In power meters frequencies up to 40 GHz are attained by this method.

20.3.3
Measurement of the Peak Value

The peak value can be measured very simply by charging a capacitor via a diode. To eliminate its forward voltage, the diode is inserted into the feedback loop of a voltage follower, as shown in Fig. 20.34. For input voltages $V_i < V_C$, the diode is reverse biased. For $V_i > V_C$ the diode conducts and, due to the feedback, $V_C = V_i$. The capacitor therefore charges to the peak input voltage. The voltage follower OA2 draws only very little current from the capacitor, so that the peak value can be stored over a long period. The push-button switch P discharges the capacitor to prepare it for the next measurement.

The capacitive load of amplifier OA1 may give rise to oscillations; this effect is eliminated by resistor R_1. However, the charging time is increased as the capacitor voltage

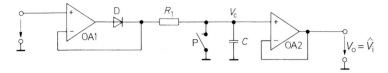

Fig. 20.34. Peak value measurement

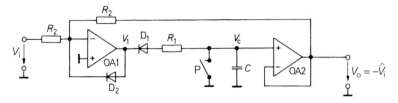

Fig. 20.35. Improved method of peak value measurement

now only approaches steady-state conditions asymptotically. A further disadvantage of the circuit is that OA1 has no feedback if $V_i < V_C$; that is, is saturated. The resulting recovery time limits the use of this circuit to low-frequency applications.

Both of these drawbacks are avoided with the peak detector shown in Fig. 20.35. Here, OA1 is operated in the inverting mode. If V_i exceeds the value of $-V_C$, V_1 becomes negative and diode D_1 becomes conducting. Because of the feedback across the two amplifiers, V_1 assumes such a value that $V_o = -V_i$. In this way, the diode forward voltage, as well as the offset voltage of the impedance converter OA2, is eliminated. If the input voltage falls, V_1 rises, thus reverse biasing diode D_1. The feedback path via R_2 is opened, but V_1 can only rise until diode D_2 becomes conducting and provides feedback for OA1. In this manner, saturation of OA1 is avoided.

The inverted positive peak value of V_i remains stored at capacitor C, as no leakage current can flow either through D_1 or through the impedance converter OA2. Before a new reading, capacitor C must be discharged by switch P. For the measurement of negative peak voltages, the diodes must be reversed.

Another method of implementing a peak voltmeter consists of using a sample-and-hold circuit and issuing the sample command at the appropriate instant. For this purpose, we can simply use a comparator, as shown in Fig. 20.36, to determine when the input voltage is larger than the output voltage, and close switch S of the sample-and-hold at this time. The output signal then follows the input signal as long as it rises, and remains stored when

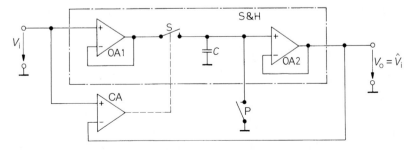

Fig. 20.36. Peak value measurement using a sample-and-hold circuit

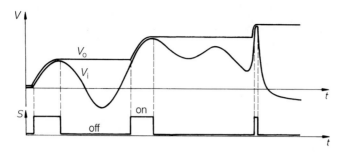

Fig. 20.37. Signal waveforms in the peak voltmeter with sample-and-hold circuit

it falls again. The output voltage only rises further when the input voltage exceeds the last maximum stored. A typical mode of operation is shown in Fig. 20.37. This circuit can be implemented using the sample-and-hold circuits of Fig. 17.32 on page 944 and the comparators from Fig. 6.39 on page 601.

Continuous Measurement of the Peak Value

The method described can be adapted for continuous peak voltage measurements if switch P is replaced by a resistor, the value of which is chosen to be large enough to insure that the discharge of the capacitor C between two voltage crests is negligible. However, this method has the disadvantage that a decrease in peak amplitude is registered only very slowly.

It is important for many applications – particularly in control circuits – for the peak amplitude to be measured with the shortest possible delay. With the methods described so far, the measuring time is at least one period of the input signal. However, for sinusoidal signals, the peak amplitude can be computed at any instant using the trigonometric equation

$$\widehat{V} = \sqrt{\widehat{V}^2 \sin^2 \omega t + \widehat{V}^2 \cos^2 \omega t} \tag{20.15}$$

We have already used this relationship for the amplitude control of the oscillator in Fig 14.30 on page 858. There, it was particularly practicable, since both the sine and cosine functions were available.

For the measurement of an unknown sinusoidal voltage, we must establish the $\cos \omega t$ function from the input signal; a differentiator can be employed for this purpose. We obtain its output voltage as follows:

$$V_1(t) = -RC\frac{dV_i(t)}{dt} = -\widehat{V}_i RC\frac{d\sin\omega t}{dt} = -\widehat{V}_i \omega RC \cos\omega t \tag{20.16}$$

If the frequency is known, the coefficient ωRC can be adjusted to 1, so that the required term for (20.15) is obtained. By squaring and adding $V_i(t)$ and $V_1(t)$, a continuous voltage for the amplitude is obtained which requires no filtering.

For variable frequencies, the circuit must be expanded as shown in Fig. 20.38, by inserting an integrator to enable provision of the $\cos^2 \omega t$ term with a frequency-independent amplitude. The output voltage of the integrator is

$$V_2(t) = -\frac{1}{RC}\int V_i(t)dt = -\frac{1}{RC}\int \widehat{V}_i \sin\omega t \, dt = \frac{\widehat{V}_i}{\omega RC}\cos\omega t \tag{20.17}$$

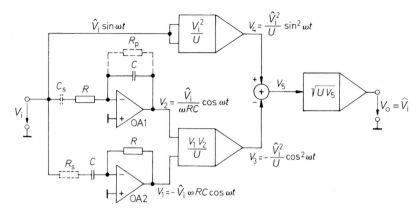

Fig. 20.38. Circuit for instantaneous amplitude measurement of sinusoidal signals

The average of the integrator output is forced to zero be means of C_s and R_p. By multiplying V_1 and V_2, the required expression,

$$V_3(t) = -\frac{\widehat{V}_i^2}{U} \cos^2 \omega t$$

is obtained. By subtracting this from V_4 and square-rooting the result, the output voltage $V_o = \widehat{V}_i$ is measured, which is equal to the amplitude of the input voltage at any instant. For steep changes in the amplitude, a temporary deviation from the correct value is incurred until the integrator output resettles to zero mean voltage. However, the output voltage will instantly change to the correct direction, so that a connected controller, for example, can obtain accurate information about the trend without delay.

20.3.4
Synchronous Demodulator

In a synchronous demodulator (synchronous detector, phase-sensitive rectifier), a rectifier is controlled by an external signal that has the same origin as the input signal.

A synchronous demodulator can be used in the arrangement shown in Fig. 20.39, to separate a sine wave from a noisy signal and to determine its amplitude. The sine wave selected has a frequency equal to that of the control signal V_{CS}, and a phase shift φ that is

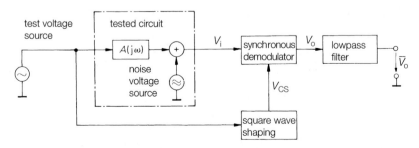

Fig. 20.39. Application of a synchronous demodulator by the measurement of noisy signals

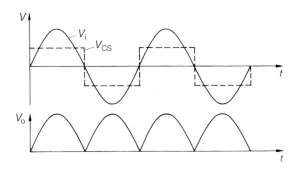

Fig. 20.40. Operation of a synchronous demodulator

constant with respect to the control signal. The special case in which $f_i = f_{CS}$ and $\varphi = 0$ is illustrated in Fig. 20.40. It can be seen that, under these conditions, the synchronous demodulator operates as a full-wave rectifier. If $\varphi \neq 0$ or $f_i \neq f_{CS}$, negative voltage–time areas occur as well as the positive areas, and reduce the mean value of the output voltage so that it is always lower than that of the example shown.

The output voltage will now be determined as a function of frequency and phase. The input voltage V_i is multiplied by $+1$ or by -1 in time with the control frequency f_{CS}, this effect being represented mathematically as

$$V_o = V_i(t) \cdot V_{CS}(t) \tag{20.18}$$

where

$$V_{CS}(t) = \begin{cases} 1 & \text{for } V_{CS} > 0 \\ -1 & \text{for } V_{CS} < 0 \end{cases}$$

By rewriting this in Fourier series form, we obtain

$$V_{CS}(t) = \frac{4}{\pi} \sum_{n=0}^{\infty} \frac{1}{2n+1} \sin(2n+1)\omega_{CS}t \tag{20.19}$$

Let us now assume the input voltage to be a sinusoidal voltage that has a frequency $f_i = m \cdot f_{CS}$ and a phase angle φ_m. Using (20.18) and (20.19), we then have the output voltage:

$$V_o(t) = \widehat{V}_i \sin(m\omega_{CS}t + \varphi_m) \cdot \frac{4}{\pi} \sum_{n=0}^{\infty} \frac{1}{2n+1} \sin(2n+1)\omega_{CS}t \tag{20.20}$$

The arithmetic mean value of this voltage is evaluated by the subsequent lowpass filter. Using the auxiliary equation

$$\frac{1}{T} \int_0^T \sin(m\omega_{CS}t + \varphi_m) = 0$$

and the orthogonal relation of sinusoidal functions, that is

$$\frac{1}{T} \int_0^T \sin(m\omega_{CS}t + \varphi_m) \sin l\omega_{CS}t \, dt = \begin{cases} 0 & \text{for } m \neq l \\ \frac{1}{2}\cos\varphi_m & \text{for } m = l \end{cases}$$

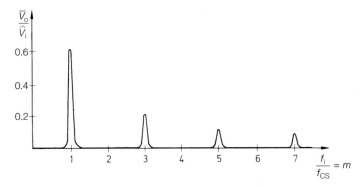

Fig. 20.41. Filter characteristic of a synchronous demodulator. A suitable chip for frequencies up to 2 MHz is the AD 630 from Analog Devices. The gain and phase detector AD 8302 is even capable up to 2.7 GHz.

and with (20.20), we obtain the final result:

$$\overline{V}_o = \begin{cases} \dfrac{2}{\pi m}\widehat{V}_i \cdot \cos\varphi_m & \text{for} \quad m = 2n+1 \\ 0 & \text{for} \quad m \neq 2n+1 \end{cases} \tag{20.21}$$

where $n = 0, 1, 2, 3, \ldots$.

If the input voltage is a mixture of frequencies, the only components that contribute to the output voltage are those the frequencies which are equal to the control frequency, or are an odd multiple thereof. This explains why the synchronous demodulator is particularly suitable for selective amplitude measurements. The synchronous demodulator is also known as the *phase-sensitive rectifier*, since the output voltage is dependent on the phase angle between the appropriate component of the input voltage and the control voltage.

For $\varphi_m = 90°$, $\overline{V}_o$ is zero even if the frequency condition is fulfilled. For our example in Fig. 20.40, we have chosen $\varphi_m = 0$ and $m = 1$. In this case, (20.21) yields

$$\overline{V}_o = \frac{2}{\pi}\widehat{V}_i$$

which is the arithmetic mean of a full-wave rectified sinusoidal voltage; a result that could have been deduced directly from Fig. 20.40.

Equation (20.21) has shown that only those voltages whose frequencies are equal to the control frequency or are odd multiples thereof contribute to the output voltage. However, this only holds if the time constant of the lowpass filter is infinitely large. In practice, this is not possible and is not even desirable, as the upper cutoff frequency would be zero; in other words, the output voltage could not change at all. If $f_c > 0$, the synchronous demodulator no longer picks out discrete frequencies, but individual frequency bands. The 3 dB bandwidth of these frequency bands is $2f_c$, and Fig. 20.41 shows the resulting filter characteristic.

The mostly unwanted contribution of the odd-order harmonics can be avoided by using an *analog multiplier* for synchronous demodulation instead of the polarity-changing switch (chips are given in Fig. 11.39). The input voltage is then multiplied by a sinusoidal function $V_{CS} = \widehat{V}_{CS}\sin\omega t$, rather than by the square wave function $V_{CS}(t)$. As this sine wave no longer contains harmonics, (20.21) only holds for $n = 0$. If the amplitude of the control

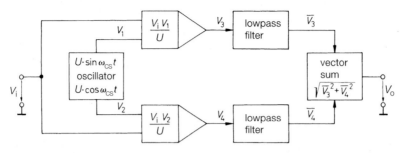

Fig. 20.42. Phase-independent synchronous demodulation

$V_o = \frac{1}{2}\widehat{V}_i$ for $f_{CS} = f_i$

voltage is chosen to be equal to the computing unit voltage U of the multiplier, we obtain, instead of (20.21), the result

$$\overline{V}_o = \begin{cases} \frac{1}{2}\widehat{V}_i \cos\varphi & \text{for } f_i = f_{CS} \\ 0 & \text{for } f_i \neq f_{CS} \end{cases} \tag{20.22}$$

According to (20.22), the synchronous demodulator does not produce the amplitude $\widehat{V}_i$, directly, but gives the real part $\widehat{V}_i \cos\varphi$ of the complex amplitude $\widehat{V}_i$. To determine the magnitude $|\underline{V}_i| = \widehat{V}_i$, the phase angle of the control voltage can be adjusted by a suitable phase-shifting network so that the output voltage of the demodulator is at a maximum. The signal $V_i(t)$ and the control voltage $V_{CS}(t)$ are then in phase, and we obtain, from (20.22),

$$\overline{V}_o = \frac{1}{2}\widehat{V}_i = \frac{1}{2}|\underline{V}_i| \quad \text{for} \quad f_i = f_{CS}$$

If a calibrated phase-shifter is employed, the phase shift φ of the tested circuit can be read directly.

Often, we are only interested in the amplitude of a spectral input component and not in its phase angle. If one wants nevertheless to use the selectivity of the synchronous rectifier two synchronous demodulators can be used in parallel that are operated by two quadrature control voltages as shown in Fig. 20.42:

$$V_1(t) = U\sin\omega_{CS}t \quad \text{and} \quad V_2(t) = U\cos\omega_{CS}t$$

where U is the computing unit voltage of the demodulating multipliers. The oscillator shown in Fig. 14.30 on page 858 is particularly suitable for generating these two voltages.

Only the spectral component of the input voltage which has the frequency f_{CS} contributes to the output voltages of the two demodulators. If it has the phase angle φ with respect to V_1, it is of the form

$$V_i = \widehat{V}_i \sin(\omega_{CS}t + \varphi)$$

According to (20.22), the upper demodulator produces the output voltage

$$\overline{V}_3 = \frac{1}{2}\widehat{V}_i \cos\varphi \tag{20.23}$$

whereas the lower demodulator gives

$$\overline{V}_4 \;=\; \frac{1}{2}\widehat{V}_i \sin \varphi \tag{20.24}$$

By squaring and adding, we obtain the output voltage

$$V_o \;=\; \frac{1}{2}\widehat{V}_i\sqrt{\sin^2\varphi + \cos^2\varphi} = \frac{1}{2}\widehat{V}_i \tag{20.25}$$

which is independent of the phase angle φ. The circuit can therefore be used as a tunable selective voltmeter. Its bandwidth is constant and equal to twice the cutoff frequency of the lowpass filters. The attainable Q-factor is considerably higher than that of conventional active filters: a 1 MHz signal can be filtered with a bandwidth of 1 Hz without any trouble. This corresponds to a Q-factor of 10^6.

If the control frequency f_{CS} of the local oscillator is made to sweep through a given range, the circuit operates as a spectrum analyzer.

Chapter 21:
Sensors and Measurement Systems

This chapter deals with circuits for measuring nonelectrical quantities. For this purpose, it must first be detected by a sensor and then converted into an electrical quantity. The interface circuit for the sensor normally converts this quantity into a voltage which, after conditioning, is then displayed or employed for control purposes.

The individual stages are shown in Fig. 21.1. A concrete example is then given in Fig. 21.2, for a humidity sensor. In this example, the sensor has a capacitance that is dependent on the relative humidity. In order to measure it, the sensor must be incorporated in a capacitance-measuring circuit, the output of which delivers a voltage that is proportional to the capacitance but in no way proportional to the humidity. Another circuit is therefore required for linearization and calibration of the sensor. There are a wide variety of sensors for the most diverse measurands and measurement ranges. An overview of the types available is provided in Fig. 21.3.

21.1
Temperature Measurement

The following subsections describe various ways in which temperature can be measured. It is evident from the overview table that metallic sensors, such as the thermocouple and the resistance thermometer, can be employed for a very large range of temperatures. Semiconductor-based temperature sensors (PTC and NTC thermistors, transistors) produce larger output signals but at a smaller temperature range.

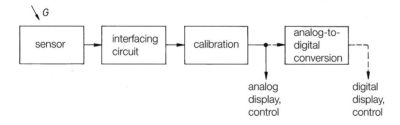

Fig. 21.1. Conversion of a physical quantity G into a calibrated electrical signal

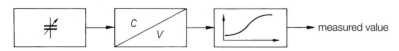

Fig. 21.2. A humidity sensor, as an example of how the measured value is obtained

Measurand	Sensor	Measurement range	Principle
Temperature	PTC metal	$-200 \ldots +800°C$	Positive temperature coefficient of the resistance of metals; e.g., platinum
	PTC thermistor	$-50 \ldots +150°C$	Positive temperature coefficient of the resistance of semiconductors; e.g., silicon
	NTC thermistor	$-50 \ldots +150°C$	Negative temperature coefficient of the resistance of metal-oxide ceramic
	Transistor	$-50 \ldots +150°C$	Negative temperature coefficient of the base–emitter voltage of a transistor
	Thermocouple	$-200 \ldots +2,800°C$	Thermo-electric voltage at contact of different metals
	Crystal oscillator	$-50 \ldots +300°C$	Temperature coefficient of the resonant frequency of specially cut quartz crystals
Temperature via heat radiation	Pyrometer	$-100 \ldots +3,000°C$	The spectral distribution of the luminance is temperature-dependent
	Pyroelement	$-50 \ldots +2,200°C$	The increase in temperature due to radiated heat generates a polarization voltage
Light intensity	Photodiode Phototransistor	$10^{-2} \ldots 10^5 \, \text{lx}$	Current increases with light intensity due to optically released charge carriers
	Photoresistor	$10^{-2} \ldots 10^5 \, \text{lx}$	Electrical resistance reduces as the illumination increases
	Photomultiplier	$10^{-6} \ldots 10^3 \, \text{lx}$	Light releases electrons from a photocathode, which are multiplied by subsequent dynodes
Sound	Dynamic microphone		The induction of a voltage by movement of a coil within a magnetic field
	Condenser microphone		The voltage of a charged capacitor varies with the distance between the plates
	Crystal microphone		The piezoelectric effect generates a voltage
Magnetic field	Induction coil		Supplies voltage if the magnetic field changes or the coil moves within the field
	Hall-effect device	$0.1 \, \text{m} \ldots 1 \, \text{T}$	Produces a voltage across the semiconductor by deflection of electrons in the magnetic field
	Magnetoresistor	$0.1 \ldots 1 \, \text{T}$	Resistance increases in the semiconductor as a function of field strength

Fig. 21.3. Overview of sensors, part 1

Measurand	Sensor	Measurement range	Principle
Force	Strain gauge	$10^{-2} \ldots 10^{7}$N	Force causes elastic elongation of a thin-film resistor, thereby increasing its resistance
Pressure	Strain gauge	$10^{-3} \ldots 10^{3}$ bar	The bridge circuit of the strain gauge on the diaphragm is detuned by pressure
Acceleration	Strain gauge	$1 \ldots 5{,}000$ g	The strain-gauge bridge is detuned by acceleration force on weighted diaphragm
Linear displacement	Potentiometric displacement transducer	μm $\ldots$ m	The potentiometer tap is shifted
	Inductive displacement transducer	μm $\ldots 10^{-1}$ m	The inductive bridge is unbalanced by displacement of a ferrite core
	Incremental displacement transducer, optical	μm $\ldots$ m	The reticle pattern is scanned. The number gives the displacement
Angle	Incremental angular displacement transducer, optical	$1 \ldots 20{,}000$ per revolution	The reticle pattern is scanned. The number gives the angle of rotation
	Incremental angular displacement transducer, magnetic	$1 \ldots 1{,}000$ per revolution	Magnetic scanning of a toothed-wheel sensor
	Incremental angular displacement transducer, capacitive	$1 \ldots 1{,}000$ per revolution	Capacitive scanning of a toothed-wheel sensor
Flow velocity	Windmill-type anemometer		The rotational speed increases with the flow speed
	Heated-wire anemometer		Cooling increases with the flow rate
	Ultrasound transceiver		The Doppler shift increases with the flow rate
Gas concentration	Ceramic resistor		The resistance changes with the adsorption of the test substance
	MOSFET		Change in threshold voltage during adsorption of the test substance under the gate
	Absorption spectrum		Absorption lines are characteristic for each gas
Humidity	Capacitor	$1 \ldots 100\%$	The dielectric constant increases due to water absorption as the relative humidity rises
	Resistor	$5 \ldots 95\%$	The resistance decreases due to water absorption as the relative humidity rises

Fig. 21.3. Overview of sensors, part 2

21.1.1
Metals as PTC Thermistors

The resistance of metals increases with temperature; that is, metals possess a positive temperature coefficient. The metals most widely used for temperature measurement are platinum and nickel–iron. To a first approximation, the resistance increases linearly by some 0.4% per degree of temperature. Thus, if the temperature increases by 100 K, the resistance increases by a factor of 1.4.

With *platinum temperature detectors*, the resistance R_0 is specified at 0°C. A usual value is 100 Ω (Pt 100), but it can also be 200 Ω (Pt 200), 500 Ω (Pt 500), or 1,000 Ω (Pt 1,000 Ω). In the range $0°C \leq \vartheta \leq 850°C$, the resistance is given by the equation (DIN43760 and IEC 570):

$$R_\vartheta = R_0\left[1 + 3.90802 \cdot 10^{-3}\vartheta/°C - 0.580195 \cdot 10^{-6}(\vartheta/°C)^2\right]$$

and in the range $-200°C \leq \vartheta \leq 0°C$ by

$$R_\vartheta = R_0[1 + 3.90802 \cdot 10^{-3}\vartheta/°C - 0.580195 \cdot 10^{-6}(\vartheta/°C)^2$$
$$+ 0.42735 \cdot 10^{-9}(\vartheta/°C)^3 - 4.2735 \cdot 10^{-12}(\vartheta/°C)^4]$$

The usable temperature range of $-200°C$ to $+850°C$ is very wide. For higher temperatures, thermocouples are employed (see Sect. 21.1.6). The nonlinearity of the equation is relatively small. Within a limited range of temperatures, linearization can therefore often be dispensed with. Examples of interfacing circuits are given in Sect. 21.1.4.

With *nickel–iron temperature detectors*, the nominal resistance R_0 is specified at 20°C. The temperature curve within the range $-50°C \leq \vartheta \leq 150°C$ is then given by

$$R_\vartheta = R_{20}[1 + 3.83 \cdot 10^{-3}(\vartheta/°C) + 4.64 \cdot 10^{-6}(\vartheta/°C)^2]$$

It can be seen that in addition to the linear term there is a quadratic component that causes a deviation of approximately 25° at 150°C. Linearization is therefore invariably necessary. Section 21.1.4 describes the operation and design of the relevant interfacing circuits.

21.1.2
Silicon-Based PTC Thermistors

The resistance of uniformly doped silicon increases with temperature. The temperature coefficient is approximately twice as large as for metals. The resistance approximately doubles for an increase in temperature of 100 K. The relevant equation takes the form

$$R_\vartheta = R_{25}[1 + 7.95 \cdot 10^{-3}\Delta\vartheta/°C + 1.95 \cdot 10^{-5}(\Delta\vartheta/°C)^2]$$

This applies only for sensors of KTS-series from Infineon or Philips and is only approximate for other manufacturers. In the equation, R_{25} is the nominal resistance at 25°C, being mainly between 1 and 2 kΩ. $\Delta\vartheta$ is the difference between the actual temperature and the nominal temperature: $\Delta\vartheta = \vartheta - 25°C$. As with nickel–iron sensors, the usable temperature range is between 50°C and +150°C. Section 21.1.4 will show how silicon PTC thermistors are used and how their characteristics are linearized.

21.1.3
NTC Thermistors

NTC thermistors are temperature-dependent resistors with a negative temperature coefficient. They are made of metal-oxide ceramic material and their temperature coefficients are very large, being between -3% and -5% per degree. NTC *power thermistors* are used for inrush current limiting. With these devices, heating due to the flow of current is desirable. When hot, they must possess a low resistance and a high current-carrying capacity. In contrast, self-heating in NTC *measurement thermistors* is kept to a minimum. What matters here is a resistance curve that is specified as precisely as possible. If the temperature of interest T is close to the nominal temperature T_N, the relationship between temperature and resistance can be approximated by the relation

$$R_T = R_N \cdot \exp\left[B\left(\frac{1}{T} - \frac{1}{T_N}\right)\right] \tag{21.1}$$

Temperatures must be inserted in Kelvin ($T = \vartheta + 273°$). Depending on the type of thermistor, the constant B is between 1,500 K and 7,000 K. To enable the resistance characteristic to be described precisely, even if the temperature differences are large, it is preferable to use the equation

$$\frac{1}{T} = \frac{1}{T_N} + \frac{1}{B}\ln\frac{R_T}{R_N} + \frac{1}{C}\left(\ln\frac{R_T}{R_N}\right)^3$$

This additionally includes the term with the coefficient 1/C, allowing an accuracy of 0.1 K to be achieved even over a temperature range of 100 K. This naturally requires that the coefficients or the resistance characteristics are specified with sufficient accuracy by the manufacturer. Interfacing circuits for NTC thermistors are dealt with in Sect. 21.1.4.

21.1.4
Operation of Resistive Temperature Detectors

With the resistive temperature detectors (RTD) described here, resistance is a function of temperature; the relationship is described by the relevant equations $R = f(\vartheta)$. The magnitude of the change in resistance with temperature is given by the temperature coefficient:

$$TK = \frac{1}{R} \cdot \frac{dR}{d\vartheta} \tag{21.2}$$

in percent per degree. This also allows the resulting temperature tolerance to be calculated from a resistance tolerance:

$$\underbrace{\Delta\vartheta}_{\text{Temperature tolerance}} = \frac{1}{TK} \cdot \underbrace{\frac{\Delta R}{R}}_{\text{Resistance tolerance}} \tag{21.3}$$

For a temperature coefficient of 0.3% per degree, a resistance tolerance of $\pm 1\%$ consequently produces a temperature tolerance of ± 3 K. The larger the temperature coefficient, the smaller is the temperature tolerance for a given resistance tolerance.

The resistance of resistive temperature detectors can be measured by making a constant current flow through the sensor. This current must be small enough to insure that no appreciable self-heating occurs, if possible keeping the heat dissipation below 1 mW. A

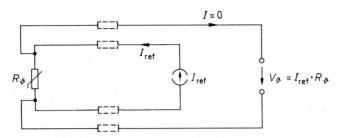

Fig. 21.4. A four-wire resistance measuring circuit that provides independence from lead resistances

voltage is then obtained at the sensor that is proportional to its resistance. Where there are long leads between the current source and the sensor, it may be useful to employ a four-wire resistance measuring circuit, such as the one shown in Fig. 21.4. Here, the lead resistances do not falsify the result when V_ϑ is measured using high- resistance instruments.

Although the voltage V_ϑ is proportional to the resistance R_ϑ, it is not necessarily a linear function of temperature due to nonlinear characteristics. However, if the measured values are digitized, the corresponding temperature can be calculated by solving the relevant characteristic curve equation for ϑ. For analog linearization, a function network of the type described in Sect. 11.7.5 can be connected at the output.

For most applications, however, adequate linearization is obtained by connecting a suitable fixed resistor R_{lin} in parallel with the sensor, as shown in Fig. 21.5a. Figure 21.6 shows the effect of R_{lin} on a silicon PTC thermistor. As the value of R_ϑ increases, the linearization resistor causes the value of the parallel circuit to increase more slowly. This largely compensates for the quadratic term in the characteristic curve equation. The quality of the linearization is basically dependent on optimizing the linearization resistance for the required measurement range. In the simplest case, this value can be obtained from the data sheet.

The question that remains, however, is how to proceed when no information is available in the data sheet for the required measurement range. The usual requirement is for a constant error limit that is as low as possible throughout the range. The linearizing resistance allows the error for three temperatures (ϑ_L, ϑ_M, and ϑ_V) to be reduced to zero. These three temperatures are now shifted, selecting an appropriate value of R_{lin}, until the maximum error between them and at either end of the range is of the same magnitude. This process is illustrated in Fig. 21.6.

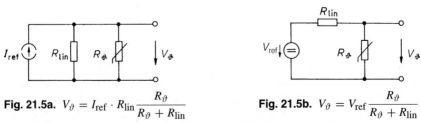

Fig. 21.5a. $V_\vartheta = I_{\text{ref}} \cdot R_{\text{lin}} \dfrac{R_\vartheta}{R_\vartheta + R_{\text{lin}}}$ **Fig. 21.5b.** $V_\vartheta = V_{\text{ref}} \dfrac{R_\vartheta}{R_\vartheta + R_{\text{lin}}}$

Fig. 21.5a,b Linearization of an NTC thermistor characteristic using R_{lin}.
For $V_{\text{ref}} = I_{\text{ref}} \cdot R_{\text{lin}}$, both circuits produce the same output signal

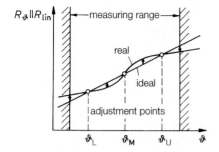

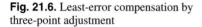

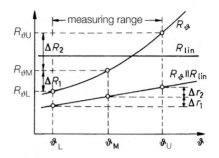

Fig. 21.6. Least-error compensation by three-point adjustment

Fig. 21.7. Simplified method of calculating the linearizing resistance

A simple approximate value for R_{lin} is obtained by placing temperatures ϑ_L and ϑ_U at the ends of the measurement range and ϑ_M in the middle. This is shown in Fig. 21.7. The linearization condition therefore follows from the requirement that the change in resistance of the parallel circuit ($R_\vartheta \| R_{\text{lin}}$) in the lower half of the measurement range must be of exactly the same magnitude as in the upper half. For R_{lin}, this gives

$$R_{\text{lin}} = \frac{R_{\vartheta M}(R_{\vartheta L} + R_{\vartheta U}) - 2R_{\vartheta L} \cdot R_{\vartheta U}}{R_{\vartheta L} + R_{\vartheta U} - 2R_{\vartheta M}} \tag{21.4}$$

where $R_{\vartheta L}$, $R_{\vartheta M}$, and $R_{\vartheta U}$ are the resistance values of the sensor at the lower (ϑ_L), middle (ϑ_M), and upper (ϑ_U) temperatures. We can see that the linearizing resistance tends to infinity – that is, can be omitted – when $R_{\vartheta M}$ is halfway between $R_{\vartheta L}$ and $R_{\vartheta U}$, since the sensor is then itself linear. If $R_{\vartheta M}$ is above the mid-point, R_{lin} becomes negative. This situation occurs when the quadratic term of the sensor characteristic is negative – as, for example, with platinum sensors.

The linearization described is also obtained if the current source shown in Fig. 21.5a is combined with the linearizing resistance and converted into an equivalent voltage source, as shown in Fig. 21.5b. The linearizing resistance R_{lin} is the same in both cases. Figure 21.8 shows the resulting measurement circuit. Voltage V_ϑ is now the linearized function of the temperature. In order not to influence it by loading, it is applied to the noninverting input of an instrumentation amplifier. Due to the configuration of R_1, R_2, and R_3 it can simultaneously provide the required gain and zero shift. It can be seen from Fig. 21.8 that the resulting circuit can also be regarded as a measuring bridge.

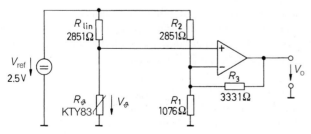

Fig. 21.8. Linearization, zero shift, and gain for a silicon PTC thermistor
$V_o = 20\,\text{mV}\,\vartheta/°\text{C}$ for $0° \leq \vartheta \leq 100°\text{C}$

ϑ	R_ϑ	V_ϑ	V_o
$\vartheta_L = \quad 0°C$	$R_{\vartheta L} = \quad 820\,\Omega$	$V_{\vartheta L} = 0.558\,V$	$V_{oL} = 0.00\,V$
$\vartheta_M = 50°C$	$R_{\vartheta M} = 1202\,\Omega$	$V_{\vartheta M} = 0.741\,V$	$V_{oM} = 1.00\,V$
$\vartheta_U = 100°C$	$R_{\vartheta U} = 1706\,\Omega$	$V_{\vartheta U} = 0.936\,V$	$V_{oU} = 2.00\,V$

Fig. 21.9. Operation of the circuit in Fig. 21.8

The selection of component values will now be explained by reference to an example. The circuit is to measure a temperature in the range from 0°C to 100°C, and to produce output voltages between 0 V and 2 V. The reference voltage is to be 2.5 V. The sensor chosen is a silicon PTC thermistor, in this case type KTY 83/110 from Philips. Linearization is to be computed for this range. To do this, we take the resistance values at the ends of the range and at mid-range, as specified in the data sheet, in accordance with Fig. 21.9. From (21.4), this gives a linearizing resistance $R_{lin} = 2,851\,\Omega$. The linearization error is greatest halfway between the predetermined accurate values; that is, at 25°C and 75°C in this case. However, it is only 0.2 K. The values for V_ϑ also given in Fig. 21.9 are then produced at voltage divider R_δ, R_{lin}. It can be seen that the differences with respect to mid-range actually become equal in size.

To calculate resistances the R_1, R_2, and R_3, one of the values can be prespecified. We select $R_2 = R_{lin} = 2,851\,\Omega$. Resistances R_1 and R_3 determine the gain and the zero point. The gain of the circuit results, on the one hand, from the output voltage required,

$$A = \frac{V_{oU} - V_{oL}}{V_{\vartheta U} - V_{\vartheta L}} = \frac{2.00\,V}{380\,mV} = 5.263$$

and, on the other hand, from the formula for the instrumentation amplifier:

$$A = 1 + R_3/(R_1 || R_2)$$

In the circuit of Fig. 21.8, the selection of the zero point $V_{oL} = 0\,V$ forces the condition that the voltage drop across the virtual-parallel resistors R_1 and R_3 is precisely V_ϑ:

$$V_{\vartheta L} = \frac{R_1 || R_3}{(R_1 || R_3) + R_2} V_{ref}$$

From these two conditional equations, we obtain

$$R_1 = 1,076\,\Omega \quad \text{and} \quad R_3 = 3,331\,\Omega$$

To implement the circuit, we select the next standard value from the E 96 series (see Sect. 29.6 on page 1515). Circuit balancing is unnecessary in most cases, provided that a closely toleranced reference voltage source is used.

Where a high degree of accuracy is required, the zero point can be adjusted by varying R_1, and the gain by varying R_3. In order to obviate the need for any iterative adjustment routine here, nulling is initially performed at a temperature at which the voltage at R_3 is zero. Then R_1 can be adjusted, regardless of the value of R_3. In our example,

$$V_o = V_\vartheta = 0.685\,V$$

for $R_\vartheta = 1,076\,\Omega$ or $\vartheta = 34.3°C$. The gain can then be adjusted for any other temperature (e.g., 0°C or 100°C) at R_3 without affecting the zero point. The general procedure for sensor circuit balancing is given in Sect. 21.5.

If an operational amplifier is selected whose common-mode voltage range and output voltage swing extend as far as the negative operating voltage (rail-to-rail output), the circuit

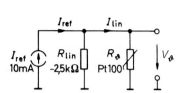

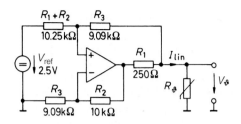

Fig. 21.10. Principle of linearized operation of Pt 100 sensors

Fig. 21.11. Implementation of a current source that has a negative output resistance for a Pt 100 sensor

can be operated from a single 5 V source. In order to achieve almost zero output, the output should be loaded with an additional 1 kΩ.

To linearize platinum temperature sensors, a negative linearizing resistance is required because of the negative quadratic term in the characteristic. For a Pt 100 sensor operated in the temperature range between 0°C and 400°C, a linearization resistor $R_{lin} = -2.5\,k\Omega$ is required in accordance with (21.4). Linearization of the type shown in Fig. 21.8 is therefore impossible. In this case, a current source with a negative internal resistance must be used. The equivalent circuit diagram is shown in Fig. 21.10. The current source shown in Fig. 12.9 is particularly suitable for implementing this circuit. If R_3 is given a slightly lower value than would be necessary for a constant-current source, a negative resistance is produced:

$$r_o = -\frac{\Delta V_\vartheta}{\Delta I} = \frac{R_1 R_3}{R_3 - R_2} = R_{lin}$$

This is a conditional equation for R_3. If we make $R_1 = 250\,\Omega$, $R_2 = 10\,k\Omega$, and $R_{lin} = -2.5\,k\Omega$, we obtain a value of 9.09 kΩ for R_3. The circuit thus dimensioned is shown in Fig. 21.11. The circuit for the gain adjustment and zero shift of sensor voltage V_ϑ can he devised analogously to that of Fig. 21.8.

In a restricted range of temperatures, and where the degree of accuracy required is not too great, the resistance characteristic of an NTC thermistor can also be linearized using a parallel resistor. Figure 21.12 shows how this arrangement operates. Optimum linearization can also be obtained here if the inflection point of $R_{lin} \| R_T$ is at the mid-point, T_M, of the

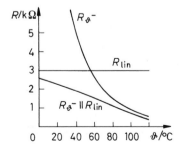

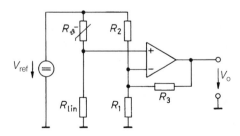

Fig. 21.12. Linearization of an NTC thermistor using a parallel resistor

Fig. 21.13. An interfacing circuit that provides linearization, zero shift, and gain for NTC thermistors

desired temperature range. Here the linearizing resistance can be calculated according:

$$R_{\text{lin}} = \frac{B - T_M}{B + 2T_M} R_{TM} \sim R_{TM}$$

where B is the B-value of the NTC thermistor in (21.1). Once again, the temperature sensor can be connected in series with the same linearizing resistor R_{lin}, giving a linearized voltage curve. In order to obtain a voltage that increases with temperature, it is advisable to take off the voltage across the linearizing resistor. This is shown in Fig. 21.13. The circuit and its design procedure are otherwise identical to the interfacing circuit for PTC thermistors shown in Fig. 21.8.

21.1.5
Transistors as Temperature Sensors

Because of its internal structure, a bipolar transistor is a heavily temperature-dependent component. Its reverse current doubles for an approximately 10 K increase in temperature, and its base–emitter voltage falls by some 2 mV/K (see Sect. 2.21). These otherwise undesirable side-effects can be utilized for temperature measurement. In Fig. 21.14, a transistor configured as a diode is operated with a constant current. This produces the temperature dependence of the base–emitter voltage shown in Fig. 21.15. At room temperature ($T \approx 300$ K), its normal value is about 600 mV. With a temperature increase of 100 K, V_{BE} falls by 200 mV, and it increases accordingly if the temperature is reduced. The temperature coefficient is therefore

$$\frac{\Delta V}{V \cdot \Delta T} = 0.3\%/\text{K}$$

Unfortunately, the dispersion of the forward voltage and of the temperature coefficient is large. For this reason, individual transistors are nowadays only used for temperature measurement where the degree of measuring accuracy required is not too great for instance in CPUs. Here the type MAX 6699 is usefull. Improved calibration can be achieved for circuits based on the difference in the base–emitter voltages of two bipolar transistors operated at different current densities. The principle is shown in Fig. 21.16. This is effectively a bandgap reference, as described in Sect. 16.4.2. The difference between the base–emitter voltages is given by

$$\Delta V_{\text{BE}} = V_T \ln \frac{I_{C2}}{I_{C0} A_2} - V_T \ln \frac{I_{C1}}{I_{C0} A_1} = V_T \ln \frac{I_{C2} A_1}{I_{C1} A_2}$$

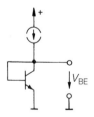

Fig. 21.14. The use of the base–emitter voltage for temperature measurement

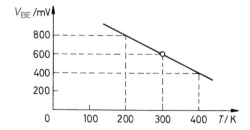

Fig. 21.15. Base-emitter voltage as a function of (absolute) temperature (typical)

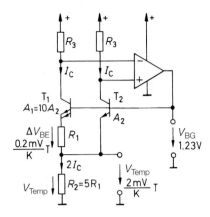

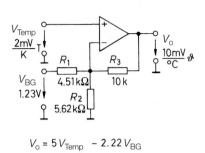

$$V_o = 5\,V_{\text{Temp}} - 2.22\,V_{\text{BG}}$$

Fig. 21.16. Use of a bandgap reference for temperature measurement (e.g., the LT 1019 from Linear Technology)

Fig. 21.17. A supplementary circuit for implementing a Celsius zero point

Since the two collector currents are the same and the surface ratio of the transistors is $A_1/A_2 = 10$, it follows:

$$\Delta V_{\text{BE}} = \frac{kT}{e}\ln 10 = 200\frac{\mu V}{K}\cdot T$$

To implement a bandgap reference, this voltage is amplified with R_2 so that a voltage $V_{\text{Temp}} \approx (2\,\text{mV}/\,K)\cdot T$ is produced, which compensates for the temperature coefficient of T_2 (see Sect. 16.4.2).

Voltage V_{Temp} can be used directly for temperature measurement: it is proportional to the absolute temperature T (PTAT). For $\vartheta = 0°C$,

$$V_{\text{Temp}} = 2\frac{\text{mV}}{K}\cdot 273\,K = 546\,\text{mV}$$

To obtain a Celsius zero point, a constant voltage of this magnitude can be subtracted from V_{Temp}. For this purpose, the subtractor in Fig. 21.17 uses the appropriately weighted voltage V_{BG}.

The principle employed in Fig. 21.16 can be modified by connecting the emitters to the same potential. The output voltage of the operational amplifier in Fig. 21.18 again assumes a value such that the two collector currents have the same magnitude. This produces the same value for ΔV_{BE}, but in this case between the base terminals. The voltage across R_1 is therefore proportional to T, (PTAT). It can be increased to any value by the series connection of additional resistors. In the example shown in Fig. 21.18, it is amplified by a factor of 50:

$$V_{\text{Temp}} = 50\Delta V_{\text{BE}} = 10\frac{\text{mV}}{K}\cdot T$$

At room temperature ($T \approx 300\,K$) this produces a voltage of $V_{\text{Temp}} \approx 3\,V$. The advantage of this variant is that V_{Temp} occurs at the output of the operational amplifier and a load can therefore be applied to it.

Temperature sensors that employ the principle shown in Fig. 21.18 are manufactured as ICs by National (LM 335). They do not have a separate supply voltage terminal and therefore behave like Zener diodes.

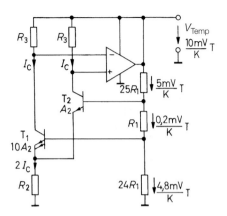

Fig. 21.18. A modified bandgap reference for direct temperature measurement (e.g., LM 335 from National)

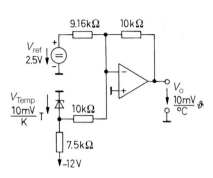

Fig. 21.19. An example of Celsius zero shift for a two-terminal temperature sensor

The operation of this type of temperature sensor with Celsius zero shift will be illustrated by the example shown in Fig. 21.19. Since the sensor behaves like a Zener diode with a low internal resistance (approximately $0.5\,\Omega$ at 1 mA), the current flowing has virtually no effect on the voltage and the sensor can be operated from any supply voltage. It is merely necessary to insure that the minimum operating current (here, 0.4 mA) is maintained. On the other hand, an unnecessarily large operating current should not be chosen, so that self-heating is kept to a minimum. If a series resistance of $7.5\,\text{k}\Omega$ is selected in Fig. 21.19, a current of about 1 mA flows through the sensor at 0°C; at 150°C it is still higher than 0.4 mA. To obtain a Celsius zero point, a current of

$$\frac{2.73\,\text{V}}{10\,\text{k}\Omega} = \frac{2.5\,\text{V}}{9.16\,\text{k}\Omega} = 273\,\mu\text{A}$$

must be subtracted. Since the operational amplifier inverts in this case, a positive Celsius scale is obtained at the output, as required.

A sensor that incorporates Celsius zero shift is also available. The LM 35 from National, for example, produces a voltage of $10\,\text{mV}/°\text{C}$. It represents a significantly simpler solution if only positive temperatures are to be measured.

Voltage ΔV_{BE}, which is proportional to temperature, can also be used to generate a *current* that is proportional to the absolute temperature. In both Fig. 21.16 and Fig. 21.18, the collector current I_C is proportional to T. In order to obtain the required current, it is therefore sufficient to replace the operational amplifier in Fig. 21.16 with the current balancing circuit shown in Fig. 21.20. The condition $I_{C1} = I_{C2}$ is then still fulfilled. The voltage

$$\Delta V_{\text{BE}} = V_T \ln \frac{A_1}{A_2} = \frac{k}{e} \ln \frac{A_1}{A_2} \cdot T = 86 \frac{\mu\text{V}}{\text{K}} \ln \frac{A_1}{A_2} \cdot T$$

then results in a current of

$$I = 2I_C = 2\Delta V_{\text{BE}}/R_1$$

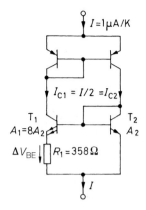

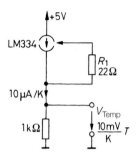

Fig. 21.20. A temperature-controlled current source using the bandgap principle (e.g., the AD 592 from Analog Devices)

Fig. 21.21. A temperature-controlled current source with a freely selectable output current (e.g., the LM 344 from National)

A surface ratio of $A_1/A_2 = 8$ and a resistance $R_1 = 358\,\Omega$ therefore produce a current of

$$I = T \cdot 1\,\mu A/K$$

An example of a sensor that employs this principle is the AD 592 from Analog Devices.

Sometimes, resistor R_1 is not incorporated in the integrated circuit but can be connected externally. This provides the option of selecting any required output signal; in other words, of programming even a relatively large current of $10\,\mu A/K$. A monolithic sensor that offers this facility is the LM 334 from National. An example of how it operates is shown in Fig. 21.21. The $22\,\Omega$ resistor sets the current to $10\,\mu A/K$. It causes a voltage drop of $10\,mV/K$ across the load resistance of $1\,k\Omega$. Since these temperature sensors constitute constant-current sources, the supply voltage has no effect on the current, provided that the voltage is no lower than a given minimum (1 V in the case of the LM 334). For this reason, a accurate voltage supply is not required.

21.1.6
Thermocouple

At the contact point of two different metals or alloys, the Seebeek effect gives rise to a voltage, in the millivolt range, which is known as the thermoelectric voltage. The principle of temperature measurement illustrated in Fig. 21.22 shows that even if one of the two metals is copper, we always obtain two thermocouples with opposite polarities. At identical temperatures $\vartheta_M = \vartheta_R$, their thermoelectric voltages therefore compensate one another.

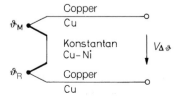

Fig. 21.22. Principle *of* temperature measurement with thermocouples, using a copper–constantan thermocouple as an example

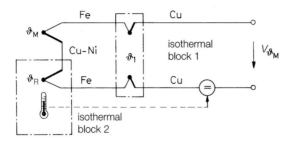

Fig. 21.23. Compensation of the reference junction temperature ϑ_R

Consequently, only the temperature difference $\Delta\vartheta = \vartheta_M - \vartheta_R$ can be measured. To measure individual temperatures, a *reference junction* with reference temperature ϑ_R is therefore required. The case is particularly simple if $\vartheta_R = 0°C$. This can be achieved by immersing one leg of the thermocouple in an ice/water mixture. The measured values then indicate by how many degrees ϑ_M exceeds $0°C$.

This method of generating the reference temperature $\vartheta_R = 0\%$ is of course simply a conceptual model, which would be difficult to implement. A practical solution is to construct an oven that can be kept at a constant temperature of, for example, $60°C$ and to use this as a reference temperature. In this case, the measured value is therefore referred to $60°C$. To convert it to $0°C$, a constant voltage that corresponds to the reference temperature of $60°C$ can simply be added.

However, it is even simpler to leave the temperature of the reference junction as it is. It will then be close to the ambient temperature. Nevertheless, if this reference temperature is not taken into account, an error of between $20°$ C and $0°$ C can easily occur, which would be too large for most applications. But if this temperature is measured (which would be a simple matter using a transistor thermometer IC, for example), the associated voltage can be added into the measuring circuit. This procedure is shown schematically in Fig. 21.23, which also shows the case in which neither of the thermocouple metals is copper. In this case, an additional unintentional thermocouple occurs at the connection point to a copper wire in the evaluation circuit. In order to insure that these two thermoelectric voltages compensate for each other, the two additional elements must be at the same temperature.

The arrangement shown in Fig. 21.23 can be simplified by combining the two isothermal blocks into one with a temperature of ϑ_R and reducing to zero the length of the joining metal lead (in this case, iron). This produces the commonly used arrangement shown in Fig. 21.24, which only requires one isothermal block.

There are various combinations of metals or alloys for thermocouples, which are standardized in IEC 584 and DIN 43710. They are listed in Fig. 21.25. It is evident that their maximum working temperatures vary widely and that the noble metal thermocouples possess markedly lower temperature coefficients. The thermoelectric voltage characteristic is

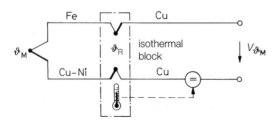

Fig. 21.24. A practical configuration for a thermocouple systems

Ty	Metal 1, positive terminal	Metal 2, negative terminal	Temperature coefficient, average	Usable temperature range
T	Copper	Constantan	42.8 μV/°C	− 200 to + 400°C
J	**Iron**	**Constantan**	51.7 μV/°C	− 200 to + 700°C
E	Chromel	Constantan	60.9 μV/°C	− 200 to +1,000°C
K	**Chromel**	**Alumel**	40.5 μV/°C	− 200 to +1,300°C
S	Platinum	Platinum− 10% rhodium	6.4 μV/°C	0 to +1,500°C
R	Platinum	Platinum− 13% rhodium	6.4 μV/°C	0 to +1,600°C
B	Platinum− 6% rhodium	Platinum− 30% rhodium		0 to +1,800°C
G	Tungsten	Tungsten− 26% rhenium		0 to +2,800°C
C	Tungsten− 5% rhenium	Tungsten− 26% rhenium	15 μV/°C	0 to +2,800°C

Fig. 21.25. Overview of thermocouples. The most widely used types, J and K, are shown in bold type. Types B and G are so nonlinear that no average temperature coefficient can be specified Constantan = copper-nickel; Chromel = nickel-chromium; Alumel = aluminum-nickel

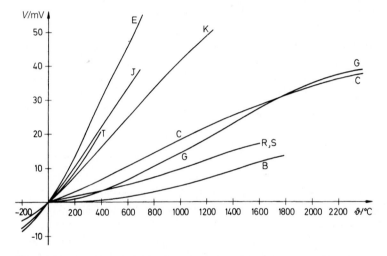

Fig. 21.26. Thermoelectric voltage versus temperature for various thermocouples, at a reference temperature of 0°C

plotted in Fig. 21.26. As we can see, none of the curves is precisely linear. Types T, J, E, and K, however, do possess a decent linearity and also deliver relatively high voltages. They are therefore preferred if the temperature range permits their use. For the other types, if it is not possible to restrict the application to a narrow temperature range, linearization is required.

To evaluate the thermoelectric voltage, a voltage corresponding to reference temperature ϑ_R must be added in accordance with Fig. 21.24, in order to refer the measurement to the "ice point"; that is, 0°C. This correction can either be made at thermocouple level or after amplification. Figure 21.27 gives a schematic representation of the second situation, using an iron–constantan thermocouple as an example. To amplify its voltage to 10 mV/K, a gain of

$$A = \frac{10\,\mathrm{mV/K}}{51.7\,\mathrm{\mu V/K}} = 193$$

is required. Then the reference temperature must be added with the same sensitivity; that is, also with 10 mV/K. Figure 21.28 shows one way of implementing this principle. Since

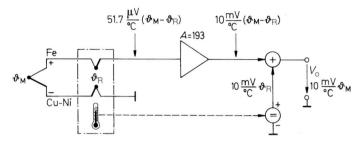

Fig. 21.27. Amplification and reference point compensation for thermocouples, using an iron-constantan thermocouple as an example

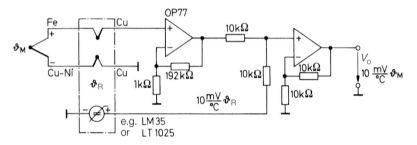

Fig. 21.28. A practical arrangement for the interfacing circuit for thermocouples, using an iron-constantan type as an example

the thermoelectric voltages are in the microvolt range, a low-drift operational amplifier is necessary. In order to obtain adequate loop gain despite the high voltage gain of 193, the operational amplifier must possess a high open loop gain A_D. Measurement of the reference junction temperature can be greatly simplified by using a ready-made temperature sensor with a Celsius zero point, such as the LM 35 from National or the LT 1025 from Linear Technology. However, any other circuit described in this chapter that provides an output signal of 10 mV/K could also be used.

Figure 21.29 shows the alternative principle whereby the ice point correction value is added to the thermocouple voltage before the signal is amplified. For this purpose, a voltage of 51.7 µV/K must be added in the case of an iron–constantan thermocouple. The circuit becomes very simple if we make use of the fact that the thermocouple is electrically

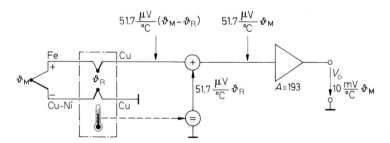

Fig. 21.29. Reference junction compensation prior to amplification of thermocouple signals, using an iron-constantan type as an example

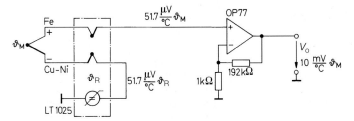

Fig. 21.30. A practical design for reference junction compensation prior to amplification, using iron-constantan thermocouples as an example

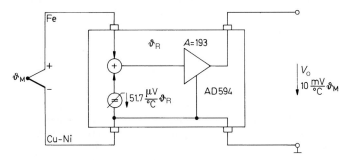

Fig. 21.31. Using integrated thermocouple amplifiers

floating and can therefore simply be connected in series with the correction voltage source, as shown in Fig. 21.30.

The simplest solution is to use specific ICs for operating thermocouples, such as the AD 594–597 series from Analog Devices. Types AD 594 and 596 of this series are calibrated for the operation of iron-constantan thermocouples (type J) and types AD 595 and 597 for Chromel-Alumel (type K). Here, the wires of the thermocouple are connected directly to the integrated circuit, as shown in Fig. 21.31. The latter constitutes the isothermal block with reference temperature ϑ_R. It is assumed that the silicon crystal is at the same temperature as the IC pins. The ice point correction is generated for the chip temperature, added to the thermoelectric voltage, and amplified. An internal zero point and gain calibration to within $1°$ C are available. If the inputs are shorted and the thermocouple omitted, only the ice point correction voltage of

$$V_o = 51.7 \frac{\mu V}{°C} \cdot \vartheta_R \cdot 193 = 10 \frac{mV}{°C} \vartheta_R$$

is produced at the output. The circuit then operates as a transistor temperature sensor with a Celsius zero point.

21.1.7
An Overview of Types

A number of representative manufacturers and products for temperature measurement are listed in Fig. 21.32. There are large variations in price. It is therefore worthwhile comparing different principles and types. In general, it can be said, however, that the more expensive a sensor is, the more accurately it has been adjusted (laser-trimmed) by the manufacturer.

Type	Manufacturer	Output signal nominal value	Temperature range
Metal PTC thermistor			
Pt 100...1000	Heraeus	100...1000 Ω	$-50...+500°C$
Fk 100...2000	Heraeus	100...2000 Ω	$-200...+500°C$
1 Pt 100...1000	Omega	100...1000 Ω	$-70...+500°C$
Pt 100...1000	Murata	100...1000 Ω	$-50...+600°C$
Pt 100...1000	Sensycon	100...1000 Ω	$-50...+600°C$
Silicon PTC thermistor			
AD 22100[1]	Analog D.	22 mV/K	$-50...+150°C$
KTY-Series	Infineon	1...2 kΩ	$-50...+150°C$
KTY-Series	Philips	1...2 kΩ	$-50...+300°C$
Metal-ceramic NTC thermistors			
M-Series	Infineon	1 k...100 kΩ	$-50...+200°C$
NTH-Series	Murata	100 Ω...100 kΩ	$-50...+120°C$
Thermistors	Philips	1 kΩ...1 MΩ	$-50...+200°C$
Bandgap sensors			
AD 7818[2]	Analog Dev.	4 LSB/K	$-55...+125°C$
TMP 04	Analog Dev.	PWM-Output	$-40...+100°C$
TMP 17	Analog Dev.	1µA/K	$-40...+105°C$
TMP 36	Analog Dev.	10 mV/K	$-40...+125°C$
LT 1025[3]	Lin. Tech.	10 mV/K	$0...+60°C$
MAX 6607	Maxim	10 mV/K	$-20...+85°C$
DS 18B 20[2]	Maxim	20 LSB/K	$-55...+125°C$
LM 45	National	10 mV/K	$-20...+100°C$
LM 60	National	6 mV/K	$-40...+125°C$
LM 74[2]	National	16 LSB/K	$-55...+150°C$
LM 134	National	0.1...10 µA/K	$-40...+125°C$
TMP 125	Texas Inst.	4 LSB/K	$-40...+125°C$
Thermocouples			
J, K, S, R, B	Heraeus	Fig. 21.26	Fig. 21.26
J, K, S, R, B, T, E, C, G	Omega	Fig. 21.26	Fig. 21.26
J, K, S	Philips	Fig. 21.26	Fig. 21.26
J, K, S, R, B	Sensycon	Fig. 21.26	Fig. 21.26
Thermocouple amplifiers			
AD 594	Analog Dev.	type J 10 mV/°C	$-55°C...+125°C$
AD 595	Analog Dev.	type K 10 mV/°C	$-55°C...+125°C$

[1] Amplifier integrated

[2] ADC integrated

[3] Additional outputs for reference junction compensation of thermocouples

Fig. 21.32. Examples of temperature sensors

21.2
Pressure Measurement

Pressure is defined as force per unit area

$$p = F/A$$

The unit of pressure is

Pressure range	Application
< 40 mbar	Water level in a washing machine, dishwashert
100 mbar	Vacuum cleaner, filtration monitoring, flow measurement
200 mbar	Blood pressure measurement
1 bar	Barometer, motor vehicle (correction for ignition and fuel injection)
2 bar	Motor vehicle (tire pressure)
10 bar	Expresso machinery
50 bar	Pneumatics, industrial robots
500 bar	Hydraulics, construction machinery
2000 bar	Car motor with fuel injection

Fig. 21.33. Pressures that occur in practical applications

$$1 \text{ Pascal} = \frac{1 \text{ Newton}}{1 \text{ Square meter}}; \quad 1 \text{ Pa} = \frac{1 \text{ N}}{1 \text{ m}^2}$$

The term "bar" is also widely used. It is related to the above:

$$1 \text{ bar} = 100 \text{ kPa} \quad \text{or} \quad 1 \text{ mbar} = 1 \text{hPa}$$

Pressure is sometimes defined as the height of a column of water or mercury. The relationships are:

$$1 \text{ cm H}_2\text{O} \; \stackrel{\wedge}{=} \; 98.1 \text{ Pa} \; = \; 0.981 \text{ mbar}$$
$$1 \text{ mm Hg} \; \stackrel{\wedge}{=} \; 133 \text{ Pa} \; = \; 1.33 \text{ mbar}$$

In English-language data sheets, pressure is usually specified in

$$\text{psi} = \text{pounds per square inch}$$

For the conversion follows:

$$1 \text{ psi} = 6.89 \text{ kPa} = 68.9 \text{ mbar} \quad \text{or} \quad 15 \text{ psi} \approx 1 \text{ bar}$$

Figure 21.33 shows a few examples of the orders of magnitude of pressures that occurs in practical applications.

Pressure sensors can be used in a wide variety of applications. They can even be used to determine flow rates and flow volumes via differences in pressure.

21.2.1
Design of Pressure Sensors

Pressure sensors record the pressure-induced deflection of a diaphragm. For this purpose, a number of strain gauges are mounted on the diaphragm and form a Wheatstone bridge. They vary their resistance as a result of the piezo-resistive effect of deflection, pressure, or tensile force. They are mainly constructed from vapor-deposited constantan or platinum–iridium layers. Nowadays, resistors implanted in silicon are generally used, with the silicon substrate simultaneously acting as the diaphragm. They have the advantage of being cheaper to manufacture and are over ten times more sensitive. Their disadvantage, however, is their higher temperature coefficient.

Figure 21.34 gives a schematic representation of a pressure sensor. In the differential pressure sensor shown in Fig. 21.34a, a pressure p_1 is exerted on one side of the diaphragm

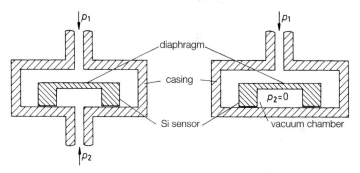

Fig. 21.34a. A differential pressure sensor

Fig. 21.34b. An absolute pressure sensor

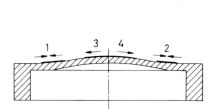

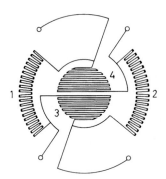

Fig. 21.35a. Expansion and compression of the diaphragm of pressure sensors

Fig. 21.35b. Arrangement of strain gauges on the diaphragm

and a pressure of p_2 on the other. Therefore, only the pressure difference $p_1 - p_2$ causes the deflection of the diaphragm. With the absolute pressure sensor shown in Fig. 21.34b, one side of the diaphragm takes the form of a vacuum chamber.

Figure 21.35 shows a typical arrangement of the strain gauges on the diaphragm. The left-hand diagram is intended to show that when the diaphragm deflects, zones are produced which are elongated and others which are compressed. It is in these areas (see the right-hand diagram) that the four bridge resistors are arranged. They are interconnected in such a way that the resistances in the bridge arms change inversely. As can be seen from Fig. 21.36, this arrangement produces a particularly large output signal, while concurring effects, such as the absolute value of the resistances and their temperature coefficient, compensate for each other. In spite of this, the output signal is low because of the very small changes in resistance ΔR. At maximum pressure and at an operating voltage of $V_{\text{ref}} = 5\,\text{V}$, it is between 25 and 250 mV, depending on the sensor. The relative change in resistance is therefore between 0.5% and 5%.

The output signal of a real pressure sensor is made up of a component that is proportional to the pressure and an undesirable offset component:

$$V_D = S \cdot p \cdot V_{\text{ref}} + O \cdot V_{\text{ref}} = V_p + V_O \qquad (21.5)$$

where p is the pressure,

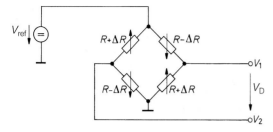

Fig. 21.36. Measuring bridge of a pressure sensor

$$\frac{V_D}{V_{\text{ref}}} = \frac{R + \Delta R}{2R} - \frac{R - \Delta R}{2R} = \frac{\Delta R}{R}$$

$$S = \frac{\Delta V_D}{\Delta p \, V_{\text{ref}}} = \frac{\Delta R}{\Delta p \cdot R}$$

is the sensitivity, and O is the offset. Both terms provide a contribution that is proportional to the reference voltage. In order not to obtain too small a signal, the reference voltage should be as large as possible. However, constraints exist due to the self-heating of the sensor. Reference voltages between 2 V and 12 V are therefore used.

21.2.2
The Operation of Temperature-Compensated Pressure Sensors

Silicon-based pressure sensors have such high temperature coefficients that some form of temperature compensation is generally required. The easiest solution is for the user to employ pressure sensors which already have temperature compensation, provided by the manufacturer. However, in some cases cost considerations may force users to implement their own temperature compensation. We shall now describe a possible approach.

There are a number of basic considerations relating to the conditioning of pressure sensor signals:

1) Although the four bridge resistors in Fig. 21.36 are well matched among each other, their absolute value exhibits a large tolerance and is also heavily temperature-dependent. For this reason, the output signals must not be loaded, and an instrumentation amplifier is normally used to provide gain.
2) Pressure sensors usually have a zero error, which is quite small in absolute terms (e.g., ± 50 mV); however, comparison with the wanted signal shows that it is usually of the order of magnitude of the measurement range. A zero adjuster covering the entire measurement range is therefore required.
3) In general, the sensitivity of a pressure sensor also exhibits significant tolerances (e.g., $\pm 30\%$), so that gain adjustment is additionally required.
4) Zero and gain adjustment must be possible without having to use iteration procedures.
5) Since the wanted signals of a pressure sensor are small, a large amount of amplification is required. This results in appreciable amplifier noise. The pressure sensor itself also produces considerable circuit noise. The bandwidth at the amplifier output must therefore be limited to the required frequency range of the pressure variations.

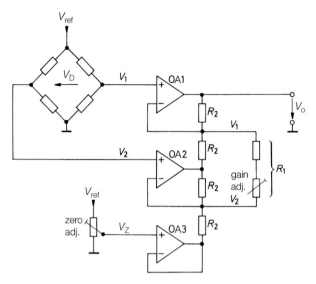

Fig. 21.37. An interfacing circuit for pressure sensors

$$V_o = 2\left(1 + \frac{R_2}{R_1}\right) \cdot (V_1 - V_2) + V_Z$$

$$V_o = \quad A \quad \cdot \quad V_D \quad -V_O$$

6) The pressure-measuring circuit will often be required to operate exclusively from a positive voltage, so the circuit should be designed accordingly. Rail-to-rail amplifiers will be useful for that purpose.

The circuit normally used for conditioning pressure sensor signals is an instrumentation amplifier. In Fig. 21.37, the asymmetrical amplifier from Fig. 20.6 is illustrated as an example. The gain is adjusted by resistor R_1 to suit the sensor. For zeroing purposes, the foot of voltage divider R_2 has not been connected to ground, but via impedance converter OA3 to the zero adjuster. This causes voltage V_Z to be added to the output voltage.

The circuit shown in Fig. 21.37 can be operated from a single positive supply voltage, since the quiescent potentials are approximately $\frac{1}{2}V_{\text{ref}}$. A serious disadvantage, however, is that it is not possible to adjust the zero and the gain noniteratively. If, for example, the output voltage of the pressure sensor is broken down in accordance with Eq. (21.5) into a pressure-dependent component V_P and the offset voltage V_O, it can be seen that both are amplified by the factor $A = 2(1 + R_2/R_1)$:

$$V_o = A(V_1 - V_2) + V_Z = AV_P + AV_O + V_Z$$

whereas the voltage for zero correction is not amplified. The zero adjustment

$$V_Z = -AV_O$$

is therefore dependent on the gain A. This means that for iteration-free adjustment, the voltage for zero correction must also be amplified by A. Consequently, zeroing must *precede* the point in the signal flow at which gain adjustment takes place. When there is no output signal ($V_o = V_{\text{ref}}$) the gain adjust has no effect on the output voltage.

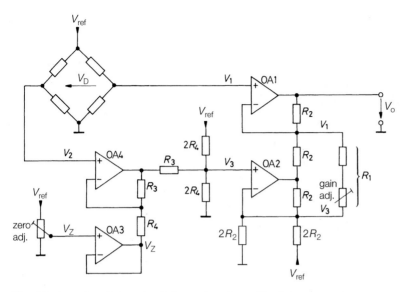

Fig. 21.38. An interfacing circuit for noniterative calibration

$$V_o = 2 \left(1 + \frac{R_2}{R_1}\right)\left[(V_1 - V_2) + \frac{R_3}{R_3 + R_4}\left(V_Z - \frac{1}{2}V_{ref}\right)\right] + V_{ref}$$

$$V_o = \underbrace{\phantom{2\left(1+\frac{R_2}{R_1}\right)}}_{A} \qquad [V_D - V_O] \qquad + V_{ref}$$

To allow zeroing to be performed before the subtractor containing the gain adjuster, operational amplifier OA4 has been added in Fig. 21.38. For potential V_3, this gives:

$$V_3 = V_2 + \frac{R_3}{R_3 + R_4}\left(\frac{1}{2}V_{ref} - V_Z\right)$$

Thus a voltage of up to $\pm \frac{1}{2}V_{ref}R_3/(R_3 + R_4)$ can be added to potential V_2, depending on the size of V_Z. The amplifier zero has not been fixed at 0 V but at V_{ref} by connecting the foot of voltage divider chain R_2 to V_{ref} instead of to ground. This shifts the zero balance point from zero pressure and zero output voltage to the pressure that corresponds to $V_o = V_{ref}$ (see Sect. 21.5.1).

An example will serve to illustrate component value selection for the circuit. An air pressure meter is to deliver an output voltage of 2.5 mV/hPa. An uncalibrated pressure sensor is to be used. At a supply voltage of $V_{ref} = 5$ V, it delivers a signal of 10...40 μV/hPa; its zero error can be as much as ± 50 mV. To calculate the zero adjustment, we choose $R_3 = 1$ kΩ. At a reference voltage of $V_{ref} = 5$ V, this then produces the required adjustment range with $R_4 = 49$ kΩ. The gain must be adjustable between 62.5 and 250, depending on the sensitivity of the sensor. If $R_2 = 10$ kΩ is specified, a minimum value of 80 Ω (fixed resistor) and a maximum value of 330 Ω is obtained for R_1. The variable resistor must therefore have a value of 250 Ω.

To calibrate the circuit, it is first necessary to adjust the zero point. To avoid an iterative process, the pressure at which no voltage is present at the gain adjuster is selected; that is, $V_1 - V_3 = 0$. This is the case when $V_o = V_{ref}/2 = 2.5$ V and corresponds to a pressure of 1,000 hPa. The zero adjuster is therefore set so that V_o is actually 2.5 V at 1,000 hPa.

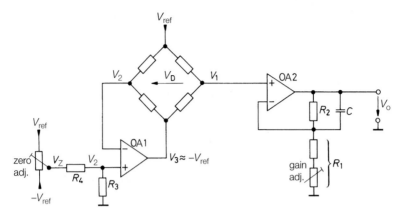

Fig. 21.39. Transfer of the bridge signal to the right-hand bridge arm

$$V_o = \left(1 + \frac{R_2}{R_1}\right)(V_D + V_n) = A(V_D - V_O)$$

Since no voltage is dropped across R_1 after the adjustment, its value has no effect on the adjustment. To calibrate the gain, we select the pressure that is as far away as possible from the zero point at 1,000 hPa – at the upper or lower end of the desired measurement range, for example – and adjust the output voltage to the nominal value using R_1. A more detailed description of sensor calibration is given in Sect. 21.5.

Since the sensor signals are in the microvolt range, it is advisable to use operational amplifiers with low offset voltages and offset voltage drifts. However, as the bandwidth requirements are small, operational amplifiers with a low power consumption can he used. A ready-made instrumentation amplifier can, of course, be used for the subtractor, although the advantage is not significant: only the four resistors R_2 are saved.

The circuit for conditioning the sensor signals can be greatly simplified if a negative voltage is additionally available or can be generated with a voltage converter. In the circuit shown in Fig. 21.39, one bridge arm of the pressure sensor is in the negative feedback path of amplifier OA1. If we imagine $V_2 = 0$, the entire bridge signal V_D will therefore be transferred to the right-hand bridge output and subtraction will no longer be required. Therefore the simple noninverting amplifier OA2 is all that is required here to provide gain. For nulling the offset voltage V_O of the sensor, voltage $V_2 = -V_O$ is applied

A lowpass filter can easily be implemented using capacitor C to limit the noise bandwidth of the circuit. A second-order lowpass filter can also be implemented by connecting a second capacitor to ground directly at the bridge output.

21.2.3
Temperature Compensation for Pressure Sensors

By their nature, the doped silicon resistors of a pressure sensor are temperature-dependent. They are even used for temperature measurement themselves (see Sect. 21.1). The typical

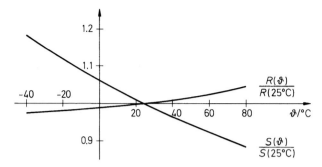

Fig. 21.40. Resistance and sensitivity of silicon pressure sensors as a function of temperature

$$TK_R = \frac{\Delta R}{R\Delta\vartheta} \approx +1350\frac{ppm}{K}, \quad TK_S = \frac{\Delta S}{S\Delta\vartheta} \approx -2350\frac{ppm}{K}$$

resistance curve is shown in Fig. 21.40. At room temperature, the resistor's temperature coefficient is

$$TK_R = \frac{\Delta R}{R \cdot \Delta\vartheta} \approx 1350\frac{ppm}{K} = 0.135\frac{\%}{K}$$

In a bridge arrangement of the type employed in pressure sensors, the temperature-induced resistance variation has no adverse effect, provided that it is the same in all of the resistors and no load is placed on the output signal. However, a problem arises due to the fact that the pressure sensitivity of the sensor is also temperature-dependent; its temperature coefficient is

$$TK_S = \frac{\Delta S}{S \cdot \Delta\vartheta} \approx -2350\frac{ppm}{K} = -0.235\frac{\%}{K}$$

Thus, for a temperature increase of 40°C, it has already dropped by 10%, as can be seen from Fig. 21.40. To prevent this from invalidating the measurement, the gain must be increased accordingly with temperature. Naturally, this must not be based on the temperature of the amplifier, but on that of the pressure sensor. The temperature detector must therefore be incorporated into the pressure sensor. Hence, the obvious solution would be to perform the temperature compensation in the sensor itself. This can be done by increasing the reference voltage V_{ref} with temperature in such a way that the reduction in sensitivity is just compensated:

$$V_D = S \cdot P \cdot V_{ref} + O \cdot V_{ref} = V_P + V_O$$

It is usually accepted that the zero point $V_o = O \cdot V_{ref}$ will shift slightly.

Temperature-compensated pressure sensors differ only in the method of temperature compensation employed. Three widely used methods are illustrated in Fig. 21.41. In Fig. 21.41a, an NTC thermistor is used to increase the bridge voltage with temperature. In Fig. 21.41b, the negative temperature coefficient of a diode, of $-2\,mV/K$, is used. The arrangement of the transistor in the circuit produces the effect of three diodes. A temperature sensor using the bandgap principle can also be incorporated; the type used in Fig. 21.41c is the LM 335 (see Fig. 21.18). It operates as shown in Fig. 21.18 and delivers a voltage of $10\,mV/K$; that is, about 3 V at room temperature. The interesting feature of this solution is that the temperature compensation circuit can be simultaneously used for temperature measurement.

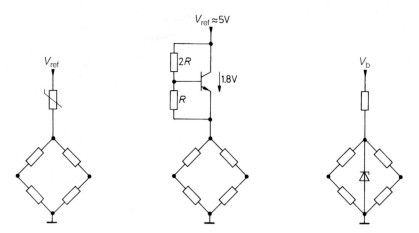

a An NTC thermistor; for example, in the SDX-series from SenSym   **b** Approximately three diodes; for example, in the KP 100 A 1 from Philips   **c** A bandgap temperature sensor; for example in the LM 335 from National

Fig. 21.41a,b,c. Temperature compensation methods for pressure sensors

One method of temperature compensation that dispenses with an additional temperature detector involved uses the temperature dependence of the bridge resistors themselves for temperature compensation. If the bridge is operated with a constant current I_{ref} instead of a constant voltage V_{ref}, the voltage across the bridge increases with temperature by the same amount as its resistance. Unfortunately, the voltage increase of $TK_R = 1350$ ppm/K is not enough to compensate for the reduction in sensitivity of $TK_S = -2350$ ppm/K. If, however, the current source in Fig. 21.42 is given a negative internal resistance, current I_B will rise as the voltage increases. The requirement that bridge voltage V_B must rise by a factor of $|TK_S/TK_R|$ more quickly than with a constant current produces the condition:

$$V_B = |TK_S/TK_R|R_B I_k = (R_i\|R_B)I_k$$

This results in the equation for determining the value of R_i:

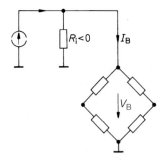

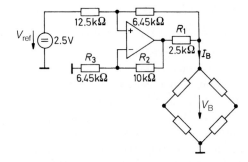

Fig. 21.42. Operation of a pressure sensor from a current source that has a negative internal resistance

Fig. 21.43. Practical implementation of the current source

$$I_k = 1\,\text{mA} \quad R_i = -7.05\,\text{k}\Omega$$

$$R_i = \frac{|TK_S|}{TK_R - |TK_S|} R_B = -2.35 R_B$$

The circuit shown in Fig. 12.9 is again ideally suited for use as a current source. Figure 21.43 shows how it can be used for temperature compensation in a pressure sensor with a bridge resistance of $R_B = 3\,k\Omega$. We select a short-circuit current of $I_k = 1\,mA$. This gives $R_1 = V_{ref}/I_k = 2.5\,k\Omega$. The rated operating voltage of the bridge is therefore

$$V_B = |TK_S/TK_R| R_B I_k = (2350/1350) \cdot 3\,k\Omega \cdot 1\,mA = 5.22\,V$$

In order to obtain the component values for the circuit, the required internal resistance must first be determined:

$$R_i = \frac{|TK_S|}{TK_R - |TK_S|} R_B = -2.35 \cdot 3\,k\Omega = -7.05\,k\Omega$$

If $R_2 = 10\,k\Omega$ is then specified, we obtain

$$R_3 = R_2 \left(1 + \frac{R_1}{R_i}\right) = 6.45\,k\Omega$$

21.2.4
Commercially Available Pressure Sensors

Figure 21.44 gives some idea of the wide range of pressure sensors available. There are not only a large number of other manufacturers, but most of the types listed are merely representative of whole families of sensors. It can be seen from Fig. 21.44 that in addition to the pressure sensors with a range of 1–2 bar, which are primarily designed for barometers, there are also types with very much smaller and greater measurement ranges. There are two designs of pressure sensors: one type measures pressure against atmospheric pressure, while the other measures the difference in pressure between two ports.

Type	Manufacturer	Pressure range	Measurement range	Zero point error	Bridge resistance	Temp. compensation
MPX 10	Freescale	0.1...0.5 bar	35 mV	50%	1 kΩ	—
MPX 2000	Freescale	0.1...2 bar	40 mV	3%	2 kΩ	internal
MPX 5000*	Freescale	40 m...10 bar	4.5 V	1%	—	internal
40PC*	Honeywell	1...17 bar	4 V	2%	—	internal
170PC	Honeywell	17...70 mbar	30 mV	10%	6 kΩ	21.41a
180PC*	Honeywell	0.3...10 bar	5 V	2%	—	internal
240PC*	Honeywell	1...35 bar	5 V	1%	—	internal
KP100*	Infineon	1 bar	14 bit	10 %	—	internal
KP120*	Infineon	1 bar	4 V	2 %	—	internal
NPC 12xx	Novasensor	70 m...7 bar	75 mV	3 %	4 kΩ	internal
NPH	Novasensor	25 m...0.4 bar	100 mV	3 %	4 kΩ	internal
ASDX*	SenSym	70 m...7 bar	4 V	2 %	—	internal
SDX	SenSym	70 m...7 bar	90 mV	3 %	4 kΩ	21.41a
SX	SenSym	70 m...20 bar	110 mV	40 %	4 kΩ	—
Sensor Amplifiers						
MAX 1450	Maxim		4.5 V	1 %		21.42

* Integrated amplifier

Fig. 21.44. Examples for piezoresistive pressure sensor families

The sensitivity of the sensors appears to vary considerably. The reason for this is the great diversity of measurement ranges. At full pressure and a nominal supply voltage, they all deliver a difference signal of $50 - 150$ mV. The only exception to this rule concerns those types that incorporate amplifiers. These provide an amplified, temperature-compensated, and calibrated output signal. With many types, the zero error is of the order of the entire measurement range. The types with internal compensation (Fig. 21.41a) perform much better in this respect, since not only the sensitivity but also the zero point is calibrated by the manufacturer. Amplifiers with built-in calibration for offset, gain, temperature coefficient and linearity are found in the family MAX 1450–1458 Maxim.

21.3
Humidity Measurement

Humidity specifies water content. Of particular interest is the water content of air. The *absolute humidity* H_{abs} is defined as the amount of water contained in a unit volume of air:

$$H_{abs} = \frac{\text{Mass of water}}{\text{Volume of air}}; \qquad [H_{abs}] = \frac{g}{m^3}$$

The maximum amount of water that can be dissolved in air is given by the *saturation humidity* H_{sat}:

$$H_{sat} = H_{abs\,max} = f(\vartheta)$$

Its magnitude is heavily dependent on temperature, as shown in Fig. 21.45. When saturation humidity is reached or exceeded, water condenses: the *dew point is* reached. Determining the dew point thus allows Fig. 21.45 to be directly employed to specify the absolute humidity.

Most of the reactions due to atmospheric humidity, which include such things as physical well-being, are dependent on the *relatire humidity* H_{rel}:

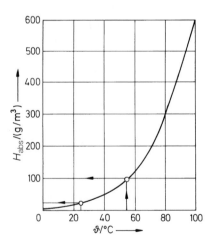
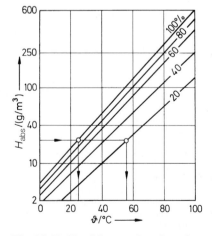

Fig. 21.45. Saturation humidity as a function of temperature

Fig. 21.46. Humidity as a function of temperature Parameter: relative humidity H_{rel}

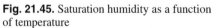

For $H_{rel} = 100\%$, the plots in Figs. 21.45 and 21.46 coincide.

$$H_{\text{rel}} = \frac{H_{\text{abs}}}{H_{\text{sat}}}$$

which indicates the percentage of the saturation humidity attained. The amount of the relative humidity can be determined with the aid of Fig. 21.45. If, for example, a dew point of 25°C is established by cooling the air, the absolute humidity $H_{\text{abs}} = 20 \text{ g/m}^3$. However, at a temperature of, for example, 55°C, the air could take up $H_{\text{sat}} = 100 \text{ g/m}^3$ of water. Therefore, at 55°C the relative humidity is

$$H_{\text{rel}} = \frac{H_{\text{abs}}}{H_{\text{sat}}} = \frac{20 \text{ g/m}^3}{100 \text{ g/m}^3} = 20\%$$

The relationship between relative air humidity and temperature can be obtained directly from Fig. 21.46.

21.3.1
Humidity Sensors

The example given above shows that the relative humidity can be determined by measuring the ambient temperature and the dew point. Although the dew point can be measured precisely and no further calibration is required, the cooling equipment required is complex. The sensors commonly used for determining humidity simplify the measurement by providing an output that is a direct function of the relative humidity – the quantity generally of interest. They consist of a capacitor with a dielectric whose permittivity is humidity-dependent.

Figure 21.47 shows this type of device. The dielectric is made from aluminum oxide or a special plastic foil. One or both electrodes consist of a metal that is permeable to water vapor. The capacitance characteristic for a typical device is shown in Fig. 21.48. It can be seen that a specific basic capacitance C_0 occurs and that the increase in capacitance is nonlinear. Within a restricted range, this nonlinearity can be largely eliminated using a series capacitor. A humidity sensor with a voltage output is the HIH 4000 family from Honeywell.

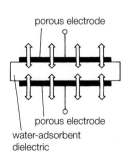

Fig. 21.47. Internal design principle of a capacitive humidity sensor

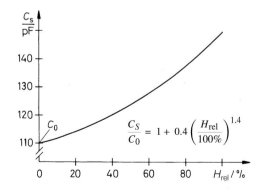

$$\frac{C_S}{C_0} = 1 + 0.4 \left(\frac{H_{\text{rel}}}{100\%} \right)^{1.4}$$

Fig. 21.48. Sensor capacitance versus relative humidity. Example: No. 2322691 90001 from Philips

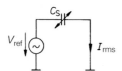

Fig. 21.49. Capacitance measurement by measuring the impedance
$I_{rms} = 2\pi I_{rms} \cdot f \cdot C_S$

21.3.2
Interfacing Circuits for Capacitive Humidity Sensors

In order to determine humidity, it is necessary to measure the capacitance of the humidity sensor. This means that any circuit for measuring capacitance could be employed here. For example, an AC voltage can be applied to the sensor and the flow of current measured, as shown schematically in Fig. 21.49. Although this method appears simple, it is actually quite complex, as it requires not only a calibrated AC meter, but also an AC voltage source with constant amplitude and frequency.

A simpler solution is to incorporate the sensor in an astable circuit, with the sensor determining its operating frequency or duty factor. Figure 21.50 shows a circuit of this type. It consists of two multivibrators of the type shown in Fig. 6.29. Multivibrator M1 oscillates at a constant frequency of about 10 kHz if CMOS gates are used. It synchronizes multivibrator M2, whose ON time is determined by humidity sensor C_S. The ON times of both multivibrators are of equal duration at zero humidity; but as the humidity increases, the ON time of M2 becomes longer, as shown in Fig. 21.51. From the difference in the ON times, a signal V_3 is obtained which is proportional to ΔC and thus approximately proportional to the humidity. The lowpass filter at the output averages the signal.

A circuit that provides a considerably higher degree of accuracy is shown in Fig. 21.52. Here, the capacitance of the humidity sensor is determined in accordance with the definition of capacitance $C_S = Q/V$. Capacitor C_S is initially charged to V_{ref} and then discharged via the summing point. The average current flowing is

$$\overline{I}_S = V_{ref} \cdot f \cdot C_S$$

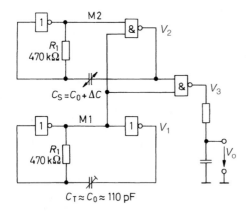

Fig. 21.50. Determining the increase in capacitance by measuring the increase in oscillation period Gartes: CMOS; for example, the CD 4001

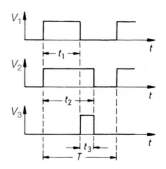

Fig. 21.51. Output signal resulting from the difference in switching times

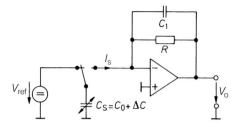

Fig. 21.52. Humidity measurement realized using switched-capacitor technology

$$V_o = -V_{ref} \cdot R \cdot f \cdot C_S$$

where f is the frequency at which the switch is operated. A DC voltage proportional to C_S is then produced at the output voltage $V_o = V_{ref} \cdot f \cdot C_S \cdot R$ due to averaging by C_1.

In order to implement a humidity-measuring circuit based on the principle illustrated in Fig. 21.52, additional devices are required for adjusting the zero point and the full-scale span. The complete circuit is shown in Fig. 21.53. Capacitor C_T is used for nulling. It is likewise charged to voltage V_{ref}, but is then applied to the summing point with reverse polarity. This produces a current given by

$$\overline{I}_T = -V_{ref} \cdot f \cdot C_T$$

The feedback resistor R has also been replaced by a switched capacitor. Its average current is

$$\overline{I}_G = V_o \cdot f \cdot C_G$$

Application of Kirchhoff's Law to the summing point – that is, $\overline{I}_S + \overline{I}_T + \overline{I}_G = 0$ yields the output voltage

$$V_o = -V_{ref} \frac{C_S - C_T}{C_G} = -V_{ref} \frac{\Delta C}{C_G}$$

It can be seen that by using switched-capacitor technology for nulling and gain adjustment, all of the currents are proportional to f, thereby canceling the switching frequency out of

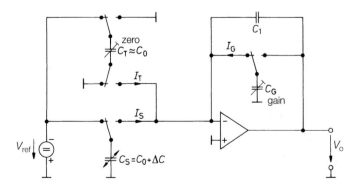

Fig. 21.53. Humidity measurement with nulling and sensitivity adjustment

$$V_o = -V_{ref} \Delta C / C_G$$

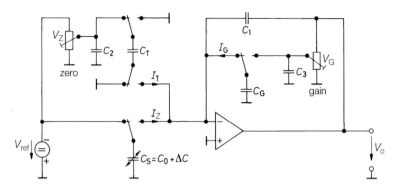

Fig. 21.54. Nulling and gain adjustment for a humidity sensor using potentiometers

the result. This benefit is lost if resistors are used, as in normal SC techniques; this is also evident from the circuit in Fig. 21.52. The LTC 1043 from Linear Technology is particularly suitable for implementing the switches, since it not only contains four changeover switches but also a clock generator that drives the switches.

It is desirable to adjust the zero point and full-scale span with potentiometers rather than trimming capacitors. A method that can be employed without losing the benefits of pure SC technology is shown in Fig. 21.54. Here, the current flowing through C_T or C_G is varied by the voltage tapped off via the potentiometers. In order to insure that this does not increase the charging time, capacitors C_2 and C_3 are additionally employed and are selected to be large in relation to C_T or C_G.

21.4
The Transmission of Sensor Signals

There is often a considerable distance, or an environment with high levels of interference, between the sensor and the point at which the signal is to be evaluated. For this reason, special action must be taken in such cases to insure that the measured values are not impaired by external effects. Depending on the field of application and the protection class required, a distinction is drawn between electrical signal transmission and a more complex method that employs electrically isolated transmission.

21.4.1
Electrical (Direct-Coupled) Signal Transmission

For long-distance transmission the line resistance R_L cannot be ignored. Even small currents necessary for the operation of the sensor result in such large voltage drops that impair the measured value. This problem can be solved by transmitting the measurement signal for evaluation via two additional lines in which no current flows. The measured quantity is then obtained by employing an instrumentation amplifier such as the one shown in Fig. 21.55. The voltage drop in the measurement circuit merely causes a common-mode signal $V_{CM} = I_0 R_L$, which disappears after subtraction.

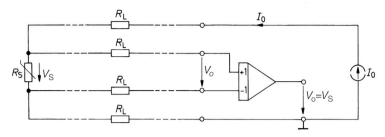

Fig. 21.55. Four-wire measurement, using a resistive temperature detector as an example

$V_o = I_0 R_S = V_S \qquad V_{CM} = I_0 R_L$

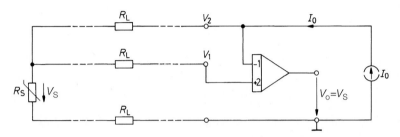

Fig. 21.56. Three-wire measurement, using a resistive temperature sensor as an example

$V_1 = I_0(R_S + R_L) \qquad V_2 = I_0(R_S + 2R_L) \qquad V_o = 2V_1 - V_2 = I_0 R_S = V_s$

One wire can be dispensed with by specifying that the resistance in all the circuits must be identical, giving the three-wire method shown in Fig. 21.56. The voltage drop across R_L can be eliminated in this case by formulating the expression

$$V_o = 2V_1 - V_2 = 2V_S + 2I_0 R_L - V_S - 2I_0 R_L = V_S$$

If the sensor signals are small as, for example, in pressure sensors and thermocouples, they must be preamplified near the sensor before the signal is transmitted over a long circuit. This principle is illustrated in Fig. 21.57. Although the output signal is affected by the voltage drop across R_L, if gain A is selected to be sufficiently large, this error becomes insignificant. It can be eliminated altogether if the four-wire method of Fig. 21.55 is additionally employed. However, one more instrumentation amplifier on the receiver side would then be required.

It is simpler in this case to convert the sensor signal into a current proportional to it. A current is unaffected by the line resistances. The principle is shown in Fig. 21.58.

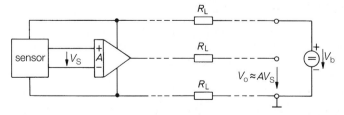

Fig. 21.57. A preamplifier for small sensor signals reduces signal transmission errors

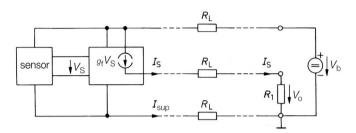

Fig. 21.58. A preamplifier with a current output at the sensor eliminates errors in signal transmission. An example of a voltage-controlled IC current source: XTR 110 from Texas Instruments
$$V_o = I_S R_1 = g_f V_S R_1 = A V_S$$

The voltage-controlled source converts sensor voltage V_S into a current $I_S = g_f V_S$. This produces a voltage drop of $V_o = g_f R_1 V_S$ across the load resistor. If $R_1 = 1/g_f$ is selected, the sensor signal is reproduced: $V_o = V_S$. However, the arrangement can be used at the same time to amplify the sensor voltage by making $A = g_f R_1 \gg 1$.

A further way of simplifying signal transmission is to insure that the current consumption of the sensor and the voltage-controlled current source are constant. In this case, signal current I_S and supply current I_{sup} can be transmitted over the same line. Only two lines are then required, as shown in Fig. 21.59. They are used both to supply the sensor and interfacing circuit and to transmit the measurement signal. If current I_{sup} or the resulting voltage $R_1 I_{sup}$ are subtracted at the load resistor, the sensor signal will remain. As with current transmission in Fig. 21.58, the line resistances R_L have no effect on the measurement result. However, the supply voltage V_b must be large enough to prevent the current sources from being driven into saturation, in spite of all the voltage drops present in the current loop.

Currents $I_{sup} + I_S$ of a current loop are standardized. Their values are between 4 mA and 20 mA. Here, 4 mA corresponds to the lower end of the range and 20 mA to the upper

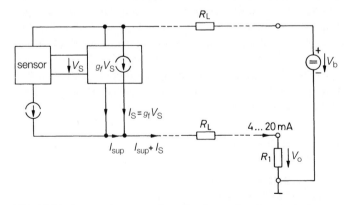

Fig. 21.59. A two-wire current loop for sensor signal transmission. IC types: XTR 101 from Burr Brown or AD 693 from Analog Devices
$$V_o = (I_{sup} + I_S)R_1 = R_1 I_{sup} + R_1 g_f V_S$$

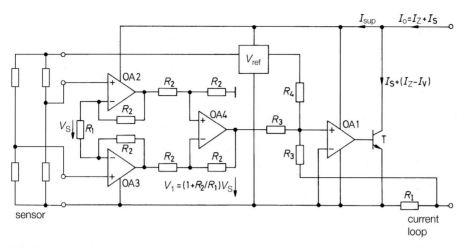

Fig. 21.60. Internal design of a current loop transmitter, using the AD 693 from Analog Devices with a resistance bridge as an example for the sensor

$$I_o = I_Z + I_S = \frac{R_3}{R_4} \frac{V_{\text{ref}}}{R_I} + \left(1 + \frac{R_2}{R_1}\right) \frac{V_S}{R_I}$$

end. For unipolar signals, the zero point is set to 4 mA. For bipolar signals it is set to 12 mA, giving a control range of ± 8 mA. If $R_1 = 250\,\Omega$ is selected, as is usual, both cases produce voltages of $V_o = 1\text{--}5\,$V at the receiver end.

Figure 21.60 shows the internal structure of a sensor interfacing circuit with a current loop output. The core of the circuit is a precision current source consisting of transistor T, operational amplifier OA1, and shunt resistor R_I. Current I_o assumes a value such that the input voltage difference of OA1 becomes zero. If R_4 is omitted for the sake of simplicity, this will occur when the voltage drop is $I_o R_I = V_1$. Resistor R_4 is merely used to add the zero current of $I_Z = 4\,$mA or 12 mA. The sensor signal is conditioned by the instrumentation amplifier and then controls the current source. The clever feature of the arrangement shown in Fig. 21.60 is that the load currents for the four operational amplifiers, the reference voltage source, and any sensors connected to it also flow through shunt resistor R_I. Their sum is thus taken into account for current measurement. Transistor T then carries only the current lacking in the wanted output current. To insure that the arrangement operates even with the smallest loop current of $I_o = 4\,$mA, the sum of the load currents I_{sup} must be less than 4 mA. In commercially available ICs, the internal current drain is less than 1 mA, so that up to 3 mA is still available for operating the sensor.

A positive side-effect of the method described above is that faults can easily be detected. If the loop current is less than 4 mA, a fault has occurred; for example, a bypass or an open circuit.

21.4.2
Electrically Isolated Signal Transmission

For transmission over long distances, and in environments with heavy electrical interference, noise signals of such magnitude can occur that the methods of signal transmission described above will not provide an adequate signal-to-noise ratio. In such cases, there is

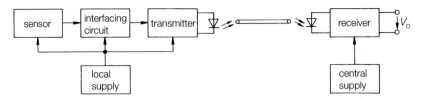

Fig. 21.61. Principle of optical transmission of sensor signals

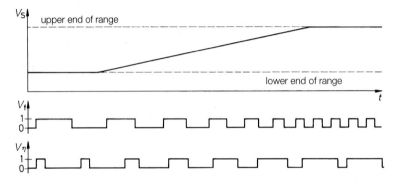

Fig. 21.62. Digital modulation of a sensor signal.
(*upper*):   The analog sensor signal
(*middle*): voltage-to-frequency conversion
(*bottom*): voltage-to-duty factor conversion

only one viable solution: the use of fiber-optic transmission. This is affected neither by electrostatic nor electromagnetic fields, and can handle almost any potential difference. Figure 21.61 shows the principle involved in the optical transmission of sensor signals.

However, analog signals are not usually transmitted over optical fibers, since the attenuation of the optical transmission paths is not well defined and is also subject to temperature variations and aging. The sensor signal is therefore converted into a serial digital signal in the transmitter. There are various ways of doing this. In the case of voltage–frequency conversion, the frequency is a linear function of the voltage; the duty factor of the output signal is a constant 1 : 1. With voltage to duty factor conversion, the frequency is constant but the duty factor is a linear function of the voltage. Figure 21.62 shows the principles of the two methods. They are particularly useful in cases in which an analog signal is to be recovered at the receiver end.

Digital processing of the received signals is also possible by measuring the frequency or the duty factor digitally. However, if high accuracy is required, it is preferable to digitize the signal using a commercially available A/D converter at the sensor end and to transmit the result serially word by word.

21.5
Calibration of Sensor Signals

Some sensors are manufactured to such tight tolerances that no calibration is required, provided that the interfacing circuit also employs components with sufficiently tight tolerances. In this case, the sensor can be replaced without recalibration becoming necessary. However, this unfortunately applies only to a few temperature sensors. The general rule is that recalibration is always necessary when a sensor is replaced. Where a high degree of accuracy is required, regular recalibration may even be necessary.

21.5.1
Calibration of the Analog Signal

For an explanation of the calibration process without reference to the specific characteristics of the sensor, we shall consider the calibration circuit to be separated from the interfacing circuit of the sensor, as shown in Fig. 21.63. We shall assume that the sensor signal is a linear function of the physical quantity G or is linearized by the interfacing circuit. The input voltage of the calibration circuit can then be expressed in the form

$$V_i = a' + m'G \tag{21.6}$$

The calibrated signal should generally be proportional to the measured quantity, in accordance with

$$V_o = mG \tag{21.7}$$

Figure 21.64 shows the voltage characteristic for a temperature measurement as an example. The calibration circuit must allow correction of the zero point and the gain. An important constraint is that calibration should be possible *without iteration*; in other words, there must be a procedure whereby one setting does not affect the other. This is possible with the arrangement shown in Fig. 21.63. Its output is:

$$V_o = A(V_i + V_Z) \tag{21.8}$$

Using (21.6) and (21.7), comparison of the coefficients produces the following calibration conditions:

Zero point: $V_Z = -a'$
Gain: $A = m/m'$

For zero adjustment, the physical quantity $G = G_0$, associated with output value $V_o = 0$ is applied to the sensor. The output voltage is then adjusted to $V_o = 0$ by varying V_Z. This adjustment is independent of any setting of gain A: the only requirement is that $A \neq 0$. In Fig. 21.64, zeroing causes a parallel shift of the input characteristic through the origin.

For gain adjustment, we apply a physical quantity G_1 and calibrate the gain A so that the nominal value of the output voltage $V_{o1} = mG_1$ is produced. In Fig. 21.64, this corresponds to a rotation of the shifted input characteristic until it corresponds to the

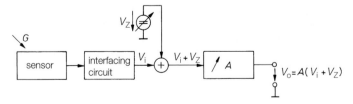

Fig. 21.63. Basic arrangement for the calibration of sensor signals by adjusting the zero point V_Z and the gain A

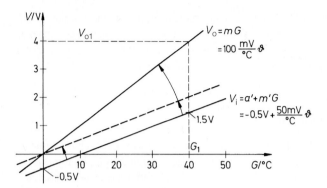

Fig. 21.64. Illustration of the calibration process – first nulling, and then gain adjustment – using the example of a clinical thermometer

required function. Nulling is not affected by this, since gain adjustment merely involves varying factor A in Eq. (21.8).

It can be seen that the reverse sequence does not result in noniterative adjustment. It is therefore absolutely essential for the zero adjuster to *precede* the gain adjuster in the signal path. The circuit arrangement in Fig. 21.63 is therefore invariable.

Calibration will now be further explained using the clinical thermometer example from Fig. 21.64. For zeroing, the sensor is set to temperature $\vartheta = 0°C$ and the output voltage adjusted to $V_o = 0$ by varying V_Z. This is the case for voltage

$$V_Z = -a' = +0.5\,V$$

To calibrate the gain, the second calibration value is applied to the sensor – for example, $G_1 = \vartheta_1 = 40°C$ – and gain A is adjusted until the wanted value of the output voltage

$$V_{o1} = mG_1 = \frac{100\,mV}{°C} \cdot 40° = 4\,V$$

is obtained. The gain is therefore

$$A = \frac{m}{m'} = \frac{100\,mV/°C}{50\,mV/°C} = 2$$

The adjustment described requires that the zero point $V_o = 0$ be initially adjusted for $G = 0$. However, the situation may arise in which physical quantity $G = 0$ cannot be implemented, or is not contained in the measure range as is the case in a clinical thermometer. We might also wish to set both calibration points close to the measurement range of interest; in other words, in the clinical thermometer example shown in Fig. 21.64, to $G_1 = 40°C$ and $G_2 = 30°C$, for example. This allows errors resulting from nonlinearity to be minimized within this range. In order to obtain a noniterative adjustment here too, the zero point of the input characteristic curve can be displaced to one of these calibration values, as shown in Fig. 21.65, and an appropriate voltage added at the output. Additional voltage V_d in Fig. 21.66 is used for this purpose. It is preferable to dimension it for the smaller of the two calibration values:

$$V_d = V_{o2} = mG_2$$

For zeroing purposes, physical quantity G_2 is applied and voltage $V_i + V_Z$ or $A(V_i + V_Z)$ is zeroed by varying V_Z. To do this, we do not need to measure into the circuit but follow

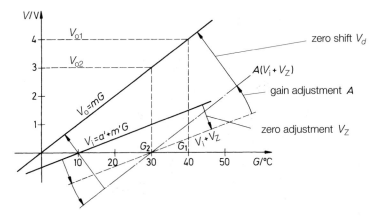

Fig. 21.65. Iteration-free adjustment procedure with two nonzero calibration points G_1, G_2

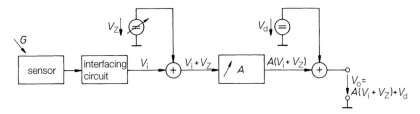

Fig. 21.66. Arrangement for noniterative calibration of sensor signals if none of the calibration points is zero

the adjustment at the output. Calibration value $V_{o2} = V_d$ must then be obtained here. Since the output voltage of the amplifier after adjustment is precisely zero, it is independent of the value of A.

The second calibration value is then applied and gain A is adjusted as previously described. The shifted input characteristic in Fig. 21.65 will rotate until it possesses the correct slope. The calibrated output signal is then obtained by voltage addition at the output side.

An example of the practical implementation of a calibration circuit is shown in Fig. 21.67. The input voltage and the voltage of the zero adjuster are added at the summing point of OA1. The gain is set at the feedback resistor. The fixed-value resistor is used to limit the adjustment range; it also prevents the gain being set to zero. Amplifier OA2 effects the output-side zero shift for the first calibration point. Since the amount of shift can be selected by a suitable choice of R_3, no adjustment is required here.

The adjustment procedure will now be further explained using the clinical thermometer example. Let the input and output characteristics be of the form

$$V_i = -0.5\,\text{V} + \frac{50\,\text{mV}}{°\text{C}}\vartheta; \quad V_o = \frac{100\,\text{mV}}{°\text{C}}\vartheta$$

and the calibration points

$$(\vartheta_2 = 30°\text{C}, \ V_{o2} = 3\,\text{V}); \quad (\vartheta_1 = 40°\text{C}, \ V_{o1} = 4\,\text{V})$$

This gives an output-side zero shift of $V_d = V_{o2} = 3\,\text{V}$. If $R_1 = 10\,\text{k}\Omega$ is specified, resistance $R_3 = 16.7\,\text{k}\Omega$ is obtained if the reference voltage is $-5\,\text{V}$. For zeroing, we

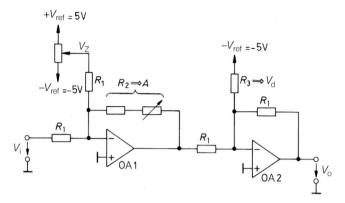

Fig. 21.67. Practical design of a calibration circuit

$$V_o = \underbrace{\frac{R_1}{R_3} V_{\text{ref}}}_{V_d} + \underbrace{\frac{R_2}{R_1}}_{A} (V_i + V_Z)$$

apply a temperature of $\vartheta_2 = 30°$ to the sensor and adjust the output voltage to $V_{o2} = 3\,\text{V}$. The voltage required for this purpose is

$$V_Z = -V_{i1} = +0.5\,\text{V} - \frac{50\,\text{mV}}{°\text{C}} \cdot 30°\text{C} = -1\,\text{V}$$

The output voltage of OA1 is then zero and the value that happens to be set for A has no effect on the zero adjustment. To calibrate the gain, the other calibration point of $\vartheta_1 = 40°\text{C}$ is set and the output voltage adjusted to $V_{o1} = 4\,\text{V}$. This is obtained for a gain of

$$A = \frac{m}{m'} = \frac{100\,\text{mV}/°\text{C}}{50\,\text{mV}/°\text{C}} = 2$$

With $R_1 = 10\,\text{k}\Omega$, the calibrated condition produces a value of $R_2 = 20\,\text{k}\Omega$.

21.5.2
Computer-Aided Calibration

If we intend to undertake further processing of a sensor signal by a microcomputer, it is advantageous to calibrate the sensor with the microcomputer as well. As can be seen from Fig. 21.68, this not only obviates the need for an analog calibration circuit, but also allows calibration to be performed more easily and improves accuracy and stability. For calibration, let us assume that the number N at the output of the A/D converter shown in Fig. 21.69 is a linear function of measured quantity G:

$$N = a + bG \tag{21.9}$$

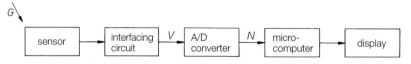

Fig. 21.68. Arrangement for computer-aided calibration of sensor signals

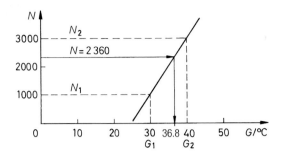

Fig. 21.69. Numerical calibration of a sensor with calibration points (G_1, N_1) and (G_2, N_2)

The calibration coefficients a and b are determined from two calibration points

$$(G_1, N_1) \quad \text{and} \quad (G_2, N_2)$$

by solving the system of equations

$$N_1 = a + bG_1 \quad \text{and} \quad N_2 = a + bG_2$$

for a and b:

$$b = \frac{N_2 - N_1}{G_2 - G_1} \tag{21.10}$$

and

$$a = N_1 - bG_1 \tag{21.11}$$

To compute the appropriate physical quantity from a measured value N, (21.9) must be solved for G

$$G = (N - a)/b \tag{21.12}$$

For practical calibration, the intended calibration values – for example, $G_1 = 30°C$ and $G_2 = 40°C$ – are stored in a table. They are then applied in turn to the sensor and the microcomputer is instructed – for example, via push-buttons – to read in the appropriate measured values – for example, $N_1 = 1{,}000$ and $N_2 = 3{,}000$ – and store them in the table. Using these entries, a program in the microcomputer can compute the calibration values in accordance with (21.10) and (21.11 and store them in the table as well:

$$b = 200/°C \quad \text{and} \quad a = -5{,}000$$

Calibration is now complete. The analysis program can then compute values G_i in accordance with (21.12). For a measured value of $N = 2{,}360$, the example gives a temperature of

$$G = \frac{N - a}{b} = \frac{2360 + 5000}{200/°C} = 36.8°C$$

For computer calibration, the hardware characteristic is therefore taken as given, its equation is formulated, and it is then used to map measured values N_i onto physical quantities G_i. It is therefore unnecessary to shift or rotate characteristic curves as in analog calibration. Any calibration point can be chosen: calibration is always noniterative, since the calibration points are determined by solving a system of equations.

A particularly difficult problem is posed by the calibration of sensors whose signals are not merely a function of the quantity sought, but of a second value as well. The most prevalent form of such unwanted dual dependence occurs in the temperature-dependence of sensor signals. Pressure sensors are an example of this, and they will be used here to illustrate the procedure involved. Measured value N comprises four components:

$$N = a + bp + c\vartheta + d\vartheta p \tag{21.13}$$

where p is the pressure, ϑ is the temperature, a is the zero error, b is the pressure sensitivity, c is the temperature coefficient of the zero point, and d is the temperature coefficient of the sensitivity.

To determine the four coefficients a, b, c, and d, four calibration measurements are performed, each of which differs in one quantity,

$$N_{11} = a + bp_1 + c\vartheta_1 + dp_1\vartheta_1 \qquad N_{21} = a + bp_2 + c\vartheta_1 + dp_2\vartheta_1$$
$$N_{12} = a + bp_1 + c\vartheta_2 + dp_1\vartheta_2 \qquad N_{22} = a + bp_2 + c\vartheta_2 + dp_2\vartheta_2$$

and we obtain

$$d = \frac{N_{22} + N_{11} - N_{12} - N_{21}}{(p_2 - p_1)(\vartheta_2 - \vartheta_1)} \qquad b = \frac{N_{22} - N_{12}}{(p_2 - p_1)} - d\vartheta_2$$

$$c = \frac{N_{22} - N_{21}}{\vartheta_2 - \vartheta_1} - dp_2 \qquad a = N_{22} - bp_2 - c\vartheta_2 - dp_2\vartheta_2 \tag{21.14}$$

Calibration is now complete and the pressure can then be computed from (21.13):

$$p = \frac{N - a - c\vartheta}{b + d\vartheta} \tag{21.15}$$

To give an example of how calibration is performed, let us assume that the four calibration values required are to be obtained at pressures of $p_1 = 900$ mbar and $p_2 = 1,035$ mbar, and at temperatures of $\vartheta_1 = 25°C$ and $\vartheta_2 = 50°C$. This produces the measured values shown in Fig. 21.70. Using Eq. (21.14), we obtain from the calibration coefficients:

$$a = -1375 \qquad b = 5.18\frac{1}{\text{mbar}}$$

$$c = 1.71\frac{1}{°C} \qquad d = -0.0119\frac{1}{\text{mbar} \cdot °C}$$

This calibration is very precise, since it not only calibrates the zero point and gain but also takes account of the temperature coefficients of the sensitivity and zero point. This method allows low-cost, uncalibrated sensors to be used for performing precision measurements.

For pressure measurement, we use (21.15). If, for example, a measured value of $N = 3351$ is obtained for a temperature of $\vartheta = 15°C$, this gives a pressure of

$$p = \frac{N - a - c\vartheta}{b + d\vartheta} = \frac{3351 + 1375 - 1.71 \cdot 15}{5.18 - 0.0119 \cdot 15} \text{ mbar} = 940 \text{ mbar}$$

Obviously, a calibrated temperature measurement is required to enable proper account to be taken of the influence of temperature. The temperature measurement will, of course,

	$\vartheta_1 = 25°C$	$\vartheta_2 = 50°C$
$p_1 = 900$ mbar	$N_{11} = 3061$	$N_{12} = 2837$
$p_2 = 1035$ mbar	$N_{21} = 3720$	$N_{22} = 3456$

Fig. 21.70. An example of pressure calibration

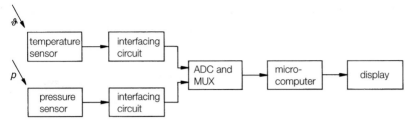

Fig. 21.71. An arrangement for computer-aided temperature and pressure calibration and measurement

also be calibrated by computer, in the same way as described above. The resulting block diagram is shown in Fig. 21.71. The pressure and temperature sensor signals are conditioned by the interfacing circuits and are fed to an analog-to-digital converter with a built-in multiplexer. The microcomputer receives the measured values N and computes from them the calibration coefficients during calibration and the measured quantities during normal operation. To enable sufficient accuracy to be attained, the A/D converter must have an accuracy of at least 12 bits. As A/D converters possessing this degree of accuracy are not available in most single-chip microcomputers, separate A/D converters which also contain an input multiplexer, have to be employed.

The sensor signal processor MSP 430 family from Texas Instruments is specifically tailored for the evaluation of sensor signals. In addition to the 14 bit AD converter with multiplexer, it contains a driver for a ten-digit liquid crystal display (LCD). It represent a particularly simple solution in cases where the measured values are to be displayed only.

Chapter 22:
Electronic Controllers

22.1
Underlying Principles

The purpose of a controller is to bring a physical quantity (the controlled variable X) to a predetermined value (the reference variable W) and to hold it at this value. To achieve this, the controller must counteract the effect of disturbances in a suitable way.

The basic arrangement of a simple control circuit is shown in Fig. 22.1. The controller influences the controlled variable X by means of the correcting variable Y, so that the error signal $W - X$ is as small as possible. The disturbances acting on the controlled system (plant) are represented formally by the disturbance variable Z, which is superimposed on the correcting variable. In what follows, we shall assume that the controlled variable is a voltage and that the system is electrically controlled. Electronic controllers can then be employed.

In the simplest case, such a controller is a circuit that amplifies the error signal $W - X$. If the controlled variable X rises above the reference signal W, the difference $W - X$ becomes negative. The correcting variable Y is thereby reduced by a factor defined by the amplifier gain. This reduction counteracts the increase in the controlled variable; that is, there is negative feedback. At steady-state, the remaining error signal is smaller the larger is the gain A_C of the controller. It can be seen from Fig. 22.1 that, for linear systems,

$$Y = A_C(W - X) \quad \text{and} \quad Y = A_S(Y + Z), \tag{22.1}$$

where A_S is the gain of the controlled system. Hence, the controlled variable X is:

$$X = \frac{A_C A_S}{1 + A_C A_S} W + \frac{A_S}{1 + A_C A_S} Z. \tag{22.2}$$

It is now obvious that the response of the control system to a reference input, $\partial X / \partial W$, approaches unity more closely the greater is the loop gain:

$$g = A_C A_S = \frac{\partial X}{\partial (W - X)}. \tag{22.3}$$

The response to a disturbance, $\partial X / \partial Z$, approaches zero more closely the larger is the gain A_C of the controller.

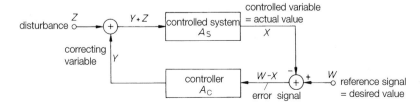

Fig. 22.1. Block diagram of a feedback control loop

It must be pointed out, however, that there is a limit to the value of the loop gain g since, if the gain is too large, the unavoidable phase shifts within the control loop give rise to oscillations. This problem has already been discussed in connection with the frequency compensation of operational amplifiers. The objective of control engineering is to obtain, despite this restriction, the smallest possible error signal and good transient behavior at the same time. For this reason, an integrator and a differentiator are added to the proportional amplifier, and the P-controller is thus turned into one that exhibits PI, or even PID, action. The electronic realization of such circuits is dealt with below.

22.2
Controller Types

22.2.1
P-controller

A P-controller (a controller with proportional action) is a linear amplifier. Its phase shift must be negligibly small within the frequency range in which the loop gain g of the control system is larger than unity. For example, an operational amplifier with resistive feedback is a P-controller of this kind.

To determine the maximum possible proportional gain A_P, we consider the Bode plot of a typical controlled system, represented in Fig. 22.2. The phase lag is 180° at the frequency $f = 3.3\,\text{kHz}$. The negative feedback then becomes a positive feedback. In other words, the phase condition (14.3) for a self-sustaining oscillation is fulfilled. The value of the proportional gain A_P determines whether or not the amplitude condition of Eq. (14.2) is also fulfilled. For the example in Fig. 22.2, the gain $|A_S|$ of the system at 3.3 kHz is about $0.01 \cong -40\,\text{dB}$. If we select $A_P = 100 \cong 40\,\text{dB}$, the loop gain at this frequency would

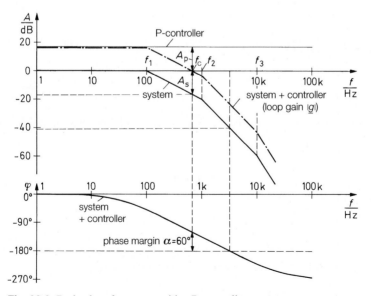

Fig. 22.2. Bode plot of a system with a P-controller

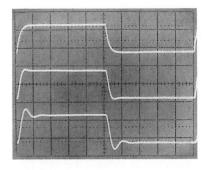

Fig. 22.3. Step response as a function of the phase margin, for the constant-gain crossover frequency f_c: $\alpha = 90°$ (*upper*), $\alpha = 60°$ (*middle*), and $\alpha = 45°$ (*lower*)

then be $|\underline{g}| = |\underline{A}_S| \cdot A_P = 1$; in other words, the amplitude condition of an oscillator would also be fulfilled and the system would oscillate permanently at $f = 3.3$ kHz. If $A_P > 100$ is chosen, an oscillation with an exponentially rising amplitude is obtained; and for $A_P < 100$, a damped oscillation occurs.

There is now the question of how much A_P must be reduced for an optimum transient behavior. An indication of the damping of the transient response can be obtained directly from the Bode diagram: the *phase margin* is the phase angle required at the *gain crossover frequency* f_c to produce a phase lag of 180°. The gain crossover frequency f_c is the critical frequency for which the loop gain is unity, and is therefore often called the unit loop-gain frequency. Hence, the phase margin is

$$\alpha = 180° - |\varphi_{\text{loop}}(f_c)| = 180° - |\varphi_S(f_c) + \varphi_c(f_c)|, \tag{22.4}$$

where φ_S is the phase shift of the system and φ_C is that of the controller. For the case of a P-controller, by definition $\varphi_C(f_c) = 0$, so that

$$\alpha = 180° - |\varphi_S(f_c)|. \tag{22.5}$$

A phase margin of $\alpha = 0°$ results in an undamped oscillation, since the amplitude condition and the phase condition for an oscillator are simultaneously fulfilled; $\alpha = 90°$ represents the case of critical damping. For $\alpha \approx 60°$, the step response of the closed control loop shows an overshoot of about 4%. The settling time is then at a minimum. This phase margin is therefore optimum for most cases. The oscillogram shown in Fig. 22.3 provides a comparison of the step responses obtained for different phase margins.

To determine the optimum P-controller gain, we examine the Bode diagram to find the frequency at which the controlled system has a phase shift of 120°. In the example shown in Fig. 22.2, that frequency is 700 Hz. This can be made the critical frequency by selecting the P-controller gain, $\underline{A}_P$, such that $|\underline{g}| = 1$. Thus, from (22.3),

$$A_C = A_P = \frac{1}{A_S} = \frac{1}{0.14} = 7$$

or

$$A_{PdB} = 20\,\text{dB} \cdot \lg 7 = 17\,\text{dB}.$$

This case is plotted in Fig. 22.2. The low-frequency limit value of the loop gain is therefore

$$g = A_S A_P = 1 \cdot 7 = 7.$$

At steady-state, this gives us, from (22.2), a relative deviation of

$$\frac{W - X}{W} = \frac{1}{1 + g} = \frac{1}{1 + 7} = 12.5\%.$$

If the controller gain is increased to obtain a smaller deviation, the transient response suffers. A proportional gain of any magnitude can only be set for systems that behave like first-order lowpass filters, because the phase margin is then greater than 90° at any frequency.

22.2.2
PI-Controller

The previous section has shown that, for reasons of stability, the gain of a P-controller must not be too large. One way of improving the control accuracy is to increase the loop gain at low frequencies, as illustrated in Fig. 22.4. In the vicinity of the critical frequency f_c, the frequency response of the loop gain thereby remains unchanged, so that the transient behavior is not affected. However, the remaining error signal is now zero, since

$$\lim_{f \to 0} |g| = \infty.$$

To implement a frequency response of this kind, an integrator is connected in parallel with the P-controller, as shown in Fig. 22.5. The Bode plot of the resulting *PI-controller* (a controller with proportional-integral action) is presented in Fig. 22.6. It can be seen that, for low frequencies, the PI-controller acts as an integrator and for high frequencies as a proportional amplifier. The changeover between the two regions is characterized by the cutoff frequency f_1 of the PI-controller. At this frequency the phase shift is 45°, and the controller gain $|A_C|$ is 3 dB above A_P.

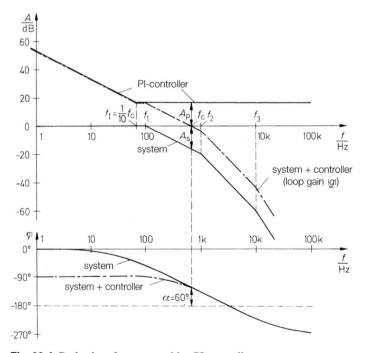

Fig. 22.4. Bode plot of a system with a PI-controller

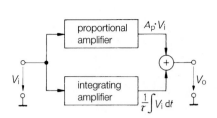

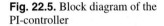

Fig. 22.5. Block diagram of the
PI-controller

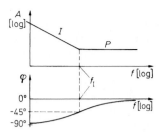

Fig. 22.6. Bode plot of the PI-controller

To determine the cutoff frequency f_1, we calculate from Fig. 22.5 the complex controller gain

$$\underline{A}_C = A_P + \frac{1}{j\omega\tau_I} = A_P\left(1 + \frac{1}{j\omega\tau_I A_P}\right).$$

Hence,

$$\underline{A}_C = A_P\left(1 + \frac{\omega_1}{j\omega}\right) \quad \text{where} \quad \omega_1 = 2\pi f_I = \frac{1}{\tau_I A_P} \tag{22.6}$$

A PI-controller can also be realized using a single operational amplifier, the appropriate circuit being shown in Fig. 22.7. Its complex gain is

$$\underline{A}_C = -\frac{R_2 + 1/j\omega C_I}{R_1} = -\frac{R_2}{R_1}\left(1 + \frac{1}{j\omega C_I R_2}\right). \tag{22.7}$$

By comparing coefficients with Eq. (22.6), we obtain the controller parameters as

$$A_P = -\frac{R_2}{R_1} \quad \text{and} \quad f_I = \frac{1}{2\pi C_I R_2}. \tag{22.8}$$

It is quite simple to design a PI-controller if we make use of the fact that the I-component does not change the phase margin. The size of the P-component (the gain) is therefore retained; that is, in the example given, $f_c = 700\,\text{Hz}$ and $A_P = 7$.

In order to insure that the I-component does not reduce the phase margin, it is necessary to select $f_I \ll f_c$. However, it is not advisable to select it to be unnecessarily low, as it will then take longer for the integrator to bring the deviation to zero. An optimum value is $f_I = 0.1\,f_c$. The I-component then reduces the phase margin by less than $6°$. Figure 22.4 has been plotted using these parameters. The appropriate transient behavior of the error signal is shown by the oscillogram of Fig. 22.8. It is obvious from the lower trace that, with these optimum parameters, the PI-controller settles to a zero error signal in the same time interval that a purely P-action controller requires to adjust to a relative error signal of $1/(1 + g) = 1/8 = 12.5\%$.

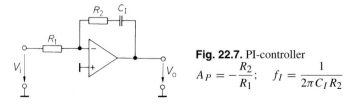

Fig. 22.7. PI-controller

$$A_P = -\frac{R_2}{R_1}; \quad f_I = \frac{1}{2\pi C_I R_2}$$

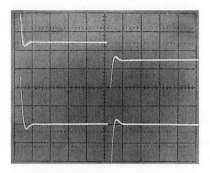

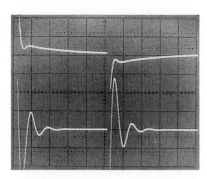

Fig. 22.8. Error signal for the P-controller (*upper*) and the PI-controller with an optimum value of f_I (*lower*)

Fig. 22.9. Error signal of the PI-controller: f_I too small (*upper*) and f_I too large (*lower*)

The effect of a less than optimum value of f_I is demonstrated by the oscillogram shown in Fig. 22.9. For the upper trace, f_I is too small: the settling time is increased. For the lower trace, f_I is too large: the phase margin is reduced.

22.2.3
PID-Controller

By connecting a differentiator in parallel as shown in Fig. 22.10, a PI-controller can be extended to become a PID-controller (a controller with proportional-integral-derivative action). Above the differentiation cutoff frequency f_D, the circuit behaves like a differentiator. The phase shift rises to $+90°$, as can be seen from the Bode plot in Fig. 22.11. This phase lead at high frequencies can be used to partially compensate for the phase lag of the controlled system in the vicinity of f_c, allowing a higher proportional gain, so that a higher gain crossover frequency f_c is obtained. The transient behavior is thereby speeded up.

The selection of the controller parameters will again be illustrated for our example system. In the first step, we raise the proportional gain A_P until the phase margin is only about $15°$. In this case, we can infer from Fig. 22.12 that $A_P = 50 \cong 34\,\mathrm{dB}$ and $f_c \approx 2.2\,\mathrm{kHz}$, as against 700 Hz for the PI-controller. If the differentiation cutoff frequency is now chosen to be $f_D \approx f_c$, the phase lag of the controller at frequency f_c is approx.

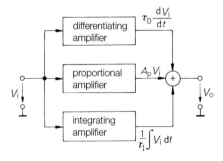

Fig. 22.10. Block diagram of the PID-controller

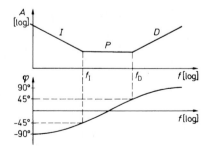

Fig. 22.11. Bode plot of the PID-controller

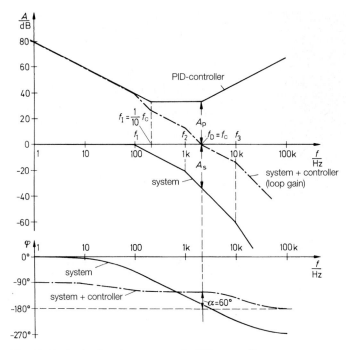

Fig. 22.12. Bode plot of a system and a PID-controller

$+45°$; in other words, the phase margin is increased from $15°$ to $60°$ and we obtain the desired transient behavior.

To determine the integration cutoff frequency f_I, the same principles apply as for the PI-controller; that is, $f_I \approx \frac{1}{10} f_c$. This results in the frequency response of the loop gain shown in Fig. 22.12.

The reduction in settling time is illustrated by a comparison of the oscillograms for a PI- and a PID-controller, shown in Fig. 22.13.

In order to implement a PID-controller, we must first determine the complex gain from the block diagram shown in Fig. 22.10:

$$\underline{A}_C = A_P + j\omega\tau_D + \frac{1}{j\omega\tau_I} = A_P\left[1 + j\left(\frac{\omega}{\omega_D} - \frac{\omega_I}{\omega}\right)\right] \tag{22.9}$$

where

$$f_D = \frac{A_P}{\pi\tau_D} \quad \text{and} \quad f_I = \frac{1}{2\pi A_P\tau_I}. \tag{22.10}$$

A circuit that has the frequency response of (22.9) can also be realized using a single operational amplifier; this is shown in Fig. 22.14. Its complex gain is given by

$$\underline{A}_C = -\left[\frac{R_2}{R_1} + \frac{C_D}{C_I} = j\omega C_D R_2 + \frac{1}{j\omega C_I R_1}\right].$$

Hence, with $\dfrac{C_D}{C_I} \ll \dfrac{R_2}{R_1}$,

$$\underline{A}_C = -\frac{R_2}{R_1}\left[1 + j\left(\omega C_D R_1 - \frac{1}{\omega C_I R_2}\right)\right]. \tag{22.11}$$

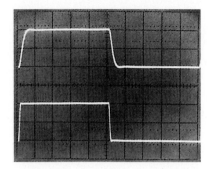

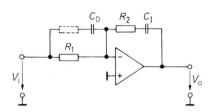

Fig. 22.13. Comparison of the transient behavior of a system with a PI-controller (*upper*) and with a PID-controller (*lower*)

Fig. 22.14. PID-controller

$$A_P = -\frac{R_2}{R_1}, \quad f_I = \frac{1}{2\pi C_I R_2}, \quad f_D = \frac{1}{2\pi C_D R_1}$$

A comparison of the coefficients with those in (22.9) yields the controller parameters

$$A_P = -\frac{R_2}{R_1}, \quad f_I = \frac{1}{2\pi C_I R_2}, \quad f_D = \frac{1}{2\pi C_D R_1}. \tag{22.12}$$

22.2.4
The PID-Controller with Adjustable Parameters

To determine the different controller parameters, we assumed that the parameters of the controlled system were known. However, these data are often difficult to measure, particularly for very slow systems. It is therefore usually better to establish the optimum controller parameters by experiment. A circuit is then required which allows *independent* adjustment of the controller parameters A_P, f_I, and f_D. It can be seen from Eq. (22.12) and Eq. (22.10) that this condition can be fulfilled neither by the circuit shown in Fig. 22.14 nor by that in Fig. 22.10 since, for any variation in A_P, the cutoff frequencies f_I and f_D will also change.

For the circuit shown in Fig. 22.15, however, the parameters are decoupled and can therefore be adjusted independently. The complex gain of the circuit is

$$\underline{A}_C = \frac{R_P}{R_I}\left[1 + j\left(\omega C_D R_D - \frac{1}{\omega C_I R_1}\right)\right]. \tag{22.13}$$

Comparing coefficients with Eq. (22.9) gives the controller parameters:

$$A_P = \frac{R_P}{R_I}, \quad f_I = \frac{1}{2\pi C_I R_I}, \quad f_D = \frac{1}{2\pi C_D R_D}. \tag{22.14}$$

Controller optimization is again illustrated with reference to our example system. In the first step, switch S is closed to render the integrator inactive. Resistor R_D is made zero; that is, the differentiator does not contribute to the output signal. The circuit is therefore a purely P-action controller.

We now apply a square wave to the reference signal input and record the transient behavior of the controlled variable X. Starting from zero, A_P is increased until the step response is only slightly damped, as in the upper trace of Fig. 22.16. The step response

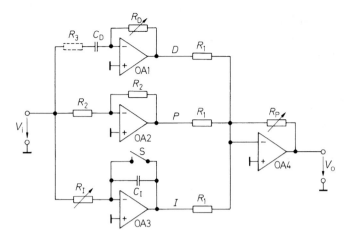

Fig. 22.15. PID-controller with decoupled parameters

$$A_P = \frac{R_P}{R_1}, \quad f_I = \frac{1}{2\pi C_I R_I}, \quad f_D = \frac{1}{2\pi C_D R_D}$$

is then that obtained for the system shown in Fig. 22.12, for a phase margin of 15° and without derivative or integral controller action.

In the second step, the differentiation cutoff frequency f_D is lowered from infinity by increasing R_D, and is adjusted to a value for which the desired damping is achieved (see the lower trace in Fig. 22.16).

In the third step, we consider the transient behavior of the error signal $W - X$. After opening switch S, the integration cutoff frequency f_I is increased until the settling time is at a minimum. The appropriate oscillograms have already been shown in Figs. 22.8 and 22.9.

The great advantage of this optimization method is that the optimum controller parameters represented by Fig. 22.12 are obtained immediately and without iteration. Using the controller parameters found in this way, the simple PID-controller shown in Fig. 22.14 can be designed.

The *oscillation test* – that is, the occurrence of a just slightly damped oscillation (Fig. 22.16) – also allows calculation of all the data required for designing the PID controller: the oscillation frequency is the critical frequency $f_{osc} = 1/T_{osc} = f_c$. The gain for which oscillation occurs yields the P-gain $A_{Cosc} = A_P$. The differentiation cutoff fre-

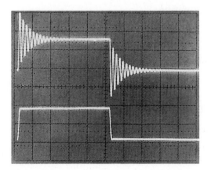

Fig. 22.16. Experimental determination of the proportional and derivative controller parameters

quency is selected to be equal to the oscillation frequency $f_D = f_{osc}$, and the integration cutoff frequency to be equal to one-tenth of the oscillation frequency: $f_I = \frac{1}{10} f_{osc}$. Taken together, these parameters produce the design parameters for a PID controller:

$$A_P \approx A_{Cosc} \qquad \tau_D \approx T_{osc} \qquad \tau_I \approx 10 T_{osc}$$

22.3
Control of Nonlinear Systems

22.3.1
Static Nonlinearity

We have assumed so far that the equation for the controlled system in Fig. 27.1 is

$$X = A_S Y$$

in other words, that the controlled system is linear. For many systems, however, this condition is not fulfilled, so that in general

$$X = f(Y).$$

For small changes about a given operating point X_0, each system can be considered to be linear as long as its transfer characteristic is continuous and differentiable in the vicinity of this point. In such cases, the derivative

$$a_S = \frac{dX}{dY}$$

is used, so that for small-signal operation

$$x \approx a_S y,$$

where $x = (X - X_0)$ and $y = (Y - Y_0)$. For a fixed point of operation, the controller can now be optimized as described above. However, a problem arises if larger changes in the reference variable are allowed: since the incremental system gain a_S is dependent on the actual point of operation, the transient behavior varies as a function of the magnitude of W.

This can be avoided by providing linearity; that is, by connecting a function network of Sect. 11.7.5 on p. 750 in front of the controlled system. The corresponding block diagram is shown in Fig. 22.17. If the function network is used to implement the function $Y = f^{-1}(Y')$, we obtain the required linear system equation

$$X = f(Y) = f[f^{-1}(Y')] = Y.$$

If, for instance, the system exhibits exponential behavior,

$$X = A e^Y,$$

the function network must be a logarithmic one, that has the characteristic

$$Y = f^{-1}(Y') = \ln \frac{Y'}{A}.$$

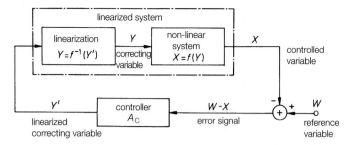

Fig. 22.17. Linearization of a system that has static nonlinearity

22.3.2
Dynamic Nonlinearity

A different kind of nonlinearity of a controlled system may arise as the result of the rate of change of some quantity in the system being limited to a value that cannot be raised by increasing the correcting variable. We have encountered this effect in operational amplifiers, in the form of limitation of the slew rate. For large input steps and controllers with integral action, the above effect leads to large overshoots that decay only slowly.

The reason for this is as follows. For optimized integral action of the controller and for a small voltage step, the integrator reaches its steady-state output voltage at the precise instant at which the error signal becomes zero. For double the input step of a linear system, the rate of change in the system, as well as that of the integrator, would double. The increased reference value would therefore be reached within exactly the same settling time.

For a system that has a limited slew rate, only the rate of change of the integrator is doubled, whereas that of the system remains unchanged. This results in the controlled system reaching the reference value very much later and the integrator overshooting. The controlled variable therefore greatly exceeds the reference value and the decay takes longer the further the integrator output voltage is from the steady-state value. The decay time constant for this nonlinear operation therefore becomes larger for increasing input steps.

The effect is avoided by increasing the integration time constant (i.e., reducing f_I) until no overshoot is incurred for the largest possible input step. However, this results in considerably prolonged settling times for small-signal operation (see the lower trace in Fig. 22.9).

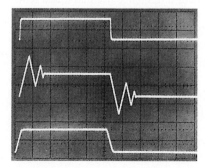

Fig. 22.18. Transient behavior of the controlled variable for a limited system slew rate.
Upper trace: small-signal characteristic.
Middle trace: large-scale characteristic.
Lower trace: large-signal characteristic for a slew-rate limited reference variable

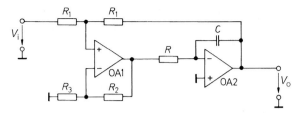

Fig. 22.19. Slew rate limiter for the reference variable. Resistors R_2, R_3 limit the gain of OA1 and provide additional frequency compensation

Steady-state output voltage: $V_o = -V_i$

Maximum slope of output voltage: $\dfrac{dV_o}{dt} = \dfrac{V_{max}}{RC}$

A much more effective measure is to limit the slew rate of the reference variable to the maximum slew rate of the controlled system. This insures linear operation throughout, and the overshoot effect is thus avoided. This will not increase the settling time for large reference signals, since the controlled variable cannot change any faster anyway. This is illustrated by the oscillogram shown in Fig. 22.18.

In principle, a lowpass filter could be used to limit the slew rate, but this would also reduce the small-signal bandwidth. A better solution is shown in Fig. 22.19. If a voltage step is applied to the circuit input, amplifier OA 1 saturates at the output limit V_{max}. The output voltage of OA 2 therefore rises at the rate

$$\frac{dV_o}{dt} = \frac{V_{max}}{RC}$$

until it reaches the values $-V_i$ determined by the overall feedback. A square-wave voltage would therefore be shaped into the required trapezoidal voltage. The signal remains unchanged if the rate of change of the input voltage is smaller than the predetermined maximum. The small-signal bandwidth is therefore not affected.

22.4
Phase-Locked Loop

A particularly important application of feedback control in communications systems is the phase-locked loop (PLL). Its purpose is to control the frequency f_2 of an oscillator in such a manner that it is the same as the frequency f_1 of a reference oscillator, and to do this so accurately that the phase shift between the two signals remains constant. The basic arrangement of such a circuit is illustrated in Fig. 22.20.

The frequency of the voltage-controlled oscillator (VCO) can be varied by means of the control voltage V_f, according to the relationship

$$f_2 = f_0 + k_f V_f . \tag{22.15}$$

Such voltage-controlled oscillators are described in Chap. 14. For low frequencies, the second-order differential equation circuit of Sect. 14.4 or the function generators of Sect. 14.5 can be employed. For higher frequencies, the emitter-coupled multi-vibrator of Fig. 6.21 is more suitable, or any *LC* oscillator if a varactor diode is connected in paral-

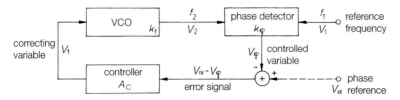

Fig. 22.20. Principle of the phase-locked loop (PLL)

lel with the capacitor of the oscillating circuit. However, the linear relationship of (22.15) then holds only for small variations around the point of operation, f_0, as the incremental control constant (the VCO "gain") $k_f = \mathrm{d}f_2/\mathrm{d}V_f$ is dependent on the operating point.

The phase detector produces an output voltage that is defined by the phase angle φ between the tracking oscillator voltage V_2 and the reference oscillator voltage V_1:

$$V_\varphi = k_\varphi \cdot \varphi .$$

The integrating property of the controlled system is of particular interest. If frequency f_2 deviates from the reference frequency f_1, the phase shift φ will increase proportionally with time, and without limit. The error signal in the closed loop therefore rises, even for a finite controller gain, until both frequencies are exactly the same. The remaining error signal of the *frequency* is thus zero.

The remaining error signal of the *phase shift*, however, does not usually become zero. From Fig. 22.20, we deduce that $V_\alpha - V_\varphi = V_f/A_C$; hence

$$\alpha - \varphi = \frac{f_1 - f_0}{A_C k_f k_\varphi} , \qquad (22.16)$$

where f_0 is the VCO frequency for $V_f = 0$. If it is important not only to keep the phase shift constant but also to hold it precisely at a predetermined value of

$$\alpha = V_\alpha/k_\varphi = -\varphi ,$$

a PI-controller must be used for which $A_C(f = 0) = \infty$. In many applications it is sufficient to control for identical frequencies ($f_1 = f_2$) – that is, for a *constant* phase shift (the angle α being unimportant) – so that the control input V_α can be omitted. Voltage V_φ is then the error signal.

To determine the controller parameters, the frequency response of the system must be known. As mentioned before, the phase detector exhibits integral behavior, so that the phase shift is given by

$$\varphi = \int_0^t \omega_2 \mathrm{d}\tilde{t} - \int_0^t \omega_1 \mathrm{d}\tilde{t} = \int_0^t \Delta\omega \mathrm{d}\tilde{t} , \qquad (22.17)$$

where $\tilde{t}$ is a dummy time variable of integration. To determine the frequency response of the controlled system, we modulate frequency ω_2 sinusoidally with a modulating frequency ω_m around the center frequency ω_1. Hence

$$\Delta\omega(t) = \widehat{\Delta\omega} \cos \omega_m t .$$

By inserting this in (22.17), we obtain:

$$\varphi(t) = \frac{\widehat{\Delta\omega}}{\omega_m} \cdot \sin\omega_m t .$$

Taking the phase lag of 90° into account, we obtain, in complex notation,

$$\frac{\underline{\varphi}}{\underline{\Delta\omega}} = \frac{1}{j\omega_m} , \tag{22.18}$$

which is the equation for an integrator. With the constants k_f and k_φ, the complex gain of the controlled system is then

$$\underline{A}_S = \frac{\underline{V}_\varphi}{\underline{V}_f} = \frac{2\pi k_f k_\varphi}{j\omega_m} = \frac{k_f k_\varphi}{jf_m} . \tag{22.19}$$

As will be seen later, the measurement of the phase shift involves a certain delay and the factor k_φ is therefore complex; in other words, the order of the system is raised.

The behavior of a phase-locked loop generally depends on the type of phase detector used. The most important circuits will now be discussed.

22.4.1
Sample-and-Hold Circuit as a Phase Detector

The phase angle φ between two voltages V_1 and V_2 can, for example, be measured with a sample-and-hold circuit by sampling the instantaneous value of V_1 at the rising edge of V_2. For this purpose, V_2 in Fig. 22.21 is used to activate an edge-triggered one-shot that produces the sampling pulse for the sample-and-hold circuit. It can be seen in Fig. 22.22 that the output voltage of the circuit is given by

$$V_\varphi = \hat{V}_1 \sin\varphi . \tag{22.20}$$

Around the point of operation ($\varphi = 0$), the detector characteristic is approximately linear:

$$V_\varphi \approx \hat{V}_1 \varphi/\text{rad} .$$

Hence, the factor of the phase detector is

$$k_\varphi = \frac{dV_\varphi}{d\varphi} = \hat{V}_1/\text{rad} . \tag{22.21}$$

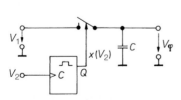

Fig. 22.21. A sample-and-hold circuit used as a phase detector

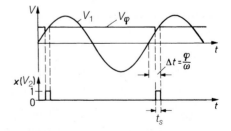

Fig. 22.22. Voltage waveform in the phase detector. The dips in V_φ disappear to a great extent if the sampling time t_s is not much larger than the time constant of the sample-and-hold circuit

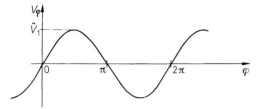

Fig. 22.23. Transfer characteristic of a sample-and-hold circuit used as a phase detector

It is obvious from Fig. 22.23 that a further possible operating point at which $V_\varphi = 0$ would be $\varphi = \pi$. Then, $k_\varphi = -\hat{V}_1/\text{rad}$. The sign of the controller gain defines which of the two operating points is stable. Further stable points of operation occur at intervals of 2π. This indicates that the phase detector does not recognize a displacement by a whole number of periods.

If a triangular wave-shape V_1 is employed rather than the sinusoidal one, a triangular detector characteristic is obtained. For rectangular input voltages V_1, the circuit is obviously unusable.

The Dynamic Behavior

The phase detector described determines a new value for the phase shift only once during a period of the input waveform and consequently has a dead time. The delay is between 0 and $T_2 = 1/f_2$, depending on the instant at which a change in phase occurs. The average delay is therefore $\frac{1}{2}T_2$. To allow for this, the "gain" k_φ of the phase detector for higher phase modulation frequencies f_m must be written in complex form, so that

$$\underline{k_\varphi} = k_\varphi e^{-j\omega_m \cdot \frac{1}{2}T_2} = \hat{V}_1 e^{-j\pi f_m/f_2} . \tag{22.22}$$

Using (22.19), we therefore obtain, for the complex gain of the entire controlled system,

$$\underline{A_S} = \frac{k_f \underline{k_\varphi}}{jf_m} = \frac{k_f \hat{V}_1}{jf_m e^{j\pi f_m/f_2}}$$

or

$$|\underline{A_S}| = \frac{|\underline{V}_\varphi|}{|\underline{V}_f|} = \frac{k_f \hat{V}_1}{f_m} \quad \text{and} \quad \varphi_m = -\frac{\pi}{2} - \frac{\pi f_m}{f_2} . \tag{22.23}$$

Controller Parameters

For the controller, it is best to choose a circuit without derivative action, since the output voltage of the sample-and-hold element changes only in steps. According to (22.23), the phase shift φ_m between $\underline{V}_\varphi$ and $\underline{V}_f$ at frequency $f_m = \frac{1}{4}f_2$, is $-135°$. The phase margin is thus $45°$ if we adjust the proportional gain A_P in such a way that the gain crossover frequency $f_c = \frac{1}{4}f_2$. By definition, at $f_m = f_c$,

$$|\underline{g}| = |\underline{A_S}| \cdot |\underline{A_C}| = 1 .$$

Using $\underline{A_C} = A_P$ and (22.23), this results in

$$A_P = \frac{f_c}{k_f k_\varphi} = \frac{f_2}{4k_f \hat{V}_1} .$$

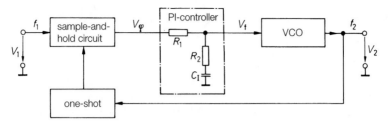

Fig. 22.24. PLL with a sample-and-hold circuit as phase detector

A typical example is $f_2 = 10\,\text{kHz}$, $k_f = 5\,\text{kHz/V}$, and $k_\varphi/ = \hat{V}_1/\text{rad} = 10\,\text{V/rad}$. Hence, $A_P = 0.05$. In this case, the controller can be a passive voltage divider.

To reduce the remaining phase error – see Eq. (22.16) – the gain at low frequencies can be increased by integral action ($f_I = \frac{1}{10} f_c = \frac{1}{40} f_2$). It is advisable, however, to limit the low-frequency value of the gain to a finite value A_I, since otherwise the integrator would drift and saturate for as long as the loop is not in the locked condition. The VCO can thereby be detuned to such an extent that the loop will not acquire lock.

The passive voltage divider can be extended in a simple way to become a PI-controller with limited gain A_I, by connecting a capacitor in series with the resistor R_2, as illustrated in Fig. 22.24. The controller parameters are then

$$A_P = \frac{R_2}{R_1 + R_2} \qquad f_I = \frac{1}{2\pi R_2 C_I} \qquad A_I = 1.$$

Pull-In

After switch-on there is usually a frequency offset $\Delta f = f_1 - f_2$. The phase shift therefore increases proportionally with time. According to Fig. 22.23, this produces an alternating voltage at the output of the phase detector, with a frequency Δf and an amplitude $\hat{V}_\varphi = \hat{V}_1$, so that the tracking oscillator is thus frequency-modulated by the voltage:

$$V_f = A_P \hat{V}_1 \sin \Delta \omega t.$$

There will therefore be an instant at which the frequencies are identical, and the loop will pull in and acquire lock. The precondition for this is that the frequency offset $\Delta f = f_1 - f_2$ is smaller than the sweep width:

$$\Delta f_{2\,max} = \pm k_f A_P \hat{V}_1. \tag{22.24}$$

This maximum permissible offset is known as the *capture range* and represents the normal range of operation of the loop. For our example, it is $\pm 2.5\,\text{kHz}$; that is, 25% of $f_1 = 10\,\text{kHz}$.

22.4.2
Synchronous Demodulator as a Phase Detector

Section 20.3.4 on page 1053 describes the application of the multiplier as a phase-sensitive rectifier. If two sinusoidally alternating voltages, $V_1 = U \cos \omega_1 t$ and $V_2 = U \cos(\omega_2 t + \varphi)$, are applied to the inputs, the output voltage will be (U = multiplier unit)

$$V_o = \frac{V_1 V_2}{U} = \frac{1}{2} U \cos[(\omega_1 + \omega_2)t + \varphi] + \frac{1}{2} U \cos[(\omega_1 - \omega_2)t - \varphi]. \tag{22.25}$$

For $\omega_1 = \omega_2$, there is an oscillating term with double the frequency, superimposed on a DC voltage of magnitude

$$V_\varphi = \overline{V}_o = \frac{1}{2} U \cos \varphi, \tag{22.26}$$

in accordance with (20.22).

This function is represented in Fig. 22.25. It becomes immediately obvious that the voltage cannot be used as a control variable in the vicinity of $\varphi = 0$, as the sign of the error signal cannot be detected. The two operating points $\varphi = \pm\pi/2$ are well suited, however, because the voltage V_φ has a zero crossing. The sign of the gain determines at which of the two points lock-in occurs. Further stable points of operation occur at intervals of 2π. This implies that this kind of phase detector also cannot recognize a phase displacement by a whole number of periods; that is, by a multiple of 2π.

Within a range of about $\pm\pi/4$ around the stable point of operation, φ_0, the characteristic of the phase detector, is approximately linear, so that, with $\varphi = \varphi_0 + \vartheta$,

$$V_\varphi = \frac{U}{2} \cos(\varphi_0 + \vartheta) = \pm\frac{U}{2} \sin \vartheta \approx \pm\frac{U}{2} \vartheta / \text{rad}. \tag{22.27}$$

The sensitivity (the phase detector "gain") is therefore

$$k_\varphi = \frac{dV_\varphi}{d\vartheta} = \pm\frac{U}{2\,\text{rad}}. \tag{22.28}$$

If, instead of the two sinusoidal voltages, two square-waves with the magnitude $\pm U$ are applied, the detector characteristic becomes triangular, as shown by the dashed lines in Fig. 22.25. The stable points of operation are again at $\varphi_0 = \pm(\pi/2)\pm n\cdot 2\pi$, the sensitivity in this case being

$$k_\varphi = \pm\frac{2U}{\pi\,\text{rad}}. \tag{22.29}$$

For square-wave input voltages, an analog multiplier is obviously no longer required. In this case, considerably higher frequencies are achieved using the sign of gain switch shown in Fig. 17.22 on page 939 or an EXCLUIV-OR gate.

If the ripple of V_φ is to be kept small, a lowpass filter must be connected at the multiplier output, the cutoff frequency f_{lp} of which is small against $2f_1$, in accordance with (22.25). This is a definite disadvantage since, unlike in the previous circuit, the proportional gain of the controller must now be chosen so low that the gain crossover frequency $f_c \approx f_{lp}$. At this frequency, the phase shift of the controlled system together with that of the lowpass filter is already $-135°$. If $f_c \approx f_{lp} \ll f_1$, the control loop obtained would be impractical,

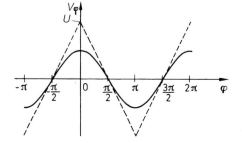

Fig. 22.25. Mean absolute value of the output voltage of a multiplier for sinusoidal input voltages of amplitude U.
Dashed line: the multiplier mean output voltage for a square-wave input signal with the levels $\pm U$

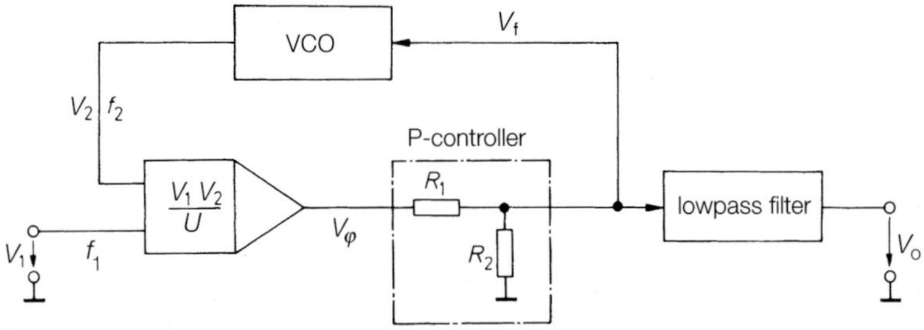

Fig. 22.26. PLL with a multiplier as phase detector for frequency demodulation

because of its very slow response. In principle, it could be speeded up by derivative action of the controller, but this would nullify the effect of the lowpass filter; that is, increase the ripple.

An increase in the control system bandwidth at the expense of the ripple of V_f can be attained very simply by using a proportional controller and omitting the lowpass filter, as shown in Fig. 22.26. A phase margin of 90° is then available for any proportional gain chosen; that is, the control loop is aperiodically damped.

Because of the feedback, the ripple of V_f causes the tracking oscillator to be frequency-modulated with twice the signal frequency. This results in a distortion of the output sine wave which is equivalent to a phase noise of the oscillator. For square waves, the mark-space ratio is changed. The proportional gain must not be too high if the distortion is to be kept within tolerable limits. The condition $f_{lp} \leq \frac{1}{3} f_1$ can be used as a rule of thumb.

The resulting arrangement from Fig. 22.26 is available as an integrated circuit PLL. Usually, the multiplier is simplified and reduced to a modulator, as shown in Fig. 17.22. The 74 HC 4046 from Philips and National, for instance, is based on this principle.

When operated without a lowpass filter, the circuit is usable only in those applications where it is important to have frequency f_2 identical to f_1, and where the shape and phase shift of the output signal are not significant, for example, as a discriminator for frequency demodulation. The FM-signal is used as input signal. If the VCO frequency f_2 is proportional to V_f, this voltage is also proportional to the input frequency f_1. The superimposed ripple can be filtered out by a lowpass filter outside the control loop.

22.4.3
The Frequency-Sensitive Phase Detector

One drawback of the phase detectors described above is that they possess only a limited capture range; in other words, they cannot pull in if the initial frequency offset exceeds a certain limit. The reason for this is that the phase-equivalent signal V_φ for a frequency deviation is an alternating voltage that is symmetrical about zero. The voltage V_f therefore effects a periodic frequency modulation of the tracking oscillator, but never a systematic tuning in the right direction; that is, toward the lock-in frequency.

The phase detector in Fig. 22.27, however, is different in this respect, in that it produces a signal that has the correct sign information for any given frequency offset. The circuit basically comprises two edge-triggered D flip-flops. For control purposes, the input voltages $V_1(t)$ and $V_2(t)$ are converted into the rectangular signals x_1 and x_2.

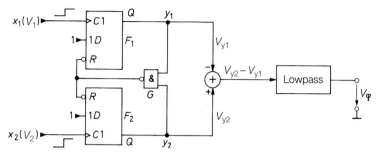

Fig. 22.27. Phase detector with memory for the sign of the phase shift

We now assume that both flip-flops are reset. If voltage V_2 leads voltage V_1 (i.e., $\varphi > 0$), we first obtain a positive edge of x_2 which sets flip-flop F_2. It remains set until the following positive edge of x_1 sets flip-flop F_1. The state in which both flip-flops are set exists only during the propagation delay time, since they are both reset subsequently by gate G. It can be seen from Fig. 22.28 that the output of the subtractor shows a sequence of positive rectangular pulses. Correspondingly, a sequence of negative pulses is obtained if the positive-going edge of x_2 occurs *after* the positive-going edge of x_1; that is, if $\varphi < 0$. This behavior is summarized in the state diagram of Fig. 22.29.

The duration of the output pulses is equal to the time interval between the positive-going zero crossings of $V_1(t)$ and those of $V_2(t)$. Hence, the mean value of the output voltage is:

$$V_\varphi = \hat{V}\frac{\Delta t}{T} = \hat{V} \cdot \frac{\varphi/\text{rad}}{2\pi} . \tag{22.30}$$

As the value of the time interval increases proportionally with φ until the limits $\pm 360°$ are reached, a range of linear phase measurement of $\pm 360°$ is obtained. When this limit is exceeded, the output voltage jumps to zero and increases again, still with the same polarity. The result is the sawtooth characteristic shown in Fig. 22.30.

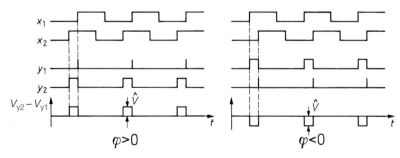

Fig. 22.28. Input and output signals of the phase detector

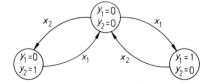

Fig. 22.29. State diagram of the phase detector

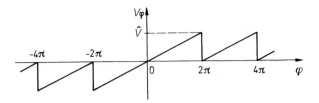

Fig. 22.30. Transfer characteristic of the phase frequency detector

The basic difference between this characteristic and all previous ones is that, for $\varphi > 0$, V_φ is always positive and, for $\varphi < 0$, always negative. This is the reason for the frequency sensitivity of the detector. If frequency f_2 is, for instance, larger than f_1, the phase shift increases continuously and proportionally with time. As shown in Fig. 22.30, we then obtain for V_φ a sawtooth voltage that has a positive mean value. If this detector is used in a phase-locked loop, it always indicates a leading phase. For a controller with integral action, the tracking freqticncy f_2 is therefore reduced until it coincides with f_1. The capture range is thus theoretically infinite, and in practice limited only by the input voltage range of the VCO.

We described in Sect. 22.4.2 how the averaging lowpass filter has a very unfavorable effect on the transient behavior. For this reason, it is usually also omitted in this circuit. If we wish to have $\varphi = 0$ (with the aid of a PI-controller), no phase distortions are incurred, since in this case $V_\varphi = 0$ even without filtering. The flip-flops in the phase detector then produce no output pulses.

One drawback of the circuit is that very small deviations in phase are not detected, since the flip-flops would then have to produce extremely short output pulses that would be lost due to the limited rise times within the circuit. This is the reason why the phase jitter (the phase noise) is somewhat larger than with the sample-and-hold detector.

If a PLL with a large capture range and a small phase jitter is required, this circuit can be combined with a sample-and-hold detector. After lock-in has been accomplished, the sample-and-hold detector is switched into the loop instead of the frequency-sensitive phase detector.

22.4.4
The Phase Detector with an Extensible Measuring Range

With the phase detectors described so far, it is not possible to detect a phase shift of more than one oscillation period, as the phase-measuring range is limited to values between $\pi/2$ and 2π, depending on the detector used. There are applications, however, for which a phase delay of several oscillations must be recovered. The phase detector shown in Fig. 22.31 is suitable for this purpose. It is based on the up–down counter as illustrated in Fig. 9.36 on page 672, which is insensitive to coincident clock pulses.

Near zero phase shift, the detector behaves in the same way as the previous circuit. If x_2 leads x_1, positive pulses of the magnitude V_{LSB} arise, the duration of which is equal to the interval between the zero crossings of the input voltages. For a phase lag, negative pulses occur. The mean value of the pulses is:

$$V_\varphi = \overline{V_D} = V_{LSB}\frac{\Delta t}{T} = V_{LSB} \cdot \frac{\varphi/\text{rad}}{2\pi}.$$

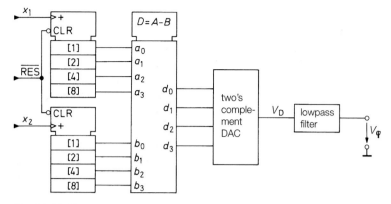

Fig. 22.31. Phase detector with an extensible measuring range.
Range in example: $+7$, -8 periods

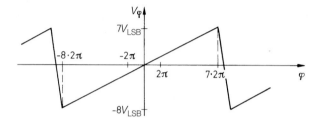

Fig. 22.32. Detection characteristic of the phase detector

If the phase displacement reaches the value 2π, the value for the interval Δt jumps from T to 0. In contrast to the previous circuit, however, the output does not assume zero voltage, but remains at V_{LSB}, as the difference D simultaneously increases by 1. To generalize, the resultant output voltage is given by:

$$V_\varphi = V_{LSB}\left(D + \frac{\Delta t}{T}\right) = V_{LSB} \cdot \frac{\varphi/\text{rad}}{2\pi}.$$

The expression $D + \Delta t/T$ indicates the number of periods by which the two signals are displaced. The resulting detector characteristic is shown in Fig. 22.32 for 4 bits. The measurement range can he increased as required by extending the counting range.

22.4.5
The PLL as a Frequency Multiplier

A particularly important application of the PLL is that of frequency multiplication. A frequency divider is connected to each of the two inputs of the phase detector, as illustrated in Fig. 22.33. The frequency of the VCO assumes such a value that

$$\frac{f_1}{n_1} = \frac{f_2}{n_2}.$$

In this manner, the frequency of the VCO,

$$f_2 = \frac{n_2}{n_1} f_1$$

can be adjusted to any rational multiple of the reference frequency f_1.

In this application, the phase detector may operate at a frequency that is considerably lower than that of the VCO. It must therefore be insured that the control voltage V_φ contains no ripple. An undesired frequency modulation and phase noise would otherwise occur instead of simply the distortion of the output waveform, as described in Sect. 22.4.2.

The frequency multiplier circuit can be used to generate frequencies above 50 MHz with crystal accuracy for which practically no crystals are available. For this purpose, we use a crystal oscillator that operates at, for example, $f_1 = 10$ MHz and select $n_2 > n_1$. If it is merely a question of obtaining an integral multiple of the crystal frequency, we can select $n_1 = 1$, thereby dispensing with the input divider. However, if we would like, for instance, to step through the frequencies from 90 to 100 MHz in 100 kHz increments, the crystal frequency must first be divided down to 100 kHz with $n_1 = 100$. We can then generate all of the desired frequencies using a divider factor of $n_2 = 900 \ldots 1000$. This is the principle on which the digital tuners widely used in today's radio and TV receivers are based. A number of integrated PLL components are listed in Fig. 22.34.

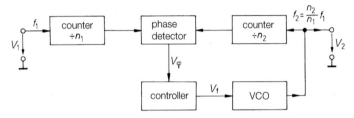

Fig. 22.33. Frequency multiplication using a PLL

Type	Manufacturer	Technology	Frequency max	Principle
Phase detectors, PD				
AD 8343	Analog D.	Bipolar	2500 MHz	analog multiplier
AD 9100	Analog D.	Bipolar	200 MHz	sample-hold
AD 9901	Analog D.	TTL/ECL	200 MHz	phase/frequ. det.
MAX 9382	Maxim	PECL	450 MHz	phase/frequ. det.
LF 398	National	Bifet	0.3 MHz	sample-hold
MC 100EP 140	On Semi.	ECL	2000 MHz	phase/frequ. det
Voltage-controlled oscillators, VCOs				
F 100	Fujitsu	CMOS	30 MHz	piezo-oscillator
VC 80	Fujitsu	Bipolar	2500 MHz	LC-oscillator
LTC 1799	Lin. Tech.	CMOS	33 MHz	multivibrator
LTC 6905	Lin. Tech.	CMOS	170 MHz	multivibrator
MAX 2609	Maxim	PECL	600 MHz	LC-oscillator
MAX 2754	Maxim	Bipolar	1200 MHz	LC-oscillator
MC 100 EL 1648	On Semi.	ECL	1100 MHz	LC-oscillator
74 LS624	Texas I.	TTL	20 MHz	Multivibrator
VFC 110	Texas I.	Bipolar	4 MHz	V → f converter

Fig. 22.34. Examples for PLLs and components

Type	Manufacturer	Technology	Frequency max	Principle
Phase-locked loops, PLLs				
74 HC 4046	many	CMOS	20 MHz	PD+VCO
AD 800	Analog D.	ECL	155 MHz	PD+VCO
AD 9540	Analog D.	PECL	655 MHz	PD+DDS+divider
CY 22394	Cypress	PECL/CMOS	400 MHz	PD+VCO+divider
CY 7B9940	Cypress	LVTTL	200 MHz	PD+VCO+divider
MPC 9331	Freescale	CMOS	240 MHz	PD+VCO+divider
ispclock 5500	Lattice	LVTTL	320 MHz	PD+VCO+divider
LMX 2325	National	BiCMOS	2500 MHz	PD+VCO+divider
NBC 12430	On Semi.	PECL	800 MHz	PD+VCO+divider
TLC 2932	Texas I.	TTL	32 MHz	PD+VCO

Fig. 22.34 (cont.). Examples for PLLs and components

Chapter 23:
Optoelectronic Components

23.1
Basic Photometric Terms

The human eye perceives electromagnetic waves in the range 400 to 700 nm as light. The wavelength produces the sensation of color, and the intensity that of brightness. In order to quantify brightness, it is necessary to define a number of photometric quantities. The *luminous flux* Φ is a measure of the number of quanta of light (photons) passing through a cross-sectional area of observation area A per time. It is expressed in lumen (lm). The luminous flux Φ is unsuitable for characterizing the brightness of a light source, as it is generally a function of the cross-sectional area A and the distance r from the light source. In the case of a spherically symmetrical point source, the luminous flux is proportional to the solid angle Ω. This is defined as $\Omega =$ surface/(radius)^2 and is actually dimensionless. However, it is generally assigned the unit *steradian* (sr). The solid angle that encloses the entire surrounding sphere is given by

$$\Omega_0 = \frac{4\pi r^2}{r^2} \text{ sr} = 4\pi \text{ sr}.$$

A circular cone of aperture angle $\pm\varphi$ encloses the solid angle

$$\Omega = 2\pi(1 - \cos\varphi) \text{ sr}. \tag{23.1}$$

At $\pm 33°$, Ω is approximately unity. For small solid angles, we can, as an approximation, replace the spherical surface by a flat surface, obtaining

$$\Omega = \frac{A_n}{r^2} \text{ sr} \tag{23.2}$$

where r is the distance of the surface from the center. As the luminous flux of a point source of light is proportional to the solid angle Ω, the brightness of the light source can be characterized by the quantity $I = d\Phi/d\Omega$, the *luminous intensity*. The unit of luminous intensity is the candela (cd). The relationship between the above units is given by 1 cd = 1 lm/sr. A light source therefore possesses a luminous intensity of 1 cd if it emits a luminous flux of 1 lm into a solid angle of 1 sr. In the case of spherical symmetry, the total emitted luminous flux is therefore $\Phi_{tot} = I\Omega_0 = 1$ cd 4π sr $= 4\pi$ lm. 1 cd is defined as the luminous intensity of a black body with a surface area of 1.6667 mm^2 at the temperature of solidifying platinum (1769°C). A large candle flame has a luminous intensity of approximately 1 cd. For incandescent lamps, the relation $I = 1\frac{cd}{W}P$ can be used as an approximation, where P is the rated power of the incandescent lamp.

In the case of extended light sources, the *luminance* $L = dI/dA_n$ is generally specified, A_n being the projection of the light source area onto the plane perpendicular to the direction of observation. If the angle between the surface normal and the specified direction is ε, then $dA_n = dA \cdot \cos\varepsilon$. The unit of luminance is the stilb (sb): 1 sb = 1 cd/cm^2.

A measure of how bright an illuminated area A appears to the observer is the *illuminance* $E = d\Phi/dA_n$. The unit is the lux (lx): 1 lx = 1 lm/m^2. A full moon gives an illuminance of 0.1 to 0.2 lx. A newspaper is just readable at an illuminance of 0.5 to 2.0 lx. At a writing

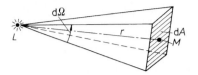

Fig. 23.1. Relationship between luminous intensity and illuminance

desk, there should be an illuminance of 500 to 1,000 lx. Daylight can produce illuminances of up to 50,000 lx.

We shall now calculate the illuminance produced by a point source of light that has a given luminance at a specified distance r (Fig. 23.1). In order to calculate the illuminance, we assume that the area term dA is small compared to r^2 and is perpendicular to the connecting line LM. From (23.2), the solid angle $d\Omega$ subtended by dA at point L is therefore given by

$$d\Omega = \frac{dA}{r^2}\,\text{sr}\,.$$

As defined, the luminous flux emitted by the lamp L is

$$d\Phi = I\,d\Omega = I\frac{dA}{r^2}\,\text{sr}\,.$$

For the illuminance, we obtain the following:

$$E = \frac{d\Phi}{dA} = \frac{I}{r^2}\,\text{sr}\,. \tag{23.3}$$

The illuminance is therefore inversely proportional to the square of the distance.

As each quantum of light possesses the energy hf, a relationship can be established between the light power P_L and the luminous flux Φ for a specific frequency. At a wavelength of 553 nm,

$$P_L = \frac{1.47\ \text{mW}}{\text{lm}}\Phi\,.$$

Thus the illuminance is given by

$$1\ \text{lx} = 1\frac{\text{lm}}{\text{m}^2}\hat{=}\frac{1.47\ \text{mW}}{\text{m}^2}$$

When giving approximate values for various luminous intensifies, we stated that an incandescent lamp of rated power $P = 10\ W$ possesses a luminous intensity of about 10 cd. It therefore radiates a luminous flux $\Phi_{tot} = 4\ \text{sr} \cdot 10\ \text{cd} = 126\ \text{lm}$ into the full solid angle; at a wavelength $\lambda = 555$ nm, this corresponds to a light power $P_L = 0.185\ W$. An incandescent lamp consequently has an efficiency $\eta = P_L/P \approx 2\%$. In addition to the photometric units given above, other units are often used, particularly in the American literature. They are listed in Fig. 23.2. An overview of the efficiency of different light sources is given in Fig. 23.3. It is remarkable that even the best high-efficiency LEDs do not reach the efficiency of flourescent lamps.

Physical quantities	Relationship	Units
Luminous flux	Φ	$1\,\text{lm} = 1\,\text{cd sr} \,\hat{=}\, 1.47\,\text{mW}\ (\lambda = 555\,\text{nm})$
Luminous intensity	$I = \dfrac{d\Phi}{d\Omega}$	$1\,\text{cd} = 1\,\dfrac{\text{lm}}{\text{sr}} \,\hat{=}\, 1.47\,\dfrac{\text{mW}}{\text{sr}}$
Luminance	$L = \dfrac{dI}{dA_n}$	$1\,\text{sb} = 1\,\dfrac{\text{cd}}{\text{cm}^2} = \pi\ \text{lambert} = \pi \cdot 10^4\ \text{apostilb}$
		$= 2919\ \text{lambert}$
Illuminance	$E = \dfrac{d\Phi}{dA_n}$	$1\,\text{lx} = 1\,\dfrac{\text{lm}}{\text{m}^2} = 0.0929\ \text{foot-candle} \,\hat{=}\, 0.147\,\dfrac{\mu\text{W}}{\text{cm}^2}$

Fig. 23.2. Table of photometric quantities

Lamp	Luminous efficiency lm/W	Optical output per Watt	Efficiency
Incandescent	14	20 mW	2%
Halogen	20	30 mW	3%
Flourescent	80	120 mW	12%
LED	20 ... 70	30 ... 100 mW	3 ... 10%

Fig. 23.3. Efficiency of common light sources compared to high-efficiency LEDs

23.2
Photoconductive Cells

Photoconductive cells are junctionless semiconductor devices whose resistance is a function of the illuminance. Figure 23.4 shows the circuit symbol and Fig. 23.5 the characteristic.

A photoconductive cell behaves like an ohmic resistor; that is, its resistance is neither a function of the voltage applied nor of its sign. With moderate illuminance the relationship $R \sim E^{-\gamma}$ applies, where γ is a constant between 0.5 and 1. With higher illuminance,

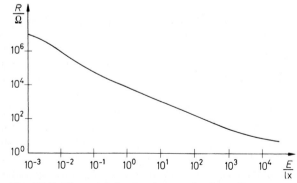

Fig. 23.4. Circuit symbol

Fig. 23.5. Characteristic of a photoconductive cell

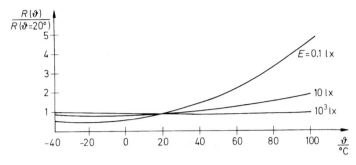

Fig. 23.6. Temperature dependence of a photoconductive cell

the resistance tends to a minimum value. At low illuminance the value of γ increases, and at very low illuminance the resistance tends to the dark resistance. The dark-to-light resistance ratio may exceed 10^6. The resistance is markedly temperature-dependent at low illuminance. This is shown by Fig. 23.6.

When the cell is illuminated, a steady-state resistance value is not established immediately. The photoconductive cell requires a certain settling time. This is in the millisecond range at illuminances of a few thousands of lux, but may exceed several seconds at values below 1 lx. The steady-state value at which the resistance settles depends not only on the illuminance but also on the preceding optical history. After comparatively long exposure to high illuminance, higher resistance values are obtained than if the photoconductive cell had been kept in the dark.

Photoconductive cells are generally made of cadmium sulfide, for which the figures quoted above apply. Photoconductive cells made of cadmium selenide are characterized by shorter settling times and a higher dark-to-light resistance ratio. However, they possess higher temperature coefficients and exhibit greater dependence on the past optical history. Cadmium-based photoconductive cells are sensitive in the spectral range from 400 to 800 nm. Some types can be used over the entire range, while others possess a quite specific color sensitivity. Photoconductive cells with high infrared sensitivity are fabricated from lead sulfide or indium antimonide. They are suitable for wavelengths up to some 3 or 7 µm, but are considerably less sensitive than cadmium-based cells.

Photoconductive cells have a sensitivity comparable to that of photomultipliers. They are therefore suitable for measuring low levels of illumination. They can also be employed as controllable resistors. As the load may be several watts, components such as relays can be directly connected without additional amplification.

23.3
Photodiodes

The reverse current of a diode increases on exposure to light. This effect can be used to measure light. For this purpose, photodiodes are provided with a glass window in the package. Figure 23.7 shows the circuit symbol, Fig. 23.8 the equivalent circuit diagram, and Fig. 23.9 the characteristics. Essentially, a short-circuit current flows that is proportional to the illuminance. Thus, in contrast to photoconductive cells, no external voltage source is required. Typical sensitivity values are of the order of $0.1 \ \mu A/lx$ for small photo diodes.

Fig. 23.7. Circuit symbol   **Fig. 23.8.** Model of a photodiode

When a reverse bias is applied, the photoelectric current remains virtually unchanged. This operating mode is useful if short response times are required, as the junction capacitance decreases with reverse bias.

As the illuminance increases, the no-load voltage rises to approximately 0.5 V in the case of silicon photodiodes. As Fig. 23.9 shows, the diode voltage decreases only slightly on load as long as the current is smaller than the short-circuit current I_P determined by the illuminance. Photodiodes are therefore suitable not only for measurement of light, but also for the generation of electrical energy. For this purpose, particularly large-area photodiodes are manufactured, which are known as photovoltaic or solar cells.

The spectral sensitivity range of silicon photodiodes is between 0.6 and 1 μm, while that of germanium photodiodes is between 0.5 and 1.7 μm. The relative spectral response is shown in Fig. 23.10.

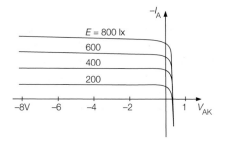

Fig. 23.9a. Characteristic of a silicon photo diode

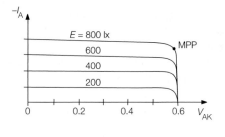

Fig. 23.9b. Extended presentation of photovoltaic range. MPP = Maximum Power Point

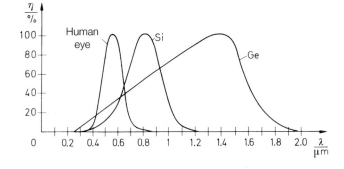

Fig. 23.10. Relative response η of germanium and silicon photodiodes

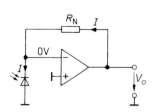

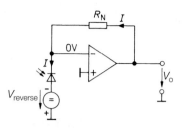

Fig. 23.11. A current–voltage converter for a particularly low dark current
Output voltage: $V_o = R_N \cdot I$
ICs: OPT 101 Texas Instr.
 ISL 29000 Intersil

Fig. 23.12. A current–voltage converter for a particularly large bandwidth
Output voltage: $V_o = R_N \cdot I$

Photodiodes have considerably shorter response times than photoconductive cells. Their cutoff frequency is typically 10 MHz, although cutoff frequencies of up to 1 GHz can be obtained using pin photodiodes.

Due to their low photoelectric current, photodiodes generally require an amplifier. In order to obtain maximum bandwidth, the voltage across the photodiodes is held constant, as the charge of their junction capacitance need not then be changed. The corresponding operational amplifier circuits are shown in Figs. 23.11 and 23.12. These are current–voltage converters, which are described in Sect. 13.2 on page 806. In the circuit in Fig. 23.11, no voltage is dropped across the photodiode – apart from the small offset voltage of the operational amplifier. Consequently, with this circuit the dark current is particularly small. In the circuit in Fig. 23.12, the photodiode is driven with negative bias. It therefore has low junction capacitance and higher bandwidths can be achieved.

The input bias current of the operational amplifiers must always be small compared to the photocurrent. The feedback resistance R_N must have a low capacitance; otherwise, it limits the circuit bandwidth. A resistance $R_N = 1$ GΩ in parallel with a capacitance $C_N = 1$ pF results in a cutoff frequency of as low as

$$f_c = 1/2\pi\, R_N\, C_N = 160\ \text{Hz}\,.$$

23.4
Phototransistors

In a phototransistor, the collector–base junction is designed as a photodiode. Figure 23.13 shows its circuit symbol and Fig. 23.14 its equivalent circuit.

The way in which a phototransistor operates can be easily explained by reference to the equivalent circuit in Fig. 23.14: the current through the photodiode causes a base current and thus an amplified collector current to flow. Whether it is preferable to connect the base or leave it open depends on the particular circuit. Phototransistors in which the base lead is not brought out are also known as photo-duodiodes.

In order to achieve a particularly high current gain, a Darlington-connected phototransistor can be used. Its equivalent circuit is shown in Fig. 23.15.

The equivalent circuits show that phototransistors have a similar performance to that of comparable photodiodes in terms of their spectral range. However, their cutoff frequency

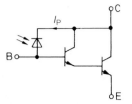

Fig. 23.13.
Phototransistor symbol

Fig. 23.14. Equivalent
circuit of a phototransistor

Fig. 23.15. Equivalent circuit of
a photo-Darlington connection

Fig. 23.16.

Fig. 23.17.

Fig. 23.16 and **23.17.** Simple photodetector circuits

is considerably lower. For phototransistors it is of the order of 300 kHz and for photo-Darlington connections some 30 kHz.

Figure 23.16 shows a phototransistor used as a photodetector. Denoting the photocurrent through the collector–base diode by I_P, we obtain an output voltage

$$V_o = V^+ - BR_1 I_P .$$

Correspondingly, for the circuit in Fig. 23.17,

$$V_o = BR_1 I_P .$$

A wide range of phototransistors and optoelectronic switches is available from Optek and Infineon.

23.5
Light-Emitting Diodes

Light-emitting diodes (LEDs) are not made from silicon or germanium, but from gallium arsenide phosphide (III-V compound). These diodes emit light when a forward current flows. The spectral range of the luminous flux emitted is quite sharply delimited, its frequencies depending on the basic material used. Figure 23.18 shows the circuit symbol, and Fig. 23.19 provides an overview of the most important characteristics.

The efficiency of high efficiency LEDs can be as high as 10 % but at older types it is less than 0.1%.The luminance is proportional to the forward current over a wide range. Currents of a few milliamperes are sufficient to provide a clearly visible display. LEDs are therefore particularly suitable as display elements in semiconductor circuits. They are also available as seven-segment or matrix displays.

A   K

Fig. 23.18. LED symbol

Color	Wave-length at maximum intensity [nm]	Basic material	Forward voltage at 10 mA [V]	Luminous intensity at 10 mA [cd]	Light power at 10 mA [mW]
infrared	900	Ga As	1.3...1.5		1...5
bright red	630	Al In Ga P	2.0...2.2	1...2	
yellow	590	Al In Ga P	2.0...2.2	1...2	
green	525	In Ga N	2.5...3.5	1...4	
blue	470	In Ga N	2.7...3.7	1...4	
white		In Ga N	2.7...3.7	1...4	

Fig. 23.19. High performance LEDs. White LEDs are blue LEDs with built-in color-converter

LEDs are supplied by most semiconductor manufacturers. A particularly wide selection is offered by Agilent, Hamamatsu, Optek and Ledtronics.

23.6
Optocouplers

If a LED is combined with a photodetector – for example, a phototransistor – it is possible to convert an input current into a floating output current at any potential. Optocoupler devices of this kind are available in standard IC packages. In order to achieve a high level of efficiency, they are generally operated in the infrared region. The most important feature of an optocoupler is the current transmission ratio $\alpha = I_o/I_i$. This is essentially determined by the detector characteristics. Typical values are listed in Fig. 23.20. We can see that the highest current gain is obtained using photo-Darlingtons, although they exhibit the lowest cutoff frequency.

Optocouplers are suitable for transmitting both digital and analog signals. The relevant circuits are described in the Figs. 23.11 and 23.16. For sensor applications, optocouplers are also designed as slotted optical switches or reflective optical switches. The main manufacturers of optocouplers are Infineon, Optek, Sharp, and Vishay.

23.7
Visual Displays

Digital information can be displayed in many ways; for example, using incandescent lamps, glow lamps, LEDs, or liquid crystals. LED and liquid crystal displays (LCDs) have

Detector	Transmission ratio $\alpha = I_o/I_i$	Cutoff frequency
photodiode	ca. 0.1%	10 MHz
phototransistor	10... 300%	300 kHz
photo-Darlington transistor	100...1000%	30 kHz

Fig. 23.20. A comparison of optocouplers

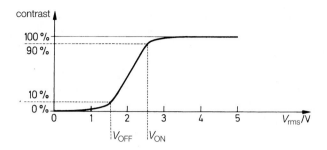

Fig. 23.21. Contrast of a LCD-Display as a function of the AC voltage applied

assumed great importance, as they can be operated with low voltages and low currents. The application is simplified by the large number of integrated drivers available.

LCDs are not semiconductor components. Unlike LEDs, they themselves generate no light, relying instead on external illumination. An optical effect is produced because a liquid crystal element is transparent when no voltage is applied and therefore appears bright, whereas it is opaque when a voltage is applied and therefore appears dark. The liquid crystal element comprises two electrodes with an organic substance sandwiched between them. This substance contains crystals whose orientation can be varied by an electric field therefore they are named liquid cristal displays. The state of the element therefore depends on the electric field strength; it behaves, in effect, like a capacitor.

The device is driven by AC voltages at a frequency high enough to insure that no flicker occurs. On the other hand, the frequency selected must be low enough to insure that the alternating current flowing through the capacitor remains small. In practice, the values selected are between 30 and 100 Hz. The driving AC voltage must not contain any DC component, as even 50 mV electrolyze the LCD and reduce lifetime.

In Fig. 23.21, the contrast is plotted as a function of the rms value of the alternating voltage amplitude applied. For AC voltages of less than $V_{OFF\ rms} \approx 1.5$ V, the display is virtually invisible; voltages of more than $V_{ON\ rms} \approx 2.5$ V produce maximum contrast.

As the capacitance of a liquid crystal element is only about 1 nF/cm², the currents required for driving the device are well below 1 μA. This extremely low current requirement represents a significant advantage over LEDs.

23.7.1
Binary Displays

LEDs require a forward current of 5...20 mA for good visibility in daylight. These currents can be provided most conveniently using gates, as in Figs. 23.22 and 23.23. In Fig. 23.22, the LED lights when an H-level appears at the gate output by applying an L-level at the input. In Fig. 23.23, the reverse is true. Current limiting is provided in each case via

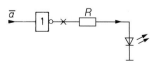

Fig. 23.22. Driving a LED connected to ground from a logic gate

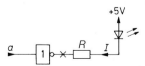

Fig. 23.23. Driving a LED connected to supply voltage from a logic gate

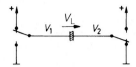   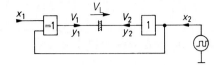

Fig. 23.24. Principle   **Fig. 23.25.** Practical implementation

Figs. 23.24 and **23.25.** The driving of a liquid crystal display from a single supply voltage, without DC offset

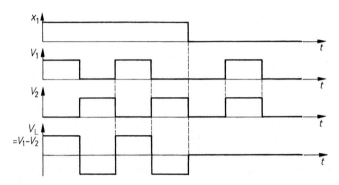

Fig. 23.26. Voltage waveform for a liquid crystal display that is switched on and off

the resistor R. Due to the relatively high load due to the LEDs, the gate outputs have no specified voltage level and must therefore not be used for logic operations. This is indicated on the circuit diagram by a cross at the gate output.

In order to control the light output, it is possible to use gates with a second input at which a square-wave AC voltage is applied. Its duty cycle then enables the average diode current to be reduced to zero. The frequency must be at least $100\,\mathrm{Hz}$ to insure that no flicker is visible.

Drive signal generation for liquid crystal displays is somewhat more complicated. An alternating voltage must be generated whose rms value is sufficiently high and whose mean value is zero. The easiest way to achieve this is to connect the display between two switches (Fig. 23.24), which are switched back and forth between ground and operating voltage V^+ either in phase or in antiphase. For in-phase operation, $V_L = 0$, while for antiphase operation, $V_{L\,\mathrm{rms}} = V^+$. This is illustrated by the waveform diagram in Fig. 23.26.

The practical implementation of this principle is shown in Fig. 23.25. When $x_1 = 0$, $y_1 = y_2 = x_2$; the two terminals of the display therefore switch in phase with the square-wave signal x_2. For $x_1 = 1$, $y_1 = \bar{x}_2$, and the display receives antiphased signals. CMOS gates are most suitable for this purpose, as their output levels with purely capacitive loading only differ by a few millivolts from V^+ or zero potential. In addition, only the use of CMOS gates fully exploits the low power requirement of liquid crystal displays.

23.7.2
Analog Displays

A quasi-analog display can be obtained using a row of indicating elements. This provides a dot-position display if only the element belonging to a particular indication value is

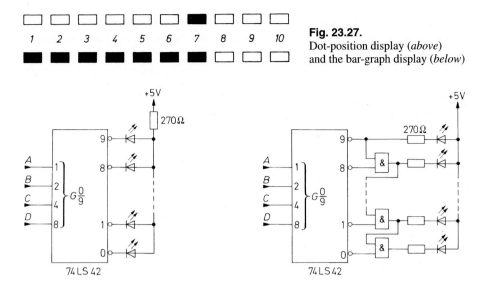

Fig. 23.27.
Dot-position display (*above*)
and the bar-graph display (*below*)

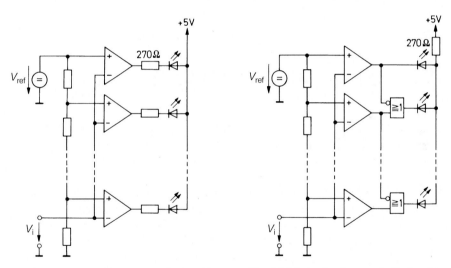

Fig. 23.28. Binary dot-position display drive

Fig. 23.29. Binary bar-graph display drive

Fig. 23.30. Analog dot-position display drive

Fig. 23.31. Analog bar-graph display drive

turned on. A bar-graph display is obtained if all the lower elements are turned on as well. Figure 23.27 compares these two alternatives.

A dot-position display can be driven by binary signals using a 1-out-of-n encoder (see Sect. 8.2.1 on page 643). In this case, only the LED connected to the output selected is turned on. The bar-graph display in Fig. 23.29 is obtained when all the LEDs below the selected output are also turned on via the additional output gates.

In order to drive a display row using analog signals, it is advisable to employ an analog–digital converter in a parallel arrangement, as the signals required for driving a bar-graph display are then produced directly. As shown in Fig. 23.30, the input voltage is compared with a reference voltage by means of a comparator chain. As a result, all the comparators whose reference voltages are smaller than the input voltage are activated. With this method, additional gates are required in order to implement a dot-position display, as shown in Fig. 23.31.

In many applications the easiest way of implementing a dot-position or bar-graph display is to use a simple microcontroller for instance the PIC-family from Microchip. The input signal can be connected to the AD-input, the LEDs to the port outputs.

23.7.3
Numerical Displays

The simplest means of representing the numerals 0 to 9 is to arrange seven indicating elements to form a seven-segment display (Fig. 23.32). Depending on which combination of segments a to g is turned on, all digits can be represented with adequate readability.

In order to drive a seven-segment display, it is necessary to assign each digit, which is normally present in binary coded form (BCD), the associated combination of segments. A circuit of this type is known as a BCD seven-segment decoder. Its truth table is shown in Fig. 23.33. The principles for connecting LED and liquid crystal displays are those of Figs. 23.23 and 23.25 respectively. The corresponding circuits are given in Figs. 23.34 and 23.35.

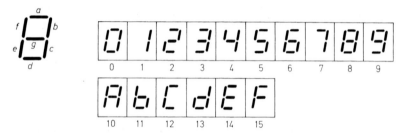

Fig. 23.32. Seven-segment display with hexadecimal extensions

Digit	BCD input				Seven-segment output						
Z	2^3	2^2	2^1	2^0	a	b	c	d	e	f	g
0	0	0	0	0	1	1	1	1	1	1	0
1	0	0	0	1	0	1	1	0	0	0	0
2	0	0	1	0	1	1	0	1	1	0	1
3	0	0	1	1	1	1	1	1	0	0	1
4	0	1	0	0	0	1	1	0	0	1	1
5	0	1	0	1	1	0	1	1	0	1	1
6	0	1	1	0	1	0	1	1	1	1	1
7	0	1	1	1	1	1	1	0	0	0	0
8	1	0	0	0	1	1	1	1	1	1	1
9	1	0	0	1	1	1	1	1	0	1	1
10	1	0	1	0	1	1	1	0	1	1	1
11	1	0	1	1	0	0	1	1	1	1	1
12	1	1	0	0	1	0	0	1	1	1	0
13	1	1	0	1	0	1	1	1	1	0	1
14	1	1	1	0	1	0	0	1	1	1	1
15	1	1	1	1	1	0	0	0	1	1	1

Fig. 23.33. Truth table for a BCD seven-segment decoder with hexadecimal extension

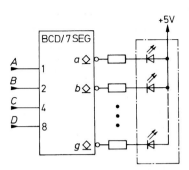

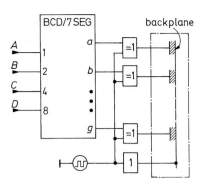

Fig. 23.34. Connection of a LED display to a seven-segment decoder

Fig. 23.35. Connection of a liquid crystal display to a seven-segrnent decoder

Type	Manufacturer	Techno-logy	Internal memory
For LED displays			
74 LS 47	Texas Instr.	TTL	no
74 LS 247	Texas Instr.	TTL	no
CD 4511	numerous	CMOS	yes
For liquid-crystal displays (LCD)			
CD 4055	Harris	CMOS	yes
CD 4056	Harris	CMOS	yes
CD 4543	Harris	CMOS	yes

Fig. 23.36. Seven-segment decoders

BCD seven-segment decoders are available as integrated circuits; examples are listed in Fig. 23.36. Some of the LED-driving types possess current source outputs, in which case the external current-limiting resistors are not required. In addition to the decoders for driving common-anode displays, types are also available for a common cathode. In the case of the liquid-crystal driving decoders, the exclusive-OR gates are already incorporated. The only external device required is therefore the square-wave generator.

In some cases the numbers 10 to 15 that are not used for numeric display are assigned to the letters A to F. The usual shape of the additional characters in a hexadecimal display is also shown in Fig. 23.32.

23.7.4
Multiplex Displays

The LED and liquid crystal displays described here can be used for visual representation of multi-digit data. However, in order to minimize the number of drivers and lines required, it is advisable in the case of multi-character displays to connect them together as a matrix and operate them on a time-division multiplex basis. Figure 23.37 shows an arrangement of this kind for an 8-character, 7-segment LED display. The corresponding segments of all the displays are connected in parallel. In order to ensure that the same segments of all

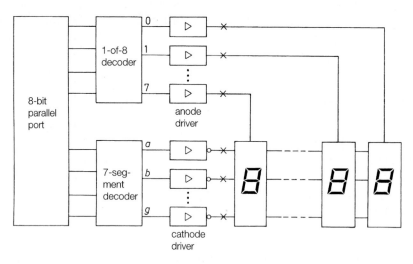

Fig. 23.37. Connection of an 8-character seven-segment display to a parallel port

the characters do not now light up simultaneously, only one character is activated via the 1-of-8 decoder at a time.

Therefore, only 15 lines are required to operate an 8-character, 7-segment display. A single 8-bit parallel interface is suitable as a microprocessor interface. A number of 7-segment decoders are shown in Fig. 23.36. Anode and cathode drivers are listed in Fig. 23.39.

Multiplex operation is performed by the microprocessor program. To do this, the position number is specified with four bits and the character to be represented in BCD code using the other four bits. The output is then repeated for the next character position. In order to ensure that a flicker-free display is produced, the complete display cycle must be executed at least 100 times a second. There are many applications, particularly in simple devices, where the processing time required for driving the display is spare. However, it may be disturbing to have the display flickering when the microprocessor is required for long periods for other tasks.

If the display is to operate stand-alone, it must possess an additional display memory and an internal multiplexing facility. The resulting circuit is shown in Fig. 23.38. The display data is written by the microprocessor into a two-port memory (see Sect. 10.2 on page 697) which is connected to the microcomputer bus like a normal RAM. The display contents are read from the two-port memory, independently of the bus operation. During this process, the binary counter issues the addresses cyclically and activates the appropriate character positions via the 1-of-8 decoder.

Display drivers operating on this principle are widely available as fully-integrated circuits. Some types are listed in Fig. 23.47 on page 1145. As well as the types with parallel data inputs, there are also versions in which the display data is stored in a shift register. They only require one serial data line and no addresses to control them.

Liquid crystal displays (LCDs) require an AC voltage of a specific amplitude. The push-pull methods described in Fig. 23.25 on page 1136 is only used for generating this voltage for a small number of segment drivers. For larger numbers of segments, liquid crystal displays are also linked together to form matrices, in order to minimize the number

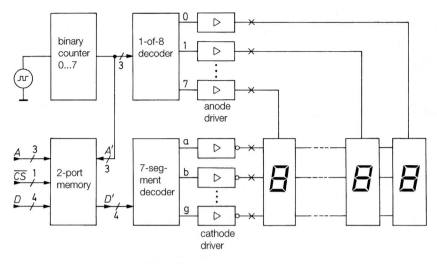

Fig. 23.38. Stand-along multiplex display with data memory

Type	Manufac- turer	Number of drivers	Current max.	Voltage drop max.	Interface
Anode drivers, current sources					
IRF 7504[1]	Intern. Rect.	2	1400 mA	0.38 V	2 bit
UDN 2981A	Allegro	8	350 mA	1.8 V	8 bit
TLC 5920[2]	Texas Inst.	16	…30 mA	const. current	seriell
Cathode drivers, current sink					
IRF 7501[1]	Intern. Rect.	2	1900 mA	0.26 V	2 bit
CA 3262	Intersil	4	600 mA	0.6 V	4 bit
A 6277	Allegro	8	…150 mA	const. current	serial
A 6801	Allegro	8	500 mA	1.0 V	8bit
TPIC 2701	Texas Inst.	7	500 mA	0.4 V	7 bit
TPIC 2810	Texas Inst.	8	100 mA	0.5 V	I^2C
TPIC 6273	Texas Inst.	8	250 mA	0.8 V	8 bit
TLC 5920[2]	Texas Inst.	8	640 mA	0.6 V	serial
TLC 5941	Texas Inst.	16	…80 mA	const. current	SPI

[1] Logic Level Mosfets   [2] Includes anode- and cathode-drivers

Fig. 23.39. High current drivers for LED displays and other applications requirung high currents

of connecting lines. However, *three* voltage levels are required (in addition to ground potential) for driving liquid crystal matrices of this kind, in order to ensure that the selected segments receive a sufficiently high and the remainder a suitable low AC voltage. This special type of multiplexing is known as the triplex method.

23.7.5
Alphanumeric Displays

Seven-segment displays only allow a few letters to be represented. In order to display the entire alphabet, a higher resolution is required. This can be achieved by using 16-segment displays or 35-dot matrix displays.

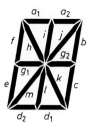

Fig. 23.40. 16-segment display

	0	1	2	3	4	5	6	7	8	9	A	B	C	D	E	F
2		!	"	#	$	%	&	'	(	)	*	+	,	--	.	/
3	0	1	2	3	4	5	6	7	8	9	:	;	<	=	>	?
4	@	A	B	C	D	E	F	G	H	I	J	K	L	M	N	O
5	P	Q	R	S	T	U	V	W	X	Y	Z	[	\	]	^	_

Fig. 23.41. Usual character set of a 16-segment display. At the border you find the ASCII equivalent. The first hex-sign is in the top row, the second sign in the column on the left. For example: $K \cong B4_{Hex}$

16-Segment Displays

The arrangement of the segments in a 16-segment display is shown in Fig. 23.40. In comparison with the seven-segment display in Fig. 23.32, segments a, d, and g are divided into two sections and segments h–m are added. This enables generation of the character set shown in Fig. 23.41. It is usually limited to 64 characters, which include the upper-case letters, the numerals, and the most important special characters. Some display drivers are listed in Fig. 23.47.

35-Dot Matrix Displays

A better resolution than that of a 16-segment display is obtained by using a 5×7 dot matrix, as shown in Fig. 23.42. This allows an approximate representation of virtually all conceivable characters. Figure 23.44 shows how all the 96 ASCII characters and 32 additional special characters can be represented using commercially available character generators.

Due to the large numbers of LEDs involved in matrix displays, they are also electrically connected as a matrix. This is shown in Fig. 23.43, using LEDs as an example. There are only 12 external connections. However, as the matrix makes it impossible to turn on all the required elements simultaneously, the display is driven on a time-division multiplex basis by selecting row by row and activating the desired combination of columns each time. If the progression is sufficiently fast, the observer has the impression that all the dots addressed are activated simultaneously. If the cycle frequency is more than 100 Hz, the display appears virtually flicker-free to the human eye.

Figure 23.45 shows a schematic illustration of a drive circuit for LED matrices. Rows are selected one by one using the straight binary counter and the 1-out-of-8 decoder. The row number, together with the ASCII code for the desired character, is fed to the

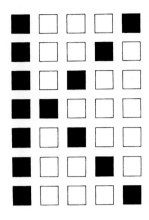

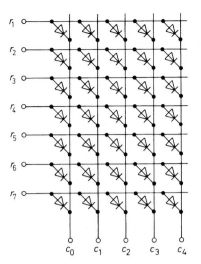

Fig. 23.42. Arrangement of dots in a 35-dot matrix, in seven rows and five columns

Fig. 23.43. Matrix arrangement of display elements, using LEDs as an example

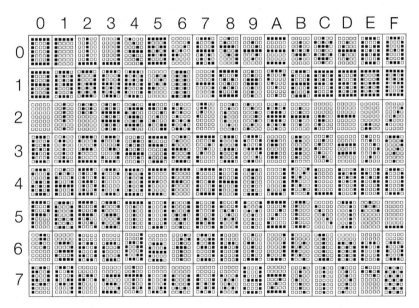

Fig. 23.44. Example for an ASCII character generator. For example: K $\hat{=}$ B4$_{Hex}$

character generator. The latter determines, in accordance with Fig. 23.43, which dots of the particular row are to be turned on. Character generators are available as mask-programmed ROMs, providing the symbols shown in Fig. 23.44. If other character sets are required, it is advisable to program an EEPROM accordingly. Figure 23.46 gives the character generator contents required for the character "K." Typical matrix display with integrated drive electronics are shown in Fig. 23.47.

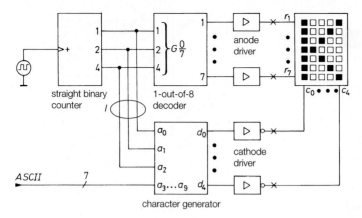

Fig. 23.45. Drive circuit for LED matrices comprising 5×7 elements

Row number	ROM address										ROM Contents				
	ASCII "K"										Column data				
								I							
I	a_9	a_8	a_7	a_6	a_5	a_4	a_3	a_2	a_1	a_0	c_0	c_1	c_2	c_3	c_4
1	1	0	0	1	0	1	1	0	0	1	1	0	0	0	1
2	1	0	0	1	0	1	1	0	1	0	1	0	0	1	0
3	1	0	0	1	0	1	1	0	1	1	1	0	1	0	0
4	1	0	0	1	0	1	1	1	0	0	1	1	0	0	0
5	1	0	0	1	0	1	1	1	0	1	1	0	1	0	0
6	1	0	0	1	0	1	1	1	1	0	1	0	0	1	0
7	1	0	0	1	0	1	1	1	1	1	1	0	0	0	1

Fig. 23.46. Character generator contents for representing the character "K"

Type	Manufacturer	Display	Technology	Characters	Data input
Drivers for 7-segment displays			common		
ICM 7212	Maxim	LED	anode	4	4 bit
MAX 7219	Maxim	LED	cathode	8	SPI
MAX 6951	Maxim	LED	cathode	8	SPI
ICM 7211	Maxim	LCD	static	4	4 bit
MAX 7231	Maxim	LCD	multiplex	8	6 bit
MAX 7232	Maxim	LCD	multiplex	10	serial
Drivers for 16-segment displays					
ICM 7243	Intersil	LED	cathode	8	6 bit
MAX 6955	Maxim	LED	cathode	8	I^2C
Drivers for 35-segment displays					
MAX 6953	Maxim	LED	cathode	4	I^2C
Smart 35-segment displays including drivers			size		
DLO 7135	Osram	LED	17 mm	1	7 bit
SCE 5780	Osram	LED	4.5 mm	8	serial
HDSP 2132	Agilent	LED	5 mm	8	8 bit
HCMS 3976	Agilent	LED	5 mm	8	serial

Fig. 23.47. Display decoders and drivers and displays with integrated drivers

Part III

Communication Circuits

Chapter 24:
Basics

24.1
Telecommunication Systems

Today, telecommunication systems are as much a part of everyday life as electrical energy. Besides the analog telephone as a conventional cable system and analog radio and TV broadcasting as classical wireless systems, there are countless more modern telecommunication systems including ISDN telephones, cordless and mobile telephones, radio and TV broadcasting via wideband cable networks or satellite transmission, PC modems, wireless PC mouses and keyboards, wireless garage door openers, and remote controlled car locks with the actuator integrated within the car key. Furthermore, heterogeneous systems such as the Internet evolve from the combination of several systems and the application of specific network procedures.

A transmission system is defined as a telecommunication system if a *modulation* is used for the interface to the transmission channel. According to this criterion, telecommunication engineering must be perceived as the theory of *modulation methods*. In contrast, there are transmission systems without modulation, e.g. computer interfaces such as V.24 and SCSI, which only provide specific lines and drivers for direct signal transmission over larger distances. Telecommunication systems are thus characterized by the use of a *modulator* in the transmitter and a corresponding *demodulator* in the receiver.

Figure 24.1 shows the components of an analog and a digital telecommunication system. Ordered top to bottom, the components form a *transmitter*; from bottom to top they form a *receiver*. The transmission medium between the transmitter and the receiver is the *channel*, which may be a cable or radio link with sending and receiving antenna.

In analog systems, the useful signal to be transmitted $s(t)$ is supplied directly to the *analog modulator*. The modulator output signal is amplified in a *transmitter amplifier* and fed to the channel. Most analog modulators generate a signal with the desired transmission frequency; in such cases the transmitter amplifier consists solely of one or more amplifiers in series. In other cases, the modulator generates a signal with an intermediate frequency which then must be converted to the transmission frequency by a mixer incorporated within the transmitter amplifier. The channel causes a signal attenuation that may be as much as 150 dB in radio transmission links (e.g. 1 kW $= 10^3$ W transmitted power $\rightarrow$ 1 pW $= 10^{-12}$ W received power). In extreme cases the power of the signal is only slightly higher than the unavoidable thermal noise. In the receiver, the *receiving amplifier* enhances the signal enough so that it can be fed to the demodulator. This requires a gain control to ensure a fixed signal level at the demodulator despite the large variations in the received signal level due to the varying distances to the transmitter. In radio and cable systems for multiple use, the receiving amplifier must also perform frequency selection by separating the desired incoming signal from signals of adjacent frequency ranges. Here, several filters as well as one or two mixers are used for frequency conversion. Finally, the *analog demodulator* generates the received useful signal $r(t)$ from the selected and amplified signal.

The digital system contains all of the same components as the analog system except that the modulator and demodulator are of digital design and are connected to the amplifiers

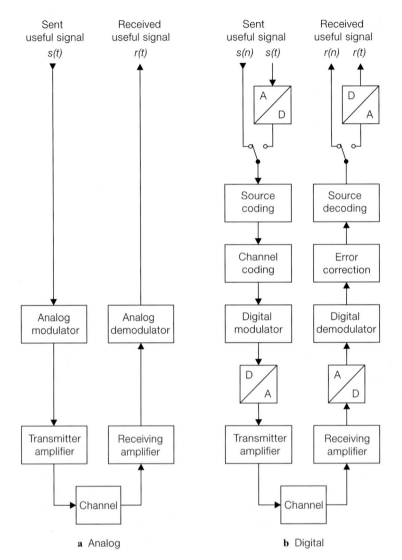

Sent useful signal s(t)   Received useful signal r(t)   Sent useful signal s(n) s(t)   Received useful signal r(n) r(t)

a Analog   **b** Digital

Fig. 24.1. Components of a telecommunication system

via D/A or A/D converters. Sometimes these converters are regarded as integral parts of the modulator or demodulator and are not shown separately; in such cases the digital modulator comprises a digital input and an analog output, while the digital demodulator has an analog input and a digital output. These components, which correspond to the analog system, are then ready for application. The system is complemented by a *channel coding circuit* in the transmitter that introduces a redundancy in terms of check bits, parity bits, checksums or a specific code; this redundancy is used for *error correction* in the receiver. Some systems use an additional *source coding* and *source decoding* in order to reduce the amount of transferred data. Source coding is usually not loss-free, i.e. the signal is not reconstructed exactly in the decoding process. Rather, source coding is based on physiological findings, asserting that the human receiver is not capable of detecting certain parts of speech or

Feature	Analog	Digital
Circuit complexity	Low	High
Bandwidth utilization	Poor	Good – very good
Complexity of modulation method	Low	High
Required signal-to-noise ratio at the receiver input	High	Low
Required transmitter power	High	Low
Transmission performance:		
– With low signal-to-noise ratio	Poor	Very high
– With high signal-to-noise ratio	High	Ideal
Accuracy of arithmetic operations	Poor	High – ideal
Temperature drift	Yes	No
Drift due to aging	Yes	No
Alignment complexity during manufacture	High	Low

Fig. 24.2. Features of analog and digital telecommunication systems

image signals. This is the level at which the digital signal $s(n)$ is sent and the signal $r(n)$ is received. For the transmission of analog signals, additional converters are required in the transmitter and the receiver; this is the case, for example, in digital telephones in which the transmitted useful signal $s(t)$ originates from a microphone and the received useful signal $r(t)$ is reproduced by a loudspeaker.

Compared to a digital system, an analog system comprises fewer components, which are often of simpler design than the corresponding components of the digital system. One disadvantage of the analog system is that noise and other interferences caused during signal transmission can no longer be separated from the signal; therefore the signal-to-noise ratio declines rapidly, especially in transmissions over considerable distances. Furthermore, analog modulation methods do not make full use of the available bandwidth and need a relatively high signal-to-noise ratio at the receiver input in order to achieve a high transmission performance.

Digital systems use sophisticated modulation methods that provide a substantially better bandwidth utilization than analog systems. Noise and other interferences are totally eliminated by threshold evaluation in the demodulator as long as a certain amplitude is not exceeded. If this amplitude is exceeded, an initial erroneous evaluation is made which can be remedied by error correction if the erroneous failure rate remains below a certain limit. Digital systems are thus capable of providing near ideal transmission performance even with a low signal-to-noise ratio at the receiver input. Better utilization of the bandwidth by using more complex modulation methods is also of high importance as the continual introduction of new systems causes an increasing shortage of transmission frequencies.

Figure 24.2 presents a comparison of the most important features of analog and digital systems; these also contain the *common* advantages of digital systems such as low drift and low alignment complexity. Some of the properties are redundant; for instance, better utilization of bandwidth in digital systems is a consequence of higher complexity of modulation methods, while a lower required signal-to-noise ratio at the receiver input enables the transmit power to be reduced.

24.2
Transmission Channels

The transmission channels are discussed in the order of their industrial use: *cable, radio link* and *fibre-optics*. Despite their technological differences, all channels have one thing in common, that is, the transmission is based on electromagnetic waves.

24.2.1
Cable

Telecommunications predominantly use *coaxial cables* and *two-wire lines*. Figure 24.3 shows a cross-section of these cables together with the E and H field lines, as well as their characteristic dimensions. The coaxial cable is a *shielded line* since the fields are restricted to the space between the inner and outer conductor; influence on adjacent components is thus excluded[1]. In contrast, the signal of an *unshielded* two-wire line may be emitted into neighboring components or other unshielded cables running in parallel due to capacitive (E field) or inductive (H field) coupling; this occurrence is called *crosstalk*.

The space between the internal and external conductor of a coaxial cable is filled by a dielectric to centre the conductors; the dielectric material is usually Teflon ($\epsilon_r = 2.05$) or polystyrene ($\epsilon_r = 2.5$). The conductors of the two-wire line each have a polyethylene sheath; the conductor and sheath are then either twisted or strapped together.

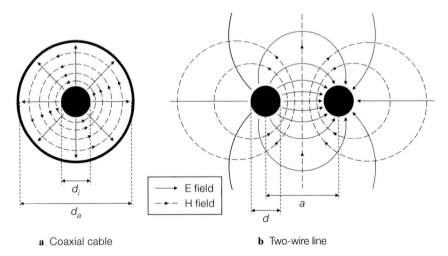

⟶ E field	
– ⋅– H field	

a Coaxial cable   **b** Two-wire line

Fig. 24.3. Cross-section and field lines of cables used in telecommunications

---

[1] In many coaxial cables used in practice the outer conductor is not perfectly *sealed* so that weak fields also exist outside the cable.

Field Characteristic Impedance and Velocity of Propagation

The ratio of E to H field strength of a propagating electromagnetic wave is defined by the *field characteristic impedance* Z_F; Maxwell's equations lead to [24.1]:

$$Z_F = \frac{|E|}{|H|} = \sqrt{\frac{\mu_0 \mu_r}{\epsilon_0 \epsilon_r}} = 120\pi\,\Omega\,\sqrt{\frac{\mu_r}{\epsilon_r}} \overset{\mu_r=1}{=} \frac{120\pi\,\Omega}{\sqrt{\epsilon_r}} = \frac{377\,\Omega}{\sqrt{\epsilon_r}}$$

Since cables do not usually contain magnetic substances one can assume $\mu_r = 1$. The *velocity of propagation* is

$$v = \frac{c_0}{\sqrt{\epsilon_r \mu_r}} \overset{\mu_r=1}{=} \frac{c_0}{\sqrt{\epsilon_r}} \tag{24.1}$$

with the free-space speed of light $c_0 = 3 \cdot 10^8$ m/s. For the common dielectrics with $\epsilon_r \approx 2 \ldots 2.5$ this amounts to approximately $2 \cdot 10^8$ m/s, or $2/3$ of the speed of light.

Line Characteristic Impedance

The ratio of voltage to current of a propagating wave is defined by the *line characteristic impedance* Z_W. It is calculated by determining the voltage via integration along an E field line from conductor 1 to conductor 2 and by determining the current via integration along an H field line [24.1]:

$$V = \int_1^2 E \, dr \quad , \quad I = \oint H \, dr$$

This leads to:

$$Z_W = \frac{V}{I} = Z_F k_g = Z_F \cdot \begin{cases} \dfrac{1}{2\pi} \ln \dfrac{d_a}{d_i} & \text{Coaxial cable} \\[4mm] \dfrac{1}{\pi} \ln \left(\dfrac{a}{d} \sqrt{\left(\dfrac{a}{d}\right)^2 - 1} \right) & \text{Two-wire line} \end{cases}$$

The line characteristic impedance consists of the field characteristic impedance Z_F and a *geometric factor* k_G that describes the line. Insertion of Z_F results in:

$$Z_W = \begin{cases} \dfrac{60\,\Omega}{\sqrt{\epsilon_r}} \ln \dfrac{d_a}{d_i} & \text{Coaxial cable} \\[4mm] \dfrac{120\,\Omega}{\sqrt{\epsilon_r}} \ln \left(\dfrac{a}{d} \sqrt{\left(\dfrac{a}{d}\right)^2 - 1} \right) & \text{Two-wire line} \end{cases} \tag{24.2}$$

In practice, cables are used with $Z_W = 50\,\Omega$ (e.g. $\epsilon_r = 2.05$, $d_i = 2.6$ mm, $d_a = 8.6$ mm) and $Z_W = 75\,\Omega$ for coaxial cables and with $Z_W = 110\,\Omega$ for twisted pair lines. The calculation of Z_W for two-wire lines is difficult since fields exist inside the cable sheath ($\epsilon_r > 1$) and in the surrounding space ($\epsilon_r = 1$); therefore it is necessary to use the *effective* value of ϵ_r in (24.2) that can be determined only by field simulation or measurement.

The line characteristic impedance is not an ohmic resistance and thus cannot be measured using an ohmmeter or impedance meter. It describes solely the ratio between the voltage and the current of *one* wave. Later we shall see that normally two waves exist on a line: the *incidental* (forward) *wave* with $V_f = Z_W I_f$ and the *reflected wave* with $V_r = Z_W I_r$. With the equations $V = V_f + V_r$ and $I = I_f - I_r$, we can calculate the voltage V that can be measured between the conductors and the current I that flows through the cable.

In practice, the prefix *line* is usually omitted and the parameter is called *characteristic impedance*. Often the symbols Z_L or Z_0 are used where Z indicates that it is a complex impedance; but sometimes the symbols R_W, R_L or R_0 are also used.

Transmission Equation

A short piece of cable can be described by an equivalent circuit with four components (see Fig. 24.4) where the following four *parameters per unit length* are used [24.1]:

- The *inductance per unit length* L' represents the energy per unit length stored in the H field. The unit is *Henry per meter*: $[L'] = \mathrm{H/m}$.
- The *capacitance per unit length* C' represents the energy per unit length stored in the E field. The unit is *Farad per meter*: $[C'] = \mathrm{F/m}$.
- The *resistance per unit length* R' takes into account the ohmic losses in the conductors. The unit is *Ohm per meter*: $[R'] = \Omega/\mathrm{m}$. At low frequencies this parameter corresponds to the DC resistance of the conductor. At frequencies above approximately $10\,\mathrm{kHz}$ its value increases in proportion to the square root of the frequency of $R' \sim \sqrt{f}$ due to the current displacement *(skin effect)*; this causes an attenuation that increases with frequency.
- The *leakage per unit length* G' takes into account the insulation conductance and the polarization losses of the dielectric. The unit is *Siemens per meter*: $[G'] = \mathrm{S/m}$. Usually the insulation conductance is negligibly low. The polarization losses increase proportionally to the frequency $(G' \sim f)$, but in technical applications are typically lower than the ohmic losses.

According to Fig. 24.4, the voltages and currents can be calculated:

$$V_2 = V_1 - \left(R'dz + j\omega L'dz\right) I_1$$
$$I_2 = I_1 - \left(G'dz + j\omega C'dz\right) V_2$$

Insertion of

$$V_2 = V_1 + dV \quad , \quad I_2 = I_1 + dI$$

and division by dz and subsequent limit transition

$$dz \to 0 \quad , \quad V_1 \to V_2 = V \quad , \quad I_1 \to I_2 = I$$

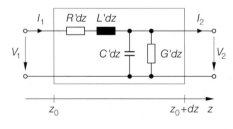

Fig. 24.4. Equivalent circuit for a short piece of cable with the length dz

leads to:

$$\frac{dV}{dz} = -(R' + j\omega L') I \tag{24.3}$$

$$\frac{dI}{dz} = -(G' + j\omega C') V \tag{24.4}$$

Differentiation of (24.3) with respect to z and insertion of (24.4) lead to the *transmission equation:*

$$\frac{d^2V}{dz^2} = (R' + j\omega L')(G' + j\omega C') V = \gamma_L^2 V \tag{24.5}$$

The general solution is

$$V(z) = V_f e^{-\gamma_L z} + V_r e^{\gamma_L z} \tag{24.6}$$

and the *propagation constant* is:

$$\gamma_L = \sqrt{(R' + j\omega L')(G' + j\omega C')} \tag{24.7}$$

Even at frequencies in the lower kHz range the following conditions apply to low-loss lines: $j\omega L' \gg R'$ and $j\omega C' \gg G'$; consequently the propagation constant [24.1] is:

$$\gamma_L \approx \underbrace{\frac{R'}{2}\sqrt{\frac{C'}{L'}} + \frac{G'}{2}\sqrt{\frac{L'}{C'}}}_{\alpha_L} + j\underbrace{\omega\sqrt{L'C'}}_{\beta_L} \tag{24.8}$$

with the *attenuation constant* α_L and the *phase constant* β_L. In a loss-free line ($R' = G' = 0$) the attenuation constant is zero.

For further clarification we take the function of time:

$$u(t,z) = \mathrm{Re}\left\{V(z)\,e^{j\omega t}\right\} \overset{(24.6)}{=} \mathrm{Re}\left\{V_f\,e^{j\omega t-\gamma_L z} + V_r\,e^{j\omega t+\gamma_L z}\right\}$$

$$= \underbrace{|V_f|e^{-\alpha_L z}\cos\left(\omega t - \beta_L z + \varphi_f\right)}_{\text{Forward wave}} + \underbrace{|V_r|e^{\alpha_L z}\cos\left(\omega t + \beta_L z + \varphi_r\right)}_{\text{Reflected wave}}$$

$$= u_f(t,z) + u_r(t,z)$$

It consists of a *forward wave* $u_f(t,z)$ and a *reflected wave* $u_r(t,z)$. Figure 24.5 shows these waves at a given time t_0 and a quarter of a period length later. The propagation in the opposite direction and the increasing attenuation in the direction of propagation can be seen. The *velocity of propagation* v is determined by the displacement of the maximum value of the cosine function. For the forward wave the following applies:

$$\omega t - \beta_L z + \varphi_f = 0 \Rightarrow \boxed{v = \frac{dz}{dt} = \frac{\omega}{\beta_L} = \frac{1}{\sqrt{L'C'}}} \tag{24.9}$$

For the reflected wave the absolute value of the velocity of propagation is the same, but the sign is negative; this again indicates that the two waves propagate in opposite directions. The *wave length* λ corresponds to the distance between two consecutive maxima; this

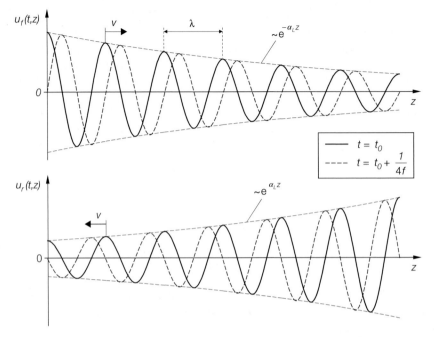

Fig. 24.5. Forward (top) and reflected (bottom) wave on a line at a given time t_0 and a quarter of a period length later

requires the z-dependant portion of the argument of the cosine function to pass through the 2π region:

$$\beta_L\lambda = 2\pi \implies \boxed{\lambda = \frac{2\pi}{\beta_L} = \frac{1}{f\sqrt{L'C'}} = \frac{v}{f}} \tag{24.10}$$

To calculate the current I on the line, (24.3) is solved for I with the value of V from (24.6) inserted:

$$I = -\frac{1}{R'+j\omega L'}\frac{dV}{dz} = -\frac{1}{R'+j\omega L'}\left(-\gamma_L V_f\, e^{-\gamma_L z} + \gamma_L V_r\, e^{\gamma_L z}\right)$$

$$= \sqrt{\frac{G'+j\omega C'}{R'+jwL'}}\left(V_f\, e^{-\gamma_L z} - V_r\, e^{\gamma_L z}\right)$$

With the line *characteristic impedance*

$$Z_W = \sqrt{\frac{R'+jwL'}{G'+j\omega C'}} \tag{24.11}$$

the following applies:

$$I = \frac{V_f}{Z_W}e^{-\gamma_L z} - \frac{V_r}{Z_W}e^{\gamma_L z} = I_f\, e^{-\gamma_L z} - I_r\, e^{\gamma_L z} \tag{24.12}$$

Here, too, we find a forward and a reflected wave, although in this case these are subtracted. The current waves are coupled to the corresponding voltage waves via the line characteristic impedance. This relationship has already been described in the previous section.

The voltages V_f and V_r, as well as the currents I_f and I_r, of the forward and reflected waves cannot be measured directly because the two waves are always superimposed on the cable, thus, only V and I can be measured. A directional coupler must be used for the measurement of the waves [24.1].

For low-loss lines the influence of R' and G' on the line characteristic impedance can be disregarded; then:

$$Z_W \approx \sqrt{\frac{L'}{C'}} \tag{24.13}$$

This applies to loss-free lines exactly.

In the previous section we calculated the line characteristic impedance for specific cables by using the geometric factor k_G from the field characteristic impedance. This, and the calculation by means of the line quantities per unit length, are identical since L' and C' are also geometric properties.

Attenuation

For lines and cables the *attenuation per unit length a'* is normally specified in decibel per meter. For a standard 50 Ω coaxial cable [24.1]:

$$\frac{a'}{\text{dB/m}} \approx 2.35 \cdot 10^{-3} \sqrt{\frac{f}{\text{MHz}}}$$

For the attenuation a of a line with length l it follows:

$$a = a'l$$

Figure 24.6 shows that the attenuation is dependent upon length and frequency. Two-wire lines have an attenuation which is two to five times higher, depending on the line type.

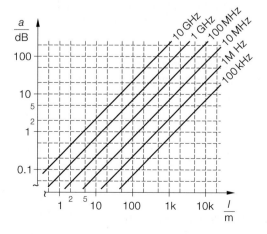

Fig. 24.6. Attenuation a of a standard 50 Ω coaxial cable with length l for several frequencies

Line characteristic impedance	$Z_W = \sqrt{\dfrac{L'}{C'}}$
Velocity of propagation	$v = \dfrac{1}{\sqrt{L'C'}} = \dfrac{c_0}{\sqrt{\epsilon_r}} = 3 \cdot 10^8 \dfrac{\text{m}}{\text{s}} \cdot \dfrac{1}{\sqrt{\epsilon_r}}$
Inductance per unit lengthg	$L' = \dfrac{Z_W}{v}$
Capacitance per unit length	$C' = \dfrac{1}{Z_W v}$
Attenuation constant	$\alpha_L = 0.115\,\text{m}^{-1} \cdot \dfrac{a'}{\text{dB/m}}$
Phase constant	$\beta_L = \dfrac{\omega}{v} = \dfrac{2\pi f}{v} = \dfrac{2\pi}{\lambda}$ with $\lambda = \dfrac{v}{f}$
Propagation constant	$\gamma_L = \alpha_L + j\beta_L$

Fig. 24.7. Cable and line parameters

The attenuation constant α_L can be calculated from the attenuation per unit length:

$$\alpha_L = 0.115\,\text{m}^{-1} \cdot \frac{a'}{\text{dB/m}} = 0.115\,\text{m}^{-1} \cdot \left(\frac{a'}{\text{dB/m}}\bigg|_{f_0}\right)\sqrt{\frac{f}{f_0}}$$

For a standard 50 Ω coaxial cable it thus follows:

$$\alpha_L \approx 2.7 \cdot 10^{-7}\,\text{m}^{-1} \cdot \sqrt{\frac{f}{\text{Hz}}} = \begin{cases} 2.7 \cdot 10^{-4}\,\text{m}^{-1} & f = 1\,\text{MHz} \\ 2.7 \cdot 10^{-2}\,\text{m}^{-1} & f = 10\,\text{GHz} \end{cases}$$

Parameters of a Cable

A cable is usually specified by the line characteristic impedance Z_W, the velocity of propagation v and the attenuation per unit length a'. The relative dielectric constant ϵ_r may also be given instead of the velocity of propagation; the velocity of propagation can then be calculated with (24.1). As an alternative to Z_W and v or ϵ_r, the inductance per unit length L' and the capacitance per unit length C' can also be specified but this is uncommon in practice. An overview of the parameters and relationships is given in Fig. 24.7.

Four-Pole Representation of a Line

Figure 24.8 shows the four-pole diagram of a line with length l together with the relevant currents and voltages. According to (24.6), voltage V_1 is shown by coordinate $z = 0$ and

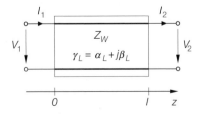

Fig. 24.8. Four-pole representation of a line

voltage V_2 by coordinate $z = l$ as the sum of a forward and a reflected wave:

$$V_1 = V_f + V_r \qquad (24.14)$$

$$V_2 = V_f e^{-\gamma_L l} + V_r e^{\gamma_L l} \qquad (24.15)$$

Similarly, the following applies to the currents:

$$I_1 = I_f - I_r = \frac{V_f}{Z_W} - \frac{V_r}{Z_W} \qquad (24.16)$$

$$I_2 = I_f e^{-\gamma_L l} - I_r e^{\gamma_L l} = \frac{V_f}{Z_W} e^{-\gamma_L l} - \frac{V_r}{Z_W} e^{\gamma_L l} \qquad (24.17)$$

Equations (24.15) and (24.17) lead to:

$$V_2 + Z_W I_2 = 2 V_f e^{-\gamma_L l} \quad , \quad V_2 - Z_W I_2 = 2 V_r e^{\gamma_L l} \qquad (24.18)$$

It follows that the reflected wave is determined by the circuitry at gate 2. No reflected wave exists for $V_2 - Z_W I_2 = 0$, i.e. with a resistance $R = Z_W = V_2/I_2$ at gate 2; this is known as the *termination of the line with its characteristic impedance*. Solving (24.18) for V_f and V_r and insertion into (24.14) and (24.16) results in:

$$V_1 = \frac{V_2}{2}\left(e^{\gamma_L l} + e^{-\gamma_L l}\right) + \frac{Z_W I_2}{2}\left(e^{\gamma_L l} - e^{-\gamma_L l}\right)$$

$$I_1 = \frac{V_2}{2Z_W}\left(e^{\gamma_L l} - e^{-\gamma_L l}\right) + \frac{I_2}{2}\left(e^{\gamma_L l} + e^{-\gamma_L l}\right)$$

With

$$\cosh(\gamma_L l) = \frac{1}{2}\left(e^{\gamma_L l} + e^{-\gamma_L l}\right) \quad , \quad \sinh(\gamma_L l) = \frac{1}{2}\left(e^{\gamma_L l} - e^{-\gamma_L l}\right)$$

the *four-pole equation of a line* is obtained:

$$\begin{bmatrix} V_1 \\ I_1 \end{bmatrix} = \begin{bmatrix} \cosh(\gamma_L l) & Z_W \sinh(\gamma_L l) \\ \dfrac{1}{Z_W}\sinh(\gamma_L l) & \cosh(\gamma_L l) \end{bmatrix} \begin{bmatrix} V_2 \\ I_2 \end{bmatrix} \qquad (24.19)$$

Line with Termination

Let us take a line with terminating impedance Z_2 and calculate the input impedance Z_1 (see Fig. 24.9). For $V_2 = Z_2 I_2$ it follows from (24.19):

$$Z_1 = \frac{V_1}{I_1} = \frac{Z_2 \cosh(\gamma_L l) + Z_W \sinh(\gamma_L l)}{\dfrac{Z_2}{Z_W}\sinh(\gamma_L l) + \cosh(\gamma_L l)} = \frac{Z_2 + Z_W \tanh(\gamma_L l)}{\dfrac{Z_2}{Z_W}\tanh(\gamma_L l) + 1} \qquad (24.20)$$

With

$$\gamma_L = j\beta_L = j\frac{2\pi}{\lambda}$$

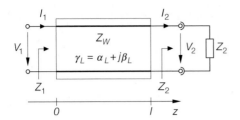

Fig. 24.9. Line with termination

and $\tanh(j\beta_L l) = j \tan(\beta_L l)$, the input impedance for a loss-free line ($\alpha_L = 0$) is:

$$Z_1 = \frac{Z_2 + j\, Z_W \tan\left(\dfrac{2\pi l}{\lambda}\right)}{1 + j\, \dfrac{Z_2}{Z_W} \tan\left(\dfrac{2\pi l}{\lambda}\right)} \tag{24.21}$$

Equations (24.20) and (24.21) show that the line produces an impedance transformation $Z_2 \to Z_1$. For further clarification let us look at some special cases.

- **Termination with the characteristic impedance:** With $Z_2 = Z_W$ it follows that $Z_1 = Z_2 = Z_W$, independent of the length of the line. It was already mentioned in the previous section that this eliminates the reflected wave. Termination with the characteristic impedance is the preferred operating mode for transmission lines since it ensures an optimum power transmission from the signal source to the load; this is further detailed in Sect. 24.3.
- **Electrically short line:** If a line is very much shorter than the wave length λ the tanh and the tan terms can be disregarded; then $Z_1 = Z_2$. This corresponds to the *normal* connecting line in low-frequency circuits which can be regarded as the ideal connection. With higher frequencies, the permissible length of an electrically short line decreases in accordance with the wave length, i.e. inversely proportional to the frequency; in the GHz range this causes a noticeable impedance transformation even at lengths of a few millimeters.
- **$\lambda/4$-line:** A loss-free line with a length equal to a quarter of the wave length λ results in $\tan(2\pi l/\lambda) = \tan(\pi/2) \to \infty$; from (24.21) it follows:

$$Z_1 = \frac{Z_W^2}{Z_2} \tag{24.22}$$

This relationship is sufficiently accurate for low-loss lines as well. The $\lambda/4$-line is often used instead of a transformer for resistance transformations; the resistance $Z_2 = R_2$ is transformed to $Z_1 = R_1$ by a $\lambda/4$-line with $Z_W = \sqrt{R_1 R_2}$. Such lines are also known as $\lambda/4$-*transformers*.
- **Open line:** A line with $Z_2 \to \infty$ is called an *open, open-ended* or *unloaded line*; if it is loss-free, it follows from (24.21):

$$Z_1 = \frac{Z_W}{j \tan\left(\dfrac{2\pi l}{\lambda}\right)} \overset{l < \lambda/8}{\approx} \frac{Z_W}{j\, \dfrac{2\pi l}{\lambda}} = \frac{1}{j\omega C' l} = \frac{1}{j\omega C} \tag{24.23}$$

An open, loss-free line acts as a reactance with a capacitive response $(\tan(2\pi l/\lambda) > 0)$ or an inductive response $(\tan(2\pi l/\lambda) < 0)$ depending on the length; for $l < \lambda/8$ the line acts as a capacitance with $C = C'l$.

- **Short circuited line:** For a short circuited $(Z_2 = 0)$ loss-free line, (24.21) leads to:

$$Z_1 = jZ_W \tan\left(\frac{2\pi l}{\lambda}\right) \overset{l<\lambda/8}{\approx} j Z_W \frac{2\pi l}{\lambda} = j\omega L'l = j\omega L \tag{24.24}$$

Thus, a short circuited, loss-free line also acts as a reactance with an inductive response $(\tan(2\pi l/\lambda) > 0)$ or a capacitive response $(\tan(2\pi l/\lambda) < 0)$ depending on the length; for $l < \lambda/8$ the line acts as an inductance with $L = L'l$.

The latter three cases play an important role in the realization of matching circuits in the upper MHz and the GHz range; these applications no longer use coaxial cables or two-wire lines but strip lines as described below. Figure 24.10 shows the various transformation properties of a given line.

Example: A 10 MHz signal is to be measured with an oscilloscope; a one meter long 50 Ω coaxial cable connects the test point to the input of the oscilloscope. Since the input impedance of the oscilloscope (1 MΩ || 20 pF $\Rightarrow$ $Z_2 \approx -j\,1.6\,\text{k}\Omega$) is significantly higher than the characteristic impedance of the line $(Z_W = 50\,\Omega)$, the line is essentially *open-ended*. With $v \approx 2 \cdot 10^8$ m/sec we have $\lambda = v/f = 20$ m so that $l < \lambda/8 = 2.5$ m; according to (24.23), $Z_1 = 1/j\omega C$ with $C = C'l = 1/Z_W v \approx 100$ pF. The input capacitance of the oscilloscope is then added to this capacitance of the nearly open line: $C = 100\,\text{pF} + 20\,\text{pF} = 120\,\text{pF}$. An accurate calculation with (24.21) yields:

$$Z_1 = \frac{- j\,1.6\,\text{k}\Omega + j\,50\,\Omega\,\tan\left(\dfrac{\pi}{10}\right)}{1 + j\,\dfrac{-j\,1.,6\,\text{k}\Omega}{50\,\Omega}\,\tan\left(\dfrac{\pi}{10}\right)} = -j\,139\,\Omega \overset{!}{=} \frac{1}{j\omega C}$$

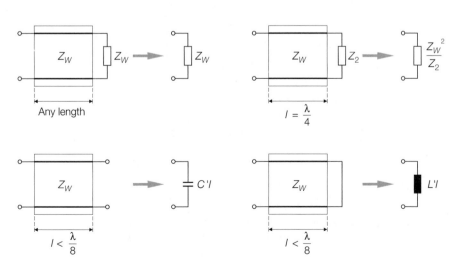

Fig. 24.10. Transformation properties of a line

It follows that $C = 114\,\text{pF}$. This means that the signal to be measured is loaded with a capacitance that is substantially higher than the input capacitance of the oscilloscope. The line of one meter in length is therefore not an electrically short line.

Strip Line

With increasing frequencies, it becomes necessary on printed circuit boards to incorporate connections as conductors with defined characteristic impedance in order to ensure the distortion-free transfer of high-frequency analog signals and fast digital signals; *strip lines* of various designs are used for this purpose [24.1].

The simplest type of strip line is the *microstrip*, as shown in Fig. 24.11. It is almost indistinguishable from normal PC board conductors and can thus be produced in the normal etching process. Owing to the uniform ground surface on the rear side of the board, circuit boards with a copper layer on both sides must be used. Boards made of *Pertinax* should not be used because of their high dielectric losses. Epoxy resin boards ($\epsilon_r \approx 4.8$) can achieve acceptable results in low-demand applications with frequencies below 1 GHz; although the scatter of ϵ_r with epoxy resin is particularly problematic. In general, however, substrates of Teflon ($\epsilon_r = 2.05$) and, especially in the GHz range, aluminum ceramics ($\text{Al}_2\text{O}_3, \epsilon_r = 9.7$) are used.

Calculating the line characteristic impedance and the quantities per unit length is only possible with very complex mathematical models. In practice, the required parameters are usually determined by field simulation. However, there are semi-empiric equations for determining the line characteristic impedance of a microstrip with the dimensions given in Fig. 24.11. These equations provide an accuracy of about 2% provided that $w/d \gg 10$, a condition which is easy to satisfy in practice [24.1]; for $w > h$:

$$\frac{Z_W}{\Omega} \approx \frac{188.5/\sqrt{\epsilon_r}}{\dfrac{w}{2h} + 0.441 + \dfrac{\epsilon_r + 1}{2\pi\epsilon_r}\left[\ln\left(\dfrac{w}{2h} + 0.94\right) + 1.451\right] + \dfrac{0.082\,(\epsilon_r - 1)}{\epsilon_r^2}}$$

and for $w < h$:

$$\frac{Z_W}{\Omega} \approx \frac{60}{\sqrt{\dfrac{\epsilon_r + 1}{2}}}\left[\ln\left(\dfrac{8h}{w}\right) + \dfrac{1}{32}\left(\dfrac{w}{h}\right)^2 - \dfrac{1}{2}\dfrac{\epsilon_r - 1}{\epsilon_r + 1}\left(0.4516 + \dfrac{0.2416}{\epsilon_r}\right)\right]$$

Figure 24.12 shows the curves for Teflon, epoxy resin and Al_2O_3.

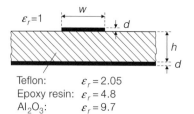

Teflon: $\varepsilon_r = 2.05$
Epoxy resin: $\varepsilon_r = 4.8$
Al_2O_3: $\varepsilon_r = 9.7$

Fig. 24.11. Cross-section of a microstrip

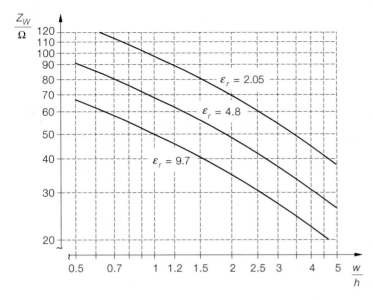

Fig. 24.12. Characteristic impedance of a microstripline for Teflon ($\epsilon_r = 2.05$), epoxy resin ($\epsilon_r = 4.8$) and Al$_2$O$_3$ ($\epsilon_r = 9.7$)

24.2.2
Radio Communication

Figure 24.13 shows the components of a radio transmission system. The output signal of the transmitter amplifier is fed to the *transmitting antenna* through a cable. As the antenna input impedance is usually not matched to the characteristic impedance of the cable, optimum power transfer requires a *matching network*. The electromagnetic wave radiated from the *transmitting antenna* is received by the *receiving antenna* erected at a distance r. The received signal passes through another *matching network* before it is fed along a cable to the receiver amplifier.

Antennas

There are many different antenna designs; an overview is given in [24.1]. They differ in terms of frequency range, bandwidth and *radiation pattern*. Normally, the transmitting

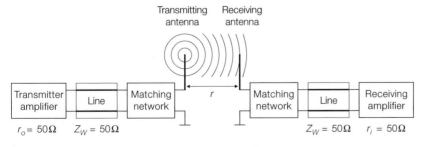

Fig. 24.13. Components of a radio transmission system

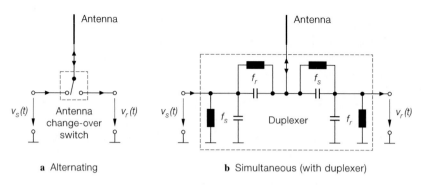

a Alternating

b Simultaneous (with duplexer)

Fig. 24.14. Operating modes using a common transmit and receive antenna

antennas of *radio and TV stations* send the signal horizontally in all directions so that it can be received by any receiver within the transmission range. The *receiving antennas* of portable radio and TV sets also have a wide radiation pattern to, if possible, avoid any alignment to the broadcasting station; this, however, allows only relatively powerful stations to be received. In contrast, stationary devices use *directional antennas* which can pick up weak stations as well but must be exactly adjusted for the given transmitter station; antenna misalignment blocks reception. An example of this are parabolic reflector antennas of satellite receivers. In *mobile communications* the mobile unit cannot be aligned because the site of the base station is generally unknown and undergoes changes in relation to the location of the mobile unit and the given propagation conditions; this is the reason that here, too, antennas with a wide radiation pattern are used. The base stations themselves use sectoring, i.e. the surroundings are divided into sectors each of which is served by one antenna of the respective directional characteristic. *Directional* (line-of-sight) *radio systems* use transmitting and receiving antennas with extremely narrow radiation patterns; this provides benefits including long distances to be covered with relatively low transmit power, the prevention of unwanted radio monitoring and the same transmit frequency to be used for transmitting in other directions. In principle, each antenna can be used for transmitting and receiving; the directional characteristic is the same.

In bi-directional transmission links with a common transmit and receive antenna, the output signal of the transmitter amplifier must be prevented from reaching the sensitive input of the receiving amplifier which would otherwise be immediately destroyed. Systems with alternating transmission and receiving modes use an antenna change-over switch (see Fig. 24.14a). Simultaneous transmitting and receiving is also possible if separate frequencies are used for the received and transmitted signals; signal separation is done by a specific filter (*duplexer*). Figure 24.14b shows a simple duplexer with parallel resonant circuitry.

Directivity: The parameter for the directional characteristic is the *directivity D*; it indicates the factor by which the transmit power is higher in the main direction, as opposed to a hypothetical antenna which has an omnidirectional radiation pattern in all directions. The reference antenna is hypothetical since there is no single antenna that features an omnidirectional radiation pattern; the directivity value of a real antenna is thus always higher than one.

The directivity relates to the *radiated* power; in practice, however, the *supplied power*, which is higher than the radiated power due to losses, is of interest.

Equivalent circuit: Figure 24.15 shows the equivalent circuit of an electrically short rod antenna (length $< \lambda/4$) including the connection to the transmitter amplifier; in this case L_A and C_A are the reactive antenna elements, R_S is the *radiation resistance* and R_V the *ohmic loss resistance* [24.1]. The operating frequency is below the resonant frequency, i.e. the antenna impedance has a capacitive part; the sum of the radiation and loss impedances is less than 50 Ω. The antenna impedance is transformed to 50 Ω by the matching network.

Figure 24.16 shows the radiation resistance R_S of a rod antenna plotted over the relative length l/λ [24.1]. The resistance becomes very low for $l < \lambda/8$; at this point matching to 50 Ω is only possible for a very narrow bandwidth. Especially favorable are rod antennas with $l/\lambda \approx 0.26 \ldots 27$. Including the antenna loss resistance, they have an overall resistance of 50 Ω and are operated slightly above the resonant frequency; matching is performed with a series capacitance.

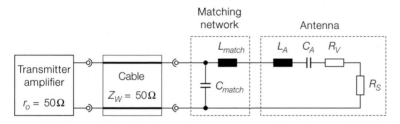

Fig. 24.15. Equivalent circuit of a rod antenna ($l < \lambda/4$) including the connection to the transmitter amplifier

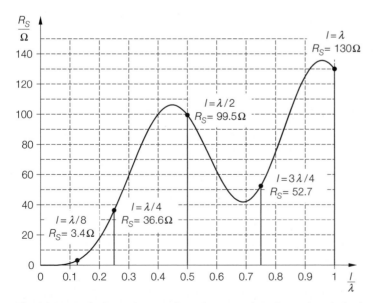

Fig. 24.16. Radiation resistance of a rod antenna plotted over the relative length l/λ

Antenna radiation efficiency: The *radiation efficiency* η can be taken directly from Fig. 24.15:

$$\eta = \frac{R_S}{R_S + R_V} < 1$$

This shows the ratio of the radiated power to the supplied power. If the antenna is used as the receiving antenna, the equivalent circuit is primarily the same except that the loss resistance does not have the same value due to a slightly different current distribution; it is therefore necessary to differentiate between the *transmission efficiency* η_S and the *receiving efficiency* η_R.

Antenna gain: The product of the directivity and the antenna efficiency is known as the *antenna gain:*

$$G = D\eta$$

The antenna gain compares the transmit power of a real, lossy antenna in the main radiation direction with the transmit power of a hypothetical, loss-free antenna with an omnidirectional radiation pattern at the *same supplied power*. Due to the different antenna efficiencies in the transmission and receiving mode, it is necessary to differentiate between the *transmit gain* and the *receive gain*; in practice, however, these differences are usually so low that there is no need for such a differentiation.

Power Transfer via a Radio Link

The relationship between the transmit power P_S and the receive power P_E of a radio link can be specified by means of the antenna gain G_S of the transmission antenna and the antenna gain G_E of the receiving antenna [24.1]:

$$P_E = P_S \, G_S \, G_E \left(\frac{\lambda}{4\pi r}\right)^2 \tag{24.25}$$

The free space wave length is

$$\lambda = \frac{c_0}{f} = \frac{3 \cdot 10^8 \, \text{m/s}}{f}$$

and r is the distance between the transmitter and the receiver. The factor

$$\left(\frac{\lambda}{4\pi r}\right)^2 = \frac{\lambda^2/(4\pi)}{4\pi r^2} = \frac{\text{Effective surface of the receiving antenna}}{\text{Spherical surface}}$$

takes into consideration that the receiving antenna covers only a portion of the evenly radiated spherical surface[2].

In practice, the *link attenuation* is specified:

$$\frac{a}{\text{dB}} = 10 \log \frac{P_S}{P_E} = 20 \log \frac{4\pi r}{\lambda} - \frac{G_S}{\text{dB}} - \frac{G_E}{\text{dB}} \tag{24.26}$$

---

[2] Please note that the transmitting and receiving antennas are now to be considered as loss-free antennas with an omnidirectional radiation pattern as any deviation is already accounted for in the antenna gains G_S and G_E.

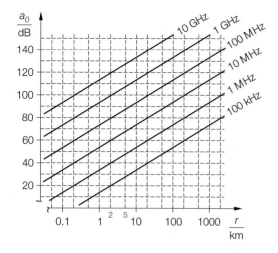

Fig. 24.17. Basic attenuation of a radio link

The *basic attenuation* is:

$$\frac{a_0}{\mathrm{dB}} = 20\log\frac{4\pi r}{\lambda} = 20\log\frac{4\pi r f}{c_0} \tag{24.27}$$

Accordingly, the attenuation rises at a rate of 20 dB per decade as the distance and frequency increase. Inserting the constants leads to:

$$\frac{a}{\mathrm{dB}} = 32.4 + 20\log\frac{r}{\mathrm{km}} + 20\log\frac{f}{\mathrm{MHz}} - \frac{G_S}{\mathrm{dB}} - \frac{G_E}{\mathrm{dB}}$$

Figure 24.17 shows the basic attenuation a_0 in its dependency on distance and frequency.

Equations (24.25) and (24.26) only apply to an ideal propagation in space. Real links have a *supplementary attenuation* that is caused by air, fog or rain and whose value is higher or lower depending on the given frequency; in addition, there are absorptions close to ground and local signal breakdowns due to multi-path propagation. For a detailed description of propagation conditions in different frequency ranges refer to [24.1].

Frequency Ranges

Frequencies are divided into ranges; Fig. 24.18 shows the ranges from 30 kHz to 300 GHz and their respective designations. The range between 200 MHz and 220 GHz is also known as the *microwave range*; it is subdivided into 12 bands (see Fig. 24.19). The names of the

Frequency	Wave length	Abbreviation	Full designation
30 kHz − 300 kHz	10 km − 1 km	LF	Low frequencies
300 kHz − 3 MHz	1 km − 100 m	MF	Medium frequencies
3 MHz − 30 MHz	100 m − 10 m	HF	High frequencies
30 MHz − 300 MHz	10 m − 1 m	VHF	Very high frequencies
300 MHz − 3 GHz	1 m − 10 cm	UHF	Ultra high frequencies
3 GHz − 30 GHz	10 cm − 1 cm	SHF	Super high frequencies
30 GHz − 300 GHz	1 cm − 1 mm	EHF	Extremely high frequencies

Fig. 24.18. Frequency ranges and wave lengths for radio links in the range from 30 kHz to 300 GHz

Designation	P	L	S	C	X	Ku	K	Ka	Q	E	F	G
from (GHz)	0.2	1	2	4	8	12	18	27	40	60	90	140
to (GHz)		1	2	4	8	12	18	27	40	60	90	140	220

Fig. 24.19. Microwave bands

Designation	Frequency	Wavelength
Long-wave radio broadcasting	148.5 ... 283.5 kHz	2.02 ... 1.06 km
Medium-wave radio broadcasting	526.5 ... 1606.5 kHz	572 ... 187 m
Short-wave radio broadcasting	3.95 ... 26.1 MHz	76 ... 11.5 m
TV broadcasting range I	47 ... 68 MHz	6.38 ... 4.41 m
Ultra-short wave (FM) radio broadcasting	88 ... 108 MHz	3.41 ... 2.78 m
TV broadcasting range III	174 ... 223 MHz	1.72 ... 1.34 m
TV broadcasting range IV+V	470 ... 790 MHz	63.8 ... 38 cm

Fig. 24.20. Frequency and wave length ranges for radio and TV broadcasting in Germany

System	Network	Frequeny range	
FM (analog)	C	U: 451 ... 455.74 MHz	D: 461 ... 465.74 MHz
GSM900	D	U: 890 ... 915 MHz	D: 935 ... 960 MHz
GSM1800	E	U: 1710 ... 1785 MHz	D: 1805 ... 1880 MHz
DECT		U: 1880 ... 1900 MHz	D = U
UMTS		U: 1920 ... 1980 MHz	D: 2110 ... 2170 MHz

Fig. 24.21. Frequency ranges for mobile communication and mobile telephones in Germany (U = uplink: mobile unit → base station, D = downlink: base station → mobile unit)

ranges and bands are often combined with the type designations of components, e.g. *UHF transistor* or *S-band FET*.

Besides this classification of frequency ranges, or bands, which is independent of the application, a particular frequency range is also assigned to each special use. Figure 24.20 contains an overview of radio and TV broadcasting ranges, while Fig. 24.21 lists the bands for mobile communication, including mobile telephones according to the DECT standard.

24.2.3
Fibre Optic Links

In addition to coaxial and two-wire cables and radio links, fibre optic links via *optical waveguides (glass fibres)* are also becoming increasingly important. This technology uses a carrier signal in the infrared range ($f = 190 \ldots 360$ THz, $\lambda = 1.55 \ldots 0.85\,\mu m$) which can be modulated by signal frequencies of up to 100 GHz; theoretically this allows transmission rates of up to 200 Gbit/s. Today's systems use 10 Gbit/s while systems with up to 40 Gbit/s are presently being tested. Due to their very small relative modulation bandwidth (signal-to-carrier frequency ratio $\ll 10^{-3}$), the transmission attenuation is constant across the transmission band; therefore, the equalization requirements in the receiver are lower than in cable links despite the substantially higher data rates.

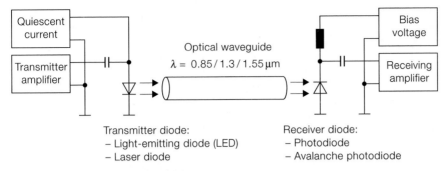

Transmitter diode:
– Light-emitting diode (LED)
– Laser diode

Receiver diode:
– Photodiode
– Avalanche photodiode

Fig. 24.22. Components of a simple fibre optic transmission system

Another advantage of fibre optics is its insensitivity to external electromagnetic interference (optimum *passive electromagnetic compatibility*) and the lack of any interference emission (optimum *active electromagnetic compatibility*); it is therefore possible to route optical *waveguides* in bundles through locations with very strong electromagnetic interferences without disturbance.

Figure 24.22 shows the components of a simple fibre optic transmission system. In the transmitter the radiation intensity of the *transmitter diode* is modulated by an electrical signal; in the receiver the incident radiation intensity is converted back into an electrical signal by the *receiver diode*. More powerful systems use additional special electro-optical components such as optical amplifiers, wave length multiplexers and optical oscillators. The following description is restricted to the properties of the different optical *waveguides* with reference to the relevant literature [24.2, 24.3].

Optical Waveguides

A high quality *waveguide* consists of a very thin fibre made of silicate glass; here, the cross-sections shown in Fig. 24.23 are used. The radiation propagates in the core with the refractive index n_{co} and the diameter d_{co}. The cladding, with its slightly lower refractive index n_{cl} and the outer diameter d_{cl}, is required for guidance only while the outer sheathing serves to protect the optical waveguide. Typical values for a step-index fibre are $n_{co} \approx 1.4$ and $n_{cl}/n_{co} \approx 0.99$, i.e. the refractive index of the cladding is only 1% lower than that of the core. Waveguides made of glass are called *glass fibres*.

Optical waveguides made of plastic are on the market. These are known as *plastic fibres*. They are lower in price and, due to their high mechanical flexibility, are easier to lay than glass fibres but have significantly poorer propagation properties so that they can only be used for short distances and low data rates. Their diameter is much bigger than that of waveguides made of glass; typical diameters are $d_{co} = 0.98$ mm and $d_{cl} = 1$ mm.

Limiting angle and acceptance angle: The signal propagation can be explained by means of radiation optics. The ray in the core is totally reflected at the interface to the cladding, i.e. it is returned to the core if the angle between the ray and the interface is smaller than the *limiting angle* β_g; this is described by[3]:

---

[3] In radiation optics the angle between the ray and the *perpendicular* to the interface is often used; in this case, it is described by $\sin \beta_g = n_{cl}/n_{co}$. We relate the angle to the fibre axis.

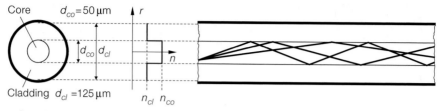

Core d_{co}=50 µm

Cladding d_{cl} =125 µm

a Step-index fibre

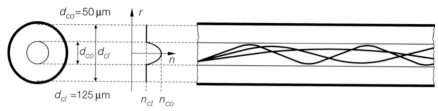

d_{co}=50 µm

d_{cl} =125 µm

b Gradient fibre

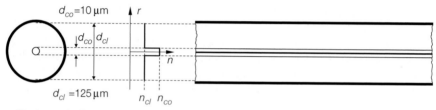

d_{co}=10 µm

d_{cl} =125 µm

c Single-mode fibre

Fig. 24.23. Cross-section, refraction index and propagation characteristics in optical waveguides made of silicate glass

$$\cos \beta_g \ = \ \frac{n_{cl}}{n_{co}} \ < \ 1$$

The typical step-index fibre values result in the angle $\beta_g \approx 8°$. To ensure that the angle within the waveguide remains below the limiting angle, the angle of incidence at the front end must be smaller than the *acceptance angle* α_A. Figure 24.24 illustrates this. From the refraction law it follows:

$$\frac{\sin \alpha_A}{\sin \beta_g} \ = \ n_{co}$$

Numerical aperture: In practice, the acceptance angle is replaced by the *numeric aperture*

$$A_N \ = \ \sin \alpha_A \ = \ n_{co} \sin \beta_g \ = \ n_{co} \sqrt{1 - \cos^2 \beta_g} \ = \ \sqrt{n_{co}^2 - n_{cl}^2}$$

A typical value is $A_N = 0.2$. The specification of β_g and A_N is equivalent to the specification of n_{co} and n_{cl}. The numeric aperture is an important parameter with regard to the

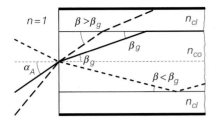

Fig. 24.24. Limiting angle β_g and acceptance angle α_A

coupling between the transmitter diode and the waveguide; a high value in combination with a correspondingly high acceptance angle is advantageous. The velocity of propagation is:

$$v = \frac{c_0}{\sqrt{\epsilon_{r,co}}} = \frac{c_0}{n_{co}}$$

Here, $\epsilon_{r,co} = n_{co}^2$ is the dielectric constant of the core material.

Modes: Applying Maxwell's equations shows that, due to boundary conditions for the fields, not all angles within $0 \leq \beta < \beta_g$ can be used for signal propagation; moreover, according to the condition

$$\sin \beta_m = \frac{\sqrt{2}\lambda m}{\pi d_{co}} \quad \text{with } m = 0, 1, 2, \ldots \text{ und } m \leq \frac{\pi d_{co}}{\sqrt{2}\lambda}$$

only discrete angles β_m are available [24.2]. The rays corresponding to these angles are known as modes; their number becomes higher with an increasing core diameter.

The core diameter of the *step-index fibre* is so large that it allows several modes to propagate (see Fig. 24.23a). Since the different modes cover different distances, a pulse originating from the transmitting diode becomes continuously wider as the length of the fibre becomes longer. This *modal dispersion* has a strong limiting effect on the bandwidth, particularly for large fibre lengths; this is the reason why step-index fibres are no longer used in long-distance communication. Step-index fibres made of plastic are used in simple systems with distances of up to 100 m and data rates of up to a maximum of 40 Mbit/s [24.4].

The *gradient fibre* uses a continuous transition of the refractive index; this causes the modes to be sent back in the form of continuous total reflection in the direction of the fibre axis (see Fig. 24.23b). Since in the outer portions of the core the velocity of propagation increases due to the declining refractive index, the *slated* modes propagate faster than the modes along the fibre axis; this strongly reduces the modal dispersion and increases the bandwidth. Even though the gradient fibre does not reach the bandwidth of the single-mode fibre described below, it still has the advantage that simpler connection systems, with higher tolerances in terms of alignment, can be used due to the larger core diameter.

In the *single-mode fibre*[4], the core diameter is so small that it allows only the basic mode to propagate (see Fig. 24.23c), thus eliminating the modal dispersion. The permissible core diameter is derived from the condition that the angle of the mode with $m = 1$ must be wider than the limiting angle:

$$\beta_1 > \beta_g \Rightarrow d_{co} < \frac{\sqrt{2}\lambda}{\pi\sqrt{1 - \left(\dfrac{n_{cl}}{n_{co}}\right)^2}} \quad \overset{n_{co}/n_{cl} \approx 0.999}{\approx} \quad 10\,\lambda$$

---

[4] This type was formerly known as the *mono-mode fibre*.

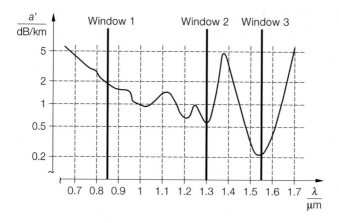

Fig. 24.25. Attenuation coefficient of a typical waveguide of silicate glass plotted over the wave length

In this case, the refractive index of the cladding is only 0.1% lower than that of the core in order to prevent the permissible core diameter from getting too small. This fibre achieves the highest bandwidth. One drawback is that complex connectors are required.

Wave Length Ranges

Three ranges, in which the attenuation is particularly low, are used for signal transmission in optical waveguides made of silicate glass *(glass fibres)* (see Fig. 24.25). These ranges are known as *windows*. The characteristic parameters of these windows are listed in Fig. 24.26. Generally, the *free-space wave length* is specified which renders the value independent of the refractive index of the waveguide.

Window 1 is often used despite its comparatively high attenuation because it enables the use of conventional infrared light-emitting diodes (IR LEDs) in the transmitter and conventional infrared photodiodes (IR photodiodes) in the receiver. The lengths of the links are shorter than 5 kilometres and the data rates are below 200 Mbit/s; gradient fibres with $d_{co} = 50\,\mu$m are used for this purpose.

Windows 2 and 3 are exclusively used for long-distance cables with maximum data rates; gradient fibres are increasingly being replaced by single-mode fibres with $d_{co} = 10\,\mu$m $< 10\lambda$. Data rates exceeding 1 Gbit/s are achieved solely with single-mode fibres. Laser diodes are used on the transmitter side and avalanche photodiodes on the receiving side.

For signal transmission with optical waveguides made of plastic *(plastic fibres)* visible radiation (light) with a wave length of $\lambda = 660\,\mu$m is often used. The attenuation is

Name	Wavelength [nm]	Frequency [THz]	Attenuation [dB/km]	Waveguide
	660	455	230 (!)	Plastic fibres
Window 1	850	353	2	Gradient fibre
Window 2	1300	231	0.6	Gradient and single-mode fibres
Window 3	1550	194	0.2	Single-mode fibre

Fig. 24.26. Wave length ranges for optical waveguides

extremely high, thus restricting the link length to 100 m. For the light spectrum, red light-emitting diodes (LEDs) are used in the transmitter and photodiodes are used in the receiver.

24.2.4
Comparison of Transmission Channels

The following description is restricted to a comparison of the attenuation because comparing data rates is only possible by taking the method of modulation into consideration. Furthermore, the data rate of radio transmission is limited by the assigned frequency range and not by the carrier frequency.

Figure 24.27 shows the superiority of optical waveguides as compared to coaxial cables. Since waveguides have a very narrowband modulation, the attenuation depends solely on the distance; with a permissible attenuation of 40 dB between transmitter and receiver it is possible to cover up to 100 km without a repeater. In the case of coaxial cables, the attenuation also depends on the frequency. The distance coverage is therefore determined by the maximum permissible attenuation at the upper frequency limit.

In radio links, distance has only a logarithmic effect on attenuation; the semilogarithmic plot in Fig. 24.27 thus shows straight lines. For the borderline case of very large distances, the radio link is superior to all other types. However, the technically available bandwidth must be distributed among a high number of systems. Due to the high sensitivity of narrowband receivers, the permissible attenuation can reach up to 150 dB. Figure 24.27 shows the basic attenuation only; the reduction in attenuation due to the gains of transmitter and receiver antennas (usually 10 ... 20 dB; more than 40 dB in large parabolic antennas) and the additional attenuation caused by air, rain, fog and absorptions close to ground are not taken into account. The main advantage of the radio link is, of course, its independence from wires and cables.

Today, almost all telephone and data traffic uses fibre optic systems with several optical waveguides in parallel. This is the basic technology used for the high transmission capabilities of public and private wide-area networks such as the *Internet*.

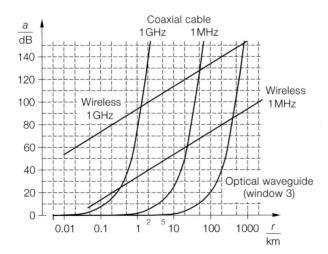

Fig. 24.27. Attenuation of several transmission channels

24.3
Reflection Coefficient and S Parameters

In Sect. 24.2.1 we saw that voltages and currents along a wire are described by forward and reflected waves, that the relationship between these waves depends on the circuitry and that an impedance transformation generally takes place; only in the case of electrically short lines may an ideal link be assumed. This description is now extended to cover any two-terminal and four-terminal network, i.e. *all* voltages and currents in a circuit are broken down into a forward and a reflected wave, allowing a uniform description of components and interconnecting lines. In this case, components are no longer characterized by impedances and admittances but by the ratio between the forward and the reflected wave. The corresponding parameters are the *reflection coefficient* and the *S parameters*.

24.3.1
Wave Parameters

The voltages of the forward (index f) and reflected (index r) waves on a line are interrelated by the respective currents via the characteristic impedance Z_W of the line:

$$V_f = Z_W I_f \quad , \quad V_r = Z_W I_r$$

For this reason, one parameter is sufficient to describe each of the two waves. The following *wave parameters* are used:

$$
\begin{aligned}
a &= \frac{V_f}{\sqrt{Z_W}} = I_f \sqrt{Z_W} & \text{Forward wave} \\[2mm]
b &= \frac{V_r}{\sqrt{Z_W}} = I_r \sqrt{Z_W} & \text{Reflected wave}
\end{aligned}
\tag{24.28}
$$

They provide a measure for the power transported by the waves and are given in units of *root of Watt*:

$$[a] = [b] = \sqrt{VA} = \sqrt{W}$$

The transported power is[5]:

$$
\begin{aligned}
P_f &= \mathrm{Re}\left\{V_f I_f^*\right\} \stackrel{Z_W \text{ real}}{=} |a|^2 \\[2mm]
P_r &= \mathrm{Re}\left\{V_r I_r^*\right\} \stackrel{Z_W \text{ real}}{=} |b|^2
\end{aligned}
\tag{24.29}
$$

The characteristic impedance Z_W of the line is real, therefore, V_f and I_f as well as V_r and I_r, are always in phase and both waves transport effective power only.

Description by Means of Voltage and Current

The voltage V and the current I are obtained by superimposing the forward and the reflected waves[6]:

$$V = V_f + V_r \quad , \quad I = I_f - I_r$$

---

[5] Root-mean-square phasors are used; consequently $P = VI$ for real phasors and $P = \mathrm{Re}\left\{VI^*\right\}$ for complex phasors where $I^* = \mathrm{Re}\{I\} - j\,\mathrm{Im}\{I\}$.

[6] These relationships result from (24.6) and (24.12) by inserting $z = 0$.

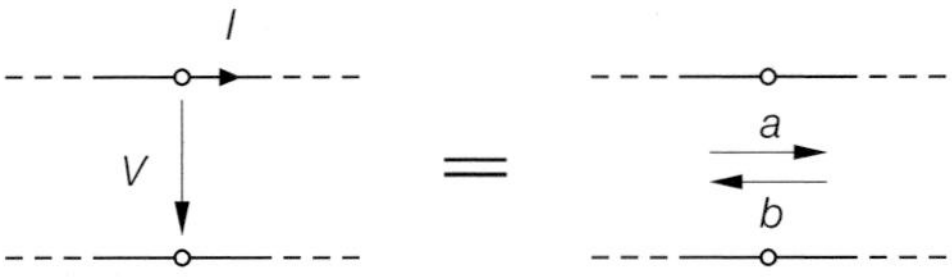

Fig. 24.28. Equivalent representations of the parameters in a circuit

Inserting the wave parameters from (24.28) leads to

$$V = \sqrt{Z_W}\,(a + b) \tag{24.30}$$

$$I = \frac{1}{\sqrt{Z_W}}\,(a - b) \tag{24.31}$$

and after inversion:

$$a = \frac{1}{2}\left(\frac{V}{\sqrt{Z_W}} + I\,\sqrt{Z_W}\right) \tag{24.32}$$

$$b = \frac{1}{2}\left(\frac{V}{\sqrt{Z_W}} - I\,\sqrt{Z_W}\right) \tag{24.33}$$

This results in equivalent representations of the parameters in a circuit as shown in Fig. 24.28.

Equations (24.30)–(24.33) are not completely clear in their given form because the principle of wave parameters, as a substitution for the voltages and currents of the forward and reflected waves, is only contained indirectly therein. Therefore, these equations must always be considered in relation to (24.28).

24.3.2
Reflection Coefficient

Converting to wave parameters means that a two-terminal network is no longer described by the impedance Z but by the forward and reflected waves (see Fig. 24.29). The forward wave is also called the *incident wave*. The ratio of the reflected wave to the incident wave is known as the *reflection coefficient* r:

$$\text{Reflection coefficient } r = \frac{\text{Reflected wave}}{\text{Incident wave}} = \frac{V_r}{V_f} = \frac{b}{a}$$

Using $Z = V/I$ it follows from (24.32) and (24.33) that:

$$r = \frac{V_r}{V_f} = \frac{b}{a} = \frac{Z - Z_W}{Z + Z_W} \tag{24.34}$$

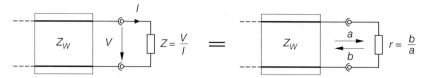

Fig. 24.29. Impedance and reflection coefficients of a two-terminal network

After solving for Z:

$$Z = Z_W \frac{1+r}{1-r} \tag{24.35}$$

Reflection Coefficient Plane (r Plane)

Equation (24.34) describes the mapping of the *impedance plane (Z plane)* on the *reflection coefficient plane* (*r plane*). The region of passive two-terminal networks with Re $\{Z\} \geq 0$ (*right Z semi-plane*) is mapped into the unit circle of the r plane, i.e. $|r| \leq 1$ applies to passive two-terminal networks (see Fig. 24.30). The passivity is demonstrated by the fact that the active power absorbed by the two-terminal network as a difference between incident and reflected active power is always positive or zero:

$$P = P_f - P_r \overset{(24.29)}{=} |a|^2 - |b|^2 \overset{(24.34)}{=} |a|^2 \left(1 - |r|^2\right) \overset{|r| \leq 1}{\geq} 0$$

Factor

$$k_P = 1 - |r|^2 \tag{24.36}$$

is called the *power transmission factor*. For active two-terminal networks we obtain Re $\{Z\} < 0$, $|r| > 1$ and $P < 0$, i.e. active two-terminal networks deliver active power.

Mapping the Z plane on the r plane creates three special points:

- **Matching:** Matching to the characteristic impedance is achieved for $Z = Z_W$. Section 24.2 has already shown that in this case the reflected wave disappears ($b = 0$);

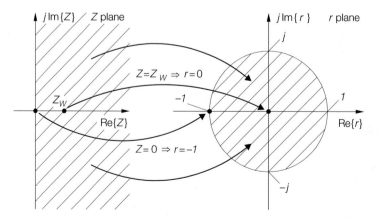

Fig. 24.30. Mapping the impedance plane (Z plane) on the reflection coefficient plane (r plane) in the case of passive two-terminal networks (Re $\{Z\} \geq 0$)

consequently it follows from (24.34) that $r = 0$. The incident active power P_f is fully absorbed by the two-terminal network.

- **Short circuit:** $Z = 0$ results in $r = -1$, i.e. the incident and reflected waves are of the same absolute size, but of opposing phase: $b = -a$. In this case, the two-terminal network does not absorb any active power; the incident active power is fully reflected: $P_r = P_f$.
- **Open circuit:** $Z \to \infty$ results in $r = 1$; the incident and reflected waves are of the same size and in phase: $b = a$. Here again, the incident active power is fully reflected: $P_r = P_f$.

The following regions also occur:

- **Ohmic resistances:** With ohmic resistances ($Z = R$) we have a real reflection coefficient in the range of $-1 < r < 1$. This region consists of a subregion for $0 < R < Z_W$ and $-1 < r < 0$, the matching point for $R = Z_W$ and $r = 0$ and a subregion for $Z_W < R < \infty$ and $0 < r < 1$.
- **Inductances:** With inductances ($\mathrm{Re}\{Z\} = 0$, $\mathrm{Im}\{Z\} > 0$) we have $|r| = 1$ and $0 < \arg\{r\} < \pi$, i.e. the upper half of the unit circle in the r plane.
- **Capacitances:** With capacitances ($\mathrm{Re}\{Z\} = 0$, $\mathrm{Im}\{Z\} < 0$) we also have $|r| = 1$ but $-\pi < \arg\{r\} < 0$, i.e. the lower half of the unit circle in the r plane.

Figure 24.31 shows these special points and regions in the r plane.

Figure 24.32 shows the absolute value of the reflection coefficient and the power transmission factor of ohmic resistances for $Z_W = 50\,\Omega$. The absolute value of the reflection coefficient increases rapidly with any deviation from the matched condition $Z = R = 50\,\Omega$ and asymptotically approaches one. The power transmission factor is less steep around the matching point; a slight mismatch is thus not critical in terms of power transmission. In the range $20\,\Omega < Z = R < 130\,\Omega$, (24.34) yields $|r| < 0.45$ and

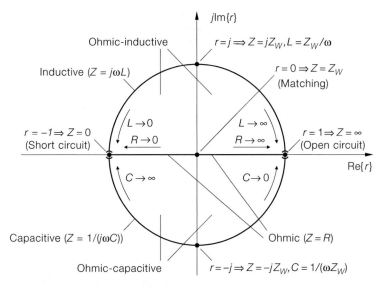

Fig. 24.31. Special points and regions in the reflection coefficient plane (r plane)

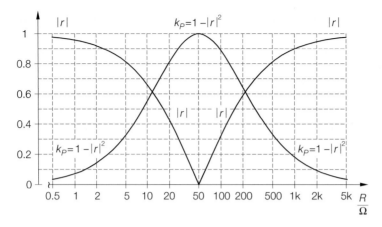

Fig. 24.32. Absolute value of the reflection coefficient and power transmission factor $kp = 1 - |r|^2$ for ohmic resistances where $Z_W = 50\,\Omega$

(24.36) yields $kp = 1 - |r|^2 > 0.8$. In this case the transmission power loss is below 1 dB ($10 \log kp = -0.97$ dB).

Influence of a Line on the Reflection Coefficient

Section 24.2.1 explained that a line causes an impedance transformation. This impedance transformation can now be explained by means of the reflection coefficient. To do so, we assume a line of the length l with a terminating impedance Z_2 and corresponding reflection coefficient r_2, we then calculate the reflection coefficient r_1 at the input of the line (see Fig. 24.33).

The voltage along the line is:

$$V(z) = V_f(z) + V_r(z) \overset{(24.6)}{=} V_f(0)\, e^{-\gamma_L z} + V_r(0)\, e^{\gamma_L z}$$

Here, $V_f(0)$ and $V_r(0)$ are the voltages of the incident and the reflected waves at the point $z = 0$. Equation (24.28) leads to the waves $a(z)$ and $b(z)$ along the line:

$$a(z) = \frac{V_f(z)}{\sqrt{Z_W}} = \frac{V_f(0)}{\sqrt{Z_W}}\, e^{-\gamma_L z} \quad , \quad b(z) = \frac{V_r(z)}{\sqrt{Z_W}} = \frac{V_r(0)}{\sqrt{Z_W}}\, e^{\gamma_L z}$$

This allows the reflection coefficients r_1 and r_2 to be calculated:

$$r_1 = \frac{b_1}{a_1} = \frac{b(0)}{a(0)} = \frac{V_r(0)}{V_f(0)} \quad , \quad r_2 = \frac{b_2}{a_2} = \frac{b(l)}{a(l)} = \frac{V_r(0)}{V_f(0)}\, e^{2\gamma_L l}$$

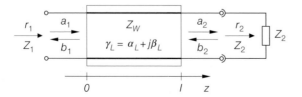

Fig. 24.33. Influence of a line on the reflection coefficient

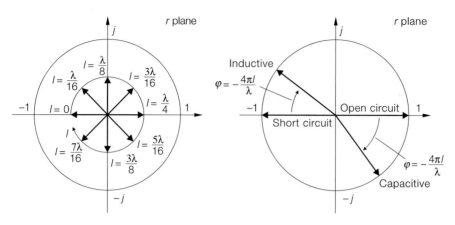

a Resistance: $Z_2 = R_2 = Z_W/3$, $r_2 = -1/2$ **b** Short circuit ($r_2 = -1$) and open circuit ($r_2 = 1$)

Fig. 24.34. Rotation of the reflection coefficient in a loss-free line

With $\gamma_L = \alpha_L + j\beta_L$ it follows:

$$r_1 = r_2 e^{-2\gamma_L l} = r_2 e^{-2\alpha_L l} e^{-2j\beta_L l} \tag{24.37}$$

This means that the line causes an attenuation of the reflection coefficient by double the value of the attenuation constant α_L and a shift double the phase constant β_L.

The case of a loss-free line is of particular importance. If $\alpha_L = 0$ it follows from (24.37):

$$r_1 = r_2 e^{-2j\beta_L l} \overset{\beta_L = 2\pi/\lambda}{=} r_2 e^{-j\frac{4\pi l}{\lambda}} \overset{\varphi = -4\pi l/\lambda}{=} r_2 e^{j\varphi} \tag{24.38}$$

In this case, the reflection coefficient is only rotated clockwise by two turns per wave length: $l = \lambda \Rightarrow \varphi = -4\pi$. Figure 24.34a demonstrates this with a resistance $Z_2 = R_2 = Z_W/3$ and $r_2 = -1/2$ and a gradual increase in the line length by $\Delta l = \lambda/16$. First, the reflection coefficient enters the ohmic inductive region. For $l = \lambda/4$ ($\varphi = -\pi$) the condition of $r_1 = -r_2 = 1/2$ is reached with $Z_1 = Z_W^2/R_2 = 3Z_W$. This property of a $\lambda/4$ line was already described in (24.22) and Fig. 24.8. With a further increase in line length, the reflection coefficient passes through the ohmic capacitive region until the starting point $r_1 = r_2$ is reached for $l = \lambda/2$ ($\varphi = -2\pi$). This means that the reflection coefficient r_1 is periodic with $\Delta l = \lambda/2$.

Figure 24.34b shows that a short short-circuited line ($r_2 = -1$) has an inductive effect and a short open-circuited line ($r_2 = 1$) has a capacitive effect; this, too, has already been described in (24.23) and (24.24) and in Fig. 24.10. With $l = \lambda/4$ the short circuit changes to an open circuit and the open circuit to a short circuit.

Termination with the characteristic impedance ($Z_2 = Z_W$) results in $r_2 = 0$. In this case, the rotation is without effect; $r_1 = 0$ and $Z_1 = Z_W$, regardless of the length of the line.

Standing Wave Ratio

Let us now look at the trend of the voltage phasor $V(z)$ along a loss-free line; when using (24.34) and (24.28) it follows from (24.30):

$$V(z) = \sqrt{Z_W}\,(a(z)+b(z)) = \sqrt{Z_W}\,a(z)\,(1+r(z)) = V_f(z)\,(1+r(z)) \qquad (24.39)$$

Here, $V_f(z)$ is the voltage phasor of the incident wave and $r(z)$ is the reflection coefficient. The waves are not attenuated on a loss-free line. Therefore, the magnitude of the voltage phasor $V_f(z)$ remains constant along the line:

$$|V_f(z)| = |V_f| = \text{const.}$$

The magnitude of the voltage phasor $V(z)$ is thus determined from (24.39):

$$|V(z)| = |V_f|\,|1+r(z)| \qquad (24.40)$$

Likewise, the magnitude of the reflection coefficient of a loss-free line is constant:

$$|r(z)| = |r| = \text{const.}$$

As the reflection coefficient undergoes a rotation, the factor $|1+r(z)|$ in (24.40) remains within the range:

$$1-|r| \leq |1+r(z)| \leq 1+|r|$$

Consequently, points with the maximum and minimum values of the voltage phasor $V(z)$ appear alternately along the line:

$$V_{max} = |V_f|\,(1+|r|) \quad , \quad V_{min} = |V_f|\,(1-|r|) \qquad (24.41)$$

This results in a standing wave with the *voltage standing wave ratio (VSWR)*:

$$s = \frac{V_{max}}{V_{min}} = \frac{1+|r|}{1-|r|} \qquad (24.42)$$

If the system is matched ($r = 0$), the VSWR evaluates to one. In this case, there is no standing wave and the magnitude of the voltage phasor $V(z)$ is constant along the entire length of the line: $|V(z)| = |V_f|$. In the purely reactive case ($|r| = 1$), the VSWR tends to infinity and $V_{max} = 2|V_f|$ and $V_{min} = 0$. The distance between the maxima and the minima is $\lambda/4$, which corresponds to a rotation of the reflection coefficient by an angle of π (180°).

Figure 24.35 shows a standing wave on a loss-free line of length $l = \lambda/2$ for $r_2 = 0.5\,e^{j\,30°}$. Thus, the magnitude of the reflection coefficients are $|r(z)| = |r| = |r_1| = |r_2| = 0.5$. According to (24.40), the magnitude of the voltage phasor $V(z)$ is proportional to the magnitude of the factor $1 + r(z)$; this factor is constructed geometrically at five locations at an interval of $\lambda/8$ in Fig. 24.35. Since the reflection coefficient of a line with length $l = \lambda/2$ is rotated by 2π (360°), there is exactly one maximum and one minimum value. From (24.42) the voltage standing wave ratio for $|r| = 0.5$ is $s = 3$.

The standing wave ratio is also of importance for the active power P. Along a loss-free line the magnitudes of the wave parameters and the reflection coefficient are constant; consequently:

$$P = P_f - P_r = |a|^2 - |b|^2 = |a|^2\left(1-|r|^2\right) = \frac{|V_f|^2}{Z_W}\left(1-|r|^2\right)$$

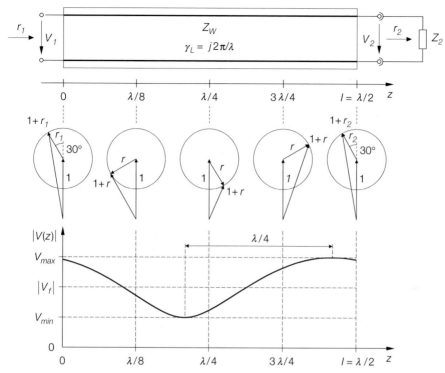

Fig. 24.35. Standing wave on a loss-free line of length $\lambda/2$ for $r_2 = 0.5\,e^{j\,30°}$

If we insert (24.41), the following is obtained:

$$P = \frac{V_{max}^2}{Z_W}\frac{1-|r|^2}{(1+|r|)^2} = \frac{V_{max}^2}{Z_W}\frac{1-|r|}{1+|r|} = \frac{1}{s}\frac{V_{max}^2}{Z_W} = \frac{P_{max}}{s}$$

Consequently, the transmitted effective power P is less than the power P_{max}, which can be transmitted in a matched system with the same maximum voltage, by a value equal to the VSWR.

In practice, the voltage standing wave ratio is of high importance as it can be measured directly by means of a voltage or E field probe moved along the line; the wave length can also be determined. The magnitude of the reflection coefficient can be calculated from the measured standing wave ratio using (24.42):

$$|r| = \frac{s-1}{s+1} \tag{24.43}$$

However, this method does not allow the phase to be determined.

24.3.3
Wave Source

A signal source with an internal resistance is called a *wave source*. A wave source emits an independent wave while the passive, two-terminal networks discussed so far only reflect incident waves. Figure 24.36 shows a wave source with the related parameters.

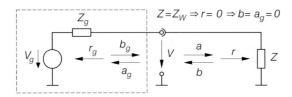

$$Z = Z_W \Rightarrow r = 0 \Rightarrow b = a_g = 0$$

Fig. 24.36. Wave source

Independent Wave from a Wave Source

The wave b_g emanating from a source comprises a portion $b_{g,0}$ generated by the source and a reflected portion $r_g a_g$:

$$b_g = b_{g,0} + r_g a_g \qquad \text{with } r_g = \frac{Z_g - Z_W}{Z_g + Z_W} \tag{24.44}$$

The portion generated by the source is called the *independent wave* as it is not dependent on the incident wave a_g. When loaded with a characteristic impedance $Z = Z_W$ the values are $r = 0$ and $b = a_g = 0$ (see Fig. 24.36); in this case we have $a = b_g = b_{g,0}$. Thus, it is possible to determine the independent wave $b_{g,0}$ by calculating the voltage V for the situation $Z = Z_W$ and converting it into a wave; with

$$V = \frac{V_g Z}{Z_g + Z} \overset{Z = Z_W}{=} \frac{V_g Z_W}{Z_g + Z_W} = \frac{V_g}{2}\left(1 - \frac{Z_g - Z_W}{Z_g + Z_W}\right) = \frac{V_g}{2}\left(1 - r_g\right)$$

and $a = b_g = b_{g,0}$ and $b = a_g = 0$ it follows from (24.30):

$$b_{g,0} = \frac{V}{\sqrt{Z_W}} = \frac{V_g}{2\sqrt{Z_W}}\left(1 - r_g\right) \tag{24.45}$$

For a matched wave source with $Z_g = Z_W$ and $r_g = 0$, we have:

$$b_{g,0} = \frac{V_g}{2\sqrt{Z_W}} \tag{24.46}$$

Available Power

For high-frequency amplifiers the *available power gain* is usually specified. The power at the amplifier output is not related to the power taken from the source but rather to the *available power* $P_{A,g}$. The available power is the maximum active power that can be taken from a source with power matching[7]:

$$P_{A,g} = \frac{|V_g|^2}{4\,\mathrm{Re}\left\{Z_g\right\}} \overset{Z_g = R_g}{=} \frac{|V_g|^2}{4 R_g} \tag{24.47}$$

For calculations with the wave parameters, a notation with $b_{g,0}$ and r_g is required. From (24.45) it follows:

$$|V_g|^2 = \frac{4 Z_W |b_{g,0}|^2}{|1 - r_g|^2}$$

---

[7] We use *root-mean-square phasors;* with peak value phasors we have $P_{A,g} = |\hat{V}_g|^2 / (8\,\mathrm{Re}\left\{Z_g\right\})$.

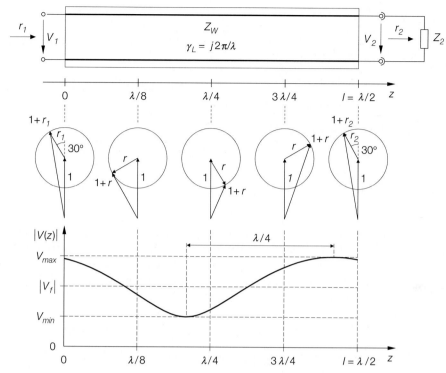

Fig. 24.35. Standing wave on a loss-free line of length $\lambda/2$ for $r_2 = 0.5\,e^{j\,30°}$

If we insert (24.41), the following is obtained:

$$P = \frac{V_{max}^2}{Z_W}\frac{1-|r|^2}{(1+|r|)^2} = \frac{V_{max}^2}{Z_W}\frac{1-|r|}{1+|r|} = \frac{1}{s}\frac{V_{max}^2}{Z_W} = \frac{P_{max}}{s}$$

Consequently, the transmitted effective power P is less than the power P_{max}, which can be transmitted in a matched system with the same maximum voltage, by a value equal to the VSWR.

In practice, the voltage standing wave ratio is of high importance as it can be measured directly by means of a voltage or E field probe moved along the line; the wave length can also be determined. The magnitude of the reflection coefficient can be calculated from the measured standing wave ratio using (24.42):

$$|r| = \frac{s-1}{s+1} \tag{24.43}$$

However, this method does not allow the phase to be determined.

24.3.3
Wave Source

A signal source with an internal resistance is called a *wave source*. A wave source emits an independent wave while the passive, two-terminal networks discussed so far only reflect incident waves. Figure 24.36 shows a wave source with the related parameters.

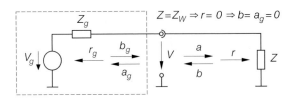

Fig. 24.36. Wave source

Independent Wave from a Wave Source

The wave b_g emanating from a source comprises a portion $b_{g,0}$ generated by the source and a reflected portion $r_g a_g$:

$$b_g = b_{g,0} + r_g a_g \quad \text{with } r_g = \frac{Z_g - Z_W}{Z_g + Z_W} \tag{24.44}$$

The portion generated by the source is called the *independent wave* as it is not dependent on the incident wave a_g. When loaded with a characteristic impedance $Z = Z_W$ the values are $r = 0$ and $b = a_g = 0$ (see Fig. 24.36); in this case we have $a = b_g = b_{g,0}$. Thus, it is possible to determine the independent wave $b_{g,0}$ by calculating the voltage V for the situation $Z = Z_W$ and converting it into a wave; with

$$V = \frac{V_g Z}{Z_g + Z} \overset{Z=Z_W}{=} \frac{V_g Z_W}{Z_g + Z_W} = \frac{V_g}{2}\left(1 - \frac{Z_g - Z_W}{Z_g + Z_W}\right) = \frac{V_g}{2}\left(1 - r_g\right)$$

and $a = b_g = b_{g,0}$ and $b = a_g = 0$ it follows from (24.30):

$$\boxed{b_{g,0} = \frac{V}{\sqrt{Z_W}} = \frac{V_g}{2\sqrt{Z_W}}\left(1 - r_g\right)} \tag{24.45}$$

For a matched wave source with $Z_g = Z_W$ and $r_g = 0$, we have:

$$b_{g,0} = \frac{V_g}{2\sqrt{Z_W}} \tag{24.46}$$

Available Power

For high-frequency amplifiers the *available power gain* is usually specified. The power at the amplifier output is not related to the power taken from the source but rather to the *available power* $P_{A,g}$. The available power is the maximum active power that can be taken from a source with power matching[7]:

$$P_{A,g} = \frac{|V_g|^2}{4\,\mathrm{Re}\left\{Z_g\right\}} \overset{Z_g=R_g}{=} \frac{|V_g|^2}{4R_g} \tag{24.47}$$

For calculations with the wave parameters, a notation with $b_{g,0}$ and r_g is required. From (24.45) it follows:

$$|V_g|^2 = \frac{4\,Z_W\,|b_{g,0}|^2}{|1 - r_g|^2}$$

---

[7] We use *root-mean-square phasors;* with peak value phasors we have $P_{A,g} = |\hat{V}_g|^2/\left(8\,\mathrm{Re}\left\{Z_g\right\}\right)$.

from Fig. 2.41. The results for a FET are almost the same because the small-signal models differ very slightly (see Fig. 3.49). For practical purposes the S parameters are always quoted for $Z_W = 50\,\Omega$.

The low-frequency values for parameters S_{11} and S_{22} can easily be determined since at low frequencies the transistor has no reverse transmission. These parameters correspond to the reflection coefficient r_1 at the input and r_2 at the output and can be calculated directly from the input resistance r_i and the output resistance r_o of the transistor at low frequencies:

$$S_{11} \overset{(24.50)}{=} r_1 \overset{(24.34)}{=} \frac{r_i - Z_W}{r_i + Z_W}$$

$$S_{22} \overset{(24.51)}{=} r_2 \overset{(24.34)}{=} \frac{r_o - Z_W}{r_o + Z_W}$$

From Fig. 24.41, we obtain for low frequencies $r_i = R_B + r_{BE}$ and $r_o = r_{CE}$; consequently:

$$S_{11} = \frac{R_B + r_{BE} - Z_W}{R_B + r_{BE} + Z_W} \approx 1 - \frac{2Z_W}{r_{BE}} \tag{24.58}$$

$$S_{22} = \frac{r_{CE} - Z_W}{r_{CE} + Z_W} \approx 1 - \frac{2Z_W}{r_{CE}} \tag{24.59}$$

For the approximations, it is assumed that $R_B < Z_W \ll r_{BE}, r_{CE}$. To determine S_{21} we first calculate the overall gain with $R_g = R_L = Z_W$:

$$A_B = -\frac{r_{BE}}{Z_W + R_B + r_{BE}}\, g_m\,(Z_W \| r_{CE}) \approx -\frac{g_m r_{BE} Z_W}{Z_W + r_{BE}} = -\frac{\beta Z_W}{Z_W + r_{BE}}$$

Again, we make the assumption that $R_B < Z_W \ll r_{BE}, r_{CE}$. From (24.54) it follows:

$$S_{21} = 2A_B \approx -\frac{2\beta Z_W}{Z_W + r_{BE}} \tag{24.60}$$

Since there is no reverse transmission, $S_{12} = 0$. This allows us to locate the low-frequency S parameters in the r plane: S_{11} and S_{22} are located close to the open-circuit point $r = 1$, while S_{12} is at the origin $r = 0$ and S_{21} on the negative real axis outside the unit circle.

Loci: The frequency response of the S parameters is presented by means of loci in the r plane. Figure 24.42 shows these for a bipolar transistor without case using the small-signal model of Fig. 24.41a and for a transistor with case using the small-signal model of

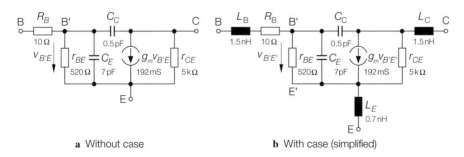

a Without case   **b** With case (simplified)

Fig. 24.41. Small-signal model of a bipolar transistor

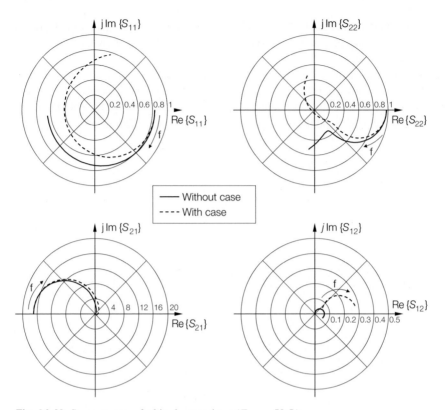

Fig. 24.42. S parameters of a bipolar transistor ($Z_W = 50\,\Omega$)

Fig. 24.41b which uses a simplified case model with three lead inductances. The small-signal parameters of the transistor are determined with the help of Fig. 2.45 on page 83 for $I_C = 5\,\text{mA}$, $\beta = 100$, $V_A = 25\,\text{V}$, $f_C = 4\,\text{GHz}$ and $C_C = 0.5\,\text{pF}$. These parameters are typical of high-frequency discrete transistors of the BFR series. By inserting

$$g_m = 192\,\text{mS} \quad , \quad r_{BE} = 520\,\Omega \quad , \quad r_{CE} = 5\,\text{k}\Omega$$

and $Z_W = 50\,\Omega$ into (24.58)–(24.60), we obtain the low-frequency values of the S parameters:

$$S_{11} = 0.83 \quad , \quad S_{12} = 0 \quad , \quad S_{21} = -16.9 \quad , \quad S_{22} = 0.98$$

With these values the loci in Fig. 24.42 begin with $f = 0$ and climb to $f = 6\,\text{GHz}$. Without case, S_{11} and S_{22} are in the ohmic-capacitive region (Im $\{r\} < 0$). Due to the lead inductances, with case there is a series resonance at both the input and output which causes the impedances to become ohmic (Im $\{r\} = 0$). Above the resonant frequencies, S_{11} and S_{22} are inductive (Im $\{r\} > 0$). The case has very little effect on S21; with increasing frequency the magnitude declines while the phase is shifted by approximately 180°. Without case, the magnitude of S_{12} remains below 0.07 even at very high frequencies, i.e. the reverse transmission remains relatively low. With case, the reverse transmission increases markedly due to the lead inductances. As the reverse transmission is a measure of stability (reverse transmission → feedback → oscillator), high-frequency discrete transistors with relatively long leads are particularly susceptible to parasitic oscillations which is the reason why, in

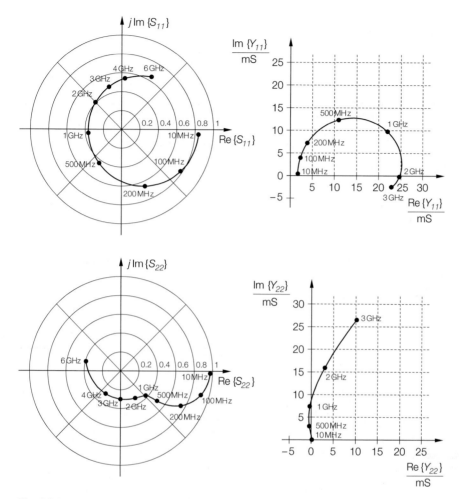

Fig. 24.43. Parameters of the high-frequency transistor BRF93 (part 1)

the GHz range, SMD cases with low lead inductances are mandatory. In integrated circuits, this problem occurs only with circuit parts connected to external leads; inside integrated circuits, the lead inductances are usually negligibly low.

Example: Figures 24.43 and 24.44 show the S and Y parameters of the high-frequency discrete transistor BRF93 for $I_C = 5\,mA$ and $V_{CE} = 5\,V$. The loci of the S parameters are determined with $Z_W = 50\,\Omega$ and, with the exception of S_{12}, correspond well to the general curves in Fig. 24.42. The deviation of S_{12} is a consequence of the simplified case design in Fig. 24.41b.

According to Fig. 24.43, the series resonance at the input occurs at 1 GHz and at the output at 5.5 GHz. Comparing the locus of S_{11} to that of Y_{11} shows the influence of the reverse transmission: for S_{11}, which is measured with a termination with Z_W at the output, the series resonance occurs at 1 GHz, whereas for Y_{11}, which is measured with an output short circuit, it occurs at 2 GHz (Im $\{Y_{11}\} = 0$). In this case, the operating conditions,

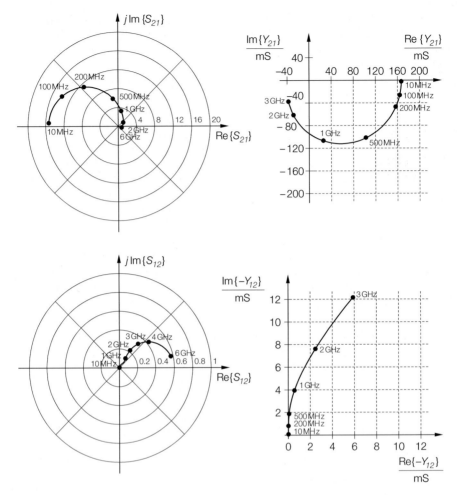

Fig. 24.44. Parameters of the high-frequency transistor BFR93 (part 2)

*termination with Z_W and *short circuit*, must be seen in terms of *small-signal conditions*, i.e. the output is connected to resistance Z_W or to ground via a sufficiently high capacitance.

The locus of Y_{22} has a negative real portion between 230 MHz and 1.09 GHz; in this region the transistor is potentially unstable. If a load Y_L with Re $\{Y_{22} + Y_L\} < 0$ and Im $\{Y_{22} + Y_L\} = 0$ is connected to the output[9], a parasitic oscillation occurs; here, this takes place with inductances between 16 nH ($Y_L = 1/(j\omega L) = -j\,9$ mS at $f = 1.09$ GHz) and 550 nH ($Y_L = 1/(j\omega L) = -j\,1.25$ mS at $f = 230$ MHz). If the input is terminated with $Z_W = 50\,\Omega$, no instability occurs since the locus of S_{22} remains entirely within the unit circle of the r plane. Consequently, the output impedance is Re$\{Z_o\} > 0$

---

[9] A transistor is unstable whenever the input and output admittance of the transistor, in combination with the admittance of the external circuitry, form a negative resistance; then Re$\{Y\} < 0$ and Im$\{Y\} = 0$ must apply. The same applies to impedances; here Re$\{Z\} < 0$ and Im$\{Z\} = 0$ must apply. These conditions result from the known oscillating conditions for the loop gain and the phase of an oscillator; here the condition for the real part corresponds to the condition for the loop gain, and the condition for the imaginary part corresponds to the condition for the phase.

and the output admittance is $\text{Re}\{Y_o\} > 0$. This performance is typical of high-frequency transistors. Thus, the S parameters can be measured without stability problems while the measurement of Y, Z or H parameters causes parasitic oscillations that render an accurate measurement impossible.

24.4
Modulation Methods

The useful signal to be transferred generally has to be converted into a signal suitable for transmission; this process is called *modulation* and the methods used are called *modulation methods*. One distinguishes between a *transmission in the baseband*, which transmits the useful signal in its original frequency range, and *carrier-frequency transmission*, which transposes the useful signal to a higher transmit frequency. Transmission in the *baseband* is typical in line transmission systems such as telephones. In its most simple form, the useful signal can be transmitted without conversion, i.e. without modulation, as is the case in analog telephone systems. However, there are also line transmission systems that use carrier-frequency transmission, for example, radio and TV signals transmitted via the broadband network. Transmission via optical waveguides is a line transmission technique considered suitable for both methods. Wireless systems, on the other hand, must use carrier-frequency transmission since the size of the required antennas is inversely proportional to the transmit frequency and the direct transmission of low-frequency signals would require extremely large antennas. Furthermore, only *one* transmission channel is available for radio transmission so that the various systems have no choice but to use different frequency ranges.

The following discussion is restricted to *carrier-frequency transmission*. The parameters *amplitude, frequency* and *phase* of a high-frequency *carrier signal*

$$s_C(t) = a_C \cos \omega_C t \qquad (24.61)$$

vary due to the useful signal $s(t)$. In *analog modulation methods* this variation is achieved directly by the useful signal:

- *Amplitude modulation* (AM): $\quad s_C(t) = [a_C + k_{AM}s(t)] \cos \omega_C t$
- *Frequency modulation* (FM): $\quad s_C(t) = a_C \cos \left[\omega_C t + k_{FM} \int_0^t s(\tau)d\tau \right]$
- *Phase modulation* (PM): $\quad s_C(t) = a_C \cos \left[\omega_C t + k_{PM}s(t) \right]$

The parameters k_{AM}, k_{FM} and k_{PM} represent the modulation depth. Figure 24.45 shows the modulated carrier signals for these methods. Since frequency is the derivative of the phase by time ($\omega = d\phi/dt$), these parameters are interdependent; frequency and phase modulation are thus combined under the term *angle modulation*. Amplitude and frequency modulation are the conventional methods for radio broadcasting where long and medium wave radio uses the AM method while VHF broadcasting uses FM. These modulation methods will be described in more detail in the following sections.

The binary transmission of digital signals is best achieved by using a two-level square-wave signal for $s(t)$, e.g. $s(t) = 0$ for zero and $s(t) = 1$ for one. A switch-over takes place between two amplitudes, two frequencies or two phases. This method is known as keying and the modulation methods are *amplitude shift keying* (ASK), *frequency shift keying* (FSK) and *phase shift keying* (PSK); Fig. 24.46 shows the modulated carrier signals.

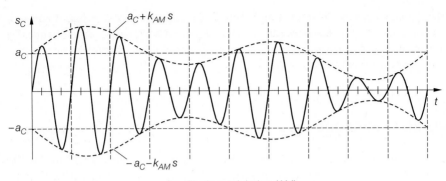

a Amplitude modulation (AM)

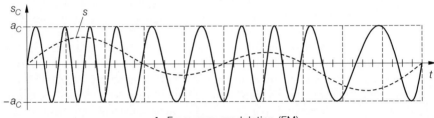

b Frequency modulation (FM)

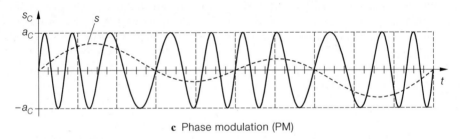

c Phase modulation (PM)

Fig. 24.45. Analog modulation methods

It is also possible to use more than two levels. In this case, the corresponding methods are called n-ASK, n-FSK and n-PSK, where n indicates the number of levels. Therefore, the two-level method is also called 2-ASK, 2-FSK and 2-PSK. Strictly speaking, the keying methods are not independent techniques because what we actually have are the common AM, FM or PM methods utilizing a specific useful signal.

There is also a multitude of other modulation methods which modulate the amplitude *and* the phase. In these methods there is no simple relationship between the useful signal $s(t)$ and the modulated carrier signal $s_C(t)$. The notation

$$s_C(t) = a(t) \cos [\omega_C t + \varphi(t)] \tag{24.62}$$

can be used for the general amplitude modulation $a(t)$ and the general phase modulation $\varphi(t)$. The relationship between a(t), $\varphi(t)$ and the useful signal $s(t)$ characterizes the given method. In most cases another notation is used that is based on a trigonometric conversion of (24.62):

$$s_C(t) = a(t) \cos [\omega_C t + \varphi(t)]$$

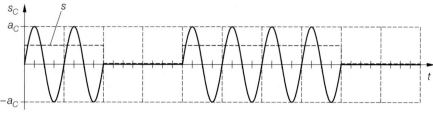

a Amplitude shift keying (ASK)

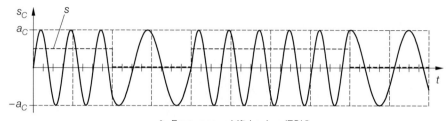

b Frequency shift keying (FSK)

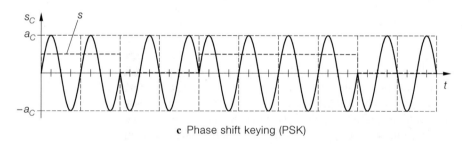

c Phase shift keying (PSK)

Fig. 24.46. Simple digital modulation methods

$$= a(t)\cos\varphi(t)\cos\omega_C t - a(t)\sin\varphi(t)\sin\omega_C t$$

$$= i(t)\cos\omega_C t - q(t)\sin\omega_C t \tag{24.63}$$

The signals

$$i(t) = a(t)\cos\varphi(t) \quad , \quad q(t) = a(t)\sin\varphi(t) \tag{24.64}$$

are known as the *quadrature components*, where $i(t)$ is the *in-phase signal* and $q(t)$ the *quadrature signal*. Thus, it is possible to interpret an amplitude-modulated and phase-modulated signal as the difference between a cosine carrier signal amplitude-modulated with $i(t)$ and a sine carrier signal amplitude-modulated with $q(t)$. The related analog modulation method is called *quadrature amplitude modulation* (QAM). Using the inversion[10]

$$a(t) = \sqrt{i^2(t) + q^2(t)} \quad , \quad \varphi(t) = \arctan\frac{q(t)}{i(t)} + \frac{\pi}{2}(1 - \operatorname{sign} i(t)) \tag{24.65}$$

the notation changes from the quadrature components to the notation for amplitude and phase modulation.

---

[10] The value of the signum function is: $\operatorname{sign} x = 1$ for $x > 0$, $\operatorname{sign} x = 0$ for $x = 0$ and $\operatorname{sign} x = -1$ for $x < 0$.

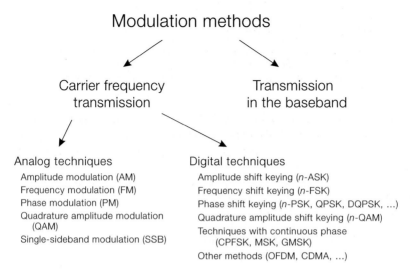

Fig. 24.47. Overview of the most important modulation methods

Figure 24.47 gives an overview of the most important modulation methods. In all modern carrier-frequency methods the modulator first produces the quadrature components $i(t)$ and $q(t)$ from the useful signal, then an *I/Q mixer* generates the modulated carrier signal according to (24.63). Section 24.4.3 will describe this in more detail. Transmission in the *baseband* is performed either directly, i.e. without modulation, or with the use of pulse modulation methods which digitize the message to be transmitted and recode it in a suitable pulse sequence (see [24.5]). This will not be elaborated on.

24.4.1
Amplitude Modulation

In *amplitude modulation* (AM), the useful signal $s(t)$ which is to be transmitted modulates the amplitude of the carrier signal $s_C(t)$ while the phase of the carrier signal remains constant. A distinction is made between *amplitude modulation with carrier* and *amplitude modulation without carrier*:

$$s_C(t) = \begin{cases} [a_C + k_{AM}s(t)]\cos\omega_C t & \text{AM with carrier} \\ k_{AM}s(t)\cos\omega_C t & \text{AM without carrier} \end{cases} \qquad (24.66)$$

Presentation in the Time Domain

For AM with carrier, the sinusoidal modulating signal

$$s(t) = a_s \cos\omega_s t$$

renders a modulated signal

$$s_C(t) = [a_C + k_{AM}a_s \cos\omega_s t]\cos\omega_C t \qquad (24.67)$$

$$= a_C \cos \omega_C t + \frac{k_{AM} a_s}{2} \cos (\omega_C - \omega_s) t + \frac{k_{AM} a_s}{2} \cos (\omega_C + \omega_s) t$$

Unmodulated	Useful signal in	Useful signal in
carrier signal	the lower sideband	the upper sideband
$s_{C,u}(t)$	$s_{LSB}(t)$	$s_{USB}(t)$

The modulated carrier signal consists of the *unmodulated carrier signal*, a useful signal at the frequency $f_C - f_s$ in the *lower sideband* and a useful signal at the frequency $f_C + f_s$ in the *upper sideband*. For AM without carrier the unmodulated carrier signal is missing. Since the useful signal appears twice, once in each sideband, AM is also called *double-sideband modulation*. Figure 24.48 shows the signal components that occur in AM as well as the modulated carrier signals with and without carrier.

The magnitude of the modulated carrier signal is called the *envelope curve* $s_{C,E}$:

$$s_{C,E} = \begin{cases} |a_C + k_{AM} s(t)| & \text{AM with carrier} \\ |k_{AM} s(t)| & \text{AM without carrier} \end{cases}$$

For AM with carrier, the envelope consists of the useful signal and the carrier amplitude providing that the *modulation depth*

$$m = \frac{k_{AM} a_s}{a_C} \tag{24.68}$$

remains below one; then:

$$a_C + k_{AM} s(t) > 0$$

Figure 24.48 shows this for $m = 0.8$. The useful signal can then be regained by a peak value rectification of the modulated carrier signal with subsequent separation of the DC portion. This type of demodulation is known as *envelope detection*. Due to the existence of this simple demodulation method, the AM broadcasting system uses exclusively AM with carrier.

Presentation in the Frequency Range

The frequency-related presentation of AM with carrier has already been described for sinusoidal modulation by (24.67); with (24.68) it follows:

$$s_C(t) = a_C \cos \omega_C t + \frac{m a_C}{2} \cos (\omega_C - \omega_s) t + \frac{m a_C}{2} \cos (\omega_C + \omega_s) t$$

Figure 24.49a shows the spectra of the useful signal for the unmodulated and the modulated carrier. Since AM is a *linear modulation method*, the sidebands for any combination of useful signals can be formed by superimposing the sidebands of the individual useful signals. The sidebands of a carrier modulated with a general signal therefore correspond to the useful signal band, whereby the upper sideband is in *noninverted mode* and the lower sideband is in *inverted mode*, i.e. with an inverted frequency sequence (see Fig. 24.49b). The bandwidth of the modulated carrier thus corresponds to double the upper cutoff frequency of the useful signal:

$$B_{AM} = 2 f_g \tag{24.69}$$

For AM without carrier, the carrier portion is missing in the modulated signal.

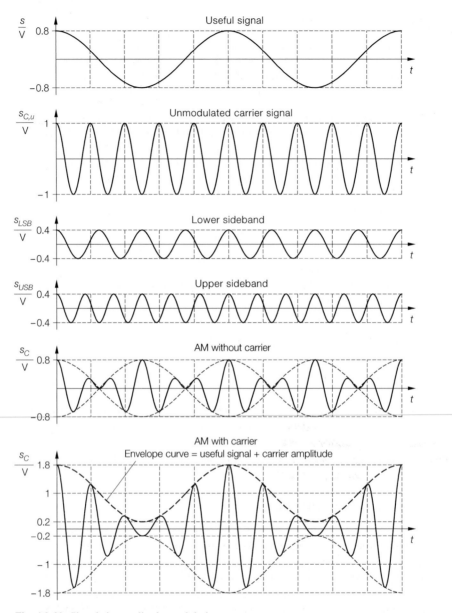

Fig. 24.48. Signals in amplitude modulation

Modulation

To produce an amplitude-modulated signal according to (24.66), it is necessary to use a multiplier and a sinusoidal carrier signal $\cos \omega_C t$. Figure 24.50 shows this for AM with carrier.

Instead of the sinusoidal carrier signal $\cos \omega_C t$, a square wave signal with amplitude levels 0 and 1 and the period length $T_C = 1/f_C$ can be used. In this case, the multiplication uses only the factors 0 and 1 so that the multiplier can be replaced by a switch. From the

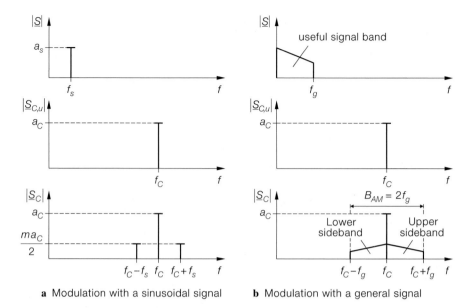

a Modulation with a sinusoidal signal   **b** Modulation with a general signal

Fig. 24.49. Amplitude modulation with carrier in the frequency range. In AM without carrier, the carrier is missing in the modulated signal.

Fourier series of the square wave signal

$$s_{C,u}(t) = \begin{cases} 1 & \text{for } nT_C \le t < (n+1/2)T_C \\ 0 & \text{for } (n+1/2)T_C \le t < (n+1)T_C \end{cases} \quad n \text{ integer}$$

$$= \frac{1}{2} + \frac{2}{\pi} \cos \omega_C t - \frac{2}{3\pi} \cos 3\omega_C t + \frac{2}{5\pi} \cos 5\omega_C t + \cdots$$

$$= \frac{1}{2} + \frac{2}{\pi} \sum_{n=0}^{\infty} \frac{(-1)^n}{2n+1} \cos (2n+1)\,\omega_C t$$

one can see that besides the desired carrier of the frequency f_C, other carrier components also occur at uneven multiples of f_C, as does a direct component. Each of these components is modulated by the useful signal and has the corresponding sidebands. The desired carrier with its sidebands is extracted from this mixture by means of a bandpass filter. Figure 24.51 shows the amplitude modulator with switch along with the time- and frequency-related presentations of the signals. If the square wave signal is not symmetrical (pulse duty ratio $\ne 50\%$), additional carrier components occur at all even multiples of f_C; at the same time the amplitude of the desired carrier decreases. The electronic switches described in Sect. 17.2 and the mixers described in Chap. 28 can be used as the switch.

   Figure 24.52 shows the sample circuit of a MOSFET used as a short circuit switch and a two-circuit bandpass filter for extracting the desired carrier and the related sidebands. Voltage V_S corresponds to the signal $a_C + k_{AM} s(t)$ in Fig. 24.51; it must be larger than zero in order to obtain an AM with carrier. An amplifier is required for decoupling the switch and the filter; Sect. 28.2 describes the dimensioning process for a two-circuit bandpass filter.

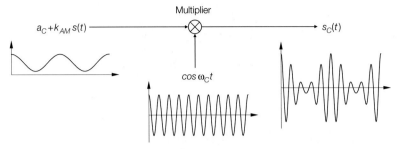

Fig. 24.50. Amplitude modulator with multiplier

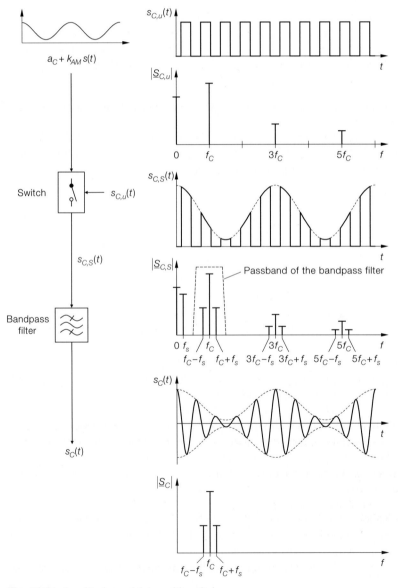

Fig. 24.51. Amplitude modulator with switch

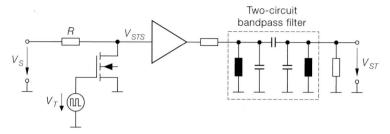

Fig. 24.52. Example of an amplitude modulator with short circuit switch

Demodulation

Envelope detector: The *envelope detector* shown in Fig. 24.53 can be used to de-
modulate an AM with carrier; this consists of a peak-type rectifier with a lossy storage
circuit (R_S, C_S) and a highpass filter (C_c, R_L) to cancel the DC component. The following
conditions must be met for accurate demodulation:

- The carrier frequency must be significantly higher than the maximum frequency of the
  useful signal.
- The minimum of the envelope must be higher than the forward voltage of the diode.
- The time constant $T_S = C_S(R_S \| R_L)$ of the storage circuit must be selected such that
  the rectified voltage can follow the envelope curve[11].

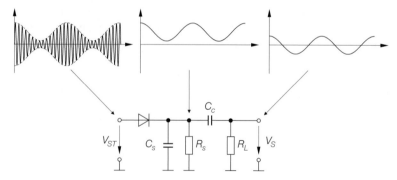

Fig. 24.53. Envelope detector

- The useful signal must be a pure alternating voltage signal as the highpass filter suppresses both the DC component of the useful signal and the DC component caused by
  the carrier.
- The cutoff frequency of the highpass filter must be lower than the minimum frequency
  of the useful signal.

The prime advantage of the envelope detector is its simple design. A drawback is the
nonlinearity due to the nonlinear characteristic of the diode, especially with smaller carrier

---

[11] The capacitance C_c can be regarded as a short circuit in the carrier frequency range; $R_S \| R_L$
thus becomes effective.

amplitudes; this produces a lower modulation limit. The envelope detector is used in simple AM radio receivers.

Synchronous demodulator: Higher quality demodulation is achieved by *synchronous demodulation*, although this requires much more elaborate circuitry. In this demodulation method, the modulated carrier signal is multiplied by an unmodulated carrier signal of the same frequency and the same phase. For sinusoidal modulation of the carrier signal this results in:

$$s_M(t) = s_C(t) \cos \omega_C t \qquad (24.70)$$

$$= [a_C + k_{AM} a_s \cos \omega_s t] \cos \omega_C t \cos \omega_C t$$

$$= [a_C + k_{AM} a_s \cos \omega_s t] \frac{1 + \cos 2\omega_C t}{2}$$

$$= \frac{a_C}{2} + \frac{k_{AM} a_s}{2} \cos \omega_s t + \frac{a_C}{2} \cos 2\omega_C t$$

$$+ \frac{k_{AM} a_s}{4} \cos(2\omega_C - \omega_s)t + \frac{k_{AM} a_s}{4} \cos(2\omega_C + \omega_s)t$$

Besides the required component

$$a_C + k_{AM} a_s \cos \omega_s t$$

the signal product $s_M(t)$ also contains additional components of 1/2 weighting in the double carrier frequency range; the latter are suppressed by a lowpass filter. Figure 24.54 shows the *synchronous demodulator* including the time- and frequency-related representation of the signals. The modulated carrier signal may also be multiplied by a square-wave signal with a period length of $T_C = 1/f_C$; in this case the multiplier can be replaced by a switch. The resulting additional components in the signal product $s_M(t)$ are also suppressed by the lowpass filter.

The synchronous demodulator with multiplier or switch largely corresponds to the amplitude modulator with multiplier or switch; they differ only with regard to the necessary filters. The use of a switch in the modulator requires an additional bandpass filter in order to suppress unwanted signal components. On the other hand, the lowpass filter in the demodulator is always required, regardless of whether a multiplier or a switch is used. Therefore, in practice, the synchronous demodulator is generally provided with an electronic switch as described in Sect. 17.2 or with a mixer as described in Chap. 28.

For AM with carrier, the sine- or square-wave carrier signal, which is required for demodulation in the synchronous demodulator and has the same frequency and phase as the carrier signal in the modulator, can be extracted from the carrier component contained in the modulated signal by means of a *phase-locked loop* (PLL) (see Fig. 24.55); this accounts for much of the complexity in the circuitry required for the synchronous demodulator. In the AM without carrier this is not possible. In this case, the useful signal itself must have a suitable characteristic to allow synchronization in the demodulator.

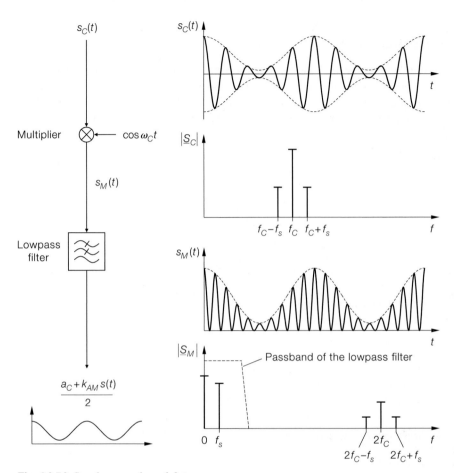

Fig. 24.54. Synchronous demodulator

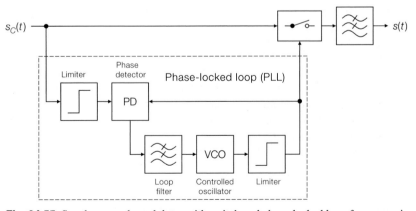

Fig. 24.55. Synchronous demodulator with switch and phase-locked loop for recovering the carrier

24.4.2
Frequency Modulation

In *frequency modulation* (FM), the *instant frequency* or *instant angular frequency*

$$\omega(t) = \frac{d\phi}{dt} \quad \Rightarrow \quad f(t) = \frac{\omega(t)}{2\pi} = \frac{1}{2\pi} \frac{d\phi}{dt}$$

is modulated by the useful signal:

$$\omega(t) = \omega_C + k_{FM} s(t) \tag{24.71}$$

To produce the modulated carrier signal, the instantaneous phase $\phi(t)$ must be generated by integrating the instantaneous angular frequency $\omega(t)$[12]:

$$s_C(t) = a_C \cos\phi(t) = a_C \cos\left[\int_0^t \omega(\tau)\, d\tau\right]$$

By inserting (24.71) and performing integration, we obtain:

$$s_C(t) = a_C \cos\left[\omega_C t + k_{FM}\int_0^t s(\tau)\, d\tau\right] \tag{24.72}$$

This means that the frequency-modulated carrier signal corresponds to a phase-modulated carrier signal

$$s_C(t) = a_C \cos[\omega_C t + \varphi(t)]$$

with the phase:

$$\varphi(t) = k_{FM}\int_0^t s(\tau)\, d\tau$$

Presentation in the Time Domain

For a sinusoidal useful signal

$$s(t) = a_s \cos\omega_s t$$

the instantaneous angular frequency is:

$$\omega(t) = \omega_C + k_{FM} a_s \cos\omega_s t$$

It varies sinusoidally in the range $\omega_C \pm k_{FM} a_s$. The maximum deviation from the carrier frequency is known as the *frequency deviation*:

$$\Delta\omega = k_{FM} a_s \quad \Rightarrow \quad \Delta f = \frac{\Delta\omega}{2\pi} = \frac{k_{FM} a_s}{2\pi} \tag{24.73}$$

For the modulated carrier signal we obtain

$$s_C(t) = a_C \cos\left[\omega_C t + k_{FM} a_s \int_0^t \cos\omega_s \tau\, d\tau\right]$$

$$= a_C \cos\left[\omega_C t + \frac{k_{FM} a_s}{\omega_s} \sin\omega_s t\right] \tag{24.74}$$

---

[12] For the general case, $-\infty$ must be used as the lower limit for the integrals since the phase at the time t depends on the entire preceding curve of the signal s. Here, only the region $t \geq 0$ is taken into consideration under the assumption that $\int_{-\infty}^0 s(\tau)d\tau = 0$; the lower limit can then be set to zero.

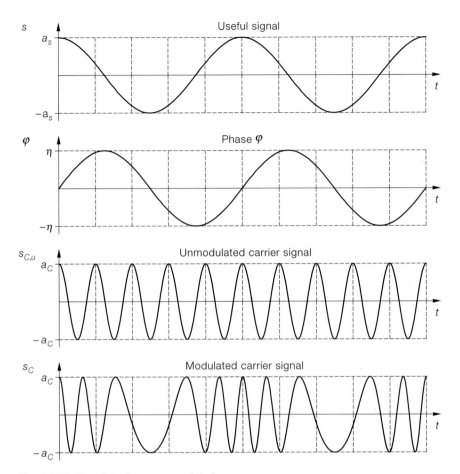

Fig. 24.56. Signals in frequency modulation

with the phase:

$$\varphi(t) = \frac{k_{FM} a_s}{\omega_s} \sin \omega_s t = \eta \sin \omega_s t$$

The phase deviation

$$\eta = \frac{k_{FM} a_s}{\omega_s} \overset{(24.73)}{=} \frac{\Delta\omega}{\omega_s} = \frac{\Delta f}{f_s} \qquad (24.75)$$

is called the *modulation index* and corresponds to the ratio of the frequency deviation Δf and the useful signal frequency f_s. Figure 24.56 shows the signals occurring during frequency modulation.

Presentation in the Frequency Domain

The frequency-related presentation of FM for sinusoidal modulation follows from the series expansion of the modulated carrier:

$$s_C(t) = a_C \cos [\omega_C t + \eta \sin \omega_s t]$$

$$= a_C\, J_0(\eta) \cos \omega_C t$$

$$- a_C\, J_1(\eta) \cos(\omega_C - \omega_s)t + a_C\, J_1(\eta) \cos(\omega_C + \omega_s)t$$

$$+ a_C\, J_2(\eta) \cos(\omega_C - 2\omega_s)t + a_C\, J_2(\eta) \cos(\omega_C + 2\omega_s)t$$

$$- a_C\, J_3(\eta) \cos(\omega_C - 3\omega_s)t + a_C\, J_3(\eta) \cos(\omega_C + 3\omega_s)t$$

$$+ a_C\, J_4(\eta) \cos(\omega_C - 4\omega_s)t + a_C\, J_4(\eta) \cos(\omega_C + 4\omega_s)t$$

$$- \cdots$$

$$= a_C\, J_0(\eta) \cos \omega_C t \qquad\qquad\qquad \text{Carrier} \qquad\qquad (24.76)$$

$$+ a_C \sum_{n=1}^{\infty} (-1)^n J_n(\eta) \cos(\omega_C - n\omega_s)t \qquad \text{Lower sideband}$$

$$+ a_C \sum_{n=1}^{\infty} J_n(\eta) \cos(\omega_C + n\omega_s)t \qquad\qquad \text{Upper sideband}$$

J_n are the *Bessel functions of first order* shown in Fig. 24.57 while η is the modulation index according to (24.75). The spectrum thus consists of an infinite number of components located on both sides of the carrier with a spacing according to the frequency of the useful signal; they form a lower and an upper sideband. Since the magnitude of the Bessel functions rapidly declines with a constant argument η and higher orders n, for practical purposes, the two series in (24.76) can be interrupted after a finite number of elements. For a better understanding, Fig. 24.58 shows the magnitude of the Bessel functions in decibel and the spectra in decibel for three values of η. It can be seen that the spectrum widens with an increase in η. Since the Bessel functions contain zeros, some individual components may be zero, for example, the carrier component at $\eta = 2.4$ is missing since $J_0(2.4) = 0$.

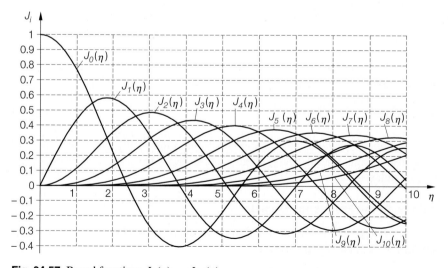

Fig. 24.57. Bessel functions $J_0(\eta) \ldots J_{10}(\eta)$

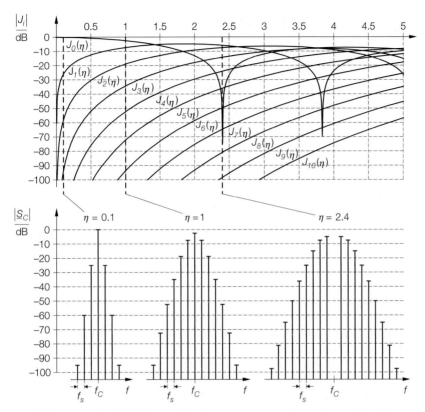

Fig. 24.58. Magnitude of the Bessel functions $J_0(\eta)\ldots J_{10}(\eta)$ in decibel and spectra of the modulated carrier signal for $\eta = 0.1/1/2.4$

The bandwidth of a frequency-modulated carrier signal cannot be specified accurately. A detailed study shows that 99% of the transmit power is contained in the carrier and in the $(\eta + 1)$ components below and above; therefore, the *Carson bandwidth*

$$B_{FM} = 2(\eta + 1) f_s \tag{24.77}$$

is given as the bandwidth of a frequency-modulated carrier which is modulated with a sinusoidal signal with the frequency f_s. Insertion of η from (24.75) results in:

$$B_{FM} = 2(\Delta f + f_s) \tag{24.78}$$

The bandwidth reaches its peak at the maximum useful signal frequency $f_{s,max}$. The depth of frequency modulation is represented by the *minimum modulation index* η_{min}, which is reached at $f_s = f_{s,max}$ and corresponds to the ratio of the frequency deviation and the maximum useful signal frequency:

$$\eta_{min} = \frac{\Delta f}{f_{s,max}}$$

Thus, the following is true:

$$B_{FM} = 2(\eta_{min} + 1) f_{s,max}$$

FM radio uses $\Delta f = 75\,\text{kHz}$ and $f_{s,max} = 15\,\text{kHz}$; consequently, $\eta_{min} = 5$ and $B_{FM} = 180\,\text{kHz}$.

FM is a *nonlinear modulation method*. For this reason, the spectrum of the carrier signal modulated by a general signal cannot be calculated by summing the spectra of the individual components. Only in exceptional cases does a general signal have a spectrum that is symmetric to the carrier. Despite these restrictions, the bandwidth formulas can also be used for the general case; $f_{s,max}$ then represents the upper cutoff frequency of the useful signal.

Modulation

A *voltage-controlled oscillator* (VCO) controlled by the useful signal $s(t)$ is used as the frequency modulator (see Fig. 24.59a). The constant k_{FM} is then determined by the tuning characteristic of the oscillator:

$$k_{FM} = \frac{d\omega}{ds}$$

Figure 24.59b shows a simple FM modulator based on a Colpitts oscillator with a tuning diode D for modulating the frequency. The slope of tuning depends on the characteristic and coupling of the diode to the resonant circuit; the latter is adjusted by means of capacitance C_c. As the oscillator output signal usually contains strong harmonics, the desired signal must be extracted by means of a bandpass filter.

FM modulators based on high-frequency oscillators are used whenever the carrier frequency should be identical to the transmit frequency. If, however, the modulator signal is generated on a lower intermediate frequency and later converted to the transmit frequency, it is also possible to use low-frequency oscillators such as the emitter-coupled multivibrator from Sect. 6.3.2.

Demodulation

Discriminator: One method of demodulating an FM signal is to convert it to an amplitude-modulated signal with subsequent envelope detection as shown in Fig. 24.60. First, the amplitude of the input signal is held constant and independent of the receiving

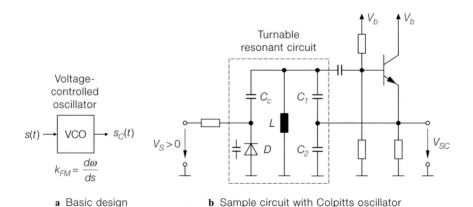

a Basic design   **b** Sample circuit with Colpitts oscillator

Fig. 24.59. Frequency modulator

conditions by means of a limiter and a bandpass filter. At the same time, any amplitude modulation that may interfere with the demodulation process is eliminated (AM suppression).
A series connection of several differential amplifiers with feedback for the DC voltage for
operating point setting is used as a limiter (see Fig. 24.61); the resistances are selected
such that the transistors are not driven into the saturation region.

A *(frequency) discriminator* with frequency-dependent gain is used for the conversion
of FM to AM. Since FM frequency deviation is generally much smaller than the carrier
frequency, the relative frequency deviation is very small. Therefore, the frequency dependence of the gain must be very high in the region of the carrier frequency in order to
obtain a sufficiently high sensitivity. For the *slope detector* (discriminator), a circuit with
resonant frequency slightly above the carrier frequency is used so that the FM-modulated
carrier signal undergoes a frequency-dependent amplification at the slope of the resonance
curve. Figure 24.62 shows the slope detector together with the following envelope detector.

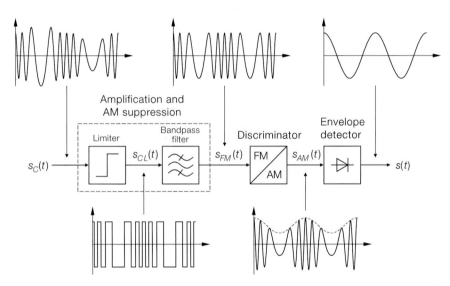

Fig. 24.60. Frequency demodulator with discriminator

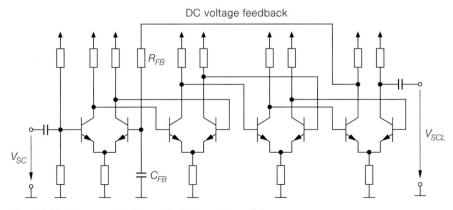

Fig. 24.61. Four-stage limiter with differential amplifiers

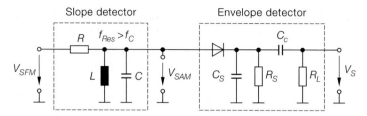

Fig. 24.62. Slope detector with envelope detector

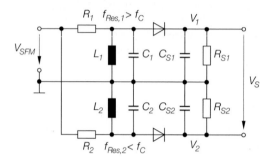

Fig. 24.63. Differential discriminator

Since the slope of the resonance curve is not constant, this simple circuit does not produce sufficiently linear characteristics, such that even with low modulations the distortion factor increases rapidly. For this reason, practical applications always use the *differential discriminator* shown in Fig. 24.63 which evaluates the difference between two shifted resonance curves and produces a region with a linear characteristic (see Fig. 24.64). With a frequency deviation of Δf, the linear portion of the characteristic must be $2\Delta f$ wide and the following must therefore be true:

$$\Delta f_{Res} = f_{Res,1} - f_{Res,2} \approx 5\Delta f$$

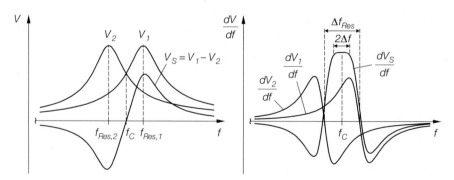

a Transfer characteristics   **b** Slope of the transfer characteristics

Fig. 24.64. Characteristic of the differential discriminator

The carrier frequency approximately corresponds to the mean value of both resonance frequencies:

$$f_C = \sqrt{f_{Res,1} f_{Res,2}} \underset{\Delta f_{Res} \ll f_{Res,1}, f_{Res,2}}{\approx} \frac{f_{Res,1} + f_{Res,2}}{2}$$

This determines the selection of the resonant frequencies:

$$f_{Res,1} = f_C + \frac{5\Delta f}{2} \quad , \quad f_{Res,2} = f_C - \frac{5\Delta f}{2}$$

The bandwidth B of the two resonant circuits must be $4\Delta f$; the qualities are thus:

$$Q_1 = \frac{f_{Res,1}}{B} \approx \frac{f_C}{4\Delta f} + 0.6 \quad , \quad Q_2 = \frac{f_{Res,1}}{B} \approx \frac{f_C}{4\Delta f} - 0.6$$

The following allows the resistances to be determined:

$$R_1 = Q_1 \sqrt{\frac{L_1}{C_1}} \quad , \quad R_2 = Q_2 \sqrt{\frac{L_2}{C_2}}$$

In practice, slightly higher resistors must be selected since the envelope detectors place an additional load on the circuits; for $C_{S1}, C_{S2} \leq C_1, C_2$ and $R_{S1}, R_{S2} \gg R_1, R_2$ this load is small. The time constant of the envelope detectors must be selected such that the detectors can follow the maximum signal frequency.

Example: In FM radio with $\Delta f = 75\,\text{kHz}$, demodulation is accomplished at the intermediate frequency $f_C = 10.7\,\text{MHz}$ where it follows that $f_{Res,1} = 10.89\,\text{MHz}$ and $f_{Res,2} = 10.51\,\text{MHz}$. The default value $C_1 = C_2 = 1\,\text{nF}$ results in $L_1 = 214\,\text{nH}$ and $L_2 = 229\,\text{nH}$. With $Q_1 = 36.2$ and $Q_2 = 35.1$, we arrive at $R_1 = 530\,\Omega$ and $R_2 = 531\,\Omega$. On the basis of these values, a fine adjustment is made that also compensates the influence of the envelope detectors, for the latter one may select $C_{S1} = C_{S2} = 1\,\text{nF}$ and $R_{S1} = R_{S2} = 10\,\text{k}\Omega$ in order to meet the conditions specified above.

PLL demodulator: The *PLL demodulator* shown in Fig. 24.65 is of high quality and very easy to integrate; it is used to make the frequency of a voltage-controlled oscillator (VCO) follow the instantaneous frequency of the modulated carrier by means of a phase-locked loop (PLL). If the VCO has a linear characteristic and the bandwidth of the loop filter is larger than the maximum frequency of the useful signal, then the output signal of the loop filter is proportional to the useful signal. In practice, the PLL demodulator usually operates on an intermediate frequency which is significantly lower than the receiving frequency; this allows the use of a VCO with a square-wave output signal and renders the subsequent limiter unnecessary.

24.4.3
Digital Modulation Methods

Digital modulation methods are used for the transmission of binary data. There are two types of digital modulation: the simple keying methods derived from the analog methods and the more sophisticated methods; they differ in both the transmission rate and the susceptibility to errors as well as in the circuit designs used.

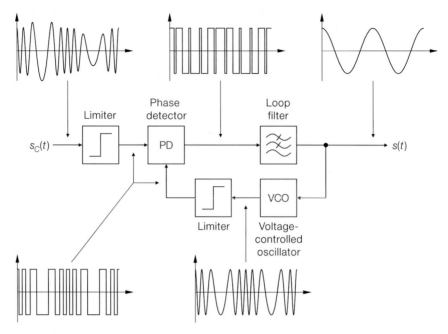

Fig. 24.65. Frequency modulation with a phase-locked loop (PLL demodulator)

Simple Keying Methods

The simple keying methods comprise the *amplitude shift keying* (ASK) and the *frequency shift keying* (FSK). They are based on the analog amplitude and frequency modulation methods and use a binary signal instead of a general useful signal. The related signals have already been shown in Fig. 24.46.

Amplitude Shift Keying (2-ASK):   Amplitude shift keying uses a switch for the modulator in order to turn the carrier signal on and off. An envelope detector with a subsequent comparator is used as the demodulator where the signal level below the switching threshold of the comparator is regarded as a binary zero and above the threshold as a binary one. As the amplitude of the received carrier signal can vary considerably, one must either use a controlled amplifier to enhance the signal to a defined level or perform a suitable adaptation of the comparator switching threshold. Adaptation of the switching threshold can be done by a second envelope detector with a considerably larger time constant which determines the amplitude $V_{s,max}$ of a binary one and maintains it in accordance with its time constant. The comparator switching threshold is then adapted to half the carrier amplitude (see Fig. 24.66).

Amplitude shift keying is only used in very simple systems which have a maximum transmission rate of up to 1.2 kBit/s. The main advantage is the simple circuit design. Amplitude shift keying with several levels (n-ASK with $n > 2$) that allows a higher transmission rate, is not used in practical applications; there are other more suitable methods, e.g. frequency shift keying.

Frequency Shift Keying (2-FSK):   Frequency shift keying uses the same components as analog frequency modulation. The FM modulator is switched between the two frequencies

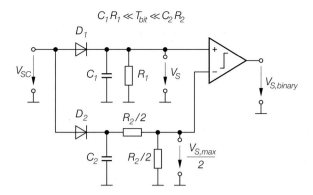

$$C_1 R_1 \ll T_{bit} \ll C_2 R_2$$

Fig. 24.66. Demodulator for amplitude shift keying with automatic adaptation of the switching threshold

f_1 and f_2 by the desired binary signal. The differential detector can be used as the demodulator while the two resonant circuits are set to the frequencies f_1 and f_2, and the output signals of the envelope detectors are compared by a comparator. A linear discriminator characteristic is not required in this case.

Integrated receive circuits for 2-FSK usually use the binary frequency discriminator with an edge triggered D flip-flop shown in Fig. 24.67. The modulated carrier signal

$$s_C(t) = \cos(\omega_C \pm \Delta\omega)t$$

with the frequencies $f_C - \Delta f$ for the binary zero and $f_C + \Delta f$ for the binary one is multiplied by a cosine and a sine trigger signal thus yielding the following components:

$$\cos(\omega_C \pm \Delta\omega)t \cdot \cos\omega_C t = \frac{1}{2}\cos(\pm\Delta\omega)t + \frac{1}{2}\cos(2\omega_C \pm \Delta\omega)t$$

$$\cos(\omega_C \pm \Delta\omega)t \cdot \sin\omega_C t = -\frac{1}{2}\sin(\pm\Delta\omega)t + \frac{1}{2}\sin(2\omega_C \pm \Delta\omega)t$$

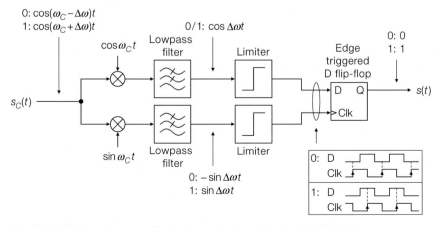

Fig. 24.67. Binary frequency discriminator for demodulating 2-FSK signals

The components at the double carrier frequency are suppressed by means of lowpass filters. If we neglect the prefactors and consider the symmetry of the cosine and sine functions, then the output of the lowpass filters yields:

$$\cos(\pm\Delta\omega)t = \cos\Delta\omega t \quad , \quad -\sin(\pm\Delta\omega)t = \mp\sin\Delta\omega t$$

After conversion to square-wave signals by means of limiters, the binary data is obtained from the time sequence of rising edges; an edge triggered D flip-flop is used for evaluating the binary data. In practice, the multipliers are replaced by two electronic switches that are actuated by two staggered square-wave signals; the resultant harmonics at multiples of the carrier frequency are suppressed by the lowpass filters. The carrier frequency in the receiver may not correspond to the carrier frequency in the transmitter exactly but must be between $f_C - \Delta f$ and $f_C + \Delta f$. In practice, the carrier frequencies in the transmitter and receiver are derived from crystal oscillators of the same nominal frequency; this usually makes the difference between them much smaller than the frequency deviation Δf.

The frequency shift keying 2-FSK is often used in simple systems with data rates of up to several kilobits per second; 4-FSK systems are also used. However, more complex methods are used for higher data rates because they allow higher rates at the same bandwidth of the transmitted signal and are less prone to errors.

I/Q Presentation of Digital Modulation Methods

As a rule, in digital modulation methods both the amplitude and the phase are modulated; this enables higher data rates at the same bandwidth. To represent the modulated carrier signal, the *quadrature components* $i(t)$ and $q(t)$ from (24.63) are used:

$$s_C(t) = a(t)\cos[\omega_C t + \varphi(t)] = i(t)\cos\omega_C t - q(t)\sin\omega_C t$$

Modulation and demodulation: Modulation is done in two steps. In the first step, a *digital modulator* generates the *in-phase signal* $i(t)$ and the *quadrature signal* $q(t)$ from the binary data signal $s(n)$. In the second step an *I/Q mixer* forms the modulated carrier signal $s_C(t)$. Figure 24.68 shows the configuration of the modulator. In practice, the I/Q mixer must be followed by a bandpass filter in order to suppress unwanted components. This particularly applies to mixers used as switches, which is almost always the case in practical solutions.

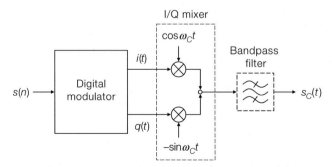

Fig. 24.68. Modulator for digital modulation methods

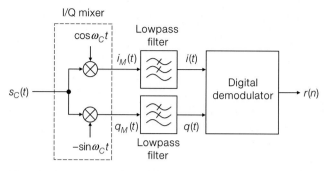

Fig. 24.69. Demodulator for digital modulation methods

Demodulation is also performed in two steps. In the first step, an I/Q mixer forms the signals

$$i_M(t) = s_C(t) \cos \omega_C t = [i(t) \cos \omega_C t - q(t) \sin \omega_C t] \cos \omega_C t$$

$$= \frac{1}{2}[i(t) + i(t) \cos 2\omega_C t - q(t) \sin 2\omega_C t]$$

$$q_M(t) = s_C(t)(-\sin \omega_C t) = [i(t) \cos \omega_C t - q(t) \sin \omega_C t](-\sin \omega_C t)$$

$$= \frac{1}{2}[q(t) - q(t) \cos 2\omega_C t - i(t) \sin 2\omega_C t]$$

After lowpass filtering, we obtain the quadrature components $i(t)$ and $q(t)$. In the second step, a *digital demodulator* determines the binary data signal $r(n)$. Figure 24.69 shows the configuration of the demodulator. Generating the unmodulated carrier signals $\cos \omega_C t$ und $-\sin \omega_C t$ of the correct frequency and phase is complicated. In practice, the carrier frequencies in the transmitter and receiver are derived from crystal oscillators of the same nominal frequency; thus the initial frequency deviation is low. The crystal oscillator in the receiver can be tuned and readjusted by phase-locked codes transmitted at given intervals. In mobile communication systems a special *pilot channel* is often evaluated in addition to the active channel; the pilot channel carries a special pilot signal that allows synchronization.

The carrier frequency f_C often corresponds to the transmit frequency; in this case the modulated signal is only amplified and fed to the transmitting antenna. However, with increasing transmit frequencies, it becomes more and more difficult to produce I/Q mixers with the same properties in the I and Q branch and to provide the unmodulated carrier signals $\cos \omega_C t$ and $-\sin \omega_C t$ with the same amplitude and exactly the same phase shift. Therefore, a low intermediate frequency is used as the carrier frequency. Conversion to the transmit frequency is done by another mixer.

Complex baseband signal: The quadrature components are combined to form a *complex baseband signal*

$$s_B(t) = i(t) + j q(t) \tag{24.79}$$

This signal corresponds to the complex phasors known from AC calculations; the following is true

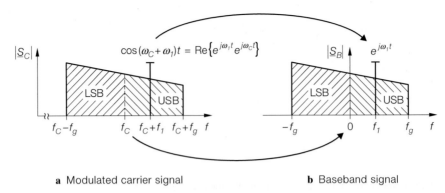

a Modulated carrier signal   **b** Baseband signal

Fig. 24.70. Spectra of the signals (LSB: lower sideband; USB: upper sideband) using the single-tone signal with baseband frequency f_1 as an example

$$v(t) = \hat{v}\cos(\omega t + \varphi) = \mathrm{Re}\left\{\hat{v}\,e^{j\varphi}e^{j\omega t}\right\} = \mathrm{Re}\left\{V\,e^{j\omega t}\right\}$$

$$\Rightarrow \quad V = \hat{v}\,e^{j\varphi}$$

with the complex phasor V. Correspondingly, for the modulated carrier signal, the following applies:

$$s_C(t) = a(t)\cos[\omega_C t + \varphi(t)] = \mathrm{Re}\left\{a(t)\,e^{j\varphi(t)}e^{j\omega_C t}\right\}$$

$$= i(t)\cos\omega_C t - q(t)\sin\omega_C t = \mathrm{Re}\left\{[i(t) + j\,q(t)]\,e^{j\omega_C t}\right\}$$

$$\Rightarrow \quad s_B(t) = a(t)\,e^{j\varphi(t)} = i(t) + j\,q(t)$$

The complex phasor is dependent on time because the amplitude and phase of the real and imaginary components are time-dependent; the result is a complex signal instead of a complex phasor. With

$$s_C(t) = \mathrm{Re}\left\{s_B(t)\,e^{j\omega_C t}\right\} \tag{24.80}$$

the modulated carrier signal is derived from the complex baseband signal. For practical purposes, the term *complex* is usually omitted and it is simply called the *baseband signal*.

In the frequency domain, the transition from the modulated carrier signal to the baseband signal corresponds to a shift in the spectrum by the carrier frequency (see Fig. 24.70). The lower sideband is then mapped on the negative baseband frequencies and the upper sideband to the positive baseband frequencies. The unmodulated carrier has the baseband frequency zero. Since the sidebands are independent of one another, the spectrum is usually asymmetrical.

The main advantages of the baseband signal are the independence of the carrier frequency and the representation of the carrier status by a signal with an amplitude and phase that correspond to the amplitude and phase of the carrier. For sinusoidal high-frequency and intermediate-frequency signals, it is not the absolute frequency that is normally specified, but the frequency deviation from the carrier. This deviation corresponds to the baseband frequency.

Examples: For an amplitude-modulated carrier signal

$$s_C(t) = [a_C + k_{AM}s(t)]\cos\omega_C t$$

the following is true:

$$s_C(t) = \text{Re}\left\{[a_C + k_{AM}s(t)]\,e^{j\omega_C t}\right\} \quad \Rightarrow \quad s_B(t) = a_C + k_{AM}s(t)$$

It follows:

$$i(t) = a_C + k_{AM}s(t) \quad , \quad q(t) = 0$$

The baseband signal is real. For the frequency-modulated carrier signal

$$s_C(t) = a_C\cos\left[\omega_C t + k_{FM}\int_0^t s(\tau)\,d\tau\right]$$

the following is true:

$$s_C(t) = \text{Re}\left\{\left[a_C\,e^{jk_{FM}\int_0^t s(\tau)\,d\tau}\right]e^{j\omega_C t}\right\}$$

$$\Rightarrow \quad s_B(t) = a_C\,e^{jk_{FM}\int_0^t s(\tau)\,d\tau}$$

It follows:

$$i(t) = a_C\cos\left[k_{FM}\int_0^t s(\tau)\,d\tau\right] \quad , \quad q(t) = a_C\sin\left[k_{FM}\int_0^t s(\tau)\,d\tau\right]$$

In this case, the baseband signal is complex.

Bandwidth: The upper cutoff frequency $f_{g,B}$ of the complex baseband signal corresponds to the maximum of the cutoff frequencies of the quadrature components. If $f_{g,i}$ is the upper cutoff frequency of the in-phase signal $i(t)$ and $f_{g,q}$ is the upper cutoff frequency of the quadrature signal $q(t)$, then:

$$f_{g,B} = \max\left\{f_{g,i},\ f_{g,q}\right\}$$

According to (24.69), the two amplitude-modulated signals $i(t)\cos\omega_C t$ and $q(t)\sin\omega_C t$ have a bandwidth corresponding to double the upper cutoff frequency:

$$B_{AM,i} = 2f_{g,i} \quad , \quad B_{AM,q} = 2f_{g,q}$$

This means that the bandwidth of the modulated carrier frequency corresponds to double the maximum of the cutoff frequencies of the quadrature components:

$$B = \max\left\{B_{AM,i},\ B_{AM,q}\right\} = 2f_{g,B} = \max\left\{2f_{g,i},\ 2f_{g,q}\right\} \tag{24.81}$$

For quadrature components, the *double-sided bandwidth* is always given in practice; it corresponds to the *single-sided bandwidth* of amplitude-modulated signals:

$$B_i = 2f_{g,i} = B_{AM,i} \quad , \quad B_q = 2f_{g,q} = B_{AM,q}$$

This eliminates the factor 2, and the (single-sided) bandwidth of the modulated carrier signal, which is identical to the required transmission bandwidth, corresponds to the maximum of the (double-sided) bandwidth of the quadrature components. This is simply referred to as the bandwidth B and is illustrated in Fig. 24.71.

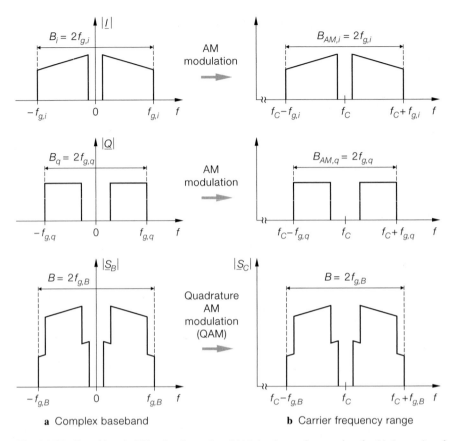

Fig. 24.71. Signal bandwidths: in-phase signal $i(t)$ (top), quadrature signal $q(t)$ (centre) and complex baseband signal $s_B(t)$ (bottom)

Constellation diagrams: For the transmission of a binary data signal $s(n)$, m bit are combined in one symbol (see Fig. 24.72); the *data rate* r_D (clock frequency f_D) is reduced to the *symbol rate* $r_S = r_D/m$ (symbol clock $f_S = f_D/m$). The digital modulator assigns a certain carrier state to each of the 2^m possible symbols and generates the associated quadrature components i and q. Mapping the 2^m carrier states, as described by the respective baseband pointers $s_B = i + j q$, in the IQ plane results in the *constellation diagram* of the modulation method. Figure 24.73 shows the constellation diagrams for 2-PSK ($m = 1$), 4-PSK ($m = 2$) and 8-PSK ($m = 3$) together with the resulting quadrature components for the data signal from Fig. 24.72. The assignment of the symbols to the carrier states is carried out according to the *Gray code* so that adjacent carrier states differ by one bit only. This results in a minimum bit error rate since, in most cases, erroneous symbol detection in the demodulator caused by interference supplies an adjacent symbol and thus generates *one* bit error only.

   The bandwidth of the modulated carrier signal is proportional to the symbol clock and, in practice, amounts to $B \approx (1.3\ldots 2) f_S$. This means that, as compared to 2-PSK for a given bandwidth, double the data rate is achieved for 4-PSK while it is tripled for 8-PSK. The ratio of the data rate to the bandwidth is known as the *bandwidth efficiency* Γ [24.6]:

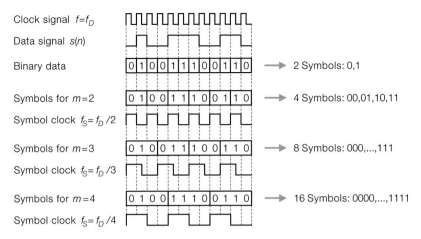

Fig. 24.72. Generation of symbols from the binary data signal

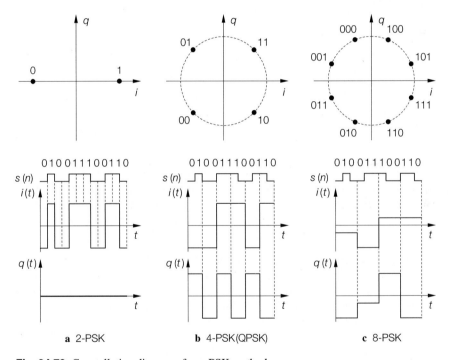

a 2-PSK   **b** 4-PSK(QPSK)   **c** 8-PSK

Fig. 24.73. Constellation diagrams for n-PSK methods

$$\Gamma = \frac{r_D}{B} \overset{\substack{r_D=mr_S \\ B=(1.3...2)\cdot f_S}}{=} \frac{m}{(1.3...2)} \frac{\text{Bit}}{\text{s} \cdot \text{Hz}} \tag{24.82}$$

As m increases, the spacing of the carrier states decreases if the power of the modulated carrier signal remains the same, thus increasing the susceptibility to errors. A measure of the susceptibility is the *power efficiency* E_b/N_0 [24.6]; this indicates by which factor the mean energy E_b per received bit must exceed the thermal noise-power density N_0 in order

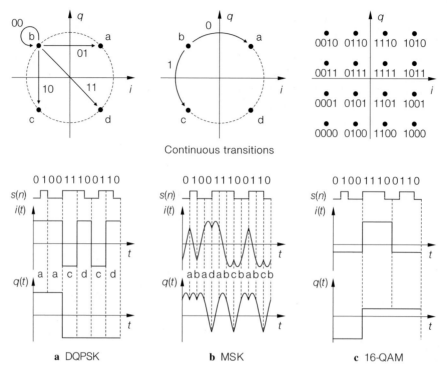

Fig. 24.74. Constellation diagrams for DQPSK, MSK and 16-QAM

to remain at or below a given bit error rate. The power efficiency corresponds to the required signal-to-noise ratio at the input of the demodulator multiplied by a given factor. With the received useful signal power $P_r = E_b f_D$ (mean energy per received bit x data rate) and the noise power $P_n = N_0 B$ (noise-power density x bandwidth), the signal-to-noise ratio is:

$$SNR = \frac{P_r}{P_n} = \frac{E_b f_D}{N_0 B} \overset{\substack{f_D = m f_S \\ B = (1.3...2) \cdot f_S}}{=} = \frac{m}{(1.3...2)} \frac{E_b}{N_0} \tag{24.83}$$

The demands for a high bandwidth efficiency (high Γ) and a high power efficiency (small $(E_b/N_0$ or SNR) are diametrically opposed. A good compromise is achieved with 4-PSK which is also known as *quadrature-phase shift keying* (QPSK); this method is used quite often.

Figure 24.74 shows the constellation diagrams of other common modulation methods. DQPSK (*differential quadri-phase shift keying*) is one of the *differential* modulation methods in which the symbols are represented by status transitions and not by the carrier states. In these methods, the demodulator can determine the binary data signal by the successive comparison of two consecutive symbols without knowing the absolute phase; this makes the demodulator comparably simple. Another differential method is MSK (*minimum shift keying*); this method continuously changes the carrier phase with each data bit by ±90°. The advantage of this method is the constant carrier amplitude which is independent of the speed of the status transitions. Here, nonlinear amplifiers can be used without causing intermodulation distortions. Likewise in n-PSK and DQPSK, all states have the same amplitude, although in practice, transitions cannot occur suddenly as we will see in the section below; this causes a change in the carrier amplitude in the transition regions. 16-QAM (*quadrature amplitude modulation*) uses a 4 × 4 constellation diagram. QAM methods feature a high bandwidth efficiency and are used whenever extremely high transmission rates are required at limited bandwidths; systems with 64-QAM (8 × 8) and 256-QAM (16 × 16) are also used. However, these methods require a high signal-to-noise ratio at the input of the demodulator.

Pulse Shaping

For the quadrature components $i(t)$ and $q(t)$, the methods n-PSK, DQPSK and 16-QAM provide a sequence of square pulses with the *symbol duration* $T_S = 1/f_S$ (see Figs. 24.73 and 24.74). In this form they are not suitable for transmission as the spectrum of a square pulse is relatively wide and decreases very slowly with increasing frequencies; the bandwidth required for the transmission would be disproportionately high. A considerable reduction in the bandwidth can be achieved by *pulse shaping* using suitable filters; for this purpose the quadrature components $i(t)$ and $q(t)$ are filtered by *pulse filters*.

Cosine roll-off pulses: *Cosine roll-off pulses* have particularly favorable characteristics

$$s_{(r)}(t) = \frac{\sin(\pi f_S t)}{\pi f_S t} \frac{\cos(\pi r f_S t)}{1 - (2 r f_S t)^2} \qquad \text{with } 0 < r \leq 1$$

with the spectrum

$$\underline{S}_{(r)}(f) = \begin{cases} 1 & \text{for } |f| < (1-r)\dfrac{f_S}{2} \\[2mm] \dfrac{1}{2}\left[1 + \cos\dfrac{\pi}{r}\left(\dfrac{|f|}{f_S} - \dfrac{1-r}{2}\right)\right] & \text{for } (1-r)\dfrac{f_S}{2} \leq |f| \leq (1+r)\dfrac{f_S}{2} \\[2mm] 0 & \text{for } |f| > (1+r)\dfrac{f_S}{2} \end{cases}$$

Parameter r is known as the *roll-off factor* and influences the (double-sided) bandwidth of the pulse:

$$B = (1+r) f_S \quad \Rightarrow \quad B T_S = 1 + r \tag{24.84}$$

Figure 24.75 shows the spectrum of the cosine roll-off pulse. A typical value in practice is $r = 0.3 \ldots 1$; consequently $B = (1.3 \ldots 2) f_S$.

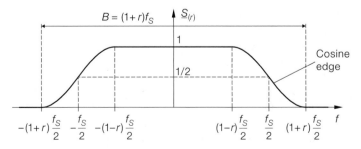

Fig. 24.75. Spectrum of a cosine roll-off pulse

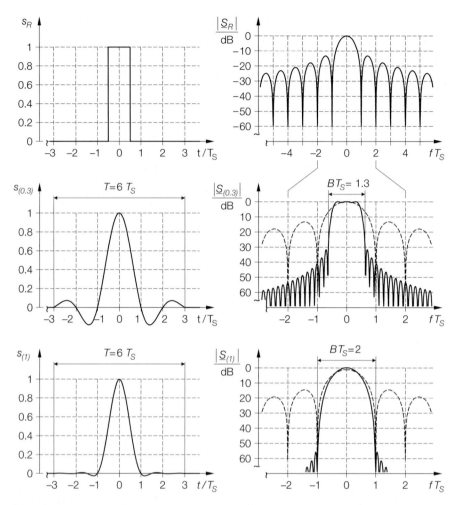

Fig. 24.76. Pulses and absolute spectra: square pulse (top), cosine roll-off pulse $s_{(0.3)}$ with $r = 0.3$ (centre) and cosine roll-off pulse $s_{(1)}$ with $r = 1$ (bottom) with the pulse duration $T = 6\,T_S$. Together with the cosine roll-off pulses the spectrum of the square pulse is shown for comparison.

Figure 24.76 shows the time signals and the spectra of the cosine roll-off pulses with $r = 0.3$ and $r = 1$ compared to a square pulse. The spectra of the cosine roll-off pulses have a much steeper trailing edge. The bandwidth corresponds to the width of the main region between the two inner zero points. The components outside the main region result from the necessary limitation of the infinitely long pulse duration; they can be reduced to any size by lengthening the duration of the pulse. In Fig. 24.76, the pulse duration is $T = 6\,T_S$ ($-3 \leq t/T_S \leq 3$). With an increasing roll-off factor the trailing edge of the pulse becomes steeper so that the limitation has less effect.

As the cosine roll-off pulses are longer than the symbol duration T_S, pulse crosstalk occurs and is known as *inter-symbol interference* (ISI). A special feature of the cosine roll-off pulse is the fact that the central maximum has the value 1, and there are zero points interspaced by T_S on both sides (see Fig. 24.76). This eliminates the inter-symbol

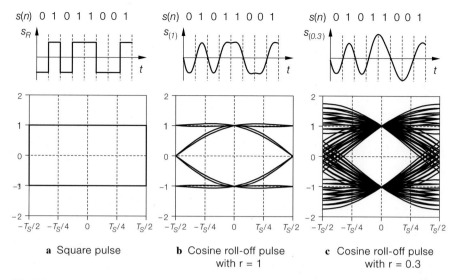

a Square pulse

b Cosine roll-off pulse
with r = 1

c Cosine roll-off pulse
with r = 0.3

Fig. 24.77. Time signals (above) and eye patterns (below)

interference if, in the demodulator, the symbols are sampled in the centre of the symbol duration. Deviation from the ideal sampling time may falsify the sampled value due to adjacent pulses to such an extent that the demodulator makes a false decision resulting in a bit error. Knowledge of the permissible shift in the sampling time and the related reduction in the signal-to-noise ratio can be gained from the *eye pattern*; for this diagram, all signal curves that are possible within one symbol duration are calculated and plotted over one common time axis $-T_S/2 < t < T_S/2$. Figure 24.77 shows the eye patterns for cosine roll-off pulses with $r = 0.3$ and $r = 1$ compared to the ideal eye pattern of square pulses. It can be seen that sampling in the centre of the pulse duration ($t = 0$) causes no reduction in the signal-to-noise ratio. For any deviation from this sampling time the signal-to-noise ratio declines; the smaller the roll-off factor r, the faster the signal-to-noise ratio declines. The region between the lowest course of the curve for level 1 and the uppermost course of the curve for the zero level is called the *eye*. With square pulses, the eye is opened to its maximum; with cosine roll-off pulses the eye closes for $r \to 0$. The eye opening is a measure of the synchronization requirements in the receiver: the smaller the eye, the more accurate the sampling instant must be.

Furthermore, the eye pattern shows that after pulse shaping the amplitude is no longer constant. This results in a situation whereby amplitude modulation also occurs for n-PSK and DQPSK, even though all states in the constellation diagram have the same magnitude. The amplitude modulation increases with a decreasing roll-off factor; this causes an increase in the crest factor (ratio of the peak value to the (rms) effective value).

When selecting the roll-off factor it is necessary to compromise between the required bandwidth and the opening of the eye. For $r \to 0$ the bandwidth assumes the minimum value $B = f_S$. For $r = 1$, the opening of the eye is at its maximum and the bandwidth is thus $B = 2f_S$.

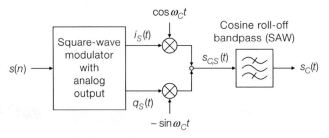

a Cosine roll-off bandpass in the carrier range

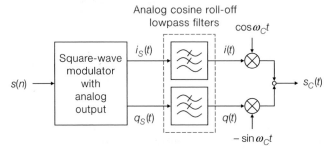

b Analog cosine roll-off lowpass filters in the baseband

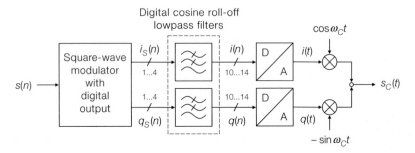

c Digital cosine roll-off lowpass filters and D/A converters in the baseband

Fig. 24.78. Pulse filters

Pulse filter: Linear-phase transversal filters with finite pulse response are used for pulse shaping. These filters contain delay elements whose output signals are weighted and added. Normally, they function as digital *FIR (finite impulse response) filters*. The delay elements can then be designed as shift registers (see Sect. 19.3). The standard design of a transversal filter is shown in Fig. 19.12 on page 995. The transversal filter can also function as an analog filter by using delay lines, sample-and-hold circuits or a *charge-coupled device (CCD)* for signal delay. Another possibility is to use *surface acoustic wave filters (SAW filters)*, which use the transit time of an acoustic wave for signal delay. However, SAW filters operate in the form of bandpass filters only.

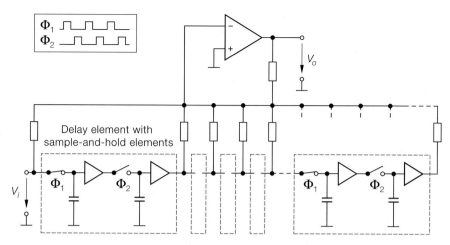

Fig. 24.79. Analog transversal filter with sample-and-hold elements

The most simple pulse filtering method is via a SAW bandpass filter with cosine roll-off characteristic in the carrier range, i.e. after the I/Q mixer; the modulator thus has the configuration shown in Fig. 24.78a. In practice, this requires no extra circuitry since a bandpass is required after the I/Q mixer in order to suppress unwanted components (see Fig. 24.68). An intermediate frequency must be used for the carrier frequency ($f_C \approx 10 \ldots 100\,\text{MHz}$) so that the SAW filter with the required bandwidth can function.

Pulse shaping in the baseband requires separate filters for the quadrature components $i(t)$ and $q(t)$. Figure 24.78b shows a modulator with analog cosine roll-off lowpass filters. The analog transversal filter with sample-and-hold circuits and an inverting operational amplifier for the weighted summation can be used as shown in Fig. 24.79. As the quadrature components in the baseband have a double-sided bandwidth $B = (1 + r)f_S \leq 2f_S$, the clock frequency of the transversal filter must be higher than the symbol frequency f_S by at least a factor of 2 in order to meet the requirements of the sampling theorem. In practice, this filter is usually clocked with four times the symbol frequency to increase the distance to the aliasing components (*oversampling*). This means that for cosine roll-off pulses with an impulse duration of $6T_S$, a filter with $6 \cdot 4 = 24$ delay elements or 48 sample-and-hold elements is required. In modulation methods with binary quadrature components, the delay elements can be replaced by D flip-flops as is the case in 2-PSK, 4-PSK (QPSK) and DQPSK. To reduce the circuitry of the filters, in simple systems the cosine roll-off lowpass filters are often realized in approximation only. If a somewhat higher bandwidth and lower eye diagram opening are acceptable, a standard lowpass filter can be used instead of the transversal filter.

In complex systems, pulse shaping is accomplished by digital FIR filters requiring additional D/A converters to generate the analog quadrature components. Figure 24.78c shows a modulator with digital cosine roll-off lowpass filters. The word length at the filter input results from the constellation diagram and is a maximum of 4 bit (256-QAM → 16 × 16 constellation diagram → 4 bit each for $i_S(n)$ and $q_S(n)$). The word length at the output corresponds to the resolution of the D/A converter and is to be selected in accordance with the required signal-to-noise ratio; $10 \ldots 14$ bit are common in practice. Filters for modulation methods with binary quadrature components (2-PSK, QPSK and

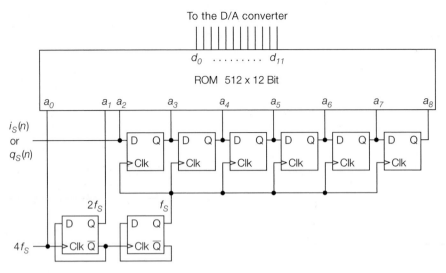

Fig. 24.80. Digital cosine roll-off filter with ROM for modulation methods of binary quadrature components $(i_R(n), q_R(n) \in [0; 1])$

DQPSK) are particularly simple as the input signal assumes the values ± 1 only and is represented by one bit. Since the output word of a filter with a pulse length of $6T_S$ depends on a maximum of 7 consecutive bits, all of the $4 \cdot 2^7 = 512$ possible output words can be stored in a ROM with a clock frequency of $4 f_S$. A shift register of length 7 and both the full and half clock frequency, i.e. $4 f_S$ and $2 f_S$, are used for addressing. Figure 24.80 shows this simple filter. Often the clock frequency is raised to $8 f_S$ or $16 f_S$ in order to enlarge the distance to the aliasing components in which case a ROM of 1024 or 2048 words is necessary.

Pulse shaping in most modern systems is performed by a digital signal processor (DSP) which also carries out all other digital functions, i.e. all functions that are shown *above* the D/A or A/D converters in Fig. 24.1b. Where the calculatory power of a standard DSP is insufficient or the power loss of a standard DSP is too high for the required computing power, customized DSPs with special digital components are used to speed up time-critical functions. Such DSPs may contain, for example, two of the filters shown in Fig. 24.80 and subsequent D/A converters.

If analog transversal filters or digital filters are used for pulse shaping, additional analog anti-aliasing filters must be used to remove the aliasing components at multiples of the clock frequencies; these filters are not shown in Fig. 24.78b/c.

A Simple QPSK Modulator

A simple modulator for a QPSK system is shown below and may be used equally well for DQPSK if the binary useful signal is coded before entering the modulator. Let us assume that the modulator generates the modulated carrier signal on an intermediate frequency that is subsequently converted to the transmit frequency.

Figure 24.81 shows the QPSK modulator with I/Q mixer and Fig. 24.82 the signal wave forms. The digital modulator consists of a 2-bit series-parallel converter that distributes the bits of the binary data signal $s(n)$ to the i and q branches. The flip-flop FF1 reduces the

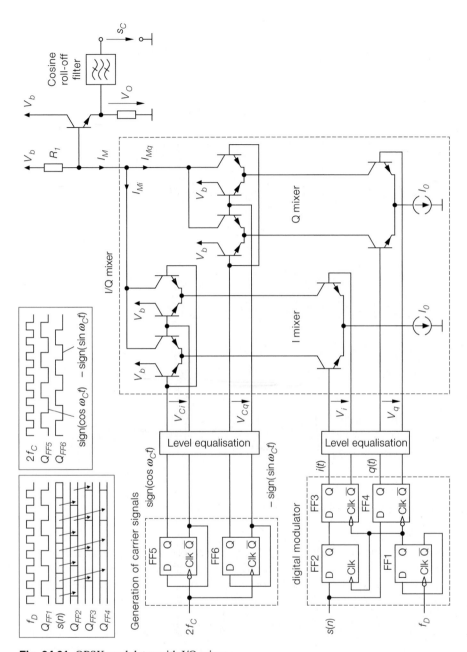

Fig. 24.81. QPSK modulator with I/Q mixer

clock frequency $f_D = 1/T_D$ by a factor of 2 to the symbol frequency $f_S = 1/T_S = f_D/2$. The i bits are buffered in the flip-flop FF2 until the related q bits are available, then both bits are taken over synchronously by the flip-flops FF3 and FF4. The level-equalized output voltages V_i and V_q of the modulator are converted to the carrier frequency by an I/Q mixer. Two square-wave signals, staggered by a quarter of a period with a carrier frequency $f_C = 1/T_C$, serve as the carrier signals. These signals are derived from a square-

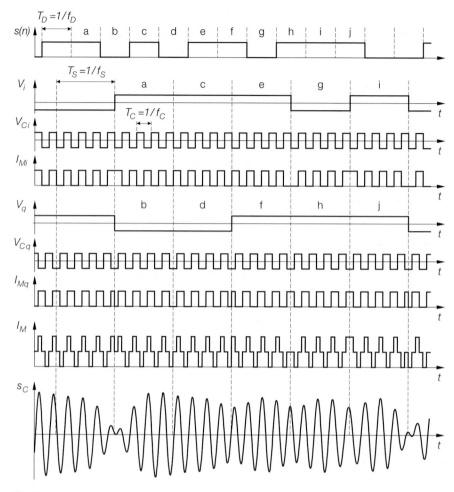

Fig. 24.82. Signals in the modulator

wave signal with double the carrier frequency by the divider flip-flops FF5 and FF6. The fundamental waves of the square-wave signals correspond to the carrier signals $\cos \omega_C t$ and $- \sin \omega_C t$ of an ideal I/Q mixer. The current switches of the mixers are activated with the level-equalized carrier voltages V_{Ci} and V_{Cq} resulting in square-wave currents I_{Mi} and I_{Mq} at the output of the mixers. The summation of the mixer output signals is done by adding the currents I_{Mi} and I_{Mq}. The summation current I_M is converted into a voltage by resistance R_1 while the common-collector circuit serves as a buffer. The modulated carrier signal $s_C(t)$ is obtained from the output voltage V_o after filtering by a cosine roll-off bandpass filter (SAW filter). Figure 24.81 presents the modulated carrier signal $s_C(t)$ without the delay caused by the filter, thus illustrating the relationship with current I_M.

Although all points of the QPSK constellation diagram have the same absolute value, there is also an amplitude modulation, in addition to the phase modulation, which is caused by the cosine roll-off filter. A diagonal transition in the constellation diagram passes the origin and, in this case, the amplitude falls briefly back to zero.

24.5
Multiple Use and Grouping of Communication Channels

A two-dimensional space defined by the frequency and time axes is available for wireless transmission of signals. The transmission channels of all data communication systems must be arranged within this space so that there is multiple utilization. The mode of subdividing this space is called *multiplex operation*.

The transmission between two communication parties can be unidirectional or bi-directional. In unidirectional transmission, one of the parties acts as the data transmitter while the other is the data receiver; typical examples are radio and TV broadcasting. Unidirectional systems usually have a distributive characteristic, i.e. one transmitter serves many receivers with the same information. Such systems are therefore called *broadcast systems* and the signal distribution is known as *broadcasting*. In bi-directional transmission, the two parties act as both data transmitter and data receiver. They can alternately use one channel or separate channels for the transmission in both directions. The first case is known as *half duplex operation* and the second as *duplex* or *full duplex operation*. An example of half duplex operation is CB radio telephony which allows only one partner to speak at any given time and requires a special change-over signal (*Over!*) for the transition to the other party. Modern systems like cordless or mobile phones use duplex operation for signal transmission requiring two channels to form one link. The method of grouping is called *duplex mode*.

24.5.1
Multiplex Operation

Frequency Division Multiplex

The most important method of dividing transmission space is called *frequency division multiple access (FDMA)* or frequency division multiplex. This approach permanently assigns a certain frequency range to each transmission channel. All channels of a certain application are grouped together to cover the frequency range available for this application; some examples are listed in Figs. 24.20 and 24.21 on page 1168. All communications systems use frequency division multiple access at the uppermost level; there is no system that utilizes the entire frequency range available. Figure 24.83a illustrates the division of the transmission space in the frequency division multiple access mode. In this context, the channels are also known as *frequency channels*. There is a frequency gap between the channels that is required as a transition region for the filter in the receiver, therefore the channel separation C is larger than the bandwidth B of the signals.

Frequency division multiple access requires no coordination between the systems in adjacent channels. Each system can use the assigned channel without limitations.

Time Division Multiple Access

The division of transmission time for individual frequency channels into *time slots* is called *time division multiple access (TDMA)* or *time division multiplex*. Figure 24.83b illustrates a situation where all frequency channels use the same time slot pattern. This is the case in many applications but is not a necessity.

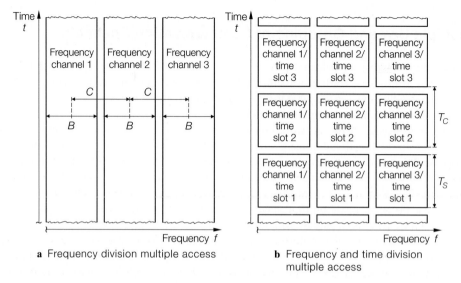

Fig. 24.83. Multiplex modes

A distinction must be made between *time division multiple access on the data level* and *time division multiple access on the transmitter level*. On the data level, several data streams are combined into one and sent by a *single* transmitter. Correspondingly, the transmitted signal is received by one receiver and the resulting data stream is split up into the original data streams. One example is the radio relay transmission of telephone conversations. Here, for instance, 30 digitized speech signals with a data rate of 64 kbit/s each are combined into one data stream of 1.92 Mbit/s for transmission. In this case, the division of the transmission time into data slots only refers to the arrangement of the data and has no influence on the transmitter or the transmit signal[13].

In the time division multiple access mode on the transmitter level, time slots are used by *different* transmitters which must be coordinated in order to avoid overlapping of the transmit times. A time gap between the time slots is needed to switch over from one transmitter to the next. The interval T_C between the beginning of two successive time slots is therefore slightly larger than the duration T_S of a time slot (see Fig. 24.83b).

The time slots are cyclically and consecutively numbered and combined into *frames* where all time slots of the same number form one *time channel*. Figure 24.84 illustrates this for the case of four time channels. The time channels can be further divided if m transmitters share a single channel so that each transmitter uses one time slot in every mth frame. This method is used, for example, in GSM radio communications (global system for mobile communications).

The time division multiple access operation is used in communication systems where several participants communicate with one common *base station (BS)* or *base transceiver station (BTS)*. Frequency division multiple access operation would make it necessary for the base station to provide each participant with one transmitter and one receiver. Time

---

[13] The term *transmitter* is not used here in the wider sense and thus only specifies the components from the modulator to the antenna. This means that the components for combining the data streams into one are not part of the transmitter.

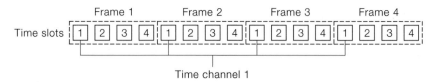

Fig. 24.84. Frames and time channels for time divison multiplex with four channels

division multiple access operation, on the other hand, can serve several participants with one transmitter and one receiver. GSM mobile communication uses a time division multiple access mode with 8 time slots enabling one GSM base station to serve a maximum of $6 \cdot 8 = 48$ participants with 6 transmitter-receiver units. Cordless telephone systems of the DECT standard use a time division multiple access mode with 24 time slots, 12 of which are intended for both transmission directions. Thus a DECT base station can serve a maximum of 12 telephones with one transmitter/receiver unit. Consequently, with respect to the connection capacity, the number of time slots should be as high as possible. This, however, is more complex in terms of coordination and reduces the efficiency because the ratio of time slot length to time gap between the time slots is less favorable.

Code Division Multiple Access

The *code division multiple access (CDMA)* or code division multiplex operation is a method for multiple use of one frequency channel by several transmitters without subdividing the transmit time. The data streams of the transmitter are coded with *orthogonal code words* and simultaneously transmitted at the same frequency by digital transmitters without any further coordination. Each receiver receives the sum of all transmitted signals and can extract the relevant data by means of the related codes. This method is also known as *direct sequence CDMA (DS-CDMA)*. Figure 24.85 shows the basic principle of DS-CDMA. In this diagram the transmitter and receiver components do not contain a specific modulator or demodulator.

In addition to the direct sequence CDMA, there are also other code division multiple access modes, such as *frequency hopping CDMA (FH-CDMA)* which changes the transmit frequency in accordance with a code pattern; this is not further detailed here but can be referred to in the relevant literature [24.7]. Since the code division multiplex mode requires a code for each connection, its connecting capacity corresponds to the number of orthogonal code words. With the use of suitable codes, this capacity is significantly higher than that of the time division multiple access mode.

Principle of direct sequence CDMA operation: The DS-CDMA performs an Exclusive-Or operation with a binary code word for every bit of the transmitted data stream. Figure 24.86 illustrates this using coding with Walsh codes of the length 8 (transmitter 6: $s_6(t) = d_6(t) \oplus c_6(t)$ as an example). Due to coding, the bit rate increases according to the length of the code word. This also enlarges the bandwidth necessary for transmission. For this reason, the coding method is also called *spreading*, the length of the code word is called the *spreading factor (SF)* and the code division multiple access operation is known as *spread spectrum modulation*. The spreading factor (SF) is derived from the bit length T_B of the uncoded data stream and from the bit length T_C of the code word:

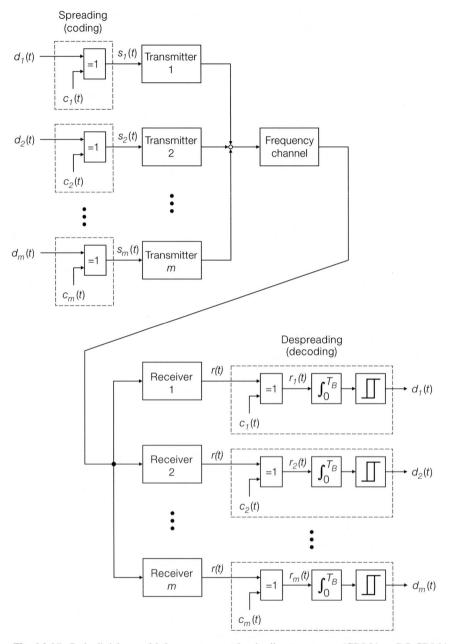

Fig. 24.85. Code division multiple access operation in direct sequence (CDMA or DS-CDMA)

$$SF = \frac{T_B}{T_C} \tag{24.85}$$

The spreading factor in Fig. 24.86 is $SF = 8$. The bits of the coded data stream and the code words are called *chips* to distinguish them from the bits of the uncoded data stream. Thus T_B is the *bit duration* and T_C is the *chip duration*.

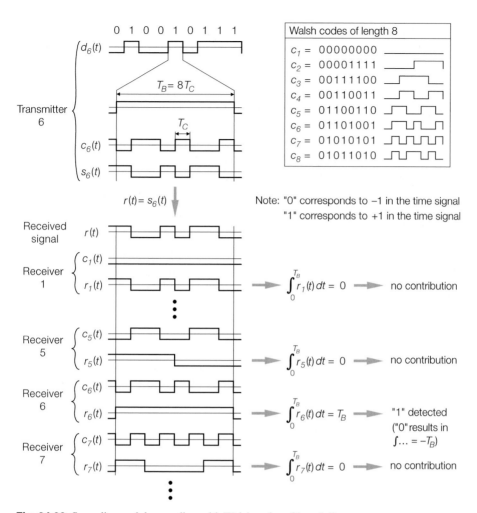

Fig. 24.86. Spreading and despreading with Walsh codes of length 8

In the receivers, the received signal undergoes an Exclusive-Or operation with the code word and is integrated throughout a bit duration. This decoding operation is called *despreading*. Owing to the orthogonality[14] of the code words the integration only yields a component not equal to zero in the receiver that uses the same code word as the transmitter. Figure 24.86 illustrates the situation where the received signal $r(t)$ is equal to the transmitted signal $s_6(t)$ of transmitter 6. Since the spreading, despreading, and addition of the transmit signals are linear operations, the separation of a received signal consisting of several transmit signals is achieved in the same fashion.

---

[14] Signals of the length T ($t \in [0, T]$) are *orthogonal* if the following is true:

$$\int_0^T c_i(t)\, c_j(t)\, dt = \begin{cases} k \neq 0 & \text{for } i = j \\ 0 & \text{for } i \neq j \end{cases}$$

Practical realization: Figures 24.85 and 24.86 show the basic principle of code division multiplex without the use of a special modulation method. In practice, however, the code division multiple access mode is always used in combination with one of the known modulation methods, most commonly QPSK or DQPSK. Figure 24.87 shows the integration of the components for code division multiple access in a system with QPSK modulation. Spreading is done after modulation but before the roll-off filtering; despreading is done before demodulation. The IF and RF components of the transmitter and receiver are not shown in Fig. 24.87. The transmitter is usually that shown in Fig. 25.6c on page 1243 with a digital I/Q mixer. In this case the components of the modulator also operate digitally and are implemented with a digital signal processor (DSP). The preferred receiver is that with IF sampling as shown in Fig. 25.23c on page 1266 or the direct conversion receiver shown in Fig. 25.33 on page 1277. The components of the demodulator are also implemented with a DSP.

When designing a transmission system with code division multiple access some additional aspects must be taken into consideration. To illustrate these aspects, let us thus look at a mobile communications system in which several mobile units communicate with one common base station (see Fig. 24.88). Here, all *downlink* channels (base station → mobile unit) are transmitted *synchronously* from the transmitter of the base station, while the *uplink* channels (mobile unit → base station) operate *asynchronously*, i.e. without coordination between the transmitters of the mobile units.

– The Walsh codes that were used in Fig. 24.86 are orthogonal only in synchronous operation; if the code words are shifted in time, an accurate separation of the channels is

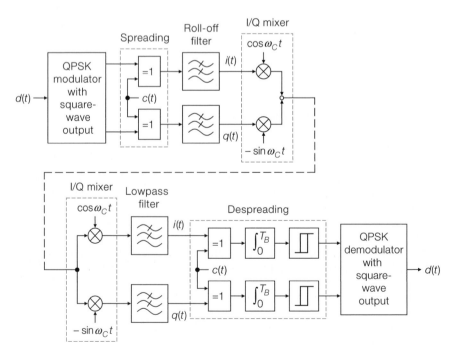

Fig. 24.87. Code division multiple access in combination with QPSK modulation: modulator (top) and demodulator (bottom)

Mobile unit 1

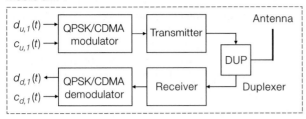

Base station

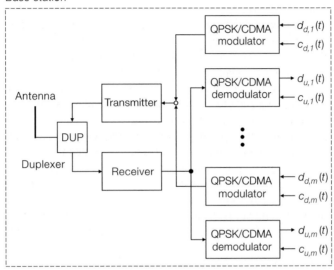

mobile unit m

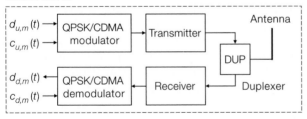

Fig. 24.88. Mobile communications system with QPSK modulation and code division multiple access with separate code words for *uplink* (mobile unit → base station, index u) and *downlink* (base station → mobile unit, index d)

no longer possible. Therefore, the use of the Walsh codes is found in *downlink* channels only. The *uplink* channels require code words that remain approximately orthogonal even with a time shift. A measure for this is the cross correlation function, which measures the similarity of signals in dependence of their time shift[15]. The absolute value must be

---

[15] The cross-correlation function of two signals with length T ($t \in [0, T]$) is:

$$R_{ij}(\tau) = \int_0^T c_i(t)\,c_j((t+\tau) \bmod T)\,dt$$

as small as possible for all code words and all time shifts. In practice, a set of binary *pseudo noise (PN)* or a *pseudo random binary sequence (PRBS)* is usually used [24.7].

The time-shifted signal $c_j(t + \tau)$ is assumed to continue periodically by regarding the argument $t + \tau$ modulo T so that it always lies in $[0, T]$. The cross-correlation function is thus also periodic with T, i.e. only the range $\tau \in [0, T]$ must be considered.

- The code words are used for separation of the channels and for spectral spreading of the transmit signal. Often this leads to the problem of code words with little cross-correlation causing an unfavorable spectral distribution of the transmission power. One way of avoiding the problem is to use *two* codes in order to decouple the properties regarding channel separation and spectral spreading. First, the channel separation is performed using *long codes* and then the spectral spreading using *short codes*. Both code words are usually of the same chip duration, where the length of the short code corresponds to the bit duration of the uncoded data stream and the duration of the long code covers several bits of the uncoded data stream [24.7].
- Since the code words used in practice are not exactly orthogonal, each transmit signal generates a noise-like interference signal in all nonrelated receivers which reduces the signal-to-noise ratio. The connecting capacity of the system is fully exploited if the number of transmit signals has increased so much that the signal-to-noise ratio has dropped to the minimum value required for an accurate demodulation. In this case, the number of transmit signals is usually clearly below the number of code words. Therefore, the connecting capacity of a practical system is not limited by the number of code words but by the interference levels which in turn depend on the distribution of the mobile units. This means that the connecting capacity is variable.
- The connecting capacity attains its maximum if the received signal level intended for that receiver is higher than all other transmit signals or if all received transmit signal levels are equal. A power control scheme must be used in order to meet this condition. The transmission power of mobile units must be adjusted such that all *uplink* channels reach the base station with the same level; the signal-to-noise ratio of all channels is thus equal. The power of the *downlink* channels must be so small that the respective mobile units are just capable of receiving, thus reducing the interfering signals in the receivers of other mobile units.

Despite these demands and the corresponding complexity involved in the functioning of such systems, code division multiple access systems are superior to systems with time division multiple access. For this reason, existing systems with time division multiplex (GSM, DECT) are increasingly being replaced by systems with code division multiplex (UMTS, IS-95).

24.5.2
Duplex Operation

Duplex operation can be explained by way of a mobile communications system. The channels for the two transmission directions are called the *uplink* channel (mobile unit $\rightarrow$ base station) and the *downlink* channel (base station $\rightarrow$ mobile unit).

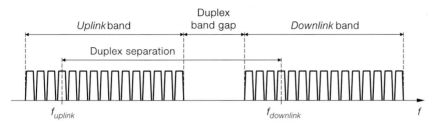

Fig. 24.89. Channel grouping for frequency division duplex operation

Frequency Division Duplex

In *frequency division duplex operation (FDD)* separate frequency channels are used for the *uplink* and the *downlink* channel of a connection. All *uplink* channels form the *uplink* band and all *downlink* channels form the *downlink* band. One downlink channel is permanently assigned to each uplink channel (see Fig. 24.89). The frequency separation between the two channels is called the *duplex separation*. In mobile units and base stations, the bands are separated by a *duplexer* (see Fig. 24.14b on page 1164 and 24.88 on page 1233). For this purpose, a *duplex band gap* that serves as the transition region for the filters of the duplexer is introduced between the *uplink* and the *downlink* bands.

In frequency division duplex operation, the transmitters and receivers are operated simultaneously. The attenuation of the filters in the duplexer must be sufficiently high to prevent the transmit signal from reaching the receiver with too high a level and from blocking the RF preamplifier. Furthermore, efficient shielding is necessary between transmitter and receiver in order to reduce cross-talk to a noncritical level.

Time Division Duplex

In *time division duplex operation (TDD)*, different time slots of a frequency channel with time multiplex are used for the *uplink* channel and for the *downlink* channel of one connection. In this case, the transmitter and receiver only operate for the duration of the given time slot, and an antenna change-over switch can be used to alternate the antenna function between transmitting and receiving (see Fig. 24.14a on page 1164).

Since in the time division duplex mode, the transmitter and the receiver of one unit are not operated at the same time, shielding between the transmitter and the receiver is not required. The required antenna change-over switch is also less complicated, cheaper and significantly smaller than the duplexer required for frequency division duplex mode. For this reason, a combination of time division multiplex and time division duplex is principally used in simple systems with only a few time slots. The advantages then outweigh the disadvantages caused by the coordination required to access the individual time slots.

Chapter 25:
Transmitters and Receivers

This chapter describes the design of transmitters and receivers for radio transmission. The terms used shall have a defined meaning such that the components from the modulator up to the transmitting antenna form the *transmitter*, while the components from the receiver antenna up to the demodulator form the *receiver*.

The demands placed on the transmitter and receiver are clearly distinct since the transmitter must process only the desired signal while the receiver must separate the desired signal from the frequency mixture received by the antenna. Furthermore, the transmitter handles signal levels which are constant or which vary very slightly, while the receiver copes with extremely large level differences that depend on the distance to the transmitter. The main challenges for the transmitter include the task of converting the useful signal into a high-frequency transmission signal with as little interference as possible, to amplify this signal with the highest possible efficiency and to minimize the transmission of undesirable interference signals generated by the conversion or amplification. The main challenges for the receiver are to filter out the desired signal even from very weak levels, while at the same time receiving very strong signals from adjacent frequency ranges, and producing a clear signal with a high signal-to-noise ratio and minimum intermodulation distortions. Thus, the main obstacle for concern in transmitters is *efficiency*, while receivers face *issues of selection, dynamics* and *noise*.

25.1
Transmitters

First we will look at the construction of transmitters with analog modulation, followed by a description of transmitters with digital modulation. These descriptions are supported by simplified block diagrams showing only the essential components.

25.1.1
Transmitters with Analogue Modulation

Transmitters with Direct Modulation

The most simple transmitter is obtained when the carrier frequency f_C of the analog modulator is identical to the transmission frequency f_{RF}. In this case, the modulator output signal only needs to be amplified and fed to the antenna. In practice, the transmission amplifier must be followed by an *output filter* that reduces the distortion products originating in the amplifier to an acceptable level. Figure 25.1a shows the construction of a transmitter with *direct modulation*. The signal spectra are shown in Fig. 25.2.

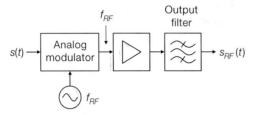

a With direct modulation

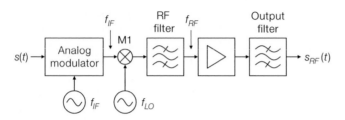

b With intermediate frequency

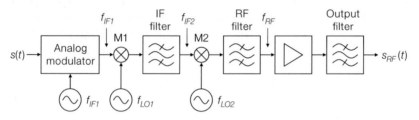

c With two intermediate frequencies

Fig. 25.1. Transmitter with analog modulation

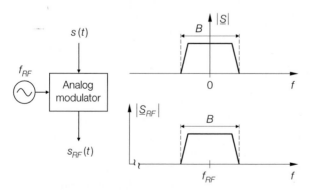

Fig. 25.2. Signal spectra in transmitters with direct modulation

Transmitters with One Intermediate Frequency

With increasing frequencies and growing demands, it becomes more and more difficult to obtain a modulator with the required accuracy. Therefore, a lower *intermediate frequency* f_{IF} with which the modulator can be easily built is used as carrier frequency f_C:

$$f_C = f_{IF} \ll f_{RF}$$

Figure 25.1b shows the construction of a transmitter with one *intermediate frequency*. Conversion to the transmission frequency f_{RF} is done by mixer M1 which is provided with the frequency

$$f_{LO} = f_{RF} - f_{IF}$$

from a *local oscillator (LO)*. The mixing process generates the sum and difference frequencies

$$f_{LO} + f_{IF} = f_{RF} \quad , \quad f_{LO} - f_{IF} = f_{RF} - 2f_{IF}$$

The portion at the transmission frequency is filtered by an *RF filter* and fed to the transmitter amplifier. Figure 25.3 shows the signal spectra.

Owing to $f_{RF} = f_{LO} + f_{IF}$, the frequency sequence is identical in the IF and RF signals, which means that a higher IF frequency results in a higher RF frequency; this is known as *noninverted mode*. It is also possible to choose $f_{RF} = f_{LO} - f_{IF}$ by filtering out the signal

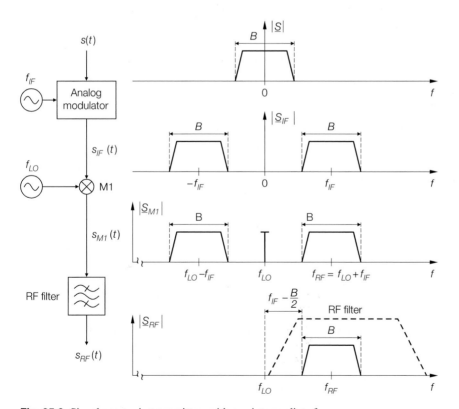

Fig. 25.3. Signal spectra in transmitters with one intermediate frequency

portion below the local oscillator frequency in Fig. 25.3. Then the frequency sequence in the transmission signal is inverted; this is known as *inverted mode*. The receiver must take the inverted frequency operation into account in order to correctly receive the desired signal. For this purpose, the receiver uses a mixer operated in inverted mode.

The mixer output signal contains a signal portion at the local oscillator frequency f_{LO} (see Fig. 25.3). Consequently, the transition region of the RF filter (transition from the pass band to the cutoff band) must not exceed the width $f_{IF} - B/2$ to ensure that the transmission signal lies fully within the pass band and the local oscillator signal is in the cutoff band. Particularly suitable are surface acoustic wave (SAW) filters with their very narrow transition region and constant group delay but whose high insertion loss ($>20\,\text{dB}$) is disadvantageous. Where no SAW filters are available for the desired transmission frequency, LC filters or filters with dielectric resonators must be used. As these filters have unwanted group delay distortion at the borders of the transition region, it is necessary to select a clearly smaller transit region in order to prevent the transmission signal from being affected. As an alternative, one may use the entire range between the portions above and below the local oscillator frequency as the transition region and suppress the local oscillator frequency by a separate serial or parallel resonant circuit (zero transmission at f_{LO}).

With rising transmission frequencies, the ratio of the transmitter frequency to the width of the transition region increases; hence, the quality of the RF filter must also increase:

$$Q_{RF} \sim \frac{f_{RF}}{f_{IF} - B/2} \overset{f_{IF}=f_C \gg B}{\approx} \frac{f_{RF}}{f_C}$$

This results in a higher filter order and increased group delay distortions. In practice, the intermediate frequency is made as high as possible so that the transition region becomes wider and the RF filter quality becomes correspondingly low.

Transmitters with Two Intermediate Frequencies

In transmitters with one intermediate frequency and high transmission frequencies, the quality of the RF filter becomes impermissibly high. A second intermediate frequency is then required that ranges between the carrier frequency of the modulator and the transmission frequency:

$$f_C = f_{IF1} < f_{IF2} < f_{RF}$$

Figure 25.1c shows the construction of a transmitter with *two intermediate frequencies*, while the signal spectra are presented in Fig. 25.4. Mixer M1 converts the modulator's output signal from the first to the second intermediate frequency. This requires a local oscillator with the frequency $f_{LO1} = f_{IF2} - f_{IF1}$. Subsequently the portion above the local oscillator frequency is filtered out by an *IF filter*. The quality of the IF filter is proportional to the ratio of the second intermediate frequency and the width of the transition region:

$$Q_{IF} \sim \frac{f_{IF2}}{f_{IF1} - B/2} \overset{f_{IF1}=f_C \gg B}{\approx} \frac{f_{IF2}}{f_C}$$

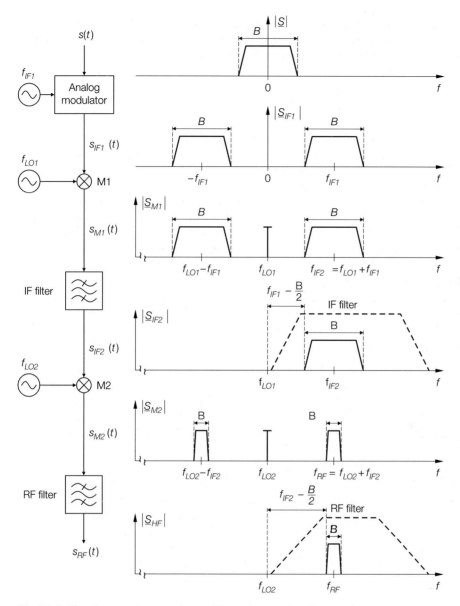

Fig. 25.4. Signal spectra in transmitters with two intermediate frequencies

The conversion to the transmission frequency is achieved with a mixer M2, which is fed by a second local oscillator with the frequency $f_{LO2} = f_{RF} - f_{IF2}$. An RF filter of the quality

$$Q_{RF} \sim \frac{f_{RF}}{f_{IF2} - B/2} \overset{f_{IF2} \gg B}{\approx} \frac{f_{RF}}{f_{IF2}}$$

is required to filter out the transmission signal.

Obviously the overall quality is $Q \approx f_{RF}/f_C$, which, in transmitters with one intermediate frequency, has to be generated by the RF filter and in transmitters with two intermediate frequencies can be distributed to two filters:

$$Q = Q_{RF}\, Q_{IF} \sim \frac{f_{RF}}{f_C}$$

The relative amounts can be controlled by the value of the second intermediate frequency, more specifically, if it is relatively high then $Q_{IF} > Q_{RF}$, if it is relatively low then $Q_{IF} < Q_{RF}$. In practice, the values selected depend on the transmission frequency and the available filters. The planned number of units also has an important influence since for high unit numbers customized dielectric or SAW filters can be used, but for mass applications such as mobile communication even the design of new filter technologies is warranted. For small batch production, on the other hand, standard filters are used. The use of LC filters with discrete components is avoided where possible for reasons of space and calibration.

In transmitters with two intermediate frequencies, one can also operate one or both mixers in inverted mode by filtering out the portions below the local oscillator frequency. If both mixers are operated in inverted mode, then the transmission signal is in noninverted mode again.

Transmitters with Variable Transmission Frequencies

In transmitters with a variable transmission frequency, the frequency of the last local oscillator is variable, thus allowing the transmission frequency to be altered without affecting the other components. Variations take place within the frequency range assigned to the specific application according to the channel spacing C. Figure 25.5 illustrates this taking a transmitter with five channels as an example. The RF filter is rated such that all channels are within the pass band and all local oscillator frequencies are within the cutoff band. Alternatively, a tuneable RF filter may be used, but only in exceptional practical cases.

For a lower number of channels and less channel spacing, the local oscillator and transmission frequencies change very little. For such applications a transmitter with one intermediate frequency can be used as long as the transition region between the highest local oscillator frequency and the lowest limit of the channel pattern is sufficiently wide. Although in most cases this requires a transmitter with two intermediate frequencies where the second intermediate frequency is selected relatively high so that the transition region becomes as wide as possible.

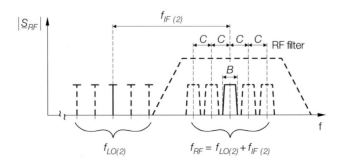

Fig. 25.5. Transmitter with variable transmission frequency

25.1.2
Transmitters with Digital Modulation

In principle, transmitters with digital modulation are of the same design as transmitters with analog modulation. The essential difference is that digital modulators primarily generate the quadrature components $i(t)$ and $q(t)$ that are combined into a modulated carrier signal by an I/Q mixer.

Figure 25.6a shows a digital transmitter with direct modulation. It corresponds to the analog transmitter with direct modulation in Fig. 25.1a if the combination of digital modulator, I/Q mixer (MI and MQ) and the subsequent filter are regarded as being equivalent to the analog modulator. The same applies to the digital transmitter with one or two in-

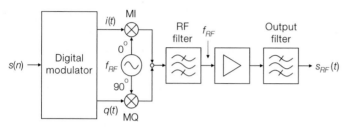

a With direct modulation

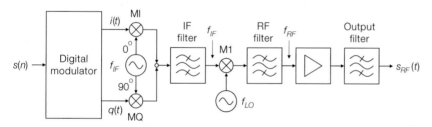

b With one intermediate frequency and an analog I/Q mixer

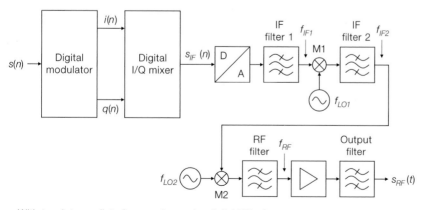

c With two intermediate frequencies and a digital I/Q mixer

Fig. 25.6. Transmitter with digital modulation

termediate frequencies. A digital transmitter with one intermediate frequency is shown in Fig. 25.6b.

If particularly high demands are made in terms of the accuracy of the I/Q mixer, a digital I/Q mixer is used to prevent amplitude and phase errors between the two branches. The output of the digital I/Q mixer provides a digital IF signal that is converted into an analog IF signal by a D/A converter and subsequent IF filter. As the frequency of the IF signal must be comparatively low due to the limited sampling rate of the digital I/Q mixer and the D/A converter, a second intermediate frequency is usually utilized. Figure 25.6c shows the resulting transmitter.

25.1.3
Generating Local Oscillator Frequencies

The required local oscillator frequencies are derived by phase-locked loops (PLL) from a crystal oscillator with reference frequency f_{REF}. Figure 25.7 depicts this for a transmitter with one intermediate frequency and variable transmission frequency. The intermediate frequency is fixed and is determined by the divider factors n_1 and n_2:

$$f_{IF} = \frac{n_2}{n_1} f_{REF}$$

The local oscillator frequency is variable in steps according to the channel spacing C. For this purpose, the reference frequency is divided to the channel distance by the divider factor n_3 and multiplied by a PLL with the programmable divider factor n_4:

$$C = \frac{f_{REF}}{n_3} \quad , \quad f_{LO} = n_4 C = \frac{n_4}{n_3} f_{REF}$$

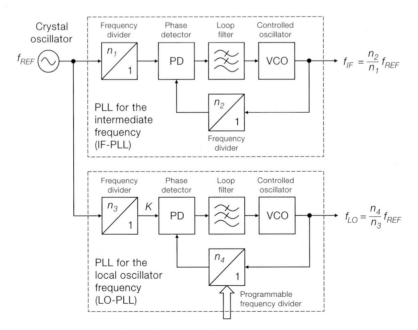

Fig. 25.7. Generation of the local oscillator frequencies

The local oscillator frequency and thus the transmission frequency is adjusted by changing the divider factor n_4. If the local oscillator frequencies are not divisible by C, then the reference frequency must be divided by means of the divider factor n_3 to the largest common divisor of C and the local oscillator frequencies and this common divisor must be multiplied by n_4.

Example: In Fig. 24.81 on page 1225, a QPSK modulator with I/Q mixer is to be converted into a transmitter with one intermediate frequency that is capable of a data rate of 200 kbit/s at a roll-off factor $r = 1$. A crystal oscillator with $f_{REF} = 10\,\text{MHz}$ is to be used as a reference. The data rate $f_D = 200\,\text{kHz}$ is obtained by division by a factor of 50. The carrier or intermediate frequency is $f_C = f_{IF} = 70\,\text{MHz}$ since inexpensive SAW filters are available for this frequency. Since the I/Q mixer in Fig. 24.81 must be driven with the frequency $2f_C = 140\,\text{MHz}$, we select $n_1 = 1$ and $n_2 = 14$ for the IF PLL in Fig. 25.7. For QPSK, the symbol frequency is equal to half the data rate $f_S = f_D/2$, resulting in a bandwidth of $B = (1 + r)f_S = 200\,\text{kHz}$. We assume that the transmitter can use 4 channels ranging from 433 to 434 MHz with a channel spacing of $C = 250\,\text{kHz}$. From the transmission frequencies $f_{RF} = 433.125/433.375/433.625/433.875\,\text{MHz}$ we obtain the local oscillator frequencies $f_{LO} = f_{RF} - f_{IF} = 363.125/363.375/363.625/363.875\,\text{MHz}$. Since these are not multiples of C, we calculate the largest common divisor: $lcd\{K, f_{LO}\} = 125\,\text{kHz}$. For the LO PLL this leads to $n_3 = 10\,\text{MHz}/125\,\text{kHz} = 80$ and $n_4 = f_{LO}/125\,\text{kHz} = 2905/2907/2909/2911$. For all channels, the RF filter must allow signal transmission without major group delay distortion and, at the same time, sufficiently attenuate the highest local oscillator frequency. The double-tuned-circuit bandpass filter described in Sect. 26.2 can be set up for a center frequency of 434.4 MHz and a bandwidth of 10 MHz. Thus the desired signal is attenuated by 6 dB, while the local oscillator frequency is reduced by more than 54 dB and the portion below the local oscillator frequency by more than 70 dB.

25.2
Receivers

The receiver has the task of filtering out the desired signal from the antenna signal and amplifying it enough to feed it to the demodulator. In most instances, the receive frequency is variable so that different channels, for example, various radio stations, can be received. As the signal level may vary widely depending on the distance between transmitter and receiver, the receiver must be provided with amplifiers of variable gain and gain control in order to compensate for the different levels of receive signals. Limiting amplifiers that convert the receive signal into a square wave signal and subsequent filtering can be used only for signals from transmitters with pure angle modulation.

First we shall describe receivers for analog modulation in which the receive signal is converted to an intermediate frequency and then demodulated in an analog demodulator (for example, detector for AM and envelope discriminator for FM). Then we shall discuss the expansions to enable the reception of digital modulated signals.

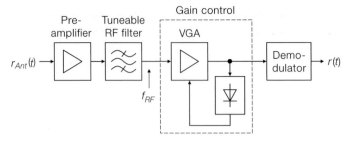

a Direct-detection receiver

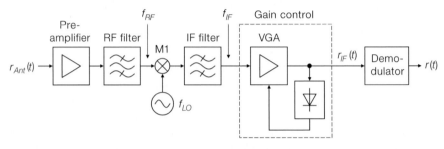

b Superheterodyne receiver (with one intermediate frequency)

Fig. 25.8. Types of receivers

25.2.1
Direct-Detection Receivers

In the pioneer days of radio engineering, the *direct-detector* receiver shown in Fig. 25.8a was used. The receive signal was filtered by an RF filter and fed directly to the demodulator after a fixed or variable amplification. The RF filter needed tuning in order to receive the signals from different radio stations. The only modulation technology that could be used was amplitude modulation since the envelope detector was the only demodulator that worked satisfactorily with a variable carrier frequency $f_C = f_{RF}$. All other demodulators must be set up for a fixed carrier frequency or require frequency-synchronous tuning according to the RF filter.

Besides being limited to amplitude modulation, the direct-detection receiver has other significant draw-backs:

- The transmission frequency must be no more than two orders of magnitude greater than the bandwidth of the signal to be received; otherwise, the quality of the RF filter becomes too high. In the early days of broadcasting systems, there were only a few stations with significantly differing transmission frequencies. A simple resonant circuit was therefore sufficient to filter out the desired station.
- Tuneable filters of high quality are expensive and can only be tuned to a very limited frequency range if the bandwidth is to be maintained. On the other hand, the resonant circuits used in the early days allowed easy tuning by means of a variable capacitor.
- The entire amplification must be done at the transmission frequency, thus high-frequency transistors with high quiescent currents and relatively low gains must be used.

– With increasing frequencies, the performance of envelope detectors decreases due to the parasitic capacitance of the rectifier diode.

With the growing density of transmitting stations and the use of higher frequencies, the direct-detection receiver soon reached its limits.

25.2.2
Superheterodyne Receivers

In the *superhet(erodyne)* receiver, the tuning of the RF filter is replaced by the frequency conversion from a mixer with variable local oscillator frequency f_{LO}. This converts the signal to be received to a fixed *intermediate frequency (IF frequency)*:

$$f_{IF} = f_{RF} - f_{LO} \ll f_{RF}$$

An *intermediate frequency filter (IF filter)* of a substantially lower quality

$$Q_{IF} \sim \frac{f_{IF}}{B} \overset{f_{IF} \ll f_{RF}}{\ll} \frac{f_{RF}}{B} \sim Q_{RF}$$

is used to filter out the signal. The variable amplification and the demodulation are also done at the IF frequency. Thus, all disadvantages of the direct-detection receiver are eliminated. Figure 25.8b shows the construction of a superhet receiver with one intermediate frequency.

RF Filters

In the process of frequency conversion, not only the desired receive frequency

$$f_{RF} = f_{LO} + f_{IF}$$

but also the *image frequency*

$$f_{RF,im} = f_{LO} - f_{IF}$$

are converted to the IF frequency (see Fig. 25.9). This causes a region located at the opposite side of the local oscillator frequency to be converted to the pass band of the IF filter. In order to prevent this, the RF filter in front of the mixer must be set up such that all desired receive frequencies are within the pass band and the related image frequencies in the cutoff region (see Fig. 25.10). The RF filter is thus also known as the *image filter*. In practice, the RF filter is designed such that the local oscillator frequencies are also in the cutoff region. This prevents the relatively strong signal of the local oscillator from moving backwards into the pre-amplifier and to the receiving antenna. This characteristic is of high importance because the undesirable emission of local oscillator signals from the

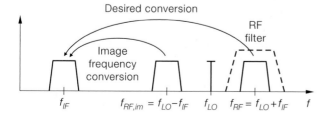

Fig. 25.9. Image frequency in the superhet receiver

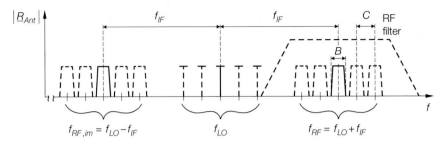

Fig. 25.10. RF filter design in the superhet receiver

receiving antenna is a major problem in the design of receivers which comply with EMC regulations.

In practice, the local oscillator signals are not sinusoidal but present strong harmonic distortions. This results in additional image frequencies of higher order on both sides of the harmonics of the local oscillator frequency which are also converted to the pass band region of the IF filter:

$$f_{RF,im(n)} = nf_{LO} \pm f_{IF}$$

These image frequencies and the corresponding harmonics of the local oscillator frequency must also be suppressed by the RF filter. The RF filter must therefore provide a high stopband attenuation even above the range of reception. LC filters or filters with dielectric resonators are used in practical applications where two to four resonant circuits are typical. These filters are called 2-, 3- or 4-pole filters. The number of poles refers to the equivalent lowpass filter and is thus equal to the number of resonant circuits[1].

With an increasing receive frequency and a constant IF frequency, the relative difference between the receive frequency and the image frequency becomes smaller and smaller; thus causing the quality

$$Q_{RF} \sim \frac{f_{RF}}{f_{IF}}$$

of the RF filter to increase. Where the separation of the receive and image frequencies can no longer be achieved by reasonable means, it is necessary to either increase the IF frequency in order to reduce the quality of the RF filter or to use a superhet receiver with two intermediate frequencies.

It is also possible to configure the RF filter such that the frequency $f_{LO} - f_{IF}$ below the local oscillator frequency is used as the receive frequency f_{RF}, while the corresponding image frequency $f_{RF,im} = f_{LO} + f_{IF}$ is suppressed. In this case, the mixer M1 operates in the *inverted mode* as the frequency sequence is inverted due to the relation $f_{IF} = f_{LO} - f_{RF}$; but, with $f_{IF} = f_{RF} - f_{LO}$, the mixer operates in the *noninverted mode* and the frequency sequence remains the same.

In noninverted mode, the image frequency is below the receive frequency, while in inverted mode, it is above. Therefore, the inverted mode is always used in cases where the frequency range above the receive frequency has clearly weaker signals than the frequency

---

[1] A simple resonant circuit has two poles: $s = \pm j\omega_0$. A filter with four resonant circuits therefore has eight poles, but is still called a 4-pole filter in practice since bandpass filters with a lowpass/bandpass transformation are calculated on the basis of an equivalent lowpass filter with half the number of poles.

range below the receive frequency; in this way it is easier to suppress the image frequency. The inverted mode must be compensated for in the modulator or by an inverted mode in the transmitter.

Pre-Amplifiers

A *low-noise amplifier* (LNA) is used in front of the RF filter to keep the noise figure of the receiver low (see Fig. 25.8b). Without a pre-amplifier, the noise figure is according to (4.201):

$$F'_r = F_{RFF} + \frac{F_{M1} - 1}{G_{A,RFF}} \overset{\substack{F_{RFF}=D_{RFF} \\ G_{A,RFF}=1/D_{RFF}}}{=} D_{RFF} F_{M1}$$

Here, F_{RFF} is the noise figure and $G_{A,RFF}$ is the available power gain of the RF filter and F_{M1} is the noise figure at the input of mixer M1. The latter is calculated with (4.201) from the noise figure of the mixer and the noise figures of the subsequent components. An overall impedance matching is assumed so that the noise figure of the filter corresponds to the power attenuation D_{RFF} in the pass band region, and the available power gain corresponds to the reciprocal value of the power attenuation. With the typical values $D_{RFF} \approx 1.6\,(2\,dB)$ and $F_{M1} \approx 10\,(10\,dB)$, the noise figure becomes unacceptably high: $F'_r \approx 16\,(12\,dB)$. Using a pre-amplifier with noise figure F_{LNA} and available power gain $G_{A,LNA}$ the noise figure is:

$$F_r = F_{LNA} + \frac{F'_r - 1}{G_{A,LNA}} = F_{LNA} + \frac{D_{RFF} F_{M1} - 1}{G_{A,LNA}}$$

With a sufficiently high gain this value is much smaller than the noise figure without a pre-amplifier and in the limiting case of a very high gain it approaches the noise figure of the pre-amplifier.

In practice, the gain of the pre-amplifier cannot be increased without limits since at this point it is still the entire receive signal of the antenna that is amplified. This means that both the signal to be received and, under good receiving conditions, the signals of neighboring channels can reach relatively high levels which may overdrive a pre-amplifier with too high a gain. In addition, a high gain in the RF range is achievable with great effort only. Therefore, the gain is selected at a level which is high enough to reduce the noise figure of the receiver to an acceptable level. Typical values are $F_{LNA} \approx 2\,(3\,dB)$ and $G_{A,LNA} \approx 10 \ldots 100$ $(10 \ldots 20\,dB)$. In the above example, these values lead to $F_r \approx 2.15 \ldots 3.5\,(3.3 \ldots 5.4\,dB)$ compared to $F'_r \approx 16\,(12\,dB)$ without pre-amplification.

IF Filters

Due to the mixer, the entire pass band region of the RF filter is shifted to the intermediate frequency range (see Fig. 25.11). Here, the channel with the desired receive frequency is filtered out by the IF filter. For this reason the IF filter is also known as the *channel filter*. It must have very steep edges since the transition region between the pass band and the cutoff band must not be wider than the region between adjacent channels. Particularly well suited are surface acoustic wave (*SAW*) *filters* which, despite extremely steep edges, have almost no group delay distortions. In contrast, the group delay distortions of LC or dielectric filters increase with rising edge steepness. Filters with ceramic resonators (*ceramic filters*) are used in applications that are relatively insensitive to group delay distortions, such is the

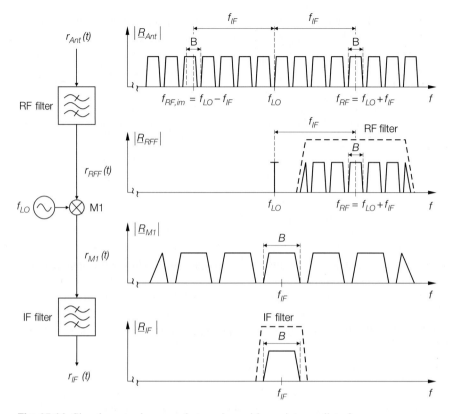

Fig. 25.11. Signal spectra in a superhet receiver with one intermediate frequency

case in AM broadcasting for example. In digital modulation modes, on the other hand, group delay distortions have to be kept as low as possible and thus the use of SAW filters is usually mandatory.

Superhet Receiver with Two Intermediate Frequencies

In the superhet receiver with two intermediate frequencies as shown in Fig. 25.12, the receive frequency is converted into a relatively high first intermediate frequency f_{IF1}, which is selected such that the separation of receive and image frequencies can occur with an RF filter of acceptable quality:

$$Q_{RF} \sim \frac{f_{RF}}{f_{IF1}}$$

Fig. 25.13 shows the signal spectra.

IF filter 1 filters out a portion that contains the desired channel. It is not possible to filter out the desired channel alone at this point because of the necessary high quality. IF filter 1 serves as the image frequency filter for the second mixer, this means that the image frequency

$$f_{IF1,im} = f_{IF1} - 2 f_{IF2}$$

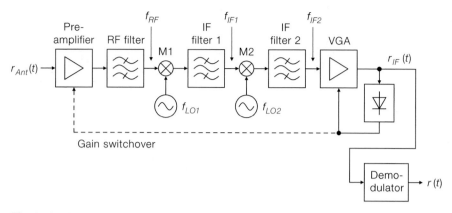

Fig. 25.12. Superhet receiver with two intermediate frequencies

must be within the cutoff band of the filter. To prevent a backwards transmission of the second local oscillator frequency

$$f_{LO2} = f_{IF1} - f_{IF2}$$

this frequency must also be within the cutoff band; consequently, the quality of the filter is:

$$Q_{IF1} \sim \frac{f_{IF1}}{f_{IF2}}$$

After conversion to the second intermediate frequency with the mixer M2, the desired channel is filtered out by means of *IF filter 2* which acts as the channel filter.

It is possible to operate one or both mixers in inverted mode by regarding the frequencies $f_{LO1} - f_{IF1}$ or $f_{LO2} - f_{IF2}$ below the local oscillator frequencies as the receive frequencies. In this case, the RF filter suppresses the image frequency $f_{RF,im} = f_{LO1} + f_{IF1}$ while IF filter 1 suppresses the image frequency $f_{IF1,im} = f_{LO2} + f_{IF2}$. If only one of the mixers is operated in inverted mode, then the frequency sequence is inverted, due to $f_{IF1} = f_{LO1} - f_{RF}$ or $f_{IF2} = f_{LO2} - f_{IF1}$. This must be taken into account in the demodulator or must be compensated via an inverted mode in the transmitter. If both mixers are operated in inverted mode, the overall receiver operates in noninverted mode.

The advantage of the superhet receiver with two intermediate frequencies is that the quality for filtering out the desired channel can be distributed to two IF filters

$$Q_{IF} \sim \frac{f_{IF1}}{B} = \frac{f_{IF1}}{f_{IF2}} \frac{f_{IF2}}{B} \sim Q_{IF1} Q_{IF2}$$

which is in contrast to the superhet receiver with one intermediate frequency where the task must be performed by one IF filter. This is required whenever the receive frequency f_{RF} is very high, meaning that a high (first) intermediate frequency f_{IF1} is required in order to limit the quality of the RF filter or if the bandwidth B of the receive signal is very low.

Generating the Local Oscillator Frequencies

The local oscillator frequencies required are derived from a crystal oscillator by means of a phase-locked loop (PLL) which has already been explained in the description of transmitters (see page 1244 and Fig. 25.7). In receivers with a variable receive frequency,

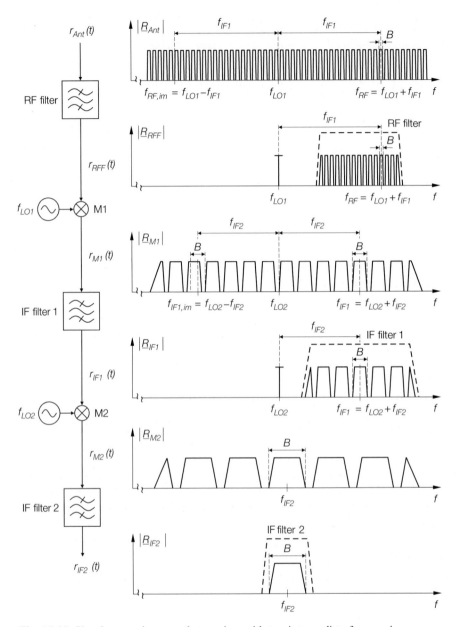

Fig. 25.13. Signal spectra in a superhet receiver with two intermediate frequencies

the frequency of the first local oscillator is varied by adapting the divider factors of the corresponding PLL.

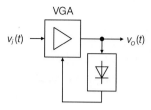

a Simplified diagram

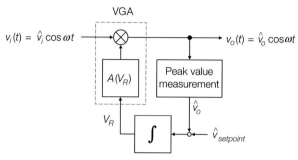

b Equivalent circuit

Fig. 25.14. Gain control

25.2.3
Gain Control

A *variable gain amplifier (VGA)* and an amplitude detector are used for gain control as shown by a simplified diagram in Fig. 25.14a. The VGA generates the voltage

$$v_o(t) = A(V_R) v_i(t) \quad \Rightarrow \quad \hat{v}_o = |A(V_R)| \hat{v}_i \tag{25.1}$$

with the variable gain $A(V_R)$ and the control voltage V_R. A peak value rectifier is usually used to determine the amplitude. By comparing the rectifier output with the setpoint value an integrator generates the control voltage V_R from the difference. Figure 25.14b shows the equivalent circuit for the gain control.

Control Characteristic

In steady state (operating point A) we have $\hat{v}_o = \hat{v}_{setpoint}$ and $V_R = V_{R,A}$ with:

$$|A(V_{R,A})| = \frac{\hat{v}_{setpoint}}{\hat{v}_i}$$

For examination of the dynamic response we linearize (25.1) at the operating point:

$$d\hat{v}_o = \left(\hat{v}_i \frac{d|A|}{dV_R} \right)\bigg|_A dV_R + |A(V_R)|\big|_A d\hat{v}_i$$

$$= \underbrace{\hat{v}_{i,A} \frac{d|A|}{dV_R}\bigg|_A dV_R}_{k_R} + \underbrace{|A(V_{R,A})| \, d\hat{v}_i}_{k_F} \tag{25.2}$$

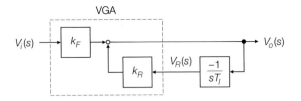

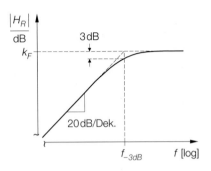

Fig. 25.15. Linear model of the gain control

Fig. 25.16. Frequency response of the gain control

Using factors k_R and k_F and the Laplace transforms

$$V_i(s) = \mathcal{L}\{d\hat{v}_i\} \quad , \quad V_o(s) = \mathcal{L}\{d\hat{v}_o\} \quad , \quad V_R(s) = \mathcal{L}\{dV_R\}$$

we obtain the linear gain control model as shown in Fig. 25.15 featuring the transfer function:

$$H_R(s) = \frac{V_o(s)}{V_i(s)} = k_F \frac{sT_I/k_R}{1+sT_I/k_R} \overset{T_R=T_I/k_R}{=} k_F \frac{sT_R}{1+sT_R}$$

Here, T_I is the time constant of the integrator and T_R the resulting time constant of the control circuit. This results in a highpass filter with gain k_F and a -3 dB cutoff frequency of:

$$f_{-3dB} = \frac{1}{2\pi T_R} = \frac{k_R}{2\pi T_I} = \frac{\hat{v}_{i,A}}{2\pi T_I} \left. \frac{d|A|}{dV_R} \right|_A \tag{25.3}$$

Figure 25.16 shows the frequency response. Changes to the input amplitude with a frequency, that is below the cutoff frequency, are better suppressed with decreasing frequency; while changes with frequencies above the cutoff frequency are amplified with $k_F = |A(V_{R,A})|$. The cutoff frequency must be less than the lower cutoff frequency of the amplitude modulation contained in the desired signal to prevent the desired signal from being invalidated.

According to (25.3) the cutoff frequency is proportional to the input amplitude $\hat{v}_i$ and to the derivative of the gain characteristic $|A(V_R)|$. In order to prevent the cutoff frequency from being dependent on the operating point, the condition

$$k_R = \hat{v}_i \frac{d|A|}{dV_R} = \frac{\hat{v}_{setpoint}}{|A(V_R)|} \frac{d|A|}{dV_R} = \text{const.}$$

must be met; it follows:

$$\frac{d|A|}{dV_R} = \frac{k_R}{\hat{v}_{setpoint}} |A(V_R)| \quad \Rightarrow \quad |A(V_R)| = A_0 \, e^{\frac{k_R V_R}{\hat{v}_{setpoint}}} \tag{25.4}$$

Therefore, the VGA must have an exponential gain characteristic. In practice, the gain is quoted in decibel, i.e. logarithmically, thus producing a linear relationship:

$$A(V_R)\,[\text{dB}] \;=\; A_0\,[\text{dB}] + \frac{k_R V_R}{\hat{v}_{setpoint}} \cdot 8.68\,\text{dB}$$

Variable Gain Amplifier (VGA)

There are several circuit designs for constructing a *variable gain amplifier* (VGA). In integrated circuits, the VGA with differential amplifiers for current distribution as shown in Fig. 25.17 is used almost exclusively. It offers a control range of approximately 60 dB with the required exponential characteristic.

The VGA cell consists of a common-emitter circuit with current feedback (T_1, R_1) and a differential amplifier (T_2, T_3). The quiescent current is adjusted with resistances R_2 and R_3, while R_7 serves as the load resistance. The output current

$$I_{C1}(t) \;=\; I_{C1,A} + i_{C1}(t) \;=\; I_{C1,A} + \frac{g_{m1}}{1 + g_{m1}R_1}\,v_i(t)$$

of the common-emitter circuit is distributed by the differential amplifier to the load resistance and the supply voltage; according to (4.61)[2] this is:

$$I_{C3} \;=\; \frac{I_{C1}}{2}\left(1 + \tanh\frac{V_R}{2V_T}\right) \;=\; \frac{I_{C1}}{1 + e^{-\frac{V_R}{V_T}}}$$

Taking the subsequent amplifier with gain A_V into account, the small-signal output voltage is:

$$v_o(t) \;=\; -A_V\, i_{C3}(t)\, R_7 \;=\; -\frac{A_V\, i_{C1}(t)\, R_7}{1 + e^{-\frac{V_R}{V_T}}} \;=\; -\frac{A_V\, g_{m1}\, R_7}{1 + g_{m1}R_1}\,\frac{v_i(t)}{1 + e^{-\frac{V_R}{V_T}}}$$

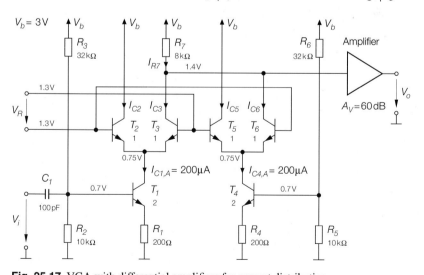

Fig. 25.17. VGA with differential amplifiers for current distribution

---

[2] Current I_{C1} corresponds to the quiescent current $2I_0$ of the differential amplifier.

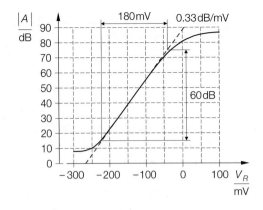

Fig. 25.18. Characteristic of the VGA of Fig. 25.17 ($f = 3$ MHz)

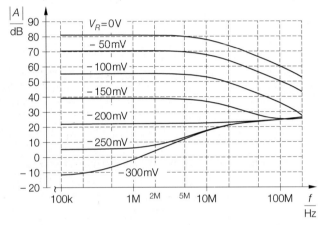

Fig. 25.19. Frequency response of the VGA of Fig. 25.17

The control range is $V_R < -2V_T$. Here, the constant value of one is negligible with respect to the exponential function, and the desired exponential gain characteristic is:

$$v_o(t) \approx -\frac{A_V g_{m1} R_7}{1 + g_{m1} R_1} e^{\frac{V_R}{V_T}} v_i(t) \quad \Rightarrow \quad A(V_R) \approx -\frac{A_V g_{m1} R_7}{1 + g_{m1} R_1} e^{\frac{V_R}{V_T}} \tag{25.5}$$

Figure 25.18 shows the characteristic of the VGA in Fig. 25.17 for a signal frequency of 3 MHz. The control range covers 60 dB with a slope of 0.33 dB/mV. It is limited upward by the deviation from the exponential shape and downward by the reverse attenuation of the VGA cell. The latter depends on the parasitic capacitances and becomes worse with a frequency increase. Figure 25.19 shows the frequency response for different control voltages. Above 10 MHz the gain drops at a rate of 20 dB/decade; thus, the control range narrows accordingly. In this region, the minimum gain increases to 25 dB due to the declining reverse attenuation of the VGA cell.

The change in the current distribution also changes the DC voltage at the output of the VGA cell making the galvanic coupling with the subsequent amplifier difficult. The change can be compensated by connecting a second VGA cell with the same quiescent current ($T_4 \ldots T_6$, $R_4 \ldots R_6$) in parallel and inversely controlling the differential amplifier. Then we have

$$I_{R7,A} = I_{C3,A} + I_{C6,A} = I_{C1,A} = I_{C4,A}$$

and the DC voltage remains constant.

Dimensioning the control circuit according to (25.3) requires factor k_R to be determined. A comparison between (25.4) and (25.5) yields:

$$k_R = \frac{\hat{v}_{setpoint}}{V_T} \tag{25.6}$$

Here, $\hat{v}_{setpoint}$ is the desired amplitude at the VGA output (see Fig. 25.14b). The time constant T_I of the integrator can be calculated from $\hat{v}_{setpoint}$ and the cutoff frequency f_{-3dB}:

$$T_I = \frac{k_R}{2\pi f_{-3dB}} = \frac{\hat{v}_{setpoint}}{2\pi f_{-3dB} V_T} \tag{25.7}$$

Localization of Gain Control in the Receiver

In the direct-detection receiver of Fig. 25.8a, the gain control must be located in the RF section. This is inconvenient because the control range decreases with rising frequencies and the RF frequency is variable. In the superhet receiver with one intermediate frequency shown in Fig. 25.8b, the gain control is located in the IF section behind the IF filter. This arrangement is compulsory since, before the IF filter, the signal contains not only the desired channel but also all the adjacent channels with frequencies in the pass band region of the RF filter.

In systems with received levels that vary extremely, the high levels require an additional gain reduction of the pre-amplifier in order to prevent the subsequent components from being overloaded. The gain switchover of Fig. 25.12 serves this purpose. However, it only works well under the presumption that the high level is caused solely by the desirable channel. Thus, overdriving of the pre-amplifier by a neighboring channel can not be prevented.

From these considerations, it follows that an optimum operating range for all components is only possible if *all* amplifiers are made controllable by the level at their own output. This provides maximum sensitivity for the desired channel independent of the levels of adjacent channels. Such an elaborate design for the gain control is used in exceptional cases only. For most applications, a control system based on the level of the desired signal, as described here, is sufficient.

Level Detection

In addition to the amplitude-controlled useful signal, many systems require a measure for the received level of the useful signal. Typical examples include the VHF broadcasting system with automatic stereo/mono switchover controlled by the received level, and mobile communication in which several base stations receive a signal transmitted from a mobile unit and then the base station with the highest received level takes over the communication.

Level detection can be based on the control voltage of the gain control. If the controllable amplifier has an exponential characteristic, the control voltage V_R is a logarithmic measure for the received level. In steady state, (25.4) provides:

$$\hat{v}_{setpoint} = |A(V_R)| \, \hat{v}_i = A_0 \hat{v}_i \, e^{\frac{k_R V_R}{\hat{v}_{setpoint}}} \quad \Rightarrow \quad V_R = \frac{\hat{v}_{setpoint}}{k_R} \ln\left(\frac{\hat{v}_{setpoint}}{A_0 \hat{v}_i}\right)$$

Using (25.6), it follows for the VGA of Fig. 25.17:

$$V_R = V_T \ln \left(\frac{\hat{v}_{setpoint}}{A_0 \cdot 1\,V} \right) - V_T \ln \left(\frac{\hat{v}_i}{1\,V} \right)$$

If $\hat{v}_i$ increases by a factor of 10 (20 dB), V_R decreases by $V_T \ln 10 \approx 60\,\text{mV}$. Therefore, the slope of the level detection is $-3\,\text{mV/dB}$.

This simple level detection is confined to the exponential portion of the characteristic and depends on the temperature. Integrated receiver circuits usually provide a temperature compensated level signal positive with a slope which is called the *received signal strength indicator (RSSI)*.

Digital Gain Control

With respect to the cutoff frequency of the gain control there are contradicting demands. On one hand, it should be as low as possible so that an amplitude modulation contained in the useful signal is not invalidated; whereas, on the other hand, it should be as high as possible so that following a channel switchover, the steady state is reached in the shortest time possible. One method of optimization is to switch over the time constant of the integrator. In normal operation a large time constant with a correspondingly low cutoff frequency is used, but in the case of large deviations, for example following a switchover to another channel, the system changes to a smaller time constant.

A more flexible and suitable solution is to use a *digital gain control* according to Fig. 25.20. Here, a microcontroller evaluates the *received signal strength indicator (RSSI)* of the last IF amplifier and performs a gain adaptation of the RF and IF amplifiers. Here, too, the majority of the control range has to be covered by the last IF amplifier because all other amplifiers also boost the neighboring channels. If, in addition to the desired channel, the neighboring channels have comparably high levels, then the risk of overdriving exists. Switching the three amplifiers on the input side of Fig. 25.20 is optional. In practice, usually only one amplifier is switched over.

Very often the digital gain control is performed in steps of $2\ldots 4\,\text{dB}$ resolution in accordance with the gain graduation of the last IF amplifier. The gain is adjusted by a binary command (n_{VGA} Bit in Fig. 25.20). The change in gain is done either by a gain switchover in the individual amplifier stages or by using programmable attenuators between the stages.

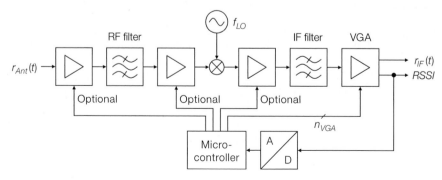

Fig. 25.20. Digital gain control

The microcontroller can evaluate the received level by averaging relatively quickly the $RSSI$ signal, while at the same time taking into account the current amplifier setting. The microcontroller can thus programme all controllable amplifiers in one step with high accuracy, thus significantly reducing the transient time. Following this pre-setting, the duration of averaging is increased so that only amplitude variations with frequencies below the lower limit of the desired signal amplitude modulation are adjusted. In practice, the gain is set by a central microcontroller that controls the overall system. Therefore, it is particularly easy to adapt the performance to the given mode of operation (normal reception, channel switching, search mode, etc.).

25.2.4
Dynamic Range of a Receiver

The dynamic range of a receiver corresponds to the difference between the maximum and minimum received level. The maximum received level is determined by the maximum permissible intermodulation distortions and depends on the *intercept point* of the receiver. The minimum received level follows from the minimum signal-to-noise ratio at the input of the demodulator and depends on the *noise figure* of the receiver. In turn, the intercept point and noise figure of the receiver are dependent on the intercept points, the noise figures and the gain factors of the individual components. Therefore, the main task in designing a receiver is the selection of components with suitable characteristics. On one hand, the performance of the signal processing chain is limited by its weakest member, while on the other hand, components with unnecessarily high characteristics are either expensive or have a high power consumption. Thus, the selection of components must be balanced between the two extremes in order to achieve an optimum result.

In the example below, the dynamic range of the receiver shown in Fig. 25.21 is calculated. It is assumed that the receiver picks up channels in the range of 434 MHz with a bandwidth of $B = 200\,\text{kHz}$ and a channel spacing of $C = 250\,\text{kHz}$. We use a receiver with one intermediate frequency $f_{IF} = 70\,\text{MHz}$. Two identical RF amplifiers with gain $A = 12\,\text{dB}$ are used in the RF stage where RF amplifier 1 corresponds to the pre-amplifier in Fig. 25.8a. The RF filter for suppressing the image frequency

$$f_{RF,im} = f_{RF} - 2f_{IF} = 434\,\text{MHz} - 2 \cdot 70\,\text{MHz} = 294\,\text{MHz}$$

is arranged between the two RF amplifiers and is designed as a two-circuit bandpass filter with an attenuation of 6 dB ($A = -6\,\text{dB}$). A programmable attenuator performs the gain switching to adapt the received level. The attenuator performance can be switched between 1 dB and 25 dB ($A_1 = -1\,\text{dB}$, $A_2 = -25\,\text{dB}$). It should be noted in this respect that the noise figures of a passive reactive filter and an attenuator correspond to the respective attenuation. A diode mixer with a conversion loss of 7 dB ($A = -7\,\text{dB}$) and a noise figure of 7 dB is used as the mixer. Two identical IF amplifiers with gain $A = 25\,\text{dB}$, and the IF filter arranged between them, follow in the IF stage. The IF filter is a surface acoustic wave (SAW) filter with a center frequency of 70 MHz and a bandwidth of 200 kHz. The attenuation is 24 dB ($A = -24\,\text{dB}$). This is followed by a variable gain IF amplifier that provides the

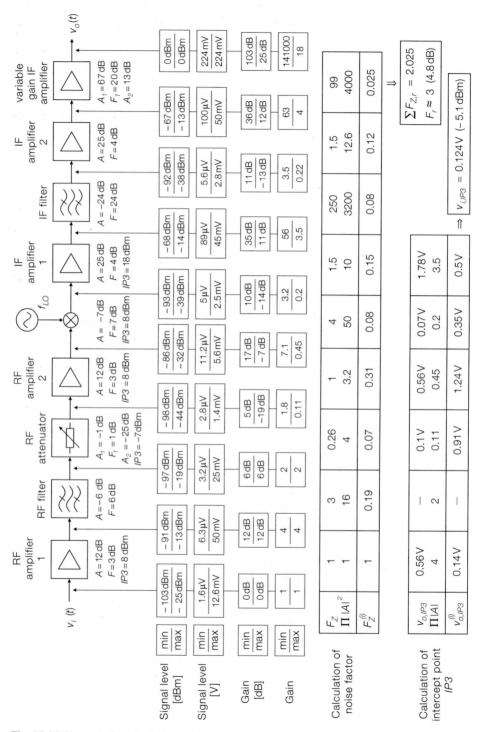

Fig. 25.21. Example for calculating the dynamic range of a receiver

subsequent demodulator with a constant output level of $0\,\mathrm{dBm}$ ($v_{\mathit{eff}} = 224\,\mathrm{mV}$)[3]. It is based on the VGA of Fig. 25.17 and has a high noise figure of $20\,\mathrm{dB}$ which is typical for VGA cells.

Noise Figure of the Receiver

To calculate the noise figure F_r of the receiver, we assume that all components are matched to the characteristic impedance and the quoted gain factors in decibel correspond to the available power gain G_A; it thus follows:

$$G_A\,[\mathrm{dB}] \;=\; A\,[\mathrm{dB}] \quad \Rightarrow \quad G_A \;=\; |A|^2$$

The noise figure can be calculated with (4.201):

$$F_r \;=\; F_1 + \frac{F_2 - 1}{G_{A1}} + \frac{F_3 - 1}{G_{A1}G_{A2}} + \cdots \overset{(4.199)}{=} 1 + F_{Z1} + \frac{F_{Z2}}{|A_1|^2} + \frac{F_{Z3}}{|A_1 A_2|^2} + \cdots$$

Here, $F_Z = F - 1$ is the supplementary noise figure of the respective component. In Fig. 25.21 the noise figures of the components are quoted in decibel, and with

$$F_Z \;=\; 10^{\frac{F\,[\mathrm{dB}]}{10}} - 1$$

we obtain the supplementary noise figures stated in the upper portion of the table. Beneath the supplementary noise figures, the power gains at the input of the receiver, up to the input of the given component, are stated ($\Pi\,|A|^2$). This allows us to convert the supplementary noise figures to the input of the receiver:

$$F_Z^{(i)} \;=\; \frac{F_Z}{\Pi\,|A|^2}$$

The supplementary noise figure and the noise figure of the receiver are obtained by arithmetic addition:

$$F_{Z,r} \;=\; \Sigma\,F_Z^{(i)} \quad \Rightarrow \quad F_r \;=\; F_{Z,r} + 1$$

For the receiver in Fig. 25.21, $F_{Z,r} \approx 2$ and $F_r \approx 3$ ($4.8\,\mathrm{dB}$).

After conversion to the input, the supplementary noise figures of the components indicate their individual contribution to the supplementary noise figure of the receiver. This shows which of the components must be of low-noise design to ensure that the noise figure of the receiver is markedly reduced and which of the components may have a higher noise figure without causing a noticeable increase in the noise figure of the receiver. In the receiver of Fig. 25.21, the contribution of the first RF amplifier dominates, followed by the contributions of the second RF amplifier and the RF filter. Under practical considerations the receiver appears to be well balanced since the noise figures of the RF amplifiers may only be decreased with a great effort. The first RF amplifier often requires a compromise between a low noise figure and a high intercept point. A high intercept point necessitates feedback and this causes an increase in the noise figure.

---

[3] The $0\,\mathrm{dBm}$ level corresponds to a power of $1\,\mathrm{mW}$ at $50\,\Omega$:

$$P \;=\; \frac{v_{\mathit{eff}}^2}{50\,\Omega} \;=\; 1\,\mathrm{mW} \quad \Rightarrow \quad v_{\mathit{eff}} = 223.6\,\mathrm{mV} \quad \Rightarrow \quad v_{\mathit{eff}}\,[\mathrm{dBm}] \;=\; 20\log\frac{v_{\mathit{eff}}\,[\mathrm{V}]}{0.2236\,\mathrm{V}}$$

Minimum Received Level

The minimum received level $P_{i,min}$ is determined by the effective noise power $P_{n,i}$ at the receiver input and the minimum required signal-to-noise ratio $SNR_{i,min}$ for an error-free demodulation of the received signal:

$$SNR_{i,min} = \frac{P_{i,min}}{P_{n,i}} \quad \Rightarrow \quad P_{i,min} = SNR_{i,min} P_{n,i} \tag{25.8}$$

The minimum received level is also called the *sensitivity*, where a lower minimum received level is the same as an increased sensitivity.

The effective noise power results from the thermal noise power density N_0, the bandwidth B and the noise figure F_r of the receiver:

$$P_{n,i} = N_0 B F_r = kT B F_r \overset{T=300\,\text{K}}{=} 4.14 \cdot 10^{-21} \frac{\text{W}}{\text{Hz}} \cdot B F_r \tag{25.9}$$

Consequently:

$$P_{n,i}\,[\text{dBm}] = -174\,\text{dBm} + 10\,\text{dB} \cdot \log \frac{B}{\text{Hz}} + F_r\,[\text{dB}] \tag{25.10}$$

Insertion into (25.8) yields the minimum received level:

$$P_{i,min}\,[\text{dBm}] = -174\,\text{dBm} + 10\,\text{dB} \cdot \log \frac{B}{\text{Hz}} + F_r\,[\text{dB}] + SNR_{i,min}\,[\text{dB}] \tag{25.11}$$

The minimum received level essentially depends on the bandwidth. Therefore, the minimum received level of a system with a high data rate and a resulting high bandwidth is greater than that of systems with a low data rate, provided the systems use the same modulation mode (same as $SNR_{i,min}$) and receivers with the same noise figure. An increase in the data rate by a factor of 10 increases the minimum received level by 10 dB.

We assume that the receiver in Fig. 25.21 receives a QPSK-modulated signal with a maximum symbol failure rate of 10^{-6}. According to [25.2] this requires a power efficiency of $E_b/N_0 = 13\,\text{dB}$. With the required power efficiency, the assumed data frequency $f_D = 200\,\text{kHz}$, and the bandwidth $B = 200\,\text{kHz}^4$, equation (24.83) provides the required signal-to-noise ratio:

$$SNR_{i,min}\,[\text{dB}] = \left(\frac{E_b f_D}{N_0 B} \right)\,[\text{dB}] = 13\,\text{dB}$$

Insertion into (25.11) with $B = 200\,\text{kHz}$ and $F_r \approx 5\,\text{dB}$ leads to the minimum received level:

$$P_{i,min}\,[\text{dBm}] = -174\,\text{dBm} + 53\,\text{dB} + 5\,\text{dB} + 13\,\text{dB} = -103\,\text{dBm}$$

This corresponds to an rms voltage of $1.6\,\mu\text{V}$.

Maximum Received Level

The maximum received level depends on the permissible intermodulation distortions. The dominating intermodulation of 3rd order ($IM3$) is as described by the intermodulation ratio

---

[4] We presume a QPSK system with a data rate $r_D = 200\,\text{kbit/s}$ and a roll-off factor $r = 1$. This results in a data frequency of $f_D = 200\,\text{kHz}$, the symbol frequency $f_S = f_D/2 = 100\,\text{kHz}$ (two bits per symbol) and the bandwidth $B = (1 + r) f_S = 200\,\text{kHz}$ (see (24.84)).

*IM*3 which is characterized by the intercept point *IP*3. These relationships are described in Sect. 4.2.3 on page 426 by way of the amplitude of sinusoidal signals. In telecommunication engineering, the levels are usually given in dBm or the corresponding rms values, but this does not influence the intermodulation ratio *IM*3. From (4.184) it follows:

$$IM3 \approx \left(\frac{\hat{v}_{i,IP3}}{\hat{v}_i}\right)^2 = \left(\frac{v_{i,IP3}}{v_i}\right)^2 \tag{25.12}$$

Here, $v_{i,IP3}$ and v_i are the rms values and $\hat{v}_{i,IP3} = \sqrt{2}v_{i,IP3}$ and $\hat{v}_i = \sqrt{2}v_i$ are the amplitudes of the intercept point *IP*3 and the received signal, both of which are related to the input of the receiver. In practice, the intermodulation ratio is quoted in decibel and the rms values of the intercept point and the received signals are quoted in dBm. It thus follows:

$$IM3\,[\text{dB}] \approx 2\left(v_{i,IP3}\,[\text{dBm}] - v_i\,[\text{dBm}]\right) \tag{25.13}$$

The intercept point is determined by means of a two-tone signal; therefore, the intermodulation ratios according to (25.12) and (25.13) are also only valid for a two-tone signal. The receiver receives a very complex signal that is composed of the desired receive signal and the signals of adjacent channels. For this reason, the intermodulation ratio cannot be given; whereas, in practice, the two-tone intermodulation ratio is used as a substitute. For this purpose, the permissible nonlinearity is determined for the case of two neighboring channels with the same level, then the relevant two-tone intermodulation ratio and the intercept point are calculated from this value. We will not go into details on this and assume that the required two-tone intermodulation ratio is known.

The intercept point $v_{i,IP3}$ of the receiver is calculated from the intercept points of the components, but only the components up to the last IF filter are taken into account since, behind this filter, the signals of all adjacent channels are suppressed. Figure 25.21 shows the output intercept points of the components in dBm. The lower portion of the table contains the resulting effective values $v_{o,IP3}$, which are converted to the input by means of the related gain factors from the receiver input to the output of each component ($\Pi\,|A|$):

$$v_{o,IP3}^{(i)} = \frac{v_{o,IP3}}{\Pi\,|A|}$$

In Sect. 4.2.3 we demonstrated that the inverse square of the intercept points of 3rd order of a series connection must be added (see page 439):

$$\frac{1}{v_{i,IP3}^2} = \Sigma\,\frac{1}{v_{o,IP3}^{(i)\,2}}$$

For the receiver in Fig. 25.21, this leads to $v_{i,IP3} = 0.124\,\text{V}$ ($-5.1\,\text{dBm}$). QPSK usually requires an intermodulation ratio of $IM3 \approx 10000$ (40 dB); the maximum received level is obtained with (25.13):

$$P_{i,max}\,[\text{dBm}] = v_i\,[\text{dBm}] = v_{i,IP3}\,[\text{dBm}] - \frac{IM3\,[\text{dB}]}{2}$$

$$= -5.1\,\text{dBm} - \frac{40\,\text{dB}}{2} \approx -25\,\text{dBm}$$

This corresponds to an rms value of 12.6 mV.

The component intercept points converted to the input indicate the contribution which the components make to the intercept point of the receiver; a higher value is better than a lower value. In Fig. 25.21, the contribution of the first RF amplifier dominates and is enhanced by squaring the values for the inverted quadratic addition. The dominance of the intercept point of the first RF amplifier is typical of receivers and an improvement at this point is difficult to achieve and only possible at the cost of the noise figure or the current consumption.

Dynamic Range

The *maximum dynamic range* of the receiver can be determined from the minimum and maximum received levels:

$$D_{max} \, [\text{dB}] \; = \; P_{i,max} \, [\text{dBm}] \, - \, P_{i,min} \, [\text{dBm}] \tag{25.14}$$

For the receiver in Fig.25.21 it follows:

$$D_{max} \; = \; -25 \, \text{dBm} - (-103 \, \text{dBm}) \; = \; 78 \, \text{dB}$$

The signal levels for borderline cases are given in Fig. 25.21 in both dBm and Volt. One should note that the signal levels are related to the portion of the received useful signal. The overall levels may be significantly higher if adjacent channels with higher levels exist. Only after the last IF filter are the levels of the useful signal and the overall system equal; in this case, the signals of all adjacent channels are suppressed.

The *available dynamic range* depends on the signal levels in the adjacent channels and may be much lower than the maximum dynamic range. We examine the case of the receiver in Fig. 25.21 receiving an adjacent channel with a maximum level of $P_{i,max} = -25 \, \text{dBm}$. In this case, there are intermodulation distortions, some of which add to the received channel and cause a noise-like interference with a level that is clearly higher than the thermal noise level. Therefore, the level of the useful signal must be above the minimum received level $P_{i,min} = -103 \, \text{dBm}$ by the same factor to guarantee the required signal-to-noise ratio. This reduction in sensitivity is particularly undesirable in radio receivers and is the reason why weak stations which are located close to powerful stations can not be received. The same problem occurs in base stations of mobile communication systems that must be capable of receiving very different signal levels from several mobile units. The mobile units themselves are less critical since they normally use the highest received level in their communication with the base station. Blocking of one mobile unit by other mobile units operating in the immediate vicinity is prevented by using a different frequency range for communication from the mobile units to the base stations (*uplink*) than from the base stations to the mobile units (*downlink*) (see Fig. 24.21). The separation of *uplink* and *downlink* ranges is achieved by a *duplexer* consisting of two bandpass filters. Figure 25.22 shows an example of a mobile unit for GSM900. The two ranges are separated by a frequency gap which is needed as the transition region for the bandpass filters of the duplexer. One disadvantage of this is the increase in the noise figure caused by the duplexer. The noise figure is increased by the power attenuation D_D of the duplexer:

$$F'_r \; \overset{(4.201)}{=} \; F_D + \frac{F_r - 1}{G_{A,D}} \; \overset{F_D = 1/G_{A,D} = D_D}{=} \; D_D + D_D \, (F_r - 1) \; = \; D_D F_r$$

Here, F_r is the noise figure of the receiver without a duplexer. Consequently:

$$F'_r \, [\text{dB}] \; = \; D_D \, [\text{dB}] + F_r \, [\text{dB}]$$

A typical value for duplexers is $D_D \approx 3 \ldots 4 \, \text{dB}$. This means that the maximum dynamic range is reduced by the factor D_D when a duplexer is used. On the other hand, the available

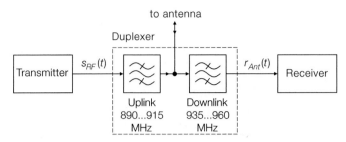

Fig. 25.22. Separation of the *uplink* and *downlink* regions by means of the duplexer in case of a mobile unit for GSM900

dynamic range increases significantly when the unit is operated in the vicinity of other mobile units since their comparably strong transmission signals can no longer reach the receiver.

The available dynamic range depends on the stop-band attenuation of the RF and IF filters. If, for example, the last IF filter has a stop-band attenuation of 50 dB, but the level of the adjacent channel is 50 dB higher, then the levels of the desired and the adjacent channels at the output of the filter are the same, and no reception is possible. The position of the image frequencies and the levels that occur at these frequencies, which are determined by the selected IF frequencies, also have an effect on the available dynamic range. Therefore, besides the above-mentioned considerations, a multitude of additional considerations must be taken into account when designing a receiver.

25.2.5
Receivers for Digital Modulation

Receivers for digital modulation methods have basically the same design as receivers for analog modulation but differ in terms of the demodulator. While analog demodulators process the IF signal directly, digital demodulators perform an additional frequency conversion by means of an I/Q mixer in order to provide the quadrature components $i(t)$ and $q(t)$.

The principle construction of a demodulator for digital modulation methods is illustrated in Fig. 24.69; Fig. 25.23a shows the same version with an additional gain control. The input signal is given by the IF signal $r_{IF}(t)$ of a superhet receiver with one or two intermediate frequencies (see Fig. 25.8b or Fig. 25.12) and corresponds to the carrier signal $s_C(t)$ of Fig. 24.69. The quadrature components $i(t)$ and $q(t)$ are derived with the help of an I/Q mixer and two lowpass filters and fed to the demodulator.

Compared to a receiver for analog modulation, the lowpass filters behind the I/Q mixer act like an additional filter. Therefore, in a receiver for digital modulation, the desired channel is normally not filtered out by the last IF filter but by the lowpass filters behind the I/Q mixer; this is the reason they are called *channel filters* in Fig. 25.23a. With respect to the filtering function, a receiver for digital modulation with one intermediate frequency already has the same characteristics as a receiver for analog modulation with two intermediate frequencies. Figure 25.24 shows the relevant signal spectra for the i branch which are the same as for the q branch.

However, channel filtering behind the I/Q mixer has two disadvantages:

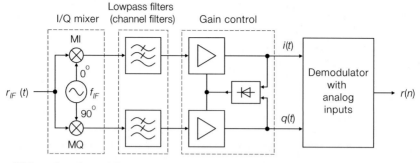

a With analog channel filters and analog gain control

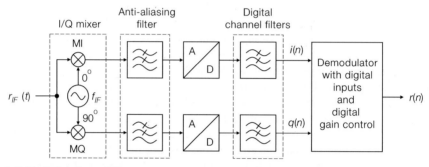

b With digital channel filters

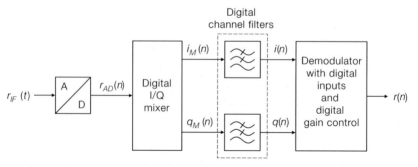

c With IF sampling and digital channel filters

Fig. 25.23. Receiver for digital modulation methods (without RF and IF components refer to Fig. 25.8b and Fig. 25.12)

- The gain control can be performed only after the lowpass filters since the IF signal may still contain adjacent channels with considerably higher signal levels. Gain control requires two variable gain amplifiers that amplify the mean of the absolute value

$$\overline{|r_{CB}(t)|} = \overline{\sqrt{i^2(t) + q^2(t)}}$$

of the complex baseband signal $r_{CB}(t) = i(t) + j\,q(t)$ to a setpoint value. An analog realization of this gain control is rather complex.
- The lowpass filters for channel filtering must have very steep edges as the frequency gap between the useful and the adjacent channels is very narrow. At the same time, the group

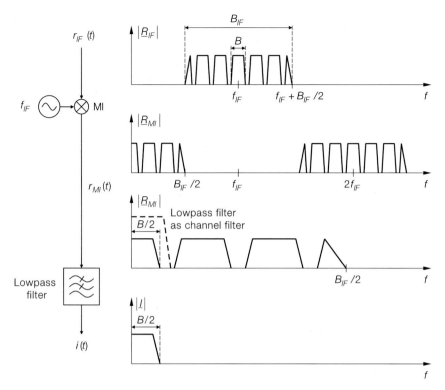

Fig. 25.24. Signal spectra for a digital receiver with analog channel filters according to Fig. 25.23a (shown for branch i only, branch q is identical)

delay in the useful channel must be as constant as possible since digital modulation is very sensitive to group delay distortions. It is difficult to meet these demands with analog lowpass filters.

Due to these disadvantages, a demodulator with analog inputs usually is used in combination with channel filtering and gain control in the IF range. In this case, the lowpass filters in Fig. 25.23a are only required to suppress the portions of the double IF frequency and the gain control for i and q is unnecessary.

Receivers with Digital Channel Filters

A version which is more suitable for practical use is obtained with digital filters as channel filters and a demodulator with digital inputs (see Fig. 25.23b). The output signals of the I/Q mixer undergo anti-aliasing filtering and are digitized by two A/D converters. The digital channel filters function as linear-phase FIR filters which prevent group delay distortions. The gain control is integrated in the demodulator and adapted to the modulation method used. Figure 25.25 shows the signal spectra for branch i which is identical to branch q.

The demands on the anti-aliasing filter are comparatively low as the range $2f_{IF} - (B_{IF} + B)/2$ is available for the transition from the pass-band to the stop-band according to Fig. 25.25; an LC filter of the 2nd or 3rd order is usually sufficient. In practice, an attenuated portion of the IF signal and the local oscillator signal are contained in the

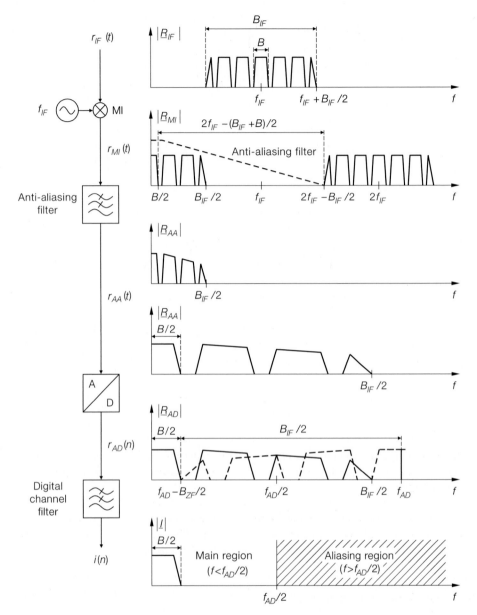

Fig. 25.25. Signal spectra of a digital receiver with digital channel filters according to Fig. 25.23b (for branch i only, branch q is identical)

output signal of the mixers which is caused by asymmetries and crosstalk. Usually the attenuation of the IF signal is high enough to prevent any interference. The level of the local oscillator signal is substantially higher and must be reduced. This can be done in two ways:

− Stop filters with a resonant frequency that is matched to the IF frequency are added to the anti-aliasing filters (see Fig. 25.26).

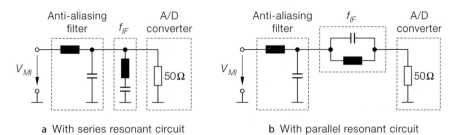

a With series resonant circuit   b With parallel resonant circuit

Fig. 25.26. Anti-aliasing filters with stop filters for the IF frequency to attenuate the local oscillator signal

– The sampling frequency of the A/D converter is selected such that the difference between the IF frequency and the harmonics of the sampling frequency is larger than half the bandwidth of the desired signal ($= B/2$). After sampling, the IF frequency is then within the stop-band of the digital channel filters.

A combination of both methods is also possible.

After anti-aliasing filtering, the signal has an upper cutoff frequency corresponding to half the bandwidth of the IF filter ($= B_{IF}/2$) (see Fig. 25.25). Therefore, a sampling frequency $f_{AD} > B_{IF}$ would be required for nonaliasing A/D conversion. Since the subsequent digital channel filter suppresses all portions above a frequency of half the bandwidth of the desired signal ($= B/2$), aliasing may be permitted in this region. Consequently the sampling frequency is:

$$f_{AD} > \frac{B_{IF} + B}{2} \tag{25.15}$$

Figure 25.25 is the borderline case of the minimum sampling frequency. The aliasing components shown as a broken line go right up to the desired channel.

The IF signal and the signals behind the mixers still contain several adjacent channels; therefore, the overall level of these signals may be significantly higher than the level of the desired channel. In order to prevent overdriving of the A/D converters, it is necessary to use a gain control for the IF signal in addition to the gain control for the desired channel that is integrated in the demodulator. For this purpose, the gain control in the IF range, which exists in the superhet receivers shown in Figs. 25.8b and 25.12, is used.

Dynamic range: The available dynamic range of the receiver depends primarily on the resolution of the A/D converter. We demonstrate this for a desired channel of power P_C and an adjacent channel of power P_{AC}. The corresponding spectrum at the output of an A/D converter is plotted in Fig. 25.27. The powers of the channels correspond to the areas below the respective curves[5]. $P_{n,Q}$ is the power of the quantization noise of the A/D converter and is evenly distributed in the frequency interval from zero up to half the sampling frequency. We assume that the power in the neighboring channel is significantly higher than the power

---

[5] The power of a signal $x(t)$ with the Fourier transform (two-sided spectrum) $X(f)$ is:

$$P_x = \int_{-\infty}^{+\infty} |X(f)|^2 \, df$$

This is called the *Parseval equation*. We use unilateral absolute spectra; this eliminates the negative frequencies, and the lower limit of the integral becomes zero.

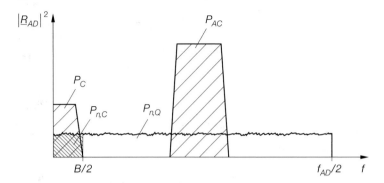

Fig. 25.27. Spectrum at the output of the A/D converter for a desired channel with the power P_C and an adjacent channel with the power P_{AC}. $P_{n,Q}$ is the power of the quantization noise, $P_{n,C}$ is the portion in the desired channel.

in the desired channel. Thus, the overall power is approximately equal to the power in the adjacent channel:

$$P = P_C + P_{AC} + P_{n,Q} \overset{P_{AC} \gg P_C, P_{n,Q}}{\approx} P_{AC}$$

When fully modulated, an ideal A/D converter with a resolution of N bits achieves the signal-to-noise ratio:

$$SNR = \frac{3 \cdot 2^{2N}}{C^2} \quad \Rightarrow \quad SNR\,[\mathrm{dB}] = N \cdot 6\,\mathrm{dB} + 4.8\,\mathrm{dB} - CF\,[\mathrm{dB}] \tag{25.16}$$

The *crest factor* of the signal is

$$CF = \frac{\text{Peak value}}{\text{rms value}} = \frac{v_{max}}{v_{eff}} \tag{25.17}$$

The crest factor ranges from $CF = 1$ (0 dB) for a square-wave signal to $CF \approx 4$ (12 dB) for a noise-like signal[6]. This means that the achievable signal-to-noise ratio depends on the signal in the adjacent channel. The power of the quantization noise can be calculated from the overall power P and the signal-to-noise ratio:

$$SNR = \frac{P}{P_{n,Q}} \quad \Rightarrow \quad P_{n,Q} = \frac{P}{SNR} = \frac{PCF^2}{3 \cdot 2^{2N}}$$

The portion is within the desired channel (see Fig. 25.27):

$$P_{n,C} = P_{n,Q} \frac{B}{f_{AD}} = \frac{PCF^2}{3 \cdot 2^{2N}} \frac{B}{f_{AD}}$$

In order to ensure correct demodulation of the desired signal, the signal-to-noise ratio SNR_C in the desired channel must be higher than the minimum signal-to-noise ratio $SNR_{i,min}$ of the modulation method used:

$$SNR_C = \frac{P_C}{P_{n,C}} > SNR_{i,min}$$

---

[6] For a sinusoidal signal with $CF = \sqrt{2}$ (3 dB) (25.16) produces the relationship $SNR = N \cdot 6\,\mathrm{dB} + 1.8\,\mathrm{dB}$ (see (18.9) on page 950).

Thus, the power in the desired channel is

$$P_C > \frac{SNR_{i,min} P C F^2}{3 \cdot 2^{2N}} \frac{B}{f_{AD}} \tag{25.18}$$

and the permitted ratio of the powers in the adjacent channel and the desired channel (available dynamic range) is:

$$\frac{P_{AC}}{P_C} \overset{P_{AC} \approx P}{\approx} \frac{P}{P_C} < \frac{3 \cdot 2^{2N}}{SNR_{i,min} C F^2} \frac{f_{AD}}{B} \tag{25.19}$$

The parameters $SNR_{i,min}$, CF and B are defined by the modulation method used. Therefore, the available dynamic range is essentially determined by the resolution N of the A/D converter and the sampling frequency f_{AD}. While the sampling rate is often increased in audio applications to achieve a better signal-to-noise ratio (*oversampling*), this is usually not possible in receivers due to their very high minimum sampling rate; here, it is the resolution that must be increased if the available dynamic range is not large enough.

Owing to a number of interferences, the signal-to-noise ratio of a real A/D converter is lower than that of an ideal A/D converter according to (25.16). In practice, one must use the *effective resolution* $N_{eff} < N$, which is quoted in data sheets, in place of the resolution N. Instead of the effective resolution, many data sheets specify the signal-to-noise ratio for a sinusoidal signal as a function of the signal and the sampling frequencies; in this case, the effective resolution is

$$N_{eff} = \frac{SNR\,[\mathrm{dB}] - 1.8\,\mathrm{dB}}{6\,\mathrm{dB}} \tag{25.20}$$

Example: We consider a receiver for a QPSK system with a data rate $r_D = 200\,\mathrm{kbit/s}$, a roll-off factor $r = 1$ and a bandwidth of $B = 200\,\mathrm{kHz}$. The bandwidth of the last IF filter is assumed to be $B_{IF} = 1\,\mathrm{MHz}$. According to (25.15) the sampling frequency is

$$f_{AD} > \frac{B_{IF} + B}{2} = 600\,\mathrm{kHz}$$

We choose $f_{AD} = 800\,\mathrm{kHz}$. At a bit error rate of 10^{-6} QPSK requires a minimum signal-to-noise ratio $SNR_{i,min} = 20$ (13 dB); with $r = 1$ the crest factor is $CF \approx 1.25$ (2 dB). Furthermore, we assume an available dynamic range of $P_{AC}/P_C = 10^6$ (60 dB). Solving (25.19) for N we obtain:

$$N > \frac{1}{2}\,\mathrm{ld}\left(\frac{P_{AC}}{P_C} \frac{SNR_{i,min} C F^2}{3} \frac{B}{f_{AD}}\right) = \frac{1}{2}\,\mathrm{ld}\left(10^6 \cdot 10.4 \cdot \frac{1}{4}\right) \approx 10.7$$

Thus, an A/D converter with an effective resolution of at least 10.7 bits at $f_{AD} = 800\,\mathrm{kHz}$ is required. If operated with a sinusoidal signal, the signal-to-noise ratio according to (25.20) is $SNR = 10.7 \cdot 6\,\mathrm{dB} + 1.8\,\mathrm{dB} = 66\,\mathrm{dB}$. In practice, this means that a 12 bit converter is required.

This example is typical of receivers with digital channel filters. A/D converters with a comparably high resolution are required despite the fact that the signal-to-noise ratio $SNR_{i,min}$ required in the desired channel is very low. This is necessary due to the high signal levels in adjacent channels.

Receivers with IF Sampling and Digital Channel Filters

Using digital I/Q mixers in addition to digital channel filters results in a receiver with IF sampling in which the IF signal is already digitized as shown in Fig. 25.23c. As the bandwidth B_{IF} of the IF signal is usually significantly lower than the *IF* frequency, *sub-sampling* is possible, which means that the sampling frequency f_{AD} is selected lower than the IF frequency without compromising the demand that $f_{AD} > 2B_{IF}$. The aliasing effect converts the IF signal to a lower frequency. Figure 25.28 illustrates this with sampling in the first, second and third aliasing region as compared to sampling in the main region.

Sampling in the main region means that it is necessary to comply with the sampling theorem in its standard form, i.e. the upper cutoff frequency must be lower than half the

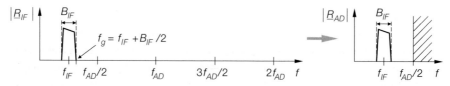

a Sampling in the main region ($m = 0$, normal mode)

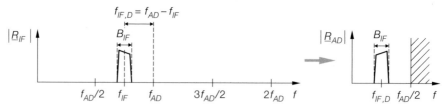

b Subsampling in the first aliasing region ($m = 1$, inverted mode)

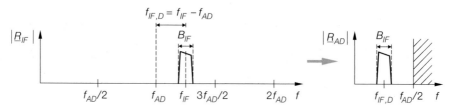

c Subsampling in the second aliasing region ($m = 2$, normal mode)

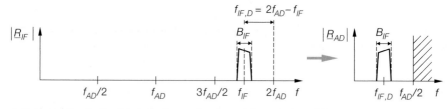

d Subsampling in the third aliasing region ($m = 3$, inverted mode)

Fig. 25.28. Frequency conversion in IF sampling

sampling frequency:

$$f_g = f_{IF} + \frac{B_{IF}}{2} < f_{AD}$$

For subsampling in the mth aliasing region, the IF signal must be fully within this frequency range[7]. Thus, at the lower limit this requires

$$f_{IF} - \frac{B_{IF}}{2} > m\frac{f_{AD}}{2}$$

and at the upper limit:

$$f_{IF} + \frac{B_{IF}}{2} < (m+1)\frac{f_{AD}}{2}$$

In summary, the general condition for the sampling frequency f_{AD} is:

$$\frac{2f_{IF} + B_{IF}}{m+1} < f_{AD} < \frac{2f_{IF} - B_{IF}}{m} \qquad \text{with } m \le \frac{f_{IF}}{B_{IF}} - \frac{1}{2} \qquad (25.21)$$

For $m = 0$ this also applies to the main region but in this case the upper limit is excluded. By inserting the maximum possible integer for m into (25.21), the minimum sampling frequency $f_{AD,min}$ is obtained. This minimum value depends on the ratio f_{IF}/B_{IF} and lies within the range:

$$2B_{IF} < f_{AD,min} < 2B_{IF}\left(1 + \frac{B_{IF}}{2f_{IF}}\right)$$

The digital IF frequency $f_{IF,D}$ at the output of the A/D converter is:

$$f_{IF,D} = \begin{cases} f_{IF} - m\dfrac{f_{AD}}{2} & m \text{ even} \\[2mm] (m+1)\dfrac{f_{AD}}{2} - f_{IF} & m \text{ odd} \end{cases} \qquad (25.22)$$

It follows that with even m values, the IF signal is converted in noninverted mode and with odd m values in inverted mode (see Fig. 25.28). The inverted mode must either be taken into account in the demodulator or compensated by an inverted mode in the transmitter or in the mixers of the previous superhet receiver.

The I/Q mixer generates the signals from the digital output signal $r_{AD}(n)$ of the A/D converter:

$$i_M(n) = r_{AD}(n)\cos\left(2\pi n\frac{f_{IF,D}}{f_{AD}}\right)$$

$$q_M(n) = -r_{AD}(n)\sin\left(2\pi n\frac{f_{IF,D}}{f_{AD}}\right)$$

The digital quadrature components $i(n)$ and $q(n)$ are obtained after channel filtering. The digital I/Q mixer becomes particularly simple if the digital IF frequency is equal to a

---

[7] This condition applies only to cases where the entire IF signal is to be processed digitally. If limited to the desired channel, aliasing can be allowed as long as the desired channel is not affected. This will be detailed further below.

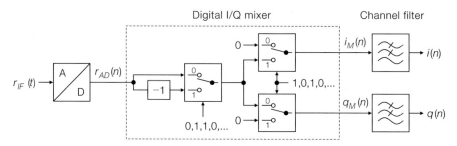

Fig. 25.29. Digital receiver with IF sampling where $f_{IF,D} = f_{AD}/4$. The switches operate in synchronization with the A/D converter.

quarter of the sampling frequency; then we have

$$f_{IF,D} = \frac{f_{AD}}{4} \quad \Rightarrow \quad \begin{cases} i_M(n) = r_{AD}(n) \cos\left(\dfrac{\pi n}{2}\right) \\[2mm] q_M(n) = -r_{AD}(n) \sin\left(\dfrac{\pi n}{2}\right) \end{cases} \tag{25.23}$$

with:

$$\cos\left(\frac{\pi n}{2}\right) = 1, 0, -1, 0, \ldots \qquad \text{for } n = 0, 1, 2, 3, \ldots$$

$$\sin\left(\frac{\pi n}{2}\right) = 0, 1, 0, -1, \ldots \qquad \text{for } n = 0, 1, 2, 3, \ldots$$

In this case, only the factors 0 (value is suppressed), 1 (value is taken over) and -1 (value is taken over with the sign inverted) occur and no multiplication is required. Equation (25.23) specifies the relationship:

$$i_M(n) = [\; r_{AD}(0), \quad 0 \quad , -r_{AD}(2), \quad 0 \quad , r_{AD}(4), \quad 0 \quad , \ldots]$$
$$q_M(n) = [\quad 0 \quad , -r_{AD}(1), \quad 0 \quad , r_{AD}(3), \quad 0 \quad , -r_{AD}(5), \ldots]$$

Accordingly, the signal sequence $r_{AD}(n)$ must pass a controlled inverter and then be distributed to the two outputs by a demultiplexer resulting in a digital receiver with IF sampling as shown in Fig. 25.29.

The condition for the sampling frequency is obtained by inserting (25.23) into (25.22):

$$f_{AD} = \frac{4 f_{IF}}{2m + 1} \qquad \text{with } m \leq \frac{f_{IF}}{B_{IF}} - \frac{1}{2} \tag{25.24}$$

It follows that $f_{AD} = 4 f_{IF}$ in the main region ($m = 0$, noninverted mode), $f_{AD} = 4 f_{IF}/3$ in the first aliasing region ($m = 1$, inverted mode), and $f_{AD} = 4 f_{IF}/5$ in the second aliasing region ($m = 2$, noninverted mode), etc. This condition is met in Fig. 25.28. Figure 25.30 lists some of the commonly used IF frequencies together with the corresponding sampling frequencies for $m = 0 \ldots 4$.

For subsampling, it is necessary to use special A/D converters suitable for subsampling because the analog bandwidth (i.e. the bandwidth of the analog input circuitry and the sample-and-hold element) must be higher than the sampling frequency.

IF-frequencies	Sampling frequencies				
	$m = 0$	$m = 1$	$m = 2$	$m = 3$	$m = 4$
455 kHz	1.82 MHz	606.67 kHz	364 kHz	260 kHz	202.22 kHz
10.7 MHz	42.8 MHz	14.267 MHz	8.56 MHz	6.114 MHz	4.756 MHz
21.4 MHz	85.6 MHz	28.533 MHz	17.12 MHz	12.23 MHz	9.511 MHz
70 MHz	280 MHz	93.33 MHz	56 MHz	40 MHz	31.11 MHz

Fig. 25.30. Sampling frequencies for some commonly used IF frequencies

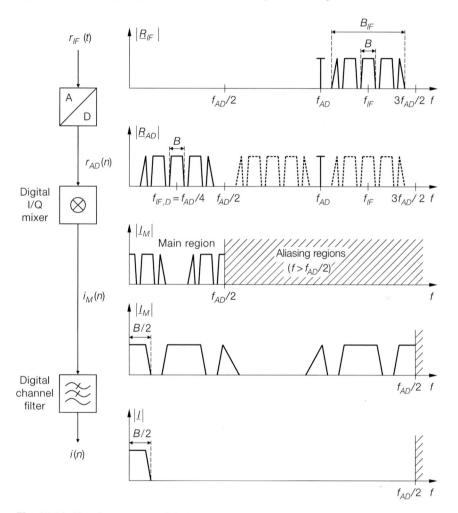

Fig. 25.31. Signal spectra in a digital receiver with IF sampling for $f_{IF,D} = f_{AD}/4$ and $f_{AD} = 4 f_{IF}/5$ ($m = 2$)

Figure 25.31 shows the signal spectra of a digital receiver with IF sampling for $f_{IF,D} = f_{AD}/4$ and $f_{AD} = 4 f_{IF}/5$ ($m = 2$). We can see that no aliasing occurs when the condition of (25.24) is met; this means that the entire IF signal is digitized without any loss. Thus, it is possible to receive adjacent channels by using bandpass filters, instead of lowpass filters, as channel filters and by once again converting the frequency of the output signal. This enables

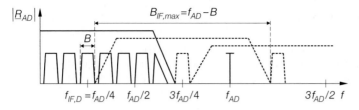

Fig. 25.32. Maximum IF bandwidth in subsampling

the reception of *all* channels that are completely within the pass band of the IF filter without changing the local oscillator frequency. Switching the channel filters is particularly easy in practice as the channel filtering is generally done by a digital signal processor (DSP); only the coefficients for the filter must be exchanged. This method is of particular importance in narrow-band systems since it enables an entire group of channels to be received with the same local oscillator frequency. In extreme cases the entire frequency band of the application is within the IF bandwidth; here, one can use a fixed local oscillator frequency and perform the channel selection solely by switching the channel filters. If, however, only the desired channel is to be processed, as in Fig. 25.31, aliasing can be permitted as long as the desired channel is not affected; thus the condition for *m* in (25.24) can then be widened. We illustrate this by enlarging the IF bandwidth in Fig. 25.31 to the point just before aliasing in the desired channel occurs (see Fig. 25.32). It thus follows:

$$B_{IF,max} = f_{AD} - B \quad \Rightarrow \quad f_{AD} > B_{IF} + B \tag{25.25}$$

Inserting (25.24) into (25.25) and solving for *m* yields the condition:

$$m < \frac{2 f_{IF}}{B_{IF} + B} - \frac{1}{2} \tag{25.26}$$

A comparison of (25.25) and (25.15) shows that with IF sampling the minimum sampling frequency is twice as high than for sampling of the quadature components after analog I/Q mixing. The reason for this is that the IF signal contains *both* quadrature components:

$$r_{IF}(t) = i(t) \cos(2\pi f_{IF} t) - q(t) \sin(2\pi f_{IF} t)$$

This shows that it is possible to perform IF sampling with *one* A/D converter and a sampling rate according to (25.25) or to perform sampling of the quadrature components with two A/D converters and *half* the sampling rate.

Comparison of Receivers with Digital Modulation Methods

The receiver with analog channel filters shown in Fig. 25.23a is not used in practive. Only the version with channel filtering and gain control in the IF range is of importance and the analog lowpass filters are only used for suppressing the components with twice the IF frequency. This version is often found in uncomplicated systems with simple modulation methods and comparably low data rates.

The receiver with digital channel filters is a widespread system. It allows a much better separation from adjacent channels thus allowing very narrow frequency gaps between the channels so that better use is made of the frequency band available for the application. Sampling of the quadrature components is possible with A/D converters with lower analog bandwidth thus keeping the power loss in the analog section of the converter low. As the modulation method becomes more complex, the disturbing effects of the unavoidable asymmetries in the analog I/Q mixer increase which in turn increases the bit error rate. Thorough tuning of the I/Q mixer, in regards to the amplitude and phase of the two signal paths, is essential if complex modulation methods are used. This adjustment must have a high degree of stability in terms of temperature sensitivity and durability in order to permanently satisfy the demands.

The digital I/Q mixer in the receiver with IF sampling works accurately. Thus, this receiver yields the best results. If the condition $f_{IF,D} = f_{AD}/4$ is met, the mixer only consists of three multiplexers and one inverter.

Direct Conversion Receiver

If, instead of the IF signal, the RF signal is used as the input signal for receivers with digital modulation as shown in Fig. 25.23 on page 1266, then the receiver becomes a *direct conversion receiver*. The previous superhet receiver is reduced to the pre-amplifier and the RF filter and all IF components are eliminated. In practice, the receiver with digital channel filters, according to Fig. 25.23b, is used almost exclusively. For optimum utilization of the A/D converter, a gain control is necessary after the I/Q mixer. The gain control for the desired channel takes place as usual in the demodulator. This gives us the direct conversion receiver, a typical version of which is shown in Fig. 25.33. Figure 25.34 shows the related signal spectra for branch i, which also apply for branch q.

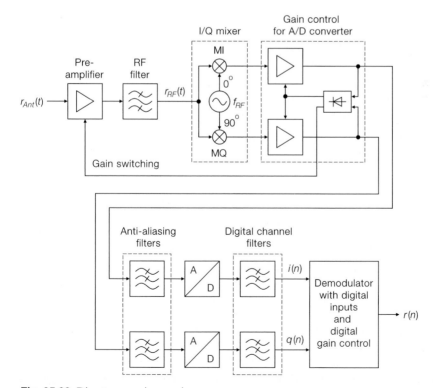

Fig. 25.33. Direct conversion receiver

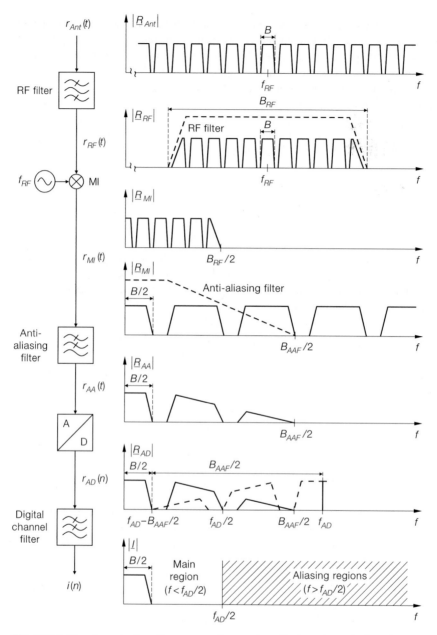

Fig. 25.34. Signal spectra in a direct conversion receiver (for branch i only, branch q is identical)

There are no image frequencies in the direct conversion receiver. Therefore, the RF filter is only required to limit the received band with the aim of limiting the received power. In the superhet receiver, the bandwidth of the RF filter must be at least as large as the frequency range to be received; it may be even larger as long as the additional received power does not restrict the dynamic range of the subsequent components too much.

In addition to the portions with differential frequencies in the range of $0 \leq f \leq B_{RF}/2$, the output signals of the I/Q mixer also contain portions of the sum frequencies in the range of $2 f_{RF}$. Furthermore, there are portions at the frequency f_{RF} generated by crosstalk in the mixers that are suppressed by the anti-aliasing filter.

The minimum sampling frequency of the A/D converter depends on the bandwidth B of the desired channel and the bandwidth B_{AAF} of the anti-aliasing filter or B_{RF} of the RF filter depending on which of the two bandwidths is lower:

$$f_{AD} > \begin{cases} \dfrac{B + B_{AAF}}{2} & \text{for } B_{AAF} < B_{RF} \\[2mm] \dfrac{B + B_{RF}}{2} & \text{for } B_{AAF} \geq B_{RF} \end{cases} \tag{25.27}$$

In both cases, the desired signal remains free of any aliasing contents. The situation of $B_{AAF} < B_{RF}$ is shown in Fig. 25.34. However, the sampling frequency can also be selected such that all channels are digitized in the pass band region of the RF filter without aliasing, and the channel selection can be achieved by switching the digital channel filters where, in this case, the following must apply: $f_{AD} > B_{RF}$. The anti-aliasing filter is then used exclusively for suppressing the signal portions in the frequency range f_{RF} and $2 f_{RF}$.

The primary advantage of a direct conversion receiver is its reduced number of filters. It is particularly suitable for monolithic integration as only the RF filter is required as an external component and RC filters are used as the anti-aliasing filters. At the same time, only one local oscillator with an RC quadrature network ($0°/90°$) is required, which may also be integrated with the exception of a resonant circuit that determines the frequency, and a variable capacitance diode for frequency tuning. The elimination of the IF components substantially reduces the current consumption of the receiver. In particular, power-consuming drivers required for the SAW-IF filters in the superhet receiver and the subsequent amplifiers for compensating the relatively high attenuation of these filters are eliminated.

Besides the advantages mentioned, the direct conversion amplifier also has three problems with negative effects that must be reduced to a noncritical level by additional circuitry:

– The local oscillator frequency corresponds to the received frequency which causes the risk of the relatively strong local oscillator signal reaching the antenna via the RF filter and the pre-amplifier and then being transmitted (see Fig. 25.35). In order to avoid this, the pre-amplifier must have a particularly low reverse transmission. As an alternative, a 3-gate circulator can be introduced between the pre-amplifier and the RF filter. This eliminates the local oscillator signal at the third gate so that it no longer reaches the output of the pre-amplifier (see Fig. 25.36).
– If the local oscillator signal reaches the RF path and is reflected, the result is a *self-mixing effect*. This results in a DC component at the outputs of the I/Q mixer which

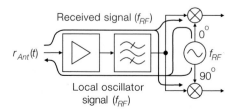

Fig. 25.35. Transmission of the local oscillator signal in direct conversion receivers

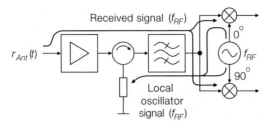

Fig. 25.36. Direct conversion receiver with circulator

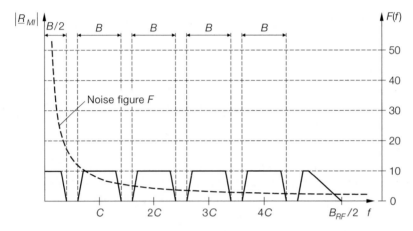

Fig. 25.37. Plot of the spectral noise figure $F(f)$ of variable gain amplifiers over the channel sequence in direct conversion amplifiers

superimposes the DC component of the desired signal. Since the removal of this disturbing DC component is not possible, the total DC component must be removed in the demodulator by a digital highpass filter with a very low cutoff frequency. This must be done in such a way that the useful signal is affected as little as possible.

- The variable gain amplifiers operate as LF amplifiers in the frequency of 1/f noise, which increases the noise figure significantly above that of an IF amplifier. The influence from the noise figure of the receiver can, of course, be reduced by making the gain of the RF pre-amplifier as high as possible, but this is limited by the fact that a high gain in the RF range is only possible with several amplifier stages and comparably high power consumption; at the same time this reduces the overload margin. Figure 25.37 shows the plot of the spectral noise figure $F(f)$ of the variable gain amplifier over the channel sequence. One possible way of improving the noise figure is to reduce the spectral noise figure by using the mth adjacent channel at $f = mC$ as the desired channel instead of the channel at $f = 0$[8].

---

[8] According to Fig. 25.37 the bandwidth of the adjacent channels is twice as high as the bandwidth of the channel at $f = 0$. However, these channels contain two RF channels, i.e. $f_{RF} + C$ and $f_{RF} - C$, which are separated by combining the quadrature components in the following digital processing; only half of the noise power becomes effective so that factor 2 in the bandwidth is compensated.

Maintaining the required amplitude and phase position accuracy of the I/Q mixer poses an additional problem. The demands on I/Q mixers in direct conversion receivers and in superhet receivers are identical, but meeting them is much more difficult in the I/Q mixer with RF input than in the I/Q mixer with IF input due to the higher frequency.

Today, the problems outlined for direct conversion receivers are dealt with well. Therefore, it may be assumed that direct conversion receivers will replace superhet receivers. In this respect, considerations arise as to whether the receiver with IF sampling, according to Fig. 25.23c, shall be also used as a direct conversion receiver by placing only an additional pre-amplifier and an RF filter in front of the A/D converter. This is known as *RF sampling*.

Chapter 26:
Passive Components

26.1
High-Frequency Equivalent Circuits

When dimensioning and simulating high-frequency and intermediate-frequency circuits, the high-frequency response of passive components must be taken into consideration. For this purpose, the high-frequency equivalent circuits shown in Fig. 26.1 are used to model resistors, inductors, and capacitors.

It is common practice to name the reactive components *inductor* and *capacitor* and their ideal values *inductance* and *capacitance*, respectively. Resistive components are called *resistors* and their value is called *resistance*.

Supplementary elements in the equivalent circuits are called *parasitic* elements or parasitics. Their values depend on the construction of the given component. One of the most important values is the parasitic inductance of the component body and the connecting leads. It is roughly proportional to the length and amounts to approximately 1 nH/mm, i.e. a standard resistor with an overall length of 15 mm (5 mm each for the body and the two leads) has an inductance of $L_R \approx 15$ nH. Wound film capacitors have even higher values since the wound film layers act as an inductance. For inductors, this portion can be neglected if their main inductance is sufficiently high. Similar considerations apply to the parasitic capacitance.

The values of parasitics can be minimized by producing miniaturized components without connecting leads, which is the case for surface-mounted components (*surface mounted devices* or SMD). Modern RF and IF circuits use SMD components only and our explanations are restricted to this type. The frequency range, in which the equivalent circuits are valid, depends on the size of the SMD component and increases with decreasing size. For components of the size 1206 (3 mm × 1.5 mm) the equivalent circuits are valid up to 1 GHz (with restrictions up to 2 GHz). We will state the impedances and reflection factors up to 5 GHz to characterize the response of the equivalent circuits in this range. The response of real components in this range is not only dependent on the characteristics

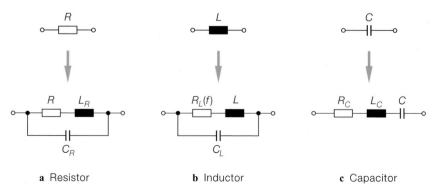

a Resistor   **b** Inductor   **c** Capacitor

Fig. 26.1. High-frequency equivalent circuit of SMD components

of the component but also on the type of mounting. Therefore, the demands in terms of mounting and soldering precision increase with frequency.

26.1.1
Resistor

Figure 26.1 a shows the equivalent circuit for an SMD resistor. It corresponds to the equivalent circuit of a parallel resonant circuit with imperfect inductance. The impedance is:

$$Z_R(s) = (R + sL_R) \,||\, \frac{1}{sC_R} = \frac{R + sL_R}{1 + sC_RR + s^2L_RC_R} \tag{26.1}$$

Consequently:

$$Z_R(j\omega) = \frac{R + j\left(\omega\left(L_R - C_RR^2\right) - \omega^3L_R^2C_R\right)}{\left(1 - \omega^2L_RC_R\right)^2 + \omega^2C_R^2R^2} \tag{26.2}$$

The dominating response of the resistance depends on the sign of the term $(L_R - C_RR^2)$ in the imaginary part of $Z_R(j\omega)$:

$$R < \sqrt{L_R/C_R} \quad \Rightarrow \quad \text{inductive response}$$
$$R > \sqrt{L_R/C_R} \quad \Rightarrow \quad \text{capacitive response}$$

For $R = \sqrt{L_R/C_R}$, the imaginary part is maximally flat and the impedance remains real for as long as possible. For very high frequencies, there is always a capacitive response because the capacitance C_R dominates. But in this region, the equivalent circuit is no longer valid.

Figure 26.2 shows the magnitude and phase of the impedance of SMD resistors of the size 1206 with $L_R = 3\,\text{nH}$ and $C_R = 0.2\,\text{pF}$. A maximally flat imaginary part in which the phase remains zero for as long as possible is obtained for $R = \sqrt{L_R/C_R} \approx 120\,\Omega$. With lower values, the resistors respond inductively (positive phase) and with higher values, capacitively (negative phase). For $R = 190\,\Omega$ the magnitude remains flat up to high frequencies.

In addition to the impedance, the reflection factor

$$r_R(j\omega) = \frac{Z_R(j\omega) - Z_W}{Z_R(j\omega) + Z_W} \tag{26.3}$$

is also of interest. Z_W is the characteristic impedance of the connecting lines. Figure 26.3 shows a plot of the reflection factor of the resistors from Fig. 26.2. The maximally flat phase of the impedance for the $120\,\Omega$ resistor also results in a maximally flat phase of the reflection factor; the curve of the reflection factor thus begins tangential to the real part axis. In contrast, the curve of the $190\,\Omega$ resistor with maximally flat magnitude runs perpendicular to the real part axis.

Furthermore, it is noticeable that a wideband $50\,\Omega$ termination is not possible with a $50\,\Omega$ resistance. This requires a capacitance $C \approx 1\,\text{pF}$ in parallel to make the imaginary part maximal flat:

$$L_R = (C_R + C)\,R^2 \quad \Rightarrow \quad C = \frac{L_R}{R^2} - C_R = \frac{3\,\text{nH}}{(50\,\Omega)^2} - 0.2\,\text{pF} \approx 1\,\text{pF}$$

In this way, it is possible to compensate all resistances with $R < \sqrt{L_R/C_R}$.

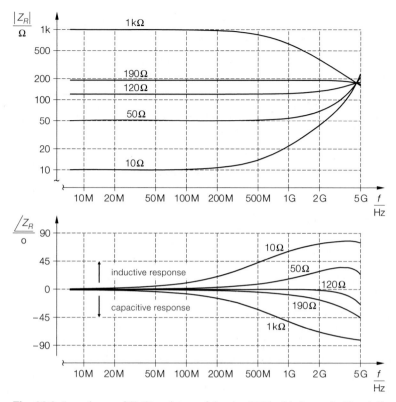

Fig. 26.2. Impedance of SMD resistors of the size 1206 with $L_R = 3\,\text{nH}$ and $C_R = 0.2\,\text{pF}$

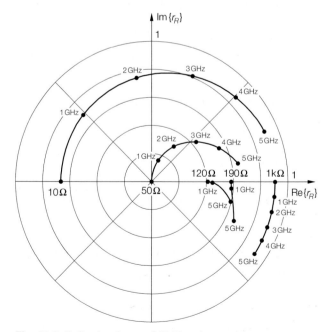

Fig. 26.3. Reflection factor of SMD resistors

26.1.2
Inductor

The equivalent circuit of an inductor, as shown in Fig. 26.1 b, is essentially similar to the equivalent circuit of a resistor, except that the magnitudes of the values are different. The parasitic resistance R_L is caused by the skin resistance (*skin* effect) of the winding and is proportional to the square root of the frequency [26.1]:

$$R_L(f) = k_{RL}\sqrt{f} \tag{26.4}$$

In SMD inductors with an inductance of up to $10\,\mu H$ the loss resistance coefficient k_{RL} with the unit $\Omega/\sqrt{Hz}$ is approximately proportional to the inductance:

$$k_{RL} \approx k_L L \tag{26.5}$$

Typical values are $k_L \approx 1200\,\Omega/(\sqrt{Hz} \cdot H)$ for the size 1206 and $k_L \approx 600\,\Omega/(\sqrt{Hz} \cdot H)$ for the size 1812 [26.2]. The following approximation applies to SMD inductors of size 1812 with an inductance above $10\,\mu H$ [26.2]:

$$k_{RL} \approx 20\,\Omega/\sqrt{Hz} \cdot \left(\frac{L}{H}\right)^{0.7}$$

In inductors, the parallel resonance is of high quality as the plot of the magnitude of the impedance in the upper part of Fig. 26.4 indicates. At a frequency of

$$\omega_r = \frac{1}{\sqrt{LC_L}} \quad \Rightarrow \quad f_r = \frac{1}{2\pi\sqrt{LC_L}} \tag{26.6}$$

the quality is:

$$Q_r = \frac{1}{R_L(f_r)}\sqrt{\frac{L}{C_L}} = \frac{\sqrt{2\pi}}{k_{RL}}\sqrt[4]{\frac{L^3}{C_L}} \tag{26.7}$$

The quality of SMD inductors of size 1206 and 1812 is $Q_r \approx 100\ldots300$. In regards to the resonant frequency, one must distinguish between the *phase resonant frequency*

$$f_{r,ph} = f_r\sqrt{1 - \frac{1}{Q_r^2}}$$

and the *magnitude resonant frequency* [26.1]

$$f_{r,max} \approx f_r\sqrt{1 - \frac{1}{2Q_r^4}}$$

At the phase resonant frequency the impedance of the inductor is real. At the magnitude resonant frequency the impedance reaches its maximum value:

$$Z_{L,max} \approx Q_r^2 R_L(f_r)$$

Due to the high quality Q_r, the frequencies f_r, $f_{r,ph}$ and $f_{r,max}$ differ only very slightly. In practice, the frequency f_r is usually called the *resonant frequency* or *self-resonating frequency (SRF)*.

The inductor *quality factor (QF)* Q_L is more important than the quality Q_r:

$$Q_L(f) = \frac{\text{Im}\{Z_L(j2\pi f)\}}{\text{Re}\{Z_L(j2\pi f)\}} \overset{f < f_r/4}{\approx} \frac{2\pi f L}{R_L(f)} = \frac{2\pi L}{k_{RL}}\sqrt{f} \tag{26.8}$$

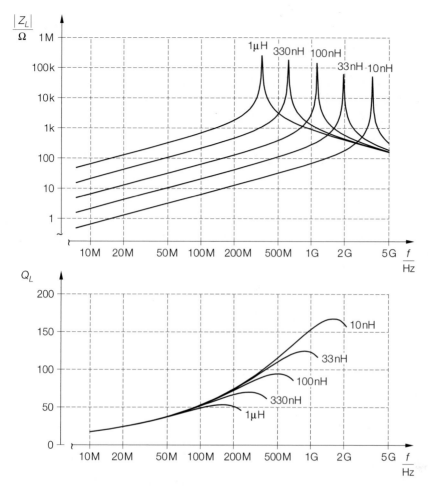

Fig. 26.4. Magnitudes of impedance and inductor quality for SMD inductors of size 1206 with $k_L = 1200\,\Omega/(\sqrt{Hz} \cdot H)$ and $C = 0.2\,pF$

This value is a measure of the losses (high $Q_L \rightarrow$ low losses) and is defined for the frequency range of inductive response only ($f < f_{r,ph}$). In the frequency range up to $f_r/4$, it is almost proportional to the root of the frequency and reaches its maximum at approximately $f_r/2$; above this maximum, it declines rapidly and becomes zero at the phase resonant frequency. This is illustrated in the lower part of Fig. 26.4. The loss resistance coefficient of SMD inductors with an inductance lower than $10\,\mu H$ is $k_{RL} \approx k_L L$. From (26.8) it follows.

$$Q_L(f) \approx \frac{2\pi}{k_L}\sqrt{f} \approx \frac{\sqrt{f/Hz}}{100\ldots200}$$

Factor 100 applies to the size 1812 and factor 200 to the size 1206.

Due to the high quality factor Q_L and the high quality Q_r, the impedance is almost purely imaginary with the exception of a small region around the resonant frequency; consequently, the magnitude of the reflection factor is approximately one:

$$r_L(j\omega) = \frac{Z_L(j\omega) - Z_W}{Z_L(j\omega) + Z_W} \approx e^{j\left(\pi - 2\arctan\frac{\mathrm{Im}\{Z_L(j\omega)\}}{Z_W}\right)}$$

The values for $\omega = 0$ are $\mathrm{Im}\{Z_L(j0)\} = 0$ and $r_L(j0) \approx -1$, i.e. the locus diagram of the reflection factor for $f = 0$ commences in the short-circuit point of the r plane. Figure 26.6 a shows the typical curve of the reflection factor taking an SMD coil with $L = 100\,\mathrm{nH}$ as an example.

26.1.3
Capacitor

The equivalent circuit of a capacitor is shown in Fig. 26.1 c which results in a lossy series resonant circuit with the impedance:

$$Z_C(s) = R_C + sL_C + \frac{1}{sC} = \frac{1 + sCR_C + s^2L_CC}{sC} \tag{26.9}$$

The *self-resonating frequency (SRF)* is

$$\omega_r = \frac{1}{\sqrt{L_CC}} \quad \Rightarrow \quad f_r = \frac{1}{2\pi\sqrt{L_CC}} \tag{26.10}$$

with the quality

$$Q_r = \frac{1}{R_C}\sqrt{\frac{L_C}{C}} \tag{26.11}$$

The phase and magnitude resonant frequencies are equal to the resonant frequency f_r; a differentiation, as in the case of an inductor, is not required. Figure 26.5 shows the plot of

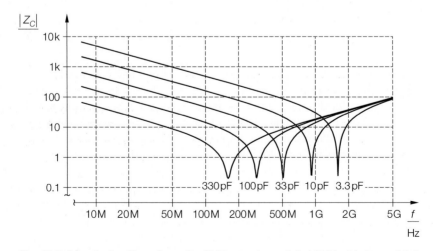

Fig. 26.5. Magnitude of impedance for SMD capacitors of size 1206 with $R_C = 0.2\,\Omega$ and $L_C = 3\,\mathrm{nH}$

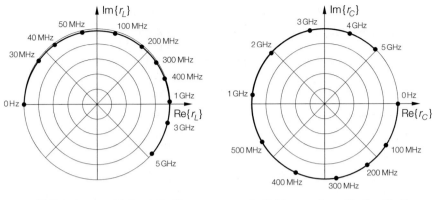

a SMD inductor with $L = 100\,\text{nH}$ **b** SMD capacitor with $C = 10\,\text{pF}$

Fig. 26.6. Typical curve of the reflection factor for SMD conductors and SMD capacitors of size 1206

the magnitude of the impedance of SMD capacitors of size 1206 with $R_C = 0.2\,\Omega$ and $L_C = 3\,\text{nH}$.

The *capacitor quality factor (QF)* Q_C is more important than the quality Q_r:

$$Q_C(f) = -\frac{\text{Im}\{Z_C(j2\pi f)\}}{\text{Re}\{Z_C(j2\pi f)\}} \overset{f<f_r/4}{\approx} \frac{1}{2\pi f C R_C} \tag{26.12}$$

This value is a measure of the losses (high $Q_C \rightarrow$ low losses) and is defined for the frequency range with capacitive response only ($f < f_r$). In the frequency range up to $f_r/4$, it is approximately inverse proportional to the frequency and for $f \rightarrow 0$, it approaches infinity.

Since the impedance is almost purely imaginary with the exception of a small region around the resonant frequency, the reflection factor has a magnitude of approximately one:

$$r_C(j\omega) = \frac{Z_C(j\omega) - Z_W}{Z_C(j\omega) + Z_W} \approx e^{j\left(\pi - 2\arctan\frac{\text{Im}\{Z_C(j\omega)\}}{Z_W}\right)}$$

The values for $\omega = 0$ are $\text{Im}\{Z_C(j\,0)\} = \infty$ and $r_C(j\,0) \approx 1$, i.e. the locus diagram of the reflection factor for $f = 0$ commences in the open-circuit point of the r plane. Figure 26.6 b shows the typical curve of the reflection factor taking an SMD capacitor with $C = 10\,\text{pF}$ as an example.

26.2
Filters

Besides amplifiers and mixers, filters are amongst the most important components of telecommunication systems. With the exception of filters in low-frequency bands, the filters used are passive because active filters are only suitable for IF and RF frequencies in exceptional cases. The standard LC filters are being increasingly replaced by dielectric filters or surface acoustic wave (SAW) filters. This is particularly the case in applications with high unit numbers in which customer-specific or application-specific filters are used.

The essential advantage of dielectric and SAW filters is that they are designed as a single component which the vendor supplies with specified tolerances, so that they can be used in very demanding applications without any adjustment. In contrast, LC filters are made up of several components and can only be used, without adjustments, in noncritical applications. SAW filters provide another very important advantage, namely an almost constant group delay that can be achieved independent of the magnitude response.

26.2.1
LC-Filters

LC filters are often designed with the aid of filter catalogues. The first step is to select a suitable filter characteristic (Butterworth, Thompsen, Chebyshev, etc.) in order to meet the demands in terms of magnitude response, group delay and steepness. Then, the desired order of the filter is determined. The filter structures and the normalized component values for filters are listed in filter catalogues as, for example, in [26.3].

In most cases lowpass filters have the branching structure shown in Fig. 26.7 a and are designed *directly*. For bandpass filters, however, an equivalent lowpass filter with the desired properties is designed and then converted into the corresponding bandpass filter by means of a lowpass/bandpass transformation. This procedure changes the lowpass structure of Fig. 26.7 a into the bandpass structure of Fig. 26.7 b [26.3]. However, it is not always successful as the lowpass/bandpass transformation is based on a nonlinear mapping of the frequency axis which changes the curve of the group delay.

Two-Circuit Bandpass Filter

When designing a transmitter or receiver, it is usually possible to use standard IF filters by selecting the IF frequencies accordingly, unlike RF filters which must be created for the specific transmission or receive frequencies. If no standard filter is available and no high unit numbers are required, the *two-circuit bandpass filter* as shown in Fig. 26.8 is used very often. In terms of its properties it can be compared to the two-circuit dielectric filter.

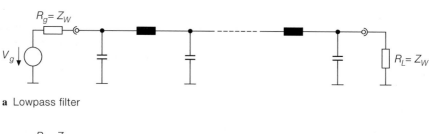

a Lowpass filter

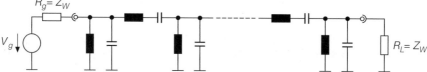

b Bandpass filter

Fig. 26.7. LC filters

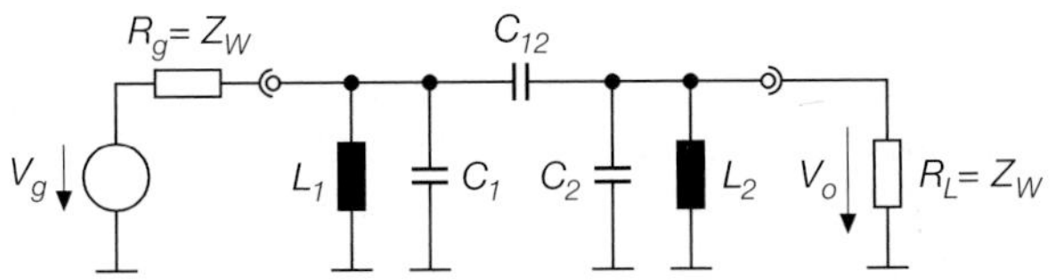

Fig. 26.8. Two-circuit bandpass filter

The following explanation is restricted to the symmetrical case with $R_g = R_L = Z_W$, $L_1 = L_2 = L$ and $C_1 = C_2 = C$. First the resonant frequency

$$f_r = \frac{1}{2\pi\sqrt{L(C + C_{12})}} \tag{26.13}$$

and the resonance quality factor

$$Q_r = Z_W \sqrt{\frac{C + C_{12}}{L}} \tag{26.14}$$

are defined. With the off-resonance factor

$$v = Q_r \left(\frac{\omega}{\omega_r} - \frac{\omega_r}{\omega}\right) = Q_r \left(\frac{f}{f_r} - \frac{f_r}{f}\right) \tag{26.15}$$

and the coupling factor

$$k = \omega_r C_{12} Z_W = 2\pi f_r C_{12} Z_W \tag{26.16}$$

we obtain the operational transfer function [26.1]

$$A_B(jv) = \frac{V_o(jv)}{V_g(jv)} = \frac{jk}{1 + k^2 - v^2 + 2jv} \tag{26.17}$$

with the squared magnitude (power transfer function):

$$|A_B(jv)|^2 = \frac{k^2}{\left(1 + k^2\right)^2 + \left(2 - 2k^2\right)v^2 + v^4} \tag{26.18}$$

Here, the off-resonance factor v takes the place of the radian frequency ω and the argument jv is therefore used instead of $j\omega$. The change-over from ω or f to v, according to (26.15), corresponds to a bandpass/lowpass transformation. $A_B(jv)$ is the transfer function of the equivalent lowpass filter.

When calculating the group delay one must start with the radian frequency ω and the nonlinear relationship between ω and the off-resonance factor v, thus the following applies:

$$\tau_{Gr}(\omega) = -\frac{d}{d\omega}\left[\arctan\frac{\text{Im}\{A_B(j\omega)\}}{\text{Re}\{A_B(jw)\}}\right]$$

$$= -\frac{d}{dv}\left[\arctan\frac{\text{Im}\{A_B(jv)\}}{\text{Re}\{A_B(jv)\}}\right]\frac{dv}{d\omega}$$

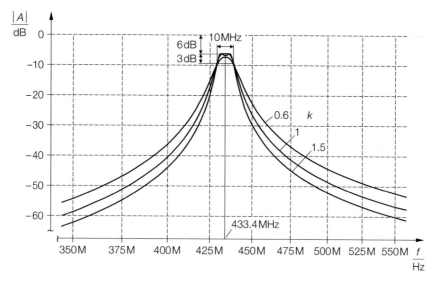

Fig. 26.9. Frequency response of a two-circuit bandpass filter with $f_C = 433.4\,\text{MHz}$ and $B = 10\,\text{MHz}$ for several coupling factors k.

Using (26.17) and (26.15) for the calculation leads to:

$$\tau_{Gr}(\omega) = \frac{2\left(v^2 + k^2 + 1\right)}{\left(v^2 - k^2 - 1\right)^2 + 4v^2} \frac{Q_r}{\omega_r}\left(1 + \left(\frac{\omega_r}{\omega}\right)^2\right) \tag{26.19}$$

According to (26.15), v also depends on ω.

Figure 26.9 shows the frequency response of a two-circuit bandpass filter with a center frequency of $f_C = 433.4\,\text{MHz}$ and a bandwidth $B = 10\,\text{MHz}$ for several coupling factors k. Fig. 26.10 shows the enlarged passband and the plot of the group delay. The relationship between the center frequency f_C and resonant frequency f_r is discussed in more detail below.

Three situations can be derived from (26.18):

– **Critically coupled (k = 1):** The bandpass filter displays a maximally flat frequency response because the equivalent lowpass filter has a Butterworth characteristic as can be seen in a comparison with (13.6) on page 791:

$$|A_B(jv)|^2 = \frac{1}{4 + v^4}$$

At the resonant frequency, the frequency response becomes maximum:

$$A_{B,max} = |A_B(j\,0)| = \frac{1}{2}$$

This corresponds to an attenuation of 6 dB. The $-3\,\text{dB}$ cutoff frequencies with an attenuation of 9 dB occur at the off-resonance factor $v = \pm\sqrt{2}$:

$$\left|A_B\left(\pm j\sqrt{2}\right)\right| = \frac{A_{B,max}}{\sqrt{2}} = \frac{1}{2\sqrt{2}}$$

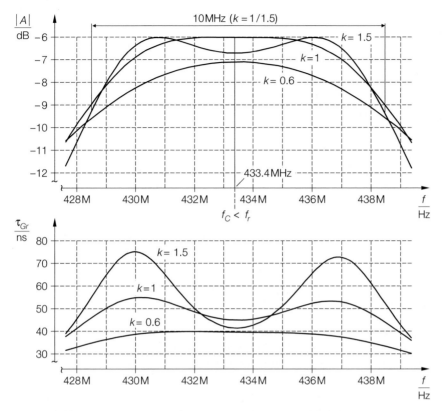

Fig. 26.10. Frequency response and group delay of a two-circuit bandpass filter with $f_C = 433.4\,\text{MHz}$ and $B = 10\,\text{MHz}$ in the passband for several coupling factors k

In practice, the critical coupling is often used as it provides a satisfactory compromise between the requirements of a high steepness at the transition to the stop-band and a reasonable flat group delay curve in the passband.

– **Overcritically coupled ($k > 1$):** The frequency response displays two maxima with

$$A_{B,max} = \left| A_B\left(\pm j\sqrt{k^2 - 1}\right)\right| = \frac{1}{2}$$

which are located at either side of a local minimum at the resonant frequency:

$$A_{B,0} = |A_B(j\,0)| = \frac{k}{1 + k^2} < \frac{1}{2} \qquad \text{for } k > 1$$

The equivalent lowpass filter has a Chebyshev characteristic with the ripple:

$$w = \frac{A_{B,max}}{A_{B,0}} = \frac{1 + k^2}{2k} > 1 \qquad \text{for } k > 1$$

Figure 26.11 shows the ripple plotted over the coupling factor. The $-3\,\text{dB}$ cutoff frequencies are reached at an off-resonance factor $v = \pm\sqrt{2}\,k$. Here, however, they are

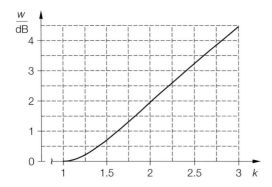

Fig. 26.11. Ripple of an overcritically coupled two-circuit bandpass filter

related to the squared mean value of the maximum and the local minimum at the center frequency:

$$\left|A_B\left(\pm j\sqrt{2}\,k\right)\right| = \frac{\sqrt{\frac{1}{2}\left(A_{B,max}^2 + A_{B,0}^2\right)}}{\sqrt{2}} = \frac{1}{4}\sqrt{1 + \frac{1}{w^2}}$$

The corresponding attenuation is therefore higher than 9 dB. In practice, overcritically coupled filters are used whenever a high steepness at the transition to the stop-band is required. However, this is achieved at the cost of the group delay which for $k > 1$ shows a distinct ripple.

– **Undercritically coupled ($k < 1$):** As in the case of critical coupling, the frequency response curve has a maximum value at the center frequency with the corresponding attenuation higher than 6 dB:

$$A_{B,max} = \frac{k}{1 + k^2} < \frac{1}{2} \quad \text{for } k < 1$$

On both sides of the maximum, the frequency response declines faster than in critically coupled filters. The equivalent lowpass filter has a Bessel characteristic at $k \approx 0.6$. Undercritical coupling is used in cases that demand a constant group delay over the entire passband. This is rarely the case as the filter is almost exclusively used as an RF filter in transmitters and receivers. The bandwidth of the filter is substantially higher than the bandwidth of one channel, and group delay variations within a channel are sufficiently low even in critically or overcritically coupled filters.

The center frequency f_C and the -3 dB bandwidth B are required to define the filter and then the resonant frequency is calculated:

$$f_r = \sqrt{f_C^2 - \frac{B^2}{4}} \tag{26.20}$$

The difference between the two frequencies results from the nonlinear relationship between the frequency f and the off-resonance factor v (26.15). This makes the filter curves symmetrical with respect to the off-resonance factor v, but asymmetrical with respect to the frequency f. For the two -3 dB cutoff frequencies (L/U: lower/upper cutoff frequency)

$$f_L = f_C - \frac{B}{2} \quad , \quad f_U = f_C + \frac{B}{2}$$

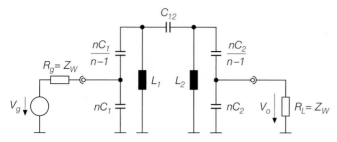

Fig. 26.12. Two-circuit bandpass filter with capacitive coupling

we obtain the following formula by inserting into (26.15) and rearranging:

$$v_L = -\frac{Q_r B}{f_r} \quad , \quad v_U = \frac{Q_r B}{f_r}$$

At the −3dB cutoff frequencies, the off-resonance factor is $v = \pm\sqrt{2}\,k$, i.e. $v_L = -\sqrt{2}\,k$ and $v_U = \sqrt{2}\,k$ which leads to the resonance quality factor:

$$Q_r = \sqrt{2}\,k\,\frac{B}{f_r} \tag{26.21}$$

By inserting f_r according to (26.20) and Q_r according to (26.21) into (26.13), (26.14) and (26.16), the values of the components can be calculated:

$$L = \frac{Z_W}{2\pi f_r Q_r} \quad , \quad C = \frac{Q_r - k}{2\pi f_r Z_W} \quad , \quad C_{12} = \frac{k}{2\pi f_r Z_W} \tag{26.22}$$

In Fig. 26.8 it thus follows that $L_1 = L_2 = L$ and $C_1 = C_2 = C$.

With a characteristic impedance $Z_W = 50\,\Omega$, the inductances L_1 and L_2 become very low at high frequencies. Then the version with capacitive coupling as shown in Fig. 26.12 can be used which requires that $n > 1$. Equations (26.20)–(26.22) are again used for calculating the values, but in (26.22), Z_W is now replaced by $n^2 Z_W$; thus, the inductances are increased by the factor n^2 and the capacitances are decreased by the same factor.

Example: Dimensioning of a two-circuit bandpass filter with center frequency $f_C = 433.4\,\text{MHz}$, a −3dB bandwidth $B = 10\,\text{MHz}$ and a coupling factor $k = 1$ for a characteristic impedance $Z_W = 50\,\Omega$ is as follows. From (26.20) and (26.21) we obtain $f_r = 433.37\,\text{MHz}$ and $Q_r = 61.29$, and with (26.22), $L \approx 300\,\text{pH}$, $C = 442\,\text{pF}$ and $C_{12} = 7.3\,\text{pF}$. The inductance of 300 pH is impractically low. Therefore we use the filter with capacitive coupling from Fig. 26.12 and select n such that L is at the standard value of 22 nH. As the inductances increase by the factor n^2, we have $n^2 \cdot 300\,\text{pH} = 22\,\text{nH}$; consequently, $n^2 \approx 73.3$ and $n \approx 8.56$. The inductances can now be implemented by using SMD inductors and the capacitances are decreased by the factor n^2 : $C \approx 6\,\text{pF}$ and $C_{12} \approx 0.1\,\text{pF}$. The capacitance C is divided into the two capacitances $nC \approx 51.7\,\text{pF}$ $= 47\,\text{pF}||4.7\,\text{pF}$ and $nC/(n-1) \approx 6.8\,\text{pF}$. The coupling capacitance C_{12} is produced by capacitive coupling between two adjacent printed circuit lines.

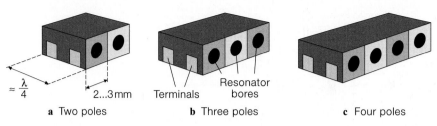

$\approx \frac{\lambda}{4}$ 2...3 mm   Terminals   Resonator bores

a Two poles   **b** Three poles   **c** Four poles

Fig. 26.13. Dielectric bandpass filters

Filters with Lines

At frequencies above 500 MHz the inductances and capacitances of an LC filter become so low that an implementation using inductors and capacitors is no longer possible. In this case, lines must be used. Since only bandpass filters are employed in this range, the implementation of the filter structure in Fig. 26.7 b requires a line element with the characteristic of a series resonant circuit and an element with the characteristic of a parallel resonant circuit. Suitable for this purpose are lines with length $\lambda/4$ which are open-ended or short-circuited at one end. The design is based on the *Richards transformation* which allows the direct calculation of a line filter from the equivalent lowpass filter. Refer to the section on *microwave filters with lines* in [26.1].

26.2.2
Dielectric Filters

In the frequency range of 800 MHz to 5 GHz, filters with coupled resonators of the length $\lambda/4$ are used where one end of the resonator remains open, and the other is short-circuited. To prevent the dimensions from becoming too large, a dielectric with a minimum of losses and maximum relative dielectric coefficient is employed to reduce the wave length from the free-space wave length $\lambda_0 = c_0/f$ to:

$$\lambda = \frac{v}{f} = \frac{c_0}{\sqrt{\epsilon_r}\, f} \qquad (26.23)$$

Such filters are called *dielectric filters*. In the frequency range up to 1 GHz, barium-titanate with $\epsilon_r \approx 90$ is used; therefore the length of the resonator is approximately 8 mm ($\lambda \approx 32$ mm at $f = 1$ GHz). For higher frequencies, dielectrics with a lower relative dielectric coefficients are employed.

A dielectric bandpass filter with n resonators is called an n-pole filter. The number of poles relates to the equivalent lowpass filter because the transfer function of the bandpass filter has $2n$ poles (two per resonator). Typical construction of standard dielectric filters is shown in Fig. 26.13 [26.4].

Figure 26.14 illustrates a cross section of a two-pole filter. It consists of two resonator bodies made of barium-titanate with an axial resonator bore and a radial bore for coupling. The resonator bodies are metalized with the exception of the no-load side, the bores for coupling and a small gap for capacitive coupling of the resonators. It should be noted that electromagnetic fields expand through the resonator bodies; no field exists in the resonator bores. The length of the physical resonators is always slightly smaller than $\lambda/4$ as the fields at the open end expand into the ambient space (stray field). Thus, the electrical length of the resonator is longer than the physical length.

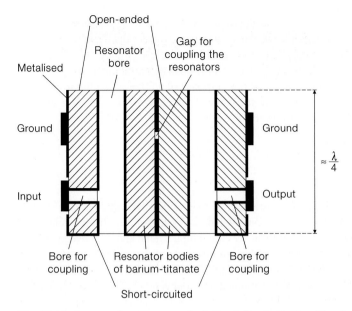

Fig. 26.14. Cross section of a two-pole dielectric bandpass filter. The electromagnetic fields spread in the hatched resonator bodies; no fields exist in the bores.

The equivalent circuit of a two-pole dielectric filter corresponds to the circuit diagram of the two-circuit bandpass filter in Figs. 26.8 and 26.12. In the case of three-pole or multi-pole filters, additional parallel resonant circuits with the same type of capacitive coupling are added. Although, this correspondence only exists for the pass band and the adjacent

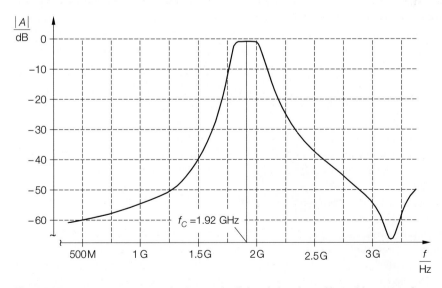

Fig. 26.15. Frequency response of a three-pole dielectric bandpass filter with a center frequency $f_C = 1.92$ GHz [26.4]

portions of the stop-band since, with all odd harmonics, the resonators also have a parallel resonance causing additional regions with lower attenuation above the desired pass band.

Dielectric filters, predominantly two-pole or three-pole, are employed as RF filters in transmitters and receivers. The small size of the filters is particularly important for mobile units. Figure 26.15 shows the frequency response of a three-pole filter used in the PCS mobile communication system with the center frequency $f_C = 1.92\,$GHz and the dimensions 6.5 mm × 4.3 mm × 2 mm [26.4].

26.2.3
SAW Filters

A *SAW* (*surface acoustic wave*) filter is a transversal filter (FIR filter) which uses the transit time of an acoustic surface wave across a piezo-electric crystal for a controlled delay. The excitation of the surface acoustic wave by an electrical input signal and its conversion back to an electrical output signal is achieved with piezo-electric transducers which are known as *interdigital transducers* due to their comb-like interlocked electrodes. Figure 26.16 shows the construction of a SAW filter with a weighted and nonweighted transducer separated by the transit distance.

An acoustic surface wave (Rayleigh wave) is an elastic wave travelling across a solid surface with a speed of $v \approx 3000 \ldots 4000\,$m/s. On smooth surfaces, the speed, or propagation, does not depend on the frequency of the wave, i.e. there is no dispersion, thus the shape of the wave remains the same and the group delay is constant. The piezo-electric crystal is mostly made of lithium-niobate ($LiNbO_3$) with $v = 3990\,$m/s. The resulting wave lengths range from $\lambda = 400\,\mu$m at $f = 10\,$MHz to $\lambda = 4\,\mu$m at $f = 1\,$GHz. The electrodes of the transducers are $\lambda/4$ wide and spaced at $\lambda/2$ making the electrode dimensions for frequencies above 1 GHz smaller than 1 µm. Thus, the maximum operating frequency of a SAW filter is determined by the minimum size allowed in the manufacturing process. Today, SAW filters with center frequencies of up to 400 MHz are available and, when frequencies are above 200 MHz, the lithium-niobate is replaced by a quartz crystal.

The transfer function of a SAW filter corresponds to that of a transversal or FIR filter (see Sect. 19.3 on page 994). The filter coefficients result from the electrode length of both transducers and as the two transducers become consecutively active, it is necessary to form the convolution product for the electrode lengths of the two transducers. In practice, one of the transducers is often *nonweighted*, i.e. all electrodes are of the same length, and

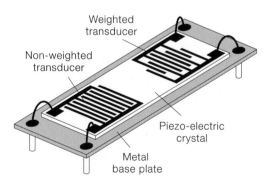

Weighted
transducer

Non-weighted
transducer

Piezo-electric
crystal

Metal
base plate

Fig. 26.16. Construction of a SAW filter

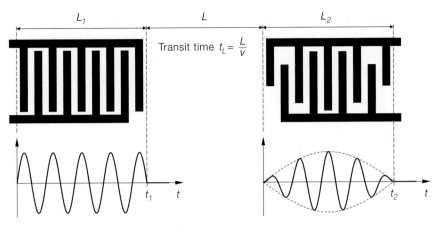

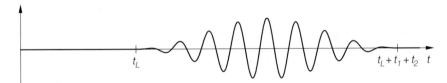

a Geometry and pulse response of the transducers

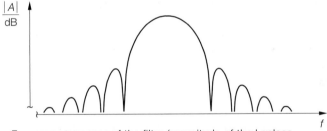

b Pulse response of the filter (convolution product of the pulse responses of the transducers + transit time

c Frequency response of the filter (magnitude of the Laplace transformation of the pulse response)

Fig. 26.17. Relationship between geometry and absolute frequency response of a SAW filter

the other is *weighted* (see Fig. 26.16). Figure 26.17 explains the relationship between the geometry and the frequency response.

The pass band attenuation of a SAW filter is relatively high. The transducer used as the transmitter emits one wave in the direction of the receiving transducer and one wave in the opposite direction. This results in a loss of half of the power, which corresponds to an attenuation of 3 dB. For symmetry reasons, the attenuation of the receiving transducer is also 3 dB so that the theoretical lower limit for the attenuation of a SAW filter is 6 dB. In practice, the attenuation must be much higher to prevent too strong a degradation of the transfer function caused by the waves reflected at the transducers and the ends of the crystal. Particularly disturbing is the *triple-transit* echo, which is reflected once at each transducer and therefore passes the transit distance three times. The *triple-transit* echo undergoes an attenuation (in decibel) that is three times higher than that of the desired

Standard filter (*D*=21.5 dB)   *Low-loss* filter (*D*=7.5 dB)

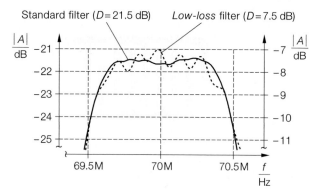

Fig. 26.18. Frequency response of a standard SAW filter and a low-loss SAW filter with a center frequency $f_C = 70\,\text{MHz}$ and a $-3\,\text{dB}$ bandwidth $B = 1\,\text{MHz}$ [26.5]

signal and must have an attenuation that is at least 40 dB higher. Thus, in standard filters the attenuation of the desired signal is at least 20 dB. If the attenuation of the *triple-transit* echo is insufficient, it will cause a ripple in the frequency response and in the group delay. This is the case in *low-loss* filters in which the ripple is accepted in favor of the lower attenuation. Figure 26.18 shows the frequency response of a standard SAW filter compared to that of a low-loss SAW filter [26.5].

Figure 26.18 shows the frequency response of a SAW filter for the situation of double-ended impedance matching to the characteristic impedance $Z_W = 50\,\Omega$. Without impedance matching, neither the specified frequency response nor the specified attenuation can be reached. The circuits for impedance matching are listed on the data sheet. The impedance of the two transducers can be described by means of the electro-mechanical equivalent circuit of a piezo-electric transducer as shown in Fig. 26.19 a. R_m, L_m and C_m are substitute elements for describing the mechanical properties; C_{stat} is the static capacitance of the transducer electrodes. At the center frequency, the impedance of the electro-mechanical portion becomes real and then only the electro-mechanical resistance R_m and the static capacitance C_{stat} are effective (see Fig. 26.19 b). The dimensions are such that the impedance of the converter is ohmic capacitive not only at the center frequency but throughout the entire pass band range and beyond. Generally the resistance R_m is higher than $50\,\Omega$ so that the impedance matching circuit consisting of the capacitance C_{stat} and

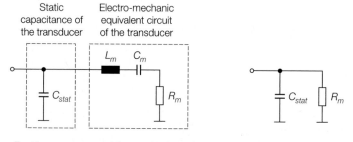

a For the range around the centre frequency   **b** At the centre frequency

Fig. 26.19. Equivalent circuit for a piezo-electric transducer

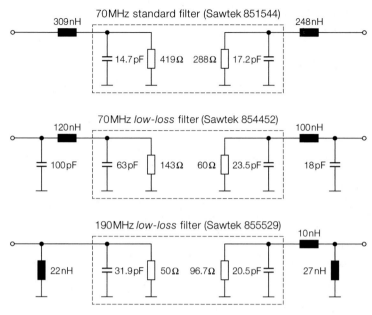

Fig. 26.20. Impedance matching of SAW filters to $Z_W = 50\,\Omega$ [26.5]

external circuitry components must therefore transform R_m to 50 Ω. Figure 26.20 shows three examples.

26.3
Circuits for Impedance Transformation

Circuits for impedance transformation are used for *impedance matching* and *coupling*. In the matching process an input or output impedance of a component is matched to the characteristic impedance of a line to prevent reflections and to maximize the transfered power. In some cases a controlled mismatch is performed. In the coupling (or interfacing) process, a load is connected to a resonant circuit with the purpose of transforming the impedance of the load such that the quality of the resonant circuit reaches a predetermined value.

26.3.1
Impedance Matching

The following section contains a description of simple reactive networks for loss-free matching of an impedance to the characteristic impedance Z_W of a line. This type of matching is known as *narrow-band* impedance matching and is accurate at one frequency only. This is sufficient for practical applications as long as the impedance matching bandwidth is larger than the bandwidth of the useful signal. The criterion is the reflection factor r, which must drop to zero (match) at the center frequency and must not exceed a certain value at the pass band borders; usually the demand is $|r| < 0.1$. The compliance check is done by circuit simulation or by measurement in a sample circuit.

The bandwidth of the matching circuit declines with an increasing transformation factor. Therefore, matching of impedances with $|Z| \ll Z_W$ and $|Z| \gg Z_W$ is possible for a very narrow band only. If the bandwidth of the simple network is not sufficient, more sophisticated networks must be used for wideband impedance matching. Often these networks are not loss-free since both the reflection factor and the broadband frequency response must be optimized. For a more detailed description refer to the literature [26.1].

Impedance Matching Networks with Two Elements

Figure 26.21 shows two networks for matching an impedance of $Z = R + jX$ to a characteristic impedance Z_W of a line. This is done using two reactive elements with the reactances X_1 and X_2 at the center frequency f_C. If the admittance $Y = G + jB$ is known instead of the impedance Z, the following conversion must be performed:

$$Z = \frac{1}{Y} = \frac{1}{G + jB} = \frac{G - jB}{G^2 + B^2}$$

$$\Rightarrow \quad R = \frac{G}{G^2 + B^2} \quad , \quad X = -\frac{B}{G^2 + B^2} \tag{26.24}$$

The following conditions apply to the network in Fig. 26.21 a:

$$jX_1 \| (Z + jX_2) = \frac{jX_1 (Z + jX_2)}{Z + j(X_1 + X_2)} \overset{!}{=} Z_W$$

By inserting $Z = R + jX$, then separating into real and imaginary parts and solving the equation for X_1 and X_2, the following conditions are derived:

$$X_1 = \pm \frac{Z_W R}{\sqrt{R(Z_W - R)}} \quad , \quad X_2 = \mp \sqrt{R(Z_W - R)} - X \tag{26.25}$$

Here, the condition $R < Z_W$ must apply to keep the term under the square root positive; thus, this network only allows a *step-up transformation* $R \to Z_W > R$. There are two solutions according to the $\pm$ signs; the positive sign must be assigned to one reactance

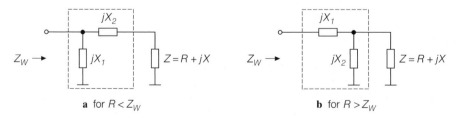

Fig. 26.21. Impedance matching networks with two elements. Dimensions are defined with (26.25) for $R < Z_W$ and with (26.27) for $R > Z_W$.

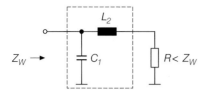

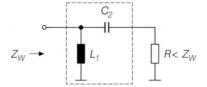

a Characteristic of a lowpass filter
   $(X_1 < 0, X_2 > 0)$

b Characteristic of a highpass filter
   $(X_1 > 0, X_2 < 0)$

Fig. 26.22. Step-up transformation of resistances. Dimensions are defined with (26.25) and (26.26).

and the negative sign to the other. A positive reactance is produced by an inductance and a negative reactance by a capacitance:

$$X_{1/2} > 0 \quad \Rightarrow \quad L_{1/2} = \frac{X_{1/2}}{2\pi f_C}$$

$$X_{1/2} < 0 \quad \Rightarrow \quad C_{1/2} = -\frac{1}{2\pi f_C X_{1/2}}$$

(26.26)

For resistances ($Z = R$, $X = 0$), the signs of X_1 and X_2 in (26.25) are different, thus resulting in the versions with an inductance and a capacitance shown in Fig. 26.22. The version in Fig. 26.22 a has the characteristic of a lowpass filter and that in Fig. 26.22 b the characteristic of a highpass filter. For general impedances ($X \neq 0$), the sign of X_2 is also influenced by the reactance X which allows versions with two inductances (X_1, $X_2 > 0$) or two capacitances (X_1, $X_2 < 0$). $X_2 = 0$ eliminates the series element and impedance matching is achieved with a parallel inductance ($X_1 > 0$) or a parallel capacitance ($X_1 < 0$).

For the network in Fig. 26.21 b we obtain the condition:

$$jX_1 + (Z \,\|\, jX_2) = \frac{jZ(X_1 + X_2) - X_1 X_2}{Z + jX_2} \overset{!}{=} Z_W$$

By inserting $Z = R + jX$, separating into real and imaginary parts and solving for X_1 and X_2, we obtain the conditions:

$$X_1 = \pm Z_W \sqrt{\frac{R^2 + X^2}{Z_W R} - 1}$$

$$X_2 = \frac{\mp \left(R^2 + X^2\right)}{R \sqrt{\dfrac{R^2 + X^2}{Z_W R} - 1} \pm X}$$

(26.27)

For resistances ($Z = R$, $X = 0$) the following applies:

$$X_1 = \pm \sqrt{Z_W(R - Z_W)} \quad , \quad X_2 = \mp \frac{Z_W R}{\sqrt{Z_W(R - Z_W)}}$$

(26.28)

Here, the condition $R > Z_W$ must apply to make sure that the term under the square root remains positive. Therefore, this network only enables a *step-down transformation*

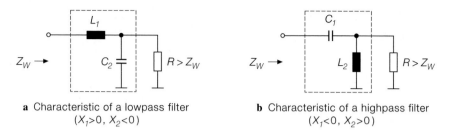

a Characteristic of a lowpass filter
$(X_1>0, X_2<0)$

b Characteristic of a highpass filter
$(X_1<0, X_2>0)$

Fig. 26.23. Step-up transformation of resistances. Dimensions are defined with (26.28) and (26.26)

$R \rightarrow Z_W < R$ for resistances. With complex impedances $(X \neq 0)$, however, a step-up transformation is also possible, provided that:

$$R^2 + X^2 > Z_W R$$

For $|X| > Z_W/2$ this is possible for all values of R. Here, again, there are two solutions and in accordance with (26.26) the elements are produced by an inductance or a capacitance.

For resistances $(Z = R, X = 0)$, the signs of X_1 and X_2 in (26.28) are different, resulting in the versions with an inductance and a capacitance shown in Fig. 26.23. The version in Fig. 26.23 a has the characteristic of a lowpass filter and that in Fig. 26.23 b the characteristic of a highpass filter. For general impedances $(X \neq 0)$, the sign of X_2 is also influenced by the reactance X which allows versions with two inductances $(X_1, X_2 > 0)$ or two capacitances $(X_1, X_2 < 0)$. If in (26.27) the term in the denominator of X_2 becomes zero, then the parallel element is eliminated and impedance matching is achieved with a series inductance $(X_1 > 0)$ or a series capacitance $(X_1 < 0)$.

The filter characteristic of the impedance matching networks can be used to suppress undesirable signal contents. If the signal contains, for example, remnants of a local oscillator signal or an undesirable side band as a result of a previous frequency conversion, a lowpass characteristic is selected when those portions are above the center frequency and a highpass characteristic when they are below the center frequency. For impedance matching in amplifiers, stability should be the first consideration.

Example: Let us consider the impedance matching at the input side of the 70 MHz *lowloss* SAW filter in Fig. 26.20 on page 1301. The equivalent circuit consists of a resistance $R_m = 143\,\Omega$ and a parallel capacitance $C_{stat} = 63\,\mathrm{pF}$, which, at the center frequency $f_C = 70\,\mathrm{MHz}$, results in the admittance

$$Y = G + jB = \frac{1}{R_m} + j\omega C_{stat} \overset{\omega=2\pi \cdot 70\,\mathrm{MHz}}{=} (7 + j\,27.7)\,\mathrm{mS}$$

where $G = 7\,\mathrm{mS}$ and $B = 27.7\,\mathrm{mS}$. Conversion with (26.24) yields the impedance Z with $R = 8.58\,\Omega$ and $X = -33.9\,\Omega$. Matching to $Z_W = 50\,\Omega$ must be achieved with the network in Fig. 26.21 a due to $R < Z_W$. From (26.25) it follows that $X_1 = \pm 22.8\,\Omega$ and $X_2 = (\mp 18.9 + 33.9)\,\Omega$. We select the lowpass characteristic with $X_1 = -22.8\,\Omega$ and $X_2 = 52.8\,\Omega$ in order to increase the attenuation at frequencies above the pass band; thus, with (26.26) we obtain:

$$C_1 = \frac{1}{2\pi \cdot 70\,\mathrm{MHz} \cdot 22.8\,\Omega} \approx 100\,\mathrm{pF} \quad, \quad L_2 = \frac{X_2}{2\pi \cdot 70\,\mathrm{MHz}} \approx 120\,\mathrm{nH}$$

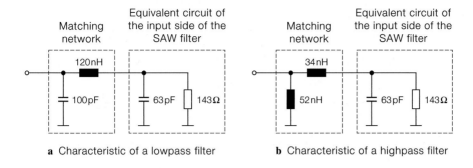

a Characteristic of a lowpass filter   **b** Characteristic of a highpass filter

Fig. 26.24. Impedance matching to $Z_W = 50\,\Omega$ at the input side of a 70 MHz *low-loss* SAW filter

For the version with the characteristic of a highpass filter, this results in two inductances: $X_1 = 22.8\,\Omega \rightarrow L_1 \approx 52\,\text{nH}$ and $X_2 = 15\,\Omega \rightarrow L_2 \approx 34\,\text{nH}$. Owing to the series inductance L_2, this causes a lowpass characteristic, meaning that the overall characteristic is that of a bandpass filter. Figure 26.24 shows both versions.

Collins Filter

In practical applications, the π network in Fig. 26.25 with two parallel capacitances and one series inductance is often used instead of the simple impedance matching network with two elements. This is known as the *Collins filter* and has the characteristic of a lowpass filter. The additional degree of freedom resulting from the third element can be used to optimize the bandwidth or to shift the values of the elements into a region better suited for the implementation.

Here we shall concentrate first on the matching of resistances. The following condition is obtained for the center frequency $\omega_C = 2\pi f_C$:

$$\cfrac{1}{j\omega_C C_1 + \cfrac{1}{j\omega_C L + \cfrac{1}{j\omega_C C_2 + \cfrac{1}{R}}}} \overset{!}{=} Z_W$$

After multiplying out the terms, separating the real and imaginary parts and using the transformation ratio

$$t = \frac{R}{Z_W} \tag{26.29}$$

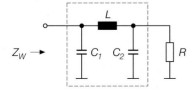

Fig. 26.25. Collins filter

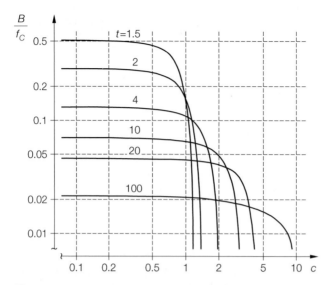

Fig. 26.26. Relative bandwidth B/f_C ($|r| < 0.1$ of a Collins filter for several transformation ratios t

and the capacitance ratio

$$c = \frac{C_1}{C_2} \tag{26.30}$$

the equations for dimensioning are obtained [26.6]:

$$C_1 = \frac{c}{2\pi f_C R} \sqrt{\frac{t(t-1)}{t - c^2}} \tag{26.31}$$

$$C_2 = \frac{1}{2\pi f_C R} \sqrt{\frac{t(t-1)}{t - c^2}} \tag{26.32}$$

$$L = \frac{R}{2\pi f_C} \sqrt{\frac{(t-1)\left(t - c^2\right)}{t(t-c)^2}} \tag{26.33}$$

The capacitance ratio must be selected on the basis of the transformation ratio so that the terms under the square roots are positive:

$$\begin{aligned} t > 1 &\Rightarrow c < \sqrt{t} \\ t < 1 &\Rightarrow c > \sqrt{t} \end{aligned} \tag{26.34}$$

The selection of the capacitance ratio c influences the values of the elements and the bandwidth. Figure 26.26 shows the relative bandwidth B/f_C for which the magnitude of reflection factor remains below 0.1 for several transformation ratios t. The bandwidth declines with an increasing transformation ratio. Figure 26.26 only presents the curves for $t > 1$. For $t < 1$ the input and output are simply interchanged by replacing t with $1/t$ and c with $1/c$.

The Collins filter can also be used for matching general impedances Z. This is done by starting with

$$Z = R \,||\, jX$$

and compensating the reactive portion with a parallel reactance $X_p = -X$:

$$Z_p = Z \,||\, jX_p = R \,||\, jX \,||\, jX_p \overset{X_p = -X}{=} R$$

The parallel reactance X_p is combined with capacitance C_2 connected in parallel to form reactance X_2:

$$jX_2 = \frac{1}{j2\pi f_C C_2} \,||\, jX_p \quad\Rightarrow\quad X_2 = \frac{X_p}{1 - 2\pi f_C C_2 X_p}$$

According to (26.26), this is accomplished by a capacitance or an inductance.

The Collins filter is predominantly used for matching amplifiers and some of the elements of the filter are implemented by the parasitic elements of the transistors (see Fig. 27.13 on on page 1341). Chapter 27 describes this in more detail.

Impedance Matching with Strip Lines

With increasing frequencies, inductances and capacitances in impedance matching networks become smaller and smaller making the implementation with conventional components increasingly difficult. Furthermore, the parasitic effects of the inductors and capacitors used become more noticeable with increasing frequencies. For this reason strip lines are often used for impedance matching in the GHz frequency range. There is a multitude of suitable structures that are well detailed in the literature [26.1]. Some of the typical structures are described below. It should be noted that the individual strip lines of one structure must be *directly* connected to one another; the spatial separation in the illustrations is for clarity only.

An important class of structures for impedance matching with strip lines is based on the $\lambda/4$ transformer described in Sect. 26.2 (see Fig. 26.8 on page 1291 and (26.22) on page 1295). A $\lambda/4$ transformer consists of a line that has a length $\lambda/4$ and a characteristic impedance Z_{W1}. If one end of the line is terminated with an impedance $Z = R + jX$, the impedance at the other end is:

$$Z_1 \overset{(24.22)}{=} \frac{Z_{W1}^2}{Z} = \frac{Z_{W1}^2}{R + jX} \overset{!}{=} Z_W$$

In the matched state it should correspond with the characteristic impedance Z_W of the connecting line.

Figure 26.27 a shows the matching of a resistance ($Z = R$, $X = 0$). Here, the line of the $\lambda/4$ transformer must have the characteristic impedance:

$$Z_{W1} = \sqrt{Z_W R}$$

The transformation range is very limited since, in practice, the characteristic impedance of a strip line can vary by a factor of 4 only (see Fig. 26.10 on page 1293); thus, for $Z_W/2 < Z_{W1} < 2Z_W$, the transformation range is $Z_W/4 < R < 4Z_W$.

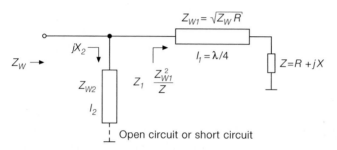

a Transformation of a resistance

b Transformation of a impedance with subsequent shunt compensation

c Transformation of a series-compensated impedance

Fig. 26.27. Examples of impedance matching with strip lines with the use of a $\lambda/4$ transformer

For a general impedance Z the structure in Fig. 26.27 b can be used. First, a $\lambda/4$ transformation to

$$
Z_1 = \frac{Z_{W1}^2}{Z} \overset{\substack{Z_{W1}=\sqrt{Z_W R} \\ Z=R+jX}}{=} \frac{Z_W R}{R+jX} = \frac{1}{\dfrac{1}{Z_W}+j\dfrac{X}{Z_W R}}
$$

is performed. The reactive portion is then compensated with a shunt reactance X_2. From the condition $Z_1 \parallel jX_2 = Z_W$, it follows:

$$
X_2 = \frac{Z_W R}{X}
$$

In the capacitive case ($X < 0 \rightarrow X_2 < 0$), the shunt reactance is implemented by a short open-ended line and in the inductive case ($X > 0 \rightarrow X_2 > 0$) by a short short-circuited line. The required length is derived from (26.23) in the capacitive case

$$
l_2 = \frac{\lambda}{2\pi} \arctan\left(-\frac{Z_{W1}}{X_2}\right) = \frac{\lambda}{2\pi} \arctan\left(-\frac{Z_{W1}X}{Z_W R}\right) \qquad \text{for } X < 0
$$

and from (24.24) in the inductive case:

$$l_2 = \frac{\lambda}{2\pi} \arctan\left(\frac{X_2}{Z_{W1}}\right) = \frac{\lambda}{2\pi} \arctan\left(\frac{Z_W R}{Z_{W1} X}\right) \qquad \text{for } X > 0$$

In the capacitive case, the characteristic impedance Z_{W1} should be as low as possible (wide strip line) and in the inductive case as high as possible (narrow strip line), thus minimizing the length. These lines are known as *capacitive* or *inductive stub lines*.

Figure 26.27 c shows another structure for matching a general impedance Z. First, the reflection factor

$$r_Z = |r_Z| e^{j\varphi_z} = \frac{Z - Z_W}{Z + Z_W}$$

is rotated with a connecting line ($Z_{W2} = Z_W$) of the length l_2 so that it becomes real (series compensation): $r_1 = \pm|r_Z|$. Then, the corresponding resistance

$$R_1 = \frac{1 \pm |r_Z|}{1 \mp |r_Z|}$$

is transformed with

$$Z_{W1} = \sqrt{Z_W R_1}$$

to the characteristic impedance Z_W by a $\lambda/4$ transformer. The rotation of the reflection factor r_Z is in accordance with (24.38) on page 1179:

$$r_1 = r_Z e^{-j\frac{4\pi l_2}{\lambda}} = |r_Z| e^{j\left(\varphi_z - \frac{4\pi l_2}{\lambda}\right)}$$

It is real for

$$\varphi_z - \frac{4\pi l_2}{\lambda} = n\pi \quad \Rightarrow \quad l_2 = \frac{\lambda}{4}\left(\frac{\varphi_z}{\pi} - n\right) \qquad n \text{ integer}$$

The following values are selected to keep the line as short as possible:

$$\varphi_z > 0 \quad \Rightarrow \quad n = 0 \quad \Rightarrow \quad r_1 = |r_Z|$$
$$\varphi_z < 0 \quad \Rightarrow \quad n = -1 \quad \Rightarrow \quad r_1 = -|r_Z|$$

The structures in Fig. 26.27 are designed so that the first step in the matching process is done with a series line. This causes a spatial distance between the impedance to be matched and the other components, thereby facilitating the arrangement of the strip lines on the substrate. On the matched side, the arrangement of other components causes no problem because a connecting line with the characteristic impedance Z_W can be used for the spatial separation.

Impedance matching with a $\lambda/4$ transformer is possible only for a very limited transformation ratio and is thus not optimum with respect to the required line length. Better results are achieved with the structures shown in Fig. 26.28. Let us first look at matching with the series line according to Fig. 26.28 a. Here, we use (26.21) in order to obtain the input impedance Z_1 of a line with characteristic impedance Z_{W1} and length l_1 if terminated with an impedance $Z_2 = Z = R + jX$; then we set $Z_1 = Z_W$:

$$Z_1 = \frac{Z + j Z_{W1} \tan\left(\frac{2\pi l_1}{\lambda}\right)}{1 + j \dfrac{Z}{Z_{W1}} \tan\left(\frac{2\pi l_1}{\lambda}\right)} \overset{!}{=} Z_W$$

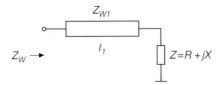

a With a series line

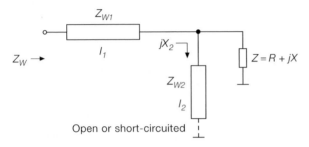

b With a series line and output compensation

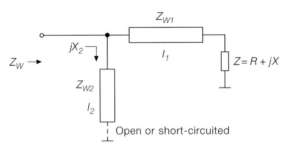

c With a series line and input compensation

Fig. 26.28. Examples of impedance matching with strip lines

Using the abbreviation

$$k_{l1} = \tan\left(\frac{2\pi l_1}{\lambda}\right) \tag{26.35}$$

multiplication and separation into real and imaginary parts lead to the conditions:

$$R = Z_W\left(1 - \frac{k_{l1} X}{Z_{W1}}\right) \quad , \quad X = k_{l1}\left(\frac{Z_W R}{Z_{W1}} - Z_{W1}\right)$$

By solving for Z_{W1} and k_{l1} the equations for defining the dimensions are derived:

$$
\begin{aligned}
Z_{W1} &= \sqrt{Z_{W1}\left(R - \frac{X^2}{Z_W - R}\right)} \\
k_{l1} &= \frac{Z_{W1}}{X}\left(1 - \frac{R}{Z_W}\right)
\end{aligned}
\tag{26.36}
$$

For $R > Z_W$ matching is possible for all values of X while for $R < Z_W$ the condition

$$|X| < \sqrt{R(Z_W - R)} \tag{26.37}$$

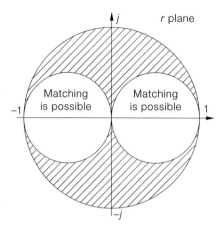

Fig. 26.29. Region in which matching with a series line is possible

must be met to ensure that the term under the square root in (26.36) is positive. This condition is quite easily represented in the r plane where matching with a simple series line is possible for all impedances with a reflection factor located in the r plane within the two circular regions shown in Fig. 26.29.

Matching of impedances for which condition (26.37) is not met, requires the structures in Figs. 26.28 b and 26.28 c. In Fig. 26.28 b, the structure's reactance X is compensated by the parallel reactance X_2 such that condition (26.37) is complied with; thus, impedance matching with a series line is possible. In the structure of Fig. 26.28 c, a parallel reactance X_1 at the input of the series line is permitted:

$$Z_1 = Z_W \, || \, jX_1$$

This is subsequently compensated by a parallel reactance $X_2 = -X_1$. This will not be described in more detail as these cases provide degrees of freedom that can be used to optimize the characteristic impedances Z_{W1} and Z_{W2} as well as the line length. In practice this is done with the aid of simulation programs for high-frequency circuits which contain suitable optimization algorithms. Both structures are often combined to obtain additional degrees of freedom for optimization.

26.3.2
Coupling

For power decoupling from a parallel resonant circuit, a load resistance must be coupled to the circuit. For the quality of a parallel resonant circuit loaded with a load resistance R_L but otherwise loss-free the following equation applies:

$$Q_r = R_L \sqrt{\frac{C}{L}}$$

Additionally, in high-frequency circuits the load resistance is usually $R_L = Z_W = 50 \, \Omega$. This means that the C/L ratio must be selected comparably high in order to achieve a sufficient quality, thus making the inductance very small at high resonant frequencies. Let

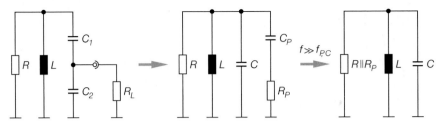

a With capacitive voltage divider

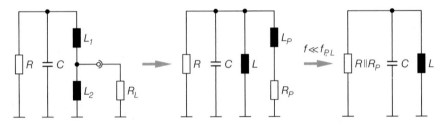

b With inductive voltage divider

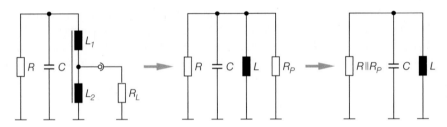

c With coupled inductive voltage divider

Fig. 26.30. Procedures for coupling a resistance R_L to a parallel resonant circuit

us take the example of a resonant circuit with a quality $Q_r = 50$ at a resonant frequency of $f_r = 1\,\text{GHz}$; it thus follows:

$$f_r = \frac{1}{2\pi\sqrt{LC}} = 1\,\text{GHz}\,, \quad Q_r = 50 \quad \Rightarrow \quad C = 159\,\text{pF}\,, \quad L = 159\,\text{pH}$$

An inductance of 159 pH is impractically low. At the same time, the capacitance is too high because generally the self-resonant frequency of a capacitor with capacitance $C = 159\,\text{pF}$ is clearly below 1 GHz (see Fig. 26.5 on page 1288). This means that sufficient quality and a practical rating for the elements can be achieved only by transforming the load resistance; the coupling procedures shown in Fig. 26.30 serve this purpose. For each of the procedures shown in Fig. 26.30 (left), the equivalent circuit (center) and a simplified equivalent circuit (right) are shown.

Coupling with a Capacitive Voltage Divider

With the divider factor

$$n_C = 1 + \frac{C_2}{C_1} \tag{26.38}$$

the elements of the equivalent circuit diagram in Fig. 26.30 a are:

$$R_P = n_C^2 R_L \quad , \quad C_P = \frac{C_1}{n_C} \quad , \quad C = \frac{C_1 C_2}{C_1 + C_2} \tag{26.39}$$

For

$$f \gg f_{P,C} = \frac{1}{2\pi C_P R_P} = \frac{1}{2\pi n_C C_1 R_L} \approx \frac{1}{2\pi C_2 R_L} \tag{26.40}$$

the capacitance C_P can be disregarded. The resonant circuit is then loaded with the transformed resistance R_P, which is arranged in parallel to the resonant impedance R.

Coupling with an Inductive Voltage Divider

With the divider factor

$$n_L = 1 + \frac{L_1}{L_2} \tag{26.41}$$

the elements of the equivalent circuit diagram in Fig. 26.30 b are:

$$R_P = n_L^2 R_L \quad , \quad L_P = n_L L_1 \quad , \quad L = L_1 + L_2 \tag{26.42}$$

For

$$f \ll f_{P,L} = \frac{R_P}{2\pi L_P} = \frac{n_L R_L}{2\pi L_1} \approx \frac{R_L}{2\pi L_2} \tag{26.43}$$

the inductance L_P can be disregarded. The resonant circuit is then loaded with the transformed resistance R_P, which is arranged in parallel to the resonant impedance R.

Coupling with Inductive Coupled Voltage Divider

If an inductive coupling of the inductances of the inductive voltage divider is provided so that the mutual inductance is

$$M = \sqrt{L_1 L_2}$$

the following divider factor is obtained:

$$n_{L,k} = 1 + \sqrt{\frac{L_1}{L_2}} \tag{26.44}$$

The components of the equivalent circuit in Fig. 26.30 c are:

$$R_P = n_{L,k}^2 R_L \quad , \quad L = L_1 + L_2 + 2M = \left(\sqrt{L_1} + \sqrt{L_2} \right)^2 \tag{26.45}$$

The resonant circuit is loaded with the transformed resistance R_P, which is arranged in parallel to the resonant impedance R. The transformation is independent of the frequency.

26.4
Power Splitters and Hybrids

When the output power of a matched amplifier is to be divided into two load resistances, a *power splitter* (power divider) is required; the splitter allows loss-free matching to the characteristic impedance Z_W. The principle of splitting the power of a matched RF amplifier as compared to an LF amplifier is shown in Fig. 26.31. Generally, LF amplifiers have a very low output resistance r_o. It is therefore possible to connect several load resistances to the output as long as the permissible output current is not exceeded. The power delivered by the amplifier depends on the load resistances. However, a matched RF amplifier must always be operated with a load resistance $R_L = Z_W$ to achieve the maximum output power and to prevent reflections which could destroy the amplifier. Consequently, the delivered power is constant and, where several load resistances are used, it must be divided by a power splitter.

The following section describes power splitters with three and four terminals. The latter are known as *hybrids* and may also be used as *power combiners*.

A typical application of power splitters and combiners are RF power amplifiers that consist of two parallel stages (see Fig. 26.32). The input power is distributed to the two stages by a power splitter, and the outputs of the two stages are added by a power combiner.

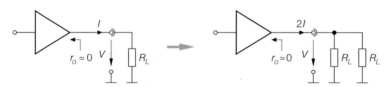

a LF amplifier with two load resistances

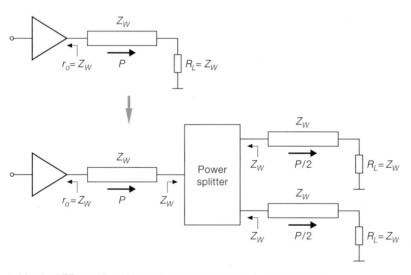

b Matched RF amplifier with two load resistances and power splitter

Fig. 26.31. Amplifier with two load resistances

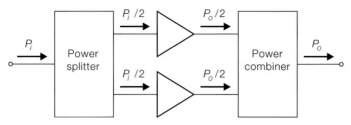

Fig. 26.32. Power splitter and power combiner in an RF amplifier with two stages in parallel

26.4.1
Power Splitter

Lossy Resistive Power Splitters

The lossy *resistive power splitters* shown in Fig. 26.33 are employed for wide-band power splitting. They are fully matched but deliver only half of the input power to the outputs while the other half is dissipated at the internal resistances. Since a quarter of the input power is available at each output, such splitters are also knows as *6 dB power splitters*. An identification of the three terminals is not needed because of the symmetry.

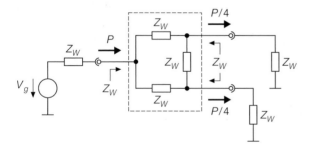

a Delta configuration

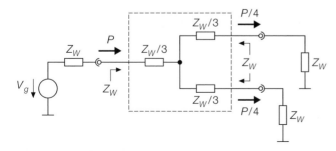

b Star configuration

Fig. 26.33. Lossy resistive power splitters

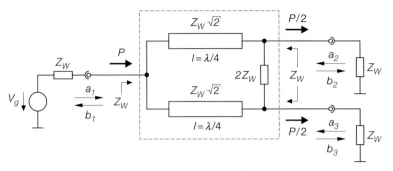

Fig. 26.34. Wilkinson splitter

Wilkinson Splitter

Fully matched and loss-free properties characterize the *Wilkinson splitter* shown in Fig. 26.34. It consists of two $\lambda/4$ lines and one resistance, thus having a narrow bandwidth. The input must be identified because the splitter is asymmetrical, and a loss-free operation is obtained only with the configuration shown in Fig. 26.34. Since half of the input power is available at each output, this splitter is also known as *3 dB power splitter*.

The response of the Wilkinson splitter is best described by means of the S parameters [26.1]:

$$
\begin{bmatrix} b_1 \\ b_2 \\ b_3 \end{bmatrix} = \begin{bmatrix} S_{11} & S_{12} & S_{13} \\ S_{21} & S_{22} & S_{23} \\ S_{31} & S_{32} & S_{33} \end{bmatrix} \begin{bmatrix} a_1 \\ a_2 \\ a_3 \end{bmatrix} = \frac{-j}{\sqrt{2}} \begin{bmatrix} 0 & 1 & 1 \\ 1 & 0 & 0 \\ 1 & 0 & 0 \end{bmatrix} \begin{bmatrix} a_1 \\ a_2 \\ a_3 \end{bmatrix} \tag{26.46}
$$

We can see that it is fully matched from the fact that the reflection factors at the three terminals are zero: $S_{11} = S_{22} = S_{33} = 0$. If a wave a_1 with the power

$$
P_1 = |a_1|^2
$$

occurs at terminal 1, then waves with the following powers are obtained at terminals 2 and 3:

$$
P_2 = |b_2|^2 = |S_{21}|^2|a_1|^2 = \frac{|a_1|^2}{2} = \frac{P_1}{2}
$$

$$
P_3 = |b_3|^2 = |S_{31}|^2|a_1|^2 = \frac{|a_1|^2}{2} = \frac{P_1}{2}
$$

It should be noted that in this case $b_1 = a_2 = a_3 = 0$ due to the fully matched state. On the other hand, if a wave a_2 with the power $P_2 = |a_2|^2$ occurs at terminal 2, we obtain $P_1 = |S_{12}|^2|a_2|^2 = |a_2|^2/2 = P_2/2$ and $P_3 = |S_{32}|^2|a_2|^2 = 0$, i.e., half the power is available at terminal 1 while the other half is dissipated at the internal resistance. The same situation arises if a wave occurs at terminal 3.

26.4.2
Hybrids

It can be demonstrated that a loss-free symmetrical power splitter that is fully matched to the characteristic impedance can only be constructed with four terminals. In a configuration with three terminals, the demands on the S parameters lead to inconsistencies [26.1]. Power splitters with four terminals are known as *hybrids* or *ring couplers*. The power fed to one terminal is divided and distributed to two of the other three terminals while the fourth terminal carries no signal.

S Parameters of a Hybrid

The properties of a hybrid can be easily described by means of the S parameters. One must distinguish between the *180° Hybrid* with

$$\begin{bmatrix} b_1 \\ b_2 \\ b_3 \\ b_4 \end{bmatrix} = \frac{-j}{\sqrt{2}} \begin{bmatrix} 0 & 0 & 1 & 1 \\ 0 & 0 & 1 & -1 \\ 1 & 1 & 0 & 0 \\ 1 & -1 & 0 & 0 \end{bmatrix} \begin{bmatrix} a_1 \\ a_2 \\ a_3 \\ a_4 \end{bmatrix} \tag{26.47}$$

and the *90° Hybrid* with:

$$\begin{bmatrix} b_1 \\ b_2 \\ b_3 \\ b_4 \end{bmatrix} = \frac{-j}{\sqrt{2}} \begin{bmatrix} 0 & 0 & -j & 1 \\ 0 & 0 & 1 & -j \\ -j & 1 & 0 & 0 \\ 1 & -j & 0 & 0 \end{bmatrix} \begin{bmatrix} a_1 \\ a_2 \\ a_3 \\ a_4 \end{bmatrix} \tag{26.48}$$

Both hybrids are fully matched: $S_{11} = S_{22} = S_{33} = S_{44} = 0$. Figure 26.35 shows the symbols of the two hybrid versions.

Let us first look at the 180° hybrid. The power of a wave a_1 occurring at terminal 1 is distributed to terminals 3 and 4. With $a_2 = 0$, it follows from (26.47):

$$b_3 = S_{31}a_1 = \frac{-j\,a_1}{\sqrt{2}} \quad \Rightarrow \quad P_3 = |b_3|^2 = \frac{|a_1|^2}{2} = \frac{P_1}{2}$$

$$b_4 = S_{41}a_1 = \frac{-j\,a_1}{\sqrt{2}} \quad \Rightarrow \quad P_4 = |b_3|^2 = \frac{|a_1|^2}{2} = \frac{P_1}{2}$$

The output waves b_3 and b_4 are in phase. The power of an incident wave a_2 at terminal 2 is also distributed to terminals 3 and 4 but the output waves b_3 and b_4 are in phase opposition. With $a_1 = 0$, it follows from (26.47):

$$b_3 = S_{32}a_2 = \frac{-j\,a_2}{\sqrt{2}} \quad \Rightarrow \quad P_3 = |b_3|^2 = \frac{|a_2|^2}{2} = \frac{P_2}{2}$$

$$b_4 = S_{42}a_2 = \frac{j\,a_2}{\sqrt{2}} \quad \Rightarrow \quad P_4 = |b_3|^2 = \frac{|a_2|^2}{2} = \frac{P_2}{2}$$

The 180° phase shift between terminals 2 and 4 is indicated in the diagram shown in Fig. 26.35 a. The 90° hybrid provides the following outputs if a wave occurs at terminal 1:

$$b_3 = S_{31}a_1 = \frac{-a_1}{\sqrt{2}} \quad \Rightarrow \quad P_3 = |b_3|^2 = \frac{|a_1|^2}{2} = \frac{P_1}{2}$$

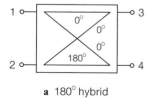

a 180° hybrid

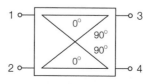

b 90° hybrid

Fig. 26.35. Hybrids

$$b_4 = S_{41}a_1 = \frac{-j\,a_1}{\sqrt{2}} \quad \Rightarrow \quad P_4 = |b_3|^2 = \frac{|a_1|^2}{2} = \frac{P_1}{2}$$

and if a wave occurs at terminal 2:

$$b_3 = S_{32}a_2 = \frac{-j\,a_2}{\sqrt{2}} \quad \Rightarrow \quad P_3 = |b_3|^2 = \frac{|a_2|^2}{2} = \frac{P_2}{2}$$

$$b_4 = S_{42}a_2 = \frac{-a_2}{\sqrt{2}} \quad \Rightarrow \quad P_4 = |b_3|^2 = \frac{|a_2|^2}{2} = \frac{P_2}{2}$$

In both cases there is a 90° phase shift between the two outputs. This is indicated in the diagram shown in Fig. 26.35 b.

Hybrids with Inductors and Capacitors

Figure 26.36 shows the three hybrids with inductors and capacitors [26.7]. In the 180° hybrid of Fig. 26.36 a the following condition must apply:

$$L = \frac{Z_W\sqrt{2}}{2\pi f_C} \quad , \quad C = \frac{1}{2\pi f_C Z_W \sqrt{2}} \tag{26.49}$$

Here, f_C is the center frequency at which the hybrid operates accurately. The bandwidth amounts to approximately 20% of the center frequency. In the 90° hybrid of Fig. 26.36 b the following condition must apply:

$$L = \frac{Z_W}{2\pi f_C \sqrt{2}} \quad , \quad C_1 = \frac{1}{2\pi f_C Z_W} \quad , \quad C_2 = \frac{\sqrt{2}-1}{2\pi f_C Z_W} \tag{26.50}$$

Here, the bandwidth amounts to only approximately 2% of the center frequency. For the 90° hybrid with inductor coupling, according to Fig. 26.36 c, the following condition must apply:

$$L = \frac{Z_W}{2\pi f_C} \quad , \quad C = \frac{1}{2\pi f_C Z_W} \tag{26.51}$$

Again, the bandwidth amounts to only approximately 2% of the center frequency.

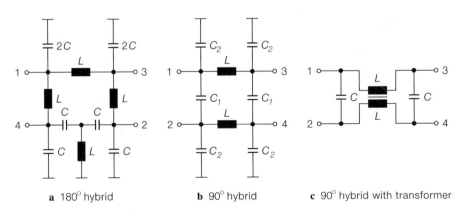

a 180° hybrid   b 90° hybrid   c 90° hybrid with transformer

Fig. 26.36. Hybrids with inductors and capacitors

Hybrids with Lines

For frequencies in the GHz range, most hybrids are provided with strip lines; Fig. 26.37 shows three versions [26.1, 26.7]. A particularly compact version with a relatively wide bandwidth of approximately 10% of the center frequency is that shown in Fig. 26.37 c with two noncoupled lines of length $\lambda/8$ and two capacitances:

$$C = \frac{1}{2\pi f_C Z_W} \tag{26.52}$$

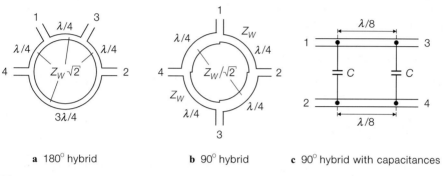

a 180° hybrid   b 90° hybrid   c 90° hybrid with capacitances

Fig. 26.37. Hybrids with lines

Chapter 27:
High-Frequency Amplifiers

Today, in the high- and intermediate-frequency assemblies of telecommunication systems, amplifiers composed of discrete transistors are still used in addition to modern integrated amplifiers. This is particularly the case in high-frequency power amplifiers employed in transmitters. In low-frequency assemblies, on the other hand, only integrated amplifiers are used. The use of discrete transistors is due to the status quo of semiconductor technology. The development of new semiconductor processes with higher transit frequencies is soon followed by the production of discrete transistors, but the production of integrated circuits on the basis of a new process does not usually occur until some years later. Furthermore, the production of discrete transistors with particularly high transit frequencies often makes use of materials or processes which are not (or not yet) suitable for the production of integrated circuits in the scope of production engineering or for economic reasons. The high growth rate in radio communication systems has, however, boosted the development of semiconductor processes for high-frequency applications. Integrated circuits on the basis of compound semiconductors such as gallium-arsenide (GaAs) or silicon-germanium (SiGe) can be used up to the GHz range. For applications up to approximately 3 GHz bipolar transistors are mainly used, which, in the case of GaAs or SiGe designs, are known as *hetero-junction* bipolar transistors (HBT). Above 3 GHz, gallium-arsenide junction FETs or *metal-semiconductor field effect transistors* (MESFETs) are used.[1] The transit frequencies range between 50 . . . 100 GHz.

27.1
Integrated High-Frequency Amplifiers

In principle, integrated high-frequency amplifiers use the same circuitry as low-frequency or operational amplifiers. A typical amplifier consists of a differential amplifier used as a voltage amplifier and common-collector circuits used as current amplifiers or impedance converters (see Fig. 27.1a). The differential amplifier is often designed as a cascode differential amplifier to reduce its reverse transmission and its input capacitance (no Miller effect). Such circuits are described in Chap. 4, Sect. 4.1. Since the transit frequency of high-frequency transistors ($f_T \approx 50 \ldots 100$ GHz) is approximately 100 times higher than that of low-frequency transistors ($f_T \approx 500$ MHz . . . 1 GHz), the bandwidth of the amplifier increases by approximately the same factor. This, however, presumes that the parasitics of the bond wires and the connections within the integrated circuit can be reduced enough so that the bandwidth is primarily determined by the transit frequency of the transistors and is not limited by the connections. This is a key problem in both the design and use of high-frequency semiconductor processes.

---

[1] The construction of an HBT corresponds to that of a conventional bipolar transistor. Here, however, different material compositions are used for the base and emitter regions in order to enhance the current gain at high frequencies. The construction of a MESFET is shown in Fig. 3.26b on page 196.

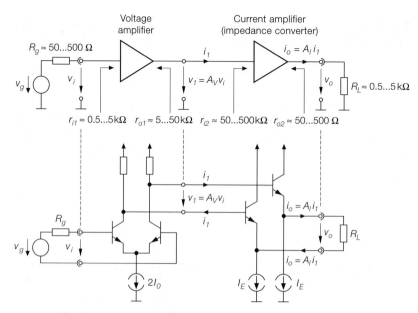

a Principle and design of an integrated amplifier

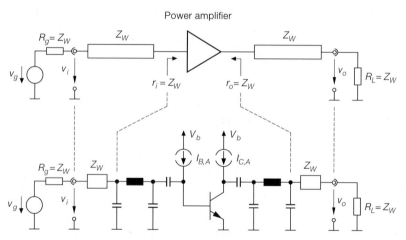

b Principle and design of a matched amplifier with one discrete transistor

Fig. 27.1. Principle construction of high-frequency amplifiers

27.1.1
Impedance Matching

Generally, the connecting leads within integrated circuits are so short that they can be considered as ideal connections even in the GHz range;[2] therefore, it is not necessary to carry

---

[2] These are *electrically short lines* (see Sect. 26.2). In this context the term *ideal* does not refer to the losses; these are relatively high in integrated circuits due to the comparably thin metal coating and the losses in the substrate.

out matching to the characteristic impedance within the circuit. In contrast, the external signal-carrying terminals must be matched to the characteristic impedance of the external lines to prevent any reflections. In the ideal case, the circuit is dimensioned such that input and output impedances, including the parasitic effects of bond wires, connecting limbs and the case, correspond to the characteristic impedance. Otherwise, external components or strip lines must be used for impedance matching (see Sect. 26.3).

Figure 27.1a shows typical values of low-frequency input and output resistances of the voltage and the current amplifier in an integrated high-frequency amplifier where it is assumed that equivalent amplifiers are employed as signal source and load.

Impedance Matching at the Input

For high frequencies, the input impedance of a differential amplifier is ohmic-capacitive due to the capacitances of the transistor. Generally, up to around 100 MHz, its value is clearly higher than the usual characteristic impedance $Z_W = 50\,\Omega$.

A rigorous impedance matching method involves inserting a terminating resistance $R = 2Z_W = 100\,\Omega$ between the two inputs of the differential amplifier (see Fig. 27.2a);

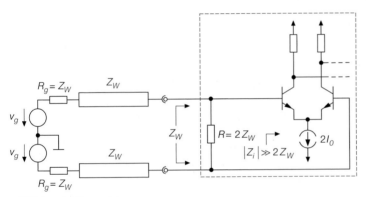

a With terminating resistance

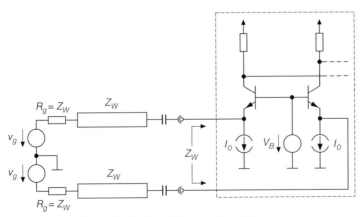

b With common-base circuits ($I_0 \approx 520\,\mu\text{A}$ for $Z_W = 50\,\Omega$)

Fig. 27.2. Impedance matching at the input side of an integrated amplifier

this matches both inputs to $Z_W = 50\,\Omega$. This method is simple, easy to accomplish with a resistor in the integrated circuit and acts across a wide band. A disadvantage is the poor power coupling due to the dissipation of the resistor and the large increase in the noise figure (see Sect. 27.1.2). Instead of placing a resistance $R = 2Z_W$ between the two inputs, each of the two inputs can be connected to ground via a resistance $R = Z_W$. However, this means that a galvanic coupling to signal sources with a DC voltage is no longer possible as the inputs are connected to ground with low resistance. The version with a resistance $R = 2Z_W$ is thus preferred.

As an alternative, common-base circuits can be used for the input stages (see Fig. 27.2b); then, the input impedance corresponds approximately to the transconductance resistance $1/g_m = V_T/I_0$ of the transistors. With a bias current $I_0 \approx 520\,\mu\text{A}$, this resistance is $1/g_m \approx Z_W = 50\,\Omega$. In this case, the power coupling is optimal. A disadvantage is the comparably high noise figure (see Sect. 27.1.2).

Both methods are suitable for frequencies in the MHz range only. In the GHz range, the influence of the bond wires, the connecting limbs and the casing have a noticeable effect. The situation can be improved by using loss-free matching networks made up of reactive components or strip lines that must be fitted externally. This will provide an optimum power coupling with a very low noise figure. In practice, impedance matching focuses less on optimum power transmission than it does on optimum noise figure, or a compromise between both optima. This is described in more detail in Sect. 27.1.2.

Impedance Matching at the Output

Wideband matching of the output impedance of a common-collector circuit to the usual characteristic impedance $Z_W = 50\,\Omega$ can be achieved by influencing the output impedance of the voltage amplifier while taking into consideration the impedance transformation in a common-collector circuit. For the qualitative aspects refer to Fig. 2.105a on page 147 and to the case shown in the left portion of Fig. 2.106 where the output impedance of a common-collector circuit has a wideband ohmic characteristic if the preceding amplifier stage has an ohmic-capacitive output impedance with a cutoff frequency that corresponds to the cutoff frequency $\omega_\beta = 2\pi f_\beta$ of the transistor. Due to secondary effects this type of matching can be achieved *quantitatively* only with the aid of circuit simulation. Again, in the GHz range, the influence of the bond wires, the connecting limb and the casing show a disturbing effect. In principle, impedance matching remains possible, but not with the wideband effect.

If impedance matching is not possible by influencing the output impedance of the common-collector circuit, external matching networks with reactive components or strip lines are used.

27.1.2
Noise Figure

In Sect. 2.3.4 we showed that the noise figure of a bipolar transistor with a given collector current $I_{C,A}$ is minimum if the effective source resistance between the base and the emitter terminal reaches its optimum value:

$$R_{gopt} = \sqrt{R_B^2 + \frac{\beta\, V_T}{I_{C,A}}\left(\frac{V_T}{I_{C,A}} + 2R_B\right)} \overset{R_B \to 0}{\approx} \frac{V_T\,\sqrt{\beta}}{I_{C,A}} \tag{27.1}$$

Here, R_B is the base spreading resistance and β the current gain of the transistor. For the collector currents $I_{C,A} \approx 0.1 \ldots 1\,\mathrm{mA}$, which are typical of integrated high-frequency circuits, the source resistance for $\beta \approx 100$ is in the region $R_{gopt} \approx 260 \ldots 2600\,\Omega$. With larger collector currents, R_{gopt} can be further reduced, e.g. to $50\,\Omega$ at $I_{C,A} = 23\,\mathrm{mA}$ and $R_B = 10\,\Omega$, but the noise figure reaches only a local minimum as shown in Fig. 2.52 on page 90. This is caused by the base spreading resistance. Very large transistors with very small base spreading resistances are used in low-frequency applications which enables the global minimum of the noise figure to be nearly reached even with small source resistances. However, in this case the transit frequency of the transistors drops rapidly; thus, in high-frequency applications, this method can be used in exceptional cases only.

In impedance matching at the input side by means of a terminating resistance as shown in Fig. 27.2a, the effective source resistance has the value $R_{g,eff} = R_g || R/2 = Z_W/2 = 25\,\Omega$ for each of the two transistors in the differential amplifier due to the parallel connection of the external resistances $R_g = Z_W$ and the internal terminating resistance $R = 2Z_W$. It is thus clearly lower than the optimum source resistance $R_{gopt} \approx 260 \ldots 2600\,\Omega$. Furthermore, the noise of the terminating resistance causes the noise figure to become relatively high. With impedance matching at the input side by means of a common-base circuit as shown in Fig. 27.2b, the effective source resistance has the value $R_{g,eff} = R_g = Z_W = 50\,\Omega$; here, too, the noise figure is comparably high.

For impedance matching with reactive components or strip lines, the internal resistance R_g of the signal source can be matched to the input resistance r_i of the transistor by means of a loss-free and noise-free matching network. If we disregard the base spreading resistance R_B, then $r_i = r_{BE}$. For the effective source resistance $R_{g,eff}$ between the base and emitter terminals this means that $R_{g,eff} = r_{BE}$. For $r_{BE} = \beta V_T/I_{C,A}$ and R_{gopt} the following relationship is obtained from (27.1) with $R_B = 0$:

$$R_{g,eff} = r_{BE} = R_{gopt}\sqrt{\beta} \tag{27.2}$$

Thus, with impedance matching, the effective source resistance is higher than the optimum source resistance by a factor of $\sqrt{\beta} \approx 10$. This might make the noise figure lower than that in the configurations with a terminating resistance or a common-base circuit, but it is still clearly higher than the optimum noise figure.

The optimum noise figure is only obtained when noise matching is performed instead of power matching. This means that the internal resistance $R_g = Z_W$ of the signal source is not matched to $r_i = r_{BE}$ but to $R_{gopt} = r_{BE}/\sqrt{\beta}$. Conversely, the input resistance of the (noise) matched amplifier is no longer Z_W but $Z_W\sqrt{\beta}$. This leads to the input reflection factor

$$r \overset{(24.34)}{=} \frac{Z_W\sqrt{\beta} - Z_W}{Z_W\sqrt{\beta} + Z_W} = \frac{\sqrt{\beta} - 1}{\sqrt{\beta} + 1} \overset{\beta \approx 100}{\approx} 0.82$$

and a standing wave ratio (SWR):

$$s \overset{(24.42)}{=} \frac{1 + |r|}{1 - |r|} = \sqrt{\beta} \overset{\beta \approx 100}{\approx} 10$$

In most applications this is not acceptable. Therefore, a compromise between power and noise matching is used in most practical cases where a low noise figure is of importance. Power matching is generally used if the noise figure is of no importance.

Above $f = f_T/\sqrt{\beta} \approx f_T/10$ the optimum source resistance decreases, as can be seen from the equation for $R_{gopt,RF}$ in Sect. 2.3.4. This does not mean that the matching methods in Fig. 27.2 can achieve a lower noise figure in this range. Factor $R_{g,eff}/R_{gopt}$ does go down but the minimum noise figure increases as the equation for $F_{opt,RF}$ in Sect. 2.3.4 shows. We will not examine this range more closely as the noise model for bipolar transistors with a transit frequency above 10 GHz as used in Sect. 2.3.4 will only allow qualitative statements in this case. The range $f > f_T/10$ is then entirely in the GHz range and some secondary effects, such as the correlation between the noise sources of the transistor, which were disregarded in Sect. 2.3.4, become significant, and the optimum source impedance is no longer real.

Example: With the help of circuit simulation we have determined the noise figure of the different circuit versions for an integrated amplifier with the transistor parameters in Fig. 4.5 on page 274. Owing to the symmetry, we can restrict the calculations to one of the two input transistors; Fig. 27.3 shows the corresponding circuits. We use a transistor of size 10 and a bias current of $I_{C,A} = 1$ mA. In the common-base circuit according to Fig. 27.3c, we reduce the bias current to 520 μA in order to achieve impedance matching to $Z_W = 50\,\Omega$.

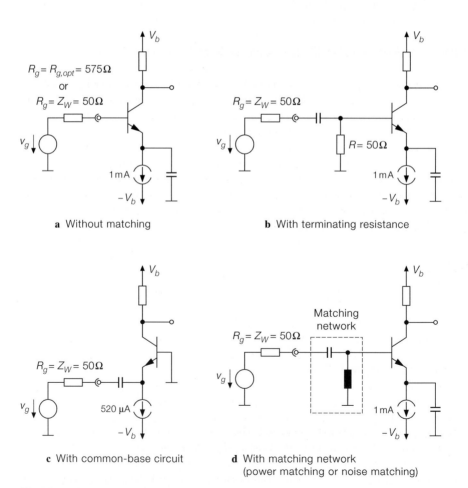

Fig. 27.3. Circuits for a noise figure comparison

The base spreading resistance is $R_B = 50\,\Omega$ and the frequency is $f = 10\,\text{MHz}$. From (27.1) it follows $R_{gopt} = 575\,\Omega$ for $I_{C,A} = 1\,\text{mA}$ and $R_{gopt} = 867\,\Omega$ for $I_{C,A} = 520\,\mu\text{A}$.

The circuit without matching in Fig. 27.3a achieves an optimum noise figure $F_{opt} = 1.12$ (0.5 dB) for $R_g = R_{gopt} = 575\,\Omega$ and $F = 1.52$ (1.8 dB) for $R_g = 50\,\Omega$. The circuit with terminating resistance in Fig. 27.3b results in the noise figure $F = 2.66$ (4.2 dB); the noise figure thus clearly increases. A more favorable value is achieved with the common-base circuit in Fig. 27.3c where $F = 1.6$ (2 dB). With power matching to $R_g = Z_W = 50\,\Omega$, according to Fig. 27.3d, the value obtained is $F = 1.25$ (0.97 dB), which is only a factor of 1.1 (0.5 dB) above the optimum value. The optimum noise figure is achieved with noise matching.

If power matching is essential in order to prevent reflections, the circuit with matching network and power matching according to Fig. 27.3d leads to the lowest noise figure, followed by the common-base circuit in Fig. 27.3c and then the circuit with terminating resistance in Fig. 27.3b. Without power matching, the circuit with matching network and noise matching according to Fig. 27.3d is clearly superior to the circuit without matching in Fig. 27.3a for $R_g = 50\,\Omega$ with regard to both the noise figure and the reflection factor.

27.2
High-Frequency Amplifiers with Discrete Transistors

Figure 27.1b shows the principle design of high-frequency amplifiers made up of discrete transistors. It is clear that the circuit design differs fundamentally from that of the integrated amplifier shown in Fig. 27.1a. The actual amplifier consists of a bipolar transistor in common-emitter configuration and circuitry for setting the operating point, which is presented in Fig. 27.1b, by the two current sources $I_{B,A}$ and $I_{C,A}$. The practical functionality will be further described below. Instead of a bipolar transistor, a field effect transistor can also be used. Coupling capacitances are used in front of and behind the transistor to prevent the operating point from being influenced by the additional circuitry. The networks for impedance matching to the characteristic impedance of the signal lines include π elements (Collins filters) with a series inductance and two shunt capacitances as shown in Fig. 27.1b.

27.2.1
Generalized Discrete Transistor

The term *discrete transistor* should not be misunderstood in a limited sense because the components used in practice often contain several transistors and additional resistances and capacitances in order to simplify the process of setting the operating point. We call these components *generalized discrete transistors*.[3]

Figure 27.4a shows the graphic symbol and the most important versions of a generalized discrete transistor without additional components for setting the operating point. A Darlington circuit is often used to enhance the current gain at high frequencies.

Figure 27.4b presents some typical designs with additions for setting the operating point. The version at the left can be used equally well for the Darlington circuits in

---

[3] This can be related to the CC operational amplifier which may also be regarded as a generalized discrete transistor (see Sect. 5.5 and Figs. 5.82 to 5.87).

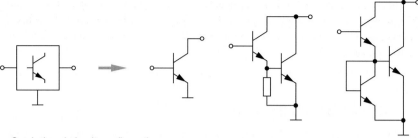

a Symbol and circuit configurations

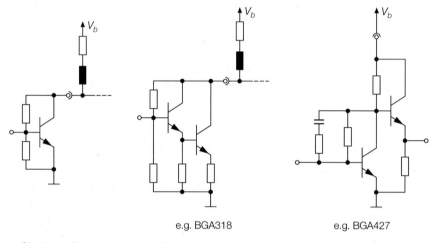

e.g. BGA318   e.g. BGA427

b Circuit configurations with additional elements for setting the operating point

Fig. 27.4. Generalized discrete transistor

Fig. 27.4a. The resistances provide a voltage feedback which, at sufficiently high-resistive dimensions, becomes virtually ineffective at high frequencies if the impedance of the collector-base capacitance falls below the value of the feedback resistor. The external element is an inductance which represents an open circuit at the operating frequency and consequently causes a separation of the signal path and the DC path. The version shown in the center of Fig. 27.4b has an additional emitter resistance for current feedback; therefore, it is particularly suitable for wideband amplifiers or amplifiers with a high demand in terms of linearity.

The version shown at the right of Fig. 27.4b consists of a common-emitter circuit with voltage feedback followed by a common-collector circuit. Strictly speaking, this does not belong to the group of discrete transistors since, like the integrated amplifier in Fig. 27.1b, it comprises a voltage amplifier (common-emitter circuit) and a current amplifier (common-collector circuit). Nevertheless, we have included it since it usually comes in a casing that is typical of discrete transistors. The voltage feedback is often operated with two resistances and one capacitance. Only the resistance, which is directly connected between the base and the collector, influences the operating point and is used for setting the collector voltage at the operating point. The capacitance is given dimensions such that it functions as a short circuit at the operating frequency, thus allowing the parallel arrangement of the two resistances to become effective.

The versions shown in Fig. 27.4 are regarded as low-integrated circuits and are termed *monolithic microwave integrated circuits (MMIC)*. They are made of silicon (Si-MMIC), silicon-germanium (SiGe-MMIC) or gallium-arsenide (GaAs-MMIC) and are suitable for frequencies of up to 20 GHz.

27.2.2
Setting the Operating Point (Biasing)

Generally, the operating point is set in the same way as for low-frequency transistors. However, with high-frequency transistors, one attempts to make the resistances required in order to set the operating point ineffective at the operating frequency otherwise they will have an adverse effect on the gain and noise figure. For this reason, the resistances are combined with one or more inductances which can be considered short-circuited with regard to setting the operating point, and nearly open-circuited at the operating frequency.

A description of how the operating point is set in a bipolar transistor is given below. The circuits described may equally well be used for field effect transistors.

DC Current Feedback

If we apply the above-mentioned principle to the operating point adjustment with DC current feedback as shown in Fig. 2.75a on page 117, we obtain the circuit design shown in Fig. 27.5a in which high-frequency decoupling is achieved for the base and the collector of the transistor by means of inductances L_B and L_C respectively. The collector resistance can be omitted in this case. Thus, there is no DC voltage drop in the collector circuit so that this method is particularly suitable for low supply voltages. In extreme situations, one may remove R_1 and R_2 and connect the free contact of L_B directly to the supply voltage; the transistor then operates with $V_{BE,A} = V_{CE,A}$. Due to the decoupled base, the noise of resistors R_1 and R_2 have only very little influence on the noise figure of the amplifier at the operating frequency; thus, this is a particularly low-noise design for setting the operating point. This is especially the case if an additional capacitance C_B is introduced which, at

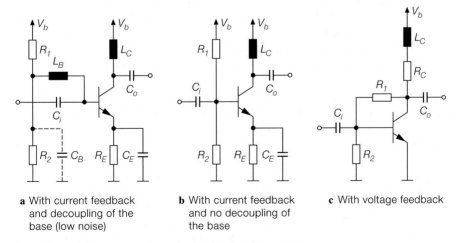

a With current feedback and decoupling of the base (low noise)

b With current feedback and no decoupling of the base

c With voltage feedback

Fig. 27.5. Setting the operating point in high-frequency transistors

the operating frequency, acts almost as a short circuit. Where a slight increase in the noise figure is not critical, it may not be necessary to decouple the base and thus the circuit shown in Fig. 27.5b may be used.

With an increase in frequency decoupling becomes more and more difficult since the characteristics of the inductors used to achieve the required inductance become less favorable. In order to make the magnitude of the impedance as high as possible, an inductor with a resonant frequency that is as close as possible to the operating frequency is used. As a result, the resonant impedance is approximately reached which, however, decreases with an increasing resonant frequency as shown in Fig. 28.4 on page 1366. For this reason, in the GHz range, the inductances are replaced by strip lines of the length $\lambda/4$. These lines are short-circuited for small signals at the end opposite the transistor by capacitance C_B or by connecting them to the supply voltage. The end closest to the transistor then acts as an open circuit.

Particularly problematic is the capacitance C_E which, at the operating frequency, must perform as a short circuit. Here, too, a capacitance with a resonant frequency as close as possible to the operating frequency is used, whereby doing so results in impedances with a magnitude close to that for the series resistance of the capacitance (typically $0.2\,\Omega$). However, with increasing resonant frequency, the resonance quality of the capacitances increases (see Fig. 28.5 on page 1366), thus making the adjustment more and more difficult. As an alternative, an open-circuited strip line of length $\lambda/4$ could be used that acts as a short circuit at the transistor end but, due to the unavoidable radiation at the open-ended side (antenna effect), this method is not practical. A short-circuited strip line must also be rejected as it provides a short circuit for the DC current and thus short-circuits the resistance R_E. Owing to these problems, the DC current feedback is used only in the MHz range while in the GHz range the emitter terminal of the transistor must be connected directly to ground.

DC Voltage Feedback

Figure 27.5c shows the method of setting the operating point by means of DC voltage feedback. This is used in many monolithic microwave integrated circuits (see Fig. 27.4b). A collector resistance R_C is essential in order to render the feedback effective and to ensure a stable operating point. The collector is decoupled by the inductance L_C so that, at the operating frequency, the output is not loaded by the collector resistance. The base can be decoupled by adding series inductances to the resistances R_1 and R_2; however, this method is not used in practice. A disadvantage is an increase in the noise figure due to the noise contributions from R_1 and R_2, but these can be kept low using high-resistive dimensioning.

Automatic Operating Point Control

Amplifiers, whether consisting of integrated circuits or discrete components, are often provided with automatic control of the operating point as shown in Fig. 27.6. Here, the collector current of the high-frequency transistor T_1 is measured from the voltage drop V_{RC} across the collector resistance R_C and compared with a setpoint value V_{D1}. Transistor T_2 controls the voltage at the base of transistor T_1 so that $V_{RC} \approx V_{D1} \approx 0.7\,\text{V}$.

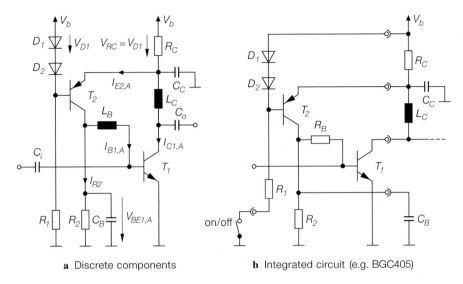

a Discrete components   **b** Integrated circuit (e.g. BGC405)

Fig. 27.6. Automatic operating point control

Let us first look at the circuit in Fig. 27.6a. It follows:

$$V_{RC} = \left(I_{C1,A} + I_{E2,A}\right) R_C \; , \quad V_{BE\,1,A} = I_{R2} R_2 \; , \quad I_{E2,A} \overset{I_{B2,A} \approx 0}{\approx} I_{B1,A} + I_{R2}$$

Consequently:

$$V_{RC} = \left(I_{C1,A} + I_{B1,A} + \frac{V_{BE\,1,A}}{R_2}\right) R_C \overset{I_{C1,A} \gg I_{B1,A}}{\approx} \left(I_{C1,A} + \frac{V_{BE\,1,A}}{R_2}\right) R_C$$

If the emitter-base voltage of transistor T_2 corresponds approximately to the voltage of diode D_2, then:

$$V_{RC} \approx V_{D1} \quad \Rightarrow \quad I_{C1,A} \approx \frac{V_{D1}}{R_C} - \frac{V_{BE\,1,A}}{R_2} \approx 0.7\,\mathrm{V}\left(\frac{1}{R_C} - \frac{1}{R_2}\right)$$

$R_2 \gg R_C$ is typically the case in practice; thus $I_{C1,A} \approx 0.7\,\mathrm{V}/R_C$.

The control circuit must have a pronounced lowpass characteristic of first order to ensure stability; capacitance C_B serves this purpose. It is selected such that the cutoff frequency

$$f_g = \frac{1}{2\pi C_B \left(R_2 \,\|\, r_{BE\,1}\right)}$$

is below the operating frequency by a factor of at least 10^4.

Figure 27.6b shows the control of the operating point for an integrated circuit where the elements L_C and C_B must be provided externally. The inductance L_B is usually replaced by a resistance which slightly shifts the operating point. Resistance R_C is usually an external component so that the bias current can be adjusted. This adjustment is necessary as the bias current, which is optimum in terms of gain and noise figure, depends on the operating frequency. Furthermore, the ground connection of resistance R_1 is usually accessible from the outside so that the amplifier can be turned on and off by a switch.

27.2.3
Impedance Matching for a Single-Stage Amplifier

Calculation of the matching networks for an amplifier with a generalized single transistor is complex because the impedances at the input and output port depend on the circuitry connected to the other port, respectively; this is due to the internal reactive feedback which also leads to a non-zero reverse transmission. The calculation is usually based on the S parameters of the transistor *including* the circuitry for setting the operating point.

Conditions for Impedance Matching

Figure 27.7 shows a transistor with matching networks and the corresponding reflection factors at various positions. Since these points are fully matched, the reflection factors at the signal source and the load are zero. The matching network at the input side transforms the reflection factor of the signal source from zero to r_g at the transistor input where it meets the input reflection factor r_1 of the transistor. Similarly, the matching network at the transistor output transforms the reflection factor of the load from zero to r_L, which meets the output reflection factor r_2 of the transistor. For two-sided impedance matching, the respective reflection factors must be conjugate complex:

$$r_g = r_1^* \quad , \quad r_L = r_2^* \tag{27.3}$$

The related impedances are also conjugate complex:

$$Z_g = Z_W \frac{1+r_g}{1-r_g} \overset{r_g=r_1^*}{=} Z_W \frac{1+r_1^*}{1-r_1^*} = Z_1^*$$

$$Z_L = Z_W \frac{1+r_L}{1-r_L} \overset{r_L=r_2^*}{=} Z_W \frac{1+r_2^*}{1-r_2^*} = Z_2^*$$

The conditions for power matching are thus met.

Reflection Factors of the Transistor

The reflection factors r_1 and r_2 of the transistor depend on r_L and r_g due to the reverse transmission (see Fig. 27.8). For the transistor, including the circuitry for setting the operating point, the following is true:

$$\begin{bmatrix} b_1 \\ b_2 \end{bmatrix} = \begin{bmatrix} S_{11} & S_{12} \\ S_{21} & S_{22} \end{bmatrix} \begin{bmatrix} a_1 \\ a_2 \end{bmatrix}$$

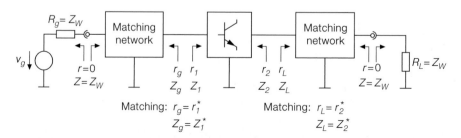

Fig. 27.7. Conditions for impedance matching on both sides

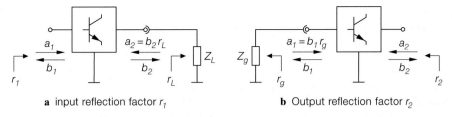

a input reflection factor r_1 **b** Output reflection factor r_2

Fig. 27.8. Calculating the reflection factors of a connected transistor

With a load with reflection factor r_L connected to the output, the input reflection factor r_1 is determined by inserting the condition $a_2 = b_2 r_L$ from Fig. 27.8a and solving the equation for $r_1 = b_1/a_1$. Similarly, the output reflection factor r_2 with a source with reflection factor r_g connected to the input is calculated by inserting the condition $a_1 = b_1 r_g$ from Fig. 27.8b and solving the equation for $r_2 = b_2/a_2$. This leads to:

$$r_1 = S_{11} + \frac{S_{12}S_{21}r_L}{1 - S_{22}r_L} \tag{27.4}$$

$$r_2 = S_{22} + \frac{S_{12}S_{21}r_g}{1 - S_{11}r_g} \tag{27.5}$$

Without reverse transmission ($S_{12} = 0$), there is no interdependence and the reflection factors are $r_1 = S_{11}$ und $r_2 = S_{22}$.

Calculating Impedance Matching

If we insert the conditions (27.3) into (27.4) and (27.5), the reflection factors r_g and r_L of the matched condition are obtained through elaborate calculations [27.1]:

$$r_{g,m} = \frac{B_1 \pm \sqrt{B_1^2 - 4|C_1|^2}}{2C_1} \tag{27.6}$$

$$r_{L,m} = \frac{B_2 \pm \sqrt{B_2^2 - 4|C_2|^2}}{2C_2} \tag{27.7}$$

The parameters are:

$$B_1 = 1 + |S_{11}|^2 - |S_{22}|^2 - |\Delta_S|^2$$

$$B_2 = 1 - |S_{11}|^2 + |S_{22}|^2 - |\Delta_S|^2$$

$$C_1 = S_{11} - \Delta_S S_{22}^*$$

$$C_2 = S_{22} - \Delta_S S_{11}^*$$

$$\Delta_S = S_{11}S_{22} - S_{12}S_{21}$$

In (27.6) and (27.7) the negative sign applies to $B_1 > 0$ or $B_2 > 0$ and the positive sign to $B_1 < 0$ or $B_2 < 0$.

Stability at the Operating Frequency

To ensure that the amplifier is stable, the following must apply:

$$|r_{g,m}| < 1 \quad , \quad |r_{L,m}| < 1$$

The real parts of the impedances are thus positive:

$$\operatorname{Re}\{Z_g\} = \operatorname{Re}\{Z_1\} > 0 \quad , \quad \operatorname{Re}\{Z_L\} = \operatorname{Re}\{Z_2\} > 0$$

It can be demonstrated that this is the case when the *stability factor (k factor)* is

$$k = \frac{1 + |S_{11}S_{22} - S_{12}S_{21}|^2 - |S_{11}|^2 - |S_{22}|^2}{2\,|S_{12}S_{21}|} > 1 \tag{27.8}$$

and the secondary conditions

$$|S_{12}S_{21}| < 1 - |S_{11}|^2 \quad , \quad |S_{12}S_{21}| < 1 - |S_{22}|^2 \tag{27.9}$$

are met [27.1].

Without reverse transmission ($S_{12} = 0$), the k factor is $k \to \infty$. In this case the secondary conditions require that $|S_{11}| < 1$ and $|S_{22}| < 1$, i.e., the real parts of the input and output impedances of the transistor, including the circuitry for setting the operating point, must be greater than zero. Therefore, a transistor without reverse transmission can be matched at both sides if the real parts of the impedances are greater than zero. If reverse transmission exists ($S_{12} \neq 0$), the secondary conditions are more stringent and thus positive real parts of the input and output impedance are no longer sufficient. In this case, however, the condition $k > 1$ is more crucial than the secondary conditions, i.e., the secondary conditions are usually met but the condition $k > 1$ is not.

Calculating Matching Networks

If the conditions (27.8) and (27.9) are met, the matching networks can be determined from (27.6) and (27.7) with the help of the reflection factors $r_{g,m}$ and $r_{L,m}$. First, the input and output impedances of the transistor, whose operating point is set for the matched condition, are calculated:

$$Z_{1,m} = Z_W \frac{1 + r_{1,m}}{1 - r_{1,m}} \overset{r_{1,m}=r_{g,m}^*}{=} Z_W \frac{1 + r_{g,m}^*}{1 - r_{g,m}^*} \tag{27.10}$$

$$Z_{2,m} = Z_W \frac{1 + r_{2,m}}{1 - r_{2,m}} \overset{r_{2,m}=r_{L,m}^*}{=} Z_W \frac{1 + r_{L,m}^*}{1 - r_{L,m}^*} \tag{27.11}$$

Using the procedure described in Sect. 26.3, it is now possible to calculate the matching networks for these impedances.

If conditions (27.8) and (27.9) are not met, a straight-forward procedure is not available. In this case a mismatch at the input or output must be accepted. A problem arises in finding suitable reflection factors r_g and r_L for which the mismatch is as small as possible while the operation of the system is sufficiently stable. [27.1] describes a procedure on the basis of stability circles which is not discussed in more detail here. A relatively easy procedure is to connect additional load resistances to the input or output of the transistor so that the S parameters meet the conditions of (27.8) and (27.9). However, it depends on the given application whether this yields a better overall result than a slight mismatch.

Stability Across the Entire Frequency Range

The stability conditions (27.8) and (27.9) ensure stability only at the operating frequency for which the matching networks are determined. However, in no way does this guarantee that the amplifier will be stable at all frequencies. This can be investigated by means of a test setup or by simulating the small-signal frequency response across the entire frequency range from zero up to and beyond the transit frequency of the transistor. When measuring the small-signal frequency response with a network analyser it should be noted that, in this case, the amplifier is connected to wide-band circuitry with $R_g = Z_W$ and $R_L = Z_W$. In the actual application, the amplifier may only have narrow-band matching that can cause instability at frequencies other than the operating frequency, i.e., the stability at the network analyser does not necessarily indicate stable operating conditions in the actual application.

Power Gain

For impedance matching on both sides with reactive, i.e., loss-free, matching networks, the *maximum available power gain (MAG)* [27.1]

$$MAG = \left| \frac{S_{21}}{S_{12}} \right| \left(k - \sqrt{k^2 - 1} \right) \tag{27.12}$$

can be determined from (27.8) with the stability factor $k > 1$. This and other power gains are described in Sect. 27.4 in more detail.

Example: The task is to design a high-frequency amplifier with transistor type BFR93 matched at both sides for an operating frequency (center frequency) $f_C = 1.88\,\text{GHz}$. The supply voltage is to be 3.3 V. We use automatic control of the operating point according to Fig. 27.6a with a bias current of $I_C = 5\,\text{mA}$. For this bias current we obtain a minimum noise figure as stated in the data sheet.[4]

Figure 27.9 shows the dimensioned components of a circuit for setting the operating point. The following aspects were taken into consideration:

- Since the input impedance of the transistor is very low (Re $\{S_{11}\} < 0 \rightarrow$ Re $\{Z_i\} <$ 50 Ω), the inductive decoupling of the base is omitted; therefore, the inductance L_B of Fig. 27.6a is replaced by a resistor $R_B = 1\,\text{k}\Omega$.
- An inductor with $L_C = 33\,\text{nH}$ and a parallel resonant frequency of approximately 1.9 GHz ($C \approx 0, 2\,\text{pF}$) is used for the inductive decoupling of the collector.
- A resistor $R_{LC} = 100\,\Omega$ is placed in series with L_C so that at frequencies below the operating frequency it causes losses which increase the k factor in the frequency range 100 MHz . . . 1.8 GHz (see Fig. 27.10). This reduces the tendency to oscillate in this frequency range.
- For capacitive blocking at the operating frequency, the capacitors C_{B1} and C_{C1}, whose series resonant frequency is approximately 1.9 GHz, are used ($C = 4.7\,\text{pF}$, size 0604: $L \approx 1.5\,\text{nH}$).

---

[4] The data sheet also specifies that the maximum transit frequency is reached with $I_C = 20\,\text{mA}$ so that $I_C = 5\,\text{mA}$ is not optimum. However, one should be careful, since the transit frequency is measured with the output short-circuited, allowing only limited conclusions to be drawn as regards to the power gain that can be achieved with impedance matching on both sides. In another design, conducted in parallel to this one, for $I_C = 20\,\text{mA}$ a power gain was achieved that was a mere 0.2 dB greater, a value which does not warrant the higher bias current, especially since the noise figure increases significantly.

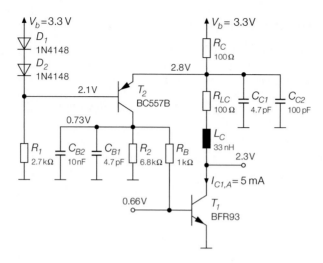

Fig. 27.9. Circuit for setting the operating point of transistor BFR93

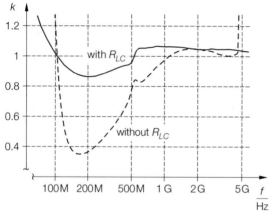

Fig. 27.10. k factor for the circuit shown in Fig. 27.9

- An additional capacitor C_{C2} with a higher capacitance is placed in parallel to C_{C1} in order to improve the capacitive blocking effect at low frequencies.
- Capacitor C_{B2} determines the cutoff frequency of the operating point control and therefore has a relatively high capacitance.

The S parameters of the transistor with operating point setting are determined by circuit simulation:[5]

$$S_{11} \;=\; -0.3223 + j\,0.2527 \quad,\quad S_{12} \;=\; 0.1428 + j\,0.1833$$

$$S_{21} \;=\; 1.178 + j\,1.3254 \quad,\quad S_{22} \;=\; 0.09015 - j\,0.249$$

---

[5] In this simulation, the high-frequency equivalent circuits of resistors and capacitors were taken into consideration. Nevertheless, the results of the simulation cannot be used for a real circuit design since the simulation model for transistor BFR93 provided by the manufacturer is not accurate enough for this frequency range. In practice, the S parameters of the transistor, including the network for setting the operating point, must be measured with a network analyser. In this example we use the S parameters from the simulation so that it can repeated with *PSpice*.

With (27.8) it follows $k = 1.05 > 1$, i.e., impedance matching on both sides is possible. The power gain to be expected is obtained with (27.12): $MAG = 5.57 \approx 7.5\,\text{dB}$. Equations (27.6) and (27.7) lead to:

$$r_{g,m} = -0.6475 - j\,0.402 \quad , \quad r_{L,m} = 0.3791 + j\,0.6$$

Then, using (27.10) and (27.11) we can calculate the input and output impedances of the transistor with operating point setting in the matched condition:

$$Z_{1,m} = (7.3 + j\,14)\,\Omega \quad , \quad Z_{2,m} = (33 - j\,80)\,\Omega$$

For both impedances, the real part is smaller than $Z_W = 50\,\Omega$ so that matching requires a step-up transformation according to Fig. 26.21a on page 1302.

For matching at the input side we obtain from (26.25) with $R = 7.3\,\Omega$ and $X = 14\,\Omega$:

$$X_1 = \pm 20.7\,\Omega \quad , \quad X_2 = \mp 17.7\,\Omega - 14\,\Omega$$

We select the highpass filter characteristic ($X_1 > 0$, $X_2 < 0$) according to Fig. 26.22b on page 1303, because then the series capacitance C_2 can simultaneously serve as a coupling capacitor. From

$$X_1 = 20.7\,\Omega \quad , \quad X_2 = -31.7\,\Omega$$

it follows with (26.26):

$$L_{1,i} = 1.75\,\text{nH} \quad , \quad C_{2,i} = 2.65\,\text{pF}$$

The additional index i refers to the *input side* matching.

For matching at the output side we obtain from (26.25) with $R = 33\,\Omega$ and $X = -80\,\Omega$:

$$X_1 = \pm 70\,\Omega \quad , \quad X_2 = \mp 24\,\Omega + 80\,\Omega$$

We now select the lowpass filter characteristic ($X_1 < 0$, $X_2 > 0$) according to Fig. 26.22a on page 1303 so that the overall characteristic is that of a bandpass filter. From

$$X_1 = -70\,\Omega \quad , \quad X_2 = 104\,\Omega$$

it follows with (26.26):

$$C_{1,o} = 1.2\,\text{pF} \quad , \quad L_{2,o} = 8.8\,\text{nH}$$

The additional index o refers to matching at the *output* side. An additional coupling capacitor is required at the output. We use a 4.7 pF capacitor with a series resonant frequency of 1.9 GHz which, at the operating frequency $f_C = 1.88\,\text{GHz}$, it acts as a short-circuit and thus has no influence on the matching effect.

Figure 27.11 shows the amplifier with the two matching networks. The elements of the matching networks are ideal; at this stage the design is not ready for practical use. It is necessary to check at which points inductors and capacitors can be connected and where strip lines may be advantageous or are mandatory for functionality of the elements. This is not discussed any further; please refer to the notes on impedance matching in multi-stage amplifiers in the next section.

Finally we present the results achieved. The upper part of Fig. 27.12 shows the magnitudes of the S parameters in the matched amplifier at the operating frequency $f_C = 1.88\,\text{GHz}$. One can see that matching covers a relatively narrow frequency band. If the requirements $|S_{11}| < 0.1$ and $|S_{22}| < 0.1$ hold for the reflection factors, then the bandwidth is approximately 53 MHz. Matching at the input covers a narrower band than

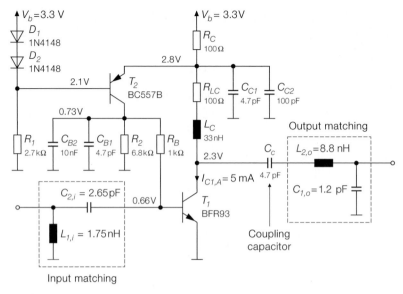

Fig. 27.11. Amplifier with matching networks

at the output since the transformation factor for the real part of the impedance is higher: $7.3\,\Omega \rightarrow 50\,\Omega$ at the input compared to $33\,\Omega \rightarrow 50\,\Omega$ at the output. In the center of Fig. 27.12 the magnitudes of the S parameters are plotted over a wider range. This shows that the output is also nearly matched ($|S_{22}| \approx 0.1$) in the range around 600 MHz. The position of this range depends on the capacitance of the coupling capacitor at the output, which can be used for adjustment. This can be useful if the amplifier is followed by a mixer for conversion to a low intermediate frequency. A suitable choice of the coupling capacitor can also provide a sufficient matching for the intermediate frequency. This indicates that high-frequency circuit engineering often takes advantage of secondary effects. The bottom diagram of Fig. 27.12 shows the gain in decibel. At the operating frequency it reaches its maximum, which we have calculated with (27.12): $MAG \approx 7.5\,\text{dB}$. The gain is comparatively low as the transistor type BFR93 has a transit frequency of only 5 GHz and is operated at its performance limit in our example. Modern circuits for the frequency range around 2 GHz use transistors with transit frequencies of about 25 GHz, resulting in gains of $20\ldots25\,\text{dB}$.

27.2.4
Impedance Matching in Multi-stage Amplifiers

Matching in multi-stage amplifiers is done in the same way as in single-stage amplifiers. Each stage is matched at both sides and then arranged in series, where the matching networks between the stages can often be simplified by combining the elements. In most cases, however, this is not the optimum procedure. In practice, it is used only if, for construction purposes, the stages are so far apart that the connections between the stages can no longer be considered as electrically short lines as is especially the case in the GHz range.

In all other cases the output of each stage is matched directly to the input of the next stage. The calculation of this type of impedance matching is complicated since an amplifier

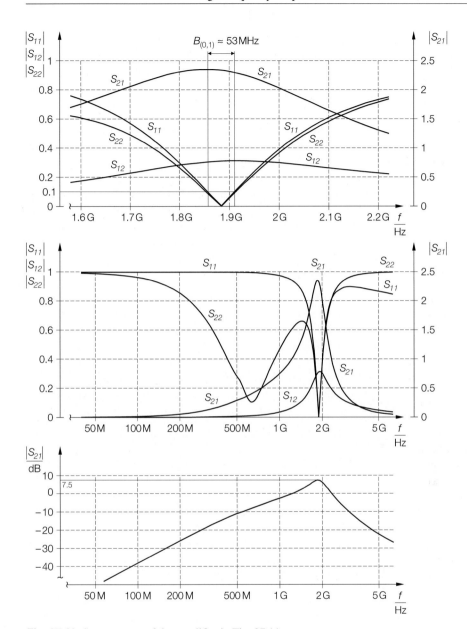

Fig. 27.12. S parameters of the amplifier in Fig. 27.11

with n stages, including $n+1$ matching networks (input side, output side and $n-1$ networks between the stages), are interdependent due to the reverse transmission of the transistors. The procedure is divided into two steps:

– In the first step, structures must be selected that, in principle, allow impedance matching on the basis of the S parameters of the individual transistors. This must include all wiring that is required for construction, i.e., the PC board layout of the amplifier must be roughly outlined.

– In the second step, the values of the elements in the individual structures must be determined by means of a simulation program. For this purpose, iterative optimization methods (*optimizers*) are used to find the ideal dimensions with regard to the criteria specified by the user. Often these criteria include maximizing $|S_{21}|$ observing the secondary conditions $|S_{11}| < 0.1$ and $|S_{22}| < 0.1$ in the specified frequency range.

If the reverse transmission of the transistors is not very high, the first run may already provide a satisfactory result. Otherwise the structures must be varied before further runs are carried out. These may become necessary solely because the established element values cannot be achieved or arranged on the predetermined layout of the PC board.

In practice, this procedure is also used for single-stage amplifiers. The ideal matching networks can, of course, be calculated directly by the procedure described in the previous section, but practical operation on the basis of the properties of the real components and the PC board layout require additional computer-aided optimization.

Impedance Matching with Series Inductance

For high-frequency bipolar transistors with a transit frequency above 10 GHz, the capacitances of the actual transistor are so low that the input and output capacitances are formed by the parasitic capacitance of the case. The equivalent circuit for these transistors with case capacitances C_{BE} and C_{CE} and case inductances L_B, L_C and L_E is shown in Fig. 27.13a where the relationships are $C_{BE} > C_{CE} > C_C$ and $L_B \approx L_C > L_E$. The equivalent circuit can be simplified due to the component dimensions. When using the simplified equivalent circuit for a multi-stage amplifier as shown in Fig. 27.13b, the circuitry between each of the stages represents a Collins filter. The capacitances of the filter are formed by the capacitances of the transistor and the inductances of the filter by the series connection of the case inductances and an external inductance. Therefore, if the dimensions are favorable, matching between the stages can be achieved with a series inductance. Similarly, the parasitic elements of the transistors at the input and output of the amplifier can be integrated into a Collins filter.

27.2.5
Neutralization

The main obstacle in impedance matching is the reverse transmission of the transistors which reduces the stability factor k and prevents matching on both sides if $k < 1$. For a transistor without reverse transmission $S_{12} = 0$ and $k \to \infty$ holds, so both sides can be matched provided the real parts of the input and output impedances are positive, i.e., $|S_{11}| < 1$ and $|S_{22}| < 1$. A transistor without reverse transmission operates *unilaterally* which means that signal transmission takes place only in the forward direction.

Circuits for Neutralization

The reverse transmission is caused by the collector-base capacitance C_C in bipolar transistors and by the gate-drain capacitance C_{GD} in FETs. It can be eliminated by connecting a *neutralization capacitance* C_n of the same value between the base and a point in the circuit that carries the inverted small-signal voltage of the collector. Such a point is created by using an inductor with center tap for decoupling the collector and connecting this tap to the supply voltage (see Fig. 27.14). The point opposite the collector then carries the in-

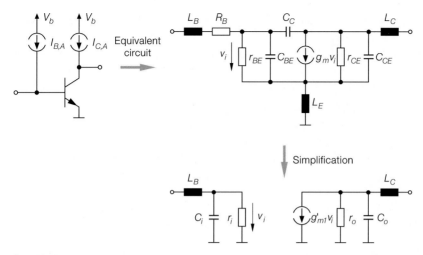

a Simplified equivalent circuit of a bipolar transistor in common-emitter configuration

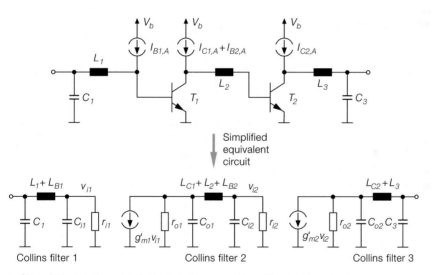

b Simplified equivalent circuit of a two-stage amplifier with matching circuitry

Fig. 27.13. Impedance matching of a two-stage amplifier with Collins filters utilizing the parasitic elements of the transistors

verted small-signal voltage. Neutralization is almost ideal up to approximately 300 MHz, but above this frequency the disturbing influence caused by the parasitic effects of the transistor (base spreading resistance and base inductance), the inductor and the capacitor becomes apparent. Amplifiers for higher output power often use two transistors in push-pull arrangement which can then be neutralized by cross-coupling with two capacitances C_{n1} and C_{n2} (see Fig. 27.15). Neutralization of a differential amplifier according to Fig. 27.16 is based on the same principle.

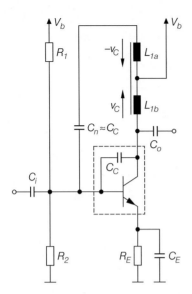

Fig. 27.14. Neutralization of a transistor

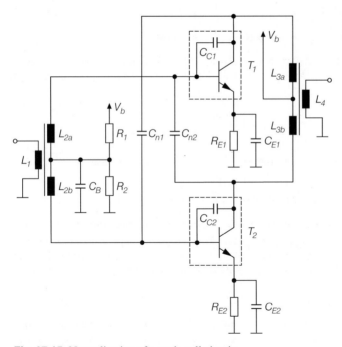

Fig. 27.15. Neutralization of a push-pull circuit

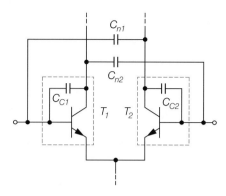

Fig. 27.16. Neutralization of a differential amplifier

Power Gain in the Case of Neutralization

Neutralization and two-sided impedance matching produce the highest possible power gain, known as *unilateral power gain* [27.1]:

$$U = \frac{\dfrac{1}{2} \left| \dfrac{S_{21}}{S_{12}} - 1 \right|^2}{k \left| \dfrac{S_{21}}{S_{12}} \right| - \mathrm{Re}\left\{ \dfrac{S_{21}}{S_{12}} \right\}} \tag{27.13}$$

Here, the S parameters of the transistor *without* neutralization and the stability factor k from (27.8) on page 1334 are to be inserted. However, the S parameters of the neutralized transistor can also be used, making $S_{12,n} = 0$ and leading to:[6]

$$U = \frac{|S_{21,n}|^2}{\left(1 - |S_{11,n}|^2\right) \left(1 - |S_{22,n}|^2\right)}$$

27.2.6
Special Circuits for Improved Impedance Matching

If the methods described so far fail to provide acceptable matching, circulators or 90° hybrids can be used to improve matching. This is the case, for example, if noise matching is carried out at the input of an amplifier in order to minimize the noise figure and at the same time the lowest possible reflection factor is required.

Impedance Matching with Circulators

A *circulator* is a transmission-asymmetric multi-port element. In practice, 3-port circulators are used exclusively, which are suitable for frequencies in the GHz range and achieve their transmission asymmetry by means of premagnetized ferrites [27.1].

---

[6] This relationship is obtained by calculating the transfer gain G_T according to (27.30) on page 1357 with matching on both sides and without reverse transmission, thus giving $S_{12} = 0$, $r_g = S_{11}^*$ and $r_L = S_{22}^*$.

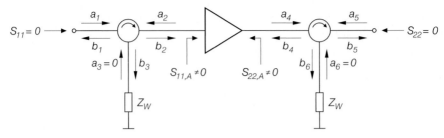

Fig. 27.17. Impedance matching with circulators

An ideal 3-port circulator is characterized by

$$
\begin{bmatrix} b_1 \\ b_2 \\ b_3 \end{bmatrix} = e^{j\varphi} \begin{bmatrix} 0 & 0 & 1 \\ 1 & 0 & 0 \\ 0 & 1 & 0 \end{bmatrix} \begin{bmatrix} a_1 \\ a_2 \\ a_3 \end{bmatrix}
\tag{27.14}
$$

where a_1, a_2, a_3 are the incoming and b_1, b_2, b_3 the reflected waves at the three ports. The circulator is fully matched, thus $S_{11,C} = S_{22,C} = S_{33,C} = 0$. The incident waves are transmitted to the next port in the order $1 \to 2 \to 3 \to 1$ and simultaneously undergo a rotation with the angle φ. Transmission asymmetry is seen in the asymmetry of the S matrix where $S_{12,C} \neq S_{21,C}$, $S_{13,C} \neq S_{31,C}$ and $S_{23,C} \neq S_{32,C}$.

Figure 27.17 shows a non-matched amplifier ($S_{11,A} \neq 0$, $S_{22,A} \neq 0$) provided with a circulator at the input and the output. The directional orientation of the circulators is indicated by arrows in the diagrams. Let us first look at the circulator at the input and assume the non-limiting condition $\varphi = 0$ which ensures that the wave a1 coming from the signal source is passed on to the amplifier unaltered:

$$
b_2 = S_{21,C}\, a_1 \overset{\varphi=0}{=} a_1
$$

Wave $a_2 = S_{11,A}\, b_2$, which is reflected at the amplifier input, is then transferred to the terminating resistance Z_W at port 3:

$$
b_3 = S_{32,C}\, a_2 = S_{32,C}\, S_{11,A}\, S_{21,C}\, a_1 \overset{\varphi=0}{=} S_{11,A}\, a_1
$$

Here, the wave is absorbed without reflection. This means that no incident wave occurs at port 3 and thus no reflected wave at port 1:

$$
a_3 = 0 \quad \Rightarrow \quad b_1 = S_{13,C}\, a_3 = 0
$$

This causes the reflection factor at the input to be zero:

$$
S_{11} = \frac{b_1}{a_1} \overset{b_1=0}{=} 0
$$

The functional principle of this matching method is based on the fact that a wave reflected at the input of the amplifier does not reach the signal source but is absorbed in the terminating resistance. In practice, this requires a circulator with very favorable characteristics and equally good termination at port 3. The circulator at the output of the amplifier operates in the same fashion.

In practice, only one circulator is normally used to improve the reflection factor of the amplifier. In low-noise amplifiers, the circulator is used at the input side to correct any existing input mismatch in the case of noise matching. The next section will describe noise matching in more detail. Power amplifiers sometimes use a circulator at the output; the circulator then performs two tasks:

- It reduces the reflection factor S_{22} at the amplifier output to zero.
- It prevents the wave reflected by the load from reaching the output of the amplifier; instead, the wave is absorbed in the terminating resistance Z_W.

The second task is of particular importance as the power amplifier can be destructed by the reflected wave.

Matching with 90° Hybrids

Matching with 90° hybrids requires two hybrids and two amplifiers with identical characteristics as shown by the circuit arrangement in Fig. 27.18. The S parameters of the 90° hybrid are determined with (28.48) on page 1419.

Let us first look at the relationships at the input. An incident wave a_1 is distributed in terms of its power to the two amplifiers. Wave b_4 at amplifier 2 leads in phase by 90°:

$$b_3 = S_{31,H}\, a_1 = -\frac{a_1}{\sqrt{2}} \quad , \quad b_4 = S_{41,H}\, a_1 = -j\,\frac{a_1}{\sqrt{2}}$$

At the inputs of the amplifiers, the waves are reflected with the input reflection factor $S_{11,A}$:

$$a_3 = S_{11,A}\, b_3 = -S_{11,A}\,\frac{a_1}{\sqrt{2}} \quad , \quad a_4 = S_{11,A}\, b_4 = -j\, S_{11,A}\,\frac{a_1}{\sqrt{2}}$$

This allows the calculation of the waves that occur at ports 1 and 2:

$$b_1 = S_{13,H}\, a_3 + S_{14,H}\, a_4 = -\frac{a_3}{\sqrt{2}} - j\,\frac{a_4}{\sqrt{2}} = 0$$

$$b_2 = S_{23,H}\, a_3 + S_{24,H}\, a_4 = -j\,\frac{a_3}{\sqrt{2}} - \frac{a_4}{\sqrt{2}} = j\, S_{11,A}\, a_1$$

One can see that the waves reflected by the amplifiers are transmitted to the terminating resistance Z_W at port 2 and that the reflection factor at port 1 becomes zero:

$$S_{11} = \frac{b_1}{a_1}\overset{b_1=0}{=} 0$$

Similarly, the value at the output is $S_{22} = 0$.

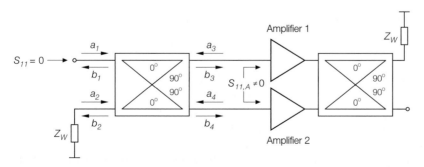

Fig. 27.18. Matching with 90° hybrids

The hybrid at the output acts as a *power combiner* and adds the output powers of the two amplifiers. Therefore, this version of matching is often used in power amplifiers despite its comparatively elaborate circuit construction.

27.2.7
Noise

In relation to the noise figure of integrated high-frequency amplifiers, we explained in Sect. 27.1 that bipolar transistors with (power) matching do not achieve the minimum noise figure since the matching network transforms the source resistance R_g to the input resistance r_{BE} of the transistor, while the optimum source resistance is $r_{BE}/\sqrt{\beta}$. In order to minimize the noise figure, power matching can be replaced by noise matching although this causes unacceptably high input reflection factors in most cases. The same applies to field effect transistors where here, too, power and noise matching differ substantially.

Noise Parameters and Noise Figure

At frequencies in the GHz range, the characteristic noise behavior of bipolar and field effect transistors can no longer be accurately described by the noise models discussed in Sects. 2.3.4 and 3.3.4. Instead, it is necessary to use the noise parameters specified in the data sheets including the minimum noise figure F_{opt}, the optimum reflection factor $r_{g,opt}$ of the signal source and the normalized noise resistance r_n. Often the noise resistance $R_n = r_n Z_W$ is quoted instead of the normalized noise resistance. The noise parameters allow the noise figure to be calculated for any given reflection factor r_g [27.2]:

$$F = F_{opt} + 4\, r_n \frac{\left|r_g - r_{g,opt}\right|^2}{\left(1 - |r_g|^2\right)\left|1 + r_{g,opt}\right|^2} \tag{27.15}$$

For $r_g = r_{g,opt}$ it follows: $F = F_{opt}$.

Design of a Low-Noise Amplifier

When designing an amplifier, the noise figure is calculated for all reflection factors with $|r_g| < 1$ and presented in the r plane, thus resulting in circles of constant noise figures. The diagram also represents the related power gain where, for the reflection factor $r_{g,m}$ with power matching, the maximum available power gain (*MAG*) is achieved if matching on both sides is possible. The power gain for other values of r_g corresponds to the gain G_T and is calculated with:

$$r_g \overset{(27.5)}{\Longrightarrow} r_2 = S_{22} + \frac{S_{12}S_{21}r_g}{1 - S_{11}r_g} \overset{\text{Matching}}{\Longrightarrow} r_L = r_2^*$$

$$\overset{(27.30)}{\Longrightarrow} G_T = \frac{|S_{21}|^2 \left(1 - |r_g|^2\right)\left(1 - |r_L|^2\right)}{\left|(1 - S_{11}r_g)(1 - S_{22}r_L) - S_{12}S_{21}r_g r_L\right|^2}$$

This results in circles with constant power gain. Normally the calculation is performed with the aid of suitable simulation or mathematical programs.

Figure 27.19 shows the noise figure and the power gain of a GaAs MESFET CFY10 at $f = 9\,\text{GHz}$. Power matching is achieved for $r_g = r_{g,m} = -0.68 + j\,0.5$ and noise

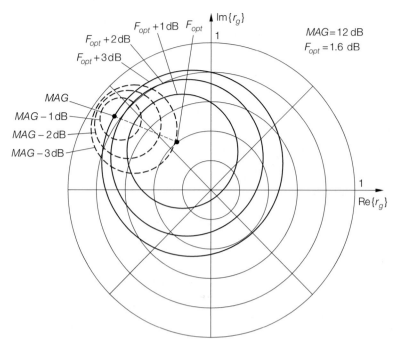

Fig. 27.19. Noise figure and power gain of a GaAs MESFET CFY10 at $f = 9\,\mathrm{GHz}$ ($I_{D,A} = 15\,\mathrm{mA}$, $V_{DS,A} = 4\,\mathrm{V}$)

matching is achieved for $r_g = r_{g,opt} = -0.24 + j\,0.33$. The circles of constant noise figures show that with power matching the noise figure is 3 dB higher than with noise matching. Likewise one can see from the circles of constant power gain that with noise matching the power gain is 3.1 dB lower than *MAG* . Now it is possible to draw a connecting line between $r_{g,m}$ and $r_{g,opt}$ and to select a reflection factor r_g for which the user-specific requirements are met.

Where matching at both sides is not possible, one can often carry out noise matching at the input and power matching at the output. For this purpose, the circles of constant noise figures are first drawn in the r plane. Then the power gain is calculated for all r_g values for which stable operation is possible. The procedure is:

– Starting with a given reflection factor r_g, the reflection factor at the output is calculated:

$$r_2 \overset{(27.5)}{=} S_{22} + \frac{S_{12}S_{21}r_g}{1 - S_{11}r_g}$$

If $|r_2| \geq 1$, stable operation with power matching at the output is not possible.
– If $|r_2| < 1$, power matching at the output is assumed: $r_L = r_2^*$.
– The corresponding reflection factor at the input is calculated:

$$r_1 \overset{(27.4)}{=} S_{11} + \frac{S_{12}S_{21}r_L}{1 - S_{22}r_L} = S_{11} + \frac{S_{12}S_{21}r_2^*}{1 - S_{22}r_2^*}$$

If $|r_1| \geq 1$, stable operation with power matching at the output is not possible.

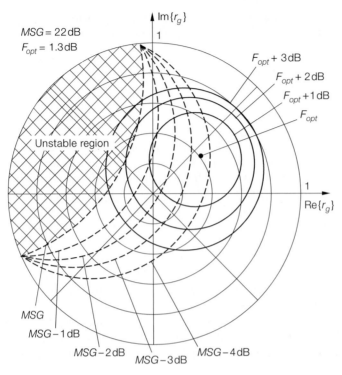

Fig. 27.20. Noise figure and power gain of a bipolar transistor BFP405 at $f = 2, 4\,\text{GHz}$ ($I_{C,A} = 5\,\text{mA}$, $V_{CE,A} = 4\,\text{V}$)

– If $|r_1| < 1$, the related transfer gain G_T is calculated:

$$G_T = \frac{|S_{21}|^2 \left(1 - |r_g|^2\right) \left(1 - |r_L|^2\right)}{\left|(1 - S_{11}r_g)(1 - S_{22}r_L) - S_{12}S_{21}r_g r_L\right|^2}$$

This results in circles of constant power gain that are limited by a stability border which is also circular. The stability border is the point at which the *maximum stable power gain MSG* is obtained; this will be described in more detail in Sect. 27.4.

Figure 27.20 shows the noise figure and the power gain of a bipolar transistor BFP405 at $f = 2.4\,\text{GHz}$. The stability factor is below one so that power matching at both sides is not possible. Noise matching is obtained for $r_g = r_{g,opt} = 0.32 + j\,0.25$. The circles of constant power gain are limited by a stability border at which the maximum stable power gain *MSG* is reached. The circles of constant power gain show that, with noise matching, the power gain is 3.5 dB below *MSG*. Likewise the circles of constant noise figure indicate that for operation with the power gain *MSG*, the noise figure is 1.8 dB above the minimum noise figure. A suitable reflection factor r_g can now be selected.

If the optimum reflection factor $r_{g,opt}$ with power matching at the output side is located inside the instable region, one must do without power matching and shift the stability border by a suitable choice of $r_L \neq r_2^*$ until $r_{g,opt}$ is within the stable region.

In practice, optimizing the parameters r_g and r_L in terms of noise, power gain and other criteria is done by means of simulation or mathematical programs with which nonlinear optimization processes can be carried out.

27.3
Broadband Amplifiers

Amplifiers with a constant gain over an extended frequency range are known as *broadband amplifiers*. High-frequency amplifiers are called broadband amplifiers if their bandwidth B is wider than the center frequency f_C thus producing a lower cutoff frequency $f_L = f_C - B/2 < f_C/2$ and an upper cutoff frequency $f_U = f_C + B/2 > 3f_C/2$ as well as a ratio $f_U/f_L > 3$. Sometimes $f_U/f_L > 2$ is used as a criterion. The term *broadband* is given to these amplifiers only because their bandwidth is clearly higher than the bandwidth of reactively matched amplifiers that are typical of high-frequency applications and in most cases have a ratio of $f_U/f_L < 1.1$. Furthermore, the wideband characteristic of high-frequency amplifiers is also related to impedance matching. Therefore, it is not the -3dB bandwidth that is used as the bandwidth, but the bandwidth within which the magnitude of the input and output reflection factors remain below a given limit. While reactively matched amplifiers usually require reflection factors of $|r| < 0.1$, broadband amplifiers accept reflection factors of $|r| < 0.2$. The less stringent demand reflects the fact that wideband matching in the MHz or GHz range is much more complicated than the narrow-band reactive matching.

27.3.1
Principle of a Broadband Amplifier

The functional principle of a broadband amplifier is based on the fact that a voltage-controlled current source with resistive feedback can be matched at both sides to the characteristic impedance Z_W. To implement the voltage-controlled current source, one of the generalized discrete transistors from Fig. 27.4 on page 1328 is used.[7] Figure 27.21 shows the principle of a broadband amplifier.

Let us first calculate the gain using the small-signal equivalent circuit shown in Fig. 27.22a. The nodal equation at the output is:

$$\frac{v_i - v_o}{R} = g_m v_i + \frac{v_o}{R_L}$$

This leads to the gain:

$$A = \frac{v_o}{v_i} = \frac{R_L (1 - g_m R)}{R + R_L} \tag{27.16}$$

The input current is

$$i_i = \frac{v_i - v_o}{R} = \frac{v_i (1 - A)}{R}$$

which leads to the input resistance:

---

[7] The version in the right portion of Fig. 27.4b cannot be used as it has no high-resistance output.

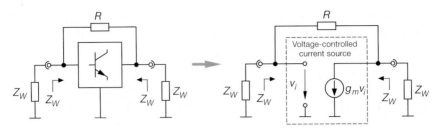

Fig. 27.21. Principle of a broadband amplifier

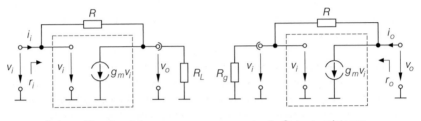

a Gain and input resistance   b Output resistance

Fig. 27.22. Equivalent circuits for calculating the gain as well as the input and output resistances of a broadband amplifier

$$r_i = \frac{v_i}{i_i} = \frac{R + R_L}{1 + g_m R_L} \tag{27.17}$$

According to Fig. 27.22b, the output current is:

$$i_o = \frac{v_o}{R + R_g} + g_m v_i = \frac{v_o}{R + R_g} + g_m \frac{R_g v_o}{R + R_g}$$

This leads to the output resistance:

$$r_o = \frac{v_o}{i_o} = \frac{R + R_g}{1 + g_m R_g} \tag{27.18}$$

We set $R_L = R_g = Z_W$ and calculate the reflection factors at the input and output:

$$S_{11} = \left.\frac{r_i - Z_W}{r_i - Z_W}\right|_{R_L = Z_W} = \frac{R - g_m Z_W^2}{R + 2Z_W + g_m Z_W^2} \tag{27.19}$$

$$S_{22} = \left.\frac{r_o - Z_W}{r_o - Z_W}\right|_{R_g = Z_W} = \frac{R - g_m Z_W^2}{R + 2Z_W + g_m Z_W^2} = S_{11} \tag{27.20}$$

The reflection factors S_{11} and S_{22} are identical and become zero for:

$$\boxed{R = g_m Z_W^2} \tag{27.21}$$

This means that both sides are matched. The forward transmission factor is:

$$S_{21} = \left. A \right|_{R_L = Z_W,\ R = g_m Z_W^2} = -\frac{R}{Z_W} + 1 = -g_m Z_W + 1 \tag{27.22}$$

This is identical to the gain in a circuit which is matched at both ends. It can be influenced only by means of the transconductance g_m as the feedback resistance is linked to the transconductance. A high transconductance results in a high gain.

27.3.2
Design of a Broadband Amplifier

Figure 27.23 shows the practical design of a broadband amplifier on the basis of an integrated Darlington transistor with resistances for the operating point adjustment. Resistances R_3 und R_4 have values in the kΩ range and are therefore negligible. This is especially
the case for the internal feedback resistance R_3 which is higher by at least a factor of 10
than the resistance R required for impedance matching. The effective feedback resistance
is thus:

$$R_{eff} = R \parallel R_3 \overset{R \ll R_3}{\approx} R$$

Resistance R_C serves to adjust the bias current. In terms of the small-signal parameters,
it is parallel to the amplifier output and acts like an additional load resistance. This means
that the amplifier no longer exactly fulfils the symmetry condition $S_{11} = S_{22}$ of an ideal
broadband amplifier, in other words, the matching condition $S_{11} = S_{22} = 0$ can only be
approximately satisfied. R_C must therefore be made as high as possible. In the region of the
upper cutoff frequency, the gain and matching can be improved by the inductances L_R and
L_C. The inductance L_R also contains the parasitic inductances of the resistance R and the
coupling capacitance C_c. Therefore, C_c can be a capacitor with a relatively high capacitance
and inductance, i.e., with a low resonant frequency, without producing any negative effect.
Capacitances C_i and C_o serve as coupling capacitances at the input and the output. These
are critical as most capacitors only achieve an impedance of $|X| \ll Z_W = 50\,\Omega$ in a
relatively narrow range around the resonant frequency (see Fig. 28.5 on page 1366). Thus,
the matching bandwidth is usually limited by the coupling capacitors.

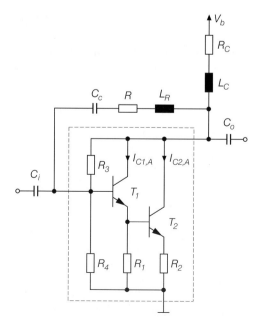

Fig. 27.23. Practical design of a broadband
amplifier

With the help of (27.22) we can use the desired gain to derive the necessary transconductance g_m of the voltage-controlled current source which corresponds approximately to the transconductance of transistor T_2 when taking the current feedback via R_2 into account:

$$g_m \approx \frac{g_{m2}}{1 + g_{m2} R_2} \qquad \text{with } g_{m2} = \frac{I_{C2,A}}{V_T}$$

The selection of the bias current $I_{C2,A}$ determines the maximum output power of the amplifier. In practice, modulation with an rms value of up to $I_{eff} \approx I_{C2,A}/2$ is useful; the distortion factor then remains below 10%. Consequently, the output power and the quiescent current are:

$$P_{o,max} = I_{eff}^2 Z_W \approx \frac{I_{C2,A}^2 Z_W}{4} \quad \Rightarrow \quad I_{C2,A} > \sqrt{\frac{4 P_{o,max}}{Z_W}} \qquad (27.23)$$

However, the bias current must be high enough to achieve the necessary transconductance: $I_{C2,A} \geq g_m V_T$. In this case, the resistance of the current feedback is:

$$R_2 \overset{I_{C,A} > g_m V_T}{=} \frac{1}{g_m} - \frac{V_T}{I_{C2,A}} \qquad (27.24)$$

The parasitic inductance of resistor R_2 must be as low as possible in order to avoid undesirable reactive feedback and is of particular importance with values below 20 Ω. If the expected bandwidth is not achieved in a broadband amplifier with current feedback, the reason is often because the parasitic inductance in the emitter circuit of T_2 is too high.

The current feedback via R_2 also influences the bandwidth by causing it to increase with increasing feedback. This is the reason why amplifiers with a particularly wide bandwidth make use of current feedback even if this is not required on the basis of the output power; typical examples are broadband amplifiers for instrumentation.

Example: In the following, a broadband amplifier is designed according to Fig. 27.23 for a 50 Ω system by using two transistors of the type BFR93 in Darlington configuration (see Fig. 27.24). A gain of $A = 16\,dB$ and a maximum output power of $P_{o,max} = 0.3\,mW = -5\,dBm$ are required. For the supply voltage we assume $V_b = 5\,V$. The gain is:

$$|A| = |S_{21}| = 10^{\frac{A\,[dB]}{20\,dB}} = 10^{\frac{16\,dB}{20\,dB}} = 6.3$$

With (27.22) we obtain the necessary transconductance:

$$S_{21} = -g_m Z_W + 1 \overset{!}{=} 6.3 \overset{Z_W = 50\,\Omega}{\Longrightarrow} g_m = \frac{7.3}{50\,\Omega} = 146\,mS$$

For the quiescent current of T_2 it follows $I_{C2,A} > g_m V_T = 3.8\,mA$. With (27.23) we obtain from the maximum output power $I_{C2,A} > 4.9\,mA$. We select $I_{C2,A} = 5\,mA$. The resistor R_2 is calculated with (27.24) to be $R_2 = 1.6\,\Omega$. The resulting small current feedback is not implemented for the moment as we must expect a loss in gain due to secondary effects.

For the bias current of transistor T_1 we select $I_{C1,A} = 2\,mA$ since, with smaller currents, the transit frequency drops rapidly. As the base-emitter voltage of T_2 is approximately 0.66 V and the base current $I_{B2,A} \approx 50\,\mu A$ (current gain approximately 100) is negligible compared to $I_{C1,A} = 2\,mA$, the value for the resistor R_1 is obtained:

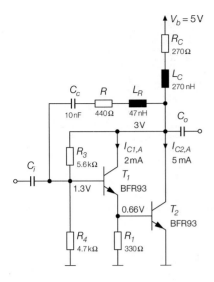

Fig. 27.24. Example of a broadband amplifier

$R_1 \approx 0,66\,\text{V}/2\,\text{mA} = 330\,\Omega$. Concerning the voltage divider for operating point adjustment we select $R_3 = 5.6\,\text{k}\Omega$ and $R_4 = 4.7\,\text{k}\Omega$ resulting in a voltage of 3 V at the collectors of the transistors (see Fig. 27.24). To ensure that the desired bias current for T_2 is achieved ($I_{C2,A} = 5\,\text{mA}$), a collector resistor $R_C = 270\,\Omega$ must be used for the supply voltage $V_b = 5\,\text{V}$.

After all resistors for operating point setting have been dimensioned, we can calculate the transconductance g_m. For this purpose we use the equation for the transconductance of a Darlington transistor with resistance R from Sect. 2.3.4 and insert $R = R_1$:

$$g_m \approx g_{m1} \frac{1 + g_{m2}\,(r_{BE\,2} \| R_1)}{1 + g_{m1}\,(r_{BE\,2} \| R_1)}$$

For $g_{m1} = I_{C1,A}/V_T = 77\,\text{mS}$, $g_{m2} = I_{C2,A}/V_T = 192\,\text{mS}$ and $R_1 = 330\,\Omega$, the transconductance is $g_m \approx 185\,\text{mS}$. From (27.21) the feedback resistance is thus $R = g_m Z_W^2 = 463\,\Omega$.

Further dimensioning is done with the aid of circuit simulation. We have used the high-frequency equivalent circuits for all resistances and inductances as well as the capacitor C_c, only for the coupling capacitances C_i and C_o have we assumed ideal capacitances. First, the reflection factors S_{11} and S_{22} are optimized at low frequencies by finely tuning the resistance R; the result is $R \approx 440\,\Omega$. Then, the gain and the impedance matching at high frequencies is optimized by adding inductors L_R and L_C. For $L_R = 47\,\text{nH}$ and $L_C = 270\,\text{nH}$, the plots of the magnitudes of the S parameters are obtained as shown in Fig. 27.25. The typical demand on broadband amplifiers of $|S_{22}| < 0.2$ is complied with up to about 1 GHz. In this range $|S_{11}| < 0.1$, i.e., the input matching is extremely good for a broadband amplifier. The desired gain $|S_{21}| = 6.3 = 16\,\text{dB}$ is reached up to approximately 300 MHz. The -3dB cutoff frequency is at 700 MHz.

The current feedback calculated for transistor T_2 with $R_2 \approx 1.6\,\Omega$ can be neglected because the amplifier achieves the desired gain. Deviations from the calculated values have two sources. First, the transconductance $g_m = 185\,\text{mS}$ of the Darlington transistor is lower than the transconductance $g_{m2} = 192\,\text{mS}$ of transistor T_2, and second, the transistor BFR93 has a parasitic emitter resistance of approximately $1\,\Omega$.

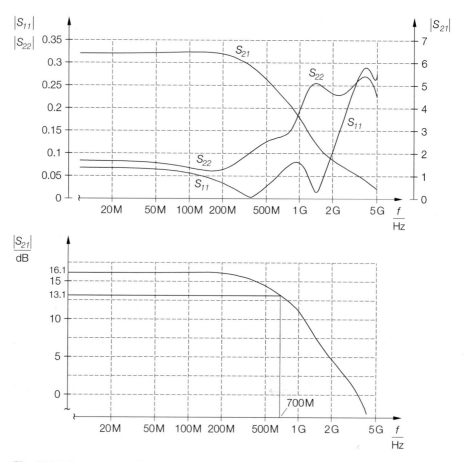

Fig. 27.25. S parameters of the broadband amplifier from Fig. 27.24

In practice, the very good overall performance of this amplifier can only be used in a comparatively small frequency band as the coupling capacitances C_i and C_o cannot be given a wide-band low-resistance characteristic. If necessary, several capacitors with staggered resonant frequencies must be used.

27.4
Power Gain

Usually the *power gain* is specified for high-frequency amplifiers. There are different definitions of *gain* which relate to different parameters. Some of the related equations on the basis of S or Y parameters are very complicated. We shall begin by explaining the definitions of gain for an ideal amplifier and then extend these to cover a more general situation. The complex equations on the basis of S and Y parameters are intended for computer-aided evaluations only as *manual* calculation is very involved.

Figure 27.26 shows the ideal amplifier with the open-circuit gain factor A, the input resistance r_i and the output resistance r_o; there is no reverse transmission. The amplifier

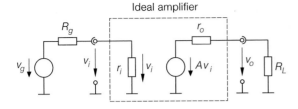

Fig. 27.26. Ideal amplifier with signal source and load

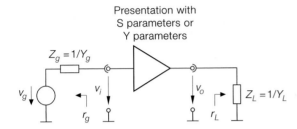

Fig. 27.27. General amplifier with signal source and load

is operated with a signal source of the internal resistance Rg and a load R_L. For further calculations we require the *overall gain*

$$A_B = \frac{v_o}{v_g} = \frac{r_i}{R_g + r_i} A \frac{R_L}{r_o + R_L}$$

and the *gain under load:*

$$A_L = \frac{v_o}{v_i} = A \frac{R_L}{r_o + R_L}$$

For the general situation, we look at an amplifier that is characterized by S and Y parameters. It is operated with a source of the impedance $Z_g = 1/Y_g$ and a load $Z_L = 1/Y_L$ (see Fig. 27.27). For presentation with the help of the S parameters we also need the reflection factors of the source and the load

$$r_g = \frac{Z_g - Z_W}{Z_g + Z_W} \quad , \quad r_L = \frac{Z_L - Z_W}{Z_L + Z_W}$$

and the determinant of the S matrix:

$$\Delta_S = S_{11} S_{22} - S_{12} S_{21}$$

It should be noted that the parameters r_g and r_L are reflection factors while r_i and r_o are the resistances of the ideal amplifier from Fig. 27.26.

27.4.1
Direct Power Gain

Direct power gain refers to the power gain in the conventional sense:

$$G = \frac{P_L}{P_i} = \frac{\text{Effective power absorbed by the load}}{\text{Effective power absorbed at the amplifier input}}$$

For the ideal amplifier from Fig. 27.26 it follows [8]:

$$P_L = \frac{v_o^2}{R_L} \quad , \quad P_i = \frac{v_i^2}{r_i}$$

This leads to:

$$G = \left(\frac{v_o}{v_i}\right)^2 \frac{r_i}{R_L} = A_L^2 \frac{r_i}{R_L} = \frac{A^2 r_i R_L}{(r_o + R_L)^2} \tag{27.25}$$

The corresponding calculation for the amplifier in Fig. 27.27 leads to:

$$G = \frac{|S_{21}|^2 \left(1 - |r_L|^2\right)}{1 - |S_{11}|^2 + |r_L|^2 \left(|S_{22}|^2 - |\Delta_S|^2\right) - 2\,\mathrm{Re}\left\{r_L\left(S_{22} - \Delta_S S_{11}^*\right)\right\}}$$

$$= \frac{|Y_{21}|^2\,\mathrm{Re}\,\{Y_L\}}{\mathrm{Re}\left\{Y_{11} - \dfrac{Y_{12}Y_{21}}{Y_{22}Y_L}\right\}|Y_{22} + Y_L|^2} \tag{27.26}$$

The direct power gain is independent of the signal source impedance and therefore contains no indication regarding the impedance matching on the input side. Comparison of, say, two amplifiers that use the same signal source, the same load, and output the same effective power to the load reveals that the amplifier with the lower effective input power has a higher direct power gain. In relation to high-frequency amplifiers, this property is not useful; therefore, the direct power gain is rarely used in high-frequency engineering.

27.4.2
Insertion Gain

Insertion gain is the ratio of the effective powers absorbed by the load with or without amplification:

$$G_I = \frac{P_L}{P_{L,wa}} = \frac{\text{Effective power absorbed by the load with amplifier}}{\text{Effective power absorbed by the load without amplifier}}$$

Consequently, $P_{L,wa}$ is the effective power which the signal source can deliver directly to the load. For the ideal amplifier from Fig. 27.26, we obtain:

$$P_L = \frac{v_o^2}{R_L} \quad , \quad P_{L,wa} = \frac{v_g^2 R_L}{(R_g + R_L)^2}$$

Consequently:

$$G_I = \left(\frac{v_o}{v_g}\right)^2 \left(\frac{R_g + R_L}{R_L}\right)^2 = A_B^2 \left(\frac{R_g + R_L}{R_L}\right)^2$$

$$= \left(\frac{r_i}{R_g + r_i}\right)^2 A^2 \left(\frac{R_g + R_L}{r_o + R_L}\right)^2 \tag{27.27}$$

---

[8] We use rms values so that $P = v^2/R$.

The corresponding calculation for the amplifier in Fig. 27.27 leads to:

$$G_I = \frac{|S_{21}|^2 |1 - r_g r_L|^2}{|(1 - S_{11}r_g)(1 - S_{22}r_L) - S_{12}S_{21}r_g r_L|^2}$$

$$= \frac{|Y_{21}|^2 \operatorname{Re}\{Y_g\} \operatorname{Re}\{Y_L\} |Y_g + Y_L|^2}{|(Y_{11} + Y_g)(Y_{22} + Y_L) - Y_{12}Y_{21}|^2 |Y_g Y_L|} \tag{27.28}$$

The insertion gain depends on the impedance of the signal source and the load and therefore takes the input and output impedance matching into account. However, the maximum gain is generally not reached with matching at both sides. This can be exemplified with the ideal amplifier. With two-sided matching, $R_g = r_i$ and $R_L = r_o$ where insertion into (27.27) leads to:

$$G_{I,match} = \left(\frac{1}{2}\right)^2 A^2 \left(\frac{R_g + R_L}{2R_L}\right)^2$$

This shows that, despite impedance matching at both sides, the insertion gain depends on the ratio R_g/R_L. A constant insertion gain is achieved only in the special case of equal resistances at the input and output, i.e., $R_g = r_i = r_o = R_L$. Owing to this characteristic, the insertion gain is hardly used.

27.4.3
Transfer Gain

Transfer gain specifies the ratio of the effective power absorbed by the load to the available (effective) power at the signal source:[9]

$$G_T = \frac{P_L}{P_{A,g}} = \frac{\text{Effective power absorbed by the load}}{\text{Available power at the signal source}}$$

For the ideal amplifier from Fig. 27.26 we obtain:

$$P_L = \frac{v_o^2}{R_L} \quad , \quad P_{A,g} = \frac{v_g^2}{4R_g}$$

This leads to:

$$G_T = \left(\frac{v_o}{v_g}\right)^2 \frac{4R_g}{R_L} = A_B^2 \frac{4R_g}{R_L} = \left(\frac{r_i}{R_g + r_i}\right)^2 A^2 \frac{4R_g R_L}{(r_o + R_L)^2} \tag{27.29}$$

The corresponding calculation for the amplifier in Fig. 27.27 leads to:

$$G_T = \frac{|S_{21}|^2 \left(1 - |r_g|^2\right)\left(1 - |r_L|^2\right)}{|(1 - S_{11}r_g)(1 - S_{22}r_L) - S_{12}S_{21}r_g r_L|^2}$$

$$= \frac{4|Y_{21}|^2 \operatorname{Re}\{Y_g\} \operatorname{Re}\{Y_L\}}{|(Y_{11} + Y_g)(Y_{22} + Y_L) - Y_{12}Y_{21}|^2} \tag{27.30}$$

---

[9] The available power is an effective power by definition and thus does not have to be explicitly specified as being effective.

The transfer gain depends on the impedance of the signal source and the load and becomes maximum with impedance matching at both sides. This can be demonstrated with (27.29):

$$\frac{\partial G_T}{\partial R_g} = 0 \quad , \quad \frac{\partial G_T}{\partial R_L} = 0 \quad \Longrightarrow \quad R_g = r_i \quad , \quad R_L = r_o$$

Thus, the transfer gain meets the reasonable demands expected of a gain definition.

27.4.4
Available Power Gain

Available power gain specifies the ratio of the available powers of the amplifier to the signal source:

$$G_A = \frac{P_{A,amp}}{P_{A,g}} = \frac{\text{Available power of the amplifier}}{\text{Available power of the signal source}}$$

For the ideal amplifier from Fig. 27.26 we obtain:

$$P_{A,amp} = \frac{(Av_i)^2}{4r_o} \quad , \quad P_{A,g} = \frac{v_g^2}{4R_g}$$

This leads to:

$$G_A = \left(\frac{Av_i}{v_g}\right)^2 \frac{R_g}{r_o} = \left(\frac{r_i}{R_g + r_i}\right)^2 A^2 \frac{R_g}{r_o} \tag{27.31}$$

The corresponding calculation for the amplifier in Fig. 27.27 leads to:

$$G_A = \frac{|S_{21}|^2 \left(1 - |r_g|^2\right)}{1 - |S_{22}|^2 + |r_g|^2 \left(|S_{11}|^2 - |\Delta_S|^2\right) - 2\text{Re}\left\{r_g \left(S_{11} - \Delta_S S_{22}^*\right)\right\}}$$

$$= \frac{|Y_{21}|^2 \text{Re}\left\{Y_g\right\}}{\text{Re}\left\{\left((Y_{11} + Y_g)\,Y_{22} - Y_{12}Y_{21}\right)\left(Y_{11} + Y_g\right)\right\}} \tag{27.32}$$

The available power gain is independent of the load and includes no indication with regard to impedance matching at the output side. It is required for noise calculations since these are based on the available power. The available power gain has already been described in Sect. 4.2.4 in connection with the calculation of the noise figure of amplifiers connected in series (see (4.200) and (4.201) on page 452).

27.4.5
Comparison of Gain Definitions

Specific properties of the various gain definitions have already been described in the relevant sections; for this reason, we shall restrict ourselves to a brief comparison here.

Direct power gain G is of no relevance in high-frequency amplifiers since the optimum use of the available power of the signal source is required and because impedance matching at the input side necessary for this purpose has no bearing on the direct power gain. In fact, it reaches its maximum if the amplifier absorbs as little power from the signal source as possible, i.e., with the poorest possible impedance matching. The direct power gain is relevant for low-frequency amplifiers since, in these cases, the aim is to achieve the highest possible voltage gain, which means a minimum load on the signal source. In high-frequency amplifiers such mismatches are undesirable because of the resulting reflections.

The insertion gain G_I is of no real significance for matched amplifiers. This will be explained for the ideal amplifier in Fig. 27.26. With matching at both sides and different resistances at the input and the output a mismatch occurs in the direct connection of the signal source and the load that, in practice, would be corrected by a matching network. For this reason, the two operating modes which are compared in the definition of the insertion gain are not *practical* but *theoretical* alternatives only. With impedance matching at both sides and equal resistances at the input and output, the matched condition ($R_g = R_L$) exists even with a direct connection between the signal source and the load, but in this case the available power of the signal source is delivered to the load and the insertion gain G_I corresponds to the transfer gain G_T.

Due to its properties, the transfer gain G_T is the preferred definition of gain in high-frequency engineering and is simply referred to as gain. However, it is not to be confused with *voltage gain* or *power gain*. Only in the case of impedance matching at both sides and the same resistances at the input and output are the voltage gain, current gain and transfer gain identical in their *decibel* values.

The available power gain G_A is required for noise calculations, as mentioned above, but beyond this it is of no importance.

27.4.6
Gain with Impedance Matching at Both Sides

With identical resistances at the input and output, matching at both sides means that, for the ideal amplifier in Fig. 27.26, $R_g = r_i = r_o = R_L = Z_W$. In this case, all gain definitions are identical:

$$G = G_I = G_T = G_A = \frac{A^2}{4} = 4A_B^2 \tag{27.33}$$

This is also true for a general amplifier which can be demonstrated by comparing the equations on the basis of the S and Y parameters, taking into account the given matching conditions. Due to the length of the required calculations, proof thereof is not given here.

Using the S parameters for an amplifier matched at both sides with $R_g = R_L = Z_W$ leads to:

$$S_{11} = S_{22} = r_g = r_L = 0 \quad \Rightarrow \quad G = G_I = G_T = G_A = |S_{21}|^2$$

This is a simple relationship because the measuring condition $R_L = Z_W$ for determining S_{21} is equal to the operating condition.

When using the Y parameters, the two-sided match to $1/Y_g = 1/Y_L = Z_W$ is reached when certain conditions are met:[10]

$$Y_{11} = Y_{22} \quad , \quad (Y_{11}Y_{22} - Y_{12}Y_{21})\, Z_W^2 = 1 \tag{27.34}$$

Then:

$$G = G_I = G_T = G_A = \frac{|Y_{21}|^2 Z_W^2}{|1 + Y_{11}Z_W|^2} \tag{27.35}$$

---

[10] These conditions are determined by calculating the Y parameters according to Fig. 24.40 on page 1186 from the S parameters while taking into account the condition $S_{11} = S_{22} = 0$.

For an amplifier without reverse transmission $Y_{12} = 0$; from the above conditions it follows $Y_{11} = Y_{22} = 1/Z_W$, i.e., the input resistance $r_i = 1/Y_{11}$ and the output resistance $r_o = 1/Y_{22}$ must be equal to the characteristic resistance Z_W. This case corresponds to the ideal amplifier in Fig. 27.26 from which the matching conditions $r_i = Z_W$ and $r_o = Z_W$ can be directly derived if $R_g = R_L = Z_W$.

27.4.7
Maximum Power Gain with Transistors

Sect. 27.2 showed that a generalized discrete transistor can be matched at both sides if the stability factor is

$$k = \frac{1 + |S_{11}S_{22} - S_{12}S_{21}|^2 - |S_{11}|^2 - |S_{22}|^2}{2\,|S_{12}S_{21}|} > 1 \tag{27.36}$$

and the secondary conditions

$$|S_{12}S_{21}| < 1 - |S_{11}|^2 \quad , \quad |S_{12}S_{21}| < 1 - |S_{22}|^2 \tag{27.37}$$

are met; here $S_{11}, \ldots, S_{22}$ are the S parameters of the transistor. The conditions for the Y parameters are

$$k = \frac{2\operatorname{Re}\{Y_{11}\}\operatorname{Re}\{Y_{22}\} - \operatorname{Re}\{Y_{12}Y_{21}\}}{|Y_{12}Y_{21}|} > 1 \tag{27.38}$$

and:

$$\operatorname{Re}\{Y_{11}\} \geq 0 \quad , \quad \operatorname{Re}\{Y_{22}\} \geq 0 \tag{27.39}$$

Maximum Available Power Gain

If matched on both sides, the transistor, *including* the matching networks, fulfils the condition $S_{11,match} = S_{22,match} = 0$ (see Fig. 27.28). The corresponding power gain is known as the *maximum available power gain (MAG)* and is given by [27.1]:

$$MAG = |S_{21,match}|^2 = \left|\frac{S_{21}}{S_{12}}\right|\left(k - \sqrt{k^2 - 1}\right) = \left|\frac{Y_{21}}{Y_{12}}\right|\left(k - \sqrt{k^2 - 1}\right) \tag{27.40}$$

At high frequencies, *MAG* is inversely proportional to the square of the frequency: $MAG \sim 1/f^2$, which corresponds to a declining rate of 20 dB/decade. This is caused by the frequency dependence of the S and Y parameters.

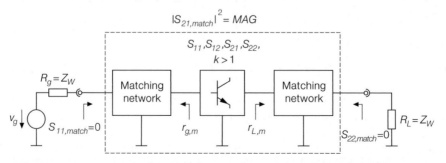

Fig. 27.28. *Maximum available power gain (MAG) of an amplifier matched at both sides*

Maximum Stable Power Gain

At frequencies above approximately a quarter of the transit frequency, the conditions for impedance matching at both sides are usually met. Below this frequency range $k < 1$, i.e., matching at both sides is no longer possible; in this case the maximum available power gain is not defined. Only the *maximum stable power gain (MSG)* can be achieved [27.1]:

$$MSG = \left| \frac{S_{21}}{S_{12}} \right| = \left| \frac{Y_{21}}{Y_{12}} \right| \tag{27.41}$$

At low frequencies it is approximately inversely proportional to the frequency: $MSG \sim 1/f$, which corresponds to a rate of decline of $10\,\text{dB/decade}$. When approaching the frequency for $k = 1$, the decline rate increases to $20\,\text{dB/decade}$ resulting in a smooth transition between *MSG* and *MAG*.

Unilateral Power Gain

The highest achievable power gain is the *unilateral power gain (U)*:

$$U = \frac{\dfrac{1}{2} \left| \dfrac{S_{21}}{S_{12}} - 1 \right|^2}{k \left| \dfrac{S_{21}}{S_{12}} \right| - \text{Re} \left\{ \dfrac{S_{21}}{S_{12}} \right\}} = \frac{|Y_{21} - Y_{12}|^2}{4 \left(\text{Re}\{Y_{11}\} \, \text{Re}\{Y_{22}\} - \text{Re}\{Y_{12}Y_{21}\} \right)} \tag{27.42}$$

This assumes that the transistor is *neutralized* by suitable circuitry, i.e., it has no reverse transmission; it then operates *unilaterally*. Circuits for neutralization are described in Sect. 27.2. At high frequencies, the unilateral power gain is approximately inversely proportional to the square of the frequency: $U \sim 1/f^2$, which corresponds to a decline rate of $20\,\text{dB/decade}$.

Limit Frequencies

The maximum available power gain (*MAG*) assumes the value 1 or $0\,\text{dB}$ at the transit frequency f_T of the transistor. The unilateral power gain (U) is higher than one even above the transit frequency since the reverse transmission is eliminated. The frequency at which U assumes the value 1 or $0\,\text{dB}$ is called the *maximum oscillation frequency* f_{max}. This represents the maximum frequency at which the transistor can be operated as an oscillator.

Example: Figure 27.29 shows the maximum power gains for transistor BFR93 at $V_{CE,A} = 5\,\text{V}$ and $I_{C,A} = 30\,\text{mA}$. The maximum available power gain (*MAG*) is only defined for $f > 500\,\text{MHz}$ as only here does the stability factor k rise above one. It declines at a rate of $20\,\text{dB/decade}$ and assumes the value 1 or $0\,\text{dB}$ at the transit frequency $f_T = 5\,\text{GHz}$. For $f < 500\,\text{MHz}$ the maximum stable power gain (*MSG*) is obtained which, at lower frequencies, declines at a rate of $10\,\text{dB/decade}$. At high frequencies the unilateral power gain U is higher than *MAG* by approximately $7.5\,\text{dB}$ and assumes the value 1 or $0\,\text{dB}$ at $f_{max} = 12\,\text{GHz}$.

In transistors with transit frequencies above $20\,\text{GHz}$, the collector-base capacitance C_C or the gate-drain capacitance C_{GD} are usually reduced to such an extent that the transistor can be regarded as having no reverse transmission even without neutralization. In this case, the maximum oscillation frequency f_{max} is only slightly higher than the transit frequency f_T.

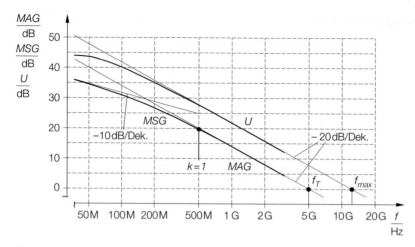

Fig. 27.29. Maximum power gains for transistor BFR93 at $V_{CE,A} = 5\,\mathrm{V}$ and $I_{C,A} = 30\,\mathrm{mA}$

Chapter 28:
Mixer

Mixers are required for *frequency conversion* in transmitters and receivers and together with amplifiers and filters are among the essential components of radio transmission systems. The following sections explain the functional principle of a mixer and then describe the circuits used in practical applications.

28.1
Functional Principle of an Ideal Mixer

An ideal mixer corresponds to a multiplier (see Fig. 28.1). The signal to be converted along with the *local oscillator signal* required for the conversion are fed into the inputs; in the ideal case the latter signal is sinusoidal. The output provides the converted signal as well as additional components generated in the conversion process. The unwanted components must be suppressed by filters in further processing. For this reason, one or two filters are required in addition to the mixer for frequency conversion. Usually the input for the signal to be converted is called the *input* and the input for the local oscillator signal is called the *local oscillator input*.

   The process of converting the input signal to a higher frequency is called *up-conversion* and the corresponding mixer is known as *up-conversion mixer*. Correspondingly, the terms *down conversion* and *down-conversion mixer* are used when the input signal is converted to a lower frequency. Figure 28.2 shows the characteristic frequencies in up-conversion and down-conversion mixers.

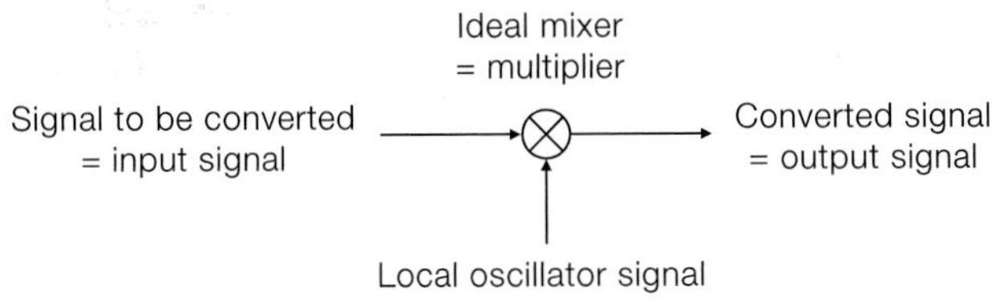

Ideal mixer
= multiplier

Signal to be converted
= input signal

Converted signal
= output signal

Local oscillator signal

Fig. 28.1. Ideal mixer

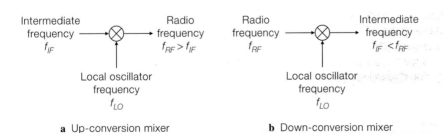

Intermediate frequency f_{IF}

Radio frequency $f_{RF} > f_{IF}$

Local oscillator frequency f_{LO}

a Up-conversion mixer

Radio frequency f_{RF}

Intermediate frequency $f_{IF} < f_{RF}$

Local oscillator frequency f_{LO}

b Down-conversion mixer

Fig. 28.2. Frequencies in mixers

- The *intermediate frequency (IF)* f_{IF} is the lower of the two carrier frequencies, i.e. the carrier frequency of the input signal of the up-conversion mixer or the carrier frequency of the output signal in the down-conversion mixer. In up conversion of the signal from the base band or down conversion of the signal into the base band, the intermediate frequency is $f_{IF} = 0$, which is the case, for example, in I/Q mixers.
- The *radio frequency (RF)* f_{RF} is the higher of the two carrier frequencies, i.e. the carrier frequency of the output signal in the up-conversion mixer or the carrier frequency of the input signal in the down-conversion mixer.
- The *local oscillator frequency (LO)* is the frequency of the required local oscillator signal and corresponds to the frequency shift achieved by the conversion.

The corresponding signals are called IF, RF and LO signals.

A distinction must be made between the frequencies related to the individual mixers and the frequencies in a specific transmitter or receiver. In a transmitter, each IF frequency of the transmitter occurs at the IF frequency of one of the mixers. Consequently, each IF frequency and the transmit frequency of a transmitter is generated by a mixer and therefore occurs as the RF frequency of the respective mixer. The same applies to the receiver. In the following we look at the frequencies of a single mixer while the meaning of these frequencies in a specific transmitter or receiver remains open.

28.1.1
Up-Conversion Mixer

An IF signal[1]

$$s_{IF}(t) = a(t) \cos [\omega_{IF} t + \varphi(t)]$$

is fed to the input of an up-conversion mixer and multiplied by the local oscillator signal

$$s_{LO}(t) = 2 \cos \omega_{LO} t$$

(see Fig. 28.3). The amplitude of the local oscillator signal is set to 2 so that the following equations contain no factors of $1/2$; this has no influence on the basic relations. The output signal is:

$$s_{RF}(t) = s_{IF}(t) \cdot s_{LO}(t) = a(t) \cos [\omega_{IF} t + \varphi(t)] \cdot 2 \cos \omega_{LO} t$$

$$= \underbrace{a(t) \cos [(\omega_{LO} + \omega_{IF}) t + \varphi(t)]}_{\substack{\text{Upper band } (f > f_{LO}) \\ \text{in noninverted mode}}} + \underbrace{a(t) \cos [(\omega_{LO} - \omega_{IF}) t - \varphi(t)]}_{\substack{\text{Lower band } (f < f_{LO}) \\ \text{in inverted mode}}}$$

The component of frequency $f_{LO} + f_{IF}$ is called the *upper band* and is of the same frequency sequence as the IF signal; this is known as *noninverted mode*. The component at frequency $f_{LO} - f_{IF}$ is called the *lower band* and is of inverted frequency sequence compared to the IF signal; this is known as *inverted mode*. Each of the two bands can serve as the output signal. The undesired band must be suppressed by means of a filter.

---

[1] Here, the amplitude modulation $a(t)$ and the angular modulation $\varphi(t)$ are used for the notation since this is more compact than the expression with the quadrature components: $s_{IF}(t) = i(t) \cos \omega_{IF} t - q(t) \sin \omega_{IF} t$.

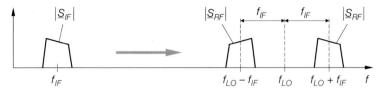

$s_{IF}(t) = a(t) \cos\left[\omega_{IF}t + \varphi(t)\right]$ $\qquad$ $s_{RF}(t) = a(t) \cos\left[(\omega_{LO} + \omega_{IF})t + \varphi(t)\right]$
$$+ a(t) \cos\left[(\omega_{LO} - \omega_{IF})t - \varphi(t)\right]$$

$2 \cos \omega_{LO} t$

$|\underline{S}_{IF}|$ $\qquad$ $|\underline{S}_{RF}|$ $\quad f_{IF} \quad f_{IF} \quad |\underline{S}_{RF}|$

f_{IF} $\qquad\qquad$ $f_{LO} - f_{IF} \quad f_{LO} \quad f_{LO} + f_{IF} \quad f$

Fig. 28.3. Time signals and spectra in an up-conversion mixer

28.1.2
Down-Conversion Mixer

An RF signal

$$s_{RF}(t) = a(t) \cos\left[\omega_{RF}t + \varphi(t)\right]$$

is fed to the input of the down-conversion mixer and multiplied by the local oscillator signal

$$s_{LO}(t) = 2 \cos \omega_{LO} t$$

(see Fig. 28.4). The output signal is:

$$s_M(t) = s_{IF}(t) \cdot s_{LO}(t) = a(t) \cos\left[\omega_{RF}t + \varphi(t)\right] \cdot 2 \cos \omega_{LO} t$$

$$= \begin{cases} a(t) \cos\left[(\omega_{RF} - \omega_{LO})t + \varphi(t)\right] & \text{Noninverted mode } (f_{RF} > f_{LO}) \\ \quad + a(t) \cos\left[(\omega_{RF} + \omega_{LO})t + \varphi(t)\right] & \\[2mm] a(t) \cos\left[(\omega_{LO} - \omega_{RF})t - \varphi(t)\right] & \text{Inverted mode } (f_{RF} < f_{LO}) \\ \quad + a(t) \cos\left[(\omega_{LO} + \omega_{RF})t + \varphi(t)\right] & \end{cases}$$

In addition to the desired components at the differential frequency, the output signal contains an extra component at the summation frequency which must be suppressed by a filter. The IF signal of interest is therefore:

$$s_{IF}(t) = \begin{cases} a(t) \cos\left[(\omega_{RF} - \omega_{LO})t + \varphi(t)\right] & \text{Noninverted mode } (f_{RF} > f_{LO}) \\ a(t) \cos\left[(\omega_{LO} - \omega_{RF})t - \varphi(t)\right] & \text{Inverted mode } (f_{RF} < f_{LO}) \end{cases}$$

If the RF frequency is higher than the LO frequency, then the result is an IF signal in noninverted mode of the same frequency sequence (see Fig. 28.4a). Otherwise we obtain an IF signal in *inverted mode* with an inverted frequency sequence (see Fig. 28.4b).

With down-conversion mixers it often occurs that the signal supplied to the RF input, in addition to the desired RF signal of the frequency $f_{RF} = f_{LO} \pm f_{IF}$, also contains an *image signal* with the *image frequency* $f_{RF,im} = f_{LO} \mp f_{IF}$, which is likewise converted to the IF frequency. In this case, the mixer operates in the noninverted mode *and* in the inverted mode. Figure 28.5 shows a down-conversion mixer with the RF frequency $f_{RF} = f_{LO} + f_{IF}$ in noninverted mode and the image frequency $f_{RF,im} = f_{LO} - f_{IF}$ in

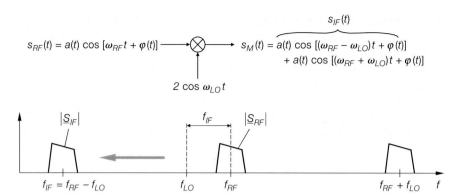

a In noninverted mode ($f_{RF} > f_{LO}$)

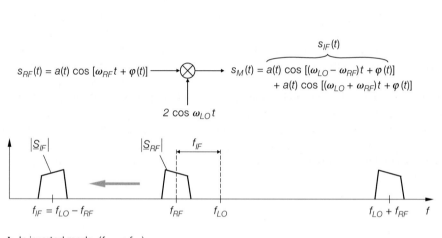

b In inverted mode ($f_{RF} < f_{LO}$)

Fig. 28.4. Time signals and spectra in the up-conversion mixer

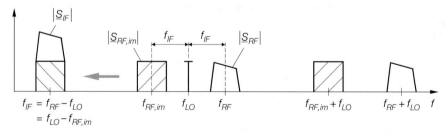

Fig. 28.5. Image frequency $f_{RF,im}$ in a down-conversion mixer in noninverted mode. The frequency sequence of the image signal $|\underline{S}_{RF,im}|$ is reversed due to the inverted mode

inverted mode where the frequency sequence of the image signal is reversed due to the inverted mode. In order to ensure that the mixer only converts the desired RF signal, the image signal must be suppressed by an *image frequency filter* arranged in front of the mixer; this will be explained in more detail in Sect. 27.2 on receivers. The existence of the image signal is a consequence of the functional symmetry of the up- and down-conversion mixers. The up-conversion mixer converts one IF signal into two RF signals, one of which must be selected *behind* the mixer. Similarly the down-conversion mixer converts two RF signals into one IF signal, meaning that one of the RF signals must be selected *in front* of the mixer.

28.2
Functional Principles of Practical Mixers

Multipliers are rarely used in practice. Practical multipliers feature a high linearity for both inputs which is not required for frequency conversion as explained below. Thus, practical multipliers used as mixers are even undesirable because they have a high noise figure due to the complex circuitry that is required to obtain the high linearity. In fact, in most cases the noise figure that occurs when a multiplier is used as a mixer is unacceptably high.

For a practical mixer it is sufficient for the signals of an ideal up- or down-conversion mixer to be *contained* in the voltages or currents of the mixer. The voltages and currents may contain other signals as long as these are separated from the useful signals with respect to the frequency so that they can be suppressed by filters at the output. Here, a distinction must be made between additive and multiplicative mixing. The two types of mixing will be described in the sections below taking an up-conversion mixer as an example.

28.2.1
Additive Mixing

In *additive mixing*, the IF and the LO signals are added together, provided with a suitable DC component V_0 and fed to a component with a nonlinear characteristic. The nonlinearity results in a multitude of mixing frequencies including the desired RF frequency, which is separated by means of a bandpass filter. Figure 28.6. shows the principle of additive mixing.

Description by Equations

In practice, the current-voltage characteristic $I(V)$ of a diode or transistor is usually used for the nonlinear characteristic, i.e. the input signal is a voltage and the output signal is a

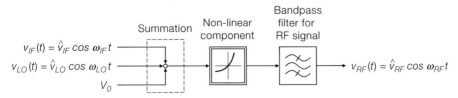

Fig. 28.6. Principle of additive mixing

current. The characteristic is expressed by a Taylor series at the operating point V_0:

$$I(V) = I(V_0) + \left.\frac{dI}{dV}\right|_{V=V_0} (V - V_0) + \frac{1}{2}\left.\frac{d^2I}{dV^2}\right|_{V=V_0} (V - V_0)^2$$

$$+ \frac{1}{6}\left.\frac{d^3I}{dV^3}\right|_{V=V_0} (V - V_0)^3 + \frac{1}{24}\left.\frac{d^4I}{dV^4}\right|_{V=V_0} (V - V_0)^4 + \cdots$$

With the small-signal parameters

$$i = I(V) - I(V_0) \quad, \quad v = V - V_0$$

and the abbreviations $a_1, a_2, \ldots$ for the derivatives we obtain:

$$i = a_1 v + a_2 v^2 + a_3 v^3 + a_4 v^4 + \cdots$$

This allows insertion of the small-signal voltage

$$v(t) = v_{IF}(t) + v_{LO}(t) = \hat{v}_{IF} \cos \omega_{IF} t + \hat{v}_{LO} \cos \omega_{LO} t$$

which results in:

$$i(t) = a_1 \left(\hat{v}_{IF} \cos \omega_{IF} t + v_{LO} \cos \omega_{LO} t\right)$$

$$+ a_2 \left(\hat{v}_{IF}^2 \cos^2 \omega_{IF} t + 2\,\hat{v}_{IF}\,\hat{v}_{LO} \cos \omega_{IF} t \cos \omega_{LO} t + \hat{v}_{LO}^2 \cos^2 \omega_{LO} t\right)$$

$$+ \cdots$$

The quadratic term

$$2\,a_2\,\hat{v}_{IF}\,\hat{v}_{LO} \cos \omega_{IF} t \cos \omega_{LO} t$$

$$= a_2\,\hat{v}_{IF}\,\hat{v}_{LO} \left[\cos (\omega_{LO} + \omega_{IF})\,t + \cos (\omega_{LO} - \omega_{IF})\,t\right]$$

contains the desired expression. The current $i(t)$ is fed to a bandpass filter that separates the signal portion at $f_{RF} = f_{LO} + f_{IF}$ (noninverted mode) or at $f_{RF} = f_{LO} - f_{IF}$ (inverted mode) in order to convert it into an output voltage

$$v_{RF}(t) = \hat{v}_{RF} \cos \omega_{RF} t = R_{BP}\,a_2\,\hat{v}_{IF}\,\hat{v}_{LO} \cos \omega_{RF}\,t$$

Here, R_{BP} is the transmission impedance of the bandpass filter in the passband.[2]

The amplitude $\hat{v}_{RF}$ of the output voltage is proportional to the coefficient a_2 of the nonlinear characteristic, where a_2 should be as high as possible so that the oscillator amplitude $\hat{v}_{LO}$ required for a certain output amplitude is kept low.

Nonlinearity

If we evaluate other terms of the current $i(t)$, we see that all coefficients a_i with an even index i contribute a component at frequency f_{RF}. For example, the term

$$a_4 v^4(t) = a_4 \left(\hat{v}_{IF} \cos \omega_{IF} t + \hat{v}_{LO} \cos \omega_{LO} t\right)^4$$

contains the components

$$\frac{3}{2}\,a_4\,\hat{v}_{IF}\,\hat{v}_{LO}^3 \left[\cos (\omega_{LO} + \omega_{IF})\,t + \cos (\omega_{LO} - \omega_{IF})\,t\right]$$

---

[2] Please note the units: $[R_{BP}] = \Omega$, $[a_2] = $ A/V^2 and $[\hat{v}_{IF}] = [\hat{v}_{LO}] = $ V; consequently $[R_{BP}\,a_2\,\hat{v}_{IF}\,\hat{v}_{LO}] = $ V.

and:

$$\frac{3}{2} a_4 \, \hat{v}_{IF}^3 \, \hat{v}_{LO} \left[\cos \left(\omega_{LO} + \omega_{IF} \right) t + \cos \left(\omega_{LO} - \omega_{IF} \right) t \right]$$

The amplitude of the first component is proportional to $\hat{v}_{IF}$ and adds to the desired output signal. In contrast, the amplitude of the second component is proportional to $\hat{v}_{IF}^3$ and is therefore nonlinear. With low IF amplitudes, the nonlinear component is negligible; however, the following condition must be met:

$$\left| \frac{3}{2} a_4 \, \hat{v}_{IF}^3 \, \hat{v}_{LO} \right| \ll \left| a_2 \, \hat{v}_{IF} \, \hat{v}_{LO} \right| \quad \Rightarrow \quad \hat{v}_{IF} \ll \sqrt{\left| \frac{2 \, a_2}{3 \, a_4} \right|}$$

By evaluating additional terms of the current $i(t)$, more conditions for $\hat{v}_{IF}$ are found that depend on the coefficients a_6, a_8, ... Normally, additive mixing is nonlinear and can be regarded as quasi linear only for low IF amplitudes. Conversely, it is strictly linear only if all coefficients with an even index $i \geq 4$ are zero as is the case with a quadratic characteristic of $i = a_2 v^2$.

The frequencies generated in additive mixing can be presented systematically in the form of a *frequency pyramid* (see Fig. 28.7). The coefficients a_i produce frequency groups (m, n) with nonnegative integer values for m and n and $m + n = i$; the coefficients a_2 thus produce the groups $(2,0)$, $(1,1)$ and $(0,2)$. The frequencies belonging to one group (m, n) are determined by calculating the sum

$$\underbrace{\pm f_{LO} \pm \cdots \pm f_{LO}}_{m \text{ addends}} \underbrace{\pm f_{IF} \pm \cdots \pm f_{IF}}_{n \text{ addends}}$$

for all possible sign configurations that evaluate to nonnegative values. For group $(1, 1)$ the addends are

$$f_{LO} + f_{IF} \quad , \quad f_{LO} - f_{IF} \quad , \quad - f_{LO} + f_{IF} \quad , \quad - f_{LO} - f_{IF}$$

and the frequencies under the assumption $f_{LO} > f_{IF}$ are:

$$f_{LO} + f_{IF} \quad , \quad f_{LO} - f_{IF}$$

In groups with higher values for m and n, the number of frequencies increases. All frequencies of the group (m, n) are also contained in the groups $(m + 2, n)$ and $(m, n + 2)$. The recursive application shows that all frequencies of one coefficient a_i are also produced by the coefficients $a_{(i+2)}$, $a_{(i+4)}$, $a_{(i+6)}$, ... Therefore, besides the coefficient a_2, the coefficients $a_4, a_6, a_8, \ldots$ are also of importance in additive mixers. The amplitudes of a group (m, n) are proportional to $\hat{v}_{LO}^m \, \hat{v}_{IF}^n$. The desired output signal with $f_{RF} = f_{LO} \pm f_{IF}$ lies in the group $(1,1)$ and is thus proportional to $\hat{v}_{LO} \, \hat{v}_{IF}$. Other components of the same frequency occur, for example, in the groups $(3,1)$ and $(1,3)$. The component in group $(3,1)$ is proportional to $\hat{v}_{LO}^3 \, \hat{v}_{IF}$ and is thus linear with respect to $\hat{v}_{IF}$. The component in group $(1,3)$, on the other hand, is proportional to $\hat{v}_{LO} \, \hat{v}_{IF}^3$ and is thus nonlinear with respect to $\hat{v}_{IF}$.

The nonlinearity of additive mixing not only results in a nonlinear relation between the IF amplitude $\hat{v}_{IF}$ and the RF amplitude $\hat{v}_{RF}$, but causes additional intermodulation distortions in modulated IF signals. For this we replace the constant IF amplitude $\hat{v}_{IF}$ by an amplitude-modulated signal without carrier of the modulation frequency f_m; then:

$$v_{IF}(t) = \hat{v}_{IF} \cos \omega_m t \cos \omega_{IF} t$$

$$= \frac{\hat{v}_{IF}}{2} \left[\cos \left(\omega_{IF} + \omega_m \right) t + \cos \left(\omega_{IF} - \omega_m \right) t \right]$$

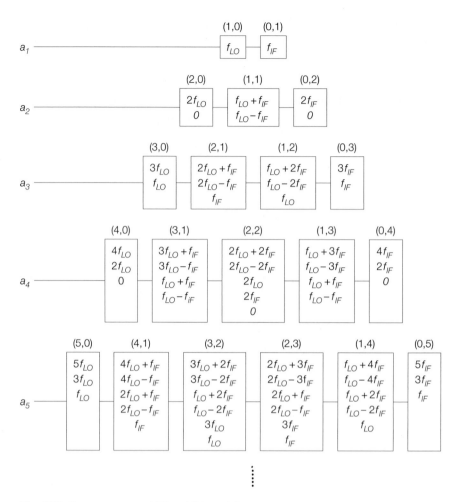

Fig. 28.7. Frequency pyramid for additive mixing

A calculation that is not shown here in detail reveals that all coefficients a_i with an even index i provide components at the desired output frequencies $f_{LO} \pm f_{IF} \pm f_m$. The coefficients a_i with $i = 4, 6, 8, \ldots$ supply additional components at the frequencies $f_{LO} \pm f_{IF} \pm 3 f_m$ and those with $i = 6, 8, \ldots$ supply components at $f_{LO} \pm f_{IF} \pm 5 f_m$, etc. These unwanted components, known as *intermodulation products*, are proportional to higher powers of the IF amplitude and must therefore also be reduced to an acceptable level by limiting the IF amplitude. For this case, too, a frequency pyramid can be built by utilizing the frequencies $f_{IF} + f_m$ and $f_{IF} - f_m$ instead of the frequency f_{IF} in group (0,1) and then calculating the other groups in the usual fashion. The number of frequencies in the groups increases compared to Fig. 28.7 because the following summations must be calculated:

$$\underbrace{\pm f_{LO} \pm \cdots \pm f_{LO}}_{m \text{ addends}} \underbrace{\pm (f_{IF} \pm f_m) \pm \cdots \pm (f_{IF} \pm f_m)}_{n \text{ addends } f_{IF} \pm f_m}$$

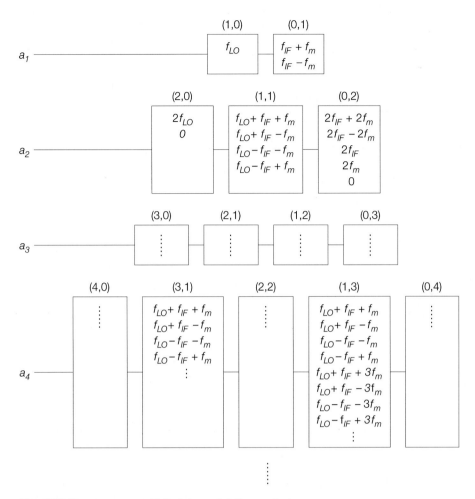

Fig. 28.8. Frequency pyramid for intermodulation products

Figure 28.8 shows a portion of the frequency pyramid for intermodulation products with the intermodulation products of third order ($f_{LO} \pm f_{IF} \pm 3 f_m$) in group (1,3) which are proportional to $\hat{v}_{LO} \hat{v}_{IF}^3$.

Basically, the relationships are the same as in a nonlinear amplifier. Therefore, the nonlinear characteristic parameters (compression and intercept points) can be used for mixers in the same way. However, the relationships between the characteristic parameters and the coefficients of the nonlinear characteristic are different because in an amplifier the coefficients a_i with an uneven index i are important unlike those with an even index i as is the case in a mixer. A nonlinear mixer can be regarded as a nonlinear amplifier with an additional frequency shift. For the *qualitative* aspects please refer to Sect. 4.2.3.

Practical Circuits

Figure 28.9 shows some typical circuits for additive mixing. In the circuit with a diode in Fig. 28.9a, the IF signal, together with the voltage V_0 for operating point setting, is supplied

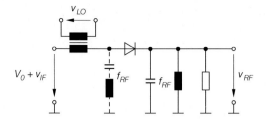

a With diode and summation by means of a transformer

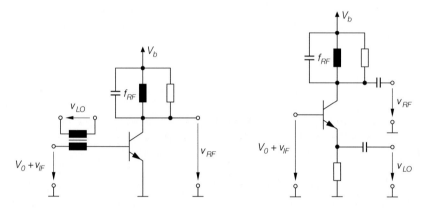

b With bipolar transistor summation by means
of a transformer

c With bipolar transistor and summation
by means of a separate supply to the
base and emitter

Fig. 28.9. Typical circuits for additive mixing. The parallel resonant circuits are tuned to the RF
frequency f_{RF}

directly whereas the LO signal is added via a transformer. A parallel resonant circuit is used
as the bandpass filter for the RF signal. The voltage at the diode corresponds to the sum
$V_0 + v_{IF} + v_{LO}$ since, at the IF and LO frequencies, the parallel resonant circuit acts as a
short circuit. At the output, all current components, with the exception of the RF component,
are short-circuited by the parallel resonant circuit. The RF component generates the RF
output voltage v_{RF} at the resistance of the resonant circuit. The disadvantage of this circuit
is that the RF current must flow through the IF and LO signal sources in order to close the
RF current loop. This can be prevented by connecting a series-resonant circuit between
the transformer and the diode to ground in order to short-circuit the RF current at this
point (see Fig. 28.9a). For optimum operation, complete decoupling of the IF, LO and RF
terminals is necessary. To do so, the IF and the LO circuit must also be provided with a
parallel resonant circuit which acts as an open circuit for the respective frequency, while
for all other frequencies it acts almost as a short circuit. Section 28.3 explains this in more
detail. Since the diode is a passive component, the RF power available at the output is
always lower than the supplied IF power, i.e. there is a *conversion loss*.

A *conversion gain* can be achieved by using a bipolar transistor (see Fig. 28.9b).
Here, the RF current does not flow through the IF or LO signal source. The RF portion is
decoupled from the other components relatively well. This decoupling can be enhanced
by introducing a cascode transistor. Since the RF current of the transistor can be picked

up at a separate connection, namely the collector, the IF and the LO signals can be added by supplying one signal to the base and the other to the emitter (see Fig. 28.6c) which eliminates the need for a transformer. However, the RF current then flows through the LO signal source.

The exponential characteristic

$$I(V) = I_S \left(e^{\frac{V}{nV_T}} - 1 \right)$$

of a diode ($V = V_D$, $I = I_D$, $n = 1 \ldots 2$) or of a bipolar transistor ($V = V_{BE}$, $I = I_C$, $n = 1$) can be used to calculate the coefficients of the nonlinear characteristic at the operating point $I_0 = I(V_0)$:

$$a_i = \frac{1}{i!} \left. \frac{d^i I}{dV^i} \right|_{V=V_0} = \frac{1}{i!} \frac{I_0}{(nV_T)^i} \quad \Rightarrow \quad a_2 = \frac{1}{2} \frac{I_0}{(nV_T)^2} \quad , \quad \ldots$$

The coefficients are proportional to the bias current such that if $I_0 = 100\,\mu A$ and $n = 1$, $a_2 \approx 74\,mA/V^2$. The bias current I_0 and the amplitudes of the IF and the LO signals must be selected such that the peak current

$$I_{max} = I_S \left(e^{\frac{V_0 + \hat{v}_{LO} + \hat{v}_{IF}}{nV_T}} - 1 \right) \approx I_S\, e^{\frac{V_0 + \hat{v}_{LO} + \hat{v}_{IF}}{nV_T}} = I_0\, e^{\frac{\hat{v}_{LO} + \hat{v}_{IF}}{nV_T}}$$

is not too high considering that $\hat{v}_{LO} + \hat{v}_{IF} = 100\,mV$ and $n = 1$ already cause a current of $I_{max} \approx 47\,I_0$. In practice, the levels are chosen so that the maximum IF signal amplitude is clearly lower than the local oscillator amplitude; then, the peak current is almost independent of the IF signal.

Instead of a bipolar transistor it is also possible to use a field effect transistor. This is even more advantageous since the field effect transistor only produces very low intermodulation distortions due to its approximately parabola-shaped transfer characteristic ($a_2 \neq 0$ and $a_i \approx 0$ for $i > 2$). Figure 28.10a shows a commonly used circuite with a junction FET which allows the bias voltage V_0 to be omitted and the operating point to be set by resistance R_S. From the transfer characteristic

$$I_D = \frac{K}{2} (V_{GS} - V_{th})^2$$

it follows:

$$a_2 = \frac{1}{2} \frac{d^2 I_D}{dV_{GS}^2} = \frac{K}{2}$$

This means that coefficient a_2 does not depend on the bias current but rather on the size of the FET represented by the transconductance coefficient K. Even in very large-sized FETs, it is clearly lower than in a bipolar transistor with typical operating point values. Typical values lie in the range of $a_2 \approx 1 \ldots 10\,mA/V^2$.

Another circuit with an approximately parabola-shaped characteristic is the current squaring circuit with bipolar transistors shown in Fig. 28.10b where $I_{C4} \sim I_1^2$. For transis-

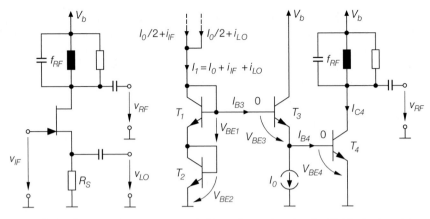

a Field-effect transistor   **b** Current squaring circuit with bipolar transistors

Fig. 28.10. Additive mixers with approximately parabola-shaped characteristic. The parallel resonant circuits are tuned to the RF frequency f_{RF}.

tors of equal size (identical saturation reverse current I_S) and if base currents are neglected, the following applies:

$$V_{BE\,1} = V_{BE\,2} = V_T \ln \frac{I_1}{I_S} \quad , \quad V_{BE\,3} = V_T \ln \frac{I_0}{I_S} \quad , \quad V_{BE\,4} = V_T \ln \frac{I_{C4}}{I_S}$$

From the loop equation

$$V_{BE\,1} + V_{BE\,2} = V_{BE\,3} + V_{BE\,4}$$

we obtain:

$$V_{BE\,4} = V_{BE\,1} + V_{BE\,2} - V_{BE\,3} = V_T \ln \frac{I_1^2}{I_0 I_S} \quad \Rightarrow \quad I_{C4} = \frac{I_1^2}{I_0}$$

The input current I_1 is composed of the bias current I_0, the IF current i_{IF} and the LO current i_{LO}. It originates from two current sources with the bias current $I_0/2$ and the small-signal currents i_{IF} and i_{LO}.

A commonly used additive mixer is the one shown in Fig. 28.11a, which is provided with two MOSFETs in cascode configuration and functions in most cases by means of a *dual-gate MOSFET (DGFET)* (see Fig. 28.11b). We assume that the MOSFETs are of equal size, i.e. both MOSFETs have the same transconductance coefficient K. The lower MOSFET is operated in the ohmic region; thus, its drain current I_{D1} not only depends on the gate-source voltage V_{GS1}, but also to a high degree on the drain source voltage V_{DS1}. For $V_1 = V_{GS1}$ the drain current is:

$$I_{D1} \overset{(3.2)}{=} K V_{DS1} \left(V_1 - V_{th} - \frac{V_{DS1}}{2} \right)$$

The channel length modulation can be neglected. The upper MOSFET is operated in the pinch-off region. With respect to the lower MOSFET the upper MOSFET operates in common drain configuration (source follower) and thus sets the voltage V_{DS1}. From

$$I_{D2} \overset{(3.3)}{=} \frac{K}{2} (V_{GS2} - V_{th})^2$$

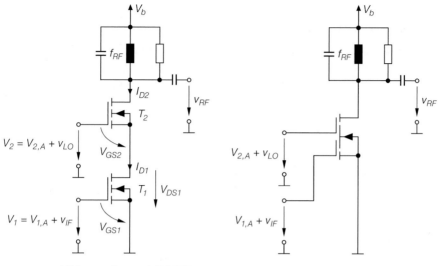

a With two discrete MOSFETs   **b** With dual-gate MOSFET (e.g. BF999)

Fig. 28.11. Additive mixers with two MOSFETs

and $I_{D1} = I_{D2}$ it follows:

$$V_{DS1} = V_2 - V_{GS2} = V_2 - V_{th} - \sqrt{\frac{2I_{D1}}{K}}$$

Insertion of this equation into the equation for I_{D1} results in, among other things, the term $K V_1 V_2$, which contains the desired product of the IF and LO signals due to $V_1 = V_{1,A} + v_{IF}$ and $V_2 = V_{2,A} + v_{LO}$. Since V_{DS1} also depends on I_{D1}, the resulting equation cannot be solved for I_{D1}; therefore, the drain current and the RF signal filtered out by the parallel resonant circuit can only be determined numerically. The LO signal and the operating point voltage $V_{2,A}$ are chosen such that the ohmic range of the lower MOSFET is used to its full extent which results in the maximum IF signal. In literature, this mixer is often listed under multiplicative mixers as the IF and LO signals are supplied to separate connections and are not added explicitly. We regard it as an additive mixer since, because of its dependence on I_{D1}, the drain source voltage V_{DS1} not only depends on V_2 but also on V_1, where $V_{DS1} = V_{DS1}(V_1, V_2)$. Thus, the drain current contains components that are nonlinear with respect to V_1.

In modern transmitters and receivers, additive mixers are seldom used. There are essentially two reasons for this:

- A practical mixer must be designed such that the inputs and outputs are decoupled as well as possible and can be matched to the characteristic impedance; also the gain should be as high as possible and the noise figure as low as possible. When using an additive mixer this is feasible with major limitations only.
- Owing to the general nonlinearity, the intermodulation distortions are comparatively high which limits the dynamic range.

28.2.2
Multiplicative Mixers

In *multiplicative mixers* the IF signal is multiplied by the LO signal. In contrast to an ideal mixer, the multiplicative mixer uses a general periodic LO signal of the fundamental frequency f_{LO} instead of a sinusoidal signal. The desired RF frequency is obtained from the output frequencies by a bandpass filter.

Multiplicative mixing becomes very simple if square-wave LO signals are used because multiplication can then be achieved with switches. Figure 28.12 shows the principle

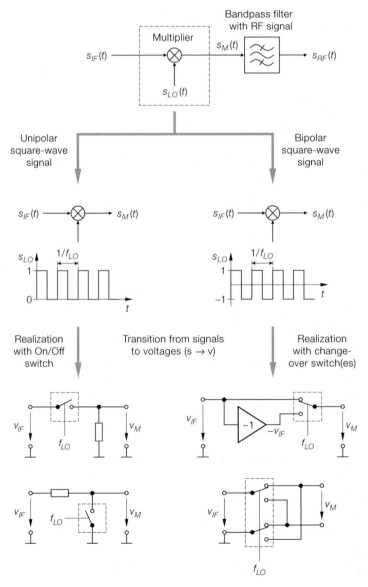

Fig. 28.12. Principle of multiplicative mixing (top) and special cases with square-wave LO signals

of multiplicative mixing and the special situations for unipolar and bipolar square-wave signals. In the case of the unipolar square-wave signal, the IF signal is only multiplied by 0 and 1 which can be implemented by an On/Off switch that operates as a series or short-circuiting switch. In the case of the bipolar square-wave signal, the IF signal is multiplied by $+1$ and -1. For this purpose one can generate $-v_{IF}$ by means of an inverting amplifier and use a change-over switch to alternate between v_{IF} and $-v_{IF}$. A two-pole switch can be used as an alternative. For this purpose electronic switches that are driven by a square-wave signal of the frequency f_{LO} are used.

Description by Equations

The signal $s_M(t)$ at the output of the multiplier in Fig. 28.12 is obtained by a Fourier series expansion of the LO signal:

$$s_M(t) = s_{IF}(t) \cdot s_{LO}(t)$$

$$= s_{IF}(t) \cdot [c_0 + c_1 \cos(\omega_{LO}t + \varphi_1) + c_2 \cos(2\omega_{LO}t + \varphi_2) + \cdots]$$

$$= s_{IF}(t) \cdot \left[c_0 + \sum_{n=1}^{\infty} c_n \cos(n\omega_{LO}t + \varphi_n) \right]$$

The IF signal is multiplied by the fundamental wave (c_1) and the harmonics ($c_2, \ldots$) of the LO signal. Furthermore, there is a direct transmission corresponding to the DC component (c_0).

The following Fourier series are obtained from the square-wave signals in Fig. 28.12:

$$s_{LO}(t) = \begin{cases} \dfrac{1}{2} + \dfrac{2}{\pi} \cos\omega_{LO}t - \dfrac{2}{3\pi} \cos 3\omega_{LO}t + \cdots & \text{unipolar} \\[2mm] \dfrac{4}{\pi} \cos\omega_{LO}t - \dfrac{4}{3\pi} \cos 3\omega_{LO}t + \cdots & \text{bipolar} \end{cases}$$

$$= \begin{cases} \dfrac{1}{2} + \dfrac{2}{\pi} \displaystyle\sum_{n=0}^{\infty} \dfrac{(-1)^n}{2n+1} \cos(2n+1)\omega_{LO}t & \text{unipolar} \\[2mm] \dfrac{4}{\pi} \displaystyle\sum_{n=0}^{\infty} \dfrac{(-1)^n}{2n+1} \cos(2n+1)\omega_{LO}t & \text{bipolar} \end{cases} \qquad (28.1)$$

This only produces odd multiples of the LO frequency. Furthermore, the bipolar square-wave signal has no DC component. With the modulated IF signal

$$s_{IF}(t) = a(t) \cos[\omega_{IF}t + \varphi(t)]$$

the output of the multiplier provides in the case of a unipolar square-wave signal:

$$s_M(t) = \frac{a(t)}{2} \cos[\omega_{IF}t + \varphi(t)]$$

$$+ \frac{a(t)}{\pi} \{\cos[(\omega_{LO} + \omega_{IF})t + \varphi(t)] + \cos[(\omega_{LO} - \omega_{IF})t - \varphi(t)]\}$$

$$- \frac{a(t)}{3\pi} \{\cos[(3\omega_{LO} + \omega_{IF})t + \varphi(t)] + \cos[(3\omega_{LO} - \omega_{IF})t - \varphi(t)]\}$$

$$+ \cdots$$

a Spectrum of the IF signal

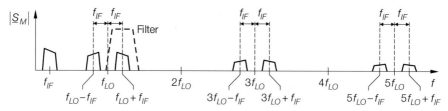

b Spectrum at the output of the multiplier in the case of a unipolar square-wave signal

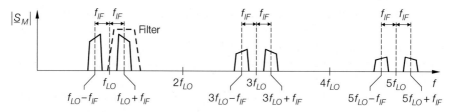

c Spectrum at the output of the multiplier in the case of a bipolar square-wave signal

Fig. 28.13. Absolute spectra in multiplicative up-conversion with square-wave LO signals

In the case of a bipolar square-wave signal there is no IF frequency component and the amplitudes of all other components are doubled. Figure 28.13 shows the corresponding spectra. At the LO frequency and all odd multiples of the LO frequency there is an upper band in noninverted mode and a lower band in inverted mode. The amplitudes decrease with increasing frequency according to the Fourier coefficients of the LO signal. The upper and lower bands of the LO frequency are used as the RF signal, i.e. $f_{RF} = f_{LO} \pm f_{IF}$ while all other components are suppressed by a filter. Generally, components of higher frequencies can also be used for the RF output signal.

If the square-wave signals are asymmetric (duty cycle $\neq 50\%$), the Fourier series also contains components at even multiples of the LO frequency. The output of the mixer then supplies upper and lower bands even at these frequencies, e.g. at $2f_{LO} \pm f_{IF}$.

Switching Characteristic of the Switch

The electronic switches used in practical mixers have no ideal switching characteristic but show a transient behavior which means that the IF signal is multiplied not by an ideal square pulse but by a square-wave signal distorted by the switching characteristic (see Fig. 28.14). Distinction must be made between the supplied LO signal s_{LO} and the LO signal s'_{LO} that is used in the multiplication. This does not affect the basic function of the

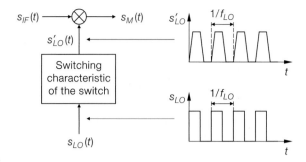

Fig. 28.14. Effect of the switching characteristic of the electronic switch

mixer as the distortion only causes a change in the Fourier coefficient of the LO signal, which can be accepted as long as the fundamental wave of the effective LO signal remains sufficiently high.

In practice, the supplied LO signal s_{LO} is also not usually a square-wave signal as the generation of high frequency square pulses is difficult and causes significant interferences; instead, the virtually sinusoidal signal of a high frequency oscillator is used. The effective LO signal then depends on the switching characteristic of the electronic switches that are controlled by sinusoidal signals.

With an increasing LO frequency the effect of the nonideal switching characteristic becomes more noticeable. With frequencies above 10 GHz even the fastest switching diodes no longer function as accurate switches but rather operate only within the transition region; thus, the multiplicative mixer changes into an additive mixer.

Image frequencies in down-conversion mixer operation: The functional symmetry between up and down-conversion mixers was explained for the ideal mixer in connection with the image frequency. The general relationship is: every band generated in up-conversion mixing acts as an image frequency band in down-conversion mixing. For a multiplicative down-conversion mixer this means that at the RF frequency $f_{RF} = f_{LO} + f_{IF}$ not only the band at $f_{LO} - f_{IF}$ but also the bands at all harmonics of the LO signal ($nf_{LO} \pm f_{IF}$ with $n = 2, 3, \ldots$) act as image frequency bands since they, too, are converted to the IF frequency. Therefore, the image frequency filter must have a sufficiently high attenuation at these frequencies as well.

Nonlinearity

The description by equations shows that multiplicative mixing is linear with regard to the relationship between IF and RF amplitudes; consequently, there are no intermodulation products. In practice, this is not the case as the required electronic switches are not exactly linear in operation and have modulation limits. This is the reason why multiplicative mixers are also characterized by nonlinear parameters (compression and intercept points). The relationships are basically the same as in additive mixers; however, the nonlinearity is usually much lower since it is caused solely by secondary effects while in additive mixers nonlinearity is a precondition for the mixer function. Therefore, the compression and intercept points achieved with multiplicative mixers are higher.

Practical Circuits

By principle, every additive mixer can also be used as a multiplicative mixer by using a square-wave LO signal and selecting the operating point and the LO signal amplitude such that the diode or transistor of the additive mixer switches between the nonconductive and the conductive state when driven by the LO frequency. At the same time, the amplitude of the input signal is selected so small that the small-signal modulation prevails; then the input signal is amplified by the respective small-signal gain, i.e. alternating between zero (= small-signal gain in the nonconductive state) and a constant value (= small-signal gain in the conductive state). The result is a multiplicative mixer with On/Off switch and, depending on the circuit, additional gain or attenuation. Figure 28.15 illustrates this using the mixer with diode from Fig. 28.9a as an example. Thus the diode acts as an electronic On/Off switch with a forward resistance corresponding to the small-signal resistance $r_D(\hat{v}_{LO})$ in the conductive state.

The multiplicative mixers used in practice are described in the following sections.

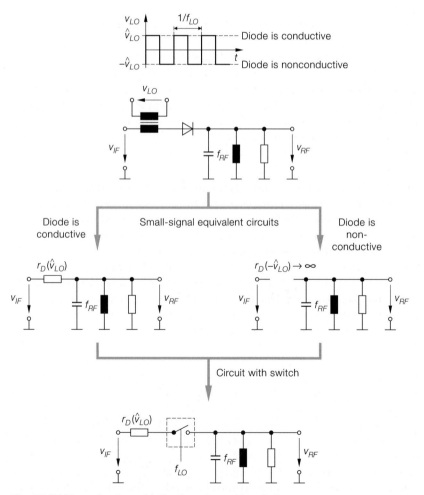

Fig. 28.15. Example of a multiplicative mixer with a diode

28.3
Mixers with Diodes

Mixers with diodes are widely used and predominantly found in circuits made up of discrete components. They operate almost exclusively as multiplicative mixers, i.e. the diodes are used as switches. Since the diode is a passive component these mixers always have a typical conversion loss of 5 . . . 8 dB. For this reason they are also known as passive mixers.

Due to the high frequencies, diodes with excellent switching performance are required. These are usually special Schottky diodes (mixer diodes) with very low junction capacitance; the diffusion capacitance of Schottky diodes is negligibly low. In order to minimize the junction capacitance, the area of the metal-semiconductor contact must be minimized and doping must be reduced in comparison to standard diodes which consequently increases the spreading resistance. Mixer diodes are thus characterized by a very low capacitance and a relatively high spreading resistance.

The following section describes the mixer with a single diode in more detail as all mixers with diodes can be reduced to this construction in terms of their transmission characteristic.

28.3.1
Unbalanced Diode Mixer

Figure 28.16a shows the circuit of a mixer with a single diode called an unbalanced (diode) mixer. The LO voltage V_{LO} is a large-signal voltage that is used to periodically switch over the operating point of the diode between the forward and reverse state. The IF voltage v_{IF} is a small-signal voltage and is transferred to the RF output according to the small-signal response of the diode. The LO and the IF voltages are added by a 1:1 transformer.

Frequency separation at the three connections is achieved by means of three narrow-band parallel resonant circuits. At the resonant frequency they act like an open circuit and are thus ineffective, while at all other frequencies they nearly cause a short circuit. Consequently, these connections only carry voltages and currents at the relevant resonant frequency. The capacitance of the diode is considered an integral part of the parallel resonant circuit which eliminates the need to account for it separately.[3]

The method for calculating the properties is basically the same as for all small-signal circuits. First, the operating point is determined and the circuit linearized, it is then possible to calculate the small-signal behavior. Unlike amplifiers, mixers do not have a constant operating point, but one that changes periodically in accordance with the LO voltage resulting in a *time-variable operating point*. The calculation of this time-variable operating point is based on the *LO circuit*.

LO Circuit

Figure 28.16b shows the LO circuit of the unbalanced diode mixer which assumes that the parallel resonant circuits at the IF and the RF terminals act as short circuits for the LO frequency and multiples thereof. The 1:1 transformer supplies the LO voltage to the

---

[3] The combined effect of capacitance and spreading resistance of the diode cause losses that are proportional to the frequency and are neglected in our simplified approach. A detailed calculation can be found in [28.1].

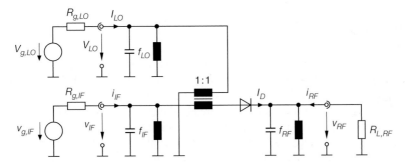

a Circuit diagram with signal sources and RF load resistance

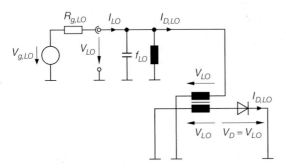

b LO circuit

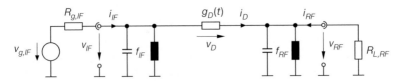

c Small-signal equivalent circuit for the RF and the IF circuit

Fig. 28.16. Unbalanced diode mixer

diode. It is sinusoidal with the LO frequency f_{LO} since the LO parallel resonant circuit suppresses all harmonics at multiples of f_{LO}:

$$V_D(t) = V_{LO}(t) = \hat{v}_{LO} \cos \omega_{LO} t$$

From the voltage the current $I_{D,LO}(t)$ of the time-variable operating point

$$I_{D,LO}(t) = I_D(V_{LO}(t))$$

is determined by means of the diode characteristic and has the maximum value:

$$I_{D,max} = I_D(\hat{v}_{LO})$$

It cannot be calculated using the simple exponential diode characteristic according to (1.1) as the mixer diodes are operated in a region in which the spreading resistance has a noticeable influence. Figure 28.17 shows the typical curves of $V_{LO}(t)$ and $I_{D,LO}(t)$. The amplitude $\hat{v}_{LO}$ must be higher than the forward voltage V_F of the diode in order to cause an accountable current flow.

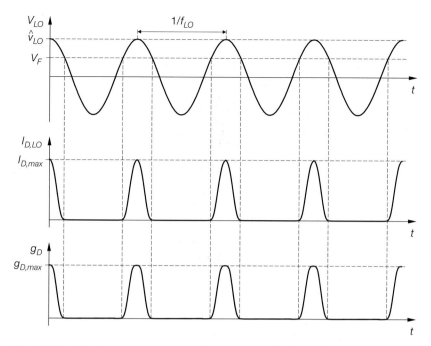

Fig. 28.17. Unbalanced diode mixer: voltage $V_{LO}(t)$ at the LO circuit, current $I_{D,LO}(t)$ of the diode and the resulting curve of the small-signal conductivity $g_D(t)$. V_F is the forward voltage of the diode.

Current $I_{D,LO}(t)$ can be expanded in the form of a Fourier series:

$$I_{D,LO}(t) = I_{D,0} + \sum_{n=1}^{\infty} \hat{i}_{D,n} \cos n\omega_{LO} t \qquad (28.2)$$

Here, $I_{D,0}$ is the DC component and $\hat{i}_{D,1}$ is the amplitude of the fundamental wave at frequency f_{LO}. The series contains only cosine components since in Fig. 28.17 the current is an even function of the time ($I_{D,LO}(-t) = I_{D,LO}(t)$). In this case, the coefficients of the Fourier series are:

$$I_{D,0} = f_{LO} \int_{0}^{1/f_{LO}} I_{D,LO}(t)\, dt$$

$$\hat{i}_{D,n} = 2f_{LO} \int_{0}^{1/f_{LO}} I_{D,LO}(t) \cos n\omega_{LO} t\, dt$$

In practice, the coefficients can be determined with the help of a circuit simulation by performing a time domain simulation for the LO circuit, presenting the current $I_{D,LO}(t)$ spectrally[4] and reading the amplitudes of the components.

Figure 28.18a shows the DC component $I_{D,0}$, the fundamental component $\hat{i}_{D,1}$ and the maximum current $I_{D,max}$ of a Schottky diode of type BAS40 dependent upon the LO amplitude $\hat{v}_{LO}$. Above $\hat{v}_{LO} = 0.3$ V, the components are no longer exponential due to the spreading resistance.

---

[4] In *PSpice* this is done using the *FFT* function of the *Probe* program.

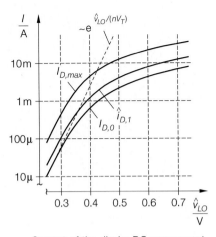

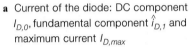

a Current of the diode: DC component $I_{D,0}$, fundamental component $\hat{i}_{D,1}$ and maximum current $I_{D,max}$

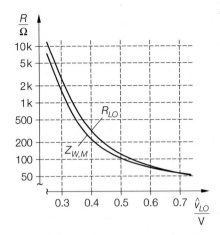

b Resistances for power matching: R_{LO} at the LO terminal and $Z_{W,M}$ at the IF and RF terminals

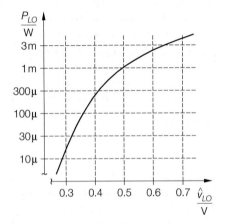

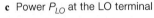

c Power P_{LO} at the LO terminal

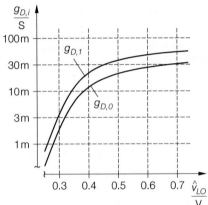

d Small-signal conductance: DC component $g_{D,0}$ and fundamental component $g_{D,1}$

Fig. 28.18. Characteristic parameters of an unbalanced diode mixer with a Schottky diode of type BAS40

The DC component and harmonics of current $I_{D,LO}(t)$ are short-circuited by the LO parallel resonant circuit where the resonant circuit is noneffective only for the fundamental wave. Consequently, the current $I_{LO}(t)$ at the LO terminal corresponds to the fundamental wave of the current $I_{D,LO}(t)$:

$$I_{LO}(t) = I_{D,LO}(t)\Big|_{f=f_{LO}} = \hat{i}_{D,1}\cos\omega_{LO}t$$

Since both $V_{LO}(t)$ and $I_{LO}(t)$ are sinusoidal, at a constant LO amplitude the LO circuit behaves like an ohmic resistance with:

$$R_{LO} = \frac{V_{LO}(t)}{I_{LO}(t)} = \frac{\hat{v}_{LO}}{\hat{i}_{D,1}} \tag{28.3}$$

Thus, if operated with a sinusoidal LO voltage source $V_{g,LO}$ with internal resistance $R_{g,LO}$, the current $I_{LO}(t)$ contains no harmonics. For $R_{LO} = R_{g,LO}$ there is power matching between the LO voltage source and the LO circuit, but if $R_{LO} \neq R_{g,LO}$ a matching network can be used or the transformation ratio of the transformer can be altered. However, power matching is only achieved for the given LO amplitude since the resistance R_{LO} decreases with an increasing LO amplitude due to the nonlinear relationship between $\hat{v}_{LO}$ and $\hat{i}_{D,1}$. Figure 28.18b shows the resistance R_{LO} dependent upon the LO amplitude for a Schottky diode of type BAS40.

In practice, the LO voltage is generated by a high-frequency oscillator; of interest is the required power at the LO terminal:

$$P_{LO} = \frac{1}{2} \hat{v}_{LO} \hat{i}_{D,1} = \frac{1}{2} \frac{\hat{v}_{LO}^2}{R_{LO}} \tag{28.4}$$

With an increasing LO amplitude, it shows a steeper rise than with an ohmic resistance since R_{LO} decreases at the same time. Figure 28.18c shows the LO power dependent upon the LO amplitude for a Schottky diode of type BAS40.

Small-Signal Equivalent Circuit

The small-signal equivalent circuit for the IF and RF circuits shown in Fig. 28.16c is derived by linearization of the diode. Since the operating point is time-variable, the diode is described by a time-variable small-signal conductivity $g_D(t)$

$$g_D(t) = g_D(V_{LO}(t)) = \left. \frac{dI_D}{dV_D} \right|_{V_D=V_{LO}(t)} \tag{28.5}$$

with the maximum value:

$$g_{D,max} = g_D(\hat{v}_{LO}) = \left. \frac{dI_D}{dV_D} \right|_{V_D=\hat{v}_{LO}}$$

The small-signal conductivity is used because, in the reverse region, the small-signal resistance $r_D(t) = 1/g_D(t)$ tends toward infinity and thus cannot be adequately shown in the figures.

Figure 28.17 shows the plot of the small-signal conductivity. With small currents it is proportional to $I_{D,LO}(t)$ since here the conductivity according to (1.3) is

$$g_D(t) = \frac{1}{r_D(t)} \approx \frac{I_{D,LO}(t)}{nV_T} \tag{28.6}$$

With large currents the spreading resistance has an influence. The conductivity no longer increases proportionally to the current; therefore, the peaks in the conductivity curve are less pronounced than in the current curve.

The small-signal conductivity is also expanded in a Fourier series:

$$g_D(t) = g_{D,0} + \sum_{n=1}^{\infty} g_{D,n} \cos n\omega_{LO} t \tag{28.7}$$

As for the current $I_{D,LO}(t)$, the coefficients can be calculated using the integral equations for the Fourier series expansion. In practice, this is not necessary as the required coefficients can be determined by means of circuit simulation which will be further described below. Figure 28.18d shows the DC component $g_{D,0}$ and the fundamental component $g_{D,1}$ for a Schottky diode of type BAS40 dependent upon the LO amplitude.

Small-Signal Response

In the following, the mixer is operated in noninverting mode with $f_{RF} = f_{LO} + f_{IF}$ and the small-signal current $i_D(t)$ of the diode is calculated. From Fig. 28.16c it follows:

$$i_D(t) = g_D(t) v_D(t) = g_D(t) (v_{IF}(t) - v_{RF}(t)) \tag{28.8}$$

The voltages $v_{IF}(t)$ and $v_{RF}(t)$ only contain components at the IF or RF frequencies since the parallel resonant circuits short-circuit all other frequencies:

$$v_{IF}(t) = \hat{v}_{IF} \cos \omega_{IF} t \quad , \quad v_{RF}(t) = \hat{v}_{RF} \cos \omega_{RF} t \tag{28.9}$$

Inserting (28.7) and (28.9) into (28.8) results in:

$$i_D(t) = \left(g_{D,0} + \sum_{n=1}^{\infty} g_{D,n} \cos n\omega_{LO} t \right) \left(\hat{v}_{IF} \cos \omega_{IF} t - \hat{v}_{RF} \cos \omega_{RF} t \right)$$

$$= \left(g_{D,0} + g_{D,1} \cos \omega_{LO} t + \cdots \right) \left(\hat{v}_{IF} \cos \omega_{IF} t - \hat{v}_{RF} \cos \omega_{RF} t \right)$$

$$= g_{D,0} \hat{v}_{IF} \cos \omega_{IF} t - g_{D,0} \hat{v}_{RF} \cos \omega_{RF} t$$

$$+ g_{D,1} \hat{v}_{IF} \cos \omega_{LO} t \cos \omega_{IF} t$$

$$- g_{D,1} \hat{v}_{RF} \cos \omega_{LO} t \cos \omega_{RF} t$$

$$+ \cdots$$

$$= g_{D,0} \hat{v}_{IF} \cos \omega_{IF} t - g_{D,0} \hat{v}_{RF} \cos \omega_{RF} t$$

$$+ \frac{g_{D,1} \hat{v}_{IF}}{2} [\cos (\underbrace{\omega_{LO} + \omega_{IF})t + \cos (\omega_{LO} - \omega_{IF})t}_{\omega_{RF}}]$$

$$- \frac{g_{D,1} \hat{v}_{RF}}{2} [\cos (\omega_{RF} + \omega_{LO})t + \cos (\underbrace{\omega_{RF} - \omega_{LO}}_{\omega_{IF}})t]$$

$$+ \cdots$$

It can be seen that the fundamental component $g_{D,1}$ of the small-signal conductivity $g_D(t)$ produces the desired frequency conversion from f_{IF} to f_{RF} by yielding, at the frequency $f_{LO} + f_{IF} = f_{RF}$, a component that is proportional to the IF amplitude $\hat{v}_{IF}$. Similarly, there is a conversion from f_{RF} to f_{IF}, i.e. at frequency $f_{RF} - f_{LO} = f_{IF}$ a component is produced that is proportional to the RF amplitude $\hat{v}_{RF}$. As a result of the harmonic components of the small-signal conductivity, additional components are generated at higher frequencies which are not relevant for further calculations.

The small-signal current $i_D(t)$ of the diode flows through the IF and RF circuits. The parallel resonant circuits short-circuit all components at $f \neq f_{IF}$ in the IF circuit and all components at $f \neq f_{RF}$ in the RF circuit; thus, only the components at the respective resonant frequencies are available at the terminals. The small-signal currents $i_{IF}(t)$ and $i_{RF}(t)$ become available by extracting the components at f_{IF} and f_{RF} from the current $i_D(t)$:

$$i_{IF}(t) = i_D(t)\Big|_{f=f_{IF}} = \left(g_{D,0}\,\hat{v}_{IF} - \frac{g_{D,1}\,\hat{v}_{RF}}{2}\right)\cos\omega_{IF}t$$

$$i_{RF}(t) = -i_D(t)\Big|_{f=f_{RF}} = \left(g_{D,0}\,\hat{v}_{RF} - \frac{g_{D,1}\,\hat{v}_{IF}}{2}\right)\cos\omega_{RF}t$$

The following relationships for the voltage and current phasors are then derived:

$$\underline{i}_{IF} = g_{D,0}\,\underline{v}_{IF} - \frac{g_{D,1}\,\underline{v}_{RF}}{2} \tag{28.10}$$

$$\underline{i}_{RF} = g_{D,0}\,\underline{v}_{RF} - \frac{g_{D,1}\,\underline{v}_{IF}}{2} \tag{28.11}$$

These equations correspond to the four-pole equations in Y notation. The small-signal response of the mixer can thus be described by a Y matrix:

$$\begin{bmatrix} \underline{i}_{IF} \\ \underline{i}_{RF} \end{bmatrix} = \begin{bmatrix} g_{D,0} & -\dfrac{g_{D,1}}{2} \\ -\dfrac{g_{D,1}}{2} & g_{D,0} \end{bmatrix} \begin{bmatrix} \underline{v}_{IF} \\ \underline{v}_{RF} \end{bmatrix} \tag{28.12}$$

All parameters of interest, e.g. the small-signal gain and the input and output resistances at the terminals, can be calculated by means of this Y matrix. The frequency conversion of the mixer is no longer explicitly apparent.

A method for determining the coefficients $g_{D,0}$ and $g_{D,1}$ by means of a circuit simulation follows directly from the Y notation of the mixer. For this method, the mixer is operated according to Fig. 28.19 with an LO voltage source with the intended amplitude $\hat{v}_{LO}$ and an IF voltage source with the small-signal amplitude $\hat{v}_{IF} \ll \hat{v}_{LO}$. In the circuit simulation, these two voltage sources can be connected directly in series and thus the transformer is no longer required. The parallel resonant circuits at the LO and IF terminals are also omitted since the voltage sources only contain components of the respective frequencies, while all other frequencies are short-circuited, thus they adopt the same function as the resonant

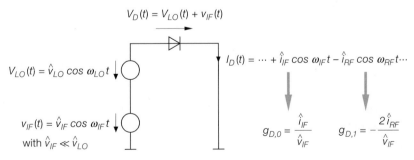

Fig. 28.19. Circuit simulation to determine the coefficients $g_{D,0}$ and $g_{D,1}$ of unbalanced mixer. Amplitudes $\hat{i}_{IF}$ and $\hat{i}_{RF}$ are obtained from the spectral display of current $I_D(t)$.

circuits. The RF output is short-circuited which renders the RF parallel resonant circuit unnecessary. In this mode of operation it follows from (28.12) with $\underline{v}_{RF} = 0$ that:

$$\underline{i}_{IF} = g_{D,0}\underline{v}_{IF} \quad , \quad \underline{i}_{RF} = -\frac{g_{D,1}}{2}\underline{v}_{IF}$$

By inserting the small-signal amplitudes we obtain the defining equations for the coefficients:

$$g_{D,0} = \frac{\hat{i}_{IF}}{\hat{v}_{IF}} \quad , \quad g_{D,1} = -\frac{2\hat{i}_{RF}}{\hat{v}_{IF}}$$

The small-signal amplitude $\hat{v}_{IF}$ is given by the IF voltage source. The current $I_D(t)$ of the diode is now determined by means of a time domain simulation. The small-signal amplitudes $\hat{i}_{IF}$ (component at f_{IF}) and $\hat{i}_{RF}$ (component at f_{RF}) can be determined from the spectral display of $I_D(t)$.

Mixer Voltage Gain

Now we can calculate the *voltage gain* of the mixer:

$$A_M = \frac{\underline{v}_{RF}}{\underline{v}_{IF}}$$

According to Fig. 28.16c the current at the RF terminal is:

$$\underline{i}_{RF} = -\frac{\underline{v}_{RF}}{R_{L,RF}}$$

Insertion into (28.11) and solving for $\underline{v}_{RF}/\underline{v}_{IF}$ leads to:

$$A_M = \frac{\underline{v}_{RF}}{\underline{v}_{IF}} = \frac{1}{2}\frac{g_{D,1}R_{L,RF}}{1 + g_{D,0}R_{L,RF}} \tag{28.13}$$

The voltage gain is proportional to the fundamental component $g_{D,1}$ of the small-signal conductivity of the diode. It reaches its maximum as $R_{L,RF} \rightarrow \infty$; however, in this case no power is provided to the RF circuit.

Conversion Gain

In most cases, mixers are used in matched systems; but, in this case, the internal resistance $R_{g,IF}$ of the IF voltage source and the RF load resistance $R_{L,RF}$ correspond to the characteristic impedance Z_W of the system: $R_{g,IF} = R_{L,RF} = Z_W$. The related power gain is called the conversion gain G_M and corresponds to the gain G_T of an amplifier. With the Y parameters

$$Y_{11} = Y_{22} = g_{D,0} \quad , \quad Y_{12} = Y_{21} = -\frac{g_{D,1}}{2}$$

of the mixer and the source and load conductivities

$$\text{Re}\{Y_g\} = \frac{1}{R_{g,IF}} \quad , \quad \text{Re}\{Y_L\} = \frac{1}{R_{L,RF}}$$

Equation (27.30) leads to:

$$G_M = \frac{g_{D,1}^2 R_{g,IF} R_{L,RF}}{\left[(1 + g_{D,0}R_{g,IF})(1 + g_{D,0}R_{L,RF}) - \frac{1}{4}g_{D,1}^2 R_{g,IF} R_{L,RF}\right]^2}$$

For $R_{g,IF} = R_{L,RF} = Z_W$ we have:

$$GM = \left[\frac{g_{D,1} Z_W}{\left(1 + g_{D,0} Z_W\right)^2 - \frac{1}{4} g_{D,1}^2 Z_W^2} \right]^2 \tag{28.14}$$

Since the conversion gain of mixers with diodes is less than one, the conversion loss $L_M = 1/G_M$ is often quoted. The quantities are usually stated in decibel:

$$G_M \,[\text{dB}] = 10 \log G_M \quad , \quad L_M \,[\text{dB}] = 10 \log L_M = -G_M \,[\text{dB}]$$

In Sect. 27.4.6 we explained that an amplifier is matched at both sides to the characteristic impedance Z_W at the point where the Y parameters meet the following conditions (see (27.34)):

$$Y_{11} = Y_{22} \quad , \quad (Y_{11} Y_{22} - Y_{12} Y_{21}) Z_W^2 = 1$$

Since a diode mixer fulfils the first condition, the second can be used to calculate the characteristic impedance necessary for matching both sides:

$$Z_{W,M} = \frac{1}{\sqrt{Y_{11} Y_{22} - Y_{12} Y_{21}}} = \frac{1}{\sqrt{g_{D,0}^2 - \frac{g_{D,1}^2}{4}}} \tag{28.15}$$

The related power gain corresponds to the maximum available power gain (*MAG*) of an amplifier and is calculated using (27.35):

$$MAG = \frac{|Y_{21}|^2 Z_{W,M}^2}{\left|1 + Y_{11} Z_{W,M}\right|^2} = \frac{1 - \sqrt{1 - \frac{1}{4}\left(\frac{g_{D,1}}{g_{D,0}}\right)^2}}{1 + \sqrt{1 - \frac{1}{4}\left(\frac{g_{D,1}}{g_{D,0}}\right)^2}} \tag{28.16}$$

The same result is achieved if we insert (28.15) into (28.14) and determine the maximum conversion gain through the condition:

$$\frac{dG_M}{dZ_W} = 0$$

The maximum available power gain depends only on the ratio of the coefficients $g_{D,1}$ and $g_{D,0}$ of the small-signal conductivity of the diode (see Fig. 28.20). The *stability condition*

$$g_{D,0} > \frac{g_{D,1}}{2} \tag{28.17}$$

must be met in order to keep the fully matched mixer stable; only then the terms under the square root in (28.15) and (28.16) are positive. The stability condition leads to the stability limit in Fig. 28.20. In mixers with diodes, the stability condition is always met due to the passivity of the diode.

Figure 28.21 shows the conversion gain $G_{M\,(50)}$ in a 50 Ω system ($R_{g,IF} = R_{L,RF} = 50\,\Omega$) and the *MAG* ($R_{g,IF} = R_{L,RF} = Z_{W,M}$) for an unbalanced diode mixer with a Schottky diode of type BAS40 dependent upon the LO amplitude. The characteristic impedance $Z_{W,M}$ for matching at both sides is shown in Fig. 28.18b on page 1384. The *MAG* reaches a maximum of approximately $-3\,\text{dB}$ for $\hat{v}_{LO} \approx 0.3\,\text{V}$ after which it decreases slowly.

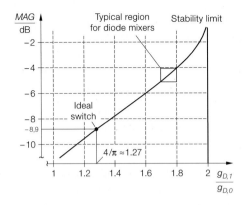

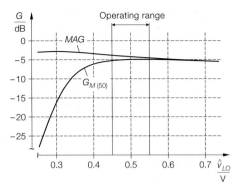

Fig. 28.20. Maximum available power gain (*MAG*) dependent upon the ratio of the coefficients $g_{D,1}$ and $g_{D,0}$ of the small-signal conductivity $g_D(t)$

Fig. 28.21. Conversion gain $G_{M\,(50)}$ ($R_{g,IF} = R_{L,RF} = 50\,\Omega$) and *MAG* of an unbalanced diode mixer with a Schottky diode of type BAS40

With small LO amplitudes, the conversion gain $G_{M\,(50)}$ is very small due to the strong mismatch ($Z_{W,M} \gg 50\,\Omega$), but increases rapidly with an increasing LO amplitude and reaches a wide maximum in the region $\hat{v}_{LO} = 0.5\ldots0.6\,\mathrm{V}$. Above this maximum $G_{M\,(50)}$ and *MAG* are almost identical since $Z_{W,M}$ tends toward $50\,\Omega$ in this region.

In most cases, mixers with diodes are operated without any specific matching circuits. The LO amplitude is selected such that the maximum conversion gain $G_{M\,(50)}$ is almost reached and thus a good compromise between conversion gain and required LO power is achieved. The reasonable operating range for the BAS40 diode is shown in Fig. 28.21; here, the conversion gain reaches $G_{M\,(50)} \approx -5\,\mathrm{dB}$. This does not warrant any matching efforts as the *MAG* is no more than approximately 1 dB higher.

The interaction of the spreading resistance and the capacitance of the diode produce additional frequency-proportional losses which were not taken into account here. Depending on the diode and frequency, this can reduce the conversion gain to $G_{M\,(50)} \approx -5\ldots-8\,\mathrm{dB}$. These losses are detailed in [28.1].

Comparison with an Ideal Switch

Figure 28.15 on page 1380 shows an unbalanced diode mixer with square-wave LO signal as an example of a multiplicative mixer with a switch. If an ideal switching characteristic is assumed, this mixer features a rectangular curve of the small-signal conductivity $g_D(t)$ with the values $g_D = 0$ (switch open) and $g_{D,max} = 1/r_D(\hat{v}_{LO})$ (switch closed). When using the series expansion, this leads to a unipolar square-wave signal according to (28.1) with:

$$g_{D,0} = \frac{g_{D,max}}{2} \quad , \quad g_{D,1} = \frac{2\,g_{D,max}}{\pi} \quad \Rightarrow \quad \frac{g_{D,1}}{g_{D,0}} = \frac{4}{\pi}$$

Insertion into (28.16) provides

$$MAG = \frac{1 - \sqrt{1 - 4/\pi^2}}{1 + \sqrt{1 - 4/\pi^2}} \approx 0.13$$

or *MAG* $\approx -8.9\,\text{dB}$. Therefore, with a square-wave LO voltage, the *MAG* is clearly lower than with a sinusoidal LO voltage (*MAG* $\approx -4\ldots-5\,\text{dB}$). At first, this seems surprising, but the reason is that *MAG* is only influenced by the ratio $g_{D,1}/g_{D,0}$. In an ideal switch, this ratio amounts to $4/\pi \approx 1.27$ and is thus lower than in typical diode mixers with a sinusoidal LO voltage ($g_{D,1}/g_{D,0} \approx 1.7\ldots1.8$) (see Fig. 28.20). For practical applications, this result is advantageous as the LO voltage is produced by a high frequency oscillator with almost sinusoidal output voltage.

Disadvantages of Unbalanced Diode Mixers

The coupling of the connections is particularly troublesome in unbalanced diode mixers. Frequency separation by means of the three parallel resonant circuits, which were assumed to be ideal in the previous calculation, is only approximately possible in practice. In particular, the RF and LO frequencies are often close together, meaning that it is only possible to prevent coupling of the strong LO signal into the RF circuit with complex filtering.

　　If used in transmitters and receivers with variable transmit and receiving frequencies, both the RF and LO frequencies are variable, which further complicates the separation of the frequencies. Frequency separation by fixed-frequency filters is no longer possible if the tuning ranges of the RF and LO frequencies overlap.

28.3.2
Single Balanced Diode Mixers

Figure 28.22a shows the circuit of a mixer with two diodes that is known as a *single balanced diode mixer*. The LO voltage V_{LO} regularly switches the operating point of the diodes alternately between the forward and the reverse region. The IF small-signal voltage v_{IF} is added to the voltage of diode D_1 and subtracted from the voltage of diode D_2 by the 1:1:1 transformer *TR1*. This means that, with respect to small signals, the diodes are driven in balanced mode, i.e. both diodes carry the same amount of small-signal current i_D, but of opposing directions. On the RF side, the small-signal current i_D is decoupled by the 1:1:1 transformer *TR2* and fed to the RF filter.

　　In the single balanced diode mixer the IF and the RF circuits are decoupled from the LO circuit. The small-signal current i_D of the diodes that flows through the IF and the RF circuit does not contain any portions of the LO frequency f_{LO} or multiples thereof. On the other hand, no currents of the IF or RF frequency flow through the LO circuit. Figure 28.23 shows this on the basis of the operating modes of transformer *TR1*. Figure 28.23a shows that, with a symmetrical load, the centre tap of the secondary side is current-free; therefore, the IF voltage v_{IF} generates no current in the LO circuit. Figure 28.23b shows that a symmetrical modulation on the secondary side has no effect on the primary side since the magnetic fluxes of the two secondary coils eliminate each other because their currents flow in opposite directions. Thus, the LO voltage V_{LO} generates no current in the IF circuit. This means that both the IF and LO circuits are decoupled. The RF and LO circuits are also decoupled in the same way. Consequently, the filters in the IF and RF circuits are only required to suppress the frequency of the other circuit and must no longer suppress the LO frequency. This reduces the requirements for the RF filter as it no longer faces the task of separating the closely adjacent frequencies f_{RF} and f_{LO}. Similarly, the demands on the LO filter are reduced as it only has to suppress the harmonics at multiples of f_{LO}. If harmonics

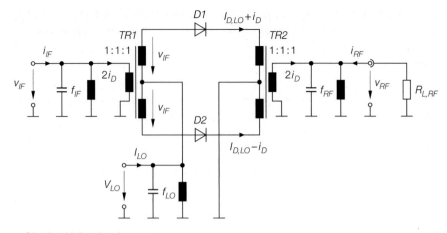

a Circuit with load resistance

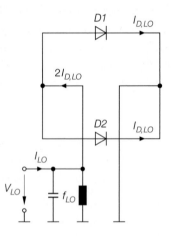

b LO circuit

Fig. 28.22. Single balanced diode mixer

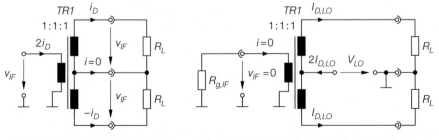

a Modulation on the primary side　　**b** Symmetrical modulation on the secondary side

Fig. 28.23. Voltages and currents of transformer *TR1* in the case of symmetrical loads on the secondary side

at the LO terminal are acceptable, the LO filter can be omitted. This, however, changes the waveforms of the LO voltage and LO current and thus the small-signal characteristics.

LO Circuit

Figure 28.22b shows the LO circuit of the single balanced diode mixer. The same current $I_{D,LO}(t)$ as in the single-phase mixer flows through both diodes with the same LO amplitude $\hat{v}_{LO}$. Since both diodes are connected in parallel, the current $I_{LO}(t)$ at the LO terminal is twice as high as in an unbalanced diode mixer:

$$I_{LO}(t) = 2 I_{D,LO}(t)\Big|_{f=f_{LO}}$$

This halves the resistance R_{LO}:

$$R_{LO} = \frac{1}{2} R_{LO\,(unbal)} \qquad (28.18)$$

Small-Signal Equivalent Circuit and Small-Signal Characteristics

Linearization of the diodes results in the small-signal equivalent circuit shown in the upper part of Fig. 28.24. The small-signal conductivities $g_{D1}(t)$ and $g_{D2}(t)$ have the same value and correspond to the small-signal conductivity of an unbalanced diode mixer with the same LO amplitude since the voltages and currents of the diodes are the same in both cases:

$$g_{D1}(t) = g_{D2}(t) = g_{D\,(unbal)}(t) \qquad (28.19)$$

The small-signal equivalent circuit of a single balanced diode mixer can be converted into the small-signal equivalent circuit of an unbalanced diode mixer by converting the small-signal conductivities of the diodes to the primary side of the transformer. To do so,

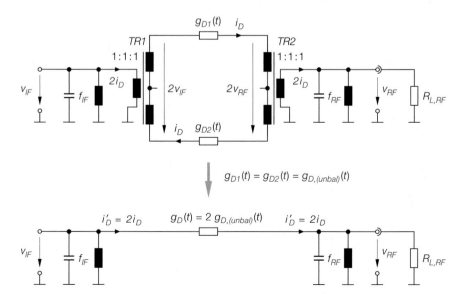

Fig. 28.24. Small-signal equivalent circuit of a single balanced diode mixer

we first calculate the small-signal current $i_D(t)$ of the diodes:

$$2v_{IF}(t) - 2v_{RF}(t) = \frac{i_D(t)}{g_{D1}(t)} + \frac{i_D(t)}{g_{D2}(t)} \overset{(28.19)}{=} \frac{2\,i_D(t)}{g_{D\,(unbal)}(t)}$$

$$\Rightarrow \quad i_D(t) = g_{D\,(unbal)}(t)\,(v_{IF}(t) - v_{RF}(t))$$

Conversion to the primary side of the transformer yields:

$$i_D'(t) = 2\,i_D(t) = 2\,g_{D\,(unbal)}(t)\,(v_{IF}(t) - v_{RF}(t))$$

This results in the small-signal equivalent circuit shown in Fig. 28.24 that corresponds to a small-signal equivalent circuit of an unbalanced diode mixer. The small-signal conductivity $g_D(t)$ is twice as high as in an unbalanced diode mixer with the same LO amplitude:

$$g_D(t) = 2\,g_{D\,(unbal)}(t) \tag{28.20}$$

This also makes the coefficients of the Fourier series of $g_D(t)$ twice as high:

$$g_{D,0} = 2\,g_{D,0\,(unbal)} \quad , \quad g_{D,1} = 2\,g_{D,1\,(unbal)} \tag{28.21}$$

The Y matrix of the single balanced diode mixer can now be determined according to (28.12) and all other parameters can be calculated by using (28.13)–(28.16) for the unbalanced diode mixer.

The MAG has the same magnitude as that of the unbalanced diode mixer with the same LO amplitude since here only the ratio of $g_{D,1}$ and $g_{D,0}$ are effective according to (28.16). However, the related characteristic impedance is lower by a factor of 2 (see 28.15):

$$Z_{W,M} = \frac{1}{2}\,Z_{W,M\,(unbal)} \tag{28.22}$$

The conversion gain $G_{M\,(50)}$ in a 50 Ω system is similar to that of an unbalanced diode mixer. Figure 28.25 shows a comparison of different mixers with Schottky diodes of type BAS40. In the single balanced diode mixer, the maximum value of $G_{M\,(50)}$ is slightly higher than in the unbalanced diode mixer and is achieved at a lower LO amplitude.

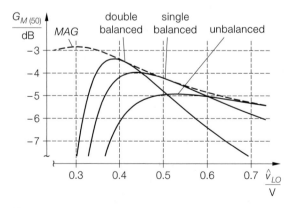

Fig. 28.25. Conversion gain $G_{M\,(50)}$ of an unbalanced, a single balanced and a double balanced diode mixer with Schottky diodes of type BAS40 in a 50 Ω system

Pros and Cons of the Single Balanced Diode Mixer

The main advantage of the single balanced diode mixer compared to the unbalanced diode mixer is that the LO and RF circuits are decoupled which prevent the strong LO signal from being coupled into the RF circuit. In practice, the degree of decoupling depends on the symmetry of the transformer. Another advantage is the low nonlinearity due to the balanced modulation of the diodes. Therefore, the values of the compression point and the intercept points are higher than in the unbalanced diode mixer.

The main disadvantage of the single balanced diode mixer is that, like the unbalanced diode mixer, it uses only one half wave of the LO voltage.

28.3.3
Double Balanced Diode Mixer

Figure 28.26 shows the circuit of a mixer with four diodes known as a *double balanced diode mixer* or *ring modulator*. It consists of the anti-parallel connection of two single balanced diode mixers (D_1/D_2 and D_3/D_4) connected crosswise (see Fig. 28.26a). A modification of the circuit diagram results in a diode ring as shown in Fig. 28.26b which led to the name ring modulator. For reasons of clarity in the following explanation, we shall use two single balanced diode mixers in our illustrations.

Owing to the anti-parallel connection of the two single balanced diode mixers, the double balanced diode mixer uses both half waves of the LO voltage where diodes D_1 and D_2 are conductive in the positive phase and diodes D_3 and D_4 are conductive in the negative phase. Figure 28.27 shows the two phases of the LO circuit. The crosswise connection causes the polarity of the small-signal current $2i_D$ on the RF side to change with each half wave of the LO voltage. As a result, the double balanced diode mixer generally operates as a multiplicative mixer with a bipolar square-wave signal.

Section 28.2.2 showed that the RF signal of a multiplicative mixer with bipolar square-wave signal has no component at the IF frequency (see Fig. 28.13c on page 1378); correspondingly, the IF signal has no part at the RF frequency. This means that the IF and RF circuits in the double balanced diode mixer are decoupled. Since the LO circuit is already decoupled from the IF and RF circuits due to the characteristics of the two single balanced diode mixers, all three circuits are decoupled. Nevertheless, the number of filters required cannot be reduced as both the IF and RF signals have components at multiples of the LO frequency.

LO Circuit

Figure 28.27 shows the LO circuit of the double balance diode mixer for the two half waves of the LO voltage. Currents $I_{D1,LO}(t), \ldots, I_{D4,LO}(t)$ of the diodes, as well as the total current $I_{LOD}(t)$, are plotted in Fig. 28.28. The same current flows through each of the diodes as in the unbalanced diode mixer with the same LO amplitude. The total current $I_{LOD}(t)$ has no DC component due to the symmetry and contains portions at odd multiples of the LO frequency:

$$I_{LOD}(t) = \hat{i}_{LOD,1} \cos \omega_{LO} t + \hat{i}_{LOD,3} \cos 3\omega_{LO} t + \hat{i}_{LOD,5} \cos 5\omega_{LO} t + \cdots$$

The harmonics of $I_{LOD}(t)$ are short-circuited by the LO filter. The portion of the funda-

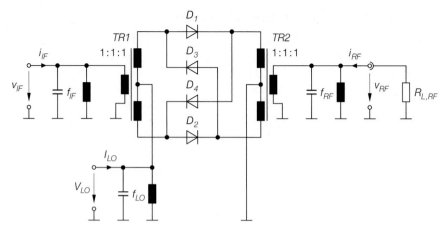

a Circuit diagram showing two single balanced diode mixers in anti-parallel connection

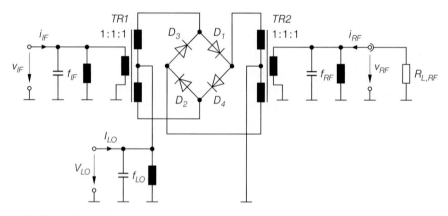

b Modified diagram showing the diode ring

Fig. 28.26. Double balanced diode mixer

mental wave corresponds to the LO current $I_{LO}(t)$:

$$I_{LO}(t) = I_{LOD}(t)\Big|_{f=f_{LO}} = \hat{i}_{LOD,1} \cos \omega_{LO} t$$

With the same LO amplitude, the amplitude $\hat{i}_{LOD,1}$ in the double balanced diode mixer is higher than in an unbalanced diode mixer by a factor of four. This is true because a current flows through a set of two diodes connected in parallel during both half waves of the LO voltage, thus reducing the resistance R_{LO} by a factor of four:

$$R_{LO} = \frac{\hat{v}_{LO}}{\hat{i}_{LOD,1}} = \frac{1}{4} R_{LO\,(unbal)}$$

Small-Signal Equivalent Circuit and Small-Signal Characteristic

Figure 28.29 shows a small-signal equivalent circuit of a double balanced diode mixer which results from the small-signal equivalent circuit of a single balanced diode mixer

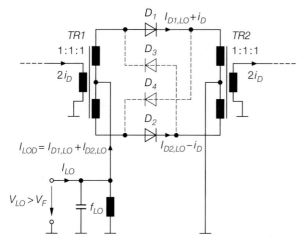

a Positive LO voltage

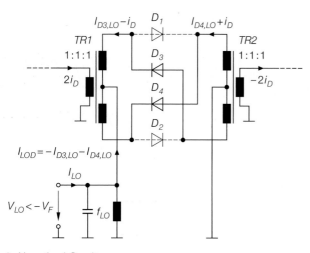

b Negative LO voltage

Fig. 28.27. LO circuit of a double balanced diode mixer. V_F represents the forward voltage of the diodes

by inserting two change-over switches to represent the polarity reversal. The small-signal conductivity $g_D(t)$ consists of the small-signal conductivities of the two single balanced diode mixers that are phase-shifted by half the LO period. The plot of $g_D(t)$ can be seen in Fig. 28.28.

Calculating the small-signal behavior is more complex than for an unbalanced or a single balanced diode mixer due to the polarity reversal. First we determine the small-signal currents $i'_D(t)$ and $i'_{D,S}(t)$ on the IF and RF side. From Fig. 28.29 it follows:

$$i'_D(t) = \begin{cases} g_D(t)\,[\,v_{IF}(t) - v_{RF}(t)\,] & V_{LO} \geq 0 \\ g_D(t)\,[\,v_{IF}(t) + v_{RF}(t)\,] & V_{LO} < 0 \end{cases}$$

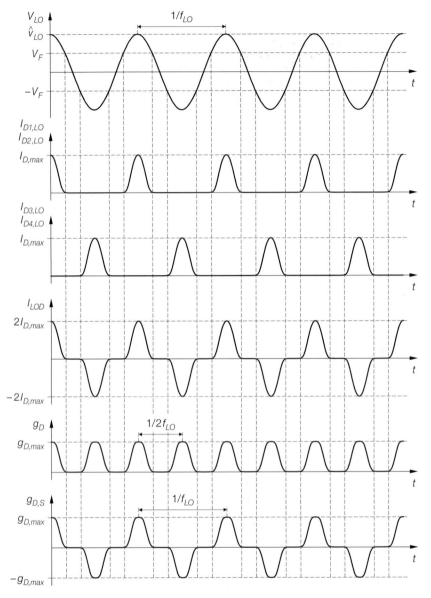

Fig. 28.28. Ring mixer with voltage $V_{LO}(t)$ of the LO circuit, currents of diodes $D_1 \ldots D_4$, total current I_{LOD} of the diodes, small-signal parameters $g_D(t)$ and $g_{D,S}(t)$. V_F represents the forward voltage of the diodes

$$i'_{D,S}(t) = \begin{cases} i'_D(t) = g_D(t)\,[\,v_{IF}(t) - v_{RF}(t)\,] & V_{LO} \geq 0 \\ -i'_D(t) = -g_D(t)\,[\,v_{IF}(t) + v_{RF}(t)\,] & V_{LO} < 0 \end{cases}$$

The additional introduction of the small-signal conductivity

$$g_{D,S}(t) = \begin{cases} g_D(t) & V_{LO} \geq 0 \\ -g_D(t) & V_{LO} < 0 \end{cases}$$

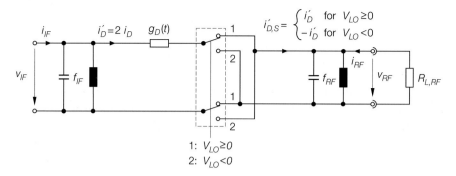

$$i'_{D,S} = \begin{cases} i'_D & \text{for } V_{LO} \geq 0 \\ -i'_D & \text{for } V_{LO} < 0 \end{cases}$$

1: $V_{LO} \geq 0$
2: $V_{LO} < 0$

Fig. 28.29. Small-signal equivalent circuit of a double balanced diode mixer

renders a case discrimination unnecessary; consequently:

$$
\begin{aligned}
i'_D(t) &= g_D(t)\, v_{IF}(t) - g_{D,S}(t)\, v_{RF}(t) \\
i'_{D,S}(t) &= g_{D,S}(t)\, v_{IF}(t) - g_D(t)\, v_{RF}(t)
\end{aligned}
\tag{28.23}
$$

The curve of $g_{D,S}(t)$ is shown in Fig. 28.28. It can be seen from the equations that the currents and voltages on the same side of the switch are interrelated by the small-signal conductivity $g_D(t)$ while the crosswise interrelation is caused by the small-signal conductivity $g_{D,S}(t)$. The small-signal conductivities $g_D(t)$ and $g_{D,S}(t)$ can be expressed by means of the small-signal conductivity of an unbalanced diode mixer at the same LO amplitude:

$$g_D(t) = 2\left[g_{D\,(unbal)}(t) + g_{D\,(unbal)}(t - T_{LO}/2) \right]$$

$$g_{D,S}(t) = 2\left[g_{D\,(unbal)}(t) - g_{D\,(unbal)}(t - T_{LO}/2) \right]$$

This takes into consideration the fact that the small-signal conductivities of the two single balanced diode mixers are both twice as high as those of the unbalanced diode mixer and are added or subtracted by the respective phase shift of half an LO period ($T_{LO} = 1/f_{LO}$). Inserting the Fourier series

$$g_{D\,(unbal)}(t) = g_{D,0\,(unbal)} + \sum_{n=1}^{\infty} g_{D,n\,(unbal)} \cos n\omega_{LO} t$$

from (28.7) eliminates all components at odd multiples of f_{LO} from $g_D(t)$ and all components at even multiples of f_{LO} including the DC component from $g_{D,S}(t)$; therefore:

$$
\begin{aligned}
g_D(t) &= 4\left[g_{D,0\,(unbal)} + g_{D,2\,(unbal)} \cos 2\omega_{LO} t + \cdots \right] \\
g_{D,S}(t) &= 4\left[g_{D,1\,(unbal)} \cos \omega_{LO} t + g_{D,3\,(unbal)} \cos 3\omega_{LO} t + \cdots \right]
\end{aligned}
\tag{28.24}
$$

These characteristics can be derived from the plots of $g_D(t)$ and $g_{D,S}(t)$ shown in Fig. 28.28. $g_D(t)$ has the fundamental frequency $2f_{LO}$ and thus contains components at frequencies $0, 2f_{LO}, 4f_{LO}, \ldots$. In contrast, $g_{D,S}(t)$ has the fundamental frequency f_{LO}, is symmetrical and has no DC component and therefore consists of components at the frequencies $f_{LO}, 3f_{LO}, 5f_{LO}, \ldots$.

Inserting the Fourier series for the small-signal conductivities from (28.24) into (28.23) for the small-signal currents and rearranging the terms according to the frequencies leads to the components at the following frequencies:

$$i'_D(t): \quad f_{IF}, \ 2f_{LO} \pm f_{IF}, \ 4f_{LO} \pm f_{IF}, \ 6f_{LO} \pm f_{IF}, \ \ldots$$

$$i'_{D,S}(t): \quad \underbrace{f_{LO} + f_{IF}}_{f_{RF}}, \ \underbrace{f_{LO} - f_{IF}}_{f_{RF,im}}, \ 3f_{LO} \pm f_{IF}, \ 5f_{LO} \pm f_{IF}, \ \ldots$$

The small-signal current $i'_D(t)$ on the IF side contains no component at the RF frequency and the small-signal current $i'_{D,S}(t)$ on the RF side contains no component at the IF frequency, i.e. the IF and RF circuits are decoupled as already mentioned.

From the small-signal currents $i'_D(t)$ and $i'_{D,S}(t)$, the respective portions of the IF and RF currents can be determined by extraction:

$$i_{IF}(t) = i'_D(t)\Big|_{f=f_{IF}} \quad , \quad i_{RF}(t) = -i'_{D,S}(t)\Big|_{f=f_{RF}}$$

All other components are short-circuited by the filters. The calculation, which will not be explained here in detail, is similar to that for the unbalanced diode mixer and also leads to a Y matrix:

$$\begin{bmatrix} \underline{i}_{IF} \\ \underline{i}_{RF} \end{bmatrix} = \begin{bmatrix} g_{D,0} & -\dfrac{g_{D,1}}{2} \\ -\dfrac{g_{D,1}}{2} & g_{D,0} \end{bmatrix} \begin{bmatrix} \underline{v}_{IF} \\ \underline{v}_{RF} \end{bmatrix}$$

Coefficient $g_{D,0}$ corresponds to the DC component in $g_D(t)$ and coefficient $g_{D,1}$ corresponds to the fundamental component in $g_{D,S}(t)$. Using (28.24), this leads to the relationship with the coefficients of an unbalanced diode mixer of the same LO amplitude:

$$g_{D,0} = 4 g_{D,0\,(unbal)} \quad , \quad g_{D,1} = 4 g_{D,1\,(unbal)} \tag{28.25}$$

This allows the Y matrix of the double balanced diode mixer to be determined and all other parameters to be calculated using (28.13)–(28.16) for the unbalanced diode mixer.

The *MAG* has the same magnitude as in the unbalanced diode mixer or the single balanced diode mixer of the same LO amplitude since only the ratio of $g_{D,1}$ and $g_{D,0}$ becomes effective according to (28.16). However, the related characteristic impedance is four times lower (see (28.15)):

$$Z_{W,M} = \frac{1}{4} Z_{W,M\,(unbal)} \tag{28.26}$$

The conversion gain $G_{M\,(50)}$ in a $50\,\Omega$ system has a similar characteristic as that of an unbalanced or a single balanced diode mixer. Figure 28.25 on page 1394 shows a comparison of mixers with Schottky diodes of type BAS40. In a double balanced diode mixer, the maximum value for $G_{M\,(50)}$ is always somewhat higher than in an unbalanced or a single balanced diode mixer and is reached at a lower LO amplitude.

Broadband Operation

Double balanced diode mixers are often operated in the mode shown in Fig. 28.30. The IF and RF filters are then separated from the mixer by amplifiers. The unwanted portions of the small-signal currents $i'_D(t)$ at the input and $i'_{D,S}(t)$ at the output of the mixer are no longer short-circuited by the filters but have an influence in accordance with the impedances $Z_{IF}(s)$ and $Z_{RF}(s)$ of the amplifiers. This is especially the case with the image frequency component at the frequency $f_{RF,im}$, which has the same size as the component at the

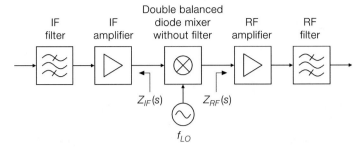

Fig. 28.30. Broadband operation of a double balanced diode mixer

RF frequency f_{RF}. Like all other interfering components, these are amplified by the RF amplifier shown in Fig. 28.30 and are suppressed only by the subsequent RF filter.

The small-signal response of the mixer depends on the values of the impedances $Z_{IF}(s)$ and $Z_{RF}(s)$ at all frequencies involved and can only be determined using numeric methods or circuit simulation. This is called the *broadband operation* of the mixer. There are qualitatively similar relationships as in narrow-band operation as discussed above, but the quantitative changes for the broadband operation are:

– The conversion gain $G_{M\,(50)}$ and the *MAG* are approximately $1\ldots 2$ dB lower.
– The characteristic impedances $Z_{W,M\,(IF)}$ and $Z_{W,M\,(RF)}$ for power matching at the IF and/or RF terminal are higher than in narrow-band operation and are no longer equal. Typical values are:

$$Z_{W,M\,(IF)} \approx (2\ldots 3)\,Z_{W,M} \quad , \quad Z_{W,M\,(RF)} \approx (1.2\ldots 1.5)\,Z_{W,M}$$

Here, $Z_{W,M}$ is the characteristic impedance in narrow-band operation (see (28.15)). This means that the characteristic impedance $Z_{W,M\,(IF)}$ at the IF terminal is twice as large as the characteristic impedance $Z_{W,M\,(RF)}$ at the RF terminal.

No separate matching circuits are used in broadband operation. The conversion gain $G_{M\,(50)}$ is optimized by a suitable choice of the LO amplitude and the transformation ratios of the transformers.

28.3.4
Diode Mixers in Practical Use

In practice, double balanced diode mixers are used predominantly. They are available as discrete components and, in addition to the four diodes, also contain two transformers. Figure 28.31 shows a common design with a total of four terminals: LO, IF, RF and common ground. In some models the ground terminals are not tied to a common ground; the double balanced diode mixer then has six terminals.

Since the terminals are decoupled and the circuit is symmetrical, the connections of the double balanced diode mixer can be interchanged. Of course, this alters the voltages and currents of the diodes, but the small-signal equivalent circuit and the operating parameters (R_{LO}, $Z_{W,M}$, *MAG*, etc.) remain the same if the transformers are symmetrical. In double balanced diode mixers used in practical applications, the LO and IF terminals are often interchanged (see Fig. 28.31). Then transformer *TR1* is no longer supplied with the IF frequency but with the significantly higher LO frequency, thus reducing the transformer

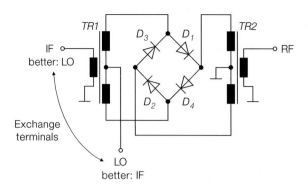

Fig. 28.31. Double balanced diode mixer as a discrete component

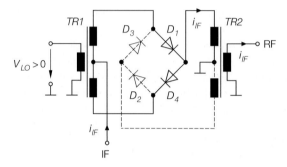

a Positive LO voltage

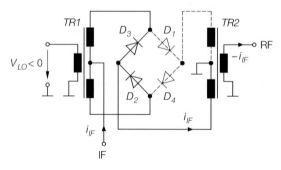

b Negative LO voltage

Fig. 28.32. Current distribution in a double balanced diode mixer with interchanged LO and IF terminals

size. In this case, a very low IF frequency can be used. Figure 28.32 shows the related current distribution. In the LO circuit, current flows alternately through diodes D_1/D_4 ($V_{LO} > 0$) and D_2/D_3 ($V_{LO} < 0$) which causes the IF current to flow alternately through both secondary windings of transformer $TR2$.

Discrete double balanced diode mixers are always designed for a certain LO power. For this power, the conversion gain $G_{M\,(50)}$ in a 50 Ω system reaches its maximum and the terminals are optimally matched to 50 Ω. This is achieved by using suitable diodes and matching the transformation ratios of the transformers. The double balanced diode mixer is then no longer symmetrical, i.e. the terminals must be used as designated by the manufacturer. Instead of the conversion gain, the data sheet specifies the *conversion loss* in dB:

$$L_{M\,(50)}\,[\text{dB}] \;=\; -G_{M\,(50)}\,[\text{dB}] \;=\; -10\log G_{M\,(50)}$$

The LO power is given in dBm:

$$P_{LO}\,[\text{dBm}] \;=\; 10\log\frac{P_{LO}}{1\,\text{mW}}$$

A mixer for an LO power of n dBm is called a *level n mixer*.

According to specifications, frequency ranges at which the mixer operates are defined for the IF and LO/RF terminals and result from the bandwidth of the diodes and the transformers. The decoupling of the terminals is not ideal due to the transformer asymmetries and due to capacitive and inductive coupling. Particularly critical is the crosstalk from the strong LO signal into the IF and RF circuits. Data sheets thus specify the *isolation* between the LO and RF terminals (*LO-RF isolation*) as well as between the LO and IF terminals (*LO-IF isolation*). The isolation decreases with an increasing LO frequency with typical values in the range of $50\ldots70$ dB for $f_{LO} < 10$ MHz and $20\ldots30$ dB for $f_{LO} > 1$ GHz. These values only apply if the IF and RF terminals are terminated with $50\,\Omega$ for the LO frequency, which means that broadband operation of the mixer is assumed. In narrow-band operation with IF and RF filters directly at the mixer, the isolation is generally significantly higher. A large selection of double balanced diode mixers can be found in [28.2].

At frequencies above 5 GHz the transformers are replaced by stripline hybrids. Figure 28.33 shows a commonly used version of a single balanced diode mixer with a 180° hybrid. The LO signal is supplied to terminal 4 of the hybrid and made available at terminals 1 and 2 with half the power. The LO signals at terminals 1 and 2 are in phase opposition due to the 180° shift in the hybrid; thus, with respect to the LO signal, the diodes are connected in series and conductive during one half wave. During this half wave, the IF and RF circuits are connected by the small-signal conductivities of the diodes. As the signals at terminals 1 and 2 are in phase with respect to the IF and RF circuits, the LO signal available in phase opposition is not transferred to the IF and RF circuits. Consequently, the IF and RF signals are not transferred to the LO circuit since co-phasal signals at terminals 1 and 2 are compensated at terminal 4. Owing to this characteristic of the single balanced diode mixer with 180° hybrid, decoupling of the LO circuit from the IF and RF circuits is achieved with one hybrid while in single balanced diode mixers with transformers, two transformers are required.

At frequencies exceeding 10 GHz, unbalanced mixers are also used quite often. The summation of the LO and IF signals, which is performed by a transformer in Fig. 28.16a, is then achieved by coupled lines.

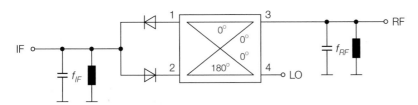

Fig. 28.33. Single balanced diode mixer with 180° hybrid

28.4
Mixers with Transistors

Multiplicative mixers with transistors are used almost exclusively in integrated circuits. In these mixers the input signal is converted to a current by a voltage-to-current converter and connected to the output by means of one or two differential amplifiers operated as change-over switches. In the following section, the two commonly used circuits are described. The first is the *single balanced mixer* and the second is the *double balanced mixer* which is also named *Gilbert mixer* after its inventor B. Gilbert. Both circuits can be made up of bipolar transistors or MOSFETs. The following explanations are based on bipolar transistors.

28.4.1
Single Balanced Mixer

Figure 28.34 shows the circuit diagram of a *single balanced mixer* operated as an up-conversion mixer. It consists of a common-emitter circuit with current feedback (T_3, R_E) that operates as a voltage-to-current converter (V/I converter) and a differential amplifier (T_1, T_2) that acts as a switch and supplies the output current alternately to the RF output and the supply voltage. The small-signal IF voltage v_{IF}, together with a DC voltage V_0, is supplied to the input to determine the operating point. The differential amplifier is switched by the LO voltage V_{LO}, which, under ideal conditions, is rectangular. From the mixed products in current I_{C2} the small-signal RF current i_{RF} is separated by an RF filter and supplied to the RF load resistance $R_{L,RF}$ via a coupling capacitance C_c.

The functional principle of the single balanced mixer is shown in Fig. 28.35. One can see that in terms of the transfer characteristic, the change-over switch functions as an On/Off switch only. In this form, the single balanced mixer works as a multiplicative mixer with a unipolar square-wave signal as shown in a comparison with Fig. 28.12 on page 1376.

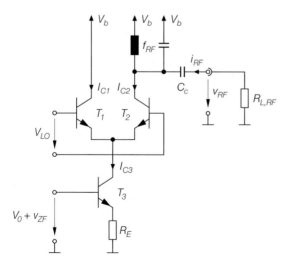

Fig. 28.34. Single balanced mixer with transistors

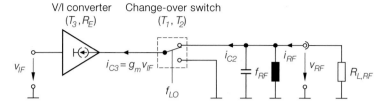

Fig. 28.35. Functional principle (= small-signal equivalent circuit) of a single balanced mixer with transistors

Calculation of the Transfer Characteristic

Current I_{C3} at the output of the common-emitter circuit comprises the bias current $I_{C3,A}$ and the small-signal current i_{C3}:

$$I_{C3} = I_{C3,A} + i_{C3} \tag{28.27}$$

The bias current $I_{C3,A}$ is determined by the DC voltage V_0 at the input. The small-signal current i_{C3} is:

$$i_{C3} = g_m v_{IF} \tag{28.28}$$

Here,

$$g_m = \frac{g_{m3}}{1 + g_{m3} R_E} \overset{g_{m3} = I_{C3,A}/V_T}{=} \frac{I_{C3,A}}{V_T + I_{C3,A} R_E} \tag{28.29}$$

is the transconductance of the voltage-to-current converter.

The collector currents of transistors T_1 and T_2 can be taken from current I_{C3} with the help of the current characteristic of the differential amplifier. For $2I_0 = I_{C3}$ and $V_D = V_{LO}$ from (4.61) on page 331, it follows:

$$I_{C1} = \frac{I_{C3}}{2}\left(1 + \tanh\frac{V_{LO}}{2V_T}\right) \quad , \quad I_{C2} = \frac{I_{C3}}{2}\left(1 - \tanh\frac{V_{LO}}{2V_T}\right) \tag{28.30}$$

Here, $V_T = 26\,\text{mV}$ is the thermal voltage. Figure 28.36 shows the currents plotted over the LO voltage. For $V_{LO} < -5V_T = -130\,\text{mV}$ and $V_{LO} > 5V_T = 130\,\text{mV}$, the differential amplifier nearly reaches its full modulation and operates like a switch as desired. The range in between is the switch-over region in which both transistors are forward biased.

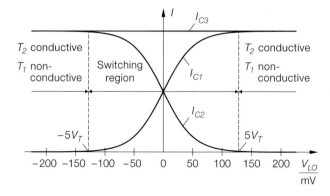

Fig. 28.36. Current characteristics of the differential amplifier

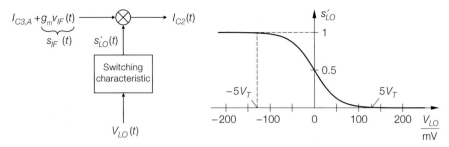

a Model of a single balanced mixer   **b** Curve of the switching characteristic

Fig. 28.37. Single balanced mixer as a multiplicative mixer taking into consideration the switching characteristic of the differential amplifier

The characteristic of current I_{C2} as a function of time can be calculated by inserting (28.27) and (28.28) into (28.30):

$$I_{C2}(t) = [\underbrace{I_{C3,A} + g_m v_{IF}(t)}_{s_{IF}(t)}] \underbrace{\left[\frac{1}{2}\left(1 - \tanh \frac{V_{LO}(t)}{2V_T}\right)\right]}_{s'_{LO}(t)} \qquad (28.31)$$

This shows that the single-balanced mixer operates as a multiplicative mixer where the IF signal $s_{IF}(t)$ is multiplied by the LO signal $s'_{LO}(t)$. In addition, a DC component occurs that corresponds to the bias current $I_{C3,A}$, which is also multiplied by $s'_{LO}(t)$. The LO signal $s'_{LO}(t)$ results from the LO voltage $V_{LO}(t)$ when taking the switching characteristic of the differential amplifier into consideration. Figure 28.37 illustrates this.

Rectangular LO Voltage

First we look at the operation with a bipolar rectangular LO voltage of the amplitude $\hat{v}_{LO}$. The LO signal $s'_{LO}(t)$ is also rectangular with the following values:

$$s'_{LO} = \frac{1}{2}\left(1 - \tanh\frac{V_{LO}}{2V_T}\right) \overset{\substack{V_{LO}=\pm\hat{v}_{LO} \\ \tanh(-x)=-\tanh x}}{=} \frac{1}{2}\left(1 \mp \tanh\frac{\hat{v}_{LO}}{2V_T}\right)$$

Figure 28.38a shows the plot of $V_{LO}(t)$ and $s'_{LO}(t)$ for different amplitudes. For $\hat{v}_{LO} > 5V_T = 130\,\mathrm{mV}$ the signal $s'_{LO}(t)$ is approximately a unipolar square-wave signal with the values 0 and 1. In this case, the mixer can be considered an ideal switch.

For further calculations, the signal $s'_{LO}(t)$ is expanded into a Fourier series:

$$s'_{LO}(t) = c_0 + c_1 \cos \omega_{LO} t + c_3 \cos 3\omega_{LO} t + c_5 \cos 5\omega_{LO} t + \cdots$$

$$= c_0 + \sum_{n=0}^{\infty} c_{(2n+1)} \cos(2n+1)\omega_{LO} t \qquad (28.32)$$

Besides the DC component c_0, the series contains only cosine components of the LO frequency f_{LO} and uneven multiples thereof since $s'_{LO}(t)$ is an *even* function of time according to Fig. 28.38 ($s'_{LO}(t) = s'_{LO}(-t)$) with a pulse duty ratio of 50 %. Signal $s'_{LO}(t)$

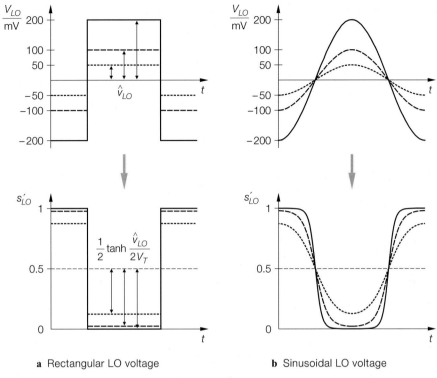

a Rectangular LO voltage   **b** Sinusoidal LO voltage

Fig. 28.38. LO voltage $V_{LO}(t)$ and LO signal $s'_{LO}(t)$ for the amplitudes
$\hat{v}_{LO} = 50\,\text{mV} \,/\, 100\,\text{mV} \,/\, 200\,\text{mV}$

can be viewed as the sum of a DC component $c_0 = 1/2$ and a bipolar square-wave signal of the amplitude:

$$\frac{1}{2}\tanh\frac{\hat{v}_{LO}}{2V_T}$$

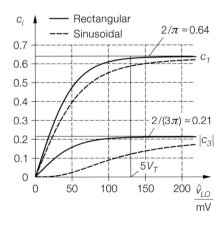

Fig. 28.39. Coefficients c_1 and $|c_3|$ of the Fourier series of the LO signal $s'_{LO}(t)$ for a rectangular and a sinusoidal LO voltage

Using the series expansion for a bipolar square-wave signal in (28.1) leads to the coefficients:

$$c_0 = \frac{1}{2} \quad , \quad c_1 = \frac{2}{\pi} \tanh \frac{\hat{v}_{LO}}{2V_T} \quad , \quad c_3 = -\frac{2}{3\pi} \tanh \frac{\hat{v}_{LO}}{2V_T} \quad , \quad \cdots \quad (28.33)$$

In Fig. 28.39 the coefficients c_1 and $|c_3|$ are plotted over the amplitude $\hat{v}_{LO}$, where, for $\hat{v}_{LO} \to \infty$, they assume the state of coefficients of a unipolar square-wave signal. Essentially this is the case at $\hat{v}_{LO} > 5V_T$.

Sinusoidal LO Voltage

Generating a rectangular LO voltage becomes more and more difficult with an increasing LO frequency. For this reason, at high frequencies the nearly sinusoidal output voltage of a high frequency oscillator is used. Figure 28.38b shows the curves of $V_{LO}(t)$ and $s'_{LO}(t)$ for this case. Here, too, with an increasing amplitude the LO signal $s'_{LO}(t)$ enters the state of a unipolar square-wave signal. With the exception of the DC component c_0 the coefficients of the Fourier series are smaller than for a rectangular LO voltage of the same amplitude. Fig. 28.39 compares the coefficients c_1 und $|c_3|$.

Small-Signal Response

Now the small-signal dynamic response can be calculated by inserting a sinusoidal IF voltage

$$v_{IF}(t) = \hat{v}_{IF} \cos \omega_{IF} t$$

and the Fourier series for $s'_{LO}(t)$ from (28.32) into (28.31):

$$\begin{aligned}
I_{C2}(t) &= \left[I_{C3,A} + g_m v_{IF}(t) \right] s'_{LO}(t) \\
&= \left[I_{C3,A} + g_m \hat{v}_{IF} \cos \omega_{IF} t \right] \left[c_0 + c_1 \cos \omega_{LO} t + c_3 \cos 3\omega_{LO} t + \cdots \right] \\
&= I_{C3,A} \left[c_0 + c_1 \cos \omega_{LO} t + c_3 \cos 3\omega_{LO} t + \cdots \right] \\
&\quad + c_0 g_m \hat{v}_{IF} \cos \omega_{IF} t \\
&\quad + \frac{c_1 g_m \hat{v}_{IF}}{2} \left[\cos(\omega_{LO} + \omega_{IF})t + \cos(\omega_{LO} - \omega_{IF})t \right] \\
&\quad + \frac{c_3 g_m \hat{v}_{IF}}{2} \left[\cos(3\omega_{LO} + \omega_{IF})t + \cos(3\omega_{LO} - \omega_{IF})t \right] \\
&\quad + \cdots
\end{aligned}$$

Figure 28.40 shows the spectrum of current I_{C2}. We presume that the mixer operates in noninverted mode, then the RF frequency is $f_{RF} = f_{LO} + f_{IF}$. The RF filter short-circuits all components with the exception of the RF component which then results in the RF current

$$i_{RF}(t) = I_{C2}(t) \Big|_{f=f_{RF}=f_{LO}+f_{IF}} = \frac{c_1}{2} g_m \hat{v}_{IF} \cos \omega_{RF} t$$

and the RF voltage:

$$v_{RF}(t) = -R_{L,RF} \, i_{RF}(t) = -\frac{c_1}{2} g_m R_{L,RF} \, \hat{v}_{IF} \cos \omega_{RF} t$$

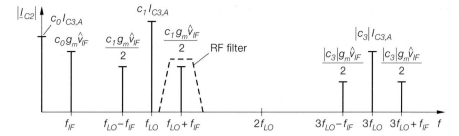

Fig. 28.40. Spectrum of current $I_{C2}(t)$ in the case of a sinusoidal IF voltage

This assumes that the output resistance of transistor T_2 can be neglected. The voltage phasors are:

$$\underline{v}_{RF} = -\frac{c_1}{2} g_m R_{L,RF} \, \underline{v}_{IF} \tag{28.34}$$

Mixer Voltage Gain

Equation (28.34) indicates that the single balanced mixer operates like an amplifier with the *mixer voltage gain:*

$$A_M = \frac{\underline{v}_{RF}}{\underline{v}_{IF}} = -\frac{c_1}{2} g_m R_{L,RF} \overset{(28.29)}{=} -\frac{c_1}{2} \frac{g_{m3} R_{L,RF}}{1 + g_{m3} R_E} \tag{28.35}$$

The frequency conversion is no longer explicitly visible.

The mixer voltage gain is lower than the gain A of an equivalent common-emitter circuit with current feedback by a factor of $c_1/2$:

$$A \overset{(2.70)}{=} -\frac{g_m R_C}{1 + g_m R_E} \overset{\substack{g_m = g_{m3} \\ R_C = R_{L,RF}}}{=} -\frac{g_{m3} R_{L,RF}}{1 + g_{m3} R_E} \quad \Rightarrow \quad A_M = \frac{c_1}{2} A$$

Coefficient c_1 results from the operational principle of a multiplicative mixer and reaches the maximum value $2/\pi \approx 0.64$ (see Fig. 28.39). The factor $1/2$ is caused by the fact that, besides the desired RF band at $f_{LO} + f_{IF}$, the mixing process produces an image frequency band of the same amplitude at $f_{LO} - f_{IF}$ which is suppressed by the RF filter (see Fig. 28.40). This means that the mixer voltage gain is lower than the gain of an equivalent common-emitter circuit with current feedback by at least a factor of $1/\pi$ (≈ 10 dB). Typical values are in the range of $|A_M| \approx 2 \dots 10$ ($6 \dots 20$ dB).

Bandwidth

We have calculated the mixer voltage gain A_M for the static condition only, i.e. without accounting for the capacitances of the transistors. Therefore, strictly speaking, it applies to low frequencies only. However, the bandwidth of the single balanced mixer is usually very wide for three reasons:

- The common-emitter circuit with current feedback, together with the transistors of the differential amplifier, forms a cascode circuit and therefore reaches a cutoff frequency that is between the transconductance cutoff frequency f_{Y21e} and the transit frequency f_T of transistor T_3.

- With respect to the small-signal current, transistor T_2 operates in common-base configuration with the α cutoff frequency $f_\alpha \approx f_T$.
- The output capacitance of transistor T_2 can be regarded as an integral part of the RF filter capacitance and thus has no interfering effect.

Therefore, even for higher frequencies, the mixer voltage gain can be evaluated by means of the static voltage gain.

Impedance Matching

For high frequencies, the single balanced mixer must be matched at all sides to the characteristic impedance Z_W of the connecting lines in order to avoid undesirable reflections and impedance transformations. This is possible with the same methods used in amplifiers:

- Circuits for impedance transformation described in Sect. 28.3.1.
- Methods for matching integrated amplifiers described in Sect. 27.1.1 (see Fig. 27.2 on page 1323).

Figure 28.41 shows a typical example:

- The common-emitter circuit at the input is replaced by a common-base circuit with bias current I_0 and input impedance:

$$\frac{1}{g_{m3}} = \frac{V_T}{I_0} \tag{28.36}$$

$I_0 \approx 520\,\mu\text{A}$ provides an impedance matching to $Z_W = 50\,\Omega$. If a current feedback is required to improve the linearity, one may select a higher bias current and introduce an

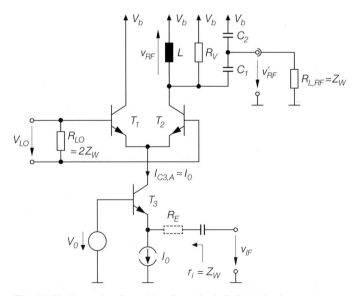

Fig. 28.41. Example of matching for a single balanced mixer

additional series resistance:

$$R_E = Z_W - \frac{1}{g_{m3}} = Z_W - \frac{V_T}{I_0} \tag{28.37}$$

This ensures that the matched condition is maintained. However, the common-base circuit has the disadvantage that the transconductance g_m of the voltage-to-current conversion is linked to the characteristic impedance Z_W. Inserting (28.36) and (28.37) into (28.29), with and without R_E, leads to:

$$g_m = \frac{1}{Z_W} \tag{28.38}$$

The conversion gain can therefore not be influenced by transconductance g_m.

- A terminating resistance R_{LO} is used at the LO input. If the LO voltage is supplied symmetrically, $R_{LO} = 2Z_W$ is required in order to terminate both inputs with Z_W. This assumes that the input impedances of transistors T_1 and T_2 are substantially higher than Z_W and are negligible.
- The load resistance $R_{L,RF} = Z_W$ at the output is coupled by a capacitive voltage divider (C_1, C_2). According to (28.38) and (28.39) the transformed load resistance is:

$$R_P = Z_W \left(1 + \frac{C_2}{C_1}\right)^2 \tag{28.39}$$

The coupling capacitance C_c from Fig. 28.34 is no longer required as the load resistance is already decoupled with respect to DC voltages by the capacitive voltage divider. Figure 28.42 shows the transformation. In the following, we assume that the loss resistance R_V represents all loss resistances of the parallel resonant circuit including the output resistance of transistor T_2.[5] The matching condition is then $R_V = R_P$.

Impedance matching at the input side is particularly important for down-conversion mixer operation because the RF signal is available at the input and often the RF components in front of the mixer are very sensitive to mismatching.

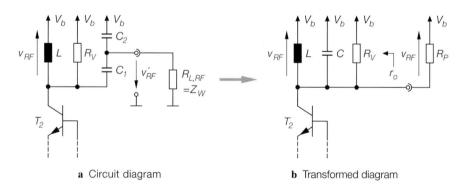

a Circuit diagram   b Transformed diagram

Fig. 28.42. Transformation of the load resistance by capacitive matching

---

[5] The output resistance of the transistor is defined here as the reciprocal of the real part of the output admittance: $r_o = 1/\mathrm{Re}\{Y_o\}$. For low frequencies, $r_o = r_{CE} = V_A/I_{C,A}$ (see (2.13)); for high frequencies the output resistance is clearly lower.

Conversion Gain

Now we can determine the *conversion gain* G_M of the single balanced mixer. It corresponds to the gain G_T of the amplifier and is calculated with (27.29):

$$G_M = G_T = \left(\frac{r_i}{R_g + r_i}\right)^2 A^2 \frac{4 R_g R_L}{(r_o + R_L)^2}$$

Here, r_i = input resistance, A = open-circuit gain, r_o = output resistance of the single balanced mixer. Additionally, R_g is the internal resistance of the signal source and R_L is the load resistance. We assume that the mixer is matched at both ends with $r_i = R_g = Z_W$ and $r_o = R_L$; then:

$$G_M = \frac{A^2 Z_W}{4 r_o} \tag{28.40}$$

Since the transformation is loss-free we can determine the open-circuit gain and the output resistance by means of the transformed diagram in Fig. 28.42b where R_P and $R_{L,RF}$ absorb the same power. As R_V represents the loss resistances by definition from Fig. 28.42b it follows:

$$r_o = R_V \tag{28.41}$$

In an open circuit, i.e. without R_P, R_V acts as the load resistance. The open-circuit voltage gain A can thus be calculated with (28.35) by replacing $R_{L,RF}$ by R_V:

$$A = -\frac{1}{2} c_1 g_m R_V \tag{28.42}$$

Here, g_m is the transconductance of the voltage-to-current converter:

$$g_m = \begin{cases} g_{m3} & \text{Without current feedback} \\ g_{m3}/(1 + g_{m3} R_E) & \text{With current feedback} \end{cases} \tag{28.43}$$

Inserting (28.41) and (28.42) into (28.40) leads to the *conversion gain of a single balanced mixer with impedance matching at both ends*:

$$\boxed{G_M = \frac{1}{16} c_1^2 g_m^2 Z_W R_V} \tag{28.44}$$

This is proportional to the loss resistance R_V. At low frequencies, R_V is very high and it may be necessary to reduce it by an additional parallel resistance in order to prevent the voltage of the parallel resonant circuit from becoming too high. R_V decreases with an increasing frequency. A benefit is that R_V only has a *linear* effect on the conversion gain (= power gain). Therefore, with a decrease in loss resistance in a matched circuit, the voltage gain only decreases in proportion to $\sqrt{R_V}$ and not in proportion to R_V as is the case for the open-circuit voltage gain in (28.42).

At the output we have $R_V = R_P$. The required capacitance ratio of the capacitive voltage divider results from (28.39):

$$R_V = Z_W \left(1 + \frac{C_2}{C_1}\right)^2 \quad \Rightarrow \quad \frac{C_2}{C_1} = \sqrt{\frac{R_V}{Z_W}} - 1 \tag{28.45}$$

Example: Let us look at a matched single balanced mixer with common-base circuit as shown in Fig. 28.41. Here, according to (28.38) the relation $g_m = 1/Z_W$ applies. For a fully modulated differential amplifier $(c_1 = 2/\pi)$ and $Z_W = 50\,\Omega$, substitution into (28.44) leads to:

$$G_M \overset{g_m=1/Z_W}{=} \frac{1}{16} c_1^2 \frac{R_V}{Z_W} = \frac{1}{16}\left(\frac{2}{\pi}\right)^2 \frac{R_V}{50\,\Omega} = \frac{R_V}{1974\,\Omega}$$

This means that for a conversion gain of $G_M = 4\,(6\,\text{dB})$, a loss resistance of $R_V \approx 7.9\,\text{k}\Omega$ is required. The capacitance ratio of the capacitive voltage divider is calculated with (28.45) which yields $C_2/C_1 \approx 11.6$. This represents the limit of what can be achieved in practical applications. The reason for this is the fundamental relation of $g_m = 1/Z_W$, which applies to matched single balanced mixers with common-base circuit. The transconductance g_m, which has a quadratic effect on the conversion gain, is thus limited to a comparably low value.

Better results are achievable with a single balanced mixer with common-emitter circuit as shown in Fig. 28.43. Here, any value can be selected for the transconductance g_m and the relatively high input resistance can be matched by a terminating resistance $R_1 \approx Z_W$ independent of g_m. The current feedback is omitted and $I_0 = 2\,\text{mA}$ is selected so that $g_m = g_{m3} = I_0/V_T \approx 77\,\text{mS}$. For $\beta_3 = 100$, the input resistance of transistor T_3 is $r_{BE3} = \beta_3/g_{m3} \approx 1.3\,\text{k}\Omega$; for $R_1 = 52\,\Omega$ we have $r_i = (R_1 \| r_{BE}) = 50\,\Omega$. By substituting into (28.44) and using $c_1 = 2/\pi$ we obtain:

$$G_M = \frac{1}{16}\left(\frac{2}{\pi}\right)^2 (77\,\text{mS})^2 \cdot 50\,\Omega \cdot R_V = \frac{R_V}{133\,\Omega}$$

We assume that the loss resistance R_V is caused by the output resistance of transistor T_2. Since the bias current is higher than in the single balanced mixer with common-base circuit, we can assume a reduced loss resistance of $R_V = 7.9\,\text{k}\Omega \cdot (520\,\mu\text{A}/2\,\text{mA}) \approx 2050\,\Omega$.

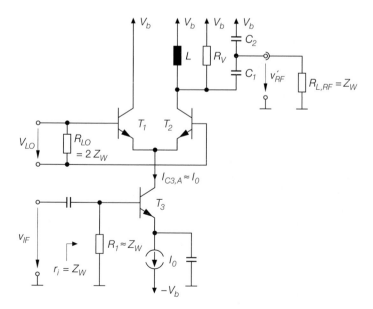

Fig. 28.43. Single balanced mixer with common-emitter circuit and impedance matching

Thus the conversion gain achieved is $G_M \approx 15$ (12 dB). From (28.45) for the capacitive voltage divider it follows $C_2/C_1 \approx 5.4$.

In this example the conversion gain of the single balanced mixer with common-emitter circuit is higher than that of the single balanced mixer with common-base circuit by a factor of 4 (6 dB). A disadvantage is the increase in the noise figure due to the terminating resistance R_1. Therefore, this design is not used for front-end down-conversion mixers in receivers.

Practical Design

Figure 28.44 shows the design of a single balanced mixer for practical applications with all components required for operating point setting and matching to $Z_W = 50\,\Omega$. The voltages V_0 and V_1 for setting the operating point are generated with resistors R_1, R_2 and R_3; C_3 and C_6 serve as blocking capacitors. Resistors R_4 and R_5 supply the voltage V_1 to the inputs of the differential amplifier and at the same time serve as LO terminating resistors $R_4 = R_5 = 50\,\Omega$. The series connection of R_4 and R_5 corresponds to resistance $R_{LO} = 2Z_W$ in Figs. 28.41 and 28.43. The LO voltage is provided via the coupling capacitors C_4 and C_5. The bias current $I_{C3,A} \approx 520\,\mu A$ required for impedance matching to $50\,\Omega$ is set with resistor R_6. No current feedback is used. The IF voltage is supplied via the coupling capacitor C_7. The circuitry at the output with capacitive voltage divider C_1, C_2 and the resonant impedance R_V is adopted from Fig. 28.41. However, in this case capacitor C_2 is not connected to the supply voltage V_b (small-signal ground) but to ground. With very high voltage divider factors this causes the RF output current which flows almost entirely through C_2 not to flow onto the supply voltage line.

The symmetric LO voltage can be generated by an oscillator with differential output. It often occurs that only an asymmetric LO voltage is available, then the supply methods shown in Fig. 28.45 can be used. In Fig. 28.45a the LO voltage is supplied asymmetrically to one of the two LO inputs while the other input is short-circuited (C_5 to ground) with respect

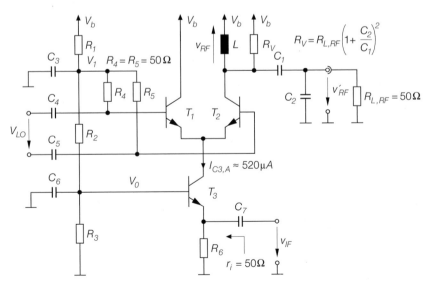

Fig. 28.44. Practical design of a single balanced mixer with matching to $Z_W = 50\,\Omega$

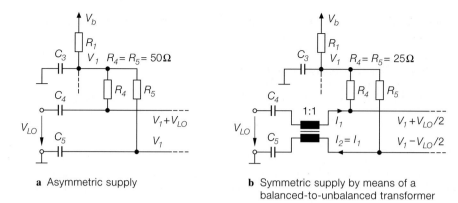

a Asymmetric supply

b Symmetric supply by means of a balanced-to-unbalanced transformer

Fig. 28.45. Use of an asymmetric LO voltage

to small signals. But the asymmetry is disadvantageous to the distortion characteristic of the mixer; therefore, the symmetric supply by means of a balanced-to-unbalanced transformer (balun) is often used in practice. The balun requires $I_1 = I_2$ so that there is pure differential modulation at the LO inputs.

In integrated circuits, balancing of an asymmetric LO voltage is achieved by means of a differential amplifier with an asymmetric input and a symmetric output. At the same time, this differential amplifier serves as an amplifier for the LO signal and is directly coupled to the differential amplifier of the single balanced mixer. Figure 28.46 shows a typical design. The inductance L and the capacitances $C_1, \ldots, C_4$ are generally not integrated, rather they are connected externally. Matching at the LO side is done with resistance $R_B \approx Z_W$. The differential amplifier T_4, T_5 is driven into saturation and generates an almost rectangular LO voltage V_{LO} of the amplitude $I_1 R_C > 5V_T$ from the sinusoidal voltage V'_{LO}. The

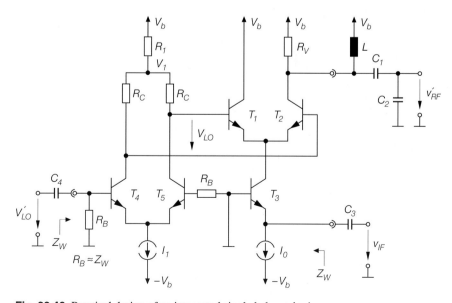

Fig. 28.46. Practical design of an integrated single balanced mixer

maximum voltage V_1 at the LO inputs is adjusted by resistance R_1 which causes the voltages at the LO inputs to alternately assume the values V_1 and $V_1 - I_1 R_C$.

Single Balanced Mixers with Transformers

Single balanced mixer designs often use transformers. Figure 28.47 shows a typical design with two transformers. The LO transformer *TR1* provides the symmetric supply of an asymmetric LO voltage and can be simultaneously used for matching if a suitable transformation ratio is selected.

The output transformer *TR2* is also symmetric which means that the current I_{C1} of transistor T_1 can be also used. The following description is based on a 1:1:1 transformer. The secondary current I_1 thus corresponds to the difference of the primary currents:

$$I_1(t) = I_{C2}(t) - I_{C1}(t)$$

According to (28.31) the current I_{C2} is:

$$I_{C2}(t) = [I_{C3,A} + g_m v_{IF}(t)] \left[\frac{1}{2}\left(1 - \tanh\frac{V_{LO}(t)}{2V_T}\right)\right]$$

Accordingly, current I_{C1} is:

$$I_{C1}(t) = [I_{C3,A} + g_m v_{IF}(t)] \left[\frac{1}{2}\left(1 + \tanh\frac{V_{LO}(t)}{2V_T}\right)\right]$$

Consequently, the secondary current of the transformer is:

$$I_1(t) = [I_{C3,A} + g_m v_{IF}(t)] \underbrace{\left[- \tanh\frac{V_{LO}(t)}{2V_T}\right]}_{s'_{LO}(t)} \tag{28.46}$$

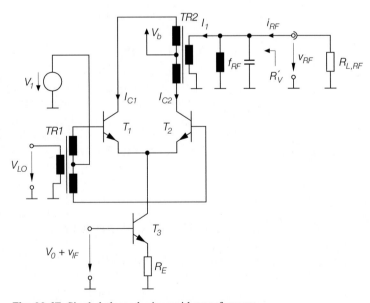

Fig. 28.47. Single balanced mixer with transformers

In this case, the LO signal $s'_{LO}(t)$ has no DC component and twice the amplitude as that of a single balanced mixer without output transformer. For the coefficients of the Fourier series of $s'_{LO}(t)$ this means that coefficient c_0 becomes zero, while all other coefficients increase by a factor of 2. Consequently, the mixer voltage gain A_M rises by a factor of 2 as well since, according to (28.35), it is proportional to the coefficient c_1. According to (28.44), the conversion gain G_M in a matched circuit is proportional to the square of coefficient c_1 and must thus increase by a factor of 4. In practice this is usually not the case since the output resistance of transistor T_1 becomes effective, too, and causes a reduction in the loss resistance R_V. In extreme cases the loss resistance is caused solely by the transistors, and then the conversion gain only rises by a factor of 2.

Transformer $TR2$ is also used for output matching. To this end, the transformation ratio n is selected such that the loss resistance $R'_V = R_V/n^2$, as related to the secondary side, is equal to the load resistance $R_{L,RF}$.

Disadvantage of Single Balanced Mixers with Transistors

The main disadvantage of single balanced mixers is that the differential amplifier switches not only the small signal current $i_{C3} = g_m v_{IF}$ but also the bias current $I_{C3,A}$ of the voltage-to-current converter. Therefore, with the differential amplifier fully modulated, the collector currents of transistors T_1 and T_2 contain a rectangular component of the amplitude $I_{C3,A}$ and the frequency f_{LO} that is significantly higher than the small-signal component. In the spectrum of the collector currents, this produces components of the LO frequency and odd multiples thereof which are proportional to $I_{C3,A}$ (see Fig. 28.40 on page 1409). Particularly problematic is the component of the LO frequency which is close to the RF frequency and must be suppressed by the RF filter; for this reason the demands on that filter are very high.

This disadvantage prevents an efficient integrated version of the single balanced mixer. In addition, it is desirable to replace the RF filter by an ohmic load resistance to match the resulting output signal to the characteristic impedance Z_W by using an integrated impedance converter (one or more common-collector circuits) and afterwards filtering the signal. Here, too, the square-wave component in the collector current of T_2 is undesirable. To avoid an overmodulation by this component, the ohmic load resistance must be chosen so small that no conversion gain is achieved.

28.4.2
Double Balanced Mixer (Gilbert Mixer)

Figure 28.48 shows the circuit of a *double balanced mixer*, which is also named *Gilbert mixer* after its inventor B. Gilbert. It is the mixer of choice in integrated circuits since it can be operated without any filters connected directly to the mixer. Suppression of unwanted components in the output voltages is then achieved in subsequent components. The following description is based on an up-conversion mixer.

A comparison of the double balanced mixer in Fig. 28.48 with the single balanced mixer in Fig. 28.34 on page 1404 shows that the double balanced mixer consists of two single balanced mixers whose outputs are connected: T_1, T_2, T_5 and T_3, T_4 and T_6. The common-emitter circuits with current feedback (T_5 and T_6) that operate as voltage-to-current converters (V/I converters) are combined to a differential amplifier with current feedback and are inversely modulated by the IF voltage v_{IF}. This makes the connection

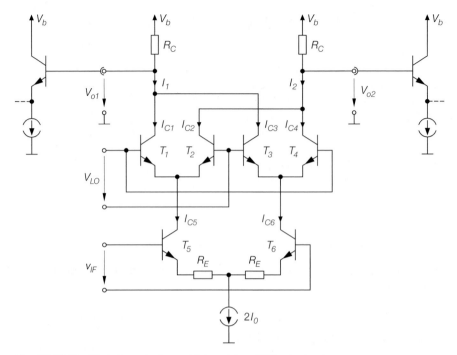

Fig. 28.48. Double balanced mixer with transistors (Gilbert mixer)

point between the two feedback resistances R_E a virtual ground (small-signal ground). The bias currents are set by means of a current source $2I_0$: $I_{C5,A} = I_{C6,A} = I_0$. Ideally, the LO voltage V_{LO} is rectangular and is supplied to the differential amplifiers (T_1, T_2 and T_3, T_4) which operate as switches with opposite polarity. This portion of the circuit is called the *Gilbert cell*. Two collector resistances R_C are used instead of the RF filter; therefore, the output voltages are not filtered at this point, and in addition to the desired RF components, they still contain all other components that are generated in the conversion process. Common-collector circuits at the output are usually used as impedance converters. Then the RF filters follow; in most cases dielectric or SAW filters are used.

The doubled balanced mixer in Fig. 28.48 corresponds to a differential amplifier with current feedback and collector resistances in which the polarity between the IF inputs and the outputs can be reversed. Similar to a differential amplifier, the double balanced mixer allows asymmetric operation by supplying a constant potential to one of the two IF inputs or using only one output or a combination of both. Also, the LO input can be operated asymmetrically, although this has a negative effect on the distortion characteristic. For this reason, asymmetrical IF or LO voltages are usually converted into symmetric voltages by a balun or an asymmetric differential amplifier before they reach the mixer. This method is shown in Figs. 28.45b and 28.46 for the example of a single balanced mixer with asymmetric LO voltages. In the case of an asymmetric output, the collector resistance usually remains at the unused output.

Figure 28.49 shows the functional principle of the double balanced mixer. It can be seen that both single balanced mixers are inversely modulated with half the IF voltage each.

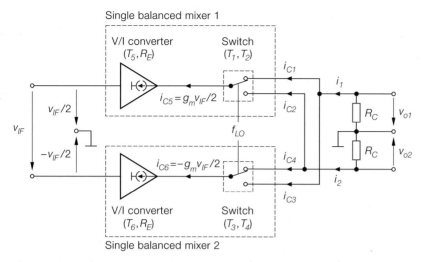

Single balanced mixer 1

V/I converter
(T_5, R_E)

Switch
(T_1, T_2)

$i_{C5} = g_m v_{IF}/2$

i_{C1}

i_1

$v_{IF}/2$

i_{C2}

R_C v_{o1}

v_{IF}

f_{LO}

$-v_{IF}/2$

$i_{C6} = -g_m v_{IF}/2$

i_{C4}

R_C v_{o2}

i_2

i_{C3}

V/I converter
(T_6, R_E)

Switch
(T_3, T_4)

Single balanced mixer 2

Fig. 28.49. Functional principle (= small-signal equivalent circuit) of a double balanced mixer with transistors

The double balanced mixer operates as a multiplicative mixer with a bipolar square-wave signal as a comparison with Fig. 28.12 on page 1376 shows.

Calculating the Transfer Characteristic

The calculation is the same as that for the single balanced mixer. The collector currents of the differential amplifier T_5, T_6 are

$$I_{C5} = I_0 + \frac{1}{2} g_m v_{IF} \quad , \quad I_{C6} = I_0 - \frac{1}{2} g_m v_{IF} \tag{28.47}$$

where

$$g_m = \frac{g_{m5}}{1 + g_{m5} R_E} = \frac{g_{m6}}{1 + g_{m6} R_E} \overset{g_{m5}=g_{m6}=I_0/V_T}{=} \frac{I_0}{V_T + I_0 R_E} \tag{28.48}$$

is the transconductance of the voltage-to-current converter. According to (28.30), the collector currents of transistor $T_1, \ldots, T_4$ are:

$$I_{C1} = \frac{I_{C5}}{2} \left(1 + \tanh \frac{V_{LO}}{2V_T} \right) \quad , \quad I_{C2} = \frac{I_{C5}}{2} \left(1 - \tanh \frac{V_{LO}}{2V_T} \right)$$

$$I_{C3} = \frac{I_{C6}}{2} \left(1 - \tanh \frac{V_{LO}}{2V_T} \right) \quad , \quad I_{C4} = \frac{I_{C6}}{2} \left(1 + \tanh \frac{V_{LO}}{2V_T} \right) \tag{28.49}$$

At the output of the Gilbert cell the currents are added:

$$I_1 = I_{C1} + I_{C3} \quad , \quad I_2 = I_{C2} + I_{C4} \tag{28.50}$$

The time responses can be calculated by inserting (28.47) and (28.49) into (28.50):

$$I_1(t) = I_0 + \frac{1}{2} g_m v_{IF}(t) \tanh \frac{V_{LO}(t)}{2V_T}$$

$$I_2(t) = I_0 - \frac{1}{2} g_m v_{IF}(t) \tanh \frac{V_{LO}(t)}{2V_T} \tag{28.51}$$

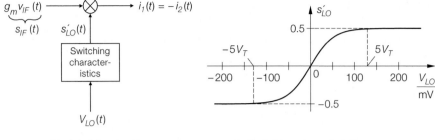

a Model of the double balanced mixer

b Switching characteristics

Fig. 28.50. Double balanced mixer as a multiplicative mixer taking into account the switching characteristics

It can be seen that in the double balanced mixer only the small-signal components are switched over while the bias current I_0 remains constant. This is an essential advantage as compared to the single balanced mixer in which the bias current is switched over as well (see (28.31) on page 1406). Due to this fact, we can restrict the following considerations to the small-signal currents:

$$i_1(t) = \underbrace{g_m v_{IF}(t)}_{s_{IF}(t)} \underbrace{\left[\frac{1}{2} \tanh \frac{V_{LO}(t)}{2V_T} \right]}_{s'_{LO}(t)} \quad , \quad i_2(t) = -i_1(t) \tag{28.52}$$

We can see that the double balanced mixer operates as a multiplicative mixer: the IF signal $s_{IF}(t)$ is multiplied by the LO signal $s'_{LO}(t)$. The LO signal $s'_{LO}(t)$ is derived from the voltage $V_{LO}(t)$ when taking the switching characteristics into consideration. Figure 28.50 illustrates the characteristic behavior.

The relationship between a rectangular or sinusoidal LO voltage $V_{LO}(t)$ and the LO signal $s'_{LO}(t)$ has already been explained for the single balanced mixer. In the double balanced mixer $s'_{LO}(t)$ contains no DC component since the plot of the switching characteristics is symmetrical to the original.[6] Thus, the coefficient c_0 of the Fourier series expansion for $s'_{LO}(t)$ becomes zero. From (28.32) it follows:

$$s'_{LO}(t) = c_1 \cos \omega_{LO} t + c_3 \cos 3\omega_{LO} t + c_5 \cos 5\omega_{LO} t + \cdots \tag{28.53}$$

Coefficients c_1, c_3, ... have the same values as in the single balanced mixer.[7] For a rectangular LO voltage of the amplitude $\hat{v}_{LO}$, the coefficients according to (28.33) are:

---

[6] In order to transpose the characteristic of the single balanced mixer in Fig. 28.37b into the characteristic of the double balanced mixer in Fig. 28.50b, it is necessary to mirror the V_{LO} axis in addition to a vertical shift by $1/2$. The reason for this is that the characteristic of the double balanced mixer is based on current I_1 of transistor T_1 while in the single balanced mixer it is based on current I_{C2} of transistor T_2.

[7] Literature often states that the coefficients c_1, c_3, ... in the double balanced mixer are twice as high as in the single balanced mixer. In this case, the factor $1/2$ in (28.52) is not considered to be a part of $s'_{LO}(t)$ but is treated separately. The coefficients are then higher by a factor of 2 but this is compensated for in the course of the calculation by the separate factor $1/2$. In this context the definition of the transconductance g_m must be thoroughly examined in order to determine whether a symmetric or asymmetric output signal is used.

$$c_1 = \frac{2}{\pi} \tanh \frac{\hat{v}_{LO}}{2V_T} \quad , \quad c_3 = -\frac{2}{3\pi} \tanh \frac{\hat{v}_{LO}}{2V_T} \quad , \quad \cdots$$

Figure 28.39 on page 1407 presents the coefficients c_1 and $|c_3|$ for a rectangular and a sinusoidal LO voltage.

Small-Signal Response

We can now calculate the small-signal output voltages

$$v_{o1}(t) = -R_C\, i_1(t) \quad , \quad v_{o2}(t) = -R_C\, i_2(t) = -v_{o1}(t)$$

for a sinusoidal IF voltage:

$$v_{IF}(t) = \hat{v}_{IF} \cos \omega_{IF} t$$

Inserting the small-signal currents from (28.52) and the Fourier series expansion from (28.53) leads to

$$v_{o1}(t) = -g_m R_C\, \hat{v}_{IF} \cos \omega_{IF} t \,[\, c_1 \cos \omega_{LO} t + c_3 \cos 3\omega_{LO} t + \cdots \,]$$

$$= -\frac{c_1}{2} g_m R_C\, \hat{v}_{IF} \,[\cos(\omega_{LO} + \omega_{IF})t + \cos(\omega_{LO} - \omega_{IF})t\,]$$

$$-\frac{c_3}{2} g_m R_C\, \hat{v}_{IF} \,[\cos(3\omega_{LO} + \omega_{IF})t + \cos(3\omega_{LO} - \omega_{IF})t\,]$$

$$-\cdots$$

with the RF component:

$$v_{RF}(t) = v_{o1}(t)\Big|_{f=f_{RF}=f_{LO}+f_{IF}} = -\frac{c_1}{2} g_m R_C\, \hat{v}_{IF} \cos \omega_{RF} t \tag{28.54}$$

Figure 28.51 shows the related spectrum which corresponds to the spectrum of a multiplicative mixer with bipolar square-wave signal in Fig. 28.13c on page 1378. It contains no undesirable components at the LO frequency f_{LO} and at multiples thereof as they are caused in the single balanced mixer by switching the bias current. This can be seen in a comparison with Fig. 28.40 on page 1409.

The maximum value of the output voltage $v_{o1}(t)$ is

$$v_{o1,max} = max|v_{o1}(t)| = \frac{1}{2} g_m R_C\, \hat{v}_{IF}$$

and with an ideal switch ($c_1 = 2/\pi$), it exceeds the amplitude of the RF component in (28.54) by a factor of $1/c_1 = \pi/2 \approx 1.57$ (4 dB) only. This allows the entire output

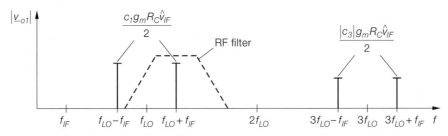

Fig. 28.51. Spectrum of the output voltage $v_{o1}(t)$ for a sinusoidal IF voltage

voltage to be further processed without major limitations to the dynamic range and the RF component to be filtered out at a later stage. The demands on the RF filter are lower than in the single balanced mixer since there is no component at the LO frequency as can be seen by comparing Fig. 28.51 with Fig. 28.40 on page 1409.

Mixer Voltage Gain

The RF voltage is obtained from (28.54):

$$\underline{v}_{RF} = -\frac{c_1}{2} g_m R_C \underline{v}_{IF} \tag{28.55}$$

This leads to the *mixer voltage gain:*

$$A_M = \frac{\underline{v}_{RF}}{\underline{v}_{IF}} = -\frac{c_1}{2} g_m R_C \overset{(28.48)}{=} -\frac{c_1}{2} \frac{g_{m5} R_C}{1 + g_{m5} R_E} \tag{28.56}$$

Only the RF component of the output voltage $v_{o1}(t)$ is taken into account when determining the mixer voltage gain, which thus corresponds to the differential gain A_D of a differential amplifier. In most cases, however, the differential output voltage $v_o(t) = v_{o1}(t) - v_{o2}(t)$ is used. The mixer voltage gain is then twice as high:

$$A_{M,diff} = 2A_M = -c_1 g_m R_C \tag{28.57}$$

In the following, we refer to A_M as the *single-ended mixer voltage gain* and to $A_{M,diff}$ as the *differential mixer voltage gain*.

The single-ended mixer voltage gain A_M of the double balanced mixer corresponds to the mixer voltage gain of the single balanced mixer in (28.35) on page 1409 if we assume that $R_C = R_{L,RF}$, i.e. equal load resistances for the RF component. Typical values are in the range of $|A_M| \approx 2 \dots 10$ ($6 \dots 20$ dB).

Bandwidth

With regard to bandwidth, basically the same considerations apply to the double balanced mixer as to the single balanced mixer. However, in a double balanced mixer with collector resistances and subsequent impedance converters, it is not possible to compensate for the output capacitances of transistors $T_1, \dots, T_4$ because, together with the collector resistances, they form lowpass filters and thus limit the bandwidth of the output. This is particularly apparent in up-conversion mixers with output signals of a high frequency f_{RF}. This negative effect is much less in down-conversion mixers whose output frequency f_{IF} is significantly lower. To solve this, one can increase the transconductance g_m and reduce the resistances R_C accordingly; however, this is at the cost of the current consumption.

As an alternative, inductances can be used to compensate the capacitances; Fig. 28.52 shows two ways of doing this. In both cases we obtain parallel resonant circuits at the outputs which are tuned to the RF frequency in up-conversion mixers so that their function corresponds to that of the RF filter in a single balanced mixer. This method is particularly interesting in integrated circuits if the required inductances are so small that they can be integrated or implemented by means of bond wires; otherwise external inductances must be used.

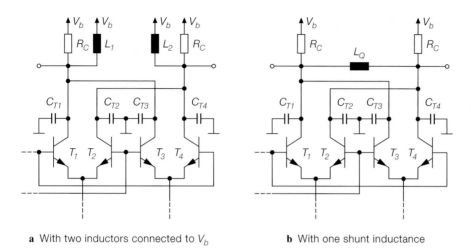

a With two inductors connected to V_b **b** With one shunt inductance

Fig. 28.52. Compensation of the output capacitances of transistors $T_1, \ldots, T_4$ by inductances tuned to resonance

Double Balanced Mixers with Integrated Circuits

In integrated circuits, the double balanced mixer is often constructed with additional amplifiers; Fig. 28.53 shows a typical model. Matching to the characteristic impedance of external lines is required only at the input and output of the integrated circuit. The mixer itself is operated without impedance matching. The conversion of asymmetric external voltages into symmetric voltages for the mixer is done by means of asymmetrically operated differential amplifier stages in the three amplifiers. Since it is necessary to dimension the input and output amplifiers for a certain frequency range, integrated circuits of this type are mostly suitable for a narrow frequency band only. This is not true for the LO amplifier which can be designed as a broadband limiting amplifier. Figure 28.54 shows an example with common-base circuits for impedance matching at the inputs.

A mixer in an integrated circuit is described by the mixer voltage gain and the input and output impedances at the three pairs of terminals. Quoting the conversion gain (power gain) is not useful due to the unmatched operating mode.

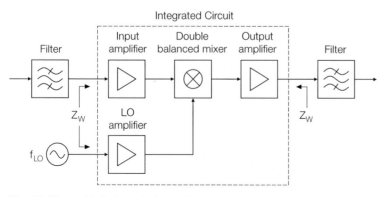

Fig. 28.53. Double balanced mixer with amplifiers in an integrated circuit

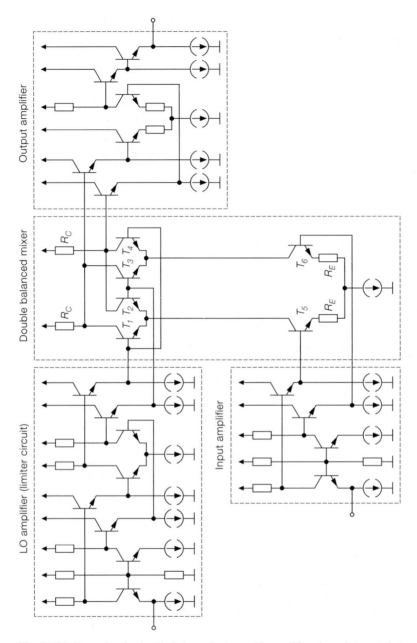

Fig. 28.54. Example of a double balanced mixer with amplifiers in an integrated circuit

Impedance Matching

In order to allow universal use of integrated double balanced mixers, they are used without input and output amplifiers. In this case, the input and output of the mixer must be matched to the characteristic impedance. To do this, the same methods are used as in single balanced mixers. Figure 28.55 shows some examples of impedance matching at the

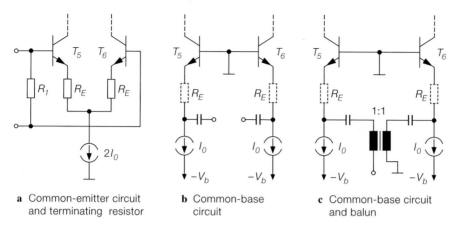

a Common-emitter circuit
and terminating resistor

b Common-base
circuit

c Common-base circuit
and balun

Fig. 28.55. Examples of impedance matching at the input of a double balanced mixer

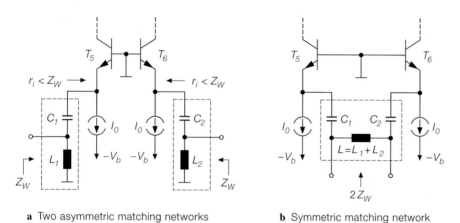

a Two asymmetric matching networks

b Symmetric matching network

Fig. 28.56. Impedance matching at the input of a double balanced mixer by means of matching networks

input side. As in the single balanced mixer, common-base circuits are often used instead of common-emitter circuits. Where an asymmetric input is required, a balance-to-unbalance transformer (balun) can be added. The matching networks described in Sect. 28.3.1 can also be used as an alternative. In the case of a symmetric input, either two asymmetric or one symmetric matching network may be applied. Fig. 28.56 illustrates the use of the matching network from Fig. 28.22b on page 1392 for a step-up transformer from $r_i < Z_W$ to Z_W.

The matching networks from Sect. 28.3.1 are also used at the output. Figure 28.57 shows the use of the matching network from Fig. 28.23b on page 1392 for a step-down transformation. In down- or up-conversion mixers with low RF frequencies, the output impedance of transistors $T_1, \ldots, T_4$ is very high at the output frequency. In this case, the collector resistances are required to limit the voltage amplitudes at the collectors; at the same time they allow a feasible transformation ratio R_C/Z_W. This is shown in Fig. 28.57a in connection with a symmetric matching network. In up-conversion mixers with high RF frequencies, the output impedance of the transistors is often so low that the

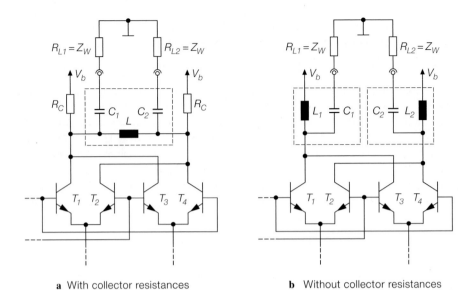

a With collector resistances   **b** Without collector resistances

Fig. 28.57. Impedance matching at the output of a double balanced mixer by means of matching networks

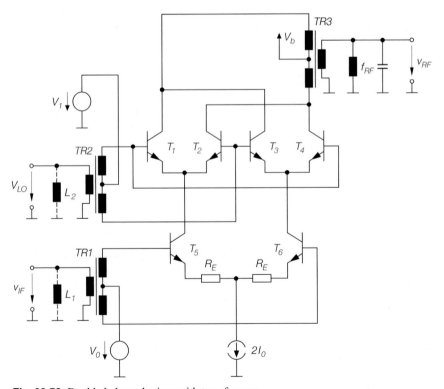

Fig. 28.58. Double balanced mixer with transformers

collector resistances can be omitted. Matching is then achieved as in Fig. 28.57b with two asymmetric matching networks since the inductances of the impedance matching networks can be used to supply the operating voltage at the same time.

Besides baluns, 1:1:n and n:n:1 transformers are also used to convert asymmetric signal sources and loads to symmetric inputs and outputs of double balanced mixers. Impedance matching can then be achieved fully or in part by a suitable choice of the transformer ratio. Figure 28.58 shows an example with three transformers. Since the input admittance of the transformers is ohmic-capacitive, the primary side of transformers $TR1$ and $TR2$ also carry capacitive admittances; this necessitates an additional compensation of the capacitive component to match the characteristic impedance. In the most simple case, this can be done by tuning the inductances L_1 and L_2 to resonance. The output admittance on the secondary side of transformer $TR3$ also has a capacitive component which, however, can be regarded as an integral portion of the RF filter.

Conversion Gain

To calculate the conversion gain of a circuit matched at both sides, we combine the collector resistances R_C and the output resistances of transistors $T_1, \ldots, T_4$ into two loss resistances R_V. The load resistances $R_{L1} = R_{L2} = Z_W$ are transformed by the matching networks into two resistances R_P which are connected in parallel to the loss resistances. In the matched state we have $R_V = R_P$. Figure 28.59 shows the transformation at one of the two outputs. At both outputs this produces the same conditions as at the output of a single balanced mixer (see Fig. 28.42 on page 1411). According to (28.40) the conversion gain is:

$$G_M = \frac{A^2 Z_W}{r_o} \overset{r_o=R_V}{=} \frac{A^2 Z_W}{R_V} \qquad (28.58)$$

Here, Z_W is the input resistance at *one* input and R_V is the transformed load resistance at *one* output. Therefore, it is necessary to use either the open-circuit gain from *one* input to *one* output or the differential open-circuit gain for the open-circuit gain A in the equation. The differential open-circuit gain is determined from the differential mixer voltage gain $A_{M,diff}$ found by replacing R_C by R_V:

$$A = A_{M,diff}\Big|_{R_C=R_V} \overset{(28.57)}{=} -c_1 g_m R_V$$

By insertion into (28.58), the *conversion gain of the double balanced mixer with impedance*

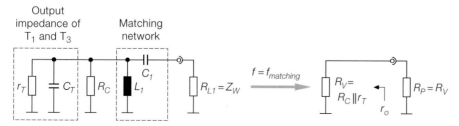

Fig. 28.59. Small-signal equivalent circuit for transformation of the load resistance at one of the two outputs

matching at both sides is obtained:

$$G_M = \frac{1}{4} c_1^2 g_m^2 Z_W R_V \tag{28.59}$$

A comparison with the conversion gain of a single balanced mixer in (28.44) shows that, with the same loss resistances R_V, the conversion gain of the double balanced mixer is four times higher. The same loss resistances, however, can be assumed only at low frequencies because only then are the output resistances of the transistors negligible and the loss resistances correspond to the collector resistances. In this case, the double balanced mixer achieves double the output voltage and four times the output power due to its differential output, but at high frequencies the output resistances of the transistors have a dominating effect. Since there are two transistors connected in parallel at each output of the double balanced mixer, the loss resistances then have half the value of that for the single balanced mixer in Fig. 28.42. In this case, the conversion gain of the double balanced mixer is only twice as high as that of the single balanced mixer.

I/Q Mixer with Double Balanced Mixers

The double balanced mixer is particularly suitable for operation as I/Q mixers in digital modulators and demodulators; in both cases two mixers are required. Figure 28.60 shows the respective location of the mixers which were adopted from Fig. 24.68 on page 1212 and Fig. 24.69 on page 1213.

In I/Q mixers, the RF and LO frequencies of both mixers are identical to the carrier frequency f_C of the carrier signal $s_C(t)$: $f_{RF} = f_{LO} = f_C$. The quadrature components $i(t)$ and $q(t)$ are baseband signals with the carrier frequency zero: $f_{IF} = 0$. In this case, there is no image frequency since the RF and image frequencies coincide due to $f_{IF} = 0$: $f_{RF} = f_{LO} \pm f_{IF} = f_{LO} \mp f_{IF} = f_{RF,im}$. An I/Q mixer only operates correctly if the mixer voltage gains of both mixers are identical and the phase shift between the two LO signals is 90°. Without balancing, these conditions can be met only if both mixers, including the components for generating the LO signals, are combined in *one* integrated circuit. For this purpose, only the double balanced mixer of Fig. 28.48 on page 1418 is used because it needs no filters directly at the mixer and can thus be used without external components.

The I/Q down-conversion mixer shown in Fig. 28.60b uses two double balanced mixers that are connected at their inputs; the output signals are processed separately. For the I/Q up-conversion mixer shown in Fig. 28.60a, the output signals of both double balanced mixers must be added. This summation can be done without additional circuitry by adding the output currents instead of the output voltages and using common-collector resistances according to Fig. 28.61. Additionally, each of the two output voltages or the differential output voltage can be considered to be the output signal $s_M(t)$ of the I/Q up-conversion mixer.

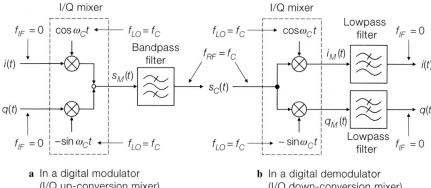

a In a digital modulator
(I/Q up-conversion mixer)

b In a digital demodulator
(I/Q down-conversion mixer)

Fig. 28.60. I/Q mixers

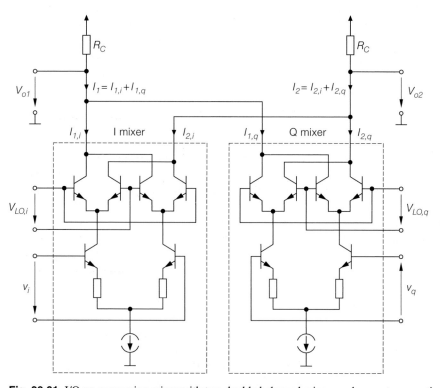

Fig. 28.61. I/Q up-conversion mixer with two double balanced mixers and current summation

Chapter 29:
Appendix

29.1
PSpice – Brief User's Guide

29.1.1
General

PSpice from *OrCAD* (previously *MicroSim*) is a circuit simulator from the *Spice* family (*Simulation Program with Integrated Circuit Emphasis*) for the simulation of analog, digital, and mixed analog/digital circuits. In 1970 *Spice* was developed at Berkely University and is available today in the version 3F5 for use without licence. On this basis, commercial offshoots evolved which contain specific expansions and additional modules for entering circuits graphically, presenting results and controlling processes. The most common ones are *PSpice* and *HSpice*. While *HSpice* from *Synopsys* (previously *Meta Software*) was designed for developing integrated circuits comprising several thousand transistors and is used in many IC design systems as the simulator, *PSpice* is a particularly well-priced and easy to operate simulation environment for developing small and medium-sized circuits on PCs with a Microsoft Windows operating system.

The following short instructions apply to the demo version of *PSpice 8* (*PSpice Eval 8*) for *Microsoft Windows 98/ME/2000/XP and Vista*.

29.1.2
Programs and Files

Spice

Every simulator in the *Spice* family uses *netlists*. A netlist is a description of a circuit prepared by an editor which contains component lists and circuit topology data augmented by simulation instructions and references to model libraries. Fig. 29.1.1 shows the programs and files involved in the circuit simulation process.

– The netlist of the circuit to be simulated is prepared by an editor and stored in the circuit file *<name>.CIR* (*CIRcuit*).
– The circuit file is read in by the simulator (*PSpice or Spice 3F5*), which then performs the simulations according to the simulation instructions; this may include the use of models from component libraries *<xxx>.LIB* (*LIBrary*).
– The simulation results and (error) messages are stored in the output file *<name>.OUT* (*OUTput*) and can be displayed and printed with the aid of an editor.

PSpice

In addition to the simulator *PSpice*, the *PSpice* software package also contains a program for the graphic input of circuit diagrams (*Schematics*) and a program for the graphic representation of the simulation results (*Probe*). Figure 29.1.2 shows the process with the programs and files involved:

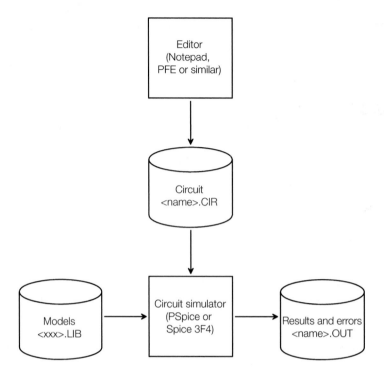

Fig. 29.1.1. Programs and files used in *Spice*

- The *Schematics* program allows you to enter the circuit diagram for the circuit to be simulated and store it in the schematics file *<name>.SCH* (*SCHematic*); this process uses schematic symbols from symbol libraries *<xxx>.SLB* (*Schematic LiBrary*).
- By starting the simulation (*Analysis/Simulate*) or creating the netlist (*Analysis/Create Netlist*), the Schematics program generates the circuit file *<name>.CIR*; at the same time, the netlist is stored in the file *<name>.NET* and incorporated by an *Include* instruction. Another file *<name>.ALS* is generated which contains a list of alias names, but is of no relevance for the user.
- The *PSpice* simulator is activated in the Schematics program by starting the simulation (*Analysis/Simulate*); as an alternative *PSpice* can be started manually and the circuit file can be selected with *File/Open*. The simulation uses models from the component libraries *<xxx>.LIB*.
- The simulation results that may be presented graphically are stored in the data file *<name>.DAT*; nongraphic results and messages are stored in the output file *<name>.OUT* and can be displayed by selecting *Analysis/Examine Output* in the Schematics program or using an external editor.
- The simulation results can be presented graphically using the *Probe* program; this allows you to directly display individual signals or to perform calculations with one or more signals. The commands required to create a graph may be stored in the display file *<name>.PRB* using the *Options/Display Control* function and are available for later retrieval. If the simulation has been started via *Analysis/Simulate* in the *Schematics* program, the *Probe* program will start automatically upon completion of the simulation;

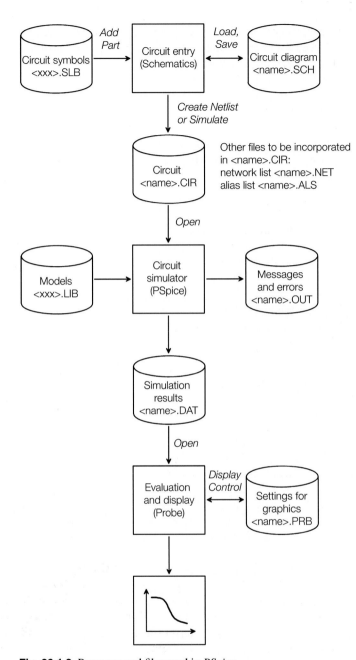

Fig. 29.1.2. Programs and files used in *PSpice*

in this case the data file *<name>.DAT* is loaded automatically. If it has been started manually, the data file must be selected via *File/Open*.

With *PSpice* you can also work with netlists directly. To do so the circuit diagram should not be entered graphically; instead, the circuit file *<name>.CIR* should be created with the help of an editor. Unlike *Spice,* this offers the advantage of being able to represent the simulation results graphically using *Probe*. This procedure is often used when creating new models since the experienced user can eliminate any errors that may occur during model testing much faster in the circuit file than via the graphic diagram entry.

29.1.3
A Simple Example

A small-signal amplifier with AC coupling is taken as an example to illustrate how a circuit diagram is entered and circuit simulation is performed; Fig. 29.1.3 shows the corresponding circuit diagram.

Entering the Circuit Diagram

The *Schematics* program is activated in order for the circuit diagram to be entered; Fig. 29.1.4 shows the program window. The tool bar contains (from left to right) the *File* operations *New*, *Open*, *Save* and *Print*, the *Edit* operations *Cut*, *Copy*, *Paste*, *Undo* and *Redo*, and the *Draw* operations *Redraw*, *Zoom In*, *Zoom Out*, *Zoom Area* and *Zoom to Fit Page*. All are used in the usual way.

The diagram is entered by performing the following steps:

- Insert components
- Configure components
- Draw connecting wires

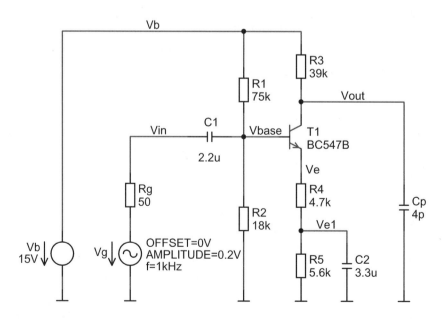

Fig. 29.1.3. Circuit diagram for our example

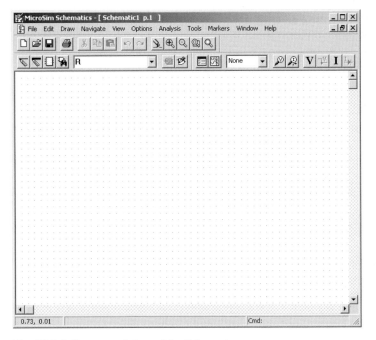

Fig. 29.1.4. Program window of the *Schematics* program

This requires the following tools:

Step		Tool	Action
1		*Get New Part*	Inserts components
2		*Edit Attributes*	Configures the components
3		*Draw Wire*	Adds connecting wires
4		*Setup Analysis*	Enters the simulation instructions
5		*Simulate*	Starts the simulation

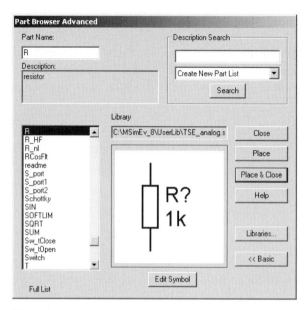

Fig. 29.1.5. *Get New Part* dialog

Inserting the components: Use the *Get New Part* tool to open the dialog window *Part Browser Basic*; use *Advanced* to open the *Part Browser Advance* dialog window shown in Fig. 29.1.5. If the component name is known, it can be entered in the *Part Name* field; the component appears in the preview window and can be accepted by clicking on the *Place* or *Place & Close* button. If the name is not known, the component list must be searched to find the desired component. The *Libraries* button opens a dialog window containing the component list arranged according to libraries; here, however, a preview is not shown until a component has been selected and confirmed with *Ok*.

After accepting the component with the *Place* or *Place & Close* button the part is inserted into the diagram by clicking the left mouse button. Prior to insertion the symbol of the component can be rotated using *Ctrl-R* and/or mirrored using *Ctrl-F*. The insert mode continues until you have clicked the right mouse button or pressed the Esc key.

The names of some important passive and active components are listed in the table on page 1437

Configuring the components: Most components have to be configured following insertion. In the case of passive components such as resistors, capacitors and inductors this means entering the value; in the case of voltage and current sources this involves entering the parameters of the signal shape (amplitude, frequency, etc.) and for controlled sources the control ratio must be entered. For integrated transistors the size of the transistor and the name of the substrate node must be entered while for operational amplifiers parameters like transit frequency and slew-rate are needed. Some components such as standard transistors (e.g. BC547B) need not be configured since they are assigned a reference to a model in a model library which contains all the data required.

The value of a passive component can be changed by double clicking on the indicated value; this opens the *Set Attribute Value* option for entering the value (see Fig. 29.1.6).

Clicking on *Edit Attributes* or double clicking the component symbol opens the *Part* dialog window shown in Fig. 29.1.7 which contains a list of all parameters. Parameters not

The names of some important passive and active components:

Name	Description	Library
R	Resistor	TSE_ANALOG.SLB
C	Capacitor	
L	Inductor	
K	Inductive coupling	
E	Voltage-controlled voltage source	
F	Current-controlled current source	
G	Voltage-controlled current source	
H	Current-controlled voltage source	
Tr	Ideal transformer	
V	General purpose voltage source	
V_AC	Small-signal voltage source	
V_bias	DC voltage source	
V_pulse	Large-signal pulse voltage source	
V_sine	Large-signal sine-wave voltage source	
V_square	Large-signal square-wave voltage source	
V_triangle	Large-signal triangle voltage source	
I	General purpose current source	
I_bias	DC current source	
GND	Ground	
1N4148	Small-signal diode 1N4148 (100mA)	TSE_BIPOLAR.SLB
1N4001	Rectifier diode 1N4001 (1A)	
BAS40	Small-signal Schottky diode BAS40	
BC547B	npn small-signal transistor BC547B	
BC557B	pnp small-signal transistor BC557B	
BD239	npn power transistor BD239	
BD240	pnp power transistor BD240	
BF245B	n-channel JFET BF245B	TSE_FET.SLB
IRF142	n-channel power MOSFET IRF142	
IRF9142	p-channel power MOSFET IRF9142	
N1	Integrated npn transistor	TSE_INTEGRATED.SLB
P1	Integrated pnp transistor	
NMOS	Integrated n-channel MOSFET	
PMOS	Integrated p-channel MOSFET	
VV	Operational amplifier	TSE_MODEL.SLB
VC	Transconductance operational amplifier	
CV	Current-feedback operational amplifier	

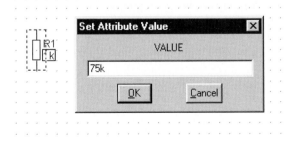

Fig. 29.1.6. Dialog *Set Attribute Value* dialog

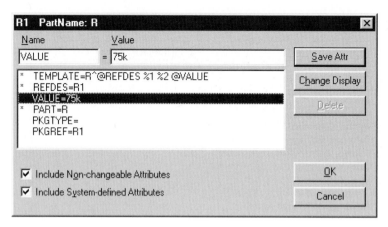

Fig. 29.1.7. *Part* dialog

marked by an asterisk can be selected, modified in the *Value* field and saved by pressing the *Save Attr* button. Pressing *Change Display* allows you to ascertain whether or not and how the selected parameters are presented in the circuit diagram; usually only the value, e.g. 1*k*, or the parameter name and the value, e.g. $R = 1k$, is displayed.

Numeric values can be entered as an exponential expression (e.g. *1.5E-3*) or can be provided with the following suffixes:

Suffix	f	p	n	u	m	k	Mega	G	T
Name	Femto	Pico	Nano	Micro	Milli	Kilo	Mega	Giga	Terra
Value	10^{-15}	10^{-12}	10^{-9}	10^{-6}	10^{-3}	10^3	10^6	10^9	10^{12}

There is no distinction between upper and lower case letters. A common error is the use of *M* for *Mega* which is then interpreted by *PSpice* as *Milli*.

Adding the connecting wires: After all the components have been placed and configured, the connecting wires must be entered using the *Draw Wire* tool; here, the cursor takes the form of a pencil. First fix the starting point of the wire with a left mouse click. The wire is then shown as a broken line and can be confirmed from point to point by clicking the left mouse button (see Fig. 29.1.8). The most simple case is a straight line between starting point and end point; in this case the course is determined automatically. By fixing intermediate points the course can be influenced. Placing a point on a component terminal or any other wire causes the program to assume that the wire entry is complete and conclude the entry. As an alternative the entry can be interrupted at any given point by clicking the right mouse button or pressing *Esc*.

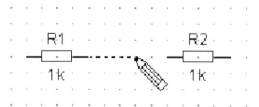

Fig. 29.1.8. Drawing a connecting wire

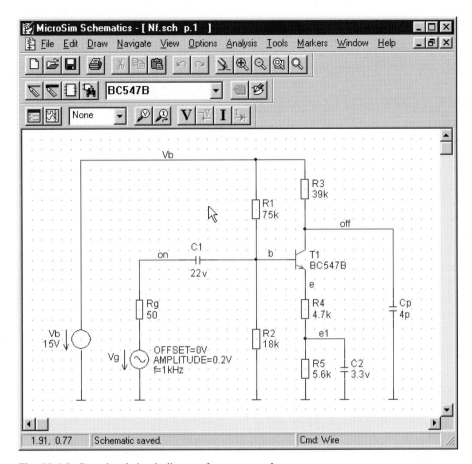

Fig. 29.1.9. Completed circuit diagram for our example

Normally ground wires are not drawn. Each point connected to ground is marked with a ground symbol GND. In the netlist the nodal name 0 is assigned to any point that is part of the GND net.

A nodal point 0 must exist in every diagram; every circuit diagram must therefore contain at least one ground symbol GND.

A name is assigned automatically to every node. These names appear in the netlist and are required in the *Probe* program to select the signals to be displayed. Since the names assigned automatically do not appear in the circuit diagram and are thus not known without referencing the netlist, an appropriate name should be given to any nodal point in the diagram that is of particular interest; this is done by double clicking on one of the wires that is connected to the node and entering an appropriate name.

After all the components have been entered, connecting wires inserted and nodal names assigned, the completed circuit diagram is presented as shown in Fig. 29.1.9; if you have not already done so, store the diagram by clicking on the *File/Save* button.

Entering Simulation Instructions

This step describes the simulations to be performed and the parameters for controlling the voltage and current sources used. There are three simulation methods that use different sources:

- *DC Sweep (DC voltage analysis):* This analysis examines the DC response of a circuit; in this process one or two of the sources are varied. The result is a characteristic curve or a family of characteristics. In this analysis only DC voltage sources and DC components of other sources (parameter *DC=*) are taken into consideration.
- *AC Sweep (Small-signal analysis):* This analysis examines the small-signal response. First the operating point of the circuit is determined with the help of the DC sources and the circuit is linearized at this operating point. Then the frequency response is determined for the given frequency range using complex-valued linear AC circuit analysis. In this second step, only the small-signal components of the sources are taken into consideration (parameter *AC=*). As the small-signal analysis is linear there is a linear relation between the result and the given amplitudes of the sources; therefore a normalized amplitude of 1V or 1A, i.e. *AC=1* is commonly used.
- Transient (Large-signal analysis): This analysis covers the large-signal response; it determines the temporal characteristic of all voltages and currents by numeric integration. This analysis only takes into consideration large-signal sources and the large-signal components of other sources.

In our example a small-signal analysis is to be performed to determine the small-signal frequency response, as well as a large-signal analysis with a sine-wave signal with an amplitude of 0.2 V and a frequency of 1 kHz. In this case a large-signal voltage source *V_sine* with the additional *AC* parameter is used at the input (see circuit diagram in Fig.29.1.9). Figure 29.1.10 shows the source parameters that result from the values specified.

Simulation instructions are required in addition to the source settings; these instructions determine the analyses to be performed and the parameters used in each analysis:

- *DC Sweep:* name and range of the source(s) to be varied
- *AC Sweep:* frequency range

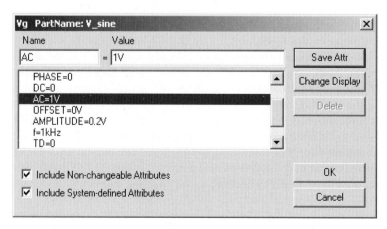

Fig. 29.1.10. Parameters of the driving voltage source

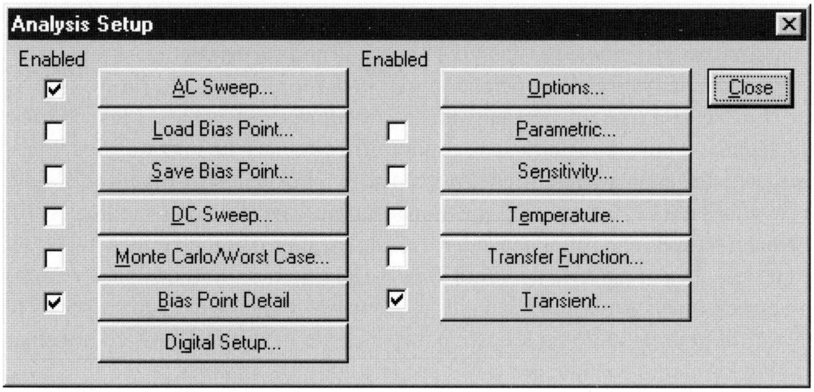

Fig. 29.1.11. List of analysis options

– *Transient:* length of the time interval to be simulated and, if applicable, the increment for numeric integration

The simulation instructions are finalized with the *Setup Analysis* tool. Here, a list of analysis set-up options appears. This list is shown in Fig. 29.1.11. In addition to the *AC Sweep, DC Sweep* and *Transient* analyses already described, there are other analysis types and capabilities which will be explained later. The *Bias Point Detail* analysis calculates the operating point on the basis of the DC sources and saves the result in the output file *<name>.OUT*; this analysis is activated by default. For our example the *AC Sweep* and *Transient* analyses must be activated.

Selecting the field *AC Sweep* opens the *AC Sweep* dialog shown in Fig.29.1.12, in which the frequency range is entered. In our example the frequency range 1 Hz to 10 MHz with 10 points per decade is to be investigated.

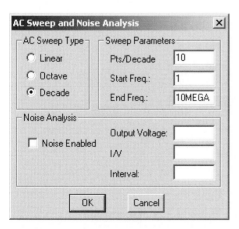

Fig. 29.1.12. Setting the frequency range for *AC Sweep*

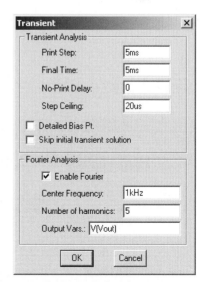

Fig. 29.1.13. Setting the parameters for *Transient*

Selecting the *Transient* field opens the *Transient* dialog shown in Fig. 29.1.13. Here, the end point of the simulation is entered in the *Final Time* field and the maximum increment for numeric integration in the *Step Ceiling* field. The start time for recording the results is to be entered in the *No-Print Delay* field; normally this is 0 to allow every value calculated to be displayed graphically. If only the steady-state condition of circuits with long transient times is to be investigated, the estimated transient time can be entered in the *No-Print Delay* field so that recording does not begin until after the transient time. The *Print Step* parameter still exists for historical reasons and is not needed. Nevertheless, this parameter must not be set to 0 and must be less than or equal to the *Final Time*. In addition, the output signal V(Vout) undergoes a Fourier analysis at a fundamental frequency of 1 kHz, which corresponds to the frequency of the source; in doing so, five harmonics are determined, which, together with the resulting harmonic content, are saved in the output file *<name>.OUT*.

After the simulation instructions have been entered, the schematics file is complete and can be stored by pressing the *File/Save* button.

Starting the Simulation

The *Simulate* tool starts the simulation; first the netlist is generated, after which the *PSpice* simulator is started. The *PSpice* window appears during the simulation process; the window shown in Fig. 29.1.14 appears when the simulation is complete.

Displaying the Results

If no errors have occurred during simulation, the Probe program will start automatically. If the simulation contains several analyses, the window shown in Fig. 29.1.15 for selecting an analysis will appear; selecting *AC* opens the *AC* window shown in Fig.29.1.16, which already contains a frequency scale corresponding to the frequency range simulated.

The *Add Trace* tool is used to select the signals to be displayed:

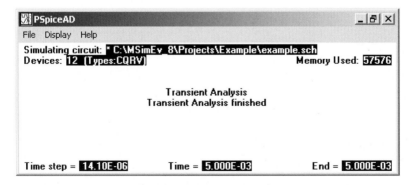

Fig. 29.1.14. *PSpice* window following completion of the simulation

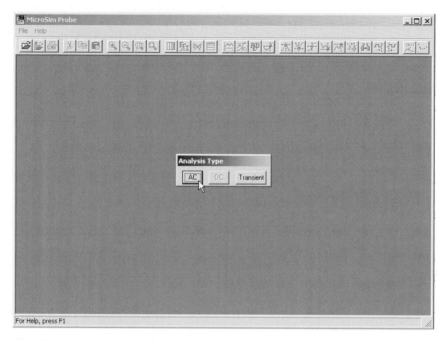

Fig. 29.1.15. Selecting the analysis type in the *Probe* start-up window

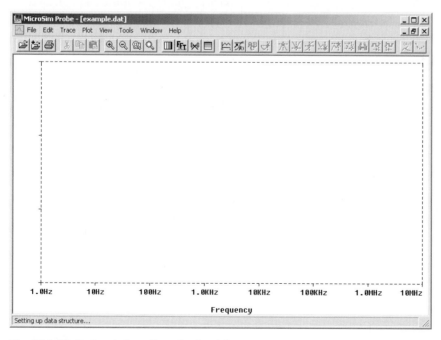

Fig. 29.1.16. *Probe* window after selecting *AC*

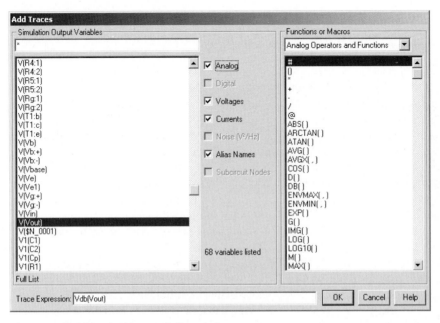

Fig. 29.1.17. Dialog *Add Traces* dialog window

Figure 29.1.17 shows the *Add Traces* dialog window displaying a list of signals on the left and a selection of mathematical functions on the right. The following designations are used:

Designation	Example	Meaning
I(<component>)	I(R1)	Current flowing through a component with two terminals, e.g. the current through resistor R1
I<terminal>(<component>)	IB(T1)	Current flowing through the terminal of a component, e.g. the base current of transistor T1
V(<node name>)	V(Vout)	Voltage between a node and ground, e.g. the voltage at the *Vout* node
V(<component:terminal>)	V(C1:1)	Voltage at the terminal of a component, e.g. the voltage at terminal 1 of capacitor C1
V<terminal>(<component>)	VB(T1)	Voltage at the terminal of a component, e.g. the voltage at the base terminal of transistor T1

Click on the desired signals or functions to select. These will then be shown in the *Trace Expression* field, where they can be edited if necessary. The following options are available for *AC* signals:

Display	Magnitude	Magnitude in dB	Phase
Example	M(V(Vout)) VM(Vout) V(Vout)	DB(V(Vout)) VDB(Vout)	P(V(Vout)) VP(Vout)

In our example *Vdb(Vout)* represents the magnitude of the output voltage (see Fig. 29.1.18). Since the driving voltage source has an amplitude of 1 V ($AC = 1$) this value represents the small-signal gain of the circuit. The scale factor of the x and y-axes can be altered via the menu options *Plot/X Axis Settings* and *Plot/Y Axis Settings*.

Other signals that use the same scale factors can be added to the display without further alteration. If signals with different scale factors, e.g. the phase *Vp(Vout)*, are to be represented in a meaningful manner, it is first necessary to create another y-axis using the *Plot/Add Y Axis* menu option. The active Y-axis is marked with a » and can be selected with a mouse click; after using *Plot/Add Y Axis* the new y-axis is automatically active. Adding the phase *Vp(Vout)* displays the window shown in Fig. 29.1.19.

Finally we also wish to display the results of the large-signal transient analysis. To do so, we must switch over using the *Plot/Transient* menu option; an empty window appears which already has a time scale corresponding to the time interval simulated. If the voltages *V(Vin)*, *V(Vbase)*, *V(Ve)* and *V(Vout)* are added via the *Add Traces* option, the display shown in Fig. 29.1.20 appears.

The settings for a particular display can be saved with the *Tools/Display Control* menu option and can be retrieved at a later date. The settings for the various analyses are saved separately so that only those settings are shown that belong to the analysis selected. The settings used last can be called up via the *Last Session* option.

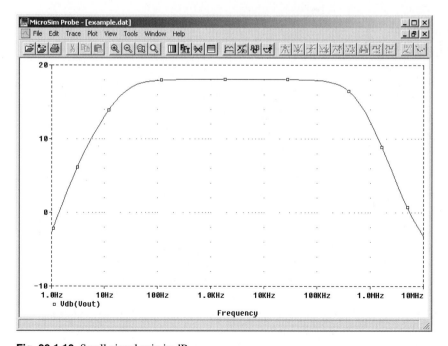

Fig. 29.1.18. Small-signal gain in dB

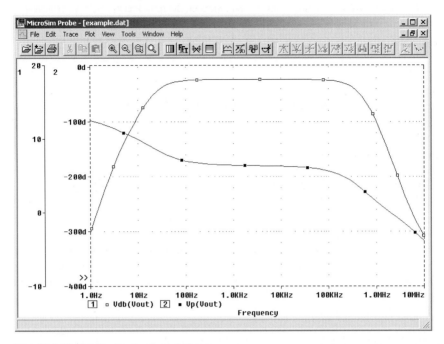

Fig. 29.1.19. Small-signal gain and phase

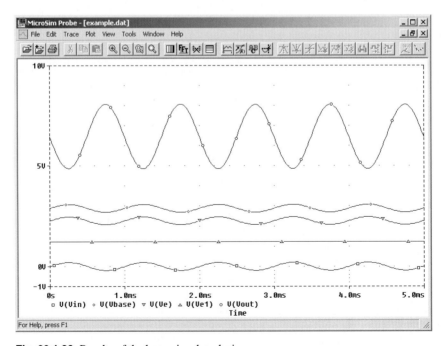

Fig. 29.1.20. Results of the large-signal analysis

The *Tools/Cursor/Display* menu option allows you to display two markers simultaneously, which are moved with the left or right mouse button; at the same time the x and y values of the marker position are shown in an additional window. For more information refer to *Cursor* in the help index. The markers can also be turned on and off with the *Toggle Cursor* tool:

|  | *Toggle Cursor* | Switches the markers on or off |

Indicating the operating point

Following a simulation the voltages and currents of the operating point can be shown in the circuit diagram (see Figs. 29.1.21 and 29.1.22); this is done with the folldue tools in the *Schematics* program:

| **V** | *Enable Bias Voltage Display* | Displays the operating point voltages |
| **I** | *Enable Bias Current Display* | Displays the operating point currents |

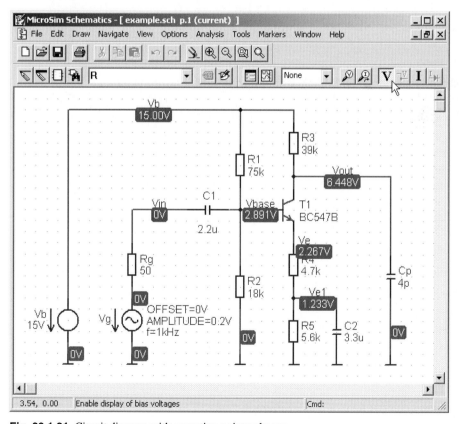

Fig. 29.1.21. Circuit diagram with operating point voltages

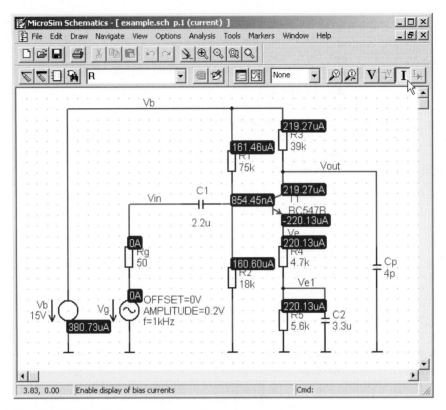

Fig. 29.1.22. Circuit diagram with operating point currents

After an extensive circuit diagram has been entered, the first step is usually to check the operating point by performing a simulation using a *Bias Point Detail* Analysis, which is activated by default, and to examine the results. This ensures that the circuit has been entered correctly and is functional, before other analyses, which may be somewhat time-consuming, are performed. When following this procedure the *Probe* display program does not start automatically because the *Bias Point Detail* analysis does not produce any graphic data.

Netlist and Output File

The files from our example include the following contents (shortened in some instances):

– **Circuit file EXAMPLE.CIR:**

```
** Analysis setup **
.ac DEC 10 1 10MEGA
.tran 5m 5m 0 20us
.four 1kHz 5 V([Vout])
.OP
* From [SCHEMATICS NETLIST] section of msim.ini:
.lib "C:\MSimEv_8\UserLib\TSE.lib"
.lib "nom.lib"
.INC "example.net"
.INC "example.als"
.probe
.END
```

This file includes simulation instructions (.ac/.tran/.four/.OP), references to the model libraries (.lib) and instructions for including the netlist and the alias file (.INC).

– **Netlist EXAMPLE.NET**:

```
* Schematics Netlist *
R_R4          Ve Ve1 4.7k
R_R5          Ve1 0 5.6k
R_R3          Vb Vout 39k
Q_T1          Vout Vbase Ve BC547B
R_R1          Vb Vbase 75k
R_R2          Vbase 0 18k
C_C2          Ve1 0 3.3u
C_C1          Vin Vbase 2.2u
R_Rg          Vin $N_0001 50
V_Vb          Vb 0 DC 15V
C_Cp          Vout 0 4p
V_Vg          $N_0001 0 DC 0 AC 1V
+             SIN 0V 0.2V 1kHz 0 0
```

– **Output file EXAMPLE.OUT:**

```
****     BJT MODEL PARAMETERS
              BC547B
              NPN
     IS     7.049000E-15
     BF   374.6
     NF     1
     VAF   62.79
     IKF     .08157
     ISE   68.000000E-15
     NE     1.576
     BR     1
     NR     1
     IKR    3.924
     ISC   12.400000E-15
     NC     1.835
     NK     .4767
     RC     .9747
     CJE   11.500000E-12
     VJE     .5
     MJE     .6715
     CJC    5.250000E-12
     VJC     .5697
     MJC     .3147
     TF   410.200000E-12
     XTF   40.06
     VTF   10
     ITF    1.491
     TR    10.000000E-09
     XTB    1.5
```

```
****    SMALL SIGNAL BIAS SOLUTION     TEMPERATURE = 27.000 DEG C

     NODE   VOLTAGE     NODE   VOLTAGE     NODE   VOLTAGE     NODE   VOLTAGE
  (   Vb)   15.0000  (    Ve)    2.2673  (   Ve1)   1.2327  (   Vin)   0.0000
  ( Vout)    6.4484  (Vbase)    2.8908  ($N_0001)  0.0000

     VOLTAGE SOURCE CURRENTS
     NAME          CURRENT
     V_Vb         -3.807E-04
     V_Vg          0.000E+00
     TOTAL POWER DISSIPATION    5.71E-03   WATTS
```

```
****    OPERATING POINT INFORMATION    TEMPERATURE = 27.000 DEG C

**** BIPOLAR JUNCTION TRANSISTORS
```

```
NAME            Q_T1
MODEL           BC547B
IB              8.54E-07
IC              2.19E-04
VBE             6.24E-01
VBC             -3.56E+00
VCE             4.18E+00
BETADC          2.57E+02
GM              8.45E-03
RPI             3.47E+04
RX              0.00E+00
RO              3.03E+05
CBE             4.02E-11
CBC             2.82E-12
CJS             0.00E+00
BETAAC          2.93E+02
CBX             0.00E+00
FT              3.13E+07


****   FOURIER ANALYSIS                    TEMPERATURE = 27.000 DEG C

FOURIER COMPONENTS OF TRANSIENT RESPONSE V(Vout)

   DC COMPONENT =    6.461795E+00

 HARMONIC   FREQUENCY    FOURIER    NORMALIZED     PHASE       NORMALIZED
   NO         (HZ)      COMPONENT   COMPONENT      (DEG)      PHASE (DEG)
    1       1.000E+03   1.598E+00   1.000E+00   -1.792E+02    0.000E+00
    2       2.000E+03   1.879E-03   1.176E-03    7.745E+01    2.567E+02
    3       3.000E+03   4.460E-05   2.791E-05    1.200E+02    1.200E+02
    4       4.000E+03   1.202E-04   7.521E-05    8.062E+00    1.873E+02
    5       5.000E+03   9.737E-05   6.093E-05    3.090E-01    1.796E+02

       TOTAL HARMONIC DISTORTION =   1.180047E-01 PERCENT
```

This file contains the parameters for the models used (here: *BJT Model Parameters*), operating point information (*Small Signal Bias Solution*) with the small-signal parameters of the components (*Operating Point Information*) and the results of the Fourier analysis (*Fourier Analysis*).

29.1.4
Further Examples

Characteristics of a Transistor

Figure 29.1.23 shows the circuit used in this example. *DC Sweep* is activated in the *Setup Analysis* dialog window (see Fig. 29.1.24). Then the parameters shown in Fig. 29.1.25 are entered:

– In the internal loop *DC Sweep* the collector-emitter voltage source VCE is varied in steps of 50 mV through a range of 0...5 V.
– In the external loop *DC Nested Sweep* the base current source IB is varied in steps of 1 μA through a range of 1...10 μA.

After the parameters have been entered, the simulation is started using *Simulate* and the collector current is displayed via *Add Traces* in the *Probe* program (see Fig. 29.1.26).

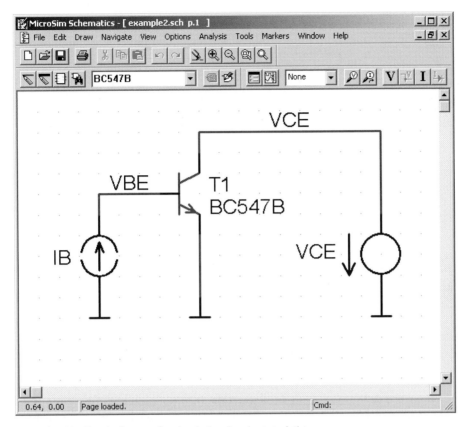

Fig. 29.1.23. Circuit diagram for simulating the characteristics

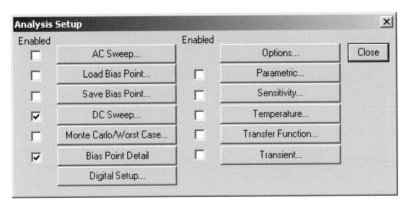

Fig. 29.1.24. Activating the *DC Sweep* analysis

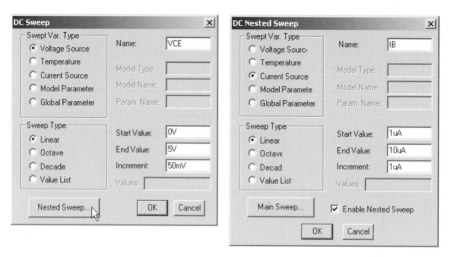

Fig. 29.1.25. Parameters for the internal and external loop

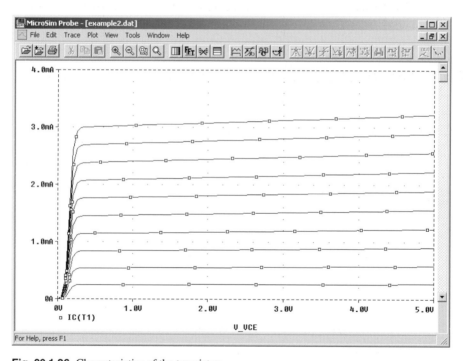

Fig. 29.1.26. Characteristics of the transistor

Using Parameters

Often it is desirable to perform the same analysis several times while varying one circuit parameter, e.g. the value of a resistor. The example in Fig. 29.1.27 shows the circuit diagram of an inverter with variable base resistance RB. This means that the RB value has to be replaced by a parameter in curly brackets (here: R), which then has to be defined. This is done by means of the *Parameters* component that has been inserted in the upper left corner of the diagram in Fig. 29.1.27. Double clicking on the *Parameter* symbol opens the *Param* dialog window shown in Fig. 29.1.28. You will then be prompted to enter the name of the parameter and its default value; the default value is used in analyses without parameter variations.

In the *Setup Analysis* dialog, *DC Sweep* must be activated in order to simulate the characteristics; similarly, *Parametric* must be activated to vary the parameter (see Fig. 29.1.29). The corresponding parameters are shown in Fig. 29.1.30. In *DC Sweep* a given parameter can also be varied via the *Nested Sweep* option; however, this is not the most flexible of methods as the *Nested Sweep* option is only available in *DC Sweep* analyses, while *Parametric* allows variation in any analysis.

After simulation using *Simulate* the *Probe* program initially displays the window shown in Fig. 29.1.31 for selecting the curves or parameter values to be indicated; in the default

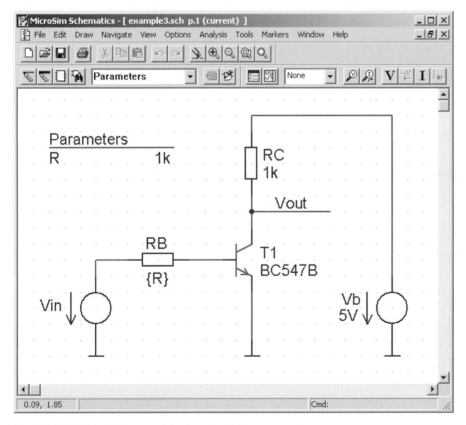

Fig. 29.1.27. Circuit diagram of the inverter with parameter R

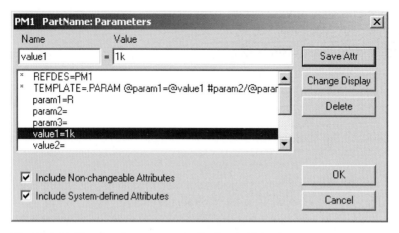

Fig. 29.1.28. Entering the parameter in the *Param* dialog box

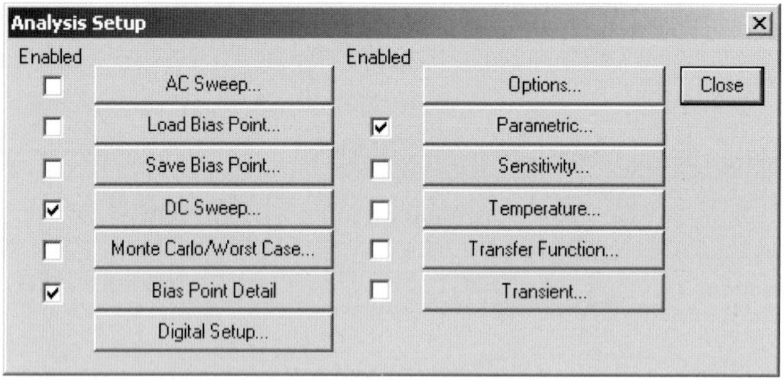

Fig. 29.1.29. Activating *DC Sweep* and *Parametric*

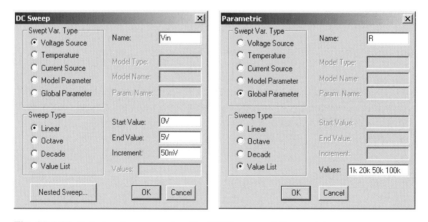

Fig. 29.1.30. Entering the parameters for *DC Sweep* and *Parametric*

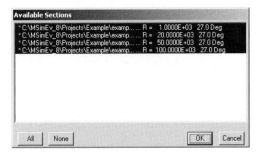

Fig. 29.1.31. Selections for curves to be shown

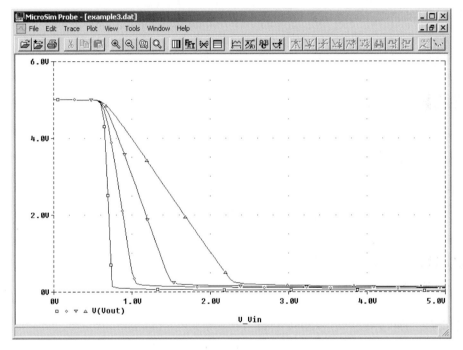

Fig. 29.1.32. Characteristics of the inverter for R = 1 k/20 k/50 k/100 k

setting all curves are selected. Entering $V(Vout)$ results in the characteristics shown in Fig. 29.1.32. The individual characteristic curves are marked by different symbols, which are shown underneath the graph in the order of the parameter values.

29.1.5
Integrating Other Libraries

A library comprises two sections (see Fig. 29.1.2):

– The *symbol library* „xxx".SLB contains the circuit symbols of the components and information on the representation of the components in the netlist.
– The *model library* „xxx".LIB contains the component models; these are either *elementary models* with the parameters given as .MODEL instructions or *macro models* which

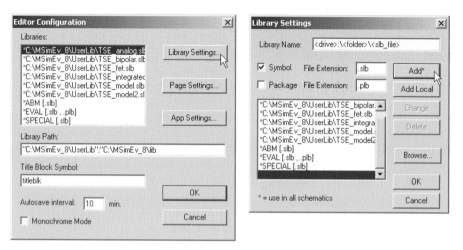

Fig. 29.1.33. *Editor Configuration* and *Library Settings* dialog windows

comprise several elementary models that are combined to a *sub-circuit* and are contained in the model library in the form of *.SUBCKT„name"„terminals"„circuit".ENDS.*

A symbol library is integrated in the *Schematics* program via the *Options/Editor Configuration* option. This opens the *Editor Configuration* dialog window shown in the left hand side of Fig. 29.1.33, which lists the existing symbol libraries and their corresponding paths. Selecting the *Library Settings* option opens the dialog box shown in the right hand side of Fig. 29.1.33. This option is used to add, edit or delete symbol libraries. It is possible

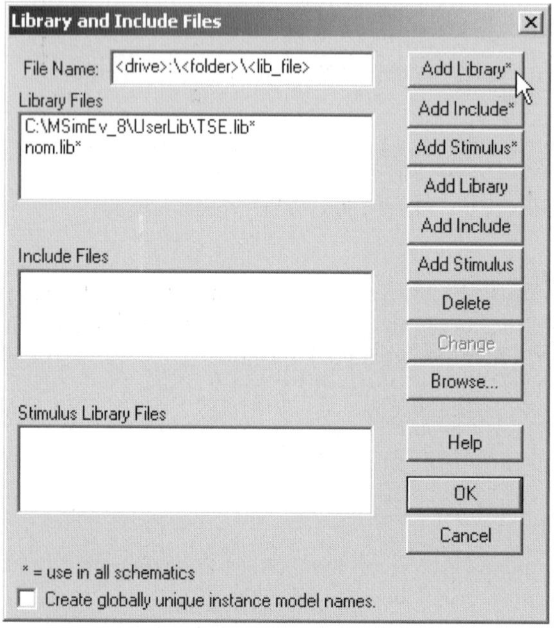

Fig. 29.1.34. *Library and Include Files* dialog window

to enter the name and path (drive and directory) of the library in the *Library Name* field or to search for the desired library using *Browse*. Clicking on *Add** adds the symbol library to the list. To close the dialog window click on *Ok.*

Similarly, a model library is added in the *Schematics* program via the option *Analysis/Library and Include Files*. The name and path of the library are entered in the same way and the library can then be added with *Add Library** (see Fig. 29.1.34).

To add libraries, always use the *asterisk* commands *Add** or *Add Library** since this adds the libraries *permanently* and makes them automatically available when the program is re-activated. Since in the demo version of *PSpice* both the number of libraries and the number of library elements are limited it is necessary to *exchange* libraries if other libraries are required for further simulations and the total capacity has been used up.

29.1.6
Some Typical Errors

The typical errors are described on the basis of the circuit diagram shown in Fig. 29.1.35 which contains several errors. If an error occurs, the *MicroSim Message Viewer* appears showing the relevant error message either before or after the simulation (see Fig. 29.1.36).

– *Floating Pin:* One of a component's terminals is not connected, e.g. *R2* in Fig.29.1.35. This error occurs during netlist generation; a dialog window with the message *ERC: Netlist/ERC errors – netlist not created* appears and, after clicking on *Ok* to confirm, the *Message Viewer* with the error message *ERROR Floating Pin: R2 pin 2* appears. As a general rule, every terminal must be connected. The only exceptions are components which have been specially configured or macro models which already have *internal* circuitry at one or more terminals so that no *external* circuit is required.

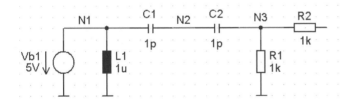

Fig. 29.1.35. Circuit diagram containing typical errors

Severity	Origin	Time	Message Text
● INFO	Schematics	04:32PM	Creating PSPICE netlist...
● ERROR	Schematics	04:32PM	Floating pin: R2 pin 2
● INFO	Schematics	04:32PM	Netlist/ERC errors - netlist not created.

3 Messages: 1 Error, 0 Warning, 2 Info

Fig. 29.1.36. *MicroSim Message Viewer* window

- *Node < node name> is floating*: The voltage of a node cannot be determined since it is indeterminate; this is the case for node *N2* in Fig. 29.1.35. This error message appears at any time when only capacitors and/or current sources are connected to a node; consequently, Kirchhoff's node law is not fulfilled. Every node must have a DC path to ground in order to clearly determine the nodal voltage. In the case of node *N2* in Fig. 29.1.35 the error can be rectified, for example, by adding a high-resistive resistor between *N2* and ground.
- *Voltage and/or inductor loop involving „component"*: A loop formed by voltage sources and/or inductors which goes against Kirchhoff's loop law exists; in the example shown in Fig. 29.1.35 the voltage source *Vb1* is short-circuited for DC voltages by the inductor *L1*.

29.2
ispLEVER – Brief User's Guide

29.2.1
Outline

Programming PLDs as described in Chap. 10.4 requires the creation of a so called fusemap containing a list of the desired connections. This may be done manually using a text editor or, more comfortably, with the aid of a design environment like the one provided by ispLEVER. The starter software of ispLEVER can be downloaded by Lattice (www.latticesemi.com) and can be licensed there at no cost.

This program supports the entry of the circuit by either using a programming language or by generating a circuit diagram. Furthermore, in order to test the functionality of the design simulation with graphic output is carried out. In addition to this, the propagation delay time can be tested with the help of timing analysis and optimised for different design goals.

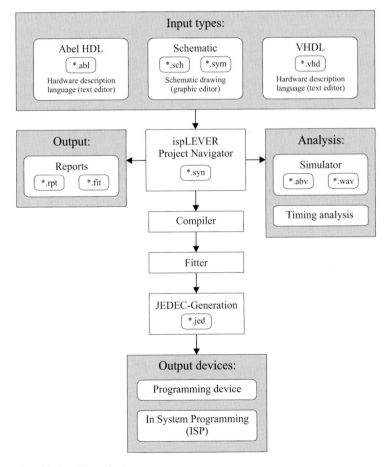

Fig. 29.2.1. Flow chart

The design environment itself has come a long way. It was initially developed by Data I/O under the name Synario. The software was then acquired by MINC, who were bought out by Vantis, who proceeded to rename the product DesignDirect. After the merger with Lattice it was named DesignExpert and finally became ispLEVER.

The flow chart in Fig. 29.2.1 shows the interconnections between and procedures carried out in the various input modes as well as the analysis process and the generation of output files. The project navigator is used to start all the activities and to setup the various options.

The hardware description language Abel HDL (High Definition Language) is entered via the text editor. In addition, circuit diagrams can be drawn with the help of the Schematic's input option. A description in the programming language VHDL is also possible, but this is not explained here. The interdependence of the source files is managed and represented in the form of a hierarchy in the source window.

The integrated compiler converts the source files independent of the source into a uniform machine format and generates a process-related report. Several reports are displayed during each phase of the programming process in the output window below. Such reports include not only design errors but also design analysis to show additional information such as timing analysis and utilisation of the chip resources.

The ispLEVER software package comprises various program areas. Their combined features enable the generation of complex designs. The integrated development environment is the project navigator shown in Fig. 29.2.2 which is started via the program ispLEVER. The project navigator shows all the files related to a given project and is used to start up all processes required for the project.

As is common with Windows, the menus and their various options and commands are activated by a left mouse click or a designated key combination. For reasons of simplicity

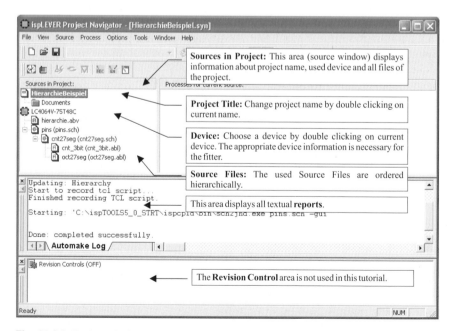

Fig. 29.2.2. Project Navigator source window

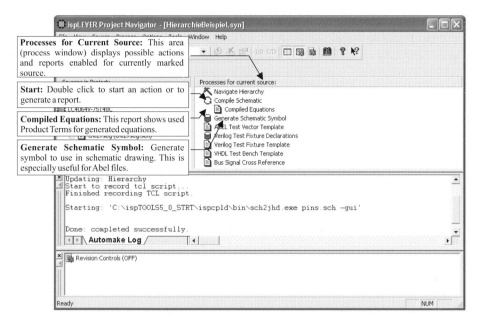

Fig. 29.2.3. Project Navigator Process window

the following description is limited to the mouse operations. In the source or the process window (see Fig. 29.2.2 and Fig. 29.2.3, respectively) you can perform a specific action or review a report by double clicking the left mouse button.

For example, a double click on the project title "Untitled" allows you to modify the project title. With a double click on the device name you can select the device, while double clicking a file name opens the file.

Both a text and a graphic editor are available for data input. The system provides text reports on the success of the individual compilation steps and the analysis results in the Automake Log window. The functional simulation results may also be presented graphically. The results of the timing analysis appear in the form of a table. Finally a fusemap is generated as a JEDEC file for device programming.

The window at the button of the Project Navigator is for revision control purpose, which is not reasonable for small designs and therefore neglected in this tutorial.

ispLEVER is a very comprehensive software package, which means that this description can only cover the most important commands and features. Detailed information is available in the Help function of each element of the software package.

29.2.2
Circuit Entry

Every new project begins with data input. In principle it is unimportant which device is used at this stage of the design process. But the chip-family should suit because this determines which libraries are available. ispLEVER supports almost every device from Lattice, beginning with the simple PLDs to the complex CPLDs and FPGAs. For our example, we have opted for the model ispM4A5-64/32 to illustrate the various design steps, starting with an empty project and proceeding to circuit input and design analysis and finishing with the creation of the JEDEC file. This device is a CPLD that consists of four PLDs of the 33V16 type and an additional programmable interconnection matrix. The programming logic is located on the chip; it is thus programmable within the circuit (ISP – In System Programmable) as is the case with all newer devices. The only thing required is a passive download cable and a download program as described in Sect. 29.2.5.

The following table shows the meaning of the file name extensions (also refer to Fig. 29.2.1). To archive a project only the files shown in bold print must be copied. The remaining files will be regenerated by the program as required.

abl	Abel HDL file	**sch**	Schematic file
abv	Abel test vectors	**sym**	Symbol for schematic
fit	Fitter report	**syn**	ispLEVER project file
jed	JEDEC file (Fusemap)	**wav**	simulation output
rpt	Report		

Hardware Description Language Abel

The process of describing a circuit function using Abel HDL is explained below using the example of a 3 bit counter that counts from zero to five and then restarts from zero. The state diagram and the schematic symbol are shown in Fig. 29.2.4.

A new project is begun by selecting "New Project ..." in the File menu of the Project Navigator. This opens the Project Wizard window (see Fig. 29.2.5a) in which you first type in the desired project name and select or create the desired directory. Furthermore the design entry type is chosen. Use the "Project type" Schematic/Abel as shown. Following the dialog a screen for selecting the desired PLD pops up (see Fig. 29.2.5b). After selecting the device's family a list with all devices of this family is displayed on the right. Don't forget to set the correct package type underneath and to check the part number.

The next dialog is for adding source files of former projects (see Fig. 29.2.6.). If you don't want to import any modules go on without any entry.

The Project Navigator does not automatically adopt the project file name as the project title; the project title can, however, be edited manually if desired by double clicking at the first entry in the source window of the project navigator.

To generate a new Abel data record use the "New ..." button from the source menu of the project navigator. Select the Abel HDL module from the list (see left figure in Fig. 29.2.7). Clicking on OK opens a dialog window which prompts the user to enter the Module Name, the File Name and the Title (see right Figure in Fig. 29.2.6). After these informations have

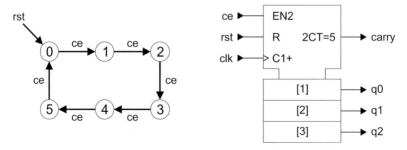

Fig. 29.2.4. State diagram and schematic symbol of the counter

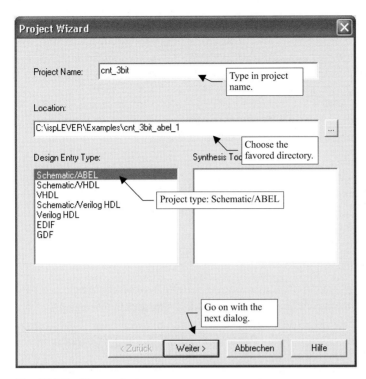

Fig. 29.2.5a. New project

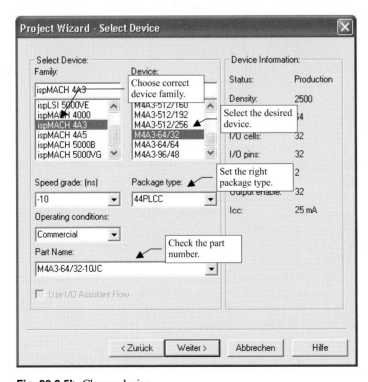

Fig. 29.2.5b. Choose device

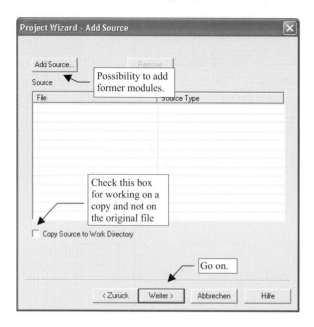

Fig. 29.2.6. File import

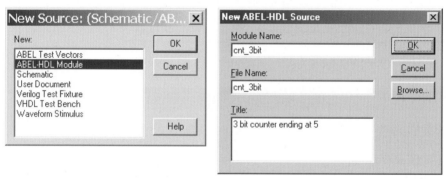

Fig. 29.2.7. New Abel source

been entered and confirmed via the OK button, the text editor for entering the Abel code will open automatically.

The module name is the identification code within each project, while the file name is the name under which the data record is saved in the directory. Both entries are mandatory; the title, on the other hand, is an optional description of the function. It makes sense to use identical names for module and file in order to facilitate later retrieval. The title should be as detailed as possible so that the function of the module can be readily identified later.

The following program sample shows the complete generation of an Abel HDL module. Of particular importance are the key words shown in bold print, which must exist in every Abel file. If you wish to format the text as shown in the example below spaces or tabs should be used since these are ignored by the compiler.

Enter the example from Fig. 29.2.9 using the names from Fig. 29.2.8 and save the project. If you want to avoid typing you can import the source files from the sample directory (my files\ispLEVER examples that the TS-installer has generated) using the

Project name:	cnt_3bit
Directory:	cnt_3bit_abel_1
Abel module name:	cmt_3bit
Abel file name:	cmt_3bit
Abel title:	3 bit counter ending at 5

Fig. 29.2.8. Suggested names for example 1

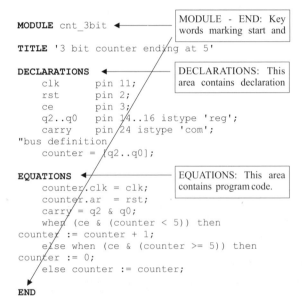

```
MODULE cnt_3bit  ◄───────   MODULE - END: Key
                            words marking start and

TITLE '3 bit counter ending at 5'

DECLARATIONS  ◄─────────    DECLARATIONS: This
    clk       pin 11;       area contains declaration
    rst       pin 2;
    ce        pin 3;
    q2..q0    pin 14..16 istype 'reg';
    carry     pin 24 istype 'com';
"bus definition
    counter = [q2..q0];

EQUATIONS  ◄────────────    EQUATIONS: This area
    counter.clk = clk;      contains program code.
    counter.ar  = rst;
    carry = q2 & q0;
    when (ce & (counter < 5)) then
counter := counter + 1;
    else when (ce & (counter >= 5)) then
counter := 0;
    else counter := counter;

END
```

Fig. 29.2.9. Example 1. Abel module for cnt_3bit

Source/Import option in the Project Navigator. The files can be used to avoid typing and to prevent possible errors. However, the project should not be opened in the existing directory as this would prevent you from being able to follow the individual steps.

Inputs and outputs in the Abel module: The description of the behaviour of a circuit (see Fig. 29.2.9) starts immediately underneath the module name and title with a list of the input and output signals and their interrelations. Pin numbers may also be assigned at this point, which will be explained further below.

Abel recognises output signals from the key word "istype". It would be more correct to speak here of a resulting signal since it is not necessary available at a pin (see Fig. 29.2.10). The outputs q2 to q0 were declared "istype 'reg' ", which means that each of these outputs is made available via a register (to build a sequential circuit). The carry signal is in the form of a combinatorial circuit which calls for a corresponding designation. The syntax here is "istype 'com' ". The "istype" command has many more variations. But only the " 'reg' " (registered) and " 'com' " (combinatorial) types are required for this brief user's guide. A list of some more variations is contained in the on-line help menu of the Text Editor. The outputs are not necessarily available at a pin of the chip; several signals are only intended for internal feedback. Therefore it would be more accurate to speak of resulting signals.

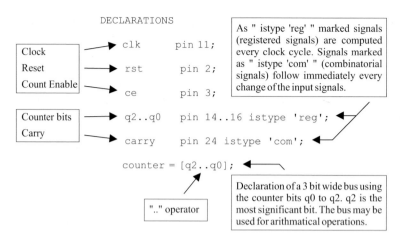

DECLARATIONS

Clock   → clk   pin 11;

Reset   → rst   pin 2;

Count Enable → ce   pin 3;

Counter bits → q2..q0   pin 14..16 istype 'reg';

Carry → carry   pin 24 istype 'com';

counter = [q2..q0];

As " istype 'reg' " marked signals (registered signals) are computed every clock cycle. Signals marked as " istype 'com' " (combinatorial signals) follow immediately every change of the input signals.

".." operator

Declaration of a 3 bit wide bus using the counter bits q0 to q2. q2 is the most significant bit. The bus may be used for arithmatical operations.

Fig. 29.2.10. Declaration of inputs and outputs

The ".." operator is an abbreviation for sequences, meaning that not every element of a sequence needs to be described in detail. In this example (Fig. 29.2.9) the benefit is minimal since only the element q1 needs not to be entered.

Behaviour of the circuit in the Abel module: The following syntax of the functional description is only one among many, but it is nevertheless a very powerful method for a counter. Figure 29.2.11 shows this method of defining the counter. The second and third program lines tell the compiler to permanently transmit the input signals "clk" and "rst" to the bus "counter". Normally a decimal value is assigned to the contents of the bus.

The next five lines contain the actual program for the desired function. The carry signal is the AND operation of the first and third bit and therefore becomes active at $5_{dec} = 101_{dual}$. Afterwards the counter reading is checked to determine whether or not it has reached the final count of 5. If it is true the counter is reset to 0 otherwise it is incremented.

The "." operator allows the sub-elements of signals and busses to be accessed. The ":=" operator assigns a signal that is synchronous to the clock (registered), while "=" sets the signal directly (combinatorial) to the given value. This works only with the appropriate variables, e.g. a combinatorial signal cannot be assigned to a registered signal (combinatorial).

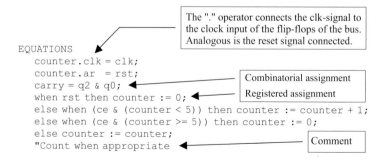

The "." operator connects the clk-signal to the clock input of the flip-flops of the bus. Analogous is the reset signal connected.

```
EQUATIONS
    counter.clk = clk;
    counter.ar  = rst;
    carry = q2 & q0;
    when rst then counter := 0;
    else when (ce & (counter < 5)) then counter := counter + 1;
    else when (ce & (counter >= 5)) then counter := 0;
    else counter := counter;
    "Count when appropriate
```

Combinatorial assignment

Registered assignment

Comment

Fig. 29.2.11. Functional description

The "when then else" expression is used to distinguish between different situations. Any logic expression can serve as a condition; ensure that parentheses are used intelligently to avoid erroneous operations. The operators are processed in the sequence given by the compiler; but parentheses make it easier to understand the intention of the programmer.

After closing the text editor you can compile the Abel module by double clicking on "Compile Logic" in the process window of the project navigator. The compiler automatically checks the syntax.

The results of the compilation can be checked in the report window by opening the desired object in the "Processes for current source" window in the Project Navigator.

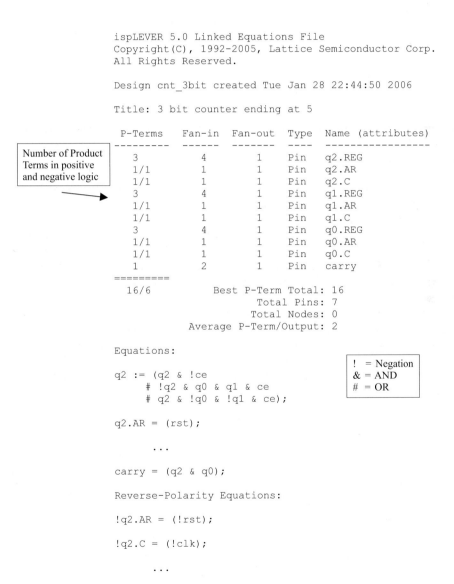

```
ispLEVER 5.0 Linked Equations File
Copyright(C), 1992-2005, Lattice Semiconductor Corp.
All Rights Reserved.

Design cnt_3bit created Tue Jan 28 22:44:50 2006

Title: 3 bit counter ending at 5

        P-Terms    Fan-in   Fan-out   Type   Name (attributes)
        ---------  ------   -------   ----   ------------------
          3          4         1      Pin    q2.REG
          1/1        1         1      Pin    q2.AR
          1/1        1         1      Pin    q2.C
          3          4         1      Pin    q1.REG
          1/1        1         1      Pin    q1.AR
          1/1        1         1      Pin    q1.C
          3          4         1      Pin    q0.REG
          1/1        1         1      Pin    q0.AR
          1/1        1         1      Pin    q0.C
          1          2         1      Pin    carry
        =========
          16/6                 Best P-Term Total: 16
                                     Total Pins: 7
                                    Total Nodes: 0
                          Average P-Term/Output: 2

Equations:

q2 := (q2 & !ce
      # !q2 & q0 & q1 & ce
      # q2 & !q0 & !q1 & ce);

q2.AR = (rst);

      ...

carry = (q2 & q0);

Reverse-Polarity Equations:

!q2.AR = (!rst);

!q2.C = (!clk);

      ...
```

Number of Product Terms in positive and negative logic

! = Negation
& = AND
# = OR

Fig. 29.2.12. Compiled equations for example 1

You can view the reports in the Automake Log window (see Fig. 29.2.2) that imforms you about the success of compiling your design. If you follow this procedure for the "Compiled Equations", you will receive a report as shown in Fig. 29.2.12. Please note that the Project Navigator automatically performs all preparatory steps. In our example the Project Navigator would cause the compiler to begin compilation (Compile Logic) as soon as the report is requested. Use "Generate Schematic Symbol" to create a symbol for later use.

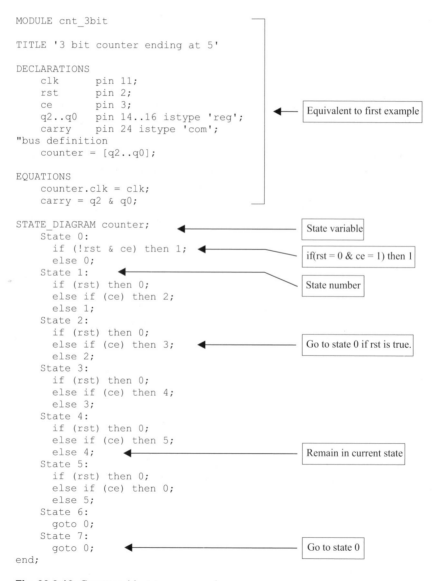

```
MODULE cnt_3bit

TITLE '3 bit counter ending at 5'

DECLARATIONS
    clk       pin 11;
    rst       pin 2;
    ce        pin 3;
    q2..q0    pin 14..16 istype 'reg';        ◄─── Equivalent to first example
    carry     pin 24 istype 'com';
"bus definition
    counter = [q2..q0];

EQUATIONS
    counter.clk = clk;
    carry = q2 & q0;

STATE_DIAGRAM counter;                 ◄───────────  State variable
    State 0:
      if (!rst & ce) then 1;    ◄──────────────  if(rst = 0 & ce = 1) then 1
      else 0;
    State 1:       ◄────────────────────────  State number
      if (rst) then 0;
      else if (ce) then 2;
      else 1;
    State 2:
      if (rst) then 0;
      else if (ce) then 3;       ◄──────────  Go to state 0 if rst is true.
      else 2;
    State 3:
      if (rst) then 0;
      else if (ce) then 4;
      else 3;
    State 4:
      if (rst) then 0;
      else if (ce) then 5;
      else 4;       ◄─────────────────────  Remain in current state
    State 5:
      if (rst) then 0;
      else if (ce) then 0;
      else 5;
    State 6:
      goto 0;
    State 7:
      goto 0;       ◄─────────────────────  Go to state 0
end;
```

Fig. 29.2.13. Counter with status commands

Project name:	ent_3bit
Directory:	ent_3bit_abel_2
Abel module name:	ent_3bit
Abel file name	ent_3bit
Abel title:	3 bit counter ending at 5

Fig. 29.2.14. Suggested names for example 2

State Diagram in Abel

The hardware description language Abel HDL also offers the possibility of creating a sequential logic as a state diagram in text form.

With the "STATE_DIAGRAM" key word (see Fig. 29.2.13) you can enter each transient after making the usual declarations of module name, inputs/outputs and buses (vectors).

Create a new project. Use the names from Fig. 29.2.14 and generate the new directory. Enter the Abel HDL module or import it with the Project Navigator: Source/Import option from my files\ispLEVER examples.

The basic difference between the two sample counter designs is obvious: When describing the counter using a loop, only the maximum counter reading needs to be changed in order to create a counter of any desired count cycle.

In the state diagram every state must be listed individually. For this reason this type of entry is suitable only for sequential logic circuits with few states. On the other hand it enables complex transient conditions to be described.

Truth Table in Abel

Another possible entry option in Abel HDL is the truth table. It is primarily intended to create combinatorial circuits. In our example we will create a seven-segment decoder. This is meant to translate the output of the counter in the previous examples into a seven-segment display (shown in Fig. 29.2.15). The suggested names for the third example are listed in Fig. 29.2.16. Figure 29.2.17 shows the data record to be entered. Again observe the "compiled equations" and generate a schematic symbol using "Generate Schematic Symbol".

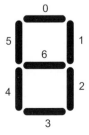

Fig. 29.2.15. Seven-segment display

Project name:	oct27seg
Directory:	oct27seg
Abel module name:	oct27set
Abel file name	oct27seg
Abel title:	octal to seven segment decoder

Fig. 29.2.16. Suggested names for example 3

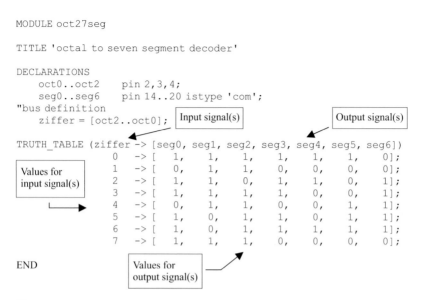

```
MODULE oct27seg

TITLE 'octal to seven segment decoder'

DECLARATIONS
    oct0..oct2    pin 2,3,4;
    seg0..seg6    pin 14..20 istype 'com';
"bus definition
    ziffer = [oct2..oct0];
```

Input signal(s)   Output signal(s)

```
TRUTH_TABLE (ziffer -> [seg0, seg1, seg2, seg3, seg4, seg5, seg6])
            0  -> [   1,    1,    1,    1,    1,    1,    0];
            1  -> [   0,    1,    1,    0,    0,    0,    0];
            2  -> [   1,    1,    0,    1,    1,    0,    1];
            3  -> [   1,    1,    1,    1,    0,    0,    1];
            4  -> [   0,    1,    1,    0,    0,    1,    1];
            5  -> [   1,    0,    1,    1,    0,    1,    1];
            6  -> [   1,    0,    1,    1,    1,    1,    1];
            7  -> [   1,    1,    1,    0,    0,    0,    0];
```

Values for input signal(s)

```
END
```

Values for output signal(s)

Fig. 29.2.17. Data record for the seven-segment decoder

Circuit Diagram with Schematic

The Schematic Editor is a tool used to produce graphic designs for digital circuits. The design process for such circuit diagrams is supported by several libraries of logic cells, registers, input/output buffers, etc. (graphic symbols with inputs and outputs) to which the designer has access. These libraries allow the user to add his own components created either in Abel HDL or with the Schematic Editor.

If you choose the Source/New command "Schematic" a dialog box opens that prompts the entry of a name for the new schematic. After a suitable name has been fed, the Schematic Editor is ready for the circuit to be entered.

When creating a circuit diagram in Schematic it is very important to know the tool box commands described in more detail in Fig. 29.2.18. Each description in the table corresponds to a button in the tool box.

Again the design example is the 3 bit counter from Fig. 29.2.9. Three toggle flip-flops with reset input are used and connected accordingly. The corresponding circuit is shown in Fig. 29.2.19.

Enter the example using the names from Fig. 29.2.20 or use the Project Navigator: Source/Import option to import the information from my files\ispLEVER examples.

	Add Symbol	Add Wire	Add Bus Tap
	Add Instance	Add Net Name	Add I/O Marker
	Edit Pin	Edit Symbol	Edit Net
	Duplicate	Move	Drag
	Rotate	Mirror	Delete
	Draw Text	Draw Line	Draw Rectangle
	Draw Arc	Draw Circle	Highlight

Fig. 29.2.18. Tool box of the Schematic Editor

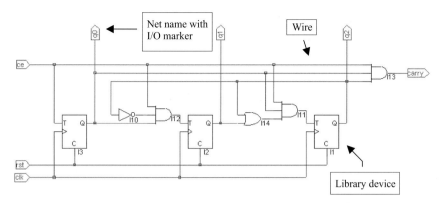

Fig. 29.2.19. Counter Schematic

Project name:	cnt_3bit
Directory:	cnt_3bit_sch
Schematic file name:	cnt_3bit

Fig. 29.2.20. Suggested names for example 4

Only the most important steps for the design of a circuit diagram will be outlined here. For further information refer to the Help menu. To perform an operation first select a command from the tool box and then the object that is to be used. If you want to terminate a command, click on the circuit diagram with the right mouse button. If an entire region is marked, the command is applied to several objects simultaneously. To do so, press and hold the left mouse button and draw a box around the desired objects. If you want to cancel the last step click on Undo (either in the Edit menu or in the symbol bar).

The following steps must be executed to create the sample design:

– Insert the toggle flip-flops (library REGS.LIB, component G_TC) selected from the Symbol Libraries (Add Symbol) into the diagram. The desired library can be selected in the upper section of the popped up window. Then mark the component in the lower section and place it three times in the desired positions in the circuit diagram.
– Add the required logic gates to the diagram as shown in Fig. 29.2.19. You find them in the GATES.LIB; select the gates G_INV, G_2OR and G_3AND.
– Use Add Wire to connect components at the respective red dots.
– Assign "net names" to all inputs and outputs so they can be addressed in the superordinate modules or for testing. After selecting the command (Add Net Name) enter the desired name at the bottom of the Schematic Editor, confirm it by clicking on Enter and click on the red dot at the end of the wire.
– Now use the I/O markers (Add I/O Marker) to define the signals as input or output signals. In the case of several I/Os, this can also be done for all of them in one step by marking the relevant region.
– Save the schematic diagram.
– Use File/Matching Symbol to create a schematic symbol.

In order to illustrate how a bus is generated the outputs q0 up to q2 are connected to one bus (see Fig. 29.2.21).

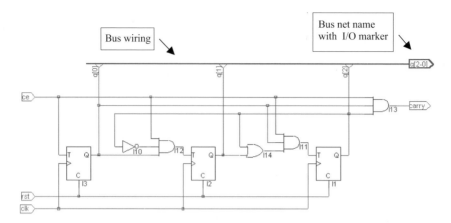

Fig. 29.2.21. Circuit diagram of counter with bus

Follow the description below:

- First remove the output wires and markers from the diagram (Delete). To do so you can either click on the element to be deleted or mark and delete an entire region completely.
- Draw a horizontal wire above the circuit.
- Designate the right end of the label with a net name called q[0-2]. This makes the line a bus which then appears as a bolder line.
- Define the bus as an output using an I/O marker.

The next step is to establish the bus connections, i.e. to connect the output signals to the bus. This is done in the following manner:

- First enter the range of bus connections via Add Net Name (q[0-2] in this example) and confirm by pressing Enter.
- Then hold Shift and press the right mouse button. The name assigned to the first connection (here q[0]) will appear at the cross-hair cursor.
- Hold down the left mouse button and draw a line **from** the desired output point to the bus. Releasing the mouse button will place a bus tap and at the cross-hair cursor appears the label of the next connection (here q[1]).
- Repeat this procedure for all other connections to be made. For necessary corrections use the Undo function.

Alternatively you can do the connections using the Bus Tap Tool from the toolbox:

- Click the bus and drag a horizontal line from the bus to the pin to be connected.
- Choose the naming tool and click with the left mouse button at the bus. The bus name will appear at your cursor tip.
- Then click right for changing the bus name into connection names and place them with a left click at the bus taps.

Component libraries

In Schematic the options File/Matching Symbol and Add Symbol make it possible to reuse an existing design. For example the 3 bit counter could be connected with the 7-segment decoder. In this context we wish to draw your attention to the existing component libraries.

Library	Description
vanprim.lib	selguide.pdf
vanttl.lib	vanttl.pdf
vanfunc.lib	vanfunc.pdf

Fig. 29.2.22. Component libraries for MACH PLDs

They contain not only simple gates but also complex components such as entire counters, multiplexers and adders. The use of these modules saves a lot of time in the designing and testing process; in this way the design process is performed on the same level as before when using complex TTL devices.

Which of the libraries are available depends on the PLD used. Figure 29.2.22 shows a brief list of the libraries that are most important for the MACH modules. Libraries for the ispLSI1k...8k family are described in the files ispmacro.pdf and 58kmcr.pdf.

In our exercise a BCD counter and a 7-segment decoder from the TTL library are used. Again we generate a counter with a 7-segment output. The corresponding circuit diagram is shown in Fig. 29.2.23.

Enter the sample component with the names from Fig. 29.2.24 or use the Source/Import option to get it from the directory my files\ispLEVER examples. Follow the steps described below to create the circuit diagram:

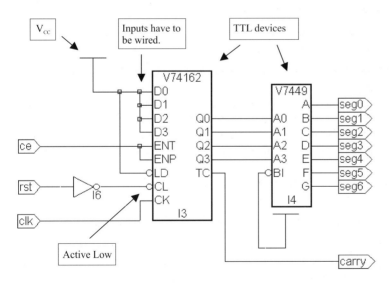

Fig. 29.2.23. TTL modules with wiring

Project name:	library
Directory:	library
Schematic file name:	library

Fig. 29.2.24. Suggested names for example 5 with component libraries

- Insert the TTL modules V74162 and V7449 (library VANTTL.LIB) selected from the Symbol Libraries (Add Symbol) into the diagram.
- Add the required logic gates, inverters and VCC (library GATES.LIB, G_INV and VCC) to the diagram according to Fig. 29.2.23.
- Draw the wires using Add Wire.
- Assign net names to all inputs and outputs. After selecting the Add Net Name option enter the desired names, confirm using Enter and click on the wire; ensure that you click onto the red dot at the end of the line. For the segment designations use the name "seg0+". This causes the program to increase the final digit by one after each designation process.
- Define the designated I/O signals as inputs or outputs using the I/O marker (Add I/O Marker). In the case of several I/Os, all of these can be defined in one step by marking an entire region.

Hierarchy

With ispLEVER you can link or interleaf several modules that have been designed in Abel or Schematic. This is similar to a C or Pascal program with various different functions and procedures. ispLEVER shows the hierarchy using a tree structure.

The simplest method of linking hierarchy levels is to develop the upper-most module as a schematic as shown in Fig. 29.2.25. The desired sub-modules are then integrated into this. In the hierarchy tree these modules appear below the main module. To do so, it may be necessary to import the Abel files used.

In this exercise we will use the counter and the octal-to-seven-segment decoder to create the design shown in Fig. 29.2.26.

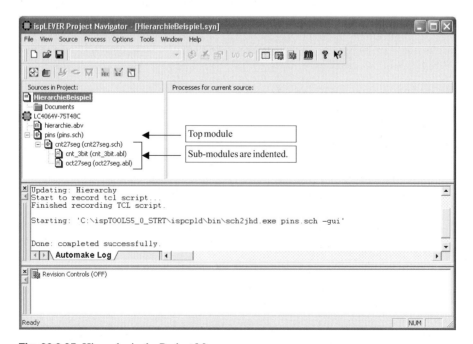

Fig. 29.2.25. Hierarchy in the Project Manager

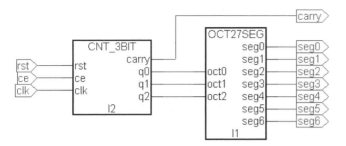

Fig. 29.2.26. Top module in Schematic

Project name:	cnt27seg
Directory:	cnt27seg
Schematic file name:	cnt27seg

Fig. 29.2.27. Suggested names for example 6

Enter the example with the name from Fig. 29.2.27 or use the Project Navigator: Source/Import option to import it from the ispLEVER examples.

Import the previously created Oct27Seg decoder and the first Abel counter (from cnt_3bit_abel_1). Use menu option Source/Import in the Project Navigator and select the corresponding Abel file. Then open a new Schematic design using Source/New…/ Schematic. Add the two symbols from the "local" library to the design.

Finally draw the necessary wires, define them as inputs/outputs and mark them with the corresponding I/O markers.

Use the File/Matching Symbol to generate a symbol for the entire design before quitting the Schematic Editor. This is effectively the same as using Generate Schematic Symbol in the Project Navigator.

29.2.3
Pin Assignment

If no pins are assigned, the Device Fitter will perform the pin assignment itself when generating the JEDEC file, so that it is best for the internal wiring (also refer to Sect. 29.2.5 Optimization). Manual pin assignment is done either directly in the source file (e.g. Abel or Schematic) as in the examples described or by entering the values in the Constraint Editor.

If the pin assignment is to be taken from a source file, the import option must be chosen in the dialog popping up during the fitting procedure. The fitter then adopts the assignments from the top-level module, i.e. all pin assignments in the sub-modules are discarded.

The information of the specific chip is first fetched from the chip-library when it is used for simulation and for converting the net lists into JEDEC format. Only in such cases does it actually make sense to assign pins.

The device selection window is opened by double clicking on the current device name in the source window. This opens the Device Selector window (Fig. 29.2.28). Don't forget to choose the correct package type.

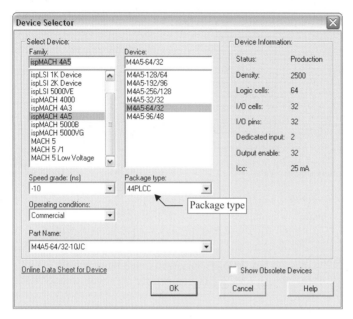

Fig. 29.2.28. Device Selector window

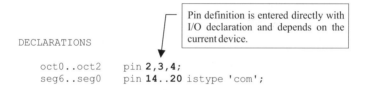

DECLARATIONS

    oct0..oct2   pin **2,3,4**;
    seg6..seg0   pin **14..20** istype 'com';

Fig. 29.2.29. Pin assignment in Abel

Abel

In Abel pins are assigned simply by entering the desired pin number after the key word "pin" (see Fig. 29.2.29). The Oct27Seg decoder is used as an example; here, the pin numbers are shown in bold print. The numbers can be entered either individually or by using the ".." operator.

By assigning the pin numbers directly in the Abel HDL module, which is also used for the function, the designer is forced to adapt the pin numbers accordingly when changing to another module or when importing the module into another project.

Schematic

Pin assignment can be done in any Schematic file. However, it is easier and more common to combine the entire circuit in a single block and to assign the pins in a top-level schematic (see Fig. 29.2.30).

Create a new project and import the files shown in Fig. 29.2.31. Generate a new Schematic module named "pins" and perform the following steps:

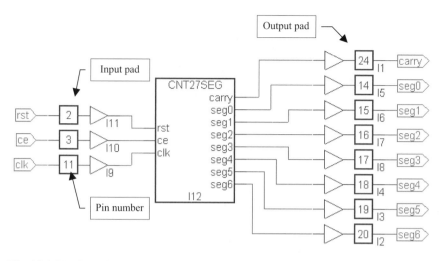

Fig. 29.2.30. Pin assignment in Schematic

Project name:	cnt27seg
Directory:	cnt27seg
Schematic file name:	cnt27seg
Import files:	cnt27seg\cnt27seg.sch
	cnt27seg\cnt_3bit.abl
	cnt27seg\oct27seg.abl

Fig. 29.2.31. Suggested names for example 7

- Insert the circuit symbol cnt27seg from the library ("local")
- Insert the I/O pads (library I/OPADS.LIB)
- Draw wires to connect the I/O pads to the module cnt27seg
- Add short wires to the outside of the I/O pads
- Add the input and output names at the end of the wiresProvide the inputs/outputs with I/O markers
- Open the Symbol Attribute Editor from the tool box (see Fig. 29.2.32)
- Mark one I/O pad after the other and assign the according pin number.

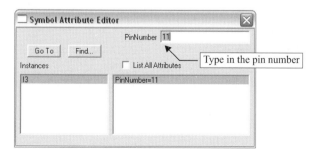

Fig. 29.2.32. Symbol Attribute Editor

Constraint Editor

By the use of the Constraint Editor the pin numbers are assigned to the signals by hand. To start this editor mark the device entry in the source window and double-click at Constraint Editor in the process window.

Double click at the "carry" signal in the left window (see Fig. 29.2.33). The selected signal appears as entry on the right. Then double click t the "pin" cell and enter the pin number 25. In the GLB cell the logic block of the device is assigned automatically, which indicates the geometrical placement of the signal on the chip.

The "carry" signal is now assigned to pin 25. In order to undo an assignment click at the signal with the right mouse button and choose "clear selected".If you have assigned the pin numbers in the schematic or ABEL file these numbers are taken for initializing the constraint editor. You are also able to change the pin numbers in this case by the help of the constraint editor.

Depending on the device selected other location assignments are possible, e.g. the allocation to a special PAL-block or the macro cell which is to calculate the signal.

Open the "pins" project and start the Constraint Editor. Click on the "Loc" button and mark the "carry" signal. Then mark pin number 25. Confirm the assignment by clicking on Add. The "carry" signal is now included in the list underneath. In order to undo an assignment use the cancel command. Alternative you mark the "carry" signal, select pin number 24 and confirm this with the "Update" button. If the Import Source Constraint option is active, the signals are shown instantly in the list underneath.

The assignment can be changed in 3 steps:

– mark the signal in the window at the bottom
– Modify transfers the signal to the upper window
– here it can be connected to the desired pin
– confirm with update

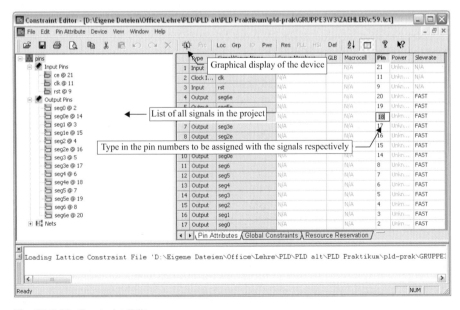

Fig. 29.2.33. Constraint Editor

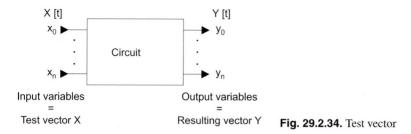

Fig. 29.2.34. Test vector

29.2.4
Simulation

Trouble shooting is the prime purpose of design analysis. The behaviour of the newly designed circuit is checked by way of simulation. On the other hand the timing/frequency analysis is suitable for checking the functionality under given conditions, such as the maximum frequency. ispLEVER features an integrated simulator which applies the input signals described in a test vector file to the design and calculates the resulting signal forms (see Fig. 29.2.34).

These input signals are the test vectors that are applied to the design one after the other. Another option is to compare the simulation results with a given result vector.

Whether the results of such simulation are useful and the extent (in terms of percentage) to which they test the design depends on the selection of suitable input signals. Therefore it is important to define suitable start conditions. For example the registers should be reset prior to initial use.

Flip-flops accept input signals only in the event of a rising clock pulse edge. For this reason, signals that occur after a positive test pulse edge and disappear before the next pulse have no effect. Furthermore, the input signals of flip-flops must not change during the positive clock pulse edge since otherwise they do not comply with the set-up and hold time, which would result in undefined conditions.

The results of the simulation can be viewed in the Waveform Viewer. All signal forms can be examined here.

Test vectors

Test of combinatorial circuits: Open the Oct27Seg project and open the Abel file. Insert the text after the truth table as shown in Fig. 29.2.35. Only one input and one output signal must be given in the test vector. All the other output signals can be viewed in the Waveform Viewer.

Now mark the Oct27Seg vectors in the Project Navigator and start the functional simulation. In the Simulator Control Panel the simulation is activated via the Simulate/Run button. The Waveform Viewer opens and shows the simulation result. A more detailed description of the operation is given further below.

Testing sequential logic: In this example a test vector file that is separate from the Abel file is to be created in order to test the counter.

Open the "cnt_3bit" project in the "cnt_3bit_abel_1" directory, create a new source test vectors module and enter the program according to Fig. 29.2.36. The test vectors are saved in a separate file, which provides the advantage that the test vector file can be used

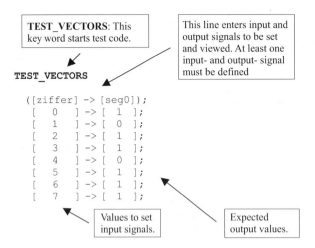

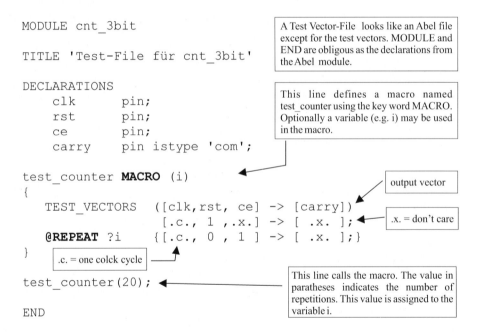

Fig. 29.2.35. Structure of test vectors

Fig. 29.2.36. Test vector file for the counter

in different projects. In other words, this test vector file can be used for all three "cnt_3bit" projects.

Test vectors are also used for stimulating counters. In this case macros are used. This allows longer simulations to be set up without every time increment having to be specified explicitly.

First you define the "test_counter" macro via the key word "macro". Please note that the entire macro is in parentheses. The variable (i in this example) is used to specify the number of repetitions. The macro itself has a structure similar to that of a common test vector. In order to avoid having to specify every single step, the repeat command ensures that the test vector is repeated i number of times. Several test vectors could also be used – the important thing is to pay close attention to the use of parentheses. To activate the macro, its designated name must be entered; the value in parenthesis specifies the number of repetitions desired.

The simulator knows how to interpret the values ".c." and ".x.". ".c." means that the simulator has to generate a clock pulse, while ".x." (don't care) instructs the simulator to ignore the values.

The test vector must again contain at least one input and one output element. Other signals can be selected and viewed in the Waveform Viewer. The test vector file created can now be used for the two other counter projects and can be imported into the library, hierarchy and pins projects; these projects are then ready for testing. Observe: there must exist only one test vector file in a project.

Waveform Viewer

Open the hierarchy project, import the test vector file of the counter and start the simulation. This opens the Simulator Control Panel (see Fig. 29.2.37), which allows several settings to be made before the actual simulation is started (Simulate/Run). The Waveform Viewer (see Fig. 29.2.38) is the most suitable tool in the ispLEVER software package for signal representation.

All signals available for viewing (see Fig. 29.2.39) can be displayed via the Show command in the Edit menu. Instances indicates the current level of signals; the corresponding signals can be found under **Nets**. Double clicking on a **net** displays the given signal. Alternatively you can mark one or more signals and view them via the **Show** option. A marked signal shown in the plot-window can be deleted when it is marked by the edit/hide command.

In order to show a summary of the corresponding output signals when viewing a counter, you have to use the bus option. In our example we wish to view the internal counter output only. This is done by double clicking on D underneath Instances. This displays additional internal signals. Expand the window by pressing the **Bus** button and enter a bus name. Then mark the **Nets** N_1, N_2 and N_3. The signals marked are added to the active bus via **Add Net(s)**. The sequence of the signals, i.e. the significance within the bus, can be reversed using the **Reverse** button. With **Save Bus** the bus can be saved and subsequently displayed via **Show**. The sate of the bus is shown in decimal notation. **New Bus** allows you to create additional buses.

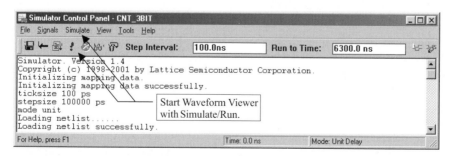

Fig. 29.2.37. Simulator Control Panel

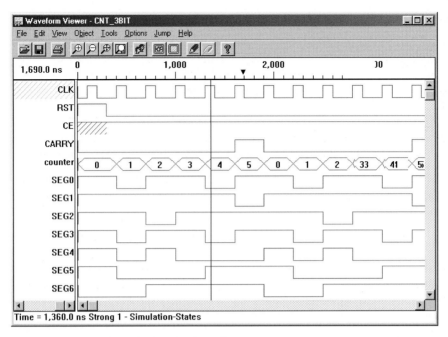

Fig. 29.2.38. Signal forms in the Waveform Viewer

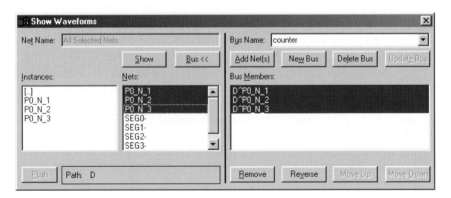

Fig. 29.2.39. Show Waveforms dialog window

The Waveform Viewer allows you to zoom in on an image. Click on the Zoom In option in the View menu. Mark the desired area in the display or simply click in the display to increase the zoom.

Functional simulation performs solely a functional analysis, while timing simulation takes the real chip and rooting into consideration. This requires that the signals are connected to pins. However, calculations become far more complex, which take much computation time in large-scale projects.

Time and Frequency Analysis

With ispLEVER you can calculate the delay times of the signals in the circuit, although this is only possible with newer components and does not work with old GALs.

Timing analysis can be started as soon as a MACH component has been selected and marked in the source window. Double click on the Timing Analysis button in the process window.

On the left hand side you can choose from several analysis types (see Fig. 29.2.40). The analysis of the maximum frequency is most important for obtaining a first impression. This illustrates the general usability of the design. If you continue to point the cursor to

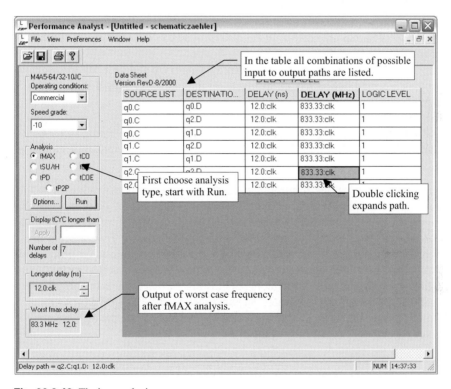

Fig. 29.2.40. Timing analysis

Fig. 29.2.41. Timing analysis, logic path information

one of the seven points for several seconds, you will receive more information. The Run option starts the analysis chosen.

The analysis result is presented in the window to the right. Detailed information on the transit times in the chip can be obtained by double clicking on one of the elements (see Fig. 29.2.41).

29.2.5
Optimization

There are different ways of improving a design. First you have to define the intended design goal. Is it to consume as few resources as possible or is it to achieve the highest possible clock frequency?

With the information obtained from the analysis you can alter the allocated modules in order to improve the design.

A simple way of increasing the permissible frequency is to choose a faster chip. However, such a chip may not be readily available on the market or may draw a too much current or may be too expensive.

Often the only remaining option is to change the existing design. This may involve changing the pin assignment (Constraint Editor) or redefining the operating conditions of the Fitter in the "Optimization Constraint Editor" in the process window of the device entry.

Maximum frequency versus minimum space requirement: By activating the "Optimization Constraint Editor" from the process window of the device entry you are able to configure the fitting constraints. By double clicking each table entry respectively you are able to select the possible values from the drop down menu. By changing line 6 ("node_collapsing_mode") from "Speed" to "Area" the fitter is optimized for bringing big designs in a small chip: The logic cells of the chips are used to full capacity. Use this option if the fitting process failes with "need more pins". Be aware that the area option optimized packing the design at the cost of signal transition time.

In small designs, as in the examples described here, these effects are not noticeable as the module capabilities are not fully utilised.

Without fixed pin assignment: If the desired maximum frequency cannot be achieved by the methods described above, you still have the possibility of not defining the pin assignment. This enables the fitter to spread the design more suitably on the chip. But this means that the printed circuit board has to be adapted to the pin assignment of the PLD.

Constraint Editor: The Constraint Editor not only allows you to change the pin assignment in some chip types but also to influence the assignment of the macrocells, blocks and segments. So you can force corresponding functions in one PAL-block and reduce transmission times on the chip.

29.2.6
Programming

Generally, the newer PLDs are programmable in the circuit (ISP – In System Programmable). This makes programming equipment unnecessary since the required programming logic is located on the chip. Sometimes "ISP" is included in the model designation,

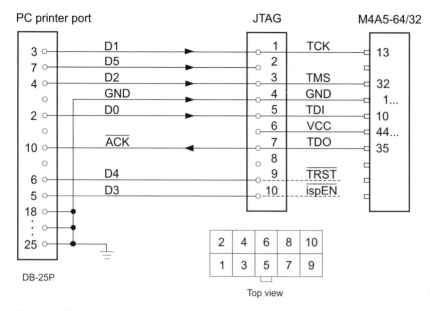

Fig. 29.2.42. Passive JTAG download cable with connections according to Lattice

as is the case with the chip from the ispM4A family used in our example. Programming is done via the standardised JTAG interface[1] which is also used to test the circuits.

To program the component all you need is a download program to transfer the JEDEC file to the chip via a download cable. The download program for Lattice-products is named ispVM. ispLEVER software contains the download program ispVM which uses the parallel port (printer interface). The necessary connections are shown in Fig. 26.2.42.

A download cable can be ordered from Lattice under the order number HW-DL-3C and is used to connect the PC parallel interface to the standardised JTAG plug on the printed board of the PLD. As an extra measure, the Lattice cable also features integrated drivers (74VHC244) to guarantee the correct levels at the PLD even in unfavourable conditions. In most cases a simple passive cable like the one shown in Fig. 29.2.42 is sufficient. For

JTAG pin	Signal name	Meaning
1	TCK	Test clock
2		not used
3	TMS	Test mode select
4	GND	Ground
5	TDI	Test data in
6	VCC	Interface supply
7	TDO	Test data out
8		not used
9	TRST	Test reset
10	ispEN	Enable programming

Fig. 29.2.43. Signals in the JTAG connector

---

[1] IEEE 1149.1 Boundary Scan Test Interface from the Joint Test Action Group (JTAG)

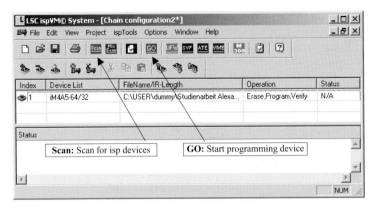

Fig. 29.2.44. ispVM-System

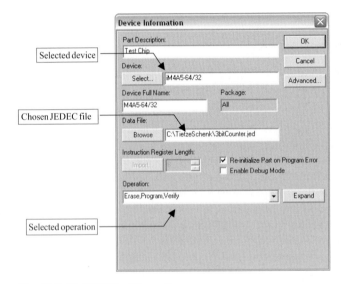

Fig. 29.2.45. JTAG Part Properties

the JTAG connection the download cable has a 10-pin socket like those commonly used on interface cables for PC mainboards. The matching 10-pin plug is located on the printed board of the PLD to be programmed. The names and meaning of the signals in the JTAG interface are listed in Fig. 29.2.43. As seen in our example, the TRST and the ENABLE signals are not required in many PLDs.

To program a device start the program ispVM (Fig 29.2.44).

− At first install the download-cable, insert the PLD and switch the power on for the board.
− Press the "Scan" button in Fig. 29.2.44. Now all PLDs in the chain of the circuit are scanned and displayed in the right order.
− Double click on the Device in the list to be changed. Insert in the dialog the information on the Device by choosing Select (resulting in the menu in Fig. 29.2.46), the Data File (JEDEC-file) and programming operation as seen in Fig. 29.2.45.

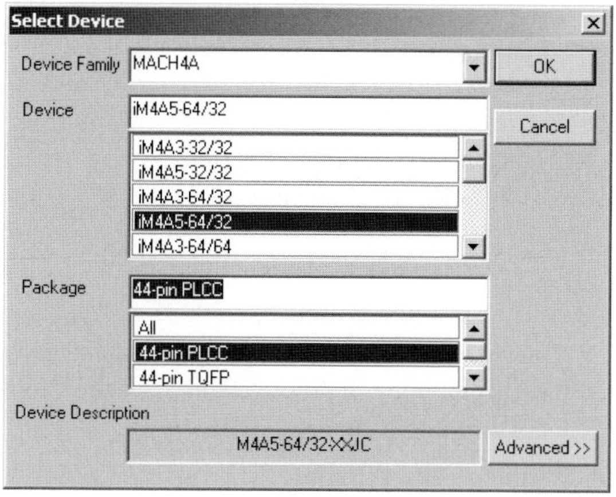

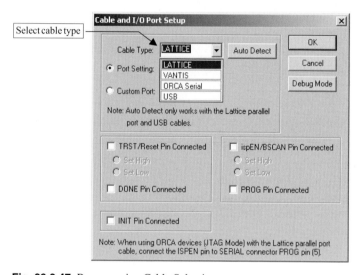

Fig. 29.2.46. Select Device

Fig. 29.2.47. Programming Cable Selection

– With Options/Cable and I/O Port Setup you select the PC port and the type of the download-cable (Fig. 29.2.47).
– Start programming with "Go" from the main window (Fig. 29.2.44)

29.2.7
Outlook

We hope that the information provided in this chapter will enable you to design your own circuits using ispLEVER and program them in a PLD. For those of you requiring extra or additional information we suggest using the help menue in ispLEVER.

If a design is not only to be simulated but also to be tested in a circuit, the evaluation board from Lattice is particularly useful. The provided configuration differs for the mounted devices CPLDs and FPGAs.

29.3
Passiv RC and LRC Networks

RC Networks are of fundamental importance to circuit design. As their effect is the same in all circuits, their operation will be described in some detail.

29.3.1
The Lowpass Filter

A lowpass filter is a circuit which passes low-frequency signals unchanged and attentuates at high frequencies, introducing a phase lag. Figure 29.3.1 shows the simplest type of *RC* lowpass filter circuit.

Frequency-domain analysis

To calculate the frequency response of the circuit, we use the voltage divider formula, written in complex notation as:

$$\underline{A}(s) = \frac{V_o}{\underline{V}_i} = \frac{1/(sC)}{R + 1/(sC)} = \frac{1}{1 + sRC} \tag{29.3.1}$$

Factoring according to

$$\underline{A} = |\underline{A}|e^{j\varphi}$$

we obtain the frequency response of the absolute value or magnitude and of the phase shift:

$$|\underline{A}| = \frac{1}{\sqrt{1 + \omega^2 R^2 C^2}} \quad , \quad \varphi = -\arctan \omega RC \tag{29.3.2}$$

The two curves are shown in Fig. 29.3.2.
  To calculate the 3 dB cutoff frequency f_c, we substitute

$$|\underline{A}| = \frac{1}{\sqrt{2}} = \frac{1}{\sqrt{1 + \omega_g^2 R^2 C^2}}$$

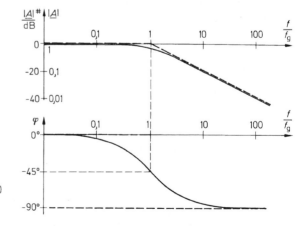

Fig. 29.3.1. Simple lowpass filter   **Fig. 29.3.2.** Bode plot of a lowpass filter

inti (29.3.2), which gives

$$f_c = \frac{\omega_g}{2\pi} = \frac{1}{2\pi RC} \qquad (29.3.3)$$

From (29.3.2) the phase shift at this frequency is $\varphi = -45°$.

As we can see from Fig. 29.3.2 the amplitude-frequency response $|\underline{A}| = \widehat{V}_o/\widehat{V}_i$ can be easily constructed using the two asymptotes:

1. At low frequencies $f \ll f_c$, $|\underline{A}| = 1 \hat{=} 0$ dB.
2. At high frequencies $f \gg f_c$, from (29.3.2) $|\underline{A}| \approx 1/\omega RC$, i.e. the gain is inversely proportional to the frequency. When the frequency is increased by a factor of 10, the gain is reduced by the same factor, i.e. it decreases by 20 dB decade or 6 dB octave.
3. At $f = f_c$ ist $|\underline{A}| = 1/\sqrt{2} \hat{=} - 3$ dB.

Time-Domain Analysis

In order to analyze the circuit in the time domain, we apply a step function of voltage to the input, as shown in Fig. 29.3.3. To calculate the output voltage, we apply Kirchhoff's current law to the (unloaded) output and obtain in accordance with Fig. 29.3.1

$$\frac{V_i - V_o}{R} - I_C = 0$$

With $I_C = C\dot{V}_o$, we obtain the differential equation

$$RC\dot{V}_o + V_o = V_i = \begin{cases} V_r & \text{for } t > 0 \text{ in Case a} \\ 0 & \text{for } t > 0 \text{ in Case b} \end{cases} \qquad (29.3.4)$$

It has the following solutions

$$\begin{array}{cc} \text{Case a:} & \text{Case b:} \\ V_o(t) = V_r\left(1 - e^{-t/RC}\right) & V_o(t) = V_r e^{-t/RC} \end{array} \qquad (29.3.5)$$

This curve is also plotted in Fig. 29.3.3. We can see that the steady-state values $V_o = V_r$ or $V_o = 0$ are only attained asymptotically. As a measure of the response time, a *time constant* τ is therefore defined. This indicates how long it takes for the deviation from the

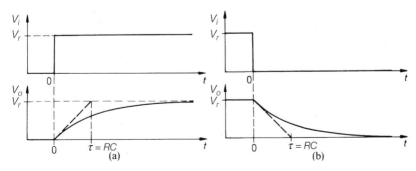

Fig. 29.3.3 a and **b.** Step-response of a lowpass filter

Response accurace	37%	10%	1%	0,1%
Response time	τ	$2, 3\tau$	$4, 6\tau$	$6, 9\tau$

Fig. 29.3.4. Response time of a lowpass filter

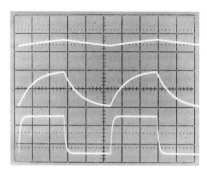

Fig. 29.3.5. Square-wave response of a lowpass filter for various frequencies

Upper trace:	$f_i = 10 f_c$
Middle trace:	$f_i = f_c$
Lower trace:	$f_i = \frac{1}{10} f_c$

steady-state value to equal $1/e$ times the step magnitude. From (29.3.5) the time constant is

$$\tau = RC \qquad\qquad (29.3.6)$$

The response time for smaller deviations can also be derived from (29.3.5). Figure 29.3.4 lists a number of important parameters.

If a square-wave voltage of period T is applied as the input signal, the e-function is truncated after time $T/2$ by the subsequent step. Which final value is obtained at the output depends on the ratio between the time $T/2$ and the time constant τ. This characteristic is clearly illustrated by the oscillogram in Fig. 29.3.5.

Lowpass filter as an integrating circuit: In the previous section we saw that the alternating output voltage is small compared with the input voltage if a signal frequency $f \gg f_c$ is selected. The lowpass filter operates then as an integrating circuit. This property can be inferred directly from differential equation (29.3.4). Assuming that $|V_o| \ll |V_i|$, it follows:

$$RC\dot{V}_o = V_i,$$

$$V_o = \frac{1}{RC} \int\limits_0^t U_e(\tilde{t})d\tilde{t} + V_o(0)$$

Lowpass filter as an averaging circuit: For unsymmetrical alternating voltages, the above condition $f \gg f_c$ is not satisfied. The Fourier expansion in fact contains a constant which is identical to the *arithmetic mean*

$$\overline{V}_i = \frac{1}{T} \int\limits_0^T V_i(t)\, dt$$

where T is the period of the input voltage. If all the higher-order terms of the Fourier series are combined, a voltage $V_i'(t)$ is obtained whose characteristic corresponds to that of the input voltage, but which is displaced from zero such that its arithmetic mean is zero. The input voltage may therefore be expressed in the form

$$V_i(t) = \overline{V}_i + V_i'(t)$$

For voltage $U ein'(t)$, the condition $f \gg f_c$ can be satisfied; it is integrated, whereas the DC component is transferred linearly. The output voltage therefore becomes

$$V_o = \underbrace{\frac{1}{RC} \int_0^t V_i'(\tilde{t}) \, d\tilde{t}}_{\text{residual ripple}} + \underbrace{\overline{V}_i}_{\text{mean value}} \tag{29.3.7}$$

If the time constant

$$\tau = RC$$

is made sufficiently large, the ripple is insignificant compared with the mean value and we get

$$V_o \approx \overline{V}_i \tag{29.3.8}$$

Rise time and cutoff frequency: Another parameter for characterizing lowpass filters is the rise time t_r. This denotes the time taken for the output voltage to rise from 10 to 90% of the final value when a step is applied to the input. From the e-function in (29.3.5)

$$t_r = t_{90\%} - t_{10\%} = \tau(\ln 0.9 - \ln 0.1) = \tau \ln 9 \approx 2.2\tau$$

Consequently, with $f_c = 1/2\pi\tau$

$$\boxed{t_r \approx \frac{1}{3 f_c}} \tag{29.3.9}$$

In approximation this relation is also true for higher-order lowpass filters.

If a number of lowpass filters with various rise times t_{ri} are connected in series, the resultant rise time is

$$t_r \approx \sqrt{\sum_i t_{ri}^2} \tag{29.3.10}$$

and the cutoff frequency is

$$f_c \approx \left(\sum_i f_{ci}^{-2} \right)^{-\frac{1}{2}}$$

Hence, for n lowpass filters having the same cutoff frequency

$$\boxed{f_c \approx \frac{f_{ci}}{\sqrt{n}}} \tag{29.3.11}$$

29.3.2
The Highpass Filter

A highpass filter is a circuit which passes high-frequency signals unchanged and attenuates at low frequencies, introducing a phase lead. Figure 29.3.6 shows the simplest form of RC

highpass filter circuit. The frequency response of the gain and phase shift is again obtained from the voltage divider formula:

$$\underline{A}(s) = \frac{V_o}{\underline{V}_i} = \frac{R}{R + 1/(sC)} = \frac{1}{1 + 1/(sRC)} \qquad (29.3.12)$$

This yields

$$|\underline{A}| = \frac{1}{\sqrt{1 + 1/\omega^2 R^2 C^2}} \quad \text{and} \quad \varphi = \arctan \frac{1}{\omega RC} \qquad (29.3.13)$$

The two curves are shown in Fig. 29.3.7. For the cutoff frequency, we obtain as with the lowpass filter:

$$f_c = \frac{1}{2\pi RC} \qquad (29.3.14)$$

At this frequency the phase shift is $+45°$.

As in the case of the lowpass filter, the amplitude-frequency response can be easily plotted on a double-logarithmic scale using the asymptotes:

1) At high frequences $f \gg f_c$, $|\underline{A}| = 1 \,\hat{=}\, 0$ dB.
2) At low frequencies $f \ll f_c$, from (29.3.13) $|\underline{A}| \approx \omega RC$, i.e. the gain is proportional to the frequency. The slope of the asymptote is therefore $+20$ dB/decade or $+6$ dB/octave.
3) For $f = f_c$, $|\underline{A}| = 1/\sqrt{2} \,\hat{=}\, -3$ dB, as with the lowpass filter.

To calculate the step response, we apply Kirchhoff's current law to the (unloaded) output:

$$C \cdot \frac{d}{dt}(V_i - V_o) - \frac{V_o}{R} = 0 \qquad (29.3.15)$$

With $\dot{V}_i = 0$, this yields the differential equation

$$RC\dot{V}_o + V_o = 0 \qquad (29.3.16)$$

the solution of which is

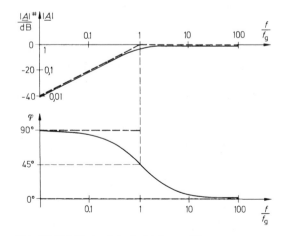

Fig. 29.3.6. Simple highpass filter   Fig. 29.3.7. Bode plot of a highpass filter

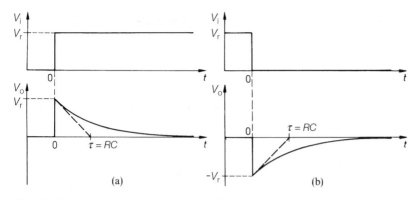

Fig. 29.3.8a,b. Step response of a highpass filter

$$V_o(t) = V_{00}e^{-\frac{t}{RC}} \tag{29.3.17}$$

The time constant is therefore $\tau = RC$, as in the case of the lowpass filter.

In order to determine the initioal value $V_{00} = V_o(t = 0)$, we have to consider that at the instant when the input voltage changes abruptly, the capacitor charge remains unchanged. The capacitor therefore acts as a voltage source of value $V = Q/C$. The output voltage accordingly shows the same stem ΔV as the input voltage. If V_i goes from zero to V_r, the output voltage likewise jumps from zero to V_r, (see Fig. 29.3.8 a) then decays exponentially to zero again in accordance with (29.3.17)

If the input voltage now goes abruptly from V_r to zero, V_o jumps from zero to $-V_r$ (see Fig. 29.3.8 b). Note that the output voltage assumes negative values even though the input voltage is always positive. This distinctive characteristic is frequently used in circuit design.

Use as an RC coupling network: If a square-wave voltage periodic in $T \ll \tau$ is applied to the input, the capacitor charge barely changes during one half-cycle; the output voltage is identical to the input voltage apart from an additive constant. As no direct current can flow via the capacitor, the arithmetic mean of the output voltage is zero. No DC component of th einput voltage is therefore transferred. It is this property which enables a highpass filter to be used as an RC coupling network.

Use as a differentiating circuit: If input voltages with frequencies $f \ll f_c$ are applied, $|\underline{V}_o| \ll |\underline{V}_i|$. Consequently, from differential equation (29.3.15)

$$V_o = RC\frac{dV_i}{dt}$$

Low-freqency input voltages are therefore differentiated.

The oscillograms in Fig. 29.3.9 summarize the transient response of a highpass filter.

Series Connection of Several Highpass Filters

If a number of highpass filters are connected in series, the resultant cutoff frequency is

$$f_c \approx \sqrt{\sum_i f_{gi}^2} \tag{29.3.18}$$

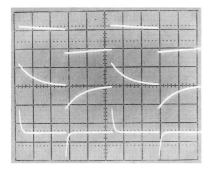

Fig. 29.3.9. Square-wave response of a highpass filter for various frequencies
Upper trace: $f_i = 10 f_c$
Middle trace: $f_i = \quad f_c$
Lower trace: $f_i = \frac{1}{10} f_c$

Consequently, for n highpass filters having identical cutoff frequencies

$$\boxed{f_c \approx f_{ci} \cdot \sqrt{n}}$$

(29.3.19)

29.3.3
Compensated Voltage Divider

It is frequently the case that a resistive voltage divider is capacitively loaded, making it a lowpass filter. The lower the resistance selected for the voltage divider, the higher the cutoff frequency of the filter. However, limits are imposed in that the input resistance of the divider should not be reduced below a specified value.

Another possible way of raising the cutoff frequency is to use a highpass filter to compensate for the effect of the lowpass filter. This is the purpose of capacitor C_k in Fig. 29.3.10. It is dimensioned such that the resultant parallel-connected capacitive voltage divider has the same division ratio as the resistive voltage divider. Consequently, the same voltage division is produced at high and low frequencies. This means that

$$\frac{C_k}{C_L} = \frac{R_2}{R_1}$$

For optimum adjustment of C_k, it is useful to test with the step function. For the correct value C_k, the step response becomes ideal.

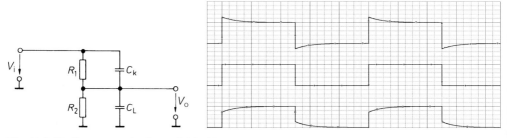

Fig. 29.3.10. Compensated voltage divider. Squarewave response:
Upper trace: overcompensated;
Middle trace: correct coompensated;
Lower trace: undercompensated

29.3.4
Passive *RC* Bandpass Filter

By connecting a highpass and a lowpass filter in series, we obtain a bandpass filter whose output voltage is zero for high and low frequencies. One widely used combination is shown in Fig. 29.3.11. We shall now calculate the output voltage at medium frequencies and the phase shifts introduced. In complex notation, the formula for the unloaded voltage divider yields:

$$\frac{V_o}{V_i} = \frac{\dfrac{1}{\dfrac{1}{R} + sC}}{\dfrac{1}{\dfrac{1}{R} + sC} + R + \dfrac{1}{sC}} = \frac{sRC}{1 + 3sRC + s^2 R^2 C^2}$$

Simplifying with $s_n = sRC$, we obtain for the gain

$$A(s_n) = \frac{V_o}{V_i} = \frac{s_n}{1 + 3s_n + s_n^2} \tag{29.3.20}$$

Hence the magnitude and phase shift are given by

$$|A| = \frac{1}{\sqrt{\left(\dfrac{1}{\omega_n} - \omega_n\right)^2 + 9}}, \quad \varphi = \arctan \frac{1 - \omega_n^2}{3\omega_n} \tag{29.3.21}$$

The output voltage is maximum for $\omega_n = 1$. The resonant frequency is therefore

$$f_r = \frac{1}{2\pi RC} \tag{29.3.22}$$

The quantity ω_n initially introduced for simplification expresses the normalized frequency

$$\omega_n = \frac{\omega}{\omega_r} = \frac{f}{f_r}$$

The phase shift at resonance is zero and the gain $A_r = \frac{1}{3}$. The frequency response of $|A|$ and φ is shown in Fig. 29.3.12.

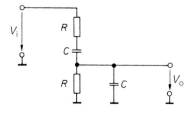

Fig. 29.3.11. Passive *RC* bandpass filter

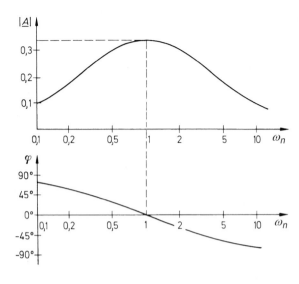

Fig. 29.3.12. Bode plot of the passiven RC bandpass filter

$$\omega_n = \frac{\omega_0}{\omega_r}; \ \omega_r = \frac{1}{RC}$$

29.3.5
Wien–Robinson Bridge

If the bandpass filter shown in Fig. 29.3.11 is modified by inserting resistors R_1 and $2R_1$, as shown in Fig. 29.3.13, a Wien–Robinson bridge is obtained. The resistive voltage divider furnishes the voltage $\frac{1}{3}V_i$ irrespective of frequency. At the resonant frequency, the output voltage is therefore zero. Unlike the bandpass filter response, the frequency response of the Wien–Robinson bridge is then minimum. The circuit is therefore useful for suppressing a given frequency band. The output voltage can be calculated from (29.3.20):

$$\frac{\underline{V}_o}{\underline{V}_i} = \frac{1}{3} - \frac{s_n}{1 + 3s_n + s_n^2}$$

Hence

$$A(s_n) = \frac{1}{3} \cdot \frac{1 + s_n^2}{1 + 3s_n + s_n^2} \tag{29.3.23}$$

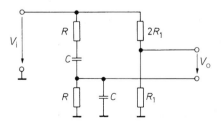

Fig. 29.3.13. Wien–Robinson bridge

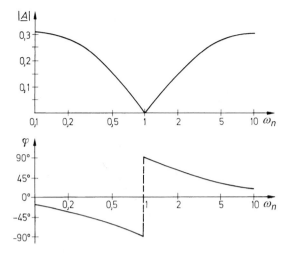

Fig. 29.3.14. Bode plot of the Wien–Robinson Bridge

$$\omega_n = \frac{\omega_0}{\omega_r}; \ \omega_r = \frac{1}{RC}$$

The magnitude of the gain and the phase shift are

$$|\underline{A}| = \frac{|1 - \omega_n^2|}{3\sqrt{(1 - \omega_n^2)^2 + 9\omega_n^2}}, \quad \varphi = \arctan \frac{3\omega_n}{\omega_n^2 - 1}$$

The frequency response $|\underline{A}|$ and φ is plotted in Fig. 29.3.14. For high and low frequencies the gain is $|\underline{A}| = \frac{1}{3}$; the Q-Factor amounts here to $Q = \frac{1}{3}$.

29.3.6
Parallel-T Filter

The parallel-T filter in Fig.29.3.15 has a very similar frequency response to that of the Wien–Robinson bridge and is therefore also suitable for suppressing a given range of frequencies. However, it differs from the Wien–Robinson bridge in that the output voltage can be measured with respect to ground. For high and low frequencies $\underline{V}_o = \underline{V}_i$. High frequencies are transferred without attenuation via the two capacitors C and low frequencies via the two resistors R.

To calculate the frequency response we apply Kirchhoff's current law to points 1, 2 and 3 in Fig. 29.3.15 and obtain, for an unloaded output:

Node *1*: $\quad \dfrac{\underline{V}_i - \underline{V}_1}{R} + \dfrac{\underline{V}_o - \underline{V}_1}{R} - \underline{V}_1 \cdot 2sC = 0$

Node 2: $\quad (\underline{V}_i - \underline{V}_2)sC + (\underline{V}_o - \underline{V}_2)sC - \dfrac{2\underline{V}_2}{R} = 0$

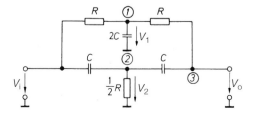

Fig. 29.3.15. Parallel-T filter

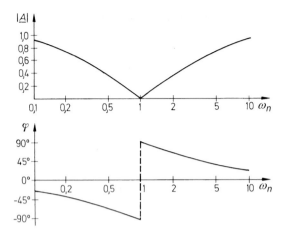

Fig. 29.3.16. Bode plot of the parallel-T filter

$$\omega_n = \frac{\omega_0}{\omega_r}; \ \omega_r = \frac{1}{RC}$$

Node 3: $(V_2 - V_o)sC + \dfrac{V_1 - V_o}{R} = 0.$

Thus, be eliminating V_1 and V_2 and with the normalization $s_n = sRC$, we obtain for the gain:

$$A(s_n) = \frac{1 + s_n^2}{1 + 4s_n + s_n^2} \tag{29.3.24}$$

Hence the magnitude and phase shift are given by:

$$|A| = \frac{|1 - \omega_n^2|}{\sqrt{(1 - \omega_n^2)^2 + 16\omega_n^2}}, \qquad \varphi = \arctan \frac{4\omega_n}{\omega_n^2 - 1}$$

The two curves have been plotten in Fig. 29.3.16. It can be seen that the gain here is also zero at the resonant frequency $f_r = 1/(2\pi RC)$. For low and high frequencies the gain is $|A| = 1$; the Q-factor here is even lower with $Q = \frac{1}{4}$.

29.3.7
Resonant Circuit

Series resonant circuit	**Parallel resonant circuit**

Impedance:

$$\underline{Z} = R + sL + \frac{1}{sC}$$

$$\underline{Z} = \frac{R + sL}{1 + sRC + s^2LC}$$

$$\underline{Z} \approx \frac{sL}{1 + sRC + s^2LC}$$

Resonant frequency:

$$f_r = \frac{1}{2\pi\sqrt{LC}}$$

$$f_r \approx \frac{1}{2\pi\sqrt{LC}}$$

Series resonant circuit

Resonant resistance:

$$\underline{Z}_r = R$$

Bandwidth:

$$B = \frac{R}{2\pi L}$$

Quality factor:

$$Q = \frac{f_r}{B} = \frac{1}{R}\sqrt{\frac{L}{C}}$$

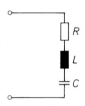

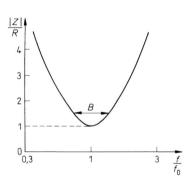

Fig. 29.3.17a. Series resonant circuit

Fig. 29.3.18a. Frequency response for $Q = 2$

Parallel resonant cicuit

Resonant resistance:

$$\underline{Z}_r \approx \frac{L}{RC} = Q^2 R$$

Bandwidth:

$$B \approx \frac{R}{2\pi L}$$

Quality factor:

$$Q = \frac{f_r}{B} \approx \frac{1}{R}\sqrt{\frac{L}{C}}$$

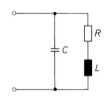

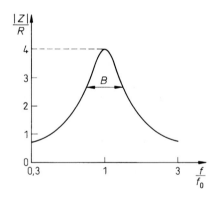

Fig. 29.3.17b. Parallel resonant circuit

Fig. 29.3.18b. Frequency response for $Q = 2$

29.4
Definitions and Nomenclature

We hope that the following list of definitions will help to avoid confusion and enable a better understanding. Where possible, the definitions are based on IEC recommendations

Voltage. A voltage between two points x and y is denoted by V_{xy}. It is defined as being positive if point x is positive with respect to point y, and negative, if point x is negative with respect to point y. Therefore, $V_{xy} = -V_{yx}$. The statement

$$
\begin{aligned}
V_{BE} &= -5\,\mathrm{V} \quad \text{or} \\
-V_{BE} &= 5\,\mathrm{V} \quad \text{or} \\
V_{EB} &= 5\,\mathrm{V}
\end{aligned}
$$

thus indicates that there is a voltage of 5 V between E and B where E is positive with respect to B. In a circuit diagram, the double indices are often omitted and the notation V_{xy} is replaced by a voltage arrow V *pointing from node x to node y*

Potential The Potential V is the voltage of a node with relation to a common reference node 0 or ground:

$$
V_x = V_{x0}
$$

In electrical circuits, the referenc epotential is denoted by a ground symbol. Often V_x is used when actually implying V_x. Specialists then speak, although not quite correctly, of the voltage of a node, e.g. the anode voltage. For the voltage between two nodes, x and y,

$$
V_{xy} = V_x - V_y
$$

Current. The current is indicated by a current arrow on the connecting line. One defines the current I as being positive if the current in its conventional sense, i.e. the transport of positive charge, flows in the direction of the arrow. I is thus positive if the arrow of the current flowing through a load points from the larger to the smaller potential. The directions of the current and voltage arrows in a circuit diagram are not important as long as the actual values of V and I are given the correct signs.

If current and voltage arrow of a circuit element have the same direction, Ohm's law, with the above definitions, is $R = V/I$. If they have opposite directions it changes to $R = -V/I$. This fact is illustrated in Fig. 29.4.1.

Resistance. If the resistance is voltage- or current-dependent, a static resistance $R = V/I$ and an incremental resistance $r = \partial V/\partial I \approx \Delta V/\Delta I$ can be defined. These formulas are valid if the voltage and current arrows point in the same direction. If the directions are opposed, a negative sign must be inserted, as in Fig. 29.4.1.

Voltage and current source. A real voltage source can be described by the equation

$$
V_o = V_0 - R_{int} I_o \tag{29.4.1}
$$

$$R = \frac{V}{I}$$

$$R = -\frac{V}{I}$$

Fig. 29.4.1. Ohm's law

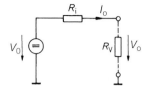

Fig. 29.4.2. Equivalent circuit of a real voltage source

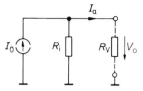

Fig. 29.4.3. Equivalent circuit of a real current source

where V_0 is the no-load voltage and $R_{int} = -dV_o/dI_o$ the internal resistance. This is represented by the equivalent circuit in Fig. 29.4.2. An ideal voltage source is characterized by the property $R_{int} = 0$, i.e. the output voltage is independent of the current.

A different equavalent circuit for a real voltage source can be deduced by rewriting (29.4.1):

$$I_o = \frac{V_0 - V_o}{R_{int}} = I_{sc} - \frac{V_o}{R_{int}} \qquad (29.4.2)$$

where $I_{sc} = V_0/R_{int}$ is the short-circuit current. The appropriate circuit is shown in Fig. 29.4.3. It is obvious that, the large R_{int}, the less the output current depends on the output voltage. For $R_{int} \to \infty$, one obtains an ideal current source.

According to Figs. 29.4.2 and 29.4.3, a real voltage source can be represented either by an ideal voltage source or by an ideal current source. Which representation is chosen depends on whether the internal resistance R_{int} is small or large in comparison with the load resistance R_L

Kirchhoff's current law (KCL). For the calculation of the parameters of many electronic circuits, we use Kirchhoff's current law. It states that the sum of all currents flowing into a node is zero. Currents fldue toward the node are counted as being positive, and currents flowing from the node are negative. Figure 29.4.4 demonstrates this fact. It can be seen that, for Node N

$$\sum_i I_i = I_1 + I_2 - I_3 = 0$$

Ohm's law states that

$$I_1 = \frac{V_1 - V_3}{R_1}$$

$$I_2 = \frac{V_2 - V_3}{R_2}$$

$$I_3 = \frac{V_3}{R_3}$$

By substitution we obtain

$$\frac{V_1 - V_3}{R_1} + \frac{V_2 - V_3}{R_2} - \frac{V_3}{R_3} = 0$$

giving the result

$$V_3 = \frac{V_1 R_2 R_3 + V_2 R_1 R_3}{R_1 R_2 + R_1 R_3 + R_2 R_3}$$

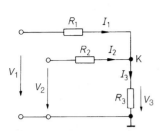

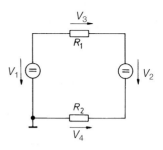

Fig. 29.4.4. Example to demonstrate Kirchhoff's current law (KCL)

Fig. 29.4.5. Example to demonstrate Kirchhoff's voltage law (KVL)

Kirchhoff's voltage law (KVL). Kirchoff's voltage law states that the sum of all voltages around any loop in an electrical network is zero. The voltage is entered in the appropriate equation with a positive sign if its arrow points in the direction in which one proceeds around the loop; the voltage is entered with a negative sign if the voltage arrow points against this direction. For the example in Fig. 29.4.5

$$\sum_i V_i \ = \ V_1 + V_4 - V_2 - V_3 = 0$$

AC circuits (alternating-current circuits). If the circuit can be described by a DC (direct-current) transfer characteristic $V_o = f(V_i)$, this relationship necessarily also holds for any time-dependent voltage, i.e. $V_o(t) = f[V_i(t)]$, as long as the changes in the input voltage are quasi stationary, i.e. not too fast. For this reason, we use upper-case letters for DC as well as for time-dependent quantities, e.g. $V = V(t)$.

However, cases exist wehre a transfer characteristic is only valid for alternating voltages without DC components and it is therefore sensible to have a special symbol to dinstinguish such alternating voltages. We use the lower-case letter u to denote their instantaneous values.

A particularly important special case is that of sinusoidally alternating voltages, i.e.

$$v(t) \ = \ \widehat{V} \ \cos(\omega t + \varphi_u) \tag{29.4.3}$$

where $\widehat{V}$ is the peak value (amplitude). Other values for the caracterization of the voltage magnitude are the root-mean-square value $V_{\text{eff}} \ = \ \widehat{V}/\sqrt{2}$ or the peak-to-peak value $V_{SS} = 2\,\widehat{V}$.

The calculus for trigonometric functions is rather involved, but that for exponential functions fairly simple. Euler's theorem

$$e^{j\alpha} \ = \ \cos\alpha + j\sin\alpha \tag{29.4.4}$$

enables a sine function to be expressed by the imaginary part of the complex exponential function

$$\sin\alpha = \text{Im}\{e^{j\alpha}\}$$

Equation (29.4.3) can therefore also be written as

$$v = \widehat{V} \cdot \text{Im}\{e^{jt\omega t+\varphi_u}\} = \text{Im}\{\widehat{U}e^{j\varphi_u}e^{j\omega t}\} = \text{Im}\{\widehat{U}e^{j\omega t}\}$$

where $\underline{V} = \widehat{V}e^{j\varphi_v}$ is the complex amplitude. Its magnitude is given by

$$|\underline{V}| \ = \ \widehat{V} \cdot |e^{j\varphi_v}| = \ \widehat{V} \ \sqrt{\cos^2\varphi_v + \sin^2\varphi_v} = \ \widehat{V}$$

i.e. it is equal to the peak value of the sine wave. Time-dependent currents are treated in an analogous way. The corresponding symbols are then

$$I, \quad I(t), \quad i, \quad \hat{I}, \quad \underline{I}$$

Arrows can also be assigned to alternating voltages and currents. Of course, the direction of the arrow then no longer indicates polarity but denotes the mathematical sign with which the values must be entered in the formulas: The rule illustratede in Fig. 29.4.2 for DC voltages also applies in this case.

In analogy to the resistance in a DC circuit, a complex resistance is defined as the impedance $\underline{Z}$

$$\underline{Z} = \frac{\underline{V}}{\underline{I}} = \frac{\hat{V}e^{j\varphi_u}}{\hat{I}e^{j\varphi_i}} = \frac{\hat{V}}{\hat{I}}e^{j(\varphi_u - \varphi_i)} = |\underline{Z}|e^{j\varphi}$$

where φ is the phase angle between current and voltage. If the voltage is leading with respect to the current, φ is positive. For a purely ohmic resistance, $\underline{Z} = R$; for a capacitance

$$\underline{Z} = \frac{1}{j\omega C} = -\frac{j}{\omega C}$$

and for an inductance $\underline{Z} = j\omega L$. Ths laws for the DC circuit quantities can be applied to complex quantities as well

One can also define a complex gain

$$\underline{A} = \frac{\underline{V}_o}{\underline{V}_i} = \frac{\hat{V}_o e^{j\varphi_a}}{\hat{V}_i e^{j\varphi_e}} = \frac{\hat{V}_o}{\hat{V}_i}e^{j(\varphi_a - \varphi_e)} = |\underline{A}|e^{j\varphi}$$

where φ is the phase angle between input and output voltage. If the output voltage is leading with respect to the input voltage, φ is positive; for a lagging output voltage, it is negative.

Power. There are different definitions of power. Their relationship is summarized here: The *instantaneous power* is defined as

$$p(t) = v(t) \cdot i(t). \tag{29.4.5}$$

The *effective power* is the mean instantaneous power. It is found by averaging over one period:

$$P = \frac{1}{T}\int_0^T v(t) \cdot i(t)\, dt. \tag{29.4.6}$$

For a *cosine*-form time behavior, this gives us

$$p(t) = \frac{1}{T}\int_0^T \hat{V}\cos(\omega t + \varphi_u) \cdot \hat{I}\cos(\omega t + \varphi_i)\, dt \tag{29.4.7}$$

$$= \tfrac{1}{2}\hat{V}\hat{I}\cos(\varphi_u - \varphi_i) = V_{eff}I_{eff}\cos(\varphi_v - \varphi_i).$$

The *apparent power* is the effective power for the case that $\varphi_v - \varphi_i$ would occur.

$$S = \frac{1}{2}\hat{V}\hat{I} = V_{eff}I_{eff}. \tag{29.4.8}$$

The *reactive power* is obtained from the equation which gives the effective power by substituting the angle $\varphi_i - 90°$ for the angle φ_i.

$$Q(t) = \frac{1}{T} \int_0^T \hat{V} \cos(\omega t + \varphi_v) \cdot \hat{I} \cos(\omega t + \varphi_i - 90°) \, dt \tag{29.4.9}$$

$$= \tfrac{1}{2} \hat{V} \hat{I} \sin(\varphi_v - \varphi_i) = V_{eff} I_{eff} \sin(\varphi_u - \varphi_i).$$

This results in the relationship

$$P^2 + Q^2 = S^2. \tag{29.4.10}$$

The power factor is a measure of how large the effective power component is:

$$PF = \frac{P}{S} \overset{(29.4.10)}{=} \frac{P}{\sqrt{P^2 + Q^2}} = \cos(\varphi_v - \varphi_i) = \cos\varphi. \tag{29.4.11}$$

The maximum power factor is given for $Q = 0$; the condition $PF = 1$ then applies, and the effective power is equal to the apparent power.

The relationship between the different power parameters can be made clear very easily with the use of phasors: from the definition of the apparent power

$$\underline{S} = \frac{1}{2} \underline{V} \underline{I}^* = \underline{V}_{eff} \underline{I}_{eff}^* \tag{29.4.12}$$

it follows with $\underline{I} = \sqrt{2} I_{eff} e^{j\varphi_i}$ and $\underline{I}^* = \sqrt{2} I_{eff} e^{-j\varphi_i}$ and $\underline{V} = \sqrt{2} V_{eff} e^{j\varphi_v}$ that

$$\underline{S} = V_{eff} I_{eff} e^{j(\varphi_u - \varphi_i)} = V_{eff} I_{eff} \cos(\varphi_v - \varphi_i) + j \, V_{eff} I_{eff} \sin(\varphi_v - \varphi_i) = P + jQ.$$

Parameter	General	Sine
Signal	$v(t)$	$v(t) = \hat{V} \cos(\omega t + \varphi_v)$
	$i(t)$	$i(t) = \hat{I} \cos(\omega t + \varphi_i)$
		$\varphi = \varphi_u - \varphi_i$
Effective value	$V_{eff} = \sqrt{\dfrac{1}{T} \int_0^T v^2(t) \, dt}$	$V_{eff} = \dfrac{\hat{V}}{\sqrt{2}}$
	$I_{eff} = \sqrt{\dfrac{1}{T} \int_0^T i^2(t) \, dt}$	$I_{eff} = \dfrac{\hat{I}}{\sqrt{2}}$
Effective power	$P = \dfrac{1}{T} \int_0^T v(t) i(t) \, dt$	$P = V_{eff} I_{eff} \cos\varphi = \dfrac{\hat{V} \hat{I}}{2} \cos\varphi$
Apparent power	$S = V_{eff} I_{eff}$	$S = V_{eff} I_{eff} = \dfrac{\hat{V} \hat{I}}{2}$
Reactive power	$Q = \sqrt{S^2 - P^2}$	$Q = V_{eff} I_{eff} \sin\varphi = \dfrac{\hat{V} \hat{I}}{2} \sin\varphi$
Power factor	$PF = \dfrac{P}{S}$	$PF = \cos\varphi$

Fig. 29.4.6. Calculation of power parameters

Linear voltage ratio A	Logarithmic voltage ratio $A[\text{dB}]$
10^{-6}	$-120\,\text{dB}$
10^{-3}	$-\ \ 60\,\text{dB}$
10^{-2}	$-\ \ 40\,\text{dB}$
10^{-1}	$-\ \ 20\,\text{dB}$
0.5	$-\ \ \ \ 6\,\text{dB}$
0.7	$-\ \ \ \ 3\,\text{dB}$
1	$0\,\text{dB}$
1.4	$3\,\text{dB}$
2	$6\,\text{dB}$
10	$20\,\text{dB}$
10^2	$40\,\text{dB}$
10^3	$60\,\text{dB}$
10^6	$120\,\text{dB}$

Fig. 29.4.7. Table for conversion of voltage ratios: $A[\text{dB}] = 20\,\text{dB lg } A$

Logarithmic absolute $P[\text{dBm}]$	Linear absolute P	Linear absolute V
$-100\,\text{dBm}$	$100\,\text{fW}$	$2.24\,\mu\text{V}$
$-\ \ 80\,\text{dBm}$	$10\,\text{pW}$	$22.4\,\mu\text{V}$
$-\ \ 60\,\text{dBm}$	$1\,\text{nW}$	$224\,\mu\text{V}$
$-\ \ 40\,\text{dBm}$	$100\,\text{nW}$	$2.24\,\text{mV}$
$-\ \ 20\,\text{dBm}$	$10\,\mu\text{W}$	$22.4\,\text{mV}$
$-\ \ 10\,\text{dBm}$	$100\,\mu\text{W}$	$70.8\,\text{mV}$
$-\ \ \ \ 6\,\text{dBm}$	$250\,\mu\text{W}$	$112\,\text{mV}$
$-\ \ \ \ 3\,\text{dBm}$	$500\,\mu\text{W}$	$159\,\text{mV}$
$0\,\text{dBm}$	$1\,\text{mW}$	$224\,\text{mV}$
$3\,\text{dBm}$	$2\,\text{mW}$	$316\,\text{mV}$
$6\,\text{dBm}$	$4\,\text{mW}$	$448\,\text{mV}$
$10\,\text{dBm}$	$10\,\text{mW}$	$708\,\text{mV}$
$20\,\text{dBm}$	$100\,\text{mW}$	$2.24\ \ \text{V}$

Fig. 29.4.8. Conversions of absolute power
$$P[\text{dBm}] = 10\,\text{dBm lg } \frac{P}{1\,\text{mW}} = 20\,\text{dBm lg } \frac{V}{0.224\,\text{V}}$$

This results in the relation

$$|\underline{S}|^2 = S^2 = P^2 + Q^2$$

in agreement with (29.4.10).

A summary of the different power parameters is given in Fig. 29.4.6. The relationships for any arbitrary time signals are also included here. Substituting cosine-form signals in these general relationships results in the familiar relationships.

Logarithmic voltage ratio. In electrical engineering, a logarithmic value $A[\text{dB}]$ is often used to express the voltage ratio $A = \widehat{V}_o/\widehat{V}_i$, i.e. the gain of a circuit. The relationship between A and $A[\text{dB}]$ is given by

$$A[\text{dB}] = 20\,\text{dB lg } \frac{\widehat{V}_o}{\widehat{V}_i} = 20\,\text{dB lg } A$$

The table in Fig. 29.4.7 lists some values.

Logarithmic power. In radio frequency circuits it is not usual to calculate with voltages but with power rations with respect to a reference level of $0\,\text{dBm}$ for a power of $1\,\text{mW}$ at a resistor of $50\,\Omega$. The "m" in dBm stands for $1\,\text{mW}$. Some examples for the conversion are given in Fig. 29.4.8.

Logarithms. The logarithm of a denominate number (e.g. $\text{lg }10\,\text{Hz}$) is not defined. We therefore write, for example, not $\text{lg } f$ but $\text{lg}(f/\text{Hz})$. The difference of logarithms is a different matter: the expressen $\Delta \text{lg } f = \text{lg } f_2 - \text{lg } f_1$ is well defined since it can bet written in the form $\text{lg}(f_2/f_1)$.

Mathematical symbols. We often use a shortened notation for the time derivatives:

$$\frac{dV}{dt} = \dot{V}, \qquad \frac{d^2V}{dt^2} = \ddot{V}$$

The symbol $\sim$ represents a *proportional* relationship; the symbol $\approx$ stands for *approximately equal to*; the symbol $\hat{=}$ means *corresponding to*. The symbol $||$ means *parallel*. We use it to indicate that resistors are connected in parallel, i.e.

$$R_1||R_2 = \frac{R_1 R_2}{R_1 + R_2}$$

List of the most important symbols

V	any time-dependent voltage, also DC voltages
v	alternating voltage without DC component
$\hat{V}$	amplitude (peak value) of a voltage
$\underline{V}$	complex voltage amplitude
V_{rms}	root-mean-square value of a voltage
U	computing unit voltage
V_T	thermal voltage kT/e_0 (k = Voltzmann's constant, T = absolute temperature, e_0 = charge of electron
V_b	supply voltage = battery voltage
V^+	positive supply potential; in circuit diagrams indicated by $(+)$
V^-	negative supply potential; in circuit diagrams indicated by $(-)$
I	any time-dependent current, also direct currents
i	alternating current without DC component
$\hat{I}$	amplitude (peak value) of a current
$\underline{I}$	complex current amplitude
I_{rms}	root-mean-square value of a current
R	ohmic resistance
r	differential resistance
$\underline{Z}$	complex resistance (impedance)
t	time
τ	time constant
T	period, cycle time
$f = 1/T$	frequency
f_g	3 dB cutoff frequency
f_{gA}	3 dB cutoff frequency of the open-loop gain $\underline{A}_D$ of an operational amplifier
f_T	gain-bandwidth product; unity-gain bandwidth
B	3 dB bandwidth
$\omega = 2\pi f$	angular frequency
$\omega_n = \omega/\omega_0$	normalized angular frequency
$s = \sigma + j\omega$	complex angular frequency
$s_n = s/\omega_0$	normalized complex angular frequency
$A = dV_o/dV_i$	small-signal voltage gain for low frequncies

$\underline{A}(j\omega)=\underline{V}_o/\underline{V}_i$	complex voltage gain; frequency response
$A(s)$	general transfer function
A_D	difference mode gain, open-loop gain of an operational amplifier
g	loop gain
G	common-mode rejection ratio, CMRR
k	feedback factor
$\beta = dI_C/dI_B$	small-signal current gain
$g_m = dI_c/dU_{BE}$	small-signal transconductance
ϑ	temperature on the Celsius scale
T	temperature on the Kelvin scale; absolute temperature
$y = x_1 \cdot x_2$	logic AND operation (conjunktion)
$y = x_1 + x_2$	logic OR operation (disjunction)
$y = \overline{x}$	logic NOT operation (negation)
$y = x_1 \oplus x_2$	logic exclusiv-OR operation
$\dot{x} = dx/dt$	first derivative of x with respect to time
$\ddot{x} = d^2x/dt^2$	second derivative of x with respect to time
$^a\log x$	logarithm to the base a
lg	logarithm to the base 10
ln	logarithm to the base e
ld	logarithm to the base 2
a^*	conjugated complex value of a
	for $a = \mathrm{Re}(a) + j\mathrm{Im}(a)$
	$\Rightarrow a^* = \mathrm{Re}(a) - j\mathrm{Im}(a)$

29.5
Types of the 7400 Digital Families

The importance of primitive logic circuits has decreased since most functions are realized by PLDs and FPGAs today. Therefore the number of manufacturers and types has decreased also. Some manufacturers that produce the 7400 family until today are Fairchild, National; On-Semiconductor; Philips, ST-Microelectronics and naturally Texas Instruments.

Type	NAND-gates	Output	Pins
00	Quad 2 input NAND	TP	14
01	Quad 2 input NAND	OC	14
03	Quad 2 input NAND	TP	14
10	Triple 3 input NAND	TP	14
12	Triple 3 input NAND	OC	14
13	Dual 4 input NAND schmitt-trigger	TP	14
18	Dual 4 input NAND schmitt-trigger	TP	14
20	Dual 4 input NAND	TP	14
22	Dual 4 input NAND	OC	14
24	Quad 2 input NAND schmitt-trigger	TP	14
26	Quad 2 input gate NAND 15V output	OC	14
30	8 input NAND	TP	14
37	Quad 2 input NAND buffer	TP	14
38	Quad 2 input NAND buffer	OC	14
40	Dual 4 input NAND buffer	TP	16
132	Quad 2 input NAND schmitt-trigger	TP	14
133	13 input NAND	TP	16
1000	Buffer '00' gate	TP	14
1003	Buffer '03' gate	TP	14
1010	Buffer '10' gate	TP	14
1020	Buffer '20' gate	TP	14

Type	NOR-gates	Output	Pins
02	Quad 2 input NOR	TP	14
23	Dual 4 input strobe expandable I/P NOR	TP	16
25	Dual 4 input strobe NOR	TP	14
27	Triple 3 input NOR	TP	14
28	Quad 2 input NOR buffer	TP	14
33	Quad 2 input NOR buffer	OC	14
36	Quad 2 input NOR	TP	14
1002	Buffer '02' gate	TP	14

Type	AND-gates	Output	Pins
08	Quad 2 input AND	TP	14
09	Quad 2 input AND	OC	14
11	Triple 3 input AND	TP	14
15	Triple 3 input AND	OC	14
21	Dual 4 input AND	TP	14
1008	Buffer '08' gate	OC	14

TP = Totem Pole, OC = Open Collector, TS = Tristate

Type	OR-gates	Output	Pins
32	Quad 2 input OR	TP	14
802	Triple 4 input OR NOR	TP	
832	Hex 2 input buffer	TP	20
1032	Buffer '32' gate	TP	14

Type	AND-OR-gates	Output	Pins
51	Dual 2 wide input AND-OR-Invert	TP	14
54	4 wide 2 input AND-OR-Invert	TP	14
64	4-2-3-2 input AND-OR-Invert	TP	14

Type	EXOR-gates	Output	Pins
86	Quad exclusive OR	TP	14
136	Quad exclusive OR	OC	14
266	Quad 2 input exclusive NOR	OC	16
386	Quad exclusive OR	TP	14
7266	'266' with totempole output	TP	16

Type	Inverters	Output	Pins
04	Hex inverter	TP	14
05	Hex inverter	OC	14
14	Hex inverter schmitt-trigger	TP	14
19	Hex inverter schmitt-trigger	TP	14
1004	Buffer '04' gate	TP	14
1005	Buffer '05' gate	OC	14

Type	Buffers	Output	Pins
34	Hex buffer	TP	14
35	Hex buffer	OC	14
125	Quad 3 state buffer	TS	14
126	Quad 3 state buffer	TS	14
1034	Hex buffer	TP	14
1035	Hex buffer	OC	14

Type	Line-drivers	Output	Pins
804	Hex 2 input NAND line driver	TP	20
805	Hex 2 input NOR line driver	TP	20
808	Hex 2 input AND line driver	TP	20
832	Hex 2 input OR line driver	TP	20

Type	Flip-Flops, transparent	Output	Pins
75	Quad D-latch	TP	16
77	Quad D-latch	TP	16
279	Hex SR-flip-flop	TP	16
375	Quad D-latch	TP	16

Type	Flip-Flops, Master-Slave		Output	Pins
73	Dual JK-flip-flop, preset, clear		TP	14
74	Dual D-flip-flop, preset, clear		TP	14
76	Dual JK-flip-flop, preset, clear		TP	16
78	Dual JK-flip-flop, preset, clear		TP	14
107	Dual JK-flip-flop, clear		TP	14
109	Dual JK-flip-flop, preset, clear		TP	16
112	Dual JK-flip-flop, preset, clear		TP	16
113	Dual JK-flip-flop, preset		TP	14
114	Dual JK-flip-flop, preset, clear		TP	14
171	Quad D-flip-flop, clear		TP	16
173	Quad D-flip-flop, clear, enable		TS	16
174	Hex D-flip-flop, clear		TP	16
175	Quad D-flip-flop, clear		TP	16
11478	Quad metastable resistant		TP	24

Type	Shift registers		Output	Pins
91	8 bit shift register		TP	14
95	4 bit shift register	PIPO	TP	14
96	5 bit shift register	PI	TP	16
164	8 bit shift register	PO	TP	14
165	8 bit shift register	PI	TP	16
166	8 bit shift register	PI	TP	16
195	4 bit shift register	PIPO	TP	16
299	8 bit shift reg. right/left	PIPO	TS	20
673	16 bit shift register	PO	TP	24
674	16 bit shift register	PI	TP	24

Type	Shift registers with output registers		Output	Pins
594	8 bit shift reg. w. output reg.	PO	TP	16
595	8 bit shift reg. w. output reg.	PO	TS	16
596	8 bit shift reg. w. output reg.	PO	OC	16
597	8 bit shift reg. w. input reg.	PI	TP	16
598	8 bit shift reg. w. input reg.	PIPO	TS	20
599	8 bit shift reg. w. output reg.	PO	OC	16
671	4 bit shift reg. w. outp. reg. right/left	PO	TS	20
672	4 bit shift reg. w. outp. reg. right/left	PO	TS	20
962	8 bit shift reg. dual rank	PIPO	TS	18
963	8 bit shift reg. dual rank	PIPO	TS	20
964	8 bit shift reg. dual rank	PIPO	TS	18

Type	Asynchronous counters	Output	Pins
90	Decade counter	TP	14
92	Divide by 12 counter	TP	14
93	4 bit binary counter	TP	14
293	4 bit binary counter	TP	14
390	Dual decade counter	TP	16
393	Dual 4 bit binary counter	TP	14

PI = parallel input, PO = parallel output

Type	Synchronous counters	Output	Pins
161	4 bit binary counter, sync. load	TP	16
163	4 bit binary counter, sync. load	TP	16
169	4 bit binary up/down counter, sync. load	TP	16
191	4 bit binary up/down counter, async. load	TP	16
193	4 bit binary up/down counter, async. load	TP	16
669	4 bit binary up/down counter, sync. load	TP	16

Type	Synchronous counters with registers	Output	Pins
590	8 bit binary counter w. output reg.	TS	16
592	8 bit binary counter w. input reg.	TP	16
593	8 bit binary counter w. input reg.	TS	20
697	4 bit binary counter w. output reg.	TS	20

Type	Bus-drivers (one direction)	Output	Pins
240	8 bit bus driver, data inverting	TS	20
241	8 bit bus driver	TS	20
244	8 bit bus driver	TS	20
365	6 bit bus driver	TS	16
366	6 bit bus driver, data inverting	TS	16
367	6 bit bus driver	TS	16
368	6 bit bus driver, data inverting	TS	16
465	8 bit bus driver	TS	20
540	8 bit bus driver, data inverting	TS	20
541	8 bit bus driver	TS	20
1240	'240' reduced power	TS	20
1241	'241' reduced power	TS	20
1244	'244' reduced power	TS	20
2240	'240' with serial damping Resistor	TS	20
2241	'241' with serial damping Resistor	TS	20
2244	'244' with serial damping Resistor	TS	20
2410	11 bit bus driver, data noninvert., ser. damp. Res.	TS	28
2541	'541' with serial damping Resistor	TS	20
2827	'827' with serial damping Resistor	TS	24
16240	16 bit bus driver, data inverting	TS	48
16244	16 bit bus driver, data noninverting	TS	48

Type	Bus-drivers with transparent latches	Output	Pins
373	8 bit latch	TS	20
533	8 bit latch, data inverting	TS	20
563	'533' bus pinout	TS	20
573	'373' bus pinout	TS	20
667	8 bit latch, data inverting, readback	TS	24
990	8 bit latch, readback	TP	20
992	9 bit latch, readback	TS	24
994	10 bit latch, readback	TS	24
16373	16 bit latch, data non inverting	TS	48
29841	10 bit latch	TS	24
29843	9 bit latch	TS	24

Type	Bus-drivers with edge-triggered D-Flip-Flops	Output	Pins
273	8 bit D-Flip-Flop with clear	TP	20
374	8 bit D-Flip-Flop	TS	20
377	8 bit D-Flip-Flop with enable	TP	20
563	8 bit D-Flip-Flop, data inverting	TS	20
564	8 bit D-Flip-Flop, data inverting	TS	20
574	'374' bus pinout	TS	20
575	'574' with syncronous clear	TS	24
576	8 bit D-Flip-Flop, data inverting	TS	20
874	8 bit D-Flip-Flop	TS	24
876	8 bit D-Flip-Flop, data inverting	TS	24
996	8 bit D-Flip-Flop, data readback	TS	24
16374	16 bit D-Flip-Flop	TS	48
29821	10 bit D-Flip-Flop	TS	24

Type	Transceivers (bidirectional)	Output	Pins
245	8 bit transceiver, bus pinout	TS	20
645	8 bit transceiver	TS	20
1245	'245' reduced power	TS	20
1645	'645' reduced power	TS	20
2245	'245' with serial damping resistor	TS	20
16245	16 bit transceiver	TS	48

Type	Transceivers with edge-triggered registers	Output	Pins
646	8 bit reg. transceiver	TS	24
16651	16 bit reg. transceiver, data inverting	TS	56
16652	16 bit reg. transceiver	TS	56

Type	Comparators	Output	Pins
85	4 bit magnitude comparator	TP	16
518	8 bit identity comparator	OC	20
520	8 bit identity comparator	TP	20
521	8 bit identity comparator	TP	20
679	12 bit address comparator	TP	20
682	8 bit magnitude comparator	TP	20
684	8 bit magnitude comparator	TP	20
688	8 bit identity comparator w. enable	TP	20

Type	Decoders, Demultiplexers	Output	Pins
42	BCD to 10 line decoder	TP	16
(45	BCD to 10 line decoder	OC	16)
137	3 to 8 line decoder w. addr. latch	TP	16
138	3 to 8 line decoder	TP	16
139	Dual 2 to 4 line decoder	TP	16
154	4 to 16 line decoder	TP	24
155	Dual 2 to 4 line decoder	TP	16
156	Dual 2 to 4 line decoder	OC	16
237	3 to 8 line decoder w. addr. latch	TP	16
238	3 to 8 line decoder	TP	16
259	3 to 8 line decoder w. output latch	TP	16
538	3 to 8 line decoder	TS	20

Type	Multiplexers, digital	Output	Pins
151	8 input multiplexer	TP	16
153	Dual 4 input multiplexer	TP	16
157	Duad 2 input multiplexer	TP	16
158	Quad 2 input multiplexer	TP	16
251	8 input multiplexer	TP/TS	16
253	Dual 4 input multiplexer	TS	16
257	Duad 2 input multiplexer	TS	16
258	Duad 2 input multiplexer	TS	16
352	Dual 4 input multiplexer	TP	16
354	8 input multiplexer w. input data latch	TS	20
356	8 input multiplexer w. data reg.+adr. latch	TS	20
398	Quad 2 input multiplexer w. data reg.	TP	20
857	Hex 2 input multiplexer, masking	TS	24

Type	Priority decoders	Output	Pins
147	10 line to binary priority encoder	TP	16
148	8 line to binary priority encoder	TP	16
348	8 line to binary priority encoder	TS	16

Type	Display decoders	Output	Pins
47	BCD to seven segment for LEDs	OC	16
49	BCD to seven segment for LEDs	OC	16
247	BCD to seven segment for LEDs	OC	16

Type	Monostables	Output	Pins
122	Monostable, retriggerable	TP	14
123	Dual monostable, retriggerable	TP	16
221	Dual monostable	TP	16
423	Dual monostable, retriggerable	TP	16

Type	Oscillators	Output	Pins
624	Voltage controlled oscillator	TP	14
628	Voltage controlled oscillator	TP	14
629	Dual voltage controlled oscillator	TP	16

Type	Phase locked loop	Output	Pins
297	Digital phase locked loop	TP	16

Type	Adders and Arithmetic Logic Units (ALUs)	Output	Pins
83	4 bit binary full adder	TP	16
181	4 bit arithmetic logic unit	TP	24
182	Carry look ahead unit for 4 adders	TP	16
183	Dual carry save full adder	TP	14
283	4 bit binary full adder	TP	16
385	Quad serial adder/subtractor	TP	20
583	4 bit BCD adder	TP	16
881	4 bit arithmetic logic unit with status check	TP	24

Type	Parity generators	Output	Pins
180	8 bit parity generator	TP	14
280	9 bit parity generator/checker	TP	14

29.6
Standard Series

E 3 ±20%	E 6 ±20%	E 12 ±10%	E 24 ±5%	E 48 ±2%	E 96 ±1%
1.0	1.0	1.0	1.0	1.00	1.00
					1.02
				1.05	1.05
					1.07
			1.1	1.10	1.10
					1.13
				1.15	1.15
					1.18
		1.2	1.2	1.21	1.21
					1.24
				1.27	1.27
			1.3		1.30
				1.33	1.33
					1.37
				1.40	1.40
					1.43
				1.47	1.47
	1.5	1.5	1.5		1.50
				1.54	1.54
					1.58
			1.6	1.62	1.62
					1.65
				1.69	1.69
					1.74
				1.78	1.78
		1.8	1.8		1.82
				1.87	1.87
					1.91
				1.96	1.96
			2.0		2.00
				2.05	2.05
					2.10
				2.15	2.15
2.2	2.2	2.2	2.2		2.21
				2.26	2.26
					2.32
				2.37	2.37
			2.4		2.43
				2.49	2.49
					2.55
				2.61	2.61
					2.67
		2.7	2.7	2.74	2.74
					2.80
				2.87	2.87
					2.94
			3.0	3.01	3.01
					3.09
				3.16	3.16
					3.24

E 3 ±20%	E 6 ±20%	E 12 ±10%	E 24 ±5%	E 48 ±2%	E 96 ±1%
	3.3	3.3	3.3	3.32	3.32
					3.40
				3.48	3.48
					3.57
			3.6	3.65	3.65
					3.74
				3.83	3.83
		3.9	3.9		3.92
				4.02	4.02
					4.12
				4.22	4.22
			4.3		4.32
				4.42	4.42
					4.53
				4.64	4.64
4.7	4.7	4.7	4.7		4.75
				4.87	4.87
					4.99
			5.1	5.11	5.11
					5.23
				5.36	5.36
					5.49
		5.6	5.6	5.62	5.62
					5.76
				5.90	5.90
					6.04
			6.2	6.19	6.19
					6.34
				6.49	6.49
					6.65
	6.8	6.8	6.8	6.81	6.81
					6.98
				7.15	7.15
					7.32
			7.5	7.50	7.50
					7.68
				7.87	7.87
					8.06
		8.2	8.2	8.25	8.25
					8.45
				8.66	8.66
					8.87
			9.1	9.09	9.09
					9.31
				9.53	9.53
					9.76

Fig. 29.6.1. Standard series of values per DIN 41426 or IEC 63

29.7
Color code

Color	1st digit	2nd digit	Multiplier	Tolerance
none				± 20 %
silver			× 0.01 Ω	± 10 %
gold			× 0.1 Ω	± 5%
black		0	× 1.0 Ω	± 20 %
brown	1	1	× 10 Ω	± 1%
red	2	2	× 100 Ω	± 2%
orange	3	3	× 1 kΩ	
yellow	4	4	× 10 kΩ	
green	5	5	× 100 kΩ	
blue	6	6	× 1 MΩ	
violet	7	7	× 10 MΩ	
grey	8	8	× 100 MΩ	
white	9	9		

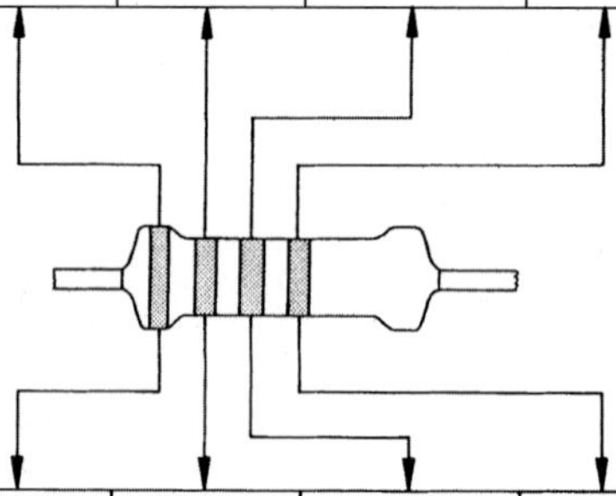

Example	yellow	violet	red	silver
4.7 kΩ	4	7	× 100 Ω	10%

Fig. 29.7.1. 4-ring color code per DIN 41429

Color	1st digit	2nd digit	3rd digit	Multiplier	Tolerance	Temperature coefficient
silver				× 0.01 Ω		
gold				× 0.1 Ω	± 5 %	
black		0	0	× 1.0 Ω		± 250 ppm/K
brown	1	1	1	× 10 Ω	± 1 %	± 100 ppm/K
red	2	2	2	× 100 Ω	± 2 %	± 50 ppm/K
orange	3	3	3	× 1 kΩ		± 15 ppm/K
yellow	4	4	4	× 10 kΩ		± 25 ppm/K
green	5	5	5	× 100 kΩ	± 0.5%	20 ppm/K
blue	6	6	6	× 1 MΩ		± 10 ppm/K
violet	7	7	7	× 10 MΩ		± 5 ppm/K
grey	8	8	8	× 100 MΩ		± 1 ppm/K
white	9	9	9			

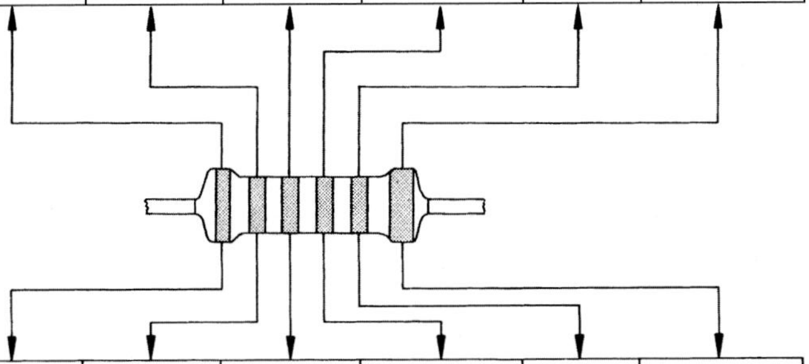

Example	yellow	violet	green	brown	brown	blue
4.75 kΩ	4	7	5	× 10 Ω	1%	10 ppm/K

Fig. 29.7.2. ring color code per IEC 62.
The temperature coefficient is usually only specified if it is less than 50 ppm K

29.8
Manufacturers

Here are listed the most important manufacturers for semiconductors and integrated circuits. The first line gives the short form notation, the second line specifies the internet
address of the homepage, the third line contains the most important products. Some older
manufacturers that are in the meantime transferred to other firms are also mentioned to
enable the reader to find older components.

Actel
http://www.actel.com/
FPGAs

Agilent
http://www.semiconductor.agilent.com/
Optoelectronics, RF-semiconductors

Allegro
http://www.allegromicro.com/
Power-drivers, Hall-Sensors

Altera
http://www.altera.com/
PLDs, FPGAs

AMD, Advanced Micro Devices
http://www.amd.com/
CPUs for PCs, Flash-memory

Analog Devices
http://www.analog.com/
OPAmps, AD-DA-converters, Signalprocessors, sensors

Apex
http://www.apexmicrotech.com/
Power-OPAmps, PWM-amplifiers

Atmel
http://www.atmel.com/
Flash memories, SRAMs, microcontroller

Benchmarq $\Longrightarrow$ Texas Instruments

Burr Brown $\Longrightarrow$ Texas Instruments

Cherry Semiconductor $\Longrightarrow$ ON Semiconductor

Coilcraft
http://www.coilcraft.com
Inductors

Cypress
http://www.cypressmicro.com/
SRAMs, PLDs

Dallas $\Longrightarrow$ Maxim

Datel
http://www.datel.com/
AD-DA-converter

Elantec $\Longrightarrow$ Intersil

Exar
http://www.exar.com/
Line Transceivers, UARTs

Fairchild
http://www.fairchildsemi.com/
Power mosfets, EEPROMs, 7400-Logic

Freescale
http://www.freescale.com/
Microcontrollers, DSPs, sensors

Fuji
http://www.fujisemiconductor.com/
Power mosfets and IGBTs, power modules

Fujitsu
http://www.fujitsumicro.com/
Memories, microcontrollers

General Electric
http://www.gesensing.com/
Sensors

Harris $\Longrightarrow$ Intersil

Honeywell
http://content.honeywell.com
Sensors

Hitachi
http://www.halsp.hitachi.com/
Memories, microcontrollers

Honeywell
www.honeywell.com/sensing
Sensors

IDT
http://www.idt.com/
SRAMs, Fifos, dual port memories

Infineon
http://www.infineon.com/
Power mosfets, memories, optoelectronics

Inmos $\Longrightarrow$ ST Microelectronics

Intel
http://www.intel.com/
CPUs for PCs, flash memories

Intermetall $\Longrightarrow$ Visahy

International Rectifier
http://www.irf.com/
Power mosfets, IGBTs, diodes

Intersil
http://www.intersil.com/
AD-DA-converters, OPAmps, MOS-drivers

IXYS
http://www.ixys.com/
Power mosfets, IGBTs, Diodes

Lattice
http://www.latticesemi.com/
PLDs, FPGAs

LEM
http://www.lem.com
Current sensors

Linear Technology
http://www.linear.com/
OPAmps, voltage regulators, SC-filters, AD-DA-converters, sensors

Maxim
http://www.maxim-ic.com/
OPAmps, voltage regulators, SC-filters, AD-DA-converters
Memories, sensors

Microchip
http://www.microchip.com/
Microcontrollers

Micron
http://www.micron.com/
DRAMs, flash memories

Microsemi
http://www.microsemi.com/
Diodes, transistors, voltage regulators, RF semiconductors

Mini-Circuits
http://www.minicircuits.com/
RF-Amplifiers, -oscillators, -mixer, -transformers, -attenuators

MIPS
http://www.mips.com/
IPs for Signal processors

Mitsubishi
http://www.mitsubishichips.com/
SRAM, DRAM, flash-memories

Monolithic Memories (MMI) ⟹ **AMD**

Motorola ⟹ **Freescale, On Semiconductor**

Murata
http://www.murata.com/
Sensors, passive components

National
http://www.national.com/
OPAmps, voltage regulators, AD-DA-converters

NEC
http://www.necel.com/
Microcontrollers, AD-DA-converter, MOS-power transistors, optoelectronics

Novasensor ⟹ **General Electric**

Oki
http://www.okisemi.com/
SRAMs. DRAMs, microcontrollers

Omega
http://www.omega.com
Temperature sensors

ON Semiconductor
http://www.onsemi.com/
Analog ICs, 7400 logic, ECL logic, discrete semiconductors

Optek
http://www.optekinc.com
Optoelectronics

Osram
http://www.osram-os.com/
Optoelectronics, LEDs

Philips
http://www.semiconductors.philips.com/
Microcontrollers, 74oo logic, sensors

Power Integrations
http://www.powerint.com/
Off-line switchers

Pulse
http://www.pulseeng.com
Inductors, pulse transformers, RF transformers

QuickLogic
http://www.quicklogic.com/
FPGAs

Raytheon $\Longrightarrow$ Fairchild

RCA $\Longrightarrow$ Intersil

Renesas
http://www.renesas.com/
SRAMs, flash-memories

Rohm
http://www.rohm.com/
Liner ICs, CMOS logic, transistors, resistors

Samsung
http://www.samsungsemi.com/
SRAM, DRAM, Flash-memories, microprocessors

Semikron
http://www.semikron.com/
Power modules

SensorTechnics
http://www.sensortechnics.com/
Pressure sensors

SGS-Thomson $\Longrightarrow$ ST Microelectronics

Signetics $\Longrightarrow$ Philips, ON semiconductor

Siemens $\Longrightarrow$ Infineon

Siliconix $\Longrightarrow$ Vishay

Sharp
http://www.sharp-sme.com/
DRAM, SRAM, Flash-memories, microcontrollers, optoelectronics

Silicon Systems $\Longrightarrow$ Texas Instruments

Sipex
http://www.sipex.com/
Power management

Silicon General $\Longrightarrow$ Microsemi

Sony
http://www.sony.com/semi
AD-DA-converters, optoelectronics, microwave ICs

Spansion
http://www.spansion.com/
Flash memories

Sprague ⟹ Allegro

SPT – Signal Processing Technologies ⟹ Fairchild

ST Microelectronics
http://www.st.com/
Memories, microcontrollers, power semiconductors

Supertex
http://www.supertex.com/
High voltage MOS components

Telefunken ⟹ Vishay

Temic ⟹ Vishay

Texas Instruments
http://www.ti.com/
OPAmps, AD-DA-converters, signal processors, 7400 logic

Thomson ⟹ ST Microelectronics

Toshiba
http://www.toshiba.com/taec/
DRAMs, SRAMs, Flash-memories, microcontrollers, 7400 CMOS logic

Triquint
http://www.triquint.com/
GaAs-semiconductors

Unitrode ⟹ Texas Instruments

Valvo ⟹ Philips

Vantis ⟹ Lattice

Vishay
http://www.vishay.com/
Mosfets, voltage regulators, diodes, Z-diodes, bipolartransistors, optoelectronics

Waferscale ⟹ ST Microelectronics

Xicor ⟹ Intersil

Xilinx
http://www.xilinx.com/
CPLDs, FPGAs

Zilog
http://www.zilog.com/
Microprozessors, signal processors

Bibliography

Chapter 1:

[1.1]   Sze, S.M.: Physics of Semiconductor Devices, 2nd Edition. New York: John Wiley & Sons, 1981.

[1.2]   Hoffmann, K.: VLSI-Entwurf. München: R. Oldenbourg, 1990.

[1.3]   Löcherer, K.-H.: Halbleiterbauelemente. Stuttgart: B.G. Teubner, 1992.

[1.4]   MicroSim: PSpice A/D Reference Manual.

[1.5]   Antognetti, P.; Massobrio, G.: Semiconductor Device Modeling with SPICE. New York: McGraw-Hill, 1988.

[1.6]   Zinke, O.; Brunswig, H.; Hartnagel, H.L.: Lehrbuch der Hochfrequenztechnik, vol. 2, 3rd Edition. Berlin: Springer, 1987.

[1.7]   Bauer, W.: Bauelemente und Grundschaltungen der Elektronik, 3rd Edition. Müchen: Carl Hanser, 1989.

[1.8]   Kesel, K.; Hammerschmitt, J.; Lange, E.: Signalverarbeitende Dioden. Halbleiter-Elektronik vol. 8. Berlin: Springer, 1982.

[1.9]   Mini-Circuits: Frequency Mixers, datasheets.

Chapter 2:

[2.1]   Gray, P.R.; Meyer, R.G.: Analysis and Design of Analog Integrated Circuits, 2nd Edition. New York: John Wiley & Sons, 1984.

[2.2]   Sze, S.M.: Physics of Semiconductor Devices, 2nd Edition. New York: John Wiley & Sons, 1981.

[2.3]   Rein, H.-M.; Ranfft, R.: Integrierte Bipolarschaltungen. Halbleiter-Elektronik vol. 13. Berlin: Springer, 1980.

[2.4]   Antognetti, P.; Massobrio, G.: Semiconductor Device Modeling with SPICE. New York: McGraw-Hill, 1988.

[2.5]   Getreu, I.: Modeling the Bipolar Transistor. Amsterdam: Elsevier, 1978.

[2.6]   MicroSim: PSpice A/D Reference Manual.

[2.7]   Hoffmann, K.: VLSI-Entwurf. München: R. Oldenbourg, 1990.

[2.8]   Schrenk, H.: Bipolare Transistoren. Halbleiter-Elektronik vol. 6. Berlin: Springer, 1978.

[2.9]   Müller, R.: Rauschen. Halbleiter-Elektronik vol. 15. Berlin: Springer, 1979.

[2.10]   Motchenbacher, C.D.; Fitchen, F.C.: Low-Noise Electronic Design. New York: John Wiley & Sons, 1973.

[2.11]   Thorton, R.D.; Searle, C.L.; Pederson, D.O.; Adler, R.B.; Angelo, E.J.: Multistage Transistor Circuits. Semiconductor Electronics Education Committee, vol. 5. New York: John Wiley & Sons, 1965.

Chapter 3:

[3.1]   Sze, S.M.: Physics of Semiconductor Devices, 2nd Edition. New York: John Wiley & Sons, 1981.

[3.2]   Hoffmann, K.: VLSI-Entwurf. München: R. Oldenbourg, 1990.

[3.3]   Antognetti, P.; Massobrio, G.: Semiconductor Device Modeling with SPICE. New York: McGraw-Hill, 1988.

[3.4]   Spenke, E.: pn-Übergänge. Halbleiter-Elektronik vol. 5. Berlin: Springer, 1979.

[3.5]   MicroSim: PSpice A/D Reference Manual.

[3.6]   Müller, R.: Rauschen. Halbleiter-Elektronik vol. 15. Berlin: Springer, 1990.

Chapter 4:

[4.1]   Gray, P.R.; Meyer, R.G.: Analysis and Design of Analog Integrated Circuits, 2nd Edition. New York: John Wiley & Sons, 1984.

[4.2]   Geiger, L.G.; Allen, P.E.; Strader, N.R.: VLSI – Design Techniques for Analog and Digital Circuits. New York: McGraw-Hill, 1990.

[4.3]   Antognetti, P.; Massobrio, G.: Semiconductor Device Modeling with SPICE. New York: McGraw-Hill, 1988.

[4.4]   Weiner, D.D.; Spina, J.F.: Sinusoidal Analysis and Modeling of Weakly Nonlinear Circuits. New York: Van Nostrand, 1980.

[4.5]   Maas, S.A.: Nonlinear Microwave Circuits. Norwood: Artech House, 1988

[4.6]   Motchenbacher, C.D.; Fitchen, F.C.: Low-Noise Electronic Design. New York: John Wiley & Sons, 1973.

[4.7]   Müller, R.: Rauschen. Halbleiter-Elektronik vol. 15. Berlin: Springer, 1979.

[4.8]   Haus, H.A.; Adler, R.B.: Circuit theory of noisy networks. New York: John Wiley & Sons, 1959.

[4.9]   Vanisri, T.; Toumazou, C.: Integrated high frequency low-noise current-mode optical transimpedance preamplifiers: theory and practice. IEEE Journal of solid state circuits, vol. 30, no. 6, June 1995, p. 677.

Chapter 24:

[24.1]   Zinke, O.; Brunswig, H.: Lehrbuch der Hochfrequenztechnik. vol. 1, 4th Edition. Berlin: Springer, 1990.

[24.2]   Ebeling, K.J.: Integrierte Optoelektronik. 2nd Edition. Berlin: Springer, 1992.

[24.3]   Grau, G.; Freude, W.: Optische Nachrichtentechnik. 3rd Edition. Berlin: Springer, 1991.

[24.4]   Weinert, A.: Kunststofflichtwellenleiter. Erlangen: Publicis MCD, 1998.

[24.5]   Pehl, E.: Digitale und analoge Nachrichtenübertragung. Heidelberg: Hüthig, 1998.

[24.6]   Huber, J.: Digitale Übertragung I & II. Script of a lecture. University Erlangen-Nürnberg, Lehrstuhl für Nachrichtentechnik II, 1999.

[24.7]   Lee, J.S.; Miller, L.E.: CDMA Systems Engineering Handbook. Boston: Artech House, 1998.

Chapter 25:

[25.1]   Pettai, R.: Noise in Receiving Systems. New York: John Wiley & Sons, 1984.

[25.2]   Huber, J.: Digitale Übertragung I & II. Script of a lecture. University Erlangen-Nürnberg, Lehrstuhl für Nachrichtentechnik II, 1999.

Chapter 26:

[26.1]   Zinke, O.; Brunswig, H.: Lehrbuch der Hochfrequenztechnik. vol. 1, 4th Edition. Berlin: Springer, 1990.

[26.2]   Coilcraft: SMD Inductors 1206 CS and 1812 CS, datasheets.

[26.3]   Saal, R.: Handbuch zum Filterentwurf. 2nd Edition. Heidelberg: Hüthig, 1988.

[26.4]   Toko: Chip Dielectric Filters, datasheets.

[26.5]   Sawtek: SAW Filters, datasheets.

[26.6]   Kupferschmidt, K.H.: Die Dimensionierung des π-Filters zur Resonanztransformation. Frequenz 24, 1970, pp. 215–218.

[26.7]   Larson, L.E.: RF and Microwave Circuit Design for Wireless Communications. Boston: Artech House, 1996.

Chapter 27:

[27.1]   Zinke, O.; Brunswig, H.: Lehrbuch der Hochfrequenztechnik. vol. 1, 4th Edition. Berlin: Springer, 1990.

[27.2]   Hewlett Packard: S-Parameter Design. Application Note 154.

Chapter 28:

[28.1]   Meinke, Gundlach: Taschenbuch der Hochfrequenztechnik. 5th Edition. Berlin: Springer, 1992.

[28.2]   Mini-Circuits: Frequency Mixers, datasheets.

Index

Printing: Krips bv, Meppel, The Netherlands
Binding: Stürtz, Würzburg, Germany